HANDBOOK OF CRYSTAL GROWTH 2

BULK CRYSTAL GROWTH

Part A: Basic Techniques

HANDBOOK OF CRYSTAL GROWTH

Volume 1: FUNDAMENTALS

A: Thermodynamics and Kinetics
B: Transport and Stability

Volume 2: BULK CRYSTAL GROWTH

A: Basic Techniques
B: Growth Mechanisms and Dynamics

Volume 3: THIN FILMS AND EPITAXY

A: Basic Techniques
B: Growth Mechanisms and Dynamics

NORTH-HOLLAND
Amsterdam – London – New York – Tokyo

HANDBOOK OF CRYSTAL GROWTH 2

BULK CRYSTAL GROWTH

Part A: Basic Techniques

Edited by

D.T.J. Hurle

H.H. Wills Physics Laboratory
University of Bristol
Bristol, UK BS8 1TL

1994
NORTH-HOLLAND
Amsterdam – London – New York – Tokyo

ELSEVIER SCIENCE B.V.
Sara Burgerhartstraat 25
P.O. Box 211, 1000 AE Amsterdam, The Netherlands

Library of Congress Cataloging-in-Publication Data
(Revised for vol. 2, pt. A-B)

Handbook of crystal growth.

Includes bibliographical references and indexes.
Contents: v.1. Fundamentals. pt. A. Thermodynamics and kinetics. pt. B. Transport and stability -- v.2. Bulk crystal growth. pt. A. Basic techniques. pt. B. Growth mechanisms and dynamics.
1. Crystal--Growth. I. Hurle, D.T.J.
QD921.H32 1993 548'.5 93-7317
ISBN 0-444-81554-6

The beautiful natural snowflake on the front cover was photographed by Yoshinori Furukawa on Mt. Asahidake in Japan. The adjacent vertical strip is from a transmission electron micrograph taken by Tony Cullis of an In-implanted Si layer exhibiting cellular microsegregation produced by pulsed laser recrystallisation from the melt. The two images were chosen to reflect the natural and man-made aspects of crystal growth, respectively.

ISBN: 0 444 81554 6 (Part A and B)
0 444 89933 2 (Series)

Printed in The Netherlands

CONTENTS OF THE HANDBOOK

Volume 1. Fundamentals

Preface

Foreword
Sir Charles Frank

Part A. Thermodynamics and Kinetics

1. Historical introduction
   H.J. Scheel
2. Phase equilibria
   R.F. Brebrick
3. Defect thermodynamics and phase diagrams in compound crystal growth processes
   H. Wenzl, W.A. Oates and K. Mika
4. Nucleation theory
   B.Mutaftschiev
5. The structure of supersaturated solutions
   A.S. Myerson and A.F. Izmailov
6. Crystal growth mechanisms
   J.P. van der Eerden
7. Growth and morphology of crystals: Integration of theories of roughening and Hartman–Perdok theory
   P. Bennema
8. Atomic-scale models of crystal growth
   G.H. Gilmer
9. Low-energy ion/surface interactions during crystal growth from the vapor phase: Effects on nucleation and growth, defect creation and annihilation, microstructure evolution and synthesis of metastable phases
   J.E. Greene

Part B. Transport and Stability

10. Description of transport processes
    A.A. Wheeler
11. Bulk models of solidification
    H. Huppert
12. Morphological stability
    S.R. Coriell and G.B. McFadden

13. Effects of flow on morphological stability
S.H. Davis
14. Pattern formation in crystal growth
B. Billia and R. Trivedi
15. The dendrite
M.E. Glicksman and S.P. Marsh
16. Kinetics of electrocrystallisation
P. Ramasamy

Volume 2. Bulk Crystal Growth

Part A. Basic Techniques

1. Investigations of crystal growth in earth and planetary sciences
I. Sunagawa
2. Bridgman and related growth techniques
E. Monberg
3. Czochralski growth
D.T.J. Hurle and B. Cockayne
4. Crystal growth by floating zone melting
J. Bohm, A. Lüdge and W. Schröder
5. Use of a magnetic field in melt growth
D.T.J. Hurle and R.W. Series
6. Fractional crystallisation
P.J. Jansens and G.M. van Rosmalen
7. Industrial mass crystallisation
A.E.D.M. van der Heijden and G.M. van Rosmalen
8. Crystallisation of biological macromolecules
A. McPherson
9. Hydrothermal growth of crystals
K. Byrappa
10. Flux growth
W. Tolksdorf
11. Bulk crystal growth by physical vapour transport
E. Kaldis and M. Piechotka

Part B. Growth Mechanisms and Dynamics

12. Segregation phenomena in crystal growth from the melt
J.P. Garandet, J.J. Favier and D. Camel
13. Convection in melt growth
G. Müller and A. Ostrogorsky
14. Stress in the cooling crystal
J. Völkl
15. Modelling Bridgman and Czochralski growth
F. Dupret and N. van den Bogaert

16. Shaped crystal growth
V.A. Tatarchenko
17. Crystallisation of eutecties, monotectics and peritectics
J.D. Hunt and S.-Z. Lu
18. Spherulitic crystallisation in macromolecules
P.J. Phillips
19. Fundamentals of aqueous solution growth
S. Sarig
20. Crystal growth in gels
F. Lefaucheux and M.C. Robert

Volume 3. Thin Films and Epitaxy

Part A. Basic Techniques

1. Halogen transport epitaxy
H. Watanabe
2. Levitation epitaxy
H.M. Cox
3. Organo-metallic vapour phase epitaxy: principles and practices
D.W. Kisker and T.F. Kuech
4. Principles of molecular beam epitaxy
C.T. Foxon
5. MBE with gaseous source
H. Hirayama and H. Asahi
6. Liquid phase epitaxy
M.B. Small, E.A. Giess and R. Ghez
7. Solid phase epitaxy
G.L. Olson and J.A. Roth
8. Rapid solidification of silicon and some silicides by pulsed laser annealing
P. Baeri and S.U. Campisano
9. Pulsed laser ablation-deposition of multi-component oxide thin films: basic laser ablation and deposition processes and influence on film characteristics
O. Auciello
10. Growth of diamond films from the vapour phase
B.V. Spitsyn

Part B. Growth Mechanisms and Dynamics

11. Kinetic processes in vapour phase epitaxy
A.A Chernov
12. OMVPE: reaction kinetics
G.B. Stringfellow
13. Transport phenomena in vapour phase epitaxy reactors
K.F. Jensen
14. Atomic layer epitaxy
T. Suntola

15. Nucleation and surface diffusion in MBE
T. Nishinaga
16. Migration enhanced epitaxy
Y. Horikoshi
17. OMVPE of ultra-thin quantum wells and heterostructures
L. Samuelson
18. Photo-assisted epitaxy
S.J.C. Irvine
19. In-situ optical study of epitaxial growth
C. Pickering
20. Atomic mechanisms of LPE
E. Bauser
21. Artificial epitaxy (graphoepitaxy)
E.I. Givargizov
22. Structural effects in coherent epitaxial films
A. Zunger

PREFACE TO VOLUME 2

As explained in the Preface to Volume 1, this Handbook seeks to cover the science underlying crystal growth processes and the application of that science to the development of the technology of these processes on both industrial and research scales. By focussing on the underlying scientific basis of the subject, it is hoped that the Handbook will help to keep theory and practice in touch with each other. In this volume bulk crystal growth processes are reviewed; in Part A the basic techniques are covered starting, in chapter 1, with nature's techniques and in Part B mechanisms and dynamical aspects of the growth processes are addressed. Unfortunately, the allocation of the chapters between the two Parts is a little arbitrary in several cases because both topics are covered.

Emphasis has been placed on those techniques which are of current importance either for the commercial production of single crystals or for the preparation of crystals for research. Techniques of largely historical interest only, such as, for example, the Verneuil technique have not been included. Thus, in melt growth only Czochralski, floating zone and Bridgman and related techniques are covered in detail. However, the underlying science, which provides a rational basis for process design, is covered in depth.

Although the focus is on the science and technology rather than on the state-of-the-art in a spectrum of materials, the design of commercial bulk growth processes is very material specific so that several authors, myself included, have found it necessary to describe the application of the basic techniques material by material.

To contain the size of the Handbook within manageable and affordable limits, the scope of the materials coverage has had to be limited. Industrial mass crystallisation of inorganic materials has been included but polymer crystallisation has not, except for the chapter by Phillips on spherulitic crystallisation in macromolecules.

A fairly wide range of solution growth techniques is covered but electrocrystallisation is excluded, the kinetics of which were comprehensively covered by Ramasamy in Volume 1B. The treatment of bulk crystal growth from the vapour has been confined to physical vapour transport techniques. Chemical vapour transport techniques—but applied to the growth of epitaxial lavers—are covered in considerable depth in Volume 3.

Part A opens with a stimulating chapter by Ichiro Sunagawa reviewing crystal growth under terrestrial and other planetary conditions (which is almost exclusively crystal growth from solution or from the vapour). He clearly shows that the time is ripe for a concerted effort to apply crystal growth science to the interpretation of the microstructure of naturally occurring minerals.

As with Volume 1, the authority of the chapters is impressive and I am very grateful to all the authors for the immense effort they have given to the project. In several cases this has been done against a background of considerable pressure and anxiety about their employment situations; I have very much appreciated their commitment.

Don Hurle
May 1993

CONTENTS OF VOLUME 2

Preface ix

Part A. Basic Techniques

1. Investigations of crystal growth in earth and planetary sciences 1
I. Sunagawa
2. Bridgman and related growth techniques 51
E. Monberg
3. Czochralski growth 99
D.T.J. Hurle and B. Cockayne
4. Crystal growth by floating zone melting 213
J. Bohm, A. Lüdge and W. Schröder
5. Use of a magnetic field in melt growth 259
D.T.J. Hurle and R.W. Series
6. Fractional crystallisation 287
P.J. Jansens and G.M. van Rosmalen
7. Industrial mass crystallisation 315
A.E.D.M. van der Heijden and G.M. van Rosmalen
8. Crystallisation of biological macromolecules 417
A. McPherson
9. Hydrothermal growth of crystals 465
K. Byrappa
10. Flux growth 563
W. Tolksdorf
11. Bulk crystal growth by physical vapour transport 613
E. Kaldis and M. Piechotka

Part B. Growth Mechanisms and Dynamics

12. Segregation phenomena in crystal growth from the melt 659
J.P. Garandet, J.J. Favier and D. Camel
13. Convection in melt growth 709
G. Müller and A. Ostrogorsky
14. Stress in the cooling crystal 821
J. Völkl
15. Modelling Bridgman and Czochralski growth 875
F. Dupret and N. van den Bogaert
16. Shaped crystal growth 1011
V.A. Tatarchenko
17. Crystallisation of eutectics, monotectics and peritectics 1111
J.D. Hunt and S.-Z. Lu

18. Spherulitic crystallisation in macromolecules 1167
P.J. Phillips
19. Fundamentals of aqueous solution growth 1217
S. Sarig
20. Crystal growth in gels 1271
F. Lefaucheux and M.C. Robert

Subject Index 1305

CHAPTER 1

Investigations of Crystal Growth in Earth and Planetary Sciences

Ichiro SUNAGAWA*

Yamanashi Institute of Gemmology and Jewelry Arts, Tokoji-machi 1955-1, Kofu, 400 Japan

Contents

1. Introduction 3
2. Growth conditions 5
   2.1. Conditions and environments of natural crystallization 5
   2.2. Methodologies 7
3. Nucleation and textures 8
4. Morphology and perfection of crystals; deductive analyses of growth and post-growth histories 14
   4.1. Problems 14
   4.2. External morphology 16
   4.3. Surface microtopography 22
   4.4. Internal morphology (perfection and homogeneity) 26
5. Experimental methods to simulate natural crystallization 33
   5.1. In-situ observation of crystallization from aqueous solution 34
   5.2. Dynamic crystallization experiments and in-situ investigation methods of magmatic crystallization at high temperatures 41
   5.3. Simulation of extra-terrestrial materials 47
6. Concluding remarks 47
References 48

* Formerly, Institute of Mineralogy, Petrology and Economic Geology, Tohoku University, Aramaki, Aoba-ku, Sendai, 980 Japan.

HANDBOOK OF CRYSTAL GROWTH, VOL. 2
edited by D.T.J. Hurle

1. Introduction

It has been the main interest of the modern science and technology of crystal growth to understand the growth mechanisms of crystals at an atomic level and to synthesize single crystals or thin films with desired properties, perfection and homogeneity. To achieve this goal, every effort has been made to realize precisely controlled growth conditions. In contrast to this, crystallization to form Earth and Planetary solid materials takes place under a wide range of fluctuating conditions, and in various environments. The conditions are exactly the opposite to those in the modern science and technology of crystal growth. The conditions are not controlled; they range from very close to equilibrium to far from it; they may fluctuate greatly during growth, or may change abruptly; the chemical systems are multi-component and impure; the process is impossible to observe directly. It appears to pose problems which are much too complicated to apply directly the achievements in modern science and technology of crystal growth. It is, however, this point that stimulates the curiosity of Earth and Planetary scientists. It is worthwhile to make an effort to build a bridge between the two disciplines: Earth and Planetary sciences and the science and technology of crystal growth. This will be mutually beneficial.

We should also recall that the science of crystal growth was born when Nicolaus Steno [1] explained the origin of a variety of external forms exhibited by natural quartz crystals in terms of different growth rates in different crystallographic directions. The concept of growth rate anisotropy still forms the basis of the modern science of crystal growth and morphology. Since the time of Steno, the morphology of crystals has become the core of the study of mineral crystallization. This has further developed to the analysis of growth and post-growth histories of mineral crystals based on the analysis of morphological characteristics.

Mineralogy and petrology are the materials science in the Earth and Planetary sciences. They developed at first as a science of description and classification. Nearly 3500 independent mineral species, i.e., those having respectively different chemical compositions, crystal structures and stability fields, are presently known. When crystals of a few mineral species, with or without noncrystalline solid, aggregate to form a solid mass, it is called a rock or an ore (or mineral) deposit. Rocks are classified into igneous, sedimentary and metamorphic rocks depending on how they are formed.

The temperature, pressure, fugacity and other conditions of mineral formation have been evaluated based on geological investigations and equilibrium thermodynamic analyses of mineral paragenesis.

It has recently been realized, however, that equilibrium thermodynamic analysis is not enough to understand how mineral crystals were formed. The importance of kinetic analysis of the processes of mineral formation has increasingly been realized in the discipline of the Earth and Planetary sciences, since such knowledge contributes much to deepening our understanding of the formation of Earth and Planetary bodies. Experimental investigations along this line, e.g., dynamic crystallization experiments of silicate systems and in-situ investigations of crystal growth processes in aqueous solution at ordinary temperature, and in silicate systems at high temperatures, have developed rapidly. Although there is still a big gap between the laboratory model systems and reality, owing to the fact that natural systems are too complicated and impure for one to apply directly the experimental results, investigations along this line have brought about a new insight into Earth and Planetary sciences.

Although there may still be a gap between the achievements of the modern science and technology of crystal growth and of Earth and Planetary sciences, it is worthwhile to summarize in this chapter how crystal growth in the mineral kingdom has been investigated, with a hope of introducing the problems which may attract the interest of the modern crystal growth community and rebuilding a bridge between the two communities. Useful information for modern science and technology of crystal growth will also be obtained from mineral crystallization, since the Earth is a ceaseless crucible growing crystals under an extremely wide range of conditions.

This chapter consists of six sections. In section 2, physical, chemical and environmental conditions encountered in the formation of Earth and Planetary solid materials are summarized. Various methodologies to understand how mineral crystals nucleated, grew or dissolved, and what sort of post-growth histories they experienced are also explained. The results of such investigations have given new insight into and enriched the understandings of Earth and Planetary solid bodies, since the previous investigation has been principally based on equilibrium thermodynamics, neglecting kinetics.

In section 3, nucleation and texture formation of rocks, ores and meteorites are discussed. Instead of giving a comprehensive explanation of nucleation, mainly topical problems relating to nucleation and texture formation of Earth and Planetary solid materials are discussed in this section, which includes the role of secondary nucleation phenomena and metastable nucleation.

Since we cannot see directly the process of natural crystallization, deductive analyses of growth and post-growth histories of mineral crystals based on the character of single crystals and crystal aggregates have been the principal methodology, a description of which is given in section 4. Morphology of crystals is at the core of this methodology. The origin of a variety of external forms and surface microtopographs are discussed. Analyses of growth and post-growth histories of minerals based on the character of the perfection and homogeneity (internal morphology) are explained using diamond and other minerals as representative examples.

Recent experimental investigations to simulate natural crystallization and to verify crystal growth mechanisms which have contributed much to deepening our understanding of kinetic problems of Earth and Planetary solid materials are reviewed in section 5. The results of in-situ observation of crystallization of simple compounds in aqueous solution at ordinary temperature and atmospheric pressure are useful in understanding the process of crystallization in surfacial precipitation, sedimentogenetic and diagenetic conditions. Dynamic crystallization experiments have contributed much to the understanding of magmatic crystallization. In-situ observation at high temperatures is a promising future methodology for studying magmatic crystallization. Other experimental methods to simulate the formation of extra-terrestrial solid materials, including lunar rocks, chondrules, comets and interplanetary dust particles are briefly explained as well.

Section 6 gives a concluding remark.

2. *Growth conditions*

2.1. *Conditions and environments of natural crystallization*

Our interest is to understand how mineral crystals which constitute solid Earth and Planetary bodies nucleated, grew, dissolved and transformed, and what sort of post-growth histories they experienced. From this understanding, we can deepen our understanding of the constitution and history of Earth and Planetary bodies. The solid materials were formed through igneous activities, sedimentogenetic and diagenetic, and metamorphic and metasomatic processes on the Earth, and through condensation, accumulation, melting, solidification and impact processes in the case of extra-terrestrial materials. The temperature, pressure, fugacity, driving force (supersaturation, supercooling, etc.), chemistry and duration of mineral formation range extremely widely, incomparable with laboratory conditions.

Figure 1a illustrates schematically the temperature, pressure and depth conditions under which various geological processes in the Earth's crust and upper mantle take place [2, 3]. Figure 1a indicates only the conditions by which mineral crystals may be brought out to our hands through the Earth's tectonic movements. Minerals formed below this depth, i.e., those formed in the lower mantle and core, seldom reach our hands. In the case of minerals of extra-terrestrial origins, such as lunar rocks, meteorites and cosmic dust (interplanetary dust particles), much more severe conditions have to be assumed. Temperatures may range from near 0 K to more than 10 000 K, pressure from 10^{-10} Torr to many GPa.

Mineral crystals are formed in a variety of environmental conditions. In igneous process, mineral crystallization occurs in a magma, a high-temperature solution which was formed by partial melting of the mantle or the crust. There are dry and wet magmas, depending on the positions of partial melting, whose melting and solidification temperatures are different (fig. 1b). At the later stage of solidification of a magma, some minerals may grow freely in an open space (in bubbles formed in the solidifying magma) forming a pegmatite or amygdal void. Supercritical vapor or hydrothermal solutions derived from igneous activities precipitate various minerals in a form of fissure-filling ore veins, or penetrate into the surrounding rocks forming hydrothermal metasomatic ore deposits. Mineral precipitation in sedimentogenetic, diagenetic, and surfacial environments takes place principally in an aqueous solution at ordinary temperature and near atmospheric pressure conditions. In the case of regional metamorphism and contact metasomatism, a confined environment, i.e., crystallization in the presence of other solid materials has to be assumed. It is still a controversial problem whether metamorphic and metasomatic processes proceed through solid-state phase transition or via a dissolution and precipitation process. The present author believes, however, that except for a few special cases, mineral crystallization in the majority of geological processes, including metamorphic and metasomatic processes, can be regarded essentially as crystal growth from solution phases [3]. This is indicated in fig. 1b, which schematically shows various geological processes in terms of depth and time. Neither crystal growth from a pure melt phase nor by pure physical vapor deposition are present in mineral crystallization in the Earth. Vapor growth in mineral

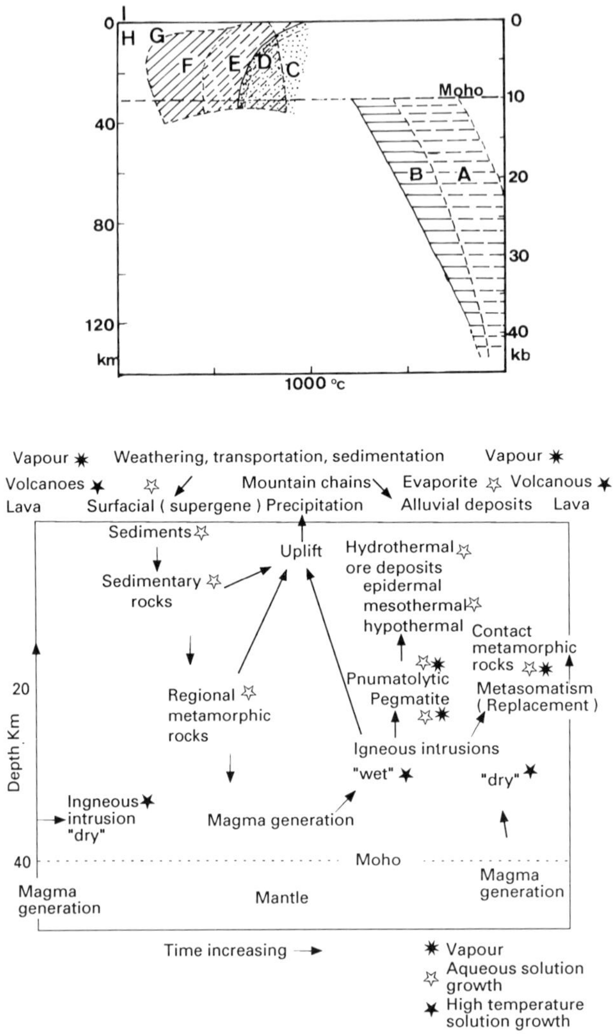

Fig. 1. (a) Temperature, pressure and depth of geological processes; A: peridotite and eclogite; B: basaltic magma; C: granitic magma; D: pegmatite; E: contact metamorphism; F: regional metamorphism; G: hydrothermal vein and metasomatism; H: diagenesis; I: weathering, transportation, sedimentation and surfacial (supergene) precipitation. (b) Types of crystal growth assumed in respective process, expressed in time and depth [3]. Modified, based on [2].

formation occurs in the majority of cases by a chemical vapor deposition process. Solid-state crystallization is limited to special cases, e.g., formation of marble from limestone, or of quartzite from sandstone.

In the case of extra-terrestrial materials, such as cosmic dusts (interplanetary dust particles) and meteorites, vapor growth and high-temperature solution growth are considered to play a more important role than in the Earth.

Through geological processes and tectonic movements of the Earth, including movement and subduction of plates, mineral crystals, once formed primarily, experience further complex post-growth histories. The Earth has experienced constant change and evolution throughout its 4.5 billion years' history. Oxygen fugacity on the Earth's surface changed drastically. Through the processes of weathering, transportation and sedimentation, and of plate movement, subduction and mantle convection, the cycle and the regeneration of Earth's materials take place. On a smaller scale, if, e.g., a magma in which crystals (phenocrysts) were already present was brought to a different position or erupted onto the Earth's surface, the phenocrysts would experience different P, T conditions. Through the process, they will experience plastic deformation, mechanical destruction, partial dissolution, transition and exsolution (precipitation of a mixed phase), or they may be quenched and survive the new conditions. The changes may be repeated several times. All such growth and post-growth histories are recorded by changes in their morphological characteristics, perfection and homogeneity in single crystals or textural characteristics of mineral aggregates, provided that they survived the new conditions. Exsolution lamellae observed in rock-forming minerals, such as feldspar, pyroxene, etc., in both terrestrial and extra-terrestrial materials, are representative examples that have been investigated in detail with the hope of deducing the cooling histories of magmas. Diamond is a unique example that survived and preserved its original growth characteristics which took place at a depth of more than 120 km, in spite of its experience of severe and drastic changes of conditions while it was being transported to the Earth's surface.

2.2. *Methodologies*

Since we are unable to observe directly the processes of mineral formation and successive changes, investigations of morphology, perfection and homogeneity of mineral crystals and deductive analyses of growth and post-growth histories based on these characteristics are traditionally adopted as the most important methodology for understanding the kinetics of mineral crystallization. Investigations of this type have recently been reactivated among Earth and Planetary scientists, since they realized the limits of the equilibrium thermodynamic approach, which had been the major methodology in these disciplines. The recent development of experimental methods to investigate imperfection and inhomogeneity of solid materials is another reason for this reactivation.

Deductive analyses should, however, be supplemented by laboratory simulation. From the beginning of this century, equilibrium phase diagrams of silicate, and later sulfide and oxide, systems were prepared. Mineral parageneses were analysed based on these data. The experimental conditions were expanded from the dry-system experiments at 1 atm pressure and high temperature of the earlier period to wet-system, and/or ultrahigh-pressure conditions. It is now possible to elucidate mineral phases expected in the lower mantle, which are impossible to obtain in our hands. The aims of experiments of this type are to analyse mineral paragenesis based on the equilibrium phase relations, assuming that naturally occurring mineral paragenesis reached the equilibrium state through a long geologic history.

Since neither nucleation, growth nor dissolution takes place at equilibrium, and textures exhibited by mineral paragenesis are the result of such kinetic processes, another type of experiment should be planned with the aim of understanding the kinetics of mineral crystallization. This was started in the 1970s, and is called "Dynamic Crystallization Experiments" (see, e.g., [4]), which deal principally with magmatic crystallization. Experiments of this type were originally aimed at reproducing igneous textures of lunar samples. The results stimulated interest in kinetic problems among Earth and Planetary scientists. They further developed a quantitative analysis of kinetics (see, e.g., [5]). However, all these are, in principle, indirect methods, in that paragenetic identification, texture observations, and even rate measurements were made on quenched samples, whose crystallization processes were concealed. An in-situ investigation method at high temperatures was thus devised, so that whole crystallization processes could be observed directly and in-situ and recorded on video tape with high visibility, with the conditions controlled precisely enough to enable kinetic data to be theoretically analysed [6–9]. This method was developed on the experience of the in-situ observation method of crystallization in an aqueous solution at ordinary temperature and atmospheric pressure. The results obtained by the latter methods can be used as a simulation of mineral formation in surfacial, sedimentogenetic and diagenetic environments. These methods will be described in more detail in section 5.

3. *Nucleation and textures*

Textures of rocks and ores are indicators of how nucleation took place. Depending on the conditions, a variety of textures appear. If a magma was solidified slowly at a depth with a small supercooling and slow cooling rate, a holocrystalline texture appears, whereas when it was quenched by eruption, a glassy rock containing only crystallites appears. If a magma which contains already nucleated crystals (phenocrysts) was transported to a different position or to the Earth's surface and solidified, we expect a porphyritic texture, consisting of phenocrysts and groundmass. The groundmass may consist of only glass or much smaller phenocrysts than older phenocrysts. The latter texture is called a hyalopyritic texture, which is commonly observed in natural basaltic rocks. In such a case, one usually assumes nucleation at two different stages and supercoolings. Representative textures of igneous rocks are illustrated in fig. 2.

Through a series of dynamic crystallization experiments, most igneous rock textures, including both terrestrial and lunar rocks, were successfully reproduced. However, experiments performed at two different supercoolings could reproduce only a texture consisting of phenocrysts and radiating dendrites in the groundmass, and not a hyalopyritic texture, fig. 3a. The reason is that the morphology of crystals changes systematically from polyhedral to dendritic with increasing supercooling (see section 4.1). Kouchi et al. [10] successfully reproduced hyalopyritic texture in the laboratory, by forced mixing of two basaltic magmas containing phenocrysts, fig. 3b. They used an infra-red focusing furnace, and realized forced mixing by rotating the upper and lower rods in opposite directions. Nucleation density increased drastically, fig. 4, and a hyalopyritic texture was formed when forced mixing was applied, as compared to

Fig. 2. Representative igneous textures. (a) Holocrystalline texture (granite), (b) porphyritic (hypocrystalline) texture (andesite), and (c) holohyaline (glassy) texture (pitchstone).

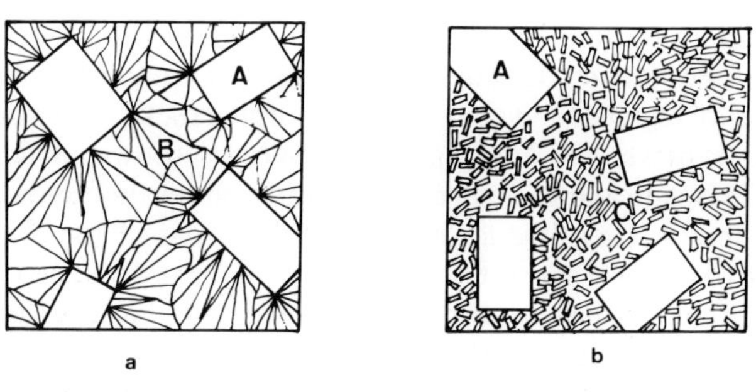

Fig. 3. (a) Texture of a basalt reproduced without forced convection by a two-step dynamic crystallization experiment, (b) naturally observed and experimentally reproduced hyalopyritic texture by forced mixing [10]. A: phenocryst; B: dendritic groundmass; C: microphenocryst.

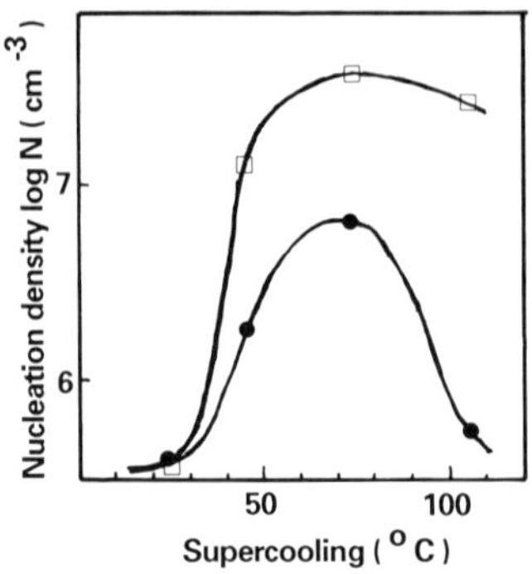

Fig. 4. Nucleation densities [vertical axis, log N (cm^{-3})] versus supercooling (horizontal axis, °C) obtained in static (solid circles) and forced (open squares, 48 rpm) mixing experiments of two basaltic melts containing artificial phenocrysts [10].

the case of a static experiment, i.e., with only a little forced convection produced by rotating two rods in the same direction. This suggests that the phenomena of secondary nucleation (growth on detached particles from the seed), which has been utilized in the field of industrial crystallization to produce a large number of equally sized crystal grains, should also be taken into consideration in magmatic crystallization. Detached particles from phenocrysts by forced mixing acted as preferential sites for the growth of microphenocrysts.

Although it has been entirely neglected in the discipline of Earth and Planetary sciences, secondary nucleation phenomena should be expected more commonly in natural crystallization, not only in magmatic crystallization but also in surfacial precipitation of minerals like those in evaporite deposits, and sedimentogenetic crystallization, since turbulent flow on a micro-scale is always present in these processes [11].

Another interesting application of a concept in industrial crystallization to geological problems has recently been initiated by Cashman [12]. She tried to evaluate nucleation rates, growth rates and growth times using quantitative measurements of crystal size distributions (CSD), which were originally developed in the industrial crystallization field to obtain kinetic information on the crystallization of industrial compounds [13]. Population densities and crystal sizes (quantitative texture analysis) of lavas, and later of regional and contact-metamorphic rocks were measured. Igneous and metamorphic rocks indicated different CSDs, as shown in fig. 5a and b. By combining the CSD data and geological information, she evaluated the growth rates of silicate and oxide crystals in the lavas to be of the order of 10^{-10}–10^{-11} cm/s, the nucleation rates 10^{-2}–10^{-3}/cm^{3}s, the growth time of the order of 10 to 100 years. For high-temperature contact-metamorphic rocks, the growth time was less than 100 years at $\Delta T = 10$°C, and the nucleation rates were between 10^{-4} and 10^{-1}/cm^{3}s. It should be noted that the nucleation, growth rate and the growth time of minerals, even in a lava, which erupted onto the Earth's surface and thus should have experienced a large supercooling, are very small. It is also interesting to note that the time required to

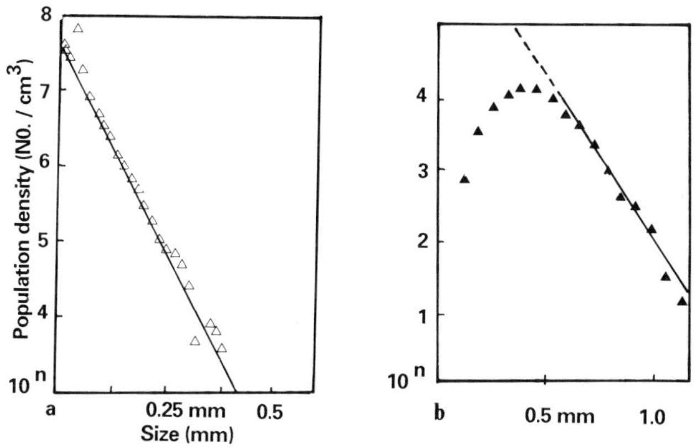

Fig. 5. Comparison of CSDs (crystal size distributions) obtained in igneous rocks (plagioclase in a lava)(a) and metamorphic rocks (garnet in a contact-metamorphic rock) (b) [12]. Vertical axis: population density (N°/cm^3); horizontal axis: crystal size in mm.

complete one geological process is rather short, a matter of years and not of millions of years. The evaluated values indicate crystallization under near-equilibrium conditions even in the case of solidification of a lava. This is probably due to the large volume of a geologic system. In the case of contact metamorphic rocks, she obtained bell-shaped CSDs (fig. 5b), which are different from the almost straight CSDs of lavas (fig. 5a). She interpreted this difference as due to modification of the original CSD by a successive annealing process involved in contact metamorphism.

Sedimentary rocks are formed by the crystallization of authigenic minerals (those precipitated in-situ) in the intergranular spaces of allogenic minerals (fragments of residual minerals survived from weathering and transportation). When they are buried, authigenic minerals transform to different minerals to meet new temperature and pressure conditions. When such a transformation takes place at lower temperatures and pressures than regional metamorphism, the process is called diagenesis. Clay minerals, carbonates, hydroxides and zeolites are formed in this process. Further increasing the depth, i.e., temperature and pressure, regional metamorphism takes place, before the solid materials are partially melted to form a new magma. Contact metasomatism takes place when a magma intrudes into pre-existing strata, giving heat and materials (volatiles and dissociated elements) to the latter. When the source of heat and materials is fumarole or hydrothermal, the term hydrothermal metasomatism is applied, which takes place at a shallower depth than ordinary contact metasomatism. In both regional metamorphism and contact metasomatism, neomineralization occurs under pressure in the presence of other solid materials. Crystal grains are small and xenomorphic, and often align in a certain direction, giving a characteristic texture called schistosity in regional metamorphic rocks. Often a few mineral grains grow much larger than the average size, which are called porphyroblasts. It was suggested that Ostwald ripening was responsible for the formation of porphyroblasts of muscovite in regional metamorphic rocks [14]. This suggestion was based on the observations of surface microtopographs of mica minerals in relation to increasing metamorphic grades. Surface microtopographs of {001} faces representing growth and dissolution were distinguished. The smaller crystals were characterized by a dissolution pattern, whereas co-existing porphyroblastic mica crystals were characterized by growth patterns, suggesting the growth of porphyroblastic crystals at the expense of smaller mica crystals (see fig. 14). In fig. 6, textures of sedimentary, regional and contact metamorphic rocks are shown.

A mica mineral, muscovite, $KAl_3Si_3O_{10}(OH)_2$, occurs in igneous rocks and pegmatites, as well as in regional metamorphic rocks and hydrothermal metasomatic deposits. In pegmatite, it can grow as meter-sized crystals, whereas only micron-sized crystals are found in regional metamorphic rocks and hydrothermal metasomatic deposits. Why is this so? It should be related to the heterogeneous nucleation problem. We may assume much higher densities of heterogeneous nucleation sites in regional metamorphic and hydrothermal metasomatic environments than in pegmatitic environments or in a magma. This is an interesting problem to be answered quantitatively.

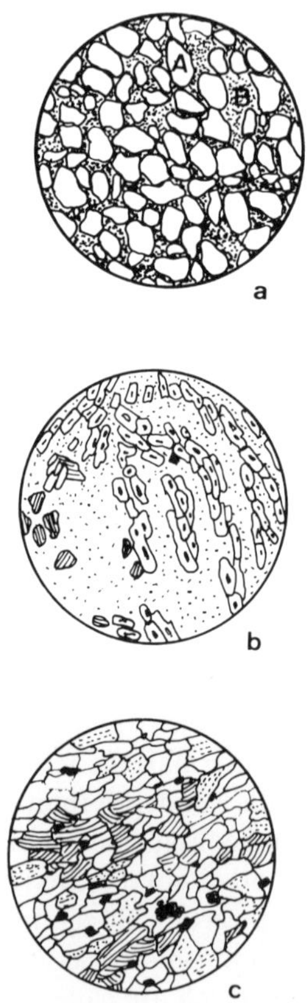

Fig. 6. Textures of (a) sedimentary (sandstone), (b) contact-metamorphic (hornfels), and (c) regional metamorphic (gneiss) rocks. A: allogenic mineral fragments: B: authigenic minerals.

A similar question is applicable to the case of nucleation of quartz crystals, which are one of the most persistent minerals. Quartz crystals as large as a meter can be found in pegmatites, but those composing agates and chalcedony are of micron or even nanometer size. The origin of this difference was accounted for [15] on the basis of the difference of the state of silica polymerization depending on temperature. At lower temperature where agates, etc., are formed, silica is more likely to be polymerized as compared to the monomeric state at higher temperature conditions such as in pegmatite. The perfection of quartz crystals is markedly different between those occurring in pegmatites and in epithermal veins and agates. On decreasing the growth temperature, we see less perfect and much smaller crystals, and eventually the amorphous state, opal.

Chondrules are spherical particles found in stony meteorites (chondrites), and consist of olivine and/or pyroxene of extra-terrestrial origin. They exhibit various textures; barred, porphyritic, radiated, etc. All naturally observed chondrule textures have been reproduced by dynamic crystallization experiments [16], which offered crucial information as to the temperature, supercooling and cooling rate of the formation of chondrules. There was once a state of silicate liquid droplets, in which crystallization of olivine and pyroxene took place to form chondrules, in the process of

Table 1
Examples of metastable nucleation phenomena observed in silicate systems by the in situ observation method at high temperatures (upper) and the result of analysis of factors causing the phenomena in the case of Di–Tp–Ak compositions (lower).

Liquid composition (wt%)	Stable phase	Phase that appeared prior to the stable phase
An_{100}	Triclinic anorthite (An) ($CaAl_2Si_2O_8$)	pseudo-hexagonal and pseudo-orthorhombic $CaAl_2Si_2O_8$ polymorphs, mullite, corundum
$An_{80}Wo_{20}$	An	two $CaAl_2Si_2O_8$ polymorphs
$Di_{40}An_{60}$	An	Diopside (Di) ($CaMgSi_2O_6$)
$Di_{50}An_{50}$	An	Di
$Di_{60}An_{40}$	An	Di
Di_{100}	Di	Forsterite (Fo) (Mg_2SiO_4)
$Di_{80}Tp_{10}Ak_{10}$	Di	Fo
$Di_{65}Tp_{25}Ak_{10}$	Fo	Perovskite (Pv) ($CaTiO_3$), Di
$Di_{55}Tp_{35}Ak_{10}$	Pv	Fo
$Di_{85}En_{15}$	Di	Fo

Tp: Ti-pyroxene; Ak: Akermanite; En: Enstatite

	Expected	Observed
Growth rate	Pv > Di > Fo	Pv < Di < Fo
Transient effect	Fo > Di > Pv	$Pv \gg Di \sim Fo$
Interfacial energy	Pv < Di < Fo	Pv < Di < Fo (from T/V_m values)

meteorite formation. Chondrules are regarded as the original source materials of the Earth.

Another interesting problem relating to nucleation is a phenomenon called metastable nucleation. This term is applied to two cases: (1) When nucleation of a phase takes place under a condition apart from that expected from the equilibrium thermodynamic relation of the phase (CVD diamond, and aragonite in living shell fish are representative examples. Both are high-pressure forms); (2) When a phase or phases not expected from the relevant phase relation nucleate prior to the stable phase. By in-situ investigation at high temperatures, it has recently been clarified that metastable nucleation occurs more commonly and universally in silicate and oxide systems. For example, forsterite always nucleates prior to diopside, the only stable phase, in a pure diopside melt, and can grow steadily as if forsterite was the stable phase. This behavior is seen before the appearance of diopside. Forsterite crystals start to dissolve in the presence of diopside. See also fig. 21, showing the observed nucleation sequence in a pure anorthite, $CaAl_2Si_2O_8$, melt, in relation to temperature and time. Table 1 summarizes metastable nucleation phenomena of category 2 so far observed in silicate systems by in-situ observation methods. The reasons why metastable nucleation can take place have been analysed based on the observation of the sequence of nucleation, the rates of nucleation and growth of metastable phases, and the analysis of nucleation behavior (steady-state or transient nucleation), etc. [17, 18]. It has been clarified that the key factor is interfacial energy between the solid and liquid. The stability of metastable phases in the presence or absence of a stable phase has also been investigated [17]. Metastable nucleation will be treated in some more detail in section 5.2.

4. *Morphology and perfection of crystals; deductive analyses of growth and post-growth histories*

4.1. *Problems*

Since the time of Steno, a variety of external forms of polyhedral crystals which are exhibited by the same mineral species has attracted the interest of mineralogists. Such interest had brought forth the concept of equilibrium form, structural form (e.g., Bravais' law, Donnay–Harker's extension of Bravais' law, Hartman–Perdok's PBC analysis, see [19]), and growth forms. It is only after the birth of atomistic models of crystal growth in the 1930s that much attention has been given to other aspects of crystal morphology, such as the relations amongst polyhedral, hopper, dendritic and spherulitic morphologies, fractal patterns, surface microtopographs of crystal faces, and internal morphologies (e.g., growth sectors, growth banding, etc.) of crystals.

According to [19], the following are the major problems of morphology of mineral crystals to be studied:

(1) What are the mutual relations between fractal patterns, spherulitic (both are morphologies exhibited by polycrystalline aggregates), dendritic, hopper and polyhedral morphologies (single crystals) in relation to growth conditions?

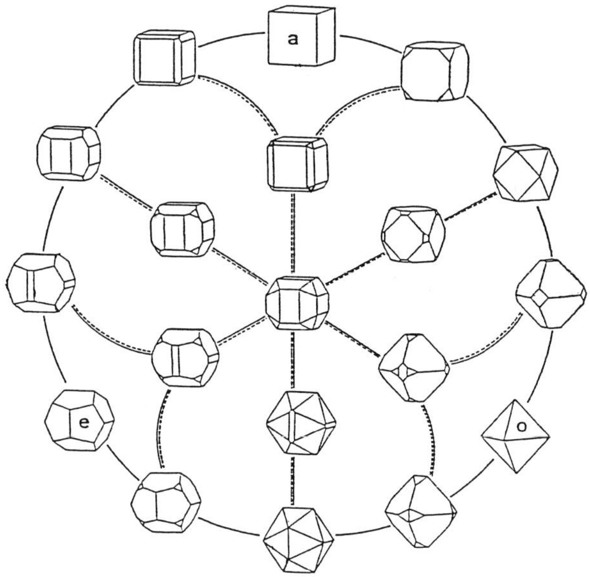

Fig. 7. By a combination of a {100}, e {210} and o {111} faces, pyrite crystals exhibit all these habits. They are commonly encountered in natural pyrite crystals [22].

(2) Why can one mineral species take a wide variety of polyhedral forms? This is further divided into the following two problems:

(a) Variation due to a combination of different faces (fig. 7), i.e., so-called habit variation.

(b) Variation due to the anisotropic development of crystallographically equivalent faces (fig. 8).

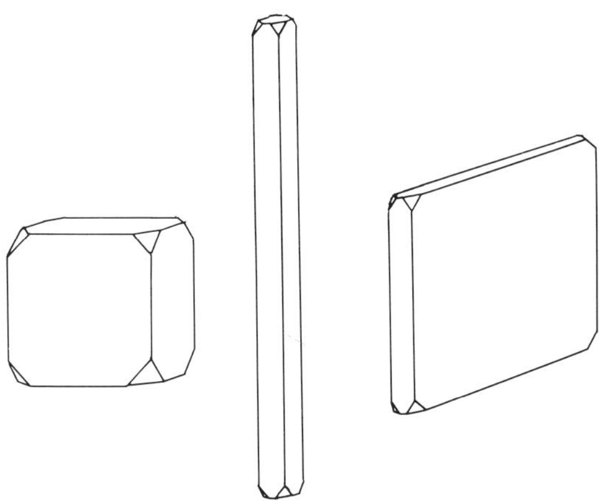

Fig. 8. By anisotropic development of crystallographically equivalent faces (in this case {100} faces of a cubic crystal), various malformed external forms appear [19].

(3) Surface microtopography of crystal faces (morphology of spiral step patterns) in relation to growth conditions and also to the external forms of polyhedral crystals.

(4) Internal morphology, e.g., morphologies of growth sectors, growth banding, and so on.

(5) Application of the understanding of morphological problems to geological and petrological problems.

We shall mainly deal with problem 5, based on the present knowledge of problems 1 to 4.

4.2. External morphology

Figure 9 shows a general view of variations of the external morphologies of crystals in relation to growth mechanisms, growth rates, interface roughness, and driving forces [19]. This figure indicates only one section of the whole relationship, and the boundaries (dotted lines) vary depending on phases (melt, solution, vapor), types of solvent (with different solute–solvent interaction), chemical composition, crystal sizes, etc. But the general tendency of variation from polyhedral, hopper, dendritic (single crystals), and further to spherulitic and fractal (polycrystals) with increasing driving force is unchanged. In fig. 10, experimentally observed morphological variations of silicate crystals depending on supercooling and liquid composition are shown. Figure 10a shows a result obtained by dynamical crystallization experiments, and fig. 10b those obtained by in-situ observation methods. In both cases, the general tendency of morphological changes from polyhedral via hopper to dendritic crystals and further spherulitic aggregates with increasing supercooling is clearly noted.

As can be judged from fig. 9, polyhedral crystals bounded by low-index smooth faces appear principally by a spiral growth mechanism at a smaller driving force. Hopper crystals appear by two-dimensional nucleation growth or spiral growth preferentially

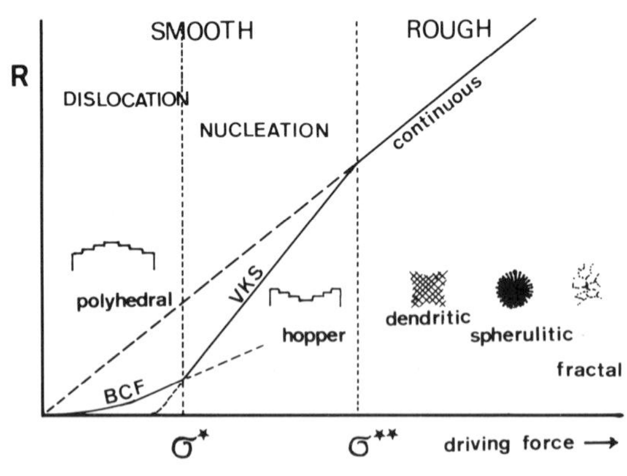

Fig. 9. Schematic diagram illustrating general relations among growth mechanisms (BCF: special growth; VKS: 2D nucleation growth; and continuous growth mechanisms), interface roughness, driving force, and external morphologies of crystals (polyhedral, hopper, dendritic, spherulitic and fractal). Modified from [19].

(a)

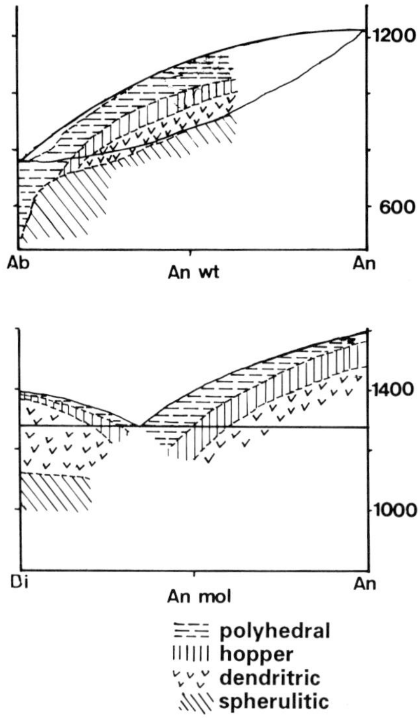

(b)

Fig. 10. Morphological changes of some silicate minerals depending on supercooling and liquid composition. (a) Albite(Ab)–anorthite(An)–H_2O system; by dynamic crystallization experiment. (b) Diopside(Di)–anorthite(An) system; by in-situ observation method [19]. Temperature in °C, composition in %.

originating along the edges of smooth faces, where higher supersaturation is expected than the central area of a growing face [4, 19]. Further increasing the driving force, one may expect dendritic crystals, whose growth is principally governed by adhesive type growth on rough interfaces. At a much higher driving force, crystal aggregates with a spherulitic morphology are expected, due to radiating growth starting from a nucleus. Spherulitic morphology may also be expected at a smaller driving force when needle, platy, or, rarely, bulky crystals grow radiating on the surface of foreign particles or on spherical aggregates of the same crystals which are formed by coalescence due to external forces such as agitation. Fractal patterns appear when nucleation occurs sporadically in a confined and static environment.

Reflecting the wide range of growth conditions, mineral crystals exhibit a full range of morphological variations, from fractal to polyhedral morphologies. A good example of fractal patterns exhibited by mineral crystals is manganese oxide or hydro-oxide minerals formed in sedimentary seams. A variety of minerals, particularly those formed under surfacial conditions, occur in the form of spherulitic aggregates, e.g., aragonite, gypsum (desert rose), diagenetic zeolites. Dendritic and hopper morphologies are also often encountered. In fig. 11, a few photographs of mineral samples exhibiting these morphologies are shown. However, it should be stressed that these morphologies, which represent growth at larger driving forces than polyhedral crystals, are only seldom encountered in rock-forming minerals of magmatic origin. They are found only in lavas erupted and solidified on the oceanic floor, such as in pillow lavas [20]. These morphologies are seldom encountered, even in the chilled margin of an igneous dyke. This indicates that, in magmatic crystallization, rather slow growth rates near equilibrium conditions have to be assumed, owing to the larger scale of the system. This is consistent with the growth rates evaluated by Cashman [12] using CSD data of lavas (see section 3).

A wide range of morphologies is seen even in one mineral species. Diamond is a representative example, whose morphologies are classified by Sunagawa [21] into (1) single-crystalline type (octahedral, dodecahedral and twinned crystals);

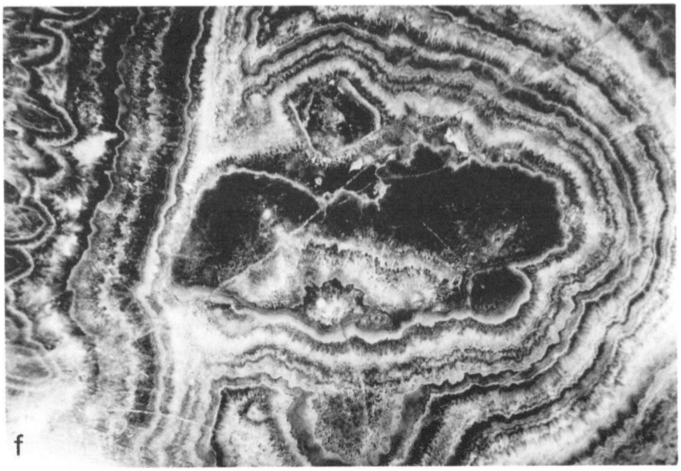

Fig. 11. Photographs of mineral specimens, exhibiting (a) fractal patterns (Mn oxide), (b) spherulite (native arsenic), (c) spherulitic aggregate of needle crystals (zeolite), (d) semi-spherulitic form (calcite), and patterns exhibited by these aggregates: (e) rhodochrosite and (f) quartz.

(2) polycrystalline type (spherulitic crystals, or randomly oriented crystal aggregates, such as borts, hailstone borts, stewartite, carbonado, etc.); and (3) a combined type with a single-crystalline core and a fibrous mantle formed by columnar or dendritic growth. The morphological variation represents the difference of growth conditions (see section 5.4). Figure 12 indicates schematically the morphologies exhibited by natural diamonds.

The origin of habit variation exhibited by polyhedral crystals of one mineral species has been the subject of investigations having a long history since the time of Steno. Figure 7 shows a representative example of habit variation due to the combination and development of three low-index faces exhibited by one particular mineral, pyrite, FeS_2. Due to the combination of three faces $\{100\}$, $\{210\}$ and $\{111\}$, in decreasing order of morphological importance, all these habits appear in nature. In addition to the variation of this type, pyrite crystals sometimes exhibit a malformed morphology (one or more pairs of faces develop much larger than other pairs of crystallographically equivalent faces, resulting in platy, prismatic or acicular habits, see fig. 8). The origin of the malformed morphology may be inferred to be due to either anisotropy in environmental factors such as flow of solution, or anisotropy in growth sites. Whiskers are a representative example of malformed morphology of crystals, resulting from anisotropy in growth sites. A eutectic liquid droplet serves as an anisotropic growth site in the case of the VLS (vapour–liquid–solid) mechanism, and outcrops of capillary play a similar role in KCl or NaCl whisker formation. Pyrite rarely occurs as a form of whisker, often kinked. Its origin has not been clarified, although it was shown that pyrite whiskers do not contain axial dislocations.

Since polyhedral morphology bounded by low-index faces are the morphology that is expected in the region where growth is principally controlled by the spiral growth

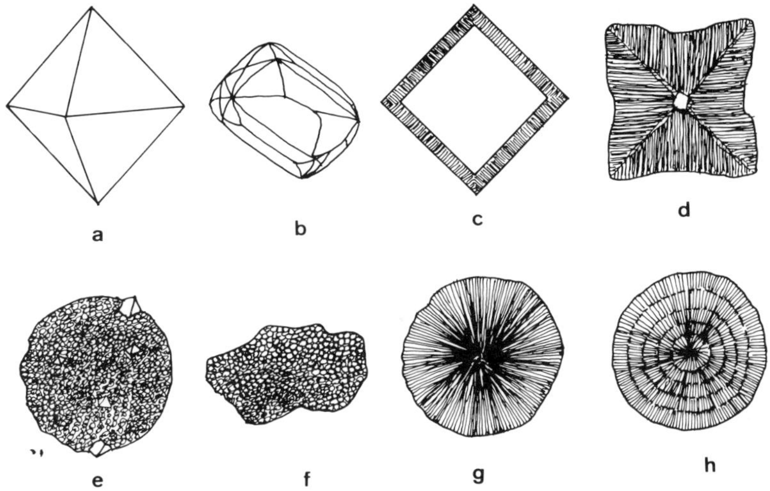

Fig. 12. Morphologies exhibited by natural diamonds. Single-crystalline type (a, b); combined type (c, d), and polycrystalline type (e–h); (e) framesite, (f) carbonado, (g) short bort and ballas, (h) hailstone bort.

mechanism (fig. 9), and the external form is determined by relative normal growth rates of faces concerned, the factors that control habit variation are those affecting the height and advancing rate of elemental spiral layers and the step separations. The main factors are now understood to be temperature, supersaturation, type of solvent, fugacity, impurity, phases and size of growth units [19]. The $\{100\}$ of pyrite is morphologically the most important face, on which layer-by-layer (spiral) growth can take place under any growth condition. Thus cubic habit appears most persistently, occurring under any conditions. The $\{210\}$ appears in most cases due to pile-up of edges of growth layers on $\{100\}$, and only under special conditions can spiral growth take place on this face. The $\{111\}$ behaves similarly to the $\{210\}$, but under different conditions. As a result of such a tendency, pyrite crystals show cubic habit when they are small or under conditions not favorable for pyrite crystallization, such as in the periphery of an ore deposit, or in country rock. Based on this knowledge, habit variation of pyrite crystals was investigated in relation to the mode of occurrence, and utilized for ore prospecting [22]. Similar types of investigation have been carried out extensively on other minerals as well, particularly in the Eastern World. The names of "typomorphology" and "ontogeny" [23] are given to such investigations.

Recent interesting observations on the morphology of crystals of extra-terrestrial materials are observations of whiskers or of disk-like morphology exhibited by cosmic (interplanetary) dust particles. Minerals of meteorites usually do not show euhedral morphology, although a recent cathodoluminescence topographic investigation of olivine crystals in a chondrule exhibited polyhedral zoning. However, Bradley et al. [24] reported whiskers and disk-like morphology of enstatite in interplanetary dust particles, and interpreted this as indicative of growth of these particles by a spiral growth mechanism from the vapor phase. Later, a similar morphology was also observed on low-Ni- pentrandite crystals of cosmic dust.

4.3. Surface microtopography

The surface microtopographs of low-index crystal faces of minerals have been investigated extensively mainly by the present author [3, 25], using sensitive optical (phase contrast, differential interference contrast, etc.) and interferometric methods, and the decoration technique of transmission and scanning electron microscopy. These methods enable one to visualize and measure the steps of growth spirals or etch features of nanometer scale in height, down to 2.3 Å [26]. Recently developed STM (scanning tunnelling microscopy) and AFM (atomic force microscopy) are also powerful techniques to visualize such small steps, together with an ability to visualizing atomic structures. Phase-shifting interferometry, which can visualize the profile of surface microtopographs of nanometer height, is another very recently developed optical method applied to surface microtopographic investigation [27]. These are, however, still at an early stage in their application to mineral crystals. In this section, we briefly summarize only the results of observations on mineral crystals made by the former methods.

The aim of surface microtopographic investigations is to verify, at an atomic level, the growth mechanism of mineral crystals, at least at the latest stage of their formation. Whether there are any post-growth events, such as etching or dissolution is another objective of these investigations. The earliest surface microtopographic investigations of mineral crystals were made in the 1920s, and attention was mainly paid to the symmetry of natural etch figures (etch pits and etch hillocks) observed on the surfaces of well-formed polyhedral crystals, mostly of pegmatite origin [28, 29]. After the birth of the BCF theory [30], growth spiral patterns have attracted much interest. Mineral crystals whose surface microtopographs were investigated in some details are hematite grown by a natural CVT process due to post-volcanic action, diamond and anorthite grown in magmas, beryl, topaz, quartz crystals occurring in pegmatites, phlogopite in druses of lavas, sphalerite, pyrite, quartz and apatite formed in hydrothermal veins, zeolites formed by diagenesis, kaolin group minerals (kaolinite, dickite, nacrite) and sericite (micron-sized muscovite) formed by hydrothermal metasomatism, and muscovite (both sericite and porphyroblastic white mica) in regional metamorphic rocks. Minerals representing most types of geologic processes have been covered.

Surface microtopographic characteristics of natural and synthetic crystals, as well as of those grown from vapor and solution phases were summarized and compared by Sunagawa [25, 31]. When growth spirals observed on the same face of crystals of the same mineral grown from a similar type of environmental phase are compared, natural crystals generally exhibit a more polygonized morphology and a wider step separation than synthetic crystals, indicating crystal growth of minerals at a smaller driving force. The ratios of step separation versus step height of mono-molecular growth spirals are 10^4–10^6 on crystals grown from vapor phases, 10^2–10^3 on those grown from aqueous and hydrothermal solution phases. The ratios are in general one or two orders of magnitude larger on natural minerals than on synthetic crystals on the same face of the same crystal species [31].

It should be stressed that, on the low-index faces (morphologically the most important faces), of all the minerals mentioned above, except muscovite of regional metamorphic origin, growth spirals of some kind have been observed, in spite of the fact that they come from a wide range of geologic origins. Beryl crystals from a pegmatite served as the first sample to prove the spiral growth theory and show mono-molecular growth spirals on the prism [32], and basal faces. The surface microtopographs of hematite of post-volcanic origin were investigated most extensively [26, 33], and growth spirals of 2.3 Å in step height (the height of an Fe_2O_3 single layer, $\frac{1}{3}$ unit cell height) were observed. Phlogopite occurring in voids of lavas, sphalerite and apatite from hydrothermal veins, kaolin group minerals formed in hydrothermal metasomatic deposits, zeolites of diagenetic origin, all exhibited growth spirals with mono-molecular step height. Since, in the case of rock-forming minerals of magmatic origin, it is difficult to secure their as-grown surfaces, the exceptional examples whose surface microtopographs were studied are diamond and anorthite megacrysts. Anorthite showed growth spirals of macro-steps. In the case of diamonds, one-to-one correlation was established, on crystals from Siberia which received only a slight dissolution, between the summits of growth hillocks on the {111} faces, consisting of triangular growth layers with step heights less than 5 Å and the outcrops of screw-type dislocations, verifying the spiral growth mechanism [34]. In fig. 13, an example of such a correlation between a surface microtopograph and an X-ray topograph of Siberian diamond is shown.

These observations demonstrate that, in most geologic processes including magmatic crystallization and metasomatic processes, vapor or solution growth at a small driving force prevails. It is to be noted that growth spirals with a mono-molecular step height have been universally observed even in the case of minerals formed by hydrothermal metasomatic processes (sericite and kaolin group minerals [35]), not only on crystals grown in environments free from vapor, aqueous, hydrothermal or high-temperature solution phases.

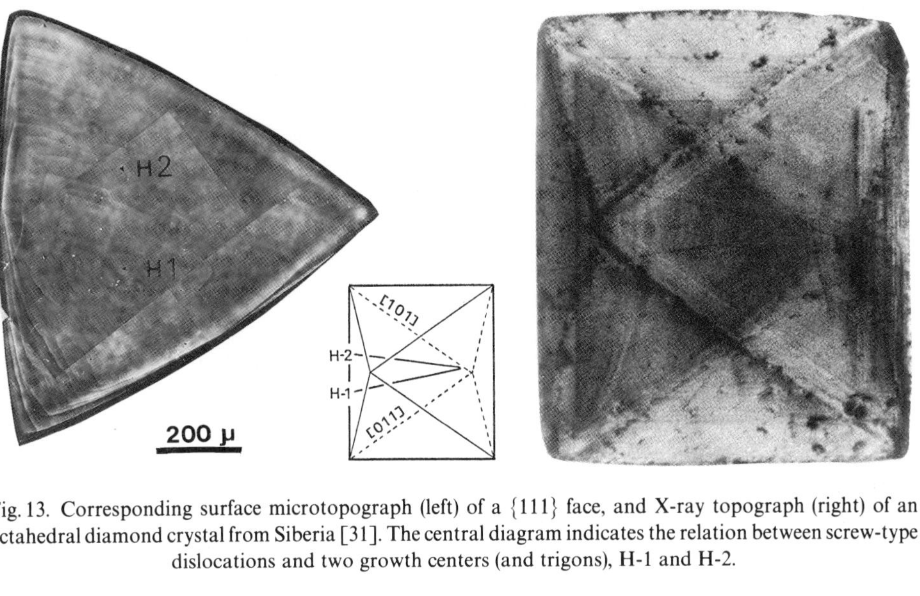

Fig. 13. Corresponding surface microtopograph (left) of a {111} face, and X-ray topograph (right) of an octahedral diamond crystal from Siberia [31]. The central diagram indicates the relation between screw-type dislocations and two growth centers (and trigons), H-1 and H-2.

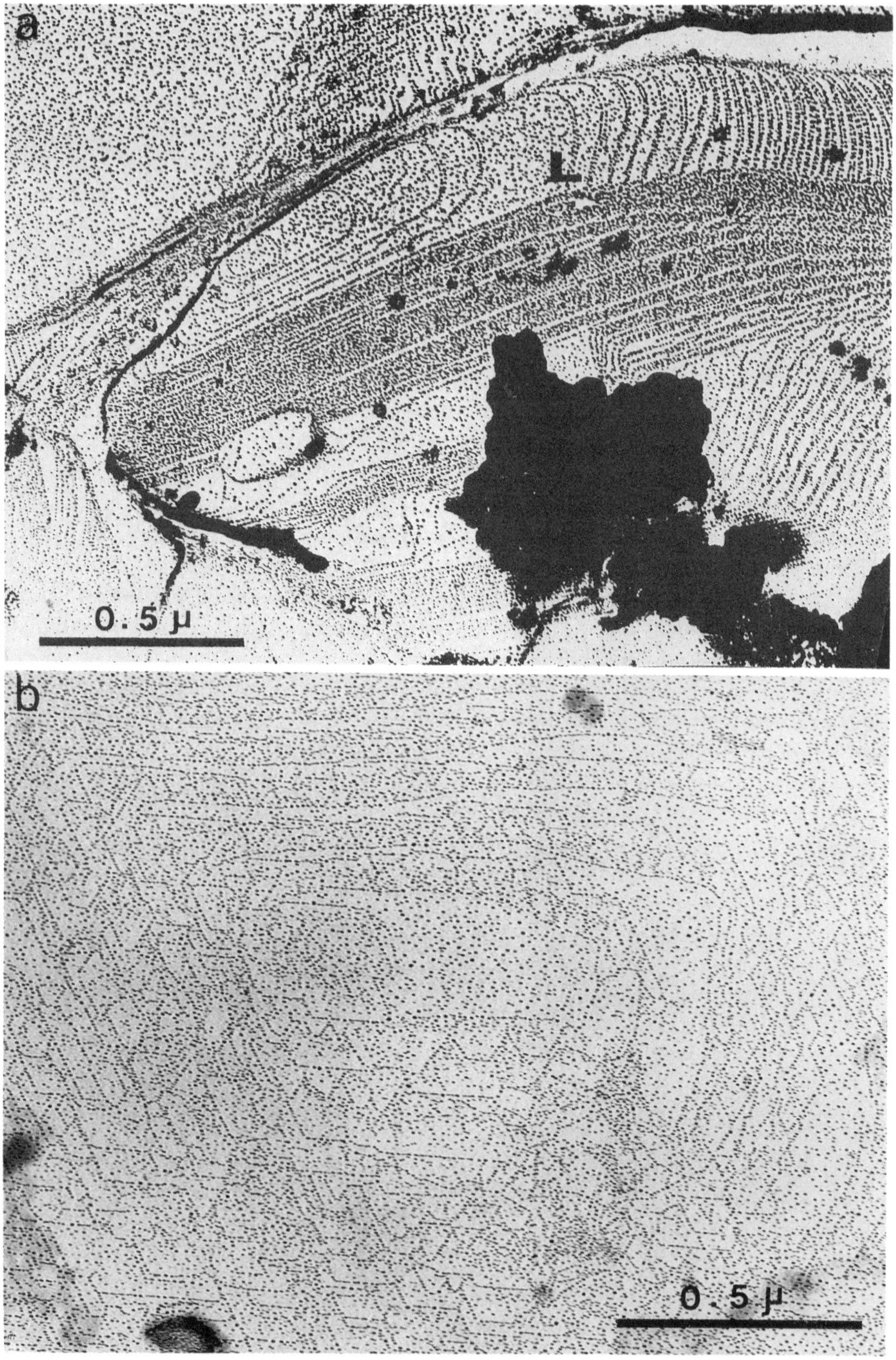
a
L
0.5 μ
b
0.5 μ

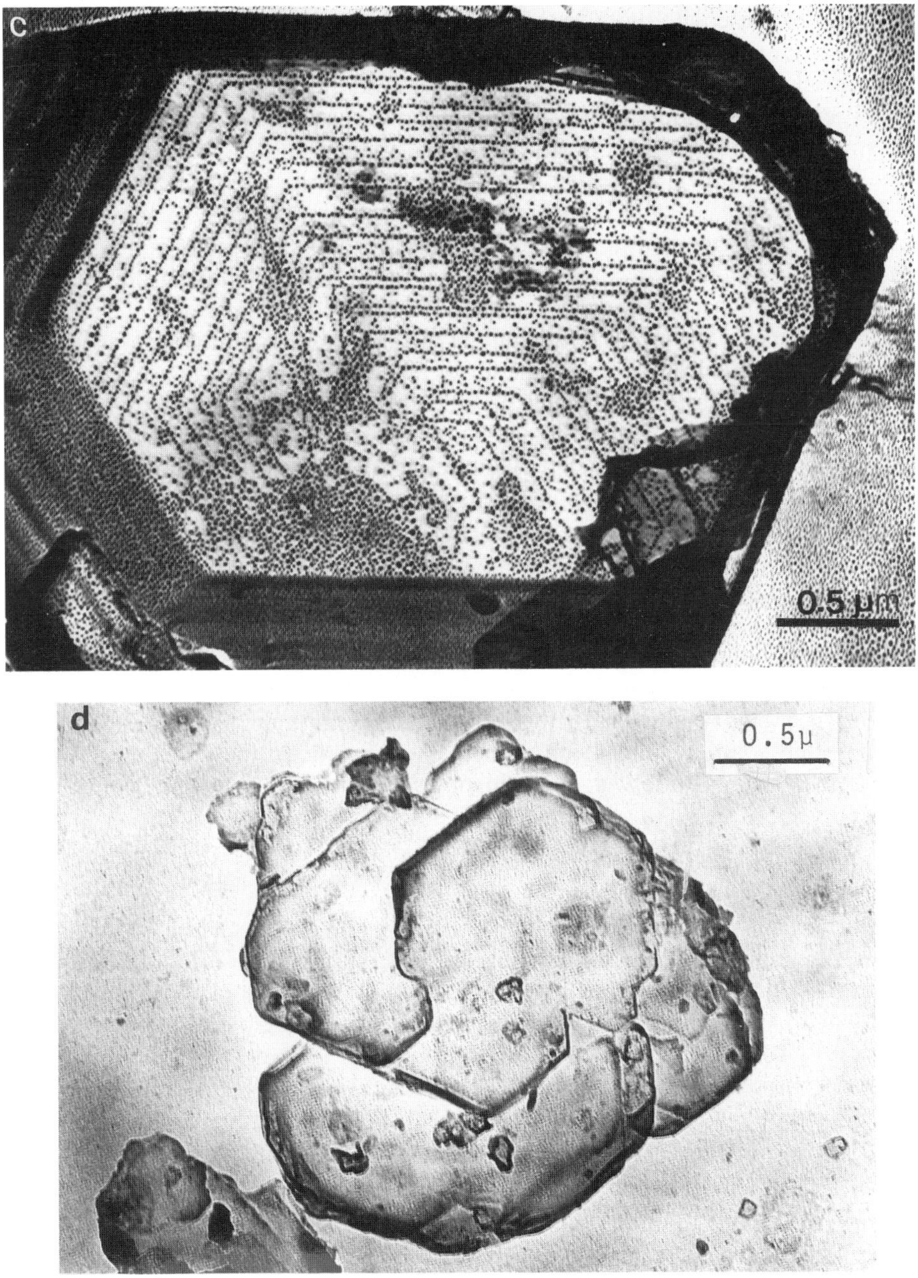

Fig. 14. Surface microtopographs of the {001} face of sericite and porphyloblastic white mica, representing growth (a) and dissolution (b) [14]. Paired steps in (a) indicate that the crystal is the 2M polytype consisting of two unit mica layers. (c) Dickite of hydrothermal metasomatic origin, and (d) chlorite from hydrothermal vein deposit.

In contrast to the above, growth spiral patterns have never been observed on the {001} face of muscovite (both sericite and porphyroblastic white mica) of regional metamorphic origin [14]. They exhibit concentric smooth or rugged step patterns with 10 Å step height, fig. 14a, b. It was even possible to identify different polytypes, e.g., 1M, 2M, 3T, etc., by the modes of pairing or tripling of the steps. The smooth concentric steps (fig. 14a) represent growth steps, whereas the rugged ones (fig. 14b) are dissolution steps. The characteristic concentric, but not spiral, patterns were inferred to be due to the development of growth or dissolution steps originating from the presence of other solid particles on the growing surfaces. In contrast to this, the same mineral, sericite (muscovite) and kaolin group minerals (fig. 14c) of hydrothermal metasomatic origin, universally exhibits growth spiral patterns with mono-molecular step height, in spite of the fact that crystallization in both regional metamorphism and hydrothermal metasomatism started from solid rocks [35]. The difference between the two types was inferred to be due to the fact that a complete dissolution of the original rock and precipitation of sericite and kaolin group minerals from a thin film of solution at least at the front of metasomatism occurs in hydrothermal metasomatism whereas, in regional metamorphism, crystallization occurs only through interstitials of solid grains. The difference in pressure conditions between regional metamorphism and hydrothermal metasomatism may be pointed out as a possible cause for this difference.

It is interesting to note that individual single crystals of sericite or kaolinite group minerals with a diameter of around a micron and nicely polygonal outlines are universally encountered in hydrothermal metasomatic deposits, whereas sericite crystals are usually coalesced in a group when they occur in hydrothermal vein deposits (fig. 14d) [35]. This suggests that growth of crystals occurs under more static conditions and in a thin solution film in the former environment. In the latter environment, crystals grow in an agitated, violently flowing solution of larger volume.

4.4. *Internal morphology (perfection and homogeneity)*

Since the internal morphologies of single crystals of minerals, such as growth banding, zonal structures, growth sector boundaries, distribution of dislocations and other lattice defects are the record of their growth histories, these characteristics have been investigated extensively, with the hope of deducing their growth and post-growth histories. Various modern techniques to characterize lattice defects, perfection, and chemical homogeneity have been applied to mineral crystals, which include polarization microscopy, X-ray topography, laser-beam tomography, cathodoluminescence topography, ESR, EPMA and SIMS. To apply these techniques, nearly perfect single crystals are desired. Thus, most previous work of this kind has been made on crystals of gem-quality minerals.

Diamond crystals survive through their severe post-growth history and preserve their original as-grown characteristics, and highly perfect crystals are obtainable. Pegmatites produce not only gigantic crystals, but also highly perfect crystals to which various modern characterization techniques are applicable; examples are beryl, topaz, spodumene, quartz, etc. They grew freely in non-confined environments, and produced highly perfect crystals with polyhedral morphology. In these cases, the knowledge and techniques aquired by modern science and technology is directly applicable. Chemical and isotopic variations associated with physical variations can be investigated with ESR, EPMA or SIMS. Combining the results of both physical and chemical characterization, the morphological and chemical evolution during growth and the events in their post-growth histories have been deduced.

As a representative example of deducing the growth and post-growth histories of minerals based on the analysis of external and internal morphological charateristics of single crystals of minerals, diamond will be described in some detail in the following, as a case study.

(1) *Morphological types*: As described in section 4.2, natural diamond crystals can be grouped into (a) a single-crystalline type, (b) a polycrystalline type, and (c) a combined type. Crystals of category (a) were formed under near-equilibrium conditions, at small driving force; those of category (b) were formed under conditions with a much higer driving force than (a); and those of category (c) experienced two different conditions, namely at first they grew under the same conditions as (a), and later experienced the conditions of (b) [21]. Crystals of category (b) usually have radiating or columnar and sometimes concentric texture, consisting of elongated diamond crystals, micron-sized in width, and interstitial foreign minerals like apatite. Crystals of category (b) and (c) are encountered more commonly in eclogite xenolith than in ultra-mafic rocks.

(2) *External morphologies*: Most single-crystalline natural diamond crystals are rounded. Dodecahedral faces of larger crystals are heavily rounded and divided into four rounded $\{hhl\}$ faces, which are evidently due to dissolution starting from simple octahedral crystals. They are dodecahedroidal rather than dodecahedral faces. Most dodecahedroidal faces are explained as due to the dissolution process. But on smaller, micron-sized crystals, the $\{110\}$ faces often appear as flat, but striated faces. These are dodecahedral faces that appeared during the growth process, due to the pile-up of growth steps developing on the $\{111\}$ faces, and not by dissolution. On the $\{111\}$ faces of crystals which received only a slight dissolution (i.e., simple octahedral crystals with only slightly rounded corners and edges), triangular growth hillocks with the same orientation as the triangle of the $\{111\}$ face are observed. At their summits, tiny trigons are sometimes observed, which could be correlated with the outcrops of screw-type dislocations, fig. 13 [34]. No growth layers have so far been observed on the $\{100\}$, or more correctly cuboid faces of natural diamond crystals which show only rugged surfaces. This indicates that, in the growth process of natural diamond, $\{100\}$ faces were not smooth interfaces and layer-by-layer growth did not take place. This is also supported by the hummocky growth banding seen in the $\{100\}$ growth sectors in center-cross diamond [36] (see point (6), below).

(3) *Dissolution features*: On $\{111\}$ faces, trigons (triangular point-bottomed and flat-bottomed etch pits) are universally observed, fig. 15. On dodecahedroidal faces, ditches, net-works, superimposed ring-patterns, fig. 16, are observed. All these surface characteristics are due to a dissolution process that natural diamond crystals experienced, most probably during the ascension from the depth to the Earth's surface.

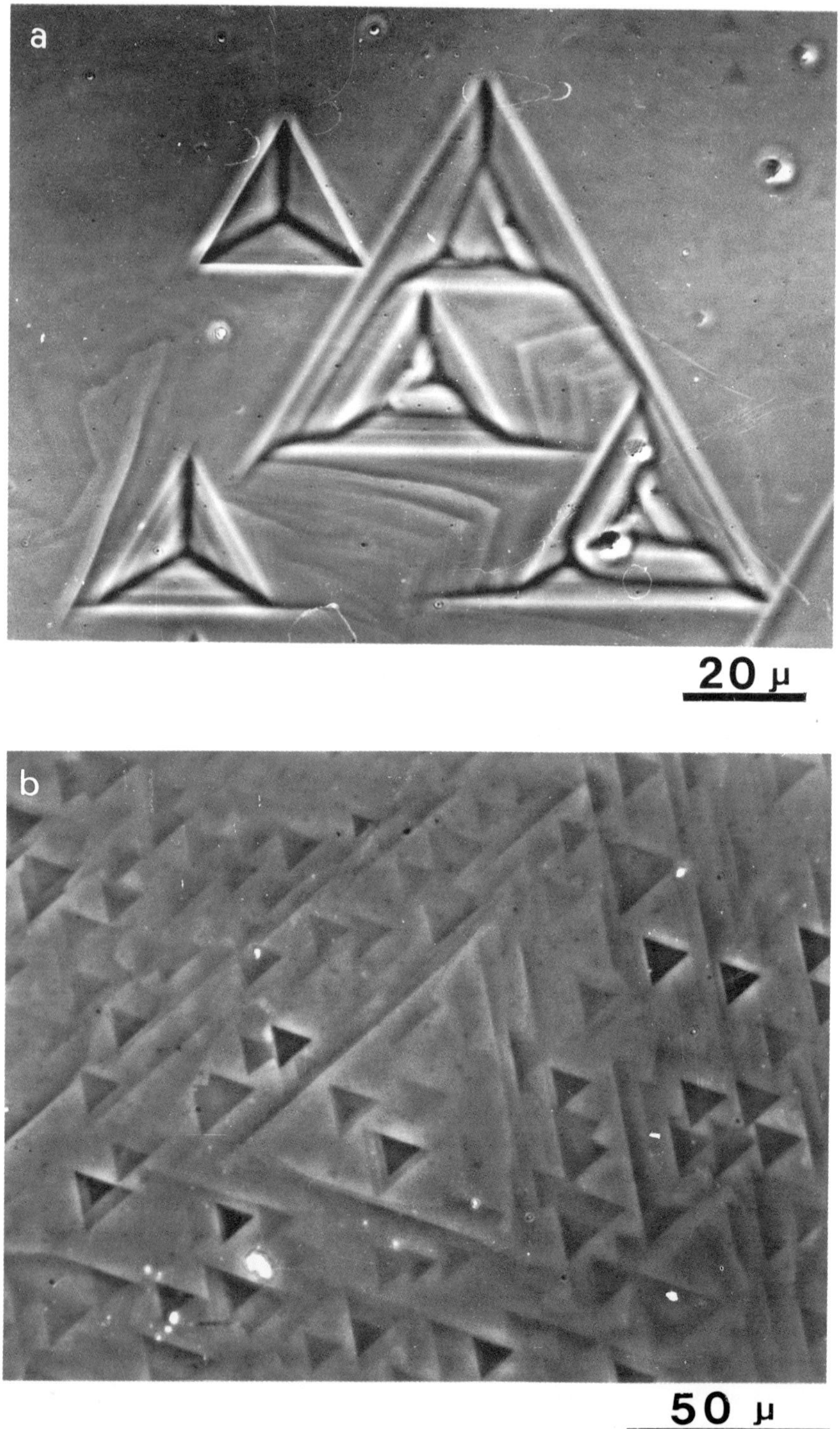

Fig. 15. Two types [(a) point bottom and (b) flat bottom] of trigons commonly observed on the {111} face of natural diamond [21]. Phase contrast.

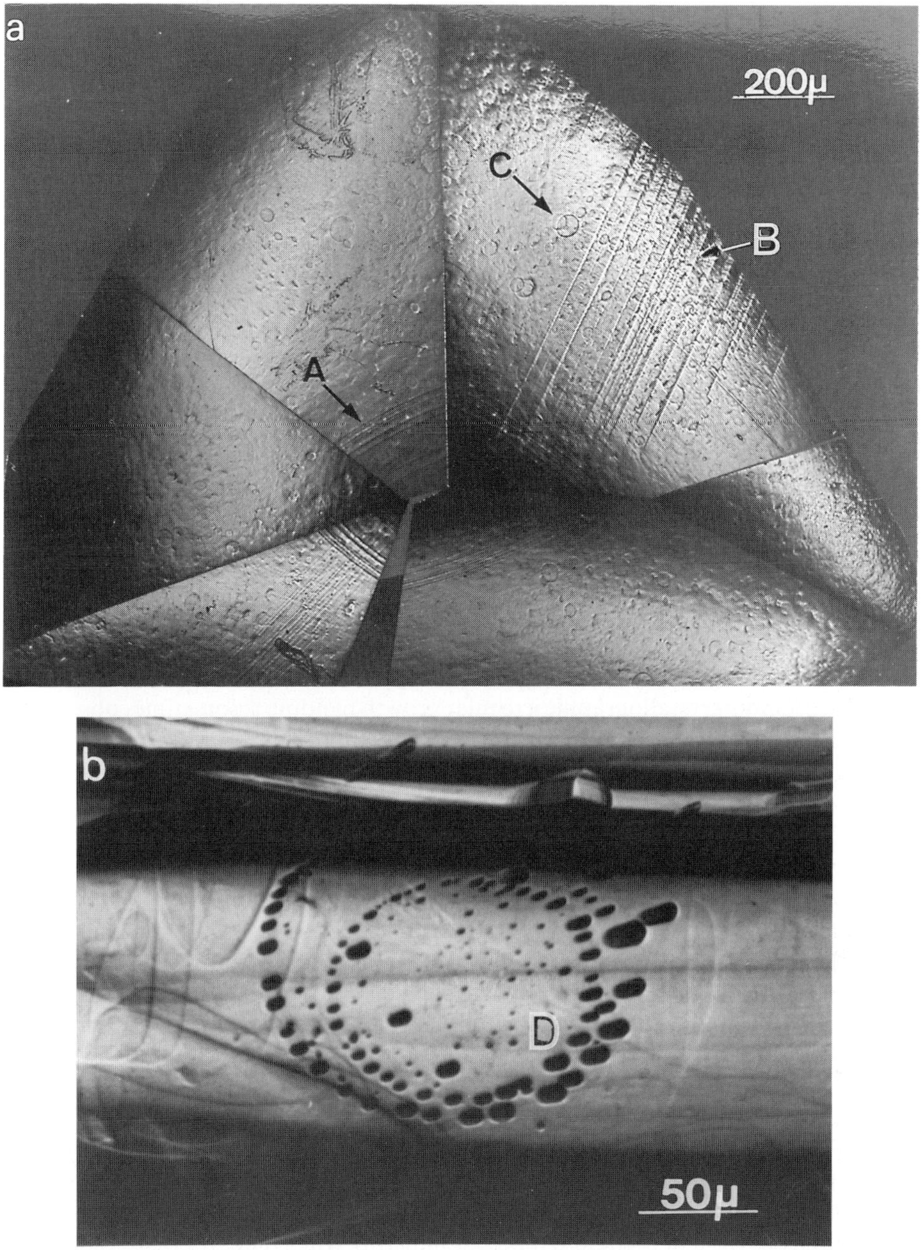

Fig. 16. Dissolution features observed on dodecahedroidal faces of natural diamond [21]. Note the occurrence of ditches (A), networks (B), superimposed circular ditches (C), and ring patterns (D).

(4) *Inclusions*: Syngenetic solid inclusions in natural diamonds have been investigated in a great detail [37]. Two different categories have been classified in respect to the types and chemistries of mineral inclusions in diamonds; ultra-mafic and eclogitic suites. This indicates there are at least two chemically different media in which natural diamonds grew.

(5) *Nitrogen platelets*: In type-Ia diamonds, nitrogen platelets, with a size of 1000 Å^2 are universally observed, the origin of which was interpreted as precipitation due to the annealing process that natural diamond crystals experienced either in the mantle or during the process of ascension. The bands with full-of- and free-of-nitrogen platelets alternate usually parallel to {111} in one single crystal, with electron-microscopically sharp boundaries.

(6) *Growth banding*: Growth banding becomes clearly discernable (a) by etching the polished surfaces, in the form of alternating heavily etched and less etched bands, (b) by X-ray topography, and particularly (c) by cathodoluminescence topography, in a form of alternating different contrasts. Straight banding parallel to {111} is universally observed in the {111} growth sectors, whereas hummocky banding always appears in the {100} growth sectors [36]. This is so when both {111} and {100} growth sectors co-exist in a single crystal, in which cases a center-cross pattern appears. This indicates that, under the conditions of growth of single-crystalline natural diamonds, the {111} interface was a smooth interface on which layer-by-layer growth occurs, whereas the {100} was a rough interface on which an adhesive-type growth mechanism operated. This makes a strong contrast to the growth of synthetic diamond crystals from metal solution. Also non-crystallographic irregular growth banding is sometimes observed in natural diamond crystals. As to the origin of irregular growth banding, an appropriate explanation has so far not been given. There is a possibility that the conditions of natural diamond growth are situated close to roughening transition points (both thermodynamic and kinetic roughening transitions). Further investigation is needed.

(7) *Dislocations and planar defects*: Growth-induced dislocations radiating in a form of bundle from a point at the center or an off-center position and running nearly perpendicular to the {111} surface are often observed in single crystalline diamond of high perfection. One-to-one correlation between the outcrops of dislocations and the summit of a growth pyramid or concentrated area of trigons was established, indicating the role of dislocations in the growth and dissolution of diamond crystals [21, 34, 36]. Dislocation tangles are also observed by transmission electron microscopy, which suggests that natural diamond crystals experienced plastic deformation during their post-growth history.

(8) *Geochemical data*: With increasing data on carbon isotope ratios of natural diamond crystals, it has become more evident that there should be both inorganic and organic sources as to the origin of carbon in diamond [38]. Geochemical data also indicate that the ages of diamond and the kimberlite which contain the former are in many cases distinctly different, indicating that kimberlite (and lamproite, another mother rock of diamond) played only the role of diamond carrier [38]. However, this argument by petrologists has very recently been questioned by the success of diamond synthesis in a mixture of kimberlite powder and graphite in the diamond-stable region. The morphological characters of these synthetic diamonds are very close to those of

natural diamonds and distinctly different from synthetic diamonds grown from metallic solutions.

Based on the above information, the following may be suggested at present as to the growth and post-growth histories of natural diamond crystals:

(1) Natural diamond crystals were formed in deep-seated silicate magmas of two distinct compositions, ultra-mafic and eclogitic, in the diamond-stable regions, at depths greater than 120 km. They grew from a high-temperature solution phase, and not by solid-state crystallization, nor from a pure melt of carbon, nor through metamorphic processes.

(2) Possible carbon sources are two-fold: an inorganic and magmatic source which originated from the beginning of the earth, and an organic source, which was subducted into the mantle through subduction of plate. In the latter case, eclogite is the most likely candidate as the environment of diamond growth, where a high concentration of carbon is expected.

(3) Two different growth conditions can be assumed for the growth of diamonds. One is the condition with smaller driving force, under which single-crystalline diamonds grew. The other is the condition with relatively high driving force, where polycrystalline diamonds are more likely formed. The latter conditions are likely in the eclogitic suite, whereas the former corresponds to the ultra-mafic suite.

(4) Diamond crystals thus formed were brought up by rapid ascent of kimberlite or lamproite magmas, which trapped solidified magma containing diamond and acted as a carrier. Such rapid ascent of deep-seated magmas is possible only in the continental shield, and not in a subduction zone or orogenic belt. This is in agreement with the distribution of diamond localities on the Earth.

(5) During the ascent, diamond crystals experienced an annealing process and precipitated nitrogen platelets (this could also occur in situ in the mantle after the magma was solidified), weak plastic deformation to create tangled dislocations, partial dissolution to result in rounded morphology and the formation of trigons and other dissolution features. Finally diamond crystals were quenched due to adiabatic explosion of the ascending magma, which contained volatile components such as CO_2.

Diamond crystals have been synthesized from metal or alloy solutions (metals are often regarded as catalysts, but they are in reality solvents) under high-temperature and high-pressure conditions. By using seed crystals and the temperature gradient method, large synthetic diamond crystals of fairly high perfection, the present record exceeding 14 carats, have been successfully grown in several laboratories. It is to be anticipated that gem-quality synthetic diamonds will appear on the gem market in the very near future, which poses problems for gem identification. However, one should not be afraid, since both natural and synthetic diamonds grew in different media, and thus exhibit different morphological characteristics. Because synthetic diamonds grow in a metal or alloy solution, layer-by-layer (and spiral) growth takes place on both $\{111\}$ and $\{100\}$ faces. Growth spirals with large step heights are observed on both faces. In contrast, in the case of natural diamond, only the $\{111\}$ face can grow by a layer-by-layer or spiral growth mechanism. The difference comes from the difference of solvents between the two, and the resultant possibility of surface reconstruction of {100} interfaces in metal solutions. As a result, the two diamond crystals exhibit different growth sectors, which can be detected easily by cathodoluminescence topography or even polarization microscopy.

Similar types of investigation as in the case of diamond have been made, although not as extensive as in the case of diamond, on quartz, beryl, topaz or spodumene crystals of pegmatite or pneumatolytic and hydrothermal vein origins, etc., which produce highly perfect crystals. It has been revealed from these observations that a sequence of growth–slight dissolution–regrowth took place in the process of formation of a pegmatite. The direction of flow of nutrient vapor or hydrothermal solutions invaded into the void was also analysed. Difference of characteristics between the same mineral crystals of different origins has also been investigated. A nice example exhibiting such differences is of beryl crystals from pegmatite and the same mineral, emerald crystals formed in metamorphic rocks (e.g., Africa, Ural, etc.) and in calcite veins (Colombian). This knowledge is useful in gem identification. So the knowledge obtained through the investigations of mineral crystal growth has made a great and essential impact to the discipline of gem identification.

Most of the above examples are on highly perfect crystals, which can be formed in rather restricted geological processes, like pegmatite formation. The crystals formed in ordinary geological processes are usually far less perfect to visualize clearly their heterogeneities and imperfections by applying very sensitive modern techniques of characterization. However, if heterogeneities and imperfections can be visualized by applying appropriate methods, even if they are less sensitive, the information contributes greatly to the understanding of geological processes. A nice example can be seen in recent investigations on the inhomogeneity of garnet crystals in regional metamorphic and contact-metasomatic rocks.

Both regional metamorphism and contact metasomatism had been analysed based on a concept that the chemical reaction was completed to its equilibrium, assuming that crystals became chemically homogeneous. Owing to the development of microanalytical techniques, such as EPMA, it gradually became clear that the crystals, even in these rocks, such as garnet are also chemically heterogeneous and exhibit a morphological evolution associated with the growth process. It was also possible to visualize morphological evolution during the growth process from smooth interface growth, morphological instability of this interface associated with inclusion trapment, cellular growth and the following stabilization of the interface [39]. Such morphological evolution during the growth process is more distinctly discernible in contact-metasomatic garnet than in regional metamorphic garnets, due to a prolonged annealing process in the latter. Although less distinct, it has recently been observed in the latter case that, on the original growth-induced heterogeneity, is superimposed a heterogeneity induced by the later diffusion of elements during annealing. However, from the difference of element distribution patterns, it was possible to deduce post-growth differential diffusion rates depending on elements during the annealing process. Distributions of less mobile elements show a polygonal distribution pattern in a single crystal, reflecting the original growth zoning, whereas distributions of more mobile elements show a less polygonal, rounded pattern.

It should be noted that there are no mineral crystals which are perfect and thoroughly homogeneous. They exhibit oscillatory changes in physical perfection and chemical homogeneity from the centre to the outside of the crystal, normal (in accordance with the expectation from the relevant phase relation) or reverse (opposite to the above) chemical zoning from the center to the periphery, or different element partitioning in different growth sectors. Each mineral crystal has its own personality, i.e., characteristic heterogeneity and imperfection, which are the record of its growth and post-growth histories.

The information obtainable through this type of investigation is essentially deductive and qualitative. We have at present no method to evaluate quantitative the absolute growth rates or the time required to complete the growth of a mineral crystal based on the data of characterization of single crystals. One exceptional example was reported by Nassau and Jackson [40] on Trapiche emerald from Colombia, which consists of a clear central core of beryl and a sectored outer area, consisting of beryl and albite. They deduced that the time required for complete changes from central clear core to sectored outer area was a matter of minutes. But this is the time required to change from one state to another, and not the time to complete whole crystallization. One may thus argue that millions of years or a matter of minutes, hours, days, months or years were required for the formation of, say, diamond crystals.

We may obtain through isotopic data the difference in the absolute ages of different geologic epochs (such as the ages of diamond crystallization and that of the magma which brought up the diamonds; the two are not the same), but isotopic age determination is not sensitive enough to determine the difference in ages between the onset and the end of crystallization of one particular diamond crystal.

Through a comparison of the surface microtopographic characteristics of as-grown faces between natural and synthetic crystals of quartz, however, the present author prefers the concept of a matter of years to that of millions of years, at least for the formation of natural quartz crystals occurring in pegmatites and hydrothermal ore veins. But this is still a qualitative argument.

Based on the CSD analysis of textures of igneous rocks and geological observations, Cashman [12] evaluated the time required to complete crystallization of rock-forming minerals in a lava and in contact-metamorphic rocks to be of the order of 10 to 100 years. Although the analyses and argument to evaluate the time required to complete the crystallization of minerals is still in an embryonic state, it is noteworthy that the time so far evaluated is a matter of years and not of a million years.

5. *Experimental methods to simulate natural crystallization*

Since we are unable to observe the growth process of mineral crystals, laboratory simulations of mineral crystallization are also important. Along this line, various experimental methods have been devised. A brief explanation will be given in this section of these methods. In section 5.1, the in-situ observation methods of crystallization taking place in aqueous and hydrothermal solutions at ordinary to moderate temperatures and atmospheric to moderate pressures are described, and geological processes to which the results are applicable are discussed. Section 5.2 deals with dynamic crystallization experiments and in-situ observation methods at high temperatures, whose aims are investigation of magmatic crystallization. In section 5.3, experiments aimed at solving the problems of planetary science are briefly reviewed.

5.1. *In-situ observation of crystallization from aqueous solution*

Experimental verification of crystal growth mechanisms has been one of the main subjects in the modern science of crystal growth. For this purpose, the following have been the main methodologies, in addition to deductive analyses of growth mechanisms based on characterization: (1) investigations of the relations between the rates of nucleation, growth or dissolution and the driving force and the analysis of the obtained data compared to theoretical expectations, and (2) in-situ observation of growth or dissolution processes. Owing to less difficulty in controlling the growth parameters (temperature, pressure, purity of solution, supersaturation, etc.), growth of simple ionic crystals, such as NaCl, $NaClO_3$, $Ba(NO_3)_2$, alum in aqueous solution (and sometimes in organic solution) has been investigated principally. Through such investigations, the presence of a diffusion boundary layer around a growing crystal, concentration gradient in the boundary layer, concentration distribution around a growing crystal (Berg effect [41]), layer and spiral growth processes on low-index, smooth interfaces, roughening transition of smooth interfaces have been visualized or verified. The relations between growth or dissolution rates and supersaturation have been measured precisely to facilitate theoretical analyses.

In the above two types of methodologies, investigation by method 1 is in principle an indirect one, since rate measurements have to be done from the outside, and neither the process of crystallization nor the perfection of the crystal are exactly known. The earlier method 2 was good only for demonstration purposes, and precise condition control and rate measurements were not possible. So, by combining the merits of the two methods 1 and 2, a new in-situ observation method of crystal growth in aqueous solution was designed in several laboratories, including Tohoku University, the author's former affiliation [8]. With these methods, a number of problems have so far been investigated at a resolution of nanometer order and a temperature control better than 0.01°C. The outline, ability, and advantages of the in-situ observation methods are summarized in table 2, together with the in-situ observation methods (and ex-situ methods) at high temperatures, a description of which will be given in the next section. The problems so far studied in relation to crystal growth mechanisms from aqueous solution include [8]:

(1) Cluster formation of nanometer size in a solution was detected by laser light scattering, and the cluster size and density and the diffusion constant were measured by Raman spectroscopy.

(2) The diffusion boundary layer around and buoyancy-driven convection from a growing or dissolving crystal was visualized by a Schlieren technique, and the concentration gradient in the diffusion boundary layer and the supersaturation very close to a growing interface were measured by the use of Mach–Zehnder interferometry.

(3) The effects of solution flow upon the growth or dissolution rates, thickness of diffusion boundary layer, behavior of buoyancy-driven convection, etc., were investigated.

(4) The advancement of spiral growth layers with heights of nanometer order was observed and the rates were measured using phase-contrast and interference-contrast microscopy and by Michelson interferometry.

(5) The in situ identification of the dislocation characteristics was accomplished by polarization microscopy and other methods.

(6) The normal growth rates of a face, of an individual growth hillock on the face, its slope and the advancing rate of the spiral layers were precisely measured by the use of Michelson interferometry.

(7) The relation between growth or dissolution rates and super- or under-saturation was thus measured precisely enough to be theoretically analysed on individual spiral growth hillocks, not simply of a bulk crystal or of a face.

Although mineral crystallization occurs in impure systems and under fluctuating growth conditions, the knowledge obtained through the above type of investigations on simple crystals growing from pure aqueous solution and under precisely controlled conditions are very useful, since mineral formation in surfacial conditions, in sedimentogenesis and diagenesis, and precipitation in hydrothermal solution and hydrothermal metasomatism occur, in principle, from an aqueous solution phase. We shall briefly summarize, in this section, essential problems involved in mineral formation in the sedimentogenetic–diagenetic processes based on the achievements obtained through the investigations of simple and pure aqueous solution systems. The readers may refer to [11] for a detailed description and discussion.

The sedimentogenetic–diagenetic processes can be classified as follows:

(1) Crystallization under surfacial conditions, such as precipitation of evaporates, etc.;

(2) precipitation of authigenic minerals in intergranular spaces of allogenic mineral fragments in sediments; and

(3) crystallization and transformation of minerals during diagenesis, catagenesis, or low-grade metamorphism.

In all these processes, precipitation of minerals occurs in an aqueous solution phase at ordinary temperature and pressure. Proceeding from (1) to (3), the volume of aqueous solution diminishes and crystallization takes place in a more confined environment. In (1), crystallization takes place in a large volume of aqueous solution when a supersaturated state is achieved principally by evaporation. So the nucleation often occurs on the meniscus surface of the aqueous solution, as well as on the surface of emerged or submerged foreign solid particles. Ceaseless movement of the solution is expected. In (2), crystallization occurs in the intergranular aqueous solution of allogenic mineral fragments, whereas in (3) crystallization proceeds through a much thinner film of aqueous solution. In (2) turbulent movement of solution on the micro-scale is expected. Temperature and pressure also increase from (1) to (3).

Under these circumstances, important points to be taken into consideration in understanding crystal growth in these processes are (1) the state of polymerization of

Table 2
In situ observation (investigation) methods of the crystallization process in an aqueous solution at ordinary temperature, in silicate liquids at high temperature, and by ex situ investigation methods. (After [9].)

	In situ				Ex situ
	aqueous solution, ordinary temperature		silicate system, high temperature		
		methods		remarks	remarks
Condition control					
T, range	$< 100°C$	micro-computer, thermocouple, thermostated	$< 1600°C$		homogeneous T distribution
T, fluctuation	$< 1/100°C$	or clean room	$\pm 0.3°C$	sharp T gradient, direct measurements of T, ΔT, R	indirect measurement of T, ΔT, R
P	1 atm		1–20 atm		
σ_b	$0.0x\%$	from solubility curve	$\Delta T(1°C)$		
σ_s	measured	Mach–Zehnder	not yet		
u	up to 40 cm/s	two-bath flow system	not yet	not controllable	
Observable phenomena in bulk solution					
Clustering	clusters of 5 nm	light scattering, Raman	observable	observable	nothing observable
Heterogeneous nucleation	observable		observable		
Convection	observed	Schlieren	observed		
Diffusion boundary layer	observed	Mach–Zehnder	observed		

Observable phenomena at the interface and in a bulk crystal					
Growth step pattern	steps of nm height are discernible	PCM, DICM	less visibility		nothing observable
Dissolution step and etch pit	steps of nm height are discernible	PCM, DICM	observed		
Morphological change	observed	PCM, DICM, PM, video	observed		
Growth sector	observed	PCM, DICM, PM, video	observed		
Growth bands	observed	PCM, DICM, PM, video	observed		
Inclusion trapment	observed	PCM, DICM, PM, video	observed		
Reaction	observed	PCM, DICM, PM, video	observed		
Dislocations	identified	PM, etch pit	not yet		
Measurable kinetic rates					
Nucleation					
incubation time	measured	light scattering	measured	direct observation on video	indirect
probability			measured		
Growth and dissolution					
R and $-R$, interface	measured	Michelson, video	measured	equal thickness fringes, video	indirect
v and $-v$, spiral steps	measured	Michelson, video	not yet		
R, v, p of individual growth hillock	measured	Michelson	not yet		

Abbreviations: T: temperature; ΔT: temperature difference between equilibrium and growth temperatures; P: pressure; σ_b: supersaturation of a solution apart from a growing crystal: σ_s: supersaturation of a solution adjacent to a growing interface: u: flow velocity of a solution; r: size of cluster; D: diffusion constant of clusters; R and $-R$: normal growth and dissolution rates of an interface or a growth hillock, or an etch pit; v and $-v$: advancing and retarding rates of a spiral layer or a dissolution step; p and $-p$: slope of a growth hillock and an etch pit; PM, PCM, DICM: various optical microscopies, polarization, phase contrast, differential interference contrast microscopy, respectively; Mach–Zehnder, Michelson are interferometries.

the crystallizing species, (2) the effect of flow and micro-convection, and of buoyancy-driven convection upon secondary nucleation phenomena. In the following, some of the results obtained by the in-situ observation methods which are useful in understanding nucleation and growth of crystals in the processes from (1) to (3) will be described.

Quartz crystals of high perfection occur in pegmatite, pneumatolytic and hydrothermal veins. The perfection and size of single crystals diminish with decreasing growth temperature, eventually becoming the micron-sized micro-crystalline quartz which composes agates and chalcedony. Upon further lowering of the temperature, SiO_2 forms amorphous silica, opal. Agates and chalcedony, and opal are representative forms of silica formed under sedimentogenetic and diagenetic conditions. Brazil twinning consisting of right-handed and left-handed quartz are persistent in these quartz crystals, but far less frequently encountered in quartz crystals grown under higher temperature. These general tendencies suggest a possibility of silica polymerization under sedimentogenetic and diagenetic conditions, whereas under pegmatitic and hydrothermal conditions, they are less polymerized [15].

By in-situ observation, Onuma et al. [42] investigated (1) the effect of solution flow upon the thickness of the diffusion boundary layer, supersaturation close to a growing

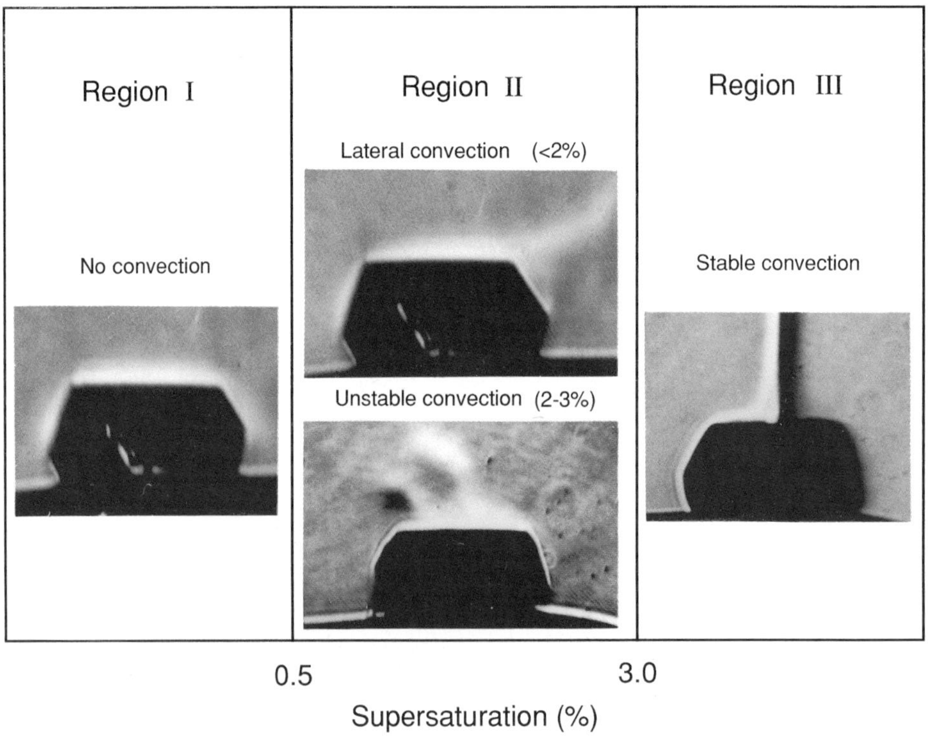

Fig. 17. Behavior of a diffusion boundary layer and buoyancy-driven convection in relation to bulk supersaturation. $Ba(NO_3)_2$ crystal growing from static aqueous solution [42]. By courtesy of K. Onuma.

interface (surface supersaturation), supersaturation distribution over the interface (Berg effect [41]), (2) behavior of buoyancy-driven convection and its effect upon surface microtopographs and perfection of crystals. In (1), it was demonstrated that, with increasing flow rate, the thickness of the diffusion boundary layer diminishes but never becomes zero; and the supersaturation distribution over the growing surface

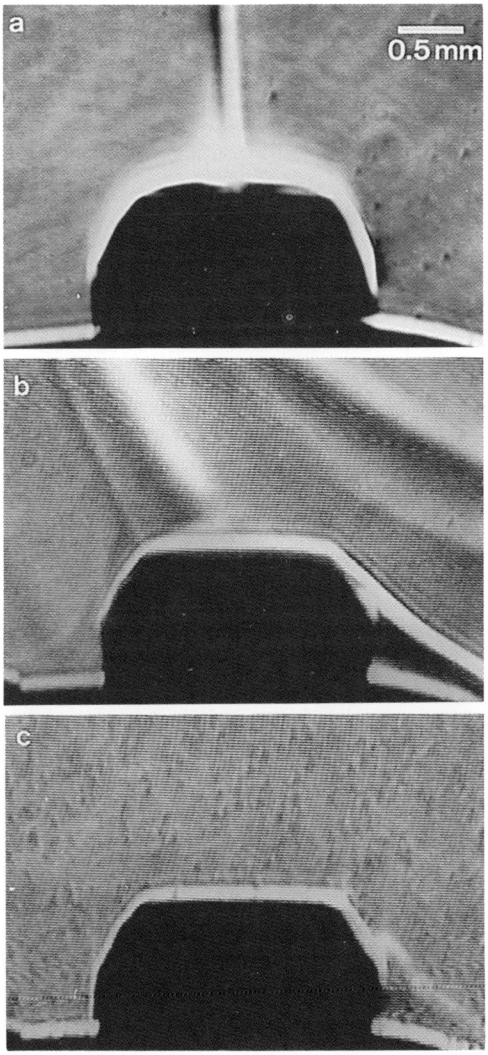

Fig. 18. Schlieren photographs showing blizzard-like concentration changes in the solution (b) by giving a stir to a static solution in which a crystal is growing with a stable buoyancy-driven convection plume (a), and the appearance of an avalanche of tiny crystalline particles due to secondary nucleation, after a few tens of seconds. $Ba(NO_3)_2$ solution [42]. By courtesy of K. Onuma.

becomes more homogeneous (the Berg effect is less enhanced). In (2), it was shown that, at low bulk supersaturation in a static solution, only the thickness of the diffusion boundary layer increases upon increasing the supersaturation but, above a critical value of supersaturation, buoyancy-driven convection starts to appear but behaves unsteadily up to the second critical supersaturation. Beyond this critical supersaturation, a buoyancy-driven convection plume rises steadily, and perpendicularly from the interface. In fig. 17, the behavior of buoyancy-driven convection in a static solution is shown in relation to bulk supersaturation.

When agitation was applied to such a system at higher supersaturation, they observed an avalanche of tiny particles in the solution after a few tens of seconds. During agitation, blizzard-like violent movement was observed on the Schlieren image, and after a few tens of seconds, an avalanche of tiny particles appeared as shown in fig. 18. This indicates that there were embryonic particles around the interface at higher bulk supersaturation which were dispersed into the bulk solution by agitation and acted as secondary nucleation sites.

Buoyancy driven-convection has a definitive effect upon the surface microtopography of the interface. Owing to the uplifting flow of buoyancy-driven convection plume, which results in depletion at the root of the plume, a hopper-like depression area appears on the surface. Figure 19 shows such a surface. The central amoeboid depression is due to the presence of a buoyancy-driven convection plume. Such a depression on a growing surface will act as preferential sites for the trapment of mother liquid inclusions. In the presence of flow along the interface, the root of the buoyancy-driven convection plume moves from the center of a face to the rear side of the face escaping from there to the bulk solution. All these sorts of phenomena should be expected to occur frequently in real systems.

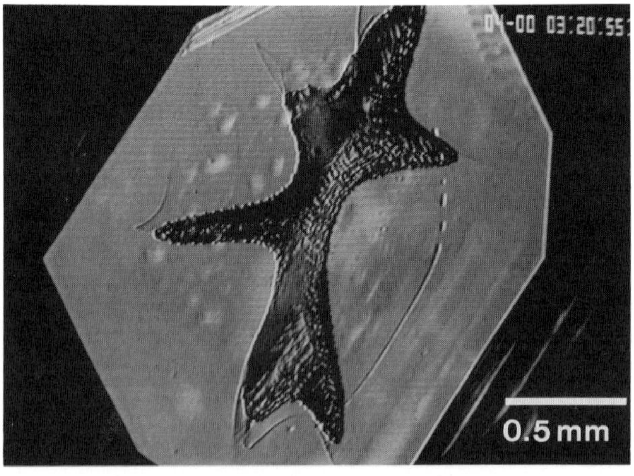

Fig. 19. Amoeboid depression formed at the root of buoyancy-driven convection plume on a {111} face of $Ba(NO_3)_2$ crystal [42]. By courtesy of K. Onuma.

5.2. Dynamic crystallization experiments and in-situ investigation methods of magmatic crystallization at high temperatures

Dynamical crystallization experiments were started in the 1970s (see, e.g., [4]) to experimentally reproduce various igneous rock textures in laboratories and to see the dependence of morphologies of silicate crystals upon the driving force (supercooling) and chemical composition. Textures of various igneous rocks and even those of chondrules in stony meteorites [16] were reproduced. In this type of experiment, either powders of simulated compositions or natural rock samples are used as nutrients. They are melted above the liquidus temperature in the form of a droplet in an electric furnace, followed by being given a desired supercooling at a desired cooling rate, and kept for a desired duration and quenching. The quenched products are investigated under optical microscopes to see the texture and by X-ray diffraction to identify the phases. Experiments of this type, however, gradually faded out because, although they were useful demonstrations, the data obtained were not quantitative or reproducible enough to be suitable for theoretical analysis.

Following the dynamic crystallization experiments, experiments on silicate systems to investigate the relationships between driving force and nucleation or growth rate were started. Conditions were well controlled, but the growth rates were measured on quenched samples, and the crystallization process itself was completely concealed. There was a trial experiment to directly observe the crystallization process at high temperature [43], but the visibility was low and the observations were limited to a narrow temperature range.

Dynamic crystallization experiments are in a sense static experiments, since no forced convection is applied. This was probably one of the reasons why naturally observed hyalopyritic textures of basaltic rocks, consisting of larger phenocrysts and micro-phenocrysts in the groundmass, were not reproduced by this method (see section 3).

An interesting breakthrough in igneous petrology was made by a simple experiment using an infra-red focusing furnace. The origin of andesite magma had been a subject of severe controversy, and a model of magma mixing was put forward, but no experimental verification supporting the model was obtained. Kouchi and Sunagawa [44] applied forced mixing of two chemically different magmas, basalt and dacite. An andesite magma was reprodced in the laboratory in a matter of an hour by this experiment. This was the first experimental verification of the magma-mixing model for the origin of the andesite.

In dynamic crystallization experiments, the process is not observable. Although the experiments were further developed to quantitatively analyse the kinetics involved, the measurements were made on the quenched products, not knowing what was happening during the process.

If the whole process of a magmatic crystallization could be observed and recorded, this would provide us with the most direct information. This was realized by the in-situ observation method at high temperatures developed at Tohoku University [6–9]. A thin film of silicate liquid of a desired composition is prepared using a heater of loop form set in an electric furnace installed on the stage of an optical microscope, and the crystallization process is observed in situ by transmitted light and recorded on video tape. This is equivalent to preparing magma and directly observing a magmatic crystallization under the optical microscope. Temperatures are controlled by personal computer with fluctuations less than 0.5°C at 1600°C. In fig. 20, a block diagram of the experimental set-up is shown.

In this in-situ observation method, the visibility is high because observation is made by transmitted light. It is further enhanced greatly by using various sensitive optical microscopies, such as polarization, phase-contrast and differential interference-contrast microscopes, and electronic methods. Even thin growth layers advancing on a crystal face of silicate crystals, and the diffusion boundary layer and convective flow around growing or dissolving crystals are discernible. Since the whole process is recorded with high visibility on video tape, and growth conditions are well controlled and precise kinetic measurements are possible. The in-situ observation method at high temperatures is a very promising one for investigating crystallization phenomena taking place in a magma, and to analyse the kinetics in a quantitative manner. Various

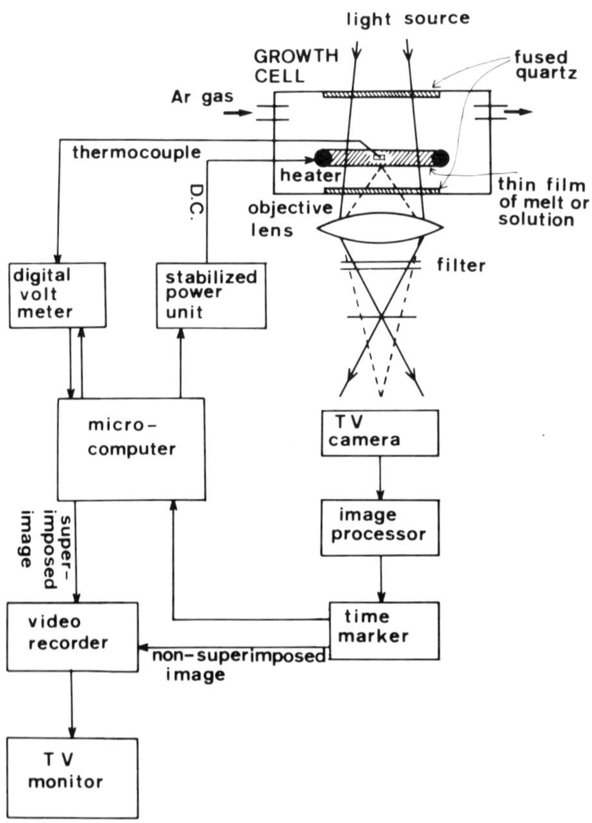

Fig. 20. A block diagram of an in-situ observation method at high temperatures [6–8].

unexpected pieces of information have been obtained, which include convective movement of crystals, growth and dissociation of spherulitic aggregates followed by re-growth in a sinlge-crystalline state, growth and dissolution processes of crystals, morphological changes depending on supercooling and liquid composition, nucleation and metastable nucleation phenomena, stability of metastably nucleated phases in the presence and absence of stable phases, growth and dissolution, etc. Only the phenomena relating to metastable nucleation will be described in the following. Readers interested in other aspects may refer to [6–9].

It is universally observed in several, and perhaps most, silicate systems that a phase or phases which are not expected from the relevant equilibrium phase relations appear prior to the appearance of the stable phase. For example, when a melt of pure anorthite composition, $CaAl_2Si_2O_8$, is supercooled, the phases appearing at first in the melt are not anorthite, but corundum, mullite, pseudo-orthorhombic and pseudo-hexagonal $CaAl_2Si_2O_8$ polymorphs, as shown in fig. 21 [17]. The only phase to be expected in this

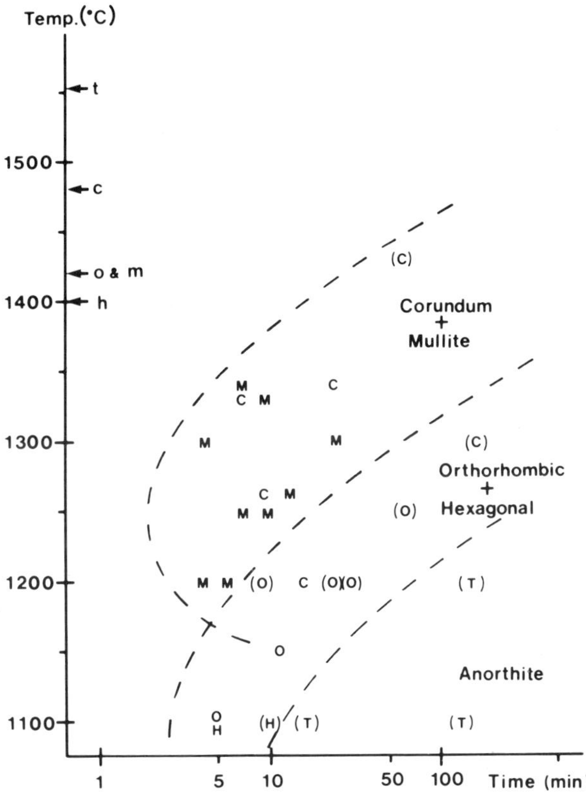

Fig. 21. Observed nucleation sequence under isothermal conditions in relation to temperature and time, in the pure anorthite melt. Phases nucleated secondarily are indicated in parentheses. The measured melting temperatures are indicated by arrows. t, T: anorthite, the stable phase; o, O: orthorhombic polymorph; h, H: hexagonal polymorph; c, C: corundum; m, M: mullite [17].

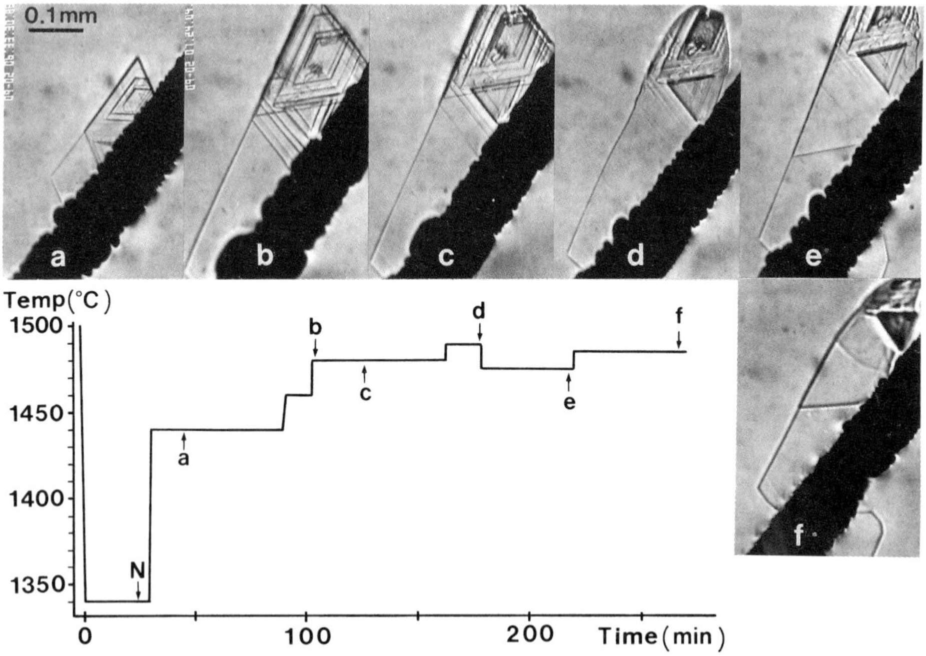

Fig. 22. An example showing how to determine the melting point of a metastably nucleated phase [17]. A series of photomicrographs of metastably nucleated corundum crystals in pure An melt and the corresponding temperature–time diagram. Note that the crystal becomes rounded in d and f, whereas in a, b, c, e, it has a straight outline. The melting point is in between d, f and c, e.

composition is anorthite, triclinic $CaAl_2Si_2O_8$. We shall call such a phenomenon "metastable nucleation", which has been observed in many silicate systems so far investigated by the in-situ observation method at high temperatures, table 1.

The reason why metastable nucleation can take place has been investigated in several silicate systems, and the role of interfacial energy difference between the liquid–stable solid phase and the liquid–metastably nucleated solid phase was pointed out as the governing factor [17, 18]. This conclusion was obtained based on the analyses of the results on (1) an in-situ check of nucleation sites, (2) the dependency of nucleation rates on the degree and duration of superheat and atmosphere compositions, (3) measurements of the sequence and rate of nucleation, (4) measurement of growth rate versus supercooling relations of both stable and metastable phases, and (5) nucleation behavior, steady state or transient.

The stabilities of metastably nucleated phases in the presence and absence of a stable phase, and the mechanism of transformation of a metastable phase to a stable phase in the presence of the latter have been investigated [17]. When crystals of the stable phase appear in the solution, metastably nucleated phases quickly disappear, transforming to the stable phase, either through a dissolution and precipitation process or through solid-state transformation depending on the temperature conditions [17].

Since the whole process of crystallization is recorded, the high temperature in-situ observation method has various advantages such as:

(1) Nucleation and growth processes are directly observable. This is a big advantage compared to ex-situ experiments, since we know whether heterogeneous nucleation on the crucible wall has taken place or not. Morphological changes of crystals can be traced exactly. This enables precise measurement of the melting points of any phases, including that of a metastable phase, by observing the transformation from polyhedral to rounded morphologies. Figure 22 is an example to show how such a determination can be done on metastably nucleated phases. The melting points of metastably nucleated phases in fig. 21 were determined by this method.

(2) Precise measurements of nucleation, growth and dissolution rates are possible, because the process is recorded on video tape, enabling repeated measurements. In figs. 23 and 24, examples of the results of measurements of nucleation and growth rates are shown, respectively.

(3) Since a diffusion boundary layer and buoyancy-driven convection, or formation of embryonic particles, can be visualized by Schlieren methods or interferometric techniques, one can obtain kinetic data, knowing the state of the environmental conditions.

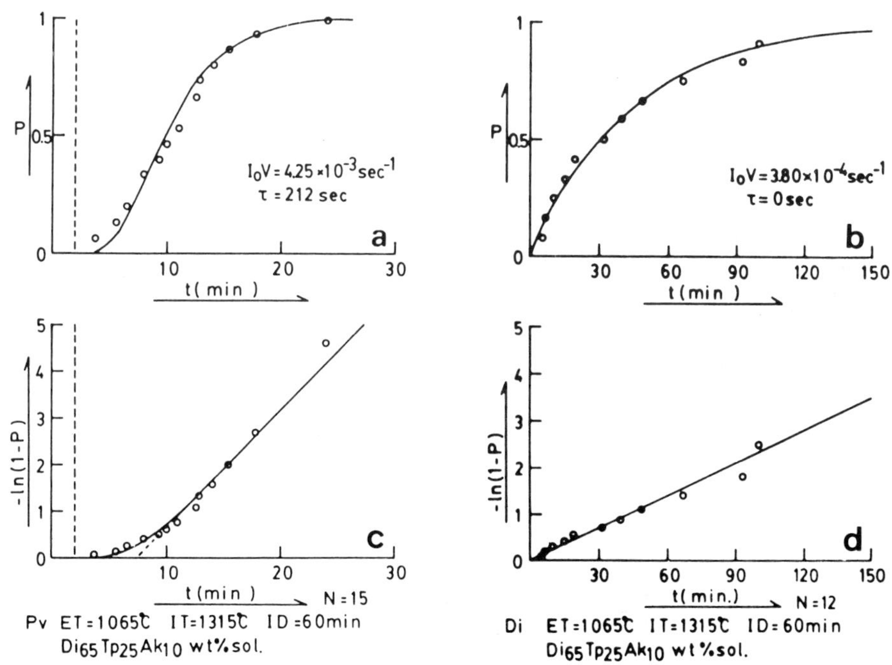

Fig. 23. Nucleation rate versus time relations of metastably nucleated phases and the analysis of nucleation characters based on Toschev's theory [18]. Pv: perovskite (a, c); Di: diopside (b, d); Tp: Ti-pyroxene; Ak: åkermanite. See also table 1.

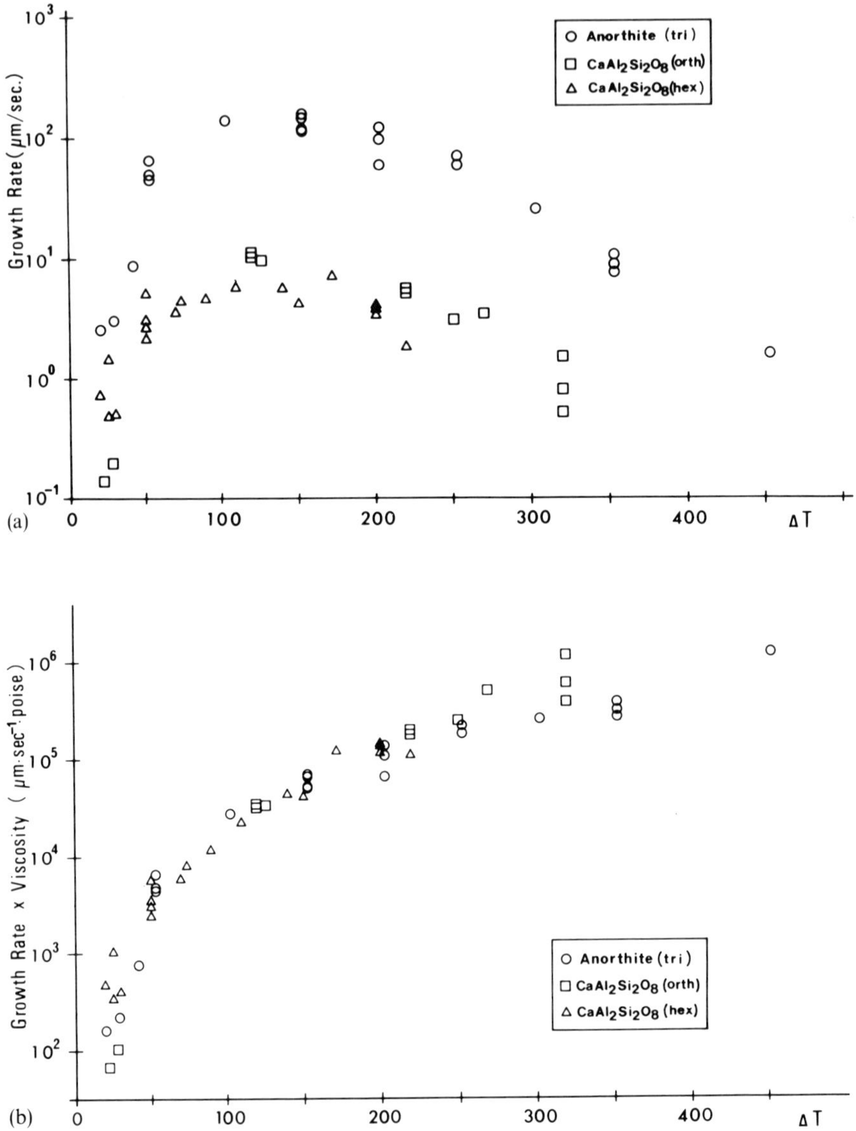

Fig. 24. Examples of growth rate measurements by the in-situ observation method at high temperatures. Growth rate versus supercooling relation for crystals of both stable and metastably nucleated phases in pure anorthite melt. The growth rates were measured along the long direction (a). (b) Indicates the growth rate × viscosity versus supercooling relation based on (a) [17].

5.3. *Simulation of extra-terrestrial materials*

Along the same lines as the experimental work described above to investigate kinetic problems of the Earth's solid materials, laboratory investigations aimed at extra-terrestrial materials has recently been started. Two examples will be cited in this section.

Comets are considered to consist of ice covering silicate dust particles. It is known that the brightness of the comet tail changes as it is approaching and departing the Sun, the origin of which was interpreted based on the equilibrium phase relation of crystalline ice. Kouchi [45] undertook laboratory investigation of the condensation and evaporation rates of crystalline and amorphous ice of H_2O and other compounds. He found a distinct difference in temperatures and rates of condensation and evaporation between crystalline and amorphous ices. Based on these results, he proposed a kinetic model to explain the changes in brightness of comet tails. He also demonstrated the effect of ultra-violet rays upon the evaporation rate of ice. Further, Kouchi et al. [46] theoretically analysed the conditions for the formation of amorphous ice on the surface of single-crystalline ice. Based on this model, they concluded that ice, condensed in a molecular cloud, is amorphous and that crystalline ice is formed around mass-loss stars and primordial solar nebulae [46]. This prediction was supported by infra-red spectral observation.

Laboratory reproduction of interplanetary dust particles is another example of the application of crystal growth ideas to solve the problems of Planetary sciences. Evaporation and condensation experiments of silicates and sulfides have been made along these lines at two laboratories. Various silicate and sulfide phases with extraordinary morphologies, including whiskers, and those which have never been synthesized by ordinary crystal growth techniques, have been synthesized by these experiments carried out under extreme conditions.

Only one example will be cited here. Kaito and Saito [47] were successful in growing the ordered $Fe_{50}Ni_{50}$ phase (tetrataenite), which was not possible by other methods, by coalescence of Fe and Ni smoke grains. The method is simple, just converging smoke streams originating from independent Fe and Ni evaporation sources. This success has an important implication in explaining a possible origin for natural remnant magnetism of chondrules, since it has not been possible to account for how tetrataenite was formed in space.

6. *Concluding remarks*

Crystal growth to form Earth and Planetary solid materials occurs under an extremely wide range and fluctuating conditions, producing a wide range of morphology, perfection and homogeneity of crystals. It appears to be a subject of too much complexity to be investigated within the scope of the modern science and technology of crystal growth. As a result, there has been a gap between Earth and Planetary sciences and the modern science and technology of crystal growth. However, the gap has recently become much narrower, through the effort of scientists in both fields, either by applying knowledge obtained in the field of crystal growth to complicated geological systems, or by developing new experimental methods to investigate relevant systems. The present author would be pleased if the reader appreciated through reading this chapter how the achievements in the modern science and technology of crystal growth can be applied to solve the problems in Earth and Planetary science, where crystallization takes place under a wide range and fluctuating conditions and on a much larger scale than in the laboratory. Very detailed investigations are not necessary to solve the problems. Experiments based on simple ideas are in most cases more helpful.

Although crystal growth in Earth and Planetary solid bodies might be very complicated, we should also note that this complexity offers the pleasure of decoding a detective story written in the Earth and Planetary bodies, which is worthwhile to read. The morphology of crystals forms the core of the studies on this topic.

References

[1] N. Steno, De Solido Intra Sodium Naturaliter Contento Dissertations Prodromus, Florence (1669); English translation by J.G. Winter (Hafner, New York, 1968).

[2] P.J. Wyllie, The Dynamic Earth (Wiley, New York, 1972).

[3] I. Sunagawa, in: Materials Science of the Earth's Interior, ed. I. Sunagawa (Terra, Tokyo/Reidel, Dordrecht, 1984) p. 63.

[4] G. Lofgren, in: Physics of Magmatic Processes, ed. R.B. Hargraves (Princeton Univ. Press, Princeton, 1980) p. 487.

[5] R.J. Kirkpatric, in: Kinetics of Geochemical Processes, eds. A.C. Lasaga and R.J. Kirkpatrick, Reviews in Mineralogy, Vol. 8 (Min. Soc. America, 1981) 321;
R.J. Kirkpatrick, Amer. Mineralogist 68 (1983) 66;
A.C. Lasaga, Amer. J. Sci. 282 (1982) 1264.

[6] K. Tsukamoto, T. Abe and I. Sunagawa, J. Cryst. Growth 63 (1983) 215.

[7] T. Abe, K. Tsukamoto and I. Sunagawa, Mineral. J., 13 (1987) 479.

[8] I. Sunagawa, in: Dynamic Processes of Material Transport and Transformation in the Earth's Interior, ed. F. Marumo (Terra, Tokyo, 1990) p. 139.

[9] I. Sunagawa, Annu. Rev. Earth Planet. Sci. 20 (1992) 113.

[10] A. Kouchi, A. Tsuchiyama and I. Sunagawa, Contrib. Mineral. Petrol. 93 (1987) 429.

[11] I. Sunagawa, in: Diagenesis IV, eds. K.H. Wolf and G.V. Chilingarian (Elsevier, Amsterdam, 1993) in press.

[12] K.V. Cashman, in: Review in Mineralogy, Vol. 24, Modern Methods of Igneous Petrology, eds. J. Nicholls and J.K. Russell (Min. Soc. America, 1989) p. 259;
K.V. Cashman, Bull. Volcanol. 50 (1988) 194;
K.V. Cashman and B.D. Marsh, Contrib. Mineral. Petrol. 99 (1988) 292;
K.V. Cashman and J.M. Ferry, Contrib. Mineral. Petrol. 99 (1988) 401.

[13] A.D. Randolph and M.A. Larson, Theory of Particulate Processes (Academic Press, New York, 1971).

[14] S. Tomura, M. Kitamura and I. Sunagawa, Phys. Chem. Miner. 5 (1979) 65.

[15] I. Sunagawa, in: Crystal Growth in Science and Technology, eds. H. Arend and J. Hulliger (Plenum Press, New York, 1989) 253.

[16] A. Tsuchiyama, H. Nagahara and I. Kushiro, Earth Planet. Sci. Lett. 48 (1980) 155.

[17] T. Abe, K. Tsukamoto and I. Sunagawa, Phys. Chem. Miner. 17 (1991) 473.

[18] H. Nakamura, K. Tsukamoto and I. Sunagawa, J. Cryst. Growth 99 (1990) 1227.

[19] I. Sunagawa, in: Morphology of Crystals, ed. I. Sunagawa (Terra, Tokyo/Reidel, Dordrecht, 1987) Part B, p. 509.

[20] I. Sunagawa, Bull. Mineral. (France) 104 (1981) 81.

[21] I. Sunagawa, in: Materials Science of the Earth's Interior, ed. I. Sunagawa (Terra, Tokyo, 1984) 303.
[22] I. Sunagawa, Habit Variation of Pyrite, Rept. Geol. Surv. Japan. No. 175 (1957).
[23] D.P. Grigoriev, Ontogeny of Minerals (Israel Prog. Sci. Tansl., Jerusalem, 1965).
[24] J.P. Bradley, D.E. Brownlee and D.R. Veblen, Nature 301 (1983) 473.
[25] I. Sunagawa and P. Bennema, in: Preparation and Properties of Solid Materials, Vol. 7, ed. W.R. Wilcox (Marcel Dekker, New York, 1982) p. 1;
I. Sunagawa, in: Morphology of Crystals, ed. I. Sunagawa (Terra, Tokyo, 1987) p. 321;
K. Sangwal and R. Rodriguez-Clemente, Surface Morphology of Crystalline Solids (Trans Tech Pub., Switzerland, 1991).
[26] I. Sunagawa, Amer. Mineralogist 46 (1961) 1216.
[27] W.J.P. van Enckevort, J. Cryst. Growth 119 (1992) 177.
[28] S. Ichikawa, Amer. J. Sci., Ser. 5, 6 (1923), 53; 137.
[29] A.P. Honess, The Nature, Origin and Interpretation of the Etch Figures of Crystals (Wiley, New York, 1927).
[30] W.K. Burton, N. Cabrera and F.C. Frank, Nature 163 (1949) 398.
[31] I. Sunagawa, J. Cryst. Growth 45 (1978) 3.
[32] L.J. Griffin, Philos. Mag. 41 (1950) 196.
[33] I. Sunagawa, Amer. Mineralogist 47 (1962) 1139; 1332.
[34] I. Sunagawa, K. Tsukamoto and T. Yasuda, in: Materials Science of the Earth's Interior, ed. I. Sunagawa (Terra, Tokyo, 1984) p. 331.
[35] I. Sunagawa and Y. Koshino, Amer. Mineralogist 60 (1975) 407;
I. Sunagawa, Y. Koshino, M. Asakura and T. Yamamoto, Fortschr. Miner. 52 (1975) 515.
[36] A.R. Lang, in: Physical Properties of Diamond, ed. R. Berman (Clarendon Press, Oxford, 1965) p. 69;
A.R. Lang, in: Properties of Diamond, ed. J.E. Field (Academic Press, London, 1979) p. 425.
[37] H.O.A. Meyer, in: Mantle Xenoliths, ed. P.H. Nixon (Wiley, New York, 1987) p. 501.
[38] M.B. Kirkley, J.J. Gurney and A.A. Levinson, Gems & Gemol. 27 (1991) 2.
[39] B. Jamtveit and T.B. Anderson, Phys. Chem. Minerals 19 (1992) 176.
[40] K. Nassau and K.A. Jackson, Amer. Mineralogist 55 (1970) 416.
[41] W.F. Berg, Proc. Roy. Soc. (London) A 164 (1938) 79.
[42] K. Onuma, K. Tsukamoto and I. Sunagawa, J. Cryst. Growth 98 (1989) 377; 384; 100 (1990) 125.
[43] R.J. Kirkpatrick, G.R. Robinson and J.F. Hays, J. Geophys. Res. 81 (1976) 5715.
[44] A. Kouchi and I. Sunagawa, Nature 304 (1983) 527; Contrib. Mineral. Petrol. 89 (1985) 17.
[45] A. Kouchi, Nature 330 (1987) 550.
[46] A. Kouchi, T. Kuroda and J.M. Greenberg, IGC-29 Abstracts Vol. 3, No. 3, 683, (1992) Kyoto.
[47] C. Kaito and Y. Saito, Proc. Japan Acad., Ser. B, 65 (1989) 125.

CHAPTER 2

Bridgman and Related Growth Techniques

E. MONBERG

AT&T Bell Laboratories, Murray Hill, NJ 07984, USA

Contents

1. Introduction . . . 53
2. Normal freezing growth techniques . . . 54
   2.1. Bridgman/Stockbarger . . . 55
   2.2. Gradient freeze . . . 58
   2.3. Dynamic gradient freeze . . . 58
3. Control of melt composition and crystal stoichiometry in III–V compound semiconductor materials. . . . 62
   3.1. Modified directional freezing . . . 62
   3.2. Boric oxide encapsulation . . . 64
4. Crucible confined growth of GaAs . . . 67
   4.1. Horizontal Bridgman growth of GaAs . . . 67
   4.2. Vertical growth of GaAs . . . 69
      4.2.1. Vertical gradient freeze growth of GaAs . . . 70
      4.2.2. Liquid encapsulated vertical Bridgman . . . 72
      4.2.3. Liquid encapsulated vertical zone melting . . . 72
5. Crucible confined growth of GaP . . . 73
   5.1. Early gradient freeze growth of GaP . . . 73
   5.2. Vertical gradient freeze growth of GaP . . . 74
6. Crucible confined growth of InP. . . . 83
   6.1. Horizontal Bridgman and LEC growth of InP . . . 83
   6.2. Vertical gradient freeze growth of InP . . . 83
      6.2.1. Dislocation reduction in VGF InP . . . 85
      6.2.2. Striations in VGF/EDGF InP . . . 88
      6.2.3. Growth twins in InP . . . 93
7. Bridgman growth of II–VI compound semiconductor single crystals . . . 94
References . . . 96

HANDBOOK OF CRYSTAL GROWTH, VOL. 2
edited by D.T.J. Hurle

1. Introduction

Almost all the elemental materials and congruently or near congruently melting compounds can be crystallized from the melt. Not surprisingly therefore, slow directional solidification of the melt is a long established method for the preparation of single crystals.

The pioneering work by Bridgman was directed toward the growth of single crystals of mainly low-melting point metals required for the high pressure studies for which he became famous. His first paper [1] described the growth of tungsten, antimony, bismuth, tellurium, cadmium, zinc and tin. To obtain single crystallinity, he arranged for the initially self-seeded growth to be propagated through a capillary section in the growth ampoule in order to select a single grain. Credit for this particular idea is probably due to Tammann [2]. Bridgman used a vertically mounted tubular electric furnace to melt the charge and lowered the ampoule containing the charge through it.

A similar vertical configuration was employed by Stockbarger to grow single crystals of lithium fluoride [3]. In one variant of the method, he utilized two furnaces mounted coaxially and separated by a baffle having a central hole of just sufficient diameter to allow the crucible to be lowered through. The upper furnace was at a temperature sufficient to melt the charge and the temperature of the lower one empirically adjusted to give the best crystal quality. Crucibles were made of platinum of up to 3 inches diameter and were double walled to avoid the possibility of leaks. They were tapered at the bottom where self seeding occurred.

The basic techniques were taken up and modified by numerous workers; the history up to the late 1940s is reviewed in Buckley's book [4]. Horizontal directional solidification has also been widely employed. It was used by Chalmers and co-workers in an extensive study of the crystallization of low melting point metals in the 1950s (see [5] for a review). This work incidentally led to the recognition of constitutional supercooling which is probably the single most important concept in melt growth [6]. Horizontal directional solidification is today known either as "horizontal Bridgman" (HB) growth or as the Chalmer technique.

Following the publication of the classic paper on zone melting by Pfann [7], both horizontal and vertical directional solidification (so-called normal freezing) techniques were extended to melt not the whole charge, but a zone which could be traversed along the charge. The choice between a normal freezing or zone melting configuration is dictated by the requirements for dopant distribution. It may be that one wishes to simultaneously purify the material in which case a zone melting configuration is required. If there is no need to purify the material and no dopants are to be added, normal freezing is the simplest option. If a dopant is added, then the macrosegregation of that dopant is different depending on whether a normal freeze or a single zone pass is performed. One way of avoiding macrosegregation is to use the zone levelling technique where a molten zone is passed repeatedly one way and then the other way along the charge [8].

Because of the simplicity of these several techniques and the ease of construction of suitable equipment from readily available components, they are very widely used to prepare research samples. The range of materials grown by the techniques is enormous and far too wide to document here. As with all crystal growth techniques, the design of the equipment is very material specific depending on melting temperature, chemical reactivity of the melt, vapor pressure etc. It would therefore be impossible to cover the topic, material by material, and this review is confined to some general comments on the base technology followed by a detailed consideration of the growth of III–V semiconductors and a brief review of II–VI compound growth.

The terminology applied to the techniques has become even more refined in the case of III–V growth depending on the means by which the crystal–melt interface is caused to traverse the length of the charge. If this is done by translating charge or furnace, the term Bridgman or Bridgman–Stockbarger is applied. However, if a temperature profile is caused to move along the charge by ramping the power supplied to a single zone furnace or a linear array of furnaces then the terms gradient freeze or dynamic gradient freeze are used. The adjective "modified" is often prefixed when describing the growth of compound materials to indicate the addition of a heated reservoir for the column V species in order to maintain the composition of the melt or to control the stoichiometry of the crystal.

The study of the growth of these materials embraces a number of materials issues of interest to crystal growers in general. These include the propensity for twinning, high temperature compound dissociation, formation of defects together with the practical aspects of designing apparatus for crystal growth at high temperature and pressure. One can apply many of the lessons learned from the practice of III–V growth to the Bridgman growth of other materials.

The chapter is organized as follows:

In section 2 the different normal freeze techniques are described with emphasis on the impact of the choice of technique on crystal properties and perfection and with special reference to III–V materials. In section 3, the problem of maintaining the desired melt composition and hence crystal stoichiometry is discussed. The use of boric oxide encapsulation and the maintenance of a partial pressure of the volatile species are the two techniques described. In sections 4, 5 and 6 the application of the technology to the growth of GaAs, GaP and InP is described in detail. Finally, in section 7, there is a brief discussion of the Bridgman growth of II–VI materials with emphasis on CdTe and ZnSe.

InP and GaAs are taken as the prototype materials demonstrating the problems (and some solutions) of seeding, crucible interaction, temperature and melt–solid interface profiles and stoichiometry control. Study of the group III phosphides additionally introduces the issues of high pressure system design and the propensity for twinning exhibited by these materials.

2. *Normal freezing growth techniques*

The growth of single crystals from the melt can be either seeded or unseeded. In many cases when investigating new materials, the initial experiments are unseeded. The preferred growth directions of the material will eventually lead to some large single crystal domains which can be "mined" out, oriented and used in subsequent seeded attempts. Initially, most seeded growth methods required a transparent furnace or a viewing port in order to avoid melting all of the seed, or none of it. As the sophistication of the apparatus for crystal growth has increased it is now possible to reproducibly melt back a portion of the seed by controlling the temperature profile and the position of the growth assembly relative to the profile. The growth orientation for single crystals used for substrates for electronic and photonic applications must be controlled. The anisotropy of parameters such as epitaxial growth morphology, etch rates, electro-optic behavior are all orientation dependent. Thus, all commercial crystal growth of these materials are seeded processes, and the following discussions will concentrate on seeded growth techniques.

2.1. Bridgman/Stockbarger

The Bridgman technique is characterized by the relative translation of the crucible containing the melt to the axial temperature gradient in a vertical furnace as shown in fig. 1. Whilst it is generally simpler in a mechanical sense to lower the crucible through the gradient, in some cases where vibration of the ampoule is a concern, the crucible is held stationary and the furnace is translated on a linear track assembly.

There are some advantages to a horizontal configuration. In some situations it is necessary to achieve a steady state equilibrium with a gaseous species, such as arsenic vapor in the growth of GaAs. This is easier to achieve in a horizontal boat configuration where the distance from the vapor to the growth interface is fairly constant over the entire length of the crystal. In the vertical mode, this distance varies with the fraction of the melt solidified. The seeding step is generally more easily observed in the horizontal mode especially where opaque crucible materials, such as pyrolytic boron nitride (PBN), are used. However, for the growth of crystals at higher temperature and pressures, it becomes increasingly more difficult to observe seeding in either the horizontal or vertical Bridgman configuration due to increased insulation and the pressure containment vessels. The removal of the ingot from the crucible is generally easier from the horizontal boat crucibles. The crystal–crucible interface is half that of vertical crucibles

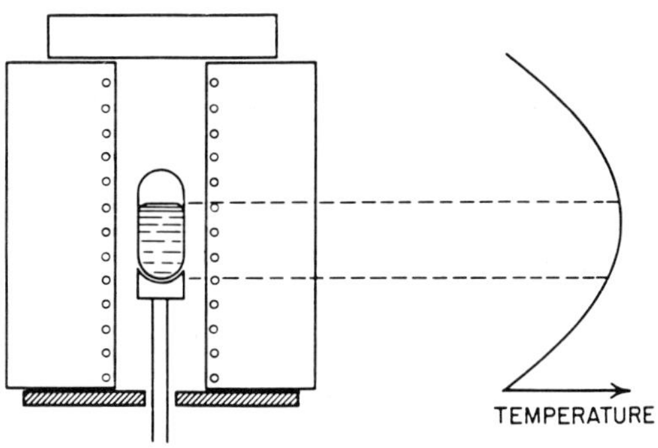

Fig. 1. Vertical Bridgman furnace configuration.

and it is easier for solvents to work their way into the annular gap between the crucible and the crystal.

There are several advantages to the vertical approach. The crucible and the furnace can be designed with axial symmetry. This results in a more homogeneous crystal where the melt-solid interface is normal to the growth axis. Heat flow, convective flow patterns, and solute distribution are all more symmetric in this case.

However, this alone will not assure radial uniformity. Control of the convective flow, by careful thermal design and possibly by introduction of forced convection by rotating the growth ampoule, may be necessary. The use of accelerated rotation of the crucible, a technique pioneered in flux growth has been reported for the growth of the incongruently melting compound Rb_2MnCl_4 [9]. This enabled the authors of this paper to increase their growth rate four fold without inducing cellular breakdown of the crystalizing interface as a result of the onset of constitutional supercooling.

Ideally, one would like to be able to grow with a planar crystal–melt interface into a quiescent melt where solute transport is by diffusion alone. This would yield a radially uniform dopant distribution and, after an initial transient, axial segregation would also be avoided. This prospect has motivated crystal growth experiments in a microgravity environment. Diffusion limited conditions during Bridgman growth of gallium-doped germanium crystals have been demonstrated [10].

In the case of commercial crystal growth, the vertical mode has distinct economic advantages. Wafers that are sliced from the axisymmetric vertical boules are circular whereas horizontally grown boules yield D-shaped wafers. Subsequent wafer polishing and epitaxial processing steps are best suited for handling circular wafers. In addition, the intra-wafer electrical and compositional uniformity for vertically grown crystals is generally better due the near coincidence of the wafer slices with the melt–solid interface traces. In the horizontal case, the non-axisymmetric isotherms lead to melt–solid interfaces that are not normal to the seeding axis. A survey of the Journal of Crystal Growth over the last decade shows that the vertical configuration was strongly favored over horizontal growth.

There is a common misconception regarding vertical Bridgman–Stockbarger growth of compound semiconductors. These materials expand several percent upon freezing. The notion is that the crucible constrains the solidifying ingot and leads to large stresses and structural imperfections. This would be true if the crucible is quench cooled. A skin of solid material would form and confine the molten core. As the core freezes, it would expand and split the outer shell. However, the situation for normal freezing is distinctly different. The polycrystalline charge and a portion of the seed is first melted. As the melt solidifies epitaxially on the seed, it occupies the same cross sectional area as the solid seed, but the level of the melt is raised slightly, due to the lower density of the solid compared to the melt. As solidification proceeds, the recently solidified material is cooled and contracts since the coefficient of thermal expansion is positive. The crucible material must be chosen so that it has a lower coefficient of thermal expansion than the solid crystal, but as long as the top of the melt is not constrained from rising, there will be no stress imposed on the solidifying crystal. This interpretation is confirmed by the exceedingly low dislocation levels and absence of slip in the vertical gradient freeze (VGF) results that are discussed in sections 4 and 6.

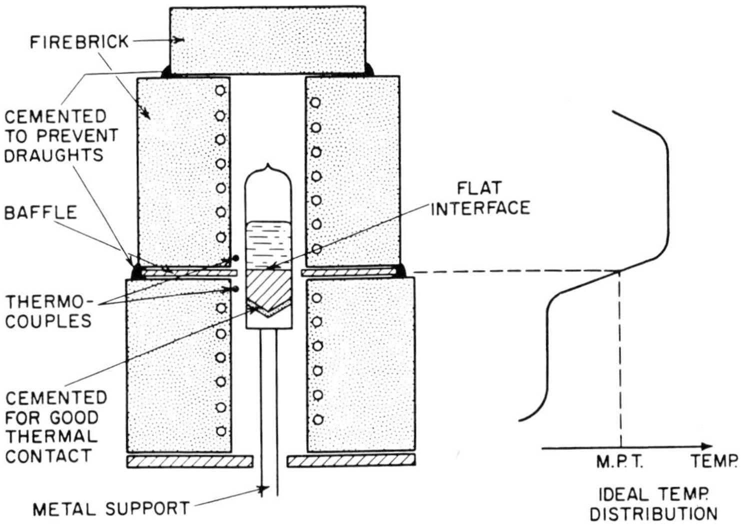

Fig. 2. Vertical Stockbarger furnace configuration.

The Stockbarger method is a more sophisticated modification of the Bridgman method and is illustrated in fig. 2. There is a high temperature zone, an adiabatic loss zone and a low temperature zone. The upper and lower temperature zones are generally independently controlled and the lossy zone is either unheated or poorly insulated. The shape of the axial temperature profile is more readily controlled and/or modified by using two independent heating zones. A further extension of this multizone approach is the dynamic gradient freeze method described in section 2.3. An additional advantage of the Stockbarger method is that the axial gradient at the melt–solid interface can be considerably larger than that which is normally achievable in Bridgman growth. This is achieved by adjusting the power levels so that the melting point of the crystal is located within the lossy region between the high and low temperature zones.

The conditions pertaining at the three-phase boundary between melt, crystal and crucible are critical to obtaining single crystals. The first issue is the avoidance of keying (or wetting) of the crystal to the crucible. This requires chemical inertness and a thermal contraction coefficient of the crystal greater than that of the crucible so that as both cool to room temperature the crystal contracts away from the crucible wall. Bridgman describes preparing his ampoule by pouring in a heavy paraffin and then washing with a light paraffin and allowing to drain. This was said to help the solidified metal to detach itself from the crucible more easily. This presumably was the consequence of the formation of a carbon film on heating which has also been used by numerous workers subsequently to obtain a roughened surface. Sand blasting of silica ampoules has also been reported to reduce wetting. The sand blasting produces a microscopically roughened surface. The surface tension of the melt causes it to sit upon the points of the fractured silica and not allow it to spread over the surface. The difficulty here is achieving a reproducible surface over the entire portion of the crucible in contact with the melt.

A novel approach is the use of a "soft" mold obtained by tamping a suitable powder around a pre-cast polycrystalline charge. This was first reported by Noggle and Kohler who used it to grow aluminum single crystals in a soft mold of alumina powder [11]. The use of boric oxide to wet the crucible is described in detail in section 3 below.

For extremely refractory materials a crucible-less variant of Bridgman growth, known as skull melting, can be deployed [12] in which a solid skin of the refractory powder is maintained around the molten charge to retain it. The material serves as its own crucible, reducing contamination and increasing the upper limit of temperature for melt growth. One of the more successful application of this technique has been the growth of cubic zirconia. The skull melting technique usually yields large-grained polycrystals rather than one large single crystal.

Suitable furnaces, complete with power supplies and control systems are commercially available but, often on grounds of cost, many researchers produce their own resistance heaters. For relatively low temperature applications Kanthal tape is used to wind furnaces (see for example [13]) which can either be uniform to obtain an isothermal environmental or graded to facilitate gradient freeze growth. At intermediate temperatures, crusilite (silicon carbide) heating elements are useful (see for example [14]) whilst at temperatures above about 1500°C precious metal wire, molybdenum disilicide (or induction heating, see for example [15]), has to be employed.

2.2. *Gradient freeze*

The third approach to normal freeze growth is that of gradient freeze. The vertical configuration of the gradient freeze method is illustrated in fig. 3. The crucible is positioned in the furnace and a monotonically increasing temperature profile is imposed along the crucible. The power to the furnace zone(s) is increased to the point where the entire polycrystalline charge and a portion of the single crystal seed is melted. Once a thermal steady state is reached, the power is controllably ramped down so that the melt is solidified starting from the seed. There need be no moving parts in this arrangement although it is usually desirable to rotate the crucible to average out temperature asymmetries. This technique has advantages in situations where it is difficult to introduce translational freedom, as in the case of growth in high pressure autoclaves. In cases where there are large temperature gradients, or long crystals, overheating of the melt can be a concern. It is also important to design the gradient freeze growth system so that the axial temperature profiles are parallel over the range of input power necessary to grow the entire crystal. This helps to maintain a constant growth rate for a given cooling rate.

2.3. *Dynamic gradient freeze*

Many of the compromises discussed in the previous two sections are a result of the limited number of independently controllable furnace zones. The difficulties that arise

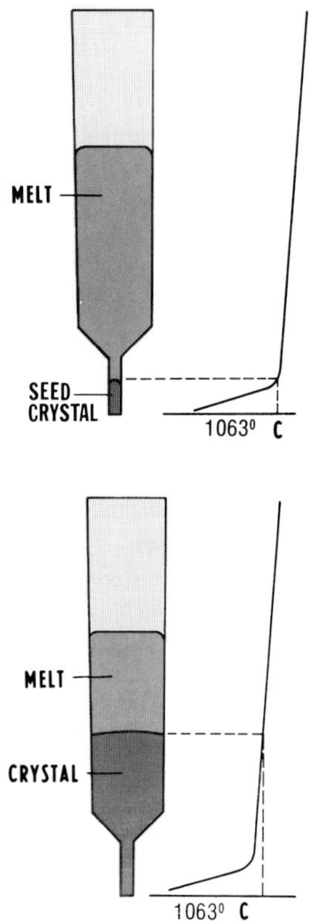

Fig. 3. Vertical gradient freeze temperature profile for InP crystal growth.

from this limitation include, but are not limited to, overheating of the melt, non-constant growth rates, end effects and complex mechanical means for translation. With the availability of low cost, multichannel, integrated control systems it has become possible to build multizone furnaces with 22–50 independently controlled heating zones of one inch or more in length [16]. Such a multizone furnace can be configured to operate as a gradient freeze or an electronic Bridgman system. The term electronic Bridgman refers to a configuration whereby a desired Bridgman-like temperature profile is established, but instead of moving the crucible or the furnace, the gradient is moved along the furnace by sequentially reducing power to the appropriate adjacent independent control zones. The smoothness of this moving profile depends upon the length of each independent zone, relative to the diameter of the tube furnace.

Figure 4 schematically compares the VGF process with the electrodynamic gradient freeze (EDGF) multi-zone process. In the VGF approach, the hot zone has a characteristic axial temperature profile and as growth proceeds from the seed into the melt, the entire profile is shifted to lower temperatures. The melt temperature increases monotonically and the maximum temperature that the melt sees is a function of the length of

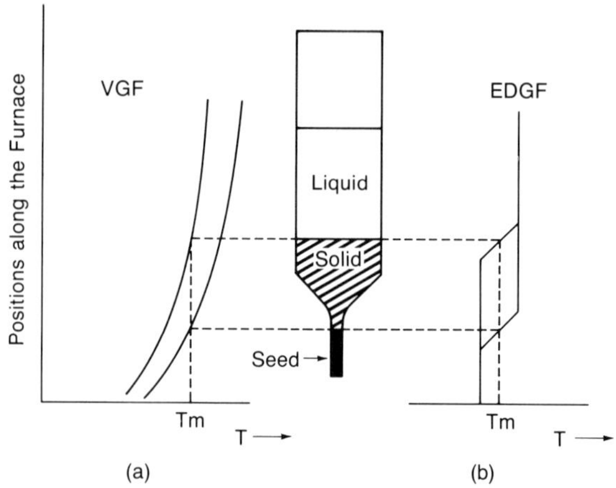

Fig. 4. Comparison of axial temperature profiles for (a) vertical gradient freeze growth to (b) electrodynamic gradient freeze growth.

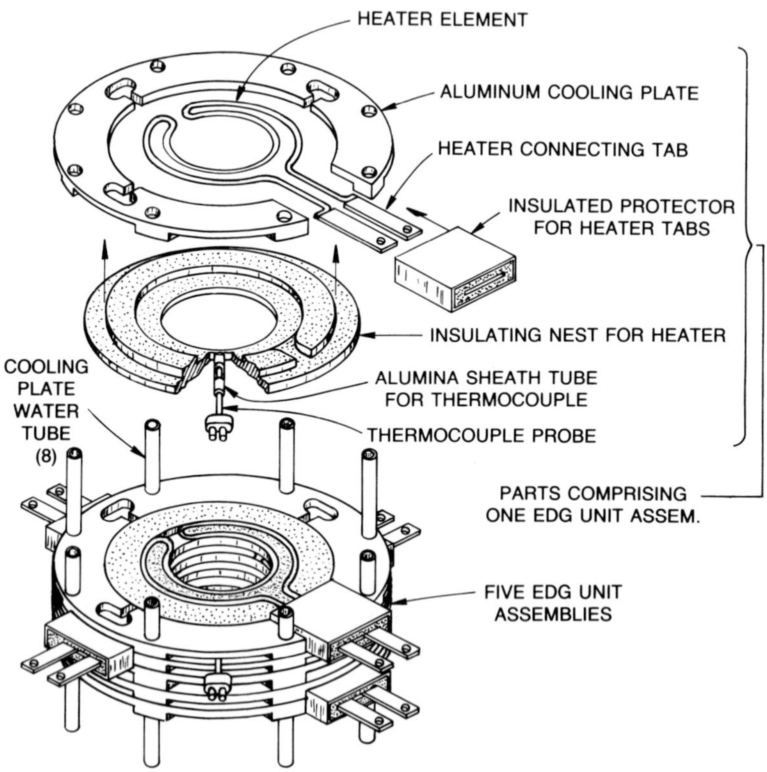

Fig 5. Electrodynamic gradient freeze growth furnace assembly.

the charge. In the case of EDGF growth, a small gradient segment is programmed to pass through an otherwise isothermal melt. There need be no moving parts involved and the smoothness of the moving gradient is limited only by the zone resolution.

Figure 5 is a representation of the dynamic gradient freeze heater assembly designed for the EDGF growth of InP [17]. Each successive 0.5 inch thick heater unit is rotated 90 degrees relative to the preceding unit. Groups of 2 or 4 adjacent stacked units are wired in series to give either 1 or 2 inch zones. Figure 6 shows the multi-zone furnace configured for 23 independently controlled zones. In order to maintain as much application flexibility as possible, the entire furnace was constructed from 58 identical EDGF units. In theory, each of these repeating units could be used as an independent zone. Practically speaking, there are many factors which limit the total number of

Fig. 6. Twenty-three-zone electrodynamic gradient freeze furnace.

zones; simplicity of design, the need to provide pressure vessel penetrations for each control thermocouple and set of power leads, and the desire to minimize the cycle time for the response of the individual control loops. The furnace configuration can be changed as needs dictate simply by rerouting the external braided power cables, relocating the control thermocouples and changing the control signal connections.

3. Control of melt composition and crystal stoichiometry in III–V compound semiconductor materials

3.1. Modified directional freezing

The term, modified growth, has been earlier designated as describing a means for maintaining a controlled vapor pressure of the V compound over the III–V melt, i.e. As, P, or Sb. The control of the vapor pressure is usually accomplished by heating a reservoir of the condensed phase of the column V elements. This reservoir must be at the coldest point of the ampoule. This approach has proven quite successful for the arsenides. This is due to the fact that the arsenic vapor pressure is a reproducible function of temperature and is little affected by the details of the arsenic preparation and purification.

Parsey and Thiel found that the density of dislocations in Bridgman-grown GaAs depends critically on the As partial pressure [18]. Crystals that were grown under an As pressure from a reservoir whose temperature was precisely maintained at 617°C were essentially dislocation free. The free carrier concentration and corresponding compensation ratio was found to vary with the arsenic pressure. In the case of LEC growth from PBN crucibles, Holmes et al. have shown that the conductivity of GaAs goes from p-type to semi-insulating as the As concentration in the melt exceeds 0.475 mole fraction [19]. Regardless of the growth technique, variations in the arsenic partial pressure are reflected in the melt composition and ultimately in the crystal stoichiometry.

In the case of red phosphorus, reproducible control of crystal stoichiometry has not been achieved with a "modified" growth system. There have been many reports of InP single crystal growth using either the modified horizontal gradient freeze or horizontal Bridgman techniques. In these examples, in-situ synthesis was followed by crystal growth. No encapsulant was usually employed because of the need to transport phosphorus to the In + P solution during synthesis. Bachman and Buehler showed that the reproducibility of the vapor pressure of red P from different commercial sources is no better than 10% [20]. This is due to the number of different allotropes of phosphorus, all with different equilibrium vapor pressures at a given temperature, that usually coexist in commercially available material. The work of Farr and of Smits and Bokhorst shows that there are slow phase transformations to the triclinic form IV and form V allotropes from the monoclinic form in the 500–550°C range [21]. This happens to be in the range where one would like to control the phosphorus reservoir temperature for the modified growth of GaP and InP.

Wardill et al., in an elegant set of experiments, have directly sensed the phosphorus pressure in a sealed ampoule during the horizontal Bridgman growth of InP [22]. They used this signal in a closed control loop to maintain the pressure by varying the power to the phosphorus heater. The external pressure was adjusted so that there was no differential pressure across the ampoule wall which could lead to explosion or implosion. Variations of 30°C in the set point of the phosphorus heater were sometimes necessary to maintain the desired phosphorus pressure.

The first report of the modified VGF growth of indium and gallium phosphides was by Gault et al. [23]. This approach was similar, in the manner of composition control, to the technique for unencapsulated GaP VGF growth reported by Blum et al. [24]. In the unencapsulated approach, melt composition is maintained by heating a separate reservoir of condensed red phosphorus to produce the desired phosphorus vapor pressure over the melt. Using the same unencapsulated modified VGF approach in the growth of InP, there was significant variation in the melt composition from run to run for the same phosphorus heater set point [25]. Figure 7 shows the last-to-freeze portions of three successive InP crystals with indium-rich, stoichiometric and phosphorus-rich tops.

Fig. 7. Composition of the last-to-freeze portions of VGF-grown InP single crystals.

The demonstrated difficulties in obtaining a reproducible vapor pressure over condensed red phosphorus within the necessary temperature range of 500–600°C has led to the introduction of B_2O_3 as an encapsulant for the VGF growth of III–V compound semiconductors.

3.2. Boric oxide encapsulation

Boric oxide plays an extremely important role in the development of compound semiconductor bulk growth processes. Its use as a liquid encapsulant made possible the growth of large single crystals of materials with volatile components by a simple modification of Czochralski technique that had been well developed for germanium growth. Metz et al. were the first to use B_2O_3 for the Czochralski growth of compound semiconductors [26]. They grew PbTe and PbSe crystals and provided the impetus for Mullin et al. to produce single crystals of InAs and GaAs by the same technique [27]. Shortly thereafter, the LEC growth of InP and GaP was also reported by Mullin et al. [28]. While the pressures involved necessitated a high strength steel enclosure, the technique was essentially the same as for the arsenide materials. Mullin et al. recognized that the B_2O_3 wetted the silica crucible and formed a liquid film between the crucible and the melt, thus the term "liquid encapsulation" was coined. They also recognized that the encapsulant preferentially dissolved surface oxides and gettered some trace impurities from the polycrystalline charge during melting, and thus, served as an additional purification step. Indeed, the most important result was that high purity III–V single crystals could be grown using B_2O_3 with no deleterious effect on their electrical properties.

In a similar manner, the use of boric oxide has made possible the production of high quality, large single crystals of InP and GaAs grown by the vertical gradient freeze and vertical Bridgman techniques. It serves as an inert barrier to the loss from the melt of the volatile arsenic or phosphorus species. At the same time, the boric oxide will wet the surface of the pyrolytic boron nitride (PBN) crucible. This reduces the interaction between the solidifying melt and the crucible and retards twinning and polycrystalline nucleation. The key to the success of the boric oxide–PBN combination is the laminar nature of the PBN. As the boric oxide solidifies in the annular space between the ingot and the crystal, it pulls away a thin layer of the PBN crucible. This sacrificial layer prevents the crystal from being put under strain due to the differences in the coefficient of thermal expansion of the crucible, the boric oxide skin and the crystal. The destructive results of this interaction can be seen in the substitution of quartz crucibles for PBN. The lower thermal conductivity of the quartz crucible is used to try to maintain a planar melt–solid interface at the boundary with the crucible, but there is no corresponding sacrificial layer in this system. Figure 8 shows a completely single InP crystal grown in a quartz crucible with boric oxide as the encapsulant. Extensive cracking of the crystal results from the mechanical stresses due to the differential thermal contraction between the crystal and the crucible with a borosilicate glass in between the two.

The first report of encapsulated normal freezing growth of InP was by Monberg et al. and described the electrodynamic gradient freeze (EDGF) growth using a total

Fig. 8. Stress cracking in a single crystal of InP grown in a silica crucible with B_2O_3 encapsulant.

encapsulation approach [17]. This work showed that it was possible to maintain composition during both the melting of the polycrystalline InP and the actual growth process without the use of a separate phosphorus source. The concept is to totally encapsulate the InP throughout the process. The boric oxide in the EDGF system was added as solid disks at the top of the polycrystalline charge in the PBN crucible. It was found that appreciable amounts of phosphorus were lost in a matter of hours at temperatures above 700°C. Boric oxide, on the other hand, begins to soften at 450°C and will flow reasonably well at 550°C. Thus, the strategy to avoid InP decomposition is to ensure that all of the InP is covered with the molten encapsulant before raising the temperature of the InP charge above 650°C. Obviously, the most advantageous situation is to use a dense polycrystalline charge that conforms to the shape of the PBN crucible. Smaller pieces of InP will not pack as compactly and will require far more B_2O_3 to achieve total encapsulation. Since there is no independent phosphorus source with which to alter the melt composition, it is desirable that the starting polycrystalline InP be stoichiometric. Under these conditions, InP single crystal boules, exceeding 1200 grams, can be grown with no discernable second phase indium in the last-to-freeze portion. Figure 9 shows two temperature profiles; one is representative of the encapsulation process and second is at the beginning of seeded growth. The ability of the independently controlled multi-zone EDGF furnace to smoothly change thermal

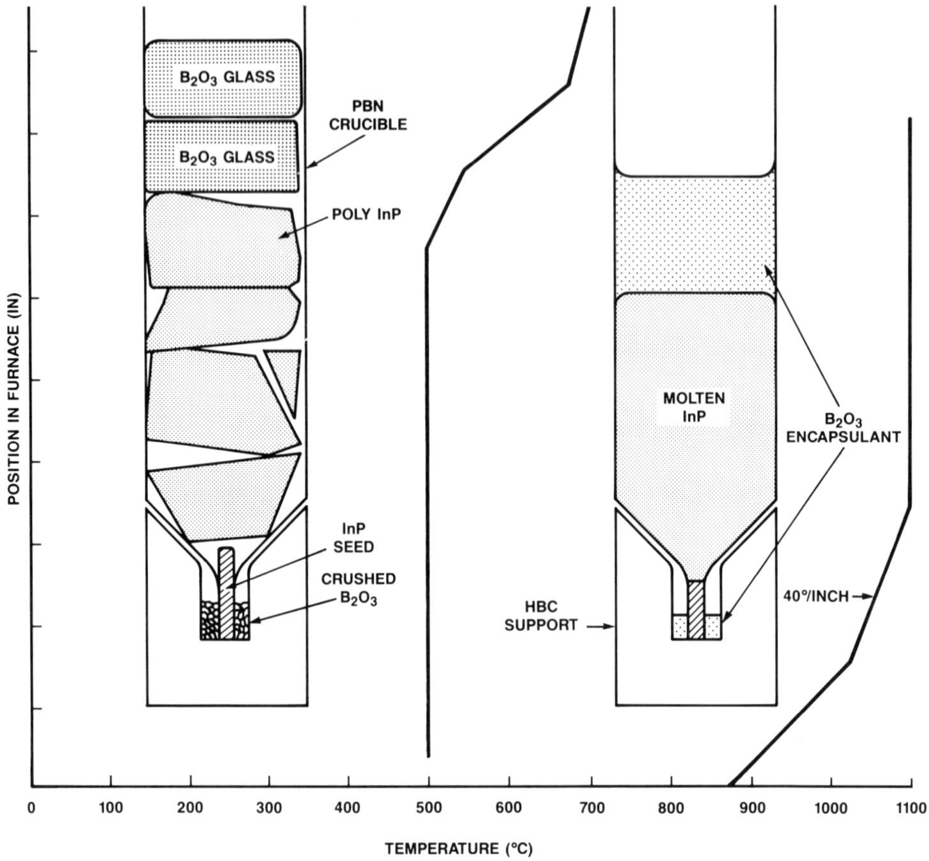

Fig. 9. Axial temperature profiles for totally encapsulated EDGF growth of InP.

profile shapes during the growth run is exploited in this instance. One can establish an encapsulating profile that preferentially melts the B_2O_3 while, at the same time, prevents significant decomposition of the polycrystalline InP. Once encapsulation is complete, the thermal profile for single crystal growth can be established. After the crystal growth run employing "total encapsulation" is complete, there is no evidence of any loss of phosphorus. The solidified B_2O_3 encapsulant is optically clear. There are some wisps of diffused phosphorus at the InP–B_2O_3 interface, but for the most part, the top of the ingot is visible through the solidified encapsulant. This indicates that there is no significant loss of phosphorus into or through the B_2O_3 encapsulant.

A second, and equally important, consequence of liquid encapsulation in the VGF/EDGF systems has been observed. When B_2O_3 is allowed to uniformly wet the inner surface of the PBN crucible, it prevents the InP from interacting with the PBN surface and thus causing polycrystalline nucleation. Prior to the implementation of boric oxide encapsulation the solidified InP ingot would adhere to and delaminate large areas of the PBN crucible. This apparent crucible–melt interaction was often accompanied by the onset of polycrystalline growth. Archer claims that an indium-rich solution of In + P will attack PBN at 1500°C within two hours [29]. Thus, it is not unlikely that over the course of a crystal growth run there is significant interaction with the crucible. Following the implementation of encapsulation, the ingot tended to release from the crucible more freely than previously; however the surface of the ingot was not uniform in appearance. The predominantly highly reflective surface would be interrupted by regions with a matte finish having the surface texture of the PBN crucible. These variation in surface finish were caused by a PBN surface that was non-uniformly wet by the B_2O_3. This problem was addressed by a more aggressive crucible cleaning procedure and the addition of a thin coating of B_2O_3 on the inner surface of the PBN crucible via a high temperature oxidation step [30]. The actual thickness of the layer for a given temperature is determined by the balance of the oxidation rate of PBN and the evaporation rate of B_2O_3. The oxidation of BN is represented by

$$2BN + \tfrac{7}{2}O_2 = B_2O_3 + 2NO_2.$$

The rate constants measured for PBN oxidative weight loss at 1100°C and 1200°C are $1.04 \times 10^{-7}\,g\,cm^{-2}$ and $5.4 \times 10^{-7}\,g\,cm^{-2}$, respectively [31]. The vapor pressure of vitreous B_2O_3 is given by the expression [32]

$$\log P(atm) = 6.742 - 16\,960/T.$$

At 1100°C, the B_2O_3 vapor pressure is only a few millitorr, but at 1300°C, the rate of evaporation of the B_2O_3 exceeds that of the PBN oxidation rate. Thus, a temperature in the range of 1100–1200°C represents an effective compromise for producing a sufficiently thick layer in a reasonable length of time.

4. *Crucible confined growth of GaAs*

4.1. *Horizontal Bridgman growth of GaAs*

There are two primary reasons why so much investment has gone into the development and production of GaAs single crystals by boat growth methods compared to Czochralski techniques. The first is that boat growth systems are relatively inexpensive and easy to build compared to Czochralski systems, whether we speak of the research or production levels. Secondly, the major commercial interest in GaAs single crystals is for substrates for optoelectronic devices. It is well known that dislocations in the substrate lead to non-radiative recombination centers; dark line and dark spot defects [33]. The large axial and radial temperature gradients that are characteristic of the Czochralski technique lead to stress levels much higher than those found in Bridgman growth and consequently larger dislocation levels [34]. Thus, almost all GaAs substrates used for optoelectronic devices are grown by low temperature gradient techniques.

There are a plethora of horizontal boat growth techniques that have been used to grow GaAs. They are all variations on the directional solidification techniques

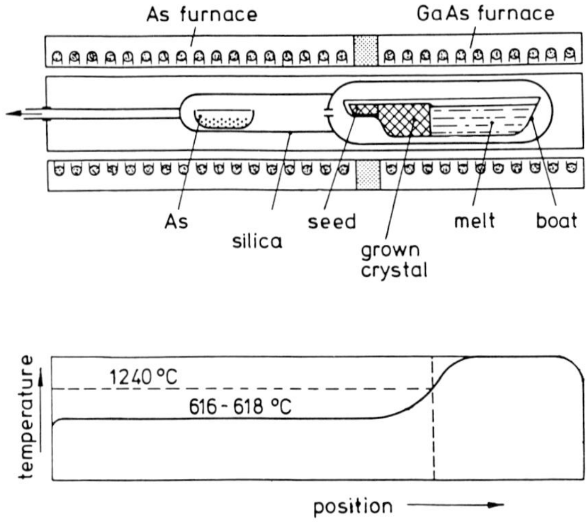

Fig. 10. Modified horizontal Bridgman furnace configuration for the growth of GaAs.

described earlier; gradient freeze, zone melting, Bridgman. They all have a low temperature arsenic zone which is maintained at the temperature necessary to provide the desired arsenic partial pressure over the melt. Figure 10 shows a typical configuration for the HB growth of GaAs. There is, by necessity, a transitional zone or zones which are intermediate in temperature between the arsenic zone and the GaAs melt zone. The design of the high temperature GaAs melt zone is the major distinction among the several techniques. In general, the sophistication of the high temperature zone(s) is directly related to the number of independently controllable heating elements. In the simplest case, that of gradient freeze growth, a monotonically increasing temperature profile is established along the length of the boat. The power to this zone is then ramped down so that the melt–solid interface is swept from the seed throughout the length of the melt. With a single zone it can be difficult to maintain a constant growth rate and a uniform temperature gradient along the length of the solidified crystal. Suzuki et al. added a third, intermediate, controllable zone which reduced the axial and radial gradient at the solidifying interface and helped to maintain the same shape profile from the initiation of seeding to the solidification of the last to freeze tail potion [35].

This approach was taken even further by Parsey and Thiel [18] and Bourret et al. [36] with the multi-zone electrodynamic gradient furnace. The basis of this approach is a stack of independently controlled axially symmetric heater zones. The repeating unit is 0.5 inch long and several elements are generally wired in series so that a 1 or 2 inch zone is the independently controlled zone. The resulting system can emulate the heat flow characteristics of the Bridgman–Stockbarger or gradient freeze growth techniques without requiring the motion of either the containment vessel, growth ampoule or furnace. The profile can be electronically modified as growth proceeds in order to compensate for end effects. Extensive profiling and iterative adjustment of the relative power levels to the zones are necessary in order to take full advantage of this approach.

Of course, the added number of control loops, heater element connections and thermocouples make it a more demanding system to maintain. Nonetheless, the flexibility of the EDGF system offers a unique experimental tool to the crystal grower to determine the optimum growth parameters for a material system. These could then be built into a more conventional furnace if the low cost production of single crystals is the goal.

Young and Caruso claimed that the use of B_2O_3 in the horizontal gradient freeze growth of GaAs significantly reduced the wetting of the PBN crucible by the melt [37]. They also showed that B_2O_3, being of significantly lower thermal conductivity than PBN in the *a* (in-plane) direction, acts as an thermal insulator between the melt and the boat surface. This leads to less radial heat loss and to a flattering of the melt–solid interface.

4.2. Vertical growth of GaAs

Various approaches have been employed to grow GaAs in a vertical, non-LEC mode. Clemans [23] was the first to utilize the VGF approach, Hoshikawa et al. [38] subsequently reported on liquid encapsulated vertical Bridgman while Swiggard [39] used yet another technique, liquid encapsulated vertical zone melting. There are several common features to all these approaches: (a) The melt is confined in a vertical, tapered, PBN crucible and is bottom seeded. (b) A low axial temperature gradient is established across the melt–solid interface. The resulting radial gradients are very small and the melt–solid interface is quite planar. (c) The initial charge is polycrystalline GaAs. In-situ synthesis has not been demonstrated because of the difficulties in protecting the seed either from dissolving in the Ga-rich melt or from decomposing due to the loss of arsenic during the heating of the charge prior to growth.

As was mentioned earlier, in the general discussion about vertical growth, there are several advantages to these vertical growth methods in comparison to the horizontal approaches. This is especially true for the growth of GaAs. The inevitable D-shaped wafer, cut perpendicular to the growth direction in the HB material produces wafers that are more difficult to handle and process in any automated fashion. The wafers are generally ground to a circular shape or are cleaved square with the unavoidable waste of material. The axially asymmetric heat flow and convection patterns in HB growth lead to a melt–solid interface that is not perpendicular to the seeding direction. Compounding this problem is the fact that a propensity for twinning makes (100) seeded growth very difficult in HB. Thus, seeding is normally along the (111) direction and the ingot is sliced at a 54.5° angle from the seeding axis to obtain the desired (100) wafers. This leads to irregular wafer shapes with non-uniform properties from edge-to-edge.

In comparison, vertical growth leads to round wafers with more uniform intra-wafer properties, especially if rotation of the crucible is practiced. Due, in part, to the symmetric heat flow patterns and the planar interface shape, (100) seeded growth is routinely achieved in the vertical methods.

At one time it was felt that a significant advantage of HB was the possibility of in-situ synthesis followed directly by single crystal growth. In practice however these process steps are almost always decoupled. This has to do with the fact that the optimized furnace configuration for synthesis is not the same as for single crystal growth. In addition, it is desirable to be able to etch away any unreacted gallium prior to crystal growth. One major practical advantage of present HB systems is the ability to view the meltback of the seed in the open-top boat. To this point there have been no reports of vertical systems where one could view the melt–solid interface. This places a premium on thermal profile reproducibility in the vertical systems.

4.2.1. Vertical gradient freeze growth of GaAs

The first report of large GaAs single crystals grown by the VGF method was by Clemans and co-workers [23]. Their approach was a variation of the earlier work on the VGF growth of GaP and InP. The details of the VGF system are given in more detail in section 5 but basically it consists of a two-zone furnace; a high power graphite resistance heater and a lower coaxial wire wound heater. The upper heater zone melts the polycrystalline GaAs charge and a portion of the single crystal seed. The power is then controllably ramped down and the melt is solidified within the PBN crucible. The lower zone heats a reservoir of arsenic to provide the requisite As vapor pressure over the molten GaAs. The system has been scaled to the point where 4 kg, 75 mm diameter crystals are routinely grown. Figure 11 shows the progress made by Clemans in scaling up of the VGF process for the growth of (100) seeded, semi-insulating crystals.

The axial temperature gradients in the VGF system are less than 5°C cm^{-1}. This is more than a factor of ten reduction from those established for the LEC growth of

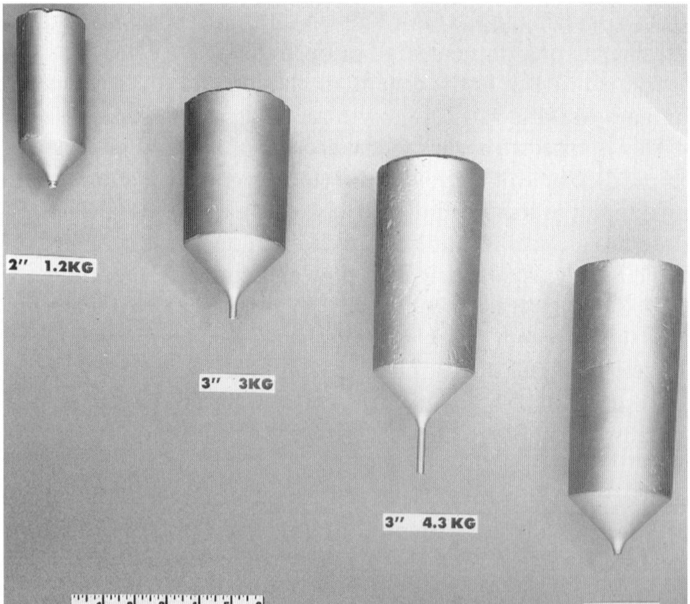

Fig. 11. Progress in the scale-up of the VGF growth of GaAs.

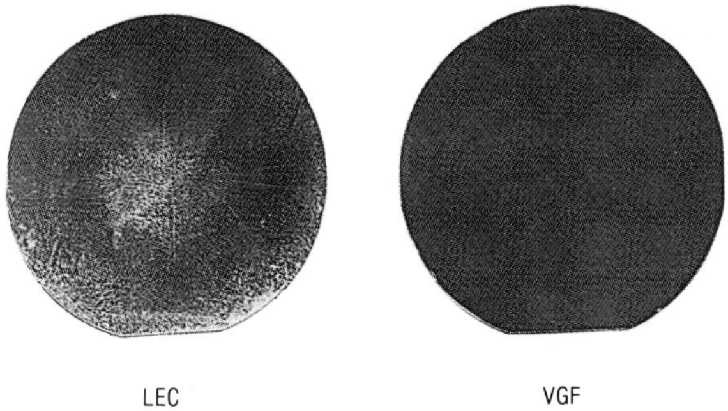

Fig. 12. Etch pit densities of KOH-etched VGF and LEC 75 mm diameter GaAs wafers.

GaAs. The resulting low level of dislocations agree well with the predictions of Jordan et al. for 50 and 75 mm diameter GaAs crystals [40]. The reduction is attributable to the lowering of the thermoelastic stress below the critical resolved shear stress. Undoped semi-insulating 75 mm diameter crystals have been grown with an average etch pit density (EPD) of 2500 cm^{-2} compared to 1500 cm^{-2} in the 50 mm diameter crystals [41]. Silicon doped, n-type crystals have even lower dislocation levels, well below 1000 cm^{-2} [23]. Figure 12 is a comparison of the defect density in 3 inch semi-insulating GaAs wafers grown by the LEC and VGF techniques. These (100) wafers have been polished and subjected to a molten KOH defect revealing etch. The LEC wafer has a significant radial variation in dislocation density with an overall level in the mid 10^4 cm^{-2} range. The VGF wafer has a considerably more uniform defect distribution with the overall level in the low 10^3 cm^{-2} range.

Abernathy and co-workers [42] repeated Clemens work on 50 mm diameter semi-insulating GaAs in a furnace of similar design, with axial gradients of less than 4°C cm^{-1}. Typical growth rates were 2–3 mm h^{-1}. After solidification was complete, the furnace was ramped to room temperature at a rate of less than 100°C h^{-1}. The entire crystal was slowly cooled after solidification in order to minimize any hoop stress. Axial and radial temperature gradients generated in the cooling solid are just as important in determining the stress levels as those at the melt–solid interface. The resulting dislocation levels ranged from 2000 cm^{-2} at the seed end to 6000 cm^{-2} at the tail. Alloying with indium at a concentration of 2×10^{19} cm^{-3} produces a lattice hardened material with dislocation levels on the order of 100 cm^{-2}.

Bourret and Merk [43] discovered that boric oxide played a similar crucial role in their growth of GaAs by the VGF technique. They employed a multi-zone furnace to grow 2 inch diameter boules that were totally encapsulated by B_2O_3. They found that the degree of wetting of the PBN was directly related to the water content of the boric oxide. However, "wet" B_2O_3 led to water vapor being trapped in the annular B_2O_3 layer between the PBN crucible and the ingot. This caused a significant degradation of the surface morphology of the crystal although no increase in polycrystalline grain nucleation. In addition to adding bulk boric oxide for composition control, they employed a high temperature oxidation cycle involving flowing O_2 (100 sccm) at 1100°C for 14 h producing a B_2O_3 layer 50 microns thick on the PBN surface.

An additional important aspect of B_2O_3 encapsulation in the growth of semi-insulating GaAs is the complex interdependence of the material resistivity on the water content of the encapsulant and the composition of the melt. Bourret and Merk have seen a sharp drop in resistivity toward the tail end of VGF GaAs ingots grown with "wet" (water content greater than 600 ppm) boric oxide in contrast to their findings with "dry" B_2O_3 [43]. The resistivity of the undoped GaAs, grown from Ga-rich melts, varied from 10^2-10^8 ohm cm as the water content increased from 100 to 2000 ppm. What is clear is that there are numerous subtle differences among LEC, VGF and horizontal Bridgman GaAs growth techniques involving melt composition, carbon, silicon and boron impurities, B_2O_3 water content and the crucible material. The resulting material properites of the material grown by each of these techniques will therefore be unique.

4.2.2. Liquid encapsulated vertical Bridgman growth of GaAs

A second approach, that of liquid encapsulated vertical Bridgman growth was first reported by Hoshikawa et al. [38] who grew 3 inch diameter GaAs crystals using the liquid encapsulated vertical Bridgman (LEVB) technique. They claim that boric oxide is critical for success in two respects: (1) The boric oxide forms a protective layer around the single crystal seed in the PBN seed well. The seed is intentionally etched to a loose fit in the seedwell and a meniscus of the GaAs melt wets the sides of the single crystal seed. The encapsulant prevents the generation of grain boundaries at the three phase junction. (2) Total coverage of the inner surface of the crucible wall with B_2O_3 prevents the melt from coming into direct contact with irregularities which can cause nucleation of grain boundaries. In addition, the LEVB approach, as practiced by Hoshikawa, relies entirely upon the encapsulant to maintain melt composition; there is no separate arsenic reservoir.

In Hoshikawa's LEVB system, the PBN crucible is lowered through a fixed temperature gradient at a rate $3\,mm\,h^{-1}$, although the average growth rate was estimated to be twice that. The crucible is rotated in the hot zone at 2 rpm in order to average out thermal asymmetries. The use of the boric oxide encapsulant simplified the furnace design by eliminating the lower arsenic heating zone. The reported dislocation density in the $\langle 100 \rangle$ seeded, 75 mm diameter crystals varied from 5000 to 40 000 cm^{-2}. While this is somewhat higher than the levels reported for the VGF approach, it is still only 20%–50% of the level obtained with the standard LEC process.

In theory, with encapsulation, one gives up the ability to fine tune the melt composition during growth. The fact of the matter is that this capability has never been exploited in any reported VGF work.

4.2.3. Liquid encapsulated vertical zone melting

Swiggard has reported on a novel method for GaAs growth; liquid encapsulated vertical zone melt growth (LEVZM) [39]. This is shown schematically in fig. 13. A molten zone is established via a narrow "spike" heater within a dense charge. The crucible is then translated through this spike zone in a manner similar to zone refining. Independently controlled heater zones above and below the spike adjust the length of

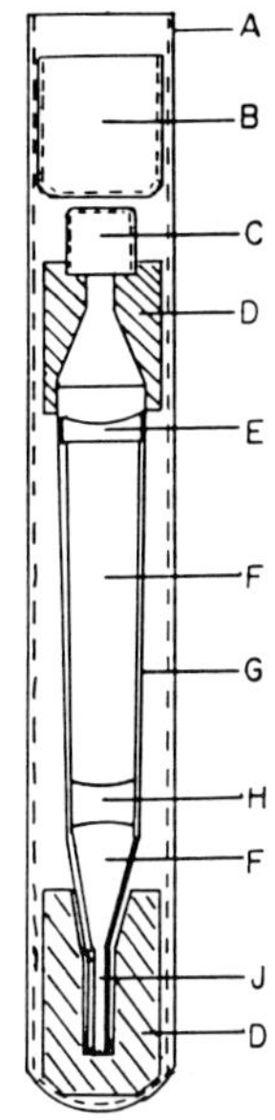

Fig. 13. Growth container schematic for the vertical zone melting (VZM) growth of GaAs. A, quartz compound; B, quartz plug; C, quartz cup; D, BN parts; E, boric oxide; F, solid GaAs; G, PBN crucible; H, liquid GaAs; J, GaAs seed.

the molten zone and can control the temparature gradient at the growth interface. Composition control is achieved through the use of boric oxide as an encapsulant. Single crystals, ⟨100⟩ seeded, 28 mm in diameter, weighing 400 g have been grown and characterized. There is nothing particularly unusual about the properties of the semi-insulating GaAs grown by the LEVZM method; what is intriguing are some of the possible directions that the development of this technique could lead to. The curvature of the melt–solid interface can be adjusted by varying the height of the molten zone. Since the heat flux to the crystal is normal to the growth axis and is concentrated in a narrow spike-heater zone, the two melt–solid interfaces should be convex. This is contrary to the normal Bridgman or VGF situation where the input heat flux has a larger axial component, with the resulting concave interface shape. Important questions, concerning the realitionship of the interface shape with edge faceting and twinning could begin to be answered with the aid of this experimental tool.

5. *Crucible confined growth of GaP*

5.1. *Early gradient freeze growth of GaP*

The need for high quality substrates for red and green LEDs drove the early development efforts in the growth of GaP. A.G. Fischer [44], in a comprehensive survey of the synthesis and growth of high melting II–VI and III–V compounds, refers to

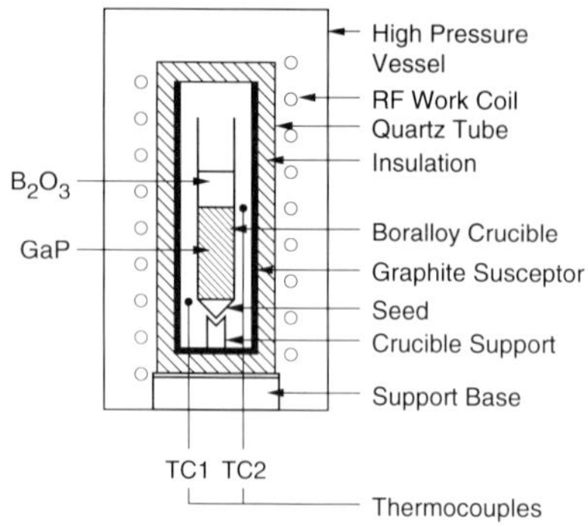

Fig. 14. Configuration for the liquid encapsulated vertical gradient freeze growth of GaP.

the Bridgman growth of GaP in PBN crucibles with B_2O_3 encapsulation, but no details of results are given. Blum and Chicotka were the first to report the results of vertical unseeded Bridgman growth in PBN(also known as Boralloy) crucibles [24]. In their initial experiments, a separately heated phosphorus reservoir maintained the melt composition as the 1.5 cm diameter crystal was grown. The results were not reproducible as only one single crystal was obtained. Several years later, Blum and Chicotka reported on the liquid encapsulated gradient freeze growth of GaP [45]. As motivation for the VGF effort, they point out five problems with the growth of GaP via an LEC approach: (i) high residual strain in the crystals, (ii) high dislocation levels, (iii) difficulty in maintaining diameter control, (iv) pronounced growth striations, and (v) the requirements for close attention during entire growth cycle.

This work featured a considerably simplified apparatus that dispensed with the separate reservoir of phosphorus entirely and is shown in fig. 14. The 1.5 cm diameter crystals were ⟨111⟩ seeded and showed considerably less strain and lower dislocation levels than comparable LEC material. They claim growth rates of 25 mm h^{-1} followed by cooling to room temperature in one hour. The resulting dislocation density was 10^3 cm^{-2}, an order of magnitude below that achieved by LEC growth.

Woodbury later attempted to repeat their work but was unsuccessful in achieving low dislocation levels [46]. He did however make several significant contributions to the understanding of LEVGF. Some of the aspects on which he focused attention were: (i) measuring the radial and axial temperature profiles of the hot zone, (ii) criteria for the proper fit of the seed crystal to the crucible seed well, (iii) scale-up to 1 inch diameter crystals and (iv) preliminary studies of the LEVGF growth of CdSe. He also introduced a "foaming" step by reducing the inert gas pressure prior to the B_2O_3 solidifying during the cooling cycle. This removed some of the encapsulant from the annular space between the ingot and crucible walls, thus facilitating removal of the ingots and preserving the expensive PBN crucible.

5.2. *Vertical gradient freeze growth of GaP*

Due, in part, to the difficulties involved with the high pressures (exceeding 40 atm), high temperatures (approximately 1500°C) and the control of the phosphorus activity, there

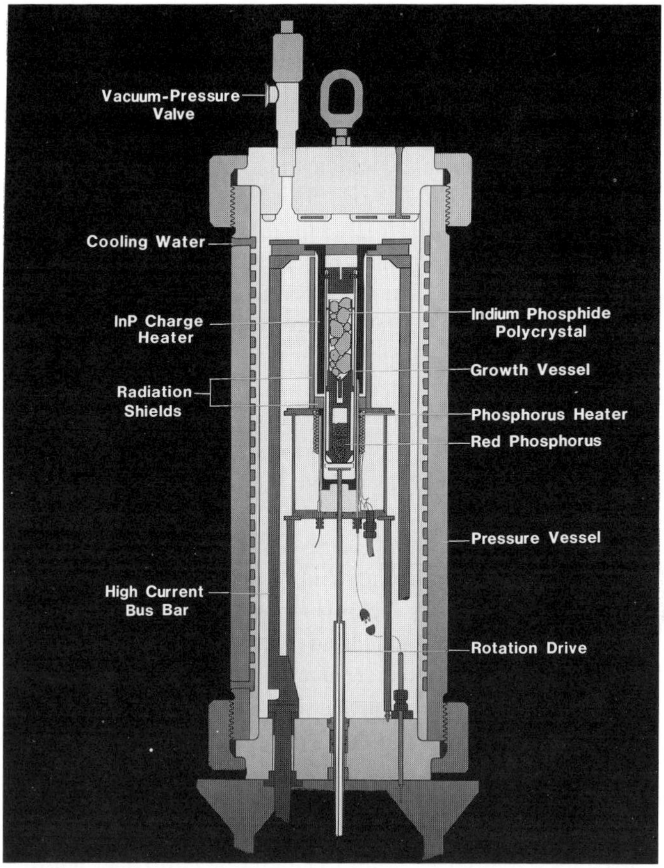

Fig. 15. Cross sectional view of a high pressure modified VGF InP crystal growth system.

was little progress in non-LEC growth of GaP until the work of Gault et al. was published in 1986 [23]. One of the major driving forces for this work was the reduction of dislocations in GaP, since the electroluminescent efficiency of the red and green LEDs grown on GaP substrates was shown by Beppu et al. to be dependent on the dislocation level [47]. The VGF method has the advantage of minimizing the thermal stress by employing low thermal gradients, low growth rates and controlling post-solidification cooling rates beyond the yield point for stress-induced dislocation generation.

There are several basic design concepts that are common to the growth of III–Vs by this process. Figure 15 is a schematic of the VGF crystal growth system minus the power supply and control system. A vertical two zone heater assembly is enclosed within a water cooled, stainless steel lined pressure vessel. Within the heater assembly is a growth vessel which contains a source of phosphorus and the charge of the polycrystalline material. The lower zone heats a reservoir of condensed phosphorus and provides a means of composition control by regulating the phosphorus partial pressure over the melt. The upper zone heats a bottom-seeded crucible and an unencapsulated charge of the III–V compound; this zone controls solidification and subsequent cooling during the growth run. The desired axial and radial temperature gradients are attained via an empirically optimized combination of insulation, heat shields and heater design. Rotation of the growth vessel averages out the thermal asymmetries of the assembly. The growth vessel can be translated within the heater in order to adjust its position in the temperature profile. There are appropriate high pressure feed-throughs for a rotation–translation shaft, various monitor and control thermocouples, and power leads. There are separate pressurizing and vent lines, and a high conductance vacuum valve. The upper zone heater is a shaped graphite "picket-fence" resistance heater. The heater resistance is altered by machining the wall, on a lathe, to the desired thickness. By such an iterative process the desired thermal profile of the charge heater can be established. The heater is supported by two stainless steel electrodes carrying the power from a low voltage, high current power supply. The heater is, in turn, surrounded by a wound laminated graphite insulation shield which extends along the length of, as well as below, the heater. The shield below the charge heater is required for adequate isolation of the low temperature zone from the high temperature zone. The hot junction of the upper zone control thermocouple is located in the bottom of the charge heater adjacent to the seedwell region of the crucible. Directly below the charge heater is the lower zone, or phosphorus control heater. The phosphorus heater consists of an insulated Nichrome wire which is wound around a cylindrical graphite core, so that there is an axial monotonic temperature increase with height. The phosphorus heater control thermocouple is housed within the hollow, rotation–translation shaft, and the junction is located in the growth vessel support pedestal just below the coldest region of the phosphorus reservoir.

Figure 16 is the temperature control system for the modified VGF system laid out in block diagram form. The control loop for the upper zone is straight-forward; a process programmer provides a remote set point to an analog controller whose output fires an SCR power supply. Closed loop control is accomplished by feeding back the output of a high temperature W5% Re/W26% Re thermocouple to the controller. The control system for the phosphorus heater is somewhat more involved and is a good example of a cascaded control system.

The temperature at the junction of the S-type thermocouple housed in the shaft below the growth assembly pedestal is to be controlled independently of the upper zone heater. However, this control point is spatially well removed from the windings of the phosphorus heater and there is a considerable time lag between the time a temperature step is made at the heater and when it is fully realized at the shaft thermocouple. A second thermocouple, embedded in the phosphorus heater, has a much smaller time constant with respect to phosphorus heater temperature changes. However, its junction is well removed from the desired phosphorus control position. The solution to this problem lay in the employment of a cascade control system. Controller #1, with a faster responding thermocouple adjacent to the heater windings, can be thought of as the coarse control. Controller #2, whose thermocouple is just below the phosphorus reservoir, is the fine control. The 0–10 V output of controller #1 is fed directly to the trigger circuitry of the SCR power supply. The output of controller #2 is level shifted to

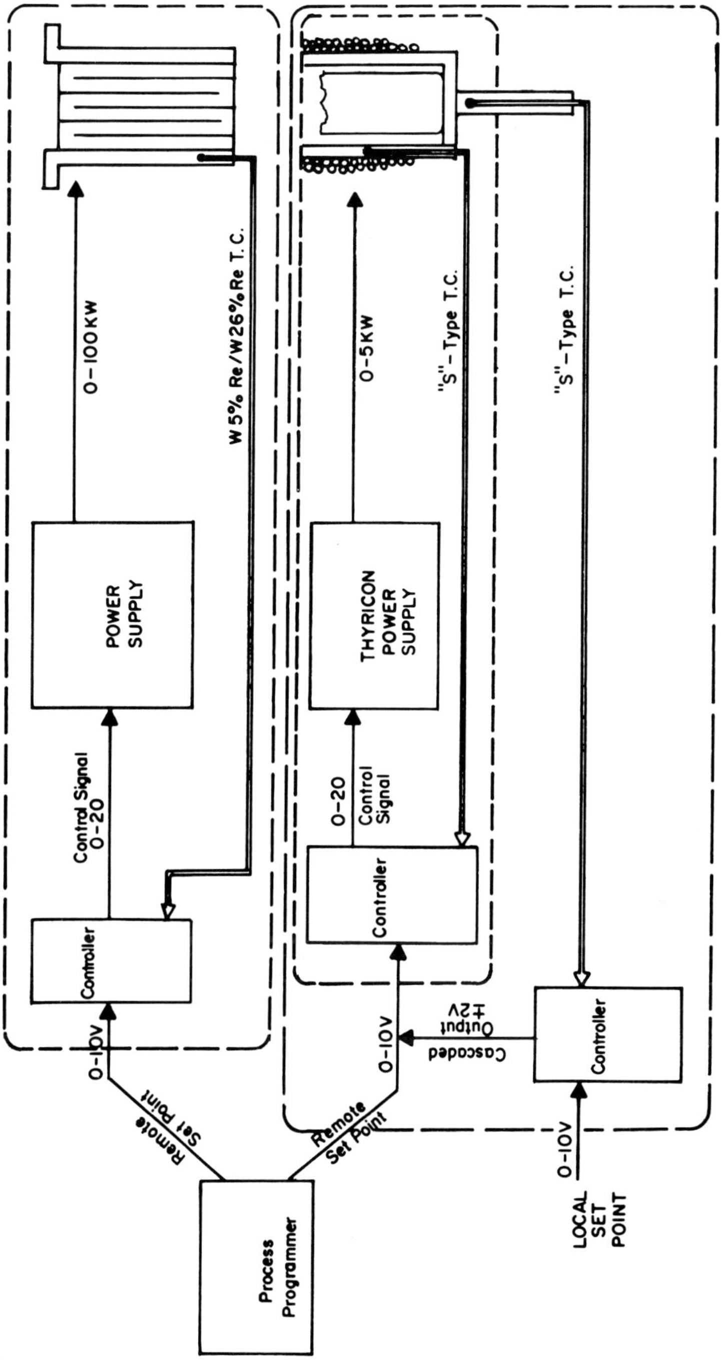

Fig. 16. Temperature control system schematic for a two-zone modified VGF crystal growth system.

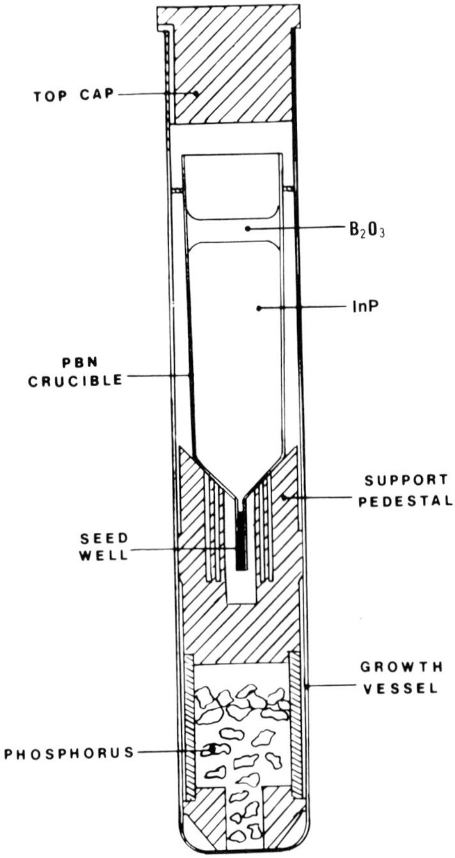

Fig. 17. Cross sectional view of the growth vessel assembly for the modified VGF growth of InP.

a range of ± 5 V and then biases the controller #1 set point, providing the fine tuning to maintain the phosphorus set point with a much faster response than would otherwise be the case in a single control loop system.

Figure 17 is a more detailed view of the pyrolytic boron nitride (PBN) growth vessel and its contents. A reservoir to contain the red phosphorus is provided below the support pedestal. The support pedestal is a cylinder of hot pressed BN in which several concentric slots are machined around the seedwell area. By proper choice of the number and dimensions of the slots, the thermal geometry and the resulting melt–solid interface shape can be altered in the seedwell region.

The PBN crucible is nominally 50 mm I.D. and is terminated in a 6.3 mm I.D. seed-well. A top cap of hot pressed BN is pinned to the top of the PBN growth vessel. The top cap confines the phosphorus vapor as well as providing a means by which to lift it in and out of the heater assembly. During growth, there is a continuous loss of phosphorus vapor through the annulus between the top cap and the wall of the

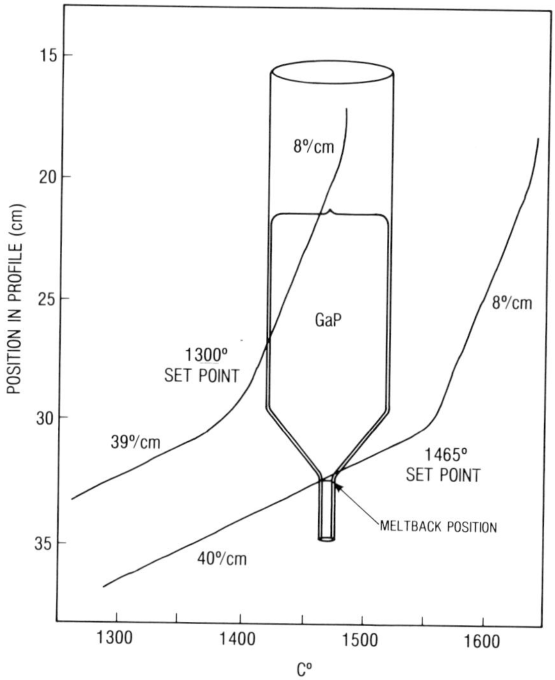

Fig. 18. Axial temperature profiles at the seeding and last-to-freeze segments of VGF growth of GaP.

growth vessel. However, sufficient condensed phosphorus remains in the reservoir to maintain the desired partial pressure for the duration of the growth run.

Figure 18 shows two axial temperature profiles of the hot zone of the GaP VGF systems. These profiles were measured with a growth vessel that was loaded with a dummy profiling charge, a shaped piece of hot pressed BN substituting for the PBN crucible and melt. A seeded crucible with the meltback position representing the melt–solid interface (for a control set point of 1465°C) is shown for reference. The axial temperature profile can be approximated by a linear gradient of 40°C cm^{-1} in the seedwell region and a linear gradient of 8°C cm^{-1} over the 50 mm diameter region of the crystal. The relatively large gradient in the seedwell facilitates the reproducible meltback of the seed. In order to achieve a flat to slightly convex interface shape, the imposition of a large thermal gradient over the solid seed, with its lower thermal conductivity relative to the melt, has been shown to be helpful. The two profiles shown in fig. 18 were taken at two power levels, corresponding to temperatures of 1465 and 1300°C in the seedwell region. The 165°C difference is the amount of cooling found to be necessary to solidify a 1000 g charge of GaP under these conditions. The growth assembly components have been designed so that the two profiles are similar in shape. Within this range, the fact that the thermal profile shape is independent of the absolute power level simplifies the determination of a cooling rate program to achieve the desired growth rate.

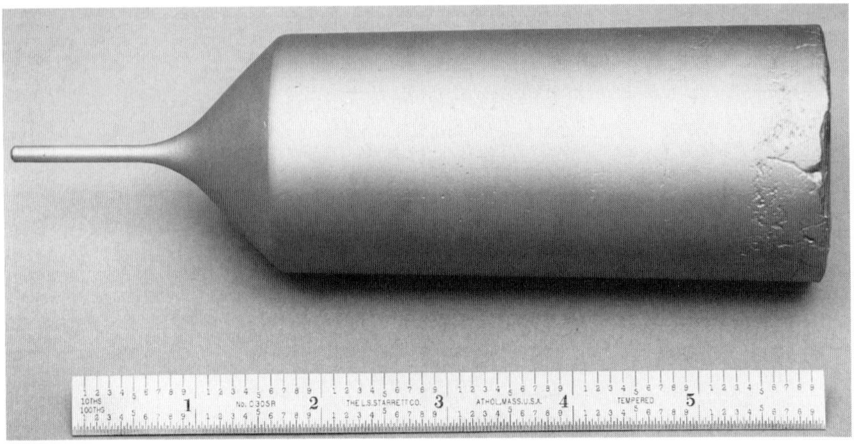

Fig. 19. Tellurium-doped, 1300 g, (111)-seeded gallium phosphide single crystal grown by the modified VGF method.

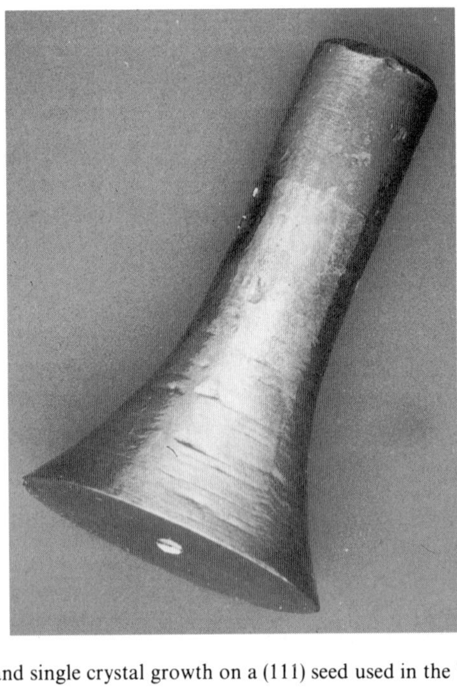

Fig. 20. Meltback and single crystal growth on a (111) seed used in the VGF growth of GaP.

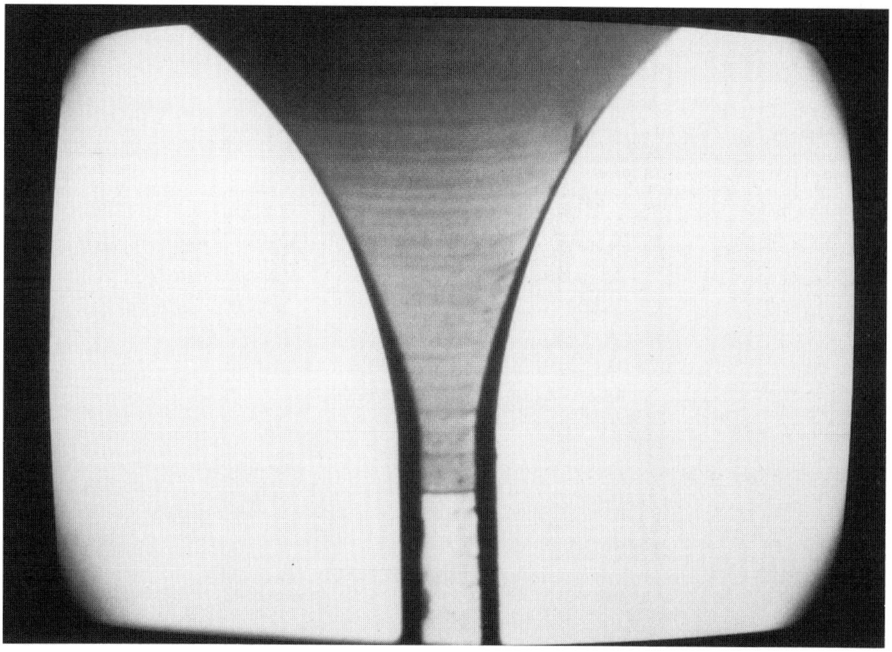

Fig. 21. Infrared transmission photograph of the seeding region of a VGF grown GaP single crystal.

Figure 19 shows a 1300 g, (111) seeded, 50 mm diameter GaP single crystal grown by this process. Both sulfur and tellurium doped n-type crystals have been grown. Small steps on the inner surface of the PBN crucible wall, due to partial delamination, are replicated by the ingot surface with no polycrystalline nucleation. Figure 20 is an example of the typical meltback of the (111) seed and the initial growth necking out towards the 50 mm body diameter. If the seed is properly sized by a dilute bromine/methanol etch to slip fit into the PBN seedwell, there is no spillover of the melt along the side of the original seed. This can lead to random nucleation and subsequent polycrystalline growth. The faceted growth visible in the neck region indicate that the surface is free and not wetting the PBN crucible. Figure 21 shows a transverse slice of the seed and transition region of a VGF-grown GaP single crystal. The seed was undoped while the polycrystalline GaP charge was sulfur-doped. The position of the meltback is clearly seen. The striations are typically slightly concave in this section of the crystal since the reducing cross sectional area acts as a thermal choke to the axial flow of heat. Efforts to closely match the thermal conductivity of the crucible support to that of the solid crystal have flattened the interface in the seeding region (see figs. 31 and 32).

Figure 22 shows photomicrographs taken in three regions of a (111) wafer sliced from a VGF-grown GaP crystal similar to the one shown in fig. 19. The wafer was polished and etched with a ferricyanide–KOH–H_2O solution to reveal dislocations. The magnification is such that 50 pits in the field of view corresponds to an EPD level of

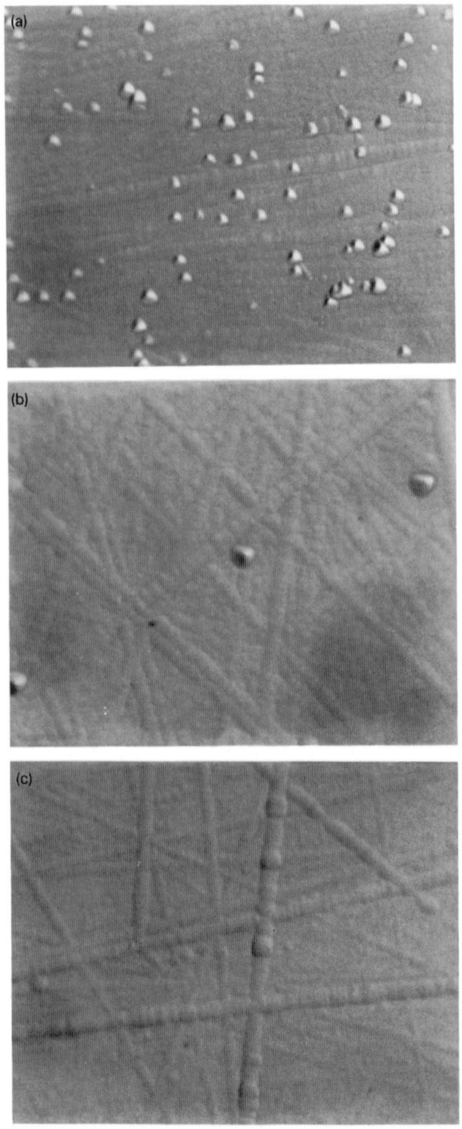

(a) (b) (c)

Fig. 22. Dislocation density of VGF-grown GaP wafers.

10^3 cm^{-2}. The etch pit count ranges from a high of mid 10^3 cm^{-2} at the very edge of the wafer (fig. 22a), to less than 100 cm^{-2} at a point 3 mm from the edge (fig. 22b), to a dislocation free area in a center 25 mm core of the wafer (fig. 22c). These results are comparable to the best published LEC results by Watanabe [48] and are one to two orders of magnitude better than the normally reported values for large diameter LEC-grown GaP.

6. *Crucible confined growth of InP*

6.1. Horizontal Bridgman and LEC growth of InP

The primary application of InP is in devices for long wavelength (1.3–1.6 microns) optical communication. Semi-insulating InP is also used as the substrate for the developing field of integrated optoelectronics. The substrates for such purposes should have the following attributes; (1) low dislocation densities and a corresponding absence of slip, (2) uniform properties and dopant distribution across a wafer and within a given boule, (3) circular wafers with a minimum 50 mm diameter, (4) low optical absorption coefficient, (5) minimal grown-in strain, and (6) low cost.

Weisberg et al. [49] improved upon the horizontal gradient freeze system for the growth of InP that was first demonstrated by Guire and Weiser [50]. It was felt that this approach "provided optimum control of the thermal decomposition equilibrium of InP during the process of crystal growth". Bachman and Buehler [51] refined this approach and used a modified growth technique where all of the phosphorus was in the vapor phase in a near isothermal ampoule in order to avoid transport of the volatile ZnP_2 or CdP_2, used for p-type dopants, to the cold end of the quartz ampoule. The technique was unseeded, due in part, to the shallow axial gradients in this system. Since the seed is not visible, it is difficult to avoid melting the entire seed or not melting it at all in a shallow gradient. The synthesis and growth of InP in these closed quartz systems was always quite exciting since the spectacle of a phosphorus conflagration was never far off.

Mullin et al. first applied the LEC process to the growth of InP [28]. Although this approach was more costly than zone melting or gradient freeze, it had the distinct advantage that oriented crystals could be grown in a single working day. Single crystals were obtained at a much higher yield when stoichiometric polycrystalline starting material was used. Constitutional supercooling in In-rich melts could only be avoided by considerably slowing down the grown rate. Nonetheless, the LEC approach has been used nearly exclusively since 1968 for the growth of single crystal InP.

6.2. Vertical gradient freeze growth of InP

The first report of the application of the VGF approach to the growth of InP was in 1986 [23]. The motivation for this effort was to produce substrates with a lower level of stress and a higher degree of uniformity than was available from LEC-grown boules. The process was very similar to that described for the VGF growth of GaP with the following noteworthy differences:

(1) The approximate 400°C decrease in melting point, 1465°C versus 1063°C, results in a lower axial gradient between the upper and lower heater zones for InP and thus affects the heat transfer between them. The design of the insulation package for the hot zone must be modified so that control of the phosphorus heater is maintained over the entire range of input power to the upper zone.

(2) The decrease (40 to 27.5 atm) in the equilibrium vapor pressure of phosphorus over the melt corresponds to a reduced phosphorus control temperature (590°C to 558°C). This lower control temperature is nearer to the temperature (550°C) of the reported phase transformation of phosphorus from tetragonal form IV to triclinic form V [21]. The slow rate (on the order of days) for this transformation to occur and the uncertainty as to which modification one initially starts with makes the control of the phosphorus pressure more difficult for InP growth.

(3) One is able to grow InP under an overall lower argon pressure than is necessary for GaP. This has the desirable effect of reducing convection-driven temperature control instability.

Figure 23 shows a typical axial temperature profile for the VGF growth of InP superimposed on the typical crucible position within this profile. The gradient is steeper in the seed region. This allows a greater axial heat flow during growth and it enhances the ability to melt back to a reproducible position on the seed. The gradient is

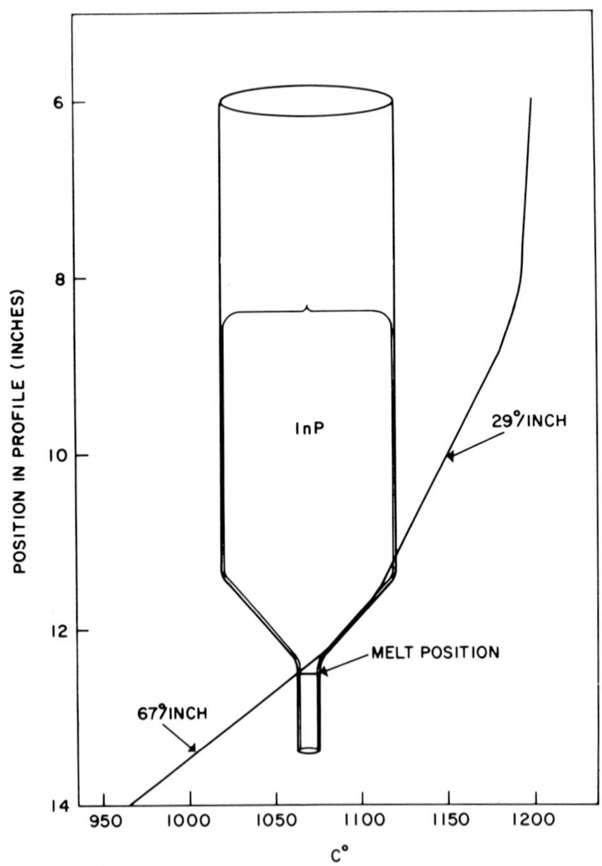

Fig. 23. Axial temperature profile for the VGF growth of InP.

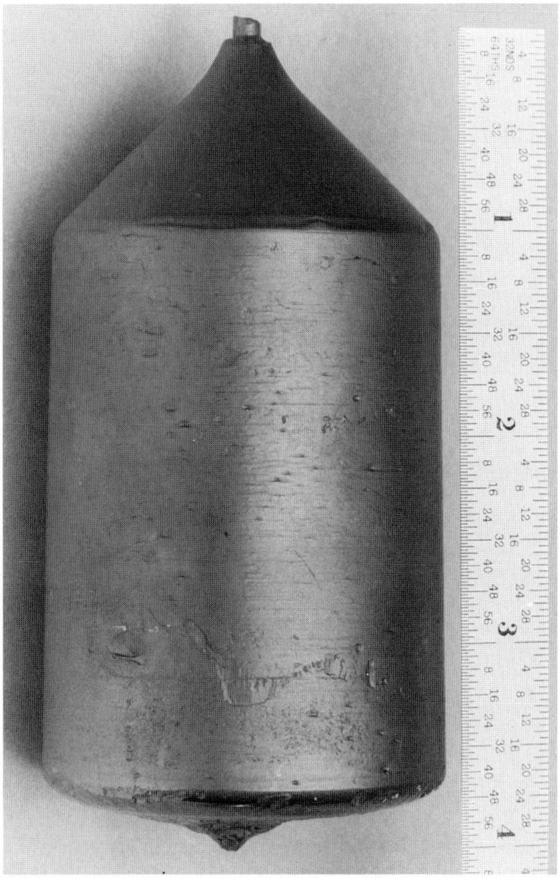

Fig. 24. Indium phosphide single crystal grown by the VGF method.

then somewhat shallower in the main body of the crystal in order to reduce the thermoelastic stress on the cooling crystal. Figure 24 is a typical 750 g, (111) seeded single crystal of InP. The last-to-freeze portion is fully composed, with no excess indium, indicating that the compositional control of the melt was effective. It can be seen that the seed-to-crystal meltback is not a smooth transition, rather it is somewhat high on the seed. In fact, this is similar to Hoshikawa's prescribed seed–crystal interface in the LEVB growth of GaAs [38]. Nevertheless, the growth from this seed is single crystalline. This is due to the planar melt–solid interface shape and the lack of interaction or wetting of the crucible.

6.2.1. Dislocation reduction in VGF InP

The low axial and radial temperature gradients associated with the vertical gradient freeze technique result in a low level of thermoelastic stress and a correspondingly low

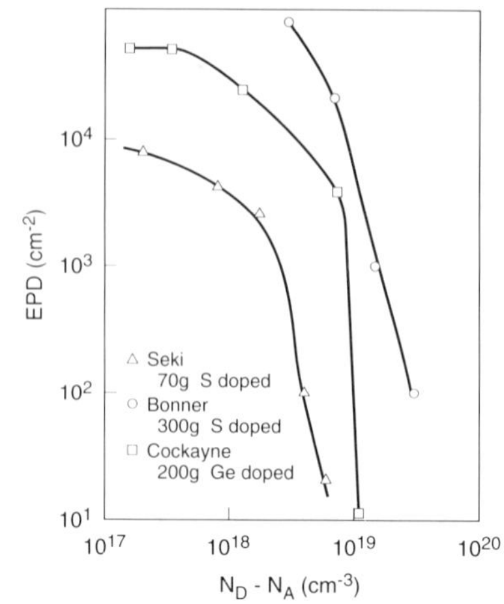

Fig. 25. Dislocation reduction by impurity hardening in LEC-InP.

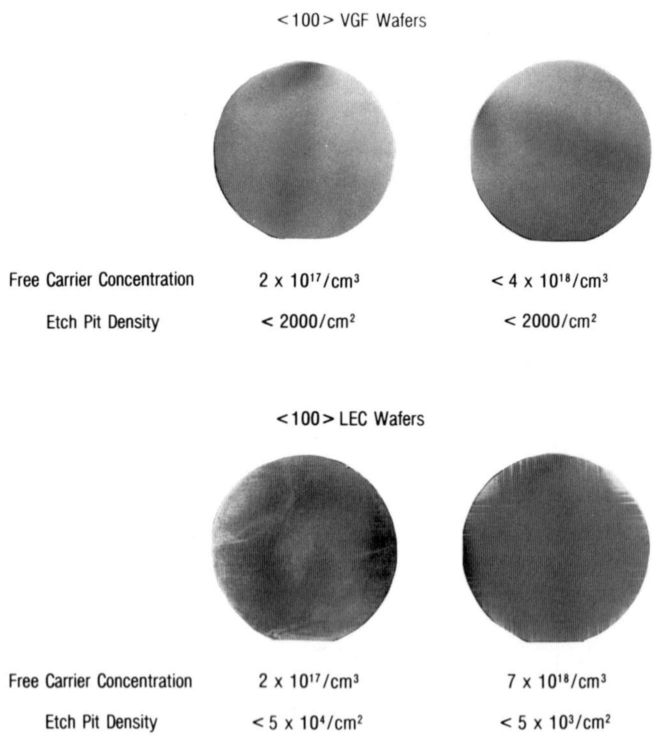

Fig. 26. Comparison of dislocation reduction by impurity hardening in LEC- and VGF-grown InP wafers.

level of defects. This is achieved without resorting to the addition of impurity hardening dopants, as is common for LEC-grown material and illustrated in fig. 25. Seki et al. first demonstrated the hardening effect of zinc, sulfur and tellurium doping of small LEC-grown InP single crystals [52]. Bonner later demonstrated the effectiveness of this approach upon scaling up to larger crystals [53]. Brown et al. showed dislocation reduction for Ge-doped InP [54]. In all the cases there is a sharp drop in EPD at a characteristic doping threshold, usually exceeding 4×10^{18} cm^{-3}.

Figure 26 shows the dependence of dislocation density on doping levels for both LEC and VGF grown InP. The wafers shown are (100), sulfur-doped, InP wafers that have been polished and subjected to a defect revealing etch [55]. The crystals were grown with doping levels both above and below the threshold for lattice hardening. The LEC wafer with the lower doping has extensive slip and a high level of dislocations. Even the highly doped LEC wafer has extensive peripheral defects that must be trimmed. In comparison, the VGF material shows very low defect densities and no slip at either level of doping.

Figure 27 shows an undoped InP single crystal grown with a seed that was cut from an updoped LEC crystal. The dislocation density in the LEC seed exceeds 10^4 cm^{-2},

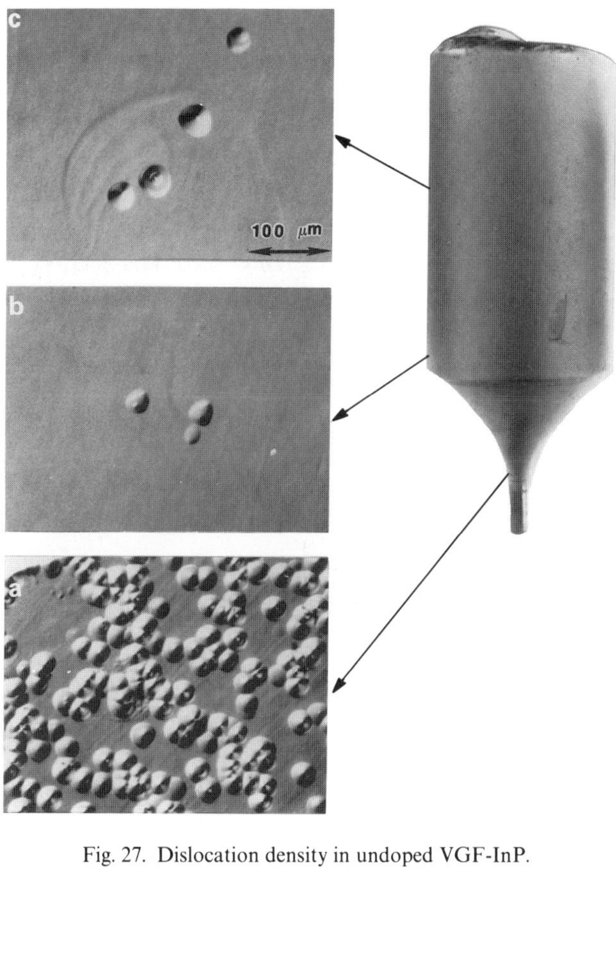

Fig. 27. Dislocation density in undoped VGF-InP.

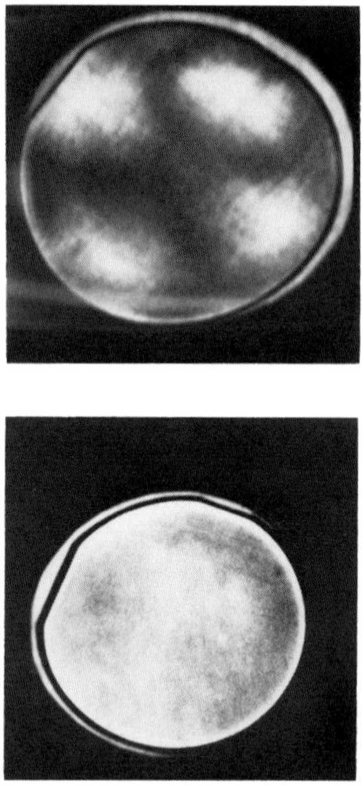

Fig. 28. Residual strain in InP substrates.

while the VGF grown material is nearly dislocation free. The striking reduction in EPD in the bulk crystal is all the more remarkable given the fact that many of the dislocations in the bulk thread up from the LEC seed. Dislocation levels well below $100\,\mathrm{cm}^{-2}$, with no peripheral slip, are routinely measured for an entire VGF undoped crystal.

Despite the dislocation reduction achieved by impurity hardening, a high level of residual strain can still be found in LEC grown material. Figure 28 compares the residual strain in low dislocation LEC and VGF wafers. Using crossed polarizers and a large field CCD camera, the stress induced birefringence contrast is compared. The LEC wafer shows a four-fold stress pattern in contrast to the VGF material. These stress patterns compare quite favorably to those calculated by Jordan et al. for (100) LEC and VGF wafers [56].

6.2.2. *Striations in VGF/EDGF InP*

Variations in growth rate result in a corresponding fluctuation in impurity incorporation. These variations can be induced by factors such as vibration, convective instability, temperature fluctuations, etc. These fluctuations are responsible for free carrier and refractive index changes and can appear as growth striations which demarcate the progress and shape of the melt–solid interface as growth proceeds. Imaging these growth striations allows one to see the balance between radial and axial heat flow, measure growth rates, and assess the relative stability of the growth process. Various etchants have enjoyed some degree of success in bringing out these striations in doped InP, Brown et al. used an A–B etch to look at striations in small sections of the crystal [54].

Chu has compared the transmission cathodo-luminescence (TCL) of striations in VGF and LEC-grown InP [25]. The luminescence efficiency for S-doped (100) InP has been found to increase with dopant concentration in the rage $1–20 \times 10^{18}\ cm^{-3}$. Therefore, the degree of variation in dopant incorporation can be determined qualitatively in the TCL contrast of the dopant striae. TCL photos are presented in fig. 29 for the LEC and VGF material. The image for the LEC material in fig. 29a (sulfur-doped, $1 \times 10^{18}\ cm^{-3}$) shows sharply contrasting bands with rather irregular spacing. The TCL image of the VGF sample taken under similar SEM conditions is shown in fig. 29b (sulfur-doped, $3 \times 10^{18}\ cm^{-3}$). There is very little contrast apparent. This indicates that the level of dopant fluctuation in the VGF material is much less than that of LEC and that the process itself is, in all likelihood, considerably more stable.

Transmission X-ray topography gives a larger scale view of the striation shape, but is quite a laborious procedure. However, Barns has developed a rather simple

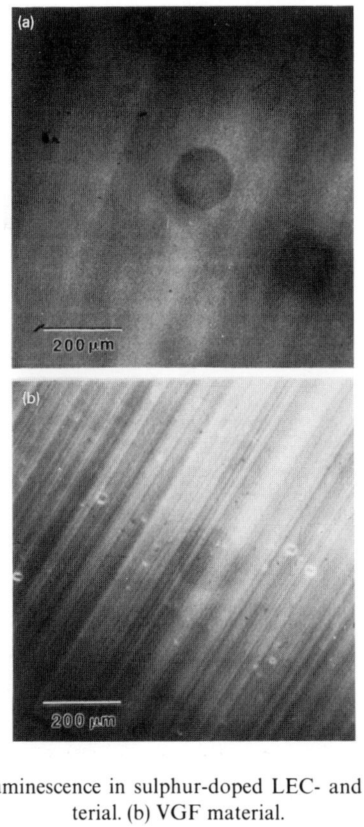

Fig. 29. Transmission cathodoluminescence in sulphur-doped LEC- and VGF-grown InP. (a) LEC material. (b) VGF material.

procedure for imaging, on high speed infrared sensitive film, the variations in transmission through polished longitudinal slabs of InP crystals [57].

The shape of the melt–solid interface at the intersection with the crucible wall influences edge faceting, twin initiation, and the nucleation of secondary growth orientations. Planar or slightly convex interface shapes are desirable. The thermal conductivity of the crucible has been shown by Jasinski, et al. [58] and Holland [59] to affect the shape at this triple point. If the thermal conductivity along the crucible wall is greater than that of the solid, it tends to make the interface shape more concave.

Figures 30a and b are striagraphs taken of InP crystals grown in silica and pyrolytic boron nitride (PBN) crucibles respectively. The in-plane thermal conductivity of the PBN is about 20 times higher than that of the quartz crucible, yet there is no discernable difference in the interface curvature, at the ingot edge, between the two. The curvature of the edge of the 2.5 mm thick slab, cut from the 50 mm diameter ingot, limits how close to the actual edge of the crystal the interface can be resolved.

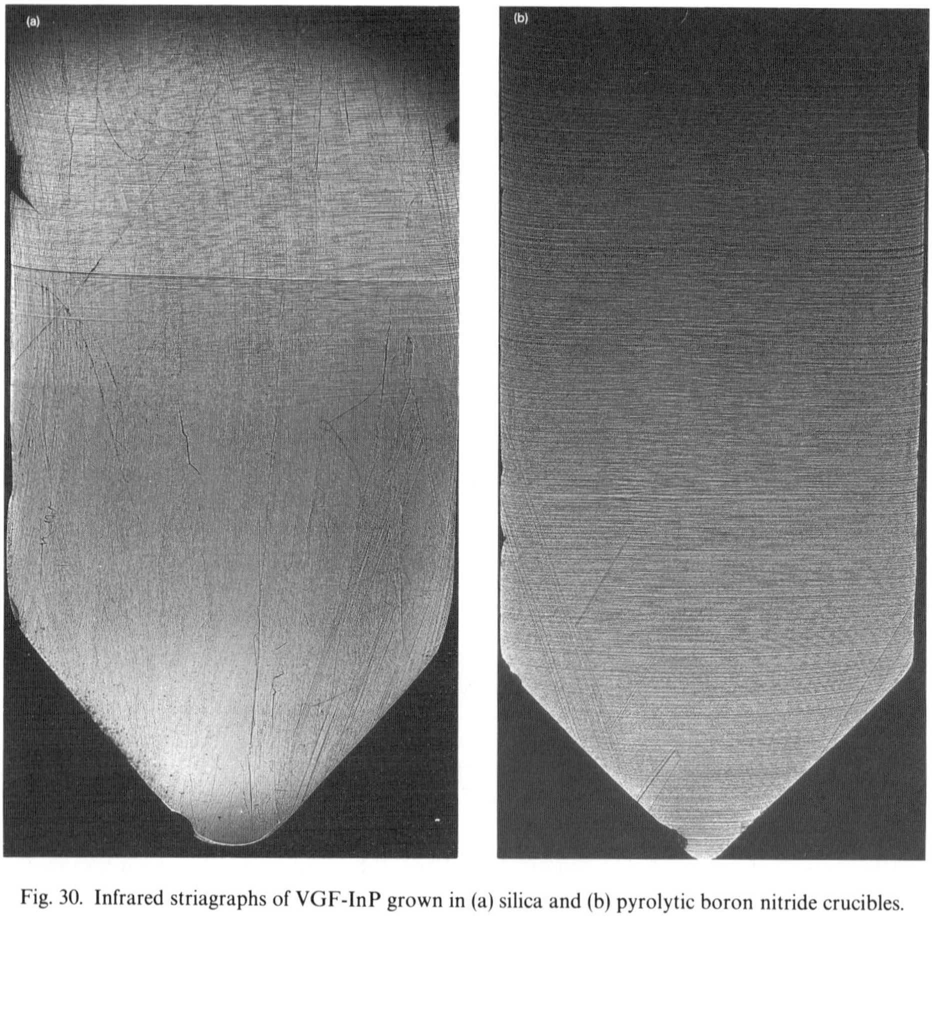

Fig. 30. Infrared striagraphs of VGF-InP grown in (a) silica and (b) pyrolytic boron nitride crucibles.

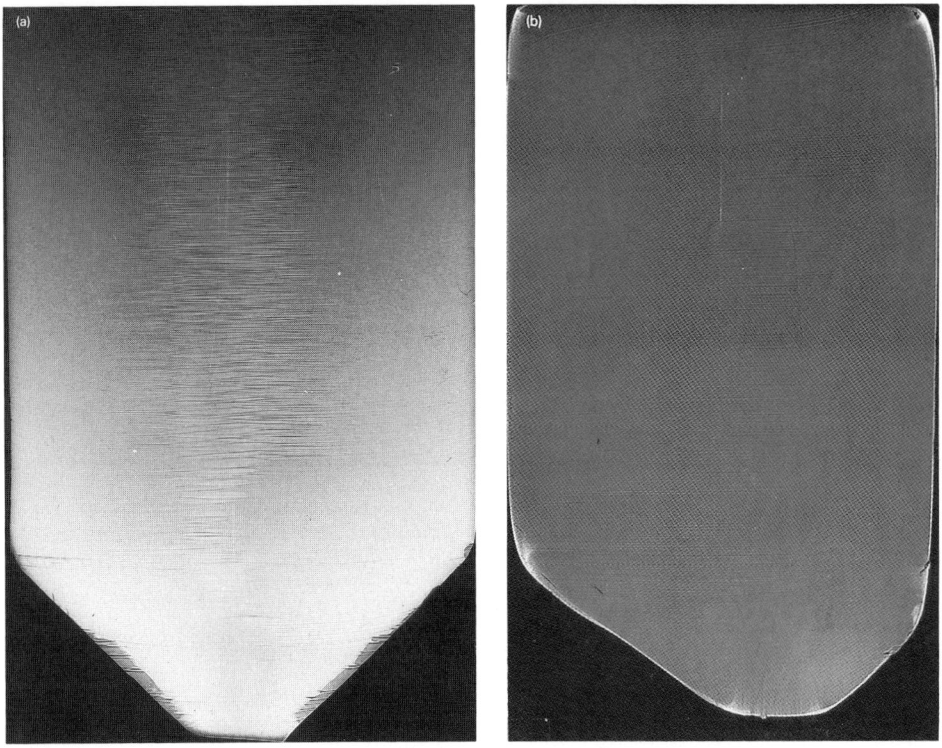

Fig. 31. Infrared striagraphs of (111)-seeded VGF-InP. (a) showing cone faceting. (b) scaled-up 63 mm ∅ crystal.

The influence of temperature stability in the melt, on striations in the grown crystal can be seen in figs. 31a and b. These striagraphs were taken from samples cut from (111) seeded VGF InP crystals where the temperature stability of the control thermocouples has been improved to better than $\pm 0.15°C$. These should be compared to the striagraphs of figs. 30a and b where the temperature fluctuations were one order of magnitude larger. There is a marked decrease in striation contrast for the growth under more stable conditions. Figure 31a shows some degree of (111) core faceting, while fig. 31b shows the interface shape in a scaled-up 63 mm diameter crystal. As can be seen, there is no significant difference in the isotherm shape in going from 50 to 63 mm in diameter.

A striking example of the diagnostic power of this striation imaging technique is shown in fig. 32. A power failure occurred during the growth run and one can easily see the effect of rapid cooling upon the interface curvature. The heat loss flux is primarily radial and the interface becomes strongly concave. In addition, the sudden increase in growth rate is reflected in the sharp increase in dopant incorporation shown by the higher infrared absorption.

Fig. 32. Infrared striagraph of an InP-VGF crystal grown with rapid cooling due to a power failure midway through solidification.

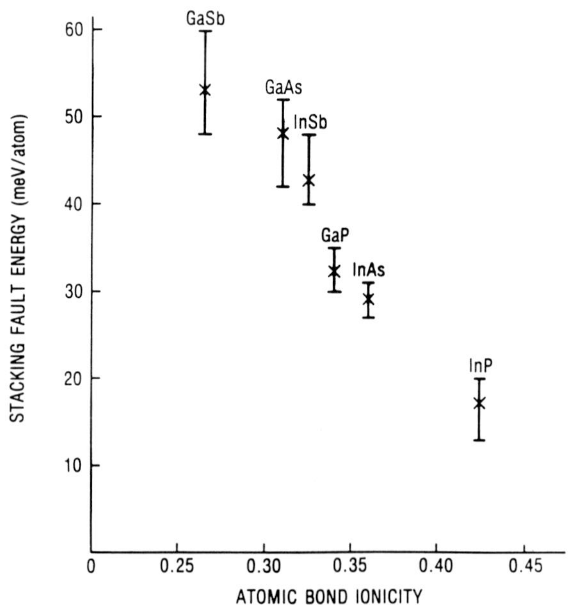

Fig. 33. Stacking fault energy as a function of the bond ionicity for a number of III–V compound semiconductors.

6.2.3. Growth twins in InP

Twinning is probably the single most persistent problem in the single crystal growth of InP, by either the VGF or LEC technique. The high degree of ionicity of the In–P bond leads to a low stacking fault energy for the zincblende to wurtzite transition that occurs on the (111) planes as a twin boundary [60]. In fact, InP has the lowest stacking fault energy of all the III–V compound semiconductors as is shown in fig. 33. The greater the bond ionicity, the lower is the energy barrier to the eclipsed stacking configuration

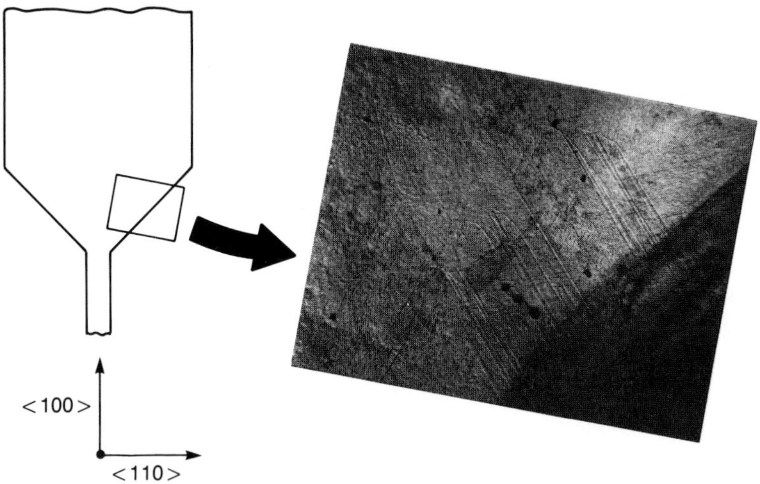

Fig. 34. ⟨111⟩ Edge facets leading to twinning in ⟨100⟩-seeded VGF-InP.

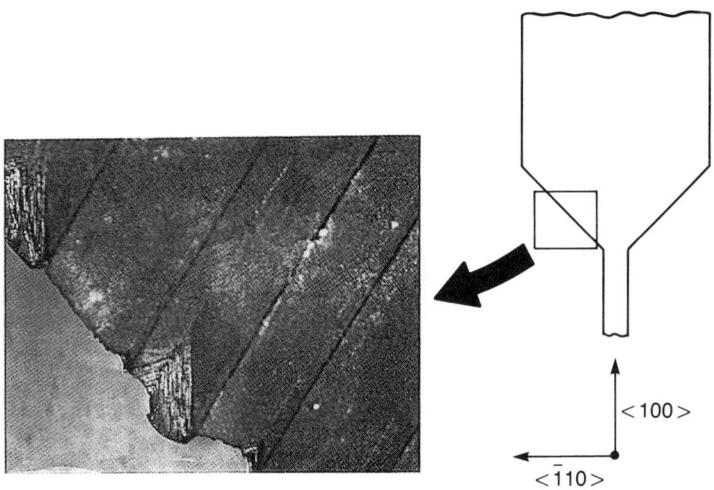

Fig. 35. Local stress induced twinning in ⟨100⟩-seeded VGF-InP.

characteristic of the wurtzite structure. The more covalent bonded semiconductors find the staggered stacking configuration energetically more favorable and remain in the zincblende structure.

Shibata et al. have observed in the LEC growth of InP that temperature fluctuations in the melt can lead to a series of irregularly spaced, (111) edge facets [61]. Occassionally these facets result in a twinning event as can be seen in fig. 34. Local stress can be generated at the melt–solid–crucible interface by the crystal sticking to the crucible. Stress induced twinning can be the result as can be seen in fig. 35. At the beginning of each and every fracture pattern, we can see the initiation of a twin lamellae. The procedure of forming a thin boric oxide layer on the inner surface of the PBN crucible, described earlier, minimizes the wetting of the crucible by the melt and concurrently, the incidence of twinning.

7. *Bridgman growth of II–VI Compound semiconductor single crystals*

The II–VI compound semiconductors and their alloys span the bandgap range from the semi-metal HgCdTe to the wide gap materials such as CdTe and ZnSe. ZnSe has recently attracted much attention because of its use as a substrate for blue lasers and LEDs. The demand for high performance infrared detector arrays has been a major driving force in the area of the growth of HgCdTe and CdTe.

Numerous growth methods have been applied to the growth of II–VI single crystals with one common tie; none of them result in large single crystal boules on a reproducible basis. There are numerous factors which account for the difficulty of single crystal bulk growth, but the most salient of these is the high ionicity of these materials. This influences the segregation, diffusion, reactivity and stacking fault energy.

The major appeal of melt growth is the speed at which crystals of reasonable size can be grown. Rates in the range of mm/h are common, although slower rates often improve the crystalline perfection, especially if constitutional supercooling is a problem. Other approaches to the growth of bulk II–VI crystals are reviewed by Fitzpatrick [62]. These include vapor growth, solution growth and solid state recrystallization. While the growth at lower temperatures can avoid the problems of phase transitions (see below) and produce crystals of very high quality, growth rates are exceedingly slow. Thus, much work continues to be done on improving the bulk growth approach.

Many II–VI compounds exhibit both the cubic (zincblende) and the hexagonal (wurtzite) form. Because of the high degree of ionicity, the energy differences between these two modifications is low and phase transformations from the hexagonal form at solidification to the cubic form at some lower temperature are a major problem in the growth of ZnS and ZnSe. This low energy barrier between the two forms results in a low stacking fault energy which often manifests itself as twinning [60].

The softness of the II–VI compounds, relative to group IV or III–V materials, makes them susceptible to strain. This is a major problem if there is any interaction between the II–VI material and the crucible. Various alternatives to the standard silica crucible have been tried with the goal of reducing this interaction. These include PBN crucibles, graphitized silica crucibles, and graphitized and non-graphitized quartz cloth liners. Yasuda et al. reported on the vertical Bridgman growth of CdTe [63]. They found that the sticking of the ingot to the silica crucible was due to a layer of TeO_2 on the CdTe and could be avoided by a hydrogen anneal of the polycrystalline source material and the use of PBN crucibles. Omino and Suzuki and Yoshida et al. both used PBN crucibles for the vertical Bridgman growth of ZnSe [64, 65]. Graphite crucibles had proven to be too porous and could not maintain the melt composition while a silica crucible supported by a graphite ampoule devitrified too rapidly at the growth temperature of 1520°C.

Most of the reported Bridgman growth of II–VI single crystals is unseeded. There are several possible explanations for this. In order to reduce decomposition of the starting material, growth is usually carried out under an inert gas overpressure. The containment vessel makes it much more difficult to see the seed and control the meltback. Thus, fairly tight requirements are placed on the thermal profile reproducibility in order to seed "blindly". Secondly, twinning and polycrystalline growth off the seed is so prevalent that it may be a waste of good single crystal material to use it as a seed. Instead, it is not uncommon to have a capillary neck in the seedwell and let the preferred orientation grow at the expense of the faster growth directions. Yoshida et al. found that a crucible cone angle of 30° in the unseeded growth of ZnSe gave the largest single crystal grains. Sen et al. have grown 2 inch diameter unseeded crystals of CdTe and $Cd_{0.96}Zn_{0.04}Te$ up to 7.5 inch in length where 75% of the boule was single crystalline [66].

In an effort to improve the quality of the available CdTe single crystals, much of the recent work in vertical Bridgman growth has involved multi-zone furnaces. Yasuda et al. used an EDG furnace with 8 independent segments having an active 14 cm growth region comprised of the center 6 segments [63]. Sen et al. built multi-zone furnace for the modified Bridgman growth of CdTe [66]. It consisted of three stacked tubular furnaces; the upper furnace contained a heat pipe and was used to maintain a constant Cd overpressure. The middle EDG furnace had 10 independently controlled zones in 1.5 inch increments. The bottom furnace is the cold zone and contains a second heat pipe that was used to maintain a flat temperature profile. Using this multizone furnace, Sen et al. were able to grow CdTe and $Cd_{0.96}Zn_{0.04}Te$ with low EPD (10^{-4}) without low angle or subgrain boundary features in regions away from the ampoule walls. This accomplishment is attributed to several factors. Raising the position of the liquid–solid interface relative to the gradient zone made it more convex and thus confined the nucleation of polycrystalline growth to the periphery. By an iterative process of temperature profiling and adjustment of the relative power input to the multiple independent zones they were able to achieve axial temperature uniformity in the melt zone (≈ 3°C maximum variation over 6 inches) and a low axial temperature gradient (< 30°C/inch).

Finally, the problem of second phase precipitates is also a significant factor in the growth of II–VI single crystals. The cationic species (column II elements) have a significantly higher dissociation pressure than do the related cations, Al, Ga, or In, in the III–V materials. This is in addition to the dissociation pressure of the anionic species in the II–VI family; S, Se and Te. Sen et al. observed tellurium precipitates in Bridgman-grown CdTe and CdZnTe materials at a density of up to $5 \times 10^3\ cm^{-2}$ [66]. This is despite the fact that the crystal was grown under a constant Cd overpressure. The precipitates ranged from 6 to 35 microns in size, with the largest in the core of the boule. Post growth thermal annealing of the boules under Cd-saturated conditions was only partially effective in reducing the size of the Te precipitates.

References

[1] P.W. Bridgman, Proc. Am. Acad. Arts Sci. 60 (1925) 303.

[2] G. Tammann, Metallography, translated by Dean and Swenson (The Chemical Catalogue, New York, 1925) p. 26.

[3] D.C. Stockbarger, Rev. Sci. Instrum. 7 (1963) 133.

[4] H.E. Buckley, Crystal Growth (Wiley, New York, 1951).

[5] D.T.J. Hurle, Prog. Mater. Sci. 10 (1962) 79.

[6] J.W. Rutter and B. Chalmers, Can. J. Phys. 31 (1953) 15.

[7] W.G. Pfann, Trans. AIME 194 (1952) 747.

[8] W.G. Pffan, Zone Melting, 2nd Ed. (Wiley, New York, 1966) p. 199.

[9] A. Horowitz, D. Gazit, J. Makovsky and L. Ben-Dor, J. Cryst. Growth 61 (1983) 323.

[10] A.F. Witt, H.C. Gatos, L. Lichtensteiger and C.J. Herman, J. Electrochem. Soc. 125 (1978) 1832.

[11] T.S. Noggle and J.S. Koehler, Acta Met. 3 (1955) 260.

[12] V.V. Osiko, Proc. 10th Int. Conf. on Crystal Growth (San Diego, 1992); J.F. Wenkusherhaal, R. Aragon, H.R. Harrison, R.C. McCallister and C.J. Sandberg, J. Cryst. Growth 61 (1983) 221.

[13] M. Cervera-Marzal, J.P. Chaminade, J. Ravez, M. Pouchard, and P. Hagenmuller, J. Cryst. Growth 64 (1983) 516;
A. Kojima, S. Hoshino, Y. Noda, K. Fujii and T. Kanashiro, J. Cryst. Growth 94 (1989) 293.

[14] M. Voda, S.V. Nistor and V. Ghiordanescu, J. Cryst. Growth 119 (1992) 339.

[15] W. Korczak and P. Mikolajczak, J. Cryst. Growth 61 (1983) 601.

[16] R.H. Mellen Sr., US Patent Nos. 4,086,424, 4,423,516, and 4,518,351.

[17] E.M. Monberg, H. Brown and C.E. Bonner, J. Cryst. Growth 94 (1989) 109.

[18] J.M. Parsey Jr. and F.A. Thiel, J. Cryst. Growth 73 (1985) 211.

[19] D.E. Holmes, R.T. Chen, K.R. Elliot and C.G. Kirkpatrick, Appl. Phys. Lett. 40 (1982) 46.

[20] K.J. Bachman and E. Buehler, J. Electrochem. Soc. 121 (1974) 835.

[21] A. Smits and S.C. Bokhorst, Z. Physik. Chem. 91 (1916); T.D. Farr, Tennessee Valley Authority, Chem. Eng. Rep. No. 8 (Mucle Shoals, AL , 1950).

[22] J.E. Wardill, D.J. Dowling, R.A. Brunton, D.A.E. Crouch, J.R. Stockbridge and A.J. Thompson, J. Cryst. Growth 64 (1983) 15.

[23] W.A. Gault, E.M. Monberg and J.E. Clemans, J. Cryst. Growth 74 (1986) 491.

[24] S.E. Blum, R.J. Chicotka and B.K. Bischoff, J. Electrochem. Soc. 115 (1968) 324.

[25] E.M. Monberg, W.A. Gault, F. Dominguez, F. Simchock, S.N.G. Chu and C.M. Stiles, J. Electrochem. Soc. 135 (1988).

[26] E.P.A. Metz, R.C. Miller and R. Mazelsky, J. Appl. Phys. 33 (1962) 2016.

[27] J.B. Mullin, B.W. Straughan and W.S. Brickell, J. Phys. Chem. Solids 26 (1965) 782.

[28] J.B. Mullin, R.J. Heritage, C.W. Holiday and B.W. Straughan, J. Cryst. Growth 3&4 (1968) 281.

[29] N.J. Archer, Special Publication of the Am. Chem. Soc. High Temp. Inorg. Chem. 30 (1977) 167.

[30] E.M. Monberg and J.E. Clemans, US Patent No. 4,923,561.

[31] V.A. Lavrenko and A.F. Alexeev, Ceram. Int. 12 (1986) 25.

[32] G.V. Samsonov, ed., Boron, its Compounds and Alloys (Acad. Sci. USSR, Kiev, 1960).

[33] P. Petroff and R.L. Hartman, Appl. Phys. Lett. 23 (1973) 469.

[34] A.S. Jordan, R. Caruso and A.R. VonNeida, Bell System Tech. J. 59 (1980) 593.

[35] T. Suzuki, S. Akai, K. Kohe, Y. Nishida, K. Fujita and N. Kito, Sumitomo Electron. Tech. Rev. 18 (1978) 105.

[36] E.D. Bourret, M.L. Galiano, R.D. Mih, J.B. Guitron and E.E. Haller, J. Cryst. Growth 102 (1990) 877.
[37] M.S.S. Young and R. Caruso, unpublished results.
[38] K. Hoshikawa, H. Nakanishi, H. Kohda and M. Saaura, J. Cryst. Growth 94 (1989) 643.
[39] E.M. Swiggard, J. Cryst. Growth 94 (1989) 556.
[40] A.S. Jordan, E.M. Monberg and J.E. Clemans, J. Appl. Phys. 73 (1993) 477.
[41] J.E. Clemans and J.H. Conway, Proc. 5th Conf. on Semi-Insulating Materials (Malmo, 1988) eds. G. Grossman and L. Lebedo (IOP Publishing, Bristol, 1988).
[42] C.R. Abernathy, A.P. Kinsella, A.S. Jordan, R. Caruso, S.J. Pearton, H. Temkin and H. Wade, J. Cryst. Growth 85 (1987) 106.
[43] E.D. Bourret and E.C. Merk, J. Cryst. Growth 110 (1991) 395.
[44] A.G. Fischer, J. Electroehem. Soc. 117 (1970) 41c.
[45] S.E. Blum and R.J. Chicotka, J. Electrochem. Soc. 120 (1973) 588.
[46] H.H. Woodbury, J. Cryst. Growth 35 (1976) 49.
[47] T. Beppu, M. Tashiro, K. Hirahara and M. Watanabe, Jpn. J. Appl. Phs. 17 (1978) 509.
[48] M. Watanabe, Jpn. Semi. Tech. New 3 (1984) 44.
[49] L.R. Weisberg, F.D. Rosi and P.G. Herkart, in: Properties of Elemental and Compound Semiconductors Vol. 5, ed. H.C. Gatos (Interscience, New York, 1959) p. 2.
[50] R.J. Guire and K. Weiser, US Patent No. 2, 871, 100 (Jan. 27, 1959).
[51] K.J. Bachman and E. Buehler, J. Electron. Mater. 3 (1974) 279.
[52] Y. Seki, H. Watanabe and J. Matsui, J. Appl. Phys. 49 (1978) 822.
[53] W.A. Bonner, Mater. Res. Bull. 15 (1980) 63.
[54] G.T. Brown, B. Cockayne and W.R. MacEwan, J. Cryst. Growth 51 (1981) 369.
[55] T.I. Ejim, F. Simchock and E.M. Monberg, Proc. 1st Int. Conf. on InP and Related Compounds, (Norman, OK), eds. R. Singh and L.J. Messick, SPIE 1144.
[56] A.S. Jordan, E.M. Monberg and J.E. Clemans, Proc. 10th Int. Conf. on Crystal Growth (San Diego), eds. J.B. Mullin and T. Surek (Elsevier, Amsterdam, 1993).
[57] R.L. Barns, J. Electron. Mater. 18 (1989) 703.
[58] T.J. Jasinski, W.N. Rohsenow and A.F. Witt, J. Cryst. Growth 61 (1983) 339.
[59] L.R. Holland, J. Cryst. Growth 96 (1989) 577.
[60] H. Gottschalk, G. Patzer and H. Alexander, Phys. Stat. Sol. (a) 45 (1978) 207.
[61] M. Shibata, Y. Sasaki, T. Inada and S. Kuma, J. Cryst. Growth 102 (1990) 557.
[62] B.J. Fitzpatrick, J. Cryst. Growth 86 (1988) 106.
[63] K. Yasuda, Y. Iwakami and M. Saji, J. Cryst. Growth 99 (1990) 727.
[64] A. Omino and T. Suzuki, J. Cryst. Growth 117 (1992) 80.
[65] H. Yoshida, T. Fujii, A Kamata and Y. Nakata, J. Cryst. Growth 117 (1992) 75.
[66] S. Sen, W.H. Konkel, S.J. Tighe, L.G. Bland, S.R. Sharma and R.E. Taylor, J. Cryst. Growth 86 (1988) 111.

CHAPTER 3

Czochralski Growth

D.T.J. HURLE

H.H. Wills Physics Laboratory, University of Bristol, Tyndall Avenue, Bristol, UK BS8 1TL

B. COCKAYNE

DRA Malvern, St. Andrews Road, Malvern, Worcestershire, UK WR14 3PS

Contents

1. Introduction 101
2. Elements of the process 102
   2.1. Introduction 102
   2.2. Thermal balance and growth rate 104
   2.3. The meniscus 105
   2.4. Crystal and crucible rotation 108
3. Basic Technology 109
   3.1. Introduction 109
   3.2. Mechanical design 110
   3.3. Heating systems 112
   3.4. Growth chamber furniture 113
   3.5. Growth chamber ambient 115
   3.6. Control systems 116
4. Growth of elemental materials 123
   4.1. Metals 123
   4.2. Germanium 126
      4.2.1. Introduction 126
      4.2.2. Crystal growth 127
      4.2.3. Cellular structure 130
   4.3. Silicon 140
      4.3.1. Introduction 140
      4.3.2. Puller design 144
      4.3.3. Thermal distribution, interface shape and dislocation generation. 148

HANDBOOK OF CRYSTAL GROWTH, VOL. 2
edited by D.T.J. Hurle

4.3.4. Convection and conservative dopant incorporation . . . 149
4.3.5. Non-conservative dopant incorporation . . . 150
4.3.6. Pulled silicon ribbons . . . 154
5. Growth of compound semiconductors . . . 155
5.1. Pulling of materials with a volatile component . . . 155
5.1.1. Hot wall technology . . . 156
5.1.2. The liquid encapsulation Czochralski technique . . . 157
5.2. Group III antimonides . . . 159
5.2.1. Indium antimonide . . . 159
5.2.2. The facet effect in pulled crystals . . . 161
5.2.3. Growth twinning . . . 164
5.2.4. Pulling from a non-congruent melt . . . 164
5.2.5. Gallium and aluminium antimonides . . . 167
5.3. Group IIIB arsenides . . . 167
5.3.1. Gallium arsenide . . . 167
5.3.2. Indium arsenide . . . 173
5.4. Group IIIB phosphides . . . 173
5.4.1. Gallium phosphide . . . 173
5.4.2. Indium phosphide . . . 174
5.5. Mixed III–V alloy single crystals . . . 177
5.6. II–VI and IV–VI compounds . . . 180
5.6.1. Cadmium telluride . . . 180
5.6.2. Lead telluride . . . 181
6. Growth of oxide crystals . . . 181
6.1. Introduction . . . 181
6.2. The basic problems . . . 184
6.2.1. Technological problems . . . 184
6.2.2. Fundamental problems . . . 185
6.2.2.1. Chemical factors . . . 185
6.2.2.2. Stress factors . . . 189
6.3. Growth of specific materials . . . 192
6.3.1. Sapphire and magnesium aluminate spinel . . . 192
6.3.2. Garnets . . . 193
6.3.3. Lithium niobate and tantalate . . . 195
6.3.4. Rare earth gallates and aluminates . . . 197
6.4. Pulling optical fibres . . . 198
6.4.1. Edge defined film fed growth . . . 198
6.4.2. Laser heated pedestal growth . . . 200
7. Growth of fluoride crystals . . . 201
8. Growth of organic crystals . . . 203
9. Concluding remarks . . . 204
References . . . 205

1. Introduction

Crystal pulling was first developed as a technique for growing single crystals of useful size and perfection, by Teal and Little who pioneered the growth of germanium single crystals at Bell Laboratories in 1949 [1]. Since that time its application has been extended to the growth first of silicon and later of a very wide range of compound semiconductors, metals, oxides and halides. It is in fact the dominant technique for the commercial production of most of these materials. Industrial scale equipment has been developed with a high degree of automation and is commercially available. Each class of material brings its own special difficulties and machines are tailored to specific ones; there is no such thing as the universal crystal pulling machine.

The technique originates, and takes its other name, from pioneering work by Czochralski in 1917 [2] who pulled single crystals wires (fibres) of low melting point metals from the melt in order to determine the maximum speeds at which each could be crystallised. The technique was modified by von Gomperz [3] and by Mark, Polanyi and Schmid [4] to provide cooling of the growing crystal. This early work was made known to the wider world by Buckley in his book *Crystal Growth*: which was the first book to comprehensively describe the various techniques of crystal growth [5].

The principal extensions to the technique made by Teal and Little, which are crucial to its use in the production of large and homogeneous single crystals, were:

(a) Provision to control carefully and adjust the heating power supplied to the melt in order that the diameter of the wire could be increased to the point that the growing crystal was a right circular cylinder of a diameter that was a significant fraction of that of the crucible.

(b) The facility to rotate the crystal as it was being pulled from the melt.

This latter feature served to ensure radial symmetry of the thermal distribution thus ensuring that the crystal, in the absence of any disturbing crystallographically-related effects, grew as a cylinder. Later it came to be realised that rotation also helped to ensure the radial uniformity of distribution of added dopants. Another important refinement introduced by Teal and Little was the provision of an inert ambient atmosphere in the growth chamber but the manner in which this was originally done [6] appears to have been both hazardous and probably unreliable!

To this core technology has been added a whole range of refinements and novel concepts to enable the technique to be used for an ever increasing range of materials. Thus the problem of the decomposition of III–V semiconducting compounds has been overcome by the development of the liquid encapsulation technique [7, 8] including its operation in high pressure growth chambers [9, 10] to facilitate the growth of the Group III phosphides. Highly reactive metals such as the rare earths have been grown using the ingenious concept of the "cold crucible". Novel heating techniques such as laser or electron beam heating have supplemented the more conventional methods of resistive or radio frequency induction heating.

Control of the growth process and, in particular, control of the diameter of the growing crystal has become the norm on all commercial equipments and this has required the development of suitable sensors capable of working in the hostile environment of a crystal puller. By the use of dies, special shapes of crystal such as tubes and polygonal shapes have been grown for various purposes [11].

In the space of the forty or so years since the work of Teal and Little, the technique has in fact come to dominate the production of bulk single crystals.

In this chapter, we first describe the elements of the technique and the underlying physical processes in section 2 and then describe the core technologies in section 3. In sections 4 to 8 specific classes of material are considered together with the modifications to the technology which are required for each of them. Growth-related defect formation and its control is also considered material by material. Finally in section 9, some reflections on the reasons for the success of the technique and its likely further development are given.

Fundamental aspects of the process are covered in other chapters in this Volume. This is also the subject of a recent book by one of us [245].

2. *Elements of the process*

2.1. *Introduction*

In its simplest form (fig. 1), the technique consists of a crucible which contains the charge material to be crystallised surrounded by a heater capable of melting the charge. A pull rod with a chuck attached to its lower end is mounted co-axially above the

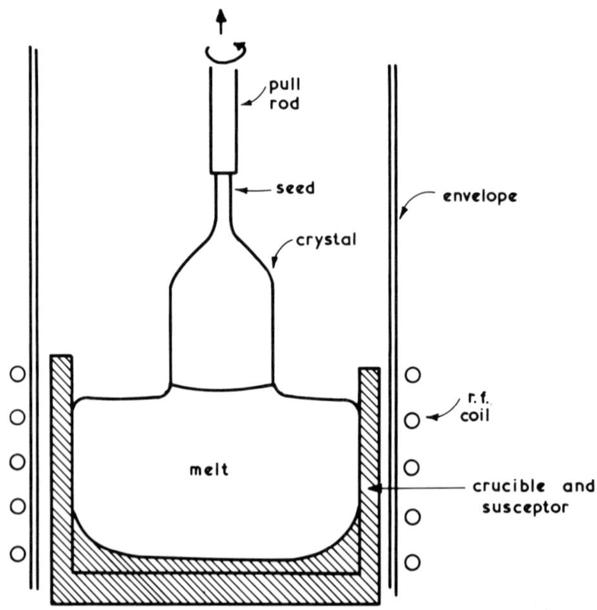

Fig. 1. Schematic representation of the elements of crystal pulling shown with rf induction heating of the crucible.

crucible. A seed crystal is attached to the chuck and the pull rod is lowered until the end of the seed crystal is dipped into the melt. The melt temperature is carefully adjusted until a meniscus is supported by the end of the seed. Once a thermal steady state has been achieved, the pull rod is slowly lifted and rotated and crystallisation onto the end of the seed occurs. Lifting rates vary with the material, the size of the crystal and the amount of dopant in the melt but are in the range of a few tenths of a mm per hour up to tens of cms per hour. Crystal rotation rates usually range from a few up to a maximum of a few hundred revolutions per minute.

To increase the diameter of the crystal from that of the seed, small and careful heater power reduction is imposed either manually or under automatic control. The whole crystal growth assembly is located within an envelope containing an inert ambient gas or, alternatively, is evacuated so that the crystal and melt are protected from attack by atmospheric gases and from ambient temperature fluctuations. The containing envelope is, in part at least, transparent so that the growth process can be continuously observed.

A typical procedure for operating such a puller might be as follows:

(a) Etch the charge material and load it into the crucible. Fit an oriented seed crystal into the chuck having first removed work damage in the cut seed by etching.

(b) Assemble the growth chamber, evacuate and then purge with ambient gas (if one is to be used).

(c) Apply crucible rotation if used. Melt the charge, holding its temperature just a little above the melting point.

(d) Lower the seed crystal to a position a little above the melt surface and allow it to come to thermal equilibrium; apply seed rotation. Then lower the end of the seed very slowly into the melt to minimise the thermal shock.

(e) Make fine adjustments to the heater power until the seed stably supports a liquid meniscus. This is a crucial step. Too high a melt temperature and the seed is lost; too low a temperature and the introduction of the seed can nucleate a supercooled melt which rapidly freezes over.

(f) Commence pulling and adjust the melt temperature and/or the pull rate so that a thin neck is grown on to the seed. This helps to prevent defects in the seed from propagating into the crystal.

(g) Increase the crystal diameter slowly and smoothly by reducing the heater power until the crystal has grown out to the required diameter. Restabilise the process at this diameter by further adjustment of the heater power.

The linear rate of growth of a cylindrical crystal is constrained to be equal to the pulling speed plus the rate of fall of melt level in the crucible. However, this is achieved only by careful adjustment of the heat flow conditions such that the rate at which the latent heat of fusion is removed is exactly equal to its rate of generation so that a thermal quasi-steady state is obtained. In general it is this removal of latent heat which is the rate-controlling process.

Where the material being crystallised is an alloy with segregation of one or more constituents, then removal of rejected components (which we shall hereafter call solutes except where the solute is deliberately added in order to achieve certain desirable properties of the crystal, when it is designated a dopant) can become rate-limiting. In these circumstances constitutional supercooling of the melt can occur which ruins the crystalline perfection.

In fact crystal pulling is useful only for dilute alloys and near-congruent compounds where the depression of the liquidus temperature at the interface due to rejected components is limited to, at most, a few tens of degrees Celsius. By and large therefore, crystal pulling is a thermally-controlled process.

Other practical requirements which have to be satisfied in order for the technique to be successfully applied are:

(a) It must be possible to achieve a sufficiently high temperature to melt the substance.

(b) Having melted it, it must be possible to contain and maintain partial pressures of any volatile constituents in equilibrium with the melt (or avoid the formation of a vapour phase – see section 5.1 below).

(c) It must be possible to contain the molten charge in a crucible without deleterious corrosion.

These criteria can be satisfied for most of the single crystals which are commercially important. The major exception is quartz.

The principal advantages of the technique include:

(a) The crystal is not constrained as it cools and a high structural perfection can be obtained.

(b) Crystal rotation ensures a radially uniform distribution of dopant.

(c) The crystal can be observed at all stages of its growth. This makes the process of "seeding on" relatively straightforward and permits melting-back to the seed if the initial growth is seen to be defective because of flaws introduced in the seeding process or as a result of twinning during "grow-out" of the crystal from the seed to its final diameter.

2.2. Thermal balance and growth rate

We have indicated that, in most circumstances, it is the removal of latent heat which controls the crystal pulling process. Immediately adjacent to the solid-liquid interface, heat is transported by conduction except in the case of materials which are self-transparent to their thermal radiation in which case radiative heat transport is also important. Excluding this latter situation, we can express the heat flux balance at a planar crystallising interface as follows:

$$k_S G_S - k_L G_L = Lv, \tag{2.1}$$

where $k_{S,L}$ are the thermal conductivities and $G_{S,L}$ the temperature gradients of solid (crystal) and liquid (melt) respectively. The gradients are normal to the plane of the interface if the latter is isothermal. L is the latent heat per unit volume of crystal and v is the normal growth rate of the interface.

For the case of the growth of a cylindrical crystal, the growth rate v is equal to the pulling speed v_p plus the rate of fall of the level of the melt in the crucible. The relationship is

$$v = v_p \rho_L R^2/(\rho_L R^2 - \rho_S r^2), \tag{2.2}$$

where $\rho_{S,L}$ are the densities of solid and liquid and R and r are the radii of the crystal and the crucible respectively. This correction to the pulling speed to obtain the true growth speed is usually quite significant. For example for a material with no volume change on crystallisation – i.e. equal solid and liquid densities – the growth speed is 33% greater than the pulling speed if the crucible radius has the typical value of twice that of the crystal. When the crystal radius is changing with time then the height of the meniscus is also changing so that there is a further contribution to the growth speed.

Thus far we have supposed that the interface is planar and the heat flow one-dimensional. In reality of course the heat flow is three-dimensional. However, to achieve a high crystal perfection by minimising thermal stress in the cooling crystal (see chapter 14 by Volkl in this Volume) we usually need to adjust the thermal conditions so that the interface is indeed nearly planar. This can be done empirically or can be aided by numerical modelling of the global process (see chapter 15 by Dupret and Van den Bogaert). Modelling requires super-computer power because it is necessary to include in the modelling both the convective flow in the melt and the complex radiative heat exchanges between the crystal, the melt and the various components of the puller "furniture".

Optically self-transparent crystals pose additional problems. Cockayne et al. [12] have shown experimentally the importance of the degree of self-transparency in determining the interface shape in DyAG–YAG solid solution crystals and, recently, Wallace et al. [13] have suggested that this may also be true for some undoped (and lightly doped) semiconductors.

2.3. The meniscus

The behaviour of the meniscus is central to the dynamics and the control of the diameter of a growing Czochralski crystal. The shape of the growing crystal is determined by the angle at which the meniscus contacts the crystal at the three-phase boundary. This is determined by the height of the meniscus. The meniscus height is, in turn, controlled by the heat transport in the melt and crystal. If the crystallising interface is essentially in thermodynamic equilibrium (i.e., the undercooling needed to drive the crystallisation process is negligibly small) then the position of the three-phase boundary is determined by the location of the freezing point isotherm.

If additional heat is supplied to the melt, this isotherm will rise thereby increasing the meniscus height and causing the meniscus to "neck-in". Conversely, if the heat input is reduced, the meniscus height is reduced and the meniscus will contact the crystal at a greater angle.

In general the crystal will grow conically with a cone angle Θ_S determined by the angle of contact of the meniscus at the three-phase boundary. Cylindrical growth will

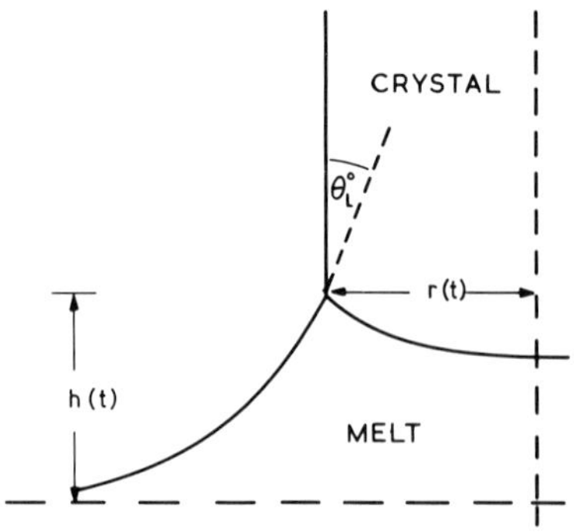

Fig. 2. Illustration of the parameters defining the meniscus during the growth of a cylindrical crystal. The contacting angle at the three-phase boundary (Θ_L^0) is a characteristic of the given material.

be achieved only when this angle of contact takes some unique value which is characteristic of the given material being crystallised.

One might suppose that, in the absence of any volume change on solidification, the conical angle of the crystal would be equal to the contacting angle of the meniscus, and this is indeed true for a certain class of materials – namely those where the melt completely wets the crystal. Most metals appear to fall within this category. However, for those materials which do not completely wet their own solids, (and this may indeed be the majority of crystalline materials), the melt–gas interface makes a finite angle with the crystal–gas interface at the three-phase boundary as shown in fig. 2. This characteristic angle has, to date, been measured for only a few materials which are listed in table 1.

The condition of incomplete wetting is expressed by the inequality

$$\sigma_{SG} < \sigma_{SL} + \sigma_{LG}, \tag{2.3}$$

where σ_{SG}, σ_{LG} and σ_{SL} are respectively the crystal–gas, melt–gas and crystal–melt interfacial energies.

The Herring condition

$$\sum(\sigma_i t_i + \partial\sigma_i/\partial t_i) = 0, \quad i = \text{SG, LG, SL} \tag{2.4}$$

Table 1
Characteristic contact angle for some crystal–melt systems.

Material	Contact angle (Θ_L^0)(degrees)	Reference
Ge	13 ± 1	[167]
Si	11 ± 1	[167]
Cu	0	[168]
$LiNbO_3$	0	[169]
LiF	19	[170]
Al_2O_3	17 ± 4	[171]

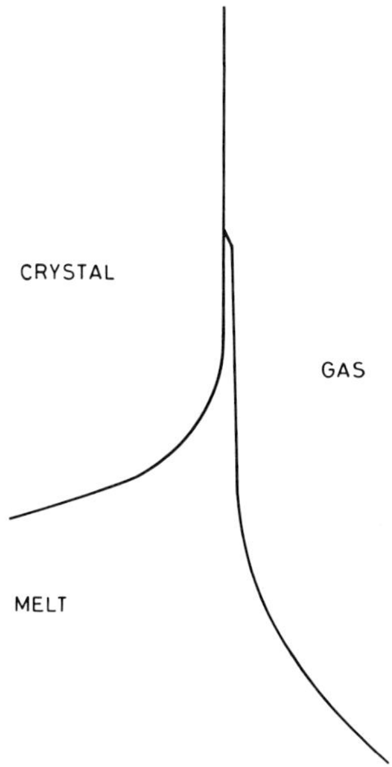

Fig. 3. Schematic representation of the three-phase boundary for a material system where the melt wets the crystal.

can be used to express this contacting angle in terms of the surface energies and their orientation dependencies. t_i is the surface normal vector [14].

For metallic materials which completely wet their own solids, a thin liquid film extends for a short distance up the crystal with the effective contacting angle being zero (see fig. 3). This phenomenon, which can be thought of as surface melting, has been observed on droplets of molten metals and a thermodynamic model has been derived which predicts the thickness and extent of the layer [15].

The profile of the meniscus for any given contacting angle Θ_L^0 is obtained by solution of the Laplace–Young equation for a holm:

$$2z = -\beta[z''/(1+z'^2)^{3/2} + z'/x(1+z'^2)^{1/2}], \tag{2.5}$$

where z and x are the vertical and radial co-ordinates respectively and the prime denotes differentiation with respect to the radial co-ordinate x.

$$\beta = 2\sigma_{LG}/\delta\rho_L g \tag{2.6}$$

is the Laplace constant, $\delta\rho_L$ being the difference between the densities of the liquid and the gas; g is the gravitational acceleration. Remote from the crystal the interface becomes horizontal for a sufficiently large crucible.

Unfortunately analytical solutions of the equation in this geometry do not appear to exist. However useful approximate solutions have been given by Hurle [16] and by Boucher and Jones [17].

2.4. Crystal and crucible rotation

A disc rotating in a fluid acts as a centrifugal fan, drawing in fluid axially, imparting angular momentum to it within a rather thin boundary layer (known as an Ekman layer) and expelling it tangentially. If both the disc and the fluid are infinite in extent so that there is no return flow at boundary walls, then the thickness of the Ekman layer is independent of radiàl or azimuthal position on the disc; i.e. it is of uniform thickness over the whole disc. If the "disc" is in fact a growing crystal rejecting solute into an even thinner solute boundary layer then this rejected solute is advected away at a rate which depends on the crystal rotation rate and which is everywhere the same thereby ensuring that the solute incorporation into the crystal is radially uniform.

Of course the crystal is not infinite in extent and neither is the melt. However, edge effects due to the finite radius of the crystal are small. The recirculation produced by the presence of the crucible wall results in an axial suction which is significantly different from that in an infinite fluid but does not seriously impair the radial uniformity as has been demostrated in numerous experimental studies of segregation in Czochralski growth. This is indeed one of the principal advantages offered by the pulling technique.

A further departure from the idealisation of the infinite disc is the fact that the crystal is positioned at the melt surface and not within it. This reduces the coupling but, again, does not appear to impair the radial uniformity.

Flow at a rotating disc has been extensively analysed following the first presentation of a similarity solution by von Karman [18] which was numerically integrated by Cochran [19]. The effect of crystallisation at the "disc" is equivalent to the existence of a suction velocity at the disc surface of magnitude equal to the growth rate. This aspect has been incorporated into the theory by Stuart [20] and by others. The literature relevant to crystal growth has been reviewed by Jones [21].

The von Karman theory has been used by Burton, Prim and Slichter (hereafter BPS) [22] in a, now classic, paper to relate the effective distribution coefficient (k_{eff}) of a solute to its interface distribution coefficient (k) as follows:

The equation of mass transport in the melt, written in a co-ordinate frame moving with the crystallising interface is

$$\mathrm{d}C/\mathrm{d}t = D\nabla^2 C + v_{\mathrm{p}}\,\mathrm{d}C/\mathrm{d}z - \boldsymbol{u}\nabla C, \tag{2.7}$$

where C is the solute concentration (assumed dilute), t is time, z the co-ordinate in the normal growth direction and $\mathbf{u}$ the flow velocity vector. The problem can be treated one-dimensionally because of the uniformity of the boundary layer and, for the z-component of $\mathbf{u}$ (denoted $H(z)$) we take a truncated form of the series expansion obtained by Cochran. It is adequate to use just the first term which is quadratic in z.

Imposing the boundary conditions that, remote from the crystal, the melt concentration takes the value C_∞ and that, at the interface, there is a flux boundary condition

$$-D\,\mathrm{d}C/\mathrm{d}z = v_{\mathrm{P}}[C_{\mathrm{S}} - C(0)], \tag{2.8}$$

where C_{s} is the solute concentration in the crystal and $C(0)$ the concentration in the melt at the interface, we obtain the following expression for the effective distribution coefficient:

$$k_{\text{eff}} = C_S/C_\infty = k/[1-(1-k)J], \tag{2.9}$$

where $k = C_S/C(0)$ and J is given by

$$J = v_p(\nu/\Omega)^{1/2}/D \int_0^\infty \exp\left[\mathrm{Sc} - \int_0^Z H(z)\,\mathrm{d}z'\right] \mathrm{d}z, \tag{2.10}$$

$\mathrm{Sc} = \nu/D$ being the Schmit number, where ν is the kinematic viscosity of the melt. Burton, Prim and Slichter introduced the concept of a boundary layer which is related to J by

$$v_p\delta/D = -\ln(1-J). \tag{2.11}$$

However, the BPS approach has been criticised by Wilson [23] and by Wheeler [24] and Wilson (see also chapter 1) has demonstrated that a more sensible definition is

$$v_p\delta'/D = J. \tag{2.12}$$

These two definitions of course approach each other as J becomes small compared to unity. Evaluating the integral in eq. (2.10) yields the following approximate expression for the BPS boundary layer thickness δ:

$$\delta = 1.6\,\mathrm{Sc}^{-1/3}(\nu/\Omega)^{1/2}. \tag{2.13}$$

Substituting into eq. (2.9) gives the famous BPS equation:

$$k_{\text{eff}} = k/[k+(1-k)\exp(-v_p\delta/D)] \tag{2.14}$$

In the limit that δ goes to zero (i.e., when the boundary layer is obliterated by a very strong centrifugal flow) k_{eff} becomes equal to k. In the other limit, as the boundary layer thickness becomes infinite (no convective mixing), k_{eff} approaches unity i.e., the concentration of solute in the crystal is equal to that in the bulk of the melt. Because of natural convective mixing in the melt this state is never obtained in practise in Czochralski growth.

In addition to crystal rotation, rotation of the crucible is simultaneously carried out in commercial growth processes for the production of semiconductor crystals usually with the crucible counter-rotated with respect to the crystal. The reasons for this appear to be, in part, historical but in the case of silicon growth, crucible rotation provides an additional control variable to tune the oxygen concentration in the crystal (see section 4.3 below). Crucible rotation also influences the shape of the crystal–melt interface, a parameter which is of considerable importance in determining crystal perfection.

3. *Basic technology*

3.1. *Introduction*

As has already been indicated, commercial crystal pullers are carefully designed for a specific material; indeed for a specific size of crystal and melt charge and a given crystal specification. Smaller scale research pullers, on the other hand, are designed to be more flexible and are usually of modular construction so that growth chambers, crucibles, heating methods are all interchangeable. In this section we outline only the basic design requirements of a crystal puller. The modifications and refinements introduced to grow specific materials are covered in sections 4–8.

3.2. Mechanical design

The principal engineering requirements of the equipments are:

(1) A rigid vibration-free frame to prevent differential movement between seed and melt surface as a result of mechanical or thermal stresses on the frame and to prevent the transmission of any building- or motor-generated vibration to the melt.

(2) Precision vibration-free lifting and rotation mechanisms. In commercial systems the crucible is usually lifted in order to maintain the melt surface level at the same vertical position with respect to the heater. Pull rates as low as 0.1 mm/h are necessary for some doped or complex oxide crystals, whilst silicon crystals of modest size can be grown at speeds in excess of 10 cm/h. Crystal rotation rates of up to 200 rpm are sometimes required for oxide growth, but crucible rotation rates seldom exceed 20 rpm. The facility to translate the pull rod rapidly is also required for speedy initial location of the seed crystal with respect to the melt surface.

Precision lead-screws are frequently used to effect the lifting motion. The screws are commonly of stainless steel running in a bronze nut. A shear pin must be fitted to prevent accidental damage as a result of a "freeze-up" of the melt. As crystal sizes have steadily increased, the height of pullers has also increased to the point that they can exceed the height of single storey factories. This problem is avoided in some models of silicon puller by the use of a ball-chain which is wound up on to a drum which is rotated to provide the crystal rotation facility. The ball-chain has sufficient resistance to torsional shear to enable the crystal to be rotated but, in the event of a heater failure following which the crystal becomes frozen to the charge, catastrophic damage to the mechanism is avoided. An inherent disadvantage is that there is inevitably some small lack of smoothness of pull as successive links of the ball-chain are wound on to the drum. More recently, suitable poly-strand cables, made of tungsten or of stainless steel have been developed which overcome this problem.

The wide range of pulling speeds required for research pullers usually necessitates the use of gearboxes with the attendant problems of backlash. DC servo-motors coupled to tacho-generators are commonly used for both lift and rotation mechanisms. All motors and gear assemblies should be carefully balanced to minimise vibration. The wide dynamic range of stepping motors makes them attractive but, again, they are not completely smooth in their operation and conceivably could introduce dopant striations into the crystals.

(3) A seal between pull rod and growth chamber. Where the ambient atmosphere is corrosive or where lubricants on the motors and gears would contaminate the crystal, the mechanisms have to be located outside the growth chamber and therefore the pull rod and crucible lift rod have to pass through seals in the chamber. A variety of designs

of such seals have been employed from the simple "O" ring to a ferrofluidic seal. In pressure pullers and where growth takes place in vacuo or at reduced pressure, the seals have to withstand the pressure differential across the chamber wall.

Evaporants in the chamber can condense on to the pull rod and, when this is pulled through the seal, can produce excessive wear of that seal. The simple "O" ring attached to the pull rod whose diameter is slightly reduced below the seal position largely overcomes this problem but requires that the "O" ring is pulled up through a precision-bore tube.

(4) A suitable growth chamber. Growth chamber design varies widely and is dictated by such diverse factors as: (a) ease of opening up to load and to remove the grown crystal and, it should not be forgotten, to clean the chamber, (b) visibility of the growing crystal, (c) maintenance of radial thermal symmetry around the crystal, (d) must not couple into a rf crucible-heating power source, (e) must not release contaminant gases during operation of the puller, (f) provision of water or gas cooling to keep chamber walls at near room temperature, (g) provision of ports for the introduction of heater and thermocouple leads, ambient gas inlet and outlet, vacuum lines, doping and viewing ports etc.

Commercial designs include:

(i) Chambers which separate at the top from the pulling head assembly and can be lowered to expose the crucible on its pedestal to facilitate loading. Hydraulic jacks are usually employed to raise and lower the chamber. Alternatively the chamber may be counter-balanced so that it can be moved freely by hand.

(ii) "Clam shells" which open laterally into two halves to expose the interior. Viewing is via a quartz plate mounted in the top of the chamber.

(iii) Simple quartz cylindrical envelopes sealed by gasket against end plates top and bottom with facility to lower the bottom plate (or raise the top one). Such quartz envelopes are sometimes in the form of a "Y-tube" to permit oblique viewing. For very high temperature materials where the radiation is too intense for direct viewing, the viewing port can be fitted with an adjustable pair of polaroids to control the transmitted intensity.

Where it is desired to recharge the puller without cooling to room temperature the crystal can be pulled into a separate, upper, openable chamber and a shutter can be provided which divides off this chamber from the main chamber containing the crucible.

Pressure chambers have to be built and tested to specific safety standards and should be provided with both pressure relief valves and bursting discs to avoid the hazard of accidental over-pressurisation. Austenitic stainless steel is used for the manufacture of pressure chambers and any welds are tested by radiography and ultrasonics. Water cooling coils are brazed to the chamber exterior. The two sections of the chamber can be sealed using high tensile bolts with a neoprene "O" ring seal backed with PTFE spacer rings. Alternatively self-sealing clamps have been employed in one commercial design.

Viewing is through thick sapphire windows fitted to an inclined side port on the chamber. If necessary, external illumination can be provided through a second side port. Careful design is necessary to avoid fogging of the windows by condensates during growth. For safety the growth process is usually viewed remotely on closed-circuit TV.

3.3. Heating systems

The choice of heating system is both very material dependent and subject to economic considerations since power consumption is a major factor in the economics of bulk crystal production. Where its use is possible, resistive heating scores on cost and low voltage. High current "picket fence heaters" (see fig. 4) are common in commercial semiconductor crystal pullers. These are made of graphite which can be coated with a refractory material such as silicon carbide or boron nitride.

Radio frequency induction heating is "cleaner" in that it does not introduce heated foreign materials except the susceptor. In fact, the susceptor is frequently also the crucible. Higher capital and maintenance costs are associated with this technique however. Its most important use is with very refractory oxide materials whose melting temperatures are above those of viable resistive heating elements.

Fig. 4. A pBN-coated "picket fence" resistance heater for a GaAs LEC puller (Photograph by courtesy of Ringsdorff).

Resistive and rf induction heating are the dominant heating methods employed, but, in specialised applications such as the pulling of fibres of refractory oxides, lasers are being used. In even fewer cases electron beam heating has been reported.

Both power control and temperature control are in common use. Power control is usually effected as a fast-acting inner servo-loop of the diameter control servo (see section 3.6 below). Alternatively, if the puller is not fitted with automatic diameter control, the heater power can be programmed to any desired schedule by the use of a digital controller obtaining reference input data from a "look-up" table.

The choice of temperature sensor depends on the environment in which it is to be placed. Precious metal thermocouples are versatile and fairly reliable and can be positioned in obscured locations which optical sensors cannot reach. However, optical pyrometers and CCD array cameras are ideal for directly observing and measuring melt temperature provided that fogging-up problems can be avoided.

Digital temperature and power control is now such a developed art with so much commercial equipment available that this is best bought "off the shelf". Accordingly we do not attempt to review this topic here.

Equally, rf induction heater design and coupling to crystal pullers has been taken to a high level of specialisation by the rf equipment manufacturers and we confine ourselves to a few basic remarks.

The need to thermally insulate the melt, crucible and susceptor from the water-cooled rf coil necessarily makes the coupling of the rf to the susceptor quite loose, resulting in a very poor power factor. In order to draw the full available power from the rf generator therefore a high tank circuit kva is required. This has to be achieved whilst keeping the coil voltage relatively low to avoid excessive ionisation of the ambient gas in the growth chamber which can result in arcing, sometimes with catastrophic effects.

Generators with frequencies in the range 200–450 kHz and power outputs in the 10–50 kW region have generally been employed. There is now an increasing tendency to use solid state inverters in the respective frequency and power ranges 10–30 kHz and 10–50 kW due to their compactness, stability and better compatibility with skin depth requirements as modern electronics have replaced the noisy motor generators which were once used to provide output in this frequency range.

Thermal design of the system needs to be carefully considered. Using an industrial standard 450 kHz frequency generator the skin depth of the coupling into a graphite susceptor at 1400°C is only 2.3 mm so that there is no penetration of the rf field through the susceptor into the melt.

3.4. Growth chamber furniture

By "furniture" is implied crucible, pedestal, susceptor (if any), radiation shields, after-heaters etc.

Chemical compatibility of crucible and melt is the crucial factor which dictates choice of material. For metals and semiconductors, quartz and graphite are common choices. Both of these materials have been developed to a very high state of purity. The purest quartz is synthetic and not made directly from the naturally occurring mineral.

Graphite has been highly purified by the nuclear industry on account of its neutron absorbing properties. In particular it is available in a high density form which is much more readily and accurately machinable. Both quartz and graphite need to be baked out prior to use to remove absorbed gases.

More recently pyrolytic vapour deposition processes have been developed for the fabrication of crucibles of boron nitride (known as pBN) and these are now in common use for the growth of III–V semiconductors. In particular, by providing an alternative to quartz for the growth of semi-insulating GaAs, the use of pBN has removed the problem of silicon contamination of that material which hitherto compromised its semi-insulating nature [25].

Few container materials are compatible with molten oxides and, in general, these are the expensive platinum group metals [188]. At temperatures up to approximately 1500°C, platinum is utilised because of its relative ductility in manufacture and oxidation resistant properties even in a 100% oxygen atmosphere. Rhodium also has excellent oxidation resistance but it tends to dissolve slightly in some oxide melts and can cause unwanted broad band absorption effects in crystals grown therefrom. At higher temperatures, up to around 2200°C, iridium is the favoured material. However, this is the least oxidation resistant of the useable platinum metals. It is capable of withstanding 1–4 vol% O_2 in an otherwise inert atmosphere at 1600°C and above but undergoes severe volatilisation in the presence of oxygen in the temperature range 1000–1500°C due to volatile oxide formation. Tungsten and molybdenum crucibles have been utilised for some aluminates and sapphire but these metals require an inert or slightly reducing atmosphere which can limit the range of composition of the gas ambient used to avoid crystal defects associated with oxygen loss [160].

Gold or silver can be used where the melt is strongly oxidising provided the growth temperature is low enough. (The melting points of gold and silver are 1063°C and 960°C respectively.) Gold has been used for the growth of lead germanate but very small inclusions of metallic gold were found in the pulled crystals [26].

The useful life of a precious metal crucibles depends on many things. Where the melting point of the charge is only a little less than the crucible melting point, inadvertent melting of the crucible as a result of non-uniform or, more particularly, too rapid, heating is an ever present risk. The growth of sapphire or spinel in iridium are such cases. A crucible can fail catastrophically in other ways also; e.g. by fracture when containing materials which cause distortion (such as the garnets). This occurs because of the different expansion coefficients and different temperature dependence of the expansion coefficients of charge and crucible. (The expansion coefficient of iridium increases from $6.24 \times 10^{-6}\,°C^{-1}$ at room temperature to $15.19 \times 10^{-6}\,°C^{-1}$ at 2225°C whereas the garnets have an intermediate, much more nearly constant, value.)

Oxidation of the crucible, when the ambient is made oxidising to maintain melt stoichiometry (as in the case of lithium tantalate for example) has already been mentioned. Flame spraying the outside of the crucible with zirconia can prolong its life significantly provided that crucible distortion can be avoided.

Crucible contamination and its control have been major areas of research into the growth of Si and GaAs and this is reviewed in sections 4.3 and 5.2 respectively.

To control the temperature gradients during cooling, after-heaters are sometimes used, especially with refractory oxides where thermal stress during cooling can result in cracking of the crystals. Such after-heaters can be either passive (e.g. just a ceramic sleeve into which the crystal is pulled or active (i.e. a secondary heater). Tungsten and molybdenum have little general compatibility with oxides and hence, when these are used as crucibles, any heat shields must be made of the same materials. The platinum group of metals are however, compatible with most oxide refractories including those readily available such as alumina, magnesia and zirconia. A variety of shapes and sizes can be obtained commercially, which permits their use as both the thermal surround to the crucible and the thermal envelope required above the melt to protect the growing and cooling crystal.

There is a two-fold purpose for such heat shields. Firstly, to prevent the crucible from melting; most oxide melts are insulating in character which means that the crucible wall must be maintained at a higher temperature than the oxide melting point so that the centre of the melt remains molten. Secondly, to reduce temperature gradients in the thermal environment above the melt so as to minimise stress effects and to prevent crack formation during crystal cooling. Heat shields of iridium or platinum, appropriate to the temperature required, are often used to line the refractory environment above the melt as an additional safeguard. The crucibles and heat shields most frequently employed are cylindrical in shape with appropriate slots cut to facilitate viewing and crystal withdrawal. Heat baffles placed within the melt, also made from the platinum metals, have also been used to minimise convective effects within molten oxide melts [161]; these do, however, cause complications in both melt preparation and melt removal subsequent to growth.

An important technological problem related directly to the high temperatures involved is the attachment of the seed crystal to the pulling mechanism. Very few materials are available which are machineable, malleable and capable of repeated use without damage in an oxidising atmosphere close to the melt surface. Iridium or refractory couplings allow only crude connections due to their intrinsically brittle character. Platinum–rhodium or platinum connections allow sufficient refinement but are generally limited to temperatures lower than 1500°C. Such components, used close to the melt surface, allow seed crystals to be generated or extended but the ultimate solution is to generate long seeds so that the necessary couplings can be effected in the cooler regions outside the refractory enclosure where components made of platinum, its alloys and stainless steel can be used with relative ease.

3.5. Growth chamber ambient

Choice of ambient is dictated by two principal needs; namely the maintenance and control of stoichiometry of the melt and avoidance of chemical reaction between crystal or melt and the ambient gas or any impurities contained in or transported by it.

A secondary issue is the thermal transport properties of the gas as this has an impact on the temperature distribution in the cooling crystal. The thermal conductivity of a gas is an inverse function of its molecular weight, helium and hydrogen having roughly an order of magnitude greater value than argon. Convective heat transport scales with a fractional power of the Grashof number of the flow [27] and, for an ideal gas, this is proportional to the square of the gas pressure. Thus, in high pressure systems violent convection occurs and convective heat transport can be dominant.

Where the material to be grown is involatile, a vacuum provides the cleanest environment and also offers the possibility of continuous removal of volatile impurities from the system. It also prevents any convective cooling of the grown crystal.

3.6. Control systems

The economic advantages which accrue from being able to reproducibly grow accurately cylindrical crystals of pre-determined diameter are obvious and can be obtained by replacing the human skills which were honed in the research phases of the development of the pulling process with automatic servo-control systems.

Servo-control engineering is, of course, a highly advanced engineering field, but crystal growth brings its own specific problems. These are principally: (a) the very slow and "noisy" nature of the process and (b) the difficulty of providing suitable sensors because of the hostile environment. Direct imaging of the crystal profile is made additionally difficult because of the location of the crystal–melt interface deep within the crucible.

The first problem makes a study of the system dynamics important because empirical optimisation of the process is too slow and costly. An experimental determination of the transfer function of the pulling process has been described by Hurle et al. [28, 29] and theoretical system identification is comprehensively covered in the chapter in this volume by Tatarchenko and so will not be further considered here. Instead we focus on the second problem, namely the provision of adequate, rugged sensors.

Techniques for automatic diameter control can be conveniently classified under four headings namely: (i) use of optical reflection from the meniscus, (ii) techniques for imaging the profile of the crystal, (iii) weighing the crystal or the crucible, (iv) the use of a die.

It might be thought that the best techniques would be those which directly measure the crystal diameter at, or as near as possible to, the crystal–melt interface. In fact however, better control of diameter can be obtained by sensing changes in the meniscus shape. The reason for this is as follows:

As was explained in section 2.3, the meniscus contacts the crystal at the three-phase boundary at a characteristic angle Θ_L^0. If the crystal is growing conically at an angle to the vertical of Θ_S then the meniscus will make an angle to the vertical at the point of contact with the crystal of

$$\Theta_L = \Theta_S + \Theta_L^0. \tag{3.1}$$

The rate of change of the crystal radius is simply given by

$$\dot{R} = v \tan \Theta_S = v \tan(\Theta_L - \Theta_L^0), \tag{3.2}$$

where v is the vertical growth speed

$$v = v_0 - \dot{h}, \tag{3.3}$$

and where v_0 is the pull rate plus the rate of fall of melt level in the crucible (as given by eq. (2.2)) and $\dot{h}$ is the rate of change of meniscus height which is obtained by integrating the Laplace–Young equation (eq. (2.5)) once with current values of R and Θ_L.

Thus Θ_L is a measure of the rate of change of R. Hence deviation of Θ_L from its desired value of Θ_L^0 (which corresponds to the growth of a cylindrical crystal) is an indicator of impending change in crystal radius. Therefore, by detecting change in meniscus angle rather than in radius directly, we are able to anticipate change in radius and take action before it has occurred to any significant extent. As we shall see, the most successful methods of automatic diameter control all use this approach but achieved in several different ways.

The first reported operational system for automatic diameter control was for the growth of silicon single crystals pulled from 1kg melts [30]. This used an optical pyrometer, working in the wavelength range 0.65–0.9 microns to monitor the "bright ring" visible around the growing crystal (see fig. 5). Originally it was thought that this bright ring was due to latent heat evolution at the crystallising interface but it is now known that it is in fact a reflection of the hot upper regions of the crucible in the meniscus.

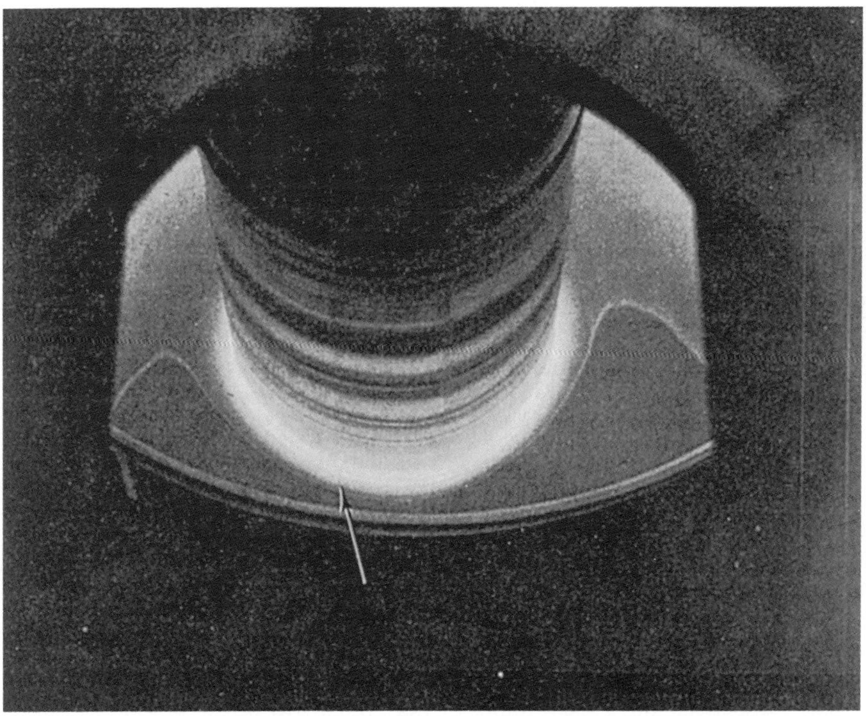

Fig. 5. Silicon single crystal being pulled. The arrow shows the meniscus "bright ring" [155].

Therefore any change in meniscus contacting angle will cause lateral motion of the bright ring image across the field of view of the pyrometer and this signal change can be used to provide a control signal to maintain constant meniscus angle and hence constant crystal radius.

Because of the relatively high speed at which silicon is pulled, the latent heat generation represents a significant contribution to the heat flux at the interface. Changing the pulling speed therefore changes the meniscus height (see section 2.2) and therefore the crystal radius. Accordingly the servo-controller can be made to respond to changes in pyrometer signal by changing either (a) the pulling speed in the approriate direction to bring the meniscus height and hence the contacting angle back to its proper position or (b) by changing the heater power to achieve the same effect. Method (a) has the advantage that it is quick acting but, by changing the growth speed, the effective distribution coefficient of any dopant is also changed. Method (b) is slower in response but leaves the growth speed unaltered after the transient has decayed. In practise, changes to both pulling speed and heater power are applied. Pulling speed changes are made by a fast-acting inner servo-loop with the power level adjusted in an outer control loop so as to maintain the mean value of the pulling speed at about its desired level.

In modern commercial equipments the simple pyrometer has been replaced by the CCD array camera and sophisticated computer algorithms are used to maximise crystal yield within specification. If the melt level is allowed to fall then the camera needs to be tracked so that it continues to image the same field of view relative to the position of the three-phase boundary.

The bright ring technique is particularly suited to the growth of silicon because of its high Laplace constant (which gives a large meniscus height) and the high reflectivity of molten silicon. An alternative “active” meniscus reflection system has been described by Gross and Kersten [31] in which a laser beam is reflected vertically from the melt surface. By having the laser beam at near normal incidence to the bulk melt surface, the problem of falling melt level is obviated but this reduces the sensitivity of the technique since the point of observation of the meniscus has to be more remote from the three-phase boundary. The technique was successfully used for the growth of potassium chloride and bismuth germanate crystals but wave motion on the melt surface due to vibration posed a serious problem.

A number of attempts have been made to directly image the crystal in the vicinity of the three-phase boundary so that a direct measure of current crystal radius might be made. A fundamental problem is that, if the technique is to work throughout the growth run as the crucible is progressively emptied, the crystal can only be observed at an oblique angle and not normally. One technological tour de force to avoid this problem was made by Pruett and Lien [32] and by van Dijk et al. [33] who used an X-ray source and an image intensifier to image a growing gallium phosphide crystal through the walls of the pressure chamber and through the crucible, from which it was being grown.

All such methods have used TV images with a computer algorithm to select the relevant frame scan-line from which to obtain a measure of the current crystal diameter. Different approaches have been made to the difficult problem of how to select the correct line. Van Dijk et al. [33] located the melt level remote from the crystal and counted a fixed number of lines, corresponding to the meniscus height, above this level. Interestingly both they and Pruett and Lien [32] found that they obtained their best results when they selected a line which was below the level of the three-phase boundary. They were therefore measuring the width of the meniscus at some fixed position and were thus obtaining a measure of meniscus contacting angle thereby reaping the benefits of anticipating radius change expounded above.

Optical techniques are not readily suited to oxide growth where melt reflectivities are, in general, very low nor to the growth of the III–V compound semiconductors grown by the liquid encapsulation technique where the three-phase boundary is immersed beneath a layer of boric oxide. Bubbles which form in the boric oxide make imaging difficult (although this problem can be largely overcome using fairly sophisticated image averaging techniques [34]). Furthermore, the boric oxide can sometimes become cloudy with included particulate matter thus obscuring a view of the crystal–melt interface. For both III–V and oxide growth, weighing has become the established technique. Indeed it is the most versatile technique and can be used for almost any material; it is therefore likely to be the best choice for a research puller.

Bardsley et al. were the first to report on the use of a servo-controller utilising crystal weighing [35]. They compared an observed weight with a demanded weight and used the error signal to modulate heater power. They demonstrated an ability to grow long single crystals of a low density material (sodium chloride) with closely controlled diameter. This work was followed by a number of publications describing the use of both crystal and crucible weighing as a sensor for automatic diameter controllers. These were almost exclusively related to the growth of oxide crystals (for a review see Hurle [36]).

There is in fact an inherent problem in the use of weighing with the common semiconductor materials as was discovered by Bardsley et al. [37] when they attempted to grow germanium at speeds of less than several cm per hour. This problem is evident from the expression for the force experienced by the crystal weighing device:

$$F(t) = m_0 g + \int_0^t \rho_S \pi R^2 g v \, \mathrm{d}t + \pi R^2 \rho_L g h + 2\pi R \sigma_{LG}(\cos \Theta_L - \Theta_S \sin \Theta_L), \tag{3.4}$$

where m_0 is the mass of the pull rod plus crystal at time $t = 0$, $\rho_{S,L}$ are the densities of crystal and melt respectively. The third and fourth terms on the right-hand side of eq. (3.4) are related to the supported meniscus. The third term is the excess force due to the supported meniscus whilst the fourth term is the vertically resolved component of the surface tension force around the three-phase boundary. If we differentiate this equation:

$$\dot{F}(t) = \rho_S \pi R^2 g v + \pi \rho_L R g(2\dot{R}h + R\dot{h}) + 2\pi\sigma_{LG}(\dot{R} \cos \Theta_L - R \sin \Theta_L / v) \tag{3.5}$$

we see that the first term on the right-hand side expresses the fact that the rate of increase in weight of the crystal is proportional to the square of the current radius value at constant growth speed. However, the sensor also records the additional forces due to the surface tension terms, and so the dependence of weight on crystal radius is in fact much more complicated. Substituting for Θ_L from eq. (3.2) we see that the relationship is given by a second-order differential equation.

In the jargon of control engineering, the use of a weight sensor contributes two "zeros" to the transfer function of the process. The problem arises because, for semiconductors, one of these "zeros" is in the right-hand plane of the Nyquist diagram [38] and this gives rise to an instability [39]. Physically the phenomenon is due to two effects which can be explained as follows:

(1) Non-zero value of Θ_L^0. Suppose a crystal, which is initially growing as a right circular cylinder, starts to decrease in diameter. The contacting angle Θ_L will decrease and so the vertically resolved component of the surface tension will increase. Ultimately, of course, if the crystal stabilises at some new but smaller diameter, Θ_L will have returned to its old value and the vertically resolved component of the surface tension will be reduced because of a reduction in the peripheral length (equal to $2\pi R$) over which it is acting. However the initial effect is to increase the apparent "weight" of the crystal and a simple servo-controller which equates weight with the square of the crystal radius will make a heater power change of the *wrong* sign causing the crystal to deviate even more from the set point. Eventually non-linear effects will take over and the crystal will probably undergo limit cycle oscillations in diameter.

(2) A small increase in melt temperature will cause the meniscus height to increase and therefore the contacting angle Θ_L to decrease resulting in a steady decrease in crystal diameter. However, the increase in meniscus height is effectively replacing a wafer of crystalline material of area πR^2 by a wafer of melt of approximately the same area. Since the density of the common semiconductors (the Group IV and III–V materials) is greater in their liquid than in their crystalline state, this rise in melt temperature produces an apparent increase in weight. Ultimately, the meniscus height will again reduce if the crystal stabilises at a new, smaller radius, but the initial effect is again to produce an error signal of the "wrong" sign.

Apparent weight is therefore an ill-conditioned measure of crystal radius for these materials. However the problem was overcome by Joyce and co-workers [39, 40] who developed a servo-controller in which the magnitudes of these "anomalous" contributions to the crystal weight were estimated and that estimate was subtracted from the raw weight signal to give a corrected signal which yielded a stable control loop. This system was subsequently incorporated on the range of high pressure liquid encapsulation pullers for the commercial growth of III–V compounds marketed by Cambridge Instruments (now Metals Research Semiconductors).

Two modes of control are possible with the weighing technique (fig. 6): either the actual weight (or the weight corrected for the anomalous terms) is compared with a demanded weight (eq. (3.4)) or the weight signal is first differentiated and compared with the demanded rate of increase of weight (eq. (3.5)), the difference forming the error signal. The former ("weight mode") seeks to keep the total integrated weight of the crystal at its demanded level and so compensates for any error in diameter at a previous stage by causing an equal error of opposite sign. Thus "start-up" errors can produce a damped oscillation in shape which is propagated down the growing crystal. However, the signal-to-noise ratio is good in this mode so that it can be deployed down to very low growth rates.

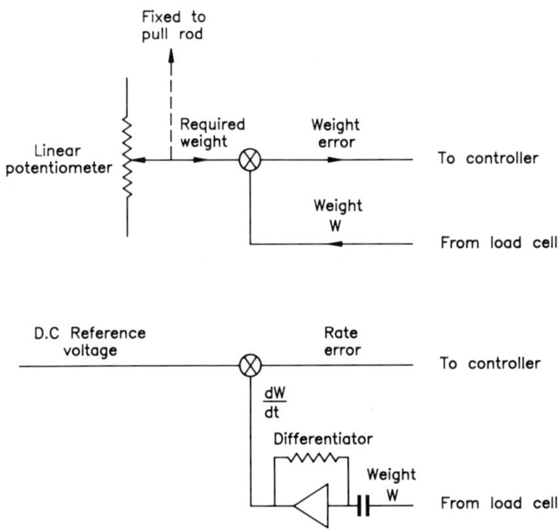

Fig. 6. Illustration of the two modes of automatic diameter control using a weight sensor. In the direct weight method (top) (referred to as the "potentiometric" method), the crystal weight is compared with a reference signal derived from a long-throw linear potentiometer coupled to the pull rod. If the pull rate is constant, this signal is a linearly ramped voltage. In the "differential" mode (bottom) the weight signal is first differentiated and then compared with a reference voltage.

The "differentiated weight" mode seeks to keep the current value of diameter at its correct value, ignoring past history. Signal differentiation increases the noise level markedly but this is a problem only at the slowest growth rates. Further, because of the phase advance provided by signal differentiation, this mode tends to provide a more stable servo-loop which is less sensitive to the thermal lags in the puller system. This latter advantage is particularly significant at high growth rates. Also the above described "anomaly" effect is less serious when the controller is operated in this mode.

The relative advantages and disadvantages of the two modes make the "weight" mode preferred for oxide growth (where growth rates tend to be very low) whilst, for "anomalous" materials, the "differentiated weight" mode is favoured. The attainable performance of weight control is shown in fig. 7 for the Y_2O_3–Al_2O_3 system.

The use of a die floating on the melt surface or rigidly attached to the crucible has been widely researched, principally for the growth of ribbon crystals. Two classes of die have been employed – wetting and non-wetting. The use of non-wetting dies has been extensively explored in the ex-Soviet Union following pioneering work by Stepanov [41]. In work reported in 1959, but performed much earlier, he described the fabrication of metal strips, tubes and bars by pulling through a die floating on, or fixed in the plane of the surface of a bath of molten metal. Subsequently he considered the use of the process to grow single crystals. Antonov [42] reported specifically on the use of the technique for the growth of cylindrical crystals of germanium and indium antimonide. The die used is shown schematically in fig. 8.

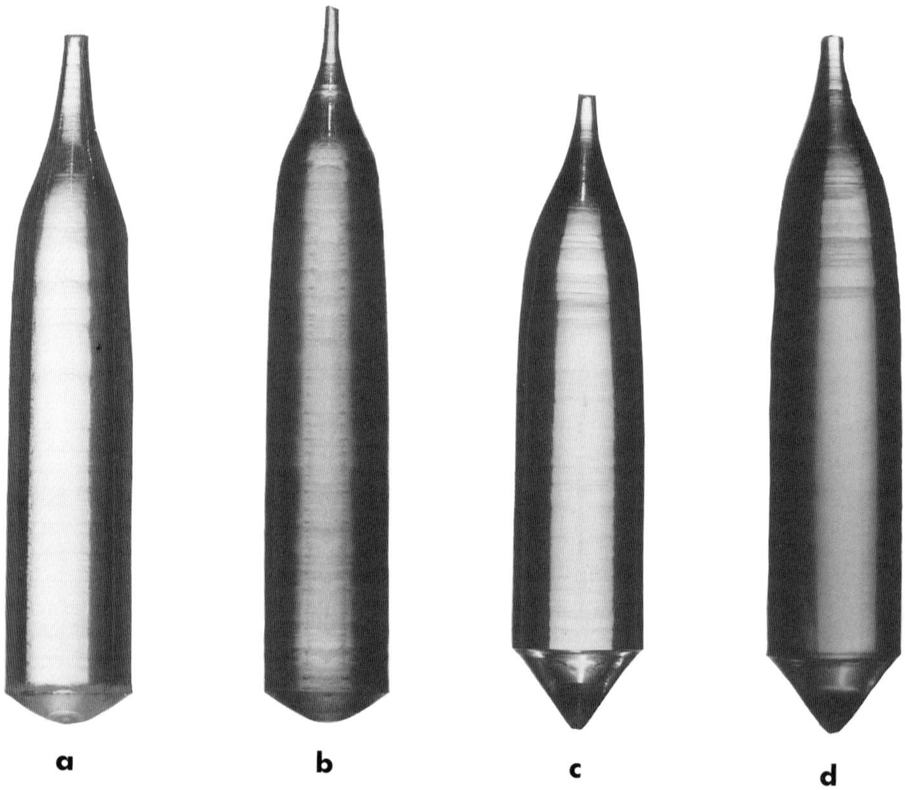

Fig. 7. Yttrium aluminate single crystals grown using automatic diameter control based on crucible weighing; (a) and (b) are $YAlO_3$, (c) and (d) are $Y_3Al_5O_{12}$.

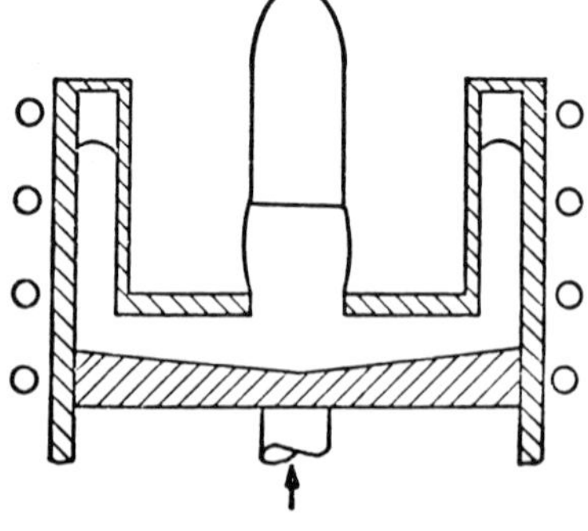

Fig. 8. Crystal pulling through a die [42].

An excess pressure is used to force the melt to the top of the die and to control the meniscus shape. It has been shown that there is an optimal pressure for which the crystal diameter is sensibly independent of small fluctuations in meniscus height. The dynamic stability of the process has been extensively analysed by Tatarchenko and is reported in his chapter in this Volume.

The use of wetted dies was pioneered by LaBelle and Mlavsky [43] with their so-called edge-defined film-fed growth (EFG) technique. This has been used to produce sapphire crystals of a variety of shapes and was subsequently developed for the growth of silicon ribbons. Although applicable to the growth of cylindrical crystals, it does not appear to have been used for this purpose.

In Stepavov's method, the crystal diameter is controlled by anchoring the lower end of the meniscus at the edge of the die. An alternative approach is to confine the lower region of the meniscus within a slightly conical surface which is not wetted by the melt. Indeed the wall of the crucible acts in this way when growth at a diameter approaching that of the crucible is performed in conventional Czochralski growth. The use of a conical die, termed a "coracle", floating at the interface between a gallium arsenide or phosphide melt and a boric oxide liquid encapsulant has been reported by Cole et al. [44]. The die was made of silicon nitride and was wetted by the boric oxide but not by the semiconductor melt. It can be shown that growth at constant diameter can taken place in a stable regime where precise control of temperature is unnecessary.

The powerful advantage of both wetting and non-wetting dies is that, used under the correct conditions, they avoid the Surek type of instability [45] which is an unwanted feature of conventional Czochralski growth.

4. Growth of elemental materials

4.1. Metals

The Czochralski technique has been used to grow a variety of metal single crystals almost exclusively for research purposes. Larger crystals of a few materials, such as bismuth [46] and copper [47], which can be used as neutron monochromators, have also been pulled from the melt. Since the objective of most of the research has been a study of some physical property, the fundamentals of the growth of these materials has been but little researched and the literature is very limited.

Low melting point metals required only simple apparatus [46, 48, 49]. Most crystals have been grown manually but Uelhoff and co-workers have developed a sophisticated optical imaging technique for the computer control of the growth of copper, gold [47] and Cu_3Au [50]. This involved imaging of the three-phase boundary which was done using a mirror system to minimise fogging of the optical components in the high vacuum puller used. Bachmann et al. [48] demonstrated open-loop control of the growth of nickel and silver crystals by imaging the meniscus. Ekhult and Carlberg [51] describe a novel infrared auxillary heating system for the manual diameter control of tin crystals being pulled from the melt. Halogen lamps were focused on to the melt surface and the crystal diameter was changed by controlling the lamp power.

Some studies of the effect of cooling conditions on the structure and perfection of the crystals has been made. Bachmann and Wakiyama [52] studied the growth of hcp cobalt and double hcp $Co_{(1-x)}Fe_x$ alloy crystals. They demonstrated how they could control the martensitic transformation on cooling. Large single crystal domains were obtained by rapid cooling ($> 20°C/min$) through the martensitic temperature.

Dislocation-free growth of copper [49, 53, 54] and silver [55] single crystals has been studied. As with the common semiconductors (see below) thermal stress appears to be the principal generation mechanism. Dislocation-free crystals can be grown up to some critical (but unfortunately rather small) diameter by careful growth of a long thin neck on to the seed followed by very narrow tapering out to a final diameter.

In a low thermal gradient puller, Sworn and Brown [49] found that the maximum achievable diameter whilst retaining dislocation-free growth depended on the mini-

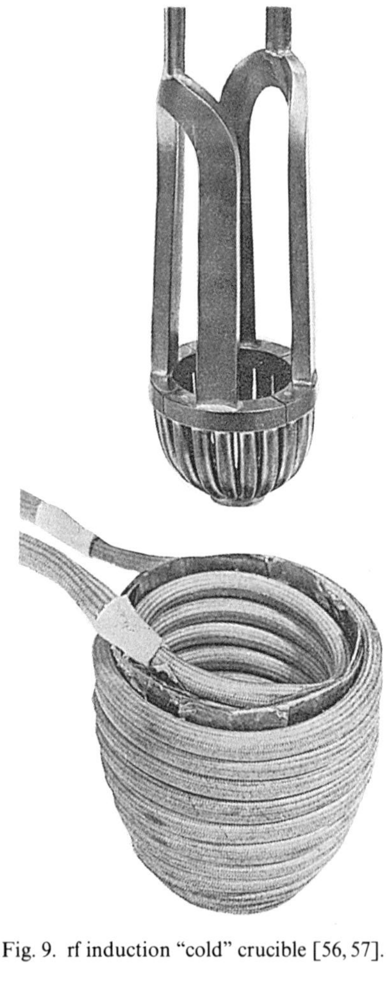

Fig. 9. rf induction "cold" crucible [56, 57].

mum thickness of the necked region. They used Bormann X-ray topography to study the perfection of their crystals. A very thin neck (0.3 mm diameter) permitted dislocation-free crystals of up to 2–3 mm diameter to be grown whereas, if the neck diameter was not less than 0.5 mm, dislocation-free growth was not obtained at any (greater) diameter. Dislocation density was weakly dependent on pulling speed, dislocations being more prone to occur at higher speeds (6 cm h^{-1} as compared to their standard speed of 3 cm h^{-1}). Crystal orientation was not a significant factor in determining crystal perfection. (They grew on [321], [110] and [100] axes.)

Collisions between the crystal and scum on the melt surface (particularly copper oxide or graphite particles) were also shown to be a potent cause of dislocation generation.

The highly reactive rare earth metals have been pulled from the melt using a novel form of crucible in which contact with the melt is avoided. This, so-called, cold crucible was pioneered by Sterling and Warren [56, 57] and by Hukin [58] who first applied it to the growth of the rare earth metals. Sterling and Warren's design was evolved from their horizontal silver boat technique [57] in which high frequency currents induced into the boat from a surrounding rf coil levitated the melt by holding it out of contact with the crucible. Heating of the melt, though relatively inefficient, was produced by these induced currents. Hence the term "cold crucible". Silver was chosen as crucible material because of its very high electrical conductivity. Their design for a Czochralski crucible together with the rf coil is shown in fig. 9. (They demonstrated the growth of oxygen-free silicon single crystals using this technique.)

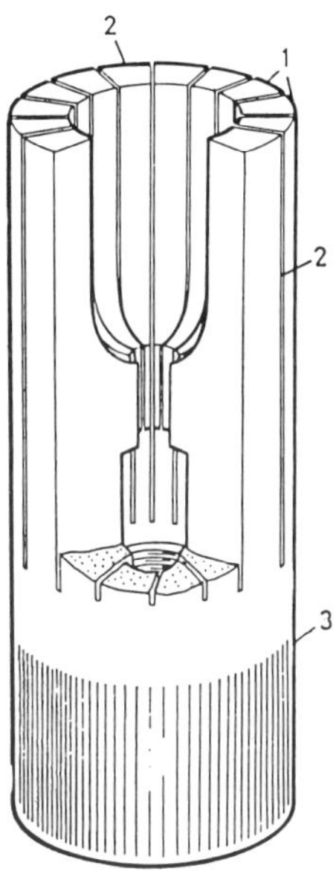

Fig. 10. Improved design of cold crucible. (1) Radial slots; (2) segmented walls; (3) housing for water cooling manifolds [58].

Hukin [58] improved upon the design of this crucible in significant ways. He fabricated the crucible from a cylindrical billet of high conductivity copper producing sixteen vertically slotted segments (fig. 10). Each segment has an integral water cooling channel and the channels are connected via manifolds in the lower portion of the body to a co-axial supply and return pipe which also serves as the support column for the crucible assembly. The cylindrical exterior facilitates close rf coupling. The vertical slots in the crucible extend well below the bottom of the containment section in order that each segment is effectively isolated electrically within the region of the field and so acts as a single turn step-down transformer between the rf coil and the charge. The crucible is highly polished, silver plated and rhodium flashed in order to produce an easily cleaned surface of high reflectivity and electrical conductivity.

A crucial refinement is the provision of a central hole at the base of the containment section where the radial slots merge. This concentrates the rf flux immediately below the charge where it is most effective in producing levitation. The Sterling and Warren design connected the segmented conductors together at this point resulting in current cancellation which produced a "cold" spot.

The choice of rf frequency is important. Low frequencies (in the 10 kHz range) produce strong levitation but the rf penetration depth is then large and excessive dissipation in the copper occurs. The optimum frequency is in the 100 kHz region. The following power levels are reported by Hukin: to melt 100 g nickel or platinum charge, 15 kW; 60 g lanthanum 5 kW; 35 g niobium 25 kW.

4.2. *Germanium*

4.2.1. *Introduction*

Germanium was in fact the material on which the pulling technique was first pioneered by Teal and Little [1]. However its use as a transistor material was soon supplanted by silicon and commercial production declined. More recently production rates have revived as the material is finding use as a passive infrared material. It is used for optical components in the thermal imaging 8–14 micron atmospheric window.

Whilst it may not now be amongst the most widely used semiconductors, it has been and remains the workhorse material in basic studies of semiconductor crystal pulling. Thus it was used (a) to verify the Burton, Prim and Slichter theory of the effect of crystal rotation on effective distribution coefficient [22], (b) to study the nature of the cellular structure occurring under conditions of constitutional supercooling [59–61] and (c) as the prototype material for the study of Czochralski dynamics [28] and the development of automatic diameter control techniques based on a weight sensor [35, 40, 62].

Melting at only 937°C, having a low vapour pressure at its melting point, being of low toxity, containable in quartz or graphite crucibles and having had its physical and physico-chemical properties more thoroughly documented than almost any other material, it is an ideal research vehicle.

4.2.2. *Crystal growth*

Puller design is straightforward with no special features required to cope with the particular properties of germanium. Automatic diameter control can be performed with a weight servo-controller provided that signal processing to cope with the "anomalous" dependence of apparent weight on crystal radius is incorporated [40].

Commercial equipment is available to grow crystals up to 30 cm diameter. Heating is by a picket fence element fabricated from electro-graphite. The growth chamber can be used in vacuo or in an ambient inert gas.

For the production of ultrapure material smaller pullers are available in which the heated components within the chamber are reduced to a minimum by the use of rf induction heating of an ultrapure graphite crucible with thermal shielding using synthetic quartz and extensive water cooling. Flowing ultrapure hydrogen is used as ambient.

A problem of yield arises with doped material since all the electrically active dopants in germanium have distribution coefficients which differ markedly from unity and therefore exhibit marked macrosegregation along the ingot. Only a small section of any given crystal will therefore be likely to be within specification.

The problem can be overcome by the use of a second, inner "floating crucible" first described by Leverton [63] and by Goorissen [64] and shown schematically in fig. 11.

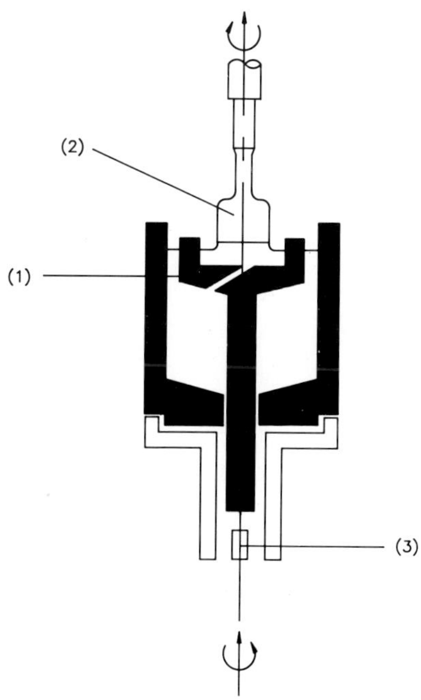

Fig. 11. Schematic representation of the floating crucible technique. The floating crucible (1) from which the crystal (2) is growing is weighted down with the counterweight (3).

The inner crucible, from which the crystal is pulled, is connected to the outer one via an orifice which is small enough to prevent significant mixing of the melt between the inner and outer crucibles, but large enough to permit Poiseuille flow from outer to inner as the melt in the inner is consumed by growth of the crystal. This condition can be expressed as $D \ll uL$ where D is the solute diffusion coefficient in the melt, u the flow velocity through the connecting orifice and L its length.

If the initial concentration of solute in both crucibles is C_0 then the initial growth will yield crystal of concentration kC_0 where k is the distribution coefficient and the inner melt will be replenished with melt of concentration C_0. The average concentration of the inner crucible melt will therefore rise during a transient phase to a concentration of C_0/k whereupon the crystal composition will have risen to C_0. At this point a steady state is reached with material leaving the crucible as crystal of composition C_0 and entering as melt of the same composition and at the same rate. If the volume of the inner crucible is small in relation to the overall melt volume, then the fraction of crystal grown during the transient phase is kept small. Alternatively, by pre-charging the inner crucible with solute at concentration C_0/k, the transient is avoided altogether.

With care, dislocation-free crystals can be obtained in undoped material. This is even easier with "p$^+$" doped material as acceptor doping seems to suppress dislocation

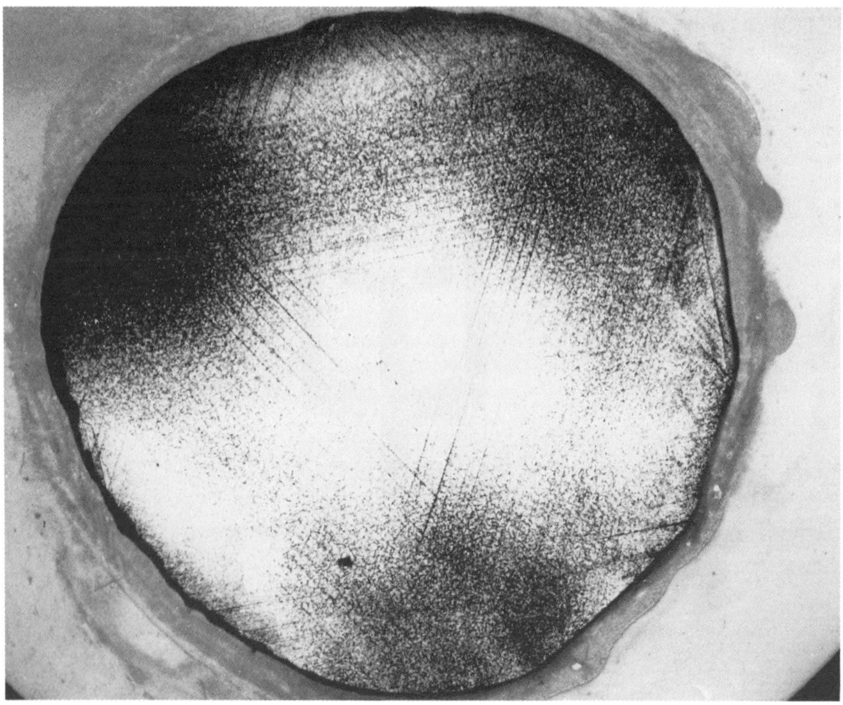

Fig. 12. Etched (111) section of a germanium single crystal showing a characteristic pattern of octahedral slip produced by thermal stress during cooling from the growth temperature. Mag. × 8.

formation [65]. On the other hand, "n$^+$" material presents more problems [61]. Dislocations, when they are formed, exhibit the pattern of octahedral slip characteristic of diamond-cubic semiconductors. (fig. 12)

The $\{111\}$ faces are below the roughening temperature at the melting point of germanium and form facets on convex-to-the-melt interfaces to which they are tangential. Solutes are incorporated at these facets in an amount which differs from the near-equilibrium segregation which occurs on the non-facetted portions of the interface. A typical facet is shown in fig. 13 and fig. 14 displays the effect of such a facet in giving a core of differing concentration down a $\langle 111 \rangle$ axis crystal. Whilst most solutes are incorporated at facets in concentrations in excess of the equilibrium concentration (which may be thought of as being due a trapping of a surface excess absorbed concentration by rapid lateral motion of growth steps – see section 5.1 below), the solute gallium is incorporated in a lower concentration on the facet in germanium [67].

Growth twins can form on $\{111\}$ facets at the three-phase boundary during growth, particularly during the "grow-out" phase as was first noted and studied by Billig [68].

Fig. 13. Interface of a $\langle 111 \rangle$ germanium single crystal snatched from the melt. The dark central region is a retained droplet of melt. This partly obscures the (111) on its right-hand side (the bright crescent). Mag. × 2.6.

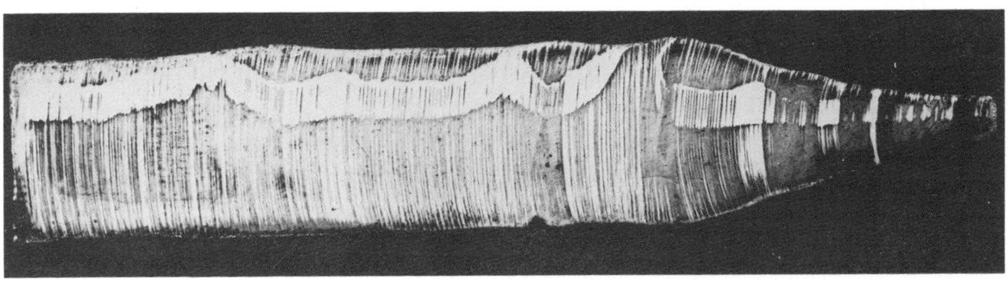

Fig. 14. Longitudinal section of a near-$\langle 111 \rangle$ oriented gallium-doped germanium single crystal. Growth striations and a meandering (111) growth facet revealed by pulse plating [155]. Mag. × 1.2.

However the incidence of twinning in a clean puller/melt environment and with a good heater power control system which avoids marked growth rate fluctuations, is extremely low.

The occurrence and morphology of cellular structure has been extensively studied in germanium and as it is a paradigm for all diamond-cubic and zinc-blende semiconductors, a summary of its main characteristics is now given.

4.2.3. Cellular structure

Conditions for the occurrence of a morphological instability leading to the formation of a finite-amplitude cellular structure in the presence of a zone of constitutionally supercooled melt ahead of the crystal have been fully described in Volume I of this Handbook to which the interested reader should refer. The special features associated with the pulling process are that crystal rotation provides an additional control parameter affecting the conditions of onset of constitutional supercooling and the hydrodynamics of the flow due to that rotation influences the evolution and form of the cellular structure.

The fact that germanium (and the other Group IV and III–V semiconductors) can form $\{111\}$ facets on the growth interface profoundly influences the morphology as we shall show.

Using the Burton, Prim and Slichter analysis, Hurle [69] showed that the gradient of constitutional supercooling at the interface of a crystal rotating at an angular rate Ω and growing at a normal growth rate v, from a melt containing a solute at concentration C_L, the solute having a diffusivity D and a distribution coefficient k is given by

$$\mathrm{d}S/\mathrm{d}z|_0 = -mvC_\mathrm{L}(1-k)\,[k+(1-k)\exp(-v\delta/D)]^{-1}/D - G_\mathrm{L} \tag{4.1}$$

$$= -mvC_\mathrm{S}(1-k)/Dk - G_\mathrm{L}, \tag{4.2}$$

where z is the co-ordinate normal to the plane of the undeformed interface, δ is the BPS boundary layer thickness (see eq. (2.13)), C_S is the solute concentration in the crystal and m is the slope of the liquidus line for the alloy system. G_L is the temperature gradient in the melt at the interface. From the two forms of the equation it can be seen that, in terms of the solute concentration in the bulk melt, increasing the crystal rotation rate (i.e. reducing the boundary layer thickness δ) decreases the gradient of constitutional supercooling. However, if the objective is to grow an alloy crystal of given compostion C_S, then crystal rotation has no effect: one can either put a relatively small amount of solute into the melt and pull the crystal fairly rapidly to achieve a high value of the effective distribution coefficient to give the required C_S or one can put more solute into the melt and pull more slowly.

As with all melt growth systems, the way to avoid constitutional supercooling (for a given melt concentration C_L) is to pull slowly with a steep temperature gradient in the melt. A steep melt gradient also requires a steep gradient in the crystal (see eq. (2.1.)) and so increases the thermal stresses on the cooling crystal (see the chapter by Volkl). Lowering the growth rate brings economic disbenefit and also requires that fluctuations in growth rate due to melt turbulence, mechanical and power controller fluctuations etc. have to be minimised. As with most aspects of crystal growth, an optimal compromise has to be sought. Avoidance of constitutional supercooling is, however, vital if homogeneous single crystals are to be produced.

The morphology of the cellular structure in germanium can be revealed in several ways: (a) by snatching the crystal from the melt, (b) by sectioning the grown crystal and etching to reveal micro-variations in solute concentration, (c) using topographic or radiographic techniques.

We give examples of each of these in the following.

The fact that a $\{111\}$ growth facet forms on the crystal–melt interface whenever it presents a convex profile to the melt has a marked influence on the development of the cellular morphology in such materials. An initial near-sinusoidal perturbation of the

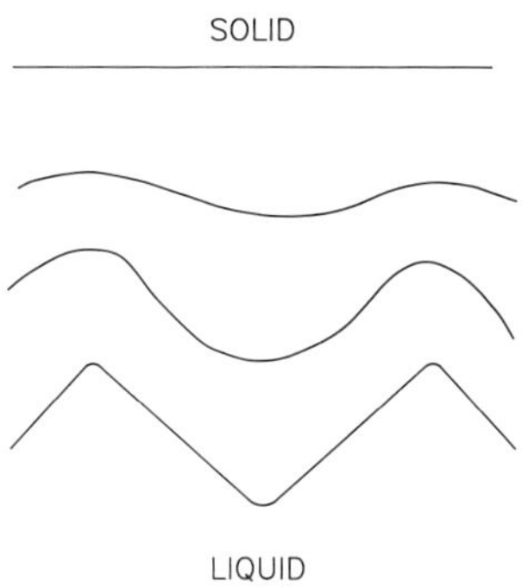

Fig. 15. Schematic representation of the development of a cellular structure in a facetting material. Micro-facets develop once the amplitude of the interfacial perturbation has produced sections of the interface which are tangential to $\{111\}$ planes.

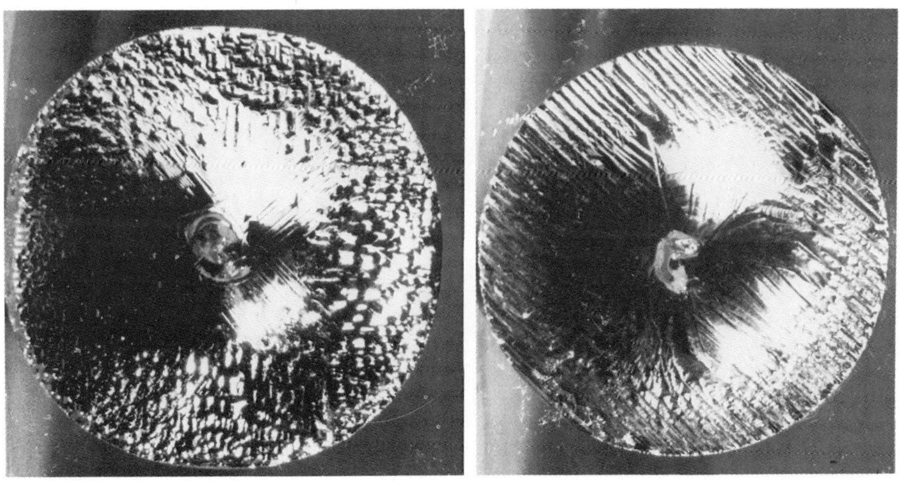

Fig. 16. Interfaces of snatched germanium crystals growing with a cellular interface. The right crystal is $\langle 110\rangle$ oriented; the left one is $\langle 100\rangle$. The cellular interfaces are composed of arrays of $\{111\}$ micro-facets. Mag. $\times$ 2.6.

interface, which forms as morphological instability sets in, grows in amplitude until it becomes tangential to one or more sets of {111} planes and micro-facets start to form. This is shown schematically in fig. 15. It causes the morphology of the cellular structure to be very orientation dependent. For growth on a ⟨110⟩ axis, the structure takes up a "factory roof" configuration of very elongated cells formed by pairs of {111} planes inclined at approximately 35° to the growth axis. For growth in the ⟨100⟩ direction, the cells are approximately equiaxed being composed of square-based pyramids formed by the four {111} planes equally inclined at approximately 44° to the ⟨100⟩ growth axis. This can be seen in fig. 16 which shows the interfaces of two crystals respectively of ⟨100⟩ and ⟨110⟩ orientation which were snatched from the melt. A magnified portion of the ⟨100⟩ crystal interface is shown in fig. 17. On the free cylindrical surface of the crystal the onset of a cellular structure is made evident by the appearance of "corrugations" (fig. 18) where pairs of cells intersect the surface. This can be seen as the crystal is

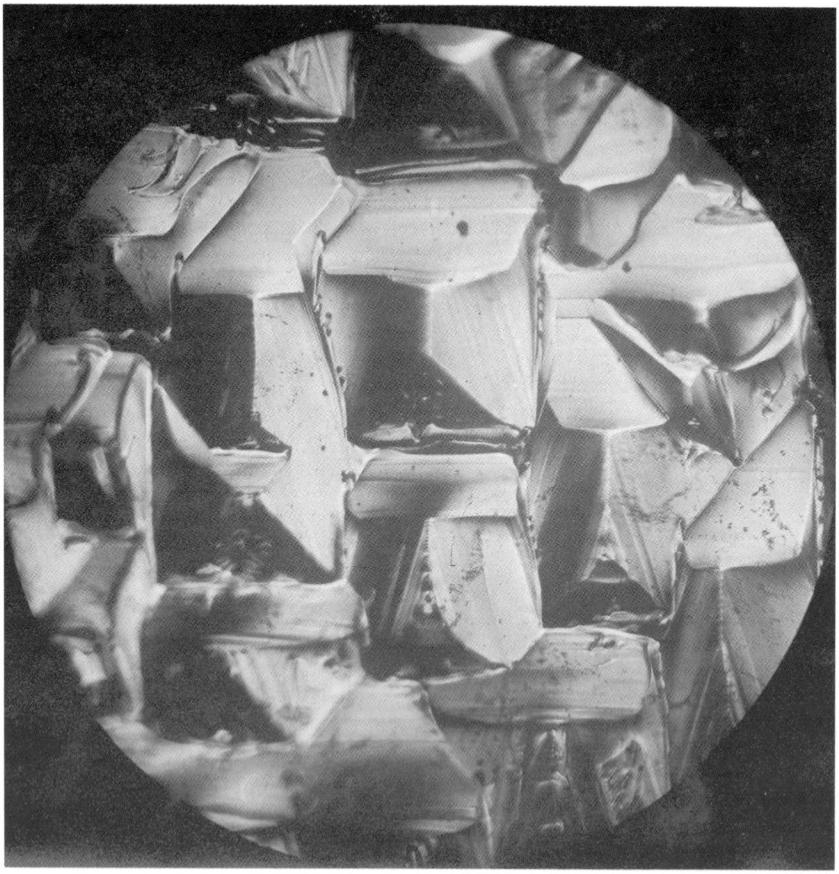

Fig. 17. Magnified portion of the interface of the ⟨100⟩ crystal shown in fig. 16. Mag. × 76.

Fig. 18. Profile of the interface of the crystal shown in fig. 16 (left) viewed normal to the growth axis showing the intersection of pairs of {111} micro-facets, which form the cellular structure, with the cylindrical surface of the crystal. The ribbed pattern is a modulation of the crystal profile produced by the crystal rotation. Mag. × 120.

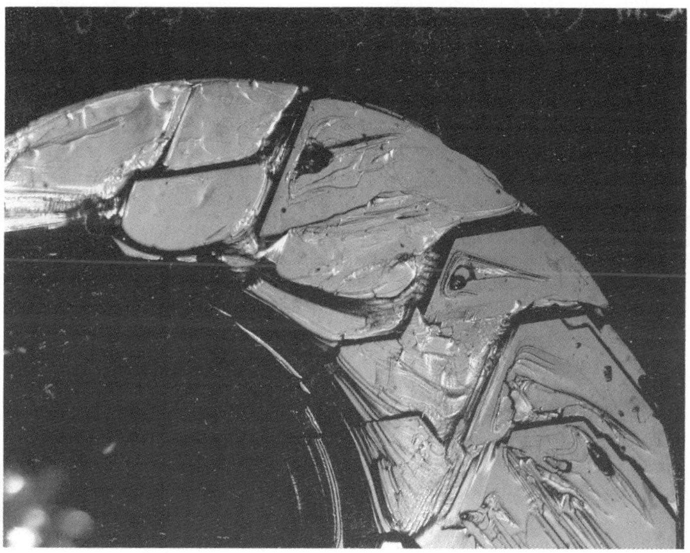

Fig. 19. {111} macro-steps, produced by constitutional supercooling, on the interface of ⟨111⟩ oriented doped Ge crystal growing under conditions of a convex-to-the-melt freezing point isotherm. The large dark region in the bottom left-hand corner is the melt droplet which adhered when the crystal was snatched from the melt. Mag. × 6.

being grown, thus permitting the operator to remelt and regrow it at a slower rate which avoids the occurrence of constitutional supercooling.

When a (111) plane is parallel to the macroscopic plane of the interface, the interface is somewhat stabilised against cellular breakdown [60] but, when breakdown does occur, it is by the formation of macro-steps as seen in fig. 19.

Rotation of the crystal imposes an overall pattern to the cellular interface. The cellular projections, which tend to form first near the axis of the rotating crystal, produce local turbulence which propagates spirally outwards and imposes a pattern on the structure at greater radial distances. This can be seen in fig. 20 which is of a {110} etched section parallel to the plane of a cellular interface. The multi-start spiral of segregation caused by the local turbulence is evident. The elongated cells are terminated at their interesections with the spiral arms. Figure 21 is a {100} etched section cut parallel to the plane of the interface of a ⟨100⟩ axis crystal which, similarly, shows the elongation of the {111} pyramids where their axes are aligned parallel to the spiral flows.

The profile of the cellular interface can be studied by etching longitudinal sections cut perpendicular to the macroscopic plane of the interface. Figure 22 is an etched (110) plane orthogonal to the ⟨110⟩ growth axis of a gallium-doped germanium crystal. The profile of the "factory roof" cellular structure is delineated by growth striations

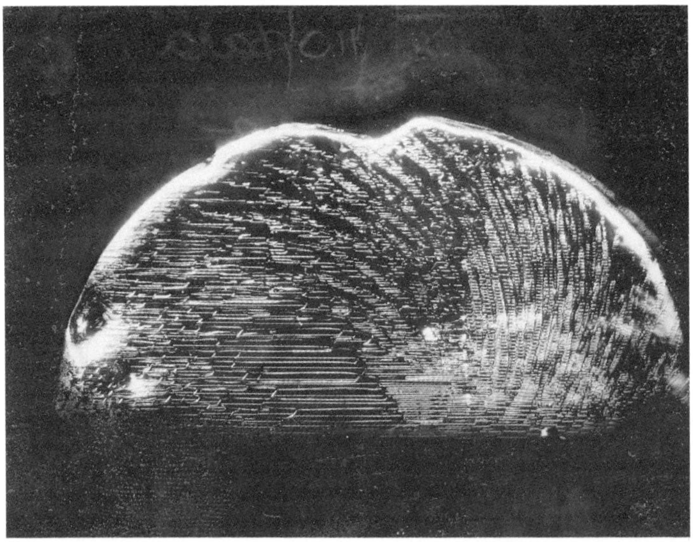

Fig. 20. Dark field macrograph of half of a wafer cut from a ⟨110⟩ oriented Ga-doped Ge single crystal growing with a cellular interface. The cellular structure has been revealed by chemical etching. Note that the cell size and aspect ratio varies markedly with position on the wafer. Cells form first in the centre and generate local turbulence which spreads radially. Because of the crystal rotation this takes the form of a multi-start spiral. The pattern of micro-segregation imprints itself on the pattern of cellular development with the long axis of the cells (which are formed by pairs of {111} planes inclined at roughly 35° to the growth axis) being terminated by the spiral arms. Where the arms are roughly parallel to the cell axis the cells are very elongated (as at lower left); where orthogonal to the cell axis their length is limited to the spacing between the spiral arms (as at upper right). Mag. × 4.5.

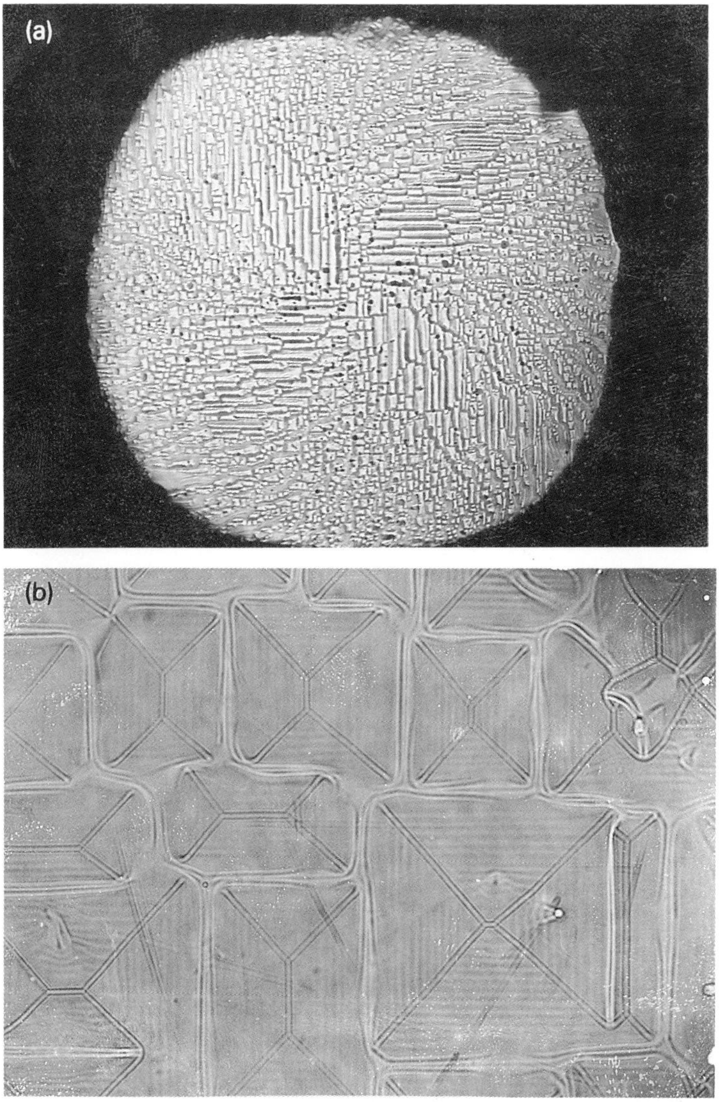

Fig. 21. Etched wafer of a $\langle 100 \rangle$ oriented Ga-doped Ge single crystal growing with a cellular interface. (a) Macro-graph of the whole wafer showing the elongation of the cells (which are composed of pyramids having sides which are $\{111\}$ micro-facets) in $\langle 110 \rangle$ directions produced by crystal rotation. Mag. × 3.2. (b) micrograph of a region near the bottom of (a). It shows (i) cell boundaries (the wavy double lines), (ii) the intersections of pairs of $\{111\}$ planes (straight double lines) and (iii) the terracing on the cells formed by the growth striations which show the amount of crystal grown on each face per crystal revolution. (The micrograph is of course a section through the growing pyramids.) Mag. × 58.

Fig. 22. Etched ⟨110⟩ longitudinal section through a ⟨110⟩ oriented Ga-doped Ge single crystal showing the profile of the cellular interface which is composed of pairs of {111} planes. The single vertical lines are the traces of the cell peaks. The vertical double lines are the cell boundaries. The herring-bone pattern is formed by the growth striations caused by crystal rotation. Mag. × 77.

produced by rotation of the crystal in a melt which lacks perfect thermal symmetry and therefore produces a periodic fluctuation in growth rate as the crystal is rotated. The spacing of these striations is equal to the amount of crystal pulled per revolution.

As the degree of constitutional supercooling increases as growth proceeds (because the solute concentration in the melt steadily rises due to macrosegregation from a finite melt for a solute with $k < 1$), the cell boundaries deepen. Solute can only be lost from deep within the liquid channel by diffusion into the crystal. Whilst this can occur in some metal systems, the diffusion rates of substitutional solutes in the diamond-cubic and zinc-blende semiconductors are so low that solute loss by this means is negligible. The cell boundaries continue to deepen therefore until either (a) solid is formed by a eutectic reaction or (b) an instability in the liquid channel occurs causing it to pinch-off leaving a liquid droplet trapped within the crystal. This latter instability is

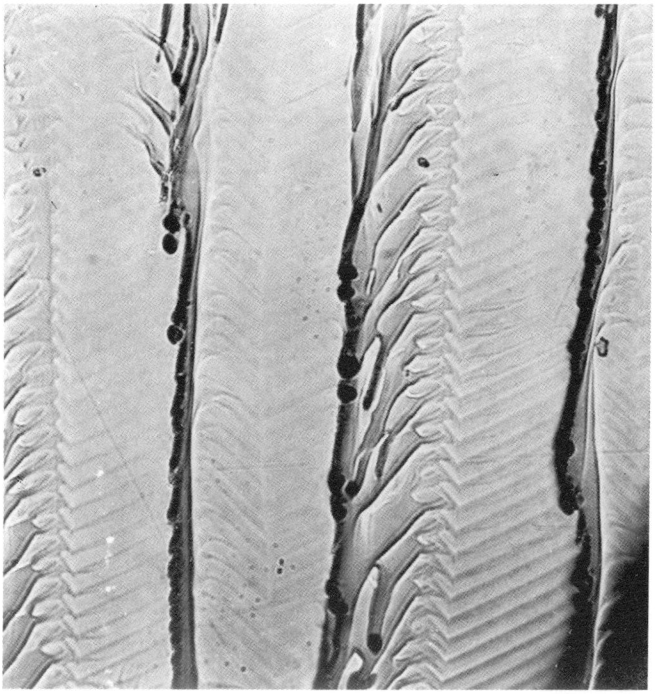

Fig. 23. Etched section, as in fig. 22 but at a slightly later stage in the growth of the crystal. Liquid droplets have been "pinched off" in the cell boundary grooves and have migrated slowly down the crystal under the influence of the axial temperature gradient finally solidifying (circular dark blobs) as the crystal is cooled to room temperature. Mag. × 77.

common in semiconductor growth as can be seen from fig. 23 which shows solute-rich droplets trapped within the germanium crystal.

Having become trapped, these droplets do not remain stationary since they are embedded within crystal on which there is an imposed temperature gradient. They therefore dissolve crystal on their hot side and the dissolved germanium diffuses across the droplet to be precipitated out on the cold side. By this process, known as temperature gradient zone melting [70], the droplet migrates down the crystal chasing, but never catching up, the main crystal/melt interface. In so doing it leaves a trail of solute-rich solid which can be revealed by etching (fig. 24) or by X-ray absorption topography (fig. 25). The droplets migrate approximately along the direction of maximum temperature gradient (which is vertically down the crystal). They therefore remain tied to cell boundaries where these are also aligned vertically, as is the case when the pairs of $\{111\}$ planes forming them are symmetrically disposed with respect to the growth axis. (This is the case for $\langle 110 \rangle$ and $\langle 100 \rangle$ growth.) For other orientations, they cut across cell boundaries as they move down the crystal. The whole process can be thought of as a "snowstorm" of droplets descending slowly downwards repeatedly regrowing any given portion of the crystal (fig. 26).

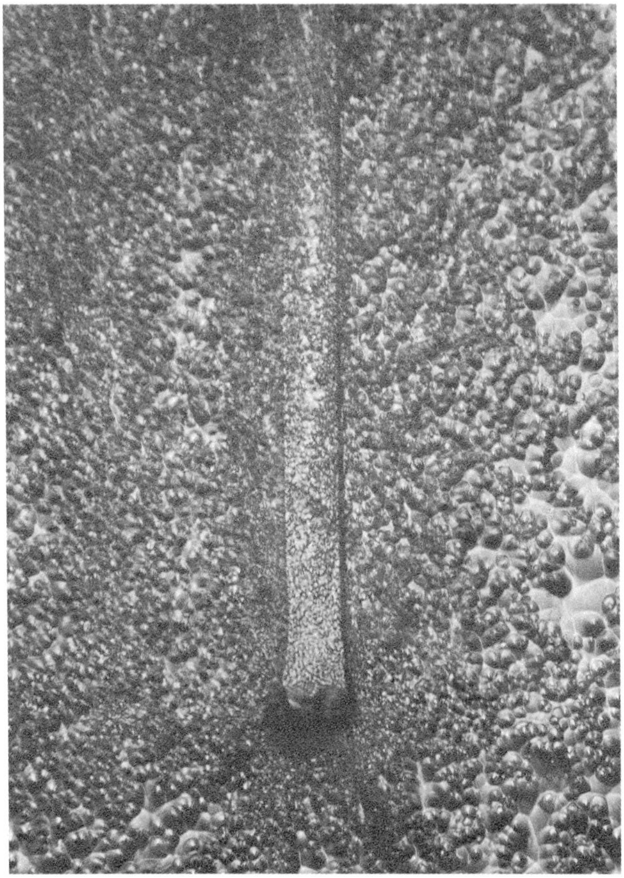

Fig. 24. Longitudinal (111) plane sectioned from a ⟨110⟩ oriented Ga-doped Ge crystal showing dislocation etch pits and the trace of the trapped liquid droplet (called a solute trail). Ga-rich material has been etched out at the termination of the trail.The dislocations are the result of the stress produced by recrystallisation of highly doped material in the solute trail which has a lattice parameter which differs significantly from that of the matrix. Mag. × 120.

Rather different behaviour occurs with solutes having a distribution coefficient greater than unity (such as silicon in germanium). Here the solute raises the freezing points and is preferentially segregated to the peaks of the cells, the cell boundaries being depleted. Since this depletion can occur only down to zero per cent solute, the depth of the cell boundary is limited and the massive droplet formation, which is characteristic of gallium dopant for example, is not seen when silicon is the dopant.

Dislocation-free growth is very rapidly lost once a cellular interface forms because the microsegregation of the solute produces stresses within the lattice which rapidly exceed the critical resolved shear stress. Massive dislocation arrays are associated with the cell boundaries (fig. 27) and with the solute trails (fig. 28). Germanium expands on

Fig. 25. X-ray absorption topograph of an In-doped Ge single crystal containing trapped solute-rich droplets which have migrated down the crystal by temperature gradient zone melting. The solute trails show dark contrast on account of the high indium content within them. The dark dots are the droplets at the end of the solute trails. They are probably composed of In–Ge eutectic [61]. Mag. × 5.2.

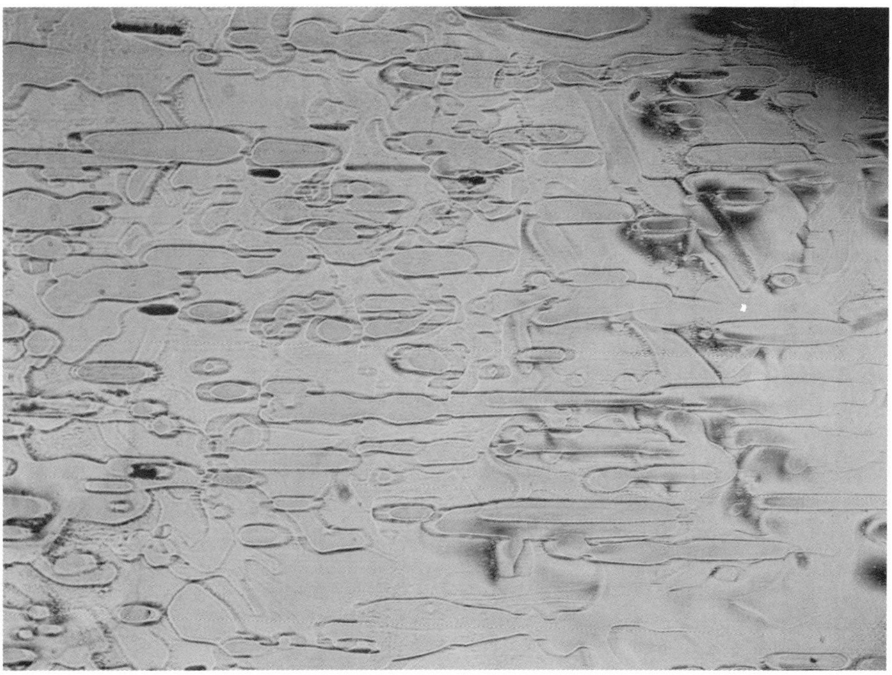

Fig. 26. Micrograph of a section of a (110) wafer cut from a Ga-doped Ge single crystal showing extensive regions of regrowth by the passage of solute trails down the crystal produced by a "snowstorm" of trapped Ga-rich liquid droplets. Mag. × 55.

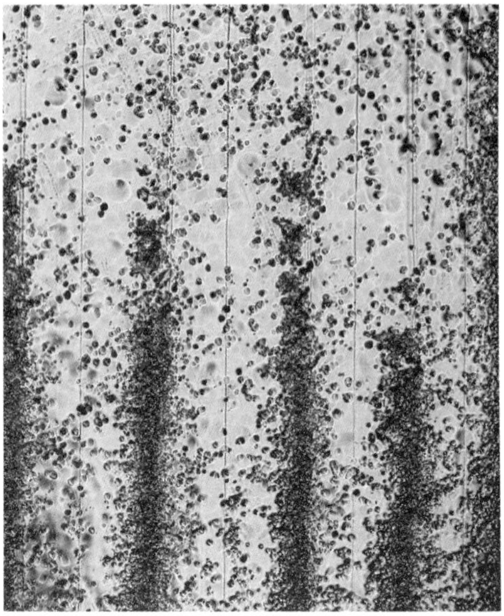

Fig. 27. Oblique {111} section taken from a ⟨110⟩ oriented Ga-doped Ge single crystal just below the point of onset of a cellular structure. Single vertical lines show the cell peaks; double lines the cell boundaries. As growth proceeds the lattice mismatch in the cell boundaries increases resulting in the formation of massive arrays of dislocations in the lower part of the micrograph. Mag. × 48.

solidification and so, when the liquid droplets finally freeze, they put the surrounding lattice in compression. If this occurs at a sufficiently high temperature then prismatic dislocation loops will be punched out (fig. 29). The correlation between the microsegregation and the dislocation pattern can be seen from a comparison of an autoradiograph of a section of crystal which had been doped with radio-gallium with an optical micrograph of the same section after etching in an etchant capable of revealing dislocation etch pits (fig. 30).

To summarise, constitutional supercooling produces a disastrous degradation in crystal quality both in terms of microsegregation and of structural perfection in crystals grown from the melt by whatever technique. The rotation imposed during crystal pulling gives a unique contribution to the pattern of the resulting cellular structure. The ability to reveal both microscale fluctuations in solute concentration and dislocations (through etch pit formation) in germanium makes this an ideal material in which to study the nature and dynamics of the evolution of cellular structure.

4.3. Silicon

4.3.1. Introduction

In 1952, Teal and Buehler [71] produced the first pulled silicon single crystals using the technology already developed in the Bell Laboratories for the growth of germanium

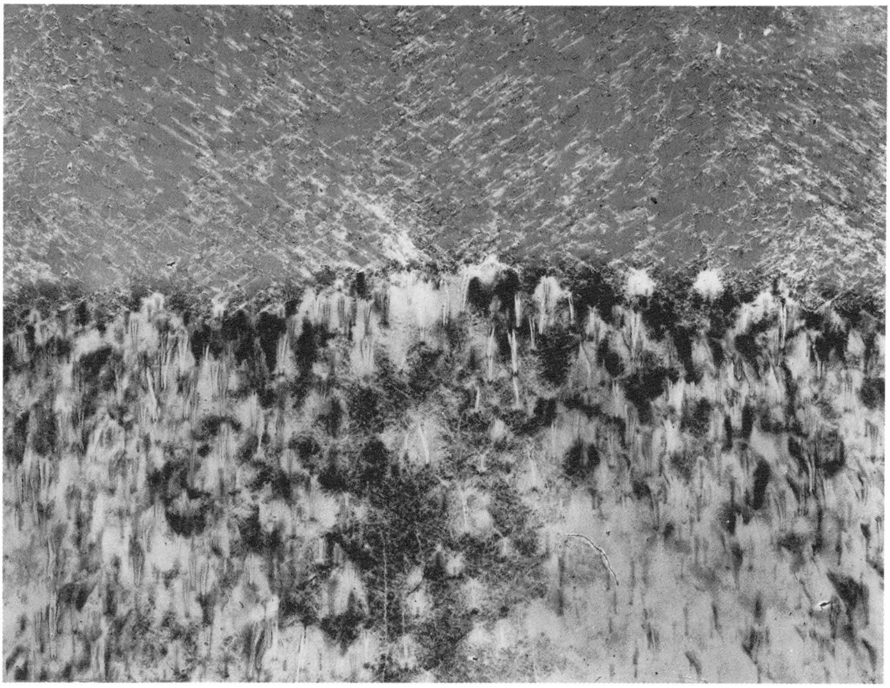

Fig. 28. X-ray diffraction topograph of a longitudinal section of a $\langle 110 \rangle$ oriented heavily Ga-doped Ge single crystal showing the abrupt onset of a cellular structure with the formation of solute trails surrounded by massive arrays of dislocations. The stress generated by the micro-segregation generates slip back into the previously grown, non-cellular, portion of the crystal (top of topograph). Mag. × 8.

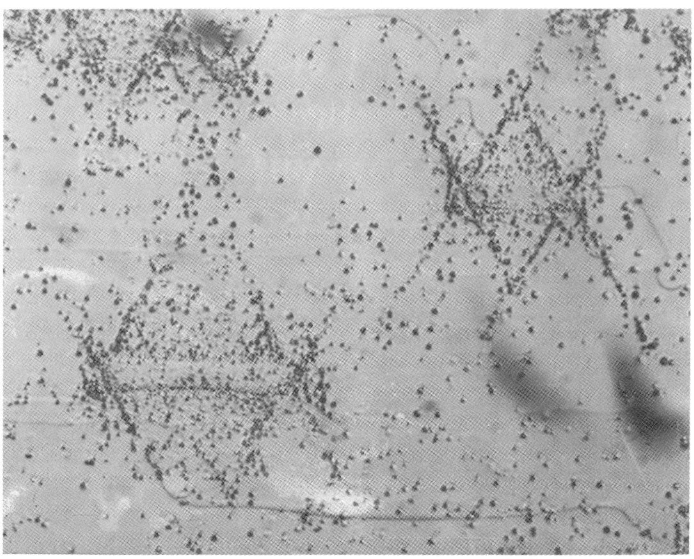

Fig. 29. Oblique $\{111\}$ section of a $\langle 110 \rangle$ oriented crystal containing solute trails. Etch pits show the dislocation loops punched out at the highly stressed ends of the solute trails. Mag. × 70.

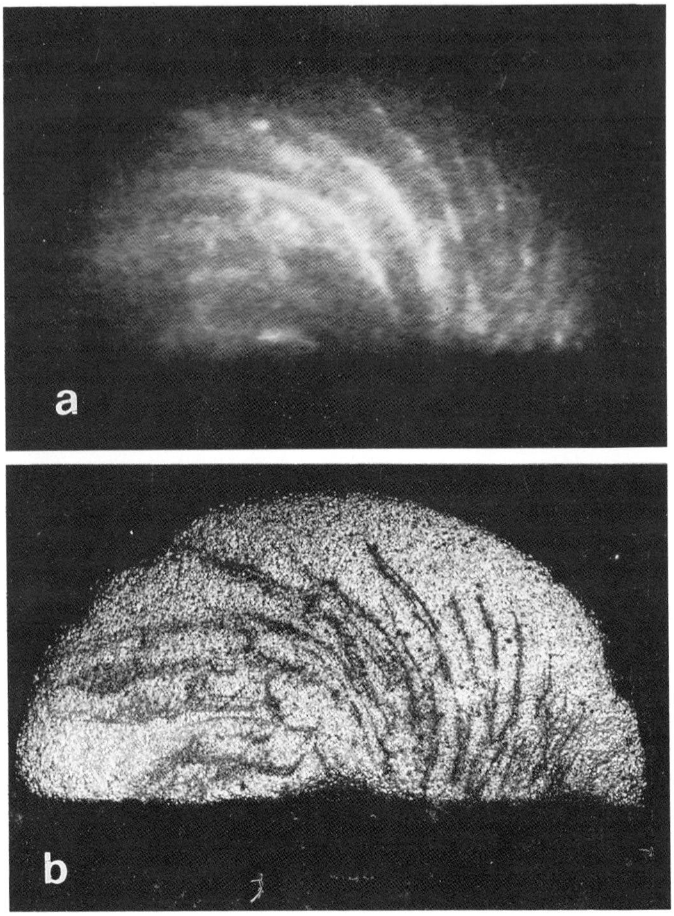

Fig. . Half-wafer from a radio-Ga-doped ⟨111⟩ oriented Ge single crystal growing under conditions of consi tutional supercooling with a concave-to-the-melt freezing isotherm producing a multi-start spiral pattern of {111} macrosteps. (a) Auto-radiograph showing the segregation of radio-Ga at the macrosteps and (b) an etched section with dislocation etch pits revealing the spiral macro-steps. Mag. × 3.8.

but with the single variant of placing within the graphite crucible, used for germanium, an inner silica crucible.

Early crystals were heavily dislocated but, in 1958, Dash [72] showed how dislocation-free crystals could be obtained by growing a long thin neck on the end of the seed crystal and growing-out the crystal at a shallow angle. Subsequently, Ziegler [73] demonstrated that by initially pulling at a high rate a long thin neck could be produced which became dislocation-free. The rapid cooling which this produced was believed to trap into the crystal a high supersaturation of vacancies which facilitated the climb out of the crystal of any dislocations propagated from the seed. Once dislocation-free, the crystal was quite resistant to the formation of new dislocations unless its surface became damaged and rapid grow-out of the shoulder region without introducing dislocations became possible.

From these early beginnings, the use of the technique has developed to dominate the production of the world's most important single crystal.

Worldwide, more than 8000 tons of silicon single crystal is produced annually for device fabrication, mostly for integrated circuits. In fact around 98% of all commercial semiconductor devices are made from silicon. About 85% of that silicon single crystal is made by crystal pulling, essentially all of which is free from extended dislocations.

Currently, melts of up to 150 kg (corresponding to a volume of 59 ℓ) are used from which single crystals of up to 20 cm in diameter are grown. Crystals of up to 30 cm diameter are in the development stage.

The charge material, known as polysilicon, is prepared by the Siemens process [74] which is the hydrogen reduction of trichlorosilane:

$$4SiHCl_3 + 2H_2 = 3Si + 8HCl + SiCl_4.$$

The $SiHCl_3$ is prepared by reaction of metallurgical grade silicon with hot HCl and subsequently purified by distillation. Deposition of the polysilicon is caused to take place on a thin, electrically heated, silicon rod. The rod is subsequently broken up into lumps of the the order of a few cm^3 volume which, after etching, are loaded into the crucible.

Grown crystals (Cz silicon, as such material is called), contain oxygen as a result of corrosion of the silica crucible by the molten silicon. With proper control of its concentration and uniformity, this oxygen confers two benefits to the integrated circuit manufacturer: (a) it hardens the crystal lattice, reducing the incidence of plastic slip during device processing and (b) by annealing the wafers in a particular way, the oxygen can be aggregated in the form of micro-precipitates which getter deleterious heavy metals which can get into the crystals at low, but significant, levels during processing. These contaminants, if not gettered, degrade the performance and the lifetime of the devices.

However, the presence of oxygen is deleterious for high power device applications and it is this segment of the market which is serviced by the floating-zone technique which can yield very pure, high resistivity material which can be subsequently doped very uniformly by nuclear transmutation (see chapter 4 by Bohm, Ludge and Schroder).

The steady increase in diameter of commercial crystals is shown in fig. 31. Each increase in crucible size increases the strength of the natural convective flow in the melt which, because the viscosity of molten silicon is low, is turbulent. The incorporation of oxygen on both macro- and micro-scales, and indeed other dopants, is very dependent on the convective behaviour of the melt. To minimise the increase in turbulence, crucibles of low height to radius aspect ratio are used [75].

Whilst there is no known fundamental limitation to the scale-up of the process, increased melt turbulence is likely to set a limit to the dopant micro-uniformity obtainable. Current interest is therefore directed to changing the process from a batch one where a crucible is charged, the charge melted and the crystal pulled from it with

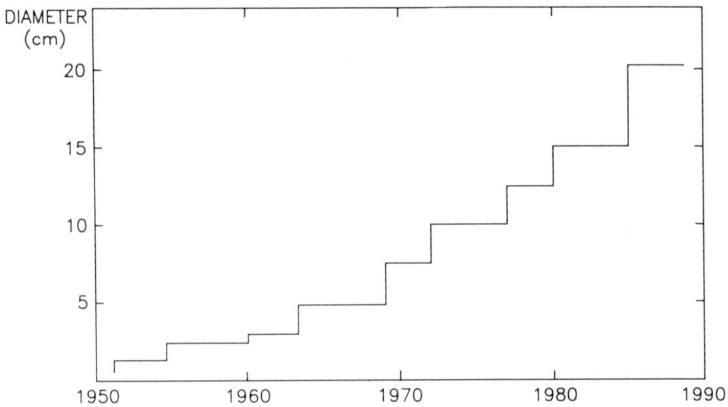

Fig. 31. Production silicon wafer diameter evolution (Siltec data).

repetition of the process, to a semi-continuous process where new charge material is continuously fed into the melt.

4.3.2. Puller design

Furnace chambers are commonly made of stainless steel welded to vacuum standards since pullers are operated under a reduced pressure of argon (see section 4.3.5 below). The chambers can be double-walled to facilitate water cooling. Gas flow is controlled by mass flow controllers. The crystals are pulled into an upper chamber separated from the main chamber by a gate valve to permit removal of the crystal and replacement of the seed without cooling down the furnace. Servo-controlled drive motors for lift and rotation of both crystal and crucible are coupled using magnetic clutches which are used to obtain high speed positioning of seed and crucible.

Both ball chain lifting mechanisms and pulling shafts are available. Some manufacturers use poly-strand stainless steel or tungsten cables instead of a ball chain. Picket fence heaters appear to be universal. Hydraulic lifting devices are used to open and close the chamber and even to remove the grown crystal. A mock-up of a crucible and heater assembly is shown in fig. 32.

Microprocessor control of all major functions is standard. These include, pump-down, leak testing, melt temperature stabilisation, seed dipping, growth of neck, shoulder, body and end-cone sections of the crystal (with automatic diameter control using CCD imaging cameras controlling the "bright ring" position), average pulling speed, heater power, gas pressure and flow rate, crucible position and crystal and crucible rotation rates and the rates of ramping these during the growth run. Fibre optics are used to achieve optimal imaging of the "bright-ring". A digital control system has been described in detail by Kim et al. [76].

The stored energy in 150 kg of molten silicon at its melting point (1420°C) is approximately 4×10^8 J. Crucible failure dumps the charge at the base of the puller where special cooled spill pans are deployed to contain the spillage.

Fig. 32. Mock-up of components for silicon crystal pulling (Photograph by courtesy of Ringsdorff).

Typical heater power consumption of a puller using a 100 kg charge is around 200 kV A. Cooling water is used at a rate of around 250 ℓ min^{-1} and argon ambient gas at 80 ℓ min^{-1}. The crucible for such a charge is approximately 60 cm in diameter and 40 cm in height. A commercial puller is shown in fig. 33 and a modest sized ⟨100⟩ oriented single crystal in fig. 34.

Several proposals for modification of the process to a continuous one have been made. Lorenzini et al. [77] described a two-crucible system, the second crucible providing a source of liquid silicon obtained by melting-in new charge material. Fiegl [78] proposed the use of a siphon to transfer the molten silicon from one crucible to the other. Such approaches appear likely to be difficult to operate reliably and, by using two crucibles, destroy the radial symmetry of the system.

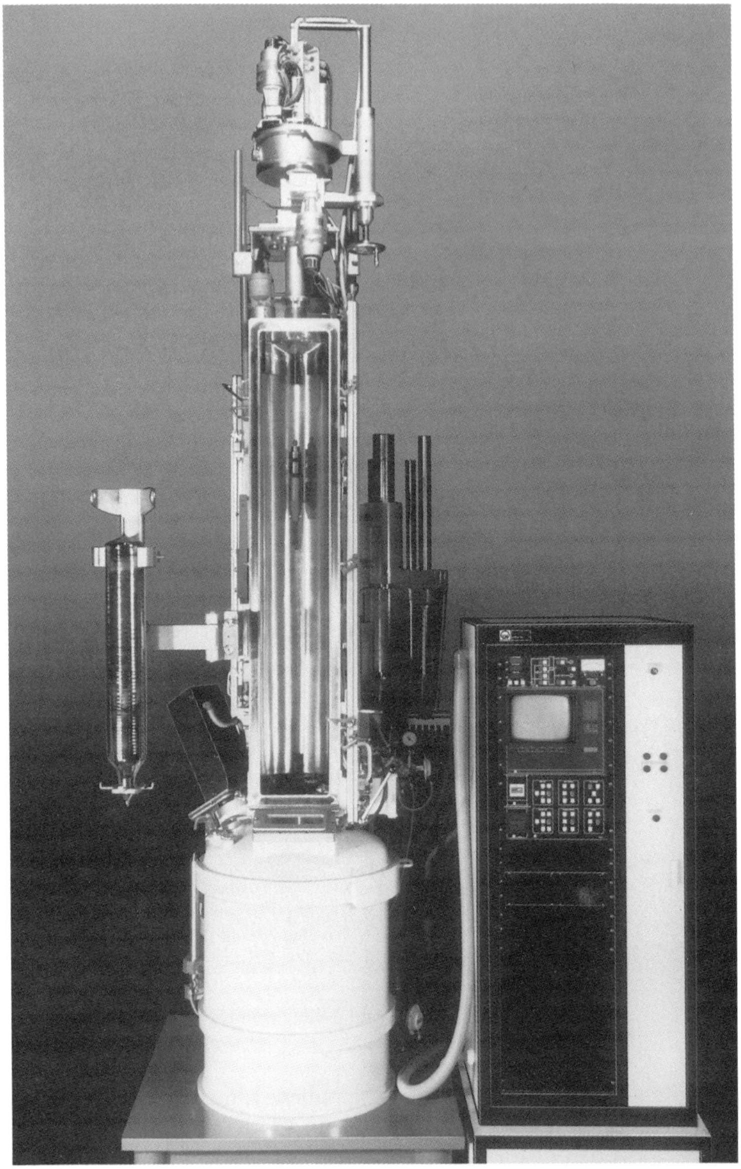

Fig. 33. Commercial silicon crystal puller with cable pulling mechanism. The puller can grow crystals of up to 205 mm diameter from a 45 kg charge. (Courtesy of Leybold AG, Alzenau, Germany.)

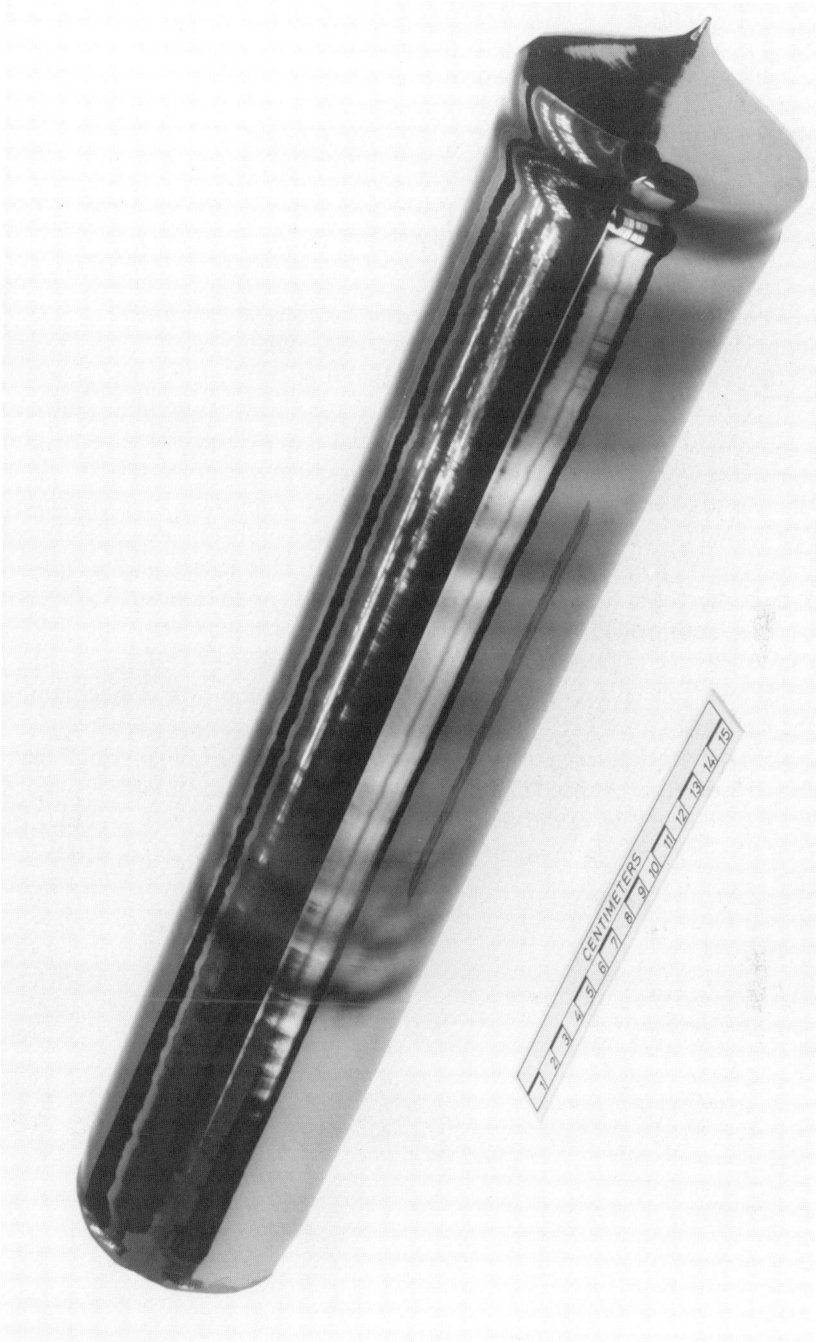

Fig. 34. 100 mm diameter ⟨100⟩ oriented single crystal of silicon pulled from the melt. (Courtesy of K.G. Barraclough.)

An alternate approach is to immerse an annular quartz baffle into the melt, which prevents polysilicon, fed into the outer annular region, from contacting the crystal and disrupting its growth. The presence of additional quartz in the system has implications for the oxygen content of the crystals of course. The recent availability of high purity polysilicon pellets makes this approach more practicable.

However, one major problem remains. It is that the operational life of a quartz crucible appears to be limited to a few tens of hours after which an instability in the erosion introduces particulate matter into the melt.

4.3.3. Thermal distribution, interface shape and dislocation generation

The design features of commercial silicon crystal pullers result from the required specification put on the product; principally that it be dislocation-free, micro-uniform to some specified degree and radially uniform across a wafer to within a per cent or so. Axial uniformity has also to be maximised in order that as large a portion of the crystal as possible falls within specification for electrical resistivity etc.

To these ends control of interface shape is very important and therefore much attention is paid to thermal design using sophisticated global modelling of the process (see the chapter 15 by Dupret and Van den Bogaert). This aspect, and the questions of dopant uniformity and oxygen uptake, depend heavily on the convective behaviour of the melt. This is controlled by the temperature gradients in the system, the melt size and aspect ratio and the rates of rotation of crystal and crucible. Additionally the incorporated oxygen concentration depends on the evaporation rate of silicon monoxide from the melt surface which is influenced by the ambient gas pressure.

To obtain dislocation-free growth throughout the length of silicon single crystals of commercial size requires special attention to the control of the interface shape as well as to the initial extension of the seed as first demonstrated by Dash and by Ziegler as described above. The requirements differ for ⟨111⟩ oriented and ⟨100⟩ oriented crystals.

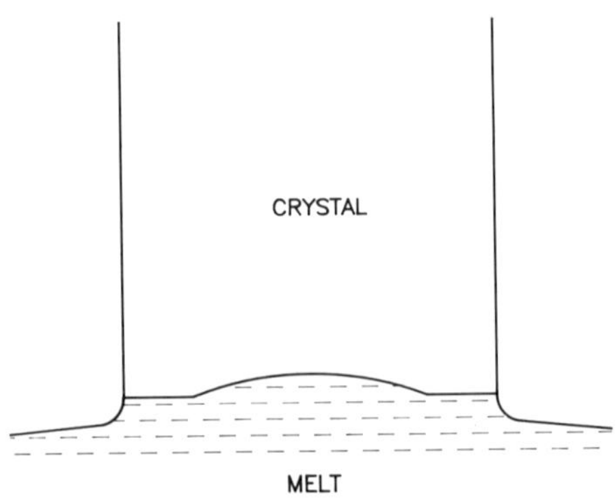

Fig. 35. Schematic profile of the required shape of the crystal–melt interface to achieve dislocation-free growth of silicon on a ⟨111⟩ axis. Note the peripheral ring facet and the concave central region.

For ⟨111⟩ crystals, it is necessary to have a crystal-melt interface shape which is concave toward the melt in its central region and which is surrounded by a annular (111) facet as shown in fig. 35. This is achieved by arranging the thermal geometry of the puller so that, without pulling, the crystal–melt interface is slightly convex to the melt. The interface will then display a central (111) facet. Increasing pull rate increases the amount of latent heat generated at the interface and this produces the desired concavity of the central region of the interface with the facet becoming annular. This requires a pulling velocity in excess of some minimum value and, provided that this is achieved, the growth of a very long thin neck on the seed is not necessary.

In contrast, growth on a ⟨100⟩ axis does require a long thin neck and a somewhat lower pulling velocity and, again, a slightly concave interface is necessary.

The onset of dislocation-free growth is evident to the operator on the free surface of the growing crystal and is known in some circles as "lock-on". "Lock-on" is characterised by the appearance of small axial ridges at the azimuthal positions corresponding to the {111} edge facets (disposed at 90° on ⟨100⟩ crystals and 120° on ⟨111⟩ ones). "Lock-on" is accompanied by a sudden small increase in crystal diameter to a new stable value which occurs without any change having been made to the heater power. This subject has been recently documented by Wilkes [75].

If "lock-on" is lost due to a reduction in growth rate resulting in the formation of a convex-to-the-melt interface shape, then small {111} edge facets form (inclined at about 71° and 44° to the (111) or (100) macroscopic plane of the interface respectively). As growth proceeds, dislocations nucleate at these facets and are propagated into the crystal. These dislocations interact and multiply giving a dislocation pattern having, in the case of ⟨111⟩ growth, a three-fold symmetry in a transverse slice (i.e., a wafer) rather than the six-fold symmetry ("star of David") which occurs when the dislocation pattern is the result of thermal stress during cooling (see chapter by Volkl).

Whilst, as has just been indicated, a slightly concave interface is required to obtain extended dislocation-free growth, a highly concave interface can introduce dislocations by plastic deformation during cooling because of the large hoop stresses which result from the high radial temperature gradient in the hotter portions of the crystal associated with a highly concave interface. To maintain the crystal entirely dislocation-free, its diameter must be carefully tapered down at the end of the growth before it is withdrawn from the residual melt.

4.3.4. Convection and conservative dopant incorporation

The natural, buoyancy-driven, convective flow in the crucible is turbulent and produces a spectrum of temperature fluctuations which are propagated through the crystal–melt interface, temporally modifying the heat fluxes and hence the instantaneous growth rate. This fluctuating growth rate produces fluctuations in the magnitude of the boundary layer of rejected solute (see chapter 12 by Garandet et al.) and hence gives rise to solute striations in the crystal. Additionally, fluctuations in the rate of melt flow past the crystallising interface partially wash away the solute boundary layer to a degree that fluctuates with the fluctuating flow. This too contributes to the fluctuations in dopant concentration in the crystal. However, if the crystal rotation rate is sufficiently high, the flow adjacent to the crystallising interface will be dominated by that rotation so that the solute boundary layer will barely "see" the turbulent convection in the bulk of the melt.

Rotation of the crucible, counter to that of the crystal, is deployed to modify the melt flow and, concomitantly, the temperature distribution in the melt. The stability of the growing crystal against heater power and other fluctuations is intimately related to the melt flow pattern. This may be connected with the occurrence of a baroclinic wave instability which has been recently observed using X-ray tomographic imaging techniques [79]. There has been extensive modelling of melt flows and this is reviewed in chapter 13 by Müller and Ostrogorsky.

Fortunately, the commonly deployed "n" and "p" dopants in silicon, namely phosphorus and boron, both have distribution coefficients near to unity (the values are 0.35 and 0.8 respectively) so that solute striations of these dopants are not very pronounced. (Incidently, it is this fact which allows the use of growth rate modulation as a means for achieving servo-control of crystal diameter in silicon.)

However, the specification on radial uniformity of dopant is demanding and, if the natural convective flow in the bulk of the melt is sufficiently strong to erode the solute boundary layer due to crystal rotation in the peripheral region of the interface, this will result in a reduction in solute concentation along a radial direction in the wafer (for a solute with $k < 1$) which is unacceptably large (more than a few per cent from centre to edge of the wafer).

There are, of course, other useful dopants of silicon which have distribution coefficients which differ markedly from unity, such as antimony ($k = 0.023$) and, with such dopants, striations and radial non-uniformity can be major problems.

Because dopants with $k \approx 1$ are available, axial dopant segregation is not such a serious problem as it is with germanium and, in consequence, there has been little effort devoted to developing double crucible techniques such as were described in section 4.2.2.

4.3.5. *Non-conservative dopant incorporation*

We have indicated that control of the deliberately added, non-volatile electrically active, dopants boron and phosphorus present no serious problems. The control of the oxygen concentration in Czochralski silicon on the other hand is a problem on which a great amount of research has been, and still is, being carried out in an attempt to improve control of its incorporation and distribution.

The benefits arising from the controlled incorporation of oxygen have already been mentioned. The oxygen originates from reaction of the molten silicon with the silica crucible. The reaction is temperature dependent [80, 81] but it is probable that the rate of erosion of the silica is controlled by the rate of transport of oxygen by diffusion through a boundary layer at the melt–crucible interface and hence is dependent on the flow conditions in the melt.

Oxygen is lost from the melt by two mechanisms: incorporation into the growing crystal at an interstitial site and evaporation from the melt surface as silicon monoxide. The latter loss mechanism accounts for about 98% of the eroded silica. The magnitude of the flux dictates that large concentration gradients must exist in the melt. It therefore follows that, unlike the conservative dopants considered above, the oxygen content in the crystal is highly influenced by the convective flows in the bulk of the melt and by boundary layers at the melt–crucible and melt–ambient gas interfaces.

Flows at the free surface are largely determined by the Marangoni effect which, because of the temperature dependence [82] (and perhaps also the oxygen concentration dependence [83]) of the surface tension of molten silicon, produces a surface traction which drives the flow.

Transport in the ambient gas phase depends on the strength of the forced and natural convective flows in the gas. This is gas pressure dependent. To achieve the required levels of oxygen in the crystal, and to purge the chamber to prevent the formation of deposits of silicon monoxide on its colder parts (which might subsequently fall into the melt and disrupt the single crystal growth by inducing twinning), argon gas is flowed through the chamber. The argon is introduced into the top of the chamber and extracted through the base and maintained at an ambient pressure of around 20 Torr. (Vacuum pumping during growth is not possible as it produces strong bubbling of the melt due to enhanced evolution of the silicon monoxide.)

Uncertainty as to the value of the distribution coefficient of oxygen in silicon exists but it probably lies within a factor of two of unity with the result that fluctuations in interstitial oxygen concentration in the crystal are due principally to fluctuations in transport rate in the bulk melt. Further, since almost all of the dissolved oxygen evaporates, small changes in that loss rate produce marked changes in the concentration in the crystal. Hence, to control the mean level of oxygen in the crystal at the required level, requires understanding and control of both the kinetics of erosion and of the transport process.

Here the batch nature of the process produces added complications because, whilst the surface area of the melt from which the SiO evaporates changes little as growth proceeds, the area of melt in contact with crucible wall, and hence the area available for erosion, steadily diminishes as the melt level in the crucible falls tending to produce a reduction in oxygen concentration in the tail of the crystal. Things are not quite that simple though, since with the falling melt level comes a reduction in the Grashof number and hence of the strength of the natural convective flow and with it the rate of transport in the melt. The complex interplay between these several factors can only be elucidated by careful experimentation coupled to realistic modelling of the flow processes.

The required oxygen levels depend on the particular device being made. Where gettering by oxide precipitates is vital to device performance, a high concentration (around $1 \times 10^{18}\,\mathrm{cm}^{-3}$) is required. Where device performance is sensitive to lattice defects, oxygen-related defect formation must be avoided and this necessitates lower oxygen levels. Thus the required oxygen concentration for dynamic RAM production decreases with increasing packing density (i.e. capacity) of the chip.

These extremely demanding requirements on oxygen level and distribution have led to experimentation with the use of a magnetic field as an additional control parameter and the use of a cusped magnetic field to achieve uniform, low oxygen level material looks promising [84, 85]. (This topic is reviewed by Hurle and Series in chapter 5.)

Whilst pulled silicon single crystals are free from extended dislocations, they nonetheless contain microdefects which can include submicron interstitial dislocation loops and microscopic chemical inhomogeneities not imaged by electron microscopy. Such defects can be revealed by high temperature annealing of a polished sample followed by chemical etching (fig. 36.) High densities of etch pits which give a cloudy appearance to the wafer are commonly called "swirl defects" because of their macroscopic configuration which is clearly related to crystal and/or crucible rotation. These micro-defects constitute the nucleation sites at which oxygen precipitates and oxygen-induced stacking faults (OSFs) form during subsequent heat treatment of the wafer in processing. Extensive studies of swirl defects have been made by Föll et al. [156].

Fig. 36. Micrograph of a longitudinal section of a pulled silicon crystal showing layers of oxygen precipitates and their secondary defects. Between these layers phosphorus doping striations are evident. (Courtesy of W. Zulehner.) Mag. × 88.

The oxygen precipitation is strongly correlated with local remelting at the crystal–melt interface resulting from turbulent temperature fluctuations in the melt. Figure 36 shows both phosphorus striations (fine dark lines) and broader bands of oxygen micro-precipitates in a longitudinal section of a pulled crystal. The broad nature of these bands is thought to be due to the nature of the fluctuation in the rate of transport of dissolved oxygen in the melt to the interface as a result of the fluctuations in the bulk flow in the melt. The phosphorus striations on the other hand, are due to boundary layer fluctuations due to the fluctuating growth rate.

The presence of carbon is believed to be implicated in the nucleation of oxygen micro-precipitates. Certainly, at relatively high concentrations, it is known to enhance micro-precipitation. Moreover, its distribution in the crystal exhibits a swirl-like pattern as shown by X-ray topography (fig. 37). Its distribution coefficient is 0.073 [86].

Fig. 37. X-ray diffraction topograph of a Czochralski silicon wafer showing the strain distribution due to micro-segregation of carbon at a mean concentration of $8 \times 10^{16}\ cm^{-3}$. (Courtesy M.J. Hill.) Mag. × 2.5.

The source of the carbon has been the subject of some study. One possible mechanism is reaction of the evaporating silicon monoxide with the graphite heater generating CO and CO_2 which is subsequently absorbed into the melt. However, Series and Barraclough [87] have shown that, at least at the ppm level, most of the contamination of the melt occurs prior to growth.

4.3.6. Pulled silicon ribbons

The quest for cheap single crystal sheets for the production of solar cell arrays has motivated numerous programmes to pull ribbons of silicon from a melt using a shaped die. Progress can be assessed from study of the Proceedings of international conferences held on shaped crystal growth [89, 90]. Pioneering work in the field was carried out by Labelle and Mlavsky with their edge-defined, film-fed growth (EFG) technique [43].

All such techniques are beset with several intrinsic problems. The first of these is the lack of inertness of any available die material in the presence of molten silicon. The use of graphite, for example, results in the formation of particles of silicon carbide which cause twinning [162]. Non-uniformity of doping is another problem. Most serious of all is the difficulty of scaling up the process whilst maintaining the flatness of the ribbon. As the ribbon width is increased, locked-in thermal stress causes the ribbons to buckle slightly which has serious consequences for subsequent processing technology which requires a flat surface against which a mask can be placed.

A novel approach which helps to minimise this problem and which increases the output per machine is to grow closed tubular or "nonagonal" shapes [163]. Yet another way of addressing the problem of flatness, but one which yields polycrystalline

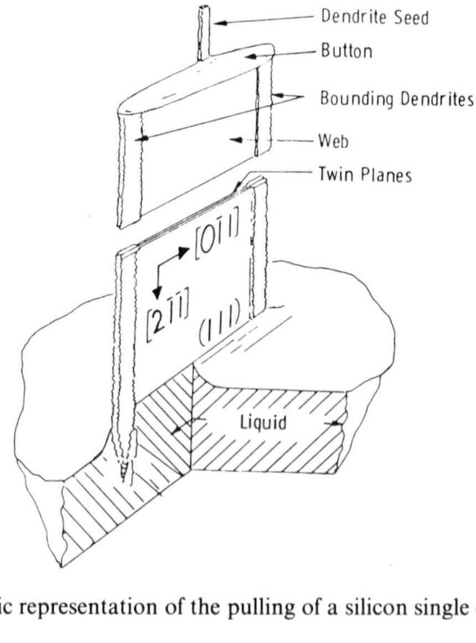

Fig. 38. Schematic representation of the pulling of a silicon single crystal web [165].

material, is to coat a soft graphite ribbon with silicon by pulling it up through a slot in the base of the crucible [164].

Of much older vintage is dendritic web growth which utilises the curious morphology of silicon dendrites [165] (see fig. 38) and which has the advantage that it does not require a die. Silicon grows pseudo-dendritically in directions in the $\{111\}$ plane if it contains, parallel to that plane, at least two closely-spaced twin planes. The presence of this pair of planes produces a permanent re-entrancy at the crystal–melt interface which facilitates rapid growth at low supercooling without the need for repeated two-dimensional nucleation [166].

If a "seed" contains two such dendrites laterally separated and with their twin planes co-planar (fig. 38) then, when these are dipped into a suitably supercooled melt and withdrawn, they pull up a film of single crystal silicon between them (rather as a soap film can be generated by withdrawing a wire frame from a dish of soapy water).

The dynamics and stability of all of these shaped crystal growth processes have been analysed by Tatarchenko and are reviewed in his chapter in this Volume.

5. *Growth of compound semiconductors*

The expanding commercial importance of a number of compound semiconductors, serving device applications which cannot be met by silicon, has led to the development of pulling techniques for the growth of these compounds. The dissociability of a number of these materials adds a significant complication and difficulty but one which has been overcome, to a major degree, by the development of the liquid encapsulated Czochralski (LEC) technique.

Provided that only crystals of near-congruent composition are required, these materials can be pulled at reasonably fast rates, though generally not as fast as germanium or silicon. Where it is necessary to grow from a melt composition significantly different from the congruent composition, then the problem of constitutional supercooling and cellular structure formation becomes serious.

In this section we describe pulling of the Group IIIB antimonides, arsenides and phosphides, and representative examples of the II–VI and IV–VI semiconductors. The Group III nitrides cannot be grown from the melt. First though, we consider the problems posed by the dissociability of these materials and describe modifications to the pulling technique that have been developed to cope with the problem. The melting points and dissociation pressures at the melting point for the more important III–V compounds are shown in table 2.

5.1. *Pulling of materials with a volatile component*

Two basic approaches are possible to the problem of the loss of one or more components of the melt by evaporation. They are either (a) to provide a gaseous ambient which is in thermodynamic equilibrium with the melt or (b) to prevent the formation of the vapour phase by sealing the melt in a medium which exerts a hydrostatic pressure

Table 2
Group V dissociation pressure of several III–V compounds at their melting points.

Compound	Melting point (°C)	Vapour pressure (atm)	Reference
InSb	525	4×10^{-8}	[172]
InAs	939	0.244	[173]
InP	1062	27.5	[174]
GaSb	712	1×10^{-6}	[172]
GaAs	1238	0.98	[175]
		0.705	[173]
		≈ 2	[176]
GaP	1465	32	[177]

on it greater than the vapour or dissociation pressure so that vapour bubbles cannot form.

Both approaches have been successfully engineered but the second approach, known as the liquid encapsulated Czochralski (LEC) technique has generally proved superior and has revolutionised the production of indium and gallium phosphides and of gallium arsenide.

5.1.1. *Hot wall technology*

Approach (a) above requires that the whole growth chamber containing the equilibrated vapour must be at a sufficiently high temperature to prevent the condensation of that vapour. Thus, in the case of gallium arsenide for example, a partial pressure of around two atmospheres is required to maintain equilibrium with a stoichiometric melt at the melting point of GaAs (1511 K) [176]. To prevent arsenic vapour at two atm pressure from condensing to form elemental arsenic requires a minimum temperature of around 630°C. Therefore the walls of the puller must be maintained at least at this temperature. Additionally the seal through which the pull rod is rotated and drawn must withstand both the high temperature and the corrosion of the arsenic vapour.

The problem of corrosion severely limits the choice of construction materials, quartz being the favoured one. The design problems have been tackled in several ways. Firstly syringe pullers have been developed in which the growth chamber is made from quartz with the pull rod seal formed by a precision turned quartz cylinder which is drawn up a precision-bored quartz tube forming part of the growth chamber. Inevitably there is some vapour loss past such a seal but this is contained in an outer chamber and the method has been successfully used for the growth of gallium arsenide.

A sophisticated refinement of this is to enclose the entire pull rod within the chamber and to lift and rotate it remotely by magnetic coupling to a high Curie temperature magnetic alloy slug sealed within the quartz pull rod. The mechanical and electromagnetic design of such a system is tricky if instabilities are to be avoided. Additionally of course, charging the crucible and removing the crystal are major exercises in silica-blowing not suited to a production environment. The techniques have been reviewed by Gremmelmaier [94] and by Fischer [95].

A somewhat simpler approach has been proposed by Mullin et al. [96] which has been called pressure balancing. This addresses the problem of the seal by having a long pyrolytic boron nitride bearing in which a screwed channel is milled forming a nut of extended thickness. A reservoir at the top of the bearing contains boric oxide as a liquid seal between the pull rod and the bearing. Rotation of the pull rod provides an Archimedes screw action which prevents the molten boric oxide from draining out of the bottom of the bearing.

For this to work, the pressures on the two sides of the bearing have to be equalised. This can be achieved with a two-way bubbler which allows any excess vapour to partition itself between inner and outer chambers. Provided that the inert gas pressure in the outside chamber is approximately correct, the loss of vapour from the growth chamber will be negligible.

This approach looks a promising one for making hot wall technology practicable, but the ease of and success achieved with the LEC technique has meant that very little work has gone into its development.

5.1.2. The liquid encapsulation Czochralski technique

A striking feature of LEC growth is its elegant simplicity. It requires no mechanical modifications to a standard crystal puller provided that elevated pressures are not required. In the case of gallium arsenide for example, it simply involves charging the crucible with a quantity of boric oxide, a low melting point glass, in addition to the polycrystalline gallium arsenide pieces.

On heating the crucible, the boric oxide softens and flows over the charge. Since it wets the silica crucible, it also totally encapsulates the charge. Now, providing that an inert gas pressure in excess of the dissociation pressure of arsenic over molten gallium arsenide is maintained in the growth chamber, no loss of arsenic will occur since arsenic bubble formation will be suppressed by the excess pressure. Being essentially insoluble in the boric oxide, arsenic cannot diffuse through it.

Crystal growth is performed by lowering the seed crystal through the boric oxide until it contacts the melt at the boric oxide/melt interface. Pulling is then initiated in the usual way. The growing crystal, as it breaks the boric oxide surface, retains a thin skin of boric oxide on it which serves to minimise dissociation of the hot crystal as it cools to room temperature. The principle is shown schematically in fig. 39.

This method was first reported by Metz et al. [7] who grew PbTe crystals of up to 1 inch diameter and 3 inch in length but this attracted little interest until Mullin et al. [8] applied the technique to the growth of InAs and GaAs and showed how the boric oxide encapsulated the melt coining the term liquid encapsulation. They followed this by developing a pressure pulling technology, which enabled them to apply the LEC principle to the growth of high quality single crystals of gallium and indium phosphides [9].

The basic requirements for an encapsulant are the following:

(1) immiscible with the melt,
(2) chemically inert to both melt and crucible,
(3) less dense than the melt so that it floats on it,
(4) wets the crystal and the crucible,

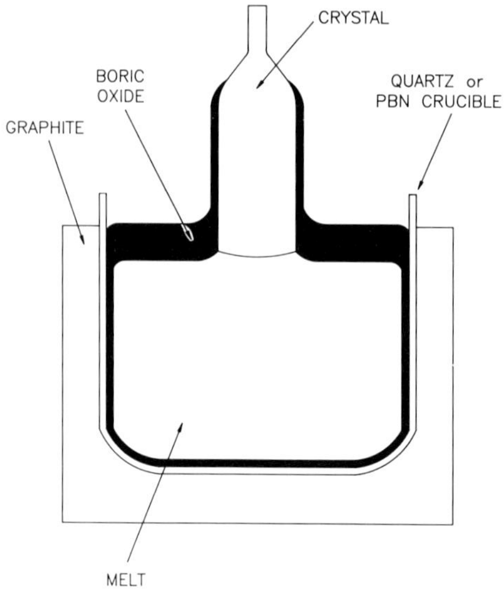

Fig. 39. Schematic representation of the liquid encapsulation principle. Note that both the melt and the crystal are encapsulated in the molten boric oxide encapsulant.

(5) has a suitable viscosity and softening temperature so that it encapsulates the charge before any significant dissociation has taken place during heating yet is viscous enough not to drain too rapidly from the emerging crystal.

The experience of many laboratories has shown that boric oxide meets these requirements better than any other material. Relevant properties of boric oxide are given in table 3. It has the additional property that, depending on its residual water content, it is a "getter" for a number of metal contaminants and thus helps to further purify the charge. Control of the water vapour content (typically in the range 100–2000 ppm) is crucial and this is discussed in more detail in respect to its use with GaAs in section 5.3.1 below.

Table 3
Some physical properties of boric oxide.

Density	$1.52\ g\ cm^{-3}$
Specific heat	$1.83\ J\ (g\ K)^{-1}$
Viscosity 750°C	436 poise
900°C	115 poise
1115°C	35 poise
Refractive index	1.463
Surface tension 860°C	80 dyne cm
1300°C	100 dyne cm
Thermal expansion coefficient 900°C	$9.8 \times 10^{-5}\,°C^{-1}$
1300°C	$4.3 \times 10^{-5}\,°C^{-1}$
Vapour pressure 1300°C	3.4 mm Hg

The limitations of boric oxide are that it is too viscous for some of the lower melting point materials. This can be partially overcome by making small additions of cryolite (Na_2AlF_6) [97, 98]. Another approach has been to use eutectic mixtures of inorganic halides but these tend not to wet the crystal. For really low melting point materials fatty acids, such as adipic acid, can be used.

The development of high pressure pullers allowed the LEC technique to be applied to the growth of first, gallium phosphide to meet the LED market and later, indium phosphide now widely used as a substrate for infrared opto-electronic devices particularly in fibre-optic communications.

Pullers developed for this purpose have in fact found their biggest use in the growth of semi-insulating gallium arsenide for substrates in integrated circuits fabricated using an ion implantation technology. This has come about because, by operating at about 50 atm pressure, the gallium arsenide melt can be formed by charging the crucible with gallium and arsenic separately rather than having first to pre-synthesise the compound in separate equipment. Such direct synthesis results in purer material which more reproducibly yields semi-insulating crystals.

The current generation of pressure pullers take up to 25 kg charges of GaAs in crucibles up to 10 inch diameter from which crystals of up to 6 inch diameter can be grown. Growth chambers are designed for working pressures up to or in excess of 1000 psi. One such system is shown in fig. 40.

Puller frames are designed to accommodate solenoids for magnetic Czochralski growth (see chapter 5). The machines are fully computer-controlled with servo-control of crystal diameter using either a weight sensor or optical diameter sensing. Weighing is performed by a rugged load cell mounted within the pressure chamber. An assembly for weighing the crucible is shown in fig. 41. The diameter sensing technique requires sophisticated signal averaging to eliminate the spurious effects caused by the formation of bubbles in the boric oxide encapsulant during growth.

Provision is made for after heaters to control the rate of cooling of the crystal and thereby, to attempt to control the dislocation density in the crystal (see chapter 14).

5.2. Group III antimonides

5.2.1. Indium antimonide

Indium antimonide received early exploitation as a photo-detector in the infrared region, with single crystals of modest size being pulled from the melt. The very low dissociation pressure of this low melting point material (m.p. 525°C) makes liquid encapsulation unnecessary. The principal problems which it presents are proneness to twinning during growth and doping inhomogeneity on a macro-scale caused by a pronounced "facet effect" (see below). These problems were first studied by Hulme and Mullin [99] using the novel and elegant research puller shown in fig. 42.

The growth chamber consists of a large quartz test tube (B) sealed with an "O" ring to a flange at the top through which the pull rod (A) passed. One novel feature is that the growth chamber is also the crucible and is heated by a cup-shaped basket resistance heater (C), made of pyrophyllite and wound with Kanthal tape, which fits around the

Fig. 40. Commercial high pressure puller for the liquid encapsulated growth of III–V compounds. The puller is capable of growing GaAs single crystals of up to 150 mm diameter from charges of up to 25 kg. (Courtesy of Metals Research Semiconductors Ltd.)

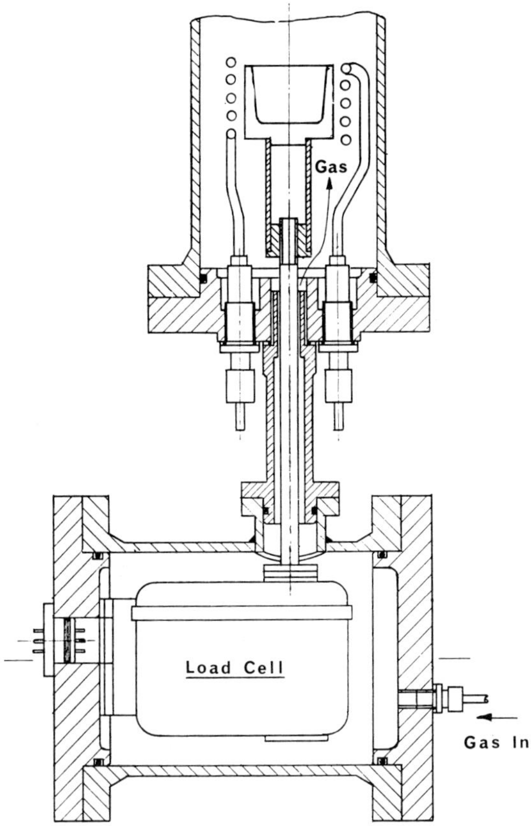

Fig. 41. Load cell attachment to a pressure puller for crucible weighing. (Courtesy J.B. Mullin.)

base of the test tube. Electro-magnetic stirring of the melt is facilitated by three symmetrically disposed water cooled copper coils (D) fed from a 0–300 A three-phase 400 Hz supply which provides a rotating magnetic field which vigorously stirs the melt.

5.2.2. The facet effect in pulled crystals

The electro-magnetic stirring device described above was first introduced [100] in an attempt to improve the radial uniformity of doping of the grown crystals but Hulme and Mullin subsequently discovered that this was caused by anomalous segregation of the dopants at $\{111\}$ facets on the crystallising interface; an effect which has become known as the "facet effect". Values of the distribution coefficient for a number of dopants in indium antimonide both on and off the facet are given in table 4. The most dramatic effect is with tellurium where, off the facet the (near-thermodynamic equilibrium) distribution coefficient (k) is 0.5 whereas, on the (111) facet, excess Te is incorporated giving a distribution coefficient on the opposite side of unity ($k = 4.0$). This eight-fold difference in doping concentration is shown by the autoradiographs taken

Fig. 42. Novel research puller for the growth of InSb developed by Hulme and Mullin [99]. The crystal (A) is grown in a test-tube crucible/growth chamber. The charge is melted by the basket resistance heater (C) and the melt is stirred by a three-phase electro-magnetic field applied through coils (D).

of longitudinal sections of a radio-Te-doped ⟨111⟩ InSb single crystal shown in fig. 43.

The proposed explanation of this facet effect supposes that a surface-excess concentration of dopant is adsorbed onto the crystal interface. In the non-faceted region which is above its roughening temperature (see chapter 6 in Volume I), atoms (or molecules)

Table 4
Facet effect ratio of solutes in InSb [104].

Solute	K_{eff} (111)	Facet ratio
Cu	6.6×10^{-4}	1.3
Zn	2.3–3.5	1.2–1.4
Cd	0.26	3.3
Si	0.1	1.7
Ge	1.2×10^{-2}	1.4–1.8
Sn	5.7×10^{-2}	3.7–4
P	0.16	4
S	0.1	3–3.3
Se	0.17–0.35 (1.9 on facet)	5.4–6.1
Te	0.47 (4.2 on facet)	5.9–8.9

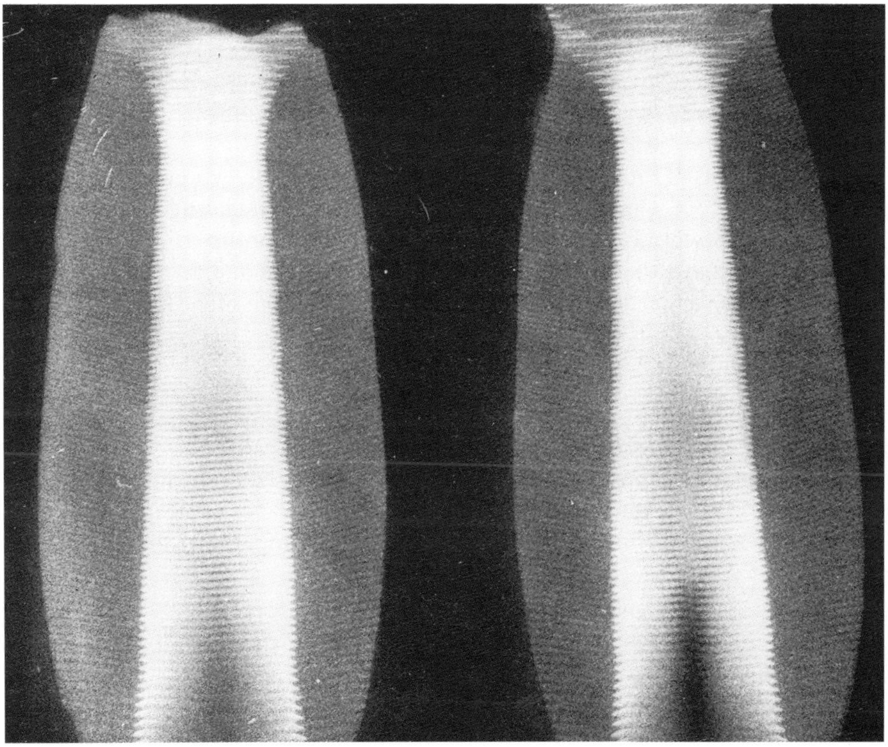

Fig. 43. Auto-radiographs of a longitudinally sectioned ⟨111⟩ oriented single crystal of InSb doped with radio-Te. The bright central column in each section is a region of enhanced Te concentration where the crystal grew with a faceted interface. The fine, near horizontal, lines are the growth striations produced by crystal rotation. (Courtesy J.B. Mullin.) Mag. × 2.1.

can be added singly and the crystal equilibrates with the bulk melt incorporating an equilibrium concentration of solute.

On the faceted region of the interface however, two-dimensional nucleation of new layers occurs followed by very rapid lateral motion of the re-entrant edges provided by the nucleus. This rapid step motion is believed to trap into the crystal some of the adsorbed excess concentration. Quantitative theories have been developed by Holmes [101] and by Trainor and Bartlett [102] but their experimental test is prevented by lack of knowledge of the basic adsorption parameters.

The equilibrium surface adsorbed concentration can of course be less than the bulk concentration when the interface is "phobic" to the dopant, in which case the distribution coefficient on the facet would be expected to be less than the equilibrium value. This is the case for Ga in germanium [67].

At high doping levels, this non-uniform distribution of dopant strains the crystal [103].

5.2.3. *Growth twinning*

The polarity of ⟨111⟩ oriented seed crystals was found early to be a critical factor in obtaining twin-free single crystals of indium antimonide [104]. Single crystal growth is relatively readily obtained when growth is propagated in the (111) indium-terminating direction but, if the seed is reversed to present the (111) antimony-terminating face then twinning during growth at the three peripheral {111} indium-terminating facets is almost inevitable.

Autoradiographs of radio-Te-doped crystals grown on the two polar directions are shown in fig. 44 from which it can be seen that the In-terminating facets tend to be larger than the Sb-terminating ones. Twinning is most prone to occur during the grow-out phase before the shoulder of the crystal is reached and, to avoid this, Hulme and Mullin employed seed crystals of a diameter equal to that of the final crystal [104]. This propensity to twin is shared with indium phosphide and a possible mechanism is discussed in section 5.4.2 below.

5.2.4. *Pulling from a non-congruent melt*

If growth is promoted from a non-congruent melt then the component in excess will be rejected by the crystallising interface forming a boundary layer in the melt ahead of the interface. This rejected component lowers the liquidus temperature of the melt just as

Fig. 44. Auto-radiographs of wafers cut from [111] oriented (left) and [100] oriented (right) single crystals of radio-Te doped InSb. The bright segments are the periphery of the wafers are the flare-out of β particles from the excess concentration of radio-Te incorporated at the {111} edge facets. A central (111) facet is also evident in the third and fourth radiographs in the left-hand sequence. There is also evidence of a small central (100) facet in the third radiograph of the right-hand sequence. Note that one of the edge facets in the fourth radiograph of the left-hand sequence has disappeared because the crystal has twinned on the facet. Note, on the {100} wafers of the right-hand sequence, that one pair of opposed facets appears more pronounced than the other pair (the latter even disappearing in the fourth radiograph). This is due to the differing polarities of the two pairs of facet (see text). (Courtesy J.B. Mullin.) Mag. × 1.4.

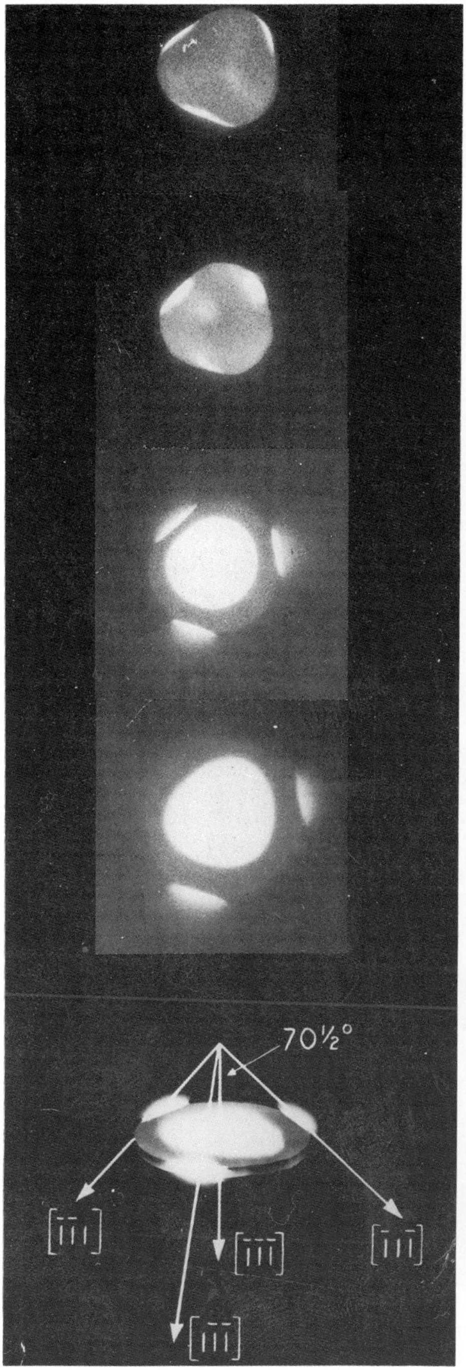
70½°
[1̄1̄1]
[1̄1̄1̄]
[1̄11̄]
[11̄1̄]

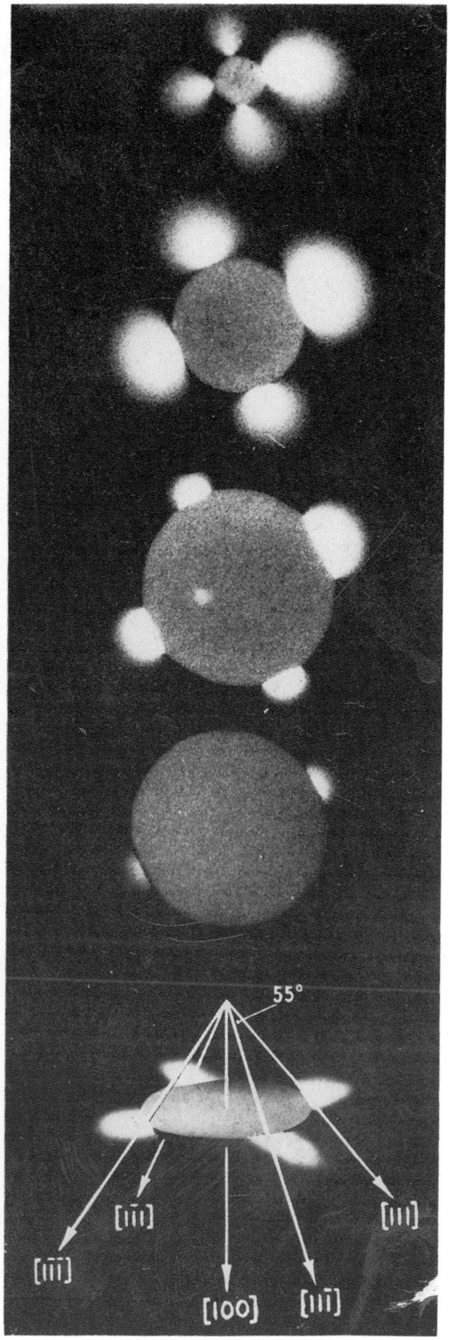
55°
[11̄1]
[111]
[11̄1̄]
[100]
[111̄]

would a dopant with distribution coefficient less than unity. In consequence, constitutional supercooling of the melt can occur leading to the formation of a cellular structure with the development of deep cell boundary grooves containing liquid which becomes progressively richer in the excess component until finally the eutectic composition is reached.

An example of such a cellular structure in a $\langle 100 \rangle$ oriented InSb crystal caused by growing it from an In-rich melt is shown in fig. 45. An elegant demonstration that the

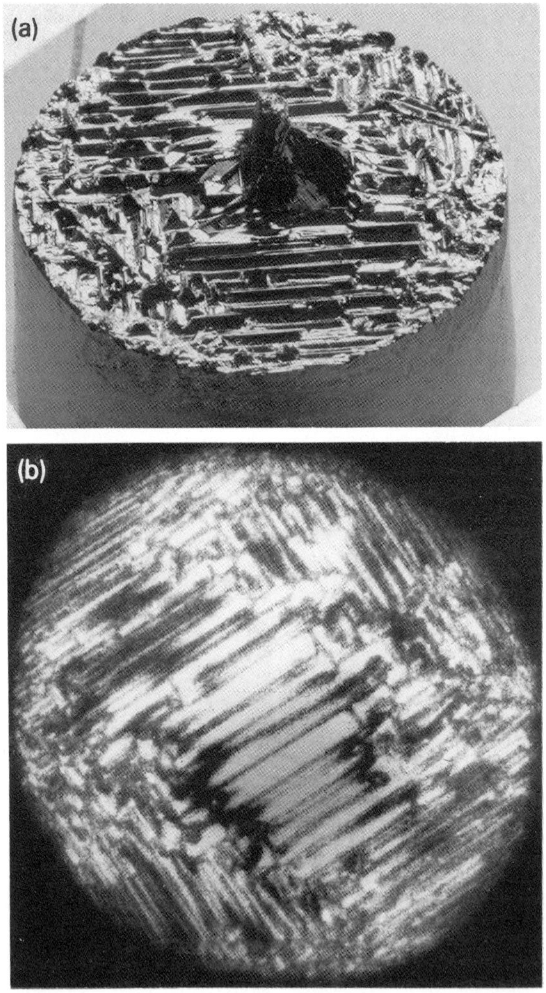

Fig. 45. Cellular structure in a $\langle 100 \rangle$ oriented single crystal of InSb produced by constitutional supercooling as a result of growing from an In-rich melt [157]. (a) Snatched crystal showing micro-faceted cellular interface. (b) Autoradiograph of a (100) section cut from the crystal showing radio-Te preferentially incorporated on the $\{111\}$ micro-facets forming the cellular structure. Mag. $\times$ 1.8.

faces of the cells are composed of microscopic facets is provided by the autoradiograph (fig. 45b). The crystal was doped with a very small quantity of radio-Te which has preferentially segregated to the faceted regions of the cellular structure produced by the excess indium.

5.2.5. Gallium and aluminium antimonides

These two materials have been much less widely researched; aluminium antimonide because of its ready oxidation in the atmosphere. However Lin et al. [105] have recently reported on the growth of AlSb single crystals of diameter up to 5 cm by pulling from a melt contained in an alumina or vitreous carbon crucible. Oxide scum on the melt surface and a tenacious oxide film on the seed presented the biggest difficulties. These were minimised by prior etching of seed and charge first in $HF:H_2O_2$ mixtures and then in HNO_3 and HCl. Finally the seed was mechanically polished until it appeared smooth and shiny. There was a small antimony loss from the melt of about $0.25\,g\,h^{-1}$ which was compensated for by addition of about 2 wt% excess Sb to the melt.

Gallium antimonide was first pulled by Allred in 1962 [106]. It has received further study in recent years on account of its possible use as a substrate for photo-detectors in fibre-optic communications at wavelengths beyond 1.5 microns. Both encapsulated and unencapsulated growth has been reported. The low melting point of the material (m.p. 712°C) means that boric oxide alone is too viscous as an encapsulant. Katsui and Uemura [98] have reported on the LEC growth under an encapsulant of boric oxide to which 3.2 mol% Na_2AlF_6 had been added. They showed that the added cryolite did not significantly contaminate the growth crystal.

Sunder et al. [107] preferred not to use any encapsulant growing instead in a flowing hydrogen ambient. They found that use of an encapsulant increased the probability of twinning. Their results suggest that the small antimony loss produced a negligible change in stoichiometry.

Their puller consisted of a cylindrical stainless steel chamber with an inserted, angled viewing port. The melt was contained in a hemispherical quartz crucible placed in a carbon susceptor which was rf heated by a 20 kW, 450 kHz generator supplied through an eight-turn water-cooled coil. Crystal diameter was controlled manually. Untwinned crystals typically 3.5 cm diameter and 6 cm long were obtained with high yield.

In accord with all previous published experience, crystals grown from a stoichiometric melt were "p"-type with a carrier concentration of around $1\text{–}2 \times 10^{17}\,cm^{-3}$ which is believed to be due to an unidentified charged native point defect. Evidence for a facet effect was obtained from studies of the macro-scale homogeneity of the crystals.

5.3. Group IIIB arsenides

5.3.1. Gallium arsenide

The commercial development of the GaAs integrated circuit became possible only after reproducible, high quality semi-insulating substrates were made available [108, 109]. This development depended crucially on the exploitation of the LEC technique without which the commercial use of GaAs today would be much more limited.

Initially, semi-insulating GaAs was made by doping the crystal with chromium which formed deep levels at about mid band gap and compensated the residual chemical donors in the crystal. The principal donor was silicon which was picked up during the synthesis of polycrystalline charge material produced by distilling arsenic into molten gallium contained in a silica boat. In the absence of chromium, this residual silicon rendered the crystal "n"-type.

Swiggard et al. [110] demonstrated that, by compounding the material in pyrolytic boron nitride, silicon contamination levels could be markedly reduced and crystals pulled from these charges were semi-insulating even without chromium doping. This was a significant advance since the chromium was unsatisfactory because, having a relatively high diffusivity in the GaAs, it spread during device fabrication. The principle of compounding III–V and II–VI materials under an encapsulant had been patented by Haggerty and Wenkus [66].

It is now known that this semi-insulating behaviour is due to the presence of a deep donor level (located at 0.75 eV below the conduction band edge) known as EL2. The precise structure of EL2 remains a matter of dispute but it is known that it contains an arsenic antisite (As_{Ga}). When the silicon concentration is sufficiently reduced, the concentration of residual chemical acceptors (notably carbon) exceeds the silicon donor concentration. If now there is a concentration of deep donors [EL2] in excess of the net (uncompensated) shallow acceptors then the Fermi level will be pinned near to the deep donor level and the material will be semi-insulating.

On thermodynamic grounds (see chapter 3 in Volume I), the concentration of EL2 will increase with increasing arsenic content in the melt. Semi-insulating material is therefore produced from slightly As-rich melts.

High pressure pullers, first developed for the growth of GaP, are used to avoid the pre-synthesis stage by loading the crucible, made of pyrolytic boron nitride, with elemental Ga and As in the correct ratio. The arsenic is placed in the bottom of the crucible, rods of gallium are laid on top and then a preformed disc of dried boric oxide is placed on top of this (fig. 46).

The crucible temperature is raised to about 450°C at which point the boric oxide softens and flows over the charge of arsenic and, now molten, gallium. The ambient

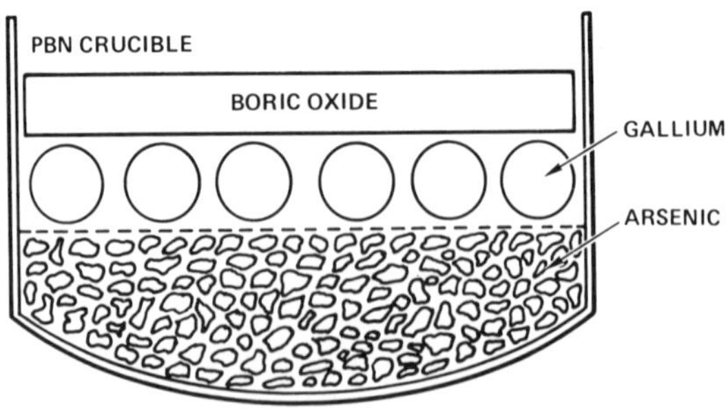

Fig. 46. Schematic representation of the loading of the crucible for direct synthesis, high pressure, LEC growth of GaAs [158].

argon pressure is then raised to around 60 atm and the crucible temperature to around 800°C. At this temperature the gallium and the arsenic react exothermically raising the temperature still further. The charge temperature is then raised progressively up to the melting point of GaAs (1238°C) until all of the arsenic has dissolved in the molten gallium. The chamber pressure can then be reduced to about 20 atm for the crystal growth phase of the process. Some small loss of arsenic inevitably occurs during synthesis. This is within acceptable limits but, more stringent specification on crystal stoichiometry in the future may require this problem to be addressed.

A commercial high pressure puller developed specifically for semi-insulating GaAs is shown in fig. 40. Control of all functions, from chamber pressurisation through "seed-on", diameter control, tail off and shut down, is fully computerised. Diameter control uses a combination of optical and weight sensing. It is designed to accommodate after-heaters and also any axially symmetric magnetic field.

The puller height is approximately 4.2 m with a crystal lift stroke of 1 m. The required heater power is about 60 kW which is servo-controlled using a W/Re thermocouple sensor. Cooling water is used at rate of 30 $\ell\,\text{min}^{-1}$.

A $\langle 100 \rangle$ oriented seed crystal, attached to the pull rod, is lowered through the boric oxide (the molten oxide layer thickness is of the order of 1 to 2 cm) to the boric oxide/melt interface and the melt temperature carefully adjusted until a meniscus is supported. A simple electrical circuit can be used to indicate when contact between the seed and the melt has been established [111]. Seed crystal and crucible are both rotated at rates in the range 0 to $\pm$ 20 rpm determined empirically by each manufacturer.

The allowable pull rate decreases with increasing crystal diameter and is typically less than 1 cm h^{-1} for 3 inch diameter crystals. Crucible lift is employed in commercial pullers to maintain the melt surface level in a fixed position with respect to the crucible heater. After-heaters can be used to minimise the thermal stress in the cooling crystal but this enhances the rate of drainage of boric oxide from the crystal and can result in surface degradation due to dissociation with the formation of gallium droplets which can migrate into the crystal in the shoulder region by the process known as temperature gradient zone melting [88].

Current production is mainly of 3 or 4 inch diameter crystals. A typical crystal and some wafers are shown in figs. 47 and 48 respectively.

Attempts have been made to avoid the need for expensive pressure pullers by compounding the charge using a reservoir of arsenic which is heated to generate arsenic vapour before being introduced into molten gallium, contained in the crucible under a layer of molten boric oxide, through a bubbler tube dipped into the melt (fig. 49). This tricky operation has been perfected but, notwithstanding, the technique has found only limited commercial use because it appears that the thermal conditions in the puller at low pressure are prone to give rise to unfavourable interface shapes, which lead to progressive degradation of crystal quality so that the maximum usable length of crystal which can be grown is much less than with the high pressure technique. An additional factor is that use of a low pressure chamber limits the extent to which the melt can be made As-rich without excessive As loss occurring. The approach is also not practicable in very large pullers since the required size of the As-reservoir becomes too large. A

Fig. 47. A three inch diameter single crystal of semi-insulating GaAs grown by the high pressure LEC technique. (Courtesy of MCP Wafer Technology.)

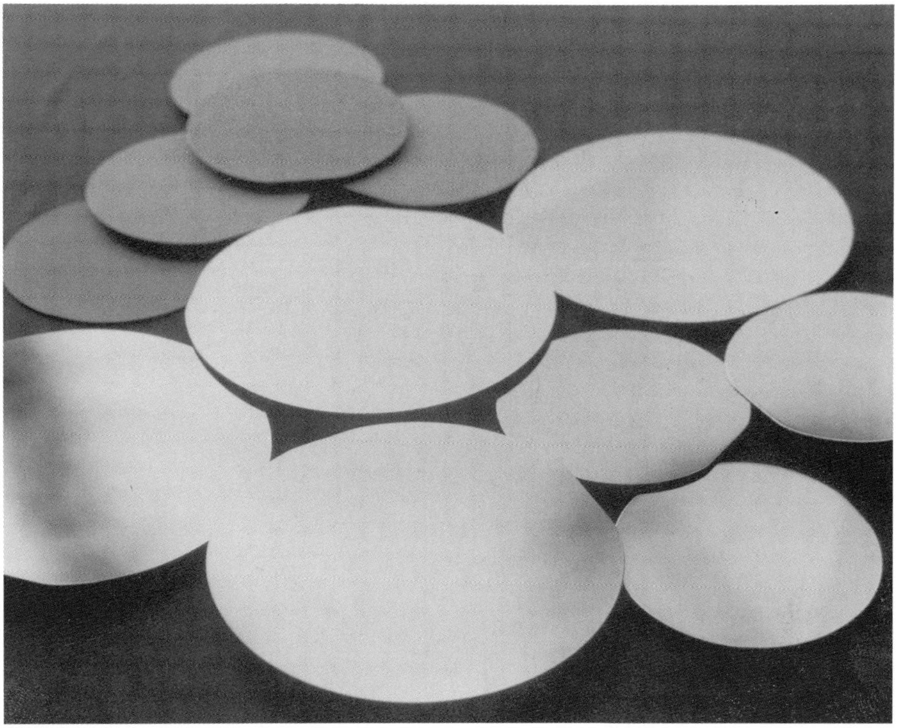

Fig. 48. Polished GaAs wafers of 2, 3 and 4 inch diameter. (Courtesy of MCP Wafer Technology.)

small reservoir can still be used however in order to adjust the stoichiometry to As-richness [112].

The water content of the boric oxide is a critical parameter in the process. A number of impurities, notably silicon and carbon, are gettered by the [OH] content of the oxide. In fact undoped semi-insulating material can be grown from a quartz crucible if the boric oxide is sufficiently "wet". However, the incidence of twinning increases at high water content. Typically the water content is controlled to be around or less than 1000 ppm, the purest material available containing around 100 ppm water. Dry boric oxide (100–150 ppm water) results in carbon concentrations as high as 8×10^{15} cm^{-3} whilst wet oxide (> 500 ppm) reduces carbon to below the detection limit of 3–5 $\times 10^{14}$ cm^{-3} [93].

The water also causes uptake of gallium from the melt as Ga_2O_3 tending to make the melt more As-rich [113]. The volatilisation of this Ga_2O_3 and its subsequent condensation on the graphite furniture in the puller provides the mechanism for the carbon uptake [114].

"Dry" boric oxide gives rise to high concentrations of boron in the crystal (in excess of 1×10^{17} cm^{-3} for the driest oxide) but this does not appear to produce any (significant) electrical effect nor to be deleterious to device performance.

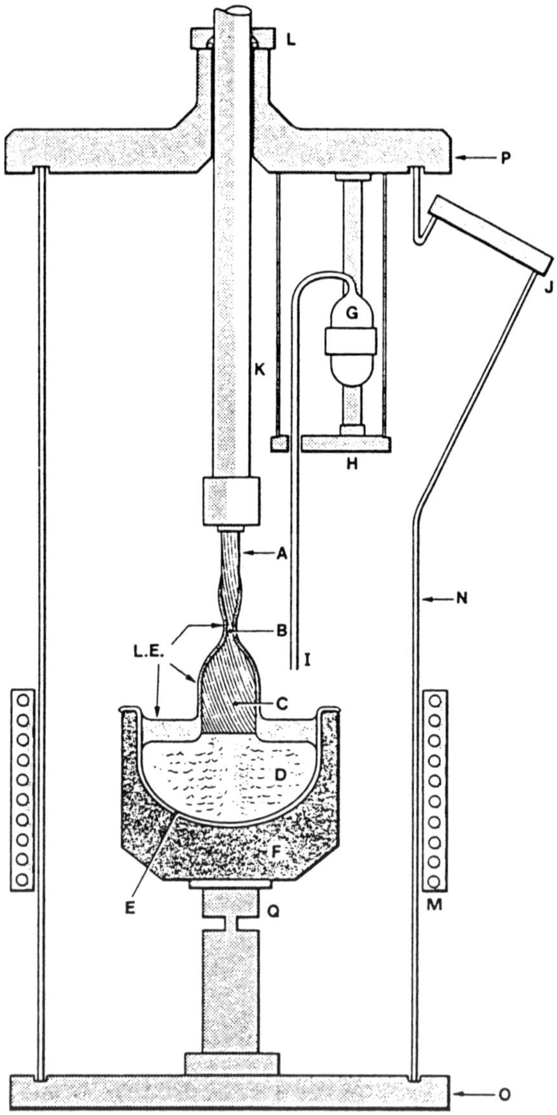

Fig. 49. Apparatus for As injection into a GaAs melt. The As is contained in a separately-heated resevoir (G) which has an outlet tube (I). The reservoir assembly can be lowered through the boric oxide encapsulant (LE) into the melt (D) with the reservoir temperature raised sufficiently to bubble As into the melt. (Courtesy of J.B. Mullin.)

Numerous refinements have been tried by various researchers, particularly in Japan. These include total immersion of the crystal in boric oxide in order to reduce the thermal stress in the cooling crystal. Reduction of the dislocation density by reducing the thermal stress imposed on the cooling crystal has been a major issue and this is reviewed by Volkl in chapter 14. To this end after-heaters have been employed although they exacerbate the problem of dissociation of the grown crystal with formation of gallium droplets on the surface and also make crystal diameter control more difficult. An alternative approach to the problem has been to seek to harden the lattice by isoelectronic doping with indium. The growth of such alloy single crystals is reviewed in section 5.5 below.

The use of a magnetic field to control micro-homogeneity is reviewed in a companion chapter (chapter 5). Doped GaAs single crystals for opto-electronic components such as LEDs and lasers are more commonly grown by Bridgman techniques which are described in chapter 2.

Finally it has to be remembered that it is polished wafers which are required industrially not just grown boules. The cutting and polishing of semi-insulating GaAs wafers to the exacting standards of integrated circuit manufacture is at least as difficult a task as growing the crystal but that topic is outside the scope of this Handbook.

5.3.2. Indium arsenide

Mullin et al. [8] first demonstrated the LEC growth of InAs. Having a melting point of 943°C and a dissociation pressure of less than one atmosphere it is readily grown under boric oxide in a conventional puller such as might be used to grow germanium.

Extensive studies of the doping of this material grown by LEC have been made by Bublik and co-workers in Russia [115–118].

5.4. Group IIIB phosphides

The development of the high pressure LEC technique has revolutionised the production of GaP and InP making them commercially useful substrate materials.

The in-situ compounding of InP and GaP by careful heating of the elements under boric oxide as is done with GaAs is not readily possible because of the very high pressure needed to avoid phosphorus loss. It is however possible to distil phosphorus into either Ga or In in the manner shown in fig. 49 and such an approach has been employed for InP.

5.4.1. Gallium phosphide

The LEC growth of GaP was first reported independently by Bass and Oliver [10] and by Mullin et al. [9]. Synthesis of charge material has been described by Ringel [119].

Commercial pullers have been fitted with automatic diameter control which has been shown to yield improved material from which yellow LEDs can be made having the highest measured external quantum efficiency [120]. First attempts at automatic diameter control of this material utilised the "die" principle where a "coracle" of silicon nitride was floated at the interface between the melt and the boric oxide (see section 3.6). However, whilst this worked well for ⟨111⟩ oriented single crystals of GaP it was found that, for ⟨100⟩ axis growth, a progressive degradation of crystal quality occurred as growth proceeded finally resulting in polycrystallinity. The mechanisms causing this peculiar effect have not been fully elucidated.

The problem has been overcome by the development of weight control servos [121] which are now used in commerical operations [120]. Diameter control proved a serious problem in early work on scale-up of the process with problems of control of the thermal environment in the high pressure chamber where the coupling between the crystal shape and the heat transfer produced by turbulent convection in the ambient gas, radiative transport at the high temperature of growth (the melting point of GaP is 1465 °C) and the "blanketing" effect of the boric oxide layer, led to instability and often to cracking of the crystals [122].

The defect structure of LEC-grown crystals has been studied because of its relevance to the efficiency of opto-electronic devices made from the crystals. Whilst it is possible to pull dislocation-free crystals of up to 15 mm diameter [123], such crystals still contain micro-defects which, when revealed by chemical etching, are termed "S" pits. The principal micro-defects are probably the result of the condensation of native point defects in the crystal during cooling [124].

Von Neida et al. [125] pulled small single crystals from Ga-rich melts (up to 64 at% Ga). By use of a low growth rate they were able to avoid constitutional supercooling. They found that these Ga-rich crystals had a lower dislocation density than equivalent crystals grown from a stoichiometric melt and, more significantly, there was a large increase in red photoluminescence quantum efficiency in the Ga-rich crystals.

5.4.2. *Indium phosphide*

The initial exploitation of InP was as a microwave oscillator [126], but, more recently, it has assumed a much greater importance for long wavelength opto-electronic devices particularly in long distance fibre-optic communications. Most crystals are grown by the high pressure LEC technique. This was first reported by Mullin and coworkers [127]. Recently low dislocation density crystals of commerical size have been produced by the vertical gradient freeze technique (see chapter 2).

Commercial LEC crystals are grown in the high pressure pullers originally developed for GaP and now used also for GaAs. Crystal sizes lag behind those of GaAs, current production being of 2 and 3 inch diameter crystals grown from melts of a few kilograms weight contained in crucibles of up to six inch diameter. Crucibles are made of pyrolytic boron nitride or quartz [128]. Operating pressures of around 40 atm are used to prevent phosphorus loss through the boric oxide encapsulant. These high pressures result in highly turbulent convection in the ambient gas and again produce problems in scaling-up as were experienced with GaP.

Purification of the elements and synthesis of stoichiometric charge material are issues which have received considerable study. State-of-the-art as-grown undoped crystals are "n"-type with carrier concentrations in the range $1\text{–}5 \times 10^{15}\,\mathrm{cm}^{-3}$. Semi-insulating material is produced by doping with iron [134]. There is some evidence that undoped material can be made semi-insulating by annealing under phosphorus-rich conditions. This could be the result of the formation of the analogue of EL2 in GaAs: viz. the phosphorus antisite P_{In}. However the position of the energy level in the band gap produced by this annealing procedure appears to be near identical to that in Fe-doped material and the possibility exists that Fe contamination during annealing is a factor. The issue is at present unresolved; it may prove worthwhile to attempt to grow crystals under phosphorus-rich conditions but this will stretch the high pressure technology to new limits.

A variety of synthesis techniques have been developed which have been reviewed by Oda et al. [129]. We limit ourselves here to mention of techniques for direct synthesis in the puller. There have been two approaches. The first is the injection of phosphorus vapour from a heated red phosphorus source in the manner illustrated for GaAs in fig. 49. First proposed by Farges [130], this method has been scaled up for the production of 10 kg ingots of polycrystalline InP [131].

The second method, developed by Inada and Fukuda [132], is quite novel and involves liquid phosphorus. Indium and phosphorus are loaded into a specially designed quartz chamber positioned within a high pressure puller and heated. The upper part of the chamber is relatively cool so that the phosphorus vapour is condensed to liquid phosphorus which runs down into the indium melt reacting with it at a temperature of around 1000–1100°C. If excess phosphorus is added to the chamber initially then, when reaction is complete, the InP melt is said to be covered with a layer of liquid phosphorus whose thickness depends on the amount of the phosphorus excess. Small single crystals (up to 2 cm diameter and 100 g weight) were pulled from such melts using the liquid phosphorus as encapsulant rather than boric oxide. The acronym LP-CZ (liquid phosphorus Czochralski) was coined to describe the process. The electrical purity of this material was no better than conventional LEC grown crystals however and it is not clear at this time that there are any benefits sufficient to make the greatly added technical difficulties worth tackling.

Donor doping with S, Se, Te, Ge or Sn and acceptor doping with Zn or Cd is readily possible in InP but, as with GaAs, there are problems of reaction with the encapsulant when silicon is used as a dopant.

Concerted attempts have been made to reduce the dislocation density in LEC crystals either by reducing the thermal gradients or by lattice hardening achieved by sulphur doping in the range 5×10^{18} to $1 \times 10^{19}\,\text{cm}^{-3}$. Reduction of the thermal gradients by the use of after-heaters in the puller brings with it the dual problems of difficulty in control of crystal diameter and dissociation of the grown crystal due to the enhanced drain-off of encapsulant resulting from the higher crystal temperature. This latter problem has been addressed by Oda and co-workers [133] who have maintained a partial pressure of phosphorus in the ambient gas above the encapsulant to minimise the dissociation. This brings mechanical problems of how to maintain the integrity of the pull rod seal in such a corrosive ambient and, for good commercial reasons, full details of how this has been achieved have not been published.

However, sulphur doping results in a crystal of low resistivity and high optical absorption in the infrared which is not acceptable in some opto-electronic applications. Further, it produces a significant increase in lattice parameter which brings problems with any subsequent epitaxy on wafers. The use of isoelectronic dopants such as Ga,

As and Sb has been explored [91, 92]. These do not contribute charged carriers but, in order to affect significantly the dislocation density, they must be incorporated in concentrations which are an order of magnitude greater than with sulphur doping.

The dominant and serious problem in the LEC growth of InP is its proneness to twin at peripheral facets during growth [137]. This significantly reduces the commercial yield of single crystals. Examples of such growth twins are shown in fig 50. In common

Fig. 50. Twinned crystals of InP grown by the high pressure LEC technique. (Courtesy of W. Bonner.)

with InSb, InP has a low stacking fault energy and so the energy required to form a twin nucleus in both materials is quite small.

Numerous lists of the factors which induce twinning in both materials have been given. Common amongst most of these lists are :

(1) Interface shape – which controls the size of the peripheral facets.
(2) The moisture content of the boric oxide.
(3) The seed orientation – which determines the angle and azimuthal disposition of the {111} facets.
(4) The angle of conical grow-out of the shoulder which is the region where twinning most readily occurs.
(5) The magnitude of axial and radial temperature gradients (which of course also covers point 1 – interface shape).
(6) Temperature and/or diameter fluctuations.

Recently Hurle [135] has shown, using thermodynamic arguments, that above some minimum supercooling on a peripheral facet, a twinned nucleus is actually thermodynamically favoured over a correctly oriented one for the particular cone angle which causes the twinned nucleus to present a low-index face to the ambient encapsulant at the three-phase boundry. This situation obtains only if, simultaneously, the peripheral facet is locked to the three-phase boudary. Hurle has derived criteria for this condition and demonstrated that they are satisfied for the materials which twin readily (InP, InSb and, to a somewhat lesser degree, GaAs) but not for Si and Ge which do not readily twin. The critical minimum undercooling is shown to be quite small for InP (only around 7°C) so that fluctuations in the local temperature which change the facet width and cause the conical angle of growth to fluctuate through a wide angular range can be sufficient to thermodynamically favour twin formation. Shibata et al. [136] have shown experimentally that twinning does indeed occur following a rapid fluctuation in facet size as shown in fig 51. A similar observation in GaAs has recently been made by Tower et al. [159].

As with GaAs, cutting and polishing of wafers from the grown ingots presents a range of difficult problems, some of which are related to the growth conditions (such as residual elastic strain in the crystal).

5.5. Mixed III–V alloy single crystals

The availability of only a few binary substrate materials is a constraint on the exploitation of the concept of band structure engineering which, with the aid of the dramatic advances made in epitaxial growth techniques, has nonetheless led to an amazing range of novel electronic and opto-electronic devices.

If ternary mixed III–V alloy single crystal substrates could be grown across the composition range, close lattice matching to hetero-epitaxial layers would be possible without the need to produce compositionally-graded layers to minimise strain and misfit dislocations. Moreover, band structure engineering could be extended to bulk semiconductors.

Fig. 51. Longitudinal section through a ⟨111⟩ oriented crystal of InP etched to show a {111} edge facet. Near the bottom of the micrograph the width of the facet increases suddenly and this is followed by twinning on the facet [136]. Mag. × 11.

Accordingly, there have, over the years, been numerous attempts to produce bulk ternary samples but the dual problems of avoiding constitutional supercooling and of obtaining good seeding have defeated all efforts. The bigger the lattice parameter difference between the two end members the "fatter" is the solid solution "lens-shaped" phase diagram (see fig. 52) and the more intractable is the problem of overcoming constitutional supercooling.

One of the most promising systems for study is the $Ga_{1-x}In_xAs$ system because of its relatively "thin" lens (The system $InAs_{1-x}P_x$ has the least segregation of all the III–V alloys). Interest in the $Ga_{1-x}In_xAs$ system was generated when it was shown [138, 139] that the addition of 1 at% In to GaAs hardened the lattice, dramatically reducing the dislocation density in pulled semi-insulating crystals. It was hoped that this would effect an improvement in GaAs MESFET integrated circuit technology but the alloyed material rapidly fell from favour once it was found that the ion implantation conditions were very sensitive to In concentration and that unavoidable variations in In content on both macro- and micro-scales were unacceptably large.

Subsequently Bonner and co-workers [140] have succeeded in growing alloy single crystals of up to 1 inch diameter with In concentrations up to $x = 0.12$. They have shown that the key to this is the use of ternary near lattice-matched seeds for all values of $x > 0.05$. Such seeds were cut from previously grown crystals so that the value of x was incrementally increased with each growth run. Bonner used an rf heated high pressure research puller growing from hemispherical bottomed silica or PBN crucibles

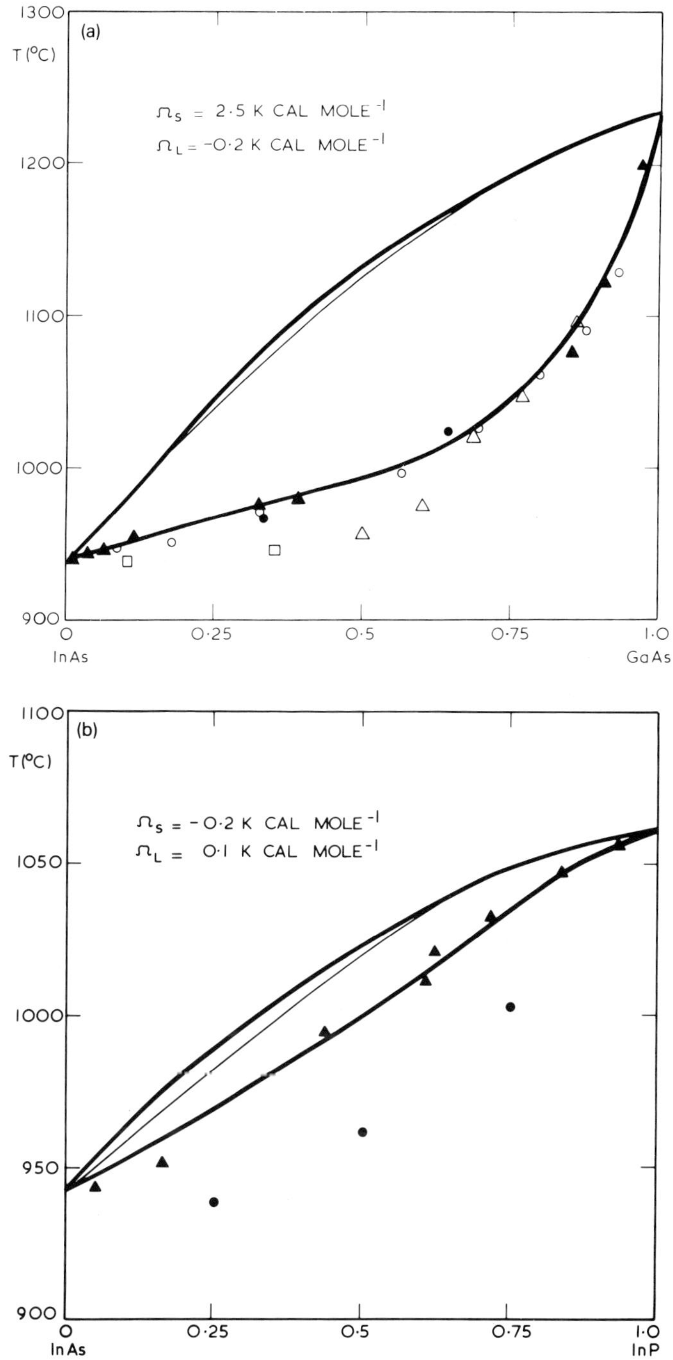

Fig. 52. Pseudo binary phase diagrams for the systems (a) $In_{1-x}Ga_xAs$ and (b) $InAs_{1-x}P_x$. The faint lines are the experimentally determined liquidus curves. The various symbols are several published experimental determinations of the solidus. The bold lines are liquidus and solidus curves obtained from regular solution theory with fitted values of the solid and liquid interaction parameters (Ω_S and Ω_L respectively) shown.

containing charges of up to 200 g weight. Very dry boric oxide was used as an encapsulant and ambient nitrogen pressures were in the range 2–20 atm. Crystal rotation rates were in the range 20–50 rpm and, to avoid constitutional supercooling, growth speeds down to as low as 1 mm h^{-1} were employed. A maximum of about 60% of the melt could be pulled before a cellular structure formed. The crystals were grown on the ⟨111⟩ B axis.

For $x < 0.05$ the crystals were semi-insulating but became progressively less resistive as x was increased. However semi-insulating behaviour could be obtained in the higher x-value crystals by chromium doping.

Kusunoki and co-workers [149, 150] have also succeeded in growing single crystals by the LEC technique obtaining a uniform composition by continuously feeding the melt with GaAs to maintain a constant melt composition. They were able to grow a uniform crystal of 4 mm length with $x = 0.14$ [150].

One other system is worth mentioning. Joukoff and Jean-Louis [141] have pulled single crystals of the alloy system $InSb_{1-x}Bi_x$ in the range up to $x = 0.026$ in an attempt to obtain material which would be suitable for the fabrication of infrared detectors for wavelengths beyond that covered by InSb. They used a resistance heated furnace, graphite crucibles and an argon/10% hydrogen flowing ambient gas. Crystals were pulled in the ⟨111⟩ direction at rates in the range 1–10 mm h^{-1}.

Some InBi precipitation was observed. InBi does not have the zinc-blende structure and so the solubility in InSb is expected to be limited.

5.6. *II–VI and IV–VI compounds*

Both of these classes of compound either dissociate at high temperature or evaporate congruently and both are potential candidates for LEC growth. This has proved to be more successful with the IV–VI compounds than with the II–VI ones overall. We briefly review work on CdTe and PbTe as illustrative examples.

5.6.1. *Cadmium telluride*

At the congruent melting point of cadmium telluride, the partial pressures of Cd and Te are respectively 0.7 atm and 5×10^{-3} atm [142]. Mullin and Straughan [143] have reported on LEC growth in a pressure puller and noted that some cadmium loss occurred at pressures below about 5 atm. Klausutis et al. [144] grew polycrystalline boules of $CdTe_{1-x}Se_x$ ($0.05 < x < 0.15$) and noted Cd loss at chamber pressures below 13 atm.

However, the real problems with the Czochralski growth of CdTe are not vapour loss but the very low thermal conductivity of the material together with its very low stacking fault energy [145]. The low thermal conductivity (only about 1/3 of that of GaAs at the melting point) produces very large thermal stresses in the cooling crystal resulting in dislocation generation and polygonisation whilst the low stacking fault energy means that twin planes can form very readily. These problems have proved insuperable and large, wholly-single, crystals have not been produced by pulling to the authors' knowledge.

Mullin and Straughan [143] grew from silica crucibles under a boric oxide encapsulant, pulling at a rate of 0.5 cm h^{-1} with a seed rotation rate of 10 rpm. They experimented with boric oxide depth compromising on a value of 7 mm. Too thin a layer resulted in too rapid drain-off from the crystal with consequent dissociation; too thick a layer made diameter control impossible. This latter problem seems to be similar to that experienced with GaP by Nygren et al. [122]. The largest single crystal grains obtained were about 10 mm × 5 mm × 50 mm.

5.6.2. Lead telluride

Growth under a layer of molten boric oxide was first demonstrated with lead telluride by Metz et al. [7]. Crystals were grown under an approximately 3 mm thick layer from crucibles of 4 cm diameter in a nitrogen gas ambient at one atmosphere pressure. No doubt because of problems due to residual water in the boric oxide, the latter was first placed in the crucible alone and raised rapidly to 1100°C and held there until frothing ceased before cooling to room temperature. Elemental Pb and Te in stoichiometric ratio were then added on top of the boric oxide and heated slowly until the lead melted followed by rapid heating to above the melting point of PbTe. The boric oxide would, by this point, have become molten and floated on top of the PbTe melt. Any PbO present in the Pb charge dissolved in the boric oxide. Good quality single crystals of up to 1 inch diameter and 3 inch length were obtained. Because of the high viscosity of the boric oxide, only a small differential between crystal and crucible rotation rates could be used.

The technique has been extended to the growth of the solid solution $Pb_{1-x}Sn_xTe$ [146, 147]. This system has a narrow "lens" (i.e., a small freezing range) and it has proved possible to grow single crystals across the whole composition range free of cellular structure. Hiscocks and West [146] pulled from 60 g melts contained in a silica crucible under boric oxide. Growth rates in the range 1–3 mm h^{-1} were used to obtain good single crystals. As seed crystals of the alloy compositions were not available, self seeding using a narrow bore tantalum tube attached to the end of the pull rod was employed. Some evaporation from the crystal was observed for PbTe-rich alloys as a result of drain-off of the boric oxide from the crystal as it emerged. The measured compositions of the crystals were used to construct an improved pseudo-binary section of the ternary phase diagram.

Similar work was carried out independently by Wagner and Willardson [147] with near identical results. In addition, they successfully grew from non-stoichiometric Pb-rich melts with up to 5 at% Pb excess. Attempts to grow from Te-rich melts were hampered by volatilisation problems which made growth impossible.

Using an electro-etch Burke et al. [148] revealed compositional striations and a cellular structure in $(Pb_{1-x}Sn_x)_{1-y}Te_y$ alloy single crystals grown by the Bell and Howell Company.

6. *Growth of oxide crystals*

6.1. Introduction

The stimulus for the single crystal growth of oxides was supplied by the discovery of pulsed laser action in ruby in 1960 [178] and both pulsed and continuous-wave laser emission in neodymium-doped calcium tungstate in the following two years [179, 180]. The subsequent three decades have seen such crystals assume increasing importance in the fields of optics and electronics. Lasers, substrates and shock-proof transparent windows are just a few examples of the many applications which have evolved. Such uses employ the principal properties of this group of materials such as fluorescence, optical transparency, electrical insulation and mechanical strength but, in addition, a vast range of specific characteristics has also been exploited including ferroelectric, piezoelectric, pyroelectric, acoustic and magnetic properties. In the majority of cases, the requisite highly specialised properties can only be obtained when the material has the high degree of structural and chemical perfection exhibited by single crystals.

Oxide single crystals have been grown from the melt, vapour and solution but, for the size and quality demanded by most specifications, only melt growth processes are generally applicable and, of these, the Czochralski technique has proved to be the most adaptable.

The temperature range encompassed by the Czochralski growth of refractory materials is typically from the melting point of lead germanate (738°C) to that of magnesium aluminate spinel (2150°C). The range of compounds which has been grown is vast and includes both single metal–oxygen and mixed metal–oxygen compounds. Both congruently melting compositions and solid solutions have been solidified as single crystals.

Comprehensive lists of compounds which are amenable to Czochralski growth can be found in the literature [181–183]; a representative list illustrating the growth conditions for the materials of major electronic and optical interest is provided in table 5.

The first detailed accounts of Czochralski growth as applied to oxides was that reported by Nassau and Broyer for the single crystal growth of calcium tungstate [184].

Table 5
Typical conditions for the Czochralski growth of oxide single crystals.

Compound	Melting point (°C)	Crucible material	Atmosphere	Pulling rate ($mm\,h^{-1}$)	Rotation rate (rpm)
$Pb_5Ge_3O_{11}$	738	Au	O_2	1–10	10–20
$Bi_{12}GeO_{20}$	930	Pt	O_2	5–15	10–50
$PbMoO_4$	1070	Pt	O_2	1–10	10–40
$ZnWO_4$	1200	Pt	air	5–15	5–50
$LiNbO_3$	1250	Pt	O_2	5–10	5–40
$Sr_xBa_{1-x}Nb_2O_6$	1400	Pt	O_2	1–10	10–20
$Ba_2NaNb_5O_{15}$	1450	Pt	O_2	1–10	10–20
$CaWO_4$	1600	Ir	A/O_2 or N_2/O_2	1–10	5–40
$LiTaO_3$	1650	Ir	A/O_2 or N_2/O_2	5–10	5–40
$Gd_3Ga_5O_{12}$	1825	Ir	A/O_2 or N_2/O_2	1–10	10–100
$YAlO_3$	1870	Ir	A/O_2 or N_2/O_2	1–10	10–20
$Gd_3Sc_2Ga_5O_{12}$	1875	Ir	A/O_2 or N_2/O_2	1–10	10–20
$Y_3Al_5O_{12}$	1970	Ir	A/O_2 or N_2/O_2	1–10	10–20
Al_2O_3	2105	Ir	A or N_2	1–10	10–20
$MgAl_2O_4$	2150	Ir	A or N_2	1–10	10–20

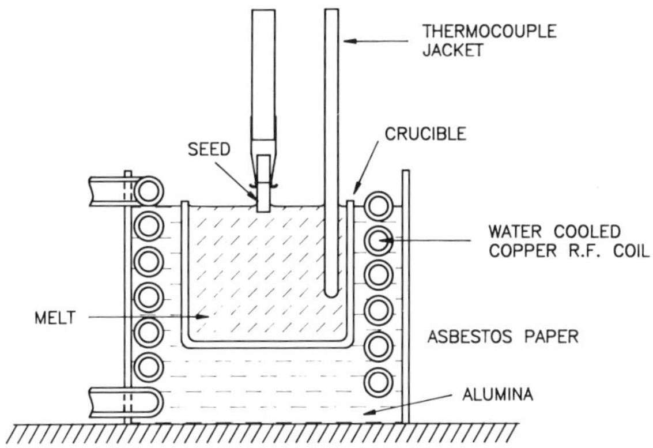

Fig. 53. A schematic representation of the apparatus used for the Czochralski growth of oxide single crystals and particularly $CaWO_4$ by Nassau and Broyer [184].

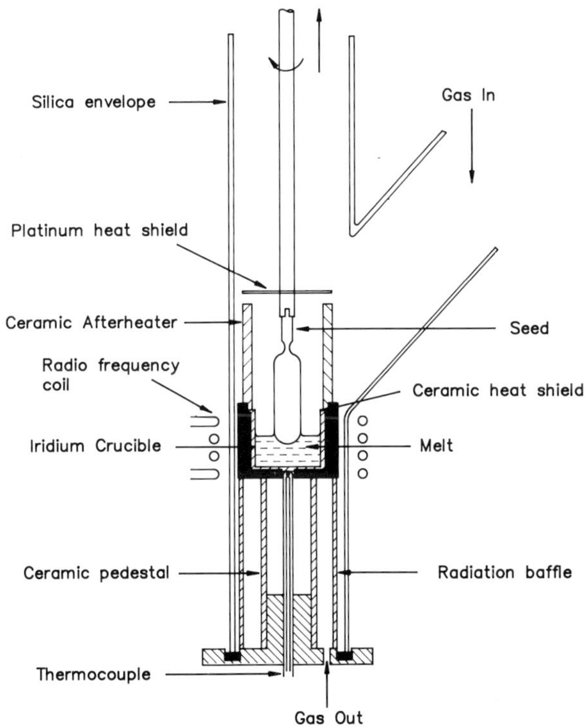

Fig. 54. A schematic representation of the apparatus used for the Czochralski growth of $CaWO_4$ and higher melting point oxide single crystals as developed by Cockayne et al. [185] incorporating heat shielding and a controlled atmosphere.

A schematic representation of their apparatus is shown in fig. 53. This illustrates the basic features of crystal pulling apparatus described earlier but also establishes the need for a refractory environment to support the crucible concentrically within the radio-frequency induction heating coil; this also provides electrical and thermal insulation in addition to reducing the rate of oxidation and evaporation of the crucible.

In subsequent work, Cockayne et al. [185, 186] developed a modified system, shown in fig. 54 which added heat shields to control the thermal environment into which the crystal grows and an enclosed growth chamber to permit control of the composition of the gas ambient so that oxidising, reducing or insert atmospheres could be utilised. Thereafter, many changes to the detail of these growth systems have been made including modifications to the type and shape of the refractories, the use of stainless steel containment vessels and the control of melt temperature by power control in response to changes detected by an rf pick-up coil and the servo-control of the crystal diameter using load cell sensors to weigh either the crystal or the crucible [37].

6.2. The basic problems

The problems associated with oxide crystal growth can be divided conveniently into two types namely, those which are technological in character and those which derive from the fundamental nature of oxides in relation to defect control. The technological problems arise from the form in which the source material is available and from the high temperatures involved. The fundamental problems are encountered because of the hard, brittle nature of many oxides, their complex crystal structure and substitutional site symmetry and from consequent growth-induced defects which can limit the quality and subsequent utility of the crystal produced.

6.2.1. Technological problems

The majority of metal oxides are available commercially as a low density powder, an exception being alumina which, as sapphire, can be obtained as small crystalline lumps produced by the Verneuil process. Several problems arise with oxide powders. Firstly the intrinsic purity, although now much improved from commercial sources, rarely exceeds 99.9999%. Secondly, such fine powders can be easily contaminated during preparation and some can absorb moisture from the atmosphere [187]. Thirdly, charging a crucible with a low density powder is time consuming. Control of water absorption and the retention of the intrinsic purity are obviously a matter of handling and storage under clean, dry conditions. The crucible charging time can be reduced by densification of the initial powder using processes such as cold pressing, sintering or continuous charging, with the crucible temperature kept above the melting point of the material concerned in the latter case. However, potential contamination or, in mixed oxide systems, preferential loss of one component must be considered in any process of densification.

A major problem arises when one or more components is volatile at the melting point of the mixed oxide compound, for example PbO in $PbMoO_4$, and, in these cases, pre-sintering to form the final compound prior to melting does improve stability during melt preparation. Some components of mixed oxide systems do not exist in a stable form, for example Li_2O in $LiNbO_3$. In these instances, the compound has to be prepared from other, more stable salts such as carbonates or nitrates, followed by subsequent thermal decomposition. For mixed oxides, many of these processes are now carried out directly by commercial manufacturers of high purity inorganic compounds. It is, however, important to recognise that lack of attention to preparative details can lead to non-stoichiometric and other compositional inhomogeneities which utimately promote defect formation in the grown crystal.

The dominant problems associated with the high temperatures involved in oxide single crystal growth are how to heat the material to its melting point, how to contain the melt and what sort of ambient atmosphere to employ. In oxide Czochralski growth, the most common practice is to use radio-frequency heating in conjunction with a crucible susceptor [182, 183]. Some details of this technology were given in section 3.3. Crucible materials and after-heaters have been discussed in section 3.4.

6.2.2. Fundamental problems

Defects which form in oxide single crystals are closely related in character to those which have already been described for semiconductor materials. The essential difference with oxides is more a matter of degree and relative importance than of concept [189]. The main defects which can degrade the quality of oxide crystals are generally recognised to be any which lead to non-uniform properties and, in many applications, optical properties in particular. Thus colour centres, facets, growth striations, dislocation low-angle boundaries, twins, individual dislocations with large strain fields, voids, cellular structures, solid state precipitates, inclusions and totally destructive effects such as cracks are all important defects which have to be eliminated or controlled. The defects can be classified into two groups according to their basic derivation from either stress or chemical effects but there exists substantial interplay between these categories.

6.2.2.1. Chemical factors. The principal problems associated with this category in oxide crystals are those apparent as facets, growth striations, cellular structures, precipitates and colour centres.

In oxide crystals with complex crystal structures, the multiplicity of close-packed low-index planes offers considerable facility for the formation of facets. These frequently control the external morphology of the crystal by their development parallel to its growth axis. However, it is those which form on the solid–liquid interface which cause problems since their formation is known to produce localised changes in lattice spacing which induce stress birefringence and impairment of optical properties. Most of the efforts attempting to understand this strain have been concerned with garnets such as $Y_3Al_5O_{12}$ and $Gd_3Ga_5O_{12}$ where the mean lattice parameter of the faceted regions has been shown to be slightly greater than non-faceted regions [190]: see fig. 55.

Three principal sources of the strain have been proposed : (i) changes in cation ratio between facet and matrix, (ii) dopant segregation and (iii) oxygen segregation. It is known that some solutes, for example Nd in $Y_3Al_5O_{12}$ and Cr in $MgAl_2O_4$ do segregate unevenly between facet and matrix and evidence also exists from lumines-

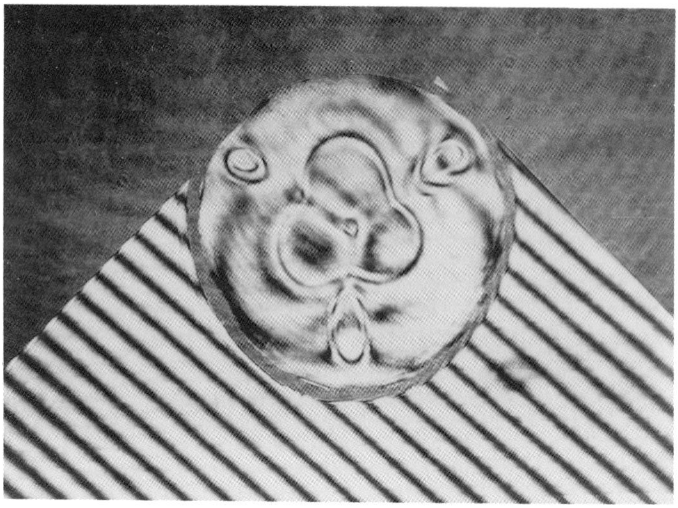

Fig. 55. The strain pattern observed along the ⟨111⟩ axis of a single crystal of $Y_3Al_5O_{12}$ showing clearly the effects induced by three symmetrically placed {211} interface facets near to the crystal centre and a further three {110} interface facets near to the crystal periphery. Mag. × 4.

cence studies to suggest that oxgen vacancies can occur in faceted regions [191]. Facet formation is generally controlled by changing the shape of the crystal–melt interface to restrict the portion which is parallel to potential faceting planes. Crystal rotation rate is used as the major control parameter. $Gd_3Ga_5O_{12}$ growth provides a good example where high crystal rotation speeds were used to induce a planar interface for ⟨111⟩ axis crystals and thereby inhibit the formation of the principal faceting planes, (110) and (211). In such instances, automatic diameter control is a distinct advantage in maintaining crystal stability.

Growth striations are particularly detrimental in oxide crystals used in optical applications since they provide an optical loss due to changes in absorption and refractive index. They can also stabilise unwanted domain structures in ferroelectric oxides such as $LiNbO_3$ and in magnetic layers grown on to substrates such as $Gd_3Ga_5O_{12}$ (see fig. 56). Such striations are known to be due to non-uniform segregation of added dopants, residual impurities or, in multicomponent systems, one of the components [192]. In some oxides, the fluctuations in growth rate which cause the non-uniform segregation evident as striations have been correlated directly with fluctuations/oscillations in temperature arising from unstable convection in the melt [193]. Striation formation is a persistent problem and is generally inhibited, although rarely eliminated, by using shallow melt depths to restrict convective instability in the melt [194]. Baffles placed within the melt to reduce the aspect ratio [161] and cooling of the crucible base [195, 196] to restrict the slow meandering of the thermal centre have also occasionally been employed.

The most apparent chemical effect in transparent oxide crystals is the formation of solute trails and cellular structures arising from constitutional supercooling

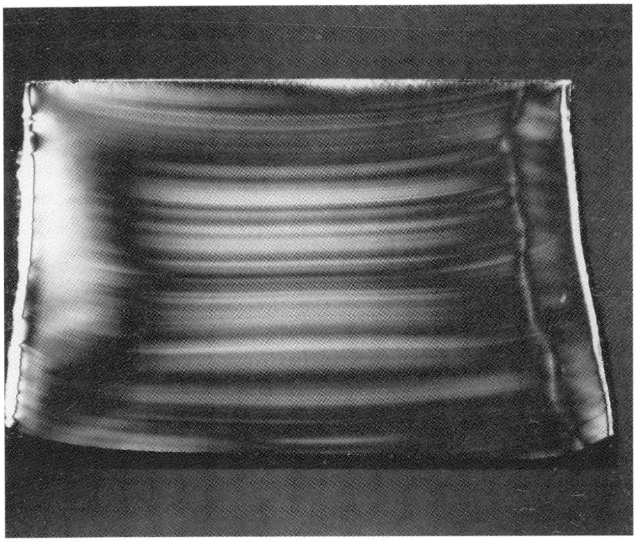

Fig. 56. A longitudinal section of a ⟨111⟩ axis single crystal of $Gd_3Ga_5O_{12}$ viewed perpendicular to the growth axis between crossed polars showing strain effects due to the presence of growth striations. Mag. × 4.

[182, 183, 186]. Under normal conditions, the full sequence of transitions from a smooth interface to a cellular form, with the development of cell boundary grooves and solute trails is rarely seen. The effect usually observed is a sudden breakdown in interface integrity with the entrapment of solute-rich material at heterogeneous nuclei [197]. Gross cellular structure in $CaWO_4$, viewed in transverse section, is shown in fig. 57. These effects can occur due to deliberately added dopants, to deviations in stoichiometry and dissolved gaseous impurities which segregate as voids.

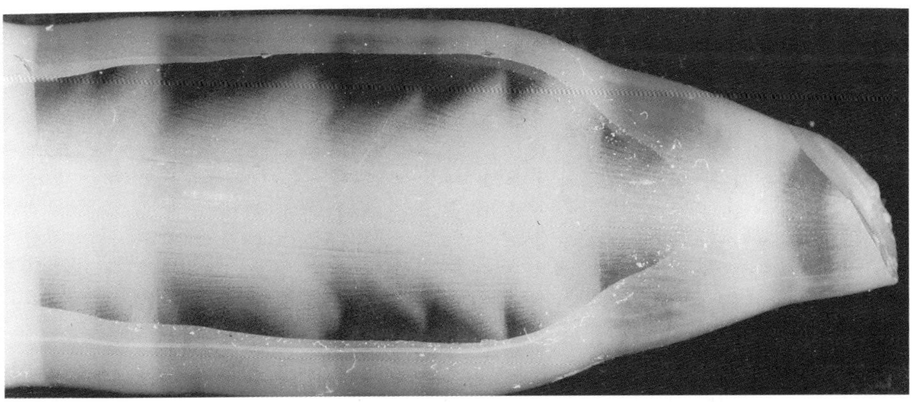

Fig. 57. The effects of constitutional supercooling in a single crystal of calcium tungstate viewed perpendicular to the [001] growth axis; note the enhancement of the structures produced where the crystal diameter changes. Mag. × 4.

The morphology of these structures depends markedly upon the types of liquid flow present in the melt. At high crystal rotation rates, when an upward hydrodynamic flow towards the interface is induced, a spiral morphology can be evident. At low crystal rotation rates, where convective flow from crucible wall to melt centre predominates, a

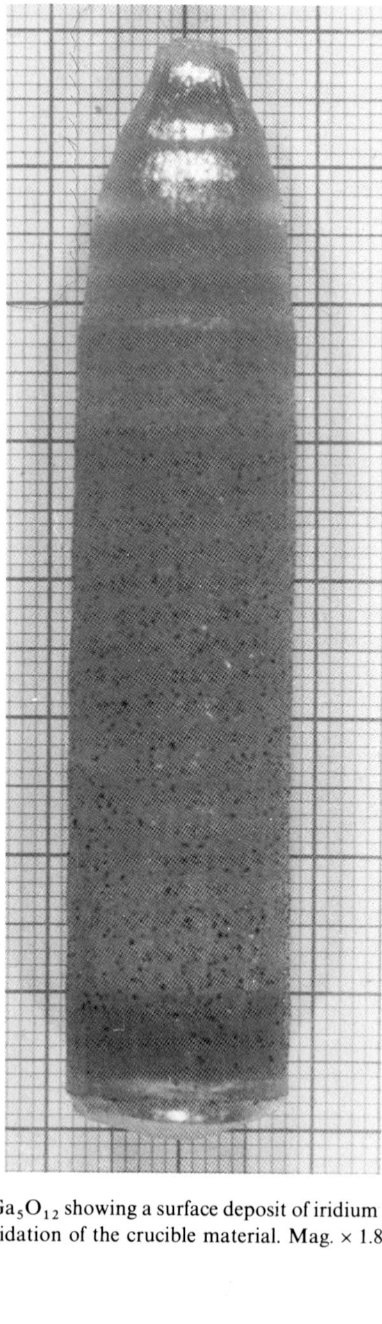

Fig. 58. A single crystal of $Gd_3Ga_5O_{12}$ showing a surface deposit of iridium platelets arising from excessive oxidation of the crucible material. Mag. × 1.8.

radial morphology is usually apparent. In between these two conditions, a wide variety of complex structures can be seen. Cellular structures are generally avoided by using low growth rates; sometimes as low as 0.5 mm h^{-1}, so that excess solute has time to diffuse away from the interface without promoting entrapment and causing interface breakdown. Steep temperature gradients also help to avoid a zone of constitutional supercooling but are incompatible with strain-free and facet-free conditions of growth.

Dopants and impurities can also cause other defects. Their concentration can exceed the solid solubility limit in the matrix on cooling leading to solid state precipitation. Such precipitates can cause small particle scattering and represent a major loss in optical transmission properties. Scatter of this type can also be caused by preferential loss of a constituent including oxygen and also by reaction between the melt or atmosphere and the crucible material. Such defects are best controlled by using very pure starting materials and by controlling the oxygen content of the gas ambient used in the growth environment.

The concentration of oxygen can be critical. For instance, in the use of gallium garnets, increasing the oxygen content of the ambient increases iridium concentration by crucible oxidation but decreases iridium entrapment by reaction of the crucible with gallium oxide. A surface deposit of iridium particles on a single crystal of $Gd_3Ga_5O_{12}$ is shown in fig. 58. A compromise at 2 vol% O_2 allows crystals free from iridium platelet inclusions to be obtained [198]. Appropriate control of oxygen levels in the ambient also produces other benefits. $YAlO_3$ crystals, grown with a small amount of oxygen present, have a smooth surface and crack less readily than those grown in inert or reducing atmospheres where induced loss of a component leads to the formation and thermomigration of liquid droplets on the crystal surface with consequent roughening [199]. An example of this is shown in fig. 59 which is of a $LiAlO_2$ crystal from which there has been superficial loss of Li_2O.

A further defect related to impurities, but usually at very low levels (< 1 ppm), is the colour centre. The general character of this defect is to provide a broad featureless absorption across the visible spectrum which imparts a brown colour to undoped crystals. Although the occurrence of colour centres and the character of the absorption is well documented, substantially less is known about their causes. In $LaAlO_3$, colour centres have been attributed to the presence of divalent impurities or vacancies on the trivalent rare earth site which couple in a complex manner with adjacent lattice sites [200]. Generally the impurity remains unidentified although Fe in Nd-doped $YAlO_3$ [201] and Mg in $Y_3Al_5O_{12}$, $YAlO_3$, $Gd_3Ga_5O_{12}$ and Al_2O_3 [203] have been established as contributing impurities. In general the problem is only eliminated by improving the quality of the starting materials, although some control can be achieved by adding other impurities to maintain charge balance; for example $Mg^{2+} + Ti^{4+} = 2Al^{3+}$ in aluminates [202].

6.2.2.2 Stress factors. Oxide crystals are usually hard materials which readily fail by brittle fracture when stressed. They can exhibit limited elastic behaviour up to quite high temperatures; $Y_3Al_5O_{12}$, for instance, obeys Hooke's law up to at least 1350°C. Most oxides have a complex crystal structure for which the deformation mechanisms are not fully established but plasticity is exhibited at high temperatures as illustrated by

Fig. 59. A [110] axis single crystal of $LiAlO_2$ showing the surface structure which occurs due to thermal migration of liquid droplets caused by the loss of a constituent, in this case Li_2O. Mag. × 1.4.

the presence of dislocations in identifiable slip systems in a number of single crystals; for example $YAlO_3$ [199] and $MgAl_2O_4$ [197]. The dislocation arrays in these and similar materials have shown that the patterns formed are consistent with the development of a hoop stress during growth and subsequent cooling; this stress is caused by the relative expansion or contraction of the crystal centre with respect to the periphery when the isotherms in the crystal are non-planar. Hoop stresses can also lead to twinning in some materials, for example $YAlO_3$ [199].

Other sources of stress can also be present. Dislocations can propagate from poor quality seeds and can be generated by thermal shock when the seed crystal is inserted into the melt [197]. Stresses associated with crystal structure type are also important. For instance, uniaxial crystals such as $LiNbO_3$, $LiTaO_3$ and $YAlO_3$ tend to be more readily stressed when the growth orientation differs from the main symmetry axis due to the dependency of thermal expansion coefficient upon crystal direction. Some crystals exhibit phase changes below the melting point. This usually prevents direct growth from the melt as such changes produce sufficient stresses to crack the crystal completely upon cooling; e.g. $Y_4Al_2O_9$ [203]. However, in a few cases, e.g. $LAlO_3$ [200] and $Ba_2NaNb_5O_{15}$ [204], the phase change involves only a slight translation of the crystal lattice and single crystals can still be produced. Additional stresses are caused by compositional changes such as facets, growth striations etc. as described in section 6.2.2.1.

The general effect of all these stresses, even in the absence of cracking or cleavage, is to induce localised changes in properties and particularly so for non-cubic crystals. For instance, in ruby crystals, dislocations which climb out of their slip planes to form low angle boundaries can cause the *c*-axis to wander and, as a result, to produce changes in refractive index. Very few oxides can be grown in dislocation-free form. A major exception is the general group of aluminium-based garnet crystals where the Burger's vector and energy associated with dislocation formation are sufficiently large to inhibit dislocation generation by slip. However, in contrast, the gallium based garnets will develop very large helical dislocations generated at heterogeneous nuclei such as voids and iridium platelets [205, 206]. An example is shown in fig. 60. These have large associated strain fields which, if propagated into magnetic layers deposited on to them when used as substrates, locally alter stress-induced magnetic anisotropy of the layer and inhibit domain motion.

The basic method of stress control is to minimise the thermal gradients which exist in the growth chamber using appropriate insulation and heat baffles; the design of these is very much a matter of empirical research for each individual group of materials since the nature of the interaction between crystal structure and crystal properties varies markedly from one such group to another. A knowledge of slip systems and thermal expansion coefficients can facilitate the use of orientations which minimise hoop stresses but this is often a minor consideration since particular orientations for growth

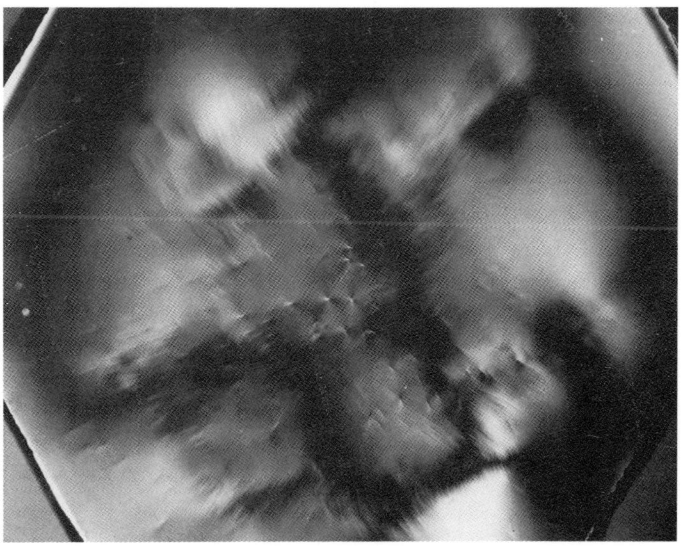

Fig. 60. The strain pattern induced by the presence of helical dislocations due to iridium platelet inclusion viewed between crossed polars along the $\langle 111 \rangle$ growth direction in a $\{111\}$ section of a $Gd_3Ga_5O_{12}$ single crystal. Mag. × 5.

are usually required to provide specific properties. Annealing subsequent to growth can also provide strain relief but significant gains are often only apparent if the heat treatment is carried out close to the melting point. The necking procedure is an important factor in controlling dislocation propagation from seed to crystal in much the same manner as for semiconductors. Current technology allows several oxides to be produced with dislocation densities $< 10^4\,cm^{-2}$ but the only significant material to be produced in dislocation-free form is $Y_3Al_5O_{12}$.

6.3. *Growth of specific materials*

To illustrate specific technological problems and problems of defect control we briefly review the growth of a few representative classes of oxide materials.

6.3.1. *Sapphire and magnesium aluminate spinel*

Cockayne et al. [207] reported on the pulling of single crystals of sapphire in 1967 using a pulling system similar to that shown schematically in fig. 54. The high melting point of sapphire (2050°C) necessitated the use of iridium as crucible material and, to avoid even this material being melted during the process, an after-heater had to be fitted above the crucible as shown in the figure. This consisted of an alumina jacket with an iridium liner. The after-heater totally enclosed the volume into which the crystal was pulled apart from a small axial hole through which the seed crystal was inserted and a small vertical slot to permit visual observation of the growing crystal. The iridium liner coupled to the rf power raising the enclosure to a suitably high temperature.

The principal defects in the crystals were microscopic particles and voids. Two types of particle were identified; one, which could be suppressed by providing an oxygen partial pressure in the ambient gas, was thought to be due to the reduction of the alumina. The second, which was made worse by the addition of oxygen, comprised iridium particles formed by reduction of volatile IrO_2 produced by attack of the crucible by the ambient oxygen. These latter particles could be prevented from becoming entrapped within the crystal by careful design of the gas flow pattern so as to sweep the IrO_2 away from the crystal and melt surfaces.

Voids were seen on the crystals when the pulling speed exceeded $12\,mm\,h^{-1}$. Similar voids were previously observed in $CaWO_4$ and shown to be the result of constitutional supercooling of the melt [208] produced by residual impurities in the charge.

Brandle [209] has shown that the crystal–melt interface is always convex to the melt but that the degree of convexity can be altered by modifying the thermal conditions and changing the crystal rotation rate. This influences the extent of facet formation on the interface.

Because it is a single metal oxide, there are no problems of non-congruency with sapphire and so growth rates can be relatively high provided that the starting charge is chemically pure. With mixed oxide systems such as $MgAl_2O_4$ (spinel) somewhat lower growth speeds are required unless exact congruency of the charge can be maintained. This material has an even higher melting point than sapphire (2105°C) making inadvertent melting of the iridium crucible an even more difficult problem. This was

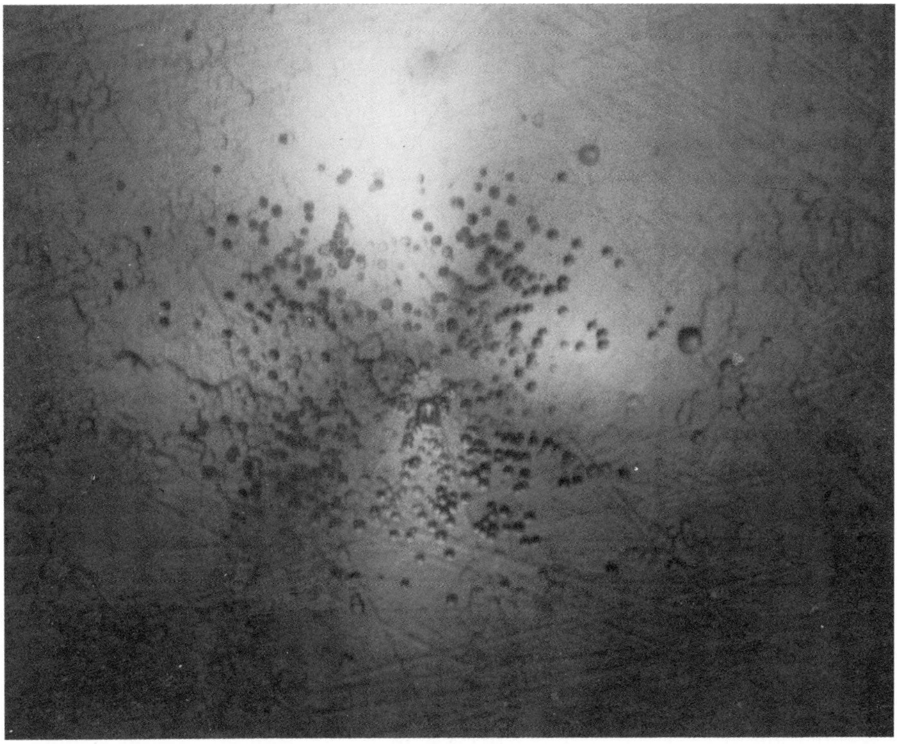

Fig. 61. Etched section showing generation of dislocations around a faceted region in a crystal of $MgAl_2O_4$. Mag. × 104.

overcome by Cockayne and Chesswas [210] by the use of iridium crucibles of double the wall thickness (0.125 inch) of those used for sapphire growth plus the use of an iridium baffle placed immediately above the crucible.

Again with spinel, the growth interface was very convex with the presence of {111} facets [211]. These authors showed, using radio-tracers, that the facets exhibit the facet effect when doped with chromium. Facets also produce elastic strain within chromium-doped crystals but not in undoped ones. This strain can be partially relieved by plastic flow giving rise to a characteristic dislocation pattern (fig. 61). This latter behaviour contrasts with the garnets for example where, because the Burgers vector of a dislocation is very large due to the large size of the unit cell, dislocations are energetically unfavourable. Typical dislocation densities in pulled sapphire are around 10^2 to 10^4 cm^{-2} [212, 207] whereas in garnets they are $< 1\ cm^{-2}$ [213].

6.3.2. Garnets

The garnet crystal composition can be written in the form $A_3(B_2C_3)O_{12}$ where the chemical elements A, B and C sit respectively on dodecahedral, octahedral and tetrahedral sites within the complex cubic unit cell. As a result, a multiplicity of compositions is possible but the majority which have been grown by the Czochralski technique are the rare earth aluminium and gallium based garnets as exemplified respectively by $Y_3Al_5O_{12}$ and $Ga_3Ga_5O_{12}$. Many crystals have been produced with the rare earths Tb, Dy, Ho, Er, Tm, Yb and Lu substituted either wholly or partially on the A-site in the aluminium based garnets whilst Nd, Sm and Eu have been added to the list of rare earth substitutions on this site for the corresponding gallium-based series. A number of partial substitutions have been made on the B-site in both series to yield important tunable laser compounds such as $Gd_3Sc_2Al_3O_{12}$ and $Gd_3Sc_2Ga_3O_{12}$ whilst a variety of coupled substitutions such as (Mg or Ca) and Zr on the B- and C-sites respectively have been made to produce single crystals with specific lattice spacings for use as substrates.

Initially garnets were grown in the type of apparatus described for $CaWO_4$ [214, 215] but the importance of this group of materials as laser hosts and substrates for magnetic bubble devices prompted the scaling-up of the process and the development of commercial pulling equipment such as the Autox system shown in fig. 62. The basic process does, however, remain the same and has been fully described by Mateika [216] and by Brandle [217] in relatively recent review articles.

Essentially, the high purity powders used as starting materials are mixed in the requisite chemical proportions, hydrostatically pressed and fired within the temperature range 1500–1600°C prior to being placed in an iridium crucible and melted. A common practice is to place the iridium crucible inside thermal insulation comprised of stabilised zirconia which can be surrounded by a containing tube of silica or alumina. Thermal shields of a suitable refractory, usually alumina or zirconia, and sometimes lined with iridium, are required above the melt in order to prevent crucible melting and to control the temperature gradients. Melting and growth are carried out under either inert or slightly oxidising atmospheres such as $99\%A/1\%O_2$, $99\%N_2/1\%O_2$, $70\%N_2/30\%CO_2$ or $69\%A/30\%N_2/1\%O_2$. Crucible volatilisation appears to be minimised by the presence of CO_2 [217]. After initial seeding, crystal shape and diameter are now normally controlled automatically by a signal derived from crucible weight changes using computer control. Growth rates vary from $< 1\ mm\ h^{-1}$ for doped crystals such as Nd–$Y_3Al_5O_{12}$ [218] to several mm h^{-1} for undoped $Gd_3Ga_5O_{12}$. When some 75% of the melt has crystallised, the crystal is usually pulled rapidly from its melt and cooled to room temperature at rates which are initially in the region of $100°C\ h^{-1}$ but ultimately become controlled by the rates of cooling of the refractory surround. The principal growth parameters are crystal rotation rate, growth rate and the composition of the gas atmosphere and melt. These parameter values have to be determined experimentally in order to control crystal defects as described in section 6.2.2.

An interesting variant of the normal Czochralski process is the double container (DC) technique developed by Mateika [219] for the growth of $Nd_3Ga_5O_{12}$ and related garnets. This uses a high frequency (1–7 MHz) rf generator to couple into the oxide melt which is contained within a water-cooled cold outer crucible. Solid oxides are too resistive to couple directly into such an rf field and so initial melting is performed by an iridium ring embedded in the base of the sintered starting charge which couples into a separate lower frequency (7–10 kHz) field supplied by a coil placed below the outer cold crucible. This produces a sufficient volume of melt for direct coupling to the higher

Fig. 62. rf-heated puller for the commercial growth of garnet and other oxide single crystals. Shown with the front half of the growth chamber removed to reveal the lift, crucible and heating assemblies. (Courtesy of Metals Research Semiconductors Ltd.)

frequency field to take over. The pronounced convection which occurs in such directly coupled melts is restricted by the use of an iridium baffle. This is a complex technique to control but reduces the use of platinum metals. It has not however become a commercially accepted process.

6.3.3. Lithium niobate and tantalate

These ferroelectric materials, which have the tetragonal tungsten-bronze structure, have been developed for electro-optic, second harmonic generation and acousto-electric applications. They illustrate problems of stoichiometry control.

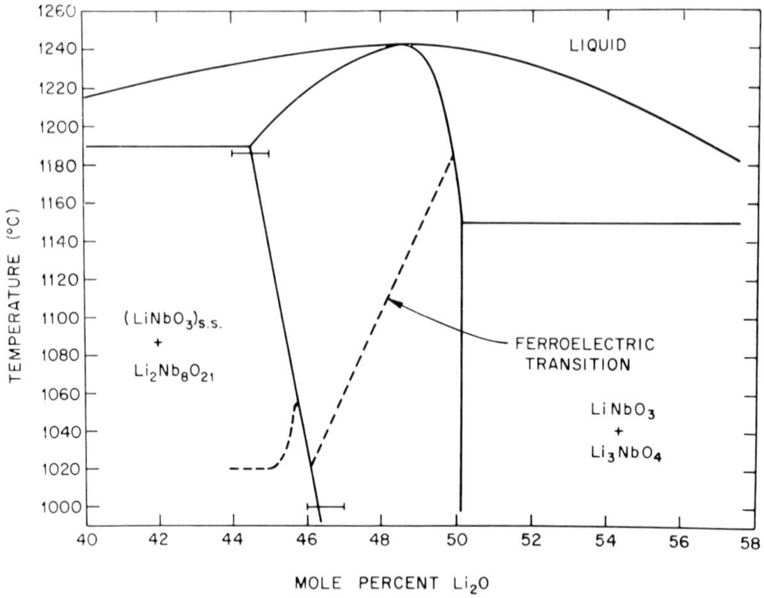

Fig. 63. Phase diagram of the system Li_2O–Nb_2O_5 in the vicinity of the compound $LiNbO_3$. Note that the congruent point corresponds to material which is Li_2O deficient [225].

The pulling of $LiNbO_3$ from an rf-heated Pt crucible was first reported by Rubin et al. [220]. Pulling at 0.25 inch h^{-1}, they obtained crystals of good optical quality with little strain. As grown, the crystals contain multiple ferroelectric domains but single domain pieces can be obtained by "poling" (i.e. heating the crystal to near its Curie temperature and applying a DC electric field along the c-axis). Bergman et al. [221] and Fay et al. [222] showed that the Curie temperature and other relevant properties were strongly dependent on melt composition implying a finite phase extent for the compound. This was studied by Lerner et al. [223] and by Peterson and Carruthers [224]. The accepted phase diagram is shown in fig. 63 from which it can be seen that the congruently melting composition is Li_2O deficient (48.6 mol% Li_2O). The segregation resulting from growth from a non-congruent melt can give rise to constitutional supercooling [225].

Macro-segregation along the length of the crystal, resulting from non-congruency, limits the length of grown crystal which can be phase-matched for second harmonic generation applications [226]. Optical damage is a problem which can be alleviated by doping the crystal with at least 4.5 mol% MgO [227].

The growth, by pulling, of lithium tantalate single crystals of the congruently-melting composition (48.8 mol% Li_2O [224]) up to 3.8 cm diameter has been reported by Brandle and Miller [227]. The crystals were grown in a nitrogen atmosphere at a rate of 8.6 mm h^{-1}. Cracking has been a persistent problem with this material. Brandle and Miller attempted to overcome this by pulling the crystals into a "muffle" furnace made of zirconia tubing. Having pulled the crystal into the muffle, the crucible temperature was lowered rapidly to about 900°C and then slowly cooled to room temperature over a 6–8 h period. After this in-situ treatment, the crystals were annealed in air and then poled.

They discovered that the cracking was associated with the development of low angle grain boundaries and that the occurrence of these boundaries was orientation dependent. *Y*-axis crystals, free of cracking, could be grown and these had a relative uniform distribution of dislocations as revealed by etching. With *x*-axis crystals large cracks which propagated throughout the crystal always formed irrespective of cool-down procedure as it did in *z*-axis crystals where etching revealed that the dislocations were aggregated into low angle boundaries.

Phase matching of $LiTaO_3$ is not possible and so only weak second harmonic generation can be obtained in this material. However, Wang et al. [228], in an elegant piece of work, have achieved quasi-phase matching by producing a periodic domain structure formed as a result of the deliberate production of pronounced striations in the crystal. They did this by offsetting the rotation axis of the crystal from that of the thermal field; the amplitude of the growth fluctuations was controlled by varying the degree of offset.

Additionally they modulated the heater power so as to produce temperature oscillations in the melt of much longer period than the rotational ones produced by the crystal offset. The latter striations were used as "rate" striations to measure the growth rate response of the crystal to the temperature oscillations. The striations were accentuated by doping the crystals with 0.1 wt% yttrium or 0.5 wt% neodymium. The grown crystals exhibited periodic laminar ferroelectric domains whose separation could be tuned to provide phase matching.

$LiNbO_3$ and $LiTaO_3$ form solid solutions cross the entire composition range. The Czochralski growth of these alloys from an rf-heated Pt crucible has been reported by Shimura and Fujino [229] who also reported on defect formation. Contrary to the results of Brandle and Miller for $LiNbO_3$ [227] they found that *z*-axis growth gave the greatest freedom from crack formation. Fukuda et al. [230] reported that avoiding cellular structure in the alloys required growth rates below 1 mm h^{-1}. It may be that the presence of a cellular structure in Shimura and Fujino's crystals contributed to the different cracking behaviour.

6.3.4. Rare earth gallates and aluminates

These materials have received recent interest as substrates for superconducting thin films. Berkstresser et al. [231] have reported on the growth by pulling of the perovskites $LaAlO_3$, $NdGaO_3$ and $LaGaO_3$. Iridium crucibles were used with furnace insulation consisting of stabilised zirconia to achieve low thermal gradients and thereby avoid melting of the iridium crucible. This is a particularly difficult problem with $LaAlO_3$ which, at 2183°C, has the highest melting point of the three. Problems of coupling of the stabilised zirconia to the rf field were experienced.

Growth speeds in the range 2–4 mm h^{-1} proved satisfactory. Low crystal rotation rates (around 10–15 rpm) were needed to give an interface shape which confined the facets to a small area of the interface. Crystals ranging up to 8 cm diameter pulled from melts of up to 3 kg in weight have been grown on ⟨100⟩ axis seeds using a nitrogen or nitrogen + 2% oxygen ambient.

These materials undergo a polymorphic phase transformation to a rhombohedral symmetry on cooling to room temperature. The phase change is accompanied by twinning.

$NdGaO_3$, melting at 1600°C, appears the most promising offering the closest lattice match to the 1:2:3 YBCO superconductor composition. It was found that the twinning could be minimised by adding 5% Sn to the melt to obtain seed crystals which are twin-free. If these seeds are then used to grow a Sn-free crystal then that too is nearly twin-free. Doping with Sm or Gd permits lattice matching. Interestingly, the growth kinetics of the material appear to change when the crystal is twin-free. One possible explanation is that the twins provide self-perpetuating re-entrancies at the growth interface and thus facilitate rapid growth at low undercooling as occurs in silicon web growth (see section 4.3.6 above).

6.4. *Pulling optical fibres*

Very small diameter crystals (fibres) can be pulled from the melt either through a die as in the edge-defined film fed growth (EFG) technique pioneered by LaBelle and Mlavsky [232] or by pulling from a small pool of melt formed on the end of a charge rod and held in place by surface tension forces (pedestal growth). This latter method can be thought of either as a crucible-less Czochralski process or as floating zone melting with different charge and crystal rod diameters. The heat source in pedestal growth must be sufficiently confinable to permit the controlled melting of the small pool of melt on the end of the charge rod. For very small diameters (down to a few microns) laser or electron beam heating sources are indicated. The latter require a high vacuum and a conducting charge and are therefore of little practical use for most optical materials. The use of lasers for the growth of fibres has been pioneered by Feigelson and co-workers [233] at Stanford University with their laser heated pedestal growth (LHPG) technique.

We briefly review the EFG and LHPG techniques as applied to the growth of optical fibres.

6.4.1. *Edge-defined film fed growth*

In 1967, LaBelle and Mlavsky reported on the growth of sapphire filaments using a modified Czochralski puller with rf heating [234]. The fibres were grown on the c-axis with diameters in the range 50–500 micron at rates of up to 150 mm min^{-1}. At these very high growth speeds the growth must almost certainly have been dendritic. If grown accurately on the *c*-axis the surface of the fibres was very smooth and they exhibited tensile strengths which were 3 or 4 times greater than that of bulk-grown crystals.

Following development of the EFG technique [235] (schematically illustrated in fig. 64), LaBelle and Mlavsky were able to grow fibres down to diameters of 100 microns in lengths of several hundred feet which could be rolled up onto a drum. By

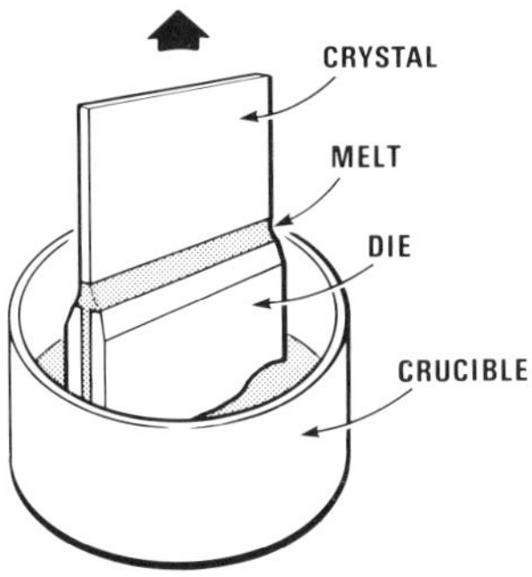

Fig. 64. Schematic representation of the EFG technique for the growth of ribbon crystals [43].

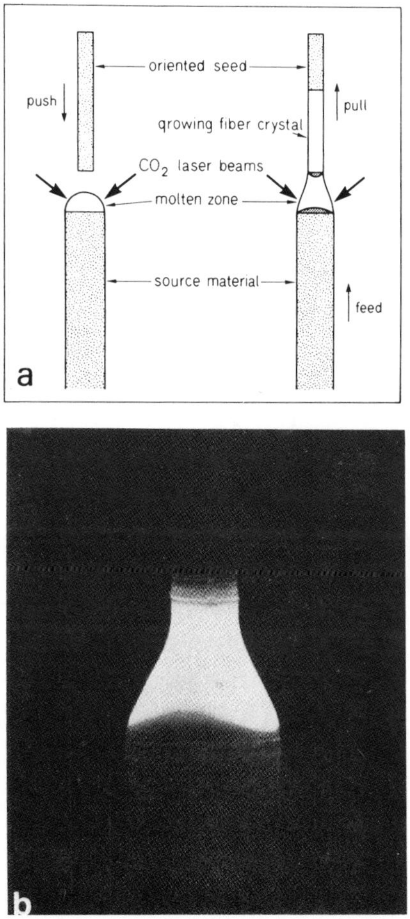

Fig. 65. (a) Schematic representation of pedestal growth and (b) a photograph of a $LiNbO_3$ fibre being grown using laser heating. (Courtesy of R.S. Feigelson [233].)

suitable design of the shape of the die a wide variety of shapes could be grown. The stability of some of these processes is reviewed in the chapter by Tatarchenko.

6.4.2. Laser heated pedestal growth

The pedestal growth concept is shown schematically in fig. 65 together with a $LiNbO_3$ fibre growing by the LHPG technique.

A CO_2 laser is an ideal heat source since it can be focused down to the size of the fibre (microns to millimetres), and can be used in any ambient atmosphere to which it is transparent and has a beam intensity capable of melting any known crystallisable material having a crystallising area comparable to the beam diameter. Because of the highly local nature of the heating it produces, it permits the design of growth equipment which has very steep temperature gradients in the melt and cooling crystal which enables growth rates to be increased significantly without the onset of constitutional supercooling. This is, of course, achieved at the price of a relatively larger amount of thermal strain. Achieving symmetrical heating of the pedestal is not simple but a very effective system, shown in fig. 66, has been developed by Fejer and co-workers [236]. Alignment of the optical system is achieved using a low power He–Ne laser.

The dynamic stability of the pedestal process has been studied by Fejer [237]. A critical parameter is the diameter reduction ratio i.e., the ratio of the diameters of the fibre and the pedestal. Below some critical value of this parameter, which is a function of the wetting angle Θ_L^0 (see section 2.3), the system is dynamically unstable. The higher the value of Θ_L^0 the greater the reduction ratio that can be stably achieved. Typical practical values of reduction ratio are 1/2 or 1/3.

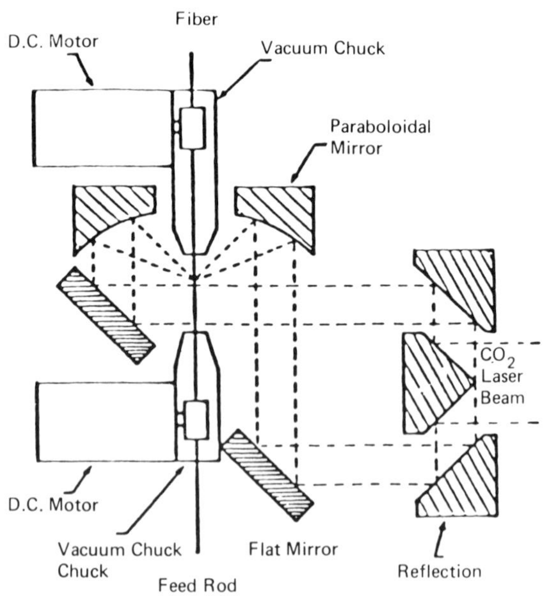

Fig. 66. Circularly symmetric laser optical system for laser-heated pedestal growth [235].

To maintain constant fibre diameter, an automatic control system is required. One published system [237] infers diameter changes by analysing interference fringes produced by a He–Ne laser. Control of diameter to 0.01% is claimed. In addition the diameter can be modulated to give a predetermined periodicity in the crystal profile.

The principal value of the technique is for rapid evaluation of new, technologically difficult, highly refractive materials. The Stanford group report the growth of over 50 different materials including, oxides, borides, halides and carbides. In addition to single phase materials, eutectic and peritectic systems have also been grown.

Defects, such as dislocations, which propagate at even a very small angle to the fibre axis will of course grow out of the crystal after a given length and, in consequence, the structural perfection of fibres tends to be higher than that of larger crystals grown by conventional pulling techniques, notwithstanding the very high thermal gradients present during fibre pulling (typically around 1000°C/cm).

The high gradients and relatively rapid growth rates offer the possibility of quenching-in high temperature phases which are metastable at room temperature. Thus fibre crystals of the high temperature non-ferroelectric hexagonal phase of $BaTiO_3$ and an entirely new phase of $ScTaO_4$ have been grown by the LHPG method [238].

7. *Growth of fluoride crystals*

The customary method used for the growth of single crystals of fluorides is the Bridgman–Stockbarger technique with the charge contained in a graphite crucible. However, in some instances, difficulties can be encountered with crystal cracking, reactivity of the melt with the graphite and control of crystal orientation; this latter factor is particularly important for crystals to be used as lasers where a close correlation can exist between optical properties and crystal direction. Some of the earliest applications of the Czochralski technique to fluoride growth were by Guggenheim [240] and Nassau [241] who grew a range of divalent metal fluorides (e.g. CaF_2, SrF_2, BaF_2, CoF_2, MnF_2) for evaluation as laser hosts.

The major problem with fluorides is the elimination of water vapour and oxygen from both the starting material and the growth environment to prevent the precipitation of oxide particles within the crystal which can cause small particle scattering. Nassau [241] attempted to overcome this problem by reacting carbonates or oxides with aqueous HF gas followed by growth in an inert argon atmosphere from a melt contained in crucibles of Pt–20% Rh, pure rhodium or iridium.

In general, the initial apparatus used for Czochralski growth of these fluorides was very similar to that described for oxide growth under a controlled ambient, the major difference being the avoidance of any oxide-containing components apart from the quartz cylinder which formed the growth chamber. Crystals up to 10 cm long and 1.5 cm diameter were grown at respective pulling and rotation rates of 0.5–5 cm h^{-1} and 20–40 rpm.

Mixed fluorides based on the schelite structure (e.g., $LiYF_4$ [242]) have also been grown by the Czochralski technique. For the $LiRF_4$ series (with R = Y, Ho, Er, Tm, Lu), Cockayne et al. [243] showed that attempts to produce single crystals directly from the molten fluorides provided only poor quality crystals due to impurities, deviations from stoichiometry and contamination during preparation. These difficulties were overcome by converting appropriate mixtures of the component fluorides into polycrystalline bars of the required compound using a zone-refining process. Single crystals were grown from the optically clear portions of such bars using rf-heated platinum crucibles and heat shields under a dry inert gas ambient, all contained within a water-cooled stainless steel chamber. Loss of the volatile LiF was restricted by raising the chamber pressure to 1.7 atm and small particle scattering, shown to be due to a solid state precipitation process, was eliminated by in-situ annealing coupled to rapid cooling. Some of these compounds, particularly those containing impurities, have been shown to melt peritectically or syntectically. In these instances, further purification or growth from melts containing a slight excess of LiF have eliminated the problem. An alternative approach to the control of the scatter and melt congruency has been reported by Pastor et al. [242] using a reactive ambient atmosphere of HF within the growth chamber which has to be constructed of either platinum or monel metal in order to withstand the corrosive chemical environment. In this case, hydrolysis of the crystalline material is prevented according to the equilibrium:

$$F^-(c) + HOH(g) = OH^-(c) + HF(g).$$

Crystals of the $LiRF_4$ series of compounds have been grown successfully in sizes up to 10 cm long and 2 cm diameter at respective crystal growth and rotation rates within the ranges 1–5 mm h^{-1} and 10–20 rpm. A similar technology has recently been applied

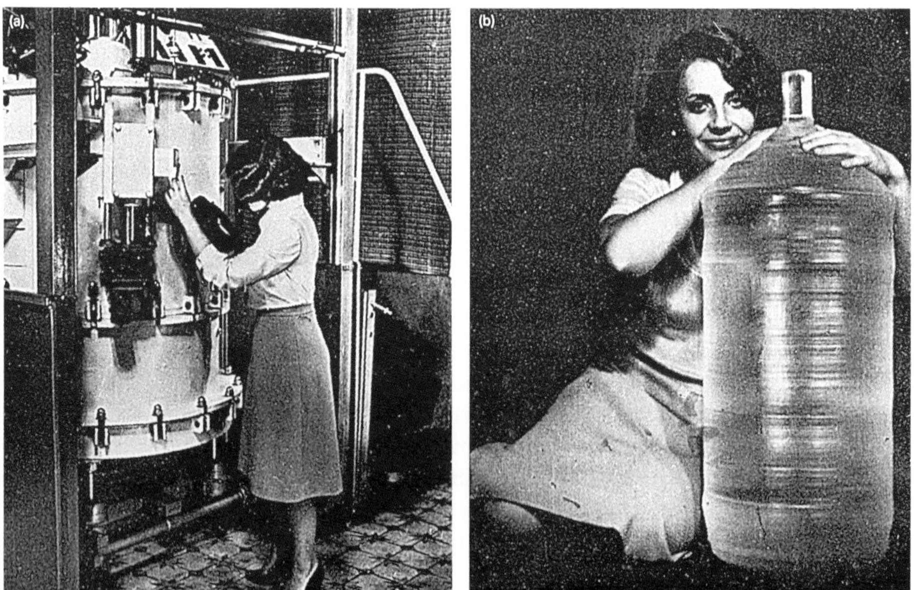

Fig. 67. (a) Automated, commercial-scale puller for the growth of alkali halides. (b) KCl single crystal grown in the apparatus [244].

to the single crystal growth of the complex mixed fluoride $LiCaAlF_6$ and its strontium analogue $LiSrAlF_6$ in both undoped form and doped with chromium for use as a tunable laser material [243]. In this case anhydrous fluorides with a purity of 99.99% or better were melted in a platinum crucible heated by a Pt–Rh resistively heated element which was electrically isolated from the pure platinum growth chamber. Flowing $HF–N_2$ gas was used as the ambient during growth and cooling of the crystals. A pulling rate of 0.75 mm h^{-1} was employed with a crystal rotation rate of 20 rpm. Typical crystals were 10 cm long and 2 cm diameter.

Some of the largest crystals grown by the Czochralski technique are the alkali halides KCl and CsI. Goriletsky et al. [244] (fig. 67) have described an automated pulling method with simultaneous feeding of starting material to the melt. The melt is contained in a platinum crucible and is supplied at the appropriate rate by molten feedstock from a adjacent reservoir to maintain a constant position for the solid–liquid interface as the crystals grows. Crystals of KCl up to 45 cm diameter and weighing 140 kg have been produced. The corresponding figures for CsI doped with Na are 30 cm and 115 kg. Crystals have been grown up to 75 cm in length.

8. *Growth of organic crystals*

The growth of a few organic single crystals of high quality has been demonstrated by several workers. The first known to the present authors is the work of Cook and Gwan [151] who grew single crystals of 12-tricosanone (laurone) from 50 g melts in a puller having a novel crucible and heater arrangement. The crucible, made of borosilicate glass, was double-walled and was heated by water circulated from a thermostat having a 24 h stability of ± 0.001 K. This proved highly suitable for the grown material (melting point 69°C) which had to be pulled at rates down to as low as 1 cm/day. A thermostatically cooled pull-rod was translated and rotated by a conventional lead screw mechanism. The ambient gas was argon at 50 kPa pressure to suppress evaporation. Initial seeding was onto a drilled stainless steel rod but subsequently, single crystal seeds were cleaved from grown crystals and oriented parallel to a fast growth direction (normal to the "*c*"-axis).

Bleay et al. [152] developed a puller suitable for the growth of benzil and benzophenone single crystals. It employed a pyrex glass crucible contained within a close fitting aluminium pot which was conventionally heated by resistive means from a thyristor controlled supply which was capable of stabilising the temperature to ± 0.02 K. To prevent condensation of volatile materials (such as benzil) on the glass growth chamber wall the latter was heated by a small independently controlled heater. Additionally, a close fitting after-heater was mounted around the pull rod thus permitting in-situ annealing of the crystals.

The pull rod was water-cooled and lead screw driven at rates of up to 5 mm h^{-1}. Seeded crystals were pulled from approximately 50 g melts at rates in the range 0.1 to 0.5 mm h^{-1}. Crystal diameter was in the range 0.5 to 2.5 cm. After-heaters were needed to reduce the thermal gradients in the case of benzil otherwise condensation on the seed occurred with concomitant polycrystallinity.

X-ray topography was used to examine the perfection of the grown crystals from which it was shown that the crystals contained a low density of dislocations over most of their cross sectional area but that a high density in the core region propagated from the seed throughout the crystal length.

Subsequently, Katoh and Kato [153] demonstrated that dislocation-free benzil single crystals could be grown in a similar puller by successively refining the quality of the seed by repeated use of the last grown crystal to obtain a seed for the current growth. They empirically found that, by growing in the "*c*"-axis direction and by carefully growing-out from the seed diameter either by lowering the pull rate or by adjusting the thermal conditions in the puller, existing dislocations could be bent outward and, in favourable cases, could be eliminated to yield a dislocation-free crystal.

Benzophenone (m.p. 48°C) and pyrene (m.p. 156°C) single crystals of several cm length have also been grown in a puller by Karl [154]. The puller was vacuum tight and of rather complex design comprised of a stainless steel chamber to which a pyrex crucible was sealed at its base. Crystal lift used a bellows system which avoided an external seal and rotating of the crystal holder was achieved remotely by the use of SmCo magnets in repulsive mode. The complete assembly was thermally shielded.

9. *Concluding remarks*

Rationally considered, crystal pulling might seem a most unpromising technique. It suffers from a range of potential problems:

(1) Growth occurs on the surface of the melt where any scum present will collect.

(2) The process is dynamically unstable, at least under some conditions (see the chapter by Tatarchenko).

(3) Control of crystal diameter requires extremely precise control of the height of the supported meniscus (typically to within better than 30 microns).

(4) The thermal geometry of the melt makes it subject to turbulent natural convection with the consequent fluctuation of temperature and hence microscopic growth rate.

(5) Particulate matter can condense in the cooler upper parts of the growth chamber and fall back into the melt disrupting the single crystallinity.

(6) It is a batch process not ideally suited to commercial production.

Given this formidable list of problems how is it that, in the last forty years, it has come to be the dominant technique for the production of bulk single crystals? One possible answer is to observe that alternative techniques which have been devised have even more, or at least more serious, disadvantages. Thus turbulent convection is a problem with all large scale growth systems, though in some cases it may by surface-tension-driven (Marangoni convection) rather than buoyancy-driven. Czochralski growth has the powerful advantage that crystal rotation largely controls the solute boundary layer thickness ahead of the crystal–melt interface giving very good radial uniformity of dopant concentration.

A second factor is the ease of seeding which can be visually observed and, if defective, can be repeated by remelting. This contributes to the very high yield which is usually obtainable with crystal pulling.

Thirdly, the absence of any ampoule around the cooling crystal obviates the problem of thermal stress due to differential expansion of crystal and ampoule during cooling which can be a problem with other techniques. However, thermal stress due to non-planar isotherms remains a problem as expounded by Volkl in chapter 14.

But perhaps most important of all is the reassurance given to the operator by being able to look into the chamber and see that all is going well!

What further developments of crystal pulling can be expected in the future? Clearly scaling-up of the process for all commercially important materials will continue since there is no obvious fundamental limit to the size of crystal that can be grown particularly if external means to damp the convective turbulence, such as the use of a magnetic field, is employed.

The dynamics of the process are only now becoming well understood and that knowledge has yet to be translated into improved control algorithms for crystal diameter control. Clearly there is some way to go in extending and refining computer control of the process. Multi-variate control, instead of just control of heater power, will become more common. The additional controlled variables could be pull rate, crystal and crucible rotation rates, crucible lift rate and, even, strength of an externally applied magnetic field. The use of adaptive control can be expected to become more common which could involve complex numerical simulation requiring supercomputing power. This will be cost effective in only a few commercial applications.

Melt replenishment, which would transform the present batch process into a semi-continuous one, offers clear commercial advantages but, notwithstanding a fairly considerable effort to date, this has not been satisfactorily achieved even for the commercially most important material silicon.

Empirical development of ever larger pullers is very uneconomic and computer-aided design of the thermal environment will become progressively more widely used. To be useful aids, this must comprise realistic global models such as are described by Dupret and Van den Bogaert in chapter 15.

The use of a magnetic field as a control variable in the growth of some semiconductor crystals holds some promise, particularly the recently developed cusped field configuration [160, 161].

Developments in oxide growth will be likely to extend the range of materials which can be grown by pulling. Going to yet higher temperatures poses formidable problems of melt containment which will require novel solutions.

Whatever the trends, the pulling technique is certain to retain its dominant role for the production of most semiconductor substrates and probably also for a wide range of other materials.

References

[1] G.K. Teal and J.B. Little, Phys. Rev. 78 (1950) 647.
[2] J. Czochralski, Z. Physik. Chem. 92 (1917) 219.
[3] E. von Gomperz, Z. Physik 8 (1921) 184.
[4] H.Mark, M. Polanyi and E. Schmid, Z. Physik 12 (1933) 3064.
[5] H.E. Buckley, Crystal Growth (Wiley, New York, 1951).
[6] G.K. Teal, IEEE Trans. Electron Devices ED-23 (1976) 621.

[7] E.P.A. Metz, R.C. Miller and R. Mazelsky, J. Appl. Phys. 33 (1962) 2016.
[8] J.B. Mullin, B.W. Straughan and W.S. Brickell, J. Phys. Chem. Solids 26 (1965) 782.
[9] J.B. Mullin, R.J. Heritage, C.H. Holliday and B.W. Straughan, J. Cryst. Growth 3 & 4 (1968) 281.
[10] J. Bass and P.E. Oliver, J. Cryst. Growth 3 & 4 (1968) 286.
[11] P.I. Antonov, J. Cryst. Growth 104 (1990) 39.
[12] B. Cockayne, M. Chesswas and D.B. Gasson, J. Mater. Sci. 4 (1969) 450.
[13] J.P. Wallace, J.K. Tien, J.A. Stefani and K.S. Choe, J. Appl. Phys. 69 (1991) 550.
[14] W. Bardsley, F.C. Frank, G.W. Green and D.T.J. Hurle, J. Cryst. Growth 23 (1974) 341.
[15] J.K. Kristensen and R.M.J. Cotterill, Phil. Mag. 36 (1977) 437.
[16] D.T.J. Hurle, J. Cryst. Growth 63 (1983) 13.
[17] E.A. Boucher and T.G.J. Jones, J. Chem. Soc. Faraday I, 76 (1980) 1419.
[18] T. von Karman, Z. Angrew. Math. Mech. 1 (1921) 244.
[19] W.G. Cochran, Proc. Camb. Phil. Soc. 30 (1934) 365.
[20] J.T. Stuart, Quart. J. Mech. Appl. Math. 7 (1954) 446.
[21] A.D.W. Jones, Prog. Cryst. Growth Charact. 9 (1984) 139.
[22] J.A. Burton, R.C. Prim and W.P. Slichter, J. Chem. Phys. 21 (1953) 1987.
[23] L.O. Wilson, J. Cryst. Growth 44 (1978) 247.
[24] A.A. Wheeler, J. Eng. Math. 14 (1980) 161.
[25] S. Makram-Ebeid, P. Langdale and G.M. Martin, in: Semi-insulating III–V Materials, eds. D.C. Look and J.S. Blakemore (Shiva Publishing, Nantwich, UK, 1984) p. 184.
[26] G.R. Jones, Gold Bull. 7 (1974) 69.
[27] M. Chesswas, B. Cockayne, D.T.J. Hurle, E. Jakeman and J.B. Mullin, J. Cryst. Growth 11 (1971) 225.
[28] D.T.J. Hurle, G.C. Joyce, G.C. Wilson, M. Ghassempoory and C. Morgan, J. Cryst. Growth 74 (1986) 480.
[29] D.T.J. Hurle, G.C. Joyce, M. Ghassempoory, A.B. Crowley and E.J. Stern, J. Cryst. Growth 100 (1990) 11.
[30] E.J. Patzner, R.G. Dessauer and M.R. Poponiak, SCP Solid State Tech. (1987) p. 201.
[31] U. Gross and R. Kersten, J. Cryst. Growth 15 (1972) 85.
[32] H.D. Pruett and H.Y. Lien, J. Electrochem. Soc. 121 (1974) 822.
[33] H.J.A. van Dijk, C.M.G. Jochern, G.J. Scholl and P. van der Werf, J. Cryst. Growth 21 (1974) 310.
[34] A.F. Witt, private communication.
[35] W. Bardsley, G.W. Green, C.H. Holliday and D.T.J. Hurle, J. Cryst. Growth 16 (1972) 277.
[36] D.T.J. Hurle, J. Cryst. Growth 42 (1977) 473.
[37] W. Bardsley, B. Cockayne, G.W. Green, D.T.J. Hurle, G.C. Joyce, J. M. Roslington, P.J. Tufton and H.C. Webber, J. Cryst. Growth 24 & 25 (1974) 369.
[38] M.A. Gevelber, G. Stephanopoulos and M.J. Wargo, J. Cryst. Growth 91 (1988) 199.
[39] W. Bardsley, G.C. Joyce and D.T.J. Hurle, J. Cryst. Growth 40 (1977) 13.
[40] W. Bardsley, D.T.J. Hurle, G.C. Joyce and G.C. Wilson, J. Cryst. Growth 40 (1977) 21.
[41] A.V. Stepanov, Sov. Phys. Tech. Phys. 4 (1959) 339.
[42] P.I. Antonov, J. Cryst. Growth 23 (1974) 318.
[43] H.E. LaBelle Jr. and A.I. Mlavsky, Mater. Res. Bull. 6 (1971) 571, 581.
[44] M. Cole, R.M. Ware and M.A. Whittaker, Proc. ECCGI (Zurich, 1976); Crystal Growth and Materials, eds. E. Kaldis and H. J. Scheel (North-Holland, Amsterdam, 1977).
[45] T. Surek, J. Appl. Phys. 47 (1976) 4384.
[46] E.M. Porbansky, J. Appl. Phys. 30 (1959) 1455.
[47] W. Uelhoff, J. Cryst. Growth 65 (1983) 278.
[48] K.J. Bachmann, H.J. Kirsch and K.J. Vetter, J. Cryst. Growth 7 (1970) 290.
[49] C.H. Sworn and T.E. Brown, J. Cryst. Growth 15 (1972) 195.
[50] Y.K. Chang, W. Uelhoff, A. Fattah and G. Hanke, J. Cryst. Growth 85 (1987) 396.
[51] U. Ekhult and T. Carlberg, J. Cryst. Growth 76 (1986) 317.
[52] K.J. Bachmann and T. Wakiyama, J. Cryst. Growth 30 (1975) 327.
[53] H. Fehmer and W. Uelhoff, J. Cryst. Growth 15 (1972) 195.
[54] I.M. Buckley-Golder, J. Cryst. Growth 40 (1977) 189.
[55] B.K. Tanner, Z. Naturforsch. 289 (1973) 676.

[56] H.F. Sterling and R.W. Warren, Nature 192 (1961) 745.
[57] H.F. Sterling and R.W. Warren, Metallurgia 67 (1963) 2.
[58] D.A. Hukin, in: Rare Earths in Modern Science and Technology, Vol. 2 (Plenum, New York, 1980) p. 25.
[59] W.Bardsley, J.M. Callan, H.A. Chedzey and D.T.J. Hurle, Solid State Electron. 3 (1961) 142.
[60] W. Bardsley, J.S. Boulton and D.T.J. Hurle, Solid State Electron. 5 (1962) 395.
[61] W. Bardsley, D.T.J. Hurle, M. Hart and A.R. Lang, J. Cryst. Growth 349 (1980) 612.
[62] D.T.J. Hurle, G.C. Joyce and Q.A.E. Vaughan, J. Cryst. Growth 132 (1993) 1.
[63] W.F. Leverton, J. Appl. Phys. 29 (1958) 1241.
[64] J. Goorissen, Philips Tech. Rev. 21 (1960) 185.
[65] J.R. Patel, R.F. Tramposch and A.R. Chaudhuri, in: Metallurgy of Elemental and Compound Semiconductors, Met. Soc. Conf. Vol. 12, ed. R.O. Grubel (Interscience, New York, 1961) pp. 45–63.
[66] J.S. Haggerty and J.F. Wenkus. US Patent No. 3704093 (Nov. 28, 1972).
[67] J.A.M. Dikhoff, Solid State Electron. 1 (1960) 202.
[68] E. Billig, J. Inst. Met. 83 (1954/5) 53.
[69] D.T.J. Hurle, Solid State Electron. 3 (1961) 37.
[70] W.G. Pfann, Zone Melting, 2nd Ed. (Wiley, New York, 1966) p. 254.
[71] G.K. Teal and E. Buehler, Phys. Rev. 87 (1952) 190.
[72] W.C. Dash, J. Appl. Phys. 29 (1958) 739.
[73] G. Ziegler, Z. Naturforsch. 16a (1961) 219.
[74] F. Bischoff, DB Patent 1, 134, 459 (filed May 1954, granted August 1962).
[75] J.G. Wilkes, Proc. 8th Int. Summer School on Crystal Growth (Palm Springs, 1992), in press.
[76] K.M. Kim, A. Kran, K. Riedling and P. Smetana, Solid State Tech. (1985).
[77] R.E. Lorenzini, A. Iwata and K. Lorenz, US Patent 4, 036, 595 (1977).
[78] G. Fiegl, Solid State Tech. (1983) 121.
[79] K. Kakimoto, M. Watanabe, M. Eguchi and T. Hibiya, J. Cryst. Growth. 126 (1993) 435.
[80] T. Carlberg, J. Electrochem. Soc. 133 (1986) 1940.
[81] H. Hirata and K. Hoshikawa, J. Cryst. Growth 106 (1990) 657.
[82] S.C. Hardy, J. Cryst. Growth. 69 (1984) 456.
[83] B.J. Keene, Surface Interface Anal. 10 (1987) 367.
[84] R.W. Series, J. Cryst. Growth, 97 (1989) 92.
[85] H. Hirata and K. Hoshikawa, J. Cryst. Growth 96 (1989) 747.
[86] T. Nozaki, Y. Yatsurugi, N. Akiyama, Y. Endo and Y. Makide, J. Radioanal. Chem. 19 (1974) 109.
[87] R.W. Series and K.G. Barraclough, J. Cryst. Growth 63 (1983) 219.
[88] W.G. Pfann, Zone Melting (Wiley, New York, 1966) p. 254.
[89] Shaped Crystal Growth. Proc. 1st Int. Conf. (Budapest 1986); J. Cryst. Growth 82 (1/2) (1987).
[90] Shaped Crystal Growth. Proc. 2nd Int. Conf. (Budapest 1989); J. Cryst. Growth 104(1) (1990).
[91] A. Katsui, S Thono. Y. Homma and T. Tanaka, J. Cryst. Growth 74 (1986) 221.
[92] R. Coquille, Y. Toudic, L. Haji, M. Gauneau, G. Moisan and D. Lecrosnier, J. Cryst. Growth 83 (1987) 167.
[93] A.T. Hunter, H. Kimura, J.P. Baukus, H.V. Winston and O. J. Marsh, Appl. Phys. Lett. 44 (1984) 74.
[94] R. Gremmelmaier, in: Compound Semiconductors, Vol. 1. Preparation of III–V Compounds (Reinhold, New York, 1962) ch. 29 p. 254.
[95] A.G. Fischer, J. Electrochem. Soc. 117 (1970) 410.
[96] J.B. Mullin, W.R. MacEwan, C.H. Holliday and A.E.V. Webb, J. Cryst. Growth 13 & 14 (1972) 640.
[97] J.B. Mullin, private communication.
[98] A. Katsui and C. Uemura, Jpn. J. Appl. Phys. 19 (1980) L318.
[99] K.F. Hulme and J.B. Mullin, Phil. Mag. 4 (1959) 1286.
[100] J.B. Mullin and K.F. Hulme, J. Electron. Control 4 (1958) 170.
[101] P.J. Holmes, J. Phys. Chem. Solids 24 (1963) 1239.
[102] A. Trainor and B.E. Bartlett, Solid State Electron. 2 (1961) 106.
[103] D.C. Miller, J. Cryst. Growth 46 (1979) 31.
[104] K.F. Hulme and J.B. Mullin, Solid State Electron. 5 (1962) 211.
[105] C.T. Lin, E. Schonherr and H. Bender, J. Cryst. Growth 104 (1990) 653.

[106] W.P. Allred, in: Compound Semiconductors, Vol. 1. Preparation of III–V compounds, eds. R.K. Willardson and H.L. Goering (Reinhold, New York, 1962) p. 187.
[107] W.A. Sunder, R.L. Barns, T.Y. Kometani, J.M. Parsey, Jr. and R.A. Laudise, J. Cryst. Growth 78 (1986) 9.
[108] R.N. Thomas, H.M. Hopgood, G.W. Eldridge, D.L. Barrett and T.T. Braggins, Solid State Electron. 24 (1981) 387.
[109] R.D. Fairman, R.T. Chen and D.R. Ch'en, IEEE Trans. Electron Devices ED-28 (1981) 135.
[110] E.M. Swiggard, S.H. Lee and F.N. von Batchelder, Inst. Phys. Conf. Ser. No. 33b (IOP, Bristol, 1977) p. 23.
[111] K. Terashima, H. Nakajima and T. Fukuda, Jpn J. Appl. Phys. 21 (1982) L452.
[112] T. Inada, T. Sato, K. Ishida, T. Fukuda and S. Takahashi, J. Electron. Mater. 15 (1986) 169.
[113] H. Emouri, T. Kikuta, T. Inada, T. Obokata and T. Fukuda, Jpn. J. Appl. Phys. 24 (1985) L291.
[114] T. Kikuta, H. Emori, T. Fukuda and K. Ishida, J. Cryst. Growth 76 (1986) 517.
[115] V.T. Bublik, V.V. Karataev, M.G. Mil'vidskii, G.A. Nemtsova, L.N. Perova, O.G. Stolyarov and T.G. Yugova, Sov. Phys. Crystallogr, 24 (1979) 302.
[116] V.T. Bublik, V.V. Karataev, M.G. Mil'vidskii, G.A. Nemtsova, L.N. Perova, O.G. Stolyarov and T.G. Yugova, Sov. Phys. Crystallogr. 24 (1979) 621.
[117] N.A. Anastas'eva, V.T. Bublik, V.V. Karataev, M.G. Mil'vidskii, O.G. Stolyarov and T.G. Yugova, Sov. Phys. Crystallogr . 27 (1982) 702.
[118] N.A. Anastas'eva, V.T. Bublik, V.V. Karataev, M.G. Mil'vidskii, L.D. Talis and T.G. Yugova, Sov. Phys. Crystallogr. 34 (1989) 912.
[119] C.M. Ringel, J. Electrochem. Soc. 118 (1971) 609.
[120] T. Fukuda, S. Washizuka, Y. Kokubun, J. Ushizawa and M. Watanabe, Int. Symp. on GaAs and Related Materials (Japan 1981); Inst. Phys. Conf. Ser. No. 63, p. 43.
[121] W. Bardsley, G.W. Green, C.H. Holliday, D.T.J. Hurle, G.C. Joyce, W.R. MacEwan and P.J. Tufton, Inst. Phys. Conf. Ser. No. 24 (1975) p. 355.
[122] S.F. Nygren, J. Cryst. Growth 19 (1973) 21.
[123] P.J. Roeksnoer, J.M.P.L. Huijbregts, W.M. van de Wijgert and A.J.R. de Kock, J. Cryst. Growth 40 (1977) 6.
[124] A.J.R. de Kock, W.M. van de Wijgert, J.H.T. Hengst, P.J. Roeksnoer and J.M.P.L. Huybregts, J. Cryst. Growth 41 (1977) 13.
[125] A.R. von Neida, L.J. Oster and J.W. Nielsen, J. Cryst. Growth 13&14 (1972) 647.
[126] C. Hilsum and H.D. Rees, Electron. Lett. 6 (1970) 277.
[127] J.B. Mullin, A. Royle and B.W. Straughan, Proc. 3rd Int. Symp. on Gallium Arsenide and Related Compounds (Aachen, 1970); Inst. Phys. Conf. Ser. No. 9 (IOP, Bristol) p. 41.
[128] K. Tada, M. Tatsumi, M. Morioka, T. Araki and T. Kawase, in: Semiconductors and Semi-metals, Vol. 31, eds. R.K. Willardson and A.C. Beer (Academic Press, New York, 1990) p. 175.
[129] O. Oda, K. Katagiri, K. Shinohara, S. Katsura, Y. Takahashi, K. Kainosho, K. Kohiro and R. Hirano, in: Semiconductors and Semi-metals, Vol. 31, eds. R.K. Willardson and A.C. Beer (Academic Press, New York, 1990) p. 93.
[130] J.P. Farges, J. Cryst. Growth 59 (1982) 665.
[131] D.J. Dowling, J.E. Wardill, R.A. Brunton, D.A.E. Crouch and J.P. Thompson, Proc. 3rd NATO Workshop on Materials Aspects of InP (Harcourt, USA, 1986).
[132] T. Inada and T. Fukuda, in: Semiconductors and Semi-metals, Vol. 31, eds. R.K. Willardson and A.C. Beer (Academic Press, New York, 1990) p. 71.
[133] O. Oda, K. Kainosho, K. Kohiro, H. Shimakura, T. Fukui, R. Hirano and S. Katsura, 6th Conf. on Semi-insulating III–V Materials (Toronto, 1990), eds. A.G. Milnes and C.J. Miner (Hilger, Bristol, 1990) p. 189.
[134] D. Rumsby, R.M. Ware and M. Whittaker, Semi-insulating III–V Materials (Nottingham, 1980), ed. G.J. Rees (Shiva Publishing, Orpington, 1980) p. 59.
[135] D.T.J. Hurle, in: Sir Charles Frank: 80th Birthday Tribute, eds. R.G. Chambers, J.E. Enderby, A. Keller, A.R. Lang and J.W. Steeds (Hilger, Bristol, 1991) p. 188.
[136] M. Shibata, Y. Sasaki, T. Inada and S. Kuma, J. Cryst. Growth 102 (1990) 557.

[137] W.A. Bonner, J. Cryst. Growth 54 (1981) 21.
[138] G. Jacob, M. Duseaux, J.P. Farges, M.M.B. van den Boom and P.J. Roeksnoer, J. Cryst. Growth 61 (1983) 417.
[139] S. McGuigan, R.N. Thomas, D.L. Barrett, G.W. Eldridge, R.L. Messham and B.W. Swanson, J. Cryst. Growth 76 (1986) 217.
[140] W.A. Bonner, R.E. Nahory, H.L. Gilchrist and E. Berry, Semi-insulating III–V Materials (Toronto, 1990), eds. A.G. Milnes and C.J. Miner (Hilger, Bristol, 1990) p. 199.
[141] B. Joukoff and A.M. Jean-Louis, J. Cryst. Growth 12 (1972) 169.
[142] A.J. Strauss, Proc. Int. Symp. on Cadmium Telluride (Strasbourg, 1971), eds. P. Siffert and A. Cornet (Centre de Reserches Nucleaires, 1971) p. 1.
[143] J.B. Mullin and B.W. Straughan, Rev. Phys. Appl. 12 (1977) 105.
[144] N. Klausutis, J.A. Adamski, C.V. Collins, M. Hunt, H. Lipson and J.R. Weiner. J. Electron. Mater. 4 (1975) 625.
[145] A.W. Vere, S. Cole and D.J. Williams, J. Electron. Mater. 12 (1983) 551.
[146] S.E.R. Hiscocks and P.D. West, J. Mater. Sci. 3 (1968) 76.
[147] J.W. Wagner and R.K. Willardson, Trans Met. Soc. AIME 242 (1968) 366.
[148] J.R. Burke, J.D. Jensen and B. Houston, J. Electrochem. Soc. 120 (1973) 431.
[149] T. Kusunoki, C. Takenaka and K. Nakajima, J. Cryst. Growth 112 (1991) 33.
[150] K. Nakajima, T. Kusunoki and C. Takenaka, J. Cryst. Growth 113 (1991) 485.
[151] J.S. Cook and P.B. Gwan, J. Cryst. Growth 33 (1976) 369.
[152] J. Bleay, R.M. Hooper, R.S. Narang and J.N. Sherwood, J. Cryst. Growth 43 (1978) 589.
[153] K. Katoh and N. Kato, J. Cryst. Growth 73 (1985) 203.
[154] N. Karl, J. Cryst. Growth 99 (1990) 1009.
[155] T.G. Digges, R. H. Hopkins and R.G. Seidensticker, J. Cryst. Growth 29 (1975) 326.
[156] H. Föll, U. Gösele and B. O. Kolbesen, J. Cryst. Growth 40 (1977) 90.
[157] D.T.J. Hurle, O. Jones and J.B. Mullin, Solid State Electron. 3 (1961) 317.
[158] Semiconductors and Semi-metals, Vol. 20, eds. R.K. Willardson and A.C. Beer (Academic Press, New York, 1984) p. 88.
[159] J.P. Tower, R. Tobin, P.J. Pearah and R.M. Ware, J. Cryst. Growth 114 (1991) 665.
[160] J. Kvapil, M. Koselja, J. Kvapil, B. Perner, V. Skoda, J. Kubelka, K. Hanal and V. Kubacek, Czech. J. Phys. B 38 (1988) 237.
[161] P.A.C. Whiffin and J. C. Brice, J. Cryst. Growth 10 (1971) 91.
[162] T.F. Ciszek and G.H. Schwuttke, J. Electrochem. Soc. 121 (1974) 207C.
[163] A.S. Taylor, B.H. Mackintosh, L. Eriss and F.V. Wald, J. Cryst. Growth 82 (1987) 134.
[164] C. Belouet, J. Cryst. Growth 82 (1987) 110.
[165] R.G. Seidensticker, J. Cryst. Growth 39 (1977) 17.
[166] R.S. Wagner, Acta Met. 8 (1960) 57.
[167] T. Surek, S.R. Coriell and B. Chalmers, J. Cryst. Growth 50 (1980) 21.
[168] K.D. Stock, Surf. Sci. 91 (1980) 655.
[169] G.A. Satunkin, B.S. Red'kin, V.N. Kurlov, S.N. Rossolenko, V.A. Tatarchenko and Yu.A. Tuflin, Cryst. Res. Technol. 21 (1986) 995.
[170] P.I. Antonov, V.M. Krymov and E. Hartmann, J. Cryst. Growth 104 (1990) 142.
[171] A.B. Dreeben, K.M. Kim and A. Schujko, J. Cryst. Growth 50 (1980) 126.
[172] G. Müller and H. Jacob, in: Numerical Data and Functional Relationships in Science and Technology, Vol. 17, eds. Landolt and Bornstein, p. 12.
[173] M. Tmar, A. Gabriel, C. Chatillon and I. Ansara, J. Cryst. Growth 69 (1984) 421.
[174] K.J. Bachmann and E. Buehler, J. Electrochem. Soc. 121 (1974) 835.
[175] J.R. Arthur, J. Phys. Chem. Solids 28 (1967) 2257.
[176] H. Wenzl, A. Dahlen, A. Fattah, S. Petersen, K. Mika and D. Henkel, J. Cryst. Growth 109 (1991) 191.
[177] S.F. Nygren, C.M. Ringel and H.W. Verleur, J. Electrochem. Soc. 118 (1971) 306.
[178] T.H. Maiman, Nature 187 (1960) 43.
[179] L.F. Johnson and K. Nassau, Proc. IRE 49 (1961) 1704.
[180] L.F. Johnson, G.D. Boyd, K. Nassau and R.R. Soden, Phys. Rev. 126 (1962) 1406.

[181] Handbook of Lasers, ed. R.J. Pressley (Chemical Rubber, 1971) p. 372.
[182] K. Nassau, in: Applied Solid State Science, Vol. 2, ed. R. Wolfe (Academic Press, New York, 1971) p. 174.
[183] B. Cockayne, J. Cryst. Growth 42 (1977) 413.
[184] K. Nassau and A.M. Broyer, J. Appl. Phys. 33 (1962) 3064.
[185] B. Cockayne, D.S. Robertson and W. Bardsley, Brit. J. Appl. Phys. 15 (1964) 1165.
[186] W. Bardsley and B. Cockayne, in: Crystal Growth Suppl. to J. Phys. Chem. Solids Proc. ICCG-1, ed. H.S. Peiser (Pergamon, Oxford, 1967) p. 109.
[187] B. Cockayne, J. Am. Ceram. Soc. 49 (1966) 204.
[188] B. Cockayne, Platinum Met. Rev. 18 (1974) 86.
[189] D.T.J. Hurle and B. Cockayne, in: Characterisation of Crystal Growth Defects by X-ray Methods eds. B.K. Tanner and D.K. Bowen (Plenum, New York, 1980) p. 46.
[190] B. Cockayne, J.M. Roslington and A.W. Vere, J. Mater. Sci. 8 (1973) 382.
[191] W.T. Stacy, J. Cryst. Growth 24&25 (1974) 137.
[192] J.R. Carruthers and A.F. Witt, in: Crystal Growth and Characterisation, eds. R. Ueda and J.B. Mullin (North-Holland, Amsterdam, 1975) p. 142
[193] B. Cockayne and M.P. Gates, J. Mater. Sci. 2 (1967) 118.
[194] B. Cockayne, M. Chesswas, J.G. Plant and A.W. Vere, J. Mater. Sci. 4 (1969) 565.
[195] B. Cockayne, J.G. Plant and A.W. Vere, J. Cryst. Growth 15 (1972) 167.
[196] J.C. Brice, O.F. Hill and P.A.C. Whiffin, J. Cryst. Growth 6 (1970) 297.
[197] B. Cockayne, J. Cryst. Growth 3 & 4 (1968) 60.
[198] C.D. Brandle, D.C. Miller and J.W. Nielsen, J. Cryst. Growth 12 (1972) 195.
[199] B. Cockayne, B. Lent, J.S. Abell and P.M. Marquis, J. Mater. Sci. 10 (1975) 1874.
[200] H. Fay and C.D. Brandle, in: Crystal Growth, Proc. ICCG-1, ed. H.S. Peiser (Pergamon, Oxford, 1967) p. 51.
[201] R.F. Belt, J.R. Latore and R. Uhrin, Appl. Phys. Lett. 25 (1974) 218.
[202] B. Cockayne, in: Crystal Growth and Materials, eds. E. Kaldis and H.J. Scheel (North-Holland, Amsterdam, 1977) p. 106.
[203] J.S. Abell, I.R. Harris, B. Cockayne and B. Lent, J. Mater. Sci. 9 (1974) 527.
[204] A.W. Vere, J.G. Plant, B. Cockayne, K.G. Barraclough and I.R. Harris, J. Mater. Sci. 4 (1969) 1075.
[205] W.T. Stacy, J.A. Pistorius and M.M. Janssen, J. Cryst. Growth 22 (1974) 37.
[206] K. Takagi, T. Fukazawa and M. Ishii, J. Cryst. Growth 36 (1976) 185.
[207] B. Cockayne, M. Chesswas and D.B. Gasson, J. Mater. Sci. 2 (1967) 7.
[208] W. Bardsley, B. Cockayne, G.W. Green and D.T.J. Hurle, Solid State Electron. 6 (1963) 389.
[209] C.D. Brandle, in: Crystal Growth. A Tutorial Approach, eds. W. Bardsley, D.T.J. Hurle and J.B. Mullin (North-Holland, Amsterdam, 1977) p. 189.
[210] B. Cockayne and M. Chesswas, J. Mater. Sci. 2 (1967) 498.
[211] B. Cockayne, M. Chesswas, P.J. Born and J.D. Filby, J. Mater. Sci. 4 (1969) 236.
[212] V.F.S. Yip and C.D. Brandle, J. Am. Ceram. Soc. 61 (1978) 8.
[213] S. Prezoisi, R.R. Soden and L.G. Van Uitert, J. Appl. Phys. 33 (1962) 1893.
[214] R.C. Linares, Solid State Commun. 2 (1964) 229.
[215] D. Mateika in: Advanced Crystal Growth, eds. P.M. Dryburgh, B. Cockayne and K.G. Barraclough (Prentice Hall, Englewood Cliffs, NY, 1981) p. 149.
[216] C.D. Brandle, in: Crystal Growth of Electronic Materials, ed. E. Kaldis (North-Holland, Amsterdam, 1985) p. 101.
[217] D. Mateika and Ch. Rusche, J. Cryst. Growth 42 (1977) 440.
[218] R.F. Belt, R.C. Puttbach and D.A. Lepore, J. Cryst. Growth 13 & 14 (1972) 268.
[219] D. Mateika, R. Laurien and M. Liehr, J. Cryst. Growth 65 (1983) 237.
[220] J.J. Rubin, L.G. van Uitert and H.L. Levinstein, J. Cryst. Growth 1 (1967) 315.
[221] J.G. Bergman, A. Ashkin, A.A. Ballman, J.M. Dziedzic, H.J. Levinstein and R.G. Smith, Appl. Phys. Lett. 12 (1968) 92.
[222] H. Fay, W.J. Alford and H.M. Dess, Appl. Phys. Lett. 12 (1968) 89.
[223] P. Lerner, C. Legras and P. Dumas, J. Cryst. Growth 3 & 4 (1968) 231.
[224] G.E. Peterson and J.R. Carruthers, J. Solid State Chem. 1 (1969) 98.

[225] P.M. Bridenbaugh, J. Cryst. Growth 19 (1973) 45.
[226] Y. Furukawa, M. Sato, F. Nitanda and K. Ito, J. Cryst. Growth 99 (1990) 832.
[227] C.D. Brandle and D.C. Miller, J. Cryst. Growth 24 & 25 (1974) 432.
[228] W. Wang, Q. Zou, Z. Geng and D. Feng, J. Cryst. Growth 79 (1986) 706.
[229] F. Shimura and Y. Fujino, J. Cryst. Growth 38 (1977) 293.
[230] T. Fukuda and H. Hirano, J. Cryst. Growth 35 (1976) 127.
[231] G.W. Berkstresser, A.J. Valentino and C.D. Brandle, J. Cryst. Growth 109 (1990) 457, 467.
[232] H.E. LaBelle and A.I. Mlavsky, Mater. Res. Bull 6 (1972) 184.
[233] R.S. Feigelson, J. Cryst. Growth 79 (1986) 669.
[234] H.E. LaBelle and A.I. Mlavsky, Nature 216 (1967) 574.
[235] M.M. Fejer, J.L. Nightingale, G.A. Magel and R.L. Byer, Rev. Sci. Instrum. 55 (1984) 1791.
[236] M.M. Fejer, Ph.D. Thesis (Stanford University, 1986).
[237] D. Elwell, W.L. Kway and R.S. Feigelson, J. Cryst. Growth 71 (1985) 237.
[238] H. Guggenheim, J. Appl. Phys. 32 (1961) 1337.
[239] K. Nassau, J. Appl. Phys. 32 (1961) 1820
[240] E.P. Chicklis, C.S. Naiman, R.C. Folweiler, D.R. Gabbe, H.P. Jenssen and A. Linz, Appl. Phys. Lett. 19 (1971) 119.
[241] B. Cockayne, J.G. Plant and R.A. Clay, J. Cryst. Growth 54 (1981) 407.
[242] R.C. Pastor, M. Robinson and W.M. Akutagawa, Mater. Res. Bull. 10 (1975) 501.
[243] R.F. Belt and R. Uhrin, J. Cryst. Growth 109 (1991) 340.
[244] V.I. Goriletsky, V.A. Nemenov, V.G. Protsenko, A.V. Radkevich and L.G. Eidelman, J. Cryst. Growth 52 (1981) 509.
[245] D.T.J. Hurle, Crystal Pulling from the Melt (Springer, Berlin, 1993).

CHAPTER 4

Crystal Growth by Floating Zone Melting*

JOACHIM BOHM, ANKE LÜDGE and WINFRIED SCHRÖDER

Institute of Crystal Growth, Berlin, Germany

Contents

1. Fundamentals of the floating zone technique 215
   1.1. Basic concepts 215
   1.2. Shape and stability of the molten zone 217
   1.3. Thermal balance and process control 221
   1.4. Fluid flow and the Marangoni effect 223
   1.5. Radio frequency heating 225
2. Crucible-free floating zone melting of silicon 228
   2.1. Technological concepts 228
   2.2. Process control 232
3. Floating zone melting of other materials 239
   3.1. Floating zone melting by electrical heating 239
   3.2. Floating zone melting by optical or thermal radiation 242
   3.3. Floating zone melting by electron bombardment 245
References 250

*Contribution based on an according chapter by the present authors in the (German) textbook of K.-Th. Wilke and J. Bohm, Kristallzüchtung, Deutscher Verlag der Wissenschaften, Berlin, 1988.

HANDBOOK OF CRYSTAL GROWTH, VOL. 2
edited by D.T.J. Hurle

1. Fundamentals of the floating zone technique

1.1. Basic concepts

Crucible-free zone melting was introduced (after a first representation in a patent of Theuerer [1]) by Keck and Golay [2] as well as by Emeis [3] for preparing high-purity silicon for semiconductor devices, to circumvent the problems resulting from the use of crucibles, especially to avoid impurities from the crucible material [4–8]. In the floating zone technique, the material is in the form of a free-standing rod clamped only at its ends, in which a relatively small zone will be melted by suitable heating equipment (fig. 1). The melt is suspended like a drop between the two parts of the rod, leading to the designation floating zone or float zone melting. This molten zone is moved through the rod over its whole length by the motion of heater or rod. A single crystal can be generated by spontaneous nucleation or using a single-crystalline seed crystal as the initial part of the rod that is kept unmolten. Mostly, the zone passage is carried out upwards, because in that case the process has a higher stability. Usually, the growing crystal and sometimes also the melting rod is rotated, mostly with different rotation rates or counterrotation.

The floating zone technique in its principle corresponds to the general zone melting set-up first invented by Kapitza [9] to grow bismuth crystals in a glass tube and afterwards developed by Pfann [10–12] for the purification of germanium. Zone melting leads to a characteristic redistribution (segregation) of impurities or components according to their distribution coefficients. Repeated zone melting can produce a substantial purifying effect, in particular for distribution coefficients $k \ll 1$ (cf. [13] or

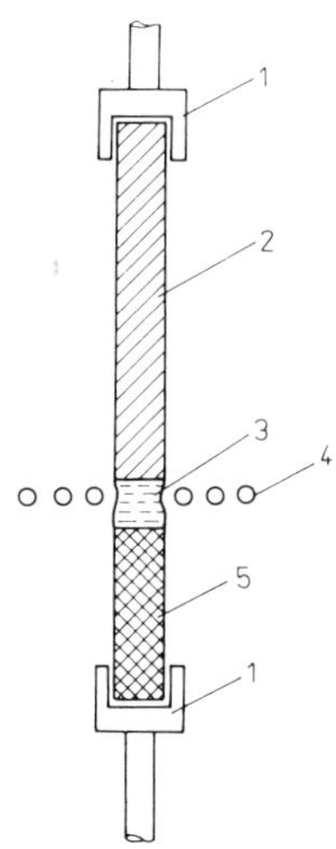

Fig. 1. Scheme of crucible-free zone melting. 1: rod holder; 2: melting (polycrystalline) rod; 3: molten zone; 4: radio frequency induction coil; 5: growing (monocrystalline) rod. (From the work of Wilke and Bohm [15].)

the according textbooks, e.g. [14,15]). Besides, this redistribution of impurities, crucible-free zone melting causes an additional purifying effect by evaporation of impurities. This can be very effective for high-melting materials, because the free surface of the molten zone is relatively large. While the segregation (for $k < 1$) is only a transportation of impurities to the rod end, evaporation means their real reduction. Certainly, the rod can be again recontaminated during the process by back-sublimation. Due to the relatively large free surface and the intensive mixing in the molten zone, the floating zone technique offers favourable conditions for doping from the gas phase [16–19].

Besides, the preparation of high-purity semiconductor silicon, floating zone melting was successfully used for the growth of materials with high or very high melting points for which no crucible material was available. The corresponding process technologies are characterized by the different heating methods to achieve melting. Most commonly used is the direct or indirect heating by radio frequency induction currents (the latter with a ring-shaped graphite or metal susceptor for converting radio frequency energy to thermal radiation). Moreover, floating zone melting can also be performed using electron bombardment (high vacuum provided), electric arc or plasma (both methods preferred for metals) as well as thermal or light radiation. Furthermore, worth mentioning is a melting technique using an electrically heated metal net that intersperses the molten zone and moves with it, of course in close contact with the melt. These various techniques are dealt with in more detail in section 3 of this chapter.

A first prerequisite to performing floating zone melting is a starting material having already the shape of a sufficiently stable and well-shaped rod. The desired rods can be obtained by casting or melting the material in moulds or by pulling from the melt using the Czochralski technique. Using the latter technique, one is faced with problems resulting from the crucible. In this respect, casting or electrical melting in a cooled crucible (skull melting) is favoured. Frequently, rods are pressed from powder or preserved by binding agents being annealed or sintered if necessary. The following method is successfully used for the production of semiconductor silicon as well as for high-melting metals. The material is chemically deposited from the gas phase (e.g. by thermal decomposition of $SiCl_4$) on a thin rod or wire of the same material producing a rod of the required size for the floating zone technique. Finally, rods were prepared for floating zone melting by the Verneuil technique and by melting or welding powder or pieces on a cold hearth.

Often a first floating zone passage is used only to melt down the material. In many cases this first zone passage is the most difficult one. Sometimes during the first passage it is helpful to leave a solid core in the centre of the rod to give a sufficient stability to the assembly. Frequently, it happens that the diameter of the crystallizing rod varies because of some irregularities. Such variations of the rod diameter can be diminished by corresponding variations of the feed rate of the upper rod to the molten zone. A general change in the rod diameter can be realized by different speeds of the upper and lower rod in relation to the molten zone ("squeezing" or "stretching" a rod). Likewise, a lower density of the rod resulting from its preparation can be compensated for by a somewhat higher feed rate. The following simple relation holds between the radii r_F of the melting rod and r_K of the crystallizing rod, their densities ρ_F and ρ_K and the related

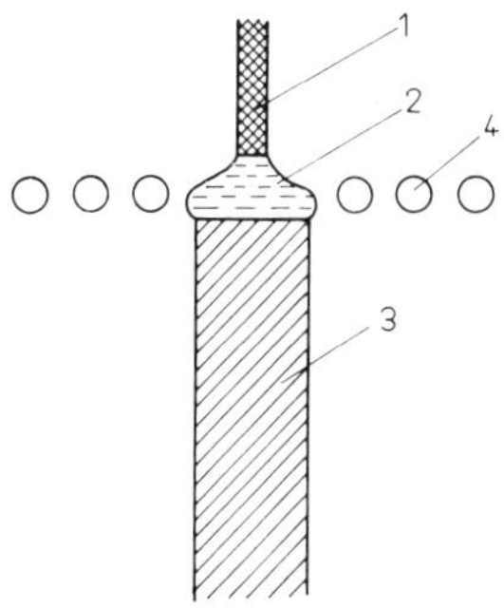

Fig 2. Pedestal technique (from the work of Dash [21]). 1: crystal; 2: melting top; 3: starting material; 4: RF coil. (From the work of Wilke and Bohm [15].)

speeds v_F and v_K:

$$v_F/v_K = (r_K/r_F)^2 \rho_K/\rho_F. \tag{1}$$

A special variant of crucible-free zone melting is the pedestal technique introduced by Dash [20, 21] as well as by Poplawsky and Thomas [22] (fig. 2): The top of a thicker (polycrystalline) rod is melted and from this melted top a (thinner) single crystal is pulled upwards using a seed crystal. The pedestal technique closely resembles the set-up of pulling crystals from the melt by the Czochralski technique.

1.2. Shape and stability of the molten zone

The shape and stability of the molten zone play an important role in floating zone melting. The melt is hanging free like a drop between the two parts of the rod and held against gravity essentially by the surface tension. In general, a molten zone is more stable the higher its surface tension is and the lower its specific weight. Besides, additional forces can act on the melt whose influence will be discussed later.

Several authors have studied the dependence of shape and stability of a molten zone on the surface tension [4, 6, 23–40]. The papers of Coriell and co-authors include an extensive bibliography concerning the mathematical aspccts of the subject.

In principle, the length up to which a molten zone can be stable is limited: A liquid cylinder held only by its surface tension between two rods will be unstable if its length exceeds its circumference. In the latter case the surface of the liquid cylinder can be reduced by a small change of its shape (perturbation) and in the further development of the perturbation the liquid cylinder breaks down into droplets. This stability limit for a molten zone is valid if the weight of the melt can be neglected, implying very small diameter and gravity-free conditions.

Under the influence of gravity the molten zone develops a meniscus-like shape (fig 3). The description of the meniscus results from the Laplace (or Laplace–Young) equation that gives the relation between the hydrostatic pressure in the melt and the capillary pressure caused by its surface tension [33, 34]:

$$\gamma_L(1/R_1 + 1/R_2) = -\rho_L gh + P. \tag{2}$$

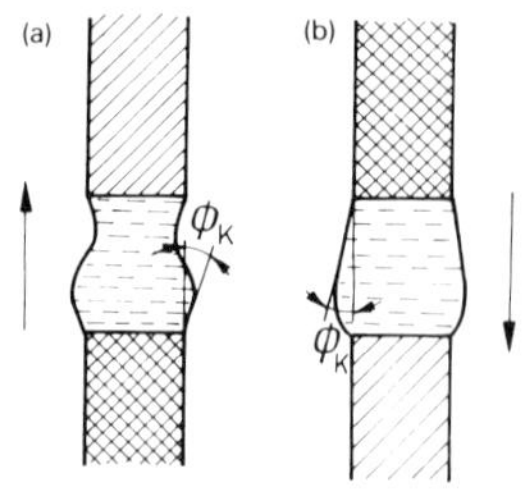

Fig. 3. Meniscus shapes of a molten floating zone. (a) Zone passage upwards. (b) Zone passage downwards; ϕ_K meniscus angle at the crystallizing rod. (From the work of Wilke and Bohm [15].)

In this equation, γ_L is the surface tension of the melt, R_1 and R_2 are the principal radii of curvature along the meniscus surface, ρ_L is the density of the melt, g the gravitational acceleration, and h the height of the melt; P is a constant not further defined at first and being determined, e.g. by the meniscus volume V_M. This equation is strictly true for stationary flow-free liquid surfaces with a constant position-independent surface tension and for some special flow patterns, e.g. a uniform rotating melt. These assumptions, however, are not generally fulfilled in floating zone melting, so some reservations have to be made concerning the applicability of the Laplace–Young equation. Coriell et al. [33, 34] and Surek and Coriell [35] gave a numerical solution of this equation for a rotationally symmetric liquid meniscus under boundary conditions applicable to the floating zone technique. They have shown that the shape of the zone can be completely described by four dimensionless variables: r_F/r_K; l/r_K; $V_M/\pi r_K^2$ and $\varepsilon = r_K^2 \rho_L g/\gamma_L$ (r_F and r_K are the radii of the melting and the crystallizing rods, l the zone length, V_M the meniscus volume, g the gravitational acceleration, ρ_L density and γ_L surface tension of the melt). The first three of these variables are geometric ones, while the variable ε (the so-called Bond number) contains the material-specific quantities ρ_L and γ_L. These four variables determine, among others, the meniscus angle ϕ_K (to the perpendicular line) which represents an important quantity for the floating zone process. To get a constant diameter of the crystallizing rod, a certain, well-defined material-specific value $\phi_K = \phi_K^0$ has to be maintained.

Depending on whether the crystallization, and therefore the meniscus angle ϕ_K^0, is realized at the lower or upper rod (that means depending on whether the zone moves upwards or downwards), different meniscus shapes and meniscus volumes are obtained (fig. 3).

For most materials, the meniscus angle required to keep a constant diameter is $\phi_K^0 = 0$. But according to Bardsley et al. [41] as well as to Surek and Chalmers [42–44] one finds for silicon $\phi_K^0 = 11°$ and for germanium $\phi_K^0 = 13°$ (Wenzl et al. [45] found $\phi_K^0 = 7° \pm 3°$ for Ge). For Al_2O_3, Dreeben et al. [46] have found $\phi_K^0 = 17° \pm 4°$ at an (atomically) rough phase boundary and $\phi_K^0 = 35° \pm 4°$ at an (atomically flat) facet.

As already mentioned, a molten zone can be stable only up to a certain limited length. Table 1 shows for a few selected values of the variable l/r_K, the maximum values ε^* of the variable $\varepsilon = r_K^2 \rho_L g/\gamma_L$ up to which, under given conditions ($r_K = r_F$; $\phi_K = 0$, a solution exists representing a meniscus stable against small perturbations. (Negative

Table 1
Stability of a liquid zone by Coriell and Cordes [33] (rod radii $r_K = r_F$; meniscus angle $\phi_K = 0$; further information the text).

l/r_K	ε	ε_{90}	λ^*	ρ^*
6.0	4.6×10^{-5}		0.041	0.007
4.0	5.24×10^{-2}		0.916	0.229
2.0	1.384		2.35	1.175
1.0	7.48	7.223	2.74	2.74
0.5	30.2	28.98	2.75	5.50
0.25	120.2	115.3	2.74	10.96
−6.0	0.1103	0.1046	1.99	0.332
−4.0	0.307	0.307	2.26	0.565
−2.0	1.605	1.543	2.53	1.265
−1.0	7.00	6.728	2.65	2.65
−0.5	29.00	27.838	2.69	5.38
−0.25	117.7	112.88	2.71	10.84

values of the variable l/r_K mean set-ups with an upper crystallizing rod). Furthermore, the values ε_{90} of the variable ε are given for which the meniscus meets the rod horizontally (meniscus angle at the melting rod $\phi_F = 90°$). At places where no values are given in table 1, ε_{90} would be greater than ε^* and could not be reached. As long as $\varepsilon_{90} < \varepsilon < \varepsilon^*$ it follows that $\varepsilon_F > 90°$ which is only possible if the liquid/solid phase boundary proceeds concavely into the melting rod. From ε^*, the quantities $\rho^* = \sqrt{\varepsilon^*}$ and $\lambda^* = (l/r_K)\sqrt{\varepsilon^*}$ are derived (see table 1). With these quantities, the material-specific values of the rod radius r_K^* and the zone length l^* corresponding to ε^* can be determined:

$$r_K^* = \rho^* \sqrt{\gamma_L/\rho_L g}, \qquad l^* = \lambda^* \sqrt{\gamma_L/\rho_L g}. \tag{3}$$

For a given rod radius r_K^*, l^* is the maximum length of a stable zone (fig. 4): For small rod diameters, l^* grows nearly proportionally to the rod radius r_K^*. For larger diameters, l^* changes over to a nearly constant value that will be reached, depending on the material, for rod diameters of the order of 20 to 30 mm. It is worth mentioning that Heywang [24] has already derived a (constant) approximate value $\lambda^* = 2.84$ for larger diameters; Green [25] (who also considered radially unsymmetrical rods and molten zones) obtained a value $\lambda^* = 2.62$ and Brice [30] a value $\lambda^* = 3.18$. All these values do not remarkably differ from each other and correspond nearly to the values given in table 1 (for large ε^*). The values in this table are valid for a meniscus angle $\phi_K = 0$ which can be assumed for most materials.

If $\phi_K > 0$ is assumed, all values in table 1 should be somewhat higher. For example, a meniscus angle $\phi_K = 11°$, relevant for silicon, leads to 10% higher values for λ^* and l^*. For silicon with $\gamma = 0.72$ N/m; $\rho_L = 2.55$ g/cm^3 and $g = 9.81$ m/s^2, e.g., $\gamma_L/\rho_L g = 0.537$, the maximum length of a stable zone can be calculated to be $l^* = 1.6$ cm for large rod diameters. In practice, a maximum zone length of 2 to 3 cm has been obtained which is clearly greater than the calculated value. This already refers to other additional

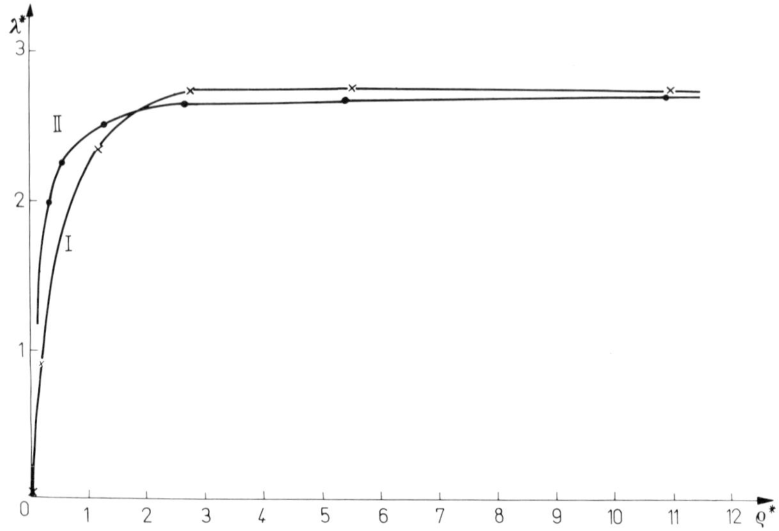

Fig. 4. Reduced maximum zone length $\lambda^* = l^* \sqrt{\rho_L g/\gamma_L}$ versus reduced rod radius $\rho^* = r_K^* \sqrt{\rho_L g/\gamma_L}$ according to table 1. (From the work of Wilke and Bohm [15].)

supporting influences, e.g., of electrodynamic forces in connection with the radio frequency heating. This will be dealt with later.

The maximum stable zone length of crystals with small diameters (0.25 to 1.5 mm) have been experimentally examined by Kim et al. [36].

If the molten zone is rotating, centrifugal forces arise that influence its shape and stability. They can be calculated taking into consideration an additional term to the Laplace–Young equation,

$$\gamma_L(1/R_1 + 1/R_2) = -\rho_L gh + \rho_L \omega^2 R_1^2/2 + P \tag{4}$$

(see above; ω is the angular rate of rotation).

According to Coriell and Cordes [33], the solution of the differential equation depends on a fifth variable $\varepsilon_R = \rho_L \omega^2 r_K^3/\gamma$ which is material-specific. The centrifugal forces cause an increase of the lower bulge of the melt meniscus and decrease its stability. Because the centrifugal forces are proportional to the radius, thicker crystal rods must be rotated slowly. This applies especially for the lower rod (for crystal diameters above 50 mm the rotation rate must be less than 10 min^{-1}).

Up to this point we have discussed the stability criterion of a molten zone in terms of whether a liquid meniscus held between two rods might be stable against some small perturbations of its shape. (This implies, to a certain extent, a statical point of view.) However, there are other (dynamic) stability problems of floating zone melting. For the success of the process, it is essential, at the crystallizing rod, to maintain a certain material-specific value ϕ_K^0 of the meniscus angle. Generally, perturbations of the zone shape (e.g. by changes of heating power) also cause changes in the meniscus angle ϕ_K. To keep the diameter of the growing crystal constant, the meniscus angle ϕ_K must return to its initial value ϕ_K^0. As Surek and Coriell [35] have shown, this is the case if both criteria $\partial \phi_K / \partial V_M > 0$ and $\partial \phi_K / \partial r_K < 0$ are simultaneously satisfied (V_M meniscus volume, r_K radius of the crystallizing rod). These conditions are fulfilled for most floating zone set-ups for both upward and downward zone motion. So, the process is dynamically stable and perturbations are self-compensating after approximately half a zone length. There are, however, exceptions. The dynamic stability, e.g. is not guaranteed for very thin rods.

Furthermore, the stability of a molten zone might be affected by the surrounding gas flow as well as the interfacial tension of the liquid/solid phase boundary and its shape. All these influences are hard to estimate. For stability it is usually favourable to have a concave shape of the lower crystallizing rod (directed to the melt). The fluid flow in the melt, though it is normally intense, only indirectly acts on the stability of the melt.

Another practically important source of instability are mechanical vibrations. They can cause considerable difficulties for the floating zone process, especially in the range of 1 to 50 Hz. Mostly, the origin is the mechanical system of the apparatus. In the case of resonance, small exciting vibrations could destabilize the zone. Those assemblies in which the seed crystal is a narrow neck carrying the growing crystal are particularly sensitive to mechanical vibrations. Therefore it is necessary to use special supporting devices for the growth of larger crystals.

1.3. Thermal balance and process control

The thermal balance of a floating zone set-up, the heater power necessary for melting, the temperature distribution in the melt and the corresponding shape of the phase boundaries have been investigated by Belk [47], Donald [48], Ratnikov et al. [49], Akiyama and Yamaguchi [50], Kuo and Wilcox [51], Arizumi and Kobayashi [52], Keller [53], Kuiken and Roksnoer [54], and Otani et al. [55]. Essential parameters are the thermal conductivities of melt and solid and the ratio of the heat dissipation via the rod ends to the heat radiation from their surface. In the literature cited, this problem is dealt with by simplified models, partly analytically, partly numerically. As an example, fig. 5 shows isotherms and heat flow lines in a cylindrical rod with a cylindrical molten zone, calculated from near-reality assumptions (without taking into consideration the heat transport by fluid flow).

Typical qualitative behaviour of floating zones is shown in fig. 6. The outer zone length l_a and the zone length l_z in the centre of the rod are plotted as a function of heater power N: Starting from a certain heater power N_a the melting of the zone begins from the outside. When reaching the heater power N_z, the zone is completely molten and the phase boundaries are convexly projecting (e.g. the zone itself is concave). With increasing heater power, l_z grows faster than l_a so that, meanwhile, a convex zone is obtained. For a certain moderate heater power N_f, the phase boundaries are nearly planar.

A more realistic calculation of the temperature distribution and heat balance, however, has to consider the real shape of the molten zone, which is sometimes far from a cylindrical one (see above). Moreover, such calculations are only valid for small

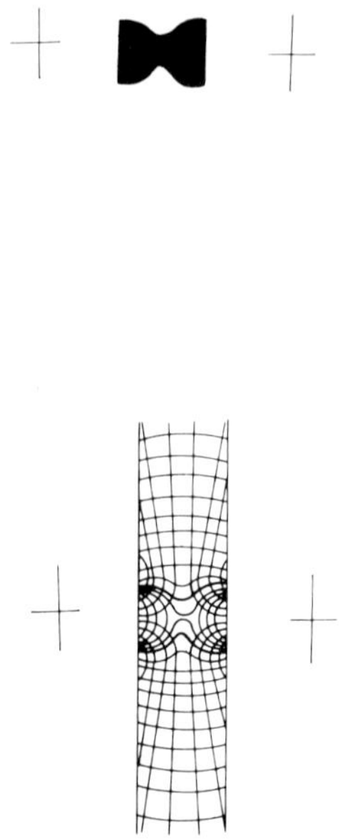

Fig. 5. Calculated isotherms and heat flow lines in a silicon rod heated by radio frequency in a floating zone set-up. (From the work of Wilke and Bohm [15] after Ratnikov et al. [49].)

Prandtl numbers ($\ll 1$) and a vanishing contribution to the heat transport by convection.

The temperature distribution and therefore the shape of the phase boundary can be affected by shape and size of the heater and by the installation of radiation shields and afterheaters. Kokorish [57] and Raab and Lebek [58] tried to get a planar phase boundary by applying a direct current using the Peltier effect.

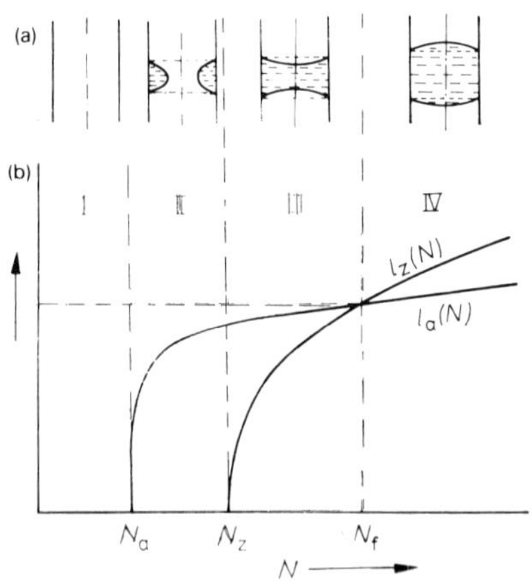

Fig. 6. Zone shapes versus heating power. l_a: outer zone length, l_z: zone length in the centre; I: no zone; II: zone with core; III concave zone; IV: convex zone. (From the work of Wilke and Bohm [15] after Kobayashi [56].)

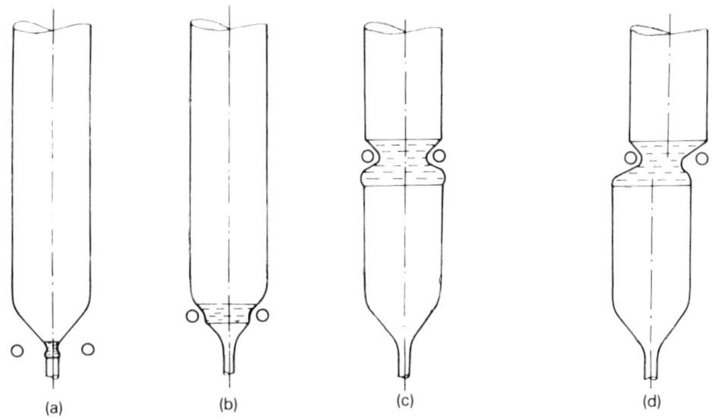

Fig. 7. Needle-eye technique. (a)–(c) Process phases. (d) Asymmetrical needle-eye technique. (From the work of Wilke and Bohm [15].)

Taking into account the limited length l^* of a stable molten zone as described in section 1.2, the rod radius of a molten zone is also limited: As already mentioned, heat is generally supplied to the molten zone from the outside, whereas the heat dissipation goes via the rods. Therefore, the solid/liquid phase boundary projects into the molten zone. In the case of rod diameter which are too large, the zone cannot be completely molten and a solid core remains in the centre between the rods.

In the case of upward zone motion, one has the advantage of having a waist-shaped melt shown in fig. 3a. This allows one to get larger rod diameters up to 150 mm. The diameter of the hole of the RF-inductor can be made smaller than the diameter of the rod (fig. 7). This technique is called the needle-eye technique. Using this technique, the influence of the electromagnetic supporting forces can be very effective. A related technique is the unsymmetrical needle-eye technique having the axes of the melting and crystallizing rods shifted (fig. 7d).

For floating zone melting, the diameter control of the growing crystal is generally not as difficult as it is for other techniques, e.g. the (dynamically unstable) Czochralski technique. This is due to the dynamic stability of the floating zone process mentioned before. There are various suggestions concerning diameter control in the patent literature. For example, Quenisset and Naslan [59] use, as a reference signal, the torque caused by inner friction in the viscous melt when the two rods are rotating with different rates, and which is dependent on the size of the molten zone. Other methods of diameter control are based on video observations of the molten zone with an (infrared-sensitive) video camera followed by electronic or computerized image processing.

1.4. Fluid flow and the Marangoni effect

In floating zone melting, generally rapid movements of the melt can be observed, indicating intense convection. This convection is driven by different sources. First of all, there are buoyancy forces, caused by large temperature differences and steep temperature gradients resulting in respective differences in melt density and leading to so-called thermal or natural convection. This convection will result in a torus-like circulation of the melt which ascends in the region of the hot surface, flows inwards at the upper liquid/solid phase boundary and descends in the centre of the molten zone.

The fluid flow in a floating zone decisively influences heat and mass transfer as well as the shape of the interfaces [29, 53, 56, 60–67]. In the case of rotation, the thermal convection is modified by centrifugal forces. Opportunities to manipulate the forced convection by a suitable choice of rotation rates of the rods, however, are limited because of stability reasons. Nevertheless, the rotation rate of the upper melting rod can be made so high that, for sufficiently large rod diameters, the thermal convection is effectively suppressed and the convection pattern gets partly inverted. Lan and Kou [66, 67] succeeded in obtaining nearly planar phase boundaries by counter-rotating the upper and lower rods or rotating the lower rod only. This allows the outer zone length to be reduced, especially in the case of large-diameter crystals and it results in an increased stability of the molten zone.

Another driving force for convection is caused by the Marangoni effect that arises at the melt surface [60, 68–78].

The surface tension of a hot melt is generally lower than that of a cold one. Thus, if there are differences in the temperature of a free melt surface, the melt tends to minimize its surface energy by extending the area of lower surface tension: The free surface is pulled from the hot regions to the colder ones like a rubber skin. Due to the spatial gradient $\partial\gamma/\partial y$ of the surface tension γ (coordinate y is tangential to the free surface), there is a tangential action force τ_y per unit area,

$$\tau_y = -\partial\gamma/\partial y, \qquad (5)$$

which drives the Marangoni convection like a shear stress. Because of the steep temperature gradients occurring in floating zone melting, Marangoni flow is expected to be considerable. For the floating zone arrangement, the highest surface temperature occurs in the centre of the zone. Accordingly, the surface melt is pulled from the middle belt of the zone to both the upper and the lower end, respectively. This should lead to a flow pattern consisting of two toroidal vortices, one in the upper half, the other in the lower half of the zone (fig. 8) – and not to a uniform torus-like vortex across the whole zone that would have resulted from natural (buoyancy-driven) convection. Especially in the lower half zone, the Marangoni flow is directed contrary to the natural convection. The flow pattern of fig. 8 has been realized in model experiments with a molten zone of sodium nitrate, $NaNO_3$, by Schwabe et al. [71], indicating the predominance of Marangoni flow compared to buoyancy-driven flow in this case.

The unique influence of Marangoni flow can be studied from experiments with zero or reduced gravity, in space experiments (cf., e.g. [79, 80]).

To describe Marangoni convection, one refers to the so-called dimensionless Marangoni number,

$$\mathrm{Ma} = (\partial\gamma/\partial T)\Delta T L/\eta\kappa, \qquad (6)$$

with the following parameters: $\partial\gamma/\partial T$ the temperature coefficient of surface tension, ΔT the temperature difference, L a characteristic length (half zone length has to be used), η the dynamical viscosity, and κ the thermal diffusivity. The Marangoni number gives a measure of the relation between heat transfer by Marangoni convection and transfer by

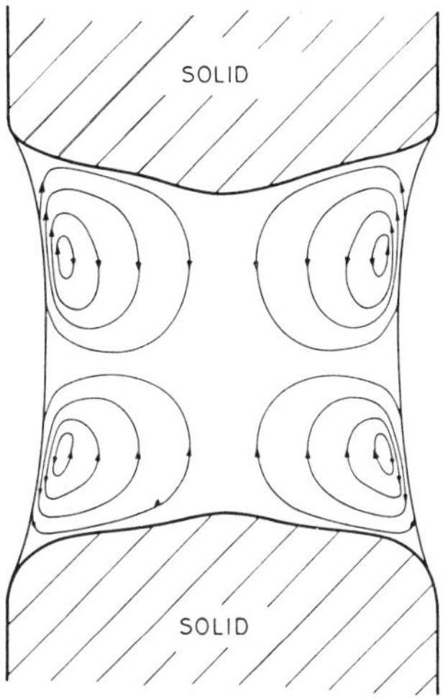

Fig. 8. Streamlines of Marangoni convection in a $NaNO_3$ floating zone. (From the work of Schwabe et al. [71].)

heat diffusion (heat conduction). The role of the Marangoni number is analogous to that of the (thermal) Rayleigh number for natural convection. If the Marangoni number exceeds some critical value with its order of magnitude in the range of 10^5, the laminar Marangoni flow becomes unstable, secondary convection cells appear and an oscillatory flow arises in the molten zone [80–82]. As is well known from other assessments, oscillatory convection causes corresponding temperature oscillations at the growing interface which result in an oscillatory growth velocity leading to fluctuations in impurity distribution, growth striae, etc. Generally, Marangoni convection will be important – excepting reduced-gravity experiments – for relatively small zones up to 10 mm in diameter. Under practical conditions and using large floating zone facilities, both the thermal (buoyancy) and rotational effects dominate the Marangoni effect by orders of magnitude [83].

1.5. Radio frequency heating

For a set-up with radio-frequency induction heating, electrodynamic forces play an important role [77, 83–86]. In the magnetic field generated by the inductor, the Lorentz force $\boldsymbol{f}_{el}$ acts on an induction current-carrying volume element of the melt,

$$\boldsymbol{f}_{el} = \boldsymbol{j} \times \mu \boldsymbol{H}. \tag{7}$$

This force is directed orthogonally to both the vectors of current density $\boldsymbol{j}$ and magnetic field strength $\boldsymbol{H}$. In melts, the permeability μ corresponds to the magnetic induction constant μ_0. Applying the Maxwell equation $\boldsymbol{j} = \text{rot}\,\boldsymbol{H} \equiv \nabla \times \boldsymbol{H}$, by means of vector analysis the following relation is obtained:

$$\boldsymbol{f}_{el} = -\mu \nabla \boldsymbol{H}^2/2 + \mu_0(\boldsymbol{H}\nabla)\boldsymbol{H}, \tag{8}$$

with ∇ the Nabla operator, $(\partial/\partial x, \partial/\partial y, \partial/\partial z)$. The first term of the electrodynamic force leads to an electrodynamically caused pressure $P_{el} = \mu H^2/2$ on the melt. The second term acts as a driving force for an electrodynamically caused convection. Because of the skin effect, the induction currents and therefore the electrodynamic forces arise only in a very thin surface layer of the melt with a thickness

$$\delta_{el} = \sqrt{\frac{1}{\pi \nu_{el} \sigma_L \mu_0}} \tag{9}$$

(ν_{el} frequency of induction current, σ_L conductivity of the melt). Using realistic values for silicon with $\nu_{el} = 2.4$ MHz and $\sigma_L = 12\,300\,\Omega^{-1}\,cm^{-1}$ a value $\delta_{el} = 0.292$ mm is obtained. Mühlbauer et al. [83], on the basis of a simplified Navier–Stokes equation, have performed numerical calculations of the electrodynamic pressure P_{el} and the electrodynamically caused flow function under the boundary condition (a) of a cylindrical zone (fig. 9), and (b) a realistic zone configuration (needle-eye technique, fig. 10). They used the following parameters, which are valid for floating zone melting of silicon: frequency $\nu_{el} = 2.4$ MHz applied voltage $U_{eff} = 600$ V, dynamic viscosity

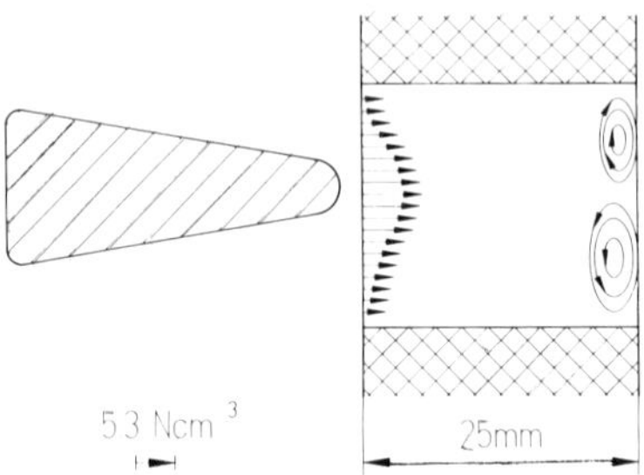

Fig. 9. Electrodynamical force (left-hand part of the picture) and electrodynamically caused flow pattern (right-hand part of the picture) in a symmetric molten zone. (From the work of Wilke and Bohm [15] after Mühlbauer et al. [83].)

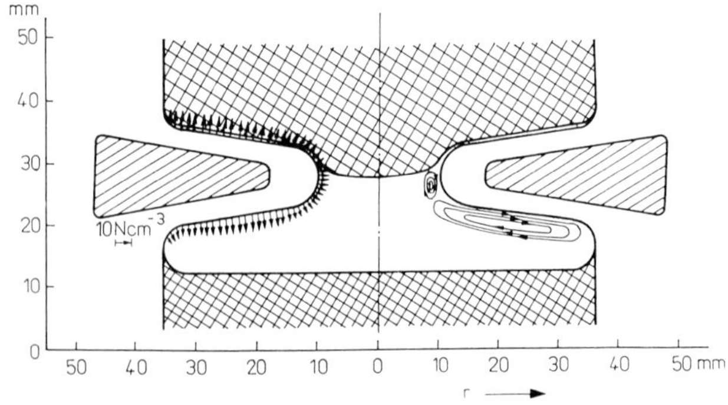

Fig. 10. Electrodynamical force (left-hand part of the picture) and electrodynamically caused flow pattern (right-hand part of the picture) using a needle-eye configuration. (From the work of Wilke and Bohm [15] after numerical results of Mühlbauer et al. [83].)

$\eta = 8.8 \times 10^{-8}$ N s/cm^2. The configuration of fig. 10 was empirically derived by directly observing and copying the zone shape from a floating zone experiment. For the two configurations, the results are quite different. For the practically relevant needle-eye configuration, the electrodynamic pressure P_{el} near to the upper rod is directed upwards, so that the melting silicon is pushed off and only a thin melting skin remains on the melting rod. Near the lower rod, the pressure is directed downwards and pushes down the molten zone resulting in an increased curvature of the melting surface at the liquid/solid boundary. Consequently, the capillary pressure is increased resulting in a stabilization of the molten zone. In the stationary case, the following equilibrium between the pressure components holds:

$$P_v + P_\rho = P_{el} + P_\gamma, \tag{10}$$

with the flow pressure $P_v = \rho_L v_L^2/2$ (ρ_L density; v_L flow rate), the gravitational pressure $P_\rho = \rho_L g h$ (g gravitational acceleration, h height of the melt), the electrodynamic pressure $P_{el} = \mu H^2/2$ and the capillary pressure $P_\gamma = \gamma_L(1/R_1 + 1/R_2)$ (γ_L surface tension, R_1 and R_2 principal radii of curvature). With knowledge of the remaining three components of the pressure, the flow pressure P_v can be calculated and, from this, the flow rate $v_L = \sqrt{2P_v/\rho_L}$ (fig. 11). According to this figure, the electrodynamic pressure clearly dominates. Under practical conditions the corresponding electrodynamical forces exceed those derived from temperature differences, rotation or surface tension by orders of magnitude [77, 83]. According to the numerical results of the latter authors, v reaches a maximum value of 100 cm/s at the surface region of the larger vortex shown in fig. 10, so that the authors assume a turbulent flow (the back flow inside has a speed of only 25 cm/s).

As the electrodynamically caused flow is limited to a near-surface region, it has no direct influence on the processes at the growth front. By this flow, however, impurities arriving at the melting surface will be brought inside the melt very rapidly. Chun [87] gives an explicit formulation of all forces acting on a molten zone. He also considers

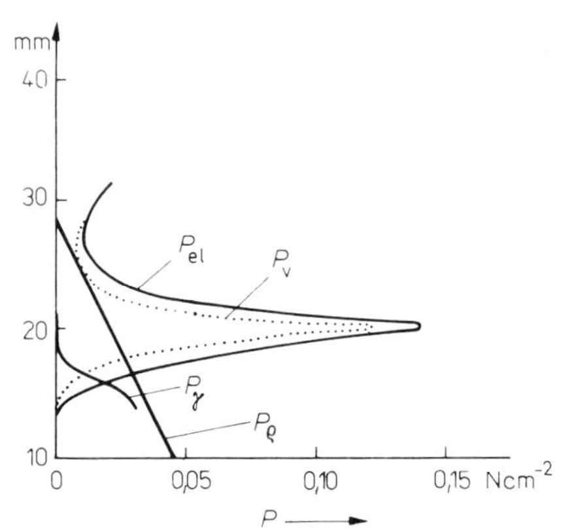

Fig. 11. Pressure components in the molten zone applying needle-eye configuration. $P_{el} + P_\gamma - P_\rho = P_v$ (dotted); the ordinate corresponds to fig. 10. (From the work of Wilke and Bohm [15] after Mühlbauer et al. [83].)

the electrical forces ("electrical pressure") possibly playing a role in zone melting of dielectrics. Considering all components acting on the flow in a molten zone, a quantitative description of the expected flow pattern in the molten zone has proved to be difficult. The convection scheme as far as it is concerned, will influence the temperature distribution in the molten zone, the shape of the liquid/solid phase boundaries and certainly the distribution of impurities in the growing crystal. Much work has been done to improve the homogeneity of the distribution of impurities or dopants in the growing crystal by acting on the convection scheme in the zone.

2. *Crucible-free floating zone melting of silicon*

2.1. *Technological concepts*

The crucible-free floating zone technique (shortly FZ) with radio frequency heating was developed particularly for the production of silicon as a material for electronic components. As floating-zone melting can be processed without a crucible, it meets one important requirement for the growth of high-purity crystals. With the avoidance of a crucible, one prominent source for contamination of molten silicon, reaction with the crucible material, has been excluded. This is an essential advantage of the FZ technique in comparison with the Czochralski technique (CZ) where the silicon is molten in quartz crucible. In FZ silicon, not only is the oxygen content (10^{16} cm^{-3}) two orders of magnitude lower than that of CZ silicon, but also the concentrations of carbon, heavy metals, alkali and alkaline earth metals are lower. Using the FZ technique, silicon crystals have been produced with a purity characterized by a specific resistance of above 100 000 Ω·cm, which corresponds to the intrinsic conductivity. Applying this method, doped crystals with an exactly adjustable dopant concentration can also be produced. Further advantages of the FZ method are better axial homogeneity of the specific resistance in the crystals and a pull rate about twice that of the CZ technique.

The following disadvantages compared with the CZ technique are to be mentioned:

- the obtainable crystal diameters are smaller;
- the crystals tend to crack while processing them;
- as starting material, smooth rods have to be prepared;
- the technique is more complicated and difficult to control;
- the equipment is more complex and expensive.

Especially the latter points are disadvantageous: The requirements on mechanism, RF technique and control raise the price of these crystals. This is the reason why the FZ technique in general is only applied to grow crystals with a specific resistance > 40 Ω·cm. These resistance values, up to now, cannot be obtained by the otherwise more economical CZ technique. Therefore, the more expensive FZ crystals are mostly applied in power electronics. Recently, world-wide, about several hundred tons a year are produced using the FZ technique, which is about 15% of the total production of silicon single crystals.

Table 2
Publications about silicon growth by the floating zone technique (selection).

Müller [7]	Dash [20,21,104][v]	Braun + [167]	Eckart + [171,172]	Green [177]
Hemmat + [159]	Ratnikow + [49]	Trainor + [111][d]	Ziegler [105]	Ban + [178]
Böttger + [160][v]	Vlasak [369]	Peibst + [168][v]	Hadamovsky [89]	King [179]
Ciszek [106][d] [107]	Spenke [91] Mühlbauer + [84][d]	Wilcox [108]	Keller + [100], [109][d], [53][v], [95]	
Schröder + [31]	de Kock + [164][s,v]	Petroff + [169][s]	Geil + [173][d,v], [174][v], [175][s,v]	
Roksnoer + [119, 120, 161][s]		Collins [93]	Doerschel + [276][v]	Föll + [117][s]
Gupta + [101]	Jakowlew + [165][s] Jastrzebski + [99]	Belov + [170]	Dietze + [96]	
Kimura + [110, 113][d]	De Leon + [112][d] Hoffmann + [180][s]	Kolbesen + [118][s], Kramer [97]		
Eyer + [162, 163][d]	Buzynin + [166]	Thomas + [116][d]		

+: Co-authors.
[d] Doping and homogeneity.
[s] Swirls and other microdefects.
[v] Dislocations.

Surveys of FZ growth of silicon have been given by Hadamovski [88, 89], Benson [90], Spenke and Heywang [91, 92], Collins [93], Silbernagel [94], Keller and Mühlbauer [95], Dietze et al. [96], Kramer [97], Zulehner [98].

A voluminous representation of the electronic data of FZ silicon was made by Jastrzebski and Zanzucchi [99]. A selection of further literature is given in table 2. Because of the high industrial relevance, many technological steps and details are published only in the patent literature.

Similar to the CZ technique, the FZ technique has achieved a high technological state: as a result of a highly sophisticated development of the apparatus and process, dislocation-free silicon crystals with a diameter of 150 mm and a length of 1.5 m can be grown. On the one hand, the development is going to larger diameters, but, on the other hand, above all to improve both homogeneity and defect structure. It is typical for the FZ technique that there are many parameters which could be varied, such as heater design, diameters of starting rod and crystal, their respective rotation rates, pull rates, the zone length, the crystallographic orientation (pulling direction), use and design of afterheaters, ambient atmosphere, and others. During process development, all these parameters have to be optimized and adjusted with respect to each other. The necessary know-how to realize reproducible growth is much more extensive than for the CZ technique.

Nowadays, modern FZ equipments exclusively use inductive heating by radio frequency fields applying the variant of internal heating which means that the radio frequency inductor is placed inside the growing chamber. The equipment combines elements of precision mechanics with those of heavy machine construction, the latter mostly to suppress vibrations. The mechanical drives must be carefully selected to avoid vibrations and the so-called stick–slip effect. The drives must work at low speed

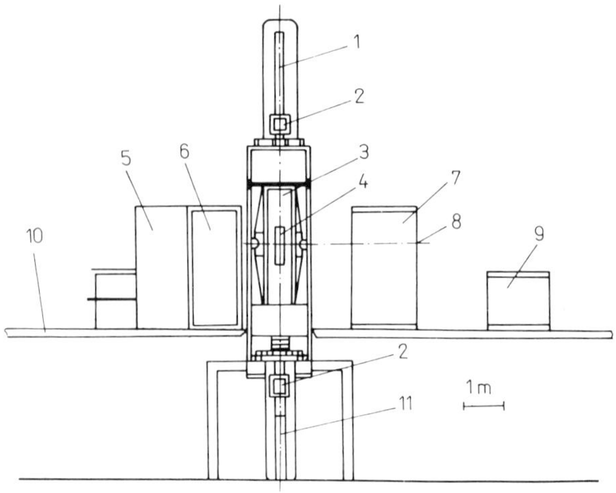

Fig. 12. Modern floating-zone equipment with 1.5 m pull length. 1: upper pull shaft; 2: drives for travel and rotation; 3: growing chamber; 4: observation windows; 5: electrical supply; 6: operation panel; 7: RF generator; 8: inductor plane; 9: hydraulics; 10: working stage; 11: lower pull shaft. (From the work of Wilke and Bohm [15].)

and should be somewhat oversized: the mechanical motions (pulling and rotation) have to be controlled to an accuracy of 5‰.

To accommodate the movements for pulling, a considerable height of the equipment is necessary, having usually the fourfold pulling length L (the so-called $4L$ variant). The equipments reach an overall height of about 10 m (fig. 12). The height could significantly be reduced by constructive solutions with the driving motors located inside the growing chamber. This causes, however, the risk of escape of impurities into the working atmosphere or failure of the drives by evaporated and precipitating particles. Therefore, metal bellows are used above and below the growing chamber to avoid bushing problems. The growth chamber itself is thereby kept as small as possible.

Frequencies of 2 to 5 MHz are applied for radio frequency induction heating (RF heating). The RF power (35 kW for large equipment) must be controlled with an accuracy of 0.5‰, the ripple of the RF current should not exceed 0.5‰ also. Superimposed frequencies below 300 Hz must be avoided because they produce vibrations. Keller and Stut [100] have discussed further points concerning the application of RF heating. The relatively high working frequency of 2 MHz and above is necessary to guarantee satisfactory melting of the feed rod using the skin effect [101]. Above 5 MHz, however, the electromagnetic supporting effect on the molten zone, which is important for its stability, is reduced considerably so this frequency should not be exceeded. Frequencies above 3 MHz cause an increasing tendency to electrical flash-overs at bushings as well as between the inductor turns and between inductor and melt. The following means are suited to reduce flash-overs:

- reduction of frequency (deterioration of the melting behaviour of the feed rod);
- inductor design as a single-turn inductor (opposite to multi-turn pipe coils);

- earthing of the inductor in the centre of the outer diameter (halves the voltage);
- application of special high-purity protective gas mixtures.

The starting material for the FZ technique is rod-shaped microcrystalline silicon manufactured already with a high purity from high-purity trichlorsilane ($SiHCl_3$) by chemical precipitation on thin electrically heated silicon rods. These thin silicon rods with a diameter of 2 to 4 mm are also obtained crucible-free before, growing them from the molten top of a thicker rod (pedestal method), so that, in this respect, high purity is ensured also. The polycrystalline silicon rods, however, must be mechanically treated to guarantee reliable melting. The contaminated surface layer is removed by etching in a HF–HNO_3 mixture. Fenzl et al. [102] produced the needed silicon rods with the desired accuracy using a graphite mould casting. The floating zone process itself causes a further purifying effect.

Earlier it was common to perform several zone passages. But it is much more economical to realize the high purification as far as possible by distillation steps when producing trichlorsilane. Therefore, nowadays, only two zone passages are performed, the first one with the rod remaining polycrystalline, for purification, homogenization and removing of disturbing particles and the second one for single-crystal growth. If the starting material is of very high purity, only one step is needed to grow the single crystal. The following variants of growing silicon crystals with large diameters (> 30 mm) have proven to be successful:

- growing under protective gas (Ar, or Ar with a small amount of other diatomic gases);
- zone passage with fixed inductor and moving rods;
- growth direction (direction of zone passage) upwards;
- needle-eye technique using a compact inductor;
- application of a seed crystal and growing a "bottle neck" for single-crystal growth;

The protective gas argon is used with a purity of 99.999%; the gas flow rate amounts to 400 to 1200 l/h. An additive of hydrogen improves the electrical breakdown strength but causes structural defects. (Helium is not only expensive but also has a lower breakdown strength.) An additive of nitrogen (0.05 to 1%) to the argon atmosphere decreases the breakdown frequency in the region of the inductor and does not cause structural defects. Growing in vacuum is favourable with regard to evaporation of impurities, but it has the disadvantage that evaporated silicon could fall back onto the melt from the cold inductor or other devices leading to generation of dislocations and to failure of the growth run. Only crystals up to 30 mm diameter can be vacuum-grown with a sufficient yield. For large-diameter crystals the first zone passage for purification and homogenization is usually performed under vacuum (polycrystalline growth of the rod), while for growing the single crystals the protective gas must be used.

An assembly with a fixed coil and moving rods has not only constructional advantages for large equipments, it also allows the parameters to be more effectively varied. A moving coil has only been applied for smaller equipments producing crystals with diameters below 50 mm.

The zone passage upwards (growing from a hanging platform) not only leads to a more stable zone in combination with the needle-eye technique, but also makes it possible to squeeze and stretch the crystal during the growing process. It is much more favourable than a downward-moving zone [31]. The latter arrangement can only be applied for diameters up to 40 mm and is nowadays used for special purposes, because the pull rate can be varied widely.

The inner diameter of the inductor is smaller than that of the crystal to be grown (see fig. 7) to ensure stabilization of the molten zone by electrodynamical forces. Only this technique enables the growth of large-diameter crystals. Design and adjustment of the RF inductor are very important for stability of the molten zone as well as the melting of the feed rod and the crystallization behaviour. The best success was reached using a compact pancake-shaped, single-turn inductor. The inner diameter should not exceed 37 mm to join the tapering feed rod with the small seed crystal by a small melt drop as well as to realize the growth of the "bottle-neck". The outer diameter should always be larger than that of the crystal to be grown and should have a value of 100–150 mm to grow crystals with a diameter of 75 mm. Keller and Stut [100] used a horizontally movable inductor to facilitate the melting. Besides, they described a separable inductor to circumvent a melting-off at the end of the zone passage, which is especially important when performing several zone passages. The asymmetrical variant of the needle-eye technique (see fig. 7d) having been successfully used by Keller [95] to grow crystals of 30 to 45 mm diameter with a homogeneous dislocation distribution, nowadays does not play any role in growing large crystals.

To get the phase boundary as flat as possible and to diminish the stress during cooling down, the growing crystal is surrounded by an afterheater or a cooling body or a combination of both. A preheater for the melting feed rod did not prove successful because it unfavourably influenced the melting behaviour [97].

2.2. Process control

The first step in successfully growing silicon crystals is to achieve crystals entirely free of dislocations. This is substantial not only with respect to their quality but also an important prerequisite to mastering the floating zone process on its own. In floating zone melting, there are steep radial temperature gradients of about 150 K/cm resulting in large thermal stresses. This is due to inductive heating supplying the heat only to a thin surface layer of the molten zone. If there are any dislocations present, the thermal stresses cause their multiplication by plastic deformation, resulting in a dislocation density of 10^3 to 10^5 cm^{-2} for crystals of about 30 mm in diameter. For larger crystals the dislocation density will become still higher even when carefully controlling the process, applying afterheaters, etc. Afterwards, subgrain boundaries will occur and finally the growth becomes polycrystalline. The growth of crystals with larger diameter was successfully performed using the bottle-neck technique proposed by Dash [20, 21, 104] and adapted to the FZ technique by Ziegler [105], to realize dislocation-free growth of crystals. In a dislocation-free crystal, even if the phase boundaries are not flat and there are large thermal stresses, no creation and multiplication of dislocations will occur.

The growth direction is given by the crystallographic orientation of the seed crystal. The usual choice is [111], or sometimes [100], depending on the demands of the subsequent device technology. Pulling into a [110] direction has proven to be difficult, because the dislocation lines mostly run in this direction and the crystal cannot be made dislocation-free by pulling a bottle-neck.

The seed crystal is a small crystal rod with a diameter of about 4 mm that is supposed to be orientated with an accuracy of $\pm 0.5°$. Usually, seed crystals are cut from optically orientated crystals and again optically re-orientated after their input into the growing chamber.

Commonly, the growing process has the following sequence: Inside the growing chamber, the seed crystal is mounted at the lower holding shaft, the feed rod to be melted at the upper one. After mounting, the growing chamber is evacuated and then filled with the protective gas. The process steps, when starting the growing process, are represented in fig. 13: The feed rod is conically tapered below with an angle of 45° to 60°. Because silicon is electrically non-conducting at room temperature, the feed rod is preheated to 400°C to enable coupling to the RF field. This is realized using a graphite ring that is heated by the RF field of the inductor and transfers the heat to the feed rod by heat radiation. When the silicon couples to the RF field, the graphite ring is moved away, the tip of the feed rod is melted and the seed crystal is approached. After joining them, a bottle-neck 2 to 3 mm in diameter and length of 50 to 70 mm is grown at a relatively high growing rate of 15 mm/min or more to eliminate the dislocations. Then, the crystal is gradually broadened to the desired diameter by changing both the RF power and pull rate. The cylindrical part of the crystal is grown at a pull rate of 2 to 4 mm/min with a slow rotation (3 to 8 min^{-1}); the feed rod is usually oppositely rotated and somewhat slower (2 to 5 min^{-1}).

The operator recognizes the dislocation-free state of the growing crystal by its cross-section not being exactly circular but more or less three-sided (when pulling in a [111] direction). Furthermore, the surface of dislocation-free crystals has characteristic ridges due to the formation of {211} facets on the rim of the liquid/solid phase boundary [106–108]. A crystal growing in a [111] direction shows three, or, for larger diameters, up to six, ridges. Not all of them are necessarily marked likewise: A deviation in orientation of only 0.7° causes only one ridge to be strongly marked. Since the shape of

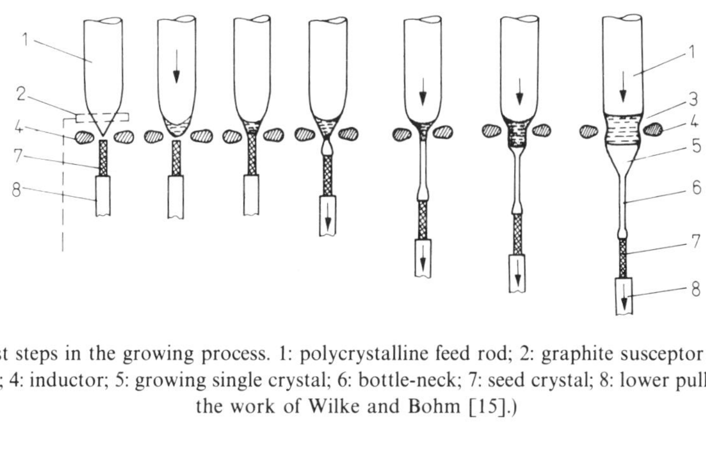

Fig. 13. First steps in the growing process. 1: polycrystalline feed rod; 2: graphite susceptor (movable); 3: molten zone; 4: inductor; 5: growing single crystal; 6: bottle-neck; 7: seed crystal; 8: lower pull shaft. (From the work of Wilke and Bohm [15].)

the phase boundary and the growth conditions for facet formation at the rim are changing, the ridges could change their width, possibly vanish and return again during further growth.

If dislocations are generated, e.g. by an impurity particle getting into the melt and to the phase boundary, they rapidly multiply and spread over the whole cross-section and it is of no use to continue the growing experiment.

Because the growing crystal is supported only by the narrow bottle-neck, the assembly is very sensitive to vibrations. The latter can lead to a leakage of the floating zone or to a crack of the bottle-neck if the assembly is resonant. It is impossible to enlarge the diameter of the bottle-neck since there is then no guarantee of dislocation-free crystals. If the mass of the crystal grown exceeds 4 kg, supporting devices become necessary. Numerous proposals for supporting devices are described in the literature. Mostly it is sufficient to damp the crystal vibrations because the bottle-neck is able to carry the mass itself if it is resting and there are no vibrations. As illustrated in fig. 14,

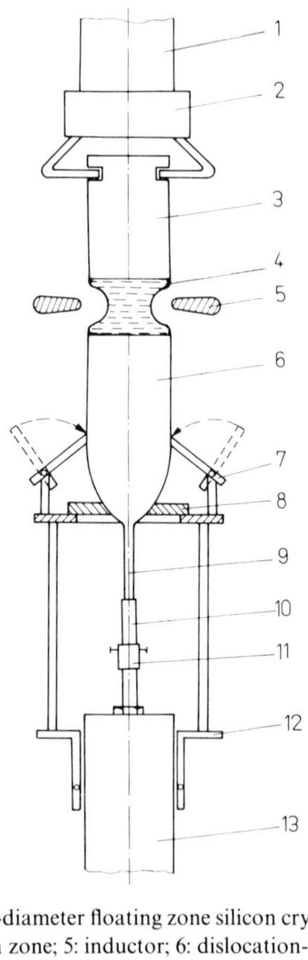

Fig. 14. Supporting device for large-diameter floating zone silicon crystals. 1: upper pull shaft; 2: rod holder; 3: polycrystalline feed rod; 4: molten zone; 5: inductor; 6: dislocation-free single crystal; 7: support finger; 8: float-ring; 9: bottle-neck; 10: seed crystal; 11: seed holder; 12: support basket; 13: lower pull shaft. (From the work of Wilke and Bohm [15].)

vibration damping is realized by three fingers made from ceramic or metal and displaced 120° from each other. They are approached to the crystal after a certain pulling length and locked in this position. Besides, the weight of the crystal can be carried by a so-called support ring that is freely movable and is also approached from below together with the supporting device after a certain pulling length.

The doping of silicon demanded for semiconductor components preferably is carried out for large charges during the chemical deposition of the starting material from trichlorsilane. When adjusting the required concentration, a change in concentration by the FZ process must be taken into consideration.

The following doping techniques by FZ melting are used:

- *gas-phase doping*: Doping gases as PH_3, B_2H_6, PO_2Cl_3, $AsCl_3$, BCl_3 are added to the protective gas or these gases are blown on the molten zone when growing in vacuum.
- *pill doping*: A "pill" of the doping element or better an alloy of it to Si is added at the beginning of the melting of the zone. This technique is especially suitable for elements having a small distribution coefficient.
- *drop doping*: Drops of a solution containing the doping element are applied to the starting rod and dried. Nowadays, this technique is seldom used because of possible impurities.

The homogeneity of the doping concentration determining the resistance homogeneity in the crystal volume is a problem. As for the other melt growth techniques, the significant inhomogeneities are striation-shaped concentration variations due to fluctations of the local growth rate. The striations follow the position of the growth front or the phase boundary, respectively. Since the phase boundary is not planar but has a more or less special shape, not only axial but also radial inhomogeneities of the doping concentration and the resistance may result from striation formation. The striation-shaped doping inhomogeneities are detected by etching and the corresponding resistance inhomogeneities by high-resolution spreading resistance measurements. Ciszek [106], Keller and Mühlbauer [53, 95, 109], Mühlbauer and Sirtl [84] and Kimura et al. [110] have described striation formation and doping inhomogeneities.

Because of the steep temperature gradient, in FZ-grown crystals the inhomogeneities are larger than in CZ-grown crystals and perturbations of the growing process play a more important role. A distinction has to be made between rotational and non-rotational striations. The first ones are strongly marked and periodic in the axial direction, spaced according to the ratio of growth and rotation rate. The non-rotational striations are caused by transient fluctuations of convective flow and superposed on the rotational ones. Non-rotational striations are non-periodic and weaker. The relative intensity of the striations is stronger the smaller the distribution coefficient k of the dopant in question is. This has importance for carbon ($k = 0.07$), which represents the most significant impurity for silicon and plays a role in the formation of microdefects.

The radial homogeneity of doping element concentration and resistivity is a superior quality parameter of the silicon crystal. The radial homogeneity in a complex way depends on the shape of the phase boundary, the fluctuations of dopant concentration inside the melt (due to convection), and the local fluctuations of the crystallization rate and the crystallization kinetics (locally changing thickness of diffusion layer, atomically rough or atomically flat growth – the latter at facets). Many technological parameters which can be (not all independently of each other) varied by the operator affect these conditions:

- diameter and rotation of the feed rod;
- outer zone length (heat power);
- inductor geometry, RF frequency;
- diameter, orientation and rotation of the crystal rod;
- eccentricity (displacement between the axes of the feed rod and the crystal rod);
- pull rate.

The rotation rate, the orientation of the crystal rod, the inductor geometry and the pull rate have the most important influence on the growing crystal. Lower rotation rates (up to 5 $\min^{-1}$) cause a concave phase boundary, the isotherms in the crystal rod acquire a simple shape and follow the shape of the phase boundary (fig 15a). If higher rotation rates are employed (above 30 $\min^{-1}$), the convection pattern is governed by centrifugal forces (fig 15b): In the centre of the molten zone a relatively cool region arises causing a convex phase boundary. Particularly when growing in a [111] direction, a (111) facet is formed and, due to the changed crystallization kinetics (atomically flat growth at the facet), a core with strongly enhanced (for a distribution coefficient $k < 1$) doping concentration is formed [111]. Mühlbauer and Sirtl [84] have shown that the formation of the (111) facet takes place not only at a convexly projecting phase boundary but also at the deepest point of a concave phase boundary. It is generated in the remelting phase during one rotation turn in an asymmetrical temperature field. The atomically rough phase boundary surrounding the facet melts back somewhat faster than the atomically flat facet – resulting in the persistence of the facet. Certainly having a concave facet, the doping concentration is only slightly increased

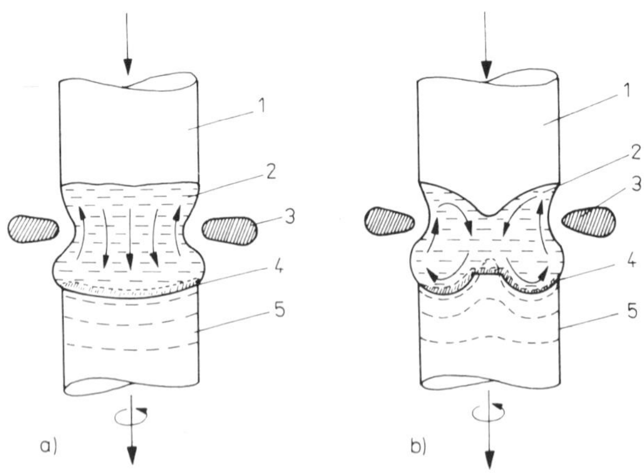

Fig. 15. Convection scheme and phase boundaries, (a) for low pull rate (< 5 $\min^{-1}$); (b) for higher pull rates (> 30 $\min^{-1}$) of the crystal rod. 1: feed rod; 2: molten zone; 3: susceptor; 4: limiting diffusion layer; 5: crystal rod with isotherms. (From the work of Wilke and Bohm [15].)

contrary to the conditions valid for convex facets. According to Keller and Mühlbauer [109], not only the facet growth plays a role in doping concentration but also the flow pattern causing a locally changing thickness of the diffusion boundary layer and the doping concentration as well. The authors mentioned above obtained a relatively low variation in doping concentration for both very low (1 to 6 min^{-1}) and very high (50 to 200 min^{-1}) rotation rates. The latter, however, are far from being realized for thicker rods. Moderate rotation rates (10 to 30 min^{-1}) result in large variations of the doping concentration.

The influence of the pull rate on the shape of the phase boundary (in the realizable range of 1 to 5 mm/min) is more complicated. For small pull rates (1 to 2 mm/min), no dislocation-free crystals can be grown. On the other hand, for thicker rods the pull rate is limited to 5 mm/min to ensure a complete melting of the zone. The best results are obtained for pull rates of 3 to 5 mm/min and rotation rates of 3 to 10 min^{-1}.

The application of magnetic fields to improve the doping and resistance homogeneities are reported by De Leon et al. [112] (axial magnetic field of 180 G), Kimura et al. [113] (transverse magnetic field of 1800 G, fig. 16) and Lie et al. [114]. Similar to the effect of applying a magnetic field using the Czochralski technique, also for the FZ technique the convection in the molten zone is suppressed and therefore the resistance homogeneity is improved especially for n-type crystals. The influence of the inductor geometry on radial doping homogeneity of dislocation-free FZ silicon crystals was investigated by Schröder et al. [115]. Thomas et al. [116] studied the effect of vertical and axial magnetic fields on the oxygen distribution in the crystal. Summarizing, the following conditions have been established to improve the resistance homogeneity of FZ-grown crystals:

- low rotation rate of the crystal rod (2 to 10 min^{-1});
- relatively high pull rate (3 to 5 mm/min);
- precisely concentric arrangement of crystal rod and inductor, no eccentricity, radially;
- symmetric temperature field;

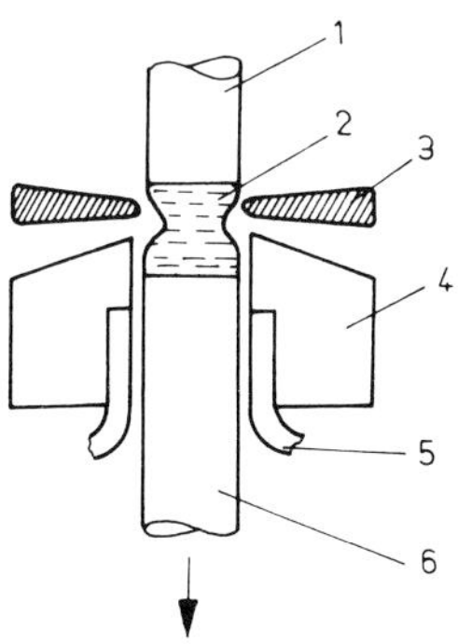

Fig. 16. FZ growth of silicon in a transverse magnetic field (from the work of Wilke and Bohm [15] after Kimura et al. [113]). 1: feed rod; 2: molten zone; 3: inductor; 4: magnetic poles; 5: cooling water; 6: single crystal.

- avoiding facet formation;
- application of a magnetic field to the molten zone.

To circumvent the problems of homogeneous doping, for the highest demands on homogeneity, the so-called transmutation doping by irradiating the silicon crystals (after growth) with thermal neutrons in a nuclear reactor was introduced. According to the nuclear reaction

$$^{30}\mathrm{Si} \xrightarrow{(n,\gamma)} {}^{31}\mathrm{Si} \xrightarrow{3.6\mathrm{h}} {}^{31}\mathrm{P} + \beta^-, \tag{11}$$

the (stable) phosphorus isotope ^{31}P is obtained from the silicon isotope ^{30}Si. Another way to obtain homogeneous doping is by means of ion implantation during the subsequent processing of the silicon crystals.

Just as with silicon crystals grown by the Czochralski technique, also those grown by the floating zone technique contain microdefects. These microdefects become visible when etching polished faces. Because of their appearance, they are called swirls. Their origin is the precipitation of point defects during the growth process or during the subsequent cooling. Depending on their size, larger A- and the smaller B-, C- and D-swirls are distinguished. Detailed investigations on the formation of A- and B-swirls have been made by Föll et al. [117] and by Kolbesen and Mühlbauer [118]. A-swirls are collapsed agglomerations of interstitial and impurity atoms surrounded by small dislocation loops with a diameter of about 300 Å as it has been proven by directly imaging A-swirls by means of an electron microscope. When slowly cooling the as-grown crystals, A-swirls can act as a source of dislocations [164].

Up to now, because of the small size, the direct proof of B-swirls has not succeeded. They are supposed to be the first step to A-swirls. When using small pull rates, B-swirls can be converted into A-swirls or dislocations. Therefore, B-swirls are supposed to consist of drop-shaped agglomerates of interstitial atoms and impurity atoms. The character of C- and D-swirls is widely unknown.

According to Föll et al. [117], the carbon content influences concentration, formation and the behaviour of B-swirls. This is demonstrated by the relation between the carbon striations and the arrangement of swirls. High carbon concentrations ($> 5 \times 10^{16}\,\mathrm{cm}^{-3}$) cause A-swirls and stacking faults. Therefore, reduction of the carbon content is a decisive means by which to grow nearly defect-free FZ crystals. The swirl concentration in FZ crystals shows a strong dependence on the pull rate (fig. 17). This behaviour is connected with their formation kinetics, not yet clarified in all details. Pull rates above 5 mm/min, where no microdefects are produced, can however only be applied for thin crystals and cannot be realized for rod diameters relevant for production.

When growing FZ silicon crystals in a hydrogen-containing atmosphere, numerous strongly formed defects arise. Their nature and formation mechanism are also largely unknown. Due to these defects, the material cannot be used in semiconductor applications. Therefore, a hydrogen atmosphere is not used for FZ growth of silicon crystals. Single crystals are usually grown in pure argon or with the addition of another diatomic gas. Wolf and Schröder [103] report on the influence of nitrogen on the

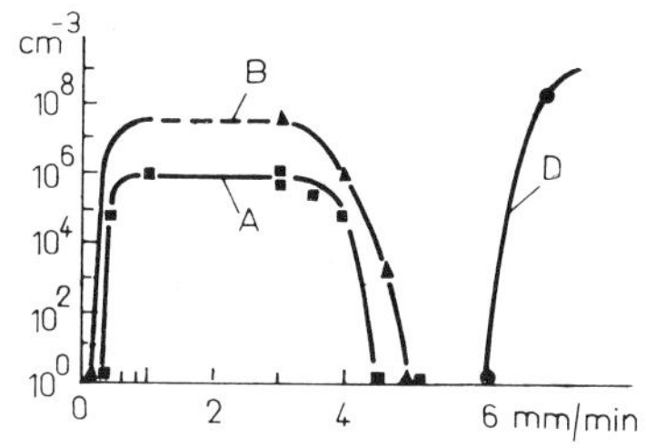

Fig. 17. Concentration of A-, B- and D-swirls in FZ silicon versus growth rate. (From the work of Wilke and Bohm [15] after Roksnoer et al. [119, 120].)

growth of silicon by the floating-zone technique. The grown crystals were free from swirls and other structural defects.

3. *Floating zone melting of other materials*

3.1. *Floating zone melting by electrical heating*

The crucible-free floating zone technique has been applied for growing many other materials, particularly for refractory ones with high or very high melting points for which no suitable crucibles are available. The various techniques and apparatus variants developed up to now can be characterized by their different heating methods.

Electrically conductive materials (at least at higher temperatures) can be melted directly by means of radio frequency induction heating (RF heating), as described for silicon in the previous section. Using this heating technique, the attainable melting temperatures are nearly unlimited. Radio frequency generators with an RF power input of 30 kW are typical for melting rods with diameters of about 10 mm.

The RF coil can be mounted outside a tubular growing chamber (so-called external heating) or inside a, correspondingly larger, growth chamber (internal heating). External heating can be realized with less expense. However, it has the disadvantage

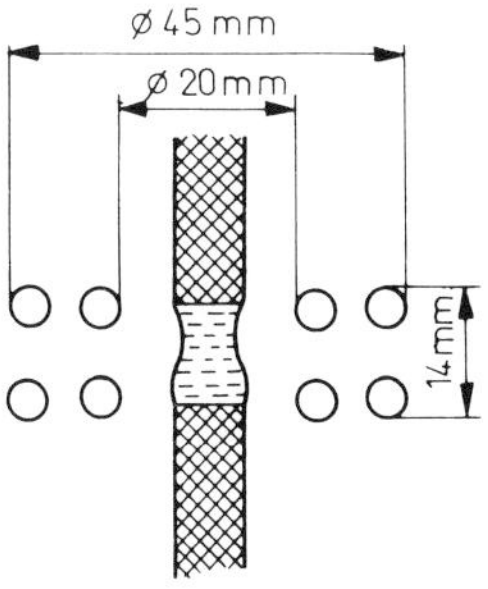

Fig. 18. Radio frequency induction coil used to melt TiC, TiN, and ZrN. Rod diameter about 10 mm. (From the work of Wilke and Bohm [15] after Christensen [121].)

that vaporized material condenses on the walls of the tube, which disturbs the growth process and limits visibility. Using internal heating, radiation shields (mostly of tantalum or molybdenum) or afterheaters can be placed inside the growing chamber. If necessary, a tubular afterheater can be inductively heated in the same circuit by suitably designing the RF inductor (e.g. by an additional turn around the afterheater). However, electrically conductive materials have a relatively good thermal conductivity so that mostly the temperature gradient is not a critical question. The shape of the RF field can be influenced, as mentioned, by the design of the RF coil and a concentrator (figs. 18, 19 and 20). Concentrators are used especially for materials having a very high melting point.

The melting of metals and other oxidizable materials are usually performed in vacuum or an inert atmosphere (argon, helium). Nevertheless, RF floating zone melting

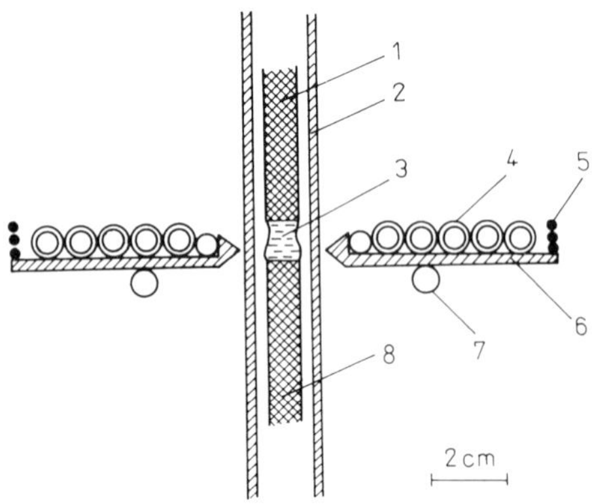

Fig. 19. RF concentrator for crucible-free zone melting. 1: melting rod (Ni); 2: protective tube; 3: molten zone; 4: RF induction coil (the most inner turn is conductively connected with the concentrator, the other turns are insulated); 5: sensor coil (for power control); 6: concentrator (copper); 7: additional cooling; 8: growing crystal. (From the work of Wilke and Bohm [15] after van Drunen and Saimoto [122].)

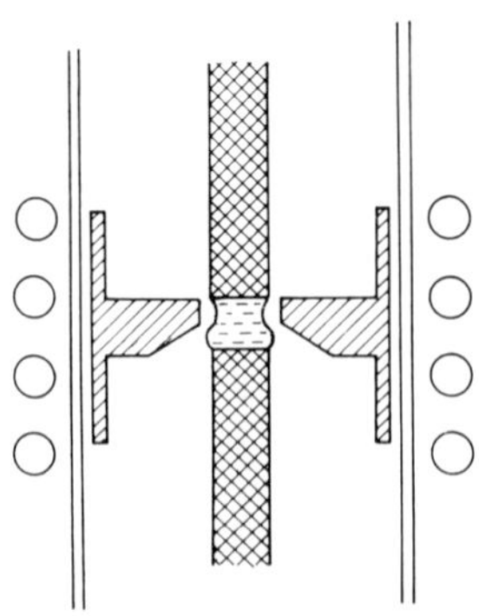

Fig. 20. RF concentrator for floating zone melting above 2000°C. The concentrator is made of copper and water-cooled (the cooling water channels are not represented). (From the work of Wilke and Bohm [15] after Johnson [123].)

offers easy possibilities to employ reducing (containing H_2 or CO) or oxidizing (containing O_2 or CO_2) or otherwise reactive atmospheres. For example, there is the opportunity to maintain a fixed oxidizing potential during the full run of the zone passage, which may be important to grow compounds (e.g. oxides) with well-defined stoichiometry.

Decomposing materials, having a noticeable vapour pressure at their melting point, have to be grown in an entirely closed growth chamber (encapsulated apparatus). The wall of the growing chamber may be additionally heated to avoid deposition of the volatile component (hot-wall technique). Johnson [124] described the floating zone melting of GaAs covered by a protective melt (liquid encapsulation) of B_2O_3 to avoid vaporization of As. Furthermore, there are equipments which allow floating zone melting (e.g. of GaP) under enhanced pressure.

When using RF heating for melting non-conductive materials, a ring-shaped susceptor (usually made of graphite or a refractory metal) surrounding the rod is needed to convert the RF energy into thermal radiation. A heated susceptor, however, is always a source of impurities.

Instead of an RF susceptor, a ring-shaped or loop-shaped resistance heating element can be used as a heat source. Resistance heating is mainly applied for materials with moderate or low melting temperatures.

An interesting variant of resistance heating uses the so-called strip heater (fig. 21): A strip-shaped heating conductor penetrates the molten zone; in the region of the molten zone, the strip is net-like perforated. Because of the immediate heat transmission, the power consumption is relatively small (<1 kW) and the strip confers additional stability to the molten zone [65, 128, 129]. Certainly, there is direct contact between melt and strip heater; to that extent it is not really a crucible-free technique.

Another variant of floating zone melting employs an electric arc burning between the sample and one or more cathodes (see, e.g. [130]). Furthermore, a gas discharge may be used [131–136]. Such a set-up is relatively simple, but it must be noted that material could be transferred from the cathodes to the molten zone. Therefore, they should be

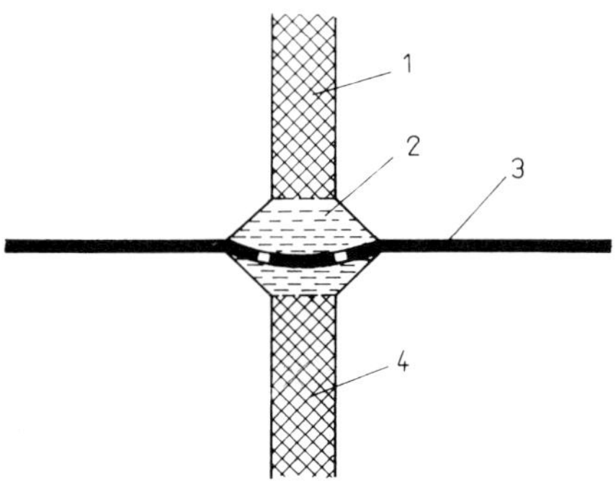

Fig. 21. Crucible-free zone melting with a strip heater. 1: melting rod; 2: melt; 3: strip heater (perforated iridium); 4: growing crystal. (From the work of Wilke and Bohm [15] after Gasson [125] and Muto and Awazu [126].)

manufactured from a material which is compatible with the sample material (e.g. from Al for melting Al_2O_3, etc).

A set-up for floating zone melting of VC, NbC and NbN using a radio frequency plasma has been described by Kumashiro et al. [135]: Four plasma torches are mounted around the sample rod. The pressure exerted by the plasma suppresses the dissociation of the sample material; furthermore, a reactive component (nitrogen) could be added to the plasma torch, which is injected into the sample.

3.2. *Floating zone melting by optical or thermal radiation*

Another technique widely used for floating zone melting, especially of dielectric substances, uses the radiation coming from an optical or thermal radiation source which is focussed onto the molten zone by means of a concave mirror. Firstly, Kooy and Couwenberg [137] have melted rods of ferrite $MnFe_2O_4$ and other oxides using the light of the carbon-arc lamp of a film projector (cf. [138]). Subsequently, optical furnaces or radiation furnaces constructed for floating zone techniques have become commercially available. Their main component is an ellipsoidal reflector (fig. 22). There is a high-power W-halogen lamp or Xe high-pressure lamp at its first focus and the sample rod is arranged at the second one. So, light radiation falls onto the molten zone from all directions. Assemblies consisting of two lamps and a double ellipsoidal reflector with a common focus are still more powerful (fig. 23). The apparatus represented in the figure is also available as a compact assembly used for space experiments. By computer simulation, Eyer et al. [140] investigated the radiation distribution as well as the available radiative power in single- and multiple-reflector assemblies. Balbashov and Egorov [141] described an apparatus for floating zone melting by radiative

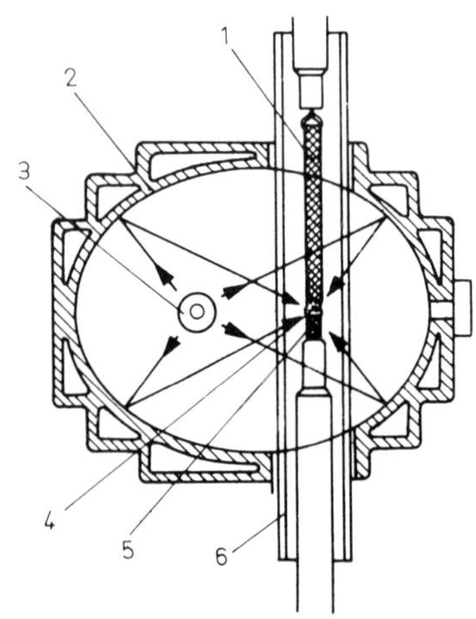

Fig. 22. Optical furnace for floating zone melting. 1: melting rod; 2: ellipsoid mirror; 3: W-halogen lamp; 4: molten zone; 5: growing crystal; 6: protective tube (quartz glass). (From the work of Wilke and Bohm [15] after Takei [127].)

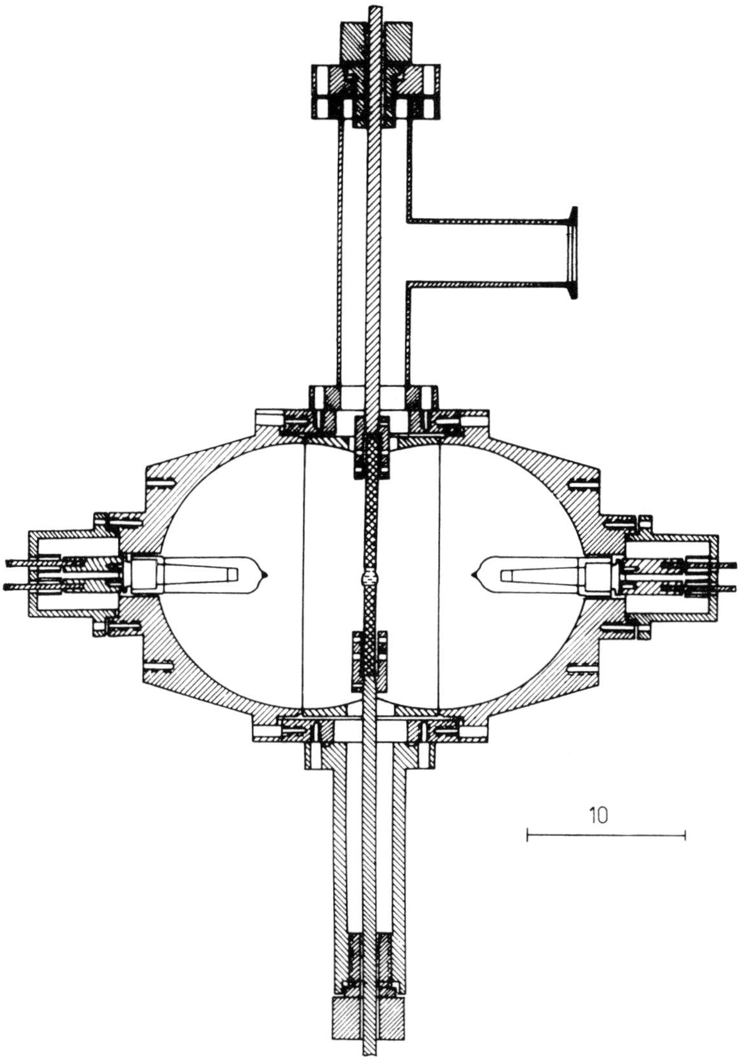

Fig. 23. Optical furnace for the floating zone melting with two lamps and a double-ellipsoidal reflector. (From the work of Wilke and Bohm [15] after Eyer et al. [139].)

heating under a pressure of 100 atm (oxygen or other gases). They have grown single crystals of numerous oxides, amongst others some incongruent melting ones (see table 3). Figure 24 shows the pressure chamber of their apparatus, being arranged in one focus of the ellipsoidal reflector. The light radiation (of a Xe lamp) enters the pressure chamber through a massive quartz tube. Additionally, a radio frequency afterheater is used.

Furthermore, laser radiation has been frequently used for floating zone melting. Mostly the infrared radiation of a CO_2 laser (with a wavelength of 10.6 μm) has been

Table 3
Single crystal growth using crucible-free zone melting (compounds).

Substance	Melting point (°C)	Heating method	Literature (+ co-authors)
Al_2O_3	2050	electron bombardment	Neumann + [157], Somov + [373]
		resistance heating	Petit-Le Du + [296]
		iridium strip	Cockayne + [297]
		gas discharge (plasma)	Class + [131]
		light furnace	Field + [298], Revcolevschi + [299], Cox + [300], Dhalenne + [301], Saito + [302, 303]
		laser radiation	Eickhoff + [142], Gasson + [304], Burrus + [144, 145], Kim + [36], Dreeben + [46]
Al_2SiO_5		light furnace	Kitazawa + [305]
BaB_2O_4		laser radiation	Tang + [306]
$BaFe_{12}O_{19}$		light furnace	Balbashov + [141]
$BaTiO_3$		gas flame	Brown + [307], Tien + [308]
		light furnace	Nakanishi + [309]
$BaTi_4O_9$	1432	light furnace	Tanaka + [310]
$Bi_2CaSrCu_2O_8$		light furnace	Takekawa + [311], Bredy + [312], Lu + [313]
$Ca_2Al_2SiO_7$	1590	light furnace	Ii + [314]
$Ca_2MgSi_2O_7$	1454	light furnace	Ii + [314]
CaO	2580	light furnace	Kitazawa + [305]
$CaO{\cdot}Al_2O_3$, $CaO{\cdot}2Al_2O_3$, $12CaO{\cdot}7Al_2O_3$		electron bombardment	Oudalov + [315]
$CaWO_4$	1570	iridium strip	Gasson [125]
$CaZrO_3$	2375	laser radiation	Gasson + [304]
CdTe	1042	light furnace	Eyer + [140]
$CaTiSiO_5$	1382	light furnace	Tanaka + [316]
$CoGaO_4$	1825	light furnace	Kramer + [317]
$Co_xNi_{1-x}O$		light furnace	Jouini + [318]
Cu_2O	1230	gas discharge (plasma)	Brower + [319]
		light furnace	Schmidt-Whitley + [320], Loison + [321]
$Fe_{1-x}O$		light furnace	Berthon + [322]
Fe_2SiO_4	1180	light furnace	Takei [127]
$FeTa_2O_6$	1500	light furnace	Tanaka + [288, 316]
$FeTiO_3$	1390	light furnace	Takei + [127]
$(FeTi)_3O_4$		light furnace	Brabers + [323]
GaAs	1238	RF induction	Whelan + [324], Cunnell + [325], Johnson [124]
GaP	1350	RF induction	Lorant [326]
GaSb	703	electron bombardment	van Mau + [327]
$GdFeO_3$		light furnace	Balbashov + [141]
$Gd_3Ga_5O_{12}$	1730	light furnace	Balbashov + [141]
$LaCoO_3$		electric arc	Gerthsen [130]
$LaMnO_3$		light furnace	Hashimoto + [328]
$La_2Ti_2O_7$	1790	light furnace	Kimura + [329]
$LnFe_2O_4$		light furnace	Iida + [330]
LiB_3O_5		laser radiation	Yangyang + [331]

Table 3 (cont'd)

Substance	Melting point (°C)	Heating method	Literature (+ co-authors)
Li_2CO_3	723	platinum strip	Palaniswamy + [332], Raju + [333]
$Lu_2Si_2O_7$	1850	light furnace	Bretheau-Raynal + [334]
$MgAl_2O_4$	2150	laser radiation	Gasson + [304]
		light furnace	Kitazawa + [305], Hanede + [335]
$MgFe_2O_4$		light furnace	Balbashov + [141]
$(MgFe)_2SiO_4$		light furnace	Kitamura + [336], Inoue + [337, 338], Hosoya + [339]
$MnFe_2O_4$	1600	light furnace	Kooy + [137], Brabers + [340]
$(MnZn)Fe_2O_4$		light furnace	Balbashov + [141]
$NaNO_3$	306	resistance heating	Schwabe + [71, 80], Lan + [65–67, 128, 129]
NbO_2	1915	light furnace	Kun + [341]
Nb_2O_5	1512	light furnace	Shindo + [342]
$NdAlO_3$		light furnace	Bednorz + [343]
$NdBO_3$		platinum strip	Kumar + [344]
$Nd_3Ga_5O_{12}$	1640	light furnace	Kitamura + [336, 345–347]
Nd_2O_3	2250	laser radiation	Cockayne + [348]
$NiFe_2O_4$		light furnace	Poplapwski + [22, 349, 350], Weaver + [351], Balbashov + [141]
NiO	1990	light furnace	Kooy + [137], Dhalenne [301, 352]
$ScTaO_4$	2340	laser radiation	Elwell + [353]
$SmFeO_3$		light furnace	Fairholme + [354]
$SrTa_2O_6$		light furnace	Tanaka + [289]
$SrTiO_3$		light furnace	Balbashov + [141], Higuchi [355]
		RF induction	Holt [356]
		gas discharge (plasma)	Class + [131]
		laser radiation	Tiller + [357]
UO_2	2730	RF induction	Chapman + [358, 359]
VO		RF induction	Billingham + [273]
$YAlO_3$	1875	gas discharge (plasma)	Class [131]
		light furnace	Balbashov + [141]
$Y_3Al_5O_{12}$	1970	light furnace	O'Bryan + [360], Balbashov + [141], Kitamura + [347]
		gas discharge (plasma)	Class [132]
		laser radiation	Burrus + [143]
$YFeO_3$	1700	light furnace	Fairholme + [354], Marsh + [361], Kitamura + [336], Balbashov + [141]
YFe_2O_4	1320	light furnace	Shindo + [342], Kitamura + [336, 362]
$Y_3Fe_5O_{12}$	1582	RF induction	Abernethy + [363], Petrov + [374], Kimura + [364, 365], Balbashov + [141]
Y_2O_3	2450	laser radiation	Cockayne + [348], Gasson + [304], Kitazawa + [305], Stone + [366]
		light furnace	Tsuiki + [367]
YVO_4	1810	iridium strip	Muto + [126]
ZrO_2	2600	light furnace	Kitazawa + [305], Yamakawa [368]

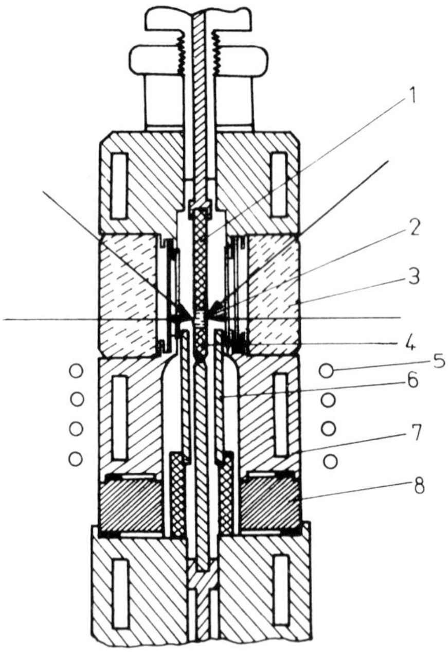

Fig. 24. High-pressure part of an optical furnace for the floating zone technique. 1: melting rod; 2: molten zone; 3: quartz glass tube; 4: growing crystal; 5: RF coil (for the afterheater); 6: RF susceptor (platinum–rhodium); 7: RF concentrator (copper); 8: insulating tube; black: silicon rubber seals. (From the work of Wilke and Bohm [15] after Balbashov and Egorov [141].)

used [46, 142–145]. To utilize the laser power (some 100 W) effectively and also to allow the melting of thicker rods, the rod is surrounded by an additional tubular heating element having a slot for laser radiation and observation. These laser apparatuses, however, could not succeed against perfectly conventional light furnaces. Gurtler et al. [146], have transferred a polycrystalline silicon ribbon into a monocrystalline one by zone melting using two CO_2 lasers directed onto both sides of the ribbon. They called it the RTR (ribbon-to-ribbon) technique.

3.3. Floating zone melting by electron bombardment

Another technique is the floating zone melting by electron bombardment mainly of refractory metals (W, Mo, Re, Nb, Ta, V, Zr, Pt, Pd and others) (see table 4). The electron radiation is emitted from a wire loop concentrically surrounding the sample rod (fig. 25). Commonly, this loop as well as the focussing shields are kept at earth potential, whereas a positive potential (up to 5 kV) is applied to the sample rod. Cylindrical focussing shields have proved to be favourable where the wire loop serving as electron source is positioned somewhat below the molten zone (fig. 26). Therefore, the influence of the molten zone on the electron source is diminished. The electron beam current (approximately 500 mA) must be kept constant.

Table 4
Single-crystal growth using crucible-free zone melting (elements, alloys, boronides, carbides, silicides, nitrides).

Substance	Melting point (°C)	Heating method	Literature (+ co-authors)
Al	659	RF induction	Albert [181], Chaudron [182], Revel [183]
		resistance heating	Albert [181]
B	2200	RF induction	Greiner [184]
		electron bombardment	Talley [185], Hood + [186]
		gas discharge (plasma)	Borchert + [133]
B_4C	2350	RF induction	Borchert + [187]
Be	1283	RF induction	Schaub + [188], Jönsson [189]
		light furnace	Jönsson [189]
CeB_4	2400	RF induction	Kawakumi + [190]
Co	1493	electron bombardment	Bachmann [155], Hayashi + [191]
CrSi	1750	RF induction	Jørgensen + [370]
CrV	1890	RF induction	Camargo + [371]
Cu	1083	electron bombardment	Calverley + [148], Hayashi + [192]
		RF induction	Le Héricy [193, 194]
EuB_6	2500	RF induction	Isikawa + [195], Tanaka + [196]
Fe	1536	RF induction	Smith + [197], Albert [181], Hillmann + [198, 199]
Fe–Si		RF induction	Swets [200], Vanek + [77, 201, 202], Kadecková + [203–206]
Gd	1312	RF induction	Whittaker [207]
HfC	3928	gas discharge (plasma)	Storey + [208]
		RF induction	Otani + [209]
La etc.		RF induction	Ford [210]
LaB_6	2400	RF induction	Curtis + [211], Tanaka + [212], Otani + [213, 214]
		electric arc	Verhoeven + [215], Noack + [216]
		laser radiation	Takagi + [217]
LiAl	700	light furnace	Kitahama + [218]
Li_3N	840	light furnace	Nishida + [219]
Mo	2620	RF induction	Calverley + [148], Smith + [197], Buehler [220], Savickij + [221–223], Otani + [55, 213]
		electron bombardment	Belk [47], Zedler + [224], Göbel + [156], Barthel + [225–227], Bas + [158], Capp + [228], Balan + [229], Jurisch + [230], Glebovsky + [231]
MoB	2200	gas discharge (plasma)	Storey + [208]
Mo_2C	2692	gas discharge (plasma)	Storey + [208]
		RF induction	Christensen [232]
Mo_3Ir	2110	RF induction	Koksbang + [233]
Nb	2468	RF induction	Buehler [220]
		electron bombardment	Savickij + [221, 222], Sell + [234], Zedler + [224], Achter + [235], Bas + [158], Reed + [236], Naramoto + [237–239], Glebovsky + [240]
NbB_2		RF induction	Nakano + [241, 242]

(cont'd)

Table 4 (cont'd)

Substance	Melting point (°C)	Heating method	Literature (+ co-authors)
NbC; NbC_3	3600	RF induction	Packer + [243], Kumashiro + [244], Otani + [245]
		gas discharge (plasma)	Kumashiro + [135, 136]
NbN		RF induction	Christensen [246], Scheerer [247], Kumashiro + [244]
		gas discharge (plasma)	Kumashiro + [135, 136]
Ni	1255	electron bombardment	Calverley + [148], Sell + [234], Hayashi + [248]
		RF induction	Smith + [197], Wernick + [249], Albano + [250], Soden + [251], van Drunen + [122]
Pb	327	resistance heating	Albert [181]
Pd	1550	electron bombardment	Sell + [234]
		RF induction	Hornfeldt + [252]
Pt		RF induction	Vuillemin + [253]
Re	3180	electron bombardment	Calverley + [148], Geach + [254], Lawley + [255], Savickij + [221, 222], Sell + [234], Soden + [256]
Rh	1960	electron bombardment	Sell + [234]
		RF induction	Hornfeldt + [252]
Sn	232	resistance heating	Albert [181]
Ta	2996	electron bombardment	Calverley + [148], Mordike [257, 258] Savickij + [221, 222], Sell + [234] Zedler + [224], Bas [158], Kralina + [259]
		RF induction	Otani + [213]
TaC	3983	gas discharge (plasma)	Storey + [208]
		RF induction	Packer + [243], Otani + [260]
Ta_2C		gas discharge (plasma)	Storey + [208]
TaN	3090	RF induction	Conroy + [261]
Ta_2N		RF induction	Conroy + [261]
Ti	1668	RF induction	Smith + [197], Wernick + [249]
		electron bombardment	Cass + [262]
Ti_3Au, Ti_3Pt,			
TiC, TiC_x	3067	gas discharge (plasma)	Storey + [208]
		RF induction	Packer + [243], Kumashiro + [263, 264], Christensen [121], Yajima + [265], Otani + [55, 245, 266 – 271]
TiN	2949	RF induction	Christensen [121]
U	1130	RF induction	Albert [181]
UC		RF induction	Beucherie + [272]
V	1890	electron bombardment	Calverley + [148], Savickij + [222], Sell + [234]
		RF induction	Wernick + [249]
VB_2		RF induction	Nakano + [241, 242]
VC, VC_x	2700	RF induction	Billingham + [273], Packer + [243], Hou + [274], Otani + [213, 267]
		gas discharge (plasma)	Kumashiro + [135, 136]
VN	2100	RF induction	Billingham + [273], Christensen + [275]

Table 4 (cont'd)

Substance	Melting point (°C)	Heating method	Literature (+ co-authors)
V_3Si	1935	RF induction	Greiner + [176]
		electron bombardment	Jurisch + [277], Callaghan + [278], Behr + [279]
W	3390	electron bombardment	Calverley + [148], Geach + [254, 372], Carlson [280], Schadler [281], Savickij + [221, 222], Schnitzel + [282], Herbst + [283], Sell + [234], Hay + [284, 285], Bas + [158], Carlson [286], Glebovsky + [231]
		RF induction	Buehler [220], Otani + [213]
W_2C	2867	gas discharge (plasma)	Storey + [208]
Y	1550	RF induction	Murphy + [287]
YB_2		RF induction	Johnson [123]
YB_{66}	2100	RF induction	Tanaka + [288, 289]
Zr	1855	RF induction	Albert [181], Langeron [290, 291], Flottmann + [292]
		electron bombardment	Sell + [234], Bas + [158]
ZrC, ZrC_x	3420	RF induction	Lee + [293], Packer + [243], Otani + [294]
		electric arc	Mackie + [295]
ZrN	2982	RF induction	Christensen [121]

It is important that there is a very fast response to fluctuations by outbursts of gas during melting. Powerful pumps are necessary to maintain a high vacuum. Equipments are described by Calverley et al. [148], Birbeck and Calverley [149], Cole et al. [150], Hurwitt and Adams [151], Brandes and Flynn [152], Ketterson et al. [153], Abdelallatif et al. [154] and Bachmann [155]. The temperature distribution in electron beam melted zones has been calculated by Donald [48] and by Göbel et al. [156].

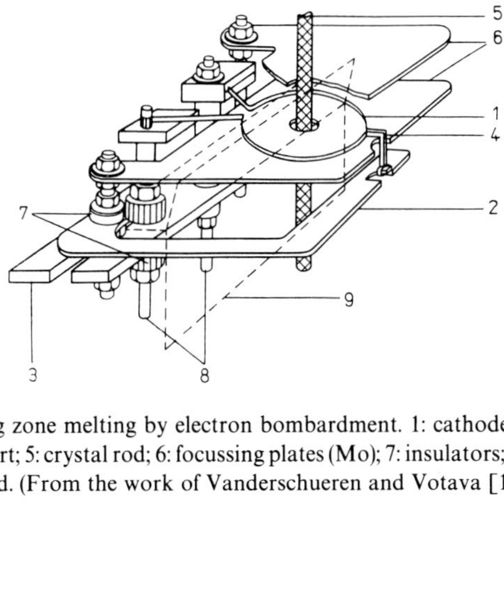

Fig. 25. Heater for floating zone melting by electron bombardment. 1: cathode (W-wire); 2: holding plate (Mo); 3: base plate; 4: support; 5: crystal rod; 6: focussing plates (Mo); 7: insulators; 8: supply lead; 9: protecting shield. (From the work of Vanderschueren and Votava [147].)

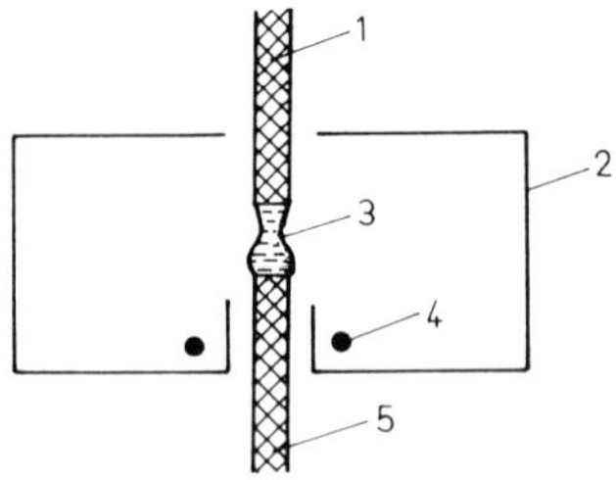

Fig. 26. Cylindrical focussing shield for floating zone melting by electron bombardment. 1: melting rod; 2: focussing shield; 3: molten zone; 4: cathode loop; 5: growing crystal. (From the work of Wilke and Bohm [15].)

Neumann and Huggins [157] published auxiliary devices for the electron beam zone melting of insulators like Al_2O_3. Bas and Stevens [158] have melted thin metal rods in a horizontal set-up using a point-shaped focussed electron beam.

The electron beam floating zone (EBFZ) technique is not only a technique for single-crystal growth but also for high-purification of the refractive metals, especially to remove traces of non-metals such as O, N, etc. To avoid impurities, the wire loop, which is the electron source should be manufactured from the same material as the substance being melted.

References

[1] H.C. Theuerer, US Patent No. 3060123 v.17.12. (1952).
[2] P.H. Keck and M.J.E. Golay, Phys. Rev. 89 (1953) 1297.
[3] R. Emeis, Z. Naturforsch. 9a (1954) 67.
[4] P.H. Keck and M. Green. J. Appl. Phys. 24 (1953) 1459.
[5] P.H. Keck, Physica 20 (1954) 1059.
[6] P.H. Keck, H. Horn, J. Van Soled and A. MacDonald, Rev. Sci. Instrum. 25 (1954) 298.
[7] S. Müller, Z. Naturforsch. 9b (1954) 504.
[8] H.C. Theuerer, Bell Labor. Rec. 33 (1955) 327.
[9] P. Kapitza, Proc. R. Soc. London A 119 (1928) 358.
[10] W.G. Pfann, Trans. AIME 194 (1952) 747.
[11] W.G. Pfann, Zone Melting (Wiley, New York, 1958) (2. ed. 1966).
[12] W.G. Pfann, Science 135 (1962) 1101.
[13] G.H. Rodway and D. Hunt, J. Cryst. Growth 97 (1989) 680.
[14] K. Hein and E. Buhrig (Eds.), Kristallisation aus Schmelzen (Dtsch. Verlag Grundstoffind, Leipzig, 1983)
[15] K.-Th. Wilke and J. Bohm, Kristallzüchtung (Deutscher Verlag der Wissenschaften, Berlin, 1988).
[16] J. van den Boomgard, J. Philips Res. Rep. 10 (1955) 319.
[17] J. van den Boomgard, J. Philips Res Rep. 10 (1956) 27.
[18] J. van den Boomgard, J. Philips Res. Rep. 11 (1956) 91.
[19] O. Madelung, Z. Physik 162 (1961) 508.
[20] W.C. Dash, J. Appl. Phys. 29 (1958) 705.
[21] W.C. Dash, J. Appl. Phys. 31 (1960) 736.
[22] R. Poplawsky and J.E. Thomas, Rev. Sci. Instrum. 31 (1960) 1303.
[23] W. Heywang and G. Ziegler, Z. Naturforsch. 9a (1954) 561.
[24] W. Heywang, Z. Naturforsch. 11a (1956) 238.

[25] E. Green. J. Appl. Phys. 35 (1964) 1297.
[26] C. Huh and L.E. Scriven. J. Colloid Interf. Sci. 30 (1969) 323.
[27] J.F. Padday, Phil. Trans. R. Soc. London A 269 (1971) 165.
[28] J.R. Carruthers and M. Grasso, J. Cryst. Growth 13/14 (1971) 611.
[29] J.R. Carruthers and M. Grasso, J. Appl. Phys. 43 (1972) 436.
[30] J.C. Brice, The Growth of Crystals from Liquid (North-Holland, Amsterdam, 1973).
[31] W. Schröder and H.-G. Dörrfeld, Kristall und Technik 9 (1974) K11–K13.
[32] S. Hartland and R.W. Hartley, Axisymmetric Fluid–Liquid Interfaces (Elsevier, Amsterdam, 1976).
[33] S.R. Coriell and M.R. Cordes, J. Cryst. Growth 42 (1977) 466.
[34] S.R. Coriell, S.C. Hardy and M.R. Cordes, J. Colloid Interf. Sci. 60 (1977) 126.
[35] T. Surek and S.R. Coriell, J. Cryst. Growth 37 (1977) 253.
[36] K.M. Kim, A.B. Dreeben and A. Schujko, J. Appl. Phys. 50 (1979) 4472.
[37] N.E. Mokrousov, A.M. Pronin and D.A. Jaškov, Izvest. AN, Neorg. Mater. 18 (1982) 1913.
[38] E.A. Brener, G.A. Satunkin and V.A. Tatarchenko, Rost Kristallov 14 (1983) 153.
[39] J. Meseguer, J. Cryst. Growth 73 (1985) 599.
[40] J. Meseguer, L.A. Mayo, J.C. Lorentente and A. Fernandez, J. Cryst Growth 73 (1985) 609.
[41] W. Bardsley, F.C. Frank, G.W. Green and D.T.J. Hurle, J. Cryst. Growth 23 (1974) 341.
[42] T. Surek and B. Chalmers, J. Cryst. Growth 29 (1975) 1.
[43] T. Surek, Scr. Metall. 10 (1976) 425.
[44] T. Surek, B. Chalmers and A.I. Mlavsky, J. Cryst. Growth 42 (1977) 453.
[45] H. Wenzl, A. Fattah et al., J. Cryst. Growth 43 (1978) 607.
[46] A.B. Dreeben, K.M. Kim and A. Schujko, J. Cryst. Growth 50 (1980) 126.
[47] J.A. Belk, J. Less-Common Metals 1 (1959) 50.
[48] D.K. Donald, Rev. Sci. Instrum. 32 (1961) 811.
[49] D.G. Ratnikov, W.J. Dobrovolskaja and J.E. Nesvedzki, Promyshlenoe primenenic tokov, vysokoj chastoty velektrotermii (Gos. Nauchotekhn. Izdat. Mash. Lit., Moscow, 1961).
[50] K. Akiyama and J. Yamaguchi, J. Appl. Phys. 33 (1962) 1899.
[51] V.H.S. Kuo and W.R. Wilcox, J. Cryst. Growth 12 (1972) 1910.
[52] T. Arizumi and N. Kobayashi, J. Cryst. Growth 13/14 (1972) 615.
[53] W. Keller, J. Cryst. Growth 36 (1976) 215.
[54] H.K. Kuiken and P.J. Roksnoer, J. Cryst. Growth 47 (1979) 29.
[55] S. Otani, T. Tanaka and Y. Ishizawa, J. Cryst. Growth 66 (1984) 419.
[56] N. Kobayashi, Jap. J. Appl. Phys. (1988) 20.
[57] E. Ju. Kokorish, Kristallografija 5 (1960) 815.
[58] S. Raab and A. Lebek, Phys. Status Solidi 12 (1965) 333.
[59] J.M. Quenisset and R. Naslan, J. Cryst. Growth 30 (1975) 169.
[60] C.E. Chang, J. Cryst. Growth 44 (1978) 168; 178.
[61] N. Kobayashi and W.R. Wilcox, J. Cryst. Growth 59 (1982) 616.
[62] Y. Kamotani, S. Ostrach and M. Vargas, J. Cryst. Growth 66 (1984) 83.
[63] J.Y. Murthy, J. Cryst. Growth 83 (1987) 23.
[64] G.W. Young and A. Chait, J. Cryst. Growth 96 (1989) 65.
[65] C.W. Lan and S. Kou, J. Cryst. Growth 108 (1991) 351.
[66] C.W. Lan and S. Kou, J. Cryst. Growth 119 (1992) 281.
[67] C.W. Lan and S. Kou, J. Cryst. Growth 118 (1992) 151.
[68] C.E. Chang and W.R. Wilcox, J. Cryst. Growth 28 (1975) 8.
[69] C.E. Chang and W.R. Wilcox, Int. J. Heat Mass Transfer 19 (1976) 355; 20 (1977) 891.
[70] C,E. Chang, W.R. Wilcox and R.A. Lefever, Mater. Res. Bull. 14 (1979) 527.
[71] D. Schwabe, A. Scharmann, F. Preisser and R. Oeder, J. Cryst. Growth 43 (1978) 305.
[72] D. Schwabe and A. Scharmann, J. Cryst. Growth 46 (1979) 125.
[73] D. Schwabe and A. Scharmann, J. Cryst. Growth 52 (1981) 435.
[74] J. Barthel, K. Eichler, M. Jurisch and W. Löser, Kristall und Technik 14 (1979) 637.
[75] P.A. Clark and W.R. Wilcox, J. Cryst. Growth 50 (1980) 461.
[76] Ch.-H. Chun and D. Schwabe, in: Convective Transport and Instability Phenomena, eds. J. Zierep and H. Oertel (Braun, Karlsruhe 1982) p. 297.

[77] P. Vanek, S. Kadeckova, M. Jurisch and W. Löser, J. Cryst. Growth 63 (1983) 191.
[78] J.L. Duranceau and R.A. Brown, J. Cryst. Growth 75 (1986) 367.
[79] G. Müller, J. Cryst. Growth 99 (1990) 1242.
[80] D. Schwabe, R. Velten and A. Scharmann, J. Cryst. Growth 99 (1990) 1258.
[81] M. Jurisch and W. Löser, J. Cryst. Growth 102 (1990) 214.
[82] M. Jurisch, J. Cryst. Growth 102 (1990) 223.
[83] A. Mühlbauer, W. Erdmann and W. Keller, J. Cryst. Growth 64 (1983) 529.
[84] A. Mühlbauer and E. Sirtl, Phys. Status Solidi (a) 23 (1974) 555.
[85] D.N. Riahi and J.S. Walker, J. Cryst. Growth 94 (1989) 635.
[86] K.H. Lie, J.S. Walker and D.N. Riahi, J. Cryst. Growth 100 (1990) 450.
[87] Ch.-H. Chun, Forschungsber W 77-42, Bundesminister. Forschung und Technik, Bonn (1977).
[88] H.-F. Hadamovsky, Neue Hütte 10 (1965) 449; 11 (1966) 28.
[89] H.-F. Hadamovsky (ed.), Halbleiterwerkstoffe (2nd ed.) (Dtsch. Verl. Grundstoffind., Leipzig, 1972).
[90] K.E. Benson, Chem. Eng. Progr. 64 (1968) 93.
[91] E. Spenke, in: Semiconductor Silicon (Electrochem. Soc., New York, 1969).
[92] E. Spenke and W. Heywang, Phys Solidi (a) 64 (1981) 11.
[93] R.L. Collins, J. Cryst. Growth 42 (1977) 490.
[94] H. Silbernagel, Technik 50 (1978) 611.
[95] W. Keller and A. Mühlbauer, (Marcel Dekker, New York, 1981).
[96] W. Dietze, W. Keller and A. Mühlbauer, in: Crystals, Properties and Application: Silicon, ed. J. Grabmaier (Springer, Berlin, 1981).
[97] H.G. Kramer, Solid State Technol. (1983) 137.
[98] W. Zulehner, Mater. Sci. Eng. B4 (1989) 1.
[99] L. Jastrzebski and P. Zauzucchi, in: Semiconductor Silicon, eds. H.R. Huff, R.J. Kriegler and Y. Takeishi (Electrochem. Soc., Pennington, NJ, 1981).
[100] W. Keller and H. Stut, Feinwerktechnik 75 (1971) 207.
[101] K.P. Gupta, R.O. Gregory and M. Rossnick, J. Cryst. Growth 44 (1978) 526.
[102] H.J. Fenzl, W. Erdmann, A. Mühlbauer and J.M. Welter, J. Cryst. Growth 68 (1984) 771.
[103] E. Wolf and W. Schröder, Cryst. Res. Technol. 22 (1987) K21.
[104] W.C. Dash, J. Appl. Phys. 30 (1959) 459.
[105] G. Ziegler, Z. Naturforsch. 16a (1961) 219.
[106] T.F. Ciszek, in: Semiconductor Silicon 1969, eds. R.R. Haberecht and E.L. Kern (Electrochem. Soc. New York, 1969).
[107] T.F. Ciszek, J. Cryst. Growth 10 (1971) 263.
[108] W.R. Wilcox, J. Cryst. Growth 7 (1970) 203.
[109] W. Keller and A. Mühlbauer, Phys. Status Solidi (a) 25 (1974) 149.
[110] H. Kimura, D.J.O'Connor et al., SPIE Infrared Detector Mater. 285 (1981) 143.
[111] A. Trainor, B.E. Bartlett, Solid State Electron. 2 (1961) 106.
[112] N. DE Leon, J. Guldberg and J. Salling, J. Cryst. Growth 55 (1981) 406.
[113] H. Kimura, M.F. Harvey et al., J. Cryst. Growth 62 (1983) 523.
[114] K.H. Lie, J.S. Walker and D.N. Riahi, J. Cryst. Growth 109 (1991) 167.
[115] W. Schröder, H.-J. Rost and E. Wolf, Cryst. Res. Technol. 24 (1989) 3.
[116] R.N. Thomas, H.M. Hobgood, P.S. Ravishankar and T.T. Braggins, J. Cryst. Growth 99 (1990) 643.
[117] H. Föll, U. Gösele and B.O. Kolbesen, J. Cryst. Growth 40 (1977) 90.
[118] B.O. Kolbesen and A. Mühlbauer, Solid State Electron. 25 (1982) 759.
[119] P.J. Roksnoer, W.J. Bartels and C.W.T. Bulle, J. Cryst. Growth 35 (1976) 245.
[120] P.J. Roksnoer and M.M.B. van den Boom, J. Cryst. Growth 53 (1981) 563.
[121] A.N. Christensen, J. Cryst. Growth 33 (1976) 99.
[122] G. van Drunen and S. Saimoto, J. Cryst. Growth 11 (1971) 151.
[123] R.W. Johnson, J. Appl. Phys. 34 (1963) 352.
[124] E.S. Johnson, J. Cryst. Growth 30 (1975) 249.
[125] D.B. Gasson, J. Sci. Instrum. 42 (1965) 114.
[126] K. Muto and K. Awazu, Jap. J. Appl. Phys. 8 (1969) 1360.
[127] H. Takei, J. Cryst. Growth 44 (1978) 629.

[128] C.W. Lan and S. Kou, J. Cryst. Growth 108 (1991) 340.
[129] C.W. Lan and S. Kou, J. Cryst. Growth 108 (1991) 541.
[130] P. Gerthsen, Z. Angew. Phys. 15 (1963) 301.
[131] W. Class, H.R. Nasor and G.T. Murray, in: Crystal Growth, ed. H.S. Peiser (Pergamon, Oxford, 1967) p. 75.
[132] W. Class, J. Cryst. Growth 3/4 (1968) 241.
[133] W. Borchert, W. Dietz and H. Kölker, Z. Angew. Phys. 29 (1970) 277.
[134] Z. Trousil, Czech. J. Phys. B 12 (1962) 227.
[135] Y. Kumashiro, A. Itoh, E. Sakuma and S. Misawa, J. Cryst. Growth 52 (1981) 495.
[136] Y. Kumashiro, E. Sakuma et al., J. Cryst. Growth 52 (1981) 597.
[137] C. Kooy and H.J.M. Couwenberg, Philips Techn. Rundsch. 23 (1961/62) 143.
[138] M.R. Null and W.W. Lozier, Rev. Sci. Instrum. 29 (1958) 163.
[139] A. Eyer, B.O. Kolbesen and R. Nitsche, J. Cryst. Growth 57 (1982) 145.
[140] A. Eyer, R. Nietsche and H. Zimmermann, J. Cryst. Growth 47 (1979) 219.
[141] A.M. Balbashov and S.K. Egorov, J. Cryst. Growth 52 (1981) 498.
[142] K. Eickhoff and K. Gürs, J. Cryst. Growth 6 (1969) 21.
[143] C.A. Burrus and J. Stone, Appl. Phys. Lett. 26 (1975) 318.
[144] C.A. Burrus and L.A. Coldren, Appl. Phys. Lett. 31 (1977) 383.
[145] C.A. Burrus and J. Stone, J. Appl. Phys. 49 (1978) 3118.
[146] R.W. Gurtler, A. Baghdadi, R.J. Ellis and J.A. Lesk, J. Electron. Mater. 7 (1978) 441.
[147] E. Vanderschueren and E. Votava, Rev. Sci. Instrum. 33 (1962) 389.
[148] A. Calverley, M. Davis and R.F. Lever, J. Sci. Instrum. 34 (1957) 142.
[149] F.E. Birbeck and A. Calverley, J. Sci. Instrum. 36 (1959) 460.
[150] M. Cole, C. Fisher and I.A. Bucklow, J. Appl. Phys. 12 (1961) 577.
[151] S.D. Hurwitt and M.A. Adams, Met. Progr., Febr. (1963).
[152] E.A. Brandes and M.J. Flynn, J. Sci. Instrum. 43 (1966) 716.
[153] J.B. Ketterson, J.S. Tait and L.R. Windmiller, J. Cryst. Growth 1 (1967) 323.
[154] A. Adbelallatif, E.B. Bas and G. Wulff, Z. Angew. Math. Phys. 18 (1967) 339
[155] K.J. Bachmann, J. Cryst. Growth 18 (1973) 13.
[156] R. Göbel, M. Jurisch, W. Löser et al., Cryst. Res. Technol. 21 (1986) 1015.
[157] L. Neumann and R.A. Huggins, Rev. Sci. Instrum. 33 (1962) 433.
[158] E.B. Bas and H. Stevens, Z. Angew. Math. Phys. 18 (1967) 747.
[159] N. Hemmat and A.L. Mac Donald, Rev. Sci. Instrum. 32 (1961) 601.
[160] O. Böttger and E. Richter, Z. Naturforsch. 17a (1962) 526.
[161] P.J. Roksnoer, J. Cryst. Growth 68 (1984) 596.
[162] A. Eyer, H. Leiste and R. Nitsche, J. Cryst. Growth 71 (1985) 173.
[163] A. Eyer and H. Leiste, J. Cryst. Growth 71 (1985) 249.
[164] A.J.R. de Kock and P.J. Roksnoer, J. Cryst. Growth 28 (1975) 125.
[165] B. Jakowlew, W. Jeske and J. Weyher, Kristall und Technik 13 (1978) 553.
[166] A.N. Buzynin, V.V. Zaichkov, V.V. Osiko and V.M. Tatarintsev, Sov. Crystallogr. 34 (1) (1989).
[167] J.H. Braun and R.A. Pellin, J. Electrochem. Soc. 108 (1961) 969.
[168] H. Peibst, Kristall and Technik 1 (1966) 459.
[169] P.M. Petroff and A.J. de Kock, J. Cryst. Growth 30 (1975) 117.
[170] S.T. Belov and V.V. Dobrovenskii, Izvest. AN, Neorg. Mater. 17 (1981) 762.
[171] F. Eckart and H. Menninger, Exp. Technik Physik 9 (1961) 253.
[172] F. Eckart and H. Lemke, Mber. Dtsch. Akad. Wiss. Berlin 4 (1962) 401.
[173] W. Geil, A. Lebek and K. Schmugge, Kristall und Technik 11 (1976) 189.
[174] W. Geil and K. Schmugge, Kristall und Technik 11 (1976) 1181.
[175] W. Geil and K. Schmugge, Kristall und Technik 14 (1979) 343.
[176] E.S. Greiner and H. Mason, J. Appl. Phys. 35 (1964) 3058.
[177] G.W. Green, J. Sci. Instrum. 38 (1961) 167.
[178] Z. Ban and M. Sikirica, Croat. Chim. Acta 34 (1962) 175.
[179] G.D. King, J. Sci. Instrum. 1 (1968) 923.
[180] K. Hoffmann, M. Pasemann and S. Sevcik, Cryst. Res. Technol. 19 (1984) 991.

[181] Ph. Albert, Chimia 14 (1960) 218.
[182] G. Chaudron, Receuil 79 (1960) 502.
[183] G. Revel, C.R. Acad. Sci. Paris C 259 (1964) 4031.
[184] E.S. Greiner, J. Appl. Phys. 30 (1959) 598.
[185] C.P. Talley, J. Appl. Phys. 32 (1961) 1787.
[186] C.B. Hood and M.O. Thursten, J. Electrochem. Soc. 109 (1962) 66.
[187] W. Borchert and E. Born, Kristall und Technik 4 (1969) 293.
[188] B. Schaub and G. Cabane. C.R. Acad. Sci. Paris C 257 (1963) 444.
[189] S. Jönsson, J. Cryst. Growth 63 (1983) 116.
[190] M. Kawakumi, S. Kunil, T. Komatsubara and T. Kasuya, Solid State Commun. 36 (1980) 435.
[191] S. Hayashi, S. Ono and H. Komatsu, Kristall und Technik 13 (1978) 263.
[192] S. Hayashi, T. Inoue and H. Komatsu, Kristall und Technik 14 (1979) K1.
[193] J. Le. Héricy, C.R. Acad. Sci. Paris C 251 (1960) 1385.
[194] J.Le. Héricy, C.R. Acad. Sci. Paris C 251 (1960) 1509.
[195] Y. Isikawa, M. Bajaja et al., Solid State Commun. 22 (1977) 573.
[196] T. Tanaka, E. Bannai et al., J. Cryst. Growth 40 (1977) 125.
[197] R.I. Smith and J.L. Rutherford, Trans AIME 209 (1957) 478.
[198] H. Hillmann and A. Mager, Z. Metallk. 51 (1960) 663.
[199] H. Hillmann, Z. Metallk. 53 (1962) 256.
[200] E.D. Swets, J. Appl. Phys. 33 (1962) 1893.
[201] P. Vanek and S. Kadeckova, J. Cryst. Growth 47 (1979) 458.
[202] P. Vanek, Crystal Res. Technol. 19 (1984) 1207.
[203] S. Kadeckova and B. Sestak, Brit. J. Appl. Phys. 14 (1963) 927.
[204] S. Kadeckova and B. Sestak, Canad. J. Phys. 45 (1967) 1041.
[205] S. Kadeckova and B. Sestak, Kristall und Technik 4 (1969) 243.
[206] S. Kadeckova, P. Toula and J. Adàmek, J. Cryst. Growth 83 (1987) 410.
[207] M. Whittaker, J. Cryst. Growth 3/4 (1968) 317.
[208] R.N. Storey and R.A. Laudise, J. Cryst. Growth 6 (1970) 261.
[209] S. Otani and T. Tanaka, J. Cryst. Growth 51 (1981) 381.
[210] D. Ford, J. Cryst. Growth 94 (1989) 85.
[211] B.J. Curtis and H. Graffenberger, Mater. Res. Bull. 1 (1966) 27.
[212] T. Tanaka, E. Bannai, S. Kawai and T. Yamane, J. Cryst. Growth 30 (1975) 193.
[213] S. Otani, T. Tanaka and Y. Ishizawa, J. Cryst. Growth 106 (1990) 498.
[214] S. Otani, T. Tanaka and Y. Ishizawa, J. Cryst. Growth 108 (1991) 425.
[215] J.D. Verhoeven, E.D. Gibson, M.A. Noack and R.J. Conzemius, J. Cryst. Growth 36 (1976) 115.
[216] M.A. Noack and J.D. Verhoeven, J. Cryst. Growth 49 (1980) 595.
[217] K. Takagi and M. Ishii, J. Cryst. Growth 40 (1977) 1.
[218] K. Kitahama, M. Hiratani and S. Kwai, J. Cryst. Growth 62 (1983) 177.
[219] K. Nishida, K. Kitahama and S. Kawai, J. Cryst. Growth 62 (1983) 475.
[220] E. Buehler, Trans. AIME 212 (1958) 694.
[221] J.M. Savickij, Ch.V. Kopeckij, A.I. Pekarev and M.I. Novosadov, Izvest. AN Ser. Techn. Nauk Metallurg., Topliva 6 (1961) 74.
[222] J.M. Savickij, Ch.V. Kopeckij, A.I. Pekarev and M.I. Novosadov, Zavod. Lab. 27 (1961) 1041.
[223] J.M. Savickij and Z.-Z. Tao, Izvest. AN Ser. Techn. Nauk Metallurg, Topliva 4 (1962) 133.
[224] E. Zedler, D. Müller and U. Wiesner, Z. Metallk. 56 (1965) 316.
[225] J. Barthel and R. Scharfenberg, in: Crystal Growth, ed. S.H. Peiser (Pergamon, Oxford, 1967) p. 133.
[226] J. Barthel and K. Eichler, Kristall und Technik 2 (1967) 205.
[227] J. Barthel, R. Göbel et al., J. Cryst. Growth 52 (1981) 369.
[228] D.J. Capp, H.W. Evans and B.L. Eyre, J. Less-Common Metals 40 (1975) 9.
[229] V.Z. Balan, J. Barthel et al., Kristall und Technik 11 (1976) 1041.
[230] M. Jurisch, W. Löser et al., Cryst. Res. Technol. 17 (1982) 963.
[231] V.G. Glebovsky, V.N. Semenov and V.V. Lomeyko, J. Cryst. Growth 87 (1988) 142.
[232] A.N. Christensen, J. Cryst. Growth 33 (1976) 58.
[233] R. Koksbang and S.E. Rasmussen, J. Cryst. Growth 72 (1985) 745.

[234] H.G. Sell and W.M. Grimes, Rev. Sci. Instrum. 35 (1964) 64.
[235] M.R. Achter, L.C. Vold and T.G. Digges, Trans. AIME 236 (1966) 1597.
[236] T.B. Reed, in: Crystal Growth, ed. H.S. Peiser (Pergamon, Oxford, 1967) p. 39.
[237] H. Naramoto and K. Kamada, J. Cryst. Growth 20 (1973) 313.
[238] H. Naramoto and K. Kamada, J. Cryst. Growth 24/25 (1974) 531.
[239] H. Naramoto and K. Kamada, J. Cryst. Growth 30 (1975) 145.
[240] V.G. Glebovsky, I.V. Kapchenko, V.V. Kiryenko, A.N. Oblivantzev and A.G.Rybasov, J. Cryst. Growth 74 (1986) 529.
[241] K. Nakano, K. Nakamura, Y. Kumashiro and E. Sakuma, J. Cryst. Growth 52 (1981) 602.
[242] K. Nakano, M. Doi, K. Kuwayama and T. Imura, J. Less-Common Metals 82 (1981) 309.
[243] M.E. Packer and M.J. Murray, J. Cryst. Growth 16 (1972) 240.
[244] Y. Kumashiro, E. Sakuma et al., J. Less-Common Metals 75 (1980) 187.
[245] S. Otani, T. Tanaka and Y. Ishizawa, J. Cryst. Growth 62 (1983) 211.
[246] A.N. Christensen, Acta Chim. Scand. A 31 (1977) 77.
[247] B. Scheerer, J. Cryst. Growth 49 (1980) 61.
[248] S. Hayashi and H. Komatsu, Kristall und Technik 14 (1979) 761.
[249] J.H. Wernick, D. Dorsi and J.J.Byrnes, J. Electrochem. Soc. 106 (1959) 245.
[250] V.J. Albano and R.R. Soden, J. Electrochem. Soc. 113 (1966) 511.
[251] R.R. Soden and V.J. Albano, J. Electrochem. Soc. 115 (1968) 766.
[252] S. Hornfeldt, J.B. Ketterson and L.R. Windmiller, J. Cryst. Growth 5 (1969) 289.
[253] J.J. Vuillemin and Ch. Y. Lee, J. Cryst. Growth 100 (1990) 279.
[254] G.A. Geach and F.O. Jones, J. Less-Common Metals 1 (1959) 56.
[255] A. Lawley and R. Maddin, Acta Metallurg. 8 (1960) 896.
[256] R.R. Soden, G.F. Brennert and E. Buehler, J. Electrochem. Soc. 112 (1965) 77.
[257] B.L. Mordike, Z. Metallk. 52 (1961) 587.
[258] B.L. Mordike, Z. Metallk. 53 (1962) 586.
[259] A.A. Kralina, T.V. Zoltikova, V.O. Esin and V.A. Sazonova, Fiz. Met. Metalloved. 25 (1968) 333.
[260] S. Otani, T. Tanaka and Y. Ishizawa, J. Cryst. Growth 55 (1981) 431.
[261] L.E. Conroy and A.N. Christensen, J. Solid State Chem. 20 (1977) 205.
[262] T.R. Cass, R.W. Quinn and W.R. Spencer, J. Cryst. Growth 2 (1968) 413.
[263] Y. Kumashiro, A. Itoh and S. Misawa, J. Less-Common Metals 32 (1973) 21.
[264] Y. Kumashiro, A. Itoh and S. Misawa, Jap. J. Appl. Phys. 15 (1976) 921.
[265] F. Yajima, T. Tanaka, E. Bannai and S. Kawai, J. Cryst. Growth 47 (1979) 493.
[266] S. Otani, T. Tanaka and Y. Ishizawa, J. Cryst. Growth 71 (1985) 615.
[267] S. Otani, T. Tanaka and Y. Ishizawa, J. Cryst. Growth 80 (1987) 241.
[268] S. Otani, T. Tanaka and Y. Ishizawa, J. Cryst. Growth 83 (1987) 481.
[269] S. Otani, T. Tanaka and Y. Ishizawa, J. Cryst. Growth 92 (1988) 8.
[270] S. Otani, T. Tanaka and Y. Ishizawa, J. Cryst. Growth 92 (1988) 359.
[271] S. Otani, T. Tanaka and Y. Ishizawa, J. Cryst. Growth 97 (1989) 522.
[272] P. Beucherie, M. Block and J.G. Wurm, Metallurgie 5 (1964) 385.
[273] J. Billingham, S.P. Bell and H.M. Lewis, J. Cryst. Growth 13/14 (1972) 693.
[274] Y. Hou, S. Otani, T. Tanaka and Y. Ishizawa, J. Cryst. Growth 68 (1984) 733.
[275] A.N. Christensen and P. Roehammer, J. Cryst. Growth 38 (1977) 281.
[276] J. Doerschel, F.G. Kirscht and R. Behr. Kristall und Technik 12 (1977) 1191.
[277] M. Jurisch, K.H. Berthel and H.-J. Ullrich, Phys. Status Solidi (a) 44 (1977) 277.
[278] T. Callaghan, J. Schwanebeck, L. Toth, M. Dayan and A.M. Goldmann, J. Appl. Phys. 49 (1978) 2523.
[279] G. Behr, K. Bartsch et al., Cryst. Res. Technol. 20 (1985) K93.
[280] R.G. Carlson, J. Electrochem. Soc. 106 (1959) 49.
[281] H.W. Schadler, Trans. AIME 218 (1960) 649.
[282] R.H. Schnitzel and G.H. Keith, J. Less-Common Metals 4 (1962) 98.
[283] E.-M. Herbst, D. Schulze and H. Wadewitz, Phys. Status Solidi 3 (1963) K151.
[284] D.R. Hay and E. Scala, Trans. AIME 233 (1965) 1153.
[285] D.R. Hay, R.K.Skogerboe and E. Scala, J. Less-Common Metals 15 (1968) 121.
[286] E.H. Carlson, J. Cryst. Growth 1 (1967) 271.

[287] J.E. Murphy and M.M. Wong, J. Less-Common Metals 40 (1975) 65.
[288] T. Tanaka, S. Otani and Y. Ishizawa, J. Cryst. Growth 73 (1985) 31.
[289] I. Tanaka, Y. Sato and H. Kojima, J. Cryst. Growth 99 (1990) 837.
[290] J.P. Langeron, C.R. Acad. Sci. Paris C 256 (1963) 4436.
[291] J.P. Langeron, C.R. Acad. Sci. Paris C 256 (1963) 5570.
[292] T. Flottmann, M. Beya, P. Pelzer, H. Wenzl and A. Heiming, J. Cryst. Growth 87 (1988) 213.
[293] D.W. Lee and J.S. Haggerty, J. Amer. Ceram. Soc. 52 (1969) 641.
[294] S. Otani, T. Tanaka and A. Hara, J. Cryst. Growth 51 (1981) 164.
[295] A. Mackie and C.H. Hinrichs, J. Cryst. Growth 87 (1988) 101.
[296] G. Petit-Le Du and J. Aubree, Rev. Phys. Appl. 2 (1967) 264.
[297] B. Cockayne, M. Chesswas and D.B. Gasson, J. Mater. Sci. 2 (1967) 7.
[298] W.G. Field and R.W. Wagner, J. Cryst. Growth 3/4 (1986) 799.
[299] A. Revcolevschi and R. Colongues, C.R. Acad. Sci. Paris C 266 (1968) 1767.
[300] R.T. Cox, A. Revcolevschi and R. Collongues, J. Cryst. Growth 15 (1972) 301.
[301] G. Dhalenne, A. Revcolevschi and R. Collongues, Mater. Res. Bull. 7 (1972) 933.
[302] M. Saito, J. Cryst. Growth 71 (1985) 664.
[303] M. Saito, J. Cryst. Growth 74 (1986) 385.
[304] D.B. Gasson and B. Cockayne, J. Mater. Sci. 5 (1970) 100.
[305] K. Kitazawa, K. Nagashima et al., J. Cryst. Growth 39 (1977) 211.
[306] D.Y. Tang and R.K. Route, J. Cryst. Growth 91 (1988) 81.
[307] D.M. Brown and F.K. Heumann, J. Appl. Phys. 35 (1964) 1947.
[308] T.Y. Tien and R.M. Garretson, J. Cryst. Growth 16 (1972) 177.
[309] K. Nakanishi, T. Mori, Y. Shintani and O. Tada, Jap. J. Appl. Phys. 14 (1975) 1231.
[310] I. Tanaka, H. Kojima, and F. Sudo, J. Cryst. Growth 76 (1986) 311.
[311] S. Takekawa, H. Nozaki, A. Umezono et al., J. Cryst. Growth 92 (1988) 687.
[312] H.D. Bredy, J.S. Haggerty, M.J. Cima et al., J. Cryst. Growth 96 (1989) 225.
[313] Z. Lu, L.V. Moulto, R.S. Feigelson, R.J. Raymakers and P.N. Peszkin, J. Cryst. Growth 106 (1990) 732.
[314] N. Ii and I. Shindo, J. Cryst. Growth 46 (1979) 569.
[315] Y.P. Oudalov and Z.S. Medvedeva, Mater. Res. Bull. 4 (1969) 887.
[316] I. Tanaka, T. Obuchi and H. Kojiama, J. Cryst. Growth 87 (1988) 169.
[317] W.E. Kramer, A.M. Stewart and R.H. Hopkins, J. Cryst. Growth 73 (1985) 329.
[318] A. Jouini and L.-C. Dufour, J. Cryst. Growth 50 (1980) 873.
[319] W.S. Brower and H.S. Parker, J. Cryst. Growth 8 (1971) 227.
[320] R.D. Schmidt-Whitley, M. Martinez-Clemente and A. Revcolevschi, J. Cryst. Growth 23 (1974) 113.
[321] J.L. Loison, M. Robino and C. Schwab, J. Cryst. Growth 50 (1980).
[322] J. Berthon, A. Revcolevschi, H. Morikawa and B. Touzelin, J. Cryst. Growth 47 (1979) 736.
[323] V.A.M. Brabers, T.E. Whall and P.S.A. Knapen, J. Cryst. Growth 69 (1984) 101.
[324] J.M. Whelan and G.H. Wheatley, J. Phys. Chem. Solids 6 (1958) 169.
[325] F.A. Cunnell and R. Wickham, J. Sci. Instrum. 37 (1960) 410.
[326] M. Lorant, Chem. Rundsch. 14 (1961) 469.
[327] A.N. van Mau, C. Ance and G. Bougnot, Mater. Res. Bull. 3 (1968) 901.
[328] T. Hashimoto, N. Ishizawa, N. Mizutani and M. Kato, J. Cryst. Growth 84 (1987) 207.
[329] M. Kimura, S. Nanamatsu et al., Jap. J. Appl. Phys. 11 (1972) 904.
[330] J. Iida, S. Takekawa and N. Kimizuka, J. Cryst. Growth 102 (1990) 398.
[331] Ji. Yangyang, Z. Shuqing, H. Yijing et al., J. Cryst. Growth 112 (1991) 283.
[332] M. Palaniswamy, F.D. Gnanam, P. Ramasamy and G.S. Laddha, Cryst. Res. Technol. 17 (1982) 911.
[333] K.S. Raju, M. Palaniswamy, P. Ramasamy and G.S. Laddha, Cryst. Res. Technol. 19 (1984) 939.
[334] F. Bretheau-Raynal, N. Tercier, B. Blanzat and M. Drifford, Mater. Res. Bull. 15 (1980) 639.
[335] H. Hanede, M. Isobe, A. Watanabe and S. Shirasaki, J. Cryst. Growth 92 (1988) 155.
[336] K. Kitamura, S. Kimura and S. Hosoya. J. Cryst. Growth 48 (1980) 469.
[337] T. Inoue, H. Komatsu, S. Hosoya and H. Takei, J. Cryst. Growth 55 (1981) 307.
[338] T. Inoue, H. Komatsu et al., J. Cryst. Growth 57 (1982) 203.
[339] S. Hosoya and H. Takei, J. Cryst. Growth 57 (1982) 343.
[340] V.A.M. Brabers, J. Cryst. Growth 8 (1971) 26.

[341] Z.K. Kun, W.E. Kramer and G.W. Roland, J. Cryst. Growth 58 (1982) 122.
[342] I. Shindo and H. Komatsu, J. Cryst. Growth 34 (1976) 152.
[343] J.G. Bednorz and H. Arend, J. Cryst. Growth 67 (1984) 660.
[344] J. Kumar, M. Thirumavalavan, F.D. Gnanam and P. Ramasamy, Cryst. Res. Technol. 23 (1988) 1337.
[345] K. Kitamura, S. Kimura, Y. Miyazawa, Y. Mori and O. Kamada, J. Cryst. Growth 62 (1983) 351.
[346] K. Kitamura, M. Tsutsumi, S. Kimura and H. Komatsu, J. Cryst. Growth 67 (1984) 656.
[347] K. Kitamura, S. Kimura et al., J. Cryst. Growth 69 (1984) 537.
[348] B. Cockayne, D.B. Gasson and N. Forbes, J. Mater. Sci. 5 (1970) 837.
[349] K.P. Poplawsky, J. Appl. Phys. 33 (1962) 1616.
[350] R.P. Poplawsky and M.R. Wright J. Cryst. Growth 1 (1967) 139.
[351] E.A. Weaver, H.D. Merchaut and R.P. Poplawsky, J. Amer. Ceram. Soc. 52 (1969) 214.
[352] G. Dhalenne, A. Revcolevschi and A. Gervais, J. Cryst. Growth 44 (1978) 297.
[353] D. Elwell, W.L. Kway and R.S. Feigelson, J. Cryst. Growth 71 (1985) 237.
[354] R.J. Fairholme, G.P. Gill and A. Marsh, Mater. Res. Bull. 6 (1971) 1131.
[355] M. Higuchi, T. Hosokawa and S. Kimura, J. Cryst. Growth 112 (1991) 354.
[356] J. Holt, Brit. J. Appl. Phys. 16 (1965) 639.
[357] W.A. Tiller and C.T. Yen, J. Cryst. Growth 109 (1991) 120.
[358] A.T. Chapman and G.W. Clark, J. Amer. Ceram. Soc. 48 (1965) 494.
[359] A.T. Chapman, G.W. Clark et al., J. Amer. Ceram. Soc. 53 (1970) 46.
[360] H.M. O'Bryan and B. O'Connor, Amer. Ceram. Soc. Bull. 45 (1966) 578.
[361] A. Marsh, R.J. Fairholme and G.P. Gill, IEEE Trans. Magn. MAG 7 (1971) 470.
[362] K. Kitamura, N. Ii, I. Shindo and S. Kimura, J. Cryst. Growth 46 (1979) 277.
[363] L.L. Abernethy, T.H. Ramsey and J.W. Ross, J. Appl. Phys. Suppl. 32 (1961) 376.
[364] S. Kimura and I. Shindo, J. Cryst. Growth 41 (1977) 192.
[365] S. Kimura, I. Shindo et al., J. Cryst. Growth 44 (1978) 621.
[366] J. Stone and C.A. Burrus, J. Appl. Phys. 49 (1978) 2281.
[367] H. Tsuiki, K. Kitazawa et al., J. Cryst. Growth 49 (1980) 71.
[368] T. Yamakawa, N. Ishizawa, K. Uematsu, N. Mizutani and M. Kato, J. Cryst. Growth 75 (1986) 623.
[369] T. Vlasak, Elektrie 16 (1962) 135.
[370] J.-E. Jørgensen and S.E. Rasmussen, J. Cryst. Growth 47 (1979) 124.
[371] P.C. Camargo, J.A. Ehlert and F.R. Brotzen, J. Cryst. Growth 57 (1982) 241.
[372] G.A. Geach and F.O. Jones, in: Proc. 3. Plansee-seminar Hochschmelzende Metalle (1958) 77.
[373] A.I. Somov, A.P. Svinarenko and A.G. Tjutjunnik, Izvest. AN Neorg. Mater. 2 (1966) 1892.
[374] D.A. Petrov and A.A. Bukhanova, Izvest. AN Neorg. Mater. 4 (1968) 1433.

CHAPTER 5

Use of a Magnetic Field in Melt Growth

D.T.J. HURLE

H.H. Wills Physics Laboratory, University of Bristol, Tyndall Avenue, Bristol, United Kingdom, BS8 1TL

R.W. SERIES

DRA Malvern, St. Andrews Road, Malvern, Worcester, United Kingdom WR14 3PS

Contents

1. Introduction 261
2. Design of equipment 261
3. The MHD equations 264
4. Czochralski growth with a transverse field 267
5. Czochralski growth with an axial field 268
   5.1. Flow and thermal fields 268
   5.2 Dopant segregation 270
      5.2.1. Dopant striations 270
      5.2.2. Extended Burton, Prim and Slichter analysis 271
      5.2.3. Experimental studies 273
6. Czochralski growth with a cusped field 278
7. Bridgman and floating-zone growth. 279
8. Concluding remarks 283
References. 283

HANDBOOK OF CRYSTAL GROWTH, VOL. 2
edited by D.T.J. Hurle

1. Introduction

Metal and semiconductor melts, having a relatively high electrical conductivity, are subject to a Lorentz force when moving in a magnetic field. This can be used to damp natural convective flows in the melt and, most particularly, to damp out turbulence and thereby remove the dopant striations caused by that turbulence. This was first practically demonstrated independently by Utech and Flemings [1] and by Chedzey and Hurle [2] in 1966.

Equally, a time-varying field can induce motion in the melt and hence can be used to provide forced convection. This concept is rather older and has been practised in the metal castings industry for some time.

This chapter focusses principally on the effects of applying a steady magnetic field to a crystal growth melt.

In the next section we consider the technology involved in generating fields of several configurations applied to the principal growth techniques. In section 3, permissible approximations to the full MHD equations for a conducting fluid are derived.

By far the most research on the use of a magnetic field to date has been on its application to the Czochralski growth of silicon and the III–V compounds. The motivation has been different for the two classes of material. In the case of silicon, the work has been directed at the control of the uptake of oxygen in the crystal whereas, in the case of the III–V compounds, the principal motivation has been the elimination of dopant striations. Both of these topics are treated in sections 4–7. The results of growth experiments in a magnetic field performed using Bridgman and floating-zone techniques are reviewed in section 7.

2. Design of equipment

Experimental studies so far performed indicate that useful field magnitudes are in the range of 0.05 to 0.5 T or perhaps somewhat higher. These fields have to be generated in a large volume. For example, in the Czochralski configuration, the bore must be large enough to embrace the crucible, heater and any necessary heat shields. Thus, for the commercial growth of 6 inch diameter silicon single crystals in the presence of an axial magnetic field, this could require a solenoid having a bore diameter approaching 1 m. Such fields can be generated either by resistive or superconducting magnets.

Conventional wire-wound solenoids are simpler to operate, easier and cheaper to fabricate and are more readily modified, e.g. by the addition of subsidiary windings. However, because of the large volume over which the field has to be generated, their power dissipation is a major problem. For a modest field value of only a few tenths of a tesla it is typically of the order of magnitude of the power supplied to the crucible to melt the charge. For commercial silicon pullers this is about 100 kW. It is therefore vital that the solenoid windings are of as low resistance as possible making high-conductivity copper the only viable choice.

The power dissipation problem forces consideration of superconducting magnet technology. This makes somewhat higher fields practical but at the cost of higher maintenance. Monetary saving is somewhat illusory since, though the cost of providing the cryogenics during the crystal growth period is small compared to the energy cost of a conventional solenoid, this is offset by the higher capital cost and by the need to keep the superconducting magnet cooled during the "down" time.

There is a considerable amount of energy stored in the windings of both types of magnet which produces large stresses on the magnet structures. The cost of manufacture reflects this and depends sensitively on the design. For a conventional solenoid the cost increases roughly as the fourth power of the magnet bore and as the square of the field.

The field also imposes stresses on the crucible heater. In order that these are not periodic which could lead to resonance and fatigue failure, the heater current must be

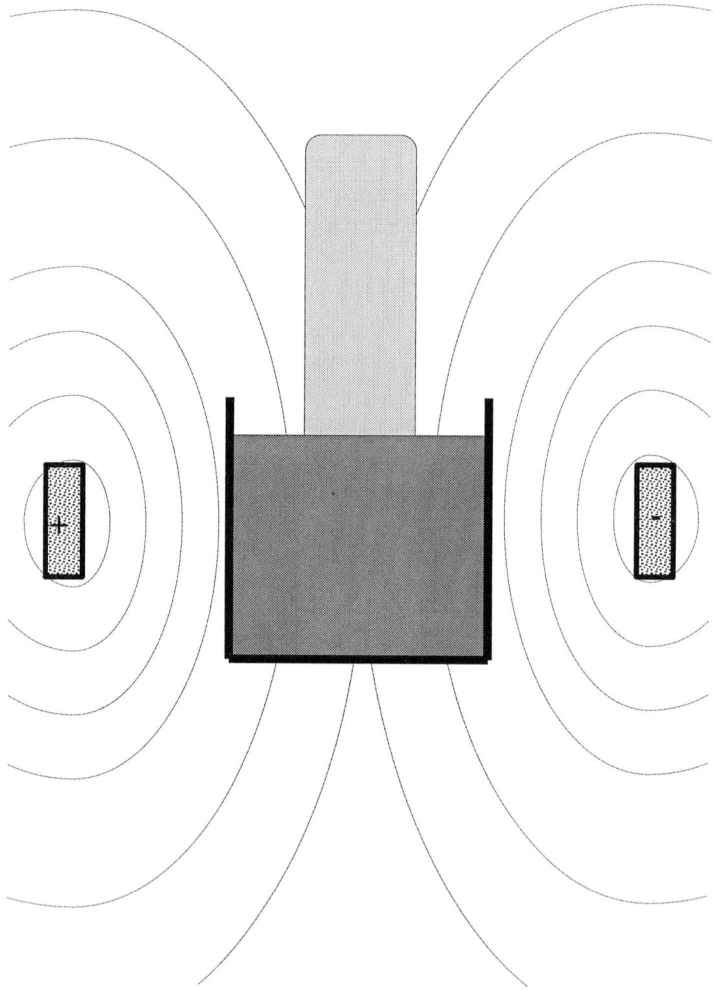

Fig. 1. Axial field configuration: schematic representation of field lines.

DC. This can be provided from simple bridge-rectified phase-angle-controlled AC or more elaborate solid state regulated systems provided always that ripple levels are kept sufficiently low.

The simplest configurations of field that can be applied are of course those provided between the pole pieces of a conventional electromagnet (fig. 1) and that of a simple, singly wound, solenoid (fig. 2) and indeed these are the two configurations which have been most studied. As applied to Czochralski growth, both have been shown to have serious limitations which are considered in sections 4 and 5. This has led to the development of a cusped field configuration, described in section 6, which largely overcomes the deficiencies of the other two configurations (fig. 3). In Bridgman and floating-zone configurations, only simple axial or transverse fields have been studied.

Axial fields are generated with air-cored solenoids. Additional windings can be used to trim the field uniformity [3]. Use of soft iron cores or pole pieces is counter-productive as they tend to reduce the field in the central region of the bore where the melt is situated.

The transverse field provided by a conventional electromagnet tends to be rather non-uniform. The large bore required means that the pole-piece-radius to bore-diameter ratio cannot be made sufficiently large. Air-cored designs are better, but, to keep power consumption within reasonable bounds, superconducting magnet designs are indicated. In the case of Czochralski growth, the current in the "picket fence" melt heater is directed vertically and, in consequence, the force experienced by the heater is much greater with a transverse field than with an axial one.

For reasons which are explained in section 6, the cusped field was designed for use in Czochralski growth to provide an essentially radial field in the plane of the crystal–melt interface, but an axial one deep in the melt. This is achieved by use of a pair of

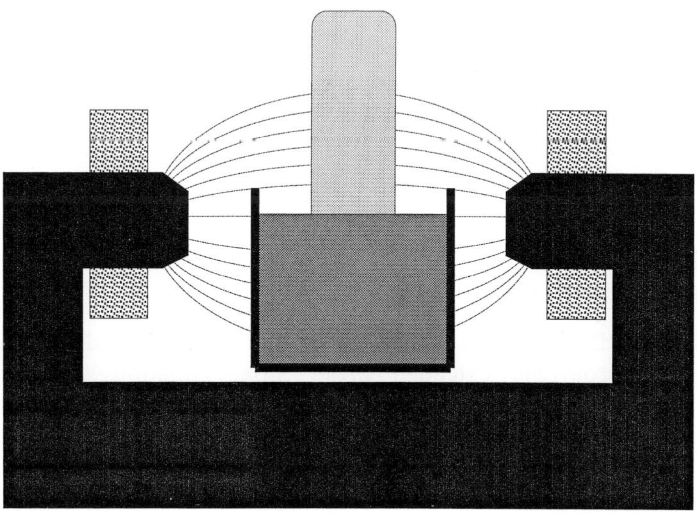

Fig. 2. Transverse field configuration using an eletromagnet: schematic representation.

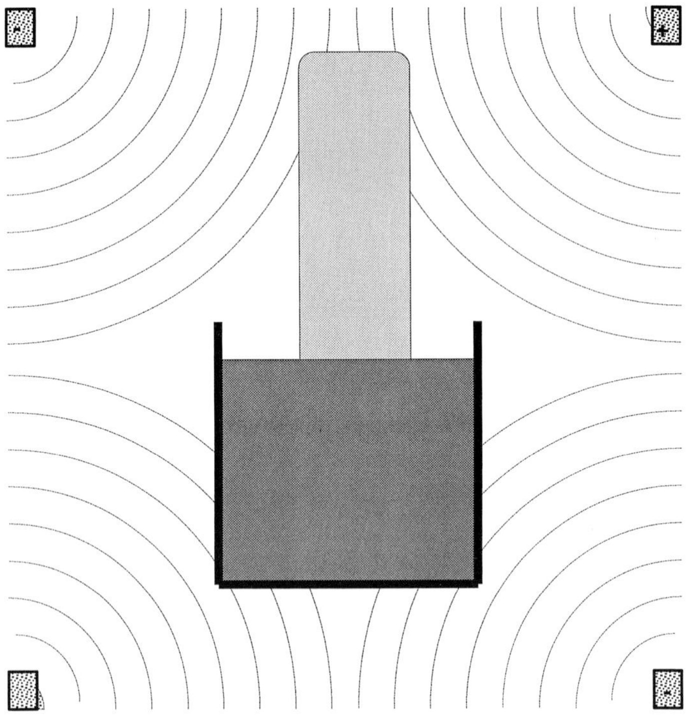

Fig. 3. Schematic representation of a cusped magnetic field formed using a pair of Helmholtz coils operated in opposed current mode.

Helmholtz coils operated in opposed current mode. It has been implemented in two different laboratories using superconducting magnets [4,5].

3. The MHD equations

The relevant considerations are based on the following two physical facts:

(1) The motion of electrically conducting fluid across magnetic lines of force generates a current.

(2) Fluid elements carrying current which traverse magnetic lines of force give rise to additional forces acting on the elements.

The relevant MHD equations result from the combination of Maxwell's equations coupled to the appropriately extended Navier–Stokes equations. The former can be written:

$$\nabla \times \boldsymbol{H} = \boldsymbol{j}, \tag{3.1}$$

$$\nabla \times \boldsymbol{E} = -\dot{\boldsymbol{B}}, \tag{3.2}$$

$$\nabla \cdot \boldsymbol{B} = 0, \tag{3.3}$$

where we have neglected the displacement current, which is a valid approximation provided that the fluid velocity remains non-relativistic. $\boldsymbol{H}$ is the magnetic field, $\boldsymbol{B}$ the magnetic induction, $\boldsymbol{E}$ the electric field and $\boldsymbol{j}$ the current density.

$$\boldsymbol{B} = \mu\mu_0 \boldsymbol{H}, \tag{3.4}$$

where $\mu\mu_0$ is the permeability.

Dissipation is taken into account through the extended form of the phenomenological Ohm's law:

$$\boldsymbol{j} = \sigma(\boldsymbol{E} + \boldsymbol{u} \times \boldsymbol{B}), \tag{3.5}$$

where σ is the electrical conductivity of the fluid (assumed isotropic) and $\boldsymbol{u}$ the vector flow velocity.

From eqs. (3.1)–(3.5) we can eliminate $\boldsymbol{E}$ and $\boldsymbol{H}$ to obtain an equation for the magnetic induction:

$$\dot{\boldsymbol{B}} = \nabla \times (\boldsymbol{u} \times \boldsymbol{B}) - (\mu_0\sigma)^{-1} \nabla \times (\nabla \times \boldsymbol{B}). \tag{3.6}$$

This describes the temporal variation of $\boldsymbol{B}$ as the result of convection (first term on the right-hand side) and of diffusion (second term on the right-hand side). Equation (3.3) remains as an initial condition on eq. (3.6). The ratio of the first to the second term on the right-hand side of eq. (3.6) is of order of magnitude

$$\mathrm{Re}_\mathrm{m} = \sigma\mu_0 U d, \tag{3.7}$$

which is designated the magnetic Reynolds number. U is a characteristic flow velocity and d a typical spatial dimension.

For large flow velocity (i.e., large Re_m), the second term on the right-hand side of eq. (3.6) (the magnetic dissipation term) can be neglected. When Re_m is very small, the first term is small and, in consequence, the flow does not distort the magnetic flux lines significantly. The quantity $(\sigma\mu_0)^{-1} = \nu_\mathrm{m}$ can be thought of as analogous to the kinematic viscosity. The latter is the diffusion coefficient for momentum in the fluid whilst ν_m is the diffusion coefficient for the magnetic field. Thus one can define a magnetic Prandtl number $\mathrm{Pr}_\mathrm{m} = \nu/\nu_m$ which is, for liquid metals, very small. For example, for liquid gallium $\nu_\mathrm{m}/\kappa \sim 10^4$ and $\nu/\kappa \sim 2 \times 10^{-2}$ so that $\mathrm{Pr}_\mathrm{m} \sim 2 \times 10^{-6}$ (κ being the thermal diffusivity).

The essential physics is in the incorporation of the Lorentz force into the equation of motion of an incompressible fluid element from which the hydromagnetic Navier–Stokes equation follows. The Lorentz force is, with neglect of excess charge, (a valid approximation for a liquid metal) [6]:

$$\boldsymbol{L} = \boldsymbol{j} \times \boldsymbol{B}. \tag{3.8}$$

Inserting eq. (3.8) into the equation of motion leads to the appropriate hydromagnetic Navier–Stokes equation.

When the magnetic Reynolds number is very small compared to unity, a situation which pertains under all crystal growth conditions, we can replace $\boldsymbol{B}$ by $\boldsymbol{B}_0$ the externally imposed field, from which it is evident that $\partial \boldsymbol{B}/\partial t = 0$. From eq. (3.2) it can then be deduced that $\boldsymbol{E}$ is restricted to the form $\boldsymbol{E} = -\nabla\Phi$. Substituting eq. (3.5) into

(3.8) yields:

$$L = \sigma(E + u \times B_0) \times B_0.$$

Some authors have neglected the term in E but Baumgartl and Müller [7] have recently shown that, except for the description of steady two-dimensional flows, this is not justified.

In the small Reynolds and Prandtl number limit a considerable simplification of the Navier–Stokes equation is also possible [7] because the flow perturbs the induction due to the imposed field by only a very small amount: i.e. $B = B_0 + b$ where b is the small field induced by electric currents generated by melt flow across the field lines. In this case the equation becomes, with the Boussinesq approximation, when suitably non-dimensionalised (length by d and time by d^2/κ):

$$\frac{\partial u}{\partial t} = -\nabla p - (u \cdot \nabla)u + \mathrm{Pr}\,\nabla^2 u + \mathrm{Ra}\,\mathrm{Pr}\,T e_z + \mathrm{Pr}\,\mathrm{Ha}^2[(u \times e_B - \nabla\Phi \times e_B], \tag{3.9}$$

where $\mathrm{Pr} = \nu/\kappa$ is the Prandtl number, $\mathrm{Ha} = d|B_0|\,(\sigma/\rho\nu)^{1/2}$ is the Hartmann number, $\mathrm{Ra} = \alpha\Delta T d^3/\nu\kappa$, p is the pressure, T the non-dimensional temperature, $e_B = B_0/|B_0|$ and Φ is the electric potential given by

$$\nabla^2\Phi = \nabla(u \times e_B) \tag{3.10}$$

and where α is the fluid expansion coefficient, ρ the fluid density and ΔT the imposed temperature difference.

To these equations are coupled the energy transport equation:

$$\frac{\partial T}{\partial t} = -(u \cdot \nabla)T + \nabla^2 T. \tag{3.11}$$

Note that the Hartmann number (proportional to H) appears squared so that most interactions will scale with the square of the field. Boundary conditions on Φ have to be specified; complications arise when the induced currents can penetrate into the crystal [8].

There is an analogue of the Taylor–Proudman theorem for inviscid flows in the presence of a steady field H_z which states that motions cannot vary in the direction of H (the z-direction) so that all steady motions in an inviscid fluid are two-dimensional [6]. However, as with the Taylor–Proudman theorem, this breaks down in the vicinity of boundary layers where dissipative effects are important.

A most important point to make is that electric currents can transport energy over much greater distances than viscous or thermal mechanisms. This is evident in shear layers close to rigid boundaries with streamlines normal to the field. For example, under an axial field, crystal rotation stirs the melt to a much greater depth than in the nonmagnetic case, while under a transverse field the crucible wall acts to damp the boundary layer flow to a much greater degree.

An additional source of forced convection in a semiconducting melt during crystal growth in a magnetic field has been pointed out by Gel'fgat and Gorbunov [9]. They suggest that the large thermo-EMF of such melts (as distinct from melts of the common crystalline semiconductors such as silicon whose melts are metallic) will, in the presence of the temperature gradients in the melt, give rise to thermo-currents which will interact with the imposed magnetic field, to give a significant additional contribution to the Lorentz force experienced by the melt. They quantify their model and report some initial experiments which appear to demonstrate the effect.

4. Czochralski growth with a transverse field

A transverse field destroys the rotational symmetry of the Czochralski configuration. It removes the axial symmetry of the natural convective flow so that the rotating crystal experiences a cyclical variation in temperature which modulates its growth rate locally resulting in pronounced rotational striations. This has been observed experimentally by Ravishankar et al. [10] for the case of silicon.

Kobayashi [11] has demonstrated that a similarity solution to the flow field in the melt is obtainable for a purely transverse field. He has shown that, in an infinite melt, the Cochran flow [12] due to the rotating crystal "disc" (the classic solution used in the Burton, Prim and Slichter [13] theory) is not affected by a transverse magnetic field provided that the electrical conductivity of the crystal is negligible compared to that of the melt. The analysis is more sensitive to this latter assumption than is the case for an axial field (see below). The transverse field affects only the pressure distribution in the melt. A more rigorous analysis, taking account of the finite size of the crucible has been carried out by Williams et al. [14]. Hence the segregation of conservative dopants should not be affected by such a field other than by the modulation of the growth rate brought about by the induced thermal asymmetry already mentioned.

In practice, field uniformity is a key issue since the generation of an accurately transverse field over the large diameter of a commercial crystal growth melt is very difficult. Even small design errors can result in a significant axial component of field in the vicinity of the crystal–melt interface which will change the effective distribution coefficient and degrade the radial uniformity (see below).

The behaviour of nonconservative dopants depends of course on the way in which the transport of the dopant in the melt is modified by the changes to the pattern of flow brought about by applying the field. Suzuki et al. [15] were the first to report on the use of a transverse field during the Czochralski growth of silicon single crystals. They demonstrated that the field markedly reduced the melt temperature fluctuations and significantly reduced the oxygen concentration in the crystals. It was this latter observation that led to widespread interest in the application of a magnetic field to Czochralski growth.

Obtaining an understanding of the behaviour of oxygen during growth with a transverse field is made difficult because it appears, from the obtained experimental results, that crucible rotation plays a dominant role, unlike the axial field case, where it is crystal rotation which appears to be the more important parameter.

Terashima and Fukuda and their co-workers [16] were the first to apply a magnetic field to the liquid-encapsulation Czochralski growth of gallium-arsenide single crystals. They used a pair of superconducting magnets to apply a transverse field of up to

0.3 T. They showed that the temperature fluctuations were markedly reduced from about 18°C down to 0.1°C as the field was increased from zero to around 0.125 T.

5. *Czochralski growth with an axial field*

A purely solenoidal field around the melt with axis coincident with the axis of the pull rod has been used in studies of the growth of silicon, gallium arsenide and indium phosphide. Its merit is that it retains the axial symmetry which is characteristic of Czochralski growth, but, as we shall see, it destroys the radial uniformity of doping which is conferred by crystal rotation in the absence of a magnetic field.

We first consider its effect on the flow and temperature fields in the melt and then its effect on segregation from which a criterion is derived for the retention of radial dopant uniformity.

5.1. *Flow and thermal fields*

An understanding of the flow field in the bulk of the melt is necessary in order to model the transport of nonconservative dopants (such as oxygen in silicon) to the crystal–melt interface. It is also necessary to establish the thermal stability of the system.

Digital simulation, using the MHD equations outlined in the previous section in either two dimensions (axial symmetry about the pull rod axis assumed) or three dimensions, has been performed by a number of workers [8,17–20]. Considerable insight into the effects of an axial field has been provided by Hjellming and Walker [21–24] who have used asymptotic analysis to deduce the nature of the boundary layers formed.

The factors contributing to the flow, in addition to the effect of the field, are:

(1) The forced flow produced by rotation of the crystal and/or the crucible.

(2) Natural convective flow due to buoyancy forces caused by the temperature gradients within the melt. (Buoyancy effects due to concentration gradients can also be of importance in some instances.)

(3) Flow induced in the melt by the traction produced at the free melt surface (Marangoni effect) resulting from the temperature (and/or concentration) dependence of the melt surface tension.

The operation and interaction of these several effects results in the presence of thin shear layers within the melt shown schematically in fig. 4. These shear layers pose serious grid resolution problems to the numerical modeller requiring considerable computing power to deal with them adequately.

Surprisingly, the addition of a magnetic field to the problem does not actually make the situation any worse since the overall damping effect of the flow can result in a thickening of the boundary layers so that the increased number of equations to be solved simultaneously is offset by a relaxation in grid mesh size.

Analytic modelling of specific aspects of the flow gives insight into mechanisms. Experimental studies in model flow systems help to build confidence in the veracity of

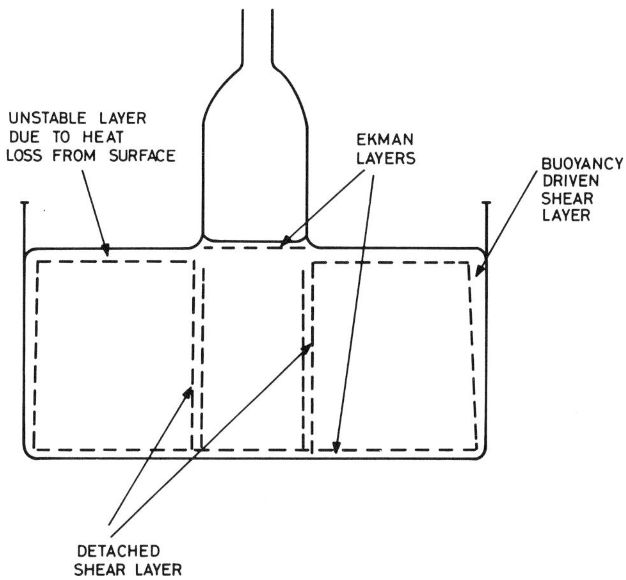

Fig. 4. Schematic representation of the momentum boundary layers present in a Czochralski melt.

the numerical simulation but the inability to scale all the relevant dimensionless parameters appropriately in the model fluid means that the limitations of these model systems must be well understood. In particular, model transparent fluids having $\mathrm{Pr} \ll 1$ do not exist so that they fail to properly represent the thermal fields in real metal or semiconductor melts.

The forced flow generated by the rotating crystal is modified by the presence of an axial magnetic field. Hydromagnetic flow due to an electrically insulating disc rotating in an infinite fluid embedded in a uniform axial magnetic field has been treated by Kakutani [25] and an equivalent analysis has also been performed by Kobayashi [11]. Similarity solutions analogous to those first obtained by von Karman [26] and evaluated by Cochran [12] for the purely hydrodynamic case were obtained. These show that the axial component of the flow is independent of radial and azimuthal position giving a uniformly accessible boundary layer across the disc surface. The flow is characterised by a magnetic interaction parameter,

$$N = B_0^2 \sigma / \rho \Omega, \tag{5.1}$$

where $B_0 = \mu \mu_0 H$ is the uniform magnetic flux density. σ is the electrical conductivity of the melt and ρ its density. Ω is the crystal (disc) rotation rate. Kakutani [25] has shown that a power series expansion in $1/N$ well describes the axial flow for $N > 2$. Hurle and Series [27] have further shown that, by defining a modified magnetic interaction parameter,

$$N' = N + 0.427, \tag{5.2}$$

the power series expansion adequately describes the flow for all N, including $N = 0$. The expression for the axial component of the flow, scaled by the Ekman layer thickness $(\nu/\Omega)^{1/2}$, is:

$$h(z) = -[1 - 2\exp(-(N')^{1/2}z)]/3(N')^{3/2}. \tag{5.3}$$

The form of eq. (5.2) is in keeping with the concept of a "magnetic viscosity". The constant 0.427 prevents N' from going to zero as B_0 becomes zero and represents the effect of the kinematic viscosity of the melt. The assumptions in this analysis have been vindicated in a more detailed study by Organ and Riley [28] and by Hicks and Riley [29]. The analysis has been extended to the case where the far field fluid is rotating with steady angular velocity (representing the effects of crucible rotation) by Cartwright et al. [30]

5.2. Dopant segregation

In conventional Czochralski growth, segregation of conservative dopants is controlled by the forced convection due principally to crystal rotation with some influence due to rotation of the crucible. This is because the solute diffusion boundary layer thickness is significantly less than the momentum boundary layer produced by the rotating crystal. This boundary layer is uniformly accessible, i.e. its thickness is independent of radial position and this confers a similar uniformity on the radial dopant profile. The application of an axial magnetic field modifies the flow profile within the momentum boundary layer thereby changing the effective distribution coefficient of a dopant. This can be described by an extension of the classic Burton, Prim and Slichter (BPS) analysis to include an axial magnetic field as an additional control parameter. As we shall see this provides a valuable approach to understanding the effect of the field on segregation but conditions can occur for which the solute boundary layer extends beyond the momentum boundary layer produced by the rotating crystal at which point the segregation becomes dependent on the details of the flow in the bulk of the melt. Under these conditions full numerical simulation is needed to predict dopant incorporation.

5.2.1. Dopant striations

Müller and co-workers [31] made a study of the temperature fluctuations in a liquid-encapsulated indium phosphide melt and of iron dopant striations in the grown crystals. Temperature measurements in the melt showed chaotic fluctuations of large amplitude in zero field which were transformed into near-periodic oscillations of diminished amplitude and period ~ 0.1 Hz upon application of the field. Irregular striations were seen in the crystals grown in zero field whilst in crystals grown with an applied field the striations were regularly spaced by a distance corresponding to the temporal frequency of the temperature oscillation.

Miyairi et al. [32] and Satoh et al. [33] have grown 50 mm diameter indium phosphide single crystals in an axial field and they too noted that the amplitude of the temperature fluctuations was decreased following application of the field from around 9°C amplitude in zero field down to < 0.3°C in a field of 0.1 T. Resistivity fluctuations in doped crystals were markedly reduced when the field was applied; there was additionally some reduction in dislocation density in the crystals.

Munakata and Tanasawa [34] have demonstrated, in model experiments and numerical simulation, a macroscopic oscillation of the flow in a Czochralski melt which they ascribe as the cause of nonrotational growth striations. The oscillatory flow is produced by an interaction between the buoyancy-driven and the forced flow due to the rotating crucible. They further modelled conditions in a gallium arsenide melt in the presence of an axial magnetic field and demonstrated, by numerical simulation, that the field was effective in suppressing the oscillatory flow.

5.2.2. Extended Burton, Prim and Slichter analysis [27]

Because the axial flow field adjacent to the crystal–melt interface is independent of radius and azimuth we can obtain the solute field in the melt by solution of the mass diffusion equation in the z coordinate alone inserting the expression for the flow velocity given in section 5.1 [eq. (5.3)] added to the suction velocity imparted to the fluid by the growth of the crystal, namely:

$$\frac{D\,\mathrm{d}^2C}{\mathrm{d}z^2}+[V_0-V(z)]\frac{\mathrm{d}C}{\mathrm{d}z}=\frac{\mathrm{d}C}{\mathrm{d}t}=0, \tag{5.4}$$

where $V(z)=(\nu/\Omega)^{1/2}\,h(z)$ and V_0 is the crystal growth rate. The relevant boundary conditions are that the far-field solute concentration is constant (C_0) and that the concentration in the melt at the interface is $C(0)=C_s/k_0$ where C_s is the concentration in the crystal at the interface and k_0 is the interface distribution coefficient. Integration of eq. (5.4) gives the following relationship between k_0 and the effective distribution coefficient $k_{\mathrm{eff}}=C_s/C_0$:

$$k_{\mathrm{eff}}=k_0/[1-(1-k_0)J], \tag{5.5}$$

where

$$J=\int_0^{\infty}\exp\left[-z+\int_0^{z'}\{V(z)/V_0\}\,\mathrm{d}z\right]\mathrm{d}z', \tag{5.6}$$

with $z'=V_0z/D$.

The entity J is related to the parameter $\Delta=V_0\delta/D$ in the BPS formulation, sometimes loosely termed the normalised boundary layer thickness δ. The relation between the two quantities is:

$$\Delta=-\ln(1-J). \tag{5.7}$$

A second significant distance is that of the diffusion profile defined by Wilson [35]:

$$\bar{\Delta}=J. \tag{5.8}$$

Expanding the logarithm in eq. (5.7) we see that the two thicknesses become nearly equal for $J\ll 1$.

To obtain an expression for k_{eff} we have to integrate eq. (5.6) after insertion of the expression for the flow field [eq. (5.3)]. The result is:

$$J=\eta\theta^{-1/2}\int_0^{\infty}\exp[-\eta\theta^{-1/2}z-(z+2\mathrm{e}^{-z}-\tfrac{1}{2}\mathrm{e}^{-2z}-\tfrac{3}{2})/3\theta^2]\,\mathrm{d}z, \tag{5.9}$$

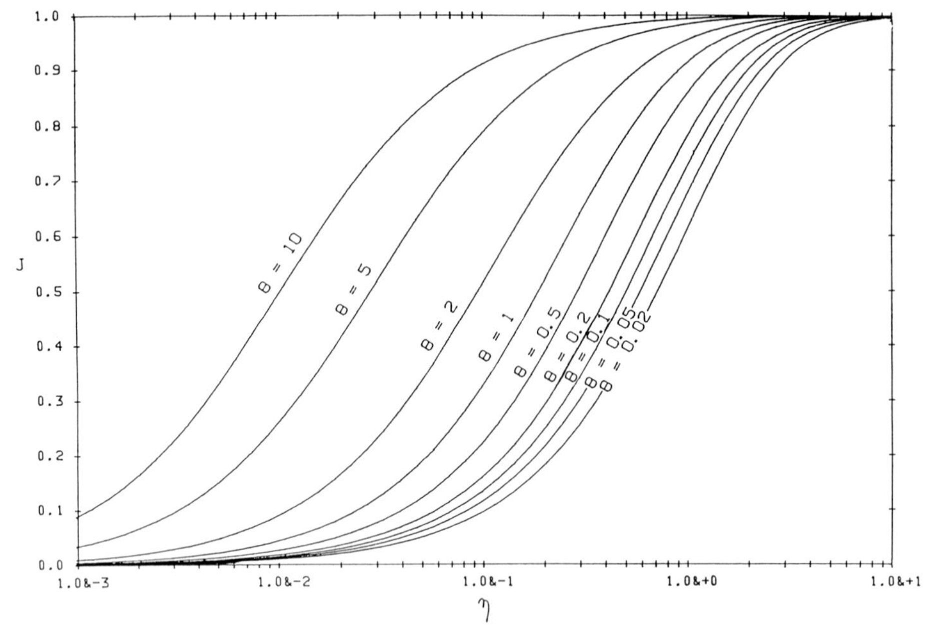

Fig. 5. Plot of the parameter J (see text) as a function of a dimensionless growth rate (η) for various values of the non-dimensional magnetic field (θ).

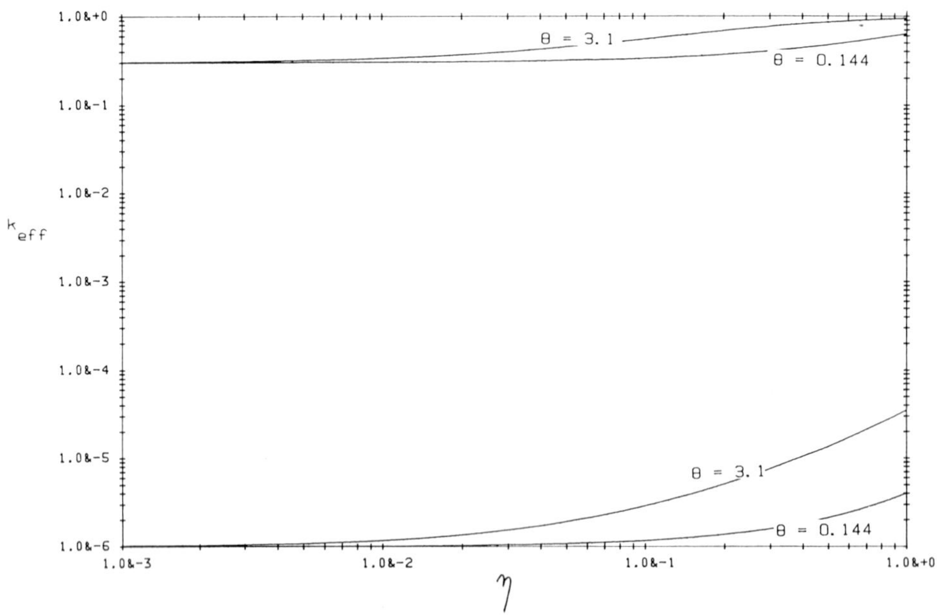

Fig. 6. Variation of the effective distribution coefficient with dimensionless growth rate (η) for two dopants having $k_0 = 0.3$ and 10^{-6}, respectively. Data are plotted for two values of $\theta = 0.144$ and 3.1 (see text).

where

$$\theta = N'/\mathrm{Sc}^{1/2} \tag{5.10}$$

is a renormalised magnetic field, $\mathrm{Sc} = \nu/D$ being the Schmidt number, and

$$\eta = V_0 D^{-1} (\nu/\Omega)^{1/2}\, \mathrm{Sc}^{-1/4} \tag{5.11}$$

incorporates the influences of growth rate and crystal rotation rate.

A plot of J as a function of η and θ is shown in fig. 5. This can be converted into a plot of k_{eff} versus η or θ for any given k_0. Such a plot is shown in fig. 6 for two very different values of k_0 and for two values of θ corresponding to the growth of silicon with a seed rotation rate of 22 rpm, with zero magnetic field and with a field of 0.2 T, respectively. From the figure, it can be seen that, for $\eta = 1$ (which corresponds in this case to a growth rate of 1.2 mm min^{-1}) the effective distribution coefficient for the solute with the larger k_0 ($= 0.3$) has approached unity whilst for the solute with the very small k_0, k_{eff} has risen by nearly an order of magnitude. It can be seen therefore that even relatively modest fields have a profound effect on segregation. The analysis has been extended by Cartwright et al. [30] to include the effects of crucible rotation.

5.2.3. Experimental studies

The theory has been validated in experimental studies of the axial distribution of phosphorus and carbon in silicon by Series [36] and by Ravishankar et al. [10] for the case of gallium in silicon. The results obtained by the last named authors are shown in fig. 7 from which it will be seen that good agreement is obtained.

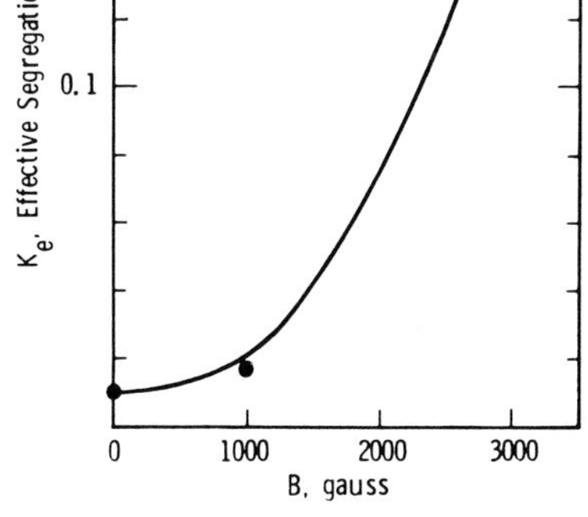

Fig. 7. Dependence of the effective distribution coefficient of gallium in silicon on axial magnetic field strength [10]. The curve is from the theory of Hurle and Series [27].

Müller and co-workers [31] have grown 3 inch diameter iron-doped indium phosphide single crystals using the high-pressure LEC technique with and without an applied axial magnetic field. The distribution coefficient of iron in indium phosphide is approximately 10^{-3} so there is a strong macrosegregation down the crystal. By applying a large magnetic field during the initial stages of growth and stepwise reducing this field as growth proceeded, Ozawa et al. [37] and, independently, Hofmann et al. [31] were able to obtain indium phosphide crystals having a near-constant axial concentration of iron up to a melt fraction solidified approaching 70% (fig. 8). Kobayashi and Muguruma [38] have shown theoretically how the magnetic field should be varied as a function of the melt fraction solidified in order to maximise the yield of crystal within some given bound on dopant concentration.

However, a difficulty is apparent in all these experimental studies: it is found that the radial uniformity of the dopant is poor compared to that obtained with zero magnetic field. The above theory, being one-dimensional, does not of course predict any radial variation. The reasons for this radial non-uniformity can be understood as follows:

The parameter J is a measure of the normalised extent of the diffusion profile. A small value of J indicates that the diffusion profile is confined close to the crystal and

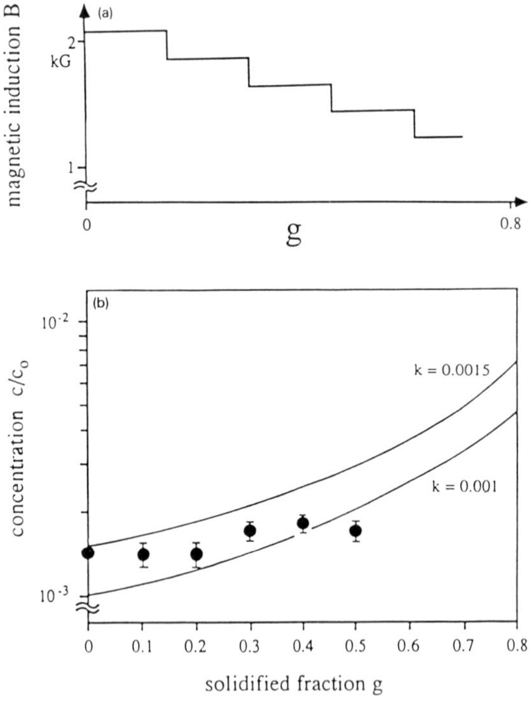

Fig. 8. Use of step-wise reduction in axial magnetic field strength to minimise macrosegregation in Fe-doped InP. (a) Applied field as a function of fraction of melt solidified; (b) measured concentration of iron normalised to initial melt concentration. The curves are the theoretical segregation curves for a solute having values of the distribution coefficient (k) shown. From Hofmann et al. [31].

most of the profile lies within the region of melt controlled by the rotating crystal. As J approaches unity the boundary layer thickens and the diffusion profile extends into the region of the melt where the flow is influenced by the far-field conditions such as convective flow driven by buoyancy generated at the heated crucible walls. If this flow has a radial dependence it will destroy the radial uniformity of the dopant. The transition occurs roughly at the point at which the solute boundary layer thickness defined by

$$L_d = JD/V_0$$

exceeds the momentum boundary layer thickness due to the rotating crystal. This is given by

$$L_m = (\nu/\Omega N')^{1/2}.$$

Note that L_d increases with increasing magnetic field whilst L_m decreases so that, with increasing field, a point of cross-over is reached. Defining the condition for radial uniformity as $L_d < L_m$, this is then:

$$\int_0^\infty \exp[-\eta\theta^{-1/2}z - (z + 2e^{-z} - \tfrac{1}{2}e^{-2z} - \tfrac{3}{2})/3\theta^2]dz < 1. \tag{5.12}$$

Analytic expression of this inequality is difficult but it can be evaluated numerically as shown in fig. 9. The regime for which the inequality (5.12) is not satisfied is marked on the figure. The horizontal line $\theta = 0.427\ (D/\nu)^{1/2}$ corresponds to zero magnetic field so that the region beneath this line is not experimentally accessible. Only growth in the unshaded region is to be expected to yield crystals having reasonable radial uniformity.

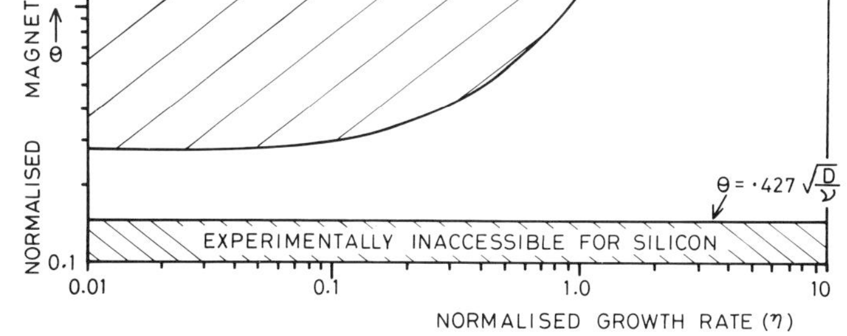

Fig. 9. Non-dimensional growth rate (η) versus non-dimensional magnetic field (θ) illustrating the region of parameter space for which radial uniformity of dopant concentration is obtained. Plotted for a Schmidt number of Sc = 8.75 appropriate to silicon.

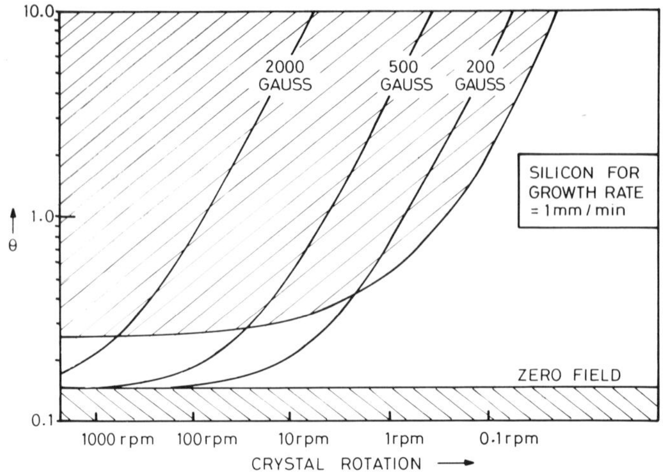

Fig. 10. As fig. 9 but re-plotted in practical units, assuming a crystal growth rate of 1 mm min^{-1}. Contours are for various values of axial magnetic field.

In fig. 10 this information is translated into practical units for parameters appropriate to silicon growing at a rate of 1 mm/min. Contours of constant magnetic field are plotted on the diagram. It can be seen that, for zero field, growth will always occur in a regime of good radial uniformity. However, the application of a quite modest field of 0.05 T will require a crystal rotation rate in excess of a 30 rpm to achieve acceptable uniformity. To obtain this with a 0.2 T field would require a crystal rotation rate in excess of 700 rpm which is not practicable. Similar graphs can be constructed from the more complex model incorporating far-field melt rotation but the effect is to reduce the size of the regime in which acceptable crystals can be grown.

Thus far we have considered only "conservative" dopants, i.e. those for which there is no exchange of dopant between the melt and the ambient or the crucible. However, the principal motivation for studying the use of a magnetic field in the Czochralski growth of silicon is to use it to control the concentration and uniformity of oxygen in the crystal. Now oxygen is a " nonconservative " dopant being continuously introduced into the melt by corrosion of the crucible walls by the molten silicon and being lost from the melt surface by evaporation as silicon monoxide.

Extensive studies have been made by Hoshikawa [39, 40], Hirata and Inoue [41], Ohwa et al. [42] and Series [36]. The situation has been numerically modelled by Organ and Riley [20] and others. The behaviour is complex but, in general terms, application of an axial field leads to reduced mixing in the bulk of the melt with poor uniformity and high oxygen concentration in the crystal.

An axial field has been applied to the LEC growth of gallium arsenide by Terashima et al. [43–45] and by Osaka and Hoshikawa [46], the former using a superconducting magnet and the latter a conventional solenoid capable of generating a field of up to 0.2 T. An interesting effect noted was the influence of the field on the semi-insulating properties of undoped gallium arsenide.

Undoped gallium arsenide single crystals are semi-insulating, having a resistivity of around $10^8 \Omega \cdot cm$ if grown under slightly arsenic-rich conditions which gives rise to a sufficient concentration of an ionised native defect which forms a deep donor level in the crystal. This defect, labelled EL2, is known to be associated with the arsenic-antisite (As_{Ga}), but its exact structure is still a matter of controversy. The crystal is semi-insulating if the EL2 concentration is greater than the net shallow acceptor concentration (due to residual chemical acceptors such as zinc and also, possibly, to some native defect acceptors). The concentration of EL2 representing, as it does, an excess arsenic concentration in the crystal, increases with increasing arsenic concentration in the melt.

If the crystal is growing from a noncongruent melt then there will exist ahead of the crystal–melt interface a boundary layer of the rejected excess component. If the melt composition is to the As-rich side of congruency then there will be a boundary layer of rejected excess arsenic. As we have already seen, the effect of an axial field is to enhance that boundary layer, i.e. to increase the "effective distribution coefficient of the *excess* As". For Ga-rich melts a similar effect will occur but now with a Ga-rich boundary layer. The result of applying the field is to markedly enhance the dependence of the EL2 concentration on the melt composition compared to that obtained with conventional LEC growth because the field acts to enhance the magnitude of the boundary layer whether it be Ga-rich or As-rich. This effect is evident in the experimental results of Terashima et al. [44] as was pointed out by Series and Hurle [47]. This explanation is given added weight by subsequent work by Kimura et al. [45] who used an axial field of up to 0.3 T generated by a superconducting solenoid. They made careful measurements

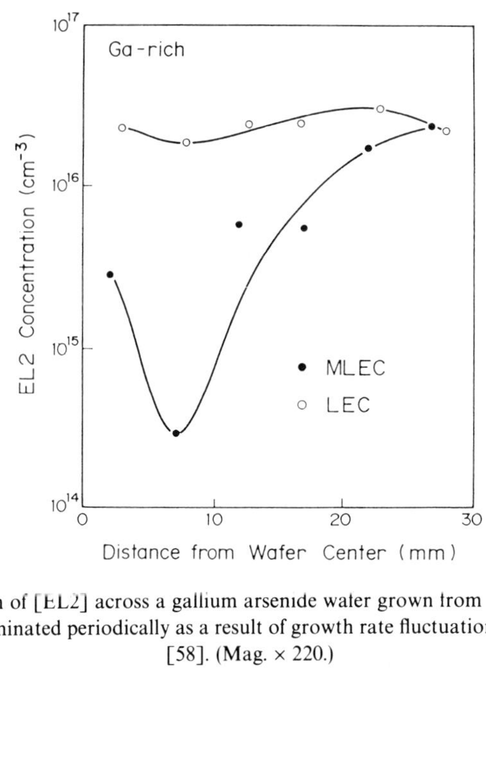

Fig. 11. Radial variation of [EL2] across a gallium arsenide wafer grown from a Ga-rich melt in an axial the eutectic reaction terminated periodically as a result of growth rate fluctuations produced by convective [58]. (Mag. × 220.)

of the EL2 distribution axially down their crystals and showed that increasing the field strength from 0.15 to 0.3 T reduced the EL2 concentration for melts to the Ga-rich side of congruency but increased it for those to the As-rich side. For a melt of 50.3 at.% no change was observed as the field was increased, suggesting that this composition is near to congruency.

Kimura et al. [45] further found that the axial uniformity of [EL2] obeyed the normal freeze equation in the presence of the field but was much less reproducible in the absence of the field. However, from the experience with silicon reported above, one would expect the radial uniformity to be degraded by the presence of the field and this was indeed observed as is shown in fig. 11 taken from the work of Terashima et al. [44] for the case of growth from a slightly Ga-rich melt.

6. *Czochralski growth with a cusped field*

We have seen above that both axial and transverse magnetic fields have serious limitations. Whilst both achieve the desirable result of damping out the turbulent convection in the melt, they have different and undesirable effects on the solute segregation: an axial field seriously degrades the radial symmetry which is a feature of conventional Czochralski growth whilst a transverse field destroys the thermal symmetry and introduces pronounced rotational striations.

An elegant solution to these problems was devised independently by Series [4] and by Hirata and Hoshikawa [5]. They proposed the use of a field configured so that, in the plane of the crystal–melt interface, it was purely radial whereas deep in the melt, it was predominantly axial. Thus, the rotational symmetry of the system is preserved whilst, close to the crystal–melt interface, there is no axial component of field to destroy the radial uniformity.

Such a configuration can be obtained by the use of a pair of axially mounted Helmholtz coils operated in opposed current mode (fig. 3). The diameter and separation of the coils can be chosen to maximise the component of the field normal to the crucible wall thus having a maximal effect on the oxygen erosion rate kinetics. Also with this configuration, a large fraction of the melt volume is subject to magnetic damping so that there is a marked reduction in turbulence and concomitantly in the dopant striations.

Both Series [4] and Hirata and Hoshikawa [5] report low oxygen concentration and good radial uniformity in silicon growth. For a wide range of growth conditions the radial uniformity is as good as, or possibly better than, that achieved under zero field conditions. The flow field and oxygen segregation in the presence of a cusped field have been modelled by Riley and co-workers [48, 49]. They show [48] that the model correctly predicts the reduction in the overall oxygen incorporation, the magnitude of that reduction being in reasonable accord with the experimental data of both research groups. Both model and experiment show the most rapid rate of reduction in oxygen concentration at low values of the field.

The desirable properties of the field can be attributed to several factors. Firstly, the crystal grows in a region of zero axial field and so none of the problems of interaction of the field with the momentum boundary layer adjacent to the crystal are experienced. Secondly, the melt close to the crystal is in a region of low overall field strength and so can become well mixed. Thirdly, the bulk of the melt is in a region of high field so that turbulence is effectively damped. Further, since the field is approximately normal to the vertical crucible wall, the boundary layer adjacent to the wall is thickened, an effect which has been analysed by Kerr and Wheeler [50]. As a result, erosion of the crucible is reduced leading to a reduction in oxygen concentration in the melt and, in turn, to the observed reduction in oxygen in the grown crystal.

In later work, Hirata and Hoshikawa [51] explored the effect of moving the neutral plane of the field away from the plane of the crystal–melt interface. They reasoned that by doing so, i.e. by applying a vertical component of the field in the plane of the free melt surface and a variable radial component of the field at the vertical crucible wall, they could independently control the erosion rate of the crucible and the rate of evaporation of oxygen as silicon monoxide at the melt surface. By reducing the rate of evaporation by applying a field having a *vertical* component of 0.055 T at the melt surface, they were able to *increase* the oxygen concentration in the crystal to a value some 35% higher than that obtained with zero magnetic field.

Although it is difficult to make totally general valid statements about the relative effects of crystal and of crucible rotation it is found that, for axial field growth, crystal rotation is the dominant control parameter, whilst for transverse and for cusped fields, it is crucible rotation which has the most significant effect. This is in line with the concept enunciated above that a field normal to a rigid (nonconducting) boundary strengthens the coupling between that boundary and the melt flow. These concepts are largely borne out by numerical simulations of the effects of axial and cusped fields on oxygen segregation in silicon performed by Sabhapathy and Salcudean [52]. These authors also showed the importance of thermocapillary convection in controlling the meridional flow in the cusped field configuration. Hjellming et al. [53] have shown, by an analytical analysis, the importance of a nonzero electrical conductivity of the crystal in determining the angular velocity of the melt in the presence of a cusped field.

As far as the authors are aware, at the time of writing, there has been no work reported on the application of a cusped field to any other material than silicon.

7. *Bridgman and floating-zone growth*

Whilst the main interest has been in the area of the commerical growth of silicon, gallium arsenide and indium phosphide by the Czochralski and LEC techniques, there have been some studies on the effects of a magnetic field on the floating-zone growth of silicon and on the horizontal and vertical Bridgman growth of several other materials. These date back to the original experiments of Utech and Flemings [1, 54], who studied horizontal Bridgman growth of tellurium-doped indium antimonide in the presence of a transverse, vertically oriented magnetic field. They performed similar experiments with a dilute aluminium–copper alloy [54] and they showed that, in both systems, the application of the field eliminated the dopant striations.

Independently, Chedzey and Hurle [2, 55] performed similar horizontal Bridgman growth experiments on indium antimonide with a transverse field, but applied it along a horizontal axis. Both configurations of field appear to be equally efficacious in removing the dopant striations. Chedzey and Hurle [2] were able to correlate the waveform of the temperature fluctuations with the spacing of the striations.

Vertical Bridgman growth in the presence of a transverse horizontal magnetic field has been used by Kim [56] to grow indium antimonide single crystals and by Sen et al. [57] to grow indium–gallium antimonide. Kim showed that striation-free growth was obtained above some critical value of the field which corresponded to the damping out of convective oscillations in the melt.

Sen et al. [57] studied a concentrated alloy growing under conditions of constitutional supercooling. Using a transverse field of 0.4 T they showed that the field produced a dramatic reduction in the density of grain boundaries and twin boundaries in the crystals. They suggested that a sudden increase in freezing rate caused by turbulence could nucleate twins at the interface before a cellular structure developed. Such twinning would be absent when the fluctuations were removed.

Hurle and Hunt [58] described the directional solidification, horizontally, of the indium antimonide–nickel antimonide eutectic. This is a rod eutectic system and, in the absence of a magnetic field, turbulence in the melt produces a banded form of eutectic in which the rods are repeatedly terminated by the temperature fluctuations. By applying a transverse horizontal magnetic field this banding was eliminated and very long rods in a uniform eutectic microstructure were obtained (see fig. 12). Similar experiments have been performed on the bismuth–manganese eutectic system by de Carlo and Pirich [59] with similar results.

Matthiesen et al. [60] grew 1.5 cm diameter germanium single crystals in a vertical Bridgman configuration in the presence of an applied axial magnetic field of the considerable magnitude of 3 T. The solute distribution was characteristic of that in diffusion-controlled growth, showing that the field had effectively suppressed convective motion in the melt. Motakef [61] has analysed the scaling laws which apply to segregation in vertically controlled Bridgman growth and applied his analysis to the Matthiesen et al. data.

Robertson and O'Connor [62] have studied the effect of both transverse and axial fields on the floating-zone melting of gallium-doped silicon single crystals grown on a (1 0 0) axis. Transverse fields of up to 0.55 T were shown to have a pronounced effect on the shape of the crystal–melt interface and, for non-rotated crystals, on the cross-sectional shape of the crystal which assumed an elliptical form with the major axis of that ellipse aligned in the field direction. Interestingly, under these conditions, pronounced dopant striations were found in the crystal spaced in time by roughly a 30 s period. The effective distribution coefficient of gallium was found to increase in the presence of the field. A marked reduction in dopant concentration fluctuations is observed for field strengths greater than 0.3 T.

However, in the presence of a large axial field, control of the crystal shape became extremely difficult, particularly with low rotation rates. The improved uniformity was confined to a central core region extending to about one-third of the crystal diameter.

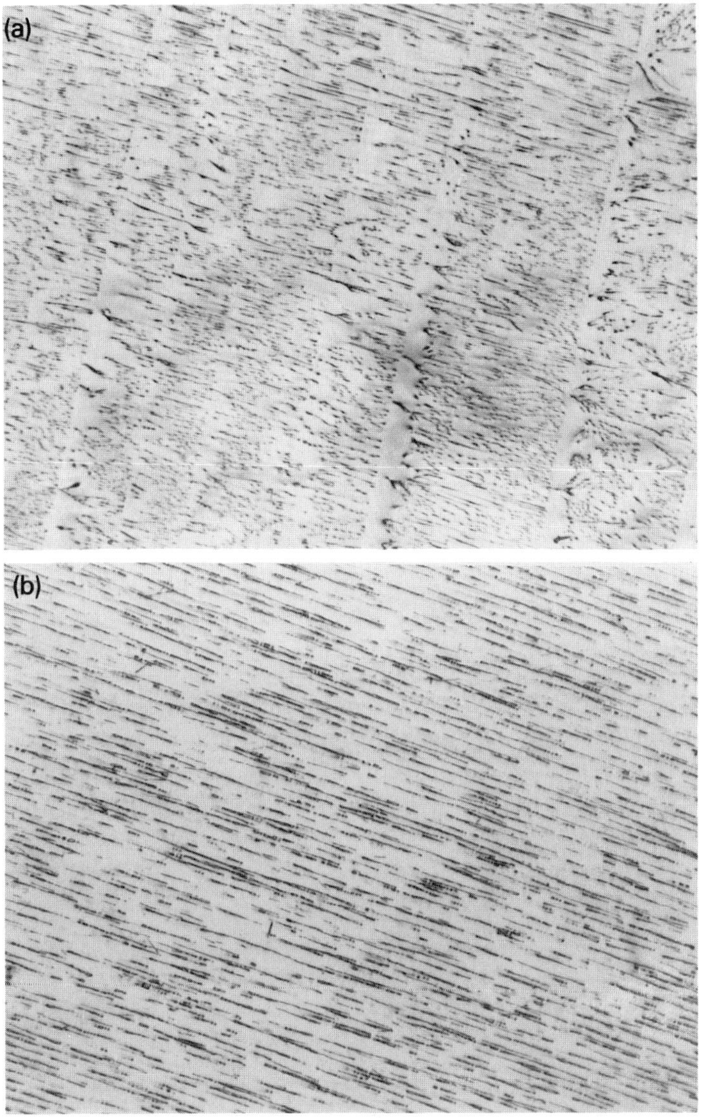

Fig. 12. Microstructures of directionally solidified InSb–NiSb eutectic. (a) Banded eutectic structure with the eutectic reaction terminated periodically as a result of growth rate fluctuations produced by convective temperature fluctuations in the melt; the growth direction is from left to right. (b) Long aligned rods of NiSb produced by growth in a transverse magnetic field which has damped the convective turbulence in the melt [58]. (Mag. × 220.)

To date there has been but little theoretical analysis of the effects of a magnetic field on floating zone growth although Wilson [63] has analysed the effect of an axial field on convective instability in a rigidly rotating cylindrical fluid column.

Solution growth, using the travelling heater technique in the presence of an axial magnetic field, has been studied experimentally by Danilewsky et al. [64]. The materials grown were S-doped indium phosphide obtained from In solution as well as Te-doped gallium antimonide and Zn-doped gallium arsenide grown from Ga solution. The crystals were grown vertically in sealed quartz ampoules of 15 mm internal diameter using a monoellipsoidal mirror furnace. Dopant striations due to natural convection were entirely suppressed by a 0.44 T field in the gallium antimonide and gallium arsenide crystals but not in indium phosphide. The latter result is explained as due to the large solutal expansion coefficient in this case which would require an applied field in excess of 2 T to suppress the time-dependent convection. Switching on and off of the field-induced morphological instability in the case of indium phosphide.

A quite different use of a steady magnetic field in horizontal zone melting has been described by Pfann and Hagelbarger [65]. These authors achieved magnetic suspension of the molten zone by passing a DC current axially along the crystal and applying a transverse magnetic field over the zone volume. The resulting vertical force suspended the molten zone against the force of gravity (fig. 13). The condition that the force exactly balances gravity is given as

$$HI = 10\rho g\pi a^2,$$

where a is the radius of the rod, ρ the melt density, g the gravitational acceleration, H the field and I the current. For the zone to remain stably suspended, additional criteria have to be satisfied, notably that the zone length does not exceed some maximum critical value. The method was demonstrated for iron, tin and aluminium but does not work for materials, such as the common semiconductors, which have very different electrical conductivities in their molten and crystalline states.

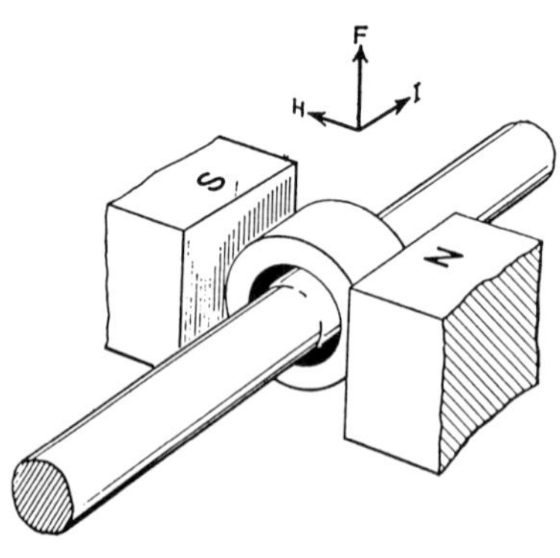

Fig. 13. Schematic representation of magnetic suspension of a molten zone (from Pfann and Hagelbarger [65]).

8. Concluding remarks

We have seen that the most universal effect of the applied field is the damping of the convective turbulence in the melt and that this imparts an improved micro-homogeneity to the dopant distribution in the crystal. It can additionally have potent effects on the macrosegregation: in Czochralski growth the effective distribution coefficient is driven strongly towards unity by the application of an axial field but, at the same time, the radial uniformity of the dopant distribution can be seriously impaired.

Nonconservative dopants are influenced in very complex ways which are impacted by changes in the uptake processes as well as in the rate of loss by evaporation from the free surface, so that, even if the distribution coefficient of the dopant is near unity (as is the case for oxygen in silicon), the incorporation of dopant into the crystal can be profoundly modified by application of the field. Here the configuration of the field is of central importance and the recent innovation of the cusped field has enabled these processes to be largely brought under control, offering the possibility of much tighter specification of oxygen distribution in silicon over a wider range of concentration values. Equally, with gallium arsenide, the microdistribution of the very important deep level EL2 is much improved by use of a magnetic field.

An alternative method for removing melt turbulence produced by buoyancy-driven convection and the concomitant striations is to grow the crystals in a gravity-free environment. This has been demonstrated in a number of space experiments [66]. It is therefore appropriate to compare the merits and de-merits of magnetic field damping as against gravity reduction. Aside from the technological aspects of generating the separate environments lies the difference that magnetic field damping depends on the action of the Lorentz force which is proportional to the vector product of the flow velocity and the field strength. It therefore has a decreasing influence with decreasing flow velocity and so is ill-fitted to the further damping of very weak flows.

By contrast, a reduction in gravity represents a reduction in the driving body force for natural convection. However, the attainable space environment does not provide zero gravity and the residual levels of micro-gravity (including so-called g-jitter) set a lower bound on the effectiveness of space processing.

The best results can be anticipated to come from the combined use of microgravity and a magnetic field, wherein the micro-gravity would reduce the slow, steady convective flows to a minimum with the magnetic field damping any g-jitter fluctuations. This concept has been discussed by Baumgartl et al. [67].

The use of a magnetic field as an additional control parameter in crystal growth has yet to find commercial use and more research is needed to elucidate its true potential.

References

[1] H.P. Utech and M.C. Flemings, J. Appl. Phys. 37 (1966) 2021.
[2] H.A. Chedzey and D.T.J. Hurle, Nature 210 (1966) 933.
[3] K.M. Kim and P. Smetana, J. Appl. Phys. 58 (1985) 2731.
[4] R.W. Series: J. Cryst. Growth 7 (1989) 92.

[5] H. Hirata and K. Hoshikawa, J. Cryst. Growth 6 (1989) 747.
[6] S. Chandrasekhar, Hydrodynamic and Hydromagnetic Stability, Clarendon Press, Oxford (1961).
[7] J. Baumgartl and G. Müller, Proc. 8th European Symp. on Materials and Fluid Science in Microgravity, Brussels (1992). ESA Publication no SP. 333.
[8] W.E. Langlois, L.N. Hjellming and J.S. Walker, J. Cryst. Growth 87 (1987) 51.
[9] Yu.M. Gel 'fgat and L.A. Gorbunov, Sov. Phys. Dokl. 34 (1989) 470.
[10] P.S. Ravishankar, T.T. Braggins and R.N. Thomas, J. Cryst. Growth 104 (1990) 617.
[11] S. Kobayashi, J. Cryst. Growth 75 (1986) 301.
[12] W.R. Cochran, Proc. Cambridge Phil. Soc. 30 (1934) 365.
[13] J.A. Burton, R.C. Prim and W.P. Slichter, J. Chem. Phys. 21 (1953) 1987.
[14] M.G. Williams, J.S. Walker and W.E. Langlois, J. Cryst. Growth 100 (1990) 233.
[15] T. Suzuki, N. Isawa, Y. Okubo and K. Hoshi, in: Semiconductor Silicon 1981, eds. H.R. Huff, R.J. Kriegler and Y. Takeishi (Electrochem. Soc., Pennington, NJ, 1981).
[16] K. Terashima and T. Fukuda, J. Cryst. Growth 63 (1983) 423.
[17] W.E. Langlois and K.E. Lee, J. Cryst, Growth 62 (1983) 481.
[18] M. Mihelcic and K. Wingerath, J. Cryst. Growth 71 (1985) 613.
[19] R. Cartwright, O.J. Ilegbusi and J. Szekely, J. Cryst. Growth 94 (1989) 321.
[20] A.E. Organ and N. Riley, J. Cryst. Growth 82 (1987) 465.
[21] L.N. Hjellming and J.S. Walker, J. Fluid Mech. 164 (1986) 237.
[22] L.N. Hjellming and J.S. Walker, J.Fluid Mech. 182 (1987) 335.
[23] L.N. Hjellming, J. Cryst. Growth 104 (1990) 327.
[24] L.N. Hjellming and J.S. Walker, J. Cryst. Growth 92 (1988) 371.
[25] T. Kakutani, J. Phys. Soc. Jpn. 17 (1962) 1496.
[26] T. von Karman, Z. Angew Math. Mech. 1 (1921) 244.
[27] D.T.J. Hurle and R.W. Series, J. Cryst. Growth 73 (1985) 1.
[28] A.E. Organ and N.Riley, J. Appl. Math. 36 (1986) 117.
[29] T.W. Hicks and N.Riley, J. Cryst. Growth 96 (1989) 957.
[30] R.A. Cartwright, D.T.J. Hurle, R.W. Series and J. Szekely, J. Cryst. Growth 82 (1987) 327.
[31] D. Hofmann, M. Mosel and G. Müller, in: Semi-insulating III–V Materials, Malmö, 1988, eds. G. Grossmann and L. Ledebo (Hilger, Bristol, 1988) p 429.
[32] H. Miyairi, T. Inada, M. Eguchi and T. Fukuda, J. Cryst. Growth 79 (1986) 291.
[33] T. Satoh, I. Hashimoto, Y. Sakaguchi and K. Matsuzawa, in: Semi-insulating Materials, Hakone, 1986, eds. H. Kukimoto and S. Niyazawa (Ohmsha, Tokyo/North-Holland, Amsterdam, 1986).
[34] T. Munakata and I. Tanasawa, J. Cryst. Growth 106 (1990) 566.
[35] L.O. Wilson, J. Cryst. Growth 44 (1978) 247.
[36] R.W. Series, J. Cryst. Growth 97 (1989) 85.
[37] S. Ozawa, T. Kimura, J. Kobayashi and T. Fukuda, Appl. Phys. Lett. 50 (6) (1987) 329.
[38] S. Kobayashi and T. Muguruma, J. Cryst. Growth 84 (1987) 559.
[39] K. Hoshikawa, Jpn. J. Appl. Phys. 21 (1982) L545.
[40] H. Hoshikawa, H. Kohda and H. Hirata, J. Appl. Phys. 23 (1984) L37.
[41] H. Hirata and N. Inoue, Jpn. J. Appl. Phys. 23 (1984) L527.
[42] M. Ohwa, T. Higuchi, M. Toji, M. Watanabe, K. Omma and S. Takasuma, in: Semiconductor Silicon 1986, eds. H. R. Huff, T. Abe and B.O. Kolbesen (Electrochem. Soc., Pennington, NJ, 1986) p. 117.
[43] K. Terashima, N. Nishio, S. Washizuka and M. Watanabe, J. Cryst. Growth 84 (1987) 247.
[44] K. Terashima, A. Yahata and T. Fukuda, J. Appl. Phys. 59 (1986) 982.
[45] T. Kimura, T. Katsumata, M. Nakajima and T.Fukuda, J. Cryst. Growth 79 (1986) 164.
[46] J. Osaka and K. Hoshikawa, In: Semi-insulating III–V Materials, Kah-nee-ta, Warm Springs, OR, 1984, eds. D.C. Look and J.S. Blakemore (Shiva, Nantwich, 1984) p. 126.
[47] R.W. Series and D.T.J. Hurle, J. Cryst. Growth 113 (1991) 305.
[48] T.W. Hicks, A.E. Organ and N. Riley, J. Cryst. Growth 94 (1989) 213.
[49] N. Riley, J. Cryst. Growth 97 (1989) 76.
[50] O.S. Kerr and A.A. Wheeler, J. Fluid Mech. 199 (1989) 217.
[51] H. Hirata and K. Hoshikawa, J. Cryst. Growth 98 (1989) 777.

[52] P. Sabhapathy and M.E. Salcudean, J. Cryst. Growth 113 (1991) 1564.
[53] L.N. Hjellming, P.A. Tolley and J.S. Walker, J. Fluid Mech. 249 (1993) 1.
[54] H.P. Utech and M.C. Flemings, in: Crystal Growth, ed. H.S. Peiser (Pergamon, Oxford, 1967) p. 65
[55] D.T.J. Hurle, in: Crystal Growth, ed. H.S. Peiser (Pergamon, Oxford, 1967) p. 659
[56] K.M. Kim, J. Electrochem. Soc. 129 (1982) 427.
[57] S. Sen, R.A. Lefever and R. Wilcox, J. Cryst. Growth 43 (1978) 526.
[58] D.T.J. Hurle and J.D. Hunt, The Solidification of Metals. ISI P110 (Iron and Steel Institute, 1968) p. 162.
[59] J.L. de Carlo and R.G. Pirich, Met. Trans. A 15 (1984) (1984) 2155.
[60] D.H. Matthiesen, M.J. Wargo, S. Motakef, D.J. Carlson, J.S. Nakos and A.F. Witt, J. Cryst. Growth 85 (1987) 557.
[61] S. Motakef, J. Cryst. Growth 104 (1990) 833.
[62] G.D. Robertson Jr. and D.J.O'Connor, J. Cryst. Growth 76 (1986) 111.
[63] S.K. Wilson, Quart. J. Mech. Appl. Math. 45 (1992) 363.
[64] A.N. Danilewsky, P. Dold and K.W. Benz, J. Cryst. Growth 121 (1992) 305.
[65] W.G. Pfann and D.W. Hagelbarger, J. Appl. Phys. 27 (1956) 12.
[66] D.T.J. Hurle, G. Müller and R. Nitsche, in: Fluid Sciences and Material Science in Space, ed. H.U. Walter (Springer, Berlin, 1987) p. 313.
[67] J. Baumgartl, M. Gewald, R. Rupp, J. Stierlen and G. Müller, in: Proc. 7th European Symposium on Materials and Fluid Sciences in Microgravity, Oxford (1989). ESA Publication No. SP 295.

CHAPTER 6

Fractional Crystallisation

P.J. JANSENS and G.M. van ROSMALEN

Delft University of Technology, Leeghwaterstraat 44, 2628 CA Delft, Netherlands

Contents

1. Introduction 289
2. Basic concepts 291
   2.1. Phase diagrams 291
   2.2. Incorporation of impurities 292
   2.3. Inclusion of impurities due to unstable growth morphologies 294
3. Crystal growth techniques 295
4. Purification mechanisms 297
5. Industrial layer growth equipment 299
6. Industrial suspension growth equipment 303
7. Industrial pressure crystallisation equipment 308
8. Evaluation 311
Acknowledgements 311
List of symbols 312
References 312

HANDBOOK OF CRYSTAL GROWTH, VOL. 2
edited by D.T.J. Hurle

1. Introduction

Fractional crystallisation (FC) is a general term for separation process which can be used for large-scale separation and ultrapurification of organic compounds from multicomponent mixtures in an economically favourable and environmentally safe way. The driving force for crystallisation is either cooling or pressurisation (fig. 1). The vast majority of the FC processes is based on cooling crystallisation but in Japan the relatively new and promising high-pressure crystallisation is already applied on an industrial scale. In a cooling crystallisation process crystals are grown either freely suspended or as a fixed layer on a cooled surface. In Europe mainly layer growth processes are applied whereas in Japan suspension growth processes are dominant. Worldwide both crystal growth techniques seem to be equally represented.

FC processes can be characterised as a multistep operation which involves a countercurrent contact between crystals and mother liquor in such a way that the crystals are grown and/or washed in mother liquor of increasing purity at ascending temperatures (fig. 2). Layer growth and pressure crystallisation processes are usually operated batchwise whereas suspension crystallisation processes mostly feature a continuous operation.

Advantages of FC in comparison to conventional separation techniques like distillation and extraction, are its superior selectivity, lower energy consumption (providing the operating temperature exceeds 273 K) and moderate operating conditions whereas addition of toxic or explosive solvents can usually be avoided. The superior selectivity stems from the fact that the molecular mobility is very restricted in the crystal lattice

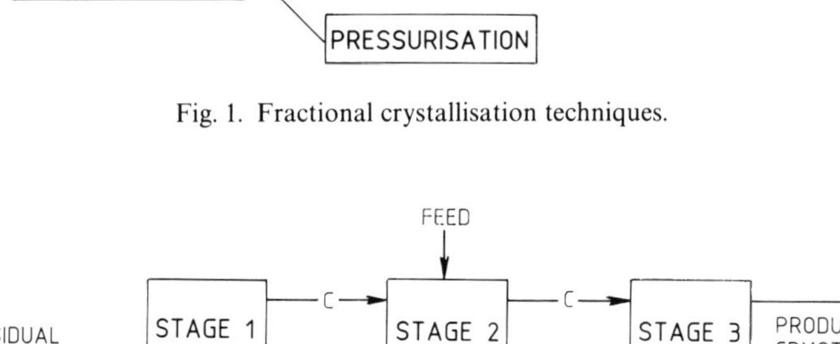

Fig. 1. Fractional crystallisation techniques.

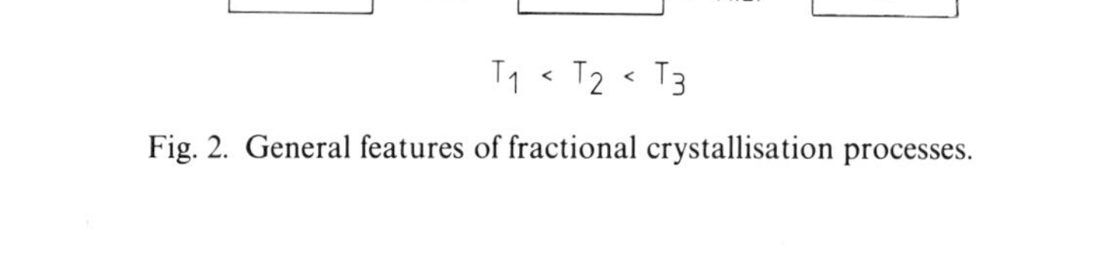

Fig. 2. General features of fractional crystallisation processes.

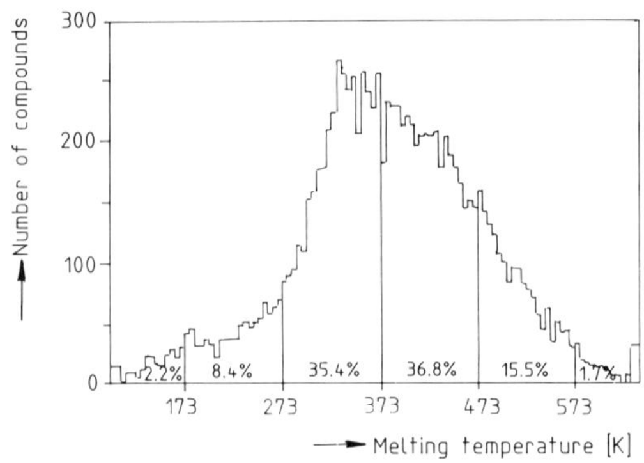

Fig. 3. Melting points of organic compounds at atmospheric pressure [1].

due to strong molecular bonds and a regular arrangement which makes it difficult for impurities to enter the lattice. The energy consumption of FC is relatively low because of its high selectivity which allows low reflux ratios and, when compared to distillation, because the latent heat of fusion of organic compounds is often two to four times smaller than their latent heat of evaporisation. Reservations have to be made for FC processes which require deep cooling to crystallise a low melting point compound and for the so-called dynamic layer growth processes which involve repeated crystallisation operations. The operating conditions for cooling crystallisation processes lie close to the melting point of the pure component at atmospheric pressure. Figure 3 [1] demonstrates that almost 90% of the organic compounds have a melting point above 273 K at atmospheric pressure. In cases where the melting point of an organic compound is considerably lower than 273 K, pressure crystallisation can be considered which allows separation above room temperature at low energy demands. Addition of a solvent may be used to reduce the viscosity of the mother liquor, to enhance the growth rate of the crystals, to improve the selectivity, to lower the operating temperature and to facilitate solid/liquid separation. If possible, solvent addition has to be avoided for safety, environmental and economic reasons.

A disadvantage of FC processes is the inherent slowness of a crystallisation process resulting in low *specific* production capacities. This can result in rather large equipment and relatively high investment cost.

Table 1
Organic compounds suited for purification with fractional crystallisation and their estimated wordwide production capacities.

Organic compound	Estimated worldwide production capacity
p-xylene	4.5×10^6 t/y (1992)
acetic acid	4.0×10^6 t/y (1989)
phenol	3.7×10^6 t/y (1989)
dimethylterephthalate	4.6×10^6 t/y (1992)
caprolactam	3.6×10^6 t/y (1992)
naphthalene	$0.6–1.0 \times 10^6$ t/y (1992)

Evaluating the merits and drawbacks of FC it can be said that application of FC is most interesting as a tool in ultrapurification where the feedstock is already 70–80% pure and a product purity of 99.99% is strived for. Such stringent purity demands are frequently needed for advanced polymer and pharmaceutical applications. Another important field of application lies in the purification of thermally unstable components which have to be treated in the modern food industry. Table 1 summarises a number of organic (bulk) compounds which can be purified with FC and their estimated world-wide production capacities.

2. *Basic concepts*

2.1 Phase diagrams

Phase diagrams are the graphical expression of phase equilibria and can be used to obtain a first indication of the separability of the components of a mixture. The complexity of phase diagrams rapidly increases with the number of components. Binary phase diagrams are simple and well documented, whereas ternary phase diagrams are often very complicated and rarely described in literature. Data on phase equilibria of systems with four components or more are almost never available. In practice, however, multicomponent systems are usually of great interest and these systems are simply treated as combinations of separate binary systems. This simplification is only valid in the absence of chemical interactions and when each impurity behaves independent of the others in its relationship to the main component. In this section only binary phase diagrams will be discussed.

Binary phase diagrams were divided by Roozeboom [2] in two categories: *eutectics* which show no solid solubility and *solid solutions* which exhibit limited or complete solid solubility. Solid solution may involve either the inclusion of guest molecules in the interstitial voids of the host lattice or the isomorphous substitution of host molecules by guest molecules. In case of organic systems only the substitutional mode is known [3]. Criteria for the formation of solid solutions are extensively described by Kitaiigorodskii [4]. It is obvious that similarity in size and shape between the host and the guest molecules is essential for the formation of solid solutions. Introduction of look-alike guest molecules will not lead to major changes in the crystal lattice of the host molecules. In this respect Kitaiigorodskii argues that host molecules are replaced more easily by smaller molecules than by larger ones. Furthermore, for solid solution to be continuous throughout the full range of compositions the structure of the components must have identical space-groups, the same number of molecules per unit cell and similar molecular packing. Matsuoka and Fukushima [1] showed that 85.9% of the binary organic mixtures belong to the eutectic category of which the "simple eutectic" is the most important representative (54.3%). Complete solid solubility occurs in about 1% of the investigated binary mixtures (fig. 4).

The composition changes of the liquid and the solid phase during a crystallisation process, at infinitely slow growth rates under ideal mixing conditions in both phases, can be read from a phase diagram as is illustrated for the cooling crystallisation of a

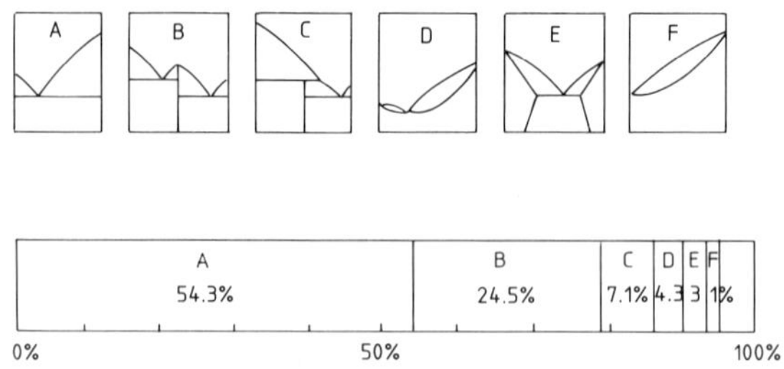

Fig. 4. Statistical probability of binary phase diagrams of organic compounds [1].

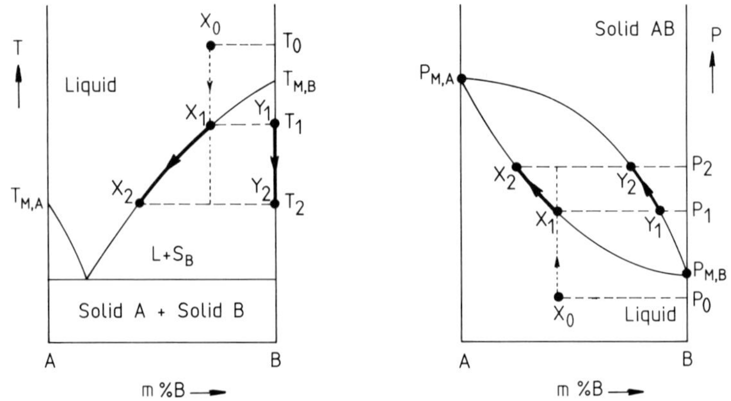

Fig. 5. Equilibrium crystallisation process by (a) cooling of a eutectic mixture and (b) pressurisation of a complete solid solution system.

eutectic mixture and for the pressure crystallisation of a solid solution mixture in fig. 5. The recovery of these crystallisation processes at temperature T_2 or pressure P_2 can be calculated using the lever rule:

$$R_{T_2/P_2} = \frac{(X_0 - X_2)}{(Y_2 - X_2)}. \tag{1}$$

2.2. *Incorporation of impurities*

The incorporation of impurities in the crystal lattice under equilibrium conditions can be quantified as the *equilibrium distribution coefficient* (K_0) which is defined as the ratio of the impurity concentration in the crystal phase (C_S) to the impurity concentration in the mother liquor (C_L):

$$K_0 = \frac{C_S}{C_L} \tag{2}$$

For the binary mixtures illustrated in fig. 5 the equilibrium distribution coefficient K_0 at temperature T_2 or pressure P_2 can be expressed as:

$$K_{0,T_2/P_2} = \frac{(1 - Y_2)}{(1 - X_2)}. \tag{3}$$

For eutectic mixtures $K_0 = 0$ and for solid solution systems $0 < K_0 < 1$. For multicomponent mixtures a multidimensional equilibrium distribution matrix is described by Tiller [5].

During a non-equilibrium crystallisation process the balance between crystal growth, which involves selective rejection of impurities from the growing crystal surface, and mass transfer of impurities away from the crystal surface results in an impurity concentration gradient in front of the growing crystal surface as is illustrated in fig. 6. It is assumed here that mass transfer within the stagnant liquid boundary layer δ_m adhering to the crystal surface is governed exclusively by diffusion. The thickness of the boundary layer depends on the hydrodynamic conditions and approaches a minimum value as mixing improves ($\delta_m \approx 10^{-6}$–10^{-4}m). Because crystal growth occurs from the impurity-rich boundary layer and not from the purer bulk of the mother liquor, Burton, Prim and Slichter [6] defined an *effective distribution coefficient* (K_{eff}):

$$K_{eff} = \frac{C_{S,i}}{C_{L,b}} = \frac{K_0}{K_0 + (1 - K_0)\exp\left(\frac{-V\delta_m}{D}\right)}. \tag{4}$$

Here $C_{S,i}$ represents the impurity concentration in a crystal just behind the interface and $C_{L,b}$ is the impurity concentration in the bulk of the mother liquor, both under non-equilibrium growth conditions. Equation (3) neglects the difference in specific weight between the crystal phase and the mother liquor but this omission can easily be repaired by substituting V' for the growth rate V [5]:

$$V' = \left(\frac{\rho_S}{\rho_L}\right) V. \tag{5}$$

Tiller [5] argues that a delay in mass transfer caused by surface integration and adsorption effects should be accounted for in the calculation of the effective distribution coefficient K_{eff}. When surface integration becomes the rate determining step of a

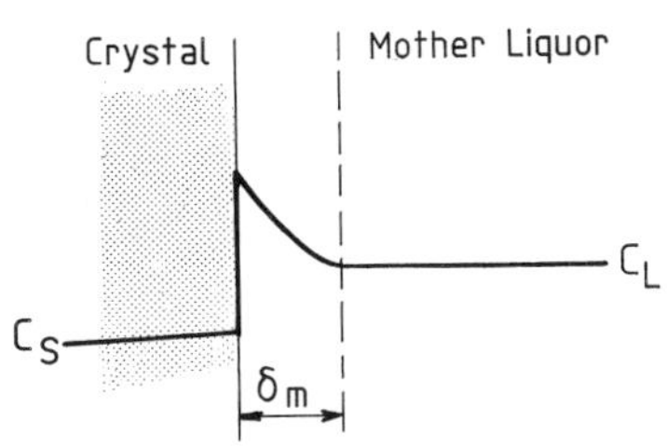

Fig. 6. Impurity concentration gradient at the crystal/mother-liquor interface.

crystal growth process, the equilibrium distribution coefficient K_0 is to be replaced by an interface distribution coefficient K_i in eq. (4). Ostrogorsky and Müller also modified the expression for the effective distribution coefficient to correct for convective solute transfer [7].

2.3. *Inclusion of impurities due to unstable growth morphologies*

Inclusion of impure mother liquor in the crystal matrix reduces the average crystal purity and, consequently, the selectivity of the crystallisation process. Inclusion of mother liquor can hardly be avoided in bulk crystallisation processes of highly concentrated solutions or near melts. It creates a major problem in large-scale ultrapurification of organic compounds and is the main reason why the design and operation of FC equipment differs significantly from the conventional industrial crystallisation processes described in the chapter by van der Heijden and van Rosmalen (Chapter 7, this volume).

An inclusion is formed when a small portion of mother liquor gets entrapped between irregular growth patterns on the crystal surface and becomes eventually enclosed by the crystal matrix. The irregular patterns, usually referred to as *unstable growth morphologies*, occur most easily at high growth rates and poor mixing conditions in the mother liquor.

Crystal growth cannot proceed without heat and mass transfer. Heat transfer was shown to be the rate determining step when crystallisation occurs from a melt by de Goede et al. [8, 9] and Matsuoka and Garside [10]. Since heat release is proportional to the heat exchanging area, a growing crystal has a tendency to increase its surface area. There are two ways available to oppose this tendency. First, there is an intrinsic stabilising effect of a crystal striving for minimisation of its surface tension and consequently a reduction of its surface. Second, an externally applied temperature gradient may reduce the driving force in a direction from the crystal interface into the mother liquor. The stabilising surface tension effect increases with stronger curvature of the crystal surface [11]. At low growth rates the effect of surface tension is dominant and well-faceted crystals can be produced, but at high growth rates the effect of heat transfer is dominant and cellular or dendritic morphologies will appear on the crystal surface [12].

An intrinsically unstable growth situation can occur when the supersaturation (undercooling or overpressure) of the mother liquor *increases* in the direction from the crystal surface towards the bulk of the mother liquor. The presence of an impurity gradient in front of the crystal surface locally lowers the equilibrium temperature. Consequently, the presence of the impurity gradient enhances the probability of the occurrence of a zone of increasing supersaturation adjacent to the crystal surface. This condition is usually referred to as *constitutional supersaturation* [11,13,14]. In the presence of constitutional supersaturation, an accidental protrusion on a flat crystal surface will enter a region of higher supersaturation and tends to outgrow the surrounding surface (fig. 7). Likewise, small dips in the crystal surface tend to become retarded even more. From fig. 7e it is easy to see how portions of mother liquor can get

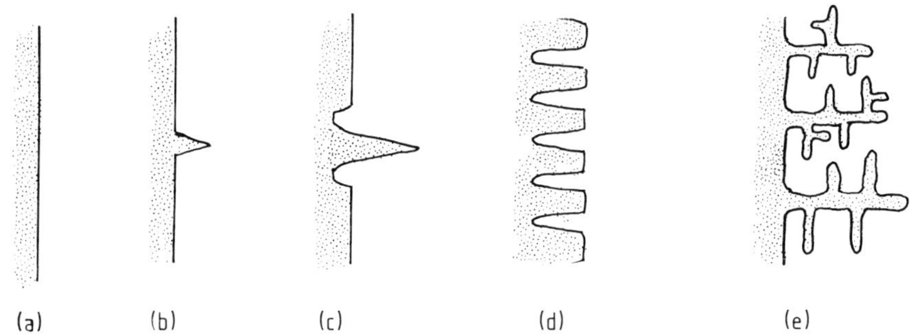

Fig. 7. Macroscale roughening of a crystal surface [9].

entrapped between the branches of the dendritic structures. The concept of constitutional supersaturation and other stability criteria which are necessary to predict stable growth morphologies are more extensively described by Tiller [15] and by Sekerka [12].

The level of constitutional supersaturation and therefore the stability of a crystal morphology under practical crystallisation conditions is to a large extent dependent on the applied crystallisation technique as will be discussed in the next section.

3. *Crystal growth techniques*

The *layer growth* technique is based on the formation of a crystal layer on a cooled heat exchanger surface. The main growth direction is perpendicular to the cooled surface into the bulk of the mother liquor. The heat of crystallisation is removed through the cooled surface via the crystal layer and consequently the bulk of the mother liquor is undersaturated. By keeping the growth rates low and providing good mixing conditions constitutional supersaturation can in principle be avoided in layer growth processes as is illustrated in fig. 8a. Unfortunately, the growth rates applied in industrial layer crystallisers have to be high ($\pm 10^{-6}$–10^{-5} m/s) to meet the specified production capacity and the mixing conditions in some equipment designs are moderate to poor. The production capacity is defined here as the product of the average linear growth rate and the surface area available for crystal growth. Since the available crystal growth area in layer crystallisers is relatively small because it is limited to the installed heat exchanger surface ($\leqslant 100\,m^2/m^3$), high growth rates are necessary to compensate for it. As a result significant constitutional supersaturation is always present in industrial layer growth processes (fig. 8b). According to the Mullins–Sekerka stability criterion the stabilising effect of surface tension on the flat surface of a crystal layer is week [11]. Consequently, unstable growth morphologies easily occur and the purity of the crystal layer is reduced significantly due to inclusions of mother liquor. In conclusion, the selectivity of most industrial layer growth processes is moderate because of the high growth rates and poor mixing conditions which promote both incorporation and inclusion of impurities.

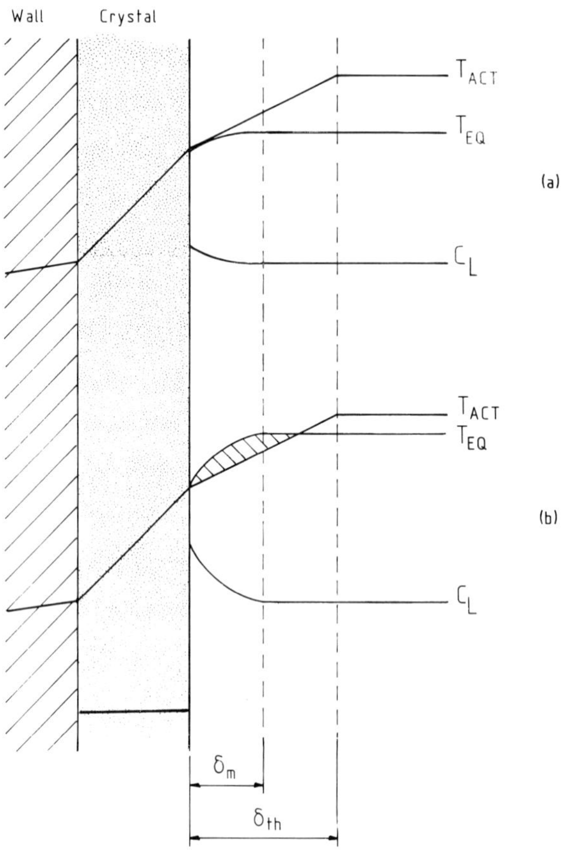

Fig. 8. (a) Stable and (b) unstable growth conditions in layer growth.

In *suspension growth* processes crystals grow freely suspended in their undercooled mother liquor. A suspension growth situation is intrinsically unstable because some degree of constitutional supersaturation cannot be avoided (fig. 9). Fortunately, the growth rates applied in suspension crystallisation equipment can be kept low ($\pm$ 10^{-8}–10^{-7} m/s) and the mixing conditions are usually good. Since the area available for crystal growth is equal to the total crystal surface area, which can measure up to 10 000 m^2/m^3, fairly high specific production capacities can already be accomplished at low growth rates. As a result the degree of constitutional supersaturation in suspension growth processes is small. Furthermore, the stabilising effect of surface tension in the case of suspension growth is strong because of the strong curvature of the surface of individual crystals. Consequently, unstable growth morphologies can usually be avoided in industrial suspension crystallisers. In conclusion, the selectivity of most suspension growth processes is very good because of the low growth rates and good mixing conditions.

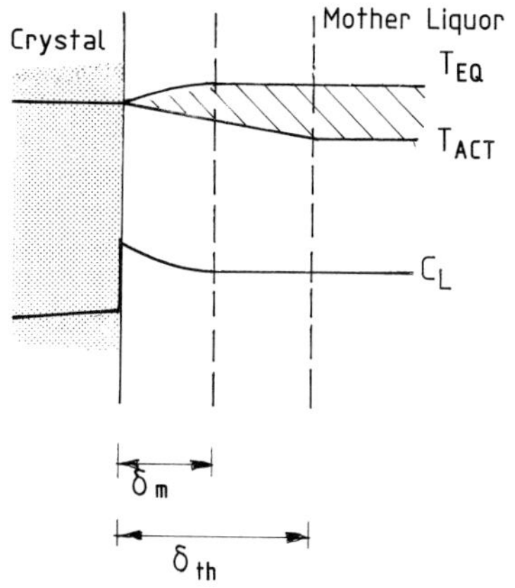

Fig. 9. Intrinsically unstable growth conditions in suspension growth.

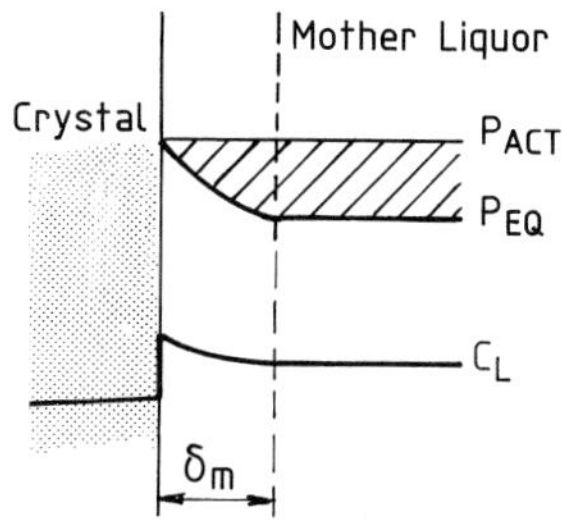

Fig. 10. Intrinsically unstable growth conditions in pressure crystallisation .

In a *pressure crystallisation* process crystals grow in an unstirred compressed compartment at extremely high growth rates ($\pm 10^{-4}$–10^{-3} m/s). The crystals grow from a supersaturated melt and this situation is intrinsically unstable (fig. 10). Even though the applied pressure distributes instantaneously and homogeneously through the crystallising mixture, considerable constitutional supersaturation is inevitable in pressure crystallisers. At the extremely high growth rates, the stabilising effect of surface tension is not strong enough to prevent unstable growth morphologies. Industrial practice shows that the selectivity of a single pressure crystallisation step is only moderate, mainly due to the inclusion of mother liquor in the product crystal cake.

4. *Purification mechanisms*

Crystal purity is hampered by impurities, which are *incorporated* in the crystal lattice, *included* in the crystal matrix, and/or *adhered* to the crystal surface. Several purification mechanisms can be applied to upgrade the crystal purity and thus to improve the overall separation efficiency of a fractional crystallisation process. Usually four purification mechanisms are distinguished: washing, sweating, recrystallisation and solid-state diffusion.

Washing involves an intensive contact between a crystal and a wash liquid. It promotes the mass transfer between a stagnant layer adhering to the crystal surface and the bulk of the liquid phase. The efficiency of washing can be improved by establishing a countercurrent transport between the crystal phase and the wash liquid. In fractional crystallisation preferably pure molten product is used as the wash liquid. Washing has only a small effect on included and no effect on incorporated impurities.

Sweating is a technical term which originates from the wax purification industry. Raw wax containing water and oils is heated slowly until droplets of the low melting components appear on the surface, resembling sweat droplets. By removing the droplets from the surface, the purity of the remaining wax is increased. In fractional crystallisation processes sweating is applied for the removal of included impurities from the crystal matrix. Sweating takes place by increasing the temperature or by reducing the pressure in order to partially melt the impure portions of the crystals thus forming an impurity-rich liquid phase on the crystal surface which is then separated. For a sweating operation to be effective, it is essential that the impurities are locally dispersed as inclusions of mother liquor in the crystal phase [16]. The inclusions, which have solidified during the progress of the crystallisation, will melt first upon gradual heating because of their lower melting point while the surrounding (purer) crystal structure remains solid. Exactly what forces cause the melt to move towards the crystal surface is still unknown but thermal gradients [17], difference in expansion coefficient between the solid and the liquid phase, capillary forces, and gravity are expected to play a role. The pathways for the transportation of the melt to the crystal surface may be pores resulting from a recrystallisation process or thermal cracks [18]. Research on sweating of crystal layers and separate crystals was performed extensively by Matsuoka [16, 19–21], Ulrich and Özoguz [22] and Toyokura et al. [23].

Recrystallisation can be used to purify crystals that contain incorporated impurities. In a complete recrystallisation operation crystals are separated from impure mother liquor, next the crystals are melted and a portion of this melt is again crystallised. In practice, this operation is carried out batchwise in layer growth and pressure crystallisers. In suspension crystallisers continuous in situ partial recrystallisation is strived for by contacting a hot and purer liquid phase with colder and impurer crystals. During this encounter an impure crystal will partially melt because of its lower melting point but only at the expense of a certain enthalpy change. Since suspension crystallisation on a microscale occurs almost adiabatically, this enthalpy change is the driving force for the crystallisation of a portion of the surrounding purer liquid either on the surface of existing crystals or via the formation of new crystals. Because the recrystallisation process is slow and residence times in industrial suspension crystallisers are relatively short, the recrystallisation process is unlikely to be complete.

Solid-state diffusion is best described as the migration of molecules through the crystal structure. The driving force for solid state diffusion is a concentration gradient inside the crystal. The purifying effect of solid state diffusion may be considered negligible in industrial practice because of the low diffusion coefficients ($< 1 \times 10^{-14}$ m^2/s) of impurities in organic crystals.

5. *Industrial layer growth equipment*

Industrial layer growth crystallisation processes can in principle be operated *continuously* but the vast majority are operated *batchwise*. Two types of batchwise layer growth processes can be distinguished: static and dynamic. In *static* equipment crystal growth occurs from a stagnant mother liquor. In *dynamic* equipment crystal growth occurs from a mother liquor in motion. The movement of the mother liquor can be achieved by *internal mixing* or by *external circulation* (see fig. 11).

A rotary-drum crystalliser [24, 25] consists of a horizontally mounted internally cooled drum which is partially immersed in a liquid feedstock. Upon rotation of the drum the crystal layer which is formed on the immersed part of the drum emerges from the liquid feedstock. The crystal layer is removed from the rotating drum with a knife. Extra mixing devices can be installed in the liquid feedstock to enhance mass and heat transfer. High growth rates are applied and thick crystal layers need to be grown to meet production capacities. Hence the single-step separation efficiency of the rotary drums is rather poor and nowadays these devices are mainly used as product flakers.

Another recent attempt to develop a continuous process based on layer growth is the Brembelt [26, 27]. In this design a conveyor belt is positioned at an angle to the horizontal (fig. 12). A coolant is sprayed in different compartments at the bottom of the belt and a liquid feedstock is supplied on top of the belt. The belt transports a growing crystal layer upwards while the mother liquor flows downwards as a film. The temperature of the coolant in every compartment is adjusted to achieve an ascending temperature profile along the belt in the direction of the crystal layer transport. The temperature rise will induce sweating and possibly recrystallisation. The Brembelt is at present being tested on pilot plant scale for its ability to combine a good separation efficiency with an acceptable specific production capacity.

The Proabd refiner of the French BEFS-technologies is the most applied representative of the *batchwise static* layer growth processes [28, 29]. The original design features

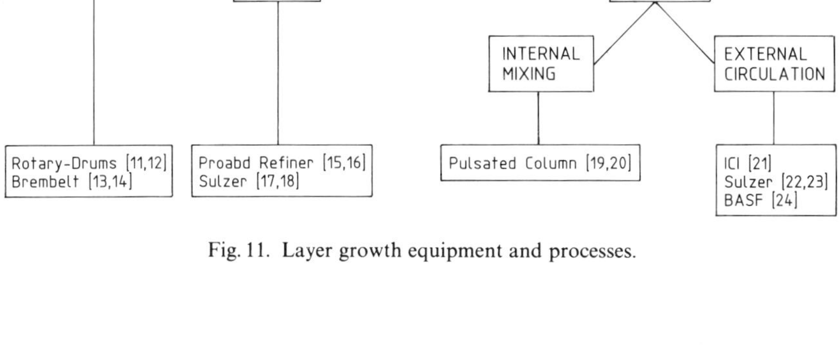

Fig. 11. Layer growth equipment and processes.

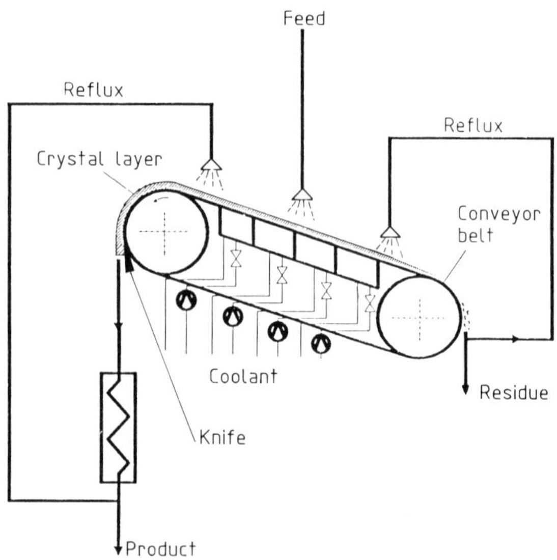

Fig. 12. Flow scheme of the Brembelt [26, 27].

a rectangular closed vessel equipped with finned heat exchange tubes. The heat exchange medium flows along the inside and crystallisation occurs on the outside of the heat exchange tubes. Other static designs differ slightly from the Proabd refiner in the shape of the heat exchange elements and the geometry of the crystalliser. The largest static crystalliser manufactured by Sulzer has a length of 6.1 m, a width of 2.0 m and a height of 2.4 m [31]. One static operation cycle comprises five phases. First, the vessel is filled with liquid feedstock. The temperature of the heat exchange medium is then carefully lowered and part of the feed is crystallised. Next, an outlet valve is opened and the residual mother liquor is allowed to drain away. In the fourth phase, the temperature of the heat exchange medium is gradually increased to induce sweating. Drainage continues during the sweating phase. Finally, the remaining crystal cake is melted completely and the product is withdrawn.

In the stagnant melt, mass and heat transfer are diffusion controlled and despite the fact that the growth rates are kept only at $\pm 10^{-6}$ m/s constitutional supersaturation is inevitable and unstable growth morphologies are bound to occur. In industrial practice networks of dendrite structures fill up the available space between the heat exchange tubes during the crystallisation process and cause entrapment of large quantities of impure residual mother liquor in the crystal layer. Long drainage and sweating times are necessary to release the entrapped mother liquor. Due to the slow crystallisation and the long sweating period, a typical operation cycle may take about one day.

Sulzer is market leader with its *batchwise dynamic* falling film system which features an *external transport loop* for the feedstock and the heat exchange medium [32, 33]. The principle of this system is explained from the simplified scheme in fig. 13. A liquid

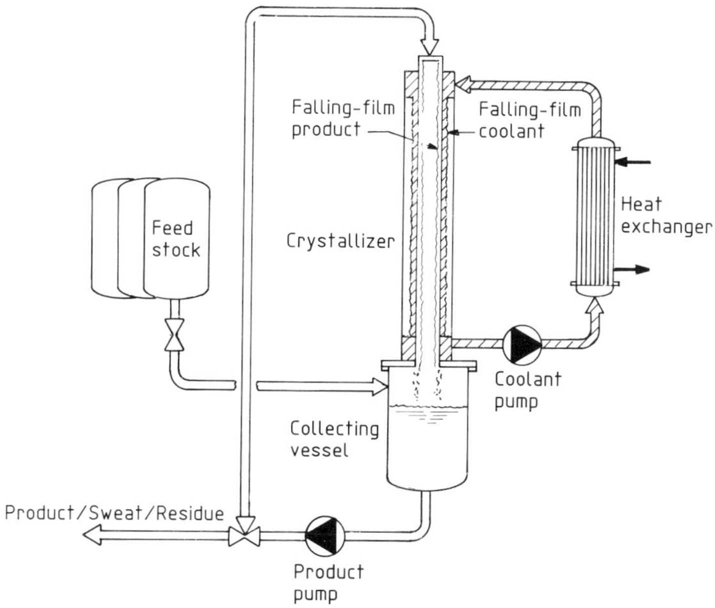

Fig. 13. Sulzer dynamic falling film system.

feedstock is supplied to a collecting vessel. When the filling of the collecting vessel is completed, the product and the coolant pump are activated at the same time, producing cocurrent falling films on both sides of the vertical tubes. The tubes are 0.07 m in diameter and 12 m long. The maximum number of tubes in one crystalliser shell is 1250. Like in the static systems, one operation cycle consists of five phases: filling, crystallisation, drainage of residual mother liquor, sweating and product melt-off [34].

The turbulent flow regime in the falling film enhances heat and mass transfer, which facilitates high growth rates (10^{-6}–10^{-5} m/s) and reduces the degree of constitutional supersaturation. Consequently, unstable growth morphologies do not occur as easily as in static layer crystallisers. Some typical values: crystallisation of a portion of the feedstock takes about 45 min and in this period a 0.015 m thick crystal layer is produced. Drainage of the residual mother liquor needs only 1–5 min, sweating takes about 15–20 min and product removal is completed in 5–10 min. Thus, one operation cycle is completed in about 1.5 h [34].

The short operation cycle time makes the dynamic system well suited for repeated processing of an ever purer feedstock. Sulzer adopted the *sequential countercurrent staging* method illustrated in fig. 14. The residual mother liquor from an arbitrary stage 2 operated at temperature T_2 is transported to an earlier stage 1 to be processed at a lower temperature T_1. The sweat obtained in stage 2 can be mixed with a new feedstock for stage 2 or with the drained residue from stage 3. The molten product from stage 2 is transferred to consecutive stage 3 and will be processed again at a higher

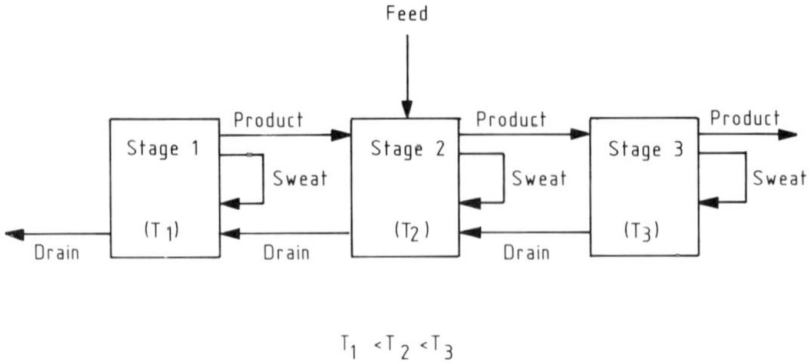

Fig. 14. Sequential countercurrent staging.

temperature T_3. Sequential countercurrent staging can be performed *intermittently* in one crystalliser providing that sufficient intermediate storage tanks are available.

Figure 15 displays the increase in crystal purity per stage for an unspecified substance obtained with a (sequential countercurrent) multistage falling film system and with a single-step static crystalliser [33]. The separation achieved in the single-stage static unit is equivalent to about two and a half dynamic stages of the dynamic multistep system. However, the static operation took about 20 h while the same separation effect was achieved with the dynamic system in only 2–3 h.

In the dynamic systems from ICI [35] and BASF [36] the mother liquor and the coolant are supplied as fully developed tubular flows. The falling film principle has distinct advantages over the fully developed flow principle: (i) lower energy consumption for transport, cooling and heating of the flows, and (ii) easy monitoring of the progression of the crystallisation via a measurement of the liquid level in the collecting vessel [37].

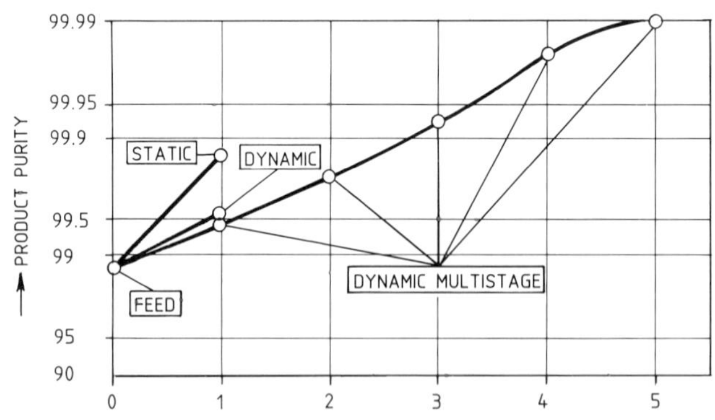

Fig. 15. Comparison of purification results obtained with static and dynamic layer growth systems [33].

Sulzer [31] claims the applicability of their static system for production capacities up to 15 000 tonnes per year and product purities up to 99.7%. Their dynamic falling film system is recommended for higher purities and production capacities. A dynamic multistage plant has been constructed with a capacity of 150 000 tonnes per year for the purification of bisphenol-A.

In recent years research has been focused on the development of *dynamic* crystallisers with *internal mixing*. Rutgerswerke [38] developed a bubble-column. In this design bubbles of inert gas are blown through the crystalliser to enhance mass transfer in the mother liquor. However, during operation, crystal deposition via sublimation occurs in the circulation circuit of the inert gas and also unwanted nucleation of suspended crystals is initiated by the gas bubbles in the crystalliser. A more successful design seems to be the pulsated column [39, 40]. In this column mass transfer in the mother liquor is promoted by pulsation which is effected by means of an external moving piston unit.

Layer crystallisers have difficulty in purifying relatively impure feedstocks. At higher impurity levels in the liquid phase, constitutional supersaturation becomes more pronounced and unstable growth morphologies occur more easily. Another phenomenon which is not yet well described in the literature concerns the attachment of a crystal layer to the cooling surface which seems to decrease when growth occurs from a more impure mother liquor.

6. *Industrial suspension growth equipment*

Industrial suspension crystallisation processes are generally operated continuously and feature a cascade of crystallisers followed by a solid/liquid separation unit (see table 2).

The selectivity of a suspension crystallisation process is usually good but the overall separation efficiency strongly depends on the way slurry handling and solid/liquid separation are performed. The slurry handling in the cascade is aimed at either a

Table 2
Suspension growth processes.

Process	Cascade	Solid/liquid separation	Refs.
Phillips Petroleum	cocurrent	wash column, mechanical bed transport (I)	[58, 60]
(I and II)		wash column, hydraulic bed transport (II)	[59, 60]
Brodie Purifier	countercurrent	wash column, gravity bed transport	[44, 45]
Niro [Grenco]	countercurrent	wash column, mechanical bed transport	[54, 55]
GMF-Gouda (CDC)	cocurrent	centrifuge	[41, 42]
Kureha (KCCP)	cocurrent	beltfilter + sweat column	[61, 62]
Newton Chambers	countercurrent	centrifuge	[63, 64]
Nippon Steel (BMC)	single step	wash column, gravity bed transport	[65, 66]
TNO-Thijssen	cocurrent	wash column, hydraulic bed transport	[49, 50]
Tsukish. Kikai (4C)	countercurrent	wash column, gravity bed transport	[46, 47]

cocurrent or a *countercurrent* contact pattern between the crystals and the mother liquor. With a cocurrent contact pattern, crystals and mother liquor are transported in the same direction and crystals are grown from a mother liquor of decreasing purity at descending temperatures. With a countercurrent contact pattern, crystals and mother liquor are transported in opposite directions and crystals are grown from a mother liquor of increasing purity at ascending temperatures.

A good example of a *cocurrent* cascade is the Cooling Disc Crystalliser (CDC) of the Gouda machine factory [41, 42]. The CDC consists of a horizontal trough which is divided into separate compartments, which are comparable to separate crystallisers, by fixed cooling elements (fig. 16). In each compartment a disc rotates which is equipped with teflon wipers and mixing blades. The discs are mounted on a longitudinal axis which rotates with 10–45 rpm. Internal slurry transport facilities are absent in a CDC since both crystals and mother liquor flow freely from one crystalliser to the next via openings at the bottom of the cooling elements. The crystal content increases in the successive compartments and may amount up to 45 vol% in the last compartment. Since the temperature difference between the product slurry and the coolant is kept approximately constant throughout the CDC and the specific crystal surface area increases from inlet to outlet, the crystal growth rates are lowest in the last compartments where the impurity concentration in the mother liquor is highest. Consequently, in a CDC a good selectivity can be maintained at high production rates whereas the minimal slurry handling contributes to preserving a good crystal quality. The largest available CDC is 8 m long, has a volume of 20 m^3, contains 19 cooling elements with a

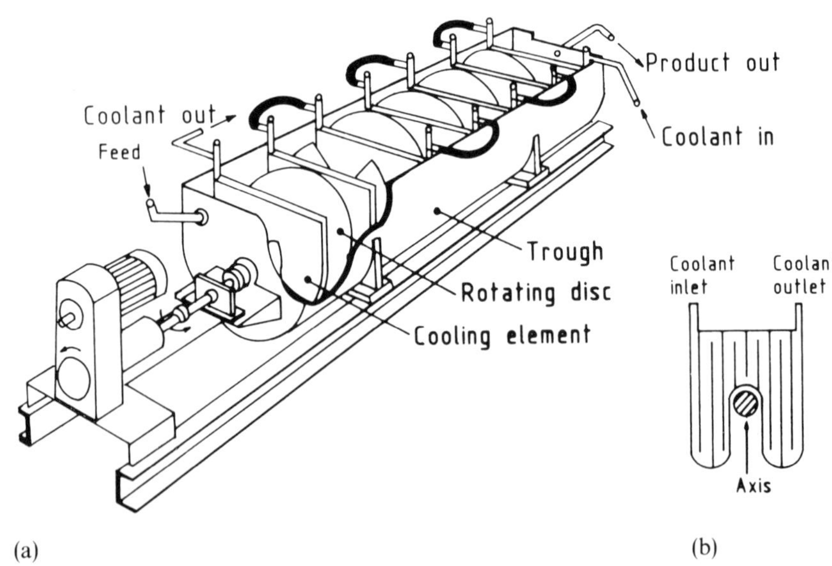

Fig. 16. (a) GMF Cooling Disc Crystalliser; (b) cooling element.

total cooling surface of 130 m^2 and has a maximum production capacity of about 30 000 tonnes per year [43].

The first industrial *countercurrent* cascade was designed by Brodie [44, 45]. This ingenuous design features two or three horizontally positioned crystallisers in which crystal transport is promoted by helix-shaped steel scrapers. The horizontal crystallisers are mounted on top of each other with a direct open connection between them. This open connection on the one hand provides a *continuous* countercurrent contact between crystals and mother liquor which has a favourable effect on crystal shape and purity, but on the other hand it reduces, because of thermal convection, the maximum temperature difference between the cold and the hot end of the cascade and consequently the maximum recovery. The complex precision design of the scrapers caused major scale-up problems and limited the capacity of an industrial Brodie purification process to approximately 10 000 tonnes per year. Industrial Brodie crystallisers are 10–15 m long and have a maximum diameter of about 1 m.

Based on their experience with the Brodie purification process, Tsukishima Kikai (TSK) developed the Continuous Counter Current Crystallisation (4C) system which features totally separated crystallisers resulting in a *discontinuous* contact between the crystals and the mother liquor [46, 47]. The crystallisers are equipped with an external cooling jacket and steel scrapers which are mounted on a rotating draft tube. Slurry transport outside the crystallisers is effected by pumps followed by concentration in cyclones. Meanwhile the residual mother liquor is withdrawn from a crystalliser via an overflow (fig. 17). The height and diameter of the industrial crystallisers amount to 5 m. The draft tube rotates at about 25 rpm and in the crystallisers the suspension density is about 25 vol%. A relatively large temperature difference of 15–20 K per stage can easily be obtained which provides a good recovery. TSK has built a cascade comprising three

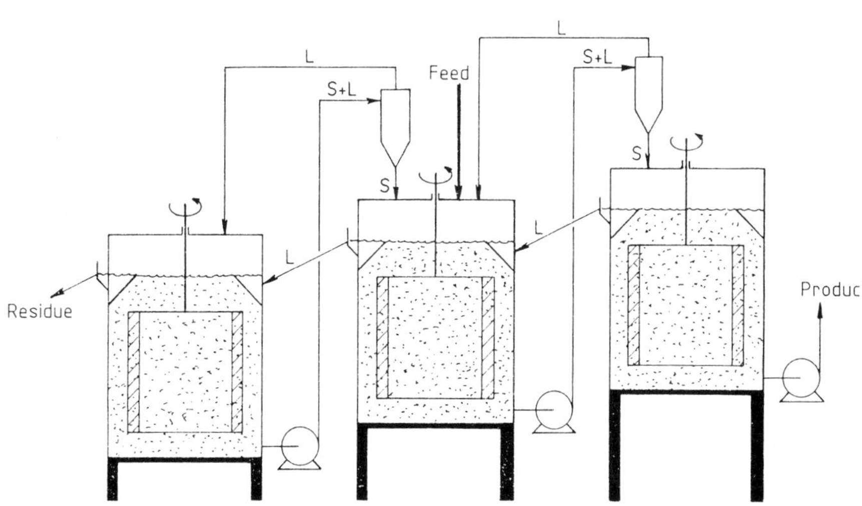

Fig. 17. Countercurrent cascade of the TSK-4C system.

crystallisers with a design capacity of 70 000 tonnes per year for the purification of p-xylene [48].

To summarise, *cocurrent* cascades are simple designs, easy to operate and require limited auxiliary equipment. These cascades are designed to produce a pure and easily filterable crystal product by maintaining low growth rates and minimisation of slurry handling. *Countercurrent* cascades feature more complex designs and usually need more auxiliary equipment. These cascades are designed to introduce extra purification by allowing washing, sweating and recrystallisation already to occur in the crystallisation section of an FC plant. So far, significant extra purification due to a countercurrent contact pattern has not been positively demonstrated on an industrial scale. A major advantage of countercurrent cascades is the fact that the product slurry is discharged from the hot end at the highest temperature. This "*hot*" *product discharge* can lead to savings in the downstream solid/liquid separation section as will be discussed in the following paragraphs.

Solid/liquid separation may be achieved by using well-known equipment such as centrifuges, beltfilters and cyclones but more and more the so-called *wash columns* are applied. These wash columns combine a virtually complete solid/liquid separation with a highly efficient crystal washing action. Basically, three types of wash columns can be distinguished after the different mechanisms employed for crystal bed transport (see fig. 18).

In the first type a crystal bed is transported by *gravity*. The feed slurry enters at the top of the wash column. The crystals settle under the influence of gravity into a loosely packed downward moving bed and the mother liquor leaves the column via an overflow. At the bottom of the column the crystal bed is melted. In industrial gravity wash columns the melter is simply positioned *inside* the wash column. Though heat

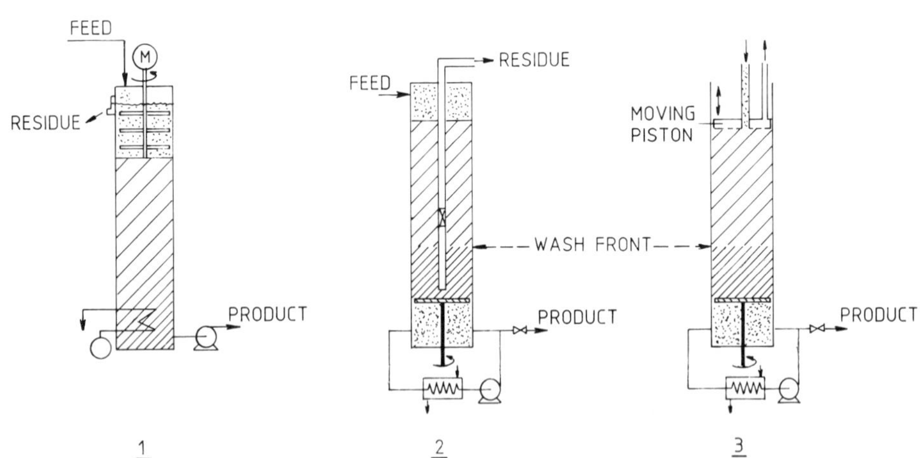

Fig. 18. Three wash column types with different mechanisms for crystal bed transport. (1) Crystal bed transport by gravity. (2) Crystal bed transport by hydraulic force. (3) Crystal bed transport by mechanical aids.

transfer is poor in these almost stagnant melting sections, it is sufficient to meet the low production capacities of this type of wash column. A portion of the molten product is withdrawn from the bottom section and the remainder ascends in the column thus providing a countercurrent washing action. A sharp wash front, as can be detected in the other two types of wash columns, does not appear because of the loose packing and the disturbance of the bed by the agitator. The agitator covers the total height of the column and serves to prevent agglomeration and/or channelling in the crystal bed. The heat input, required to overcome the difference in temperature between the slurry feed and the molten product, is produced mainly by the crystallisation of the wash liquid. The mentioned temperature difference should not exceed 15–20 K because crystallisation of large amounts of wash liquid will lead to serious agglomeration problems. When larger temperature differences have to be overcome, another type of separator (e.g. a centrifuge) has to be used as a pre-separation step. In view of this, the advantage of a *hot product discharge* from the crystallisation section, as is the case in a countercurrent cascade, becomes obvious. An industrial TSK-4C gravity wash column is 8 m high, has a maximum diameter of 2.2 m and is filled for about 6 m with a crystal bed. The revolution rate of the agitator is low ($\leq$ 1.5 rpm). The capacity of gravity wash columns of course depends on physical properties of the solid and the liquid phase, the crystal size distribution and the wall friction but it does not exceed 1500 kg/m^2 h for organic compounds [46–48].

In the second type of wash column a dense crystal bed is transported by means of *hydraulic force*. A slurry is fed under pressure at the top of the column. After a short zone of rapid thickening, the crystals are caught in a densely packed downward moving bed. The residual mother liquor leaves the column via filters which are positioned either in separate filter tubes [49, 50] or in the wall of the column [51]. From a scale-up point of view the version with separate filter tubes, which is known as the TNO-Thijssen column, is favourable. In this version the ratio of the total surface area of the filters to the cross-sectional surface area of the column can be kept constant and the radial distance which has to be overcome by the mother liquor before it reaches the filter surface can be minimised by installing more filter tubes as the column diameter increases. At the bottom of the column the crystals are scraped off, suspended in the wash liquid and subsequently melted in an external melter. The wash liquid (refluxed molten product) flows upwards in the column due to the overpressure in the reslurry chamber below the scraper and thus provides a countercurrent washing action. The ascending wash liquid does not leave the column through the filters but crystallises on the cold crystal bed. The perfect plug flow of the wash liquid through the packed bed results in a sharp wash front. The wash front marks a steep temperature gradient and a sharp decrease in the porosity and permeability of the bed which are caused by the crystallisation of the wash liquid. The permeability decrease underneath the wash front improves the washing efficiency and is essential for a stable washing action [52]. The maximum temperature difference which can be overcome between the slurry feed and the molten product is 10–15 K. A larger temperature difference will make the permeability underneath the wash front extremely low. Consequently, the production capacity will be reduced drastically and/or the washing action will be hampered. At present, the TNO-Thijssen wash column has been tested extensively on a pilot plant scale and the positive results encourage further scale-up. The maximum production capacity of hydraulic wash columns will probably be 5000–10 000 kg/m^2 h [53].

In the third type of wash column, a dense crystal bed is transported by *mechanical force* created by a moving piston or a screw. The mother liquor leaves the column via pores in the mechanical aid. Otherwise this type of wash column is operated identically to the hydraulic wash column. An industrial Niro (Grenco) mechanical wash column is about 1.5 m high and has a maximum diameter of 1.2 m. The capacity of mechanical wash columns is limited to 10 000 kg/m^2 h [54, 55].

A comparison of the three types of wash columns shows that gravity columns have significantly smaller capacities than hydraulic and mechanical columns. Gravity columns require, as a rule of thumb, an average crystal size larger than ±500 μm to achieve an economically viable operation [48, 56], whereas the mechanical and hydraulic columns can already be successfully operated with much smaller crystals (100–200 μm) [53, 56]. The requirement of a larger average crystal size for the gravity column has an important impact on the design of the preceding cascade of the crystallisers: the residence times need to be longer, equipment volume will become larger and investment cost will increase. The residence times of crystals in gravity columns may amount to several hours which allows extra purification by sweating and recrystallisation to occur. The residence times of crystals in columns with hydraulic or mechanical bed transport is about 10 min which is too short for extra purification effects in addition to the washing action. Finally, the compressive stresses exerted on the crystal bed during operation are negligible for gravity bed transport, somewhat higher (0–2 bar) for hydraulic bed transport [57], and very high (3–10 bar) for mechanical bed transport. The level of these compressive stresses is a relevant parameter in the choice of a particular type of wash column in a fractional crystallisation process because various organic compounds (e.g., cyclohexane) form a mechanically weak crystals.

7. *Industrial pressure crystallisation equipment*

In pressure crystallisation the driving force for crystallisation is created by an increase of pressure. A pressure crystallisation process under near equilibrium conditions is illustrated in fig. 5b.

The most lucrative aspect of pressure crystallisation might be the fact that phase equilibria can often be manipulated through pressure variations. This makes it possible to process compounds with melting points well below 273 K at 1 bar, at about room temperature at elevated pressures or to increase the maximum recovery by shifting the eutectic composition just by raising the pressure. Figure 19 represents the binary phase diagram of p- and m-cresol at various pressures. From the phase diagram at 1 bar, it follows that p-cresol crystals can only be recovered from feedstocks containing at least 58 m% p-cresol corresponding with the composition at the eutectic. Moritoki [67] observed that this eutectic shifts to lower p-cresol mass percentages at higher pressures. p-Cresol crystals can still be recovered from feedstocks containing 40 m% p-cresol at 3000 bar. This demonstrates that the maximum recovery for some organic mixtures

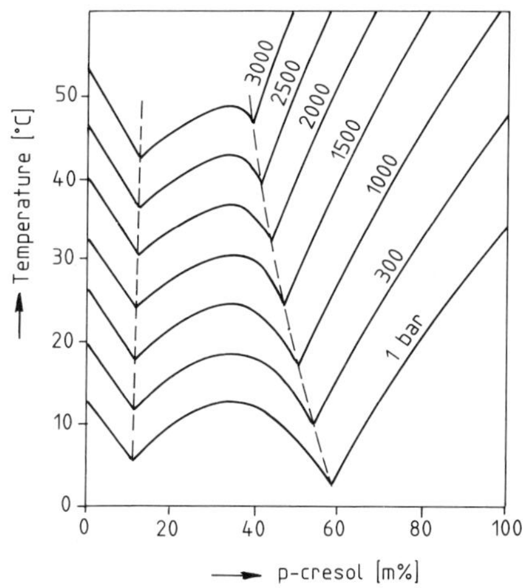

Fig. 19. Phase diagram of p- and m-cresol.

depends on the pressure level and for the case of m- and p-cresol can be increased through pressurisation. Figure 19 also shows higher melting points for p- and m-cresol at increased pressures.

The pressure crystallisation process was developed by Kobe Steel [67–73] to the large-scale adiabatically operated plant design illustrated in fig. 20. The pressure variations in the crystalliser during one operation cycle are presented in fig. 21. The plant is operated as follows: A liquid feedstock is supplied to a chiller vessel, where it is kept at a fixed temperature and some seed crystals are formed or added. The feed is transferred to the crystalliser by means of a slurry pump. When the filling is completed, the piston moves stepwise into the crystalliser thus building up pressure and triggering crystallisation. The pressure in the crystalliser is gradually increased to a predetermined

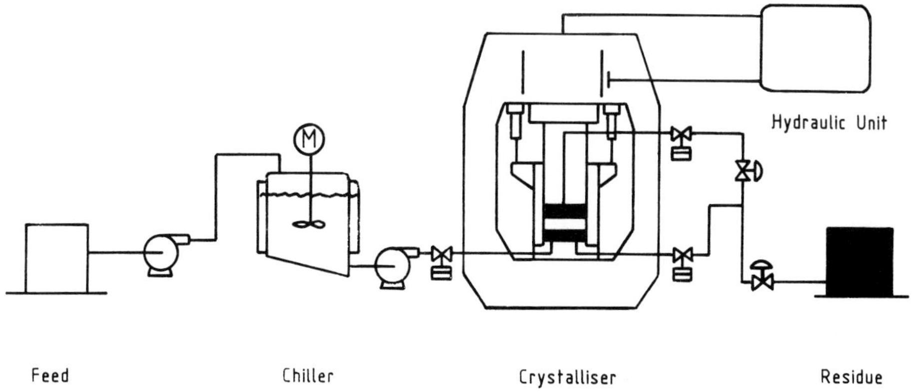

Fig. 20. Flow scheme of the adiabatic Kobe Steel high-pressure process [70].

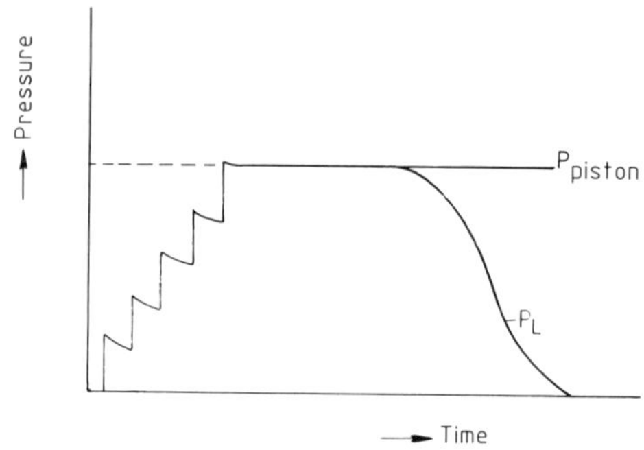

Fig. 21. Pressure variations during a pressure crystallisation cycle.

level (P_0) to obtain the desired recovery. The residual mother liquor is drained from the crystalliser by opening the valve on the right. The remaining crystal cake is squeezed out during this drainage by the continuing inward motion of the piston. Eventually, the liquid pressure gradually decreases and sweating occurs. Nitrogen gas is supplied at the top of the crystalliser to remove the residual mother liquor remaining in the dead spaces of the crystalliser. Finally, the piston and the wall of the crystalliser are lifted and the remaining crystal cake is discharged. Contrary to the cooling crystallisation processes the product is obtained as *solid* material.

In 1984 the first commercial Kobe Steel high-pressure crystalliser was taken into operation at Sumitomo Chemical Co. in Japan to purify p-cresol. In one operation cycle 15 kg feedstock (80 m% p-cresol) is treated and 5 kg product (99.5 m% p-cresol) is obtained. This operation cycle is completed in 2 to 4 min. The maximum operation pressure is about 2000 bar. The production capacity of this plant does not exceed 1000 tonnes per year [74–76].

A practical problem in pressure crystallisation is the removal of residual mother liquor and sweat from the product crystal cake. Experimentally it was found [77] that the centre part of the product cake is the most impure due to entrapped mother liquor which cannot be squeezed out. The commercial plant of Kobe Steel is equipped with filters in the wall of the crystalliser to improve the efficiency of the solid/liquid separation.

Advantages of the pressure crystallisation process are:

- the easy and accurate control of crystallisation and melting rates due to the instantaneous and homogeneous distribution of the driving force (pressure);
- the absence of auxiliary solid/liquid separators, stirring and scraping devices;
- low-energy consumption since only mechanical energy for compression of 10–20% of the volume is required.

The major disadvantages of this process are the high investment and maintenance costs relative to its low production capacity [75, 76].

8. Evaluation

All three crystallisation techniques described in the preceding chapters being: layer growth, suspension growth and (high-)pressure crystallisation, can be successfully applied for ultrapurification of organic compounds on an industrial scale. Each technique has, however, its own advantages and disadvantages which must be taken into consideration when designing a purification plant.

The *layer growth* technique is a proven technology which is usually applied batchwise. Scale-up of layer growth processes is very easy to perform. Operation is simple and flexible for limited variations in the feed conditions and purity demands. Slurry handling can be avoided and there is no need for auxiliary solid/liquid separation units. However, the single-step separation efficiency and the specific production capacity are moderate in comparison to suspension growth processes. The investment costs and energy demands for the dynamic systems are relatively high because, in most cases, a multistage operation is needed. The maintenance costs, on the other hand, are low. Finally, this technique is well suited for ultrapurification purposes but it is less well applicable for separation of more impure feedstocks.

The *suspension growth* technique is also a proven technology and is usually operated continuously. The single-step separation efficiency is excellent and the specific production capacity is high. Slurry handling is, however, inevitable and the overall separation efficiency of a suspension growth process is to a large extent determined by the efficiency of the downstream solid/liquid separation unit. Investment and maintenance costs are moderate and the energy demands of suspension processes are low. Scale-up requires extensive testing on laboratory and pilot-plant scale. The suspension growth technique is applicable for ultra-purification of rather pure feedstocks but also for separation of impure feedstocks. The presence of impurities in the feedstock often has a positive side effect because it reduces encrustation problems that might occur.

The *pressure crystallisation* technique must be considered as a technology in development. Pressure crystallisation is always performed batchwise. Scale-up beyond productions of 1000 tonnes per year will be difficult. The single-step separation efficiency is moderate but one operation cycle is very short which makes the technique ideal for multistage operation. This technique is particularly well suited to treat low-melting-point compounds at about room temperature. Furthermore, the recovery of a crystallisation step may improve at higher operating pressures. Energy consumption is low. Slurry handling is avoided and there is no need for auxiliary solid/liquid separation equipment. Investment and maintenance cost are at present still high.

Reviews on fractional crystallisation and/or comparisons between the three crystal growth techniques are presented in [78–82].

Acknowledgements

The authors are most indebted to M. Matsuoka of Tokyo University of Agriculture and Technology, who organised and coached the study tour of P.J. Jansens through

Japan. They sincerely thank A.E.D.M. van der Heijden of Delft University of Technology for his useful suggestions and comments regarding the section on basic concepts and M. Moritoki of Kobe Steel for his valuable contributions to the section on pressure crystallisation. The authors are grateful to G.J. Arkenbout of TNO-IMET, Y.M. Özoguz of Bremen University, K. Takegami and K. Nakamaru of Tsukishima Kikai, A.N. de Moet of GMF-Gouda, O. Fischer and A.K. Kuszlik of Sulzer Chemtech, J.P. Coutie of BEFS Technologies, K. Sakuma of Nippon Steel Chemical, for their information and in-depth discussions.

List of symbols

C_L = impurity concentration in the liquid phase [m%]
C_S = impurity concentration in the solid phase [m%]
D = diffusion coefficient [m^2/s]
K_0 = equilibrium distribution coefficient [–]
K_i = interface distribution coefficient [–]
K_{eff} = effective distribution coefficient [–]
P_{act} = actual pressure [N/m^2]
P_{eq} = saturation pressure under equilibrium conditions [N/m^2]
P_i = pressure at time i or stage i [N/m^2]
R = recovery of the main component [–]
T_{act} = actual temperature [K]
T_{eq} = saturation temperature under equilibrium conditions [K]
T_i = temperature at time i or stage i [K]
T_M = melting-point temperature [K]
V = linear growth rate [m/s]
X_i = composition of the mother liquor at T_i or P_i [m%]
Y_i = composition of the crystals at T_i or P_i [m%]
δ_m = thickness of stagnant mass transfer boundary layer at the crystal surface [m]
δ_{th} = thickness of stagnant heat transfer boundary layer at the crystal surface [m]
ρ_L = specific weight or density of the liquid phase [kg/m^3]
ρ_S = specific weight or density of the solid phase [kg/m^3]

References

[1] M. Matsuoka and H. Fukushima, Bunri Gijutsu (Separation Process Engineering) 16 (1986) 4.
[2] H.W.B. Roozeboom, Z. Phys. Chem. 10 (1892) 145.
[3] G.J. Sloan and A.R. McGhie, Techniques of Melt Crystallisation (Wiley, New York, 1988).
[4] A.I. Kitaiigorodskii, Molecular Crystals and Molecules (Academic Press, New York, 1973).
[5] W.A. Tiller, The Science of Crystallisation, Microscopic Interfacial Phenomena (Cambridge University Press, Cambridge, 1991).
[6] J.A. Burton, R.C. Prim and W.P. Slichter, J. Chem. Phys. 21 (1953) 1987.
[7] A.G. Ostrogorsky and G. Müller, J. Cryst. Growth 128 (1992) 207.
[8] R. de Goede and G.M. van Rosmalen, J. Cryst. Growth 104 (1990) 392.
[9] R. de Goede, Ph.D. thesis, Delft University of Technology, Delft (1988).
[10] M. Matsuoka and J. Garside, Chem. Eng. Sci. 46 (1991) 183.
[11] W.W. Mullins and R.F. Sekerka, J. Appl. Phys. 34 (1963) 323; 52 (1964) 444.
[12] R.F. Sekerka, in: Proc. ICCG-10, J. Cryst. Growth 128 (1992) 1.
[13] J.W. Rutter and B. Chalmers, Can. J. Phys. 31 (1975) 15.
[14] J.S. Langer, Rev. Mod. Phys. **52** (1980) 1.
[15] W.A. Tiller, The Science of Crystallization, Macroscopic Phenomena and Defect Generation (Cambridge University Press, Cambridge, 1991).

[16] M. Matsuoka, Bunri Gijutsu (Separation Process Engineering) 14 (1984) 113.

[17] R.V. Scholz, J. Ulrich, and K. Genthner, in: Proc. 12th Symp. on Industrial Crystallization, ed. Z.H. Rojkowski (Zaklad Poligraficzny Wlodarski, Warsaw, 1993) pp. 1–25.

[18] M. Stepanski, O. Fischer and R. Stadler, in: Abstr. G.V.C. Working Party on Crystallisation (Delft University of Technology, Delft, 1992) p. 14.

[19] M. Matsuoka and A. Sumitani, in: Industrial crystallisation 87, eds. J.N. Nyvilt and S. Zacek (Bechyne, Czechoslovakia, 1987) p. 57.

[20] M. Matsuoka, M. Ohishi and S. Kasama, J. Chem. Eng. Jpn. 19 (3) (1986) 181.

[21] M. Matsuoka and A. Sumitani, J. Chem. Eng. Jpn. 21 (1) (1988) 6.

[22] J. Ulrich and M.Y. Özoguz, Chem. Ing. Tech. MS 1725/89; Summary, in: Chem. Ing. Tech. 61 (1) (1989) 76.

[23] K. Toyokura, K. Yamada, T. Kajiwara, J. Sato, I. Hirasawa, M. Moritoki and N. Nishiguchi, in: Proc. 4th World Congr. of Chemical Engineering (1991).

[24] S.C. Carney US 2622114 (1948).

[25] J.C. Chaty, in: Fractional Solification, eds. M. Zief and W.R. Wilcox (Marcel Dekker, New York, 1967) p. 409.

[26] J. Ulrich, M. Stepanski and Y. Özoğuz, Swiss patent application 03 750/90–0 (1990).

[27] I. Hünken, Y. Özoğuz and J. Ulrich, in: BIWIC 1991, ed. J. Ulrich (Bremen University, Bremen, 1991) p. 42.

[28] C. Ab-Der-Halden, U.K. patent 837295 (1959); UK 899799 (1961).

[29] J.G.D. Molinari, in: Fractional Solidification, eds. M. Zief and W.R. Wilcox (Marcel Dekker, New York, 1967) p. 393.

[30] S.J. Jancic, Chem. Prod. 5 (1988) 90.

[31] Sulzer Chemtech, Fractional Crystallisation: Falling Film, Static and Scraped Wall, Brochure from Sulzer Chemtech e/28.56.06.40-IV.91.100 (1991) p. 9.

[32] K. Saxer, CH 501421 (1968); US 3621664 (1968).

[33] S.J. Jancic, in: Industrial Crystallisation 87, eds. J.N. Nyvilt and S. Zacek (Bechyne, Chechoslovakia, 1987) p. 57.

[34] Private communication to Dr. O. Fischer and Dr. A.K. Kuszlik of Sulzer (1992).

[35] E. Barton, UK 1083850 (1963).

[36] K. Wintermantel, D. Stockburger and W. Fuchs, DE 2 606 364 (1976).

[37] M.Y. Özoguz, K. Wagnick, A.K. Kuszlick and J. Ulrich, in: BIWIC 1990, ed. J. Ulrich (Bremen University, Bremen, 1990) p. 42.

[38] K. Stolzenberg, Chem. Ing. Tech. 55 (1) (1983) 45.

[39] S.J. Jancic, Swiss patent application 1631/90 (1990).

[40] A.K. Kuszlik, Ph.D. thesis, Bremen University, Bremen (1990).

[41] J.F. Witte and J.W.L.M. Horsmans, EP 0031134 (17-12-1980); US 4486395 (26-05-1982).

[42] A.N. de Moet, in: Industrial Crystallisation 84, eds. S.J. Jancic and E.J. de Jong (Elsevier, Amsterdam, 1984) p. 223.

[43] Private communication to Ir. A.N. de Moet of the Gouda machine factory (1992).

[44] D.L. McKay, in: Fractional Solidification, eds. M. Zief and W.R. Wilcox (Marcel Dekker, New York, 1967) 427.

[45] J.A. Brodie, DE 1947251-C3 (1965).

[46] K. Takegami, M. Morita, K. Nakamaru and K. Miwa, EP 0105524 B1 (1983).

[47] K. Takegami, N. Nakamaru and M. Morita, in: Industrial Crystallisation 84, eds. S.J. Jancic and E.J. de Jong (Elsevier, Amsterdam, 1984) p. 143.

[48] Private communication to Dr. K. Takegami and Dr. K. Nakamaru of Tsukishima Kikai Company (1991).

[49] H.A.C. Thijssen and G.J. Arkenbout, US 4666456 (1983); US 4735781 (1985).

[50] G.J. Arkenbout, A. van Kuijk and L.J.H.M. Schneiders, in: Industrial Crystallisation 84, eds. S.J. Jancic and E.J. de Jong (Elsevier, Amsterdam, 1984) p. 137.

[51] E.D. Brennan, US 4309878 (1980).

[52] M.A.G. Vorstman and H.A.C. Thijssen, in: Proc. Symp. on Heat and Mass Transfer Problems in Food Engineering, Wageningen (1972) D4.

[53] Private communication to Dr. G.J. Arkenbout of TNO-Apeldoorn (1990).
[54] H.A. Jansen, in: BIWIC 1991, ed. J. Ulrich (Bremen University, Bremen, 1991) p. 58.
[55] Private communication to Ir. H.A. Jansen of Niro (Grenco) (1991).
[56] G.J. Arkenbout, M. Nienoord and E.J. de Jong, in: Industrial Crystallisation 90, ed. A. Mersmann, Garmisch-Partenkirchen, Germany (1990) p. 715.
[57] P.J. Jansens, O.S.L. Bruinsma, G.M. van Rosmalen and de R. de Goede, in: BIWIC 1991, ed. J. Ulrich (Bremen University, Bremen, 1991) p. 59.
[58] J. Schmidt, DBP 900209 (1950); US 2617274 (1950).
[59] P.M. Arnold, US 2540083 (1945).
[60] J.G.D. Molinari and B.V. Dodgson, The Chemical Engineer July/August (1974) 460.
[61] J. Yamada, C. Shimizu and S. Saitoh, in: Industrial Crystallisation 81, eds. S.J. Jancic and E.J. de Jong, (North-Holland, Amsterdam, 1982) p. 265.
[62] M. Shimada, S. Kumagai, S. Sagawa and F. Iwaki, DE 3043823 A1 (1980).
[63] D. Bradley and M.J. Shires, UK 908302 (1960).
[64] J.G. Molinari, in: Fractional Solidification, eds. M. Zief and W.R. Wilcox (Marcel Dekker, New York, 1967) p. 401.
[65] K. Sakuma and J. Ikeda, in: Industrial Crystallisation 84, eds. S.J. Jancic and E.J. de Jong (Elsevier, Amsterdam, 1984) p. 147.
[66] K. Sakuma JP 58-463222 (1974); JP 55-18521 (1974).
[67] M. Moritoki, Int. Chem. Eng. 20(3) (1980) 394.
[68] M. Moritoki, Japan 0970170 (1969); DE 2135539 C3 (1971).
[69] M. Moritoki and K. Kitagawa, US 4784766 (1985).
[70] M. Moritoki and T. Fujikawa, in: Industrial Crystallisation 84, eds. S.J. Jancic and E.J. de Jong (Elsevier, Amsterdam, 1984) p. 369.
[71] M. Moritoki, K. Kitagawa and K. Kaneko, in: Industrial Crystallisation 84, eds. S.J. Jancic and E.J. de Jong (Elsevier, Amsterdam, 1984) p. 377.
[72] M. Moritoki, M. Ito, T. Sawada, M. Ishiyama, Y. Yamazaki, T. Fukutomi and K. Toyokura, in: Industrial Crystallisation 87, eds. J.N. Nyvilt and S. Zacek (Bechyne, Chechoslovakia, 1987) p. 485.
[73] M. Moritoki, in: Industrial Crystallisation 84, eds. S.J. Jancic and E.J. de Jong (Elsevier, Amsterdam, 1984) p. 373.
[74] Private communication to Dr. M. Moritoki from Kobe Steel (1990, 1991).
[75] Private communication to Dr. S. Arakawa, Dr. F. Horiuchi, Dr. N. Kanda and Dr. T. Umemura of Sumitomo Chemical (Ohita Works) (1991).
[76] S.J. Jancic, in: BIWIC 1990, ed. J. Ulrich (Bremen University, Bremen, 1990) p. 42.
[77] K. Toyokura, H. Narahara, M. Yokota, K. Yamada, M. Moritoki and N. Nischiguchi, in: Industrial Crystallisation 90, ed. A. Mersmann, Garmisch-Partenkirchen, Germany (1990) p. 721.
[78] G.J. Arkenbout, Improvement of melt crystallisation's efficiency for industrial applications, Paper presented at the EC seminar: Energy Efficiency in Industrial Processes – Future R&D Requirement, Brussels (1980).
[79] R. Rittner and R. Steiner, Chem. Ing. Tech. 57(2) (1985) 91.
[80] P.J. Jansens, Fractional crystallisation, in: Post Academic Course on Industrial Crystallisation, Delft University of Technology and Nijmegen University, Delft (1990, 1991).
[81] M.Y. Özoguz, Ph.D. thesis, Bremen University, Bremen (1991).
[82] N.P. Wynn, Chem. Eng. Prog. 3 (1992) 52.

CHAPTER 7

Industrial Mass Crystallisation

A.E.D.M. van der HEIJDEN and G.M. van ROSMALEN

Delft University of Technology, Leeghwaterstraat 44, 2628 CA Delft, Netherlands

Contents

1. Scope of industrial crystallisation 318
2. Crystallisation techniques 321
   2.1. Driving force for crystallisation from solution 321
   2.2. Classification of crystallisation methods 322
      2.2.1. Cooling crystallisation 323
      2.2.2. Evaporative crystallisation 323
      2.2.3. Flash evaporative crystallisation 324
      2.2.4. Precipitation 324
      2.2.5. Salting-out crystallisation 325
3. Nucleation 325
   3.1. Nucleation mechanisms 325
      3.1.1. Primary nucleation 326
      3.1.2. Secondary nucleation 326
   3.2. Metastable zone width 334
4. Growth 336
   4.1. Aspects of crystal growth 336
   4.2. Crystal growth mechanisms 336
      4.2.1. Rough growth 337
      4.2.2. Layer growth 337
   4.3. Simplified growth rate models 339
      4.3.1. Mass transfer and heat transfer 339
      4.3.2. Size-dependent growth (SDG) 341
      4.3.3. Growth rate dispersion (GRD) 342
      4.3.4. Aggregation, attrition and breakage 343
   4.4. Additives as habit modifiers 346
5. Crystal size distribution 346

HANDBOOK OF CRYSTAL GROWTH, VOL. 2
edited by D.T.J. Hurle

5.1. CSD properties . . . 347
5.2. CSD measuring techniques . . . 349
5.2.1. Sieve analysis . . . 350
5.2.2. Coulter counter . . . 350
5.2.3. Forward light scattering . . . 352
5.2.4. Quasi-elastic light scattering . . . 355
5.2.5. Image analysis . . . 355
5.2.6. Backward light scattering . . . 356
5.2.7. Acoustic spectroscopy . . . 357
6. Continuous crystallisation models . . . 358
6.1. Mass, heat and population balances . . . 358
6.1.1. Mass balance . . . 359
6.1.2. Heat balance . . . 359
6.1.3. Population balance . . . 360
6.2. Steady-state operation . . . 362
6.2.1. The MSMPR model . . . 362
6.2.2. Mixed suspension with fines destruction . . . 365
6.2.3. Mixed suspension with classified product removal . . . 367
6.2.4. The *R–z* model . . . 369
6.2.5. Two-stage crystalliser . . . 370
6.2.6. Steady-state CSD including size-dependent birth and death rate functions . . . 371
6.3. Dynamic operation . . . 372
6.3.1. Outline of simulation procedure . . . 372
6.4. Scale-up of industrial crystallisers . . . 377
7. Batch crystallisation models . . . 379
7.1. Population, mass and heat balances during batchwise operation . . . 379
7.1.1. Batchwise operation . . . 379
7.1.2. Population balance . . . 381
7.1.3. Mass balance and moment equations . . . 382
7.2. Control strategies of batch processes . . . 382
7.2.1. Seeded batch . . . 383
7.2.2. Unseeded batch . . . 384
7.2.3. Batch crystallisation with fines destruction . . . 385
8. Industrial crystallisers . . . 385
8.1. Selection of crystallisation method . . . 386
8.2. Selection of crystalliser . . . 387
8.2.1. Solar ponds . . . 387
8.2.2. Thermo-syphon crystalliser . . . 388
8.2.3. Stirred draft-tube baffle crystalliser . . . 388
8.2.4. Forced circulation crystalliser . . . 391
8.2.5. Cascade of evaporative crystallisers . . . 391
8.2.6. Cooling disc crystalliser . . . 394
8.2.7. Oslo crystalliser . . . 395
8.3. Solid–liquid separation . . . 396
8.3.1. Sedimentation . . . 396

8.3.2. Filtration 397

9. Precipitation 397

9.1. General aspects of precipitation 397

9.2. Solution chemistry and supersaturation 399

9.3. Recrystallisation 400

9.4. Mixing 402

9.4.1. Introduction 402

9.4.2. Macromixing 402

9.4.3. Micromixing 403

9.4.4. Mixing times 406

9.4.5. Process parameters 407

9.4.6. Eulerian models for mixing 408

Acknowledgements 409

List of symbols 409

References 412

1. Scope of industrial crystallisation

Industrial crystallisation, sometimes also referred to as bulk or mass crystallisation, concerns the nucleation and growth of a huge number of small crystals (~ 200–1000 μm) either from solution or from the melt. The target in industrial crystallisation is the production and separation of compounds as a solid phase as well as their simultaneous purification. Numerous examples of solid products can be found in the chemical, polymer, pharmaceutical, semiconductor and food industry.

The advantage of crystallisation compared to other separation techniques like distillation, adsorption and extraction is that it can be applied at low temperatures and mostly without an auxiliary solid (adsorbent) or liquid (extractant) phase. Furthermore, during growth from solution or melt, one crystallisation step often suffices to reach a high-purity product since crystallisation is a selective technique with respect to the incorporation of foreign ions, atoms or molecules into the crystalline solid. A drawback is the low rate at which the crystals increase their dimensions, implying a low yield rate. Growth rates of well-soluble compounds are typically in the range of 10^{-8} to 10^{-7} m s^{-1}.

After the crystallisation process is completed, further downstream product handling is required to remove the mother liquor in which the crystallisation took place, from the product particles. This solid-liquid separation can be achieved by either sedimentation, filtration, centrifugation or with help of a wash column. This puts restrictions upon the achieved crystal size distribution (CSD). The CSD is usually characterised by the width of the distribution and by its d_{50} value, being the size where 50% of the total crystal volume consists of crystals below this size. Separation units are usually capable of handling particles ranging in size from 200 to 1000 μm. To reach these sizes, generally a residence time of crystals in the crystalliser of 0.5 to 5 h is needed. The CSD also affects the washability and drying of the crystals as well as the storage (high density) and transport (non-caking) behaviour of the product. In all these aspects of the general product demands, control of the shape of the crystals is of importance as well. Also the purity of the crystals can have a great influence on the performance of the product. For instance, inclusions of mother liquor or gas bubbles in crystals may lead to cracking of the crystals during mechanical handling and in the case of liquid inclusions also to caking of the crystals induced by temperature and pressure gradients during storage.

To obtain a wide variety of solid products which fulfill the general product demands mentioned above, several crystallisation techniques have been developed and various types of crystallisers have been designed.

Industrial crystallisation is still a field of intensive research, since a large number of physical, chemical and hydrodynamical processes meet simultaneously in a crystalliser. Hence, to give a thorough description of the performance of an industrial crystalliser is a difficult task. In the past, a process engineering approach proved to be quite satisfactory. Under the pressure of ever more stringent product demands, however, not only for specialties but also for bulk products in a competitive market, a better insight in the underlying mechanisms of the crystallisation processes becomes a requisite. Such knowledge opens ways for a better process and product control and can also be used for more reliable scale-up purposes.

In this chapter only an overview will be given of the classical field of industrial crystallisation. Some subjects will be addressed in more detail than others, but this is compensated by providing the reader with literature references that can be consulted for further information.

Besides the classical field of industrial crystallisation, however, other fast developing fields can be recognised. Fractional crystallisation of organic compounds can be identified as such. Nowadays the demand for ultrapure organic bulk products with purities between 99 and 99.99%, increases rapidly. Here the final product is usually an ultrapure molten phase and the crystals only serve as an intermediate, although their role is essential in the purification process. This counts for, e.g. organics in the food industry but also for monomers which serve as intermediates for the production of polymers and fibers. Further processing of these intermediates to the final product can often be quite sensitive to minor impurities and to quality fluctuations of the starting materials. Because of this increasing demand for ultrapure organic compounds the field of high-performance separation techniques had to be developed and improved further. A promising technique is multi-stage melt or fractional crystallisation. The superior selectivity characterising the crystallisation process and the low temperature of operation are the main reasons to apply this method. The method can handle relatively impure input materials (70–80% pure) and normally only a few crystallisation steps are required before the ultrapure product composition is reached. Supersaturation can be established either by cooling or by pressurisation of the liquid and the mode of crystallisation can be layer or suspension growth. For layer growth, crystalline material is deposited on a cooled surface. In order to obtain an acceptable yield rate, the growth rate has to be rather high, up to 10^{-5} m s^{-1}. For growth from suspensions, growth rates are in the range of 10^{-8} to 10^{-7} m s^{-1}. If a countercurrent contact between crystals and liquid is established the crystals are grown in and washed by mother liquor of ascending purity and temperature. The technique of in particular layer growth suffers from morphological instabilities, cellular and dendritic growth [1] owing to constitutional supercooling, and recrystallisation of impure inclusions and washing of the crystals with their purer melt is needed to reach the final goal. In the suspension growth technique adequate separation and washing of the crystals is essential and can form a major problem in the operation. Consequently, the design, modelling and performance of the equipment used (crystalliser, solid–liquid separation units) have developed into their own field of research. A separate chapter is therefore devoted to this subject [2].

Apart from the demand for ultrapurification of organic compounds, other specific product demands can acquire their own tailor-made crystallisation method and equipment. Such demands can follow from a typical application of the product or from inherent physical, chemical or mechanical properties of the compound. Another illustrative example in this respect is supercritical crystallisation as a technique to obtain very tiny ultrapure crystals or particularly shaped polymer particles. This and other nonclassical techniques will not be treated in this chapter.

Since the classical industrial crystallisation processes are mostly restricted to crystallisation from solution, cooling and pressure crystallisation from the melt will not be treated here, but in relation to fractional crystallisation [2].

2. Crystallisation techniques

2.1. Driving force for crystallisation from solution

The solubility of a substance in a solvent generally depends on temperature, pressure and the presence of other dissolved substances. For most substances the solubility increases with temperature – exceptions are, e.g. sparingly soluble, inorganic salts like $CaCO_3$ in water – and decreases with pressure. The solubility of a substance leading to a saturated ideal solution can thermodynamically be shown to depend on the molar enthalpy of melting ΔH_M, of the solid material and the difference between the inverse of the melting temperature T_M (of the solid) and that of the actual temperature T of the solution. No dependence is found on properties of the solvent, implying that for ideal situations the solubility of a given substance in any solvent as a function of temperature is the same. This is of course hardly ever encountered in practice, since solute–solvent interactions (accounted for by the activity a related to the concentration via the activity coefficient γ where $a = \gamma c$) may greatly affect the ideality of the solution.

A necessary requirement for crystallisation is the creation of a non-equilibrium situation during which a driving force is established. From a thermodynamic point of view the fact that the chemical potentials of the solid (μ_S) and liquid (μ_L) phase are different reflects the non-equilibrium situation. Thus $\Delta\mu = \mu_L - \mu_S$ is a measure for the supersaturation. $\Delta\mu$ is called the growth affinity or driving force for crystallisation. Another way of expressing the departure from equilibrium is with a dimensionless parameter. The advantage of a dimensionless parameter is that it can be used to compare driving forces for different crystallising substances. The relative supersaturation σ defined as

$$\sigma = \exp(\Delta\mu/kT) - 1 \approx \Delta\mu/kT \quad \text{for } \Delta\mu \ll kT, \tag{1}$$

can be used as such, where k is Boltzmann's constant. For $\Delta\mu$ small in comparison to kT, this may be approximated by $\sigma \approx \Delta\mu/kT$. Another frequently adopted dimensionless parameter is S, defined as $1 + \sigma$. The value of $\Delta\mu$ is directly related to Δc, as will be demonstrated below for the various crystallisation methods.

In practice, the driving force $\Delta\mu$ cannot be determined directly but usually the concentration c can be measured (whether or not directly). For solution growth, the difference Δc between the actual concentration c (at temperature T or pressure p) and the equilibrium (saturated) concentration c^* at the equilibrium or saturation temperature T^* (or equilibrium pressure p^*) provides an absolute value of the supersaturation expressed either as kg per kg solvent (mass ratio) or kg per m^3 solution (mass concentration). Dividing Δc by c^* gives a relative supersaturation, usually expressed as a percentage. For cooling crystallisation from solution, the undercooling $\Delta T = T^* - T$ is also used to express the supersaturation, where T is the actual solution temperature and T^* the equilibrium temperature. From the solubility curve of c^* versus T^*, ΔT can be easily related to Δc. When describing the different crystallisation methods in the next subsections, it will be shown how $\Delta\mu$ relates to Δc or ΔT [3].

An accurate measurement of the supersaturation is usually difficult. If the solubility curve $c^*(T^*)$ of a substance in a certain solvent is known, the Δc value can be calculated by measuring the temperature of the solution and its equilibrium temperature. The concentration of a solution in the case of medium to highly soluble compounds can also be determined by measuring either the density, the refractive index, the viscosity or any other physical parameter which varies in a distinct way with concentration. Yet another, new, way of measuring the supersaturation is the use of a fluorescence technique, developed by Chakraborty and Berglung [4] for sucrose in water. This technique uses a trace fluorescent probe, added to the solution at ppm level, which can exist in a protonated or deprotonated form depending on the amount of free water available. Although the method still needs more development and refinement to become more generally applicable, it is a promising alternative for determination of supersaturations of easily soluble compounds. In the case of slightly soluble substances, the supersaturation can be determined by measuring the electrical conductivity of the solution. Also a measurement of the activities a_i of the ionic species present in the solution can be used.

The answer to the question as to which of the techniques available for measuring the imposed supersaturation should be used, depends strongly on both material and solution properties.

2.2. *Classification of crystallisation methods*

Crystallisation processes applied in the field of industrial crystallisation generally proceed via cooling, evaporation, flash evaporation, precipitation or salting-out techniques. Each of these techniques will be treated shortly and the conditions will be indicated when a particular crystallisation method is used. For quick reference, fig. 1 can be used. The concentration versus temperature plots (i.e. solubility diagrams) summarise the crystallisation methods discussed here together with the principle to

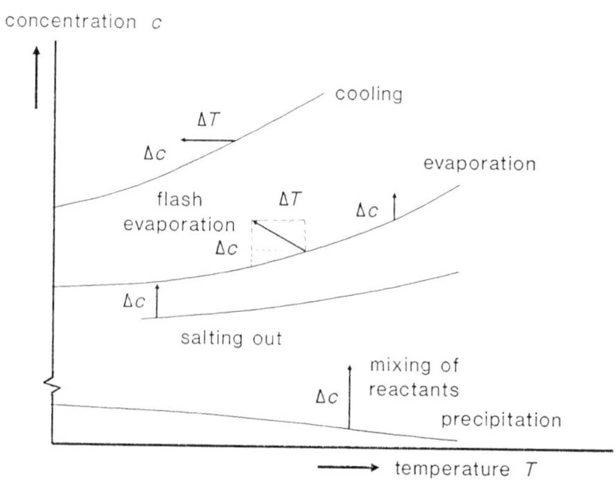

Fig. 1. Several solubility plots (concentration versus temperature) indicating the method of crystallisation and the principle used to establish the supersaturation.

establish the driving force. The position of the plots is arbitrarily chosen. Later in this chapter the crystalliser types required for crystallisation by the evaporative/cooling principle (section 8) and by precipitation (section 9) will be treated.

2.2.1 Cooling crystallisation

The supersaturation of a solution can be increased by decreasing the temperature in cases where the solubility of the substance is an increasing function of temperature. The cooling crystallisation method is usually applied for highly soluble materials exhibiting a steep (positive) solubility curve (i.e. a large, positive dc^*/dT). Owing to the large number of solute entities available in solution, sufficiently large crystals (up to 1000 μm) are formed. The supersaturations required to obtain a satisfactory product yield are usually rather low (σ in the order of $10^{-3} - 10^{-2}$). At these low supersaturations primary nucleation does not play a role and the prevailing mechanism for new phase formation is secondary nucleation (see section 3).

The driving force for crystallisation $\Delta\mu$ can be calculated as follows: at the equilibrium temperature T^* the chemical potentials μ_L and μ_S of the liquid and the solid phase are the same. In a non-equilibrium situation at temperature T we can write:

$$\Delta\mu = \mu_L(T) - \mu_S(T) = \mu_L(T) - \mu_L(T^*) + \mu_S(T^*) - \mu_S(T) \quad [\text{J mol}^{-1}], \tag{2}$$

or, using one of the Maxwell relations $s = -\partial\mu/\partial T$ (in $\text{J K}^{-1}\,\text{mol}^{-1}$ and at constant pressure and composition):

$$\Delta\mu = -\int_{T^*}^{T} (s_L - s_S)\,dT = -\Delta s_{eq}(T - T^*) = \Delta H_{eq}\Delta T/T^*, \tag{3}$$

where we assumed that the molar entropy s does not vary with temperature between T and T^*. $\Delta\mu$ is thus proportional to the undercooling ΔT ($= T^* - T$), which can be directly translated into Δc via the solubility diagram.

2.2.2. Evaporative crystallisation

A supersaturated solution can be obtained by evaporation of the solvent. This technique is applied for easily and medium-soluble substances with a weakly increasing solubility curve, which do not decompose at the boiling temperature of the solvent. Similar to crystallisation by cooling, relatively low supersaturations ($\sigma < 0.01$) suffice to obtain adequately large crystals. At these low supersaturations, secondary nucleation is the dominant mechanism whereas primary nucleation is operative only at a minor level or can be completely neglected.

In this case the driving force $\Delta\mu$ can be calculated as follows. The chemical potential μ_L of the liquid is given by $\mu_L^0 + RT\ln a$, with μ_L^0 the standard chemical potential, R the ideal gas constant and a the activity of the solute entities in the solvent. The chemical potential μ_S of the solid equals $\mu_{L,eq} = \mu_L^0 + RT\ln a_{eq}$. Since $\Delta\mu = \mu_L - \mu_S$, it also equals:

$$\Delta\mu = RT\ln(a/a_{eq}) = RT\ln(\gamma c/\gamma_{eq}c^*). \tag{4}$$

If the ratio of the activity coefficients $\gamma/\gamma^* \approx 1$, eq. (4) can be expressed as a function of $\Delta c/c^*$:

$$\Delta\mu \approx RT \ln(c/c^*) = RT \ln(1 + \Delta c/c^*) \approx RT\, \Delta c/c^* \quad \text{for } \Delta c/c^* \ll 1. \tag{5}$$

This further reduces to $kT\Delta c/c^*$ if we express $\Delta\mu$ in J per solute entity (instead of $J\,mol^{-1}$) by dividing eqs. (4) and (5) by Avogadro's number N_A. Comparing this with eq. (1), it is found that the relative supersaturation σ equals $\Delta c/c^*$ while $S = c/c^*$.

In the case of – at least – two ions, A and B, reacting to form the crystal C, e.g., $\nu_A A + \nu_B B + \cdots \rightarrow \nu_C C$, where ν_i (i = A, B, C) are the stoichiometric coefficients, the activity a must be replaced by the product summation of the activity a_i of the ith component to the power ν_i. Generally, it is necessary to write down the complete expression for $\Delta\mu$ including all participating ions during the reaction. Approximations can be made only after careful examination of the individual importance of each term contributing to the crystallisation process and even then misinterpretations concerning the definition behind the reported supersaturations are easily introduced [3].

2.2.3. *Flash evaporative crystallisation*

In the case of flash evaporation mostly a cascade of crystallisers is used, operated at ever lower pressures and temperatures. This procedure enables an optimal utilisation of energy. Two causes contribute to the driving force: a temperature and a concentration difference (see fig. 1). This technique is applicable to highly and medium soluble compounds and sufficiently large crystals are generally obtained at relatively low supersaturations, where again secondary rather than primary nucleation represents the dominating nucleation mechanism.

2.2.4. *Precipitation*

By mixing two feeds containing the separate components A and B (e.g. as ions) constituting the crystal structure, the desired AB compound forms once its solubility product is exceeded. This method is applied for sparingly soluble substances, where sufficient production levels can only be obtained by creating high supersaturation levels ($\sigma \gg 1$). High supersaturations for precipitation cause primary nucleation to be the dominant nucleation mechanism. A large number of primary nuclei is mostly formed which, in spite of the high supersaturation, grow to considerably smaller sizes (1–10 μm) than those of easily soluble compounds obtained by cooling or evaporative crystallisation. In fact they are so small that no secondary nuclei can be bred by the growing crystals. To avoid high local supersaturations upon mixing and thus uncontrollable excessive primary nucleation, intensive mixing is strived for. On the other hand, the large supersaturations often cause aggregation of crystals, which may be further stimulated by the forced hydrodynamics. For these reasons precipitation is in general much harder to control than the other crystallisation methods.

The driving force for crystallisation by means of evaporation, as derived in section 2.2.2, also applies to precipitation. However, the approximation made in eq. (5), $\sigma \ll 1$, is not valid in the case of precipitation and hence the exact result, i.e. the logarithmic dependence on S or σ, has to be used.

2.2.5. Salting-out crystallisation

In salting-out or drowning-out crystallisation an anti-solvent is added to a saturated, mostly aqueous solution which reduces the solubility of the highly or medium soluble compound or salt, thus creating a supersaturation. To avoid locally high supersaturations good mixing is essential just as it is in the case of precipitation.

By applying this technique, both organic (by adding, e.g. water) and inorganic substances (by addition of an organic solvent) can be crystallised. The second solvent is selected on account of its good miscibility with the original solvent. Salting-out processes are sometimes also carried out by addition of a common ion salt. To a solution of, e.g. the salt AB, the salt AC is added which has a higher solubility than AB and causes the crystallisation of AB.

The method is applied for substances with a solubility that hardly changes with temperature. An advantage is that the method can be operated at ambient temperatures, which is advantageous for heat-sensitive substances. Although in most cases relatively high supersaturations are used, this is not a stringent demand. Depending on the supersaturation and the size of the developed crystals, primary and/or secondary nucleation may occur. An example of this technique is the salting-out of potash alum with aqueous acetone [5].

Salting-out coupled with thermal regeneration of the solvent is called extractive crystallisation. In this technique, which is still at a very early stage of development, a phase separation of the solvent, which remains after filtration of the crystals, is provoked by altering the temperature. The solvent phase is subsequently recycled in the process [6].

3. Nucleation

3.1. Nucleation mechanisms

The formation of solid material, or nucleation, from a liquid (either solution or melt) is only possible if this liquid is supersaturated or undercooled. The mechanisms by which nucleation is able to occur will be treated here. They are twofold: primary and secondary nucleation. Primary nucleation is new phase formation from a clear liquid. This type of nucleation can, in turn, be subdivided into homogeneous and heterogeneous nucleation. In the latter situation a foreign substrate, e.g. in the form of dust particles, is present in the solution on which nucleation starts. In the former case such a substrate is absent and new phase formation has to proceed by statistical fluctuations of solute entities clustering together. Secondary nucleation, on the contrary, only happens because of the prior presence of the material being crystallised, a definition given by Botsaris [7].

Both primary and secondary nucleation mechanisms will be treated in this section together with a summary of modelling results on the latter subject. nucleation strongly determines the final CSD and is an important factor in the control of industrial crystallisers. Since primary nucleation is discussed by Mutaftschiev [8] as well as by Van der Eerden.[3] and Sarig [9] here only an overview is given, considering its most important features.

3.1.1. Primary nucleation

Classical theories describing new phase formation due to primary (or spontaneous) homogeneous nucleation start from the assumption that in supersaturated solutions, solute atoms, ions or molecules have to combine to form a cluster. As a result of statistical fluctuations numerous clusters are produced but fall apart as well – it is a reversible process. In undersaturated and just saturated solutions the system is in equilibrium and stable against new phase formation. In supersaturated solutions, however, there will be a competition between two opposing terms constituting the free energy of a nucleus: each solute entity which is added to a cluster will on the one hand reduce its free energy by an amount $\Delta\mu$ (which is taken here as the chemical potential difference per solute entity, instead of per mole, between fluid and solid) while on the other hand the surface increment leads to a positive contribution to the free energy. Both terms obviously depend on the number of solute entities n building up the cluster: the former term (being related to the volume of the nucleus) is proportional to n, while the latter is proportional to $n^{2/3}$, since this is a term related to the nucleus' surface area. Since the terms are opposite in sign the free energy of the nucleus exhibits a maximum at a certain critical size. Above this size the nucleation process becomes irreversible, i.e. the cluster will only grow from this size on and a nucleus is formed. This critical size can be shown to be inversely proportional to the supersaturation which – in turn – is related to $\Delta\mu$.

The primary nucleation rate J is defined as the number of primary nuclei appearing per unit volume and unit time. Its dependence on the supersaturation S is given by:

$$J = AS\exp(-\Delta G^*/kT) \quad [\# \text{ m}^{-3}\text{s}^{-1}] \tag{6}$$

where A is (practically) S-independent and ΔG^* is still a function of S. Since the nucleation rate is dominated by the exponential term in eq. (6), ΔG^* is given explicitly:

$$\Delta G^* = a_4\gamma^3\Omega^2/[\delta kT \ln S]^2 \quad [\text{J}]. \tag{7}$$

Here a_4 is a numerical shape factor (e.g. $16\pi/3$ for spheres), Ω the molecular volume (m^3 per solid entity), γ the specific surface excess free energy (J m^{-2}) of the crystal surface in contact with the solution and δ the number of ions in the formula unit. ΔG is the work of cluster formation which reaches a maximum, ΔG^*. Thus, in a supersaturated solution a nucleation barrier exists, the height of which is given by eq. (7). The height of this barrier is inversely proportional to $\ln^2 S$ and proportional to γ^3. This means that, in principle, every supersaturated solution is unstable against new phase formation and that, sooner or later, crystallisation is bound to occur.

The occurrence of homogeneous nucleation in practice is rare and generally heterogeneous nucleation will take place, i.e. nucleation induced by the presence of an interface different from that of the solute. The wetting conditions (the interactions across the interface between substrate, solute and solution) will be altered and are responsible for lowering the nucleation barrier. This can be interpreted as an effective specific surface excess free energy, γ_{eff}, which is lower than γ due to the presence of the foreign substrate. Volmer [10] showed that, in effect, the nucleation barrier is modified by a multiplication factor Ψ, which varies between 0 and 1. By replacing $\Psi\gamma^3$ by γ_{eff}^3, it turns out that in most practical cases γ_{eff} is smaller than γ by a factor of 0.3 to 0.5.

Knowledge of γ is of major importance for understanding not only primary nucleation processes, but, e.g. crystal growth kinetics as well. Nielsen and Söhnel [11] were the first to use primary nucleation data (induction time measurements) to determine specific free surface energies of crystals grown from solution. However, in order to obtain a reliable measurement of the primary nucleation rate, the supersaturated solution has to be free of foreign particles which may induce heterogeneous nucleation, especially at low supersaturations, or very high supersaturations must be created, where homogeneous nuclei outnumber heterogeneous nuclei. Mersmann [12] derived a simple equation based on work of Guggenheim [13] for determination of γ from solubility data. This comprises a more accessible type of measurement compared to the measurement of primary nucleation rates. In approximately two out of three cases considered in [12], the calculated values of γ are underestimated with respect to those determined from primary nucleation data [11]. The discrepancy between the two sets of data might be even larger if it is assumed that the data of Nielsen and Söhnel [11] already yielded lower values of γ due to heterogeneous nucleation.

From a fundamental point of view on cluster formation it is interesting to gain more insight in to the structure of clear, supersaturated solutions. Mullin and Leci [14] measured the concentration of citric acid as a function of height in a column (~ 0.5 m) containing its supersaturated, aqueous solution and found an increasing concentration going from the top to the bottom of the column. In saturated and undersaturated solutions this concentration gradient did not develop. Similar results were found later for other substances as well (sucrose [15], $NaNO_3$, K_2SO_4, citric acid and urea [16]). Hussmann et al. [17] carried out Raman spectroscopic studies in $NaNO_3$ aqueous solutions and found evidence for the existence of solute clusters. No information was gained, however, on the size of these clusters.

Recently, Veverka et al. [18] provided a much more rigorous version of the work of Larson and Garside [16], based on non-equilibrium thermodynamics.

Measurements of the solute diffusivity as a function of concentration were performed by Chang and Myerson [19–21]. Their results are also indicative of the presence of solute clusters. These as well as more recent attempts to unravel the structure of supersaturated solutions are reviewed by Myerson and Izmailov in [22].

3.1.2. Secondary nucleation

Nucleation mechanisms. Secondary nucleation is a phenomenon encountered during many crystallisation processes. It is considered as the main source of the production of new crystals in particular in cooling and (flash) evaporative crystallisation and it occurs at already moderate values of the supersaturation, i.e. much smaller than that required for primary nucleation.

It is generally accepted that secondary nucleation can be divided into four main breeding sources:

(1) *Initial* (or apparent) *breeding*, which usually occurs if dry crystals are introduced into a solution. Small fragments adhering to the crystals are washed from the surface and become suspended in the solution. During seeding of a crystalliser with dry seed

crystals this plays a role. It can be eliminated by introducing the fresh seeds into an undersaturated solution where the initial, secondary nuclei are dissolved before feeding the seed slurry into the crystalliser.

(2) *Dendritic breeding*, which only occurs at such high values of the supersaturation that facet instabilities occur. The corners and edges of the crystals experience a higher supersaturation than the flat faces and in the extreme even dendrites are formed. Forced hydrodynamic flow may be capable of removing some of the protruding crystal edges or dendrite branches, which act as secondary nuclei.

(3) *Contact nucleation*, which occurs as a result of collisions between two crystals, between a crystal and an impeller blade, and between a crystal and the walls of the vessel. The secondary nuclei are formed by attrition of so-called parent crystals.

(4) *Fluid shear breeding*, which, due to hydrodynamic instead of mechanical forces, causes attrition of the parent crystals.

Of the four sources, contact nucleation is considered to make the largest contribution to the secondary nucleation rate in crystallisers, although experimental work [23, 24] and recent theoretical estimations [3] show that the fluid shear frictional forces exerted on surfaces of crystals passing each other very closely should not be underestimated.

Pioneering work on the *contact nucleation* mechanism was carried out by Clontz and McCabe [25]: a $MgSO_4 \cdot 7H_2O$ crystal surface present in a supersaturated solution was contacted with a rod and the resulting secondary nuclei were counted after outgrowth. They established an increasing amount of secondary crystals with increasing supersaturation and increasing impact energy. Another result was that at a given impact energy the number of secondary crystals per unit contact area is inversely proportional to approximately the square root of the contact area (see also [26]). Visual inspection of the contacted crystals after removal from the solution did not reveal any damage of the crystal surface. To explain their results they applied the idea of Powers [27], who proposed that the interface between crystal surface and bulk solution is separated by an intermediate layer consisting of solute clusters of sub- and near-critical size. While contacting the crystal, the rod was supposed to enter this interfacial layer during which clusters were dislodged from it. This kind of nucleation was named *catalytic breeding*. The higher the supersaturation, the smaller the critical nucleus size and the more nuclei survive. The initial size distribution of the surviving secondary nuclei was therefore thought to be peaked at approximately the critical nucleus size corresponding to the imposed supersaturation. This model became known as the "survival theory" and was first introduced by Lal et al. [28]. The hypothesis concerning the monodispersity of the initial size distribution could, however, not be tested at that time because not the initial size distribution but the distribution after a certain growth time was observed.

Cerreta and Berglund [29] designed and carried out experiments allowing discrimination between the mechanisms of catalytic breeding and attrition by applying a laser Raman microprobe technique. Their results on α-dextrose, which is known to exist in hydrated and anhydrous form, indicate that under certain experimental conditions, attrition is the most likely mechanisms, although in a specific temperature range their results could not be explained properly with either the attrition or the catalytic breeding mechanism. Earlier, Denk and Botsaris [30] showed a 100% correspondence between the enantiomorphic form of the nuclei and that of the parent crystal of $NaClO_3$ in stirred solutions. Surprisingly and still unexplained is that both L- and D-nuclei were found in unstirred solutions at low supersaturations.

The first in situ microscopic observations of contact breeding were conducted by Garside and Larson [31]. Their observations clearly revealed damaging of a potash alum crystal after contacting its surface, even at the slightest touch. In a complementary paper [32], the initial size distribution of secondary crystals was reported. The measurements indicated a subdivision over two size ranges: particles between 1 and 4 μm, the number of which was independent of supersaturation (the particles were found to be formed in undersaturated solutions as well), and particles of 4–50 μm, the number of which increased with supersaturation. Furthermore, these secondary nuclei displayed different growth characteristics: many of the nuclei smaller than about 10 μm appeared not to grow or to grow very slowly, whereas the larger ones showed growth rates comparable to macroscopic crystals. This might be the first, concrete indication for two types of populations of nuclei of different origin – although both origins have to be related to the crystal surface – and different growth rates. The observation of secondary nuclei being much larger than the critical nucleus size just after their formation, clearly contradicted the "survival theory". The term "fragment" instead of "nucleus" may therefore be less confusing. Together with the results of Cerreta and Berglund [29] and Denk and Botsaris [30] these findings were the main reasons for disregarding the mechanism of catalytic breeding. Nevertheless, a recent paper by Tai et al. [33] demonstrates that models concerning the catalytic breeding mechanism are still developed further.

Subdivision of contact nucleation. For the purpose of modelling secondary contact nucleation kinetics in an industrial crystalliser, Daudey [34] (also [35]) suggested the following scheme (see table 1): secondary nucleation can be subdivided into a *mechanical* and a *surface breeding* mechanism. These names are quite arbitrarily chosen, since both sources of fragments are generated from the crystal surface due to a mechanical action. Nevertheless, mechanical breeding relates to the formation of secondary nuclei as a result of abrasion of the parent crystals which occurs even in a nonsolvent and hence is not primarily related to the growth of these crystals. Surface breeding generates secondary nuclei due to growth-related features at the crystal surface. This breeding mechanism is therefore directly dependent on growth rate or supersaturation. Clearly,

Table 1
Divison of secondary nucleation rate into mechanical and surface breeding. The table shows which of the steps (may) depend on the supersaturation σ.

Mechanism	Formation	Removal	Outgrowth
mechanical breeding	$\neq f(\sigma)$	$=f(\sigma)$	$=f(\sigma)$
surface breeding	$=f(\sigma)$	$=f(\sigma)$	$=f(\sigma)$

the crystal surface structure as a result of the growth kinetics is of major importance here.

Each of the two mechanisms comprises three successive steps, namely (i) formation of secondary nuclei or fragments, (ii) removal of the nuclei from the parent crystal from which they are generated, and (iii) subsequent growth of the nuclei to crystals. The removal step is a competition between the removal rate R of nuclei from the parent crystal surface and their incorporation rate I through overgrowth by propagating growth layers [35]. The removal efficiency is given by the fraction $R/(R + I)$. For $R \gg I$, the removal efficiency tends to unity, while in the opposite limit, $R \ll I$, the removal efficiency can be approximated by R/I. I is likely to be growth-rate (G) dependent, e.g. $I \sim G$. The third step can also be looked upon as an efficiency step which depends on supersaturation.

In accordance with this framework, Daudey [34] further assumed a subdivision in growth characteristics, i.e. slow growth for more mechanically damaged secondary nuclei originating from mechanical breeeding and more normal growth rates for nuclei produced via surface breeding.

With this concept in mind, several experimental results can be found which are indicative of this subdivision of the secondary nucleation rate into mechanical and surface breeding. The work of Garside and Larson [31] and Garside et al. [32], already mentioned earlier, serves as one of the first examples.

The work of Wissing et al. [36] pointed to the existence of surface breeding and revealed the important role of the surface structure. They showed, using an in situ microscopic technique, that the roughness of the crystal surface structure of KH_2PO_4(KDP), which is correlated with the supersaturation and the amount of impurities (Fe^{3+}) present in the solution, influences the value of a critical impact energy E_c. This critical value is defined as the impact energy above which value damage to the crystal surface and high nucleation rates are observed, while below E_c no damage was visible and nucleation rates were a factor of 10 to 100 times smaller. By "rough" we mean a flat crystal surface with a relatively high macrostep density, and the rougher the surface, the smaller the value of E_c.

The work of Van der Heijden et al. [37] showed that the initial size distribution (i.e., within 10 s after formation) of secondary crystals of potassium hydrogen phthalate (KAP), can be described by a log-normal distribution. Fragments smaller than 0.5–1 μm were not found; the mean size of the secondary fragments ranged from 5–10 μm. Using simple fracture-mechanical considerations, minimal and mean sizes could be understood quantitatively. This would suggest that the secondary nuclei were formed by the mechanical breeding mechanism. Scanning electron microscopic observations of the contacted area, however, showed regions with clearly stepped structures, indicating that the secondary nuclei might also be due to partially removed surface growth layers (i.e. macrosteps), which suggests a surface breeding mechanism. It is thus not clear which of these two mechanisms applies or whether a combination of these two has to be assumed.

The formation step during mechanical breeding has no direct dependence on the supersaturation; this dependence is introduced by the second and/or third step in the process, depending on whether or not the removal step is rate-limiting. The formation step during surface breeding on the other hand, depends on the supersaturation, which dependence may be further influenced by the supersaturation dependencies of the removal and subsequent growth steps, as in the case of mechanical breeding.

Modelling of secondary contact nucleation. The derivation of crystallisation kinetics is based on knowledge of the secondary nucleation rate B_0 (i.e. the number of secondary crystals produced per unit of crystalliser volume and unit time) and therefore on its dependence on accessible process variables being the supersaturation σ (or growth rate G), the power input ε (or related to this the stirrer rotation rate N) and the total mass per m^3 slurry M_T or some other moment of the CSD such as the total surface area per unit crystalliser volume, A_T (for an explanation of CSD properties, see section 5). Most generally, B_0 can be expressed as a power law in which we do not specify yet which of the CSD moments applies. This moment will be denoted by M_p, with p still unspecified:

$$B_0 \sim \sigma^b \varepsilon^k M_p^j \sim G^i N^h M_p^j. \tag{8}$$

The secondary nucleation rate is expressed as a function of G instead of σ, because for medium and highly soluble substances the supersaturation is very low and thus difficult to measure in a crystalliser. The power input ε is related to N^3 (see, e.g. Jančić and Grootscholten [38]; hence $h = 3k$. Crystal growth kinetics from an engineering point of view are often written as $G \sim \sigma^g$, where g is the growth order (simplified growth rate models are discussed in section 4). This means that $i = b/g$. From crystal growth theory [3] it follows that g is usually of the order of 1–2. The experimentally determined ranges of the powers b, k, and j for a wide variety of high and medium soluble substances are: $1 < b < 3$; $0.6 < k < 0.7$ and $j = 1$ for crystal–impeller and $j = 2$ for crystal–crystal collisions. Generally the third moment of the CSD ($p = 3$) must be substituted in eq. (8). This moment can be expressed as the total mass M_T per m^3 of slurry if the volumetric shape factor k_v and the solid density ρ_c are known. The proportionality factor in eq. (8) relating B_0 to these process variables is a function of the geometry and design of the crystalliser and the type of impeller or pump used.

The crystallisation kinetics, expressed as a power law of several experimentally accessible parameters, eq. (8), are a result of numerous experiments carried out in various kinds of crystallisers. In order to predict the influence of changes in crystalliser design, scale of operation or operating conditions on the crystallisation kinetics, an attempt has to be made to understand the physical aspects behind this empirically derived power law. This can be achieved by modelling of the processes taking place in a crystalliser. Mechanistic modelling of the secondary nucleation rate B_0 started with the work of Ottens et al. [39], followed by Botsaris [7], Mersmann et al. [40] and Ayazi Shamlou et al. [24]. All the models assume a dependence of B_0 on the supersaturation σ introduced only via the growth of the secondary fragments formed. Thus, the results of these studies are only applicable to the mechanical breeding mechanism. Van der Heijden et al. [41] recently extended the modelling with a description of the surface breeding mechanism thereby including the σ-dependence of the formation step. If the removal step is rate-limiting, the power b or i in eq. (8) will be reduced since the removal efficiency can then be written as R/I where the incorporation rate I is growth-rate dependent. The secondary fragments which are formed and removed, subsequently grow, thus populating the larger size classes. Often it is found that these small fragments grow with a lower rate compared to larger, macroscopic crystals due to size-dependent growth (SDG) and growth rate dispersion (GRD) effects which will be treated in section 4. For modelling purposes this growth step is usually considered as an efficiency factor in which only a certain fraction of the formed fragments survives. This fraction is also assumed to be a function of the supersaturation. In section 4.3.3 a physical model is treated concerning SDG/GRD effects. Incorporation of this result in the efficiency factor for growth, effectively leads to a lowering of the powers b or i in eq. (8).

The modelling of mechanical and surface breeding will now be treated separately in the following two subsections.

Modelling of mechanical breeding. The work of Ottens et al. [39] was based on the kinetic energy transferred to a crystal during a collision with either another crystal, a moving object like an impeller blade or a stationary surface such as a vessel wall. Evans et al. [42] showed that the contribution of crystal–wall contacts to the secondary nucleation rate can be neglected. Therefore, this type of collision is not considered further here. For all of these situations a linear relationship between B_0 and the power input ε was found. Depending on the collision type either a quadratic (crystal–crystal) or linear (crystal–impeller) dependence on M_T resulted. The σ-dependence could not be quantified within the scope of their description as can be seen in table 2 which (amongst others) summarises the results of Ottens et al. [39].

Table 2
List of the parameters constituting the dependence of the secondary nucleation rate on supersaturation, power input and moment of the CSD [see eq. (8)] for crystal–crystal (c–c) and crystal–impeller (c–i) contacts. The results for mechanical and surface breeding are represented separately.

References	Power	c–c contacts	c–i contacts
mechanical breeding			
Ottens et al. [39]	b	–	–
	k	1	1
	j	2	1
	p	3	3
Ayazi-Shamlou et al.	b	–	–
[24]	k	1	1.33
	j	2	1
	p	3	3
Mersmann et al. [40]	b	–	–
	k	0.73	0.73
Pohlisch and	j	2	1
Mersmann [43]	p	3	3
surface breeding			
Van der Heijden et al.	b	0–3	0–3
[41]	k	(>)0.47–0.60	(>)0.47–0.60
	j	~2	~1
	p	~2	2 or 3

Botsaris [7] represented the secondary nucleation process as the product of the rate of energy transfer to the crystals due to contacting and two efficiency functions, one describing the production of secondary nuclei per unit of transferred energy (production function) and the other accounting for the fraction of nuclei actually growing (survival function). This product determined the secondary nucleation rate B_0.

Mersmann et al. [40] and Pohlisch and Mersmann [43] modelled mechanical breeding as the rate of abraded volume from a crystal surface layer due to collisions between either two crystals or a crystal and an impeller blade. They assumed that 100% of the secondary fragments which were formed was removed from the surface. They thus derived a power law for the secondary nucleation rate B_0 comprising N (or ε) and M_T. Again, the σ-dependence was not quantified, although they define an effective mass fraction which grows to populate the size distribution, while the remaining fraction is lost, e.g. through dissolution. In the formation step, leading to an abraded volume, the dimensions of the impeller blade (e.g. width, thickness and pitch angle) were incorporated, which were supposed to be useful for scale-up considerations. The results on the mechanical breeding mechanism according to Mersmann and co-workers are summarised in table 2.

Ayazi Shamlou et al. [24] not only modelled mechanical breeding, the results of which are incorporated in table 2, but also fluid shear breeding. Their experimental results on potassium sulphate suspended in a nonsolvent were indicative of attrition of crystals by purely hydrodynamic effects.

Modelling of surface breeding. Contrary to mechanical breeding, surface breeding strongly depends on the parent crystal surface features as a result of growth. This implies that the formation step during surface breeding depends on the supersaturation. If it is accepted that the surface breeding mechanism is operative as a source of secondary nuclei, it is evident that more knowledge has to be gained on the crystal surface structure due to growth and on its response to imposed mechanical forces. Suppose the crystals grow according to a spiral growth mechanism (see section 4). Generally, macrosteps are formed due to bunching effects of monatomic steps induced by unspecified, but ever present, impurities. The height and density of the macrosteps are both functions of the supersaturation. These macrosteps are, from a fracture-mechanical point of view, weak points of the crystal surface and during a collision these parts will be broken off relatively easily and, consequently, may grow with a higher rate than fragments generated by mechanical breeding. Upon this idea a model was developed for the formation step of the surface breeding mechanism [41]. This model thus gives a physical explanation of the supersaturation dependence of the formation step in B_0. The results of this model are summarised in table 2. For crystal–crystal contacts, e.g. a dependence of the formation step on the square of the total surface area A_T^2 (related to the second moment of the CSD) is found, which can be understood from the fact that surface breeding is directly related to the surface of the crystals.

The ranges given for the power b of σ also cover the fact that the removal efficiency can be σ-dependent. Such a situation occurs in the case where the removal rate R is the rate-limiting step. The removal efficiency $R/(R+I)$ can then be approximated by R/I. Since the incorporation rate I is likely to be growth-rate dependent (e.g. $I \sim G$), this removal efficiency may lead to a *reduction* of the power of σ found in the formation step. The removal rate R can be a function of N and it is reasonable to assume that it increases with N. So, for a removal efficiency smaller than 1, the power h of N (and accordingly k of ε) can be larger than the value given in table 2. This is indicated by the $>$ sign in table 2.

Such a difference in removal efficiencies has been found experimentally in the case of contact nucleation experiments on KDP [36] and on KAP [37]: under similar hydrodynamic conditions practically all secondary fragments became suspended in the solution in the former case while in the latter situation the majority of secondary crystals was not removed from the parent crystal surface. This effect is very likely related to differences in material properties.

One of the major problems is how to discriminate experimentally between surface and mechanical breeding. A suggestion was put forward to study attrition of crystals in a nonsolvent [34, 44, 45], but Chianese et al. [46] pointed out that the mechanical breeding rate and the abraded fragments in a supersaturated solution and in a nonsolvent may differ drastically (apart from differences introduced by different fluid properties like viscosity), simply because crystals with sharp edges and corners are subject to attrition in a different way than crystals which are rounded off by continued abrasion. In a supersaturated solution an abraded area on a crystal surface will heal itself as a result of growth and this is clearly not the case in a nonsolvent. Extrapolating this line of reasoning, it seems likely that a newly grown piece of crystalline material onto a parent crystal can be broken off more easily compared to the first time it was removed. This idea is supported by work of, e.g. Bhat et al. [47] and Sherwood et al. [48]. Upon regrowth of as-grown crystals, damaged on purpose, stressed and probably mechanically less strong areas were found at the regrown interface. The effects described here will hamper a distinct, experimental discrimination of the surface and mechanical breeding mechanisms on the basis of attrition experiments only.

3.2. Metastable zone width

For most of the easily and medium soluble substances the solubility is an increasing function of temperature (see fig. 2). The curve describes the concentration where the solution is just saturated at a given temperature. For practical reasons, the region of supersaturated solutions is often divided into a metastable and a labile zone and the boundary between them indicates the maximum supersaturation at which the excess solute does not crystallise via primary nucleation under normal operating conditions of a crystalliser. Much experimental work (e.g. [49]) has been carried out to determine the metastable zone widths for a number of materials. The practical importance of knowing the metastable zone width for a certain crystallising substance is provided by the fact that it limits the operational supersaturation. Exceeding this maximum supersaturation is accompanied by excessive nucleation and, consequently, appearance of enormous numbers of fines which decrease the d_{50} of the final product.

On theoretical grounds such a boundary line does not exist, since even at the slightest supersaturation crystallisation will occur, although the process may be slow. So, in the

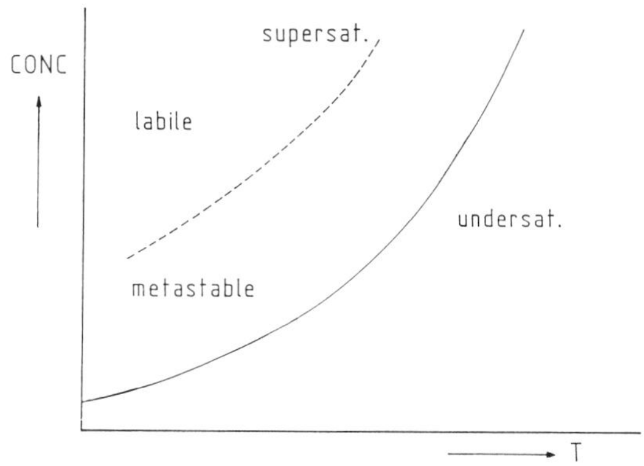

Fig. 2. Solubility plot indicating the undersaturated and supersaturated regions. The latter region is subdivided into a labile and metastable zone.

definition of the metastable zone width implicitly a time scale is introduced which can be related to the residence time of the solution in the crystalliser. This residence time should not exceed the time delay between the moment that a given supersaturation of the solution is reached and the moment that the metastability is observably broken by primary nucleation. If this happens by the formation of the first nucleus, as is the case in the mononuclear (MN) model, the time delay, called induction time t_i is given by [50]:

$$t_i = (JV)^{-1} \quad [\mathrm{s}], \tag{9}$$

where J is the primary nucleation rate, given by eq. (6), and V the volume. Since J is a steep, increasing function of S, t_i obviously depends on this parameter as well. Hence a certain imposed supersaturation corresponds to an induction time and vice versa. Kashchiev et al. [51] showed that eq. (9) is valid only for very small volumes V or in cases where the growth rate $G \to \infty$. Otherwise, detection of the first supercritical nucleus somewhere in the solution volume is impossible even with the techniques nowadays available. These restrictions are in industrial practice hardly ever satisfied hence the definition for the induction time given above is purely academic.

A more generally applicable definition of the induction time t_i is the time required for the formation of a detectable macroscopic volume V_{macro} of the new phase which is a specified fraction $\alpha = V_{macro}/V$ of the total volume V. Now, more nuclei are allowed to develop while the first nucleus grows out to a macroscopic size. Here a polynuclear (PN) model is thus operative and the induction time is given by:

$$t_i = (\alpha / a_4 J G^3)^{1/4} \quad [\mathrm{s}], \tag{10}$$

where a_4 is a shape factor for the crystals formed (the dimensionality of growth is assumed to be 3, i.e. nuclei grow isotropically in all directions).

The general expression for t_i, which is valid irrespective of whether one, several or many nuclei cause the loss of metastability of the supersaturated parent phase, can be attained simply by adding the induction times in eqs. (9) and (10) [51].

The metastability limit S_c can be calculated by selecting a certain time scale τ which is supposed to be equal to the induction time t_i. This time scale together with the specific dependence of J and G on S, the scale of the experiment characterised by the volume V and the detectable volume fraction α, then determine S_c. In by far the most cases the pure PN model applies and there is no V-dependence. The relation between G and S depends on the active growth mechanism, such as two-dimensional nucleation, spiral growth or normal growth. The complete description of t_i as a function of S for each different growth mechanism is given by Kashchiev et al. [51].

4. Growth

4.1. Aspects of crystal growth

The relevance of the crystal surface structure for the surface breeding mechanism has already been emphasised in the previous section. This crystal surface structure is indissolubly connected to its growth mechanism [3] and the shape of the crystal through the role of the growth rates of the individual crystal faces [52]. Both properties (surface structure and crystal shape) play an important role in the collision processes during secondary nucleation. But also for the downstream processing of the produced crystals and their final performance, the shape of the crystals and their uptake of inclusions during growth are essential. Besides the influence of crystal growth also aggregation, attrition and breakage processes affect the CSD. Therefore, these phenomena are included in this section on growth as well.

For most industrial processes, in particular for cooling and (flash) evaporative crystallisers, secondary nucleation is the dominating process for the production of new crystals. The growth behaviour of these fines can be rather different from that of the larger crystals. Especially in case of mechanical breeding, either due to impacts between crystals and crystalliser parts or due to hydrodynamic forces, the major part of the fines may grow very slowly or may not show any growth at all. In the past this phenomenon has been attributed to either size-dependent growth (SDG) or growth rate dispersion (GRD). The physical mechanism underlying this phenomenon is the strain content within the small particles which can be built up during the refaceting step after their formation or which is a result of the formation step itself because of the mechanical and/or hydrodynamic forces involved. Consequently, the fines – though large in number – do not all contribute to the final product yield. Understanding and modelling of SDG and GRD provides more insight in to the relation between growth behaviour and formation mechanism (either mechanical or surface breeding) of secondary nuclei and finally into the resulting CSD.

4.2. Crystal growth mechanisms

Crystal growth processes can be subdivided roughly into two subsequent steps: volume diffusion of growth units towards the crystal surface and incorporation of these units at kink sites on the surface, including surface diffusion of these units. For crystallisation from very concentrated solutions and from the melt, transport of heat through the bulk phases becomes a third step. In this section only the surface incorporation step of growth units will be discussed. In section 4.3 a bit more attention will be given to the diffusion-controlled process as well as to heat transport.

4.2.1. Rough growth

The atomic or molecular structure of a growing crystal surface is influenced by several factors such as driving force, temperature, solvents, impurities and hydrodynamics of the solution. Depending on these factors smooth or rough crystal surfaces can be discerned. Each crystal surface has an absolute critical temperature T_R, the roughening temperature, which value is determined by the interaction of surface ions, atoms or molecules mutually and with the mother phase. The transition from smooth to rough surfaces occurs at T_R (e.g. [3, 52, 53]). The order parameter for this phase transition is the edge free energy which is nonzero below T_R and zero at and above this temperature. This implies that for $T > T_R$ a growth unit can be incorporated anywhere on the crystal surface, in contrast to smooth growth for which energy is gained only if a growth unit is incorporated at an edge of a step. For rough growth the crystal faces tend to become rounded while a smooth surface results if layer growth prevails. The so-called continuous growth of rough surfaces turns out to be directly proportional to $\Delta\mu$ or σ [see eq. (1)] for small deviations from equilibrium and is therefore also referred to as linear growth.

Besides roughening of a crystal surface during growth at $T > T_R$, a surface can also become rough even at temperatures below T_R if the driving force for crystallisation, $\Delta\mu$, exceeds a value for which the corresponding two-dimensional critical nucleus size becomes of the order of a growth unit or, more precisely, for $\Delta\mu$ values where the activation energy required for the formation of a two-dimensional nucleus of critical size becomes of the order of kT. In this situation the growth rate is also linearly proportional to the supersaturation $\Delta\mu$, which for most solution growth processes may be approximated by σkT [see eq. (1)]. The process described here is called kinetic roughening rather than the thermal roughening which was discussed previously. The main difference between the two roughening mechanisms is that during kinetic roughening the edge free energy is still nonzero, while it is zero for thermodynamic roughening. For more detail we refer to Van der Eerden [3] and Bennema [52].

4.2.2. Layer growth

For layer growth of a smooth crystal surface, two mechanisms can be discerned: a spiral growth and a two-dimensional growth mechanism. The latter mechanism comprises the formation of two-dimensional islands on the crystal surface which are assumed to spread laterally in an isotropic manner. An island can either grow and cover the whole surface before a new island is formed on top of the previously formed layer or, more realistically, islands are allowed to nucleate all over the surface including that of the new layers formed by laterally spreading islands. This is also referred to as the "birth-and-spread" model (see fig. 3). Two-dimensional growth occurs either if no structural defects like dislocations or grain boundaries emerge at the crystal surface, or

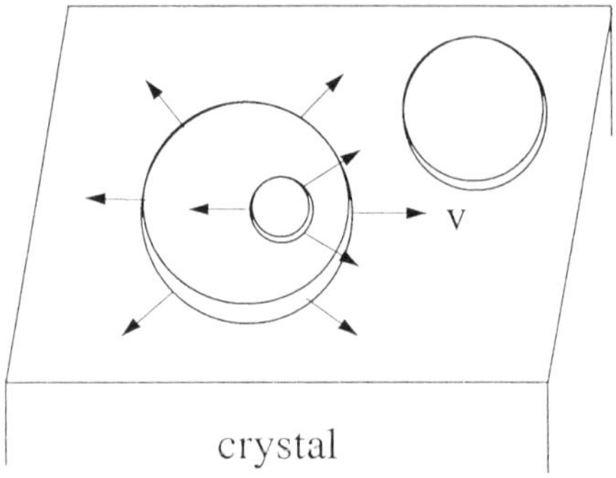

Fig. 3. Visualisation of two-dimensional nucleation model: two-dimensional nuclei are born and subsequently grow, forming a new layer on the crystal.

if the supersaturation is so high that the nucleation barrier for the formation of a two-dimensional nucleus is relatively low.

For two-dimensional nucleation the growth rate normal to the surface equals [3]:

$$G = 1.137\, h(v^2 J_{2D})^{1/3} \sim (\Delta\mu/kT)^{5/6} \exp(-\Delta G^*_{2D}/3kT) \quad [\mathrm{m\,s^{-1}}], \tag{11}$$

where h is the height of the two-dimensional islands, v the lateral spreading velocity of an island, J_{2D} the two-dimensional nucleation rate (in $\#\mathrm{m^{-2}s^{-1}}$) and ΔG^*_{2D} the activation energy for creation of a two-dimensional critical nucleus. Since J_{2D} is a steep function of the supersaturation (since $\Delta G^*_{2D} \sim 1/\Delta\mu$ and $v \sim \Delta\mu$), G also shows an abrupt increase if the supersaturation exceeds a certain critical value. For $\Delta G^*_{2D} \ll 3kT$ (large $\Delta\mu$), G is proportional to $(\Delta\mu)^{5/6}$ [3].

The former mechanism, spiral growth, occurs due to the presence of screw dislocations, the formation of which can hardly be avoided during a crystal growth process. These defects lead to a faster growth at low supersaturations than can be achieved via two-dimensional nucleation. This is due to the fact that a screw dislocation (or a dislocation with a screw component) emerging at the surface serves as a permanent step source (see fig. 4). Steps are preferred sites for growth units to become incorporated in the surface. Around the defect line the step will curve and a growth spiral is formed. These ideas were developed by Frank [54]. The growth rate normal to the crystal surface as a consequence of the spiral mechanism was derived by Burton et al. [55].

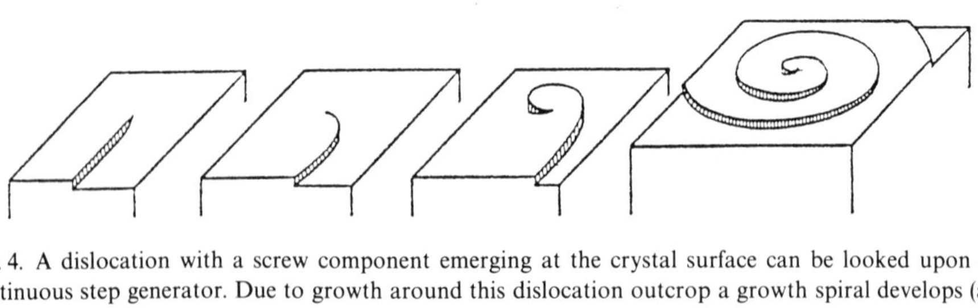

Fig. 4. A dislocation with a screw component emerging at the crystal surface can be looked upon as a continuous step generator. Due to growth around this dislocation outcrop a growth spiral develops (after [65]).

This growth rate is given by:

$$G = hv/19r_c = (hv_0/19r_c)\tanh(19r_c/2\lambda) \sim (\Delta\mu^2/\Delta\mu_1)\tanh(\Delta\mu_1/\Delta\mu), \qquad (12)$$

where h is the height of the spiral step, v the step velocity (far away from the centre of the growth spiral), r_c the critical nucleus size and $\Delta\mu_1$ a constant. v_0 is the velocity of one, infinitely long, straight step and it is proportional to $\Delta\mu$. λ is the diffusion length which is the mean displacement of adatoms during their stay at the terrace between two steps. For $\Delta\mu \ll \Delta\mu_1$, $G \sim \Delta\mu^2$ which may be further approximated to $G \sim \sigma^2$ if $\Delta\mu \ll kT$ (parabolic law) while for $\Delta\mu \gg \Delta\mu_1$, the diffusion fields of the individual steps start to overlap, leading to a linear dependence of G on $\Delta\mu$.

Upon gradually increasing $\Delta\mu$ (starting at equilibrium), the normal growth rate G passes first the regime of spiral growth (assuming that screw dislocations emerge at the surface) going from a quadratic to a linear dependence on $\Delta\mu$ and then either enters the regime of two-dimensional nucleation where $G \sim (\Delta\mu)^{5/6}$ followed by the regime of kinetic roughening in which $G \sim \Delta\mu$, or directly enters the latter regime.

Over the years, both theory and experiment have been extended including the influence of (i) impurities on crystal surface structure and shape and (ii) solvents on crystal shape, subjects which are partly addressed in [3, 52]. As illustrative examples, we mention growth of m-chloronitrobenzene from binary [56] and ternary [57] melts of its para- and ortho-isomers and solvent-dependent growth shapes of m-chloronitrobenzene, 2-methyl-4-nitroalinine [58] and of ε-caprolactam crystals [59, 60].

4.3. Simplified growth rate models

This section is devoted to growth models derived from an engineering point of view. In this approach the precise growth mechanism active for the surface integration step of growth units is of less importance and only an approximate proportionality or power law is used to describe crystal growth processes, including the diffusion process preceding the incorporation step. Also some attention is paid to heat transfer processes. In general the crystal growth rate from an engineering point of view depends on the supersaturation, volume or surface-area-related shape factors of the crystals, their size, and mass and heat transfer coefficients.

4.3.1. Mass transfer and heat transfer

Mass transfer. The rate of change per unit surface area of a crystal mass due to growth is assumed to be directly proportional to the concentration difference $(c_b - c_i)$ (see fig. 5), where c_b is the bulk concentration (far away from the surface) and c_i the concentration near to the interface (both in kg m^{-3}). This step accounts for the volume diffusion of growth units through the stagnant liquid layer at the crystal surface. The proportionality constant k_d (in m s^{-1}) is the mass transfer coefficient. For dilute solutions, $k_d = D/\delta_m = D\,\mathrm{Sh}/x$, in which D is the diffusion constant, δ_m the boundary layer thickness related to mass transfer, x a characteristic dimension of the crystal and Sh the Sherwood number defined as $k_d x/D$. For concentrated solutions at low

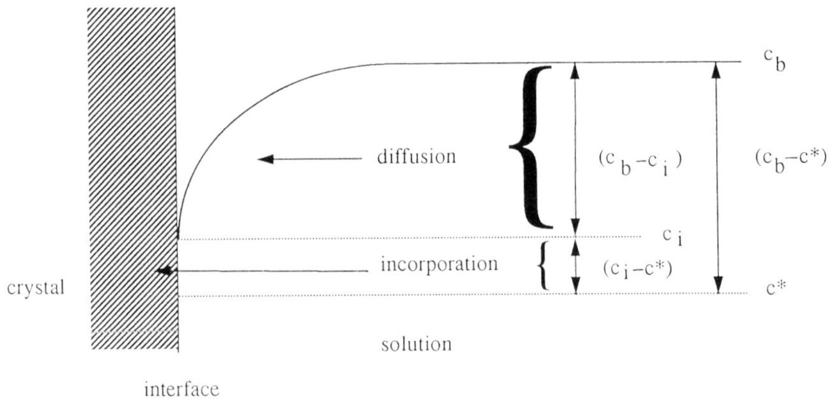

Fig. 5. The concentration profile perpendicular to the crystal surface during mass transfer.

supersaturations, however, k_d includes a correction factor, i.e., $k_d = D/(\delta_m[1 - m_s])$, where m_s is the mass fraction of solute in the bulk solution. The term $(1 - m_s)$ is a correction for convective transport.

The rate of incorporation of growth units into the crystal surface is determined by a power of the concentration difference near the interface: $(c_i - c^*)^n$, with c^* the equilibrium concentration at the interface and n ranging from 1 to 4. The mass transfer coefficient involved in this step is denoted as k_r. Note that this formulation serves as a simplified version of eq. (11) or (12). Due to conservation of mass, the incorporation flux has to equal the diffusion flux of mass to the crystal surface. The case $n = 1$ provides a particularly simple situation in which the rate of change in mass, $(dm/dt)_{mass}$ (in kg s^{-1}), can be directly related to the imposed concentration difference $(c_b - c^*)$:

$$(dm/dt)_{mass} = k_t a(c_b - c^*) \quad \text{and} \quad k_t^{-1} = k_d^{-1} + k_r^{-1}, \tag{13}$$

where a is the crystal surface area. For highly soluble substances [like NaCl or $(NH_4)_2SO_4$] which give rise to rather concentrated solutions, but also in highly viscous solutions (e.g. sucrose), volume diffusion is often the rate-determining process. On the other hand, for sparingly soluble materials (precipitation), often the integration step is the most important one. In these examples where one of the two steps is rate-limiting, the highest value of the two mass transfer coefficients can be neglected. The general case (arbitrary n, k_d and k_r) cannot be solved explicitly and one has to use an implicit expression relating growth to supersaturation.

Heat transfer. During crystal growth from solution, heat transfer is unimportant at low temperatures or for low solubilities [61, 62]. Thus for such situations, crystal growth is, to a great extent, determined by mass transfer only. During crystal growth from a melt or from solution at elevated temperatures or high concentrations, however, heat transfer comes into play. For instance, at the crystal interface growing from its melt a liquid/solid phase transition has to occur during which the latent heat of solidification has to be removed.

For heat-transfer-controlled crystallisation $(\mathrm{d}m/\mathrm{d}t)_{\mathrm{heat}}$ is given by:

$$(\mathrm{d}m/\mathrm{d}t)_{\mathrm{heat}} = -\alpha a \Delta T/\Delta H_{\mathrm{cr}}, \tag{14}$$

where α is the heat transfer coefficient (in $\mathrm{W\,m^{-2}K^{-1}}$), $\Delta T = T_{\mathrm{b}} - T^*$ is the difference between the bulk liquid temperature and the equilibrium temperature and ΔH_{cr} the enthalpy of crystallisation (in $\mathrm{J\,kg^{-1}}$). α equals $\lambda/\delta_{\mathrm{h}}$, where λ is the thermal conductivity (in $\mathrm{W\,m^{-1}K^{-1}}$) and δ_{h} the boundary layer thickness related to heat transfer. The heat transfer coefficient α can also be expressed in terms of the Nusselt number (Nu), i.e. $\alpha = \lambda\,\mathrm{Nu}/x$, with x the size of the crystal.

Simultaneous mass and heat transfer. Situations exist where both mass and heat transfer determine the increase of mass in time. The combination of eqs. (13) and (14) has to result in an overall mass rate $(\mathrm{d}m/\mathrm{d}t)_{\mathrm{m\&h}}$. Since mass and heat transfer occur in parallel, the following expression can be derived [63]:

$$(\mathrm{d}m/\mathrm{d}t)_{\mathrm{m\&h}} = [(\mathrm{d}m/\mathrm{d}t)_{\mathrm{mass}}^{-1} + (\mathrm{d}m/\mathrm{d}t)_{\mathrm{heat}}^{-1}]^{-1} = k_{\mathrm{m\&h}}\, a \Delta c, \tag{15}$$

where $k_{\mathrm{m\&h}}$ is the overall mass transfer coefficient including the contributions of both mass and heat transfer. It is given by:

$$k_{\mathrm{m\&h}} = [k_{\mathrm{t}}^{-1} + (\Delta H_{\mathrm{cr}} M c)/(\alpha R T^2)]^{-1} \quad [\mathrm{m\,s^{-1}}], \tag{16}$$

with M the molecular weight ($\mathrm{kg\,mol^{-1}}$) of the crystallising substance and R the ideal gas constant ($\mathrm{J\,K^{-1}mol^{-1}}$).

4.3.2. Size-dependent growth (SDG)

In section 4.3.1 the mass transfer coefficient k_{t} was treated [see eq. (13)]. It consists of a coefficient describing the volume diffusion (k_{d}) and one taking into account the surface integration part of the growth process (k_{r}). k_{d} is inversely proportional to the size of the crystal suspended in a fluid growing as a result of molecular diffusion. This size-dependency of k_{d} [and thus also of k_{t}, eq. (13)] is most profound at crystal sizes of the order of a few micron or less [3, 64]. However, the value of k_{d} approaches infinity for sizes going down to zero, which is obviously a nonphysical situation. In this size region the crystal growth process is taken over by the surface integration coefficient k_{r} which becomes the rate-limiting step. The size-dependence of the overall mass transfer coefficient k_{t} can thus hardly be an explanation for the observed SDG of crystals up to 200 µm.

Another size-dependency of the growth rate can be introduced if the equilibrium concentration changes with size. This effect is called the Gibbs–Thomson effect, which is a shift of phase equilibrium at a curved surface. For a spherical interface this shift, which enters the expression for the imposed $\Delta\mu$ via μ_{S} as an additional term, is given by $2\Omega\gamma/R$ (see, e.g., [65]), where Ω is the molecular volume, γ the specific surface excess free energy and R the radius of curvature, which can be positive (bump) or negative (indentation). This expression neglects the anisotropy of γ. Since the Gibbs–Thomson effect becomes significant only for sizes below ~ 1 µm, it does not seem a solid explanation for SDG. Another reason for disregarding the Gibbs–Thomson effect as a mechanism for SDG is the observation (ex situ, using scanning electron microscopy,

SEM) that secondary nuclei (~ 1–50 μm) generated by crystal–rod contacts during growth, are all faceted, implying growth rather than dissolution, within at least 10 s after formation [37].

So, no physical explanation existed for why in the intermediate size range (say, 1–200 μm), crystals generally show size-dependent growth. To describe the measurements obtained during runs in an industrial crystalliser, size-dependent growth rate functions were postulated. The ones most often applied were reviewed by Garside [66], Tavare [67] and Berglund [68] but also more recently, new empirical functions were introduced [69].

An attempt to understand SDG from a physical point of view resulted in connection with growth rate dispersion, which is discussed below (section 4.3.3).

For still larger crystal sizes, an effective growth rate can be introduced which is defined as the sum of the kinetic growth rate, determined by the super-saturation, and a negative attrition rate. The attrition rate is assumed to increase with crystal size, in this way leading to an effective, size-dependent growth rate [43, 70].

4.3.3. *Growth rate dispersion (GRD)*

The phenomenon of growth rate dispersion is the observation that crystals of the same size and under identical conditions like temperature, supersaturation and hydrodynamics show a different growth rate.

GRD can be understood by assuming a dislocation content which varies from crystal to crystal [71]. Especially if a (small) crystal contains only few dislocations contributing to growth, it is possible that at a certain moment during growth a dislocation initially emerging on a fast growing face, ends in the corner of two adjacent faces. Consequently, the growth rate of that particular face is strongly reduced and thus the overall growth rate of the crystal. A direct verification of this mechanism will be difficult to obtain because of the small sizes of the crystals involved which makes them unsuitable for Lang topographic examination.

Measurements by Ristić et al. [72], later followed by Mitrović et al. [73] and Ristić et al. [74] with several crystallising substances, indicate that secondary nuclei, ranging in size from about 1 to 200 μm, are disordered to a much higher degree than can be expected from variations in low dislocation densities as proposed by Chernov [71]. Therefore, another approach for an explanation of GRD, is to assume that elastic stress is built up in the crystal due to the presence of dislocation networks and/or grain boundaries. A measure for the total disorder inside a crystal is given by the mosaic spread which can be determined by the Laue X-ray diffraction technique [72–74]: an X-ray passes the crystal along its characteristic dimension x during which the beam is deflected over a certain angle due to slight misorientations or stress fields within the crystal. This technique is not sensitive to the individual defects but to the overall disorder. Modelling GRD [75] taking into account that the elastic stress increases the chemical potential of the solid μ_S, and thus decreases $\Delta\mu$ with respect to the chemical potential of the fluid, leads to a growth rate depending on the mosaic spread in the crystal (being the source for growth dispersion) as well as on the size x of the crystal. For a constant mosaic spread this expression becomes purely size-dependent. The most important conclusion of this model is that GRD and SDG have the same origin, contrary to the normally separate description of the two phenomena.

A formula, in terms of the so-called grain boundary model, relating the growth rate G to the supersaturation σ and the mosaic spread η (in rad) present in a crystal of size x was derived by Van der Heijden and Van der Eerden [75]. Here a simplified expression is given:

$$G(\eta, x) = G_0 [1 - (A_1 \eta / \sigma \sqrt{x}) \ln (A_2 \sqrt{x} / \eta)]^g, \quad (17)$$

where g is in the range $1-2$. G_0 is the growth rate for the case of zero elastic strain. A_1 (in $m^{1/2} rad^{-1}$) and A_2 (in $rad\, m^{-1/2}$) are constants related to material parameters such as the shear modulus of the material, Burgers' vector of the dislocations present in the crystal, the molecular volume, the Poisson ratio, the dislocation core radius and the length scale of the mosaic spread l (in m) and further Boltzmann's constant and the temperature. Equation (17) is valid for crystals with sizes x larger than l. Generally, l is of the order of 0.5 to 1.0 µm. It is easily confirmed that if $A_1 \eta / \sigma \sqrt{x} \ll 1$, $G \to G_0$, i.e., the growth rate becomes constant. This condition cannot be simply related to a certain size x since it also depends on the values of σ and η. This implies that for small supersaturations, some GRD can be present also among crystals larger than, say, 200 µm. If η is constant, G as given by eq. (17), is purely size-dependent.

The term $G(\eta, x)/G_0$ in eq. (17) can be thought of as an efficiency factor for growth of secondary fragments (see section 3.1.2). Including this efficiency in the modelling of the secondary nucleation rate B_0 [eq. (8)], will, in effect, lead to a reduction of the power of σ (or G).

4.3.4. *Aggregation, attrition and breakage*

Up to now we have considered the influence of crystal growth on the crystal size distribution. To complete this, rate processes such as aggregation but also attrition and breakage have to be considered as well, although attrition and in particular breakage decrease the size of a crystal, contrary to growth and aggregation processes. In this section these phenomena will be considered because they are often observed during crystallisation processes and because of their, usually, considerable influence on the CSD.

Throughout this chapter the term "agglomerate" is used to describe a loosely bounded conglomerate of individual particles, whereas the term "aggregate" is preserved for a number of particles which have grown together during or after agglomeration. So, an aggregate is a more tightly bound conglomerate than an agglomerate.

Aggregation. Aggregation preferentially occurs in the presence of a large number of small particles combined with a high supersaturation; sometimes it is mistaken for growth. For instance, during start-up of a crystallisation process, aggregation may easily occur because, as supersaturation increases, an explosion of small crystals due to primary nucleation is initiated. During precipitation processes aggregation may occur throughout a large part of the total duration of the experiment because, in spite of the high supersaturation, the crystals grow slowly and remain rather small. In some cases, aggregation can be advantageous because it may improve the filterability; on the other hand, liquid inclusions between the particles building up the aggregate will be formed quite easily, which reduce the purity of the product. Some illustrative examples can be found in the literature [5, 76–78].

Formally, aggregation can be divided into primary and secondary aggregation. Primary aggregation is defined as an apparent conglomerate which has developed as a result of growth, e.g. by (multiple) twinning or by dendritic growth of a single crystal. Secondary aggregation on the other hand is the formation of conglomerates as a result of collisions between individual particles already present in the solution which remain grown together owing to the presence of a supersaturation. In most practical cases we are dealing with the latter form of aggregation, which will be described in more detail below.

The conditions for (secondary) aggregation depend on a large number of variables. In addition to the size of the crystals and the supersaturation also the viscosity, temperature, pH, solid content of the slurry, surface double layer potential and stirring rate play an important role. If conditions for aggregation are favourable, it can be a rather fast process. All these dependencies make aggregation a difficult phenomenon to model adequately. Modelling work on aggregation therefore simplifies it to a second-order process, i.e., only binary collisions are assumed to occur. Furthermore – and this is a rather crude approximation – the shape of the aggregates is assumed to be conserved with respect to the shape of the individual particles constituting the agglomerate. Usually the particles are assumed to be spherical. Aggregation can also be described with the constraint that the total volume of the crystals is conserved.

Secondary aggregation can be subdivided into perikinetic (particles smaller than 1 μm) and orthokinetic (particles larger than 1 μm) aggregation. Perikinetic aggregation results from Brownian motion of particles caused by random fluctuations in temperature in a stagnant liquid. Perikinetic aggregation was already described in 1917 by Von Smoluchowski [79] who showed that the collision frequency factor $K(u, v)$ depends on the particle volumes u and v, the temperature T and the fluid viscosity η (in $\mathrm{kg\,m^{-1}s^{-1}}$):

$$K(u,v) = (2kT/3\eta)(u^{1/3} + v^{1/3})(u^{-1/3} + v^{-1/3}) \quad [\mathrm{m^3s^{-1}}]. \tag{18}$$

For a monodisperse sample ($u = v$) this, so-called, aggregation kernel is independent of particle volume. This leads to the most simple aggregation kernel possible: a constant ($8kT/3\eta$). Note, however, that as soon as some aggregates have been formed, the sample is not monodisperse anymore and the assumption of a constant $K(u, v)$ is, in principle, in error.

Orthokinetic aggregation is enhanced by stirring, which introduces velocity gradients within the liquid thereby increasing the collision rate. Very high stirring rates, however, tend to disrupt the aggregates or agglomerates formed previously. Orthokinetic aggregation was described by Von Smoluchowski [79] as well and assuming a perfect collection efficiency, the corresponding kernel for particles larger than 1 μm becomes:

$$K(u,v) = (4G'/3)(u^{1/3} + v^{1/3})^3 \quad [\mathrm{m^3\,s^{-1}}], \tag{19}$$

where G' (in $\mathrm{s^{-1}}$) is the mean velocity gradient in the liquid.

The problem of aggregation is of increasing interest for crystallisation and product handling and over the years, other, both (semi-) empirical and theoretical kernels for orthokinetic aggregation were proposed in the literature. A short survey can be found in the paper of Hartel and Randolph [80]. These models, however, do not consider supersaturation and its effect on the stability of these aggregates. Since the supersaturation has a large influence on the aggregation properties of crystals, a more complete description of aggregation kinetics should, at least, include this parameter.

An attempt to do this has been made by David et al. [81] who defined a sticking probability accounting for the building up of a crystalline bridge between two neighbouring particles. This necessitates the assumption that the particles remain close to each other for a sufficiently long time. Their result, an aggregation rate r between two particles of different size (x_1 and x_2), can be condensed into one, simple expression:

$$r = K(x_1, x_2, T, D_t, D_s, \varepsilon, \ldots) N \Delta c^j n_1 n_2 \quad [\# \mathrm{m}^{-3} \mathrm{s}^{-1}], \tag{20}$$

where $K(x_1, x_2, T, D_t, D_s, \varepsilon, \ldots)$ is the aggregation kernel, N the stirrer rotation speed, Δc the solute concentration difference and n_1 and n_2 the particle concentrations of the two different sizes taking part in the aggregation process. T, D_t, D_s and ε are the temperature, tank diameter, stirrer diameter and power input, respectively. The main conclusions of the model of David et al. [81] are (i) that aggregation cannot be considered as a process independent of size, (ii) that aggregation may occur within a size range and that the ratio of the sizes of the aggregating crystals plays an important role. With respect to the latter conclusion the model shows that small crystals stick more easily to larger ones – particles approximately equal in size tend to aggregate much less easily. The model includes crystalliser dimensions and power input allowing scale-up of the aggregation process. The model was successfully applied to describe aggregation of adipic acid crystals growing from an aqueous solution [78].

In a recent study by Wachi and Jones [82] the particle size distribution and degree of aggregation during precipitation was dynamically modelled, thereby including a mean primary crystal size, or – in other words – the number of (primary) crystals within an aggregate. This attempt allows for an explicit formulation of aggregate disruption by primary particle attrition in the population balance. It accounts for the formation of a bimodal CSD as well as the prediction of secondary particle characteristics such as the specific surface area.

Experimental results of Melikhov et al. [83] indicate that particles of similar size show a higher tendency to aggregate than particles of different size. Clearly, work within this field still needs more attention before a complete understanding of aggregation is obtained.

Attrition and breakage. Although attrition and breakage both decrease the crystal size undergoing one of these fracture mechanisms while new fragments are generated, they differ in this respect that breakage occurs for energies considerably higher than are required for fracture of a particle, whereas attrition is fracture of a small volume of the particle by local stresses insufficient to break the particle as a whole [84]. During breakage a large number of fragments is formed within a relatively wide size distribution. A particle undergoing attrition remains approximately the same size, but produces a number of much smaller fragments.

Breakage of particles requires rather high energies and it seems questionable whether this phenomenon can be a dominant process unless under conditions of, e.g. (extreme) high stirring rates. Attrition, on the other hand, can be a more easily occurring process since smaller impact energies are involved. Experimental results point to the existence of a critical size x_c above which attrition occurs and below which the parent crystals remain intact. Therefore, attrition is most realistically described by an S-shaped curve, e.g. according to a dimensionless classification function of the form $s^p/(1 + s^p)$, where s is a size ratio (x/x_c) and p a power influencing the steepness of the S-curve. Depending on stirrer rotation rate N and solid content M_T of the slurry, x_c is usually of the order of 500 μm. Other functions describing attrition can be found in [43, 70, 85].

In section 6 it will be shown how attrition and breakage (as well as aggregation) can be incorporated in the modelling of the development of the CSD in time.

4.4. *Additives as habit modifiers*

The presence of impurities in a crystallisation process can hardly be avoided. Fortunately, crystallisation is a highly selective method compared to other techniques like distillation, adsorption and extraction, i.e. it is rather difficult for an impurity to become incorporated in the crystal – although there are, of course, exceptions to this rule. As a result of the ability of impurities to adsorb at specific sites on the crystal surface, the impurities in this way block the possibilities for incorporation of growth units. Accordingly, the growth of that specific face will be reduced. If the retardation effect varies for the different faces, the habit of the crystal will be changed with respect to the pure case. This habit modification can take drastic forms if only a few faces are (completely) blocked by the impurities: the morphology may change from, e.g. a needle to a block-shaped crystal.

Although impurities sometimes cause undesired effects, intentional use of additives can be of major importance, e.g. to improve the filterability of the product. A review of "tailor-made" auxiliaries for the control of nucleation, growth and dissolution of crystals was given by Lahav and Leiserowitz [86]. In their paper they give numerous examples and references of, e.g. inhibitor- or solvent-induced morphology changes.

5. *Crystal size distribution*

One of the main characteristics of a crystalline product is its size distribution. The CSD is important for the product quality and performance, but also influences the separation of the crystals from the mother liquor and the subsequent drying of the crystals. It is also essential for storage and handling of the final product. For instance, small crystals contain more adhering mother liquor after filtration due to the relatively large surface area compared to a similar mass comprising larger crystals. This results in a less pure product after drying and a higher tendency towards caking. Another example concerns the dissolution time, since in many cases crystals are dissolved or digested in their final use and the size of the crystals determines, to a great extent, the time required for dissolution. As a final example we mention that many inorganic pigments obtained by means of precipitation exhibit a dependence of colour on CSD.

The dominant properties of the product will usually be the mean size and the width of the distribution, but the zeroth to third moments of the CSD are also related to important physical parameters of the product. These CSD properties will be treated first. After that, a classification of the most currently applied CSD measuring techniques will be given.

5.1. *CSD properties*

In order to describe the size distribution of a population of crystals, three methods can be used:

(1) The *cumulative size distribution* $N(x)$ (#m^{-3}), which equals the total number of crystals between size 0 and x per unit of slurry volume.

(2) The *population density* $n(x)$ (#$\mathrm{m}^{-3}\,\mathrm{m}^{-1}$), which is defined as $\mathrm{d}N(x)/\mathrm{d}x$, so $n\mathrm{d}x$ is the number of crystals with size between x and $x + \mathrm{d}x$.

(3) The *cumulative mass density distribution* $M(x)$ ($\mathrm{kg\,m^{-3}}$), defined as:

$$M(x) = \int_0^x \rho_c k_v n(x) x^3 \,\mathrm{d}x, \tag{21}$$

which gives the weight of crystals per unit of slurry volume of size smaller than x; ρ_c is the density of the solid (in $\mathrm{kg\,m^{-3}}$) and k_v is a volumetric shape factor (e.g. equal to $4\pi/3$ for the case of spherical particles).

Figure 6 shows a typical example of a cumulative size distribution $N(x)$ and its related population density $n(x)$ and the cumulative mass density distribution $M(x)$. The mass density distribution $m(x)$ is obtained through $\mathrm{d}M(x)/\mathrm{d}x$.

The *i*th moment of the distribution, m_i, is defined as:

$$m_i = \int_0^\infty n(x) x^i \,\mathrm{d}x \quad [\mathrm{m}^i\,\mathrm{m}^{-3}]. \tag{22}$$

The definition in eq. (22) can be used to represent some relevant properties of the crystalline product. The zeroth moment m_0 of the distribution is the total number N_T of crystals per unit of crystalliser volume; m_1 is the total length L_T of the crystals per m^3; m_2 (together with a surface area shape factor k_a) is the total surface area of the crystals per m^3 and, finally, m_3 (together with the density of the crystals ρ_c and a volumetric shape factor k_v) describes the total mass M_T of crystals per m^3.

Several average crystal sizes can also be defined: $x_N \equiv m_1/m_0$ is the number averaged or mean crystal size; $x_M \equiv m_4/m_3$ is the mass-averaged crystal size; x_D is the dominant crystal size for which $\mathrm{d}M(x)/\mathrm{d}x = 0$. The d_{50} value is defined as the crystal size for which 50 wt.% of the solid is smaller than d_{50}. All these definitions are summarised in table 3. Finally, the coefficient of variation (CV), which is a measure of the width of the crystal size distribution, is defined as $[(m_3 m_5/m_4^2) - 1]^{1/2}$. It is a definition on the basis of mass.

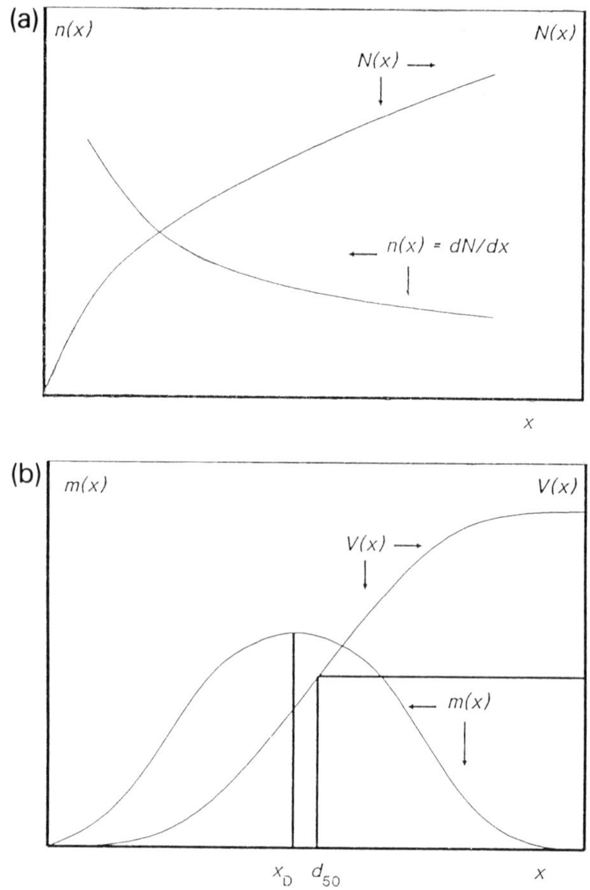

Fig. 6. (a) Example of a typical cumulative size distribution, $N(x)$; the population density $n(x)$ is also shown. (b) Cumulative particle volume $V(x)$ and mass density distribution $m(x)$, indicating the location of x_D and d_{50}.

Table 3
Definition, description and dimension of several important properties describing a crystal size distribution.

Definition	Description	Dimension
$N_T = m_0$	total number of crystals per m^3 of slurry	$\# m^{-3}$
$L_T = m_1$	total length of crystals per m^3 of slurry	$m\,m^{-3}$
$A_T = k_a m_2$	total surface area of crystals per m^3 of slurry	$m^2 m^{-3}$
$M_T = k_v \rho_c m_3$	solid content	$kg\,m^{-3}$
$x_N = m_1/m_0$	number-averaged or mean crystal size	m
$x_M = m_4/m_3$	mass-averaged crystal size	m
x_D	x at which $dM/dx = 0$; dominant crystal size	m
d_{50}	x at which $\int_0^{d_{50}} n(x)x^3\,dx = \int_{d_{50}}^{\infty} n(x)x^3\,dx$	m
$CV = [(m_3 m_5/m_4^2) - 1]^{1/2}$	coefficient of variation	–

Solving the population balance equation (see section 6) with the proper boundary conditions at $x = 0$, leads to the population density $n(x)$. This will be treated in more detail in sections 6 and 7, which show the derivation of the CSD for specific industrial continuous and batch crystallisation processes, respectively. Often, however, this balance equation cannot be solved analytically and for empirical purposes, several mass density functions $m(x)$ have been proposed over the years. These comprise normal (Gaussian), log-normal, gamma, Rosin–Rammler and Gaudin–Melloy distribution functions (see, e.g. Randolph and Larson [87]).

5.2. *CSD measuring techniques*

The techniques for measuring the particle size distribution can be divided in three groups; *off-line*, *on-line* and *in-line*. In off-line techniques, like the sieve analysis, the size distribution becomes available after taking a slurry sample, pretreatment of the sample and performance of the analysis. It can be used for specification of an existing product, but the time lag between production and actual measurement has the disadvantage that it cannot be used for continuous monitoring and control of the crystallisation process. With on-line CSD measurement units, this time lag caused by sample pretreatment has been substantially reduced, e.g. the time needed for a dilution step. In the case of in-line detection, sampling is not needed and the CSD can be monitored directly during the process. In table 4, the CSD measurement techniques already available and currently under development are summarised. Generally, the CSD can be represented either on a number or volume distribution base. A review can be found in [88].

An important point during sampling is that the sample has to be representative of the total distribution [89]. So, at the sampling point, classification by size has to be avoided. The sample has to be treated carefully in order to avoid attrition, breakage and agglomeration during handling. Several techniques require dilution with saturated mother liquor to give a reliable result. Shape, and for some techniques also the refractive index of the crystals, can influence the finally measured CSD. Often, the dynamic range of the technique is insufficient to cover the whole relevant crystal size

Table 4
CSD measuring techniques classified according to type (off-line, on-line and in-line).

Name	Type	Size range [μm]	Classes [#]	CSD result
sieve analysis	off-line	30–2000	15–20	weight
Coulter counter	off-line	1–150	(≥) 64	number
forward light scattering	on-line/dil.	1–2000	~ 30	weight
quasi-elastic light scattering	on-line/dil.	0.005–3	–	volume
image analysis	on-line/dil.	> 5	–	number/shape
backward light scattering	in-line	1–1000	38	number
acoustic spectroscopy	on-line	5–3000	31	weight

range (below 1 μm up to a few mm) without intermediate adjustment of the measuring device.

5.2.1. *Sieve analysis*

An off-line technique to obtain a particle size distribution is by means of a sieve analysis. A sieve is an open, usually cylindrical container, having equally sized apertures that are uniformly distributed over the base. The openings are either square (wire mesh) or circular (punched holes). Mounting the sieves in order of increasing aperture size, loading the top sieve with the dried crystals and agitating the unit (manually or mechanically), the powder will be classified according to the sieve fractions used. At the bottom of the stack, the fines are collected in the so-called receiver. The result of this method is usually, but not necessarily since it is an absolute method, presented as a cumulative undersize percentage distribution.

The sieve analysis is, like any other size measurement technique, subject to errors. Pretreatment of the sample is crucial since the product has to be dried during which agglomeration and/or caking have to be minimised as much as possible. The probability that a particle presents itself at a pore depends on the size distribution of the crystals, the number of particles on the sieve (load), physical properties (like the particle's surface roughness), tendency towards agglomeration or caking, method of agitating the sieve, aspect (length-over-width) ratio of the particle and the geometry of the sieving surface (ratio of open area to total area). The particle size distribution which results from the sieve analysis depends on the duration of the sieving, variation of sieve apertures, wear of particles, errors of sampling (classification during sampling), and observational and experimental errors.

The size range covered by a sieve analysis starts at several tens of microns going up to mm's. Since sieving is subject to a variety of errors it is generally not worthwhile to calibrate sets of sieves.

The introduction and development of new, on-line and even in-line particle size measurement techniques has led to considerably less use of sieve analysis in industry. One of the other reasons that sieving is no longer popular, is that the method is rather time consuming.

5.2.2. *Coulter counter*

The Coulter counter makes use of the electrical sensing zone method. Number and size of particles suspended in an electrolyte are measured by passing them through a small orifice. Two electrodes on both sides of the orifice measure a decrease in current due to the passage of a particle through the orifice. The pulse results from an increase in resistance over the orifice. The amplitude of this pulse is proportional to the volume of the particle. The particles are counted in 64 classes. By interpolation the number of size classes can be increased. For each specific application, the device should be calibrated using, e.g. latex particles suspended in the same electrolyte and under similar conditions as the subsequent experiments. Nevertheless, errors may be introduced since the lattices may shrink or swell due to the electrolyte solution. Therefore, if possible, an in situ optical microscopic verification of latex sizes is preferred for calibration.

The Coulter counter technique has found a wide range of applications from counting and sizing red blood cells to analysing crystal size distributions [39, 90]. The (super) saturated solution itself can be used as the electrolyte medium [32].

In order to obtain a reliable CSD measurement, an electrolyte medium has to be chosen in which the crystalline substance is stable, i.e. does not dissolve. The particles should have a high specific resistance compared to the liquid; if not (i.e. the particles are electrically conductive), corrections have to be applied, because the device will underestimate the actual size. The number of particles per unit of volume has to be low (smaller than $\sim 10^5$ particles per ml) in order to avoid simultaneous passage of two or more particles through the sensing zone (coincidence measurement). In general, this implies a considerable dilution of the sample prior to measurement. The particle size should not be too large compared to the orifice diameter otherwise the device yields a nonlinear response which is not accounted for during the conversion of the measured data to the CSD. Moreover, a small size of the particles avoids blockage of the sensing zone.

A cylindrical orifice of width L, cross-sectional area A submerged in an electrolyte medium with resistivity ρ_f (see fig. 7) has a resistance R. With a nonconducting particle of volume V_p present within the orifice opening of volume V, the resistance changes to R'. The difference in resistance, $\Delta R = R - R'$, to first order in V_p can be shown to be [88]:

$$\Delta R \approx -\rho_f V_p / A^2, \quad \text{for } V_p / V \ll 1 \quad [\Omega]. \tag{23}$$

The measurement results in a volume-based number distribution which can be transformed to a number distribution as a function of size by calculating the particle radius from the volume V_p on the basis of spherical particles; volumetric shape factors

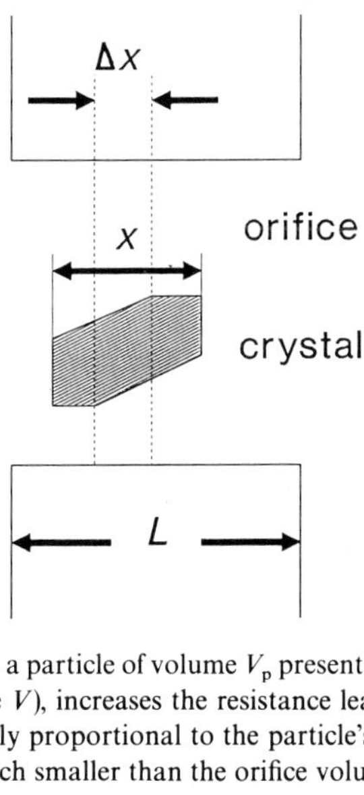

Fig. 7. The Coulter counter principle: a particle of volume V_p present within the electric sensing zone (width L, cross-sectional area A and volume V), increases the resistance leading to a voltage pulse which can be measured. The output signal is directly proportional to the particle's volume provided that this volume is much smaller than the orifice volume.

accounting for nonsphericity of the crystals can be incorporated. No errors are expected due to the fact that, generally, crystals are faceted.

The dynamic range of the Coulter counter covers 2–60% of the orifice diameter D. This maximum attainable size stems from the fact that higher-order terms in V_p/V are neglected in eq. (23). If the CSD to be measured comprises sizes that cannot be measured within one run using one orifice diameter tube (which is often the case), a separation of the coarser particles from the remaining particles is required. The small dynamic range and the required dilution, make the Coulter counter less suitable for on-line CSD analysis.

5.2.3. *Forward light scattering*

Instruments based on the principle of forward light scattering (FLS) to measure particle size distributions have been developed in the last decades and are applied in many fields [91–96].

The application of the FLS method can be twofold. First, the method can be of the optical sensing zone type (comparable to the electric sensing zone principle of the Coulter counter) in which a single particle passes through a focused beam of light. During passage, the particle scatters light and the total scattered light intensity is measured on a detector. Secondly, an ensemble of particles scatter an incident beam of light, the intensity of which is detected. The first method is an absolute technique in the sense that particles are both counted and sized, whereas the second method is a relative one. Upon applying the optical sensing zone technique, the same problems are encountered as with the Coulter concerning blockage of the sensing zone, dynamic range and dilution. Additionally, the optical properties of the crystals play an important role. The principle of FLS will be shortly outlined below. This description holds for both the optical sensing zone devices and the technique by which the scattered light by an ensemble of particles is detected. Attention is focused, however, on the latter technique.

The intensity pattern of light scattered by a particle depends on its size, its shape and its optical properties. The intensity of light scattered by an ensemble of particles suspended in a liquid flowing through an optical cell is detected on a spatial detector (see fig. 8) after which a deconvolution of the scatter pattern yields the particle size distribution. This method is characterised by the measurement of a relatively large number of particles with a low resolution.

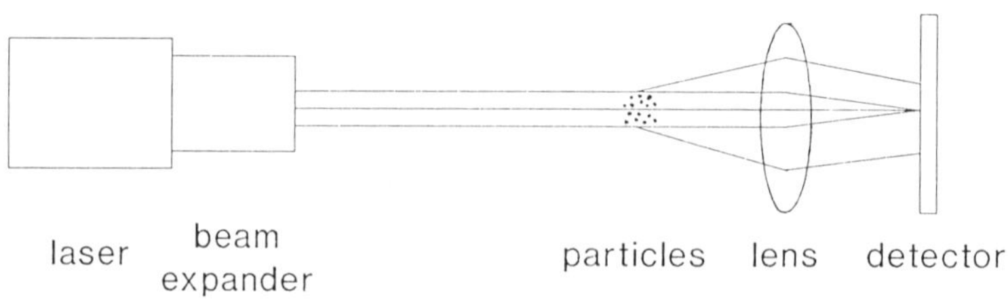

Fig. 8. The detector of a forward light scattering based device collects the light intensity due to scattering of an incident beam by an ensemble of particles.

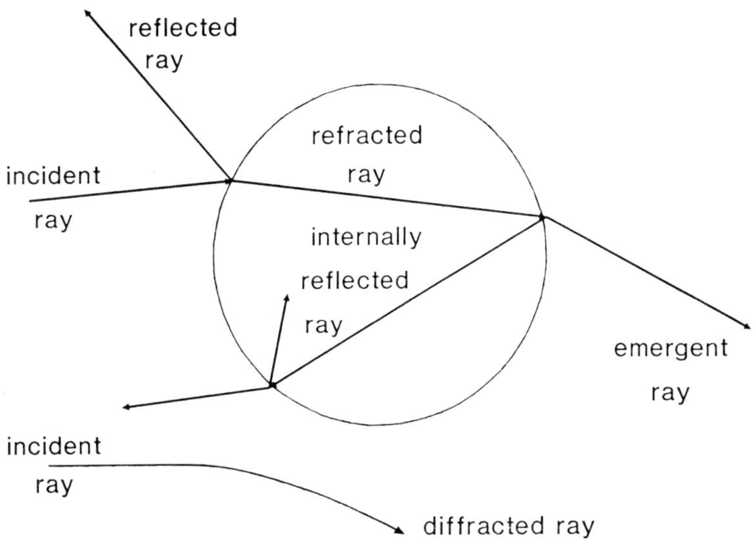

Fig. 9. Rays of light interact with a particle in several ways: reflection, refraction, diffraction and absorption. Geometric ray optics includes reflection and refraction and neglects the other two phenomena.

The theory of light scattering of a plane wave by a dielectric, spherical particle is given by Van de Hulst [97]. The beam of light is disturbed upon passing a particle (see fig. 9): (i) by reflection, both at and within the particle; (ii) by refraction at both interfaces (i.e. medium/particle and particle/medium); (iii) by diffraction around the particle; and (iv) by absorption within the particle.

The rigorous solution of scattering of light by a particle of any (spherical) size and isotropic optical properties was given by Mie in 1908. The complexity of the Mie theory arises from interference effects between diffracted light (around the particle) and refracted light (by the particle), an effect which is most prominent for particle sizes of the order of the wavelength λ. The scattered light intensity by one particle strongly depends on size as well as the refractive index m relative to that of the surrounding medium and, as a function of the scatter angle, it shows several maxima and minima. Although the theory developed by Mie is generally applicable, it is rather complicated to use, especially for large particles. Recently, it has been shown that for broad distributions, which are encountered during most crystallisation applications, also the approximative Fraunhofer theory (see below) can be used [96].

In table 5 a number of other light scattering theories are summarised together with an indication of the relative refractive index m in combination with particle size range for which the theory, if valid, yields reliable results. Rayleigh derived the intensity of scattered light by particles much smaller than the wavelength λ of the light used. Fraunhofer studied the opposite limit: particles much larger than λ. Only light diffracted around the particle is taken into account. The intensity of the refracted light is assumed to be negligible compared to the diffracted light, and the optical properties of the particle itself are not important, only the refractive index of the surrounding

Table 5
Light scattering theories and the reliability for application with wavelengths of the order of 0.5 μm (e.g., He–Ne laser with 632.8 nm). Notation: + = reliable; − = unreliable; n.v. = theory not valid.

Theory	m	$x < 20\,\mu m$	$20 < x < 500\,\mu m$	$x > 500\,\mu m$
Mie	all	+	+/−	unstable
Rayleigh	all	+	n.v.	n.v.
Fraunhofer	all	n.v.	+/−	+
ray optics	> 1.1	n.v.	+/−	+
anomalous diffr.	~ 1	n.v.	+	+

medium. For very large particles, relatively simple ray optics can be applied to derive the scattered intensity pattern. The anomalous diffraction theory applies for "invisible" particles ($m \approx 1$) with sizes much larger than the wavelength used. In this case only the phase shift of the light beam caused by diffraction is responsible for the scattering. All these examples can be shown to be special cases of the full Mie theory by making the appropriate approximations.

In summary, the detector measures an integrated light intensity vector $\boldsymbol{L}$, the components of which correspond to different size classes. $\boldsymbol{L}$ results from scattering of light by an ensemble of particles distributed over size according to a vector $\boldsymbol{q}$. The relation between $\boldsymbol{L}$ and $\boldsymbol{q}$ is given by the scatter matrix $\boldsymbol{A}$. The matrix elements depend on the scatter model used. Thus:

$$\boldsymbol{L} = \boldsymbol{A}\boldsymbol{q}. \tag{24}$$

Once a scatter pattern is detected, a deconvolution from the measured scatter pattern to a distribution comprising the fraction of particles in each size class has to be performed. Simple inversion of the scatter matrix $\boldsymbol{A}$ [according to the scheme $\boldsymbol{q} = (\boldsymbol{A}^{\mathrm{T}}\boldsymbol{A})^{-1}\boldsymbol{A}^{\mathrm{T}}\boldsymbol{L}$] often fails due to measurement errors included in the measurement of $\boldsymbol{L}$ and systematic errors in $\boldsymbol{A}$. Numerical methods, like the Phillips–Twomey smoothing technique, have been developed to overcome the ill-conditioning of the inverse problem [96]. This procedure results in stable solutions, although one should be aware of a possible broadening of the distribution and the introduction of spurious particles particularly in the wings of the distribution. Boxman [96] recently introduced yet another method to carry out the deconvolution step which eliminates the effects of smoothing and simultaneously enhances both speed and stability.

The advantages of FLS are: fast measurement, nondestructive, wide dynamic size range (0.5–2000 μm), reproducible, easy operation, and calibration is, in principle, not required. The FLS technique also allows one to measure the obscuration which is the intensity of the light scattered by the particles relative to the incident beam intensity. The obscuration is a measure of the total surface area of the particles. A comment on the wide dynamic size range is in place here. FLS claims a range of, say, three orders of magnitude in size, corresponding to nine orders of magnitude by mass, but not within one measurement run. In order to obtain a CSD covering the entire size range, lenses of different focal length have to be installed subsequently or several instruments have to be applied in series.

The two main disadvantages of the FLS technique are (i) the low resolution and accuracy in the small size range, and (ii) the low particle concentration (smaller than 1 vol.%) in order to avoid multiple scattering. This means that a dilution step has to be performed for most industrial slurries.

The FLS method yields a volume or mass distribution in percent on basis of spherical and optically isotropic particles; simple shape factors may be introduced accounting for, e.g. an aspect (length-over-width) ratio of the particles, but not for the fact that the crystals are faceted. To extend the scatter matrix A in order to account for optical anisotropy and sharp corners and facets is an extremely difficult task and it is, in fact, questionable if this extension is necessary. Due to the measurement of a relatively large number of particles, which are assumed to be randomly orientated in the beam of light, the errors introduced by anisotropic optical properties and the presence of facets might be small.

5.2.4. *Quasi-elastic light scattering*

To measure the size distribution of particles smaller than 1 μm another technique has been developed: quasi-elastic light scattering (QELS). Information on the submicron range can be important for obtaining more insight into the formation of clusters and nuclei and their growth behaviour.

QELS, also known as photon correlation spectroscopy (PCS) makes use of the rapid fluctuations (up to 10^6 Hz) of the scattered light intensity caused by the Brownian motion of small particles. To measure these signals, high time resolution instrumentation is necessary and the interpretation of the signals is complicated, especially when a polydisperse sample is studied.

The technique is mainly characterised by its sensitivity for sizes from 5 nm to 3 μm. It merely gives an indication of the mean particle size in a more or less monodisperse system. The fact that the lower size limit of this technique is in the range of the critical nucleus size (a few nm), opens another field of interest, namely the investigation of the structure of supersaturated solutions in order to establish whether clusters are present or not. With respect to this last subject, QELS has been applied to investigate the interfacial layer adjacent to a crystal surface growing from its melt. This solid–liquid interfacial layer is thought to consist of density fluctuations [98, 99] or of small gas bubbles segregated by the solid and accumulating ahead of the growth front [100–102]. Recently, Williams et al. [103] reported results which substantiate the gas bubble hypothesis.

5.2.5. *Image analysis*

Often, not a distribution on volumetric or weight basis is wanted, but a distribution by number. FLS and QELS are techniques which are not able to give a number distribution and, moreover, neither these two nor the Coulter counter and the sieve analysis give information on particle shape. For this purpose, an on-line, but often also an off-line image analysis technique can be very helpful: with a CCD (charged coupled device) camera (on-line) or a microscope (off-line), particles can be imaged, recorded on video, digitised and analysed afterwards. In this way direct visual information is gained on particle size and shape. Software is available with which surface area, perimeter, size and shape factor of the particles can be quantified.

The combination of FLS with image analysis might turn out to be a very successful concept, because the additional visual information which is obtained on particle habit and shape, allows one to identify and follow processes like agglomeration or attrition directly [104].

5.2.6. Backward light scattering

Besides forward light scattering, also the light scattered backwards by a particle can be detected and used for size determination. The principle is as follows: a focused laser beam rotates (by means of a moving lens) with a high frequency (see fig. 10). The back-scattered signal at the moment a particle passes the focus of the laser beam is

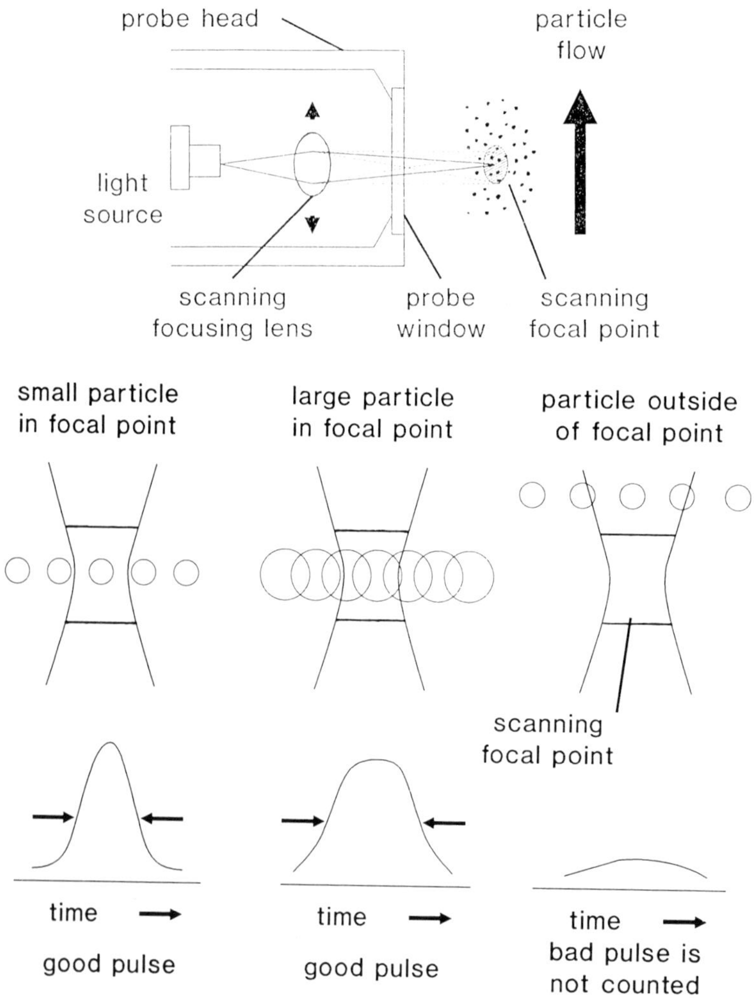

Fig. 10. Set-up and principle of the backward light scattering method.

detected within the same optical system. The width of the back-scatter pulse is, via a cord length distribution which accounts for the probability that a particle is scanned over a certain fraction of its diameter, a measure of the size of the particle. A (customer adjustable) threshold allows one to discard "bad" signals caused by particles that are out of focus, but still give a back-scatter pulse. The result is a size distribution by number which can be transformed to a volume-based distribution assuming that the particles are spherical. Nonsphericity can be accounted for by introducing a volumetric shape factor. The detectable sizes range from a few microns to a few mm's.

The main advantages compared to FLS are: (i) higher solid contents (up to 30 vol.%) can be handled so sampling and dilution are usually not necessary; (ii) a more reliable number distribution is obtained in the smaller size range because it is a counting technique; (iii) the instrument is robust so it can be used in an industrial environment; and (iv) the size of the particle is calculated from the width of the back scattered light pulse which is hardly affected by optical properties like colour, relative refractive index (unless $m \approx 1$ at which value particles become "invisible") and absorption.

A serious problem is the underestimation of particle sizes in the coarser part of the distribution, either because the crystals exhibit a certain surface roughness or because the crystals are faceted, both giving rise to more than one surface reflection from the same particle. This immediately implies an over-estimation of the number of smaller particles. The accuracy of the method will be low for particle velocities of the same order of magnitude as the scanning speed of the laser, by particles out of focus, by classification at the detector head due to the disturbed flow pattern in the vicinity of the device and by increasing coincidence measurements in slurries with increasing solid content.

The technique can be applied to the in-line detection of changes in the CSD [104].

5.2.7. Acoustic spectroscopy

This technique is, to a large extent, comparable to the forward light scattering principle, except for the fact that ultrasonic sound waves are used to interact with the particles to be measured [105].

In analogy with FLS, where the light waves respond to the optical properties of a particle, the sound waves are sensitive to the acoustic impedance of the particle. Within a measurement cell (see fig. 11), sound waves are generated by means of a piezoelectric device (transmitter). These waves are scattered by the particles present in the observation cell and the attenuated signal is detected by a second piezoelectric element (receiver). By measuring the attenuation at different frequencies (during one run) and by application of scatter theories, the particle size distribution can be calculated by deconvolution of the scattered intensity data. The frequency of the acoustic waves can be varied between 1 and 100 MHz, which means that the corresponding wavelengths are roughly in the range of a mm down to several microns. The particle sizes covered by these frequencies range from $\sim 5\ \mu m$ to 3 mm. Suspension densities from 5–10 up to 30 vol.% can be measured, hence in-line operation is possible [105]. The device yields a distribution by weight (in %). The suspension containing the particles may be regarded as a continuum, but the wave propagation velocity in such a continuum may differ considerably from the velocity in the pure liquid (without any particles) and can be used

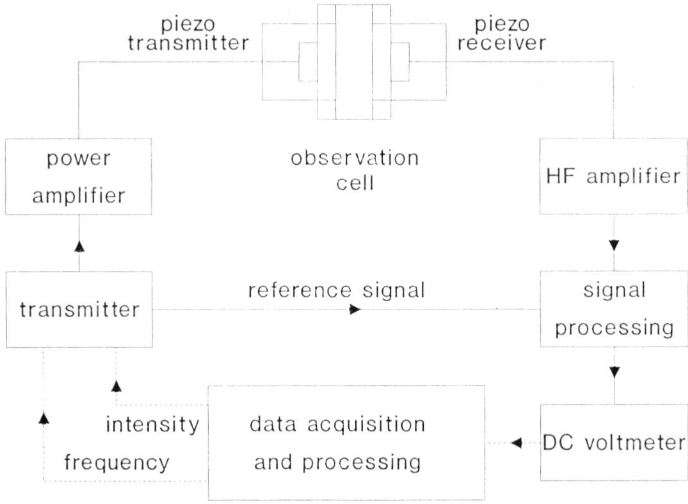

Fig. 11. Experimental set-up of the acoustic size measuring principle. A piezoelectric device (transmitter) generates an acoustic wave which is scattered by the presence of particles in the observation cell. The attenuated wave is detected by a second piezoelectric device (the receiver) and further processed to yield the size of the particle (after [105]).

as a measure of the volume concentration of particles. For particles large compared to the wavelength, the attenuation can be described with geometrical scattering (compare with Fraunhofer diffraction and ray optics, section 5.2.3).

The interaction between the acoustic waves and the particles may result in: (i) (partial) entrainment of small particles in a viscous fluid below a critical frequency leading to no or little attenuation of the incoming acoustic signal. At higher frequencies and with larger particles, but still small compared to the wavelength, the entrainment will only be partial and some scatter will occur; (ii) resonance effects in a deformable particle which may change the volume or the shape.

Although this technique is still in a state of early development, commercial instruments are already available.

6. *Continuous crystallisation models*

In this section several models of ideal types of crystallisers will be treated. The emphasis is put on the performance of the crystalliser for which three criteria can be identified: (i) CSD quality in terms of an average crystal size and distribution width; (ii) crystal production yield rate; and (iii) reactor stability.

6.1. Mass, heat and population balances

The most general way to describe the overall processes occurring in a crystalliser containing a large number of crystals (distributed over size) suspended in a supersaturated solutions is to write down overall conservation and continuity equations concerning mass, heat and size distribution of crystals. In this section these balances will be given explicitly.

6.1.1. Mass balance

The overall mass balance is governed by the requirement that the volumetric flow rate (m^3s^{-1}) multiplied by the density of the slurry (kg per m^3 slurry) entering the vessel has to equal the terms leaving the vessel:

$$Q_i\rho_i = Q_o(1 - M_T/\rho_c)\rho_o + Q_o M_T + Q_v\rho_v \quad [\mathrm{kg\,s^{-1}}], \tag{25}$$

with Q_i, Q_o and Q_v the volumetric flow rates of the inlet, outlet and vapour and ρ_i, ρ_o and ρ_v the density of the slurries and vapour, respectively. M_T is the solid content of the slurry in $\mathrm{kg\,m^{-3}}$. Under the assumption that the crystallisation process introduces negligible volume changes we have $\rho_i = \rho_o$. For cooling crystallisation the term $Q_v\rho_v$ is zero.

The mass balance equation equals the rate at which mass is gained through the production of solids with the rate at which solute is consumed from the mother liquor:

$$Q_i c_i = Q_o M_T + Q_o(1 - M_T/\rho_c)c \quad [\mathrm{kg\,s^{-1}}], \tag{26}$$

where ρ_c is the density of the solid; c_i and c are the solute concentration of the inlet feed and of the solution in the crystalliser, respectively (in $\mathrm{kg\,m^{-3}}$). Equation (26) is built up from a term describing the mass flow rate of the clear liquid feed ($Q_i c_i$), the mass flow rate of solid material ($Q_o M_T$) and the mass flow rate for the liquid corrected for the volume of solid [remaining term in eq. (26)]. Combining eqs. (25) and (26) with $\rho_i = \rho_o$ and, for cooling $Q_v = 0, c_i$ follows from:

$$[1 + (M_T/\rho_o) - (M_T/\rho_c)]c_i = (1 - M_T/\rho_c)c + M_T \quad [\mathrm{kg\,m^{-3}}], \tag{27a}$$

which can be approximated by:

$$c_i \approx c + M_T \quad \text{for } M_T \ll \rho_o, \rho_c. \tag{27b}$$

The residence time τ is determined by the total slurry discharge rate Q_o (m^3s^{-1}) and the volume of the crystalliser, V (m^3):

$$\tau = V/Q_o \quad [\mathrm{s}]. \tag{28}$$

The production rate P is given by $Q_o M_T$ (in $\mathrm{kg\,s^{-1}}$).

Two types of crystallisation systems can be identified: class I and II. For a class I system, the product yield depends on the residence time τ, while for a class II system, this is not the case. For a class II system, c in eq. (27) can be replaced by the equilibrium value c^* for a sufficiently accurate calculation of M_T from the mass balance equation. For a class I system M_T has to be measured explicitly, which requires sampling, filtration, drying and finally weighing, unless c is accurately known (and in particular the difference between c_i and c) which is generally not the case.

6.1.2. Heat balance

The energy balances for cooling and evaporative crystallisation are usually rather complicated equations. Here we will confine ourselves to simplified expressions, which disregard, e.g. energy losses due to non-optimal isolation of the crystallising equipment as well as energy input of the stirrer or pump.

For cooling crystallisation the energy E_{cool} withdrawn per unit mass of crystals reads:

$$-E_{\text{cool}} = \rho_o c_p \Delta T/c + \Delta H_{\text{cr}} \quad [\text{J kg}^{-1}], \tag{29}$$

where c_p is the specific heat ($\text{J K}^{-1}\text{kg}^{-1}$) of the mother liquor, ΔT the temperature difference between inlet feed and slurry and ΔH_{cr} (in J kg^{-1}) the enthalpy of crystallisation.

The energy consumption E_{evap} per unit mass of crystals during evaporative crystallisation can be written as:

$$E_{\text{evap}} = \rho_o c_p \Delta T/c + \rho_o \Delta H_{\text{evap}}/c - \Delta H_{\text{cr}} \quad [\text{J kg}^{-1}], \tag{30}$$

with ΔH_{evap} (in J kg^{-1}) the enthalpy of evaporation of the mother liquor. For most practical cases of evaporative crystallisation the contribution of ΔH_{evap} to the heat balance is dominant with regard to the other terms in eq. (30).

Despite the fact that generally $E_{\text{cool}} < E_{\text{evap}}$, it cannot simply be concluded that the overall costs per unit of produced mass for cooling crystallisation are always less than those for evaporative crystallisation.

6.1.3. Population balance

The population balance equation [87] is expressed in terms of the population number density $n(x)$ (see section 5.1). $n(x)\,\mathrm{d}x$ is the number of crystals per unit volume within the size range from x to $x + \mathrm{d}x$. For crystals, four ways are available to enter and to leave this infinitesimally small size range (see fig. 12), namely (i) by growth of smaller crystals into this size range or by growth of crystals out of this range; (ii) by volumetric input and output flow streams concerning this size range; (iii) by birth of crystals due to breakage or aggregation into the given size range; and (iv) by death of crystals due to aggregation or breakage out of this size range. The macroscopic population balance equation for a crystalliser with constant volume V, as derived by Randolph and Larson [87], reads:

$$\frac{\partial n(x,t)}{\partial t} + \frac{\partial [G(x,t)n(x,t)]}{\partial x} + D(x,t) - B(x,t) = -\sum_k \frac{n_k(x,t)Q_k}{V}. \tag{31}$$

The terms in eq. (31) are expressed in $\#\,\text{m}^{-1}\text{m}^{-3}\text{s}^{-1}$. The right-hand side of eq. (31) represents the inflow and outflow of slurry to the system, where Q_k is (by convention) taken positive for a flow Q_{out} out of V and negative for a flow Q_{in} into V. SDG or GRD functions for the growth rate G, as given by eq. (14), can be implemented in the population balance equation. $D(x,t)$ and $B(x,t)$ are the death and birth functions respectively. Equation (31) allows – through the functions D and B – the incorporation of, e.g. the aggregation kernels referred to in section 4.3.4. In principle the birth function could also comprise the formation of secondary nuclei. It is important, however, to realise that the terms in eq. (31) are all expressed in $\#\,\text{m}^{-1}\text{m}^{-3}\text{s}^{-1}$, whereas nucleation rates (like J and B_0) are expressed in $\#\,\text{m}^{-3}\text{s}^{-1}$. Suppose $B(x,t)$ describes an attrition process during which particles are generated in a relatively wide range of sizes. If we think of secondary nucleation as an attrition process (see section 3.1.2) being

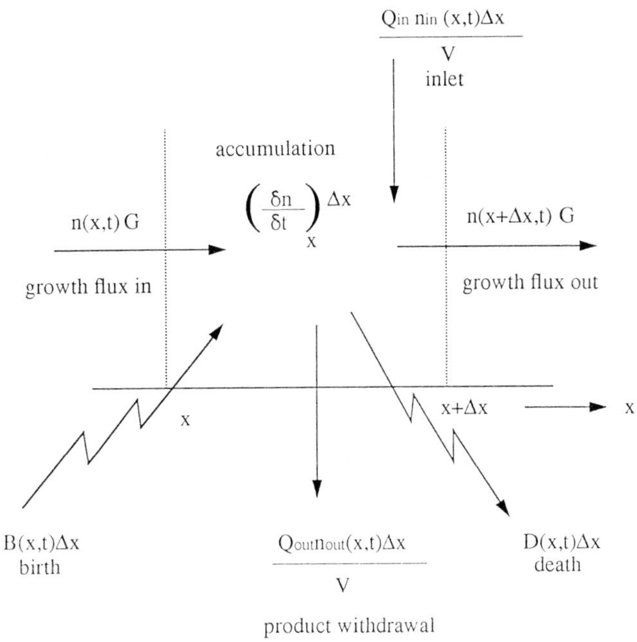

Fig. 12. Schematic depiction of an infinitesimal size range from x to $x + dx$ which crystals may enter and leave due to growth, aggregation, attrition, breakage and volumetric input and output flow streams.

part of the overall attrition process described by B, the secondary nucleation process can be assumed to generate particles smaller than an arbitrarily chosen size y. The relation between the secondary nucleation rate $B_0(y,t)$ and the birth function $B(x,t)$ is then given by the following integral:

$$B_0(y,t) = \int_0^y B(x,t)\,\mathrm{d}x \quad [\#\,\mathrm{m}^{-3}\mathrm{s}^{-1}]. \tag{32}$$

In this case crystals appear only at sizes $x > 0$ and $n(x=0,t) \equiv n_0 = 0$ (an example of such a situation is given in section 6.2.6). Usually, however, secondary nuclei are rather small. In fact, they are assumed to enter the size distribution at zero size. Their number density is therefore imposed as a boundary condition at $x = 0$ to the solution of eq. (31). The relation between the population density at zero size and the secondary nucleation rate B_0 is given by:

$$n_0 = \left.\frac{\mathrm{d}N}{\mathrm{d}x}\right|_{x=0} = \left.\left[\frac{\mathrm{d}N}{\mathrm{d}t}\right]\cdot\left[\frac{\mathrm{d}x}{\mathrm{d}t}\right]^{-1}\right|_{x=0} = \frac{B_0(t)}{G(t)}. \tag{33}$$

The term $J(t)$, accounting for primary nucleation also at zero size, can be incorporated in the boundary condition at $x = 0$ simply by adding $J(t)$ to $B_0(t)$ in eq. (33). Except for the boundary condition with respect to size, also a boundary condition with respect to time is required to solve eq. (31). This latter boundary condition is given by the initial population density distribution $n(x, t = 0)$.

The general expression for the population balance given in eq. (31) cannot be solved analytically, except for a few, rather simplified cases (section 6.2). Often, solutions for $n(x, t)$ have to be found via numerical methods. This is the subject of section 6.3.

6.2. *Steady-state operation*

An already considerable simplification of the population balance equation is obtained under steady-state operation, i.e. the system has reached an equilibrium situation in which there is no time dependence. For steady-state processes eq. (31) reduces to:

$$\frac{\partial[G(x)n(x)]}{\partial x} + D(x) - B(x) = -\sum_k \frac{n_k(x)Q_k}{V} \quad [\#\mathrm{m}^{-1}\mathrm{m}^{-3}\mathrm{s}^{-1}]. \tag{34}$$

Nevertheless, more assumptions are needed before an analytical solution for $n(x)$ can be found. In the next few sections several examples of steady-state situations will be treated.

6.2.1. *The MSMPR model*

In fig. 13 a mixed suspension, mixed product removal (MSMPR) crystalliser is depicted schematically. The MSMPR concept assumes that a clear liquid is used for feeding, that rate processes such as secondary nucleation (except at size zero), attrition, breakage and aggregation can be neglected, that G is size-independent and that unclassified withdrawal of the product slurry occurs. Furthermore, it is assumed that the suspension is ideally mixed. Under these restrictions eq. (34) simplifies to:

$$G\,\mathrm{d}n(x)/\mathrm{d}x + n(x)/\tau = 0, \tag{35}$$

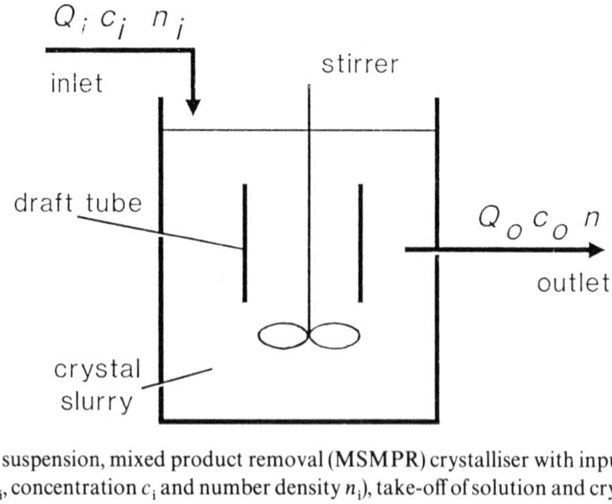

Fig. 13. The mixed suspension, mixed product removal (MSMPR) crystalliser with input feed (characterised by mass flow rate Q_i, concentration c_i and number density n_i), take-off of solution and crystals (Q_o), stirrer and draft tube.

Table 6
The main CSD characteristics for the MSMPR calculated with help of the definitions in table 3.

N_T [$\#\mathrm{m}^{-3}$]	$n_0 G\tau$
L_T [$\mathrm{m\,m}^{-3}$]	$n_0(G\tau)^2$
A_T [$\mathrm{m^2 m}^{-3}$]	$2k_a n_0(G\tau)^3$
M_T [$\mathrm{kg\,m}^{-3}$]	$6\rho_c k_v n_0(G\tau)^4$
x_N [m]	$G\tau$
x_D [m]	$3G\tau$
x_M [m]	$4G\tau$
d_{50} [m]	$3.67G\tau$
CV [–]	0.5

with boundary condition as given by eq. (33). The solution to eq. (35) now becomes:

$$n(x) = n_0 \exp(-x/G\tau) \quad [\#\mathrm{m}^{-1}\mathrm{m}^{-3}]. \tag{36}$$

The main characteristics of the steady-state CSD for the MSMPR, like the first four moments of the distribution and several average crystal sizes, can be easily calculated from eq. (36), see section 5. These are listed in table 6. The coefficient of variation, CV, being a measure for the relative width of the CSD, equals 0.5 and does not depend on the parameters n_0, G and τ.

The exponential relation between population density and crystal size is illustrated in fig. 14. A semi-logarithmic plot ($\ln n$ versus x) yields a straight line with a slope of $-(G\tau)^{-1}$. G can be determined from the slope of the steady-state CSD given in fig. 14 since the residence time τ is known. The intercept at the vertical axis gives the value of

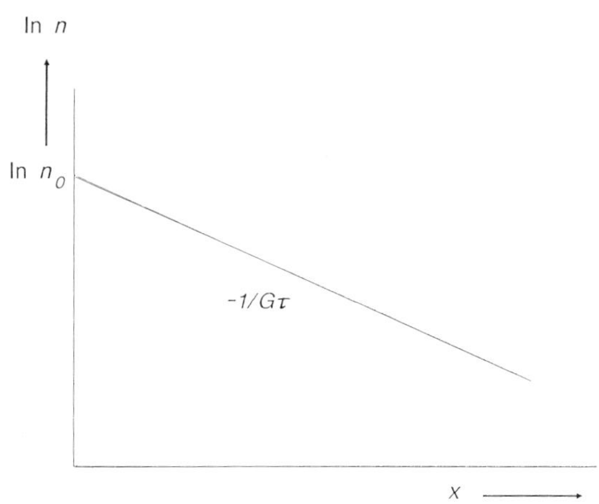

Fig. 14. The steady-state MSMPR population density distribution $n(x)$ plotted semi-logarithmically versus length x of the crystals. The MSMPR solution predicts a straight line with a slope of $-(G\tau)^{-1}$ and n_0 the intercept of the line with the vertical axis.

n_0. In practice, however, the population density $n(x)$ is often found to deviate from the straight line predicted by eq. (36). Moreover, $n(x)$ cannot be measured reliably (or even not at all) for $x \to 0$. Thus, finding n_0 from an extrapolation of the CSD is often not very reliable. Determination of n_0 from a calculated (for class II systems) or measured (for class I systems) M_T value is therefore preferred. With the help of the third moment m_3 of the CSD the solid content M_T of the slurry can be related to n_0 (see section 5). It is given by:

$$n_0 = M_T/[6k_v\rho_c(G\tau)^4] = B_0/G \quad [\# \, m^{-1} m^{-3}]. \tag{37}$$

In this way the value of B_0 can be calculated as well. The values of n_0 and B_0 are "effective" values under the assumption that G is constant, i.e., $G \neq G(x)$.

In section 3.2 the secondary nucleation rate was written as a power law in terms of the power input ε, the solid content M_T of the slurry and growth rate G. By performing experiments under MSMPR conditions with different stirrer rotation rates, residence times and solid contents, the overall dependence of B_0 on G, ε and M_T can be determined [34].

In an attempt to improve the dominant size of the CSD under MSMPR conditions one could consider the influence of a different residence time τ. The effect of τ on, e.g. the dominant crystal size $x_D \equiv m_3/m_2 = 3G\tau$ (see section 5.1 and table 6) is not straightforward since a change in τ and thus in supersaturation affects the nucleation and the growth rate as well [87]. In order to establish the effect of a different residence time on x_D, while maintaining the same solid content M_T, the expression for the solid content M_T of the slurry as derived from the steady-state population density is used [see eq. (37)]. The ratio of the dominant crystal sizes in the two crystallisers can then be expressed in terms of the ratio τ_1/τ_2:

$$x_{D,1}/x_{D,2} \equiv (3G_1\tau_1)/(3G_2\tau_2) = [\tau_1/\tau_2]^{(i-1)/(i+3)}, \tag{38}$$

under the assumption that $B_0 = n_0 G = k_N G^i$ for constant M_T and stirrer speed N, and that the proportionality constant k_N is the same for both experiments. From the result in eq. (38), it can be concluded that the dominant crystal size may either increase ($i > 1$) or decrease with $\tau (i < 1)$. Usually i is found to be in the range of 0 to 3. Accordingly, the dependence of x_D on τ is weak. Hence, attempts to improve the performance of the MSMPR crystalliser with respect to the dominant crystal size by changing the residence time is not very effective. Besides, increasing the residence time increases capital investment and/or decreases production rate. Since the width of the CSD, characterised by CV, is constant under MSMPR conditions (namely 0.5), variation of τ has no influence on this CSD property.

Similarly, the effect of an MSMPR operated at different solid contents $M_{T,1}$ and $M_{T,2}$ but at the same residence times can be investigated by using different feed concentrations and the same feed flow. Also temperature and hydrodynamics are similar in both cases and again $B_0 \sim G^i M_T^j$. This leads to the ratio:

$$x_{D,1}/x_{D,2} = [M_{T,1}/M_{T,2}]^{(1-j)/(i+3)}. \tag{39}$$

For $j = 1$, which is a realistic example, the ratio is unity and the dominant crystal size is independent of solid content. However, if $j > 1$, which is also not unrealistic, the

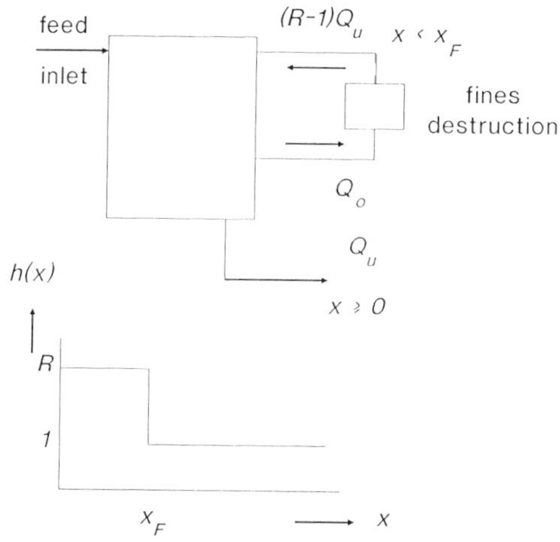

Fig. 15. Schematic set-up of the mixed suspension with fines destruction (MSFD) crystalliser.

dominant size slightly decreases (since always $i > -3$) with increasing solid content because of increased nucleation.

Another approach to influence the CSD quality is the development of different types of crystallisers. For this purpose the principles of fines removal and classified product removal will be discussed in the next sections for a few simple cases [87].

6.2.2. *Mixed suspension with fines destruction*

The accelerated removal of fine crystals below a size x_F is a way to influence the CSD of a crystalliser. In the most simple way this is established in a mixed suspension, fines destruction (MSFD) crystalliser which is depicted schematically for the case of a total fines destruction system in fig. 15. The fines ranging from 0 to x_F are withdrawn from the crystalliser and dissolved after which the clear liquid is fed back to the system. For an ordinary MSMPR, $n(x)/\tau$ particles of size x are withdrawn per unit of time. In the case of an MSFD crystalliser an excess amount of crystals is removed per unit of time for $0 \leq x < x_F$ at a volumetric flow rate of $(R-1)Q_u$, where $R > 1$ and Q_u the product flow by which crystals are removed over the entire size range (note that the two volumetric flow rates have different solid contents). This mode of operation implies a residence time of τ/R for crystals smaller than x_F and τ for those larger than x_F. The steady-state population balance, eq. (33), becomes:

$$G\,\mathrm{d}n(x)/\mathrm{d}x + h(x)n(x)/\tau = 0, \tag{40}$$

where the function $h(x)$ is given by:

$$h(x) = R \quad \text{for } 0 \leq x < x_F, \tag{41a}$$

$$h(x) = 1 \quad \text{for } x \geq x_F. \tag{41b}$$

For each size range the population balance equation, eq. (40) can be solved. The

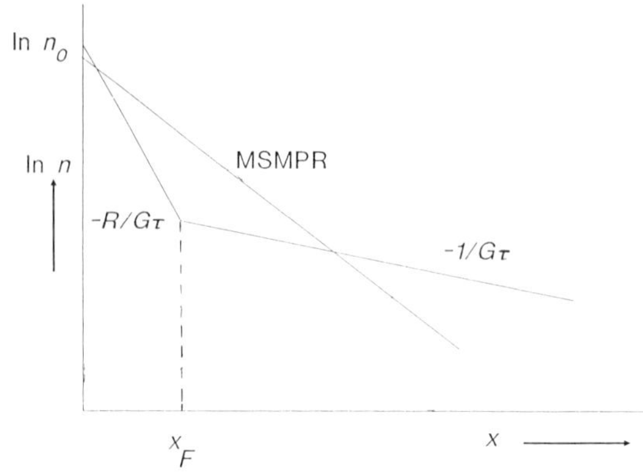

Fig. 16. The steady-state CSD obtained with an MSFD.

requirement that at $x = x_F$ the two expressions for $n(x)$ are equal because of continuity, leads to the final solution for the MSFD crystalliser:

$$n(x) = n_0 \exp(-Rx/G\tau) \quad \text{for } 0 \leq x < x_F, \tag{42a}$$

$$n(x) = n_0 \exp(-[R-1]x_F/G\tau)\exp(-x/G\tau) \quad \text{for } x \geq x_F. \tag{42b}$$

This distribution is shown in fig. 16, together with the steady-state CSD for an MSMPR with which the MSFD is compared. In practice such a sharp cut size can never be achieved. The comparison is based on a constant m_3 of the product flow withdrawn from the crystalliser in both the MSMPR and MSFD case. The intercepts with the vertical axis for both distributions do not coincide – in fact the intercept for the MSFD case is located at a higher position compared to the MSMPR case. This might be surprising, since fines are removed at an increased rate. However, due to the fact that the growth rate is increased to a higher level in the MSFD crystalliser, the value of $n_0 \sim G^{i-1}$ (assuming $i > 1$) for the MSFD crystalliser is larger than the population density at zero size for the MSMPR case. For $i < 1$, this is reversed, while for $i = 1$ the two intercepts coincide. The mass-averaged crystal size x_M, defined as m_4/m_3 (section 5), can be shown to increase compared to the MSMPR mass-averaged crystal size ($= 4G\tau$). A simulation study of a class II system in a 970 l cooling crystalliser with $R = 5$, $x_F = 100$ μm and $\tau = 4500$ s and with secondary nucleation kinetics according to $B_0 \sim G^2 M_T$ shows that $x_M \approx 960$ μm while, in this example, it is ~ 690 μm for the MSMPR case [106]. The net effect of fines removal is that the growth rate in an MSFD ($\sim 5.3 \times 10^{-8}$ m s^{-1} in this example) is increased to a higher level compared to an MSMPR ($\sim 3.8 \times 10^{-8}$ m s^{-1}), while the same production of crystals is obtained of larger size and less total surface area. For $R = 20$, $x_M \approx 1540$ μm and the growth rate is $\sim 9.4 \times 10^{-8}$ m s^{-1}. It is, however, important to realise that fines destruction in principle has a destabilising effect on the operation of the crystalliser, but often facilitates filtration of the product since the coefficient of variation (CV) slightly decreases (in this example from 0.5, being the MSMPR value, to 0.49 and 0.41 for $R = 5$ and 20, respectively).

6.2.3. Mixed suspension with classified product removal

Figure 17 shows the so-called mixed suspension, classified product removal (MSCPR) crystalliser in which crystals of size x larger than x_P, are withdrawn, by means of a classifying unit, at an accelerated rate compared to the MSMPR rate. For a perfect classifier, the separation between crystals smaller and larger than the cut-size x_P would be ideal, i.e. only crystals larger than x_P would leave the vessel, while all crystals smaller would be fed back. In this ideal situation the residence time for crystals in the range $0 \leq x < x_P$ would be infinite and the population density would result in a horizontal line (a constant) for $x < x_P$.

In practice, however, the horizontal line in the CSD for sizes smaller than x_P can never be achieved since always some particles in this size range will slip through together with the product flow as a result of the non-deal behaviour of the classification unit. The function $h(x)$ introduced in section 6.2.2 describing the size-dependent withdrawal, will be some smooth function of x. Usually, it can be represented by an S-shaped curve. So, in order to obtain a more realistic picture still allowing an analytical solution of the CSD, this S-curve is approximated by a step function assuming that crystals smaller than x_P are removed at a removal rate which is arbitrarily set at the MSMPR level, in this way accounting for the so-called dead flux of small particles in the flow carrying the large ones. The classification function $h(x)$ is depicted in fig. 17 as well.

Note that the three flows entering or leaving the classifier (i.e., the classifier feed, the product flow, which leaves the classifier and the overflow, which is fed back to the crystalliser) have different solid contents, in increasing order: overflow < feed flow < product flow. Therefore, it is more convenient to characterise the residence times of the crystals distributed over size: crystals smaller than x_P are withdrawn at a rate which is arbitrarily set to MSMPR level, implying a residence time of τ; crystals larger than x_P have a residence time of τ/z with z larger than unity, because these crystals are

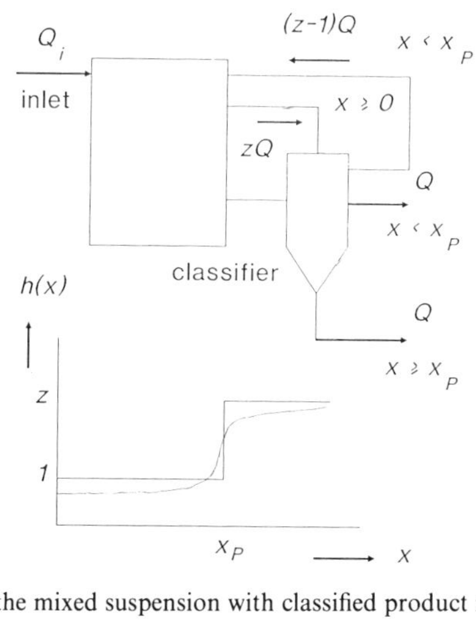

Fig. 17. Schematic set-up of the mixed suspension with classified product removal (MSCPR) crystalliser.

withdrawn at a higher rate compared to the MSMPR case. The population balance under these highly idealised conditions is then given by:

$$G dn(x)/dx + h(x)n(x)/\tau = 0, \tag{43}$$

where the classification function $h(x)$ is defined as:

$$h(x) = 1 \quad \text{for } 0 \le x < x_P, \tag{44a}$$

$$h(x) = z \quad \text{for } x \ge x_P. \tag{44b}$$

The differential equations resulting for the two size ranges can now be solved. The result is:

$$n(x) = n_0 \exp(-x/G\tau) \qquad \text{for } 0 \le x < x_P, \tag{45a}$$

$$n(x) = n_0 \exp(-[1-z]x_P/G\tau)\exp(-zx/G\tau) \quad \text{for } x \ge x_P. \tag{45b}$$

The CSD is plotted in fig. 18 together with the CSD for the MSMPR situation. The value of the intercept of the CSD for the MSCPR crystalliser is smaller in comparison with the MSMPR case when operated at the same m_3 as the MSCPR in the product flow. The CSD is shifted to the left compared to the MSMPR implying a lower mass-averaged crystal size for the MSCPR. An advantage can be that the resulting product has a more narrow CSD than found for crystallising under MSMPR conditions.

Simulation of a cooling crystalliser with classified product removal with $z = 5$ and further similar assumptions with respect to B_0, x_P and τ as mentioned at the end of section 6.2.2, shows that x_M of the product is decreased from $\sim 690\,\mu m$ to $\sim 640\,\mu m$. Just as for the MSFD case, the growth rate is slightly increased, in this example from $\sim 3.8 \times 10^{-8}$ (MSMPR) to $\sim 4.0 \times 10^{-8}\,m\,s^{-1}$ (MSCPR) [106]. Analogous to fines destruction, classified product removal will generally destabilise crystalliser operation, but it results in a product with an improved filterability (CV = 0.40). For $z = 20$, x_M is even further reduced to $\sim 630\,\mu m$ while the CV is 0.39.

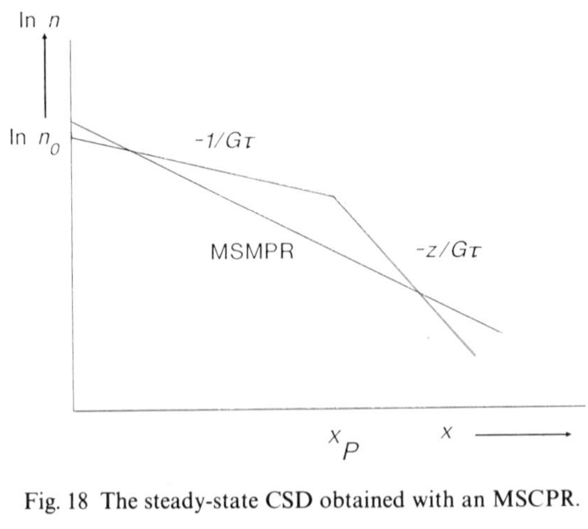

Fig. 18 The steady-state CSD obtained with an MSCPR.

6.2.4. The R–z model

A combination of the MSFD and MSCPR crystalliser leads to the so-called R–z crystalliser after the parameters R and z comprised by the classification function $h(x)$ (see fig. 19). This type is discussed here since it provides the calculation of a CSD obtained under conditions which are often close to those found in real crystallisers.

For crystals with sizes smaller than x_F, a residence time of τ/R is assumed, while it is τ/z for crystals larger than x_P. In the intermediate size range, from x_F to x_P, crystals are removed at MSMPR rate, i.e. the residence time for these crystals is τ. Under these conditions, the CSD is given by [87]:

$$n(x) = n_0 \exp(-Rx/G\tau) \qquad \text{for } 0 \le x < x_F, \tag{46a}$$

$$n(x) = n_0 \exp(-[R-1]x_F/G\tau)\exp(-x/G\tau) \qquad \text{for } x_F \le x < x_P, \tag{46b}$$

$$n(x) = n_0 \exp(-[R-1]x_F/G\tau)\exp(-[1-z]x_P/G\tau)\exp(-zx/G\tau) \qquad \text{for } x \ge x_P. \tag{46c}$$

For $R, z \neq 1$, fines destruction and classified product removal are simulated. If $R = z = 1$ the result in eq. (46) coincides with the result obtained for the MSMPR crystalliser. Figure 20 shows the steady state CSD obtained with the R–z crystalliser.

Simulation of an R–z crystalliser with $R = z = 5$ and otherwise similar to the previous two examples (sections 6.2.2 and 6.2.3) shows that x_M of the product is $\sim 780\,\mu m$, which is larger compared to the MSMPR case where $\sim 690\,\mu m$ was found [106]. The growth rate increases from $3.8 \times 10^{-8}\,m\,s^{-1}$ (MSMPR case) to $5.8 \times 10^{-8}\,m\,s^{-1}$ (R–z case). What is usually gained is an improved filterability of the product corresponding to a smaller coefficient of variation, namely CV = 0.32, as compared to the MSMPR situation [107]. Depending on the product demands, the advantage of an improved filterability should be outweighed with the smaller mass-averaged crystal size of the product and the destabilising effect which fines destruction

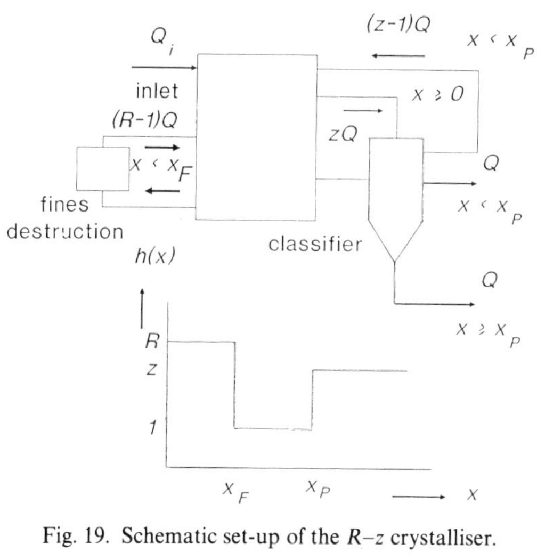

Fig. 19. Schematic set-up of the R–z crystalliser.

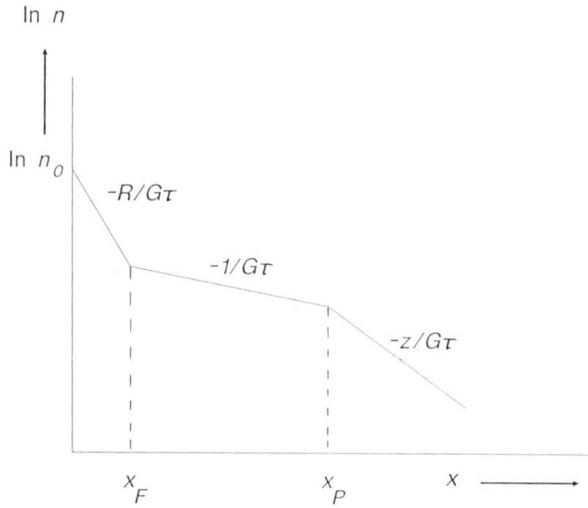

Fig. 20. The steady-state CSD obtained with the R–z crystalliser.

and classified product removal impose on crystalliser operation. Simulation of an R–z crystalliser with $R=z=20$ leads to $x_M \approx 910\,\mu m$ and CV = 0.21.

An additional remark with respect to the filterability of the product crystals is that Ouchiyama and Tanaka [108, 109] have given a procedure for estimating the porosity of a random packing of spherical particles of various sizes. This quantitative approach is a first attempt to link the experimentally found CSD with the porosity of a random packing of these crystals under the assumption of spherical particles.

6.2.5. Two-stage crystalliser

CSDs can be altered by crystallising in a series of vessels, each operated at MSMPR conditions, with the suspension flowing from one vessel into the next, in other words, a cascade of MSMPRs with forward feeding. Here the only case to be considered is that of two equally sized vessels and a fresh feed entering only into the first vessel. The result will be compared with a single MSMPR of the same volume. Unequally sized vessels and especially a distributed feeding system will lead to a totally different final CSD.

The streams in a two-stage crystalliser (fig. 21) are characterised by (c_i, $n=0$, i.e. a crystal-free flow) for the feed inlet, by (c_1, n_1) for the intermediate flow from the first to the second vessel and by (c_2, n_2) for the outlet of the second crystalliser. In order to compare the results for the cascade with a single-stage MSMPR (described in section 6.2.1) we assume that the residence time in each vessel of the cascade is $\tau_1 = \tau_2 = \tau/2$, where τ is the residence time for a single MSMPR. Further we assume that the production in each of the two vessels equals one half of the production obtained with the single stage crystalliser. The corresponding population balance for each vessel is given by [87]:

$$G_1 \mathrm{d}n_1(x)/\mathrm{d}x + n_1(x)/\tau_1 = 0, \tag{47a}$$

$$G_2 \mathrm{d}n_2(x)/\mathrm{d}x + [n_2(x) - n_1(x)]/\tau_2 = 0, \tag{47b}$$

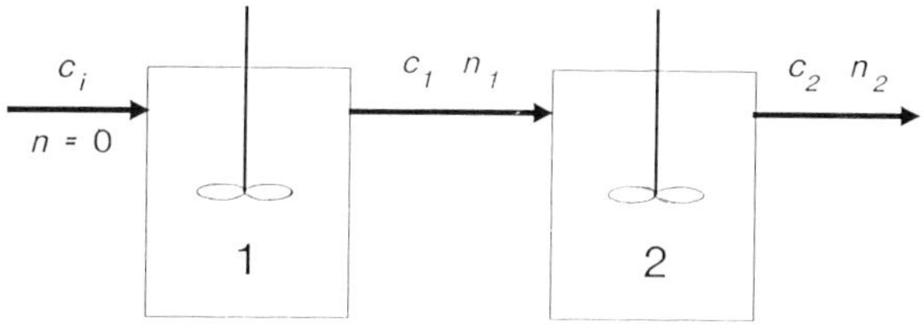

Fig. 21. The two-stage cascade crystalliser with forward feeding.

with solutions:

$$n_1(x) = n_{0,1} \exp(-x/G_1\tau_1), \tag{48a}$$

$$n_2(x) = n_{0,2} \exp(-x/G_2\tau_2) + n_{0,1}[G_1\tau_1/(G_1\tau_1 - G_2\tau_2)] \times [\exp(-x/G_1\tau_1) - \exp(-x/G_2\tau_2)], \tag{48b}$$

where nucleation is assumed to occur in both vessels.

For nucleation in the first stage only, the CSD is obtained simply by putting $n_{0,2} = 0$ in eq. (48b). From the requirement $M_{T,1} = M_{T,2} = M_T/2$ under the assumption that $n_{0,1} = B_{0,1}/G_1 = k_N G_1^{i-1}$ and $n_0 = k_N G^{i-1}$ (for the moment we neglect the M_T dependence), the growth rate G_1 in the first vessel can be expressed in terms of the growth rate G for the single-stage case. The relation between G_1 and G is [87]:

$$G_1/G = 2^{3/(i+3)}, \tag{49}$$

which implies that G_1 is larger than G. A more realistic point of view is to include an M_T^j term in the secondary nucleation rate B_0. Calculations show that the dominant crystal size obtained with a cascade of two crystallisers is smaller than the x_D obtained with a single-stage system [87]. The growth rate G_1 in the first vessel relative to G in the single-stage system then modifies to:

$$G_1/G = 2^{(j+3)/(i+3)}. \tag{50}$$

This ratio is, like eq. (49), also larger than unity for all typical values of i and j.

More complicated configurations of forward feed stage systems may improve the average crystal size of the CSD, but the complexity of such systems often requires a numerical solution of the process.

6.2.6. Steady-state CSD including size-dependent birth and death rate functions

As an example of calculating the CSD from the steady state population balance, eq. (33), where no crystals are assumed to appear at $x = 0$, only at sizes larger than zero, the birth and death rate functions $B(x)$and $D(x)$, respectively, are not neglected. The birth and death rate functions can be any continuous function of size. The boundary condition with respect to size is given by $n(x = 0) \equiv n_0 = 0$. The population balance now reads:

$$G\,dn(x)/dx + n(x)/\tau = B(x) - D(x), \tag{51}$$

which is a standard differential equation of the Bernoulli type. Its solution is given by (with x' the integration variable):

$$n(x) = \exp(-x/G\tau) \int_0^x [\{B(x') - D(x')\}/G] \exp(x'/G\tau)\,\mathrm{d}x'. \tag{52}$$

This solution already incorporates the boundary condition with respect to size. For simple birth and death rate functions (e.g. polynomials in x) the integral in eq. (52) can be determined explicitly. For more difficult (and probably more relevant) functions, integration must be performed numerically. Anyhow, in most cases, the solution will be rather different from the CSD for the MSMPR case as given by eq. (36).

6.3. *Dynamic operation*

Systems operated at steady state can be described with time-independent equations, see eq. (32), which have an analytical solution only in some ideal situations (section 6.2.). A system that varies in time, e.g. during start-up or subsequent changes in process conditions (transient response) or perturbations (such as feed stream interruptions or fluctuations) is a dynamic system and the equations describing its behaviour in time are, in general, complicated. The equations can no longer be solved analytically so numerical solutions are required. A numerical solution of the process equations can be obtained by simulation of the process with the help of a computer. Calculation of the population balance equation for complex steady-state situations is also feasible by a simulation procedure using sufficiently long run times.

Process optimisation and control strategies in the case of dynamic behaviour of a continuous crystalliser can also be obtained by simulation of the process. Once the proper data are implemented in the program, the simulation of processes can be used as a predictive tool, e.g. for some characteristic values (d_{50}, CV) of the resulting CSD. Also, simulations could give an indication about how to optimise these characteristic values. Finally, control algorithms can be designed and validated by which perturbations can be compensated as smoothly as possible by temporary adjustments of externally accessible process conditions.

6.3.1. *Outline of simulation procedure*

Design of dynamic models usually consists of a combination of a mathematical and physical description of the process under consideration. The former is related to a system identification approach [110] which gives a mathematical description of (parts of) the process in which the parameters used have no obvious link to physical process characteristics. The latter is based on physical insight in (part of) the different mechanisms of the processes occurring in the crystalliser. Roughly speaking, the more physical knowledge that is incorporated in the computer program, the better is the simulation and its predictive value. Consequently, less information (data) is needed in order to obtain a good model.

The CSD of a certain substance with its physical and chemical properties is determined by process conditions such as the volumetric flow rate of the feeds, the concentration of the feed solution, the mean residence time, the temperature, the stirrer

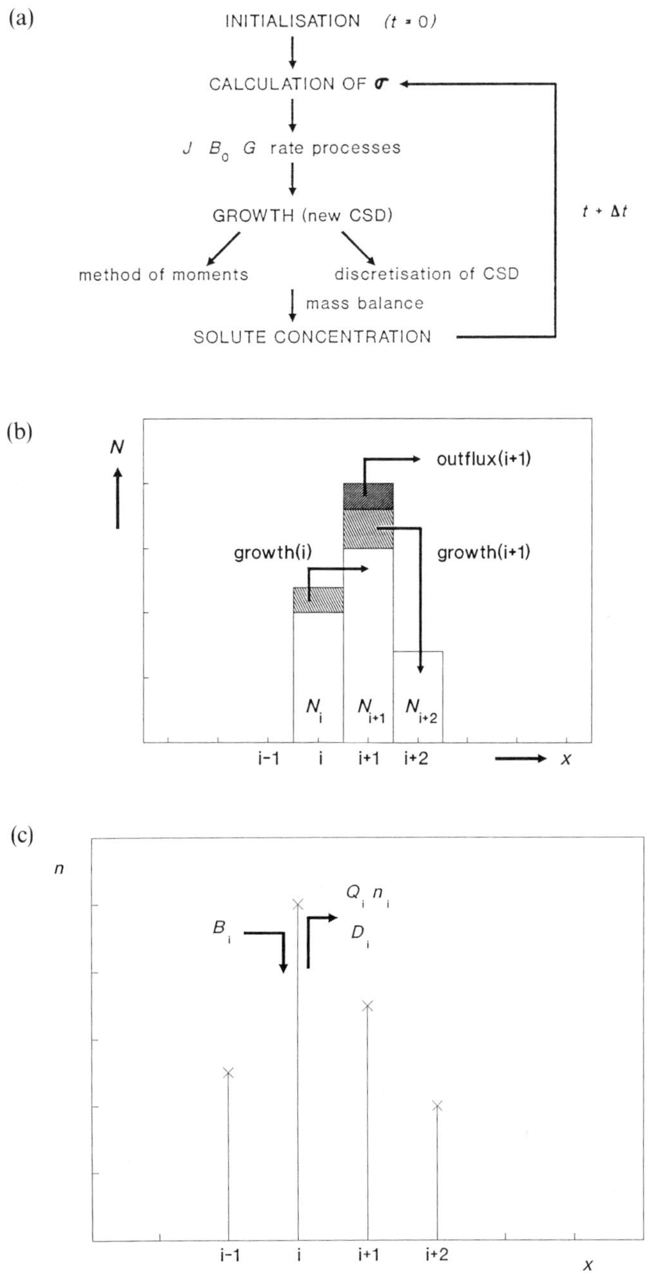

Fig. 22. (*a*) Schematic visualisation of simulation steps during dynamic modelling of crystallisation processes. (*b*) Scheme of the change in the discretised CSD in one cycle of a simulation. In a time interval Δt the occupation of class $i+1$ changes due to growth and due to the fact that a small portion of the entire population is removed from the reactor (outflux). (*c*) Example of a discrete population density in terms of $n(x_i, t)$ with the different terms contributing to the shift in CSD.

rotation rate and the geometry and design of the equipment. These parameters impose their influence on the CSD through the growth rate G and the secondary nucleation rate B_0 and, in some processes, also through the primary nucleation rate J. The CSD may be further subject to changes due to attrition, breakage, dissolution of particles and aggregation processes.

The procedure of a computer simulation is depicted schematically in fig. 22a. The initialisation comprises the input of the parameters fixing the initial size distribution in terms of its first few moments or in its discretised form, the residence time τ, the initial concentration (together with the solubility curve, which allows calculation of the supersaturation), the stirring rate and the crystalliser configuration (such as application of fines removal, product classification, multi-stage crystallisers). If one starts off with a clear liquid, crystals have to appear via a primary nucleation rate J. Furthermore, relations describing the mass and heat balances as well as the occurring rate processes (like growth, attrition, etc.) have to be incorporated. General expressions for G can be found in eqs. (11) and (12) or a simplified version as described in section 4.3 can be used. Equations (6) and (8) give the relations for J and B_0, respectively. After initialisation, the crystallisation process is represented by a loop during which the contributions of the occurring rate processes to the development of the CSD at the prevailing supersaturation are calculated in time. Time evolves with a constant time interval Δt each time the loop is passed. During each cycle a new CSD is found, shifted with respect to the previous one. With the new CSD, a new value for M_T can be calculated which is inserted into the mass balance in order to calculate the new value of the solute concentration c and thus of the new supersaturation. For evaporative and cooling crystallisation strictly speaking also the solute concentration balance, which depends on the mass and heat balances, must be incorporated for a proper calculation of c as a function of time, but often one assumes that the reduction in supersaturation is a much faster process than the overall CSD dynamics (class II system, see section 6.1.1), which in fact implies that the concentration c and the temperature T are assumed to be independent of time. The cycle is repeated a number of times yielding the development of the CSD in time. In the case of steady-state operation, the cycle has to be repeated until convergence is reached, i.e. no detectable changes in CSD occur anymore.

The subsequent growth of the crystals and the corresponding shift in the CSD can be described either by a change in moments of the CSD or by determining the changes in the CSD written in a discrete (rather than a continuous) form. Both approaches will be treated below.

Method of moments. The method of moments represents the CSD in its first (usually) four moments m_0 to m_3, related to total number N_T, total length L_T, total surface area A_T and total volume V_T of the crystals, all expressed per unit of crystalliser volume (see section 5). After initialisation of the process, the supersaturation σ is calculated using the initial solute concentration and the solubility curve. The subsequent growth of the crystals after a small time interval Δt is calculated by determining the new values of N_T, L_T, A_T and V_T as follows (disregarding influx terms in this example):

$$N_{T,new} = N_{T,old} + (J + B_0)\Delta t - N_T\Delta t/\tau, \tag{53a}$$

$$L_{\mathrm{T,new}} = L_{\mathrm{T,old}} + N_{\mathrm{T}} G \Delta t - L_{\mathrm{T}} \Delta t / \tau, \tag{53b}$$

$$A_{\mathrm{T,new}} = A_{\mathrm{T,old}} + 2k_{\mathrm{a}} L_{\mathrm{T}} G \Delta t - A_{\mathrm{T}} \Delta t / \tau, \tag{53c}$$

$$V_{\mathrm{T,new}} = V_{\mathrm{T,old}} + 3(k_{\mathrm{v}}/k_{\mathrm{a}}) A_{\mathrm{T}} G \Delta t - V_{\mathrm{T}} \Delta t / \tau, \tag{53d}$$

where τ is the residence time. In those places in eq. (53) where $N_{\mathrm{T}},\ldots,V_{\mathrm{T}}$ are given without the index "old" or "new", the simulation is not affected by whether the old or new value is taken. This approximation is only valid if the time step Δt of the simulation process is sufficiently small. The difference between $V_{\mathrm{T,new}}$ and $V_{\mathrm{T,old}}$ multiplied by the density ρ_{c} of the crystals equals the increase in crystal mass. Substitution of this value into the mass balance equations allows one to calculate the amount of solute left. Together with the equilibrium concentration a new value for the supersaturation is found and the loop, as depicted in fig. 22a, is completed. It will be clear that the accuracy of the numerical solution is increased if a smaller time interval Δt is chosen.

The numerical calculations involved in the method of moments can be easily implemented. This method can be applied successfully for describing dynamic continuous or batch (which is intrinsically dynamic) crystallisation processes with "straightforward" growth and nucleation kinetics. If, however, the crystallisation process shows size-dependent processes (e.g. SDG or GRD, breakage, attrition, dissolution, aggregation or agglomeration) the method of moments breaks down, because these processes cannot, or only in a very limited way, be incorporated (an example to incorporate aggregation was given by Seckler et al. [111]; Witkamp et al. [112] made an attempt to describe dissolution of crystals with this method). Furthermore, one cannot reconstruct, in general, the population density $n(x, t)$ from the time-dependent moments of a distribution, except for specific conditions [87].

Discretisation of CSD. The second method discussed here is the transormation of the CSD from a continuous into a discrete form. This alternative approach is characterised by writing the CSD as a set of classes containing the number (population density, length, surface area or volume can be used equally well) of crystals in a given size range of width Δx. This method has several advantages compared to the method of moments described previously: (i) the CSD is obtained directly instead of only derived characteristics (i.e. its moments); and (ii) size-dependent growth, dissolution of crystals, aggregation, classification, in fact any process involving size can be implemented much easier. Illustrative examples can be found in [76, 80, 82, 113–115].

The fact that this method yields the CSD is further advantageous since direct information is gained on, e.g. bimodality of the distribution, information which is lost with the method of moments. Besides, it is very easy to calculate directly any arbitrary moment of a known CSD.

The CSD can be represented as a plot of the number of crystals per unit of crystalliser volume $N_i(t) = N(x_i, t) = N(i\Delta x, t)$ with sizes between $i\Delta x$ and $(i+1)\ \Delta x$ for $i = 0,\ldots, i_{\max}$. A reasonable value for $i_{\max}$ is 500. The initialisation step requires information on earlier mentioned parameters at the beginning of this section and additionally on the maximum crystal size $x_{\max}$ or the number of classes $i_{\max}$ for a given Δx of, typically, a few microns. The initial CSD is fixed by giving the values $N_i(t = 0)$. For a clear liquid these values are obviously zero, and a primary nucleation rate J or another source of crystals, e.g. an influx, has to be incorporated to account for the appearance of crystals.

In the loop only the step describing the shift in CSD is essentially different from the one used by the method of moments. At any time t the change in the number of crystals N_{i+1} per unit volume in class $i+1$ is given by (disregarding any influx terms):

$$N_{\text{new},i+1} = N_{\text{old},i+1} + \text{growth}(i) - \text{growth}(i+1) - \text{outflux}(i+1), \tag{54}$$

with growth(i) (in $\# \text{m}^{-3}$) the number of crystals from class i entering class $i+1$ by growth; growth($i+1$) the number of crystals leaving class $i+1$ by growth, and outflux($i+1$) the number of crystals in class $i+1$ leaving the reactor (see also fig. 22b). For $i=0$, the number term growth(i) has to account for particles entering the volume by primary and/or secondary nucleation if it is assumed that these particles are produced at smallest size x_0. Thus, growth($i=0$) could read:

$$\text{growth}(i=0) = (J + B_0)\,\Delta t \tag{55}$$

A birth and death number term (also expressed in $\# \text{m}^{-3}$) can be incorporated into eq. (54) for particles generated or destroyed at sizes larger than x_0. The following assumptions are made with respect to the "growth" and "outflux" number terms:

$$\text{growth}(i) = N_i G\,\Delta t/\Delta x, \tag{56a}$$

$$\text{growth}(i+1) = N_{i+1} G\,\Delta t/\Delta x, \tag{56b}$$

$$\text{outflux}(i+1) = N_{i+1}\,\Delta t/\tau. \tag{56c}$$

The dimensionless term $G\Delta t/\Delta x$ describes the fraction of a certain size class which leaves one class to populate the next. In order to allow $G\Delta t$ to stay well below Δx, it is necessary to choose Δt sufficiently small. As an extension of this method, G can be written with an index i as well, which allows a dependence of growth rate on size x_i.

The above-described method was based on a discrete description of $N(x,t)$, but an alternative approach is to discretise the population density $n(x,t)$ into $n_i(t) = n(x_i,t) = n(i\Delta x,t)$ fractions and to solve the time-dependent population balance equation, not in its continuous form [see also eq. (33)]:

$$\partial n(x,t)/\partial t = -\partial[G(x,t)n(x,t)]/\partial x - Q(x,t)n(x,t)/V + B(x,t) - D(x,t), \tag{57a}$$

but in its discrete form in both the variables x and t according to the so-called *method of lines* visualised in fig. 22c [116–118]:

$$\partial n_i/\partial t = -[G_{i+1}n_{i+1} - G_{i-1}n_{i-1}]/(2\Delta x) - Q_i n_i/V + B_i - D_i, \tag{57b}$$

$$\partial n_i/\partial t = [n_i(t_{j+1}) - n_i(t_j)]/\Delta t, \tag{57c}$$

where $Q(x,t)$ (or Q_i) describes outflux ($Q>0$) or influx ($Q<0$) processes which include the possibility of an enhanced removal (and destruction) of fines and/or classified product as described for steady state situations in sections 6.2.2 to 6.2.4; $B(x,t)$ (or B_i) and $D(x,t)$ (or D_i) are the birth and death rate, respectively, expressed in $\# \text{m}^{-1}\text{m}^{-3}\text{s}^{-1}$ [70]. In this way the partial differential equation (57a) is transformed into a set of ordinary differential equations (57b) and (57c) that can be integrated with a standard first-order Euler approximation algorithm. The interval Δx is constant and given by $x_{i+1} - x_i$, where i ranges from 0 to i_{max}, and similarly for the discretisation of t into steps of $\Delta t = t_{j+1} - t_j$, with j ranging from 0 to j_{max}. At the edges of the intervals slightly adjusted schemes have to be used. Equation (57b) can be described mathematically in a matrix structure which is convenient to handle. It turns out that the first-order Euler integration scheme, which is in fact a quite "rough" integration method, is already sufficient to obtain an acceptable accuracy-to-computation-time ratio. This, of course, also depends on the chosen time and size intervals.

Besides the fact that the method based on discretisation of either $N(x, t)$ or $n(x, t)$ can be extended relatively simply to include size-dependent processes, also the effects of perturbations, like the temporarily addition of a growth inhibitor and failures or fluctuations in feed streams, on the CSD can be investigated. If the behaviour of the crystallising system is completely known, as far as J, B_0, G and its size-dependent rate processes are concerned, through previously performed experiments, this information can be used to construct control strategies for processes performed under different conditions. Often, a relatively simple process control on the basis of mass balance equations can be a valuable method to compensate for, e.g. (temporary) interruptions in one of the feed streams during precipitation, where several components are added simultaneously. The discretisation method is also suitable to obtain information on, e.g. secondary nucleation kinetics with respect to its dependence on supersaturation, power input and solid content of the slurry for a set of data obtained under given experimental conditions. This can be achieved by a nonlinear optimisation routine which minimises the difference between the experimental data of the measured $N(x, t)$ or $n(x, t)$ and the distribution as calculated by the model. The minimisation proceeds by adjusting a set of fitting parameters in B_0, comprising of e.g. the powers b, k and j of σ, ε and the total solid content M_T, respectively, [see eq. (8)] until convergence is reached, e.g. by the simplex method. It is important to take into account the physical relevance of of the values of the powers b, k and j in order to avoid convergence with respect to a local minimum rather than the absolute minimum of the parameter set.

6.4. *Scale-up of industrial crystallisers*

In order to produce at full scale at the lowest possible cost per unit of mass produced, large crystallisers are required. Generally, scale-up has to be performed on the basis of results obtained with laboratory-scale equipment. The work of Mersmann [119] gives information on how to calculate the crystalliser volume for a given crystalliser design starting with data determined by means of laboratory or even bench-scale crystallisers usually operated at MSMPR conditions. It requires measurement of (i) the metastable zone width for a certain induction time; (ii) the growth rate, whether it is volume diffusion or surface integration controlled; (iii) the secondary nucleation rate as a function of operating conditions; and (iv) the attrition rate of large crystals, not only in nonsolvents but in supersaturated solutions as well. Knowledge of these parameters allows one to calculate the crystalliser volume for a given mean crystal size to be obtained with it.

Scaling up of industrial crystallisers, however, is a procedure which cannot yet be regarded as a mature science, since too many features of it are still incompletely understood. Ideally, scaling up would mean that all characteristics of a produced crystalline substance, as obtained in a small-scale crystalliser, would be equal when the substance was produced in a full-size crystalliser.

In order to realise this, the following demands should be fulfilled in theory: (i) identical flow conditions for liquid and solid; (ii) identical supersaturation in all equivalent regions; (iii) identical initial seed sizes for seeded batch processes; (iv) identical magma densities; and (v) identical contact times between growing crystals and supersaturated liquors.

Dimensionless numbers are frequently applied in many scale-up processes, as a tool to maintain the same conditions in equipment of different geometries. The flow in a crystalliser can be characterised most simply by the Reynolds number (Re = ratio of inertial force to viscous force) and the Froude number (Fr = ratio of inertial force to gravitational force). In crystallisation processes, however, they are of limited use. This is a consequence of the fact that keeping Re and Fr simultaneously constant when changing either the stirrer diameter or the stirrer speed is not physically feasible.

Dimensional analysis enables one to determine a third dimensionless group, the Power number Po:

$$\mathrm{Po} = P/(\rho N^3 D_s^5) = f(\mathrm{Re}, \mathrm{Fr}), \tag{58}$$

where P is the power input (in W), ρ the density of the liquid, N the stirrer rotation rate and D_s the stirrer diameter. This number, which is a power-law function of Re and Fr, should be maintained at a constant level in order to maintain the same stirring conditions for different N, D_s and P (via the stirrer). The stirrer not only plays an important role with respect to the flow conditions for liquid and solid, but also affects the nucleation rate, which determines, together with the growth rate, the final CSD of the product. The major parameters for scale-up of secondary nucleation rate are (i) a constant power input ε; (ii) a constant stirrer speed N; and (iii) a constant stirrer tip speed v_t. This stirrer speed should at least suffice to keep all crystals in suspension (off-bottom criterion). Scale-up to a larger crystalliser volume is often done on the basis of a constant specific power input ε.

In section 3.1.2, the secondary nucleation rate B_0 was discussed and in eq. (8) an (empirical) expression was given, showing the dependence of B_0 on σ (or G), N (or ε) and M_T. Little attention, however, was paid to the proportionality factor k_N. From experimental work it turns out that this factor depends strongly on crystalliser design: type of impeller (turbine, propeller, number of blades, blade thickness, blade width), vessel diameter, distance between draft tube and impeller blade tips, in other words, the proportionality "constant" is a function of geometry and size of both crystalliser and stirrer. The behaviour of the secondary nucleation rate during scale-up is mainly determined by target efficiencies for collisions between crystals with the edges and widths of the impeller blades. In large crystallisers the contribution of crystal–crystal contacts dominates that of crystal–impeller contacts. Pohlisch and Mersmann [43] developed a first attempt to incorporate the characteristics of the stirrer (number of blades, pitch angle, blade thickness and width) into the proportionality constant k_N.

Contrary to results of other models (e.g. [24, 120]), Pohlisch and Mersmann [43] found B_0 to be proportional to $\varepsilon^{0.6-0.7}$ rather than to ε. Also for surface breeding a similar range for the power of ε was found [41].

For crystallisers of similar geometry (i.e. constant ratio of impeller diameter to draft tube diameter) and of size characterised by a dimension λ, the empirical relation given in eq. (8) can be extended with a factor λ^d. If we simply write $B_0 \sim \lambda^q$, the value of q can be determined on the basis of the three different criteria mentioned above, i.e. constant ε, N or v_t [66]. Results obtained in practice with this rather pragmatic approach are as follows. With a constant power input a slight increase of secondary nucleation is observed. With a constant tip speed and a slight decrease of secondary nucleation is observed. Scale-up on the basis of a constant stirrer rotation rate strongly enhances the secondary nucleation rate. In general, scaling-up problems are approached by using the constant power input method.

Scaling up is still more art than science, since each specific crystalliser configuration has its own difficulties due to the fact that the complex flow patterns are usually hard to quantify and to reproduce on another scale. Nevertheless, more detailed information can be found in [38, 43, 119, 121–124] and references therein.

It will, regardless of the type of equipment, remain essential in scaling-up procedures to assess the sensitivities of the crystallising system towards the factors which may change during scaling up in advance, in order to prevent undesirable surprises.

7. Batch crystallisation models

The most striking difference between batch and continuous crystallisation modes is that in batch crystallisation no steady state will be reached. Thus, contrary to continuous crystallisers, batchwise operation is intrinsically a dynamic process. The dynamic modelling procedures outlined in section 6.3 are therefore also well suited to describe batch processes.

7.1. Population, mass and heat balances during batchwise operation

7.1.1. Batchwise operation

In a vessel containing a clear mother liquor a supersaturation is created which gradually increases with time. At a certain moment, crystals appear due to primary nucleation (unseeded batch). Alternatively, seed crystals can be added. During growth of these crystals, more crystals may be formed by primary and/or secondary nucleation and finally a product is formed with either a wanted or unwanted CSD. After crystallisation, the vessel is emptied and a solid/liquid separation is performed.

During *seeded batch* operation, initially added seed crystals grow, thereby populating the larger size classes. The major problem during seeded batch operation is that the seed crystals have to be added at the right moment: if they are added too early the solution may not yet be (super) saturated and the crystals dissolve; when the seeds are added too late, primary nucleation might already have occurred and the, thus obtained, initial CSD will grow together with the seeds to an often bimodal CSD which usually does not satisfy the demands.

Seeded batch experiments with dry seed crystals often suffer from initial breeding (see section 3.1.2) due to dislodging of macroscopic fragments adhering to the dry seed crystals. These small fragments lead to a bimodal initial distribution which, at the end of the batch run, results in a broader CSD. Initial breeding can be avoided by adding the seeds to the crystalliser as a slurry, which is obtained by premixing of the dry seeds with an undersaturated solution, so that the smaller fragments are dissolved.

One of the most important objectives of a batch crystallisation process is to produce a product with a narrow CSD. After the addition of the seeds nucleation must thus be suppressed, since any subsequent nucleation broadens the CSD. This nucleation will be mostly secondary nucleation, unless the seeds are added too late and some primary nucleation has already occurred [125]. Also the addition of a too small amount of seed crystals, especially under controlled operation of the batch run, promotes the development of secondary nucleation, since the seed surface does not suffice to consume the created supersaturation [125]. Controlled operation is therefore focused on a control of the supersaturation. Secondary nucleation is suppressed when the supersaturation is kept at a constant, sufficiently low, level.

Since also SDG/GRD effects tend to widen the CSD, the quality of the seed crystals and their pretreatment may play a role in the final width of the CSD.

During *unseeded batch* runs, the moment at which primary (either homogeneous or heterogeneous) nucleation occurs, cannot be predicted. Nucleation occurs in an explosive way if a certain supersaturation level is exceeded. The crystals subsequently grow and the final CSD is obtained. The initial size distribution of the crystals will be rather monodisperse if the majority of the particles is formed at approximately the same time, while no progressive nucleation occurs afterwards. Progressive primary nucleation will be of no importance if the supersaturation directly after the nucleation explosion drops considerably since the occurrence of primary nucleation can be neglected at these more moderate values of the supersaturation. If size-dependent processes, like aggregation, are operative at only a minor level, this rather narrow distribution moves along the size axis in time thereby more or less preserving its shape. In practice, however, the distribution will broaden with time since new fines appear due to secondary nucleation, unless the supersaturation is kept at a constant, low level. Also SDG/GRD effects contribute to this broadening. Consequently, the d_{50} value of the final CSD will decrease.

During uncontrolled seeded or unseeded batchwise operation the supersaturation changes with time and so does the growth rate. The size of a crystal at an arbitrary time $t > 0$ is given by:

$$x(t) = x_0 + \int_0^t G(t)\,\mathrm{d}t = x_0 + k_G \int_0^t \sigma^g(t)\,\mathrm{d}t, \tag{59}$$

where the initial crystal size x_0 is zero for the unseeded case and where k_G is a kinetic coefficient and g the growth order. For each batch run, the time dependence of σ (and G) can be different since the time t of addition of the seeds to the crystalliser content at supersaturation $\sigma(t)$ as well as spontaneous primary nucleation are statistical events while the subsequent nucleation processes are also related to $\sigma(t)$. As a consequence, it will be very difficult to obtain reproducible CSDs at the end of each batch run.

Obviously, uncontrolled operations is not the most efficient way to perform batch crystallisation. The two statistical events cannot be circumvented, but any subsequent primary and/or secondary nucleation can be suppressed by following an appropriate evaporation or cooling strategy in order to keep the supersaturation at a constant, low level. This requires insight into the population and mass balances governing the crystallisation process and in the moment equations needed for dynamic simulation. The control strategies are part of section 7.2.

7.1.2. Population balance

A batch system has no net inflow (except for the addition of seed crystals) or outflow of crystals. During a semibatch run, a continuous feed enters the volume, but withdrawal takes place only at the end of the run. During evaporative crystallisation, the total operating volume also changes in time but here due to evaporation of the solvent. It is, therefore, more convenient to use $\tilde{n}(x,t) = n(x,t)V(t)$, instead of $n(x,t)$, for batchwise operated processes, since $\tilde{n}(x,t)$ takes changes in $V(t)$ into account.

The population balance equation for a (semi)batchwise operated crystalliser is, expressed in terms of $\tilde{n}(x,t)$ rather than $n(x,t)$, given by [87]:

$$\partial\tilde{n}(x,t)/\partial t + \partial[G(x,t)\tilde{n}(x,t)]/\partial x - B(x,t)V(t) + D(x,t)V(t) = 0 \quad [\# \, \mathrm{m}^{-1}\mathrm{s}^{-1}] \tag{60}$$

Equation (60) includes birth and death rate functions, $B(x,t)V$ and $D(x,t)V$, respectively, in order to cover size-dependent processes like aggregation, breakage and dissolution. The boundary condition with respect to size is introduced by defining a time-dependent nucleation rate, i.e., the number of nuclei produced per unit volume and time at size zero. This nucleation rate includes any progressive primary nucleation J as well as secondary nucleation B_0, and thus:

$$\tilde{n}(x=0,t) \equiv \tilde{n}_0 = (J + B_0)V/G \quad [\# \, \mathrm{m}^{-1}]. \tag{61}$$

To solve the population balance equation, the initial population density is also needed as a boundary condition with respect to time.

A simple solution of the population balance method can be analytically obtained for the following situation. Imagine a seeded batch run under conditions of constant supersaturation (constant G) and zero birth and death rate functions ($BV = DV = 0$). The population balance, eq. (60), then reduces to:

$$\partial\tilde{n}(x,t)/\partial t + G\partial\tilde{n}(x,t)/\partial x = 0 \quad [\# \, \mathrm{m}^{-1}\mathrm{s}^{-1}]. \tag{62}$$

The general solution of this, particularly simple, partial differential equation is given by:

$$\tilde{n}(x,t) = f(y) \quad [\# \, \mathrm{m}^{-1}], \tag{63}$$

where f is any function of $y \equiv x - Gt$. For any given initial population density $\tilde{n}(x,t=0) \equiv \tilde{n}_0(x)$, the population density varying with size and time is easily found from the substitution $x \rightarrow y = x - Gt$ in the expression for $\tilde{n}_0(x)$. The solution given by eq. (63) shows that the crystal sizes are simply shifted in time; the absolute value of the shift at time t is determined by the product Gt. Of course, this example is a very unrealistic situation. A more realistic situation of seeded, batchwise operated processes, including a birth rate function $B(x,t)$ which varies with the total surface area of the crystals present in the crystalliser batch, was provided by Pan and Mersmann [126], who solved this problem analytically by application of the population balance method.

7.1.3. *Mass balance and moment equations*

The mass balance equation for batch operation is determined by the rate of change of the solute concentration which matches the rate of change of the mass of the solids:

$$\mathrm{d}(cV)/\mathrm{d}t + \mathrm{d}(M_\mathrm{T}V)/\mathrm{d}t = 0 \quad [\mathrm{kg\,s^{-1}}] \tag{64}$$

where $M_\mathrm{T}V$ is the total mass of solids in suspension and cV the total mass of dissolved solute. For an evaporative class II system, the time derivative of cV becomes $c\mathrm{d}V/\mathrm{d}t$ (T is constant), whereas for a cooling batch system this is $V\mathrm{d}c/\mathrm{d}t$ (V is constant). In the latter case, $\mathrm{d}c/\mathrm{d}t$ can be expressed in terms of $\mathrm{d}T/\mathrm{d}t$ if c as a function of temperature T is known (for $\mathrm{d}c/\mathrm{d}t = [\mathrm{d}c/\mathrm{d}T][\mathrm{d}T/\mathrm{d}t]$). Semibatch operation requires an additional term describing the input of solute which equals $Q_\mathrm{in}c_\mathrm{in}$ where c_in represents the solute concentration.

The time-dependent moments for batch and semibatch operation, which are needed to calculate changes in V from B_0 and G are given by:

$$\mathrm{d}(N_\mathrm{T}V)/\mathrm{d}t = (J + B_0)V \quad [\#\mathrm{s^{-1}}], \tag{65a}$$

$$\mathrm{d}(L_\mathrm{T}V)/\mathrm{d}t = N_\mathrm{T}VG \quad [\#\mathrm{m\,s^{-1}}], \tag{65b}$$

$$\mathrm{d}(A_\mathrm{T}V)/\mathrm{d}t = 2k_\mathrm{a}L_\mathrm{T}VG \quad [\#\mathrm{m^2s^{-1}}] \tag{65c}$$

$$\mathrm{d}(M_\mathrm{T}V)/\mathrm{d}t = 3\rho_\mathrm{c}(k_\mathrm{v}/k_\mathrm{a})A_\mathrm{T}VG \quad [\#\mathrm{m^3s^{-1}}] \tag{65d}$$

The initial moments of the distribution, together with eqs. (60), (61), (64) and (65) and the appropriate equations for J, B_0 and G determine the new moments of the CSD of the product. In general the model must be solved numerically. The difference between $M_\mathrm{T,new}$ and $M_\mathrm{T,old}$ equals the increase in mass. Substitution of this value into the mass balance equation allows one to calculate the amount of solute left. Together with the equilibrium concentration a new value for the supersaturation is found and the next step in the process can be calculated.

7.2. *Control strategies of batch processes*

In principle, batch crystallisation is an adequate method to obtain a coarse product with a narrow CSD, if the initial CSD could be controlled and if the nucleation in time could be suppressed. In this section some control strategies will be discussed for both unseeded and seeded batchwise operations based on modelling procedures which either apply the method of moments or just solve the population balance equation. Most of the examples which are discussed, apply the former method.

7.2.1. Seeded batch

The strategy for a narrow CSD is, as was said before, based on operating the entire batch process at constant supersaturation (thus G is constant, disregarding SDG/GRD) which is chosen at such a level that at least primary, but preferably also secondary, nucleation will not occur: $d(N_T V)/dt = 0$ [eq. (65a)]. So, $N_T V$ is constant and equal to the total number of seeds N_0 (# and not $\# m^{-3}$) added to the vessel at $t = 0$. For an evaporative system the four moment equations, given by eq. (65), combined with the mass balance [eq. (64)] merge into [87]:

$$c d^3 V/dt^3 + 6\rho_c k_v G^3 N_0 = 0 \quad [kg\,s^{-3}], \tag{66}$$

which is simply integrated to:

$$c[V(t) - V(0)] = p_1 t + p_2 t^2 + p_3 t^3 \quad [kg]. \tag{67}$$

The integration constants p_1, p_2 and p_3 can be determined from the initial conditions imposed by the initial values of the first and second derivative of V. So, in the case of a seeded batch, operated at constant supersaturation and zero nucleation rate (both primary and secondary), the corresponding evaporative program is given by the time-derivative of eq. (67), which equals the net rate of evaporation of the solvent (m^3 solvent per unit of time), as a function of time for a chosen low value of G. This rate must be low at the start because of the small crystal surface available for growth, and high at the end of the process to keep G constant.

For a cooling batch run the variables c and V in eq. (66) may be inter-changed leading to an expression for $c(t)$ similar to the one for $V(t)$ in eq. (67). In case of cooling, Nývlt et al. [127] derived the temperature program for the case of a seeded batch operation under the assumptions of a constant supersaturation and zero nucleation rate. The expression, derived for the addition of a fixed number of seeds having zero initial size, area and volume, reads:

$$[T(0) - T(t)]/[T(0) - T(t_{end})] = (t/t_{end})^3, \tag{68}$$

where $T(t)$ defines the temperature program starting at $T(0)$ at $t = 0$ with a supersaturated solution and finishing the batch run at $t = t_{end}$ at the corresponding temperature $T(t_{end})$. Equation (68) assumes a linear dependence of c on temperature T. In these cooling programs where a narrow CSD is strived for, the decrease in temperature must be slow at the start and fast at the end of the batch process.

In actual operation nucleation cannot be totally suppressed. Upon including primary nucleation (occurring in combination with seeding), under the assumption that $d(N_T V)/dt = JV$ = constant and supersaturation is constant, the derivation by Baliga [128] – actually Baliga solved the set of equations for constant secondary nucleation B_0 – can be used to solve this problem. For a seeded cooling batch run operated at constant supersaturation assuming a constant nucleation rate (i.e. $dN_T V/dt = (J + B_0)V$ = constant) a similar expression is found as given by eq. (68) except for a power of four instead of three of the ratio t/t_{end}. This is due to the fact that the general solution of the set of differential equations, eqs. (64) and (65), contains an additional t^4 term in eq. (67). Again, this result was derived for the addition of a fixed number of seeds with zero initial size, area and volume. However, a system with a constant nucleation rate is not a very realistic situation.

A more likely situation during cooling and evaporative modes of crystallisation is that secondary nucleation contributes to the development of new crystals during the course of the process. Putting $B_0 = k_N G^i \varepsilon^k M_T^j$ according to eq. (8), the following, coupled differential equations are obtained for constant supersaturation:

$$d^4(M_T V)/dt^4 = 6\rho_c k_v k_N G^{i+3} \varepsilon^k M_T^j V \quad [\mathrm{kg\,s^{-4}}], \tag{69a}$$

$$d(M_T V)/dt = -d(cV)/dt \quad [\mathrm{kg\,s^{-1}}]. \tag{69b}$$

In general, eq. (69) cannot be solved analytically. Numerically, e.g., by application of the methods outlined in section 6.3, an expression can be obtained for the rate at which solvent must be evaporated (dV/dt) or at which the crystalliser should be cooled [dT/dt deduced from dc/dt and the solubility curve $c(T)$].

Recently, Nývlt [129] analytically modelled batch crystallisation according to the salting-out principle. An expression was derived describing the rate of addition of the second solvent for a seeded salting-out batch process under conditions of constant supersaturation and constant nucleation rate. The initial size, area and volume of the fixed number of added seeds were chosen to be zero.

The assumption that the initial size, area and volume of the seeds are zero is, for most practical cases, not very realistic. For modelling purposes it is necessary, therefore , to know the initial size distribution explicitly, or derived properties like the first four moments, which serve as the boundary conditions for the (analytical or numerical) solution of the population and mass balances in order to obtain either the evaporation or the cooling rate as a function of time. During seeded batch, the initial size distribution can be measured prior to the addition of the seeds to the crystalliser.

7.2.2. *Unseeded batch*

For unseeded batch operation the set of differential equations, governing the process and given in section 7.1 [in particular eqs. (64) and (65)], are still applicable, but the initial conditions required to solve the problem, either analytically or numerically, differ from those applied for seeded batch crystallisation. During unseeded batch, crystals appear in the initially clear, supersaturated solution due to homogeneous or heterogeneous primary nucleation. After this "shower" of prímary crystals, supersaturation drops considerably and the subsequent nucleation mechanism is of the secondary rather than primary type.

The condition of operating at a constant, low supersaturation, imposed by an evaporative or cooling program, cannot be fulfilled here since this will lead to a mostly inacceptable long induction period prior to nucleation and to progressive primary nucleation afterwards, and not to an outburst of nuclei at the start. A nonconstant supersaturation implies that the primary and secondary nucleation rate J and B_0 have to be used with their explicit dependence on σ [see eqs. (6) and (8)]. Hence, a process control strategy has to be deduced from dynamic simulation of the batch process. As a consequence, the problem has to be treated numerically. Since nucleation starts from a clear liquid, the initial values of the first four moments (number, size, area and volume) are all zero, which is a rather inconvenient starting point for a numerical approach. Moreover, primary nucleation is a purely statistical event, which is hard to model adequately. In reality, however, an initial distribution of crystals exists, although it cannot be measured. Therefore, a function for the initial CSD can be selected [87] and the simulation procedure can be started. Another procedure is to make an assumption concerning the shape of the initial distribution (e.g. initial mean size and initial coefficient of variation, CV) or concerning its first four moments. Once this assumption has been made, the problem in fact "reduces" to the seeded batch situation, where it is here further assumed that supersaturation is so low that secondary nucleation becomes the prevailing mechanism. Obviously, the specific choice of the initial size distribution will affect the finally obtained CSD.

An example of such an initial distribution of primary crystals could be a distribution of nuclei with sizes approximately equal to the critical nucleus size r_c corresponding to a specific supersaturation. A possible subsequent procedure is then to use this distribution as initial condition for the seeded batch modelling described in section 7.2.1, under the assumption of, e.g. constant supersaturation and constant secondary nucleation.

7.2.3. Batch crystallisation with fines destruction

Broadening of the CSD during batch operation is caused by size-dependent processes like aggregation, SDG/GRD, breakage, progressive primary and/or secondary nucleation. Destruction of fines is usually a good method to improve the mean size as well as the width of the distribution given by the coefficient of variation, CV, defined in section 5. This can be achieved (i) by introducing one or more periods of undersaturation during the batch cycle, or (ii) by withdrawal of fines from the crystalliser, by dissolving these fines and by returning the clear liquor to the vessel. The former method, although relatively easy, has the disadvantage that the larger crystals (partly) dissolve as well. The latter method, on the other hand, requires a classifying system, a fines pump and a dissolution vessel. It has the advantage that fines are removed and subsequently dissolved in a continuous manner and has the possibility to control the product quality on-line. The continuous removal/dissolution can be readily modelled in terms of the population balance given in eq. (60) analogously to the MSFD crystalliser in the continuous crystallisation case (see section 6.3.1.).

8. *Industrial crystallisers*

The previous sections reviewed several methods of crystallisation as well as the theoretical background of kinetics (e.g. primary and secondary nucleation, growth, aggregation) and modelling of the rates of these processes. This section will be devoted to selection of a certain crystallisation method, and of appropriate crystallisers. The list of available types of crystallisers has no pretence to be complete and only serves as a means to describe important features of and differences between several crystallisers some of which are commonly used in industry. For more detailed equipment descriptions we refer to [38, 119, 123, 130–133] and references therein.

8.1 Selection of crystallisation method

As already pointed out in section 2.2, there are basically five ways of establishing a supersaturation required for crystallisation from solution: cooling, evaporation, flash evaporation, precipitation and salting out.

Evaporative and cooling crystallisation are the most frequently applied methods in large-scale industrial plants, since salting out – although it can be a reasonable alternative and is a widely used method, e.g. in the production of proteins – requires an auxiliary phase (usually an alcohol) that has to be recovered afterwards. Precipitation is only applied for the production of slightly soluble salts. Precipitation and, to some extent, also salting out require intensive mixing of the two feed streams to avoid locally high supersaturations, which puts restrictions on the suitability of a specific crystalliser for this type of application.

For industrial purpose, the choice between evaporative and cooling crystallisation is dictated by several criteria: (i) the energy demand per unit amount of crystals produced; (ii) the mean crystal size; (iii) encrustation; and (iv) equipment costs. These criteria are strongly influenced by the material to be processed and by the solvent. Therefore, each of the different types of crystalliser has features that must be suited to the material as well as to the specific product demands.

The maximum attainable crystal size appears to increase with the solubility of the compound, at least for solubilities up to $\sim 10\,\mathrm{mol\,l^{-1}}$ (see fig. 23, after [134]). Above this value, the mean crystal size levels off at ~ 1 mm. Apart from the solubility, also the

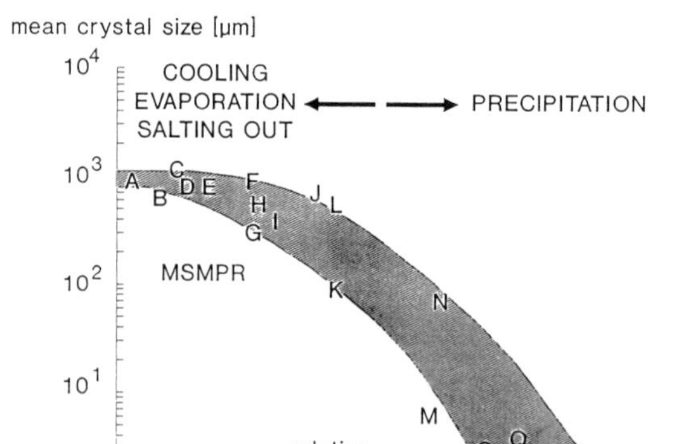

Fig. 23. The mean attainable crystal size as a function of the relative supersaturation and the solubility for seventeen systems (after [134]).

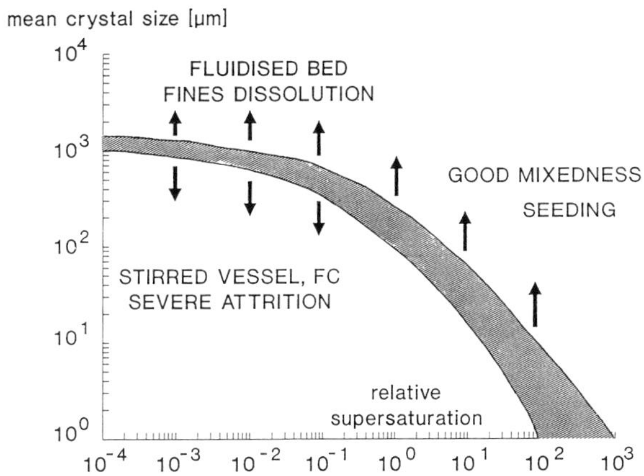

Fig. 24. Influence of crystalliser type and hydrodynamics on mean crystal size (after [134]).

hydrodynamics of the slurry and the type of crystalliser may affect the mean crystal size, as is indicated in fig. 24 (after [134]).

Locally, supersaturations can become quite high, e.g. at the cooling surface in a cooling crystalliser or in the boiling zone of an evaporative crystalliser. This may induce encrustation or scaling. In the case of cooling this can be avoided by reducing the temperature difference over the cooling surface and by increasing the flow rate in the heat exchanger tubes. During evaporative crystallisation encrustation often occurs at the vessel walls of the boiling zone thus reducing the effective evaporative area. The stirrer shaft may also suffer from scaling within this region leading to an imbalance of the stirrer. Fragments of scale deposits can break away from the encrusted layers causing blocked pipe lines. This emphasises the need to avoid, or at least reduce, the occurrence of scaling. In some cases continuous spraying of the vessel walls and the stirrer shaft just above the boiling zone with water or an undersaturated feed is applied to reduce scaling.

8.2. Selection of crystalliser

After a method of crystallisation has been chosen which should satisfy the product requirements, the equipment has to be selected or designed which meets the desired product demands. In this section the attention is mainly focused on evaporative crystallisers, although similar equipment can often be used for cooling crystallisation by replacing the evaporation unit by a cooling device.

8.2.1. Solar ponds

One of the simplest applications of evaporative crystallisation is by using solar ponds. In this way NaCl can be gained from seawater at relatively low cost. This process lacks, of course, any control of production rate. Obviously, higher yield rates can be obtained by heating the solution by means of heat exchangers. To improve heat transfer and to keep the crystals suspended, the vessel should be equipped with a stirrer or a pump. The disadvantage of a stirred open vessel is that a comparatively high energy demand for vaporisation of water at atmospheric pressure is required. Therefore, industrial evaporative crystallisers are usually closed and operated at reduced pressure. The stirrer speed and, related to this, the nucleation rate due to crystal/impeller collisions as well as the energy demand for mixing can be significantly reduced by use of a draft tube, which changes the hydrodynamic flow pattern.

8.2.2. *Thermo-syphon crystalliser*

The simplest evaporative crystalliser operating sub-atmospherically, is the thermo-syphon crystalliser, fig. 25. A schematic visualisation of a thermo-syphon crystalliser is also depicted in fig. 25. It consists of two plugflow reactors (PF) and two mixed vessels. PF I represents the draft tube through which the supersaturated crystal slurry flows downward into a poorly mixed vessel where the fresh feed (F) is added. At the bottom, large crystals settle and are removed as a dense product slurry (P). The rest of the slurry is transported upwards by natural convection through the heat exchanger tubes, represented by PF II, where reheating takes place. Within these tubes, the supersaturation will decrease partly because of continued crystal growth and partly due to the usually increasing solubility with temperature. PF II may serve as an internal fines destructor if the solution concentration falls below its saturation level, which is beneficial both for increasing mean crystal size and for reducing the width of the CSD. The less supersaturated or even undersaturated solution now enters the top section of the crystalliser where mixing will be rather intense due to boiling. Part of the solvent is removed as vapour (V) and, consequently, supersaturation increases. Within this volume where the crystal–crystal collision rate is the largest and where the highest supersaturation is built up, most of the secondary nucleation takes place. The cycle is completed with the entrance of the supersaturated slurry with the newly born nuclei into the draft tube (PF I).

Although mixing through natural convection seems to be elegant, it limits the heat transfer and, related to this, the production rate per unit of crystalliser volume. Moreover, the critical size of crystals which can be transported by the natural flow up into the boiling zone, where both supersaturation and growth rate are high, may remain below the desired product size. This implies that the maximum attainable crystal size is, in a way, fixed by the process of natural convection.

8.2.3. *Stirred draft-tube baffle crystalliser*

The stirred draft-tube baffle (DTB) crystalliser is basically similar to the previously described thermo-syphon crystalliser but now both mixing and heat exchange are improved by means of a slowly rotating stirrer. First, its application as an evaporative crystalliser will be discussed. If the stirrer shaft enters the crystalliser at the top, an

---

Fig. 25. Evaporative crystalliser (thermo-syphon) consisting of a draft tube and internal heat exchanger. The corresponding flow diagram is also shown.

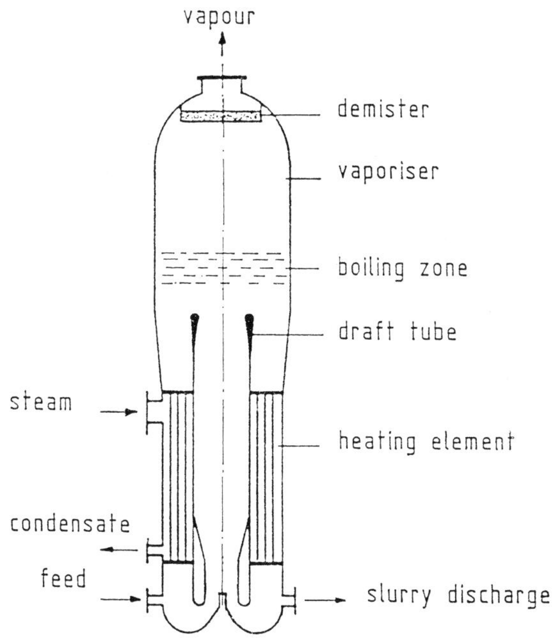
vapour
demister
vaporiser
boiling zone
draft tube
steam
heating element
condensate
feed
slurry discharge

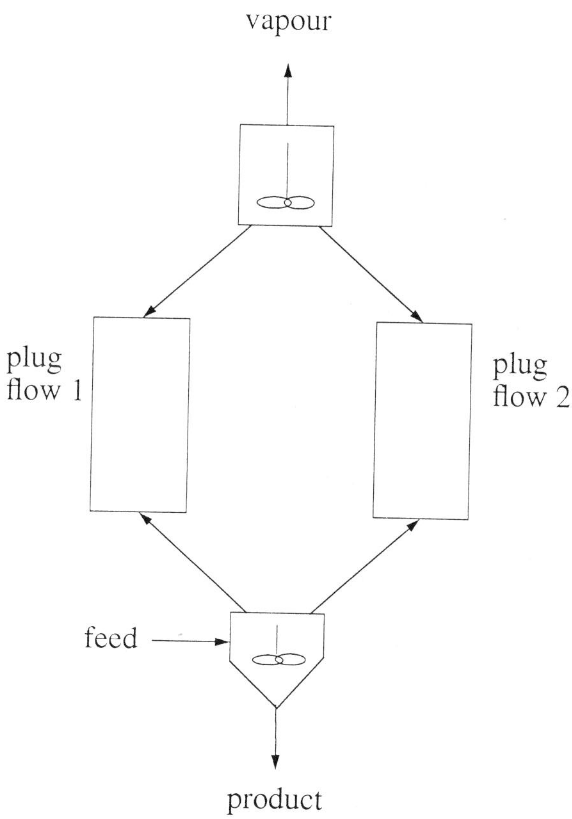
vapour
plug
flow 1
plug
flow 2
feed
product

adequate sealing is required to maintain vacuum. Since encrustation particularly occurs at the walls in the boiling zone, also the stirrer shaft in this region is a likely candidate. In spite of the rather low rotating rate, this may imbalance the shaft because of its length. This encrustation problem does not exist in crystallisers where the stirrer shaft enters at the bottom, and thus remains below the boiling zone. In this case sealing of the shaft is, however, even more complicated, since it is in contact with the supersaturated solution and crystals may enter or even grow into the sealing.

Although the stirred draft-tube crystalliser has the stirring rate as an extra degree of freedom compared to the thermo-syphon crystalliser, the hydrodynamics of both crystallisers are in essence similar. The advantage of the DTB crystalliser is only that it allows for an externally adjustable flow rate in the heat exchanger pipes, and thus for a better heat transfer, as well as for an enhanced stirring intensity in the bottom region.

The flexibility of this DTB crystalliser design with an internal heating system remains, however, limited, since the size of the heater (the size of the tube plate) and the boiling surface area are coupled because both functions have to be accommodated in the same vessel. Another drawback is that the total crystal slurry is circulated through the impeller zone, which could cause considerable attrition by crystal–impeller collisions. Special attention must therefore be given to the impeller design to minimise attrition.

The draft-tube baffle crystalliser design, shown in fig. 26, separates the crystalliser from the heat exchange facilities. Fines removal and destruction can be realised by installing a skirt baffle. In this way a settling zone is created which facilitates the removal of a fines stream. In the design with a separate heat exchanger this liquid passes, after mixing with the feed stream, through a normal tube heat exchanger outside the crystalliser and returns to the crystalliser vessle for partial evaporation. Upon heating, the fine particles may dissolve and the heater thus serves as an external fines destructor. Fines destruction causes a larger mean crystal size to develop in the crystalliser. To ensure that the fines up to a desired critical size will indeed be destroyed, the fines stream can be diluted by the solvent, e.g. by water or by dilute feed before mixing with the (under)-saturated feed stream.

At the bottom of the crystalliser, product removal occurs via an elutriation leg. Simultaneously, part of the feed (a clear liquid flow) enters the elutriation leg at the bottom and flows counter-currently to the crystal mass. This causes a classification of larger crystals which are removed as a dense slurry from the side of the elutriation leg. The slurry is transported, e.g., to a filtration unit. The filtration liquid can be returned to the crystalliser. In theory, the crystal yield could be 100% of the solute in the feed flow, assuming no losses through the vapour flow. Generally, the feed contains impurities which leads to a deviation from this ideal situation. If these do not evaporate, a gradual increase of the impurity concentration will build up within the crystalliser that may be detrimental to the performance of the crystalliser in two ways: (i) different CSD with more fines present because of a change in growth and nucleation rates; and (ii) a lesser purity of the crystals. To avoid these negative influences, the purge of mother liquor has to be sufficient in order to maintain an acceptable level of impurities. The DTB crystalliser can also be applied in both designs as a cooling crystalliser by using the heat exchanger as a cooling system. In this application the fines in the stream from

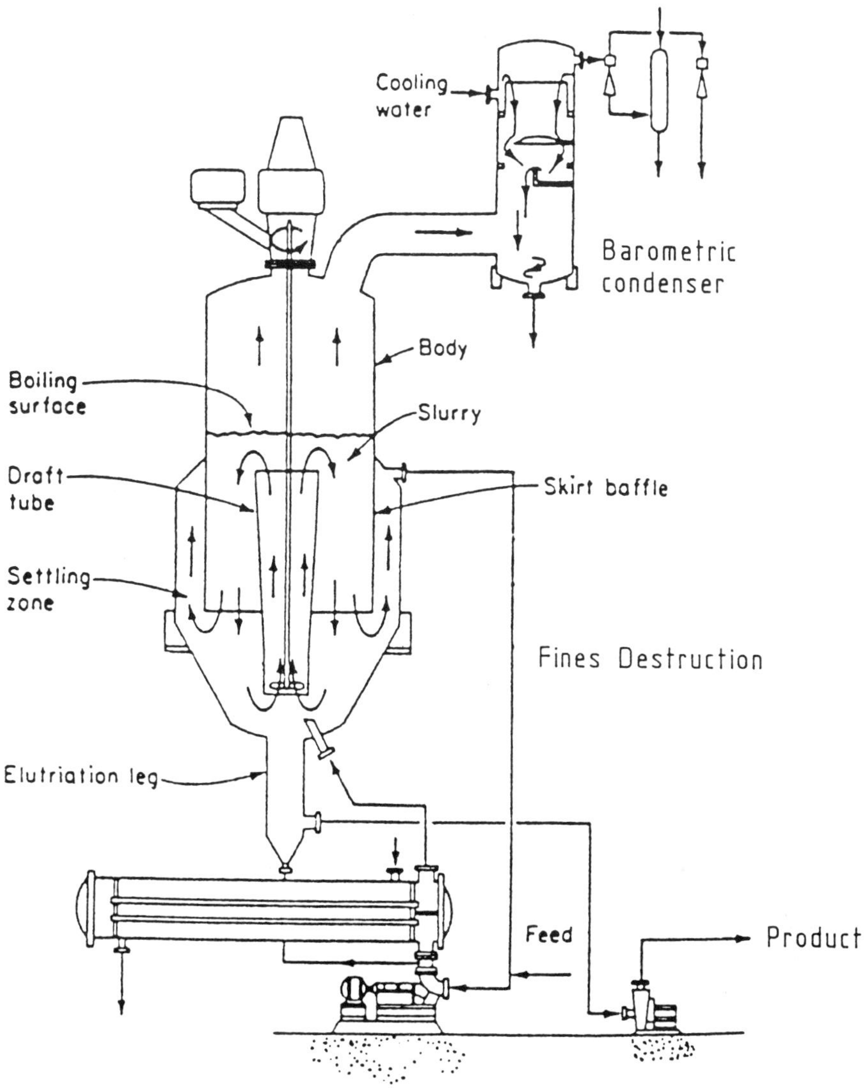

Fig. 26. Set-up of a draft-tube baffle crystalliser with an elutriation leg for product classification (after [123]).

the annular settling zone are destroyed by heating before the stream is returned with the feed stream to the cooler.

8.2.4. Forced circulation crystalliser

The forced circulation (FC) crystalliser is the most widely used crystalliser. It is depicted in fig. 27 for evaporative (vacuum) crystallisation, and comprises of two separate compartments which can be designed independently for crystallisation and heat input.

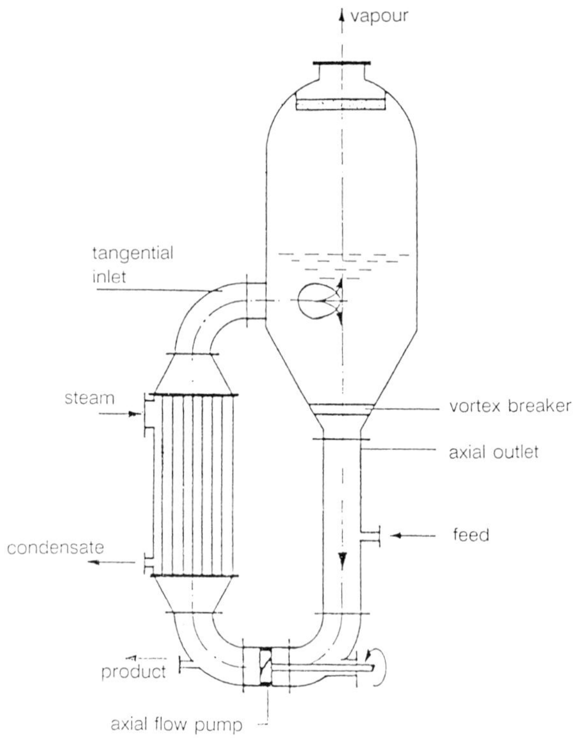

Fig. 27. A forced circulation crystalliser with a tangential inlet (after [38]).

The crystalliser body should be large enough for vapour release, to accommodate the boiling zone and to maintain a sufficiently high liquid level for retaining the growing crystals within the vessel until the supersaturation is consumed. A slurry pump is required to circulate the slurry through these two compartments. The problems outlined in section 8.2.3 concerning the sealing and imbalance of the stirrer shaft are of no importance here. On the other hand, the attrition and nucleation rate in an FC pump will usually be larger than with a DTB stirrer although special pumps can be applied to minimise this negative influence on mean crystal size. This crystalliser comes close to a mixed suspension, mixed product removal system (see section 6).

The mechanical construction of the FC crystalliser requires a more sophisticated and hence a more expensive support system because the centre of gravity is located outside the apparatus, due to its asymmetrical design. The construction has to withstand the vacuum as well as temperature changes, hence it cannot be too rigid.

The hydrodynamics in the FC crystalliser differ from those in the DTB crystalliser with the external heating system since here the reheated slurry directly enters the boiling zone, i.e. the region of highest supersaturation. Feeding of the boiling zone can proceed in three ways: (i) vertically; (ii) tangentially; and (iii) radially (the tangential possibility is shown in fig. 27). Each method affects the performance of the crystalliser in several ways [135]. The flow patterns induced by the incoming slurry flow affect the residence time in the boiling zone for crystals of various size. This classification will strongly influence the final size distribution, because it affects the nucleation rate and the average crystal growth rate which shows an apparent size dependency due to the hydrodynamic characteristics. An advantage of the FC crystalliser is that it is the least expensive vacuum crystalliser especially when substantial evaporation is required.

The FC crystalliser can also be used as a cooling crystalliser, especially when flash cooling suffices to produce crystals. In that case heat exchange is not even needed, only a circulating system.

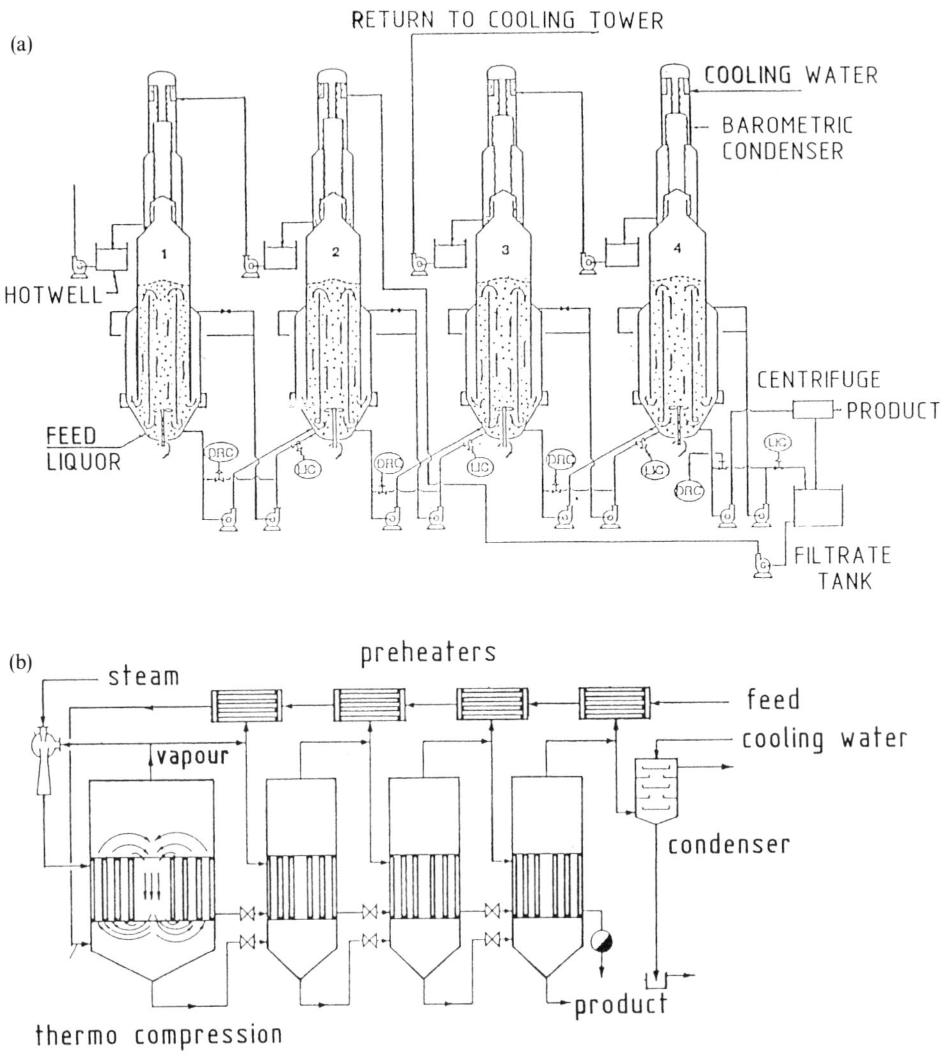

Fig. 28. (a) Cascade of four DTB crystallisers for KCl crystallisation. (b) Cascade of four FC crystallisers for NaCl crystallisation.

A major disadvantage of an FC crystalliser compared to a DTB crystalliser is encountered in scale-up. Apart from toroidal circulation for tangential inlet, vortexing also occurs, which does not occur in a DTB crystalliser.

8.2.5. Cascade of evaporative crystallisers

In order to minimise operating (mainly energy consumption) and investment costs, the evaporative DTB and FC vacuum crystallisers discussed above can be operated in series. Usually, a cascade of three or four crystallisers are applied, each controlled at lower pressure and temperature. In the case of, e.g. KCl, where the crystal yield is obtained by flash cooling (see fig. 28a) the vapour is mainly used for heating of the feed stream. In the case of, e.g. NaCl, where the crystal production only follows from vacuum evaporation of the water, the heat of condensation of the vapour is partly reused to evaporate the solvent in the next stage and for countercurrent heating of the feed stream (see fig. 28b). In principle, the total energy consumption is minimal for an infinite number of crystallisers in series, but, in practice, the optimal number is determined by the effects of cascade operation on product quality (CSD and crystal purity) and investment costs as well.

8.2.6. Cooling disc crystalliser

A cooling disc crystalliser (fig. 29) can be considered as a cascade of cooling crystallisers although they are less well separated compared to the cascade of evaporative crystallisers. The cooling disc crystalliser consists of a trough containing a number of hollow discs, mounted horizontally, through which a cooling liquid is circulated. The discs build up a set of compartments containing part of the suspension which is enabled to

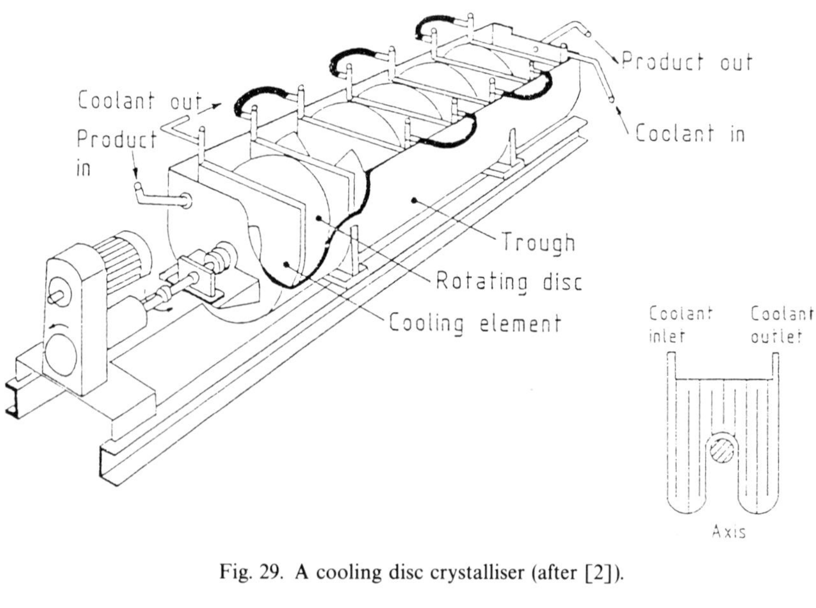

Fig. 29. A cooling disc crystalliser (after [2]).

flow from one compartment to the next through a segmental opening present in each disc. The suspension flow is countercurrent to the coolant. A gentle agitation of the slurry is provided by slow rotation of the discs. Applications of the cooling disc crystalliser for fractional crystallisation are described in [2].

8.2.7. Oslo crystalliser

The Oslo crystalliser is specially designed to produce large and uniformly sized crystals. Supersaturation is created in a vaporiser chamber after which the liquid comes down and enters a densily packed fluidised bed at the bottom of the so-called suspension chamber (see fig. 30). The supersaturation is consumed on its way up through the bed and crystalline product is removed continuously from the bottom of the classified bed. At the top of the bed the crystals are settled, and only the fines leave the crystalliser with the exhausted mother liquor. These fines could be redissolved with water before mixing this liquor with the feed stream. The mixed streams are raised in temperature in a heat exchanger which is part of the recirculation line.

A major problem in the design of the Oslo crystalliser is the close clearance through which the supersaturated liquid enters the fluidised crystal bed at the bottom of the

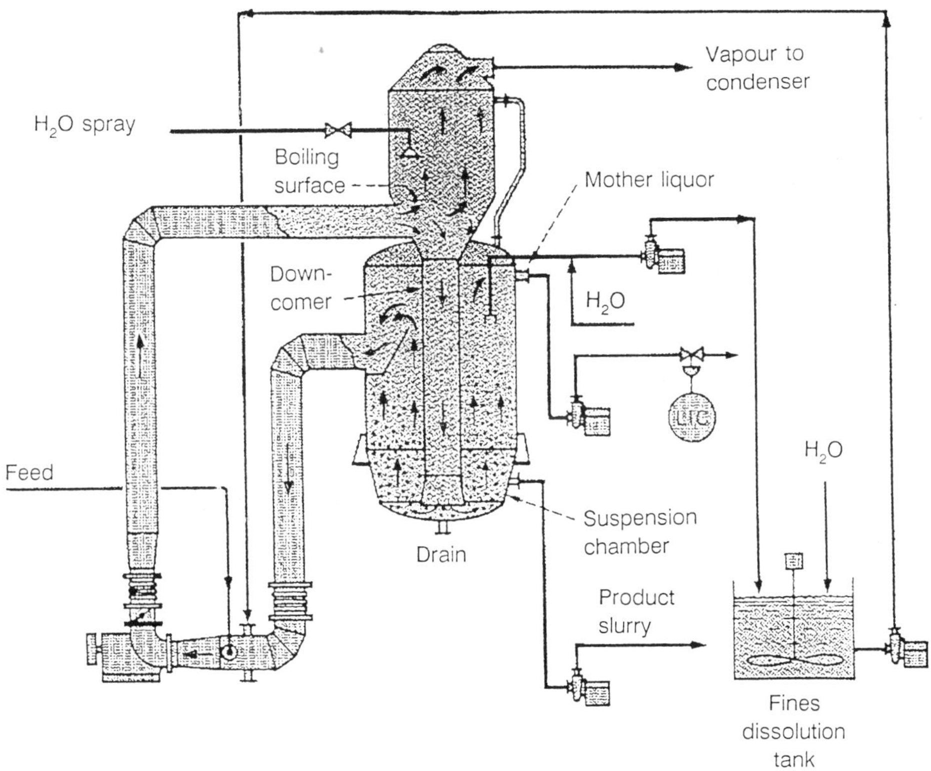

Fig. 30. Set-up of an Oslo or growth crystalliser (after [123]).

reactor. Such close clearance is needed to let the incoming liquid agitate the crystals into a fluidised bed. Together with the recirculation rate this clearance determines the local linear suspension velocity. Partial blockage due to crystal growth (especially in the case of a slightly excentric positioning of the downcomer) or due to larger agglomerates of crystals falling down from the draught tube will disturb the fluidisation of the bed and consequently the performance of the crystalliser. To minimise occasional production stops for cleaning of the crystalliser, this velocity of the incoming liquid has to be maintained above a certain lower limit.

The mean residence time of the crystals can be increased considerably in the Oslo crystalliser compared to the FC and DTB crystallisers. Besides, attrition of crystals is avoided, since no crystals are recirculated through the pump. Thus, if operated correctly, the Oslo crystalliser yields a much coarser product – a few mm's is no exception. For this reason it is sometimes also referred to as the "classifying growth"-type crystalliser [132].

8.3. Solid–liquid separation

Once the crystallisation process is completed, the liquid has to be removed from the solid material. Solid–liquid separation processes can be classified according to the principles involved [136]. Depending on whether the particles are allowed to move freely (due to acceleration fields) while the liquid is constrained or vice versa, we either have sedimentation or filtration. For sedimentation a density difference between solids and liquid is necessary, in the latter case such a difference is not required.

In this section sedimentation and filtration will be specified in some more detail since these are the most commonly applied techniques for solid–liquid separation. These topics are extensively discussed in Svarovski [136] from a theoretical, practical and design point of view.

8.3.1. Sedimentation

Solid–liquid separation by means of sedimentation can be further subdivided into gravitational and centrifugal sedimentation. Gravity sedimentation uses the gravitational field for settling of particles. Obvious parameters describing this process are particle size, particle density and liquid viscosity, but also particle shape and orientation, distortion of deformable particles, mutual interference at high concentrations, vicinity of the container wall and convection influence the performance of this technique.

Particles of the order of several microns settle very slowly. The addition of a flocculent or coagulant establishes the formation of relatively large flocs which settle more rapidly. However, flocs at the bottom will be compressed by the weight of others that settle upon them. This leads to entrapment of considerable amounts of liquid resulting in a varying density of the sediment.

Centrifugal sedimentation comprises the techniques which use a hydrodynamic flow to separate a solid fraction from a liquid. This flow can either be a result of the flow itself entering the separator or results from rotating parts within the equipment. An example of the first type of separation unit is a hydrocyclone, whereas a centrifuge is an example of the latter type. In a hydrocyclone, a vortex is produced by feeding the two-phase flow through a tangential inlet. This is accompanied by high velocity gradients, causing shear which may break possible agglomerates. It is a straightforward separation technique which is most usefully applied for particles in the 5 to 200 μm size range. An improved efficiency can be reached either by combining several cyclones in a cascade or by circulating the feed several times through the cyclone. In a sedimenting centrifuge, the flow is fed through a rotating bowl. The shearing effect is of minor importance here compared to that in a hydrocyclone. Generally, high separation efficiencies are achieved with sedimenting centrifugation.

8.3.2. Filtration

During filtration the liquid moves freely through a porous medium, whereas the particles are retained by it. The liquid is forced to move through the porous medium by applying either a vacuum at the opposite end of the filter, a pressure or centrifugal filters. Also gravitational forces can be used as a driving force.

Basically two types of filtration are available: cake filtration and deep bed filtration. The former refers to solids which are deposited in the form of a cake on the surface of a relatively thin filter, whereas the latter applies to deposition of solid material within the (thick) porous medium and cake deposition on the surface should be avoided. Surface filters are used for suspensions with concentrations of solids above $\sim 1\%$ by volume. Depth filters, on the other hand, are commonly applied for the separation of fine particles from very dilute suspensions ($< 0.1\%$ by volume). In this case the particles are smaller than the medium openings and they are attached to the medium by electrostatic and molecular forces.

Examples of filtration units are the rotary drum filter, the pan filter and the belt filter.

9. *Precipitation*

9.1. General aspects of precipitation

Precipitation is widely applied in the chemical industry to manufacture pigments, ceramic precursors, catalytic supports, pharmaceutical products, mineral salts and to separate valuable or noxious compounds from waste streams.

A precise and generally accepted definition of the term precipitation is not available [137], but often precipitation refers to a subclass of the field of solution crystallisation. Precipitation is recognised by one or more of the following characteristics: (i) low solubility of the crystallising compound; (ii) fast process; (iii) small crystal sizes; and (iv) irreversibility of the process [134, 137].

The way to achieve supersaturation for a sparingly soluble substance is in general by chemical reaction since the yield of such substances on cooling and evaporative crystallisation would be negligible. Precipitation is therefore often identified with reaction crystallisation, although many cases are known where reaction crystallisation shows more resemblance with evaporative and cooling crystallisation than with a precipitation process. This is, e.g. the case for the crystallisation of $(NH_4)_2SO_4$ from NH_3 gas and H_2SO_4.

The basic steps in any precipitation process are, just as with crystallisation, the creation of supersaturation, the generation of nuclei and the growth of these nuclei. Precipitation discriminates itself from crystallisation with respect to the low solubility involved, the use of more than one feed stream supplying the components to be precipitated and, directly related to this, mixing conditions. Contrary to crystallisation of intermediate to highly soluble substances where, due to the use of a one-feed stream, stoichiometric conditions are always present, precipitation often deals with either astoichiometric feed streams or local astoichiometry as a result of non-ideal mixing. This non-ideal mixing together with the low solubility of the precipitating compound, leads to high local supersaturations. This implies that *primary nucleation* plays a predominant role which leads to the appearance of a large number of crystals.

Although precipitation is characterised as a fast process, this does not refer to the *growth rate* of these crystals: in spite of the high supersaturations generally found, the kinetic coefficient, linking growth rate to σ, is relatively small because this coefficient is proportional to the equilibrium solubility which is generally several orders of magnitude smaller for precipitating compounds than for easily soluble substances. As a result of the high nucleation and the low growth rate, the crystals remain rather small ($1-10\,\mu m$). Therefore, secondary nucleation can be neglected (see section 3.1.2).

The small average size of the crystals present in large amounts combined with the high supersaturations and the often intensive mixing conditions, favour another size increasing mechanism, namely *agglomeration* of, in the first instance, loosely bound crystals which subsequently stick together due to the formation of crystalline bridges between the individual crystals leading to more stable aggregates. The occurrence of this process increases the average crystal size and may improve the downstream handling such as filtration. On the other hand, it can also have a negative effect concerning the purity of the final product because of liquid inclusions in the voids between the individual crystals constituting the aggregate.

Sometimes (in particular for submicron particles) *Ostwald ripening* plays a role. During this process larger crystals increase their size at the expense of smaller ones which dissolve.

During precipitation an *intensive mixing* of the crystalliser content with the two feed streams containing the components constituting the desired compound is required. The components, whether or not at stoichiometry, can be present in the solution simply as free ions, but also as ion pairs, as hydrated ions, complexed ions or as molecules. The presence of all these different species will affect the value of the supersaturation (and thus the growth rate). This means that the supersaturation is closely related to the chemistry in the solution (see section 9.2).

For slightly soluble substances *polymorphism* is a quite common feature, as is the occurrence of *hydrated forms*. This implies that, under the given process conditions, not only the most stable phase with the lowest solubility, but also metastable phases with a higher solubility could in principle develop. Since high (local) supersaturations frequently occur in precipitation processes, situations are imaginable where the solubility of a metastable phase is exceeded. According to the Ostwald rule of stages the metastable phase will then nucleate and grow unless sufficient crystals of the stable phase are already present. This implies that, for unseeded batch operation or in clear liquor regions in the reaction vessel, a metastable phase may be formed due to the high supersaturation. Depending on the operating conditions this metastable phase may transfer into the stable one, or may be withdrawn as a product. In this way products with a different performance are obtained depending on the operation of the process.

This often unwanted feature of polymorphism or hydrate formation can be turned into an advantage when *recrystallisation* of the product is required as a purification step. For slightly soluble salts recrystallisation cannot simply be realised via repeated dissolution and growth, as is done for easily soluble salts. An example where polymorphism is used for this purpose will be discussed in section 9.3. Mixing will be treated in section 9.4.

Many of the characteristic features of precipitation treated here hold for salting-out processes (see section 2.2.5) as well, provided that salting-out is accompanied by a strong suppression is solubility of the compound upon addition of the second or anti-solvent to the first solvent. A difference between precipitation and salting-out is that, during salting-out, the compound will dissolve if part of the anti-solvent is removed from the mixture (by means of evaporation), whereas extreme dilutions have to be applied during precipitation in order to dissolve only a small part of a compound.

9.2. *Solution chemistry and supersaturation*

In the previous section the entanglement between the chemistry in the solution and the resulting (local) supersaturation was mentioned. Therefore, these two subjects are treated simultaneously in this section.

The solubility of sparingly soluble compounds is expressed in terms of the solubility product K_s. For a binary electrolyte A_mB_n:

$$K_s = a^m_{A,eq}\, a^n_{B,eq}, \tag{70}$$

where A^{+n} and B^{-m} are the chemical species constituting the electrolyte and $a_{A,eq}$ and $a_{B,eq}$ are the activities of the chemical species A^{+n} and B^{-m} free in solution and in equilibrium with the solid phase.

The growth of sparingly soluble substances is mostly surface reaction controlled, and the driving force for crystallisation can be described by the supersaturation ratio defined as [138]

$$S = a^m_A\, a^n_B/K_s, \tag{71a}$$

or by:

$$S = [a^m_A\, a^n_B/K_s]^{1/(m+n)}, \tag{71b}$$

in which a_A and a_B are the activities of the species A^{+n} and B^{-m} in the supersaturated solution. The definitions of the supersaturation slightly differ from those given in section 2. It should be noted that a_A and a_B in eq. (71) are the activities of the free ions A^{+n} and B^{-m} in solution. More than for easily soluble salts, the solution may contain

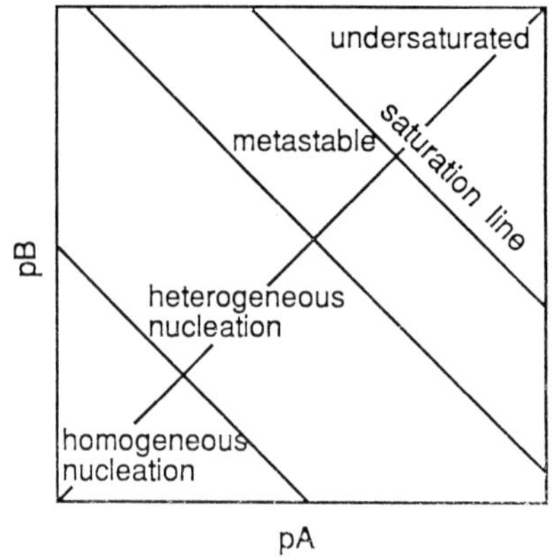

Fig. 31. Precipitation diagram in which $pB = -\log a_B$ is plotted versus $pA = -\log a_A$.

other dissolveld species such as ion pairs, hydrated ions (e.g., HB^{-m+1}) complexated ions (e.g. $\{A(HB)\}^{+n-m+1}$) and molecules (e.g., A_mB_n), the concentrations of which, although comprising the chemical species A and B, do not contribute to the supersaturation. The total, analytical concentration of A, designated by A_t is the sum of the concentrations of all the other species present in the solution and containing A. The same holds for B_t. These sums of concentrations can be used together with the value of K_s and the equations for chemical equilibrium and electroneutrality of the system to determine the concentrations of free ions $[A^{+n}]$ and $[B^{-m}]$, respectively.

Another aspect of reaction crystallisation is that crystallisation can occur from a nonstoichiometric solution. The state of the system (either under- or supersaturated) can therefore be analysed by constructing a plot of pB versus pA, where $A = [A^{+n}]$, a_A or A_t, and $pA = -\log A$ (similar for B). For $A = a_A$ and $B = a_B$, the saturation or solubility line is obtained by rearranging eq. (70) into:

$$m\,pA + n\,pB = pK_s, \tag{72}$$

where $pK_s = -\log K_s$. The solubility curve in fig. 31 is thus a straight line with slope $d(pB)/d(pA) = -m/n$. Also, lines can be drawn separating regions of metastable, heterogeneous and homogeneous nucleation. For systems in which the formation of complexes or ion pairs is negligible, the solubility is a straight line also with $A = [A^{+n}]$ or $A = A_t$ (similar for B). But in the presence of ion pairs and complexes the solubility plot expressed in concentrations will be a curved line (see fig. 32). In the precipitation diagram of a specific system, also regions can be indicated in which precursors are likely to be formed or in which there is a tendency towards formation of colloidal suspensions [139].

9.3. Recrystallisation

If the Ostwald's rule of stages holds for a given system and a batchwise precipitation is carried out, the first phase to precipitate is the most soluble one. The solute concentra-

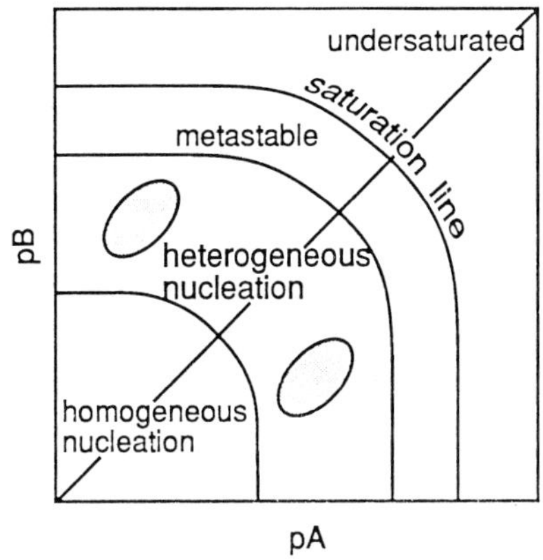

Fig. 32. Precipitation diagram in which pB = − log[B] is plotted versus pA = − log[A].

tion decreases from an initial value c_i and approaches a concentration $c_{1,\mathrm{eq}}$ (fig. 33) representing the equilibrium with respect to the first phase. The solution is still supersaturated with respect to a second, less soluble phase. After a time, exceeding the induction time of the second modification, this phase will precipitate so the solute concentration will again drop to a level $c_{2,\mathrm{eq}}$. In principle this process can be repeated with a third, less soluble, substance, etc.

Such a recrystallisation step has the advantage that the process is conducted at low supersaturation since the maximum supersaturation is the difference in solubility between the dissolving phase and the developing one.

In the case of a continuous process where the desired product is the metastable phase, the residence time has to be made sufficiently short in order to avoid significant nucleation and growth of the most stable phase.

The kinetics of recrystallisation processes can be described by coupling the dissolution rate of the less stable phase and the precipitation kinetics of the more stable one (e.g. [139, 140]).

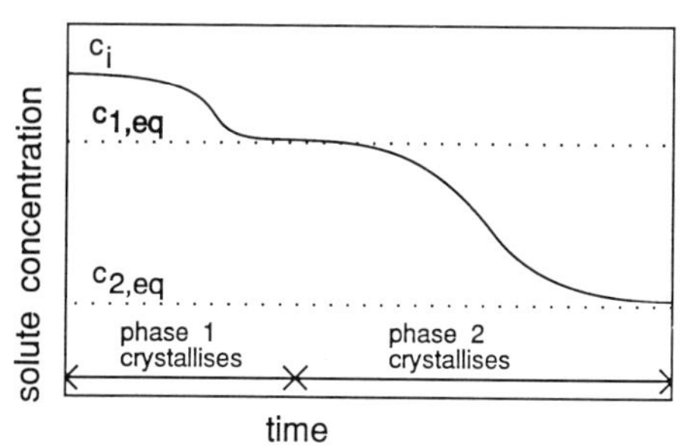

Fig. 33. Evolution of solute concentration during batch recrystallisation. At $t = 0$, a supersaturated solution with concentration c_i is fed to the reactor.

9.4. Mixing

9.4.1. Introduction

Realisation of precipitation processes in practice always involves the physical process of mixing different components. For precipitation the reactants have to be mixed and for salting-out it is the solvent and the anti-solvent that have to be mixed prior to formation of solid material.

If the precipitation rate is the rate-limiting step compared to the mixing rate, the reactor content can be considered homogeneous and mixing does not play any role. However, if the precipitation rate is faster than the mixing process, the overall precipitation rate will be controlled by the conditions of mixing and the two processes will occur simultaneously. In this case the mixing conditions are of importance for the final characteristics of the product. They can influence the product size distribution, crystal shape, product flow characteristics and sometimes even the crystal phase which is formed. Therefore, a thorough understanding of how mixing occurs and of the interaction between hydrodynamic conditions and the chemical processes involved in precipitation is of paramount importance if stringent product specifications are to be met.

The mixing process can be described either from an Eulerian (fixed frame) or a Lagrangian (material frame) point of view. In the Lagrangian frame the history of a fluid element is followed and the following *stages of mixing* are identified [141]: (i) the process of spatially distributing blobs or lumps of one fluid through the other; if completed, the lumps are evenly distributed throughout the reactor, so that all lumps of the reactor have the same average composition, but without local mixing; (ii) reduction of the scale of unmixed lumps by breakage and deformation; and (iii) mixing by molecular diffusion, whereby a homogeneous mixture on a molecular scale is obtained.

These three stages can occur either simultaneously or one after another, but in any case molecular diffusion is the last step. The first stage above is called *macromixing* and the other two stages constitute the process of *micromixing*. Ottino [142] published a detailed study of *chaotic mixing*: it turns out that mixing rates can be controlled and rapidly increased by forcing the flow into a regime of unsteady chaotic mixing. Doherty and Ottino [143] reviewed the application of chaos in dynamical systems to processes in the field of chemical engineering.

In the Eulerian description of mixing processes, conservation equations such as the mass, momentum and population balances are used to describe system properties in a spatially distributed way. Here mixing is described in terms of the theory of turbulence [144–147]. This approach has become of particular interest with the rapid improvements in the field of *computational fluid dynamics* and of experimental methods for measuring local velocities and concentrations.

9.4.2. Macromixing

In the Lagrangian approach macromixing is studied by using the concept of residence time distribution (RTD). Fluid elements enter the system, e.g. a continuous-flow vessel, remain in it for some period of time and then leave the vessel. The normalised density function describing the distribution of the residence times of the fluid elements leaving the vessel is called the residence time distribution. The RTD concept combines elegant theoretical methods with practical utility by reducing the complex flow and mixing processes to an easily manageable form without discarding their essential and complex features. Furthermore, the RTD can be obtained experimentally using stimulus–response techniques [148].

The two extremes of macromixing are represented by the plug flow reactor (no mixing whatsoever) and the ideally mixed reactor. The majority of actual crystalliser systems exhibit macromixing behaviour between these two extreme cases. Deviations from these extreme cases may be caused by: i) nonuniform liquid velocity profile; (ii) velocity fluctuation due to turbulent diffusion; (iii) short circuiting, bypassing and channelling of suspension; (iv) the presence of stagnant regions of fluid caused by the crystalliser design, internals or the configuration of addition of reactants; and (v) recycling of magma within the crystalliser as a result of agitation.

Models describing the flow pattern in a reactor vessel have been given by several authors; Tavare [148] and Westerterp et al. [149] provided a review of these models. The most popular one is based on the tanks-in-series concept, according to which the reactor is divided into a number of sections of equal volume, each of which is assumed to be perfectly mixed. The model parameter to be adjusted experimentally is the number of sections.

Another approach in describing non-ideal flow behaviour is to use dispersion models, which are based on a superposition of axial dispersion (i.e. exchange between volumes of fluid) on plug flow. The extent of axial dispersion is quantified by the Péclet number, which is also the model parameter.

When the flow behaviour within the crystalliser strongly deviates from the ideal case (i.e. an ideally mixed reactor), multi-parameter modelling is needed to represent reactor performance more closely. Usually, one or more well-mixed sections are combined with a plug flow section or with recycle streams between the sections. The volume of the sections or the value of the recycle streams are the model parameters.

9.4.3. Micromixing

In order to completely characterise the state of mixing according to the Lagrangian point of view, both macro- and micromixing have to be considered. It has been shown how the macromixing process can be characterised be means of the concept of the residence time distribution. The RTD is derived independent of the micromixing process but the RTD defines the environment surrounding the fluid elements during their presence in the reactor and therefore defines the limitations for the mixing on a molecular scale.

Characterisation of micromixing requires, strictly speaking, that both the *state of aggregation* and the *earliness of mixing* can be described [149]. Nevertheless, models for micromixing are usually based on the description of only one of these processes. In this section the concepts of state of aggregation and earliness of mixing are first described after which the most relevant micromixing models are presented. Finally, the influence of some process parameters on precipitation processes are described.

State of aggregation and earliness of mixing. The state of aggregation of a fluid element refers to the description of its size and its homogeneity with time. Some mechanisms describing changes in the state of aggregation have been proposed, which are amenable for a quantitative description [150]: (i) *Purely convective mixing* which, even in the absence of turbulence, occurs under the influence of a convective velocity field; fluid elements are stretched out and folded up, and the mixture finally exhibits a lamellar or striated structure characterised by a striation thickness; (ii) *erosive or dispersive mixing* during which fluid elements are "peeled off" by frictional forces exerted on their surfaces; (iii) *mixing by coalescence-dispersion* (see fig. 34) where the fluid elements merge and redisperse randomly with a constant frequency; and (iv) *molecular diffusion* which takes place when the fluid elements are sufficiently small; reduction of scale of fluid elements continues via a laminar stretching process. The first three mechanisms refer to the second stage of mixing presented in section 9.4.1, while the last one refers to the third stage of mixing.

For a given RTD the earliness of mixing determines the environment of the fluid elements. There are an infinite number of possibilities of how early mixing occurs. For instance, the conversion of the reaction (except for first-order reactions) will change in a chemical reactor with a plug flow followed by an ideally mixed reactor in series if the sequence of the two reactors is reversed, although the RTD is not affected by this change in sequence (see fig. 35). This is due to fact that the earliness of mixing is different for the two configurations.

Two extreme cases can be distinguished for any RTD: one of maximum mixedness, if mixing occurs at the earliest moment, and one of minimum mixedness, if mixing occurs at the latest moment. Zwietering [151] has proposed a method to calculate these two extremes of micromixing for reactors of arbitrary RTDs with premixed feed or with

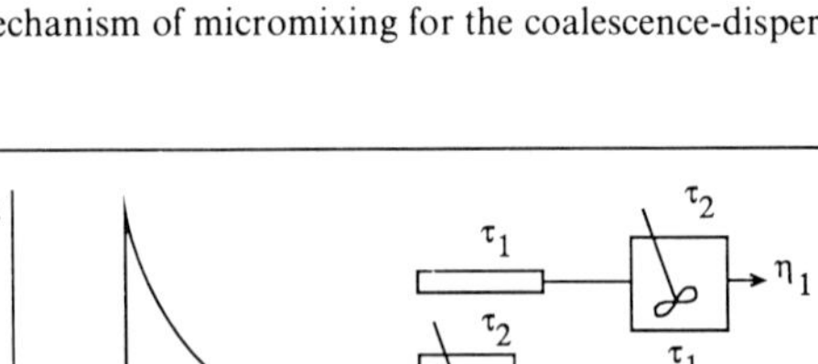

Fig. 34. Mechanism of micromixing for the coalescence-dispersion model.

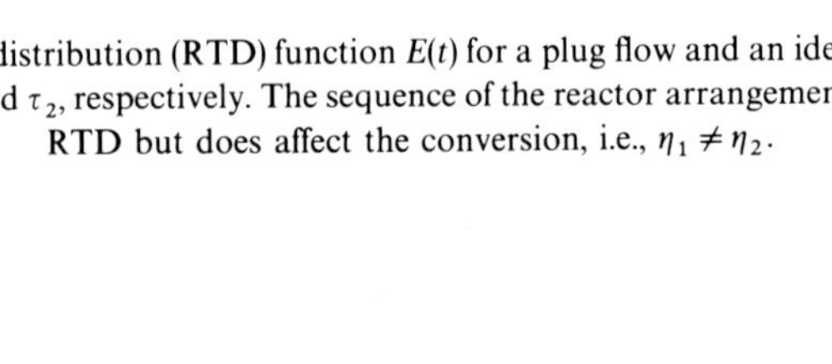

Fig. 35. Residence time distribution (RTD) function $E(t)$ for a plug flow and an ideally stirred tank in series with residence time τ_1 and τ_2, respectively. The sequence of the reactor arrangement has no influence on the RTD but does affect the conversion, i.e., $\eta_1 \neq \eta_2$.

separate feed points. The boundaries for conversion in a system with a known RTD can therefore be predicted. The boundaries lie farther apart for systems with a larger spread in the RTD function.

Micromixing models

Direct modelling of the state of aggregation. In this approach the decay of a completely segregated volume is related to the theory of turbulent mixing. The erosive or dispersive mechanism is the peeling-off of a fluid element, which is characterised by a mass transfer coefficient, the same as in the case of mass transfer between small solid particles in turbulent media [150]. A characteristic erosion time, t_e, was calculated from the power input ε and fluid properties as:

$$t_e \sim l_0 \nu^{5/12} D^{-2/3} \varepsilon^{-1/4}, \tag{73}$$

in which l_0 is the initial size of the fluid element, ν the kinematic viscosity of the liquid (in m^2s^{-1}) and D the diffusion constant (in m^2s^{-1}). Pohorecki and Baldyga [152] proposed a model which resulted in an expression for the decay of the volume V_s of a completely segregated fluid element:

$$dV_s/dt = V_s/t_{ms}, \tag{74}$$

in which t_{ms} is a characteristic time scale of the process determined by the system geometry. It is given by:

$$t_{ms} = \frac{3}{2} k_{oc}^{-3/2} \varepsilon^{-1/3}. \tag{75}$$

k_{oc} is the characteristic wave number of large eddies.

These models can be applied to continuous, semibatch or batch reactors to describe the mixing process in situations of practical interest. The latter model was applied to describe the precipitation of $BaSO_4$ in a continuous reactor for both a premixed and an unpremixed feed [153].

Diffusion models. Diffusion models describing the final stage of micromixing are based on the mechanism of molecular diffusion. The characteristic time t_d for the process is defined by:

$$t_d = \mu \lambda^2 / D, \tag{76}$$

where μ is a shape factor and λ a characteristic dimension of the fluid element. Each diffusion model assumes a specific microstructure for the fluid element, so that a specific description of the characteristic dimension λ results. Several representative models are described in [141].

Single-parameter models. In this category of models perfect macromixing is assumed and the intensity of micromixing is the same throughout the tank.

In the coalescence-dispersion model (CD model) the content of a reactor is divided into a large number of fluid elements of equal volume. These fluid elements coalesce with neighbouring elements, where mixing on a molecular scale and subsequent redispersion occur instantaneously (see fig. 34). The CD process occurs with a constant frequency, which is the model parameter. The model is mathematically described by a set of integro–differential equations which can be solved with probability functions. If two or more reacting species are present, the complexity of the formulation increases and solution by Monte Carlo simulation is necessary.

The model describing interaction by exchange with the mean (IEM model) is based on the mass exchange between a fluid element and the average concentration in the vessel. The mass transfer rate is described by a micromixing time t_{micro} which is also the model parameter. This formulation is much simpler to solve than the CD model since only a simple iterative procedure for the value of the mean concentration in the reactor is required. This model can be used for multiphase systems (e.g. solid particles in a liquid).

Models based on the earliness of mixing. These models are based on the recognition that real systems show an intermediate behaviour somewhere between the state of minimum and maximum mixedness. The fluid elements entering the reactor gradually change from a state of minimum towards a state of maximum mixedness which is assumed to be reached before the fluid element leaves the reactor. Models are characterised by a particular choice of the kinetics of transition from one state to the other. These models can handle reactors with arbitrary RTD with premixed and unpremixed feeds. The disadvantage is that they do not give a clear picture of the physical phenomena in the reactor, but adjust the observed behaviour with one or two experimental parameters.

9.4.4. Mixing times

The question whether mixing is of importance for a particular system can be assessed by comparing characteristic times for macromixing (t_{macro}), micromixing (t_{micro}) and for the chemical process involved (t_{reac}).

The macromixing time can be taken as the mean residence time or the circulation time in a draft tube crystalliser. For baffled tanks with a height equal to the tank diameter D_t in the turbulent regime, the macromixing time t_{macro} can be expressed as [134]:

$$t_{\text{macro}} = 7.3\,(D_t^2/\varepsilon_{\text{avg}})^{-1/3}, \tag{77}$$

where ε_{avg} is the average specific power input (W kg^{-1}). A similar correlation has also been proposed for the laminar flow regime [134].

Estimation of the micromixing times can be made by any of the micromixing models presented. The characteristic reaction time depends on the specific precipitation process involved. The induction time can be taken as a characteristic nucleation time. If the chemical reactions A + B → AB(liq) and AB(liq) → AB(sol) hold, and the first one is the rate limiting step (with reaction constant k), the reaction time for stoichiometric reagent addition may be defined as:

$$t_{\text{reac}} = k^{-1}(c_A c_B)^{-1/2}. \tag{78}$$

Following [141], comparison of micromixing and reaction times leads to the identification of three regimes. The reaction is therefore instantaneous if $t_{\text{micro}} \gg t_{\text{reac}}$, fast if $t_{\text{micro}} \approx t_{\text{reac}}$, and slow if $t_{\text{micro}} \ll t_{\text{reac}}$. Incorporation of the macromixing time t_{macro} (taken

Table 7
Mixing regimes as a function of characteristic times. $t_{macro} = \tau =$ residence time; MM = ideal micromixing; PS = partial segregation; S = complete segregation.

	$t_{micro} \ll t_{reac}$	$t_{micro} \approx t_{reac}$	$t_{micro} \gg t_{reac}$
$t_{micro} \ll \tau$	MM	PS	S
$t_{micro} \approx \tau$	PS	PS	S
$t_{micro} \gg \tau$	S	S	S

as the residence time τ) in the comparison of the three timescales allows one to classify regimes of ideal micromixing (MM), partial segregation (PS) or complete segregation (S) as indicated in table 7 [149].

9.4.5. *Process parameters*

Gösele and Kind [154] proposed a practical way to investigate both macro- and micromixing in a double-jet continuous reactor by considering the reactor as a system consisting of three zones. The macromixing in the reactor can be adjusted by changing the recirculation rates between the three separate zones. The micromixing in each zone, in its turn, can be adjusted by changing the mixing intensity in each of the three zones. This seems to be a suitable approach for scale-up.

The influence of micromixing upon precipitation has been verified experimentally for several systems under batch, semi-batch and continuous operation. No data are available, however, for large-scale reactors. The influence of some process parameters on the product average size and on the coefficient of variation (CV) of the distribution follows.

Effect of mixing intensity. An increase in mixing intensity always improves micromixing (since t_{micro} is proportional to ε^{-p}, with $p > 0$, section 9.4.3) but the result on the average size may vary strongly depending on the specific situation. It may enhance micromixing of reactants with the bulk, i.e. it may improve the dilution of reactants, so lower local supersaturations result and the corresponding crystal sizes are large. In other situations, micromixing between fresh reactants will be improved and small crystal sizes result. For an MSMPR configuration the former phenomenon was found [153,155], while for a plug flow configuration the latter effect was observed [156]. For batch and semi-batch configurations both tendencies have been found [156-158]. For semi-batch precipitation of a single compound ($BaSO_4$ [157]), the resulting effect depends on the addition mode (e.g., single or double jet) and the position of the feed points (e.g. close to each other or far apart).

The effect of mixing intensity can be even less straightforward due to the influence this process condition may have on agglomeration and attrition phenomena, implying that these latter effects are superimposed onto that of micromixing.

The coefficient of variation, CV, of the size distribution usually decreases as mixing intensity increases [156,159].

Effect of location of feed nozzles. Addition of reactants near the impeller outlet improves micromixing of the reactant with the bulk, thus improving dilution, corresponding to an increase in the average size of the crystals. Addition near the impeller inlet, on the other hand, results in smaller crystals [160,161].

When reactants are added by means of two nozzles, the distance between the nozzles in the Lagrangian frame has the largest effect on product size [157]: the larger the distance, the larger the average crystal size. Tosun [156] found a difference of a factor of ten in average size even when the measured intensity of micromixing was similar for both cases.

For continuous reactors, a premixed feed of reactants produces smaller crystals than an unpremixed one. Pohorecki and Baldyga [153] found a factor of two in average size during precipitation of $BaSO_4$.

The CV of the size distribution usually decreases as the distance in a Lagrangian frame between addition points (double jet) decreases [156]. When the inlet nozzle is placed at the liquid level (i.e. in a region of low turbulence), a high CV has been observed [156].

Effect of reactant concentration. Normally, higher reactant concentrations produce higher localised supersaturations and therefore smaller crystal sizes [153,156,161], although exceptions have been observed due to the interfering effect of agglomeration [157,160]. Reactant concentration, of course, also affects the mechanism of nucleation and the total conversion in the system.

9.4.6. Eulerian models for mixing

The Eulerian description, an infinitesimal control volume fixed in space, is used and conservation equations for mass, momentum and population are required in combination with the kinetics of the processes occurring in the reactor. In this way system properties can be described in a spatially distributed way. With recent advances in computational fluid dynamics, complete spatially distributed descriptions of flow fields for three-dimensional reactors of complex geometry have become available. Solutions for nonreacting systems as well as for homogeneous reaction systems have been incorporated into commercially available codes.

A distributed description of precipitation processes has been reported [111]. The precipitation of calcium phosphate in a continuous double-jet reactor in the absence of mechanical stirring has been simulated. The phenomena of macromixing, chemical equilibrium, kinetics of nucleation and growth of calcium phosphate have been coupled to the solution of the flow field. All local properties of the system could be predicted, like local flow velocities, local mixing intensities and local supersaturations. The local crystal size distribution was obtained in terms of the moments of the distribution. The influence of feed position on the product average size and on the CV of the distribution was studied. The processes governing the production of undesired particle sizes or CVs could be locally assessed, allowing a systematic and comprehensive optimisation of feed location as well as reactor geometry.

The main difficulty in the Eulerian approach is that the instabilities of the turbulent flow are still not well understood. Oversimplified assumptions are proposed with the consequence that the quantitative character of the predictions is lost. Nevertheless, it is a powerful tool for the understanding of the interactions between physical and chemical processes in reactors of complex geometry.

Acknowledgements

The authors wish to thank P. Bennema, O.S.L. Bruinsma, R.A. Eek, J.P. van der Eerden, F. Elgersma, J.H. Gerla, C.M.G. Heffels, P.J. Jansens, H.J.M. Kramer and G.J. Witkamp for helpful suggestions, discussions and contributions to the manuscript. Section 9 on precipitation was authored by M.M. Seckler, to whom we are especially grateful. B.A. van den Bosch is acknowledged for drawing several figures.

List of symbols

$a(a_{eq})$	= (equilibrium) activity [mol l^{-1}]
a	= crystal surface area [m^2]
a_4	= numerical shape factor [–]
$\boldsymbol{A}$	= dimensionless scatter matrix [–]
A	= constant [# $m^{-3}s^{-1}$]
A	= cross-sectional area [m^2]
A_1	= constant [$m^{1/2}rad^{-1}$]
A_2	= constant [rad $m^{-1/2}$]
A_t	= sum of concentrations of all other species in solution containing A [mol l^{-1}]
A_T	= total surface area of crystals per unit of crystalliser volume [m^2m^{-3}]
b	= constant [–]
B	= birth rate function [# $m^{-1}m^{-3}s^{-1}$]
B_0	= secondary nucleation rate [# $m^{-3}s^{-1}$]
B_t	= sum of concentrations of all other species in solution containing B [mol l^{-1}]
c	= concentration [mol l^{-1}]
c	= solute concentration of solution in crystalliser [kg m^{-3}]
c^*	= equilibrium concentration [kg m^{-3}]
Δc	$= c - c^*$ = solute concentration difference [kg m^{-3}]
c_b	= bulk liquid concentration [mol l^{-1}]
c_i	= interface concentration [mol l^{-1}]
c_i	= solute concentration of solution at inlet [kg m^{-3}]
c_p	= specific heat of mother liquor [J $kg^{-1}K^{-1}$]
d_{50}	= median [m]
D	= death rate function [# $m^{-1}m^{-3}s^{-1}$]
D	= diffusion constant [m^2s^{-1}]
D_s	= stirrer diameter [m]
D_t	= tank diameter [m]
E	= residence time distribution [–]
E_c	= critical impact energy [J]
E_{cool}	= energy withdrawal per unit mass of crystals during cooling [J kg^{-1}]
E_{evap}	= energy input per unit mass of crystals during evaporation [J kg^{-1}]
g	= growth order [–]
G	= growth rate [m s^{-1}]
G	= Gibbs' free energy [J]
G'	= velocity gradient in a stirred liquid [s^{-1}]
G_0	= zero elastic strain growth rate [m s^{-1}]
ΔG	= work of cluster formation [J]
ΔG^*	= nucleation barrier [J]
ΔG^*_{2D}	= activation energy for creation of a two-dimensional critical nucleus [J]
h	= height of (macro) step or of two-dimensional island [m]; constant [–]
ΔH_M	= enthalpy of melting [J mol^{-1}]
ΔH_{cr}	= enthalpy of crystallisation [J kg^{-1}]
ΔH_{evap}	= enthalpy of evaporation [J kg^{-1}]
i	= constant [–]
j	= constant [–]
I	= incorporation rate of secondary nuclei on parent crystal surface [# s^{-1}]
J	= primary nucleation rate [# $m^{-3}s^{-1}$]
J_{2D}	= 2D nucleation rate [# $m^{-2}s^{-1}$]
k	= shape factor [–]
k	= Boltzmann's constant (1.38×10^{-23} J K^{-1}) [J K^{-1}] constant [–]
k	= reaction constant (reaction order $n+1$) [L^n mol^{-n} s^{-1}]
k_a	= surface area shape factor [–]

k_d	= mass transfer coefficient via diffusion [m s^{-1}]
$k_{m\&h}$	= overall mass transfer coefficient including mass and heat transfer [m s^{-1}]
k_{oc}	= characteristic wave number of large eddies [m^{-1}]
k_r	= mass transfer coefficient via surface integration [m s^{-1}]
k_t	= overall mass transfer coefficient [m s^{-1}]
k_v	= volumetric shape factor [–]
k_G	= growth coefficient [m s^{-1}]
k_N	= nucleation coefficient [# m^{-3}s^{-1}]
K	= aggregation kernel [m^{3}s^{-1}]
K_s	= solubility product [mol^{m+n} l^{-m-n}]
l	= length scale of mosaic structure [m]
l_0	= initial size of eddy [m]
$\boldsymbol{L}$	= vector containing relative intensity of scattered light [–]
L	= orifice width [m]
L_T	= total length of crystals per unit of crystalliser volume [m m^{-3}]
m	= refraction index of particle relative to surroundings [–]; mass [kg]
m	= dM/dx = mass density distribution [kg m^{-1}m^{-3}]
m_i	= ith moment of CSD [m^{i}m^{-3}]
m_s	= mass fraction of solute in bulk solution [–]
M	= molecular weight [kg mol^{-1}] cumulative mass density distribution [kg m^{-3}]
M_T	= solid content in crystal slurry [kg m^{-3}]
n	= dN/dx = population number density [# m^{-1} m^{-3}]
n	= number of solute entities in a nucleus [#]
n	= constant [–]
$\tilde{n}$	= nV [# m^{-1}]
$\tilde{n}_0$	= n_0V [# m^{-1}]
n_0	= population number of density at zero size [# m^{-1}m^{-3}]
n_1, n_2	= particle concentrations of aggregating bodies of size x_1, x_2 [#m^{-3}]
N	= cumulative size distribution [#m^{-3}]
N	= stirrer rotation rate [rpm]
N_0	= total number of seeds [#]
N_A	= Avogadro's number (6.02 × 10^{23} # mol^{-1}) [# mol^{-1}]
N_T	= total number of crystals per unit of crystalliser volume [# m^{-3}]
p	= subscript indicating moment of distribution [–]
p_i	= integration constant [kg s^{-1}]
P	= production rate [kg s^{-1}]
P	= power [W]
$\boldsymbol{q}$	= vector containing fraction of particles in each size class [–]
q	= constant [–]
Q or Q_u	= volumetric (product) flow rate [m^{3}s^{-1}]
Q_i or Q_{in}	= volumetric inflow rate [m^{3}s^{-1}]
Q_k	= volumetric flow rate [m^{3}s^{-1}]
Q_o or Q_{out}	= volumetric outflow rate [m^{3}s^{-1}]
Q_v	= volumetric vapour flow rate [m^{3}s^{-1}]
r	= aggregation rate [#m^{-3}s^{-1}]
r_c	= critical nucleus size [m]
R	= radius of curvature [m]
R	= accelerated rate at which fines are removed [–]
R	= ideal gas constant (8.31 J K^{-1}mol^{-1}) [J K^{-1}mol^{-1}]
R	= removal rate of secondary nuclei from parent crystal surface [# s^{-1}]
R	= resistance [Ω]
R'	= resistance [Ω]
ΔR	= $R - R'$ = difference in resistance [Ω]
s	= molar entropy [J K^{-1}mol^{-1}]
s	= x/x_c = size ratio [–]
s_L	= molar entropy of liquid [J K^{-1}mol^{-1}]
s_S	= molar entropy of solid [J K^{-1}mol^{-1}]
Δs_{eq}	= molar entropy of melting [J K^{-1}mol^{-1}]
S	= c/c^* = relative supersaturation [–]
S_c	= critical supersaturation [–]
t	= time [s]
t_e	= characteristic erosion time [s]
t_i	= induction time [s]
t_{macro}	= macromixing timescale [s]
t_{micro}	= micromixing timescale [s]
t_{ms}	= characteristic timescale [s]
t_{reac}	= reaction timescale [s]
Δt	= time interval [s]
T	= temperature [K]
T_M	= melting temperature [K]
T_R	= roughening temperature [K]
ΔT	= undercooling [K]
u	= volume of an aggregating crystal [m^3]
v	= volume of an aggregating crystal [m^3]
v	= lateral spreading velocity of step [m s^{-1}]
v_0	= velocity of one infinitely long straight step [m s^{-1}]
v_t	= stirrer tip speed [m s^{-1}]

V = crystalliser volume [m^3]
V = orifice volume [m^3]
V_{macro} = detectable macroscopic volume [m^3]
V_p = volume of particle [m^3]
V_s = volume of completely segregated fluid element [m^3]
V_T = total volume of crystals per unit of crystalliser volume [m^3m^{-3}]
x = crystal size [m]
x_c = cut size [m]
x_D = dominant crystal size [m]
x_F = fines cut size [m]
x_M = m_4/m_3 = mass-averaged crystal size [m]
x_N = m_1/m_0 = number-averaged or mean crystal size [m]
x_P = product classification cut size [m]
x_1, x_2 = crystal size [m]
Δx = size interval [m]
y = $x - G\tau$ = crystal size [m]
z = accelarated rate at which product is removed [–]

Dimensionless numbers
Fr = Froude number
Nu = Nusselt number
Po = power number
Re = Reynolds number
Sh = Sherwood number

Greek letters
α = V_{macro}/V = volume fraction [–]
α = heat transfer coefficient [$W\ m^{-2}K^{-1}$]
γ = activity coefficient [–]
γ = specific surface excess free energy [$J\ m^{-2}$]
γ_{eff} = $\Psi^{1/3}\gamma$ = effective specific surface free energy [$J\ m^{-2}$]
δ = number of ions in formula unit [#]
δ_h = boundary layer thickness related to heat transfer [m]
δ_m = boundary layer thickness related to mass transfer [m]
ε = power input [$J\ kg^{-1}$]
ε_{avg} = average power input [$J\ kg^{-1}$]
η = mosaic spread [rad, °]
viscosity [$kg\ m^{-1}s^{-1}$]
conversion [–]
λ = wavelength of light [nm]
λ = characteristic length scale crystalliser [m]
λ = diffusion length of growth unit at crystal surface [m]
λ = thermal conductivity [$W\ m^{-1}K^{-1}$]
λ = characteristic size of eddy [m]
μ_L = chemical potential of liquid [$J\ mol^{-1}$]
μ_S = chemical potential of solid [$J\ mol^{-1}$]
μ_L^0 = standard chemical potential of liquid [$J\ mol^{-1}$]
$\Delta\mu$ = $\mu_L - \mu_S$ = chemical potential difference [$J\ mol^{-1}$]
$\Delta\mu_1$ = constant [$J\ mol^{-1}$]
ν = η/ρ = kinematic viscosity [m^2s^{-1}]
ν_i = stoichiometric coefficient [–]
ρ = liquid density [$kg\ m^{-3}$]
ρ_c = solid density [$kg\ m^{-3}$]
ρ_i = slurry density of inflow [$kg\ m^{-3}$]
ρ_f = resistivity of fluid [Ω m]
ρ_o = slurry density of outflow [$kg\ m^{-3}$]
ρ_v = vapour density [$kg\ m^{-3}$]
σ = $(c - c^*)/c^*$ = relative supersaturation [–]
τ = residence time [s]
Ψ = multiplicative factor for γ^3 in case of heterogeneous nucleation [–]
Ω = molecular volume [m^3 per solid entity]

Abbreviations
CCD = charged coupled device
CD = coalescence-dispersion
CSD = crystal size distribution
CV = coefficient of variation
DTB = draft-tube baffle
F = feed
FC = forced circulation
FLS = forward light scattering
GRD = growth rate dispersion
IEM = interaction by exchange with the mean
KAP = potassium hydrogen phthalate
KDP = potassium dihydrogen phosphate
MN = mononuclear
MSCPR = mixed suspension with classified product removal
MSFD = mixed suspension with fines destruction
MSMPR = mixed suspension, mixed product removal
P = product
PCS = photon correlation spectroscopy
PF = plug flow
PN = polynuclear
QELS = quasi-elastic light scattering
RTD = residence time distribution
SDG = size-dependent growth
SEM = scanning electron microscopy
V = vapour

References

[1] M.E. Glicksman and S.P. Marsh, in: Handbook of Crystal Growth, Vol. 1, ed. D.T.J. Hurle (North-Holland, Amsterdam, 1993) p. 1075.
[2] P.J. Jansens and G.M. Van Rosmalen, in: Handbook of Crystal Growth, Vol. 2, ed. D.T.J. Hurle (North-Holland, Amsterdam, 1993) Ch. 6, this volume.
[3] J.P. Van der Eerden, in: Handbook of Crystal Growth, Vol. 1, ed. D.T.J. Hurle (North-Holland, Amsterdam, 1993) p. 307.
[4] R. Chakraborty and K.A. Berglund, AIChE Symp. Ser No. 284 (87) (1991) 114.
[5] J. Mydlarz and A.G. Jones, Powder Technol. 65 (1990) 187.
[6] D.A. Weingärtner, S. Lynn and D.N. Hanson, Ind. Eng. Chem. Res. 30 (1991) 490.
[7] G.D. Botsaris, in: Industrial Crystallisation 75, ed. J.W. Mullin (Plenum, New York, 1976) p. 3.
[8] B. Mutaftschiev, in: Handbook of Crystal Growth, Vol. 1, ed. D.T.J. Hurle (North-Holland, Amsterdam, 1993) p. 187.
[9] S. Sarig, in: Handbook of Crystal Growth, Vol. 2, ed. D.T.J. Hurle (North-Holland, Amsterdam, 1993) Ch. 19, this volume.
[10] M. Volmer, Kinetik der Phasenbildung (Steinkopf, Dresden, 1939).
[11] A.E. Nielsen and O. Söhnel, J. Cryst. Growth 11 (1971) 233.
[12] A. Mersmann, J. Cryst. Growth 102 (1990) 841.
[13] E.A. Guggenheim, Trans. Faraday Soc. 36 (1940) 397.
[14] J.W. Mullin and C.L. Leci, Philos. Mag. 19 (1969) 1075.
[15] A.T. Allen, M.P. Mcdonald, W.M. Nicol and R.M. Wood, Nature 235 (1972) 36.
[16] M.A. Larson and J. Garside, Chem. Eng. Sci. 41 (1986) 1285.
[17] G.A. Hussmann, M.A. Larson and K.A. Berglund, in: Industrial Crystallisation 84, eds. S.J. Jančić and E.J. De Jong (Elsevier, Amsterdam, 1984) p. 21.
[18] V. Veverka, O. Söhnel, P. Bennema and J. Garside, AIChE J. 37 (1991) 490.
[19] Y.C. Chang and A.S. Myerson, in: Industrial Crystallisation 84, eds. S.J. Jančić and E.J. De Jong (Elsevier, Amsterdam, 1984) p. 27.
[20] Y.C. Chang and A.S. Myerson, AIChE J. 31 (1985) 980.
[21] Y.C. Chang and A.S. Myerson, AIChE J. 32 (1986) 1567.
[22] A.S. Myerson and A.F. Izmailov, in: Handbook of Crystal Growth, Vol. 1, ed. D.T.J. Hurle (North-Holland, Amsterdam, 1993) p. 249.
[23] P. Ayazi Shamlou, Z. Liu and J.G. Yates, Chem. Eng. Sci. 45 (1990) 809.
[24] P. Ayazi Shamlou, A.G. Jones and K. Djamarani, Chem. Eng. Sci. 45 (1990) 1405.
[25] N.A. Clontz and W.L. McCabe, Chem. Eng. Progr. Symp. Ser. No. 110 (67) (1971) 6.
[26] R.T. Johnson, R.W. Rousseau and W.L. McCabe, AIChE Symp. Ser. 68 (1972) 31.
[27] H.E.C. Powers, Ind. Chemist. 39 (1963) 351.
[28] D.P. Lal, R.E.A. Mason and R.F. Strickland-Constable, J. Cryst. Growth 5 (1969) 1.
[29] M.K. Cerreta and K.A. Berglund, J. Cryst. Growth 102 (1990) 869.
[30] E.G. Denk and G.D. Botsaris, J. Cryst. Growth 13/14 (1972) 493.
[31] J. Garside and M.A. Larson, J. Cryst. Growth 43 (1978) 694.
[32] J. Garside, I.T. Rusli and M.A. Larson, AIChE J. 25 (1979) 57.
[33] C.Y. Tai, J.-F. Wu and R.W. Rousseau, J. Cryst. Growth 116 (1992) 294.
[34] P.J. Daudey, Ph.D. thesis, Delft University of Technology, The Netherlands (1987).
[35] P.J. Daudey, G.M. Van Rosmalen and E.J. De Jong, J. Cryst. Growth 99 (1990) 1076.
[36] R. Wissing, M. Elwenspoek and B. Degens, J. Cryst. Growth 79 (1986) 614.
[37] A.E.D.M. Van der Heijden, M. Elwenspoek and J.P. Van der Eerden, J. Cryst. Growth 98 (1989) 398.
[38] S.J. Jančić and P.A.M. Grootscholten, in: Industrial Crystallisation (Delft University Press, Delft, 1984).
[39] E.P.K. Ottens, A.H. Janse and E.J. De Jong, J. Cryst. Growth 13/14 (1972) 500.
[40] M. Mersmann, R. Sangl, M. Kind and J. Pohlisch, Chem. Eng. Technol. 11 (1988) 80.
[41] A.E.D.M. Van der Heijden, J.P. Van der Eerden and G.M. Van Rosmalen, Chem. Eng. Sci. (1993) submitted.
[42] T.W. Evans, A.F. Sarofim and G. Margolis, AIChE J. 20 (1974) 959.

[43] J. Pohlisch and A. Mersmann, Chem. Eng. Technol. 11 (1988) 40.
[44] J. Pohlisch, Ph.D. thesis, University of München, Germany (1987).
[45] R. Sangl, Ph.D. thesis, University of München, Germany (1987).
[46] A. Chianese, F. Di Berardino, A.G. Jones and J. Mydlarz, Proc. I. Chem. E/BACG Joint Meeting, London (1988).
[47] H.L. Bhat, J.N. Sherwood and T.S. Shripathi, Chem. Eng. Sci. 42 (1987) 609.
[48] J.N. Sherwood, R.I. Ristić and T.S. Shripathi, in: Industrial Crystallisation 90, ed. A. Mersmann (1990) p. 349.
[49] J. Nývlt, R. Rychlý, J. Gottfried and J. Wurzelová, J. Cryst. Growth 6 (1970) 151.
[50] S. Toschev, A. Milchev and S. Stoyanov, J. Cryst. Growth 13/14 (1972) 123.
[51] D. Kashchiev, D. Verdoes and G.M. Van Rosmalen, J. Cryst. Growth 110 (1991) 373.
[52] P. Bennema, in: Handbook of Crystal Growth, Vol. 1, ed. D.T.J. Hurle (North-Holland, Amsterdam, 1993) p. 477.
[53] P. Bennema and J.P. Van der Eerden, in: Morphology of Crystals, ed. I. Sunagawa (Terra Scientific, Tokyo, 1987) p. 1.
[54] F.C. Frank, Disc. Farad. Soc. 48 (1949) 67.
[55] W.K. Burton, N. Cabrera and F.C. Frank, Philos. Trans. Proc. Roy. Soc. London A 243 (1951) 299.
[56] M. Matsuoka, J. Garside, J.E. Rout, S.N. Black and R.J. Davey, J. Cryst. Growth 99 (1990) 1138.
[57] B.D. Chen, K. Iwamoto, J. Garside and M. Matsuoka, in: Industrial Crystallisation 90, ed. A. Mersmann (1990) p. 727.
[58] Y. Tanaka and M. Matsuoka, J. Cryst. Growth 99 (1990) 1130.
[59] A.E.D.M. Van der Heijden, R.M. Geertman and P. Bennema, J. Phys. D Appl. Phys. 24 (1991) 123.
[60] R.M. Geertman and A.E.D.M. Van der Heijden, J. Cryst. Growth 125 (1992) 363
[61] M. Matsuoka and J. Garside, Chem. Eng. Sci. 46 (1991) 183.
[62] J. Garside, in: Crystal Growth and Materials, eds. E. Kaldis and H.J. Scheel (North-Holland, Amsterdam, 1977) p. 483.
[63] O.S.L. Bruinsma, in: Industrial Crystallisation, post-academic course, part I, Amsterdam (1991).
[64] J. Garside, A. Mersmann and J. Nývlt (eds.), Measurement of Crystal Growth Rates (European Federation of Chem. Eng./Working Party on Crystallisation; Druckhaus Deutsch, Germany, 1990).
[65] A.A. Chernov, Modern Crystallography III – Crystal Growth (Springer, Berlin, 1984).
[66] J. Garside, Chem. Eng. Sci. 40 (1985) 3.
[67] N.S. Tavare, J. Chem. Eng. 63 (1985) 436.
[68] K.A. Berglund, Chem. Eng. Commun. 41 (1986) 357.
[69] J. Mydlarz and A.G. Jones, in: Industrial Crystallisation 90, ed. A. Mersmann (1990) p. 417.
[70] S. De Wolf, Ph.D. thesis, Delft University of Technology, The Netherlands (1990).
[71] A.A. Chernov, in: Industrial Crystallisation 90, ed. A. Mersmann (1990) p. 343.
[72] R.I. Ristić, J.N. Sherwood and K. Wojciechowski, J. Cryst. Growth 91 (1988) 163.
[73] M.M. Mitrović, R.I. Ristić I. Ćirić, Appl. Phys. A 51 (1990) 374.
[74] R.I. Ristić, J.N. Sherwood and T.S. Shripathi, in: Advances in Industrial Crystallisation, eds. J. Garside, R.J. Davey and A.G. Jones (Butterworth-Heinemann, Oxford, 1991) p. 77; also: R.I. Ristić, J.N. Sherwood and T.S. Shripathi, J. Cryst. Growth 102 (1990) 245.
[75] A.E.D.M. Van der Heijden and J.P. Van der Eerden, J. Cryst. Growth 118 (1992) 14.
[76] R.W. Hartel, B.E. Gottung, A.D. Randolph and G.W. Drach, AIChE J. 32 (1986) 1176.
[77] R.C. Zumstein and R.W. Rousseau, Chem. Eng. Sci. 44 (1989) 2149.
[78] R. David, J. Villermaux, P. Marchal and J.P. Klein, Chem. Eng. Sci. 46 (1991) 1129.
[79] M. Von Smoluchowski, Z. Phys. Chem. 92 (1917) 129.
[80] R.W. Hartel and A.D. Randolph, AIChE J. 32 (1986) 1186.
[81] R. David, P. Marchal, J.P. Klein and J. Villermaux, Chem. Eng. Sci. 46 (1991) 205.
[82] S. Wachi and A.G. Jones, Chem. Eng. Sci. 47 (1992) 3145.
[83] I.V. Melikhov, I.E. Mikheeva and V.N. Rudin, Soviet Phys.: Crystall. 34 (1989) 1272.
[84] E.G. Kelly and D.J. Spottiswood, in: Introduction to Mineral Processing (Wiley, New York, 1982) p. 113.
[85] B. Mazzarotta, Chem. Eng. Sci. 47 (1992) 3105.
[86] M. Lahav and L. Leiserowitz, in: Industrial Crystallisation 90, ed. A. Mersmann (1990) p. 609.

[87] A.D. Randolph and M.A. Larson, Theory of Particulate Processes (Academic Press, San Diego, 1988).
[88] T. Allen, in: Particle Size Measurement, Powder Technology Series (Chapman and Hall, London, 1974).
[89] J. Jager, Ph.D. thesis, Delft University of Technology, The Netherlands (1990).
[90] J. Garside and S.J. Jančić, AIChE J. 22 (1976) 887.
[91] J. Swithenbank, J.M. Beer, D. Taylor, D. Abbot and G.C. McGreath, Prog. Astronaut. 53 (1977) 421.
[92] H.G. Barth, Modern Methods of Particle Size Analysis (Wiley, New York, 1985).
[93] D.J. Brown and P.G. Felton, Chem. Eng. Res. Des. 63 (1985) 125.
[94] D.J. Brown, Powder Technol. 49 (1986) 71.
[95] A. Boxman and B. Scarlett, in: Proc. Laser Technologies in Industry, ed. O.D.D. Sores, SPIE 952 (1988) p. 520.
[96] A. Boxman, Ph.D. thesis, Delft University of Technology, The Netherlands (1992).
[97] H.C. Van de Hulst, Light Scattering by Small Particles (Wiley, New York, 1957).
[98] J.H. Bilgram, H. Güttinger and W. Känzig, Phys. Rev. Lett. 40 (1978) 1394.
[99] P. Böni, J.H. Bilgram and W. Känzig, Phys. Rev. A 28 (1983) 2953.
[100] O.N. Mesquita and H.Z. Cummins, Physico-Chem. Hydrodyn. 5 (1984) 389.
[101] O.N. Mesquita, D.G. Neal, M. Copic and H.Z. Cummins, Phys. Rev. B 29 (1984) 2846.
[102] G. Livescu, M.R. Srinivasan, H. Chou, H.Z. Cummins and O.N. Mesquita, Phys. Rev. A 36 (1987) 2293.
[103] L.M. Williams, H.Z. Cummins, L.O. Ladeira and O.N. Mesquita, Phys. Rev. A 45 (1992) 3880.
[104] C.M.G. Heffels, private communication (1992).
[105] U. Riebel, Ph.D. thesis, University of Karlsruhe, Germany (1988).
[106] H.J.M. Kramer, private communication (1992).
[107] J.H. Gerla, private communication (1992).
[108] N. Ouchiyama and T. Tanaka, Ind. Eng. Chem. Fundam. 20 (1981) 66.
[109] N. Ouchiyama and T. Tanaka, Ind. Eng. Chem. Fundam. 23 (1984) 490.
[110] L. Ljung, System Identification, Theory for the User (Prentice-Hall, Englewood Cliffs, NJ, 1987).
[111] M.M. Seckler, S. Nÿssen, O.S.L. Bruinsma and G.M. Van Rosmalen, Proc. 12th Symposium on Industrial Crystallisation, ed. Z.H. Rojkowski, Warsaw, Poland (1993) p. 2.
[112] G.J. Witkamp, O.S.L. Bruinsma and G.M. Van Rosmalen, Proc. Int. Symp. on Preparation of Functional Materials and Industrial Crystallisation, Osaka, Japan (1989) p. 144.
[113] M.J. Hounslow, R.L. Ryall and V.R. Marshall, AIChE J. 34 (1988) 1821.
[114] J. Hostomsky and A.G. Jones, J. Phys. D Appl. Phys. 24 (1991) 165.
[115] S. Wachi and A.G. Jones, Chem. Eng. Sci 46 (1991) 1027.
[116] D.J. Cooper and D.E. Clough, AIChE J. 31 (1985) 1202.
[117] D.J. Cooper and D.E. Clough, AIChE J. 32 (1986) 389.
[118] S. Tsuruoka and A.D. Randolph, AIChE Symp. Ser. No. 253 (83) (1987) 104.
[119] A. Mersmann, Chem. Eng. Proc. 23 (1988) 213.
[120] E.P.K. Ottens and E.J. De Jong, Ind. Eng. Chem. Fundam. 12 (1973) 179.
[121] J.W. Mullin, Crystallisation, 3rd ed. (Butterworth, London, 1963).
[122] J.Y. Oldshoe, (1983) Chem. Eng. 13 (1983) 3.
[123] R.H. Perry and D. Green, Perry's Chemical Engineers Handbook, 6th ed. (McGraw-Hill, New York, 1984).
[124] J. Nývlt, Design of Crystallisers (CRC Press, Boca Raton, 1992).
[125] J.W. Van Put, G.J. Witkamp, F. Elgersma and G.M. Van Rosmalen, Hydrometallurgy (1993) to be submitted.
[126] D. Pan and A. Mersmann, Appl. Math. Modeling 16 (1992) 214.
[127] J. Nývlt, H. Kocova and M. Cerny, Collect. Czech. Chem. Commun. 38 (1973) 3199.
[128] J.B. Baliga, thesis, Iowa State University (1970) (see also [87]).
[129] J. Nývlt, Chem. Eng. Process. 31 (1992) 39.
[130] A.W. Bamforth, Industrial Crystallisation (Lenoard Hill, London, 1965).
[131] W. Wöhlk and G. Hofmann, Int. Chem. Eng. 27 (1987) 197.
[132] R.C. Bennett, Chem. Eng. 23 (1988) 119.

[133] S.J. Jančić, in: Industrial Crystallisation 87, eds. J. Nývlt and S. Žaćek (Bechyne, Chechoslovakia, 1987) p. 57.

[134] A. Mersmann and M. Kind, Chem. Eng. Technol. 12 (1989) 414.

[135] C.J. Asselbergs, Ph.D. thesis, Delft University of Technology, The Netherlands (1978).

[136] L. Svarovski (ed.), Solid–liquid Separation (Butterworth, London, 1991).

[137] W. Gerhartz, in: Ullmans Encyclopedia of Industrial Chemistry, Vol. B2, 5th ed. (VHC Verlagsgesellschaft, Weinheim, Germany, 1988) p. 3.

[138] A.E. Nielsen and J.M. Toft, J. Cryst. Growth 67 (1984) 278.

[139] O. Söhnel, in: Advances in Industrial Crystallisation, eds. J. Garside, R.J. Davey and A.G. Jones (Butterworth-Heinemann, Oxford, 1991).

[140] G.J. Witkamp, M.M. Seckler, O.S.L. Bruinsma and G.M. Van Rosmalen, J. Cryst. Growth 99 (1990) 1117.

[141] J. Baldyga and J.R. Bourne, in: Encyclopedia of Fluid Mechanics, Vol. 1 (Gulf Publ., Houston, 1986) p. 147.

[142] J.M. Ottino, The Kinetics of Mixing: Stretching, Chaos and Transport (Cambridge University Press, Cambridge, 1989).

[143] M.F. Doherty and J.M. Ottino, Chem. Eng. Sci. 43 (1988) 139.

[144] R.S. Brodkey, Turbulence in Mixing Operations (Academic Press, New York, 1975).

[145] R.S. Brodkey, Chem. Eng. Commun. 8 (1981) 1.

[146] J. Villermaux, ACS Symp. Ser. Vol. 226 (1983) 135.

[147] G.K. Patterson, in: Handbook of Fluids in Motion, eds. N.P. Cheremisinoff and R. Gupta (Ann Arbor Sci. Publ., Ann Arbor, 1983) Ch. 5.

[148] N.S. Tavare, AIChE J. 32 (1986) 705.

[149] K.R. Westerterp, W.P.M. Van Swaaij and A.A.C.M. Beenackers, Chemical Reactor Design and Operation (Wiley, Chichester, 1990).

[150] J. Villermaux, in: Encyclopedia of Fluid Mechanics, Vol. 2 (Gulf Publ., Houston, 1986) p. 707.

[151] T.N. Zwietering, Chem. Eng. Sci. 39 (1984) 1765.

[152] R. Pohorecki and J. Baldyga, Chem. Eng. Sci. 38 (1983) 79.

[153] R. Pohorecki and J. Baldyga, Chem. Eng. Sci. 43 (1988) 1949.

[154] W. Gösele and M. Kind, Chem.-Ing.-Technol. 63 (1991) 59 (in German).

[155] D.E. Fitchett and J.M. Tarbell, AIChE J. 36 (1990) 511.

[156] G. Tosun, Proc. 6th European Conference on Mixing, Pavia, Italy, May 1988 (BHRA, Bedford, UK, 1988) p. 161.

[157] R. Kuboi, M. Harada, J.M. Winterbottom, A.J.S. Anderson and A.W. Nienow, Proc. World Congres III of Chem. Eng., Tokyo, September (1986) p. 1040.

[158] B. Marcant and R. David, AIChE J. 37 (1991) 1698.

[159] J. Stávek, I. Fořt, J. Nývlt and M. Šipek, Proc. 6th European Conference on Mixing, Pavia, Italy, May 1988 (BHRA, Bedford, UK, 1988) p. 171.

[160] J. Baldyga, R. Pohorecki, W. Podgorska and B. Marcant, in: Industrial Crystallisation 90, ed. A. Mersmann (1990) p. 175.

[161] G. Tovgista and H.-P. Wirges, in: Industrial Crystallisation 90, ed. A. Mersmann (1990) p. 169.

CHAPTER 8

Crystallization of Biological Macromolecules

ALEXANDER McPHERSON

Department of Biochemistry, University of California, Riverside, CA 92521

Contents

1. Introduction . . . 419
2. Proteins present special problems for crystallographers . . . 420
3. Properties of macromolecular crystals . . . 423
4. Crystallization strategy . . . 430
5. Methods for promoting supersaturation . . . 432
6. Some additional notes on the crystallization of proteins by variation of pH or temperature . 441
7. Precipitating agents . . . 444
8. Factors influencing protein crystal growth . . . 447
9. Some useful considerations . . . 453
10. Seeding . . . 457
11. Concluding remarks . . . 460
References . . . 461

HANDBOOK OF CRYSTAL GROWTH, VOL. 2
edited by D.T.J. Hurle

1. Introduction

Macromolecular crystallization, which includes the crystallization of proteins, nucleic acids, and larger macromolecular assemblies such as viruses and ribosomes, is based on a rather diverse set of principles, experiences and ideas. There is no comprehensive theory, or even a very good base of fundamental data, to guide us in our efforts. As a consequence, macromolecular crystal growth is largely empirical in nature, and demands both patience and intuition.

Complicating the entire process, in addition to our limited understanding of the phenomena involved, is the astonishing complexity and range of the macromolecules themselves. Even in the case of rather small proteins, such as, e.g. ribonuclease or lysozyme, there are roughly a thousand atoms with hundreds of bonds and thousands of degrees of freedom. For viruses of weights measured in the millions of daltons, the possibilities for conformation, interaction, and mobility are almost unimaginable.

We are only now beginning to attack the development of rational approaches to macromolecular crystallization based on an understanding of the fundamental properties of the systems [1, 2]. We are only now using, in a serious and systematic manner, the classical methods of physical chemistry to determine the characteristics of those mechanisms responsible for the self-organization of large biological molecules into crystal lattices.

As a substitute for the precise and reasoned approaches that we commonly apply to scientific problems, we must rely, for the time being at least, on what is fundamentally a trial-and-error methodology. Macromolecular crystallization is generally a matter of searching, as systematically as possible, the ranges of the individual parameters that impact upon crystal formation, finding a set, or multiple sets, of these factors that yield some kind of crystals, and then optimizing the individual variables to obtain the best possible crystals. This is achieved by carrying out an extensive series, or establishing a vast array, of crystallization trials, evaluating the results, and using what information is obtained to improve conditions in successive rounds of trials. Because the number of variables is so large, and their ranges so broad, experience and insight in designing and evaluating the individual and collective trials becomes an important consideration.

To grow crystals of any compound, a solution of the molecule must by some means be transformed or brought into the supersaturated state. Its return to equilibrium entails segregation of solute molecules into the solid state. If, from a saturated solution, e.g. solvent is gradually withdrawn by evaporation, temperature is lowered or raised appropriately, or if some other property of the system, such as pH, is altered, then the solubility limit may be exceeded and the solution is then said to be supersaturated. If a solid phase is present, or introduced, then strict saturation will be re-established as molecules are recruited into the solid phase.

If no solid is present as conditions are changed, then solute will not immediately partition into two phases, and the solution will remain in the supersaturated state. The solid state does not necessarily develop spontaneously as the saturation limit is exceeded because energy, analogous to the activation energy of a chemical reaction, is required to create the second phase, the stable nucleus of a crystal or a precipitate.

Thus, a kinetic or energy barrier allows conditions to proceed further from equilibrium and further into the zone of supersaturation.

Once a stable nucleus appears in a supersaturated solution, however, it will proceed to grow until the system regains equilibrium. So long as non-equilibrium forces prevail and some degree of supersaturation exists to drive events, a crystal will grow or precipitate will continue to form.

2. *Proteins present special problems for crystallographers*

In a physical sense, perhaps, the crystallization of protein, nucleic acid, or virus is quite similar to the crystallization of conventional small molecules. It requires the creation of a supersaturated solution of the macromolecule followed by the appearance of crystal growth centers or nuclei. Once growth has commenced, emphasis shifts to maintenance of virtually invariant conditions so as to sustain continued, ordered addition of single molecules, or perhaps ordered aggregates, to surfaces of developing crystals.

Some perplexing difficulties arise in the crystallization of macromolecules in comparison with conventional small molecules due to the greater complexity, lability and dynamic properties of proteins and nucleic acids. Figure 1 shows the backbone structure of a protein molecule. The polypeptide chain has two dihedral angles that may vary per amino acid residue. The side chains of each amino acid may be any one of twenty, each with its own conformational possibilities. Thus a protein polymer may have literally thousands of degrees of rotational freedom. The problem of complexity

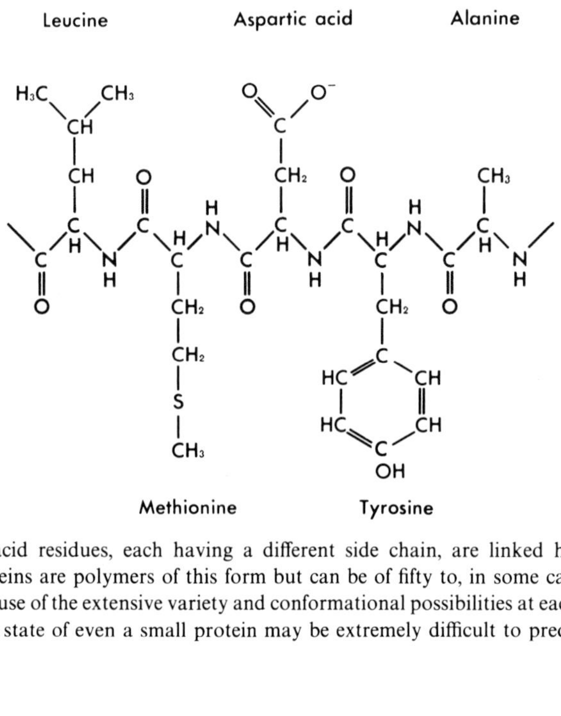

Fig. 1. Five amino acid residues, each having a different side chain, are linked here to form a short polypeptide. All proteins are polymers of this form but can be of fifty to, in some cases, over a thousand monomer units. Because of the extensive variety and conformational possibilities at each residue, the overall conformational state of even a small protein may be extremely difficult to predict or describe.

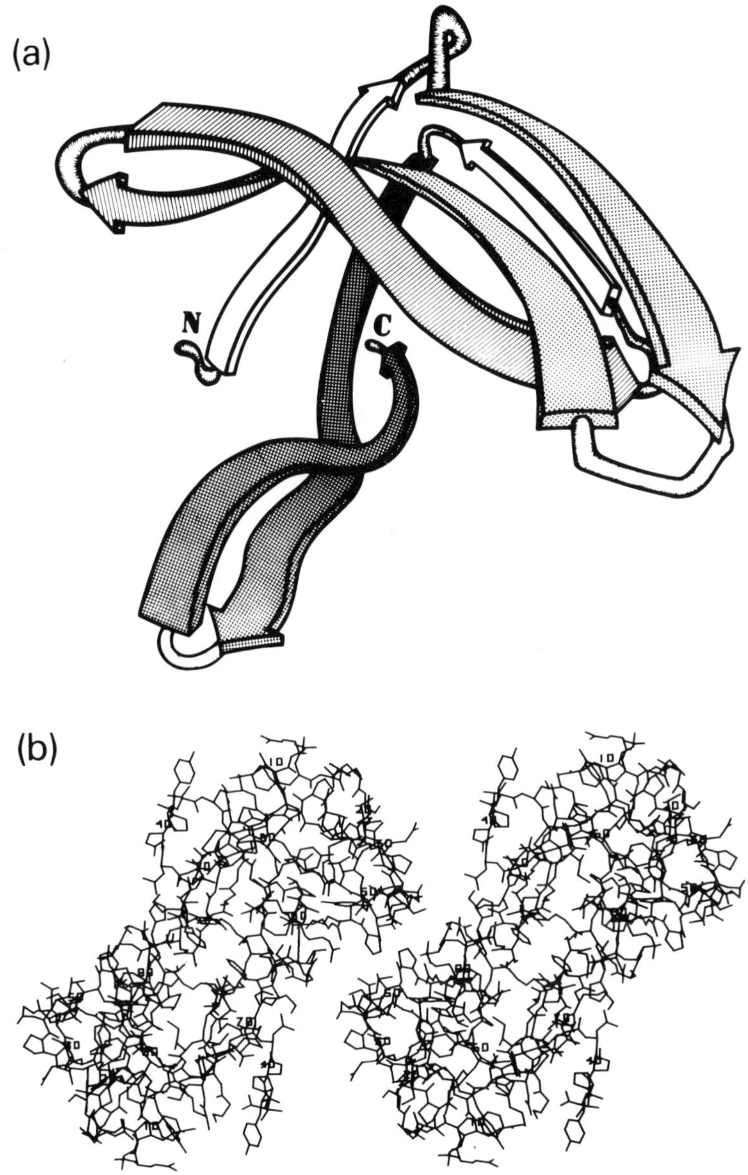

Fig. 2. (a) A schematic drawing of the course of the polypeptide comprising the relatively small Gene 5 protein of 87 amino acids. In spite of its small size, when all atoms of the Gene 5 dimer are presented as a stereodiagram in (b) the complexity is daunting.

and mobility is increased when the three-dimensional possibilities for structure are considered. In fig. 2a, e.g. is a simplified schematic diagram showing the three-dimensional folding of the Gene 5 protein [3]. This is a relatively small protein of only 87 amino acids. Even so, when one views the entire molecule in fig. 2b, even though hydrogen atoms have been omitted, the impression is one of overwhelming complexity. This, however, is typical of what one must contend with when crystallizing proteins, and in general, much larger and more varied structures than this.

The classic phase diagram of saturation characterized by the labile and metastable regions of supersaturation is still applicable to macromolecules, but it must be borne in mind that as conditions are adjusted to transport the solution away from equilibrium by alteration of its physical and chemical properties, the very nature of the solute molecules may change as well. As temperature, pH, pressure or solution are altered, so may be the conformation, charge distribution or mobility of the macromolecules.

Proteins and nucleic acids furthermore are remarkably sensitive to their environment. If exposed to sufficiently severe conditions they may denature, degrade or randomize in a manner that ultimately precludes any hope of their forming crystals. They must always be maintained in a thoroughly hydrated state, at or near physiological pH and temperature. Thus common methods for the crystallization of conventional molecules such as evaporation of solvent, extreme temperature variation, or addition of strong organic solvents are for the most part unsuitable.

To simplify the situation as much as possible, it is essential to practice whatever measures possible to purify the protein free of macromolecular contaminants and to do what is necessary to insure a state of maximum structural and chemical homogeneity. Unfortunately, we are often misled by our standard analytical approaches, such as PAGE or IEF [4,5], into believing that a specific protein preparation is completely homogeneous. This is often illustrated by distinctive differences in the crystallizability of several preparations even when all analyses indicate they are identical. Imperceptible differences can be attributed to various degrees of microheterogeneity within preparations that lie at the margin of our ability to detect them. Table 1 lists some possible causes for microheterogeneity in proteins. Others could well be added.

Table 1
Sources of microheterogeneity

Presence, absence, or variation in a bound prosthetic group, coenzyme, or metal ion.
Variation in the length or composition of the carbohydrate moiety on a glycoprotein.
Proteolytic modification of the protein during the course of isolation.
Oxidation of sulfhydryl groups during isolation.
Reaction with heavy metal ions during isolation or storage.
Presence, absence, or variation in post-translational side-chain modifications such as methylation, amidination, and phosphorylation.
Microheterogeneity in the amino or carboxy terminus or modification of termini.
Variation in the aggregation or oligomer state of the protein, association/dissociation.
Conformational instability due to the dynamic nature of the molecule.
Microheterogeneity due to the contribution of multiple but non-identical genes to the coding of the protein.
Partial denaturation of sample.
Different animals or sources of enzyme preparations.

The acute effects of microheterogeneity on protein crystallization have received increased attention from investigators. Giege et al. [6] have discussed this in detail and provided extensive evidence that purification plays a crucial role in successful crystal growth. Bott et al. [7] similarly showed the pronounced beneficial effects of isoelectric focusing on an otherwise "pure" protein.

3. Properties of macromolecular crystals

Some examples of macromolecular crystals are shown in figs. 3–7. Macromolecular crystals are composed of approximately 50% solvent on average, though this may vary from 25–90% depending on the particular macromolecule. Protein or nucleic acid occupies the remaining volume so that the entire crystal is in many ways an ordered gel with extensive interstitial spaces through which solvent and other small molecules may freely diffuse. This is seen quite dramatically in electron micrographs of protein crystals such as those in figs. 8 and 9.

In proportion to molecular mass, the number of bonds (salt bridges, hydrogen bonds, hydrophobic interactions) that a conventional molecule forms in a crystal with its neighbors, far exceeds the very few exhibited by crystalline macromolecules. Since these contacts provide the lattice interactions essential for the integrity of crystals, this

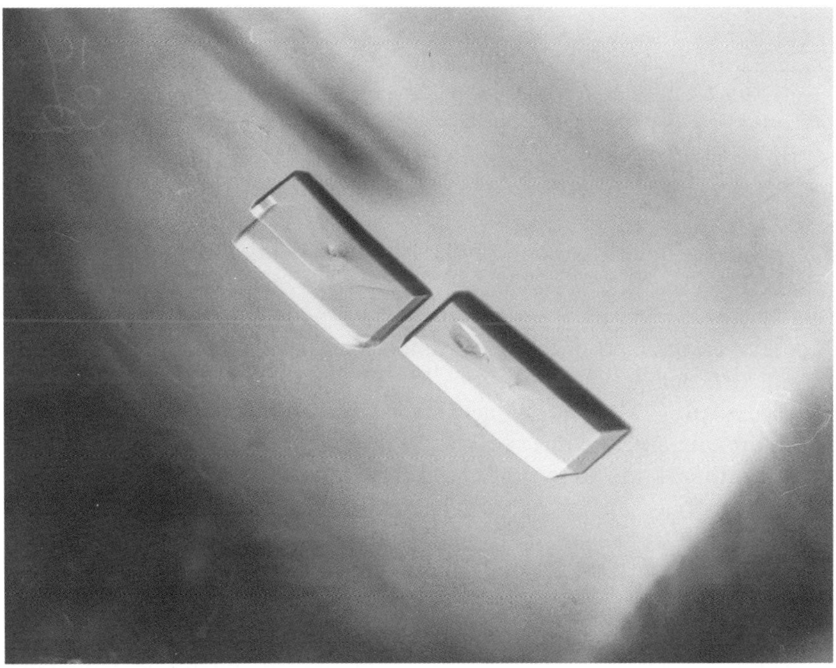

Fig. 3. Crystals grown from polyethylene glycol solution of an Fab fragment of an antibody molecule against a common feline virus. The triclinic crystals, about 0.7 mm in length were grown by vapor diffusion.

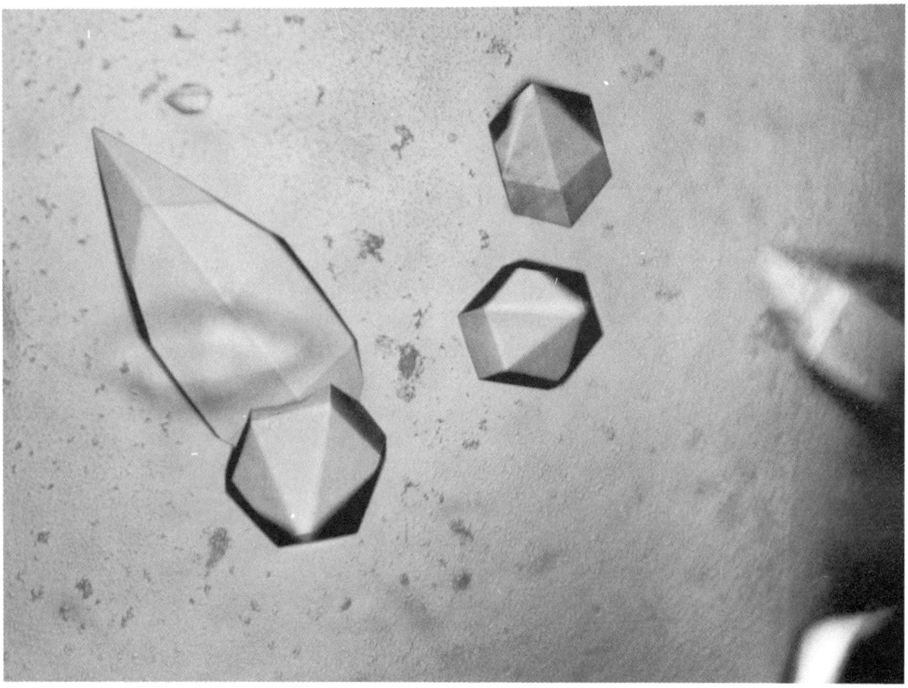

Fig. 4. Large hexagonal crystals, over a millimeter in length, of a carbohydrate-containing protein (approximately 45% sugar residues) from human serum called alpha-1-acid glycoprotein. The crystals were grown by vapor diffusion from about 65% saturated ammonium sulfate solutions in the presence of a ligand and are composed of a large amount of solvent, nearly 70%.

largely explains the difference in properties between crystals of salts or small molecules and macromolecules.

Living systems are based almost exclusively on aqueous chemistry within narrow ranges of temperature and pH. All macromolecules have, thus, evolved compatibility with that milieu and serious deviations or perturbations are rarely tolerated. As a consequence, all protein and nucleic acid crystals must be grown from a solution, one to which they are tolerant, and this is called the mother liquor. Macromolecular crystals have, to this authors knowledge, never been grown except from such aqueous solutions.

Although generally similar in their habits and appearance, there are important practical differences between crystals of low-molecular-mass compounds and crystals of proteins and nucleic acids. Crystals of conventional molecules are characterized by firm lattice forces, are highly ordered, generally physically hard and brittle, easy to manipulate, usually can be exposed to air, have strong optical properties, and diffract X-rays intensely. Macromolecular crystals are by comparison usually more limited in size, are very soft and crush easily, disintegrate if allowed to dehydrate, exhibit weak optical properties and diffract X-rays poorly. Macromolecular crystals are temperature

Fig. 5. Hexagonal prismatic crystals of the pigmented protein catalase from beef liver. This is one of the easiest proteins to crystallize and has a molecular weight of about 250 000 daltons.

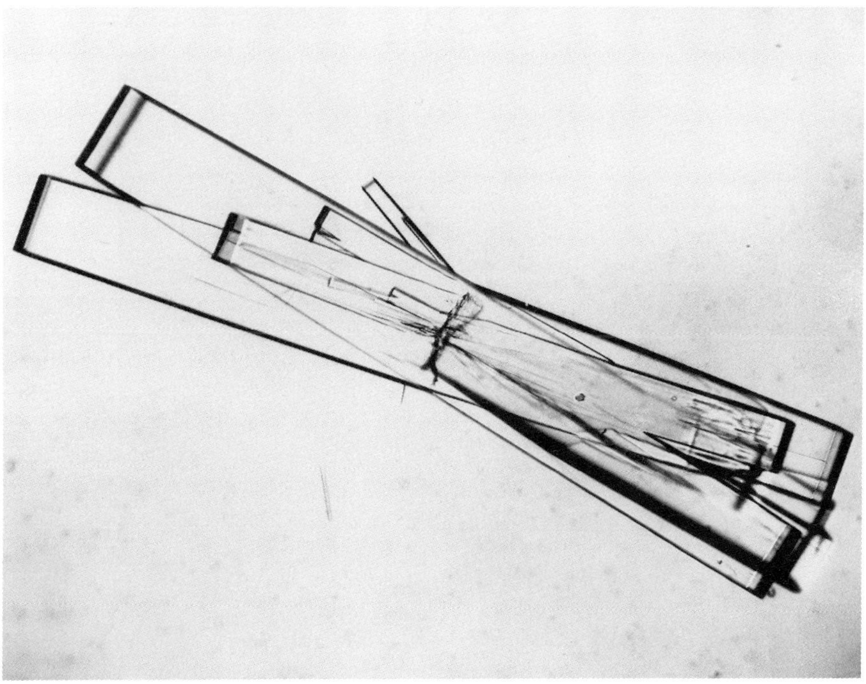

Fig. 6. The crystals seen here are not of protein, but of another macromolecule, transfer RNA, a nucleic acid of about 25 000 daltons, from yeast. Nucleic acid crystals can often be grown using procedures, approaches, and precipitants that are used for proteins.

Fig. 7. Shown here are examples of the largest kind of macromolecular assemblage that has been successfully crystallized, a virus. These crystals of satellite tobacco mosaic virus, a plant pathogen of about two million daltons molecular weight, were grown by vapor diffusion against solutions of 15 to 20% saturated ammonium sulfate using methods essentially identical to those used for crystallizing proteins.

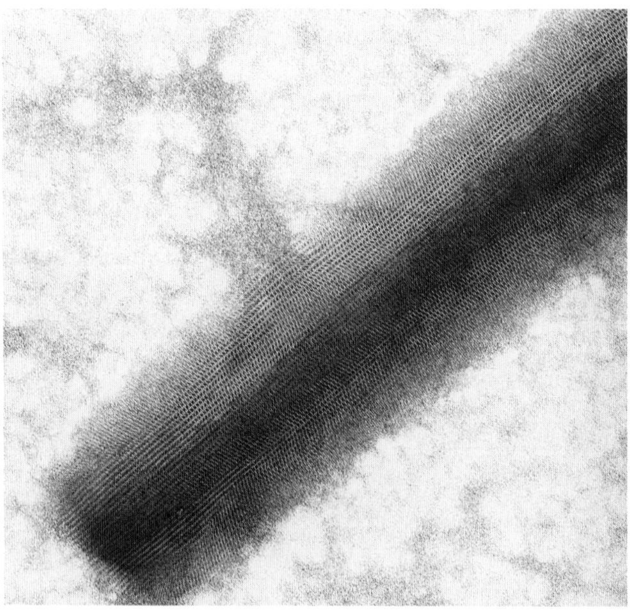

Fig. 8. An elecron micrograph of a microcrystal of the enzyme alpha amylase from a bacteria that has been negatively stained with uranyl acetate. The dark regions of the crystal are solvent channels and interstices that have taken up the heavy atom stain.

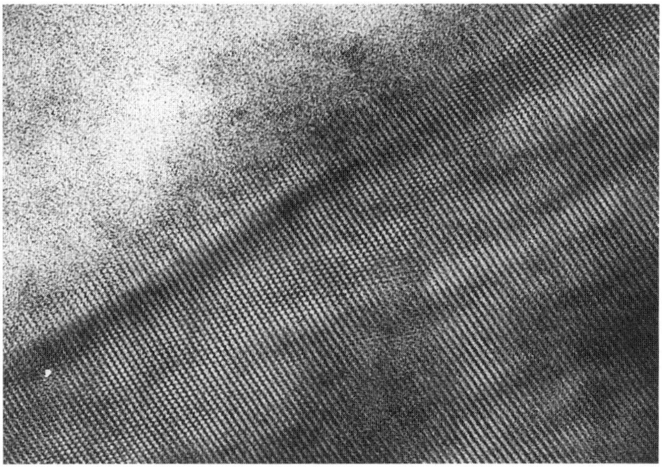

Fig. 9. A microcrystal of the same enzyme, alpha amylase, but isolated from pig pancreas. Again the crystal has been negatively stained with uranyl acetate and the long heavy traces of stain represent solvent channels that permeate the crystal and pass from edge to edge.

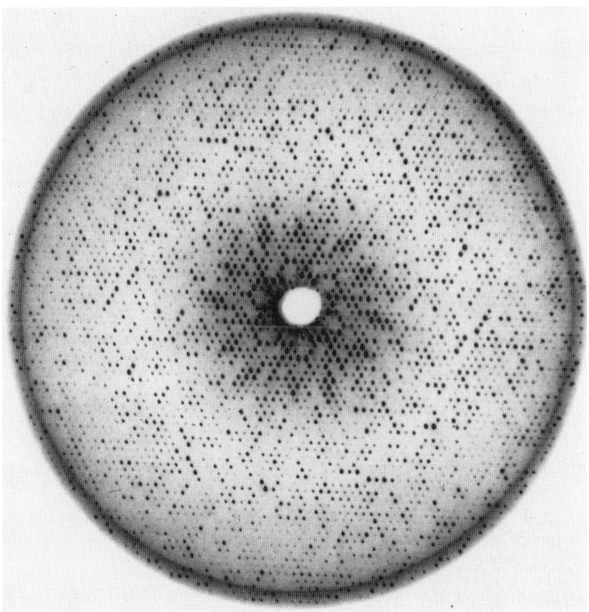

Fig. 10. An X-ray diffraction photograph using a precession-type camera of a crystal of the protein concanavalin B, a storage protein of the Jack Bean. The reflections seen near the edge of the film represent diffraction from planes in the crystal that have separations of about 2.5 Å. Note the very regular lattice on which the reflections appear and the symmetrical distribution of the reflections.

sensitive and undergo extensive damage after prolonged exposure to radiation. In general, at least several crystals must be analyzed for a structure determination to be successful although the advent of area detectors and high-intensity X-ray sources has greatly lessened this constraint.

The extent of the diffraction pattern from a crystal is directly correlated with its degree of internal order. The more vast the pattern, or the higher the resolution to which it extends, the more uniform are the molecules in the crystal and the more precise is their periodic arrangement. The level of detail to which atomic positions can be determined by crystal structure analysis corresponds closely with that degree of crystalline order. While conventional crystals often diffract to their theoretical limit of resolution, protein crystals, by comparison, produce diffraction patterns of only very limited extent. Examples of X-ray diffraction patterns are seen in figs. 10 and 11.

The liquid channels and solvent-filled cavities that permeate macromolecular crystals are primarily responsible for the limited resolution of the diffraction patterns. Because of the relatively large spaces between adjacent molecules and the consequent weak lattice forces, all molecules in the crystal may not occupy exactly equivalent orientations and positions but may vary slightly from lattice point to lattice point. Furthermore, because of their structural complexity and their potential for conformational dynamics, protein molecules in a particular crystal may exhibit slight variations in the course of their polypeptide chains or the dispositions of side groups.

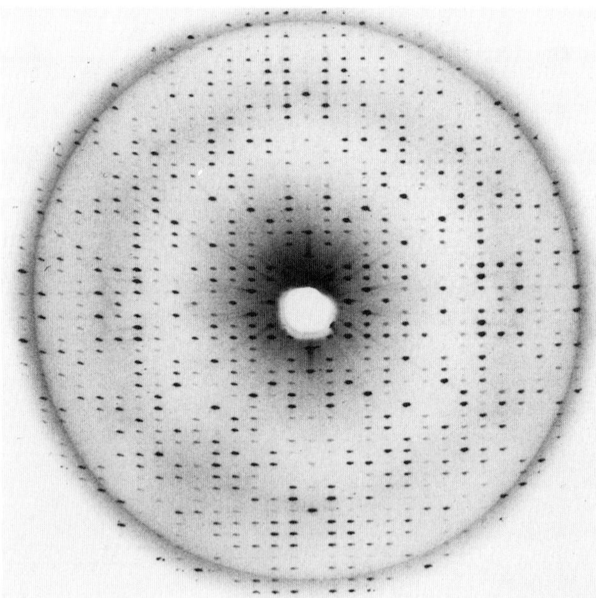

Fig. 11 An X-ray diffraction photograph from a crystal composed of both protein and nucleic acid, the enzyme RNase complexed with oligonucleotides. These crystals yield diffraction intensities to a resolution limit beyond 1.8 Å.

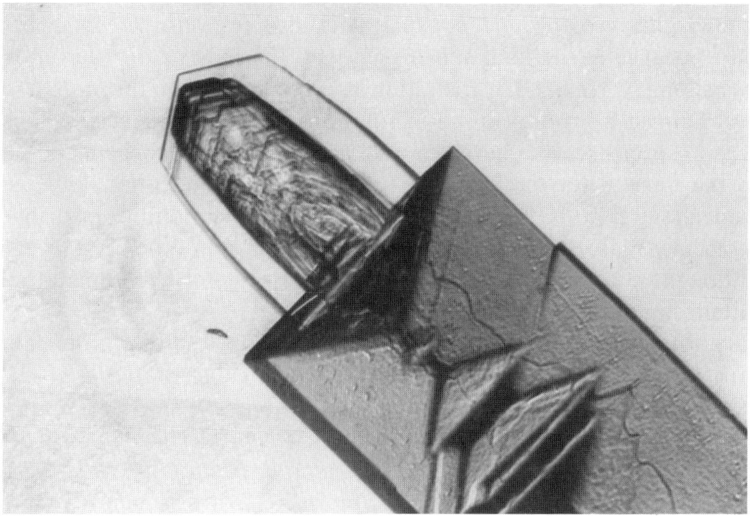

Fig. 12. The thin bladelike crystals of porcine alpha amylase are of space group $P2_12_12$ and have entirely different cell dimensions than the thick block crystal from which it is growing that has space group $P2_12_12_1$.

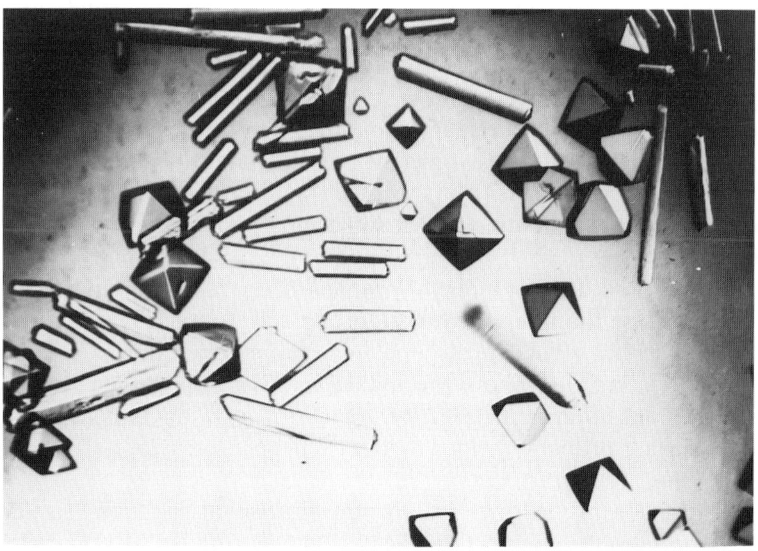

Fig. 13. A mixture of two entirely different crystal forms, hexagonal and cubic, of yeast phenylalanine tRNA are seen growing together in the same sample of mother liquor. This is but one example of the common occurrence of crystal polymorphism in macromolecules.

Although the presence of extensive solvent regions is a major contributor to the generally low quality of protein crystals, it is also responsible for their value to biochemists. Because of the high solvent content, the individual macromolecules in protein crystals are surrounded by hydration layers that maintain the protein's structure virtually unchanged from that found in solution. As a consequence, ligand binding, enzymatic, and spectroscopic characteristics, and most other biochemical features are essentially the same as for the fully solvated molecule. Conventional chemical compounds, which may be ions, ligands, substrates, coenzymes, inhibitors, drugs or other effector molecules may, e.g. be freely diffused into and out of the crystals. Crystalline enzymes, though immobilized, are completely accessible for experimentation simply through alteration of the surrounding mother liquor.

Polymorphism is a particularly common observation with both protein and nucleic acid crystals. Presumably this is a consequence of their conformational dynamic range and the sensitivity of the lattice contacts involved. Thus, different habits and different unit cells may arise from what, by most standards, would be called identical conditions. In fact, multiple crystal forms are sometimes seen coexisting in the same sample of mother liquor. Figures 12 and 13 are examples of a protein and a nucleic acid, respectively, that demonstrate this tendency.

4. Crystallization strategy

The strategy employed to bring about crystallization is to direct the macromolecular solution slowly toward a state of supersaturation by modifying the properties of the solvent. This is brought about by increasing the concentration of some precipitating agent or by altering some physical or chemical property, such as pH. This is illustrated by a version of the classical diagram for supersaturation, modified somewhat for the case of proteins, seen in fig. 14.

In concentrated solutions the macromolecules may aggregate as an amorphous precipitate. This result should be avoided. It is usually indicative that supersaturation has proceeded too far or too swiftly. One must slowly exceed the point of insolubility and thereby allow the macromolecules sufficient opportunity to assemble into a crystalline lattice.

The traditional procedure for inducing proteins to separate from solution as a solid is to gradually increase the level of saturation of a salt. Classically, the salt has been ammonium sulfate, but others are also in common use (see section 7). Frequently, the protein separates as an amorphous precipitate, but with appropriate care, manipulation of salt concentration is an effective means for growing protein crystals. This approach has, in fact, probably yielded more varieties of protein crystals than any other [8].

An explanation for this "salting-out" phenomenon (see also section 7) is that the salt ions and macromolecules compete for chemical bonds with the solvent molecules, i.e. water. Both the salt ions and the protein molecules require hydration layers to maintain their solubility. When competition between ions and proteins becomes sufficiently intense, the protein molecules begin to self-associate in order to satisfy, by intermolecu-

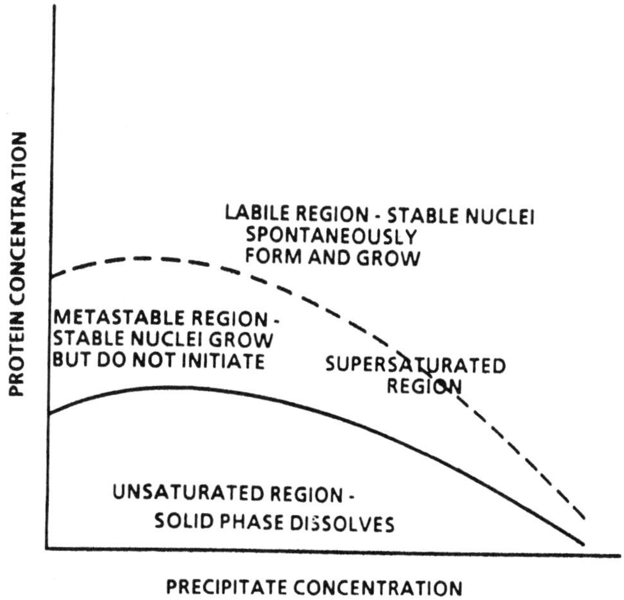

Fig. 14. A phase diagram for a hypothetical protein showing its solubility as a function of precipitant concentration. The solid line represents the maximum solubility or saturation curve for the protein. Note that the protein is less soluble at very low and high concentrations of the precipitant, corresponding to the "salting-in" and "salting-out" regions. The supersaturated region lies above the maximum solubility curve and is, in turn, demarcated by a boundary discriminating the metastable region of supersaturation from the labile region. In the labile region, crystal nuclei can both spontaneously form and grow, while in the metastable region they can only grow.

lar interactions, their electrostatic requirements. Thus dehydration, or the elimination and perturbation of solvent layers around protein molecules induces insolubility. For a specific protein, the solubility minima are usually quite dependent on the pH, temperature, the chemical composition of the precipitant as well as the properties of both the protein and the solvent (for more extensive discussions, see [9–13]).

At very low ionic strength a phenomenon known as "salting-in" occurs in which the solubility of the protein increases as the ionic strength increases from zero. Solubility at very low ionic strength is diminished by the removal of ions essential for satisfying the electrostatic requirements of the protein molecules. As ions are removed, and in this region of low ionic strength cations are most important [9,10], the protein molecules again balance their electrostatic requirements by forming interactions among themselves.

The "salting-in" effect can be applied in the opposite sense of reduced ionic strength and it can be used as a crystallization tool. In practice, a protein that is soluble at moderate ionic strength is extensively dialized against distilled water. Many proteins such as catalase, concanavalin B, immunoglobulins and plant seed proteins have been crystallized by this means [14,15].

Proteins may be driven from solution, at constant pH and temperature, by increases in ionic strength [12], or they can be crystallized or precipitated at constant ionic strength by changes in pH or temperature. This occurs because the charge distribution on the macromolecule, its surface features, or its conformation may change as a function of these variables [9]. By virtue of its ability to inhabit a range of states, proteins may exhibit a number of environmentally dependent solubility minima and each of these minima may afford the opportunity for crystal growth.

Similar effects may be achieved as well by the slow addition to the mother liquor of certain organic solvents such as ethanol or methylpentanediol. The only essential requirement for the precipitant is that, at the specific temperature and pH of the experiment, the additive does not adversely affect the structure and integrity of the protein. This is sometimes a stringent requirement and deserves some consideration. An organic solvent will compete to some extent, like salt, for water molecules. It also reduces the dielectric screening capacity of intervening bulk solvent. Reduction of dielectric shielding increases the effective strength of the electrostatic forces responsible for one protein molecule being attracted to another.

Polymers such as polyethylene glycol also serve to dehydrate macromolecules in aqueous solution, as do salts, and they furthermore alter the dielectric properties in a manner similar to organic solvents. They produce, however, an additional important effect. Polyethylene glycol perturbs the natural structure of the solvent and creates a more complex network having both water and itself as structural elements. A consequence of this restructuring of solvent is that macromolecules, particularly proteins, tend to be excluded and phase separation is promoted [16, 17]

Combinations of approaches using both precipitating agents and physical parameters are frequently effective in promoting the crystallization of macromolecules. Increasing the concentration of a precipitating agent to a point just below supersaturation and then adjusting the pH or temperature to reduce the solubility of the protein is a common method. Modification of pH can be brought about quite well with the vapor diffusion technique (see below) with volatile acids and bases such as acetic acid and ammonium hydroxide.

Because of the complicity of interactions between solute and solvent, and the shifting character of the protein, the methods of crystallization must usually be applied over a broad set of conditions with the objective of discovering the particular minimum (or minima) that yield crystals. In practice, one determines the precipitation points of the protein at sequential pH values with a given precipitant, repeats the procedure at different temperatures, and then examines the effects of different precipitating agents.

5. *Methods for promoting supersaturation*

The oldest technique of those presented in table 2, used for the crystallization of proteins and nucleic acids, having a history of over 150 years [18], is what is commonly called batch crystallization [11, 14, 15, 19–21]. It is the simplest, most thoroughly tested, and when optimized, the most reliable of all crystallization methods. Lysozyme, e.g. is most generally crystallized by the batch method. It relies on the mixing of a

Table 2
Methods for attaining a solubility minimum

(1)	Bulk crystallization.
(2)	Batch method in vials.
(3)	Evaporation.
(4)	Bulk dialysis.
(5)	Concentration dialysis.
(6)	Microdialysis.
(7)	Liquid bridge.
(8)	Free interface diffusion.
(9)	Vapor diffusion on plates or slides.
(10)	Vapor diffusion in hanging drops.
(11)	Sequential extraction.
(12)	pH-induced crystallization.
(13)	Temperature-induced crystallization.
(14)	Crystallization by effector addition.

protein solution at subsaturated concentration with other components. These alter the protein solubility or electrolyte properties of the solution to produce a mother liquor that is supersaturated with respect to protein. From such a solution the solid state may be expected to eventually form, and if the conditions and physical parameters have been chosen correctly, this solid state will be crystalline rather than amorphous. Alternatively, the physical state of a subsaturated protein solution may be altered by, e.g. lowering or raising the temperature. The supersaturated solution produced as a result yields a concomitant formation of crystals.

This technique is attractive because of its inherent simplicity and reproducibility. It requires nothing more than the combination of two or more solutions into one and a period of time until nucleation commences and growth follows. The concept is appealing because it can be carried out with an absolute minimum of human intervention. Because the apparatus is nothing more than a container for a small quantity of fluid, which can be varied as circumstances require from a few microliters to hundreds of milliliters, it is fully amenable to interface with both simple and complex observational systems designed to monitor and record the process of protein crystal growth. Thus the technique is very useful, e.g. for inelastic light scattering studies of nucleation [22–30] or time-lapse-photography investigations of protein crystal growth [31, 32].

Batch crystallization requires no complex arrangements to produce equilibration, either through the liquid or the vapor phase. It relies strictly on the energetics of the system, the activation barriers to nucleation, to regulate the process. While it has not in recent years been extensively used to screen protein crystallization conditions in most laboratories, that reflects as much a prejudice of the investigators and the popularity of other methods as any deficiency in the batch method itself. Because it was commonly used in the past with fairly large volumes of protein solution investigators have mistakenly come to believe that it is inappropriate for small samples, which it most certainly is not. It can even be carried out using microliter amounts in glass capillaries like those shown in fig. 15. It can be a useful screening technique as well as a good

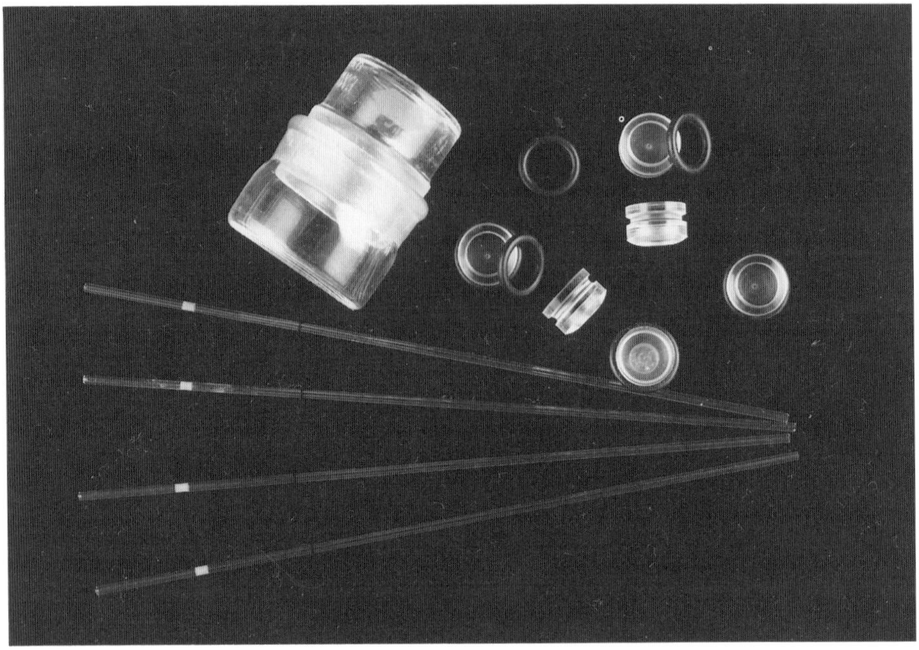

Fig. 15. The thin glass capillaries shown here may be used for either batch crystallization or crystallization by dialysis with volumes in the microliter range. The dialysis buttons seen both face-on and on-edge can be closed by dialysis membrane held in place by the rubber O-rings. Once filled with protein solution they can be submerged in a larger volume of solution contained, e.g. in the small weighing bottle.

method for the preparation of large quantities of protein crystals once optimal values for all parameters have been established.

There are a number of ingenious devices, procedures and methods for dynamically altering the degree of saturation of a protein solution. Generally, these rely on the slow increase in concentration of some precipitant such as salt or polyethylene glycol. These same approaches can of course be used as well for salting-in, modification of pH, and the introduction of materials that alter protein solubility. These techniques have been reviewed elsewhere [14, 15, 33–36] and only some of them will be dealt with here.

Dialysis is familiar to nearly all biochemists as a means of modifying the components of a protein solution. The macromolecule solution is enclosed in a membrane casing or in a container having a semi-permeable membrane partition. The membrane selectively allows the passage of small molecules and ions, but pore size prevents passage of the much larger protein molecules. The vessel or dialysis tube containing the protein is submerged in a larger volume of liquid which has the desired properties of pH, ionic strength, ligands, etc. With successive changes of the exterior solution, and concomitant equilibration of small molecules and ions across the semi-permeable membrane, the protein solution gradually acquires the desired properties.

This procedure, in some manifestation, has been used to crystallize a number of proteins [20, 21, 33–36]. It is generally applicable on a large scale, however, only when

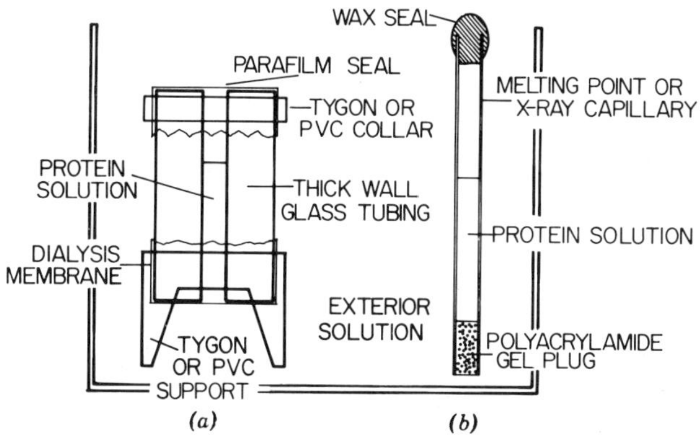

Fig. 16. (a) A microdialysis cell for crystallization of small protein samples closed at one end by a section of dialysis membrane held in place by a tygon ring that also serves as a support. (b) A similar arrangement but the membrane has been replaced by a polyacrylamide plug that is permeable to small molecules and ions.

substantial amounts of the protein are available. It has the advantage that liquid–liquid diffusion through a semi-permeable membrane permits a protein solution to be exposed to a continuum of potential crystal-producing conditions without physically perturbing the mother liquor. Diffusion through the membrane is slow and controlled. Because the rate of change of substituents in the mother liquor is a function of the concentration gradient across the membrane, the nearer the system approaches equilibrium, the more slowly it changes.

This method has been adapted to much smaller amounts of protein by crystallographers who use almost exclusively microtechniques. These involve no more than 5 to 50 μl of protein solution in each trial. First described by Zeppenzauer and Zeppenzauer [37, 38] and subsequently modified and refined by numerous others, the method confines a protein solution within a glass capillary, or the microcavity of a small plexiglass button. The cavity of the button or the ends of the microcapillary tube are closed off by a dialysis membrane. The whole arrangement, charged with protein solution, is then submerged in a much larger volume of an exterior liquid which is held within a closed vessel such as a test tube or vial. The microdyalysis method is shown schematically in fig. 16.

If the exterior solution is at an ionic strength or pH that induces the mother liquor to become supersaturated, crystals may grow. If not, the exterior solution may be exchanged for another and the experiment repeated.

The dialysis buttons, seen in fig. 15, are particularly ingenious. Not only are they compact and easy to examine, but they have a shallow groove about their waist. After a section of wet dialysis membrane is placed over the mother liquor filled cavity, it can be held securely in place by simply slipping a common rubber O ring over the top of the button and seating it in the groove. These buttons are now in wide use and have proven quite successful. Their cavities range in size from 5 to 50 μl and they can be re-used.

The technique of liquid–liquid diffusion, also known as free interface diffusion, for the crystallization of macromolecules [33–36, 39, 40], illustrated in fig. 17, is in principle

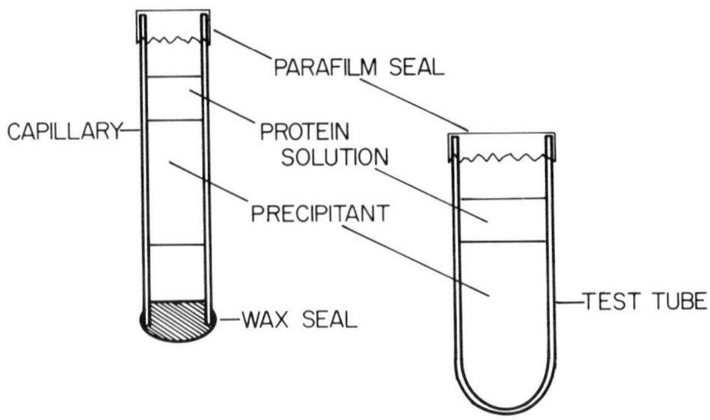

Fig. 17. Diagram illustrating the free interface diffusion technique for the crystallization of microliter amounts of protein solution on the left and milliliter amounts on the right. The method relies on the direct diffusion of precipitant and small molecules across the liquid–liquid interface.

simple and straightforward. It relies on the direct interdiffusion of a precipitant solution with a protein solution across their common liquid–liquid interface. When the two solutions are juxtaposed, precipitant gradients form in the region of the interface and produce local supersaturation. This promotes the appearance of crystal nuclei and subsequent growth to sizes useful for analysis.

The basic idea of liquid–liquid diffusion of a solvent or solute diffusing into a macromolecular solution is really quite similar to dialysis. The method is particularly applicable in microgravity because the necessity of a membrane is removed. The two liquids can, in addition, be maintained relatively stable with respect to one another for long periods of time. The diffusion rate of macromolecules is so much less than that of the small molecules of most precipitants that there is relatively little penetration of protein into the precipitant volume during the course of most experiments.

The technique of liquid–liquid diffusion has been applied successfully in laboratories to grow macromolecular crystals and with considerable success. The data base of Gilliland [8] shows 35 entries due to this technique with another 193 ascribed to various kinds of dialysis. Direct liquid diffusion has not, however, been as extensively utilized in conventional laboratories as some other techniques such as batch or vapor diffusion because of some inherent disadvantages. Significantly, most of these methodological deficiencies are relieved or eliminated in the absence of gravity [41, 42].

On earth there are density-driven convective currents that produce turbulence at the liquid–liquid interface. Transient mixing, if layering of solutions is employed, or the necessity of a membrane as in dialysis, lessens the methods' effectiveness. Sedimentation produces a particularly deleterious effect with the free interface method because nuclei and small growing crystals under the influence of gravity begin falling from the region of the liquid–liquid interface. Thus as a crystal forms, gravity increasingly pulls it downward and out of the local region of the solution where growth might favorably continue. Gravity essentially removes growing crystals from an optimal growth environment and thereby imposes a self-limiting character on the process, in turn, restricting crystal size.

Free interface diffusion, by its very nature, is certain to be at least one method of protein crystal growth that clearly will gain technical advantages from a microgravity environment. Not only will turbulent mixing due to density differences be eliminated at the interface between diffusing liquids, but the absence of gravity will prevent the removal of crystals from zones of optimal growth due to sedimentation. The additional advantages of convection free growth and the minimization of aggregation and consequent defects due to sedimentation that is enjoyed by other methods, however, would be fully maintained. The gains in efficiency resulting from microgravity would, in addition, be volume independent and obtained equally by small microsamples measured in microliters or large vessels containing many milliliters.

Currently, the most widely used method for bringing about supersaturation in microdrops of mother liquid is vapor diffusion [43, 44, 33–36]. This approach enjoys a

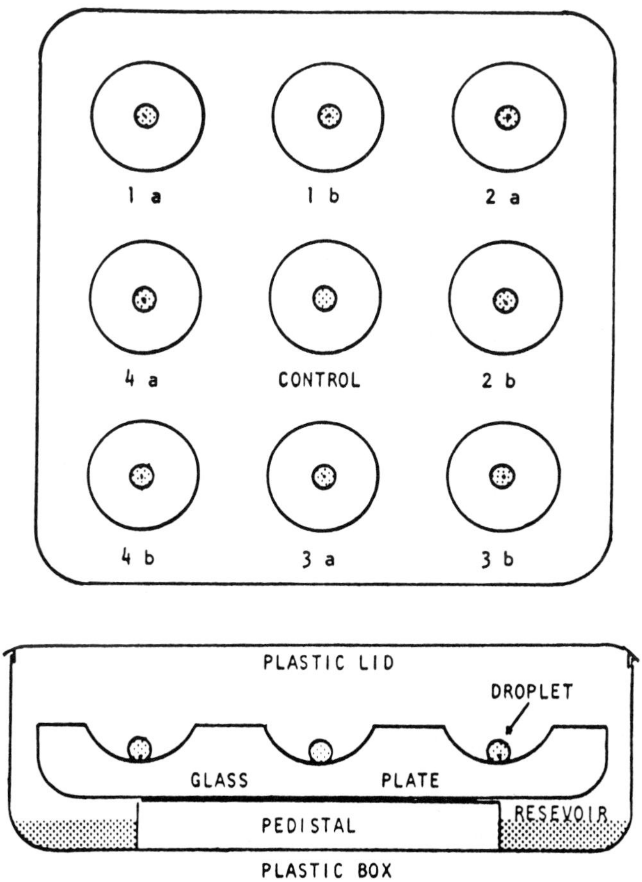

Fig. 18. Schematic diagram of a nine-well depression spot plate enclosed in a plastic sandwich box used for the crystallization of macromolecules by the vapor diffusion method. Microdrops of 10 to 25 μl can be placed in the depressions with 25 ml reservoirs of precipitating solution in the bottom of the box. Equilibration proceeds through the vapor phase.

wide diversity of applications and may be divided into those procedures that use a "sitting drop", which is illustrated in fig. 18, and those employing a "hanging drop". In either form, the method relies on the transport of water or some volatile agent between a microdrop of mother liquor, generally 5 to 25 μl volume, and a much larger reservoir solution of 0.75 to 25 ml in volume. Through the vapor phase the droplet and reservoir come to equilibrium. Because the reservoir is of much larger volume, the final conditions at equilibrium are essentially those of the initial reservoir state. A variety of devices currently in use for protein crystal growth by vapor diffusion are shown in fig. 19.

Through the vapor phase water can be removed slowly from the droplet of mother liquor, its pH may be changed, or volatile solvents such as ethanol may be gradually introduced. As with the liquid–liquid dialysis and diffusion methods, the procedure may be carried out at a number of different temperatures to gain further advantage.

According to a popular procedure [14, 15], droplets of 10 to 20 μl are placed in the nine wells of glass depression plates. The samples are then sealed in transparent containers, such as Pyrex dishes or plastic boxes which hold, in addition, reservoirs of 20 to 50 ml of the precipitating solution. As shown in fig. 18, the plates bearing protein or nucleic acid samples are supported above the reservoirs by the inverted halves of disposable Petri dishes.

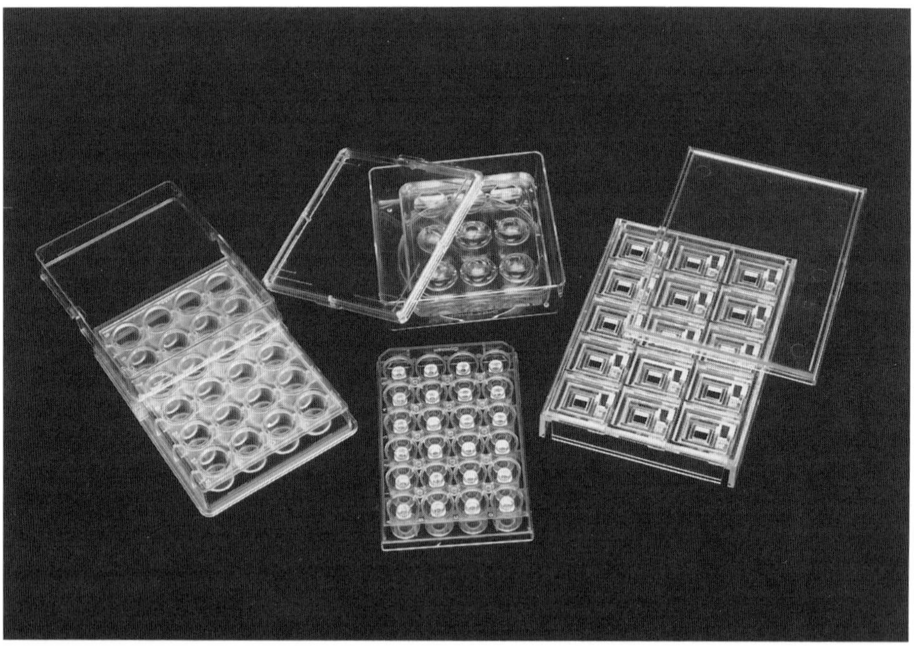

Fig. 19. Four examples of trays, plates or combinations used for crystallization by vapor diffusion. All are commercially available and can be employed for either the hanging-drop or the sitting-drop approach. That at back and center is the arrangement shown schematically in fig. 18.

Through the vapor phase, the concentration of salt or organic solvent in the reservoir equilibrates with that in the sample. In the case of salt precipitation, the droplet of mother liquor must initially contain a level of precipitant lower than the reservoir, and equilibration proceeds by passage of water out of the droplet and into the reservoir. This holds true for nonvolatile organic solvents, such as methylpentanediol and for polyethylene glycol as well. In the case of volatile precipitants, none need be added initially to the microdroplet, as distillation and equilibration proceed in the other direction.

The method described here has the advantage that it requires only micro-amounts of material and is ideally suited for screening large numbers of conditions. A disadvantage is that all the samples in a single box must be equilibrated against the identical reservoir solution. It does, however, permit some flexibility in varying conditions once the samples have been dispensed, this by modification of the concentration or pH of precipitants in the reservoir. When clear plastic boxes are used, large numbers of samples can be quickly inspected under a dissecting microscope and conveniently stored.

The disadvantage of identical reservoir conditions throughout a single box was overcome to a great extent by the introduction of two plastic plates specifically designed for protein crystallization. One of these, sponsored by the American Crystallographic Association, is a plastic plate having accommodation for fifteen protein samples. Each chamber has a separate reservoir compartment and the mother liquor microdroplet may be either suspended from the underside of a glass cover slip, as in the "hanging drop" method, or sandwiched between two glass cover slips. Sealing of the chambers from air requires silicon grease or oil between cover slips and the plastic rims of the chambers. With these plates, the optical properties are very good but equilibration tends to be slow.

A second crystallization plate [36,44], also shown in fig. 19 is produced by Cryschem Inc. With these plates, the drop is sitting on the top of a clear support post that protrudes upward from a circular cavity containing the reservoir solution. The chambers can be rapidly and conveniently sealed from air by clear plastic tape pressed onto the upper surface of the plate after the reservoirs have been filled and the drops of mother liquor dispensed. Equilibration, as with the other plate, is through the vapor phase. While the optical properties are somewhat less favorable with the Cryschem devices, they are convenient, compact and can be utilized for rapid screens of crystallization conditions.

The "hanging drop" procedure also uses vapor phase equilibration but, with this approach, a microdroplet of mother liquor (as small as 3 μl) is suspended from the underside of a microscope cover slip, which is then placed over a small well containing a milliliter of the precipitating solution. The wells are most conveniently supplied by disposable plastic tissue culture plates that have 24 wells with rims that permit sealing by application of silicone vacuum grease around the circumference. These plates provide the further advantages that they can be easily examined under a dissecting microscope and they allow compact storage. The hanging-drop technique can be used both for the optimization of conditions and for the growth of large single crystals.

While the principle of equilibration with both the "sitting drop" and the "hanging drop" are essentially the same, they frequently do not give the same results even though the reservoir and protein solutions are identical. Presumably because of differences in the apparatus used to achieve equilibration, the path is different, even though the end point may be the same. In some cases there are striking differences in the degree of reproducibility, final crystal size, morphology, required time, or degree of order. These observations illustrate the important point that the route leading to supersaturation, the kinetics of the process, may be as important as the end point achieved.

There are, of course, a nearly limitless number of variations on these themes for liquid–liquid or vapor diffusion ranging from the very simple to the unnecessarily complex. One need only use a bit of imagination to devise a novel apparatus for oneself. In fig. 20, e.g. are two very easy arrangements requiring only a depression microscope slide.

As noted earlier, one of the most powerful techniques for producing a supersaturated protein solution is adjustment of pH to values where the protein is substantially less soluble. This may be done in the presence of a variety of precipitants so that a spectrum of possibilities are created whereby crystals might form. The gradual alteration of pH is particularly useful because it may be accomplished by several gentle approaches that do not otherwise perturb the system.

Although microdialysis is probably equally suitable, more success has been achieved with the vapor diffusion method using "sitting" microdroplets on spot plates or in one of the plastic plates available for protein crystallization. The ambient salt, effector, or buffer conditions are established prior to dispensing the microdroplets in the depressions on the plate. The pH is then slowly raised or lowered by adding a small amount of volatile acid or base to the reservoir. Diffusion of the acid or base then occurs from reservoir to sample, just as for a volatile precipitant.

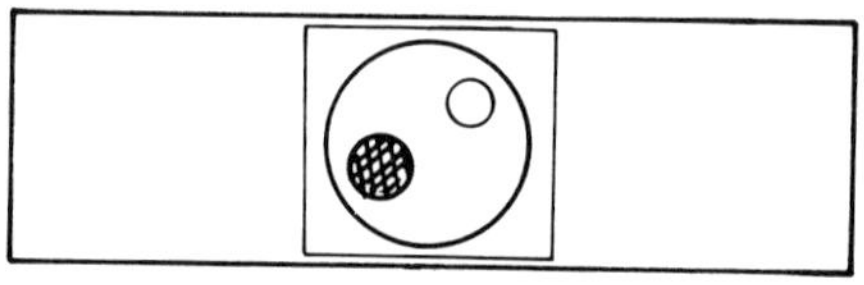

(a)

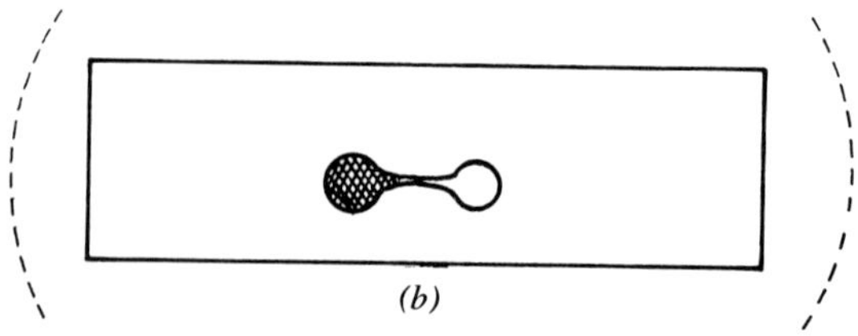

(b)

Fig. 20. (a) The well of a single-depression microscope slide, in which a droplet of protein solution can be set near a larger drop of a precipitating solution and sealed from the air. (b) A droplet of protein and precipitating solution are connected by a fine liquid bridge drawn with a needle between them. Free diffusion of the precipitant through the bridge induces crystal growth in the protein.

If the pH is to be raised, for example, a small drop of concentrated ammonium hydroxide can be added to the reservoir; a drop of acetic acid may be used to lower it. The pH can also be gradually lowered over a period of days by simply placing a tiny chip of dry ice in the reservoir. The liberated CO_2 diffuses and dissolves in the mother liquor to form weak carbonic acid. When a specific pH endpoint is required, the mother liquor may be buffered with suitable compounds at that point and then moved significantly away by addition of acid or base. The microdroplets of mother liquor may then be returned to the buffer point by addition of an appropriate volatile acid or base to the reservoir.

6. *Some additional notes on the crystallization of proteins by variation of pH or temperature*

As shown in fig. 21, individual proteins display distinct solubilities at varying levels of salt concentration and can be selectively separated from solution as amorphous precipitate or crystals. At very low ionic strengths where "salting-in" occurs, the solubility of the protein increases with the ionic strength. In this range the cations are most important, and the solubility increase occurs because of a decrease in the activity coefficient of the protein.

The "salting-out" effect, which Hofmeister [12] suggested was caused by a reduction of the chemical activity of water by salt, involves principally the anions. It occurs as the ionic strength increases. Because ionic strength is a function of the second power of the

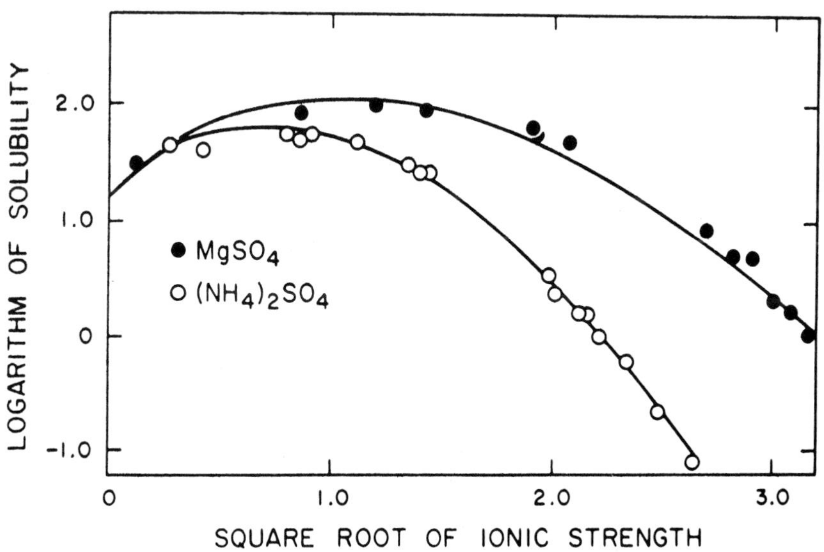

Fig. 21. The solubility of the protein enolase at constant pH and temperature as a function of ionic strength produced by two separate salts. The two ends of the curves correspond to the "salting-in" and the "salting-out" regions of the solubility diagram.

valence, multivalent ions are most efficient. In the "salting-out" region of a protein's solubility profile, the logarithm of protein solubility as a function of the ionic strength is linear and can be expressed by

$$\log S = \beta - K_S(I/2),$$

where I is the ionic strength in moles per kilogram of water, S is the solubility of the protein in grams per kilogram of water, and β and K_S are constants. K_S serves as a measure of the slope of the solubility extrapolated to zero ionic strength and β is the logarithm of the solubility at that point [9].

In principle, the precipitation or crystallization of a protein is effected by manipulating the factors that appear in the above equation. One can crystallize a macromolecule by increasing the ionic strength at constant pH and temperature, a K_S fractionation procedure, or it can be brought about at constant ionic strength by manipulating the pH and temperature, which in most cases strongly influence β. This is illustrated for two typical proteins in fig. 22.

With all techniques the objective is to produce a supersaturated protein solution by very gradual and nonperturbing adjustment of conditions. Since both pH and temperature can be accurately and sensitively altered without direct contact with the protein solution, they provide an attractive approach. In practice, a solubility minimum is determined for the protein at some specific pH, temperature, and ionic strength. The

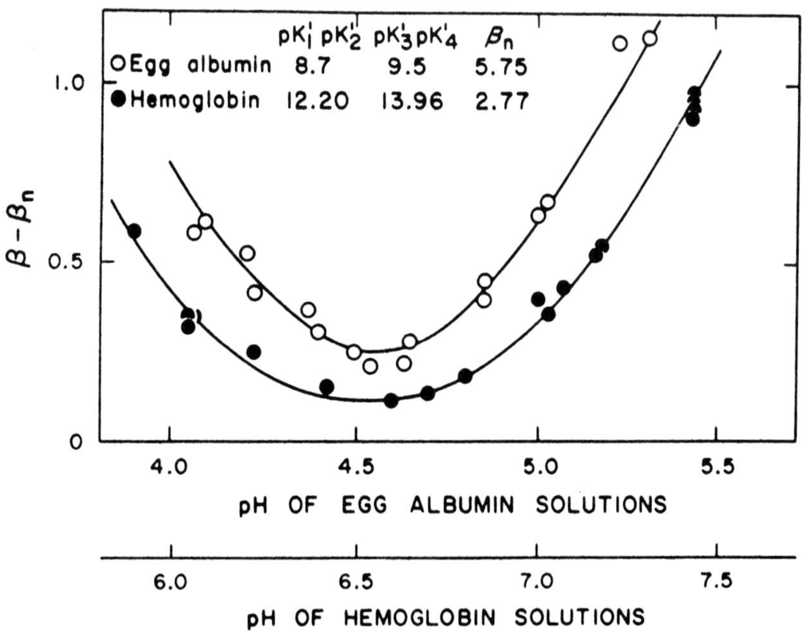

Fig. 22. The solubility behavior of two proteins, hemoglobin and egg ovalbumin, at constant ionic strength as a function of pH. The solubility minima often are good targets for the crystallization of proteins by variation of pH.

protein solution is set initially at the requisite salt concentration for this solubility minimum but either the pH or temperature is set so as to maintain protein solubility. The temperature is then slowly allowed to relax to a value that minimizes protein solubility, or the pH is allowed to gradually equilibrate via the liquid or vapor phase to an appropriate value.

In the case of temperature shifts, the protein is generally raised to a higher temperature as is done, e.g. with squash globulin, insulin, or excelsin [14, 15, 18], a process which increases its solubility. The solution is then allowed to cool slowly to a point at which the protein is no longer soluble. This can be accomplished by placing the warm solution in an insulated or thermally controlled box, or by placing the sample in a Dewar flask. Because of the thermal stability, the approach to supersaturation occurs at an almost imperceptible rate. Exactly the same technique can be applied if the protein is more soluble at low than at high temperature. Under these circumstances, the protein solution is cooled under a given set of ionic strength conditions and then allowed to warm very slowly. This was used, for example, to grow crystals of DNase by Kunitz [45] and is the basis of Jacoby's sequential extraction procedure [46] using decreasing concentrations of ammonium sulfate.

In the case of pH-induced crystallization, the protein, buffered appropriately for the final pH and made essentially insoluble by some level of salt or other precipitant, is raised to a high pH of about 9.0 with ammonium hydroxide to achieve solubilization. Alternatively, it may in some cases be resolubilized by the addition of acetic acid, which lowers the pH to the region of 3.0–4.0. In the first case, NH_3 leaves the sample through the vapor phase; in the second case, acetic acid directly evaporates from the sample. By vapor equilibration, the protein solution returns to neutrality, or whatever final point was specified by the included buffer. The protein becomes insoluble and, hopefully, crystallizes.

The technique has a number of variations; it can be used in association with direct liquid dialysis or with vapor diffusion methods as well. It can, furthermore, be employed in conjunction with a host of different precipitating agents. Temperature can be utilized in an analogous manner [47].

Proteins and nucleic acids are conformationally flexible and generally exist in several conformational equilibrium states. In particular, they may assume a substantially different conformation when they have bound coenzyme, substrate, or some other ligand. As a consequence, a protein with bound effector may exhibit appreciably different solubility properties than the native. In addition, if many conformational states are available, the presence of effector may be used to select for only one of these, thereby imposing a degree of conformity of structure that would otherwise be absent.

The addition of ligands can sometimes be employed to induce supersaturation and crystallization in those cases where its binding to the protein produces solubility differences under a given set of ambient conditions. The effector may be slowly and gently combined with the protein, e.g. by dialysis so that a resulting complex is supersaturated. The addition of ligands, substrates, and other small molecules has seen widespread use in protein crystallography, since it provides useful alternatives if the apoenzyme itself cannot be crystallized.

7. *Precipitating agents*

Precipitants of macromolecules fall into four broad categories: (1) salts; (2) organic solvents; (3) long-chain polymers; and (4) low-molecular-weight polymers and non-volatile organic compounds. The first two classes are typified by ammonium sulfate and ethyl alcohol, respectively, and higher polymers such as polyethylene glycol 4000 are characteristic of the third. In the fourth category we might place compounds such as methylpentanediol and low-molecular-weight polyethylene glycol. Commonly used members of the first two groups are presented in tables 3 and 4.

As described above, salts exert their effect principally by dehydrating proteins through competition for water molecules. Their efficiency in this is a proportional

Table 3
Salts used in crystallization of proteins

(1)	Ammonium or sodium sulfate.
(2)	Lithium sulfate.
(3)	Lithium chloride.
(4)	Sodium or ammonium citrate.
(5)	Sodium or ammonium phosphate.
(6)	Sodium or potassium or ammonium chloride.
(7)	Sodium or ammonium acetate.
(8)	Magnesium sulfate.
(9)	Cetyltrimethyl ammonium salts.
(10)	Calcium chloride.
(11)	Ammonium nitrate.
(12)	Sodium formate.
(13)	Sodium-potassium tartrate.

Table 4
Organic solvents used in crystallization of proteins

(1)	Ethanol.
(2)	Isopropanol.
(3)	1,3-Propanediol.
(4)	2-Methyl-2,4-pentanediol (MPD).
(5)	Dioxane.
(6)	Acetone.
(7)	Butanol.
(8)	Acetonitrile.
(9)	Dimethyl sulfoxide.
(10)	2,5-Hexanediol.
(11)	Methanol.
(12)	1,3-Butyrolactone.
(13)	*n*-Propanol.
(14)	Pyridine.

function of the square of the valences of the ionic species composing the salt [9–12]. Thus, multivalent ions, particularly anions, are the most efficient precipitants. Sulfates, phosphates and citrates have traditionally been employed with success.

One might expect little variation among different salts so long as the valences of their ions are the same, thus there should be little variation between two different sulfates such as Li_2SO_4 and $(NH_4)_2SO_4$. This, however, is often observed not to be the case. In addition to salting-out, which is a general dehydration effect, or reduction of the chemical activity of water, there are also specific protein–ion interactions that may have other consequences. This is perhaps not unexpected given the unique polyvalent character of individual proteins, their structural complexity, and the intimate dependence of their physical properties on their surroundings. It is inadequate, therefore, when attempting to crystallize a protein to examine only one or two salts and ignore a broader range. Changes in salt can sometimes produce crystals of varied quality, morphology, and, in some cases, diffraction properties.

It is usually not possible to predict the degree of saturation or molarity of a precipitating agent required for the crystallization of a particular protein or nucleic acid without some prior knowledge of its behavior. In general, however, it is a concentration just a few percent less than that which yields an amorphous precipitate, and this can be determined for a macromolecule under a given set of conditions using only minute amounts of material.

To determine the approximate insolubility points with a particular precipitant a 10 µl droplet of a 5 to 15 mg/ml protein solution can be placed in the well of a depression slide and observed under a low-power light microscope as increasing amounts of saturated salt solution or organic solvent (in 1- or 2-µl increments) are added. If the well is sealed between additions with a cover slip, the increases can be made over a period of many hours. Indeed, a droplet should be allowed to equilibrate for 10 to 30min after each addition, and longer in the neighborhood of the precipitation point. If larger amounts of material are available samples may be dialyzed in standard $\frac{1}{4}$in. celluloid tubing or in microdialysis buttons against precipitant solutions that are incremented over a period of time until crystallization or precipitation occurs.

The volatile organic solvents most commonly utilized (see table 4) have been ethanol, acetone, butanol and a few other common laboratory reagents. Organic solvents have been of even more general use for the crystallization of nucleic acids, particularly tRNA and oligonucleotides. This in part stems from the greater tolerance of polynucleotides to organic solvents and their polyanionic surfaces which appear to be even more sensitive to dielectric effects than are proteins.

Organic solvents should be used at a low temperature, at or below 0°C, and they should be added very slowly with good mixing. Since they are usually volatile, vapor diffusion techniques are equally applicable for both bulk or micro-amounts. Ionic strength should, in general, be maintained low and whatever means are otherwise available should be pursued to protect against denaturation.

Polyethylene glycol is a polymer produced in various lengths, from several to many hundreds of units. It exhibits as its most conspicuous feature a regular alteration of either oxygen atoms and terminal glycols. In addition to its volume exclusion property, it shares some characteristics with salts in that it competes for water and produces dehydration, and with organic solvents which alter the dielectric properties of the medium.

Aside from its general applicability and utility in obtaining crystals for diffraction analysis [48], PEG also has the advantage that it is most effective at minimal ionic strength. In addition, polyethylene glycol provides a low-electron-density medium. The first feature is important because it promotes higher ligand binding affinities than does a high ionic strength medium. As a consequence there is greater ease in obtaining isomorphous heavy-atom derivatives and in forming protein–ligand complexes for study by difference Fourier techniques. The second characteristic, a low electron-dense medium, implies a generally lower background or noise level for protein structures derived by X-ray diffraction and presumably, therefore, a more ready interpretation.

A large number of protein structures have now been solved using crystals grown from polyethylene glycol. These confirm that the protein molecules are in as native a condition in this medium as in any otherwise used. This is probably so, because the larger-molecular-weight polyethylene glycols probably do not even enter the crystals and therefore do not directly contact the interior molecules. In addition, it appears that crystals of many proteins when grown from polyethylene glycol are essentially isomorphous with, and exhibit the same unit-cell symmetry and dimensions, as those grown by other means.

PEG sizes from Mr = 400 to 20 000 have successfully provided protein crystals, but the most useful are those in the range 2000–8000. A number of cases have appeared, however, in which a protein could not easily be crystallized using this range but yielded in the presence of PEG 400 or 20 000. The molecular-weight sizes are generally not completely interchangeable for a given protein even within the mid range. Some produce the best-formed and largest crystals only at, say, Mr = 4000 and less perfect examples at other weights. This is a parameter which is best optimized by empirical means along with concentration and temperature. It might be noted that the very-low-molecular-weight PEGs, such as 200 and 400, are rather similar in character to MPD and hexanediol.

There does not appear to be any correlation between the molecular weight of a protein and that of the PEG best used for its crystallization. It is however, certainly the case that the higher-molecular-weight PEGs have a proportionally greater ability to force proteins from solution, and really obstinate polypeptides might best be approached with PEG 20 000 at the outset.

A distinct advantage of polythylene glycol over other precipitating agents is that most proteins crystallize within a fairly narrow range of PEG concentration; this being from about 4 to 18%. In addition, the exact PEG concentration at which crystals form is rather insensitive. If one is within 2 or 3% of the optimal value, some success is likely to be achieved. With most crystallizations from high-ionic-strength solutions or from organic solvents, one must be within 1 or 2% of an optimum lying anywhere between 5 and 90% saturation. The great advantage of PEG is that when conducting a series of initial trials to determine what conditions will give crystals, one can use a fairly coarse selection of concentrations and over a rather narrow total range. This means fewer trials with a corresponding reduction in the amount of protein expended.

The time required for crystal growth with PEG as the precipitant is also generally shorter than with ammonium sulfate or MPD but occasionally longer than required by volatile organic solvents such as ethanol. Although equilibration times will depend on the differential between starting and target concentrations, if this is no more than 3 or 4% then crystallization may occur within a few hours or a few days. It seldom requires more than 2 weeks. Thus evaluation of results can be made without undue demands on patience. It should be noted that protein–PEG solutions are excellent media on which to grow microbes, particularly molds, and if crystallization is being attempted at room temperature or over extended periods of time, then some retardant such as azide or chlorobutanol (commonly 0.1%) must be included in the protein solutions.

Since PEG solutions are not volatile, PEG must be used like salt or MPD [39] and equilibrated with the protein by dialysis, slow mixing, or vapor equilibration. This latter procedure, utilizing either 10 μl hanging drops over 0.5 ml reservoirs or 20 μl drops on multidepression glass plates in a sealed chamber, has proven most popular. The author has found that when the reservoir concentration is in the range of 5–12%, the protein solution to be equilibrated should be at an initial concentration of about half. That is conveniently obtained by adding 10 μl of the reservoir to 10 μl of the protein solution. When the final PEG concentration to be attained is much higher than 12%, it is advisable to start the protein at no more than 4–5% below the final value. This reduces unnecessary delays during which time the protein could denature.

Crystallization of proteins with PEG has proved most successful when the ionic strength is low. Tris or cacodylate buffer at 10 to 20 mM are good buffers in this regard. If crystallization proceeds too rapidly, addition of some neutral salt may be used to slow growth and better effect crystal form. PEGs are useful over the entire pH range and over a broad temperature range. PEG appears to be an excellent crystallization agent over the whole spectrum of proteins, although in many specific cases other precipitants may be superior.

8. *Factors influencing protein crystal growth*

Table 5 lists physical, chemical and biological considerations that may, to a greater or lesser extent, be important to the crystallization of macromolecules. Arriving at a just assignment of importance for each factor is difficult for several reasons. Every protein varies in its properties and, surprisingly perhaps, this applies even to proteins that differ by no more than one or just a few amino acids. It occasionally happens that the identical protein prepared by different procedures or at different times may show significant variations. In addition, a factor may differ in importance for each individual protein. For example, α-amylase and catalase are quite sensitive to temperature, while ovalbumin and ferritin show little, if any, alteration in crystallization properties as a function of that variable [14, 15].

Because each protein species is unique, it is virtually impossible to predict the specific values of a variable, or combinations of conditions that might be most profitably explored. Finally, the array of parameters under one's control are not independent of one another and their interrelations may be complex and ambiguous. It is, therefore,

Table 5
Factors affecting protein crystal growth

(1)	pH.
(2)	Ionic strength.
(3)	Temperature.
(4)	Concentration of precipitant.
(5)	Concentration of macromolecule.
(6)	Purity of macromolecules.
(7)	Additives, effectors, and ligands.
(8)	Organism source of macromolecule.
(9)	Substrates, coenzymes, inhibitors.
(10)	Reducing or oxidizing environment.
(11)	Metal ions.
(12)	Rate of equilibration.
(13)	Surfactants or detergents.
(14)	Gravity.
(15)	Vibrations and sound.
(16)	Volume of crystallization sample.
(17)	Presence of amorphous material.
(18)	Surfaces of crystallization vessels.
(19)	Proteolysis.
(20)	Contamination by microbes.
(21)	Pressure.
(22)	Electric and magnetic fields.
(23)	Handling by investigator.

not easy to elaborate rational guidelines relating to physical factors or ingredients in the mother liquor that can increase the probability of success in crystallizing a particular protein. The specific components and conditions must be carefully researched and refined for each individual.

Because temperature may be of great importance or have little bearing at all, in general, it is wise to conduct parallel investigations at 4°C and at 25°C. Even if no crystals are observed at either temperature, differences in the solubility behavior of the protein as a function of precipitant and effector molecules may give some indication as to whether temperature is likely to play an important role. If crystals are observed to grow at one temperature and not, under otherwise the same conditions, at the other, then further refinement of this variable is necessary. This is accomplished by conducting the trials under the previously successful conditions over a range of temperatures centered on that which initially yielded crystals.

Proteins in a high salt solution are usually more soluble at low than at higher temperatures. Proteins, however, generally precipitate or crystallize from a lower concentration of polyethylene glycol, methylpentanediol or organic solvent at low than at higher temperature. Diffusion rates, however, are less and equilibration occurs more slowly at low than higher temperature, so that the time required for precipitation or crystal formation may be longer at lower temperatures.

Following precipitant concentration, the next most important variable in protein crystal growth appears to be pH. This is a consequence of the charge distribution on a protein. All of its attendant physical and chemical properties are intimately associated with the ionization state of the amino acids or chemical groups that comprise the macromolecule. Not only does the net charge on the protein change with pH, but so do the arrangement of those charges, the dipole moment of the protein, its conformation, and in some cases its aggregation state. Thus, an investigation of the behavior of a specific protein as a function of pH is perhaps the single most essential analysis that should be carried out in attempting to crystallize the macromolecule.

As with temperature, the procedure is to first conduct multiple crystallization trials at coarse intervals over a broad pH range and then repeat the trials over a finer matrix of values in the neighborhoods of those that initially showed promise. The only limitations on the width of the initial range screened are the points at which the protein exhibits denaturation. In refining the pH for optimal growth, it should be recalled that the difference between amorphous precipitate, microcrystals, and large single crystals may be only a few tenths of a pH unit [36–38].

In addition to utilizing pH for the optimization of crystzal size, it is sometimes useful as well to explore variation of pH as a means of altering the habit or morphology of a

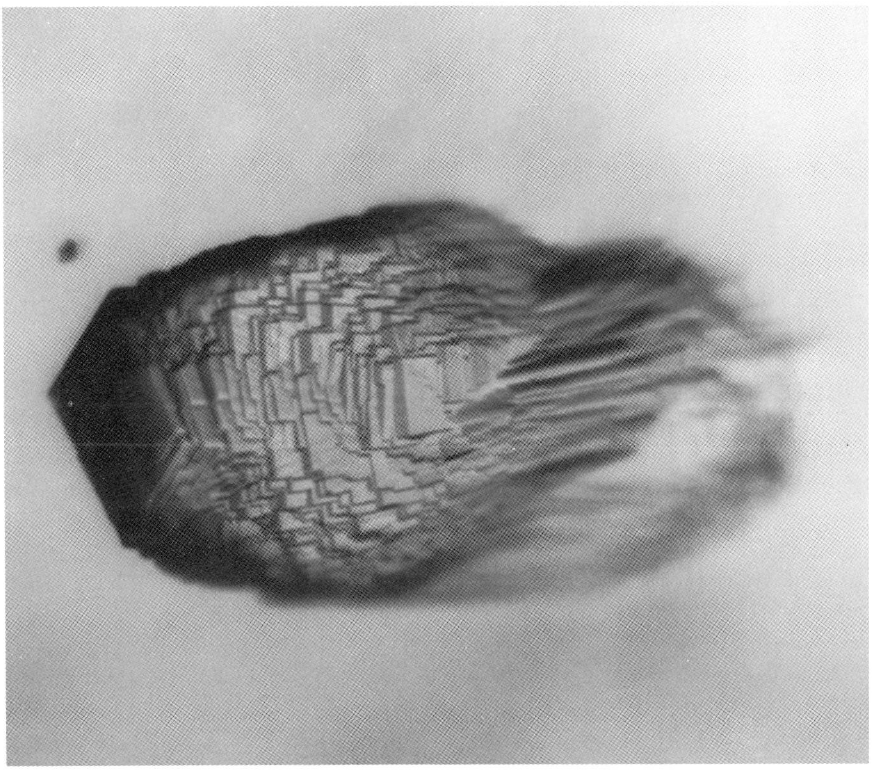

Fig. 23. A tetragonal crystal of a plant protein toxin, abrin, grown from about 50% saturated ammonium sulfate. The crystal seen here is severely twinned about the 4_1 axis of the crystal and is virtually useless for X-ray diffraction purposes.

crystalline protein. This is occasionally necessary if the initial crystal form is not amenable to analysis because it grows as fine needles or thin plates, or if it demonstrates some other unfavorable tendency such as striation or twinning. Some typical examples of this are shown in figs. 23–28.

There currently exist no systematic studies of pressure, sound, vibrations, electrical and magnetic fields, or viscosity on the rate of growth, or final quality, of protein crystals. Similarly, investigations are only now being undertaken to evaluate the effects of gravity [41, 50–52], convection, and fluid flow [53–56] on protein crystal growth, final size, and perfection. Thus it is not possible at this time to definitively evaluate their influence.

It is often, with a new and untried macromolecule, difficult to know exactly where to begin and where to go once experimental searches for successful conditions are under

Fig. 24. Long thin needles and thin plate crystals are characteristic of this protein, bromelin, isolated from the stems of pineapples. Although it is easy to obtain these crystals, none has yet been grown to a size suitable for X-ray analysis. These morphologies are, unfortunately, very common to protein crystals and are frequently a major problem to overcome.

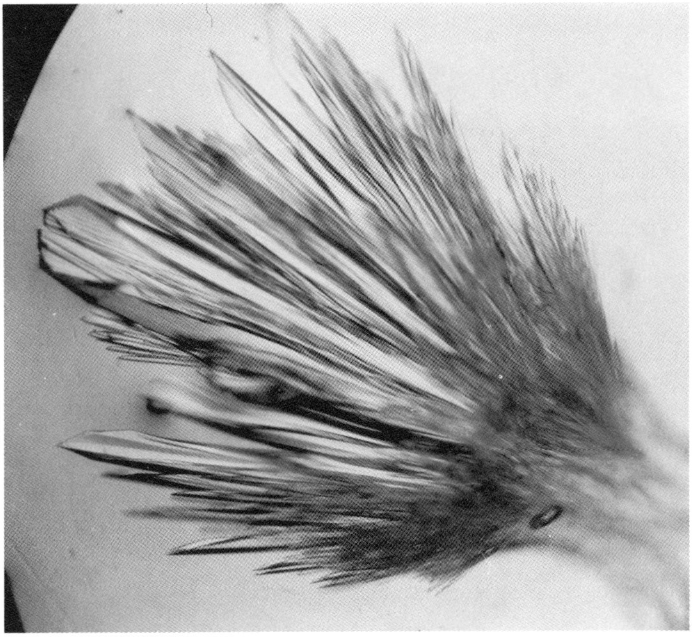

Fig. 25. Hexagonal prismatic crystals of the plant seed reserve protein concanavalin B sprouting from a common growth center. Such sprays of needle crystals in tight bunches are again common sights to protein crystallographers.

Fig. 26. Seen here is a spiral distribution of platelike microcrystals originating from a common growth center. The protein is an intact mouse monoclonal antibody against canine lymphoma and the crystals are of triclinic symmetry.

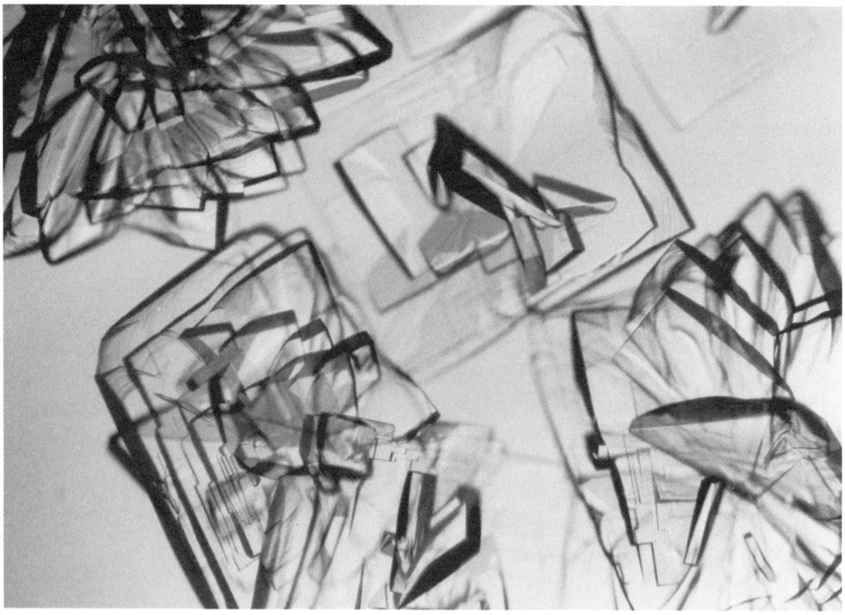

Fig. 27. The surfaces of orthorhombic crystals of the reserve storage protein canavalin often give rise to secondary growth so that new crystals appear at many angles to the first in spirals and fans.

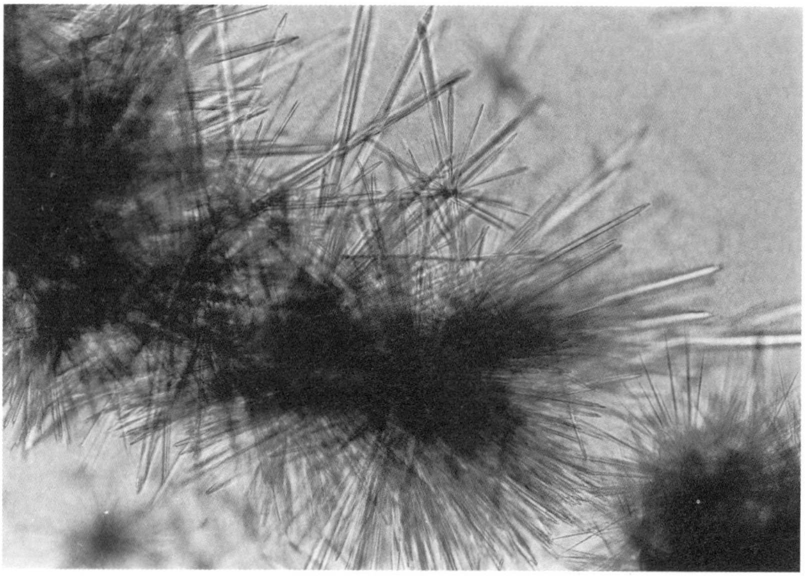

Fig. 28. The single most common habit observed by macromolecular crystallographers are bunches and clusters of fine crystalline needles like those of bacterial alpha amylase seen here. The crystals are a result of strongly preferred intermolecular interactions along one particular direction.

way. The best strategy is to begin with the most significant variables such as precipitant type and concentration, pH and temperature, and to identify limits within which one could expect crystals to grow. Because the number of parameters are potentially so great, it is essential to reduce the ranges as quickly as possible in order to concentrate one's efforts in those areas most promising of success.

Once the basic lines of pursuit are decided upon, then the other effectors such as detergents, ions, ligands, etc. can be more intelligently evaluated. Some initial screens of conditions have been described by McPherson [36] and flow diagrams for thorough and systematic investigations of crystallization variables have been described.

It should be emphasized that insight into the physicochemical properties and behavior of the macromolecule of interest is perhaps the most valuable tool that one has. Knowing the kinds of things that produce precipitation and at what concentrations, pH ranges in which the protein is viable, ligands that stabilize or control the aggregation state, all of these kinds of information are invaluable when one begins a crystallization study. These are the features that permit the design of unique, clever, and individualized approaches that are often required by particularly refractile macromolecules.

9. Some useful considerations

The earliest investigations of protein crystallization [18] suggested that the concentration of protein in the mother liquor should be as high as possible, 10 to 100 mg/ml. This is particularly true if one is attempting to grow the first crystals of a protein. The probability of obtaining crystals is certainly enhanced by increasing the concentration of protein. Concentration alone, in fact, is sometimes sufficient to drive the system into a state of supersaturation so that stable nuclei can form. High protein concentration may, however, not be the best approach for growing large, perfect crystals once optimal values for other parameters have been established.

Once conditions for nucleation and growth have been identified and the investigation of variables more or less complete, the concentration of protein should be gradually reduced in increments. This will moderate the rate of growth of the crystals. As a general rule, the largest and most perfect crystals result when the rate of accretion of molecules is slow and orderly. Reduction of concentration is a useful mechanism for effecting this.

The period necessary to grow protein crystals is quite variable and may range from a few hours in the best cases to several months in others. The actual rate of growth may not be reflected at all in the total amount of time required to obtain crystals adequate for analysis [31, 32]. The total time includes that required for achievement of solvent equilibration, formation of crystal nuclei, and maturation to occur.

When screening variables to establish optimal parameters, the practical objective is to promote crystallization at the greatest possible speed in order to expedite determination of most probable conditions. When optimizing and refining crystallization parameters, however, time itself becomes a parameter and long periods of slow growth are often an objective.

Caution should be exercised, however, if a long period elapses without the formation of crystals and then, well beyond the time required for solvent equilibration, crystals begin to appear. Some possible causes should be sought. The likelihood is that the protein has, over the time period, undergone some physical or chemical change. It may have undergone limited proteolysis, disassociated from a coenzyme or metal ion, or experienced a slow conformational change. By promoting this same event before crystallization trials are carried out, the time required for growth may be substantially reduced.

Another good possibility is that the apparatus in which the crystallization experiments were carried out was leaking and that very slow evaporation occurred. Thus the final concentration of precipitant may have reached levels appreciably higher than intended. Yet another possibility is a change in the ambient temperature. This is particularly likely when crystallization is carried out at room temperature and heating or air-conditioning systems are activated as the seasons change.

The most intriguing problem, or opportunity depending on one's perspective, is what additional components or compounds should comprise the mother liquor in addition to solvent, protein and precipitating agent. The most probable effectors are those which maintain the protein in a single, homogeneous, and invariant state. Reducing agents such as glutathione or β-mercaptoethanol are useful to preserve sulfhydryl groups and prevent oxidation. EDTA and EGTA are effective if one wishes to protect the protein from heavy or transition-metal ions. Inclusion of these components may be particularly desirable when crystallization requires a long period of time to reach completion.

When crystallization is carried out at room temperature in polyethylene glycol or low-ionic-strength solutions, then attention must be given to preventing the growth of microbes. These generally secrete proteolytic enzymes that may have serious effects on the integrity of the protein under study. Inclusion of sodium azide or thymol or chlorobutanol at low levels may be necessary to suppress invasive bacteria and fungi.

Substrates, coenzymes and inhibitors often serve to fix an enzyme in a more compact and stable form. Thus a greater degree of structural homogeneity may be imposed on a population of macromolecules and a reduced level of statistical variation achieved by complexing the protein with a natural ligand before attempting its crystallization.

In some cases an apoprotein and its ligand complexes may differ significantly in their physical behavior and can, in terms of crystallization, be treated as almost entirely separate problems. This may provide additional opportunities for growing crystals if the native apoprotein is refractile. It is worthwhile, therefore, when searching for crystallization conditions, to explore complexes of the macromolecule with substrates, coenzymes, analogues and inhibitors at an early stage. Such complexes are, in addition, inherently more interesting in a biochemical sense than the apoprotein.

Microbial growth almost invariably results in proteolysis of protein samples, something generally, but not always to be avoided. In a number of instances [57–61] limited and controlled proteolytic cleavage of a protein rendered it crystallizable while in the native state it was not. In some other cases [62], limited proteolysis resulted in a change of crystal form to a more suitable and useful habit. It should be emphasized, however, that these are examples of controlled proteolysis where the endproduct was an essentially homogeneous population of molecules, albeit cleaved molecules.

Proteolytic enzymes, it seems, occasionally trim off mobile polypeptides and degrade macromolecules to stable, compact domains. These abbreviated proteins are, as a result, more invariant, less conformationally flexible and they often form crystals more readily than does the native precursor. Although one might prefer, from a biochemical perspective, the intact protein, a partially degraded form often exhibits the activity and physical properties that are of primary interest. If a molecule can undergo limited digestion, this form may also be included in the crystallization strategy.

Various metal ions have occasionally been observed to promote the crystallization of proteins and nucleic acids. In some instances these ions were essential for activity. It was, therefore, reasonable to expect that they might aid in maintaining certain structural features of the molecule. In other cases, however, metal ions, particularly divalent metal ions of the transition series, were found that encouraged crystal growth but played no known role in the macromolecules' activity.

One of the oldest examples of a crystalline animal protein is horse spleen ferritin which forms perfect octahedra when a solution containing the protein is exposed to Cd^{2+} ions [63]. α-Lactalbumin was similarly shown to crystallize in the presence of Cd^{2+} ions [64] and several varieties of α-amylase crystallize spontaneously when presented with Ca^{2+} [65, 66]. Metal ions should be included for investigation in that class of additives that for any reason might tend to stabilize or engender conformity by specific interaction with the macromolecule.

Proteins that are naturally membrane-associated or otherwise unusually hydrophobic or lipophilic in nature present some unusual problems. Such proteins are, in general, only sparingly soluble in normal aqueous media, some virtually insoluble, and this, in turn, makes the application of conventional protein crystallization techniques very problematic. Such cases are difficult but not intractable.

To address these difficulties, the use of detergents, particularly non-ionic detergents, has been developed [67–69]. No attempt will be made here to describe the various techniques or the combinations of detergents and accessory molecules that have been used, as that involves a number of complexities and considerations that are inappropriate here. It should be noted, however, that the most popular detergent, both alone and in combination with others, was β-octylglucoside. This was extremely effective in the successful crystallization of a number of membrane proteins and has been shown to be useful, in fact, for a broad range of different macromolecules.

Detergents may be of value in the crystallization of otherwise soluble macromolecules as well [70]. Many proteins and nucleic acids, particularly when they are highly concentrated and in the presence of precipitating agents such as polyethylene glycol or methylpentanediol, tend to form transient, sometimes metastable, nonspecific aggregates. The existence of a spectrum of varying sizes, shapes and charges presents problems not appreciably different from the crystallization of a protein from a heterogeneous mixture or an impure solution composed of dissimilar macromolecules. An objective in crystallizing proteins is to limit the formation of nonuniform states and reduce the population to a set of standard individuals that can form identical interactions with one another.

As with membrane and lipophilic proteins, a means for limiting nonspecific aggregation is the inclusion of mild detergents in the crystallization mother liquor. McPherson et al. [70, 71] showed that for a fairly wide range of proteins the neutral detergent β-octylglucoside was a positive factor in obtaining crystals useful for diffraction analysis. In addition, it was demonstrated that other detergents also exhibit helpful properties in altering crystal morphology, decreasing microcrystal formation, or improving growth patterns. One example of this is shown in fig. 29.

Upon entering the field of macromolecular crystallography, one is struck by the extraordinary range of molecules and their properties that one must contend with, and the extensive variety of techniques and conditions that must be tested in order to grow crystals suitable for X-ray diffraction analysis. It would indeed be useful if some comprehensive data base existed that at least contained the experiences accumulated over the years. Indeed, such a knowledge base, combined with an excellent system to search and sift for all kinds of relevant information regarding protein crystal growth, has been compiled and is readily available. This is the crystallization data base devised by Gilliland [8, 72] and distributed through the National Institute of Standards and Technology (NIST).

Fig. 29. The large blocklike crystals distributed over the field of this photograph are the typical tetragonal modification of the protein lysozyme. In the presence of 3% (w/v) of a non-ionic detergent, as shown here, a second modification, very much different from the first, also appears. This illustrates that levels of detergent, often quite low, can significantly change the form or character of protein crystals.

This data base provides an extremely useful tool for the novice as well as the experienced crystallographer. It includes virtually all of those conditions used to grow crystals of individual proteins, and it provides innumerable ideas regarding procedures and techniques. With the current explosion in macromolecular crystal growth technology, this data base will soon be an essential resource. It should be further noted, that sophisticated mathematical approaches to the analysis of crystallization trials have been introduced [73, 74]. With these approaches, and with a realistic scoring scheme, presumably a more rapid route can be found to useful results. These methods are, however, still under development and their value under discussion.

An even more empirical approach to protein crystal growth has been suggested in which a set of more or less arbitrary conditions was applied to a macromolecule and the results inspected in the usual way. From these essentially random trials, and with some luck, leads might be obtained that could direct the investigator toward truly successful conditions.

Clearly, this procedure could be made much more effective if the collection of initial trial conditions included those mostly responsible for past successes. This information is, in fact, available in the Crystallization Data Base compiled by Gilliland [8, 72]. Subsequently, a number of laboratories have devised schemes for screening forty to sixty prospective conditions of buffer, precipitant, temperature, and so forth, and sometimes these lists have been widely circulated and in some cases published [75].

While these sets of initial trial conditions do not always yield success, and in some cases are not even very informative, they often do produce crystals and for a rather small investment of time and effort, it might be added. Thus, they are strongly recommended as a good place to start if one is at a loss on how to proceed or needs a fairly well defined jumping-off point.

10. Seeding

Often it is desirable to reproduce crystals of a protein where either the formation of nuclei is limiting, or spontaneous nucleation occurs at such a profound level of supersaturation that poor growth patterns follow. In such instances it may be desirable to promote growth in a directed fashion at low levels of supersaturation. This can sometimes be accomplished by seeding a metastable, supersaturated protein solution with crystals from earlier trials.

Seeding techniques fall into two categories, those employing microcrystals as seeds and those using larger macroseeds. In both methods, the fresh solution to be seeded should be only marginally supersaturated so that controlled, slow growth will occur. Approaches have been described in some detail by Stura and Wilson [76], by Fitzgerald and Madsen [77] and by Thaller et al. [78, 79].

In seeding with microcrystals, the danger is that too many nuclei will be introduced into the fresh supersaturated solution and masses of crystals of will result, none of which are suitable for diffraction analysis. To overcome this, a stock solution of microcrystals is serially diluted over a broad range. Some dilution sample in the series will, on average, have no more than one microseed per microliter. Others will have several fold more or none at all. A microliter of each sample in the series is then added to fresh protein crystallization solution under what are judged to be the optimal conditions for growth.

This empirical test should, ideally, identify the correct sample to use for seeding by yielding only one or a small number of single crystals when growth is complete. Seeding solutions containing too many seeds will yield additional showers of microcrystals and seeding solutions containing too low a concentration of seeds will produce nothing at all. The optimal seeding concentration as determined by the test can then be used to seed many additional samples.

Another approach involves crystal seeds large enough to be manipulated and transferred under a microscope. Again the most important consideration is to avoid spurious nucleation by transfer of too many seeds. Even if a single large crystal is employed, microcrystals adhering to its surface may be carried across to the fresh solution. It is, therefore, recommended that the macroseed be thoroughly washed by passing it through a series of intermediate transfer solutions. In doing so, not only are microcrystals removed, but if the wash solutions are chosen properly, some limited dissolution of the seed may take place. This has the effect of freshening the seed crystal surfaces and promoting new growth once it is introduced into the fresh protein solution. Again, the new solution must be at least saturated with respect to protein, but not extremely so, to insure slow and proper growth.

Recent evidence from crystallization experiments has suggested that surfaces in contact with the mother liquor may influence the formation of crystal nuclei and even the eventual habit of the crystals. Heterogeneous nucleation has long been recognized as useful in the case of conventional molecules, but has only recently been shown for macromolecules. This emphasizes further the importance of the surfaces of crystallization vessels and containers and how they are treated. At the same time it has also suggested that foreign surfaces might be useful as heterogeneous or epitaxial nucleants for macromolecules. In at least one case, the use of human hair has even been successfully employed [80].

Examples have been discovered of protein crystal nucleation on epoxides, polystyrene, and a number of other synthetic membranes. Recently, McPherson and Schlichta [81] have shown, in a series of experiments using 50 different water-insoluble minerals and five different proteins, that both heterogeneous nucleation and epitaxial growth of protein crystals from mineral faces do indeed occur. For each of the five proteins, certain specific sets of minerals were empirically identified that promoted nucleation and growth at earlier times and lower levels of supersaturation than occurred through spontaneous processes. Figure 30 shows an example of this phenomenon.

Another approach to promoting the formation of crystal nuclei has been described by Ray [82]. He introduced microdroplets of various concentrations of polyethylene glycol into protein solutions that were also sufficiently high in salt concentration to support crystal growth once stable nuclei were formed. He was able to show that protein left the salt-dominated phase of the mixture and concentrated itself in the polyethylene-glycol-rich microdroplets, sometimes reaching effective concentrations in the polyethylene glycol droplets of several hundred milligrams per milliliter.

By light-microscopic techniques it was demonstrated that crystal nuclei appeared first at the surface of the polyethylene glycol droplets and then proceeded to grow into

Fig. 30. The large crystal seen here is of the protein lysozyme that has nucleated and grown from the smaller, more irregular crystal of the mineral apophalite at the lower left. This is but one example of the heterogeneous nucleation of a protein crystal on a foreign surface.

the supersaturated salt solution that surrounded them, finally reaching a terminal size appropriate for X-ray analysis. In the absence of the polyethylene glycol droplets, no crystals were ever observed to form.

For those inexperienced in biochemical techniques, it is often difficult to find a starting point. Perhaps the best way is to grow one or two kinds of protein crystals that are relatively easy and straightforward, and from proteins that are readily accessible. Below are given the procedures for three common proteins: hen egg lysozyme, beef liver catalase and horse spleen ferritin. All three proteins can be purchased at relatively low cost from Sigma Biochemical Co., St Louis, Mo. and most other suppliers of biochemicals.

Lysozyme (white powder)
Solutions:

Stock solution–0.1 M sodium acetate/HCl, pH 4.8.

(A) 5% (w/v), equal to 50 mg/ml, of lysozyme in 0.1 M sodium acetate stock solution adjusted to pH 4.8.

(B) 6.5% (w/v) of NaCl in 0.1 M sodium acetate stock solution adjusted to pH 4.8.

Procedure:

(1) Heat (B) (NaCl solution) to 40°C.

(2) Place 1 ml of (B) in a well of a 24-well Linbro tissue culture plate.

(3) On a clean silated cover slip, create a drop of equal volumes (10 to 15 μl) of solutions (A) and (B).

or

(4) Mix equal volumes of (A) and (B) in a microfuge tube and centrifuge to remove any insoluble material, then place a 10 to 15 μl droplet on the cover slip.

(5) Place cover slip over the well.

(6) Place tray at 4 to 20°C.

Large single crystals should grow overnight.

Catalase (solution is dark green or brown in color)

(1) Shake container gently to resuspend any crystals present in the catalase solution.

(2) Remove 0.5 ml from container into a glass test tube and dilute with an equal volume of H_2O. Add 10 μl of 1.0 M NH_4OH.

(3) Hold test tube containing sample under warm water and gently shake while warming until any remaining catalase microcrystals redissolve.

(4) To crystallize:

(a) combine equal volumes of catalase solution and 0.05 M sodium phosphate adjusted to pH 6.4 in small containers or vials and place immediately at 4°C.

or

(b) dialize aliquots of the catalase solution against distilled water for 12 to 24 h at 4°C.

or

(c) combine 10 μl of the catalase solution with an equal volume of 0.05M sodium phosphate and, using vapor diffusion, equilibrate the droplets against 20% ethanol in 0.05M sodium phosphate at pH 6.4 and 4°C.

Hexagonal or orthorhombic crystals will grow in from 12 to 48 h. To slow the process reduce the catalase concentration or add increasing amounts of NaCl.

Ferritin (dark red solution)

(1) Combine equal volumes of horse spleen ferritin (as it is shipped) and 0.5% cadmium chloride or cadmium sulfate dissolved (w/v) in H_2O. Place at 4 to 20°C.

Octahedral crystals will form in from 1 h to two days. To slow the process reduce the concentration of cadmium ion. If no crystals form gradually increase the cadmium concentration.

11. Concluding remarks

The literature describing protein crystallization techniques and strategies, sparse only a few years ago, is rapidly growing. There are numerous reviews of protein crystallization [14, 15, 33, 36, 83], collections of articles [84], the proceedings of a number of symposia on macromolecular crystal growth [85] and even new textbooks [86]. While one might fear an information overload in this area of research, the contrary is in fact still true. As is probably evident from the review presented here, there are still imposing questions to be answered and important variables yet to be investigated. The results of recent years have, however, been enormously encouraging, and it is sincerely hoped, indeed expected, that by the end of the century macromolecular crystal growth can receive the same systematic and precise treatment as those other areas of crystal growth contained in these volumes.

References

[1] R. Boistelle and J.P. Astier, J. Cryst. Growth 90 (1988) 14.
[2] F. Rosenberger, J. Cryst. Growth 76 (1986) 618.
[3] G.D. Brayer and A. McPherson, J. Mol. Biol. 169 (1983) 565.
[4] J.F. Robyt and B.J. White, Biochemical Techniques, Theory and Practice (Waveland Press, Prospect Heights, Il, (1990).
[5] T.G. Cooper, The Tools of Biochemistry (Wiley-Interscience, New York, 1977).
[6] R. Giege, A.C. Dock, D. Kern, B. Lorber, J.C. Thierry and D. Moras, J. Cryst. Growth 76 (1986) 554.
[7] R.R. Bott, M.A. Navia and J.L. Smith, J. Biol. Chem. 257 (1982) 9883.
[8] G.L. Gilliland, J. Cryst. Growth 90 (1988) 51.
[9] E.J. Cohn and J.D. Ferry, in: Proteins, Amino Acids and Peptides, eds. E.J. Cohn and J.T. Edsall (Reinhold, New York, 1950).
[10] R. Czok and Th. Bucher, Adv. Protein Chem. 15 (1960) 315.
[11] J.H. Northrop, M. Kunitz and R.M. Herriott, Crystalline Enzymes (Columbia University Press, New York, 1948).
[12] T. Hofmeister, Arch. Exptl. Pathol. Pharmakol. Naunyn-Schmiedeberg's 24 (1887) 274.
[13] S.N. Timasheff and T. Arakawa, J. Cryst. Growth 90 (1988) 39.
[14] A. McPherson, in: Methods of Biochemical Analysis, Vol. 23, ed. D. Glick, (Academic Press, New York, 1976) p. 249.
[15] A. McPherson, The Preparation and Analysis of Protein Crystals (Wiley, New York, 1982).
[16] J. Hermans, J. Chem. Phys. 77 (1982) 2193.
[17] J.C. Lee and L.L.Y. Lee, J. Biol. Chem. 256 (1981) 625.
[18] A. McPherson, J. Cryst. Growth 110 (1991) 1.
[19] J.B. Sumner and G.F. Somers, The Enzymes (Academic Press, New York, 1943).
[20] K. Bailey, Trans. Faraday Soc. 38 (1942) 186.
[21] K. Bailey, Nature 145 (1940) 934.
[22] C.W. Carter, E.T. Baldwin and L. Frick, J. Cryst. Growth 90 (1988) 60.
[23] W. Kadima, A. McPherson, M.F. Dunn and F.A. Jurnak, Biophys. J. 57 (1990) 125.
[24] A. Malkin, J. Cheung and A. McPherson, J. Cryst. Growth 126 (1993) 544.
[25] A. Malkin and A. McPherson, J. Cryst. Growth 126 (1993) 555.
[26] L. Betts, L. Frick, R. Wolfenden and C.W. Carter, J. Biol. Chem. 264 (1989) 6737.
[27] E.T. Baldwin, K.V. Crumly and C.W. Carter, Biophys. J. 49 (1986) 47.
[28] G. Feher and Z. Kam, in: Methods in Enzymology, Vol. 114, eds. H.W. Wyckoff, M.F. Hirs and S. Timasheff, (Academic Press, New York, 1985) p. 77.
[29] S.D. Durbin and G. Feher, J. Cryst. Growth 76 (1986) 383.
[30] Z. Kam, A.B. Shore and G. Feher, J. Mol. Biol. 123 (1978) 539.
[31] S. Koszelak and A. McPherson, J. Cryst. Growth 90 (1989) 340.
[32] S. Koszelak, D. Martin, J. Ng and A. McPherson, J. Cryst. Growth 110 (1991) 177.
[33] A. McPherson, in: Crystallization of Membrane Proteins ed. H. Michel (CRC Press, Boca Raton, FL, 1990).

[34] A. McPherson in: Methods in Enzymology: Diffraction Methods, Vol. 114 eds. M.F. Hirs, S. Timasheff and H.W. Wyckoff (Academic Press, New York, 1985) p. 112.
[35] T.L. Blundell and L.N. Johnson, Protein Crystallography (Academic Press, New York, 1976).
[36] A. McPherson, Eur. J. Biochem. 189 (1990) 1.
[37] M. Zeppenzauer, H. Eklund and E. Zeppenzauer, Arch. Biochem. Biophys. 126 (1968) 564.
[38] M. Zeppenzauer, in: Methods of Enzymology Vol. 22, ed. W.B. Jacoby (Academic Press, New York, 1971) p. 253.
[39] F.R. Salemme, Arch. Biochem. Biophys. 151 (1972) 533.
[40] B.A. Weber and P.E. Goodkin, Arch. Biochem. Biophys. 141 (1970) 489.
[41] J. Day and A. McPherson, Protein Science 1 (1992) 1254.
[42] A. McPherson, A. Greenwood and J. Day, Adv. Space Res. 11 (1991) 343.
[43] A. Hampel, M. Labanauskas, P.G. Conners, L. Kirkegard, U.L. RajBhandary, P.B. Sigler and R.M. Bock, Science 162 (1968) 1384.
[44] D. Morris, C.Y. Kim and A. McPherson, Biotechniques 7 (1989) 522.
[45] M. Kunitz, J. Gen. Physiol. 35 (1952) 423.
[46] W.B. Jacoby, Anal. Biochem. 26 (1968) 295.
[47] F. Rosenberger and E.J. Meehan, J. Cryst. Growth 90 (1988) 74.
[48] A. McPherson, J. Biol. Chem. 251 (1976) 6300.
[49] M.V. King, J. Bello, E.H. Pagnatano and D. Harker, Acta Cryst. 15 (1962) 144.
[50] L.J. DeLucas, C.D. Smith, H.W. Smith, V.K. Senadhi, S.E. Senadhi, S.E., Ealick, C.E. Bugg, D.C. Carter, R.S. Snyder, P.C. Weber, F.R. Salemme, D.H. Ohlendorf, H.J. Einspahr, L. Clancy, M.A. Navia, B.M. McKeever, T.L. Nagabhushan, G. Nelson, Y.S. Babu, A. McPherson, S. Koszelak, D. Stammers, K. Powell and G. Darby, Science 246 (1989) 651.
[51] W. Littke and C. John, Science 225 (1984) 203.
[52] W. Littke and C. John, J. Cryst. Growth 76 (1986) 663.
[53] M. Pusey, W.K. Witherow and R. Naumann, J. Cryst. Growth 90 (1988) 105.
[54] M. Pusey, R.S. Snyder and R. Naumann, J. Biol. Chem. 261 (1986) 6524.
[55] M.L. Grant and D.A. Saville, J. Cryst. Growth 108 (1991) 8.
[56] M.L. Pusey and R. Nauman, J. Cryst. Growth 76 (1986) 593.
[57] H.W. Wyckoff, D. Tsernoglou, A.W. Hanson, J.R. Knox, B. Lee and F.M. Richards, J. Biol. Chem. 245 (1970) 305.
[58] A. McPherson and R. Spencer, Arch. Biochem. Biophys. 169 (1975) 650.
[59] J.P. Waller, J.L. Risler, C. Monteilhet and C. Zelwer, FEBS Lett. 16 (1971) 186.
[60] S.P.L. Sorensen and M. Hoyrup, Studies on proteins, C.-R. Lab. Carlesberg 12 (1915-1917) 12.
[61] A. Solomon, C.L. McLaughlin, C.H. Wei and J.R. Einstein, J. Biol. Chem. 245 (1970) 5289.
[62] F.A. Jurnak, A. McPherson, A.H.J. Wang and A. Rich, J. Biol. Chem. 255 (1980) 6751.
[63] S. Granick, J. Biol. Chem. 146 (1941) 451.
[64] D.W. Green and R. Aschaffenburg, J. Mol. Biol. 1 (1959) 54.
[65] A. McPherson and A. Rich, Biochim. Biophys. Acta 285 (1972) 493.
[66] A. McPherson and A. Rich, J. Ultrastruct. Res. 44 (1973) 75.
[67] M. Zulauf, in: Crystallization of Membrane Proteins, ed. H. Michel (CRC Press, Boca Raton, FL, 1991) Ch. 2, p. 53.
[68] H. Michel, in: Crystallization of Membrane Proteins, ed. H. Michel (CRC Press, Boca Raton, FL, 1991) Ch. 3, p. 73.
[69] R.M. Garavito, Z. Markovic-Honsiev and J.A. Jenkins, J. Cryst. Growth 76 (1986) 701.
[70] A. McPherson, S. Koszelak, H. Axelrod, J. Day, R. Williams, L. Robinson, M. McGrath and D. Cascio, J. Biol. Chem. 261 (1986) 1969.
[71] A. McPherson, S. Koszelak, H. Axelrod, J. Day, M. McGrath, R. Williams and D. Cascio, J. Cryst. Growth 76 (1986) 547.
[72] G.L. Gilliland and D.M. Bickham, in: Methods: A Campanion to Methods in Enzymology, Vol. I, ed. C.W. Carter (Academic Press, New York, 1990) p. 6.
[73] C.W. Carter Jr. and C.U. Carter, J. Biol. Chem. 254 (1979) 12219.
[74] M.J. Cox and P.C. Weber, J. Cryst. Growth 90 (1988) 318.
[75] J. Jancarik and S.H. Kim, J. Appl. Cryst. 24 (1991) 409.

[76] E.A. Stura and I.A. Wilson, in: Methods: A Companian to Methods in Enzymology, Vol. I, ed. C.W. Carter (Academic Press, New York, 1990) p. 38
[77] P.M.D. Fitzgerald and N.B.J. Madsen, J. Cryst. Growth 76 (1987) 600.
[78] C. Thaller, G. Eicher, L.H. Weaver, E. Wilson, R. Karlsson and J.N. Jansonius, Methods Enzymol. 115 (1985) 132.
[79] C. Thaller, L.H. Weaver, G. Eichele, E. Wilson, R. Karlsson and J.N. Jansonius, J. Mol. Biol. 147 (1981) 465.
[80] C.J. Leung, B.T. Nail and G.D. Brayer, J. Mol. Biol. 206 (1989) 783.
[81] A. McPherson and P. Schlichta, Science 2398 (1988) 385.
[82] W. Ray and C. Bracker Jr., J. Cryst. Growth 76 (1986) 562.
[83] P.C. Weber, Adv. Prot. Chem. 41 (1991) 1.
[84] C.W. Carter, Ed., Methods: A Companion to Methods in Enzymology, Vol. I, Protein and Nucleic Acid Crystallization (Academic Press, New York, 1990).
[85] The proceedings of four international conferences on Macromolecular Crystal Growth have been published in J. Cryst. Growth as special volumes: 76 (1986); 90 (1989); and 110 (1991). The fourth volume is in press for 1992.
[86] A. Ducruix and R. Giege, Crystallization of Nucleic Acids and Proteins, A Practical Approach (Oxford Univ. Press, Oxford, 1992).

CHAPTER 9

Hydrothermal Growth of Crystals

K. BYRAPPA

Mineralogical Institute, University of Mysore, Manasagangotri, Mysore 570 006, India

Contents

1. Introduction . . . 467
2. Historical development and trends in the modern hydrothermal method . . . 468
3. Physico-chemical and hydrodynamic principles of the hydrothermal growth of crystals . . . 470
4. Basic principles of phase formation under hydrothermal conditions . . . 473
5. Solutions, solubility and kinetics of crystallization . . . 476
   5.1. Thermodynamic principles of solubility . . . 478
   5.2. Experimental investigations of solubility. . . . 483
      5.2.1. Solubility of quartz . . . 485
      5.2.2. Solubility of berlinite . . . 486
      5.2.3. Solubility of malachite . . . 489
6. Apparatus . . . 490
   6.1. Selection of autoclaves and autoclave materials . . . 491
   6.2. Liners . . . 492
   6.3. Temperature and pressure measurements . . . 496
   6.4. Glass vessels . . . 497
   6.5. Steel autoclaves . . . 497
      6.5.1. Flat plate closure autoclave . . . 498
      6.5.2. Cone closure . . . 499
      6.5.3. Welded closure . . . 500
      6.5.4. Modified Bridgman autoclave . . . 501
   6.6. Novel equipment design . . . 503
      6.6.1. Horizontal autoclaves for controlled growth of crystals . . . 503
      6.6.2. Pendulum apparatus . . . 504
      6.6.3. Hydrothermal autoclaves for visual examination . . . 505
      6.6.4. $P-V-T$ apparatus . . . 507
      6.6.5. Apparatus for solubility determination . . . 507
   6.7. Safety and maintenance of autoclaves . . . 509

HANDBOOK OF CRYSTAL GROWTH, VOL. 2
edited by D.T.J. Hurle

6.8. Adaptability of the hydrothermal method . . . 510
6.8.1. Separated charge method . . . 511
6.8.2. Hydrothermal epitaxial growth . . . 511
6.8.3. Hydrothermal reaction sintering . . . 511
6.8.4. Redox potential . . . 511
6.8.5. Isotope doping . . . 512
6.8.6. Hydrothermal dehydration process . . . 512
7. Hydrothermal growth of some selected crystals . . . 512
7.1. Quartz . . . 512
7.1.1. Growth rate . . . 516
7.1.2. Seed effect . . . 516
7.1.3. Nutrient effect . . . 517
7.1.4. Sweeping . . . 518
7.2. Oxides . . . 518
7.2.1. Zincite . . . 518
7.2.2. Cuprite . . . 520
7.2.3. Bromelite . . . 520
7.2.4. In_2O_3 . . . 520
7.3. Phosphates . . . 520
7.3.1. Berlinite . . . 520
7.3.2. Potasium titanyl phosphate . . . 523
7.3.3. Superionic phosphates . . . 528
7.4. Borates . . . 531
7.4.1. Lithium tetraborate . . . 531
7.4.2. Superionic borates . . . 532
7.4.2.1. $Li_3B_5O_8(OH)_2$ . . . 532
7.4.2.2. $LiH_2B_5O_9$ . . . 533
7.5. Carbonates . . . 534
7.6. Growth of tellurium dioxide monocrystals . . . 536
7.7. Sulphides . . . 537
7.7.1. Sphalerite . . . 538
7.7.2. Cadmium sulphide . . . 540
7.7.3. Cinnabar . . . 540
7.7.4. Galena . . . 541
7.8. Zirconium silicates . . . 541
7.9. Sodalite . . . 541
7.10. Gemstones . . . 542
7.10.1. Emerald . . . 542
7.10.2. Coloured quartz . . . 544
7.10.3. Corundum . . . 545
7.10.4. Ruby . . . 547
Appendix . . . 548
References . . . 556

1. Introduction

The largest single crystal formed in nature (beryl crystal of more than 1000 g) and some of the larger single crystals created by man (quartz crystals of a few 100 g) are both of hydrothermal origin. The hydrothermal technique is very important for its technological efficiency in developing bigger, purer and dislocation free single crystals. The method has been widely accepted since the 1960s and practically all inorganic species, starting from native elements to the most complex silicates, germanates, phosphates and others have been obtained by this method. The technique is being employed on a large scale to prepare piezoelectric, magnetic, optic, ceramic, etc. materials. The hydrothermal technique, in contrast to the conventional techniques, offers some advantages:

(1) Compounds with elements in oxidation states that are difficult to obtain; especially important for transitional metal compounds which can be obtained in a closed system by the hydrothermal method (e.g. ferromagnetic chromium (IV) oxide).

(2) The hydrothermal method is also useful for the so-called low temperature phases e.g. α-quartz, α-berlinite and others.

(3) For the synthesis of metastable compounds, such as subiodides of tellurium, Te_2I, the hydrothermal method is quite unique.

The term "hydrothermal" is purely of geological origin. Minerals formed in the postmagmatic and metasomatic stages in the presence of water at elevated temperatures and pressures are said to be of hydrothermal origin. Hydrothermal synthesis or growth is the use of aqueous solvents or mineralizers under high temperature and high pressure to dissolve and recrystallize materials that are relatively insoluble under ordinary conditions. Usually in hydrothermal growth (excluding the crystals with negative solubility), the dissolution is carried out at the lower hotter zone (nutrient zone) and crystallization is carried out at the upper cooler zone (growth zone). The convection currents created by the temperature gradient carry the material from the nutrient zone to the growth zone. There are hydrothermal experiments carried out under extreme pressure conditions: less than 1 bar and greater than 10 kbar. Hence, hydrothermal synthesis can be defined as "any heterogeneous reaction in an aqueous media above room temperature and at pressure greater than 1 atm". The technique has advanced much in the last three decades and today it is being widely employed by physicists, chemists (both inorganic and organic), materials, scientists, ceramists, materials engineers, geoscientists, etc. The scope of the present chapter has been limited only to the hydrothermal growth of crystals instead of the hydrothermal reactions or hydrothermal treatment of various organic and inorganic materials frequently employed on a large scale. Since the number of compounds obtained in the form of single crystals by a hydrothermal technique exceeds several hundreds, for the benefit of the reader emphasis is given only to some selected crystals of interest. The table given in the appendix represents more or less a complete list of the crystals obtained by hydrothermal methods. In this chapter we discuss the following aspects of the hydrothermal growth of crystals:

(i) Historical development of the method and its present perspective.

(ii) Physico-chemical and hydrodynamic principles of the hydrothermal growth of crystals.
(iii) Principles of phase formation under hydrothermal conditions.
(iv) Solutions, solubility and kinetics of crystallization.
(v) Hydrothermal apparatus and preparation of experiments.
(vi) Hydrothermal growth of some selected crystals.

While discussing the hydrothermal growth of crystals, emphasis is made on the experimental aspects and results rather than on theoretical aspects of crystal growth due to the fact that the theory of hydrothermal crystal growth has not yet been precisely understood.

2. *Historical development and trends in the modern hydrothermal method*

There is no systematic review which can give an exact account of the historical development of hydrothermal reséarch. The hydrothermal technique was, on the whole, started by the geoscientists in the previous century in order to understand the genesis of rocks and minerals by simulating the natural conditions existing under the earth's crust and crystallizing them in the laboratory. The size of the crystals or minerals obtained usually did not exceed thousandths or hundredths of a millimeter. The earliest work on hydrothermal growth was carried out by French mineralogists who adopted this method for the modelling of the natural processes of mineral formations [1–4]. But the technique was popularized by the materials' scientists, particularly with the beginning of the growth of large size quartz crystals in the 1940s. The earliest hydrothermal studies were made in synthesizing minerals like quartz, feldspar, wollastonite, analsite and others. The first hydrothermal synthesis was carried out in 1845 by Schafhautl, who obtained quartz crystals upon transformation of freshly precipitated silicic acid in Papin's digestor [5]. Subsequently, many other workers have carried out hydrothermal synthesis of various minerals. Bunsen [6] first used thick-walled glass tubes to contain high temperature–high pressure liquids and prepare strontium and barium carbonates. The greatest contribution in the previous century was by De Senarmount, the founder of hydrothermal synthesis in geoscience. He synthesized mineral carbonates, sulphates, sulphides, fluorides using glass liners in autoclaves [7]. Von Chrustschoff (1873) first proposed the noble metal lining of autoclaves to prevent corrosion [8]. Morey and Niggli (1913) surveyed the literature and made a list of 80 mineral species of which hydrothermal synthesis was claimed before 1900 [9]. They quote an experiment carried out by one of the earliest workers, who found that by reacting, in glass for 3 months at 550°C, a mixture of colloidal silicon, colloidal ferric hydroxide, colloidal ferrous oxide, lime water, magnesium hydroxide and potassium hydroxide, some crystals resembling hornblende were obtained [10]. So, before the end of 19th century, a large number of minerals were synthesized and experiments were carried out on a wide variety of geological phenomenon ranging from the origin of ore deposits to the origin of meteorites.

The impetus for the experimental investigations during the 19th century was provided not only by a desire to explain geological phenomena but also by greatly improved equipment and techniques fostered by the industrial revolution. Such investigations helped in unravelling the hitherto unknown natural geological processes of mineral formation on one hand, and finding uses for the artificially synthesized single crystals for example, in the gemstone industry (ruby, sapphire, topaz, emerald, quartz, diamond, etc.) on the other hand. The earliest hydrothermal experiments on quartz were carried out by De Senarmount (1851) and Spezia (1905) [11]. They were able to increase the thickness of a seed plate by depositing α-quartz from an aqueous medium. In 1880, Hannay claimed to have synthesized artificial diamond by the hydrothermal technique. Similarly, Moissan in 1893 also claimed to have produced diamonds as large as 0.5 mm from charcoal by the same technique [12, 13]. But the success of these experiments is dubious. However, they provided further stimuli for hydrothermal research, particularly the development of high pressure techniques. The founding of the Geophysical Laboratory at the Carnegie Institute of Washington in 1907 probably marked the most important milestone in the history of hydrothermal research. Although, a great deal of research was carried out on hydrothermal research in the previous century, the facilities for large-scale hydrothermal research before the end of the World War II were virtually non-existent, with the exception of the University of Chicago where Bowen worked between 1937 and 1945, and Bridgman's high pressure laboratory at Harvard University during the 1940s and 1950s. This situation dramatically changed with the development of the test tube pressure vessel by Tuttle in which high pressures up to 5 kbar could be maintained at temperatures of 750°C for long periods of time. The actual impetus for the hydrothermal growth of crystals began during the 2nd World War with the growth of quartz. Here, in contrast to the very slow rate of growth achieved by the earliest workers like Spezia, the captured German reports have shown that Nacken, using natural α-quartz as seed crystals and vitreous silica as nutrient, had grown quartz crystals in an isothermal system and succeeded in obtaining a large quartz crystal from a small seed [14]. Almost at the same time, Russian workers attempted to grow large single crystals of quartz using old cannon barrels as autoclaves. Unfortunately, these works were not revealed. However, when Nacken's work was published after World War II, several laboratories began studying quartz crystal growth almost simultaneously. In the course of time, the method has evolved greatly, especially during the late 1940s and 1950s. During the 1960s, interest in the hydrothermal method was boosted in connection with the industrial growth of good quality quartz single crystals and a successful growth of most complex inorganic compounds. At this time an intensive study of the hydrothermal process of synthesis and growth of single crystals which did not have analogues in nature began. During the 1970s, there was a question about the search and growth of hitherto unknown compounds of photo-semiconductors, ferromagnets, lasers, piezo-and ferrielectrics and the hydrothermal method attracted great attention. Suddenly lots of research groups cropped up throughout the world. Among them, the earstwhile Soviet groups, particularly at the Institute of Crystallography, Moscow have played a key role in developing this method. Even today, the Hydrothermal Laboratory at the Institute of Crystallography, Moscow, is the largest group in the world working on the various aspects of hydrothermal growth of crystals such as physico-chemical principles, kinetics, designing of new apparatus, growing new compounds, etc. The laboratory has the apparatus to grow crystals starting from the laboratory scale to the industrial scale. Thus specific aspects of hydrothermal method as a modelling tool to understand the natural processes of mineral formation changed dramatically into an important method characteristic for inorganic chemistry.

One of the current problems in hydrothermal chemistry for growing various inorganic compounds as single crystals is connected with the study of the physical-chemistry and kinetics of their crystallization, recrystallization, seed-induced growth, determination of their growth mechanism and it's relation to the quality and physical properties of crystals and, on the basis of such a complex approach, to formulate a general conception of the hydrothermal method and the search for new inorganic single crystals with very important technological applications.

3. Physico-chemical and hydrodynamic principles of the hydrothermal growth of crystals

This is perhaps the least known aspect in the hydrothermal growth of crystals. No book on crystal growth gives a complete picture of the theoretical aspects of hydrothermal growth. Similarly, no author has followed a definite approach to discuss these physico-chemical aspects. There are two aspects to be discussed here:

(i) The physico-chemical studies on various aqueous solutions studied within a wide range of P, T conditions.
(ii) The physical chemistry of the hydrothermal growth of crystals. Each one of these has been studied extensively. However, the correlation of the two works has not been done with enough accuracy.

During the crystallization process from saturated solutions the mol number of the solute compounds change and therefore, the free energy is represented as a function of not only temperature and pressure but also the mol numbers :

$$G = f(T, P, n_1, n_2, \ldots n_i), \tag{1}$$

$$\mathrm{d}G = (\delta G/\delta T)_{P,n}\mathrm{d}T + (\delta G/\delta P)_{T,n}\mathrm{d}P + (\delta G/\delta n_1)_{P,T,n}\mathrm{d}n_1 + (\delta G/\delta n_1)_{P,T,n}\mathrm{d}n_2, \tag{2}$$

where

n_i = constant mol number of all components,
n_j = constant number of components except one.

The variation in the free energy can be expressed in terms of the chemical potential

$$(\delta G/\delta n_i)_{P,T,n} = \mu_i. \tag{3}$$

When P and T are kept constant, eq. (2) becomes

$$\mathrm{d}G_{P.T} = \mu_1\,\mathrm{d}n_1 + \mu_2\,\mathrm{d}n_2 + \cdots, \quad \text{or} \quad \mathrm{d}G_{P.T} = (\textstyle\sum \mu_i \mathrm{d}n_i)_{P.T}. \tag{4}$$

The integration of the eq. (4) gives

$$G_{P,T} = \sum \mu_i n_i. \quad (5)$$

Differentiation of eq. (5) gives

$$\begin{aligned} dG_{P,T} &= n_1 d\mu_1 + n_2 d\mu_2 + \cdots + n_i d\mu_i + \mu_1 dn_1 + \mu_2 dn_2 + \cdots + \mu_i dn_i \\ &= \sum n_i d\mu_i + \sum \mu_i dn_i. \end{aligned} \quad (6)$$

By equating eqs. (4) and (6) we get

$$\sum n_i d\mu_i = 0. \quad (7)$$

Equation (7) is considered as the overall equilibrium condition in the system with variable mol numbers under constant P and T.

Let us consider a hydrothermal system containing n_A mol of solid A and partially soluble (n_{1A} mol) in n_B mol solvent. Then the free energy of the system corresponds to eq. (5):

$$G = n_A \mu_A + n_{1A} \mu_{1A} + n_B \mu_B. \quad (8)$$

If the process of dissolution of the solid components continues, then the free energy dG becomes

$$G + dG = (n_A - dn)\,\mu_A + (n_{1A} + dn)\,\mu_{1A} + n_B \mu_B. \quad (9)$$

The change in the free energy corresponding to this, under constant P and T becomes

$$dG = (\mu_{1A} - \mu_A)\,dn, \quad (10)$$

or

$$dG/dn = \mu_{1A} - \mu_A. \quad (11)$$

If $\mu_{1A} < \mu_A$, then dG/dn is negative and the process occurs on its own, since the free energy decreases. When $\mu_{1A} > \mu_A$ the crystallization of the substance from the solution B takes place and the chemical potential decreases, and when $\mu_{1A} = \mu_A$ the process reaches the equilibrium state. Consequently the solute component changes over from the phase wherein its chemical potential is higher into the phase where its chemical potential is lower. The phase transition stops when the chemical potential of the component becomes equal in both phases. It is observed that the above considered mechanism is realized only when, in the non-equilibrium process, the changes in dG do not depend upon the chemical potential and concentration of other components of the solution. However, the consideration of the above relationship may help in understanding the physico-chemical principles of the hydrothermal process.

Most hydrothermal crystal growth experiments are carried out under the conditions of temperature gradient in standard autoclaves. The growth of a single crystal into the seed can be carried out in two ways:

(a) recrystallization of the solid substance, including its dissolution in the liquid phase, convective mass transfer of the dissolved part of the substance to the growth zone or seed.

(b) dissolution of the mixture of the nutrient components with the help of their convective mass transport into the growth zone and interaction of the dissolved components on the seed surface.

The methodology of the growth of single crystals on a seed is the establishment of growth conditions in which the process is represented by the sum of macro- and micro-processes occurring between the interface boundary of the solution and the crystal. The composition and concentration of the solution, temperature and pressure, hydrodynamic conditions and surface contact of the phases are some of the basic physical and chemical parameters which determine the regime and rate of dissolution of the nutrient, mass transport and possibility of the formation of new phases.

The isothermal chemical reaction can be expressed in terms of the van't Hoff equation

$$\Delta G_{P,T} = RT \ln \pi a' - RT \ln K_{ai}, \tag{12}$$

where

$\pi a' = \pi'(a_i', \gamma_i)$,
$K_a = \pi(a_i, \gamma_i)$, is the thermodynamic equilibrium constant,
γ_i = activity coefficient,
a_i' = initial non-equilibrium activity of the nutrient component in the solution,
a_i = equilibrium activity of the nutrient components in the solution.

From eq. (12) it is observed that ΔG depends upon the relative values of $\pi a'$ and K_a when

$$\pi a' < K_a; \quad \Delta G < 0, \tag{13}$$

$$\pi a' = K_a; \quad \Delta G = 0, \tag{14}$$

$$\pi a' > K_a; \quad \Delta G > 0. \tag{15}$$

For eq. (13) the process is irreversible and the reaction takes place from left to right. In case eq. (14) the reaction takes place under equilibrium condition and for eq. (15) the process becomes reversible. As a result the changes in ΔG in eq. (12) can be represented as follows:

$$\Delta G = RT \ln (a_i/a_i'),$$

where a_i/a_i' is the analogue of the relative supersaturation.

If the activity coefficient $\gamma \approx 1$ or, in the given region, γ does not depend upon the concentration then, in eq. (12), the activity coefficient can be replaced by the mol fraction. Then the ratio a_i/a_i' is replaced by S_i/S_i', where S_i' = undersaturated solubility equilibrium concentration of the supersaturated solution. The ratio S_i/S_i' is called the relative supersaturation or degree of supersaturation.

In recent years there has been good progress achieved on the experimental side. Much credit goes to the Russian hydrothermal crystal growers. Balitsky and Bublikova (1991) have studied the physico-chemical foundation of malachite synthesis. Similarly Kuznetsov (1991) has reviewed the physical chemistry of hydrothermal crystal growth and the crystallization of some A2B6 crystals. Further, Popolitov (1991) has reviewed the physical chemistry of the hydrothermal growth of tellurium dioxide crystals. The reader can get valuable information in the book *Hydrothermal Growth of Crystals* [15].

4. *Basic principles of phase formation under hydrothermal conditions*

A phase is defined as a part of the system that is homogeneous throughout and is physically separable from other phases by distinct boundaries. The number of components in a system is the minimum number of chemical constituents that must be specified to describe the composition of all phases present. Phase relationships are conveniently represented by means of phase diagrams. Phase diagrams and the phase rule apply only to systems at equilibrium. These data are extremely important to any crystal grower for a successful growth of crystals. Gibbs deducted a quantitative relationship for phase rule:

$$F = C - P + 2, \quad \text{where } F = \text{degree of freedom},$$
$$C = \text{number of components},$$
$$P = \text{number of phases}.$$

The above relation holds good if pressure, temperature and composition are the only

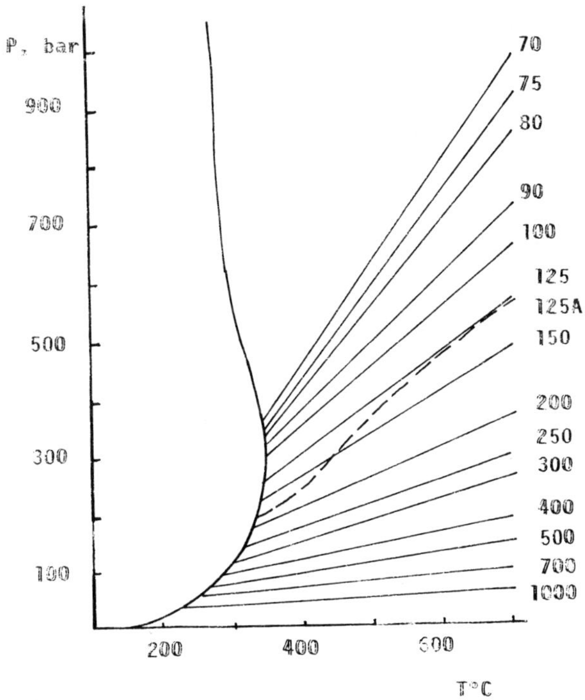

Fig. 1. PVT diagram for the system H_2O–CO_2 [16].

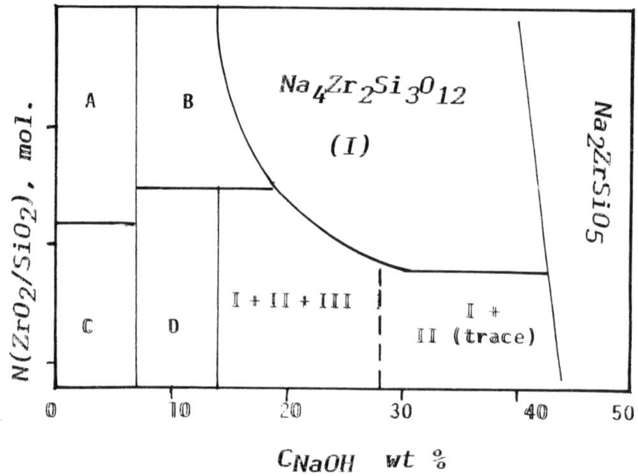

Fig. 2. NC diagram of crystallization for the system Na_2O–ZrO_2–SiO_2–H_2O [17]. (A) $ZrSiO_4$, $Na_2ZrSi_3O_9 \cdot H_2O$, $Na_4Zr_2Si_5O_{16} \cdot H_2O$, (B) $Na_2ZrSi_2O_7$, $Na_4Zr_2Si_3O_{12}$, (C) $ZrSiO_4$, $Na_2ZrSi_3O_9 \cdot H_2O$, $Na_2ZrSi_6O_{15} \cdot 3H_2O$, (D) $Na_8ZrSi_6O_{18}$, $Na_2ZrSi_2O_7$, $Na_2ZrSi_3O_9 \cdot 2H_2O$, II-$Na_8ZrSi_6O_{18}$, III-$Na_3HZrSi_2O_8$.

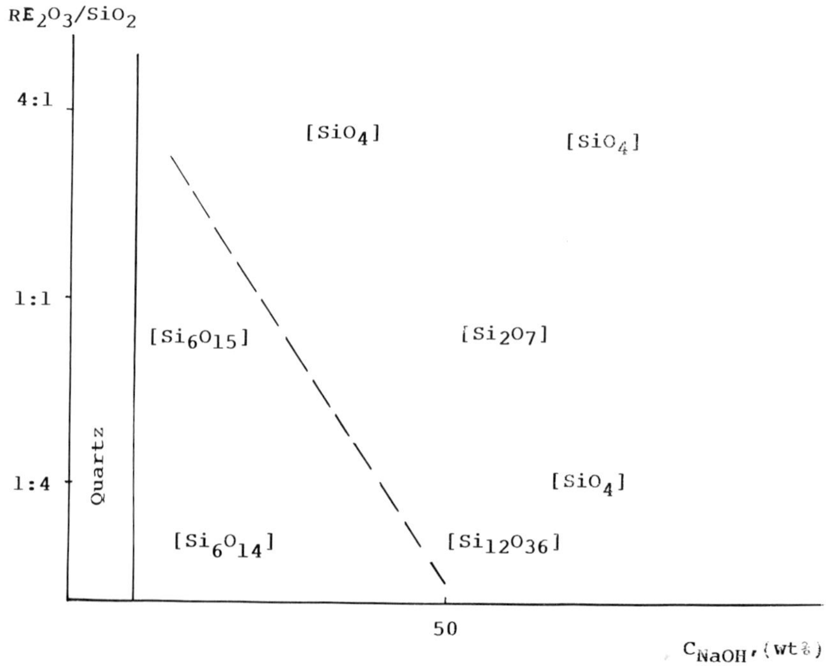

Fig. 3. NC diagram of [SiO] radical distribution in the system Na_2O–RE_2O_3–SiO_2–H_2O [18].

variables. If other variables such as magnetic, electric or gravitational fields are considered, the numerical value 2 on the right-hand side of the equation will increase by the number of new variables considered.

The hydrothermal method is considered as one of the best methods to study phase relationships in numerous systems, and the PTX diagrams thus obtained are published from time to time. Phase equilibria is considered in detail in a chapter by Brebrick in Vol. 1 of this Handbook. Hence, we restrict our discussion to some basic systems and non-conventional phase diagrams.

Among the binary systems, the H_2O–CO_2 system plays an important role with regard to the role of H_2O and CO_2 vapour pressures in native mineral formation and in the hydrothermal synthesis of inorganic compounds. Figure 1 shows the phase diagram for the H_2O–CO_2 system [16]. It is found that below the critical temperature of water the dependence V versus P has a gap at the pressure corresponding to the pressure of water at its critical density. The isochores of real mixtures are linear up to the immiscibility dome that gives a possibility to extrapolate scarce experimental data in the calculation of pressures of mineral formation from the gas–liquid inclusions.

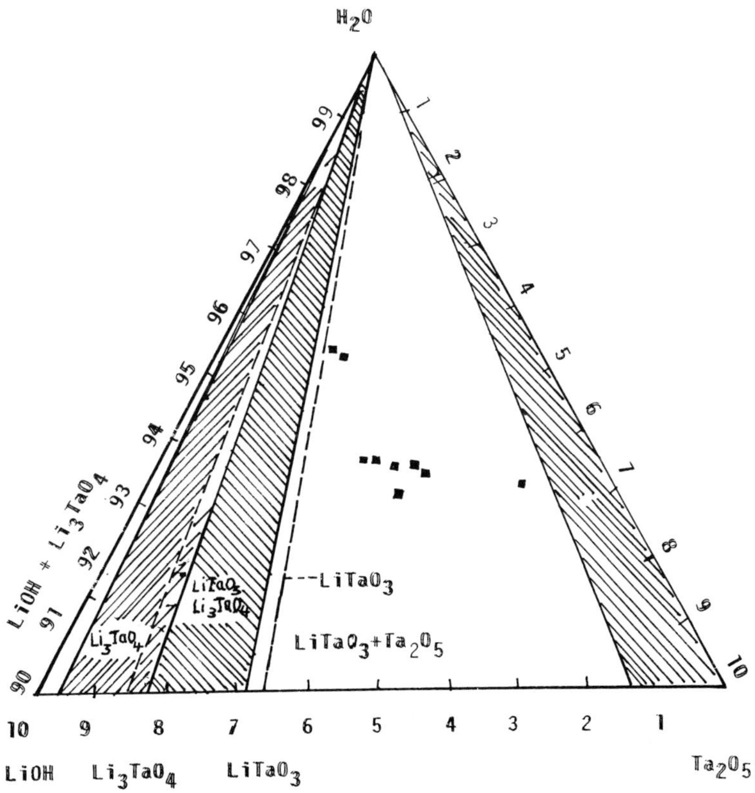

Fig. 4. Phase diagram of crystallization for the system LiO_2–Ta_2O_5–H_2O at 750°C and 1700 atm [19].

The non-conventional phase diagrams plotted for equilibrium conditions are popular especially among Russian workers. These diagrams are popularly known as composition diagrams or NC diagrams, where *N* is the nutrient composition and *C* is the concentration of the solvent mineralizer. There are hundreds of phase diagrams of composition available especially in the Russian literature. An example of such a diagram is given in fig. 2 corresponding to the system Na_2O–ZrO_2–SiO_2 [17]. Similarly, the distribution of the silicon–oxygen radical in the rare earth silicates is shown in the NC diagram (fig. 3) corresponding to the system Na_2O–RE_2O_3–SiO_2–H_2O under hydrothermal conditions [18]. The solid lines indicate the regions of monophase crystallization and dashed lines indicate the beginning of crystallization of the excess component. These non-conventional diagrams are relatively easy to obtain and are highly useful for the growth of single crystals as they clearly depict the growth conditions. Similarly, just as in the case of conventional phase diagrams – even in the phase diagrams of crystallization, the components can be represented for example, in a ternary system as a triangle. Figure 4 shows the phase diagram of crystallization in the system LiO_2–Ta_2O_5–H_2O at 750°C and 1700 atm [19].

5. *Solutions, solubility and kinetics of crystallization*

This is one of the most important aspects of the hydrothermal growth of crystals. Initial failures in the hydrothermal growth of a specific compound is usually the result of lack of proper data on the type of solvents, the solubility and solvent–solute interaction. Despite the large amount of literature data already available, it is still a highly attractive field in hydrothermal research, not only for crystal growers, but also for corrosion scientists, hydrometallurgists, geoscientists and so on. For example, geologists are now in a position to tell us much about the conditions of formation of hydrothermal ore deposits (e.g. temperature, pressure, oxidation potential, pH, fluid salinity, etc.) but, in order to fully understand the mechanisms of transport and deposition of ore minerals, information about metal complex stability and stoichiometry in high temperature–high pressure aqueous electrolyte solutions is necessary. Unfortunately, most of the studies have been performed only at the saturated vapour pressure of the system and very few combined high temperature–high pressure experiments have been designed to yield thermodynamic data.

A hydrothermal solution is generally considered as a thermodynamically ideal one, yet in the case of strong and specific interaction between the solute and the solvent or among the components of the soluble substance in them, significant deviations from Raoult's law occur. Consequently, real hydrothermal solutions differ from ideal solutions and their understanding requires a knowledge of the influence of the solvent in the process of dissolution and crystallization of various compounds. Obviously, as shown in most of the experiments, the type of solvent and its concentration determine a specific hydrothermal process and its important characteristics such as the solubility of the starting materials, quantity of the phases, their composition, output of the phases, kinetics and growth mechanism of single crystals. The value of the changes in the Gibbs free energy in reaching an equilibrium condition varies with the transition from one solvent to another and it can be shown as

$$\Delta G = \Delta H - T\Delta S = RT \ln K, \quad \text{or} \quad K = \exp(-\Delta H/T)\exp(\Delta S/R), \tag{16}$$

where K = equilibrium constant.

The above expression shows the influence of enthalpy and entropy on the equilibrium constant, so that the enthalpy and entropy of the solubility (at constant P and T) are different for different solvents. Also, the solubility of one and the same solid substance changes with the solvent.

At the moment there is no theory which can explain and estimate the solubility in real solutions. However many of the problems connected with the solubility can be explained on the basis of overall physico-chemical principles or laws.

In some cases, it is always better to use the empirical rule which agrees with the fact that solubility becomes high in solvents with higher dielectric constant (ε) and types of chemical bond which are close to those of the solute substance. Deviation from this group has a place in the case where specific interaction between solid substance and the solvent occurs.

The synthesis and recrystallization of a specific compound and the growth of single crystals on the seed are all carried out using different solvents on the basis of physico-chemical considerations. The following conditions are adopted in selecting the most suitable mineralizers:

(i) Congruence of the dissolution of the test compounds.

(ii) A fairly sharp change in the solubility of the compounds with changing temperature or pressure.

(iii) A specific quantitative value of the absolute solubility of the compound being crystallized.

(iv) The formation of readily-soluble mobile complexes in the solution.

(v) A specific redox potential of the medium ensuring the existence of ions of the required valence.

Additionally the solvent should have the desired viscosity, insignificant toxicity and very weak corrosion activity with respect to the apparatus.

These factors fulfil the requirement of the hydrothermal mineralizer in addition to determining the values of solubility of the compound under investigation.

The interaction between the solute and solvent is called solvation. Solvent molecules become linked to the dissolved molecules or ions and the resulting species containing solvent molecules are solvates. Depending upon the reactants, the linkage between the solvent and the solute may exhibit different specific (coordination, hydrogen bond formation etc.) and non-specific (electrostatic) factors or solvation process of different strength and hence solvates include very varied formations. Therefore it is not possible to find a single physical parameter characterizing the solvent which, in itself, could rationalize the solvation process. Simple ions may differ considerably in their electronic structures, charges and radii, and even larger differences are exhibited by solvent molecules representing different types of chemicals. Hence solvated ions may also be very different as regards the number of solvent molecules bound to the ion (the solvation number), the forces giving rise to the binding and the resulting physical and chemical properties like spatial requirements (there are various ways of determining the solvation number – viscosity measurements, pH, conductivity measurement, density measurements, etc.). The specific solvation process is the result of various types of complex formation between the solvent–solute components. In the case of homogeneous aqueous solutions of specific solvation, activated complex theory can be considered [20], according to which the rate constant ξ of the reaction type $A + B \rightleftharpoons X^{\neq} \rightarrow$ reaction products ($X^{\neq}$ is the activated complex) is determined by

$$\xi = (kTK^{\neq}/h)(\nu_A \nu_B/\nu^{\neq}). \tag{17}$$

The concentration of the activated complex X may be determined by the following equation:

$$[X^{\neq}] = K^{\neq}(C_A)(C_B)\,\nu_A\nu_B/\nu^{\neq}, \tag{18}$$

where h is the Planck constant, $K^{\neq}$ is the equilibrium constant of the activated complex, C_A, C_B is the concentration of the components A and B, ν_A, ν_B, $\nu^{\neq}$ is the activity coefficient of the nutrients and active complex, respectively.

In the non-specific solvation when the medium is not homogeneous – i.e. the solvent–solute interaction is not homogeneous and continuous, such an interaction plays a significant role only for ionic compounds [21]. The classical electrostatic model links the rate of reaction constant ξ (in a given known dielectric value ε), with the rate constant (in the medium with $\varepsilon = \infty$) [22]:

$$\ln \xi = \ln \xi_0 + \frac{e^2}{2\varepsilon kT}[Z_A^2/r_A + Z_B^2/r_B - (Z_A + Z_B)/r_{\neq}] + \tfrac{3}{4}\varepsilon KT(\mu_A^2/r_A^3 + \mu_B^2/r_B^3 - \mu_{\neq}^2/r_{\neq}^3), \tag{19}$$

where r_i is the radius of the particle, μ is the dipole moment of the particles, Z is the charge number, k is the Boltzman constant, T the temperature, e the electron charge.

This equation shows that the log of the rate constant depends linearly upon the reverse (–ve) values of the dielectric constants.

Let us consider briefly the role of water as a solvent in the hydrothermal growth of crystals. Water is the major component of the hydrothermal solutions always with a varied chemical composition in the laboratory and in nature. All the solutions used in hydrothermal experiments vary from one another in their properties, ability to dissolve and crystallize, in the nature of the linking between water and electrolyte. Moreover, the properties of each solution depend upon physico-chemical aspects and the structure of the pure water. The formation of associates and complexes in the aqueous solutions of electrolytes is possible because of the presence of structural water, i.e. consisting of water molecules with directional hydrogen bonding. Therefore, it is better to understand the characteristics of the pure water under hydrothermal conditions. The hydrogen ions show an influence on the solubility of various compounds under hydrothermal conditions.

5.1. Thermodynamic principles of solubility

It is well known that

$$\Delta G^0_{sol} = \Delta H^0_{sol} - T\Delta S^0_{sol}, \tag{20}$$

where ΔG^0_{sol}, ΔH^0_{sol} and ΔS^0_{sol} are the changes in the corresponding free energy, enthalpy and entropy of dissolution.

If $\Delta G^0_{sol} < 0$, then dissolution takes place. When $\Delta G^0_{sol} = 0$, the system exists in the equilibrium state, when $\Delta G^0_{sol} > 0$, crystallization is thermodynamically possible.

With a change in the enthalpy, it is observed that either:

(i) In the entire temperature interval $\Delta H^0_{sol} < 0$,

(ii) In the low temperature region $\Delta H^0_{sol} > 0$, whereas in the high temperature region $\Delta H^0_{sol} < 0$.

(iii) In the entire temperature region $\Delta H^0_{sol} > 0$.

If entropy is considered in eq. (20) then with an increase in the temperature, just like ΔH^0_{sol}, the ΔS^0_{sol} may also change. Here also we can expect three possibilities:

(i) In the entire temperature region, $T\Delta S^0_{sol} < 0$.

(ii) In the low temperature region, $T\Delta S^0_{sol} < 0$, whereas in the high temperature region, $T\Delta S^0_{sol} > 0$.

(iii) In the entire temperature interval $T\Delta S^0_{sol} > 0$.

In the first case the entropy helps in the dissolution process. In the second case it can be both helping and reducing, and in the third case it reduces the dissolution process.

In the majority of cases, the solubility of the solid substances in a liquid phase is limited. Solubility is an important physico-chemical and technological parameter, whose estimation determines any dissolution process. It acts as a factor which strongly influences the kinetics or rate of dissolution. Solubility depends upon the nature of the substance, its aggregate state, temperature, pressure and a series of other factors. Quantitative dependence of the solubility of the solid substance on the temperature at a given constant pressure is given by

$$(\delta \ln S/\delta T)_P = (Q_2/RT^2)(\delta \ln a_2/\delta \ln S_{P,T}), \tag{21}$$

where

S = solubility of the solid substance,
a_2 = activity of the solid substance,
Q_2 = heat of dissolution of one mol of the substance in the saturated solution,
R = gas constant.

The solubility of the solid substance may be expressed as a set of two processes: (a) melting of the solid substance, (b) mixing of two liquids.

In the case of ideal solubility of the solid substance, the heat of the second process is equal to zero and $\bar{Q}_2$ is equal to the heat of melting of the solid substance: $\bar{Q}_2 = \lambda$ melting and does not depend upon the selection of the solvent, whereas if $\delta \ln a_2/\delta \ln S = 1$, then eq. (21) can be written as

$$(\delta \ln S/\delta T)_P = \lambda_{melting}/RT^2 \tag{22}$$

such that the right-hand side is > 0, i.e. with an increase in the growth temperature ideal solubility of the solid substance in an ideal solution increases. The dependence of the solubility of any substance in the solution and the pressure is expressed as

$$(\delta \ln S/\delta T)_T = (V^0_{solid} - V_{liquid}/RT)(\delta \ln a_2/\delta \ln S)_{P,T}, \tag{23}$$

where

V_{liquid} = partial molar volume of the soluble substance in the solution,
V^0_{solid} = molar volume of that substance in the solid phase.

In the case of an ideal solution $\ln a_2 = \ln S$, $\delta \ln a_2/\delta \ln S = 1$, and eq. (23) becomes

$$(\delta \ln S/\delta P)/T = (V^0_{\text{solid}} - \bar{V}_{\text{liquid}})/RT. \tag{24}$$

Thus, in the case of ideal solutions of the solid substance in the investigated media, the solubility decreases with an increase in the pressure, if the molar volume of the soluble substance increases during its melting. Similarly, if $V^0_{\text{solid}} > \bar{V}_{\text{liquid}}$, then the solubility should increase with an increase in the pressure (in the case when the partial volume of the liquid is equal to the volume of the melting salt).

In recent times, many aspects of hydrothermal chemistry has been experimentally studied, particularly in the past decade. The credit goes to workers like Franck,

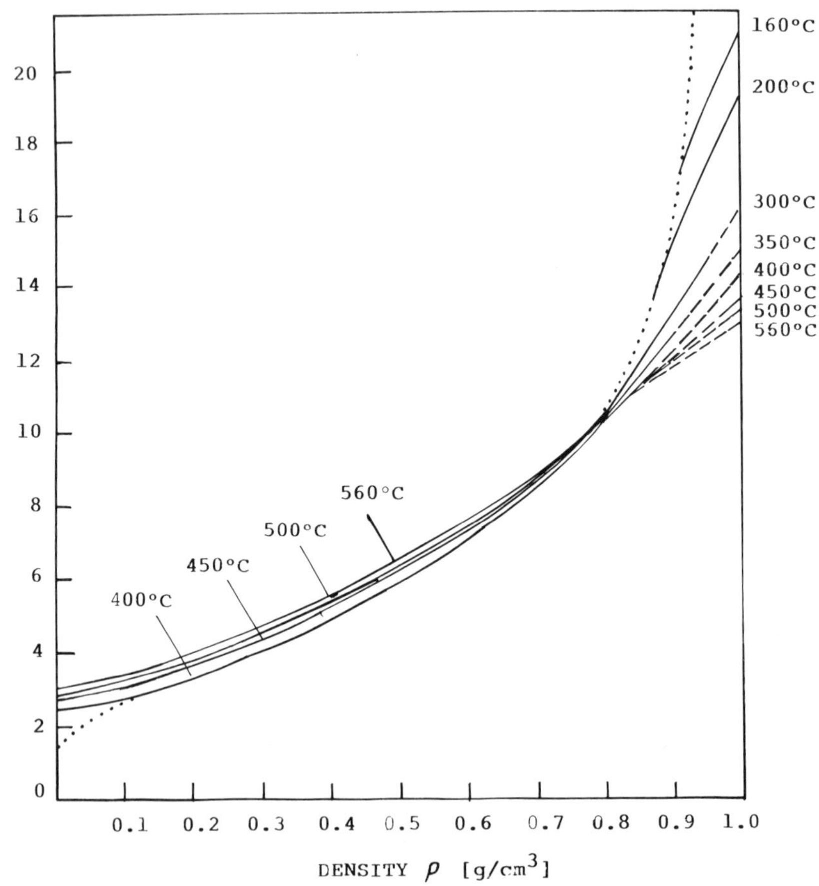

Fig. 5. Viscosity of water as a function of density and temperature [24].

Seeward, Hegleson and others [23–28]. However, most of these studies do not give an insight into the complete physical chemistry of the hydrothermal media as they represent mainly the lower P, T conditions. Helgeson has reviewed exhaustively the thermodynamic properties of electrolytes at high pressures and temperatures.

An understanding of theory and the experimental results on the hydrothermal physical chemistry is very essential to the crystal grower. Here in the experimental work some of the basic properties of the hydrothermal medium like viscosity, dielectric constant, compressibility and coefficient of expansion, etc. are discussed briefly in the crystal growth context. Since diffusion is inversely proportional to solvent viscosity, we would expect very rapid diffusion in hydrothermal growth.

This leads to the growth of perfect single crystals with well developed morphology. We can expect higher growth rates, a narrower diffusion zone close to the growing interface and less likelihood of constitutional supercooling and dendritic growth. It is thus no wonder that quartz growth rates as high as 2.5 mm/day without faults or dendritic growth have been observed [29]. Figure 5 shows the viscosity of water as a function of density and temperature. It has been demonstrated for many cases that the mineralizer solutions (typically 1 M NaOH, Na_2CO_3, NH_4F, K_2HPO_4, etc.) are close to the properties of water. For 1 M NaOH at room temperature, $\eta_{solution}/\eta_{H_2O} = 1.25$ [30] one can expect that the viscosity of hydrothermal solutions can be as much as two orders of magnitude lower than "ordinary" solutions. The mobility of molecules and ions in the supercritical range is much higher than under normal conditions. Also electrolytes which are completely dissociated under normal conditions tend to associate with rising temperature [31].

Dielectric constant is one of the important properties of a solution. More than 40 years ago, Hasted et al. (1948) observed that the dielectric constant of electrolyte solutions (ε) can be regarded as a linear function of molarity up to $\approx$ 1–2 M depending on the electrolyte [32]. Franck [24] has discussed ionization under hydrothermal conditions and made most careful and complete conductivity studies, and shown that the conductance of hydrothermal solutions remains high in spite of a decrease in ε, because that effect is more than compensated for by an increase in the ion mobility brought about by decreased viscosity under hydrothermal conditions. Figure 6 shows the dielectric constant of water [33]. Largely as a consequence of the dramatic decrease in the dielectric constant of water with increasing temperature at constant pressure and/or decreasing pressure at constant temperature "completely" dissociated electrolytes at low temperatures and pressures may become highly associated in the supercritical region. Virial expansion of Debye–Hückel theory affords a basis for extrapolating thermodynamic observations of "completely" dissociated electrolytes to high pressures and temperatures, where they may be highly associated. However, the Debye–Hückel equation fails to represent adequately observed activity coefficients for dilute solutions of 2:2 electrolytes. Landolt–Börnstein (1980) have estimated the compressibility of water, from the PVT curves [34]. The hotter and less dense the fluid is the larger is its compressibility. It is an important factor used in the calculation of density change with pressure. The coefficient of thermal expansion is also an important parameter which helps in the growth of flawless and strain free crystals. It can be calculated from PVT data. Water has a much increased coefficient of expansion under hydrothermal con-

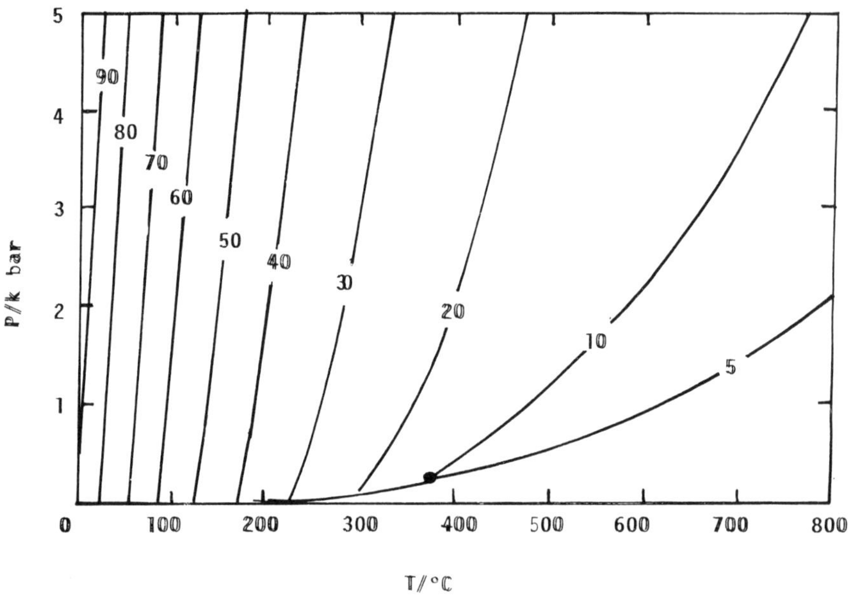

Fig. 6. Dielectric constant of water [33].

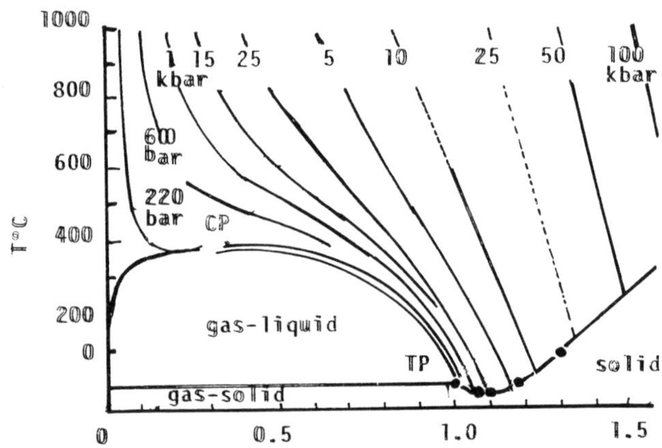

Fig. 7. Temperature–density diagram of water with pressure as a parameter [37].

ditions. PVT measurements for water and other solvents are rather rare. The PVT data for water up to 1000°C and 10 kbar are known accurately enough up to within 1% error [35, 36]. Figure 7 shows the temperature–density diagram of water, with pressure as a parameter [37]. At very high *P*, *T* conditions (1000°C and 10 kbar), water is completely dissociated into H_3O^+ and OH^- behaving like a molten salt and has a higher density of the order of 1.7–1.9 g/cm^3. In the hydrothermal growth of crystals, the PVT diagram of water proposed by Kennedy (1950) is very important (fig. 8) [38]. Usually, in most routine hydrothermal experiments, the pressure prevailing under the working

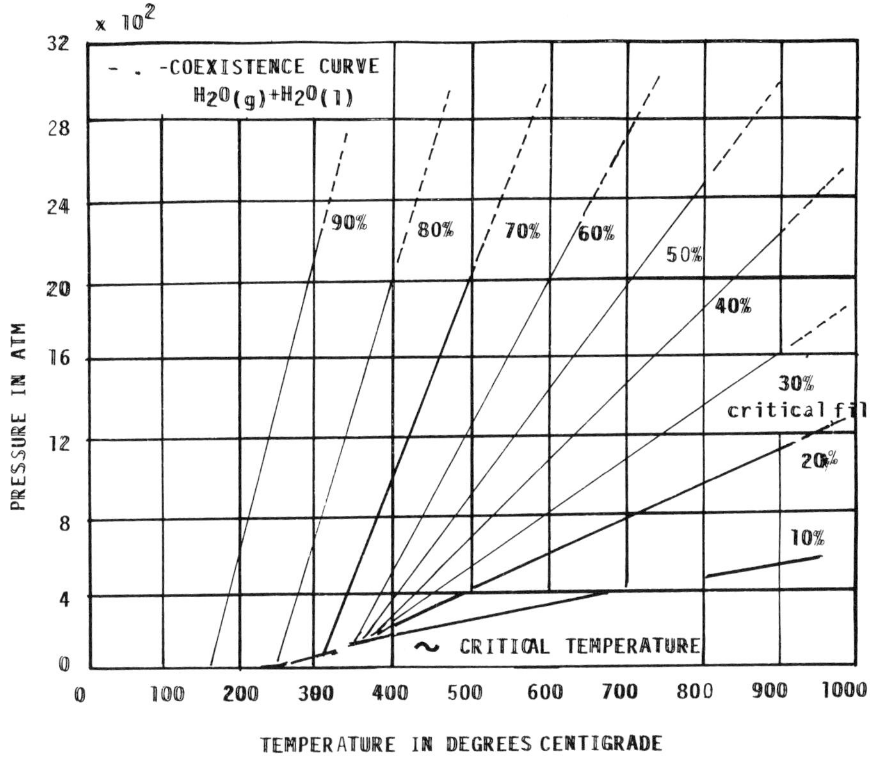

Fig. 8. PVT diagram of water [38].

conditions is determined by the degree of filling and the temperature. When concentrated solutions are used, the critical temperature can be several hundred degrees above that of pure water [39]. The critical temperatures are not known for the, usually complex, solutions at hand; hence one cannot distinguish between sub- and supercritical systems for reactions below 800°C. Although the temperature in the growth zone and the actual vapour pressure are not known to 100% accuracy, the PVT diagram of Kennedy is used by most hydrothermal crystal growers routinely. The PVT relations in $AlPO_4$ and SiO_2 systems have been reviewed in [40–42]. Also the $P-T$ behaviour of the quartz–water system has been well discussed by Brebrick in Vol. 1 of this Handbook.

5.2. Experimental investigations of solubility

The selection of a suitable solvent and the theoretical aspects of solubility have been discussed in the previous section. Here, let us consider the experimental aspect, its determination and role in growing single crystals. There are two principal methods for determining the solubility under hydrothermal conditions. The first one is a sampling technique and has been described in detail by Morey and Heiselgesser [43]. It is especially useful where two fluid phases are present. In this technique a suitably designed valve is arranged so that a fluid sample small enough so as not to perturb equilibrium can be withdrawn and chilled from an isothermal hydrothermal autoclave at operating conditions. If only one fluid phase is present a weight loss method has been proved to be somewhat less cumbersome [44]. It is quite difficult to study the solubility of crystals under hydrothermal conditions if the compound does not form any metastable solutions over a wide temperature range. If there is a considerable metastable range of this kind, however, we may use the least troublesome method of studying solubility at high temperatures and pressures – the method of weight loss. If a weighed single crystal or crystals are placed in the capsule under conditions where the phase is stable and where the fluid phase is liquid alone or gas alone, the weight loss will be a measure of solubility. However, special care must be exercised to avoid overshooting, to ensure that the desired temperature is reached during warm up, to avoid losing material before weighing, or weighing material that has precipitated during the quenching. In many cases the solubility obeys the van't Hoff equation and is linearly dependent on solution density [45]. The ratio of solubility to mineralizer concentration often gives clues to the species present [46]. Several additional difficulties were encountered with the silicates, alumino-silicates, phosphates, etc. They were first noticed while studying the solubility of quartz. These difficulties principally arise from phase separation of the solution under certain conditions, and the formation of a "heavy" phase enriched with silicon [47]. For example, it was shown by the authors of [48], in connection with the solubility of albite in H_2O at 200–350°C, that under these conditions there was a partial decomposition of the albite forming analcime. Hence, in this case, the true solubility of albite in water was not really being determined. Therefore, to obtain the true solubility of any compound which forms the heavy phases or other metastable phases, it has to be determined under conditions similar to its conditions of synthesis, so as to determine its optimum growth conditions. Russian workers at the Institute of Crystallography have designed an apparatus to suite the solubility study of various compounds.

Let us consider the solubility studies of some selected compounds like quartz, berlinite, malachite, etc. Solubility of a compound may be positive or negative. If the solubility of a compound increases with increasing temperature, it is positive. If the solubility of the compound decreases with increasing temperature, it is negative. Some compounds show positive solubility up to a certain temperature and then show negative solubility and vice versa. Quartz and berlinite form the best examples. The earliest hydrothermal experiments, carried out in the previous century, had pure water as the mineralizer. Today we use a great variety of mineralizers especially aqueous solutions of salts, acids, bases, etc. at very low concentrations. Likewise, there are a great numbers of non-aqueous solvents. Hannay and Hogarth used an alcoholic solution as a mineralizer a century ago [49]. Usually non-aqueous solvents are preferred when the required solubility cannot be reached in an aqueous medium. The table given in the Appendix shows a wide range of mineralizers used in the hydrothermal growth of crystals. However, the composition of the effective solvents used in growing crystals is fairly standard. Pure water can be a good solvent at very high temperature and pressure conditions. The use of various acids, salts and bases considerably reduce the P, T

conditions. The commonly used non-aqueous solvents are NH_3, HF, HCl, HBr, Br_2, S_2Cl_2, S_2Br_2, $SeBr_2$, $H_2S+(C_2H_5)_3$, NH_4Cl, C_2H_5OH, CS_2, CCl_4, C_6H_6, CH_3NH_2, etc. In the past decade the growth conditions have been further reduced with the use of some acid solvents. For example, high molar acids like HCl, H_2SO_4, H_3PO_4, HNO_3, HCOOH are freely used in the growth of various compounds and this has brought down the growth temperature to below 300°C. In recent years mixed solvents have been used in the growth of some compounds like $AlPO_4$. The mixed solvents are found to be highly effective. The reduction in the growth temperature considerably reduces the working pressure which in turn helps in the use of simple apparatus. Even silica autoclaves can be used, which facilitates direct visibility of the growth medium just as in any other low temperature solution growth. This has greatly attracted the attention of crystal growers in the last two years and a great variety of compounds which hither to were produced only at high temperature and pressure have been today obtained at lower *P*, *T* conditions. Hence the mineralizer plays an important role in the hydrothermal growth of crystals and it has changed the tendency and perspective of the method. All these aspects can be very well described with reference to quartz, berlinite, sodalite and α-TeO_2. The reader is advised to refer the works of Pearson [50, 51] regarding the empirical selection of solvents based on the hard and soft acids and bases. Similarly the recent literature available on berlinite and also the reviews of Laudise and others would be of great use [52–54].

5.2.1. Solubility of quartz

Much of our knowledge on the hydrothermal growth of crystals on the whole has come from the study of various aspects of the growth of quartz.

The solubility of quartz in pure water was found to be too small for crystal growth (0.1–0.3 wt%), but the solubility could be markedly increased by the addition of OH^- which acts as a mineralizer. For example, the reactions

$$SiO_2 + 2OH^- \longrightarrow SiO_3^{2-} + H_2O,$$

$$2OH^- + 3SiO_2 \longrightarrow Si_3O_7^{2-} + H_2O,$$

show the formation of various complexes or species formed during the hydrothermal crystallization of quartz. Hosaka and Taki have used Raman spectra to identifiy and quantify such species [55].

In pure aqueous solutions (even at 400°C and 25 000 psi) the solubility of quartz is too low to allow growth to take place in any reasonable time. Alkaline additions, such as NaOH, Na_2CO_3, KOH and K_2CO_3 are all effective as mineralizers in this pressure and temperature range. A small increase in molarity results in only a slight increase in the growth rate, whereas large increases begin to produce additional phase along with quartz. The minimum molarity for good growth is about 0.25 M for NaOH. Concentrations of about 4.0 M for NaOH and 2.0 M for KOH form sodium or potassium silicates along with α-quartz. Laudise and Nielson (1961) have measured the solubility of quartz in 0.5 M NaOH as a function of temperature [56]. The log of the solubility was found to be a linear function of the reciprocal of the absolute temperature at 80% and 85% fill as shown in fig. 9.

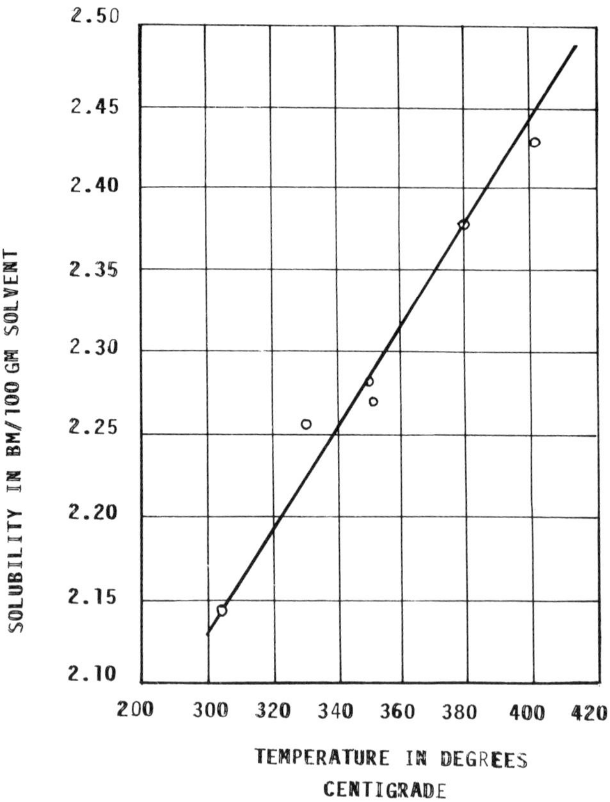

Fig. 9. Solubility of quartz as a function of temperature [56].

An important result of solubility determinations is the delineation of the pressure, temperature and composition regions where the temperature coefficient of solubility is negative. These regions are to be avoided in the growth of quartz, since it requires a different set up otherwise it will result in the loss of seed crystals.

In recent years mixed solvents are being used even in the growth of quartz. High quality large quartz crystals have been obtained in NaCl and KCl solutions (10%) [57].

5.2.2. Solubility of berlinite

Berlinite is isostructural and isoelectronic with α-quartz. Although it was grown in 1953 by Jahn and Kordes [58], there was a long gap of nearly 25 years in further growth of berlinite crystals, because of the difficulties associated with the material. These problems are directly related to the lack of a systematic study of the solubility and also partly due to the P_2O_5 presence. The solubility of berlinite was first determined by Jahn and Kordes (1950) in orthophosphoric acid above 300°C and it was found to be positive (fig. 10) [58] and subsequently Stanley (1954) reported a negative solubility for berlinite in 6.1 M H_3PO_4 [59]. The solubility of $AlPO_4$ varies widely with the type of

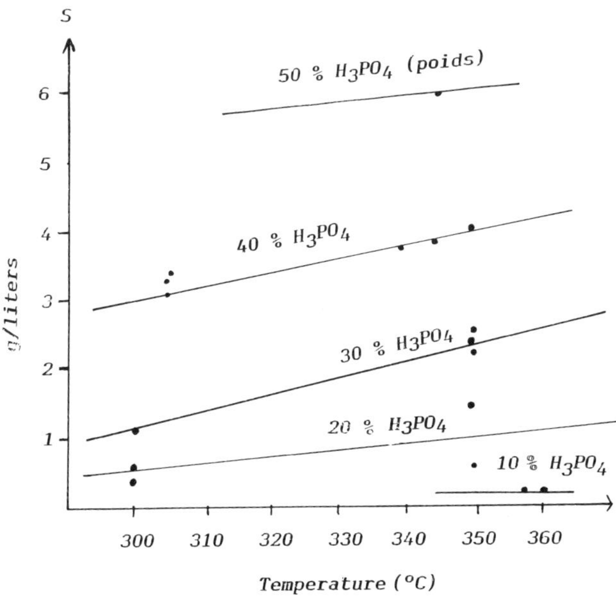

Fig. 10. Solubility of berlinite in H_3PO_4 [58].

solvent used. Some authors claim that solubility of $AlPO_4$ in HCl is similar to that in H_3PO_4. The most important difference is the higher solubility at comparable mineralizer concentration. This is an advantage in that the mineralizer concentration, being less for a given solubility, will result in less undesired mineralizer interaction in HCl; therefore, the crystals should be pure without contamination.

The authors of [53] have studied the solubility of $AlPO_4$ in some new solvents like HCOOH, NH_4Cl, Na_2CO_3, $NH_4H_2PO_4$, NaF, KF, LiF, etc. The solubility of $AlPO_4$ (in wt%) as a function of temperature and at a pressure of 2 Kpsi, in 2 M HCOOH solution is shown in fig. 11. The slope of the solubility curves for $AlPO_4$ in HCOOH is similar to the solubility curve for $AlPO_4$ in HCl and H_3PO_4 [53]. It was found that, above 300°C, the solubility does not alter much which suggests that crystal growth beyond this temperature is impracticable. Hence, the solubility data obtained for $AlPO_4$ in 2 M HCOOH suggests that the temperature range between 180 and 300°C is the most ideal and in particular 250°C, is the best growth temperature. Recently it has been shown that the solubility of $AlPO_4$ in H_2SO_4 is retrograde with respect to temperature and higher than in H_3PO_4 and HCl at comparable temperature, pressure and acid molarity. However, the viscosity of sulphuric and phosphoric acid solutions are greater than that of HCl. Hence, the addition of HCl to H_2SO_4 would form a more efficient solvent for the growth of $AlPO_4$ [60]. Figure 12 shows solubility as a function of temperature in $H_2SO_4 + HCl$ mixtures. The authors have also carried out thermodynamic interpretation of solubility data. The dissolution of berlinite in mixtures of the acids H_2SO_4 and HCl is a process of complex formation between Al^{3+} and the ligands present in the solution: OH^-, H_2O, Cl^-, HSO_4^-, SO_4^-, $H_2PO_4^-$. At low

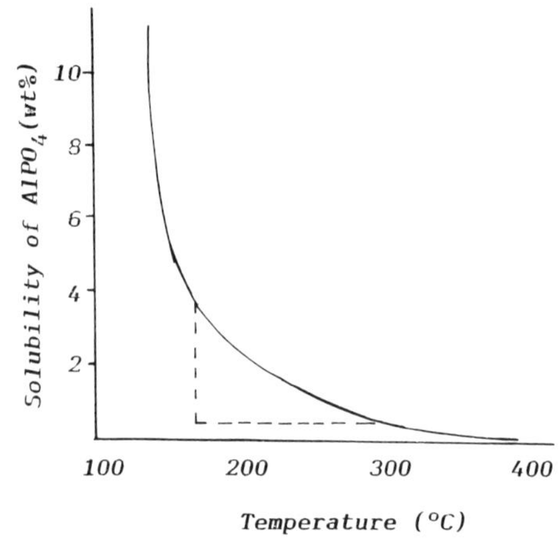

Fig. 11. Solubility of berlinite as a function of temperature in 2 M HCOOH at 2 kPsi [53].

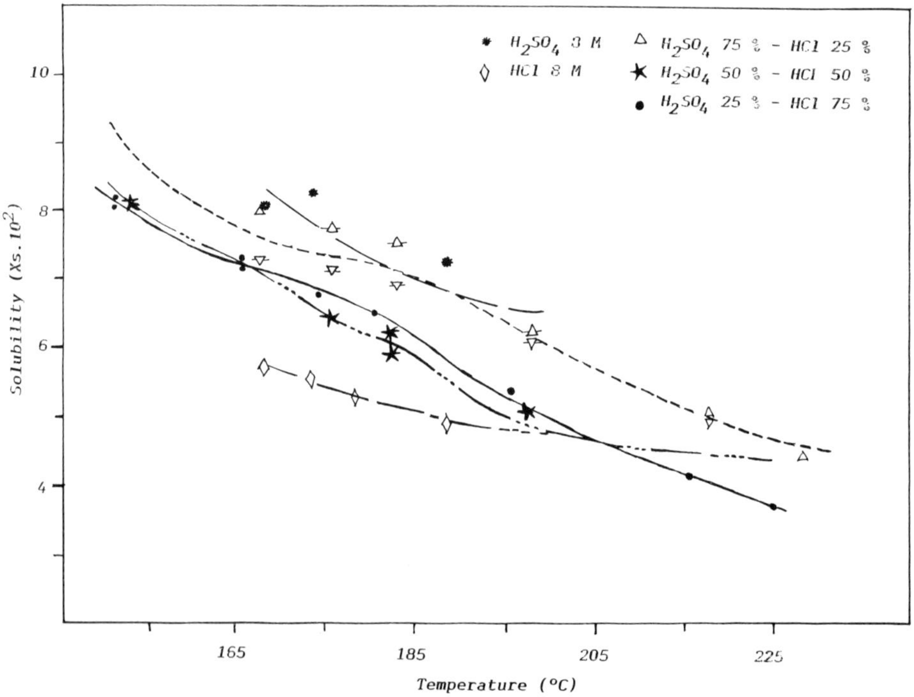

Fig. 12. Solubility of berlinite as a function of temperature in $H_2SO_4 + HCl$ [60].

temperature the complexes are similar to those of $AlPO_4$ in HCl; that is with Al–Cl complexes predominating. Even polynuclear complexes might form a broader metastable zone. With the rise in temperature, the association of Cl^- with H^+ increases the yield of HCl and destabilizes the $[Al^{3+}]$ complex. Under these circumstances, sulphate and monohydrogen phosphates replace chloride ligands and force a change in the coordination number of Al from 6 to 4 due to the steric effect. Similarly, H^+ also tends to destabilize the complex with sulphate and becomes $H_2PO_4^-$ hegemonic ligand, as its association constant with H^+ does not vary significantly with a rise in temperature. Finally, the nucleation takes place by polymerization of tetrahedral $[Al^{3+}]$ complexes.

5.2.3. Solubility of malachite

The growth of malachite single crystals has posed a challenge in the past decade. Several original methods which allow nearly all textural varieties found in natural malachite have been divised. The malachite solubility in pure water and electrolytic aqueous solutions has not been understood precisely. Symes and Kester (1984) studied the peculiarities of malachite dissolution in pure and sea water in order to reveal the

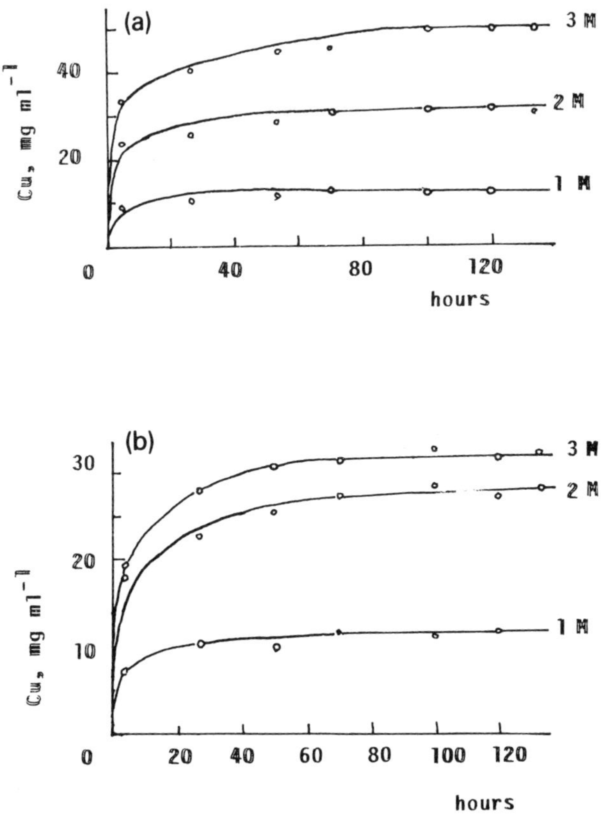

Fig. 13. Solubility curves for malachite in NH_4OH at (a) 20°C; (b) 75°C [63].

conditions necessary for it's precipitation [61]. Sulyapnikov and Shtern [62] determined the maximum copper concentration during malachite dissolution in solutions of $NaHCO_3$ and $KHCO_3$ of a different concentration at temperatures up to 200°C and different pressures of CO_2. Their data has shown that, under normal conditions, the copper content in these solutions changes strongly depending on the time of the solution mixing with the powder of the basic copper carbonate. Balitsky and Bublikova have investigated malachite solubility in 1, 2, 3 M solutions of ammonium hydroxide at temperatures 20, 25, 50 and 75°C ($P = 1$ bar). Figure 13 shows kinetic curves of malachite dissolution in Ammonia aqueous solutions at 20°C (1–10 M NH_4OH; 2 M NH_4OH; 3–3.0 M NH_4OH). As can be seen from fig. 13, equilibrium copper contents in the ammonia solutions increase as ammonia concentration increases and they decrease as temperature rises [63].

These examples give a more or less complete picture with regard to the type of solvent, the P, T conditions and the progress made in the search for new solvents for various inorganic compounds under hydrothermal conditions.

6. *Apparatus*

Crystal growth under hydrothermal conditions requires a pressure vessel capable of containing the highly corrosive solvent at high temperature and pressure and a means of producing, controlling and recording temperature and pressure. Designing a suitable hydrothermal apparatus popularly known as an autoclave, is the most difficult task associated with hydrothermal crystal growth. A literature survey clearly indicates two schools of autoclave design: (i) Western school, (ii) Russian school. The former is very important and it dominates most of the earliest designs and these designs later became standard world wide. Most of these designs are relatively simple and quite old (designed prior to the 1970s), whereas the Russian school has dominated this field from the 1970s and has come out with many new designs. The special construction Bureau at the Institute of Crystallography, Academy of Sciences, Moscow, Russian Federation, has designed several new autoclaves to suit a specific task of either growing single crystals or studying physico-chemical aspects such as solubility, PVT-behaviour, etc. Besides, several earstwhile Soviet Hydrothermal Laboratories have produced their own designs to suit their requirements and the type of crystal under investigation. Hence, most of these autoclaves are only local and have not been adopted world-wide. Hence, for the benefit of the reader, we consider only the standard autoclaves used world-wide and some important recent designs from the Russian school.

Any ideal hydrothermal autoclave should have the following characteristics: (a) inertness to acids, bases and oxidizing agents, (b) easy to assemble and dissemble, (c) should have a sufficient length to obtain a desired temperature gradient, (d) should be leak-proof with unlimited capabilities to the required temperature and pressure, (e) should be rugged enough to bear high pressure and temperature experiments for long duration, so that no machining or treatment is needed after each experimental run.

Table 1 gives a list of the autoclaves with their P, T characteristics [64]. The autoclaves listed in table 1 are externally heated pressure vessels and their PT range can not

Table 1
Autoclaves.

Type	Characteristic data
Pyrex tube 5 mm i.d. 2 mm wall thickness	6 bar at 250°C
Quartz tube 5 mm i.d. 2 mm wall thickness	6 bar at 300°C
Flat plate seal, Morey type	400 bar at 400°C
Welded Walker–Buehler closure 2600 bar at 350°C	2000 bar at 480°C
Delta ring, unsupported area	2300 bar at 400°C
Modified Bridgman, unsupported area	3700 bar at 500°C
Full Bridgman, unsupported area	3700 bar at 750°C
Cold-cone seal, Tuttle–Roy type	5000 bar at 750°C
Piston cylinder	40 kbar, 1000°C
Belt apparatus	100 kbar, > 1500°C
Opposed anvil	200 kbar, > 1500°C
Opposed diamond anvil	up to 500 kbar, > 2000°C

be extended further due to the lack of suitable refractory alloys. However, internally heated pressure vessels are now commercially available up to 10 kbar and 1400°C. S. Hirano et al. (1974) have designed such an apparatus with an inert gas as medium [65]. The list given in the table does not include some of the recent designs and modifications of the Russian workers.

6.1. Selection of autoclaves and autoclave materials

In the selection of a suitable autoclave, the first and foremost parameter is the experimental temperature and pressure followed by the solvent or the growth medium and, if the reaction is taking place directly in the vessel, the corrosion resistance is of course a prime factor in the choice of autoclave material. Laudise and Nielsen (1961) have reviewed the properties of some alloys used in the making of hydrothermal autoclaves. Table 2 gives certain properties of commonly used alloys [56]. The autoclaves can be selected accordingly with the compound to be grown. Some crystals can be grown readily within the autoclave without any lining or liners or cans. For example, the growth of quartz can be carried out in low carbon steel autoclaves. The low carbon steel is corrosion resistant in systems containing silica and NaOH, because, the relatively insoluble NaFe–silicate forms and protectively coats the ground vessel. In contrast, the growth of berlinite crystals requires a teflon lining or beakers because phosphorus is highly corrosive; it can even corrode platinum if used for a long time. Therefore, the corrosion resistance of any metal under hydrothermal condition is very important. For example, turbine engineers have long known that boiler water with $pH > 7$ is less corrosive than slightly acidic water, especially for alloys containing Si. Several conventional methods of studying corrosion are known. Usually low carbon steel is used in the fabrication of simple acid digestion bombs, Morey autoclaves, modified Bridgman, and other related autoclaves. Inconel, Udimet and Stellite are highly useful because of their high stength at high temperature. These alloys are used in the fabrication of cold cone-seal bombs used for phase equilibria studies at pressures up

Table 2
Properties of certain alloys[a].

	Resistance to aqueous OH	Rupture stress (atm) 1000 h, 800°C	Rupture stress (atm) 1000 h, 600°C
Low carbon steel	+		170
Tool steel	+		
4140 (similar to EN19)	+		170
Stainless type 304	–	270	1100
19-9-DL (Universal Cyclops Steel Co.)	?	680	2200
Croloy 15-15N	+	610	2260
Timken 17-22-A	+	–	–
Inconel X	?	1200	4500
Stellite	?	–	–
Udimet 500	?	2100	–
Silver	+[b]		
Platinum	+[b]		
Platinum–10% iridium	+[b]		

[a]Taken in part from J.W. Freeman and H.R. Voorhees, Ind. Eng. Chem. 48 (1956) 861.
[b](Most acids also).

to 5.4 kbar and temperatures up to 950°C. The recent TZM autoclaves are used at pressures up to 10 Kbar.

For low temperature and very low pressure experiments thick glass vessels can be used. Recently, Popolitov [66] has used a thick walled silica vessel with a teflon capping for experiments up to 200°C and < 500 bar pressure. The details of autoclaves are discussed in detail in the forthcoming section.

6.2 Liners

In the majority of hydrothermal experiments, the mineralizer used is highly corrosive and it can attack the vessel which is inimical to obtaining high purity crystals for devices. It requires a suitable lining for the inner wall of the autoclave or separate liners are placed in the autoclave. Even in the case of quartz growth where an acmite protective layer is found all along the inner wall of the autoclave, the purity of the substance is doubtful. Therefore for the growth of high purity quartz a silver lining for the autoclave is provided. Similarly, other metal linings are provided depending upon the solvent medium. For example, a copper lining is given to many of the borate systems. Copper partly dissolves in dilute bases at 400°C and copper crystals are deposited in the cooler region of the autoclave, likewise, nickel linings are also readily attacked. Hence noble metal lining or liners or capsules are used successfully for alkaline and neutral media. Titanium is much more corrosion resistant, but it is difficult to handle. Litvin and Tules (1971) have summarized the applicability of various liners [67] and these are listed in table 3.

The use of glass tube as a liner was first applied by De Senarmont as early as AD 1851 for the growth of carbonates, sulphates and sulphides. Figure 14 shows the sealed glass

Table 3
Materials used as reactor linings.

Material	T(°C)	Solutions	Remarks
Titanium	550	chlorides hydroxides sulphates sulphides	Corrosion in NaOH solution > 25%, in NH_4Cl solution > 10% (at 400°C)
Armco iron	450	hydroxides	Gradual oxidation producing magnetite
Silver	600	hydroxides	Gradual recrystallization and embrittlement, partial dissolution
Platinum	700	hydroxides chlorides sulphates	Blackening in chlorides in the presence of sulphur ions; partial dissolution in hydroxides
Teflon	300	chlorides hydroxides	Poor thermal conduction
Tantalum	500	chlorides	Begin to corrode in NH_4Cl solution 78%
Pyrex	300	chlorides	
Copper	450	hydroxides	Corrosion reduced in the presence of fluoride ions and organic compounds
Graphite	450	sulphates	Pyrolytic graphite most suitable for linings
Nickel	300	hydroxides	
Quartz	300	chlorides	
Gold	700	hydroxides sulphates	Partial dissolution in hydroxides

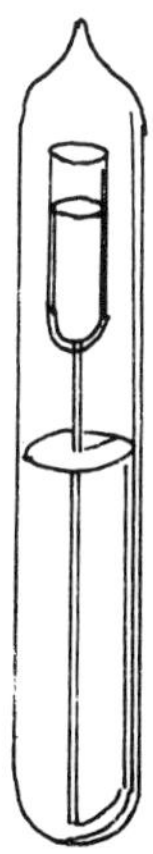

Fig. 14. Schematic diagram of the sealed glass tube used by De Senaramont [7].

tube used by De Senarmont [7]. The tubes were included in water-filled fused gun barrels and heated up. An improved version using pyrex glass and again water as pressure transmitter was reported by Allen et al. (1912) [68]. The quartz glass tube liners of 15 mm diameter can easily withstand the outside pressure of 3 kbar at 500°C. However, the internal pressure has to be kept a little below the outside value. Rabenau and Rau

(1973) have described this technique in detail [69]. For larger autoclaves, like the Morey flat plate and the modified Bridgman, noble metal or other suitable metal lining is provided. The main difficulty in such a design is that the lining should be free of cracks, pin holes or ridges and it should be honed to near mirror finish. For Tuttle cold-cone seal autoclaves capsules made of noble metals are used. The capsule tube charged with a desired amount of nutrient is clamped tightly at the top and welded shut. During the experiment the capsule collapses but does not rupture. The vessel is then heated to the desired temperature while kept at constant pressure. When the operating temperature is reached the capsule expands to its original volume and is not subjected to tension and virtually no capsule failures occur. For some experiments the capsule needs only to be pressed together; capsule length will be short for phase equilibria studies and longer where a temperature gradient for crystal growth is desired. In large diameter vessels Pt liners are usually used. These liners have silver caps to avoid Pt to Pt seizing. The liner is filled to a given percent fill and closed and the space between the liner and the inner wall of the autoclave is filled to the same percent of fill or slightly higher; this provides a pressure balance and the liner supports virtually no pressure. Pressure imbalance can cause rupture of the liner. For the growth of high purity crystals in solution in highly corrosive media, teflon beakers or liners are used. The teflon liner or beaker should sit exactly inside the autoclave without leaving any gap; alternatively pressure balancing technique can be used. A slightly higher pressure of nearly 1 to 2 percent of fill outside the teflon liner helps in the sealing of teflon liners since the cap and liner can not be welded. Also, in some cases where the experiment is carried out for longer duration, effective sealing of the teflon liner can be obtained with a sharp ring-like mechanism on the liner as shown in fig. 15. As the temperature rises the teflon expands and hermitic sealing can be obtained. The greatest disadvantage with liners or cans is the difficulty associated with measuring of the actual temperature inside. Certainly temperature gradients are appreciably lower in the liners than in the autoclaves without liner. It becomes more prominent when teflon liners are used. Despite all these difficulties single crystal growth of several materials is being carried out routinely at various laboratories.

Let us consider briefly the internal arrangement of the liner in some typical hydrothermal crystal growth experiments. In the growth of quartz crystals the liner has an assembly shown in fig. 16 [70]. The lower portion of the liner is filled with the

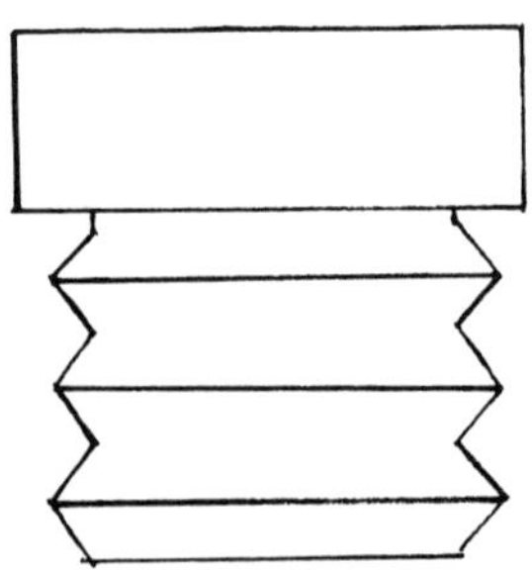

Fig. 15. Teflon liner capping mechanism for effective sealing.

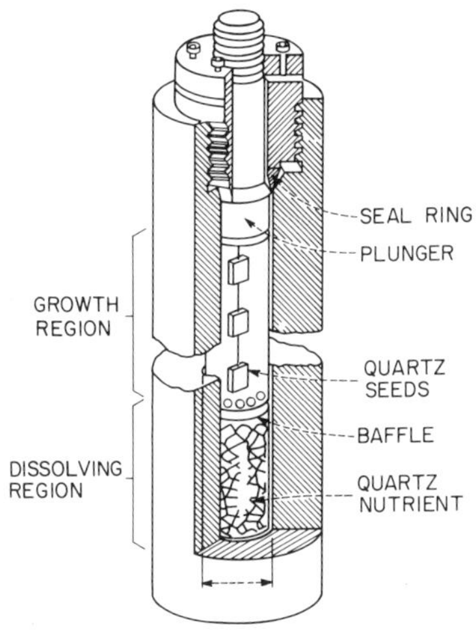

Fig. 16. Liner assembly for the growth of quartz [70].

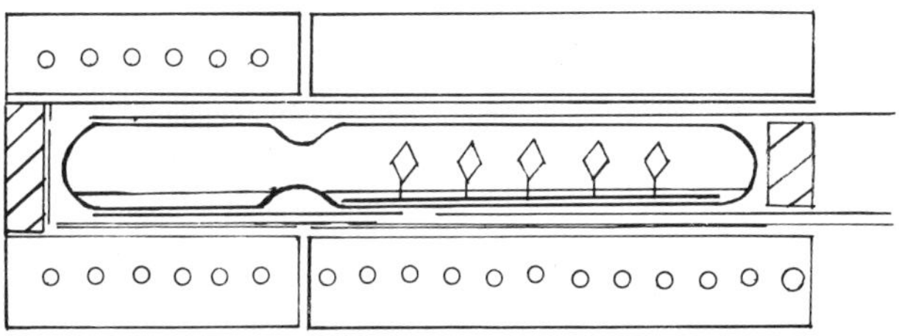

Fig. 17. Liner assembly for the growth of berlinite.

required nutrient material and, in the upper portion, a seed holder is placed and this is the growth zone (cooler zone). In between the growth zone and nutrient zone a baffle having a definite percent of opening is placed. The baffle is placed at an appropriate position. Therefore, the positioning of the baffle and the percent opening together play an important role in the growth kinetics.

In the hydrothermal growth of berlinite the internal assembly of the liner differs from that for quartz growth and is shown in fig. 17. The liner material is usually teflon. Since berlinite has a negative temperature coefficient of solubility, the nutrient material is placed in the upper portion (cooler zone) and the seed holder placed at the lower portion (hotter zone). In between the two zones a baffle with a desired percent opening is placed in addition to the vertical positioning of these liners. There are several modifications especially in horizontally mounted autoclaves. Detaint et al. (1985) proposed the liner internal assembly shown in fig. 18 [71] for the composite gradient method of growing berlinite crystals. Hence, internal assembly of the liner is very important in crystal growth.

Fig. 18. Liner assembly for the growth of berlinite [71].

6.3. Temperature and pressure measurements

Several modern electronic controllers are commercially available today to regulate temperature and pressure in hydrothermal experiments. The externally heated pressure vessels can be placed in a suitable isothermal zone or multiple heating zone furnaces controlled by thyristor-based controllers. The highly sophisticated temperature programmers commercially available would be useful in programming the experimental temperature, temperature gradient, heating and cooling rates, etc. Depending upon the

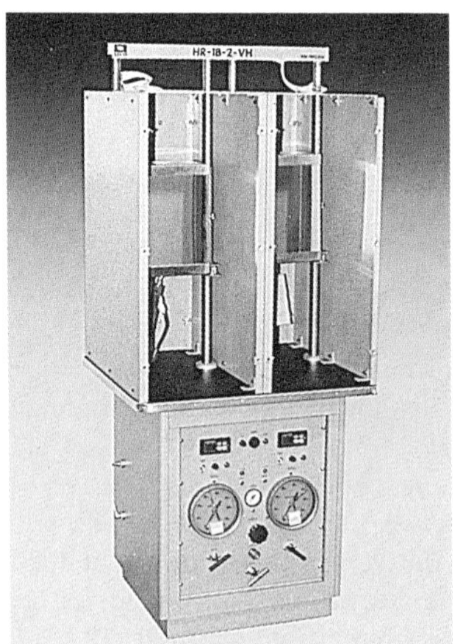

Fig. 19. Hydrothermal unit (Tempress, Leco Corp. USA).

temperature to be produced, a suitable heating element made of Kanthal or Nichrome is wound around a ceramic core. Similarly, depending upon the temperature to be measured suitable thermocouples made of Cr–Al or Pt–Rh are used. Sheathed thermocouples of small heat capacity are used to measure the internal temperatures.

The pressure in hydrothermal experiments is measured by different means. In the initial experiments in a given system, standard calibrated bourdon gauges are used and are recalibrated at regular intervals against either a dead weight tester or master gauge. Subsequently, for routine pressure measurements, Kennedy's PVT diagram (fig. 8) is used. Automatic pressure controllers and recorders are also available commercially. For example, in the Tuttle cold-cone seal autoclaves unit (Tempress, Leco Corp. USA, fig. 19) the required pressure is fixed in the gauge which will maintain the pressure at that level for fairly long duration. However, pressure measurements with a temperature gradient have not been studied in detail, but it is certain that, under hydrothermal conditions, there can be no pressure gradient. Hence, the local temperature and local density anywhere in the system would determine the pressure of the entire system.

Let us consider the design and the principle involved in the construction of various autoclaves used in hydrothermal crystal growth.

6.4. *Glass vessels*

The glass vessel in hydrothermal research has the advantages of ease of observation and resistance to acid solution (except HF acid) but the disadvantages of low P, T conditions and ease of attack by basic solutions. Hence its use has been generally neglected in hydrothermal work mainly, because the early interest concerned materials requiring basic media and P, T conditions beyond the limits of glass. When the solvent and P, T conditions permit and when the visual observation enables, glass vessels have much to recommend them and should be used more frequently in hydrothermal work. Before the design of Morey autoclaves (1913), the hydrothermal vessels consisted of tubes loaded with solid and water and sealed either by welding both ends of the tube or by means of a flat washer or screw. However, such vessels are very difficult to maintain and dangerous to operate. Croxall et al. (1979) from General Electric Company, UK, have successfully grown berlinite crystals using glass vessels [72]. With the recent discovery of new mineralizers (mixed solvents, high molar acid solutions, organic solvents, etc.) there is a growing tendency for using the glass or thick-walled silica tubes in several hydrothermal crystal growth experiments. The new mineralizers have brought down the crystallization temperature with increased solubility for compounds hitherto obtained only at high P, T conditions.

6.5. *Steel autoclaves*

The success of the earliest experiments in the hydrothermal growth of minerals using very simple autoclaves during the previous century provided an impetus for Morey at the Geophysical Laboratory of Carnegie Institute, Washington, to design a highly versatile autoclave to work in the medium pressure range [73]. This was further developed by him and his co-workers to operate at 500 bar and 400°C [74]. Following on from this, several other designs appeared and today there are many designs to suit any particular aspect of hydrothermal crystallization. All the conventional autoclaves can be divided into four types and are popularly known as: (i) flat-plate closures developed by Morey and co-workers; (ii) cold-cone seal closures designed by Tuttle [75] and popularized by Roy and co-workers; (iii) welded closures designed by Walker and Buehler [76]; (iv) unsupported area closures developed by Bridgman and modified by several commercial autoclave vendors. These closures have been described in great detail by earlier authors in several reviews [56, 64, 77]. Therefore, we restrict ourselves to a very brief survey of this and concentrate more on the recent designs.

6.5.1. *Flat plate closure autoclave*

This is popularly known as a Morey autoclave and is widely used for both synthesis and crystal growth purposes. Figure 20 shows a cross-section of a typical Morey autoclave. The usual dimensions of a Morey autoclave are 10–20 cm length and 2.5 cm inner diameter. The problems encountered by Morey and co-workers in the initial experiments were alleviated in a later design by Morey (1913) in which the closure is made by a Bridgman unsupported area seal gasket made of copper or silver. Higher *P*, *T* conditions may be obtained with an inconel vessel. At low pressures sealing is achieved by compression when the main nut or closure nut is tightened. The pressure can be directly measured and adjusted during an experiment by providing an axial hole through the closure nut. A thermocouple is inserted in the well close to the sample and the vessel is

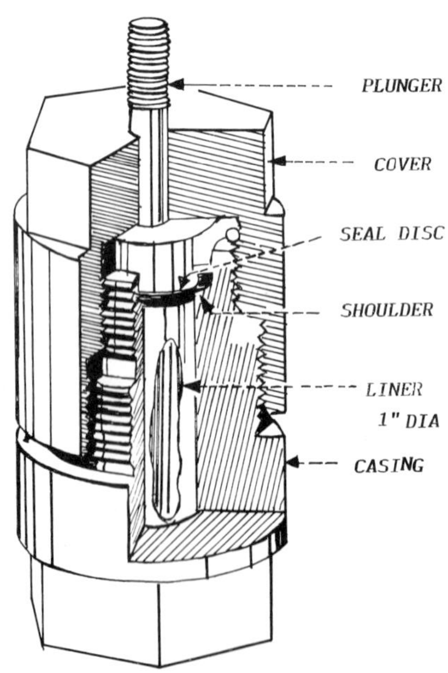

Fig. 20. Cross-section of the Morey autoclave [56].

placed inside a suitable furnace such that the entire Morey autoclave and closure lie within the element of the furnace. At the end of the run the vessel is quenched in a jet of air followed by dipping in water and the closure seal is broken.

6.5.2. Cone closure

This is known as a Tuttle cold cone seal autoclave. These autoclaves are much smaller in volume, were designed by Tuttle and are popularly known as Tuttle vessels (fig. 21) [78]. A Tuttle bomb consists of a stainless steel or low carbon steel or more exotic alloy such as inconel, udimett or stellite (Co alloy) test tube closed by a cone-in-cone seal. As with the Morey vessels, here also the entire vessel assembly is kept inside the furnace. The seal is closed by weights acting through heat resistant alloys above and below the vessels. The cone at the lower end of the support rod is machined to 59°, the cone or the vessel to 60°, this results in a ring seal. At pressures in excess of 2 kbar and above 800°C, these cones have to be machined after each experiment. Pressure is transmitted to the sample through an axial hole in the lower support rod, and temperature is measured by placing a thermocouple in a well of the vessel within 3 mm of the sample. A vertical split

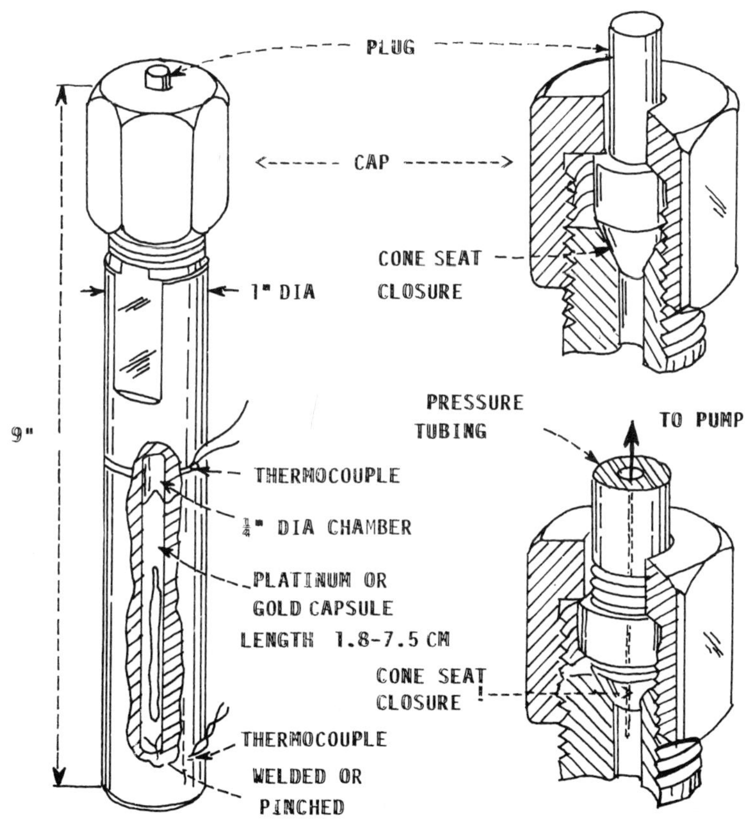

Fig. 21. Tuttle autoclave [78].

furnace is mounted on a hinge allowing it to be swung clear of the vessel during loading or quenching. This vessel can be used up to 800°C at 4 kbar.

In 1949, Tuttle described a modification to his earlier design which is much easier to use. This consists of a longer vessel in which the open end and seal are outside the furnace; hence the term "cold-cone seal" (although in actual fact the seal is far from cold). Pressure is transmitted to the sample, which is contained in a sealed capsule, through a hole in the closure. The capsules are normally made up of noble metals. These vessels may be operated closure up, closure down, or horizontally. The ratio of the vessel diameter to hole diameter (wall diameter) determines the strength of the vessel. In most standard Tuttle vessels this ratio is 4%. Using stellite 25, experiments can be carried out at 900°C and 1 kbar or 750°C and 3 kbar for long term use.

Tuttle's original design underwent several improvements in closure design using new alloys which have extended the P–T range of cold seal vessels. Luth and Tuttle (1963) [79] described a modified Tuttle-type closure which permits vessels up to 8 inch length, 1.25 inch outer diameter, and 0.25 inch inner diameter to be used up to 10 kbar pressure at temperatures as high as 750°C. Rene 41 (Ni based alloy) is used for this vessel.

The advent of new materials, particularly molybdenum alloys, has extended the temperature capabilities of cold-seal vessels to about 1150°C at a pressure above 4 kbar (the so-called TZM vessels) and the prospects of new refractory alloys to extend this range are encouraging. These cold-seal vessels are safe, inexpensive, simple and operationally routine.

All these above-described autoclaves are externally heated ones. In Morey and modified Bridgman-type autoclaves, the pressure created inside the vessel is calculated through PVT relations and in only a few cases is an external pressure gauge used. In Tuttle cold-seal vessel, the pressure is built up by an external pump and the pressure medium usually is distilled water with a little glycol added to inhibit corrosion. For higher fluid pressure a hydraulic intensifier or air-driven pump is used with argon or nitrogen as the pressure medium. Normally several vessels are run from a single pump, each vessel being connected to a high pressure line by a valve and T-junction. The valve (2 way) isolates the vessel from the line. Each vessel is connected to a separate Bourdon gauge with a safety glass dial and blow-out block. An auto-controller is normally used. The types of pressure tubing, valves and fittings used depend on the operating pressures. Normally stainless steel or steel alloy is used up to 13 kbar and size is recommended to be 1.4 inch outer diameter, 1/16 inch inner diameter.

6.5.3. Welded closure

This is a "work horse" vessel in most hydrothermal laboratories because the disposable liner permits a good deal of exploratory work to be done in systems where corrosion might ruin a more complex vessel. It is a high pressure vessel designed by Walker and Buehler (1950) for the temperature range up to 400°C and a pressure of 3 kbar [76]. The vessel consists of an inner liner welded at both ends and an outer tube of heavier construction. Figure 22 shows the construction of a welded closure.

At the conclusion of an experiment one end of the welded liner is sawn off to remove the charge. The liner is then driven out of the outer casing and discarded and a new liner

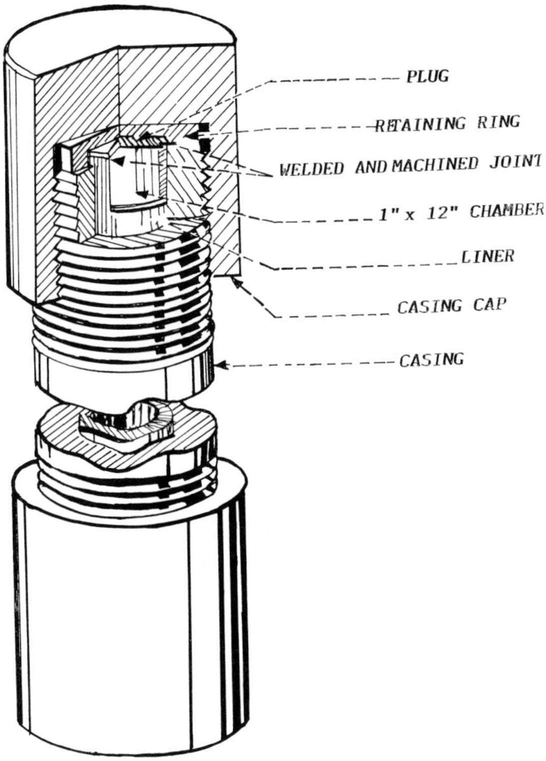

Fig. 22. Schematic diagram (cross-section) of a welded closure [76].

is inserted. Thus, for a sequence of experiments, rather extensive machine shop services are needed. However, welded liners are still relatively inexpensive compared with an entire vessel.

6.5.4. *Modified Bridgman autoclave*

A further advancement in *P*, *T* capabilities, size and design among the steel autoclaves is the modified Bridgman high pressure autoclave. This is embodied in several commercially available autoclaves. Figure 23 shows a cross-section. The initial seal that closes the vessel is made by mechanically tightening the plunger against the deformable gasket (fig. 24 shows a deformed copper ring) by the set screws in the head. The unsupported area closure makes its seal. The force of the generated pressure acts upon the piston, moving it upward against the deformable gasket and creating a perfect sealing. Such vessels can be used quite successfully in the temperature and pressure range 500°C and 3.7 kbar. The construction materials such as stainless steel or inconel, etc. are chosen according to the requirements of the particular experimental conditions. A low-carbon steel or mild steel liner or noble metal liner is normally used in experiments. The pressure inside the liner is counter-balanced with pressure between a

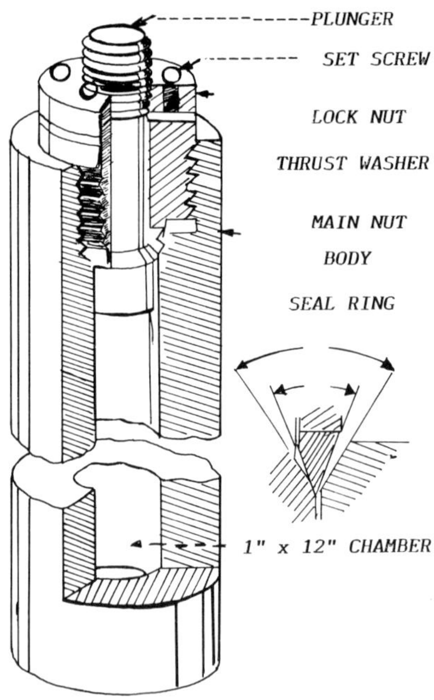

Fig. 23. Cross section of a modified Bridgman autoclave [56].

Fig. 24. Deformed copper ring used for sealing the modified Bridgman autoclave.

required amount of the nutrient material and the solvent is introduced into the liner with a definite percent fill. Then the pressure inside the autoclave is calculated through PVT relations. The modified Bridgman vessels are commercially used for crystal growth and they are further modified to grow very large single crystals of industrial importance. Sealing is the most important aspect in these autoclaves and it consumes much time and effort in the initial runs.

Most of the above closures, which are called self-energized closures, have sealing gaskets generally made of plastic materials and most frequently made of annealed copper. At temperatures below 300°C teflon can be used. If the closure is provided with a cooling system, rubber gaskets are used and sealing of such autoclaves does not require as much mechanical effort as with a copper gasket. In fact, much of the success in hydrothermal growth experiments is due to the equipment. A dramatic advance in upgrading the capacity of the autoclave (for example, as in the case of quartz growth) is the result of new sealing systems from the original flange type seals to the modified Bridgman seal which in turn has been superseded by Gray–Loc seals. The large capacity autoclaves of 1000 to 5000 liter volume are in use in Japan with this type of sealing system. In most hydrothermal experiments corrosive solutions are used and special protective inserts are used nowadays to prevent contamination of the growth zone by corrosion products. A floating-type insert or variable-volume insert has been developed at the Institute of Crystallography, Moscow. Such an insert occupies only a part of the autoclave interior while the remaining space is filled with water. These inserts are usually made of carbon-free iron, copper, gold (for alkaline media), titanium, platinum various grades of glass, molten quartz (for acid media) teflon, etc.

6.6. Novel equipment design

All the earlier mentioned designs are conventional ones. There are several novel equipment designs in recent years, particularly from the Special Construction Bureau, Institute of Crystallography, Moscow. However, most of these novel designs are made to suit a specific purpose like crystal growth, solubility measurements, PVT studies, or visual examination of the hydrothermal growth, physico-chemical investigations, kinetic studies, controlled growth, etc. The actual motive behind the designing of such novel equipment is that, in the hydrothermal method, the main difficulty is that but little information is available as to the processes taking place in the autoclave, and it is extremely difficult to observe these processes inside metal reactors or autoclaves. Hence, several workers call hydrothermal vessels/reactors "black boxes". Here let us discuss only a few of the novel autoclaves, since not all of them play a major role in the hydrothermal growth of crystals.

6.6.1. Horizontal autoclaves for controlled growth of crystals

This autoclave is highly sensitive to the displacement of mass from the nutrient zone to the growth zone through the arrangement of a highly sensitive balance system. Figure 25 shows the fundamental arrangement. The main part of the unit is a horizontal autoclave mounted at the centre of gravity on the vertical axle. The autoclave freely swings in the furnace and the deflection of the axle from the vertical indicates a shift of the mass inside the autoclave to one side or the other. Special weighing scales make it possible to estimate the amount of displaced material and additional weights help to return the autoclave to a horizontal position.

Thus, on carrying out experiments in the balance system, we obtain continuous information regarding the motion of mass in the autoclave, i.e. we know when

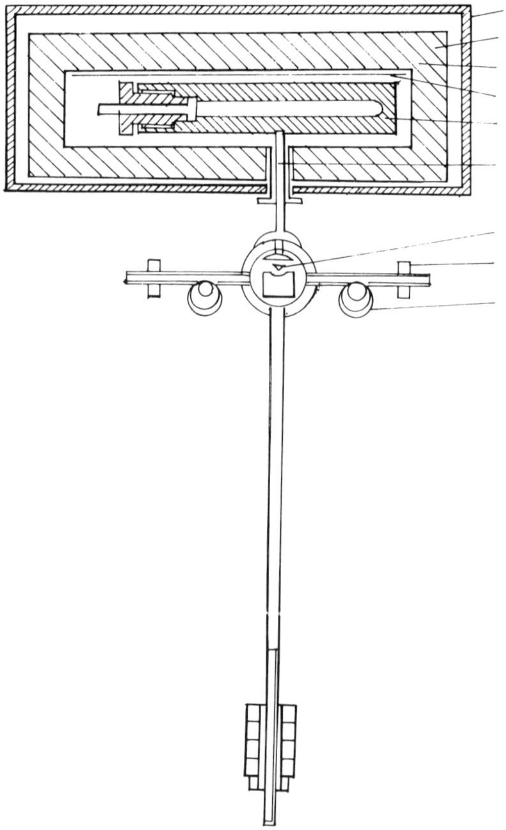

Fig. 25. Fundamental arrangement of the horizontal autoclave for controlled growth of crystals. (1) Horizontal autoclave, (2) furnace, (3) vertical axle, (4) metallic safe, (5) triangular prism, (6) ring, (7) counter weight, (8) horizontal axles, (9) balance weights, (10) cross bars, (11) sectioned heaters, (12) thermocouples, (13) oil brake.

homogenization of the medium sets in, whether recrystallization of the material is taking place, and at what velocity the recrystallization is proceeding. If crystals are being grown on seeds with a prespecified area, we shall quite easily be able to calculate the rate of growth of the crystals and to regulate this by varying the temperature difference between the dissolution and growth zones.

Shternberg (1968) [80] designed, fabricated this autoclave. He has also standardized this system to study the spontaneous crystallization of sphalerite, recrystallization of sodalite, solubility isochore of potassium niobate crystals, etc.

6.6.2. Pendulum apparatus

Mass transfer of material under hydrothermal conditions may be determined continuously or periodically from the motion of the centre of gravity of a vertical autoclave mounted in the device shown schematically in fig. 26 [80]. The oscillations of the

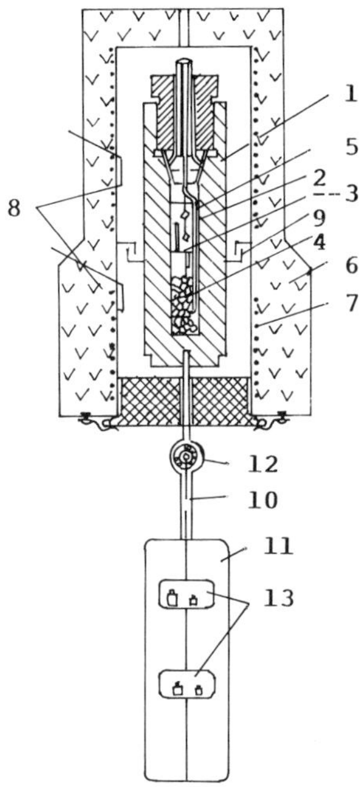

Fig. 26. Schematic diagram of the pendulum apparatus [80]. (1) Autoclave, (2) thermocouple, (3) baffle provided with tubes, (4) nutrient, (5) seed holder, (6) furnace, (7) electric winding, (8) two sets of thermocouples, (9) bellows-type barrier (10) axle, (11) counter weight, (12) roller bearing, (13) two balance shelves.

autoclave are recorded and counted by means of a photo-relay and pulse counter. Crystal growth can be monitored by observing the change in the oscillation period of the system. If the crystals are growing, the centre of gravity of the autoclave will move upward and the oscillation period of the system will increase. The pendulum apparatus designed by Shternberg is highly useful for kinetic studies.

6.6.3. Hydrothermal autoclaves for visual examination

The hydrothermal process cannot be monitored by "looking" into a steel autoclave. Systematic errors appear in the analysis of the crystallization conditions, because the post factum information is difficult to interpret and is mainly based on consecutive approximations which are only sufficient if the crystal growth is reproducible. Nevertheless, in this case, it is difficult to point out a parameter that controls the entire process, because the greater the number of non-measured parameters in the method, the more difficult it is to optimise the solution. In order to facilitate the direct observation and control of dissolution of solids, their synthesis and recrystallization and growth on seeds, several hydrothermal autoclaves with quartz and sapphire windows have been designed both in the West and in Russia, but their potential has not been fully exploited as the designs are highly complex and especially the earliest designs of Kuznetsov and Lobachev (1973) are quite cumbersome to operate [81]. Recently, Popolitov has designed, fabricated and tested experimentally an apparatus for visual

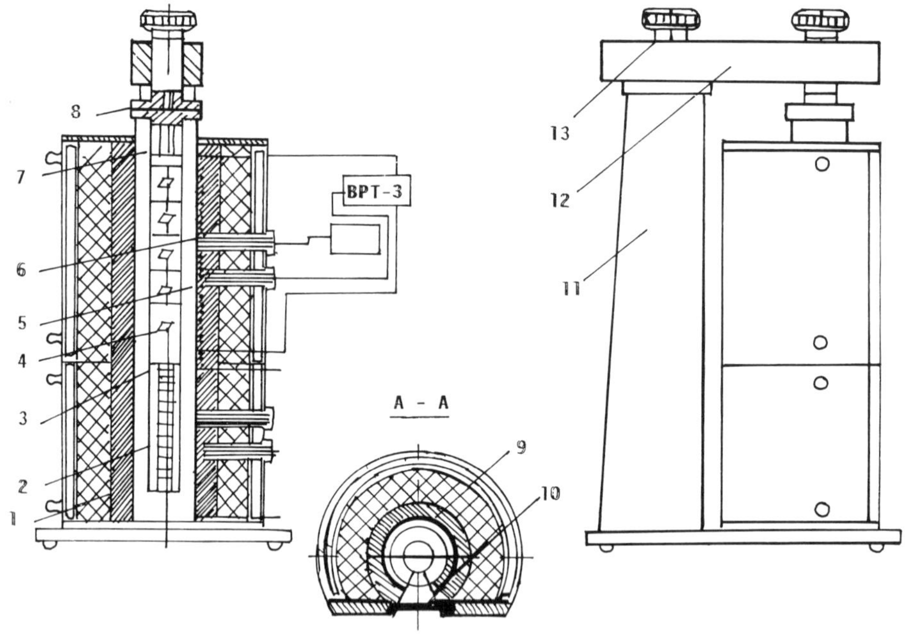

Fig. 27. Hydrothermal apparatus for visual observation [82]. (1) Quartz reactor, (2) nutrient container, (3) baffle, (4) seed holder, (5) seeds, (6) teflon tube, (7) solid cylinder, (8) sealing disc, (9) heater, (10) thermocouple, (11) autoclave frame, (12) upper mount, (13) adjustable nuts for pushing down the upper mount.

examination (fig. 27) [82]. The apparatus consists of a cylindrical reactor made of transparent quartz glass with the following parameters: compression strength = 6556 kg/cm^2, tensile strength = 2734 kg/cm^2, bending strength = 1134 kg/cm^2, shock bending strength = 1.08 kg/cm^2. These data show that the quartz glass is a good material for a transparent reactor, provided that no shocks can occur during its operation. This condition requires certain precautions to be followed during charging, discharging and sealing of the reactor.

The reactor was fabricated in the following way: a cylindrical block of quartz glass of special purity having a diameter 85–1000 mm was drilled using a diamond bore with the diameter 50 to 55 mm. Then the inner and outer walls were ground to the specified dimensions and polished. The hydrostatic pressure admissible for this reactor was not calculated before hand, but its strength was checked. To do so, the reactor was filled with water, sealed by a special valve, placed into a furnace and heated. The growing pressure due to the heating was measured with a manometer. The reactor was heated until the pressure inside it achieved 100 atm but the reactor did bear this pressure. The external heating of steel autoclaves introduces stresses on its walls due to inhomogeneous heating. By contrast, quartz glass has a very small coefficient of thermal expansion and hence the stresses induced by temperature gradients are negligible. To make the quartz reactor tight, a valve which meets the following requirements should be used:

(a) The valve must be self-sealing, i.e. stress in the seal must grow as the pressure increases, and no strong force that could damage the brittle reactor must be exerted on it when sealed.

(b) The sealing material must be plastic; if so, the initial stress needed for self-sealing is low and the reactor is safe.

(c) The valve must be chemically resistant with respect to the chemical agents or the solvents used.

The teflon ring was placed in an annular groove whose diameter is greater than that of the ring. Also, the ring's outer diameter is greater than the inner diameter of the reactor block. Hence, stresses appear in the ring due to its deformation, this being sufficient for the initial sealing.

The main advantage of quartz autoclaves is their high stability with respect to acid media which permits us to use a condensed solution, i.e. the concentration of substance to be crystallized may be enhanced without increasing the temperature. In this case a considerable gain can be obtained for the processes which run under lower temperature and pressure, this being cheaper and safer. The apparatus can also be used in alkali solutions and hydrofluoric acid media by providing a thin teflon coating to inner walls of the quartz glass. This quartz vessel can be used to synthesize and grow single crystals at temperatures up to 300°C and pressures up to 100 atm. The greatest advantage is the facility to observe all the processes taking place inside the autoclave including dissolution of the mixture, movement of the solution and growth of the crystal. By employing the appropriate thermocouple sheath the temperature distribution in the autoclave can be measured accurately. It is convenient to study the physico-chemical processes and hydrodynamics of the solution both in the homogeneous and binary phase "gas–liquid" reagents.

6.6.4. *P–V–T apparatus*

Several designs have been made for P–V–T measurements in highly corrosive systems. These devices usually consist of standard pressure manometers. We describe the P–V–T apparatus designed by Kolb and Laudise (1982) for highly corrosive fluids (fig. 28) (e.g. Berlinite systems) [83]. Also this has been used to measure the pressure in H_2O–SiO_2–NaOH systems. The pressure take-off is lined with platinum and connected to a water capillary of small capacity which goes to the pressure gauge. Several other designs have been proposed by the Special Construction Bureau, Institute of Crystallography, Moscow.

6.6.5. *Apparatus for solubility determination*

Solubility and phase equilibria studies are usually carried out in Tuttle cold-cone sealed autoclaves or small autoclaves without a pressure gauge. Sometimes other available autoclaves are also used if the quenching can be carried out effectively. The quenching ensures maintenance of conditions in the autoclave corresponding to equilibrium conditions at the test temperature. The Special Construction Bureau, Institute of Crystallography, Moscow, has designed several solubility apparatus con-

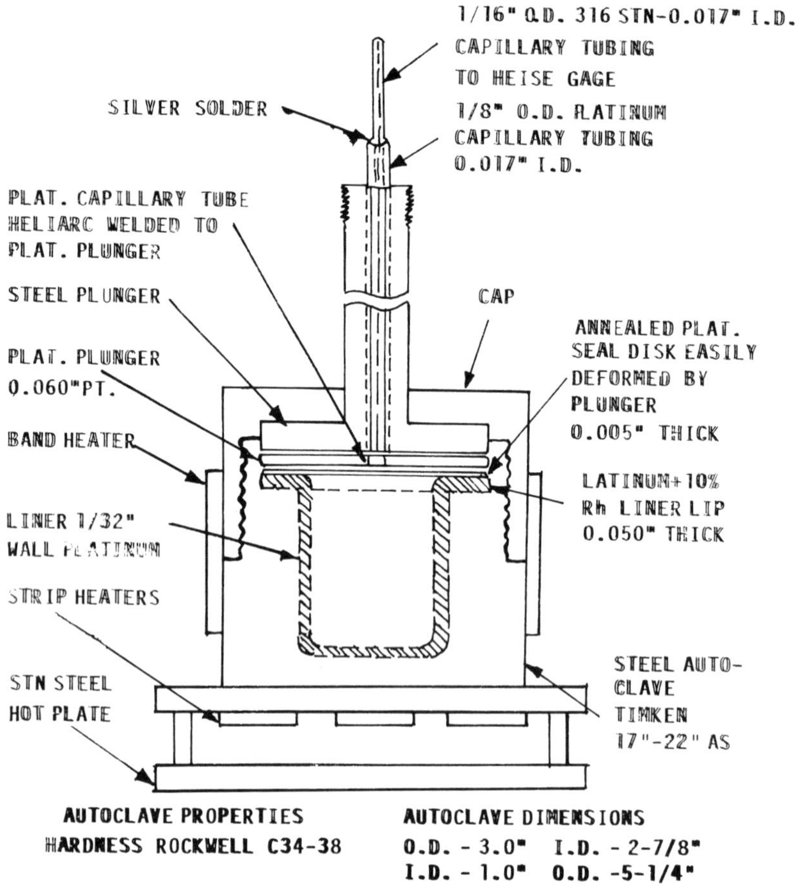

Fig. 28. PVT apparatus [83].

sisting of single autoclave or multiple autoclaves (by creating a liquid thermostat). Figure 29 shows an assembly for studying solubility simultaneously in two autoclaves [84].

There are several other designs of hydrothermal autoclaves to suit a specific purpose. For example, Strubel (1975) has designed a teflon-lined autoclave with a stirrer [85]. Similarly, Shternberg (1971) has designed a double autoclave (exoclave) in order to avoid having to have a mechanical compression system for producing a desired pressure in the reaction chamber [86]. Also this permits independent selection of temperature and pressure. The exoclave is designed for use at temperature up to 700°C and pressures up to 3000 kPa/cm^2.

In recent years high pressure systems (about 40 kbar temperatures up to 1500°C) have been used in the synthesis of new compounds. These high pressure systems are popularly known as Belt apparatuses [87]. Cylindrical platinum cells of 4 mm diameter

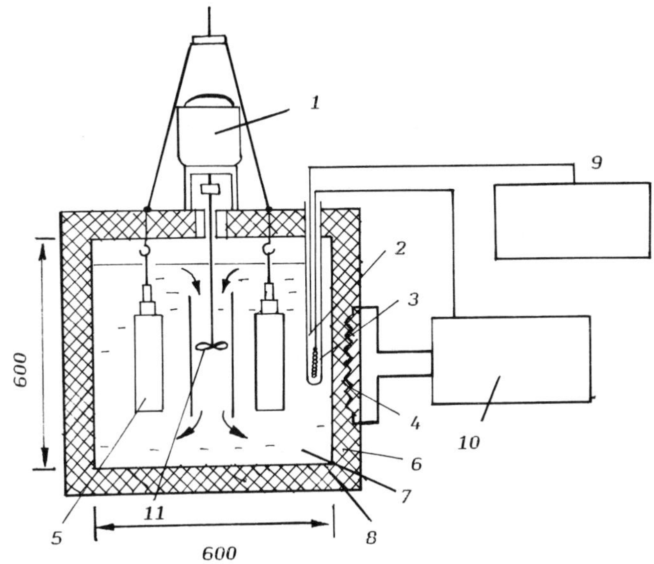

Fig. 29. Apparatus for solubility measurements [84]. (1) Motor, (2) thermocouple, (3) sensor, (4), heater, (5) autoclave, (6) body of furnace, (7) salt solution, (8) thermal insulation, (9) potentiometer, (10) electronic regulator, (11) stirrer.

and 6 mm height serve as the reaction vessels. But these high pressure vessels are not very important for the growth of crystals.

6.7. *Safety and maintenance of autoclaves*

This is the prime factor one has to bear in mind in carrying out experiments under hydrothermal conditions. It is estimated that for a 100 cm^3 vessel at 20 000 Psi the stored energy is about 15 000 foot-lb. The hydrothermal solutions – either acidic or alkaline – at high temperatures are hazardous to human beings nearby if exploded. Figures 30a and b show stellite Tuttle-cold cone sealed autoclaves and stainless steel acid digestion bombs which ruptured at high pressure. Therefore, the vessels should have rupture discs calibrated to burst above a given pressure. Such rupture discs are commercially available for various ranges of bursting pressure. The most important arrangement is that provision should be made for venting the live volatiles in the event of rupture. Proper shielding of the autoclave should be given to deviate the corrosive volatiles away from the personnel. In the case of a large autoclave, the vessels are placed with proper shielding in a pit.

The hydrothermal laboratory should have a well equipped workshop for the proper maintenance of the autoclaves. After each run the autoclave should be properly treated before using it for the next run. The contacts, especially at the sealing sections, should be kept sharp by polishing the surfaces at the contact after each run to avoid scratches.

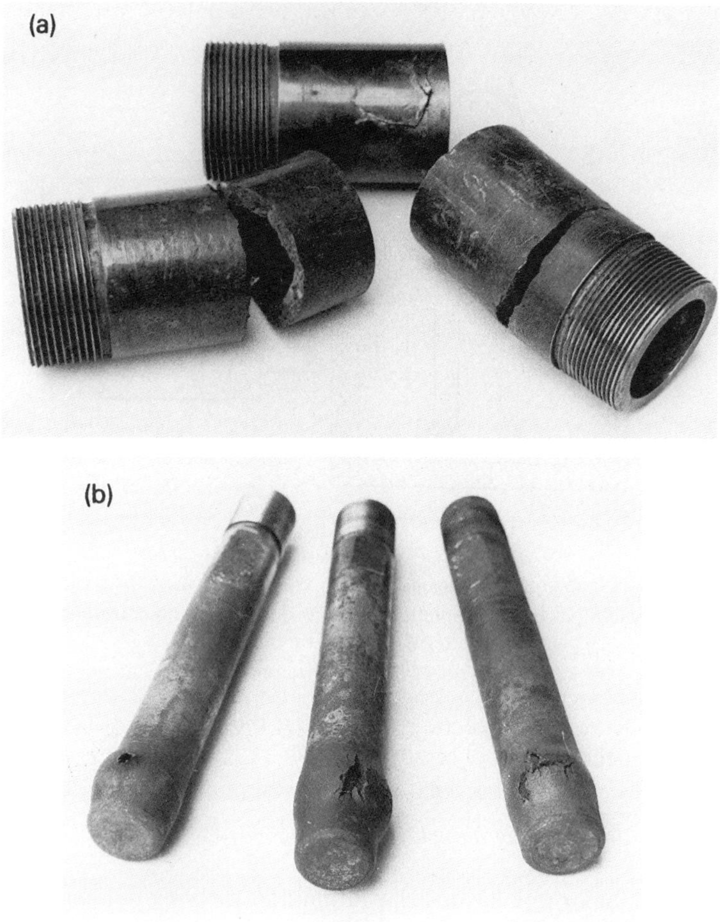

Fig. 30. Ruptured autoclaves. (a) Acid digestion bombs, (b) Tuttle autoclaves.

The autoclave threads should be maintained carefully and are always painted with pipe thread lubricant to reduce friction and prevent seizing of the autoclaves at high temperature. Some workers even place a light coating on the seal rings to insure that they are properly sealed. Graphite grease or graphite in water medium, and suspended molybdenum sulphide or copper lubricants are used. For runs above 200°C, graphite lubricant is more effective because the other lubricants may decompose. Therefore autoclave maintenance and safety make hydrothermal experiments quite expensive.

6.8. Adaptability of the hydrothermal method

The hydrothermal method is highly versatile and can be easily adopted to suit a specific requirement. Thus, in recent year, the method is becoming extremely popular among organic chemists, ceramists, metallurgists, corrosion scientists, hydrometallurgists, etc. Here we discuss briefly how the hydrothermal method can be adapted.

6.8.1. Separated charge method

Shternberg and Kuznetsov (1969) have obtained $PbTiO_3$ crystals by placing TiO_2 in the lower part and PbO in the upper part of the vessel [88]. $PbTiO_3$ crystals grew in the upper part. The autoclave can be arranged either vertically or horizontally as the case may be. The growth of garnets, for example, can be carried out in a multiple zone arrangement as reported by Puttbach et al. (1969) wherein a YAG crystal seed was placed in the middle, Y_2O_3 powder in the lower, and sapphire crystal in the upper part of the autoclave.

6.8.2. Hydrothermal epitaxial growth

Several attempts have been made to crystallize single crystals of a particular compound on some other substrate. For example, sodalite single crystals have been grown on seeds coated with layers such as amorphous silicon monoxide and carbon. Gold or silver about 100 Å units, and polycrystalline gold or silver about 300 Å units [89] have similarly been grown. Kolb and Laudise (1975) have deposited magnetic garnets on substrate material [90]. Quite recently Byrappa et al. (1984) have deposited berlinite on α-quartz on a substrate material [91].

6.8.3. Hydrothermal reaction sintering

Hydrothermal reaction sintering is a recently developed technique to prepare dense sintered oxide ceramics on a large scale. Encapsulated metal (powders, chips or even grains) and water or solution are heated under pressure. Metal is oxidized to form oxide at high temperatures–high water pressures as follows:

$$M + xH_2O \longrightarrow MOx + xH_2.$$

Ultra-fine grains are formed in this process where hydrogen and oxygen break the metal–metal bond to provide pulverization of grains. By this technique, preparation of various highly dense metal oxides of Cr, Zr, Hf, Fe and mixed oxides of La and Cr on an industrial scale is very popular.

6.8.4. Redox potential

The growth of stoichiometric magnetite in the presence of hydrogen in gold capsules has been reported [92]. The hydrogen was formed by the reaction of iron and supercritical water. Sodium iron titanium bronzes have been grown under similar conditions starting with Fe and Ti powders in NaOH [93]. The contact of the solution with the Fe of the autoclave wall at high temperature results in the formation of H and undesired reducing condition. The introduction of oxygen gas into the reaction chamber has been used to avoid this effect in the growth of $BaMnO_3$ [94]. Kuznetsov and Lobachev (1973) have discussed the redox conditions in hydrothermal autoclaves [95].

6.8.5. *Isotope doping*

Oxygen isotope doping was demonstrated on α-quartz grown in a hydrothermal solution made from water enriched in ^{18}O and ^{19}O isotopes [96].

6.8.6. *Hydrothermal dehydration process*

Water is removed from the solution by using, in the nutrient, a metal such as Mg or Zn with higher ionization tendency than hydrogen. They are oxidized during the process forming the respective hydroxides.

7. *Hydrothermal growth of some selected crystals*

Here it is extremely difficult to discuss the growth of each and every compound given in the appendix. Hence, we shall discuss the hydrothermal growth of only some selected and technologically important compounds.

7.1. *Quartz*

Quartz, SiO_2, exists both in crystalline and amorphous forms in nature. The crystalline form of quartz has over 22 polymorphic modifications; out of these, α-quartz which is stable below 573°C at atmospheric pressure, is the most popular and technologically very important among all the hydrothermally-grown crystals or materials. Quartz is an important piezoelectric crystal which has the ability to convert electric waves into mechanical waves and to reverse the process; because of this property it is widely used in filters, timing and frequency control applications. The principal source of electronic grade natural quartz is Brazil. Today the electronic industries are largely inclined to use synthetic quartz because natural quartz crystals are generally irregular in shape, automatic cutting is cumbersome and the yield is low. Over one million pounds of synthetic quartz are produced annually for a variety of applications. These applications range from optical components due to its high transparency to precise time and frequency oscillators based on its piezoelectric properties. Figure 31 shows a photograph of synthetic crystals of quartz pulled out from the largest autoclave in Japan. The inner diameter of the autoclave is 65 cm. The hydrothermal growth of quartz has been described and reviewed in several earlier works [56, 97–99]. Hence, we will only briefly consider the conventional method of the growth of α-quartz and its kinetics data. Instead we discuss more the recent developments in the growth of high purity and low dislocation quartz.

In the growth of α-quartz by the conventional method the autoclaves used by most of the workers were modified Bridgman autoclaves. The design and construction of modified Bridgman autoclaves have been described earlier in section 6.

In the growth of α-quartz, available nutrient material such as small particle size α-quartz, silica glass, high quality silica sand or silica gel is placed in a liner made up of iron or silver with a suitable baffle and a frame holding the seed plates. A mineralizer solution with a definite molarity is poured into the liner to make the required percent

Fig. 31. Photograph of synthetic crystals of quartz pulled out from largest autoclave in Japan [97].

fill. The increased solubility in the presence of mineralizer increases the supersaturation without spontaneous nucleation and consequently allows more rapid growth rates on the seeds. Figure 23 is the cross section of the modified Bridgman autoclave used in the growth of quartz crystals. The commercial autoclaves used are 10 inch inner diameter, 10 feet long unlined. These autoclaves can work at conditions up to 30 000 psi and 400°C. Most of these experiments are carried out for 25 to 90 days to obtain full size (1.5 inch) in the Z-direction and 5–6 inch in Y-direction crystals. Temperature gradient is varied according to the nutrient used. About 1 N NaOH or Na_2CO_3 is the most commonly used mineralizer. The solubility change with temperature is smaller in NaOH and slightly larger in Na_2CO_3. The temperature of the autoclave at the nutrient zone is usually kept at 355–369°C and, in the growth zone, it is kept at 350°C. The addition of lithium improves the growth rates and small amount of Li salts are routinely added to the solution [29]. The solubility is to some extent also a function of increasing pressure. The pressure is controlled by the percent fill in the autoclave, and it is usually about 80% for hydroxyl mineralizer (20 000 psi internal pressure). In most of the experiments the percent opening of the baffle is 20%; even lower percents are used by several workers. However, the actual percent open of baffle area and their geometry are not disclosed especially by commercial growers.

The quality of the grown crystals is also a function of the seed orientation and its quality. Strained seeds generally produce strained growth regions [100]. The seeds are

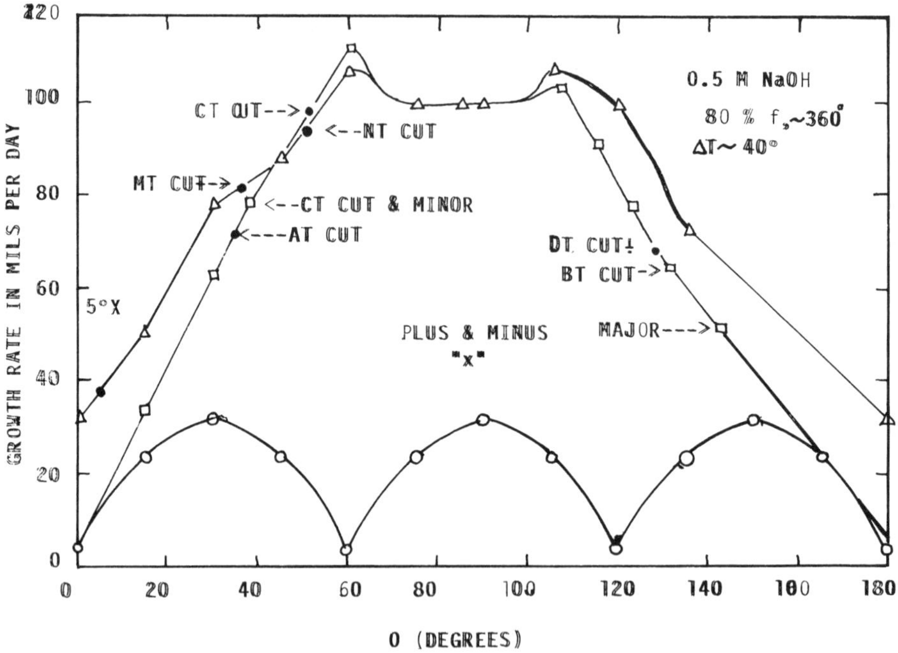

Fig. 32. Growth rate of quartz as a function of seed orientation [56].

polished to a very fine finishing before use. Most high quality crystals are grown using seeds, the surface of which is perpendicular to the Z-direction since the Z-growth region is the lowest in aluminium concentration. Earlier, most of the seeds used were natural quartz cut into a definite orientation, but in recent years this practice is only used when a high quality crystal is desired.

The growth of quartz crystals has been understood precisely with reference to the growth temperature, temperature gradient, percent of fill, solubility, percent of baffle open, orientation and nature of seed and type of nutrient. Also many kinetic studies have been carried out [56, 99]. Figures 9, 32, and 33 show the solubility of quartz with temperature, growth rate as a function of seed orientation, growth rate as a function of percent of fill. Figure 34 shows hydrothermally grown quartz crystals.

The type of crystal to be grown depends on the application, since different properties are required in each case. For optical use, high uniformity, low strain and low inclusion counts are needed since all of these can affect the transparency. For surface acoustic wave devices, large pieces are needed which can take a very high quality surface finish. The quality of the material required for resonators used in time and frequency devices varies with the application. The more precise the need, the more stringent are the requirements. For most applications, a truly high quality material is not needed. For high precision uses, such as in navigational devices and satellites, a very high quality material must be used. Most of the recent research on quartz growth is for improved resonator performance which requires the growth of high quality and low dislocation quartz. Figure 35 shows the fabrication resonators from a single crystal.

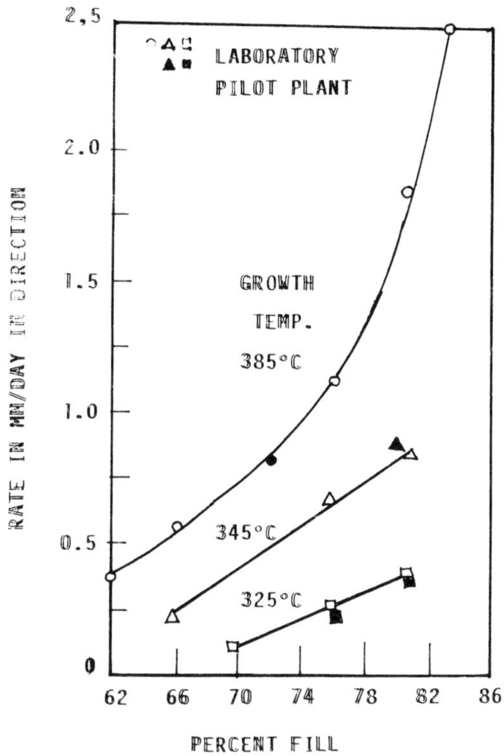

Fig. 33. Growth rate of quartz as a function of percent fill [56].

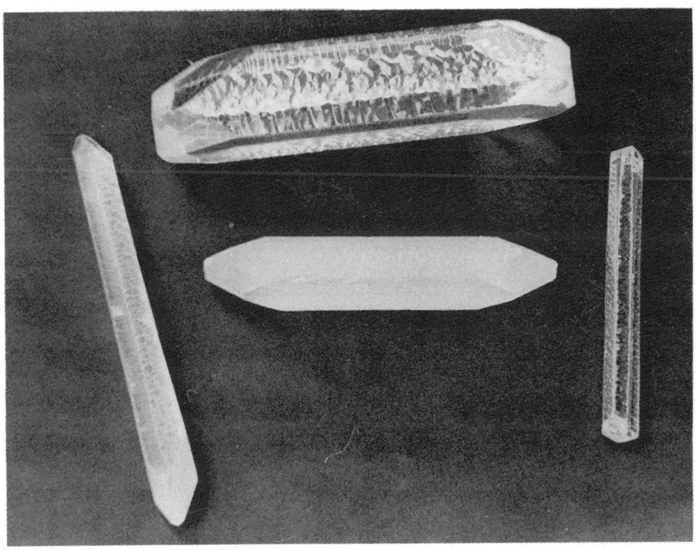

Fig. 34. Hydrothermally grown quartz crystals.

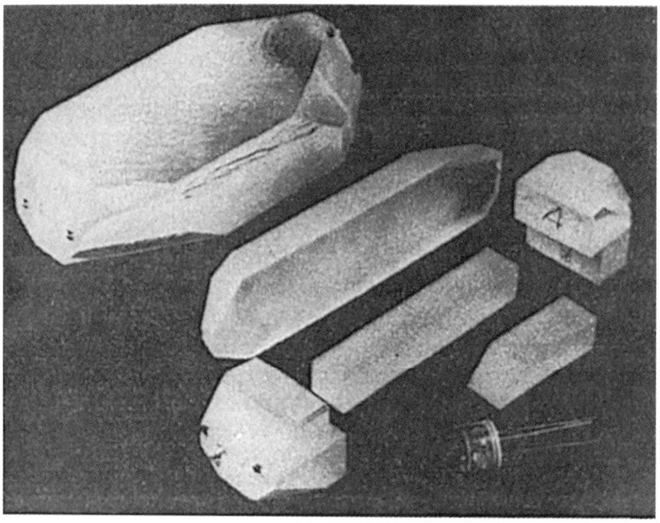

Fig. 35. The fabrication of resonators from a quartz single crystal [98].

We shall discuss the growth of high quality quartz crystals through the study of growth rate, mineralizers, liner material, seed effect, nutrient effect and sweeping.

7.1.1. *Growth rate*

Martin and Armington (1983) have reported the effect of growth rate on the aluminium content at low growth rates [101]. Aluminium is the most deleterious impurity in quartz used in frequency and timing applications. The other most important impurity is H^+, which easily fits interstitially in the large (1 Å) channels which lie parallel to the *c*-axis to charge compensate for Al^{3+} which goes to a Si^{4+}. It enters the lattice from the growth solution, as OH^-, the O^{2-} being incorporated in the SiO_2 lattice. Figure 36 shows the effect of growth rate on aluminium concentration in the Z region of grown crystals using hydroxide mineralizer. Aluminium concentration was determined by EPR. Similarly Laudise has studied the dependence of the effective partition coefficient for OH^- impurity in quartz on growth rate [102]. Lithium salt is added to the mineralizer to check the aluminium concentration in the grown crystal. As mentioned earlier the two important mineralizers used in quartz growth are NaOH and Na_2CO_3 at concentrations in the vicinity of 1 N. The entry of aluminium into the crystal occurs in both cases. The carbonate mineralizer has an advantage that it can be run at lower pressures (lower percent fill) and thus the autoclaves used for the carbonate can be constructed more inexpensively. Also the carbonate does not seem to attack the seed surface as much as the hydroxide.

7.1.2. *Seed effect*

The seed plays a predominent role in the quality of the resulting crystal. For higher frequency applications a smaller *X*-axis dimension is needed. *Z*-growth material is

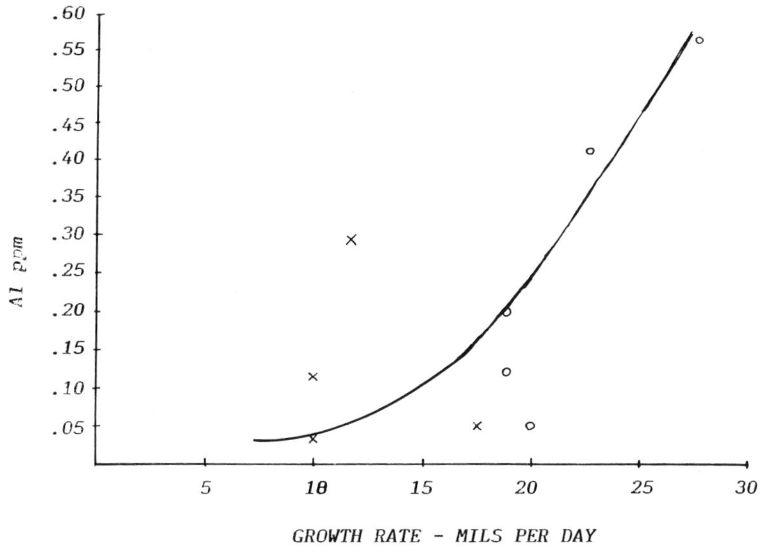

Fig. 36. Effect of growth rate on aluminium concentration in quartz [98].

desired for resonators since it has been shown that this material is about an order of magnitude lower in aluminium [103]. The thickness of the seed is usually between one or two mm. Most crystals grown from synthetic seeds however, contain large numbers of dislocations of the order of several hundred per square centimeters. This results in the formation of etch channels in the resonator which weaken it mechanically and can be a problem when electronic devices are deposited on the surface. The recent studies of Armington and Larkin (1985) have shown that, when a seed perpendicular to the Z-axis (but from the X-growth region) is used, the dislocation density can be reduced to below ten and sometimes to zero dislocations per square centimeter [104]. Although the reasons for this are not entirely understood, it may be related to the fact that, unlike dislocations in the Z-region where they form at the seed and grow close to the Z-axis during growth, most dislocations in the X-region grow at an angle of at least 45°, and usually close to 90° from the Z-axis. Studies have shown that aluminium in the seed does not migrate into the growth regions. Thus the purity of the crystals produced using seeds cut from the X-region is as good as that from crystals produced from the usual Z-seed.

7.1.3. *Nutrient effect*

The quality of the nutrient has a profound effect on the purity of the grown crystal. It appears that most of the impurities incorporated into the crystal come from the nutrient and not the autoclave walls. Therefore, in the growth of high quality quartz crystals, Z-growth material and recrystallized glass are used as the nutrient material. If future requirements become more stringent, particularly for radiation damage, it will be necessary to even further reduce the aluminium content. In this case recrystallized glass will be the leading candidate since higher purity glass than used in this study is available. This generally requires two runs, since the glass must first be converted to α-quartz.

There are several laboratories throughout the world working on finding out a suitable nutrient for high quality quartz growth. Hosaka (1991) has obtained high quality α-quartz crystals using cristabolite as the nutrient. However, the solubility of cristabolite has yet to be understood clearly [105].

7.1.4. Sweeping

Sweeping or solid state electrolysis or electrodiffusion is generally performed when high quality resonators are desired. It reduces the formation of etch tunnels. The effect of sweeping is to remove lithium and sodium deposited interstitially in the lattice during growth. These ions are usually trapped along an ångstrom wide tunnel which is parallel to the Z-axis in the quartz crystal lattice. These ions, in an interstitial position, interact with aluminium impurities substitutional in the lattice to form Al–Li centres which have been shown to be weakly bonded and are the cause of low radiation tolerance in a resonator [106]. In the sweeping process, these are replaced by Al–OH or Al–hole centres which have a much higher radiation tolerance.

Sweeping is influenced by the electric field, strength, electrode type and atmosphere. It takes less time in air due to the availability of water in the atmosphere [107]. Some research has been done in vacuum or hydrogen [108] usually for studies on point defects in quartz rather than for resonator fabrication.

Sweeping has an additional advantage that has been confirmed in recent work [109]. Since this process eliminates lithium and sodium from the crystal; it is also effective in reducing the concentration of these ions at dislocations.

7.2. Oxides

Several oxides, both simple and complex, have been obtained by the hydrothermal technique.

7.2.1. Zincite

The hydrothermal method is the most promising method for producing isometric zinc oxide crystals of good quality. Zincite is a very important material because it has the strongest piezoelectric property among semiconductor materials. It is an n-type semiconductor due to the excess zinc atoms in the compound. The solvents usually employed are 2–15 M NaOH or KOH [110]. A lot of work has been done on the hydrothermal growth of zincite covering almost all aspects. The growth is carried out within a wide range of temperature 200–500°C and pressure 150–600 atm, and percent of fill 70 to 90. The addition of lithium salts greatly improve the quality of the crystals. The solubility of zincite has been studied in great detail under hydrothermal conditions. Solubility of zincite is less dependent upon temperature at solvent concentration <3 M, but it becomes distinct for more concentrated solutions. Most of the earlier publications on hydrothermal growth of crystals deal with the growth and solubility of zincite in great detail.

The growth of zincite is carried out in 1 litre capacity autoclaves provided with silver, or copper lining [111]. The seed crystals are usually of flux growth origin. The hydrothermally-grown crystals have (0001), (0001), (1010) and (1011) faces. The growth rate does not depend upon the internal pressure. The rise in growth temperature increases the growth rate, $V(0001) > V(000\bar{1}) > V(10\bar{1}0)$ without much depending upon the solution concentration. But the rate of growth falls with an increase in the concentration of LiOH. Zincite has a greater tendency than quartz to nucleate on the walls of the autoclave. Furthermore, it shows an anisotropy in growth rate which is apparently as great as that of quartz. The zincite crystals usually show dark colour caused by an iron impurity from corrosion of the autoclave. Since ZnO crystals are usually grown for the purpose of making electrical measurements, the purity and stoichiometry of the crystals are of paramount importance. Addition of lithium salt into the solvent changes the growth kinetics. The solubility also increases from 3.7 to 5.6 wt%. Sakagami (1990) has obtained high purity ZnO crystals under a partial pressure of oxygen using a platinum-lined cold seal pressure vessel [112]. In order to provide a strong oxidizer and to obtain an adequate partial pressure of oxygen, H_2O was added to the growth solvent. The stoichiometry was measured by coulometric analysis. Figure 37 shows hydrothermally-grown crystals of ZnO. These high purity crystals could be obtained at a growth rate of about 0.2 mm per day in the direction perpendicular to the face $(10\bar{1}0)$. The different degree of colouration in picture (fig. 37) clearly shows the purity of the grown crystals. They have no major impurities > 1 ppm except for Li as detected by AAS [112].

The concentration of excess Zn in the crystals grown under excess partial pressures of oxygen decreased to a minimum of 0.8 ppm and their resistivity increased to about $10^8\ \Omega$ cm. It is considered that Li^+ acts as an acceptor compensating the substitutional zinc because 0 atomic vacancies are decreased under the oxidizer-added conditions.

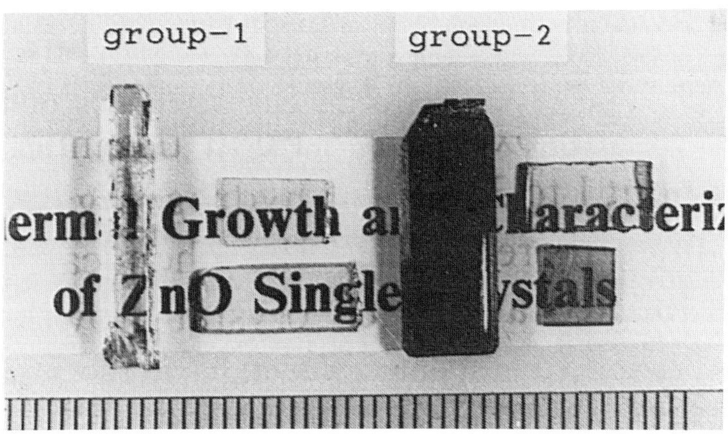

Fig. 37. Hydrothermally-grown crystals of ZnO [112].

7.2.2. *Cuprite*

Cuprite single crystals can be grown under hydrothermal conditions using hydroxides of Li, Na or Ca and aqueous solutions of alkali halogenides and NH_4. A detailed process for obtaining cuprite single crystals using such solutions is described in [113]. The process of dissolution of Cu_2O in alkali aqueous solutions can be written as

$$Cu_2O + (2a - 2)\,OH^- \rightleftharpoons 2CuO^{2a-1} + (a - 1)\,H_2O$$

i.e. transport of Cu_2O takes place with the help of CuO^{2a-1} whose solubility determines the cuprite growth conditions. With a slight change in the temperature fall, the spontaneous crystallization of cuprite takes place and the growth of single crystals from the seeds become difficult. The crystallization of cuprite in alkaline solution has been carried out in standard autoclaves with Cu or Ag lining within the temperature range of 250–450°C, and $\Delta T = 25$–75°C. For chloride solutions teflon closures have been used and the crystallization has been carried out in the alkaline solution at nearly 300°C and $\Delta T = 22$–45°C. In fluoride and chloride solutions, growth can be carried out at much lower temperatures (180–300°C, $\Delta T = 20$–40°C) with pH = 5.0–5.5. The crystals obtained are 4–5 mm in size and octahedral or dodecahedral shape.

7.2.3. *Bromelite*

The crystallization of bromelite by a hydrothermal technique is very safe compared to other techniques because of the very high toxicity of Be. The growth of bromelite can be carried out at 500–530°C and 80% fill using NaOH solutions in a silver or copper-lined standard autoclave and the percent opening of the baffle is usually 10%. BeO is used as the nutrient. Colourless single crystals of hexagonal habit with 3–4 mm size can be obtained. The most common faces are (0001) followed by (1011) and (1010) [114]. The rate of growth of the pinachoidal and prism faces are basically same but, with an increase in the supersaturation, the rate of growth of (0001) distinctly increases.

7.2.4. *In_2O_3*

Crystals of 3–5 mm size could be obtained from 5–25 wt% NaOH solutions at a temperature of 450–500°C with a temperature differential of 20–30°C [115]. Medium rate of growth in the direction of ⟨001⟩ at 450°C is about 0.5 mm/day. Crystals are transparent and cubic in habit.

7.3. *Phosphates*

7.3.1. *Berlinite*

Berlinite replaces quartz in electronic devices because of its large mechanical coupling factor (greater than α-quartz) and because its resonant frequency is nearly independent of temperature for certain orientations. The first successful growth of berlinite crystals was by Jahn and Kordes (1953) [58] followed by Stanley (1954) [59]. Stanley carried out the experiments in a sealed boro-silicate glass vessel placed in low pressure.

The main problem connected with the growth of berlinite is the negative temperature coefficient of solubility which dictates some special growth conditions. One of the

Table 4
Comparative data of solubility of berlinite.

Solvent	T(°C)	P(kpsi)	Time (days)	Remarks
3 M HCOOH	300	1	5	Good solubility
3.05 M NaOH	170	10	3	Residue not $AlPO_4$
2 M NH_4Cl	300	1	5	High solubility
4.62 M Na_3PO_4	300	10	3	Residue not $AlPO_4$
2 M HCOOH	300	11	5	Good solubility
3 M Na_2HPO_4	300	10	5	1% solubility
2.5 M NH_4Cl	300	1	5	Fairly good solubility
3.05 M CH_2COOH	170	10	3	Negligible solubility
2 M $NaHCO_3$	300	1	5	Fairly good solubility
2 M NaCl	300	1	5	Negligible solubility
3.05 M NaCl	170	10	3	4% solubility
2 M NaF	300	1	5	Good solubility
2 M KF	300	1	5	Good solubility
2 M LiF	300	1	5	Good solubility
3.05 M NH_4HF_2	300	10	5	Residue not $AlPO_4$
2 M $NH_4H_2PO_4$	300	1	5	Negligible solubility
2 M HCl	300	1	5	Good solubility
3.05 M NF_4HF	170	10	5	Residue not $AlPO_4$
3.05 M HNO_3	170	10	5	Negligible solubility
3.05 M NaCl	300	10	4	Negligible solubility
1.0 M Na_2CO_3	300	1	5	Negligible solubility

principal problems in the systematic growth of $AlPO_4$ crystals is the growth rate relationship in H_3PO_4 and the lack of good solubility data (because of the negative temperature coefficient of solubility) at the constant fill conditions which are essential for successful growth. The solubility of berlinite has been discussed in detail earlier in section 5, and table 4 gives comparative data on solubility.

Berlinite could best be grown only by a hydrothermal technique. α-berlinite is stable up to 584°C. Below 150°C the hydrates and $AlPO_4 \cdot H_2O$ are stable. Therefore, the growth temperature should be greater than 150°C but less than 584°C. However, the solubility and PVT studies have clearly shown that the crystal growth must be carried out at less than 300°C because, of the reverse solubility. Several versions of the hydrothermal growth of berlinite single crystals have been tried to suit the solubility.

Crystal growth by slow heating method. The method is mainly used for improving the quality of seeds and to enhance the size of the crystal with conditions similar to those used to obtain nucleation.

Synthesis of berlinite crystals under lower temperature and pressure conditions using a slow-heating method has been carried out using flat closure steel autoclaves or acid digestion bombs ($T = 100$–250°C; $P = 50$–100 atm) or smaller Morey autoclaves ($T = 200$–250°C; $P = 300$–1000 atm) by several workers. The nutrient is usually contained in teflon liners. The experiments are prepared either by the dissolution of P_2O_5 in water followed by the addition of respective carbonates or oxides of aluminium or by

Table 5
Experimental conditions for the growth of berlinite [53].

Nutrient composition	Fill (%)	Size of crystals (mm)	Remarks[a]
$Al_2O_3 + H_3PO_4$ mineralizer NH_4Cl	82	0.5–1.5	Well developed hexagonal crystals
$Al_2O_3 + H_3PO_4$ mineralizer NaCl	78	0.5–2.0	Large hexagonal crystals
$Al(OH)_3 + H_3PO_4$ mineralizer HCl	80	1.0–3.0	Rod-shaped needles and acicular crystals
$Al_2O_3 + H_3PO_4$ mineralizer HCOOH	80	0.5–2.5	Hexagonal and equi-dimensional crystals
$Al_2O_3 + H_3PO_4$ mineralizer $NaHCO_3$	80	0.5–1.5	Rhombohedral and hexagonal crystals

[a] Experimental duration 7 days, growth temperature 300°C.

the introduction of aluminium hydroxides or sodium aluminate into the H_3PO_4 in various proportions (maximal filling 70%). The slow raising of temperature is found to be very effective. Table 5 gives experimental conditions for the crystallization of $AlPO_4$. By this technique berlinite crystals of small size could be obtained. In some cases, experiments are carried out for 2 to 3 weeks uninterruptedly in order to obtain berlinite crystals up to 4–5 mm [116].

Crystal growth by composite gradient method. Berlinite crystals have been obtained by the composite gradient method [117]. The schematic diagram of the autoclave is shown in fig. 38. The experimental procedure is as follows: the nutrient is put in the shorter part (cold region), while seeds are installed in the largest part and these two regions are separated by a construction of the glass vessel. A horizontal temperature gradient is established in the growth zone to improve fluid circulation and to avoid parasitic crystallization.

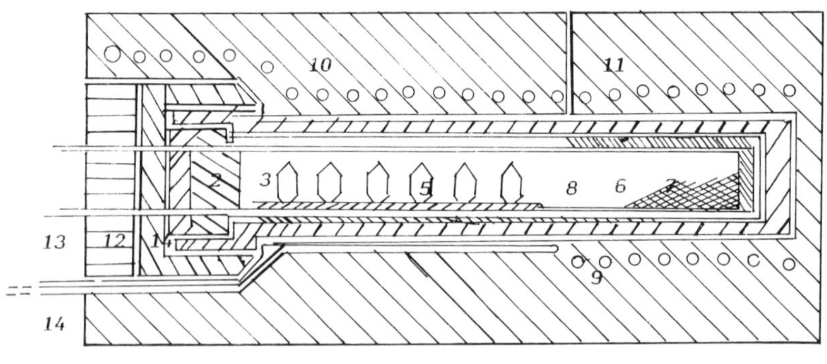

Fig. 38. Schematic diagram of the autoclave for composite gradient method. (1) Steel autoclave, (2) PTFE disc, (3) teflon liner, (4) plunger, (5) seeds, (6) pulverulent nutrient, (7) nutrient basket, (8) screen between cold and hot zones, (9) thermal insulators, (10) half band heater, (11) cold zone furnace, (12) cap (thermal insulator), (13) pyrometric rods, (14) copper circuit allowing the vertical temperature gradient of hot zone [117].

Crystal growth by reverse temperature gradient method. This is more popular and widely used in the growth of berlinite crystals. Here, the nutrient is kept in a perforated teflon gasket at the upper portion of the autoclave, which is cooler. The object is to produce growth with supersaturation provided by a temperature difference under conditions where the temperature coefficient of solubility is expected to be negative. Figure 17 shows a schematic diagram of the autoclave assembly. Fills between 60 and 83% have been investigated [52, 117, 118]. The growth rate is reduced at 83% fill and it seemed to be optimum. The growth temperature is usually between 200 and 300°C. The temperature gradient is maintained between 5 and 50°C. When growth is carried out with a low temperature gradient, the seeds are dissolved during the warm-up. The growth rates varied between 0.08 and 0.18 mm/day at 250 and 300°C and are much reduced at 200°C. The growth of berlinite crystals of sufficiently large size is possible only by this technique and this method is being adopted in the large scale production of berlinite.

Crystal growth on seeds. Spontaneously nucleated seeds and oriented seeds cut from grown crystals are mounted on a platinum frame which is placed in the bottom (hotter) region of an autoclave and the nutrient ($\sim$60 mesh particle size) prepared as described earlier, is placed above in a platinum gasket in the upper region (or in a teflon gasket). Thus, the seeds are in the warmer supersaturated region at the bottom of the autoclave and the temperature gradient is achieved by cooling the top allowing proper convection.

In the growth of berlinite on seeds, the orientation of the seed is very important because the growth rate and perfection depend upon the seed orientation. According to Laudise, the growth rates obtained for seeds with various face are [119]:

(0001) (basal plane): 0.25–0.50 mm/day,
(1021) (*X* cut): 0.23–0.30 mm/day,
(1010) (*Y* cut): 0.12–0.15 mm/day,
(0111) (minor rhombohedral face): 0.12 mm/day,
(1011) (major rhombohedral face): 0.15 mm/day.

The relative growth rates are in general agreement with the morphology of the equilibrium forms as judged from observing spontaneously nucleated crystals which are bounded by small prism (1010) faces and terminated by minor and major rhombohedral faces. Figures 39a–c show berlinite crystals obtained hydrothermally. It is to be expected that the rates on prism, major and minor faces are smaller than for (1120) and (0001) faces, since the latter two do not exist as bounding equilibrium surfaces. In general the morphology of the spontaneously nucleated crystals depends upon various parameters like pressure, temperature, volume, composition of the nutrient and mineralizer [118, 120]. For example, impurities in the powder come into the corresponding nutrient and cause twinning.

7.3.2. *Potassium titanyl phosphate*

Potassium titanyl phosphate, $KTiOPO_4$ (KTP), is one of the best non-linear optical materials increasingly being used commercially for the second harmonic generation

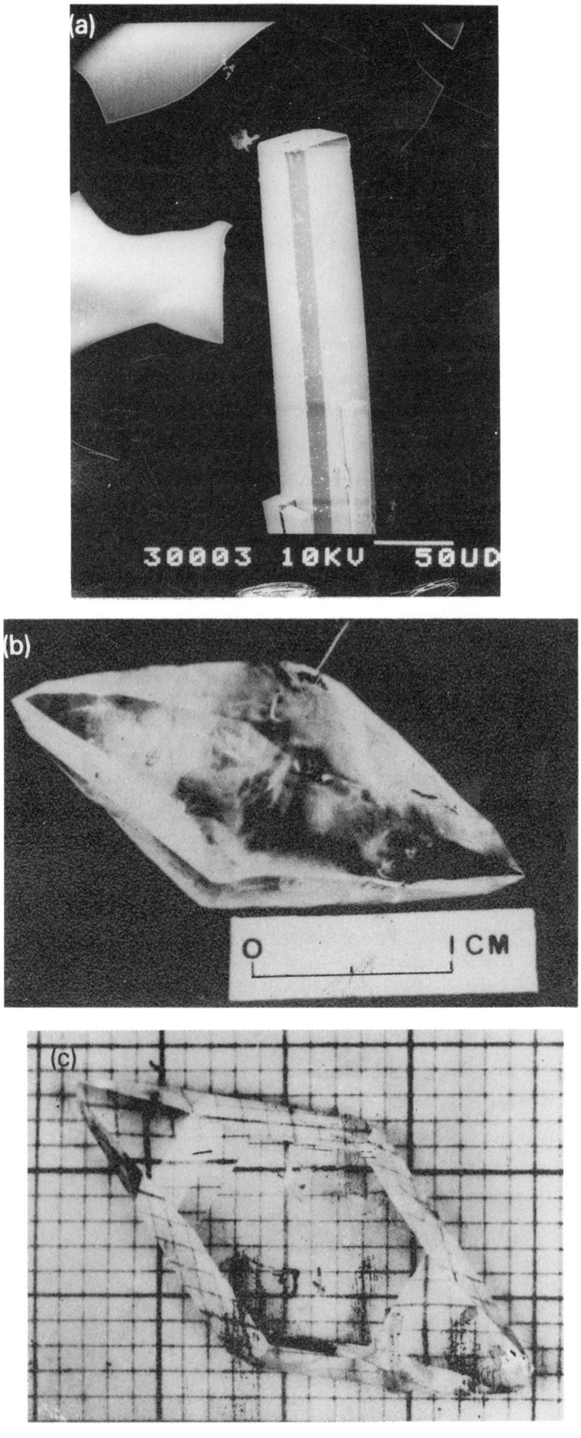

Fig. 39. Hydrothermally grown berlinite crystals: (a) rod like berlinite crystals obtained with $Al(OH)_3 + H_3PO_4$ as nutrient [53], (b) and (c) berlinite crystals grown on the seed [54].

(SHG) Nd–YAG laser. It may also be attractive in electro-optical applications such as modulators and Q-switches because of its large linear electro-optic coefficient and low dielectric constant a high optical damage threshold, a large hyper-polarizability, a wide wavelength range for phase matching and a high thermal stability. It is rather difficult to obtain good single crystals. Incongruent melting and glass formation make it impossible to grow directly from the melt. Growth from aqueous solution at low temperature ($< 300°C$) is impossible because of low solubility of KTP. Therefore, large single crystals of KTP can be grown both by hydrothermal and flux techniques. It is believed that the flux-grown crystals often have relatively poor perfection, purity and homogeneity. According to Tebo [121], the damage threshold for flux-grown KTP is at least an order of magnitude lower than that of hydrothermal material.

The first KTP crystals were obtained by Zumsteg and co-workers [122]. At present, several variants of hydrothermal growth of KTP single crystals are known [42, 123, 124]. Recently Laudise et al. (1990) [42] have carried out *P–V–T*–measurements on 2 M K_2HPO_4 and 2 M K_2HPO_4 saturated with KTP (fig. 40). It is important

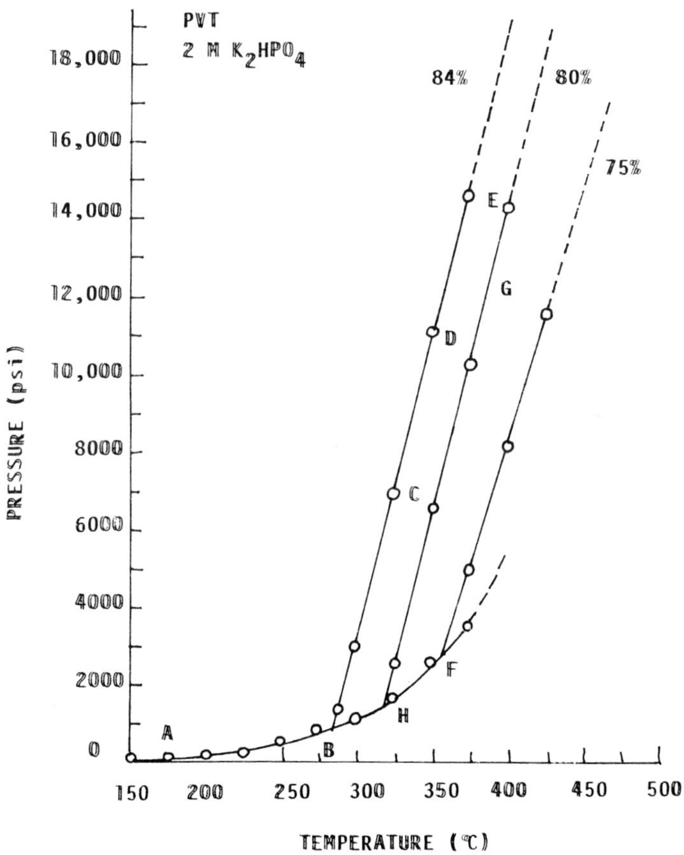

Fig. 40. PVT behaviour of K_2HPO_4 [42].

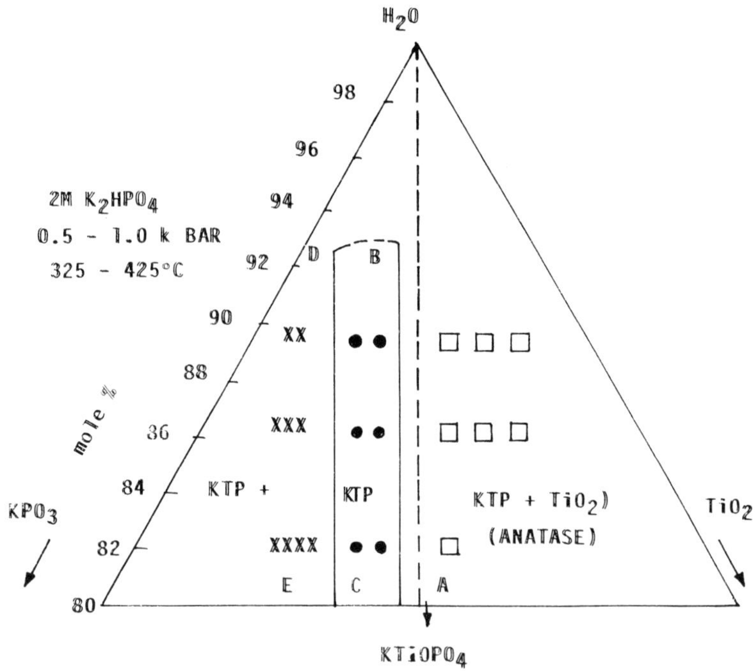

Fig. 41. Part of the ternary phase diagram KPO_3–TiO_2–H_2O in 3 M K_2HPO_4 [42].

that growth be conducted in the absence of a gas–liquid interface, otherwise seeds in the gas phase will not grow and bubbling, boiling and bumping will contribute to poor quality deposition. They made nutrient by the reaction of KH_2PO_4 and TiO_2 in platinum at 1250°C. The hydrothermal "flux" was 1.5 KH_2PO_4–1.0 TiO_2 in gold or silver lined autoclaves at $P = 1.66$–1.93 kbar, $T = 520$–560°C and $\Delta T = 30$–70°C. Growth rates on (011) were 0.2–1.8 mm/week. Laudise and co-workers have discovered much lower P, T conditions for KTP growth, permitting the use of ordinary, unlined steel autoclaves in a K_2HPO_4 solvent. By using the reaction $KH_2PO_4 + TiO_2 \rightarrow KTiOPO_4 + H_2O$ to form KTP, they carried out a phase stability study, the results of which are shown in fig. 41.

Typical conditions used in the growth runs are as follows:

Autoclave:	Plantinum lined, 1 inch i.d. × 6 inch i.e., (2.5 cm × 15 cm)
Mineralizer:	2 M K_2HPO_4 + 0.5 M KPO_3
Percent of fill:	75%
Crystallization temperature:	375°C
Nutrient (dissolving) temperature:	425°C
Temperature differential:	50°C
Seed orientation (201), (011):	Rates (~ 0.07–0.14 mm/day), (010) rate generally higher, quality good
Seed orientation (001):	Rate 27 mm/day (~ 0.68 mm/day) flawed.

The growth rates are of the total increase in crystal thickness or time of the runs. Increasing ΔT to 75°C increased wall nucleation but did not significantly increase growth rate. Increasing KPO_3 decreased rate, but growing without KPO_3 i.e. with the mineralizer (2 M K_2HPO_4) alone, did not markedly effect rate or quality. Growth in KH_2PO_4 or KPO_3 alone was very slow.

The solubility of KTP in 1 M $K_2HPO_4 + 0.5\,KPO_3$ solutions has been studied in detail [42]. Figure 42 shows the solubility of KTP (in wt%) in 1 M $K_2HPO_4 +$ $0.5\,KPO_3$ at 10 kPsi. The solubility data indicate good solubility for growth at 600°C and a positive temperature coefficient.

KTP crystals using a complex mixture of water soluble phosphate glasses, KTP, solvents and water have been obtained by several researchers. The glass composition is K_2O (42–70), $(TiO_2)_2$ (1–18), P_2O_5 (26–55) (mol%). The optimum ratio K/P is equal to 0.8–2.5, optimum glass composition is K_2O (55–60), P_2O_5 (38–42), $(TiO_2)_2$ (1–5) (mol%).

Most of these methods described above require high temperature growth with growth rates more than 0.1–0.2 mm/day. Shou-quan et al. (1990) have succeeded in growing KTP single crystals at medium temperature with high growth rates in KF solutions [125]. Belt et al. (1989) have reported the growth of KTP single crystals in 4 ℓ capacity autoclaves at temperatures of 475°C and 23 400 psi. Gold liners were used here exclusively for corrosion resistance to the phosphates. Baffle openings were adjusted in the range of 20% with $T = 30$–40°C and the total run durations were 40–50 days. The growth rates sustained were 0.5–1.0 mm/side/week on (011) seed crystals [126].

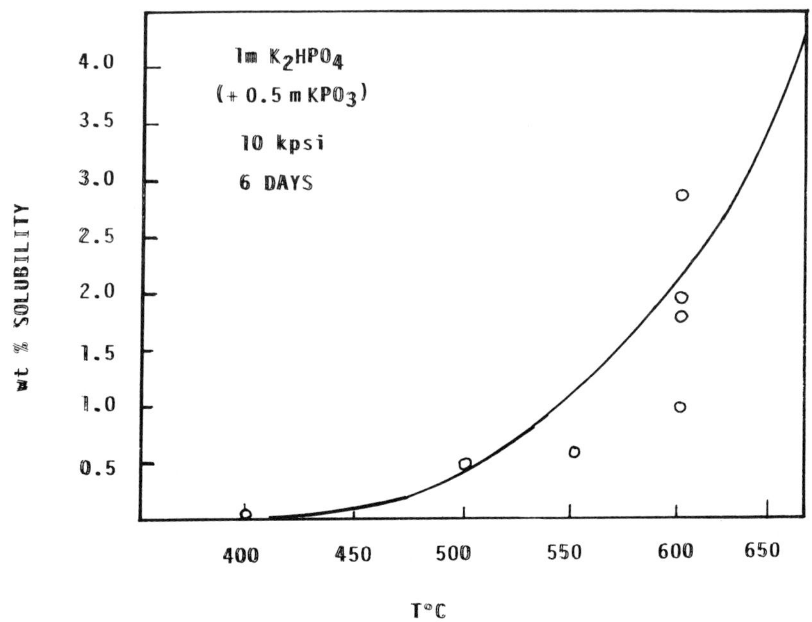

Fig. 42. Solubility of KTP (in wt%) in 1 M $K_2HPO_4 + 0.5\,KPO_3$ at 10 kpsi [42].

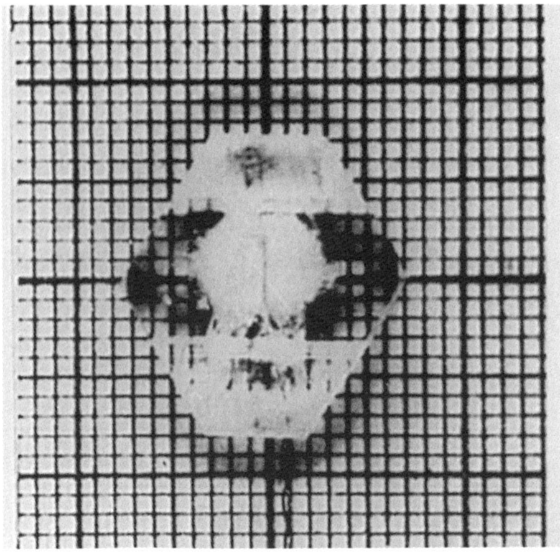

Fig. 43. Hydrothermally grown KTP crystals [125].

The darkening of KTP crystals caused by the lack of oxygen in the growth process can be overcome by introducing an oxidizing agent, for example 1–5 wt% of H_2O_2, into the mineralizer solution. Figure 43 shows KTP crystals obtained hydrothermally.

The properties of KTP crystals vary with the growth method. Hydrothermally-grown crystals generally show an absorption band between 3550–3600 cm^{-1} due to the presence of OH^-. But the linear, non-linear and electro-optical properties of KTP crystals grown by flux and hydrothermal techniques are similar. However, hydrothermally grown crystals show a somewhat higher damage threshold than flux-grown material [127].

7.3.3. Superionic phosphates

Phosphates form a major group of superionic conductors. It all began with the discovery of three-dimensional high ionic conductivity in the framework structures – NASICON ($Na_{1+x}Zr_2Si_xP_{3-x}O_{12}$, $0 < x < 3$) whose ionic conductivity is equal to that of Na-β-alumina [128]. NASICON poses a challenge to materials scientists in understanding its structure and conduction mechanism due to the lack of single crystals, non-stoichiometry in the composition, Zr deficiency, etc. Stoichiometry, structure and conductivity of NASICON are reported on by several authors from time to time. Since the structure of NASICON is highly complicated, materials scientists continue to strive to develop new superionic conductors with simple structure. Phosphates are found to be the most suitable ones. The structures of these phosphates consist of a three-dimensional framework. A series of new framework phosphates were synthesized with an intention to develop simple compounds with stoichiometric composition and relatively simple structure. These compounds were

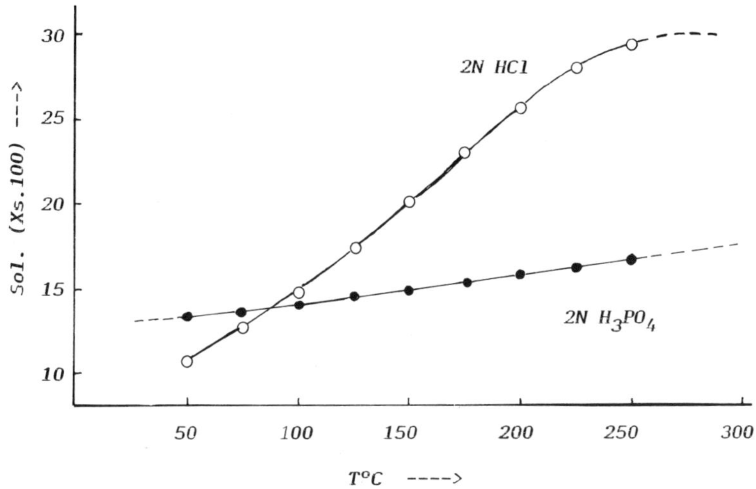

Fig. 44. Solubility curve for $Na_2H_3Al(P_2O_7)_2$.

obtained mainly by flux or solid state reaction growth. Although some of the earliest attempts made by several authors to obtain these framework phosphates were by hydrothermal techniques, the success was limited. A series of superionic phosphates have been obtained in the form of single crystals by hydrothermal technique and the majority of them are stoichiometric in composition and exhibit relatively simple structures compared to NASICON [129, 130]. The hydrothermal method was hitherto not popular for phosphates except for $AlPO_4$ and rare earth phosphates during the early 1950s and early 1980s respectively [55, 56, 131, 132]. The Appendix gives a list of superionic phosphates obtained by hydrothermal technique. We have studied various aspects of the hydrothermal growth of these materials. They include solubility, solvent–solute interaction, complex formation, growth perfection, etc. Figure 44 shows the solubility curve for $H_3Na_2Al\ (P_2O_7)_2$.

The hydrothermal experiments have been carried out in Morey type autoclaves using teflon liners within a temperature range of 200–300°C and a pressure range of 100–250 atm, 85% H_3PO_4 and NaOH solutions have been used as the solvents. Furthermore, some experiments have been carried out in other acidic solvents such as HNO_3, HCl and H_2SO_4. The crystallization has been carried out by spontaneous nucleation controlled by a systematic rate of heating. The crystals obtained were good in quality, transparent and varied in size up to 6 mm. Figures 45a–c show some of the superionic phosphates obtained by hydrothermal technique. The crystals, e.g. pyrophosphates, show prismatic, rhombohedral, columnar and tabular habits. The prism faces dominate over the others. The variation in the morphology is controlled by the degree of supersaturation, concentration of H_2O and Na_2O in the system and the concentration of cations and dopants. The ionic conductivity in these superionic phosphates varies from 10^{-4} to 10^{-1} ohm^{-1} cm^{-1} at about 300°C.

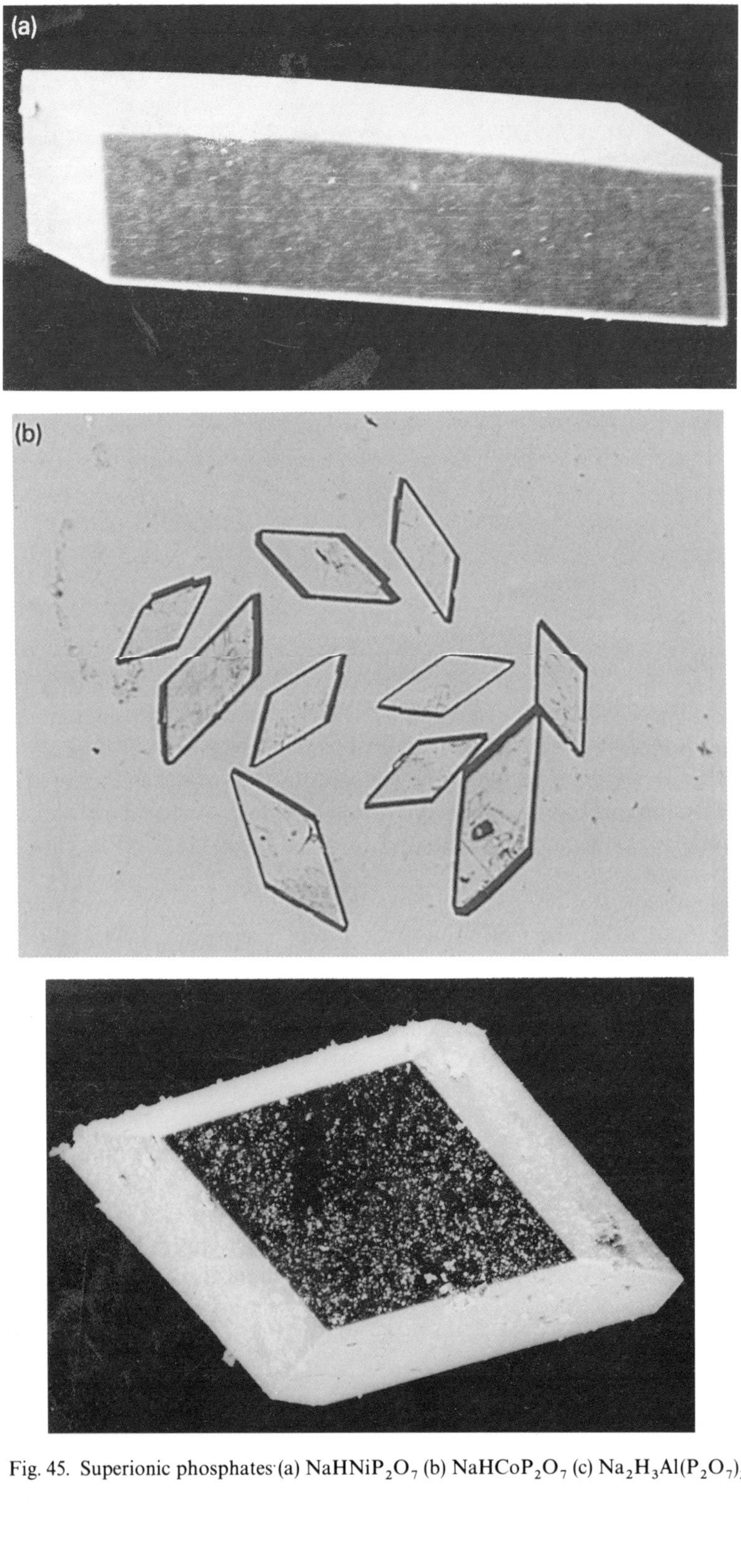

Fig. 45. Superionic phosphates (a) $NaHNiP_2O_7$ (b) $NaHCoP_2O_7$ (c) $Na_2H_3Al(P_2O_7)_2$.

7.4. Borates

Borates are becoming materials of great attraction owing to their variety of applications in modern technology, particularly after the discovery of β-barium borate [133], $Li_2B_4O_7$ [134], LiB_3O_5 [135], etc. Borates were almost ignored by hydrothermal crystal growers. It was only during the last decade that our knowledge essentially broadened about the behaviour of boron in hydrothermal solution and its interaction with various metal cations. Today we have data available on several borate systems such as Na_2O–CdO–B_2O_3–H_2O, Li_2O–CdO–B_2O_3–H_2O [136], Na_2O–ZnO–B_2O_3, Li_2O–CdO–B_2O_3–H_2O, K_2O–ZnO–B_2O_3–H_2O, ZnO–B_2O_3–H_2O [137], MgO–CdO–B_2O_3–H_2O [138] under hydrothermal conditions. More than 25 varieties of borates have been obtained by hydrothermal technique and many of them show interesting luminescence properties. Here we consider the growth of some lithium borates.

7.4.1. Lithium tetraborate

Lithium tetraborate (LBO) is an important piezoelectric and non-ferroelectric crystal used in surface acoustic wave (SAW) and bulk acoustic wave (BAW) devices. It has a lower temperature coefficient of delay than lithium niobate and tantalate and a higher electro-mechanical coupling constant than α-quartz. Since, berlinite poses several major problems in its growth as large single crystals, lithium tetraborate (LBO) is becoming an attractive prospect as a piezoelectric crystal. The raw materials for the growth of LBO in large quantities are abundant, and the ready availability of isotopically pure ^6Li, ^{10}B and ^{11}B as starting materials make LBO growth both practical and economical.

The earliest experiments on the growth of LBO were carried out by melt techniques (Czochralski and Bridgman–Stockbarger) [139, 140]. The LBO crystals obtained from the melt always show "coring" as the main macrodefect thereby hindering wide industrial application of LBO in the manufacture of SAW and BAW devices. Furthermore, Li_2O–B_2O_3 melt shows a corrosive nature towards platinum crucibles. All these problems could be avoided under hydrothermal conditions using Morey-type autoclaves, provided with teflon liners. The use of teflon liners helps to overcome the corrosive nature of the Li_2O–B_2O_3 melt. The starting material such as $LiBO_2$ or LiOH and boric acid are contained in a teflon liner. A suitable mineralizer solution of a specific molarity is added and the entire mixture is stirred well. The percent fill is varied in the range 50–60%. Several mineralizers, e.g. HCl, HCOOH, H_2SO_4 and HNO_3 have been used either independently or mixed. According to [141] the ideal experimental condition in the hydrothermal synthesis of LBO crystals could be written as: $LiBO_2$, 1.7 g; B_2O_3, 2.3 g; 1.5 mol dm^{-3}, 8 cm^3, $T = 250$°C; $P = 100$ bar, 60% fill, duration = 8 days. The quality of the crystals varies widely depending upon the starting materials and the percent fill. Figure 46 shows the solubility curve for $Li_2B_4O_7$ crystals in HCOOH mineralizer. As is evident from the solubility curve, the ideal temperature range for the growth of LBOcrystal is 200–260°C and the ideal molarity range of HCOOH as a mineralizer is 1.5–1.8 M. Figures 47a and b show representative photographs of $Li_2B_4O_7$ crystals.

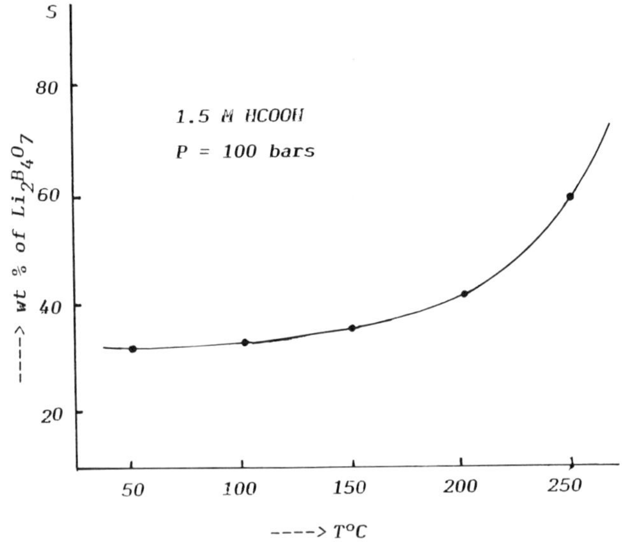

Fig. 46. Solubility curve for $Li_2B_4O_7$ in HCOOH [141].

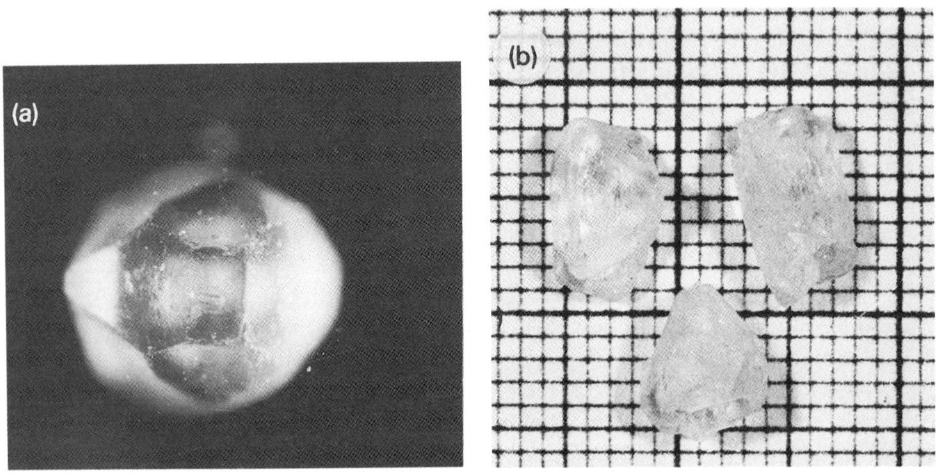

Fig. 47. Hydrothermally grown $Li_2B_4O_7$ crystals [141].

7.4.2. Superionic borates

While studying the Li_2O–B_2O_3 system under hydrothermal conditions, several new borates whose structures are suitable for fast ionic conduction have been obtained. Crystal growth of some of these borates is described below.

7.4.2.1. $Li_3B_5O_8(OH)_2$. Crystals of this material were first obtained by Melnikov et al. (1978) at high *P*, *T* conditions [142]. It has two polymorphic modifications $Li_3B_5O_8$

Fig. 48. $Li_3B_5O_8(OH)_2$ crystals [144].

$(OH)_2$-I and $Li_3B_5O_8(OH)_2$-II [137, 143]. These crystals can be obtained at fairly low P, T conditions ($T = 240°C$, $P = 100$ bar) in a pure Li_2O–B_2O_3–H_2O system [144]. The earlier systems studied by Melnikov et al. invariably contained MnO or CuO as additional components in the system. Experiments have been carried out in Morey autoclaves provided with teflon liners. The starting materials, such as LiOH and H_3BO_3, and a suitable mineralizer solution (1.5–5 M HCOOH) are contained in a teflon liner. The percent fill is varied from 50–60%. A change in the concentration of LiOH in the system changes the resultant lithium compound. The presence of additional mineralizer (e.g. HCOOH) enhances the solubility of this compound and the solvation process. Figure 48 shows a representative photograph of $Li_3B_5O_8(OH)_2$ crystals. The crystals obtained in some of the experiments are even 15 mm in size in an experimental run of 10 days. A careful study of growth parameters and the introduction of a seed should enable us to grow bigger crystals.

7.4.2.2. $LiH_2B_5O_9$. This new compound, belonging to the monoclinic system space group (P21/a), has been obtained under hydrothermal conditions at 250°C and 100 bar using $LiBO_2$, B_2O_3 and 1.5 M HCOOH solvent [145]. Figure 49 shows a representative photograph of $LiH_2B_5O_9$ crystals obtained hydrothermally. During the crystallization process hydrogen from HCOOH has entered the structure indicating a hydrolysis process. The crystals are colourless with vitreous lustre and pseudo-orthorhombic symmetry. It is a potential superionic conductor showing Li conductivity up to 300°C and at higher temperatures shows proton conductivity.

Fig. 49. $LiH_2B_5O_9$ crystal [145].

7.5. *Carbonates*

Carbonates of bivalent metals carry a great significance in the polarized devices such as optical isolators and Q-switch. They also exhibit antiferromagnetic properties. The hydrothermal growth of carbonates has been reviewed by Ikornikova [146, 147]. Earlier interest in carbonates was only amongst mineralogists. Now, with a shortage of optical-quality natural crystals of calcite, their artificial growth is becoming very popular. Among the divalent carbonates, calcite is very important.

Many artificial methods are in use for the growth of optical-grade calcite crystals [148, 149]. Usually, in the hydrothermal growth of calcite single crystals, chloride, hydroxide and carbonate solvents are employed. There are two variants of hydrothermal growth of calcite. In the first case, calcite dissolves in carbon dioxide solution in the cooler part of the autoclave, and in such a case calcium bicarbonate forms which decomposes in the hotter zone with the formation of calcite crystals. In the second variant, growth is carried out in solution, in which calcite shows a positive temperature coefficient of solubility, particularly in the aqueous solutions of chloride (NH_4Cl, LiCl) and potassium carbonate.

Solubility of calcite has been studied in great detail within a wide range of *P*, *T* conditions [147]. At a given constant pressure, solubility of calcite decreases with growth temperature. With a rise in the pressure, the solubility of calcite increases. Thus, within a wide range of *P*, *T* conditions, the temperature coefficient of solubility becomes negative. Therefore, in the growth of calcite single crystals, the nutrient is kept in the cooler zone of the autoclave and the seed in the hotter zone. Despite a large number of reports concerning calcite growth, hydrothermal reaction of calcite with the solvent has not been studied in detail except by Ikornikova.

In order to grow calcite crystals, an isothermal method in which the crystals grew with changing CO_2 concentration as the pressure was reduced at $T=$ constant was devised by Ikornikova. The charge is dissolved in a horizontal, rocking autoclave until equilibrium is achieved. After that the pressure is slowly released through a valve. The liner is made of teflon, glass or platinum. The growth of calcite can be carried out on the seed in aqueous solution of NH_4Cl at temperatures 250–350°C and pressures 50–100 atm. Calcite is being obtained today in very large autoclaves of 10 ℓ capacity provided with glass liners.

Recently Kikuta and Hirano have obtained good quality calcite single crystals from nitrate solutions [150, 151]. They have confirmed that nitrate solvents are more effective than chloride solvents from the viewpoint of solubility of calcite. Calcite crystals dissolve in the ammonium nitrate solution accompanied by the evolution of ammonia. This reaction increases the dissolution of calcite in the hydrothermal solution which gives an advantage of lowering the P, T conditions of the growth. Such a moderate growth condition facilitates the growth of optical-grade calcite single crystals using teflon liners. Rhombohedral calcite crystals spontaneously grow in NH_4NO_3 solvents even at 150°C. They have used a Ti liner to avoid contamination from the vessel material. The concentration of NH_4NO_3 in the growth experiment was selected in the range of 0.005–0.02 M to avoid heterogeneous nucleation. Great improvements could be achieved by using a baffle plate and band heaters to control the temperature gradient and the transport behaviour. Figures 50a–c show calcite single crystals grown on a (0001) and a $(10\bar{1}1)$ seed crystal. The growth rate is usually about 25μm/day. No chemical additives to change the growth direction of calcite have been found yet.

There are several reports available on the growth of $MnCO_3$, $FeCO_3$, $NiCO_3$and $CdCO_3$. The reader can refer to the work of Ikornikova [146].

Fig. 50. Calcite single crystals [146].

7.6. Growth of tellurium dioxide monocrystals

The tetragonal modification of tellurium dioxide – paratellurite – is an important piezoelectric and acousto-optic crystal. In particular, paratellurite crystals can be used in acousto-optic modulators of light, deflectors and delay lines. There are several methods of growing these crystals. The most popular is the Czochralski method, where in the growth temperature is high and there are strong temperature gradients in the growth region. Besides, high vapour pressure of tellurium dioxide often results in the production of defective and stressed crystals. Also the method is rather cumbersome and only about 20% of crystals produced have acceptable quality.

The hydrothermal method of crystallization of paratellurite started almost at the same time as that of the Czochralski method [152]. However, the effects of various physical and chemical parameters on the growth of paratellurite crystals were not properly studied. Hence, this technology is not advantageous because of its poor reproducibility and poor quality of crystals.

Paratellurite single crystals could be obtained through spontaneous nucleation from solutions of hydrohalogenic acids and the mass crystallization decreases in the following sequence HCl > HBr > HI > HF, while all other parameters are the same. The experiments carried out in platinum-lined autoclaves showed that platinum dissolves in hydrohalogenic acids and produces PtHal complexes which contaminate the system TeO_2–HHal–H_2O and make halogenide complexes. Hence the process of dissolution of TeO_2 in the acid, deteriorates and goes out of control. Thus, effective usage of hydrothermal technology of paratellurite could not be developed, owing to the

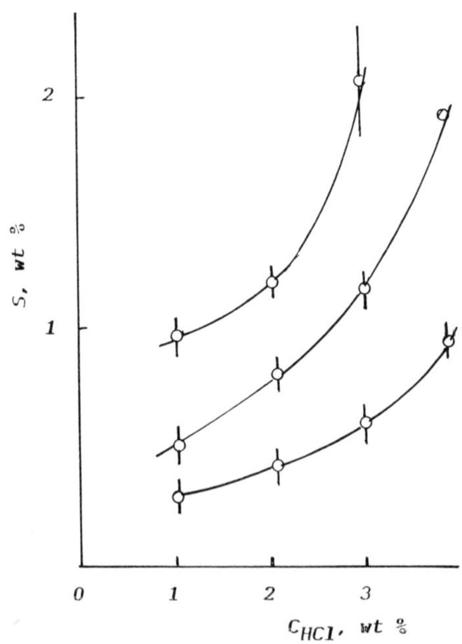

Fig. 51. Solubility of tellurium dioxide in HCl [82].

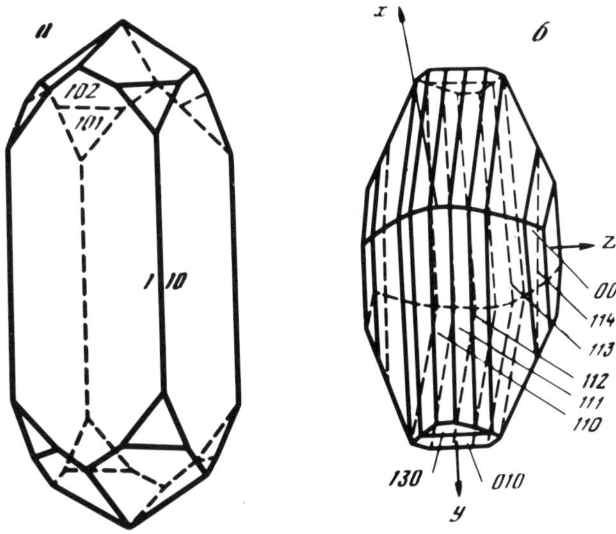

Fig. 52. Schematic diagram of α-TeO_2 crystal [82].

difficulties in observing and controlling parameters which determine the crystallization of α-TeO_2 in platinum-steel autoclaves. Paratellurite possesses a series of valuable properties but its growth technology is only now being developed.

The experiments are carried out at temperatures of 100–200°C, the filling factor being 85%. Under these conditions the pressure of the HCl solution saturated with tellurium dioxide corresponds to the saturated vapour pressure at below 20 atm. This permitted the use of sealed 5 cm^3 quartz glass ampoules. The author has studied the growth of tellurium dioxide in great detail and it is reviewed in [82]. Figure 51 shows the solubility of tellurium dioxide in HCl. Several growth aspects such as mass transfer of TeO_2 during its crystallization on the seeds, the effect of the velocity of gas bubbles on mass transport of tellurium dioxide and the effect of temperature on mass transfer of tellurium dioxide, dependence of mass transport and the solution concentration were studied. Typical conditions are: $\Delta T = 1$–4°C, percent fill = 70, mineralizer concentration = 3.65 wt% HCl, percent opening of baffle = 7 and dissolution temperature 150°C. Figure 52 shows the schematic diagram of a TeO_2 crystal. It is experimentally shown that the faces (102), (101) and (110) of TeO_2 crystals in 3.65 wt% HCl solution grow faster at higher temperature. The author has also compared the properties of hydrothermally-grown and Czochralski-grown tellurium dioxide crystals and found them to be practically similar.

7.7. *Sulphides*

Conditions for the synthesis of sulphides in hydrothermal solutions are determined by the reaction

$$S^{2-} + 4H_2O \rightleftharpoons SO_4^{2-} + 4H_2\uparrow.$$

The growth of simple sulphides has been studied in great detail by Malimin and Khitorov (1969) [153]. In the growth of simple sulphides in alkaline solution, the dissolution transport and crystallization on the seed under the temperature gradient takes place as

$$MeS + 2ROH + H_2O \rightleftharpoons Me(OH)_2 + R_2S + H^+ + OH^-.$$

The transfer of hydrolysis in this reaction leads to the possibility of the appearance of insoluble metal oxide and hydroxide on the right-hand side with the release of hydrogen:

$$MeS + 2ROH + 3H_2O \rightleftharpoons MeO\downarrow + R_2SO_4 + 4H_2\uparrow.$$

In acid solution (HCl, H_3PO_4, NH_4Cl) the dissolution process, mass transport and crystallization occur

$$MeS + 2NH_4Cl \rightleftharpoons MeCl_2 + (NH_4)_2S,$$

$$3MeS + H_3PO_4 \rightleftharpoons MeHPO_4 + 2MeHS + S,$$

$$3MeS + 2HCl \rightleftharpoons MeCl_2 + 2MeHS + S.$$

The rate of reaction in both alkali and acid solution increases with an increase in the growth temperature and the concentration.

7.7.1. Sphalerite

The cubic modification of sphalerite, ZnS, is of great interest as a semiconducting piezoelectric and electro-optical material. Sphalerite crystals obtained from the vapour phase or the melt at a growth temperature which is close to that of the phase transition temperature are characterized by the presence of interlayers of hexagonal form in a matrix of the cubic crystal. However, under hydrothermal conditions, the crystallization takes place at lower temperatures and it enables one to grow the pure cubic phase. Laudise and Ballman (1959) have investigated the systems ZnS–H_2O–Na_2S and ZnS–H_2O–NaOH [110], Laudise and Kolb (1965) have obtained the cubic form of sphalerite at temperature 350–360°C, $\Delta T = 8$–10°C and $P = 550$ atm in 5.4–8.4 M KOH solutions with growth rate of 0.032–0.4 mm/day along the $\langle 110 \rangle$ direction [154]. Kuznestov has studied various aspects of the single crystal growth of sphalerite on a seed using alkaline solution and found that the growth rate of the (111) face increases in the order KOH > CsOH > NaOH. Figure 53a shows the solubility of ZnS in KOH solution at 360°C [155].

The crystallization of sphalerite in alkaline solution was mainly carried out 360–420°C, $\Delta T = 2$–12°C, solvent concentration up to 10 M. In more concentrated KOH solutions the formation of complex sulphide takes place which separates as lamellar crystals.

The investigation of the sphalerite growth in alkaline solutions showed that:

(a) The order of the growth rates of faces is: V(111) > V(001).

(b) The growth rates of faces increase linearly with the supersaturation in the range $\Delta T = 2$–12°C. At $\Delta T > 12$°C, intensive spontaneous crystallization and dissolution of

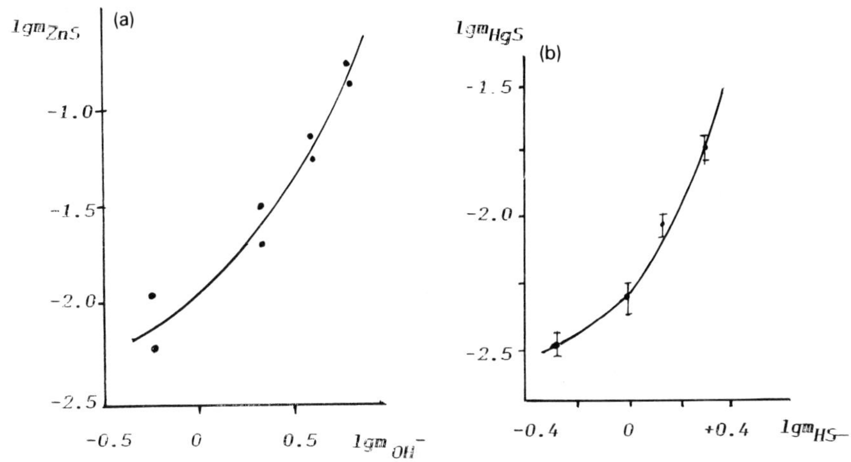

Fig. 53. Solubility curves for (a) sphalerite in KOH, (b) cinnabar in NaHS [155].

the seed are observed, because of a large radial temperature gradient, arising in the growth zone of the autoclave.

(c) The increase of the growth temperature up to 420°C slightly increases the growth rate (by 25%–40%).

The growth of sphalerite crystals on a seed was mainly performed at 370–385°C, $\Delta T = 6$–10°C, KOH and CsOH concentration 8 M. Under these conditions the growth rates of (111) and (011) faces reached 0.1–0.15 mm/day in KOH solutions and 0.1 mm/day in CsOH solutions. In all cases the surface of the (111) face is rough and the (111) growth pyramid contained many defects, principal of which were the deep cavities going from the seed to the growing surface. The (011) surface is smoother than the (111), and material of higher quality was observed in the (011) growth pyramid. Crystals grown from CsOH solution are more perfect.

Fig. 54. Hydrothermally-grown sphalerite crystal [155].

As a rule sphalerite crystals grown in alkaline solution are colourless (unlike the crystals, grown in H_3PO_4 solutions) and absolutely transparent to grey. Figure 54 shows sphalerite crystals obtained from alkaline solution.

7.7.2. *Cadmium sulphide*

Cadmium sulphide is an important semiconductor material. The growth of the material is carried out in either acidic (H_3PO_4 and NH_4Cl) or alkaline (KOH and NaOH) solutions.

In acid and chloride solutions very intensive transfer of cadmium sulphide takes place at temperatures above 400°C and grown crystals are about 3–5 mm in size [155]. In alkaline solutions, intensive transfer of the sulphide has been observed at higher temperatures (480–500°C), only at lower temperatures does partial hydrolysis take place in KOH and NaOH concentrations of 30–50 wt%. Spontaneously grown cadmium sulphide crystals are usually 10–20 × 0.5–1.0 mm in size.

The properties of cadmium sulphide crystals are important and depend on the growth conditions. Normally the crystals grown in acid and chlorine solutions are dark brown coloured, just as crystals obtained in alkaline solutions are light yellow coloured. The resistance of the crystals also depends on the growth conditions. Possibilities exist to change the resistance of growing crystals over a wide range by the variation of the solvent and its concentration. Cadmium sulphide crystals grown in alkaline solutions are characterized by the highest values of the resistance.

7.7.3. *Cinnabar*

The narrow stability field of cinnabar substantially limits the solvent for the growth of crystals. Special experiments confirmed that cinnabar is most stable against decomposition only within the pH interval 9–13 and, in this pH range, the most intensive crystallization of HgS was found to occur in NaHS and KHS solutions. Figure 53b shows the solubility of cinnabar in NaHS solution at 250°C. The growth of cinnabar in NaHS solution has been investigated in more detail.

The experiments showed that, under conditions of high supersaturation, the order of the growth rate of cinnabar faces was $V(10\bar{1}1) \approx V(0001) > V(10\bar{1}0)$. The maximum growth rate of HgS did not exceed ~ 0.05 mm/day. At the same time the as-grown crystals usually have a large number of inclusions and show mosaic structure. Moreover, the partial decomposition of HgS was sometimes observed under these conditions. The increase of pH up to 12–13 is accompanied by an increase of growth rate of the faces. Simultaneously the concentration of the inclusions and also the visible traces of mosaic structure in the crystals vanish. However, it was not always possible to avoid partial HgS decomposition in Na_2S solutions saturated with H_2S. To prevent the decomposition of cinnabar, the pH is regulated by adding NH_4C1 to Na_2S solutions. The growth rates of crystals are reduced and high magnitudes of ΔT are required to reach a rate of ~ 0.1 mm/day for a fast-growing (0001) face. At such high supersaturation, the crystals grown are, generally, poor in quality. By contrast, the addition of sodium bicarbonate to Na_2S solutions results in an increase in the growth rate of some faces (for example, $V(0001)$ reaches ~ 0.2 mm/day at $\Delta T \sim 10$°C) but cinnabar decomposes often and metallic mercury appears, indicating the considerable change in redox conditions inside the autoclave in the presence of $NaHCO_3$.

The best crystallization of cinnabar was obtained in complicated $Na_2S + NH_4Cl +$ $NaHCO_3$ solutions (pH ~ 12.5). The growth rates of (0001) in these solutions can reach 0.15 mm/day without cinnabar decomposition. The cinnabar crystals grown this way are 1.5–2.5 cm in size.

7.7.4. Galena

The crystallization of lead sulphide in alkaline solutions (pH ~ 14) is also accompanied by its partial hydrolysis and metallic lead sometimes appears in the autoclave. Simultaneously oxidized sulphur is discovered in the solution at lower temperature (400–450°C). The size of PbS crystals obtained at such temperatures was ~0.5 mm. The main experiments on the crystallization of PbS were carried out at temperatures of 500–520°C and cubic PbS crystals up to 3–4 mm in size were obtained from 20–50 wt% KOH and NaOH solutions [155].

7.8. Zirconium silicates

Among zirconium-containing phases, oxides and silicates are the most studied. The latter are important for solving many general geophysical and mineralogical problems. They are also used as solid electrolytes. Today, the number of known natural and synthetic sodium and zirconium silicates is about twenty. Amongst them, only six minerals have been studied structurally. Therefore, there is no detailed picture of their formation under hydrothermal conditions. Study of their hydrothermal crystallization in zirconium-containing systems allows one to obtain a model of the natural processes of Zr–silicate formation and transformations and the data on crystal chemistry of these compounds. The Na_2O–ZrO_2–SiO_2–H_2O system has been studied over the wide range of a nutrient ratios: $ZrO_2/GeO_2 = 3{:}1$–$1{:}9$ (mol); NaOH concentration 0–50%, temperature 300–550°C, pressure 40–300 Mpa.

The face composition of the zirconium silicates formed in Na-containing media and the crystallization region is shown in fig. 2 [17].

Analogues of zirconium minerals and $Na_4Zr_2Si_5O_{16} \cdot H_2O$ silicate are formed in the regime of solutions with the lowest concentrations and are not reported in nature. In highly concentrated solutions, silicates with the lowest silicon content are crystallized: $Na_4Zr_2Si_3O_{12}$ and Na_2ZrSiO_5. This system illustrates the role of the solvent in the formation of crystal phases. With an increase in the NaOH concentration, the process of the formation of alkali–zirconium silicates proceeds with the depolymerization of the Si–O radicals: infinite $[Si_6O_{15}]$ chains in elpidite → infinite $[Si_4O_{11}]$ chains in valsovite → island $[Si_3O_9]$ rings in catapleite → $[Si_2O_7]$ diorthogroups in parakeldyshite → $[SiO_4]$ orthogroups in the synthetic $Na_4Zr_2Si_5O_{16}$ and $Na_3HZrSi_2O_8$ phases [156].

7.9. Sodalite

Sodalite is an interesting material, because of its piezoelectric and Photochromic properties. The varieties of sodalite are chloresodalite, Lazurite, hydrosodalite and hackmanite. Sodalite contains OH^- groups and water. It can be obtained only under

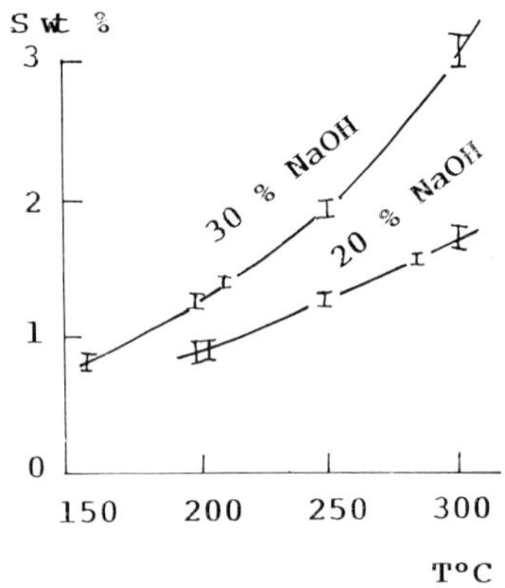

Fig. 55. Solubility curve for sodalite in NaOH solution [158].

hydrothermal conditions as single crystals of several cm^3 in volume. The growth can be carried out within a wide range of PT conditions ($T = 100–800°C$, $P = 1–3.5$ kbar). A lot of work on various aspects of the growth of sodalite has been carried out by Melnikov et al. [157]. The optimum conditions for the growth of sodalite single crystals can be determined from the values of supersaturation in the growth zone. A novel design to determine the absolute solubility of sodalite has been developed at the Institute of Crystallography, Moscow. To study the solubility of sodalite, concentrated solutions of NaOH were used since the field of stability of sodalite depends substantially on the NaOH concentration (more than 20 wt%). The solubility curve of natural sodalite (hackmanite) lies rather below the curve plotted for the solubility of artificial sodalite. Figure 55 shows the temperature dependence of solubility of sodalite in NaOH solutions. An increase in the temperature causes an increase in the solubility of sodalite. Figure 56 shows the dependence of the growth rate of sodalite (111), (100) and (110) faces on supersaturation at constant temperature. The rate of growth varies in the order (111) > (100) > (110) up to 200°C and above 200°C this order changes (100) > (111) > (110). The temperature gradient also plays a major role in determining the growth rate and it increases with rising ΔT [158]. Figure 57 shows a representative photograph of a sodalite crystal obtained by hydrothermal growth.

7.10. Gemstones

Gemstones have attracted hydrothermal crystal growers since the beginning of hydrothermal growth in the previous century. A wide range of gemstones are being grown by this technique. Here we shall discuss the hydrothermal growth of the following: corundum, ruby and emerald.

7.10.1. Emerald

Methods for the synthesis of emerald, which melts incongruently, include a hydrothermal method. Emerald synthesized by this method was first introduced onto the

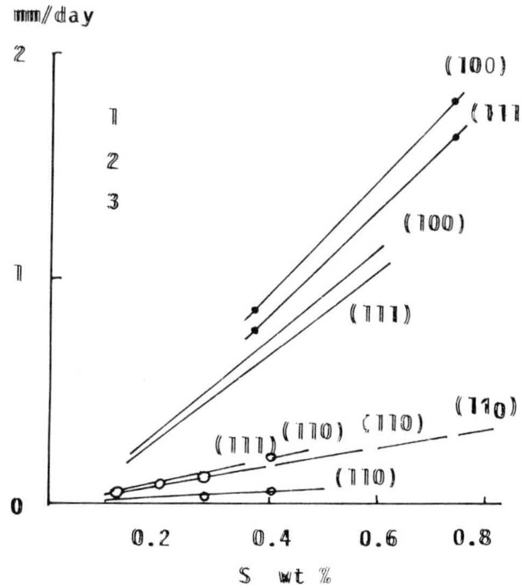

Fig. 56. Dependence of growth rate and supersaturation of sodalite [158].

Fig. 57. Hydrothermally grown sodalite crystal [158].

market in the 1960s by Johann Lechleitner of Insbruck and was given the trade name "Emerita" or "Symerald". It is presumed that Lechleitner emerald can be synthesized under the following growth conditions [159]:

Growth temperature:	300–400°C,
Pressure:	1020 kg/cm^2,
Solvent:	weak alkaline solution,
Nutrient:	silica, alumina, beryllium oxide.

Using beryl as a seed, a thin emerald layer is grown on the seed surface. A faceted beryl seed was used at first, but later a sandwich-type of beryl emerald composite stone with a more thickly grown emerald layer was introduced [160]. The deep green colour of the emerald growth layer is produced by a high concentration of chromium of the order of 7.64–13.20% Cr_2O_3. Therefore, Lechleitner emerald has a higher refractive index than natural emerald which usually contains a few percent of Cr_2O_3 and the refractive index varies with the Cr content.

In 1961 the Linde Division of the Union Carbide Company in the USA started studies on the artificial synthesis of emerald on a commercial basis and succeeded in 1964. It was introduced onto the market in 1965 but the production was suspended in 1970. The method of synthesis used for Linde hydrothermal emerald has been described by Nassau [161]. The synthesis is based on three patents and is carried out under the following conditions:

Growth temperature:	500–600°C,
Temperature gradient:	10–25°C,
Pressure:	700–1400 kg/cm^2,
Solvent:	$NH_4F + NH_4OH$ or KF, $NH_4Cl + HCl$; pH = 7–12.5, pH = 0.2–0.45,
Nutrient:	$Be(OH)_2$, $Al(OH)_3$, SiO_2,
Vessel:	lined with platinum (or silver or gold).

Using an acidic solution at a pH level below 0.1, Yancey has succeeded in obtaining a growth rate of 0.8 mm/day under the above experimental conditions [162]. The materials dissolved at the upper and lower areas of autoclaves are carried by diffusion and convection to seeds placed in the middle. The growth rate of seeds is influenced by a variety of factors such as temperature, pressure, temperature gradient, mineralizer concentration and crystallographic orientation of the seed. Linde selected the *c*-axis direction for the crystallographic orientation of a seed and obtained a maximum growth rate under the above described experimental conditions. Since the synthetic emerald layer grown on a natural beryl seed is not thick enough, strictly speaking it can not be defined as a synthetic stone. Therefore, the emerald layer alone was separated from the natural beryl seed and the separated crystal part was reused as a seed for successive growth. The formation of phenakite is affected by the growth conditions, but monocrystals are observed at the bottom of the autoclave.

There are several varieties of synthetic emerald with different trade names like Regency emerald, Biron emerald, Russian emerald, etc. However, the growth method is the same except for a slight variation in the P, T conditions, solvent and nutrient and ΔT, but the most important aspect is the dopant.

7.10.2. Coloured quartz

We have earlier discussed the hydrothermal growth of electronic grade quartz. Here we discuss the hydrothermal growth of some coloured quartz, amethyst, blue, grey, and smoky quartz.

The synthesis of amethyst was first described by Tinober and Chentsova [163]. After the growth of quartz on positive (r) and negative (z) seeds in an iron-containing alkaline solution, an amethystine colour is produced by irradiation. The colour intensity is high from the (*z*) to the (*r*) directions. This is due to the difference of growth rates between the positive and negative rhombohedra. In alkaline solutions the growth rate of the positive rhombohedron is lower than that of the negative rhombohedron. Then, the content of entrapped iron is higher in the positive rhombohedron than in the negative rhombohedron. This colour inhomogeneity with the amount of entrapped iron is distinctly discerned at areas where Dauphine twin quartz exists. A brown or green colour is emitted from the basal surface of crystals which is caused by non-trapping of structural iron in the basal surface and also a cut surface close to the basal surface [164]. The following are the growth conditions for amethyst in ammonium fluoride solvent:

Growth temperature:	210–400°C,
Temperature gradient:	30–50°C,
Pressure:	20–1000 kg/cm^2,
Solvent:	5%–15% ammonium fluoride,
Dopant:	Fe in the form of oxides or hydroxides,
Vessel:	copper or polyfluortetraethlylene,
Irradiation:	$(5–6) \times 10^6$ röntgen (^{60}Co).

An amethystine colour occurs at a growth rate of less than 0.3 mm/day at a growth temperature of 320°C, or less than 0.8 mm/day at a growth temperature of 400°C. Hosaka and Taki have synthesized purple quartz under the following conditions [165]:

Growth temperature:	400–450°C,
Temperature gradient:	25–50°C,
Pressure:	800–1300 kg/cm^2,
Solvent:	10 wt% NaCl; 10 wt% KCl,
Dopant:	Fe in the form of a metal; α-Fe_2O_3 in the form of a powder,
Vessel:	Platinum (16.5 ml),
Irradiation:	$(2–15) \times 10^6$ röntgen (^{60}Co).

When the amount of impurity exceeds a certain level the growth is inhibited and pits or crevice-flaws are induced. The degree of growth inhibition varies with the kind of impurity. In the synthesis of quartz using a solution containing titanium and iron ions, Hosaka et al. have obtained crystals showing a very faint pink colour, and found that heat treatment of the crystals in an Fe-containing vessel intensified that colour [166]. The deepening of the pink colour seems to indicate that the Fe^{2+}, required for charge transfer, was supplied. The generation of a rose colour in Ag^+-doped crystals has been reported but colour producing factors are not known.

7.10.3. Corundum

Corundum is stable under hydrothermal conditions at temperatures above 400°C and dissolves well in alkaline and carbonate solutions [167]. Laudise and Ballman [168] investigated the system $NaOH$–H_2O–Al_2O_3 and Na_2CO_3–H_2O–Al_2O_3 and succeeded in finding suitable conditions for the growth of corundum. Solubility varies little and it is very low. This is one of the reasons for the unsuccessful growth of corundum under these conditions. To obtain an economic growth rate (more than 0.1 mm/day) it requires a very high temperature gradient which is technically less feasible. According to [169], in alkaline solutions corundum dissolves and forms sodium aluminate:

$$Al_2O_3 + 2NaOH + 3H_2O \rightleftharpoons 2Na[Al(OH)_2] + 4OH^-.$$

Under higher concentrations of these alkaline solutions, the formation of heavy phases is often noticed and it can be avoided in aqueous solutions of K_2CO_3 and Pb_2CO_3. Therefore, the growth of corundum crystals is carried out in 2 M Na_2CO_3, K_2CO_3 and Pb_2CO_3 solutions [168]. The growth is usually carried out on (100), (111) and (101) seeds at a temperature of 450°C, a temperature differential of 30°C and a percent fill of 76.5%. An increased rate of growth from 1.5–14 mm/day can occur depending upon the seed orientation and the solvent. Figure 58 shows the dependence of the rate of growth and the temperature gradient for corundum crystals on different seeds. The growth kinetics have been studied for corundum in KOH solutions. With change in the concentration of mineralizer the growth rate increases and reaches a maximum in 4 M solutions and beyond that it falls. Similarly, with an increase in the growth temperature, the growth rate also increases, particularly in the temperature interval 400–500°C. With a change in supersaturation the rate of growth also increases, particularly on the pinacoidal face. High quality corundum crystals have been obtained on seeds oriented in the (122) direction. Good quality colourless and transparent crystals with (211), (311) and (111) faces have been obtained in 4 M K_2CO_3 at 480°C with a pressure 1.3 kbar, percent fill of 85% and a temperature differential of 30°C. Hydrothermally-grown corundum as a gemstone is not yet popular commercially.

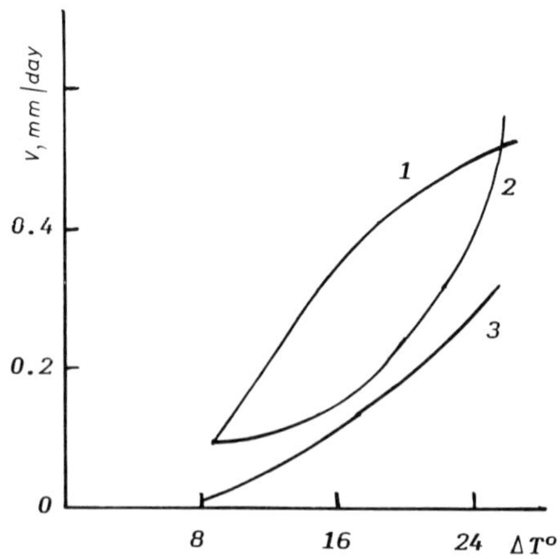

Fig. 58. Dependence of growth rate on temperature differential for corundum crystals: (1) along [122], (2) along [112], (3) along [100].

7.10.4. Ruby

Recrystallization of ruby is carried out using carbonate solutions in cold seal autoclaves [170]. The solubility of ruby is similar to that of corundum in Na_2CO_3, K_2CO_3 and $KHCO_3$ solutions and it increases with temperature. The most significant influence in ruby crystal growth is the Cr_2O_3 which is introduced in the nutrient and it strongly interacts in Na_2CO_3 solution. For example, with a concentration of Cr_2O_3 of 1.6% in the nutrient in a 10% Na_2CO_3 solution, ruby crystals do not grow. Growth of ruby crystal is controlled more by the concentration of Cr_2O_3 in the nutrient than by the growth temperature and, similarly the growth rate also varies with the concentration of Cr_2O_3 in the nutrient.

Appendix

Experimental conditions for the hydrothermal growth of crystals.

Compound	Mineralizer (s)	Temperature (°C)	Pressure (kbar)	Reference
Native elements				
Au, Ag, Pt, Co, Ni, Te, As, Bi	NaOH, KOH, HCl, HI, HBR	400–600	> 0.60	[171]
Se	NaOH, KOH	–	–	[172]
Cd	NaOH, KOH	–	–	[173]
Pb, Cu	NaOH, KOH	450–550	> 0.05	[174]
Oxides				
CuO, Cu_2O	NaOH, KOH	350–450	0.65–0.8	[175]
BeO	NaOH, KOH	450–550	0.8	[114]
ZnO	NaOH, KOH, NH_4Cl	350–600	0.78–1.0	[112]
CdO	NaOH, Na_2CO_3, $(NH_4)_2SO_4$	250–500	0.3–2.0	[176]
NiO	NH_4Cl	300	0.3	[177]
HgO	NaOH	200–300	0.1	[178]
PbO	LiOH	450	0.6	[179]
Al_2O_3, emerald	NaOH, KOH, Na_2CO_3, K_2CO_3	300–620	0.5–0.7	[161]
RE_2O_3	H_2O	> 500	0.2–0.9	[180]
In_2O_3	NaOH, NHCl	350–600	0.4–3.0	[113]
V_2O_3	NaOH	550–700	1.3	[181]
Sb_2O_3	KF, Na_2CO_3	460–600	1.0–1.5	[182]
Bi_2O_3	NaOH, KOH	360–600	0.4–1.2	[183]
Au_2O_3	Na_2CO_3	300	3.0	[184]
Fe_3O_4	NaOH	350–405	0.75–0.8	[91]
MnO_2	NaOH	450	0.75–0.8	[185]
TiO_2	KF, NaF, NH_4F, NaOH, KOH	100–700	0.5–3.0	[186]

ZrO_2, HfO_2	NaF, KF, NH_4F	520–690	1.0	[15]
SnO_2	NaOH, KOH	450–600	0.7	[187]
TeO_2	HCl, HF, HI, HBr, HNO_3	250–385	0.05–0.8	[152]
PbO_2	NH_4OH	230–290	0.5	[188]
CrO_2	$HClO_4$, HIO_4, NH_4ClO_4	350–400	2.0	[189]
SiO_2, quartz	NaOH + Na_2CO_3, NaCl + KCl	350–370	1.5	[56]
Amethyst	NH_4F	210–450	–	[164]
GeO_2	NaOH, KOH	350	1.0	[190]
$K_2Te_4O_9 \cdot 3H_2O$	$K_3PO_4 + H_2O$	380	–	[191]
$Y_3Fe_5O_{12}$	KOH	420	2.0	[192]
$Y_3Al_5O_{12}$	K_2CO_3, Na_2CO_3, NaOH	370	2.0	[89]
$ABi_{12}O_{20}$, A = Ti, Si, Ge	NaOH	500–570	1.5–2.0	[193]
$NiFe_2O_4$	NH_4Cl	475	–	[99]
Hydroxides				
$M(OH)_2$, M = Be, Mg, Ca, Cd, Mn, Ni, Co	NaOH, H_2O	100–550	0.6–4.0	[183]
$RE(OH)_3$, Re = La–Gd, Dy, Er, Yb	NaOH	350–450	0.6	[183]
$M(OH)_3$, M = Cr, In	NaOH	300–450	0.4–1.5	[183]
$Al(OH)_3$	H_2O	< 150	0.04	[194]
MOOH, MOOD, $In(OD)_3$, M = In, Cr, Gd, Sm	H_2O, D_2O	180–600	0.02–1.4	[195]
α-ScOOH, β-ScOOH	H_2O	162–350	0.006–0.17	[196]
$Sr_3Cr_2(OH)_{12}$	H_2O	150–200		[197]
Malachite, azurite	–	–		[63]
Garnets				
$A_3B_2Si_3O_{12}$, A = Mn, Ca, Fe, Cd, B = Al, Fe, In, V, Cr	CaCl, CdCl, NH_4Cl	400–900	0.4–30.0	[199,200]

(cont'd)

Appendix (cont'd)

Compound	Mineralizer (s)	Temperature (°C)	Pressure (kbar)	Reference
Silicates				
Li_2SiO_3	LiOH	450	0.15	[201]
$KHSi_2O_5$	KOH	300	–	[202]
Al_2SiO_5	H_2O	900	20.0	[203]
$CaBSiO_4(OH)$	$CaCl_2 + HCl$	300–400	0.3–0.35	[203]
Zn_2SiO_4	HCl	400–500	0.4	[204]
$R_2Si_2O_7$, R = La–Lu, Sc, Y, Cd	H_2O	700	2.5	[205]
$Na_2Fe_5TiSiO_{20}$	H_2O	500–700	0.5	[206]
$Na_2R_6Si_6O_{24}(OH)_2$, R = La, Y	H_2O	500–700	2.0	[207]
$M_4R_6Si_6O_{24}(OH)_2$, M = Ba, Ca, Sr, Pb, Mn, R = La, Nd, Sm, Gd, Dy, Y	H_2O	500–700	2.0	[207]
$MPbSiO_4$, M = Mn, Fe, Co, Ni, Zn, Mg, Cd, Be	H_2O	300–500	2.0	[208]
$MBe_2Si_2O_7$, M = Ba, Sr, Pb	H_2O	450–750	1.5–3.0	[209]
$Na_2Be_2Si_6O_{14}$	NaOH	400–600	1.5–3.0	[210]
$Na_2BeSi_2O_6$	NaOH, H_2O	400–600	–	[210]
$Ca_2MgSi_2O_7$	H_2O	798	4.06	[211]
$Na_2TiZnSiO_7$	NaOH	350–550	> 3.0	[212]
$Ba_2TiSi_2O_8$	$Ba(OH)_2$	200–500	> 2.0	[213]
$Li_4Sn_2Si_{12}O_{30}$	LiOH	500–600	> 2.0	[214]
$K_2SnSi_3O_9$	KOH	400–650	> 1.5	[215]
$RbFeSi_3O_8$	RbOH	480	1.6	[216]
$K_2CeSi_6O_{15}$	KOH	450	2.0	[217]
$K_3NdSi_6O_{15}$	$KOH + K_2CO_3$	400–600	2.0	[218]

$K(Mn, Fe)_2(Zn, Mn)_3Si_{12}O_{30}$	$KOH + KCl$	580–600	0.8	[219]
$K_2Mn_2Zn_2Si_4O_{15}$	$KOH + KCl + MnCl_2$	580–600	0.8	[220]
$FeMgAl_4Si_2O_{10}(OH)_4$	H_2O	650	9.2	[221]
$Na_2MnZn(SiO_3OH)_2(OH)_2$	NaOH	450	> 0.5	[222]
$Na_4Fe_{10}Ti_2Si_{12}O_{40}$	H_2O	700	1.0	[223]
$Na_3CaMg_2Si_8O_{22}(OH)_2$	Na_2CO_3	400–970	0.1–0.8	[224]
$LiAlSi_2O_6 \cdot H_2O$	H_2O	300–350	2.0	[225]
$Na_8SnSi_6O_{18}$	NaOH	600	1.0	[226]
$K_2ZrSi_2O_7$	KOH	310	0.5–1.0	[227]
$RBSiO_5$, R = La, Ce, Pr, Sm	NaOH + HCl	135–700	1.0	[228]
$K_2ZrSi_3O_8$	KOH	350–450	0.5	[229]
$(Na_2Ca)ZrSi_3O_9 \cdot 2H_2O$	Na_2CO_3	350–530	–	[230]
$PbCa_2Si_3O_9$	H_2O	350	2.0	[231]
$KLi_2AlSi_4O_{10}F_2$	H_2O	400–670	2.0–5.0	[232]
$K_3Li_3Al_3Si_6O_{20}(F, OH)_4$				
Sodalite				
Mica	$KOH, K_2CO_3, KHCO_3$			[233]
$Na_8Al_6Si_6O_{24}(OH)_2 \cdot nH_2O$	NaOH	200–450	0.5	[158]
Na_2MgSiO_4	NaOH	700	3.0	[234]
$Na_4Zr_2Si_3O_{12}$	NaOH	600	3.0	[235]
$NaAlSiO_4, Na_2Zn_2Si_2O_7, Na_2ZnSi_3O_8, Na_2ZnSiO_4, Na_2Mn_2Si_2O_7$	$NaOH, NaOH + Na_2CO_3$	375–450	1.0–2.0	[27]
Germanates				
$MGeO_4$, M = Zr, Hf, U	H_2O	150–300	1.0	[236]
Al_2GeO_5	H_2O	400–700	0.5	[237]
Pb_3GeO_5	KOH	450–500	0.6–1.2	[238]
$Sb_2Ge_2O_7$	KF	400–550	0.6–1.2	[239]
$Sb_4Ge_3O_{12}$	K_2CO_3	450–550	0.6–1.2	[240]

(cont'd)

Appendix (cont'd)

Compound	Mineralizer (s)	Temperature (°C)	Pressure (kbar)	Reference
$Bi_4Ge_3O_{12}$	NaOH	350–450	0.6–0.8	[241]
$CaGeO_3$	H_2O	150–450	0.8–1.8	[242]
$SrGeO_3$	NaOH	300–600	> 1.0	[243]
$Mg_5GeO_6(OH)_2$	H_2O	470–570	> 0.3	[244]
$Y_2GeO_6(OH, F)_4$	KF	450	1.5	[245]
$Yb_4Ge_3O_9(OH, F)_4$	KF	450	1.0	[246]
$Er_4Ge_3O_9(OH, F)_4$	CsF	450	0.6	[247]
$M_6Ge_4O_{10}(OH)_4$	H_2O	180–500	0.1	[248]
$Na_2LaGe_4(OH)$	NaOH	450	1.0	[249]
Na_2TiGeO_5, $Na_2Zn_2TiGeO_7$	NaOH	400–550		[250]
$Na_3ZrHGe_2O_8$	NaOH	500	1.2–2.0	[251]
Na_2ZrGeO_5, $Na_2ZrGe_2O_7$				
$Na_4Zr_2Ge_3O_{12}$				
Na_2BeSiO_2	NaOH	450–500	1.0	[252]
$KAlGe_2O_6$, $NaFeGe_2O_6$	KOH, NaOH	450	1.0	[253]
$NaAlGe_3O_9$	H_2O	800	10.0	[254]
$AgAlGe_3O_8$	$H_2O + O_2$	600	1.5	[255]
$TlAlGe_3O_8$	H_2O	1000	2.5	[255]
$MAlGe_3O_8$, M = Li, Na, K, Rb, Cs	H_2O	800–900	13.0–25.0	[255]
$Na_6Sn_4Ge_5O_{20}(OH)_2$		450	1.0	[256]
$Na_2SnGe_2O_6(OH)_2$		450	0.5	[257]
$CaREGe_2O_7$, RE = La, Nd, Eu		800	1.5	[258]
$Y_7Ge_2O_{12}(F, OH)_5$	KF, CsOH, RbOH, NaOH	450–500	1.5–2.0	[259]
$Sm_4Ge_3O_9(OH)_6$, $Y_4GeO_6(OH, F)_4$, Nd_4GeO_8, $Yb_2Ge_2O_7$, $RE_3GeO_5(OH)_3$, RE = Gd, Sm, Dy, Er				

Phosphates				
$AlPO_4$	H_3PO_4, H_2SO_4, HCl, HCOOH	150–300	0.03–0.2	[260]
ABO_4,	NaOH, H_3PO_4	200–900	0.03–50.0	[261, 262]
A = Fe, Bi, Mn, Cr, Al, B				
B = P, V, As				
$Ce(HPO_4)_2 \cdot 33H_2O$	H_3PO_4	160		[263]
$CuTi_2(PO_4)_3$	H_3PO_4	500	3.0	[264]
$AbNbOPO_4$	H_3PO_4	600	1.8	[265]
$Cu_3(PO_4)_2 \cdot H_2O$	$H_3PO_4 + H_2O$	220		[266]
$NH_4Zr_2(PO_4)_3$	NH_4Cl	250–300	0.1	[267]
$(H_3O)Zr_2(PO_4)_3$	H_2O	200	< 0.1	[268]
$KMnO_2 \cdot O(PO_4)(HPO_4)$	H_2O	400	3.0	[269]
NdP_5O_{14}, TmP_5O_{14}	H_3PO_4, HCl + H_3PO_4	240–300	0.05–0.1	[270]
$MREP_4O_{12}$,	H_3PO_4	300–500	0.1–0.5	[271]
M = Li, Na, K, Rb, Cs				
RE = La–Nd				
$(Na_{2/3}Zr_{1/3})_2P_2O_7$	H_3PO_4 + HCl	250	< 0.01	[272]
$Na_2H_3Al(P_2O_7)_2$	H_3PO_4	200–250	0.01	[273]
$NaHMP_2O_7$, M = Ni, Co	H_3PO_4	200–250	0.01	[274]
KH_2MoPO_7	H_3PO_4	180	< 0.01	[275]
$NaH_2(MoO)_2(PO_4)_3 \cdot 1/2H_2O$	H_3PO_4	600	1.8	[276]
$NbOPO_4$	H_3PO_4			
KTP	$K_2HPO_4 + KPO_3$	425–600	10 kpsi	[42]
Borates				
$Zn_4B_6O_{12}$	NaOH	500	2.4	[156]
$Li_2B_4O_7$	HCl	250	0.12	[141]
$Li_3B_5O_8(OH)_2$				[144]
$LiBO_2$	NaOH, KOH	> 300	1.0	[156]
$LiZnBO_3$	NaOH, KOH	> 300	1.0	[156]
$LiH_2B_5O_9$	HCOOH	> 240	0.01	[145]

(cont'd)

Appendix (cont'd)

Compound	Mineralizer (s)	Temperature (°C)	Pressure (kbar)	Reference
Carbonates				
MCO_3	NH_4Cl, LiCl, H_2CO_3, HNO_3	200–450	0.6–0.8	[146]
Chalcogenides				
Cu_9S_5, Ag_2S		180–343	0.01–0.15	[277]
Cu_2S		312	0.1	[278]
Ag_2Se		180–343	0.01–0.5	[279]
MSe, M = Zn, Cd, Hg, Co, Pb		180–500	0.01–1.5	[280]
MTe, M = Zn, Cd, Pd, Hg, Co		180–400	0.05–2.0	[281]
Ms, M = Cu, Zn, Hg, Pb, Fe, Cd, Mn, Co, Ni		190–640	0.05–2.0	[282]
Bi_2Se, Sb_2Se		180–343	0.01–1.4	[280]
Sb_2S_3, In_2S_3, Bi_2S_3		180–343	0.01–0.15	[283]
Bi_2Te_3, Sb_2Te_3		180–343	0.01–0.15	[284]
$AgMS_2$, M = Bi, Ga, In, Sb, Be		180–343	0.01–0.15	[283]
$CuMS_2$, M = Ga, In, Fe				
Cu_8MS_6, M = Ge, Sb		312	0.1	[278]
$CuInTe_2$		265–343	0.05–0.15	[285]
$Cu_6ZnAs_2S_7$, $Cu_6FeAs_2S_7$		180–343	0.01–0.05	[286]
Titanates				
$PbTiO_3$, $PbTi_3O_7$, $PbZrO_3$	KF	450–700	0.8–3.0	[287]
$MTiO_3$, M = Ca, Co, Mn, Ba, Sr	NaOH, KOH, KF	200–700	0.3–2.0	[288]
$Na_2Ti_2O_5$, $Na_2Ti_6O_{13}$, $Na_2Ti_3O_7$, $Na_2Ti_3O_7 \cdot 2H_2O$	NaOH, HCl, HCOOH	200–500	0.08–3.0	[289, 290]
Molybdates				
$MMoO_4$, M = Cd, Sr, Pb, Ba	NH_4Cl, NaOH, KOH	350–500		[291]
$KLn(MoO_4)_2$, Ln = La–Yb	K_2MoO_4	525–600	–	[292]
$K_5Ln(MoO_4)$, Ln = La–Yb				
Ln_2MoO_6, Ln = Pr–Er				

Tungstates				
MWO_4, M = Ba, Sr, Cd	NH_4Cl	450	–	[291]
$Li_2Fe(WO_4)_2$	$LiCl + Na_2WO_4$	575–600	1.0–1.5	[293]
$MLn(WO_4)_2$	LiCl, KCl, NH_4Cl	400–720	1.2–1.7	[294]
$LnWO_4 \cdot OH$,	$LiCl + NH_4Cl$			
M = Li, Na, K, NH_4, Ln = Ce, Pr, Nd				
Niobates, tantalates, arsenates, gallates, etc.				
ABO_4	NaOH, NaF, KF, HBr, HI	> 500	2.0	[66]
$LiNbO_3$	LiOH	650	2.0	[295]
$LiGaO_2$	NaOH	360–425	–	[296]
$Gd_3Ga_5O_{12}$	NaOH, Na_2CO_3, K_2CO_3	350–605	1.5	[89]
Chlorides, bromides, fluorides				
$Te_6O_{11}Cl_2$, $TeOCl_2$	HCl, HBr, HI	100–200	0.1	[66]
CuBr, PbBr				
$AREF_4$, $AREF_6$, A_2REF_5,	KF, H_2O	450–500	–	[297]
A = Li, Na, K, Rb, Cs				
RE = Nd, Sm, Eu, Gd, Tb, Yb, Ho, Y				
Miscellaneous				
α-ZrP	H_3PO_4, H_2O	120–280	–	[298]
β-TiP	H_3PO_4	300	–	[299]

References

[1] W. Eitel, Fortschr. Mineral. 10(1925)157.
[2] F. Fouqe and M. Levy, Synthese des Mineraux et des Roches (1882) 395.
[3] P.N. Chirvinskii, Artificial Growth of Minerals in XIX Century (Kiev, 1903–1906) p. 638.
[4] J. Koenigsberger, Fortschr. Mineral. 11 (1927) 617.
[5] K.F.E. Schafhautl, Gelehrte Anzeigen Bayer, Akad. 20(1845) 557.
[6] R. Bunsen, Annalen 65(1845) 557.
[7] H. De Senarmont, Ann. Chim. Phys. 32(1851) 129.
[8] K. Von Chrustschoff, Am. Chem. 3(1873) 281.
[9] G.W. Morey and P. Niggli, J. Am. Chem. Soc. 35(1913) 1086.
[10] P. Niggli and G.W. Morey, Z. Anorg. Chem. 83(1913) 369.
[11] G. Spezia, Atti Accad. Sci. Torino Cl. Sci. Fis. Nat. 40(1905) 254.
[12] F.Y. Loewinson-Lessing, A Historical Survey of Petrology (Oliver and Boyd, Edinburgh, 1954).
[13] N.L. Bowen, Am. J. Sc. 33(1912) 551.
[14] R. Nacken, Captured German Reports, RDRC/13/18(February 28, 1946).
[15] K. Byrappa, ed., Hydrothermal Growth of Crystals, Prog. Cryst. Growth Charact. Mater. 21(1991) 1.
[16] I.V. Zakirov, Experimental Investigations of PVT Relations in the H_2O–CO_2 System, PhD Thesis (Moscow State University, Moscow, 1987).
[17] G.D. Iljushin, L.N. Demianets, V.V. Iljukhin and N.V. Belov, Dokl. Akad. Nauk SSSR 271 (1983) 1133 [in Russian].
[18] O.V. Dimitrova, Investigations of the Phase Formations in the System Na_2O–RE_2O_3–SiO_2–H_2O under Hydrothermal Conditions, PhD Thesis (Moscow State University, Moscow, 1985)
[19] E.D. Kolb and R.A. Laudise, J. Cryst. Growth 33(1976) 145.
[20] P.W. Atkins, Physical Chemistry (ELBS, Oxford Univ. Press, Oxford, 1986) p. 745.
[21] G.A. Krestov, Thermodynamics of Ionic Processes in Solutions Chemistry (1973) p. 116.
[22] K. Deidler, Kinetics of Organic Reactions M (Mir, Moscow, 1966) p. 237.
[23] E.U. Franck, Angew. Chem. 73(1961) 309.
[24] E.U. Franck, Phys. Chem. Earth 13 & 14 (1981) 65.
[25] H.C. Helgeson, in: Chemistry and Geochemistry of Solutions at High Temperatures and Pressures. (Proc. Nobel Symp. organized by the Royal Swedish Academy of Sciences and held at Bjorkborns Herrgard, Karlskoga, Sweden, 17–21 September, 1979), eds. D.T. Rickard and Fr.E. Wickman (Pergamon Press, New York, 1981) p. 133.
[26] T.M. Seward, in: Chemistry and Geochemistry of Solutions at High Temperatures and Pressures. (Proc. Nobel Symp. organized by the Royal Swedish Academy of Sciences and held at Bjorkborns Herrgard, Karlskoga, Sweden, 17–21 September, 1979), eds. D.T. Rickard and Fr.E. Wickman (Pergamon Press, New York, 1981) p. 113.
[27] J.W. Cobble, Rapp. Tech. Cent. Belge Etude Corros. 142(1982) 119.
[28] I.L. Khodakovsky, in: Proc. 1st Int. Symp. Hydrothermal React., Gakujutsu Bunken Fukyu-Kai, 1982, ed. S. Somiya (1983) p. 76.
[29] R.A. Laudise, A.A. Ballman and J.C. King, J. Phys. Chem. Solids 26(1965) 1305.
[30] Handbook of Physics and Chemistry, 64th Ed., ed. R.C. Weast (CRC Press, Boca Raton, 1983) D261.
[31] H.C. Helgeson, Phys. Chem. Earth 13 & 14(1981) 133.
[32] J.B. Hasted, D.M. Ritson and C.H. Collie, J. Chem. Phys. 16(1948) 1.
[33] M. Uematsu, W. Harder and E.U. Franck, The Dielectric Constant of Water, Technical Paper 38p (WLKg. Gp. 3), Int. Assoc. for Prop. of Stern. Inter. Mtg. Kyoto Mtg. Japan (1976).
[34] Landolt–Bornstein, Numerical Data and Functional Relationships, Gr. IV, Vol. 4. High Pressure Properties of Matter, eds. G. Beggerow and K.L. Schafer (Springer, Berlin, 1980) 91.
[35] S.D. Haman, Phys. Chem. Earth 13 & 14(1981) 89.
[36] K. Todheide, Ber. Bunsenges. Physik. Chem. 86 (1982) 1005.
[37] E.U. Franck, Int. Corros. Conf. Ser. (1973) 109.
[38] G.C. Kennedy, Am. J. Sci. 248(1950) 540.
[39] V.I. Popolitov, DSc Thesis (Inst. of Crystallography, Russian Acad. of Sciences, Moscow, 1990).
[40] E.D. Kolb, P.L. Key, R.A. Laudise and E.E. Simpson, Bell. Sys. Tech. J. 61(1983) 639.

[41] W.L. Marshall and E.V. Jones, J. Inorg. Nucl. Chem. 36 (1974) 2313.
[42] R.A. Laudise, W.A. Sunder, R.F. Belt and G. Gashurov, J. Cryst. Growth 102 (1990) 427.
[43] G.W. Morey and J.M. Hesselgesser, Am. J. Sci. Bowen 367 (1952).
[44] R.A. Laudise, J. Am. Chem. Soc. 81 (1959) 562.
[45] R.A. Laudise, in: Progress in Inorganic Chemistry. Vol. 3, ed. F.A. Cotton (Wiley–Interscience, New York, 1962) p. 1.
[46] D.J. Marshall and R.A. Laudise, in: Crystal Growth, ed. H.S. Peiser (Pergamon, New York, 1966) p. 557.
[47] L.V. Bryatov, Some Problems Relating to the Solubility and Growth of Quartz Crystals, PhD Thesis (Inst. of Crystallography, Acad. of Sciences, Moscow, 1955).
[48] A.P. Frederickson and I.E. Cox, Am. Mineral. 39 (1954) 738.
[49] J.B. Hannay and J. Hogarth, Proc. R. Soc. London 30 (1880) 178.
[50] R.G. Pearson, J. Chem. Education 45 (1968) 581.
[51] R.G. Pearson, J. Chem. Education 45 (1968) 645.
[52] E.D. Kolb, J.C. Grenier and R.A. Laudise, J. Cryst. Growth 51 (1981) 178.
[53] K. Byrappa, V.U. Venkatachalapathy and B. Puttaraj, J. Mat. Sci. 19 (1984) 2855.
[54] M. Ouammou, Hydrothermal Synthesis of $AlPO_4$ in the HCl–H_2SO_4–H_3PO_4 mixture, PhD Thesis (ENSCM Montpellier, 1988).
[55] M. Hosaka and S. Taki, J. Cryst. Growth 100 (1990) 343.
[56] R.A. Laudise and J.W. Nielsen, Solid State Phys. 12 (1961) 149.
[57] M. Hosaka and S. Taki, J. Cryst. Growth 52 (1981) 837.
[58] V.W. Jahn and Kordes, Chem. Earth 16 (1953) 75.
[59] J.M. Stanley, Ind. Eng. Chem. 46 (1954) 1684.
[60] J. Gomez Morales, R. Rodriguez Clemente, J. Durand and L. Cot, Solubility of Berlinite in H_2SO_4–HCl Mixtures under Hydrothermal Conditions, J. Cryst. Growth 116 (1992) 427.
[61] J.L. Symes and D.R. Kester, Geochim. Cosmochim. Acta 48 (1984) 2219.
[62] D.S. Shlyapnikov and E.K. Shtern, Dokl. Acad. Nauk SSSR 207 (1972) 966.
[63] V.S. Balitsky and T. Bublikova, in: Hydrothermal Growth of Crystals, ed. K. Byrappa; Prog. Cryst. Growth Charact. Mater. 21 (1991) 139.
[64] A. Rabenau, Angew. Chem. English Ed. 24 (1985) 1026.
[65] S. Hirano, K. Nakamura and S. Somiya, Proc. 4th Int. Conf. High Pressure, Jpn. Ed. (Osugi, 1974) p. 418.
[66] V.I. Popolitov and B.N. Litvin, Growth of Crystals under Hydrothermal Conditions (Nauka, Moscow, 1986) p. 38.
[67] B.N. Litvin and D.A. Tules, in: Crystallization Processes Under Hydrothermal Conditions, ed. A.N. Lobachev under Studies in Soviet Science (Consultant Bureau, New York, 1973) p. 139.
[68] E.T. Allen, J.L. Crenshaw and J. Johnston, Am. J. Sci. 4th Ser. 33 (1912) 167.
[69] A. Rabenau and H. Rau, in: Inorganic Synthesis, Vol. 14, eds. A. World and J.K. Ruff, (McGraw–Hill, New York, 1973) Chap. 35.
[70] R.A. Laudise, Am. Assoc. Cryst. Growth 21 (1) (1991) 6.
[71] J. Detaint, E. Phillippot, J.C. Jumas, J. Schwartzel, A. Zarka, B. Capelle and J.C. Doukhan, 39th Annual Frequency Control Symp. (Philadelphia, 1985).
[72] D.F. Croxall, L.R.A. Christie, B.J. Isherwood, A.G. Todd and J. Brich, Proc. 2nd European Conf. Crystal Growth (Lancaster, 1979).
[73] G.W. Morey and P. Niggli, J. Am. Chem. Soc. 35 (1913) 1086.
[74] G.W. Morey and E. Ingerson, Am. Mineral. 22 (1937) 1121.
[75] O.F. Tuttle, Bull. Geol. Soc. Am. 60 (1949) 1727.
[76] A.C. Walker and E. Buehler, Sci. Monthly 69 (1949) 148.
[77] R.A. Laudise, in: Advanced Crystal Growth, eds. P.M. Dryburgh, B.Cockayne and K.G. Barraclough (Prentice–Hall, London, 1987) p. 267.
[78] O.F. Tuttle, Am. J. Sci. 246 (1948) 628.
[79] W.C. Luth and O.F. Tuttle, Am. Mineral. 48 (1963) 1401.
[80] A.A. Shternberg, in: Hydrothermal Synthesis of Crystals (Nauka, Moscow, 1968) p. 203 [in Russian].
[81] V.A. Kuznetsov and A.N. Lobachev, Sov. Phys. Cryst. 17 (1973) 775.

[82] V.I. Popolitov, in: Hydrothermal Growth of Crystals ed. K. Byrappa; Prog. Cryst. Growth Charact. Mater. 21 (1991) 255.
[83] E.D. Kolb and R.A. Laudise, J. Cryst. Growth 56 (1982) 83.
[84] N.Yu. Ikornikova, A.N. Lobachev, A.R. Vasenin, V.M. Egrov and A.V. Antoshin, in: Crystallization Processes under Hydrothermal Conditions, ed. A.N. Lobachev (Consultant Bureau, New York, 1973) p. 241.
[85] G. Strubel, Z. Dt. Gemmol. Ges. 24 (1975) 138.
[86] A.A. Shternberg, in: Hydrothermal Synthesis of Crystals, ed. A.N. Lobachev (Consultant Bureau, New York, 1971) p. 147
[87] H.T. Hall, Rev. Sci. Instrum. 31 (1951) 125.
[88] A.A. Shternberg and V.A. Kuznetsov, Sov. Phys. Cryst. 13 (1969) 647.
[89] G.I. Distler, S.A. Kobzareva, A.N. Lobachev, O.K. Melnikov and N.S. Triodina, Krist. Tech. 13 (1978) 1025.
[90] E.D. Kolb and R.A. Laudise, J. Cryst. Growth 29 (1975) 29.
[91] K. Byrappa, S. Srikantaswamy, G.S. Gopalakrishna and V. Venkatachalapathy, J. Mater. Sci. 21 (1986) 2202.
[92] S. Hirano and S. Somiya, J. Cryst. Growth 35 (1976) 273.
[93] S. Hirano and M.G.M. Ismail and S. Somiya, Mater. Res. Bull. 11 (1976) 1023.
[94] A.N. Christensen, Rev. Chim. Mineral. 6 (1969) 1187.
[95] V.A. Kuznetsov and A.N. Lobachev, Sov. Phys. Cryst. 17 (1973) 775.
[96] B. Sawyer, J. Cryst. Growth 36 (1976) 345.
[97] S. Taki, in: The Role of Crystal Growth for Device Development, ed. N. Nizeki; Prog. Cryst. Growth Charact. Mater. 23 (1992) 313.
[98] A.F. Armington, in: Hydrothermal Growth of Crystals, ed. K. Byrappa; Prog. Cryst. Growth Charact. Mater. 21 (1991) 97.
[99] A.A. Ballman and R.A. Laudise, in: Art and Science of Growing Crystals, ed. J.J. Gilman (Wiley, New York, 1963) p. 231.
[100] R.L. Barns, E.D. Kolb, R.A. Laudise, E.E. Simpson and K.M. Kroupa, J. Cryst. Growth 34 (1976) 189.
[101] J.J. Martin and A.F. Armington, J. Cryst. Growth 62 (1983) 203.
[102] R.A. Laudise, in: Materials for Nonlinear Optics, Chemical Perspectives, eds. S.R. Marder, J.E. Sohn and G.D. Stucky (American Chemical Society, Washington, DC, 1991) 411.
[103] H.G. Lipson and A.F. Armington, J. Cryst. Growth 80 (1987) 51.
[104] A.F. Armington and J.J. Larkin, J. Cryst. Growth 71 (1985) 799.
[105] M. Hosaka, private communication (1991).
[106] L.E. Halliburton, M.G. Jani and R.B. Bossoli, Nucl. Instrum. Methods, Phys. Res. B 1 (1984) 192.
[107] B. Fraser, Physical Accoustics, ed. W.P. Mason, 5 (1968) 54.
[108] M. Jalilian-Nosraty and J.J. Martin, J. Appl. Phys. 52 (1981) 785.
[109] A.F. Armington and J.F. Balascio, Proc. 40th Ann. Freq. Control Symp. (IEEE, Washington, DC, 1980) p. 230.
[110] R.A. Laudise and A.A. Ballman, J. Phys. Chem. 64 (1960) 668.
[111] J.W. Nielson and E.E. Dearborn, J. Phys. Chem. 64 (1960) 1762.
[112] N. Sakagami, J. Cryst. Growth 99 (1990) 905.
[113] I.P. Kuzmina and N.M. Khaidukov, Crystal Growth from High Temperature Aqueous Solutions (Nauka, Moscow, 1977) p. 178.
[114] P. Hartman, Phys. Stat. Sol. 2 (1962) 585.
[115] B.N. Litvin and L.I. Ivanova, Kristallographia 14 (1969) 1122.
[116] K.B. Byrappa and S. Srikantaswamy, in: Hydrothermal Growth of Crystals, ed. K. Byrappa; Prog. Cryst. Growth Charact. Mater. 21 (1991) 199.
[117] J.C. Jumas, A. Goiffon, B. Capelle, A. Zaraka, J.C.Doukhan, J.Schwartzel, J. Detaint and F. Phillipot, J. Cryst. Growth 80 (1987) 133.
[118] K. Byrappa, S. Srikantaswamy and K. Sangwall, Extended Abstracts ICCG-9 (Sendai, 20–25 August 1989) 25PDO5.
[119] R.A. Laudise, in: Crystal Growth of Electronic Materials, ed. E. Kaldis (Elsevier, Amsterdam, 1985) p. 159.

[120] K. Nagai, T. Okaula, J. Okuda and J. Asathara, First Int. Symp. Hydrothermal Reactions Tokyo, 1982) p. 496.
[121] A. Tebo, Laser Focus/Electro-Optics (September 1988) 28.
[122] F.G Zumsteg, J.D. Bierlein and T.E. Gier, J. Appl. Phys. 47 (1976) 4980.
[123] R.F. Belt, G. Gashurov and Y.S. Liu, Laser Focus/Electro-Optics 21 (1985) 110.
[124] R.A. Laudise, R.J. Cava and A.J. Caporaso, J. Cryst. Growth 74 (1986) 275.
[125] J. Shou-Quan, N. Hong-DA, T. Jin-Ge, X. Yan-Ping and T. Ying, J. Cryst. Growth 99 (1990) 900.
[126] R.F. Belt, Extended Abstract ICCG-9 (Sendai, 20–25 Aug. 1989) 25PDO3.
[127] G.D. Stucky, M.L.F. Phillips and T.E. Gier, Chem. Mater. 1 (1989) 492.
[128] J.B. Goodenough, H. Y-P. Hong and J.A. Kaflas, Mater-Res. Bull. 11 (1976) 203.
[129] S. Gali, K. Byrappa and G.S. Gopalakrishna, Acta Crystallogr. C 45 (1989) 1667.
[130] S. Gali and K. Byrappa, Acta Crystallogr. C 46 (1990) 2011.
[131] M. Yoshimura, K. Fuji and S. Somiya, Mater. Res. Bull. 16 (1981) 327.
[132] K. Byrappa and B.N. Litvin, J. Mater Sci. 18 (1983) 703.
[133] C.T. Chen, B.C. Nu, A.D. Jiang and G.M. You, Sci. Sinica B 28 (1985) 235.
[134] R.W. Whatmore, N.M. Shorrocks, C.O, Hara and F.W. Ainger, Electron. Lett. 17 (1981) 11.
[135] H. Koning and R. Hoppe, Z. Anorg. Allg. Chem. 439 (1978) 71.
[136] E.V. Sokolova, Hydrothermal Synthesis and Crystal Structures of Cd- and Li–Cd borates, PhD Thesis (Moscow State University, Moscow, 1980) [in Russian].
[137] N.V. Belov, A.I. Ivaschenko, O.S. Bondareva, A.N. Lobachev, Yu.A-Malinovskii, O.K. Melnikov and M.A. Simonov, in: Hydrothermal Synthesis and Growth of Single Crystals, ed. A.N. Lobachev (Nauka, Moscow, 1982) p. 158.
[138] O.V. Yakubovich, Crystal Structures of the Group of Ca, Mg and Ca, Mg-borates, PhD Thesis (Moscow State University, Moscow, 1978) [in Russian].
[139] J.D. Garrett, M.N. Iyer and J.E. Greedan, J. Cryst. Growth 41 (1977) 225.
[140] Fan Shi-ji, Shen Guan-Shun, Wang Wen, Li Jin-Long and Le Xiu-Hong, J. Cryst. Growth 99 (1990) 811.
[141] K. Byrappa and K.V.K. Shekar, J. Mater. Chem. 2 (1992) 13.
[142] O.K. Melnikov, O.S. Bondareva and Y.A. Malinovsky, 1st Int. Symp. Hydrothermal Reactions (Tokyo, Japan, 1982).
[143] E.V. Sokolova, N.A. Yamnova and N.V. Belov, Kristallografiya 25 (1980) 716.
[144] K. Byrappa, K.V.K. Shekar and S. Gali, Cryst. Res. Tech. 6 (1992) 768.
[145] A. Cardenas, J. Solans, K. Byrappa and K.V.K. Shekar, Structure of $LiH_2B_5O_9$, Acta. Cryst. C49 (1992) 645.
[146] N.Yu. Ikornikova, in: Hydrothermal Synthesis of Crystals in Chloride systems (Nauka, Moscow, 1975) p. 126.
[147] N.Yu. Ikornikova, in: Crystallization Processes under Hydrothermal Conditions, ed. A.N. Lobachev under Studies in Soviet Science (Consultant Bureau, New York, 1973) p. 93
[148] J.F. Nester and J.B. Schroeder, Am. Mineral. 52 (1967) 276.
[149] C. Belin, J. Cryst. Growth 34 (1976) 341.
[150] K. Kikuta and S. Hirano, J. Cryst. Growth 99 (1990) 895.
[151] S. Hirano, T. Yogo and K. Kikuta, in: The Role of Crystal Growth for Device Development, ed. N. Niizeki; Prog. Cryst. Growth Charact. Mater. 23 (1992) 341.
[152] V.I. Popolitov and A.N. Lobachev, Izv. Akad. Nauk USSR; Inorg. Mater. 8 (1972) 960.
[153] S.D. Malinin and N.I. Khitarov, Geochemistry 11 (1969) 1312.
[154] R.A. Laudise and E.D. Kolb, Am. Mineral. 50 (1965) 381.
[155] V.A. Kuznetsov, in: Hydrothermal Growth of Crystals, ed. K. Byrappa; Prog. Cryst. Growth Charact. Mater. 21 (1991) 163.
[156] L.N. Demyanets, in: Hydrothermal Growth of Crystals, ed. K. Byrappa; Prog. Cryst. Growth Charact. Mater. 21 (1991) 299.
[157] O.K. Melnikov, B.N. Litvin and N.S. Triodina, in: Crystallization Processes under Hydrothermal Conditions, ed. A.N. Lobachev (Consultant Bureau, New York, 1973) p. 151.
[158] L.N. Demiyanets, E.N. Emelyanova and O.K. Melnikov, in: Crystallization Process under Hydrothermal Conditions, ed. A.N. Lobachev (Consultant Bureau, New York, 1973) p. 125.
[159] E.J. Gubelin, Gems Gemmol 10 (1960) 105.

[160] H. Bank, Z. Dt. Gemmol. Ges. 25 (1976) 107.
[161] K. Nassau, J. Cryst. Growth 35 (1976) 211.
[162] P.J. Yancuj, US Patent No. 3 723 337 (1973).
[163] L.I. Tinober and L.G. Chentsova, Kristallografiya 4 (1959) 633.
[164] V.S. Balitsky, J. Cryst. Growth, 41 (1977) 100.
[165] M. Hosaka and S. Taki, J. Cryst. Growth 64 (1983) 572.
[166] M. Hosaka, T. Miyata, Y. Shimizu and O. Okuyama, J. Cryst. Growth 78 (1986) 561.
[167] G. Yamaguchi, H. Yanagida and S. Somiya, Bull. Soc. Chem. Jpn. 35 (1962) 1789.
[168] R.A. Laudise and A.A. Ballman, J. Am. Chem. Soc. 80 (1958) 2655.
[169] V.N. Rumyanstev, I.G. Ganeev and I.S.Rez, Crystal Growth, Vol. 9 (Nauka, Moscow, 1972) p. 51.
[170] R.L. Barns, R.A. Laudise and R.M. Shields, J. Phys. Chem. 67 (1963) 835.
[171] A. Rabenau and H. Rau, Naturwissenschaften 55 (1968) 336.
[172] J.F. Balasico, R.B. White and R. Roy, Mater. Res. Bull. 2 (1967) 913.
[173] L.N. Demyanets, in: Hydrothermal Synthesis of Crystals [in Russian] (Nauka, 1968).
[174] L.N. Demyanets, L.S. Garashina and B.N. Litvin, Kristallografiya 8 (1963) 800.
[175] I.P. Kuzmina, Geology Ore Deposits 6 (1963) 101.
[176] J. Bauer and P. Kaczerovsky, VSCHT Praze. G. 12 (1970) 153.
[177] E. Procharkova, D. Rykl, and V. Seidl, Patent No. 170391, Czechoslovakia, Method of obtaining monocrystals of nickel oxide, 27 August 1973, No. 5939-73.
[178] I.P. Kuzmina and B.N. Litvin, in: Crystal Growth, Vol. 4 (Nauka, Moscow, 1964) p. 160.
[179] C.J.M. Rooymans and W.F. Langenhoff, J. Cryst. Growth 3 & 4 (1968) 411.
[180] V.B. Glushkova, Polymorphism of Rare Earth Oxides (Nauka, Leningrad, 1967) p. 133.
[181] H. Guggenheim, Solid State Phys. 12 (1961).
[182] V.I. Popolitov, A.N. Lobachev and V.F. Perkin, Kristallografiya 17 (1972) 436.
[183] B.N. Litvin and V.I. Popolitov, Hydrothermal Synthesis of Inorganic Compounds (Nauka, Moscow, 1984).
[184] E.D. Kolb, A.J. Caporaso and R.A. Laudise, J. Cryst. Growth 19 (1973) 242.
[185] S.P. Fedoseeva, in: Crystal Growth, Vol. 8 (Nauka, Moscow, 1972) p. 59.
[186] R.D. Schuilling and B.W. Vink, Geochim. Cosmochim. Acta 31 (1967) 2399.
[187] I.P. Kuzmina and B.N. Litvin, Kristallografiya 7 (1963) 478.
[188] T. Eiiti, K. Iodji, Kh. Kendzi and S. Tasturo, Patent No. 52-48959 (Japan), 22-08-1975, No 50-102484.
[189] G. Demazeau and P. Maestro, Proc. 7th Int. AIRAPT Conf. (Le Creusot) 1 (1979) 572.
[190] I.P. Kuzmina, Author's abstract [in Russian] IKAN SSSR (Moscow, 1968).
[191] Jia-Shan-Quan, Jiang-Pei-Zhi, Fan-Xian-He, Niu-Hong-Da and Li-De-Zhou (1968).
[192] R.A. Laudise and E.D. Kolb, J. Am. Ceram. Soc. 45 (1962) 51.
[193] S.C. Abraham, P.B. Jameson and J.L. Bernstein, J. Chem. Phys. 47 (1967) 4034.
[194] G.Ervin and E.F. Osborn, J. Geol. 59 (1951) 381.
[195] A.N. Christensen, Acta Chem. Scand. 20 (1966) 896.
[196] A.N. Christensen, Acta Chem. Scand. 30 (1976) 133.
[197] H. Schwarz, Z. Naturforsch. 226 (1967) 554.
[198] M.C. Michel-Levy, Bull. Soc. France Mineral. Crystallogr. 79 (1956) 124.
[199] B.V. Mill, Dokl. Akad. Nauk SSSR 156 (1964) 814.
[200] J. Ito and C. Frondel, Am. Mineral. 53 N. 7/8 (1968) 1276.
[201] B.A. Maksimov, Yu. A. Kharitonov, V.V. Ilyukhin and N.V. Belov, Dokl. Akad. Nauk SSSR 178 (1968).
[202] M. Bihan, Th.A. Katt and R. Wey, Bull. Soc. France Mineral. Crystallogr. 941 (1971) 15.
[203] D.E. Appleman and J.R. Clark, Am. Mineral. 50 (1965).
[204] K. Kodaira, S. Ito and T. Masushit, J. Cryst. Growth 29 (1975) 123.
[205] Z. Ito and I. Harold, Am. Mineral. 53 (1968).
[206] D.H. Lindsley, Carnegie Inst. Annu. Rep. (Direct. Geophys. Lab. 1969, 70, Washington, 1971) p. 188
[207] I. Ito, Am. Mineral. 53 (1968).
[208] I. Ito and C. Frondel, Am. Mineral. 52 (1967) 1077.
[209] I. Ito and C. Frondel, Ark. Miner. Och. Geol. (1968) 391.
[210] G.V. Bukin, Mineralogical Museum, Akad. Nauk SSSR 19 (1969) 131.

[211] R.J. Harker and O.G. Tuttle, Am. J. Sci. 254 (1956) 468.
[212] P.A. Sandomirskii, M.A. Simonov, A.V. Arakcheeva and N.V. Belov, Dokl. Akad. Nauk SSSR 227 (1976) 856.
[213] B.Yu. Kornilovich, N.A. Ovramenko and F.D. Ovcharenko, Dokl. Akad. Nauk SSSR 261 (1981).
[214] I.Ya. Nekrasov and T.P. Dadze, J. Mineral. 3 (1980).
[215] I.Ya. Nekrasov, T.P. Dadze and N.V. Zayakina, Dokl. Akad. Nauk SSSR 261 (1981) 479.
[216] G.D. Brunton, L.A. Harris and O.C. Kopp, Am. Mineral. 57 (1972) 1720.
[217] E.E. Strelkova, O.G. Karpov and B.N. Litvin, Kristallografiya 22 (1977) 174.
[218] D.Yu. Puscharovskii, O.G. Karpov, E.A. Pobedimskaya and N.V. Belov, Dokl. Akad. Nauk SSSR 234 (1977) 1323.
[219] D.Yu. Puscharovskii, T. Baatarin, E.A. Pobedimskaya and N.V. Belov, Kristallografiya 16 (1971).
[220] D.Yu. Puscharovskii, E.A. Pobedimskaya and N.V. Belov, Dokl. Akad. Nauk 185 (1969).
[221] L.B. Halferdahl, J. Petrol. 2 (1961) 49.
[222] D.Yu. Puscharovskii, E.A. Pobedimskaya, B.N. Litvin and N.V. Belov, Dokl. Akad. Nauk SSSR 214 (1974) 91.
[223] R.N. Thompson and J.E. Chisholm, Mineral. Mag. 37 (1969) 253.
[224] C.F. Warren, Am. Mineral. 56 (1971) 997.
[225] D.J. Drysdale, Am. Mineral. 56 (1971).
[226] A.N. Safronov, N.N. Nevskii, V.I. Ilyukhin and N.V. Belov, Dokl. Akad. Nauk SSSR 255 (1980).
[227] R. Caruba, A. Baumer and G. Turco, Donnour Redlocrystallographiquel G.M. Acad. Sci 270 (1970).
[228] V.G. Chykhlanstev and K.V. Alyamovskaya, Izv. Akad. Nauk SSSR Inorg. Mater. 6 (1970) 1639.
[229] N.G. Shymyaskaya, V.A. Blinov and A.A. Voronkov, Dokl. Acad. Nauk SSSR 208 (1973).
[230] M.C. Michel-Levy, Bull. Soc. France Mineral. Crystallogr. 84 (1967).
[231] I. Ito, Am. Mineral. 53 (1968) 231.
[232] J.L. Munoz, Am. Mineral. 53 (1968) 1490.
[233] A. Baronnet, M. Amouric, B. Chabot and F. Corny, J. Cryst. Growth 43 (1978) 255.
[234] R.D. Shannon, Phys. Chem. Mineral. 4 (1979) 139.
[235] D. Tranqui, J.J. Capponi, J.C. Joubert and R.D. Shannon, J. Solid State Chem. 73 (1988) 325.
[236] R. Caruba, A. Baumer and G. Turco, Geochim. Cosmochim. Acta 39 (1975) 11.
[237] A. Baumer and G. Turco, C.R. Acad. Sci. D 270 (1970) 1197.
[238] A.A. Bush, S.A. Ivanov and S.Yu. Stefanovich, Izv. Akad. Nauk SSSR Inorg. Mater. 13 (1977) 1656.
[239] M.N. Tseitlin, G.F. Plakhov and A.N. Lobachev, Kristallografiya 18 (1973) 836.
[240] V.I. Popolitov, G.F. Plakhov and S.Yu. Stefanovich, Izv. Akad. Nauk SSSR Inorg. Mater. 17 (1981) 1841.
[241] B.N. Litvin, DSc Thesis (Institute of General and Inorganic Chemistry, Academy of Sciences, USSR, Moscow, 1978).
[242] A.N. Lazarev and A.K. Shirvinskaya, Izv. Akad. Nauk SSSR Inorg. Mater. 12 (1976) 771.
[243] T.N. Nadezhdina, E.A. Pobedimskaya, V.V. Ilyukhin and N.V. Belov, Dokl. Akad. Nauk SSSR 223 (1977) 1086.
[244] S.R. Lyon and E.S. Ehlers, Am. Mineral. 56 (1970) 118.
[245] B.A. Maksimov, Yu.V. Nikolskii and Yu.A. Kharitonov, Dokl. Akad. Nauk SSSR 239 (1978) 87.
[246] A.M. Dago, D.Yu. Pushcharovskii, E.A. Pobedimskaya and N.V. Belov, Dokl. Akad. Nauk SSSR 250 (1980) 857.
[247] D.Yu. Pushcharovskii, A.M. Dago, E.A. Pobedimskaya and N.V. Belov, Dokl. Akad. Nauk SSSR 251 (1980) 354.
[248] D.M. Roy and R. Roy, Am. Mineral. 39 (1954) 957.
[249] D.Yu. Pushcharovskii, E.A. Pobedimskaya O.V. Kudryasteva and B. Hettash, Kristallografiya 1 (1976) 1126.
[250] I.P. Kuzmina, O.K. Melnikov and B.N. Litvin, in: Hydrothermal Synthesis of Crystals (Nauka, Moscow, 1968) p. 41.
[251] N.A. Nosirev, PhD Thesis (Institute of Crystallography, Moscow, 1975).
[252] Yu.K. Egorov-Tismenko, M.A. Simonov and N.V. Belov, Dokl. Akad. Nauk SSSR 227 (1976) 2.
[253] L.I. Soloveva and V.V. Bakanin, Kristallografiya 12 (1967) 591.
[254] S. Kume, S. Ueda and M. Koizumi, J. Geophys. Res. 74 (1969) 2145.

[255] N. Kinomura, J. Am. Ceram. Soc. 56 (1973) 344.
[256] N.V. Belov, Dokl. Akad. Nauk SSSR 268 (1983) 360.
[257] A.N. Christensen, Acta Chem. Scand. 24 (1970) 1287.
[258] O.Jarchow, K.H. Kalska and H. Schenk, Naturwissenschaften 68 (1981) 475.
[259] L.N. Demianets, A.N. Lobachev and G.A. Emelchenko, Rare Earth Germanates (Nauka, Moscow, 1980).
[260] E.D. Kolb and R.A. Laudise, J. Cryst. Growth 43 (1978) 313.
[261] F. Dachille and R. Roy, Z. Kristallogr. 111 (1959) 451.
[262] S. Srikantaswamy, Synthesis and Characterization of ABO_4 Crystals, PhD Thesis (University of Mysore, Mysore, 1988).
[263] R.G. Hermann, PhD Thesis (Ohio University, Ohio, 1972).
[264] E.M. McCarron, J.L. Calabrene and M.A. Subramanian, Mater. Res. Bull. 22 (1987) 1421.
[265] J.M. Congo and P. Kierkegaard, Acta Chem. Scand. 20 (1966) 72.
[266] H. Effenberger, J. Solid State Chem. 57 (1985) 1240.
[267] A. Clearfield, B.D. Roberts and M.A. Subramanian, Mater. Res. Bull. 19 (1984) 219.
[268] M.A. Subramanian, B.D. Roberts and A. Clearfield, Mater. Res. Bull. 19 (1984) 1417.
[269] P. Lightfoot, A.K. Cheetam and A.W. Sleight, J. Solid State Chem. 73 (1988) 325.
[270] K. Byrappa. S. Srikantaswamy and S. Gali, J. Mater. Sci. Lett. 9 (1990) 235.
[271] K. Byrappa and B.N. Litvin, J. Mater. Sci. 18 (1983) 703.
[272] S. Gali and K. Byrappa, Acta Crystallogr. C 46 (1990) 2011.
[273] S. Gali, A. Cardenas, K. Byrappa and G.S. Gopalakrishna, Acta Cryst. C48 (1992) 1650.
[274] K. Byrappa, B.V. Umesh Dutt, D. Podjary and A. Clearfield, J. Mater. Res., in press.
[275] A. Clearfield, in: Hydrothermal Growth of Crystals, ed. K. Byrappa; Prog. Cryst. Growth Charact. Mater. 21 (1991) 1.
[276] R. Peascoe, Ph.D. Thesis, Texas A. & M. University, USA (1989).
[277] M. Elli, L. Cambi, Att. Accad. Naz. Lincei, Rend. Cl. Si. Fiz. Mat. Natur. 39 (1965) 87.
[278] L. Cambi and M. Elli, Chim. et. Industr. [Ital.] 47 (1968) 136.
[279] L. Cambi and M. Elli, Chim. et. Industr. [Ital.] 51 (1969) 3.
[280] L. Cambi and M. Elli, Att. Accad. Naz. Lincei. Rend. Cl. Sci. Fis. Mat. Natur. 40 (1966) 553.
[281] E.D. Kolb, A.J. Caporaso and R.A. Laudise, J. Cryst. Growth 3/4 (1968) 422.
[282] N.Kh. Abrikosov, V.F. Bankina and L.V. Porestkaya (Nauka, Moscow, 1967) p. 176.
[283] L. Cambi and M. Elli, Chim. et. Industr. [Ital.] 48 (1966) 944.
[284] L. Cambi and M. Elli, Att. Accad. Naz. Lincei. Rend. Cl. Sci. Fis. Mat. Natur. 41 (1966) 241.
[285] L. Cambi and M. Elli, Chim. et. Industr. [Ital.] 50 (1968) 94.
[286] V.S. Balitsky, V.V. Komova and N.A. Ozerova, Izv. Akad. Nauk SSSR Ser. Geol. 12 (1971) 93.
[287] V.A. Kuznetsov, in: Crystal Processes under Hydrothermal Conditions, ed. A.N. Lobachev (Consultant Bureau, New York, 1973).
[288] V.B. Mironov, F.Ya. Kulba and V.A. Nazarov, Zh. Neorg. Khim. 7 (1962) 1154.
[289] I. Keeman, Z. Anorg. Allg. Chem. 346 (1966) 30.
[290] K. Byrappa, B.V. Umesh Dutt, R.R. Clemente, S. Gali and A.B. Kulkarni, in: Current Trends in Crystal Growth and Characterization, ed. K. Byrappa (MIT, India, 1991) p. 272.
[291] L.N. Demianets, in: Hydrothermal Synthesis of Crystals (Nauka, Moscow, 1968) p. 93 [in Russian].
[292] L.Yu. Kharchenko, V.I. Protasova and P.V. Klevstov, J. Inorg. Chem. 22 (1977) 986 [in Russian].
[293] P.V. Klevstov, N.A. Novgorodsteva and L.Yu. Kharchenko, Dokl. Akad. Nauk USSR 183 (1968) 1313.
[294] R.F. Klevstova, L.Yu. Kharchenko, S.V. Borisov, V.A. Efremov and P.V. Klevstov, Kristallografiya 24 (1979) 446.
[295] V.G. Hill and K. Zimmerman, J. Electrochem. Soc. Solid State 115 (1968) 978.
[296] E.D. Kolb, A.J. Caporoso and R.A. Laudise, J. Cryst. Growth 8 (1971) 354.
[297] A.V. Novoselova, Izv, Akad. Nauk USSR, Inorg. Mater. 20 (1984) 967.
[298] A. Clearfield and G.D. Smith, Inorg. Chem. 7 (1969) 431.
[299] S. Alluli, C. Ferraggiva, A. Laginestra, M.A. Massucci and N. Tomassini, J. Inorg. Nucl. Chem. 399 (1977) 1043.

CHAPTER 10

Flux Growth

WOLFGANG TOLKSDORF

Höhenweg 3, D-63691 Ranstadt, Germany

Contents

1. Introduction . . . 565
   1.1. Literature . . . 566
      1.1.1. Books . . . 566
      1.1.2. Review articles . . . 566
2. Fluxes for oxide materials . . . 567
3. Crucibles . . . 571
4. Equipment . . . 572
5. Methods . . . 572
   5.1. Slow cooling (SC) . . . 573
      5.1.1. Slow cooling bottom growth (SCBG) . . . 574
         5.1.1.1. Iron garnets by SCBG . . . 574
         5.1.1.2. Other oxides by SCBG . . . 582
      5.1.2. Top-seeded solution growth (TSSG) . . . 584
   5.2. Vertical temperature gradient transport (VTGT) . . . 589
      5.2.1. Top-seeded VTGT growth with a nutrient . . . 589
      5.2.2. Bottom growth with a nutrient . . . 590
      5.2.3. Growth by a travelling solvent zone (TSZ) . . . 592
   5.3. Growth by flux evaporation . . . 594
   5.4. Special growth methods . . . 595
6. Growth phenomena . . . 596
References . . . 607

HANDBOOK OF CRYSTAL GROWTH, VOL. 2
edited by D.T.J. Hurle

1. Introduction

"Flux growth", the most commonly used term for crystal growth from high temperature solutions (HTS), is a method which allows the growth of single crystals of a wide range of materials. The components of the desired materials are dissolved in a solvent, the so-called "flux". The use of the word "flux" for the solvent is derived from metallurgy and ceramic technology where substances are used to assist the melting procedure or sintering. The term is used, too, for substances like welding or soldering aids. In some papers the term "flux" is used for the high temperature solution, but in this paper it will be strictly used as the high temperature solvent; in some cases this might be an excess of a compound of the solute.

The flux growth process is analogous to crystal growth from aqueous solutions, but the solvent solidifies before reaching room temperature. The main advantage of this method is that the crystals are grown below the melting temperature. If the material melts incongruently, i.e. decomposes before melting, or exhibits a phase transition below the melting point, one has indeed to look for growth temperatures lower than these phase transitions.

Flux growth might be used, when the melting temperature is very high, or too high for the facilities available for the operator. It is useful, when the vapour pressure at the melting temperature is too high. Thermal strain is minimized due to the relatively low growth temperature, the mostly very small temperature gradients used in the HTS and the free growth into a liquid, allowing the formation of growth facets.

The method is very versatile since a solvent may be found for any material required in the form of a single crystal. Flux growth is suitable for the growth of layers on single crystal substrates, the so-called liquid phase epitaxy, LPE.

The main disadvantage is the low growth rate. From the flux the growth rate of crystals is generally more than hundred times smaller than in melt growth. Thus crystals from the flux are in general of small size. Since crystals grown from HTS normally develop facets, the linear growth rate is difficult to determine and in most cases it is even not constant and might be different for each facet, which leads to inhomogeneities in the crystal. For industrial production the facetted form is a severe disadvantage compared to the round boule one can get by pulling a seed crystal from the melt as is possible e.g. by the Czochralski technique.

A further disadvantage is the unavoidable presence of ions of the flux as impurities in the crystals, if the solvent contains additional elements. Frequently there are flux inclusions found in the crystal.

In general, crystal growth from the melt is preferable whenever possible. However, since often only small crystals are needed for basic investigations or measurements, the effort to grow these from HTS might be much smaller than that of melt techniques. Furthermore, with some experience and with the choice of a suitable flux composition and growth method and automation of the growth procedure, one might be able to minimize the disadvantages of flux growth.

This review will be restricted to oxide crystals since the author, who was responsible for a crystal growth group within the Philips research laboratory in Hamburg, has, in this field, most of his personal experience. For personal reasons there will be a more detailed description of the growth of garnets and because garnets are one of the oxide materials grown nearly by all flux methods. Since there are congruently melting garnets, a comparison of both techniques is possible.

1.1. Literature

This review of flux growth technology widely refers to the literature. Of course, the selection of quoted papers from the literature is rather subjective and by far not complete or exhaustive.

1.1.1 Books

The only monograph dealing exclusively with flux growth is the book *Crystal Growth from High-Temperature Solutions* by Elwell and Scheel [1]. The work covers in detail all areas of HTS crystal growth. In particular it gives a comprehensive description of the experimental aspects such as crucibles, thermocouples and growth techniques together with experimental data. About 2500 references form a valuable literature source covering the period prior to 1975.

There are many books dealing with crystal growth which include a chapter on flux growth. One of the first is a small chapter in *Einkristalle* by Smakula [2]. In Gilman's book [3] *The Art and Science of Growing Crystals* you find a chapter on *Molten Salt Solvents* [4]. This chapter was updated in the book *Growth of Single Crystals* by Laudise [5]. In *Preparation and Properties of Solid State Materials*, Vol. 1, [6], chapter 3 is about *Exploratory Flux Crystal Growth* [7]. Those who can read German will appreciate a 75-page chapter about HTS growth in *Kristallzüchtung* by Wilke and Bohm [8]. Winkler's book *Magnetic Garnets* gives, in the chapter *Garnet Technology*, detailed descriptions of various flux growth techniques for garnets [9].

1.1.2. Review articles

There are many review articles published about flux growth. Nielsen, one of the pioneers of flux growth, describes the improved techniques used in flux growth including accelerated crucible rotation and seeding processes and he shows, in his phase equilibrium discussions, the strong relation between bulk growth and liquid phase epitaxy of garnets [10]. This review was presented already in 1972 at a conference in the USSR. Review papers by Elwell prove his interest and activity in flux growth. In 1971 he published a review about mechanisms of crystal growth from fluxed melts [11]. It was followed by another one containing a list of commonly used fluxes, a discussion of phase diagram determination including the use of a thermogravimetric apparatus and differential thermal analysis, growth mechanism and stability and general practical considerations [12]. He documented the state of technology and knowledge of growth from HTS in 1980 [13] and 1987 [14]. A very useful review paper is that on *Practical Aspects of Flux Growth by Spontaneous Nucleation* [15]. Hints to literature and practical tricks like flux draining by various methods are given, nucleation aspects are discussed and long lists of crystals and the respective fluxes are presented. A review of the present status of flux growth was given by Wanklyn [16]. Investigations of heterogeneous equilibria and data on measurements of the density, volatility, viscosity and electrical conductivity of molten salt solutions are reviewed for crystal growth of yttrium iron garnet by Timofeeva and Bykov [17]. A short review was published on flux growth technology by Tolksdorf [18].

A comparison between melt and flux growth is presented in the review article *Crystal Growth from Liquids at High Temperatures* by Brice [19]. The technical need for ferrimagnetic garnet single crystals and layers has led to an increasing understanding of the problems concerned with growth from HTS. Excellent reviews on this topic, presented by Görnert, are *High Temperature Solution Growth of Garnets: Theoretical Models and Experimental Results* [20], *Fundamental Aspects of Flux Growth* [21] and *Kinetics and Mechanisms of Flux Crystal Growth* [22] all showing the strong relation between bulk growth and layer growth.

2. *Fluxes for oxide materials*

The following properties are desirable for the high temperature solution and the flux:

(1) The material to be grown as a single crystal must be the only stable solid phase under the growth conditions.
(2) The solubility of the crystal or its components in the flux should not be too low; in general between 5 and 50 wt%, and should decrease with temperature.
(3) The flux should have a low melting point.
(4) The flux or its components should have a very low solubility in the grown crystal.
(5) The flux should have a low vapour pressure.
(6) The viscosity of the solution should be low, in the range of 1 to 10 centipoise (cP).
(7) The solution should not attack the growth container.
(8) The solution should be easily prepared.
(9) The residual melt should be easily separated from the crystals.

Summaries of fluxes used and the oxide crystals grown from them are given e.g. by Elwell [12], Wanklyn [15, 23, 24] and Elwell and Scheel [1]. The commonly used components of the flux are the basic oxides or fluorides: PbO, PbF_2, BaO, BaF, Bi_2O_3, Li_2O, Na_2O, K_2O, KF and the acidic oxides B_2O_3, SiO_2, P_2O_5, V_2O_5, MoO_3; in most cases a mixture consisting of two or three of them.

The selection of the optimum solvent for the growth of a given material is a difficult task frequently based on trial and error. Detailed investigations of the complex multicomponent phase relations for the material to be grown are in most cases not possible. One can get some initial information from existing phase diagrams of the oxides of the complex compound. Very helpful is a compilation of *Phase Diagrams for Ceramists* from 1964 with supplements to 1984 [25].

The *Use of Phase Diagrams in Crystal Growth* [26] is a chapter in a treatise in three volumes on phase diagrams for material science and technology. Wanklyn and her colleagues in the Clarendon Laboratory in Oxford have grown, from HTS, an enormous number of crystals with various compositions for investigation of their physical properties. From this experience she tried to formulate some empirical rules for the starting composition for the growth of many complex oxide crystals [23, 24]; Fewer, larger and more nearly equi-dimensional crystals are formed when (a) the excess of the acidic oxide component needed to produce the required phase in the flux is reduced to a minimum and (b) when a part of the basic oxide component is replaced by the corresponding fluoride and (c) when a mixture of two different acidic oxides is used. The fluoride increases the solubility. A higher molar ratio of basic to acidic oxides will favor the growth of fewer, and therefore, larger crystals [16]. A publication on application of the Wanklyn model to the $BaO-BaF_2-B_2O_3-Fe_2O_3$ flux system was presented by Paroli [27]. The effect of lead ions was discussed by Wanklyn and Watts [28] using WO_3 as a model sulute in flux growth.

The role of B_2O_3 was discussed by Elwell and Coe [29]. Using the simplified "hard sphere" liquid model they found, from measured expansivity and surface tension data, strong PbO/B_2O_3 complexing with the F^- ions bonded less strongly, and a high degree of solvation. Coe and Elwell [30, 31] published measurements of the viscosity and the volatility of $PbO/PbF_2/B_2O_3$ solvents and $Y_3Al_5O_{12}$ solutions.

Solubility models has been proposed by van Erck [32, 33] for garnet solutions in PbO/B_2O_3. Nitsch et al. [34] checked the validity of these models and came to the conclusion that a satisfactory explanation of the behaviour of B_2O_3 in this system has not yet been achieved. Fundamental aspects of the flux for garnet growth are described in [21]: In mixtures the basic oxide loses oxygen ions and the acidic oxide adopts them (association). A classification of typical high temperature solvents for oxide materials is presented. The solute oxides can form both bare cations and complex anions depending on the basicity of the melts. As an example the melt of a mixture of Fe_2O_3, PbO and B_2O_3 is discussed in [21]. The basic PbO will dissociate into Pb^{2+} and O^{2-}, the acidic B_2O_3 associate three O^{2-} forming two BO_3^{3-} anions. The amphoteric Fe_2O_3 might associate oxygen ions forming FeO_m^{3-2m} complexes. The free oxygen ions clearly influence the concentration and the coordination number of cation oxygen complexes. Hence the free oxygen ion concentration $[O^{2-}]$ should be a key parameter for the species actually present in the melt and for the phase boundaries. For $Y_3Fe_5O_{12}$ in a flux of PbO, PbF_2 and B_2O_3, a dissociation into three Y^{3+}, two Fe^{3+} and three FeO_4^{5-}

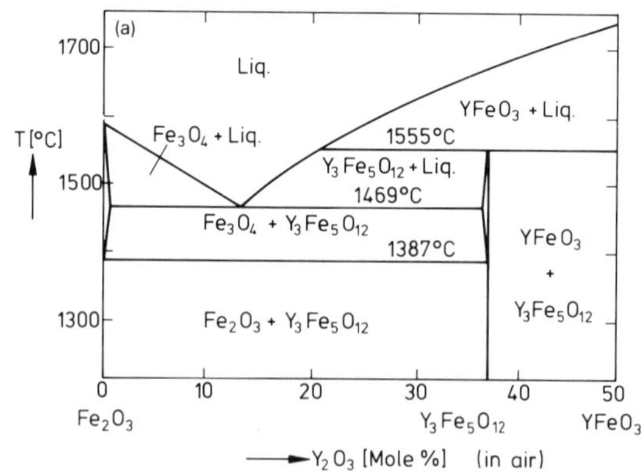

Fig. 1. (a) Simplified iron-rich part of the phase diagram of Fe_2O_3 and Y_2O_3 [42].

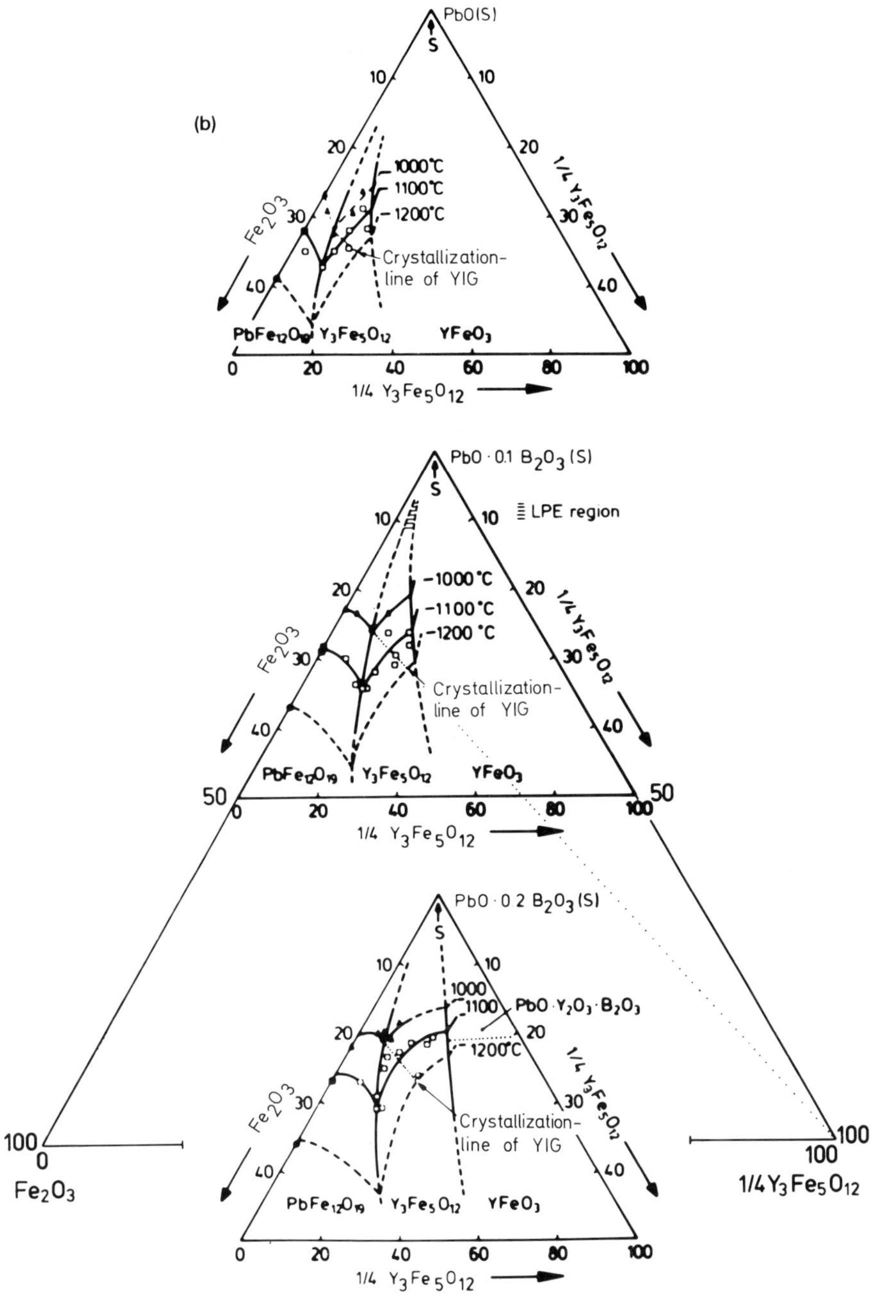

Fig. 1. (b) Pseudo-ternary phase diagram of the system. $PbO \cdot nB_2O_3–Fe_2O_3–Y_2O_3$ for YIG crystallization [45].

is suggested [22]. But even in the case of the very well studied garnets this dissociation and association example is still a suggestion and further experimental studies are required to definitely determine the species actually present in the melt. The correlation between the crystallization field of YIG and the oxygen ion concentration of HTS has been confirmed by Fischer and Sinn [35]. They found that, at a given temperature, the oxygen ion concentration essentially determines the nature of the crystallization phase and its solubility.

However, for a new material to be grown, the best way is to look in the literature to find an analogous compound. But sometimes even small changes make a big difference. When the author tried to grow $Yb_3Fe_5O_{12}$ instead of $Y_3Fe_5O_{12}$ (YIG) from the same flux, it was successful in small unstirred crucibles, but using five times larger melt volumes and forced convection no garnet crystals grew but crystals of an unknown compound were obtained instead. As in this example, such failures are mostly not published. The historical development of the flux for the incongruently melting ferrimagnetic YIG might be of interest. Nielsen and Dearborn [36] published some phase investigations of the pseudo-binary system Fe_2O_3 and Y_2O_3 with respect to YIG. They found that $Y_3Fe_5O_{12}$ melts incongruently and $YFeO_3$ congruently at temperatures above 1580°C and 1720°C, respectively. They knew that their colleague Remeika [37] had grown single crystal rare earth orthoferrites e.g. $YFeO_3$ from molten lead oxide. So they selected lead oxide as a solvent for yttrium iron garnet (YIG). Now they had to look into the pseudo-ternary system with PbO and found that a higher iron to yttrium ratio than that of garnet had to be used. In the melt composition Nielsen [38] used, magneto-plumbite crystals grew together with garnets. He showed that YIG incorporated a substantial amount of bismuth, when Bi_2O_3 was used instead of PbO. He published an improved flux in which a part of the PbO is replaced by PbF_2 [39]. An important improvement of this flux was achieved by adding a small amount of B_2O_3 as reported in [40, 41]. Only the technical interest in the garnets allowed this development.

Frequently excess of one component of the compound is used as a flux. In the case of YIG the growth temperature in air is between the eutectic (1469°C) and the peritectic (1555°C) temperature [42]. The iron-rich part of the simplified pseudo-binary phase diagram is shown in fig. 1a. Such high temperatures are not easy to control for crystal growth. Abernethy et al. [43] reported, rather soon after the proposed phase diagram for YIG [36], the growth of YIG single crystals using excess iron oxide and a travelling solvent zone. The concentration of the solute in a melt with excess of one component is mostly much higher than in other fluxes. Consequently the maximum growth rate can be higher too, reaching in some cases that of melt growth.

Nair [44] used lead sulphate instead of lead oxide in the flux growth of YIG crystals. He reported that lead sulphate decomposes during the growth and the yield is nearly double that as compared to when $PbO/PbF_2/B_2O_3$ flux is used.

The influence of the addition of B_2O_3 to PbO as a flux for YIG has been investigated by Jonker [45]. Figure 1b shows his results in the form of pseudo-ternary phase diagrams. For YIG one quarter of the formula unit was chosen as a molar concentration to have the same cation to oxygen ratio as in Fe_2O_3. It is clear from fig. 1b that excess Fe_2O_3 is necessary for being in the crystallization field of YIG. The primary crystallization field of YIG is enlarged by the addition of B_2O_3. The addition of PbF_2 enlarges this field much further. Even with the ratio $Fe/Y = 5/3 \approx 1.67$ in the melt, garnet is the only phase to precipitate in a certain temperature range [46].

3. Crucibles

Elwel and Scheel [1] present a list of, together with the noble metals, other crucible materials such as silica glass, graphite, nickel, molybdenum and tantalum. In most cases a reducing, or at least an inert, atmosphere has to be used. In the flux growth of most oxides and oxide compounds, platinum has found by far the widest application for use as a container material, despite its high price and softness, because of its high melting point and high resistance to the, mostly necessary, oxidizing atmosphere and to most of the flux components. It is reasonable to use a wall thickness of 1 mm for better handling and longer lifetimes. The capital investment as well as the refabrication costs for the expensive noble metal are very high but, with some care, the lifetime can be long. Pure grade platinum seems to be more resistent to attack by fluxes and is recommended if welding is necessary. Since this material is rather soft at high temperatures, it is very helpful to embed the platinum crucible in refractory materials. This might be e.g. an alumina crucible or pipe with end plates, which can be fastened by splints. The gap between the platinum crucible and the alumina support should be filled with alumina powder or fiber material. An alternative support is a brittle refractory brick in which hollows have been made [15]. A second brick, with matched hollows, can be used as a lid. In addition to the mechanical stability, the thermal capacity of the ceramic damps temperature fluctuations. Flux vapour or escaped melt from the platinum crucible is absorbed by the surrounding ceramic, protecting the furnace to some extent.

Fluxes containing PbO or Bi_2O_3 attack platinum severely. Certain precautions have to be taken. Non-stoichiometry of the oxides or contamination with organic compounds, such as lead acetate, can form metallic bismuth or lead which causes damage by forming a low melting alloy with the crucible. Slow heating is recommended [47]. Platinum transport in PbO or PbO/PbF_2 containing fluxes is observed quite frequently forming a kind of sponge at the bottom or sheets at the side wall of the crucible. Some crucibles are fabricated from metal with some additions such as very small particles of high melting oxides added to increase high temperature stability and prevent grain growth. These are good for BaO/B_2O_3 melts but, in combination with more corrosive melts like those containing PbO or Bi_2O_3, these additions tend to dissolve and cause leakage of the crucible. Alloys of rhodium and platinum make the crucible mechanically more stable at higher temperatures. But rhodium contaminates the melt and hence the crystal and this is unacceptable e.g. in YIG for microwave applications. Unstable oxides like iridium oxide contaminate the crucible and consequently the crystals grown in such a contaminated crucible [48]. Platinum crucibles should be cleaned in hot dilute hydrochloric or nitric acid to remove all particles from polishing and then preheated to about 1200°C. After use, the residual melt can be drained by heating the crucible upside down with a proper container to collect the melt. Adhering materials, that cannot be removed otherwise, may be dissolved in melts of potassium pyrosulfate at $\approx$ 700°C or borax at $\approx$ 1000°C.

With the necessary care, platinum crucibles can have a very long lifetime. For liquid phase epitaxy of garnets, a 12 kg melt with a flux of $PbO/Bi_2O_3/B_2O_3$ was kept for more than 12 months at temperatures above 900°C, except for the growth period when the growth temperature was 780°C. At the surface region of the melt the wall of the crucible was rough and noticeably thinner than the 1 mm thick wall outside the melt. Temperature changes and above all, frequent solidification of the melt, reduces the lifetime of a platinum crucible. Fluxes containing PbF_2 are much less corrosive. Crucibles have been used for garnet growth with a $PbO/PbF_2/B_2O_3$ flux for more than 10 runs, each of them lasting for about three weeks at high temperature.

4. *Equipment*

The equipment has to be selected according to the material to be grown and the technology to be applied. First of all the maximum temperature is determined by the heating elements, which for oxides should preferentially work in an oxidizing atmosphere. The temperature stability around the growing crystals should be as high as possible, certainly better than ± 0.05 K. A flexible programmer is necessary. Historically very often the total equipment had to be built by the operator with the help of a workshop. Elwell and Scheel [1] give a detailed overview on furnaces, heating elements, properties of firebrick and thermocouples. For slow cooling growth, mostly rectangular muffle furnaces and vertical tube furnaces are used. Today there are many kinds of these commercially available. Good temperature control and program units have been developed for these furnaces and they are available on separate order, too. The furnace must be easy to repair, since there is very often damage to bricks or heating elements by the melt which escapes out of the crucible by evaporation, creeping or leakage.

5. *Methods*

As in aqueous solutions, there are three ways to achieve the required supersaturation for crystal growth, when one ignores some special methods like electrolytic growth or solid state reactions. Supersaturation and supercooling are defined by the phase diagram. Figure 2 shows schematically a pseudo-binary system of flux and solute and the three main ways for crystal growth.

(1) Crystal growth by slow cooling (SC) when the temperature is decreased from A to B and growth occurs along the arrow to C at varying growth temperature T_g.

(2) Crystal growth by flux evaporation (Evap) when the solvent or a part of it is evaporating and the solute concentration increases from D to E. T_g is constant.

(3) Crystal growth by vertical temperature gradient transport (VTGT) when the crystal grows in a cooler region G and the crystallized material is replaced at a hotter region F generally from a polycrystalline nutrient. T_g is constant.

Sometimes a combination of the three possibilities is applied, e.g. when the flux has a high vapour pressure and part of it evaporates during slow cooling or a crystal is

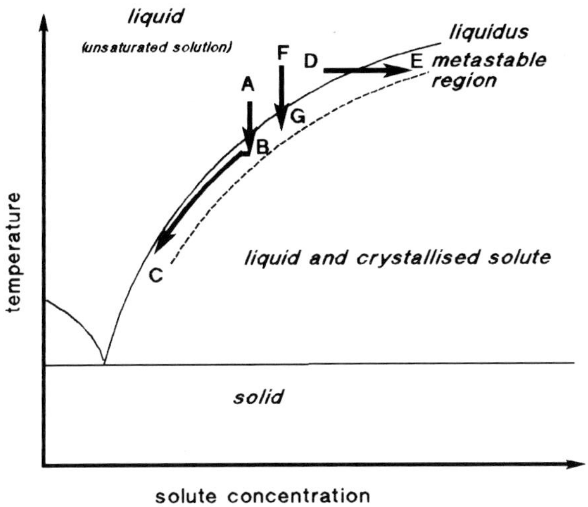

Fig. 2. Binary phase diagram showing schematically three possibilities for crystal growth from HTS. Slow cooling: A–B–C, flux evaporation: D–E and temperature gradient transport: F–G.

growing by slow cooling in a melt in which source material is present. There is no general rule which method is the best for a given material. The selection of the method depends strongly on the facilities available to the operator and the required quality – size, purity, homogeneity – of the crystals. Very frequently various methods have been applied for the same material as it will be shown in detail for garnets. Elwell and Scheel [1] present in their book a long list in alphabetical order of the crystals which have been grown together with the method used and the reference of the source publication.

5.1. Slow cooling (SC)

One possibility to create the required supersaturation for the growth of crystals is to cool a melt from point A slowly below the liquidus temperature (T_l) into the metastable region (Ostwald–Miers region) as demonstrated in fig. 2. To reach the temperature for spontaneous – three-dimensional – nucleation (T_n) at point B a larger supersaturation or supercooling ($\Delta T = T_l - T_n$) is necessary than for growth on a nucleus. Neuhaus [49], as an example, showed results for the organic compound "salol" (salicylic acid phenyl ester) which melts at 41.7°C. First nuclei are formed about 40 K below the liquidus temperature with the maximum number of nuclei at 65 K. The maximum of the linear growth rate is at 15 K below the melting temperature. For low temperature solutions the required supercooling is generally less than for melts. However, supercooling of 25 to 50 K was measured by differential thermal analysis (DTA) and thermogravimetric analysis (TGA) [50]. Wanklyn and Garrard [51] have shown that the number of spontaneous nuclei mainly depend on the acid–base ratio in the melt. Nucleation in flux growth is usually heterogeneous at the wall of the platinum crucible. The number of nucleation sites on platinum can be reduced by a long soak period above the liquidus temperature [52]. Under favourable conditions only a few nuclei are formed, since, even for heterogeneous nucleation, a larger supercooling is necessary than for growth on existing crystal faces. For atomically rough faces this supercooling is even lower. It was e.g. observed, that YIG layers grew on (111) gadolinium gallium garnet substrates by LPE from PbO/B_2O_3 fluxes even when crystals of yttrium orthoferrite had nucleated because of insufficient excess of iron oxide. Crystals grow at the lowest temperature, so it depends on the unavoidable temperature gradient, whether the cooler region is at the bottom of the crucible or near the surface of the melt.

5.1.1. *Slow cooling bottom growth (SCBG)*

This is the most commonly applied method. Crystals grow at the bottom, or on the lower side wall, of the crucible when its temperature (T_{bot}) is lower than the temperature at the surface (T_{surf}). The crucible is placed in a muffle furnace or a vertical pipe furnace in a position with the necessary temperature gradient, $\delta T = T_{surf} - T_{bot}$. The temperature is increased well above the liquidus temperature and kept there for a while (soaking time) to achieve complete dissolution of all components. The temperature is lowered to a temperature just above the liquidus temperature and then very slowly further to get supersaturation for nucleation. in some cases the temperature is increased again somewhat to reduce the number of nuclei. Then growth occurs by continuous cooling with very low rates because at the beginning of the growth the crystal surface area is small. When the crystal surface has become larger the cooling might be enhanced even more than proportional to the surface area, since usually the liquidus curve is not linear but the material precipitating from a given volume decreases with T_g at constant cooling rate.

Growth should be terminated at least before the solution solidifies. This is usually done by fast cooling. The crystals must be separated from the solidified melt as far as possible mechanically and the residual melt might be leached off with a suitable solvent. It is better to separate the crystals from the solvent whilst it is still molten at the termination of growth. To avoid a thermal shock this is best done within the furnace by inverting the crucible, since mostly the crystals stick to the platinum wall. When sealed containers of a sufficient size are used, the residual melt is saved and can be analyzed or used again after correction of the concentration. As an example, growth of garnet crystals will be described in more detail, followed by other materials where only the specialities will be mentioned.

5.1.1.1. Iron garnets by SCBG. YIG ($Y_3Fe_5O_{12}$) melts in air incongruently via a peritectic decomposition into yttrium orthoferrite ($YFeO_3$) and an iron-rich liquid at 1555°C (see fig. 1). As mentioned in section 2, lead oxide based fluxes have been used for crystal growth of YIG, preferentially by slow cooling as indicated in table 1 in which a selection of publications using various fluxes is given. Tables 1 and 2 show PbO and BaO based fluxes, respectively. In table 3 excess Fe_2O_3 is used as a flux.

The commonly used flux is a mixture of $PbO/PbF_2/B_2O_3$ with an excess of Fe_2O_3. It was used for technical growth in the Bell Laboratories [41]. A large (8.4 ℓ) platinum crucible with a diameter of about 20 cm and wall thickness of 1.5 mm was filled to about

Table 1
Garnets, PbO-based fluxes.

Composition	PbO	PbF_2	A	M/R	Remarks	Reference	Method
$Y_3Fe_5O_{12}$	52.00		–	12.30	PbO evap.	[38]	SCBG
$Y_3Fe_5O_{12}$	30.00	40.00	–	2.80	PbF_2 evap.	[39]	SCBG
$Y_3Ga_5O_{12}$	39.40	40.80	–	2.40	PbF_2 evap.	[39]	SCBG
$Y_3Fe_5O_{12}$	28.38	25.83	–	1.67	surface crystals	[53]	SCBG
$Gd_3Fe_5O_{12}$	45.90	–	5.67	9.00	skeletal	[54]	SCBG
$Nd:Y_3Al_5O_{12}$	29.70	33.06	8.34	3.03	5 kg	[40]	SCBG
$Y_3Fe_5O_{12}$	36.62	27.27	11.74	2.00	15 kg	[41]	SCBG
$Nd:Y_3Al_5O_{12}$	29.70	33.04	8.34	1.60	10 kg	[41]	SCBG
$Y_3Fe_5O_{12}$	38.00	50.00	1.32	1.67	4.2 μm/min	[55]	SCBG
$Y_3Fe_5O_{12}$	36.62	27.27	11.74	2.00	ACRT, 15 kg	[56]	SCBG
$Y_3Fe_5O_{12}$	48.81	29.85	12.64	0.68	PbF_2 evap.	[57]	SCBG
$Y_3Fe_{5-x}Ga_xO_{12}$	35.00	30.00	–	2.72	ACRT, seed	[58]	VTGT
$Y_3Fe_5O_{12}$	36.70	27.30	11.85	2.00	ind. stria.	[59]	SCBG
$Y_3Fe_5O_{12}$	74.04	–	11.85	15.30	ind. stria.	[59]	VTGT
$Y_3Fe_5O_{12}$	36.70	27.30	11.85	2.00	ACRT, local cooling	[60]	SCBG
$Y_3Fe_5O_{12}$	68.63	–	5.00	5.03		[45]	SCBG
$Nd:Y_3Al_5O_{12}$	29.70	33.04	8.39	1.60	5 kg	[61]	SCBG
$Y_3Fe_5O_{12}$	38.25	28.45	12.35	1.67	ACRT, seed	[62]	SCBG
$Y_3Al_5O_{12}$	49.68	36.96	8.50	1.67	ACRT	[63]	SCBG
$Y_3Ga_5O_{12}$	38.25	28.45	12.35	1.67	ACRT, seed	[64]	SCBG
$Mn:Y_3Fe_5O_{12}$	64.45	–	5.00	5.00	ACRT, $T_g = 1436°C$	[65]	TSSG

PbO and PbF_2 in mol%; A = $(PbO + PbF_2)/B_2O_3$, molar ratio; M/R = molar ratio, R = Y or (Y + Nd), M = Fe, Al or Ga. SCBG = slow cooling bottom growth; TSSG = top-seeded solution growth: VTGT = vertical temperature gradient transport.

Table 2
Garnets, BaO-based fluxes.

Composition	BaO	BaF_2	A	M/R	Remarks	Reference	Method
$Y_3Fe_5O_{12}$	47.71	–	1.59	2.27	c.r. = 1	[66]	SC
$Y_3Fe_5O_{12}$	47.71	–	1.59	2.27	$f = 0.9$	[67]	VTGT
$Y_3Fe_5O_{12}$	42.5	–	1.60	1.99	$f = 0.7$	[68]	VTGT
$Y_3Fe_5O_{12}$	42.5	–	1.60	1.99	$f = 0.8$	[69]	VTGT
$Y_3Fe_5O_{12}$	47.71	–	1.59	2.27	$f = 0.9$	[70]	TSZ
$Y_3Fe_5O_{12}$	47.71	–	1.59	2.27	$f = 0.7$	[70]	VTGT
$Y_3Ga_5O_{12}$	28.35	12.44	1.44	1.67	$f = 12$	[71]	VTGT
$Y_3Fe_{4.5}Ga_{0.5}O_{12}$	45.53	–	1.61	2.10	rotating seed	[72]	TSSG
$Y_3Fe_5O_{12}$	62.00	–	1.63	1.67	$f = 17$	[72]	VTGT

BaO and BaF_2 in mol%; A = $(BaO + BaF_2)/B_2O_3$, molar ratio; M/R = molar ratio, R = Y or (Y + Nd), M = Fe, Al or Ga; c.r. = cooling rate (K/h), f = growth rate (μm min).

Table 3
Garnets, excess Fe_2O_3, travelling solvent zone (TSZ)

Composition	Heating	Diam.	f	Remarks	Reference
$Y_3Fe_5O_{12}$	RF	6	42	$p_{O_2} = 3–7$, $\Delta H = >1$	[43]
$Y_3Fe_4Ga_1O_{12}$	rad.	8	17	length > 7 cm	[94]
$Y_3Fe_{5-x}Ga_xO_{12}$	rad.	8	8	$x = 0.8$, $\Delta H = 1.6$	[95]
$Y_3Fe_{5-x}Ga_xO_{12}$	rad.	8	25	$p_{O_2} = 20–50$, $x = 0$, $\Delta H = 1$, $0 < x < 1$	[96]
$GdFe_5O_{12}$	rad.	6	25	$p_{O_2} = 50$	[96]
$Y_3Fe_5O_{12}$	rad.	10	17	$T_g = 1500°C$, $p_{O_2} = 1$, $\Delta H = 0.4$	[97]
$Y_3Fe_5O_{12}$	RF		16	TSSG, 1 K/h, no facets	[91]
$Y_3Fe_5O_{12}$	rad.	1	250	laser-heated fiber growth	[98]

Diam = diameter of the single crystal rod (mm), f = growth rate (μm/min),
RF = radio frequency coil, rad. = radiation heating by the focus of a light source,
p_{O_2} = oxygen pressure (atm), ΔH = ferromagnetic resonance line width at 9.3 GHz (Oe).

one quarter of its height of 25.4 cm with more than 15 kg melt having a molar composition: PbO 36.62, PbF_2 27.27, B_2O_3 5.44, Fe_2O_3 20.38, Y_2O_3 10.19, and CaO 0.10. CaO was added for higher optical transmittance, but it turned out that nucleation of garnet was hampered without CaO. The crucible was closed but not sealed with a lid and placed in a vertical pipe furnace on a pedestal through which air was blown to cool the bottom of the crucible. To enhance mixing the crucible was rotated during the soak period at about 1300°C. Garnets were grown from 1300°C with a cooling rate of 0.5 K/h. To drain the flux at 950°C from the crucible a hole was punched in the bottom with a spear through the pedestal. They obtained as many as five 200-g bottom-grown crystals and a total yield of 2800 g. Since the vapour pressure of PbF_2 at 1297°C is 760 Torr there was a substantial loss of flux by evaporation. The majority of the crystals nucleated at the junction between surface and crucible wall. The large bottom-grown crystals had flux entrapments in the center indicating dendritic growth after nucleation at high supersaturation. Many of the smaller bottom-grown crystals were completely inclusion free and probably nucleated at a lower supersaturation.

This method was copied by a Russian team [56]: During the growth period they rotated the crucible at 25 min^{-1} for 25 s and a pause of 5 s and reversed the sense of rotation and they reported that the temperature at the bottom was higher than without the reversed rotation indicating forced convection. They found improved crystal quality [74] as far as dislocation density is concerned. The bottom-grown crystals had a dislocation density of about 10^2 cm^{-2} whereas crystals grown at the surface as a result of evaporation of the flux had a dislocation density of about 10^4 cm^{-2}.

Without knowledge of this work of the Russian group Scheel and Schulz-DuBois [75] developed a stirring technique which can be applied in conjunction with localized cooling and enables control of nucleation in closed crucibles for crystal growth from HTS. Forced convection is achieved by periodically accelerated and decelerated rotation of the crucible. Detailed information about the stirring effect of this accelerated crucible rotation technique (ACRT), the kinetics of fluid flow, material and heat transport (hydrodynamics) has been presented by Schulz–DuBois [76] and Elwell and Scheel [1]. Due to the temperature gradient for bottom growth and when a flux with a high specific weight is used, natural convection is suppressed, because the cooler and less concentrated melt at the bottom has a higher density than the melt at the surface. In this case ACRT is certainly very profitable. Brice et al. [77] published a review of models for ACRT.

Bennett [69] and Tolksdorf [78] have shown that crystals can be separated from the flux at elevated temperature by inverting the crucible within the furnace using a volume of the crucible twice as large as that of the melt. The residual melt can be used again. By doing the inverting twice even a seed crystal can be used.

In a crystal growth group of the Philips Research Laboratory in Hamburg an economical technique was developed combining ACRT, local cooling, separation of the crystals from the flux and the use of a seed crystal as shown in fig. 3 [62]. Fig. 3a is a schematic drawing of the crystal apparatus. It is a top-heated chamber furnace controlled by three thermocouples at the SiC heating elements. With a temperature-stabilized air flow around the outside of the brickwork achieving a temperature

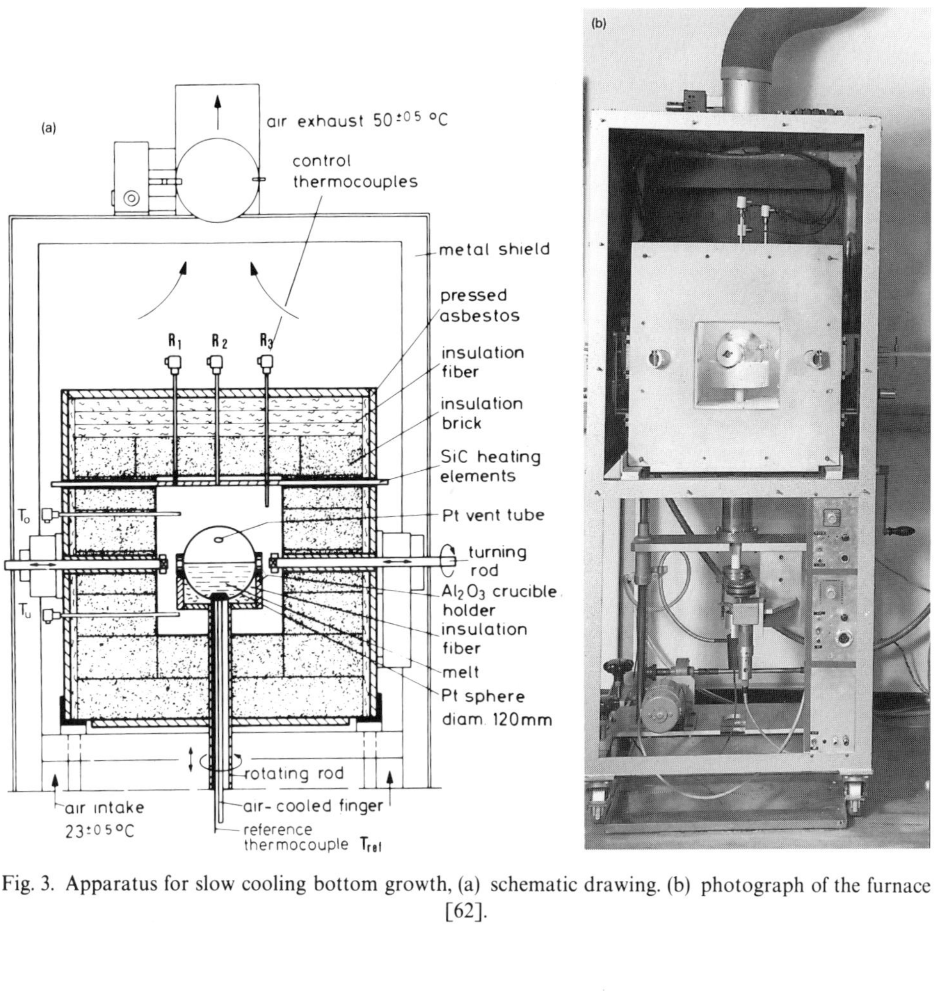

Fig. 3. Apparatus for slow cooling bottom growth, (a) schematic drawing. (b) photograph of the furnace [62].

stability of ± 0.02 K over long periods (> 5 h) at the bottom of the crucible, cooling rates of less than 0.1 K/h were possible. An air-cooled finger consists of a closed outside alumina tube and a smaller inside pipe through which air with a controlled velocity and constant temperature is blown. This air stream flows out through the outer tube, without causing turbulence in the furnace chamber. The temperature at the bottom of the crucible near the point of contact of the air-cooled finger is monitored by the reference thermocouple T_{ref} even when the crucible is rotated around its vertical axis. The reading of this thermocouple is considered as T_g. The drawing shows a spherical platinum crucible sitting on an alumina support fixed to a vertical rotating ceramic rod. The crucible and the support can be exchanged when other melt volumes should be used. The diameter of the crucible is 12 cm, the wall thickness is 1 mm. With 3 kg melt e.g. of the molar composition, PbO 38.25, PbF_2 28.45, B_2O_3 5.40, CaO 0.10 as a flux and Fe_2O_3 17.38 and Y_2O_3 10.42 (note the ratio 5:3) the crucible is just half filled. A photograph of the furnace is shown in fig. 3b and the arrangement inside the furnace is shown in fig. 4. The bearings of the horizontal axes, made from 1 mm thick platinum welded onto the crucible, are supported by ceramic. Two viewports through the sidewalls of the furnace make it possible to see whether the axes fit into the bearings; however a marker on the cold outside wall of the furnace and on the vertical axis is necessary. The small filling tube (12 mm diameter) of the crucible is completely closed by a welded lid bearing a 70 mm long spiral tube with an inner diameter of 1.6 mm as a vent for pressure release and as a diffusion barrier. Loss of flux by evaporation was not detectable (< 0.1 wt%) after closing the crucible. An airstream of 170 ℓ/h through the cooling finger results in a temperature decrease of 20 K at T_{ref} at a soaking temperature of about 1200°C. To apply ACRT, the rotation of the crucible is increased for 11 s to a

Fig. 4. A spherical platinum crucible at the growth position in a chamber furnace [62].

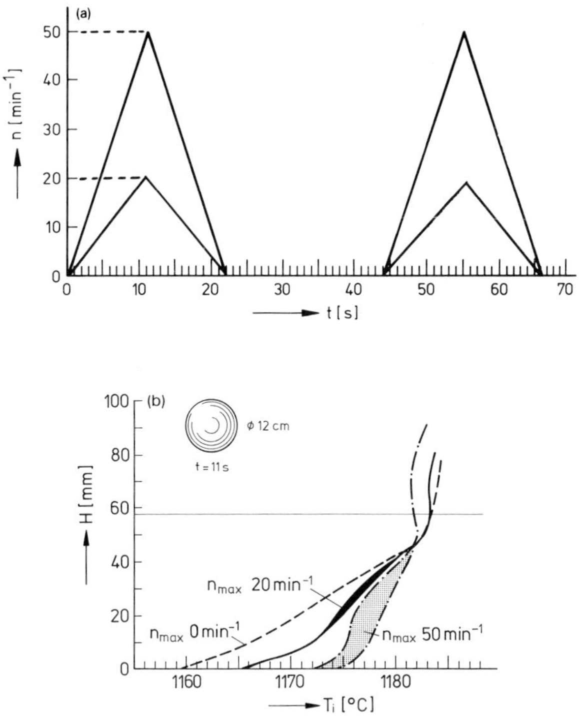

Fig. 5. ACRT with a spherical crucible of 12 cm diameter. (a) ACRT program, (b) vertical, central temperature distribution in the crucible. The horizontal line marks the melt surface [81].

maximum e.g. $n_{max} = 20$ or $50\,\text{min}^{-1}$, decreased for 11 s and kept completely stationary for another 22 s. After this pause rotation is started again in the same direction. The period of ACRT thus is 44 s as shown in fig. 5a. The vertical, central temperature distribution for these two ACRT programs in comparison with the static one are given in fig. 5b. The increased heatflow by forced convection results in a temperature increase of about 7 or 15 K, respectively. The variation of T_i is synchronous with the period of ACRT. The temperature oscillation at T_{ref} is ± 0.15 K for $n_{max} = 20\,\text{min}^{-1}$ getting smaller during the growth run. This oscillation might be reduced by reducing the stationary temperature gradient in the melt e.g. by additional heating elements at the lower side of the furnace chamber. The ACRT has to be optimized for any changes of the crucible. Details of the influence of the crucible forms on the temperature distribution are published by Tolksdorf [46, 62, 78, 79].

The spherical shape of the crucible is not only optimal for forced convection by ACRT but also for the use of a seed and the separation of the crystal from the melt. A seed crystal can be fixed just above the melt surface. After spontaneous nucleation has occurred by slow cooling – the melt is now saturated – the turning rods are introduced from the side walls of the furnace and the crucible is turned around the horizontal axis about 120° so that the seed crystal is on top of the cooling finger. Simultaneously the crystals grown by spontaneous nucleation are taken out of the melt. With a slow

Fig. 6. Seed crystal (9 g), before the growth run, screwed onto the flat area of the crucible segment [64].

turning velocity of e.g. 25°/min the liquid does not move, so that the temperature is not changed. It is obvious that, during the turning operation, it is impossible to rotate the crucible around the vertical axis. By turning the crucible after growth back into the starting position the grown crystal is separated from the melt, whilst the material crystallized by spontaneous nucleation is brought back into the melt. In the growth and starting positions the filling tube is above the melt level.

The fastening of the seed crystal is very important. To avoid any sharp points which may cause nucleation and to get good contact for heat flow, a crystal with undisturbed natural facets from spontaneous nucleation from a former run in milled flat on its lower side. The upper side has only growth facets. With a threaded hole made by ultrasonic drilling, the crystal is screwed tightly on a 3 mm diameter and 3 mm long platinum screw in the center of a flat area of a cap-like crucible segment which has been cut off from the filled platinum crucible as shown in fig. 6. This segment is again butt-welded into the crucible. The platinum screw acts as a heat sink within the seed crystal. Figure 7 shows the YIG crystal of 96 g weight grown on the seed shown in fig. 6. It should be noted, that the introduction of the seed crystal into the melt, even by the rather elegant method described above, always causes disturbance of the growth conditions. To turn the crucible, the vertical support has to be lowered by some mm, interrupting the thermal contact for about 15 min and stopping the accelerated crucible rotation. To minimize this disturbance a rather complicated program related to T_{ref} for the heating power, the air stream through the cooling finger and the crucible rotation has to be used. Such a program is shown in fig. 8. In this program standard conditions for the cooling rate of the furnace (0.2 K/h) were reached 74 h after the introduction of the seed. For the crystal shown in fig. 7 an improved program had been used shortening the interim time to 7 h [64]. Seeding temperature was 1080°C. The maximum rotation rate n_{max} was continuously increased reaching $n_{max} = 20\,\mathrm{min}^{-1}$ within 2 h. The cooling airstream was programmed to avoid any increase in T_{ref}. The cooling air stream was further increased to reach a final value of 170 ℓ/h after 5 h. During this period of 5 h the heating of the furnace was programmed to keep T_{ref} constant. Seven

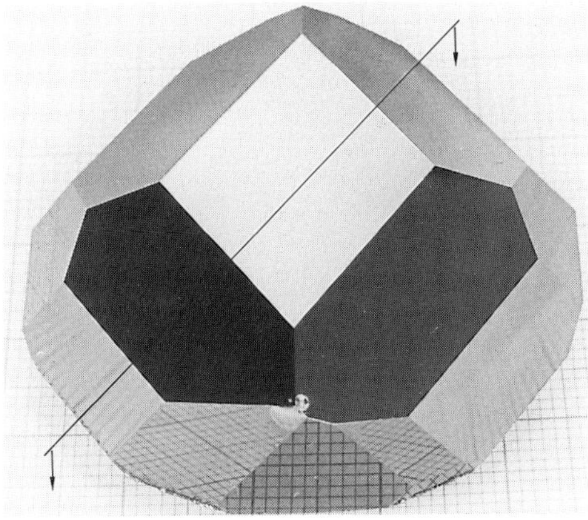

Fig. 7. YIG crystal of 96 g grown on the seed shown in fig. 6.

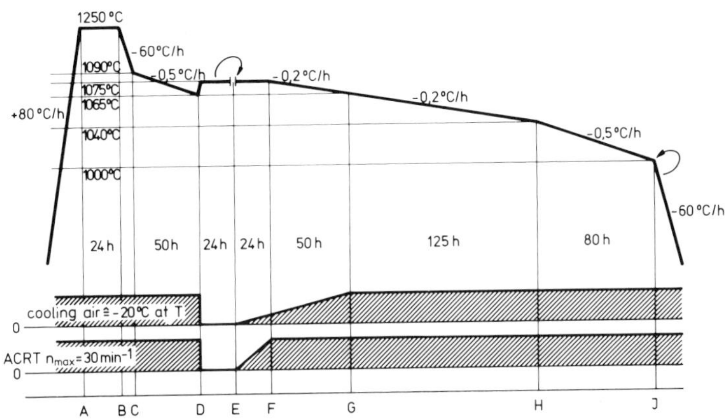

Fig. 8. Temperature program for SCBG using a seed [62]

hours after seeding standard conditions were reached again and slow cooling started at 0.15 K/h to 1060°C followed by 0.2 to 1000°C, when the crucible was turned back to its starting position. The flux used for this example was the same as mentioned above. The yield of 96 g was small compared to the yield of about 250 g usually achieved from a 3 kg melt as shown in fig. 9. In this example two smaller crystals nucleated and grew together with a large one. The lowest part of this crystal (fig. 9 at the right side) was still in contact with the melt after turning the crucible out of the growth position showing disturbed growth. A solidified droplet of the residual melt is seen near a [100] direction in fig. 9 at the lowest part of this crystal. Only in {100} directions is the crystal wetted by the flux as found by Tolksdorf and Bartels [80] using spherically shaped seed crystals.

Fig. 9. 250 g YIG crystal from a 3 kg melt [62].

As mentioned above, other crucibles can be used in the furnace shown in fig. 3 [81]. In a crucible of 160 mm diameter with an 8 kg melt of the already described composition a yield of more than 600 g has been achieved. In a cylindrical shaped crucible with a diameter of 7 cm, the bottom being bent to a radius of 6 cm, a 1 to 1.2 kg melt has been used. An assembly of four crucibles, each with a capacity of 150 g of melt, has been applied for substitution experiments.

To force nucleation on a small area, a cooled tip at the center of the bottom can be used as shown in fig. 10 for the growth of YIG. A tip was used by van Uitert et al. [61] for the growth of large yttrium aluminium garnet crystals. Scheel and Schulz-DuBois [75] grew a crystal of $GdAlO_3$ weighing 210 g in a 500 cm^3 crucible from a $PbO/PbF_2/B_2O_3$ solution on a cooled tip.

Reversing the crucible can be used to determine the saturation concentration (liquidus) at various temperatures. For this the yield has to be taken out of the crucible and the run is started again without refill. To be sure that no crystals have floated to the surface, Jonker [45] used a perforated platinum sheet above the melt surface for a PbO/B_2O_3 flux for YIG. This flux, without fluorine, might be used for substitutes forming gaseous compounds like SiF_4 or GeF_4. The formation of these gases causes irreproducible losses during melting.

Small and short temperature changes can be used to vary the growth rate for a short period. This induces compositional striations (markers) in the crystals which can be made visible and allow the determination of the growth rate as described for the growth of $YFeO_3$ [82] and for YIG [59 and 20]. Kvapil [55] minimized flux evaporation by a hydraulic seal.

Because of its low vapour pressure, BaO/B_2O_3 has been investigated for the growth of garnets by Linares [66].

5.1.1.2. Other oxides by SCBG. Due to its relative simplicity, SCBG is usually the method to start with. More than one crucible might be put simultaneously in the furnace. If one component of the melt has a high vapor pressure, sealed crucibles should be used. Otherwise competing crystals might grow at the surface due to supersaturation

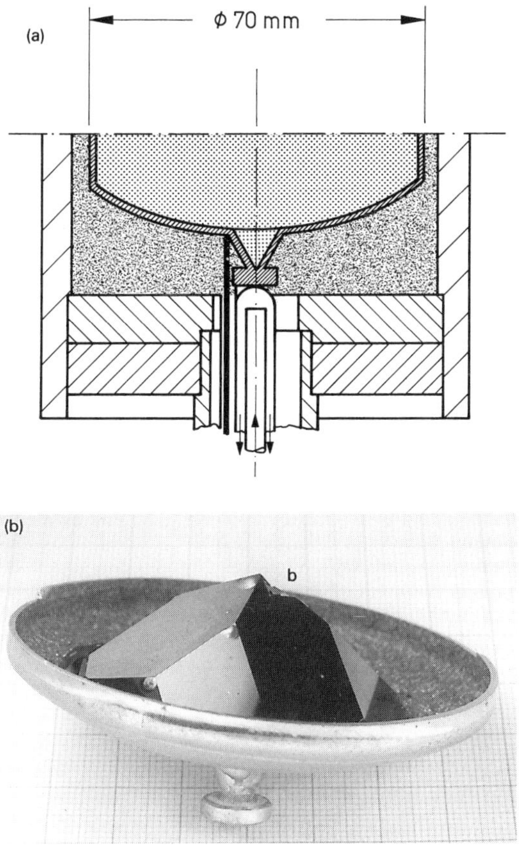

Fig. 10. Cooled tip at the bottom of a crucible for nucleation and selection of nuclei [46]. (a) Schematic with the support and air-cooled finger. (b) Photograph of the cut-off bottom with a YIG crystal.

by flux evaporation. Timofeeva [53] sealed the melt by first growing a surface crystal and then changing the temperature gradient for bottom growth. Wanklyn [83] described another sealing technique by flux evaporation. A crucible is placed upside down in a crucible of twice the size, which is closed by a crimped platinum lid. Both crucibles have a conical form. The amount of melt in the larger crucible is chosen so that it fills a part of the smaller one. By evaporation, growing crystals will seal the gap between the two crucibles and prevent further evaporation. After the run, flux can be separated from the crystals by inverting the crucibles. The melt will be drained into the smaller one, thus avoiding damage to the furnace.

In table 4 some examples of oxide crystals, in alphabetical order of the formula, are compiled together with the flux used, the reference of the source publication and some remarks. It might be of interest that Remeika [37] already used ACRT but only as a stirring method explicitly prior to the slow cooling growth of rare earth orthoferrites from a PbO flux which has a rather high viscosity. At that time the development of

Table 4
SCBG of oxides other than garnets

Composition	Flux	Remarks	Reference
$\alpha-Al_2O_3$	PbF_2	IS	[99]
$\alpha-Al_2O_3$	PbO/B_2O_3	4 K/h, 6 kg	[100]
$BaTiO_3$	KF	20 K/h, "butterfly twins"	[101]
$GdAlO_3$	$PbO/PbF_2/B_2O_3$	ACRT, 0.3 to 0.6 K/h	[75]
$FeBO_3$	B_2O_3 excess	too high viscosity	[102]
$KTa_{1-x}Nb_xO_3$	K_2O excess, Li_2O	2 K/h	[103]
$KTa_{1-x}Nb_xO_3$, $0 < x < 0.04$	K_2O excess	0.15 K/h, loss 0.1 g/h	[104]
$KTa_{1-x}Nb_xO_3$, $x \approx 0.26$	K_2O excess	ACRT, 0.13 K/h	[105]
$KTiOPO_4$	$K_2WO_4 \cdot P_2O_5$	1 K/h, 1000–700°C	[106]
$KTiOPO_4$	$K_6P_4O_{13}$	ACRT, 0.1–0.5 K/h,	[107]
$PbFe_{0.5}Nb_{0.5}O_3$	PbO excess	ACRT, 0.25 K/h	[108]
$PbMn_{0.5}Nb_{0.5}O_3$	PbO excess	ACRT, 0.25 K/h	[108]
$PbFe_{7.8}Al_{4.2}O_{19}$	$PbO/PbF_2/B_2O_3$	0.5 to 5 K/h, loss of PbF_2	[109]
$PbFe_{12-x}Ga_xO_{19}$, $0 < x < 11$	PbO excess	0.5 K/h, local cooling	[110]
$RFeO_3$, R = rare earth	PbO	30 K/h	[37]
$RFeO_3$, R = rare earth	$PbO/PbF_2/B_2O_3$	2 K/h, striations	[111]
$RFe_3B_4O_{12}$, R = rare earth	$Bi_2O_3/B_2O_3 = 1.5$	too high viscosity	[112]
$YAlO_3$	$PbO/PbF_2/B_2O_3$	3 K/h, flux evap.	[113]
Co–Ti:$YAlO_3$	$PbO/PbF_2/B_2O_3$	3 K/h, flux evap.	[114]
$YFeO_3$	$PbO/PbF_2/B_2O_3$	ACRT, 1 K/h, IS	[82]

furnaces and temperature control for crystal growth was just beginning and viscosity measurements played an increasing role. The temperature gradient for ACRT has to be as small as possible, to minimize striations. Tables of crystals grown by SCBG and the respective fluxes have been compiled by Wanklyn [15]; for complex oxides [84, 85] and for perovskites ($CaTiO_3$, $CdTiO_3$, $SrZrO_3$ and $LaGaO_3$, $PrGaO_3$, $NdGaO_3$) [86].

5.1.2. *Top-seeded solution growth (TSSG)*

For fluxes of low volatility allowing the use of open crucibles, crystals can be grown at the surface of the melt when the temperature T_{surf} is lower than T_{bot}. It is necessary, as in melt growth using the Czochralski method, to have the coldest spot at the center of the surface. A seeding rod or a seed crystal can be introduced from above at the saturation temperature. Contrary to the situation in melt growth, the temperature gradient should be small and the growth temperature T_g has to be continuously lowered by slow cooling of the melt as for SCBG. Convection plays an important role in avoiding nucleation elsewhere. A schematic drawing of a typical furnace is shown in fig. 11 for $T_g < 1200°C$ with an additional heating coil adjacent to the lower part of the platinum crucible to improve thermal convection [115]. Frequently the crystal is cooled by an air-cooled finger, but the heatflow changes during growth when the holder is slowly pulled upwards.

Smith and Elwell [87] described a thermographic method for refractory materials of density different from that of the melt to determine nucleation at a platinum wire

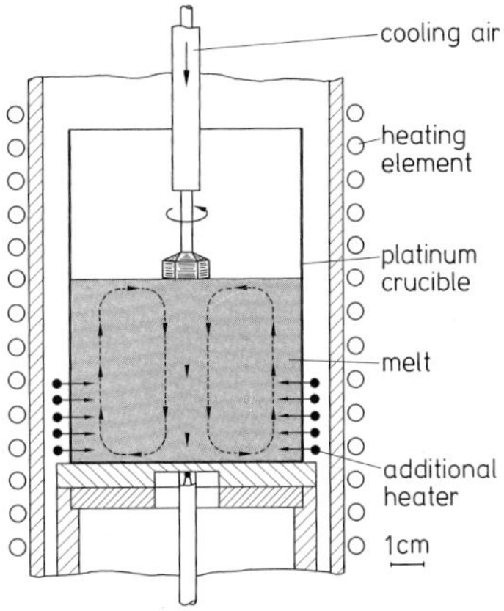

Fig. 11. Schematic of a furnace for TSSG of $BaFe_{12}O_{19}$ from a BaO/B_2O_3 flux [115].

suspended from a balance so that the free end is immersed in the center of the melt to a depth of about 5 mm. The temperature is reduced at a linear rate of 2 to 5 K/h until a change in weight of the wire is observed when crystals nucleate. If the melt is heated after crystals have been formed, the saturation temperature is indicated by a weight change at which the last traces of crystals disappear. Bruton and White [88] used this thermogravimetric method for a solution of $PbTa_2O_6$ in a flux of Bi_2O_3/B_2O_3 with a ratio of 1.2 or in a flux of PbV_2O_6 or $Pb_2V_2O_7$ to measure solubility, viscosity, density and diffusivity. The solubility and supercooling was determined by this method for $DyKMo_2O_3$ in K_2O/MoO_3 flux [89] and for rare-earth disilicates in a Bi_2O_3/V_2O_5 flux [90].

Because of the rather well developed methods for the volatile PbO based flux little work has been published using TSSG for garnets. Durcok and Pollert [65] described the growth of manganese substituted YIG. They developed an apparatus allowing ACRT with a temperature oscillation of about 1 K. To minimize radiation problems at the high temperature ($T_g = 1436°C-1390°C$) platinum reflectors were used above the melt. Oka and Unoki [91] used excess iron oxide in the ratio Fe/Y = 3.8, 4.0 and 4.6 in a RF-heated platinum crucible with a cooling rate of 0.5 to 1 K/min. The crystals had asymmetric shapes and no natural facets.

Capper and Elwell [92] investigated, with a thermobalance, a flux of $BaO/BaF_2/B_2O_3$, used for LPE of iron garnets [93], for the growth of $Y_3Al_5O_{12}$. BaF_2 was found to increase the volatility of the flux but evaporation from the solution is tolerable and the viscosity is relatively low. They believed that growth of $Y_3Al_5O_{12}$ in an unstirred solution is volume diffusion controlled.

A particular investigation has been published by Bezmaternykh et al. [72]. They studied the nucleation and growth on seeds of $Y_3Fe_{4.5}Ga_{0.5}O_{12}$ on a rapidly rotating crystal carrier in BaO/B_2O_3 fluxes with excess Fe_2O_3 and Ga_2O_3. The crystals were growing on eccentric arms of a rotating central axis in the bulk of the melt rather than near the surface as is usually done by analogy with Czochralski growth. The temperature gradient in the melt of 8.5 cm height was 1.8 K/cm. The stirring effect of the rotating

Table 5
TSSG

Composition	Flux	Remarks	Reference
$BaTiO_3$	TiO_2 excess	$T_g < 1400°C$, 0.5 K/h, $f = 2$	[117]
$Ba_{1-x}Sr_xTiO_3$, $0.02 < x < 1$	TiO_2 excess	$T_g < 1400°C$, 0.5 K/h, $f = 2$	[118]
$BaFe_{12-x}M_xO_{19}$, $x < 3$	BaO/B_2O_3 or	0.2 K/h, $f = 2$	[115]
$SrFe_{12-x}M_xO_{19}$, $x < 3$	SrO/B_2O_3 or	1200–1080°C	
M = Al, Ga or Sc	BaO or SrO excess	1400–1350°C	
$Bi_{12}TiO_{20}$	Bi_2O_3 excess	0.5 K/h, $f = 6$	[122]
$Bi_{40}Fe_2O_{63}$	Fe_2O_3 excess	0.5 K/h, $f = 6$	
$Bi_{40}Ga_2TiO_{63}$	Ga_2O_3 excess	0.5 K/h, $f = 6$	
$Bi_4Ti_3O_{12}$	$Bi_2O_3/B_2O_3 = 3$	$\delta T = 20$ K/cm, surface	[131]
$CaCO_3$	Li_2CO_3	$p_{CO_2} = 15$ atm, RF, $f = 0.5$	[132]
$KTiOPO_4$	$K_6P_4O_{13}$	seed rot., heat-pipe	[125]
$KTiOPO_4$	$K_2WO_4 \cdot P_2O_5$	seed rotation	[130]
$KNbO_3$	K_2O excess		[117]
$KNbO_3$	K_2O excess	0.2–0.5 K/h, $f = 7$, ACRT	[120]
$KNbO_3$	K_2O excess	0.3 K/h, $\delta T = 33$ K, $f = 5$	[119]
$KTa_{1-x}Nb_xO_3$	K_2O excess	add. CaO, PbO	[103]
$KTa_{1-x}Nb_xO_3$, $0.3 < x < 1$	Rb_2O	seed: 50 Hz, $f = 0.2$	[121]
RPO_4, R = Y, Dy	$Pb_2P_2O_7$	spheric. seeds, $\Delta T = 1$ K	[124]
$Mn{:}Y_3Fe_5O_{12}$	$PbO/B_2O_3 = 5$	excess Fe_2O_3, ACRT	[65]
$Zn_2Ba_2Fe_{12}O_{22}$	$Bao/B_2O_3 = 2$	0.1 K/h, $f = 1.2$	[123]

f = linear growth rate (μm/min), δT = temperature gradient (K/cm).

sample holder gave rise to temperature fluctuations of 2 to 5 K. They found an effective distribution coefficient:

$$K_{eff} = [Fe/Ga]_{crystal}/[Fe/Ga]_{melt} \approx 1.3.$$

The cooling rate was increased from 0.04 to 0.25 K/h and the corresponding growth rates were determined by induced striations to be 0.62 and 0.14 μm/min, respectively. Striations were induced by pulsed supercooling of 2 K/10 min.

Table 5 compiles some materials grown by TSSG. As an example, the growth of $BaFe_{12}O_{19}$ is described [115]. The flux is a mixture of BaO/B_2O_3 with a molar ratio of 1.8 and 35 to 50 wt% ferrite material in the temperature range of 1200°C to 1080°C. The schematic of the furnace is shown in fig. 11. Generally the seed crystal is rather small because it is taken from spontaneous nucleation. In this case large seed crystals are cut from Czochralski grown strontium hexagallate crystals in which the gallium ions were partly substituted by magnesium and zirconium [116]. For a typical composition $Sr_{1.03}Ga_{10.81}Mg_{0.58}Zr_{0.58}O_{18.99}$ the value of the lattice parameters are measured to be $a = 0.5822$ nm and $c = 2.26031$ nm. These are 1.15% and 0.51%, respectively, smaller than those of $BaFe_{12}O_{19}$. With a cooling rate of 0.2 K/h, the pulling rate is about 1 mm per day, which might be compared to a growth rate $f \approx 0.7$ μm/min. Figure 12 shows a $BaFe_{12}O_{19}$ crystal with the air-cooled platinum seed holder. The platinum double pipe is isolated by alumina fiber materials in an alumina pipe. The conical lower part is seen in fig. 12. The seed crystal is fixed in a platinum pipe by a splint, which is shown in fig. 12. The seed might equally be screwed onto a threaded platinum tip [115].

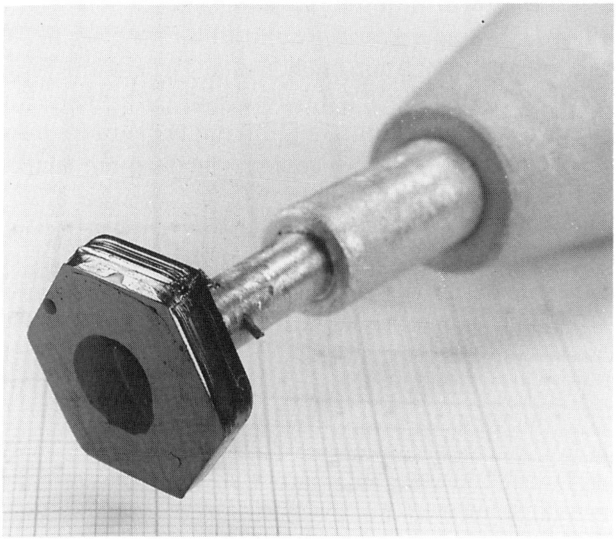

Fig. 12. $BaFe_{12}O_{19}$ crystal grown by TSSG on an air-cooled platinum holder [115].

The crystals were rather imperfect but good enough since inclusion and crack-free volumes of less than 1 mm^3 were necessary for making spheres of about 0.3 mm diameter for microwave measurements. The TSSG method had been chosen, because many compositions of $BaFe_{12-x}M_xO_{19}$ and $SrFe_{12-x}M_xO_{19}$ with M = Al, Ga or Sc and $x < 3$ had to be grown. SCBG only occasionally gave platelets which were thick enough for the measurements. Some crystals were grown by TSSG using either BaO or SrO excess as a flux. In this case T_g is between 1400°C and 1350°C and a furance with $MoSi_2$ heating elements was used.

TSSG of crystals of $BaTiO_3$, $KNbO_3$, $Bi_{12}TiO_{20}$, $SrTiO_3$ and $La_{2-x}Ba_xCuO_4$ is described by Rytz et al. [117]. They present a schematic of their furnace with hairpin shaped $MoSi_2$ heating elements for a high T_g and they compiled a reference list for the growth of $BaTiO_3$. Crystals of cubic $BaTiO_3$ are of interest for studies and application of the photorefractive effect. TSSG from a TiO_2-rich solution is able to overcome the problems associated with the hexagonal-to-cubic phase transition occurring at 1432°C and avoids incorporation of flux ions e.g. KF as in the "butterfly" type twins of Remeika [101]. As a flux, an excess of 30 mol% TiO_2 is used, with T_g from 1400°C to 1335°C. Maximum size of the crystals is in the order of 1 cm. Mixed crystals of $Ba_{1-x}Sr_xTiO_3$ with $0.02 < x < 1$ have been grown by Bethe and Welz [118]. Flückiger and Arend [119] reported on the preparation of pure and Fe-doped $KNbO_3$ crystals by TSSG with a starting composition of 51.7 to 52.8 mol% K_2O. Progress in TSSG of $KNbO_3$ is reported by Xing et al. [120] together with a compilation of earlier publications. They give a detailed schematic of the growth apparatus using a furnace in a water-cooled stainless steel vessel, ACRT and growth control by weighing the crystal. The cooling rate is 0.2 to 0.5 K/h; the corresponding growth velocity (pulling rate) is 2.5 to 6.7 μm/min, crucible size 50 cm^3 with a diameter of 4 cm, crystal size

$1.5 \times 2 \times 2.4\,cm^3$. The growth of mixed crystals of $KTa_{1-x}Nb_xO_3$ is described by Bonner et al. [103]. They used excess K_2O and some additives like CaO or PbO to influence the free carrier concentration in the crystal. Hellermann et al. [121] used, for TSSG of $KTa_{1-x}Nb_xO_3$, $0.3 < x < 1$, a flux of 12 mol% Rb_2O and forced convection by vibrating the growing crystal parallel to [100] or [111] at a frequency of 50 Hz and an amplitude of 0.3 mm. In crystals, which are cubic at room temperature ($x < 0.4$), faint striations almost perpendicular to the growth directions are observed in polarized light. Several crystals with $0.90 < x < 0.98$ have been grown. In contrast to crystals grown from melts containing excess K_2O, single domain regions can be prepared suitable for optical second harmonic generation experiments. Sections of tetragonal crystals with $x = 0.49$ and geometries of $3 \times 4 \times 2\,mm^3$ are reported to be nearly free of striations and can be completely polarized.

Bruton et al [122] published on the TSSG of $Bi_{12}TiO_{20}$, $Bi_{40}Fe_2O_{63}$ and $Bi_{40}Ga_2O_{63}$ from solution in Bi_2O_3 in the cubic γ-form.

TSSG of complex hexagonal ferrites like $Zn_2Ba_2Fe_{12}O_{22}$ (ZnY) were reported by AuCoin et al. [123] with a cooling rate of 0.1 K/h and a growth rate of 1.2 μm/min. TSSG with a weighing method for the determination of the linear growth and dissolution rates of seed crystals of YPO_4 and $DyPO_4$ in $Pb_2P_2O_7$ flux was published by Eigermann et al. [124]. Growth rates of {100} faces of YPO_4 were found to be linear with the relative supersaturation ranging from 0.004 to 0.025 for different temperatures and the dissolution rates were nearly 30% larger than the corresponding growth rate, giving evidence for an additional impedance in the growth process. Seed crystals of YPO_4 with a spherical shape were held in a melt with a low supersaturation of 1 to 2 K. Facets of the orientation {100}, {101} and {112}, the latter having a smaller radius, grew after 5 h indicating the directions of slowest growth.

A new variation of TSSG has been developed for non-linear optical crystals of $KTiOPO_4$ (KTP) from a $K_6P_4O_{13}$ flux using a heat pipe providing for a high degree of spatial temperature uniformity (± 1.9 K) throughout the growth solution [125]. The growth temperature is in the range of 970°C to 900°C with cooling rates ranging from 0.5 to 10 K/h. Initial surface dissolution appears necessary to achieve flaw-free growth immediately on a seed. The seed crystals are mounted on the end of a platinum seed rod bent 12 mm from its end. The crystals are submerged roughly midway between the melt surface and the crucible bottom. They are rotated with periodic reversal, thereby traversing through the melt at a distance from the crucible centerline. Reversal of the sense of rotation avoids asymmetrical flow rates and consequently growth rates relative to the growth facets. This effect of stirring on the crystalline quality has been demonstrated in the case of solution grown potash alum by Gits-Leon et al. [126]. Explanations of this effect are given by Chernov [127] and Wilcox [128]. Viscosity and density of solutions of KTP in $K_6P_4O_{13}$ have been measured by Bordui and Jacco [129]. Centric and eccentric rotation of the seed are reported by Shumov et al. [130] for the TSSG of KTP from melts containing WO_3 with axial and radial temperature differences varying from 0.5 to 10 K and a cooling rate from 0.3 to 1.5 K/h.

Bruton [131] reported the spontaneous nucleation of crystals of $Bi_4Ti_3O_{12}$ at the surface of a solution in B_2O_3 flux using a cooling rate of 1 K/h. The surface temperature was about 50 K lower than the temperature at the bottom.

5.2. Vertical temperature gradient transport (VTGT)

Gradient transport for crystal growth is based upon the transport of solute from a hotter (point F) to a cooler region (point G) as indicated in fig. 2. The material needed for the growth is continuously supplied by dissolution of a nutrient source at the hotter region. Since T_g is constant this steady state method is particularly suitable for the growth of solid solutions. For technical reasons and because of convective problems a vertical temperature gradient is generally applied.

5.2.1. Top-seeded VTGT growth with a nutrient

For VTGT generally $T_{surf} < T_{bot}$ and a nucleation tip or a seed is introduced from above into the melt. At the bottom of the crucible is non-dissolved material which is continuously being dissolved, replenishing the material grown on the seed crystal. Because of their high vapour pressure PbO-based fluxes are not very suitable in an open system and such melts are rarely used for garnets. Görnert and Hergt [59] placed a perforated container with polycrystalline YIG on the bottom of a crucible filled with saturated YIG solution in a PbO/B_2O_3 flux with excess iron oxide (see table 1). For spontaneous nucleation at the cooler surface they observed a supersaturation of > 10 K, but for nucleation on YIG or YAG seed crystals only < 1 K. They used positive and negative temperature "jumps" of 3 K to compare the striations so induced into the crystals with those caused by other factors such as convection, imperfect temperature control, ambient room temperature fluctuations etc.

BaO based fluxes having low volatility are favoured for top-seeded VTGT (see table 2) for the growth of garnet single crystals. Linares [67] applied this method for YIG using BaO/B_2O_3 with excess iron as a flux. The pulling rate, which might be considered as the average growth rate, was $f \approx 0.9$ μm/min, which is rather low. YIG crystals with a diameter of 9.5 mm and a length up to 45 mm have been grown by Kestigian [68]. A temperature difference of approximately 25 K was maintained over the height of the melt with T_g about 1080°C. The crystal was rotated at 60 min^{-1} and the growing interface of the crystal was approximately 3 mm below the melt surface. Bennett [69] used a seed cut from a Czochralski grown yttrium aluminium garnet (YAG) and found a junction region of 25 μm width at the interface consisting of a transient from YAG through aluminium substituted YIG to YIG. The problem of the BaO/B_2O_3 flux is the strong tendency for polynucleation and cracking [70]. The melt wets the crystal and a glassy layer of solidified melt is formed around it during cooling to room temperature.

Elwell et al. [71] used a thermobalance to determine the growth rate (f) of a rotating YAG crystal growing in a [211] direction in a $BaO/BaF_2/B_2O_3$ flux. They reported, for example, that at supercoolings $\Delta T = 2$ and 8 K, f increased from about 0.5 and 5 μm/min without rotation of the crystal to about 4 and 12 μm/min with a rotation rate of 50 min^{-1}, respectively, showing the influence of forced convection. Similar experiments have been described for YAG and YIG by Lawrence and Elwell [133]. They report a decrease of f at constant T_g due to poor mixing of the melt even when the crystal is rotated with a rate of 45 min^{-1}. Nikolov et al. [73] optimized the hydrodynamic conditions for the growth of $Y_3Fe_{5-x}M_xO_{12}$, M = Al or Ga, $x = 0$, 0.5 and

Table 6
VTGT

Composition	Flux	Remarks	Reference
BeO	K_2MoO_4/MoO_3	rot. nutrient and seed	[139]
$BaMoO_4$	Li_2MoO_4	RF, $f=8$	[134]
	Bi_2O_3/MoO_3		
$KTa_{1-x}Nb_xO_3$, $x \approx 0.26$	K_2O excess	ACRT, autonutrient	[105]
$KTa_{1-x}Nb_xO_3$, $x \approx 0.35$	K_2O excess	$f=1$	[135]
$KTIOPO_4$	$K_2WO_4 \cdot P_2O_5$	rot. seed, continuous refeeding	[130]
RVO_4, R = Tb, Dy, Tm	$Pb_2V_2O_7$	$\delta T=5$, $f=0.7$	[136]
$Zn_{1.9}Fe^{2+}_{0.1}Ba_2Fe_{12}O_{22}$	$BaO/B_2O_3=2$	$T_g=1000°C$, $f=1.2$	[137]
$Zn_{1.9}Mn^{2+}_{0.1}Ba_2Fe_{12-x}Mn_xO_{22}$		$0.1<x<0.3$	

RF = RF heating, f = growth rate (μm/min), δT = temperature gradient in the melt (K/cm).

1 from various BaO-based fluxes having kinematic viscosities from 8 to 100 m^2/s at 1100°C. They reached growth rates up to 17 μm/min. Crystals with a flat interface in the absence of temperature fluctuations proved to have minimum levels of striations, inclusions and dislocations.

For the growth of $BaMoO_4$ from a Li_2MoO_4 flux and for $Bi_2Mo_3O_{12}$ and $Bi_2Mo_2O_9$ from excess Bi_2O_3 by VTGT and top-seeding, Chen [134] used RF-heating of the platinum crucible. He observed strong convective flow such as is frequently seen in the Czochralski growth of oxides. The pulling rate ($\approx f$) was about 8 μm/min. Since no automatic diameter control had been applied, the shape of the crystals was rather irregular. Whiffin and Brice [135] pulled $KTa_xNb_{1-x}O_3$ crystals ($x \approx 0.35$) from a melt with excess K_2O and a nutrient. The growth rate varied between 0.5 and 1 μm/min depending on the supersaturation imposed by the temperature gradient. Substantial flux evaporation was reported. The growth of RVO_4 crystals (R = Tb, Dy, Tm) by pulling from a $Pb_2V_2O_7$ flux with a nutrient was described by Smith et al. [136]. With a temperature gradient (δT) of 2 to 5 K/cm and a pulling rate of 0.7 μm/min the crystal grew until it was pulled out of the melt although the nutrient had not been exhausted. The growth rate decreased as the length of the crystal increased. To get longer crystals of $TmVO_4$ they used a cooling rate of only 0.1 K/h after a length of 8 mm had grown. Tolksdorf [137] reported the growth of complex hexaferrites with Y structure from a BaO/B_2O_3 flux with a nutrient and the intergrowth of other hexagonal structures.

The problem with top-seeding is the temperature control at the growing interface, which is very sensitive to temperature changes in the, rather open, ambient atmosphere. δT has to be smaller as has the growth rate f compared to melt growth by the Czochralski technique.

5.2.2. *Bottom growth with a nutrient*

For the growth of gallium substituted iron garnets ($Y_3Fe_{5-x}Ga_xO_{12}$), a method has been developed by Tolksdorf and Welz [58] with a growth apparatus as shown in fig. 3. In this case a cylindrical crucible with a chamber for the polycrystalline nutrient was used as shown in fig. 13. The starting procedure is identical with the one described above for the SCBG of YIG. It is clear from the schematic shown in fig. 13a, that in the starting position of the crucible, the melt is neither in contact with the nutrient in the inner chamber nor with the seed fixed to the lid. After a slow cooling period to get

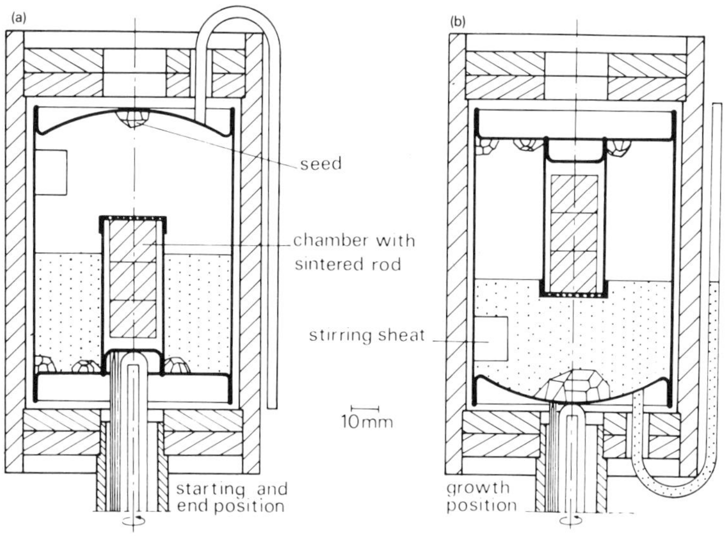

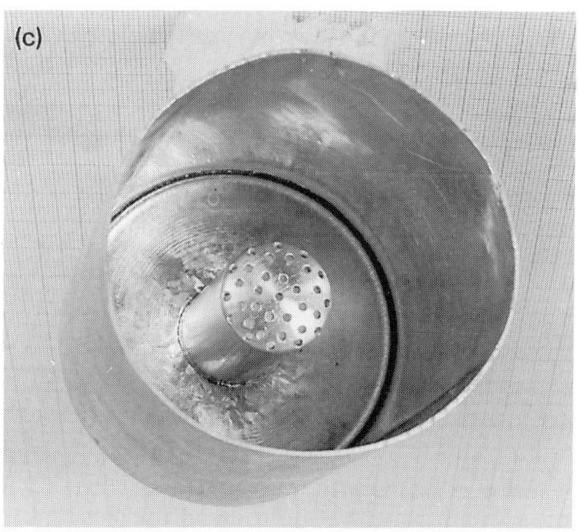

Fig. 13. Form and position of the crucible and the ceramic support for isothermal VTGT growth of $Y_3Fe_{5-x}Ga_xO_{12}$ [58]. (a) Starting and end position, (b) growth position, (c) photograph of the open crucible before filling.

spontaneous nucleation, the crucible is turned into the growth position. As shown in fig. 13b, the seed is placed above the air-cooled probe. The nutrient is in close contact with the melt through the perforated lid of the inner chamber at a somewhat higher temperature than the seed, whereas the crystals grown by spontaneous nucleation are no longer in contact with the melt. Forced convection by ACRT enhances the material transport from the nutrient to the growing crystal at constant T_g. The spread of the Ga substitution within the crystal is much less than that in crystals grown by SCBG; e.g. with a starting composition of $x = 0.8$ the measured value in the crystals was 0.776 ± 0.015, which means that the molar variation of Ga was less than $\pm 2\%$. The distribution of Ga on tetrahedral and octahedral sites in garnet depends on T_g or the annealing conditions as reported by Röschmann et al. [138].

Scheel and Sommerauer [105] developed a method for bottom growth of crystals of $KTa_{1-x}Nb_xO_3$ ($x \approx 0.26$) as already mentioned in table 4. To achieve a small concentration gradient within the crystal they used VTGT bottom growth with ACRT. In both cases spherical crucibles were used, made from two half shells by argon-arc welding along rather extended rims. For VTGT the rim is the lowest part increasing the temperature gradient and by slow cooling at a rate of 0.18 K/h crystals nucleate in the rim. At a given temperature the crucible is rotated around the horizontal axis and the temperature is kept constant at $1190 \pm 0.1°C$. The previously grown crystals are now in a hotter region of the melt forming a nutrient for the crystals to nucleate in this position at the lowest and coolest part of the crucible. From concentration fluctuations in the crystals the temperature fluctuations at the growth interface were estimated to be about 3 K.

Growth of beryllia by VTGT has been described by Austermann [139] in potassium polymolybdate, K_2MoO_4/MoO_3 in a molar ratio ranging from 0.2 to 1.0, the most commonly used ratio being 0.44. Inclusion-free BeO crystals could not be grown from a highly purified flux. Only with added oxides of e.g. Al, Mg, B, Si, ("flux conditioners") with concentrations < 0.1 wt% could virtually inclusion-free crystals be grown on polished $(10\bar{1}0)$ seed plates. Seed crystals are mounted on the lower structure of a rig above which a basket is fixed for the BeO nutrient material. The basket has a solid floor and a screen side-wall with about 50% open area, to allow saturated flux to flow to the outside regions of the crucible. The basket/rig assembly is lowered to flux submergence and rotation is started for the growth run of about six weeks. At termination the rig is lifted from the flux.

Axial rotation of the seed is reported by Shumov et al. [130] for VTGT of KTP from melts containing WO_3 with axial temperature differences varying from 5 to 30 K and radial temperature differences varying from 2 to 20 K. KTP refilling is carried out at the periphery of the surface.

5.2.3. Growth by a travelling solvent zone (TSZ)

A great variety of terminology has been used to describe this technique in which a zone of solution is made to travel through a solid in a vertical temperature gradient so that, at the hotter side, a feed rod is dissolved and a crystal is grown on the cooler side. In the examples discussed here the feed rod is on top of the hot zone and the solution zone is moved by relative motion of the heater, which may be a resistence heater strip, an RF-coil or the focus of an infrared (IR) light source.

For YIG, excess of iron oxide generally has been used as a flux (see table 3). Abernethy et al. [43] used a $MoSi_2$ susceptor heated by an RF-coil and an oxygen pressure of 3 to 7 atm to move the solvent zone at a rate of 42 μm/min through a polycrystalline feed rod.

Mirror furnaces, as described by Eyer et al. [140], can be used for local heating. The simplest geometry of the mirror is a closed ellipsoid of rotation with the radiation source (IR bulb) in one and the sample in the other focus. The sample, usually within a quartz tube for atmosphere control, is rotated and moved perpendicularly to the long (rotational) axis of the ellipsoid. YIG crystals, with low values of the ferromagnetic resonance linewidth $\Delta H = 0.4$ Oe at microwave frequency of 9.3 GHz in the core region, have been grown in excess iron oxide flux in oxygen [97]. There is a steep temperature gradient [141] which causes strong convection in the melt zone leading to growth striations. By using hollow cylinders of alumina around the crystal, striations can be minimized at least in the core region. The diameter of the crystal is limited to the order of 10 mm. The growth rate is rather large of the order of 17 μm/min. This method is particularly suited to the growth of mixed single crystals e.g. Ga-substituted YIG as demonstrated by Shindo et al. [94]. They analyzed the variation of the Ga concentration along the growth direction of a $Y_3Fe_4Ga_1O_{12}$ crystal and found, after a transient of 2 cm with higher x values, a crystal with a length of more than 5 cm which had the preset value of $x = 1.0$. Torii et al. [95] reported for $Y_3Fe_4Ga_1O_{12}$ a value of $\Delta H = 1.6$ Oe (9.3 GHz) which is comparable to crystals of the same composition grown by SCBG from a PbO-based flux. Pressures of oxygen from 20 to 50 atm have been applied using this method for the growth of crystals of $Y_3Fe_{5-x}Ga_xO_{12}$, $0 < x < 1$, by Balbashov and Egorov [96]. They demonstrated the versatility of this method by the growth of various oxide compounds. They reported the growth of $Gd_3Fe_5O_{12}$ from excess Fe_2O_3 at 50 atm oxygen pressure with a growth rate of 25 μm/min. Because they were able to apply high oxygen pressure they could grow compounds congruently without a flux which otherwise would decompose. So they reported the growth of crystals of $BaFe_{12}O_{19}$, and $SrFe_{12}O_{19}$ without a flux at $p_{O_2} = 70$ atm with a growth rate of $f = 133$ μm/min having a diameter of 8 mm. For barium hexaferrite they measured $\Delta H = 40$ Oe at 75 GHz compared to 23 Oe measured in TSSG-grown crystals [142]. For high oxygen pressure strontium hexaferrite they even measured ΔH as low as 18 Oe. They grew $Ni_{0.5}Zn_{0.5}Fe_2O_4$ crystals without a flux at $p_{O_2} = 50$ atm with $f = 167$ μm/min and a diameter of 8 mm. Evaporation of ZnO was reduced substantially for oxygen pressures above 50 atm. Furthermore, they reported the growth of mixed cobalt nickel iron spinels $Co_{1-x}Ni_xFe_2O_4$, $x = 0, 0.25, 0.5, 0.75, 1.0$, without a flux at $p_{O_2} = 10$–15 atm with $f = 100$–167 μm/min, 10 to 12 mm in diameter. They assumed that the low electric resistance of 10^2–10^3 Ω cm was due to the presence of Fe^{2+}. They also reported the growth of $Mn_{0.5}Zn_{0.5}Fe_2O_4$ without a flux at $p_{N_2} = 1$ atm with $f = 167$ μm/min having a diameter of 20 mm and a length of 50 mm. Evaporation of ZnO was compensated by the composition of the feeding rod of $Mn_{0.5}Zn_{0.7}Fe_2O_4$. The permeability of the crystals was 5000 Gauss/Oe. Shindo et al. [143] reported the growth of Mg_2TiO_4 single crystals by TSZ using a mirror furnace.

With the laser-heated pedestal growth method as developed by Feigelson [98], it is possible to grow single crystal fibres of compounds which are incongruently melting. As an example he described the growth of YIG from an Fe_2O_3-rich solution. The heat source is a 50 W CO_2 laser, operating at a wavelength of 10.6 μm, split into two beams which are focussed onto the end of a rod of the material to be grown. Starting with a YIG seed and a source rod, steady state growth can be readily achieved. Because of the high temperature gradient and the small zone, the growth rate is very high. A YIG fiber with a diameter of less than 1 mm has been grown at a rate of 250 μm/min. This method is very versatile producing fibers 3 to 1700 μm in diameter and up to 20 cm in length.

Brissot and Bellin [144] developed a method for the growth of calcite ($CaCO_3$) single crystals by TSZ using Li_2CO_3 as a flux and $p_{CO_2} = 1$ to 2 atm. In order to obtain an interface as flat as possible, a perforated platinum heater strip is embedded in the molten zone which moves from the seed to the lower end of a sintered rod with a speed of about 3 μm/min at a temperature of 700°C to 800°C. The growth of large single crystals of optical quality with a diameter of 20 mm and the shape of the heater strip has been published by Belin [145]. This method was tried for YIG with a BaO/B_2O_3 flux [70], but the results were not encouraging; the problem was the creeping flux and the porosity of the feed rod which had a density of only ≈99% of the X-ray density. The travelling heater strip has been successfully applied by Turner et al. [146] to the growth of $BaTiO_3$ with excess TiO_2 at 1350°C with a moving rate of 1.5 μm/min. They used a background temperature of 1290°C.

5.3. *Growth by flux evaporation*

As shown schematically in fig. 2 by the arrow D to E, supersaturation at constant T_g can be continuously maintained for the growth of crystals by evaporation of the solvent. This method is frequently used for aqueous and organic solutions. The advantage over growth by other techniques are that it needs no temperature programming and can use a low temperature gradient. But there are only a few examples of use of this method for oxides. Control of evaporation and thus of the growth rate might be a problem. Earlier work has been compiled in a table in the book of Elwell and Scheel [1].

Isothermal growth of YIG crystals by PbF_2/B_2O_3 flux evaporation at a T_g of 1150°C, 1200°C and 1250°C has been described by Antonini et al. [147]. Excess of Fe_2O_3 was necessary at any temperature. A hole of about 4 mm^2 was left in the lid of a 100 ml crucible. During six days at temperature with fluctuations within 0.05 K, crystals resulted of the same size as by SCBG with similar sized crucibles. PbF_2 evaporation had not caused any serious inconvenience. A ventilation system was used and only slight attack of the refractory bricks in the furnace was observed. The simplicity of this method is emphasized in comparison with bottom growth by VTGT which indeed needs sophisticated apparatus.

Wood and White [148] published on the growth of stoichiometric magnesium aluminate spinel crystal by evaporation of a PbF_2/B_2O_3 flux with a molar ratio of 60:1. They measured the weight loss with a balance. Since MgF_2 precipitates below 1150°C, they grew above 1155°C and steadily increased the temperature to get a linear evaporation rate. The crucibles were sealed by "crimping-on" the lids, lightly spot welding at intervals around the rim or by argon-arc welding. In the latter case a hole of 6 to 9 mm diameter was made in the lid to control the escape of the fluoride vapour. Crystals of $MgAl_2O_4$ over 1 cm in size were grown, doped and undoped with first period transition elements. Crystals of $ZnAl_2O_4$–$MgAl_2O_4$ and $MgGa_2O_4$–$MgAl_2O_4$ solid solutions were also prepared. The growth of large, laser quality $Nd_xR_{1-x}P_5O_{14}$, R = La, Y, $0.1 < x < 1$, crystals by flux evaporation is described by Plättner et al. [149]. They used excess $P_2O_3 + H_2O$ as a flux. Supersaturation was achieved by dehydration of phosphoric acid, $n H_3PO_4 \rightarrow H(HPO_3)_n OH + (n-1)H_2O$, at constant T_g of 550°C. The $Nd_2O_3 - P_2O_3 - H_2O$ phase diagram, solubilities, evaporation rates, seeding procedures and $\{hkl\}$ growth rates (e.g. 1.4 g/day for NdP_5O_{14}) were determined. The largest size was $15 \times 30 \times 35\ mm^3$.

5.4. *Special growth methods*

A review of electrolytic growth from high temperature solutions has been given by Elwell [150]. Electrocrystallization is used to grow crystals of conducting compounds e.g. tungsten bronzes of general formula M_xWO_3, M = Cu, alkali or rare-earth ions.

Synthesis of magnetic and non-magnetic garnets from molten salts at high pressure and high temperature was published by Mroczkowski [151]. He described a high pressure autoclave system, that is in principle capable of achieving temperatures of 1400°C and pressures of 1500 atm. In this YIG single crystals were grown by SCBG from a flux of $PbO/B_2O_3 = 5.74$ and Y/Fe = 7.36 from 1300°C–950°C with a cooling rate of 5 K/h and $100 < p_{air} < 1200$ atm. Below 150 atm the crystals showed lower IR transparency, above 150 atm they exhibited a habitus change from typical $\{110\} + \{211\}$ to dominant $\{211\}$, whilst above 500 atm the typical garnet morphology disappeared. Fe^{2+} was present in all samples. The platinum crucibles used for high pressure runs wear out much faster indicating an active role of platinum in the iron–oxygen equilibria during the growth process. Rudness and Kebler [152] described the growth of YIG by a flame fusion process, but the crystals were of poor quality.

Nomura et al. [153] described the preparation of garnet single crystals by anomalous grain growth. By solid phase reaction YIG crystals of 15 mm in diameter and 4 mm thick were prepared. Some pores were observed in the crystals but the optical absorption was about 0.4 at a wavelength of 1.55 μm, with a (100) seed f was 15 to 30 μm/min.

Görnert et al. [154] and Gořnert [20] reported the preparation and characterization of very small hexaferrite crystals by the glass crystallization method. Crystals of $BaFe_{12-2x}Co_xTi_xO_{19}$ were grown from a flux of $BaO/B_2O_3 = 1.2$ for $x = 0$ to 0.9. The melt was quenched from temperatures between 1200°C and 1400°C with a cooling rate of 10^4–10^5 K/s between twin rollers to form amorphous flakes with a thickness of 30 to 100 μm. Subsequent annealing of the glass flakes between 600°C and 790°C leads to nucleation and Oswald ripening of the hexaferrite phase, where only diffusion and coalescence play a role in the transport process. The residual melt was dissolved in dilute nitric acid; the remaining slurry was carefully rinsed and dried. Hexaferrite crystals were obtained with a diameter that can be continuously adjusted between 100 and 400 nm for unsubstituted ($x = 0$) and between 10 and 70 nm for Co–Ti substituted ($x > 0$) barium hexaferrite.

In situ observation of anorthide ($CaAl_2Si_2O_8$) and diopside ($CaMgSi_2O_6$) crystals growing in high temperature melts is reported by Tsukamoto et al. [155]. A thin film of uniform thickness (< 1 mm) of high temperature solution was suspended by a loop of Pt–Rh(20%) wire. The loop was also used as a control heater and mounted on a hot stage for high temperature microscopy. The crystals were nucleated at the tip of a thermocouple. The nucleated seed could be dissolved by increasing the temperature, so nucleation, growth and dissolution of the crystals could be observed. A TV camera was used for the observation and to document the experiment on tape. The propagation of steps of growing layers was observed.

6. Growth phenomena

Crystals grow from solution after nucleation generally in a low temperature gradient at very low supersaturation which is, in most cases, not known. According to the BCF theory [156] they develop facets at which the two-dimensional nucleation energy is reduced by the presence of nucleation centers continuously supplied by a growth spiral originated from one or a cluster of dislocations having a screw component. This theory was discussed in particular for growth from liquids by Bennema and Gilmer [157]. A review of the theoretical and experimental aspects on the morphology of growth spirals has been given by Sunagawa and Bennema [158]. Growth spirals have been observed on facets of many natural and synthetic crystals. Execpt in the case of higher-order polytypic crystals having large unit cell heights like hexaferrite [159], the monomolecular step heights of the observed growth spirals are on the order of 10 Å or less and can only made visible by special techniques. Optical methods are described e.g. by Sunagawa and Bennema [158] or by Komatsu [160]. Very often growth spirals are observed with greater step heights, sometimes even visible by inclined reflection without optical magnification as shown for a (110) garnet face [81]. These macrosteps are due to bunching of the edges of monomolecular layers. By "monomolecular layers" is meant layers having a step height which either monoatomic or of the dimension of a unit cell in the growth direction. Figure 14 shows a growth center and macro-spiral on a (110) face of a rare-earth orthoferrite ($Sm_{0.55}Tb_{0.45}FeO_3$) [162]. The width between the steps is about 10 μm, the height of the bunched steps is about 250 A. Figure 14a is a micro-photograph with differential interference contrast (DI) after Nomarski (see [160]). Note the interlacing pattern of the second and smaller spiral center which are present even at a distance of > 1 mm. The photograph was made near one edge of the

Fig. 14. Growth center and macrospiral on a (110) face of a rare-earth orthoferrite. The width between the steps is about 10 μm, the height of the bunched steps is about 250 Å [162]. (a) Micro-photograph using differential interference contrast (DI). (b) SEM photograph of the larger center as in (a), the facet was strongly tilted.

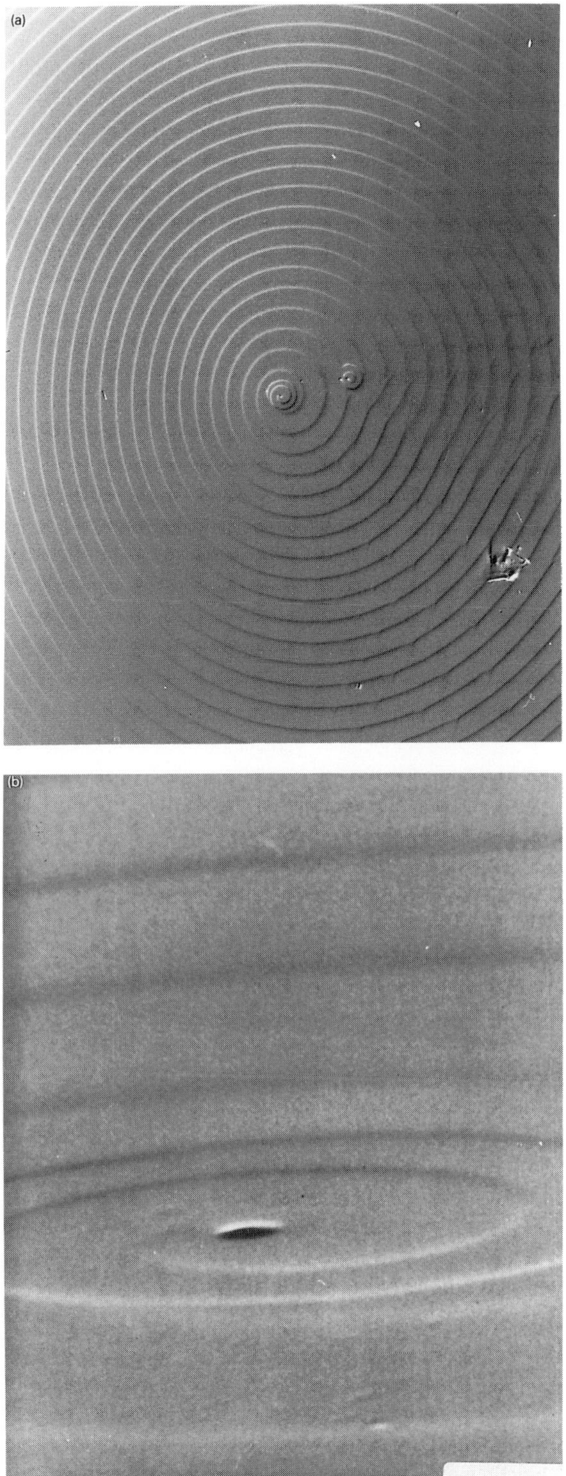
(a)
(b)
1um

crystal face which had a length of about 5 mm and which was dominated by this growth spiral. The height of the bunched steps decreased at the other end to 140 Å. A photograph of the larger center made by scanning electron microscopy is shown in fig. 14b. A very flat observation angle was used to get contrast at the steps; note the distortion of the distances in the direction of inclination. The dark contrast in the center of the spiral might be interpreted as a hollow core formed to minimize the elastic stress field associated with large step heights as discussed by Sunagawa and Bennema [161].

Growth features of a (110) face of YIG are shown in fig. 15 [162]. At a lower magnification (see fig. 15a) four segments can clearly be seen meeting at the growth center. The nearly horizontal contrast line is pointing in the [$\bar{1}$10] direction, the nearly vertical contrast line is pointing in the [001] direction. This contrast is lost at higher magnification as can be seen in fig. 15b which shows the same center with the same orientation. The height of the step near the center is about 50 Å, the width between the

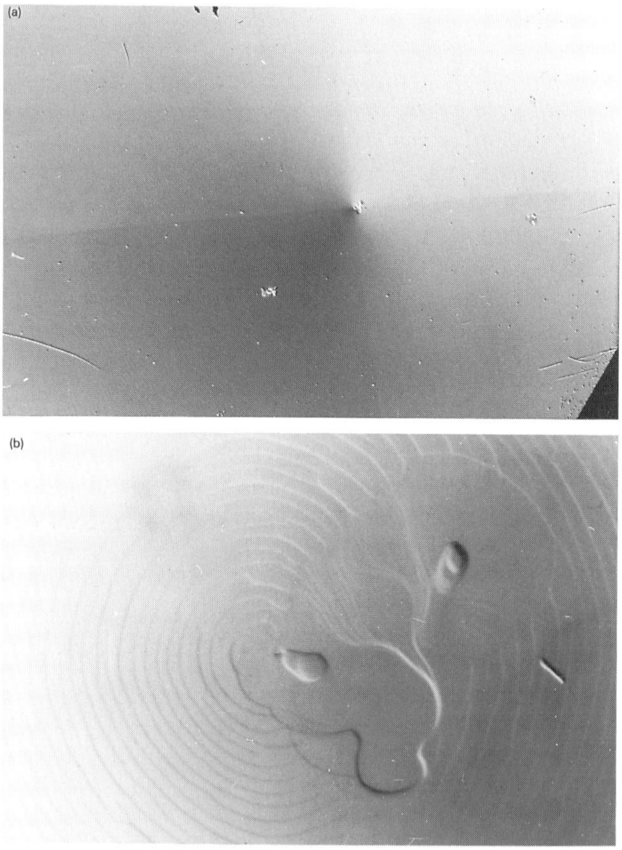

Fig. 15. Growth center of a (110) face of YIG [162]. (a) Micro-photograph using DI, total width 2.3 mm. Contrast line nearly vertical: [001] direction, contrast line nearly horizontal: [$\bar{1}$10] direction. (b) Micro-photograph with DI, larger magnification. The width between the steps is about 2 μm.

steps is about 2 Å. The contrast lines in fig. 15a are explained by Bennema et al. [163] as due to occurrence of interlaced spirals. The smallest Burgers vector of a screw dislocation giving rise to a monomolecular spiral on (110) faces has a height of 9 Å resulting in a minimum step height of 9 Å of the spiral. They assumed that the (110) face is built by double layers AB′ and BA′ of 9 Å height. The AB′ is stable but the BA′ decomposes and the B′ layer escapes rapidly to catch up with the next layer to form a stable AB′ layer. From the symmetry of (110) it can be seen, that indeed interlacing occurs in [001] and [$\bar{1}$10] with 4.5 Å which is half the spacing of the unit cell in the [110] direction. The alternating layer growth of two, in principle equal, ⟨111⟩ directions on the (110) face explains why, in the ⟨110⟩ growth sectors of a (111) wafer, no growth induced magnetic anisotropy (see [9]) is observed contrary to the adjacent ⟨211⟩ growth sectors where only one {111} in a growing {211} layer is present [81, 164]. The vicinal faces originating from a growth center on garnets have been described already by Lefever and Chase [165] and thier influence on the magnetic domains was investigated by Lefever et al. [166]. A correlation between growth hillocks and dislocations on the {110} faces of flux-grown YAG was published by Roberts and Elwell in [167], and on crystal habit and surface morphology in [168]. The formation of facets on YIG was studied by Tolksdorf and Bartels [80]. They used spherical seeds of gadolinium gallium garnet (GGG) and a dilute solution of YIG in a PbO/B_2O_3 flux with excess iron oxide indicated as "LPE region" in fig. 1b. An example is shown in fig. 16. A 4.5 μm thick layer was grown without forced convection with $f \approx 0.23$ μm/min on a GGG sphere of 2 mm diameter at a supercooling of 27 K which corresponds to a relative supersaturation $\Delta c/c_e$ of 0.3 related to the yttrium concentration as a practical approximation. For the micro-photograph in fig. 16a a non-magnifying objective lens was used for transmitting light. The illumination was made, with a condenser lens for high magnification, with the effect that a point light source was focused onto the center of the sphere. Even very small deviations from spherical shape are made visible by this simple technique. The most dominant facets were {110} and {211}; {321} facets were smaller and {210} and {332} facets were found to be very small and not completely flat. In fig. 16b a standard stereogram for [111] is shown for orientation. Figure 17 shows a typical (211) facet of about 370 μm diameter between two adjacent {321} facets. The facets frequently exhibit flat spiral-type growth hillocks as can be seen in fig. 17. The three larger facets have also been observed in YIG growth from a $PbO/PbF_2/B_2O_3$ flux on spherical YIG seeds with thicker layers. The endform corresponds to that of the YIG crystal in fig. 7. Bennema et al. [163] made a PBC (periodic bond chain) analysis of the garnet structure. They predicted facets in the order: {110}, {211}, {321}, {100}, {210} and {332} as confirmed by the sphere experiments. An exception is {100} which did not appear, but the areas near [100] showed a different growth mechanism than the atomically rough growing surface of the sphere away from the facets and in particular near [111], the direction generally used for LPE. Sticking of melt-droplets to the ⟨100⟩ directions on spheres, wafers and crystals proved the presence of wetting in these directions. The author would like to take the privilege to remark that the late John Robertson [169] confirmed the observations of YIG-LPE on spheres [80] by similar experiments. But contrary to YIG, he observed by LPE of yttrium aluminium garnet (YAG) on YAG spheres clearly

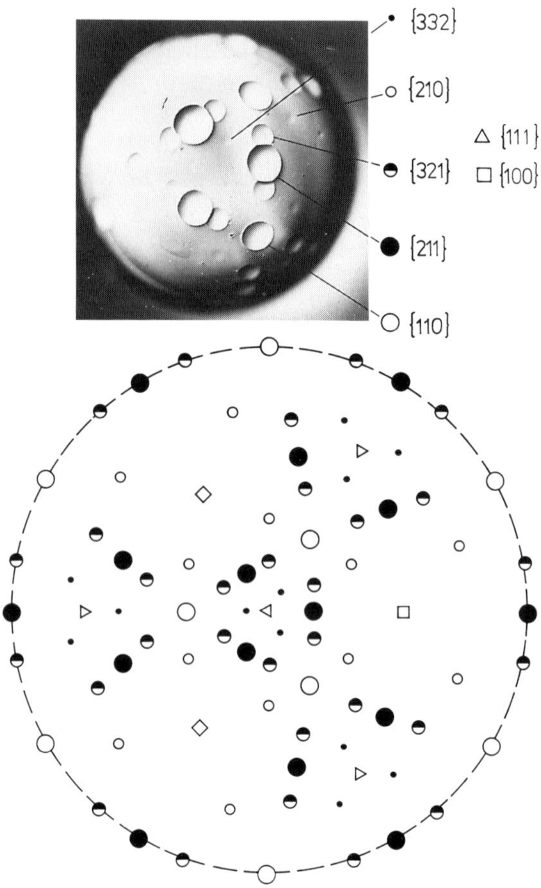

Fig. 16. YIG layer with an average thickness of 4.5 μm grown on a sphere of GGG with a diameter of 2 mm [80]. (a) Micro-photograph in transmitted light with a point light source focused onto the center of the sphere. (b) Standard stereogram on [111].

the {100} facets in a size larger than {321} and somewhat smaller than {110} and {211}. The growth rate f of the facets e.g. {211}, growing by a spiral-controlled monomolecular layer mechanism is much smaller than the atomically rough growing regions such as {111}. In the example of fig. 17 the ratio is $f_{\{211\}}/f_{\{111\}} \approx 0.1$, which is in good agreement with measurements of LPE wafers [170, 171]. Experimental evidence for a step-growth mechanism has been published by Hergt and Bürger [172], by Hergt and Pfeifer [173] and by Tsukamoto and van der Hoek [174].

Many investigations to study growth mechanisms and kinetics of flux growth have been made by LPE generally using PbO/B_2O_3-based fluxes e.g. [175, 176]. The advantages of LPE compared to SC bulk growth are evident [171]: T_g is easy adjustable and constant; supersaturation $\Delta c = c - c_e$ and supercooling $\Delta T = T_s - T_g$ generally can be kept constant. (The change during growth of a layer of 5 μm is

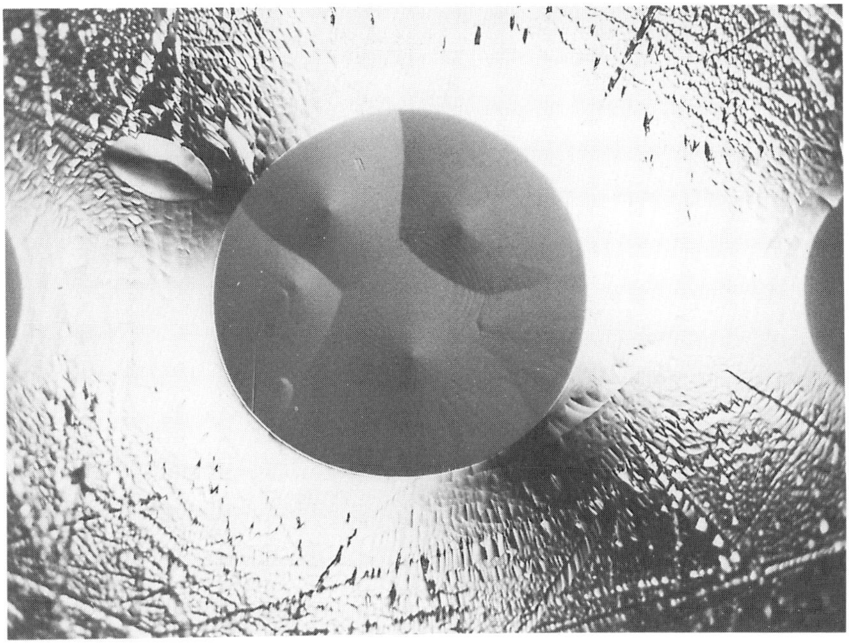

Fig. 17. Micro-photograph of a [211] facet with a diameter of about 370 μm in reflected light with DI [80].

generally less than 1%.) The linear growth rate f is, except for a short transient region, constant and easy to measure with a free choice of orientation of the substrate. It is rather easy to change f as a function of the supersaturation at the growing interface Δc_i by three parameters:

$$\Delta c_i = c - c_e(T) - \Delta c_D(\omega).$$

(1) Changing the concentration c and thereby the saturation temperature T_s and keeping T_g and the rotation rate ω constant.

(2) Changing T_g and thereby the equilibrium concentration c_e and keeping c and ω constant.

(3) Changing ω and hereby change of Δc_i by diffusion and keeping c and T_g constant.

The composition of the melt is rather easily changed, f is of the order of 1 μm/min and, even with all the procedures for homogenization and supersaturation, many growth runs can be performed in a time short compared to SCBG or TSSG. Due to the flow in the melt, the layer thickness is nearly constant over the cross section of the wafer.

For details the reader is referred to the chapter on LPE in vol. 3. Just one example is presented in fig. 18, which shows the dependence of f from the relative supersaturation. Layers of YIG were grown by LPE on a (111) GGG substrate rotating at 90 min^{-1} in three melts, which differed essentially only in the concentration of Y_2O_3 and thereby in the saturation temperature T_s [171]. For melt Y_1 and Y_2 T_g was constant, the relative supersaturation (driving force) and thereby the supercooling was increased by increas-

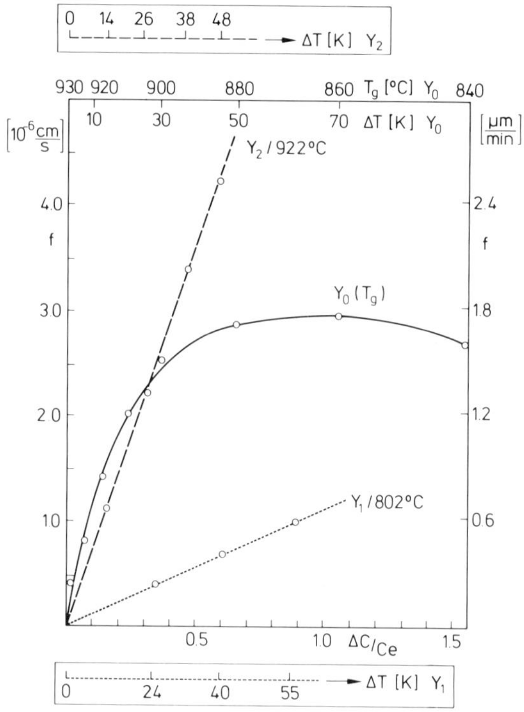

Fig. 18. The dependence of growth rate f on relative supersaturation $\Delta c/c_e$ of YIG layers grown with a rotation rate of $90\,\mathrm{min}^{-1}$ by LPE from three melts [171].

ing the Y_2O_3 concentration. There is a linear dependence of f on $\Delta c/c_e$. For melt Y_0, T_g was decreased and thereby ΔT (driving force) and c_e and hence $\Delta c/c_e$ increased. In this case f versus $\Delta c/c_e$ is not linear as shown also for order garnets [177]. This means that, at constant driving force, T_g is influencing f. At lower temperatures the incorporation of species from the melt is hampered, since viscosity and diffusivity are not changing very much in this temperature range. This example is given to demonstrate that for slow cooling growth, even when all other parameters could be kept constant, T_g is influencing the growth and thereby possibly the properties of the crystals as far as substitutions or impurities are concerned.

Contrary to the situation with LPE, f is difficult to measure: for e.g. SCBG crystals generally exhibiting facets, f is generally not constant with time or temperature. Even at an identical furnace temperature facets of the same kind, e.g. {110} for YIG on the same crystal or on a neighbouring one, generally have different growth rates even with ACRT due to complex convective flows influenced by the shape of the crucible and the shape of the growing crystal as shown by Görnert and Wende [178]. Short periodical distortions of the growth by temperature pulses induce striations which can be used as time markers. Damen and Robertson [82] and Görnert and Hergt [59] were the first to use these induced striations (IS) in flux growth, followed by Lillicrap and White [99]. Lawrence and Elwell [133] used the change of an electrical potential to induce striations in chromium-substituted YAG. The most intensive investigations using IS were made by a crystal growth group in Jena. Reviews of this method together with studies of the flux, bulk transport processes, interfacial growth and kinetics, including comparison to LPE applied to garnets, have been given by Görnert [179], by Görnert and Voigt [20] and by Görnert and Sinn [21]. By applying IS the effect of ACRT on the growth rate f was measured by Görnert [180]. He found e.g. an increase of f for Ga-substituted YIG from about 0.3 μm/min with natural convection to about 0.6 μm/min at a cooling rate of 0.7 K/h. The power of information of IS is demonstrated by Görnert and Hergt [181] in fig. 19, which shows an optical micrograph of a polished and chemically etched (110) cross section of a garnet crystal grown by SCBG marked by IS using short 2 K temperature distortions with a periodicity of $t = 54$ min. The striations at the right side are parallel to a $(1\bar{1}2)$ growth facet and on the left side parallel to a $(1\bar{1}0)$ growth facet. Growth and dissolution mechanisms are different and lead to different forms. Growth–dissolution kinetics of garnets are reported by Giess et al. [182]. Growth and dissolution rates as a function of the relative positive or negative supersaturation have been considered by Chernov and Malkin [183]. The growth form, e.g. of garnets as shown for YIG in fig. 7, which is essentially the same as for GGG, is quite different from the dissolution form in an undersaturated PbO/B_2O_3 solution [184]. When during growth, due to temperature increase, undersaturation occurs dissolution of the faces will cause a rounded form as shown for YIG [81]. A crystal plate cut parallel to the growth direction exhibits parallel striations in the growth sectors indicating the shape of the interface of the growing crystal as shown for a (110) cross section of YIG in fig. 19. When the temperature is raised above the saturation temperature the shape of the interface of the partly dissolved crystal will form bent lines at the growth sector boundaries cutting the parallel growth striations at the edges of the growth sector. The change of the growth striation from sharp edges to rounded lines by stepwise increase of temperature allows one to determine ΔT. Using this technique, Görnert et al. [185] measured ΔT of 1.2, 3 and 4 K for their growth conditions at T_g of about 980°C with a cooling rate of 0.5, 1 and 1.6 K/h, respectively. Without ACRT they measured $\Delta T > 5.7$ K. Tolksdorf et al. [64] measured $\Delta T \approx 3$ K using ACRT and a cooling rate of 0.2 K/h in their system at T_g of about 1065°C. The time to grow a certain length of crystal is given by the number of striations between two selected striations. The average growth rate f is determined by the distance of the two striations divided by the time. For example, in the lowest region of the crystal section in fig. 19 at the right side, f in the $[1\bar{1}2]$ direction is about 0.30 μm/min compared to 0.28 μm/min in the $[1\bar{1}0]$ direction at the left side. After further growing at higher f the temperature had been raised by 22 K far more than the supercooling of about 3 K necessary for the growth, causing a typical dissolution striation as described above. This formed a rounded line at the sector boundary on the trace of the interface at the time when the dissolution had stopped and growth continued due to further slow cooling. Due to an atomically rough growth mechanism f at the curved surface is much larger than f at the surface parallel to $[1\bar{1}2]$ or $[1\bar{1}0]$, as has been demonstrated with the sphere experiment [80], and the growth in the direction of the $[1\bar{1}1]$ sector boundary had become unstable as can be seen by the wavy form of the striations. This wavy form of instability is discussed by Chernov and

Fig. 19. Optical micrograph of a polished and chemically etched (110) cross section of a garnet crystal grown by SCBG marked by IS using short 2 K temperature distortions with a periodicity of $t = 54$ min. The striations at the right side are parallel to a $(1\bar{1}2)$ growth facet and on the left side parallel to a $(1\bar{1}0)$ growth facet [181].

Nishinaga [186]. Frequently this instability leads to flux inclusions as can be seen at the top of the dissolution striation in fig. 18. The shape of dissolution striations and the observation of inclusions before layer growth occurs again is reported also by other authors [64, 185, 187–189].

At the top part of fig. 18 in the layer-growth region at the right side, f in the $[1\bar{1}2]$ direction is about 0.22 μm/min compared to 0.14 μm/min in the $[1\bar{1}0]$ direction at the left side showing that not only the absolute values of f had changed but, more surprisingly, the ratio of f in the two directions had changed also. Due to different f in the striation regions the composition is different, too. Strocka and Willich [189] measured, by electronprobe microanalysis (EPMA), the lead content x along the growth direction $[1\bar{1}0]$ of a (111) wafer cut from a $Y_{3-z}Pb_zFe_5O_{12}$ crystal. In the dissolution striations z is about 0.008, whereas in the layer-growth region x is 0.014, Tolksdorf et al. [190] have shown that, with LPE of $Y_{3-x-z}Bi_xPb_zFe_{5-y}Al_yO_{12}$ on a (111) GGG substrate, x and z increase proportional to f. The effective segregation coefficients K_{eff} of Bi and Pb are much smaller than unity. The decrease of z in the

Table 7
Impurities in YIG grown from a $PbO/PbF_2/B_2O_3$ flux

Impurity	Molar units	wt%	Source
Pb	0.012–0.019	0.34–0.53	flux
Ca	< 0.001	< 0.005	flux
Fe^{2+}	0.005–0.010	0.04–0.08	p_{O_2}
B	0.002–0.006	0.003–0.009	flux
Si	0.001–0.003	0.004–0.012	oxides
F	0.010–0.016	0.026–0.042	flux
Pt	< 0.00001	< 0.00026	crucible

dissolution striations confirms that there is a different growth mechanism. The growth rate of the layers parallel to the surface (e.g. of (110)) is much larger than the linear growth rate f parallel to [110]. Willich and Strocka [188] found, in the case of $Y_{3-z}Pb_zFe_{5-y}Ga_yO_{12}$ in $[1\bar{1}0]$, that y decreases when z increases and vice versa, as has been shown also for LPE in [111] [190]. K_{eff} of Ga and of Al are larger than unity. As long as a crystal is growing with the same growth mechanism with increasing f the incorporation of substituents or impurities, having $K_{eff} < 1$ is increased and for $K_{eff} > 1$ it is decreased, as shown for Bi, Pb and Al in $Y_{3-x-z}Bi_xPb_zFe_{5-y}Al_yO_{12}$ [190] and for La, Pb and Ga in $Y_{3-x-z}La_xPb_zFe_{5-y}Ga_yO_{12}$ [191].

Typical impurities in SCBG-YIG are given in table 7 for crystals from a large number of growth runs from $PbO/PbF_2/B_2O_3$ fluxes [81]. The Pt content in these crystals is much smaller compared to values of 0.009 to 0.017 per formula unit (pfu) for LPE layers [192]. Lead will probably be incorporated generally as Pb^{2+}; roughly the charge compensation is given by Pt^{4+}, Si^{4+} and F^- and oxygen vacancies. Trivalent lead ions, Pb^{3+}, occupying dodecahedral sites, have been detected by electron-spin resonance (ESR) to occur in rare earth gallium and aluminium garnets [193]. The Pb content in YIG grown without PbF_2 is 0.01 pfu [45] a little less than the average value given in table 7. The incorporation of B in YIG LPE layers is of the same order of magnitude as in table 7 [194] and its value changes by analogy with Pb as f changes as measured by Tolle [195]. Local vibration modes in the IR spectrum of flux-grown iron and non-iron garnets have been identified and arise from a planar complex formed by a boron–oxygen group $[BO_3]^{3-}$ associated with the incorporation of boron, lead and oxygen vacancies [196]. These local modes have not been found in YIG LPE layers even with a thickness of 30 μm. This might be caused by the different growth mechanism. The values for the divalent iron (Fe^{2+}) have been calculated from the results of oxidimetric analysis. All iron garnet crystals show n-type behaviour, except those grown with a high amount of CaO in the flux, which are p-type and the presence of Fe^{4+} is assumed. Antonini et al [197] detected tetrahedral Fe^{2+} in Nb substituted yttrium iron garnet by angular variation of induced anisotropy.

To study the growth history of flux-grown garnets a cut parallel to (111) as indicated in fig. 7 is advisable because the growth facets are parallel to the zone axis [111] and separations of the striations are undistorted. X-ray reflection topography is very sensitive to changes in the lattice e.g. such as are caused by striations. Figure 20 shows

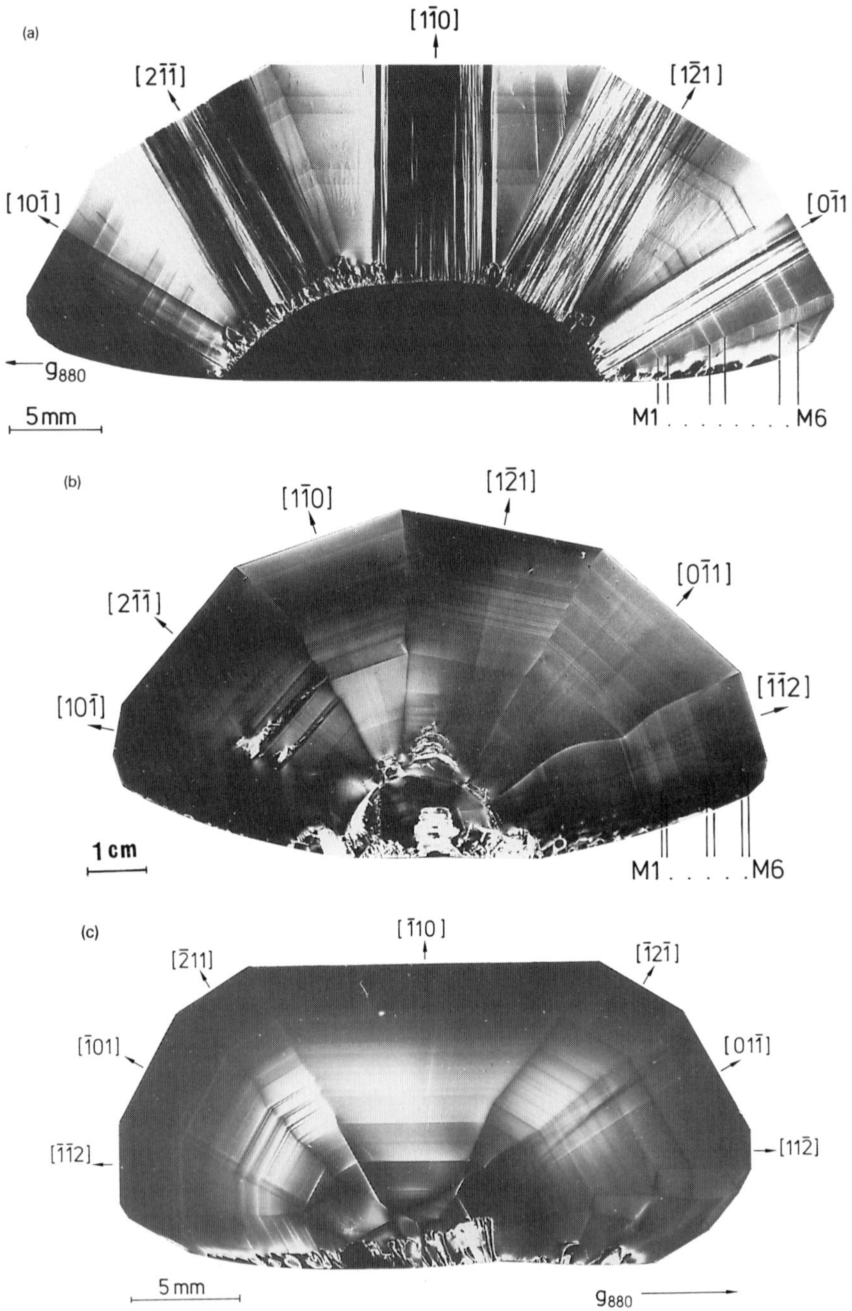

Fig. 20. X-ray topographs of 1 mm thick chemically polished (111) garnet wafers, Cu Kα1, (880) [64]. (a) $Y_3Fe_5O_{12}$ crystal with IS grown on a seed, the crystal and the location of the cut are shown in fig. 7. (b) $Y_3Ga_5O_{12}$ crystal with IS grown on a seed. (c) $Y_3Fe_{4.2}Ga_{0.8}O_{12}$ crystal grown by spontaneous nucleation.

X-ray reflection topographs of 1 mm thick chemically polished (111) garnet wafers, using Cu Kα1, (880) reflection, indicating the growth history and the imperfections. Figure 20a is a topograph of a wafer from the YIG crystal shown in fig. 7, where the location of the cut is indicated. The crystal was grown with IS on a seed. The striations had been induced by raising T_g by 0.5 K at a cooling rate of 0.15 K/h to 1060°C and 0.2 K/h to 1000°C. This abrupt temperature increase, being small compared to the supercooling of about 3 K, will only temporarily reduce f without otherwise disturbing the growth process. The seed crystal is not imaged, because its lattice constant is smaller by $\Delta a = -0.0087$ Å, due to its composition of $Y_3Fe_{4.5}Ga_{0.5}O_{12}$. From the contours of the image of fig. 20a it can be concluded that substantial dissolution of the seed had occurred during the seeding procedure, leading to an inclusion zone where bunches of dislocations were generated and propagated perpendicular to the interface similar to those described for YIG [62, 64], for potash alum [126], for $KNiF_3$ and $KCoF_3$ [187] or in other seeded solution-grown crystals [198]. The numbers of dislocations can be reduced by optimizing the seeding procedure as shown in fig. 20b for a $Y_3Ga_5O_{12}$ crystal grown, with IS, on a seed. The numbers of dislocations is generally less in spontaneously nucleated crystals, in spite of distortions at the first growth region and at the bottom of the crucible as it is shown for a $Y_3Fe_{4.2}Ga_{0.8}O_{12}$ crystal grown by spontaneous nucleation in fig. 20c. This distorted zone is reported also by Wende and Görnert [199] and by Bornmann et al. [60]. Only few dislocation lines are to be seen parallel to $[\bar{1}10]$ and $[01\bar{1}]$. This crystal was grown without IS but the contrast of the unintended striations is increased by the opposite enhancement of incorporation with varying f of Pb and Ga [189]. In fig. 20a f can be determined from the IS. f in the $[1\bar{1}0]$ direction varied from 0.67 μm/min at $T_g = 1051$°C (M4) to 0.27 μm/min at 1007°C. Contrary to the example given in fig. 19, the ratio of f in $[1\bar{2}1]$ to f in $[1\bar{1}0]$ varied only slightly in this growth period from 1.02 to 1.10. In the case of fig. 20b, this ratio was unity within the measurement error of 5%. At $T_g = 1152$°C (M2) f was 0.27 μm/min and decreased to 0.14 μm/min at $T_g = 1058$°C (M5). For unknown reasons occasionally flux inclusions occur as can be seen in the $[\overline{211}]$ growth sector. This disturbance of growth occurs preferentially in wetted areas near [100].

References

[1] D. Elwell and H.J. Scheel, Crystal Growth from High Temperature Solution (Academic Press, London, 1975).

[2] A. Smakula, Einkristalle (Springer, Berlin, 1962).

[3] J.J. Gilman, The Art and Science of Growing Crystals (Wiley, New York, 1963).

[4] R.A. Laudise, in: The Art and Science of Growing Crystals, ed. J.J. Gilman (Wiley, New York, 1963) p. 252.

[5] R.A. Laudise, The Growth of Single Crystals (Prentice-Hall, Englewood Cliffs, NJ, 1970).

[6] R.A. Lefever, Aspects of Crystal Growth (Dekker, New York, 1971).

[7] A.B. Chase, in: Aspects of Crystal Growth, ed. R.A. Lefever (Dekker, New York, 1971) p. 183.

[8] K.Th. Wilke and J. Bohm, Kristallzüchtung (Harri Deutsch, Thun, Frankfurt/Main, 1988).

[9] G. Winkler, Magnetic Garnets (Vieweg, Wiesbaden, 1981).

[10] J.W. Nielsen, in: Growth of Crystals Vol. 12, ed. A.A. Chernov (Nauka, Moscow, 1977; Translation in: Plenum Press, New York, 1984) p. 143.

[11] D. Elwell and Neate, J. Mater. Sci. 6 (1971) 1499.
[12] D. Elwell, in: Crystal Growth, Vol. 1, ed. B.R. Pamplin (Pergamon, Oxford, 1974) p. 185.
[13] D. Elwell, in: Crystal Growth, ed. B.R. Pamplin (Pergamon, Oxford, 1980) p. 463.
[14] D. Elwell, in: Crystal Growth in Science and Technology, eds. H. Ahrend and J. Hulliger (Plenum Press, New York, 1989) p. 133.
[15] B. Wanklyn, in: Crystal Growth, Vol. 1, ed. B.R. Pamplin (Pergamon, Oxford, 1974) p. 217.
[16] B. Wanklyn, J. Cryst. Growth 65 (1983) 533.
[17] V.A. Timofeeva and A.B. Bykov, Prog. Cryst. Growth Charact. 2 (1979) 377.
[18] W. Tolksdorf, in: Crystal Growth of Electronic Materials, ed. E. Kaldis (Elsevier, Amsterdam, 1985) p. 175.
[19] J.C. Brice, Prog. Cryst. Growth Charact. 1 (1978) 255.
[20] P. Görnert and F. Voigt, in: Current Topics in Material Science, Vol. 11, ed. E. Kaldis (North-Holland, Amsterdam, 1984) p. 1.
[21] P. Görnert and E. Sinn, in: Crystal Growth of Electronic Materials, ed. E. Kaldis (Elsevier, Amsterdam, 1985) p. 23.
[22] P. Görnert, Prog. Cryst. Growth Charact. 20 (1990) 263.
[23] B. Wanklyn, J. Cryst. Growth 37 (1977) 334.
[24] B. Wanklyn, J. Cryst. Growth 43 (1978) 336.
[25] E.M. Levin, C.R. Robbins and H.F. McMurdie, Phase Diagrams for Ceramists (Am. Ceram. Soc. Ohio, 1964; and Supplements 1969, 1975, 1981, 1983, 1984).
[26] J.F. Nielsen and R.R. Monchamp, in: Phase Diagrams, Vol. 3, ed. A.M. Alper (Academic Press, New York, 1970) p. 1.
[27] P. Paroli, Mater. Res. Bull. 18 (1983) 835.
[28] B. Wanklyn and B.E. Watts, J. Cryst. Growth 88 (1988) 143.
[29] D. Elwell and I.M. Coe, J. Cryst. Growth 44 (1978) 553.
[30] I.M. Coe and D. Elwell, J. Cryst. Growth 23 (1974) 345.
[31] I.M. Coe and D. Elwell, J. Cryst. Growth 33 (1976) 155.
[32] W. van Erk, J. Cryst. Growth 46 (1979) 539.
[33] W. van Erk, J. Cryst. Growth 57 (1982) 71.
[34] K. Nitsch, M. Matyás and J. Bok, Phys. Stat. Sol. (a) 98 (1986) 465.
[35] K. Fischer and E. Sinn, Cryst. Res. Technol. 23 (1988) Kl.
[36] J.F. Nielsen and E.F. Dearborn, J. Phys. Chem. Solids 5 (1958) 202.
[37] J.P. Remeika, J. Am. Chem. Soc. 78 (1956) 4259.
[38] J.W. Nielsen, J. Appl. Phys. Suppl. 29 (1960) 390.
[39] J.W. Nielsen, J. Appl. Phys. Suppl. 31 (1960) 51S.
[40] L.G. van Uitert, W.H. Grodkiewicz and E.F. Dearborn, J. Am. Ceram. Soc. 48 (1965) 105.
[41] W.H. Grodkiewicz, E.F. Dearborn and L.G. van Uitert, in: Crystal Growth, ed. H.S. Peiser (Pergamon, Oxford, 1967) p. 441.
[42] H.J. van Hook, J. Am. Ceram. Soc. 45 (1962) 369.
[43] L.L. Abernethy, T.H. Ramsay Jr. and J.W. Ross, J. Appl. Phys. 32 (1961) 376S.
[44] K.R. Nair, Mater. Res. Bull. 15 (1980) 1621.
[45] H.D. Jonker, J. Cryst. Growth 28 (1975) 231.
[46] W. Tolksdorf and F. Welz, J. Cryst. Growth 35 (1976) 285.
[47] J.M. Robertson, J.P.M. Damen, H.D. Jonker, M.J.G. van Hout, W. Kamminga and A.B. Voermans, Platinum Met. Rev. 18 (1974) 15.
[48] P. Hansen, J. Schuldt and W. Tolksdorf, Phys. Rev. B 8 (1973) 4274.
[49] A. Neuhaus, Chem. Ing. Tech. 28 (1956) 350.
[50] B. Wanklyn, B.E. Watts and V.V. Fenin, J. Cryst. Growth 70 (1984) 459.
[51] B. Wanklyn and B.J. Garrard, Mater. Res. Bull. 17 (1982) 903.
[52] B. Wanklyn and B.E. Watts, Mater. Res. Bull. 19 (1984) 711.
[53] V.A. Timofeeva and R.A. Voskanyan, Sov. Phys. Cryst. 6 (1961) 641.
[54] E.A. Giess, J. Am. Ceram. Soc. 45 (1962) 53.
[55] J. Kvapil, Krist. Tech. 1 (1966) 97.
[56] B.V. Zaitsev, S.Sh. Gendelev, A.G. Titova and W.G. Karilenko, Élektronnaya Tekhnika Ser. 7. No. 4 (1968) 10.

[57] H.H. Quon and A.J. Potvin, J. Cryst. Growth 10 (1971) 124.
[58] W. Tolksdorf and F. Welz, J. Cryst. Growth 20 (1973) 47.
[59] P. Görnert and R. Hergt, Phys. Stat. Sol. (a) 20 (1973) 577.
[60] S. Bornmann, E. Glauche, P. Görnert, R. Hergt and C. Becker, Krist. Tech. 9 (1974) 895.
[61] L.G. L.G. van Uitert, G. Zydzik, W.H. Grodkiewicz and W.A. Bonner, Mater. Res. Bull. 10 (1975) 473.
[62] W. Tolksdorf, J. Cryst. Growth 42 (1977) 275.
[63] A.W. Morris and D. Elwell, J. Mater. Sci. 14 (1979) 2139.
[64] W. Tolksdorf, B. Strocka and F. Welz, J. Cryst. Growth 65 (1983) 549.
[65] S. Durcok and E. Pollert, J. Cryst. Growth 82 (1987) 501.
[66] R.C. Linares, J. Am. Ceram. Soc. 45 (1962) 307.
[67] R.C. Linares, J. Appl. Phys. 35 (1964) 433.
[68] M. Kestigian, J. Am. Ceram. Soc. 50 (1967) 165.
[69] G.H. Bennett, J. Cryst. Growth 3 & 4 (1968) 458.
[70] W. Tolksdorf, Acta Electronica 17 (1974) 57.
[71] D. Elwell, P. Capper and M. D' Agostino J. Cryst. Growth 29 (1975) 321.
[72] L.N. Bezmaternykh, V.G. Maschenko, N.A. Sokolova and V.L. Temerov, J. Cryst. Growth 69 (1984) 407.
[73] V. Nikolov, K. Iliev and P. Peshev, J. Cryst. Growth 75 (1986) 269.
[74] S.Sh. Gendelev, L.D. Fedorovich and B.V. Zaitsev, Inorg. Mater. 8 (1972) 746.
[75] H.J. Scheel and E.O. Schulz-DuBois, J. Cryst. Growth 8 (1971) 304.
[76] E.O. Schulz-DuBois, J. Cryst. Growth 12 (1972) 81.
[77] J.C., Brice, P. Capper, C.L. Jones and J.J.G. Gosney, Prog. Cryst. Growth Charact. 13 (1986) 197.
[78] W. Tolksdorf, J. Cryst. Growth 3 & 4 (1968) 463.
[79] W. Tolksdorf, in: Current Topics in Material Science, Vol. 2, ed. E. Kaldis (North-Holland, Amsterdam, 1977) p. 639.
[80] W. Tolksdorf and I. Bartels, J. Cryst. Growth 54 (1981) 417.
[81] W. Tolksdorf and F. Welz, in: Crystals: Growth, Properties, and Applications, ed. C.J.M. Rooijmans (Springer, Berlin, 1978) p. 1.
[82] J.P.M. Damen and J.M. Robertson, J. Cryst. Growth 16 (1972) 50.
[83] B. Wanklyn, J. Cryst. Growth 54 (1981) 610.
[84] B. Wanklyn, F.R. Wondre and W. Davison, J. Mater. Sci. Lett. 3 (1984) 539.
[85] B. Wanklyn, B.E. Watts, F.R. Wondre and W. Davison, J. Mater. Sci. Lett. 5 (1986) 499.
[86] B.E. Watts, H. Dabkowska and B. Wanklyn, J. Cryst. Growth 94 (1989) 125.
[87] S.H. Smith and D. Elwell, J. Mater. Sci. 2 (1967) 297.
[88] T.M. Bruton and E.A.D. White, J. Cryst. Growth 19 (1973) 341.
[89] A. Maqsood, J. Mater. Sci. Lett. 1 (1982) 303.
[90] A. Maqsood and B. Wanklyn, J. Mater. Sci. 15 (1980) 405.
[91] K. Oka and H. Unoki, J. Appl. Phys. 56 (1984) 436.
[92] P. Capper and D. Elwell, J. Cryst. Growth 26 (1974) 65.
[93] R. Hiskes, J. Cryst. Growth 27 (1984) 287.
[94] I. Shindo, K. Kitamura and S. Kimura, J. Cryst. Growth 46 (1979) 307.
[95] M. Torii, U. Kihara, H. Goto and H. Yuri, IEEE Trans. Magn. MAG-16, No 5 (1980) 678.
[96] A.M. Balbashov and S.K. Egorov, J. Cryst. Growth 52 (1981) 498.
[97] S. Kimura, K. Kitamura and I. Shindo, J. Cryst. Growth 65 (1983) 543.
[98] B. Feigelson, in: Crystal Growth of Electronic Materials, ed. E. Kaldis (Elsevier, Amsterdam, 1985) p. 127.
[99] B.J. Lillicrab and E.A.D. Withe, J. Cryst. Growth 32 (1976) 250.
[100] D.F. Nelson and J.P. Remeika, J. Appl. Phys. 35 (1964) 522.
[101] J.P. Remeika, J. Am. Chem. Soc. 76 (1954) 940.
[102] H. Makram, L. Touron and J. Loriers, J. Cryst. Growth 13 & 14 (1972) 585.
[103] W.A. Bonner, E.F. Dearborn and L.G. van Uitert, in: Crystal Growth, ed. H.S. Peiser (Pergamon, Oxford, 1967) p. 437.
[104] D. Rytz and H.J. Scheel, J. Cryst. Growth 59 (1982) 468.
[105] H.J. Scheel and J. Sommerauer, J. Cryst. Growth 62 (1982) 291.
[106] A.A. Ballmann, H. Brown, D.H. Olson and C.E. Rice, J. Cryst. Growth 75 (1986) 390.

[107] J.C. Jacco, G.M. Loiacona, M. Jaso, G. Mizell and B. Greenberg, J. Cryst. Growth 70 (1984) 484.
[108] I.H. Brunskill, R. Boutellier, W. Depmeier, H. Schmid and H.J. Scheel, J. Cryst. Growth 56 (1982) 541.
[109] L.G. van Uitert, D.H. Smith, W.A. Bonner, W.H. Grodkiewicz and G.J. Zydzik, Mater. Res. Bull. 5 (1970) 455.
[110] E. Pollert, M. Nevrira, L. Matejková and J. Novák, Mater. Res. Bull. 16 (1981) 1499.
[111] E.A. Giess, D.C. Cronemeyer, L.L. Rosier and J.D. Kupsis, Mater. Res. Bull. 5 (1970) 495.
[112] L.I. Maltseva, N.I. Leonyuk and T.I. Timchenko, Krist. Tech. 15 (1980) 35.
[113] Y. Arai, M. Satou and K. Tsushima, J. Cryst. Growth 36 (1976) 255.
[114] Y. Arai, M. Satou and K. Tsushima, J. Cryst. Growth 36 (1976) 358.
[115] P. Röschmann, M. Lemke, W. Tolksdorf and F. Welz, Mater. Res. Bull. 19 (1984) 385.
[116] D. Mateika and R. Laurien, J. Cryst. Growth 52 (1981) 566.
[117] D. Rytz, B.A. Wechsler, C.C. Nelson and K.W. Kirby, J. Cryst. Growth 99 (1989) 864.
[118] K. Bethe and F. Welz, Mater. Res. Bull. 6 (1971) 209.
[119] U. Flückiger and H. Arend, J. Cryst. Growth 43 (1978) 406.
[120] Wu Xing, H. Losser, H, Wüest and H. Arend, J. Cryst. Growth 78 (1986) 431.
[121] B. Hellermann, F. Baller, B. Gather, H. Hesse and R. Pankrath, Ferroelectrics 124 (1991) 67.
[122] T.M. Bruton, J.C. Brice, O.F. Hill and P.A.C. Whiffin, J. Cryst. Growth 23 (1974) 21.
[123] T.R. AuCoin, R.O. Savage and A. Tauber, J. Appl. Phys. 37 (1966) 2908.
[124] W. Eigermann, G. Müller-Vogt and W. Wendl, J. Cryst. Growth 69 (1984) 605.
[125] P.F. Bordui, J.C. Jacco, G.M. Loiacono, R.A. Stolzenberger and J.J. Zola, J. Cryst. Growth 84 (1987) 403.
[126] S. Gits-Leon, F. Lefaucheux and M.C. Robert, J. Cryst. Growth 44 (1978) 345.
[127] A.A. Chernov, in: Lecture Notes ISSCG-7 (Zao, Yamagata, Japan, 1989).
[128] W.R. Wilcox, in: Lecture Notes ISSCG-7 (Zao, Yamagata, Japan, 1989).
[129] P.F. Bordui and J.C. Jacco, J. Cryst. Growth 82 (1987) 351.
[130] D.P. Shumov, V.S. Nikolov, K.N. Iliev and A.L. Aleksandrovsli, Cryst. Res. Technol. 25 (1990) 1245.
[131] T.M. Bruton, J. Cryst. Growth 35 (1976) 36.
[132] J.F. Balascio and W.B. White, J. Cryst. Growth 23 (1974) 101.
[133] C.M. Lawrence and D. Elwell, J. Cryst. Growth 32 (1976) 287.
[134] Tu. Chen, J. Cryst. Growth 20 (1973) 29.
[135] P.A.C. Whiffin and J.C. Brice, J. Cryst. Growth 23 (1974) 25.
[136] S.H. Smith, G. Garton and B.K. Tanner, J. Cryst. Growth 23 (1974) 335.
[137] W. Tolksdorf, J. Cryst. Growth 18 (1973) 57.
[138] P. Röschmann, W. Tolksdorf and F. Welz, IEEE Trans. MAG-14 (1978) 704.
[139] S.B. Austermann, J. Cryst. Growth 42 (1977) 248.
[140] A. Eyer, R. Nitsche and H. Zimmermann, J. Cryst. Growth 47 (1979) 219.
[141] K. Kitamura, S. Kimura and K. Watanabe, J. Cryst. Growth 57 (1982) 475.
[142] M. Lemke, W. Hoppe. W. Tolksdorf and F. Welz, Microwaves Opt. Acoust. 3 (1979) 253.
[143] I. Shindo, S. Kimura and K. Kitamura, J. Mater. Sci. 14 (1979) 1901.
[144] J.J. Brissot and C. Bellin, J. Cryst. Growth 8 (1971) 213.
[145] Ch. Belin, J. Cryst. Growth 34 (1976) 341.
[146] C.E. Turner, N.H. Mason and A.W. Morris, J. Cryst. Growth 56 (1982) 137.
[147] B. Antonini, A. Paoletti and P. Paroli, J. Cryst. Growth 54 (1981) 586.
[148] J.D.C. Wood and E.A.D. White, J. Cryst. Growth 3 & 4 (1968) 480.
[149] R.D. Plättner, W.W. Krühler, W.K. Zwicker, T. Kovats and S.R. Chinn, J. Cryst. Growth 49 (1980) 274.
[150] D. Elwell, in: Current Topics in Material Science, Vol. 2, ed. E. Kaldis (North-Holland, Amsterdam 1977) p. 639.
[151] S. Mroczkowski, High Temp. Sci. 19 (1985) 29.
[152] R.G. Rudness and R.W. Kebler, J. Am. Ceram. Soc. 43 (1960) 17.
[153] S. Nomura, T. Yamashita and S. Shimanuki, J. Magn. Soc. Jpn. 11 Suppl. S1 (1987) 187.
[154] P. Görnert, E. Sinn and M. Rösler, Key Eng. Mater. 58 (1991) 129.
[155] K. Tsukamoto, T. Abe and I. Sunagawa, J. Cryst. Growth 63 (1983) 215.
[156] W.K. Burton, N. Cabrera and F.C. Frank, Phil. Trans. R. Soc. 243 A (1951) 299.

[157] P. Bennema and G.H. Gilmer, in: Crystal Growth: An Introduction, ed. P. Hartmann (North-Holland, Amsterdam, 1973) p. 263.
[158] I. Sunagawa and P. Bennema, in: Preparation and Properties of Solid State Materials. Vol. 7, ed. W.R. Wilcox (Dekker, New York, 1982) p. 235.
[159] J.A. Kohn, D.W. Eckart and C.F. Cook Jr., Mater. Res. Bull. 2 (1967) 55.
[160] H. Komatsu, in: Crystal Growth of Electronic Materials, ed. E. Kaldis (Elsevier, Amsterdam, 1985) p. 359.
[161] I. Sunagawa and P. Bennema, J. Cryst. Growth 53 (1981) 490.
[162] W. Tolksdorf and F. Welz, J. Cryst. Growth 13 & 14 (1972) 566.
[163] P. Bennema, E.A. Giess and J.E. Weidenborner, J. Cryst. Growth 62 (1983) 41.
[164] W.T. Stacy and W. Tolksdorf, AIP Conf. Proc. 5 (1971) 185.
[165] R.A. Lefever and A.B. Chase, J. Am. Ceram. Soc. 45 (1962) 32.
[166] R.A. Lefever, K.A. Wickersheim and A.B. Chase, J. Phys. Chem. Solids 26 (1965) 1529.
[167] K.J. Roberts and D. Elwell, J. Cryst. Growth 62 (1983) 449.
[168] K.J. Roberts and D. Elwell, J. Cryst. Growth 53 (1981) 249.
[169] J.M. Robertson, private communication (1983).
[170] W. van Erk, H.J.G.J. Van Hoek-Martens and G. Bartels, J. Cryst. Growth 48 (1980) 621.
[171] W. Tolksdorf, in: Crystal Growth in Science and Technology, eds. H. Ahrend and J. Hulliger (Plenum Press, New York, 1989) p. 397.
[172] R. Hergt and W. Bürger, Cryst. Res. Technol. 16 (1981) 761.
[173] R. Hergt and H. Pfeiffer, Phys Stat. Sol. (a) 74 (1982) 97.
[174] K. Tsukamoto and B. van der Hoek, J. Cryst. Growth 57 (1982) 131.
[175] R. Ghez and E.A. Giess, Mater. Res. Bull. 8 (1973) 31.
[176] E.A. Giess, in: Advanced Crystal Growth, eds. P.M. Dryburg, B. Cockayne, and K.G. Barraclough (Prentice-Hall, London, 1987) p. 245.
[177] R. Ghez and E.A. Giess, J. Cryst. Growth 27 (1974) 221.
[178] P. Görnert and G. Wende, Phys. Stat. Sol. (a) 37 (1976) 505.
[179] P. Görnert, Krist. Tech. 15 (1980) 627.
[180] P. Görnert, J. Cryst. Growth 52 (1981) 88.
[181] P. Görnert and R. Hergt, Phys. Stat. Sol. (a) 47 (1978) 99.
[182] E.A. Giess, M.M. Factor, R. Ghez and C.F. Guerci, J. Cryst. Growth 56 (1982) 576.
[183] A.A. Chernov and A.I. Malkin, J. Cryst. Growth 92 (1988) 432.
[184] R.B. Heimann and W. Tolkdorf, J. Cryst. Growth 62 (1983) 75.
[185] P. Görnert, S. Bornmann and R. Hergt, Phys. Stat. Sol. (a) 35 (1976) 583.
[186] Chernov A.A. and T. Nishinaga, in: Morphology of Crystals, ed. I. Sunagawa (Terra Scientific (TERRAPUP), Tokyo, 1987) p. 211.
[187] M. Safa, B.K. Tanner, H. Klapper and B.M. Wanklyn, Phil. Mag. 35 (1977) 811.
[188] P. Willich and B. Strocka, Beitr. Elektronenmikroskop. Dierektabb. Oberfl. 13 (1980) 23.
[189] B. Strocka and P. Willich, J. Cryst. Growth 56 (1982) 605.
[190] W. Tolksdorf, H. Dammann, E. Pross, B. Strocka, F. Welz and P. Willich, J. Cryst. Growth 99 (1990) 616.
[191] W. Tolksdorf, G. Bartels and H.J. Tolle, J. Cryst. Growth 52 (1981) 722.
[192] W. Tolksdorf, H. Dammann, E. Pross, B. Strocka, H.J. Tolle and P. Willich, J. Cryst. Growth 83 (1987) 15.
[193] B. Andlauer, J. Schneider and W. Tolksdorf, Phys. Rev. B 8 (1973) 1.
[194] W. Tolksdorf, H.J. Tolle and C.P. Klages, J. Cryst. Growth 56 (1982) 217.
[195] H.J. Tolle private communication (1985).
[196] B. Andlauer and W. Tolksdorf, J. Appl. Phys. 50 (1979) 7986.
[197] B. Antonini, S. Geller, A. Paoletti, P. Paroli and A. Tucciarone, Phys. Rev. Lett. 41 (1978) 1556.
[198] M. Safa, B.K. Tanner, H. Klapper and B.M. Wanklyn, Phil. Mag. 35 (1977) 811.
[199] H. Klapper, in: Characterization of Crystal Growth Defects by X-Ray Methods, eds. K. Tanner and D.K. Bowen (NATO Advanced Study Institutes Series, Serie B: Physics, Vol. 63 (Plenum Press, New York, 1980)) p. 1.
[200] G. Wende and P. Görnert Phys. Stat. Sol. (a) 41 (1977) 263.

CHAPTER 11

Bulk Crystal Growth by Physical Vapour Transport

E. KALDIS and M. PIECHOTKA

Laboratory for Solid State Physics, ETH-Hönggerberg, CH-8093 Zürich, Switzerland

Contents

1. Introduction . . . 615
2. Principles of mass transport by PVT . . . 616
   2.1. Introduction . . . 616
   2.2. Stefan flow . . . 617
   2.3. Diffusive/advective flow in cylindrical ampoules . . . 617
   2.4. Free convection in cylindrical ampoules . . . 621
   2.5. Interplay between surface kinetics and transport . . . 623
   2.6. Dissociative sublimation . . . 629
   2.7. Morphological stability . . . 632
3. Some experimental growth techniques . . . 634
   3.1. Introduction . . . 634
   3.2. Sealed ampoule method . . . 634
   3.3. Nearly-sealed ampoule method . . . 635
   3.4. Temperature oscillation method . . . 636
4. Mercury iodide: a case study . . . 638
   4.1. A model substance for vapour growth investigations . . . 638
   4.2. Experimental: control of nucleation and growth of large crystals . . . 640
   4.3. Growth rate measurements, the rate-determining step . . . 642
   4.4. Crystal size, habit and orientation . . . 643
   4.5. Evidence for convective contributions . . . 645
      4.5.1. Nucleation and growth experiments in the presence of an inert gas . . . 645
      4.5.2. Striations in vapour-grown crystals . . . 648
   4.6. Structural defects in HgI_2 crystals . . . 650
5. II–VI compounds: model substances for high-temperature vapour growth . . . 654
6. The potential of bulk vapour growth . . . 655
References . . . 656

HANDBOOK OF CRYSTAL GROWTH, VOL. 2
edited by D.T.J. Hurle

1. Introduction

The advantage of vapour growth techniques is that the crystals tend to have a low concentration of point defects and low dislocation densities compared to crystals grown from the melt, the reason being that the temperatures employed are usually considerably lower than the melting temperature. Moreover, if the material undergoes a phase transformation, or if it melts incongruently, vapour growth can be the only choice for the growth of single crystals.

Historically, vapour growth methods in use up until the 1960s were mainly for the preparation of bulk crystals. With the enormous increase in the importance of thin films in electronic and metallurgical applications, CVD methods have now become predominant in vapour growth applications. However, as interest in more complex materials for science and technology has developed, growth of bulk crystals from the vapour, e.g. for substrates for thin films, has gained appreciable importance. Basically, the higher the perfection of the substrate, the higher is the perfection of the epitaxial layer.

Of course, for Si and GaAs, melt growth is the preferred method for the production of bulk crystals. Presently however, several compounds with interesting electronic properties have been developed which, due to their limited thermal stability, cannot be grown from the melt. Compared to melt growth, a disadvantage of vapour growth is the relatively low growth rate typically encountered. Linear growth rates from the melt range from millimeters to centimeters per hour, whereas vapour growth rates, which are similar to those for solution growth, are commonly of the order of millimeters per day. The growth of bulk crystals from the vapour for high-technology applications is, therefore, of practical interest only if the material cannot be grown from the melt. Also, as we shall see, the mass gain per unit time which is the practically important parameter increases strongly with the crystal volume. Thus long-term growth experiments can produce very large average mass uptakes. However, to reach these conditions a very high degree of control of the vapour growth process is necessary which, in most cases, has not yet been achieved technologically.

Several general classes of techniques exist for the growth of bulk single crystals from the vapour phase. They have been reviewed in detail in the past [1–3]. They are differentiated by the nature of the source material (e.g. whether it is solid or vapour) and by the means and mechanisms by which it is transported to the growing crystal surface. The conceptually simplest technique is that of sublimation where the source material is placed at one end of a sealed tube and heated so that it sublimes and is then transported to the colder region of the tube where it crystallises. If the source material is a dissociable compound then we have dissociative sublimation, the transport processes are more complex and crystal stoichiometry becomes an issue to be considered.

To further control the transport processes, an inert carrier gas can be introduced into the sealed tube. Again transport can be facilitated by means of a reversible chemical reaction where an active vapour species (often a halogen) reacts with the charge to produce a volatile species (a halide) which is transported to the crystallisation zone and where, because of the temperature dependence of the free energy of the reaction, the reaction is reversed leading to the deposition of the element or compound to be crystallised with release of the carrier (halogen) which is recycled. The van Arkel process for the deposition of tungsten is a familiar example but, in this case, the product is not required to be a single crystal.

Thus far only transport in a closed tube has been mentioned. A central advantage of this approach is that there is no loss of charge material; the conversion to single crystal is in principle 100%. Additional advantages are that only simple furnaces and instrumentation are required and hazards in the use of toxic gases are reduced. The disadvantage is that all impurities and residual gases are retained within the system and these can inhibit the growth process. However, we shall see that this deficiency can be mitigated by the use of "nearly-closed" systems. For ultimate purity though, open tube flow systems are chosen but these are largely confined to thin film deposition where the loss of charge material to the system exhaust is not an important economic factor. However, mention can be made of two examples where high-quality single crystals have been obtained in flow tube systems. The first is the work of Savage and Dobson who used hydrogen to reduce a zinc oxide furnace liner. The zinc vapour so produced was carried downstream to a region where an oxygen–nitrogen mixture was introduced which re-oxidised the zinc resulting in the deposition of rod-shaped single crystals of ZnO of several cm length. The second example is the work of Schafer who prepared large alumina single crystals by reacting aluminium trichloride gas with hydrogen in the presence of carbon dioxide. Under reduced pressure (< 50 Torr) he was able to promote growth onto a single crystal seed at temperatures in excess of 1550°C. Both methods are discussed in [1].

This chapter is confined to consideration of physical vapour transport (hereafter referred to by the acronym PVT) in closed and nearly-closed systems since this is the most widely exploited technique for the production of electronic and electro-optical quality single crystals in bulk form. Chemical vapour deposition of epitaxial thin films will be considered in much detail in Volume III of this Handbook. In section 2 the underlying principles of transport in PVT are enunciated. Nucleation and growth kinetics have been comprehensively reviewed in Volume I. In section 3 the main techniques of PVT are described. Section 4 is a detailed case study of a single material – mercury iodide – a technologically important substance for which PVT is the most promising growth technique. The growth of II–VI compounds is briefly considered in section 5 and the chapter concludes with a summary of the potential of PVT for the production of bulk crystals.

2. *Principles of mass transport by PVT*

2.1. *Introduction*

In an isothermal gas in a gravitationally-free environment the mechanisms of transport are by Fickian diffusion and advection. However, if a gravitational field is imposed then compositional gradients in the gas produce buoyancy and result in solutal convection. Further, in the presence of an imposed temperature gradient, thermal convection will also be present. Finally, it is known that the Soret effect (sometimes called, rather confusingly, thermal diffusion) – the diffusion of matter induced by an imposed temperature gradient – is important in many situations.

2.2. *Stefan flow*

In studies in 1882 on the evaporation of liquids from pipes Stefan [4] concluded that vapour is not transported by diffusion alone as had been previously thought. The large volume change on evaporation produced an advective flow (which we now call Stefan flow [3]). This is central to an understanding of PVT.

If this advective velocity is denoted by U, then, in a one-dimensional system (coordinate x) the flux of a vapour having a vapour pressure p_A and a diffusion coefficient D_{AB} will be:

$$J_A = Up_A/RT - (D_{AB}/RT)\,dp_A/dx. \tag{1}$$

where R is the gas constant and T the temperature. If "A" is the species which is crystallising then J_A represents the growth rate. The subliming species A is deemed to be diffusing into an inert gas (B) at a pressure p_B. The flux of this gas will be:

$$J_B = Up_B/RT - (D_{BA}/RT)\,dp_B/dx = 0. \tag{2}$$

The total pressure in the system is $p = p_B + p_A$. Further, $dp/dx \approx 0$, so that:

$$U \approx J_A RT/p. \tag{3}$$

Substituting eq. (3) into eq. (1), we obtain:

$$J_A = -[(D_{AB}/RTp)/(p - p_A)]\,dp_A/dx, \tag{4}$$

which is the relationship first proposed and experimentally verified by Stefan. The physical mechanism causing this flow is the volume increase on sublimation, an effect equivalent to pumping.

2.3. *Diffusive/advective flow in cylindrical ampoules*

A one-dimensional model has been used for the calculation of the advective–diffusive flux. Figure 1 shows the assumed idealised ampoule geometry, with parallel flat walls

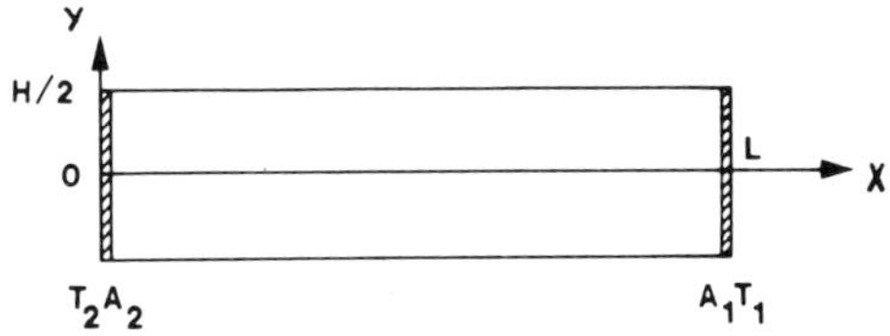

Fig. 1. Cross section of a cylindrical ampoule for steady-state PVT growth. As both source and sink are covered with the compound A, transients due to nucleation on the sink substrate do not have to be considered. Hot wall, T_2, source of the A vapour molecules (A_2). Colder wall, T_1, crystal of A growing at the sink (A_1).

covered with the material A (A_2 is the source and A_1 the crystal sink). For endothermic processes like sublimation (PVT), the source temperature T_2 is higher than the sink temperature T_1. Although only molecules of A need to be assumed for sublimation of the pure material, we are interested in the transport process when A molecules must be transported through an inert gas or a gaseous impurity B, which is rejected by the sink. For the case that advective–diffusive transport prevails over free convection the molar flux (moles per unit area per unit time) of the species A, ignoring wall effects, is given by eq. (4). The temperature and pressure dependence of D_{AB} is given approximately by:

$$D_{AB} = D_0(T/T_0)^n(p_0/p), \tag{5}$$

with $0.05 < D_0 < 2\,cm^2s^{-1}$, $n = 1.8$, $T_0 = 273\,K$, $p = 1$ atm and $p = p_A + p_B$ the total vapour pressure [14]. The local coordinate x is located at the source. In terms of molar concentrations, this flux can also be written as [5]:

$$J_A = -D_{AB}(dC_A/dx)/(1 - X_A), \tag{6}$$

with $X_A = C_A/C = p_A/p$ the mole fraction and $C_A = p_A/RT$, $C = p/RT$ the molar concentration (moles per unit volume).

When L is the distance between the source and sink, where the partial pressures of A are p_{A2} and p_{A1}, respectively, the molar flux of the species A becomes:

$$J_A = (pD_{AB}/RTL)\ln\{(p - p_{A1})/(p - p_{A2})\}, \tag{7}$$

or, in terms of the temperature difference between source and sink, $\Delta T = T_2 - T_1$:

$$J_A = (pD_{AB}/RTL)\ln(1 + p_{A2}\,\Delta H_e\,\Delta T/(p - p_{A2})\,RT^2). \tag{8}$$

Several formulations are possible for the pressure ratio in eq. (7):

$$\frac{p - p_{A1}}{p - p_{A2}} = \frac{p_{B1}}{p_{B2}} = \frac{C - C_{A1}}{C - C_{A2}} = \frac{1 - x_A(L)}{1 - x_A(0)} = \frac{x_B(L)}{x_B(0)}.$$

The total molar flux can be written as:

$$J_A + J_B = UP/RT = UC, \tag{9}$$

with U the average velocity of the molecules [3]

Reference [6] introduces the Péclet number as the figure of merit of the advective flux relative to the diffusive flux, i.e.

$$Pe = UL/D_{AB}. \tag{10}$$

The characteristic length L is taken to be equal to the distance between source and crystal. The Péclet number then becomes:

$$Pe = \ln\{[1 - x_A(L)]/[1 - x_A(0)]\}. \tag{11}$$

When the temperature dependence of the vapour density and diffusion coefficient is neglected, the molar concentration C_A of the vapour species A changes along the coordinate x from the source as [6]

$$C_A(x) = C - [C - C_A(0)]\exp(Pe\,x/L). \tag{12}$$

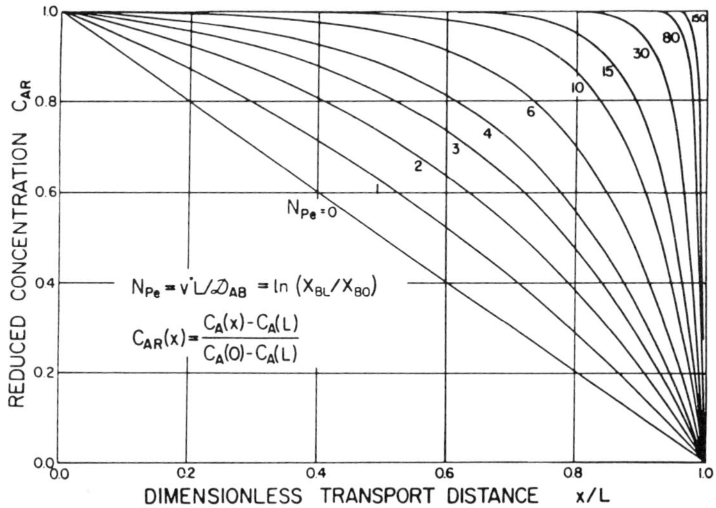

Fig. 2. PVT of A diffusing through a stagnant impurity B. Reduced concentration versus dimensionless transport distance for various Péclet numbers illustrating the increasing B accumulation at the crystal with increasing transport rate [6].

The concentration distribution is shown in normalised form in fig. 2.

For $Pe < 1$ (diffusive transport dominating) the partial pressure $p_A = C_A/RT$ of the A molecules is expected to decrease linearly from the source to the sink [6] as shown in fig. 2. This linear variation, first derived analytically by Faktor et al. [3], has been proved experimentally by Omaly et al. [15] in their work on HgI_2.

For $Pe > 1$ the one-dimensional model shows a build-up of the B impurity in front of the crystal. B is pushed towards the crystal by the "Stefan wind" and rejected from the growing crystal. This results in the deviation from linearity of the partial pressure field [6] as shown in fig. 2.

Additional important information has been derived from numerical solution of the corresponding two-dimensional model [6]. It was found that the viscous interaction of diffusive–advective flows with the ampoule wall are very important. Taking the no-slip condition into consideration, Greenwell et al. [6] confirmed, using numerical modelling, the previously suggested [16] core flow of the impurity component B, which is considered as stagnant in the one-dimensional model.

Stefan tube experiments at 100 mbar pressure (diffusion-induced binary flow in capillaries) have shown that the mass-average velocity at the walls (gas velocity component tangential to the walls) is zero, i.e.

$$V = \sum_i \rho_i U_i = 0. \tag{13}$$

ρ_i is the vapour density and U_i the average velocity of the component i. Since, however, $U_A \neq 0$ in a PVT crystal growth ampoule, we must also have $U_B \neq 0$ in order that $V = 0$ (no-slip condition). The impurity is therefore not stagnant. Since no net

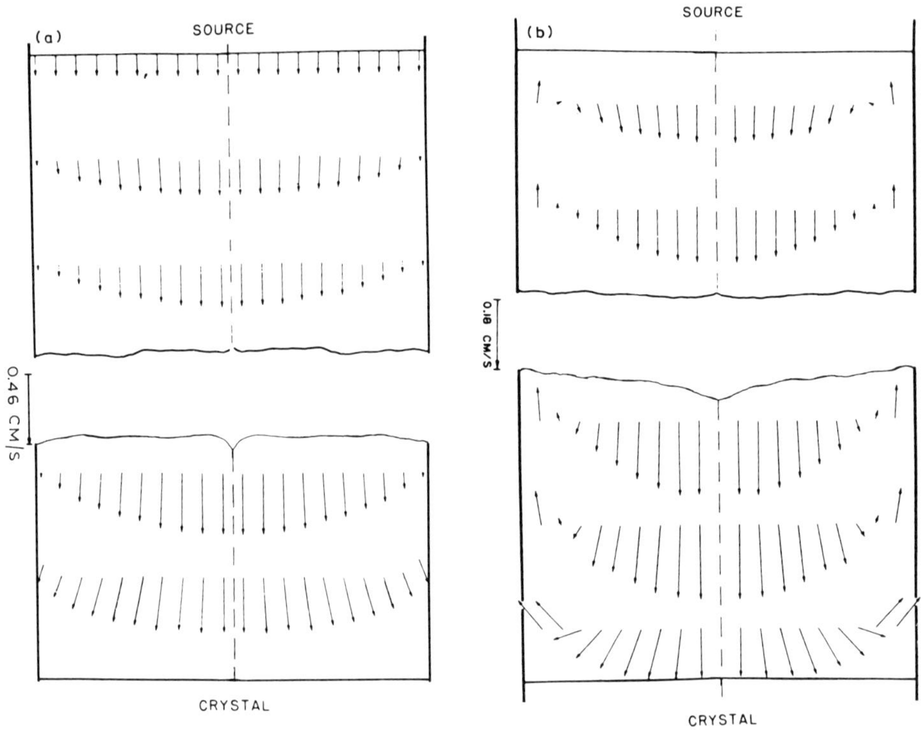

Fig. 3. Calculated velocity fields under zero gravity for PVT of the heavy component A (M_A, through a light impurity B, $M_B = 2$). Ampoule radius = 1 cm, aspect ratio $L/H = 10$. (a) Mass-average velocity which, at the crystal interface where B is rejected, is proportional to the growth flux. (b) Velocity distribution of B reflecting diffusion-induced recirculation [6].

transfer of B occurs, recirculation must take place, which leads, in the core of the ampoule, to axial flow of B opposite to its flow along the wall. Figure 3 shows numerical results under microgravity conditions [6]. Unfortunately, no experimental confirmation exists as yet.

The numerical solution of this two-dimensional diffusive–advective PVT was applied for the cases of quite different molecular weights [$M_A = 254$ (iodine), $M_B = 2$ (hydrogen)] and quite different partial pressures of the impurity component B ($p_A \approx p_B$ and $p_B < p_A$). The recirculation of the B component shows the following characteristics for each of these cases:

- For $p_B = p_A$, B accumulates at the growing interface and diffuses back along the wall in a region of vanishing velocity. This results in a radial concentration gradient of 1.0–4.5% and therefore a decrease of the deposition rate close to the cylindrical walls ($x = L$, $y = \pm\frac{1}{2}H$, fig. 1). The overall transport flux is practically identical to that of the one-dimensional solution. The variation of total pressure is $\approx 10^{-6}$ mbar for a total pressure of 20 mbar.

- For $p_B \ll p_A$, the difference in growth fluxes between the centre ($x=0$) and the periphery ($y = \pm\frac{1}{2}H$) of the sink can reach up to 15%. The transport rate tends to approach the high values obtained with Hagen–Poiseuille flow. Recirculation of the B molecules yields again a radial concentration gradient of the A species.

Ignoring gravity effects by restricting the numerical solution only to diffusive transport shows that the concentration of the component B will increase radially from the centre of the crystal A (at the sink) to the walls. It seems possible therefore to reduce the increase in the doping concentration in the outer regions of the crystal, provided that the container has a much larger radius than the crystal.

2.4. *Free convection in cylindrical ampoules*

The complexity of the quantitative theoretical description of a closed vapour growth system is due to the great difficulty in solving simultaneously the coupled conservation equations of momentum, energy and mass. Several groups have presented analytical or numerical solutions for the two-dimensional PVT problem [6, 17–24] and for the three-dimensional pure convection (without crystal growth) problem, with different aspect ratios $l = H/L$. Traditionally, the Boussinesq approximation of the Navier–Stokes equation is used for the calculation of the velocity field. This means that the density variation in the fluid is only taken into account in the buoyancy term and not in the intertial term. This is not necessarily a good approximation in vapour or gas mixtures with species of large disparity in molecular weight and has to be carefully discussed for each case.

It is useful to base discussion of convective flows on dimensionless Grashof numbers. They describe the ratio of the buoyancy force to the viscous force for free convection and are defined by the general equation

$$\mathrm{Gr} = \tilde{\rho} g H^3 (\Delta\rho)/\nu^2, \tag{14}$$

where $\tilde{\rho}$ is the average density, defined as

$$\tilde{\rho} = \frac{m_A p_A + m_B p_B}{RT(m_A/M_A + m_B/M_B)}, \tag{15}$$

as compared to the total density

$$\rho = \rho_A + \rho_B = (M_A p_A + M_B p_B)/RT, \tag{16}$$

with m the mass and M the molar mass of the species. $\Delta\rho$ is the density variation across the characteristic length H (the radius for a horizontal ampoule, fig. 1); g is the acceleration of gravity and ν the kinematic viscosity. An average value of ν is of the order of $10^{-4}\,\mathrm{g\,cm^{-1}s^{-1}}$.

For density variations due to thermal expansion only, we have

$$\Delta\rho/\rho = \Delta T,$$

where ideal gas behaviour of the vapour is assumed. The thermal expansion coefficient b is proportional to $1/T$.

For density variations solely due to concentration variations, we have

$$\Delta\rho/\rho = \alpha\Delta C,$$

where α is the solutal expansion coefficient [12]. Correspondingly one obtains the thermal Grashof number,

$$\mathrm{Gr_T} = b\rho^2 g\Delta TH^3/v^2, \tag{17}$$

and/or the solutal Grashof number,

$$\mathrm{Gr_C} = \alpha\rho^2 g\Delta CH^3/v^2, \tag{18}$$

in a given system. Note that, in real systems, $\mathrm{Gr_T}$ and $\mathrm{Gr_C}$ can either have the same or opposite signs.

In the various theoretical analyses, the thermal convection was computed by considering an imposed uniform temperature gradient (one-dimensional solutions) with a horizontal or vertical direction. In the horizontal case, two regimes appear, depending on the value of the thermal Grashof number:

- At low values of $\mathrm{Gr_T}$ the flow is governed by the temperature gradient in the centre of the closed system. This is the core-driven regime (CDR) or conduction regime (fig. 4a). The internal temperature gradient is the same as the external gradient imposed on the furnace.
- At high values of $\mathrm{Gr_T}$ the flow is governed by the gas circulation along the vertical walls. This is the boundary-layer-driven regime (BLDR), (fig. 4b). The transition occurs (for $l = 0.2$) at $\mathrm{Gr_T} = 1.1 \times 10^4$ for the two-dimensional solution and at $\mathrm{Gr_T} = 4.5 \times 10^4$ for the three-dimensional solution.

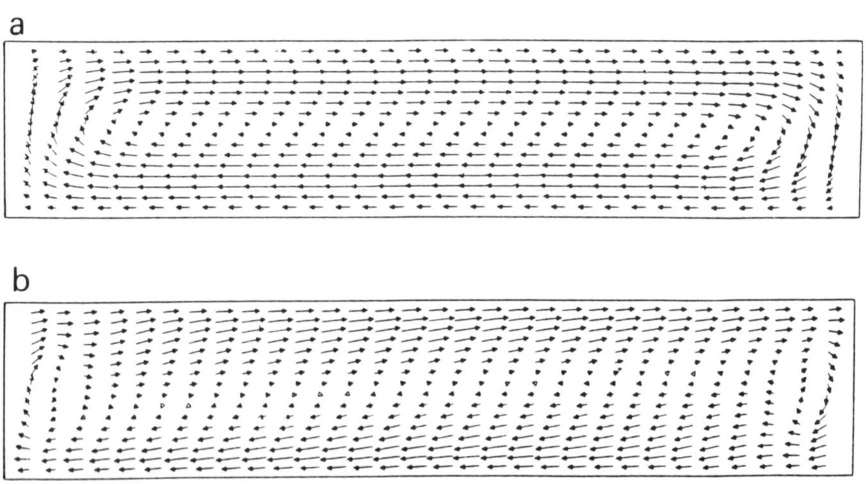

Fig. 4. (a) Core-driven regime; velocity field in the mid-plane of a horizontal cylinder of aspect ratio 10, Pr = 0.3, GrPr = 74. The flow is parallel to the axis. (b) Boundary-layer-driven regime: The same as (a) but with GrPr = 18 720. Note the inclination of the flow in the core and the formation of vortices in the vicinity of the vertical hot (left) and cold (right) walls [24].

The experimental results of Schiroky and Rosenberger [17] obtained in cylindrical ampoules are in good agreement with the three-dimensional solutions of Bontoux et al. [18]. The maximum flow velocity in the two-dimensional analytical solution is 33% higher than in the three-dimensional numerical solution in the CDR. The two-dimensional numerical solutions of Markham and Rosenberger [19], using compressible gas theory, are in good agreement with the two-dimensional numerical solutions using the Boussinesq approximation.

Markham et al. [20], using the two-dimensional compressible model, have investigated vertical ampoules taking into account the diffusive fluxes of the A species at the interfaces. For strongly different molecular weights of the components A ($M_A = 254$ (I_2)) and B ($M_B = 2$ (H_2)) an aspect ratio $l > 1$ and the hot surface up, only the inert component B is recirculating, while the transported species A condenses at the sink. An increase in the solutal Grashof number, where the radius H is replaced by the distance L between the source and sink, results in an increase of the A mass fraction and the temperature gradient towards the sink (shift of the concentration and temperature profiles). With decreasing aspect ratio ($l = 0.26$) increase of Gr_C leads to recirculation not only of B but also of A along the axis. When the hot surface is down, at low Gr_C, thermal convection cells prevail over diffusive–advective flow.

Elie [21] analysed the case of horizontal ampoules using two-dimensional modelling in the Boussinesq approximation. When the interfacial flux is high, only recirculation of the B species is obtained. At low interfacial flux a competition between solutal and thermal convection occurs. Recently, Extremet et al. [22, 23] have applied two-dimensional modelling to PVT in horizontal ampoules with a three-zone temperature profile. Most recently, the importance of radiative heat transport has been realised (see section 4).

2.5. *Interplay between surface kinetics and transport*

The kinetics and transport are a coupled process and, in the steady state, surface kinetics matches completely the vapour transport, i.e., the interfacial (crystal) flux J_C is equal to the vapour flux J_A. In the case of congruent sublimation, transport and growth in one dimension, the normal growth rate R is:

$$R = M_A J_A / \rho_C,$$

where ρ_C is the crystal density. The interplay between growth rate (interfacial kinetics) and transport rate is illustrated in fig. 5 where different rates are schematically drawn as a function of the chemical potential (the driving force). The straight line represents the rate of mass transfer and the curved one that of incorporation. The intersection of the two curves defines the growth rate of the crystal. The chemical potential difference $\Delta\mu_C = -|\mu^* - \mu_1|$ is needed for the interfacial kinetics and $\Delta\mu_A = -|\mu_2 - \mu^*|$ for the transport.

Careful observations of crystal growth in closed containers [25, 26] show that a very small amount of impurity vapour reduces the growth rate dramatically. Therefore, in closed systems, the concept that the growth might, in general, be determined by the mass transport rate, i.e. $|\Delta\mu_A| \gg |\Delta\mu_C|$ might be rather unrealistic. The experimental

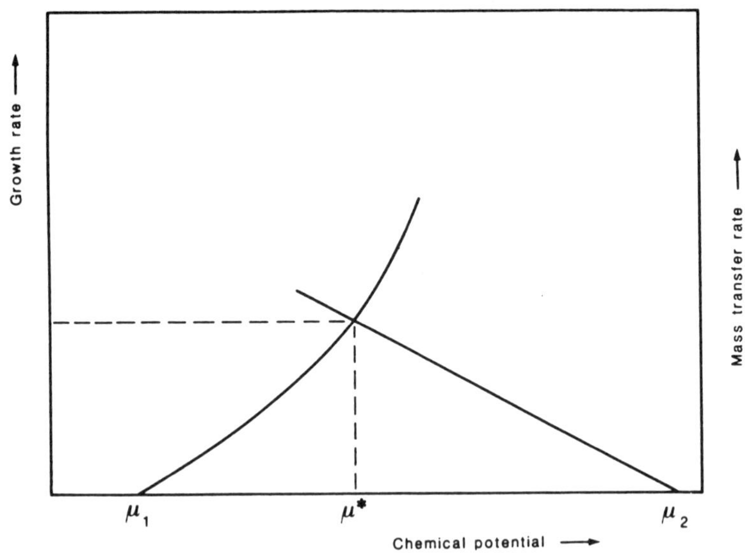

Fig. 5. Schematic illustration of the steady growth rate R as a sequential process between surface kinetics and mass transfer. The potential differences $\mu^* - \mu_1$ and $\mu_2 - \mu^*$ drive the surface kinetics and mass transfer, respectively. The inverse slopes of the curves can be interpreted as growth and transport resistances.

confirmation of theoretical growth relations is then difficult since in fact the potentials μ_2 and μ_1 can be easily obtained via the source and sink temperature but the determination of the intermediate potential μ^* needs some modifications to the method. A practical modification is to redesign the growth container in such a way that the advective flux (Up_A/RT) or the convective flux overrides the diffusive flux (D/RT) (dp_A/dx). These conditions may appear near the tip of a conically shaped ampoule [26], for the three-zone technique [15] and for the bottom seeding technique in large vessels. The latter method, where the vapour flows preferentially in a radial direction, is discussed in the following.

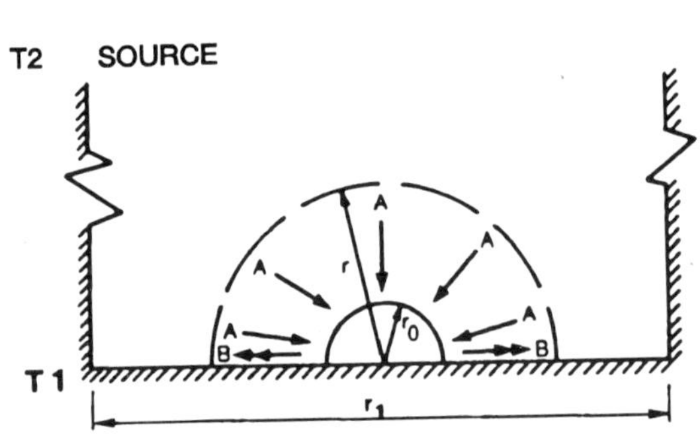

Fig. 6. Transient stage of growth with the crystal not fully covering the sink surface. Hemispherical crystal of radius r_0 and its diffusion field of A molecules. The side walls of the growth vessel are at a distance $r \gg r_0$ from the centre of the crystal. The recirculation of the impurity component B is the result of the vanishing mass velocity on the substrate (no-slip condition).

An order of magnitude estimate of the convective effects involved allows some predictions about an optimal ampoule geometry. We consider a hemi-spherical crystal of radius r_0 growing on a plane substrate with $r_0 \ll r_1$; r_1 being the ampoule radius (fig. 6). Under microgravity conditions the diffusive flux will have the hemispherical symmetry of the crystal. From the results presented in fig. 3b one can expect that the recirculation of B will only be significant in the vicinity of the substrate (see B arrows in fig. 6). As the crystal grows, the radial inhomogeneity of B-doping will increase.

To describe the concentration field we consider the vapour flux J_A of A about the hemi-spherical crystal. The vapour flux (diffusive + advective) in the radial direction $J_A(r)$ at a distance r from the centre of the crystal, is expressed by applying eq. (4) to the hemi-spherical symmetry of the concentration field [27]:

$$J_A(r) = CD_{AB}\,\mathrm{d}\ln[1 - x_A(r)]/\mathrm{d}r, \tag{19}$$

with

$$J_A(r)r^2 = J_A(r_0)r_0^2, \tag{20}$$

from the equation of continuity in the vapour phase. The flux then becomes

$$J_A(r) = [r^2(1/r_1 - 1/r)]^{-1} C\,D_{AB} \ln[1 - x_A(r)]/[1 - x_A(r_1)], \tag{21}$$

where $x_A(r_1)$ and $x_A(r)$ are the mole fractions of the vapour species A at the distance $r = r_1$ and $r_0 < r < r_1$. It is assumed that $x_A(r_1)$ is determined by the saturation pressure $p_{A2} = p_{A0}(T_2)$ of the source at temperature T_2.

For the partial pressure $p_A^* = X_A(r_0)p$ in front of the crystal interface at $r = r_0$ and for the total molar concentration $C = p/RT$, the vapour flux expressed in terms of the pressure becomes

$$J_A(r_0) = [r_0^2(1/r_1 - 1/r_0)]^{-1}(D_{AB}p/RT)\ln(p - p_A^*)/(p - p_{A2}). \tag{22}$$

Thus the flux increases with decreasing crystal size as $J_A(r_0) \approx 1/r_0$ for $r_0 \ll r_1$at constant p_A^*.

The condensation (incorporation) flux J_C is a function of the difference between the vapour pressure p_A^* at the interface and the saturation pressure $p_{A1} = p_{A0}(T_1)$ of the crystal at temperature T_1, i.e.

$$J_C \;=\; K_c(p_A^* - p_{A1})^{\mathrm{m}}. \tag{23}$$

This is shown schematically in fig. 6.

In the present case of radial mass transfer, the rate-controlling process depends on the crystal size r_0. For a crude estimation of the concentration field induced by the interplay of kinetics and mass transfer, we assume that the fastest growth kinetics possible is the one corresponding to an atomically rough interface, with $\mathrm{m} = 1$ in eq. (23). The maximal condensation coefficient $K_c = (2\pi MRT)^{-1/2}$ is given by the Hertz–Knudsen equation. For convenience of discussion, the total pressure is given as a function of

$$n = p/p_{A2} = (p_{A2} + p_{B2})/p_{A2}, \tag{24}$$

which increases with the impurity partial pressure ($n - 1 \approx p_B$).

From eqs. (19)–(22) we obtain an expression for the partial pressure as a function of distance r from the centre of the crystal:

$$p_A(r) = p - (p - p_A^*)\exp[Pe(r - r_0)/r], \tag{25}$$

where the Péclet number Pe is given by

$$Pe = J_A(r_0)r_0RT/pD_{AB} \tag{26}$$

and

$$J_A(r_0) = J_C = (p_A^* - p_{A1})/(2\pi MRT_1)^{1/2}. \tag{27}$$

It is convenient to use the reduced (normalised) concentration:

$$C_{re}(r) = [p_A(r) - p_{A1}]/(p_{A2} - p_{A1}), \tag{28}$$

with p_{A1} and p_{A2} the saturation pressures at the sink and source as in [6] to describe the partial pressure field along the radius r. By this one avoids the change of several orders of magnitude in the pressure. Using the $C_{re}(r)$ function one can describe the interplay between mass transport and growth kinetics, which leads to the following conclusions:

- when mass transport is rate-determining, the actual vapour pressure at the interface $p_A^*(r_0)$ is close to the saturation pressure p_{A1} at the crystal (sink). This is the result of depletion of the condensing species near the crystal;
- when the growth kinetics are rate-limiting, the actual vapour pressure at the interface $p_A^*(r_0)$ remains close to the saturation pressure at the source p_{A2}.

We are interested in defining the critical radius r_0^c which separates the two regimes. We define it as the average partial pressure,

$$p_A(r_0) = (p_{A1} + p_{A2})/2.$$

Substituting this value into eq. (24) we obtain for the crystal of critical size:

$$C_{re}(r_0^c) = (p_A^* - p_{A1})/(p_{A2} - p_{A1}) = 0.5,$$

if the following values are used: HgI_2 as component A, $D_0 = 0.05\,cm^2s^{-1}$, saturation pressures $p_{A0}(378\,K) = 5.22 \times 10^{-5}$ atm (crystal) and $p_{A0}(388\,K) = 1.07 \times 10^{-4}$ atm (source), $M_{HgI_2} = 454.4\,g\,mol^{-1}$ and $T_2 = 388$ K. The Péclet number is given by

$$Pe = 7.68 \times 10^5 C_{re}(r_0)p_{A0}(T_1)r_0,$$

with $p_{A0}(T_1)$ expressed in atmospheres. For $C_{re}(r_0) \ll 0.5$, the mass transport is growth-rate-limiting. For $C_{re}(r_0) \gg 0.5$, the growth is limited by surface kinetics.

Figure 7 shows the dependence of the reduced interface concentration function $C_{re}(r_0)$ on the crystal radius r_0, for different pressure ratios n.

Three conclusions can be derived from this figure:

- At a given impurity pressure (n value) the variation of $C_{re}(r_0)$ reflects the variation of the growth rate with crystal radius;
- the separation of the fields with mass transfer or kinetics as limiting process can be achieved for different impurity pressures with increasing crystal radius. Thus,

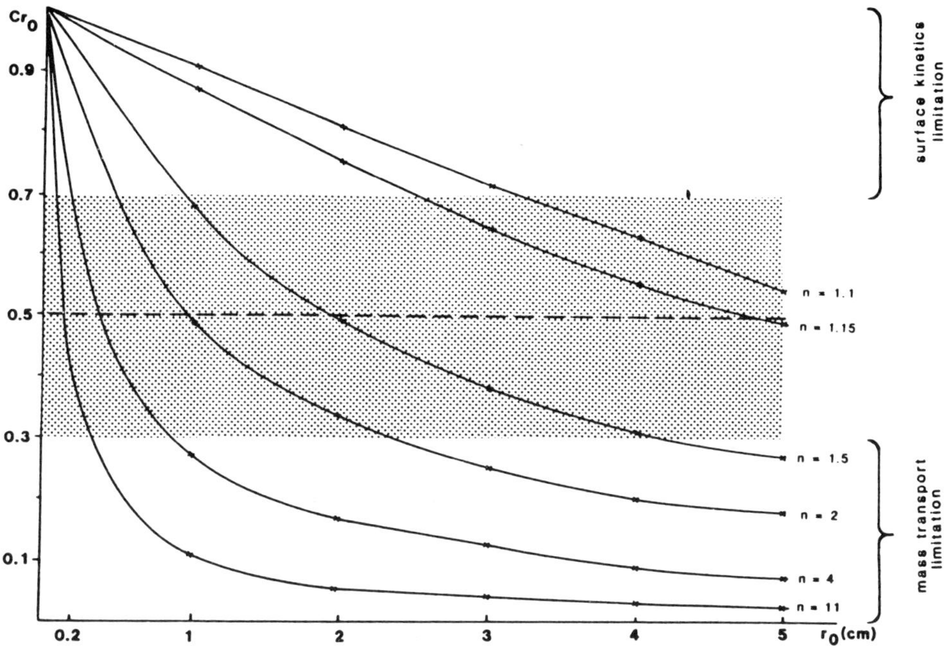

Fig. 7. Reduced concentration $C(r_0)$ of A molecules near the crystal surface versus the crystal radius for various values of the parameter n [see eq. (24)] corresponding to different pressures of the impurity component B. The following numerical values were used: $p_0(T_2) = 1.07 \times 10^{-4}$ bar, $p_0(T_1) = 5.22 \times 10^{-5}$ bar, $T_1 = 378$ K, $T_2 = 388$ K. Component A is HgI_2. The grey zone indicates schematically the transition area between mass transport and surface kinetic limitation.

whereas for $n = 1.15$ the transition starts at large radii (e.g. $r_0 > 3$ cm), for $n = 4$ it will start at a much smaller value (e.g. $r_0 \approx 1$ cm);

- whatever the limiting process, the growth rate decreases with increasing crystal radius or n value. The reason is the increase of the crystal surface at constant flux. This decrease is stronger (particularly in the kinetic region) at high impurity pressures, because impurities reduce the condensation flux on the crystal;
- for very small crystals ($r_0 < 0.2$ cm) and for all impurity concentrations, $n < 11$, surface kinetics are rate-limiting.

We recall that the curves of fig. 7 have been calculated for an atomically rough interface [a linear growth rate, $m - 1$ in eq. (23)]. For $m = 2$, the kinetic coefficient will be much lower and, therefore, the concentration gradient will be lower for the same crystal radius.

Figure 8 shows the relative partial pressure difference $[p_A(r) - p_A(r_0)]/p_A(T_1)$ as a function of the distance $(r - r_0)$ from the crystal interface for a crystal of radius $r_0 = 1$ cm. Figure 9 shows the change of the average gradient of the partial pressure of A with the impurity concentration (n) at various distances $(r - r_0)$ from the growth interface.

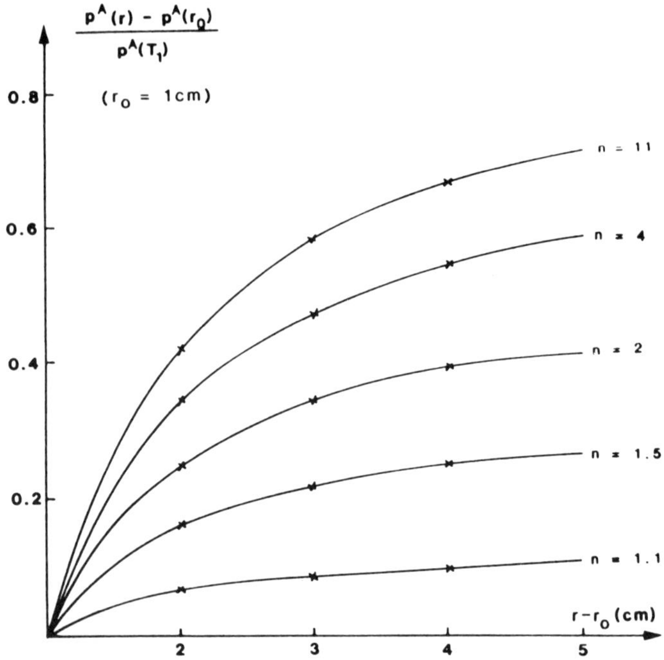

Fig. 8. Fractional partial pressure difference for component A as a function of vessel diameter r for different values of the pressure function n of the impurity component B and a constant crystal size ($r_0 = 1$ cm).

The most striking feature of these curves (fig. 9) is the high value of the partial pressure gradient of the component A just above the growth interface ($r - r_0 = 2$ mm) and its strong increase with high impurity concentrations ($n = 2$) and with crystal size for $r_0 < 2$ cm (fig. 7). The figure illustrates clearly the tremendous influence of impurities in vapour growth.

Interesting conclusions can be reached by comparing systems with radial flow (fig. 6) with one-dimensional flow systems (fig. 1). In a rectangular cavity, one-dimensional calculations for $L = 20$ cm result in a reduced concentration of $C_{re}(r_0) = 0.15$ for $n = 1.15$ and 0.07 for $n = 2$. The growth rate is always dependent on volume diffusion; the Grashof number is 1.7. Contrary to that, in the hemispherical concentration field of fig. 6 for $n = 1$, the Grashof number deduced from the concentration gradient can reach a value of 1×10^4. Of course these values are only semi-quantitative. However, they clearly underline the great difference between the two systems and the suitability of the latter for investigation in space under microgravity conditions.

In microgravity the radial transport should lead to high growth rates in spite of volatile impurities. This is due to the recirculation of the impurity component far from the crystal. On the ground such a high growth rate would result in solutal convection, which does not appear in cylindrical ampoules for steady-state growth (fig. 1) in the temperature range considered here. Solutal convection should increase with the total pressure, whilst the growth rate decreases as shown in figs. 7–9.

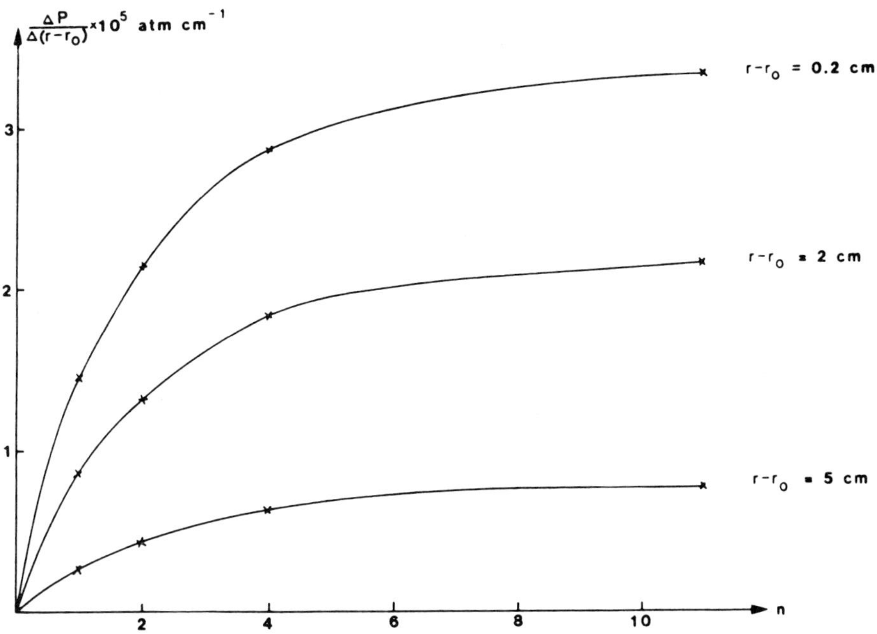

Fig. 9. Average gradient of the partial pressure of A versus the pressure function n for constant crystal size ($r_0 = 1$ cm).

In conclusion, the above estimates show that solutal convection in the concentration field surrounding the growing crystal can cause growth inhomogeneities and defects. Of course, such estimates only show the trends and application to a real crystal growth system has to be experimentally investigated. Recent studies of the growth of HgI_2 (see section 4) have shown that convective effects remain important down to very low gas pressures.

2.6. *Dissociative sublimation*

Consider a binary compound AB which dissociates entirely according to the reaction:

$$\mathrm{AB\,(solid)} \rightarrow \mathrm{A(vapour)} + 1/m\, \mathrm{B}_m\mathrm{(vapour)}. \tag{29}$$

Faktor and Garrett [3] have derived expressions for the transport rate of such a dissociated compound both in the absence or presence of a third (inert) species as follows.

Taking our one-dimensional model (fig. 1) again, and considering first the case of no inert gas, the fluxes of the two components are

$$J_\mathrm{A} = (U/RT)\,p_\mathrm{A} - (D/RT)\,\mathrm{d}p_\mathrm{A}/\mathrm{d}x,$$

and

$$J_{\mathrm{B}_m} = (U/RT)\,p_{\mathrm{B}_m} - (D/RT)\,\mathrm{d}p_{\mathrm{B}_m}/\mathrm{d}x. \tag{30}$$

Adding these equations:

$$J_A + J_{B_m} = Up/RT, \tag{31}$$

where p is the total pressure, equal, in this case, to the sum of the two partial pressures. The ratio of the fluxes must be such as to yield stoichiometric solid so that:

$$J_A = mJ_{B_m} = J, \tag{32}$$

which is the growth rate of the compound AB. From these last two equations we see that

$$Up/RT = J(1 + 1/m). \tag{33}$$

This relation can be used to eliminate U from eq. (30) which can then readily be integrated, if we ignore the very small dependence of p on x and also the temperature dependence of the diffusion coefficient, to give:

$$\delta p_A(x) = [p_A(L) - pm/(1+m)]\exp\{JRT(1+m)(x-L)/mDp\}, \tag{34}$$

$$\delta p_{B_m}(x) = [p_{B_m}(L) - p/(1+m)]\exp\{JRT(1+m)(x-L)/mDp\}. \tag{35}$$

We see that there is an exponential variation in δp_A and δp_{B_m}, where

$$\delta p_A = p_A - pm/(1+m), \tag{36}$$

and

$$\delta p_{B_m} = p_{B_m} - p/(1+m). \tag{37}$$

The latter quantities are the departures from stoichiometry in the vapour.

The physical interpretation of these partial pressure profiles is as follows: Consider first the equilibrium between solid and vapour in an isothermal closed ampoule. In general, the vapour composition will not be stoichiometric (except at one unique temperature) so that the ratio of the partial pressures,

$$\alpha = p_A/p_{B_m},$$

will not be equal to m. But if we now introduce a seed crystal at some distance and impose a small temperature difference to promote growth on the seed, then vapour will be removed in a stoichiometric ratio. This removal will bring up new vapour from the source region having ratio α and, since this differs from the value m, a boundary layer will be formed in which one component accumulates (component A in the case that $\alpha > m$) whilst the other becomes depleted.

The crystal growth rate can be expressed in terms of the vapour compositions over source and seed by re-arrangement of eq. (34) or (35):

$$J = [Dpm/RT(1+m)L]\ln\{[p_A(L) - pm/(1+m)]/[p_A(0) - pm/(1+m)]\}$$
$$= [Dpm/RT(1+m)L]\ln\{[p_{B_m}(L) - p/(1+m)]/[p_{B_m}(0) - p/(1+m)]\}. \tag{38}$$

If the growth and evaporation occur close to equilibrium conditions the partial pressures of either A or B_m at seed and source required to evaluate eq. (38) can be obtained from the equilibrium constant for the dissociation reaction.

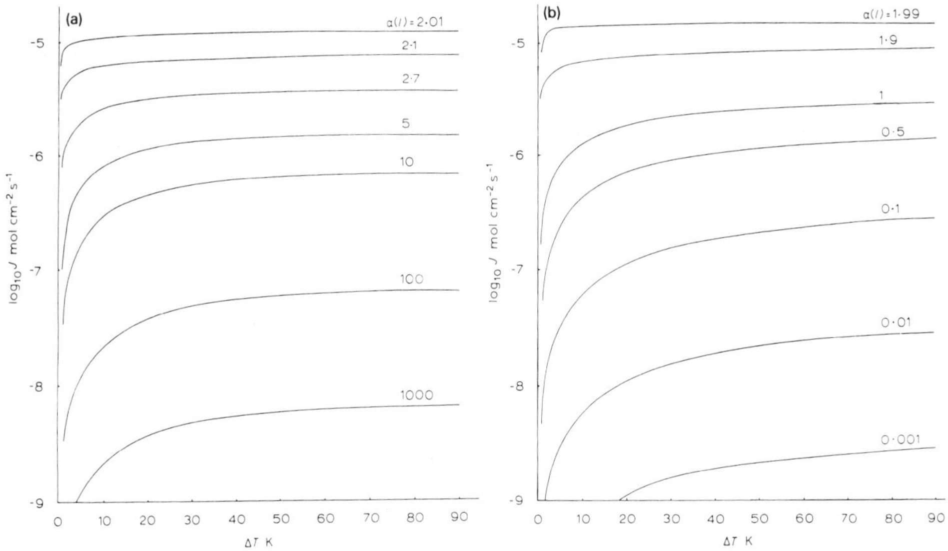

Fig. 10. Transport rate as a function of temperature difference between source and crystal for growth by dissociative sublimation of CdS. (a) With Cd-rich vapour ($\alpha > 2$). (b) With sulphur-rich vapour ($\alpha < 2$).

Numerical evaluation for parameters appropriate to the growth of CdS as a function of the driving force ΔT (the temperature difference across the ampoule) are shown in fig. 10a and b for the cases of $\alpha > m$ and $\alpha < m$, respectively. (For the temperature range 1100–1400 K sulphur exists predominantly as S_2 so that $m = 2$). For small ΔT, the growth rate is a strong function of that parameter but saturates at larger values. The most striking effect however is the catastrophic reduction in growth rate as α deviates from the "stoichiometric" value m. This is the result of the dramatic fall in the concentration of the minor component adjacent to the growing crystal face when the source vapour deviates progressively from stoichiometric composition.

Faktor and Garrett [3] demonstrate that, at 1200 K, the pressure ratio at the source $\alpha(L) \approx 1000$. In an ampoule of 30 cm^3 capacity this represent about 0.3 mg of excess cadmium or, equivalently, ten monolayers of oxide covering 100 cm^2 of source surface which, by removing free sulphur as SO_2, would give an equivalent cadmium excess. One can now begin to see why closed-tube PVT is a process that is very difficult to control when non-stoichiometry is involved. The problem is overcome by the use of an effusion hole in the ampoule as described in section 3.3 below.

Inspection of eq. (38) shows that the diffusion limitation disappears when α is exactly equal to m so that the growth rate is predicted to become very large indeed, limited only perhaps by the rate of removal of latent heat. In practice, however, the presence of a small amount of residual inert gas will have a controlling influence as Faktor and Garrett have shown. They have demonstrated that eqs. (34) and (35) still describe the distribution of the active species but with the following crucial difference. With no inert gas, the total pressure along the ampoule was equal to the sum of the partial pressures of the two active constituents. Now, with an inert gas, the total pressure is

$$p = p_A + p_{B_m} + p_z, \tag{39}$$

where p_z is the partial pressure of the inert gas which varies exponentially along the ampoule and is given by

$$p_z(x) = p_z(L)\exp\{JRT(x-L)(1+m)/mDp\}. \tag{40}$$

This re-imposes a diffusional limitation on the growth rate even at $\alpha = m$. A second effect of this inert gas is upon the morphological stability of the growing crystal surface which we now consider.

2.7. *Morphological stability*

The concept of constitutional supersaturation, analogous to that of constitutional supercooling in the melt, has been introduced by Reed and LaFleur [34] for the case of a single-component substance growing from a vapour in the presence of an inert gas. The depletion boundary layer ahead of the growing crystal of the crystallising component gives rise to an exponential partial pressure profile. The equilibrium partial pressure corresponding to a just-saturated vapour will also increase with distance from the growing face because the temperature increases toward the source region and the equilibrium vapour pressure is exponentially dependent on the temperature. If the actual partial pressure increases more rapidly with distance than the equilibrium pressure and if, at the interface, conditions are close to thermodynamic equilibrium, then a zone of supersaturated vapour will be present ahead of the crystallising interface which may induce morphological instability of that interface in exactly the manner that occurs in growth from alloy melts [35]. Further, if the magnitude of the supersaturation is sufficiently great, it could even lead to spurious nucleation in that zone and therefore polycrystalline growth. Reed and LaFleur were able to demonstrate interface instability under conditions predicted by their theory, during the growth of iodine single crystals by sublimation.

These arguments have been extended to binary compounds by Faktor et al. [36–39]. For our binary dissociating AB compound the condition for a thermodynamically stable interface (i.e. marginal or no constitutional supersaturation) is shown by them to be:

$$\{[p_A(0) - pm/(1+m)]/p_A(0) + [p_{B_m}(0) - p/(1+m)]/p_{B_m}(0)\}\, JRT(1+m)/mDp$$
$$\leqslant \Delta H/RT^2\,(\mathrm{d}T/\mathrm{d}x)_{x=0}, \tag{41}$$

where ΔH is the enthalpy of the dissociation reaction [eq. (29)]. $(\mathrm{d}T/\mathrm{d}x)_{x=0}$ is the temperature gradient in the vapour at the crystal/vapour interface. This implies a maximum critical growth rate (J_{crit}) above which instability is to be expected. In turn this implies a maximum allowable temperature difference between source and seed crystal (ΔT_{crit}). This critical state also depends strongly on α for values of $\alpha \approx m$ as shown in fig. 11 but saturates for extreme values of α. Faktor and Garrett evaluate this

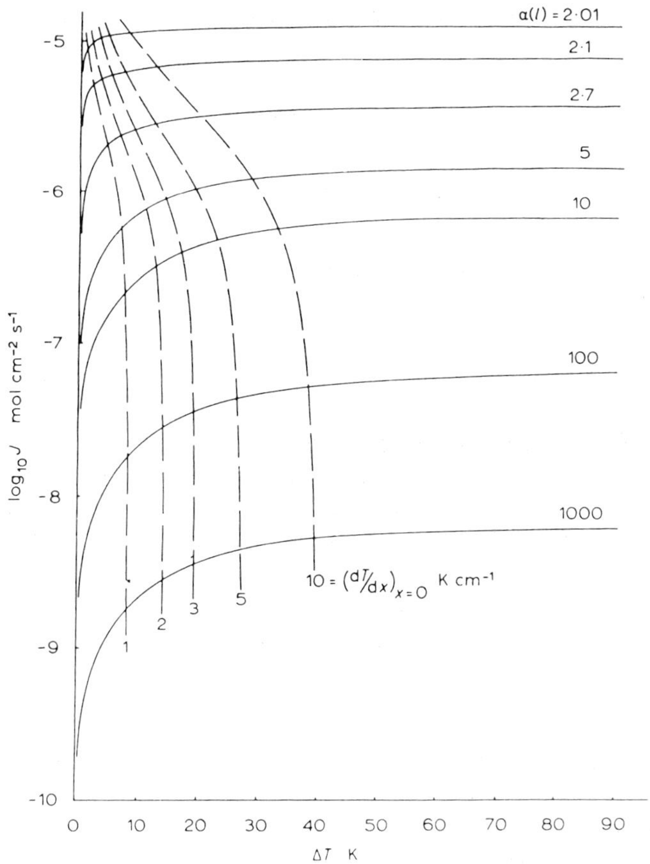

Fig. 11. Critical growth rates for CdS with various temperature gradients over the crystal.

saturation value of ΔT_{crit} for the growth of CdS at 1400 K to be, approximately,

$$\Delta T_{crit} \approx 24 \ln \{0.4(dT/dx)_{x=0} + 1\},$$

where T and ΔT are in kelvin and x in cm.

For $\alpha \approx m = 2$, needed in order to achieve a reasonable growth rate, ΔT_{crit} is very small and controlled stable growth is difficult. However, Faktor and Garrett have shown that the presence of an inert gas not only decreases J_{crit} but it also increases markedly the value of ΔT_{crit} for near-stoichiometric values of α, so that stable growth is much more readily obtained. These authors therefore posit that, for successful growth by dissociative sublimation, two conditions must be met: (i) α must differ only a little from its stoichiometric value, and (ii) a small quantity of inert gas must be added to avoid constitutional supersaturation. The latter provides a buffer against the effect of the inevitable small fluctuations in ΔT caused by convective turbulence, limitations of servo-controllers, etc.

3. Some experimental growth techniques

3.1. Introduction

A major problem encountered in the growth of bulk single crystals from the vapour is control of nucleation. This is particularly true of closed tube transport systems. Three rules need to be followed:

(1) The supersaturation must be controlled to be sub-critical so that only a few active, especially favoured sites on the ampoule wall initiate nucleation. The number of nuclei can be further reduced by careful design of a tapered tip to the ampoule;
(2) the ampoule must be transported through the imposed temperature gradient at a rate equal to the linear growth rate of the crystal. This ensures that the supersaturation conditions remain constant throughout the growth run and avoids the occurrence of spurious nucleation in the later stages of growth;
(3) the ampoule should have a large diameter and have cylindrical symmetry to stimulate convective mixing.

In some cases the provision of seed crystals has proved possible. A novel approach first introduced by Scholz [52], called the temperature oscillation method, cycles the temperature so as to periodically grow and re-sublime the crystals thereby removing all but the largest and most favoured seeds which have been formed by spontaneous nucleation.

The control of purity and, in the case of compound growth, of stoichiometry are further important issues.

3.2. Sealed ampoule method

This method was used in the earliest experiments employing cylindrical glass or quartz ampoules. A small quantity of a transporting agent gas (e.g. iodine) was added so that the technique was chemical vapour tansport (CVT) rather than PVT. The ampoule was placed in a gradient furnace and left for an extended period. The outcome of such experiments was all too frequently unpredictable and some reasons for this have been indicated in section 2. Poisoning of the growth surface by residual impurities, some of which were desorbed from the ampoule, was probably a common cause. The development of high-purity synthetic quartz only partially alleviated the purity problem.

Where growth occurred by dissociative sublimation, variation in the ratio of the vapour pressures of the components could produce orders of magnitude variation in growth rate. The amounts of excess component which needed to be added to control vapour stoichiometry are so small (see Faktor and Garrett [3]) as to make accurate control not possible. Prior [42] attempted to overcome this problem in the growth of lead selenide by connecting the ampoule to a separate reservoir containing selenium whose temperature was independently controlled. Cutter and Woods [43] employed a similar technique for the growth of zinc selenide.

Nucleation control is effected by defining the nucleation region within the ampoule by, e.g. the attachment of a cooled finger. Nitsche [44] successfully grew large single crystals of zinc sulphide by cooling one end of the ampoule with cold air passed through an attached tube. Honigmann [45] has grown large single crystals of HMTA in an ampoule in which source material was placed at both ends with a cooled finger mid-way along the length serving as a nucleation point. This idea has been extended to have axial symmetry in the technique used by Kaldis and co-workers to grow HgI_2 which is described in section 4. The ability to observe the initial stages of nucleation and growth is desirable in order to ensure that multiple nucleation is avoided. A new method is described in section 4.2. Transparent gold-coated silica furnaces pioneered by Reed et al. [46] are useful to this end.

In general, fewer and better crystals are obtained when the driving force (the temperature difference) is kept small. This was demonstrated by Butler and Harman [47] who grew crystals of $Pb_{1-x}Sn_xTe$ in a sealed system with a temperature differential of only 1°C. The crystals were used to fabricate diode infra-red lasers.

Where possible, the provision of a seed crystal is to be recommended as it affords the ability to control the orientation of the grown crystal. Kaldis and Widmer [48] grew zinc selenide by closed-tube iodine transport using this technique.

3.3. Nearly-sealed ampoule method

This concept was pioneered by Piper and Polich [49] for the growth of cadmium sulphide by dissociative sublimation. Their apparatus is shown schematically in fig. 12. The ampoule was initially open at the end where the crystal was to grow so that any volatile impurities liberated during the inital heating could escape. Either a stopper was then inserted or the growing crystal itself used to provide a self-sealing system. Because the ampoule is initially left open, the technique cannot readily be used for CVT. For PVT however, it has proved far more reliable than a fully-closed tube technique and has been used to grow large single crystals of a number of materials but especially cadmium sulphide.

Translation of the ampoule through the furnace to maintain constant growth conditions (the second criterion – section 3.1) has also been adopted. A triangular shaped thermal profile in the furnace (see fig. 12) ensures that the source material is moving toward a hotter zone in the centre of the furnace whilst the growing crystal is moving away from it. This helps to achieve a situation in which the rate of crystallisation is approximately equal to the rate of sublimation which is, in turn, equal to the rate of translation of the furnace (Schieber et al. [50]).

A refinement of the technique was introduced by Faktor and colleagues [36, 38, 39] who developed a perforated ampoule in which a 0.2 mm diameter hole was drilled near to the position of the source material providing continuous contact with the exterior environment which was maintained in vacuo (10^{-5} Torr). This hole permitted the continuous removal of impurities generated from reaction with the ampoule and sublimation from the source material but also served a second purpose. It allowed a very small amount of the charge material to escape also, thereby fixing the vapour composition and hence the crystal stoichiometry. The hole was sufficiently small that the leak rate was determined by molecular flow. If the material sublimes congruently then the components will escape in the stoichiometric ratio. This means that their

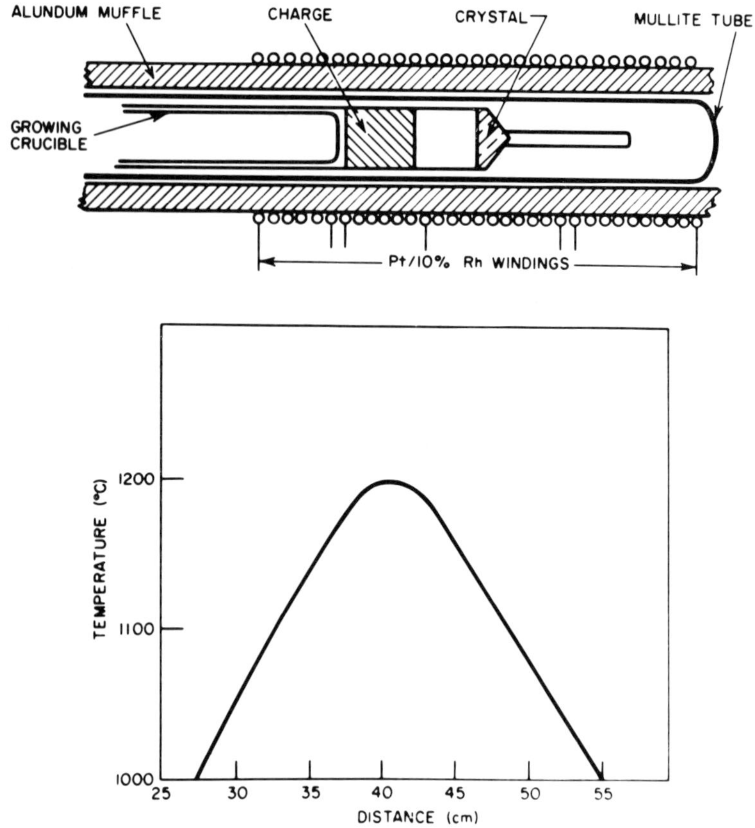

Fig. 12. Schematic representation of the Piper–Polich technique for the growth of cadmium sulphide. Also illustrated is the furnace temperature profile.

partial pressures within the ampoule will be in the ratio of their molecular weights. In the case of CdS this is $p_{cd}/p_{S_2} \approx 2.7$ which means that a crystal having a constant composition corresponding to this ratio can be grown. The rate of loss of material through the orifice is negligible (Faktor and Garrett [3]). Differing crystal stoichiometries can be obtained by replacing the ambient vacuum with a controlled vapour pressure of the component required in excess.

3.4. Temperature oscillation method

A significant deficiency of all cylindrical ampoule techniques is that the crystal is constrained by the ampoule walls during cooling so that it is compressed if the expansion coefficient of the ampoule exceeds that of the crystal. Since quartz has a very low expansion coefficient this is unlikely with ampoules of this material; more probable is that the crystal will "key" to the ampoule and thereby be placed in tension during cooling. This can be overcome by growing the crystal on a pedestal as described in section 4 for the growth of HgI_2 and as used in the temperature oscillation (or temperature reversal) method developed by Scholz [52].

The technique has been developed and exploited for the growth of HgI_2 by Schieber and co-workers [50]. Their equipment is characterised by the following features:

(1) An axial temperature gradient which is periodically reversed.

(2) An azimuthal or circumferential temperature gradient which is produced by using either an axial heater or an axial cooler which heats or cools the circumferential walls of the growth ampoule.

(3) A radial temperature gradient over the glass substrate on which the crystal grows with the coldest point – the site of nucleation – at the centre.

The apparatus can be mounted vertically or horizontally (fig. 13 shows the vertically oriented configuration). The axial temperature gradient drives the transport of HgI_2

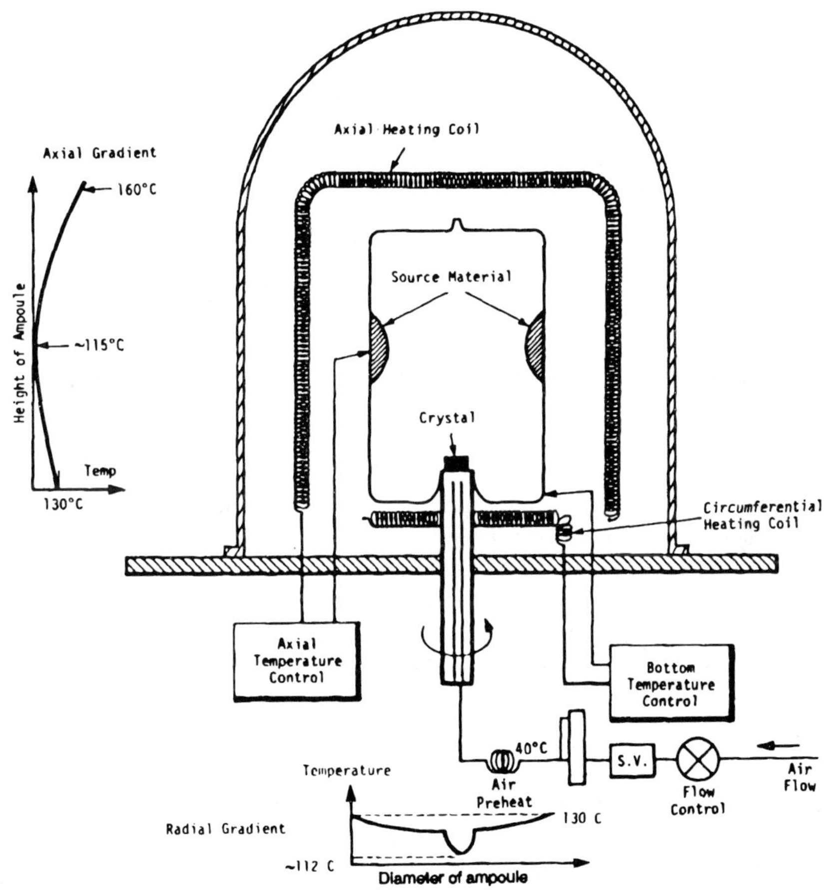

Fig. 13. Schematic representation of a vertical furnace for the growth of HgI_2 which can be operated in a temperature oscillation method (TOM) mode. For details of the cold finger construction see [50].

from the source to the crystal substrate whilst the azimuthal gradient provides a favoured site for the location of the source material on the ampoule wall. The latter gradient and the radial gradient across the glass substrate are crucial to the stability of the process (Akutagawa and Zanio [40]).

The equipment can be operated in either of two modes:

(1) The crystal temperature T_c can be kept constant whilst there is periodic oscillation of the source temperature between the limits T_{s1} and T_{s2} (POST mode).

(2) The source temperature T_s is fixed and the crystal temperature is oscillated between the limits $T_{c1} < T_s < T_{c2}$ so that the crystal periodically grows and then partially re-sublimes (POCT mode).

The driving force for evaporation in the POST mode is $\Delta T_e = T_c - T_{s2}$ and for growth $\Delta T_g = T_{s1} - T_c$. Similar quantities can be defined for the POCT mode. If the time for which $\Delta T_g > 0$ is τ_g and that for which $\Delta T_e > 0$ is τ_e, then the ratio of the integrated amounts of growth to re-evaporation is:

$$X = \int_0^{\tau_g} \Delta T_g \, \mathrm{d}t \bigg/ \int_0^{\tau_e} \Delta T_e \, \mathrm{d}t.$$

Overall growth occurs if this ratio is greater than unity.

The vertical furnace is operated in the following manner. The growth ampoule is charged with purified HgI_2, evacuated to about 10^{-6} Torr and sealed. The ampoule is placed within the furnaces which are set to the evaporation mode where the pedestal is relatively hot so that the charge material sublimes on to the upper part of the circumferential wall.

When the source material has stabilised, cooling air is introduced into the cavity beneath the pedestal to induce nucleation. After several visible nuclei have formed the undercooling and period ratios are adjusted to select a single seed nucleus. This is most readily achieved in the POCT mode since this provides the fastest response of the system.

Once this single seed nucleus has been obtained, growth is promoted by gradually increasing the air flow, and thereby increasing ΔT_g. Alternatively the period ratio is increased until all the material has been transported from source to substrate.

The ability to observe and control the nucleation process is an advantage of the technique. However, one price to be paid for the cycled growth is the presence of growth striations within the crystal – see section 4.5.2. In the case of HgI_2 it is thought that the periodic reversal of the temperature gradient may also lead to dislocation generation [50].

4. *Mercury iodide: a case study*

4.1. *A model substance for vapour growth investigations*

The vapour growth of α-HgI_2 crystals has been investigated by several laboratories in the last twenty years. The main impetus for these investigations came from the discovery of its properties as a room-temperature semiconductor detector for X- and γ-rays [51]. At about the same time the material has been crystallized by Scholz [52] who succeeded to grow single crystals up to 50 g by the temperature oscillation method. Unfortunately, a systematic characterization of these crystals has not been done. The combination of these two factors led the EG&G Energy Measurements Inc., Santa Barbara, USA to start a large detector programme with this material, supported by the US government, in order to monitor radiation in the various nuclear projects for the Department of Energy (DOE) and the Department of Defense (DOD). After almost twenty years of production of crystals and detectors, a lot of progress has been made, but also many problems concerning the growth conditions leading to defects of the α-HgI_2 crystals are still open, challenging to our fundamental knowledge in the field of vapour growth.

There have been several reasons to select α-HgI_2 as a model substance for the general study of the crystal perfection as a function of the growth conditions. An ideal substance for such investigations should grow at relatively low temperature allowing optical monitoring of the growth process without strong disturbance of the temperature field and should be transparent, so that in situ diagnostics with laser light scattering would be possible. α-HgI_2 meets these requirements: it forms red transparent crystals which grow at temperatures below 120°C, so that transparent furnaces with adequate optical windows can be used. Thus, exact measurements of the growth rate as a function of the supersaturation are possible with the crystallographic orientation as a parameter. Growth rate is a very sensitive indicator for the formation of defects and for changes in the growth mechanism. Systematic in situ growth rate measurement on single crystals from the vapour have been performed only once in the past [53].

For the study of the formation of defects, mercuric iodide has also some "negative" features as a model substance in the sense that, due to its highly antisotropic layer structure, its weak Van der Waals bonding between the layers and a low melting point, deterioration of the crystal perfection readily occurs by the incorporation of impurities and external thermal or mechanical stresses.

Another problem which might be addressed using mercuric iodide as a model material is the mechanism of the mass transport in the vapour phase. Although its equilibrium pressure is very small at the growth temperature (0.2mbar), the question arises whether convective contributions are present and to what extent they affect the crystal growth.

Ultimately, such a study is expected to yield some of the important parameters for growth from the vapour of as-perfect-as-possible single crystals. Based on our early work on this compound [28], the above-mentioned properties of mercuric iodide and the crystal growth work reported in the literature [55–57] four major sources of defects have been identified:

(1) purity of the starting material for crystal growth;
(2) structural deformation during growth as a result of thermal stresses and impurity incorporation or, for large crystals, due to mechanical deformation under their own weight;
(3) convective contributions leading to the formation of growth defects;
(4) structural deformation during the post-growth handling of the crystal.

Purification of the starting mercuric iodide material is very difficult. Due to the layer structure and the high chemical reactivity of both iodine and mercury a large amount of impurity is usually present. In order to eliminate organic (hydrocarbon) and inorganic impurities, purification of the elements and subsequent synthesis from the vapour phase has been used [58]. Although very complex and time consuming, this procedure results in material leaving no residue upon evaporation, contrary to the starting materials purified using any other method.

The mechanical deformation during growth is a complex phenomenon which is currently being investigated. Studies on convective contributions are in progress. However, results to date indicate the existence of convection even at very low total pressures (see section 4.5).

4.2. Experimental: control of nucleation and growth of large crystals

Very little is known about the nucleation of mercuric iodide. Thus, in order to produce a satisfactory nucleus, the experimenter has to proceed by trial and error frequently spending a lot of time just to obtain satisfactory nucleation. The nucleation of the red phase is complicated by the concurrent formation of metastable nuclei of the yellow phase (stable above 126°C). Already in 1933 Stranski and Totomanov [59] showed that within the undercooling range of $\Delta T = 9$–28°C the nucleation rate of the yellow and red nuclei obeys the Oswald stage law, i.e., the number of red nuclei did not exceed 21% of the total number of nuclei in each of their experiments. For nucleation from purer starting materials, much smaller undercoolings (0.2–2.0°C) are necessary [60]. This imposes rather stringent requirements on the temperature stability and distribution along the nucleation plate. The temperature reversal method is most frequently used to control nucleation of mercuric iodide. The idea behind is that perfect nuclei should evaporate more slowly than the imperfect ones; thus reversing the process from growth to evaporation should result in a natural selection of the best nuclei. It sometimes works, but due to small critical undercoolings, other factors can prevail, such as temperature inhomogeneities along the nucleation plate and the anisotropy of the thermal conductivity of the crystal (orientation of the nucleus). Thus, in the authors' laboratory another method is used [61], in which bad nuclei are removed with an argon laser beam.

Due to the mechanical fragility of mercuric iodide crystals no seeding is possible and both nucleation and growth are performed in the same ampoule. An experimental arrangement used for kinetic investigations of vapour growth is shown in fig. 14. The crystals grow on a flat pedestal without contact with the side walls (unconstrained). A five- or six-zone furnace with a temperature control of ± 0.03 K is used. All zones are kept at temperatures higher than that of the cooling finger, the latter being used to control the undercooling. Because the cooling finger, on the top of which a tiny PT-100 RTD sensor is placed, is in contact with the pedestal, internal temperature calibration is possible. After some small nuclei have appeared, one increases the temperature of the cooling finger up to the point where evaporation sets in. Because the temperature gradient across such small nuclei is negligibly small, one can assume this temperature

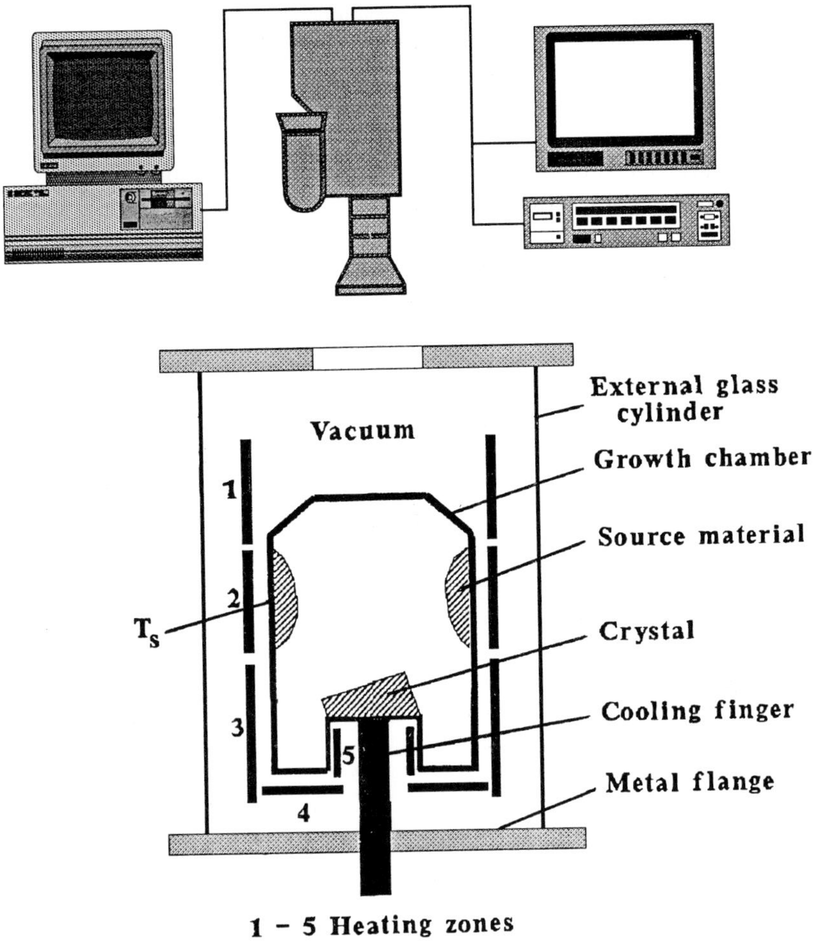

Fig. 14. Schematic representation of the experimental arrangement used for kinetic investigations of vapour growth of mercuric iodide.

to be that corresponding to equilibrium between the source and the crystal, so that any decrement of the temperature of the cooling finger may now be directly referred to as undercooling and supersaturation can be calculated.

For the measurements of the growth kinetics, an optical port is integrated into the upper part of the ampoule. A high-resolution (756 × 581 pixels), three-chip colour CCD camera (Sony DXC-730P) is used for top view observation of the pedestal, where the nucleation takes place. The camera is equipped with a 32-fold zoom objective which gives a linear resolution of ± 0.02 mm at the highest magnification. Thus, nuclei with size well below 1 mm can be visualized. Images of the growing crystals are digitised and dedicated software measures the size of the crystal along several predefined lines.

4.3. Growth rate measurements, the rate-determining step

Kobayashi et al. [28] have found for the first time, that the growth rate of α-HgI_2 declines by up to one order of magnitude in a few days under constant apparent supersaturation. A similar saturation effect of the growth rate was later reported for PVT of Hg_2Cl_2 by Singh et al. [62]. The phenomenon has been investigated by Isshiki et al. [63] in more detail. The apparent supersaturation was stepwise increased or decreased in 0.5°C increments and the change in lateral and vertical dimensions of the crystal measured. A stepped curve was obtained, with the evaporation and growth steps overlapping (fig. 15). However, the initial high growth rate after each increase of the apparent undercooling proved to be independent of previous changes (fig. 16). This implies that the actual supersaturation remains the same in spite of the stepwise increase of the apparent supersaturation. A mechanism of annihilation of the driving force must therefore exist which increases the temperature of the growing crystal interface. A detailed discussion has been given in [63].

Assuming that the thermal resistance of the crystal is responsible for this phenomenon and that only the heat of condensation has to be dissipated through the crystal, one can calculate the growth curve which fits the experimental data only up to the onset of the plateau. An important clue comes from the fact that, for crystals a few cm in size, an apparent undercooling of several degrees is still necessary to keep the crystal at zero

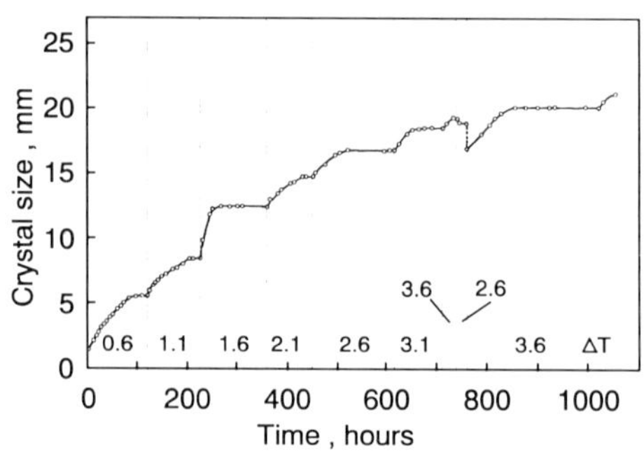

Fig. 15. Change of the crystal size with growth time. The apparent undercooling is varied in 0.5 K steps.

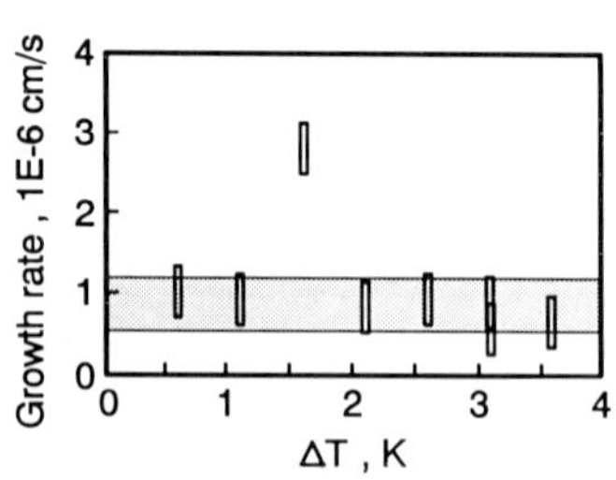

Fig. 16. Initial growth rate (at the beginning of each step on the growth curve from fig. 15) as a function of the apparent undercooling.

growth/evaporation rate. Thus, a heat flux is steadily supplied to and dissipated through the crystal in addition to the heat of condensation. Consequently, when the growth rate of the crystal approaches zero under constant apparent undercooling, the habit of the crystal tends to a rounded form which follows the isotherms in the gas phase instead of approaching the equilibrium form (a faceted polyhedron) [31, 63].

The additional heat flux is supplied to the crystal by radiation. Using a spherically symmetrical model combining conductive and radiative heat transport, mass transport and interface kinetics, Chernov et al. [65] showed that both the decrease in the growth rate and the rounding of the polygonal habit can be explained. Parallel to this investigation, numerical modeling [66] of the thermal field inside the growth furnaces used for PVT of mercuric iodide showed that, if one takes into account only thermal conductivity in both vapour and solid phases, the crystal should grow undisturbed until it reaches the interface of the source material. On the other hand, if also radiative heat transfer to the crystal is considered, the observed cessation of growth with increasing crystal dimensions at constant undercooling can be simulated rather accurately. This unexpected result leads also to the conclusion that the geometrical arrangement and the temperature of the heating zones surrounding the growing crystal play an important role in the kinetics of the growth and influence the final habit of large crystals.

In conclusion it can be said that, similar to the observations in melts, the thermal field in the growth ampoule determines the shape of large crystals growing from the vapour. As curved interfaces introduce defects, the form of the isotherms becomes very critical for large crystals.

From a practical point of view, a strong decrease of growth rate with time can be compensated for by a continuous increase of the apparent undercooling, either by decreasing the temperature of the cooling finger or increasing the temperature of the source. However, the temperature control is usually not better than 0.1°C and temperature changes made in 0.1°C steps lead to the formation of striations (see next section). The recently released temperature controllers Eurotherm series 900 allow steps as small as 0.01°C and have been succesfully used to improve the temperature control during PVT [67, 68].

Ideal control of the growth process would be achieved if the temperature could be measured and controlled with high accuracy directly at the growing crystal interface. Unfortunately, due to its spectral characteristics (transparency in the red and infra-red) this is not possible for mercuric iodide using commercially available IR detectors [69].

4.4. Crystal size, habit and orientation

The maximum crystal size in a given growth system is limited by the geometry of the growth ampoule and the temperature distribution, especially that imposed along the crystal by the cooling finger. Whereas crystals up to 500 g in weight could be grown in ampoules of 100 mm diameter and 150 mm height (0.5 l volume), much larger ampoules (approx. 2 l volume) are necessary to achieve crystals of 1 kg weight [70]. In order to show the influence of the crystal orientation on the growth kinetics and habit of mercuric iodide, two cases are discussed: (a) *c*-axis parallel to the pedestal plane, and (b)

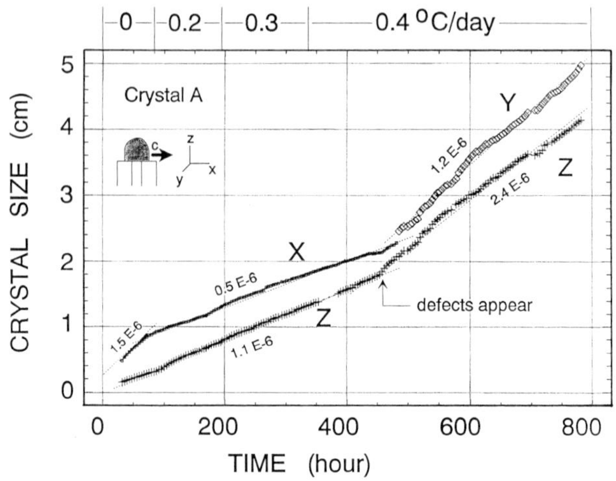

Fig. 17. Change in crystal size parallel and perpendicular to the pedestal plane as a function of growth time. The *c*-axis of crystal A is oriented almost parallel to the pedestal. Note the intentional change in measurement direction from the slow (*x*-direction) to the fast (*y*-direction) after defects appeared.

c-axis perpendicular to the pedestal plane. Accordingly, the crystal is oriented with its direction of highest thermal conductivity (2 W/m K; that along the *c*-axis is 5 times lower) (a) parallel, or (b) perpendicular to the direction of the heat dissipation through the cooling finger. Figure 17 shows changes in the lateral (x or y) and vertical (z) dimensions with time for crystal A. A similar plot for crystal B is shown in fig. 18. Due to the daily stepwise increase in apparent undercooling, the growth rate perpendicular to the pedestal, $R(z)$, remained constant in both cases (the more than twofold increase in the growth rate at around the 460th hour of growth in fig. 17 has been caused by the

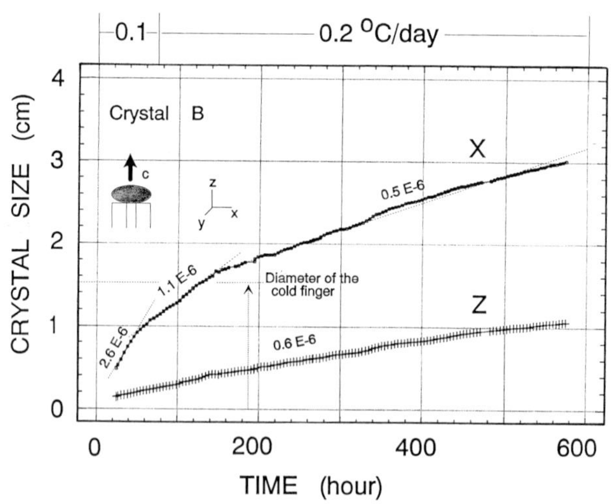

Fig. 18. As figure 17 but for crystal B having its *c*-axis oriented almost perpendicular to the pedestal plane.

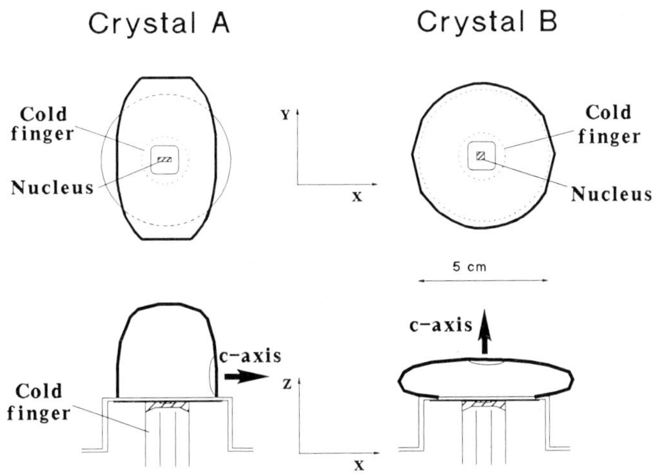

Fig. 19. Evolution of crystal shape during growth for a nucleus with its *c*-axis parallel (A) or perpendicular (B) to the pedestal plane. Note the change in growth rate ratio in the *XY*-plane when the crystal expands beyond the cold-finger area in case (A).

sudden appearance of defects on the previously very flat top face of the crystal. These quickly developed into two hillocks overgrowing the facet). A greater growth rate for crystal A [$R(z) = 1.1 \times 10^{-6}$ cm/s] compared to crystal B[$R(z) = 6.2 \times 10^{-7}$ cm/s] is easily explained by the anisotropy of the thermal conductivity mentioned above.

On the other hand, the growth rate parallel to the pedestal plane could not be kept constant despite the increasing undercooling. There are either one (crystal A) or two (crystal B) changes in the lateral growth rate $R(x)$. The first change, present on both plots, is most probably the effect of decreasing actual undercooling. In this stage of growth, the undercooling is either not being changed at all or changed only slightly in order to prevent formation of parasitic nuclei. The second decrease in $R(x)$ is clearly seen at approximately $x = 15$ mm for crystal B, but absent for crystal A. At this time, both crystals were passing the border of the direct contact area between the cold finger (diameter 16 mm) and pedestal (diameter 50 mm). However, again for the reason of the thermal conductivity anisotropy, only $R(x)$ with x coinciding with the direction of the larger thermal conductivity (crystal B, x perpendicular to c-axis) has been affected. Consequently, as it is shown in fig. 19, the initially more or less similar crystal nuclei develop to quite different final shapes.

4.5 Evidence for convective contributions

As mentioned earlier, the main question here is whether convective contributions exist in low-pressure sublimation at all. Until the middle 1980s, only one-dimensional fluid dynamics models of such systems were available, which predicted no convection, i.e. only Stefan flow and diffusion. At about this time, Cadoret et al. [71] performed the first α-HgI_2 experiments in space showing that nucleation is affected by gravity. Also the space experiment of van den Berg and Schnepple [72] resulted in a small α-HgI_2 crystal which proved to be much more homogeneous than the best earth-grown crystals. In the meantime, with improving sophistication of numerical modelling (two- and three-dimensional – see, e.g. Nadarajah et al. [73]) on one side and extensive investigations of the growth kinetics on the other [67, 78, 75], it became clear that, even in diluted gas systems, convection can contribute in an essential way to the overall mass transport and affect the crystal growth.

4.5.1. Nucleation and growth experiments in the presence of an inert gas

Although techniques exist to visualise and to measure convective fluid motions in gases under atmospheric pressure (such as laser Doppler anemometry, see, e.g. Schiroky and Rosenberger [17]) it is very difficult to apply them in closed PVT systems under a pressure of only a few millibars. Nevertheless, convective motions have been visualised during the PVT of mercuric iodide in ampoules containing apparently less pure starting material. Such a material leaves very light carbonaceous residues which become steadily captured in a convective roll located near to the pedestal on which the crystal is growing [74].

In order to enhance and investigate convection in pure systems, a small amount (1 mbar at room temperature) of helium has been added to the growth ampoules [75]. Introduction of the inert gas produced pronounced instabilities during both nucleation and growth of the crystals. Two examples are illustrated in figs. 20 and 21. Figure 20 shows the lateral dimensions of a small crystal as a function of time. Decreasing undercooling step by step by a total of 0.45 K resulted at first in a decrease in the growth rate as expected. However, after approximately 3 h (from point 3 to 4), the crystal started to grow quickly again and numerous additional nuclei appeared. No temperature variations of any of the actively controlled heating zones could be assigned

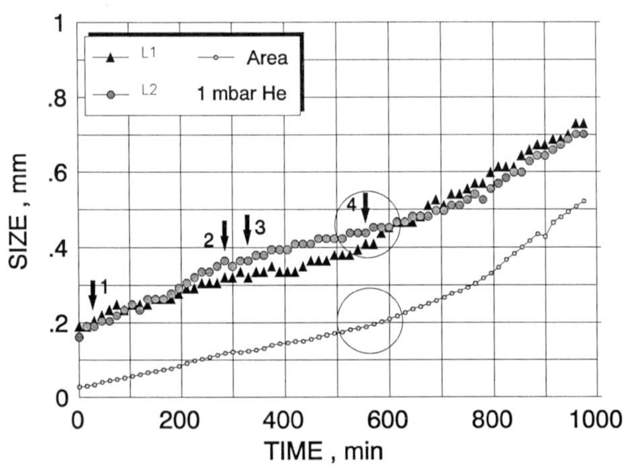

Fig. 20. Size versus time of a crystal growing in presence of helium (1 mbar at room temperature). The apparent undercooling has been decreased by 0.1 K at points 1 and 2 and by 0.25 K at point 3. Numerous additional nuclei appeared at point 4 despite decreased undercooling.

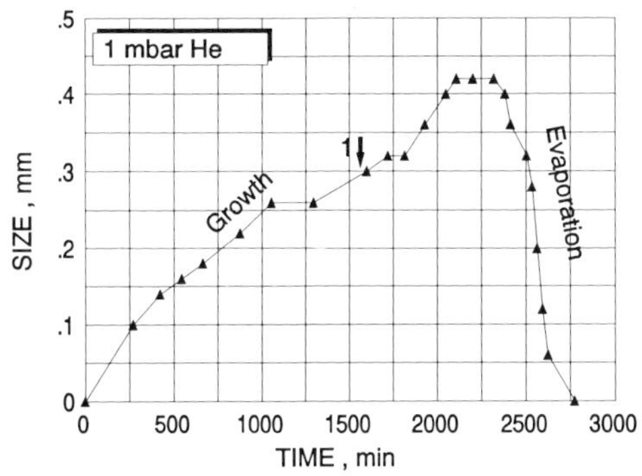

Fig. 21. Size versus time of another crystal growing in presence of helium (1 mbar at room temperature). The apparent undercooling has been increased by 0.05 K at point 1. After a period of growth, the crystal suddenly evaporated.

to this phenomenon. Temperature was controlled with ± 0.05 K accuracy. Short-term (1–2 min) spikes of ± 0.15 K are present on the temperature record of the cold finger, but are randomly distributed over the whole period of measurement. An opposite effect has been also observed (fig. 21). Increasing undercooling at point 1 resulted in growth of the crystal over the next 8 h. Then an arrest of the growth took place followed by a quick evaporation and disappearance of the crystal. Also in this case, no temperature variations of the heating zones can be traced to the kinetic instability.

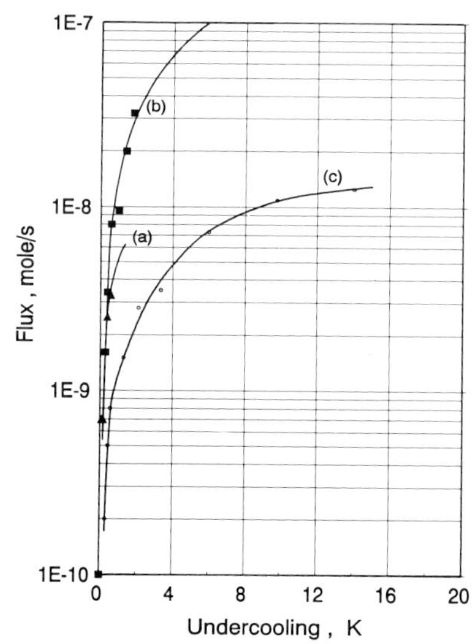

Fig. 22. Flux as a function of undercooling. Curves (a) and (b): crystal growing in presence of helium (1 mbar at room temperature), growth ampoule with 100 mm diameter. Curve (c): no inert gas added, ampoule diameter 7 mm.

Mercuric iodide crystals grow usually with a linear rate of several times 10^{-6} cm/s at an actual undercooling in the range of 0.1 – 0.5 K. Introduction of helium should decrease the growth rate if the latter is limited by the mass transport in the gas phase and no convection occurs. Figure 22 shows the growth rate as a function of the actual undercooling for a small crystal growing in helium atmosphere in an ampoule with 100 mm diameter. At first, curve (a) has been measured by varying the temperature of the cooling finger. Then the temperature of the pedestal, T_p, (heating zone 5 in fig. 14) has been raised by 26 K in one step and curve (b) was measured. For comparison, curve (c), measured in a much smaller ampoule (diameter 7 mm) in the absence of helium, is also shown [76]. An increase of T_p should increase the temperature of the crystal growing on the pedestal, i.e. decrease the undercooling and the growth rate. Instead, the growth rate increased by an order of magnitude; a very surprising effect. It can be plausibly explained however if we consider convection as a major mechanism producing mass transfer in the vapour phase. Indeed, an increase of T_p by 26 K imposed a strong horizontal temperature gradient between two vertical cylindrical walls (side wall of the pedestal and that of the ampoule, cf., fig. 14) – a classical situation to enhance a convective flow there, as well as the convective flow above the horizontal part of the pedestal (i.e. above the growing crystal).

Thus, convection seems to contribute to the vapour growth of mercuric iodide even though the total pressure in the ampoule does not exceed a few millibars. Since the molecules of mercuric iodide are very heavy (molecular weight 455 g/mol) as compared either to helium or to any residual gas likely to be present in the ampoules (CO_2, H_2O, N_2, O_2), solutal convection most probably prevails.

4.5.2. Striations in vapour-grown crystals

Striations in vapour-grown α-HgI_2 crystals were observed for the first time by Schieber et al. [50] and were attributed to the temperature oscillation method, TOM, which was used at that time. This was also the reason why this technique has been largely abandoned. Clearly, as in melt growth, a periodic reversal between growth and evaporation must produce striations. However, striations are found also in crystals grown under non-oscillating temperature conditions [75, 77]. The network of striations might be so dense (spacing 50 – 130 μm) that the crystals become “hazy” [77] or quite sparse (spacing up to 0.5 mm) [67, 75], so that, in the latter case, formation of each striae can be observed in situ during the growth and systematic investigations are possible to elucidate the mechanism of their formation.

As mentioned above (section 4.3), even if the temperature of the growth environment (ampoule and furnace) remains constant, gradual changes of temperature of the cold finger are necessary with the increasing dimensions of the crystal to keep the actual supersaturation and the growth rate constant. The rate of change of undercooling varies from 0.1 to 0.5 K/day. Zha et al. [67] showed that, if the temperature of the cooling finger is changed in 0.1 K steps, striations are formed, the distribution of which correlates with the temperature changes performed. However, despite the recent improvement in control of the growth temperature (± 0.03 K) and cooling ramps (0.005 K/h), striations have still been observed [75]. This kind of striation cannot be anymore attributed to temperature variations; therefore other sources of the kinetic instabilities must be operating. A comparison of the striation patterns with the kinetic data of growth reveals that the duration of the instabilities producing striations varies between 5 and 20 h [75]. Furthermore, whatever the origin of these disturbances, they affect all growing facets of the crystal at the same time, so that faster-growing facets leave broader striations behind. A slight decrease in the growth rate takes place at the onset of a disturbance and the accompanying striation [75]. All these observations lead to the conclusion that slow changes in the mass transport regime in the gas phase occur during the growth. Indeed, even longer transients (up to several days) in convective flow pattern have been recently measured, using laser Doppler velocimetry, by Paradies et al. [78] in PVT of mercurous chloride.

Very recently, Remy et al. [79] succeeded in obtaining X-ray transmission topographs on thin (approx. 0.1 mm) mercuric iodide samples using synchrotron radiation. Figure 23 shows a topograph [(200) reflection] measured on a (001) wafer produced from a vapour-grown crystal by solution sawing and subsequent thinning by etching in potassium ioide solution. Three grains are clearly visible, each of them containing striations running along the [110] direction. The high contrast is due to a high dis-

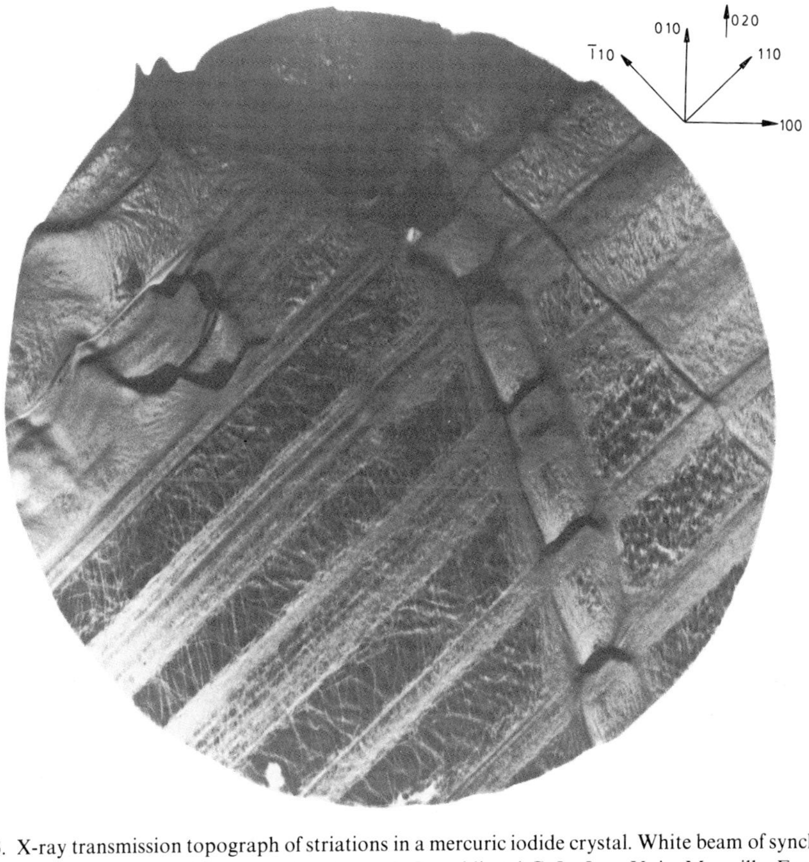

Fig. 23. X-ray transmission topograph of striations in a mercuric iodide crystal. White beam of synchrotron radiation has been used. (Courtesy of F. Remy, J. Gastaldi and G. Le Lay, Univ. Marseille, France).

location density within the striations ($>10^4\,cm^{-2}$). Thin white lines between the striations correspond to single dislocations. Interesting, within a striation, sub-striations can be seen. In terms of the nature of the disturbances, from which striations originate, this would mean that the disturbances are unsteady in time. Further investigations are in progress to understand this phenomenon.

4.6. *Structural defects in HgI_2 crystals*

High attenuation of X-rays in mercuric iodide – a reason for the very good performance of this material as a detector – makes structural investigations of the crystals by conventional X-ray diffraction methods difficult [64, 80, 84] and provides information only from the first 1–3 μm below the crystal surface. As mentioned above, first successful attempts have been made to use the synchrotron radiation topography [79, 81] for mercuric iodide samples with the thickness of a fraction of a millimeter. On the other hand, several investigations on vapour-grown crystals by means of γ-ray rocking curves have been reported in the past [82–85]. The technique seems to be particularly useful because high-energy γ-rays go through crystals of several cm thickness making investigations of large samples possible. However, this capability has never been exploited to understand the crystal growth mechanism. No attempts have been made in the past to correlate the structural perfection of the as-grown crystals with their growth history. Such investigations are important, not only for elucidating the growth mechanism but also for the technology of the X- and γ-ray spectrometers based on mercuric iodide. The as-grown crystals are undoubtedly affected by the subsequent mechanical and chemical processing during manufacturing of the spectrometers (cutting, applying electrodes, encapsulation) in a way which has become only partially understood in recent years [84]. The lack of clear differentiation between the defects introduced during growth and those produced by the subsequent processing might be the reason why the relationship between the crystal properties and the performance of the spectrometers has not been fully clarified yet.

Recently, Piechotka et al. [68] and Rossberg et al. [86] used γ-ray rocking curves to characterise the entire volume of large as-grown mercuric iodide crystals. At first, they found extended areas in the crystals, where the structural homogeneity was almost perfect, as concluded from the very small FWHM values (0.024 deg arc at instrument resolution of ± 0.015 deg arc) for the (200) reflections (tilt mosaics). A systematic mapping of the crystal perfection along both (100) and (001) planes revealed, however, that areas with destorted structure also exist.

Figure 24a shows a crystal of $4 \times 2.5 \times 2\,cm^3$ in size grown with the {001} facet almost perpendicular to the supporting glass plate. Two scans were made along the [100] direction (series A and B in fig. 24a) and the corresponding curves for (200) reflection are shown in fig. 24b,c. The narrowest peaks (FWHM 0.023 deg) have been

Fig. 24. (a) Schematic drawing of a mercuric iodide crystal grown with the *c*-axis parallel to the pedestal. Tracks of the γ-ray beam in two scans are shown. (b, c) Rocking curves for (200) reflection (tilt mosaics) measured along the scans.

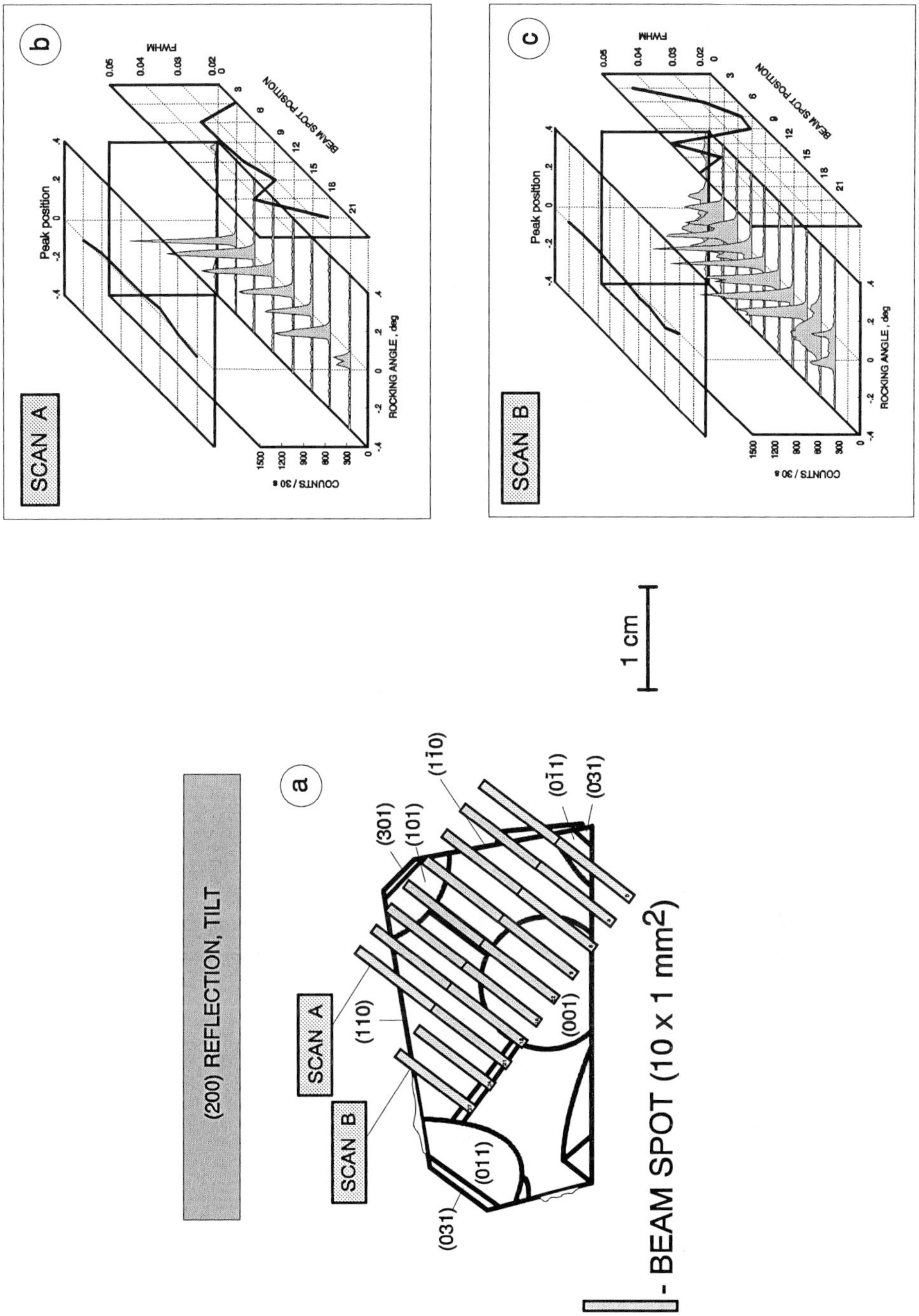
SCAN A
b
Peak position
FWHM
BEAM SPOT POSITION
ROCKING ANGLE, deg
COUNTS / 30 s
SCAN B
c
1 cm
(200) REFLECTION, TILT
a
SCAN A
SCAN B
(110)
(301)
(101)
(1$\bar{1}$0)
(0$\bar{1}$1)
(031)
(001)
(011)
(031)
- BEAM SPOT (10 x 1 mm^2)

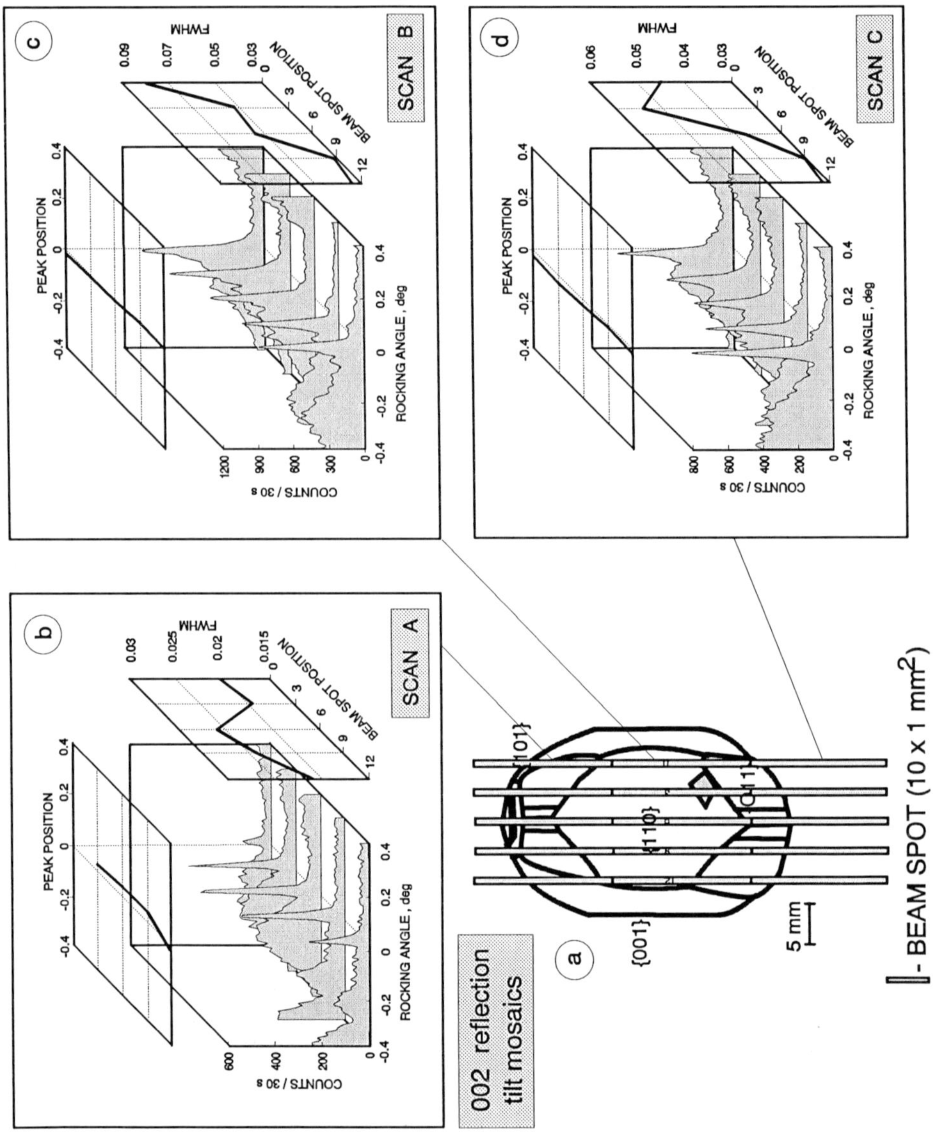
002 reflection
tilt mosaics
SCAN A
SCAN B
SCAN C
PEAK POSITION
ROCKING ANGLE, deg
COUNTS / 30 s
BEAM SPOT POSITION
FWHM
{001}
{110}
{101}
5 mm
- BEAM SPOT (10 x 1 mm²)

found around the center of the rounded {001} facet of the crystal. The crystal was scanned also in the [001] direction along three lines indicated in fig. 25a and rocking curves for the (002) reflection (tilt mosaics) were measured (fig. 25b–d). Sharp peaks of the (002) reflection (FWHM $\geq$ 0.025 deg) are accompanied by additional broad peaks indicating degradation of the crystal structure.

The red phase of mercuric iodide has a layered structure. The (001) layers are built of [HgI_4] tetrahedra with a strong, predominantly covalent, bonding within the layers whereas a weak Van der Waals bonding exists between the layers. The anisotropy of the structure is reflected in the γ-ray diffraction results: rocking curves obtained for tilt mosaics are generally better (sharp, single peaks) for (200) reflections than those for (002) reflections, which means that the crystal integrity within the layers is superior to that between the layers. Such conclusion could not be clearly drawn from the previously published data [82–84] due to the scarcity of the measurements on the (00*l*) reflections. Nevertheless, Yelon et al. [82] obtained FWHM of 0.9 deg as a typical value for the (002) rocking curves on cleaved wafers as compared to FWHM well below 0.75 deg typically for (*hk*0) reflections. A very good FWHM of 0.06 deg has also been reported by these authors at one location of an as-grown crystal, obtained, however, by rocking the crystal within a very narrow angle only ($\pm$ 0.09 deg) so that it is not clear whether or not there were additional peaks near this reflection. As the example in fig. 25 shows, a very narrow (002) peak (FWHM = 0.025 deg) might be accompanied by additional broad maxima. Multiple peaks are a quite common feature of any reflections from mercuric iodide crystals indicating the formation of subgrain boundaries and mosaics. The mosaicity seems to be non-uniformly distributed within the as-grown crystals. On both crystals investigated in [86], the lowest mosaicity was found within the regions confined between two round {001} facets.

Another clear tendency is that regions close to the bottom of the crystals are more defective than those near the top. Multiple peaks at locations 0, 3, 6 in fig. 24c were indeed measured with the beam spot close to the bottom of the crystals. By contrast, scan "a" on the crystal of fig. 24 showed very good single peaks (fig. 24b) except for locations 0 and 21, i.e. exclusively on or near the {110} facets on which parasitic nuclei with different orientation had been formed during the growth. The sharpest, even though they do not stand alone, (002) peaks (fig. 25b), have been also measured in scan "a" (fig. 25a), i.e. including mainly the region identified as the most uniform one from the scan for the (200) reflections on the same crystal. Yelon et al. [82] obtained the best (002) rocking curve also close to the top of their crystal. The improved perfection towards the top of the crystals indicates that they suffer from the contact with the glass plate on which they grew. Indeed, Burger et al. [87] estimated that the tension at the crystal/glass interface produced during the post-growth cooling might several times exceed the critical resolved shear stress of the mercuric iodide. Thus, it seems that some plastic deformation occurs at the bottom interface of the crystals, the amount of which depends on how much a particular crystal adheres to the pedestal.

Fig. 25. (a) Plan view of the same crystal as in fig. 24. Tracks of the γ-ray beam in three scans are shown. (b, c, d) Rocking curves for (002) reflection (tilt mosaics) measured along the scans.

5. *II–VI compounds: model substances for high-temperature vapour growth*

Many of these compounds are very important for the IR technology and other electronic applications, therefore, appreciable effort has been financed all over the world for the growth of large and as-perfect-as-possible single crystals. These compounds are also model substances for high-temperature vapour growth ($T < 1100^{\circ}C$). Within the frame work of this chapter it is not possible to discuss all the growth experiments which have been performed in the last 30 years. We refer, therefore, only to a recent publication which, similar to the work on HgI_2, shows the importance of radiative heat transfer in the vapour growth of large single crystals (up to 1 kg).

This method has been developed after years of investigations at the ELMA (Electronic Materials Research and Production Association, Selenograd, Moscow) by Davydov et al. [88]. As shown schematically in fig. 26, crystals are grown on a sapphire rod (4) with a diameter of up to 10 cm. As the main heat transport takes place axially (trough the polished flat front sides) and much less through the cylindrical surface, it acts like a light and heat pipe cooling the seed wafer (7) and the growing crystal (8)

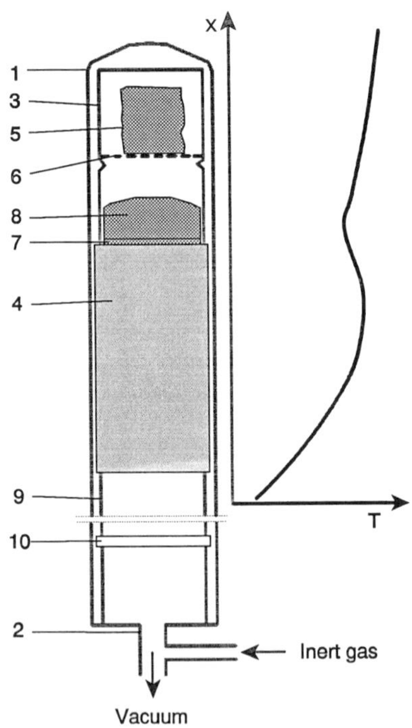

Fig. 26. Schematic drawing of the apparatus for PVT growth of II–VI crystals by the radiative cooling method. (1) Outer glass cylinder, (2) flange to vacuum system, (3) growth chamber, (4) sapphire rod, (5) source material, (6) silica glass mesh, (7) seed wafer, (8) growing crystal, (9) silica glass cylinder (support), (10) thermal shield (silica glass plate). (After Davydov et al. [88]).

relative to the source (5). This cooling is possible although the sapphire rod passes a hot zone of the furnace. The source material (5) lying on a quartz grid (6) above the seed is transported through a rather close space to the growing crystal. The crystal and source occupy a semi-closed chamber (3) inside the reactor, which is evacuated and back-filled with 0.3 bar of helium and then finally filled with 1 bar of helium. Small amounts of the source material and some volatile impurities may diffuse out to the colder parts of the reactor (9, 10). Using this method, Davydov et al. [88] have grown large single-crystalline discs up to 15 mm thickness and 20–50 mm diameter (CdTe, CdZnTe), 40 mm (ZnS), 50 mm (ZnCdS, ZnSe) and 100 mm (CdS). The growth rate achieved was of the order of 0.15 mm/h (4×10^{-6} cm/s). The optical spectra showed very sharp absorption edges but some residual absorption in IR (Davydov et al., private communication). Crystals of CdTe and GaAs grown in space with a similar method by Markov et al. have been found to have high purity and crystal perfection [54]. This method seems very promising for further development.

6. *The potential of bulk vapour growth*

The continuous search for homogeneity and perfection in the field of electronic materials gives a great impetus to investigations of the crystal perfection not only from the melt but also by alternative methods like vapour growth. Due to its well-known advantages (low temperatures, lack of other molecules, etc.), this method is expected to produce crystals of high perfection. The idea, that only thin films are needed in the future is erroneous because a thin film is seldom significantly more perfect than its substrate. Therefore, perfect substrates are in increasing demand and vapour growth of large bulk single crystals is a promising choice for substrate preparation of certain electronic materials such as the family of CdTe and other II–VI compounds, HgI_2, organic materials, etc. Also the idea that vapour growth is limited by very low growth rates can be shown to be incorrect. Large crystals can have very high mass uptake per unit time. The real problem is how to control the process in order to reach large crystal dimensions without deteriorating the perfection. The work with HgI_2 illustrates how to proceed in order to solve this problem. As fundamental experimental research in bulk vapour growth has barely begun, there are many topics still to be investigated. The main effort should go in the direction of in-depth structural and electronic characterization and the understanding how defects and traps are formed during growth. To this end in situ diagnostic methods should be further developed. The most important parameters to be studied in situ are:

- measurement of the surface temperature of the growing crystal and source material, with an accuracy 0.001 K. This is applicable to crystals with appreciable absorption coefficients within the wavelength range measured (not possible for HgI_2). It will give the true supersaturation of the system, which has, up to the present, never been measured;
- control of the temperatures of the growth ampoule with an accuracy 0.001 K;
- measurement of the growth rate of the various crystallographic orientations not only of small but also of large crystals. Very sensitive methods (interferometry etc.) can be

used only for small crystals in special cells, but measurements of larger crystals are absolutely necessary, because, with increasing dimensions, the disturbances and instabilities set-in;

- light scattering to show the growth of striations and other defects, provided that a degree of purity is achieved which allows the growth of non-scattering areas also (for transparent crystals only).

Last, but not least, the presently existing experiments show that microgravity is a very good environment for bulk vapour growth, particularly when large crystals are desired. Even under only a small vapour pressure, convection is present on earth.

The alternative to the closed systems are the open or semi-open systems (described in section 3) which have hardly been investigated for bulk growth. In principle, they give an alternative to control convection, but the ultimate degree of perfection attainable is not yet known.

Acknowledgements

The authors acknowledge with pleasure the great help of the editor in compiling parts of this manuscript. Without this help the manuscript would have not be ready in due time. Many thanks are due also to A. Rossberg, G. Lelay and J. Castaldi for communication of unpublished results. Without high purity starting material no good HgI_2-crystals could be grown. The authors are indebted to G. Wetzel for this excellent work.

The authors are also indebted to Prof. Rosenberger for pointing out that the idea of the constitutional undercooling in the gas phase introduced by Reed and Lefleur [34] has still to be confirmed experimentally. The experimental evidence provided in [34] cannot be taken into account due to an error in calculations by one order of magnitude made by these authors.

References

[1] E. Kaldis, in: Crystal Growth, ed. C.H.L. Goodman (Plenum, London, 1974) p. 70.

[2] E. Kaldis, R. Cadoret and E. Schönhert, in: Fluid Sciences and Materials Science in Space, ed. H.U. Walter (Springer, Berlin, 1987) p. 355.

[3] M.M. Faktor and T. Garrett, Growth of Crystals from the Vapour (Chapman & Hall, London, 1974).

[4] J. Stefan, Ann. Phys. Chem. 17 (1890) 550.

[5] F. Rosenberger, Fundamentals of Crystal Growth, Vol. 1 (Springer, Berlin, 1979).

[6] D.W. Greenwell, B.L. Markham and F. Rosenberger, J. Cryst. Growth 51 (1981) 413.

[7] M. Piechotka and E. Kaldis, Study of Technical Requirements for Vapour Growth Experiments under Microgravity, ESA Contract No. 5943/84/F/FL (1986).

[8] B.K. Chakraverty, in: Crystal Growth, ed. P. Hartmann (North-Holland, Amsterdam, 1973) p. 50.

[9] R. Niedermayer, in: Advances in Epitaxy, eds. H.G. Schneider and V. Ruth (VEB Verlag für Grundwissenschaften, Leipzig 1971) p. 21.

[10] F. Rosenberger, Interfacial Aspects of Phase Transformation, Proc. NATO Adv. Study Inst., Erice, Sicily, 1981, eds. R. Kern and M. Mutaftschiev (Reidel, Dordrecht, 1982).

[11] B. Bennema and C. van Leeuwen, J. Cryst. Growth 31 (1975) 3.

[12] T. Gutzow and S. Toschev, J. Cryst. Growth 7 (1970) 215.

[13] G. Perrier, C. Belouet, J. Omaly and R. Cadoret, Adv. Space Res. 19 (1979) 531.
[14] H. Schäfer, H. Jakob and K. Etzel, Z. Anorg. Allgem. Chem. 286 (1956) 27.
[15] J. Omaly, M. Robert and R. Cadoret, Mater. Res. Bull. 16 (1981) 785; 1261.
[16] S.S. Rao and C.O. Bennet, Ind. Eng. Chem. Fundamentals 6 (1967) 477.
[17] G.H. Schiroky and F. Rosenberger, Int. J. Heat Mass Transfer 27 (1984) 587.
[18] G.P. Extremet, P. Bontoux and B. Roux, Adv. Space Res. 6 (1986) 147.
[19] B..L. Markham and F. Rosenberger, J. Cryst. Growth 67 (1984) 241.
[20] B.L. Markham, D.W. Greenwell and F. Rosenberger, J. Cryst. Growth 51 (1981) 413.
[21] F. Elie, Doctorate thesis, Université Aix-Marseille 2 (1984).
[22] G. Extremet, P. Bontoux and B. Roux, Adv. Space Res. 6 (1986) 147.
[23] G. Extremet, B. Roux, P. Bontoux and F. Elie, J. Cryst. Growth 82 (1986) 761.
[24] P. Bontoux, C. Smutec, B. Rouse and J.M. Lacroix, J. Fluid Mech. 169 (1986) 211.
[25] J.R. Abernathy, D.W. Greenwell and F. Rosenberger, J. Cryst. Growth 47 (1979) 145.
[26] E. Schönherr, E. Winckler and R. Lauck, J. Cryst. Growth 75 (1986) 633.
[27] R.F. Strickland-Constable, Kinetics and Mechanisms of Crystallisation (Academic Press, London, 1968).
[28] T. Kobayashi, J.T. Muheim, P. Waegli and E. Kaldis, J. Electrochem. Soc 130 (1983) 1183.
[29] J. Omaly, M. Robert, P. Brisson and R. Cadoret, Nucl. Instr. Meth, 213 (1983) 19.
[30] M. Schieber, R.C. Calston, H.A. Lamonds, P.T. Randtke, W.F. Schnepple and J. Llacer, J. Cryst. Growth 24/25 (1974) 205.
[31] V.M. Zaletin, N.V. Lyakh and N.V. Ragozina, Cryst. Res. Technol. 20 (1985) 307; 321.
[32] M. Piechotka, unpublished results (1984).
[33] L. van den Berg and C. Ortale, private communication (1985).
[34] T.B. Reed and W.J. LaFleur, Appl. Phys. Lett. 5 (1964) 191.
[35] J.W. Rutter and B. Chalmers, Canad. J. Phys. 31 (1953) 15.
[36] M.M. Faktor, R. Heckingbottom and I. Garrett, J. Chem. Soc. A 16 (1970) 2657.
[37] M.M. Faktor, R. Heckingbottom and I. Garrett, J. Chem. Soc. A 16 (1971) 1.
[38] M.M. Faktor and I. Garrett, J. Chem. Soc. A (1971) 934.
[39] M.M. Faktor, I. Garrett and R.M. Moss, J. Chem. Soc. Faraday Trans. 1 (1973) 1915.
[40] W. Akutagawa and K. Zanio, J. Cryst. Growth 11 (1971) 191.
[41] I.F. Nicolau, private communication, published in [7].
[42] A.C. Prior, J. Electrochem. Soc. 108 (1961) 82.
[43] J.R. Cutter and J. Woods, J. Cryst. Growth 47 (1979) 405.
[44] R. Nitsche, Fortschr. Mineral. 44 (1967) 231.
[45] B. Honigmann, Z. Elektrochem. 58 (1954) 322.
[46] T.B. Reed, MIT Rep. 1 (1969) 21.
[47] J.F. Butler and T.C. Harman, Appl. Phys. Lett. 12 (1968) 347.
[48] E. Kaldis and R. Widmer, J. Phys. Chem. Solids 26 (1965) 1697.
[49] W.W. Piper and S.J. Polich, J. Appl. Phys. 32 (1961) 1278.
[50] M. Schieber, I. Beinglass, G. Dishon and A. Holzer, in: Crystal Growth and Materials (1976), eds. E. Kaldis and H.J. Scheel (North-Holland, Amsterdam, 1977) p. 280.
[51] W.R. Willig, Nucl. Instr. Meth. 96 (1971) 615.
[52] H. Scholz, Acta Electron. 17 (1974) 69.
[53] H. Heyer, J. Phys. Chem. Sol. 1 (1967) 256.
[54] H.-U. Walter, private communication.
[55] M. Schieber, Nucl. Instr. Meth. 144 (1977) 469.
[56] M. Schieber, W.F. Schnepple and L. van den Berg, J. Cryst. Growth 33 (1976) 125.
[57] L. van den Berg, W.F. Schnepple, C. Ortale and M. Schieber, J. Cryst. Growth 42 (1977) 160.
[58] M. Piechotka and E. Kaldis, Monatsh. Chemie 118 (1987) 137.
[59] I.N. Stranski and D. Totomanov, Z. Physik. Chem. A 163 (1933) 399.
[60] M. Isshiki, M. Piechotka and E. Kaldis, J. Cryst. Growth 102 (1990) 344.
[61] M. Piechotka, M. Zha and E. Kaldis, J. Cryst. Growth 113 (1991) 251.
[62] N.B. Singh, R.H. Hopkins, R. Mazelsky and M. Gottlieb, J. Cryst. Growth 85 (1987) 240.

[63] M. Isshiki, M. Piechotka and E. Kaldis, in: Proc. 7th European Symp. Materials Sci. in Microgravity, Oxford, UK, 1989, ESA SP-295 (1990) 13.
[64] S. Leon-Gits, Nucl. Instr. Meth. 285 (1989) 543.
[65] A.A. Chernov, E. Kaldis, M. Piechotka and M. Zha, J. Cryst. Growth 125 (1992) 627.
[66] A. Roux, Study on Low Temperature Vapour Growth of Mercuric Iodide. Numerical Modelling, Final Report, ESA Contract No. 8431/89/F/BZ (SC), Marseille (1992).
[67] M. Zha, M. Piechotka and E. Kaldis, J. Cryst. Growth 115 (1991) 43.
[68] M. Piechotka, G. Wetzel, A. Rosseberg, H.-J. Schwer, M. Zha and E. Kaldis, J. Cryst. Growth 128 (1993) 1040.
[69] CARRAR/MATRA, High resolution diagnostics study, ESA Contract (1991).
[70] L. van den Berg, Nucl. Instr. Meth. A 322 (1992) 453.
[71] R. Cadoret, P. Brisson and A. Magnan, Nucl. Instr. Meth. A 283 (1989) 339.
[72] L. van den Berg and W.F. Schnepple, Nucl. Instr. Meth. A 283 (1989) 335.
[73] A. Nadarajah, F. Rosenberger and J.I. Alexander, J. Cryst. Growth 118 (1992) 49.
[74] XSirius Inc., Santa Barbara, USA, video film presented at the 10th Int. Conf. on Crystal Growth, 1992, San Diego, USA.
[75] Technical Assistance in the Development of Facilities for Vapour Growth Investigations in Microgravity, Final Report, ESA Contract 9990/92/NL/GS, ETH Zurich (1993).
[76] M. Piechotka, E. Kaldis, P. Behrmann and H. Stanna, in: Proc. 7th European Symp. Materials Sci. in Microgravity, Oxford, UK, 1989, ESA SP-295 (1990) p. 587.
[77] K. James, V. Gerrish, E. Cross, M. Markakis, J. Marschall and F. Milstein, Nucl. Instr. Meth. A 322 (1992) 390.
[78] C.J. Paradies, M.E. Glicksman, O.C. Joness, G.T. Kim and J.T. Lin, in: SPIE Vol. 1561, Inorg. Crystals for Optics, Electrooptics and Frequency Conversion, (1991) p. 2.
[79] F. Remy, J. Gastaldi, G. Le Lay, A. Rossberg, M. Piechotka and E. Kaldis, to be published.
[80] S. Gits, Nucl. Instr. Meth. 213 (1983) 473.
[81] B. Steiner, R.C. Dobbyn, D. Black, H. Burdette, M. Kuriyama, R. Spal, L. van den Berg, A. Fripp, R. Schimchik, R.B. Lal, A. Batra, D. Mathiessen and B. Ditchek, J. Res. NIST 96 (1991) 305.
[82] B. Yelon, R.W. Alkire, M.M. Schieber, L. van den Berg, S.E. Rasmussen, H. Christensen and J.R. Schneider, J. Appl. Phys. 52 (1981) 4604.
[83] A. Levi, A. Burger, M. Schieber, L. van den Berg, W.B. Yelon and R.W. Alkire, Nucl. Instr. Meth. A 213 (1983) 31.
[84] M. Schieber, C. Ortale, L. van den Berg, W. F. Schnepple, L. Keller, C.N.J. Wagner, W. Yelon, F. Ross, G. Georgeson and F. Milstein, Nucl. Instr. Meth. A 283 (1989) 172.
[85] L. van den Berg and W.F. Schnepple, Nucl. Instr. Meth. A 283 (1989) 335.
[86] A. Rossberg, M. Piechotka, E. Kaldis and A. Magerl, to be published.
[87] A. Burger, S.H. Morgan, E. Silberman, D. Nason and A. Y. Cheng, Nucl. Instr. Meth. A 322 (1992) 427.
[88] A.A. Davydov, V.N. Ermolov, S.W. Neustroev and L.P. Pavlova, Neorg. Mater. 28 (1992) 42.

HANDBOOK OF CRYSTAL GROWTH 2

BULK CRYSTAL GROWTH

Part B: Growth Mechanisms and Dynamics

HANDBOOK OF CRYSTAL GROWTH

Volume 1: FUNDAMENTALS

A: Thermodynamics and Kinetics
B: Transport and Stability

Volume 2: BULK CRYSTAL GROWTH

A: Basic Techniques
B: Growth Mechanisms and Dynamics

Volume 3: THIN FILMS AND EPITAXY

A: Basic Techniques
B: Growth Mechanisms and Dynamics

NORTH-HOLLAND
Amsterdam - London - New York - Tokyo

HANDBOOK OF CRYSTAL GROWTH 2

BULK CRYSTAL GROWTH

Part B: Growth Mechanisms and Dynamics

Edited by

D.T.J. Hurle

H.H. Wills Physics Laboratory
University of Bristol
Bristol, UK BS8 1TL

1994
NORTH-HOLLAND
Amsterdam - London - New York - Tokyo

ELSEVIER SCIENCE B.V.
Sara Burgerhartstraat 25
P.O. Box 211, 1000 AE Amsterdam, The Netherlands

Library of Congress Cataloging-in-Publication Data
(Revised for vol. 2, pt. A-B)

Handbook of crystal growth.

Includes bibliographical references and indexes.
Contents: v.1. Fundamentals. pt. A. Thermodynamics and kinetics. pt. B. Transport and stability -- v.2. Bulk crystal growth. pt. A. Basic techniques. pt. B. Growth mechanisms and dynamics.
1. Crystal--Growth. I. Hurle, D.T.J.
QD921.H32 1993 548'.5 93-7317
ISBN 0-444-81554-6

The beautiful natural snowflake on the front cover was photographed by Yoshinori Furukawa on Mt. Asahidake in Japan. The adjacent vertical strip is from a transmission electron micrograph taken by Tony Cullis of an In-implanted Si layer exhibiting cellular microsegregation produced by pulsed laser recrystallisation from the melt. The two images were chosen to reflect the natural and man-made aspects of crystal growth, respectively.

ISBN: 0 444 81554 6 (Part A and B)
0 444 89933 2 (Series)

Printed in The Netherlands

CONTENTS OF VOLUME 2

Preface ix

Part A. Basic Techniques

1. Investigations of crystal growth in earth and planetary sciences 1
   I. Sunagawa
2. Bridgman and related growth techniques 51
   E. Monberg
3. Czochralski growth 99
   D.T.J. Hurle and B. Cockayne
4. Crystal growth by floating zone melting 213
   J. Bohm, A. Lüdge and W. Schröder
5. Use of a magnetic field in melt growth 259
   D.T.J. Hurle and R.W. Series
6. Fractional crystallisation 287
   P.J. Jansens and G.M. van Rosmalen
7. Industrial mass crystallisation 315
   A.E.D.M. van der Heijden and G.M. van Rosmalen
8. Crystallisation of biological macromolecules 417
   A. McPherson
9. Hydrothermal growth of crystals 465
   K. Byrappa
10. Flux growth 563
    W. Tolksdorf
11. Bulk crystal growth by physical vapour transport 613
    E. Kaldis and M. Piechotka

Part B. Growth Mechanisms and Dynamics

12. Segregation phenomena in crystal growth from the melt 659
    J.P. Garandet, J.J. Favier and D. Camel
13. Convection in melt growth 709
    G. Müller and A. Ostrogorsky
14. Stress in the cooling crystal 821
    J. Völkl
15. Modelling Bridgman and Czochralski growth 875
    F. Dupret and N. van den Bogaert
16. Shaped crystal growth 1011
    V.A. Tatarchenko
17. Crystallisation of eutectics, monotectics and peritectics 1111
    J.D. Hunt and S.-Z. Lu

18. Spherulitic crystallisation in macromolecules 1167
P.J. Phillips
19. Fundamentals of aqueous solution growth 1217
S. Sarig
20. Crystal growth in gels 1271
F. Lefaucheux and M.C. Robert

Subject Index 1305

CHAPTER 12

Segregation Phenomena in Crystal Growth from the Melt

J.P. GARANDET, J.J. FAVIER and D. CAMEL

Commissariat à l'Energie Atomique,
DTA/CEREM/DEM/SMGP,
Centre d'Etudes Nucléaires de Grenoble, 85X 38041 Grenoble Cedex, France

Contents

1. Introduction 661
2. General formulation 662
   2.1. Species conservation 663
   2.2. The convecto–diffusive fluxes 664
   2.3. The boundary conditions 665
   2.4. The dilute alloy approximation 666
3. Boundary layer concept and scaling analysis 668
   3.1. Formulation 668
   3.2. The reference diffusive case 668
   3.3. The boundary layer concept 669
   3.4. The BPS solution 670
   3.5. Scaling analysis 671
      3.5.1. Principles 671
      3.5.2. The boundary layer extent 672
      3.5.3. Verification 673
   3.6. Discussion 675
4. Axial segregation 676
   4.1. Statement of the problem 676
   4.2. Complete mixing 677
      4.2.1. Normal freezing 677
      4.2.2. Zone melting 678
   4.3. Diffusive transport 680
      4.3.1. Normal freezing 680

HANDBOOK OF CRYSTAL GROWTH, VOL. 2
edited by D.T.J. Hurle

4.3.2. Zone melting . . . 681
4.4. Partial mixing . . . 682
4.4.1. Quasi-steady-state analysis . . . 682
4.4.2. Transient analysis . . . 683
4.4.3. Connection with the boundary layer approach . . . 683
4.4.4. A word on the stagnant film model . . . 684
4.5. Experimental and numerical data . . . 685
4.6. Concluding remarks . . . 687
5. Radial segregation . . . 688
5.1. Statement of the problem . . . 688
5.2. The convective regime limit . . . 688
5.3. The purely diffusive case . . . 690
5.4. The effect of partial mixing . . . 691
5.5. Numerical and experimental results . . . 694
5.6. Concluding remarks . . . 695
6. Microsegregation . . . 696
6.1. Statement of the problem . . . 696
6.2. Quasi-steady-state models . . . 696
6.3. Theoretical analysis . . . 698
6.4. Numerical approaches . . . 700
6.5. Experimental results . . . 701
6.6. Concluding remarks . . . 703
7. General conclusion . . . 703
Acknowledgements . . . 704
References . . . 705

1. Introduction

Segregation, both at the macroscopic and the microscopic scale, is one of the major factors limiting the yield of bulk crystals grown from the melt. For instance, in electronic or optoelectronic applications, the quality of the active epilayers often depends directly on the chemical homogeneity of the substrates. Severe requirements are thus imposed by the potential users and segregation predictions would certainly be valuable assets for the practician. As the solidification conditions (e.g. thermal gradients, interface velocities, etc.) may vary significantly from one experiment to another, these predictions should be applicable to a wide range of growth parameters.

The problem is in principle tractable since the equations governing mass transfer were discovered more than a century ago. Briefly stated, solute transport is split into diffusive (often related to concentration gradients) and convective contributions. In practice, however, we are left with a formidable task: the system is often multidimensional and time dependent; moreover, the Navier–Stokes equations are generally very difficult to solve. It is only in highly idealized cases that an analytical solution can be found.

An alternative approach relies on a complete numerical modelling of the growth configuration. Accurate estimates of segregation can thus be derived but the method has its limitations in terms of calculation time and memory space. Moreover, the complex codes that perform the calculations are rather difficult to handle, making it hardly possible for the practician to use them, at least in the near future.

A third approach is based on a physical understanding of the transport phenomena. In that respect, one of the key features of the problem is the existence of a solute boundary layer, i.e. a region where all the significant composition changes take place, in the vicinity of the growth interface. We shall see that the boundary layer concept finds extensive use in a large variety of cases in the fields of axial, radial and microsegregation. It should also be said that this kind of approach can be applied to related problems, for instance in eutectic solidification.

Another important point to be stressed is that a simple mathematical analysis can be fruitfully used to obtain the orders of magnitude of the relevant variables (e.g. the maximum concentration difference across the interface) and to identify scaling laws between them. Once the method has been validated by comparison with either numerical or experimental data, predictions can be made at the expense of very simple algebra.

As for the contents of our contribution, we shall first recall (section 2) the equations governing mass transport in a crystal growth melt, outlining the usual approximations made to simplify the problem. Section 3 is devoted to a presentation of the boundary layer concept and of the scaling analysis. We then focus on the applications in terms of axial (section 4), radial (section 5) and micro (section 6) segregation.

In each case, the physical mechanisms are first identified, a brief account of the major theoretical, numerical and experimental works in the field is then given and we conclude by offering a few tips on how to improve crystal homogeneity. It should also be said that our contribution was organized to ensure that each section can, to a large extent, be read independently from the others. Finally, we would like to mention a few reference books [1–4] where the reader can find valuable information on segregation problems.

2. General formulation

2.1. Species conservation

This section is dedicated to a derivation of the equations governing mass transport in a crystal growth melt. We start with a general formulation and then proceed with some of the approximations usually made to simplify the problem. Our presentation is based on the text books by Bird, Stewart and Lightfoot and by Rosenberger [2, 5]. The interested reader can also find valuable information in the work of Wheeler in Vol. 1 of this handbook.

For each constituent j in a p-component system, let us define its mass concentration ρ^j as the mass of species j per unit volume of solution. If we denote $\boldsymbol{v}^j$ the mean velocity of the molecules of species j in a small volume element of the fluid, the mass-average velocity $\boldsymbol{V}$ in this element is given by

$$\boldsymbol{V} = \frac{\sum_1^p \rho^j \boldsymbol{v}^j}{\sum_1^p \rho^j}, \tag{2.1}$$

and $\rho_{\mathrm{L}} = \sum_1^p \rho^j$ represents the total mass density of the liquid.

As noted in [5], with such a definition $\boldsymbol{V}$ can be identified with the velocity appearing in the Navier–Stokes equation that governs momentum transfer. For our present purposes, we need only consider this mass-average velocity, but one should keep in mind that other definitions can be used (e.g. molar average, volume average, etc.) in multicomponent solutions. Now the total mass flux of constituent j with respect to the laboratory frame,

$$\boldsymbol{J}_{\mathrm{T}}^j = \rho^j \boldsymbol{v}^j, \tag{2.2}$$

can be split in two parts:

$$\boldsymbol{J}_{\mathrm{T}}^j = \rho^j(\boldsymbol{v}^j - \boldsymbol{V}) + \rho^j \boldsymbol{V} = \boldsymbol{J}_{\mathrm{D}}^j + \boldsymbol{J}_{\mathrm{C}}^j. \tag{2.3}$$

$\boldsymbol{J}_{\mathrm{D}}^j$ measures the intensity of mass transfer with respect to a frame moving with the fluid (diffusive flux) whereas $\boldsymbol{J}_{\mathrm{C}}^j$ represents the contribution of the overall velocity to mass transport (convective flux).

Let us now consider a volume element in the liquid, fixed with respect to the laboratory frame. Assuming no chemical reaction in the mixture, mass conservation of constituent j in this element requires

$$\frac{\partial \rho^j}{\partial t} + \nabla \cdot \boldsymbol{J}_{\mathrm{T}}^j = \frac{\partial \rho^j}{\partial t} + \nabla \cdot (\boldsymbol{J}_{\mathrm{D}}^j + \rho^j \boldsymbol{V}) = 0, \tag{2.4}$$

where ∇ is the Nabla operator. Upon summation over all the constituents, we obtain using eqs. (2.1) and (2.2),

$$\frac{\partial \rho_{\mathrm{L}}}{\partial t} + \nabla \cdot \rho_{\mathrm{L}} \boldsymbol{V} = 0, \tag{2.5}$$

which is the well known continuity equation of fluid mechanics. We now have to propose expressions for the convecto–diffusive fluxes as inputs to the mass conservation equations.

2.2. *The convecto–diffusive fluxes*

In order to estimate the convective contributions $\boldsymbol{J}_{\mathrm{C}}^{j} = \rho^{j}\boldsymbol{V}$ to the overall mass transfer, it is necessary to solve the Navier–Stokes equation. Convection in melt growth being the subject of another chapter in this handbook, we will not go into details, but a few points are worth making.

If we consider momentum conservation in a multicomponent system, the velocity in the fluid $\boldsymbol{V}$ is governed by the equations of continuity [eq. (2.5)] and Navier–Stokes:

$$\rho_{\mathrm{L}}\left(\frac{\partial \boldsymbol{V}}{\partial t} + (\boldsymbol{V}\cdot\nabla)\boldsymbol{V}\right) = -[\nabla\cdot\boldsymbol{\tau}] + \sum_{1}^{p}\rho^{j}\boldsymbol{f}^{j}. \tag{2.6}$$

$\boldsymbol{\tau}$ is the stress tensor that includes both pressure and viscosity effects and $\boldsymbol{f}^{j}$ the external force per unit mass on component j. For instance, an applied electric field can act differently on the various constituents of the solution.

But even if gravity is the only external force to consider, thus implying $\sum\rho^{j}\boldsymbol{f}^{j} = \rho_{\mathrm{L}}\boldsymbol{g}$, we are left with a formidable task: the species and momentum conservation equations [see eqs. (2.4) and (2.6)] cannot generally be treated separately. Furthermore, since ρ depends also on temperature, one has, in addition to consider the heat transfer phenomena, to solve the mass transfer problem. In practice, further simplifying assumptions often have to be made and we will present at the end of this section the commonly used dilute alloy approximation.

Let us now turn to the diffusion terms $\boldsymbol{J}_{\mathrm{D}}^{j}$, again avoiding all unnecessary details. Quite generally, $\boldsymbol{J}_{\mathrm{D}}^{j}$ can be split into:

$$\boldsymbol{J}_{\mathrm{D}}^{j} = \boldsymbol{J}_{\mathrm{M}}^{j} + \boldsymbol{J}_{\mathrm{F}}^{j} + \boldsymbol{J}_{\mathrm{P}}^{j} + \boldsymbol{J}_{\mathrm{H}}^{j},$$

meaning that the diffusive flux of constituent j can be due to concentration gradients ($\boldsymbol{J}_{\mathrm{M}}^{j}$), external field gradients ($\boldsymbol{J}_{\mathrm{F}}^{j}$), pressure gradients ($\boldsymbol{J}_{\mathrm{P}}^{j}$) or temperature gradients ($\boldsymbol{J}_{\mathrm{H}}^{j}$).

In the field of crystal growth, the most important contribution often comes from the ordinary concentration diffusion and we will consider only the $\boldsymbol{J}_{\mathrm{M}}^{j}$ term in the following. However, temperature gradients can sometimes lead to measurable effects.

Let us define the mass fraction of component j as the ratio $\omega^{j} = \rho^{j}/\rho_{\mathrm{L}}$. We can now proceed with a statement of Fick's law, that gives the diffusive fluxes with respect to the mass-average velocity:

$$\boldsymbol{J}_{\mathrm{D}}^{j} = -\rho_{\mathrm{L}}D_{\mathrm{m}}^{j}\nabla\omega^{j}. \tag{2.7}$$

In the above equation, D_{m}^{j} is defined as the effective binary diffusivity of constituent j in the mixture. Other equivalent expressions of the diffusive fluxes involving, e.g. molar concentration gradients, can be found elsewhere [2, 5].

Considering the true driving forces for diffusion (i.e. chemical potential gradients), one can obtain a more general – but also more complex – expression for $\boldsymbol{J}_{\mathrm{D}}^{j}$ [5]. Fortunately, D_{m}^{j} can be identified with the usual diffusion coefficient for binary systems or for dilute solutes in a nearly pure solvent. In the following, we will always assume eq. (2.7) to be valid, but one would have to go back to the textbooks when dealing with diffusion in a concentrated multicomponent solution.

Using eq. (2.7), eq. (2.4) becomes:

$$\frac{\partial \rho^j}{\partial t} + \nabla \cdot (-\rho_L D_m^j \nabla \omega^j + \rho^j \boldsymbol{V}) = 0. \tag{2.8}$$

This is the fundamental equation governing species repartition in a crystal growth melt, but we now have to find a set of boundary conditions to fully specify the problem.

2.3. The boundary conditions

Let us consider a small interface element. When the mass densities of the liquid and the solid around this element (denoted, respectively, $\rho_{L,I}$ and $\rho_{S,I}$) differ, the crystal growth rate and the fluid advection rate due to volume changes (hereafter referred to as $\boldsymbol{V}_{S,I}$ and $\boldsymbol{V}_{L,I}$) are related through [6]

$$\rho_{L,I} \boldsymbol{V}_{L,I} \cdot \boldsymbol{n} = \rho_{S,I} \boldsymbol{V}_{S,I} \cdot \boldsymbol{n}, \tag{2.9}$$

$\boldsymbol{n}$ being a unit vector normal to the boundary. Quite generally, solute incorporation in the solid depends on the interface velocity; however, in view of the low rates used for crystal growth, the undercooling is often very small and it appears reasonable to assume a thermodynamic quasi-equilibrium at the solid–liquid front. The phase diagram can then yield a relation between the composition of the crystal formed and that of the liquid that gave it birth (see fig. 1). For each constituent j, let us consider the ratio k_0^j,

$$k_0^j = \frac{\omega_{S,I}^j}{\omega_{L,I}^j}. \tag{2.10}$$

In the following, k_0^j will be termed distribution, segregation or partition coefficient. At this point, we shall not make any further hypothesis, but one should keep in mind that the k_0^j are functions of composition in the case of concentrated systems. Moreover, the actual partition coefficient sometimes differs significantly from the value predicted by

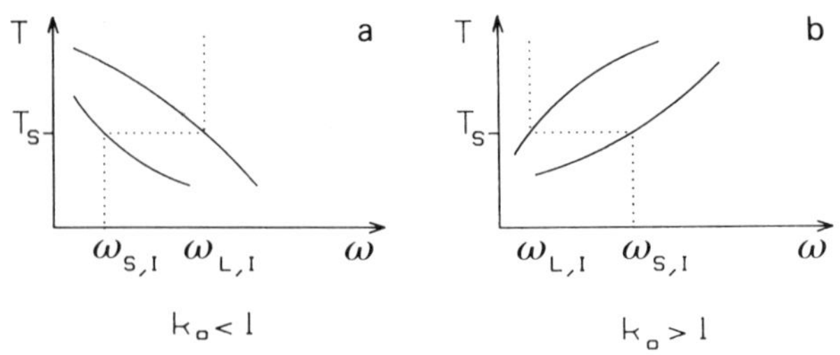

Fig. 1. Phase diagram showing solute rejection (a) or preferential incorporation (b) at the interface.

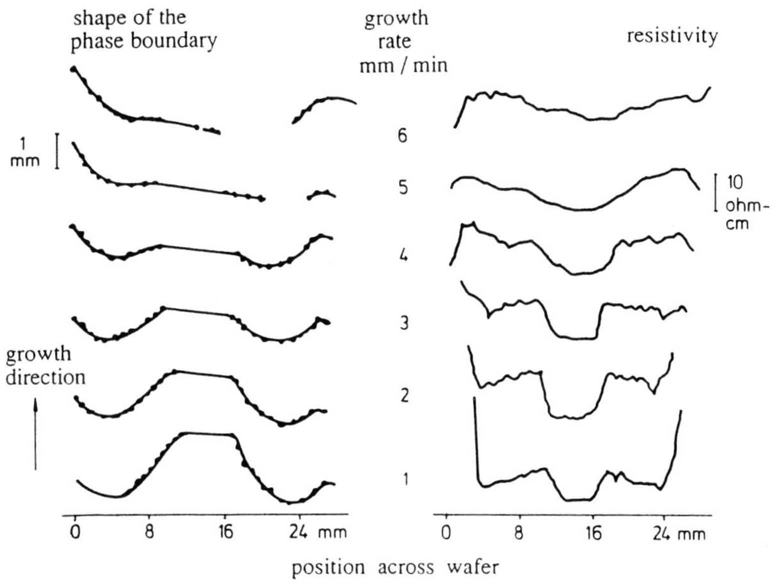

Fig. 2. Different shapes of the phase boundary of Si crystals grown at different rates (left) and the corresponding radial variations of the resistivity (right) across the wafers (from [4]).

the phase diagram when the solid–liquid front is faceted (see fig. 2); the reader is referred to [7, 8] for a detailed description of these phenomena.

When k_0^j differs from unity, i.e. when the composition in the liquid and that of the crystal are not the same, species conservation requires some mass transfer to occur. There will be a flux of matter away from ($k_0^j < 1$, solute rejection) or towards ($k_0^j > 1$, preferential incorporation) the interface. Furthermore, as the mass average velocity vanishes at the solid–liquid front, due to the no-slip condition, only diffusive transport is to be considered:

$$\boldsymbol{J}_{\mathrm{T}}^j = \boldsymbol{J}_{\mathrm{D}}^j = -\rho_{\mathrm{L,I}} D_{\mathrm{m}}^j \nabla\omega^j. \tag{2.11}$$

Introducing a solid-state diffusion coefficient D_{S}^j, species conservation at the interface becomes

$$-\rho_{\mathrm{L,I}}(\boldsymbol{V}_{\mathrm{L,I}}\cdot\boldsymbol{n}\,\omega_{\mathrm{L,I}}^j + D_{\mathrm{m}}^j\nabla\omega_{\mathrm{L,I}}^j\cdot\boldsymbol{n}) = -\rho_{\mathrm{S,I}}(\boldsymbol{V}_{\mathrm{S,I}}\cdot\boldsymbol{n}\,\omega_{\mathrm{S,I}}^j + D_{\mathrm{S}}^j\nabla\omega_{\mathrm{S,I}}^j\cdot\boldsymbol{n}).$$

$\rho_{\mathrm{L,I}}\boldsymbol{V}_{\mathrm{L,I}}\cdot\boldsymbol{n} = \rho_{\mathrm{S,I}}\boldsymbol{V}_{\mathrm{S,I}}\cdot\boldsymbol{n}$ is the total mass flux coming to the interface from the liquid. Using eqs. (2.9) and (2.10), the above expression becomes:

$$\omega_{\mathrm{L,I}}^j(1-k_0^j)\boldsymbol{V}_{\mathrm{L,I}}\cdot\boldsymbol{n} = [(\rho_{\mathrm{S,I}}/\rho_{\mathrm{L,I}})D_{\mathrm{S}}^j\nabla\omega_{\mathrm{S,I}}^j + D_{\mathrm{m}}^j\nabla\omega_{\mathrm{L,I}}^j]\cdot\boldsymbol{n}. \tag{2.12}$$

This is the fundamental equation accounting for mass transfer phenomena at the solid–liquid front. In the general case, as the k_0^j depend on composition, eq. (2.12) is not linear in terms of ω^j. It should also be noted that the effect of diffusion in the crystal is often negligible, at least when no closely spaced concentration variations (striations) are expected. From now on, we shall thus omit this contribution to the overall mass transport.

Let us now turn to the other boundary conditions. In crystal growth processes, the melt is often confined in a crucible, but contact with a surrounding atmosphere is also possible. If no exchange occurs between the fluid and the outside (closed system), species conservation requires:

$$\boldsymbol{J}_{\mathrm{T}}^{j} \cdot \boldsymbol{n} = \boldsymbol{J}_{\mathrm{D}}^{j} \cdot \boldsymbol{n} = 0.$$

Since the normal component of velocity necessarily vanishes at the boundary, only the diffusive contribution is to be considered in the total mass flux. If the above condition does not hold (e.g. growth of crystals with volatile constituents, contamination from the crucible, etc.), a law governing mass transfer in or out of the melt has to be specified.

Finally, we shall show in the following sections that it is interesting to use a coordinate system moving with the solid–liquid front for a formulation of the problem. However, when the shape of the interface changes during solidification, as can be seen for instance in the case of concentrated systems, it is not always possible to find a convenient reference frame.

Our purpose so far was to derive a set of equations and boundary conditions applying quite generally to crystal growth situations, allowing, e.g. a complete numerical modelling of the problem. We have seen that a number of difficulties arose in this approach and we shall now see in detail how the dilute alloy approximation makes matters more tractable.

2.4. *The dilute alloy approximation*

In this section, we will consider the case that the mixture is formed of traces of species $j = 2, 3, \ldots, p$ in a melt consisting of nearly pure constituent 1; our first assumption is to suppose that the total mass density of the solution does not depend on composition anymore:

$$\rho_{\mathrm{L}}(\omega^{j}, T) = \rho^{1}(T). \tag{2.13}$$

If eq. (2.13) holds, the equations of momentum and mass conservation [eqs. (2.6) and (2.8)] can now be treated separately, which is a tremendous simplification. Of course, the coupling with the heat transfer phenomena remains, and it is never an easy task to solve the Navier–Stokes equation; but what is interesting for our present purposes is that the overall velocity $\boldsymbol{V}$ appears as a sort of boundary condition for the segregation problem.

If we suppose, in addition, that the fluid is quasi-incompressible (the widely used Boussinesq approximation of fluid mechanics), the continuity equation (2.5) reduces to

$$\nabla \cdot \boldsymbol{V} = 0.$$

The divergence of the convective flux $\boldsymbol{J}_{\mathrm{C}}^{j} = \rho^{j} \boldsymbol{V}$ then becomes

$$\nabla \cdot \boldsymbol{J}_{\mathrm{C}}^{j} = \boldsymbol{V} \cdot \nabla \rho^{j} + \rho^{j} \nabla \cdot \boldsymbol{V} = \boldsymbol{V} \cdot \nabla \rho^{j}.$$

Assuming that the $D_{\rm m}^j$ do not depend on both composition and temperature, eq. (2.8) becomes, after simplification by $\rho_{\rm L} = \rho^1$,

$$\frac{\partial \omega^j}{\partial t} + V \cdot \nabla \omega^j = D_{\rm m}^j \nabla^2 \omega^j. \tag{2.14}$$

As for the associated boundary condition [eq. (2.12)], the main change is that the k_0^j can generally be considered to be constant, thus yielding a linear equation in terms of ω^j. However, let us recall that the k_0^j still depend on the structure of the interface, i.e. rough or faceted. Moreover the proportionality assumed should be taken as empirical; even at very low concentrations, there is, as stated by Müller [4], no clear theoretical evidence supporting this linear relationship.

Let us now mention another common definition of composition, in terms of the mole fractions η^j. If we assume a number of moles h^j for each species j in solution, η^j is given by

$$\eta^j = \frac{h^j}{\sum_1^p h^i}. \tag{2.15}$$

As pointed out by Wilcox [6], eqs. (2.12) and (2.14) governing species conservation at the interface and in the melt can be used equivalently in terms of ω^j and η^j. Indeed, let M^j be the molar mass of constituent j; the mass fractions ω^j are given by

$$\omega^j = \frac{M^j h^j}{\sum_1^p M^i h^i}.$$

Provided that both

$$\sum_2^p h^i \ll h^1 \quad \text{and} \quad \sum_2^p M^i h^i \ll M^1 h^1, \tag{2.16}$$

we are left with the simple proportionality relationship

$$\omega^j = \frac{M^j}{M^1} \eta^j. \tag{2.17}$$

A direct consequence of eq. (2.17) is that the k_0^j can be defined equivalently in terms of mass and moles fractions. Let us also notice that relations (2.16) can be used to test the range of validity of the dilute alloy approximation.

In the following theoretical developments, we shall restrict ourselves to the limiting case of nearly pure solutions. We will in addition always assume the segregation coefficients to be constant; we will thus have to deal with a linear problem in terms of segregation [see eqs. (2.12) and (2.14)]. However, from an experimental point of view, we will also mention some of the scarse results obtained in concentrated systems.

3. Boundary layer concept and scaling analysis

3.1. Formulation

For the readers that skipped section 2, we here briefly present the simpler formulation to be used in the following. For the sake of clarity, only dilute binary alloys are considered; however, there is no intrinsic difficulty in extending the results to the case of multicomponent systems, provided the total mass concentration of the dopants remains low enough.

We shall also assume that the mass densities of the fluid and the crystal are equal, a reasonable assumption in melt growth situations; it is then possible to define the interface velocity $\boldsymbol{V}_\mathrm{I}$ without ambiguity as can be seen from eq. (2.9). Some of the results of this section can be extended to the case where ρ_L and ρ_S differ significantly, but it is then necessary to keep in mind that solidification rate and fluid advection rate are not equal.

Solute composition, diffusion and partition coefficients are denoted C, D and k; recalling that $\boldsymbol{V}$ is the hydrodynamic velocity given by the Navier–Stokes equation and using a coordinate frame moving at a rate $\boldsymbol{V}_\mathrm{I}$ with respect to the laboratory, the governing equations for mass conservation in the bulk and at the solid–liquid front are, respectively,

$$\frac{\partial C}{\partial t} + (\boldsymbol{V} \cdot \nabla)C = D\nabla^2 C + (\boldsymbol{V}_\mathrm{I} \cdot \nabla)C, \tag{3.1}$$

$$-D\nabla C \cdot \boldsymbol{n} = C(1-k)\boldsymbol{V}_\mathrm{I} \cdot \boldsymbol{n}. \tag{3.2}$$

The $(\boldsymbol{V}_\mathrm{I} \cdot \nabla)C$ term added in eq. (3.1) comes from the change in reference frame. When k differs from unity, the composition of the fluid is not uniform, but it is the purpose of this section to show that all the significant variations occur in a boundary layer of extent δ in the vicinity of the solid–liquid front.

We shall later prove that this – so far ill defined – quantity is a measure of the convecto–diffusive state of the melt and that, as such, it will be very useful as an input in the segregation models. In this section, we shall present in some detail this boundary layer concept and propose a method to estimate δ.

Finally, we shall sometimes assume that sufficiently far away from the interface the liquid composition is constant and equal to C_∞. As noted in section 2, one would rather expect a flux condition for mass transfer out of the melt but, for modelling purposes, the $C = C_\infty$ condition is often much easier, as will be seen in the following sections.

3.2. The reference diffusive case

In this paragraph, convection is assumed to have no influence on mass transport; an additional hypothesis is that the interface lies in the plane $Z = 0$, with growth proceeding at a constant rate along the Z-direction. Considering that composition was a function of Z alone, Tiller et al. [9] treated the steady-state problem, written as:

– bulk mass transport:

$$D\frac{\partial^2 C}{\partial Z^2}+V_{\rm I}\frac{\partial C}{\partial Z}=0, \tag{3.3}$$

– boundary conditions:

$$Z=0\ \text{(interface)}: \quad -D\left(\frac{\partial C}{\partial Z}\right)_0=V_{\rm I}(1-k)C(0), \tag{3.4}$$

$$Z\to\infty: \quad C=C_\infty. \tag{3.5}$$

The solution of the above system is given by

$$C(Z)=C_\infty\left[1+\frac{1-k}{k}\exp(-ZV_{\rm I}/D)\right]. \tag{3.6}$$

The composition varies from its value at the interface C_∞/k to C_∞ over a length scale $D/V_{\rm I}$ (see fig. 3); at distances of a few times $D/V_{\rm I}$, the concentration in the liquid is quasi-uniform and equal to C_∞. Following Wilson [10], we can define the size of the solute boundary layer as

$$\delta=\frac{C(0)-C_\infty}{-(\partial C/\partial Z)_0}. \tag{3.7}$$

It is easy to check that, in this reference diffusive problem, we get $\delta=\delta_{\rm D}=D/V_{\rm I}$. Since we have here the exact solution at hand [eq. (3.6)], it can seem surprising at this point to bother with a boundary layer formalism. The reason is that such a formalism can be extended to the cases where it is necessary to consider convective transport.

3.3. The boundary layer concept

The introduction of convection in the mass transfer equation makes matters much more complex, and it is often impossible to derive analytical solutions in realistic growth cases. However, eq. (3.7) can again be used to define a length scale δ, characteristic of the convecto–diffusive state of the melt. Of course, δ will generally differ from $\delta_{\rm D}$; moreover, composition and – as a consequence – boundary layer thickness are now

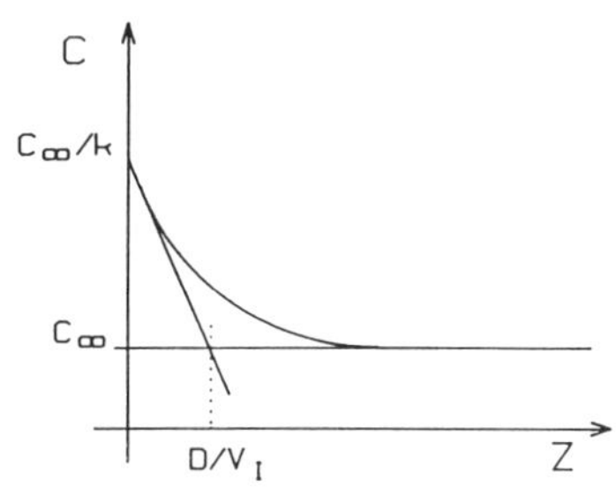

Fig. 3. Concentration profile in the fluid near the interface for purely diffusive conditions.

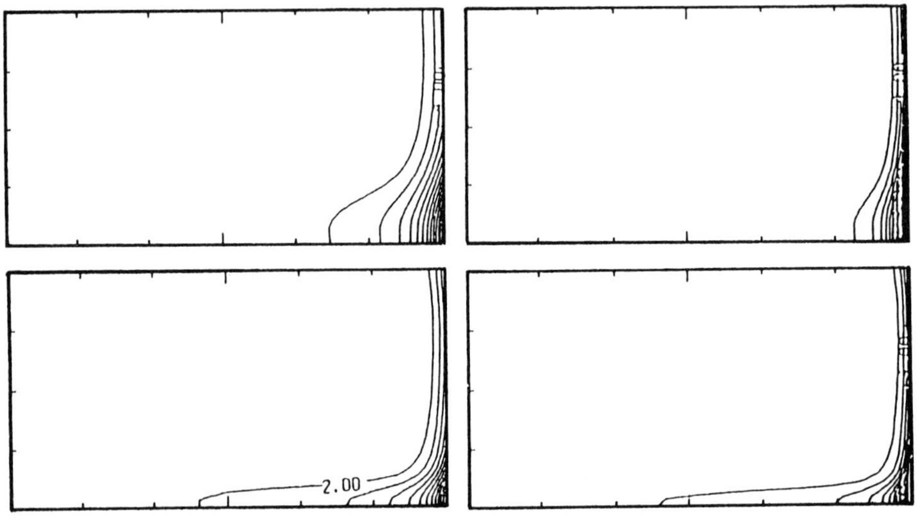

Fig. 4. Typical isoconcentration lines in a horizontal Bridgman configuration (from [15]).

allowed to vary along the interface. But the key point is that the δ thus defined can be used as inputs in almost all the physical segregation models to be presented in the frame of this chapter.

Let us now turn to the experimental data supporting the existence of a solute boundary layer in solidification configurations. Very interesting in that respect is the work of Kakimoto et al. [11] who carried out in situ X-ray topography during the growth of In-doped Si crystals by the Czochralski method. The larger absorption observed near the interface was interpreted supposing locally a higher In concentration there.

Quenching techniques have also been used, both in diffusion controlled situations [12] and in cases where thermosolutal convection was certainly present [13]. As expected, the samples exhibit an exponential-like profile at the growth front.

The boundary layer concept was also checked by means of computer simulations in the horizontal Bridgman configuration [14, 15]. As seen in fig. 4, the isoconcentration lines are much denser near the interface, indicating again a higher solute composition there; a variation of δ with position along the interface is also clearly apparent. Similar results were obtained for the vertical Bridgman case [16].

3.4. The BPS solution

We can now turn to the pioneering work of Burton, Prim and Slichter, hereafter referred to as BPS [17]; they were the first to point out the existence of a solute boundary layer at a solidification interface. They also effectively calculated its thickness in an idealized Czochralski experiment that of an infinite disk rotating over a semi-infinite liquid. We present here only their steady-state results, the transient analysis being studied in section 4.4.

In their configuration, a one-dimensional treatment of the mass transfer equation is appropriate since the isoconcentration lines are virtually normal to the growth direction, in our case the Z-axis. Assuming the hydrodynamic problem independent of the temperature and concentration fields, the normal component of the velocity profile near the interface is given by [18]:

$$W(Z) = -0.51\,\gamma^{3/2}\nu^{-1/2}Z^2. \tag{3.8}$$

It is then used as an input in the mass transfer equation:

$$D\frac{\partial^2 C}{\partial Z^2} + [V_\mathrm{I} - W(Z)]\frac{\partial C}{\partial Z} = 0. \tag{3.9}$$

The boundary conditions are still given by eqs. (3.4) and (3.5). In eq. (3.8), γ is the angular velocity of crystal rotation and ν the kinematic viscosity of the fluid. With our Z-axis oriented from the interface into the liquid, W is necessarily negative; please note that we do not use the notations of the original paper. In the low V_I range, the semi-analytic solution obtained leads to a boundary layer thickness

$$\delta = 1.6\,D^{1/3}\nu^{1/6}\gamma^{-1/2}. \tag{3.10}$$

In the high V_I range, the result would be again $\delta = \delta_\mathrm{D} = D/V_\mathrm{I}$, but this point was not outlined by the authors. Using a velocity profile as an input to the mass transfer equation, BPS were thus able to derive an expression giving δ in their idealized Czochralski configuration. Unfortunately, it is not always easy to follow the same method in real growth situations but, as we shall now see, it is often possible to find a rough estimate of δ with an order of magnitude procedure.

3.5. *Scaling analysis*

3.5.1. *Principles*

Solution of the mass transport problem as defined by eqs. (3.1) and (3.2) is in general only possible through a computer simulation of the growth configuration. Numerous interesting results were thus obtained in the past few years but even the numerical methods have their limitations in terms of calculation time and costs; moreover, the necessary memory space can also be a factor.

That is the reason why another method, involving a physical understanding of the growth phenomena, was developed in our laboratory [19, 20]. Its purpose is to identify scaling laws between the relevant groupings of variables, generally in the form of non-dimensional parameters, and to obtain order of magnitude estimates of these parameters. It should thus be used in parallel with numerical techniques, as both approaches comfort each other.

In the following, this method will be referred to as either "order of magnitude" or "scaling" analysis. Let us now illustrate its mathematical basis on the model problem of a free-fall experiment: knowing the height H of a drop tube, we shall try to estimate the time t_F needed for a particle with no initial velocity to cover the distance H. Assuming no other force than gravity, the governing equation can be written as

$$\mathrm{d}^2Z/\mathrm{d}t^2 = g. \tag{3.11}$$

Integration is obvious and the result is $t_F = (2H/g)^{1/2}$. We shall now see how our scaling procedure will lead us to an order of magnitude estimate of t_F. The principles of the method are as follows [21, 22]:

(1) Identify the ranges of variation of the different variables; in our case the relevant scales are the height H and the unknown fall time t_F.

(2) Make the equations – and if necessary the boundary conditions – non-dimensional using these scales. Setting $Z = Hz$ and $t = t_F\tau$, we get

$$(H/t_F^2)\,\mathrm{d}^2z/\mathrm{d}\tau^2 = g.$$

(3) Assume that all the non-dimensional derivatives are of the order of magnitude unity. Our estimate t_S of the free fall time is thus given by

$$H/t_S^2 = g, \qquad t_S = (H/g)^{1/2}.$$

Comparison with the exact solution shows that our order of magnitude analysis yields the right scaling law, the overall error being limited to roughly 40%. A similar procedure will now be applied to the mass transport problem to get an estimate of the boundary layer thickness.

3.5.2. *The boundary layer extent*

Let us again assume a planar interface, with growth proceeding at a constant rate V_I along the Z-axis (see fig. 5). We shall for the moment leave aside the radial segregation aspects of the problem and, as done by BPS, suppose that the concentration variations over the X- and Y-axes can be neglected.

A look at fig. 4 shows that this hypothesis is far from obvious, as the composition field is significantly distorted around the bottom left corner. But, to a first approximation, the isoconcentration lines can be considered parallel to the growth interface, at least in the upper part of the cavity. The steady state mass transfer equation thus becomes

$$D\frac{\partial^2 C}{\partial Z^2} + (V_I - W(X, Y, Z))\frac{\partial C}{\partial Z} = 0.$$

At this point, we need to get rid of X and Y to make the problem purely one-dimensional. What was done in [23] for the horizontal Bridgman problem was to take the

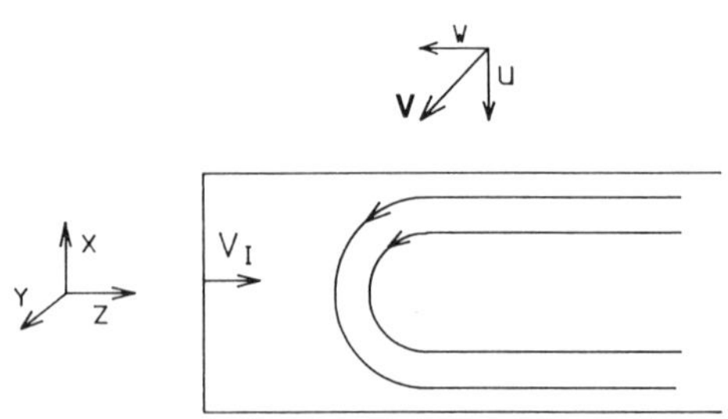

Fig. 5. Crystal growth configuration, showing the geometry of the problem and the coordinates axes.

average of W, denoted $\bar{W}$, over the upper half of the melt; any other physically sound choice (e.g. the maximum of W) would have given similar results in terms of scaling law. We shall thus write

$$D\frac{\partial^2 C}{\partial Z^2}+(V_1-\bar{W}(Z))\frac{\partial C}{\partial Z}=0. \tag{3.12}$$

We can now proceed with the order of magnitude analysis; Z ranges from 0 at the interface to a distance scaling with δ, and C from $C(0)$ (assumed constant for our present purposes) to C_∞. Setting

$$Z=\delta z, \qquad C=C_\infty+[C(0)-C_\infty]c.$$

After division by $[C(0)-C_\infty]/\delta$, eq. (3.12) becomes, in terms of the non-dimensional variables:

$$\frac{D}{\delta}\frac{\partial^2 c}{\partial z^2}+(V_1-\bar{W}(\delta z))\frac{\partial c}{\partial z}=0.$$

At the boundary layer scale $Z\simeq\delta$ (or equivalently at $z\simeq 1$) both derivatives are expected to be of the order of magnitude unity. Since their signs are opposite ($\partial^2 c/\partial z^2>0$, $\partial c/\partial z<0$), the characteristic equation for δ reads [23]

$$D/\delta=V_1-\bar{W}(\delta). \tag{3.13}$$

Let us recall again that, with our choice of axis orientation, $\bar{W}(\delta)$ is negative. At the boundary layer edge, a typical diffusive velocity D/δ is thus seen to balance the overall convective motion of the fluid $V_1-\bar{W}(\delta)$ towards the front. Moreover, depending on the relative magnitudes of V_1 and $\bar{W}(\delta)$, transport will be said to be diffusion controlled $[V_1\gg|\bar{W}(\delta)|]$ or convection controlled $[V_1\ll|\bar{W}(\delta)|]$.

3.5.3. *Verification*

Comparison of the predictions of eq. (3.13) using the velocity profile of eq. (3.8) with the results of BPS in their idealized Czochralski configuration is shown in fig. [23]. The data is presented in non-dimensional form, $\Delta=\delta/\delta_D=\delta V_1/D$ is the convecto–diffusive parameter and $B=0.17\,V_1^{-3}D^2\nu^{-1/2}\gamma^{3/2}$ – as defined by BPS – is a measure of the intensity of convection in the mass transfer problem.

The two asymptotic scaling laws ($\Delta\sim B^{-1/3}$ and $\Delta=1$) are identified through the order of magnitude procedure; the differences between the exact and the approximate solutions are always less than 25%. It is also apparent in fig. 6 that our analysis correctly predicts the location of the transition between the two regimes.

Similar results were obtained in an idealized two-dimensional horizontal Bridgman configuration [24], using the velocity profile of Batchelor–Birikh [23] as an input in the steady-state transport equation (see fig. 5 for notations):

$$D\left(\frac{\partial^2 C}{\partial X^2}+\frac{\partial^2 C}{\partial Z^2}\right)+V_1\frac{\partial C}{\partial Z}-U(X,Z)\frac{\partial C}{\partial X}-W(X,Z)\frac{\partial C}{\partial Z}=0. \tag{3.14}$$

When convection is not too strong, U and W are proportional to the Grashof number Gr that scales the driving force of natural buoyancy in the melt. But it is the

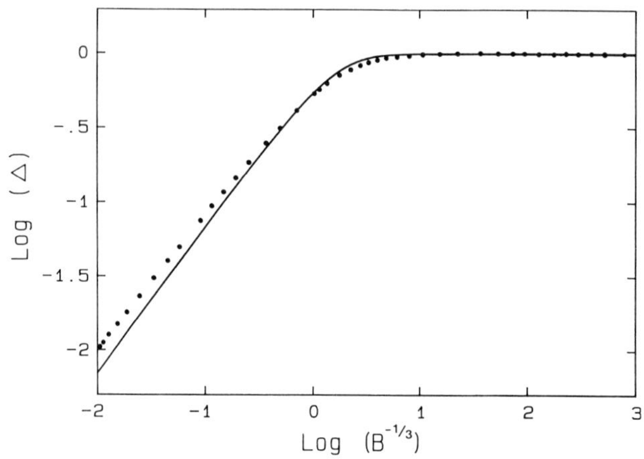

Fig. 6. Variation of Δ with the parameter B measuring the intensity of convection in the BPS problem (symbols: analytical solution; full line: scaling analysis).

product of the Grashof and Schmidt (Sc) numbers that measures the convective contribution to mass transfer. Let us also notice that with respect to the BPS problem, we have, in addition, to consider the Peclet number (Pe) that characterizes the crystal growth rate.

Finite-element computer simulations were carried out to solve eq. (3.14) along with boundary conditions (3.4) and (3.5) [24]; the results were expressed using the non-dimensional parameters GrSc and Pe. Denoting $\tilde{\delta}$ the minimum of δ along the interface, two scaling laws were identified with our numerical procedure in terms of $\tilde{\Delta} = \tilde{\delta} V_1 / D$:

– convection-controlled transport:

$$\tilde{\Delta} = 4.5 \text{ Pe GrSc}^{-1/3} \tag{3.15a}$$

– diffusion-controlled transport:

$$\tilde{\Delta} = 1. \tag{3.15b}$$

Similar scaling laws were obtained from eq. (3.13) for the asymptotic regimes [23]. Moreover, as can be seen on fig. 7, the agreement between order of magnitude estimates and numerical values of $\tilde{\Delta}$ were within 30% over a large span of the GrSc product, again justifying the validity of the analysis presented in section 3.5.2.

This choice of $\tilde{\delta}$ as a measure of the boundary layer thickness is certainly physically sound in the two asymptotic regimes, since in both cases, assuming a planar growth front, radial segregation is limited [24] and δ thus remains close to $\tilde{\delta}$ almost everywhere along the interface. It is only at the transition between these two domains that a purely one-dimensional approach is not appropriate.

Experimental evidence supporting the validity of our approach was obtained in the frame of the Mephisto programme [25] where growth takes place in a sophisticated horizontal Bridgman furnace. On-line analysis of the Seebeck voltage allows for a

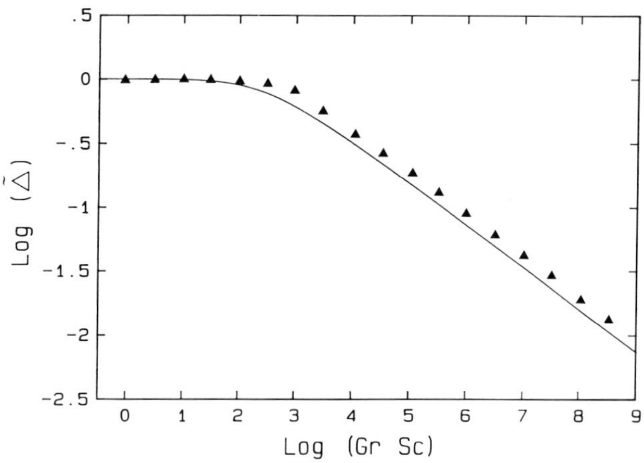

Fig. 7. Variation of the minimum of Δ across the interface with the product Gr × Sc measuring the intensity of convection in an idealized horizontal Bridgman configuration for the case $Pe = 2$ (symbols: numerical solution; full line: scaling analysis).

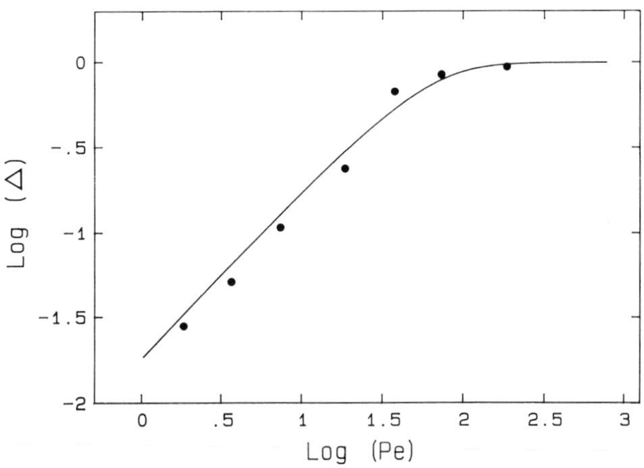

Fig. 8. Variation of Δ with Pe measuring the interface velocity (symbols: experimental data from the MEPHISTO programme; full line: scaling analysis).

precise determination of the interface composition and thus of the convecto–diffusive parameter. The results of numerous solidification runs featuring Sn:Bi and Sn:Pb alloys, presented in fig. 8 are again in very good agreement with the theoretical predictions.

3.6. Discussion

One of the apparent weaknesses of the above analysis is that it does not take into account the boundary condition at the solid–liquid front. We tested the validity of this simplification by simulating a eutectic growth in the same two-dimensional horizontal Bridgman configuration [24]. The mass transfer equation (3.14) is unchanged, but the boundary condition at the interface can now be written as [26]:

$$C = C_{\mathrm{I}}. \tag{3.16}$$

The key result is that the numerical scaling laws of eqs. (3.15a) and (3.15b) remain the same, with only a minor difference being in terms of the prefactor. What appears important is thus the existence – and not the magnitude – of the concentration gradient at the interface. This was further confirmed when we saw that, in the monophase case, a variation of the partition coefficient did not affect $\tilde{\delta}$, but modified, of course radically, the composition step near the front.

As noted in [23], it is important to keep in mind that the precision obtained is surprisingly good; scaling techniques are known to give the right orders of magnitude, but the differences with the numerical results are very often larger than 30%. However, in boundary layer problems, the ranges of variations are well defined and this could explain the accuracy of our estimates.

A word on the convecto–diffusive parameter is also necessary; as seen in figs. 6–8, Δ is close to 1 when solute transport is principally diffusive and tends towards zero with increasing buoyancy. Δ thus quantifies the respective importance of diffusion and convection on mass transport; as such, it can be used as an input in the growth models to be presented in the next section.

To sum up, our purpose here was to show that a solute boundary layer existed near the growth interface and to propose a method to estimate its thickness. So far, we only considered a steady state approach and we now have to build the link with solidification problems.

4. Axial segregation

4.1. Statement of the problem

The problem of solute distribution along melt-grown crystals was already identified in the 1920s. The first approaches assumed complete mixing in the liquid phase, whereas solutions in purely diffusive mass transport conditions were proposed in the 1950s. We shall first focus on these limiting cases (sections 4.2 and 4.3, respectively) before addressing the general problem in section 4.4.

A difference will be made between the "normal freezing" (where the whole feed is actually molten as in Czochralski or Bridgman growth) and zone melting configurations. In all the theoretical developments, the solid–liquid front will be considered perfectly planar; the effect of curvature will be thoroughly studied in section 5.

Moreover, only one-dimensional approaches will be carried out and we shall also suppose that the crystals are grown from feeds of constant composition C_0, keeping in mind that an extension to the case of inhomogeneous feeds is sometimes possible. Finally, numerical and experimental data taken from the various growth techniques will be presented in section 4.5.

4.2. Complete mixing

The first analytical works on segregation assumed the composition of the melt to be constant, which is a priori a rather surprising hypothesis. Indeed, a look at eq. (3.2) will convince the reader that there is always a concentration gradient at the interface during solidification. Except in the particular case $k = 1$, irrelevant for our present purposes since no segregation occurs, complete mixing will only be achieved when the solute diffusion coefficient is large enough to homogenize the liquid.

This was the explanation proposed by the early workers in the field to justify their assumption. Seen from another point of view, one could also say that the complete mixing hypothesis is likely to be valid when the growth velocity is very low. In practice, we shall later see that such a simple argument is not unrealistic in a variety of growth configurations.

4.2.1. Normal freezing

Let us consider a molten feed of length L and area A, having a uniform composition C_0; let us also assume that the mass densities of the solid and the liquid are equal $\rho_L = \rho_S = \rho$. When the growth front moves from z to $z + dz$, the mass of solute actually absorbed in the crystal is $\rho AkC(z)\,dz$, C being the concentration in the fluid expressed in weight fraction. Conversely, the mass of solute in the melt is always given by $Q = \rho A(L - z)C(z)$; equating dQ and $-\rho AkC(z)\,dz$, we get:

$$(1 - k)C(z)\,dz = (L - z)\,dC.$$

After division of both sides by L and integration from $C = C_0$ at $z = 0$, the above equation becomes:

$$C(z) = C_0(1 - z/L)^{k-1}, \tag{4.1a}$$

or equivalently,

$$C(\mathrm{fs}) = C_0(1 - \mathrm{fs})^{k-1}. \tag{4.1b}$$

The solidified fraction fs being defined as $\mathrm{fs} = z/L$. The concentration in the solid is thus given by

$$C_S(\mathrm{fs}) = kC_0(1 - \mathrm{fs})^{k-1}. \tag{4.2}$$

The concentration profiles of eq. (4.2), often referred to as Scheil's law, are shown in fig. 9 for various values of the partition coefficient. When k differs markedly from unity, an important segregation is to be expected; at the end of the ingot, the concentration variations are generally too high for application purposes, and part of the crystal has to be discarded.

If we now suppose that the densities of the liquid and the solid differ, a similar argument leads to a modified form for eq. (4.1a):

$$C(z) = C_0[1 - (\rho_S/\rho_L)(z/L)]^{k-1}. \tag{4.3}$$

It is interesting to notice that eq. (4.1b) [and thus eq. (4.2)] remain unchanged in terms

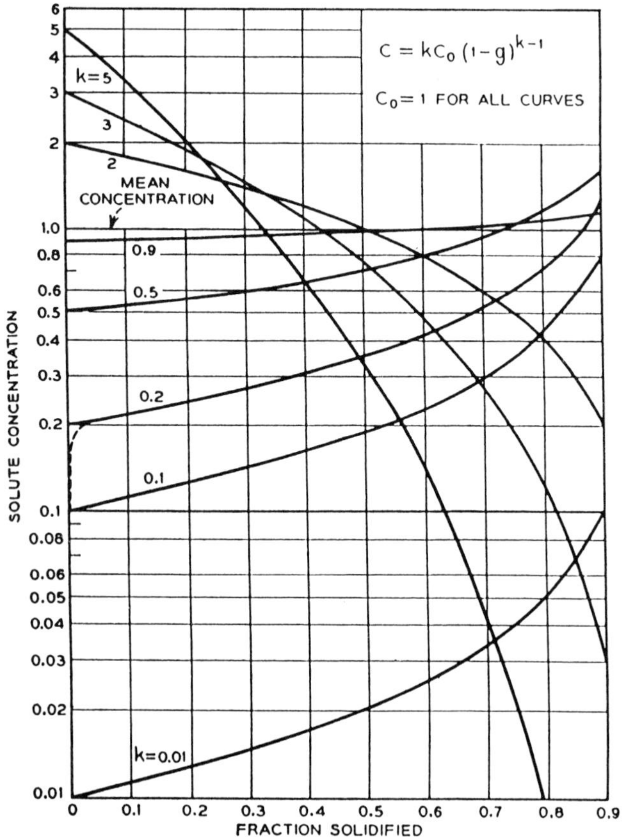

Fig. 9. Solute concentration along the crystals for various values of the partition coefficient k in the normal freezing configuration (from [27]).

of solidified fraction [27]. Indeed, the total length of the grown crystal being given by $(\rho_L/\rho_S)L$, we have $fs = (\rho_S/\rho_L)(z/L)$.

Let us also keep in mind that eq. (4.2) cannot be valid through the entire fs range in the case $k < 1$, since composition would tend to infinity when fs reaches unity. In practice, the assumption of constant k does not hold at high concentrations; besides, there is often a eutectic compound in the phase diagram, meaning that single-phase solidification cannot proceed indefinitely.

4.2.2. *Zone melting*

We now consider a molten zone of length l and area A of initial composition C_0. Let us first assume equal mass densities for the solid and the liquid. When the growth interface moves from z to $z + dz$, the solute accumulation ($\rho A l \, dC$) is equal to the difference between the fluxes in ($\rho A \, dz \, C_0$) and out ($- k\rho A \, dz \, C$). We thus get:

$$l \, dC = (C_0 - kC) \, dz.$$

Integrating from $C = C_0$ at $z = 0$, the composition in the liquid is given by

$$C = \frac{C_0}{k}[1 - (1 - k)\exp(-kz/l)]. \tag{4.4}$$

Equation (4.4), often referred to as Pfann's law, ceases to be valid when the melt reaches the end of the ingot, as the subsequent solidification proceeds by normal freezing. Using k as a parameter, a number of calculated concentration profiles are shown in fig. 10 assuming the dimension of the crystal is ten times that of the liquid zone. Segregation is again seen to be high when k differs markedly from unity.

It is clear that the purity of the material can be increased by iterating the process (multipass refining), and this result finds extensive use in the industry. Analytical efforts aiming at the derivation of the concentration profile as a function of the number n of iterations are presented in ref. [27] along with the ultimate distribution curve.

A limitation of the method is that the unwanted solutes should all have distribution coefficients less than unity. It should also be noted that eq. (4.4) holds when the liquid

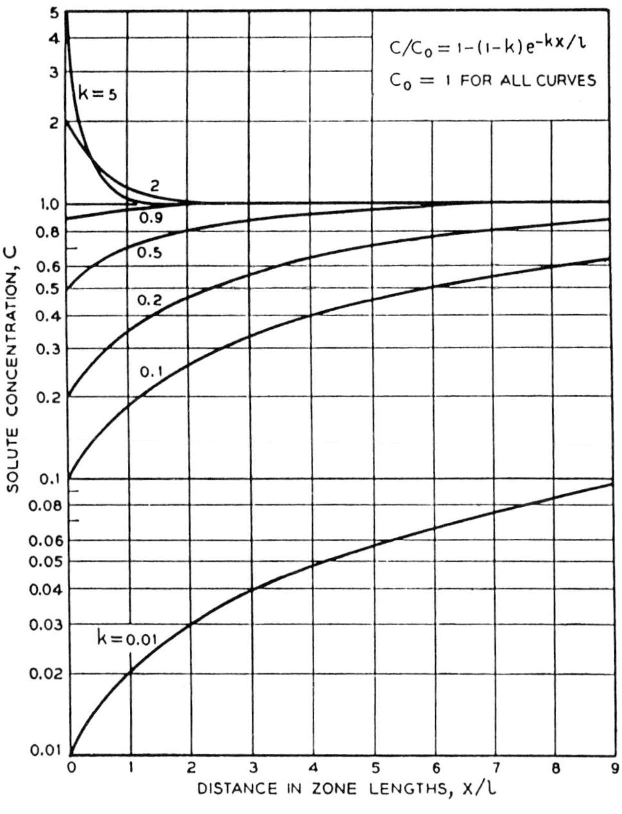

Fig. 10. Solute concentration along the crystals for various values of the partition coefficient k in the zone melting configuration (from [27]).

and solid densities differ, provided that l is defined as the length of feed molten to form the zone [27], the dimension of the liquid bridge thus being $\rho_S l/\rho_L$.

4.3. Diffusive transport

A totally different situation occurs when diffusion alone accounts for all the mass transport in the melt. Again, this appears as a rather surprising hypothesis since convection is present in all growth processes. But if it is not allowed to act at the scale $\delta = D/V_I$ – the thickness of the solute boundary layer for purely diffusive conditions – our assumption will certainly be valid.

The easiest way to avoid convective transport is to increase the growth velocity V_I, keeping in mind that the morphological stability threshold should not be exceeded. In practice, the diffusive conditions can only be reached in special cases, e.g. for Bridgman growth in microgravity [28] or for ground based devices with a magnetic field to damp buoyancy [29].

However, the study of this limiting case provides considerable physical insight into the growth problems. Besides, we shall later see that the segregation models are based on a stagnant film concept, where diffusion alone is assumed to act in a region of extent δ, generally smaller than D/V_I, in front of the growth interface.

4.3.1. Normal freezing

Let us again consider a molten feed of length L and area A, having an initial composition C_0, the mass densities in the solid and in the liquid being supposed equal. Assuming a perfectly planar interface, a one-dimensional approach is possible; in a frame moving with the interface, the problem thus becomes:

$$\partial C/\partial t = D\,\partial^2 C/\partial Z^2 + V_I\,\partial C/\partial Z, \tag{4.5a}$$

$$Z = 0: \quad D(\partial C/\partial Z)_0 = -V_I(1-k)\,C(0), \tag{4.5b}$$

$$Z \to \infty: \quad C = C_0. \tag{4.5c}$$

We have supposed here that the melt is sufficiently large to ensure a $C = C_0$ condition far away from the interface. Doing so, we of course put aside the final transient, but we shall come back later to this difficulty.

We showed in section (3.2) that a steady state regime could exist, and that the solution exhibited a solute boundary layer of extent D/V_I near the growth front [eq. (3.6)]. However, we yet have to see how this boundary layer builds up from an initially homogeneous liquid during the initial transient necessary to reach the steady state.

This problem was first treated by Tiller et al. [9]. Assuming that the rate of approach of the steady state was proportional to the difference between the actual solid concentration $C_S(Z)$ and C_0, the following approximate expression was proposed:

$$C_S(Z) = C_0(1-(1-k)\exp(-kV_I Z/D)). \tag{4.6}$$

An exact solution to the same problem was later obtained by Smith et al. [30]; the interested reader is referred to the original paper for the details of the calculations. The key result is that the difference with respect to eq. (4.6) is always less than 20%, meaning that the approximate expression can be confidently used in practice.

It is clearly seen from eq. (4.6) that the length of the initial transient scales with $D/(kV_I)$ and can often be longer than the crystal itself, precluding the establishment of a steady state regime. This is a major problem in microgravity crystal growth, and it is often necessary to use a pre-established boundary layer to allow for a steady state to be reached [31].

The uniform $C_S = C_0$ profile observed after the initial transient can be perturbed by sudden changes of the interface velocity. The ensuing segregation increases with decreasing k and with the difference between the two growth rates [30]. These results can also be applied to the study of microsegregation phenomena.

The composition profile of eq. (4.6) cannot be valid indefinitely, as the solute rejected in the melt during the boundary layer buildup must somehow be included into the crystal. Smith et al. showed that this occurred during a final transient of extent scaling with D/V_I [30]; in practice, this part of the crystal often has to be discarded, since the concentration variations there are generally too large.

Finally, it is often not possible in practice to extend these results to the case $\rho_S \neq \rho_L$; indeed, let us consider a frame moving with the solid–liquid front, i.e. at a velocity V_I with respect to the laboratory. The local fluid advection rate should then be $\rho_S V_I/\rho_L$, but it is necessarily V_I at the crucible walls due to the no-slip condition. Thus the interface cannot remain planar, meaning that a one-dimensional approach is not appropriate.

4.3.2. Zone melting

Let us now consider a molten zone of length l and area A, the feed composition being equal to C_0. The relevant mass repartition equation is again eq. (4.5a) and the solute transport at the solid–liquid front is still eq. (4.5b). However, the boundary condition at the edge of the zone now becomes

$$Z = l: \quad D(\partial C/\partial Z)_l = V_I(C_0 - C(l)). \tag{4.7}$$

The steady state solution to the system of eqs. (4.5a), (4.5b) and (4.7) is given by

$$C(Z) = C_0\left[1 + \frac{1-k}{k}\exp(-ZV_I/D)\right]. \tag{4.8}$$

Equation (4.8) coincides with the solution obtained in the normal freezing case [eq. (3.6)], the only difference being that the molten zone is now of finite length l.

As the boundary layer thickness is now limited by the dimension of the zone, the time necessary for its buildup should somehow be reduced. Using again the assumption that rate of approach of the steady state was proportional to the difference between the actual solid concentration $C_S(Z)$ and C_0, Tiller et al. [9] proposed the following approximate expression for the initial transient:

$$C_S(Z) = C_0[1 - (1-k)\exp(-\beta Z)], \tag{4.9}$$

where

$$\beta = kV_I/D(1 - \exp(-V_I l/D))^{-1}.$$

No analytical solution is available in the literature to check the validity of eq. (4.9). However, we think that it can be confidently used in practice, since the agreement between the exact and approximate solutions was good in the normal freezing case.

The results obtained in this subsection can in principle be extended to the case $\rho_S \neq \rho_L$, since the problem of the velocity difference in the vicinity of the crucible walls does not arise. Moreover, it is also interesting to note that when the parameter $V_I l/D$ is large, eq. (4.7) [respectively eq. (4.9)] reduces to eq. (4.5c) [respectively (4.6)]; the normal freezing and zone melting configurations thus become fully similar.

4.4. Partial mixing

4.4.1. Quasi-steady-state analysis

In all real cases, the boundary layer thickness takes an intermediate value between zero and D/V_I. Let us assume that, after an initial transient necessary for its formation, a quasi-steady-state is reached; it is then possible to define an effective partition coefficient k_{eff} as the ratio $kC(0)/C_B$ where $C(0)$ and C_B are, respectively, the concentrations at the interface and in the bulk of the fluid. k_{eff} thus ranges from k (complete mixing) to 1 (diffusive regime) and depends of course on the intensity of convection in the melt.

A very important result is that segregation will again be governed by eqs. (4.1) and (4.2), provided the true partition coefficient k is changed to k_{eff}. However, we yet have to see more precisely how k_{eff} can be determined in real growth experiments. A first answer was given by BPS [17] in their idealized Czochralski configuration; along with their analytical solution, they also introduced a stagnant film model that allows the computation of k_{eff} from an externally supplied parameter Δ_{SF}.

Indeed, even if we know that convection is present, let us consider a diffusion-controlled problem, where mass transfer in the bulk of the fluid and at the solid–liquid front are governed by eqs. (3.3) and (3.4), respectively. The key input is that we now impose a given concentration at a distance δ_{SF} from the interface:

$$Z = \delta_{\text{SF}}: \quad C = C_\infty. \tag{4.10}$$

In this stagnant film model, mass transfer is thus supposed to be purely diffusive in a region of extent δ_{SF} in front of the interface, the bulk of the liquid being assumed to be homogeneous. As stated by BPS, "the somewhat arbitrary quantity δ_{SF} may be characterized by defining it so that it yields the same dependence of the composition upon the growth parameters that is given by the exact solution". From eqs. (3.3), (3.4) and (4.10), we get

$$C(Z) = \frac{C_\infty}{k + (1-k)\exp(-\Delta_{\text{SF}})}[(1-k)\exp(-ZV_I/D) + k], \tag{4.11}$$

where we set $\Delta_{\text{SF}} = \delta_{\text{SF}} V_I/D$. We thus get in terms of k_{eff}:

$$k_{\text{eff}} = kC(0)/C_\infty = \frac{k}{k + (1-k)\exp(-\Delta_{\text{SF}})}. \tag{4.12}$$

We shall see in section 4.4.3 that a relevant value for Δ_{SF} can often be supplied externally and it is then possible to derive an effective partition coefficient from eq. (4.12). The simple model proposed here thus allows for macrosegregation predictions to be made, using the modified form of eq. (4.1).

4.4.2. *Transient analysis*

The problem of the solute boundary layer buildup from an initially homogeneous fluid was first treated by Favier [32], who showed that all growth techniques could be studied in a unified frame. In his work, the interface composition is given by an analytical solution depending on only two parameters, Δ_{SF} and the concentration at the edge of the stagnant film $\phi(t)$.

The initial segregation profiles are shown in fig. 11 for the limiting case where the amount of melt is supposed to be very large, so that $\phi(t)$ remains always equal to the nominal feed composition C_0. It appears clearly that the higher Δ_{SF} is, the longer it takes to reach the quasi-steady state. This was to be expected since the amount of solute present in the boundary layer increases with Δ_{SF}.

A general equation governing the behaviour of $\phi(t)$ in melts of finite extent was proposed in the same paper [32]; expressions for ϕ to be used in practice are given for a few traditional solidification configurations. A key result is that segregation is expected to be the same in the early stages of solidification for all the classical growth techniques.

Sudden changes of melt composition (addition of dopants) can also be accounted for in Favier's theoretical frame. Finally, it should also be noted that his approximate solutions for the initial transients in purely diffusive conditions are more accurate than those of Tiller et al. [see eqs. (4.6) and (4.9)].

4.4.3. *Connection with the boundary layer approach*

All the results presented in the frame of the stagnant film model assume that a relevant value of Δ_{SF} is somehow supplied externally. A way to do so in practice is to build a link

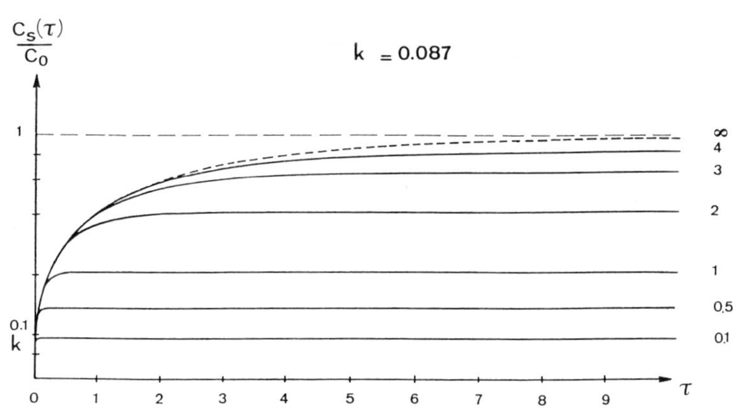

Fig. 11. Analytical solution for the initial transient and the quasi-steady-state limit for various values of the Δ_{SF} parameter.

with the boundary layer approach of section 3. Indeed, we saw there that the effect of diffusion *and* convection was to force all the significant composition variations to take place at a scale δ in the vicinity of the interface. Using the definition eq. (3.7), we find for $C(0)$:

$$C(0) = \frac{C_\infty}{1-(1-k)\Delta}, \tag{4.13}$$

Δ being the convecto–diffusive parameter, $\Delta = \delta V_1/D$. Meanwhile, setting $Z = 0$ in eq. (4.11), we get

$$C(0) = \frac{C_\infty}{k + (1-k)\exp(-\Delta_{SF})} \tag{4.14}$$

A sufficient condition for eqs. (4.13) and (4.14) to coincide can be written as

$$\Delta = 1 - \exp(-\Delta_{SF}). \tag{4.15}$$

As seen in section 3, it is generally possible to estimate the boundary layer thickness δ (and thus the convecto–diffusive parameter Δ) from order of magnitude or from numerical techniques using the definition eq. (3.7). A relevant value of Δ_{SF} can then be derived from eq. (4.15).

It is interesting to note that in the developed convective regime ($\Delta \ll 1$), $\Delta \simeq \Delta_{SF}$ or equivalently $\delta \simeq \delta_{SF}$. Under this restriction, δ_{SF} is thus the characteristic length scale of the mass transport problem, but this result certainly does not hold when diffusion becomes important.

4.4.4. *A word on the stagnant film model*

As pointed out by Wilson [10], a lot of confusion can be found in the literature on the stagnant film model, and a word of discussion is perhaps necessary at this point. To our minds, this concept does not stand on stable physical ground; for instance, the validity of eq. (4.11) away from the interface is more than questionable, since the solute flux is discontinuous at $Z = \delta_{SF}$.

However, for crystal growth purposes, we are mainly interested in the concentration at the solid–liquid front, and Δ_{SF} is by definition selected to match the exact solution there. The stagnant film model can thus be safely used to gain information on segregation matters, if the somewhat arbitrary parameter Δ_{SF} is supplied externally.

Finally, even though it may be quite useless to try to rewrite history, it is interesting to see that the results of the preceding sections could have been obtained using a physically sound approach. Indeed, let us consider the simplified mass transfer equation:

$$D\frac{d^2C}{dZ^2} + V_{eff}\frac{dC}{dZ} = 0, \tag{4.16}$$

where a constant effective velocity V_{eff}, yet to be defined, has been substituted to the $[V_1 - W(Z)]$ term in eq. (3.9). The boundary conditions being once more eqs. (3.4) and

(3.5), the solution is:

$$C(Z) = \left(\frac{(1-k)V_1}{V_{\text{eff}} - (1-k)V_1} \exp(-ZV_{\text{eff}}/D) + 1 \right) C_\infty .$$

So far we merely changed the arbitrary parameter necessary to match the exact solution of BPS at the interface, but a first advantage is that there is no discontinuity of the concentration derivative in this formulation. However, the key point is that it is now possible to make a physically sound choice for V_{eff}.

Indeed, if we set $V_{\text{eff}} = D/\delta$, we see that the characteristic length in the exponential is – as it should be – the boundary layer thickness. Moreover, according to the characteristic equation (3.13), D/δ is expected to be close to $[V_1 - W(Z)]$ at a scale δ, thus justifying the substitution in eq. (4.16).

A lot of misunderstanding could probably have been avoided, at no additional cost in terms of mathematical complexity, if BPS had introduced an effective velocity V_{eff} instead of Δ_{SF}. It would then be interesting to adapt Favier's transient analysis in this new frame for the sake of intellectual satisfaction, but no major change in segregation results is expected provided Δ and Δ_{SF} are related through eq. (4.15).

4.5. *Experimental and numerical data*

The relative simplicity of the stagnant film model induced a lot of interest in the scientific community soon after the publication of the original paper by BPS. Various

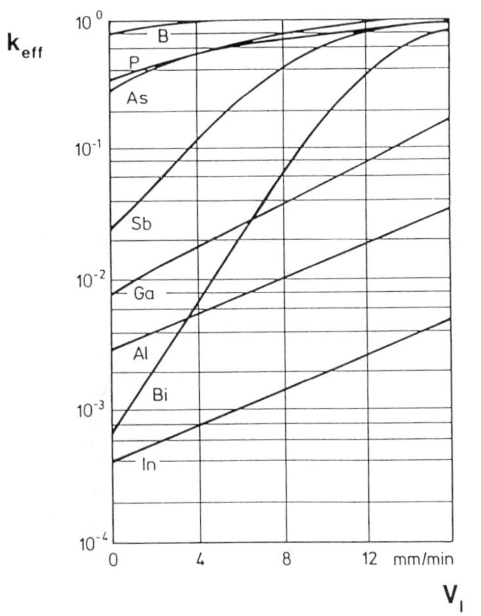

Fig. 12. Variation of the effective partition coefficient k_{eff} with growth rate for a number of dopants in Si crystals (from [4]).

authors tried to correlate the effective partition coefficient to the solidification conditions both in Czochralski [33, 34] and Bridgman-like [35] configurations.

In Czochralski growth, the observed variation of k_{eff} with interface velocity (see fig. 12) and rotation rate is generally in agreement with the theoretical predictions [33, 34]. Recent work in vertical Bridgman [36] compares well with the results of the scaling analysis [20]. However, one has to keep in mind that the precision in terms of concentration measurements is often limited.

A convincing numerical analysis was carried out by Wilson [37, 38] in the early eighties. Using the Von Karman assumption, she proceeds to solve the Navier–Stokes equations in the bulk of the melt and to determine the composition field in the cavity. Her results basically confirm and extend those of BPS; moreover, as shown in fig. 13, the agreement with the transient analysis of Favier [32] is remarkable.

The possibility of achieving diffusive transport to grow homogeneous single crystals in Bridgman furnaces led to the development of scientific programs in the space

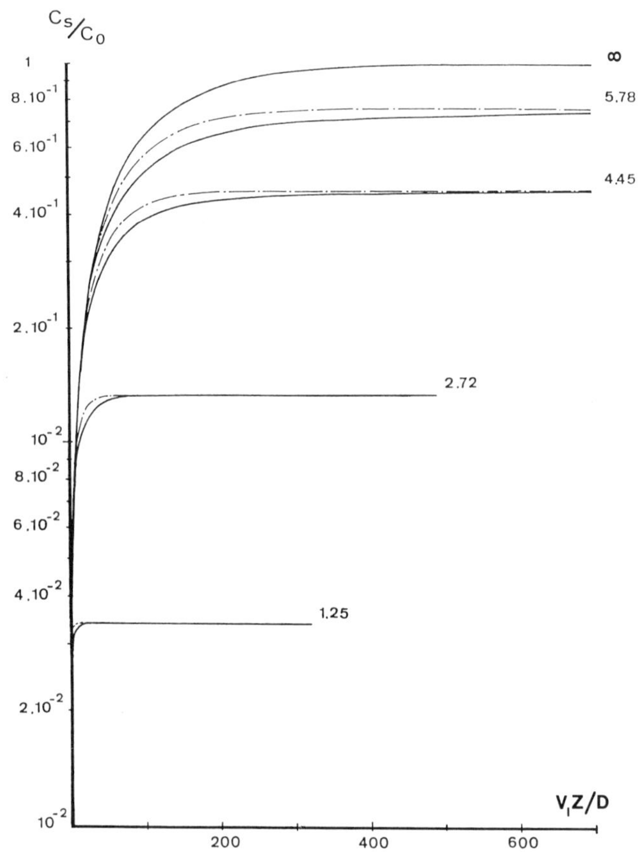

Fig. 13. Comparison between numerical (full line) and analytical (dash–dotted line) solutions for the initial transient problem (from [38]).

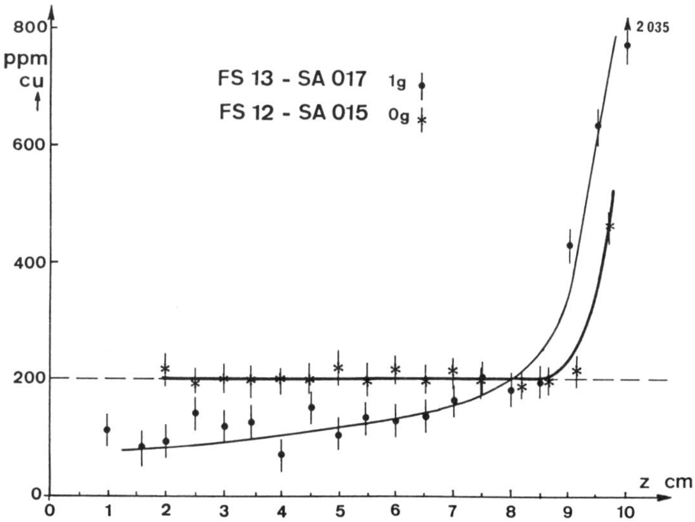

Fig. 14. Copper concentration along space-grown (*) and ground reference (·) samples.

environment. It was clearly demonstrated experimentally (see fig. 14) that the expected concentration plateau could indeed be reached in practice [28, 39, 40].

However, a number of numerical works [14, 15, 41] showed that buoyancy-driven convection due to the residual gravity could significantly perturb the diffusive regime. Moreover, one often has to reduce the initial transient necessary to reach the steady state [31], especially when the interface velocity is low.

4.6. Concluding remarks

Our purpose in this section was to show that the mechanisms leading to segregation along melt-grown crystals are now quite well understood. It should also be said that the ideas presented here can be fruitfully adapted to related solidification problems. For instance, in eutectic growth, the experimental data can often be interpreted using a similar theoretical frame [42, 43].

In the Czochralski configuration, axial segregation can be limited if the melt is large with respect to the crystal. Indeed, it is clear from eq. (4.1) that the effect of confinement only becomes important when the solidified fraction approaches unity. But the material left over often has to be recycled, since it is generally very expensive.

As for the zone melting technique, the length of the initial transient can be a problem, specially when the partition coefficient k is small. But if initially the compositions of the molten zone and of the feed are, respectively, C_0/k and C_0, it is in principle possible to grow homogeneous crystals [44]. In practice, however, the establishment of a steady state is postponed by the boundary layer buildup.

More generally, the various methods proposed to avoid axial segregation can be classified into two groups [4]. In the first class, the key idea is to ensure that the composition of the liquid in the vicinity of the interface remains constant all through solidification; this can be done using, e.g. the floating crucible or the continuous liquid feeding technique (see [2] for more details).

A second approach consists in adapting the growth conditions to balance the variations in bulk concentration. The most accessible parameter in this respect is the interface velocity [33]; another possibility is to act on convection by means of a magnetic field to change the boundary layer thickness during solidification [4, 45].

Although these methods look very promising, they are still at the laboratory stage. Industrial users have been afraid so far of either the investment or the complexity of the process, but it may be reassuring for them to know that a solution to their segregation problems exists.

5. *Radial segregation*

5.1. *Statement of the problem*

The composition of wafers cut from melt-grown semiconductor crystals has to be radially uniform (typically within a few percent) to meet the requirements of the device manufacturers. Similar demands exist for other types of materials, and we shall try in this section to identify the mechanisms leading to radial segregation, and to relate its magnitude to the solidification conditions.

An obvious origin for this problem is a curvature of the solid–liquid front, hardly avoidable in practice. We shall see theoretically that both in convective (section 5.2) and diffusive (section 5.3) mass transport conditions, an estimate of the radial segregation can be derived if the interface curvature is known. But a perfectly planar growth front does not necessarily mean lateral homogeneity, as will be shown from scaling analysis arguments in section 5.4.

Another possible origin is a difference between the apparent partition coefficients at the center and at the edge of the interface when solidification proceeds with a partially faceted front. This point will not be modelled here, but section 5.5 will be devoted to a survey of examples taken from the various growth techniques, both from a numerical and an experimental point of view.

5.2. *The convective regime limit*

Assuming both a uniform composition field in the melt and a planar growth front, we derived in section 4.2 the axial concentration profile in the solidified crystal. However, due to the one-dimensional treatment, radial segregation could not be accounted for with this simple model. If we now relax the planar interface hypothesis, we shall show that an estimate of radial segregation can be found.

Let us consider fig. 15; when the interface is at position 1, the solidified fraction is f_1 and the melt composition $C_1(f_1)$. After some time, the front reaches position 2, the solidified fraction being now f_2 and the melt composition $C_2(f_2)$. We thus get an

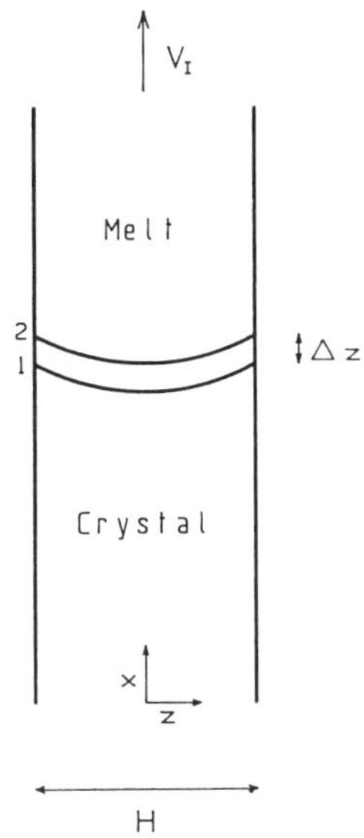

Fig. 15. Geometry and coordinates for the radial segregation problem.

estimate of the maximum radial segregation in the solid:

$$\Delta C = k[C_2(f_2) - C_1(f_1)]. \tag{5.1}$$

Keeping in mind that $C(f) = C_0(1-f)^{k-1}$ and assuming that the interface curvature is small, we can write:

$$\Delta C = k(1-k)C_0(1-f)^{k-2}\Delta f. \tag{5.2}$$

$\Delta f = f_2 - f_1$ is related to the dimensional growth front deflection ΔZ by $\Delta f = \Delta Z/L$, where L is the total length of the crystal. In the low curvature range, ΔC thus depends linearly on ΔZ and increases as solidification proceeds.

Let us say a word about a difficulty we deliberately overlooked. When the interface curvature is high, it is not obvious how to define unambiguously the solidified fraction. Eq. (5.1) could lead to significant errors, and it is only eq. (5.2) that can be used safely when the front deflection is limited.

Another difficult point is that it is often not possible to adapt the general treatment of section 4.4 to the study of radial segregation. A requirement for this approach to be valid would be that the boundary layer thickness δ remains the same all along the interface. This is an unlikely fact except in the total mixing case we just considered.

However, assuming the variations of δ to be accounted for by different effective partition coefficients between the edge k_{ee} and the center k_{ec} of the crystal, Lee [46] was able to derive an approximate radial concentration profile. His analysis could probably be extended to the case where a partially faceted interface leads to a difference between k_{ee} and k_{ec}.

5.3. The purely diffusive case

We shall here briefly recall the results of Coriell et al. [47, 48], who developed both analytical and numerical tools to relate interface curvature and radial segregation during crystal growth. The main assumptions are that mass transport proceeds only by diffusion and that the problem is two-dimensional (planar) and time-independent (quasi-steady-state). The relevant equations are thus:

- mass transfer:

$$D\nabla^2 C + (V_1 \cdot \nabla) C = 0; \tag{5.3}$$

- boundary conditions:

$$\text{interface:} \quad D\,\nabla C \cdot \boldsymbol{n} + C(1-k)\,V_1 \cdot \boldsymbol{n} = 0, \tag{5.4a}$$

$$Z \to \infty: \quad C = C_\infty, \tag{5.4b}$$

$$X = \pm\tfrac{1}{2}H \quad \partial C/\partial X = 0. \tag{5.4c}$$

For the sake of internal consistency, we do not use the notation of the original papers. Assuming first that the interface curvature was limited, they were able to express the concentration field in the liquid using a Fourier expansion series. In the low growth

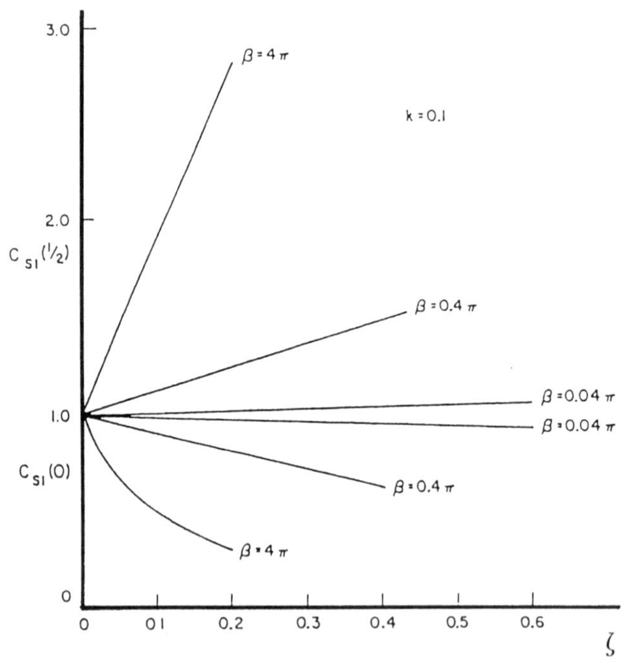

Fig. 16. Solute concentration at the interface both at the edge $[C_{SI}(\frac{1}{2})]$ and at the center $[C_{SI}(0)]$ of the crystal versus the amplitude of the interface deflection (from [48]).

velocity range, they obtain a particularly simple result for the maximum radial segregation:

$$\Delta C = \frac{V_I}{D}(1-k)\Delta Z C_\infty. \tag{5.5}$$

ΔZ is again the dimensional interface deflection. Another analytical approach involves a series expansion in terms of the Péclet number ($Pe = HV_I/D$) in the high Pe range but in the general case a full numerical solution is necessary.

A typical result of the computation is shown in fig. 16, where β is the Péclet number and ζ the non-dimensional amplitude of a sinusoidal interface shape [$z = \zeta\cos(2\pi x)$]. The concentration difference in the solid between the edge and the center of the interface [$C_{SI}(\frac{1}{2}) - C_{SI}(0)$] is seen to increase rapidly with both β and ζ.

5.4. *The effect of partial mixing*

We here want to show that even if the interface is perfectly planar, partial solute mixing in the melt can lead to a significant radial segregation. Indeed, if the boundary layer thickness changes along the solid–liquid front, eq. (4.13) shows that the interface composition cannot remain constant. It can thus be said that radial segregation is governed by the variations of the boundary layer thickness.

This point can be clearly illustrated in the case of eutectic solidification. Assuming the concentration in the liquid at the growth front equal to $\bar{C}_I$, a simple mass balance along with the definition equation of δ yield [43]:

$$C_S(X,Y) = \bar{C}_I - \frac{\bar{C}_I - C_\infty}{\Delta(X,Y)}, \tag{5.6}$$

X, Y being the coordinates of a given point on the interface; at the two extrema locations in terms of Δ, eq. (5.6) becomes:

$$C_{S,M} = \bar{C}_I - \frac{\bar{C}_I - C_\infty}{\Delta_M}, \qquad C_{S,m} = \bar{C}_I - \frac{\bar{C}_I - C_\infty}{\Delta_m},$$

where the subscript M (respectively m) refers to the maximum (respectively minimum) values of C_S and Δ. Subtracting $C_{S,m}$ from $C_{S,M}$, we get for the maximum radial segregation in the solid:

$$\Delta C = (\bar{C}_I - C_\infty)(\Delta_m^{-1} - \Delta_M^{-1}). \tag{5.7}$$

If we further assume that $\Delta_M \gg \Delta_m$ (a reasonable hypothesis when convection dominates mass transfer [43]), eq. (5.7) becomes:

$$\Delta C = (\bar{C}_I - C_\infty)/\Delta_m. \tag{5.8}$$

The maximum radial segregation in the solid thus scales with the inverse of the minimum of the convecto–diffusive parameter. This was checked numerically in the idealized two-dimensional horizontal Bridgman configuration using the Batchelor–Birikh velocity profile. The results presented in fig. 17 clearly show that the curves

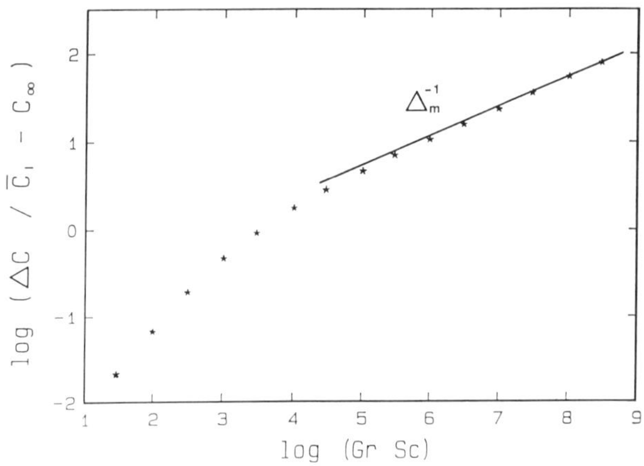

Fig. 17. Numerical values of the normalized radial segregation and Δ_m^{-1} in an idealized horizontal Bridgman eutectic configuration.

$\Delta C/(\bar{C}_l - C_\infty)$ and Δ_m^{-1} are indeed asymptotes in the high GrSc range, i.e. in the convective regime limit.

It should also be said that we observed good agreement with the experimental data, which supported the validity of a two-dimensional modelization. We thus have here a particularly simple instance where the radial segregation can be related to the boundary layer thickness.

Similar findings were obtained in the mono-phase problem, at the expense of a little more algebra. Indeed, let us again consider the idealized horizontal Bridgman configuration of section 3.4. Under a quasi-steady-state hypothesis, the solute repartition equation becomes:

$$U(X, Z)\frac{\partial C}{\partial X} + W(X, Z)\frac{\partial C}{\partial Z} = D\left(\frac{\partial^2 C}{\partial X^2} + \frac{\partial^2 C}{\partial Z^2}\right) + V_1\frac{\partial C}{\partial Z}. \tag{5.9}$$

The boundary conditions are again given by eq. (5.4). At this point, it is necessary to add some physics to the description of the problem; more precisely, we have to identify the leading terms in the above equation. We shall present here briefly the main results of our investigations [49, 50]; the interested reader is referred to the original publications for more information on the subject.

When the overall mass transport proceeds mainly by convection, it is the convective solute flux down the interface $U\,\partial C/\partial X$ (see fig. 5 for notation) that accumulates solute in the bottom left corner of the cavity, thus resulting in radial segregation. Depending on the length scale considered, this leading term is balanced either by the diffusive contributions ($D\,\partial^2 C/\partial X^2$ and $D\,\partial^2 C/\partial Z^2$) or by convection away from the front ($W\,\partial C/\partial Z$).

On the other hand, when diffusion dominates mass transfer, radial segregation is caused by a perturbation of the purely diffusive solute field that scales with the convective term $W\,\partial C/\partial Z$. Depending on the interface velocity, this perturbation can

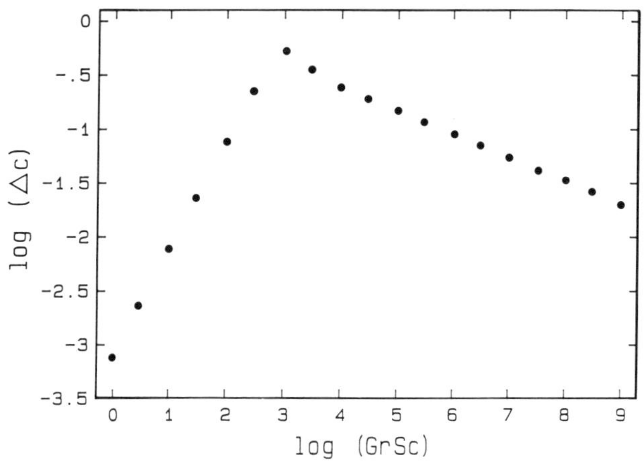

Fig. 18. Maximum radial segregation (normalized by $Pe(1-k)C_\infty$) versus the Gr × Sc product.

be balanced by the advective flow ($V_1 \partial C/\partial Z$) and/or by the diffusive contributions ($D\,\partial^2 C/\partial X^2$ and $D\,\partial^2 C/\partial Z^2$).

There is a good agreement between the order of magnitude analysis and the two-dimensional numerical simulations performed assuming the velocity field was given by the approximate Batchelor–Birikh solution [49, 50]. The theoretical approach leads to the right scaling laws in terms of the relevant non-dimensional parameters GrSc (a measure of the intensity of convection, see section 3.4) and Pe (a measure of the interface velocity).

Shown in fig. 18 is the maximum radial segregation ΔC as a function of GrSc for the case $Pe = 2$, as obtained in the computer simulations. Two different types of behaviour are clearly identified: in the low GrSc range (diffusion-controlled transport) ΔC increases linearly with GrSc, whereas in the high GrSc range (convection-controlled transport) ΔC decreases as $\mathrm{GrSc}^{-2/9}$. The associated asymptotic numerical scaling laws are:

– convective regime limit

$$\Delta C = 2Pe(1-k)\mathrm{GrSc}^{-2/9}; \tag{5.10}$$

– quasi-diffusive limit:

$$Pe \ll 1: \quad \Delta C = 5.5 \times 10^{-4}\mathrm{GrSc}\,Pe(1-k)k^{-1}C_\infty, \tag{5.11a}$$

$$Pe \gg 1: \quad \Delta C = 0.25\mathrm{GrSc}\,Pe^{-3}(1-k)k^{-2}C_\infty. \tag{5.11b}$$

Similar approaches were followed by Harriot and Brown [51] and by Naumann and Baugher [52], respectively, in idealized zone melting and vertical Bridgman configurations. In both cases, an approximate solution to the Navier–Stokes equation, valid when convection is not too strong, is first derived and then used as an input in the solute repartition problem. Scaling laws for the maximum radial segregation are proposed, in good agreement with the numerical data available.

Our purpose in this section was twofold: first, to show that convection can result in radial segregation even if the interface is perfectly planar. But we also wanted to stress that when the velocity profile in the liquid close to the interface is known, it is often possible to obtain valuable information on the solute repartition from an order of magnitude analysis.

5.5. *Numerical and experimental results*

The pioneering numerical work in the field is due to Polezhaev and coworkers [14, 53]. They solved simultaneously the temperature, momentum and concentration repartition equations in an idealized horizontal Bridgman cavity, under the restriction that the interface remained perfectly planar. Both quasi-steady state and transient analyses were carried out, thus allowing for the simulation of a complete solidification run.

They were the first to point out that radial segregation ran through a maximum at the transition between the diffusive and the convective solute transport regimes. Similar calculations by Ben Hadid [15] confirmed this result and, in addition, quantified the effect of crystal growth rate on species repartition. The observed segregation values were found to be quite low for usual solidification conditions.

In connection with the Bridgman technique in space, we tried to quantify the respective influences of curvature and convection on segregation [54]. Using the scarce experimental data available, both contributions were seen to be of the same order of magnitude for samples of circa one centimeter diameter. However, for larger crystals, convection is expected to be the leading mechanism, specially when the residual gravity is parallel to the growth front [55].

The vertical Bridgman configuration was studied in detail by Brown and his coworkers [16, 56], again from a numerical standpoint. The key innovation is that the shape of the solidification front is determined from the temperature field [16]; the effect of solutal convection in non-dilute alloys is also considered [56]. Radial segregation was found to be quite sensitive to the growth conditions, ranging from almost zero to relative values as high as 60% [16].

Even though the growth front is now curved, a maximum in radial segregation was again observed in the intermediate convective range. A similar finding was obtained by Motakef [57], who also presents scaling laws governing the interference of buoyancy-driven convection with mass transfer during the solidification of dilute alloys.

On the whole, the numerical results are in good agreement with the experimental data of Wang [58] and Rouzaud et al. [59]. Wang's experiments were carried out in a classical three-zone furnace, where the heat flux generated on the hot side passes through an adiabatic region and is finally extracted on the cold end. On the opposite, Rouzaud's design aimed at maintaining a constant gradient both in the melt and the crystal (gradient furnace).

In both experimental programs, the effect of interface velocity was investigated. Even though the convection is much weaker, the observed radial segregation was larger in Rouzaud's gradient furnace, specially when the growth conditions were close to the transition between the diffusive and the convective solute transport regimes.

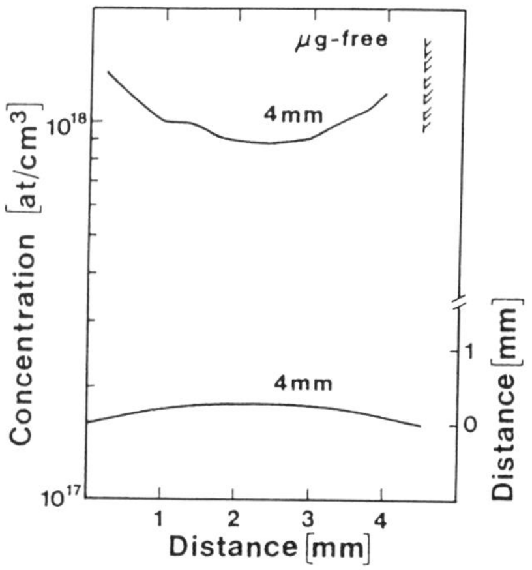

Fig. 19. Radial concentration profile and interface shape in a space-grown zone melting sample (from [63]).

In addition, Rouzaud's work featured the solidification of concentrated Ge–Si alloys and a comparison with the dilute Ge:Ga reference samples. The preferential incorporation of the lighter solute (Si) led to the formation of a Ge-rich layer which further damped the convection in the melt. However, it is interesting to note that, even with an optimum thermal design and this damping effect, the purely diffusive mass transport regime could not be reached.

Radial segregation does not appear to be a major problem in Czochralski growth, at least when there are no facet effects. Indeed, solidification generally proceeds under developed convective conditions; it is thus enough to control interface curvature (as shown in [60, 61] for the case of oxide crystals) to guarantee lateral homogeneity. It is only when the velocity field is dramatically changed by means of an applied magnetic field that radial segregation can become important [62].

As for the zone melting configuration, it was clearly shown [63, 64] that both interface curvature and convection were necessary to account for the experimentally observed radial segregation [63]. It should be noted that it is very difficult to obtain laterally homogeneous crystals with this method, relative differences ranging from 50% (see fig. 19) to more than 150% are not uncommon.

5.6. Concluding remarks

As was done for the axial problem, we should like to conclude this section with a few tips for the practician on how to avoid radial segregation. The first idea is to try to find growth parameters leading to a solidification front as flat as possible, since segregation was seen to increase always with interface deflection.

In terms of partial mixing, the problem is not so easy; indeed, mass transfer should be either principally diffusive ($\Delta \simeq 1$) or principally convective ($\Delta \ll 1$), since we saw that a maximum existed at the transition between these two regimes. The difficulty is that the flow strongly depends on the thermal design of the furnace, specially in the vertical Bridgman and zone melting configurations.

From a practical point of view, two complementary approaches may be advised: the first is to limit convection, using a stabilizing vertical Bridgman configuration or solidification in microgravity. Even then, a magnetic field might be necessary to further damp the hydrodynamic motions [4, 29] if the interface velocity is not high enough.

On the other hand, it is often easier to try to enhance convection using, e.g. higher thermal gradients or lower solidification rates. However, to avoid solute striations (see next section), the fluid flow should remain stationary. A good dose of empiricism is thus often necessary in order to find the optimal growth parameters in terms of radial segregation.

6. Microsegregation

6.1. Statement of the problem

Solute segregation at the microscopic scale is often a painful problem for the crystal grower; it may be due to time-dependent variations of the interface and/or the convection velocity. In the first class of phenomena, we find apparatus-related mechanisms, such as pulling device instabilities, or other types of vibration. Rotation of the sample in an asymmetrical thermal field also leads to striations.

Unsteady convection in the melt is another cause of microsegregation. It can act directly on the solutal field, as in the case of g-jitter during the solidification of low-Prandtl-number fluids in space. But the time-dependent temperature fluctuations associated with the flow can induce variations of the growth velocity and thus of the composition of the alloy.

Kinetic effects on faceted interfaces are also to be considered [65]. High lateral growth velocities at macrosteps or local interface breakdown due to an excessive undercooling can result in transient solute incorporation [66, 67]. A significant microsegregation is also met at cellular or dendritic solidification fronts, but a detailed study on these microscopic problems is outside the frame of this work.

We shall start with a presentation of the quasi-steady state models in section 6.2 before turning, in section 6.3, to the more complex theoretical approach that features a coupled solution to the heat and mass transfer equations. Sections 6.4 and 6.5 will be dedicated to a survey of the numerical and experimental results obtained in the field.

6.2. Quasi-steady-state models

In the simplest approach to the microsegregation problem, it is assumed that the boundary layer thickness does not depend on the unsteady velocity variations. The time-dependent convecto–diffusive parameter is thus

$$\Delta(t) = V_{\mathrm{I}}(t)\delta/D. \qquad (6.1)$$

Using this value as an input in the effective distribution coefficient equation, it is easy to

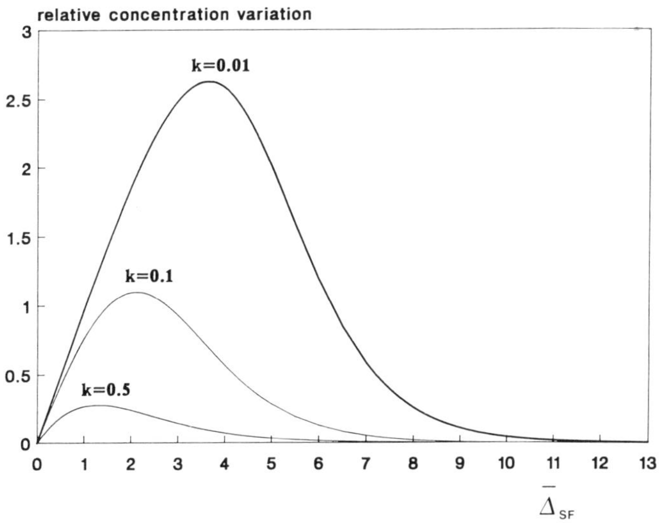

Fig. 20. Relative microsegregation versus $\bar{\Delta}_{SF}$ in the quasi-steady-state model for various values of the segregation coefficient (redrawn from [68]).

follow the composition variations at the interface. Let us now suppose that the growth rate varies as

$$V_I(t) = \bar{V}_I[1 + m(t)], \tag{6.2}$$

$m(t)$ being a modulation around the average value $\bar{V}_I$, assumed smaller than unity to avoid backmelting. In the stagnant film frame studied by Carruthers and Witt [68], the time-dependent variations of the interface composition are given by

$$C_I(t) - C_{SS} = \frac{(1-k)\bar{\Delta}_{SF}\exp(-\bar{\Delta}_{SF})}{k + (1-k)\exp(-\bar{\Delta}_{SF})} C_{SS} m(t), \tag{6.3}$$

where C_{SS} is the average interface composition derived using $\bar{\Delta}_{SF} = \bar{V}_I \delta_{SF}/D$ in eq. (4.14). The above expression, plotted in fig. 20, was derived assuming low-amplitude velocity changes, so that the problem becomes linear in $m(t)$. If mass transport in the melt is mainly convective, $\bar{\Delta}_{SF} \ll 1$, and eq. (6.3) becomes:

$$C_I(t) - C_{SS} = (1-k)\bar{\Delta}_{SF} C_{SS} m(t). \tag{6.4}$$

These results are expected to be valid if the variations of $m(t)$ are slow enough, i.e. for fluctuation periods τ larger than a characteristic diffusion time $2\pi\delta_{SF}^2/D$ [68]. Under this assumption, eqs. (6.3) and (6.4) are certainly useful for prediction purposes.

However, since δ_{SF} is assumed to be constant, the above approach cannot account for the effect of unsteady fluid flow on microsegregation. Besides, since δ_{SF} has little physical meaning except in the developed convective solute transport regime, it is probably not correct to define a diffusion time scaling with δ_{SF}^2/D.

We shall now see that a modified quasi-steady-state model, based on the solutal boundary layer concept, yields new physical insight into the problem at no additional cost in terms of mathematical complexity. Indeed, if both growth rate and convection velocity are unsteady, the time-dependent characteristic equation for δ is:

$$D/\delta(t) = V_{\rm I}(t) - \bar{W}[\delta(t), t]. \tag{6.5}$$

$\bar{W}[\delta(t), t]$ stands for the effective convection velocity at the time-dependent boundary layer scale $\delta(t)$ [see eq. (3.13)]. To obtain eq. (6.5), we had to drop the $\partial C/\partial t$ term in the mass conservation equation, which is only valid if the concentration variations take place over a time scale τ much larger than δ^2/D. Under this assumption, once $\delta(t)$ has been derived, it is easy to define a convecto–diffusive parameter $\Delta(t)$ as

$$\Delta(t) = \delta(t)\, V_{\rm I}(t)/D. \tag{6.6}$$

The interface composition can then be obtained using $\Delta(t)$ as an input in the effective distribution coefficient formula [eq. (4.13)]. For purposes of comparison, let us assume that the convective contribution does not depend explicitly on time $\{\bar{W}[\delta(t), t] = \bar{W}[\delta(t)]\}$ and that $V_{\rm I}(t)$ is given by eq. (6.2). The modified quasi-steady-state model then yields

$$\Delta(t) = \bar{\Delta}\frac{1 + m(t)}{1 + m(t)\bar{\Delta}}, \tag{6.7}$$

where $\bar{\Delta} = \bar{V}_{\rm I}\bar{\delta}/D, \bar{\delta}$ being the boundary layer thickness of the steady-state problem. With respect to the previous approach, the variations in terms of $\Delta(t)$ are reduced by a factor $1 + m(t)\bar{\Delta}$ [see eqs. (6.1) and (6.2)] which can be important for large $m(t)$. Using the same notation, the concentration fluctuation for low-amplitude growth rate changes is:

$$C_{\rm I}(t) - C_{\rm SS} = \frac{(1-k)\bar{\Delta}(1-\bar{\Delta})}{1-(1-k)\bar{\Delta}}\, C_{\rm SS}\, m(t). \tag{6.8}$$

Recalling that $\bar{\Delta} = 1 - \exp(-\bar{\Delta}_{\rm SF})$, eqs. (6.3) and (6.8) have the same form, except for a factor $\bar{\Delta}$ instead of $\bar{\Delta}_{\rm SF}$. When mass transport is mainly convective $\bar{\Delta} = \bar{\Delta}_{\rm SF}$, and the two models predict the same transient behaviour, the composition at the interface being given by eq. (6.4).

In conclusion, the modified quasi-steady-state model accounts for more physics in the problem, and the expected concentration variations, generally overestimated in the classical approach, may be substantially reduced. But the boundary layer is again assumed to react immediately to velocity changes, an hypothesis certainly incorrect in the high frequency range.

Moreover, it is not always possible, from an experimental standpoint, to quantify the growth rate variations; in that respect, an analysis based on thermal fluctuations is more interesting. It is the purpose of the following section to propose theoretically a complete solution to the microsegregation problem.

6.3. *Theoretical analysis*

We here face a formidable task, since the time-dependent heat and mass transfer equations we now have to consider both feature a nonlinear term ($V_{\rm I}\cdot\nabla T$ and $V_{\rm I}\cdot\nabla C$, respectively); that is why the two outstanding efforts in the field [69, 70] both used a perturbation approach. With respect to the pioneering work of Hurle et al. [69], the more recent approach [70] has the advantage of ensuring the continuity of the temperature field at the interface, and we shall mainly rely on it for the presentation.

However, as the algebra involved is quite tedious, we shall only outline the procedure and the main results of the analysis, the interested reader being again referred to the original publications for more details on the subject. Briefly stated, the main variables of the problem (interface velocity, composition and temperature fields) are expanded into Fourier series up to the order one. For instance, in terms of concentration, we look for a solution of the form:

$$C(Z,t) = C^0(Z) + C^1(Z)\exp(\mathrm{i}\omega t), \tag{6.9}$$

ω being the frequency of the perturbation, assumed small with respect to the $C^0(Z)$ contribution. The one-dimensional analysis is carried out in the frame of the stagnant film model, an additional hypothesis being that δ_{SF} does not depend on V_I, an assumption expected to be valid when convection dominates mass transfer ($\Delta_{SF} = \Delta \ll 1$)*.

A solution to the zeroth-order composition and temperature problem is proposed first, and the effect of a fluctuation $\phi\exp(\mathrm{i}\omega t)$ of the thermal gradient in the liquid at the interface is derived supposing all the second-order terms to be negligible. The boundary condition relating interface concentration and temperature is then:

$$T_I = T_M - m_L C_I, \tag{6.10}$$

T_M being the melting temperature of the pure solvent and m_L the slope of the liquidus line, assumed constant in the case of dilute alloys. The solute problem can thus be addressed first and its results used as an input in the heat transfer equation. Briefly stated, the results, in terms of the perturbation of the interface velocity (V_I^1) and concentration (C_I^1), are:

$$V_I^1 = \frac{\alpha_L \phi}{\alpha_S\left(\eta - \dfrac{G_S}{\mathrm{i}\omega}\right)\left(\mu - \dfrac{V_I^0}{2\alpha_S}\right) - L}, \qquad C_I^1 = (\eta/m_L)V_I^1, \tag{6.11}$$

η, λ and μ being complex parameters given by:

$$\begin{aligned}\eta = {} & \{(1-k)m_L[1 + (V_I^0)^2/2\mathrm{i}\omega D]\sinh(\lambda\delta_{SF}) \\ & - (\lambda V_I^0/\mathrm{i}\omega)[\cosh(\lambda\delta_{SF}) - \exp(-V_I^0\delta_{SF}/2D)]\} \\ & \times \{D[\lambda\cosh(\lambda\delta_{SF}) - (V_I^0/2D)(1-2k)\sinh(\lambda\delta_{SF})]\}^{-1}, \\ \lambda = {} & [(V_I^0/2D)^2 + \mathrm{i}\omega/D]^{1/2}, \qquad \mu = [(V_I^0/2\alpha_S)^2 + \mathrm{i}\omega/\alpha_S]^{1/2}.\end{aligned}$$

In the above expressions, α_S and α_L stand, respectively, for the thermal diffusivity of the solid and the liquid and V_I^0 for the unperturbed growth velocity. L denotes the ratio of the latent heat of solidification to the calorific capacity. As noted in [69], the occurrence of imaginary terms defines the phase difference between the perturbed quantities and the driving force $\phi\exp(\mathrm{i}\omega t)$.

*In a recent paper [92], a similar problem was solved in the frame of the boundary layer model to remove this limitation.

For purposes of comparison with the preceding subsection, it is fitting to note that the results in terms of concentration/velocity take a simple form in two extreme cases:

$$\delta_{SF} \to 0,\ \omega \to 0,\ \lambda\delta_{SF} \to 0; \quad C_I^1 = (1-k)\bar{\Delta}_{SF} C_I^0 V_I^1 / V_I^0. \tag{6.12}$$

We fall back on eq. (6.4), which means that the quasi-steady-state models are indeed appropriate for low-frequency perturbations.

$$\delta_{SF} \to 0,\ \omega \to \infty,\ \lambda\delta_{SF} \to \infty: \quad C_I^1 = (1-k)(\omega D)^{-1/2} V_I^1 C_I^0. \tag{6.13}$$

The amplitude of C_I^1 scales with $\omega^{-1/2}$ in the high frequency range, a result already proposed in [71]. A $\pi/4$ phase shift is observed between the fluctuations of interface velocity and concentration and C_I^1 is seen to be independent of the stagnant film thickness.

The original paper [70] considers in addition the possibility of backmelting and of kinetic undercooling at the front. However, it should be said that the validity of the model is doubtful near the diffusive regime limit since δ_{SF} may then depend on the interface velocity. The interested reader in these theoretical problems can also find valuable information in the works of Wheeler [72] and Aleksandrovskii [73].

6.4. *Numerical approaches*

The effect of fluctuating growth rate in a Czochralski configuration was thoroughly studied by Wilson [74] in the early eighties. A fully time-dependent analysis was carried out, assuming an interface velocity of the form

$$V_I(t) = \bar{V}_I[1 + A\sin(\omega t)], \tag{6.14}$$

ω being here the angular velocity of crystal rotation. The average concentration in the crystal was seen to be slightly lower than the value achieved under steady-state conditions [$V_I(t) = \bar{V}_I$]; however, this deviation can often be considered negligible for practical purposes.

Shown in fig. 21 is a comparison between the predictions of the quasi-steady-state model of section 6.2 and the numerical results. Large discrepancies are observed, both in terms of amplitude and phase shift, but it should be noted that, in the present case, the frequency ω is large with respect to D/δ^2; the quasi-steady-state approach could not be expected to be valid.

On the other hand, the agreement with the theoretical model of Hurle et al. [69] is significantly better, even though not perfect. An interesting numerical result was that the microsegregation scaled linearly with the amplitude of the fluctuation even for values of A close to unity. Thus theoretical perturbation approaches may be justified even out of their application range stricto sensu.

For the cases where backmelting occurs [amplitude $A > 1$ in eq. (6.14)], a discontinuity of the composition field in the crystal was always observed; in practice, solid-state diffusion, neglected in the formulation of the problem, would lead to a redistribution of solute and to a smoothing of the profiles. In contrast, the solute boundary layer was seen to be relatively insensitive to backmelting, and it is certainly wrong to believe that it disappears during the negative velocity period of the cycle.

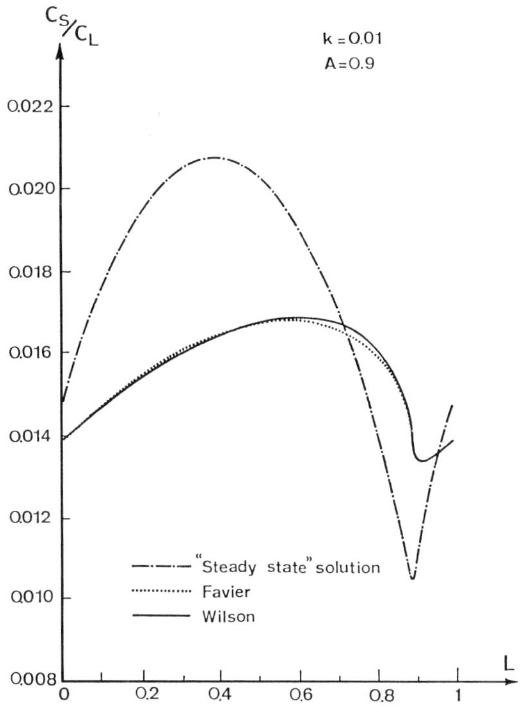

Fig. 21. Axial solute profile in a crystal grown during one rotational cycle of interface velocity.

Numerical analysis on a hybrid stagnant film model was carried out by Van Run [75] and Favier [38]. The idea was again to get rid of the hydrodynamics of the melt by considering a diffusion equation in a finite region in front of the interface; the solution to the time-dependent problem was then found using either finite differences [75] or finite elements [38]. Both approaches lead to similar results, that compare very well with Wilson's full numerical data (see fig. 21).

The coupled heat and mass transfer Stefan problem was treated by Rouzaud [76], under the assumption of purely diffusive transport. Temperature fluctuations at the boundary conditions result in microsegregation, the predictions of the model being in fair agreement with the experimental test. It should also be said that Rouzaud's work can be used in a variety of transient cases, e.g. growth rate changes in concentrated systems.

The effect of impulse, transient and periodic accelerations in Bridgman–Stockbarger microgravity growth was studied by Alexander et al. [55, 77]. The largest segregations were observed for perturbation frequencies below 10^{-2} Hz; e.g. accelerations of $10^{-5} g_0$ at 10^{-3} Hz may lead to composition variations of up to 60%.

6.5. *Experimental results*

Microsegregation phenomena were often subject to controversial interpretations in early experimental work, one of the main reasons being that the etching procedures used then did not allow for precise composition measurements. A key improvement in that respect was achieved by Witt et al. [78], who were able to measure simultaneously the dopant heterogeneity (via spreading resistance) and the microscopic interface velocity in Czochralski-grown gallium-doped germanium.

To do so, they observed the striations induced by current pulses sent at a given frequency, which allowed for quantitative correlations to be made. For instance, they showed that growth rate changes do not generally exhibit the expected sinusoidal behaviour and that marked rotational striations could be ascribed to backmelting, a result supporting the predictions of Wilson's numerical work [74].

They also found a coincidence between composition and interface velocity minima (see fig. 22). Another interesting point is that the steady-state model appeared valid only in the increasing-growth-rate part of the cycle. The electric pulse technique was thoroughly studied by Wargo [79], which allowed for the determination of the Peltier coefficient of InSb [80].

Further work by Carlson and Witt on the subject [81, 82] featured computer-based video processing techniques to determine the free charge carrier distribution in doped Czochralski-grown GaAs [81]. A striking result was that the magnitude of the fluctuations in dopant concentration were more than an order of magnitude higher than in the case of silicon. Moreover, Fourier analysis of the striations reveal random patterns which show no relationship with either seed or crucible rotation [82].

Concerning the zone melting configuration, periodic nonrotational striations in W-doped Mo were associated with unsteady thermocapillary flow [83], the characteristic frequencies observed being around 1 Hz. However, the influence of buoyancy-driven convection should not be ruled out in general, since striations in some cases disappear under microgravity conditions [84]. Another mechanism to be considered is unsteady motion in the ambient gas, even though it was shown to play no role in Fe 3%wt Si crystals [85].

The influence of various hydrodynamic regimes in the melt on a solidification front was thoroughly studied during horizontal Bridgman growth of a dilute Sn:Bi alloy using the Seebeck effect [25]. A direct relationship between measured fluctuations in the liquid near the interface and crystal defects is always observed. Such microsegrega-

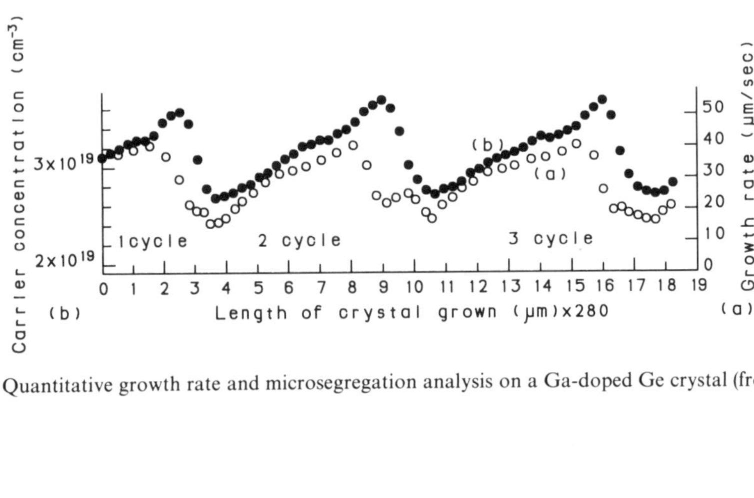

Fig. 22. Quantitative growth rate and microsegregation analysis on a Ga-doped Ge crystal (from [78]).

tion problems are not expected in the vertical Bridgman configuration, since convection is usually much weaker.

A number of papers focus on the effect of a static magnetic field in electrically conducting fluids both in Czochralski [86, 87] and zone growth [88]. For fields higher than say 0.2 T, a significant damping of the oscillatory flow is observed, but the rotational striations are of course unaffected. Besides, the application of a magnetic field may in some cases lead to unexpected results [89]. For more information on this topic, the reader is referred to the chapter of Hurle and Series.

6.6. *Concluding remarks*

To sum up, it should be said that the microsegregation problem is certainly very complex, due to its essentially transient nature. Besides, from a practician's point of view, a simple solution is not easy to find since a number of mechanisms may lead to striations; in that respect, the vertical Bridgman is probably the safest of all the growth configurations, but even in that case, there is no guarantee that one can achieve microscale homogeneity.

To start with, common sense indicates that a good design of the growth furnace is always a valuable asset. For instance, thermal asymmetry and pulling device instabilities should be limited, but this is often in practice a formidable task. Similarly, the temperature gradients leading to oscillatory buoyancy flow should also be reduced as much as possible. As for Marangoni instabilities, they can be often avoided with a coating of the free surfaces, e.g. with oxides [84].

When this is not enough, unsteady convection can be significantly damped using static magnetic fields; however, as mentioned earlier, radial segregation may then be increased [62, 89]. Another way to shift the stability threshold is to grow crystals in a centrifuge. It was demonstrated both numerically and experimentally [90] that a higher gravity led to an increased but steady convection.

To conclude, the positive influence of solid-state diffusion was asserted by Gray et al. [91] in an interesting theoretical and numerical study. Post-growth annealing can thus improve the compositional homogeneity of the crystals, especially for closed space striations.

7. *General conclusion*

One of the purposes of our contribution was to outline the existence of a solute boundary layer, i.e. of a region where all the significant concentration variations take place, in the vicinity of the solidification interface. It is then possible to define an effective diffusion velocity that can account for the effect of convection in the melt; in comparison, the stagnant film hypothesis introduced by Burton, Prim and Schlichter has little physical meaning.

Another point to be made was to show that a scaling analysis allowed for the influence of the experimental parameters to be determined through relevant non-dimensional numbers. The predictions of the method were found to be in good agreement with the existing numerical and experimental data. It can thus be used to extend the results of the numerical simulations in a wide range of parameters for any crystal growth or solidification problem when an appropriate velocity profile is available.

Our opinion is that segregation phenomena in dilute alloys are relatively well understood, even though a few problems wait for a solution. The situation is quite different for concentrated systems, mainly because the coupling with the Navier–Stokes equation makes matters much more complicated. Again, we think that future work in the field will certainly benefit from a boundary layer/scaling analysis approach.

A final point to be stressed is that there is no miracle solidification configuration. For instance, we saw that the microgravity environment or the application of magnetic fields offered interesting potential to reduce the intensity of convection. This may lead to an improvement in terms of axial and micro-homogeneity, but also to an increase in radial segregation. Practical solutions exist for each type of problem, but their overall impact on crystal quality should always be assessed.

Acknowledgements

The authors are indebted to Drs. T. Duffar and T. Alboussiere for a careful proofreading of the manuscript and to Ms. C. Barat and Mr. P. Dusserre for their assistance in the preparation of the drawings, J.P. Garandet would like to thank his friends O., Z., J. and F. for their help in improving the presentation of the manuscript. Fruitful discussions on the topic with Drs. B. Drevet and A. Rouzaud are also gratefully acknowledged. The present work was conducted within the frame of the GRAMME agreement between the CNES and the CEA.

List of symbols

General

C	= solute composition in a dilute binary alloy (mass or mole fraction)
$\bar{C}_I$	= average interface concentration in a eutectic alloy (mass or mole fraction)
D	= solute diffusion coefficient in a dilute binary alloy (m^2s^{-1})
$\mathbf{g}$	= gravity vector ($m\ s^{-2}$)
k	= solute partition coefficient in a dilute binary alloy
$\mathbf{n}$	= unit vector normal to a melt boundary
U, V, W	= dimensional velocity components on the X, Y, and Z axis ($m\ s^{-1}$)
u, v, w	= associated non-dimensional velocity components
$\mathbf{V}$	= mass-average velocity, solution of the Navier–Stokes equation ($m\ s^{-1}$)
$\mathbf{V}_I$	= interface velocity (dilute binary alloy case, $m\ s^{-1}$)
V_I	= norm of $\mathbf{V}_I$ for unidirectional growth models ($m\ s^{-1}$)
X, Y, Z	= dimensional space coordinates (m)
x, y, z	= associated non-dimensional coordinates
γ	= angular velocity of crystal rotation (s^{-1})
δ	= solute boundary layer thickness (m)
δ_D	= solute boundary layer thickness in the diffusive case ($\delta_D = D/V_I$, m)
δ_{SF}	= stagnant film thickness in the BPS model (m)

Δ = convecto-diffusive parameter ($0 \leq \Delta \leq 1$)
Δ_{SF} = parameter to be used in the stagnant film model
ν = kinematic velocity of the melt ($m^2\,s^{-1}$)
ρ_L = mass density of the liquid ($kg\,m^{-3}$)
ρ_S = mass density of the solid ($kg\,m^{-3}$)
∇ = nabla operator

Section 2

D^j_m = effective binary diffusivity of the mixture ($m^2\,s^{-1}$)
$\boldsymbol{J}^j_C$ = convective mass flux of constituent j ($\boldsymbol{J}^j_C = \rho^j \boldsymbol{V}$, $kg\,m^{-2}\,s^{-1}$)
$\boldsymbol{J}^j_D$ = diffusive mass flux of constituent j ($kg\,m^{-2}\,s^{-1}$)
$\boldsymbol{J}^j_T$ = total mass flux of constituent j with respect to a stationary frame ($kg\,m^{-2}\,s^{-1}$)
k^j_0 = equilibrium partition coefficient of constituent j
M^j = molar mass of constituent j (kg)
p = number of constituents in solution
$\boldsymbol{V}_{L,I}$ = fluid advection rate ($m\,s^{-1}$)
$\boldsymbol{V}_{S,I}$ = crystal growth rate ($m\,s^{-1}$)
$\boldsymbol{v}^j$ = mean velocity of the molecules of species j in a fluid element ($m\,s^{-1}$)
η^j = molar fraction of component j in solution
$\rho_{L,I}$ = mass density of the liquid at the interface ($kg\,m^{-3}$)
$\rho_{S,I}$ = mass density of the solid at the interface ($kg\,m^{-3}$)
ρ^j = mass concentration of constituent j, i.e., mass of species j per unit volume of solution ($kg\,m^{-3}$)
ω^j = mass fraction of component j in solution ($\omega^j = \rho^j/\rho_L$)
$\omega^j_{L,I}$ = mass fraction of component j in the liquid at the interface
$\omega^j_{S,I}$ = mass fraction of component j in the solid at the interface

Section 6

C_{SS} = steady-state interface composition (mass or mole fraction)
C^1_I = fluctuation of the interface composition (mass or mole fraction)
L = ratio of the latent heat of solidification to the calorific capacity (K)
$m(t)$ = modulation of the growth rate
m_L = slope of the liquidus (K at%$^{-1}$ or K wt%$^{-1}$)
$\bar{V}_I$ = average interface velocity ($m\,s^{-1}$)
V^1_I = fluctuation of the interface velocity ($m\,s^{-1}$)
α_L = thermal diffusivity of the liquid alloy ($m^2\,s^{-1}$)
α_S = thermal diffusivity of the solid alloy ($m^2\,s^{-1}$)
$\bar{\Delta}$ = quasi-steady-state convecto–diffusive parameter
$\bar{\Delta}_{SF}$ = parameter to be used in the quasi-steady stagnant film model
ϕ = fluctuation of the thermal gradient at the interface (liquid side)
ω = frequency of the perturbation (s^{-1})

References

[1] R. Ueda and J.B. Mullin, eds., Crystal Growth and Characterization (North-Holland, Amsterdam, 1975).
[2] F. Rosenberger, Fundamentals of Crystal Growth I. Macroscopic Equilibrium and Transport Concepts (Springer, Berlin, 1979).
[3] H.U. Walter, ed., Fluid Sciences and Materials Science in Space (Springer, Berlin, 1987).
[4] G. Müller, Crystals, Growth, Properties and Applications, Vol. 12 (Springer, Berlin, 1988).
[5] R.B. Bird, W.E. Stewart and E.N. Lightfoot, Transport Phenomena (Wiley, New York, 1960).
[6] W.R. Wilcox, J. Cryst. Growth 12 (1972) 93.
[7] J.B. Mullin and K.F. Hulme, J. Phys. Chem. Solids 17 (1960) 1.

[8] P.J. Holmes, J. Phys. Chem. Solids 24 (1963) 1239.
[9] W.A. Tiller, K.A. Jackson, J.W. Rutter and B. Chalmers, Acta Metall. 1 (1953) 428.
[10] L.O. Wilson, J. Cryst. Growth 44 (1978) 247.
[11] K. Kakimoto, M. Eguchi, H. Watanabe and T. Hibiya, J. Cryst. Growth 99 (1990) 665.
[12] R.M. Sharp and A. Hellawell, J. Cryst. Growth 5 (1969) 155.
[13] B. Billia, Ph.D. Thesis, Université Aix-Marseille (1982).
[14] S.A. Nikitin, V.I. Polezhayev and A.I. Fedyushkin, J. Cryst. Growth 52 (1981) 471.
[15] H. Ben Hadid, Ph.D. Thesis, Université Aix-Marseille (1989).
[16] C.J. Chang and R.A. Brown, J. Cryst. Growth 63 (1983) 343.
[17] J.A. Burton, R.C. Prim and W.P. Slichter, J. Chem. Phys. 21 (1953) 1987.
[18] W.G. Cochran, Proc. Cambridge Philos. Soc. 30 (1934) 365.
[19] D. Camel and J.J. Favier, J. Cryst. Growth 61 (1983) 125.
[20] D. Camel and J.J. Favier, J. Physique 47 (1986) 1001.
[21] C.C. Lin and L.A. Segel, Mathematics Applied to Deterministic problems in the Material Sciences (Macmillan, New York, 1974).
[22] A. Bejean, Convection Heat Transfer (Wiley, New York, 1984).
[23] J.P. Garandet, T. Duffar and J.J. Favier, J. Cryst. Growth 106 (1990) 437.
[24] J.P. Garandet, A. Rouzaud, T. Duffar and D. Camel, J. Cryst. Growth 113 (1991) 587.
[25] J.J. Favier, A. Rouzaud and J. Comera, Rev. Phys. Appl. 22 (1987) 713.
[26] K.A. Jackson and J.D. Hunt, Trans. Metall. Soc. AIME 236 (1966) 1129.
[27] W.G. Pfann, Zone Melting (2nd ed.) (Wiley, New York, 1966).
[28] A.F. Witt, H.C. Gatos, M. Lichtensteiger, M.C. Lavine and C.J. Hermann, J. Electrochem. Soc. 122 (1975) 276.
[29] D.H. Matthiesen, M.J. Wargo, S. Motakef, D.J. Carlson, J.S. Nakos and A.F. Witt, J. Cryst. Growth 85 (1987) 557.
[30] V.G. Smith, W.A. Tiller and J.W. Rutter, Can. J. Phys. 33 (1955) 723.
[31] W. Weber, T. Duffar and J.J. Favier, Microgravity Sci. Tech. 4 (1991) 166.
[32] J.J. Favier, Acta Metall. 29 (1981) 197; 205.
[33] J.A. Burton, E.D. Kolb, W.P. Slichter and J.D. Struthers, J. Chem. Phys. 21 (1953) 1991.
[34] H. Kodera, Jpn. J. Appl. Phys. 2 (1963) 212.
[35] F.V. Dean, J.R. Kerr and A. Hellawell, J. Inst. Met. 90 (1962) 234.
[36] I.M.S. Sidawi and S.N. Tweari, J. Cryst. Growth 131 (1993) 230.
[37] L.O. Wilson, J. Cryst. Growth 44 (1978) 371.
[38] J.J. Favier and L.O. Wilson, J. Cryst. Growth 58 (1982) 103.
[39] A.F. Witt, H.C. Gatos, M. Lichtensteiger and C.J. Hermann, J. Electrochem. Soc. 125 (1978) 1832.
[40] J.J. Favier, J. Berthier, P. Arragon, Y. Malmejac, V.T. Khryapov and I.V. Barmin, Acta Astronautica 9 (1982) 255.
[41] D. Thevenard, Ph.D. Thesis, Institut National Polytechnique, Grenoble (1991).
[42] J.D. Verhoeven and R.H. Homer, Met. Trans. 1 (1970) 3437.
[43] B. Drevet, Ph.D. Thesis, Institut National Polytechnique, Grenoble (1991); see also B. Drevet, J.P. Garandet, D. Camel and J.J. Favier, J. Cryst. Growth 129 (1993) 549.
[44] J.P. Dismukes and L. Ekstrom, Trans. Metall. Soc. AIME 233 (1965) 672.
[45] S. Osawa, T. Kimura, J. Kobayashi and T. Fukuda, Appl. Phys. Lett. 50 (1987) 329.
[46] H.H. Lee, J. Cryst. Growth 83 (1987) 610.
[47] S.R. Coriell and R.F. Sekerka, J. Cryst. Growth 46 (1979) 479.
[48] S.R. Coriell, R.F. Boisvert, R.G. Rehm and R.F. Sekerka, J. Cryst. Growth 54 (1981) 167.
[49] J.P. Garandet, J. Cryst. Growth 114 (1991) 593.
[50] J.P. Garandet, J. Cryst. Growth 125 (1992) 112.
[51] G.M. Harriot and R.A. Brown, J. Cryst. Growth 69 (1984) 589.
[52] R.J. Naumann and C. Baugher, J. Cryst. Growth 121 (1992) 751.
[53] V.I. Polezhaev, Crystals, Growth, Properties and Applications, Vol. 10 (Springer, Berlin, 1984).
[54] J.P. Garandet, T. Duffar and J.J. Favier, Proc. 8th Symp. Mat. Fluid Sci. Microgravity, Brussels (1992), ESA SP-333, 545.
[55] J.I.D. Alexander, J. Ouazzani and F. Rosenberger, J. Cryst. Growth 97 (1989) 285.

[56] P.M. Adornato and R.A. Brown, J. Cryst. Growth 80 (1987) 155.
[57] S. Motakef, J. Cryst. Growth 102 (1990) 197.
[58] C. Wang, Sc.D. Thesis, Massachusetts Institute of Technology, Cambridge, MA (1984).
[59] A. Rouzaud, D. Camel and J.J. Favier, J. Cryst. Growth 73 (1985) 149.
[60] V. Nicolov, K. Iliev and P. Peshev, J. Cryst. Growth 89 (1988) 313; 324.
[61] M. Berkowski, K. Iliev, V. Nicolov, P. Peshev and W. Piekarczyk, J. Cryst. Growth 108 (1991) 225.
[62] R.W. Series and D.T.J. Hurle, J. Cryst. Growth 113 (1991) 305.
[63] T. Carlberg, J. Cryst. Growth 79 (1986) 71.
[64] M. Levenstem, G. Amberg, E. Tillberg and T. Carlberg, J. Cryst. Growth 104 (1990) 641.
[65] J.J. Favier, J. Electrochem. Soc. 129 (1982) 2355.
[66] T. Nishinaga, T. Kazuno, T. Tanbo, J. Koide, K. Pak, T. Nakamura and Y. Yasuda, J. Cryst. Growth 65 (1983) 607.
[67] A.N. Danilewsky and K.W. Benz, J. Cryst. Growth 97 (1989) 571.
[68] J.R. Carruthers and A.F. Witt, in: Crystal Growth and Characterization (North-Holland, Amsterdam, 1975) p. 107.
[69] D.T.J. Hurle, E. Jakeman and E.R. Pike, J. Cryst. Growth 3,4 (1968) 633.
[70] D. Thevenard, A. Rouzaud, J. Comera and J.J. Favier, J. Cryst. Growth 108 (1991) 572.
[71] D.T.J. Hurle and E. Jakeman, J. Cryst. Growth 5 (1969) 227.
[72] A.A. Wheeler, J. Cryst. Growth 56 (1982) 67.
[73] A.L. Aleksandrovskii and P. Posmykiewicz, Phys. Stat. Sol. (a) 74 (1982) K23.
[74] L.O. Wilson, J. Cryst. Growth 48 (1980) 435; 451.
[75] A.M.J.G. Van Run, J. Cryst. Growth 47 (1979) 680.
[76] A. Rouzaud, J. Cryst. Growth 104 (1990) 701.
[77] J.I.D. Alexander, S. Amiroudine, J. Ouazzani and F. Rosenberger, J. Cryst. Growth 113 (1991) 21.
[78] A.F. Witt, M. Lichtensteiger and H.C. Gatos, J. Electrochem. Soc. 120 (1973) 1119.
[79] M. Wargo, Sc.D. Thesis, Massachusetts Institute of Technology, Cambridge, MA (1982).
[80] M. Wargo and A.F. Witt, J. Cryst. Growth 66 (1984) 289; 541.
[81] D.J. Carlson and A.F. Witt, J. Cryst. Growth 91 (1988) 239.
[82] D.J. Carlson and A.F. Witt, J. Cryst. Growth 108 (1991) 508.
[83] M. Jurisch and W. Löser, J. Cryst. Growth 102 (1990) 214.
[84] A. Cröll, W. Müller-Sebert and R. Nitsche, Mater. Res. Bull. 24 (1989) 995.
[85] P. Vanek and S. Kadeckova, Cryst. Res. Tech. 16 (1981) 911.
[86] T. Munakata and I. Tanasawa, J. Cryst. Growth 106 (1990) 566.
[87] G. Müller, Proc. Europ. Mater. Res. Soc. Conf. 16 (1987) 117.
[88] A.N. Danilewsky, P. Dold and K.W. Benz, J. Cryst. Growth 121 (1992) 305.
[89] D.J. Carlson and A.F. Witt, J. Cryst. Growth 116 (1992) 461.
[90] W. Weber, Ph.D. Thesis, Universität Erlangen (1990).
[91] R.T. Gray, M.F. Larrousse and W.R. Wilcox, J. Cryst. Growth 92 (1988) 530.
[92] J.P. Garandet, J. Cryst. Growth 131 (1993) 431.

CHAPTER 13

Convection in Melt Growth

GEORG MÜLLER

Institut für Werkstoffwissenschaften, Universität Erlangen-Nürnberg, D-91058 Erlangen, Germany

ALEKSANDAR OSTROGORSKY

Rensselaer Polytechnic Institute, Troy, NY 12180, USA

Contents

1. Introduction . . . 711
2. Origins and effects of convection in various melt growth configurations . . . 712
    2.1. Origins of convective phenomena . . . 712
    2.2. Effects of buoyancy convection . . . 716
    2.3. Natural and forced convection in Czochralski configurations . . . 720
    2.4. Forced convection in other melt growth configurations . . . 728
    2.5. Capillary convection . . . 730
3. Experimental tools for detection and visualization of convective flows in melt growth configurations . . . 735
    3.1. Similarity and dimensionless variables . . . 735
    3.2. Semiconductor and metallic melts (low-Prandtl number fluids) . . . 740
    3.3. Transparent melts (high-Prandtl number fluids) . . . 743
        3.3.1. Parametric studies of melt convection . . . 744
        3.3.2. Studies of instabilities and interface transitions . . . 745
    3.4. Molten salts ($0.2 < Pr < 1$) at high temperatures . . . 746
    3.5. Induced magnetic fields . . . 748
4. Experimental and numerical results of convective flows . . . 749
    4.1. Buoyancy convection in vertical cylindrical melt containers . . . 751
        4.1.1. Fundamental considerations . . . 751
        4.1.2. Vertical Bridgman/gradient freeze configuration with top seeding . . . 755
        4.1.3. Vertical Bridgman/gradient freeze configuration with bottom seeding . . . 758
            4.1.3.1. Flow configurations and transition to unsteady convection . . . 758
            4.1.3.2. Shape of the solid–liquid interface . . . 763

HANDBOOK OF CRYSTAL GROWTH, VOL. 2
edited by D.T.J. Hurle

4.2. Thermocapillary (Marangoni) convection in cylindrical floating zones 765
4.3. Buoyancy and thermocapillary (Marangoni) convection in horizontal Bridgman configurations 774
4.4. Convective flows in the Czochralski configuration 775
5. Influence of convection on solute segregation in front of a growth interface and in the crystal . . 779
5.1. Steady convection (of solute at the solid/liquid interface) 780
5.1.1. Lateral variation of solute concentration 781
5.1.2. Axial solute concentration in the crystal 783
5.1.3. Parameters influencing k_{eff} 785
5.1.4. k_{eff} for Czochralski growth 786
5.1.5. The case of weak convection 787
5.1.6. Maximum convective velocity compatible with no macrosegregation 789
5.2. Unsteady convection 789
5.2.1. Instabilities induced by convection 789
5.2.2. Instabilities carried by convection 794
6. Control of convection by external measures or variation of parameters 795
6.1. General considerations 795
6.2. Suppression of unsteady convection 796
6.2.1. Magnetic fields 797
6.2.2. Reduction of gravity 800
6.2.3. Centrifugal and Coriolis forces 803
7. Concluding considerations for an optimization of convective effects in the melt 803
Appendix 804
A. Basic equations of convection and principles of solution 804
A.1 Conservation of energy, species, mass and momentum 804
A.2 Non-dimensional form of the conservation equations 805
A.3 Analytical solutions 806
A.4 Numerical solutions 806
B. Numerical model for the simulation of thermocapillary (Marangoni) convection in cylindrical floating zones 807
C. Turbulent melt convection 810
C.1 Relative magnitude of diffusive transport properties in turbulent flow 811
C.2 Turbulent flow driven by a rotating disk 814
D. Material parameters 814
References 814

1. Introduction

Macroscopic heat and mass transport are playing a central role in crystal growth. Especially in melt growth, interface kinetics is so fast, that the rate of the growth process is controlled by the heat transport. The heat fluxes in the growth interface are driven by temperature gradients, which are needed to stabilize the solid–liquid interface. Additionally the growth process itself creates a considerable amount of heat of solidification, which must be removed from the progressing interface. Melt convection is a key transport mechanism, which controls the important growth features, like shape of the solid–liquid interface and isotherms in the crystal. Convective transport also determines the concentration distribution of rejected species in front of the growth interface and hence the composition of the crystal on the macro- and microscale. This article treats the various aspects of convection in melt growth.

In section 2 we first consider very briefly the different origins of convection mechanisms in the most important melt growth configurations. Several typical examples of the occurrence of convective flow are shown and illustrated by pictures of convective flow in melt growth configurations. Section 3 gives a survey of the various experimental tools for the detection and visualization of convective flow both for transparent and opaque melts. In section 4 a quantitative discussion of experimental and numerical results is made for different types of growth configurations: vertical Bridgman, floating zone, horizontal Bridgman and Czochralski. Here we treat systematically the action of buoyancy, capillary and rotation of the crystal/crucible and its interaction. The strong influence of convection on solute transport and segregation is discussed in section 5, since the formation of macroscopic and microscopic non-uniformity (i.e. the formation of striations) originates from convective flows. Finally, in section 6, we discuss different approaches to convection control, that can be used to optimize the conditions for melt crystal growth.

2. Origins and effects of convection in various melt growth configurations

2.1. Origins of convective phenomena

In this section we introduce the various convective phenomena and effects occurring in typical melt growth configurations. The treatment will be purely phenomenological and is illustrated by several experimental observations. The mathematical treatment of the subject is discussed in Appendix A. In section 4 quantitative experimental and theoretical results on convective flow will be discussed.

In crystal growth we are especially interested in convective transfer at the melt boundaries which are the source or sink of heat and impurities. Most important is convection at the growth interface since it controls the chemical composition (uniformity) of the growing crystal.

Melt convection may be classified by the physical mechanism driving convective currents. It includes buoyancy-driven convection (i), forced convection (ii) and capil-

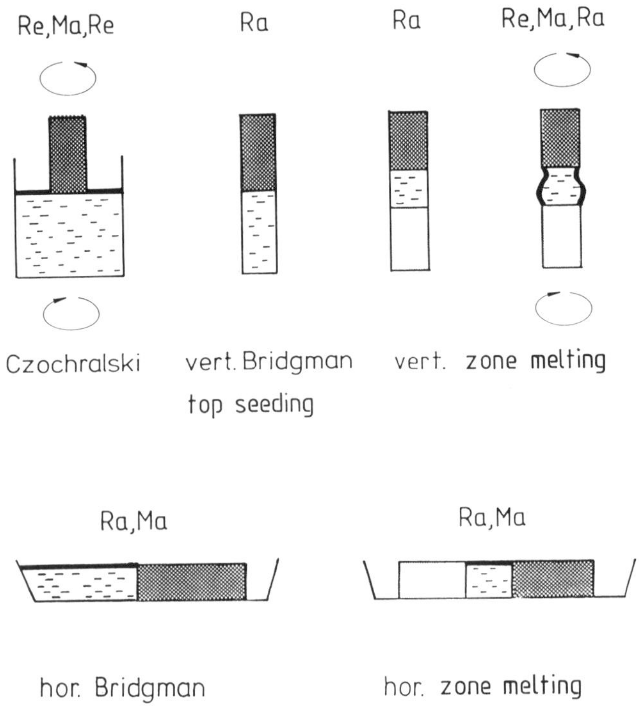

Fig. 1. Principle of important melt growth configurations showing the boundary conditions of the melt (dotted) in contact with the crystal (hatched, grey) and (in zone configurations) with the polycrystalline feed-stock (dark). The free melt surface is marked by a strong line. The corresponding mechanisms driving convective flows are denoted by the dimensionless numbers Ra (buoyancy), Ma (capillary) and Re_Ω (forced convection by rotation). From [1].

lary (Marangoni) convection (iii). The forces driving convection are not present to the same extend in each melt growth process but depend on the various principles and on the boundary conditions.

In fig. 1 we have sketched typical melt growth configurations. The mechanisms driving convective flows are denoted by the dimensionless numbers Ra (Rayleigh number), Ma (Marangoni number) and Re_Ω (Reynolds number, due to rotation Ω). The physical meaning of these parameters is that they are a dimensionless measure of the driving forces of convection. Their definitions will be given in section 3.1; their usefulness is demonstrated in section 4 where we compare experimental and theoretical results.

Buoyancy (i) occurs practically in every melt growth system, as melt growth is only possible if a temperature difference (supercooling) exists between the bulk melt and the adjacent crystal interface. To maintain this growth-driving non-equilibrium situation, it is necessary to cool the crystal and heat the melt. Melts have generally a temperature dependent density. Additionally the melt density varies with the concentration of dopants and constituents which are non-uniformly distributed within the melt due to segregation effects. Any gradients of the melt density (with the exception of those which are exactly parallel to the gravity vector) have the potential to include the onset of buoyancy flow. Therefore, we have to expect buoyancy (Ra $\neq 0$) in each melt growth configuration with increasing flow rates, when increasing the melt volume and the driving temperature gradients. Convection in any region of the melt will cause flow in the whole melt volume due to continuity.

Other body forces inducing buoyancy-driven convection in the melt include centrifugal and Coriolis forces. The effect of pressure on melt density can generally be neglected.

Typical experimental observations of buoyancy-driven flows will be shown in section 2.2 and also in section 2.3 for Czochralski configurations.

Forced convection (ii) is caused in melt growth by any mechanical part in contact with the melt, which has a relative motion with respect to the melt. The most important case is the rotation of the growing crystal relative to the crucible in the Czochralski configuration or relative to the polycrystalline feed in the floating zone technique. The effects of the rotational motion on the melt is similar to surface tension which imposes shear stress on the melt. The intensity of the shear stress is characterized by the angular velocity of rotation. In section 2.3 we illustrate this phenomenon by some typical experimental observations on Czochralski growth.

Capillary convection or Marangoni convection (iii) occurs if free melt surfaces exist in the growth configuration. The surface tension of melts (σ) is temperature and concentration dependent; σ decreases with increasing temperature. In a free melt surface with a temperature (or concentration) gradient, surface tension forces drive the flow from regions of low surface tension (hot) to areas of high surface tension (cold) (see fig. 2). The surface tension forces are balanced by viscous shear which transfers momentum to neighboring liquid layers because of fluid viscosity. Similar to buoyancy-driven convection, continuity causes the development of a bulk flow in the whole melt volume. A detailed description of the phenomenon of Marangoni convection and its investigation by model experiments is given in Chapter 4 in this handbook by Bohm et al.

The influence of capillary-driven convection is of importance in melt growth configurations when the free surface area is large in relation to the melt volume. This holds especially for the floating zone configuration, and to a certain degree, for

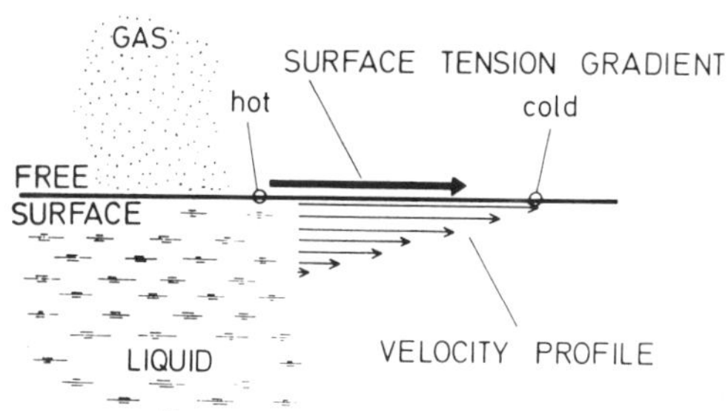

Fig. 2. Sketch of the generation of capillary (Marangoni) convection by a gradient of the surface tension. From [2].

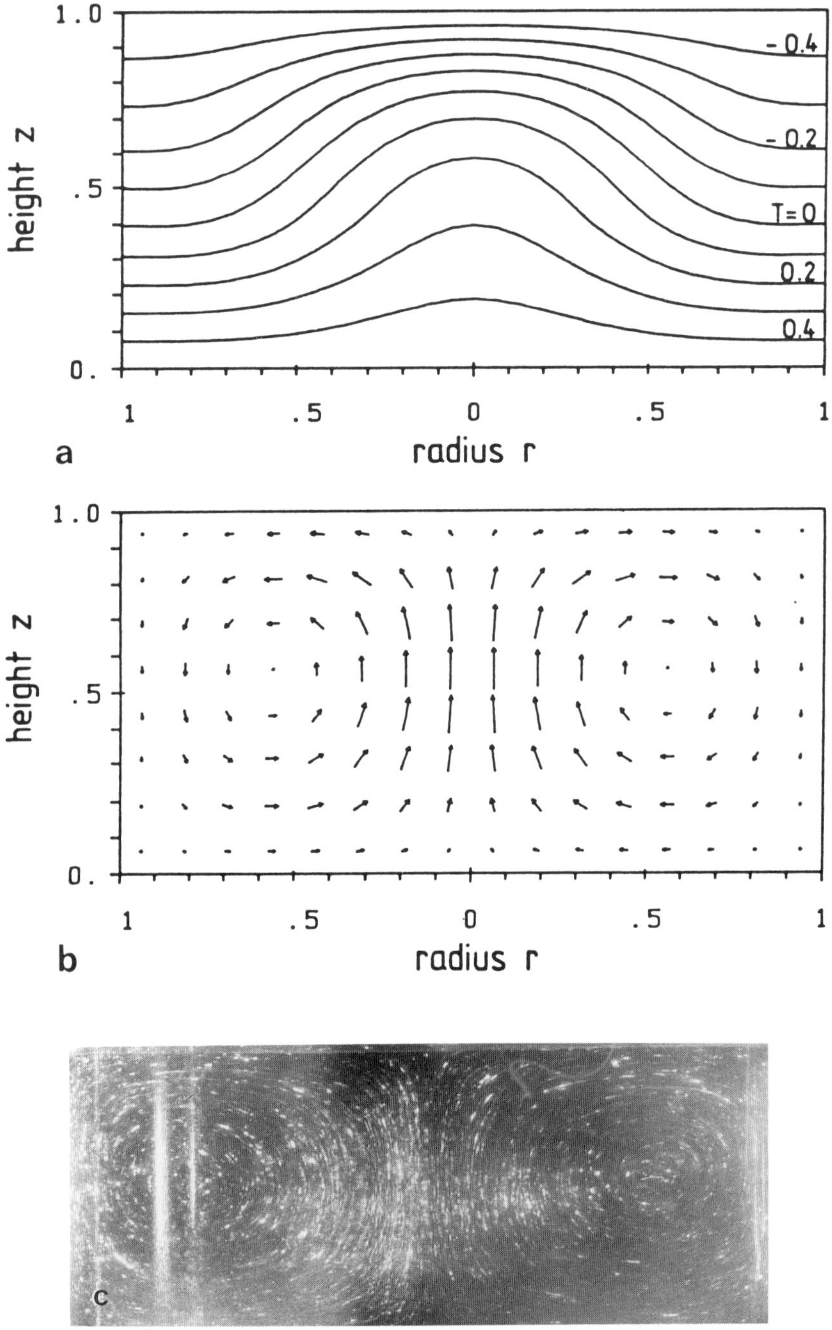

Fig. 3. Axial sections of an axisymmetric basic flow (upflow in the centre) for $h/d = 0.5$, $Pr = 6.7$ (H_2O) at $Ra = 2800$. (a) Calculated isotherms. (b) Calculated velocity field. (c) Experimentally observed flow, by light cut (H_2O). From [1].

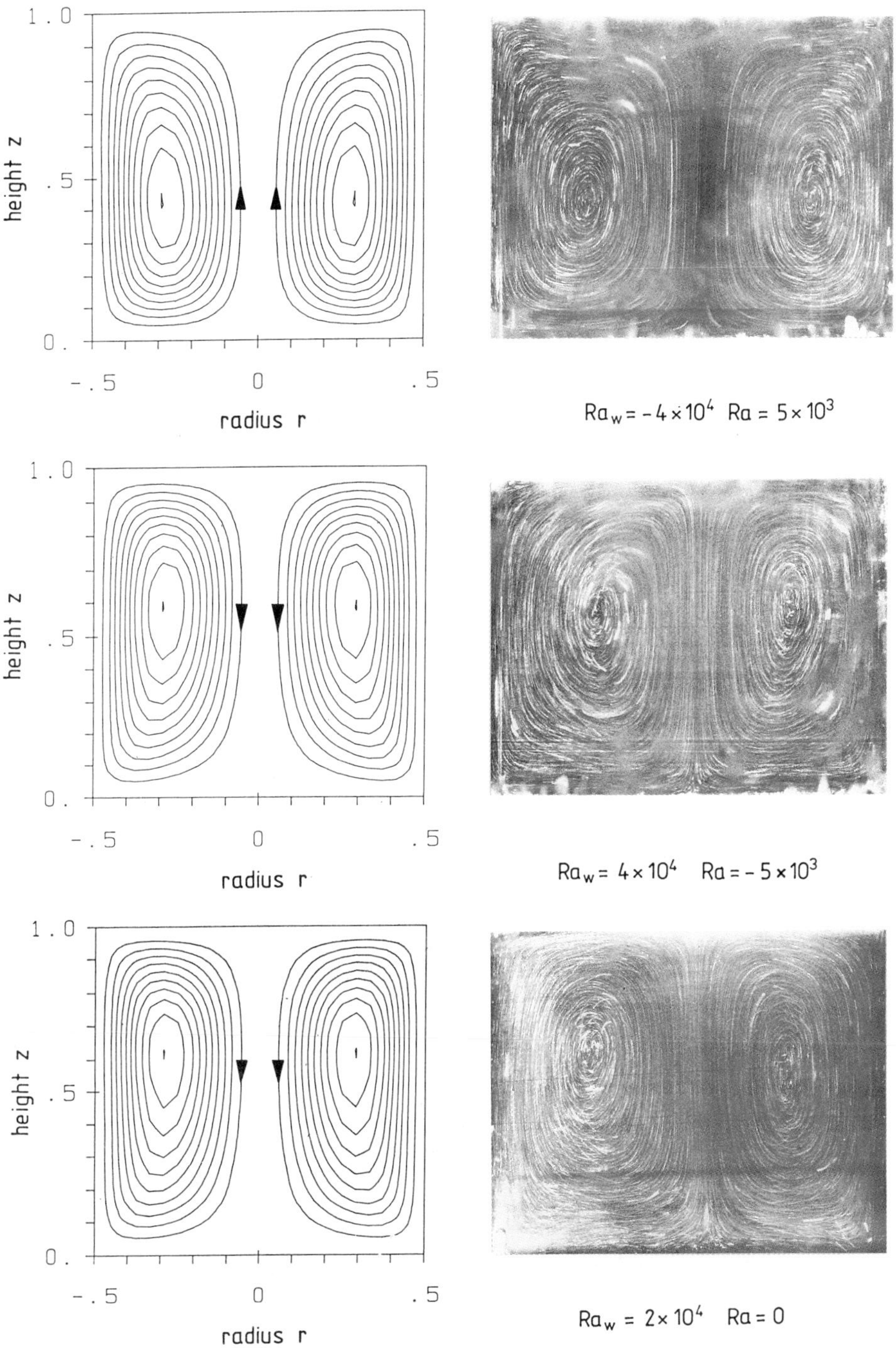

Fig. 4. Experimentally observed (right) and numerically simulated axisymmetric flows (left) in water ($Pr = 6.7$) in a cylindrical container with isothermal boundaries for $h/d = 1$ and various values of Ra and Ra_w (explained in sections 3 and 4). From [1].

horizontal open boat techniques. It can also be of importance in the Czochralski configuration (see also [2]). It should also be noted that Marangoni flows occur in the melt surface adjacent to gas bubbles e.g. in non-wetting containers.

2.2. *Effects of buoyancy convection*

The basic phenomenon of buoyancy-driven flows can be observed most clearly in crystal growth arrangements having a vertically oriented cylindrical crucible with heated sidewalls or bottom. Buoyancy is very often studied in this type of configuration as it is similar to the classical Rayleigh–Benard cell configuration (see e.g. [3]).

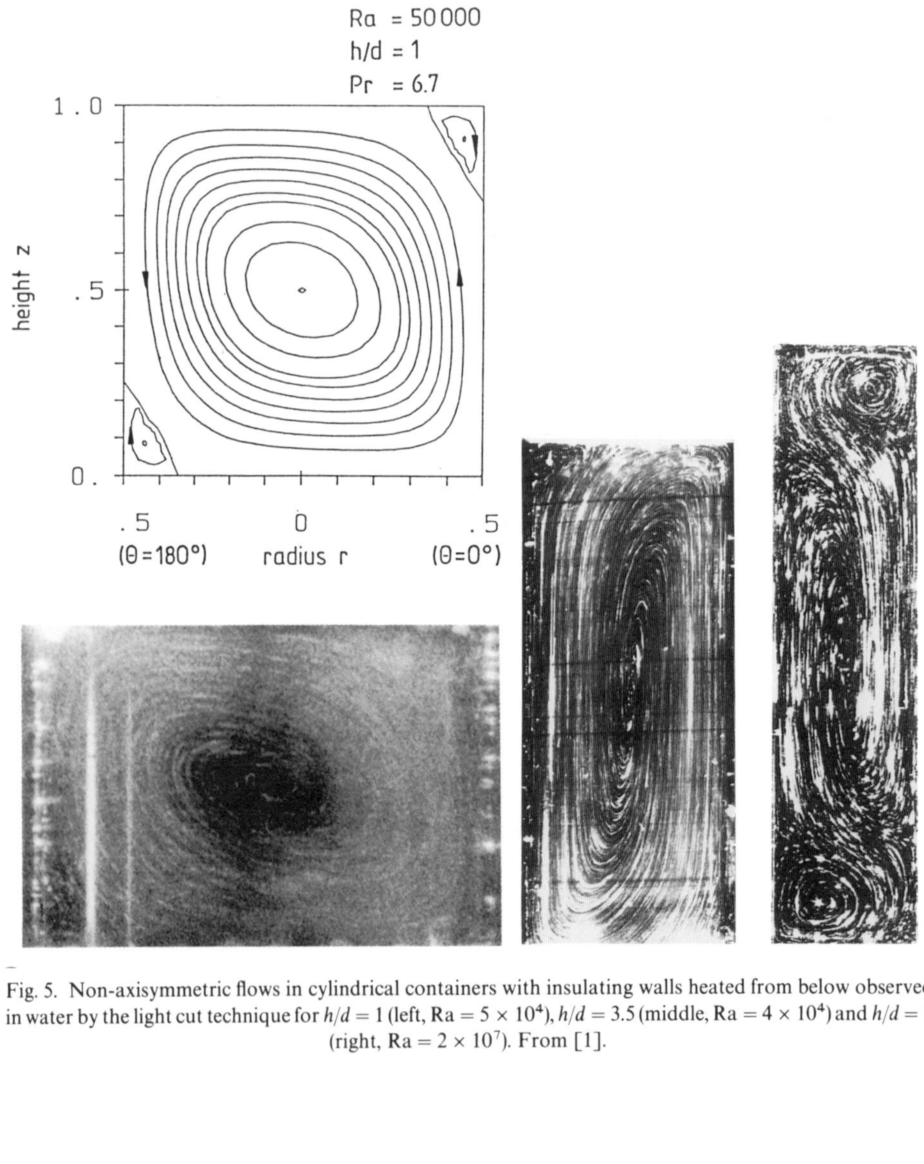

Fig. 5. Non-axisymmetric flows in cylindrical containers with insulating walls heated from below observed in water by the light cut technique for $h/d = 1$ (left, $Ra = 5 \times 10^4$), $h/d = 3.5$ (middle, $Ra = 4 \times 10^4$) and $h/d = 5$ (right, $Ra = 2 \times 10^7$). From [1].

Firstly, we consider the occurrence of convection in transparent fluids because it can be visualized. Figure 3 illustrates a typical buoyancy flow occurring in flat crucibles, i.e. with a small aspect ratio $h/d < 1$ (h = height, d = diameter). The technique for observation of such flow patterns is described in section 3.3.

For clarity the calculated flow pattern, and the resulting isotherms are given in fig. 3. The methods for such calculations are discussed later.

The axial symmetry with upflow in the center and downflow at the periphery is characteristic for this flow occurring also in opaque semiconductor and metal melts with $h/d < 1$ (compare section 4.1). The direction of the flow becomes opposite, when the transfer of heat from the sidewalls or from the upper boundary of the cylinder becomes dominant, e.g. in the vertical Bridgman and gradient freeze growth techniques with bottom seeding. This is illustrated in fig. 4.

The axial symmetry is lost if the transfer of heat from the bottom is increased or if the aspect ratio is increased $h/d > 1$ (fig. 5).

Similar behavior to that described so far exists in non-transparent melts of semiconductors and metals. But the measurement and visualization of the flow are more difficult. One elegant method is the use of X-rays and tracer particles giving pictures of flow patterns as shown in fig. 6. It is also possible to determine the flow patterns from temperature measurements with arrays of thermocouples. A typical example, showing

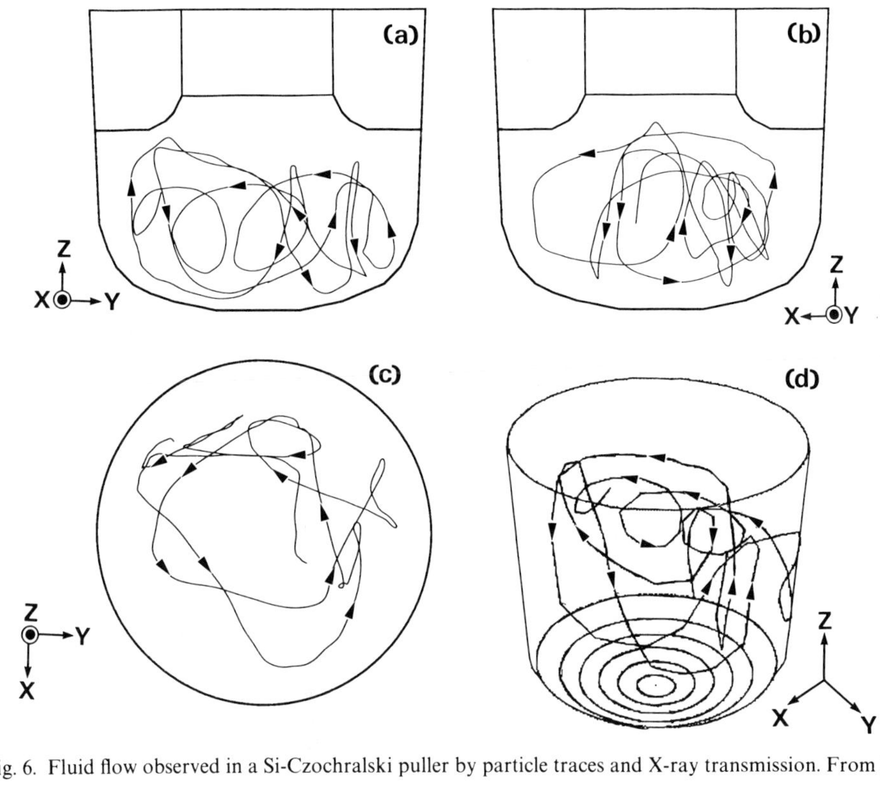

Fig. 6. Fluid flow observed in a Si-Czochralski puller by particle traces and X-ray transmission. From [4].

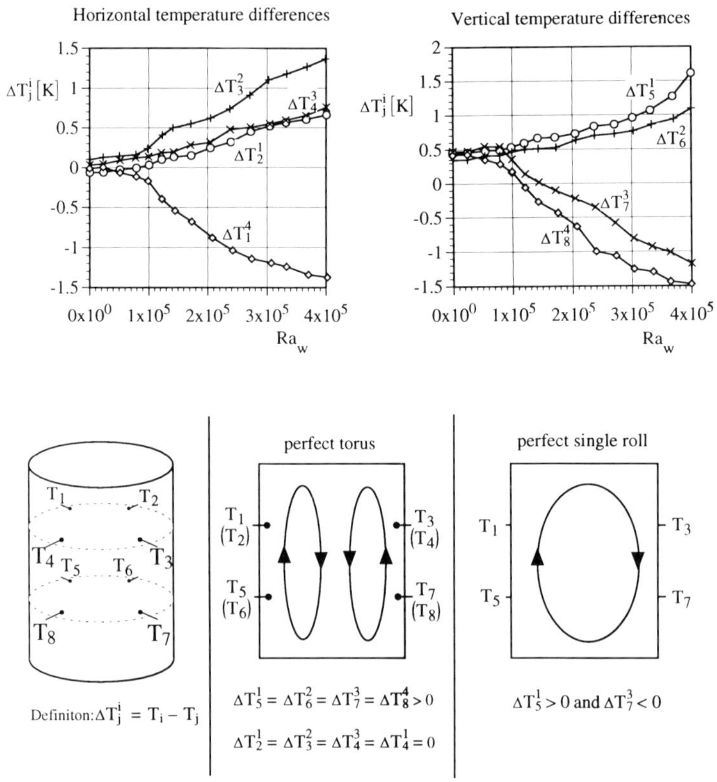

Fig. 7. Flow patterns in liquid Ga in a vertical cylindrical test cell obtained from the measurement of temperature differences between pairs of thermocouples (denoted in the sketch). $Ra = -2.5 \times 10^4$, Ra_w is varied up to 4×10^5. From [5].

the transition from a single roll for small driving forces ($Ra_w < 10^5$) to an axisymmetric torus ($Ra_w > 2 \times 10^5$) is given in fig. 7.

It is a common feature of all of these convective flows that beyond a certain flow velocity, i.e. the ratio between buoyancy and viscous forces, the flow becomes unsteady, i.e. time-dependent. This threshold depends on the geometry, the thermal boundary conditions and the physiochemical properties of the fluid. Typical examples of flow patterns of an unsteady flow are shown in fig. 5.

The transition from steady to unsteady convection can easily be detected from the signal of a thermocouple as shown in fig. 8. In this case the driving temperature difference between top (cold) and bottom (hot) of a melt (Ga) cylinder was slowly but continuously increased. The occurrence of unsteady flow is correlated to the occurrence of fluctuating temperatures (and also velocity) in the melt. Frequencies of these temperature fluctuations are in the range of 10^{-3} to 10^{-1} Hz. It depends on the geometry and the thermal boundary conditions. In a given configuration the frequency typically increases with increasing driving force for convection (see fig. 9). Such temperature fluctuations cause a fluctuating incorporation of constituents, especially

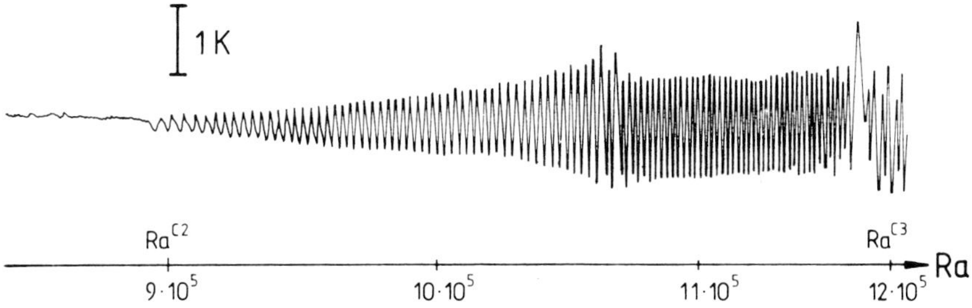

Fig. 8. Plot of the temperature measurement (thermocouple) in a Ga melt in a cylindrical container ($h/d = 5$) heated from below. The driving temperature difference between bottom and top is proportional to Ra. Ra^{c2} denotes the transition from steady to unsteady flow. From [1].

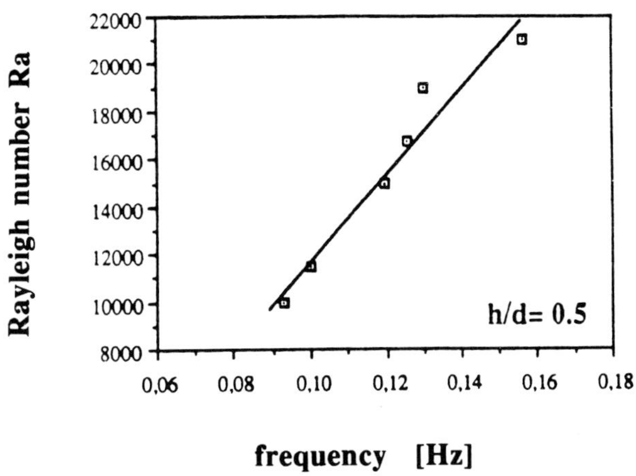

Fig. 9. Frequency of temperature fluctuations of an unsteady flow in a Ga melt contained in a vertical cylinder ($h/d = 0.5$) heated from below. From [6].

dopants in semiconductor growth. This effect, which is illustrated by fig. 10, is discussed in detail quantitatively in section 5.2.

The amplitudes of the temperature fluctuations caused by buoyancy convection are increasing (see fig. 8) and the frequency spectrum becomes more complex (see fig. 11) with increasing intensity of the flow.

In the strict sense purely buoyancy-driven convection only exists if the melt volume is completely enclosed by solid boundaries. This is only the case in the vertical Bridgman configuration with top seeding and in the vertical zone melting configuration (compare fig. 1). But also in configurations having free surfaces like the open boat and Czochralski, buoyancy is often the dominating mechanism. This has been proved for example for the horizontal zone melting configuration by the correlation of experimental results of Hurle et al. [7] and calculations by Gill [8]. Other examples will be treated below in section 4.3.

So far we have only given examples of thermally-driven buoyant flows. But density gradients, generated by segregation at the crystal–melt interface can also induce

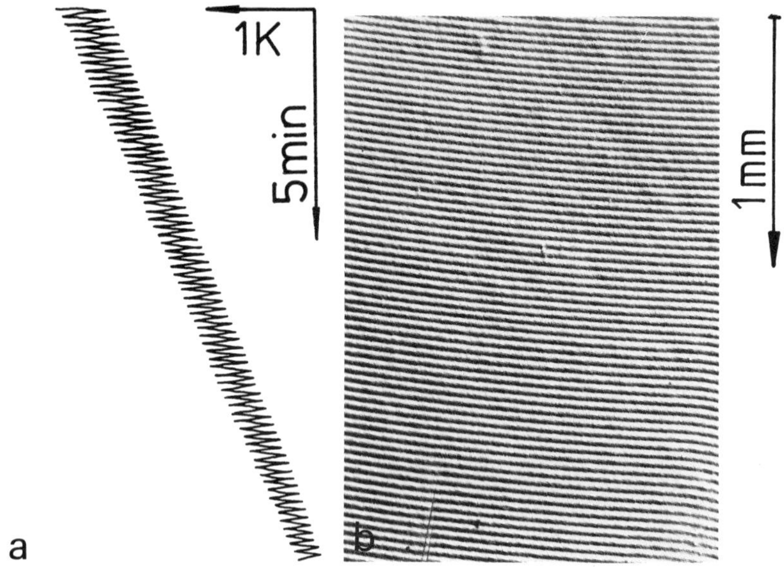

Fig. 10. Longitudinal section (interference contrast microscopy) of a Te-doped InSb crystal (b) and temperature fluctuations measured in the InSb melt (a) recorded during that time where the depicted crystal was grown. Growth method: vertical gradient freeze with top seeding. From [1].

convective flow. Their effect is, of course, significant (and eventually dominant) in solution growth. But Hurle states in his review article [9], that even the small solute concentrations – or deviations from melt congruency in the case of compound growth – associated with the melt growth can be of significance. Especially the coupling between thermal and solutal buoyancy-driven flows (so-called double diffusive convection) [10] leads to complex flows.

Also the coupling between buoyancy-driven flow and the deformation of the crystal–melt interface (morphological instability [11–13]) can be of significance. For melt growth, the likely effect of the convective flow is to sweep out the solute from the cell boundaries in the growth interface thus tending to stabilize the interface against morphological breakdown. Such effects are discussed in more detail by Hurle [9] and in Coriell and McFadden (Chapter 12, in Vol. 1 of this Handbook).

Of course buoyancy convection has also a strong effect on the heat and species transport in the Czochralski configuration. But this will be treated separately in section 2.3 as the flows often result from complex interaction of several mechanisms, such as buoyancy, rotation and capillarity.

2.3. *Natural and forced convection in Czochralski configurations*

In this section we consider typical fluid flow phenomena and flow patterns occurring in the most important of all melt growth techniques – the Czochralski configuration. Hurle characterizes in his review article [9] the flow in a Czochralski crucible (of a

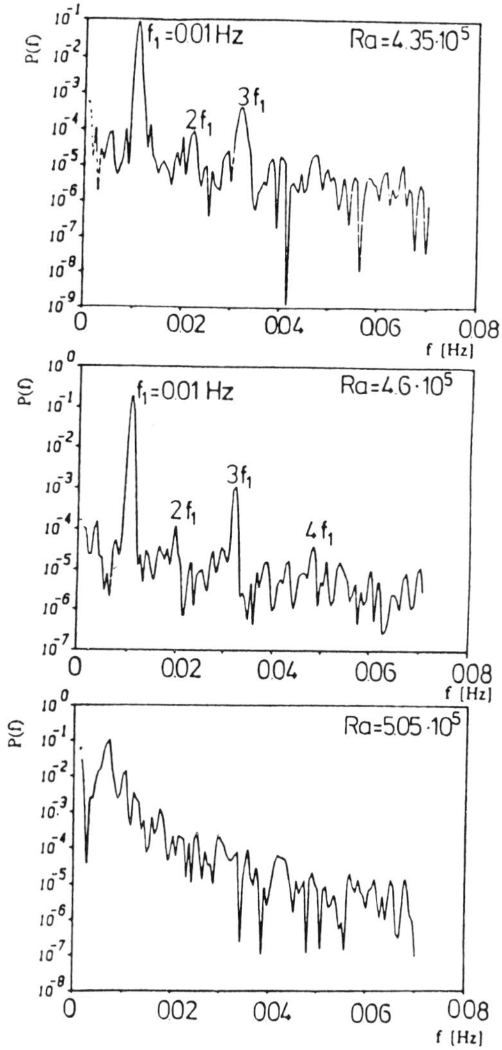

Fig. 11. Frequency spectra of unsteady flow in Ga in a vertical cylindrical container ($h/d = 4$) with increasing temperature difference between bottom and top (Ra). From [6].

technically interesting size) to be "dauntingly complicated". What is the reason for such a statement? It is the complex interaction of buoyancy, forced convection and capillary flows. A rough imagination of the resulting flow can be obtained if one considers the driving forces according to Hurle [9] as follows. At the heated vertical wall of the crucible the melt is made buoyant, rises and is then turned radially inwards at the melt surface. The rotating crystal acts as a centrifugal fan sucking up fluid axially spinning it up in a boundary layer of thickness $(\nu/\omega)^{1/2} \leq 1$ mm and ejecting it tangentially (ν is the kinematic viscosity of the melt and ω the crystal rotation rate). These two typical flows

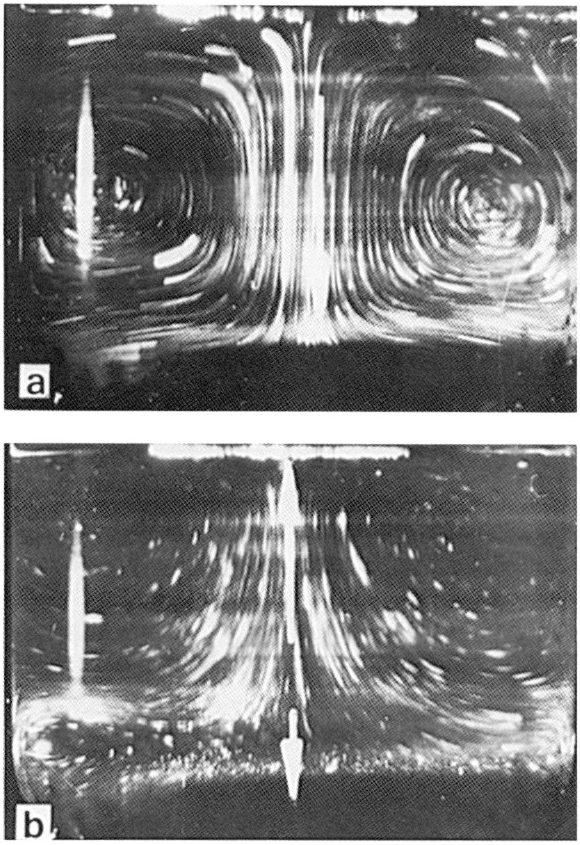

Fig. 12. Flows observed in a test cell simulating the Czochralski configuration under conditions without rotation (a), i.e. only buoyancy-driven convection ($Gr = 1.4 \times 10^4$), and forced convection (b) by crystal rotation ($Re_\Omega = 90$) in a glycerol–water solution. From [14].

(buoyancy and forced convection by crystal rotation) can be seen independently in fig. 12. This result is taken from a study of Peshev et al. [14] who used a Czochralski model with glycerol–water solutions.

The interaction of these flows causes a downflow at a certain radial distance which depends upon the relative strength of the crystal rotation and the buoyancy-driven convection.

Typical flows resulting from this interaction were studied by Carruthers [15] also in a transparent Czochralski model which is depicted in fig. 13. In this figure it can be seen that increasing crucible rotation reduces the influence of buoyancy at the growth interface. Furthermore, we have to consider that both rotation of the crystal and of the crucible are commonly employed in the growth of semiconductor crystals. Their combined effects depend on whether the two are iso-rotated or counter-rotated. With iso- (but differential) rotation under some conditions the so-called Taylor–Proudman

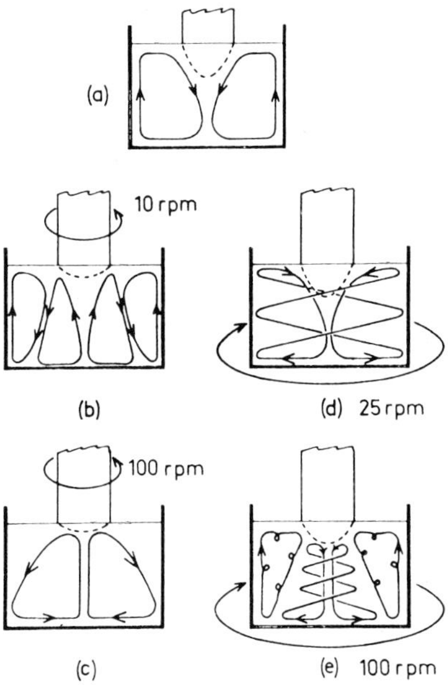

Fig. 13. Schematic flow patterns in a transparent liquid model. (a) Thermal convection only; (b) low crystal rotation rate; (c) high crystal rotation rate; (d) low crucible rotation rate; (e) high crucible rotation rate. From [15].

cell (shown in fig. 14) can be established beneath the crystal with a detached shear layer separating it from the outer region which rotates (approximately) as a solid body with the crucible.

Counter-rotation yields a more complex flow pattern beneath the crystal with a stagnation layer separating the flows driven by crystal and crucible rotation respectively (see also fig. 14 and [17–21]).

Several other factors complicate the flow behavior additionally. Firstly, the radial temperature gradient across the melt surface will lead to a thermocapillary component of the flow (see section 2.4). This effect is additionally influenced by the non-planar geometry of the melt surface, due to the meniscus at the growth interface. Secondly, the loss of heat from the melt surface produces a destabilizing vertical temperature gradient in the upper region of the melt which can give rise to the formation of convection cells [15]. Figures 15 and 16 show flows in a Czochralski model using a molten salt at high temperature obtained by Krzyminski and Ostrogorsky [22]. Their measuring technique is described in section 3. In the free melt surface, so-called spoke patterns were observed in other model experiments [23–25] and in real melt growth of Si and of garnets (fig. 17). These effects are interpreted by periodic rolls aligned parallel to the interface [18, 23].

Up to this point we considered the flow to be steady and axisymmetric, although this is not the case in reality. In Czochralski melts, a characteristic instability causes a change in angular velocity, resulting in unsteady flow. This feature of the Czochralski

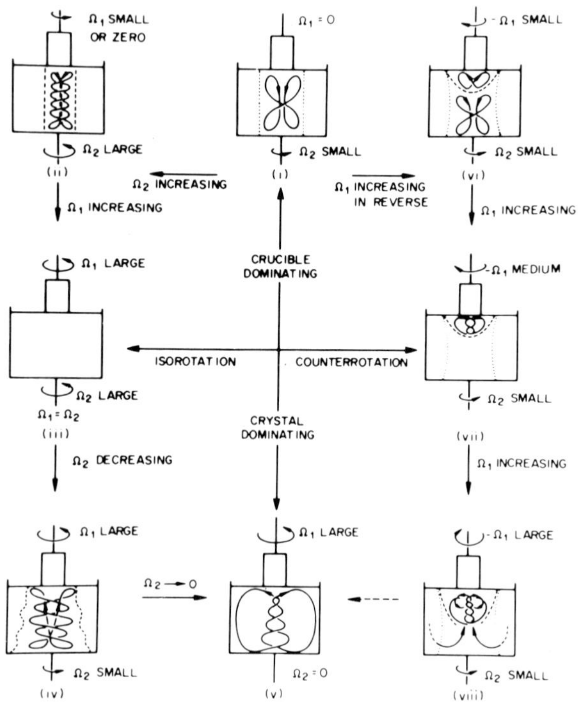

Fig. 14. Variation of flow patterns with the relative rotation directions and magnitudes of crystal and crucible. From [16].

flows is especially of importance in the region close to the growth interface [9, 20]. Typical flow patterns causing instabilities are depicted in figs. 18 and 19. Such instabilities are a source of temperature fluctuations leading to striations in the grown crystal (compare fig. 20).

Transition mechanisms of flow modes and temperature fluctuations have been studied systematically by Ozoe et al. [27]. In model experiments flow modes were visualized with liquid crystal powder in addition to temperature measurements by thermocouples. These liquid crystals depict the fluid temperature by different color. It was found that periodic changes in the convection mode are very likely to occur. When natural convection is dominant cold plumes from the rotating periphery of the cold crystal merge toward the central cold plume descending from the bottom center of the cold crystal. When forced convection becomes dominant the cold plume from the rotating edge constitutes a cold ballon-shaped region, which the hot fluid coming from the bottom of the crucible breaks periodically. This hot plume is considered as a mechanism causing a periodic remelting of the crystal, which also causes striations [28].

For a critical rotation rate, the coupling of flows causes an effect resembling a baroclinic instability. Such effects have been observed especially in the Czochralski

Fig. 15. Side view of the asymmetric, unsteady natural convection in $NaCl-CaCl_2$ melts, without crystal or crucible rotation. $Ra = 2.3 \times 10^5$. Photographs were taken by rapidly lowering the tubular heater, in 10 s intervals. From [22].

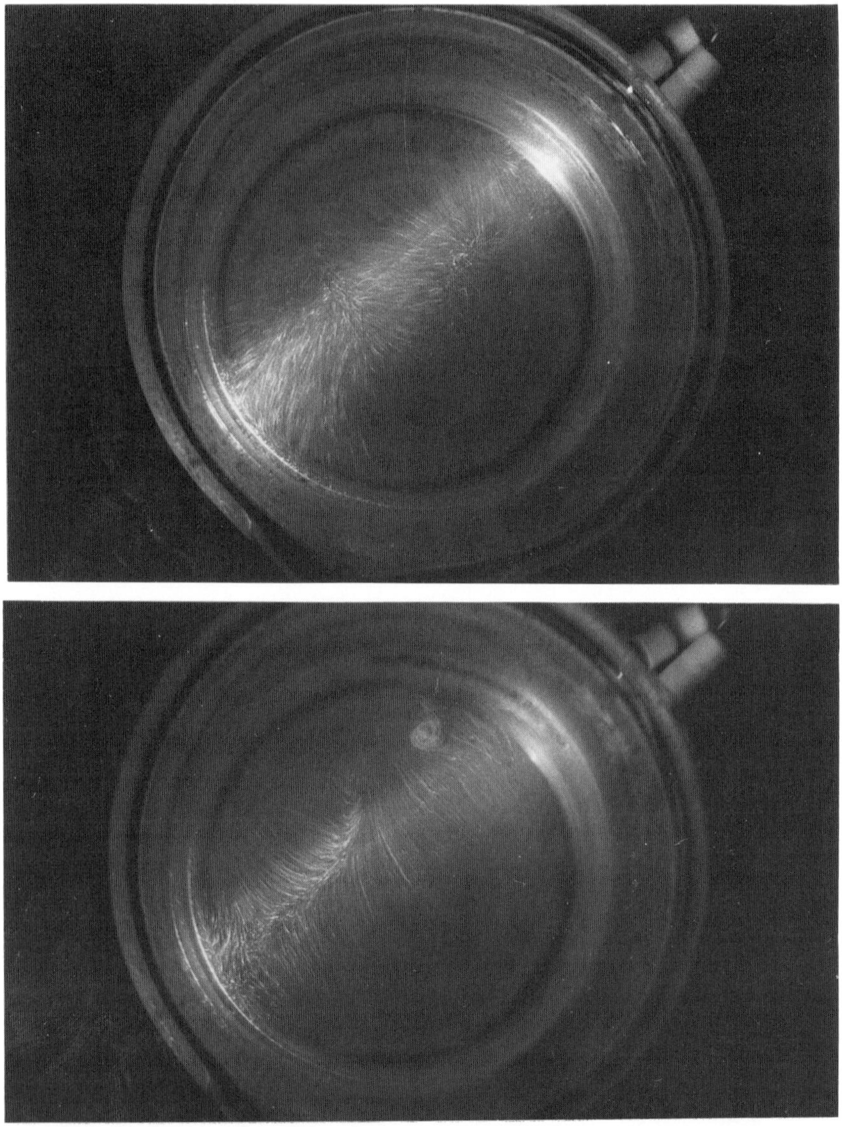

Fig. 16. Top view of asymmetric, unsteady natural convection in $NaCl-CaCl_2$ melts, without crystal or crucible rotation. From [22].

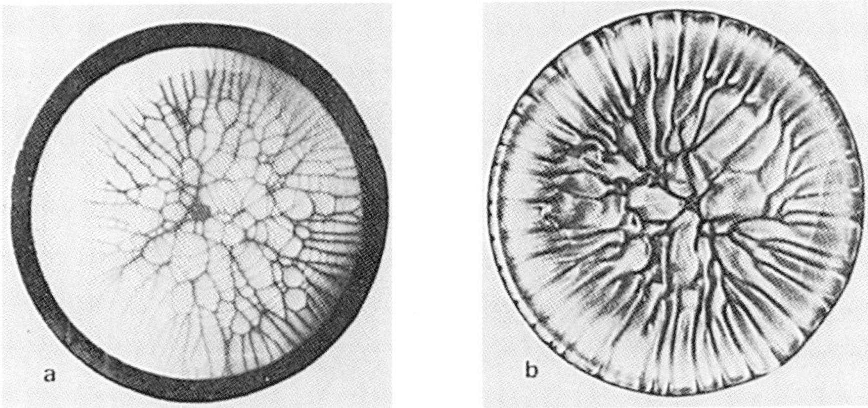

Fig. 17. Flow pattern visible on the surface of a garnet melt (a), and a water bath (b). From [18].

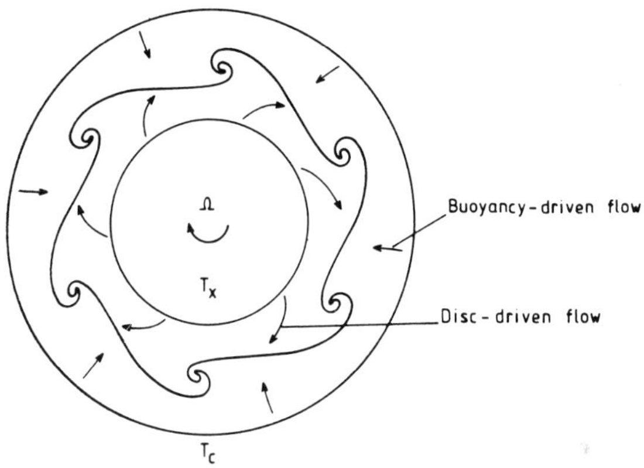

Fig. 18. Schematic drawing of typical wave patterns observed in a model system of the Czochralski configuration using a kaliroscopic fluid for visualization. From [20].

growth of oxides e.g. by Whiffin et al. [29], Cockayne and Gates [30], Brandle [31], but are now also discussed for semiconductors (Si) by Watanabe et al. [32].

The shape of the interface can be changed continuously by a variation of the rotation rates. This has been demonstrated by Hirata et al. [33] and is shown quantitatively in fig. 21. Thus crystal rotation is often used in the Czochralski configuration as the control parameter to establish a certain shape of the interface. In oxide growth, e.g. $LiNbO_3$, a flat interface is a prerequisite for the growth of large high quality crystals. Lee et al. [34] demonstrated a correlation between the interface shape and the crystal rotation rate (see fig. 22). The observed critical value of the rotation rate for achieving a flat interface was theoretically confirmed by the authors using an empirical equation based on a hydrodynamic analysis of buoyant and forced convection. But dramatic

Fig. 19. Streak photograph of vortices occurring in the configuration of fig. 18. From [20].

changes of the interface shape due to flow transitions were also observed especially in the oxide growth reported by several authors [31, 35–38].

Finally it should also be mentioned that convection (forced and buoyant) coming from the gas atmosphere (inert gas or air) within the furnace influences considerably the heat transfer between heaters, cold walls, crystal and melt. Gas convection caused by buoyancy is relevant in all growth configurations having a gaseous ambient. In gases, the Rayleigh number is proportional to the square of the gas density or pressure, respectively. Therefore, gas convection is most intense in high pressure vessels. For example in high pressure LEC and VB/VGF autoclaves, gas convection is clearly turbulent, since $Ra \approx 10^9$ [39]. At high pressures the density difference between the gas and the melt is reduced, so that coupling between the convective flow in the melt and the ambient gas can be significant [40].

2.4. Forced convection in other melt growth configurations

An interesting effect of forced convection has been recently demonstrated in the floating zone technique. A shortening of the zone height without a reduction of the zone or crystal diameter respectively would be highly desirable because of the increased stability under the action of earth gravity. Lan and Kou [41] have demonstrated, in the

Fig. 20. Longitudinal section of a 6 inch diameter Czochralski-grown p-type Si crystal (Wacker Chemitronic); annealed at 780°C (0.5 h) and 1000°C (16 h), etched (Secco-etch) to reveal the striated oxygen distribution. From [26].

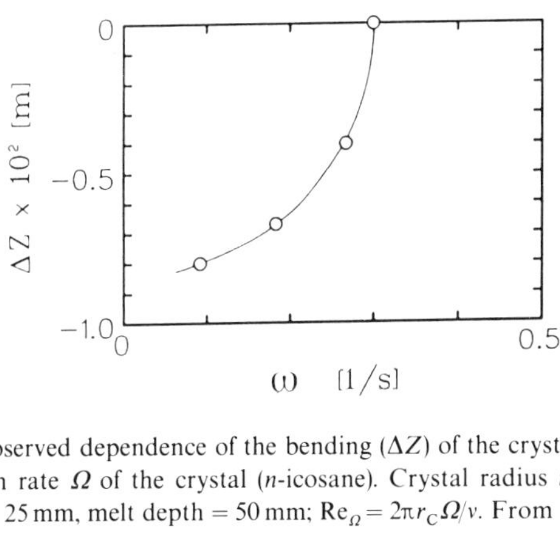

Fig. 21. Experimentally observed dependence of the bending (ΔZ) of the crystal–melt (n-icosane $C_{20}H_{42}$) interface from the rotation rate Ω of the crystal (n-icosane). Crystal radius $r_C = 11$ mm, crucible radius $R = 25$ mm, melt depth $= 50$ mm; $Re_\Omega = 2\pi r_C \Omega/\nu$. From [33].

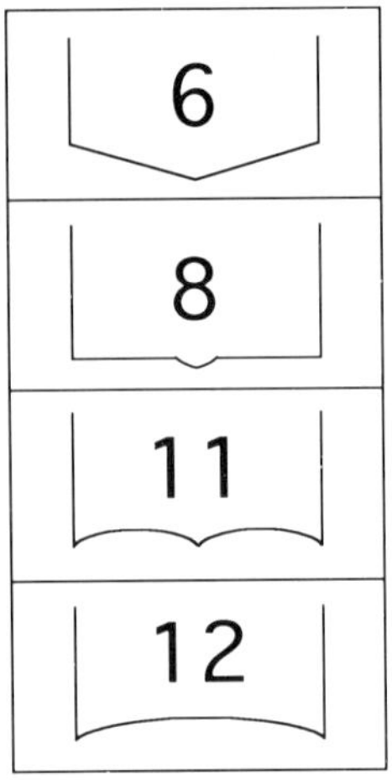

Fig. 22. Variation of the interface shape with crystal rotation rate (numbers in rpm) during the growth of $LiNbO_3$ single crystals (80 mm diameter, 132 mm length) by the Czochralski technique. From [34].

floating zone growth of $NaNO_3$, that the melt zone can be considerably shortened and stabilized by using relatively high (200 and 400 rpm) rotation rates of feed and crystal (single- and counter-rotation).

The favorable effect of the forced convection is explained (with the aid of numerical modelling) by the strongly increased heat transfer radially inward flattening both interfaces.

Forced convection is also applied in the accelerated crucible rotation technique (ACRT) firstly introduced by Scheel and Schulz-Dubois [42]. In this technique, which is mainly used in solution growth, the crucible is not rotated with a constant rate but with a periodically increased and decreased speed. The flow is thereby caused by the inertia of the fluid. Typical convection flow patterns and temperature behavior have been observed in a transparent model system (glycerin–water) by Masalov and Emel'chenko [43] depicted in fig. 23. These authors reported also the main drawback of ACRT which is the creation of temperature oscillations in the melt caused by the fluctuating heat transport. The problem with fluctuations of the heat and mass transport is the resulting fluctuating growth rate causing non-uniformity of the crystal (compare fig. 10 and for further details see sections 4 and 5.2).

2.5. *Capillary convection*

Capillary (Marangoni) flow has been mainly studied in floating zone configurations with transparent fluids. But since its importance for the growth of semiconductors was postulated by Chang and Wilcox [44], several groups are now also investigating the action of capillary flows in the growth of semiconductor and metal melts (see section 4.2.)

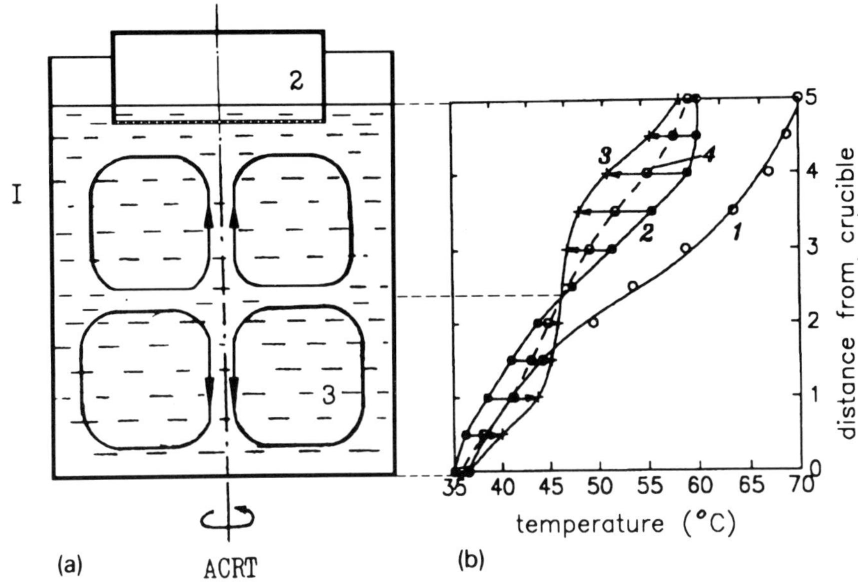

Fig. 23. Fluid flow (a) and axial temperature distribution (b) in a model system of flux growth by the accelerated crucible rotation (ACRT, 2 = feeding device). Temperature profile along crucible wall (curve 1) and axis (curve 2) with crucible at rest; curve 3 gives the maximum deviation from curve 2 under ACRT with $\Omega_{max} = 30/\text{min}$, acceleration time 6 s; hold time at zero 12 s. Curve 4 is the average solution temperature at ACRT. From [43].

Experimental studies of model systems are often conducted using zones of silicone oil [2, 45] and molten salts (e.g. sodium nitrate) [2] suspended between pairs of coaxial cylindrical thermally-conducting rods maintained at different temperatures to produce a temperature gradient along the free surface of the liquid. These systems are simulating half of a crystal growth floating zone (so-called half-zone models; compare section 4.2).

In more realistic systems the rods have the same temperature and a ring heater is used to generate the temperature gradients.

For relatively small driving temperature gradients, i.e. Marangoni number, a single axisymmetric vortex (torus) is observed in the half-zone or a pair of vortices in the full-zone correspondingly. Such a flow is shown in fig. 24a, where the flow in the $NaNO_3$ is made visible by tracer particles and the light cut technique [2]. In this case the half-zone was heated from above to suppress unsteady buoyancy-driven convection. Nevertheless there exists a weak buoyant flow driven by the horizontal (radial) temperature gradient. This flow remains, if the free surface of the zone is covered by a thin viscous oil film (fig. 24b). This flow is much weaker, as can be seen from the shorter length of the illuminated particle paths. The strong contribution of the thermocapillary flow in experiment of fig. 24a has been demonstrated by the authors by measuring the temperature distribution (fig. 25) and the flow velocity in the zone (fig. 26).

With further increase of the driving temperature gradient, thermocapillary flow becomes time-dependent i.e. unsteady, as illustrated in fig. 27.

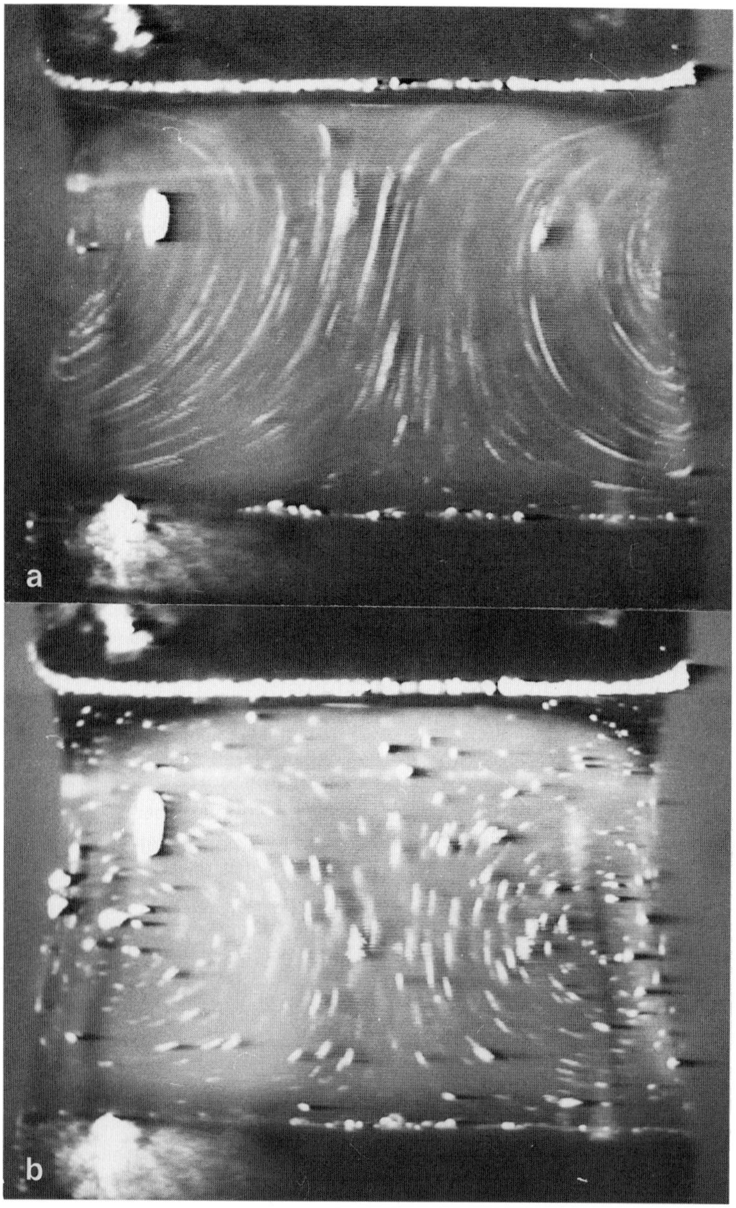

Fig. 24. Comparison of thermocapillary and buoyancy-driven flow in a $NaNO_3$ zone heated from above ($\phi = 6$ mm, length = 4 mm, $T_{upper} = 360$°C, both exposure times = 1/8 s). The flow is upwards in the middle in both cases. Compare the streak length of the tracer particles. (a) Thermocapillary forces and buoyancy forces in a zone with clean surface (velocity in the free surface about 15 mm/s). (b) Same heating conditions but thermocapillary forces suppressed by an oil film of approximately 50 μm thickness (maximum velocity about 3 mm/s). From [2].

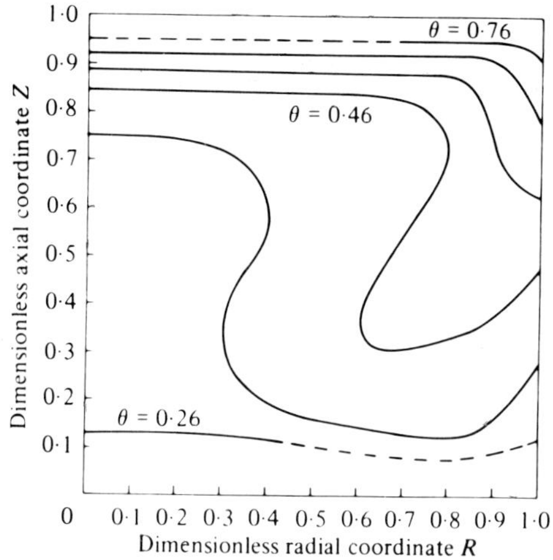

Fig. 25. Isotherms in the right half of the zone shown in fig. 24a. From [2].

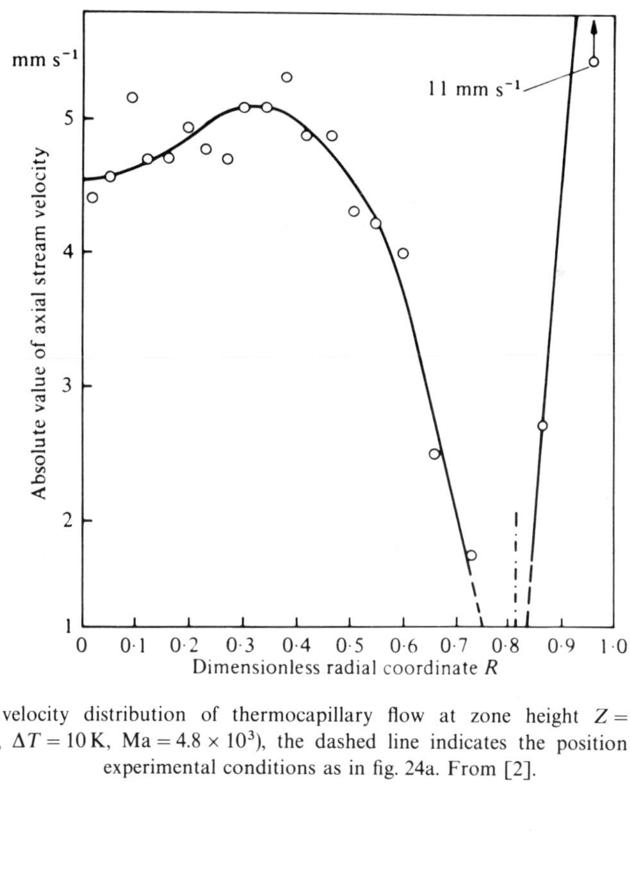

Fig. 26. Radial velocity distribution of thermocapillary flow at zone height $Z = 0.68$ ($l = 4.1$ mm, radius = 3.0 mm, $\Delta T = 10$ K, Ma = 4.8×10^3), the dashed line indicates the position of the roll axis; experimental conditions as in fig. 24a. From [2].

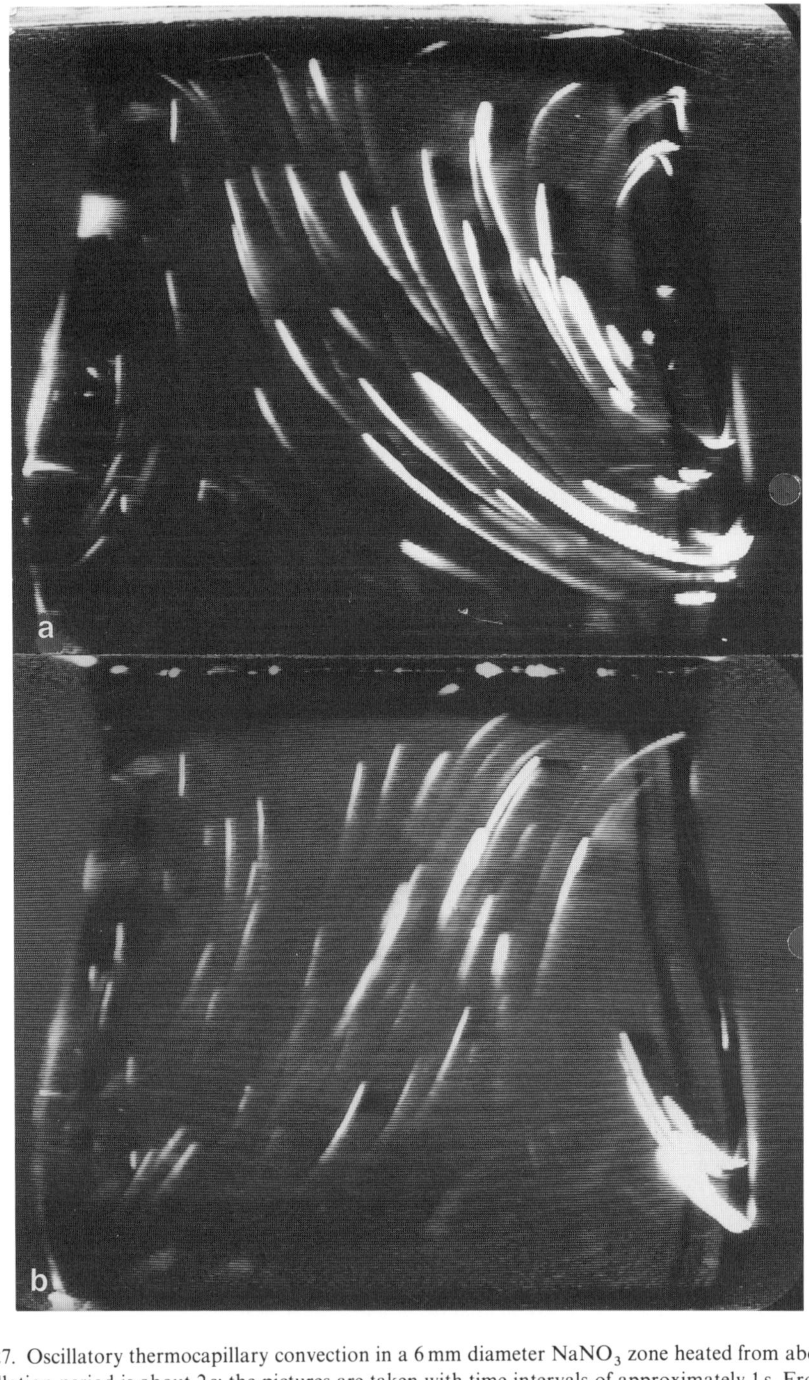

Fig. 27. Oscillatory thermocapillary convection in a 6 mm diameter $NaNO_3$ zone heated from above. The oscillation period is about 2 s; the pictures are taken with time intervals of approximately 1 s. From [2].

For more details the reader is referred to the review articles of Chun [46], Schwabe [2], Bohm et al. (Chapter 4, this volume) and section 4.2 of this article.

3. *Experimental tools for detection and visualization of convective flows in melt growth configurations*

Because of high purity requirements and the hostile environment in the furnace, little instrumentation is available during crystal growth. Instrumentation is usually limited to thermocouples or optical pyrometers needed to measure and control the temperature of the heater (or heaters) and, in Czochralski growth, the instrumentation needed for diameter control. As a result, most of the existing experimental evidence of convective flows in the melt was not obtained during actual growth experiments. Instead, model experiments were performed based on dimensional analysis and similarity.

Most model experiments were conducted for Czochralski growth, because of its importance and complexity.

3.1. Similarity and dimensionless variables

Similarity and dimensional analysis provide the key to relating the data obtained in model experiments to the unknown variables of actual problems [47]. Dimensionless analysis reduces the number of experimental variables, which is most useful in organizing experiments and interpreting the results (see the Buckingham pi theorem [47]). It is indispensable in problems which are too complex to allow for an analytical or numerical solution.

Melt convection is fully described by the set of governing differential equations and corresponding boundary conditions (see Appendix A). When these are non-dimensionalized, dimensionless parameters appear as multipliers to different terms containing non-dimensional variables and their derivatives (table 1). According to the principle of similitude, convective transport occurring in a model experiment will have the same dimensionless variables (e.g. non-dimensional velocity and temperature) as the real growth experiment if geometric similarity, kinematic similarity and dynamic similarity exist between the model and the actual growth experiment. The similarity is ensured if all relevant non-dimensional parameters describing the model are set equal to non-dimensional parameters of the real growth.

(i) *Geometric similarity* requires that all lengths and angles of the model must be in the same ratio as in the actual experiment. For the simplest melt geometry, a cylinder having height H and radius R_C, melt aspect ratio H/R_C has to be set equal. For Czochralski growth, the ratio of crystal to crucible radius, R_X/R_C, also has to be equal.

(ii) *Kinematic similarity* requires that all velocities of the model must be in the same ratio as in the actual experiment. For Czochralski growth, equal ratio of crystal to crucible rotation rates Ω_X/Ω_C ensures kinematic similarity in boundary conditions and presents a necessary requirement for full kinematic similarity. Equal Ω_X/Ω_C does not

Table 1
Dimensionless similarity parameters for melt convection.

Parameter	Definition
Geometric	
Ratio of radii	R_X/R_C
Melt aspect ratio	$AR = H/R_C$ or h/d
Kinematic	
Ratio of rotation rates	Ω_X/Ω_C
Scaled material properties	
Prandtl number	$\mathrm{Pr} \equiv \frac{\nu}{\kappa} = \frac{\text{momentum diffusivity}}{\text{heat diffusivity}}$
Schmidt number	$\mathrm{Sc} \equiv \frac{\nu}{D} = \frac{\text{momentum diffusivity}}{\text{species diffusivity}}$
Equilibrium segregation coeff.	$k = \frac{C_S}{C_L} = \frac{\text{solute concentration in solid}}{\text{solute concentration in melt}}$
Thermal-solutal	
Dimensionless temperature difference	$\frac{\Delta T}{T_a} = 2\frac{T_C - T_X}{T_C + T_X} = \frac{\text{temperature difference in melt}}{\text{absolute temperature}}$
Dimensionless radiation	$q_r^* = \frac{\varepsilon\sigma T^3}{k/H} = \frac{\text{radiation from melt surface}}{\text{conduction to melt surface}}$
Thermal Péclet number	$\mathrm{Pe}_T \equiv \mathrm{Re} \times \mathrm{Pr} = \frac{\text{convective heat transport}}{\text{diffusive heat transport}}$
Solutal Péclet number	$\mathrm{Pe}_S \equiv \mathrm{Re} \times \mathrm{Sc} = \frac{\text{convective species transport}}{\text{diffusive species transport}}$
Dynamic	
Reynolds numbers	$\mathrm{Re} \equiv \frac{VL}{\nu} = \frac{\text{inertia forces}}{\text{viscous forces}}, \quad \mathrm{Re}_X \equiv \frac{R_X^2 \Omega_X}{\nu}$
Grashof number	$\mathrm{Gr} \equiv \frac{g\beta\Delta T H^3}{\nu^2} = \frac{\text{buoyancy forces}}{\text{viscous forces}}$
Solutal Grashof number	$\mathrm{Gr}_S^i = \frac{g\beta_S^i \Delta C^i L^3}{\nu^2} = \frac{\text{buoyancy forces}}{\text{viscous forces}}$
Rayleigh number	$\mathrm{Ra} = \mathrm{Gr} \times \mathrm{Pr} = \frac{g\beta\Delta T H^3}{\nu\kappa} = \frac{\text{buoyancy forces}}{\text{viscous forces}}$
Marangoni number	$\mathrm{Ma} \equiv \frac{(\partial\sigma/\partial T)\Delta T l}{\eta\kappa} = \frac{\text{force due to surface tension}}{\text{viscous forces}}$
Bond number	$\mathrm{Bo} \equiv \frac{g\Delta\rho L^2}{\sigma} = \frac{\text{gravitational forces}}{\text{surface tension forces}}$
Hartman number	$\mathrm{Ha} \equiv \frac{\sigma B_0^2 L}{\rho V} = \frac{\text{Lorentz force}}{\text{inertia forces}}$
Ekman number	$\mathrm{Ek} \equiv \frac{\nu}{R^2\Omega_X} = \frac{\text{viscous force}}{\text{Coriolis force}}$
Rossby number	$\mathrm{Ro} \equiv \frac{\nu}{R \cdot \Omega_X} = \frac{\text{convective force}}{\text{Coriolis force}}$

ensure similarity of the velocity fields in the melts. Full kinematic similarity in velocity fields exists only if the ratios of all forces driving the flow in the model experiment and actual growth are equal.

(iii) *Dynamic similarity* requires that all forces in the model melt must be in the same ratio as in the actual experiment. The most basic requirement for dynamic similarity in laminar and turbulent flows is equal Reynolds numbers, which ensures equal ratio between inertia and viscous forces

$$\mathrm{Re} \equiv \frac{VL}{\nu} = \frac{\text{inertia forces}}{\text{viscous forces}}, \tag{3.1}$$

where V is the velocity field in the melt. For Czochralski growth, a Reynolds number based on crystal rotation rate ensures similarity in the ratio between rotational (centrifugal or Coriolis) and viscous forces

$$\mathrm{Re}_{\mathrm{X}} \equiv \frac{R_{\mathrm{X}}^2 \Omega_{\mathrm{X}}}{\nu}. \tag{3.2}$$

The Reynolds number may be replaced by the Ekman number

$$\mathrm{Ek} = \frac{\nu}{R^2 \Omega_{\mathrm{X}}} = \frac{\text{viscous force}}{\text{Coriolis force}}, \tag{3.3}$$

which is, in essence, the inverse Reynolds number [48].

The Marangoni number ensures that forces due to surface tension and viscous forces are in the same ratio

$$\mathrm{Ma} \equiv \frac{(\partial\sigma/\partial T)\Delta T l}{\eta\kappa} = \frac{\text{force due to surface tension}}{\text{viscous forces}}, \tag{3.4}$$

where l is the characteristic length of the free surface, η is the dynamic viscosity and κ is the thermal diffusivity. Since the surface tension σ is a function of the temperature, the ratios of temperatures at the free surface of the model must be set equal to the ratios of temperatures of the growth experiment.

Equal Grashof numbers are required for similarity in the ratio between buoyancy and viscous forces

$$\mathrm{Gr} \equiv \frac{g\beta\Delta T H^3}{\nu^2} = \frac{\text{buoyancy forces}}{\text{viscous forces}}, \tag{3.5}$$

where g is the gravitational acceleration, β is the coefficient of volumetric expansion, ΔT is the characteristic temperature difference imposed on the melt boundaries and ν is the kinematic viscosity.

(iv) *Similarity in thermal boundary conditions and heat transfer mechanisms in the melt.* The buoyancy forces providing the pumping power for the natural convection are concentrated inside the thermal boundary layers in the melt. In the well mixed bulk melt, the buoyancy effects are weak. Since buoyancy forces arise from temperature differences in the boundary layers, similarity in the ratio between buoyancy and viscous forces requires similar temperature fields in the melt (i.e. equal ratios of all melt temperatures are required for dynamic similarity). The temperature fields in the melt will be similar if:

(a) thermal boundary conditions are similar, and
(b) in the melt, thermal energy is transferred by similar mechanisms.

Uniform crystal temperature T_X and crucible temperature T_C are often used in model experiments and numerical simulations, because the actual temperature field in the crucible is complex and not known. Furthermore, the temperature of the free melt surface, which controls the Marangoni convection, is a priori unspecifiable [49]. The amount of radiative cooling from the free surface and through transparent crystals depends on crystal and furnace geometry, i.e. view factors, which change during growth.

Buoyancy forces (i.e. Gr) depend on the temperature difference $\Delta T = T_C - T_X$, while radiative cooling depends on the absolute temperature $T_a = (T_C + T_X)/2$. Note that T_a is approximately equal to the melting point, T_{mp}. For uniform crystal and crucible temperatures, the similarity in thermal boundary conditions (equal ratios of all temperatures at melt boundaries) is reduced to maintaining equal the dimensionless temperature difference

$$\frac{\Delta T}{T_a} = 2\frac{T_C - T_X}{T_C + T_X} = \frac{\text{temperature difference}}{\text{absolute temperature}}. \tag{3.6}$$

The (thermal) Rayleigh number Ra_T ensures similarity in heat transfer mechanisms in the melt

$$\mathrm{Ra}_T \equiv \frac{g\beta \Delta T H^3}{\nu\kappa} = \frac{\text{convective heat transport}}{\text{diffusive heat transport}}, \tag{3.7}$$

where κ is the thermal diffusivity. Note that $\mathrm{Ra}_T/\mathrm{Gr}$ is equal to the Prandtl number

$$\mathrm{Pr} \equiv \frac{\nu}{\kappa} = \frac{\text{momentum diffusivity}}{\text{heat diffusivity}} \sim \left(\frac{\delta_v}{\delta_T}\right)^3, \tag{3.8}$$

where δ_v is the thickness of momentum layer and δ_T the thickness of the thermal boundary layer (in which buoyancy forces act). Similar temperature fields have equal δ_v/δ_T (i.e. geometrically similar boundary layers). Therefore, full similarity in buoyancy-driven convection requires that the model fluid has Pr equal to that of the actual melt. Note that Pr is a fluid property which can not be adjusted (see section 3.4 for a potential exception).

In a dielectric fluid, heat is conducted by collisions of molecules which move randomly. In electrically conducting fluids, free electrons transport both thermal energy and electrical charge. Since diffusion of free electrons is a more effective mechanism for heat conduction than molecular collisions, metallic and semiconductor melts have substantially higher thermal conductivity (and lower Pr) than dielectric melts. Appendix D gives a list of important thermophysical properties and corresponding Prandtl numbers of selected melts commonly used in crystal growth and flow visualization.

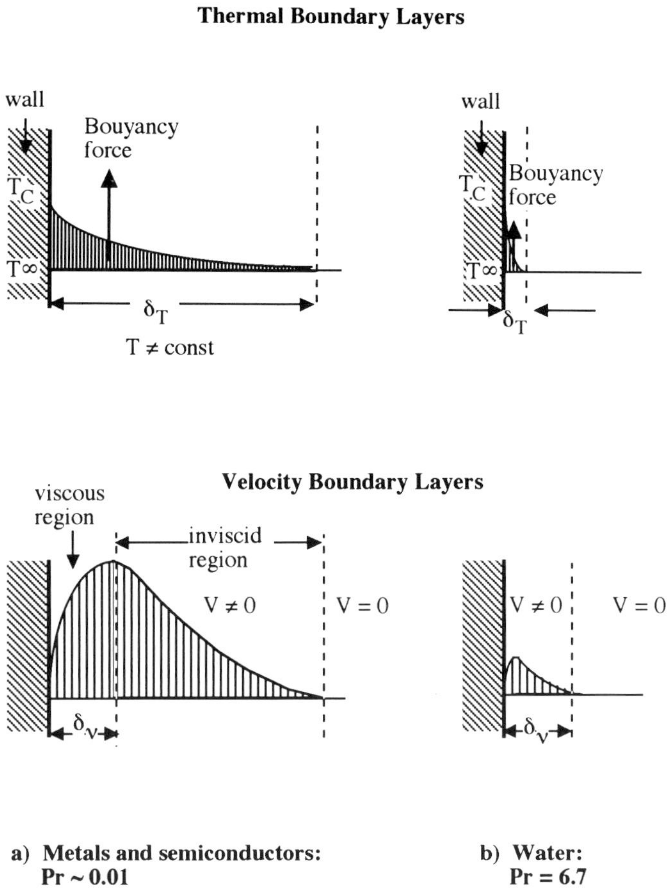

Fig. 28. Structure of boundary layers in low- and high-Prandtl number fluids.

Figure 28 shows the difference in the boundary layer structure in low- and high-Prandtl number fluids. In high-Pr number fluids, thermal boundary layers are thin. Buoyancy forces, acting close to the wall are balanced primarily by viscous forces. In contrast, in electrically conducting low-Pr number fluids, thermal boundary layers extend deep, beyond δ_v, into the inviscid flow. Buoyancy forces, which act primarily outside of the momentum boundary layer, are restrained mainly by inertial forces. Figure 29 shows the effect of Prandtl number on the flow in a square cavity heated from the side ($Gr = 10^7$). It can be seen that the non-dimensional velocity strongly depends on Pr.

Here we have discussed the key dimensionless parameters, needed even for a crude dynamic similarity. The list of non-dimensional parameters becomes longer if physically more complex situations are considered. Table 1 contains a list of non-dimensional parameters which may be of key importance in particular model experiments. For example, in studies of segregation and mass transfer in the melt, the Schmidt number

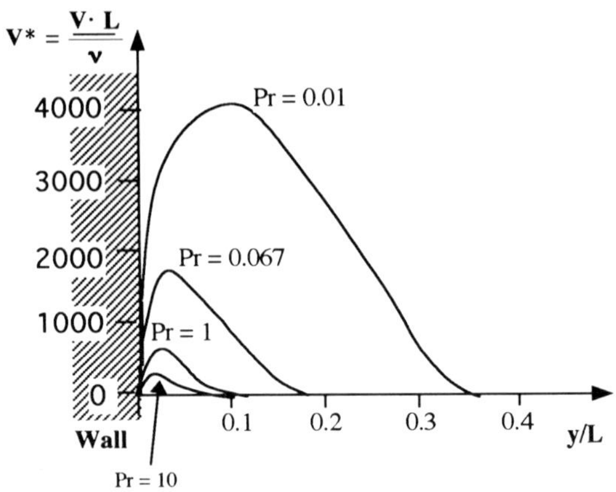

Fig. 29. Dimensionless velocity v^* close to the heated wall in a differentially heated square cavity as a function of the dimensionless distance y/L for various Prandtl numbers. The Grashof number is fixed, $Gr = 10^7$.

and the solutal Rayleigh numbers are of great importance. These non-dimensional parameters will not be discussed here because, up to the present segregation studies were always performed in real growth experiments. For more details on dimensionless groups in crystal growth see for example the textbook of Rosenberger [50].

3.2. Semiconductor and metallic melts (low-Prandtl number fluids)

Conventional flow visualization of semiconductor and metallic melts is not possible because of the presence of free charge carriers which absorb infrared, visible and ultraviolet radiation. Furthermore, the melt rapidly dissolves metals preventing the use of metallic probes.

Data obtained in model experiments with dielectric high-Pr fluids, can not be related to convection in low-Pr semiconductor melts (because of the lack of dynamic similarity). Therefore, our knowledge of free convection in semiconductor melts is based on the results of model experiments with low-Pr melts using alternative methods for flow detection, which are summarized next.

Kakimoto et al. used X-ray radiography and a tracer to visualize the convection in Si melts [51, 52]. The tracer was made out of a 0.5 mm diameter, 0.5 mm long tungsten cylinder, coated with fused silica and carbon. The 300 g silicon melt was contained in a 7.5 cm diameter crucible (melt height was $H = 4.5$ cm). The observed tracer path as shown in fig. 6 was time-dependent and asymmetric, with and without crystal/crucible rotation. Without rotation, the average velocity of the tracer projected on the vertical plane was 1 to 2 cm/s (corresponding to $Re = H\,V/\nu = 1280$ to 2500), see also section 4.4. The Grashof number of the melt was $1.\hat{6} \times 10^8$. Using X-rays, Kakimoto et al. were able to observe the solid-liquid interface during growth [53, 54].

Hirata and Inoue [55] used three 0.3 mm Pt–13% PtRh thermocouples, covered with 1.5 mm diameter quartz tubes, to measure the temperature field in a 9 cm high, 15 cm diameter silicon melt. They were able to determine the flow pattern and velocities by measuring the time of transfer of temperature fluctuations between the thermocouples. The measured spectrum of temperature fluctuations and high asymmetric flow velocities, $V = 5$–30 cm/s ($\mathrm{Re} = VH/\nu = 12\,800$–$77\,000$) suggest that the flow in the melt must have been turbulent. A typical experimental set up for studies of flow and temperature distribution in a production Czochralski apparatus is shown in fig. 30.

Müller et al. [57] and Baumgartl et al. [58] used eight 0.8 mm diameter thermistors to study the temperature and the flow field in gallium melts ($\mathrm{Pr} = 0.02$). The

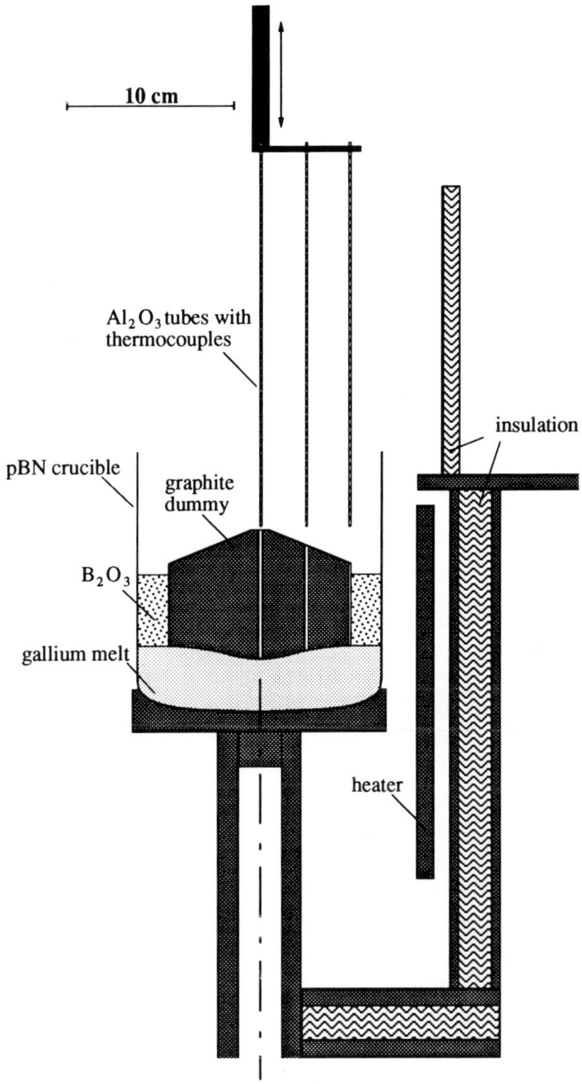

Fig. 30. Experimental set up for thermocouple measurement of temperature distribution in a melt and a dummy crystal, in a commercial production LEC puller. From [56].

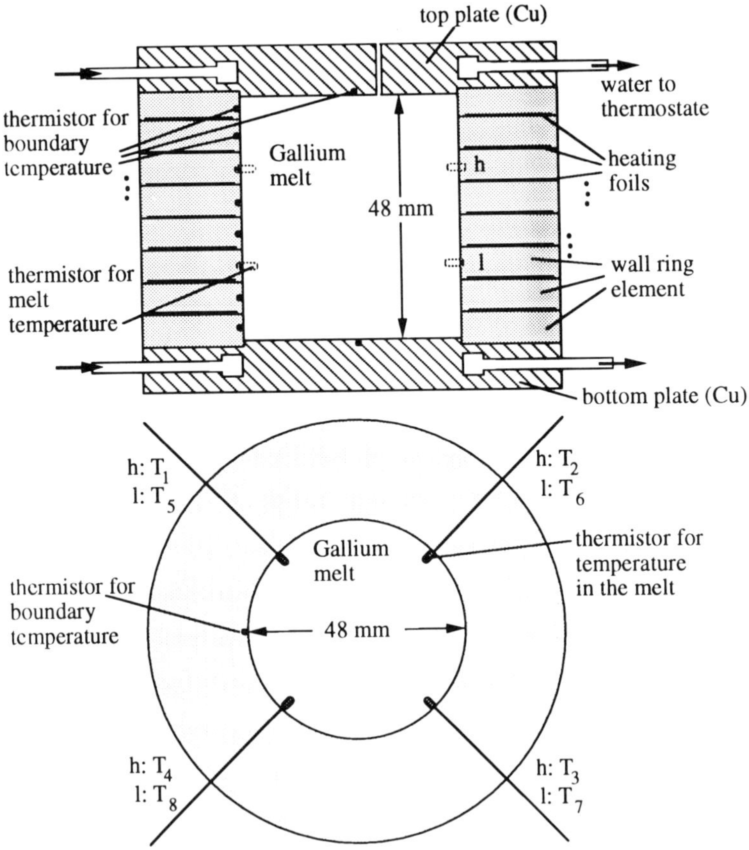

Fig. 31. Schematic drawing of a test cell for fluid flow studies in opaque metal melts by arrays of temperature probes (thermistors), with isothermal top and bottom; the temperature profile of the cylinder wall can be varied by a set of ring heaters. From [5].

thermistors were mounted in the melt, 2 mm from the wall of the vertical cylindrical test cell (fig. 31). Their model experiment was characterized by four non-dimensional groups: the melt aspect ratio, the Prandtl number, the Rayleigh number based on the vertical temperature difference, and the wall Rayleigh number Ra_W based on the radial temperature difference. Different axial and radial Rayleigh numbers imposed at the melt boundaries, allowed for a comparative study of buoyancy-driven convection in various crystal growth configurations. More results of this investigations are described in section 4. For $Ra = 0$, in the investigated range of wall Rayleigh numbers, $0 < Ra_W < 4 \times 10^5$, the flow was always non-axisymmetric, consisting of a single roll. The transition to unsteady convection occurred at $Ra_W = 1.9 \times 10^5$.

To avoid perturbing convection in molten gallium, Hurle et al. used very fine thermocouple probes (0.25 mm o.d. silica capillaries, through which 0.05 mm chromel–alumel thermocouple wires were passed) [7]. Gallium was contained in a horizontal rectangular boat, heated from the side. Temperature measurements were made at 128 positions on a three-dimensional grid, with and without the presence of a transverse magnetic field. The spatial variation of the phase of oscillations in the melt was recorded. Hurle et al. related the frequency of oscillations to the melt dimensions. They concluded that a critical velocity of circulation has to be established in order to achieve oscillations in the melt. The critical temperature difference (i.e. the critical Ra number) at which oscillations first appeared, was found to increase linearly with the square of the transverse magnetic field.

Wallace et al. demonstrated that some properties of electrically conducting fluids can be measured using eddy current techniques [59, 60]. They successfully applied this technique to measure temperatures and local thermal conductivities in semiconductor melts and crystals. The most recent experiments of Wallace et al. indicate that eddy currents may be used effectively to monitor the growth interface [61], and to measure the convective velocities in the melt [62].

3.3. Transparent melts (high-Prandtl number fluids)

Melt flow in high temperature semiconductor and oxide melts has been simulated and visualized in a number of model experiments with dielectric transparent liquids at low temperatures. The frequently used transparent liquids include: water ($Pr \approx 6$), sugar solutions of water ($6 < Pr < 200$), water and glycerol mixtures ($6 < Pr$), and tetradecane

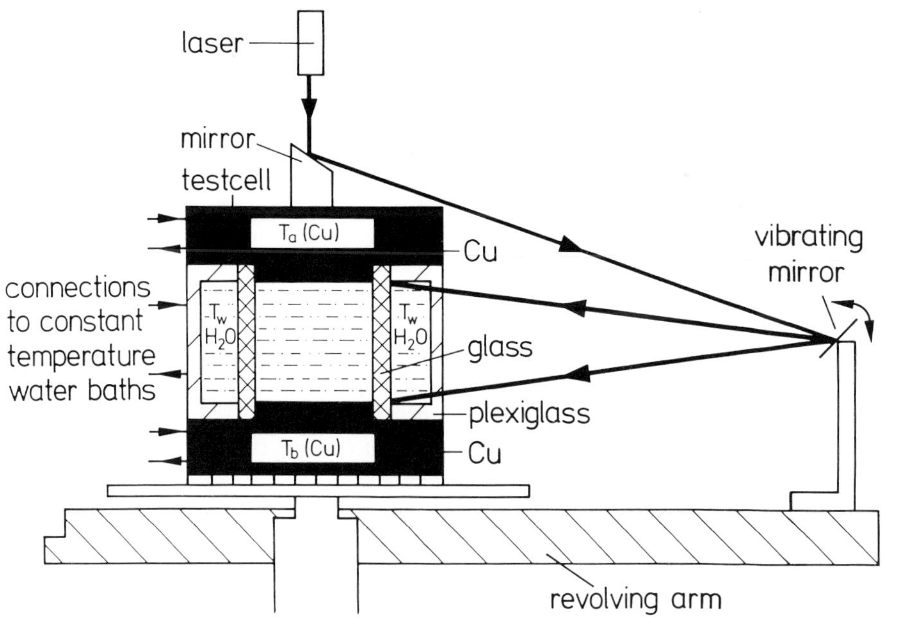

Fig. 32. Sketch of a test cell for flow visualization using a light cut technique in vertical cylindrical configurations with transparent fluids. From [63].

(Pr ≈ 26). The Prandtl number of these fluids matches well with the Prandtl number of important oxide melts at high temperatures (Pr > 1), so that similarity in thermal boundary conditions and radiative heat transfer in the melt remain as the main challenge in model experiments of oxide growth.

Flow visualization is often carried out by the light cut technique. The principle together with a test cell is shown by the sketch of fig. 32. The light cut is produced by the aid of a laser beam which is deflected by a vibrating mirror. The light is scattered by small particles (≈ 200 μm) which are suspended in the test fluid (e.g. polystyrol particles in water [63]).

Because of the inability to match the Prandtl number of semiconductor melts (0.1 < Pr < 0.01), model experiments with dielectric liquids give at best a visual, non-quantitative, indication of flows in semiconductor melts. In dielectric melts, the non-dimensional velocities (i.e. Re) are lower, while stability and symmetry of the flow is higher.

Since melt convection is governed by a system of coupled non-linear differential equations and boundary conditions, it should be sensitive to the dimensionless parameters. Ristorcelli and Lumly identified at least ten sources of possible instabilities arising from interactions between various body forces in the melt [64]. Therefore, the observed flow pattern and the velocity field in a Czochralski melt, is the outcome of a non-linear combination of ten instability mechanisms. Even small deviations from similarity may cause large differences in flow behavior.

3.3.1. *Parametric studies of melt convection*

Jones [19, 23] performed a model experiment on convection in Czochralski melts based on rigorous similarity considerations. He used a solution of water and glycerol to model silicon (Pr = 0.014) and bismuth–silicon–oxide melts (Pr = 26). He considered five non-dimensional groups: the aspect ratio R_X/H, the Eckman number ($0.9 \times 10^{-4} < E < 26.7 \times 10^{-4}$), the Grashof number ($0 < \mathrm{Gr} < 4.9 \times 10^{8}$), the Prandtl number ($4.5 < \mathrm{Pr} < 7$) and a dimensionless parameter scaling crystal to crucible rotation rates

$$\mathrm{Ro} = 2\frac{\Omega_X - \Omega_C}{\Omega_X + \Omega_C} \tag{3.9}$$

introduced by Hide and Titmen [65] as the Rossby number ($-3.25 < \mathrm{Ro} < 10.4$). Conventionally, the Rossby number is defined as the ratio of convective acceleration to the Coriolis force [48]

$$\mathrm{Ro} = \frac{V}{\Omega_X R} = \frac{\text{convective acceleration}}{\text{Coriolis force}}. \tag{3.10}$$

The flow was visualized by neutrally buoyant polystyrene beads, 0.5 mm diameter. Jones studied six flow regimes occurring at different crystal and crucible rotation rates and heat flux loads. Some of the visualized flow patterns were axisymmetric, some were not. He observed the spoke pattern and attributed it to a buoyancy-driven instability. Jones concluded [23] that the experiment cannot model exactly the melt in a crystal puller because of the difference in the Prandtl number, and because of the lack of similarity in thermal boundary conditions (the temperature of the crucible wall was kept uniform while the base was insulated; because of the low temperature of the model fluid, there was no cooling of the melt surface by thermal radiation).

Müller et al. [63] developed a model experiment for vertical Bridgman growth which accounted for both bottom and side heating (or cooling). The hydrodynamic state of the flow in the cylindrical crucible was described by four non-dimensional groups: Prandtl number (Pr $= 6.7$), radial and axial Rayleigh numbers and the aspect ratio $H/R_C = 1$.

The flow was visualized using a light cut technique, and 200 µm particles of polystyrol. The light cut was produced by deflecting the laser beam with a vibrating mirror. Müller et al. obtained the stability diagrams for a wide range of horizontal and vertical Rayleigh numbers, while the aspect ratio and the Prandtl number were kept constant (see section 4). For zero or for a stabilizing axial temperature gradient, the flow remained steady and axisymmetric up to Ra $\sim 10^7$. Comparison of these findings obtained at Pr ≈ 6 with Müller's and Baumgartl's findings obtained with Ga in a similar experimental configuration (see the previous section 3.2), demonstrates the powerful influence of Pr on the state of melt convection.

3.3.2. *Studies of instabilities and interface transitions*

Brice et al. grew bismuth oxide stabilized by small additions of SiO_2 ($Bi_{12}SiO_{20}$) and reported that above a critical rotation rate, stable growth was impossible, because the diameter of the crystal varied rapidly [66]. Similar but less catastrophic, instabilities were reported for gadolinium gallium garnets ($Gd_3Ga_5O_{12}$) [35, 67]. As the crystal rotation rate or the crystal diameter was increased, the interface, initially very convex, melted back to become nearly planar. The interface transition was accompanied by a substantial rise in melt temperature, below the crystal. This interesting instability, related to transients in melt convection, became the object of most model experiments with dielectric fluids. Most reports agree that interface instabilities result from changes in flow patterns caused by shift in relative importance of buoyancy and rotational forces. Still the detailed nature of interface transitions is not yet understood [9].

The Prandtl number of $Gd_3Ga_5O_{12}$, has not been measured. It is expected to be in the range of 1 to 30, closer to 30 [18]. Brandle [31] demonstrated that $Gd_3Ga_5O_{12}$, has similar viscosity and therefore Prandtl number as $Bi_{12}SiO_{20}$ and a number of other oxides. According to Kimura $Bi_{12}SiO_{20}$ has Pr ≈ 26 [68, 69].

The Prandtl number of high temperature oxide melts is closely matched by that of dielectric fluids. However, the strong radiative cooling from the melt surface of oxide melts (melting point $T_{mp} \sim 1000$ to 2000°C) cannot be matched in low temperature experiments. Furthermore, in most model experiments, only the value of the crystal rotation rate was reported. The non-dimensional parameters were not reported or matched.

Miller and Pernell [18] photographed the "wheel spoke" pattern, using water and an AQ-1000 reoscopic fluid. Their photographs (fig. 17) show flow patterns similar to the ones observed in $Gd_3Ga_5O_{12}$ melts. Kimura used tetradecane (Pr ≈ 30) [68] and water [69] to model the flow in $Gd_3Ga_5O_{12}$ and $Bi_{12}SiO_{20}$ (Pr ≈ 26) melts. The flow was visualized by illuminating 200 mesh polyethylene particles. The particles were illuminated by a vertical, plane of light, 3 mm wide and photographed. At the same time, the melt temperature was measured. He detected transients in melt temperature provoked by the increase in crystal rotation. The critical rotation rate was $\Omega_X \sim 60$ rpm. Grashof numbers of these experiments were not reported.

Carruthers [70] suggested that the transient state occurs when buoyancy and rotational forces have equal importance

$$\mathrm{Gr} = \mathrm{Re}^2, \tag{3.11}$$

where the characteristic length scale of Gr and Re are the crucible radius and crystal diameter respectively. Using this simple model, Carruthers predicted the critical diameter, observed in Takagi's growth experiments with $Gd_3Ga_5O_{12}$, as a function of the critical rotation rate. The effect of the capillary forces on transitions in melt convection (and thus interface instabilities) was not considered.

The model experiments of Lamprecht et al. performed with $NaNO_3$ ($\mathrm{Pr} = 9.25$, $T_{mp} = 306.8°\mathrm{C}$), shows that Marangoni convection should not be ignored a priori [71]. In this experiment the flow in the 35 mm diameter model melt was visualized using hollow spherical quartz particles (o.d. 20–70 μm) as tracer particles. The temperature of the melt surface was measured contactlessly, by an IR bolometer. A half of meridional plane was illuminated with a 1 mm thick light sheet. Convection was observed using an inclined video camera or stereo microscope. Because of the action of capillary forces, the flow velocity at the free melt surface was 18 mm/s, approximately five times higher than near the hot crucible walls ($\mathrm{Gr} = 2 \times 10^6$). This demonstrates that thermocapillary forces are dominant at the melt surface even at high Gr. The value of the Marangoni number was not reported; it can be estimated from the given dimensional parameters as $\mathrm{Ma} \sim 6.5 \times 10^4$.

3.4. *Molten salts ($0.2 < Pr < 1$) at high temperatures*

Using a special procedure schematically shown in fig. 33, Krzyminski and Ostrogorsky [22] were able to visualize the flow in Czochralski melts under realistic high temperature boundary conditions.

In Czochralski furnaces, the tubular heater surrounds the crucible, preventing side viewing the melt. This obstacle was overcome by using a "sliding" tubular resistance heater that could be rapidly lowered for the short time needed to take a photograph, as shown in fig. 33. Flow visualization was realized using streak photography. The flow was visualized using the light-cut technique. The tracer particles (fine aluminum powder) were illuminated by a vertical plane of laser light.

The best results were obtained using a mixture of sodium chloride and calcium chloride at 900 K, having $\mathrm{Pr} = 0.6$. At $\mathrm{Ra} = 2.3 \times 10^5$ and no rotation (fig. 33), the maximum melt velocity was 0.23 cm/s corresponding to $\mathrm{Re} = 64$. Despite the relatively low Re, the flow was non-symmetric and unsteady. This can be seen from the four photographs in fig. 15, which were taken by rapidly lowering the tubular heater, in 10 s intervals. Figure 16 shows the top view of the melt, which also shows asymmetric flow.

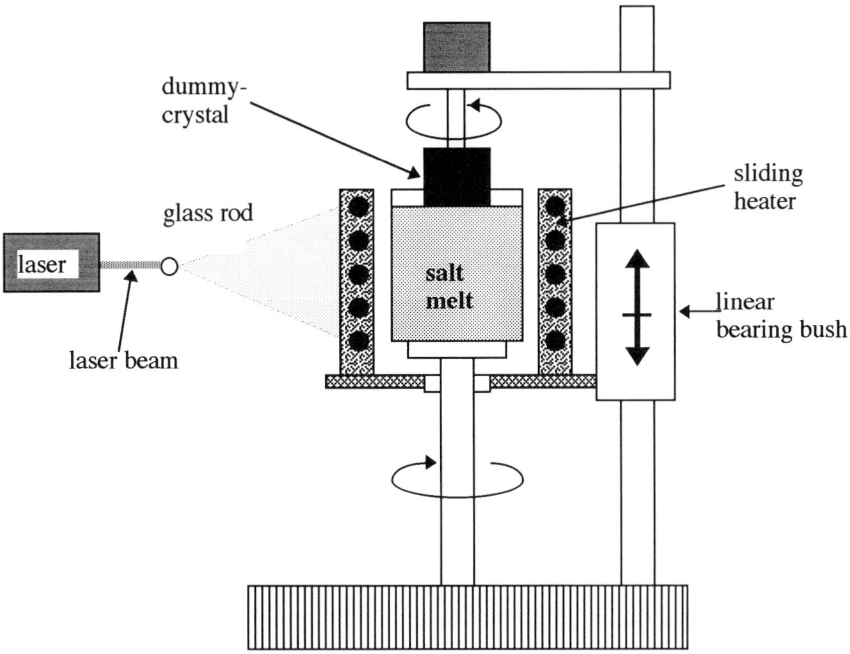

Fig. 33. Schematic representation of experimental apparatus that provides a side view of a transparent Czochralski melt at high temperatures. From [22].

While the Prandtl number of molten salts is more than one order of magnitude lower than that of water, it still remains one order of magnitude higher than the Pr of semiconductor melts. According to preliminary experiments, molten salts are semi-transparent to infrared radiation at high temperatures, and may be made optically thick by addition of impurities. In optically thick materials, the photons transfer thermal energy in a similar way as electrons in electrically conducting melts. According to the Rosseland diffusion approximation, radiative conductivity can be defined as [72]

$$k_r = \frac{16\sigma T^3}{3a_r},$$

where σ is the Stefan–Boltzmann constant and a_r is the absorption (or extinction) coefficient of the optically thick material. Therefore, the similarity in thermal energy transport may be used in order to achieve dynamic similarity with semiconductor melts, by increasing the effective conductivity of the salt melts (see Appendix A.1),

$$k_{eff} = k + k_r,$$

where k is the molecular conductivity of a dielectric fluid and thereby effectively reduce the Prandtl number. This approach may improve the dynamically similarity of model experiments of semiconductor Czochralski melts [73].

3.5. Induced magnetic fields

It can be concluded from the preceding discussion, that the detection of flow in opaque semiconductor and metal melts is difficult. Recently a new method has been studied in the group of one of the authors, which uses magnetohydrodynamics [74]. If one applies an external static magnetic field B to an electrically conducting melt volume, a small magnetic field b is induced in the melt caused by the convective flow due to the magnetohydrodynamic coupling. B must be small enough in order not to damp the flow. At the same time B should be as large as possible to allow experimental detection of the induced field b, since b is only of the order of 10^{-6} B.

This effect was demonstrated by the authors for the case of unsteady buoyancy-driven convection in a vertical cylindrical test cell containing liquid gallium (fig. 34). Firstly it was shown by a three-dimensional numerical simulation, that the temporal variation of the induced magnetic field b correlates with the flow field and the

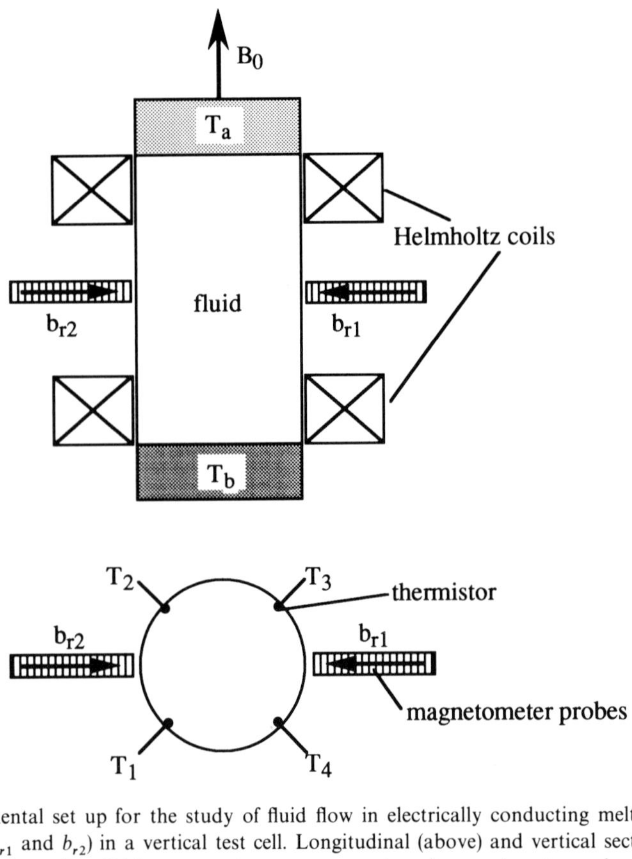

Fig. 34. Experimental set up for the study of fluid flow in electrically conducting melts by an induced magnetic field (b_{r1} and b_{r2}) in a vertical test cell. Longitudinal (above) and vertical section (below) with positions of thermistors (T_1–T_4) for temperature measurements and magnetometer probes for measurement of the radial components (b_{r1}, b_{r2}) of the induced fields. From [74].

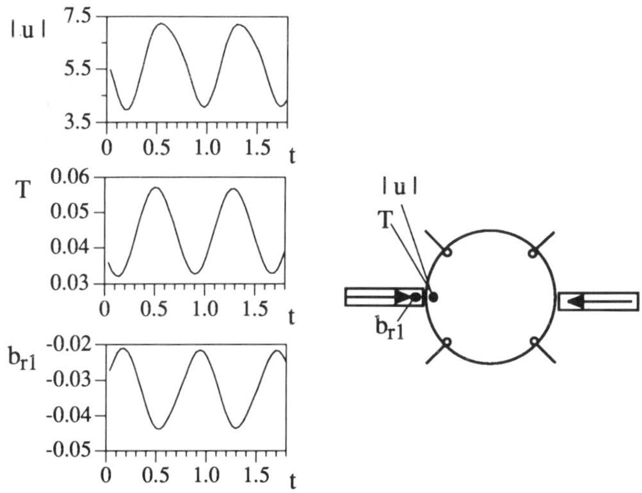

Fig. 35. Calculated flow velocity u and temperature T versus time for a monitor point inside the melt cylinder and induced field b_{r1} outside the melt for a given unsteady hydrodynamic state ($Ra = 6 \times 10^4$, $h/d = 1.9$); applied external axial field $Ha = 5.8$. From [74].

temperature in the melt (fig. 35). Then experimental proof was obtained by measuring the temperature fluctuations in the melt (by thermistors) and the induced magnetic field b by magnetometer probes outside the test cell (see fig. 36). The excellent agreement of the flow behavior (temperature signals) and detected field b is also demonstrated in fig. 36 by the corresponding Fourier spectrum. Furthermore, the authors discuss in their paper the possibilities of determining the whole flow field, e.g. in a Czochralski melt from magnetic measurements.

In any case this method has the advantage that neither electrical nor mechanical contacts with the fluid are necessary. These advantages could be particularly important in flow investigations involving high temperature or corrosive melts which are typical for semiconductor crystal growth. As a typical application in crystal growth the onset of time-dependence in the melt flow could be detected by the reaction field b. Then the crystal growth parameters (including applied magnetic fields) could be changed and the instability could be counteracted in order to maintain stationary convection. Thus a steady melt flow with optimum mixing conditions could be obtained.

4. *Experimental and numerical results of convective flows*

This section gives an extended review of the status of knowledge of convective flows in the various types of melt growth configurations by comparing quantitatively experimental and numerical results. The outline is as follows: firstly (section 4.1) only buoyancy-driven convection is treated, i.e. melt growth configurations of the vertical Bridgman (VB)/gradient freeze (VGF) type; secondly (sections 4.2 and 4.3) thermocapillary convection in floating zone and horizontal open boat type configurations;

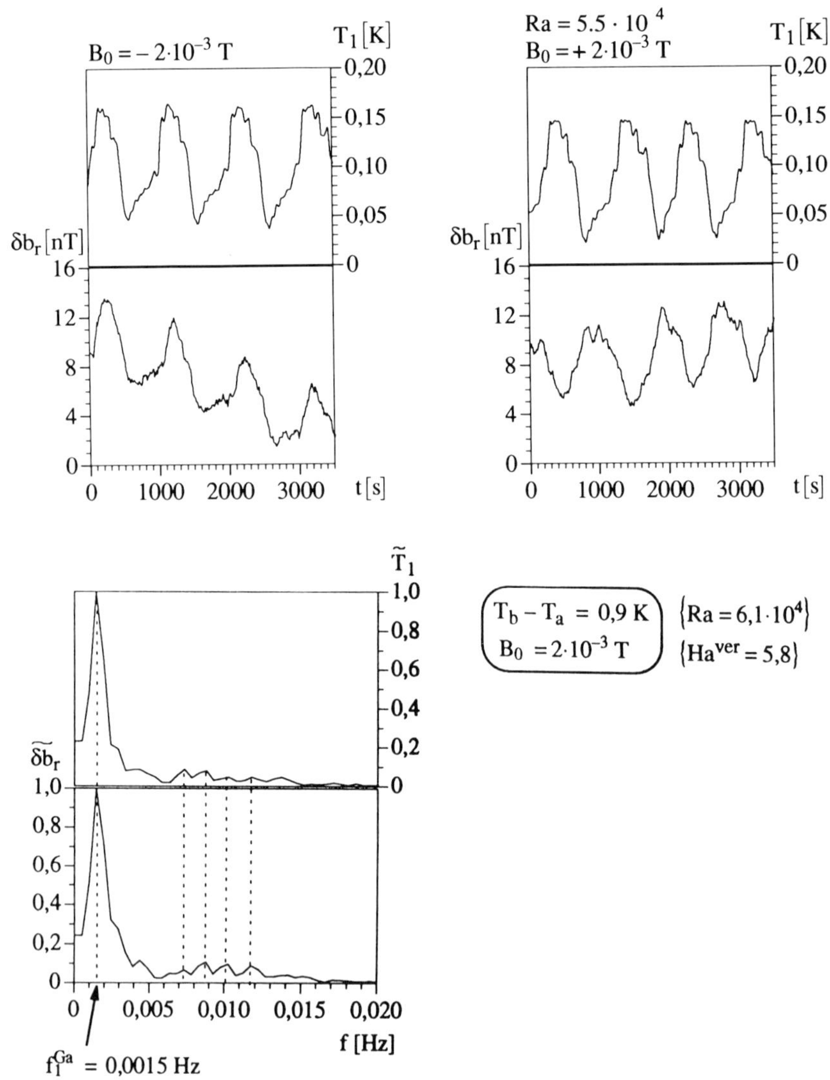

Fig. 36. Experimental result with the conditions of fig. 35, i.e. temperature difference between bottom and top 0.9 K (Ra = 6.1×10^4) and external axial magnetic field $B_0 = 2 \times 10^{-3}$ tesla (Ha = 5.8). From [74].

finally buoyancy, thermocapillary and forced convection (rotation) which are present in the Czochralski configuration (section 4.4) are discussed. The reader is also referred to the special articles in this handbook on modelling of the Czochralski configuration by Dupret and Van den Bogaert (chapter 15) and on the floating zone configurations by Bohm et al. (chapter 4).

4.1. Buoyancy convection in vertical cylindrical melt containers

4.1.1. Fundamental considerations

Buoyancy is present in all terrestrial melt growth configurations as introduced in section 2.2. It is very often the dominating mechanism driving the flows and determining the flow character. Its behavior is systematically studied in vertical cylindrical configurations without any free surface and forced convection. The fundamental question whether flows in such a configuration become unsteady can be discussed qualitatively by using the results of the pioneering work of Krishnamurti [75] on the Rayleigh–Benard [76] configuration. This model is extended by considering the finite geometry of the cylinder in terms of the aspect ratio h/d (see fig. 37). The state of convection is completely characterized in such a configuration by the three dimensionless numbers Ra, Pr and the aspect ratio h/d (for a definition of these see table 1). It can therefore be represented in a three-dimensional coordinate system as shown in fig. 38. In the Ra–Pr-plane of this figure the experimental results of the investigation in a Benard cell ($h/d \ll 1$) are plotted according to Krishnamurti [75]. The Pr-axis considers the various relevant thermophysical properties of the melts. It shows clearly that buoyancy-driven convection readily becomes unsteady or even turbulent with small driving forces in the melt growth of semiconductors and metals because of their low Prandtl number. The influence of the aspect ratio will be quantitatively discussed in section 4.1.2.

The model of fig. 37 can only be applied for $\mathrm{Ra} > 0$ i.e. melt growth configurations with predominantly vertical temperature gradients being anti-parallel to the direction of gravity (e.g. vertical Bridgman/gradient freeze with top seeding, Czochralski with bottom heating). This restriction does not exist for a model which uses a parabolic temperature profile [58] along the cylindrical sidewall, (shown in fig. 39). As the parabola is determined by the three values of the temperature of the bottom (T_b), top (T_a) and in the half of the height (T_w) of the cylinder wall, we can define an additional

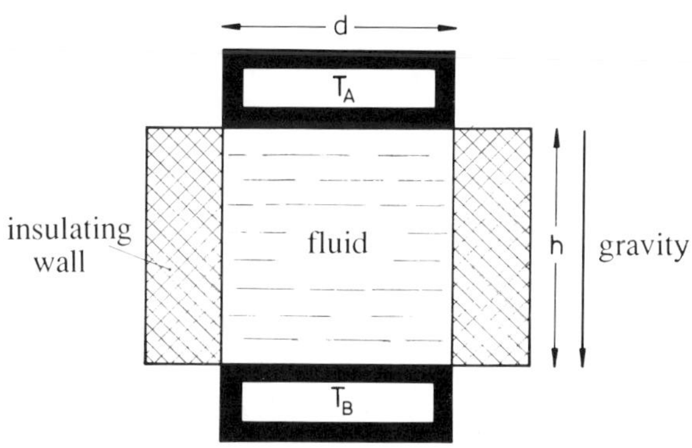

Fig. 37. Sketch of the model of vertical cylindrical melt configurations with heated/cooled end plates (T_A, T_B) and insulating wall.

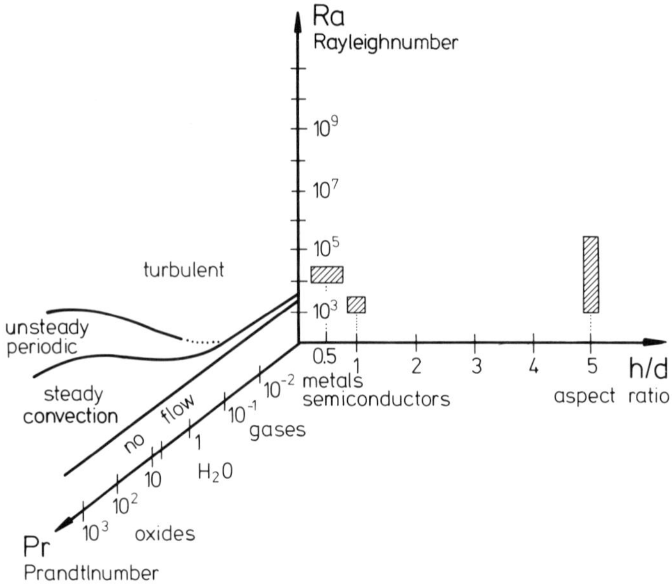

Fig. 38. Three-dimensional representation of Ra versus h/d and Pr. The results plotted in the Ra–Pr-plane belong to the limit $h/d \ll 1$, i.e. the Rayleigh–Bernard configuration. From [1].

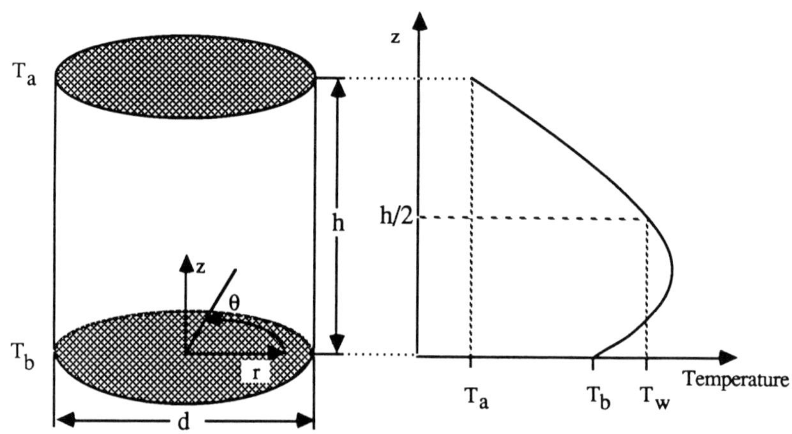

Fig. 39. Sketch of the model of vertical cylindrical melt configurations with heated/cooled end plates (T_a, T_b) and a parabolic temperature profile along the side wall. From [58].

dimensionless group called the "wall Rayleigh number Ra_w" [58, 63]:

$$Ra_w = \frac{\alpha g 2h^4(T_w - T_m)/d}{\nu\kappa}, \quad \text{with } T_m = \tfrac{1}{2}(T_a + T_b).$$

Using this definition of Ra_w and of the earlier Ra one can plot schematically the thermal boundary conditions of important melt growth configurations with regard to buoyancy-driven convection, as shown in fig. 40:

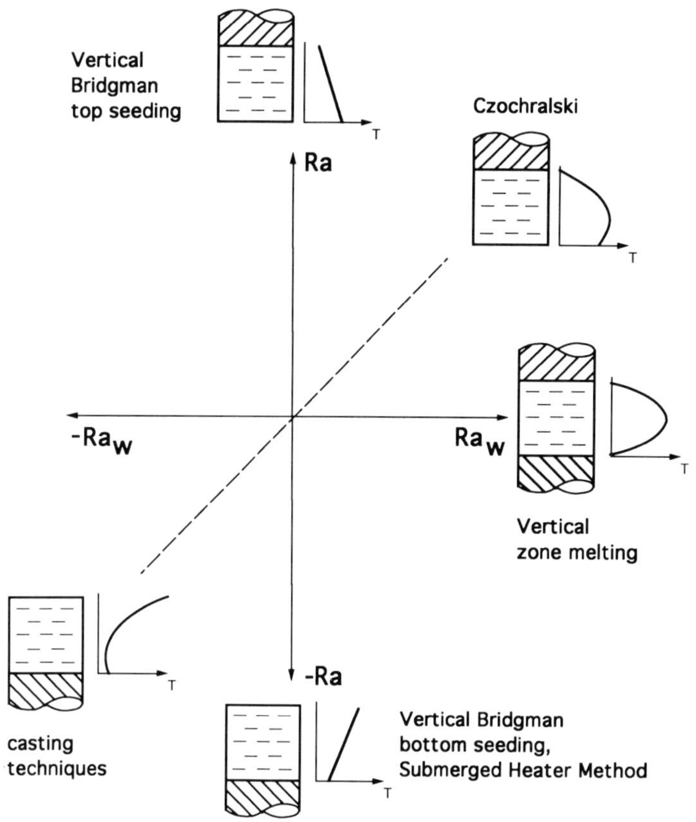

Fig. 40. Representation of different vertical crystal growth configurations in a plot of Ra versus Ra_w for the model configuration of fig. 39. From [58].

- *vertical Bridgman/gradient freeze* $|Ra_w| < |Ra|$; $Ra > 0$ means top seeding, $Ra < 0$ bottom seeding,
- *Czochralski* and *Kyropulos* technique: $Ra_w \approx Ra > 0$,
- *vertical zone melting* with top or bottom seeding: $Ra_w > 0$, $Ra = 0$,
- *block casting* (e.g. used for fabrication of Si for solar cells [77]): $Ra_w \approx Ra < 0$; $Ra < 0$,
- *heat exchanger* [78] and submerged heater [79] method: $|Ra_w| \ll |Ra|$; $Ra < 0$; $Ra_w > 0$.

The representation of fig. 40 contains some simplifications which should be kept in mind, when using it for a discussion of convective behavior in melt growth:

- The liquid–solid phase boundary is represented by an isotherm which is not moving, i.e. the Stefan problem is neglected.
- Only thermal buoyancy convection exists, effects of rotation and Marangoni convection (originating from free surfaces) are neglected.
- The aspect ratio h/d is fixed, here for example $h/d = 1$.

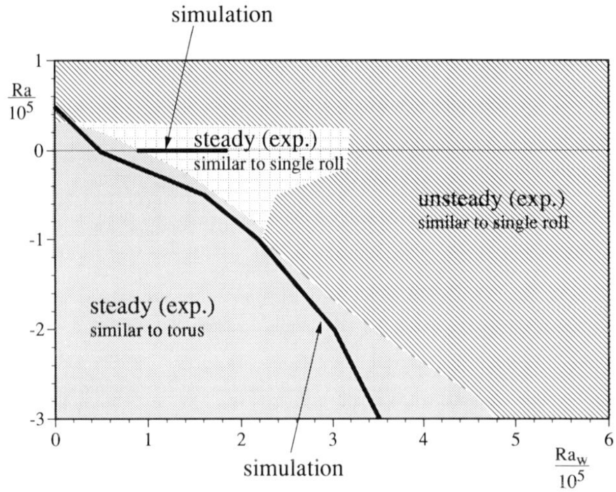

Fig. 41. Experimentally observed flow regimes (shaded areas) and numerically determined separation lines (solid). From [58].

The transition to unsteady flow has been studied for various ranges of the parameters Ra and Ra_w both by numerical simulations and experimental investigations. Typical results for the flow configurations and the boundary for the transition to unsteady flow are given in dependence of Ra and Ra_w in fig. 41.

As an example we discuss the results of numerical and experimental modelling and crystal growth for the case of vertical zone melting [58] ($Ra = 0$; $Ra_w > 0$). For any imposed horizontal temperature gradient, i.e. $Ra_w \neq 0$, convection sets in without any threshold. An axisymmetric steady convective flow (torus) was found for $0 < Ra_w < 3 \times 10^4$. A transition to a time-dependent (periodic) flow behavior was calculated for $Ra_w = 40\,000 \pm 300$. The toroidal character of the basic flow is still dominating but the flow pattern is slightly deformed and the deformation rotates around the cylinder axis with a period of $t^* = 0.45$. (The asterisk means non-dimensionalized by scaling the real time t with κ/h^2; κ is the thermal diffusivity and h is the cylinder height). This oscillatory mode becomes unstable when Ra_w is increased further and disappears for $Ra_w = 8 \times 10^4$. The re-laminarisation of the flow is connected with a fundamental change of the configuration. Laminar convection consisting of a non-axisymmetric single roll, is found for $8 \times 10^4 < Ra_w < 1.85 \times 10^5$. Unsteady convection with oscillations (frequency $f^* = 5.9$) appears if Ra_w exceeds the value of 1.85×10^5. This flow very soon becomes turbulent with a further increase of Ra_w. A typical pattern of the isotherms caused by the roll can be seen in the vertical cross section depicted in fig. 42.

In model experiments with a test cell using liquid Ga [58] only non-axisymmetric flows consisting of one main roll were observed. A transition to unsteady temperatures was measured for $Ra_w = 1.9 \times 10^5$ with a basic frequency $f = 0.027$ Hz (corresponding to a dimensionless frequency $f^* = 2.86$). Chaotic temperature signals (i.e. turbulence) was found for $Ra_w > 2.5 \times 10^5$. The transition to unsteady flow and the flow symmetry determined by numerical simulation and model experiments are in good qualitative

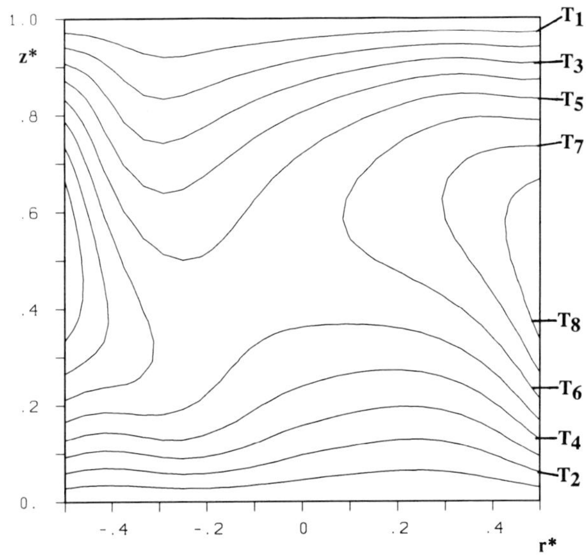

Fig. 42. Calculated isotherms in an axial section of a vertical cylindrical zone melting configuration with an axisymmetric parabolic temperature profile along the sidewall ($Ra_w = 2 \times 10^5$, $Ra = 0$, $Pr = 0.02$, $h/d = 1$). The isotherms are equally spaced between maximum (T_8) and minimum (T_1) temperature. From [58].

agreement with the results of vertical zone melting of GaAs in a closed ampoule [58]. The curves of the striation pattern depicted in fig. 43 which represent shapes of the isotherms defining the phase boundary, are very similar to the calculated ones given in fig. 42. The value of $Ra_w = 4 \times 10^5$ for this experiment corresponds to the condition for unsteady convection, which is $Ra_w > 2 \times 10^5$.

For more details on this subject see [1, 58, 63]. The influence of deviations from axisymmetric boundary conditions is discussed in [58]. For the influence of magnetic fields on the convection in this configuration see [80] and section 6.3.

4.1.2. Vertical Bridgman/gradient freeze configuration with top seeding

In this section we discuss convective flows in the vertical Bridgman (VB)/gradient freeze (VGF) configuration with top seeding. This configuration gives the optimum conditions to study quantitatively buoyancy-driven convection and also has a certain similarity to the Czochralski method.

Bottom seeding is mostly preferred in the vertical gradient freeze configuration instead of top seeding for several practical reasons including the lower potential for driving buoyant convective flows (see section 4.1.3).

The simplest approach for defining the thermal boundary conditions in a top seeding arrangement is to assume insulating (adiabatic) side walls. Then the driving buoyancy can be described simply by using the vertical (axial) temperature gradient or a temperature difference divided by the melt height as sketched in fig. 37. The corresponding dimensionless group is Ra which is often used in practice to characterize the intensity of convective flows in VB/VGF and Czochralski configurations.

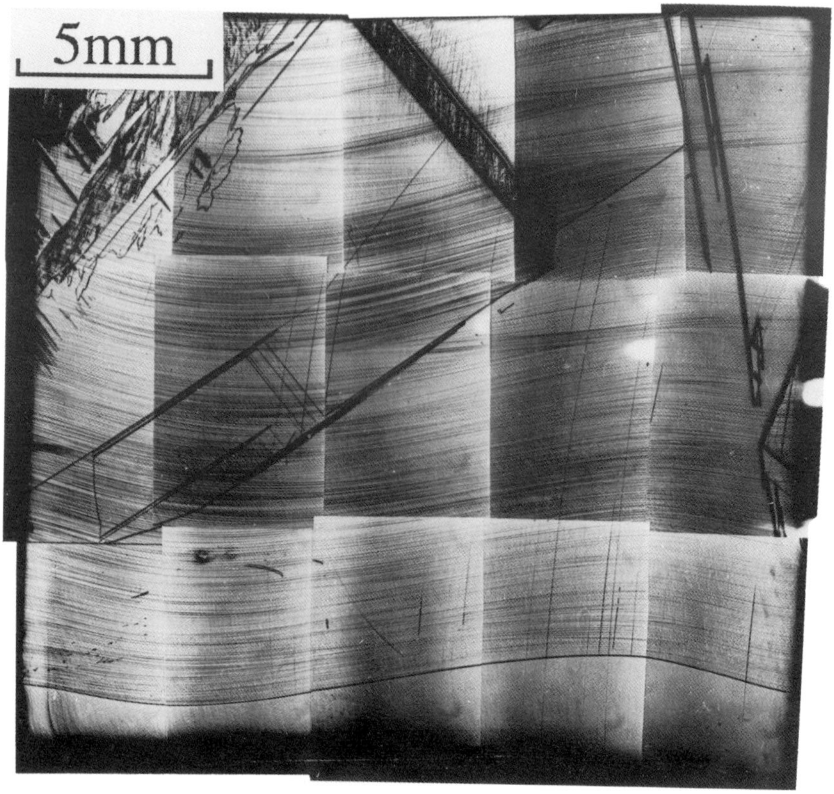

Fig. 43. Longitudinal section through a Si-doped GaAs rod after vertical zone melting. The striation pattern is revealed by the AB etch. From [58].

A more flexible consideration of the thermal boundary conditions is possible by using the approach with a parabolic temperature profile in the side walls by using Ra_w according to section 4.1.1. The stability diagram for $h/d = 1$ is given in fig. 41.

Extensive studies on convection in this configuration have been carried out by Müller and co-workers [1, 57, 58]. Their results on flow configurations and the transition from steady to unsteady flow will be briefly reviewed here. Experimental results were obtained both for metal and semiconductor melts ($Pr \sim 0.01$) and for water ($Pr = 6.7$) [63] for various aspect ratios $0.5 \leq h/d \leq 5$. In parallel with these experiments flows have been calculated numerically [58, 81]. Both experimental and numerical results agree well in showing that axisymmetric (torus-like) rolls exist only for $h/d < 1$. For all larger aspect ratios ($1 \leq h/d \leq 5$) only non-axisymmetric flows are found. These flows reveal small counter-rotating rolls in the region of the phase boundary and the bottom for increasing Ra.

Figure 44 gives an example of a typical three-dimensional flow. It can be recognized in the shape of the calculated isotherms (fig. 45), that a crystal would grow under such

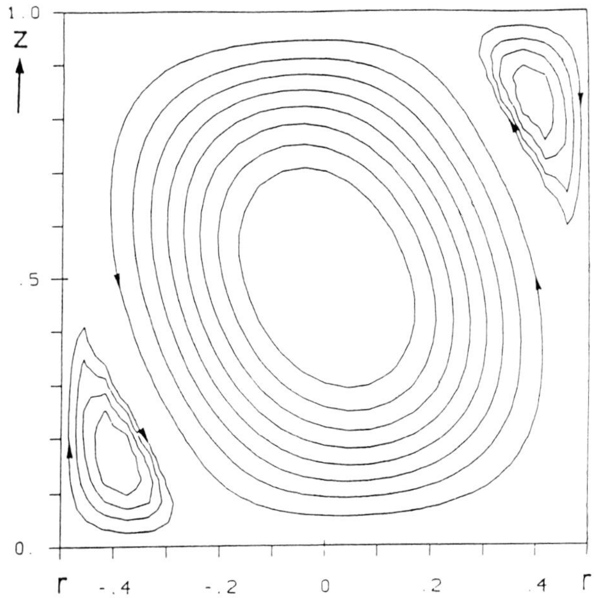

Fig. 44. Calculated flow (streamlines) in an axial section of a vertical cylindrical melt growth configuration with top seeding ($Ra = 10^4$, $Ra_w = 0$, $Pr = 00.2$, $h/d = 1$, steady flow). From [81].

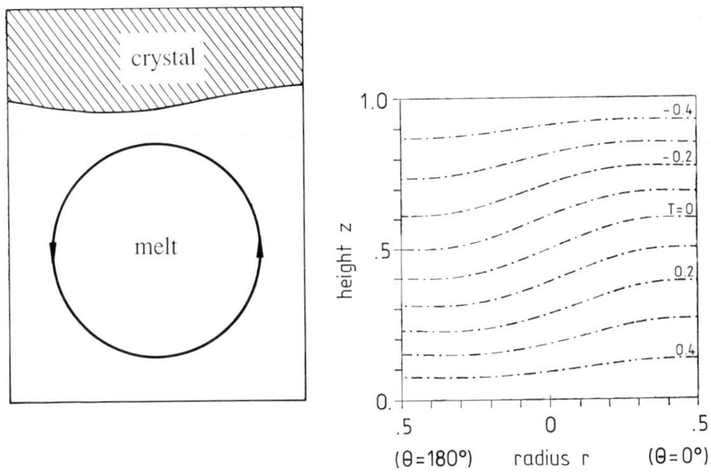

Fig. 45. Left: experimentally determined shape of the phase boundary during directional solidification of Si ($Ra \approx 4000$, $h/d = 1$ and $Pr = 0.023$) and schematic representation of the flow (according to fig. 44). Right: calculated isotherms in a melt for $Ra = 6000$, $h/d = 1$, $Pr = 0.02$. From [1].

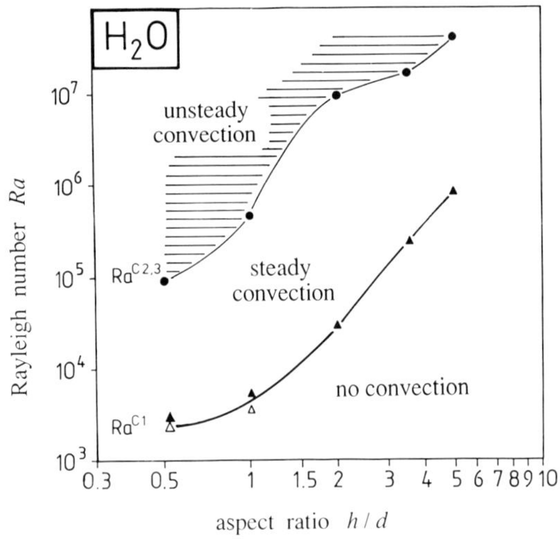

Fig. 46. Flow behavior (stability diagram) of H_2O in vertical cylindrical containers like fig. 37, Ra versus h/d. $Ra < Ra^{c1}$: no convection (open triangles: theoretical, full triangles: experimental). $Ra^{c1} < Ra < Ra^{c2}$: steady convection with flows shown in figs. 3 and 5 (full circles: experimental). $Ra > Ra^{c2}$: unsteady (turbulent) convection (shaded area: experimentally investigated). From [1].

flow conditions with a non-axisymmetric bowed and slightly inclined phase boundary which indeed has been observed in the growth of Si [82].

The stability diagrams for water ($Pr = 6.7$) and liquid Ga ($Pr = 0.02$) are shown in figs. 46 and 47. One recognizes clearly, that in contrast to water, convection flow in metal and semiconductor melts become time-dependent (unsteady) for only a very small increase of $Ra(Ra^{c2} \leq 2 \times Ra^{c1})$. The use and application of the stability diagram for predicting the occurrence of doping striations is described in section 5.1.2.

Studies of phase relations of temperature fluctuations in the vertical Bridgman configuration with top seeding were made by Knuteson et al. [83, 84]. Fourier analysis of the temperature signals shows the increasing complexity of the flow with an increase of the axial temperature gradient.

4.1.3. Vertical Bridgman/gradient freeze configuration with bottom seeding

The bottom seeding configuration is preferred in VB/VGF because of its advantages with respect to the seeding process and less intensive convective flows. In the ideal case of planar isotherms in the melt no buoyancy would result (hydrodynamic stability). But in reality radial/horizontal temperature differences always exist which drive flows without any threshold. Next, results will be discussed with respect to flow configurations, shape of the solid–liquid interface and transition to unsteady convection.

4.1.3.1. Flow configurations and transition to unsteady convection. The symmetry of the flow and its dependence on the thermal boundary conditions together with the critical values for the transition from axisymmetric two-dimensional to non-axisymmetric

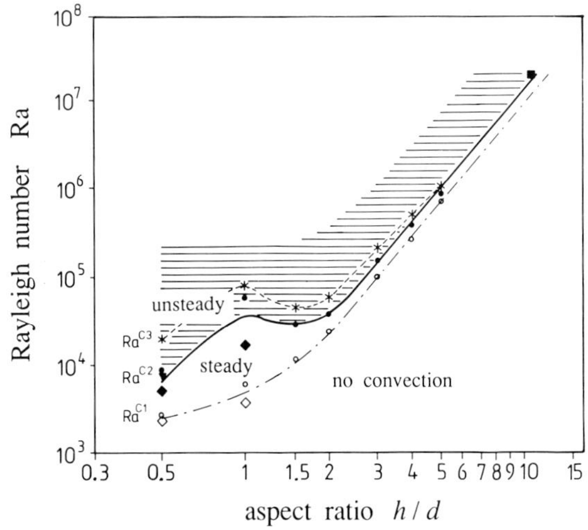

Fig. 47. Stability diagram of buoyancy convection in metal and semiconductor melts ($Pr \approx 0.01$) in vertical cylindrical model configurations with insulating side walls. Experimentally and numerically determined critical Rayleigh numbers Ra^{c1}, Ra^{c2} and Ra^{c3} versus h/d. From [1].

three-dimensional flows determined by experimental studies in a test cell with liquid Ga, is shown in fig. 41 for the aspect ratio $h/d = 1$. The numerical calculations (three-dimensional) confirm the experimental observations and give some more details of the flow patterns [85]. Axisymmetric toroidal flows are found under smaller driving forces, which turn into single roll cells with small eddies in the corners for higher values of Ra_w. The increase of negative Ra stabilizes the flow; increasing Ra_w destabilizes it. Similar behavior is found for $h/d = 2$ as shown in fig. 48, but the flow is more stable [85].

The range of stability is even more extended for $h/d = 4$. Baumgartl [85] calculated for a value of $Ra = -5 \times 10^6$, which is typical for crystal growth (20 mm diameter,

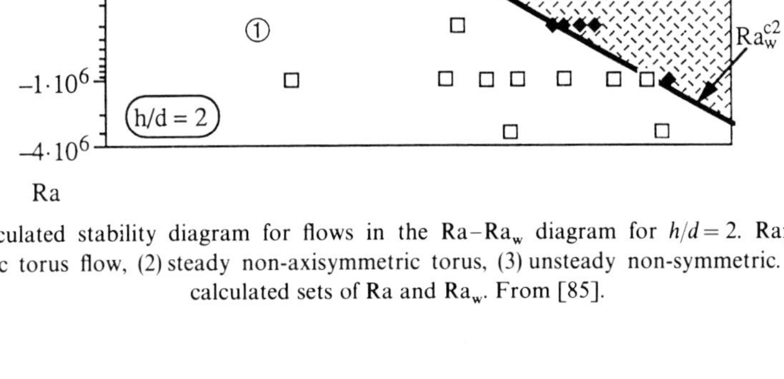

Fig. 48. Calculated stability diagram for flows in the Ra–Ra_w diagram for $h/d = 2$. Range (1) steady axisymmetric torus flow, (2) steady non-axisymmetric torus, (3) unsteady non-symmetric. Signs denote calculated sets of Ra and Ra_w. From [85].

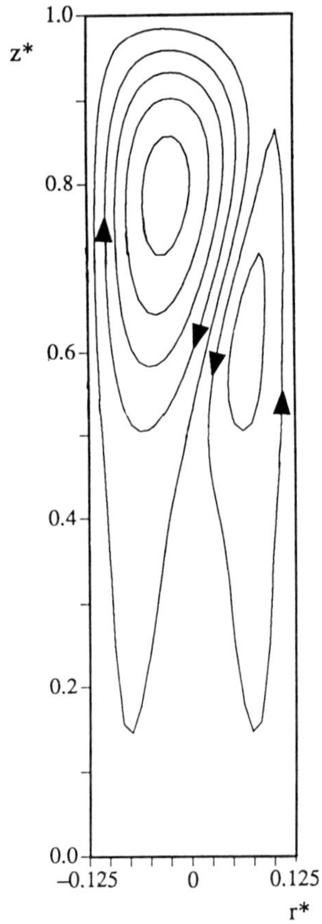

Fig. 49. Calculated convective steady flow in a vertical cylindrical melt growth configuration with bottom seeding ($Ra = -5 \times 10^6$, $Ra_w = 2 \times 10^7$, $Pr = 0.01$, $h/d = 4$). From [85].

80 mm length and axial gradient of melt temperature about 5 K/cm), that the transition to unsteady flow occurred at $Ra_w = 2.5 \times 10^7$. The flow pattern is illustrated by fig. 49. Similar results are obtained if the thermal boundary condition of the cylinder wall is defined by a heat flux $\delta T/\delta n$ instead of the parabolic temperature profile. A corresponding dimensionless number can also be defined. Figure 50 shows a calculated result of the flow pattern obtained by this heat flux-based boundary condition [85] which corresponds to $Ra_w = 2 \times 10^7$. Under these conditions the transition to unsteady flow occurs.

The results discussed so far were based on the assumption of a rigid planar solid–liquid phase boundary. Let us now consider the deformation of the solid–liquid interface due to the heat transport by solving the Stefan problem. Most of the existing literature on numerical solutions presents two-dimensional results [86–89]. Figure 51 gives a typical result of such two-dimensional calculations. The flow in these VB/VGF configurations depends strongly on the thermal boundary conditions. This is illus-

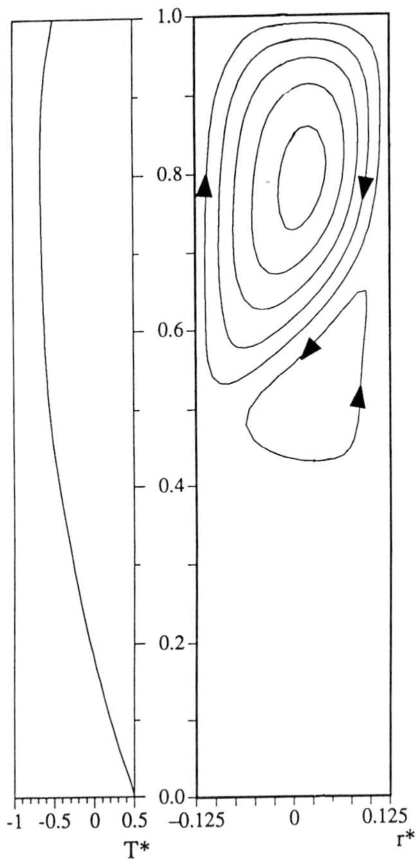

Fig. 50. Calculated convective unsteady flow in a vertical cylindrical melt growth configuration with bottom seeding under conditions of a given parabolic heat flux (corresponding to $Ra_w = 2 \times 10^7$, $Ra = -5 \times 10^6$, $Pr = 0.02$, $h/d = 4$). The resulting temperature profile along the cylinder wall is shown. From [85].

trated by fig. 52 where the temperature gradient along the crystal (2 inch diameter Ge) is varied from 10 to 40 K/cm, whereas the gradient along the melt is kept constant (5 K/cm). The resulting flow velocities are in the range of 0.5 mm/s [90].

The problem with these two-dimensional solutions is that these results cannot give the transition to three-dimensional flows and hence can also not give correct results for the transition to unsteady convection. This is demonstrated by two- and three-dimensional calculations for the same boundary conditions in fig. 53. The comparison of the results in fig. 53 shows that axisymmetric flows obtained from two-dimensional calculations are not too far from the three-dimensional results. However, the transitions to unsteady flows, which are always three dimensional, cannot be predicted. Of course two-dimensional simulations also can not correctly predict the transition to three-dimensional flows or treat the problem of deviations from axisymmetric boundary conditions which is, to a certain degree, always present in real crystal growth configurations.

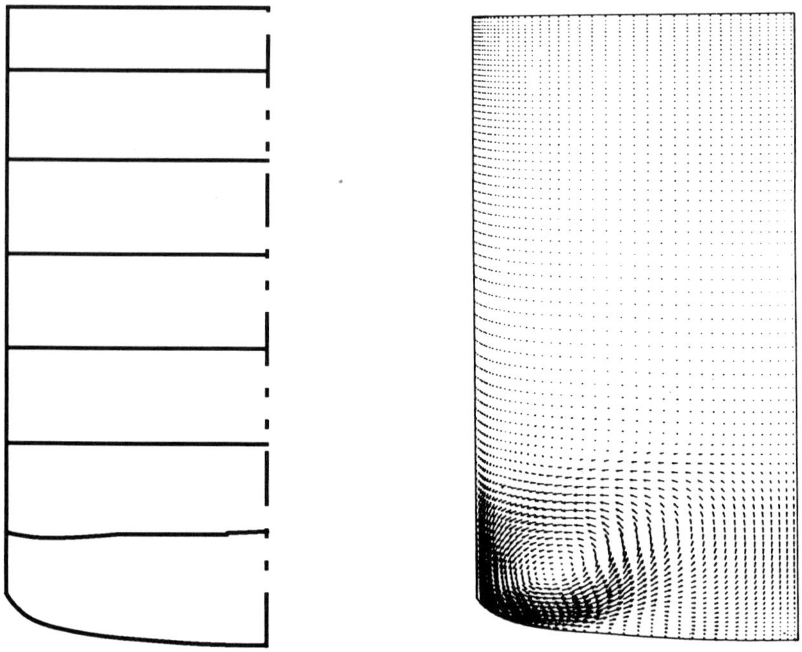

Fig. 51. Calculated (2D) convective flow in a vertical cylindrical melt growth configuration (VB/VGF) with calculated shape of the solid–liquid boundary (bottom) under conditions of a linear temperature profile along the crucible wall. Left: isotherms, right: flow. From [90].

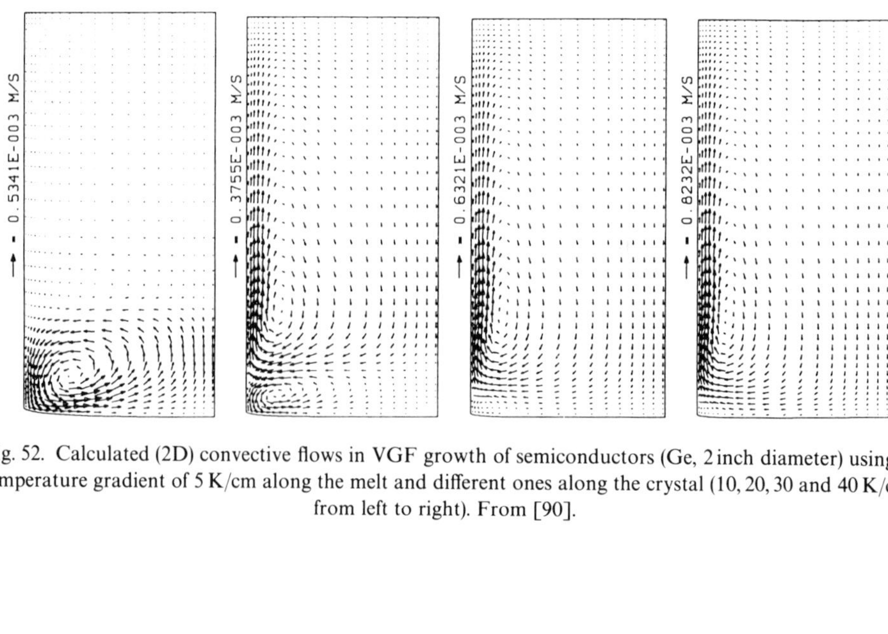

Fig. 52. Calculated (2D) convective flows in VGF growth of semiconductors (Ge, 2 inch diameter) using a temperature gradient of 5 K/cm along the melt and different ones along the crystal (10, 20, 30 and 40 K/cm from left to right). From [90].

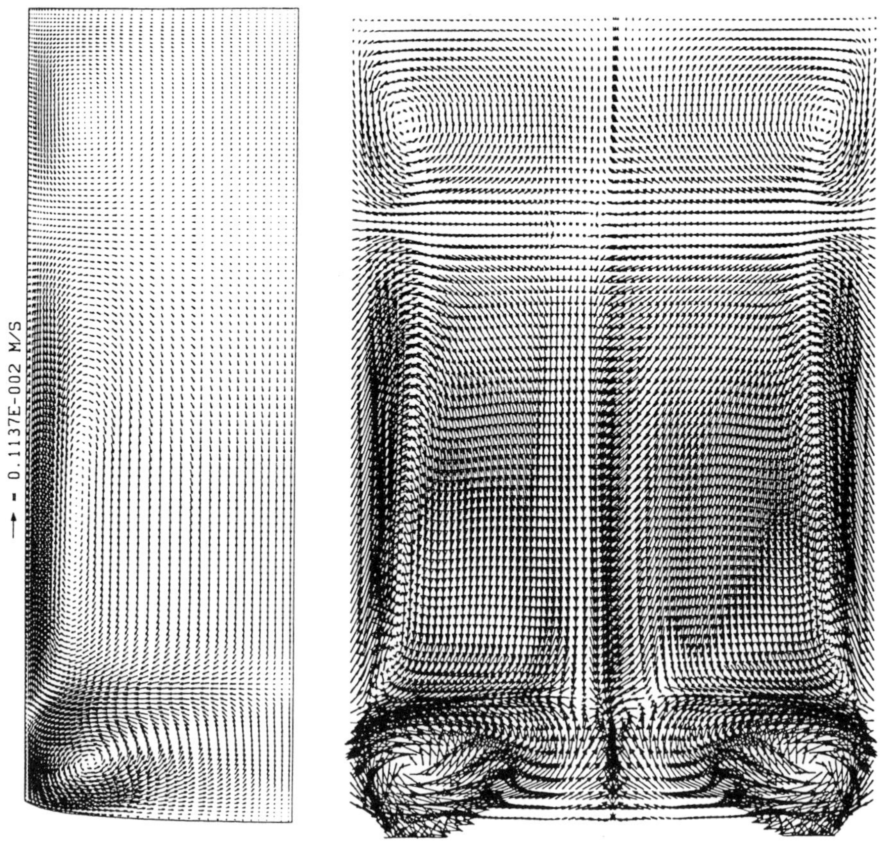

Fig. 53. Calculated (2D and 3D) convective flows in a vertical gradient freeze configuration for the growth of 2 inch diameter Ge crystals with the same non-linear temperature profile along the cylinder wall. From [90].

Jung [90] calculated also the behavior of a moving boundary under the conditions of unsteady convection. A typical result for the temporal fluctuation of the flow velocity and the temperature in the melt close to the interface are plotted together with the growth rate for the conditions corresponding to the growth of a 2 inch diameter GaAs crystal by the vertical gradient freeze technique (fig. 54). The resulting fluctuations of the concentration of the dopant Si incorporated in the growing crystal, which is also shown in this figure, is treated in more detail in section 5.2.

Brown and co-workers [87, 91] calculated the influence of density variations with concentration due to segregation effects for the cases of Ge:Ga and HgCdTe, where rejection of the heavier component at the growth interface leads to a damping of the thermally-induced convection, but does not change the basic flow structure.

4.1.3.2. Shape of the solid–liquid interface. Naumann and Lehoczky [92] were the first to show analytically that under "normal conditions" the different thermal conductivities

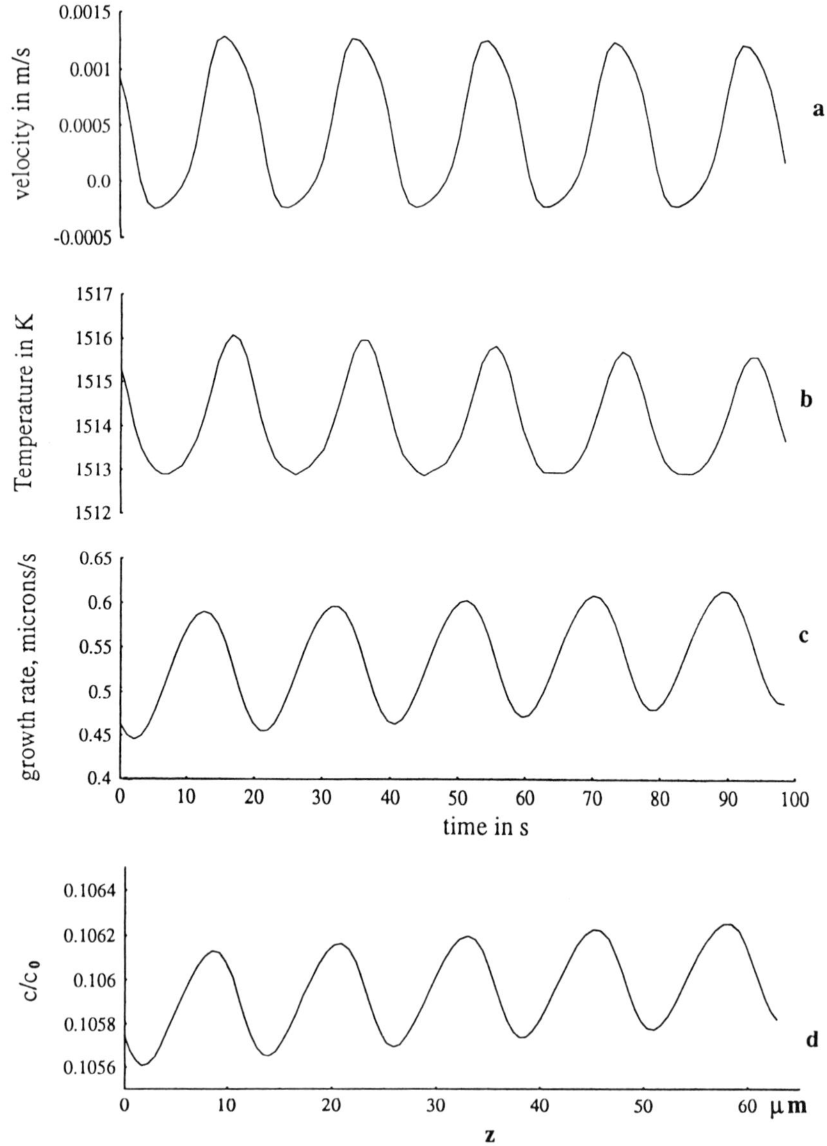

Fig. 54. VGF growth of Si-doped Ge (2 inch diameter) under conditions of unsteady buoyancy convection ($Ra = 1.1 \times 10^6$, $Ra_w = 5 \times 10^5$). Numercial calculation of: maximum velocity (a), temperature fluctuations close to the interface (b), growth rate (c) versus time and Si concentration in the growing crystal (d) versus axial growth direction. From [90].

in the liquid and solid states will control the shape of the growth interface. In semiconductors, the thermal conductivity of the melt is generally a factor of about 2 or 3 higher than of the solid. Consequently, the heat flux from the melt meets a higher resistance at the growth interface and part of it has to be conducted away laterally by radial gradients, resulting in isotherms which are concave towards the melt. Jasinski and Witt [93, 94] showed that, even inside an ideal adiabatic zone, where no radial gradients are possible, the non-zero thermal conductivity of a crucible makes the achievement of a flat interface very difficult. Experimentally this effect was observed for example in the works of Holmes and Gatos [95], Wang et al. [96], Rouzaud et al. [97], Witt et al. [98], Szofran and Lehoczky [99], where always a concave interface has been observed. On the other hand, El Mahallawy and Farag [100] reported a convex interface in solidifying aluminium, where the heat conductivity in the solid is greater than in the liquid.

Flat or convex interfaces have been achieved:

(i) In adiabatic zone furnaces, positioning the interface closer to the cold zone, e.g. Huang et al. [101].

(ii) Using "tailored" furnace-temperature profiles, applying a large temperature gradient along the crystal and a low one along the melt, compensating thereby the difference in the thermal conductivities (e.g. Monberg et al. [102, 103], Crochet [89], Look et al. [104], Kremer et al. [105] and Müllers group [106]).

Hofmann et al. [106] have demonstrated by numerical modelling and experiments with Ga-doped 2 inch diameter Ge crystals, that concave, planar and even convex shapes of the interface can be established depending on the thermal boundary conditions (see fig. 55).

Besides the effects of steady-state heat conduction, the shape of the solid–liquid interface can be determined by additional heat transfer due to convection in the melt, by changes in the melt temperature with composition and by the growth-rate dependent release of latent heat.

Semiconductor melts generally are characterized by a small Prandtl number (~ 0.01) so that the effect of convection on heat transfer is small, at least for the cases of relatively small levels of convection calculated up to now [86, 87].

4.2. Thermocapillary (Marangoni) convection in cylindrical floating zones

This subject is treated in detail in a separate article by Bohm et al. in this book (chapter 4). We will present only some results on the study of thermocapillary flows in GaAs floating zones which were obtained by the group of one of the authors.

The numerical results were obtained by using a model which is described in Appendix B.

The results of a three-dimensional and time-dependent simulation of Marangoni convection show clearly that an axisymmetric flow occurs only for small values of the Marangoni number Ma (see fig. 56a). As soon as Ma exceeds a certain value Ma^T the flows of the upper and lower part of the zone are interacting to form a new type of flow which is non-axisymmetric (see fig. 56b). To our knowledge this effect was reported for

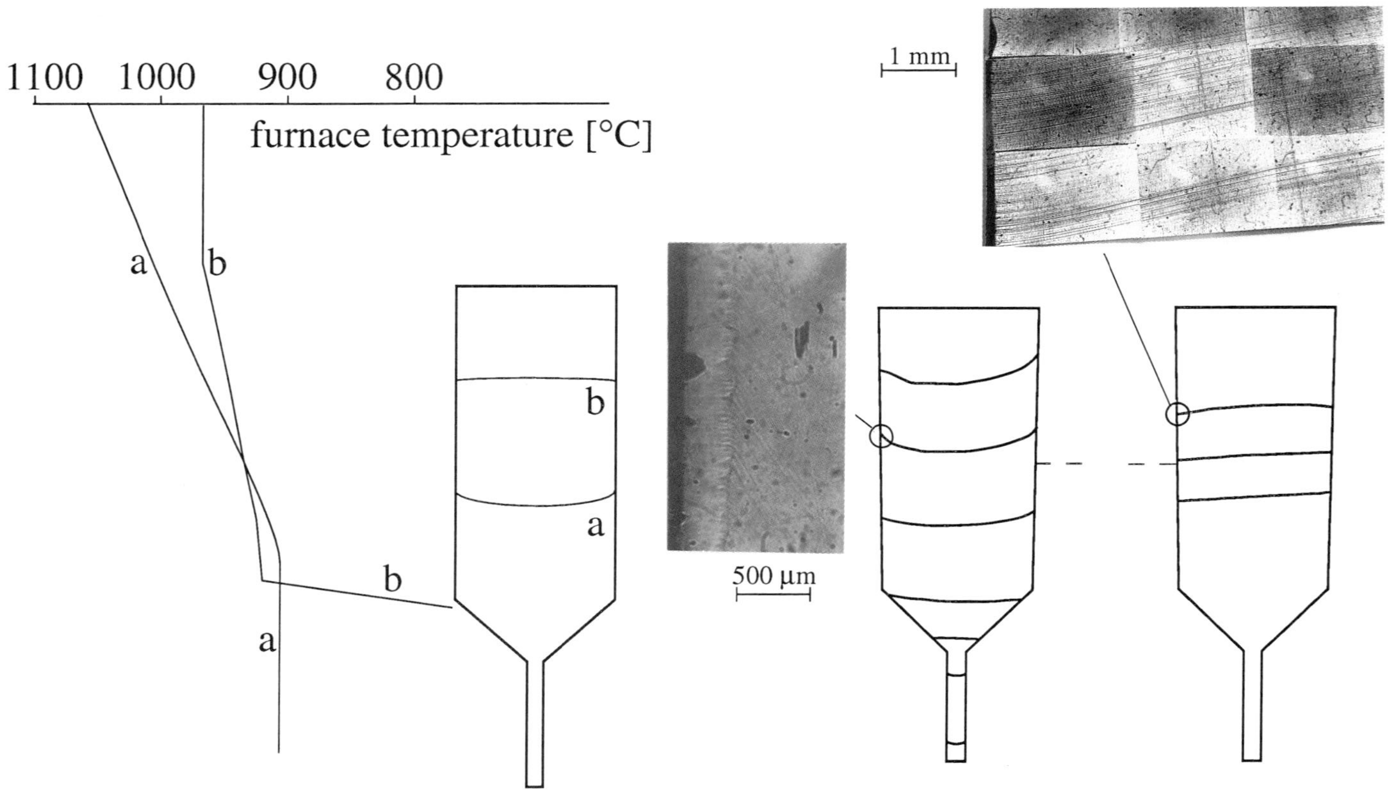

A) B) C)

Fig. 55. Numerically calculated shapes of the solid–liquid phase boundary with corresponding axial furnace profiles for the VGF growth of 2 inch diameter Ge crystals (A). Experimental verifications of concave (B) and convex (C) shapes of the growth interface using the calculated conditions (a) and (b) of (A). From [106].

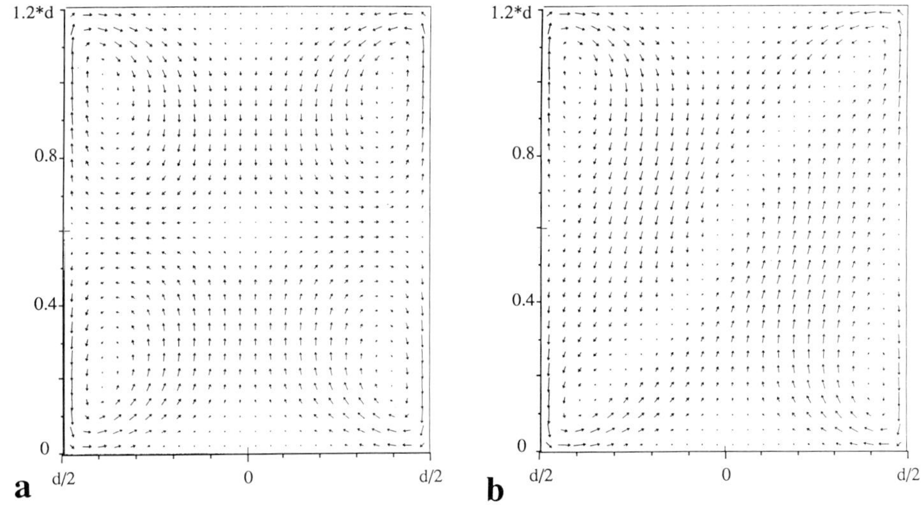

Fig. 56. Calculated thermocapillary flow in a vertical section of a cylindrical floating zone for liquid GaAs (Pr = 0.068, $h/d = 1.2$) for Ma = 50 (a) and 200 (b). From [107].

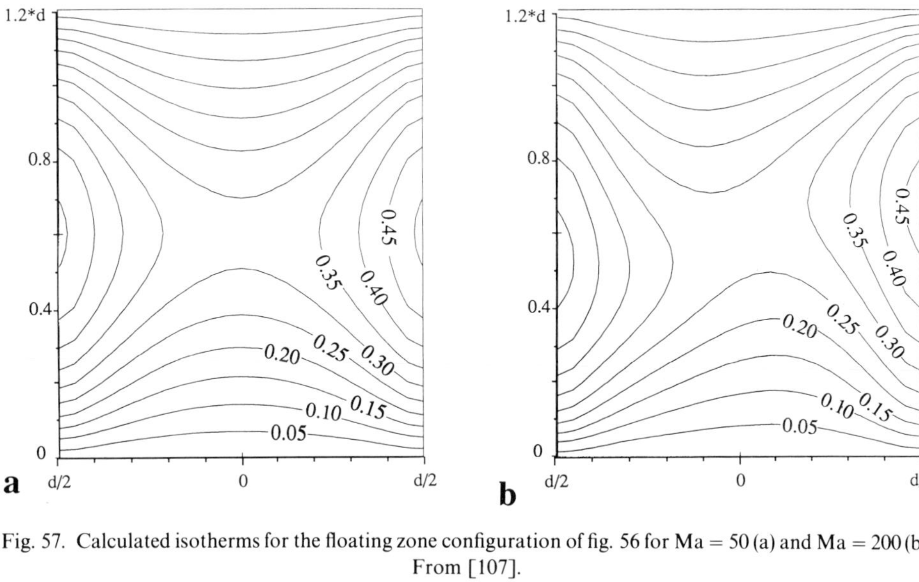

Fig. 57. Calculated isotherms for the floating zone configuration of fig. 56 for Ma = 50 (a) and Ma = 200 (b). From [107].

the first time in the literature by the group of Müller [108]. Obviously the results of earlier two-dimensional calculations given in the literature [109–114] do not correctly describe reality. Müller and Rupp's [107] results demonstrate the necessity for three-dimensional calculations for modelling Marangoni convection in semiconductors. Figures 57a and b show clearly the distortion compared to an axisymmetric flow if

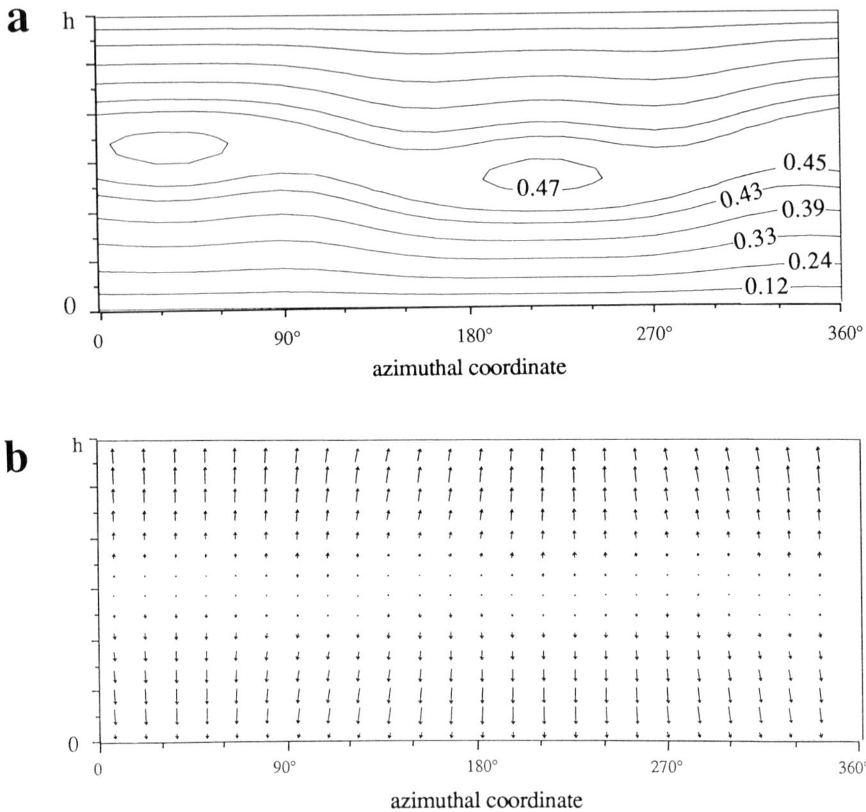

Fig. 58. Calculated non-axisymmetric flow (b) and isotherms (a) in the free melt surface of the floating zone configuration of figs. 56b and 57b. The temperatures (numbered isotherms) are non-dimensionalized by $(T - T_m)/\Delta T$. From [107].

$\mathrm{Ma} > \mathrm{Ma}^T$ by displaying the isotherms in an axial cut. Figure 58 demonstrates that, even in the free melt surface one finds azimuthal temperature gradients and components of the flow although the boundary conditions are totally axisymmetric. The transition between the two flow states at Ma^T is reversible and occurs abruptly when Ma exceeds Ma^T.

An unsteady state of Marangoni convection occurs by increasing Ma at a critical value Ma^c for the transition. These time-dependent fluctuations of the flow can principally be caused by the following three mechanisms:

(i) Axial "hydrothermal" waves, i.e. an exchange of kinetic energy between the flow in the upper and lower half of a liquid column, as postulated by Kazarinoff and Wilkowsky [115] on the basis of two-dimensional simulations.

(ii) Azimuthally travelling "hydrothermal" waves, i.e. a rotation of the whole non-axisymmetric flow configuration around the cylinder axis, as has been observed in various model experiments with transparent fluids in "half-zones" (see e.g. [2]).

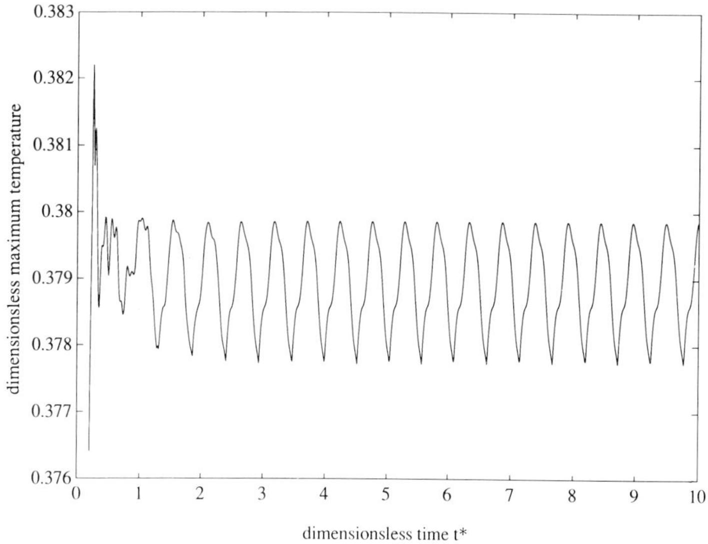

Fig. 59. Calculated temperature fluctuations in the free melt surface of a GaAs floating zone (Pr = 0.068, $h/d = 1.2$) for Ma = 720. Non-dimensionalization of temperature by $(T - T_m)/\Delta T$ and time $t\kappa/h^2$. From [107].

(iii) Fluctuations of the total thermal field of the melt zone, i.e. a kind of breathing. In this case, the maximum total temperature difference in the free melt surface is time-dependent.

Müller and Rupp's [107] results give no evidence for mechanism (i). However, they found both of the other two types: travelling waves (ii) and fluctuations of the maximum temperature (iii). Figure 59 shows that even the temperature difference ΔT across the cylinder surface which drives the Marangoni flow is fluctuating. The reason for this may be explained as follows. A small increase in ΔT increases the heat transported by the Marangoni flow which reduces the driving ΔT. But it takes some time before the increased, convective heat transport is reduced. Therefore, ΔT falls below its earlier value causing the heat transport driven by the Marangoni flow to fall below its earlier value. This leads, after some time lag, to an increase of ΔT and the cycle starts again. We found that the Prandtl number of the fluid and the aspect ratio determine whether the mechanism (iii) or travelling waves (ii) occur first when Ma^c is reached. Normally a mixture of both effects has to be expected as shown in fig. 60. This result shows that the curve of the temperature variation with time at the control point (continuous line) in the surface is similar to the curve along the circumference of the cylinder. That means that the rotation of the flow (lower curve) is modulated by the temporal fluctuation of the maximum temperature in the melt (upper curve). Obviously the characteristics of the travelling wave (ii) are large amplitudes and low frequencies. For type (iii) the opposite is true.

A summary of the relevant results for floating zone (FZ) crystal growth of GaAs and Si is given in table 2. Rupp [116] was able, by using this model, to explain quantitatively

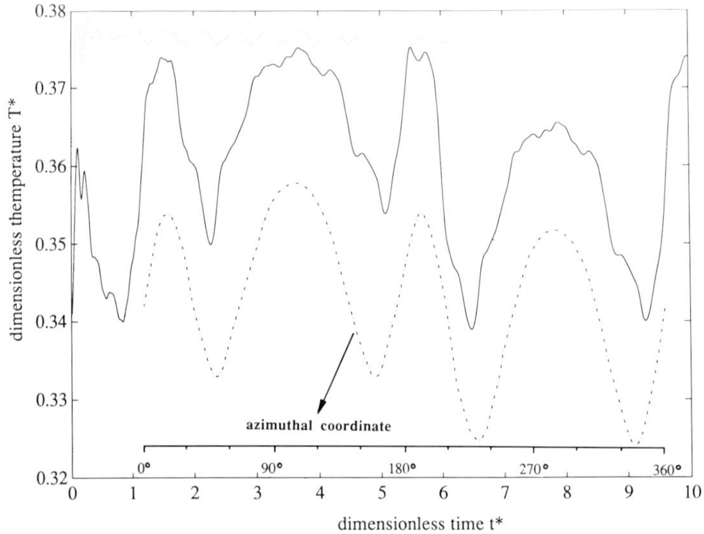

Fig. 60. Calculated fluctuations of the temperature of the thermocapillary flow (Ma = 610) in a GaAs floating zone ($h/d = 1.2$) for a control point in the free surface at half height (strongly varying solid curve) and azimuthal temperature variation lower curve. Non-dimensionalization see fig. 59. From [107].

Table 2

Values of the critical Marangoni numbers (Ma^T, Ma^c) for thermocapillary convection in GaAs and Si floating zones calculated by a full-zone and half-zone (HZ) model. Ma^T transition from 2D to 3D flow, Ma^c transition from steady to unsteady flow, used temperature coefficients of the surface tension $d\sigma/dT = -2.8 \times 10^{-4}$ (Si) and -1.8×10^{-4} (GaAs); oscillation mechanisms (ii) and (iii) see text. From [107].

Substance	GaAs, Pr = 0.068					Si, Pr = 0.02	
h/d	0.8	1.0	1.2	1.5	0.6 (HZ)	1.2	0.6 (HZ)
Ma^T	260	208	178	195	–	45	–
Ma^c	420	437	606	334	1800	115	290
Oscillation mechanism	ii(+iii)	ii(+iii)	ii(+iii)	iii	ii	iii	ii

the results of Cröll et al. [117] who have grown Si by the FZ technique with a partially covered surface of the zone.

In table 2 we have presented for comparison, the results of our earlier half-zone (HZ) calculations [108] which result in higher values of Ma^c by about a factor of 3. This difference is clearly understood as a symmetry plane is introduced in the middle of a floating zone with a no-slip condition for all velocities in the half-zone calculations. This restriction has to be considered if one compares the result obtained from experiments and simulations. We have, therefore, carried out in our group [118] computations with a half-zone configuration by assuming isothermal top and bottom planes having a temperature difference ΔT and an adiabatic side wall. The results of

Table 3
Numerical and experimental data of Ma^c and oscillation frequency of thermocapillary flow in floating zones of various transparent melts; non-dimensionalization of frequency $f^* = fh^2/\kappa$, $h = 3.5$ mm. From [107].

Pr (substances)	Ma^c numerical	Ma^c experimental	Frequency f^* numerical	Frequency f(1/s) numercial	Frequency f(1/s) experimental
1 (KCl)	2700	1470	11	0.61	0.9
7 ($NaNO_3$)	10500	5370	34	0.56	0.6
49 ($C_{24}H_{50}$)	45000	88000	135	0.70	0.6

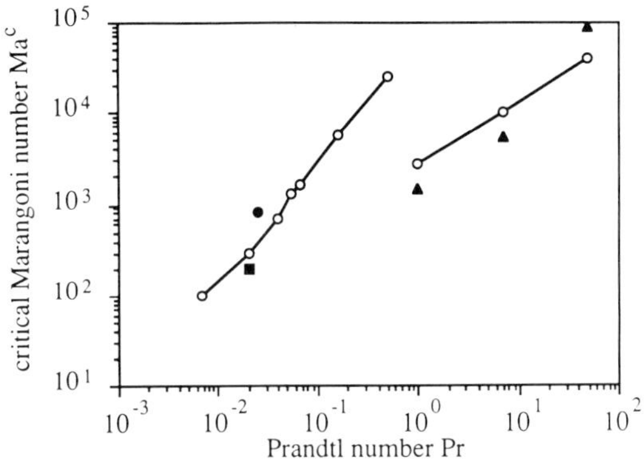

Fig. 61. Critical Marangoni number Ma^c versus Prandtl number Pr: open circles are numerical, filled symbols are experimental results. From [108].

these calculations are compared in table 3 with results of experiments from the literature [119] using transparent model fluids with $Pr \geq 1$. Numerical results concerning the critical Marangoni number Ma^c for the onset of time-dependent flow and the frequency of temperature oscillations near Ma^c for this "half-zone" configuration are in good accord with the experimental results for different fluids. Further results on this subject (half-zones) are plotted in fig. 61. The calculations of Rupp [116] give relations for half-zones and $Pr \geq 1$ of

$$Ma^c = 2.6 \times 10^3 \times Pr^{0.69}$$

and for $Pr \ll 1$ of

$$Ma^c = 6.9 \times 10^4 \times Pr^{1.36}$$

with results for various semiconductors given in table 4.

Systematic variations of the geometrical aspect ratio ($h/d = 0.8$–1.5) have only been carried out for GaAs. Ma^c is found to be maximum for $h/d = 1.2$.

Jurisch [120] studied Marangoni convection in the FZ electron beam melting of Mo and Nd by measuring the surface temperature of the melt. He was using a rapid-response micropyrometer having a response time of about 100 ms and a resolution of 0.04–0.08 K. Increasing amplitudes of these temperature fluctuations (0.25–2 HZ) were

Table 4
Numerically determined half-zone model values of Ma^c for floating zones ($h/d = 1.2$) of various semiconductor melts. From [116].

Pr	0.007	0.02	0.04	0.054	0.068	0.16	0.5
	Ge	Si	InAs	AlSb	GaAs	–	–
Ma^c	1.0×10^2	2.9×10^2	7×10^2	1.3×10^3	1.8×10^3	6.2×10^3	3.0×10^4

observed with increasing heating power, i.e. zone length. The general feature of the temperature oscillations is described as a travelling wave disturbance of the steady temperature field in the azimuthal direction of the zone. Rupp [116] compares the experimental results of Jurisch with numerical results for Ma^c and the frequencies of temperature oscillation.

It can readily be demonstrated that the Marangoni convection becomes turbulent at higher values of Ma. The route to turbulence is illustrated in fig. 62 for a Si zone with $h/d = 1.2$ by plotting the temperature of a certain control point in the zone versus the temperature in a diametral position. The graphs for various values of $115 < Ma < 160$ show closed curves i.e. fixed phase relations. For $Ma > 2 \times Ma^c$ ($= 254$) no phase relations exist, i.e. the flow is turbulent.

Influence of asymmetry of the thermal boundary conditions: Deviations of the rotational symmetry and the symmetry with respect to the middle plane of the heating flows have to be taken into account in practice. Rupp [107, 116] carried out some

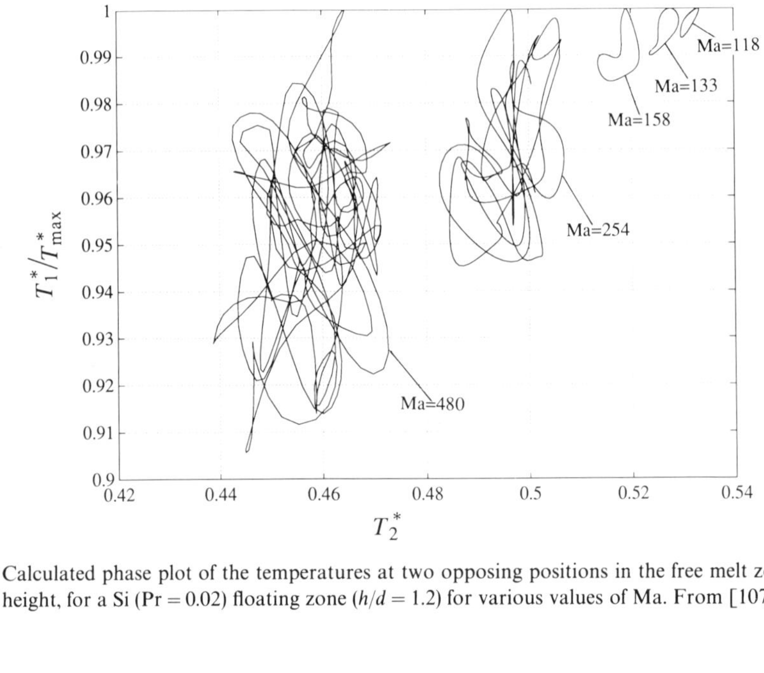

Fig. 62. Calculated phase plot of the temperatures at two opposing positions in the free melt zone at half height, for a Si ($Pr = 0.02$) floating zone ($h/d = 1.2$) for various values of Ma. From [107].

calculations by considering such deviations from the ideal symmetry of the thermal boundary conditions:

(i) Periodic azimuthal variations of the heat flux profile with an amplitude of $\pm 10\%$ of the average value: The resulting Ma^c is increased from 606 to 710, showing that this type of non-symmetry stabilizes the flow. The flow is stabilized because the rotation mechanism (ii) is suppressed by asymmetry.

(ii) Use of a realistic experimentally determined axial heat flow profile; for details see [107].

Influence of gravity, i.e. additional buoyancy convection: It is necessary to study the influence of gravity, i.e. of buoyancy-driven convection on Marangoni flow to enable quantitative comparison between the results of terrestrial FZ experiments and results obtained in microgravity. Buoyancy-driven convection can be described in vertical zone configurations in terms of a dimensionless wall Rayleigh number Ra_w defined in section 4.1. But Ma and Ra_w cannot be varied independently because they are correlated via the Bond number Bo

$$\mathrm{Bo} = \frac{\alpha g \rho h^2}{-\,\delta\sigma/\delta T}$$

by the expression

$$\mathrm{Ra_w} = T^*_{\max}\,\mathrm{Bo}\,\mathrm{Ma}_Q\,2\frac{h}{d}.$$

Using typical values, $h/d = 1.2$, $Ma_Q = 2000$ ($Ma \approx 620$), $h = 0.6\,\mathrm{cm}$, $T^*_{\max} = 0.3$ gives $Bo = 4.18 \times h^2$ (for GaAs) and results in $Ra_w = 3120$. The value of Ra_w for the transition to unsteady flow is about 40 000 (compare section 4.1) which was calculated and measured for buoyancy-driven convection in vertical zone configurations without free walls of the melt. Thus one can neglect buoyancy as a source of unsteady flows in the GaAs FZ experiments discussed in this section.

Furthermore, we studied the influence of steady buoyancy flow on Marangoni convection by several numerical simulations. It shows that e.g. for $Ma = 620$ (which is close to Ma^c) the maximum flow velocity is changed by less than 4% compared to pure Marangoni flow.

We therefore can neglect the influence of gravity for the small dimensions of terrestrial GaAs FZs. This conclusion is confirmed by the results of Cröll et al. [121] with Si. These authors demonstrated, under normal gravity conditions, that steady flow occurred when the Si melt zone ($\phi = 10\,\mathrm{mm}$) was covered by an SiO_2 skin. This flow became unsteady under the same conditions if the SiO_2 layer was absent. The revealed striation pattern was not substantially changed when this experiment was carried out under microgravity.

Velten et al. [122] have studied experimentally the influence of buoyant forces on Marangoni convection in FZs of liquids with $Pr = 1, 7$ and 49 and different aspect ratios, h/d. The zones ("half"-zone type) were heated from above or from below to discriminate the influence of buoyancy. The results are given in terms of maps in amplitude–Ma–Pr representations indicating the influence of Pr as well as that of the buoyant forces on the preferred spatial structure of periodic flows. The results are discussed in the light of existing theory and show a good agreement with the calculations of Rupp et al. [108].

Surprisingly Velten et al. found a stabilization of stationary Marangoni convection by buoyant forces which is not yet fully understood by the authors [122].

4.3. Buoyancy and thermocapillary (Marangoni) convection in horizontal Bridgman configurations

Buoyancy convection in horizontal open boat configurations has been studied experimentally by a systematic variation of the geometry of the melt (Ga) and the driving temperature difference between the horizontal end planes by Hurle et al. [7]. The resulting time averaged temperature distributions are given in three dimensions by perspex models. Furthermore, the occurrence of unsteady convection in terms of temperature oscillations was investigated. The period of oscillation was found to vary with the length of the melt, but did not depend on the depth. Additional experiments were later carried out by Pratte and Hart [123].

The correlation of the experimental results of Hurle et al. with those of a theoretical treatment by Gill [8] confirms the conclusion that buoyancy is the dominating convection mechanism. Nevertheless it has to be noted that surface tension driven convection can have a considerable effect in open boat configurations.

Okano et al. [124] have studied theoretically the action of both buoyancy-driven and Marangoni convection, on the flow in the melt inside a two-dimensional rectangular open boat. Their study is based on an order of magnitude analysis as well as on numerical simulations using a finite difference method. This yields a quantitative correlation between the velocity in the free melt surface and the velocity distribution within the melt on one side and the Grashof, Marangoni, Prandtl numbers as well as the melt depth on the other side. Okano et al. [124] concluded that Marangoni convection is affecting or even dominating the flow if the melt height is small. For deep melts, buoyancy-driven convection is dominant. The quantitative correlations were found by these authors to be rather complex ratios of the Marangoni, Grashof numbers and the aspect ratio.

A GAMM workshop held in 1988 [125] was devoted to the numerical simulation of oscillatory convection in low-Prandtl number fluids in differentially heated cavities of the horizontal boat type. Several test cases based on two-dimensional models and various aspect ratios were studied in the range of $10^4 \leq Gr \leq 4 \times 10^4$ of the Grashof number. Roux et al. [126] reported that the calculated critical values of Gr and the critical frequencies "qualitatively agree with the experimental results obtained by Hurle et al. [7] in the case of narrow open boats and for horizontal temperature gradients relevant for crystal growth applications in the horizontal Bridgman technique". However the authors stated also that, due to the complex stability diagrams and the occurrence of three-dimensional modes presently one can not expect a quantitative agreement of numerical and experimental results. The computation of the oscillatory flow modes is considered to be so time-consuming even for the two-dimensional models, that the development of more efficient algorithms and computers (e.g. vector, parallel or transputer machines) is a necessary prerequisite for further progress.

4.4. Convective flows in the Czochralski configuration

Quantitative studies of convective flows in Czochralski configurations, both calculations and experiments, are among the most challenging tasks in the science of crystal growth. This is mainly due to the combined interaction of buoyancy, forced convection and capillary flows – as already stressed in section 2.3. Furthermore, the typical melt dimensions in Si or GaAs production facilities lead to very large buoyant forces ($Ra \geq 10^8$), which is far above the critical value $Ra^c \approx 10^4-10^5$ for the transition to unsteady flow. This critical value can be estimated from the solutions of the Czochralski benchmark problem [223] and from numerical results of three-dimensional time-dependent simulations [127, 128]. Measurements of temperature in Si melts during Czochralski growth [129, 130] reveal random time traces of temperature fluctuations and broad-banded featureless Fourier power spectra suggesting chaotic flow behavior. This statement is confirmed by a recent theoretical analysis of Ristorcelli and Lumley [64]. They concluded that in either case of buoyancy, surface tension driven and rotating disc flow the non-dimensional parameters indicate that the flow is at least unsteady and three-dimensional with respect to each individual instability, considering the typical size of an industrial melt crucible for Si or even GaAs. They suggested that the Czochralski flow is a "non-linear combination of approximately ten different instabilities", i.e. strongly three-dimensional and time-dependent. The more manageable averaged momentum equations of turbulence models should, therefore, be applicable according to these authors.

Kobayashi et al. [131] carried out temperature measurements in a Si Czochralski production puller and discussed the results in terms of analytical solutions based on a $k-\varepsilon$ turbulence model.

The results of the numerical simulation of flows according to the boundary conditions defined in the international benchmark problem [222] can only be used as a check of numerical codes, but not for predictions in realistic growth configurations. Since the geometry and the thermal boundary conditions are much more complex compared to the benchmark problem [222].

Most of the presently used numerical codes can consider complex shaped geometries of the melt and locally depending temperature distributions. Curved crucibles, meniscus shapes (melt–gas interface) and the solid–liquid interface can be used as fixed boundary conditions. Some groups have also demonstrated the calculation of the shape and position of the growth interface as a free boundary problem [132, 133]. But the diameter of the crystal is kept constant. Equilibrium shapes of the meniscus have been calculated by Mika and Uelhoff [134] and Hurle [135] but not in conjunction with convective flows.

There is no special problem in introducing the realistic rotation rates of crystal and crucible. But Kakimoto et al. [136] report on convergence problems in their calculation when introducing rotation. Nevertheless this group was able to present a semi-quantitative comparison of convective flows observed by in-situ detection using double X-ray radiography (see section 3.2) and three-dimensional numerical simulation [4]. Their results for Si Czochralski growth from a 3.75 cm deep melt and 3.8 cm crystal radius (rotation 1 rpm) gives a non-axisymmetric flow having the character of a baroclinic instability.

The solution of the free boundary problem of the crystal–melt interface can only be achieved, if realistic thermal boundary conditions are imposed on the free melt surface. This can be done by global modelling of the Czochralski process [132, 136] which includes the effects of the heater, shields, growth vessels, etc.

The problem of defining realistic thermal boundary conditions is more easily solved in the global models as the corresponding temperatures at the crucible walls and interfaces are part of the solution. Nevertheless they can differ from realistic values considerably, if material parameters (e.g. emissivity) are incorrect. It is therefore more favorable to use measured temperatures to support the numerical model. This problem was studied by the groups of Brown [132] and Müller [56]. The experimental set up which was used for measurement of the temperature distribution in the melt is shown in fig. 30. In this case the GaAs crystal is replaced by a graphite dummy of the same shape and a Ga melt is used instead of GaAs. The resulting temperature distribution measured in the melt is depicted in fig. 63. Figure 64 shows a result, obtained from the global modelling with a commercial code considering conductive and radiative heat transfer in all parts [56]. But figs. 63 and 64 show a considerable disagreement between simulation and experimental data. Thus we have to conclude, that the convective heat transport in the melt has to be included. This problem was solved by combining the global conductive–radiative heat transfer model (finite element code) with a three-dimensional time-dependent finite volume code [137] which considers only the melt volume. The thermal boundary conditions from the global model were used for this calculation of convection in the melt as a first initial guess [56]. After that the flow is

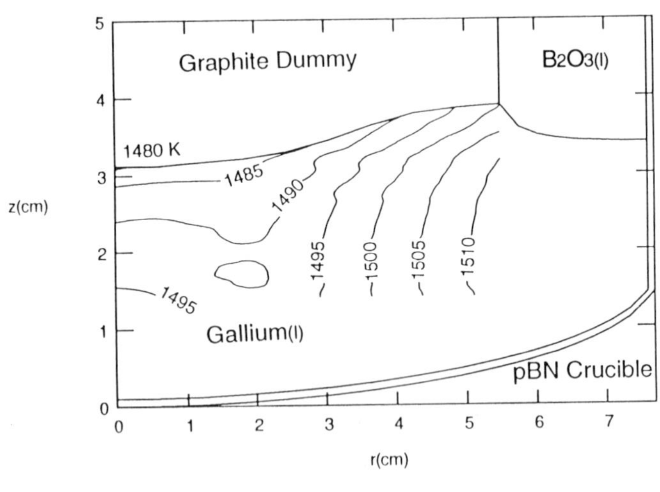

Fig. 63. Measured isotherms in the LEC configuration depicted in fig. 30 (isotherms: $\Delta T = 5$ K). From [56].

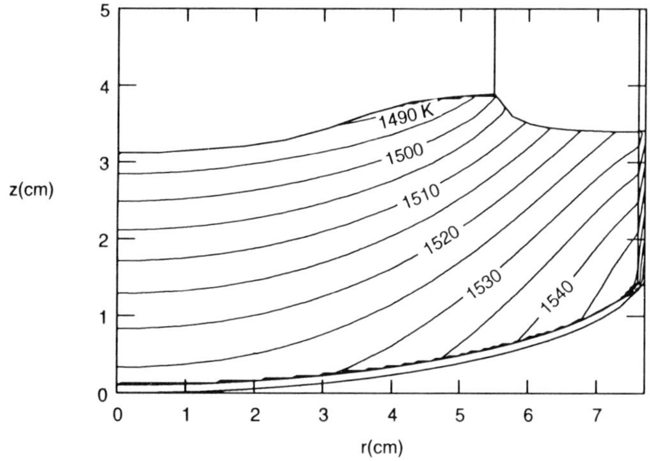

Fig. 64. Isotherms: predicted by a conduction–radiation model (isotherms: $\Delta T = 5$ K). From [56].

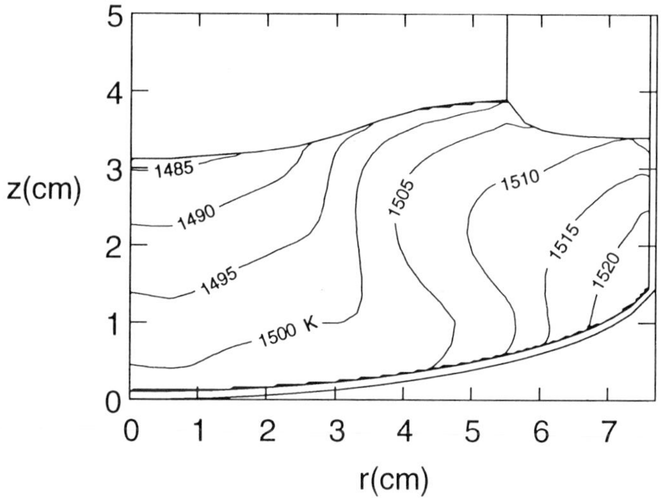

Fig. 65. Numerical simulation of the isotherms under the thermal boundary conditions of fig. 64, but now with three-dimensional time-dependent modelling of convection. The plot gives the situation at one instant of time. From [56].

computed, the temperatures of interior points in the melt were held constant for the next step of the global model, which again gives improved boundary conditions for calculation of the melt convection [138]. Figure 65 shows the result of such an iterative calculation, which fits much better to the experimental result (fig. 63), and proves the necessity of considering convective heat transport in the melt.

Convective transport in the melt has to be considered in any case, if the incorporation of dopants (see e.g. [139]), especially of oxygen is to be considered (compare e.g.

[140]). However, it has to be stated that the numerical simulation of flows under realistic conditions consumes an immense amount of computer memory space and computing time. Even the use of supercomputers gives presently not enough capacity to solve such problems with existing numerical codes. To overcome these problems researchers often introduce several artificial constraints into their models, e.g. limitation to two-dimensional flows, reduction of the melt height (small crucibles), reduction of the thermal volumetric expansion coefficient of the melt, increase of the melt viscosity.

Several attempts are presently undertaken to overcome this problem. One is the use of improved numerical methods, for example the multi-grid technique [141]. Leister and Peric [137] have recently presented the results of a three-dimensional and time-dependent calculation using a finite volume numerical code and the multi-grid technique. They simulated the experimetal configuration of a Czochralski model of Jones [20] who used water as a test fluid. They successfully calculated a time-dependent three-dimensional flow configuration which is in good qualitative correlation with the experimental results.

The next step is the use of combined interacting models, as already demonstrated above [56, 138]. In this case already a vectorized computer code and a multi-grid technique has been used for a CRAY XMP 48 computer in the vector mode – but still consuming about 5 h of CPU time per run.

It is now generally accepted that the required computing power for the simulation of convective transport in the Czochralski configuration can only by provided in the future by parallel computing.

This is the reason why many research groups worldwide are working on parallelization of the solution methods. The domain decomposition technique and MIMD (multi instruction, multiple data) parallel computers are mostly employed in conjunction with implicit solution methods. The efficiency of parallel computing, which is affected by the reduction of convergence rate due to domain decomposition and loss of CPU time due to communication, is a function of the grid size and number of processors. The above mentioned benchmark test case was solved on a transputer system with 48 processors using a grid of 512 × 348 control volumes in 1357 s, with an estimated efficiency of 80%. The same calculation took 1190 s on a CRAY YMP computer. For a given alogrithm, it is possible to create a model for the prediction of parallel efficiency as a function of the number of processors, the setup time and data transfer rates [142].

Müller and Peric [143] discussed the efficiencies of parallel computers. They came to the conclusion that up to 500 transputers would offer an efficiency of about 80% and a real computing power of over 200 MFlops for type T800 and about 4 GFlops for type T9000. The parallel processing can be achieved not only in space but also in time; this enables an overall efficiency increase [144]. It is thus expected that more accurate simulations of three-dimensional unsteady flows will soon become possible, at least for laminar regimes. The large scale simulations of turbulent flows, which is nowadays only being undertaken at an academic level, will also become a plausible means for solving engineering problems [145].

5. *Influence of convection on solute segregation in front of a growth interface and in the crystal*

Macroscopic and microscopic inhomogeneities in crystals grown from the melt are caused by segregation of solutes at the growth interface (see Garandet et al. Chapter 12, this volume). In this section, the effect of melt convection on segregation is discussed.

In equilibrium, which is established when gradients of concentration and temperature are negligibly small, the solute concentration in the solid (crystal) is related to the solute concentration in the melt (L) as

$$k = \frac{C_s}{C_L}, \tag{5.1}$$

where k is the equilibrium segregation coefficient, which can be taken as constant at low impurity concentrations ($C_L \leq 1\%$). Values of k can be found in Appendix D (table 7).

During an actual freezing process which proceeds at non-zero velocity, solute is rejected (for $k < 1$) or preferentially absorbed (for $k > 1$) by the advancing solid–liquid interface, forming an enriched or depleted solute layer in front of it, fig. 66. At low freezing rates used in crystal growth, the solute concentration in the growing crystal may be related to the melt concentration at the interface, $C_L(0)$,

$$C_s = kC_L(0), \tag{5.2}$$

where k is still the equilibrium segregation coefficient. During growth, $C_L(0)$ depends on the level of melt convection, which controls the transport of solute atoms in the melt. For the purpose of clarity, convection in solute–melt systems with $k < 1$ is discussed in the next section.

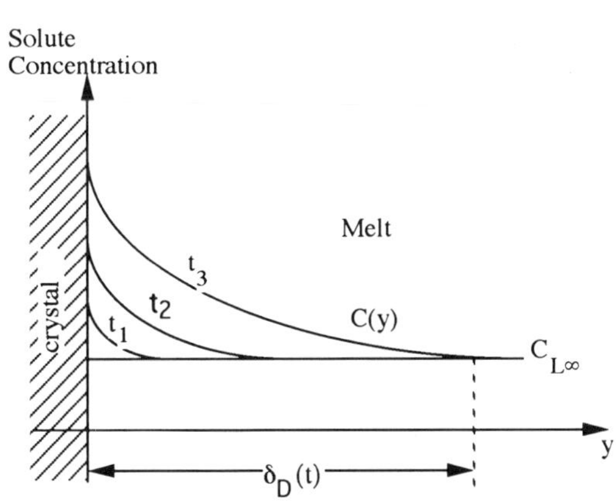

Fig. 66. Development of a solute ($k < 1$) boundary layer $\delta_D(t)$, caused by initiation of growth. Concentration profiles $C(y, t)$ are shown at different time steps, $t_1 < t_2 < t_3$.

5.1. *Steady convection (of solute at the solid/liquid interface)*

Typical flow patterns occurring in crystal growth configurations are shown schematically in fig. 67 (see also sections 2 and 4). The flow approaches the growth interface (along the wall and/or at the melt centerline), makes a 90° turn and then flows along the growth interface. The flow departs at the end of the growth interface by making another 90° turn (figs. 67b and c). In the case of Czochralski growth (fig. 67a), the flow remains tangential to the growth interface at the point of departure.

Steady convection of solute at the phase boundary is shown schematically in fig. 68. Solute atoms are rejected at the growth interface, which therefore is a source of solute. Driven by the steep axial concentration gradient at the interface, solute atoms diffuse away axially, while at the same time convection sweeps them tangentially downstream. The more the solute moves away from the wall, the higher is the convective velocity. Because typical values of diffusion coefficients are $D \sim 10^{-4}\,\mathrm{cm}^2/\mathrm{s}$, (see values of D in

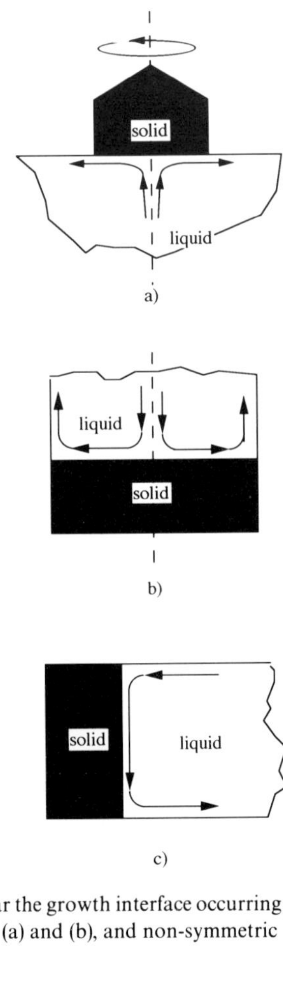

Fig. 67. Schematic flow patterns near the growth interface occurring during crystal growth for axisymmetric flows (a) and (b), and non-symmetric flows (c).

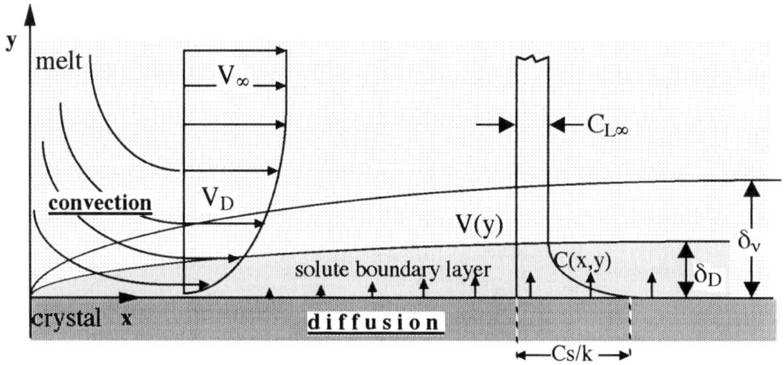

Fig. 68. Competition between diffusion and convection in the solute boundary layer: diffusion transfers the solute away from the wall, while at the same time convection sweeps the solute atoms downstream.

Appendix D, table 7), diffusion is important only very close to the growth interface (and crucible walls) where the lateral melt velocity approaches zero.

The concentration in the solute layer $C(x, y)$ approaches asymptotically the bulk melt concentration $C_{L\infty}$ for $y \geq \delta_D$. There is practically no net solute transport (by diffusion or convection) through the edge of the solute layer since, at this point, the concentration gradient $\partial C/\partial y \approx 0$. This implies that all rejected solute is removed from the growth interface laterally, by convection which carries the solute in the downstream.

5.1.1. Lateral variation of solute concentration

Analytical solutions of convective mass and momentum transfer exist only for simple flows and geometries. Blasius [146] obtained an exact analytical solution for steady laminar flow, parallel to a flat plate, which may be used as a model of the lateral convection in Bridgman growth. Defining the thickness of the momentum boundary layer δ_v as the distance from the wall to the point where velocity differs by 1% from the bulk velocity, he obtained

$$\delta_v = 5\sqrt{\frac{xv}{V_\infty}}, \tag{5.3}$$

where v is the kinematic viscosity and x is the distance along the plate (fig. 68). The analogy between momentum and mass transfer relates the thickness of solute layer to the thickness of the momentum (i.e. velocity) boundary layer [147]

$$\frac{\delta_D}{\delta_v} \approx \left(\frac{D}{v}\right)^n = \mathrm{Sc}^{-n}, \tag{5.4}$$

where $\mathrm{Sc} \equiv v/D$ is the Schmidt number. The exponent n is weakly dependent on Sc ($\frac{1}{2} \leq n \leq \frac{1}{3}$). Combining eqs. (5.3) and (5.4) results in

$$\delta_D(x) = 5\sqrt{\frac{vx}{V_\infty}}\left(\frac{D}{v}\right)^n. \tag{5.5}$$

Since the thickness of the solute layer grows as $\delta_D \sim \sqrt{x}$, the concentration $C_L(0)$ and C_S also grow, causing lateral inhomogeneity.

The flow at the growth interface of Czochralski melts is traditionally modelled as a flow driven by a rotating disk, which acts as a centrifugal pump. Von Karman [148] and later Cochran [149] obtained the exact solution of the axisymmetric flow driven by an infinite disk, surrounded by a stagnant semi-infinite fluid. According to this solution, the distance from the disk to the point where the circumferential velocity has dropped to 1% of its wall value is equal to

$$\delta_v = 5.4 \sqrt{\frac{\nu}{\omega}}, \tag{5.6}$$

where ω is the rotation rate and δ_v is the thickness of the momentum layer [146]. It is interesting to note that eq. (5.6) can be obtained from eq. (5.3), by approximating the melt velocity as $V_\infty \approx \omega R$, where R is the disk radius.

Equation (5.6) predicts uniform lateral thickness of the momentum boundary layer. Levich applied the Cochran flow solution to the concentration boundary layer at the surface of the infinite rotating disk, and used the Nernst's stagnant film theory to define the thickness of the solute boundary layer as [147]

$$\delta_D = \frac{D\,[C_L(0) - C_{L\infty}]}{j}, \tag{5.7}$$

where j is the solute flux at the disk surface, $C_L(0)$ is the concentration at the disk surface and $C_{L\infty}$ is the concentration in the bulk melt. For $\mathrm{Sc} \to \infty$, Levich obtained

$$\delta_D = 1.6 \sqrt{\frac{\nu}{\omega}} \left(\frac{D}{\nu}\right)^{1/3}, \tag{5.8}$$

which predicts uniform thickness of the solute layer, and thus no lateral segregation.

Sparrow and Gregg carried on numerical studies of heat [150] and mass [151] transfer from the infinite rotating disk for a variety of values of Prandtl/Schmidt numbers. From their results, it can be deduced that the thickness of the solute layer (i.e. the distance from the wall to the point where solute concentration differs by 1% from the bulk concentration $C_{L\infty}$) is equal to

$$\delta_D \approx 4 \sqrt{\frac{\nu}{\omega}} \left(\frac{D}{\nu}\right)^n, \tag{5.9}$$

where n is in the range $1 \leq n \leq \frac{1}{3}$ for fluids having $0 \leq \mathrm{Sc} \leq \infty$.

At low levels of convection (V_D is less or of the order of the growth rate R), the advancing solid–liquid interface can be modelled as a flat plate with uniform suction [152]. Because of the uniform suction, after an entrance length in which boundary layers grow, both δ_v and δ_D approach constant values [146, 152–154]

$$\delta_v = 4.6 \frac{\nu}{R}, \tag{5.10}$$

$$\delta_D = 4.6 \frac{D}{R}. \tag{5.11}$$

Therefore, at very low levels of convection, the thickness of the solute layer is controlled by the growth rate R.

Equation (5.8) is widely used to model segregation in Czochralski growth (see the Burton, Prim and Slichter (BPS) model [155, 156]). The main limitation of eqs. (5.8) and (5.9) stems from the infinite disk assumption, which does not permit increase of properties towards infinity. As a result, the lateral (radial) solute distribution must be constant. This implies that the physics of lateral solute transport (which is the main mechanism of solute removal from the growth interface) is not present in the infinite rotating disk model.

5.1.2. *Axial solute concentration in the crystal*

The longitudinal (radially averaged) distribution of the solute in a crystal also depends on the solute transport at the interface. The concentration of solute in the solid, C_S can be related to the bulk melt concentration using an integral control-volume analysis proposed by Ostrogorsky and Müller [154].

The boundaries of the control volume (fig. 69) are selected in such a manner as to allow easy computation of the solute transfer through its boundaries. The details of the velocity and solute fields within the control volume do not need to be known.

The control volume includes the part of the growth interface exposed to the tangential flow and the solute boundary layer attached to it. The boundaries BD and AC have length $L \gg \delta_D$. BD extends into the growing crystal. AC is just outside the solute layer. Boundary AB is at the line of symmetry or crucible wall, preventing solute transfer. At the end of the region of tangential flow, CD cuts across the solute layer. The short exit region, having length of the order of $\delta_D \ll L$, is not included in the control-volume analysis. Diffusion is neglected at all boundaries of the control volume:

(i) Lateral diffusion is not accounted for, since the concentration gradient is significant only in the growth direction ($\partial C/\partial y \gg \partial C/\partial x$).

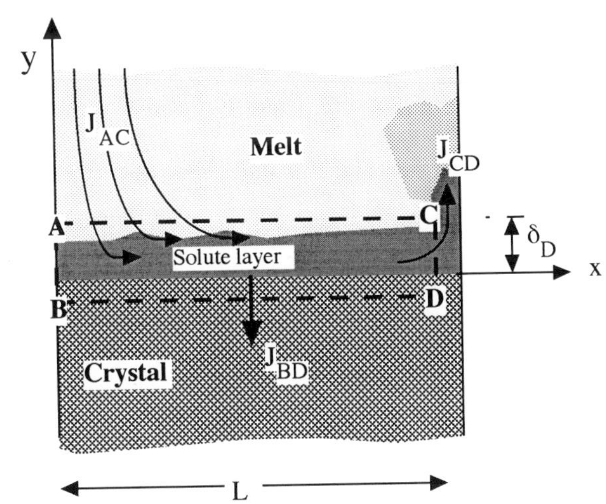

Fig. 69. Sketch of the control volume (ABCD), used to model axial segregation. From [154].

(ii) $\partial C/\partial y$ is negligible at AC, because the concentration asymptotically approaches $C_{L\infty}$.

(iii) Diffusion in the solid crystal, through BD is neglected.

Furthermore, because of the one-dimensional treatment, it is assumed that C_S does not change substantially along BD. The steady state conservation of solute in the control volume (calculated per unit depth) requires that the inflow through AC is equal to the outflow through BD and CD

$$J_{AC} = J_{BD} + J_{CD}, \tag{5.12}$$

where

$$J_{BD} = C_S RL \tag{5.13}$$

is the outflow of the solute incorporated in the solid (through BD); R is the growth rate (cm/s). The outflow of solute through CD due to convection is

$$J_{CD} = \int_0^{\delta_D} C(y)V(y)\,\mathrm{d}y. \tag{5.14}$$

J_{AC} is computed assuming that the solute concentration at AC is uniform and equal to $C_{L\infty}$

$$J_{AC} = C_{L\infty}Q_{in} = C_{L\infty}(Q_{BD} + Q_{CD}) = C_{L\infty}\left(RL + \int_0^{\delta_D} V(y)\,\mathrm{d}y\right), \tag{5.15}$$

where the volumetric inflow into the control volume per unit depth, Q_{in} (cm^2/s) is equal to the outflow $Q_{BD} + Q_{CD}$ if the density change during solidification is neglected. $V(y)$ and $C(y)$ are approximated using simple linear profiles, with the condition at the interface $C_L(0) = C_S/k$

$$V(y) = V_D \frac{y}{\delta_D}, \tag{5.16}$$

$$C(y) = C_{L\infty} + \left(\frac{C_S}{k} - C_{L\infty}\right)\left[1 - \left(\frac{y}{\delta_D}\right)\right], \tag{5.17}$$

where V_D is the convective velocity at the edge of the boundary layer, fig. 68. Integration of eqs. (5.14) and (5.15), substitution of eqs. (5.13), (5.14) and (5.15) into eq. (5.12), rearranging results in an explicit equation for C_S and $C_{L\infty}$ (the concentration in the bulk melt) with

$$\frac{C_S}{C_{L\infty}} = k_{eff}, \tag{5.18}$$

as defined in the BPS model [155], gives

$$k_{eff} = \frac{C_S}{C_{L\infty}} = \left(1 + a\frac{V_D\delta_D}{R\;L}\right)\left(1 + a\frac{V_D\,\delta_D}{R\;L}\frac{1}{k}\right)^{-1}, \tag{5.19}$$

where a is equal to $a = \frac{1}{6}$ for linear approximations of $V(y)$ and $C(y)$. When linear profiles are replaced by cubic polynomials, a becomes equal to $39/280 \approx 1/7.2$.

Table 5
δ_D and V_D for several important melt growth techniques (CZ = Czochralski, FZ = floating zone, Gr = Grashof number).

Driving mechanism	Growth method	δ_D [Reference]	V_∞ [Reference]
Crystal rotation	CZ, FZ	$4\sqrt{\frac{\nu}{\omega}}\left(\frac{D}{\nu}\right)^n$ [151, 154]	$\sim \omega L$ [154]
Natural convection	Bridgman	$5\sqrt{\frac{\nu x}{V_\infty}}\left(\frac{D}{\nu}\right)^n$ [146, 154]	$\sim (\mathrm{Gr})^{1/2}(\nu/L)$ [196]
Weak convection	Bridgman in microgravity	$4.6\,D/R$ [152, 146]	$\sim \mathrm{Gr}(\nu/L)^{1/2}$ [196]

Expressions for δ_D and V_D for several important melt growth techniques are given in table 5. δ_D represents the real physical thickness of the solute boundary layer, and therefore can be evaluated using the existing analytical solutions for momentum and mass transfer in boundary layers. V_D is the convective velocity at the edge of the boundary layer, which can be related to the bulk velocity in the melt through a linear approximation

$$V_D = V_\infty \frac{\delta_D}{\delta_\nu} = V_\infty \left(\frac{D}{\nu}\right)^n. \tag{5.20}$$

5.1.3. Parameters influencing the effective segregation coefficient

Low k has an adverse effect on k_{eff} because the flow leaving the control volume is carrying along solute having an average concentration of the order of C_S/k. The lower the value of k, the more solute is removed by the flow.

δ_D/L represents the aspect ratio of the layer. The thinner the solute layer, the less solute is removed by convection. The longer is the interface, the more the solute has to travel along the interface in order to leave the control volume. Therefore, the smaller δ_D/L, the higher is k_{eff}.

The constant $a \approx \frac{1}{6}$ accounts for the distribution of solute with respect to the velocity field: the difference in solute concentration $C(y) - C_{L\infty}$ is highest at the growth interface where $V(y) = 0$. As a result, the net convective transport of solute is lower by approximately a factor of 6 compared to the case of uniform velocity and concentration profiles.

V_D/R represents the ratio between the convective velocity removing solute and the growth rate bringing solute into the control volume. The lower V_D, the less solute is removed by convection. The higher R, the more solute enters the solute boundary layer; as a result, more removal of solute by convection can be tolerated. Therefore, the smaller the term V_D/R, the higher is k_{eff}.

5.1.4. Effective segregation coefficient for Czochralski growth

The average fluid velocity tangential to the growth interface is of the order

$$V_\infty \sim \omega L, \tag{5.21}$$

where L is the crystal radius and ω (1/s) is the angular velocity. The convective velocity at the edge of the boundary layer, V_D can be obtained using eq. (5.20)

$$V_D \approx \omega L \left(\frac{D}{\nu}\right)^n. \tag{5.22}$$

Substitution of eqs. (5.9) and (5.22) in eq. (5.19) gives the effective segregation coefficient for the Czochralski process

$$k_{\text{eff}} = \left(1 + 4a\frac{\sqrt{\omega D}}{R}\left(\frac{D}{\nu}\right)^{2n-1/2}\right)\left(1 + 4a\frac{\sqrt{\omega D}}{R}\left(\frac{D}{\nu}\right)^{2n-1/2}\frac{1}{k}\right)^{-1} \tag{5.23}$$

or, using $n = \frac{1}{2}$ for semiconductor melts having $5 < \text{Sc} < 50$

$$k_{\text{eff}} = \left(1 + 4a\sqrt{\frac{\omega}{\nu}\frac{D}{R}}\right)\left(1 + 4a\sqrt{\frac{\omega}{\nu}\frac{D}{R}\frac{1}{k}}\right)^{-1}. \tag{5.24}$$

The model for k_{eff} (5.24), is compared in fig. 70 to the segregation data obtained by

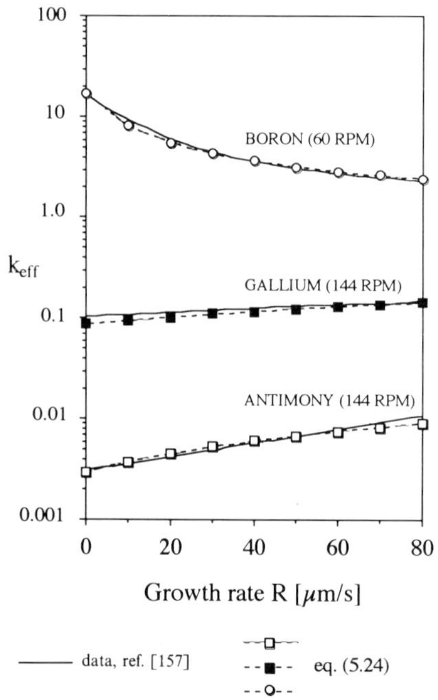

Fig. 70. Effective segregation coefficient k_{eff} versus growth rate R for Czochralski-grown Ge crystals. The experimental data (full lines) are from Bridgers [157, 158]. Dashed lines with symbols are the present model of Ostrogorsky and Müller [154], eq. (5.24).

Bridgers [157] for B, Ge and Sb-doped Ge crystals, grown by the Czochralski method. The model shows an excellent agreement with the segregation data, for both $k < 1$, and $k > 1$.

5.1.5. The case of weak convection

For conditions of very low convective velocity, V_D less or of the order of R, (e.g. growth in microgravity) substitution of eq. (5.11) in eq. (5.19) results in a model of k_{eff} for the case of weak convection

$$k_{\text{eff}} = \left[1 + 4.6a\left(\frac{V_D}{R^2LD^{-1}}\right)\right]\left[1 + 4.6a\left(\frac{V_D}{R^2LD^{-1}}\right)\frac{1}{k}\right]^{-1}. \tag{5.25}$$

5.1.6. Maximum convective velocity compatible with no macrosegregation

Theoretically, any non-zero velocity will cause lateral removal of solute, resulting in $C_S < C_{L\infty}$ (i.e. $k_{\text{eff}} < 1$). Considering the uncertainty in the parameters R, D, and k, a value of $k_{\text{eff}} = 0.95$ can be taken as a criterion for convection-free growth. For $k_{\text{eff}} = 0.95$, eq. (5.19) is transformed into the criterion for the maximum velocity, allowing diffusion-controlled segregation

$$V_{D,\max} = 0.05\left(\frac{R^2L}{D}\right)\frac{k}{0.95 - k} \tag{5.26}$$

or, using eq. (5.11)

$$V_{\infty,\max} = 0.05\left(\frac{R^2L}{D}\right)\frac{k}{0.95 - k} \times \text{Sc}^{1/3}. \tag{5.27}$$

$V_{\infty,\max}$ is plotted in fig. 71, as a function of k and the parameter R^2L/D [152]. The same figure shows the maximum allowable velocity for several solvent–solute systems, previously solidified in space.

Significant improvement in axial segregation was reported on space-grown Te-doped InSb crystals (Skylab-III mission) [159]. Figure 71 shows that the $V_{\infty,\max}$ is relatively high for the Te-doped InSb ($V_{\infty,\max} \approx 10\,\mu\text{m/s} \approx 3R$), because of the high equilibrium segregation coefficient, $k_{\text{Te}} \sim 0.5$. and low $D \sim 2 \times 10^{-5}\,\text{cm}^2/\text{s}$. The fact that $k_{\text{eff}} \approx 1$ was achieved indicates that convective velocities in the melt were $V_{\infty,\max} \leq 10\,\mu\text{m/s}$. This conclusion agrees with the results of Alexander et al. [162] who found that the steady residual acceleration of $10^{-5}\,g_0$ (= earth gravity), acting parallel to the 1 cm long growth interface, will induce melt velocities up to twice the growth rate $R = 6.5\,\mu\text{m/s}$.

$V_{\infty,\max} < 1\,\mu\text{m/s}$ is required for diffusion-controlled segregation of Ga-doped Ge, because of the low equilibrium distribution coefficient $k_{\text{Ga}} = 0.087$ and high diffusion coefficient, $D = 1.5 \times 10^{-4}\,\text{cm}^2/\text{s}$, fig. 71. Accordingly, most of the space-grown Ga-doped Ge crystals exhibited only a modest improvement in homogeneity. Only in the crystal growth experiment carried out aboard the Apollo–Soyuz mission by Witt et al. [160], a rapid increase in axial concentration of Ga was observed, attributable to the diffusion-controlled mass transfer in the melt. Yue and Voltmer grew Ga-, B- and

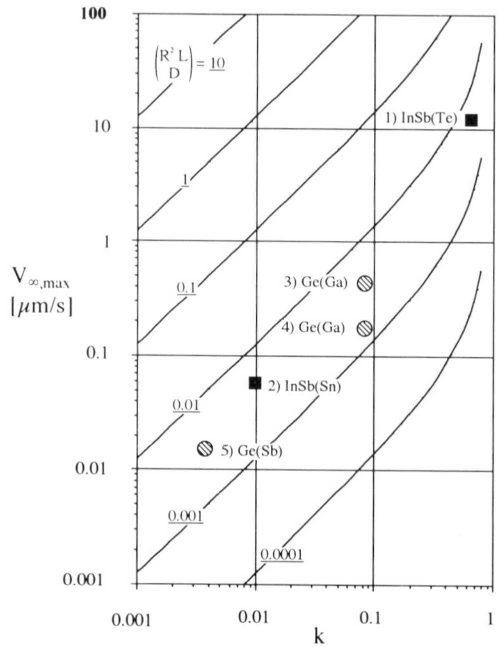

Fig. 71. Maximum tolerable melt velocity $V_{\infty,\max}$ as a function of k and parameter (R^2L/D), resulting in $k_{\text{eff}} = 0.95$. Experimental parameters of space grown crystals are: (1) Te- and (2) Sn-doped InSb; $R \approx 3.4\,\mu\text{m/s}$, $L \approx 1$ cm, $D = 2 \times 10^{-5}\,\text{cm}^2/\text{s}$ [159]; (3) Ga-doped Ge; $R \approx 8\,\mu\text{m/s}$, $L \approx 1$ cm, $D = 1.5 \times 10^{-4}\,\text{cm}^2/\text{s}$ [160]; (4) Ga-doped Ge; $R \approx 5\,\mu\text{m/s}$, $L \approx 1$ cm, $D = 1.5 \times 10^{-4}\,\text{cm}^2/\text{s}$ [161]; (5) Sb-doped Ge; $R \approx 5\,\mu\text{m/s}$, $L \approx 1$ cm, $D = 5.5 \times 10^{-5}\,\text{cm}^2/\text{s}$ [161].

Sb-doped Ge crystals aboard Skylab Mission III [161]. They reported an improvement of radial and micro segregation of Ga in Ge, while the effective distribution coefficient was $k_{\text{eff}} \sim 2k \sim 0.2$. The difference in segregation in space- and earth-grown Ge crystals doped with B and Sb was not reported. Other experiments [163–166] resulted in significant asymmetric lateral segregation, indicating convective redistribution of the dopant along the growth interface. Zemeskov et al. [164] grew Sb-doped Ge–1%Si aboard the Apollo–Soyuz mission and reported enhanced macro- and micro-inhomogeneity compared to the crystals grown under the same conditions on the ground. The experiments were performed at $4 \times 10^{-3}\,g_0$. From our model, the maximum velocity allowing diffusion-controlled segregation of Sb-doped Ge is $V_{\infty,\max} \approx 0.01\,\mu\text{m/s}$, fig. 71.

Based on fig. 71, diffusion-controlled segregation in single crystal growth from melt–solute systems with $0.1 < k < 1$ (e.g. Te in InSb) is achievable in microgravity because, in this system, significant convection can be tolerated ($V_{\infty,\max} \sim R$). On the other hand, diffusion-controlled segregation from melt–solute systems with $k \ll 0.1$ (e.g. Sn in InSb, Sb-doped Ge) is not feasible even under microgravity conditions, because the maximum velocity allowing segregation without convective interference is $V_{\infty,\max} \ll R$.

5.2. *Unsteady convection*

5.2.1. *Instabilities induced by convection*

It has been demonstrated in sections 3 and 4 that, after a flow transition, convective flows become time-dependent (unsteady), if the flow velocities are increased beyond a certain value. The transition occurs when the relevant dimensionless parameter exceeds a critical value (compare section 4). Considering the influence of flow on the species transport in the melt and the composition of the growing crystal, one would expect the compositional variations to result from time-dependent flows.

On the other hand, it has also been proved both theoretically and experimentally, that convective instabilities result in fluctuations of growth rate, which also causes compositional fluctuations in the crystal. These compositional fluctuations are generally called "striations" (see for example [1, 167]).

A direct correlation between flow instabilities, i.e. temperature fluctuations and doping striations was first experimentally shown by Hurle [168] and Müller and Wilhelm [169]. In fig. 10 Müller and co-workers [1] demonstrate a similar result. They measured the temperature by a thermocouple positioned in the melt in the front of the interface, during growth of a Te-doped InSb crystal by the top seeded vertical gradient freeze technique.

The correlation between fluctuations of the growth rate and the occurrence of doping striations has been shown by Murgai et al. [170]. These authors measured the growth rate on a microscale by interface demarcation, with current pulses (Peltier effect) during the growth of Sb-doped Si. The compositional variation of the Sb-dopant was quantitatively evaluated by electrical probing (spreading resistance technique).

The correlation between flow instabilities, (i.e. temperature fluctuations) and fluctuations of growth rate exists because the growth rate is controlled by the heat flux at the interface. At the same time, the rate of convective solute transport to and from the interface fluctuates under unsteady flow conditions. Hence it cannot be decided a priori which of the two mechanisms is responsible for solute microsegregation, i.e. the formation of striations.

The process of microsegregation can be modelled as a growth interface, oscillating due to unsteady convection. This is a rather complex task as both the heat and species transport are time-dependent and are coupled to the unsteady flow. A further complication comes from the fact that the freezing temperature $T_m(C)$ depends on the interfacial solute concentration $C(R)$ which in turn is a function of the growth rate (R). Present existing models for non-facetted growth of semiconductors are based on the assumption that growth is controlled by heat transport. The interface kinetics is ignored, i.e. the position of the interface is given by the condition $T\,(\text{interface}) = T_m(C_0)$ with $C_0 = \text{constant}$.

Several groups studied quantitative and semi-quantitative theoretical models of the variation of crystal composition due to fluctuations of growth rate. It should be mentioned in advance, that no quantitative results can be expected from using the Burton, Prim and Slichter (BPS) [155] or the Ostrogorsky and Müller (section 5.1) model of the effective segregation coefficient k_{eff}, as it has only been derived for stationary conditions. Nevertheless, eq. (5.19) for k_{eff}, correctly suggests that variations of the growth rate R and/or the thickness of the relevant boundary layer δ_D can cause compositional variations. But no information can be obtained about the dynamics of this effect, i.e. the amplitudes of the concentration variations and the frequency dependence.

Hurle et al. [171] studied the segregation using BPS-k_{eff} for low fluctuation periods. They obtained the "expected" proportionality between concentration variation and the amplitudes of the growth rate. Later on, the boundary layer model was improved by a semi-quantitative treatment by the group of Favier [172–176]. They extended the BPS model by introducing a time-dependent growth rate. The transport equation of the species containing a sinusoidal variation of the growth rate is only solved numerically in the linear term.

A similar model has been developed by Van Run [177] for modelling striation formation in the Czochralski process by adding an expression for the axial velocity taken from Cochran's [149] analysis of a rotating disk. Van Run considers also diffusion of the striated solute in the as-grown crystal.

A more basic treatment of microsegregation based on the solution of the transport equations without any boundary layer considerations is given in several papers by Wilson [178–183]. Her model of the Czochralski process is based on the Cochran analysis [149] of a rotating disk considering the solute segregation only in the axial direction. In contrast to the boundary layer models it can be used to discuss the effects of backmelting.

A further generalization of Wilson's model was made by Wheeler [184–186]. He considered limiting cases of various dimensionless groups (Re $\to \infty$, Sc $\to \infty$, Pr $\to 0$) by an analytical expansion of the governing equations.

A comparison of the results of these models with those of real crystal growth was made by Müller [1]. He considered the segregation of O and C in the Czochralski growth of Si using the experimental data of Murgai et al. [170]. The amplitudes of the concentration fluctuations are found to agree quantitatively with experiment and the three models of Favier, Wilson and Van Run.

Recently Thevenard et al. [187] studied theoretically the influence of convection-induced thermal oscillations in the melt on the growth rate. In their analytical treatment of the problem they use a linear approximation (first order Fourier series) in one dimension considering a boundary layer problem. The result shows clearly that temperature fluctuations with frequencies below 0.1 Hz give strong fluctuations of the growth rate whereas higher frequencies are damped out. In the same paper these results are checked by experimental studies of the reaction of the growth interface to convection-induced thermal fluctuations showing a good qualitative agreement.

But it also has to be considered, how far from the growth interface is the source of oscillation (compare section 5.2.2). Very recently in our group an extensive numerical

---

Fig. 72. Occurrence of doping (Te) striations as a function of the state of convective flow for three characteristic stages (a, b, c) in the crystal growth of GaSb by the VGF technique with top seeding. Interference microphotographs of etched axial crystal sections (middle) and temperature plots (right) recorded by the thermocouples $T_1 - T_2$ in the GaSb melt during that time where the depicted crystal part has been grown. From [1].

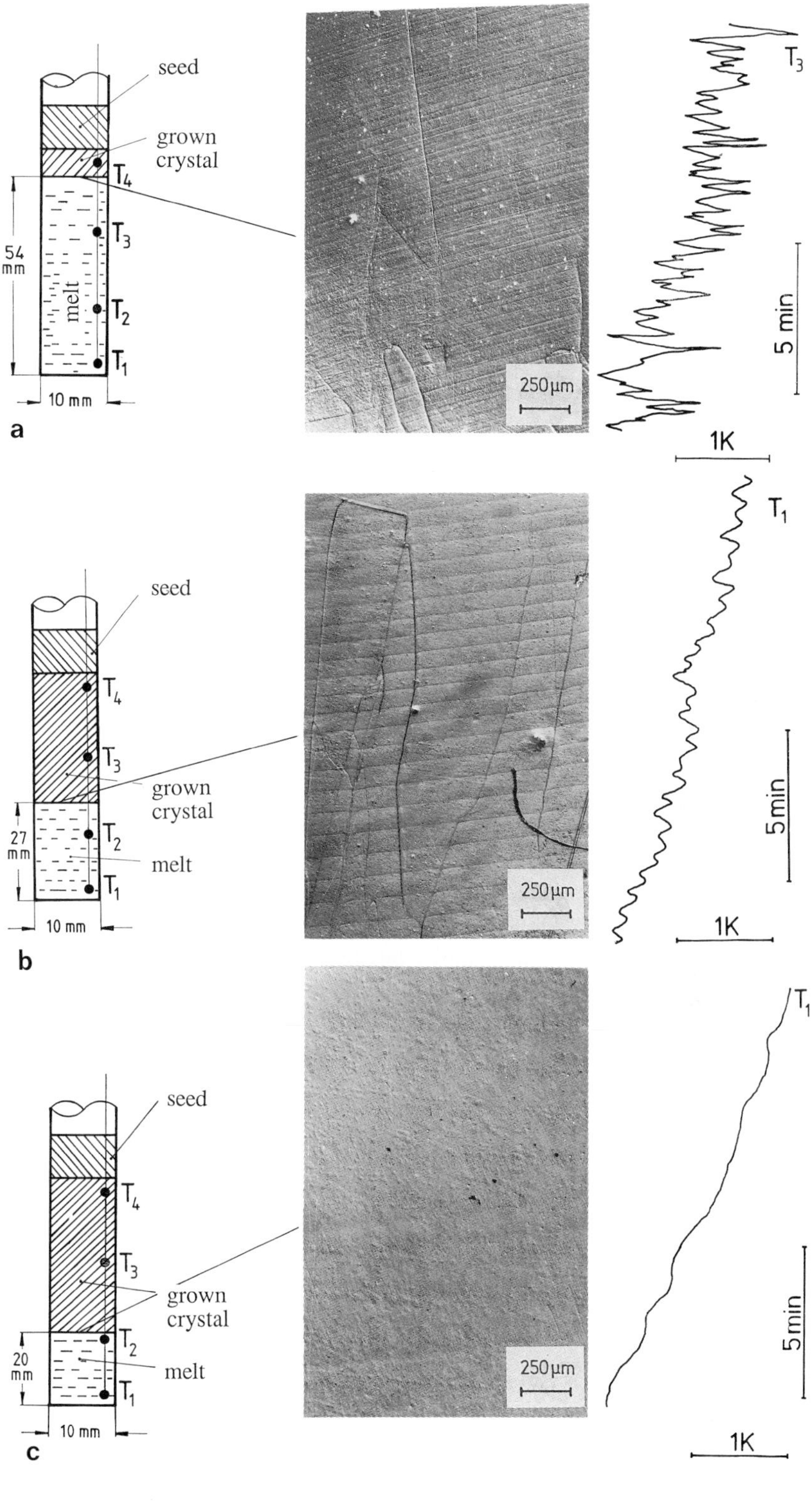

seed
grown crystal
T_4
T_3
54 mm
melt
T_2
T_1
10 mm
a
250 µm
T_3
5 min
1K
seed
T_4
T_3
grown crystal
27 mm
T_2
melt
T_1
10 mm
b
250 µm
T_1
5min
1K
seed
T_4
T_3
grown crystal
20 mm
T_2
melt
T_1
10 mm
c
250 µm
T_1
5min
1K

treatment of time-dependent segregation during crystal growth from a melt under conditions of unsteady convection has been carried out by Jung [90]. He used a finite volume code to solve the time-dependent set of transport equations including species transport for crystal growth of Si-doped GaAs by the vertical Bridgman/gradient freeze configuration (bottom seeding) in a state of unsteady convective flow. First results showing the variations of the temperature, flow rate, resulting growth rate and dopant variation are depicted in fig. 54. Jung was able to calculate also the influence of the unsteady species transport separately from the variation of the growth rate, by using a fixed (i.e. constant) growth rate under the same unsteady flow conditions as used for fig. 54. It shows, that variations in species transport rate contributes in this case about 10% to the striation amplitude.

There is a variety of experimental results concerning the occurrence of growth striations in the different melt growth configurations. However, it is rather difficult to single out from these results those cases in which the striations are clearly attributable to the occurrence of unsteady convection. According to the considerations in section 5.2.1.1, striations can be caused by each of the three driving mechanisms if it causes unsteady convection.

Historically, the occurrence of doping striations is connected with the works of Ueda [188], Gatos et al. [189], Müller and Wilhelm [169], Hurle [190], Utech and Flemings [191].

Later quantitative studies were carried out in the group of Müller mainly based on the vertical Bridgman/gradient freeze configuration with top seeding. A typical experimental result is depicted in fig. 72, which shows three different phases (a, b, c) of a crystal growth experiment. On the left-hand side there is a sketch of the set up, showing the melt heights and the positions of the thermocouples. On the right-hand side the diagram of the simultaneously recorded temperature is shown. Additionally, a longitudinal section of that part of the crystal which grew at that time is also shown. The time scale of the temperature recording and that of the grown length divided by the growth rate are identical in the figure.

Starting with fig. 72a it is notable that non-periodic temperature fluctuations are observed in the melt for large melt heights. These correspond to a non-regular pattern of striations in the grown crystal. Periodic temperature oscillations are found at a smaller melt height, which corresponds to a regular pattern of striations (fig. 72b). If the melt height falls short of a critical value, the temperature oscillations and also the striations disappear (fig. 72c). The crystal grows microscopically homogeneous, i.e. free of striations for all melt heights below that value.

A quantitative comparison of various growth results from the literature with the result of the occurrence of unsteady convection (section 4.1.2) is given in fig. 73. The course of the critical Rayleigh number Ra^{c2} is plotted in the representation versus the aspect ratio taken from section 4.1.2. The signs in the figure denote the occurrence of striations in the experiments. Striations should occur in only those cases where $\mathrm{Ra}(h/d) > \mathrm{Ra}^{c2}$. This theoretical prediction of unsteady convection (fig. 47) is confirmed by fig. 72, especially if the difficulty of a precise determination of Ra is taken into consideration (for more details see [1]). A clear correlation between the period of temperature fluctuations (caused by unsteady convection) and the distance of striations can be seen in fig. 74.

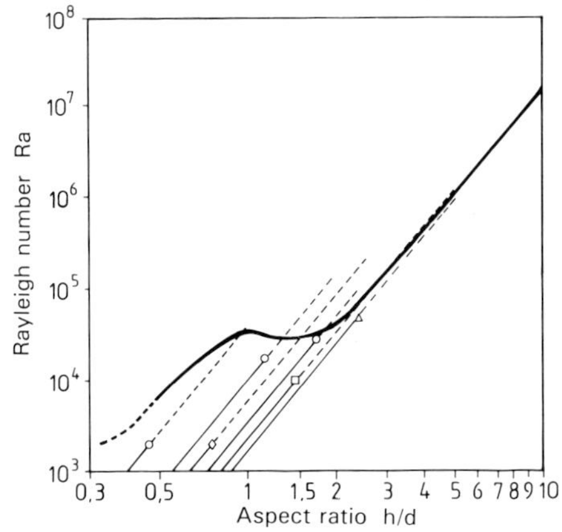

Fig. 73. Ra versus h/d for different crystal growth experiments in the literature, calculated by using the published experimental data. Solid straight lines: growth without striations, dotted lines: growth with striations. The position of the sign gives the value of Ra^{c2} estimated from the literature data. The thick curve $Ra^{c2}(h/d)$ is taken from fig. 47. From [1]. New experimental data from Grasza and Jedrzejczak [192] also fit well on the curve Ra^{c2}.

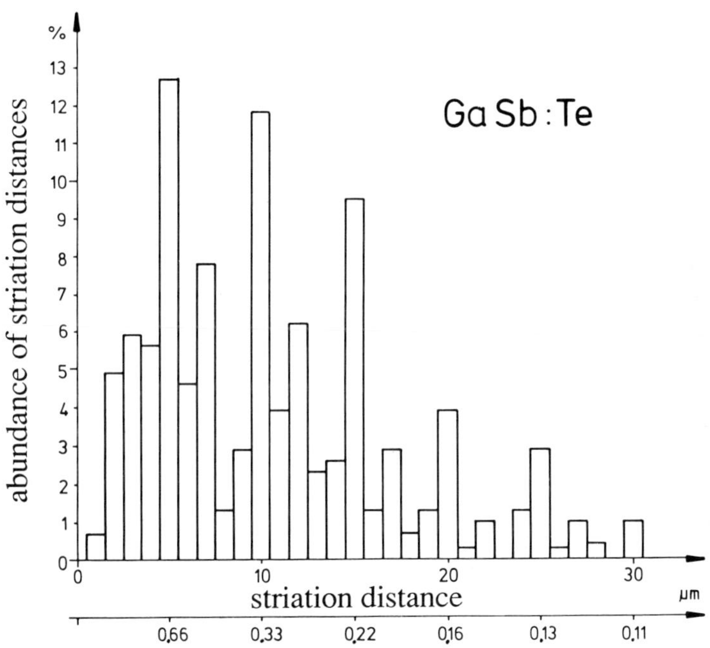

Fig. 74. Histogram of the separation distance of striations of a Te-doped GaSb crystal grown by the horizontal Bridgman technique. The occurrence of the maxima 5, 10, 15 and 30 µm correlates to a basic frequency of 0.11 Hz by using the pulling rate of 0.2 mm/min. From [1].

A corresponding result of the correlation of unsteady buoyancy-driven convection for $Ra_w = 4.2 \times 10^5$ (compare section 4.1) and the occurrence of doping striations was demonstrated by Baumgartl et al. [80] for the growth of GaAs by the vertical zone melting technique.

The occurrence of doping striations in the floating zone growth of GaAs and Si (compare section 4.2) can also quantitatively be correlated with unsteady thermocapillary convection by using the corresponding critical Marangoni numbers, see [107].

Since the temperature field in Czochralski furnaces and melts is not fully axisymmetric, the occurrence of doping striations in crystals grown by the Czochralski method is mainly caused by temperature fluctuations caused by the crystal rotation [28]. But additional to these so-called rotational striations one often finds also a fine substructure in the striation etch pattern caused by unsteady convective flow. This can be proved by suppressing these "non-rotational striations" through the action of a static magnetic field, causing convective flow to become steady (compare section 6 and the article of Hurle and Series in chapter 5 of this handbook). It should also be mentioned, that Carlson and Witt [193] concluded from a quantitative analysis of striations in LEC-grown GaAs crystals, that vortices and turbulence should exist in the melt, causing the striations.

Although the main drawback of convective instability is the occurrence of doping striations, it should be mentioned that temperature fluctuations could also cause crystallographic defects during melt growth. Under conditions where interface kinetics plays an important role, e.g. in facetted growth, generally a high degree of undercooling is needed. It should, therefore, be expected that temperature fluctuations can cause the nucleation of disoriented crystal blocks as proposed in the literature by Hurle [194]. Müller [167] gives an example of a twin in the center of a 2 inch Ge crystal (Ga-doped) grown by the vertical Bridgman technique in the ⟨111⟩ direction. The twinning occurs exactly at a position of a strong striation presumably caused by a temperature variation.

5.2.2. Instabilities carried by convection

Temperature oscillations originating in the melt, furnace atmosphere or caused by crystal rotation are carried to the growth interface mainly by melt convection. Temperature oscillations get rapidly damped out in the solid. At a distance x from the disturbance at the solid surface, the temperature oscillates with an amplitude $\Delta T(x)$, relative to that of the disturbance [195]

$$\frac{\Delta T(x)}{\Delta T(0)} = \exp\left[-\left(\frac{\omega}{2\kappa}\right)^{1/2} x\right], \tag{5.28}$$

where ω (1/s) is the frequency of the disturbance, and κ (cm^2/s) is thermal diffusivity. This one-dimensional heat conduction model is accurate if the thickness of the solid is $L > 3(2\kappa/\omega)^{1/2}$. Equation (5.28) is plotted in fig. 75. For example, in GaAs ($\kappa_{solid} = 0.035\,\mathrm{cm^2/s}$), the amplitude of temperature oscillations, caused by crystal rotation at 30 rpm ($\omega = 2\,\mathrm{s^{-1}}$, $\omega/\kappa \approx 15$) would be reduced to 10% of their surface value, within $x = 1$ cm, and could not cause striations.

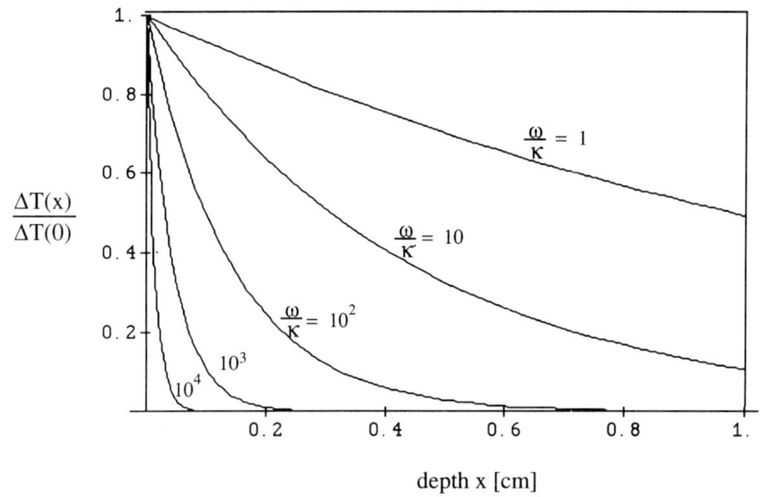

Fig. 75. Penetration depth of temperature oscillations in solids, as a function of (ω/κ). ω (1/s) is the frequency of the disturbance, κ (cm^2/s) the thermal diffusivity of the solid and x is the distance of the solid surface where temperature oscillations occur. The local amplitude of temperature oscillations $\Delta T(x)$, scaled with the disturbance amplitude $\Delta T(0)$.

6. *Control of convection by external measures or variation of parameters*

6.1. *General considerations*

The aim of any melt growth processing control is to provide conditions which give a crystal with well defined compositional and structural properties. In section 5 we have discussed correlations between convective flows and crystal properties. From this we can deduce the following criteria for strategies to control convection in melt growth:

(i) *Unsteady convection* is always undesirable and should be avoided in all cases.

(ii) *Steady convective* flow is useful in many cases of practical melt growth (e.g. production), as it increases the transport of heat (i.e. growth rate) and of species (i.e. reduces the constitutional supercooling). But with respect to the shape of the solid–liquid interface and the isotherms (thermal stress) in the growing crystal, only certain flow modes are advantageous.

Concerning (i) the goal of convection control is clear: the occurrence of unsteady flow has to be suppressed by measures which do not cause side-effects which disturb the crystal growth process. Such measures will be discussed in section 6.2.

Regarding (ii), general strategies are hard to develop, as the net result depends on special growth configurations and material systems, for which the convective flow condition is optimum at every moment of the whole growth run. In principle the same measures are useful for the control of steady flow velocity and its configuration as will

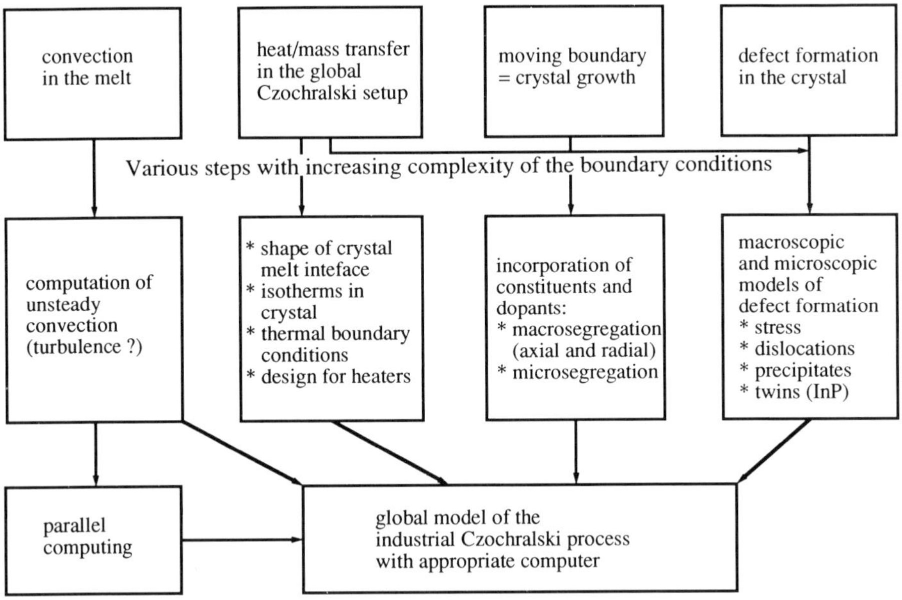

Fig. 76. Strategies for global modelling of the complex Czochralski process.

be discussed for the suppression of unsteady flow. To ensure uniformity of chemical composition of the crystal by quantitative control of flow velocities, it seems to be necessary to use the possibilities of a numerical process control, based on global models, which consider the relevant physico-chemical processes of segregation and defect formation (see fig. 76). Jung [90] gives a typical example of a numerical process modelling for the vertical gradient freeze technique [106]. His process controller minimizes the deviation from a constant value of a set growth rate ($R = 5$ mm/h) in the growth of 2 inch diameter InP crystals, by adjusting the heat transport via the variable boundary conditions (temperature gradients G_S and G_L along solid (S) and liquid (L) part). Figure 77 shows the resulting growth rates calculated for the "controlled" process and for a fixed cooling rate with constant G_S and G_L.

6.2. *Suppression of unsteady convection*

The criteria for the suppression of unsteady convection flow can easily be given by using the formalism of the relevant dimensionless numbers Ra or Gr (buoyancy), Ma (capillary) and Re (forced convection). Conditions of steady flow are achieved, if Ra(Gr), Ma and Re are smaller than the relevant critical values Ra^c, Ma^c and Re^c, as discussed in section 4. Therefore measures have to be considered, which allow for reduction of Ra(Gr), Ma and Re without disturbing the growth process by side-effects. A detailed discussion on this subject is made in the book of Müller [1]; here we only mention some important aspects.

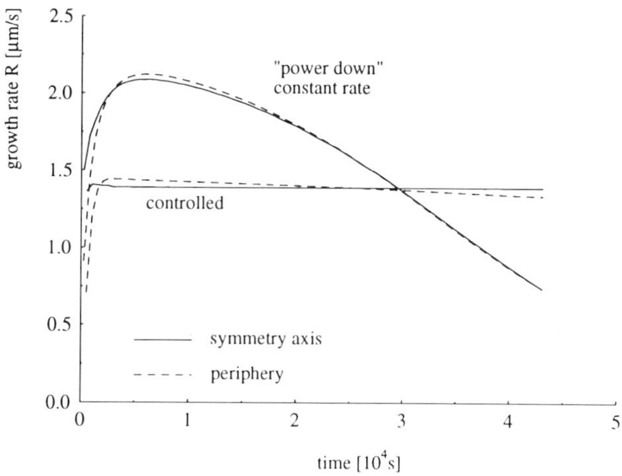

Fig. 77. Calculated growth rates in the VGF process by using a controlled adjustment of the temperature profile compared to a not-controlled power-down-method with constant temperature gradients. From [90].

From the diagrams of figs. 40 and 41 it follows clearly that the vertical Bridgman configuration with bottom seeding and the submerged heater method (SHM) are most favorable crystal growth techniques to avoid convective instability. For other melt growth configurations having a relatively large melt size (e.g. Czochralski), convective instability can only be avoided by a favorable variation or additional application of certain body forces.

6.2.1. Magnetic fields

The most widely used method of stabilizing convective flow in semiconductor melt growth is the application of magnetic fields as reviewed by Hurle and Series (Chapter 5, this Handbook). It has been shown by numerical calculation and experimental analysis that a certain strength of the magnetic induction must be applied in the various melt growth configurations to avoid convective instability in the (conductive !) semiconductor and metal melts. The results can be presented in terms of stability diagrams of the critical parameters Ra^c and Ma^c versus the Hartmann number Ha. Figure 78 shows the reduction of the flow velocity and the transition to steady flow with increasing strength of the magnetic field in the melt cylinder of fig. 39. A typical experimental result demonstrating the useful action of the magnetic field is shown in fig. 79 where the unsteady buoyancy-driven flow is clearly transformed into a steady state.

The favorable action of the magnetic field for stabilizing thermocapillary flow in floating zones is shown in fig. 80 by a plot of Ma^c versus the dimensionless magnetic field strength Ha. Typical changes of the flow patterns with increasing field are shown in fig. 81.

The effects of a magnetic field on Marangoni and buoyancy-driven convection in horizontal open boat configurations are studied numerically by Hirata et al. [197]. These authors came to the conclusion that, in open boats, buoyancy-driven convection

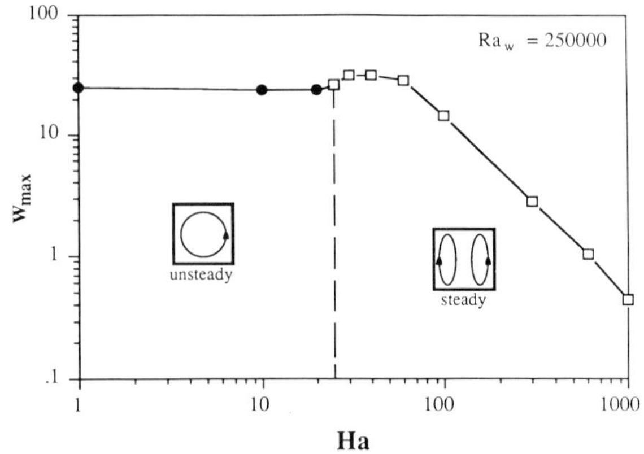

Fig. 78. Numerically calculated maximum value of the vertical (axial) velocity component w_{max} versus the Hartmann number Ha. Full symbols: unsteady convection with a single roll flow pattern. Open symbols: steady convection with axisymmetric toroidal flow. From [80].

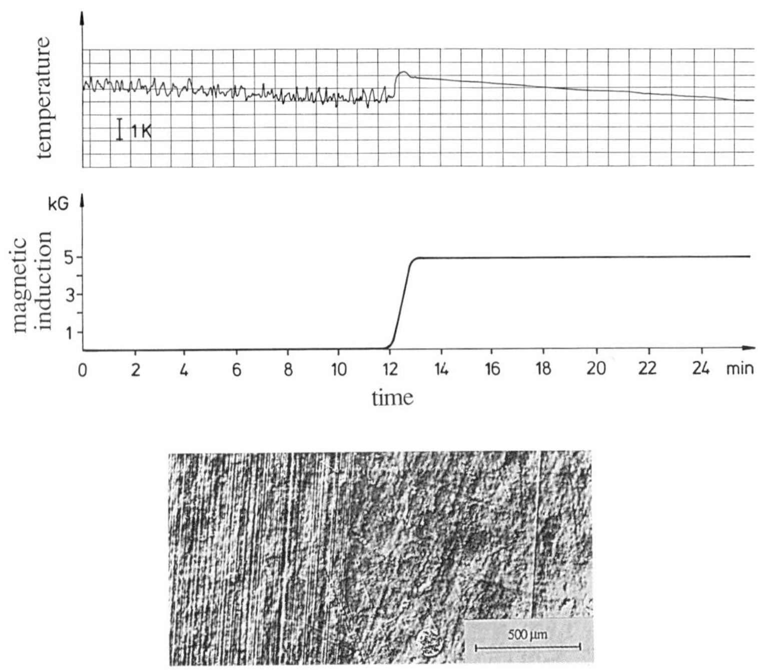

Fig. 79. Horizontal zone melting of Te-doped InSb under the action of a transverse magnetic field, switched on during growth. Longitudinal section (interference contrast microscopy), growth direction from left to right (below). Plot of temperature recording in the melt zone (above, measured during the growth of the depicted crystal part. From [1].

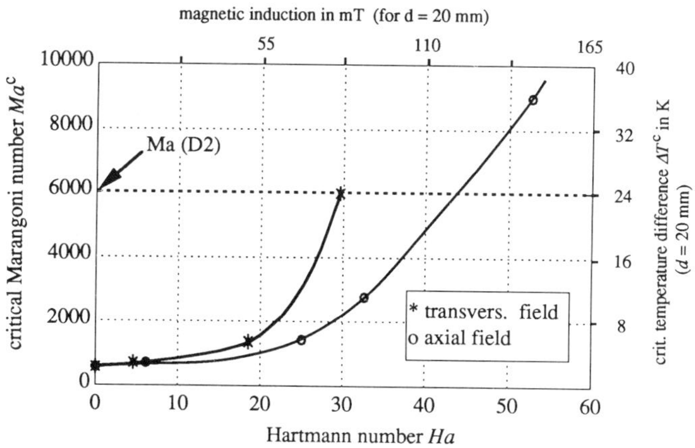

Fig. 80. Dependence of the critical Marangoni number Ma^c for the transition from steady to unsteady thermocapillary flow and of ΔT^c (the temperature difference along the free melt surface) from the strength of an axial or transverse magnetic field (expressed by Ha). The upper and the left axis give the values for relevant variables for a FZ configuration of GaAs to be flown under microgravity (Spacelab D2, 1993). From [107].

can be suppressed effectively both by horizontal and vertical fields, whereas Marangoni convection is more effectively suppressed by a vertical field.

The influence of magnetic fields on convective flow in Czochralski configurations is not so clear because of the complex interaction of the various convective mechanisms. Nevertheless some factors can be identified.

- Horizontal (transversal) fields increase the deviation of the flow field (and hence thermal field) from axial symmetry (see e.g. [198]).
- Horizontal (transversal) fields are more effective in reducing convective transport (especially flow components parallel to the crucible wall) which is important for control of the oxygen incorporation in Si crystals (for details, see Hurle and Series chapter 5, this Handbook).
- Typical values of fields to suppress unsteady convective flow in semiconductor melts under production conditions are 0.2 tesla.

But it should be noted that magnetic fields can also cause convective instability. Figure 82 shows a longitudinal section of a semiconductor eutectic InSb–NiSb sample grown by the vertical Bridgman technique with bottom seeding [1]. The part grown under the influence of a transversal magnetic field reveals strong striations, while the parts without magnetic field are free of striations. Temperature measurements confirm that there are no temperature fluctuations in the melt if the magnetic field is not applied while strong fluctuations are generated by induction fields of 0.25 to 0.45 tesla [1]. A similar result was obtained by Walter [199] with the growth of Ga-doped Ge by the vertical Bridgman configuration in a horizontal magnetic field. One possible explanation is, that the transverse field breaks the axial flow symmetry causing an asymmetric flow, which is unsteady under certain thermal boundary conditions [200]. The electrical currents induced by thermoelectric power could also cause instabilities.

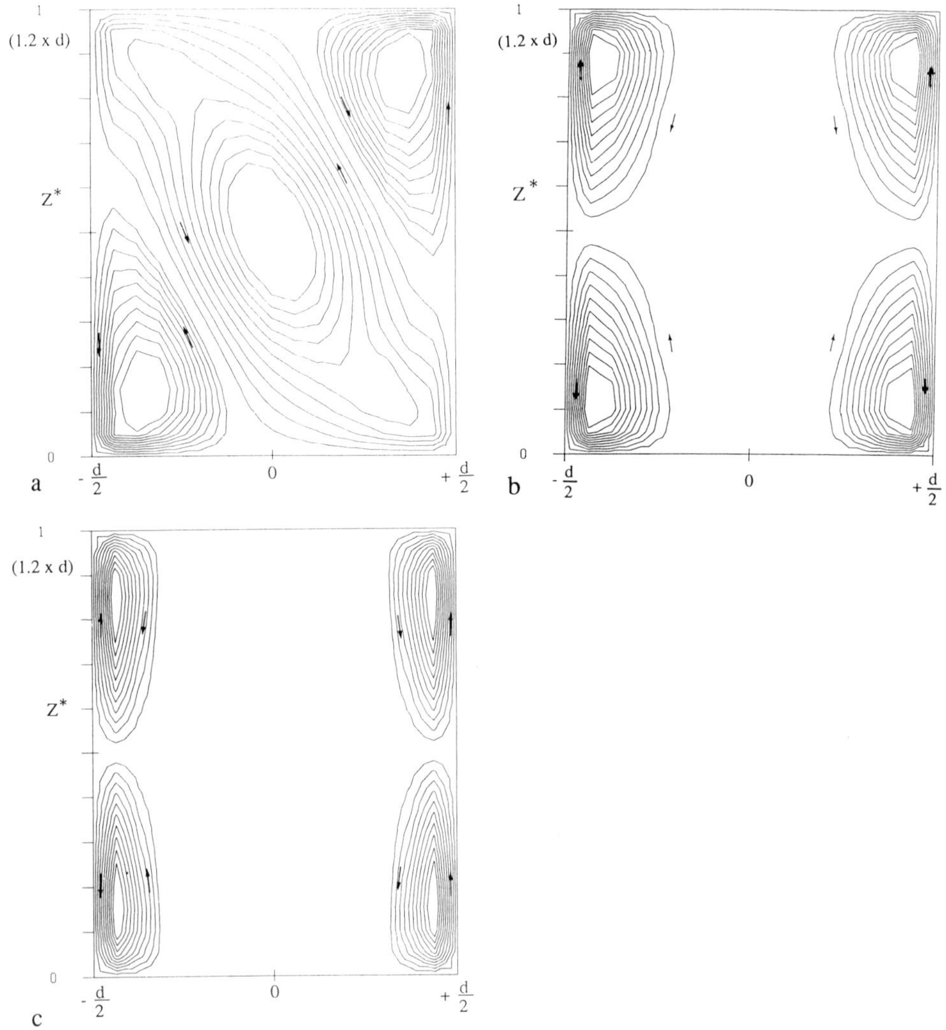

Fig. 81. Calculated flow patterns of thermocapillary convection in floating zone configuration of GaAs (Ma = 720, Pr = 0.068, $h/d = 1.2$), without magnetic field (a) and with axial magnetic field Ha = 50 (b) and 100 (c). From [107].

Baumgartl [85] has shown by numerical calculations, that the interaction of these thermoelectrically induced currents with the static magnetic field B_0 can lead to unsteady convective flows under typical growth conditions of doped semiconductors from the melt.

6.2.2. Reduction of gravity

Theoretically the simplest way to reduce buoyancy convection is to reduce gravity. However, practically this is the most expensive means, since the necessary microgravity

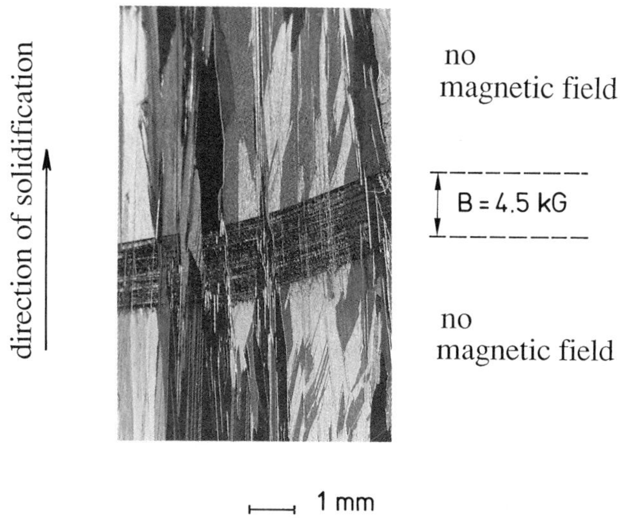

Fig. 82. Longitudinal section of a Te-doped InSb–NiSb eutectic sample solidified directionally with bottom seeding by the vertical Bridgman technique. The striations occurring in the middle part of the sample have been generated during that time when the magnetic field was switched on ($B = 4.5$ kG). From [1].

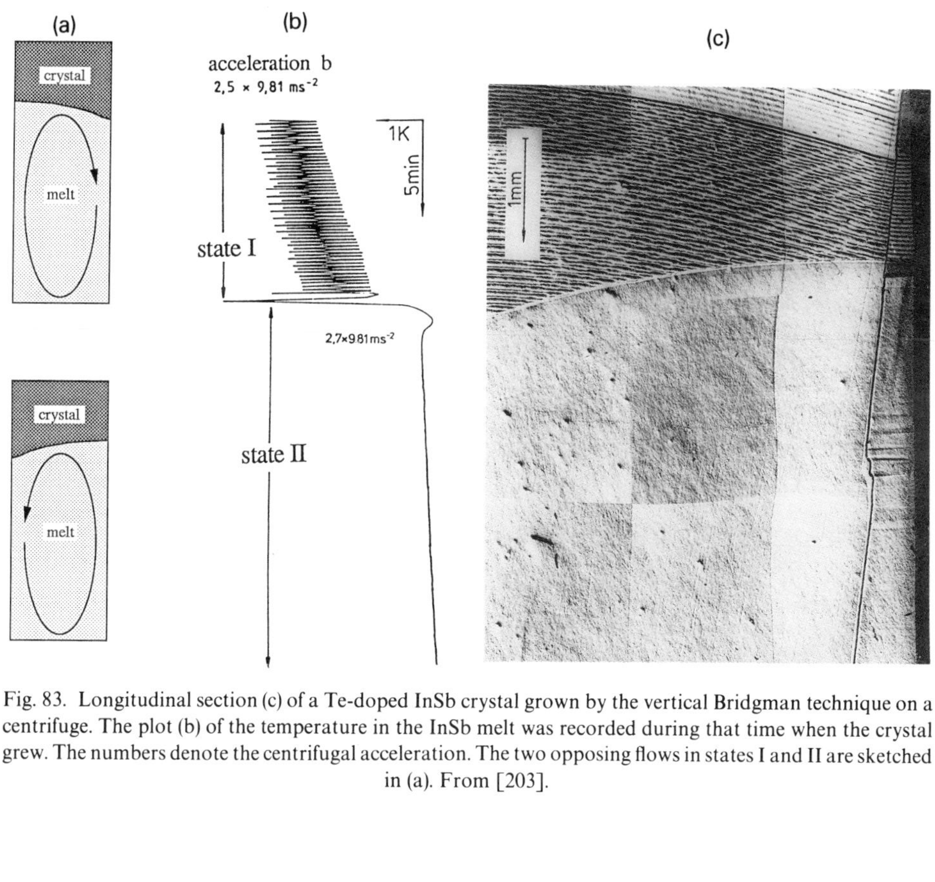

Fig. 83. Longitudinal section (c) of a Te-doped InSb crystal grown by the vertical Bridgman technique on a centrifuge. The plot (b) of the temperature in the InSb melt was recorded during that time when the crystal grew. The numbers denote the centrifugal acceleration. The two opposing flows in states I and II are sketched in (a). From [203].

conditions can only be achieved for melt growth during space missions [201]. But even in space laboratories unsteady buoyancy-driven convection can occur due to the changes of residual gravity in time and direction [202].

Microgravity conditions are useful for the study of capillary convection in floating zones without disturbances from buoyancy (see for example [2, 107]).

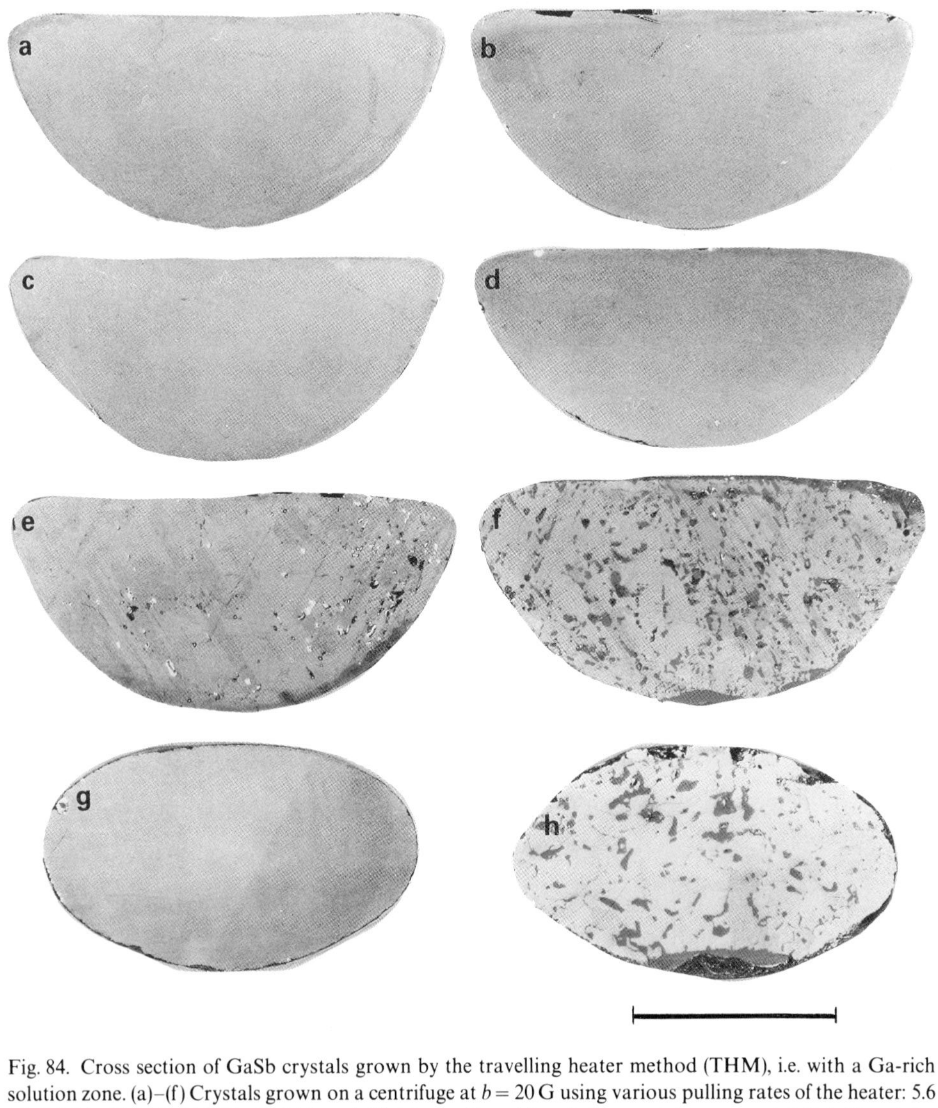

Fig. 84. Cross section of GaSb crystals grown by the travelling heater method (THM), i.e. with a Ga-rich solution zone. (a)–(f) Crystals grown on a centrifuge at $b = 20$ G using various pulling rates of the heater: 5.6 (a), 9.1 (b), 10.7 (c), 14.0 (d), 20.0 (e) and 30.0 (f) mm/day; (g) and (h): crystals grown at 1 G (without the centrifuge) with pulling rates 1.5 (g) and 4.9 (h) mm/day. Marker represents 5 mm. From [203].

6.2.3. *Centrifugal and Coriolis forces*

Extensive experimental studies and numerical simulations of the influence of centrifugal and Coriolis forces on convective flows in various melt growth configurations were carried out by Müller and co-workers [203]. It has been demonstrated that the flow in Bridgman and zone melting configurations can be changed by the combined action of the centrifugal and the Coriolis force to have an increasing transport rate as well as causing the flow to become steady. It was demonstrated that this effect can be used to suppress the formation of convection-induced striations in the growth of semiconductor crystals (see fig. 83).

The effect of the increased convective transport by an increased centrifugal force was used to improve the growth conditions in the THM (travelling heater method). The growth rate in THM is normally limited to values of a few millimeters per day, because of the limited transport rate of the species in the solution and because of the formation of solution precipitation. THM experiments at accelerations of 20 times gravity with GaSb showed that the maximum growth rate of inclusion-free crystals could be increased by nearly one order of magnitude, compared with the results obtained without centrifuge [204] (fig. 84).

7. *Concluding considerations for an optimization of convective effects in the melt*

In this final section we want to comprehensively consider convective effects towards optimization with respect to the macroscopic and microscopic homogeneity and crystallographic defects in the different melt growth methods.

The method most often used at present is the crystal pulling technique named after Czochralski. Its optimization is the most difficult one because one has to take into consideration the interaction of three different convection mechanisms (buoyancy, rotational and capillarity) and additionally crucible reactions (e.g. for Si). Nevertheless, one can expect a distinct improvement of the homogeneity with respect to buoyancy convection, by using the following measures:

(i) Use of smaller effective melt height, e.g. by the use of continuous liquid feed.
(ii) Wall heating instead of bottom heating.
(iii) Configured axial magnetic fields with $B \geq 0.2$ tesla.

The rotation of crystal and crucible seems to be necessary in order to get a cylindrical shape of the crystal and to reduce the influence of the buoyancy convection by producing an axisymmetric boundary flow. The rotation only dominates the convective flow if $\mathrm{Re} \gg (\mathrm{Ra})^{1/2}$. Under normal growth conditions, the buoyancy convection has a strong "thermal transconductance" through the boundary flow of the rotating crystal. An improvement of the homogeneity under the conditions of a rotating crystal and/or crucible is possible by:

(i) Improvement of the thermal axial (rotational) symmetry.
(ii) Co-axial adjustment of the temperature distribution in the crucible and the rotation axis of the crystal

The action of capillary convection (Marangoni effect) is generally undesirable because the transition to unsteady behavior is hard to control. Capillary convection can be reduced by a reduction of the temperature and concentration gradients along the free melt surfaces and by magnetic fields.

A real alternative to the Czochralski process, especially for the growth of compounds, is the vertical Bridgman configuration with bottom seeding. The method offers, from a theoretical point of view as well as from an experimental one, better conditions for the growth of crystals with better uniformity and less crystallographic defects. Only buoyancy convection has to be controlled, as no rotation and no free surface is necessary. The buoyancy convection can be adjusted to have a favorable steady flow. In principle it is possible to reduce buoyancy so far that conditions of a purely diffusive heat and mass transport can be approximately reached, which corresponds to microgravity conditions. This state is reached by a purely axial heat flux, which necessarily corresponds to a plane shape of the interface. This second effect of the axial heat flux has a further advantage for the growth of compounds compared with the LEC method because it reduces the thermal stress.

It is now generally accepted in the scientific community, that further progress in growing low-defect and uniform crystals by controlling convection effects in melt growth can only be achieved if the possibilities of numerical modelling are used. This depends mainly on the future progress in speeding up numerical codes and using the advantages of parallel computing.

Appendix A. Basic equations of convection and principles of solution

A.1. Conservation of energy, species, mass and momentum

If heat transfer by radiation is not accounted for, heat, mass and momentum are transferred by the same physical mechanisms: advection (i.e. bulk fluid motion) and diffusion (random motion of molecules and charge carriers). Therefore, the conservation equations, describing the transport of heat and solutes are similar. The conservation of energy presents the balance between storage, advection and diffusion of heat

$$\underset{\text{(storage)}}{\frac{\partial T}{\partial t}} + \underset{\text{(advection)}}{(V\cdot\nabla)T} = \underset{\text{(diffusion)}}{\kappa\nabla^2 T}, \tag{A.1}$$

where $\kappa = k/\rho c$ is thermal diffusivity, ρ is density and c is specific heat. Similarly, the conservation of mass (i.e. a chemical species) presents the balance between storage, advection and diffusion of each of the chemical species

$$\frac{\partial C^i}{\partial t} + (V\cdot\nabla)C^i = D_i\nabla^2 C^i. \tag{A.2}$$

C^i is the concentration of the species i per unit volume and D_i is the diffusion coefficient.

Because of advection, temperature and concentration fields in the melt are coupled to the velocity field. Since all melt flows are incompressible, the velocity field in the melt is governed by the incompressible form of the continuity equation

$$\nabla \cdot \boldsymbol{V} = 0 \tag{A.3}$$

and the Navier–Stokes equations

$$\frac{\partial \boldsymbol{V}}{\partial t} + (\boldsymbol{V} \cdot \nabla)\boldsymbol{V} = -\frac{\nabla \cdot p}{\rho} + \nu \nabla^2 \boldsymbol{V} + \boldsymbol{b}, \tag{A.4}$$

where p is pressure and $\boldsymbol{b}$ is the body force per unit mass. The buoyancy effects are usually accounted for using the Boussinesq approximation

$$\frac{\partial \boldsymbol{V}}{\partial t} + (\boldsymbol{V} \cdot \nabla)\boldsymbol{V} = -\frac{\nabla \cdot p}{\rho} + \nu \nabla^2 \boldsymbol{V} + g\beta(T - T_{\min}) + g \sum_{i=1}^{n} \beta_S^i (C^i - C^i_{\min}), \tag{A.5}$$

where g is the gravitational acceleration, $T_{\min}$ and $C^i_{\min}$ are respectively the lowest (reference) temperature and concentration of species i in the melt. β is the coefficient of thermal volumetric expansion

$$\beta = -\frac{1}{\rho}\left(\frac{\partial \rho}{\partial T}\right) \tag{A.6}$$

and β_S^i is the coefficient of volumetric expansion due to change in concentration of a species C^i

$$\beta_S^i = -\frac{1}{\rho}\left(\frac{\partial \rho}{\partial C^i}\right). \tag{A.7}$$

At low concentrations of impurities, buoyancy forces due to species transport in the melt are much smaller than the buoyancy force due to heat transfer

$$\beta(T - T_{\min}) \gg \sum_{i=1}^{n} \beta_S^i (C^i - C^i_{\min}), \tag{A.8}$$

so that the last term in eq. (A.5) is ignored.

In many cases heat transfer by radiation has additionally to be considered [205].

A.2. Non-dimensional form of the conservation equations

Studies of melt convection are usually carried on using non-dimensional forms of the above conservation equations. Combining the dimensional parameters into non-dimensional parameters (starred)

$$x_j^* = x_j/L \quad (j = 1, 2, 3),$$

$$t^* = t\nu/L^2,$$

$$V^* = VL/\nu = \mathrm{Re},$$

$$T^* = (T - T_{\min})/(T_{\max} - T_{\min}),$$

$$C^{*i} = (C^i - C^i_{\min})/(C^i_{\max} - C^i_{\min}),$$

$$p^* = p(L^2/\nu^2)/\rho,$$

reduces the total number of independent parameters. As a result, the complexity of experiments and numerical studies is reduced. At the same time, the non-dimensional results obtained from a particular experiment are more general. The conservation equations based on above non-dimensional parameters are:

$$\nabla \cdot V^* = 0, \tag{A.9}$$

$$\frac{\partial V^*}{\partial t^*} + (V^* \cdot \nabla) V^* = -\nabla p^* + \nabla^2 V^* + \mathrm{Gr} \times T^* + \sum_{i=1}^{n} \mathrm{Gr_S} \times C^*, \tag{A.10}$$

$$\frac{\partial T^*}{\partial t^*} + (V^* \cdot \nabla) T = \frac{1}{\mathrm{Pe}} \nabla^2 T^*, \tag{A.11}$$

$$\frac{\partial C^{*i}}{\partial t^*} + (V \cdot \nabla) C^{*i} = \frac{1}{\mathrm{Sc}} \nabla^2 C^{*i},$$

where

$$\mathrm{Gr} \equiv \frac{g\beta(T_{\max} - T_{\min}) L^3}{\nu^2}, \tag{A.12}$$

and $\mathrm{Gr_S^i}$ is the solutal Grashof number

$$\mathrm{Gr_S^i} \equiv \frac{g\beta_S^i (C^i_{\max} - C^i_{\min}) L^3}{\nu^2}. \tag{A.13}$$

In above equations, kinematic viscosity ν and characteristic length scale L, were used to non-dimensionalize time and velocity. An equally common approach is to use the κ and L as scaling parameters. In that case, $V^* = \mathrm{Re} \times \mathrm{Pr} = \mathrm{Pe}$ while the dimensionless time is equal to the Fourier number, $t^* = t\kappa/L^2 = \mathrm{Fo}$.

A.3. *Analytical solutions*

Because of the complexity of the conservation equations (system of coupled non-linear partial differential equations), a general analytical solution describing convection of heat, mass and momentum does not exist. For idealized initial and boundary conditions, and simple flows, a number of particular solutions has been found in terms of well-known mathematical functions. These solutions can be found in classic textbooks on fluid mechanics (e.g. [206, 207, 146]), convective heat and mass transfer [208, 209] and in the text on physicochemical hydrodynamics [147].

For melt growth, most relevant are the solutions describing transport in boundary layers along flat surfaces [147, 209] and near the infinite rotating disk [147]. From these solutions, the thickness of momentum, thermal and solute boundary layers is determined. The results, relevant for crystal growth are given in section 5.1.1.

A.4. *Numerical solutions*

Using numerical techniques on digital computers, numerical solutions of the complete set of equations governing convective transport can be obtained. For initial and boundary conditions that yield laminar flows, numerical solutions providing any desired level of accuracy can be found. Solutions of turbulent convection have taken two distinct paths:

(i) Solution of Reynolds time averaged equations, in which the eddy transport must be accounted for through a model (e.g. Prandtl's mixing length model, $k-\varepsilon$ model).

(ii) Direct simulation: the flow domain is divided into a mesh which is much finer than the size of the small eddies; the set of unsteady transport equations is solved to duplicate all the details of the flow and eddy mixing.

The state of the art of numerical solutions of convection can be found in textbooks [210, 211] and review articles [212] devoted entirely to computational fluid mechanics, heat and/or mass transfer [213]. The contribution by Dupret and Bogaert in chapter 15 of this handbook is devoted to numerical solutions of melt convection.

The numerical simulations are performed by using finite differences [214], finite elements [215], or finite volumes [211, 216]. The finite difference method is based on approximation of the partial derivatives by algebraic difference between discrete mesh points. Partial differential equations are replaced by their algebraic approximation based on mesh-points models. In the finite element method, the flow domain is divided into a number of finite elements. The finite element method is especially useful for complicated geometries and boundary conditions.

Computer programs for numerical simulations of convection have been developed during the past one or two decades. The state of the art computer codes include: NEKTON [217], STAR [216], FIDAP [218], FLUENT [219], and PHOENICS [220]. Given the level of sophistication of these programs which have been developed by teams of professionals, and the fact that these programs are user-friendly, development of "in-house software" has become increasingly less popular.

Appendix B. Numerical model for the simulation of thermocapillary (Marangoni) convection in cylindrical floating zones

The numerical modelling of the floating zone (FZ) configuration is based on a fixed cylindrical shape of the zone and the slip condition in the free surface of the cylinder mantle. Neglecting gravity, the basic transport equations for mass (continuity), momentum and heat in a fluid with constant kinematic viscosity ν thermal diffusivity κ and density ρ (Boussinesq approximation) can be written in dimensionless form as follows (compare also Appendix A):

$$\nabla \underline{v}^* = 0, \tag{B.1a}$$

$$\frac{\partial \underline{v}^*}{\partial t^*} + \underline{v}^* \cdot \nabla \underline{v}^* = -\nabla p^* + \mathrm{Pr} \times \nabla^2 \underline{v}^*, \tag{B.1b}$$

$$\frac{\partial T^*}{\partial t^*} + \underline{v}^* \cdot \nabla T^* = \nabla^2 T^*. \tag{B.1c}$$

Herein ∇ represents the gradient operator in the cylindrical coordinate system $r = (r, \theta, z)$. The Prandtl number Pr is defined as usual by $\mathrm{Pr} = \nu/\kappa$. For the non-dimensionalization of these equations the following conventions are used (the asterisk marks the dimensionless variables):

radial coordinate $$r^* = \frac{r}{h}, \tag{B.2a}$$

axial coordinate $$z^* = \frac{z}{h}, \tag{B.2b}$$

time $$t^* = t\frac{\kappa}{h^2}, \tag{B.2c}$$

velocity $$\underline{v}^* = \underline{v}\frac{\kappa}{h}, \tag{B.2d}$$

pressure $$p^* = p\frac{h^2}{\rho\kappa^2}, \tag{B.2e}$$

temperature $$T^* = \frac{T - T_\mathrm{m}}{\Delta T_Q^{\max}}. \tag{B.2f}$$

This non-dimensionalization of the temperature is related to the special thermal boundary condition which is used at the cylinder mantle. This boundary condition is implemented in the numerical model as a radial heat flux by defining a radial temperature difference $\Delta T_Q = T_{\mathrm{L}+1} - T_\mathrm{L}$ between the last points of the numerical grid outside (L + 1) and inside (L) of the melt in radial direction. The radial heat flux $Q(z)$ is defined by $Q(z) = \text{const.} \times \Delta T_Q/\Delta R$, with ΔR is the width of the numerical grid in radial direction. $\Delta T_Q^{\max}$ is then the maximum of the ΔT_Q parabola (shown in fig. 85 for $z = h/2$). $\Delta T_Q^{\max}$ is kept constant in the numerical procedure as a fixed boundary condition. This

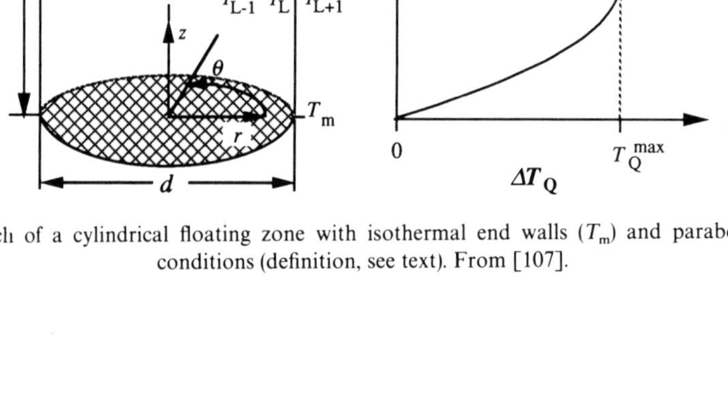

Fig. 85. Sketch of a cylindrical floating zone with isothermal end walls (T_m) and parabolic heat flux conditions (definition, see text). From [107].

parabolic heat flux boundary condition fits quite well to the experimental conditions of a mirror furnace which is used by several groups in terrestrial and microgravity FZ experiments (especially in the D2-Spacelab mission for FZ growth of GaAs by the group of Müller).

This new definition of the thermal boundary conditions seems to be necessary as the temperature differences in the free surface are the only driving force for the convective flow in the case of assumed pure thermocapillary convection. The heat transport, therefore, is most intense in the surface region. It tries, for energetic reasons, to change the driving temperature gradients in the surface by itself. Therefore, a fixed temperature distribution in the free surface given as a boundary condition would lead to unrealistic results of the numerical modelling.

With the imposed heat flux boundary condition, a temperature distribution $T(r=d/2, z, \theta)$ in the free surface is generated leading to the following equilibria for the shear tension by assuming a constant value of $\mathrm{d}\sigma/\mathrm{d}T$ ($\sigma=$ surface tension):

$$\frac{\partial T}{\partial z}\frac{\partial \sigma}{\partial T} = -v\left(\frac{\partial v_z}{\partial r}\right), \tag{B.3}$$

$$\frac{1}{r}\frac{\partial T}{\partial \theta}\frac{\partial \sigma}{\partial T} = -v\left[r\frac{\partial}{\partial r}\left(\frac{v_\theta}{r}\right)\right]. \tag{B.4}$$

Making this shear stress boundary conditions dimensionless, eqs. (B.3) and (B.4) can be written in the following form:

$$\frac{\partial T^*}{\partial z^*}\times \mathrm{Ma}_Q = \frac{\partial v_z^*}{\partial r^*}, \tag{B.5a}$$

$$\frac{\partial T}{\partial \theta}\times \mathrm{Ma}_Q = r^{*2}\frac{\partial}{\partial r^{*2}}\left(\frac{v_\theta^*}{r}\right), \tag{B.5b}$$

by defining a new Marangoni number Ma_Q based on the given heat flux Q

$$\mathrm{Ma}_Q = \frac{(\partial\sigma/\partial T)\Delta T_Q^{\max} h}{\kappa v \rho}. \tag{B.6}$$

Ma_Q is different from the normally used Ma for floating zone configurations (see for example section 5 of this handbook)

$$\mathrm{Ma} = \frac{-(\partial\sigma/\partial T)\Delta T\, h}{\kappa v \rho}. \tag{B.7}$$

Ma is based on a temperature difference ΔT which would be $\Delta T = T_{\max} - T_{\mathrm{m}} = T(h=z/2) - T_{\mathrm{m}}$ in our FZ configuration. Using the dimensionless T^* (eq. (B.2f) we obtain

$$\Delta T = T_{\max}^* \Delta T_Q + T_{\mathrm{m}} - (T_{\mathrm{m}}^* \Delta T_Q + T_{\mathrm{m}}). \tag{B.8}$$

By using $T_{\mathrm{m}}^* = 0$ it follows:

$$\Delta T = T_{\max}^* \Delta T_Q^{\max}. \tag{B.9}$$

This ΔT is a result of the Marangoni flow itself and follows from the calculations. The resulting ΔT can be used according to eq. (B.7) to calculate Ma. We have, therefore, a relation between Ma and Ma_Q.

To solve the eq. (B.1a–c) for conditions defined by eqs. (B.5a and b) a numerical code was used, which has been developed in our group by Neumann [81] and modified by Rupp et al. [108]. Additional details are described in a paper by Müller and Rupp [107].

Appendix C. Turbulent melt convection

For the past twenty five years, an increasing number of experimental [50, 54] and analytical studies [64] suggests that melt convection in large semiconductor melts should be turbulent. For example, Ristorcelli and Lumley [64] recently identified more than ten fluid mechanics instabilities associated with buoyancy, surface tension and rotational forces which should lead to turbulence in large Czochralski melts. Nevertheless, in part because of the great complexity of turbulent flows, virtually all analytical and numerical simulations of transport processes in the melt were performed assuming laminar flow in the melt.

Turbulent convection is very effective in transferring heat, mass and momentum. In contrast to the laminar flow where the fluid moves along smooth streamlines, turbulent flow is characterized by a spectrum of coexisting vortices or eddies, leading to intense mixing. Because of the mixing, the velocity, temperature and concentration fields fluctuate randomly. Figure 86 shows schematically the mechanism of eddy transport. The fluid lump A moving with velocity V_A, temperature T_A, and impurity concentration C_A is transferred by an eddy from the position y_A to position $y_B = y_A - l$, where l is the so called "mixing length". At the same time the fluid lump B (having velocity V_B, temperature T_B, and impurity concentration C_B) takes the initial position of the lump A.

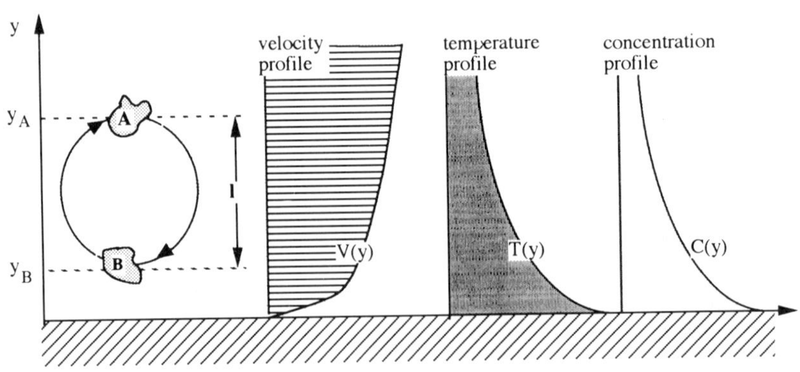

Fig. 86. Mechanism of heat, mass and momentum transfer by eddies in turbulent flow. l is the mixing length (i.e., characteristic eddy size).

There is a noticeable similarity between the mechanism of eddy mixing described above, and the molecular diffusion. In turbulent flow, instead of molecules, the fluid lumps are the main carriers the heat, mass and momentum, while the eddy size or the mixing length are equivalent to the main free path of the molecules. In laminar flow, a temperature gradient $\partial T/\partial x$ induces diffusive heat flux

$$q_{\text{molec}} = -k\frac{\partial T}{\partial y} = -\rho c\kappa\frac{\partial T}{\partial y}. \tag{C.1}$$

In the turbulent flow the same gradient would induce an apparent flux

$$q_{\text{app}} = q_{\text{molec}} + q_{\text{eddy}} = -\rho c(\kappa + \varepsilon_h)\frac{\partial T}{\partial y}, \tag{C.2}$$

where ε_h is the turbulent eddy heat diffusivity, while $(\rho c\varepsilon_h)$ is referred to as "eddy conductivity". Similarly, the apparent mass flux j_{app} and the apparent momentum flux (called the turbulent shear stress) τ_{app} are respectively

$$j_{\text{app}} = j_{\text{molec}} + j_{\text{eddy}} = -(D + \varepsilon_D)\frac{\partial T}{\partial y}, \tag{C.3}$$

$$\tau_{\text{app}} = \tau_{\text{molec}} + \tau_{\text{eddy}} = -(\nu + \varepsilon_\nu)\frac{\partial V}{\partial y}, \tag{C.4}$$

where ε_D is the eddy diffusivity and ε_ν is the eddy diffusivity of momentum, or eddy viscosity.

Since fluid lumps contain a very large number of molecules, and since the mixing length is many orders of magnitude larger than the molecular main free path, it is easy to see that the apparent diffusive (i.e. non-advective) transport in turbulent flow is controlled by the eddy size, rather than by fluid transport properties κ, D and ν. Only very close to the wall, in the "laminar sublayer", eddy mixing is inhibited. The fact that the eddy structure depends on the geometry of the flow domain and the distance from the wall, greatly complicates modelling.

Only metallic and semiconductor melts, which have very high thermal diffusivity due to diffusion of free charge carriers ($\kappa \sim 0.1\ \text{cm}^2/\text{s}$), may have the same order of magnitude as ε_h. In contrast, the diffusion coefficient of impurities in melts is extremely low, $D \sim 10^{-4}\ \text{cm}^2/\text{s}$. Therefore, ε_D is always much larger than D, and may not be ignored under any circumstances.

C.1. Relative magnitude of diffusive transport properties in turbulent flow

In laminar flow, the Prandtl number Pr shows the relative magnitude of momentum and heat transfer by diffusion. Similarly, the Schmidt number Sc scales the relative magnitude of diffusive momentum and mass transfer (see table 1).

Analog dimensionless parameters are used to characterize the eddy transport in turbulent flow. The turbulent Prandtl number Pr scales the relative magnitude of

Table 6
Thermophysical properties of common melts and model fluids.

	T_{mp} (°C)	ρ (g/cm^3)	ν (cm^2/s)	κ (cm^2/s)	k (W/cm K)	C (J/g K)	β (1/K)	Pr	σ (N/m)	dσ/dT (N/m K)	Ref.
Semiconductors											
Si	1410	2.53	0.0035	0.255	0.67	1.037	1.32×10^{-4}	0.014	0.865	-0.13×10^{-3}	[224, 227]
Ge	936	5.51	0.00135	0.175	0.39	0.404	1.11×10^{-4}	0.008	0.621	-0.26×10^{-3}	[224, 227]
GaAs	1238	5.72	0.00488	0.072	0.178	0.434	1.87×10^{-4}	0.068	0.6	-0.1×10^{-3}	[240, 234]
GaSb	712	6.03	0.00383	0.087	0.171	0.328	9.58×10^{-5}	0.044	0.44		[225, 234]
InP	1062	5.05	0.00162	0.103	0.228	0.424	4.44×10^{-4}	0.015			[225]
InSb	525	6.48	0.0033	0.06	0.0923	0.263	1.0×10^{-4}	0.055	0.29		[225, 226, 234]
Metals											
Ga	29.8	6.09	0.0032	0.105	0.255	0.398	9.85×10^{-5}	0.03[a]	0.718	-0.1×10^{-3}	[227]
Sn	232	7.0	0.0026	0.171	0.30	0.250		0.015	0.544	-0.07×10^{-3}	[227]
Zn	420	6.575	0.0058	0.1565	0.495	0.481		0.027	0.782	-0.17×10^{-3}	[227]
Oxides											
$Gd_3Ga_5O_{12}$	1740	7.6	~0.002					~15			
$Bi_{12}SiO_{20}$	895	7.63	0.03	0.00115			7×10^{-4}	26			[228, 231]
$LiNbO_3$	1260		0.115				1.5×10^{-4}				[233]
Salts											
NaCl	801	1.55	0.0103	0.0525	0.099	1.21	3.38×10^{-4}	0.2	0.116		[227]
$CaCl_2$	509	2.195	0.037	0.046	0.098	0.97	1.92×10^{-4}	0.8	0.148		[227]
$NaNO_3$	306.8	1.91	0.015	0.00158			6.6×10^{-4}	9.25	0.120	-0.07×10^{-3}	[227, 229]
Model fluids											
Water	0.0	0.998	0.01	0.00144	0.006	4.181	3×10^{-4}	6.9	0.07	-0.17×10^{-3}	[230, 241]
Tetradecane		0.76	0.03	0.001			7×10^{-4}	30			[231]
Silicon oil		0.971	5.0	0.00109			9.5×10^{-4}	4582			[232]

[a] 0.03 is for 300 K. 0.01 is for 1500 K.

Table 7
Diffusion coefficient D, equilibrium segregation coefficient k and tetrahedral covalent radii of impurities in silicon and germanium.

	D (cm^2/s)	k	Tetrahedral covalent radius (Å)
Impurities in Si			
O		1.4	0.66
C		0.07	0.77
B	$(3.3 \pm 0.4) \times 10^{-4}$ [a]	0.8	0.88
	$(2.4 \pm 0.7) \times 10^{-4}$ [b]		
P	$(2.7 \pm 0.3) \times 10^{-4}$ [a]	0.35	1.10
	$(5.1 \pm 1.7) \times 10^{-4}$ [b]		
As	$(3.3 \pm 0.9) \times 10^{-4}$ [c]	0.3	1.18
	$(3.3 \pm 0.9) \times 10^{-4}$ [b]		
Ge		0.33	1.22
Al	$\sim 2.8 \times 10^{-4}$ [c]	0.002	1.26
	$(7.0 \pm 3.1) \times 10^{-4}$ [b]		
Ga	$(6.6 \pm 0.5) \times 10^{-5}$ [a]	0.008	1.26
	$(4.8 \pm 1.5) \times 10^{-4}$ [b]		
Sb	$(1.4 \pm 0.5) \times 10^{-4}$ [a]	0.023	1.36
In	$(1.7 \pm 0.3) \times 10^{-5}$ [a]	0.0004	1.44
	$(6.9 \pm 1.2) \times 10^{-4}$ [b]		
Bi	1×10^{-4} [g]	0.0007	1.46
Impurities in Ge			
B	3.0×10^{-4} [e]	17	0.88
	2.1×10^{-4} [f]		
Si		5.5	1.17
As	1.03×10^{-4} [c]	0.04 [d]	1.18
Ga	2.1×10^{-4} [f]	0.087 [d]	1.26
	7.5×10^{-5} [d]		
	1.5×10^{-5} [e]		
	1.03×10^{-4} [b]		
Sb	5.5×10^{-4} [d]	0.003 [d]	1.36
	5.5×10^{-5} [e]		
	0.67×10^{-4} [b]		

[a] Shaskov and Gurevich [235]; direct measurement by the capillary-reservoir method.

[b] Kodera [236]; impurities in Si: D is computed from measured k_{eff}, using the BPS model [155] and estimated kinematic viscosity $\nu_{Si} = 0.0106$ (cm^2/s); impurities in Ge: recalculated D using the BPS model, Burton et al.'s data on k_{eff} [156] and $\nu_{Ge} = 0.0055$ (cm^2/s).

[c] Turovski [237]; measured k_{eff}; D is computed using the BPS model and kinematic viscosity $\nu_{Si} = 0.0027$ (cm^2/s).

[d] Burton et al. [156]; measured k_{eff}; D is computed using the BPS model and kinematic viscosity for germanium $\nu_{Ge} = 0.0025$ (cm^2/s).

[e] Ostrogorsky and Müller [154]; computed using the model presented in section 5.1, k_{eff} measured by Bridgers [157] and kinematic viscosity for germanium $\nu_{Ge} = 0.0013$ (cm^2/s).

[f] Bourret et al. [238]; measured D using shear-cell technique.

[g] Keller and Mühlbauer [239]; k_{eff} is measured; D is computed using the BPS model.

Note: According to [224], realistic values of kinematic viscosity are: $\nu_{Si} = 0.0035$ (cm^2/s) and $\nu_{Ge} = 0.00135$ (cm^2/s).

momentum and heat transfer by eddies [208]

$$\mathrm{Pr}_T = \frac{\varepsilon_v}{\varepsilon_h}. \tag{C.5}$$

The famous Reynolds analogy of heat and momentum transfer in highly turbulent flow ($\nu \ll \varepsilon_v$ and $\kappa \ll \varepsilon_h$) is equivalent to $\mathrm{Pr}_T = 1$ [208]. Indeed, for fluids having molecular $\mathrm{Pr} > 0.5$, the turbulent Prandtl number is $\mathrm{Pr}_T \approx 1$, outside of the viscous sublayer [208]. For liquid metals and semiconductors having molecular $\mathrm{Pr} \sim 0.01$, the turbulent Prandtl number is in the range $1 < \mathrm{Pr}_T < 2$.

The turbulent Schmidt number scales the relative magnitude of momentum and species transfer by eddies [208]

$$\mathrm{Sc}_T = \frac{\varepsilon_v}{\varepsilon_D}. \tag{C.6}$$

Since the molecular Schmidt number is $\mathrm{Sc} > 0.5$, one can expect $\mathrm{Sc}_T \approx 1$.

C.2. Turbulent flow driven by a rotating disk

The turbulent flow in the neighborhood of a rotating disk, was first investigated by von Karman in 1921 [148]. He established that the flow becomes turbulent at $\mathrm{Re} = VR_{\mathrm{disk}}/\nu > 3 \times 10^5$, where V is the tip velocity of the disk, $V = \omega R_{\mathrm{disk}}$. The thickness of the turbulent boundary layer depends on the radial position r,

$$\delta = 0.526 \left(\frac{\nu}{\omega}\right)^{1/5} r^{3/5}. \tag{C.7}$$

Appendix D. Material parameters

Lists of important thermophysical properties and other material parameters of various melts and model systems are given in tables 6 and 7.

For further details on viscosity, thermal conductivity and surface tension of molten semiconductors see a recent review article of Nakamura and Hibiya [234].

References

[1] G. Müller, Convection and Inhomogenities in Crystal Growth from the Melt, Crystals Vol. 12 (Springer, Berlin, 1988).
[2] D. Schwabe, in: Crystals Vol. 11, ed. H.C. Freyhardt (Springer, Berlin 1988) pp. 75–112.
[3] S. Chandrasekhar, Hydrodynamic and Hydromagnetic Stability (Clarendon, Oxford, 1961).
[4] K. Kakimoto, M. Watanabe, M. Eguchi and T. Hibiya, J. Cryst. Growth 126 (1993) 435.
[5] J. Baumgartl, R. Eckstein and G. Müller, to be published.
[6] W. Weber, PhD Thesis (University Erlangen-Nürnberg, 1988).
[7] D.T.J. Hurle, E. Jakeman and C.P. Johnson, J. Fluid Mech. 64 (1974) 565.
[8] A.E. Gill, J. Fluid Mech. 64 (1974) 577.

[9] D.T.J. Hurle, J. Cryst. Growth 65 (1983) 124.
[10] H.H. Huppert and J.S. Turner, J. Fluid Mech. 106 (1981) 299.
[11] S.R. Coriell, M.R. Cordes, W.J. Boettinger and R.F. Sekerka, J. Cryst. Growth 49 (1980) 13.
[12] D.T.J. Hurle, E. Jakeman and A.A. Wheeler, J. Cryst. Growth 58 (1982) 163.
[13] M. Glicksman, in: Proc. 4th Int. Conf. on Physico-Chemical Hydrodynamics, ed. R. Pfeffer, Ann. NY Acad. Sci. 404 (1983) 269.
[14] P. Peshev, V. Nikolov and K. Ilief, J. Cryst. Growth 65 (1983) 173.
[15] J.R. Carruthers, J. Electrochem. Soc. 114 (1967) 959.
[16] J.R. Carruthers and K. Nassau, J. Appl. Phys. 39 (1968) 5205.
[17] D.S. Robertson, Brit. J. Appl. Phys. 8 (1967) 81.
[18] D.C. Miller and T.L. Pernell, J. Cryst. Growth 57 (1982) 253.
[19] A.D.W. Jones, J. Cryst. Growth 61 (1983) 235.
[20] A.D.W. Jones, J. Cryst. Growth 94 (1989) 421.
[21] X. Ruiz, J. Massons, F. Diaz, N. Agulio and S. Gali, J. Cryst. Growth 79 (1986) 92.
[22] U. Krzyminski and A. Ostrogorsky, J. Cryst. Growth, submitted.
[23] A.D.W. Jones, J. Cryst. Growth 63 (1983) 70.
[24] D.C. Miller, in: Materials Processing in Reduced Gravity Environment of Space, ed. G.F. Rindone (Elsevier, Amsterdam, 1982) p. 373.
[25] A.D.W. Jones, Prog. Cryst. Growth Charact. 9 (1984) 139.
[26] D. Zemke, private communication.
[27] H. Ozoe, K. Totz and T. Inoue, J. Cryst. Growth 110 (1991) 472.
[28] J.R. Carruthers and A.F. Witt, in: Crystal. Growth and Characterization, eds. R. Ueda and J.B. Mullin (North-Holland, Amsterdam, 1975) p. 107.
[29] P.A.C. Whiffin, T.M. Bruton and J.C. Brice, J. Cryst. Growth 32 (1976) 205.
[30] B. Cockayne and M.P. Gates, J. Mater. Sci. 2 (1967) 118.
[31] C.D. Brandle, J. Cryst. Growth 57 (1982) 65.
[32] M. Watanabe, M. Eguchi, K. Kakimoto, Y. Baros and T Hibiya, J. Cryst. Growth 128 (1993) 288.
[33] A. Hirata, M. Tachibana, Y. Okano and T. Fukuda, J. Cryst. Growth 128 (1993) 195.
[34] S.H. Lee, Y.J. Kim, S.H. Cho and E.P. Yoon, J. Cryst. Growth 125 (1992) 175.
[35] B. Cockayne, B. Lent and J.M. Roslington, J. Mater. Sci. 11 (1967) 118.
[36] K. Takagi, T. Fukazawa and M. Ishii, J. Cryst. Growth 32 (1976) 89.
[37] J.R. Carruthers, J. Cryst. Growth 36 (1976) 212.
[38] Y. Miyazawa, Y. Mori, S. Homma and K. Kitamura, Mater. Res. Bull. 13 (1978) 675.
[39] J.A. Rocket, J. Fluid Mech. 36 (1969) 21.
[40] B. Cockayne, M. Chesswas, J.G. Plant and A.W. Vere, J. Mater. Sci. 4 (1969) 565.
[41] C.W. Lan and S. Kou, J. Cryst. Growth 119 (1992) 281.
[42] H.J. Scheel and E.O. Schulz-Dubois, J. Cryst. Growth 8 (1971) 304.
[43] V.M. Masalov and G.A. Emel' chenko, J. Cryst. Growth 119 (1992) 297.
[44] C.E. Chang and W.R. Wilcox, Int. J. Heat Mass Transfer 19 (1976) 355.
[45] Ch.-H. Chun and W. Wuest, Acta Astronautica 6 (1979) 1073.
[46] Ch.-H. Chun, J. Cryst. Growth 48 (1980) 600.
[47] P.M. Gerhart and R.J. Gross, Fundamentals of Fluid Mechanics (Addison-Wesley, Reading, MA, 1985).
[48] H.P. Greenspan, The Theory of Rotating Flows (Cambridge Univ. Press, Cambridge, 1980).
[49] S.M. Pimputkar and S. Ostrach, J. Cryst. Growth 55 (1981) 614.
[50] F. Rosenberger, Fundamentals of Crystal Growth I, Springer Series in Solid-State Sciences Vol 5. (Springer, Berlin, 1979).
[51] K. Kakimoto, M. Eguchi, H. Watanabe and T. Hibya, J. Cryst. Growth 88 (1988) 365.
[52] K. Kakimoto, M. Eguchi, H. Watanabe and T. Hibya, J. Cryst. Growth 94 (1989) 412.
[53] K. Kakimoto, M. Eguchi, H. Watanabe and T. Hibya, J. Cryst. Growth 91 (1988) 509.
[54] K. Kakimoto, M. Eguchi, H. Watanabe and T. Hibya, J. Cryst. Growth 102 (1990) 16.
[55] H. Hirata and N. Inoue, Jpn. J. Appl. Phys. 23 (1984) L527.
[56] K. Koai, A. Seidl, H.-J. Leister, A. Köhler and G. Müller, Proc. American Conf. on Crystal Growth, August 1993, J. Cryst. Growth, in press.
[57] G. Müller, G. Neumann, and W. Weber, J. Cryst. Growth 70 (1984) 78.

[58] J. Baumgartl, W. Budweiser, G. Müller and G. Neumann, J. Cryst. Growth 97 (1989) 9.
[59] J.P. Wallace, J.K. Tien, J.A. Stefain and K.S. Choe, J. Appl. Phys. 69 (1991) 550.
[60] J.A. Stefain, J.K. Tien, K.S. Choe and J.P. Wallace, J. Cryst. Growth 106 (1990) 611.
[61] D.L. Larkin, J. Kobayashi, B.S. Ashok, J.K. Tien and J.P. Wallace, AIME Proc. (October 1991).
[62] J.P. Wallace, private communication (1992).
[63] G. Müller, G. Neumann and H. Matz, J. Cryst. Growth 84 (1987) 36.
[64] J.R. Ristorcelli and J.L. Lumly, J. Cryst. Growth 116 (1992) 447.
[65] R. Hide and W.C. Titman, J. Fluid Mech. 29 (1967) 39.
[66] J.C. Brice, T.M. Bruton, O.F. Hill and P.A.C. Whiffen, J. Cryst. Growth 24 & 25 (1974) 429.
[67] K. Takagi, T. Fukazawa and M. Ishii, J. Cryst. Growth 32 (1976) 89.
[68] H. Kimura, J. Cryst. Growth 78 (1986) 19.
[69] H. Kimura, J. Cryst. Growth 78 (1986) 322.
[70] J.R. Carruthers, J. Cryst. Growth 36 (1976) 212.
[71] R. Lamprecht, D. Schwabe, A. Scharman and E. Schultheiss, J. Cryst. Growth 65 (1983) 143.
[72] R. Siegel and J.R., Howell, Thermal Radiation Heat Transfer, 2nd ed. (McGraw-Hill, New York, 1981).
[73] A.G. Ostrogorsky, to be published.
[74] J. Baumgartl, A. Hubert and G. Müller, Physics of Fluids, submitted for publication.
[75] R. Krishnamurti, J. Fluid Mech. 60 (1973) 285.
[76] E.L. Koschmieder, in: Convective Transport and Instability Phenomena, eds. J. Zirep and H. Oertel (Braun, Karlsruhe, 1982) p. 39.
[77] J. Dietl, D. Helmreich and E. Sirtl. in: Crystals Vol. 5, ed. H.C, Freyhardt (Springer, Berlin, 1981) pp. 43–107.
[78] F. Schmid, German patent, DE 30 01 815, filed 18.1.80.
[79] A. Ostrogorsky, J. Cryst. Growth 104 (1990) 233; US patent, No. 5, 047, 113 (1991).
[80] J. Baumgartl, M. Gewald, R. Rupp, J. Stierlen and G. Müller, Proc. VIIth European Symp. on Materials and Fluid Sciences in Microgravity, Oxford, UK, 10–15 September 1989, ESA SP 295 (1990) p. 47
[81] G. Neumann, J. Fluid Mech. 24 (1990) 59.
[82] H. Kölker, J. Cryst. Growth 50 (1980) 852.
[83] D.J. Knuteson, A.L. Fripp, G.A. Woodell, W.J. Debnam Jr. and R. Narayaman, J. Cryst. Growth 109 (1991) 127.
[84] D.J. Knuteson, A.L. Fripp, G.A. Woodell, W.J. Debnam Jr. and R. Narayaman, J. Cryst. Growth 110 (1991) 348.
[85] J. Baumgartl, PhD Thesis (University Erlangen-Nürnberg, 1990).
[86] Ch.J. Chang and R.A. Brown, J. Cryst. Growth 63 (1983) 343.
[87] P.M. Adornato and R.A. Brown, J. Cryst. Growth 80 (1987) 155.
[88] D.H. Kim and R.A. Brown, J. Cryst. Growth 109 (1991) 66.
[89] M.J. Crochet, F. Dupret, Y. Ryckmans, F.T. Geyling and E.M. Monberg, J. Cryst. Growth 97 (1989) 173.
[90] T. Jung, PhD Thesis (University Erlangen-Nürnberg, 1993).
[91] D.H. Kim and R.A. Brown, J. Cryst. Growth 114 (1991) 411.
[92] R.J. Naumann and S.L. Lehoczky, J. Cryst. Growth 61 (1983) 707.
[93] T. Jasinski and A.F. Witt, J. Cryst. Growth 71 (1985) 295.
[94] T. Jasinski, A.F. Witt and W.M. Rohsenow, J. Cryst. Growth 67 (1984) 173.
[95] D.E. Holmes and H.C. Gatos, J. Electrochem. Soc. 128 (1981) 429.
[96] C.A. Wang, A.F. Witt and J.R. Carruthers, J. Cryst. Growth 66 (1984) 299.
[97] A. Rouzaud, D. Camel and J.J. Favier, J. Cryst. Growth 73 (1985) 149.
[98] A.F. Witt, H.C. Gatos, M. Lichtensteiger and C.J. Hermann, J. Electrochem. Soc. 125 (1978) 1832.
[99] F.R. Szofran and S.L. Lehoczky, J. Cryst. Growth 70 (1984) 349.
[100] N.A. El-Mahallawy and M.M. Farag, J. Cryst. Growth 44 (1978) 251.
[101] Y. Huang, W.J. Debnam and A.L. Fripp, J. Cryst. Growth 104 (1990) 315.
[102] E.M. Monberg, W.A. Gault, F. Dominguez and F. Simchock, J. Cryst. Growth 83 (1987) 174.
[103] W.A. Gault, E.M. Monberg and J.E. Clemans, J. Cryst. Growth 74 (1986) 491.
[104] D.C. Look, D.C. Waiters, M.G. Mier, J.S. Sewell, J.S. Sizelove, A. Akselrad and J.E. Clemans, J. Appl. Phys. 66 (1989) 1000.

[105] R.E. Kremer, D. Francomano, G.H. Beckart and K.M. Burke, J. Mater. Res. 5 (1990) 1468.
[106] D. Hofmann, T. Jung and G. Müller, J. Cryst. Growth 128 (1993) 213.
[107] G. Müller and R. Rupp, in: Crystal Properties and Preparation, Vol. 35 (Trans Tech, Switzerland, 1991) pp. 138–154.
[108] R. Rupp, G. Müller and G. Neumann, J. Cryst. Growth 97 (1989) 34.
[109] C.E. Chang and W.R. Wilcox, J. Cryst. Growth 28 (1975) 8.
[110] C.E. Chang, W.R. Wilcox and R. A. Lefever, Mater. Res. Bull. 14 (1979) 527.
[111] Y. Kamotani, S. Ostrach and M. Vargas, J. Cryst. Growth 66 (1984) 83.
[112] Z. Kozhoukarova and S. Slavchev, J. Cryst. Growth 74 (1986) 236.
[113] N. Kobayashi, Jpn. J. Appl. Phys. 27 (1988) 20.
[114] M.Z. Saghir, Appl. Microgravity Tech. 2 (1988) 85.
[115] N.D. Kazarinoff and J.S. Wilkowski, Phys. Fluids A 1 (4) (1989) 625.
[116] R. Rupp, PhD Thesis (University Erlangen-Nürnberg, 1990).
[117] A. Cröll, W. Müller-Sebert and R. Nitsche, J. Mater. Res. Bull. 24 (1989) 995.
[118] R. Rupp. S. Auerochs, G. Müller, C. Weyrich and S. Leibenzeder, Adv. Space Res. 11 (1991) (7) 297.
[119] D. Schwabe and A. Scharmann, Adv. Space Res. 4 & 5 (1984) 43.
[120] M. Jurisch, J. Cryst. Growth 102 (1990) 223.
[121] A. Cröll, W. Müller and R. Nitsche, J. Cryst. Growth 79 (1986) 65.
[122] R. Velten, D. Schwabe and A. Scharmann, Phys. Fluids A 3 (1991) 267.
[123] J.M. Pratte and J.E. Hart, J. Cryst. Growth 102 (1990) 54.
[124] Y. Okano, M. Itoh and A. Hirata, J. Chem. Eng. Jpn. 22 (1989) 275.
[125] B. Roux, ed. GAMM Workshop on Numerical Simulation of Oscillatory Convection in Low Prandtl Number Fluids, Notes on Numerical Fluid Mechanics (Vieweg, Baunschweig, 1989).
[126] B. Roux, H. Ben Hadid and P. Laure, J. Cryst. Growth 97 (1989) 201.
[127] M. Mihelcic and K. Wingerath, J. Cryst. Growth 97 (1989) 42.
[128] A. Bottoro and A. Zebib, J. Cryst. Growth 97 (1989) 50.
[129] D. Elwell, E. Andersen and R.R. Dils, J. Cryst. Growth 98 (1989) 667.
[130] A. Seidl, G. McCord, G. Müller and H.-J. Leister, J. Cryst. Growth, submitted.
[131] S. Kobayashi, S. Miyahara and H. Fujiwara, J. Cryst. Growth 109 (1991) 149.
[132] T.A. Kinney, D.E. Bornside, R.A. Brown and K.M. Kim, J. Cryst. Growth 126 (1993) 413.
[133] H. Kopetsch, J. Cryst. Growth 88 (1988) 71.
[134] K. Mika and W. Uelhoff, J. Cryst. Growth 30 (1975) 9.
[135] D.T.J. Hurle, J. Cryst. Growth 63 (1983) 13.
[136] K. Kakimoto, P. Nicodème, M. Leconte, F. Dupret and M.J. Crochet, J. Cryst. Growth 114 (1991) 715.
[137] H.-J. Leister and M. Peric, J. Cryst. Growth 123 (1992) 567.
[138] J. Baumgartl, A. Bune, K. Koai, G. Müller and A. Seidl, E-MRS Conf. Strasbourg, May 1993, Materials Science and Engineering A (1993), in press.
[139] K.M. Kim and W.E. Langlois, J. Electrochem. Soc. 136 (1989) 1137.
[140] N. Kobayashi, J. Cryst. Growth 108 (1991) 240.
[141] M. Hortmann, M. Peric and G. Scheurer, Int. J. Numer. Methods Fluids 11 (1990) 189.
[142] M. Peric, M. Schäfer and E. Schreck, in: Parallel Computational Fluid Dynamics ed. K.G. Reinsch (Elsevier, Amsterdam, 1991) p. 297.
[143] G. Müller and M. Peric, Proc. Int. Conf. on Technical Hydrodynamics, Sofia, August 1991.
[144] G. Horton, in: Parallel Computational Fluid Dynamics, ed. K.G. Reinsch (Elsevier, Amsterdam, 1991) p. 243.
[145] M. Hussain, J. Fluid Mech. 173 (1986) 303.
[146] F.M. White, Viscous Fluid Flow, 2nd Ed. (McGraw-Hill, New York, 1991).
[147] V.G. Levich, Physicochemical Hydrodynamics (Prentice-Hall, Englewood Cliffs, NJ, 1961).
[148] T. von Karman, Z. Angew. Math. Mech. 1 (1921) 233.
[149] W.G. Cochran, Proc. Cambridge Phil. Soc. 30 (1934) 365.
[150] E.M. Sparrow and J.L. Gregg, J. Heat Transfer 81 (1959) 249.
[151] E.M. Sparrow and J.L. Gregg, J. Heat Transfer 82 (1960) 294.
[152] A.G. Ostrogorsky and G. Müller, J. Cryst. Growth 128 (1993) 207.
[153] R. Iglisch, NACA RM 1205 (1949).
[154] A. Ostrogorsky and G. Müller, J. Cryst. Growth 121 (1992) 587.

[155] J.A. Burton, R.C. Prim and W.P. Slichter, J. Chem. Phys. 21 (1953) 1987.
[156] J.A. Burton E.D. Kolb, W.P. Slichter and J.D. Struthers, J. Chem. Phys. 21 (1953) 1991.
[157] H.E. Bridgers, J. Appl. Phys. 27 (1956) 686.
[158] M.C. Flemings, Solidification Processing (McGraw-Hill, New York, 1974).
[159] A.F. Witt, H.C. Gatos, M. Lichtensteiger, M.C. Lavine and C.J. Herman, J. Electrochem. Soc. 122 (1975) 276.
[160] A.F. Witt, H.C. Gatos, M. Lichtensteiger and C.J. Herman, J. Electrochem. Soc. 125 (1978) 1832.
[161] J.T. Yue and F.W. Voltmer, J. Cryst. Growth 29 (1975) 329.
[162] J.I.D. Alexander, J. Ouazzani and F. Rosenberger, J. Cryst. Growth 97 (1989) 285.
[163] R.L. Parker, in: Crystal Growth and Materials, eds. E. Kaldis and H.J. Scheel (North-Holland, Amsterdam, 1976) p. 852.
[164] V.S. Zemeskov, I.L. Shul'pina, A.N. Titkov, I.N. Belokurova, N.B. Guseva, and V.I. Safarov, Sov. Phys. Solid State 21 (1979) 576.
[165] T. Carlberg, J. Cryst. Growth 79 (1986) 71.
[166] D. Camel and P. Tison, Proc. VIth European Symp. on Materials and Fluid Sciences in Microgravity, Oxford ESA SP-295 (1989) p. 63.
[167] G. Müller, J. Cryst. Growth 128 (1993) 26.
[168] D.T.J. Hurle, J. Cryst. Growth, Proc. Int. Conf. on Crystal Growth, Boston, 1966 (Pergamon, New York, 1967) p. 659.
[169] A. Müller and M. Wilhelm, Z. Naturforsch. 19a (1964) 254.
[170] A. Murgai, H.C. Gatos and W.A. Wesdorp, J. Electrochem. Soc. 126 (1979) 2240.
[171] D.T.J. Hurle, E. Jakeman and E.R. Pike, J. Cryst. Growth 34 (1968) 633.
[172] J.J. Favier and L.O. Wilson, J. Cryst. Growth 58 (1982) 103.
[173] J.J. Favier, Acta Met. 29 (1981) 197.
[174] D. Camel and J.J. Favier, J. Cryst. Growth 61 (1983) 125.
[175] J.J. Favier, Acta Met. 29 (1981) 205.
[176] J.J. Favier, J. Cryst. Growth 49 (1980) 373.
[177] A.M.J.G. Van Run, J. Cryst. Growth 47 (1979) 680.
[178] L.O. Wilson and N.L. Schryer, J. Fluid Mech. 85 (1978) 479.
[179] L.O. Wilson, J. Cryst. Growth 44 (1978) 371.
[180] L.O. Wilson, J. Cryst. Growth 48 (1980) 363.
[181] L.O. Wilson, J. Cryst. Growth 48 (1980) 435.
[182] L.O. Wilson, J. Cryst. Growth 48 (1980) 451.
[183] L.O. Wilson, J. Cryst. Growth 56 (1982) 219 (erratum).
[184] A.A. Wheeler, J. Cryst. Growth 56 (1982) 67.
[185] A.A. Wheeler, Proc. R. Soc. Lond. A 379 (1982) 305.
[186] A.A. Wheeler, Proc. R. Soc. Lond. A 379 (1982) 325.
[187] D. Thevenard, A. Rouzaud, J. Camera and J.J. Favier, J. Cryst. Growth 108 (1991) 572.
[188] H. Ueda, J. Phys. Soc. Jpn. 16 (1961) 61.
[189] H.C. Gatos, A.J. Strauss, M.C. Lavine and T.C. Harmon, J. Appl. Phys. 32 (1961) 2057.
[190] D.T.J. Hurle, Phil. Mag. 13 (1966) 305.
[191] H.P. Utech and M.C. Flemings, in: Crystal Growth, ed. H.S. Peiser (Pergamon, Oxford, 1967) p. 651.
[192] K. Grasza and A. Jedrejczak, J. Cryst. Growth 110 (1991) 867.
[193] D.J. Carlson and A.F. Witt, J. Cryst. Growth 108 (1991) 508.
[194] D.T.J. Hurle, Sir Ch. Frank Anniversary (Hilger, Bristol, 1991) pp. 189–206.
[195] V.S. Arpachi, Conduction Heat Transfer (Addison-Wesley, Reading, MA, 1966).
[196] S. Ostrach, Ann Rev. Fluid Mech. 14 (1982) 313.
[197] A. Hirata, M. Tachibana, Y. Okano and T. Fukuda, J. Chem. Eng. Jpn. 25 (1992) 62.
[198] M. Mihelcic and K. Wingerath, J. Cryst. Growth 82 (1987) 318.
[199] H.U. Walter, private communication.
[200] J. Baumgartl, M. Gewald, R. Rupp. J. Stierlen and G. Müller, Proc. VIIth European Symp. on Materials and Fluid Sciences under Microgravity, Oxford, UK 10–15 September 1989, ESA SP 295 (1990) p. 47.

[201] D.T.J. Hurle, G. Müller and R. Nitsche, in: Fluid Sciences and Materials Sciences in Space ed. H.U. Walter (Springer, Berlin, 1987) p. 313.

[202] J.D. Alexander, S. Amiroudine, J. Quazzani and F. Rosenberger, J. Cryst. Growth 113 (1991) 21.

[203] G. Müller, G. Neumann and W. Weber, J. Cryst. Growth 119 (1992) 8.

[204] G. Müller and G. Neumann, J. Cryst. Growth 63 (1983) 58.

[205] R. Siegel and J.R. Howell, Thermal Radiation Heat Transfer, 2nd Ed. (Hemisphere, Washington, DC, 1981).

[206] L. Prandtl, Essentials of Fluid Dymanics (Blackie, London, 1963).

[207] H. Schlichting, Boundary Layer Theory (McGraw-Hill, New York, 1979).

[208] W.M. Kays and M.E. Crawford, Convective Heat and Mass Transfer, 2nd Ed. (McGraw-Hill, New York, 1980).

[209] W.M. Rohsenow and H. Choi, Heat Mass and Momentum Transfer (Prentice, Englewood Cliffs, NJ, 1961).

[210] S.V. Patankar, Numerical Heat Transfer and Fluid Flow (Hemisphere, Washington, DC 1980).

[211] A.C. Fletcher, Computational Techniques for Fluid Dynamics, Vol. 1, 2nd. Ed. (Springer, Berlin, 1991).

[212] R.S. Rogallo and P. Moin, Ann. Rev. Fluid Mech. 16 (1984) 99.

[213] W.J. Minkowycz, E.M. Sparrow, G.E. Schneider and R.H. Pletcher, Handbook of Numerical Heat Transfer (Wiley, New York, 1988).

[214] B. Carnahan, H.A. Luther and J.O. Wilkes, Applied Numerical Methods (Wiley, New York, 1969).

[215] O.C. Zenkiewicz and R.L. Taylor, The Finite Element Method (McGraw-Hill, New York, 1988).

[216] M. Peric, R. Kessler and G. Scheuerer, Comput. Fluids 16 (1988) 389.

[217] NEKTON User's Manual 2.6 (Nektonics, Cambridge, MA, 1990).

[218] FIDAP General Introduction Manual 4.0 (Fluid Dynamics International, Evanston, IL, 1987).

[219] FLUENT User's Manual (Fluent, Lebanon, NH, 1987).

[220] D.B. Spalding, J. Math. Comput. Simulation 23 (1981) 267.

[221] G. Müller, D.T.J. Hurle and H. Wenzl, eds., Proc. 1st Workshop on Computer Modelling in Crystal Growth from the Melt, J. Cryst. Growth 97 (1989).

[222] A.A. Wheeler, J. Cryst. Growth 102 (1990) 691.

[223] U. Bückle and M. Schäfer, J. Cryst. Growth 126 (1993) 682.

[224] V.M. Glasov, S.N. Chizhevskaya and N.N. Glagoleva, Liquid Semiconductors (Plenum, New York, 1969).

[225] A.S. Jordan, J. Cryst. Growth 71 (1985) 551.

[226] T. Duffar, C. Potard and P. Dusserre, J. Cryst. Growth 92 (1988) 467.

[227] C.J. Smithells, Smithells Metals Reference Book, 6th ed. (Butterworth, London, 1983).

[228] P.A. C. Whiffen, T.M. Bruton and J.C. Brice, J. Cryst. Growth 32 (1976) 205.

[229] R. Lamprecht, D. Schwabe, A. Scharman and E. Schultheiss, J. Cryst. Growth 65 (1983) 143.

[230] P.M. Gerhart and R.J. Gross, Fundamentals of Fluid Mechanics (Addison-Wesley, Reading, MA, 1985).

[231] H. Kimura, J. Cryst. Growth 78 (1986) 19.

[232] H. Ozoe, K. Toh and T. Inoe, J. Cryst. Growth 110 (1991) 472.

[233] J. Trauth and B.C. Grabmeier, J. Cryst. Growth 112 (1991) 451.

[234] S. Nakamura and T. Hibiya, Int. J. Thermophys. 13 (1992) 1061.

[235] Yu.M. Shaskov and V.M. Gurevich, Russian J. Phys. Chem. 24 (1968) 1082.

[236] H. Kodera, Jpn. J. Appl. Phys. 2 (1963) 212.

[237] B.M. Turovskii, Russian J. Phys. Chem. 36 (English Trans.) (1962) 983.

[238] E.D. Bourret, J.J. Favier and O. Bourrel, J. Electorchem. Soc. 128 (1981) 2438.

[239] W. Keller and A. Mühlbauer, Floating Zone Silicon (Dekker, New York, 1981).

[240] R. Rupp and G. Müller, J. Cryst. Growth 113 (1991) 131.

[241] A.F. Mills, Heat Transfer (Irwin, Homewood, IL, 1992).

CHAPTER 14

Stress in the Cooling Crystal

JOHANNES VÖLKL

Siemens AG, ZFE BTPE 61, D-91050 Erlangen, Germany

Contents

1. Introduction . . . 823
2. Origin of stress in the crystal during cooling . . . 824
   2.1. Gravity . . . 824
   2.2. Interaction with containers . . . 826
   2.3. Stress induced by inhomogeneities in composition . . . 827
      2.3.1. Striations . . . 827
      2.3.2. Inclusions . . . 829
   2.4. Thermally induced stress . . . 830
3. Behaviour of material under stress . . . 836
   3.1. Stress and strain in elastic crystals . . . 836
   3.2. Plasticity in crystals . . . 838
      3.2.1. Model of plastic deformation . . . 839
      3.2.2. Deformation experiments . . . 840
   3.3. Orientation of plastic deformation – slip systems . . . 843
   3.4. Modeling of deformation experiments . . . 844
4. Modeling of stress and strain during crystal growth experiments . . . 846
   4.1. Principle of the model . . . 847
   4.2. Transfer of the model into a growing crystal . . . 848
   4.3. Evaluation of $\varepsilon_{pl}(t)$ and correlation to the dislocation density $N(t)$ . . . 851
   4.4. Numerical procedure . . . 852
   4.5. Modeling of dislocation formation and stress in real growth systems . . . 853
      4.5.1. Comparison of numerical and experimental results . . . 853
      4.5.2. Elastic–plastic interaction and the effect on elastic stress in the crystal . . . 854
      4.5.3. Influence of the pull rate on the dislocation density . . . 856
      4.5.4. Modified stress field for the growth of crystals with low dislocation density . . . 859
      4.5.5. Formation of dislocation patterns . . . 861

HANDBOOK OF CRYSTAL GROWTH, VOL. 2
edited by D.T.J. Hurle

4.6. Stress and dislocations in the literature 864
5. Residual stress in the crystal after cooling 867
5.1. Origin of residual stress 867
5.2. Comparison of residual stress and calculated plastic strain 868
6. Concluding remarks 869
References 871

1. Introduction

The final quality of a crystal in industrial crystal growth is a complicated function of the history it has experienced – starting with the solidification atom by atom and layer by layer near the melting temperature, continued by exposing this sequence of layers to a changing environment of inhomogeneously decreasing temperature and finally, after growth has been terminated, cooling the crystal down to room temperature in a more or less controlled manner. During this process, the aim of innumerable crystal growers has been and still is the perfect, especially dislocation-free crystal, which demands the crystallization and cooling process to be nearly perfectly adjusted to the properties of the growing material. But, for a long time, the approach to these perfect conditions has been investigated on an empirical basis to yield improved results.

It was Billig [1] in 1956 who discovered that the dislocation density of germanium crystals was correlated to the imposed temperature gradient. Thus, the first indication of a relation between temperature conditions during growth and final crystal quality was established. Soon afterwards Bennett and Sawyer [2] identified the $\{111\}$–$\langle 110\rangle$ slip systems in germanium. They recognized that the heat losses through the surface of the crystal could produce tensile and compressive stresses in the crystal. If these stresses were high enough to activate the $\{111\}$–$\langle 110\rangle$ slip systems, the resulting plastic deformation would be of the same symmetry as that of the dislocation pattern observed in the grown crystal. Similar observations have been made by Penning [3] in 1958 for radial heat flow in germanium, by Nygren et al. in 1971 [4] for LEC growth of GaP, and by Plaskett et al. in 1971 [5] for Bridgman growth of GaAs.

In view of the economic importance of the problem, it is surprising that it took about a quarter of a century after Billig's discovery until Jordan et al. [6] made a first attempt toward a quantitative description of the relationship between thermal conditions, thermally induced stress and resulting dislocation density in 1980.

This work presented an analytical solution for radial heat flow in a cylindrical crystal of infinite length. From this, the resolved shear stresses in the $\{111\}$–$\langle 110\rangle$ slip systems were calculated. The dislocation density was estimated to be a linear function of the resolved shear stresses reduced by a certain critical resolved shear stress, below which no dislocation generation was assumed to occur. The good qualitative agreement of this model with experimentally observed dislocation patterns in GaAs crystals demonstrated the importance of the thermal conditions during the cooling process of the crystal with respect to the dislocation density.

During the 1980s, improving computing facilities combined with the progress in software programmes enabled most of the groups engaged in crystal growth to calculate the temperature distribution and relate elastic stresses in different growth configurations. These calculations could handle real crystal shapes and even the thermal interactions of the crystal with the surrounding set-up of the growth equipment. All these calculations were "photographs" of the conditions at a certain moment of the growth process excluding dynamic influences of the process like growth rate and, especially with respect to the prediction of dislocation formation, excluding the time dependence of elastic–plastic interaction.

Such dynamic influences were incorporated in 1987 and 1988 independently by De Angelis and co-workers [7–9] for web growth of silicon and Völkl and Müller [10, 11] for bulk growth of InP. Both analyses were based on the excellent experimental and theoretical analysis of the dynamic deformation behaviour of diamond-like crystals by Alexander and Haasen in 1968 [12].

This breakthrough enabled crystal growers to investigate the influence of different growth and cooling conditions on the final "Stress-induced" dislocation density of the crystals and thus to understand the complex influence of dynamic processes during crystal growth.

The purpose of the following sections is not to reflect all those calculations but to demonstrate the fundamental aspects as well as selected problems during the cooling of crystals grown in different environments. Special emphasis is put on the elastic–plastic behaviour of single crystals and its dependence on time, temperature and material properties. The sequence will hopefully provide a better understanding of residual stresses introduced by cooling during the growth of crystals.

2. *Origin of stress in the crystal during cooling*

A homogeneous body of any shape being at homogeneous temperature and free of interactions with external forces is free of stress.

This fundamental statement is as simple to understand as it is difficult to fulfill in the real world of crystal growth. With respect to external forces the existence of the crystal itself gives rise to stress due to its own mass exposed to gravitational forces. Furthermore, some crystal growth methods use containers of different material with different thermal expansion coefficient in contact with the crystal, which leads to increasing interaction forces during the cooling of the crystal. The crystal enclosed in a crucible also is the most simple example of the cooling of an inhomogeneous body giving rise to the generation of stresses, if crystal and crucible are considered as a unit. Additionally, the crystal structure itself is far from homogeneity. Sources of stress can be introduced by the agglomeration of point defects, inhomogeneous stoichiometric composition or inclusions. The most interesting origin of stress, however, is introduced by the fundamental demand of extracting the heat out of the crystal surface. This process necessarily results in inhomogeneous temperature distributions in crystals and hence induces stresses due to the inhomogeneous thermal expansion within the body.

All of these origins of stress can be present in the crystal at the same time resulting in a highly complex stress pattern. Their significance will be discussed in the following.

2.1. *Gravity*

Although gravity is not a special consequence of the cooling of a crystal, it should briefly be considered due to its steady presence in all systems on earth. For example, a typical Czochralski crystal as shown in fig. 1 is fixed at the seed holder at D–D′ and in contact with the melt at A–A′. Except for minor effects of wetting and buoyancy-driven

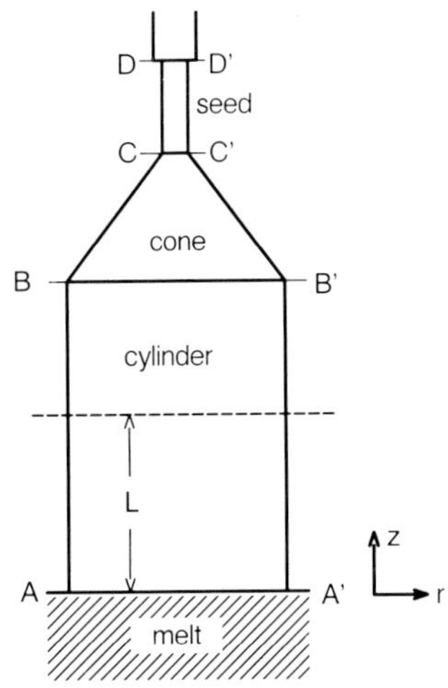

Fig. 1. Schematics of a crystal grown by the Czochralski method.

forces due to differences in the density of melt and solid, the surface A–A′ should be free of gravitational forces. Within the cylindrical part from A–A′ to B–B′ then the mass M is linearly increasing as a function of the distance l from the interface:

$$M = r^2 \pi l \rho_s. \tag{2.1.}$$

The resulting force per unit area, which is identical to the axial stress (tensional in the Czochralski set-up), is then given by

$$\sigma_{zz} = \rho_s g l. \tag{2.2}$$

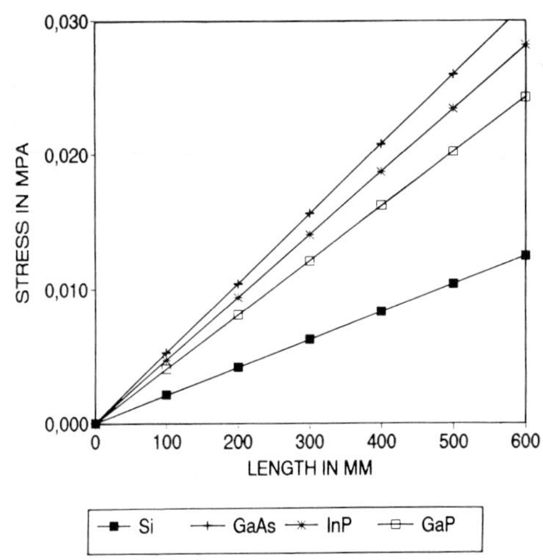

Fig. 2. Stress at the end of the cylindrical portion of a growing crystal due to gravity as a function of length for different semiconductors.

The resulting stresses for different semiconductor materials are presented in fig. 2. Even for rather long crystals the tension in the highest stressed area of the cylindrical section is very small.

In the conical part the stresses increase drastically due to the reduction of the cross section. The stress pattern within that volume is depending on the cone angle, the demonstration of which is not the target of this section. The highest stresses occur at C–C′ and within the seed, where the whole crystal is held by a small cross section of usually 15 to 40 mm^2. The average stresses in that portion thus are in the range of 0.2 to 0.6 MPa per kilogram depending on the cross section area of the seed. This small cross section limits the maximum size of the crystal in dependence of the behaviour of fracture of the material as has been discussed by Kim and Smetana [13] for silicon.

2.2. *Interaction with containers*

Certain growth methods, like the vertical Bridgman (VB) technique or horizontal boat techniques, demand that the crystal is growing in a container made from, e.g. quartz, boron nitride or graphite. Good results can be achieved if conditions to suppress wetting between boat and crystal can be established. However, very often the crystal sticks to the container. The different expansion behaviour then gives rise to enormous forces within both materials, which can even cause the complete destruction of the grown crystal. The stress situation may become quite complicated as depicted schematically in fig. 3. The thermal expansion coefficients of the crystal are α_1 in the radial and α_2 in the axial direction; those of the crucible are α_3 (radial) and α_4 (axial). If the system is cooling down the crucible is shrinking slower (or faster) than the crystal in the radial direction and thus tension (compression), is built up radially within the crystal. In the same time interval the crystal and crucible also shrink in the axial direction at different rates leading to shear stresses at the crucible–crystal interface and tensile or compressive stress components in the axial direction within both materials. All of these expansion coefficients are known to be temperature dependent leading to a complicated variation of the stress pattern during the cooling from the melting point down to

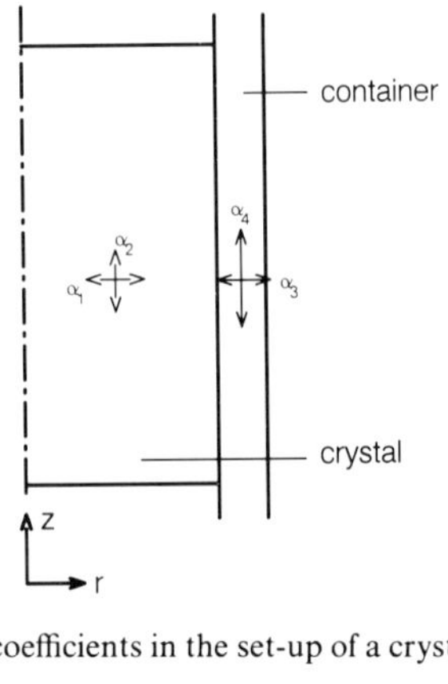

Fig. 3. Different thermal expansion coefficients in the set-up of a crystal growing in contact with a crucible.

room temperature. The magnitude of such stresses may be of the order of tens or hundreds of MPa and sometimes exceed the strength of the crystal if strongly differing material combinations are used. A typical example for such an inferior combination is the growth of InP or GaAs in quartz containers using B_2O_3 as a separation fluid. Indeed, the solidification of single–crystalline material improves due to the reduction of heterogeneous seed formation at the crucible wall as B_2O_3 is liquid at the melting point. Furthermore, B_2O_3 remains liquid down to around 480°C and hardly any forces may be generated because crystal and crucible can shrink independently in that regime, the crystal at a higher rate than the quartz crucible. At 480°C B_2O_3 solidifies and crystal and crucible stick together. As cooling proceeds down to room temperature, tensile strains are generated in the crystal. Timan and Fesenko [14] and Rosch and Carlson [15] have computed stress fields generated during the growth of GaAs by the vertical Bridgman method using sticking conditions between crystal and crucible. The calculations indicate that fracture may occur due to the enormous forces induced by the differential thermal expansion.

2.3. *Stress induced by inhomogeneities in composition*

2.3.1. *Striations*

Inhomogeneities in composition are predominantly a consequence of instabilities at the growth interface (e.g. temperature fluctuations, melt composition). Due to such instabilities, layers of different composition may grow as is schematically shown in fig. 4. The etching of such layers results in striations as depicted in fig. 5. The composition of each layer differs slightly by Δx_i from the average composition x. Even if these layers would be free from lattice mismatch (they are indeed not), they may shrink at different rates during cooling and tensile and compressive strains may locally alternate in the different layers according to the difference in their expansion coefficients. This effect is present especially in highly doped semiconductor crystals such as GaAs: In, InP:Ga, InP:As or multiple doped crystals. The intention of such doping is to get dislocation-free material due to lattice hardening. However, due to the stresses between the layers, the crystals become more brittle (e.g. GaAS:In) or the original intention of lowering the dislocation density is even converted into the opposite by dislocation formation within the layers. This is shown in fig. 6, where an InP wafer doped with Ga, Sb and As in the range of 10^{18} to $10^{19}\,\mathrm{cm}^{-3}$ exhibits a lot of dislocated circles at the

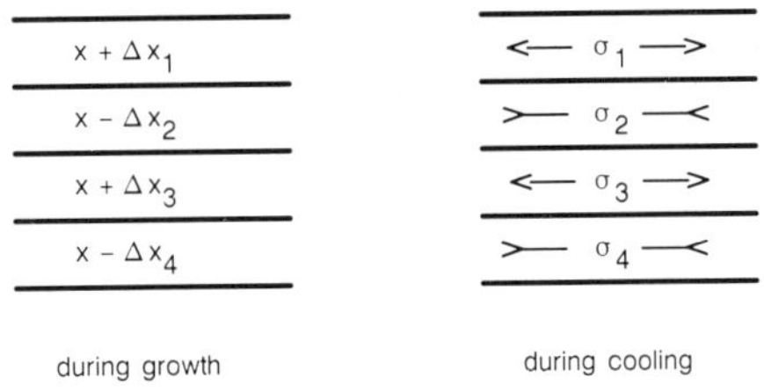

Fig. 4. Schematic representation of inhomogeneities in a growing crystal (left) and resulting situation of stress during cooling (right). x: composition; σ: stress.

Fig. 5. Striations due to inhomogeneities of composition in a sulphur-doped InP crystal.

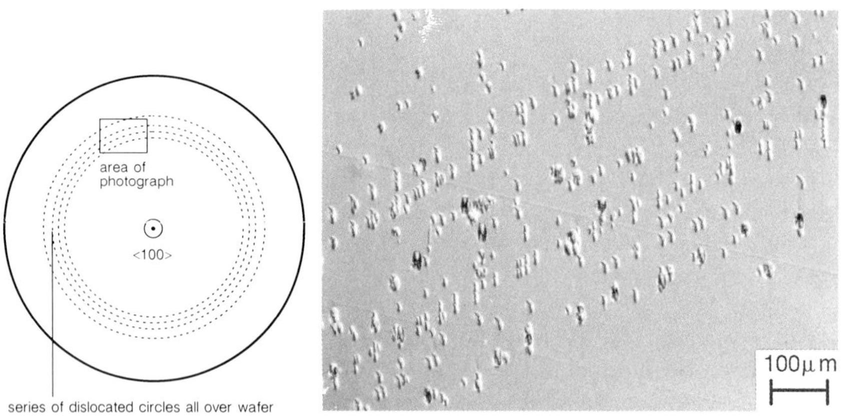

Fig. 6. Circular arrangement of dislocations on a wafer due to inhomogeneities in composition. The curved and inhomogeneous layers penetrate the surface in circles (similar to cutting an onion), where dislocations can be made visible by etching.

surface. The magnification in fig. 6 shows that local dislocation densities well above 10^4 cm^{-2} may be generated due to such inhomogeneities in composition. The fluctuations in concentration of the different layers has been determined to be of the order of up to 10% [16].

The effect described above acts in addition to the formation of dislocations due to the lattice mismatch present between such layers. However, the misfit dislocations are expected to be generated early during the solidification and may provide the initial dislocation sources for further stress-induced multiplication mechanisms during cooling. The magnitude of such stresses can become as high as several MPa.

2.3.2. Inclusions

This type of defect is frequently observed in semiconductor crystals and may be of various origins:

- SiO_2 precipitates in Czochralski silicon due to hardly avoidable residual oxygen impurities and particles from the SiO_2 crucible [17];
- arsenic precipitates in GaAs due to the growth from slightly arsenic-rich melts in order to achieve the desired electrical properties;
- precipitation of the dopant or formation of a new phase between dopant and host at concentrations above the solubility limit (Zn clusters in InP, iron phosphide needles in InP:Fe crystals).

The stresses from such precipitates can be estimated in a similar manner similar to that described in section 2.2, where two different materials with different expansion coefficients stick to each other. However, these precipitates are only a few nanometers in extent in most cases and thus are of minor importance with respect to stresses. However, the inhomogeneous distribution of such precipitates can induce long-range stress effects which may lead to the warpage of wafers as has been shown for silicon by Shimizu et al. [18].

The growth of compound semiconductors from non-stoichiometric melts can result in much bigger inclusions of 20 μm or more due to instabilities at the growth interface.

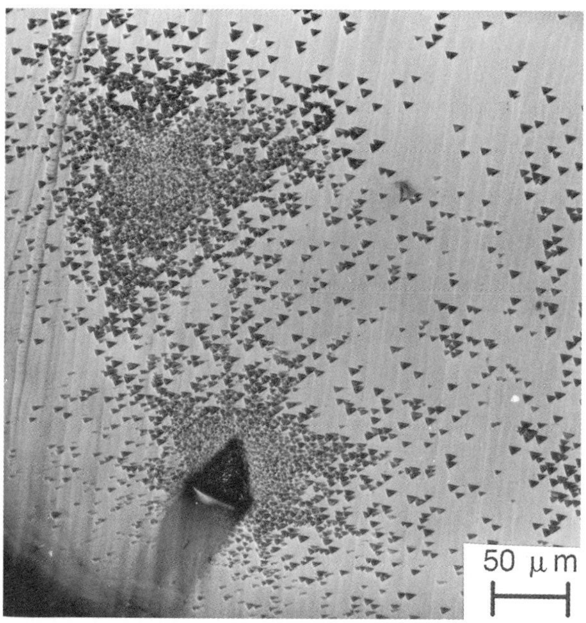

Fig. 7. Indium inclusion in an InP crystal, which has generated a high density of dislocations due to different rates of contraction during cooling. In the upper left part of the picture a further inclusion is hidden below the surface where the triangular structure of the dislocation is visible (courtesy by Dr. H.-J. Sell and Dr. R. Stein).

Typical examples are the solution growth of III–V and II–VI compounds or even Czochralski growth from initially stoichiometric melts experiencing losses of a volatile component during the growth run. Stress then is generated by the difference in volume changes of the metal inclusion and the semiconductor crystal. It can easily be estimated that a temperature difference of 1000 K during cooling and a difference in expansion coefficient of some 10^{-6} will yield a difference in strain in the range of 10^{-3} to 10^{-4}. The situation gets drastically worse if a certain amount of the second component (e.g. P in In, As in Ga) is diluted in the inclusion. According to the phase diagram, binary material will solidify within the inclusion or at its boundary during cooling which is accompanied by a volume change in the range of 5 to 10% for III–V materials. Thus a content of 10% of a second component (P, As) in the inclusion results in a volume increase of about 1% with respect to the volume before cooling. This creates enormous stresses and results in a large density of dislocations of the order of 10^6 cm^{-2} around the inclusion as depicted in fig. 7 for an In inclusion in InP. The triangular shape of the inclusion is a clear indication that the above mechanism has been present during cooling by solidification at the $\{111\}$ facets.

2.4. *Thermally induced stress*

One of the principal characteristics of solidification and cooling during crystal growth is the need for temperature gradients. These gradients are the starting-point for a heat flux which tries to compensate the temperature differences and to minimize the thermal gradients within the system. Thus a heat flux balance is evident in each system which is determined by the kind and strength of the heat sources and heat sinks. Heat sources are the furnace or the latent heat during solidification, heat sinks are the (sometimes cooled) walls to which heat is transported by conduction, radiation or convection. Thus the temperature distribution in a crystal determined by the heat flux is a complicated function of the heat exchange with its environment. Figure 8 indicates the basic mechanisms of heat exchange in a typical liquid–encapsulated Czochralski (LEC) environment.

Additionally, the crystal is being moved along a certain coordinate in several growth methods introducing changes to the steady-state heat flux by the amount of specific heat transported by this movement. The calculation of the resulting temperature distributions in the crystal has been the topic of many papers in the past [19–79].

A typical result of such calculations is shown in fig. 9. This example will serve to demonstrate the typical effects in the growing and cooling crystal in the following sections. The temperature distribution of two different LEC growth conditions is calculated at a certain moment of growth. In fig. 9a there is a small boron oxide layer thickness of about 18 mm. In fig. 9b the layer thickness is about 44 mm corresponding to nearly fully encapsulated growth of InP. The figures are taken from [58]. Such calculations reveal typical radial temperature differences in the range of 5 to 50 K between central and peripheral regions of the crystal. The axial gradient is strongly varying along the z-axis and may rise from a few degrees per centimeter up to values of more than 200 K/cm depending on the growth conditions. The results of these calcul-

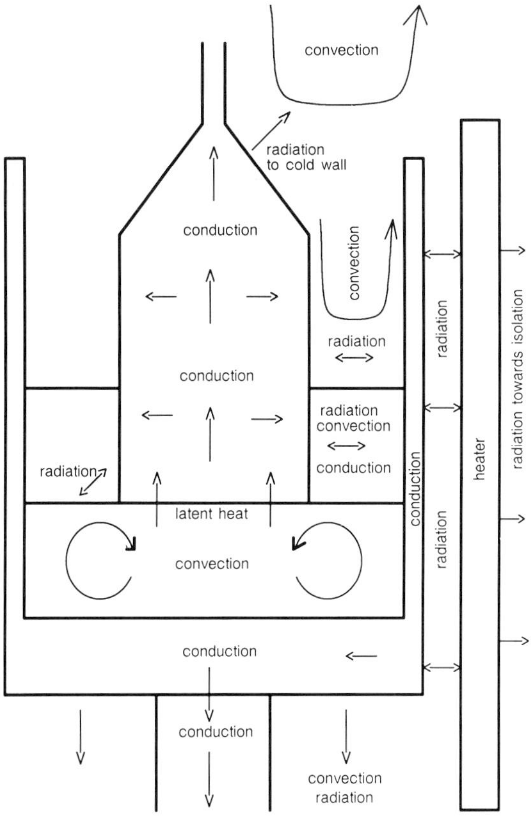

Fig. 8. Schematical visualization of heat exchange processes during LEC growth.

ations have been confirmed by an experimental determination of the thermal history in these LEC crystals carried out by Müller et al. [80]. Thermocouples have been placed in the center and at the periphery of a crystal plane approximately 15 mm below the "seed-on" position. During growth these thermocouples moved from a position 15 mm below the melt surface to the interface, grew into the crystal and moved along the z-axis recording the thermal history of two points of that crystal plane.

Figure 10 shows the temperature along the z-axis for two different growth conditions. The numbers 12/18 and 30/44 denote the initial/final boron oxide heights in mm changed by the growing volume of the crystal (e.g. 12 mm boron oxide height before growth starts increasing to a maximum height of 18 mm with crystal). During the growth run with the higher boron oxide of 30/44 mm an additional passive afterheater was used to increase the temperature of the gas atmosphere above the boron oxide and thus to adjust the temperature and viscosity of the boron oxide at the surface. Considerable changes of the growth conditions can be established by such means as is evident from fig. 10.

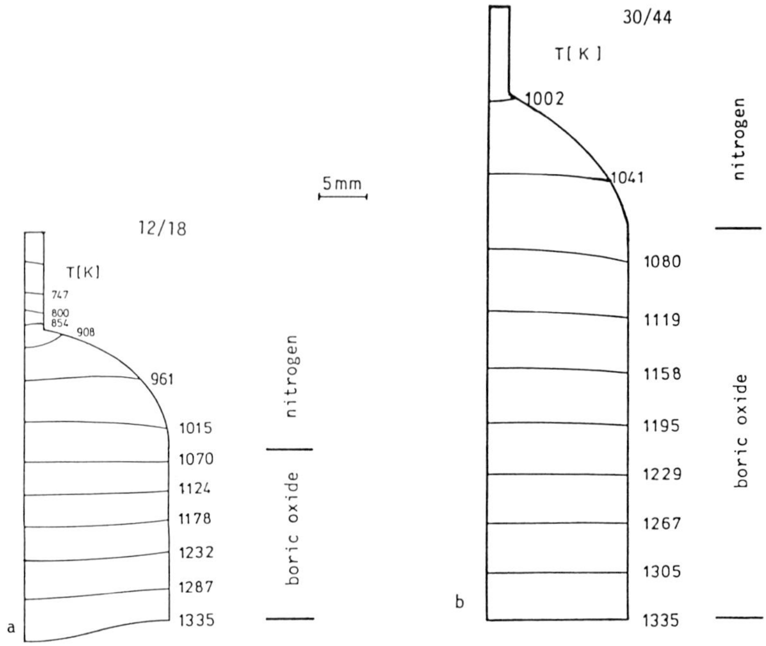

Fig. 9. Calculated isotherms in InP crystals grown according to conditions 12/18 (a) and 30/44 (b).

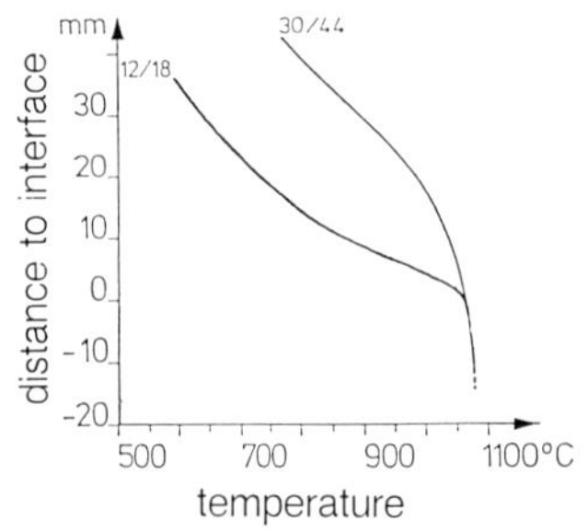

Fig. 10. History of temperature of a peripheral element during its movement along the z-axis for two different LEC growth configurations.

The resulting axial gradients for both growth conditions are shown in fig. 11. The maximum values in the crystal grown with conditions 12/18 exceed 200 K/cm in a region rather near to the interface, whereas a gentle increase of the temperature gradient along the z-axis can be established to much lower maximum values in the range of 100 K/cm with conditions 30/44.

The radial temperature difference for both conditions is shown in fig. 12. Conditions 12/18 show a radially inward heat flux in the region near the interface indicating convex isotherms as is depicted in fig. 9a. The heat flux changes to radially outward as the

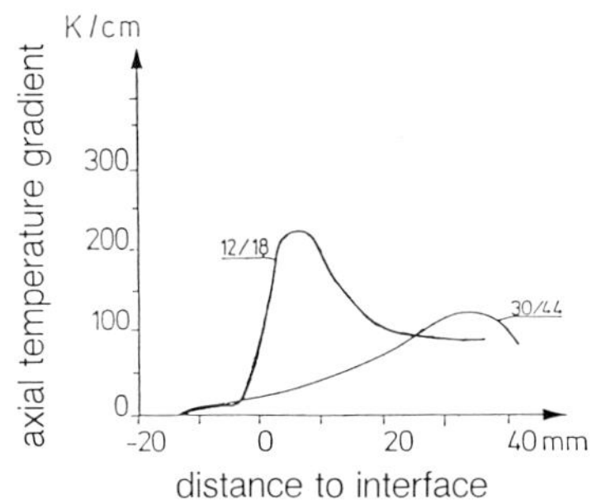

Fig. 11. Axial gradient as a function of the distance to the interface, deduced from the thermal profile of fig. 10.

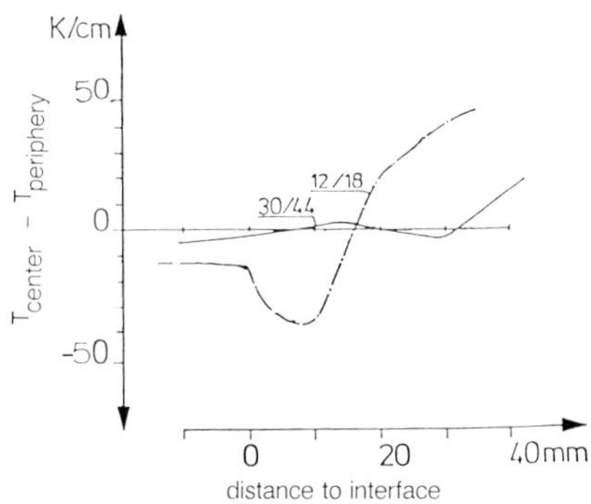

Fig. 12. Normalized temperature difference between a central and a peripheral element of a crystal plane as the plane moves away from the liquid–solid interface.

crystal plane approaches the boron oxide surface and the gas atmosphere due to the increased heat transport by the turbulent gas. The growth conditions 30/44 permit the establishment of nearly flat isotherms for a distance of approximately 30 mm from the growth interface, followed by concave isotherms near the boron oxide/gas transition region. However, the isotherms show much lower temperature differences between center and periphery, which is a result of the reduction of the heat losses into the gas atmosphere due to the passive after-heater arrangement. The thermal differences between the two growth configurations will result in important differences of dislocation patterns and residual stresses as will be shown later.

As a result of the thermal expansion coefficient characteristic for every material, the inhomogeneous temperature distribution is coupled with locally different expansion/contraction in the crystal. The continuity of the lattice combines all those partial volumes, which causes tensile/compressive strains within the volumes to fulfill the continuity condition. Of course, the amounts of tensile and compressive strain within the whole crystal have to be equal to ensure mechanical equilibrium.

These strains can be converted into the corresponding stresses by assuming completely elastic behaviour of the material all over the whole temperature range and applying Hooke's law,

$$\sigma = \varepsilon E, \tag{2.3}$$

where σ is stress, ε is strain and E is the elastic modulus.

This conversion into stresses is the most critical point of all discussions of stresses in a growing crystal, because there is no material known which keeps the same elastic behaviour and strength from room temperature up to the melting point.

Evaluations of stresses from temperature distributions are extensively treated in the literature, especially in the field of thermoelasticity. The fundamentals may be found in, e.g. [81–83].

The visualization of the real state of stress is difficult because of the manyfold directions of the stress components and their different signs indicating tensile or compressive status. Therefore, mostly stress criteria like the "von Mises" or the "Tresca" criterion are used, which allow one to calculate one single value of stress per coordinate. This embraces the disadvantage that information on directions and sign of the stress get lost. Using the von Mises criterion and assuming an elastic material, the temperature distributions in fig. 9 result in stress fields as shown in fig. 13. It is obvious that the highly inhomogeneous temperature distribution in fig. 9a results in stresses of higher magnitude than that of fig. 9b. Regions of high stresses appear at the boron oxide/gas transition and at the melt/crystal interface if the isotherms are curved as is the case in fig. 9a. Extended stress calculations of different materials and growth conditions may be found in [19–79].

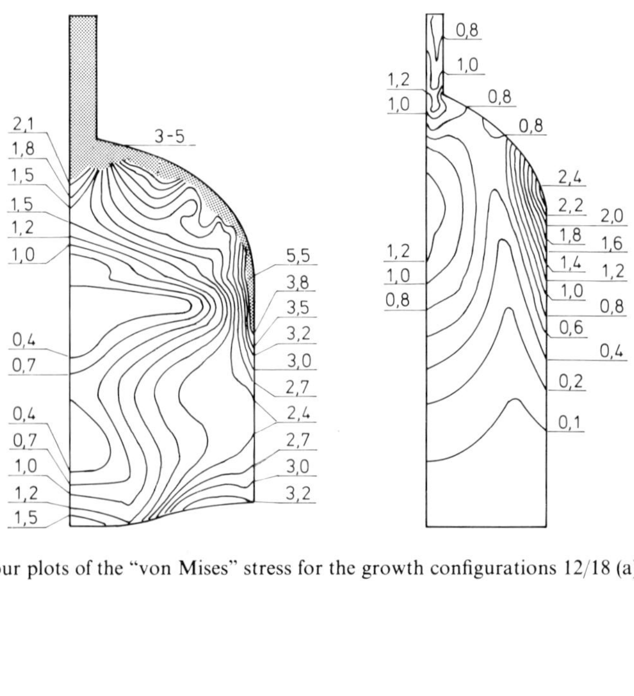

Fig. 13. Contour plots of the "von Mises" stress for the growth configurations 12/18 (a), and 30/44 (b).

In practice, such calculations are very time consuming, especially if they have to be confirmed by experimental investigations. Thus, for a rough estimation of thermally induced stresses some generally valid rules are given in the following according to the discussion of Indenbom [84] in 1979.

(a) A temperature field with linear behaviour in all coordinates results in a stress-free state, as the body only experiences free temperature bending.

This is an important rule, because it makes it impossible to estimate stress from temperature gradients as can be found in many publications for axial gradients in the past. Mathematically the formulation for stress-free conditions in a crystal is

$$\nabla^2 T = \partial^2 T/\partial x^2 + \partial^2 T/\partial y^2 + \partial^2 T/\partial z^2 = 0, \tag{2.4}$$

where x, y and z are the coordinates and T is the temperature.

In cylindrical crystals it is obvious that this condition can only be fulfilled if the radial temperature differences are equal to zero and the axial gradient is linear. This is equivalent to radially flat and axially equidistant isotherms.

(b) In bodies of the same material, mathematically similar shapes and similar temperature distributions, the stress fields are related to the temperature differences and are independent of the dimensions of the body. Thus the stresses in a crystal of 8 inch diameter are the same as those in a crystal of 1 inch diameter, if the dimensions are scaled linearly in all dimensions and the temperature differences are the same. From

$$T_{K1}(x, y, z) = a\, T_{K2}(bx, by, bz), \tag{2.5}$$

it follows that

$$\sigma_{K1}(x, y, z) = a\sigma_{K2}(bx, by, bz). \tag{2.6}$$

In the practical growth of crystals of different size, equal temperature differences and similar shapes can seldom be achieved due to dynamic influences during the solidification (latent heat, mass and heat transport due to growth rate, changes of shape with time), which results in increasing problems with increasing diameter of the crystals. However, this rule can be a very helpful support for efficient upscaling of a crystal growth apparatus.

(c) Incompatibilities of the temperature field are sources of stress. This allows one to estimate stress from the curvature of a temperature field by multiplying the curvature with the square of a characteristic dimension W:

$$\sigma \approx \alpha E \partial^2 T/(\partial w_1 \partial w_2)\, W^2, \tag{2.7}$$

where α is the linear expansion coefficient, E is the elastic modulus, T is the temperature, W is a characteristic length and w_1 and w_2 are the coordinates determining W.

The characteristic axial dimension W, taken for the estimation of stress, is in the range of 20% of the diameter of the crystal.

The application of this equation to the axial temperature profile in fig. 9b predicts a stress of about 1.5 to 2.5 MPa, which is in good agreement with the numerical calculations. For the axial temperature profile in fig. 9a values of 10 MPa or more are obtained from eq. (2.7). This exceeds the values of the numerical solution by a factor of two, which still seems to be a quite acceptable solution when considering the speed and simplicity of the estimation.

Indenbom [84] gives an impressive example of the meaning of eq. (2.7) using a simple plane plate with a width-to-length ratio of 1:10. A parabolic temperature profile in the direction of width having a temperature difference of 1 K induces the same stress field as a parabolic temperature profile in the direction of length having a temperature difference of 100 K. From that it follows that the curvature of the axial temperature distribution in a crystal may create similar magnitudes of stress as the curvature of the isotherms in the radial direction.

3. Behaviour of material under stress

One of the problems of calculating proper stress distributions in a cooling crystal is given by the dependence of many material properties on temperature. Especially at high temperatures, where the most important processes are expected to occur, very little is known about the thermal, elastic and plastic properties of semiconductor materials. During the last decade, the most comprehensive data on thermal and elastic properties have been given by the evaluations of Jordan et al. [85–87], who obtained high-temperature values by extrapolating the few low-temperature data in the literature. Most of the currently available evaluations of thermally induced stresses in crystal growth are based on these data. The following sections will not try to evaluate or improve these data but will present the principal connections of elastic and plastic behaviour in semiconductor crystals.

3.1. Stress and strain in elastic crystals

The state of stress in a homogeneous body is described by resolving the stress vector, which may point in any arbitrary direction of the coordinate system, into its principal stress components, the directions, of which are identical to the axes of the coordinate system. In fig. 14 these components are shown for a volume element in a crystal in a cylindrical coordinate system, which is the most used system for describing stresses in crystals due to its correspondence to the crystal's shape. It should be noticed, for further discussion, that the mathematical treatment of stress at a point demands the volume element to be infinitesimally small which is in contrast to the definition of stress as force per area. The stresses acting in the directions of the three basic coordinate axes are called normal stresses and represented by the character σ. Shear stresses are acting within a basic plane and denoted as τ. Additionally, two subscripts are necessary to characterize the normal of the stressed plane and the direction of stress. Thus, in a system with axes $\boldsymbol{i}, \boldsymbol{j}, \boldsymbol{k}$, τ_{ij} represents a shear stress within the plane $\boldsymbol{i}$ (set up by the vectors $\boldsymbol{j}$ and $\boldsymbol{k}$) pointing in the direction $\boldsymbol{j}$. σ_{ii} stresses the same plane but without any shear component due to the force vector pointing in the direction $\boldsymbol{i}$.

The existence of stresses in a body is coupled to the displacements of the elements of that body. The volume element experiences a strain ε (tensile or compressive), which is related to the stress in pure elastic bodies by the elastic modulus E according to eq. (2.3). A longitudinal expansion of a three-dimensional element furthermore causes a trans-

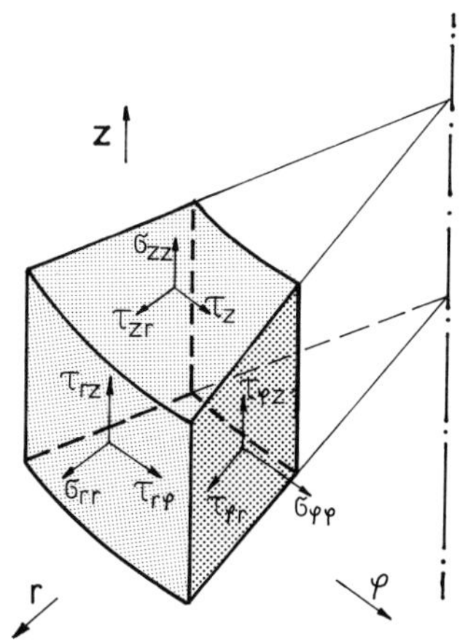

Fig. 14. Components of stress in a volume element in the cylindrical system.

verse contraction, the amount of which is material dependent and correlated with the longitudinal expansion by Poisson's ratio, ν. For absolutely homogeneous materials this ratio should be 0.25 whilst, for many existing materials, values of up to 0.35 have been measured. In cartesian coordinates, the relationship is given by the equation

$$\varepsilon_{zz} = \varepsilon_{yy} = \nu(-\varepsilon_{xx}) = -\nu/E\sigma_{xx}, \tag{3.1}$$

which can easily be transformed into three dimensions resulting in the following equations

$$\varepsilon_{xx} = \frac{\sigma_{xx} - \nu(\sigma_{yy} + \sigma_{zz})}{E}, \tag{3.2}$$

$$\varepsilon_{yy} = \frac{\sigma_{yy} - \nu(\sigma_{zz} + \sigma_{xx})}{E}, \tag{3.3}$$

$$\varepsilon_{zz} = \frac{\sigma_{zz} - \nu(\sigma_{xx} + \sigma_{yy})}{E}. \tag{3.4}$$

The shear stresses τ_{ij} in fig. 14 are correlated to the shear strains Γ_{ij} by the shear modulus G:

$$\Gamma_{ij} = \tau_{ij}/G. \tag{3.5}$$

The three constants E, G and ν characterize the principal stress–strain relationship of homogeneous isotropic elastic material.

The crystalline structure of single crystals causes a certain anisotropy of the elastic constants. An exact calculation of elastic stresses thus should be carried out in tensor notation leading to very time-consuming computations. However, the results of such calculations only exhibit minor differences to the results of isotropic calculations as has been shown by Miyazaki et al. [88].

3.2. Plasticity in crystals

In the beginning of the 1950s, Gallagher [89] and Graf et al. [90] discovered that the semiconductors silicon and germanium, which were known to be extremely brittle at room temperature, can be deformed by external mechanical forces at higher temperatures. This caused many researchers in the following years to investigate the principal mechanisms of such a change in material properties with temperature. For the first time, the availability of pure single crystals with diamond-like structure and low defect densities promised the possibility of a quantitative comparison between theory and experimental results. Soon it was established that the dominant mechanism responsible for the deformation was the generation and movement of dislocations. Other mechanisms like vacancy transport and twin formation could not explain the large degrees of deformation and were considered to play a minor role.

Furthermore, it was found that the application of a constant load does not result in a constant rate of deformation. The degree of deformation increases with time surprisingly at continuously increasing rate. This time-dependent behaviour of materials was first described by Andrade in 1910 [91]. Today it is well known as creep of a material.

Weertman and Weertman [92] discuss several types of creep mechanisms which may occur as a function of the actual conditions of stress and temperature. According to them, it is possible to handle different materials like indium, tungsten, silicon or gallium arsenide within one single diagram by normalizing the applied stresses and temperatures to the material's shear modulus G and its absolute melting point T_m. The corresponding diagram is shown in fig. 15. Additionally, the range of stress (elastic assumption) covered during crystal growth by most of the current growth processes is marked in fig. 15.

According to Weertman and Weertman no noticeable movement of dislocations is evident below a certain "critical shear stress τ_{crss}". The possible degree of deformation is very small and dominated by processes like atomic rearrangements (e.g. carbon in

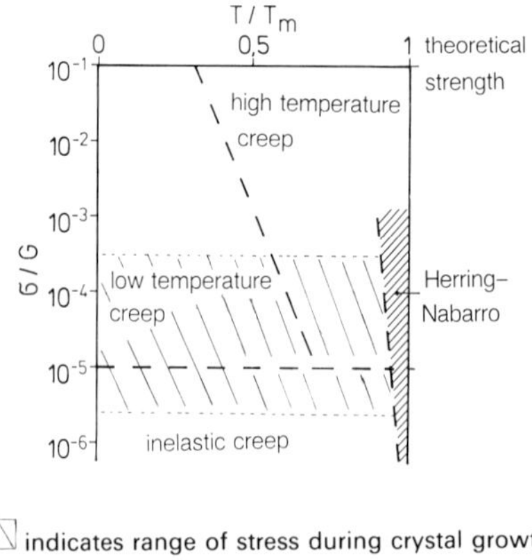

Fig. 15. Normalized diagram showing the dominating creep mechanisms as a function of temperature and load. Redrawn from Weertman and Weertman [92].

iron). The amount of this anelastic creep is negligible small compared to that generated by the movement of dislocations which is predominantly responsible for the creep above τ_{crss}. At this point, it should be critically observed that the τ_{crss} introduced by this representation has never been defined accurately in the literature. Furthermore, the movement of dislocations has been found to be a result of thermally activated processes like kink formation and propagation which do not allow one to define a fixed value of τ_{crss}. Thus, for the process of crystal growth and cooling, this τ_{crss} can be considered as a transition range of stress below which no deformation can be found within the limited duration of an experiment and finite resolution capabilities.

Above that "critical shear stress" the diagram shows the regions of low-temperature and high-temperature creep. In low-temperature creep the deformation is guided by pure slip of dislocations in their slip planes, because no climb processes are present due to the lack of vacancies at that low temperature. In the region of high-temperature creep, climbing of dislocations becomes more and more important with increasing temperature thus introducing a further degree of freedom to the movement of dislocations. The transition between low- and high-temperature creep is continuous, and certainly not as sharp as indicated in fig. 15. The "Herring–Nabarro creep" at very high temperature is dominated by vacancy diffusion in small volumes (grains). It is assumed to play a minor role in crystal growth due the dimensions of the crystal and the time scale of the growth process.

3.2.1. *Model of plastic deformation*

Most of the investigations in the literature try to reconstruct the macroscopically measurable deformation behaviour from the microscopic processes within the lattice of the crystal. Important contributions have been given by Chaudhuri et al. [93, 94], Hughes [95] and Haasen [12, 96].

The most comprehensive model for plastic deformation in semiconductors is that of Alexander and Haasen [12]. It is based on the Orowan equation [97], which gives a relationship between the macroscopic strain ε_{pl} and the microscopic process of dislocation motion, i.e. the mobile dislocation density N moving a distance dx in the crystal,

$$d\varepsilon_{pl} = \Phi b N \, dx, \tag{3.6a}$$

the integration of which results in the strain rate

$$d\varepsilon_{pl}/dt = \Phi b N v, \tag{3.6b}$$

where N is the mean mobile dislocation length per unit volume cm/cm^3, b is the Burgers vector, x is the mean free distance passed by the dislocations, v is the mean velocity of the dislocations, and Φ is a geometrical factor relating the direction of true shear strain within the slip system to the direction of the strain generated by a load.

The difficulties of solving the Orowan equation arise from the interaction of N and v and their dependence on parameters like shear stress τ and temperature T. The dislocation density N is a function of τ and, due to the process of dislocation multiplication, also of the velocity v. The velocity v is influenced by τ and temperature T: additionally it depends on N due to the effect of strain hardening of the material. The relations will be discussed shortly in the following as they are important equations to understand the elastic-plastic processes and thus the stresses during the cooling of a crystal.

The multiplication of dislocations is assumed to be a function of the dislocation density itself and of the distance covered by the moving dislocation [12, 95],

$$dN = æ N dx, \tag{3.7}$$

where æ represents the multiplication parameter, which depends on the effective stress τ_{eff} at the dislocation according to Peissker et al. [98],

$$æ = K \tau_{eff}. \tag{3.8}$$

The multiplication constant K can be evaluated from the degree of deformation at the inflection point of the creep curves [12]. The effective stress τ_{eff} is given by the applied stress τ, which is counteracted, and thus reduced, by an internal stress $A\sqrt{N}$ representing the strain hardening effect by already present dislocations. The constant A may be calculated from the elastic interaction of dislocations:

$$A = Gb/[2\pi(1 - \nu)], \tag{3.9}$$

where G is the shear modulus and ν the Poisson number.

The effective stress at the dislocation,

$$\tau_{eff} = \tau - A\sqrt{N}, \tag{3.10}$$

causes the dislocation to move at a velocity v [12]:

$$v = B_0(\tau_{eff})^m \exp(-E/kT). \tag{3.11}$$

The constant B_0 can be determined by measurements of the dislocation velocity or by fitting macroscopic deformation experiments; the activation energy E and stress exponent m are obtained directly from deformation experiments.

From eqs. (3.6) to (3.11) a system of equations follows which is called the equation of state:

$$d\varepsilon/dt = \Phi b N B_0 \exp(-E/kT)(\tau - A\sqrt{N})^m, \tag{3.12a}$$

$$dN/dt = K N B_0 \exp(-E/kT)(\tau - A\sqrt{N})^{m+1}. \tag{3.12b}$$

This more or less phenomenological description has turned out to be appropriate for a wide range of tetrahedrically coordinated semiconductors like Si, Ge, InSb, InP, GaP and GaAs.

3.2.2. *Deformation experiments*

The quantitative evaluation of the deformation behaviour can be performed by three types of deformation experiments:

(a) relaxation;
(b) deformation at constant strain rate;
(c) deformation at constant load.

Relaxation experiments are providing the most realistic experimental conditions with respect to the deformation behaviour during crystal growth, where thermally induced stresses relax by generation, multiplication and movement of dislocations. However, such experiments are very difficult to perform and to evaluate, because only very small strains can be covered by this method. Therefore, most of the deformation literature deals with the following two types fo test:

In the first one, $d\varepsilon/dt$ is kept constant. $d\varepsilon/dt$ is the compression or elongation rate, dl/dt, normalized by the specimen length, l. Starting from nearly dislocation-free material, the shear stress τ necessary to establish constant $d\varepsilon/dt$ decreases, passes through a minimum called the lower yield point τ_{ly} and finally reaches a steady state or saturation value at large strains (fig. 16a). In the second type of test, often called a creep test, the stress τ is held constant and the plastic strain rate $d\varepsilon/dt$ ("creep rate") is measured (fig. 16b). For an initially nearly dislocation-free material, the creep rate increases and passes through a maximum, until a steady state of creep is reached at large strains. Thus in both types of testing, the deformation behaviour is characterized by a minimum resistance against plastic deformation at small to intermediate strains (τ_{ly} and ε_w) and a maximum at large strains. In the context of evaluation of the parameters necessary to solve eqs. (3.11) and (3.12), this minimum deformation resistance is most important. At this minimum the softening (constant strain rate at reducing stress or increasing creep rate at constant stress) of the material due to an increasing number of dislocations is balanced by work hardening (interaction of dislocations resulting in a reduced number of mobile dislocation and reduced dislocation velocity). It has been shown by Alexander and Haasen [12] that the lower yield stress τ_{ly} and the maximum creep rate $d\varepsilon_w/dt$ both obey the same mechanical equation of state $f(d\varepsilon_w/dt, \sigma, T) = 0$, independent of deformation history, which can be expressed

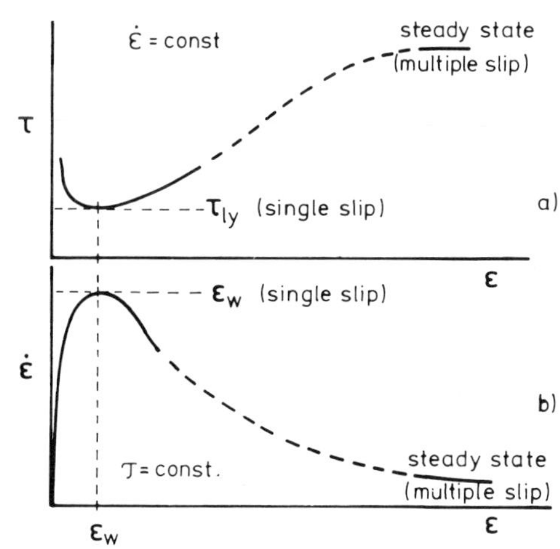

Fig. 16. Schematical representation of the material response (ε) on tests at constant strain rate (a) and tests at constant load (b).

in two equivalent ways:

$$d\varepsilon_w/dt = D\tau^{2+m}\exp(-E/kT), \tag{3.13a}$$

$$\tau_{ly} = C(d\varepsilon_w/dt)^{1/(2+m)}\exp[E/(2+m)kT], \tag{3.13b}$$

where D and $C = D^{-1/(2+m)}$ are constants.

The physical meaning of eq. (3.13) [see also eq. (3.12)], according to the model of Alexander and Haasen, is that the dislocation density varies as τ^2 while the dislocation velocity varies as τ^m. The stress exponent m has been found to be in the range of 1 to 1.5 for most tetrahedrally coordinated semiconductor materials.

In table 1 a list of parameters is given for different undoped materials. These figures represent only a small fraction of the data in the literature, especially with respect to silicon and germanium. An excellent overview on deformation of silicon and germanium has been given by Alexander [99].

The data, especially on III–V compounds, are rather incomplete at present. The figures given by different authors may show remarkably differences. One reason for this may originate from the low number of available deformation experiments due to the time-consuming deformation procedure which prevents sufficient statistics. On the other hand the origins of the basic elemental mechanisms of plastic deformation and their interaction in macroscopic deformation [e.g. generation mechanisms of new dislocations, movement of existing dislocations, processes of dislocation multiplication, type and portions of moving dislocations (60°, screw, partials) etc.] are still not fully understood. Thus a general and consistent theory is not yet available. However, the current data are of sufficient accuracy to enable an, at least semi-quantitative, modeling of deformation dynamics during the growth and cooling of crystals as will be discussed in section 4.

An impressive example of the deformation behaviour of different semiconductors is given in fig. 17. Silicon [12] and InP [100] have been deformed in constant load experiments at a temperature of 0.7 times the absolute melting point T_m ($0.7\,T_m$: Si: 1173 K; InP: 934 K) and comparable stress levels. It is evident from fig. 17 that the principal deformation behaviour is the same for both semiconductors. However, the

Table 1
Material parameters of plastic deformation for different tetrahedrically coordinated semiconductors.

Parameter	Symbol	Si	Ge	InP	InSb	GaAs	Unit
Strain-hardening factor	A	8.6	5.5	3.0	30	–	$N\,m^{-1}$
Prefactor of dislocation velocity	B_0	2.1×10^{-2}	4.6×10^{-3}	1.4×10^{-8}	9.8×10^{-4}	–	$m^{2+m}\dot{N}^{-m}S^{-1}$
Multiplication constant	K	5.7×10^{-4}	2.5×10^{-3}	1.2×10^{-2}	8.8×10^{-5}	–	$m\,N^{-1}$
Activation energy	E	2.2	1.6	1.0	0.7	1.5	eV
Stress exponent	m	0.9–1.4	1.1	1.4	1.1	1.7	

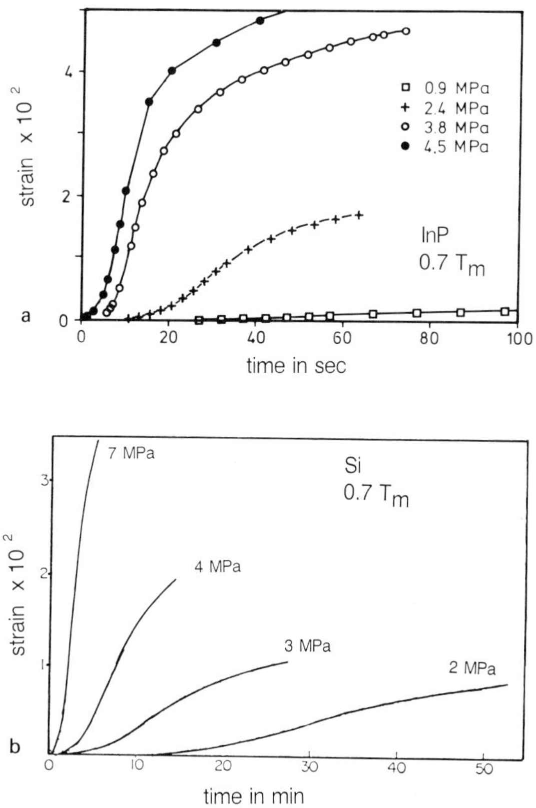

Fig. 17. Comparison of the deformation behaviour of InP (a) and Si (b) at comparable loads and a deformation temperature of 0.7 T_m, respectively. The diagram on the deformation behaviour of silicon has been redrawn from [12]. Please note different time scaling and slightly differing strain scaling for both figures.

decisive difference is given by the timescale needed to achieve similar degrees of deformation. At a constant load of about 3 MPa, a deformation of 1% is reached within less than 20 s in InP, whereas it takes about 1200 s to get the same degree of deformation in silicon!

3.3. Orientation of plastic deformation – slip systems

It is well established nowadays that the dislocation-based deformation of diamond-like semiconductors takes place along certain crystal planes and directions, i.e. slip systems. These slip systems are determined by the [111] planes and $\langle 1\bar{1}0 \rangle$ directions resulting in 24 permissible operations, which are reduced to 12 principal operations as counterwise slip within the same slip system (e.g. into $\langle 1\bar{1}0 \rangle$ and $\langle \bar{1}10 \rangle$–direction) gives equivalent results. Table 2 lists all of the 12 slip systems.

Table 2
Slip systems in tetrahedrically coordinated semiconductors.

Plane	Direction
111	$\bar{1}01$
111	$1\bar{1}0$
111	$01\bar{1}$
$1\bar{1}1$	$\bar{1}01$
$1\bar{1}1$	110
$1\bar{1}1$	011
$11\bar{1}$	$1\bar{1}0$
$11\bar{1}$	011
$11\bar{1}$	101
$\bar{1}11$	$01\bar{1}$
$\bar{1}11$	101
$\bar{1}11$	110

The meaning of these slip systems is that plastic deformation does not occur homogeneously within a single crystal. Elastic forces of any direction which can cause plastic deformation have to be projected into the appropriate slip planes and directions to find the real active force operating the slip system. This was first formulated by Schmid and Boas [101]. Consequently, a certain stress vector will induce shear stresses in all slip systems except in those perpendicular to the stress vector. Thus the different slip systems will experience different amounts of shear stress. At first the highest stressed slip system will become active and reduce the elastic stresses according eq. (3.12) simultaneously taking away part of the driving force of the other slip systems. The active slip system will carry most of the deformation, and dislocation density will increase until the rising back-stress $A\sqrt{N}$ generated by these dislocation is so high that another slip system, having a lower number of dislocations and thus higher effective stress, will operate.

3.4. Modeling of deformation experiments

Deformation experiments are carried out in a way that, according to Schmid's law, only one slip system dominates the deformation. In cubic crystals this can be achieved by choosing the $\langle 3\bar{2}\bar{1} \rangle$ orientation as a direction of the external stress resulting in a Schmid factor of $\phi = 0.467$ for the highest stressed slip system. This simplifies the evaluation of experiments as well as the modeling of the deformation due to single slip conditions.

The modeling of plastic deformation demands the simultaneous solution of eqs. (3.12a) and (3.12b). Analytically this can only be done by integrating eq. (3.12) introducing certain simplifications which are the assumption of a constant stress τ, constant dislocation velocity v, negligible backstress $A\sqrt{N}$ and constant dislocation density N. This will lead to an overestimation of the value of deformation compared to experiments where the rising back-stress $A\sqrt{N}$ reduces the effective stress and thus influences dislocation velocity [eq. (3.11)] and multiplication behaviour [eq. 3.8)]. Furthermore, the analytical solution is not able to implement a time dependence of the external stress which is necessary for a correct calculation of deformation behaviour and dislocation prediction during the process of crystal growth.

A method of solving the equation of state numerically as applied by Völkl and Müller [10, 11] is given in the following. It is based on the principle of all numerical integration methods, which is to integrate the equations within a small time interval t_i to $t_i + \Delta t_i$ and add up all intervals. For such integrations a great number of numerical methods exist. Here a simplified analytical solution is taken to predict dislocation density N and strain ε within a sufficiently small interval starting from t_i and ending at $t_i + \Delta t_i$. The simplifications necessary for the analytical solution introduce deviations of the calculated values from the real values, which can be evaluated and corrected by a suitable additional numerical procedure.

The process of the evaluation of the errors is based on a proposal of Gragg [102]. Following his ideas, extrapolation methods have been developed (see, e.g. [103]), which are based on the process of integrating a small part of the whole integration domain and to repeat that integration for the same part with a series of decreasing step widths. Finally an extrapolation procedure is used to calculate a new approximated value, which then is used as a starting value for continuing the procedure. This method has been evaluated by Hall et al. [104] to be extremely precise and stable compared to most other methods.

This procedure allows one to trace dislocation density N and velocity v as well as the amount of plastic deformation and elastic stress over the whole integration domain (period of deformation in creep tests or period of growth during crystal growth) as it is designed to follow the history of temperature, external and internal stress variations and dislocation formation during the dynamic process of crystal growth.

The simulation of deformation experiments under constant external stress σ_{ex} and the good correlation with experimental results is demonstrated in fig. 18. An estimation

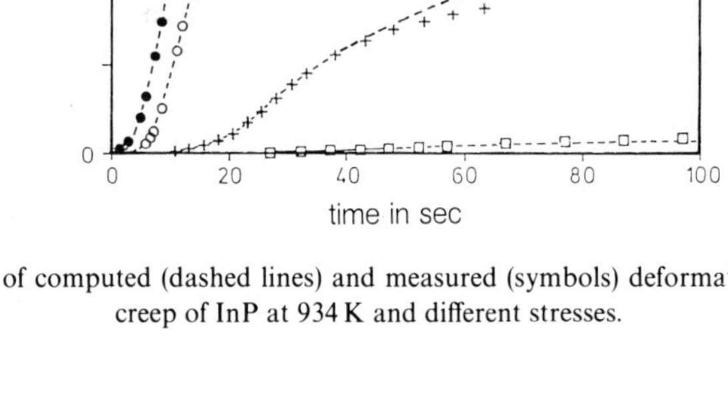

Fig. 18. Comparison of computed (dashed lines) and measured (symbols) deformation behaviour for the creep of InP at 934 K and different stresses.

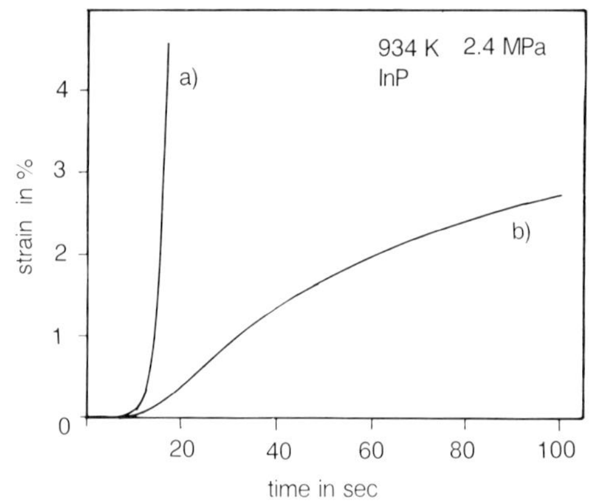

Fig. 19. Influence of neglecting the back-stress parameter (a) during the modeling of the deformation of InP. For the calculation of curve (b) a value of $A = 0.03\ \mathrm{N\,cm^{-1}}$ has been used for the back-stress parameter.

of the dislocation density at the inflection point of the curves according to the treatment of Alexander and Haasen [12] results in values above $10^6\ \mathrm{cm^{-2}}$ which is assumed to be the lower limit for obtaining reliable results by using this modeling procedure. Figure 19 reveals the enormous influence of neglecting the back-stress $A\sqrt{N}$ for the modeling of a creep experiment assuming deformation of InP at a constant load of 2.4 MPa and a constant temperature of 934 K. Considerable overestimation of strain can be observed in curve (a) if the backstress or strain hardening parameter $A\sqrt{N}$ is not considered. During crystal growth, where stresses lower than 2.4 MPa can be present due to relaxation processes, the influence of neglecting the strain-hardening parameter may cause considerable errors.

4. *Modeling of stress and strain during crystal growth experiments*

In this section the elastic and plastic properties of a material will be combined to develop a model for the evaluation of the time dependent deformation behaviour during crystal growth. In contrast to many other models, it will be designed to handle the process of the prediction of dislocation density in the final crystal without using the concept of a critical resolved shear stress. Thus it has to take into account the thermally activated nature of plastic deformation processes, i.e. the movement and multiplication of dislocations. Additionally it will be able to involve dynamic growth parameters, e.g. the influence of growth rate on final dislocation density. The model will also take into account the nonlinearities of plastic deformation. Thus it will allow one to solve any given stress–temperature–time function during the growth of a crystal using any suitable growth process. The extraordinary meaning of the model will be in the evaluation of thermal growth conditions and dynamic growth parameters to provide a prediction and engineering of crystal quality with respect to final dislocation density and residual stresses in the grown crystals.

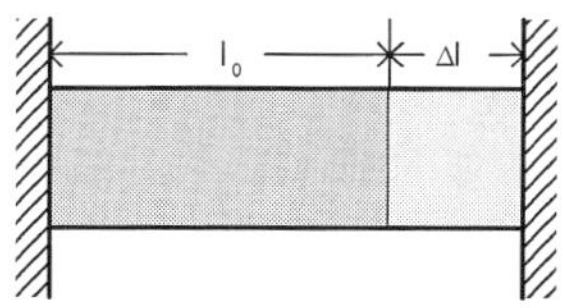

Fig. 20. Schematic drawing to describe the elastic–plastic deformation in a crystal element. For better visualisation strains are assumed to occur in the horizontal direction only.

4.1. Principles of the model

The basic problem of the evaluation of stress and strain in the cooling crystal is to solve the relaxation of an alternating elastic strain existing in a solid body. The relaxation (plastic deformation) process depends on temperature and the applied level of stress, which is induced thermally in the growing crystal and varies due to the time-dependent experimental conditions, e.g. thermal environment, as well as due to the continuously acting relaxation mechanism during the growth run.

The principles of the model can clearly be understood if a simple one-dimensional problem is assumed. A bar with an initial length l_0 is fixed at both ends as indicated in fig. 20. The bar is being elongated by an instantaneous shift of fixing 2 by a length Δl. The total strain $\varepsilon_t = \Delta l / l_0$ results which is purely elastic at time $t_0 = 0$. Because of the plasticity of the material, the initial elastic strain is partly transformed into a plastic deformation (plastic strain ε_{pl}) with increasing time. As both ends in fig. 20 are fixed, the sum of the resulting plastic strain ε_{pl} (t) and the residual elastic strain $\varepsilon_{er}(t)$ is equal to the total strain ε_t, which is constant at any time t:

$$\varepsilon_{er}(t) + \varepsilon_{pl}(t) = \varepsilon_t. \tag{4.1}$$

It should be pointed out straight away that the driving force for plastic deformation is given by the actual value of $\varepsilon_{er}(t)$ and not by the total deformation ε_t. The left-hand side of eq. (4.1) describes the state of strain of a realistic elastic–plastic body. ε_t in eq. (4.1) has to be replaced by $\varepsilon_t(t)$ if the position of the fixing of one end of the bar in fig. 20 is moving with time. This corresponds to the situation during the growth of a crystal, where an element of the crystal is exposed to a continuously varying strain by the surrounding material and is trying to relax elastic strain by transforming it into plastic strain:

$$\varepsilon_{er}(t) + \varepsilon_{pl}(t) = \varepsilon_t(t). \tag{4.2}$$

The time derivative of the strain leads to the strain rate. Consequently a correlation for the strain rates is obtained from eq. (4.2):

$$\mathrm{d}\varepsilon_{er}(t)/\mathrm{d}t + \mathrm{d}\varepsilon_{pl}(t)/\mathrm{d}t = \mathrm{d}\varepsilon_t(t)/\mathrm{d}t. \tag{4.3}$$

In the case of rigid fixings the right-hand side of eq. (4.2) is a constant resulting in the following relation for the strain rates:

$$\mathrm{d}\varepsilon_{pl}(t)/\mathrm{d}t = -\,\mathrm{d}\varepsilon_{er}(t)/\mathrm{d}t. \tag{4.4}$$

This means that each increase in plastic strain $d\varepsilon_{pl}$ results in a decrease of the residual elastic strain $d\varepsilon_{er}$ of the same amount. For time extending to infinity, ε_{pl} tends towards the total deformation ε_t. At the same time the driving force ε_{er} for the plastic deformation vanishes. Then the body of length $l_0 + \Delta l$ rests unchanged in length if the fixings are removed.

4.2. *Transfer of the model into a growing crystal*

The model now has to be applied to a growing crystal. The aim is to be able to trace the history of dislocation formation in a thin slab embedded elsewhere in the growing crystal. The initial calculation of thermal stresses assuming elastic material properties in the whole crystal has to be combined with the plastic relaxation behaviour of the material, which surely is the dominating mechanism at elevated temperatures. Several questions arise from the fact that elasticity assumed during stress modeling has to be replaced by elastic–plastic behaviour at high temperatures:

(a) What is the meaning of elastic stress calculations as shown in fig. 13, which only represent a "snapshot" at a certain moment of growth, for the calculation of the history of dislocation formation of a slab or wafer?

First of all, it is clear that these elastic stresses will not exist in the crystal at elevated temperatures due to relaxation mechanisms. However, the stresses are a measure for the local thermally induced strain due to inhomogeneous thermal expansion. The inhomogeneity of the thermal expansion is imprinted by the thermal characteristics of the whole growth configuration. In a first-order approximation, the thermal environment of the crystal may be considered to be stable during the growth run. This is because the temperature field of the crystal is a result of the heat exchange with its environment and the environmental set-up remains nearly unchanged during the growth run. Such characteristic positions in a growth set-up are, e.g. the boron oxide/gas transition in LEC growth [80, 35], the location of heat shields [60] or the imprint of gradients by a multiheater technique in Bridgman–Stockbarger or gradient freeze set-ups. For a sufficiently long crystal, each element of that crystal, except those at the very ends, can be considered to be pulled through a fixed temperature field as is indicated in fig. 21. This fixed temperature field induces characteristic strains at each local position, the derivative $d\varepsilon/dz$ of which gives the changes in strain of an element being pulled along the z-axis and thus allows one to reconstruct a load history of the element. The reconstruction of the load history can be directly transferred to the right-hand side of eq. (4.3) in conjunction with the pull rate of the crystal. This simplified approximation is advantageous as only one thermo-elastic calculation is necessary for the input of the modeling of the dislocation formation. It has been shown to give satisfying agreement with experimental results [11].

A much better but also more time-consuming method to reconstruct the history of strain would be to make a series of thermo-elastic calculations at increasing length of the crystal. Then the slab to be considered may be traced from its first formation at the growth interface through its way to the growth set-up and the cooling process to room temperature. Such calculations are extremely complex, especially with respect to the

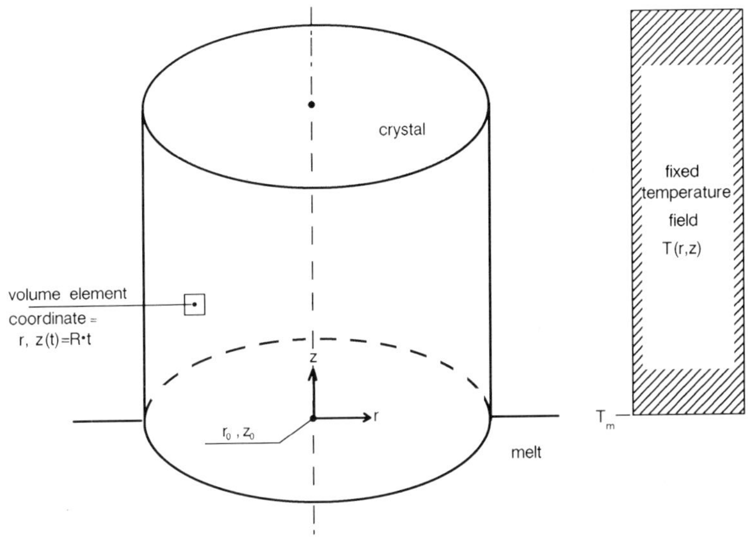

Fig. 21. Schematic drawing of a volume element within a crystal. The coordinate system is fixed at the crystal–melt interface. A volume element of the crystal moves at a constant radius r along the axial coordinate z with $z(t) = Rt$ (R = pull rate, t = time). The temperature field $T(r, z)$ inside the crystal is assumed to be stationary.

correct incorporation of dynamic effects so that, in most cases, the simplified method described above is used [10, 53, 105] in the literature.

(b) What is the state of elastic and plastic stress/strain when the slab is just created, i.e. when it is undergoing liquid–solid transformation?

Most analyses in the past have used the absolute stress/strain values of thermo-elastic calculations directly for the calculation of dislocation formation at the interface [7, 10, 11, 63, 105, 106]. Especially at and near the solid–liquid interface the stresses must be much lower than the elastically calculated ones due to the relaxation of elastic stresses by plastic deformation. The consequence was an overestimation of, e.g. the multiplication factor leading to a sudden dislocation multiplication near to the solid – liquid interface particularly under conditions of high (elastic) stress accumulation in that area as is shown in fig. 13a. The knowledge of the initial value of the stress at the solid–liquid interface is a major factor for the correct evaluation of the dislocation formation in a growing crystal. Some ideas to find out the real stress at the interface will now be given.

In principle, the solid–liquid interface itself represents an isothermal body. According to thermo-elastic theory, no stresses exist in a body of any shape being at homogeneous temperature. Thus a very thin interface layer of any curvature, constant temperature T_m and without connection to a crystal should be free of stress. Stresses at the growth interface, as shown in fig. 13a, must be introduced by the coupling to the bulk of the crystal, i.e. the introduction of divergent heat flow adjacent to the interface creating long-range stress effects due to the assumed elasticity of the material. The extension of such long-range elastic stresses can clearly be deduced from fig. 13b, where the strongly divergent heat flux at the boron-gas transition introduces elastic stresses spreading out several centimeters into the bulk of the crystal. Especially at the high temperatures near the interface such long-range effects are not expected due to the immediate (at least within seconds) relaxation of elastic stresses. The value of elastic stress at the interface is given by the counter-action and balance of long-range elastic effects and local plastic relaxation. Consequently, a method to evaluate the maximum possible elastic stresses near the interface has to be developed to determine the initial value of stress for the modeling of dislocation formation. At present this seems only to be possible by extrapolating the results of creep experiments to the melting point and by performing computerized deformation experiments able to find out the finite value of shear stress τ_i necessary to initiate dislocation movement and multiplication. In the case of InP a value of 0.1 ± 0.05 MPa has been found for τ_i using the modeling procedure described in section 3. Values for other semiconductors have been reported by Jordan et al. [107], which are extrapolated from the yield point of deformation experiments resulting in some degree of overestimation due to the strain-hardening effect generated by the high number of dislocations at the yield point (generally $\gg 10^6\ \mathrm{cm}^{-2}$) leading to an additive factor according to the corresponding values of $A\sqrt{N}$.

(c) Which magnitude of elastic shear stress τ has to be used in the equations describing dislocation formation, e.g. for the evaluation of the multiplication factor and the velocity of dislocations?

The elastic shear stress representing the driving force for dislocation formation is a function of the increase of the elastic distortion of the lattice and the plastic release of that distortion. According to elastic theory the shear stress τ within a slip plane is coupled to any arbitrarily directed strain ε by

$$\tau = \varepsilon G / \Phi, \tag{4.5}$$

where G is the shear modulus, Φ is a geometrical factor related to the directions of strain and shear.

Within each element of the crystal the shear stress is given by the actual residual elastic stress τ_{er} which is continuously changed by the varying distortions due to the movement of the element through the temperature field and the plastic response of the element to these changes. Thus an equilibrium depending on the dynamics of growth is established between the residual elastic strain and the actual plastic strain represented by both expressions on the left-hand side of eq. (4.3). This interaction represents the critical point for the solution of dislocation dynamics, because the value of both expressions depends on each other according to eq. (4.4), respectively, and the only known parameter is given by the right-hand side of eq. (4.3). It goes without saying that, in addition to that process, the equilibrium of forces has to be ensured with respect to the elastic residual stresses (tensile and compressive) within the whole crystal and each plane at any instant of growth.

(d) Which slip systems are activated by the thermally induced stress?

Within each element of the crystal all slip systems exist and are loaded differently according to their geometrical relation (Schmid factor) to the direction of the stress. The exponential dependence of plastic deformation from stress will cause the highest loaded slip system to act and thus to release the stresses on all other slip systems. A second slip system or also parallel plane of the same slip system will become activated if its effective stress $(\tau_2 - A\sqrt{N_2})$ exceeds that of the first one $(\tau_1 - A\sqrt{N_1})$.

The processes and conditions within a real crystal which is deformed plastically are described by the left-hand side of eqs. (4.2) and (4.3). The major problem to be solved in this equation is the evaluation of $\varepsilon_{pl}(t)$.

4.3. *Evaluation of $\varepsilon_{pl}(t)$ and correlation to the dislocation density $N(t)$*

The deformation rate $d\varepsilon_{pl}/dt$ is related to the movement and density of dislocations by the Orowan relation [97],

$$d\varepsilon_{pl}/dt = \Phi b N(t) v(t), \tag{4.6}$$

with the Burgers vector b, the actual average velocity v of the dislocations of density N (length of mobile dislocation lines per unit volume).

According to the treatment of Alexander and Haasen [12], in diamond-like crystals this fundamental equation for plastic deformation can be transformed into

$$d\varepsilon_{pl}(t)/dt = \Phi b N(t) B_0 \exp(-E/kT)\, \tau_{eff}^m, \tag{4.7}$$

with the average dislocation velocity

$$v = B_0 \exp(-E/kT)\, \tau_{eff}^m. \tag{4.8}$$

The pre-exponential factor B_0 can be determined from measurements of the dislocation velocity; the activation energy E of the microscopic kink processes and the stress exponent m can be obtained from macroscopic deformation experiments. τ_{eff} is given by

$$\tau_{eff} = \tau - A\sqrt{N}, \tag{4.9}$$

and takes into account the fact that already existing dislocations "harden" the lattice against further deformation, as these dislocations represent obstacles for other moving dislocations reducing their velocity. This effect is expressed in terms of an opposing stress or back-stress $A\sqrt{N}$ reducing the driving stress τ. N is the total density of dislocations which may be assumed to be equal to the number of moving dislocations in the case of crystal growth, i.e. in the range of less than about 10^6 cm^{-2} dislocations. The parameter A is called the strain-hardening factor. In the actual terminology the external stress τ in eq. (4.9) is the residual elastic stress $\tau_{er}(t)$ resulting from eq. (4.2) in connection with eq. (4.5) representing the driving force for the plastic deformation during crystal growth.

The second fundamental equation in addition to eq. (4.7) describes the successive variation of the dislocation density N during the deformation process [12]:

$$dN/dt = K N(t) B_0 \exp(-E/kT)\, \tau_{eff}^{m+1}. \tag{4.10}$$

The multiplication constant K can be obtained from the inflection point of creep curves [12] in deformation experiments with constant load.

The treatment of Alexander and Haasen, i.e. eqs. (4.7) and (4.10), is now inserted into eq. (4.3) giving a set of differential equations for the successive plastic deformation in the volume element of the crystal:

$$\frac{d\tau_{er}(t)}{dt} + \frac{G}{\Phi}\frac{d\varepsilon_{pl}(t)}{dt} = \frac{d\tau_t^{gs}(t)}{dt}, \tag{4.11}$$

$$\frac{d\varepsilon_{pl}(t)}{dt} = \Phi b N(t) B_0 \exp\left(\frac{-E}{kT}\right)(\tau_{er} - A\sqrt{N})^m \tag{4.12}$$

$$\frac{dN(t)}{dt} = K N(t) B_0 \exp\left(\frac{-E}{kT}\right)(\tau_{er} - A\sqrt{N})^{m+1}. \tag{4.13}$$

The total dislocation density in the volume element of the crystal is obtained by integrating eqs. (4.11), (4.12) and (4.13) simultaneously over the period of the whole growth process. This can be done by a numerical procedure which is described in the next section.

4.4. Numerical procedure

The system of differential equations cannot be solved analytically. In order to prepare a numerical procedure, they are integrated assuming certain simplifications which are only justified for small intervals of time t. At first, eq. (4.13) is integrated by assuming the velocity v of the dislocations and the effective stress τ_{eff} to be constant according to ref. [12], giving the solution

$$N(t) = N(t_0)\exp(K\tau_{eff} v t). \tag{4.14}$$

Equation (4.14) is inserted into eq. (4.12) which then is also integrated analytically by using the same simplifications. This results in:

$$\varepsilon_{pl}(t) = \Phi b N(t_0)\frac{\exp(K\tau_{eff} v t) - 1}{K\tau_{eff}}. \tag{4.15}$$

Due to the approximations for the integration, the resulting eqs. (4.14) and (4.15) can only be used in a stepwise numerical integration process to obtain the solution at $t_i + \Delta t_i$ when starting at a time t_i. The new dislocation density at time $t_i + \Delta t_i$ is calculated by using τ_{eff} and v from the preceding time step t_i:

$$N(t_{i+1}) = N(t_i)\exp[K\tau_{eff}(t_i)v(t_i)\Delta t_i] \tag{4.16}$$

In the same manner, the new amount of plastic deformation $\varepsilon_{pl}(t_{i+1})$ is calculated by the equation

$$\varepsilon_{pl}(t_{i+1}) = \Phi b N(t_i)\frac{\exp[K\tau_{eff}(t_i)v(t_i)\Delta t_i] - 1}{K\tau_{eff}(t_i)} + \varepsilon_{pl}(t_i). \tag{4.17}$$

The new value $\tau_{er}(t_i + \Delta t_i)$ for the next time step is expressed by using eq. (4.11):

$$\tau_{er}(t_i + \Delta t_i) = \tau_{er}(t_i) - G/\Phi[\varepsilon_{pl}(t_i + \Delta t_i) - \varepsilon_{pl}(t_i)] + \tau_t^{gs}(t_i + \Delta t_i) - \tau_t^{gs}(t_i), \tag{4.18}$$

which contains no unknown variables on the right-hand side and can now be solved.

This procedure generates a deviation of the calculated value from the exact solution within each time step due to the simplifications for the analytical integration. These deviations are minimized by a suitable method, which is to use an additional subdivision of the time interval Δt_i by a variable integer n. The new time steps are now $\Delta t_i/n$. These smaller time steps are used to find the solution for the period Δt_i in the same manner as described above. The integer n is increased to n_i to get a series of solutions with decreasing step widths and increasing precision for the same interval Δt_i. The difference between two sequential solutions for the interval Δt_i may be used as the criterion of approach; still higher precision and convergence can be achieved applying a suitable extrapolation method as described by Bulirsch and Stoer [103].

Additionally, the computation time is optimized by varying the length of the next time interval Δt_{i+1} depending on the number of calculations needed to fulfill the criterion of convergence for the actual time step Δt_i (automatic time step adjustment).

4.5. Modeling of dislocation formation and stress in real growth systems

The model is applied to calculate the formation of dislocations and the residual elastic stress during the growth of InP crystals by the LEC technique. The material parameters used for the numerical modeling are given in table 3. The numerical results are compared to dislocation densities and patterns of as-grown crystals.

4.5.1. Comparison of numerical and experimental results

The growth runs of two InP crystals, which have been grown under different, precisely determined, thermal boundary conditions labelled as 12/18 and 30/44, have been simulated by the method described above. As indicated in fig. 9, 12/18 corresponds to a rather low and 30/44 to a large boron oxide layer thickness i.e. nearly fully encapsulated LEC growth. The axial history of the temperature and the stress in the highest stressed slip system for ten equally spaced radial positions of a crystal plane 10 mm below the cone end in the cylindrical part is taken to calculate the dislocation formation. A growth rate of 16 mm/h has been applied to conditions 12/18, one of 10 mm/h to conditions 30/44. The resulting dislocation density N, giving the length of dislocation

Table 3
Material parameters of InP used for the numerical modeling of creep experiments.

Parameter	Symbol	Value for simulation
Strain-hardening factor	A	0.03 N/cm
Prefactor of dislocation velocity	B_0	$0.54\,cm^{3.8}/N^{1.4}\cdot s$
Multiplication constant	K	1.2 cm/N
Activation energy	E	1.0 eV
Stress exponent	m	1.4

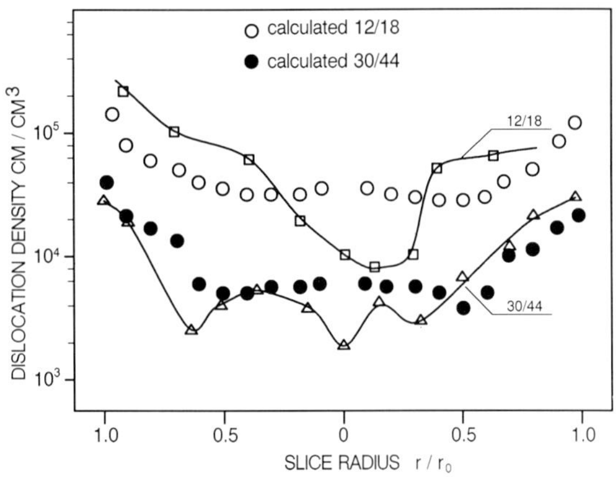

Fig. 22. Measured EPD across (111) wafers along the $\langle\bar{1}\bar{1}2\rangle$ direction in InP crystals (30 mm in diameter) grown in the $\langle 111\rangle$ direction by the LEC technique under different thermal boundary conditions 12/18 and 30/44 (solid lines). Full (30/44) and empty (12/18) circles denote the corresponding calculated values of the EPD using the model described in sections 3 and 4.

lines per cm^3, is transformed into the dislocation etch pit density (EPD) value by assuming straight dislocation lines $N = \text{EPD}$.

Figure 22 shows two curves of the EPD measured along the $\bar{1}\bar{1}2$ direction across (111) wafers cut from the crystals. The crystals were grown in a $\langle 111\rangle$ direction. The results of the calculations, indicated as full and empty circles in fig. 22, are compared to the experimental results. They show a surprisingly good agreement along the whole diameter, especially if the possible statistical variations of the EPD within different growth runs under the same conditions and the statistical deviations during dislocation counting are considered. These experimental fluctuations are about $\pm 40\%$. An estimation of the errors resulting from the calculations gives minor errors of less than 1% due to the numerical procedure itself when evaluating the dislocation formation. The contribution due to the extrapolation of property data from intermediate temperatures up to the melting point can only be estimated roughly to be in the range of about 10%. However, the errors introduced by the thermal stress analysis influencing the dislocation density calculations may rise up to about 25%.

4.5.2. Elastic–plastic interaction and the effect on elastic stress in the crystal

The relaxation of elastic stress $\tau_{gs}(t)$ in a slip system by plastic deformation according to eq. (4.3) leads to an increasing amount of plastic deformation $\varepsilon_{pl}(t)$ in the crystal and simultaneously to new and continuously changing values of the elastic residual stress $\tau_{er}(t)$. For a given stress–temperature–time function in a slip system, the elastic residual stress in a crystal element is also a function of the velocity at which the element is passing through the stress field (i.e. the growth rate). This is because of the dynamics of plastic deformation. The knowledge of $\tau_{er}(t)$ is one of the essential points of the whole procedure as discussed in section 4.2.

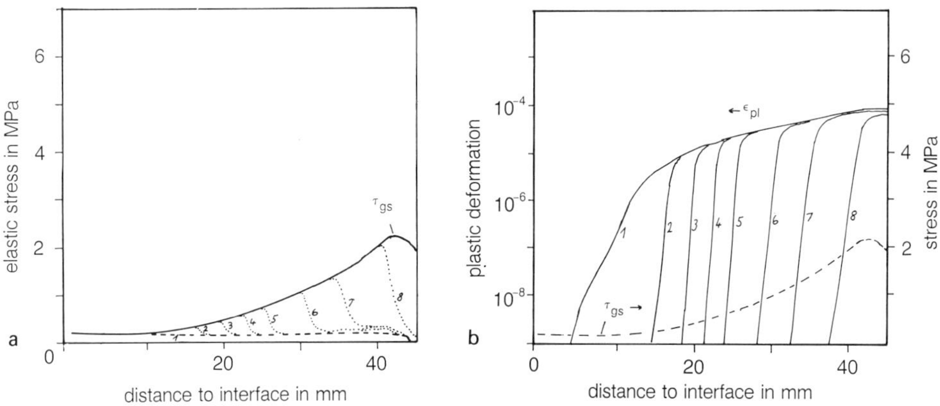

Fig. 23. Calculated residual elastic stress and plastic deformation in an elasto-plastic body using the Alexander–Haasen model compared to the calculated elastic stress in a purely elastic body (τ_{gs}). The calculations were performed using different growth rates: 1: 1 cm/h; 2: 2 cm/h; 3: 5 cm/h; 4: 10 cm/h; 5: 20 cm/h; 6: 50 cm/h; 7: 100 cm/h; 8: 200 cm/h. The relaxation of elastic stress (a), which is identical to the onset of plastic deformation (b), is shifted away from the interface with increasing growth rate showing the "equilibrium" conditions of the dynamics of deformation and the dynamics of growth.

In fig. 23 the elastic residual stress, plastic deformation as calculated and the stress function of a hypothetical elastic crystal are compared. The elastic stress data correspond to the shear stress history of a peripheral slip system under conditions 30/44. For a better comparison of the influence of the pure plastic deformation behaviour, the temperature and stress field changes in the crystal due to growth rate variations are not considered here, i.e. the stress–temperature function is kept constant for all assumed growth rates. At a growth rate of 10 mm/h (as in the experiment, the onset of plastic deformation occurs rather early in the process at about 4 mm behind the interface). Considerable amounts of plastic deformation are generated at increasing $d\tau_{gs}/dt$ about 10–15 mm behind the interface. During increasing plastic deformation the elastic residual stress in the slip system remains nearly constant at a level of about 0.15 to 0.20 MPa. Only a very small increase of the elastic residual stress can be noticed at an increasing distance from the interface which is related to the work-hardening term $A\sqrt{N}$ due to the increasing number of dislocations. The meaning of that calculation is that the real elastic stress in the growing and cooling crystal is nearly constant at a low level! As soon as an increase of the elastic stress is introduced by thermal strain, plastic deformation will occur and relax the additional elastic stress. At first sight this result is somewhat surprising. However, at low growth rates (→ low strain rates) there is enough time left for relaxation to a state of equilibrium in the whole volume of a crystal.

It follows from the low value of the elastic residual stress that the strain-hardening factor $A\sqrt{N}$ may not be neglected, even for small dislocation densities of about $10^4\,cm^{-2}$. At a dislocation density of $10^4\,cm^{-2}$ the backstress $A\sqrt{N}$ amounts to 0.03 MPa representing about 20% of the elastic residual stress during the plastic deformation! At a dislocation density of $10^5\,cm^{-2}$ the back-stress is about 0.1 MPa and thus in the same range as the elastic residual stress. Neglecting the strain-hardening factor would lead to considerable calculation errors concerning the multiplication behaviour and the dislocation velocity due to an overestimation of the effective stress parameter.

With increasing growth rate, the onset of plastic deformation is shifted more and more away from the interface. For a while, the plastic deformation is delayed until enough dislocations have been generated to facilitate plastic deformation. During that time the elastic residual stress is nearly identical to the stress in a fully elastic body. At the onset of plastic deformation the elastic residual stress is reduced to low values in the range of 0.2 MPa. The onset of plastic deformation at higher growth rates shows a yielding behaviour such as is observed from deformation experiments at constant strain rate, where an upper yield stress and a lower yield stress occur during the deformation. This behaviour is already observable at a growth rate of 20 mm/h. With increasing growth rate it becomes more and more distinct together with a steep onset of deformation.

From these calculations plastic strain rates may be deduced. These strain rates during crystal growth have been inaccessible for a long time. The question of the strain rate during crystal growth is important with respect to deformation experiments, because comparable rates have to be chosen in the deformation experiments to provide adequate conditions for valid conclusions on the yield stress of the material to be reached. At low growth rates of 10 and 20 mm/h the highest strain rate occurring during the calculations was about $1 \times 10^{-8}\,s^{-1}$. At the highest growth rates the maximum strain rate did not exceed $1 \times 10^{-5}\,s^{-1}$. Such low strain rates of $\leq 10^{-8}\,s^{-1}$ in the growing crystal can not be achieved during deformation experiments, where strain rates of $\geq 10^{-5}\,s^{-1}$ must be employed for experimental reasons. These higher strain rates lead to higher values of the lower yield stress, which often was used to evaluate a "critical resolved shear stress" for the onset of plastic deformation. In the case of undoped InP values of > 0.6 MPa have been found from such experiments [100, 107, 108]. The calculations give rise to the assumption that all of these conclusions from deformation experiments overestimate the yield stresses considerably due to the unsuitable range of strain rates used during the experiments.

4.5.3. Influence of the pull rate on the dislocation density

It is well known that dislocation-free silicon can be grown using high pull rates in the range of 3 to 10 mm/h during the necking procedure. This means that the crystal is moved quickly through the stress field. Therefore, the dislocations have only a very short time for their movement and the multiplication is then negligible. Additionally, the investigations of Dash [109] have shown that already existing dislocations (e.g. introduced from the seed or thermal shocks during the seed-on procedure) grow out towards the surface at these high pull rates. So both the propagation of existing dislocations and the generation of new ones can be suppressed by using high pull rates. Comparable successful results have not been reported for III–V semiconductors up to now. Experimentally, high growth rates are hard to achieve in III–V growth due to the low thermal conductivity and the large amount of latent heat during solidification. Thus the model presented above provides an easy and cheap way of estimating the probability of getting dislocation-free III–V crystals at increased pull rates. A variation of the pull rate results in a variation of $d\tau_{gs}/dt$ and hence influences the dynamics of dislocation formation via the onset and dynamics of plastic deformation in conjunction with the corresponding elastic residual stress value, which is determining the multiplication behaviour of the dislocations via the multiplication parameter æ [eq. (3.7)].

The stress–temperature distribution in the highest stressed slip system in the peripheral area of crystals 12/18 and 30/44 is used to calculate the final values of the released plastic strain ε_{pl} and the dislocation density N after a complete growth run in dependence of the pull rate. The difference between both stress functions is the existence of two maxima for crystal 12/18, one located very near to the interface and the second at about 18 mm from the interface and the presence of only one maximum for crystal 30/44 at a distance of about 44 mm from the interface. The results are shown in fig. 24.

An increasing dislocation density up to a pull rate of 200 cm/h results for conditions 30/44. The amount of plastic strain remains approximately constant within this range of the pull rate variation until a sudden decrease of the plastic strain and the dislocation density can be noticed at a pull rate of about 400 cm/h. With conditions 12/18 the dislocation density remains constant over a wide range of the pull rate. The tendency for a reduction can only be found at very high pull rates (> 5000 cm/h, not shown in fig. 24).

The results show that the influence of the pull rate on the dislocation density depends very much on the stress and the temperature conditions of the growth run. Under conditions 30/44 the single maximum of stress occurs at relatively low temperatures of about 0.75 to 0.8 T_m (T_m = 1335 K). At these temperatures the interaction of the dynamics of the dislocation movement and the dynamics of pulling is obvious. As has already been shown in fig. 23, the elastic residual stress at the onset of plastic

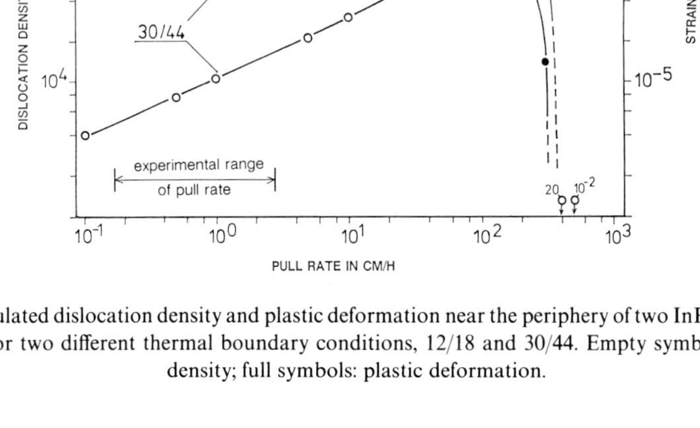

Fig. 24. Calculated dislocation density and plastic deformation near the periphery of two InP crystals versus pulling rate for two different thermal boundary conditions, 12/18 and 30/44. Empty symbols: dislocation density; full symbols: plastic deformation.

deformation increases and is shifted to lower temperatures with increasing pull rate. This results in enhanced dislocation multiplication with increasing pull rate. Microscopically, at low pull rates, the deformation ε_{pl} is achieved by a few dislocations moving large distances and having a very small tendency for multiplication. With increasing pull rate the faster variation of the stress at first induces a higher velocity of the few existing dislocations. This results in a higher internal friction (resistance). To reduce the dislocation velocity, a marked multiplication of the dislocations must occur. The same amount of plastic deformation, ε_{pl}, is achieved by many dislocations moving short distances due to their low velocity. Experimental evidence for an increasing dislocation density with increasing pull rate has been published by Elliot et al. [110] in the LEC growth of GaAs. They found a doubling of the dislocation density by increasing the pull rate from 7 to 17 mm/h. Although they suggested the increase to be a vacancy effect, above calculations also allow the conclusion that the dominating effect could be due to the increase in $d\tau/dt$ similar to the results in fig. 24.

At about 400 cm/h, the dynamics of pulling are faster than the dynamics of the dislocation processes resulting in a drastic decrease in dislocation density. A good comparative picture to imagine this process is to watch the thermal response of a solid body being pulled through a hot furnace. At low rates, the body takes on the temperature of the furnace resulting in a small temperature gradient in the body (near to equilibrium). With increasing rate the inner portion of the body remains more and more cool, resulting in an increasing gradient within the body. At very high rates the whole body remains cold because the process of pulling is much faster than the process of heating up the body. In this picture, an extremely large conductivity and very small heat capacity would lead to a complete heat-up of the body independent of the pull rate. With respect to the dislocation formation under conditions 12/18 this means a low resistance against plastic deformation at high temperatures and thus a fast response to load variations within a wide range of the pull rate.

The value of 400 cm/h for suppressing dislocation formation is about an order of magnitude larger than that for the growth of dislocation-free silicon. An explanation of this difference can be obtained from fig. 25, which shows the dependence of the strain rate on the shear stress measured by creep experiments at a temperature of 0.7 of the absolute melting temperature T_m for Si [111], Ge [112] and InP [113]. The strain rate of InP is about 1.5 orders of magnitude higher than that of Si and Ge at the same stress level. Hence, the period of loading of InP has to be shorter by a factor of 15 in order to get a deformation behaviour similar to that of Si. This correlation shows that the calculated value of the pull rate for the growth of InP is in good correlation with the growth rates used to obtain dislocation-free silicon.

Such a drastic increase in the pull rate to values of several hundred centimeters per hour to get dislocation-free III–V crystals is far away from experimental reality due to the thermal properties of these materials. The combination of a small thermal conductivity with a large amount of latent heat generated during solidification results in an upper limit for the growth rate of about 20 to 40 mm/h.

A reduction of the dislocation density is also observed with decreasing pull rate under conditions similar to 30/44. However, this method is limited by commercial considerations and technical problems (long-term stability, control problems, decom-

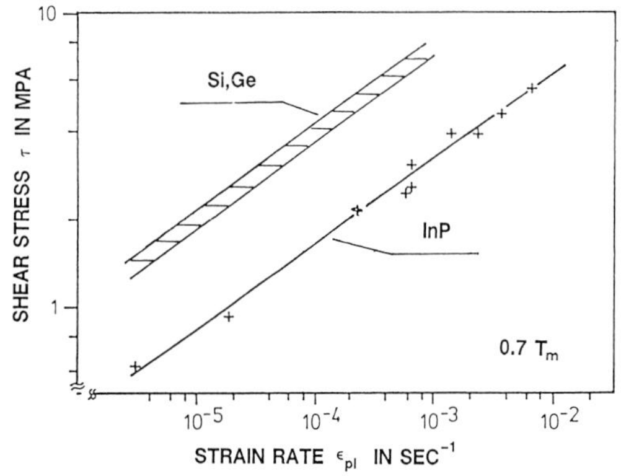

Fig. 25. Comparison of the maximum strain rates found during plastic deformation experiments (creep using constant shear stress τ) of Si [111], Ge [112] and InP [113]. The temperature of 0.7 T_m is related to the melting points of the different materials, respectively.

position of III–V compounds). A realistic lower limit to the pull rate is about 2 mm/h, which would allow the growth of crystals with a dislocation density of some $10^3\ cm^{-2}$ at the highest stressed peripheral areas.

A further reduction of dislocation density seems only to be possible by improving the growth conditions in such a way that the thermally induced stress is considerably reduced.

4.5.4. Modified stress field for the growth of crystals with low dislocation density

The model discussed above can be applied to construct thermal boundary conditions in order to grow dislocation-free III–V crystals. The dislocation density is calculated by assuming a hypothetical reduction of the stress field leaving all other parameters constant. Finally, the temperature conditions which are correlated to that stress field are discussed.

The influence of a reduction of the stress is shown in fig. 26. The calculations again are based on the highest stressed peripheral parts of the crystals 12/18 and 30/44. The stress field is reduced linearly by multiplying the stress values by a factor of between 0.01 and 1. The results in fig. 26 show clearly that the stress-induced dislocation density can only be reduced to values below $10^3\ cm^{-2}$ if the maximum stress values are less than 0.2 MPa. A dislocation density below $10^2\ cm^{-2}$ can only be achieved in InP crystals by a further reduction of the maximum stress below 0.15 MPa for both conditions 12/18 and 30/44.

The temperature conditions for the growth of such low dislocation density crystals are deduced from the following calculations. A linear temperature gradient is assumed in the axial direction allowing one to reduce the discussion to be radial temperature conditions. The radial profile is assumed to be parabolic according to eq. (4.19):

$$T_0 - T(r) = (r/r_0)^2 \Delta T, \tag{4.19}$$

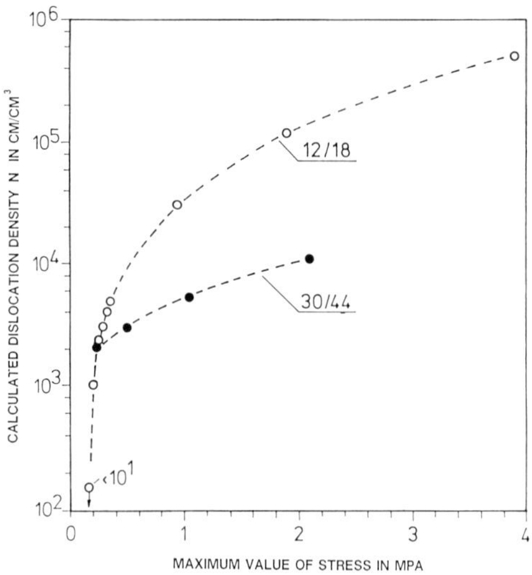

Fig. 26. Calculated values of the dislocation density versus maximum stress. The axial stress distributions, which were obtained from thermo-elastic modeling, were multiplied for these calculations by a factor between 0.01 and 1.

where ΔT is the temperature difference between center and periphery and r the radial position related to r_0, the total radius.

The highest stress then appears at the periphery of the crystal, given by the treatments of Carslaw and Jaeger [83], Schmidt and Boas [101] and Wolf [114]:

$$\tau_g^{max} = (\alpha E)/[4(1-\nu)]\,\Delta T F, \tag{4.20}$$

where τ_g^{max} is the shear stress in the relevant slip system, α the thermal expansion coefficient $= 7.4 \times 10^{-6}\,K^{-1}$ (1200 K) [86], E the shear modulus $(111) = 78.6$ GPa (1200 K), based on [86, 115–117], ν Poisson's ratio $(111) = 0.365$ (1200 K) and F the maximum efficiency factor, depending on the type of the slip system and the growth direction: for growth in $\langle 111 \rangle$ $F_{max} = 0.9428$.

According to eq. (4.20), the growth of crystals with dislocation densities below $100\,cm^{-2}$ requires temperature differences between the center and periphery of not more than 1 K, if the stress τ_g^{max} is to be kept below 0.15 MPa. Due to the uncertainty of some of the data these values can vary at most by a factor of two to three. One can hardly imagine that such a small value can be maintained reproducibly during the process of LEC growth. This value may even cause problems during the growth of crystals by the Bridgman technique, because reasonable solidification and cooling rates require higher radial temperature differences than 1 to 3 K. According to this result it is not surprising that dislocation-free undoped III–V material is not supplied on the commercial market and has even not been grown in research laboratories except for small diameters, where the stress can be kept small.

4.5.5. Formation of dislocation patterns

As has been discussed in the very beginning of this chapter, the correlation of dislocation patterns in the grown crystals to stress fields during growth has found great interest amongst many crystal growers. In this context Jordan's work [6, 46, 47, 107, 118] represents the most comprehensive investigation of the generation of stress patterns in III–V crystals, which also may be applied to all other diamond-like semiconductor lattices. The patterns exhibit a fourfold symmetry for ⟨100⟩ and a sixfold symmetry for ⟨111⟩ grown crystals. However, sometimes also a preferentially threefold symmetry may be observed in ⟨111⟩ crystals after growth, which cannot be explained by the model of Jordan due to some simplifications made during the analytical stress calculations concerning the axial temperature gradient.

Typical three- and sixfold dislocation patterns for ⟨111⟩ grown crystals (taken from [80]) are shown in fig. 27. The patterns are a result of the different growth conditions 12/18 and 30/44. In the following, the model calculations will be used to trace the history of the generation of both patterns during growth. The slice or crystal plane to be traced is divided into a grid as shown in fig. 28. Ten equidistant nodes on the radius are rotated by steps of ten degrees around the center axis of the crystal resulting in 360

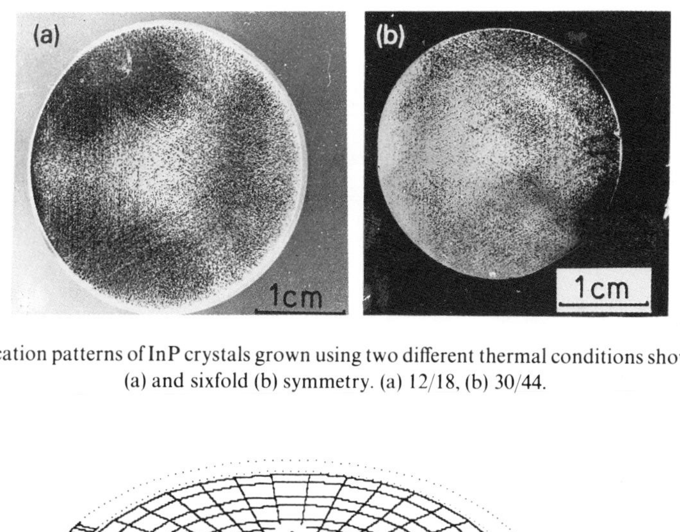

Fig. 27. Dislocation patterns of InP crystals grown using two different thermal conditions showing threefold (a) and sixfold (b) symmetry. (a) 12/18, (b) 30/44.

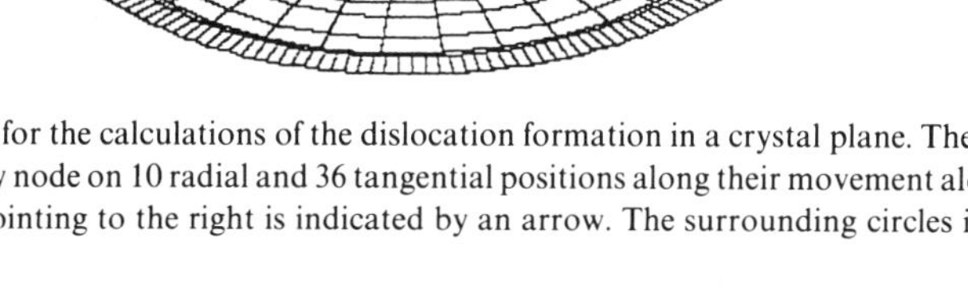

Fig. 28. Basic grid for the calculations of the dislocation formation in a crystal plane. The calculations were performed for every node on 10 radial and 36 tangential positions along their movement along the z-axis. The $\langle\bar{1}\bar{1}2\rangle$ direction pointing to the right is indicated by an arrow. The surrounding circles indicate EPD = 0.

nodes per plane. The dislocation formation and the active slip system assuming single slip is calculated during the movement of the plane from the interface along the growth direction for these nodes. For reasons of symmetry of the sphalerite lattice in the (111) plane, the calculations can be reduced to the nodes within the first 120 degrees followed by a 120 and 240 degree rotation. The $\langle\bar{1}\bar{1}2\rangle$ direction is chosen as the radial direction of reference.

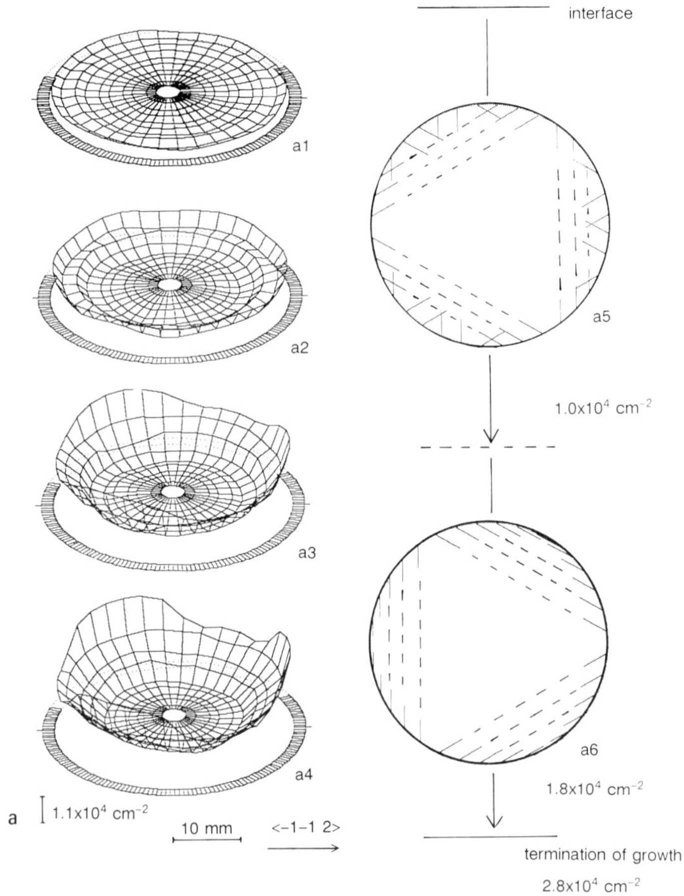

Fig. 29. History of dislocation generation in a plane of a growing InP crystal (left side of figure) and calculated projection of expected slip lines (right side of figure). Growth proceeds from the top to the bottom of the figure. (a) InP crystal grown under conditions 30/44. (a1) Distance to interface 3 mm; only a few dislocations generated; (a2) Distance to interface 39 mm; dislocation density in the peripheral region approaches values of about $10^4\,cm^{-2}$; (a3) Distance to interface 43 mm; plane is beginning to leave B_2O_3 and dislocation density increases; (a4) Distance to interface 49 mm; a dislocation density of about 2 to $3\times10^4\,cm^{-2}$ is achieved representing the final dislocation distribution; (a5) Calculated projection of slip lines at the peripheral regions. The dashed lines indicate the slip lines the most active representing a triangle pointing to the left. A dislocation density of about $1.2\times10^4\,cm^{-2}$ is generated dominated by this pattern. (a6) Calculated projection of slip lines at the peripheral regions. The dashed lines indicate the slip lines the most active representing a triangle pointing to the right. The transition between the slip line pattern of (a5) and (a6) is calculated to occur at a distance of about 40 mm to the interface, i.e. in between figs. (a2) and (a3).

A sequence of the dislocation formation is shown in fig. 29a for conditions 30/44. The left-hand side of the figure represents three-dimensional plots of the dislocation density of the slice, the right-hand side indicates the traces of the highest loaded slip systems at the periphery which are expected to be activated. Starting from the interface to a distance of about 37 mm from the interface, a triangle pointing to the left (into the

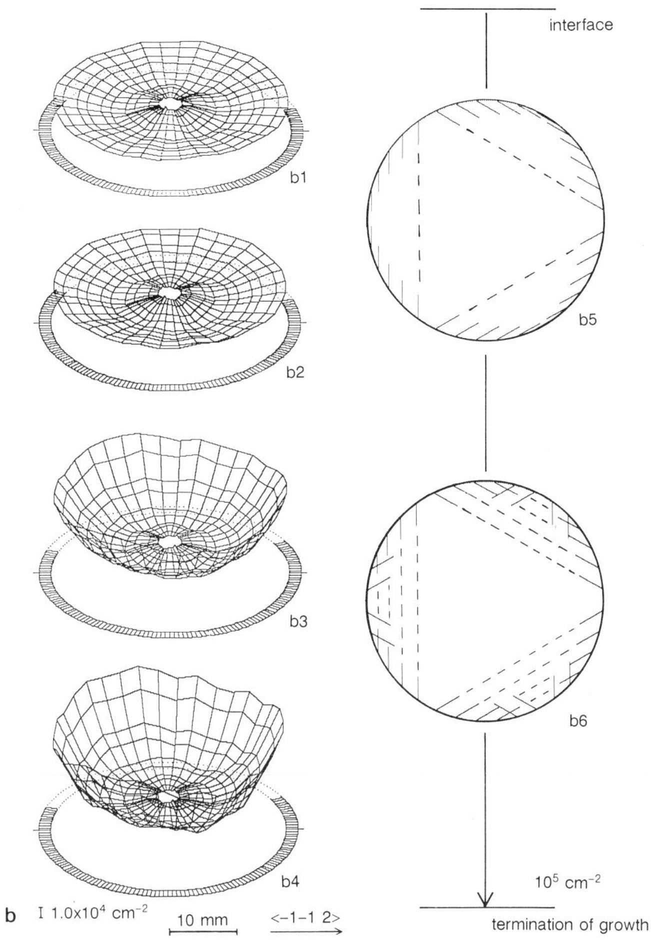

Fig. 29. (b) InP crystal grown under conditions 12/18. Please note changed scaling of dislocation density compared to fig. 29a. (b1) Distance to interface 3 mm; a dislocation density of about $3 \times 10^4\,cm^{-2}$ has been generated at the peripheral regions. (b2) Distance to the interface 15 mm; the dislocation density shows a slight increase while the plane is approaching the B_2O_3 surface being located about 18 mm from the interface. (b3) Distance to the interface 22 mm; the plane just has passed the B_2O_3 surface and the dislocation density in the peripheral regions has increased drastically to values of about $10^5\,cm^{-2}$. (b4) Distance to the interface 25 mm; the dislocation density reaches its final value slightly above $10^5\,cm^{-2}$. (b5, b6) During the whole run, the calculated projection of the slip lines exhibits a triangle pointing to the right similar to the pattern observed during the real growth run in fig. 27a. Only a few slip lines indicating other directions have been calculated at the final stage of growth, but they are not becoming dominating.

$\langle 11\bar{2} \rangle$ direction) should predominantly be operating generating a dislocation density of about $10^4\,cm^{-2}$. A change of the active slip systems occurs beginning at 38 mm from the interface until at 40 mm from the interface the slip lines form a triangle pointing to the right. This arrangement remains constant until the termination of growth, another 1.5 to $2 \times 10^4\,cm^{-2}$ of dislocations are generated in this configuration. This indicates that the sixfold symmetry of the dislocation pattern is obviously generated by adding two triangular configurations which are formed at different stages of growth.

During the growth of crystal 12/18 the traces of the slip systems result in a triangular configuration pointing to the right, which is dominating from the very beginning until the termination of growth. The calculated slip line pattern can also be observed in the etched wafer in fig. 27 as well as the remarkable correspondence of the inhomogeneity of dislocation formation. Only a few slip lines in fig. 29b are found to have other directions. They are probably not detected in the experimental picture because they were not activated for the reason that they would have had to cross already heavily dislocated slip systems in other directions. This is known as a "multiple slip condition" and would need a locally increased stress for activation which is not expected to be present during this experiment.

4.6. *Stress and dislocations in the literature*

The interaction of elasticity and plasticity in the real crystal during cooling within a wide range of temperature and its dependence on dynamic influences, e.g. growth rate, implies that the expression "stress" has to be considered as an imprecisely defined state during the cooling of a crystal. In most cases, stress is understood to be the equivalent of thermal strain in a completely elastic body. As can clearly be deduced from fig. 23, the real elastic stress in the elastic-plastic crystal is much lower than the total elastic stress found under fully elastic assumptions. The difference between the two is transformed into plastic strain, the calculation of which is especially difficult due to its dependence on dynamically varying parameters and the elastic–plastic interaction, e.g. the rate of local changes of the total thermal strain and their separation into reversible elastic and irreversible plastic parts, the volume effect given by the fact that short-range plastic deformation may release long-range elastic stresses, the inhomogeneity of plastic deformation due to the existence of slip systems and the unknown interaction parameters between different slip systems (multiple glide) or the complication that the dynamics of plastic deformation are related to the actually existing residual elastic stress and not to the total thermal stress calculated using elastic assumptions. The prediction of the dislocation density in the growing crystal, representing the key interest of the crystal grower, is only one, but essential, part of the calculation of the elastic–plastic interaction. The enormous difficulties have prevented the generation of a comprehensive model covering all of those aspects in a fully dynamic and three-dimensional way. Currently, several models are under development, i.e. those of Tsai et al. [7, 9, 119], Maroudas and Brown [106], Motakef [120, 121], Lambropoulos [122] and Nicodeme et al. [123]. The analysis of these models shows that the model of Alexander and Haasen [12] for plastic deformation in diamond-like semiconductors, sometimes also called the Haasen–Sumino model recognizing the extensive work of Sumino et al. on deformation of semiconductors, has gained world-wide acceptance. However, the aims of application and the implementation into computer programs are different.

Tsai et al. were the first to apply the Alexander and Haasen model of plastic deformation to calculate the dislocation dynamics during web growth of silicon [7–9]. Tsai [119] has extended this model to the growth of silicon bulk crystals. One of the essential differences between that model and others is the application of a three-dimensional response function generalizing the Alexander and Haasen model assuming isotropic elastic and plastic response of the material and thus neglecting the existence of slip planes. The rate of dislocation generation and deformation is linearly related to the growth rate v of the crystal. This correlation seems somewhat surprising because, even if growth is stopped, relaxation of residual elastic stresses may occur generating an increase in dislocation density and deformation. Furthermore, the model needs an additional "back-stress" of 3.6 MPa (comparable to a critical shear stress) to get reasonable results for the dislocation density. Especially in silicon web growth, where dislocation dynamics and growth dynamics (high growth rate) may compete in a way similar to the calculated suppression of dislocation generation in fig. 24, such a critical value is expected to be a dynamic parameter varying with growth rate, temperature and period of loading according to fig. 23 rather than to be a constant.

The way of calculating the plastic deformation and dislocation generation by Maroudas and Brown [106] is dominated by the need of integrating a suitable method for the evaluation of dislocation formation into an "integrated analysis of crystal growth". This integrated analysis intends to provide a fast computer modeling of the whole growth process including, e.g. melt convection, heat transfer due to thermal conduction and radiation, thermoelastic stress evaluation and estimation of the corresponding dislocation density [124]. Such an extensive analysis cannot be oriented to the exact needs of the perfect evaluation of plastic deformation and dislocation dynamics but has to be optimized for the performance of the whole model. Consequently, Maroudas and Brown handle the dynamics of dislocation formation similar to the modeling of the dynamics of an exothermal chemical reaction. A limited stock of initial materials providing the driving force (thermo-elastic stress) experiences a thermodynamically controlled reaction (multiplication of dislocations) tending to an asymptotic approach towards the final state of equilibrium of the products of the reaction (minimum of thermo-elastic stress, maximum deformation or dislocation density). The reaction is controlled by the amount of available material (maximum thermal stress) and the dynamics of consumption due to the characteristics of the reaction (plastic deformation process and related parameters). Thus the characteristic parameters of the reaction can only be evaluated and only are valid for a certain temperature–stress–time field. Furthermore, the results represent an upper bound for the number of dislocations generated under certain conditions, because the dynamics of relaxation as well as the exact time-dependent calculations of the back-stress due to generated dislocations are treated in a more general way. The effect of different stress profiles on dislocation density can be taken into consideration by a scaling of the resulting dislocation density according to the specifics of the stress profile considered.

The authors claim this model to be an "order of magnitude analysis". Indeed, the model seems to provide reasonable results [106] if carefully measured experimental data are used to set up the characteristics of the "reaction".

The model of Motakef [120, 121] tries to combine the thermo-elastic calculation and the elasto-plastic calculation by the introduction of a division of the growth process into two stages. Motakef considers this division as a simplified model for plastic deformation being more realistic than pure thermo-elastic calculations. According to an arbitrary definition, stage I represents a strain-free isothermal layer immediately upon solidification. During stage II, the evolution of that layer in the time-dependent stress–temperature field is calculated. The focus for criticism of that model is given by the arbitrary definition of two deformation stages without equivalent physical background. Furthermore, the deformation behaviour of stage I is evaluated assuming no plastic relaxation in the bulk, i.e. the total thermo-elastic stress is used for the calculation of deformation and dislocation generation at the interface. According to the preceding discussion in this chapter this leads to an overestimation of the dislocation density due to the overestimation of the multiplication parameter and the power law of creep. The overestimation of the dislocation density is enforced by neglecting the back-stress induced by the generated dislocations according to fig. 19 of this chapter and related discussions. Thus, the results of the model of Motakef are comparable to that of an asymptotic analysis with decreasing stress as Maroudas and Brown [106] have demonstrated.

Lambropoulos [122] evaluates three different models representing elastic–perfectly plastic behaviour, steady-state power law creep and the Alexander and Haasen model allowing for the multiplication of dislocations. Results of the modeling of sheet growth is compared to the growth of tubes. Again, the handling of back-stresses due to dislocations and critical thermo-elastic stress levels may lead to the conclusion that the calculation of relaxation in the bulk and especially the elastic–plastic behaviour at the interface have to be improved. Then the proposed derivation of non-dimensional groups may be a valuable contribution to a faster and integrated computation of deformation during crystal growth.

Nicodeme et al. [123] have developed a time-dependent and global model for an LEC furnace. This model so far seems to be the most comprehensive treatment of the dynamics of growth including the pulling of a crystal with respect to temperature and stress distributions. The calculation of the dislocation density is adjusted to the thermo-elastic calculations in a similar manner as in the model of Dillon et al. [7, 9]. The crystal is assumed to be isotropic in both its elastic and plastic responses, which means that the concept of a resolved shear stress and the existence of slip systems are not taken into consideration. Nevertheless, the calculations impressively show the development of an inhomogeneous dislocation distribution in the growing GaAs crystal similar to the observations in real crystals.

The advances in computing facilities and pioneering work of different groups have created enormous advances in modeling of the state of stress in the cooling crystal during the past few years. Models implementing the dynamics of growth are appearing allowing a more realistic modeling of the "never-static" crystal growth process. However, the quality of the results concerning the evaluation of the dislocation density still has to be improved in the different models, especially at the point were plasticity is introduced into initially thermo-elastic calculations. And, last but not least, the results and conclusions from the calculations have to be transformed into the real world of crystal growth. Perhaps, this transfer is the most difficult process to do which would explain the existence of numerous of calculations without the implementation of experimental proof.

5. *Residual stress in the crystal after cooling*

The meaning of "residual stress" in this section is completely different from the elastic residual stresses discussed in the preceding sections 3 and 4. Here residual stress describes the stress observed in a crystal which has been cooled down to room temperature. Then the crystal can be considered as a totally elastic body of homogeneous temperature.

5.1. Origin of residual stress

According to thermo-elastic theory, the perfect crystal at a homogeneous temperature should be free of stresses. Residual stress may only occur if some kind of deviation from the ideal crystal structure, i.e. lattice defects, have been introduced into the crystal. Such defects causing stresses are inhomogeneous distributions of point defects or dopants, dislocations or precipitates. In the following, the consequences of plastic deformation at high temperatures on the residual stress at room temperature are discussed following the arguments of Noyan and Cohen [125].

A cylindrical crystal is cooled down from a temperature T_0 to room temperature with a different cooling rate for the core and the peripheral region according to radially outward heat flux. Consider two elements of equal length l_0 at T_0 in the core and near the surface; they will experience different thermal expansion during the cooling procedure. If the crystal is purely elastic, the two pieces will have identical length when homogeneous temperature conditions are finally achieved at room temperature as may be taken from fig. 30a. No stresses will then be present in the crystal.

Now, for the purpose of better demonstration let the crystal become plastic at a certain intermediate temperature. Elastic behaviour above and below that temperature is assumed. The response of such a crystal is graphically shown in fig. 30b. Generally the surface is contracting faster than the core thus generating a shrinking of the periphery onto the core and inducing compressive stress in the core and tensile stress at the periphery. At t_x the stress is relaxed instantaneously by expanding the surface regions and/or contracting the core regions, the latter being less likely due to the considerations of volume effects on plastic deformation. The crystal is going to become elastic again by further lowering the temperature. Now the plastic changes in length at t_x are irreversible, i.e. "frozen in", thus leading to a difference in length of inner and outer element as both approach room temperature. However, the formerly expanded surface region is now going to be compressed and the core experiences tensile strain, because

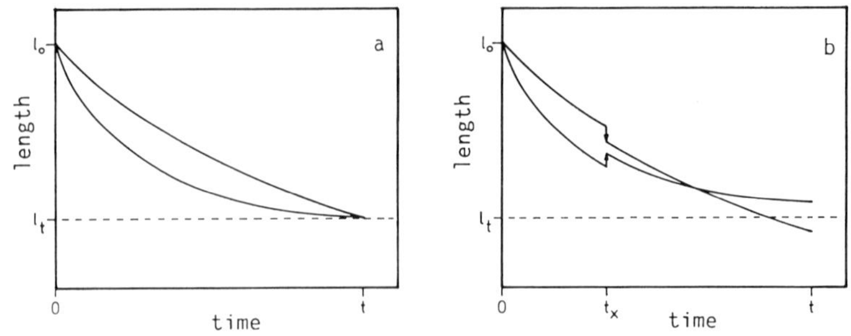

Fig. 30. Behaviour of two elements of equal finite length l_0 at the core and the surface of a cylinder, which are cooled down to a lower temperature with different cooling rates at the center and at the surface. (a) Elastic body; (b) Elastic body with plastic behaviour at time t_x.

the remaining temperature difference and hence the amount of contraction between the temperatures at t_x and room temperature is larger for the core. The remaining difference in length must be equal to the length changes at t_x. Thus the crystal now is elastically strained which is equivalent to the existence of a residual stress.

The same situation is valid during the cooling of a real crystal except that the plastic relaxation does not occur instantaneously but over a wider range of temperature and time. Thus the amount of the residual stress or strain measured in as-grown crystals should be comparable to the amount of plastic strain ε_{pl} relaxed at elevated temperatures during growth.

5.2. *Comparison of residual stress and calculated plastic strain*

The residual stress in wafers grown using conditions 12/18 and 30/44 has been investigated using a linear and circular polariscope as described in [58]. At the periphery of the wafers stresses of about 5 MPa have been measured for conditions 12/18 and of about 2 MPa for conditions 30/44. The measurements along the circumference are shown in fig. 31. These stresses correspond to an elastic deformation of 1.8 to 2.2×10^{-4} for 12/18 and 6 to 9×10^{-5} for 30/44. The values are compared to the calculated total plastic deformation in table 4. The values correspond surprisingly well, thus indicating the direct coupling of high-temperature plastic deformation and residual stresses at room temperature. The correspondence also may point to a reasonable evaluation of the elastic residual stresses in the elasto-plastic modeling of deformation at high temperatures resulting in low values of less than 0.2 MPa, the proof of which may be impossible due to the accumulated errors of the whole procedure. In addition, it should be noted that dislocation-free crystals (low amount of plastic deformation) grown by the addition of high concentrations of dopants have been found to be free of long-range residual stress, which again supports the correlation of plastic deformation and residual stress in crystals.

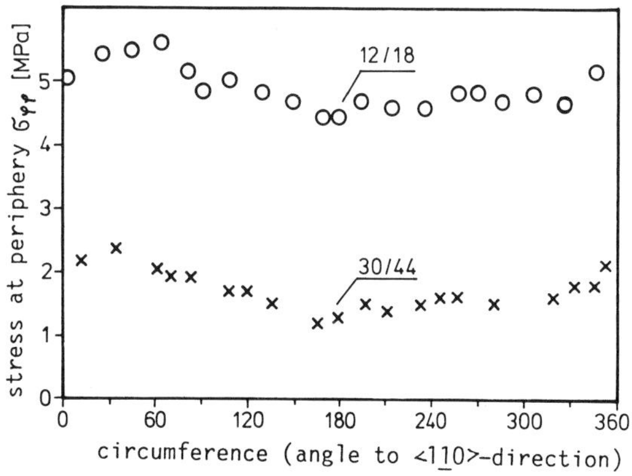

Fig. 31. Stress distribution around the circumference of ⟨111⟩ oriented crystal slices cut out of the crystals 12/18 and 30/44.

Table 4

Comparison of the strain in the peripheral region of InP crystals grown under two different thermal boundary conditions. The calculated values are results of the modeling of plastic deformation during growth, the elastic strain has been evaluated by optical birefringence measurements at room temperature using 1.05 μm wavelength. The range given represents the minimum and maximum values found at the circumference of the crystals.

Growth conditions	Residual elastic strain at room temperature	Calculated plastic deformation during cooling
12/18	1.8 to 2.2×10^{-4}	1.45×10^{-4}
30/44	6 to 9×10^{-5}	8.4×10^{-5}

6. *Concluding remarks*

The control of stress in the cooling crystal is one of most important factors with respect to the final crystal quality. The difficulty of evaluating and engineering these stresses is given by the missing comprehensive experimental access to all of the dynamically varying processes. Sufficient understanding can only be gained by a complete modeling of the growth experiments with sufficient accuracy. This however, would require incredible amounts of manpower and computation time if all possible three-dimensional, time- and temperature-dependent parameters as well as all dynamical interactions were to be considered. Therefore, the existing models up to now use more or less rigorous simplifications and extrapolations to be able to make some generally valid predictions for further improvements of the growth conditions.

From this point of view, there are many questions left to be answered for improving the situation. One of the most important is correlated to the dynamics of plasticity and dislocation generation over the whole temperature range from the melting point to room temperature at low strain rates comparable to those calculated in section 4. Improved knowledge of the process dynamics responsible for plastic deformation within a wide range of parameters, especially near the melting point, is required for the necessary increase in precision of the calculations.

Furthermore, the process of generation of the first dislocations and the process of multiplication is far from being understood or even covered by a closed theory. It is necessary to use semi-empirical models to address these problems resulting in an unknown precision. At least, an experimental proof of the existence of multiplication mechanisms in the inside of an InP crystal has been detected to be the Frank–Read mechanism as shown in fig. 32. This picture demonstrates that several dislocation loops are generated by one single source. The dislocation lines have been made visible by a copper decoration technique in InP. However, the initial source for the first dislocation is unknown, although it probably will be correlated with the imperfectness of the real crystal lattice.

Another task is to improve all data on material properties needed for the thermo-elastic calculations, especially their variation with temperature. Very little is known about the elastic behaviour of single crystals at elevated temperatures or the material properties when approaching the melting point.

At present, there is also no satisfactory model for replacing the thermo-elastic calculations by thermo-plastic calculations on a three-dimensional basis including the dynamics of growth and deformation. This will become one of the major factors in progress toward a correct evaluation of stress and strain in the cooling crystal.

The model of Alexander and Haasen [12] for dislocation formation has been shown to be able to describe the dynamics of deformation very well. At present this semi-empirical model represents the ultimate method for modeling dislocation formation, although the missing material related, theoretical background of parameters like the multiplication constant K or the stress exponent m is rather unsatisfying.

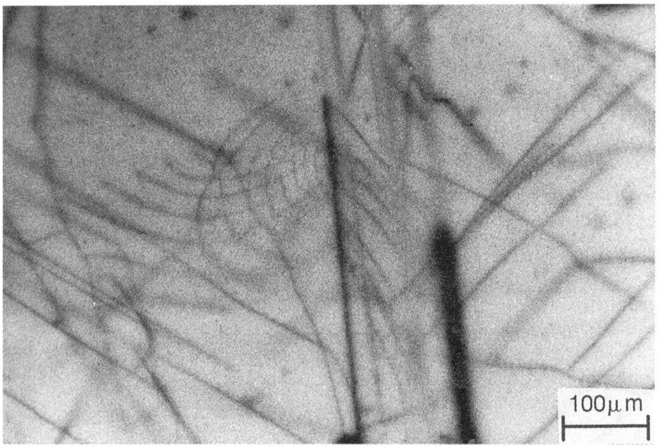

Fig. 32. Infrared transmission microphotograph of an as-grown crystal showing dislocations created by the Frank–Read mechanism. The dislocation lines have been made visible by a copper decoration technique.

Nevertheless, crystal growers nowadays have available a procedure to understand what is likely to happen in the crystal during cooling and what conclusions should be drawn from the results of their task. It will hopefully lead to a further improvement of the quality of single crystals resulting in a well-defined performance of substrates during device processing.

Acknowledgements

The author wishes to thank the University of Erlangen-Nuremberg for the unique support of experimental and computational facilities during his Ph.D. thesis. Special thanks are to the Institut für Werkstoffwissenschaften VI (Crystal growth labs, Prof. Dr. Müller), Institut für Werkstoffwissenschaften I (plastic deformation, Prof. Dr. Blum) and the Lehrstuhl für Technische Mechanik (computation of thermo-elastic stresses, Prof. Dr. Kuhn) as well as to many former colleagues for their valuable support. The investigations have been continued at the SIEMENS research laboratories in Erlangen. The excellent support of all coworkers of Dr. Raab's group is gratefully acknowledged.

References

[1] E. Billig, Proc. Roy. Soc. A 235 (1956) 37.
[2] D.C. Bennett and B. Sawyer, Bell Syst. Techn. J. 35 (1956) 637.
[3] P. Penning, Philips Res. Rept. 13 (1958) 79.
[4] S.F. Nygren, C.M. Ringel and H.W. Verleur, J. Electrochem. Soc. 118 (1971) 306.
[5] T.S. Plaskett, J.M. Woodall and A. Segmüller, J. Electrochem. Soc. 118 (1971) 115.
[6] A.S. Jordan, R. Caruso and A.R. Von Neida, Bell Syst. Techn. J. 59 (1980) 593.
[7] O.W. Dillon, C.T. Tsai and R.J. De Angelis, in: Proc. 1st Int. Symp. on Shaped Crystal Growth, Budapest, Hungary, 1986, J. Cryst. Growth 82 (1987) 50.
[8] Y.K. Kim, R.J. De Angelis, C.T. Tsai and O.W. Dillon, Acta Met. 35 (1987) 2091.
[9] C.T. Tsai, O.W. Dillon and R.J. De Angelis, J. Cryst. Growth 82 (1987) 695.
[10] J. Völkl and G. Müller, in: Semi-Insulating III–V Materials, Malmö, 1988, eds. G. Grossmann and L. Ledebo (Hilger, Bristol, 1988) p. 489.
[11] J. Völkl and G. Müller, in: Proc. 1st NATO Worksh. on Computer Modelling in Crystal Growth from the Melt, Parma, Italy, April 1989, J. Cryst. Growth 97 (1989) 136.
[12] H. Alexander, and P. Haasen, in: Solid State Physics, eds. F. Seitz, D. Turnbull and H. Ehrenreich (Academic Press, New York, 1968) 27.
[13] K.M. Kim and P. Smetana, J. Cryst. Growth 100 (1990) 527.
[14] B.L. Timan and V.M. Fesenko, Fiz. Khim. Obrab. Mater. 6 (1983) 139.
[15] W. Rosch and F. Carlson, in: Proc. 8th Amer. Conf. on Crystal Growth, Vail, CO, July 1990, J. Cryst. Growth 109 (1991) 75.
[16] T. Vetter, R. Treichler and A. Winnacker, Semicond. Sci. Technol. 7 (1992) 150.
[17] J.C. Mikkelson, J. Met. 37 (1985) 51.
[18] H. Shimizu, T. Watanabe and Y. Kakui, Jpn. J. Appl. Phys. 24 (1985) 815.
[19] N.A. Avdonin, S.S. Vakhrameev, M.G. Mil' vidskii, V.B. Osvenskii, B.A. Sakharov, V.A. Smirnov and Yu.F. Shchelkin, Dokl. Akad. Nauk SSSR 200 (1971) 316 [Tech. Phys.].
[20] S.S. Vakhrameev, V.B. Osvenskii and V.A. Smirnov, Vses. Soveshch. Rostu Krist., 4th, 2, 81-4, From: Ref. Zh., Fiz., E. (1973) Abstr. No. 1E533.
[21] S.V. Tsivinskii, Izv. Akad. Nauk SSSR, Ser. Fiz. 37 (1973) 2451.

[22] E.M. Andreev, P.I. Antonov, S.I. Bakholdin, E.V. Galaktionov and V.S. Yuferev, Izv. Akad. Nauk SSSR, Ser. Fiz. 40 (1976) 1426.
[23] P.I. Antonov, E.V. Galaktionov, V.M. Krymov and V.S. Yuferev, Izv. Akad. Nauk SSSR, Ser. Fiz. 40 (1976) 1419.
[24] E.V. Galaktionov, V.M. Krymov, E.N. Kolesnikova, S.I. Bakholdin, N.B. Guseva, I.L. Shul'pina and P.I. Antonov, Tezisy Dokl. Vses. Soveshch. Rostu Krist., 5th Vol. 2, Akad. Nauk. Gruzinskoi SSR, Tiflis, USSR (1977) p. 240.
[25] S.S. Vakhrameev, M.G. Mil'vidskii, V.B. Osvenskii, V.A. Smirnov and Yu.F. Shchelkin, Rost Krist. 12 (1977) 287.
[26] W. Geil and K. Schmugge, Krist. Tech. 13 (1978) 195.
[27] E. Zalewski and J. Zmija, Acta Phys. Pol. A 53 (1978) 503.
[28] P.I. Antonov and V.S. Yuferev, Probl. Prochn. Plast. Tverdykh Tel, ed. S.N. Zhurkov (Izd. Nauka, Leningrad, USSR, 1979) p. 236.
[29] L.A. Goncharov, V.A. Smirnov and L.N. Tityunik, in: Proc. 8th All-Union Conf. on the Production of Shaped Crystals and Components by Stepanov's Method and their Applications in the National Economy, Leningrad, Izv. Akad. Nauk SSSR Ser. Fiz. 44 (1980) 286.
[30] S.S. Vakhrameev, V.B. Osvenskii and S.S. Shifrin, Izv. Akad. Nauk SSSR, Ser. Fiz. 44 (1980) 289.
[31] I.I. Ignatov, Yu.K. Lingart, N.V. Marchenko, N.A. Tikhonova and Ya.I. Shtipel'man, J. Cryst. Growth 52 (1981) 411.
[32] T. Iwaki and N. Kobayashi, Trans. ASME J. Appl. Mech. 48 (1981) 866.
[33] M. Duseaux and G. Jacob, Appl. Phys. Lett. 40 (1982) 790.
[34] P.D. Kervalishvili and Yu.F. Shchelkin, Fiz. Khim. Obrab. Mater. 5 (1982) 70.
[35] M. Duseaux, J. Cryst. Growth 61 (1983) 576.
[36] R.H. Hopkins, R.G. Seidensticker and J. Schruben, in: Proc. 7th Int. Conf. on Crystal Growth (ICCG-7) Stuttgart, Germany, September, 1983, J. Cryst. Growth 65 (1983) 307.
[37] V.L. Indenbom, V.M. Kaganer and A.G. Frolov, in: Proc. 9th Conf. on the Preparation of Shaped Crystals and Parts by the Stepanov Method and their Application in the Economy, Leningrad, USSR, Izv. Akad. Nauk SSSR, Ser. Fiz. 47 (1983) 254.
[38] B.L. Timan and V.M. Fesenko, Fiz. Khim. Obrab. Mater. 2 (1983) 139.
[39] I.Yu. Vandakurov and E.V. Galaktionov, in: Proc. 9th Conf. on the Preparation of Shaped Crystals and Parts by the Stepanov Method and their Application in the Economy, Leningrad, USSR, Izv. Akad. Nauk SSSR, Ser. Fiz.47 (1983) 279.
[40] C.E. Huang, D. Elwell and R.S. Feigelson, J. Cryst. Growth 69 (1984) 275.
[41] A.S. Jordan, G.T. Brown, B. Cockayne, D. Brasen and W.A. Bonner, J. Appl. Phys. 58 (1985) 4383.
[42] N. Kobayashi and T. Iwaki, J. Cryst. Growth 73 (1985) 96.
[43] G. Szabo, J. Cryst. Growth 73 (1985) 131.
[44] P.I. Antonov and S.I. Bakholdin, Mater. Vses. Soveshch. Poluch. Profilirovannykh Krist. Izdelii Sposobom Stepanova Ikh Primen. Nar. Khoz., Meeting Date 1985, Akad. Nauk SSSR, Fiz.-Tekh. Inst., Leningrad, USSR (1986) p. 202.
[45] T. Iwaki and N. Kobayashi, Int. J. Solids Struct. 22 (1986) 307.
[46] A.S. Jordan and J.M. Parsey, in: Proc. 8th Int. Conf. on Crystal Growth, York, England, July 1986, J. Cryst. Growth 79 (1986) 280.
[47] A.S. Jordan, A.R. Von Neida and R. Caruso, in: Proc. 8th Int. Conf. on Crystal Growth, York, England, July 1986, J. Cryst. Growth 79 (1986) 243.
[48] V.M. Krymov, E.V. Galaktionov and S.P. Kuznetsov, Mater. Vses. Soveshch. Poluch. Profilirovannykh Krist. Izdelii Sposobom Stepanova Ikh Primen. Nar. Khoz., Meeting Date 1985, Akad. Nauk SSSR, Fiz.-Tekh. Inst., Leningrad, USSR (1986) p. 188.
[49] T. Kemin and X. Jiachi, Res. Prog. SSE 6 (1986) 347.
[50] J.C. Lambropoulos, J. Cryst. Growth 80 (1987) 245.
[51] S. Motakef, Int. J. Heat Mass Transfer 30 (1987) 1487.
[52] S. Motakef, Trans. ASME, J. Appl. Mech. 54 (1987) 813.
[53] S. Motakef and A.F. Witt, J. Cryst. Growth 80 (1987) 37.
[54] S.K. Ray, S. Utku and B.K. Wada, Comput. Struct. 25 (1987) 725.
[55] C. Schvezov, I.V. Samarasekera and F. Weinberg, J. Cryst. Growth 84 (1987) 219.

[56] C.E. Schvezov, I.V. Samarasekera and F. Weinberg, in: Proc. 7th Amer. Conf. on Crystal Growth, Monterey, CA, July 1987, J. Cryst. Growth 85 (1987) 142.
[57] E.A. Tropp, I.Yu. Vandakurov and E.V. Galaktionov, 1st Int. Symp. on Shaped Crystal Growth, Budapest, Hungary, July 1986, J. Cryst Growth 82 (1987) 65.
[58] J. Völkl and G. Müller, in: Proc. E-MRS Meeting, Strasbourg, 1987, eds. Y.I. Nissim and P.A. Glasow (Editions de Physique, Les Ulis, 1987) p. 141.
[59] G.O. Meduoye, D.J. Bacon and K.E. Evans, J. Cryst. Growth 88 (1988) 397.
[60] P. Nicodeme, F. Dupret and M.J. Crochet, in: Semi-Insul. III–V Mater., Proc. Conf., 5th, eds. G. Grossmann and L. Ledebo (Hilger, Bristol, 1988) p. 465.
[61] P. Nicodeme, F. Dupret, M.J. Crochet, J.P. Farges and G. Nagel, in: ESPRIT '88; Putting the Technology to Use; Proc. 5th Annual ESPRIT Conf. Brussels, Belgium, 1988 (North-Holland, Amsterdam, 1988) 314.
[62] P. Nicodeme, F. Dupret, M.J. Crochet, G. Grossman and L. Ledebo, J. Cryst. Growth 94 (1988) 75.
[63] C.E. Schvezov, I.V. Samarasekera and F. Weinberg, J. Cryst. Growth 92 (1988) 479.
[64] C. E. Schvezov, I.V. Samarasekera and F. Weinberg, J. Cryst. Growth 92 (1988) 489.
[65] I.Yu. Vandakurov, E.V. Galaktionov, V.S. Yuferev, V.M. Krymov and C. Barta, in: Proc. All-Union Conf. on Making Profiled Crystals by Stepanov's Method and Applications, Leningrad, USSR March 1988, Bull. Acad. Sci., USSR, Phys. Ser. 52 (1988) 6.
[66] A. Cazin-Bourguignon and S. Motakef, J. Cryst. Growth 96 (1989) 390.
[67] A.B. Crowley, E.J. Stern and D.T.J. Hurle, J. Cryst. Growth 97 (1989) 697.
[68] F. Dupret, P. Nicodeme and Y. Ryckmans, in: Proc. 1st NATO Worksh. On Computer Modelling in Crystal Growth from the Melt, Parma, Italy, April 1989, J. Cryst. Growth 97 (1989) 162.
[69] J.P. Garandet, J. Cryst. Growth 96 (1989) 680.
[70] S. Motakef, J. Cryst. Growth 96 (1989) 201.
[71] S. Motakef, J. Cryst. Growth 98 (1989) 711.
[72] C. Schvezov, I.V. Samarasekera and F. Weinberg, in: Proc. Ist NATO Worksh. on Computer Modelling in Crystal Growth from the Melt, Parma, Italy, April 1989, J. Cryst. Growth 97 (1989) 146.
[73] E. Dobrocka and J. Sladek, J. Cryst. Growth 104 (1990) 419.
[74] M. Ishida, K. Katano, S. Kawabata, Y. Higuchi, F. Orito, Y. Yamaguchi, F. Yajima and T. Okano, in: Proc. 9th Int. Conf. on Crystal Growth, Sendai, Japan, Aug. 1989, J. Cryst. Growth 99 (1990) 707.
[75] C. Parfeniuk, F. Weinberg, I.V. Samarasekera, C. Schvezov and L. Li, in: Proc. F. Weinberg Int. Symp. Solidif. Process., eds. J.E. Lait and I.V. Samarasekera (Pergamon, New York, 1990) p. 369.
[76] T. Tsukada, M. Hozawa and N. Imaishi, J. Chem. Eng. Jpn. 23 (1990) 186.
[77] K.W. Kelly, K. Koai and S. Motakef, J. Cryst. Growth 113 (1991) 254.
[78] G.O. Meduoye, D.J. Bacon and K.E. Evans, J. Cryst. Growth 108 (1991) 627.
[79] N. Miyazaki, H. Uchida, T. Munakata, K. Fujioka and Y. Sugino, J. Cryst. Growth 125 (1992) 102.
[80] G. Müller, J. Völkl and E. Tomzig, J. Cryst. Growth 64 (1983) 40.
[81] S.P. Timoshenko, Theory of Elasticity (McGraw-Hill, New York, 1951).
[82] B.A. Boley and J.H. Weiner, Theory of Thermal Stresses (Wiley, New York, 1960).
[83] H.S. Carslaw and J.C. Jaeger, Conduction of Heat in Solids, 2nd ed. (Clarendon Press, Oxford, 1959).
[84] V.L. Indenbom, Kristall und Technik 14 (1979) 493.
[85] A.S. Jordan, J. Cryst. Growth 49 (1980) 631.
[86] A.S. Jordan, J. Cryst. Growth 71 (1985) 559.
[87] A.S. Jordan, V. Swaminathan, E. Suhir, R.C. Cammarata, D.D.L. Chung and M. Jono, in: Proc. Mechanical Behavior of Materials and Structures in Microelectronics, Symposium, Anaheim, CA, April-3 May 1991 (Mater. Res. Soc, Pittsburgh, PA, 1991) 117.
[88] N. Miyazaki, H. Uchida, S. Hagihara and T. Munakata, J. Cryst. Growth 113 (1991) 227.
[89] C.J. Gallagher, Phys. Rev. 88 (1952) 271.
[90] L. Graf, H.R. Lacour and K. Seiler, Z. Metallk. 44 (1953) 113.
[91] E.N. da C. Andrade, Proc. Roy. Soc. A 84 (1910) 1.
[92] J. Weertman and J.R. Weertman, in: Physical Metallurgy, ed. R.W. Cahn (North-Holland, Amsterdam, 1970) 983.
[93] A.R. Chaudhuri, J.R. Patel and L.G. Rubin, J. Appl. Phys. 33 (1962) 2736.
[94] A.R. Chaudhuri, J.R. Patel and L.G. Rubin, J. Appl. Phys. 34 (1963) 240.

[95] D. Hughes, IBM J. Oct. (1961) 279.
[96] P. Haasen, Acta Met. 5 (1957) 598.
[97] A. Seeger, in: Handbuch der Physik, Vol. VII (Springer, Berlin, 1955) p. 1.
[98] E. Peissker, P. Haasen and H. Alexander, Phil. Mag. 7 (1962) 1279.
[99] H. Alexander, in: Dislocations in Solids, Vol. 7, ed. F.R.N. Nabarro (North-Holland, Amsterdam, 1986) 113.
[100] J. Völkl, G. Müller and W. Blum, J. Cryst. Growth 83 (1987) 383.
[101] E. Schmid and W. Boas, Plasticity of Crystals (Chapman and Hall, London, 1950).
[102] W.B. Gragg, J. SIAM Numer. Anal. (B) 2 (1965) 384.
[103] R. Bulirsch and J. Stoer, Num. Math. 8 (1966) 1.
[104] T.E. Hall, W.H. Enright, B.M. Fellen and A.E. Sedgwick, J. SIAM Numer. Anal. 9 (1972) 603.
[105] J.C. Lambropoulos, J. Mater. Res. 3 (1988) 531.
[106] D. Maroudas and R.A. Brown, J. Cryst. Growth 108 (1991) 399.
[107] A.S. Jordan, R. Caruso, A.R. VonNeida and J.W. Nielsen, J. Appl. Phys. 52 (1981) 3331.
[108] G. Müller, R. Rupp, J. Völkl, H. Wolf, W. Blum, J. Cryst. Growth 71 (1985) 771.
[109] W.C. Dash, J. Appl. Phys. 30 (1959) 459.
[110] A.G. Elliot, C.-L. Wei and D.A. Vanderwater, J. Cryst. Growth 85 (1987) 59.
[111] B. Reppich, P. Haasen and B. Ilschner, Acta Met. 12 (1964) 1283.
[112] S. Schäfer, H. Alexander and P. Haasen, Phys. Stat. Sol. 5 (1964) 247.
[113] J. Völkl, Ph.D. Thesis, University Erlangen-Nürnberg (1988) 138.
[114] H. Wolf, Diploma Thesis, University Erlangen-Nürnberg (1983) (unpublished).
[115] W.A. Brantley, J. Appl. Phys. 32 (1973) 534.
[116] F.S. Hickernell and W. Gayton, J. Appl. Phys. 37 (1966) 462.
[117] D. Gerlich and M. Wolf, in: High Pressure Science and Technology, Vol. 1, eds. B. Vodar and P. Marteau (Pergamon, Oxford, 1980) p. 506.
[118] A.S. Jordan, A.R. VonNeida and R. Caruso, in: Proc. 6th Int. Conf. on Vapor Growth and Epitaxy and 6th Amer. Conf. on Crystal Growth, Atlantic City, NJ, July 1984, J. Cryst. Growth 70 (1984) 555.
[119] C.T. Tsai, J. Cryst. Growth 113 (1991) 499.
[120] S. Motakef, J. Cryst. Growth 108 (1991) 33.
[121] S. Motakef, J. Cryst. Growth 114 (1991) 47.
[122] J.C. Lambropoulos, J. Cryst. Growth 104 (1990) 1.
[123] P. Nicodeme, N. Van den Bogaert and F. Dupret, in: Proc. Int. Symp. GaAs and Rel Comp., Jersey, 1990, Inst. Phys. Conf. Ser. 112 (1990) 43.
[124] R.A. Brown, T.A. Kinney, P.A. Sackinger and D.E. Bornside, J. Cryst. Growth 97 (1989) 99.
[125] I.C. Noyan and J.B. Cohen, in: Residual Stress and Stress Relaxation, eds. E. Kula and V. Weiss, Proc. Sagamore Army Materials Res. Conf., 1981, Lake Placid, NY (1981) 1.

CHAPTER 15

Modelling Bridgman and Czochralski Growth

F. DUPRET and N. VAN DEN BOGAERT

Centre d'Ingénierie des Systèmes, d'Automatique et de Mécanique Appliquée (CESAME), Université Catholique de Louvain, B-1348 Louvain-la-Neuve, Belgium

Contents

1. Introduction 877
   1.1. The global problem 878
   1.2. Objectives of numerical simulation 882
   1.3. Modelling of crystal growth 883
   1.4. Plan of the chapter 887
2. Physics of the growth process 887
   2.1. The crystal 887
   2.2. The melt 890
   2.3. The interfaces 891
   2.4. External agents 895
3. Finite element solution of the local Stefan problem 896
   3.1. Construction of a moving mesh 897
      3.1.1. Explicit techniques 898
      3.1.2. Implicit techniques 899
   3.2. Discretization of the thermal problem 899
      3.2.1. Energy equation 900
      3.2.2. Interface conditions 903
      3.2.3. Summary 909
   3.3. Examples of local dynamic simulations without convection 911
   3.4. Calculation of melt convection and segregation 914
   3.5. Examples of local simulations with melt convection and segregation 918
      3.5.1. Examples with melt convection 918
      3.5.2. Melt convection and segregation 919

HANDBOOK OF CRYSTAL GROWTH, VOL. 2
edited by D.T.J. Hurle

4. Radiative heat transfer in crystal growth 928
4.1. The physics of thermal radiation 928
4.1.1. Governing phenomena 929
4.1.2. Simplifying assumptions 931
4.2. Simplified models 933
4.2.1. Global surface radiation model 933
4.2.2. Gray model for the equation of transfer 936
4.2.3. Milne–Eddington differential model 939
4.3. Numerical solution of radiative transfer in Cz of VB growth 941
4.3.1. Radiative transfer in a transparent enclosure 941
4.3.2. Radiative transfer in a semi-transparent medium 950
4.4. Examples and discussion 950
4.4.1. Influence of furnace geometry on thermal stresses in the crystal for LEC growth . 950
4.4.2. Influence of internal radiation on temperature and velocity fields in a VB silicon furnace 953
4.4.3. Influence of internal radiation on the temperature field and interface configuration in a VB oxide furnace 956
5. The global problem 959
5.1. Solution of the quasi-steady global problem 959
5.1.1. Assembling technique 960
5.1.2. Iterative strategies 964
5.2. Examples of quasi-steady global simulations 965
5.2.1. Global heat transfer model for Cz growth 965
5.2.2. Modelling an induction heating system 968
5.2.3. Global heat transfer model for DGF growth 970
5.2.4. Calculation of thermoelastic stresses in the crystal 972
5.3. Solution of the dynamic global problem 976
5.3.1. Dynamic problems 977
5.3.2. Direct problem: use of Gebhart method for solving radiative transfer 980
5.3.3. Direct problem: use of Galerkin method for solving radiative transfer 980
5.3.4. Inverse problem 981
5.3.5. Dynamic simulations with melt convection 982
5.4. Examples of dynamic global calculations 982
5.4.1. Direct problem: stability analysis 983
5.4.2. Inverse problem: pulling of the head of a crystal 987
5.4.3. Direct problem based on an inverse simulation: comparison between heater power and pull rate perturbations 993
5.4.4. Dynamic global calculations including convection in the melt 998
5.4.5. Calculation of dislocation density 1000
6. Conclusions 1002
Acknowledgements 1002
Notations 1002
References 1002

1. Introduction

Major progress has been observed during the last decade in the numerical simulation of crystal growth. The constant and extraordinarily fast increase of power of available computers has generated a fierce competition in the scientific community, with the aim of developing more and more efficient algorithms and of solving problems of higher and higher complexity. Whereas our goal in the present chapter is to analyze these developments, we wish however to point out the importance of previous studies, which have been the basis of subsequent work. General textbooks on crystal growth fundamentals, and process technique and modelling, have marked the continual advancement in the field [1–10].

One of the most important objectives of early studies was to understand the mechanisms of convection [11–17]. The result of this effort, which was based mainly on using finite difference methods, has been reviewed by Pimputkar and Ostrach [18], Polezhaev [19], and Langlois [20]. The influence of magnetic fields in Czochralski semi-conductor growth was addressed by Langlois and Lee [21], Oreper and Szekely [22, 23], and Mihelčić and Wingerath [24]. Segregation was another domain where modelling was expected to provide information on important phenomena, such as crystal striations. Simplified models were developed by Scheil [25], Burton et al. [26], Hurle et al. [27], Carruthers and Witt [28], Wilson [29–34], and Favier [35], while Fourier series were used by Coriell and Sekerka [36], and finite differences by Coriell et al. [37]. The influence of magnetic fields on segregation was investigated by means of analytical models by Hurle and Series [38], and Kobayashi [39].

In view of the reduced power of past computers, process modelling was only partly addressed. The stability of meniscus shape in Czochralski growth was investigated by Mika and Uelhoff [40], while the stability of crystal radius was analysed by Surek [41], using analytical methods and simplifying the problem (which is in fact coupled and very complex). A review of the questions related to growth process stability has been given by Hurle [10]. Concerning bulk growth simulations, it is important to recall the pioneer contributions of Kobayashi and Arizumi [42–44], and Arizumi and Kobayashi [45], who introduced finite difference methods to calculate heat transfer and interface shape in Czochralski growth, taking some influence of melt convection into account. Calculation of interface shape was performed by Fu and Wilcox [46] in the case of vertical Bridgman growth. Finally, first models for the calculation of radiative exchanges in Czochralki bulk growth (taking the algebra of view factors into account) were proposed by Rea [47], Williams and Reusser [48], Ramachandran and Duduković [49], and Srivastava et al. [50].

The essential change we have faced since has been the introduction of appropriate numerical techniques in order to solve more and more non-linear and coupled problems. In addition, the trend has been clearly to develop time-dependent and, when dealing with melt convection, three-dimensional models.

A first step was taken when the finite element technique was adopted by Ettouney and Brown [51], and Crochet et al. [52] for solving the Stefan problem using the isotherm method on the solidification front. The same approach was used by Chang and Brown [53] to calculate melt convection and segregation in vertical Bridgman growth. Another important step has been to introduce in Czochralski growth the shape of all the interfaces of the crystal and melt (i.e. the solidification front, the meniscus, and the crystal wall) as unknowns of the problem, in order to perform time-dependent simulations and to analyze the dynamics of the system. This method was first used by Crowley [54], and subsequently improved and adapted to quasi-steady and dynamic finite element models by Derby et al. [55], and Derby and Brown [56], respectively.

The development of accurate numerical techniques for the calculation of radiative transfer in bulk growth has represented another domain where significant progress was observed. A first global model of Czochralski growth was introduced by Wouters [57], and Dupret et al. [58], where heat transfer is calculated in the overall furnace, while the geometry is completely arbitrary. (The numerical algorithm is detailed in [59].) Alternative methods for taking account of radiative exchanges in bulk growth simulations were developed by Atherton et al. [60], and Bornside et al. [61]. The coupling of flow and temperature calculations for quasi-steady global models was analyzed by Dupret et al. [62], Nicodème et al. [63], Sackinger et al. [140], and Derby et al. [64]. Time-dependent global models were developed by Atherton et al. [60] (using an explicit integration technique), and by Van den Bogaert and Dupret [65], and Van den Bogaert [66] (using an implicit technique). Internal radiation was introduced in the simulation of vertical Bridgman and Czochralski oxide growth by Brandon and Derby [67].

A very important result was also obtained when Crochet et al. [52, 68] demonstrated that numerical methods can be used to simulate the onset of oscillations in the flow of low-Prandtl number fluids under the action of a horizontal temperature gradient. This work and other contributions in the same field (including, in particular, the analysis of Langlois [13], and McLaughlin and Orszag [69]) have given rise to a considerable research activity, with a view to explain the periodic or chaotic behaviour of the flow which is observed in many practical situations. Examples of such investigations may be found in [70]. Today, a main difficulty remains to couple melt convection with global and time-dependent calculations.

A last domain where numerical simulation has been shown to provide a new insight into the physical understanding of the process is the calculation of thermally generated stresses in the growing crystal. A first finite element thermoelastic model, but reduced to the solid region, was developed by Duseaux [71]. The coupling of stress and global thermal calculations was introduced by Motakef and Witt [72], Motakef [73, 74], and Nicodème et al. [75]. New models for the calculation of visco-plastic stresses and dislocation density were presented by Dillon et al. [76], and Völkl and Müller [77, 78].

This list of references is of course not complete, and our aim in this chapter will be to examine other contributions in more detail, in order to highlight the advances made in bulk growth modelling. Let us mention that much information on these topics may be found in the review of Brown [79].

1.1. The global problem

A general sketch of the geometry of a typical LEC Czochralski (Cz) puller [80] is shown in fig. 1. The heater – which is often an Ohmic resistor – radiates power towards a

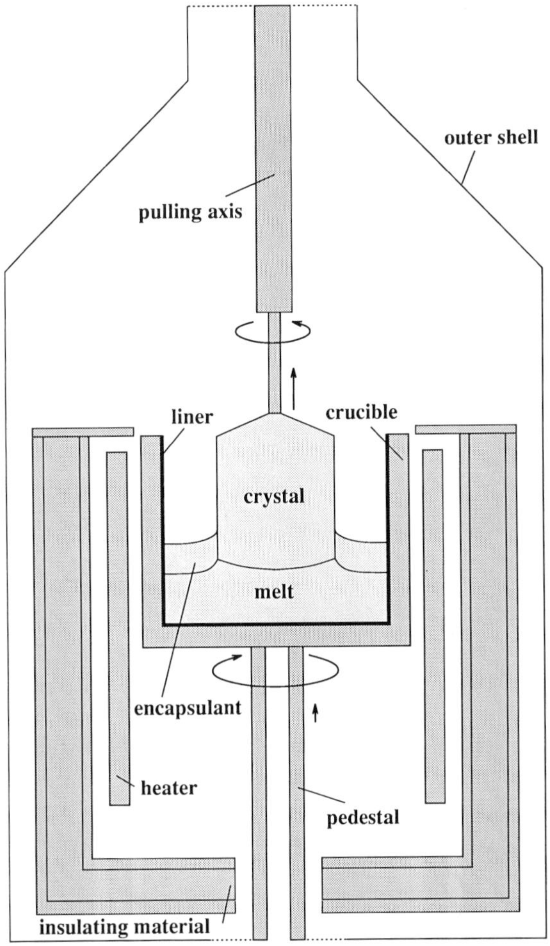

Fig. 1. Schematic diagram of a LEC furnace.

crucible containing the molten material, hence maintained above the solidification temperature. A liner of selected properties may separate the crucible from the melt. In the case of liquid encapsulated growth, a particular encapsulant covers the melt to prevent evaporation of volatile components. An alternative method to supply energy to the crucible and the melt – which is frequently used to grow oxide crystals, and sometimes to grow semiconductors – is to resort to an induction heating system, in which applying a strong alternating current to the coil generates heat directly within the crucible and/or the melt by electrical resistance to induced currents.

At the outset of the process, a crystal seed is dipped into the liquid, and then slowly pulled while a cylindrical crystal is progressively grown from the melt. During growth, the crystal radiates heat towards the surrounding environment which consists of the outer shell of the furnace (cooled by water circulation) and, in some cases, of a high pressurized gas. In this latter situation, the release of heat by conduction and convection in the gas may be important. It is in practice necessary to let both crystal and crucible rotate around the vertical axis, in order to ensure axisymmetry of the overall process, and to control buoyancy-driven flow in the liquid.

The solidification front between crystal and melt plays a major role in the process, since it is the place where interactions between the liquid and solid phases take place. In particular, the composition of the crystal may be strongly affected by the problem of segregation. In general, a major difficulty of growing high quality crystals is the dependence of interface shape and segregation on the melt flow pattern. Melt convection can be induced by temperature or solute gradients within the liquid (this effect is called natural or thermosolutal convection), by crystal and crucible rotations (forced convection), or by surface tension gradient in the liquid–environment interface (Marangoni convection). In semiconductor growth, the Grashof number of the flow is fairly high, which may give rise to an oscillating or even turbulent behaviour in the melt [81–83, 261, 278, 279]. These effects may have severe drawbacks, such as crystal

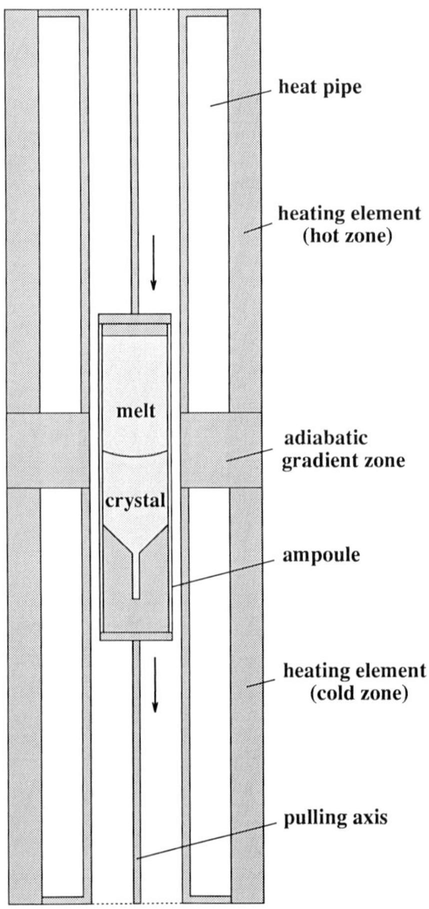

Fig. 2. Schematic diagram of a VB furnace.

striations [27, 84], and a solution to reduce their importance is to generate a strong magnetic field, which plays a role as a vigorous brake in the flow [85, 86]. The growth of oxide crystals [87] presents some important differences with respect to semiconductors. While typical Grashof numbers are lower, which means that flow behaviour is simpler, heat transfer in the melt becomes convection dominated in view of the higher Prandtl number of molten oxides. Another peculiar feature is the importance of internal radiation, which may be a significant mode of heat transfer in the crystal, as first demonstrated by Abrams and Viskanta [88].

An alternative solution to bulk crystal growth is to resort to the Bridgman technique. A typical sketch of a vertical Bridgman (VB) furnace is shown in fig. 2. Here, the molten material is contained in a cylindrical ampoule at the bottom of which solidification starts from a crystal seed (after back-melting). The outer thermal environment of the ampoule consists of a hot, an insulation, and a cold zone. During growth, the ampoule is slowly shifted downwards, and the temperature distribution induced by the outer environment in the solid and molten material shifts upwards accordingly, while keeping more or less the same axial profile during the process. Hence, melt solidifies progressively until the crystal is completely formed.

A major advantage of vertical Bridgman growth is that the solid phase is located under the liquid phase and, consequently, that the temperature gradient is pointing essentially upwards in the melt. If the isotherms were exactly horizontal, the layered fluid would be perfectly stable with respect to thermal convection. The problems related to the onset of a buoyancy-driven flow in the melt are, therefore, strongly reduced (in practice, however, a destabilizing radial temperature gradient is generally present). The situation is more complex when considering solutal convection, which may be important if the material is a non-dilute alloy. Indeed, while the density of the molten material normally decreases with increasing temperature, there is no such general rule concerning the dependence of density with respect to solute concentration, and there may be solute rejection or incorporation at the interface. Hence, the solutal field can have a priori a stabilizing or destabilizing effect. In reality, in view of the coupling between thermal and solutal convection, a flow damping is observed only for high values of the solutal Grashof number [89], and complex flow structures may be found [90–92]. It is worth noting that melt convection outside the diffusion-controlled boundary layer can have a beneficial mixing influence on crystal uniformity [93].

An efficient way to control accurately the temperature gradient at the outer surface of crystal and melt is to replace the classical vertical Bridgman apparatus by a dynamic gradient freeze (DGF) apparatus [94, 95] (fig. 3). Basically, the heating system consists of a set of controlled annular elements monitored by appropriate thermocouples, which are placed mostly around the periphery of the growth vessel. It is no longer necessary to let the ampoule move downwards in the vessel, since the evolution of the outer temperature profile is achieved by a slow vertical shift of power supply in the heating elements.

The present chapter of the Handbook is devoted to the modelling of Bridgman and Czochralski growth. Detailed practical information on these techniques may be found in Chapters 2 and 3, respectively. Although the governing equations of the model will be presented in a general and concise form, we shall in the sequel focus on axisymmetric problems (with a view to introducing radiative transfer in the system). Hence, the

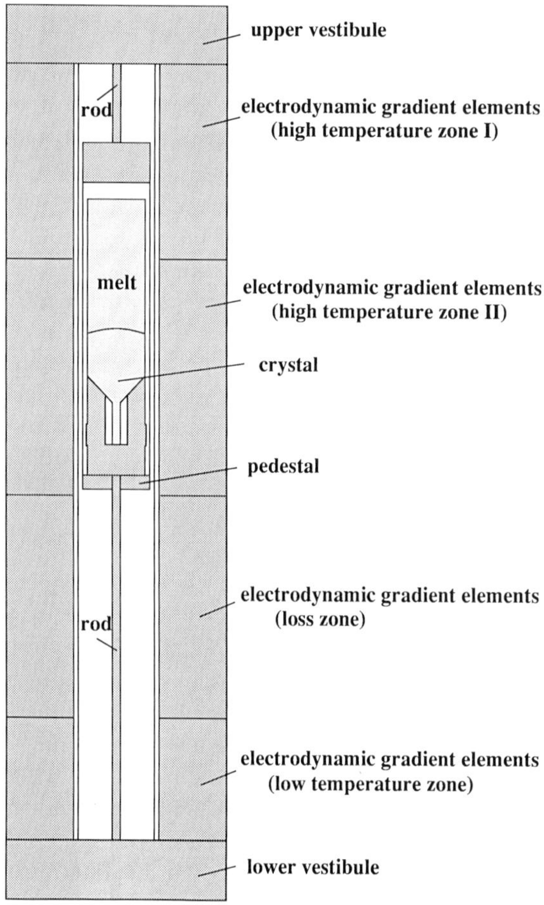

Fig. 3. Schematic diagram of a DGF furnace.

horizontal Bridgman technique will not be investigated in detail, except when particular contributions will be highlighted. The floating zone process, while involving axisymmetric calculations, will not be examined for it is the object of Chapter 4.

1.2. Objectives of numerical simulation

The first aim of modelling the growth process is to provide a better understanding of the various factors which may affect the quality of the crystal. Subsequent objectives are to predict and to control the process.

From among the numerous phenomena influencing the crystal properties, which are described in detail in other chapters of the Handbook, let us here point out a handful of effects which will be assumed to be representative of the problems to be solved. First, except in the case of shaped growth (see Chapter 16), the crystal must be cylindrical in view of industrial requirements. Second, a perfect crystal is expected to be obtained; however, the regularity of the lattice is never ideal, since dislocations, interstitials, and vacancies are generated during growth. Third, the concentration of the dopants which are transferred from melt to crystal should be as regular as possible; concentration of impurities in the crystal should be as low as possible; in the case of a pseudo-binary alloy, the composition of the crystal should be uniform; however, this is never completely obtained in view of the complexity of the transfer phenomena taking place within the melt and, in particular, in the vicinity of the solidification interface.

This rapid analysis indicates that an ideal numerical model should have the following capacities:

(i) To predict the shape of the grown crystal, and of the successive solidification fronts, from essentially a heat transfer analysis performed in the overall furnace.

(ii) To relate the quality of the lattice to the thermo-elastic and/or plastic stress generated in crystal during growth.

(iii) To take into account the phenomena governing the melt flow, and the influence of the flow patterns on the crystal composition.

(iv) To predict the stability of the process and, if necessary, the morphological stability of growth (which depends on the melt flow).

From this, we understand that numerical simulation of bulk growth requires to put together precise models associated with a high diversity of physical phenomena. Many of these models are themselves very difficult to establish and their solution may, therefore, represent per se a complex objective. In this Handbook, the process stability is analyzed in Chapter 16, stress generation in crystal is addressed in Chapter 14, the problem of melt convection is examined in Chapter 13, the peculiar aspects of segregation in the melt are analyzed in Chapter 12, while the use of magnetic fields to reduce melt convection is the subject of Chapter 5. Having in mind that the objective of the present chapter is the integrated modelling of the Bridgman and Czochralski processes, we will use all these contributions as a starting point for analyzing the problem of global simulation.

Hence, in the next sections, a general model will be introduced, as represented by a set of equations associated with many aspects of the problem. The underlying physics will however be strongly summarized, except in the case of radiative transfer, and we refer the reader to the above mentioned chapters for more detail. Moreover, the problems related to morphological stability will not be addressed, as having been analyzed in Volume 1, Chapters 12 and 13 (see also [96–98]).

1.3. Modelling of crystal growth

Fundamentally, modelling the growth process leads us to solve a three-dimensional transient problem presenting a high diversity of space and time scales. Indeed, the batchwise nature of Bridgman or Czochralski growth leads to a time-dependent system, with changing free boundaries. Moreover, even if the geometry is axisymmetric, the temperature field in a Cz furnace is often three-dimensional and time-dependent within the melt. Transient effects are also sometimes possible in VB growth in view of the complexity of thermosolutal convection [99] or the inertia of interface shape [100].

Space scales go from the very small lattice constant (which is however implicitely involved in the crystal material properties, since one is dealing with models of continuum mechanics) to the large geometrical scales, such as the changing crystal length, and the radii of crystal, crucible (in Cz growth), and overall furnace. However, smaller scales are in addition generated by the multiple boundary layers which are present in the vicinity of the solidification front and at other boundaries of the fluids in the system. In particular, one must consider viscous, thermal, and solute boundary layers in the melt (see Volume 1, Chapter 10), a dislocation boundary layer in the crystal (as demonstrated by Maroudas and Brown [101]), several "surface" boundary layers in the vicinity of the tri-junction (as related to the capillary effects which are present in the crystal–melt–outer environment separating surfaces, see Chapter 16 and the analysis of Bardsley et al. [102]), electric current boundary layers when magnetic fields are used (as shown by Hjellming and Walker [103, 104], and Hicks and Riley [105]), and finally radiative boundary layers when internal radiation is non-negligible in crystal, melt, or encapsulant (see the work of Howell and Goldstein [106]).

The situation is similar when considering time scales: we have the long characteristic time which is required to obtain a significant change of crystal length, the short time scales related to convection in melt and flow oscillations, and the very short characteristic time associated with dislocation dynamics. We have also to consider the characteristic times induced by any change of operating conditions and, in particular, the rather long response delay of the solid–liquid interface and growth angle [107], the faster response times related to solute diffusion and conductive heat transfer in the constituents [107, 89], and the rather short time which is necessary for a transition of flow regime or pattern in the melt. It is beyond the scope of this chapter to investigate the problem of scaling with more detail, and we refer to the work of Camel and Favier [108–110], and to Chapter 12 for a complete discussion.

The first objective of the modelling task is, therefore, to define precise problems from the scaling analysis. As we are clearly confronted by several singular perturbation situations [111, 112], one of the main questions is to determine whether an asymptotic model can provide satisfactory results or, equivalently, whether some equations may be simplified without a significant loss of accuracy. This approach is not only useful to save computing time, it may also be necessary when boundary layers are very thin, or when transients are very short. It is obvious that a numerical solution will never capture an effect which takes place in a subregion, or during a time, which is smaller than the mesh (or grid) size, or the time step, respectively. What is however also possible is that a coarse discretization induces oscillations in the numerical solution, which can be avoided by using an asymptotic model instead of the complete one.

The most important simplification in modelling bulk crystal growth is to introduce the *quasi-steady* [113], or *pseudo-steady* [90] assumption. In both cases, the global temperature field in the furnace is supposed to be in a steady state, while a release of latent heat of fusion is maintained at the solidification front in proportion to the pull rate. All the time derivatives are however dropped in the differential system; any external input is kept constant with time, and the evolution of the geometry is neglected. In the quasi-steady model, which is essentially used in Cz growth, the heat advection due to crystal lift is neglected, and the normal velocity component in the melt is set to zero at the liquid–solid interface. In the pseudo-steady model, which is used in VB growth, a heat advection term is added in the energy equation in both crystal and melt, which is defined exactly from the pull rate. This means that the molten and solid material is assumed to be continually injected or removed, respectively, at the top and bottom extremities of the cylindrical ampoule.

The validity of the quasi- and pseudo-steady models is based on the fact that, in many situations, the crystal lengthening and, in general, the evolution of the operating conditions are much slower than the transients associated with crystal–melt interface deformation, with the dynamics of conductive heat transfer in the various constituents, and with melt convection. This is not always valid and, in particular, poor information would be provided by quasi-steady calculations during cone formation and the tail-end stage of the Cz process. Comparisons between results of quasi-steady (or pseudo-steady) and dynamic simulations have been performed by Brown and co-workers [93, 99, 100, 107], and some examples will be presented in sections 3.3, 5.2 and 5.4.

A second important simplification, which has played a major role in the past, is to perform *local* simulations, where all the complex thermal influence of the external environment of crystal, melt, encapsulant (in LEC growth), and crucible or ampoule is supposed to be known. (Computational cost is, therefore, strongly reduced.) More precisely, the term "local" will, in the sequel, define a simplified model where the radiative exchanges in the enclosures of the furnace are not introduced in the calculations, except possibly internal radiation within crystal, melt, and encapsulant (if any).

Various methods may be used in order to obtain approximate thermal boundary conditions for the local system. In particular, imposing experimental boundary temperatures is feasible, but very difficult [71]. Imposing radiative heat flux conditions by means of appropriate external ambient temperatures has been the approach of Derby et al. [55, 114, 115] in Cz growth, and of Adornato and Brown [90, 91] and others, in VB growth. Let us emphasize however that local problems are not based on a scaling analysis, but must simply be considered as attempts to approximate the effective global system. Hence, the simplification must be used very carefully in view of the strong coupling between all the heat transfer modes in the overall furnace. In general, much better results will be provided by local simulations in VB or DGF growth than in Cz growth, since the crystal surface is not directly in contact with the radiative surrounding environment.

A more severe simplification is also often introduced by performing local simulations reduced to the melt domain (generally called *bulk flow* simulations [20]). In this case, the solidification front is assumed to be flat and at the melting temperature, while thermal boundary conditions are fixed with respect to time. The influence on interface shape of the heat transfer in the crystal, and the release of latent heat of fusion, is, therefore, completely neglected. The interest of this simplification is to facilitate the analysis of melt convection and all related effects, since the computational domain is no longer deformable. Although exact results cannot be obtained, essential information on flow patterns and regimes can be brought out in this way.

A general scheme of the effects which play a significant role in bulk growth and their relation with the simplified models is given in fig. 4.

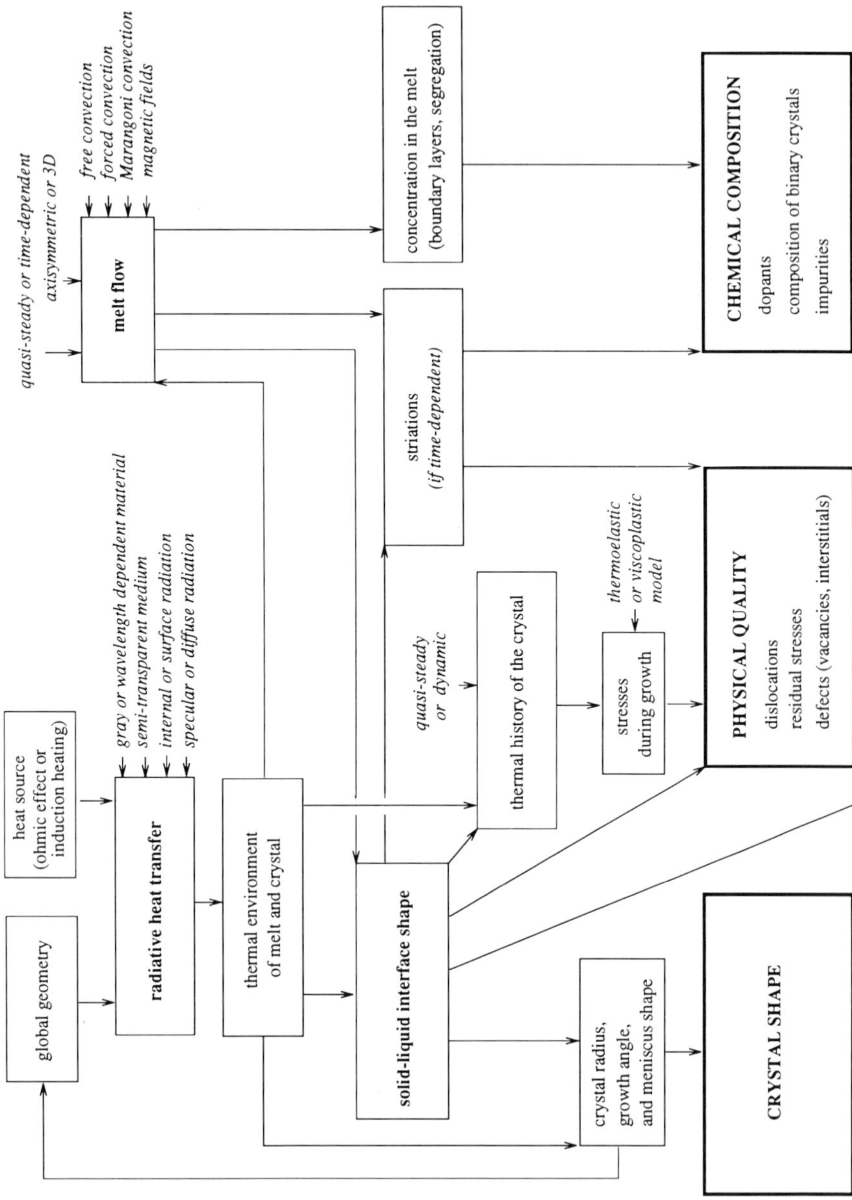

Fig. 4. Schematic representation of the physical phenomena related to bulk growth and affecting crystal quality.

1.4. Plan of the chapter

In section 2, the physics of the phenomena which play a significant role in bulk growth is presented in the form of the equations associated with heat and mass transfer in the two-phase medium. We review the existing models of stress generation in the cooling crystal, as related to its thermal history, together with the laws governing melt flow, segregation, and the shape of all the interfaces of crystal and melt. The influence of external agents, such as magnetic fields and the heating system, is also examined.

The numerical solution of the local problem by means of the finite element method is addressed in section 3. We focus first on the coupling between heat transfer and calculation of the free interfaces. The introduction of additional unknown fields in the melt, such as velocities, pressure, solute concentration, and electric potential is analyzed in a second step.

Section 4 is fully devoted to the modelling and numerical simulation of radiative transfer, whose role is two-fold in bulk growth. On the one hand, radiation in the (transparent) ambient gas couples strongly heat exchanges in all the constituents of the furnace and is, therefore, an essential ingredient for global calculations. On the other hand, when the crystal and/or melt are participating media, internal radiation must be introduced both in local and global simulations.

The global problem is the object of section 5. Quasi-steady and time-dependent models are analyzed separately. The coupling between heat exchanges in the overall furnace is examined in detail, and we highlight the difficulties related to taking into account, in unsteady calculations, the continual evolution of the geometry.

Throughout sections 3–5, examples from the literature are presented in order to illustrate the present capabilities of modelling.

2. Physics of the growth process

Recall from section 1.2 that the objective of the present section is to introduce briefly a model for most of the physical phenomena involved in Cz or VB growth. Numerical solution of these equations by means of the finite element method will be addressed in sections 3 and 5. For the sake of conciseness, a unique set of equations will be established, which will be common to all the peculiar applications we wish to consider (Cz or VB growth, pure materials or alloys, semi-conductor or oxide crystals). The complete system should not, therefore, be considered as representative of a real problem.

2.1. The crystal

Crystal production is the aim of the growth process. Several complex phenomena occurring in the solid phase must be taken into account in order to understand the relationship between crystal quality and the processing conditions. In particular, modelling all the reactions taking place in the crystal, and the formation of point defects, dislocations, or precipitates is a very difficult task, which is addressed in Chapter 14, and Volume 1, Chapter 3. These phenomena will not be analyzed in this chapter for the very reason that calculation of heat transfer, crystal shape, and segregation during growth represents a problem which is almost completely decoupled from the equations governing the reactions in the solid. (Results of local or global simulations may then be used as data for simulating the behaviour of the cooling crystal.) In what follows, we will however examine concisely the problem of stress generation during growth, since results of numerical simulations are often presented in such terms.

The heat transfer in the crystal in governed by the energy equation†,

$$\rho c \frac{\mathrm{D}T}{\mathrm{D}t} = \nabla \cdot (K \nabla T - \mathbf{q}_{\mathrm{R}}) + w. \tag{1}$$

In eq. (1), the symbols $\mathrm{D}/\mathrm{D}t$ and ∇ denote the material derivative and gradient operators, T is temperature, ρ, c and K stand for the specific mass, specific heat, and thermal conductivity of the crystal (which may depend on T), while $\mathbf{q}_{\mathrm{R}}$ is the radiative heat flux vector, and w is the heat supply per unit volume due to induction heating (if such a technique is used and a non-negligible power is generated in the crystal). These two effects will be examined with more detail in section 2.4.

Two basic methods may be used for calculating stress and evaluating dislocation formation in the crystal. The first approach is based on a thermoelastic approach [117, 118], the second one on using thermoplastic models [119–122]. In the first case, the following main hypotheses are introduced: crystal is assumed (i) to behave like an elastic body, and (ii) to be stress free at uniform temperature, i.e. to have a perfect lattice. Boundary conditions are easily identified and, in particular, the stress vector may be set to zero at the solid–liquid interface without significant loss of precision. The weight of the crystal plays some role only when very large crystals are grown. In order to provide accurate results, calculations should be performed taking into account the cubic symmetry of the crystalline material [123–125]. The method of Jordan et al. [117, 118] for evaluating dislocation formation from thermoelastic stresses is based on the observation that glide occurs at some given location when the resolved shear stress σ^i_{RS} (i.e. the tangential stress) acting along one of the four $\{111\}$ slip planes, in one of the three associated $\langle 1\bar{1}0 \rangle$ directions, exceeds the so-called critical resolved shear stress σ_{CRS}, which is a material function of temperature. Hence, a criterion for the onset of creep and generation of dislocations may be defined by comparing the maximum value (or the average) of the 12 resolved shear stresses σ^i_{RS} with σ_{CRS}. In practice, a common way to estimate an average value of the resolved shear stresses from isotropic simulations is to calculate the von Mises invariant σ_{VM}, which is defined by the formula:

$$2\sigma^2_{\mathrm{VM}} = (\sigma_1 - \sigma_2)^2 + (\sigma_2 - \sigma_3)^2 + (\sigma_3 - \sigma_1)^2, \tag{2}$$

where σ_1, σ_2, and σ_3 are the eigenvalues of the stress tensor.

For the last five years, an important effort has been paid to introducing more realistic models of dislocation formation [76–78, 101, 126–132]. Basically, they are

† Notations are explained at the end of the chapter.

defined by an ad hoc extension to multi-axial plastic strain of the uni-axial model of Alexander, Haasen, and Sumino [119–121] (as first proposed by Dillon et al. [76], and Völkl and Müller [77, 78]). These dynamic models take into account relaxation of thermal stresses in the crystal and strain hardening by dislocations. A detailed discussion is given in Chapter 14. The governing equations may be summarized as follows: the total deformation tensor consists of an elastic component ($\boldsymbol{\varepsilon}^{el}$) and a plastic component ($\boldsymbol{\varepsilon}^{pl}$); the elastic component is directly related to the stress tensor $\boldsymbol{\sigma}$ and the temperature by the equations of linear thermoelasticity; the rate of plastic strain obeys an Orowan equation,

$$\frac{\mathrm{D}\boldsymbol{\varepsilon}^{pl}}{\mathrm{D}t}=\frac{\beta v_{\mathrm{d}} N_{\mathrm{m}}}{\sigma_{\mathrm{VM}}}(\boldsymbol{\sigma}-\tfrac{1}{3}\,\mathrm{Tr}\,(\boldsymbol{\sigma})\mathbf{I}), \tag{3}$$

where v_{d} and N_{m} are the speed and density of mobile dislocations, β is a material constant, $\mathbf{I}$ is the identity tensor, and σ_{VM} has been defined in eq. (2). The rate of multiplication of the dislocations and the dislocation speed are calculated as follows:

$$\frac{\mathrm{D}N_{\mathrm{m}}}{\mathrm{D}t}=\gamma N_{\mathrm{m}} v_{\mathrm{d}} f_1(\tau_{\mathrm{eff}}), \tag{4}$$

$$v_{\mathrm{d}}=B_0 \exp\left(-\frac{Q}{kT}\right) f_2(\tau_{\mathrm{eff}}), \tag{5}$$

where γ, B_0, and Q are material constants, k is the constant of Boltzmann, while f_1 and f_2 are power-law-like material functions of the effective stress τ_{eff} experienced by the dislocations. The definition of τ_{eff}, f_1, and f_2 is different in the models of Nicodème et al. [126], Tsai et al. [132], Maroudas and Brown [101], and Motakef et al. [130]. Briefly, the effective stress is always defined as being the difference between some average scalar measure τ of the shear stress, and a back-stress due to strain hardening and impurity concentration in the crystal. There is first a subtle, but significant, difference in the definition of τ between [126, 132], on the one hand, and [101, 130], on the other hand. Indeed, in [126, 132], the scalar level of shear stress τ is defined as being the von Mises invariant of the stress tensor $\boldsymbol{\sigma}$, while, in [101, 130], a concept of scalar plastic deformation ε^{pl} is introduced, and τ is defined as being proportional to the difference between the shear strain level (measured by the Mises invariant of the strain $\boldsymbol{\varepsilon}$, multiplied by a shape factor) and this scalar plastic deformation. When dealing with plasticity, the role of the reference configuration with respect to which deformation is defined must be analyzed carefully. In [126] and [132], equations are invariant with respect to any infinitesimal change of this reference configuration which is, therefore, arbitrary. Hence, the local state of the solid medium is characterized only by its temperature T, stress $\boldsymbol{\sigma}$, and density of dislocations N. In [101] and 130], the equations may be shown not to be invariant with respect to such an infinitesimal change of reference configuration, except if the scalar plastic deformation ε^{pl} is considered as an additional intrinsic property of matter. A second, and important, difference is found between [130] and the other references in the definition of the functions f_1 and f_2. In the former case, f_1 has the same sign as τ_{eff}, while f_2 is always positive; in the latter cases, f_1 and f_2 are both positive if τ_{eff} is positive, and set to zero if

τ_{eff} is negative. The underlying problem is to know whether some creep and dislocation multiplication continue when the effective stress becomes negative (a strong reduction of these effects is certainly expected).

We consider that these problems indicate that the modelling of plastic strain and dislocation generation has not achieved a complete maturity in crystal growth simulation. The above mentioned models have many more or less important drawbacks:

- Isotropy of the crystal is assumed.
- Orientation of the dislocations is not taken into account. A tensor should be used to represent dislocation density.
- Isotropic plasticity is assumed. The extension from a uni-axial to a multi-axial creep model is based on a simplification of the physics of the problem. In particular, the concept of effective stress is not really clear.
- The motion of the dislocations is not taken into account in eq. (4), which has no diffusion-like term.
- When it is used, the concept of scalar plastic strain ε^{pl} is not clear, since no reference configuration (i.e. the configuration with respect to which deformation is defined) is specified. In fact, a natural reference configuration does not exist, since crystal grows from the melt under stress and with dislocations. While eqs. (3) to (5) are invariant with respect to infinitesimal superposed rigid body motion, i.e. objective [133, 134], a clear intrinsic definition of ε^{pl} is necessary for giving a meaning to the boundary conditions at the solidification front.
- In general, the boundary conditions to introduce at the solidification front must be clarified in accordance to the mathematical nature of the system and the rules governing constitutive equations (taking, in particular, the role of the reference configuration into account).

Nevertheless, although the modelling effort must be pursued, the introduction of thermoplastic models in the numerical simulation of bulk growth must be considered as significant progress in view of the agreement obtained between numerical and experimental results [127, 129, 130, 132]. The existence of a dislocation boundary layer located just over the solidification front has been demonstrated in [101], and represents an important concept for future developments.

2.2. *The melt*

The equations governing motion, heat transfer, and segregation in the melt are well-known. For the sake of conciseness, we shall use a unique symbol C to denote the concentration of any dopant or impurity, or the melt composition (when a pseudo-binary alloy is grown), with the convention that C denotes the mass fraction of the solute.

The Boussinesq approximation is based on the hypotheses that the fluid is Newtonian and that variations of melt density are negligible, except in the calculation of the body forces induced by thermal and solutal buoyancy. Letting $\mathbf{v}$ and p denote the velocity and pressure fields in the melt, the mass and momentum equations are written as

$$\nabla \cdot \mathbf{v} = 0, \tag{6}$$

$$\rho \frac{\mathrm{D}\mathbf{v}}{\mathrm{D}t} = -\nabla p + \nabla \cdot [\tfrac{1}{2}\mu(\nabla \mathbf{v} + \nabla \mathbf{v}^{\mathrm{T}})] + \mathbf{j} \times \mathbf{B} + \rho \mathbf{g}[1 - \beta_T(T - T_0) + \beta_C(C - C_0)], \tag{7}$$

where ρ is a reference value of the specific mass of melt, μ is the shear viscosity, $\mathbf{g}$ is the acceleration of gravity vector, β_T and β_C are the thermal and solutal expansion coefficients, T_0 and C_0 are reference values for temperature and concentration, and $\mathbf{j} \times \mathbf{B}$ is the body force per unit volume generated by the applied magnetic field, if any. In some particular cases, the coefficients μ, β_T, and β_C depend on temperature and/or solute concentration. The equations governing the calculation of the magnetic induction $\mathbf{B}$ and the electric current $\mathbf{j}$ will be summarized in section 2.4.

The energy equation in the melt has the same form (1) as in the crystal, and the symbols ρ, c, and K will keep the same meaning, while taking generally different values in the solid and the liquid. Heat supply by induction heating and internal radiation are possible effects in the melt.

The diffusion equation will be written in the self-adjoint form:

$$\rho \frac{\mathrm{D}C}{\mathrm{D}t} = \nabla \cdot \left[\rho \frac{D}{\alpha} \nabla(\alpha C) \right], \tag{8}$$

where D is normally the Fick diffusion coefficient (which may in some cases depend on temperature and/or concentration), while α is a material function of T and C, which is introduced for the sake of generality as an integrant factor in order to take thermodiffusion into account, if the Soret effect is non-negligible [135]; in most practical applications, D is a constant and α has the value 1.

The boundary conditions of eqs. (1), and (6) to (8) are examined in the next section.

2.3. The interfaces

With respect to the melt, the crystal may be assumed to behave as a rigid body (since visco-plastic deformations are infinitesimal). Hence, along the solidification front, the mass conservation and no-slip conditions are written as [96]

$$\rho_1(\mathbf{v}_1 \cdot \mathbf{n} + v_g) = \rho_1 \mathbf{v}_s \cdot \mathbf{n} + \rho_s v_g, \tag{9}$$

$$\mathbf{v}_1 - \mathbf{n}(\mathbf{v}_1 \cdot \mathbf{n}) = \boldsymbol{\omega}_s \times \mathbf{x}, \tag{10}$$

where $\mathbf{n}$ is the unit outgoing normal from the melt, v_g is the growth speed (i.e. the normal component of the velocity of the interface with respect to the solid, which is positive if the crystal is growing), $\mathbf{x}$ is the current position vector, $\boldsymbol{\omega}_s$ is the rotation rate vector of the crystal (which may be non-vanishing in Cz growth), and subscripts l and s refer to the liquid and solid phases, respectively (these subscripts are used only when necessary). Note that, in the case of quasi-steady calculations, mass conservation reduces to imposing a vanishing normal velocity component in the melt.

Two thermal conditions are prescribed along the solid–liquid interface, which is a free surface. First, the kinetics of solidification is expressed in theory through a non-isotropic relation between the growth speed and the local kinetic undercooling – i.e. the difference between the equilibrium value T_e of interface temperature and its actual value (which is the same in the liquid and solid phases). In principle, T_e is given by the Gibbs–Thomson relation [96]

$$T_e = T_m\left(1 + \frac{\sigma_f \chi_f}{\rho_s \Delta H}\right), \tag{11}$$

where T_m is the equilibrium temperature for a flat interface (which may depend on the concentration C_l in the liquid phase), while σ_f, ΔH and χ_f denote the surface free energy, the latent heat of fusion per unit mass, and the interface curvature (i.e. the sum of its principal curvatures), which is taken as positive when melt is locally convex. In bulk growth, the orders of magnitude of crystal diameter and growth rate allow one to simplify the crystallization kinetics condition by neglecting all undercooling effects and prescribing that the interface be at the temperature T_m. The second thermal condition holding on the solidification front expresses heat flux balance between the two phases, taking release of heat of fusion into account,

$$-K_l \nabla T_l \cdot \mathbf{n} = -K_s \nabla T_s \cdot \mathbf{n} - \rho_s v_g \Delta H. \tag{12}$$

In addition, when solute distribution is part of the problem, the concentration field C is governed by two conditions on the liquid–solid interface. First, the kinetics of solidification involves a relation between growth speed, interface temperature, and the local concentrations C_l and C_s in the liquid and solid phases. Again, in bulk growth, the assumption of local thermodynamic equilibrium allows one to simplify the kinetics condition, which may be put in the form of a relation between the concentrations C_l and C_s. In the case of a dilute solution, this relation is simply

$$C_s = k C_l, \tag{13}$$

where k is the segregation coefficient. The second condition for the concentration field on the solidification front expresses species conservation [96],

$$-\frac{\rho_l D}{\alpha} \nabla(\alpha C_l) \cdot \mathbf{n} + \rho_s C_l v_g = \rho_s C_s v_g, \tag{14}$$

where diffusion in the solid phase has been neglected.

In bulk growth modelling, the boundary conditions on the crystal wall are thermal only (except imposing the pull rate), since stresses do not influence heat transfer in the solid. These conditions – which relate the boundary temperatures and heat fluxes to thermal exchanges in the surrounding environment – will be developed in sections 3 to 5. (In Cz growth, it is worth noting that the coupling between all heat transfer modes in the overall furnace is so important that performing global (or integrated) simulations is the sole way to obtain a realistic temperature field in crystal and melt. Local simulations are much more significant in VB or DGF growth.) On the melt–crucible (or ampoule) interface, thermal boundary conditions are of a similar complexity. In addition, in order to perform flow calculations in Cz growth, one has to prescribe a melt–crucible (or ampoule) no-slip condition which will involve the lift and the rotation rate vector $\boldsymbol{\omega}_c$ of the crucible.

The meniscus separating the melt from its environment (ambient gas or encapsulant layer) is a supplementary interface in Cz furnaces. While thermal boundary conditions are the same as above, this interface is another free surface, whose shape is, therefore, part of the problem unknowns. It may however be observed that meniscus shape calculation can be decoupled from convection in the melt. Indeed, for most molten materials, the order of magnitude of melt density is high, and the forces governing meniscus shape are thus essentially surface tension, gravity, and hydrostatic pressure (see the analysis of Cuvelier and Driessen [136]). Hence, letting v_{m} stand for the meniscus speed (i.e. the normal component of its velocity), the kinematic condition holding at the melt boundary is

$$\mathbf{v}\cdot\mathbf{n} = v_{\mathrm{m}}, \tag{15}$$

while tangential conditions may be of two kinds. On the one hand, in the absence of an encapsulant, the shear contact forces must be imposed. In this case, the surface tension gradient can induce Marangoni convection. In general, letting $\mathbf{t}$ denote the stress vector,

$$\mathbf{t} = -p\mathbf{n} + \mu(\nabla\mathbf{v} + \nabla\mathbf{v}^{\mathrm{T}})\cdot\mathbf{n}, \tag{16}$$

shear force boundary conditions are written as

$$\mathbf{t} - \mathbf{n}(\mathbf{t}\cdot\mathbf{n}) = \gamma_T \nabla_{\mathrm{m}} T + \gamma_C \nabla_{\mathrm{m}} C, \tag{17}$$

where ∇_{m} denotes the tangential component of the gradient operator, while γ_T and γ_C are the derivatives of surface tension with respect to temperature and concentration, respectively. The material parameters γ_T and γ_C are often nearly constant, and surface tension is then of the form

$$\sigma_0 + \gamma_T(T - T_0) + \gamma_C(C - C_0),$$

where T_0 and C_0 are reference values for temperature and concentration, while σ_0 is a constant. On the other hand, when an encapsulant such as boric oxide is used, its viscosity is in practice much higher than that of the melt. Hence, encapsulant flow is mostly azimuthal and governed by crystal and crucible rotation. It is thus expected that the conditions holding at the melt–encapsulant interface may be simplified by imposing that encapsulant flow be free from shear contact forces, while constraining melt to stick on it. This approach permits one to decouple melt and encapsulant flow calculations. However, as the Prandtl number of oxides is in reality very high, this method is not satisfactory, since the slow radial and axial motion induced in an encapsulant by its contact with the melt may have a strong influence on its thin thermal boundary layers, and thus on the general heat transfer. In this case, flow calculations in the melt and encapsulant must be coupled, as shown by Fontaine et al. [137, 138].

When the distribution of a given solute must be calculated, diffusion of the species from the solid container into the melt, and from the melt across the meniscus, may in some cases be important (as in the case of oxygen transport in Cz Si growth). In practice, such diffusion boundary conditions at the surface of melt, crucible or ampoule will not be found in the situations where temperature and concentration calculations are coupled (through (i) the solutal expansion coefficient β_C, (ii) the surface tension coefficient γ_C, or (iii) the dependence of the melting temperature T_m on solute concentration), i.e. for non-dilute solutions. In this case, boundary conditions impose simply a vanishing normal flux of the species.

Under the above mentioned hypotheses, meniscus shape is governed by the Young–Laplace equation,

$$\sigma_0 \chi_m + (\rho_e - \rho_l) g (z_m - z_0) = 0, \tag{18}$$

where z_m is the vertical position of the interface (as a function of radial coordinate and time), χ_m is the interface curvature (which is taken as positive when the melt is locally convex), ρ_e is the environment specific mass, and z_0 is the theoretical axial position at which χ_m vanishes. (Using σ_0 in eq. (18) is sufficient, since the other terms defining surface tension are of the same order of magnitude as the stresses due to melt flow, which are neglected in meniscus shape calculation.) A main difficulty in solving eq. (18) is to define an appropriate boundary condition at the tri-junction. (Note that contact angle is simply imposed on the interface–crucible, or liner, contact line.) In principle, the wetting angle ϕ_0 between melt surface and crystal wall (fig. 5) is a complex function of (i) the surface free energies in the three adjacent interfaces, (ii) the orientation of these interfaces with respect to crystallographic axes, (iii) the derivatives of surface free energies with respect to interface orientation, (iv) the elastic properties of the crystal at its melting temperature, and (v) the growth velocity at the tri-junction. A detailed discussion may be found in Chapter 16 and in [102, 139, 140]. It is clear that the solution may be non-axisymmetric, and depend on pull rate and growth angle (i.e. the orientation of the crystal wall with respect to the vertical direction). However, in practice, this dependence is observed to be weak [141–145] and we will, in the sequel, assume that the wetting angle is a constant.

It is important to note that the angle formed by the solidification front and the crystal wall may not be imposed at the tri-junction when Gibbs–Thomson undercooling is neglected, because the mathematical nature of the melting temperature condition no longer requires one to introduce an interface extremity boundary condition. According to Bardsley et al. [102], the influence of this simplification on the interface shape

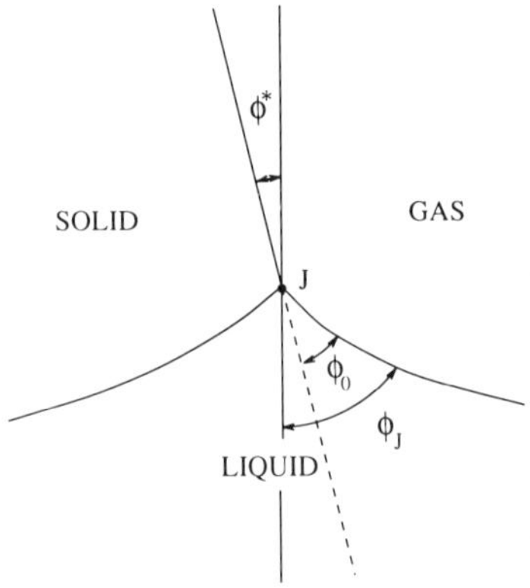

Fig. 5. Detail of the tri-junction region: ϕ_0 = wetting angle, ϕ_J = meniscus angle, ϕ^* = growth angle.

may be assumed to be non-negligible only very close to the tri-junction, which permits one to use the simplified model in bulk growth simulations.

At some stages of growth, mainly when the crystal wall is locally convex (e.g. at the end of cone formation), a short remelting may occur in the vicinity of the tri-junction. It must be clearly understood that the onset of remelting is governed by the local dynamics of heat and mass transfer at the tri-phase line, even when undercooling effects are taken into account. A careful analysis of the balance between equations and unknowns shows that the wetting angle may no longer be imposed in this situation, since the contact line is constrained to be located on the crystal wall, whose shape is known when back-melting occurs. The physics of dynamic wetting requires one to take into account deformation and stresses in both the liquid and the solid phase. When the growth speed is positive, the crystal may solidify in such a way that elastic stresses remain at a rather reduced level, for the adjacent motion in the melt is almost free, and can thus adapt to yield this "low" stress level in the crystal. When the growth speed is negative, i.e. in the case of back-melting, elastic stresses and deformation in the crystal will be much higher in the vicinity of the tri-phase line. As a consequence, crystal shape will be locally deformed, and the apparent value of the wetting angle will be changed. Again, this deformation will be significant only very close to the tri-junction, which means that using a rigid-body model for the crystal in bulk growth simulations (but not for stress calculations) is allowed, but involves dropping the wetting angle condition when remelting occurs. (This method is exactly the one which is followed in float-zone simulations for the melt–polycrystalline solid contact line condition [146, 147].) When growth restarts after back-melting, we may observe that the wetting angle will instantaneously recover its growth value, and that a sharp edge will be formed on the crystal wall.

2.4. External agents

In order to complete the model, we need to introduce the equations governing (i) the heating system, and (ii) the magnetic induction and electric current, if a magnetic field is applied to decrease the intensity of melt convection.

The heating system consists usually of one or several Ohmic resistors. If the electric conductivity of the heater were independent of temperature, it would be possible to calculate a priori the distribution of power in the resistor(s) and to express the – axisymmetrically averaged – heat supply as the product of the total power W by a known function $g(\boldsymbol{x}, t)$. In some cases, this approach (which will be followed in the sequel) is not completely satisfactory, since power and temperature distributions are coupled when electrical conductivity is thermally dependent. It will however be shown that performing coupled calculations does not involve major difficulties.

Another way to supply power to the melt is to use an induction heating system [148, 149]. The apparatus consists of an inductor, which is made of a water-cooled coil to which a strong alternating current is applied, and a conducting susceptor (often the crucible), where electric current is induced and generates power by Ohmic dissipation. The modelling of induction heating is a difficult problem, which has been addressed by Derby and co-workers [149, 64, 262] in the case of Cz or VB oxide growth. It is shown that, when the electrical conductivity of the melt is low, the induced alternating magnetic field is essentially a function of the relative position of the susceptor with respect to the inductor. In this case, the dependence of heat supply in the melt with respect to the melt–environment and melt–crystal interface shapes can be very easily characterized, since heat supply vanishes in the gas and the crystal. This simplification strongly facilitates obtaining a solution. In semiconductor growth, the electric current induced in the melt may be more important. In Cz growth, the meniscus shape can even depend on the electromagnetic field if a significant Lorentz force is exerted on the surface of the melt, due to the skin effect. Nevertheless, neglecting this latter force, we will in general state that, when induction heating is used, the heat supply distribution in any constituent of the furnace is a function of the total power supply and the geometry of the system, as represented by the vertical position of the susceptor, and the actual shapes of crystal and melt. The functional dependence of heat supply with respect to geometry may be rather complex, since it involves solving the Maxwell equations on a domain having deforming boundaries. Dependence of heat supply with respect to total power is linear if the electrical conductivity does not depend on temperature.

Using magnetic fields in order to reduce melt convection in semiconductor growth (see Chapter 5, the textbook of Moreau [150], or the review of Series and Hurle [151]) is another situation requiring solution of the Maxwell equations. In this chapter, we will only consider axial or cusp fields, since a transverse configuration gives rise to a non-axisymmetric problem, except if the field is rotating at a sufficiently high rate [150]. According to the theory developed in [150], we will assume that (i) the magnetic field is steady, or quasi-steady (rotating fields are thus not analyzed, although this would not involve major difficulties); (ii) the small currents due to advection of electric charges in the melt are negligible; (iii) the induced currents in the furnace do not affect significantly the magnetic field (inductionless approximation is used). As a consequence of these hypotheses, it is possible to show that calculation of the magnetic field is decoupled from the other unknowns and can be performed at the outset of the simulation (by solving the Laplace equation). An efficient way to pose the problem is then to introduce as a basic unknown the electric potential ϕ, and to relate the electric current to the sum of the electric field and the electromotive force by Ohm's law,

$$\mathbf{j} = \sigma_c(-\nabla\phi + \mathbf{v} \times \mathbf{B}), \tag{19}$$

where σ_c is the electrical conductivity of the medium, while the continuity equation takes the form

$$\nabla \cdot \mathbf{j} = 0. \tag{20}$$

The coupling between melt flow calculations, involving or not a magnetic field, and heat transfer simulations in Cz or VB furnaces will be analyzed in sections 3.4, 5.1 and 5.3.

3. Finite element solution of the local Stefan problem

In this section, we wish to address the numerical solution of the local problem, as defined in section 1.3. For the sake of simplicity, calculations will be limited to the domain occupied by melt and crystal. In practice, heat transfer in crucible or ampoule, and encapsulant (if any) is normally included in local simulations, but it is clear that this extension does not represent any problem and may be omitted in theoretical developments.

In order to facilitate the transition from a local to a global model, equations will be made as general as possible. In particular, the geometrical unknowns will have the same organization, although their governing equations are more complex in the global case. The main differences between local and global problems lie in the thermal boundary conditions and the power supply. These particular ingredients of the system will be redefined in section 5, when the global system is built. Another difference between local and global models resides in the selection of the system of axes, which will, in the sequel, be attached to the crucible or the ampoule in local simulations, and to the furnace in global simulations. Moreover, the axis of symmetry will be oriented upwards in Cz growth, and downwards in VB and DGF growth, since this provides a unique discrete expression for the solidification heat supply and the growth speed. On the other hand, in the present section, the time-dependent problem will be considered as representing the basic situation, from which the quasi- and pseudo-steady models can be obtained by means of simplifying assumptions whose effect will be explicitly analyzed.

3.1. Construction of a moving mesh

The problem of solving heat transfer within a domain having an unknown interface represents an important difficulty in the numerical simulation of bulk crystal growth, for a fully coupled system – involving unknown nodal interface positions, temperatures, velocities in the melt, and possibly concentrations of solute – is generally obtained when establishing the approximate equations. According to the approach of Ettouney and Brown [51], Crochet et al. [52], and Derby et al. [55], the most successful technique has been to use meshes whose elements are deformable, and to locate the melt–crystal interface on inter-elements boundaries. Development of iterative schemes based on this method is fortunately facilitated by the absence of any real geometrical difficulty to solve when implementing the numerical algorithm. The fact is that an a priori guess of the interface configuration is always roughly possible, which permits one to build rather easily a suitable deformable mesh. In a similar manner, when dealing with dynamic simulations, the change of shape of the grown crystal and the melt is not so important as to forbid one to use simple moving mesh techniques. Hence, the difficulty resides mostly in the programming task.

The concept of a moving mesh may be introduced as follows. Let, in general, in an axisymmetric geometry, the set of all the coordinates of the nodes defining the mesh (r_i, z_i) consist of two classes: $\boldsymbol{X}^{\mathrm{u}}$ is the column vector[†] of imposed nodal coordinates (such as the ones defining the crystal shape, in quasi- or pseudo-steady simulations), and of the nodal coordinates which form additional unknowns of the model (such as the vertical positions z_i of crystal–melt interface nodes, or the nodal coordinates characterizing the crystal shape, in dynamic simulations); $\boldsymbol{X}^{\mathrm{a}}$ is the column vector of the

[†] Notations are explained at the end of the chapter.

remaining nodal coordinates, which are essentially arbitrary, but must be selected in order to provide a well-behaved mesh. Hence, to any nodal coordinate belonging to X^u, corresponds a given equation of the problem (for instance, imposing the melting temperature on a given interface node is the equation associated with the axial position of this node).

3.1.1. *Explicit techniques*

A simple way to build a moving mesh is to subdivide X^u and X^a into a hierarchy of disjoined column vectors $(X^u_{(1)}, X^a_{(1)}, X^u_{(2)}, \ldots X^u_{(N)}, X^a_{(N)})$, and to define a sequence of linear relationships of the form

$$\begin{aligned}
X^a_{(1)} &= A_{(11)}X^u_{(1)},\\
X^a_{(2)} &= A_{(21)}X^u_{(1)} + B_{(21)}X^a_{(1)} + A_{(22)}X^u_{(2)},\\
&\ldots\\
X^a_{(N)} &= A_{(N1)}X^u_{(1)} + B_{(N1)}X^a_{(1)} + A_{(N2)}X^u_{(2)}\ldots\\
&\quad + B_{(NN-1)}X^a_{(N-1)} + A_{(NN)}X^u_{(N)},
\end{aligned} \tag{21}$$

where $A_{(11)}, A_{(21)}, B_{(21)}, \ldots A_{(NN)}$ are the matrices characterizing the moving mesh. It is clear that eq. (21) may formally be expressed by an equivalent linear relation of the form

$$X^a = A\,X^u, \tag{22}$$

where A is a sparse matrix. Practical calculations are of course performed using the developed form, eq. (21), in order to save computing time.

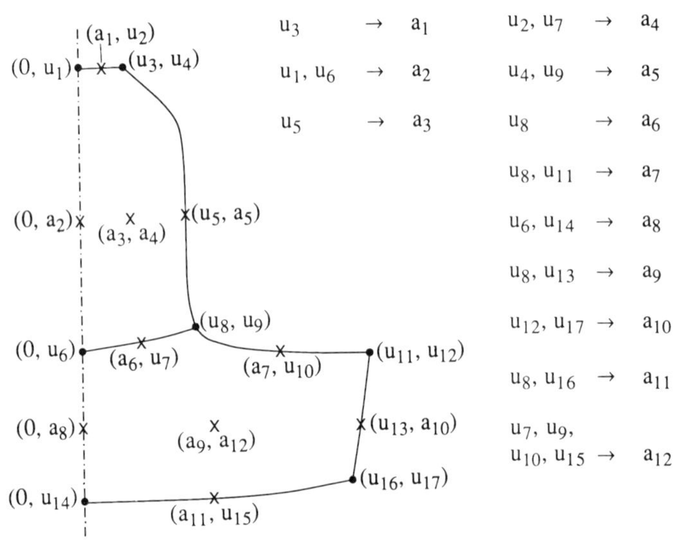

Fig. 6. General method used for mesh deformation. Definition of the vector X^u of the nodal coordinates imposed at the outset, or related to geometrical unknowns of the problem, and definition of the vector X^a of the remaining arbitrary nodal coordinates. Note that all the $B_{(ij)}$ submatrices vanish. For the sake of conciseness, the symbols u_i and a_i denote $X^u_{(i)}$ and $X^a_{(i)}$, respectively.

This technique, which is very often used, may be illustrated by means of an example used in the dynamic simulation of Cz growth [66]. Crystal and melt can be covered by a moving mesh organised according to the sequence depicted in fig. 6. We observe that nodes are arranged in different classes (17 in total), in such a way that, in any class, the radial and axial nodal coordinates belong each to a given vector $\boldsymbol{X}^{\mathrm{u}}_{(i)}$ or $\boldsymbol{X}^{\mathrm{a}}_{(i)}$, as indicated on the figure. It is easy to understand that simple interpolation rules of the form (21) can define this moving mesh (see also figs. 8 and 61).

3.1.2. Implicit techniques

An alternative approach – which is required when mesh deformation is important, and difficult to predict – is to consider all the nodal coordinates ($\boldsymbol{X}^{\mathrm{u}}$ and $\boldsymbol{X}^{\mathrm{a}}$) as coupled unknowns of the system. Here again, a well-behaved mesh is obtained by associating an appropriate equation with every component of the vector $\boldsymbol{X}^{\mathrm{a}}$. However, these equations are now implicit, and relate the components r_i or z_i of $\boldsymbol{X}^a$ to the coordinates of the nodes which are adjacent to node (i) (i.e. the nodes belonging to a same element as (i)).

A practical way to define such relations is to use coordinate transformations, among which the procedure of Thompson et al. [152] seems to be the most popular. Briefly, in this case, curvilinear coordinates (ξ and η) are defined which fit the boundary (and, in particular, all the interfaces) of the domain. Nodal coordinates (r_i, z_i) are calculated in order that the functions $\xi(r, z)$ and $\eta(r, z)$ characterizing the transformation satisfy selected partial differential equations. Hence, in the original form of the equations, independent variables are the physical coordinates r and z.

The technique for reversing, discretizing and solving the Thompson transformation is well-established, and details may be found in [152]. Its use in the case of numerical simulations of crystal growth is of particular interest in floating-zone calculations [153], in view of the importance of interface deformation. In general, using coordinate transformations suggests the following comments:

- The equations related to mesh deformation and to the physical problem represent a coupled system; hence, the computational cost will be increased with respect to using explicit techniques (this may be important in global dynamic calculations).
- Structured meshes are required by the Thompson transformation, which means that they are limited to simple topologies, such as rectangular-, L-, or T-shaped undeformed domains; however, this is exactly what is found in most cases of crystal growth simulations.

As a conclusion, coordinate transformations are expected to be more and more used in the future, in view of the continual power increase of computers (see also [154, 155]).

3.2. Discretization of the thermal problem

Let us now consider the particular case of heat transfer within the crystal and the melt in Cz or VB growth, in order to illustrate how to couple a moving mesh technique with the Galerkin finite element method. Thermosolute and forced melt convection may be taken into account according to eqs. (6)–(10), (13)–(17), (19) and (20). In a similar way, internal radiation may be introduced in the form of an additional flux term $\mathbf{q}_R$, except when using the Rosseland approximation, which simply involves a modification of the coefficient of thermal conduction $K(T)$. In view of the complexity of radiative transfer in bulk growth, the problems related to internal radiation are addressed in section 4. In the present section, we will develop the system of equations as if the radiative flux and velocity vectors were known functions of position and time. The coupled problems will be examined in sections 3.4 and 4.3. It must be emphasized that introducing convection or internal radiation in the model will always mean a very important increase of difficulty in solving the discrete equations.

3.2.1. *Energy equation*

The first step of discretization is to put eq. (1) into a weak form. Letting $T'(\mathbf{x})$ stand for an arbitrary test-function defined on the domain Ω (as occupied by the crystal or the melt), integration by parts over Ω yields the weak formulation:

$\forall\ T'(\mathbf{x})$ belonging to the Sobolev space $H_1(\Omega)$ of functions whose gradient has a summable square over Ω,

$$\left\langle \left(\rho c \frac{\mathrm{D}T}{\mathrm{D}t} - w\right) T' + (K\nabla T - \mathbf{q}_R)\cdot\nabla T' \right\rangle_{\Omega} + \langle\langle qT' \rangle\rangle_{\partial\Omega} = 0, \tag{23}$$

where symbols $\langle\ \rangle_\Omega$ and $\langle\langle\ \rangle\rangle_{\partial\Omega}$ denote integrations performed over the domain Ω or its boundary $\partial\Omega$ (using an appropriate weight r in axisymmetric calculations), while q stands for the total outgoing heat flux, as defined on the boundary of Ω (which comprises the solidification front F) (fig. 7).

In the case of a local problem, we will consider that the heat supply w, which is present in the case of induction heating, is a given function depending on the domain occupied by crystal and melt (as explained in section 2.4, this dependence may be expressed in a simple way if the electrical conductivity of the melt is low). Boundary conditions will be defined by imposing everywhere except on the crystal–melt interface a suitable relation between heat flux and temperature,

$$q = h(T - \bar{T}), \quad \mathbf{x} \in \partial\Omega \setminus \mathrm{F}, \tag{24}$$

where h is a heat transfer coefficient (which may depend on temperature) and then represent radiation and/or conduction towards a simplified environment, while $\bar{T}(\mathbf{x}, t)$ is an imposed function of position and time on the boundary. (Imposing temperature as an essential boundary condition on a part of $\partial\Omega\setminus\mathrm{F}$ is also possible, and straightforward, but this case will not be detailed here.) The weak formulation (23) must be closed by substituting $h(T - \bar{T})$ for q on $\partial\Omega\setminus\mathrm{F}$.

In order to take into account the release of latent heat of solidification W_{F} at the interface, condition (12) must be introduced in the system. A common weak formulation for crystal and melt is obtained by imposing continuity of any test-function $T'(\mathbf{x})$ on the interface, and by adding the contributions of the crystal (Ω_s) and the melt (Ω_l) in eq. (23). Boundary consists now of the outer surfaces of the crystal and the melt ($\partial\Omega_s$ and $\partial\Omega_l$), where conditions of the form (24) are imposed, and of the interface **F**, where the total outgoing flux q is the sum of the crystal and melt contributions. As explained

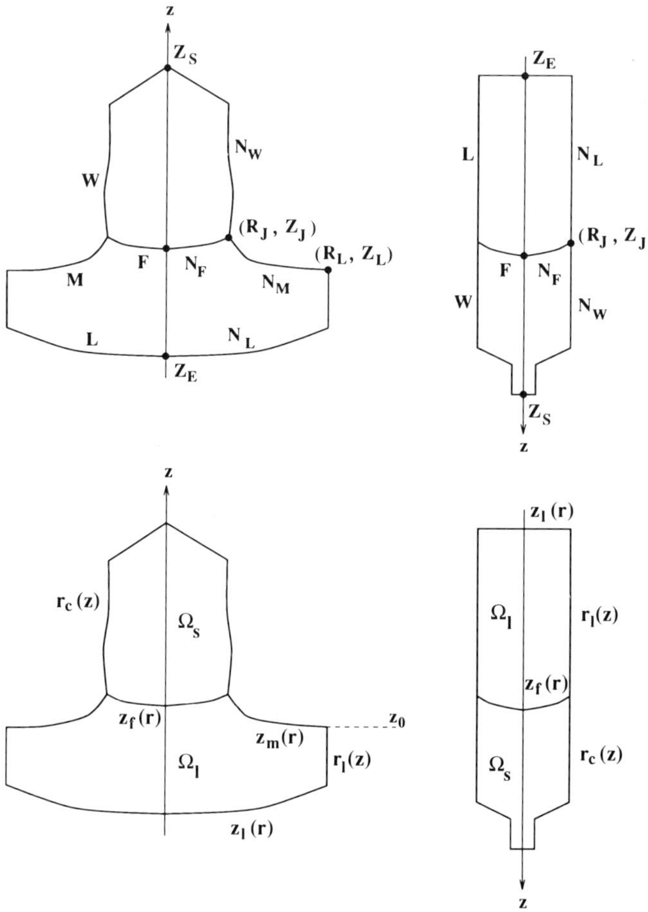

Fig. 7. Geometrical unknowns for the Cz and VB configurations.

in section 2.3, interface flux condition is written as

$$q + W_F = 0, \quad \mathbf{x} \in F, \tag{25}$$

where W_F is the product of the latent heat of fusion per unit mass ΔH by the specific mass ρ_s and the growth speed v_g of the solid phase. Note that, when using the numerical procedure developed by Brandon and Derby [67, 156], radiative fluxes are included in q, since integration by parts is carried out in eq. (23) on the total flux.

The next step of making equations discrete is space discretization, which is simply performed, when using the finite element method, by expressing the unknown temperature field as a sum,

$$T = \sum_{j=1}^{N_T} T_j(t)\phi_j(\mathbf{x}), \tag{26}$$

where symbols $T_j(t)$ and $\phi_j(\mathbf{x})$ stand for the N_T unknown nodal temperatures, which depend on time, and the associated shape functions, which depend both on position $\mathbf{x}$ and on the vector formed by the nodal coordinates defining the moving mesh (i.e. the union of X^u and X^a, as related by eq. (22)). Semi-discrete equations are easily obtained by introducing approximation (26) in the weak formulation (23) (taking eqs. (24) and (25) into account), and by applying Galerkin's method, which is exactly defined by using the shape functions as test functions,

$$T'(\mathbf{x}) = \phi_i(\mathbf{x}). \tag{27}$$

In most practical applications [51, 55, 62], shape functions have been selected as bilinear or biquadratic polynomials on the parent element, and Gaussian quadrature was used for evaluating the integrals [157, 158].

When performing space discretization, a particular treatment must be given to the solidification heat supply W_F, as defined in eq. (25). This problem will be analyzed in detail in the next section, where discretization of the interfaces will be examined. Let us however observe that growth speed is directly related in dynamic models to the motion of the nodes of the crystal mesh. In quasi- or pseudo-steady models, on the contrary, growth speed is related to the pull rate (v_{pul}) or to the heater translation rate ($-v_{pul}$), which will, at this stage, be considered as additional variables of the system.

In the case of dynamic simulations, the material derivatives involved in eq. (23) must be expressed taking into account the dependence with respect to time of both the nodal temperatures $T_i(t)$, and the nodal coordinates (which are components of X^u or X^a). This calculation is easily performed by using the formula of Lynch and Gray [159],

$$\frac{DT}{Dt} = \sum_i (\dot{T}_i(t)\phi_i(\mathbf{x}) + T_i(t)(\mathbf{v} - \mathbf{w})\cdot\nabla\phi_i), \tag{28}$$

where $\dot{T}_i(t)$ denotes the time derivative of the nodal temperature $T_i(t)$, while $\mathbf{v}$ and $\mathbf{w}$ stand for the velocity fields of the liquid or solid material, and of the deforming mesh, respectively, as calculated in a reference frame which is fixed with respect to the furnace, in global simulations, or with respect to the crucible or the ampoule, in local simulations. It is clear that $\mathbf{w}$ depends linearly on the components of $\dot{X}^u$ and $\dot{X}^a$.

The last step to execute in order to make equations discrete, in the dynamic case, is time discretization, which may be performed implicity or explicitly [160, 161]. Implicit methods are generally preferred to explicit ones in view of their better numerical stability, which allows one to use much longer discrete time intervals, although implicit systems are much more difficult to solve than explicit ones. The advantage of using a stable method usually compensates for these difficulties, because time step control is no longer related to complex numerical constraints (due to the presence of critical mesh dimensionless numbers). The most classical method is the backward Euler implicit scheme, which is strongly A-stable [160], and may be summarized as follows: if $\boldsymbol{F}(\dot{\boldsymbol{Y}}, \boldsymbol{Y}, t) = \boldsymbol{0}$ is a system of ordinary differential equations, with $\boldsymbol{Y}(t^0) = \boldsymbol{Y}^0$ as initial conditions, the algorithm consists in solving at each discrete time t^{n+1} the system

$$\boldsymbol{F}((\boldsymbol{Y}^{n+1} - \boldsymbol{Y}^n)/(t^{n+1} - t^n), \boldsymbol{Y}^{n+1}, t^{n+1}) = \boldsymbol{0}. \tag{29}$$

The Crank–Nicolson scheme has a higher precision, but lesser stability properties.

An efficient way to find the solution is provided by using a predictor–corrector technique, such as in [162]: the predictor scheme is (i) explicit (and so cheap), is (ii) of equal or higher order than the implicit scheme, and is (iii) unstable in the actual range of time steps used in the simulation; the predictor provides an initial guess $\hat{\boldsymbol{Y}}^{n+1}$ of the solution, which permits one to start Newton–Raphson iterations for finding the exact solution $\boldsymbol{Y}^{n+1}$ of system (29); control of the time step $(t^{n+2} - t^{n+1})$ is easy to perform by comparing the predicted and corrected solutions ($\hat{\boldsymbol{Y}}^{n+1}$ and $\boldsymbol{Y}^{n+1}$), in view of the known precisions of the predictor and corrector schemes (see also [93]).

At this stage, the situation may be summarized as follows: using the finite element technique, the equations governing heat transfer in the melt and the crystal have been put in the general form

$$\mathscr{T}(\dot{\boldsymbol{T}}, \boldsymbol{T}, \dot{\boldsymbol{X}}^{\mathrm{u}}, \boldsymbol{X}^{\mathrm{u}}, \dot{\boldsymbol{X}}^{\mathrm{a}}, \boldsymbol{X}^{\mathrm{a}}, \boldsymbol{V}, v_{\mathrm{pul}}) = \boldsymbol{0}, \tag{30}$$

where, for the sake of brevity, symbols $\boldsymbol{T}, \boldsymbol{X}^{\mathrm{u}}, \boldsymbol{X}^{\mathrm{a}}$, and $\boldsymbol{V}$ stand for the unknown nodal temperatures, coordinates, and velocities in the melt at time t^{n+1}, while $\dot{\boldsymbol{T}}, \dot{\boldsymbol{X}}^{\mathrm{u}}, \dot{\boldsymbol{X}}^{\mathrm{a}}$ stand for the discretized time derivatives of $\boldsymbol{T}$, $\boldsymbol{X}^{\mathrm{u}}$, and $\boldsymbol{X}^{\mathrm{a}}$, as obtained by using the Euler implicit method (e.g. $\dot{\boldsymbol{T}} = (\boldsymbol{T}^{n+1} - \boldsymbol{T}^{n})/(t^{n+1} - t^{n})$, etc.).

The following comments are important:

- There are exactly as many equations in system (30) as nodal temperatures (N_{T}). Hence, additional equations are necessary to close the discrete problem. The latter are associated with the unknowns $\boldsymbol{X}^{\mathrm{u}}$ and $\boldsymbol{V}$, and possibly v_{pul} – because nodal coordinates $\boldsymbol{X}^{\mathrm{a}}$ depend on $\boldsymbol{X}^{\mathrm{u}}$ through a relation such as eq. (22) – and correspond to the discrete fomulation of the melting temperature condition, of the relations providing the crystal and meniscus shapes, and of the equations governing convection in the melt. As explained in section 1.3, different cases must be considered, depending on whether the problem is quasi-steady, pseudo-steady or time-dependent, or whether VB, Cz or DGF growth is analyzed.
- In the case of quasi- or pseudo-steady calculations, all the time derivatives are set to zero in eq. (30). When calculating the material derivative $\mathrm{D}T/\mathrm{D}t$ defined by eq. (28), this means that both $\dot{T}_i(t)$ and the deforming mesh velocity $\mathbf{w}$ vanish. On the other hand, the velocity field $\mathbf{v}$ is evaluated as follows: in quasi-steady calculations, $\mathbf{v}$ is set to zero in the crystal while, in the melt, the normal component of $\mathbf{v}$ is constrained to vanish on the interface and the crucible, or the ampoule (where no-slip conditions hold). In pseudo-steady calculations, an additional convective term $(\mathbf{v}_{\mathrm{pul}} \cdot \nabla T)$ is introduced in $\mathrm{D}T/\mathrm{D}t$ in the crystal and the melt.
- In both quasi- and pseudo-steady calculations, the pull rate represents the axial velocity of the crystal with respect to the solidification front – and not with respect to the system of axes, as in dynamic simulations. This approach is the one which provides the most accurate results, as will be seen in the next section.

3.2.2. Interface conditions

As explained in section 2.3, interfaces consist of (i) the crystal–melt solidification front, (ii) the melt–environment meniscus, (iii) the wall of the crystal, and (iv) the melt–liner (or crucible, or ampoule) contact surface. Interface (ii) is not present in the case of VB or DGF growth. An additional interface, in the case of LEC growth, is the meniscus separating the encapsulant from the ambient gas. Calculation of this latter interface represents however essentially a decoupled problem, which is much simpler to solve than in case (ii) and will, for the sake of simplicity, not be analyzed in this section.

The *distinguished* solidification condition [51] states that crystal–melt interface is at the melting temperature (T_m), which can depend on solute or dopant concentration. The weighted residuals formulation of this equation has the form

$$\langle\langle (T - T_m)\chi_i \rangle\rangle_{F'} = 0, \tag{31}$$

where integration is performed on the projection F′ of the interface F onto the $z = 0$ plane, while symbol $\chi_i(\mathbf{x}')$ stands for the N_F shape functions of interface axial position (whose nodal values are components of the vector $\boldsymbol{X}^u$ – together with the crystal radius). Hence, the approximate solidification front has a discrete representation of the form

$$z_f = \sum_{j=1}^{N_F} z_{fj}(t)\chi_j(\mathbf{x}'). \tag{32}$$

It is worth noting that the shape functions $\chi_i(\mathbf{x}')$ may themselves depend on the crystal radius, and so on the vector $\boldsymbol{X}^u$. This particular Galerkin formulation, where test functions are selected as being the shape functions of interface configuration, permits one to use different discretizations $\phi_i(\mathbf{x})$ and $\chi_i(\mathbf{x}')$ for temperature field and interface shape (quadratic polynomials on the parent element were used in [55] for the $\chi_i(\mathbf{x}')$'s, while linear polynomials were used in [62]). Let us note that performing integration of eq. (30) on the interface F rather than on the projection F′ is also feasible and, in some sense, better, but not usual.

The general form of the discrete solidification condition (32) may be written as

$$\mathscr{F}(\boldsymbol{T}, \boldsymbol{X}^u, \boldsymbol{X}^a, \boldsymbol{C}) = \boldsymbol{0}, \tag{33}$$

where $\boldsymbol{C}$ stands for the column vector of unknown nodal concentrations in the melt, which may in this way be involved in the coupled system, even when solute distribution has no influence on melt convection. It is important to observe that eq. (33) represents a system having as many equations as nodal crystal–melt interface axial coordinates (N_F).

At this stage, it is also possible to calculate in detail the contribution of the solidification heat supply W_F to the weak formulation of the energy equation (23). In a first step, from eqs. (25) and (27), we write

$$\langle\langle qT' \rangle\rangle_F = \langle\langle q\phi_i \rangle\rangle_F = \langle\langle -W_F\phi_i \rangle\rangle_F = \langle\langle -\rho_s \Delta H v_g \phi_i \rangle\rangle_F. \tag{34}$$

In a second step, taking into account that the axial component of the growth rate is exactly the difference between the pull rate (v_{pul}) and the axial interface velocity ($\sum_{j=1}^{N_F} \dot{z}_{fj}\chi_j - z_{fj}\mathbf{w}\cdot\nabla\phi_j$), we obtain the searched term

$$\langle\langle qT' \rangle\rangle_F = \langle\langle -\rho_s \Delta H \sum_{j=1}^{N_F} (v_{pul} - \dot{z}_{fj}\chi_j + z_{fj}\mathbf{w}\cdot\nabla\phi_j) \rangle\rangle_{F'}, \tag{35}$$

where integration is carried out on the projection F′, instead of F. It is useful to recall that, in dynamic simulations, both the pull rate v_{pul} and the nodal interface axial velocities $\dot{z}_{fj}$ are components of the vector $\dot{\boldsymbol{X}}^{\mathrm{u}}$ while, in quasi- or pseudo-steady problems, $\dot{z}_{fj}$ and $\mathbf{w}$ vanish, and v_{pul} is an additional unknown. It is clear that, in all cases, the meaning of v_{pul} is exactly obtained by comparing the right-hand sides of eqs. (34) and (35).

A second interface, in Cz growth, is the meniscus M separating the melt from the ambient gas (or the encapsulant), where the influence of surface tension is essential. Following the approach of Orr et al. [163, 164], it is possible to express eq. (18) in a weak form. Let J and L denote the crystal–melt–environment, and liner–melt–environment contact lines, and $z'_{\mathrm{m}}(r)$ stand for a test-function defined on M. Letting also M′ denote the projection of M on the $z=0$ plane, we may obtain the following formulation:

$$\forall z'_{\mathrm{m}}(r) \text{ whose derivative has a summable square,}$$

$$\langle\langle \sigma_0(\boldsymbol{\nabla}_{\mathrm{m}} z_{\mathrm{m}} \cdot \boldsymbol{\nabla}_{\mathrm{m}} z'_{\mathrm{m}})\rangle\rangle_{\mathrm{M}} + \langle\langle(\rho_l - \rho_{\mathrm{e}})g(z_{\mathrm{m}} - z_0)z'_{\mathrm{m}}\rangle\rangle_{\mathrm{M}'} + \langle\langle\langle \sigma_0 z'_{\mathrm{m}} \cos\phi\rangle\rangle\rangle_{\mathrm{J}\cup\mathrm{L}} = 0, \tag{36}$$

where the symbol $\boldsymbol{\nabla}_{\mathrm{m}}$ denotes the gradient along the surface of the meniscus, while ϕ stands for the angle formed by the tangent to the meniscus with the vertical direction (fig. 5) (recall that z_0 is the theoretical axial position at which meniscus curvature vanishes). In the case of an axisymmetric problem, calculation of the different terms of eq. (36) must be performed carefully:

(i) The integral $\langle\langle\rangle\rangle_{\mathrm{M}}$ is carried out on the surface M of the meniscus, which means that a weight $r[1+(\mathrm{d}z_{\mathrm{m}}/\mathrm{d}r)^2]^{1/2}$ must be used.

(ii) The expression $\boldsymbol{\nabla}_{\mathrm{m}} z_{\mathrm{m}} \cdot \boldsymbol{\nabla}_{\mathrm{m}} z'_{\mathrm{m}}$ is equal to

$$\frac{\mathrm{d}z_{\mathrm{m}}/\mathrm{d}r}{[1+(\mathrm{d}z_{\mathrm{m}}/\mathrm{d}r)^2]^{1/2}}\,\frac{\mathrm{d}z'/\mathrm{d}r}{[1+(\mathrm{d}z_{\mathrm{m}}/\mathrm{d}r)^2]^{1/2}}.$$

(iii) An expression such as $\langle\langle\sigma_0 z'_{\mathrm{m}} \cos\phi\rangle\rangle_{\mathrm{J}}$ reduces to the product of the radius of the contact line R_{J} by the value of $\sigma_0 z'_{\mathrm{m}} \cos\phi$ on **J**.

Boundary conditions are defined by imposing the angle ϕ_{L} (whose value is generally set to 90°, since its influence is minor), and by relating the value of the angle ϕ_{J} to the growth angle ϕ^* and the wetting angle ϕ_0 (as defined in section 2.3). From fig. 5, the growth angle also obeys the condition

$$\tan\phi^* = \left.\frac{\mathrm{d}r_{\mathrm{c}}}{\mathrm{d}z}\right|_{z=Z_{\mathrm{J}}}, \tag{37}$$

where, at a given time, the curve $r=r_{\mathrm{c}}(z)$ represents the shape of the grown crystal, while $(R_{\mathrm{J}}, Z_{\mathrm{J}})$ are the actual coordinates of the tri-junction (which are components of the vector $\boldsymbol{X}^{\mathrm{u}}$). Hence, according to this discussion, contact line conditions are written as

$$\phi_{\mathrm{L}} = \tfrac{1}{2}\pi, \qquad \phi_{\mathrm{J}} = \phi_0 + \tan^{-1}\left(\left.\frac{\mathrm{d}r_{\mathrm{c}}}{\mathrm{d}z}\right|_{z=Z_{\mathrm{J}}}\right). \tag{38}$$

The weak formulation (36) must be closed by substituting the right-hand sides of eqs. (38) for ϕ_L and ϕ_J in $\langle\langle\langle\sigma_0 z'_m \cos\phi\rangle\rangle\rangle_{J\cup L}$, and by imposing the total mass M_0 of crystal and melt, in order to determine the Lagrange multiplier z_0,

$$\langle\rho\rangle_{\Omega_s\cup\Omega_l} = M_0. \tag{39}$$

The right-hand side of eq. (39) must be calculated from the knowledge of the three interfaces, as represented at a given time by the curves $z = z_f(r)$, $z = z_m(r)$, and $r = r_c(z)$.

Again, the next step is space discretization, which, applying the finite element method, yields the following approximation for meniscus shape:

$$z_m = \sum_{j=1}^{N_M} z_{mj}(t)\chi_j^*(\mathbf{x}'), \tag{40}$$

where the N_M axial positions $z_{mj}(t)$ are components of the vector X^u, while the shape functions $\chi_j^*(\mathbf{x}')$ depend implicitly on the crystal radius R_J (and so on X^u). As in the case of the energy equation, the Galerkin technique is defined by selecting the shape functions as test functions,

$$z'_m = \chi_i^*(\mathbf{x}'). \tag{41}$$

In the case of a time-dependent Cz problem, the distinction between a growing or a remelting crystal is critical (recall that, even during growth, this latter situation may occur during short time intervals when curvature of crystal shape – as represented by the function $r_c(z)$ – is high, while the crystal is locally convex).

Let us first consider the normal situation, i.e. an instant when the crystal is growing at the tri-junction. The technique developed by Derby et al. [116] is based on discretizing eq. (38) after having performed the following substitution:

$$\left.\frac{dr_c}{dz}\right|_{z=Z_J} = \frac{\dot{R}_J}{\dot{Z}_S - \dot{Z}_J}, \tag{42}$$

where Z_S stands for the axial position of the seed extremity of the crystal, which means that $\dot{Z}_S$ is simply the pull rate, and that $\dot{R}_J$ and $\dot{Z}_S - \dot{Z}_J$ are the radial and axial growth velocities at the tri-junction. Equation (42) is made discrete by using the Crank–Nicolson implicit scheme. In order to maintain a sufficient accuracy, it is essential to control strictly the axial lengthening of the element sides of the crystal which are located just above the tri-junction. This control is achieved in [56] by reducing the lift of a given number of "layers" of nodes situated just above the solidification front. A significant axial lengthening is allowed only for the other elements of the crystal. An example of such mesh deformation is shown in fig. 8(a). The importance of substitution (42) may be explained as follows: let us suppose that a given time step is such that the increase of axial length $(Z_S^{n+1} - Z_J^{n+1}) - (Z_S^n - Z_J^n)$ is clearly lower than the length of the element side which separates the crystal from its environment at the tri-junction; in this case, it is clear that calculation of the growth angle would be strongly falsified if $(dr_c/dz)|_{z=Z_J}$ were simply calculated from the mesh positions at time t^{n+1} (explicit calculations would of course be incorrect, since the growth angle ϕ^* could not change with time).

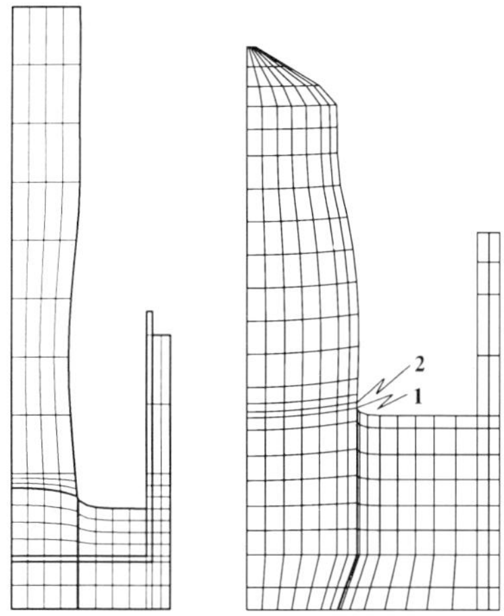

Fig. 8. Different techniques used for finite element mesh deformation: (a) as defined from [60]; (b) as defined from [65] (1. actual tri-junction location; 2. tri-junction location on the crystal wall for the previous discrete time).

The strategy of Van den Bogaert [66], and Van den Bogaert and Dupret [65, 165] is slightly different. With a view to capture accurately the radiative transfer taking place in the vicinity of the tri-junction – which is expected to play an important role – and in order to develop an implicit algorithm for radiative calculations in global simulations (as will be explained in section 5.3), the method is based on using a scheme where space and time discretizations are coupled: the material point located in the solid phase at the tri-junction at time t^n is constrained to occupy at time t^{n+1} the nodal position which is situated just above the tri-junction (fig. 8(b)). As a consequence, the implicit calculation of $(dr_c/dz)|_{z=Z_J}$ at time t^{n+1} provides exactly the same result as formula (42). It is essential to observe that this approach means that the algorithm is no longer based on integrating with respect to time a so-called semi-discrete scheme (i.e. a scheme discretized with respect to space only). The use of a coupled space–time discretization has important numerical effects, which are analyzed in [66]. The keypoint is to forbid during simulation any change of time step which would not be progressive in time (or smooth). Indeed, in formula (28), the nodal mesh velocities $\mathbf{w}$ must remain well-defined. Global results obtained by using method of [66, 65, 165] are given in section 5.4.

The situation is completely different when a short remelting occurs at the tri-junction. In this latter case, we have explained in section 2.3 that the wetting angle ϕ_0 may no longer be imposed on the contact line, whose position (R_J, Z_J) is constrained to be located on the known crystal–environment interface. The numerical technique to employ in the case of back-melting is however quite simple: the natural boundary condition holding on the contact line J must be replaced by an essential condition;

according to classical Galerkin technique, the latter will be defined by letting the test-function z'_{m} vanish at J,

$$z'_{\mathrm{m}}(\mathrm{J}) = 0, \tag{43}$$

and by forcing the contact line J to be located on the wall of the crystal,

$$R_{\mathrm{J}} = r_{\mathrm{c}}(Z_{\mathrm{J}}). \tag{44}$$

This last condition will be examined more precisely when we will analyze how to impose a discrete shape to the crystal. Let us also observe that remelting requires a modification of the space–time coupled technique described above (this is in fact very easy).

In summary, from eqs. (36) to (44), the $N_{\mathrm{M}} + 1$ equations governing the meniscus shape and imposing the total mass of the load may be written in the general form

$$\mathscr{M}(\dot{\boldsymbol{X}}^{\mathrm{u}}, \boldsymbol{X}^{\mathrm{u}}, \boldsymbol{X}^{\mathrm{a}}, z_0) = \boldsymbol{0}. \tag{45}$$

The use of r as the independent variable for representing the meniscus shape by a function $z_{\mathrm{m}}(r)$ imposes a restriction to the available range of growth angles ϕ^* (as defined in eq. (37)) for, in eq. (38), the angle ϕ_{J} must remain positive – otherwise, $z_{\mathrm{m}}(r)$ would be a multi-valued function [66]. This is however no problem, since the weak formulation (36), with extremity conditions (38), or (43) and (44), allows one to make an arbitrary selection of the independent variable for representing the configuration of the meniscus z_{m}. (Actually, this formulation may be obtained by applying the Ritz principle of minimum potential energy to the search of the meniscus shape. Potential energy consists of energy within the meniscus, due to surface tension, gravitational energy, and energy within the surface of the crystal and liner walls, again related to surface tension effects [166].)

The third interface to be discretized is the wall of the crystal. In Cz growth, we must impose that crystal shape does not change with time. The solution of Derby and Brown [56] is based on establishing an equation of motion for the crystal radius $r_{\mathrm{c}}(z, t)$,

$$\frac{\partial r_{\mathrm{c}}}{\partial t} + \dot{Z}_{\mathrm{J}} \frac{\partial r_{\mathrm{c}}}{\partial z} = 0, \tag{46}$$

which is easily integrated using a weighted residuals approach. The solution of Van den Bogaert [66] (which is slightly more accurate) is to record during simulation the successive pairs $(Z_{\mathrm{S}}^{\mathrm{i}} - Z_{\mathrm{J}}^{\mathrm{i}}, R_{\mathrm{J}}^{\mathrm{i}})$, which characterize the shape of the grown crystal, and to reconstruct the crystal shape at each discrete time t^{n+1} by interpolating between the previously collected values (except at the tri-junction, when crystal is growing). In both cases, the knowledge of the axial position of the seed extremity of the crystal Z_{S} (or the pull rate $\dot{Z}_{\mathrm{S}}$) is required. Hence, in summary, if the number of nodes $(r_{\mathrm{w}i}(t), z_{\mathrm{w}i}(t))$ located on the wall of the crystal is N_{W} (including the seed extremity of the crystal, on the axis, and the tri-junction), we have obtained a set of $N_{\mathrm{W}} - 1$ equations of the general form

$$\mathscr{W}(\dot{\boldsymbol{X}}^{\mathrm{u}}, \boldsymbol{X}^{\mathrm{u}}, \boldsymbol{X}^{\mathrm{a}}) = \boldsymbol{0}. \tag{47}$$

A peculiar situation is found when crystal surface presents a sharp edge. In this case, an additional equation will be introduced in system (47), since the edge must be the location of a given node. Similar, but much simpler relations, imposing that the crystal remains in contact with the ampoule, are easily found in VB or DGF growth. Note that eqs. (47) include the relation imposing a given value to Z_S or $\dot{Z}_S$, but not any relation holding at the tri-junction. It is clear that the radii $r_{wi}(t)$ and the axial position of the seed extremity of the crystal $Z_S(t)$ are components of $\boldsymbol{X}^u$.

The last geometrical relations we have to consider are the equations imposing that the melt remains in contact with the upward moving crucible (in Cz growth), or the ampoule (in VB or DGF growth),

$$\mathscr{L}((\dot{\boldsymbol{X}}^u), \boldsymbol{X}^u, \boldsymbol{X}^a) = 0. \tag{48}$$

Their number is exactly the number N_L of nodes located on the melt–liner (or crucible, or ampoule) interface. Again, if this surface presents a sharp edge, an additional equation will be introduced in the system. Equations (48) include a relation imposing the axial position Z_E of the interface, or its derivative $\dot{Z}_E$.

3.2.3. Summary

The best way to summarize the situation is to consider the purely conductive model, where melt convection, segregation,... are supposed to play a negligible role.

Consider first the dynamic local problem for a Cz furnace. This case may serve as a starting point for the discussion. Assuming that calculations have been performed up to a discrete time t^n, the basic unknowns are the N_T nodal temperatures and the components of the vector $\boldsymbol{X}^u$ at time t^{n+1}. The latter consists of $N_F + N_M - 1$ nodal interface and meniscus axial coordinates (since the tri-junction is common to the solidification front and meniscus), of the vertical position of the seed extremity of the grown crystal, of $N_W - 1$ nodal radii on the wall of the crystal (including the radius at the tri-junction), and of N_L nodal, either axial or radial, coordinates on the liner, that is to say a total of $N_F + N_M + N_W + N_L - 1$ components for $\boldsymbol{X}^u$. All the remaining nodal coordinates of the moving mesh belong to $\boldsymbol{X}^a$. Clearly, we have a single unknown belonging to $\boldsymbol{X}^u$ at the top of the crystal, at the crystal–melt-axis intersection, and at the bottom of the melt, a pair of unknowns belonging to $\boldsymbol{X}^u$ both at the tri-junction and on the meniscus–liner contact line, and a pair of unknowns, one belonging to $\boldsymbol{X}^u$ and the other to $\boldsymbol{X}^a$ (or both to $\boldsymbol{X}^u$ at sharp edges), on any other node of the four interfaces. In addition, we have the unknown theoretical position z_0 at which meniscus curvature vanishes. In order to find these $N_T + N_F + N_M + N_W + N_L$ unknowns, we have exactly the same number of equations given by relations (30), (33), (45), (47), and (48). The moving mesh provides an equal number of additional unknowns (i.e. the components of $\boldsymbol{X}^a$) and equations (from relations (22)). The algebraic discrete problem is, therefore, closed.

Although many unknown nodal coordinates may (and should) be calculated at the outset, such as several node positions at the melt–liner surface, solving this highly non-linear system represents of course a complex task. The best technique is probably to iterate on successive approximations of the solution by means of the Newton–Raphson or the modified Newton–Raphson [167] schemes (in this latter case, the Jacobian matrices are updated regularly, but not at each iteration, which may result in a significant gain of computing time [57, 59]). It is worth observing that the usual band-like structure of the Jacobian matrices obtained when using the finite element method is lost here, as a consequence of the presence of free boundaries. Solving the successive linear systems provided by the Newton algorithm is, therefore, more expensive, especially when additional unknown fields are involved. In particular, using a band matrix method demands that one develops a peculiar numbering strategy in order to maintain the computational cost at a reasonable level. (To this end, it was observed by Derby and Brown [114] that Jacobian matrices arising from free-boundary problems have a typical arrow-like structure provided the unknowns related to nodal coordinates be numbered after the nodal temperatures. It is essential in practice to take full account of such structure of the Jacobian matrix.) Let us note that interesting alternatives to band matrix methods are the frontal method [168], which requires to select an appropriate numbering of the elements before the calculations, and the minimum degree ordering (MDO) algorithm [169, 170].

The quasi-steady local Cz problem is very similar to the dynamic one. However, an additional unknown – the pull rate – is present in the system and requires a particular treatment. Moreover, the shape of the grown crystal which, in dynamic calculations, is provided by the past history of growth is now given at the outset. Hence, imposing simply the pull rate without other modifications of the equations (where nodal time derivatives are set to zero) provides poor information, since there is a strong inconsistency in imposing the crystal shape everywhere, but not the growth angle at the tri-junction. A solution to this problem [42–45] is obtained by letting the pull rate be an unknown of the system, while imposing the growth angle at the tri-junction. In particular, when the grown crystal is of cylindrical shape, this means that the radius at the tri-junction R_J is given. Hence, the shape of the melt–environment meniscus can be calculated at the outset of the simulation. The sole remaining geometrical unknowns are then the nodal interface vertical positions, except at the tri-junction, which strongly facilitates the solution of the problem. It is worth mentioning that, when local quasi-steady simulations are coupled with the calculation of melt convection, their interest remains very high in order to provide a better understanding of the physics of the melt flow and related effects (flow patterns, transition to periodic or chaotic behaviour, segregation, etc...).

The quasi-steady local model of Derby et al. [55] is slightly different. While pull rate is imposed, crystal shape is given only up to one degree of freedom. (When crystal shape is cylindrical, this means that the cylinder radius is an unknown.) This approach has the disadvantage that meniscus shape may no longer be calculated at the outset of the simulation. There is however an obvious interest in obtaining some information on the radius of the grown crystal.

In the case of VB or DGF furnaces, local problems are very similar to that of Cz growth, but much simpler. As a matter of fact, the only nodal coordinates which cannot be imposed at the outset are the crystal–melt interface axial coordinates – including the tri-junction – and the related nodal positions of the moving mesh. The balance between equations and unknowns is the same for the dynamic and the pseudo-steady models, and is straightforward. In the dynamic model, the axial motion of the ampoule (VB growth), or the evolution of the power supply to the heating elements (DGF growth) are taken into account by introducing appropriate thermal boundary conditions of the form (24) on the outer surface of the ampoule. In the pseudo-steady model, the above dynamic effects are modelled (i) through the solidification heat supply W_F, eqs. (34) and (35), and (ii) by means of introducing an additional convective term in the energy equation. (Recall that, in a pseudo-steady model, melt is supposed to enter one extremity of the ampoule at a uniform velocity, while the solidified material is continually removed from the ampoule at the other extremity. When available, pseudo-steady models provide better results than quasi-steady ones, for the behaviour of the solution in the vicinity of the solidification front takes into account the motion of the material with respect to this interface.)

Examples of dynamic solutions of the local Stefan problem will be presented in the next section, while sections 3.4 and 3.5 will be devoted to introducing melt convection and related effects in the local models. Let us however observe that the balance between equations and unknowns will remain equilibrated, since any additional field will bring exactly an equal number of supplementary equations and unknowns.

3.3. Examples of local dynamic simulations without convection

The first model for analyzing the dynamics of Cz growth, including calculation of crystal radius, was developed by Crowley [54] for the case of a germanium crystal. A local finite difference model was used to predict temperature field, solid–liquid interface shape, meniscus shape and crystal radius evolution as a function of an imposed meniscus base temperature. The results suggested process instability, since crystal radius was found to increase steadily outwards. This first time-dependent model was recently used by Hurle et al. [171] for studying the crystal radius response to a modulation of the heater power and compare it with experimental results and analytical calculations.

The first comprehensive local model, including time-dependent heat transfer by conduction through the crystal, melt, liner and crucible, and calculation of all the interfaces, was developed by Derby and Brown [56] on the basis of previous quasi-steady analyses [55, 114, 115]. Details of radiative heat transfer from the crystal, the melt and the crucible were still omitted, as an idealized radiative transfer from all surfaces towards a uniform ambient temperature was assumed. Boundary conditions along the crucible external wall were however derived from radiation towards a heater with a different temperature. In a first stage, a quasi-steady-state process (with long crystals and a constant melt volume) was considered. Solutions were then calculated by setting all time derivatives to zero in the time-dependent model and computing crystal radius so that tri-junction temperature equals the melting temperature. Interesting examples of quasi-steady state results concern the dependence of crystal radius on the imposed pull rate. This is represented in fig. 9 for the case of a silicon crystal. We observe that crystal radius decreases with increasing pull rate. For values of the non-dimensional pull rate exceeding a critical value of 2.85, no steady-state solution can be found, meaning that solidification heat at the solid–liquid interface becomes too large to be removed by conduction through the crystal. For some values of the pull rate, two different crystal radii could be found.

The stability of the process was examined. The quasi-steady states associated with

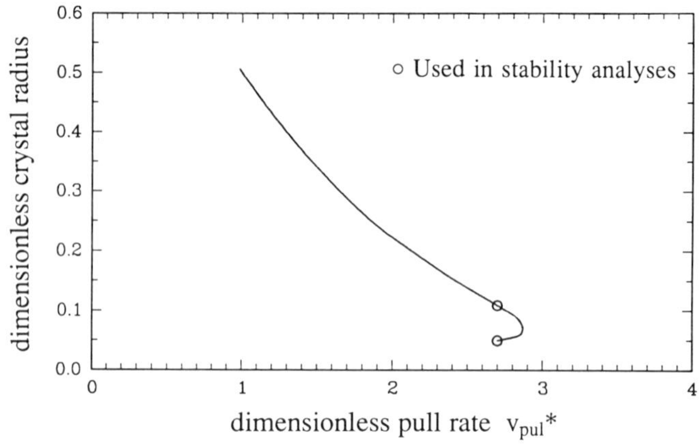

Fig. 9. Local dynamic simulation of Cz silicon growth. Crystal radius as a function of the pull rate, as predicted by a quasi-steady state model. We observe the existence of a critical value of the dimensionless pull rate (2.85) beyond which no solution can be found (From [56]).

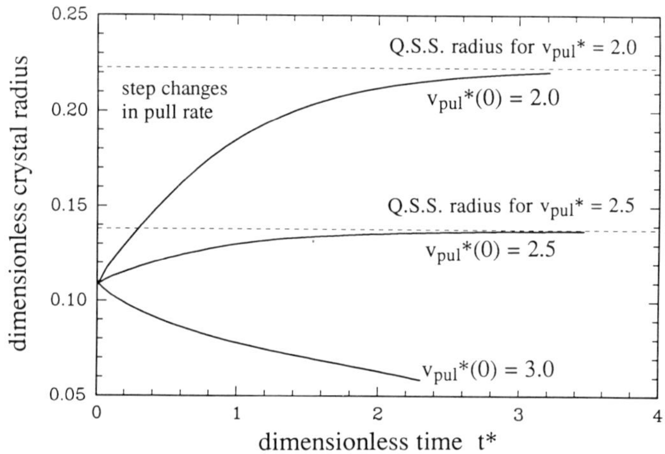

Fig. 10. Same problem as in fig. 9. Evolution of the crystal radius as a result of a step change in pull rate for a continuous-processing Cz system (constant melt volume). For pull rates higher than the critical one, the crystal radius decays towards zero (From [56]).

the radii located on the lower branch of the curve were proved to be unstable, while the states on the upper branch were proved to be stable, as shown by a linear stability analysis. The non-linear stability was investigated as well, by imposing a step change in the pull rate and analyzing the time evolution of the resulting crystal radius (fig. 10). Calculations were performed by using the dynamic model with a constant melt volume and a fixed crystal length, in order to factor out the transient behaviour of the system arising from the changing heat transfer in the melt. An upper branch steady-state corresponding to a pull rate of 2.7 (which was proved to be stable by the linear stability analysis) was chosen as the initial condition. When the pull rate is decreased, the radius evolves towards the new quasi-steady state defined by the new pull rate. When the pull rate is increased towards a value exceeding the critical value defined here above, the radius decreases towards zero, confirming the non-existence of any quasi-steady state for such a pull rate. Transient calculations performed from an initial state located on the lower branch of the curve always converge towards the corresponding solution located on the upper branch, confirming instability of the operating states on the lower branch.

Surek [41] originally showed that the system is unstable if one considers the effects of crystal shape related to meniscus shape only, without coupling the problem to the heat transfer conditions. He also suggested [144] that constraints due to heat transport would be able to stabilize the system if the crystal radius is not too small. The present results suggest that the Cz process is inherently stable for most cases. Let us note that the question of stability of the entire process is better addressed if the effects of geometrical changes are eliminated from the system. The solution of Derby and Brown [56] was to consider an ideal case where melt volume is constant and crystal length is fixed. Another method is suggested in section 5.4. If geometrical changes are not factored out, the radius cannot stabilize, and results can hardly be interpreted from a stability point of view. This is the reason for the apparent instability resulting from the Crowley [54] calculations. As a matter of fact, radius evolution was merely dictated by batch effects of crystal lengthening and was independent of any unstable behaviour.

The first batchwise simulations of Cz growth, taking actual melt volume decrease and crystal length increase into account, also appear in the work of Derby and Brown [56]. Figure 11 shows the resulting crystal shapes and temperature fields for three cases: without control, with an integral controller, and with a proportional/integral controller. We observe that uncontrolled growth leads to a steadily increasing radius as a

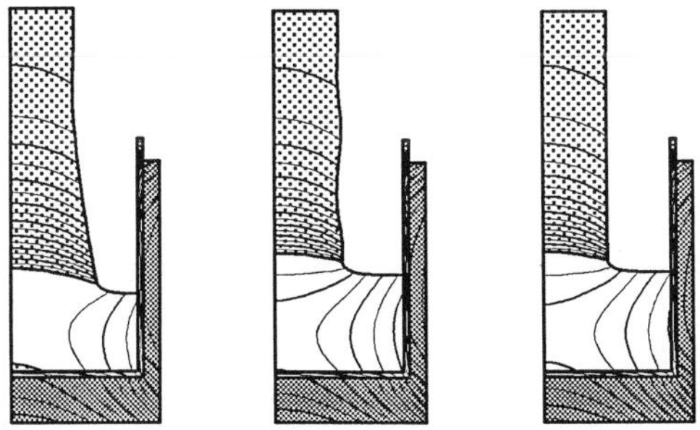

$t^* = 1.28$ $t^* = 1.28$ $t^* = 1.27$

(a) (b) (c)

Fig. 11. Same problem as in fig. 9. Crystal shape and temperature field for a batchwise simulation of Cz growth: (a) without control; (b) with integral control; (c) with proportional/integral control. Without control, a steadily increasing radius is found (From [56]).

result of the continually changing configurations of melt and crystal (remember that this cannot be interpreted as a system instability). Addition of a proportional/integral controller succeeded in obtaining an almost cylindrical crystal shape. The oscillating radius observed with integral control only was found to be due to the controller and to be independent of any batch effects. Extension of this local model with more detailed radiation calculations will however prove that, without any control, natural oscillations may still exist, as a result of the interactions between radiative heat transfer and crystal shape (see section 5.4).

Quasi-steady and time-dependent batchwise results can be compared for any given stage, as was done by Derby and Brown [107]. The validity of the quasi-steady state assumption is shown to be limited when significant changes of crystal shape occur, since the transient behaviour of the system is important under such conditions. On the other hand, even when geometrical changes are slow enough to allow for a quasi-steady assumption, apparent differences between the two approaches may arise from the fact that solidification heat at the solid–liquid interface is proportional to the "pull rate" in the quasi-steady model, while it depends on the actual growth rate in the time-dependent model. The latter may differ considerably from the real pull rate in a system where the crucible is held at a fixed position, as the effect of melt level decrease must be considered. Hence, according to the discussion of the previous section, the concept of "pull rate" must be interpreted carefully in quasi-steady Cz simulations. This problem is not found in the case of VB or DGF growth.

3.4. *Calculation of melt convection and segregation*

The numerical simulation of melt convection has represented a major objective in bulk growth modelling, in view of the complexity of the non-linear and, in some cases, unsteady and/or three-dimensional behaviour of the flow, and its important influence on stress generation and species distribution in the crystal. It is beyond the scope of this chapter to address this problem completely, and we refer the reader to Chapters 10, 13 and 14 of Volume 1, to relevant chapters in this volume, and to the reviews of [18, 20, 79, 172] and the literature therein, for additional detail.

The easiest (but not necessarily the most efficient) way to couple convection, segregation, and heat transfer calculations for a problem presenting several free boundaries and involving many constituents, is to use the finite element method [52, 53, 63, 64, 173], in view of its high geometrical flexibility. We will here follow this approach, but the interest of alternative methods will be stressed. Even in the frame of finite elements, numerous techniques have been developed in order to solve the fundamental difficulty of simulating convection at high Reynolds and Grashof numbers. Indeed, in Cz or VB growth, momentum equations are dominated by advection and have, therefore, a hyperbolic-like behaviour at the actual values of the dimensionless numbers. Moreover, heat transfer in the melt is itself advection-dominated in oxide growth, in view of the high value of the Prandtl number. The role of viscosity is essential in bulk growth for the precise reason that melt flow takes place in a limited domain. This means that, in the laminar case, the boundary layers which develop on the walls (and where viscous forces are non-negligible) have a strong influence up to within the core of the melt (see the work of Batchelor [174]). When flow is turbulent, this effect is further increased, since turbulence forms mainly on the solid walls. Similar problems are found in the energy and diffusion equations when the Péclet numbers are high. (It is worth noting that, in external flows without recirculation, the situation is completely different, for the influence of the incoming fluid is dominant.) As a consequence of these problems, melt flow calculations are very difficult to perform in bulk growth simulations, and major errors may occur when mesh refinement is not sufficient (especially in the vicinity of the walls).

In order to solve the problem by means of the finite element method, the first step is to put the system into a weak form. According to the classical technique, it is easy, for a laminar flow, to obtain from eqs. (6) and (7) the formulation

$\forall \mathbf{v}'(\mathbf{x})$ whose gradient has a summable square,

$\forall p'(\mathbf{x})$ whose square is summable,

$$\left\langle \left\{\rho\frac{\mathrm{D}\mathbf{v}}{\mathrm{D}t} - \mathbf{j}\times\mathbf{B} - \rho\mathbf{g}[1-\beta_T(T-T_0)+\beta_C(C-C_0)]\right\}\cdot\mathbf{v}' - p\nabla\cdot\mathbf{v}' - p'\nabla\cdot\mathbf{v} + \tfrac{1}{4}\mu(\nabla\mathbf{v}+\nabla\mathbf{v}^{\mathrm{T}}):(\nabla\mathbf{v}'+\nabla\mathbf{v}'^{\mathrm{T}})\right\rangle_{\Omega_1} - \langle\langle \mathbf{t}\cdot\mathbf{v}'\rangle\rangle_{\partial\Omega_1} = 0, \tag{49}$$

where the contact force $\mathbf{t}$ is defined in eq. (16). The natural boundary condition (17) must be introduced in the system by substituting in eq. (49) the right-hand side of eq. (17) for the tangential component of $\mathbf{t}$. All the remaining boundary conditions, such as eqs. (9), (10), and (15), are essential and must be imposed, while the corresponding components of the test-function $\mathbf{v}'$ are constrained to vanish on $\partial\Omega_1$. In the case of LEC growth, the coupling between melt and encapsulant flows must be performed by imposing contact and surface tension force balance on the separating interface, instead of condition (17).

When the distribution of a solute must be calculated, the weak formulation of eq. (8) is easily found to be

$\forall C'(\mathbf{x})$ whose gradient has a summable square,

$$\left\langle \rho\frac{\mathrm{D}C}{\mathrm{D}t}C' + \rho\frac{D}{\alpha}\nabla(\alpha C)\cdot\nabla C'\right\rangle_{\Omega_1} + \left\langle\left\langle\left(\rho\frac{D}{\alpha}\nabla(\alpha C)\cdot\mathbf{n}\right)C'\right\rangle\right\rangle_{\partial\Omega_1} = 0, \tag{50}$$

where all the boundary conditions are natural or mixed, except when the solute concentration is imposed on the wall of the crucible (such as in the case of oxygen diffusion in Cz Si growth). In particular, eq. (14) must be introduced in formulation (50), taking eq. (13) into account in order to eliminate the concentration in the solid C_s.

When a magnetic field is used, the weak formulation of eqs. (19) and (20), which govern the electric potential, is

$\forall \phi'(\mathbf{x})$ whose gradient has a summable square,

$$\langle \sigma_c(\nabla\phi - \mathbf{v}\times\mathbf{B})\cdot\nabla\phi'\rangle_\Omega - \langle\langle\sigma_c(\nabla\phi - \mathbf{v}\times\mathbf{B})\cdot\mathbf{n}\phi'\rangle\rangle_{\partial\Omega} = 0. \tag{51}$$

Here, calculation of the electric potential must be performed in all the constituents where electrical conductivity is non-negligible. Boundary conditions are always natural [157], and take place on the surfaces where the normal component of electric current vanishes. According to Hjellming and Walker [103, 104], the low, but non-vanishing electrical conductivity of the crystal must be taken into account in the model. Let us also observe that flow calculations are much easier to perform when a magnetic field is used, since the importance of advective terms in the equations is strongly reduced.

From weak formulations, equations are easily made discrete in space and time. Basically, the procedure is the same as for the energy equation and will thus not be detailed. Several kinds of elements may be introduced [175]. When using the Galerkin method, test-functions are selected as being the shape functions. However, this approach, while providing accurate results when it works, has the strong drawback that extremely refined meshes are required in order to perform simulations at the actual values of the dimensionless numbers [90, 91, 176, 177], since boundary layers become extremely thin (assuming again that flow is laminar). The streamline-upwinding techniques (SU or SUPG) [178] seemed to be a solution to this problem, but it was found that numerical viscosity can induce dramatic changes in the result (as was observed by Adornato and Brown [90, 91], and Ryckmans [176]). This problem may be related to the fact that, in internal flows, boundary layers influence the whole solution [174].

Examples of finite element local and global simulations will be analyzed in sections 3.5, 4.4, 5.2, and 5.4. Let us mention here the work of Crochet et al. [52, 179], Dupont et al. [180], and Wouters et al. [173] on horizontal Bridgman growth, of Crochet et al. [68] on Cz growth, and of Fontaine and Randriamampianina [181], and Fontaine et al. [137, 138] on LEC growth (where particular attention was turned to the influence of the encapsulant flow). All these contributions address the problem of the onset of an oscillatory two-dimensional or axisymmetric flow in a simplified geometry, except [180], which is devoted three-dimensional steady calculations.

In other respects, in order to solve the difficulties related to simulating high Reynolds and/or Grashof number flows, control volume methods [182] were introduced by Sabhapathy and Salcudean [183–185], and Salcudean et al. [186], for analyzing steady axisymmetric LEC problems with or without magnetic fields (including calculations in the encapsulant and the crystal). Finite volume methods were also developed by Bottaro and Zebib [187], and Leister and Perič [188], in order to simulate the unsteady three-dimensional flow in a simplified Cz geometry.

The principal effort in bulk flow calculations was however directed to the development of finite difference methods. In particular, Langlois et al. [189], Kim and Langlois [190, 191], and Williams et al. [192] used a Svanberg vorticity/stream function/ swirl/ temperature formulation to analyze the influence of axial, cusp, or weakly transverse magnetic fields on axisymmetric flow oscillations, and dopant or oxygen segregation in Cz growth (it is worth noting that appropriate electrical boundary conditions on the solidification front were introduced in the above mentioned contributions). A vorticity/ stream function/swirl/temperature formulation was developed by Organ and Riley [193], and Hicks et al. [194] for a steady axisymmetric analysis of oxygen transport in magnetic Cz growth, while a velocity/pressure/temperature formulation was used by Hjellming [195] in the case of isorotation of crystal and crucible. Let us also mention the contributions of Mihelčić et al. [196], Mihelčić and Wingerath [24], and Hirata and Hoshikawa [197], who developed variants of the MAC method [198] in order to perform transient three-dimensional simulations of Czochralski silicon melt flow with an axisymmetric magnetic field. In addition, Crespo del Arco and Bontoux [199], Ermakov et al. [200], and Ben Hadid et al. [201–203] developed accurate ADI methods in order to analyze the transient flow structure in two-dimensional rectangular, or three-dimensional cylindrical geometries without a magnetic field. Similar methods were used by Kaddeche et al. [263] to investigate segregation in a horizontal Bridgman configuration.

Spectral methods [204, 205] were also investigated by Crespo del Arco et al. [206–208], Pulicani et al. [209], and Fikri et al. [210]. A benchmark comparison between the performances of various methods was drawn at a GAMM workshop [70],

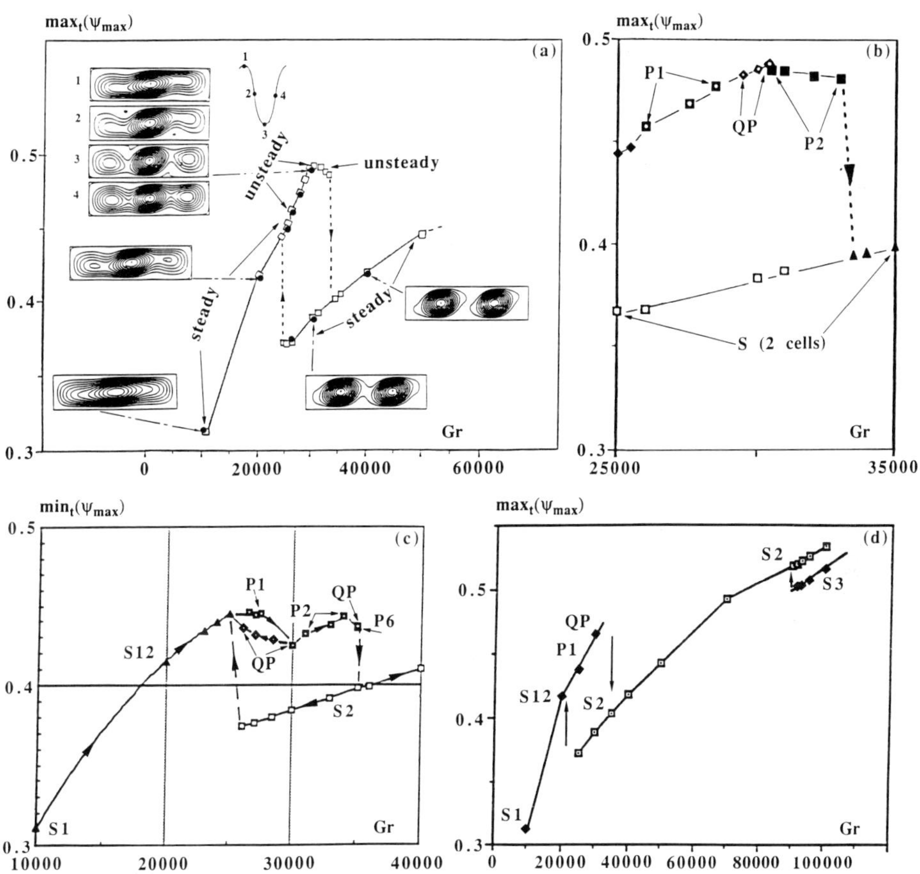

Fig. 12. Maximum value of the stream function, as a function of the Grashof number, for the case of a two-dimensional flow in a HB rectangular cavity: (a) finite element solution by Le Garrec and Magnaud, together with spectral solution by Pulicani and Peyret; (b) detail of the solution of Pulicani and Peyret; (c) spectral solution by Randriamampianina et al.; (d) finite difference solution by Ben Hadid and Roux. Different steady (Sn), oscillating (Pn), and quasi-periodic (QP) flow regimes, with hysteresis cycles, are observed (From [70]).

from which we may very roughly conclude that the most accurate results were provided by spectral methods, closely followed by the other schemes (provided strongly refined meshes or grids are used). The complexity of the flow structure which was found to occur at these rather low values of the Grashof number (2×10^4 to 4×10^4) must be stressed. In particular, different steady, oscillating, or quasi-periodic flow regimes, with hysteresis cycles, were observed. The case of the two-dimensional flow of a zero Prandtl number liquid in a rectangular cavity (supposed to represent a horizontal Bridgman boat) submitted to a horizontal temperature gradient is illustrated in fig. 12. When the Grashof number is increased, the steady one-cell initial structure of the streamlines gives rise to an alternatively oscillating or quasi-periodic flow. A sudden change of flow regime is observed at a critical Grashof number of about 3.5×10^4, and a steady two-cell flow is formed. When the Grashof number is decreased, this structure persists down to a Grashof number of about 2.5×10^4, and a rapid switch to the initial flow regime is found. More details are given in [208, 209]. Such results are essential, since we must conclude that, even at a moderate value of the Grashof number, a complex flow behaviour will be the rule. Hence, any simulation dealing with melt flow calculations must be examined very carefully.

The coupling between melt flow and heat transfer remains, therefore, a very complex problem when realistic results are expected (see also [278, 279]). It must however be taken into account that the objective of bulk growth simulations is not necessarily to capture the details of the flow structure, but to provide average values, when this is possible [211]. Here, the modelling task becomes essential. In particular, turbulent models should be developed, such as undertaken by Kobayashi [212], Kobayashi et al. [213], Ristorcelli and Lumley [83], and Kinney and Brown [264] (see also section 5.2.3).

3.5. *Examples of local simulations with melt convection and segregation*

3.5.1. *Examples with melt convection*

Taking into account melt convection is difficult to achieve in semi-conductor growth, since the Grashof number is often very high. The problem is somewhat easier to handle for oxides. In this regard, besides the work of Sackinger et al. [140], and Derby and Xiao [214], and the global analysis of Xiao and Derby [265], a very interesting local time-dependent model was developed by Kopetsch [215, 216], which allowed him to fully couple heat conduction in melt and crystal, and convection within the melt. The technique was based on a finite difference axisymmetric method associated with a coordinate transformation in order to treat moving boundaries. The melt level could be set to either drop or remain constant with time. Crystal height was taken to be fixex. Surface tension effects and meniscus curvature at the melt surface were neglected. Heat loss from the melt surface and crystal wall was modelled as radiation towards an environment at constant temperature, and a constant temperature was imposed both on the crucible bottom and the sidewalls, and at the top of the crystal. As thermal boundary conditions were all fixex, full rate was calculated so that a constant crystal radius could be imposed during growth.

The model was used to simulate the well-known inversion of the solid–liquid interface (from a convex to a concave shape with respect to the solid) occurring as a result of a transition in melt flow from free to combined free and forced convection. This can be observed for example when crystal rotation rate is increased. The phenomenon is particularly important for oxide growth, since a strong coupling between temperature field and melt flow is found due to the large Prandtl number characteristic of molten oxides. Time evolution of the system was analyzed for a progressively increasing crystal rotation rate, starting from an initial condition corresponding to a vanishingly small rotation rate. Calculations were performed for Al_2O_3, a material whose physical properties are well known. Figure 13 shows the solid–liquid interface shape and the flow pattern (as represented through velocity vectors and streamlines) for several rotation rates. The initial state, associated with free convection only, is characterized by a Grashof number of 10^7, and exhibits a single rotating cell with the fluid rising at the crucible wall and descending along the axis of symmetry, thereby pushing the isotherms near the interface down towards the melt so that a convex interface results.

When crystal rotation rate is increased, a counter-rotating vortex develops beneath the crystal, in which the fluid flows upwards along the axis of symmetry. As a matter of fact, crystal rotation forces the flow in a direction opposite to free convection. The up-flow on the axis of symmetry pushes the isotherms towards the crystal, thereby increasing the local concavity of the interface. A wavy interface shape is then obtained, i.e. slightly convex with respect to the crystal above the free convection vortex, and concave above the forced convection vortex. As crystal rotation is further increased, forced convection occupies a broader place in the melt, so that the concave part of the interface eventually extends up to the crystal periphery, yielding a totally concave shape. The transition towards a fully concave interface occurs when the boundary between free and forced convection cells is located beneath the crystal periphery.

Looking at the evolution of some critical variables as represented in fig. 14, one observes that interface transition occurs when melt temperature at a point located just beneath crystal periphery reaches a minimum. At the same time, the pull rate reaches a maximum value since solidification heat needs to be increased to maintain the melting temperature at the tri-junction. The critical rotation rate (transition occurs at about 22 rpm for this example) was found to agree fairly well with analytical predictions and experimental observations. From fig. 14, it is interesting to observe the existence of some oscillatory states for specific ranges of rotation rates (from 7 to 12 rpm for example). The nature of these oscillations, either physical or numerical, could not be identified, as a careful analysis of the time-dependent flow behaviour demonstrated that period, amplitude and spectral composition of the oscillations were dependent upon mesh refinement.

The same analysis carried out for a smaller crystal radius showed that higher rotation rates were needed to produce the interface inversion in this case. Moreover, while for the larger radius the inversion produced a wavy interface shape for a range of rotation rates, the transition from convex to concave shape occurred at a specific rotation rate associated with a virtually flat interface in the case of the smaller crystal radius.

3.5.2. Melt convection and segregation

A time-dependent finite element scheme, combining calculation of temperature, veloc-

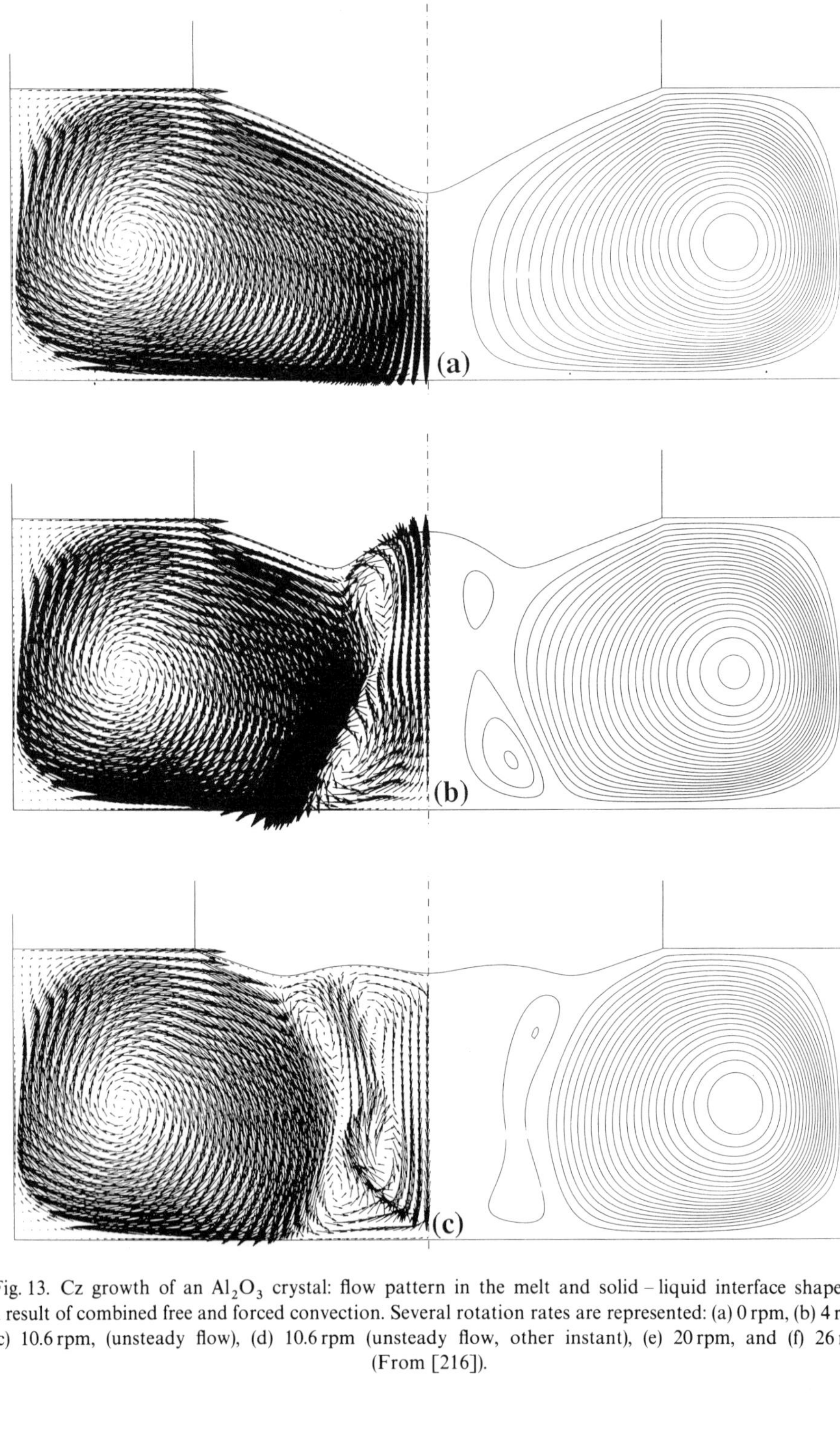

Fig. 13. Cz growth of an Al_2O_3 crystal: flow pattern in the melt and solid – liquid interface shape, as a result of combined free and forced convection. Several rotation rates are represented: (a) 0 rpm, (b) 4 rpm, (c) 10.6 rpm, (unsteady flow), (d) 10.6 rpm (unsteady flow, other instant), (e) 20 rpm, and (f) 26 rpm (From [216]).

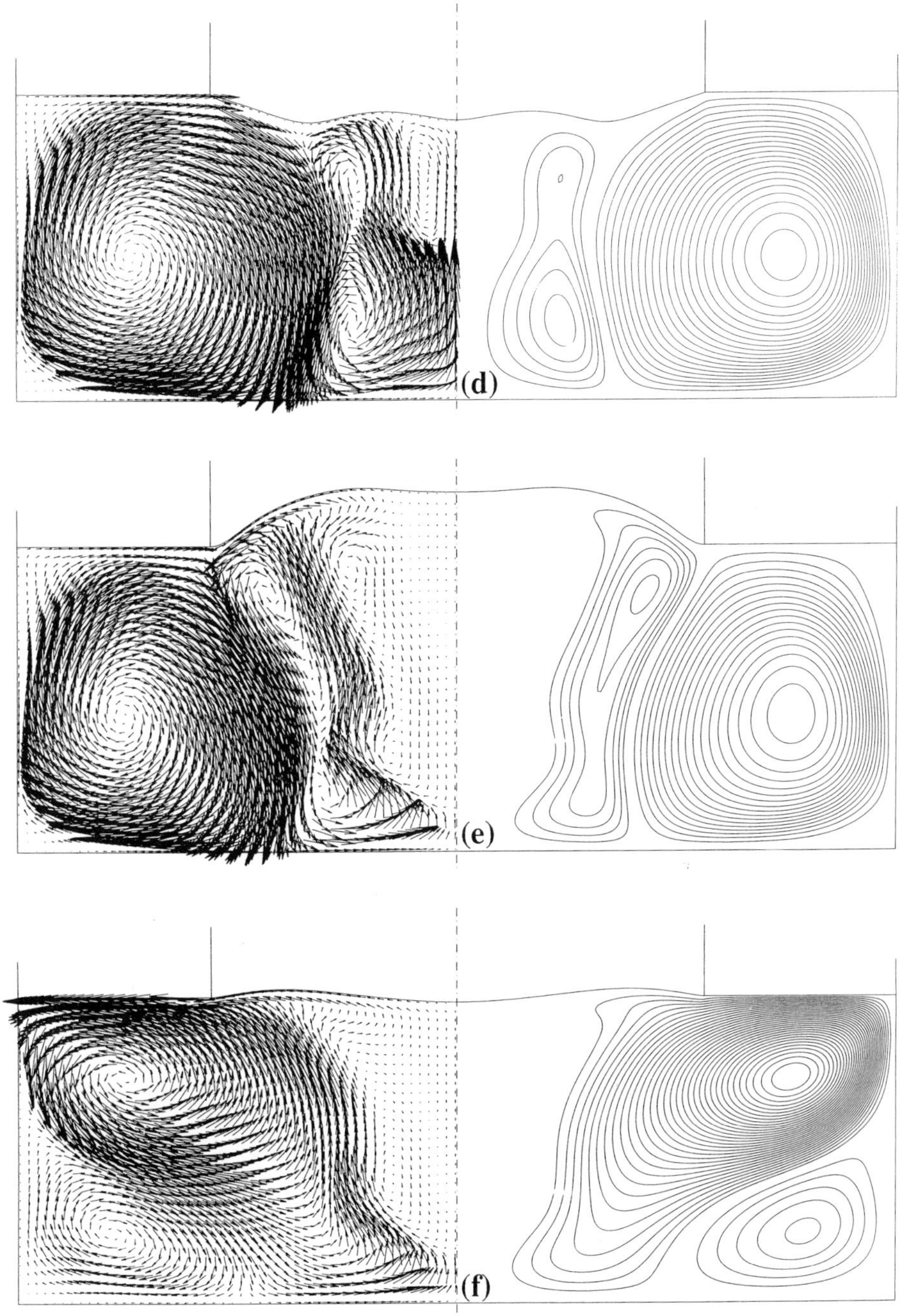

Fig. 13. (Cont'd)

(a)

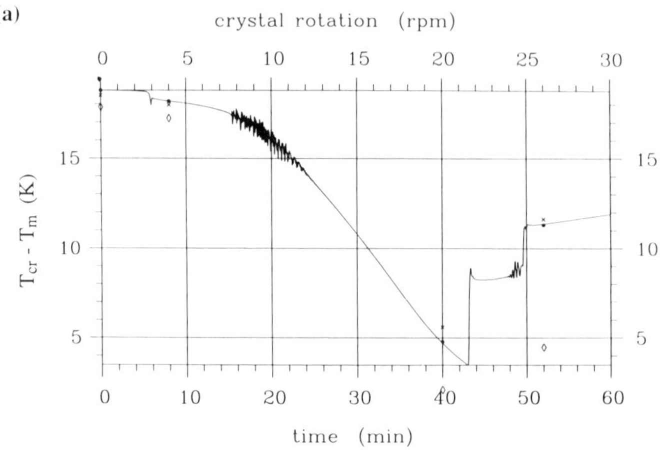

(b)

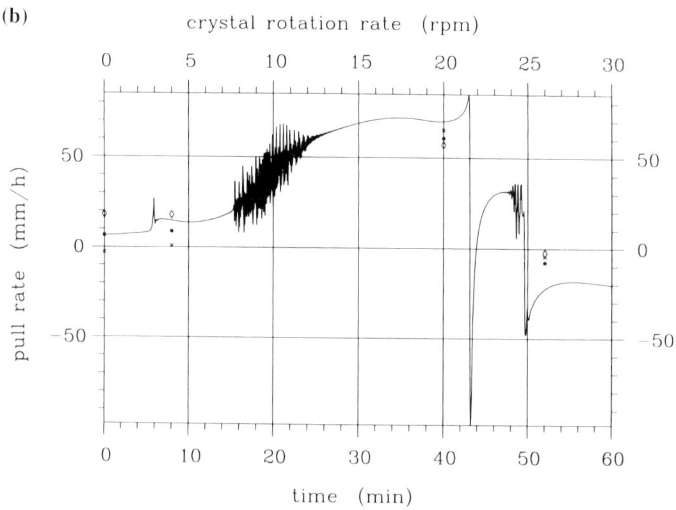

Fig. 14. Same problem as in fig 13: evolution (a) of the melt temperature beneath the crystal periphery (T_{cr}), and (b) of the pull rate as a function of time and crystal rotation rate (extreme values are found for a critical value of the rotation rate (22 rpm)) (From [216]).

ity, and concentration fields in a VB furnace, was developed by Kim and Brown [99, 100], and Brown and Kim [93], on the basis of the pseudo-steady model of [53, 90–92, 217]. The model is local, as temperatures along the ampoule are derived from a simple heat flux boundary condition which does not include calculation of view factors. Local simulations for a VB configuration are often satisfactory thanks to the simple geometry of such furnaces, as already pointed out in section 1.3.

We focus here on the growth of HgCdTe [99], a non-dilute alloy including a light (CdTe) and a heavy (HgTe) component. A schematic diagram of a similar growth system is represented in fig. 2. Melt is located in the upper part of the ampoule, and

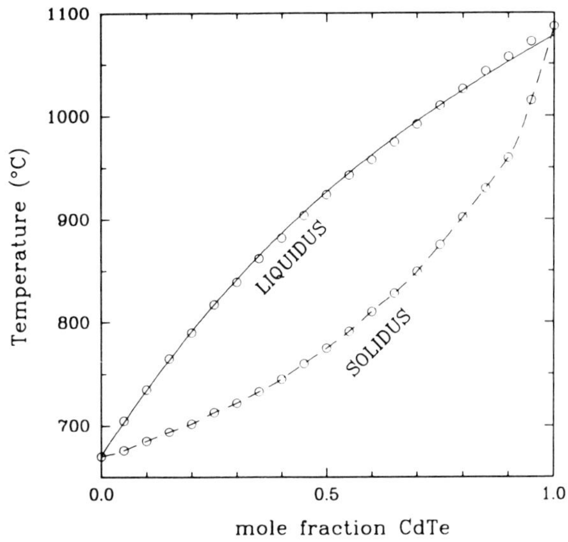

Fig. 15. Phase diagram for a CdTe-HgTe pseudo-binary alloy (From [99]).

crystal in the lower part. The phase diagram of such alloy shows a large separation between the liquidus and solidus curves (fig. 15). It can be easily deduced from this phase diagram that the lighter component, CdTe, is preferentially incorporated in the solid phase, while the heavier component, HgTe, accumulates in the melt near the solid–liquid interface. The equilibrium segregation coefficient k reaches a value of about 3 for the chosen melt concentration and dictates the ratio between the crystal composition C_s and the melt composition C_l at the solid–liquid interface (see section 2.3). Due to segregation, the composition of the melt is non-uniform, and given by the solute field. As a result of the progressive decrease of melt concentration during the process, the crystal composition is non-uniform, and the melting temperature decreases as growth proceeds. Crystal composition is measured by the axial segregation (composition along the axial coordinate), and by the radial segregation, i.e. the difference between the maximum and minimum values of composition along the interface with respect to the radially averaged interface composition. Let us recall that growing compositionally uniform crystals is a major objective for these systems.

In order to describe segregation as related to melt convection, the concept of an effective segregation coefficient was established, measuring the ratio between solute concentration in the solid and in the bulk melt away from the solid–liquid interface. The most extensively used model for correlating the effective segregation coefficient to convection is the stagnant film model developed by Burton et al. [26]. This one-dimensional model assumes that the solute is everywhere well mixed, except in a thin "stagnant layer" of thickness δ adjacent to the melt–crystal interface. This simple approach provided for a successful description of Ge(Ga) growth, as found in [93] but was not able to correctly describe HgCdTe growth, nor Se-doped GaAs growth [100]. Numerical simulation is very useful for predicting segregation in such cases.

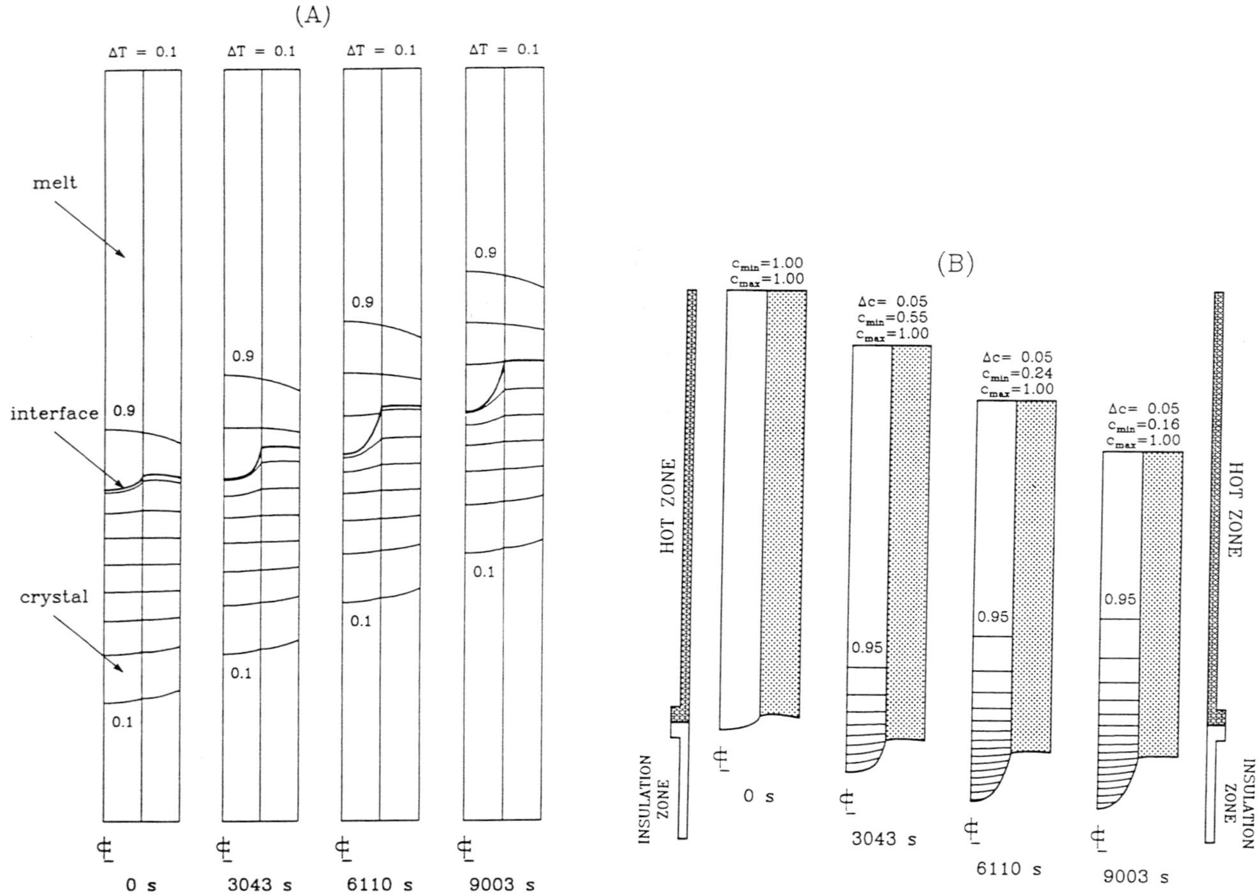

Fig. 16. VB growth of a HgCdTe crystal: evolution of the solid–liquid interface shape, temperature field (A) and solute field (B) during the initial transient of the process. The steady-state solution at 0 s corresponds to a stationary ampoule (composition throughout the melt is then uniform). No buoyancy-driven convection is considered (From [99]).

Concentration and temperature fields are highly coupled in HgCdTe growth through the dependence of melting temperature upon composition, and through the thermal and solutal expansion coefficients, β_T and β_C. Moreover, both thermal and solutal expansion coefficients depend on temperature and concentration. This coupling has an important influence on interface shape. For a pure material or a dilute alloy, interface shape is simply determined by the melting isotherm. The case of a non-dilute alloy is more complicated, since melting temperature is given by the liquidus curve and depends on composition. Interface shape is then coupled to the solute field. Figure 16 shows the time-dependent evolution of interface shape, temperature and solute fields for different instants, the initial solution corresponding to a vanishing growth rate and a uniform melt composition (initial composition of the melt is 0.2 mole fraction of CdTe or, equivalently, 1 in dimensionless form). We may observe the development of a diffusion solute layer depleted in CdTe. This solute layer reaches a quasi-steady state after a few hours. One may also observe how interface curvature progressively increases as uniformity of the melt is lost due to segregation. Melt concentration is lower on the axis than at the contact with the ampoule, which leads to a lower melting temperature on the axis. The axial interface position is then shifted towards the solid phase and leads to an increased interface concavity with respect to the crystal.

Buoyancy forces in the melt, due to density variations, are generated from the dependence of melt density on temperature and concentration, leading to thermal and solutal convection, respectively. Thermal convection is caused by the large radial temperature gradients close to the interface and leads to two axisymmetric toroïdal flow cells stacked axially in the melt. It was shown in [92] that a damping of thermal convection through solutal convection occurred in the VB system considered here. This is possible because we consider the case of a very concentrated alloy. In fig. 17, we observe how isotherms in the melt are highly deformed if only thermal convection is considered, whereas they become flatter with the addition of solutal convection. This may be understood as follows: due to the segregation effect, the heavier component of the alloy is rejected in the melt near the interface, and has a stabilizing effect on the convective flow, thereby reducing the effect of thermal convection. This damping leads to an apparent diffusion-controlled axial segregation, which is confirmed by experimental observations. Near the interface however, partial damping only is observed.

An interesting point regarding convection is to observe that melt density plotted as a function of temperature shows a maximum value, since the thermal expansion coefficient β_T exhibits a change of sign around 750°C, varying from a negative (thermally destabilizing) to a positive (thermally stabilizing) value. This phenomenon leads to a melt that is vertically unstable just above the interface, and becomes stable higher in the ampoule. In order to analyze the effect of density inversion with temperature, convection was computed for both a constant and a variable thermal expansion coefficient β_T. The resulting flow pattern was quite different, but the solute field remained almost unaffected.

The effect of parameters such as the ampoule translation rate was investigated. A reduction in growth rate causes solute diffusion to become easier. The solute diffusion layer extends then further into the melt and the axial solute gradient in the melt decreases (fig. 18). In addition, interface deflection decreases, and so does radial

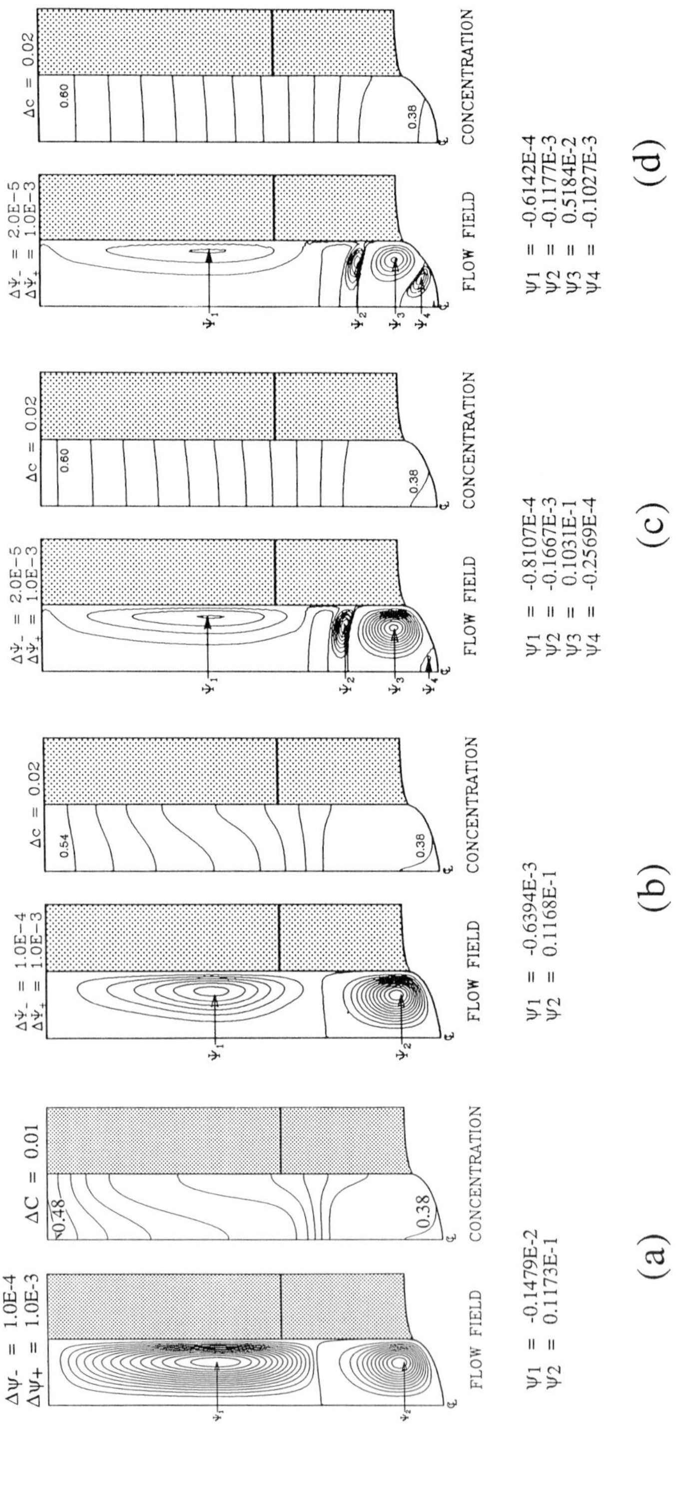

Fig. 17. VB growth of a HgCdTe crystal: damping of thermal convection by solutal convection. Streamlines and isoconcentration curves (in non-dimensional form) for increasing solutal Rayleigh numbers: (a) $Ra_s = 0$; (b) $Ra_s = 1 \times 10^6$; (c) $Ra_s = 1 \times 10^7$; (d) $Ra_s = 1.565 \times 10^7$ (From [92]).

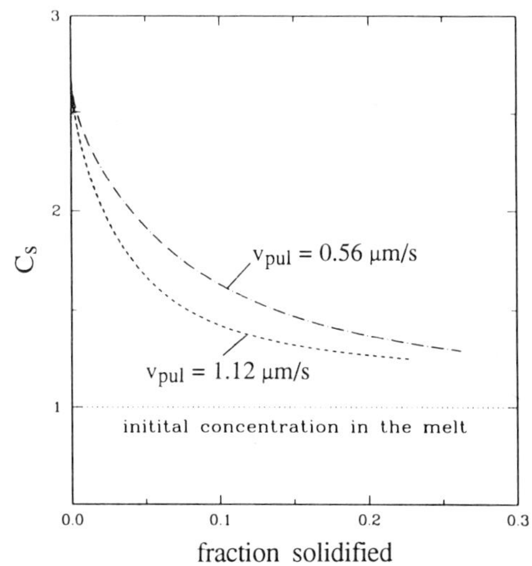

Fig. 18. Same problem as in fig. 16: axial segregation (CdTe concentration) as a function of the fraction solidified for two different growth rates (From [99]).

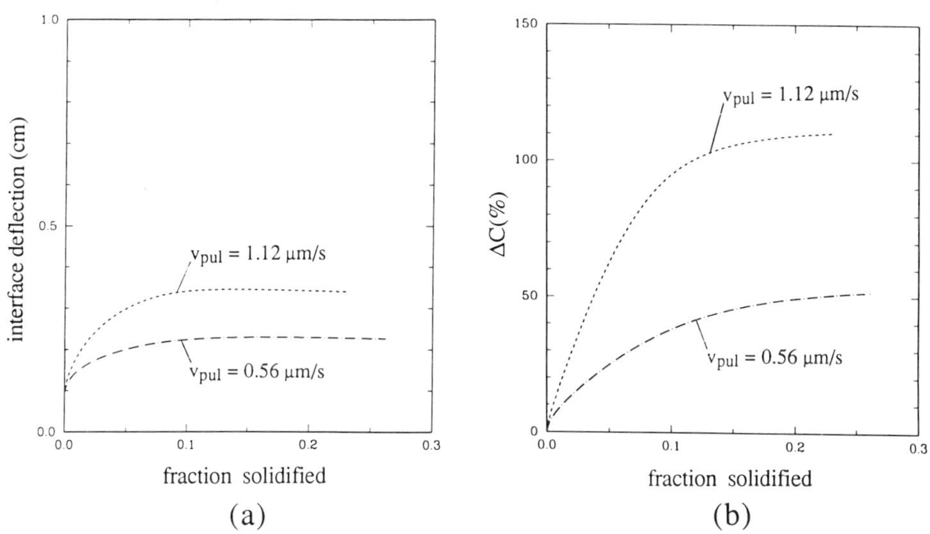

Fig. 19. Same problem as in fig. 16: (a) solid–liquid interface deflection, and (b) radial segregation, as functions of the fraction solidified for two different growth rates (From [99]).

segregation, which results mainly from interface curvature (fig. 19). Because of the damping of thermal convection by solutal convection, a decrease in growth rate was expected to have an important effect on the flow field. However, the global effect on melt flow was found to be mild as the reduction of solutal convection related to a lower axial concentration gradient was compensated by the reduction of thermal convection due to the smaller interface deflection.

4. Radiative heat transfer in crystal growth

4.1. The physics of thermal radiation

Radiation is an essential mode of heat transfer in the growth process in view of the high melting temperature of most crystalline materials. At these temperatures, radiative transfer takes place essentially in the near infrared, and weakly in the red range of the spectrum (0.6–25 μm)[218]. Indeed, letting $e_{b\lambda}$ denote the spectral emissive power of the blackbody in vacuum, i.e. the power emitted by the body per unit surface and wavelength at absolute temperature T and wavelength λ, Planck's law [219] is written as follows:

$$e_{b\lambda}(T) = \frac{2\pi h c_0^2}{\lambda^5[\exp(hc_0/k\lambda T) - 1]}, \tag{52}$$

where c_0, h and k are the speed of light in vacuum, and Planck's and Boltzmann's constants. In fig. 20, curves of $e_{b\lambda}(T)$ as a function of λ, for different values of T, show that a maximum of emissive power is found between 500 and 2500 K in the wavelength range 1.16–5.8 μm.

Radiation is governed by quantum mechanics. Hence, both wave and particle behaviour are exhibited by the photon, which is the vehicle of electromagnetic interaction. As a wave, the photon presents non-local properties, such as interference (with itself) or radiative tunnelling. As a particle, the photon shows strong local properties, i.e. seems to have a given position at a given time. In both cases, the speed c

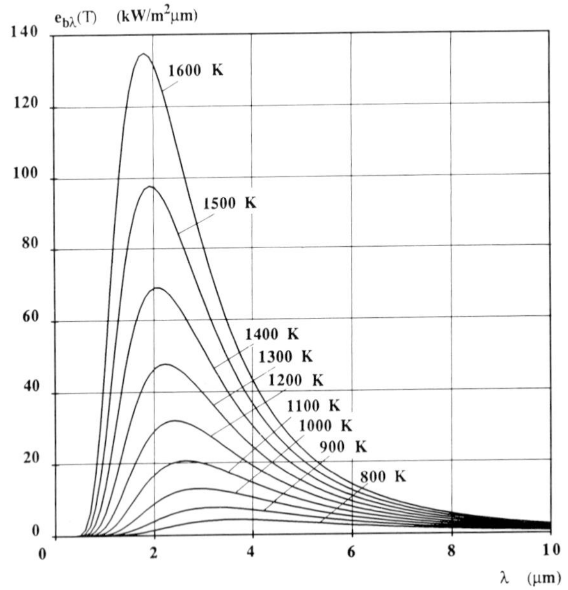

Fig. 20. Spectral emissive power of a blackbody for several temperatures.

of a photon of wavelength λ is

$$c = c_0/n, \tag{53}$$

where n is the index of refraction of the medium (which may depend on λ), and its energy e is

$$e = hc/\lambda. \tag{54}$$

Generally, the index of refraction of the medium is only slightly higher than 1, which means that c is very close to c_0. In any case, the spectral emissive power of a blackbody inside the medium is obtained by replacing c_0 with c in eq. (52).

The typical wavelengths of radiative exchanges involved in crystal growth are very small compared to the characteristic dimensions of any practical apparatus, and the emitted radiation is non-coherent. Wave effects may therefore be neglected in the modelling of the process (except for explaining the value of material parameters such as emissivities, reflectivities...). In a similar way, radiative transfer may be approximated as to be instantaneous. Lastly, energies of quanta are so small that continuity of the model may be assumed (except at the boundaries of the bodies). Hence, the *continuum model of radiative transfer* is defined by considering infinitesimal amounts of radiative energy (which will be called photons, with some abuse because they are not quantified) travelling across the medium, while experiencing absorption, scattering... Basic unknows of this model are the temperature field and the polarized spectral intensity i^*_λ, which is defined as follows: consider a given location **x** and time t, a given orientation **m** and polarization **p** (as represented by perpendicular unit vectors), and a given wavelength λ; $i^*_\lambda(\mathbf{x}, t, \mathbf{m}, \mathbf{p})$ is the flux of energy per unit surface, wavelength, solid angle, and polarization angle, of the photons crossing at time t an element of area perpendicular to **m** and centred at **x**, these photons having a direction, polarization and wavelength comprised in infinitesimal vicinities of **m**, **p** and λ (note that i^*_λ has the same value for polarizations **p** and $-$ **p**). This means that we are dealing with a problem of no less than eight independent variables (three for coordinates, one for time, two for direction of motion of the photons, as represented by two angles on the unit sphere, one for polarization, which is constrained to be perpendicular to orientation, and one for wavelength). Additional simplications are essential.

4.1.1. Governing phenomena

The physical phenomena "experienced" by the photons may be summarized as follows.

Internal natural emission. Within the medium, the density of power $r^*_{e\lambda}$ emitted at wavelength λ in a given direction and for a given polarization is

$$r^*_{e\lambda} = e_\lambda i^*_{b\lambda}, \tag{55}$$

where the "emission" coefficient e_λ has the dimension of the inverse of a length, while the polarized spectral intensity of the blackbody $i^*_{b\lambda}$ is proportional to its spectral emissive power $e_{b\lambda}$, as given by eq. (52) (taking eq. (53) into account):

$$i^*_{b\lambda} = \frac{1}{2\pi^2} e_{b\lambda}. \tag{56}$$

Integrating eq. (56) over all polarization angles, i.e. from 0 to 2π, provides the spectral intensity of the blackbody $i'_{b\lambda}$,

$$i'_{b\lambda} = \frac{1}{\pi} e_{b\lambda}. \tag{57}$$

The latter formula is obvious when observing that $e_{b\lambda}$ integrates the contributions normal to the blackbody surface of all intensities at the given location.

Internal absorption. Along the path of a "photon", the polarized spectral intensity absorbed per unit length $r^*_{a\lambda}$ is

$$r^*_{a\lambda} = a_\lambda i^*_\lambda, \tag{58}$$

where a_λ is the absorption coefficient (including both true absorption and induced emission), and i^*_λ is the polarized spectral intensity inside the body.

Scattering. Along the path of a photon, part of the beam may also be scattered, which means that it may undergo changes in direction and possible variations of intensity. Scattering does not play a significant role in the growth process, and will not be examined further.

Reflection. At the surface separating two different materials, part of the incident intensity $i^*_{i\lambda}$ on one given side (as dependent on the incident direction $\mathbf{m}_i$ and polarization $\mathbf{p}_i$) is reflected, while the remaining part is transmitted (and, generally, refracted) inside the body. The reflected intensity $i^*_{r\lambda}$, in a given direction $\mathbf{m}_r$ and for a given polarization $\mathbf{p}_r$, is the sum of all the contributions from incident beams:

$$i^*_{r\lambda} = \frac{1}{2\pi^2} \int \rho_\lambda i^*_{i\lambda} (-\mathbf{m}_i \cdot \mathbf{e}) \, dm_i \, dp_i, \tag{59}$$

where ρ_λ is the surface reflectivity, which may depend on $\mathbf{m}_i$, $\mathbf{p}_i$, $\mathbf{m}_r$, $\mathbf{p}_r$, while $\mathbf{e}$ is the unit outward normal on the surface. For a perfectly smooth (or mirror-like) surface, reflection is specular and depends on polarization, and ρ_λ takes the form of a Dirac distribution. For real surfaces, reflection is more diffuse.

Transmission and refraction. At the surface separating two materials, the transmitted part of the incident intensity will be refracted if the refraction indices of the materials are different. This means that both direction and wavelength (but not frequency) of the transmitted beam will be changed. Consequently, transmission obeys rules which are similar to those of reflection, but slightly more complex if refraction is present. In particular, for a perfectly smooth surface, transmission is specular, while it is more diffuse across a real surface. In the case of specular transmission, refraction depends on polarization. In order to avoid the difficulties related to wavelength dependence upon refraction index, the term wavelength and the symbol λ will, in the sequel, stand for the equivalent wavelength of the beam in vacuum. The calculation of $e_{b\lambda}$ from eqs. (52) and (53) must be modified accordingly.

The physics of radiative transfer may strongly be simplified by introducing the concept of *opaque* bodies, within which photons can neither be transmitted, nor emitted. Opaque bodies may be considered as a limiting case of optically very thick materials, which absorb the transmitted part of incident beams over an infinitely short path. The additional phenomena to take into account with opaque bodies are the following.

Surface emission. At the surface of an opaque body, the polarized spectral intensity $i^*_{e\lambda}$ of the emitted radiation is given by the formula

$$i^*_{e\lambda} = \varepsilon_\lambda i^*_{b\lambda}, \tag{60}$$

where the emissivity ε_λ may in particular depend on the direction and polarization of the emitted beam.

Surface absorption. Part of the intensity $i^*_{i\lambda}$ of a beam incident on an opaque surface will be absorbed by the body according to the rule

$$i^*_{a\lambda} = \alpha_\lambda i^*_\lambda, \tag{61}$$

where $i^*_{a\lambda}$ is the fraction of $i^*_{i\lambda}$ which is not reflected, while the absorptivity α_λ may again depend on the direction and polarization of the incident beam.

This general continuum model of radiative transfer must be completed by taking into account the requirements of the *second law* of thermodynamics. Under a "non far from equilibrium" assumption, these requirements involve simple rules, which may be stated as follows: (i) emitted, absorbed, scattered, reflected, and transmitted intensities are always positive; (ii) when any other source of heat but natural emission is absent, the net radiative transfer at any given frequency between any pair of infinitesimal emitting surfaces or volumes in an isothermal enclosure vanishes, for any selected couple of beam directions and polarizations. According to these rules, it is easy to demonstrate the following relations by placing arbitrary blackbodies in a given enclosure or at its boundary:

- The "emission" and absorption coefficients, eqs. (55) and (58), are equal and positive,

$$e_\lambda = a_\lambda \geqslant 0. \tag{62}$$

- The emissivity and absorptivity of any surface, eqs. (60) and (61), are equal and comprised between 0 and 1,

$$0 \leqslant \varepsilon_\lambda = \alpha_\lambda \leqslant 1. \tag{63}$$

- The reflectivity of any surface, eq. (59), is positive and symmetric with respect to incident and reflected beam direction and polarization,

$$0 \leqslant \rho_\lambda(\mathbf{m}_r, \mathbf{p}_r, \mathbf{m}_i, \mathbf{p}_i) = \rho_\lambda(\mathbf{m}_i, \mathbf{p}_i, \mathbf{m}_r, \mathbf{p}_r), \tag{64}$$

  and its average value with respect to reflected directions and polarizations is comprised between 0 and 1.
- Transmission (involving possible refraction) and scattering obey similar, but slightly more complex rules than eqs. (62), (63) and (64).

4.1.2. Simplifying assumptions

At this stage, it is possible to introduce the simplifications which are classically introduced in the modelling of radiative heat transfer in crystal growth.

I Diffuse radiation. This hypothesis assumes that absorption coefficients, emissivities, and reflectivities are non-directional and non-polarized material properties. Hence, in eqs. (55) to (61), a_λ, ε_λ, and ρ_λ depend only on $\mathbf{x}$, t, T, and λ. This means that, at a given frequency, emitted and reflected intensities are exactly proportional to those of a blackbody. The diffuse model can easily be extended to situations where scattering or non-specular transmission effects are non-negligible. In these cases, part of the beam is assumed to be exactly transmitted, while the remaining part is diffusely scattered or refracted. Such effects will no longer be considered here.

II Surface radiation. This model assumes that all the effects influencing radiative heat transfer take place at the surfaces of the bodies. Equivalently, letting h stand for a characteristic dimension of the enclosure, the model neglects the ranges of wavelengths for which the spectral optical thickness τ_λ defined by

$$\tau_\lambda = h a_\lambda, \tag{65}$$

is of the order of unity. When τ_λ is large compared to 1, the material is assumed to be opaque for the wavelength $\lambda (\tau_\lambda \cong \infty)$, while it is assumed to be transparent ($\tau_\lambda \cong 0$) when τ_λ is small compared to 1.

III Gray materials. The gray assumption is defined by supposing that material properties ($a_\lambda, \varepsilon_\lambda, \rho_\lambda, n$) do not depend on wavelength. This model requires to introduce the total emissive power of the blackbody,

$$e_\mathrm{b} = \int_0^\infty e_{\mathrm{b}\lambda}\,\mathrm{d}\lambda = n^2 \sigma T^4, \tag{66}$$

where the Stefan–Boltzmann constant σ is given by

$$\sigma = \frac{2}{15}\pi^5 \frac{k^4}{c_0^2 h^3}, \tag{67}$$

and to integrate the intensity i^* over the whole range of wavelengths,

$$i^* = \int_0^\infty i_\lambda^*\,\mathrm{d}\lambda. \tag{68}$$

IV Diffusion approximation. When the optical thickness of a material, as defined in eq. (65), is larger than 1, but not sufficiently high to permit one to introduce the opaque body approximation (see [218, 220, 106]), diffusion methods may be used to obtain better results. The Rosseland approximation is defined by replacing the total heat flux in the range of wavelengths Λ by an equivalent conductive flux:

$$\mathbf{q}_{\mathrm{R}\Lambda} = -\left(\frac{4}{3}\int_\Lambda \frac{1}{a_\lambda}\frac{\mathrm{d}e_{\mathrm{b}\lambda}}{\mathrm{d}T}\,\mathrm{d}\lambda\right)\nabla T = -\left(\frac{4}{3a_{\mathrm{R}\Lambda}}\frac{\mathrm{d}e_{\mathrm{b}\Lambda}}{\mathrm{d}T}\right)\nabla T, \tag{69}$$

where $e_{\mathrm{b}\Lambda}$ is the emissive power of the blackbody in the range Λ, while the Rosseland mean absorption coefficient $a_{\mathrm{R}\Lambda}$ is defined by

$$\frac{1}{a_{\mathrm{R}\Lambda}} = \int_\Lambda \frac{1}{a_\lambda}\frac{\mathrm{d}e_{\mathrm{b}\lambda}}{\mathrm{d}T}\,\mathrm{d}\lambda \Bigg/ \int_\Lambda \frac{\mathrm{d}e_{\mathrm{b}\lambda}}{\mathrm{d}T}\,\mathrm{d}\lambda. \tag{70}$$

In practice, Rosseland's approximation means simply that the material is treated as opaque in the range Λ, and that an additional term $(4/3a_{\mathrm{R}\lambda})\;\mathrm{d}e_{\mathrm{b}\lambda}/\mathrm{d}T$ is included in the coefficient of heat conduction. However, at the boundary of a body where Rosseland's approximation is used, a temperature jump must be introduced in order to account for peculiar effects taking place in a small boundary layer of the order of magnitude $1/a_{\mathrm{R}\Lambda}$, which is neglected in the diffusion approximation. The jump condition is written in the form

$$e_{\mathrm{b}\Lambda}(T(\mathbf{x})) - e_{\mathrm{b}\Lambda}(T(\mathbf{x}^+)) = -\psi(N_\Lambda, \Lambda)\mathbf{q}_{\mathrm{R}\Lambda}\cdot\mathbf{n}, \tag{71}$$

where $T(\mathbf{x})$ and $T(\mathbf{x}^+)$ denote temperatures calculated at the wall and in the adjacent medium, respectively, and $\mathbf{n}$ is the unit outward normal; the Stark conduction–radiation parameter in the range Λ is defined by

$$N_\Lambda = K a_{\mathrm{R}\Lambda}\left(\frac{\mathrm{d}e_{\mathrm{b}\Lambda}}{\mathrm{d}T}\right)^{-1}, \tag{72}$$

where K is the thermal conductivity of the medium; ψ is the so-called slip coefficient, whose maximum value is about $\frac{1}{2}$, but for which a general expression is not available at the present time. More details may be found in [106, 220].

A more accurate approximation is provided by the Milne–Eddington differential model, which will be examined in section 4.2.

4.2. *Simplified models*

The importance of radiative transfer in crystal growth is various. On the one hand, calculation of the global heat transfer must absolutely take radiation into account, but in a way which should be as simplified as possible, while sufficient to obtain significant heat fluxes and temperatures at the boundary of the crystal and melt. On the other hand, within the crystal, too simplified a model may be dangerous if it is a participating medium (as shown by Abrams and Viskanta [88], Yuferev and Vasil'ev [221], and Chan and Hsu [222]). Hence, different levels of accuracy are necessary depending on the location where radiative transfer takes place. In what follows, three models developed in the literature will be presented: the global surface radiation model, the gray model for integrating the equation of transfer, and the Milne – Eddington differential model.

4.2.1. *Global surface radiation model*

Assuming diffuse surface radiation (i.e. the approximations I and II of section 4.1), it is easy to integrate an accurate model for the global radiative transfer in Cz or VB growth. First developments were based on the gray assumption (see the work of Wouters [57], Dupret et al. [58], Atherton et al. [60], Tsukada et al. [223], which remains very useful in view of its strongly simplified formulation. The model we present here was developed by Dupret et al. [59] and is based on a band-energy approach, with a view to extending the gray model to more general applications, including materials having wavelength-dependent properties, or some classes of semi-transparent bodies.

Let us start by considering a range $\Lambda = [\lambda_1, \lambda_2]$ of wavelengths over which material properties are constant in a given enclosure. The total power $e_{\mathrm{b}\Lambda}$ emitted by the blackbody over this range is obtained by integrating eq. (52) with respect to λ, which yields

$$e_{b\Lambda} = \int_{\Lambda} e_{b\lambda}\, d\lambda = \gamma_{\Lambda}(T) n^2 \sigma T^4, \tag{73}$$

where the Stefan–Boltzmann constant σ was introduced in eq. (67), and γ_Λ is defined by the expression

$$\gamma_\Lambda(T) = [F(\lambda T)]_\Lambda = F(\lambda_2 T) - F(\lambda_1 T), \tag{74}$$

with

$$F(\xi) = \frac{15}{\pi^4} \int_0^{\xi nk/hc_0} \frac{d\eta}{\eta^4 (e^{1/\eta} - 1)}. \tag{75}$$

The curve $F(\xi)$ is given in fig. 21 for $n = 1$. As $\gamma_\Lambda(T)$ is the ratio between the powers emitted by the blackbody in the range Λ and in the whole spectrum, γ_Λ lies between 0 and 1. Moreover, it is important to observe, from fig. 21, that γ_Λ depends but little upon T as long as

$$0.2 \times 10^4\ \mu\text{mK} < \lambda_1 T < \lambda_2 T \leq 10^4\ \mu\text{mK}. \tag{76}$$

Hence, provided either eq. (76) holds, or Λ is the whole spectrum, or temperature may be estimated everywhere on the enclosure, γ_Λ may be assumed to be a known function $\gamma_\Lambda(\mathbf{x})$, where $\mathbf{x}$ is a generic point on the surface of the enclosure ∂V.

In surface radiation, from eqs. (59) to (61), integration of the normal contributions of intensities $(\pm\, \mathbf{m}\cdot\mathbf{e})\, i^*_\lambda$ over all incident or reflected directions $\mathbf{m}$, over all polarizations $\mathbf{p}$,

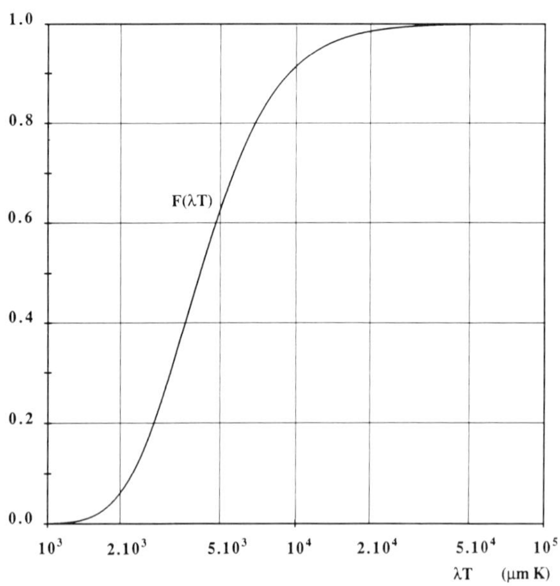

Fig. 21. Fractional blackbody emissive power in the range 0 to $\lambda T (n = 1)$.

and over the wavelength range Λ, provides easily the basic relations between the emitted ($q_{e\Lambda}(\mathbf{x})$), absorbed ($q_{a\Lambda}(\mathbf{x})$), incident ($q_{i\Lambda}(\mathbf{x})$), and reflected fluxes ($q_{r\Lambda}(\mathbf{x})$) in the range Λ at location $\mathbf{x}$:

$$\begin{aligned} q_{e\Lambda}(\mathbf{x}) &= \varepsilon_\Lambda(\mathbf{x}) e_{b\Lambda}(T(\mathbf{x})), \\ q_{a\Lambda}(\mathbf{x}) &= \varepsilon_\Lambda(\mathbf{x}) q_{i\Lambda}(\mathbf{x}), \\ q_{r\Lambda}(\mathbf{x}) &= [1-\varepsilon_\Lambda(\mathbf{x})] q_{i\Lambda}(\mathbf{x}). \end{aligned} \tag{77}$$

Here, $\varepsilon_\Lambda(\mathbf{x})$ is the surface emissivity in the range Λ at location $\mathbf{x}$, while the surface reflectivity is $1-\varepsilon_\Lambda(\mathbf{x})$. The total flux outgoing from the wall (or the radiosity) $q_{o\Lambda}(\mathbf{x})$ is the sum of the emitted and reflected fluxes:

$$q_{o\Lambda}(\mathbf{x}) = \varepsilon_\Lambda(\mathbf{x})\gamma_\Lambda(\mathbf{x}) n^2 \sigma T^4(\mathbf{x}) + [1-\varepsilon_\Lambda(\mathbf{x})] q_{i\Lambda}(\mathbf{x}). \tag{78}$$

On the other hand, the total incident flux $q_{i\Lambda}(\mathbf{x})$ is the sum of the contributions of the radiosities $q_{o\Lambda}(\mathbf{x}^*)$ from all other points on the enclosure:

$$q_{i\Lambda}(\mathbf{x}) = \int_{\mathbf{x}^* \in \partial V} K(\mathbf{x}, \mathbf{x}^*) q_{o\Lambda}(\mathbf{x}^*)\, \mathrm{d}S^*, \tag{79}$$

where $K(\mathbf{x}, \mathbf{x}^*)$ is the surface view factor between $\mathbf{x}$ and $\mathbf{x}^*$. It is easy to calculate $K(\mathbf{x}, \mathbf{x}^*)$ from Lambert's cosine law for diffuse radiation: letting $\mathbf{e}$ and $\mathbf{e}^*$ stand for the unit inward normals to the enclosure at $\mathbf{x}$ and $\mathbf{x}^*$ (fig. 22), one obtains

$$K(\mathbf{x}, \mathbf{x}^*) = \begin{cases} -\dfrac{1}{\pi} \dfrac{[(\mathbf{x}^* - \mathbf{x})\cdot\mathbf{e}][(\mathbf{x}^* - \mathbf{x})\cdot\mathbf{e}^*]}{[(\mathbf{x}^* - \mathbf{x})\cdot(\mathbf{x}^* - \mathbf{x})]^2}, & \text{if } \mathbf{x} \text{ and } \mathbf{x}^* \text{ "see" each other,} \\ 0, & \text{otherwise.} \end{cases} \tag{80}$$

Let us point out that relation (80) may easily be obtained from the rules governing transfer in a non-emitting and non-attenuating medium (as it is the case in surface radiation). The latter state that, within the enclosure, the intensity of a beam remains unchanged along its path. Hence,

$$i^*_{i\Lambda}(\mathbf{x}) = i^*_{o\Lambda}(\mathbf{x}^*), \tag{81}$$

if $\mathbf{x}$ and $\mathbf{x}^*$ see each other. Relations (79) and (80) are then obtained by appropriate

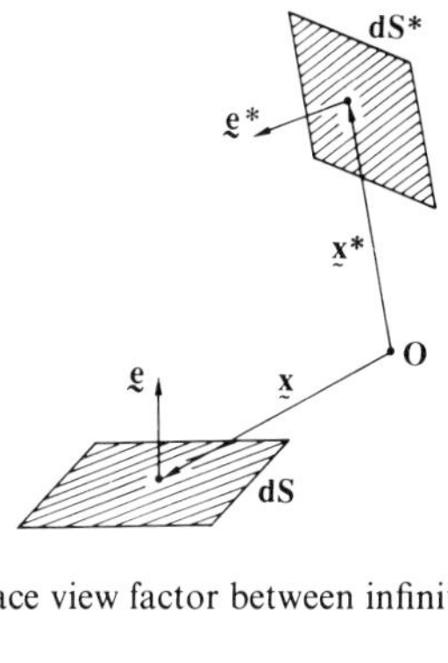

Fig. 22. Surface view factor between infinitesimal areas.

integration over all polarizations **p** and locations $\mathbf{x}^*$ on ∂V, and by taking into account the hypothesis of diffuse radiation.

The next step is to introduce the net flux at $\mathbf{x}$ in the range Λ, assumed to be positive when energy enters the enclosure:

$$q_\Lambda(\mathbf{x}) = q_{o\Lambda}(\mathbf{x}) - q_{i\Lambda}(\mathbf{x}). \tag{82}$$

Eliminating $q_{o\Lambda}(\mathbf{x})$ and $q_{i\Lambda}(\mathbf{x})$ from eqs. (78), (79) and (82) provides the integral relationship governing radiative transfer within the enclosure in the range Λ

$$\frac{q_\Lambda(\mathbf{x})}{\varepsilon_\Lambda(\mathbf{x})} - \int_{\mathbf{x}^*\in\partial V} K(\mathbf{x},\mathbf{x}^*)\frac{1-\varepsilon_\Lambda(\mathbf{x}^*)}{\varepsilon_\Lambda(\mathbf{x}^*)} q_\Lambda(\mathbf{x}^*)\,\mathrm{d}S^*$$
$$= \gamma_\Lambda(\mathbf{x})n^2\sigma T^4(\mathbf{x}) - \int_{\mathbf{x}^*\in\partial V} K(\mathbf{x},\mathbf{x}^*)\gamma_\Lambda(\mathbf{x}^*)n^2\sigma T^4(\mathbf{x}^*)\,\mathrm{d}S^*. \tag{83}$$

Extension of the above model is straightforward when using the band-energy method. The spectrum is divided in several ranges of wavelengths Λ_K, over which emissivities are assumed to be frequency-independent, whence equations similar to eq. (83) hold. The total radiative flux is then simply obtained by adding up all the contributions calculated over these different ranges [59].

In a similar way, semi-transparent materials, such as quartz tubes, or the boric oxide layer in GaAs growth, may often be considered as transparent within some range Λ_I of wavelengths, and as opaque within the remaining range Λ_II of the spectrum (this method was introduced by Nicodème et al. [75], Crochet et al. [224], Dupret et al. [225], and Thomas et al. [226]). It is then possible to define two different enclosures, where radiative exchanges take place with different wavelengths (provided the index of refraction n of the medium is not too different from 1, which allows one to neglect refraction). Enclosure I corresponds to waves for which the layer is transparent, while enclosure II corresponds to waves for which it is opaque. This model is illustrated in the simplified example of fig. 23. The total radiating flux $q_\Lambda(\mathbf{x})$ is thus obtained by adding up the contributions $q_{\Lambda\mathrm{I}}(\mathbf{x})$ and $q_{\Lambda\mathrm{II}}(\mathbf{x})$ on boundaries common to both enclosures, while a single contribution is taken on boundaries which belong to enclosures I or II only. The importance of encapsulant semi-transparency in LEC growth was first demonstrated by Motakef [73], and Motakef and Witt [72].

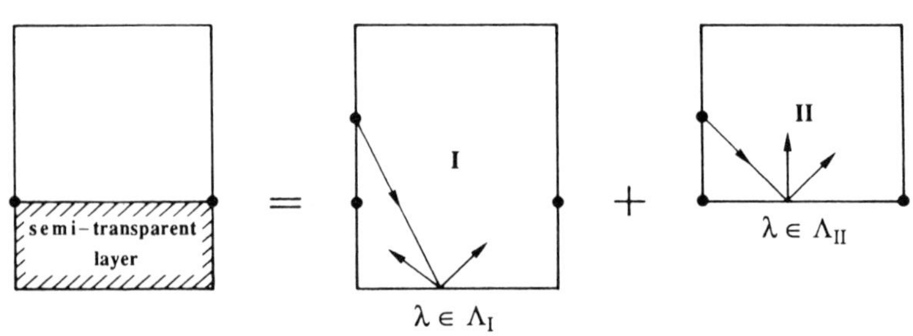

Fig. 23. Application of the band-energy method for calculating heat exchange in a semi-transparent medium.

4.2.2. Gray model for the equation of transfer

While surface radiation models are appropriate for a global modelling of the growth process, provided calculation of the view factors be accurately performed, they are not sufficient when internal radiation is non-negligible in the melt or the grown crystal. This is expected to occur within the crystal in the infrared range of the spectrum for many oxides, such as gadolinium gallium garnet, yttrium alluminium garnet, and sapphire [67, 156], but also to have some importance within both crystal and melt for semiconductors such as silicon, or gallium arsenide, or indium phosphide [227]. In this latter case, the molten or solid material is assumed to be almost opaque to infrared radiation, but to be much less absorbing at the extremities of the [1–16 μm] band.

The presence of a non-negligible internal radiation renders the problem so complex that additional simplifications are required when the final objective is numerical simulation. A first class of simplifications is found when the optical thickness of the material τ_λ, as defined in eq. (65), is clearly higher than 1 in some range of wavelengths, but not so high as to permit using the opaque body model. This situation allows one to use the Rosseland approximation, which has been introduced in section 4.1, or the differential approximation, which will be examined in the next section. The most difficult case is found when the optical thickness is of the same other of magnitude as 1, or slightly lower, which means that neither the diffusion, nor the transparent body approximations are correct.

The problem of solving the equation of transfer within the crystal has been addressed by Brandon and Derby [67, 156, 266] under the following hypotheses:

(i) radiation is assumed to be diffuse;
(ii) scattering is neglected;
(iii) optical properties are supposed to be uniform;
(iv) the boundary of the crystal is supposed to consist of opaque surfaces;
(v) the participating medium is considered as being gray.

While hypotheses (i) to (iii) are essential and should not be removed, hypothesis (iv) limits the range of applications of this model to VB growth. As a matter of fact, Cz calculations would require a coupling between radiation within the crystal and radiative transfer in the surrounding enclosure, which represents an extremely complex and expensive problem (considering the particular difficulties related to calculation of viewed and hidden parts of the enclosure). Moreover, it should be kept in mind that, when internal radiation is present, but diffusion approximations are not valid, simplification of the boundary conditions imposed at the surface of the crystal – such as assuming diffuse transmission and refraction – is not possible. Indeed, this approach would introduce errors of the same order of magnitude as the ones which are involved in using diffusion models. This shows the difficulty of extending the methods developed in [67, 156, 266] (note however that Xiao and Derby [265] consider that the crystal surface is mainly opaque in Cz growth). On the other hand, it is clear that hypothesis (v) can easily be relaxed by using a band-energy approach (with however a strong increase of computational cost).

Under hypotheses (i) to (iii), and (v), and taking into account eqs. (55) to (58), and (62), the equation of transfer may be written as follows:

$$\nabla i'(\mathbf{m}) \cdot \mathbf{m} + a[i'(\mathbf{m}) - i'_b] = 0, \tag{84}$$

where $i'(\mathbf{m})$ is the intensity in direction $\mathbf{m}$, as obtained by integrating the spectral polarized intensity $i^*_\lambda(\mathbf{m},\mathbf{p})$ over all polarization directions and on the full spectrum, while a is the absorption coefficient of the medium, which is assumed to be a constant $(\forall\lambda\in(0,\infty), a_\lambda = a)$, and i'_b is the intensity of the blackbody on the whole spectrum,

$$i'_\mathrm{b} = \frac{1}{\pi} e_\mathrm{b}. \tag{85}$$

(The total emissive power of the blackbody e_b is defined in eq. (66).)

According to [67, 156, 266], it is easy to integrate the equation of transfer (84) and to complete the model as follows: let $\mathbf{x}$ be a given point inside the medium, and $\mathbf{x}^*$ be the point of the boundary ∂V such that the beam of direction $\mathbf{m}$ arriving at $\mathbf{x}$ originates from $\mathbf{x}^*$; let also $\mathbf{x}'$ be a current point on the segment $\overline{\mathbf{x}^*\mathbf{x}}$, and let symbols l and l' denote the distances between $\mathbf{x}^*$ and $\mathbf{x}$ or $\mathbf{x}'$, respectively. From eqs. (84) and (85), it is easy to obtain the following integral equation:

$$i'(\mathbf{x},\mathbf{m}) = i'(\mathbf{x}^*,\mathbf{m}) \exp(-al) + a\int_0^l e_\mathrm{b}(T(\mathbf{x}')) \exp[-a(l-l')]\,\mathrm{d}l'. \tag{86}$$

The key point is here to introduce the radiative flux vector $\mathbf{q}_\mathrm{R}$, with the aim of coupling the model to other heat transfer modes,

$$\mathbf{q}_\mathrm{R} = \int_{m\in\Sigma} \mathbf{m}\, i'(\mathbf{m})\,\mathrm{d}m, \tag{87}$$

where Σ is the unit sphere. At the surface of the crystal, boundary conditions are easily expressed in terms of this radiative flux $\mathbf{q}_\mathrm{R}$, of the surface emissivity ε (which is assumed not to depend on wavelength), and of the unit inward normal $\mathbf{e}^*$ to ∂V. Letting q_o and q_i denote the radiosity and the incident radiative flux, respectively, the following relation may be derived by using similar methods as in the previous section (eqs. (77) and (78)):

$$q_\mathrm{o}(\mathbf{x}^*) = \varepsilon(\mathbf{x}^*) e_\mathrm{b}(T(\mathbf{x}^*)) + [(1-\varepsilon(\mathbf{x}^*)]q_\mathrm{i}(\mathbf{x}^*). \tag{88}$$

On the other hand, the total radiative flux normal to $\mathbf{e}^*$, i.e. $\mathbf{q}_\mathrm{R}\cdot\mathbf{e}^*$, and the intensity $i'(\mathbf{m},\mathbf{x}^*)$ obey the relations

$$\mathbf{q}_\mathrm{R}(\mathbf{x}^*)\cdot\mathbf{e}^* = q_\mathrm{o}(\mathbf{x}^*) - q_\mathrm{i}(\mathbf{x}^*), \tag{89}$$

and

$$i'(\mathbf{x}^*,\mathbf{m}) = \frac{1}{\pi} q_\mathrm{o}(\mathbf{x}^*), \tag{90}$$

where it has been taken into account that the intensities entering the crystal are diffuse. Eliminating $q_\mathrm{o}(\mathbf{x}^*)$ and $q_\mathrm{i}(\mathbf{x}^*)$ from eqs. (88) to (90) provides the sought after boundary condition:

$$i'(\mathbf{x}^*,\mathbf{m}) = \frac{1}{\pi}\left(e_\mathrm{b}(T(\mathbf{x}^*)) - \frac{1-\varepsilon(\mathbf{x}^*)}{\varepsilon(\mathbf{x}^*)} \mathbf{q}_\mathrm{R}(\mathbf{x}^*)\cdot\mathbf{e}^* \right). \tag{91}$$

As a conclusion, the radiative flux $\mathbf{q}_\mathrm{R}$ inside the crystal is related to the internal and boundary temperatures $T(\mathbf{x}')$ and $T(\mathbf{x}^*)$ and to the normal boundary radiative flux

$\mathbf{q}_R(\mathbf{x}^*)\cdot\mathbf{e}^*$ by relations (86), (87), and (91). The use of the radiative flux in the equations permits one to couple internal radiation in the weak formulation of the problem (section 3.2). (Note that a different approach is used in [228, 229].)

4.2.3. *Milne–Eddington differential model*

Assuming diffuse radiation, the use of matched asymptotic expansions of the spectral intensity $i'_\lambda(\mathbf{m})$ in terms of Legendre spherical harmonics, with introduction of the moments of this intensity in the differential form of the equation of transfer (as given by eq. (84) in the case of a gray material), leads to successive approximations of the spectral intensity. Among these methods, the P_1, or Milne–Eddington, approximation has been of particular interest for the modelling of internal radiation in crystal growth when the optical thickness of crystal and/or melt τ_λ, eq. (65), is higher than 1 (see the work of Amlin and Korpela [230], and Matsushima and Viskanta [227, 267]). In view of the complexity of this approach, we will however restrict ourselves to the basic aspects of the theory and refer the reader to detailed literature (see e.g. [268] for a general analysis, or [231, 269] for a review of the methods used for solving internal radiation problems).

The zeroth-order moment of the spectral intensity is the irradiance J_λ,

$$J_\lambda = \int_{\mathbf{m}\in\Sigma} i'_\lambda(\mathbf{m})\,\mathrm{d}m, \tag{92}$$

which represents the first basic unknown of the problem (the second one being temperature), and depends on wavelength, position, and time, but no longer on the beam direction $\mathbf{m}$. It is clear that irradiance is 4π times the average intensity at the given location and time. The first-order moment of the spectral intensity is the spectral radiative flux vector $\mathbf{q}_{R\lambda}$,

$$\mathbf{q}_{R\lambda} = \int_{\mathbf{m}\in\Sigma} i'_\lambda(\mathbf{m})\mathbf{m}\,\mathrm{d}m, \tag{93}$$

which is related to the irradiance by the relation

$$\mathbf{q}_{R\lambda} = -\frac{1}{3a_\lambda}\nabla J_\lambda, \tag{94}$$

where a_λ is the spectral absorption coefficient of the medium, eqs. (58) and (62). The second-order moment of the intensity is related to the irradiance by the relation.

$$\int_{\mathbf{m}\in\Sigma} i'_\lambda(\mathbf{m})\,\mathbf{m}\mathbf{m}\,\mathrm{d}m = \tfrac{1}{3}\mathbf{I}J_\lambda, \tag{95}$$

where $\mathbf{I}$ is the identity tensor. The approximate form of the intensity provided by the above moment is

$$i'_\lambda(\mathbf{m}) = \frac{1}{4\pi}\left(J_\lambda - \frac{1}{a_\lambda}\nabla J_\lambda\cdot\mathbf{m}\right). \tag{96}$$

The equation governing irradiance represents an approximation of the equation of transfer in a non-scattering medium, and is written as follows:

$$\nabla\cdot\mathbf{q}_{R\lambda} = a_\lambda(4e_{b\lambda} - J_\lambda), \tag{97}$$

where the spectral emissive power of the blackbody in vacuum is defined in eq. (52). When the refraction index n is not 1, a factor n^2 must be introduced in the calculation of $e_{b\lambda}$ in order to account for the actual speed of light in the medium, eq. (53), and the use of the concept of equivalent wavelength in vacuum. It is easy to introduce eq. (94) into eq. (97), in order to obtain an equation for J_λ. In practice, using a band-energy approach, i.e. assuming that the spectral absorption coefficient is constant over a family of wavelength ranges covering the spectrum, provides an efficient way to solve the problem [227].

The boundary conditions for irradiance associated with the Milne–Eddington model depend on the radiative properties of the adjacent medium. In any case, the spectral incident flux $q_{i\lambda}(\mathbf{x}^*)$ and the radiosity $q_{o\lambda}(\mathbf{x}^*)$ obey a relation similar to eq. (89),

$$\mathbf{q}_{R\lambda}(\mathbf{x}^*)\cdot\mathbf{e}^* = q_{o\lambda}(\mathbf{x}^*) - q_{i\lambda}(\mathbf{x}^*), \tag{98}$$

while integrating eq. (96) for all incident directions permits one to relate these fluxes to the irradiance,

$$q_{i\lambda}(\mathbf{x}^*) = \tfrac{1}{4}J_\lambda(\mathbf{x}^*) - \tfrac{1}{2}\mathbf{q}_{R\lambda}(\mathbf{x}^*)\cdot\mathbf{e}^*. \tag{99}$$

For an opaque surface, relations similar to eqs. (77) hold, and we may write

$$q_{o\lambda}(\mathbf{x}^*) = \varepsilon_\lambda(\mathbf{x}^*)e_{b\lambda}(T(\mathbf{x}^*)) + [1 - \varepsilon_\lambda(\mathbf{x}^*)]q_{i\lambda}(\mathbf{x}^*), \tag{100}$$

where $e_{b\lambda}(T(\mathbf{x}^*))$ denotes the spectral blackbody emissive power at the wall, as given by eqs. (52) and (53). The sought after boundary condition is easily obtained by eliminating $\mathbf{q}_{R\lambda}(\mathbf{x}^*)$, $q_{i\lambda}$ and $q_{o\lambda}$ between eqs. (94), and (98) to (100).

Consider now the case of a transparent adjacent medium, and let it be referred to by superscript e in the equations. The balance between reflected and transmitted fluxes in both senses yields the relations

$$\begin{aligned} q_{o\lambda}(\mathbf{x}^*) &= \rho_\lambda(\mathbf{x}^*)q_{i\lambda}(\mathbf{x}^*) + [1 - \rho^e_\lambda(\mathbf{x}^*)]q^e_{i\lambda}(\mathbf{x}^*), \\ q^e_{o\lambda}(\mathbf{x}^*) &= \rho^e_\lambda(\mathbf{x}^*)q^e_{i\lambda}(\mathbf{x}^*) + [1 - \rho_\lambda(\mathbf{x}^*)]q_{i\lambda}(\mathbf{x}^*), \end{aligned} \tag{101}$$

where it must be taken into account that, from the second law of thermodynamics, the product $n^2(1-\rho_\lambda)$ is the same on both sides of the surface. The boundary condition for irradiance is then obtained by eliminating $\mathbf{q}_{R\lambda}(\mathbf{x}^*)$, $q_{i\lambda}$, $q_{o\lambda}$, and $q^e_{o\lambda}$ between eqs. (94), (98), (99), (101), and the additional relations

$$-\mathbf{q}_{R\lambda}(\mathbf{x}^*)\cdot\mathbf{e}^* = q^e_{o\lambda}(\mathbf{x}^*) - q^e_{i\lambda}(\mathbf{x}^*) \tag{102}$$

In order to couple the calculation of temperature and irradiance – which is required because the blackbody emissive power, and even the spectral absorption coefficient, can depend on temperature – it is necessary to introduce the radiative flux vector $\mathbf{q}_R$ in both the energy equation and the heat flux boundary conditions. When the differential approximation is used on the whole spectrum, the radiative flux vector is given by the integral

$$\mathbf{q}_R = \int_0^\infty \mathbf{q}_{R\lambda}\,d\lambda, \tag{103}$$

where $\mathbf{q}_{R\lambda}$ is given by eq. (94). Using different kinds of approximations in the spectrum (and, in particular, the Rosseland approximation, eqs. (69) and (70)) is also feasible and straightforward. It is worth noting that a temperature jump boundary condition needs to be imposed only when the Rosseland approximation is introduced. When this approximation is not used, calculations will exactly yield either a temperature jump, or a radiative thermal boundary layer, depending on whether the coefficient of heat conduction is, or is not, vanishing. Let us in addition mention here the interest of the P_3 approximation (see e.g. [232, 268, 269]), although radiative boundary conditions become more complex with this approach.

4.3. Numerical solution of radiative transfer in Cz or VB growth

In this section, the different techniques developed in the literature to discretize radiative transfer in Czochralski or vertical Bridgman furnaces will be investigated. As a first step, we will consider radiation in a transparent enclosure. As a second step, the solution of the equation of transfer within the crystal, considered as being a participating medium, will be examined briefly.

4.3.1. Radiative transfer in a transparent enclosure

Up to now, all the methods presented in the literature are based on the assumptions of diffuse radiation and axisymmetric geometry. It must be emphasized that the role of calculating radiation is here to provide appropriate thermal conditions at the surface of crystal and melt, in order to permit an accurate simulation of the phenomena taking place within the molten and solidified material. Hence, all available simplifications may be retained with a view to reducing the computational cost, provided a sufficient accuracy is attained at the surface of melt and crystal. Czochralski and vertical Bridgman furnaces are essentially axisymmetric. Moreover, in the former case, axisymmetry is reinforced by letting the crystal and (or) crucible rotate around the axis. The hypothesis of diffuse radiation is more difficult to validate, for specular effects cannot be neglected at the surfaces of melt and crystal. However, as (i) these surfaces are always perturbed by small undulations, and as (ii) axisymmetric treatment means that contributions to heat flux of all the points of the enclosure located on any given circle centred on the axis are integrated, the hypothesis of diffuse radiation remains acceptable.

In the sequel, two different approaches will be presented. The first one was investigated by Dupret et al. [58, 59], and Van den Bogaert [66], and is based on an axisymmetric integration of the radiative integral equation, and a discretization by means of the Bubnov–Galerkin finite element technique. The second one was investigated by Atherton et al. [60], and Bornside et al. [61], and consists in dividing the surface of the enclosure into a set of planar or axisymmetric patches, which are each assumed to be approximately isothermal, and in calculating the total-exchange areas (or the Gebhart factors) between every pair of such patches. Gebhart factors produce the relationships between net fluxes and radiative emitted fluxes (which are proportional to fourth powers of temperature when assuming gray radiation).

First method: axisymmetric Galerkin integration. The axisymmetric form of the radiative integral equation is obtained as follows. If (r, θ, z) denote the cylindrical

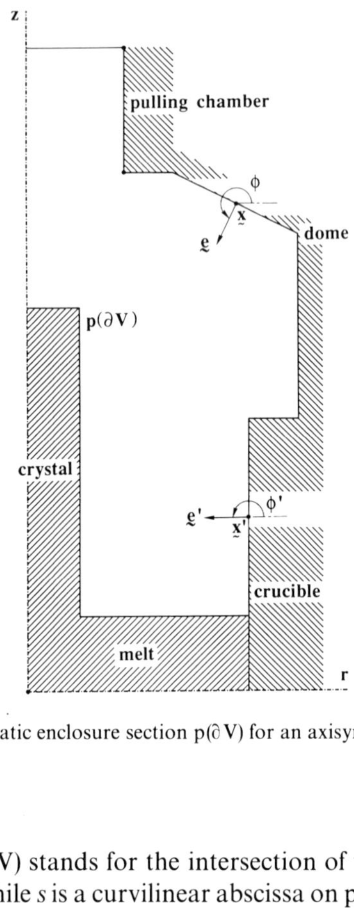

Fig. 24. Schematic enclosure section p(∂V) for an axisymmetric furnace.

coordinates of **x**, and p(∂V) stands for the intersection of the enclosure and the mid-halfplane $\theta = 0$ (fig. 24), while s is a curvilinear abscissa on p(∂V), eq. (83) is rewritten in the form

$$\frac{q_\Lambda(\mathbf{x})}{\varepsilon_\Lambda(\mathbf{x})} - \int_{\mathbf{x}'\in \mathrm{p}(\partial \mathrm{V})} r' K_c(\mathbf{x}, \mathbf{x}') \frac{1 - \varepsilon_\Lambda(\mathbf{x}')}{\varepsilon_\Lambda(\mathbf{x}')} q_\Lambda(\mathbf{x}')\, \mathrm{d}s'$$

$$= \gamma_\Lambda(\mathbf{x}) n^2 \sigma T^4(\mathbf{x}) - \int_{\mathbf{x}'\in \mathrm{p}(\partial \mathrm{V})} r' K_c(\mathbf{x}, \mathbf{x}') \sigma T^4(\mathbf{x}')\, \mathrm{d}s'. \tag{104}$$

Here, the axisymmetric view factor $K_c(\mathbf{x}, \mathbf{x}')$ is defined by the integral

$$K_c(\mathbf{x}, \mathbf{x}') = 2 \int_0^\pi K(\mathbf{x}, \mathbf{x}^*)\, \mathrm{d}\theta^*, \tag{105}$$

with

$$\mathbf{x} = (r, 0, z) \in \mathrm{p}(\partial \mathrm{V}),$$

$$\mathbf{x}' = (r', 0, z') \in \mathrm{p}(\partial \mathrm{V}),$$

$$\mathbf{x}^* = (r' \cos\theta^*, r' \sin\theta^*, z') \in \partial \mathrm{V}. \tag{106}$$

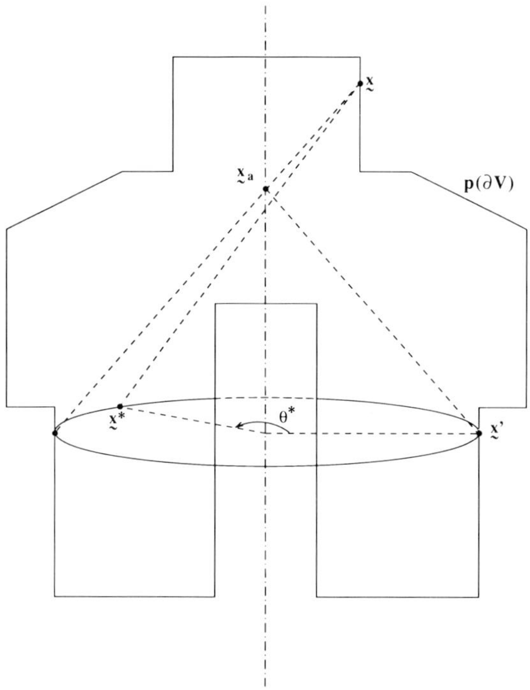

Fig. 25. Definition of the point $\mathbf{x}'$, as the circular projection of point $\mathbf{x}^*$ on the mid-halfplane $\theta^* = 0$.

The calculation of $K_c(\mathbf{x}, \mathbf{x}')$ from eqs. (105) (106) and (80) is very complex, and has represented one of the major difficulties to be solved in order to perform global simulations of the growth process, since viewed and hidden parts of the enclosure must be taken into account [59].

The procedure may be summarized as follows: for any given pair $(\mathbf{x}, \mathbf{x}')$ belonging to the section of the enclosure, $\mathbf{x}^*$ stands for a generic point of azimuthal coordinate θ^*, located on a horizontal circle centred on the axis and passing by $\mathbf{x}'$ (fig. 25). When letting $\mathbf{x}^*$ turn around the axis, the keypoint is to determine on the latter circle the set of intervals which are such that the beam $\overline{\mathbf{x}^* \mathbf{x}}$ is not intercepted by the enclosure boundary (indeed, this means that "photons" may travel freely from $\mathbf{x}^*$ to $\mathbf{x}$, and reversely). Calculation of the axisymmetric view factor $K_c\,(\mathbf{x}, \mathbf{x}')$ is thereby reduced to summing the contributions of non-hidden points $\mathbf{x}^*$, which may easily be calculated analytically.

The method used by Dupret et al. [59] in order to determine whether a given beam $\overline{\mathbf{x}^* \mathbf{x}}$ is intercepted or not by the enclosure boundary consists in defining the circular projection $\mathrm{p}(\overline{\mathbf{x}^* \mathbf{x}})$ of the line $\overline{\mathbf{x}^* \mathbf{x}}$ as being the set of points of the mid-halfplane having the same radial and axial coordinates as one of the points of $\overline{\mathbf{x}^* \mathbf{x}}$ (fig. 26). It is easy to prove that any projected line $\mathrm{p}(\overline{\mathbf{x}^* \mathbf{x}})$ forms a segment of a hyperbola. Clearly, since, for any line $\overline{\mathbf{x}^* \mathbf{x}}$, interception occurs if, and only if, the projected hyperbola segment $\mathrm{p}(\overline{\mathbf{x}^* \mathbf{x}})$ crosses the enclosure section, a geometrical algorithm may be built, which provides the analytic value (up to roundoff errors) of the axisymmetric view factor $K_c(\mathbf{x}, \mathbf{x}')$ everywhere on the enclosure section.

The result of this calculation is expressed in the form of a sum:

$$K_c(\mathbf{x}, \mathbf{x}') = 2 \sum_{i=1}^{l_v} [I(\theta_{Mi}) - I(\theta_{mi})], \tag{107}$$

where the l_v intervals $(\theta_{mi}, \theta_{Mi})$ characterize the range of values of the azimuthal coordinate θ^* of $\mathbf{x}^*$ for which $K(\mathbf{x}, \mathbf{x}^*)$ does not vanish, while $I(\theta^*)$ is a known function, which depends on the coordinates of $\mathbf{x}$ and $\mathbf{x}'$ (an equivalent shadowing algorithm was also developed by Garot and Gendre [233]).

In order to discretize the radiative integral equation (104), the Bubnov–Galerkin technique is based on putting first this latter equation into a weak form: letting $q'(\mathbf{x})$ stand for an arbitrary test-function defined on the enclosure section $\mathrm{p}(\partial \mathrm{V})$, we write:

$$\forall q'(\mathbf{x}) \quad \text{of summable square,}$$

$$\left\langle\!\left\langle \left(\frac{1}{\varepsilon_A(\mathbf{x})} q_A(\mathbf{x}) - \left\langle\!\left\langle K_c(\mathbf{x}, \mathbf{x}') \frac{1 - \varepsilon_A(\mathbf{x}')}{\varepsilon_A(\mathbf{x}')} q_A(\mathbf{x}') \right\rangle\!\right\rangle_{\mathbf{x}' \in \mathrm{p}(\partial \mathrm{V})} \right) q'(\mathbf{x}) \right\rangle\!\right\rangle_{\mathbf{x}' \in \mathrm{p}(\partial \mathrm{V})}$$
$$= \langle\langle(\gamma_A(\mathbf{x}) n^2 \sigma T^4(\mathbf{x}) - \langle\langle K_c(\mathbf{x}, \mathbf{x}') \gamma_A(\mathbf{x}') n^2 \sigma T^4(\mathbf{x}') \rangle\rangle_{\mathbf{x}' \in \mathrm{p}(\partial \mathrm{V})}) q'(\mathbf{x}) \rangle\rangle_{\mathbf{x} \in \mathrm{p}(\partial \mathrm{V})}, \tag{108}$$

where the integrals $\langle\langle\rangle\rangle_{\mathbf{x}\text{ or }\mathbf{x}' \in \mathrm{p}(\partial \mathrm{V})}$ are performed on the section of the enclosure boundary, using an appropriate weight r in view of the axisymmetry of the problem†.

In a next step, one-dimensional finite element approximations are introduced for both the fourth power of temperature $T^4(\mathbf{x})$ and the net flux $q_A(\mathbf{x})$ on $\mathrm{p}(\partial \mathrm{V})$,

$$T^4(\mathbf{x}) = \sum_{j=1}^{N_E} T_j^4 \zeta_j(\mathbf{x}),$$
$$q_A(\mathbf{x}) = \sum_{j=1}^{N_E} Q_{Aj} \zeta_j(\mathbf{x}), \tag{109}$$

where the $\zeta_j(\mathbf{x})$ are selected as quadratic polynomials on each – assumed rectilinear – segment of $\mathrm{p}(\partial \mathrm{V})$, while T_j and Q_{Aj} stand for the N_E nodal values of temperature and net flux. The latter field is not constrained to be continuous at the edges of the enclosure section, while temperatrue continuity is imposed at a later stage.

Finally, a weighted residuals formulation of the integral equation (104) is established by selecting as test-functions the N_E shape functions $\zeta_i(\mathbf{x})$,

$$q'(\mathbf{x}) = \zeta_i(\mathbf{x}), \tag{110}$$

in such a way that, from eqs. (109), the system (108) can be put after some operations in the matrix form

$$\boldsymbol{Q}_{\mathrm{R}A} = \boldsymbol{F}_A \boldsymbol{T}^4, \tag{111}$$

† Notations are explained at the end of the chapter.

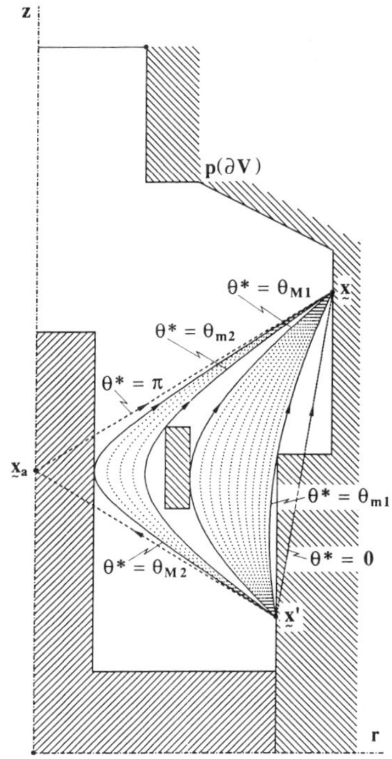

Fig. 26. Projected lines passing through **x** and **x**′ for a multi-connected enclosure (From [59]).

where $\boldsymbol{F}_{\Lambda}$ is a known matrix, while $\boldsymbol{Q}_{\mathrm{R}\,\Lambda}$ and $\boldsymbol{T}^4$ are the column vectors of nodal radiative fluxes (in the range Λ) and fourth powers of temperature. Imposing tempera-continuity along the enclosure means that $\boldsymbol{F}_{\Lambda}$ becomes a rectangular matrix. Relation (111) is the discrete form of the radiative integral equation which is used in global calculations (section 5).

Performing an efficient and accurate calculation of the matrix $\boldsymbol{F}_{\Lambda}$ is a very difficult task, for it involves the development of geometrical algorithms. The main problem is not to evaluate $K_c(\mathbf{x},\mathbf{x}')$, but to calculate the double integrals in eq. (108). Using pure 3×20 Gaussian quadrature for each pair of elements of $\mathrm{p}(\partial\,\mathrm{V})$ has been shown in [59] to give accurate results in most situations, but to provide poor values in some cases. The origin of this problem is that the axisymmetric view factor $K_c(\mathbf{x},\mathbf{x}')$ is not a *regular* function, in the sense that it may be discontinuous, or present derivative discontinuities, when regularity of the integrand is required for performing Gaussian quadrature. The solution of [59] has been to introduce division points when performing the inner integrals of eq. (108), i.e. the integrals in $\mathbf{x}'$. For a given integration node $\mathbf{x}$, division points were placed at locations where $K_c(\mathbf{x},\mathbf{x}')$ was known not to be regular in $\mathbf{x}'$, and so twenty-node Gaussian quadrature was performed separately on both sides of the division. The identity

$$\langle\langle K_c(\mathbf{x},\mathbf{x}')\rangle\rangle_{\mathbf{x}'\in \mathrm{p}(\partial \mathrm{V})} = 1, \tag{112}$$

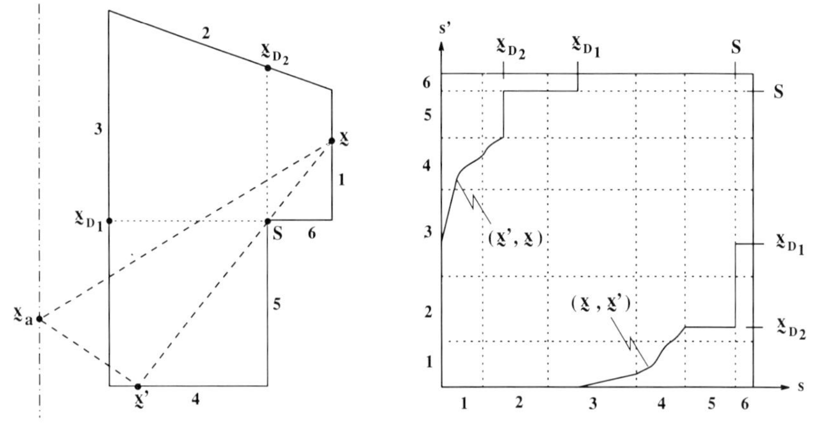

Fig. 27. Analysis of the discontinuities of the axisymmetric view factor: (a) schematic enclosure section; the salient edge S induces a discontinuity of the derivative of K_c in the (s, s') plane; (b) the discontinuity lines where K_c is non-regular have the shape of hyperbolae in the (s, s') plane.

was supposed to represent a stringent way for checking the quality of the results and for indicating whether a mesh refinement was necessary.

While several numerical tests have shown that quasi-steady solutions provided by the above algorithm seemed to be very accurate, a further analysis performed in the case of dynamic simulations has led Van den Bogaert [66] to strongly improve this method, both in efficiency and accuracy. First, the couples of locations $(\mathbf{x}, \mathbf{x}')$ where the function $K_c(\mathbf{x}, \mathbf{x}')$ is non-regular were determined carefully. Figure 27 depicts a simplified geometry and the corresponding lines where $K_c(\mathbf{x}, \mathbf{x}')$ is non-regular. (Note that, in fig. 27(b), coordinates s and s' denote the curvilinear abscissa of $\mathbf{x}$ and $\mathbf{x}'$, as measured along the enclosure section $p(\partial V)$.) Second and third improvements have consisted in introducing additional division lines where steep variations of $K_c(\mathbf{x}, \mathbf{x}')$ are found, and in performing a more accurate quadrature in the vicinity of the re-entrant edges of the enclosure, on the sides of which radiative interchange is very high. The fourth, and most important, modification of the algorithm has borne on taking into account the axisymmetric view factor symmetry,

$$K_c(\mathbf{x}, \mathbf{x}') = K_c(\mathbf{x}', \mathbf{x}), \tag{113}$$

by performing simultaneously the integrals in symmetric patches of the domain (s, s'): when a couple of symmetric patches is considered, integration is carried out on the patch which provides the best accuracy. This method (which is illustrated in fig. 28) has permitted in [66] to reduce the number of integration nodes by using a 5×5 (instead of 3×20) Gaussian quadrature on each patch. A strong gain of accuracy (with respect to the method of Dupret et al. [59]) was observed, with a simultaneous reduction of the computational cost by a factor of about 3 (a weighted residuals formulation of eq. (112) was shown to provide a more reliable check of precision than this latter equation).

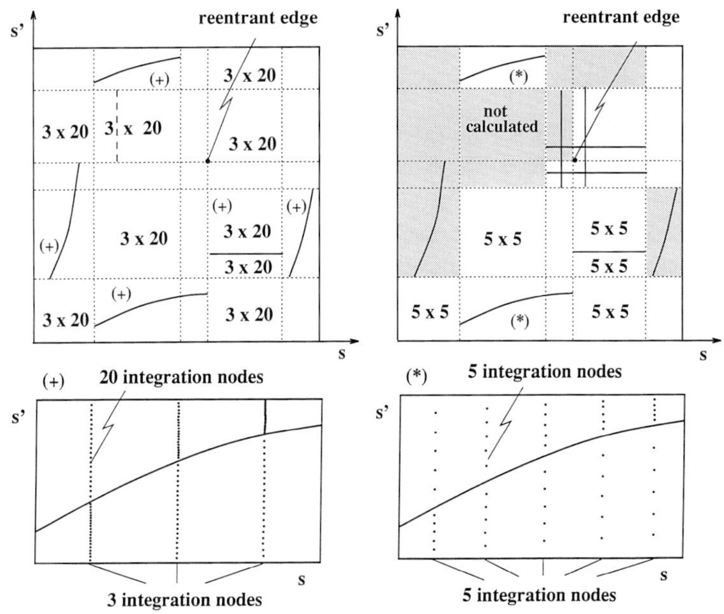

Fig. 28. Location of the integration points in the (s, s') plane for calculating the axisymmetric view factor: (a) method of Dupret et al. [59]; (b) method of Van den Bogaert [66].

It is worth noting that all these improvements have been demonstrated to be necessary for a precise evaluation of the heater power in the case of inverse dynamic simulations (where crystal shape is imposed). Hence, in general, we must conclude that global simulations are extremely sensitive to radiative transfer calculations.

Second method: calculation of the Gebhart factors between planar or axisymmetric patches. The approach of Atherton et al. [60], and Derby et al. [116] is based on subdividing the total area of the enclosure ∂V into a set of $N_s \times N_\theta$ facets F_A, and on assuming, for radiative transfer calculations, that temperature is constant on each such polygon. When using finite elements, the average temperature T_A can be selected as being the temperature of the centroid of the facet.

The method (which is based on the FACET algorithm [270]) may be summarized as follows: in a first step, the view factor $\mathscr{F}_{AB}$ between each pair of facets, i.e. the fraction of radiant energy leaving facet F_A which arrives directly at facet F_B, is evaluated, both facets being assumed as being blackbodies; in a second step, the radiative integral equation (104) is made discrete by introducing the concept of effective ambient temperature T_{aA}, which represents an equivalent temperature of the (assumed black) environment of the facet F_A; the discrete formulation of radiative transfer consists of a relation linking these effective ambient temperatures to the average temperatures T_B everywhere on the enclosure.

The calculation of view factors is based on integrating eq. (80) over a given pair of facets F_A and F_B – which are assumed to "see" each other – and on using the theorem of

Stokes, which yields the elegant formula:

$$\mathscr{F}_{AB} = \frac{1}{\bar{F}_A} \int_{F_A} \int_{F_B} K(\mathbf{x}_A, \mathbf{x}_B)\, dS_A\, dS_B$$

$$= \frac{1}{\pi \bar{F}_A} \int_{\partial F_A} \int_{\partial F_B} \ln(\mathbf{x}_A - \mathbf{x}_B)^2\, d\mathbf{x}_A \cdot d\mathbf{x}_B. \tag{114}$$

where $\bar{F}_A$ is the area of F_A. Equation (114) can easily be reduced to a line integral, which may be evaluated by means of numerical quadrature. The difficulty is again to take into account viewed and hidden parts in the calculations. In [270], a shadowing algorithm is developed, which is based on deciding whether a particular pair of facets (F_A, F_B) "view" each other by analyzing the line joining their centroids. If interception occurs, $\mathscr{F}_{AB}$ is set to zero.

The axisymmetry of Cz or VB furnaces has allowed Atherton et al. [60] to develop a particular strategy. The total enclosure boundary ∂V is subdivided into a set of N_s axisymmetric strips – which are generated by the elements forming the enclosure section (assumed as being rectilinear). These strips are basically themselves subdivided into a given number N_θ of quadrilateral facets, whose view factors $\mathscr{F}_{AB}$ are calculated. The view factor between a pair of axisymmetric strips is easily obtained by adding up the weighted contributions of their facets. A major observation in [60] is the high sensitivity of the solution to the values of N_s, and especially N_θ. As in the case of the method developed in [59, 66], the precision of view factors can easily be checked, since their sum must be 1, for a given facet or a given strip.

When the view factors $\mathscr{F}_{AB}$ between any pair of axisymmetric strips have been evaluated, every Gebhart factor $\mathscr{G}_{AB}$–i.e. the total fraction of energy emitted by strip A and absorbed by strip B–can be calculated. Assuming that the enclosure consists of strips of constant emissivity ε_A in the range Λ, it is possible to obtain from eqs. (77) and (78) the following balance:

$$(\mathscr{F}(\boldsymbol{I} - \boldsymbol{\varepsilon}) - \boldsymbol{I})\mathscr{G} + \mathscr{F}\boldsymbol{\varepsilon} = \boldsymbol{0}, \tag{115}$$

where $\mathscr{F}$ and $\mathscr{G}$ are the matrices of view and Gebhart factors $\mathscr{F}_{AB}$ and $\mathscr{G}_{AB}$, while $\boldsymbol{I}$ is the identity matrix and $\boldsymbol{\varepsilon}$ is the diagonal matrix whose elements are the emissivities ε_A (note that the matrices $\mathscr{F}(\boldsymbol{\varepsilon} - \boldsymbol{I})$ and $\mathscr{G}(\boldsymbol{I} - \boldsymbol{\varepsilon}^{-1})$ commute). It is easy to solve the system (115) in order to find the Gebhart factors.

Finally, the net radiative flux $q_\Lambda(\mathbf{x})$ leaving a given strip A can, from eqs. (73), (77), and (82), be put in the form

$$q_\Lambda(\mathbf{x}) = \varepsilon_A \gamma_A n^2 \sigma (T_A^4 - T_{aA}^4), \tag{116}$$

where the ambient temperature T_{aA} is defined by the implicit relation

$$\varepsilon_A \gamma_A \bar{F}_A T_{aA}^4 = \sum_B \varepsilon_B \gamma_B \bar{F}_B \mathscr{G}_{BA} T_B^4. \tag{117}$$

The N_s equations (117) introduce in the system N_s additional unknowns (T_{aA}), which themselves provide in eq. (116) the correct fluxes at the boundary of ∂V when performing global calculations.

When the coefficients γ_A in eqs. (116) and (117) are not set to 1, semi-transparent materials can be treated exactly as when using the first method [226].

The extremely high computational cost of the method developed in [60], as based on subdividing the surface of the enclosure into planar facets, has led Brown et al. [234], and Bornside et al. [61, 271] to propose important improvements in order to adapt the method of Gebhart to axisymmetric problems. On the basis of the work of Rea [47], Williams and Reusser [48], Ramachandran and Duduković [49], and Srivastava et al. [50, 235], these authors introduce a view factor algebra, in such a way that view factors for unknown geometries are determined as algebraic functions of view factors for known geometries. The axisymmetric strips covering the surface of the enclosure are no longer subdivided in planar facets, but view factors between every pair of axisymmetric strips are directly evaluated. The view factor algebra is based on two principles, the first one being symmetry (as easily demonstrated from eqs. (80) and (114)), and the second one conservation of energy. This method is illustrated in fig. 29, where the view factors $\mathscr{F}_{AB}$ and $\mathscr{F}_{BA}$ between two annular rings F_A and F_B are calculated from the view factors between the inner ring F_A and two hollow disks F_C and F_D, whose inner and outer radii are those of F_A and F_B, respectively. It is clear that the radiant energy leaving F_A and attaining directly F_B is the difference between the radiant energies leaving F_A and reaching F_C and F_D. View factor algebra is a general extension of such technique.

This method permits one to replace a complex shadowing algorithm, as based on geometrical and tests logical decisions, by a less expensive algebraic calculation of view factors. In this case, using Newton's method by means of a perturbation technique

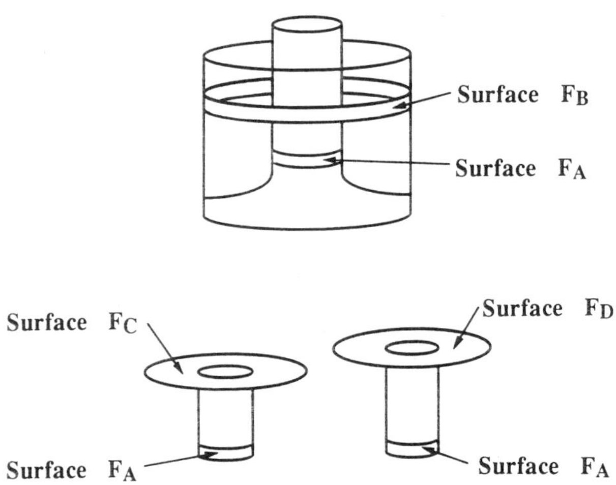

$$\mathcal{F}_{AB} = \mathcal{F}_{AC} - \mathcal{F}_{AD}$$

$$\mathcal{F}_{BA} = (\bar{F}_A/\bar{F}_B)\,\mathcal{F}_{AB}$$

Fig. 29. Use of view factor algebra in order to calculate radiation from the crystal to the wall of the quartz liner (From [61]).

becomes feasible. It must be clearly understood that this approach does not require to predict the occurrence of shadowing, but cannot be extended to an exact Newton method (with an algebraic calculation of the Jacobian matrices). Indeed, in the general case (and, in particular, dynamic simulations, where several constituents of the furnace are moving with respect to each other, or when complex reflectors are used – see fig. 26), this would require to develop an extremely complex shadowing algorithm, based on logical decisions. The numerical solution of the global problem will be examined in section 5.1.

4.3.2. Radiative transfer in a semi-transparent medium

When internal radiation is taken into account in the model, particular numerical techniques are required in order to evaluate the radiative fluxes $\mathbf{q}_R$ everywhere in the participating medium. These techniques will only be outlined here.

In order to solve the equation of transfer in the crystal, Brandon and Derby [67, 156, 266] calculate the intensities and the randiative fluxes, eqs. (86) and (87), by defining at each integration point of the mesh a system of local spherical coordinates, and by using multiple Gaussian quadratures in order to find the solution (by means of the Newton–Raphson scheme). These multi-dimensional integrals are found to be computationally very expensive. The radiative and conductive fluxes are added in the weak formulation of the energy equation (section 3.2).

The numerical technique developed by Matsushima and Viskanta [227, 267] for solving the differential approximation of the equation of transfer, eqs. (94), (97), and (103), is not based on using the finite element technique – on which we wish to focus in the present chapter – but on a control volume scheme [182]. Nevertheless, it is worth observing that putting eq. (97) into a weak form would be very easy, and would facilitate the coupling of internal radiation with the global model.

4.4. Examples and discussion

Several examples showing the importance of radiative heat exchanges in crystal growth processes will be considered in this section. We first study the case of a LEC growth system and show how modifications of furnace geometry can considerably improve crystal quality. Internal radiation in a VB configuration will then be analyzed through two other examples. It will be shown that the temperature field and the interface position are highly dependent upon optical properties such as the absorption coefficient.

4.4.1. Influence of furnace geometry on thermal stresses in the crystal for LEC growth

As a first example, we will refer to the numerical experiments carried out by Nicodème et al. [75] and Dupret et al. [225] for a GaAs LEC furnace (fig. 30), which illustrate the influence of several factors on the resulting quality of the crystal measured in terms of thermoelastic stresses. Calculations are carried out with a quasi-steady global finite element model, where radiative heat exchanges are modelled as a surface phenomenon.

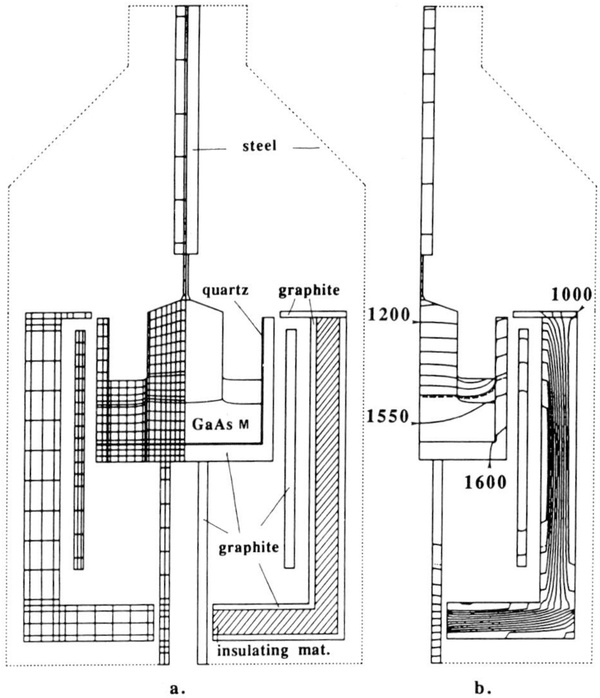

Fig. 30. LEC growth of GaAs: (a) sketch of the furnace and finite element meshes; (b) typical calculated temperature field in the furnace ($\Delta T = 50$ K) (From [225]).

The influence of pull rate and of geometrical configuration on thermoelastic stresses is investigated. Pull rate affects thermal stresses through modulation of the latent heat of fusion (wich appears in the local heat flux balance at the solidification front) and of the resulting interface shape. It is well known, for instance, that curved interfaces result in an increase of grown defects within the crystal. On the other hand, the geometry of the furnace strongly affects the temperature field and the thermal stresses through radiative heat transfer from the crystal wall.

As a first step (fig. 31), the authors have shown that a higher pull rate induces a more concave interface and a higher stress level, as represented by the von Mises invariant (see section 2.1). Peak stress values are always found at the interface and at the point where the crystal emerges from the encapsulant layer. As a second step, the addition of a heat shield located on the top of the crucible was shown to reduce the stress level in the crystal. The resulting temperature field and thermal stresses are represented in fig. 32. We observe that temperature gradients in the crystal decrease with the presence of a heat shield, as less heat is released by radiation from the crystal wall, thus leading to a general decrease of thermal stresses. While maximum stress values decrease with the presence of a heat shield, stresses in the top-end of the crystal are however increased. This is not really prejudicial, since the temperature is much lower in that part of the

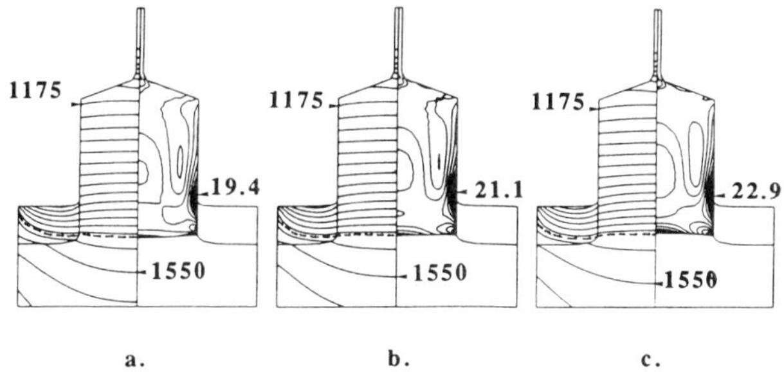

Fig. 31. Same problem as in fig. 30. Temperature field ($\Delta T = 25$ K) in the crystal, melt and encapsulant, and von Mises invariant in the crystal ($\Delta\sigma_{VM} = 2$ MPa) for different pull rates: (a) 0 cm/h; (b) 1 cm/h; (c) 2 cm/h (From [225]).

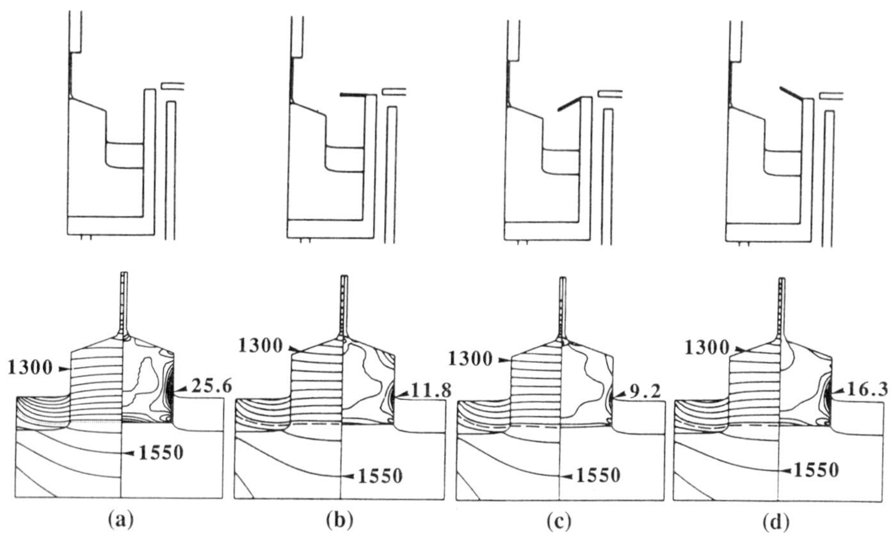

Fig. 32. Same problem as in fig. 30. Temperature field ($\Delta T = 25$ K) in the crystal, melt and encapsulant, and von Mises invariant in the crystal ($\Delta\sigma_{VM} = 2$ MPa) for different geometries: (a) without reflector; (b) with a horizontal reflector; (c) with a downwards oriented reflector; (d) with an upwards oriented reflector (From [225]).

crystal than close to the tri-junction, so that the critical resolved shear stress (CRSS) is higher. It may thus be concluded, on the basis of the ratio of the von Mises invariant to the CRSS, that crystal quality is generally improved with the addition of a heat shield. It could be demonstrated furthermore in [225] that inclining the heat shield downwards allows one to reduce the stress level even more, while inclining it upwards leads to poorer results.

It should be recalled that the conclusions drawn in [225] are only valid for the particular LEC furnace considered there and that no general rule should be inferred.

Moreover, the process should be viewed in its entirety. Hence, it must be realized that including a heat shield can be useful for reducing thermoelastic stresses, but could be disastrous for the stability of the process at the same time. As a matter of fact, addition of a heat shield causes radial thermal gradients in the melt to decrease due to the reduction of radiative heat loss from the melt surface. This is undesirable if a good control of crystal radius is to be achieved.

This example shows the possibilities of numerical simulation in order to optimize the geometry and the operating conditions in a furnace in pursuit of a particular objective. The strong influence of radiative effects was emphasized, indicating the necessity for a detailed radiation model if the temperature field and the quality of the crystal are to be calculated with accuracy. More examples related to thermal stress calculation in Cz growth can be found in section 5.2.

4.4.2. Influence of internal radiation on temperature and velocity fields in a VB silicon furnace

Most models which calculate the temperature field inside a crystal growth furnace consider radiation as a surface phenomenon. However, under high temperature conditions, radiative heat transfer inside the crystal and melt may be very important. Our second example refers to the work of Matsushima and Viskanta [227, 267], who showed the importance of internal radiation on a simple test problem representing a silicon melt and crystal in a VB furnace. The physical model, represented in fig. 33, consists of a rectangular domain including the melt at the bottom of the geometry and

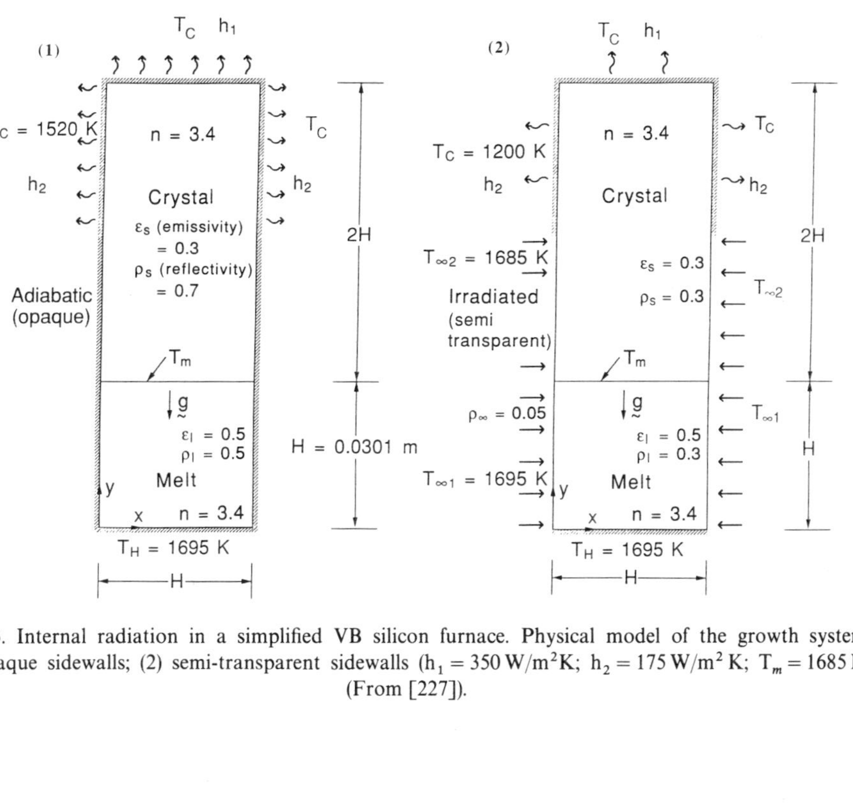

Fig. 33. Internal radiation in a simplified VB silicon furnace. Physical model of the growth system: (1) opaque sidewalls; (2) semi-transparent sidewalls ($h_1 = 350\ W/m^2K$; $h_2 = 175\ W/m^2\ K$; $T_m = 1685\ K$) (From [227]).

the crystal at the top. The solid–liquid interface is assumed to be flat, and its position is fixed. Semi-transparency of both crystal and melt is considered through the Milne–Eddington differential approximation, which has been described in section 4.2. For taking internal radiation into account, both a spectral-band model and a gray model for the absorption coefficient were tested. The results were similar enough to encourage the use of the simpler gray model. The crucible walls were choosen to be either insulated or semi-transparent and irradiated from the heater, this latter boundary conditions being more realistic. Natural convection was characterized by a Rayleigh number of 10^5. The influence of internal radiation on conduction and convection was analyzed through the use of different optical thicknesses, see section 4.1, and different boundary conditions on the crucible wall. In particular, it was found that internal radiation has a great influence on the temperature distribution close to the solid – liquid interface. Let us finally point out that though the Milne – Eddington differential model is not very well adapted for opacity coefficients clearly smaller than 1, small values were used in these calculations.

Figure 34 shows the temperature field calculated (a) in the absence, and (b–e) in the presence of internal radiation. Cases (a), (b) and (c) were obtained for insulated crucible

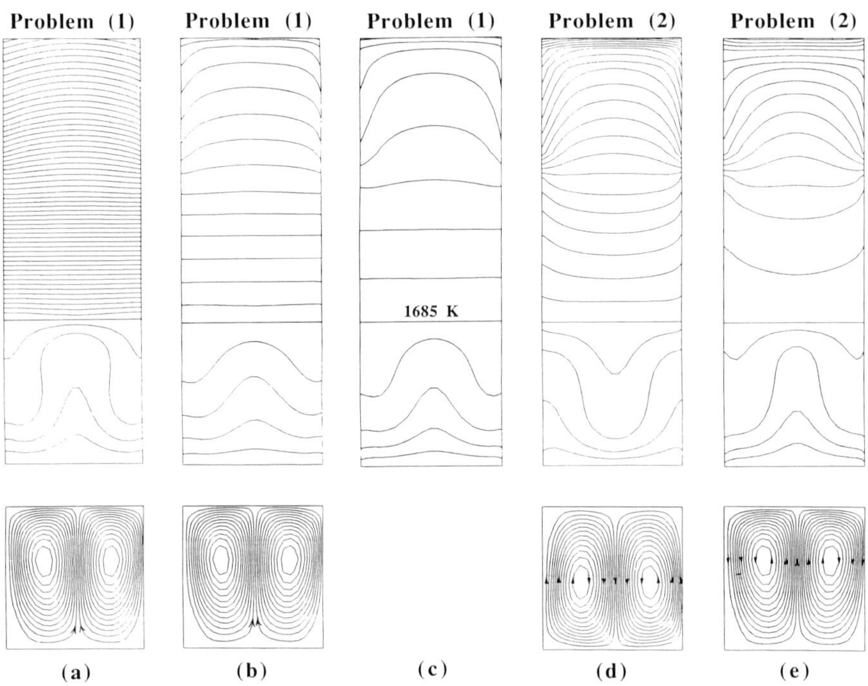

Fig. 34. Same problem as in fig. 33. Temperature field (isotherm spacing in 2 K) and melt flow for different radiative models: (a) in the absence of internal radiation; (b) with internal radiation ($\tau = 2.8$) and insulated crucible walls; (c) with internal radiation ($\tau = 1.0$) and insulated crucible walls; (d) with internal radiation ($\tau = 1.0$) and irradiated crucible walls; (e) with internal radiation ($\tau = 0.1$) and irradiated crucible walls (From [227]).

walls, while cases (d) and (e) correspond to irradiated crucible walls. Different optical thicknesses (τ) were tested. We may observe that the temperature gradient is of course highly reduced in the presence of internal radiation. This effect becomes more important as the transparency of the crystal increases, as can be observed from cases (b) and (c), where a reduction of the temperature gradient is observed when a lower opacity coefficient is used. When crucible walls are assumed to be irradiated from the surroundings (cases (d) and (e)), the temperature field in the melt is highly dependent upon the value of the opacity coefficient, which was not the case for adiabatic walls. For $\tau = 1.0$, radiant energy incident on the sides of the crucible is absorbed in the melt near the edges, resulting in a higher temperature near the wall; for a more transparent material ($\tau = 0.1$), radiant energy is absorbed in the center of the melt and increases the temperature in this region. This difference has a great impact upon the flow structure.

The flow patterns corresponding to the different cases are also represented in fig. 34. In the absence of internal radiation (a), the flow consists of two symmetric cells with the fluid ascending in the central region of the melt. With internal radiation and adiabatic crucible walls, the flow structure remains similar (case (b)). In the case of irradiated crucible walls together with $\tau = 1.0$ (d), the flow still consists of two symmetric cells, but the direction of circulation is reversed due to the higher temperature along the crucible walls (fluid is now ascending along the crucible walls). If the transparency is further increased ($\tau = 0.1$, case (e)), the flow pattern is again reversed because more radiant energy is absorbed in the central region of the melt. Interestingly, the authors could identify a critical value for the optical thickness ($\tau = 0.175$) at which the transition in the flow structure occurs. For this value, the flow pattern was not symmetric any more and a single rotating cell was found.

An interesting point was the comparison of the local conductive and radiative heat fluxes at the crystal – melt interface (fig. 35) for an opacity coefficient of 2.8. The maximum conductive flux was found to be three times smaller than the maximum radiative flux, clearly showing the importance of internal radiative transfer by comparison to diffusive transfer.

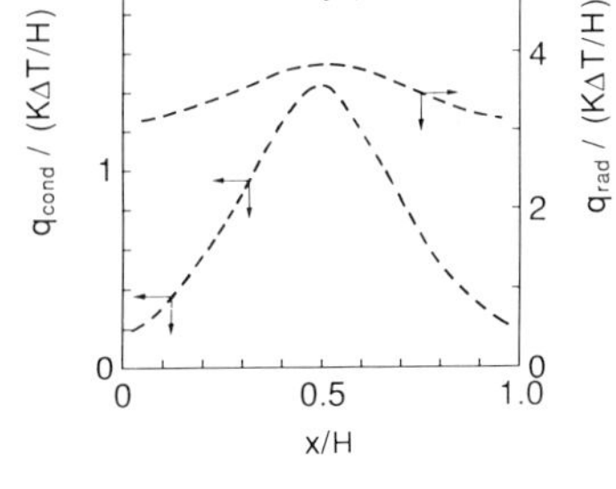

Fig. 35. Same problem as in fig. 33. Comparison of the local conductive and radiative heat fluxes at the crystal–melt interface ($\tau = 2.8$). Note that different scales are used for both variables. The maximum conductive flux is about three times smaller than the maximum radiative flux (From [227]).

4.4.3. *Influence of internal radiation on the temperature field and interface configuration in a VB oxide furnace*

Our third example deals with internal radiation in a VB system also and focuses on the model developed by Brandon and Derby [67, 266]. Calculation of the temperature field in the melt and the crystal is performed by means of a quasi-steady axisymmetric finite element model. The semi-transparent oxide crystal is considered to be a participating medium and the melt is assumed to be opaque. The model is based on the gray approximation for the equation of transfer (see section 4.2.2). No melt convection was taken into account at this stage, due to the high viscosity characteristic of oxides. Solid–liquid interface shape and position is a result of the calculations.

The growth system is represented in fig. 36. Local calculations are performed, as the ampoule is assumed to be in close proximity to the furnace, so that view factors due to complex geometrical effects need not be taken into account, and radiation toward a simplified environment may be introduced as an outer boundary condition. As a consequence of the quasi-steady-state assumption, the crystal growth rate is equal to the ampoule translation rate (1 mm/h). Material properties of melt and crystal were taken to be similar to those of YAG or GGG crystals.

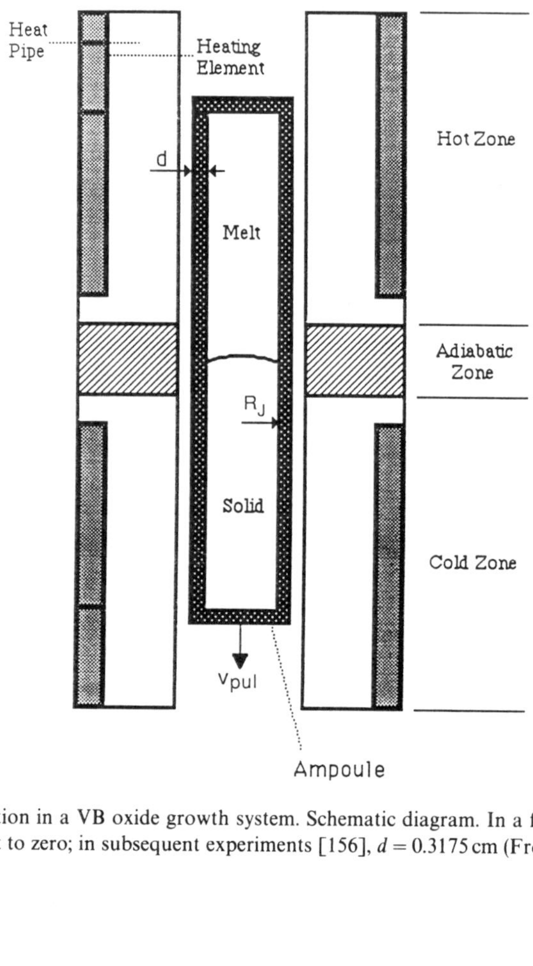

Fig. 36. Internal radiation in a VB oxide growth system. Schematic diagram. In a first set of experiments [67], d was set to zero; in subsequent experiments [156], $d = 0.3175$ cm (From [67, 156]).

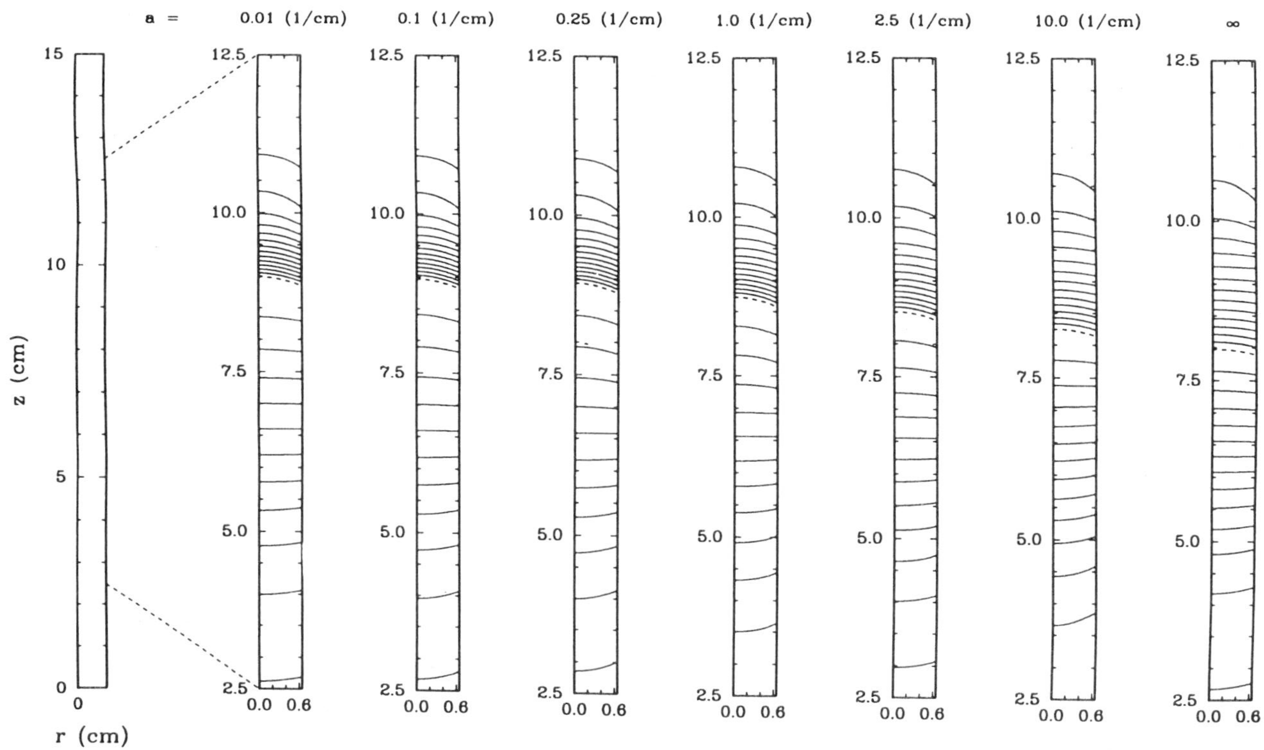

Fig. 37. Same problem as in fig. 36. Temperature field ($\Delta T = 15$ K) and solid–liquid interface configuration for increasing values of the optical absorption coefficient a in the solid. Only the middle section of the ampoule is shown, and the melt – crystal interface is represented by a dashed line (From [67]).

The effect of different optical properties was analyzed. Hence, the absorption coefficient a was varied from 0.01 to 10 cm^{-1}. The limit case of $a \to \infty$, corresponding to heat transfer occurring by conduction only was also considered. An important shift of interface position towards the hot zone is observed as the transparency of the crystal increases (fig. 37). This may be understood in the following manner. As the absorption coefficient decreases, heat transfer by radiation in the crystal is more important, and the total heat flux through the crystal is, therefore, increased. In order to satisfy the heat flux balance at the interface, thermal gradients in the melt (where heat transfer is assumed to occur by conduction only) become steeper as can be seen in fig. 38, and interface position is shifted upwards. Let us note that, despite the overall increase of axial heat flux, the conductive heat flux in the crystal is decreased, resulting in a reduced temperature gradient (fig. 38).

Other interesting simulations were related to letting the ampoule emissivity vary. These effects may be summarized as follows. Increasing the ampoule inner and outer emissivities results in an increased heat transfer between the furnace hot zone and the ampoule. Hence, curvature of the isotherms and, in particular, curvature of the solid–liquid interface is increased. In addition, due to the non-linear dependence of heat flux with respect to temperature, this increased heat transfer is favored for the hot zone and drives the interface towards the cold zone. Changing the ampoule inner emissivity only proved to have some impact on interface shape. However, a precise description of this effect could not be obtained with this simplified geometry.

As a matter of fact, the model was extended by Brandon and Derby [156] in order to take melt convection and thermal mass of the ampoule into account. Influence of melt

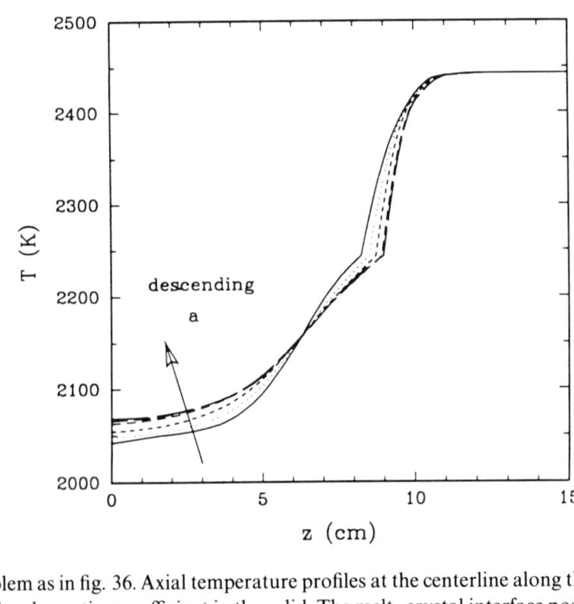

Fig. 38. Same problem as in fig. 36. Axial temperature profiles at the centerline along the crystal and melt for different values of the absorption coefficient in the solid. The melt–crystal interface position is identified by a discontinuity in the slope (From [67]).

convection proved to be unimportant for this small scale system, but coupling between heat conduction inside the ampoule and internal radiation within the crystal led to highly deflected interfaces near the ampoule wall that were not found with the simpler model. As a result, the effects of varying ampoule emissivities appeared to be quite different in [67] and [156]. It remains obvious however that emissivities play an important role regarding interface shape and that the choice of a suitable material and surface coating could help improve the process so that optimal interface morphology could be obtained. Minimizing interface curvature is usually desired, as high deflection induces diffusive radial solute segregation.

An interesting question put forth by the authors is whether Bridgman growth for participating media can attain the quasi-steady state achieved for numerous opaque crystals. Indeed, internal radiation can become very sensitive to geometry for highly transparent materials, so that solid–liquid interface configuration could continually change as growth proceeds.

We close this section by noting that the extreme sensitivity of temperature and velocity fields, and of interface position, with respect to thermal conditions at the crucible or ampoule walls that was observed in the last two examples indicate that boundary conditions have to be choosen very carefully if one wishes to obtain realistic results with local models.

5. *The global problem*

5.1. *Solution of the quasi-steady global problem*

The numerical solution of a global problem necessitates putting together all the equations of the system. The latter are furnished by the local melt–crystal model (section 3), the radiative enclosures (section 4), and the other constituents of the furnace (i.e the heater, the crucible or the ampoule, the pull rod and the pedestal in Cz growth, the insulator…) which, according to the terminology introduced in [57, 58], will all be called *macro-elements*. Briefly, the global method is based on separately analyzing the heat transfer within each macro-element, and on imposing continuity of temperature and heat flux on their interfaces, in order to couple the calculations.

The usefulness of a quasi-steady model – where the dynamics of the system geometry is assumed to be much slower than the transients associated with heat transfer within the furnace – remains very high, even at a time when dynamic global models become available [60, 65, 66, 116, 165, 273]. As a matter of fact, quasi-steady calculations are much less expensive than dynamic ones, and may provide important information on the sensitivity of the system to its different material and control parameters (such as emissivities, thermal conductivities, and also crystal diameter, pull rate…). In practice, many quasi-steady numerical experiments should always be performed before any dynamic simulation. A situation which is often encountered is the case when one or several material parameters are not well known. Here, the quasi-steady model provides an efficient way to compare the calculations with real experiments, in order to determine realistic values for the parameters.

Quasi-steady Cz problems are *inverse*, in the sense that a natural input of the system, which is generally selected as being the heater power, is calculated, while a natural output, i.e. the radius at the tri-junction, is imposed [59]. Instead of the heater power, alternative strategies have been to select as an unknown the pull rate [62], or the crystal radius (assuming a cylindrical shape) [60], or the melting temperature, but these approaches are of lower interest. In the case of VB global simulations, the problem is normally *direct*, while crystal radius must adapt to the shape of the ampoule. It is however possible to proceed as in Cz simulations, by imposing the location of the crystal–melt–ampoule contact line, and solving for the heater power.

In summary, in quasi-steady global calculations, the geometrical unknowns are strongly simplified, and reduce to the solid–liquid interface nodal vertical positions (except at the tri-junction, whose location is imposed in Cz, but not in VB quasi-steady models). Indeed, in Cz growth, the meniscus shape can be determined at the outset, and the crucible lift is imposed. Hence, using the notations of section 3.2, we have N_T unknown nodal temperatures in the so-called crystal – melt macro-element, together with N_F (in Cz simulations) or $N_\mathrm{F} - 1$ (in VB simulations) unknown nodal interface axial coordinates, and N_U additional unknowns which form a column vector $\boldsymbol{U}$; in the most general case, the latter could consist of the melt nodal velocities $\boldsymbol{V}$ (section 3.4), the nodal values of the electric potential $\boldsymbol{\Phi}$ (section 3.4), when a vertical or cusp magnetic field is used, the nodal concentrations $\boldsymbol{C}$ (section 3.4), if thermosolutal convection or segregation plays a non-negligible role, or even the nodal values of the irradiance $\boldsymbol{J}$ (section 4.3), if internal radiation is taken into account by means of a differential model, etc. It is clear that developing a general algorithm should not hide the particular difficulties related to introducing any of the above effects in the system, and to coupling a very high number of equations and unknowns.

5.1.1. Assembling technique

Different ways to perform global calculations have been presented in the literature [58–64, 72, 73, 75, 116, 224–226, 234, 236, 237]. Since most of these procedures are basically similar, we shall, in the sequel, follow the method of [59]. Particular aspects of alternative approaches will be examined at a later stage.

Let us start by recalling from section 3.2 that the solution of the local thermal problem – as defined in the crystal–melt macro-element – was found by resorting to a weighted residuals technique, which provided suitable relations, eqs. (23), (25), (31), (31), and (34), between the nodal temperatures and interface vertical positions. The system was closed by introducing appropriate boundary conditions, eqs. (24), supposed to represent the ambient environment of the crystal and the melt.

When dealing with global simulations, the treatment of natural boundary conditions in eq. (23) is quite different. At the boundary of the crystal–melt macro-element, the column vector $\mathbf{Q}^+$ of generalized nodal fluxes,

$$Q_i^+ = \langle\langle q\phi_i \rangle\rangle_{\partial\Omega}, \tag{117}$$

becomes unknown. We will also assume that, when induction heating is used, the heat supply w is proportional to the total power input in the system W, which we have explained to be an unknown in most cases of quasi-steady global Cz simulations.

Hence, we write

$$w = g(\mathbf{x}, X^{u})W, \tag{119}$$

where the function g is given and vanishes when induction heating is absent. It must be kept in mind that the dependence of g with respect to geometry can be strongly simplified in oxide growth, but may remain much more complex in semi-conductor growth. Let us on the other hand observe that conductive heat transfer in any other solid macro-element is governed by relations similar to eq. (23), as discretized from eqs. (26), (27), (32), (35), (118) and (119), except that boundary conditions of the form (24) remain imposed on the outer shell of the furnace. This analysis will permit us to express heat transfer in any solid macro-element of the overall furnace in a unique discrete formulation.

The gist of this method is to perform a *static condensation* on every macro-element, which means that internal nodal unknowns are eliminated in terms of the heater power W, and of the nodal temperatures $\boldsymbol{T}^{+}$ and generalized fluxes $\boldsymbol{Q}^{+}$ at the boundary of the macro-elements. In order to explain the procedure, let us first consider the simplest possible kind of opaque solid macro-element – which will be called *linear* – where the material properties (ρ, c, K) involved in eq. (23) are assumed to be constant and the internal radiative flux $\mathbf{q}_{R}$ is assumed to vanish. As discrete equations are linear, eliminating the internal nodal temperature $\boldsymbol{T}^{*}$ from the system readily gives a heat balance relation of the form:

$$\boldsymbol{A}\boldsymbol{T}^{+} + \boldsymbol{Q}^{+} = \boldsymbol{D}W + \boldsymbol{E}, \tag{120}$$

where the matrix $\boldsymbol{A}$ and the column vectors $\boldsymbol{D}$ and $\boldsymbol{E}$ are constant and may be calculated from eqs. (23), (24), (26), (27), (118), and (119). The right-hand side of eq. (120) relates the heat balance to the input power W, and to the external flux boundary conditions (through the vector $\boldsymbol{E}$, which originates from the conditions holding on the outer shell, eq. (24)) (note that $\boldsymbol{Q}^{+}$ concerns only internal nodes of the furnace). A careful analysis shows that, in the case of a *non-linear* macro-element (except a radiative enclosure), a similar relation to eq. (120) can be obtained, although $\boldsymbol{A}$, $\boldsymbol{D}$ and $\boldsymbol{E}$ will depend on some of the unknowns associated with the macro-element: in particular, as explained above, these hidden variables may involve, in the case of the crystal–melt macro-element, the following unknowns:

- The N_{F} or $(N_{F} - 1)$ unknown interface axial positions (which are comprised in X^{u}), depending on whether a VB or Cz configuration is considered.
- The N_{V} nodal velocities in the melt.
- The N_{T} internal and boundary nodal temperatures ($\boldsymbol{T}^{*}$ and $\boldsymbol{T}^{+}$), when material properties are temperature dependent.
- Nodal unknowns such as boundary temperatures or irradiance related to internal radiation, if it is present in the crystal and/or the melt.

Moreover, these additional variables may themselves be related to supplementary unknowns which are not explicitly present in the energy equation:

- The nodal concentrations in the melt, in the case of thermo-solutal convection.
- The nodal values of the electric potential, when dealing with magnetic fields.

Nevertheless, there are exactly as many additional equations in the crystal–melt macro-element as additional unknowns, plus one in Cz simulations; indeed, in this latter case, we must recall that melting temperature is imposed as a distinguished condition at the tri-junction, whose position is known.

In summary, static condensation provides a relation of the form (120) for any macro-element, except for *radiative enclosures* at the boundary of which relations of the form (111) hold. When several ranges of wavelength must be considered because radiative material properties are frequency dependent, it is very easy to add up their respective contributions in order to obtain a unique relation of the form

$$\boldsymbol{Q}_{\mathrm{R}} = \boldsymbol{F}\boldsymbol{T}^{4}, \tag{121}$$

at the boundary of the enclosure. In order to couple radiation with other modes of heat transfer, a vector of generalized nodal radiative fluxes $\boldsymbol{Q}_{\mathrm{R}}^{+}$ is defined by a relation similar to eq. (118),

$$Q_{\mathrm{Ri}}^{+} = \langle\langle q_{\mathrm{R}}\phi_{\mathrm{i}}\rangle\rangle_{\partial\mathrm{V}}, \tag{122}$$

where $q_{\mathrm{R}}(\mathbf{x})$ is the sum of the net radiative fluxes $q_{\Lambda}(\mathbf{x})$ in the different bands of wavelength Λ, eq. (82), as discretized using eq. (109). It is worth noting that the shape functions $\zeta_j(\mathbf{x})$ and the test-functions $\phi_i(\mathbf{x})$ are not identical, since the former are not continuous at the edges of the enclosure (because heat flux may be discontinuous), while the latter are. However, this is no problem and, eventually, from eqs. (122), (109) and (121), the generalized nodal heat fluxes can be expressed in the following form:

$$\boldsymbol{Q}_{\mathrm{R}}^{+} = \boldsymbol{B}\boldsymbol{Q}_{\mathrm{R}} = \boldsymbol{B}\boldsymbol{F}\boldsymbol{T}^{4}, \tag{123}$$

where $\boldsymbol{BF}$ is a square matrix.

Assembly of all the macro-elements which are not radiative enclosures can be performed in a next step provided the global mesh is consistent, i.e. that adjacent macro-elements have exactly the same nodes on their common boundary (fig. 39). Assembly means simply that all the balance equations of the form (120) which involve a given node on the *skeleton* of the furnace are added up (by skeleton, we mean the set of borders which connect the macro-elements of the overall furnace). According to basic finite element theory, this results in the annihilation of the generalized nodal fluxes Q_i^{+}

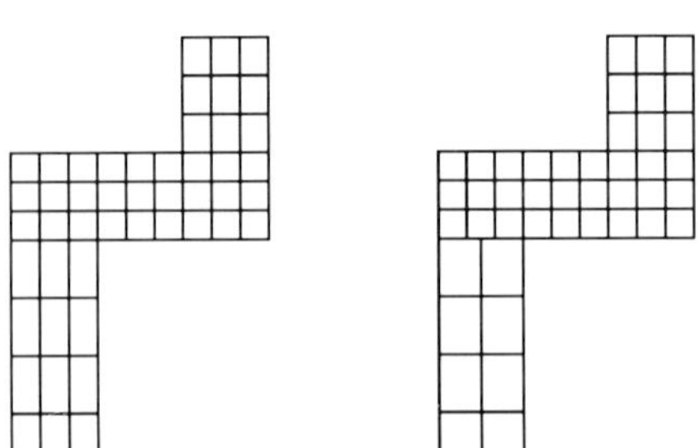

(a) (b)

Fig. 39. Example of (a) a consistent, and (b) a non-consistent mesh.

at any node which does not belong to a radiative enclosure. On the other hand, along the borders of radiative enclosures, the sum of the generalized nodal fluxes arising from balance eqs. (120) is precisely the radiative generalized nodal flux Q_{Ri}^{+}, eq. (122). Hence, substituting $\boldsymbol{BFT}^4$ for $\boldsymbol{Q}_R^{+}$ in the assembled balances, a relation of the form

$$\boldsymbol{A}^{(s)}\boldsymbol{T}^{(s)} + \boldsymbol{B}^{(s)}\boldsymbol{F}^{(s)}\boldsymbol{T}^{(s)4} = \boldsymbol{D}^{(s)}W + \boldsymbol{E}^{(s)}, \tag{124}$$

may be obtained on the skeleton of the furnace. Here, the vectors $\boldsymbol{T}^{(s)}$ and $\boldsymbol{T}^{(s)4}$ are defined by using a common numbering of nodal temperatures on the overall skeleton. The matrices $\boldsymbol{F}^{(s)}$ and $\boldsymbol{B}^{(s)}$ are rectangular, for (i) the number of nodal fluxes and temperatures on the enclosures is not exactly the same, and (ii) the nodal heat fluxes have been eliminated along adjacent conductive macro-elements (governed by eq. (120)).

In conclusion, the assembly procedure provides as many equations as unknown nodal temperatures on the skeleton. Additional unknowns are hidden in the non-linear macro-elements. The corresponding equations are provided by local analysis (sections 3.4 and 4.3). In the case of Cz growth, as heater power is normally unknown, the corresponding distinguished condition consists in imposing that the tri-junction is at the melting temperature. Hence, the global system is closed.

Many particular variants of macro-elements may be imagined. In [75, 226], semi-transparent materials are treated by superposing two radiative enclosures and a conductive macro-element, as explained in section 4.2. In [224, 225], the conductive/convective heat transfer in the ambient gas of a high pressurized enclosure is modelled by assuming that the core gas temperature T_c is almost constant, while appropriate coefficients characterize the heat transfer in the thin turbulent boundary layers which are present along the border of the enclosure; the additional unknown T_c is found by imposing heat flux balance on the macro-element. In [58], very thin shells on the outer boundary of the furnace are treated as one-dimensional macro-elements. The unknown distribution of power in an Ohmic resistor may be introduced in the system by considering the heater as a non-linear macro-element. In [58, 236], short one-dimensional segments considered as opaque blackbodies (or "twin" macro-elements) are introduced in order to subdivide the enclosure into smaller entities, with the aim of reducing the total computational cost. In [66], it is shown that setting the emissivity of twin macro-elements to a value of 2 is equivalent to modelling diffuse transmission across these segments.

The coupling strategy developed in [60, 116], although almost equivalent to the one presented above, deserves particular attention. Recall from section 4.3, eq. (117), that an effective ambient temperature is calculated everywhere on the overall enclosure, which is used to evaluate the radiative flux $q_{\Lambda}(\mathbf{x})$, eq. (116). Hence, the net radiative flux $q_R(\mathbf{x})$ is again obtained by summing all the contributions from the different wavelength ranges, the generalized nodal fluxes are calculated from eq. (122), and the subsequent coupling equations are essentially the same as above, except the solution technique. It is very useful to understand the role of the effective ambient (or sink) temperatures, in order to compare their influence with respect to that of an assumed known and constant environment.

At this stage, the difficulty remains to solve the global system, which may be highly non-linear. This problem is addressed in the next section.

5.1.2 Iterative strategies

Solving a global problem represents a very complex task in view of the high number of equations and unknowns the system involves. In this section, we shall only summarize some of the available techniques, while referring to the existing literature for more detail.

Alternate local – global iterations. In this scheme developed by Ryckmans et al. [177], each iteration consists of a local and a global step. During each local step, a local problem of the form described in sections 3.2 and 3.4 is solved. Thermal boundary conditions are furnished to the non-linear crystal – melt macro-element by using the results obtained from the previous global step. Pull rate is imposed, which means that, in Cz growth, an additional unknown must be introduced in the local system (as explained in section 3.2). In [177], this unknown is selected to be the melting temperature T_m involved in eq. (31). (When the thermodynamic melting temperature depends on concentration, as in the case of a non-dilute binary alloy, it would be very easy to select the melting temperature at vanishing concentration to be this additional unknown.)

The subsequent global step is performed by assuming that all local matrices $\boldsymbol{A}$, and the local vectors $\boldsymbol{D}$ and $\boldsymbol{E}$ involved in the heat balances (120) associated with the different macro-elements are constant. These matrices and vectors are provided by the previous local step. The global step consists in solving by Newton's method the assembled system (124) on the skeleton of the furnace. It is worth noting that the system geometry is fixex when performing global iterations, whence the radiative matrices $\boldsymbol{F}$ and $\boldsymbol{B}$ involved in eqs. (121) and (122) are known.

Convergence of the local – global scheme is achieved when the melting temperature condition (31) remains exactly satisfied after a given local iteration. The use of an incremental technique may be required when performing the local iterations if melt convection is taken into account in the model (section (3.4)).

Decoupled strategy. The method of Atherton et al. [60] and Derby et al. [116] is equivalent to performing a succession of local and global iterations. Each "local" step is defined by solving by means of Newton's method the complete system, except that the effective ambient temperatures T_{aA}, which provide the net radiative flux $q_R(\mathbf{x})$ from eq. (116), are given by the previous "global" step. It is important to observe that the geometrical unknowns of the system and, in particular, the melt – environment interface shape, are found during the "local" iterations. The "global" iterations have the simple role to update the effective ambient temperatures T_{aA} from eq. (117). A relaxation factor must be introduced in both the "local" and the "global" iterations in order to guarantee convergence.

Coupled strategies. The coupled strategy of Derby et al. [64] differs from the previous decoupled method in an essential respect. What we called "local" iterations are replaced by the solution (by Newton's scheme) of the fully coupled system, where view and Gebhart factors are however fixed. The "global" iterations update these view and Gebhart factors. (Again, the geometrical unknowns are found during the "local" iterations.) This method provides better results than the decoupled approach. Introducing appropriate continuation methods to find the solution may be extremely useful, for the global system is highly non-linear.

The coupled strategy of Brown et al. [234], and Bornside et al. [61] is based on using a full Newton scheme. The sensitivities of the Gebhart factors to geometrical unknowns are evaluated by a finite difference perturbation technique. This method is efficient, in view of its quadratic convergence, but is limited to particular geometries, such as cylindrical crystals (see the discussion in section 4.3.1).

5.2. Examples of quasi-steady global simulations

In this section, we present some results of global quasi-steady simulations in Cz and DGF systems. As a first example, we refer to the global model developed by Virzi [236] for the calculation of the temperature field and interface shape in a Cz furnace. In a second example, results from Derby et al. [64] show how heat distribution in an induction heating system can be calculated. A third example is devoted to the case of a DGF furnace, as analyzed by Crochet et al. [224]. Such a furnace has a complex structure, since many different heaters are used in order to control the temperature field in the load. We finally consider simulations which focus on crystal quality in terms of thermoelastic stresses. The selected example from Cazin-Bourguignon and Motakef [238] shows how different modifications to the furnace design can be useful in order to minimize the stress level.

5.2.1. Global heat transfer model for Cz growth

The quasi-steady finite element model developed by Virzi [236] was used for analyzing the sensitivity of the temperature field in a Cz silicon furnace with respect to pull rate, gas cooling regime, and material properties. The model is global, as it takes into account the comprehensive geometry of the puller. Conductive and radiative exchanges are taken into account, but melt convection is neglected. Prediction of the temperature field and of the solid–liquid interface position is achieved. The heater power is calculated so that the tri-junction temperature matches the melting temperature, the pull rate being imposed. A typical mesh and result are presented in fig. 40.

As a first result, the classical dependence of interface shape with respect to pull rate was displayed: solid–liquid interface was found to become more concave with respect to the crystal as the pull rate was increased. In a second step, several material emissivities were modified and a dramatic influence on the temperature field was observed (similar results were obtained in [176]). Most important were the influence of the emissivities of the silicon melt surface and of the steel outer shell. The temperature field in the melt and crystal is presented in fig. 41 for these conditions. It is easily observed that decreasing melt emissivity from 0.3 to 0.05 greatly reduces thermal gradients in the melt, as can be expected. Isotherms in the melt appear to be more perpendicular with respect to melt surface due to a decreased heat transfer towards the surroundings. The reduction of heat emitted from the liquid surface has an additional effect on the crystal: the region close to the interface receives less heat from the melt, so that the axial temperature gradient in that part is increased. The altered environment in melt and crystal leads, moreover, to a less concave solid–liquid

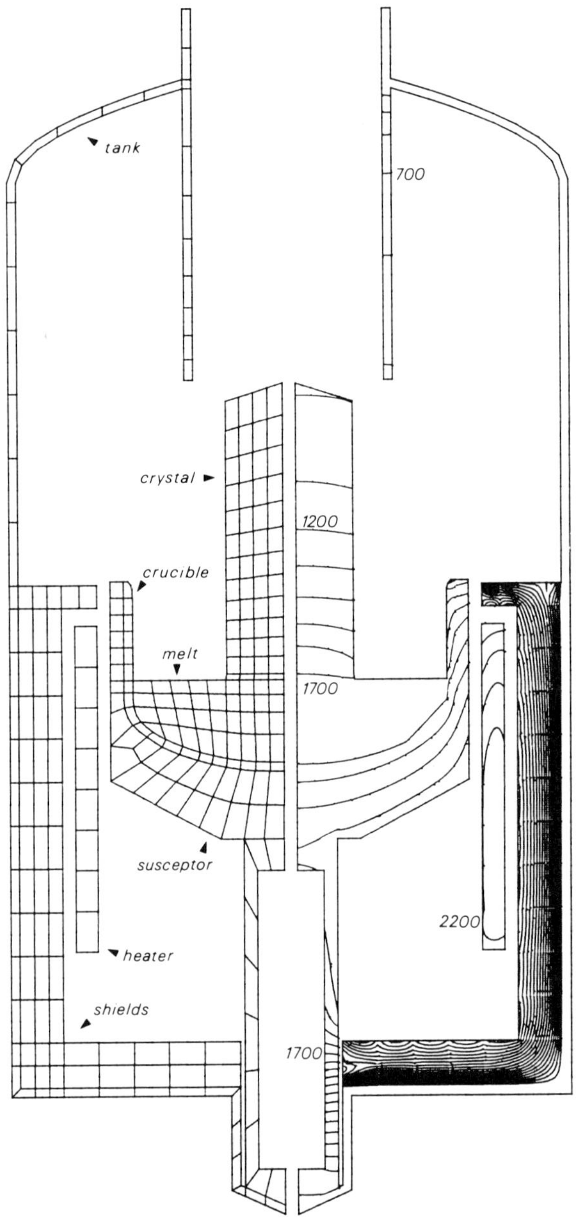

Fig. 40. Quasi-steady analysis of a Cz silicon furnace. Finite element mesh and temperature field ($\Delta T = 100$ K) in the furnace (From [236]).

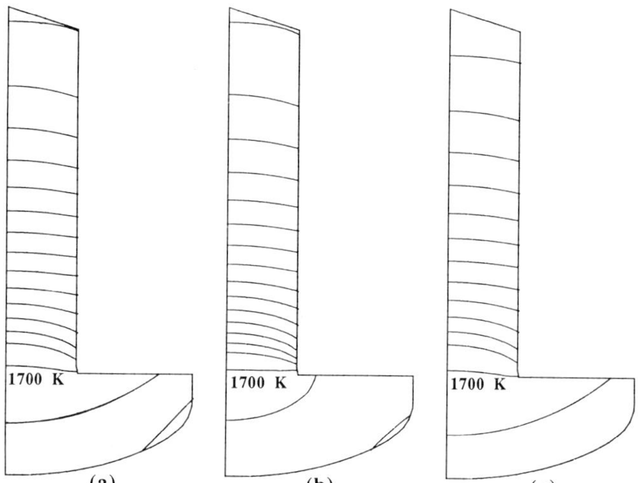

Fig. 41. Same problem as in fig. 40. Temperature field ($\Delta T = 50$ K) in the melt and crystal for (a) the reference case (melt emissivity = 0.3, outer shell emissivity = 0.4), for (b) a reduced melt emissivity (0.05), and for (c) a reduced outer shell emissivity (0.2) (From [236]).

interface, which results from the lower heat flux from the melt. On the other hand, decreasing the steel emissivity on the outer shell from 0.4 to 0.2 results in reduced radiative heat losses from both melt and crystal (through an increased reflectivity): highly reduced thermal gradients are thus observed. These results demonstrate the dramatic importance of radiative heat exchanges in a Cz puller, and stress the need for a

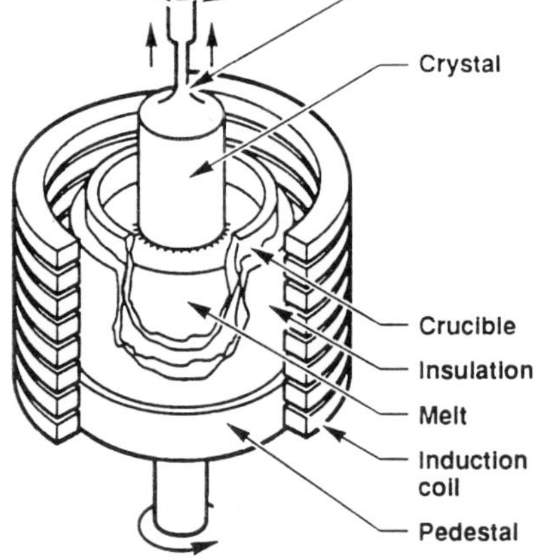

Fig. 42. VB growth of oxide crystals using an induction heating system. Schematic diagram of the growth configuration (From [64]).

better knowledge of material emissivities. The use of different convection coefficients for modelling heat transfer to the ambient gas showed that the temperature field was very little influenced by the gas cooling regime.

5.2.2. Modelling an induction heating system

In some Cz pullers, energy is delivered to the melt through an induction heating system. An alternating current through an inductive coil (fig. 42) generates magnetic and electric fields in the crucible, thereby inducing an electric current. Heat is then produced through Ohmic dissipation. The modelling of such a heating system was originally achieved by Gresho and Derby [149], and integrated in a global thermal model by Derby et al. [64]. The method may be summarized as follows. Calculation of the electric and magnetic fields is performed by solving the Maxwell equations.

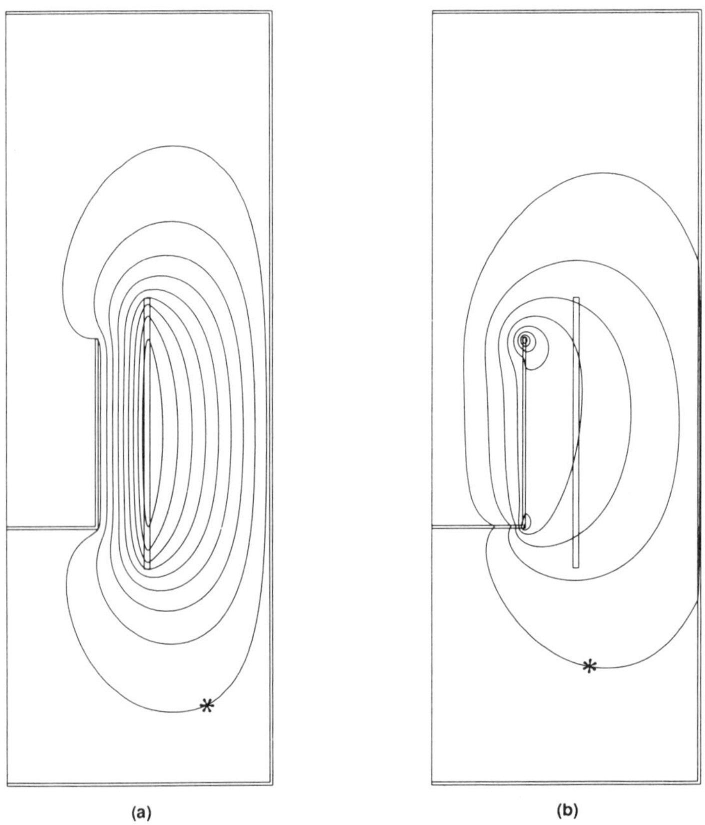

Fig. 43. Same problem as in fig. 42. Spatial distribution of the magnetic field: (a) the in-phase component C of the magnetic stream function is centred on the coil and excluded by the crucible ($\Delta C = 2 \times 10^{-5}$ weber, $C_{min} = 10^{-1}$ weber, as indicated by a $*$); (b) the out-of-phase component S originates from the crucible and exhibits maxima on the top and bottom corners of the crucible ($\Delta S = 1.5 \times 10^{-6}$ weber, $S_{min} = 7.5 \times 10^{-7}$ weber, as indicated by a $*$) (From [64]).

Electrical conductivity is supposed to be independent of temperature, so that calculation of power distribution can be decoupled from the solution of the energy equation in the overall furnace. The heat supply to the crucible can therefore be calculated a priori up to a constant. Only one induction analysis needs to be performed for, with a constant electrical conductivity, heat supply to the crucible is linearly proportional to the total electric power delivered to the coil, eq. (119). A change of generator power is then easily reflected by multiplying the heat source in the crucible by a tuning parameter, such as W. This latter can be considered as an unknown provided an additional constraint is integrated in the system (such as imposing the melting temperature at the tri-junction).

The resulting time-harmonic magnetic field in a typical Cz oxide growth system can be represented in terms of the magnetic stream function, whose in-phase and out-of-phase components are shown in fig. 43. The magnetic field is everywhere tangential to the contour lines. Figure 44 shows the resulting time-averaged power distribution inside the crucible. It can be observed that intense heating appears at the top and at the bottom of the crucible. Moreover, heating is rapidly decreasing from the outer wall to the inner wall. This effect is clearly apparent in the bottom part of the crucible. The results are in good agreement with experimental data. Note that induction heating of the melt or the crystal was disregarded, and only crucible heating was considered when solving the Maxwell equations. This assumption is justified by the small electrical conductivities of oxide crystal and melt, which are at least four orders of magnitude smaller than that of the iridium crucible (the susceptor). Replacing the oxide melt by a

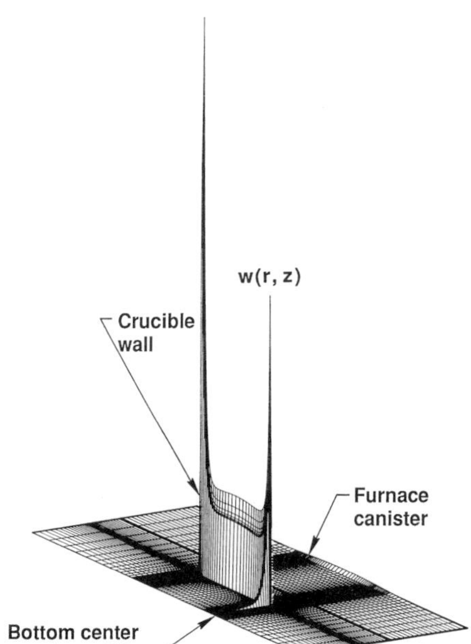

Fig. 44. Same problem as in fig. 42. Time averaged power distribution w in the crucible. Most of the power is deposited to the outer crucible wall, with local maxima at the top and bottom corners (From [64]).

material whose conductivity is two orders of magnitude higher showed virtually no change in the calculated fields in the crucible.

5.2.3. Global heat transfer model for DGF growth

As a third example, we consider the case of a DGF system, as modelled by Crochet et al. [224]. All heat transfer modes in the furnace, including convection in the ambient gas and in the melt, were taken into account. A schematic view of the furnace, together with the associated finite element mesh, is represented in fig. 45. This particular VB apparatus, the so-called Mellen electro-dynamic gradient furnace, includes a set of 23 individually controllable heating elements, thus allowing one to generate arbitrary temperature fields within the ampoule. The load consists of InP, with the melt located under the solid phase. One finds a Grashof number of 10^6 characterizing melt convection. It should be recalled that no forced convection is present in this problem.

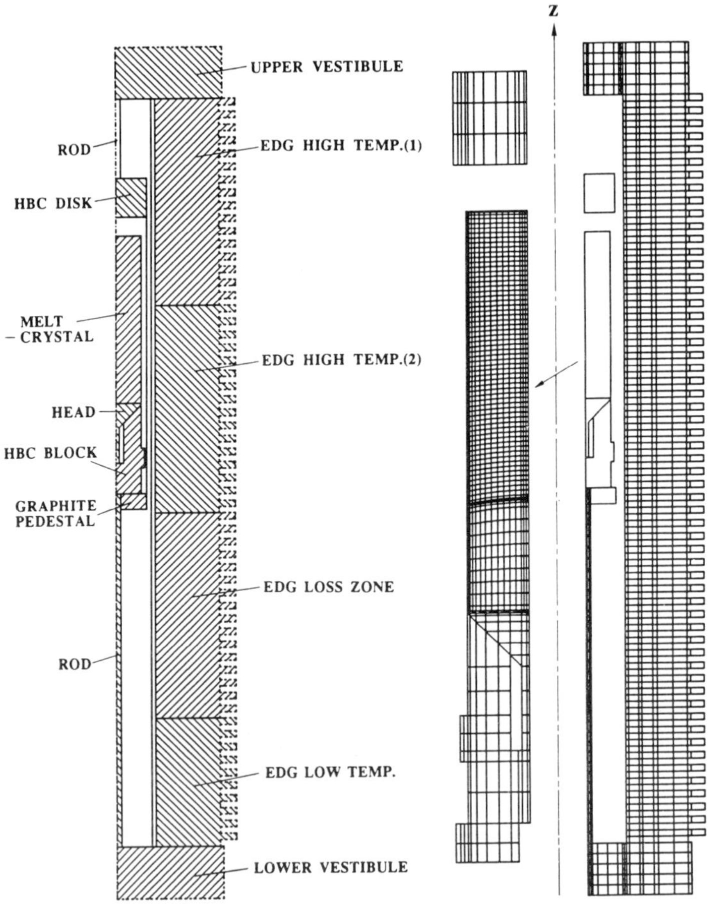

Fig. 45. DGF growth of InP. Sketch and finite element mesh representation of the Mellen furnace (From [224]).

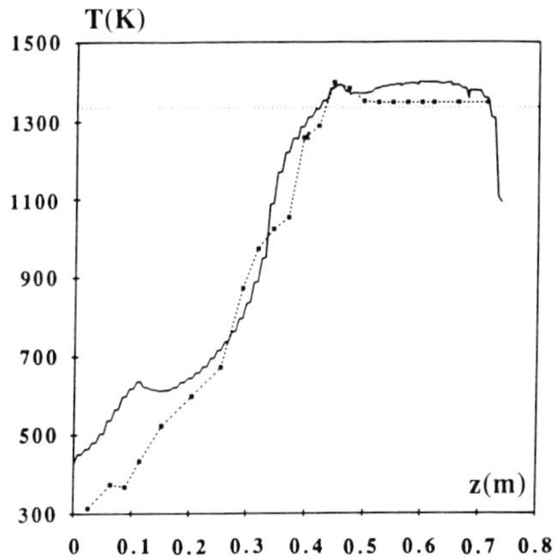

Fig. 46. Same problem as in fig. 45. Experimental (dashed line) and calculated (solid line) temperature profile along the inner wall of the furnace (From [224]).

The total heater power and its distribution among the 23 heat sources was calculated in order to obtain a temperature profile along the wall of the heater which would be similar to the measured one. The total power was such that the melting temperature of InP was found at a desired vertical position in the load. The power distribution among the 23 heat sources could in principle be calculated from 23 temperature constraints (derived from measurements, for example). However, such a problem appeared to be extremely difficult to solve as it is an inverse problem, meaning that minor errors on the experimental data lead to strong oscillations in the calculated heater powers. In practice, successive adjustments of the 23 parameters were thus performed in order to obtain the final result represented in fig. 46.

Numerical simulation of melt convection in crystal growth processes is often difficult, because the actual values of the Grashof and Reynolds numbers are usually very high. It must be noted that, in most VB systems, the cooler temperatures are located at the bottom of the melt, so that a stable configuration is found for convection. Calculation of the flow field does not present major difficulties under such conditions, and is in particular much easier to perform than for a Cz configuration (where the cold interface is always located on the top of the melt). Natural convection could be calculated in [224] up to the actual Grashof number, and fully coupled to the global thermal problem. The resulting temperature field in the whole furnace is represented in fig. 47, while the isotherms in the crystal–melt subdomain are represented in fig. 48 in the absence and in the presence of convection. The influence of this latter effect is mild: isotherms are modified in the upper part of the melt, but little change is observed in the temperature field close to the solidification front. Convexity of the interface goes however from 3.4 mm without convection to 2.0 mm with convection.

It is worth mentioning here that global modelling of a Cz system taking melt convection into account was achieved by Dupret et al. [62], Nicodème et al. [63], Ryckmans et al. [177], Kinney et al. [274], Xiao and Derby [265], and Kakimoto et al. [239, 240]. In [240], local convection calculations were done by imposing melt

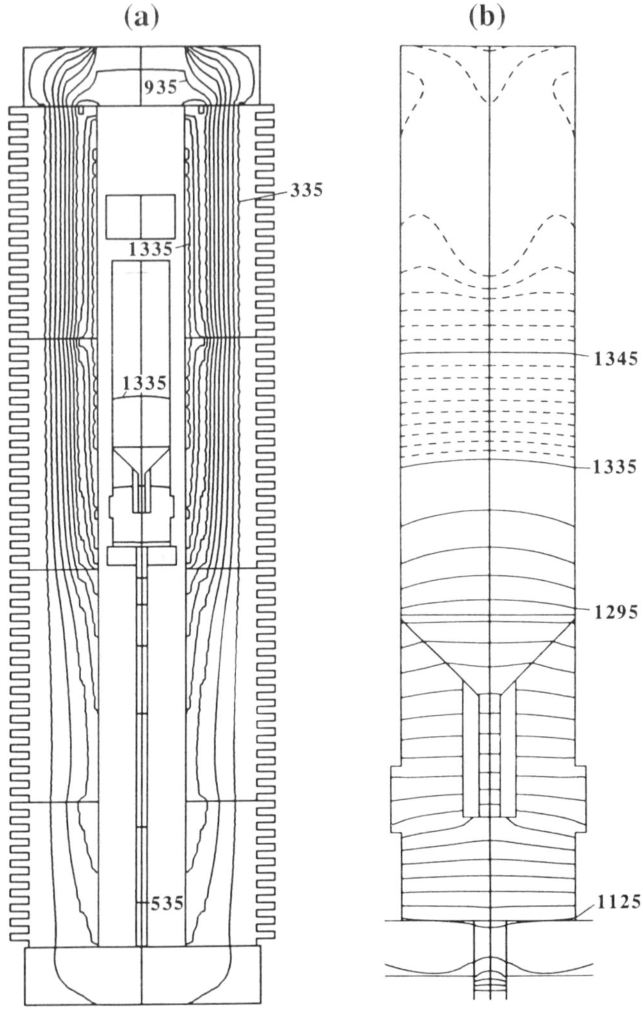

Fig. 47. Same problem as in fig. 45. Temperature field throughout the furnace: (a) global view ($T_{min} = 335$ K, $\Delta T = 100$ K), and (b) enlarged view of the load ($\Delta T = 10$ K for solid lines, $\Delta T = 1$ K for dashed lines) (From [224]).

boundary temperatures obtained by means of a global simulation performed without convection. (The importance of the role played by boundary conditions in bulk flow calculations has been demonstrated by Xiao and Derby [275].) Turbulent convection models have been introduced in global calculations by Kinney and Brown [264], in order to improve the agreement between calculated and experimental results.

5.2.4. Calculation of thermoelastic stresses in the crystal

As crystal quality is the final objective of any crystal growth process, it is interesting to be able to accurately measure this quality once the temperature field has been calculated by a suitable heat transfer model. An example of calculating thermal stresses

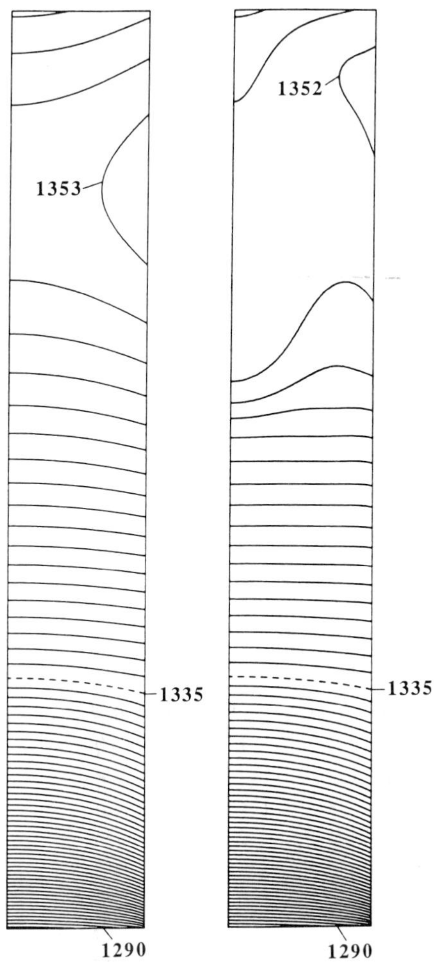

Fig. 48. Same problem as in fig. 45. Temperature field in the crystal and the melt in the absence and in the presence of convection in the melt ($\Delta T = 1$ K) (From [224]).

in the crystal was already given in section 4.4. The first investigations related to this field were carried out by Motakef [73, 74], and Motakef and Witt [72], who focused on the influence of encapsulant transparency, convection within the encapsulant layer, and on the analysis of different growth stages in LEC growth systems. Extensive analysis of thermal stresses was pursued by Motakef and co-authors in [238, 241–244]. Let us also mention the work of Meduoye et al. [245–247], of Nicodème et al. [75] and Dupret et al. [225], of Bornside et al. [248], and of Kinney et al. [274].

We would like to focus here on the results obtained by Cazin-Bourguignon and Motakef [238] for LEC GaAs growth, where different methods of reducing thermal stresses were suggested. A global model was used for the calculation of the temperature field in the furnace. Some important assumptions, such as considering a simplified furnace geometry, a uniform heat generation within the resistor, a uniform temperature in the lower part of the heater and a perfect insulation for the heater at its back side, were however introduced. Heat transfer within the ambient gas, in the melt and in the encapsulant was considered to be primarily governed by conduction, so that the velocity field representing thermal convection did not need be calculated. The semi-transparency of the boric oxide layer to thermal radiation was taken into account through a suitable band model. Thermoelastic stresses were easily calculated from the temperature field in the crystal, and were represented in the form of the excess stress field, defined as the excess of elastic shear stresses over the CRSS. (This is often a good measure of the magnitude of the driving forces for plastic deformation. Indeed, creep is supposed to occur if shear stress exceeds the critical resolved shear stress. Models for viscoplastic deformation have been detailed in section 2.1 and Chapter 14.) The excess stress field is thus used as an indicator for the onset of plastic deformation.

The influence of two different changes in the furnace design was investigated in [238]. Installation of a top assembly, leading to a higher temperature on the furnace wall (heat shield-like effect), and of a bottom heater (located under the crucible support) were considered (fig. 49). The temperature field in the furnace and the excess stress field were calculated for various combinations of top heat shield temperature, thermal conditions for the bottom of the crucible support (adiabatic or constant bottom heater temperature), and pull rate. Furthermore, the same analysis was carried out for three

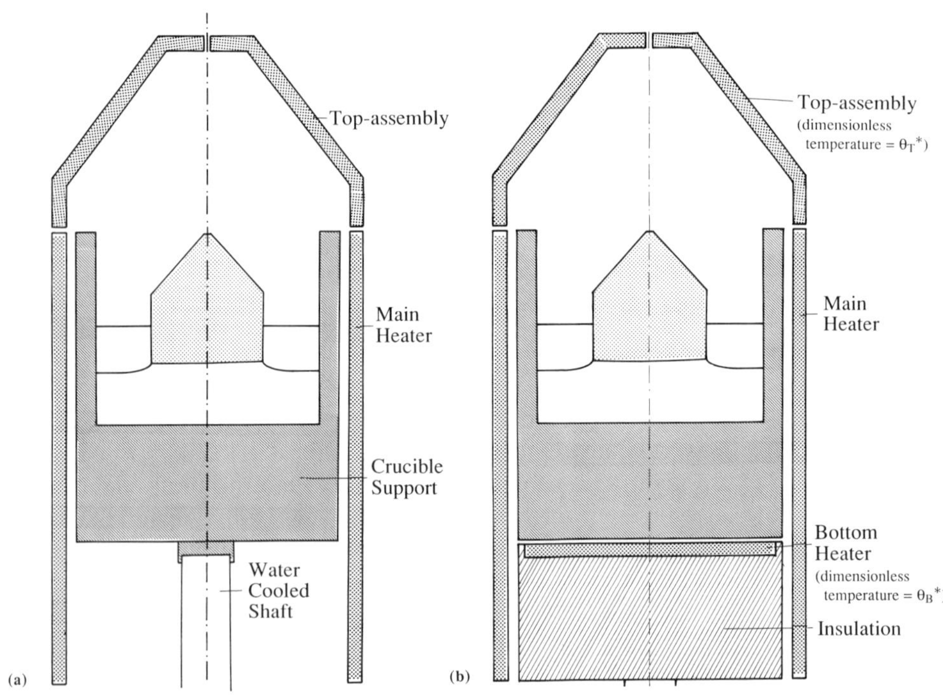

Fig. 49. LEC growth of GaAs. Modifications of furnace design in order to lower the thermal stress level: (a) addition of a top assembly; (b) addition of a top assembly together with a bottom heater (From [238]).

different stages of growth. As a general rule, very large stresses were always found close to the solid–liquid interface and at the point where the crystal emerges from the encapsulant. Typical results corresponding to an intermediate stage of growth are represented in fig. 50.

An interesting observation concerns the influence of adding a top assembly at a given temperature. The effect is twofold. First, increase of top assembly temperature leads to

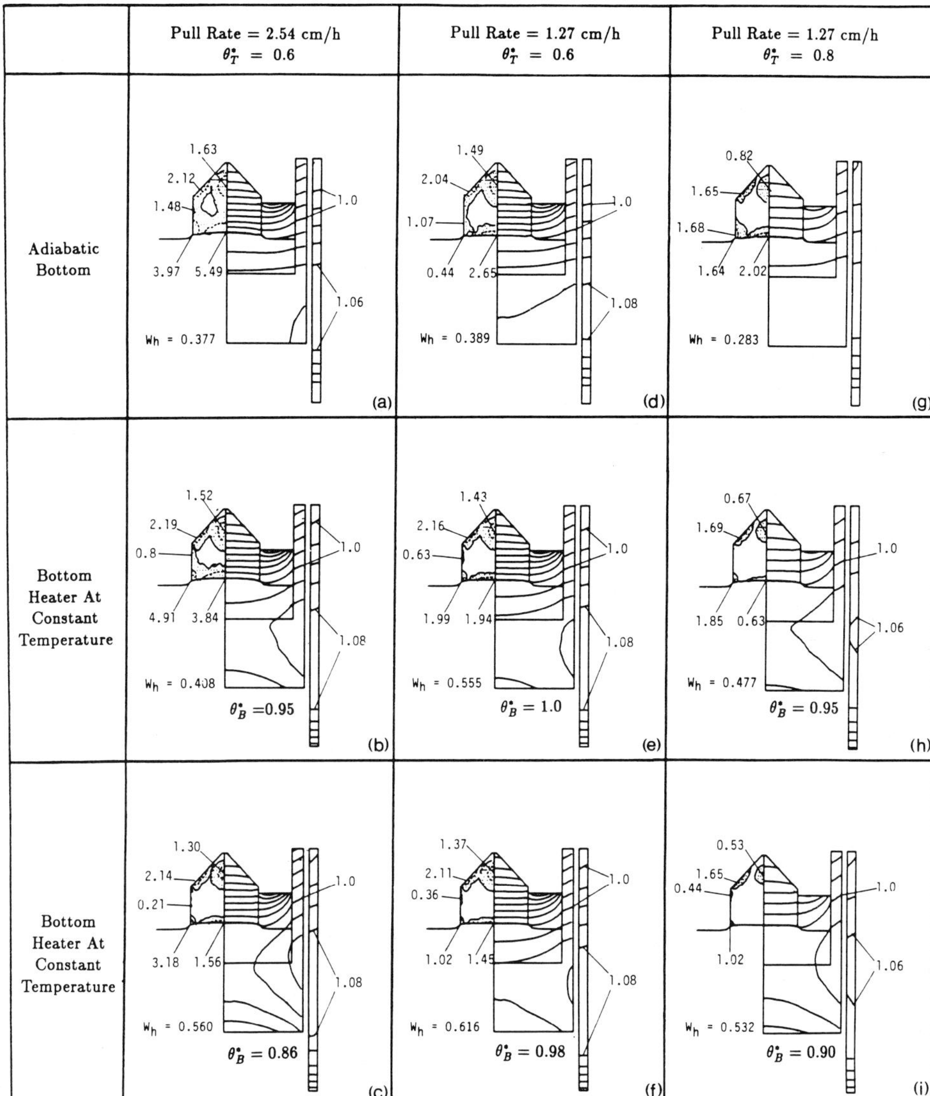

Fig. 50. Same problem as in fig. 49(b). Temperature and excess stress distributions for various bottom heater temperatures θ_B^*, pull rates, and top assembly temperatures θ_T^* during an intermediate stage of growth (From [238]).

a general reduction of thermal gradients in the crystal, because radiative heat loss from the crystal is minimized; second, as less heat is conducted towards the crystal wall, an enhanced concavity of the solid–liquid interface (with respect to the crystal) is observed. The first effect appears to reduce thermal stresses, while the second effect appears to enhance thermal stresses. As thermal stresses are the result of the combination of the influence of interface planarity and thermal gradients, the benefit of using a top assembly heat shield is strongly limited, due to the concomitant change of interface shape. Such a conclusion is also valid for the use of other heat shields or additional heaters. Bornside et al. [248], for instance, have observed the same increase of interface concavity in a silicon furnace, when using two separately controlled heaters in an attempt to reduce thermal gradients (the heat source in the upper heater was increased with respect to a reference case).

Interestingly, the use of a bottom heater was shown in [238] to be useful for counterbalancing the negative effect on interface shape induced by the addition of the top assembly. The bottom heater has a dramatic influence on the temperature field in the growth system, and particularly close to the interface, so that the concavity of the latter could be highly reduced by application of a suitable temperature at the bottom heater. In comparison with the adiabatic case, the use of a bottom heater induces a cooler melt, so that interface position can be shifted downwards.

Analysis of the influence of the pull rate was also performed. As a classical result, a pull rate reduction was proven to be associated with a lower level of thermal stresses due to a decreased interface concavity. However, this well-known effect cannot be considered as an actual tool to reduce thermal stresses, since reasonable pull rates must be maintained for economic reasons.

It clearly appears in fig. 50, which shows results obtained for different combinations of growth conditions, that the use of suitable temperatures for the top assembly and the bottom heater may provide a crystal where excess stresses have been almost totally eliminated. This example demonstrates the efficiency of numerical simulations, which may help define conditions associated with a low level of thermal stresses.

In a recent paper, Kelly et al. [244] computed stresses in the crystal for different stages of growth as a result of three different temperature evolutions in the heating system. A comparison of the calculated stresses with the dislocation density measured in corresponding real crystals demonstrated that the excess stress field could be a good indicator of viscoplastic deformation.

It is worth noting that a more realistic way to approach viscoplastic deformation is to actually consider stress relaxation and strain hardening by dislocation multiplication through a dynamic model (see section 2.1). An example of such calculation is given in section 5.4, devoted to global time-dependent problems.

5.3. Solution of the dynamic global problem

Compared with quasi-steady (or pseudo-steady) models, the interest of dynamic simulations in crystal growth is twofold. On the one hand, one wishes to capture the essential transient effects which take place during critical periods of the process, such as the initial or final stages of growth (see the work of Van den Bogaert and Braet [249]). On the other hand, developing efficient control strategies requires one to analyze the dynamic response of the system to stepwise perturbations of the input parameters (recall from section 1.1 that, in Cz growth, the latter are essentially the crystal pull and rotation rates, the crucible lift speed and rotation rate, the heater power(s), and the outer coolant temperature; in DGF growth, the most important input parameters are the power supply to the different heating elements).

Basically, much important information can be obtained on the system dynamics by means of a local analysis (i.e. an analysis reduced to crystal, melt and possibly crucible, given the radiative environment – as developed in section 3.3). Nonetheless, the coupling of time-dependent and geometrical effects may be very strong in Cz growth, in view of the complexity of radiative transfer. In the case of DGF growth, the relation between the ampoule boundary temperature and the power supply to the heating elements is also difficult to evaluate and may require one to perform integrated calculations, taking radiation into account. In the present section, we shall however restrict ourselves to Cz integrated dynamic simulations, since DGF results are not yet available in the literature.

In order to understand the coupling between the growth dynamics and the global radiative transfer in the furnace, it is worth recalling that many different time scales are identified in Cz growth. Over the short and very short scales related to the melt flow and to dislocation dynamics, we find the much slower transients related to conductive heat transfer (i) in the solid constituents (essentially crystal, crucible, and heater), and (ii) in the melt if the Prandtl number is low. Still longer is the characteristic time which is required to get a significant response of the solid–liquid interface and the growth angle to a change of operating conditions. The longest time scales are related to the effects of crystal lengthening on the overall radiative transfer in the furnace. Clearly, when the growth angle changes, the effective ambient (or sink) temperature at the meniscus can change dramatically. One of the objectives of transient global simulations is to take full consideration of such effects.

5.3.1. Dynamic problems

Whereas most quasi-steady models are inverse (section 5.1), a dynamic model may be *direct* or *inverse*. In order to develop an algorithm for the direct problem, where the heater power is imposed as a function of time, while crystal shape is integrated from the evolution of the radius at the tri-junction, it is necessary to solve the very difficult problem of the dependence of view factors on geometrical changes. This difficulty is a consequence of (i) the fact that calculation of the meniscus shape and the view factors becomes coupled, of (ii) the absence of time derivatives in the equations governing radiative transfer, and of (iii) the inherent complexity of geometrical algorithms. Problem (i) means that the nodal coordinates defining each discrete radiative enclosure may no longer be evaluated before the calculation of view factors. Problem (ii) involves that the so-called mass matrix of any semi-discrete formulation of the global problem has vanishing rows, for radiative heat exchanges play a role as constraints in the model (recall that "photons" are assumed to travel at infinite speed). As a consequence, that dynamic system is very stiff. Problem (iii) necessitates a few explanations. A geometrical algorithm will always involve several relations (such as parallelism, perpendicularity, or interception) between so-called "objects" (such as points, segments, facets). In practice, this results in introducing into the code a high number of logical decisions (e.g. "Does a given facet or a given conic frustrum intercept a given line? If yes, execute this statement, else go to this other statement.") which may render the program very intricate. Moreover, geometric relations must always be defined up to small tolerances in view of the finite floating-point precision of computers (for instance, a point is considered as a very small sphere). Unfortunately, this means that the transitivity of discretized geometrical relations is definitely lost (two straight lines which are parallel to a third one, up to some given tolerance, are not necessarily themselves parallel, up to this tolerance). The logical complexity of geometrical algorithms is, therefore, extremely increased. Another related problem is the loss of regularity of the functions defined by means of these algorithms: in view of the logical decisions involved in the calculations, functions present a very high number of discontinuities. In particular, it is almost impossible to calculate correctly a Jacobian matrix involving view or Gebhart factors (if this matrix exists mathematically), except by means of perturbation techniques [61, 273]. It is clear that difficulties of the same kind were already encountered when developing quasi-steady models, but that problems become tremendous when the objective is dynamic Cz simulation. In the second part of this section, two solutions will be discussed. For details on geometrical algorithmics, see e.g. [250, 251].

When dealing with inverse dynamic simulations, where the real input (the heater power) becomes an unknown and the real output (the radius at the tri-junction) is imposed, the geometrical problems are exactly the same as in quasi-steady calculations. Hence, the above main difficulty (i) is suppressed. However, inverse dynamic problems themselves present an important drawback, for they are *ill-posed* [252]. A well-posed problem has a solution which (i) exists, (ii) is unique, and (iii) is continuous with respect to the data [253]. An ill-posed problem violates one of the above conditions, usually condition (iii), and sometimes condition (i). In the domain of heat transfer, inverse problems are often found when experimental data are used to reconstruct a given solution in order to obtain information where direct measurements are impossible. As it is beyond the scope of this Handbook to enter into details of this mathematical problem, a simplified example will be used to explain the situation. Consider the one-dimensional linear problem of heat transfer inside a uniform layer of constant thickness (fig. 51). If temperature is imposed on both sides of the layer, the problem is direct, whereas it is inverse if both temperature and heat flux are imposed (e.g. as

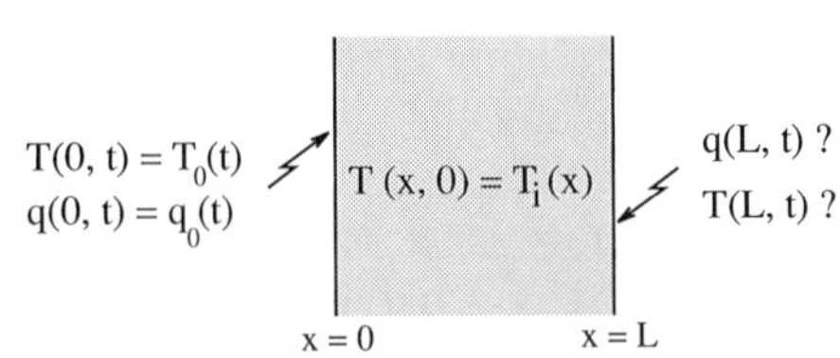

Fig. 51. One-dimensional time-dependent inverse heat conduction problem.

obtained from measurements) on one given side of the layer, in order to reconstruct the temperature field. In the former case, the left ingoing heat flux is continuous with respect to time even if the data (i.e. the boundary temperatures) are not, while being constrained to be integrable. In the latter case, it is therefore clear that a continuous left ingoing heat flux may give rise to a non-continuous temperature on the opposite side; it is possible to show that, in some cases, this flux will exist only in the sense of the distribution theory [254], or even will not exist at all. Hence, inverse problems involve a strong loss of smoothness of the solution with respect to the data. In Cz growth, the inverse dynamic problem presented above has been shown by Van den Bogaert [66] to consist of a series of inverse subproblems. At the end of this section, some peculiar aspects of solving the inverse Cz problem will be examined.

Let us first turn our attention back to the construction of the discrete global system representing the dynamic model (direct or inverse). To this end, consider the equations arising from the macro-elements which are not radiative enclosures. These relations may be subdivided into three classes: (i) the N_T equations governing heat transfer, which are obtained by discretizing eqs. (23) and (25); (ii) the equations associated with the unknown interfaces of the crystal–melt macro-element, which are summarized by eqs. (33), (45), (47), and (48); (iii) the equations governing the additional fields, essentially in the melt, and which have been analyzed in section 3.4. Time-dependent terms are now present in these relations, which must be made discrete by means of the technique described in section 3.2, eqs. (28), (29), (35), and (42). This seems to render the integrated system very complicated. Nevertheless, two major observations will stongly facilitate its construction.

- On the one hand, a careful analysis shows that the discrete equations governing heat transfer in every macro-element may be put after some operations in the same form as in the quasi-steady case, eqs. (120), provided the symbol $\boldsymbol{T}^{+}$ denotes the boundary nodal temperatures at time t^{n+1}, and implicit dependence with respect to temperature field at time t^n is assumed in the equations.
- On the other hand, it must be clear that the equations associated with the crystal and melt interfaces (of the above class (ii)) are local to the crystal–melt macro-element, in the sense that they do not require a further assembly in order to be complete, exactly as the additional field equations (of the class (iii)) did in the quasi-steady case (and still do).

From the first observation, we conclude that assembly of the discrete equations governing heat transfer in the macro-elements can be performed exactly as when building the quasi-steady global system. From the second observation, we understand that the alternate local–global solution strategy developed in section 5.1.2 for the quasi-steady problem [177] may still be used in the dynamic case, while involving more geometrical unknowns. Moreover, local iterations can be performed in the fashion of solving the local dynamic problem (sections 3.2.3 and 3.4). In a similar way, the decoupled [60, 116] and the first coupled [64] solution strategies of section 5.1 remain feasible.

The role of the geometrical variables of the system requires however a more detailed analysis. A first question arises when considering the upward lift of the crucible, which becomes an unknown in dynamic simulations. Generally, Cz growth is performed by constraining the melt surface to remain at a given level (except in the vicinity of the contact lines). The necessity of selecting this particular constraint will be understood in sections 5.4.2 and 5.4.3. All the N_L geometrical unknowns associated with the melt–liner (or melt-crucible) interface (which are components of the vector $\boldsymbol{X}^u$ of sections 3.1 and 3.2) can very simply be related to the motion of the bottom of the melt on the axis, for the crucible is rigid. Hence, expressing $(N_L - 1)$ of eqs. (48) is straightforward. The last equation of this system is obtained in [66] by imposing a given value to the theoretical level z_0 at which meniscus curvature vanishes. In order to facilitate the solution of the system, the first natural condition (38) (which is imposed at the meniscus–liner contact line L) is replaced by the essential condition constraining this contact line to be at the same axial position z_0. Hence, when using the alternate local–global iterative scheme of section 5.1.2, the successive locations of the crucible are found when performing the local iterations associated with the crystal–melt macro-element. On the other hand, as pull rate is imposed, the upward motion of the crystal is given.

The main difficulty related to the presence of geometrical unknowns in the global system is however the considerable computational expense related to view factor calculations. This forbids one to use normal implicit methods in direct dynamic simulations, which would require one to update the view factors at each iteration. In what follows, two strategies for solving this problem will be presented.

5.3.2. Direct problem: use of Gebhart method for solving radiative transfer

The method of Atherton et al. [60] is simple and robust: at each time step, the view and Gebhart factors of the previous discrete time t^n are used to perform the calculations. The decoupled strategy of section 5.1.2 is used to find the solution. This method has the disadvantage of being semi-implicit, which means that stability of the integration scheme imposes severe limits on the time step, with an important increase of computational cost. An implicit version of this scheme (based on a method of successive approximations) was therefore developed by Derby et al. [116]. On the other hand, Zhou et al. [273] have used the perturbation method of Bornside et al. [61] to perform time-dependent simulations. In general, these methods remain computationally expensive. Using efficient algorithms adapted to parallel computing is a possible remedy.

5.3.3. Direct problem: use of Galerkin method for solving radiative transfer

In order to develop an implicit integration scheme, the approach of Van den Bogaert [66], and Van den Bogaert and Dupret [65, 165] is based on a deep analysis of the geometrical evolution of the system. Recall first from section 3.2.2 that the material point located at the tri-junction at discrete time t^n is constrained to occupy at time t^{n+1} the nodal position situated just above the tri-junction (fig. 8(b)). Hence, given the time step duration $(t^{n+1} - t^n)$ and the pull rate, the crystal shape may be determined at the outset of the calculations associated with this time step, except the tri-junction location (R_J^{n+1}, Z_J^{n+1}). On the other hand, meniscus and melt–liner interface shapes are characterized

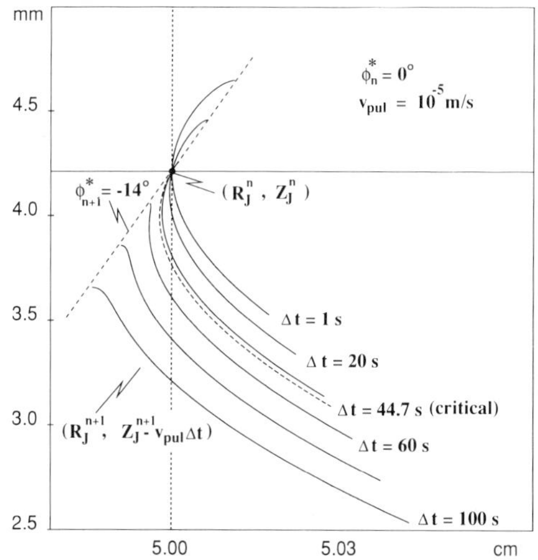

Fig. 52. Geometrical locus of the possible positions of the tri-junction for different time steps. The crystal has grown as a cylinder during the previous time steps (From [66]).

by $(N_M - 1) + N_L$ additional unknown nodal coordinates (which are components of $\boldsymbol{X}^u$), while there are $(N_M + 1) + N_L$ equations in the synthetic systems (45) and (48). The keypoint of the method is to observe that these systems involve a single additional unknown (z_0) provided the mass conservation equation (39), which is comprised in the system (45), be approximated by neglecting its slight dependence with respect to the crystal–melt interface configuration. (This is no longer an approximation when crystal and melt densities are equal, and is approximately true in the general case, since (i) the latter densities are generally very close together, and (ii) the solidification front deforms rather slowly with time.) The balance between equations and unknowns in the decoupled subsystem we have just established shows immediately that the tri-junction (R^{n+1}_J, Z^{n+1}_J) is constrained to be situated on a precise geometrical locus – given the time step duration – which may easily be determined using the growth angle ϕ^* (see fig. 5) as a parameter (fig. 52). The whole geometry of the enclosures is, therefore, a function of this single parameter. Figure 52 indicates also that, when the crystal wall is locally convex for a given growth angle, short numerical remeltings may occur, which can be avoided only by imposing that the time step duration be selected above a critical value.

A second important aspect of the method of [65, 66, 165, 249] originates from the observation that the radiative matrices $\boldsymbol{F}_A$ involved in eqs. (111) and (121) depend on the growth angle ϕ^* by a relation which may very well be approximated by a quadratic function (fig. 53). It is, therefore, very easy to calculate at every time step these radiative matrices for three selected values of the growth angle ϕ^*, and to perform the subsequent iterations (associated with the time step) by interpolating between the calculated matrices. Examples of results obtained by this method will be shown in section 5.4.3.

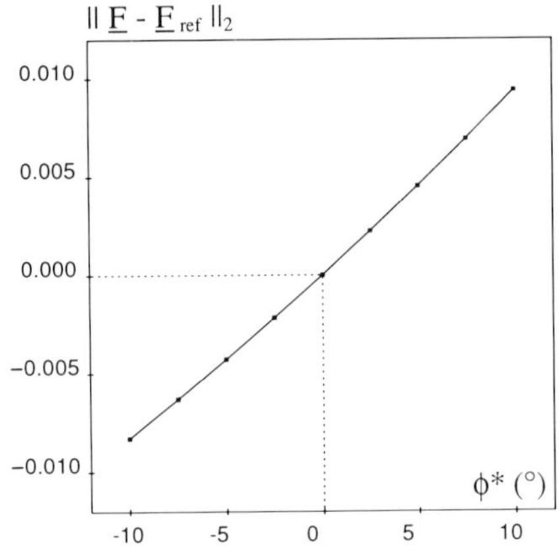

Fig. 53. Analysis of the $\| \ \|_2$ norm of the difference between the radiative matrix $\boldsymbol{F}$ and a reference matrix $\boldsymbol{F}_{\text{ref}}$ corresponding to a vertical crystal wall ($\phi^* = 0^0$). A simple dependence of $\boldsymbol{F}$ with respect to ϕ^* is observed (From [66]).

5.3.4. Inverse problem

All the geometrical difficulties of the direct problem are strongly reduced when dealing with the inverse problem, since radiative matrices may be calculated at the outset of the iterations associated with any given time step. As the problem is inverse, the calculated heater power history exhibits strong oscillations (see the results of section 5.4.2) whenever the imposed crystal shape is non-analytic [66]. However, this does not forbid one to pursue the simulation.

When inverse simulations are performed before direct simulations (as defined by perturbing the pull rate or heater power history), a smoothing of the calculated heater power – which is shown not to affect the result – is very useful.

5.3.5. Dynamic simulations with melt convection

As melt convection may not be neglected in the global heat transfer, though its calculation is extremely expensive, a simplified model has been developed by Lecomte [255], and Lecomte et al. [256], in order to take this effect into account in dynamic simulations. Briefly, quasi-steady global calculations are performed at some specific stages of growth, taking melt convection into account (these calculations may in some cases be simplified by using the extrapolation technique developed by Lecomte et al. [257]). The results of quasi-steady calculations are collected in the form of the Stokes stream function from which the velocity field derives. Interpolation between these collected values provides a velocity field for any time step of the dynamic simulation (fig. 54). The use of the Stokes stream function ensures that the velocity field represents the flow of an incompressible fluid.

This approach is based on the assumption that the characteristic time related to establishing a flow regime is much lower than those associated with the dynamic heat transfer problem, and that dynamic effects do not influence strongly the melt flow. Some results are given in section 5.4.

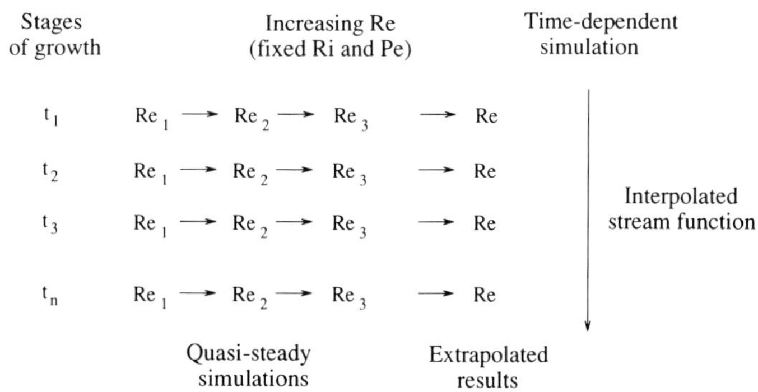

Fig. 54. General method for taking into account melt convection in global time-dependent calculations (From [255]).

5.4. *Examples of dynamic global calculations*

As pointed out in section 5.3, global time-dependent calculations for Cz growth are useful for analyzing stages of the growth where geometrical changes are important (such as the pulling of the top or bottom-end of the crystal) or for analyzing the system response (crystal radius and solid–liquid interface responses mainly) to changes in the operating conditions. The dynamic behaviour of the Cz process is very complex due to the existence of different characteristic times in the system. Control of the process is therefore very difficult. These characteristic times are related to the thermal inertia of the constituents, to the inertia of the solid–liquid interface configuration and to the batchwise transients due to melt volume decrease [56]. In practice, the crystal diameter is controlled via the heater power or the pull rate. Increasing one of these parameters makes the radius to decrease. Changes of heater power affect crystal diameter with a lag because they must propagate through the melt, while changes of pull rate work instantaneously. If no control is applied, the crystal shape evolves continuously in time, due to the geometrical effects related to crystal lengthening and melt volume decrease.

A careful analysis of the dynamic behaviour of the furnace should be very useful for developing suitable control algorithms. The basic idea for investigating the dynamics of the system, including the stability of the process, is to analyze the response to perturbations of the control parameters. Applying a step change of any of these parameters is the most simple numerical experiment and has been performed by several authors [56, 60, 116, 226, 273, 65, 165, 249]. However, in order to analyze the process stability, the batch transients due to geometrical changes must be removed; otherwise, a change in operating conditions will lead to a steadily increasing or decreasing radius, which could be interpreted as a system instability, when this change of shape would be due to geometrical effects only. Different solutions have thus been adopted. The first solution consists in simplifying the problem into a continuous processing system by assuming a constant melt volume together with a very long crystal [56, 60, 116, 226, 273]. A quasi-steady-state process can then be assumed. The second solution consists in applying a perturbation that includes actions to compensate for geometrical changes [65, 165, 249]. This technique will be described in section 5.4.3.

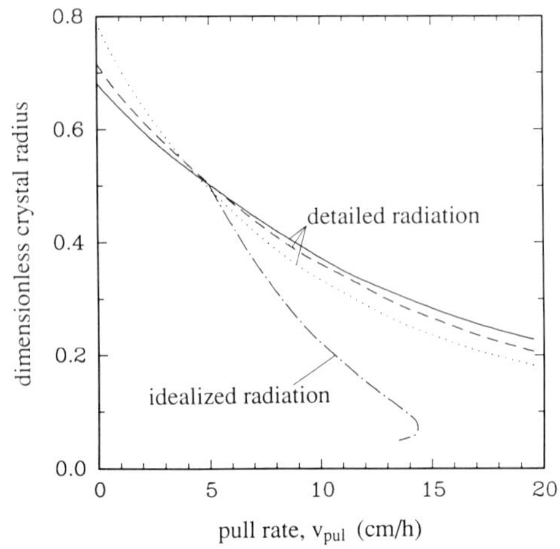

Fig. 55. Dynamic simulation of Cz silicon growth. Dimensionless crystal radius as a function of the pull rate, with detailed (the three curves correspond to different set point temperatures and thermal boundary conditions) or idealized radiation (From [60]).

5.4.1. Direct problem: stability analysis

The first global time-dependent calculations were carried out by Atherton et al. [60], who extended the local time-dependent idealized radiation model developed by Derby and Brown [56] (see section 3.3) with a detailed radiation model based on a diffuse-gray approximation. A satisfactory representation of radiative exchanges was obtained thanks to calculation of the Gebhart factors within the geometry. The global furnace was however considered with some simplifications. A spatially uniform temperature was assumed for the heater and for the furnace surroundings. In addition, the finite element description did not include all the geometrical details which are found in a real furnace. Time integration was carried out through a semi-implicit method. (The method is implicit in calculating the temperature field and interfaces shapes, but explicit in calculating the effects of radiation (section 5.3.2).) The method was applied to a silicon Cz system. Due to calculating radiative exchanges in an explicit way, the time step for transient simulations with detailed radiation was about 20 times smaller than in the initial fully implicit local model of [56].

Differences between results obtained from an idealized and a detailed radiation model can be exhibited from simple quasi-steady calculations. Let us consider in fig. 55 the crystal radius as a function of an imposed pull rate. While a critical pull rate beyond which no quasi-steady state exists was found in the case of idealized radiation (section 3.3, [56]), no such critical value was found with a detailed radiation model.

As in [56], time-dependent calculations can be performed either by assuming a constant melt volume, so that a quasi-steady state can be obtained, or by allowing the melt volume to decrease. The first possibility is used for analyzing the process stability by calculating the crystal radius evolution as a response to a step change of pull rate, while the latter possibility is used for simulating a real batchwise behaviour. In all cases, calculations are termed "direct", since the crystal radius is an unknown.

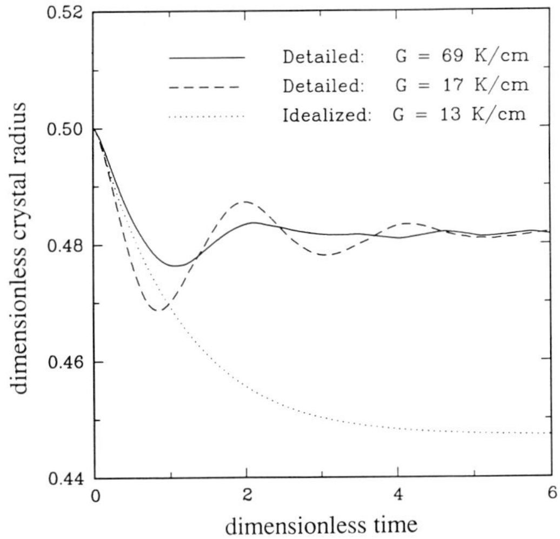

Fig. 56. Same problem as in fig. 55. Evolution of the dimensionless crystal radius after a 10% increase in pull rate with detailed or idealized radiation and different axial temperature gradients (G) in the crystal (From [60]).

With regard to the inherent stability of the process, it was shown how a detailed model for radiation leads to very different results as compared to the local model based on idealized radiation [56]. Results showing the evolution of the crystal radius as a result of a 10% step increase in pull rate are represented in fig. 56. In all cases, the radius converges towards a new value, thereby indicating the stability of the process. However, with idealized radiation, the crystal radius decreases monotonically towards its new value, while, with detailed radiation, oscillations of the crystal radius appear. Such oscillations are induced by the interaction between radiative heat exchanges and crystal shape, and are the result of the modified radiative heat transfer taking place near the tri-junction, resulting from changes in the crystal and melt surface shapes. As a matter of fact, radiative exchanges accentuate the effect of a decreasing or increasing radius. Indeed, when the crystal grows inward, a small axisymmetric surface of the crystal views the hot melt surface, thereby absorbing more heat and causing the radius to decrease even faster. This effect is eventually compensated for by a decrease in the latent heat release and in the temperature gradient at the melt surface, so that the crystal radius can increase again. The oscillations appear to be more important in systems whose axial temperature gradient is low.

As the oscillations are damped, the system is considered as stable. But, in fig. 56, we observe that the period of the oscillations is of about 3 h (real time), or 15 cm of grown crystal, which is quite large. As a result, the decay time is too long to be seen on the time scale of a batchwise decrease of the melt volume, so that the real Cz system will appear to have an oscillatory instability. In fact, the response in a real system will always interact with the batch effects of melt volume decrease and crystal length increase due to the long decay time. This is illustrated on fig. 57, showing the evolution of the crystal radius in low-gradient heat transfer conditions, for batchwise growth without control.

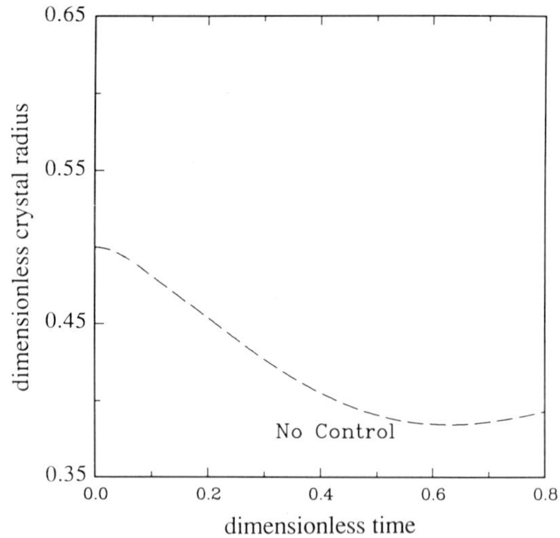

Fig. 57. Same problem as in fig. 55. Evolution of the dimensionless crystal radius predicted for diffuse-gray radiation in uncontrolled batchwise growth (From [60]).

We observe that the radius does not stabilize. In other words, the time constants related to the evolving crystal radius response to a disturbance, and to batch effects of decreasing melt volume, are not sufficiently different and the two effects cannot be decoupled. In fig. 56, the analysis of stability was performed for a hypothetical quasi-steady-state process, where very long crystals and a fixed melt level were considered (a crystal of 45 cm length was used in the simulations). Comparing the results of figs. 56 and 57, we may conclude that the results of numerical experiments based on a fixed melt volume will generally be somewhat artificial. In particular, they could not be used for comparison with experimental data, since in this latter case the geometrical changes due to the batchwise effects will prevent the radius from stabilizing. More realistic techniques can be used, as will be shown in section 5.4.3.

Global time-dependent calculations were also performed by Thomas et al. [226] for the case of GaAs LEC growth. The model used for the calculation is basically an extension of the global dynamic model developed in [60, 116]. Here, the particular radiative properties of the semi-transparent encapsulant were taken into account. The bottom of the crucible support was assumed to be insulated. The analysis of the natural dynamic response of the system through the simulation of a step change in heater temperature from a quasi-steady-state initial condition was achieved by assuming automatic melt replenishment and a fixed crystal length. A constant pull rate of 1 cm/h was considered. The heater temperature was decreased by about 20°C (i.e. about 1%) towards a new fixed value. (In a real system, the heater temperature would need to evolve continuously in order to adapt to the changing melt volume and crystal length. As a consequence, decreasing the heater temperature towards a fixed value only makes sense when melt depletion and crystal length increase are neglected.) In the solution obtained by Thomas et al. [226], the system response shows an increasing crystal radius with time (fig. 58). The evolution is different depending on the values

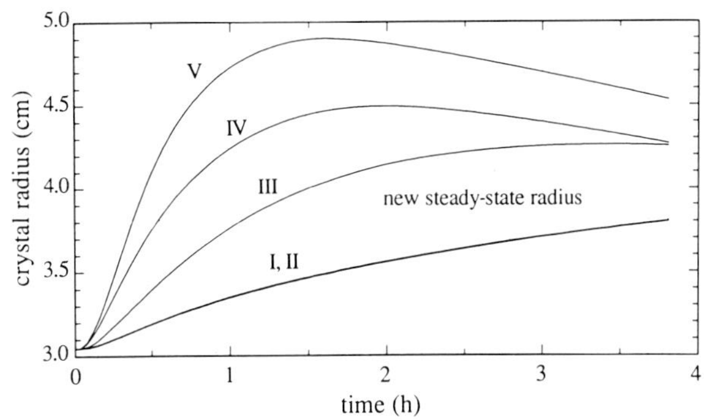

Fig. 58. Dynamic simulation of LEC GaAs growth. Evolution of the crystal radius as a result of a step change in heater temperature for different growth conditions. The conditions of curve (V) are the closest to the real situation (From [226]).

used for crystal emissivity and for conductivity of the encapsulant, and on whether the encapsulant is modelled as a semi-transparent or opaque body. The conditions corresponding to curve (V) were found to be the closest to the real case. The crystal radius was able to join a new quasi-steady state level, because crystal length was assumed to be fixed and melt depletion was neglected. Crystal radius was however very slow to reach the new quasi-steady level, because the evolving crystal shape has a long term influence on heat transfer through radiation effects. Again, the period of the oscillations is so long that, in the actual process, transient effects dominate and a quasi-steady state is never seen experimentally. The assumption of melt replenishment makes it difficult to compare numerical results with experimental data.

Similar results were obtained by Zhou et al. [273] in the case of Si growth, using the method of [61].

5.4.2. Inverse problem: pulling of the head of a crystal

Predicting a processing history of heater power and/or pull rate for the growth of a crystal of specified shape is also one of the major objectives of global dynamic simulations. This problem may be illustrated by the pulling of the head of a crystal whose shape is specified from the seed to the nominal radius. The question is to determine when, and by how much, the heater power has to be increased in order to stop radial growth, if a constant pull rate is considered. The large time constants relating the response of the crystal radius to a change in heater power render the problem very difficult. Calculating the heater power required to obtain a given crystal shape is an inverse problem, in the sense that a natural output of the system (the crystal shape) is imposed, while a natural input of the system (the heater power) is unknown. Such kinds of problem are mathematically ill-posed, as discussed in section 5.3. As a consequence, even if the imposed crystal shape is a continuous function, the resulting heater power may be discontinuous.

This inverse problem was solved in [226] by augmenting the dynamic model with a simple proportional/integral control equation, as described in [56]. The technique is thus based on direct calculations, so that the problems of oscillations due to the ill-posed nature of the inverse problem were not found. Using optimal control

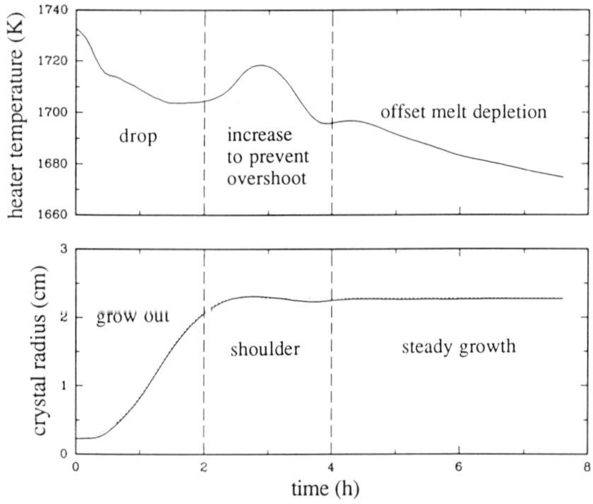

Fig. 59. Same problem as in fig. 58. Evolution of the crystal radius and heater temperature, as determined with a PI controller, for the growth of a crystal from the seed to a constant radius using a cone angle of 55° (From [226]).

constants provided results almost equivalent to those that would be obtained by means of a real inverse simulation. Direct calculations are however more expensive and difficult to implement than inverse ones, since more equations have to be coupled (see section 5.3).

The method was applied to the calculation of the heater temperature required to obtain a crystal whose cone angle and nominal radius are 55° and 2 cm, respectively. Melt depletion was taken into account. The result is represented in fig. 59, where we observe that, during the shouldering process, i.e. when switching from conical to cylindrical growth, an increase of heater temperature is necessary to stop radial growth. We may observe that the simple control algorithm worked quite well, as the calculated crystal shape was very close to the reference shape used by the controller. A typical evolution of crystal shape and temperature field for a similar problem is represented in fig. 60.

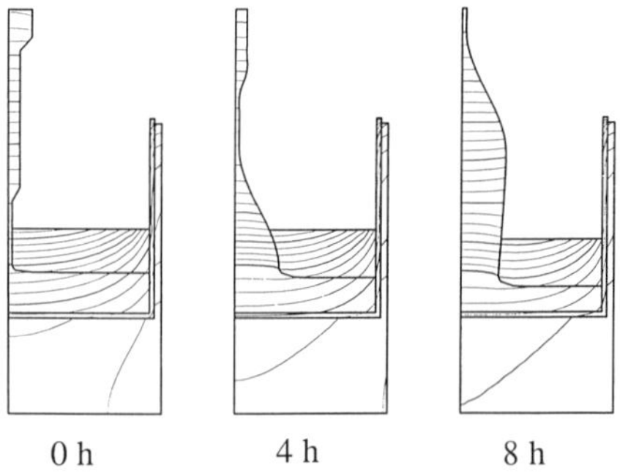

Fig. 60. Same problem as in fig. 58. Crystal shape, solid–liquid interface shape and temperature field at different instants of growth in a typical batchwise simulation ($\Delta T = 25$ K) (From [226]).

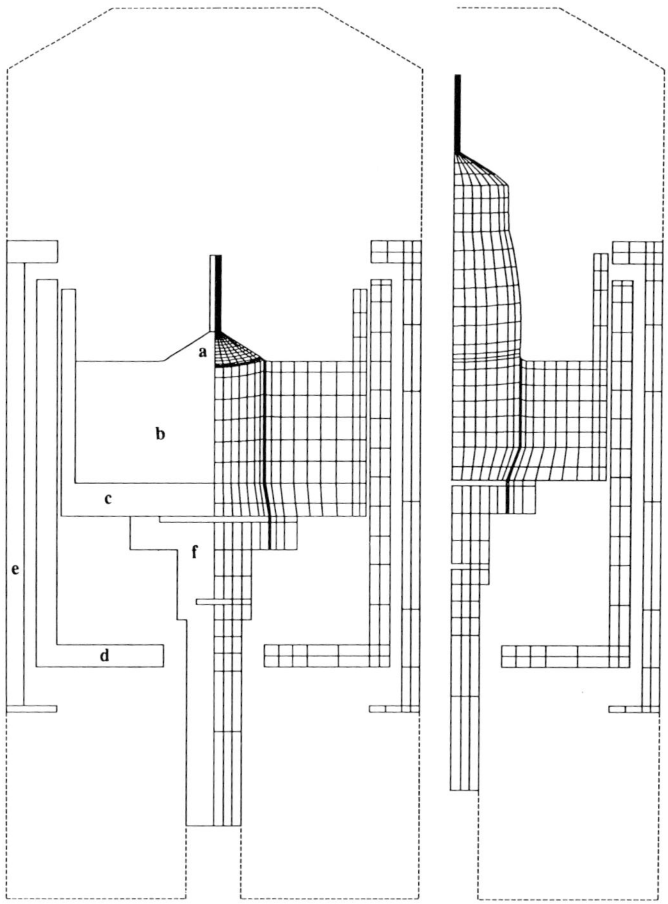

Fig. 61. Sketch and finite element representation of a Cz furnace for growing germanium crystals: (a) crystal; (b) melt; (c) crucible; (d) deater; (e) insulator; (f) pedestal (From [65]).

The inverse problem of the pulling of the head of a crystal was also addressed by Van den Bogaert and Braet [249] for the case of a germanium crystal. In this case, a fully implicit global time-dependent model was used (section 5.3.4), and inverse calculations were performed. The model considers melt volume decrease together with crucible displacement, in order to maintain the melt level at a constant axial position with respect to the heater. The furnace is represented in fig. 61. The imposed crystal shape, represented in fig. 62, includes three different parts: the conical part, with a growth angle of 60° (I), the transition from a cone to a cylinder (II), and the cylindrical part, where the radius is constant and equal to 5 cm (III).

The initial solution is calculated by means of a corresponding quasi-steady model. The pull rate is set to a constant value of 2 cm/h. The calculated heater power is represented in fig. 63, and appears to be highly discontinuous at two precise stages: first, at the beginning of the simulation, when the solution evolves from a quasi-steady to a

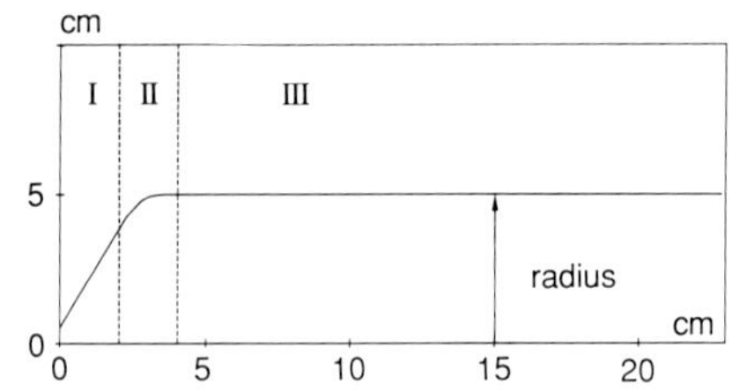

Fig. 62. Same problem as in fig. 61. Prescribed crystal shape for the inverse problem (From [249]).

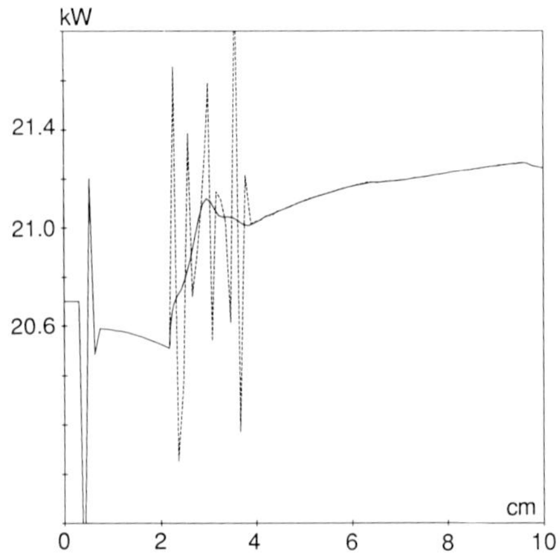

Fig. 63. Same problem as in fig. 61. Heater power as a function of crystal length. Smoothed (solid line) and non-smoothed (dotted line) curves (From [249]).

time-dependent regime, and second, during the shouldering. As a matter of fact, due to the ill-posed nature of the problem, the solution of this inverse problem may exist only in the form of a distribution. In this case, the numerical solution approximates the distribution by means of some oscillations, whose frequency and amplitude increase with a decreasing time step. The oscillating result may be easily smoothed (fig. 63), so that a legible curve is obtained (the crystal shape calculated as a response to this smoothed power fits the initial prescribed shape almost perfectly). Obtaining a heater power completely free from the oscillations related to inverse simulations is also possible by choosing an analytical expression for representing the imposed crystal shape, as demonstrated by Van den Bogaert [66].

We observe from fig. 63 that a sharp rise of heater power during the shouldering process is necessary to stop radial growth. This is immediately followed by a decrease of power, in order to avoid that the radius of the crystal starts decreasing. The results are qualitatively similar to those obtained by Thomas et al. [226]. A comparison with corresponding quasi-steady results (fig. 64) shows large differences during the shouldering, but almost no difference once constant radius is achieved, meaning that the rapid geometrical changes during shouldering induce a real transient behaviour at that stage, while constant radius growth allows for the quasi-steady assumption. In this

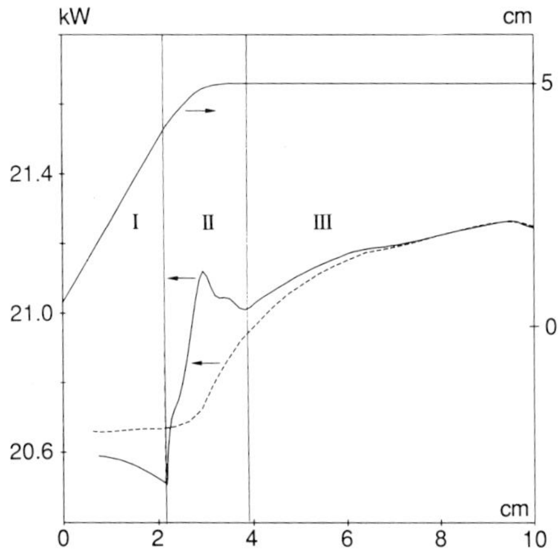

Fig. 64. Same problem as in fig. 61. Heater power as a function of crystal length. Comparison between time-dependent (solid line) and quasi-steady (dashed line) results. The imposed crystal shape is also shown (From [249]).

model, the actual growth rate is very similar to the pull rate, since crucible position is modified in order to maintain melt surface at a given position. Quasi-steady-state and time-dependent solutions very close to each other can therefore be obtained (this is not the case in the model of [107], where the crucible was held fixed, so that the crystal growth rate was always higher than the pull rate).

Let us note that the choice of a given crystal shape in inverse problems has to be done very carefully, as some configurations may be related to remelting situations (see sections 2.3, 3.2.2 and 5.3.3). As a matter of fact, when the growth angle decreases rapidly, meniscus height may increase at a higher rate than the pull rate, thereby causing the

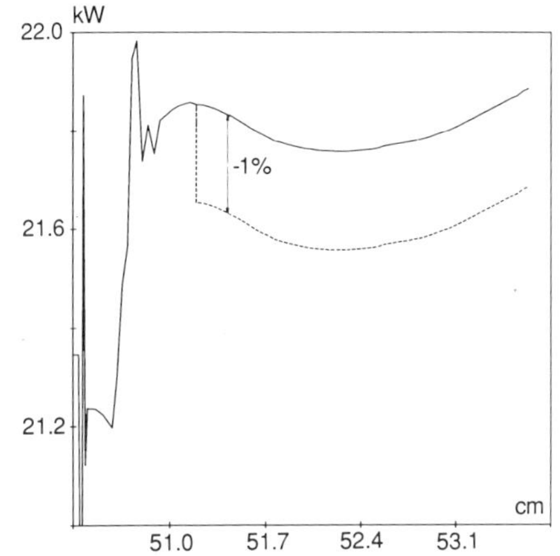

Fig. 65. Same problem as in fig. 61. Heater power as a function of the crucible vertical position, such as obtained by an inverse calculation (solid line); imposed 1% step decrease in heater power for a direct calculation (dashed line) (From [65]).

crystal to remelt at the tri-junction. In [249], local curvature during the shouldering had to be limited in order to avoid numerical remelting.

An inverse simulation calculating heater power for a longer crystal (25 cm length) was performed by Van den Bogaert and Dupret [65]. We observe how the resulting heater power (fig.65) varies only about 1% around the nominal value, once constant radius growth is achieved. These small variations compensate for the continual geometrical changes which do not allow the existence of any steady state.

Figure 66 represents the associated evolution of the solid–liquid interface position with respect to melt level, on the axis and at the tri-junction. The interface is convex with respect to the crystal during the whole process, since axial position is always lower than tri-junction position. This latter is an immediate result of the imposed growth angle. We observe a sharp increase in axial position during the shouldering,

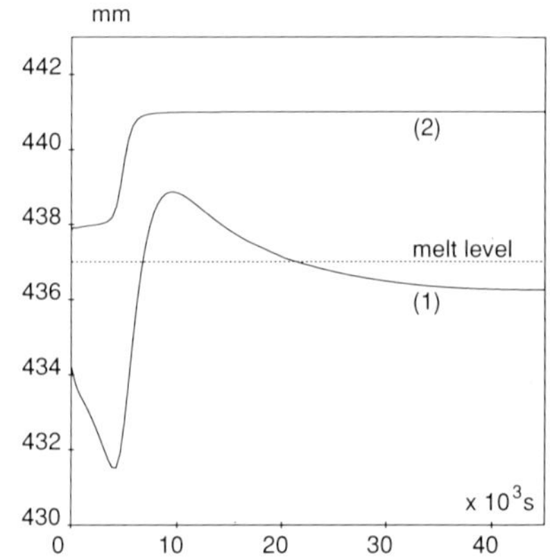

Fig. 66. Same problem as in fig. 61. Evolution of the solid–liquid interface shape (1) on the axis, and (2) at the tri-junction. The dashed line indicates the melt level (From [65]).

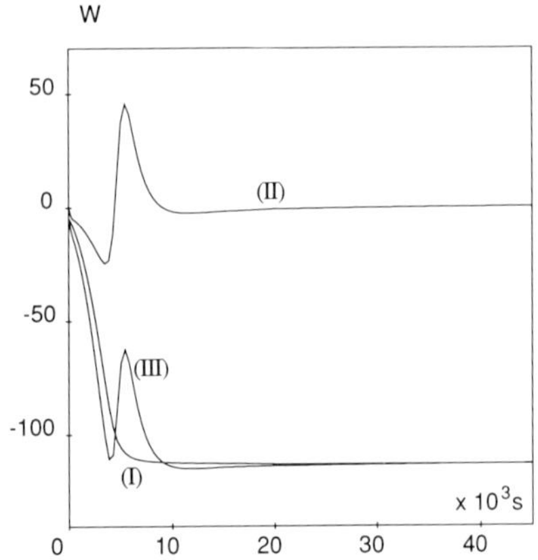

Fig. 67. Same problem as in fig. 61. Evolution of the solidification heat: (I) quasi-steady contribution, (II) time-dependent contribution, and (III) total amount (From [65]).

which matches experimental observations. The solidification heat is represented in fig. 67, where negative values stand for heat release and positive values for heat absorption. Solidification heat includes a quasi-steady contribution, due to the pull rate, and a time-dependent contribution, due to the change of configuration of the solid– liquid interface shape (see section 3.2.2). The steady contribution varies as the square of the radius and is, of course, constant for cylindrical growth. The time-dependent contribution reaches high values during the shouldering (about 40% of the quasi-steady contribution), revealing the presence of important transient effects induced by geometrical changes at that stage. Once cylindrical growth is reached, this time-dependent contribution almost vanishes, since the solidification front always remains close to the same position.

5.4.3. *Direct problem based on an inverse simulation: comparison between heater power and pull rate perturbations*

One of the main interests of inverse calculations is the possibility of solving direct problems subsequently, as has been done by Van den Bogaert and Dupret [65, 165]. The heater power evolution obtained as a result of an inverse calculation can be used as a reference evolution. Perturbations of heater power to be used in direct problems are then applied with respect to this reference curve. In fig. 65, we observe how the heater power is decreased by 1% at a given stage. The curve is vertically translated, so that variations of the heater power due to geometrical changes are taken into account. This approach allows one to eliminate the batchwise transients, in order to avoid radius oscillations due to geometrical changes. The technique is more realistic than that of fixing the melt volume, as was done in [60, 226].

It should be noted that applying a heater power step decrease as a function of crucible or crystal vertical position does not yield the same radius response. This was demonstrated in [258], where the heater power, which is related to the geometrical evolution of the system, was found to be more closely related to crucible axial position (or, equivalently, to melt volume) than to crystal axial position. As a consequence, heater power disturbances were defined on the basis of the crucible position.

A heater power step decrease, as represented in fig. 65, will cause the crystal radius to increase, since solidification obviously takes place more readily when the system is cooler. Typical finite element meshes and temperature fields for such a situation are represented in fig. 68. An interesting comparison between the system responses to a reduction of heater power and to a reduction of pull rate was drawn in [65]. It was observed that the respective responses of the system to equivalent perturbations were similar when considering long time scale effects, but strongly different when considering small time scale effects. As a first experiment, reductions of 1% of the heater power and of 10% of the pull rate were applied. Both perturbations lead to a radius increase (in the case of a reduction of pull rate, the cooling of the system is achieved through a reduction of the solidification heat at the solid – liquid interface). Both responses are very similar in shape (fig. 69), provided scales are adapted for comparison (fig. 70). The maximum radius (6 cm for 1% power reduction, and 5.3 cm for 10% pull rate reduction) is reached in both cases after growing a given crystal length (9.3 cm), but not after a given time lapse (4.5 and 5.2 h, respectively). This observation fits the idea of a time

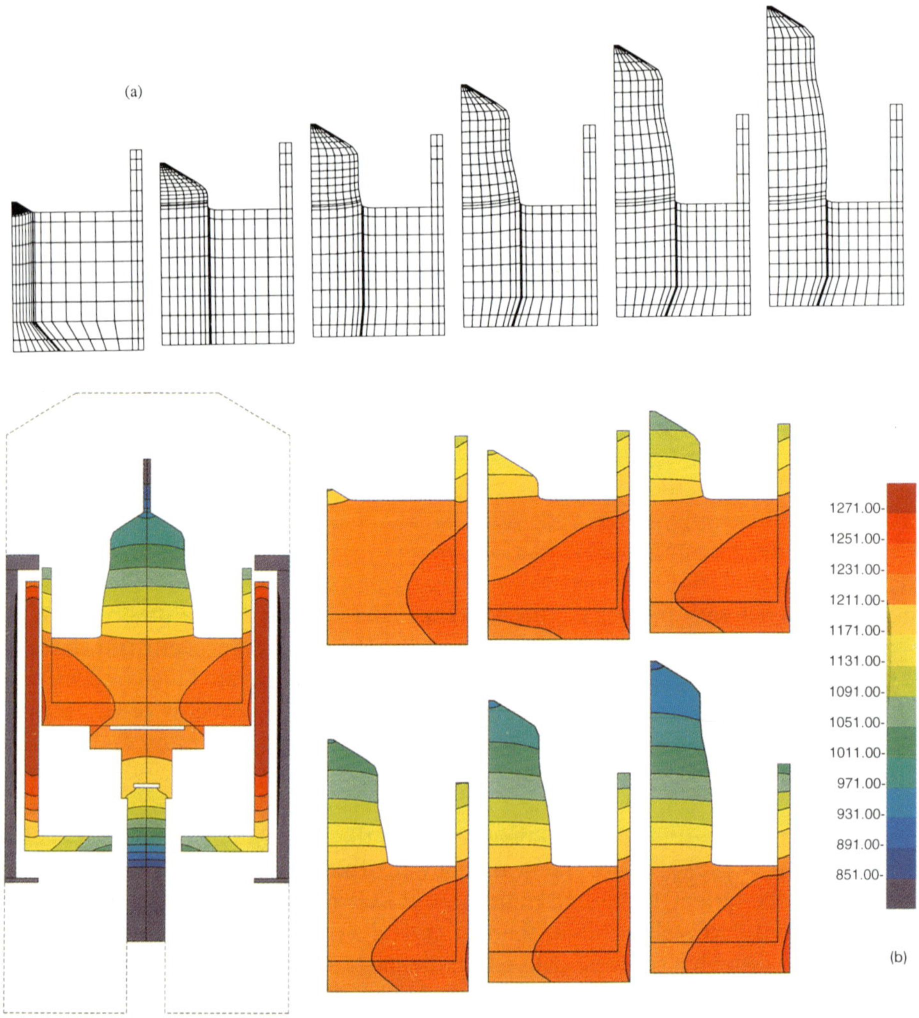

Fig. 68. Same problem as in fig. 61. Response to a 1% step decrease in heater power: (a) finite element meshes for various stages of growth; (b) temperature field in the global furnace and temperature field in the melt, crystal, and crucible for various stages of growth (color plot) (From [65]).

constant proportional to R_J/v_{pul}(where R_J is the crystal radius and v_{pul} the pull rate), as suggested in [107].

The system response to both perturbations may be investigated in terms of solid – liquid interface behaviour. Figure 71 represents the solid – liquid interface position with respect to melt level, on the axis and at the tri-junction. It is clear that, in both cases, the solidification front reacts to the perturbations by moving downwards. After a few hours, the solid–liquid interface rises back to the non-perturbed level in the case of a heater power decrease, but remains at a lower level in the case of a pull rate decrease.

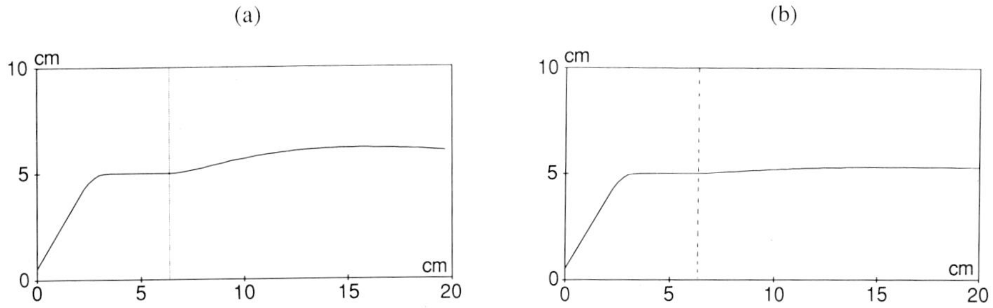

Fig. 69. Same problem as in fig. 61. Calculated crystal shape as a result of (a) a 1% step decrease in heater power, and (b) a 10% step decrease in pull rate (From [65]).

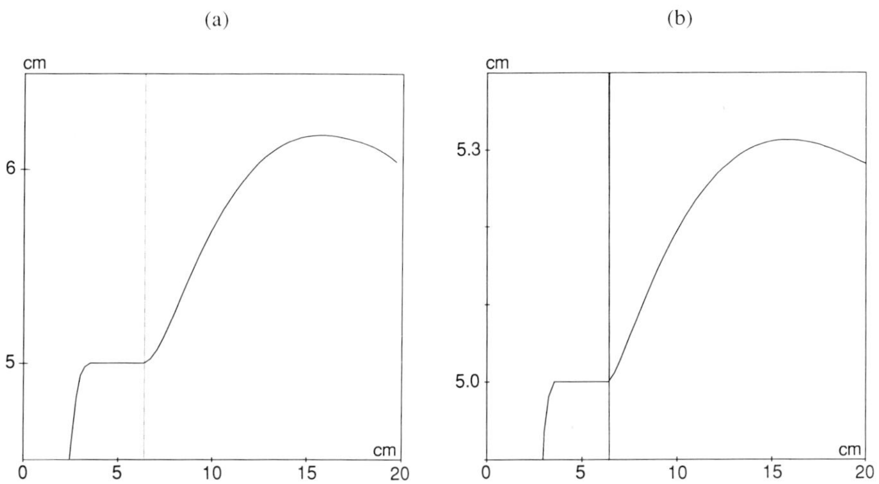

Fig. 70. Same problem as in fig. 61. Calculated crystal shape (enlarged view) as a result of (a) a 1% step decrease in heater power, and (b) a 10% step decrease in pull rate (From [65]).

Indeed, it has already been observed in quasi-steady simulations that the solid–liquid interface position is strongly related to the pull rate, in the sense that a higher pull rate yields a higher interface position, and vice versa. At the tri-junction, the interface position response is directly related to the growth angle through the contact angle constancy condition: when crystal radius increases, the growth angle is positive, and the meniscus height is lower than the steady-state value.

Due to the perturbations, an important change in the solidification heat released at the solid–liquid interface occurs, as shown in fig. 72. On the one hand, the quasi-steady contribution to the solidification heat (proportional to both pull rate and square of the radius) is modified. When decreasing the power, this contribution increases, as defined by the changing crystal radius. Decreasing the pull rate leads to an immediate decrease of this quasi-steady contribution. After a few hours, however, this quasi-steady contribution returns to the initial non-perturbed value. It is interesting to note that the

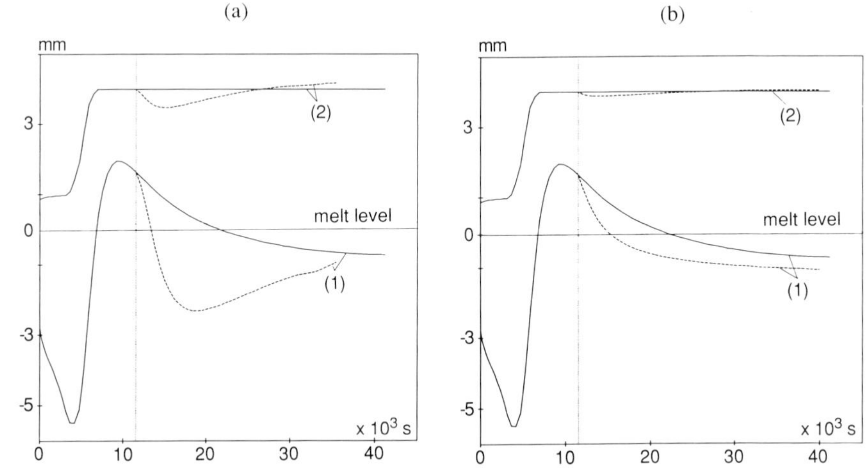

Fig. 71. Same problem as in fig. 61. Evolution of the solid–liquid interface shape (1) on the axis, and (2) at the tri-junction, as a result of (a) a 1% step decrease in heater power, and (b) a 10% step decrease in pull rate. Non-perturbed results are indicated by solid lines (From [65]).

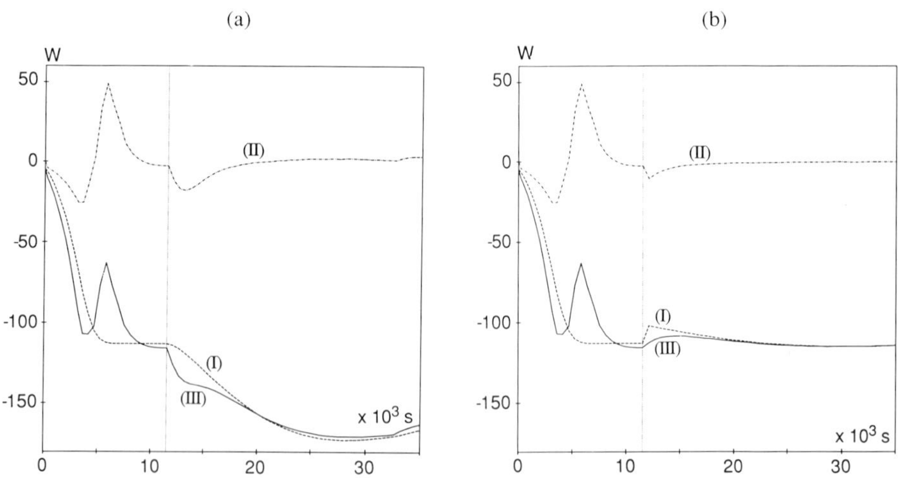

Fig. 72. Same problem as in fig. 61. Evolution of the solidification heat: (I) quasi-steady contribution, (II) time-dependent contribution, and (III) total amount, as a result of (a) a 1% step decrease in heater power, and (b) a 10% step decrease in pull rate (From [65]).

pull rate reduction is exactly compensated by the radius increase, so that the global solidification heat is maintained at the same level as prior to the perturbation.

On the other hand, the downwards motion of the interface causes a release of extra solidification heat (time-dependent contribution). In the case of a pull rate reduction, a step change of the time-dependent contribution is observed, showing an immediate interface reaction. This means that the sudden decrease of the quasi-steady contribu-

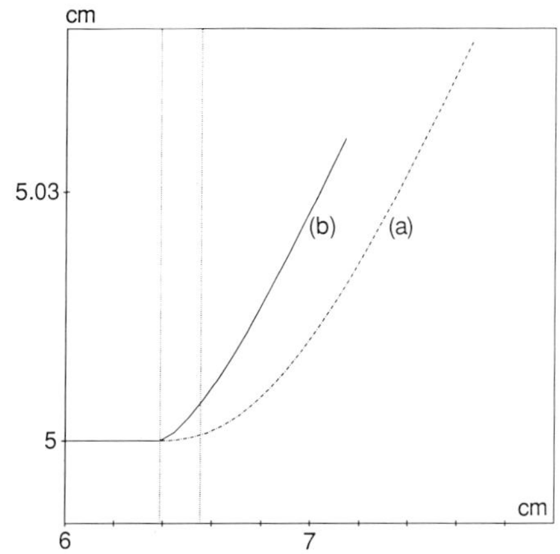

Fig. 73. Same problem as in fig. 61. Early system response. Crystal shape as a result of (a) a 0.25% step decrease of heater power, and (b) a 10% step decrease of pull rate. The vertical lines relate to the instant of the perturbation and to the situation 5 min later (From [65]).

tion caused by the pull rate reduction is immediately compensated by an increase of equal magnitude of the time-dependent contribution. At the instant the pull rate is decreased, the system reacts so that the global solidification heat is not discontinuous, or so that the local growth rate is not discontinuous.

While the effects on crystal shape for the two different perturbations are quite similar (fig. 70), it was shown in [65] that the system responses considering a small time scale (a few minutes) are quite different. The analysis of the initial system response could be achieved by repeating the simulations with a reduced time step. Furthermore, in order to render heater power and pull rate changes comparable in a quantitative way, the amplitudes of the perturbations were adapted. The heater power was decreased by 0.25%, and the pull rate by 10%, so that both perturbations yield the same final radius increase. It is clear from fig. 73 that the crystal radius responds more rapidly to a pull rate decrease than to a heater power decrease. After five minutes, for example, the radius increase is five times larger in the case of a pull rate reduction than in the case of a heater power reduction. A close view to the time-dependent part of the solidification heat (fig. 74) shows that interface reaction to pull rate decrease is felt at once, since solidification heat is directly affected by the perturbation, while reaction to power decrease is slower. In this latter case, the maximum value for time-dependent solidification heat is felt about 30 min after the perturbation. Moreover, changes in heater power must first be transported through the melt to subsequently affect the crystal radius. This difference in the small time scale responses that was exhibited accounts for the fact that Czochralski furnaces are more easily controlled by acting on the pull rate than on the heater power.

The influence of different parameters on the dynamic behaviour of a germanium Cz system was investigated by Van den Bogaert [66]. It was shown, for example, how transient effects observed during the shouldering process are increased with higher pull rates. It was observed that the sensitivity of the radius to a decrease of heater

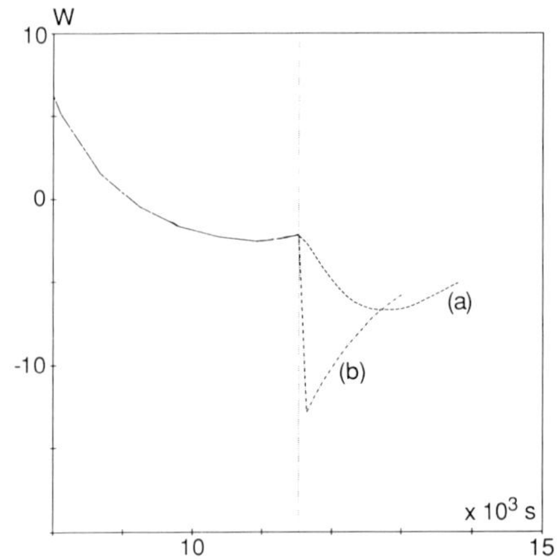

Fig. 74. Same problem as in fig. 61. Early system response. Time-dependent contribution to the solidification heat as a result of (a) a 0.25% step decrease of heater power, and (b) a 10% step decrease of pull rate (From [65]).

power is higher when the thermal conductivity of the liquid phase is lower, or when the pull rate is lower. The influence of the amplitude of the perturbation was also examined, and the response of the system was found to be almost linear. The influence of crystal radius and of crystal length on radius increase proved to be negligible.

Similar experiments comparing the actions of a power decrease or pull rate decrease were performed by Van den Bogaert and Dupret [259] for the case of a GaAs furnace.

5.4.4. Dynamic global calculations including convection in the melt

All results from time-dependent global simulations described so far have been obtained by neglecting the melt flow. In fact, one main obstacle to obtaining realistic information from global models of the Cz process is the presence of natural and forced convection in the melt. As typical Reynolds and Grashof numbers are very high for semiconductor furnaces, the computational cost for calculating melt convection using the classical finite element technique is too high to allow for dynamic simulations. In order to solve this problem, an original approach was developed by Lecomte et al. [255, 256], for computing the approximate velocity field in the melt and for performing a time-dependent simulation using this velocity field as a parameter. Convection in the melt is approximated in the laminar range by means of an extrapolation technique [255, 257], based on the method of matched asymptotic expansions: velocities within the melt for high values of the dimensionless numbers are obtained by extrapolating the velocities calculated for lower values of those numbers. A quasi-steady approximation of the flow field can be obtained by this method for any stage of growth. A subsequent interpolation of the Stokes stream function between different stages provides a flow field for any time step of a dynamic simulation, as represented in fig. 54.

Inverse simulations without convection, with convection up to 8% of the actual Reynolds number and with convection up to 50% of the actual Reynolds number have

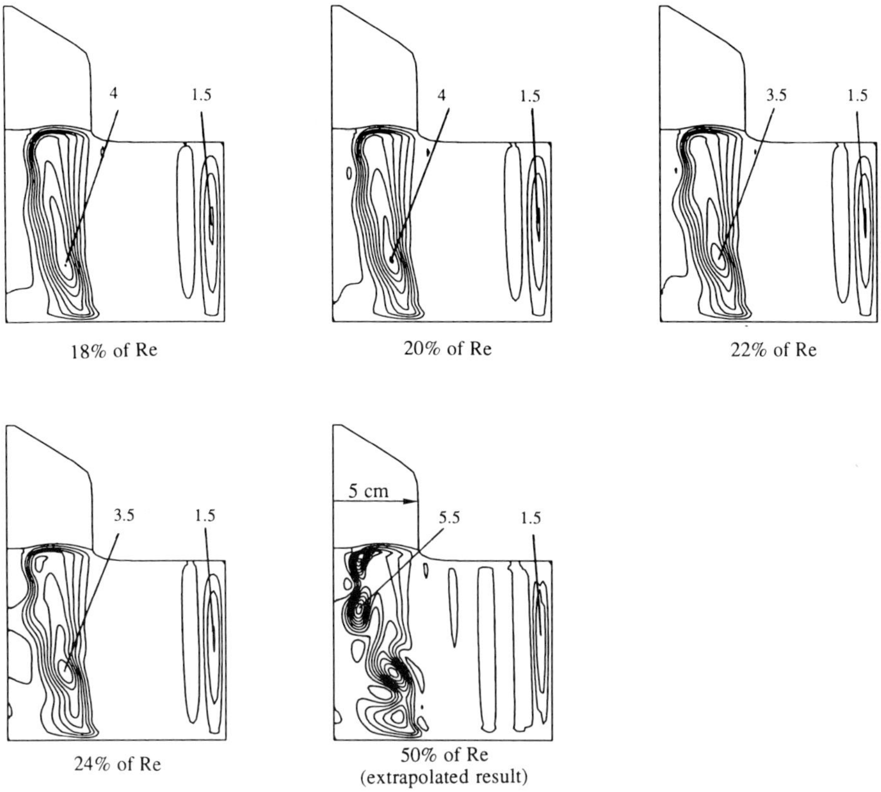

Fig. 75. Dynamic simulations of silicon growth taking melt convection into account. Extrapolated Stokes stream function (in cm^3/s) at a high Reynolds number (50% of the actual one), as calculated from Stokes stream functions at lower Reynolds numbers.

been carried out for the case of a silicon furnace. Two stages of growth were selected for quasi-steady calculations. Figure 75 shows, for one of the selected stages, the extrapolated flow field at 50% of the actual Reynolds number together with the flow fields for the lower Reynolds numbers, which were used for extrapolation. It is clear that the hypothesis of a laminar steady flow behaviour at the actual values of the dimensionless numbers is here subject to question. Nevertheless, the influence of this calculated convection on the dynamic behaviour of the system is of major interest. The calculated heater power appears to be highly dependent upon the intensity of convection (fig. 76): it is of course decreased when convection is taken into account. It is interesting to observe that taking this effect into account causes an instantaneous perturbation which is highly amplified due to the inverse nature of the problem, and which generates oscillations of the resulting heater power, just in the same way as observed in [249].

Let us point out that the influence of melt flow upon temperature field is important in this growth system. In particular, convection in the melt modifies the thermal boundary conditions on the melt and crystal, meaning that calculation of the velocity and

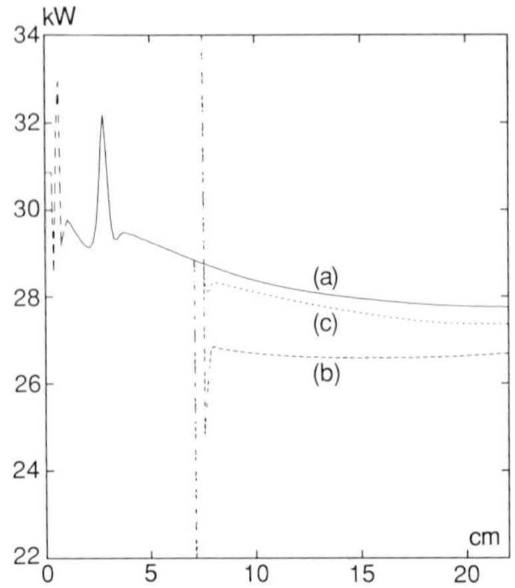

Fig. 76. Same problem as in fig. 75. Heater power as a function of crucible axial position (a) in the absence of convection, (b) for an instantaneous start of convection up to 8% of the actual Reynolds number, and (c) for an instantaneous start of convection up to 50% of the actual Reynolds number.

global temperature fields cannot be decoupled (as demonstrated by Xiao and Derby) [275]. Indeed, natural convection in the melt is dictated by the thermal boundary conditions on the melt and the crystal. Arbitrary boundary conditions on this subdomain would lead to unrealistic results. An accurate representation of the flow in the melt can, therefore,.only be achieved with a coupled global model where convection is taken into account at least by a quasi-steady approach.

5.4.5. Calculation of dislocation density

The resulting temperature evolution computed from a global time-dependent model can be used as data for subsequent calculations of dislocation density in the crystal. Such a calculation was performed by Nicodème et al. [126, 260] for the case of a GaAs LEC-grown crystal, using the model of Alexander and Haasen [119], as adapted by Dillon et al. [76]. As a first step, a global time-dependent heat transfer model was used to provide for the temperature history in the crystal. As a second step, dislocation density was calculated from this thermal history. Simulations were performed for a prescribed constant crystal radius. Results include the overall temperature field and the dislocation density field (fig. 77). Typical values for dislocation density vary between 10^4 and $10^6\,cm^{-2}$, and match experimental observations. Maximum values are located on the crystal periphery and on the axis. Because the motion of dislocations is not considered in this model, some dislocation patterns are simply convected upwards, as if they were attached to the crystal, while dislocations are observed to multiply.

Fig. 77. Calculation of dislocation density in LEC GaAs growth. Temperature field in the global furnace (in K), and decimal logarithm of the dislocation density field in the crystal (in cm^{-2}), represented for different crystal lengths: (a) 2 cm; (b) 6 cm; (c) 10 cm; (d) 14 cm (From (126]).

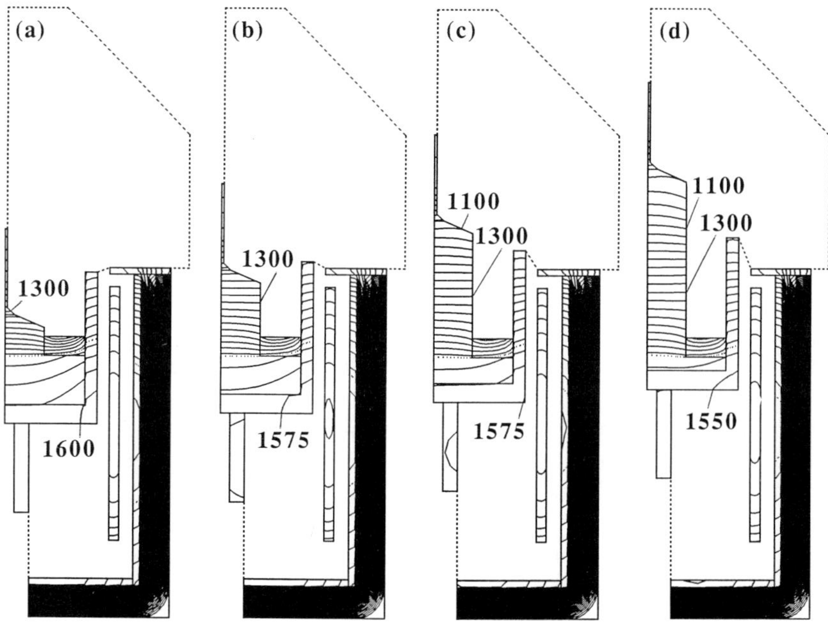
(a)
(b)
(c)
(d)
1300
1600
1300
1575
1100
1300
1575
1100
1300
1550

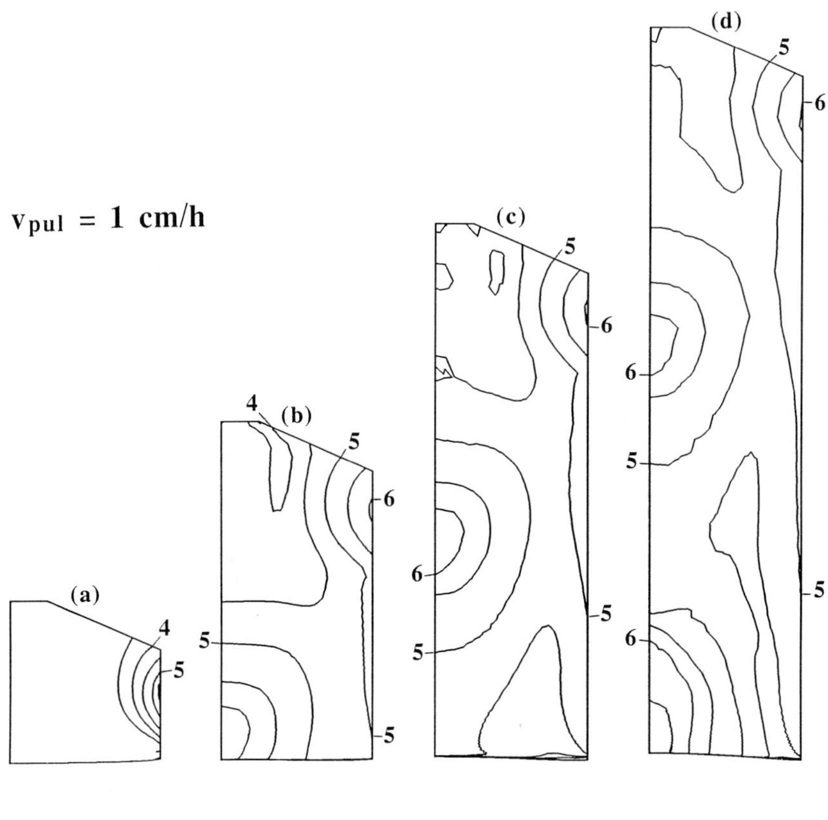
v_{pul} = 1 cm/h
(a)
4
5
(b)
4
5
6
5
5
(c)
5
6
6
5
5
(d)
5
6
6
5
5
6

Similar models for the calculation of viscoplastic deformation were presented in section 2.1. In a general manner, however, present models are rather too simplified with regard to the complexity of the phenomenon, and need to be improved in order to provide accurate predictions.

It is worth mentioning here the work of Virzi and Porrini [276], and Virzi [277], who performed dynamic simulations of the cool-down stage of the process, and demonstrated its influence on crystal quality.

6. Conclusions

In view of the results which have been presented in this chapter, we could conclude that numerical methods have reached most of their objectives and that a significant part of the development task is now behind us. Nonetheless, there is no doubt that more and more complex, accurate, and integrated models will continue to be developed in the future. The emergence of vectorized and (more) parallel computers on the market represents indeed a major event, since new classes of problem become conceivable to address and to solve at a continuously decreasing cost. Hence, a new generation of simulation software can be expected to become available soon.

The modelling effort remains however essential. In particular, we need appropriate models for calculating the melt flow in coupled global and time-dependent simulations. This question is difficult to address, since the characteristic time scales of the growth and the melt flow have different orders of magnitude. On the other hand, the problem of representing a periodic, quasi-periodic or turbulent flow behaviour is certainly one of the most attractive challenges posed to the forthcoming generation of scientists. It must be kept in mind that simplified models are required, which capture the effects taking place at the scale of the simulation, while details taking place at lower scales must be averaged. As long as accurate flow models are missing or too expensive, some of the results obtained by means of bulk growth simulations will remain subject to question.

Other domains where improved models are necessary are the calculation of the reactions in the cooling crystal, the prediction of dislocation formation, and the modelling of internal radiation in the liquid and/or the solid phase. In all these cases, a very important problem is to obtain accurate laws and material data in order to be able to perform realistic simulations. We must emphasize here that inverse numerical methods will become an essential tool for solving the important difficulties related to high temperature experiments, since this approach may provide information at locations where direct measurements are impossible. Concerning dislocation generation, it is clear that new models, taking non-isotropic effects into account by means of orientation tensors, will be required. Concerning internal radiation in Cz growth, a difficulty remains to couple this effect with the radiative heat transfer in the outer environment.

When predicting such a continual increase of the role of numerical methods with a view to obtaining high quality crystals (in particular, by means of developing control algorithms), we must also have in mind that the complexity of the software will become a more and more stringent problem. While using simplified models is, and will always remain very useful, developing integrated models gives rise to very long and intricate computer codes, which cannot be maintained by a single person. At this level, new developments will take place within organized teams, and the informatic – or software engineering – tasks will become very important.

We may, therefore, conclude in general that modelling and numerical simulation will continue to represent a fundamental research activity in crystal growth, while new structures will be required to handle the increasing difficulties of the problems.

Acknowledgments

This work was possible only owing to the contribution of several people to the research activity in numerical simulation of crystal growth in Louvain-la-Neuve. Marcel Crochet initiated this project and supervised the group for several years; his impulse was essential for us. Paul Wouters, Yves Ryckmans, Pierre Nicodème, and Michaël Lecomte participated to the development of the FEMAG software, and brought us their ideas and methods in the friendly, while rigorous, atmosphere of our group. It is impossible to describe here in detail their personal contributions, in spite of the importance of their work.

Discussions with external people were also very useful for our team. In particular, we wish to mention a fruitful collaboration with Claude Coekelbergs and Janic Braet, of Metallurgie Hoboken, with Gérard Martin, of the Laboratoire d'Electronique Physique, with Gert Nagel and Erich Dornberger, of Wacker Chemitronic, and with Koichi Kakimoto, of Nippon Electric Corporation.

Notations

In this chapter, vectors and tensors are represented by means of bold roman characters, using the *Gibbs polyadic notation*. In particular, any product involving a contraction (which yields a summation over repeated indices, in a given system of axes) is indicated by a dot ($\cdot$), such as for the scalar product.

In vector and tensor fields, for the sake of conciseness, the independent variables are not explicitly written, except when necessary. Similarly, subscripts such as l or s, which denote the liquid or solid phase, are only used on the interface, where both phases are present.

The integrations associated with the weak formulation of the equations of the problem are represented by means of single ($\langle\ \rangle$), double ($\langle\langle\ \rangle\rangle$), or triple ($\langle\langle\langle\ \rangle\rangle\rangle$) brackets, depending on whether a volume, surface, or line integral is carried out. Except for radiative enclosures, weak equations are written in a form which is valid for general three-dimensional problems, including the axisymmetric case. Radiative enclosures are considered only for axisymmetric problems.

Discrete equations are expressed in matrix form, where column vectors and matrices are represented by bold italic capital symbols.

References

[1] W.R. Wilcox, in: Preparation and Properties of Solid State Materials, Vol. 1, ed. R.A. Lefever (Dekker, New York, 1971).
[2] W.R. Wilcox, in: Preparation and Properties of Solid State Materials, Vol. 2, ed. W.R. Wilcox (Dekker, New York, 1976).
[3] D.P. Woodruff, The Solid–Liquid Interface (Cambridge Press, Cambridge, 1973).
[4] M.C. Flemings, Solidification Processing (McGraw-Hill, New York, 1974).
[5] J.C. Brice, The Growth of Crystals from Liquids (North-Holland, New York, 1976).
[6] J.R. Carruthers, in: Preparation and Properties of Solid State Materials, Vol. 3, eds. W.R. Wilcox and R.A. Lefever (Dekker, New York, 1977).
[7] F. Rosenberger, in: Springer Series in Solid-State Sciences, Vol. 5 Fundamentals of Crystal Growth I, eds. M. Cardona, P. Fulde and H.J. Queisser (Springer, Berlin, 1979).
[8] N. Kobayashi, in: Preparation and Properties of Solid State Materials, Vol. 6, ed. W.R. Wilcox (Dekker, New York, 1981).
[9] W. Zulehner and D. Huber, in: Crystal Growth, Properties and Applications, Vol. 8, ed. J. Grabmaier (Springer, New York, 1982).
[10] D.T.J. Hurle, in: Crystal Growth of Electronic Materials, ed. E. Kaldis (Elsevier, New York, 1985).
[11] M.J. Stewart and F. Weinberg, J. Cryst. Growth 12 (1972) 217.
[12] Ch. E. Chang and W.R. Wilcox, Int. J. Heat Mass Transfer 19 (1976) 355.
[13] W.E. Langlois, J. Cryst. Growth 42 (1977) 386.
[14] W.E. Langlois, J. Cryst. Growth 56 (1982) 15.
[15] W.E. Langlois, J. Cryst. Growth 63 (1983) 67.
[16] N. Kobayashi, J. Cryst. Growth 43 (1978) 357.
[17] N. Kobayashi, Comput. Methods Appl. Mech. Eng. 23 (1980) 21.
[18] S.M. Pimputkar and S. Ostrach, J. Cryst. Growth 55 (1981) 614.
[19] V.I. Polezhaev, in: Crystals: Growth, Properties, and Applications, Vol. 10. Hydrodynamics of Heat and Mass Transfer during Crystal Growth, eds. H.C. Freyhardt, A.A. Chernov and H. Müller-Krumbhaar (Springer, Berlin, 1984).
[20] W.E. Langlois, Ann. Rev. Fluid Mech. 17 (1985) 191.
[21] W.E. Langlois and K.J. Lee, J. Cryst. Growth 62 (1983) 481.
[22] G.M. Oreper and J. Szekely, J. Cryst. Growth 64 (1983) 505.
[23] G.M. Oreper and J. Szekely, J. Cryst. Growth 67 (1984) 405.
[24] M. Mihelčić and K. Wingerath, J. Cryst. Growth 71 (1985) 163.
[25] E. Scheil, Z. Metallk. 34 (1942) 70.
[26] J.A. Burton, R.C. Prim and W.P. Slichter, J. Chem. Phys. 21 (1953) 1987.
[27] D.T.J. Hurle, E. Jakeman and E.R. Pike, J. Cryst. Growth 34 (1968) 633.
[28] J.R. Carruthers and A.F. Witt, in: Crystal Growth and Characterization, eds. R. Ueda and J.B. Mullins (North-Holland, Amsterdam, 1975).
[29] L.O. Wilson, J. Cryst. Growth 44 (1978) 247.
[30] L.O. Wilson, J. Cryst. Growth 44 (1978) 371.
[31] L.O. Wilson, J. Cryst. Growth 48 (1980) 435.
[32] L.O. Wilson, J. Cryst. Growth 48 (1980) 451.
[33] L.O. Wilson, J. Cryst. Growth 48 (1980) 363.
[34] L.O. Wilson, J. Cryst. Growth 56 (1982) 219.
[35] J.J. Favier, J. Electrochem. Soc. 129 (1982) 2355.
[36] S.R. Coriell and R.F. Sekerka, J. Cryst. Growth 46 (1979) 479.
[37] S.R. Coriell, R.F. Boisvert, R.G. Rehm and R.F. Sekerka, J. Cryst. Growth 54 (1981) 167.
[38] D.T.J. Hurle and R.W. Series, J. Cryst. Growth 73 (1985) 1.
[39] S. Kobayashi, J. Cryst. Growth 75 (1986) 301.
[40] K. Mika and W. Uelhoff, J. Cryst. Growth 30 (1975) 9.
[41] T. Surek, J. Appl. Phys. 47 (1976) 4384.
[42] N. Kobayashi and T. Arizumi, Jpn. J. Appl. Phys. 8 (1969) 1091.

[43] N. Kobayashi and T. Arizumi, Jpn. J. Appl. Phys. 9 (1970) 361.
[44] N. Kobayashi and T. Arizumi, Jpn. J. Appl. Phys. 9 (1970) 1255.
[45] T. Arizumi and N. Kobayashi, J. Cryst. Growth 13 & 14 (1972) 615.
[46] T.W. Fu and W.R. Wilcox, J. Cryst. Growth 48 (1980) 416.
[47] S.N. Rea, J. Cryst. Growth 54 (1981) 267.
[48] G. Williams and R.E. Reusser, J. Cryst. Growth 64 (1983) 448.
[49] P.A. Ramachandran and M.P. Duduković, J. Cryst. Growth 71 (1985) 399.
[50] R.K. Srivastava, P.A. Ramachandran and M.P. Duduković, J. Cryst. Growth 73 (1985) 487.
[51] H.M. Ettouney and R.A. Brown, J. Comput. Phys. 49 (1983) 118.
[52] M.J. Crochet, F.T. Geyling and J.J. Van Schaftingen, J. Cryst. Growth 65 (1983) 166.
[53] Ch.J. Chang and R.A. Brown, J. Comput. Phys. 53 (1984) 1.
[54] A.B. Crowley, IMA J. Appl. Math. 30 (1983) 173.
[55] J.J. Derby, R.A. Brown, F.T. Geyling, A.S. Jordan and G.A. Nikolakopoulou, J. Electrochem. Soc. 132 (1985) 470.
[56] J.J. Derby and R.A. Brown, J. Cryst. Growth 83 (1987) 137.
[57] P. Wouters, Simulation numérique des échanges thermiques et application à la croissance des cristaux semi-conducteurs, PhD Dissertation (Université Catholique de Louvain, Louvain-la-Neuve, 1985).
[58] F. Dupret, Y. Ryckmans, P. Wouters and M.J. Crochet, J. Cryst. Growth 79 (1986) 84.
[59] F. Dupret, P. Nicodème, Y. Ryckmans, P. Wouters and M.J. Crochet, Int. J. Heat Mass Transfer 33 (1990) 1849.
[60] L.J. Atherton, J.J. Derby and R.A. Brown, J. Cryst. Growth 84 (1987) 57.
[61] D.E. Bornside, T.A. Kinney and R.A. Brown, Int. J. Numer. Methods Eng. 30 (1990) 133.
[62] F. Dupret, Y. Ryckmans, P. Wouters and M.J. Crochet, in: Proc. 6th Int. Symp. Finite Element Methods Flow Problems (June 1986, Antibes, France), eds. INRIA (1986).
[63] P. Nicodème, F. Dupret, M.J. Crochet, J. P. Farges and G. Nagel, in: Semi-Insulating III–V Materials, Proc. 5th Conf. Semi-Insulating III–V Materials (Malmö, Sweden, June 1988), eds. G. Grossman and L. Ledebo (Hilger, Bristol, 1988), pp. 471–476.
[64] J.J. Derby, J. Atherton and P.M. Gresho, J. Cryst Growth 97 (1989) 792.
[65] N. Van den Bogaert and F. Dupret, in: The Institute of Mathematics and its Applications Conference Series, New series No.12. Mathematical Modelling for Materials Processing, Proc. IMA Conf. Math. Mod. Mat. Proc. (Bristol, UK, September 1991), eds. M. Cross, J.F-T. Pittman and R.D. Wood (Clarendon Press, Oxford, 1993), pp. 493–515.
[66] N. Van den Bogaert, Simulation dynamique de la croissance dans un four Czochralski: Modèle global de calcul des échanges thermiques et de la forme du cristal sur une géométrie variable, PhD Dissertation (Université Catholique de Louvain, Louvain-la-Neuve, 1993).
[67] S. Brandon and J.J. Derby, J. Cryst. Growth 110 (1991) 481.
[68] M.J. Crochet, P.J. Wouters, F.T. Geyling and A.S. Jordan, J. Cryst. Growth 65 (1983) 153.
[69] J.B. Mc. Laughlin and S.A. Orszag, J. Fluid Mech. 122 (1982) 123.
[70] B. Roux, ed., in: Notes on Numerical Fluid Mechanics, Vol. 27. Numerical Simulation of Oscillatory Convection in Low-Pr Fluids. A GAMM Workshop (Vieweg, Braunschweig, 1990).
[71] M. Duseaux, J. Cryst. Growth 61 (1983) 576.
[72] S. Motakef and A.E. Witt, J. Cryst. Growth 80 (1987) 37.
[73] S. Motakef, Int. J. Heat Mass Transfer 30 (1987) 1487.
[74] S. Motakef, J. Cryst. Growth 88 (1988) 341.
[75] P. Nicodème, F. Dupret and M.J. Crochet, in: Semi-Insulating III–V Materials, Proc. 5th Conf. Semi-Insulating III–V Materials (Malmö, Sweden, June 1988), eds. G. Grossman and L. Ledebo (Hilger, Bristol, 1988), pp. 465–470.
[76] O.W. Dillon Jr., C.T. Tsai and R.J. De Angelis, J. Appl. Phys. 60 (1986) 1784.
[77] J. Völkl and G. Müller, in: Semi-Insulating III–V Materials, Proc. 5th Conf. Semi-Insulating III–V Materials (Malmö, Sweden, June 1988), eds. G. Grossman and L. Ledebo (Hilger, Bristol, 1988), pp. 489–494.
[78] J. Völkl and G. Müller, J. Cryst. Growth 97 (1989) 136.
[79] R.A. Brown, AIChE J., 34 (1988) 881.
[80] J.B. Mullin, B.W. Straughan and W.S. Brickell, J. Phys. Chem. Solids 26 (1965) 782.

[81] A.D.W. Jones, in: Progress in Crystal Growth and Characterization, Vol. 10. Hydrodynamics of Czochralski Growth. A Review of the Effects of Rotation and Buoyancy Forces (1984).
[82] G. Müller, in: Crystals, Properties, and Applications, Vol. 12. Convection and inhomogeneities in Crystal Growth from the Melt (Springer, Berlin, 1988).
[83] J.R. Ristorcelli and J.L. Lumley, J. Cryst. Growth 116 (1992) 447.
[84] J.R. Carruthers, J. Cryst. Growth 32 (1976) 13.
[85] H.A. Chedzey and D.T.J. Hurle, Nature 210 (1966) 933.
[86] H.P. Utech and M.C. Flemings, J. Appl. Phys. 37 (1966) 2021.
[87] B. Cockayne, J. Cryst. Growth 42 (1977) 413.
[88] M. Abrams and R. Viskanta, J. Heat Transfer 96 (1974) 184.
[89] J.J. Favier, J. Cryst. Growth 99 (1990) 18.
[90] P.M. Adornato and R.A. Brown, J. Cryst. Growth 80 (1987) 155.
[91] P.M. Adornato and R.A. Brown, Int. J. Numer. Methods Fluids 7 (1987) 761.
[92] D.H. Kim and R.A. Brown, J. Cryst. Growth 96 (1989) 609.
[93] R.A. Brown and D.H. Kim, J. Cryst. Growth 109 (1991) 50.
[94] J.M. Parsey Jr. and F.A. Thiel, J. Cryst. Growth 73 (1985) 211.
[95] E.M. Monberg, H. Brown and C.E. Bonner, J. Cryst. Growth 94 (1989) 109.
[96] M.E. Glicksman, S.R. Coriell and G.B. Mc Fadden, Ann. Rev. Fluid Mech. 18 (1986) 307.
[97] S.H. Davis, J. Fluid Mech. 212 (1990) 241.
[98] S.A. Forth and A.A. Wheeler, J. Fluid Mech. 236 (1992) 61.
[99] D.H. Kim and R.A. Brown, J. Cryst. Growth 114 (1991) 411.
[100] D.H. Kim and R.A. Brown, J. Cryst. Growth 109 (1991) 66.
[101] D. Maroudas and R.A. Brown, J. Cryst. Growth 108 (1991) 399.
[102] W. Bardsley, F.C. Frank, G.W. Green and D.T.J. Hurle, J. Cryst. Growth 23 (1974) 341.
[103] L.N. Hjellming and J.S. Walker, J. Fluid Mech. 164 (1986) 237.
[104] L.N. Hjellming and J.S. Walker, J. Fluid Mech. 182 (1987) 335.
[105] T.W. Hicks and N. Riley, J. Cryst. Growth 96 (1989) 957.
[106] J.R. Howell and M.E. Goldstein, J. Heat Transfer 91 (1969) 165.
[107] J.J. Derby and R.A. Brown, J. Cryst. Growth 87 (1988) 251.
[108] D. Camel and J.J. Favier, J. Cryst. Growth 67 (1984) 42.
[109] D. Camel and J.J. Favier, J. Phys. (Paris) 47 (1986) 1001.
[110] J.J. Favier and D. Camel, J. Cryst. Growth 79 (1986) 50.
[111] M. Van Dyke, in: Applied Mathematics and Mechanics, Vol. 8. Perturbation Methods in Fluid Mechanics (Academic Press, New York, 1964).
[112] J.D. Cole and J. Kevorkian, Perturbation Methods in Applied Mathematics (Springer, New York, 1981).
[113] E. Billig, Proc. Roy. Soc. London A 229 (1955) 346.
[114] J.J. Derby and R.A. Brown, J. Cryst. Growth 74 (1986) 605.
[115] J.J. Derby and R.A. Brown, J. Cryst. Growth 75 (1986) 227.
[116] J.J. Derby, L.J. Atherton, P.D. Thomas and R.A. Brown, J. Sci. Comput. 2 (1987) 297.
[117] A.S. Jordan, R. Caruso and A.R. Von Neida, Bell Syst. Tech. J. 59 (1980) 593.
[118] A.S. Jordan, A.R. Von Neida and R. Caruso, J. Cryst. Growth 70 (1984) 555.
[119] H. Alexander and P. Haasen, in: Solid State Physics, Vol. 22, eds. F. Seitz, D. Turnbull and H. Ehrenreich (Academic Press, New York, 1968) pp. 27–158.
[120] I. Yonenaga and K. Sumino, Phys. Stat. Sol. (a) 50 (1978) 685.
[121] M. Suezawa, K. Sumino and I. Yonenaga, Phys. Stat. Sol. (a) 51 (1979) 217.
[122] J.C. Lambropoulos, J.W. Hutchinson, R.O. Bell, B. Chalmers and J.P. Kalejs, J. Cryst. Growth 65 (1983) 324.
[123] J.C. Lambropoulos, J. Cryst. Growth 84 (1987) 349.
[124] E. Dobročka, J. Cryst. Growth 104 (1990) 428.
[125] N. Miyazaki, H. Uchida, S. Hagihara, T. Munakata and T. Fukuda, J. Cryst. Growth 113 (1991) 227.
[126] P. Nicodème, N. Van den Bogaert and F. Dupret, in: Inst. Phys. Conf. Ser. No. 112: Chapter 2. Proc. 17th Int. Symp. GaAs and Related Compounds (Jersey, September 1990) (IOP, 1990) pp. 43–48.

[127] C.T. Tsai, J. Cryst. Growth 113 (1991) 499.
[128] S. Motakef, J. Cryst. Growth 108 (1991) 33.
[129] S. Motakef, J. Cryst. Growth 114 (1991) 47.
[130] S. Motakef, K.W. Kelly and K. Koai, J. Cryst. Growth 113 (1991) 279.
[131] J.C. Lambropoulos, J. Cryst. Growth 104 (1990) 1.
[132] C.T. Tsai, M.W. Yao and A. Chait, J. Cryst. Growth 125 (1992) 69.
[133] J. Kratochvil and O.W. Dillon Jr., J. Appl. Phys. 40 (1969) 3207.
[134] J. Kratochvil and O.W. Dillon Jr., J. Appl. Phys. 41 (1970) 1470.
[135] D.T.J. Hurle and E. Jakeman, J. Fluid Mech. 47 (1971) 667.
[136] C. Cuvelier and J.M. Driessen, J. Fluid Mech. 169 (1986) 1.
[137] J.P. Fontaine, A. Randriamampianina, G.P. Extrémet and P. Bontoux, J. Cryst. Growth 97 (1989) 116.
[138] J.P. Fontaine, A. Randriamampianina and P. Bontoux, Phys. Fluids A 3 (1991) 2310.
[139] D.T.J. Hurle, J. Cryst. Growth 42 (1977) 473.
[140] P.A. Sackinger, R.A. Brown and J.J. Derby, Int. J. Numer. Methods Fluids 9 (1989) 453.
[141] T. Surek and B. Chalmers, J. Cryst. Growth 29 (1975) 1.
[142] H. Wenzl, A. Fattah and W. Uelhoff, J. Cryst. Growth 36 (1976) 319.
[143] H.Wenzl, A. Fattah, D. Gustin, M. Mihelčić and W. Uelhoff, J. Cryst. Growth 43 (1978) 607.
[144] T. Surek, S.R. Coriell and B. Chalmers, J. Cryst. Growth 50 (1980) 21.
[145] G.A. Satunkin, V.A. Tatarchenko and V.I. Shaitanov, J. Cryst. Growth 50 (1980) 133.
[146] J.L. Duranceau and R.A. Brown, J. Cryst. Growth 75 (1986) 367.
[147] G.W. Young, A. Chait, J. Cryst. Growth 106 (1990) 445.
[148] A. Mühlbauer, W. Erdmann and W. Keller, J. Cryst. Growth 64 (1983) 529.
[149] P.M. Gresho and J.J. Derby, J. Cryst. Growth 85 (1987) 40.
[150] R. Moreau, Magnetohydrodynamics (Kluwer, Dordrecht, 1990).
[151] R.W. Series and D.T.J. Hurle, J. Cryst. Growth, 113 (1991) 305.
[152] J.F. Thompson, F.C. Thames and C.W. Mastin, NASA Report NASA-CR-2729 (July 1977).
[153] C.W. Lan and S. Kou, J. Cryst. Growth 108 (1991) 351.
[154] K. Tsiveriotis and R.A. Brown, Int. J. Numer. Methods Fluids 14 (1992) 981.
[155] D. Berghezan and F. Dupret, J. Comput. Phys. 109 (1993), in press.
[156] S. Brandon and J.J. Derby, J. Cryst. Growth 121 (1992) 473.
[157] O.C. Zienkiewicz, The Finite Element Method, 3d Ed. (McGraw-Hill, London, 1977).
[158] G. Strang and G.J. Fix, An Analysis of the Finite Element Method (Prentice-Hall, Englewood Cliffs, NJ, 1973).
[159] R. Lynch and W.G. Gray, J. Comput. Phys. 36 (1980) 135.
[160] W. Gear, Numerical Initial Value Problems in Ordinary Differential Equations (Prentice-Hall, Englewood Cliffs, NJ, 1971).
[161] M.K. Jain, Numerical Solution of Differential Equations (Wiley Eastern Line, New Delhi, 1979).
[162] P.M. Gresho, R.L. Lee and R.L. Sani, in: Recent Advances in Numerical Methods in Fluids (Pineridge Press, Swansea, 1980).
[163] F.M. Orr Jr., L.E. Scriven and A.P. Rivas, J. Colloid Interface Sci. 52 (1975) 602.
[164] F.M. Orr Jr., R.A. Brown and L.E. Scriven, J. Colloid Interface Sci. 60 (1977) 137.
[165] N. Van den Bogaert and F. Dupret (1993), in preparation.
[166] Ch. Huh and L.E. Scriven, J. Colloid Interface Sci. 30 (1969) 323.
[167] J.M. Ortega and W.C. Rheinboldt, in: Computer Science and Applied Mathematics. Iterative Solution of Nonlinear Equations in Several Variables (Academic Press, New York, 1970).
[168] B.M. Irons, Int. J. Numer. Methods Eng. 2 (1970) 5.
[169] A. George and J. Liu, Computer Solution of Large Sparse Positive Definite Systems (Prentice-Hall, Englewood Cliffs, NJ, 1981).
[170] D. Berghezan, Simulation numérique d'écoulements en couche intervenant dans l'industrie photographique, PhD Dissertation (Université Catholique de Louvain, Louvain-la-Neuve, 1991).
[171] D.T.J. Hurle, G.C. Joyce, M. Ghassempoory, A.B. Crowley and E.J. Stern, J. Cryst. Growth 100 (1990) 11.
[172] H. Ben Hadid and B. Roux, J. Fluid Mech. 235 (1992) 1.

[173] P. Wouters, J.J. Van Schaftingen, M.J. Crochet and F.T. Geyling, Int. J. Numer. Methods Fluids 7 (1987) 131.
[174] G.K. Batchelor, J. Fluid Mech. 1 (1956) 177.
[175] P.S. Huyakorn, C. Taylor, R.L. Lee and P.M. Gresho, Comput. Fluids 6 (1978) 25.
[176] Y. Ryckmans, Simulation numérique de la croissance des cristaux semi-conducteurs, PhD. Dissertation (Université Catholique de Louvain, Louvain-la-Neuve, 1989).
[177] Y. Ryckmans, P. Nicodème and F. Dupret, J. Cryst. Growth 99 (1990) 702.
[178] A.N. Brooks and T.J.R. Hughes, Comput. Methods Appl. Mech. Eng. 32 (1982) 199.
[179] M.J. Crochet, F.T. Geyling and J.J. Van Schaftingen, Int. J. Numer. Methods Fluids 7 (1987) 29.
[180] S. Dupont, J.M. Marchal, M.J. Crochet and F.T. Geyling, Int. J. Numer. Methods Fluids 7 (1987) 49.
[181] J.P. Fontaine and A. Randriamampianina, Physico-Chem. Hydrodyn. 11 (1989) 693.
[182] S.V. Patankar, Numerical Heat Transfer and Fluid Flow (Hemisphere/McGraw-Hill, New York, 1980).
[183] P. Sabhapathy and M.E. Salcudean, J. Cryst. Growth 97 (1989) 201.
[184] P. Sabhapathy and M.E. Salcudean, J. Cryst. Growth 104 (1990) 371.
[185] P. Sabhapathy and M.E. Salcudean, J. Cryst. Growth 113 (1991) 164.
[186] M.E. Salcudean, P. Sabhapathy and F. Weinberg, J. Cryst. Growth 94 (1989) 522.
[187] A. Bottaro and A. Zebib, J. Cryst. Growth 97 (1989) 50.
[188] H.J. Leister and M. Perič, J. Cryst. Growth 123 (1992) 567.
[189] W.E. Langlois, L.N. Hjellming and J.S. Walker, J. Cryst. Growth 83 (1987) 51.
[190] K.M. Kim and W.E. Langlois, J. Electrochem. Soc. 136 (1989) 1137.
[191] K.M. Kim and W.E. Langlois, J. Electrochem. Soc. 138 (1991) 1850.
[192] M.G. Williams, J.S. Walker and W.E. Langlois, J. Cryst. Growth 100 (1990) 233.
[193] A.E. Organ and N. Riley, J. Cryst. Growth 82 (1987) 465.
[194] T.W. Hicks, A.E. Organ and N. Riley, J. Cryst. Growth 94 (1989) 213.
[195] L.N. Hjellming, J. Cryst. Growth 104 (1990) 327.
[196] M. Mihelčić, K. Wingerath and Ch. Pirron, J. Cryst. Growth 69 (1984) 473.
[197] H. Hirata and K. Hoshikawa, J. Cryst. Growth 125 (1992) 181.
[198] J.E. Welch, F.H. Harlow, J.P. Shannon and B.J. Daly, Los Alamos Report LA-3425 (1966).
[199] E. Crespo del Arco and P. Bontoux, Phys. Fluids A 1 (1989) 1348.
[200] M.K. Ermakov, V.L. Gryaznov, S.A. Nikitin, D.S. Pavlovsky and V.I. Polezhaev, Int. J. Numer. Methods Fluids 15 (1992) 975.
[201] H. Ben Hadid and B. Roux, J. Cryst. Growth 97 (1989) 217.
[202] H. Ben Hadid, P. Laure and B. Roux, Physico-Chem. Hydrodyn. 11 (1989) 625.
[203] H. Ben Hadid and B. Roux, J. Fluid Mech. 221 (1990) 77.
[204] D. Gottlieb and S.A. Orszag, in: CBMS Regional Conf. Series in Appl. Math., Numerical Analysis of Spectral Methods: Theory and Application (SIAM, Philadelphia, 1977).
[205] C. Canuto, M.Y. Hussaini, A. Quarteroni and T.A. Zang, in: Springer Series in Computer Phys., Spectral Methods in Fluid Dynamics (Springer, New York, 1988).
[206] E. Crespo del Arco, A. Randriamampianina and P. Bontoux, in: Proc. Conf. on Synergetics, Order and Chaos (October 1987, Madrid, Spain), ed. M. G. Velarde (World Scientific, NJ, 1988), pp. 244–257.
[207] E. Crespo del Arco, J.P. Pulicani and P. Bontoux, Physico-Chem. Hydrogen. 11 (1989) 681.
[208] E. Crespo del Arco, J.P. Pulicani and A. Randriamampianina, C.R. Acad. Sci. Paris 309, Sér. II (1989) 1869.
[209] J.P. Pulicani, E. Crespo del Arco, A. Randriamampianina, P. Bontoux and R. Peyret, Int. J. Numer. Methods Fluids 10 (1990) 481.
[210] M.A. Fikri, G. Labrosse and M. Betrouni, J. Cryst. Growth 119 (1992) 41.
[211] R.A. Brown, in: Heat and Mass Transfer in Materials Processing, Proc. Oji Int. Sem. Adv. Heat Transf. in Manufacturing and Processing of New Materials (Oct. 1990, Tomakomai, Hokkaido, Japan), eds. I. Tanasawa and N. Lior (Hemisphere, New York, 1992), pp. 137–153.
[212] S. Kobayashi, J. Cryst. Growth 99 (1990) 692.
[213] S. Kobayashi, Sh. Miyahara, T. Fujiwara, T. Kubo and H. Fujiwara, J. Cryst. Growth 109 (1991) 149.
[214] J.J. Derby and Q. Xiao, J. Cryst. Growth 113 (1991) 575.

[215] H. Kopetsch, J. Cryst. Growth 88 (1988) 71.
[216] H. Kopetsch, J. Cryst. Growth 102 (1990) 505.
[217] D.H. Kim, P.M. Adornato and R.A. Brown, J. Cryst. Growth 89 (1988) 339.
[218] R. Siegel and J.R. Howell, Thermal Radiation Heat Transfer, 2nd Ed. (McGraw-Hill, New York, 1981).
[219] M. Planck, The Theory of Heat Radiation, 2nd. Ed., translated by M. Masius (Dover, New York, 1959).
[220] M.E. Goldstein and J.R. Howell, NASA TN D-4618 (1968) 1.
[221] V.S. Yuferev and M.G. Vasil'ev, J. Cryst. Growth 82 (1987) 31.
[222] H.S. Chan and K.Y. Hsu, J. Heat Transfer 110 (1988) 260.
[223] T. Tsukada, N. Imaishi, M. Hozawa and K. Fujinawa, J. Chem. Eng. Jpn 20 (1987) 146.
[224] M.J.Crochet, F. Dupret, Y. Ryckmans, F.T. Geyling and E.M. Monberg, J. Cryst. Growth 97 (1989) 173.
[225] F. Dupret, P. Nicodème and Y. Ryckmans, J. Cryst. Growth 97 (1989) 162.
[226] P.D. Thomas, J.J. Derby, L.J. Atherton, R.A. Brown and M.J. Wargo, J. Cryst. Growth 96 (1989) 135.
[227] H. Matsushima and R. Viskanta, Int. J. Heat Mass Transfer 33 (1990) 1957.
[228] T.J. Chung and J.Y. Kim, J. Heat Transfer 106 (1984) 448.
[229] M.M. Razzaque, J.R. Howell and D.E. Klein, J. Heat Transfer 106 (1984) 613.
[230] D.W. Amlin and S.A. Korpela, J. Heat Transfer 101 (1979) 76.
[231] J.R. Howell, J. Heat Transfer 110 (1988) 1220.
[232] M.P. Mengüç and R. Viskanta, J. Heat Transfer 108 (1986) 271.
[233] C. Garot and P. Gendre, in: Proc. 1st Int. Conf. Numerical Methods in Thermal Problems, eds. R.W. Lewis and K. Morgan (Pineridge Press, Swansea, 1979), pp. 99–108.
[234] R.A. Brown, T.A. Kinney, P.A. Sackinger and D.E. Bornside, J. Cryst. Growth 97 (1989) 99.
[235] R.K. Srivastava, P.A. Ramachandran and M.P. Duduković, J. Cryst. Growth 74 (1986) 281.
[236] A. Virzi, J. Cryst. Growth 112 (1991) 699.
[237] A.B. Crowley, E.J. Stern and D.T.J. Hurle, J. Cryst. Growth 97 (1989) 697.
[238] A.M. Cazin-Bourguignon and S. Motakef, J. Cryst. Growth 96 (1989) 390.
[239] K. Kakimoto, P. Nicodème, M. Lecomte, F. Dupret and M.J. Crochet, J. Cryst. Growth 114 (1991) 715.
[240] K. Kakimoto, M. Watanabe, M. Eguchi and T. Hibiya, J. Cryst. Growth 126 (1993) 435.
[241] S. Motakef, J. Cryst. Growth 96 (1989) 201.
[242] S. Motakef, J. Cryst. Growth 98 (1989) 711.
[243] K.W. Kelly, K. Koai and S. Motakef, J. Cryst. Growth 113 (1991) 254.
[244] K.W. Kelly, S. Motakef and K. Koai, J. Cryst. Growth 113 (1991) 265.
[245] G.O. Meduoye, D.J. Bacon and K.E. Evans, J. Cryst. Growth 88 (1988) 397.
[246] G.O. Meduoye, K.E. Evans and D.J. Bacon, J. Cryst. Growth 97 (1989) 709.
[247] G.O. Meduoye, D.J. Bacon and K.E. Evans, J. Cryst. Growth 108 (1991) 627.
[248] D. E. Bornside, T. A. Kinney and R.A. Brown, J. Cryst. Growth 108 (1991) 779.
[249] N. Van den Bogaert and J. Braet, in: EMC'91: Non-Ferrous Metallurgy, Present and Future, Proc. 1st European Metals Conference (Brussels, Belgium, September 1991) (Elsevier, Amsterdam, 1991) p. 483.
[250] C. Hoffmann, J. Hopcroft and M. Karasick, in Proc. 4th Ann. ACM Symp. Computational Geometry (ACM, New York, 1988), pp. 106–117.
[251] A. Couniot, Développement d'un algorithme de remaillage automatique pour la simulation du moulage par injection, PhD Dissertation (Université Catholique de Louvain, Louvain-la-Neuve, 1991).
[252] J.V. Beck, B. Blackwell and Ch. R. Saint-Clair, Inverse Heat Conduction Problems (Wiley–Interscience, New York, 1985).
[253] J. Hadamard, La Théorie des Équations aux Dérivées Partielles (Editions Scientifiques, Pékin, 1964).
[254] L. Schwartz, Théorie des Distributions (Hermann, Paris, 1966).
[255] M. Lecomte, Étude de la convection laminaire dans un four Czochralski par méthode d'extrapolation, Masters Dissertation (Université Catholique de Louvain, Louvain-la-Neuve, 1992).

[256] M. Lecomte, E. Olivari, N. Van den Bogaert and F. Dupret, Euromech Colloquium No. 284, Fluid Flow Phenomena in Crystal Growth (Aussois, Modane, France, October 1992).
[257] M. Lecomte, K. Kakimoto and F. Dupret, in: Proc. ECCG-3 3rd European Conf. on Crystal Growth (Budapest, Hungary, May 1991) (Trans. Tech. Publications, 1991).
[258] N. Van den Bogaert, F. Dupret and M.J. Crochet, in: Proc. ECCG-3 3rd European Conf. on Crystal Growth (Budapest, Hungary, May 1991) (Trans. Tech. Publications, 1991).
[259] N. Van den Bogaert and F. Dupret, in: Inst. Phys. Conf. Ser. No. 129: Chapter 12, Proc. 19th Int. Symp. GaAs and Related Compounds (Karuizawa, Japan, October 1992) (IOP, 1993), pp. 917–918.
[260] P. Nicodème, Simulation numérique des échanges thermiques dans un four et calcul du taux de dislocations dans un cristal d'arséniure de gallium, PhD Dissertation (Université Catholique de Louvain, Louvain-la-Neuve, 1990).
[261] J.J. Derby, in: Proc. Eight International Summer School on Crystal Growth, ISSCG-8 (Palm Springs, CA, August 1992), to be published by the American Institute of Physics.
[262] S. Brandon, J.J. Derby, L.J. Atherton, D.H. Roberts and R.L. Vital, J. Cryst. Growth 132 (1993) 261.
[263] S. Kaddeche, H. Ben Hadid and D. Henry, to be published in J. Cryst. Growth (1994).
[264] T.A. Kinney and R.A. Brown, J. Cryst. Growth 132 (1993) 551.
[265] Q. Xiao and J.J. Derby, J. Cryst. Growth 128 (1993) 188.
[266] S. Brandon and J.J. Derby, Int. J. Num. Meth. Heat Fluid Flow 2 (1992) 299.
[267] H. Matsushima and R. Viskanta, in: Collected Papers in Heat Transfer 1988, Vol. 1, ed. K.T. Yang (ASME, New York, 1988), pp. 151–160.
[268] M.P. Mengüç and R. Viskanta, J. Quant. Spectrosc. Radiat. Transfer 33 (1985) 533.
[269] R. Viskanta and M.P. Mengüç, Prog. Energy Combust. Sci. 13 (1987) 97.
[270] A.B. Shapiro, UCID-1987 report, Lawrence Livermore Laboratories.
[271] D.E. Bornside and R.A. Brown, J. Thermophysics and Heat Transfer 4 (1990) 414.
[273] W. Zhou, D.E. Bornside and R.A. Brown, to be published in J. Cryst. Growth (1993).
[274] T.A. Kinney, D.E. Bornside, R.A. Brown and K.M. Kim, J. Cryst. Growth 126 (1993) 413.
[275] Q. Xiao and J.J. Derby, J. Cryst. Growth 129 (1993) 593.
[276] A. Virzi and M. Porrini, Materials Science and Engineering B17 (1993) 196.
[277] A. Virzi, to be published in the Institute of Physics Journal (1993).
[278] K. Kakimoto, M. Watanabe, M. Eguchi, T. Hibiya, Int. J. Heat Mass Transfer 35 (1992) 2551.
[279] M. Watanabe, M. Eguchi, K. Kakimoto, Y. Baros and T. Hibiya, J. Cryst. Growth 128 (1993) 288.

CHAPTER 16

Shaped Crystal Growth

V.A. TATARCHENKO

Institute for High Temperature, Russian Academy of Sciences, 127412 Moscow, Russia

Contents

1. Introduction: capillary shaping techniques (CST) . . . 1015
2. The dynamic stability of crystallization – the basis of shaped crystal growth . . . 1016
   2.1. Lyapunov set of equations . . . 1016
   2.2. The melt column shaping conditions . . . 1017
   2.3. The meniscus surface equation . . . 1018
   2.4. Boundary conditions of the capillary problem . . . 1019
   2.5. Growth angle . . . 1019
   2.6. The equation of crystal dimension change rate . . . 1021
   2.7. The equation of the crystallization front displacement rate . . . 1022
   2.8. Convection influence on the capillarity and liquid-phase heat-transfer processes . . . 1023
3. The realization of dynamic stability in the Czochralski technique (CzT) . . . 1024
   3.1. The Lyapunov set of equations . . . 1025
   3.2. Analysis of capillary coefficients . . . 1026
      3.2.1. Capillary stability . . . 1027
   3.3. The heat transfer problem . . . 1027
      3.3.1. Thermal stability . . . 1028
      3.3.2. A semi-infinite crystal where heat removal from the crystal lateral surface is allowed for, while that from the melt is not taken into consideration, i.e. $\mu_S \neq 0$; $\mu_L = 0$. 1028
      3.3.3. A crystal of finite length, l, where the temperature of the crystal upper end is equal to the environmental temperature, i.e. $\mu_L = 0$, $\mu_S \neq 0$, $T_1 = T_e$ . . . 1029
      3.3.4. A semi-infinite crystal where heat removal from the melt-column lateral surface is allowed for, i.e. $\mu_L \neq 0$, $\mu_S \neq 0$ . . . 1029
      3.3.5. Quantitative stability criteria . . . 1030
4. Techniques of pulling from a shaper (TPS) . . . 1031
   4.1. Boundary conditions of the capillary problem [37, 38] . . . 1035
      4.1.1. The catching conditions . . . 1035
      4.1.2. The angle fixing condition (the wetting condition) . . . 1036

HANDBOOK OF CRYSTAL GROWTH, VOL. 2
edited by D.T.J. Hurle

4.1.3. The wetting-to-catching transition . . . 1036
4.1.4. Influence of wetting-angle hysteresis on the capillary effects . . . 1037
4.1.5. The boundary conditions at the crystal free edges . . . 1037
4.2. Liquid meniscus heights and shapes . . . 1037
4.3. The capillary problem for systems with large Bond number [14] . . . 1039
4.4. The capillary coefficients A_{RR} and A_{Rh} . . . 1041
4.4.1. Small Bond numbers, the catching boundary condition, zero pressure . . . 1042
4.4.2. Small Bond numbers, the catching boundary condition, non-zero pressure . . . 1042
4.4.3. Small Bond numbers, the angle-fixing boundary condition . . . 1043
4.4.4. Large Bond numbers, the catching boundary condition . . . 1043
4.4.5. Large Bond numbers, the angle-fixing boundary condition . . . 1043
4.5. The conditions for crystallization stability . . . 1044
4.6. Stationary values of R^0 and h^0 as simultaneous solutions of the capillary and heat transfer problems . . . 1045
4.7. Influence of melt pressure variations on crystallization stability . . . 1047
4.8. Growth of tubular crystals . . . 1048
4.9. Impurity distribution . . . 1048
4.10. TPS growth of sapphire crystals . . . 1052
4.10.1. Morphological features of sapphire tubes . . . 1052
4.10.2. Voids . . . 1053
4.10.3. Low-angle grain boundaries . . . 1058
4.11. TPS growth of silicon crystals . . . 1059
4.11.1. Silicon tapes . . . 1059
4.11.2. Silicon tube growth . . . 1060
4.11.3. Growth of thin-walled polyhedral prism-shaped crystals . . . 1060
4.11.4. Profiled silicon structure . . . 1062
4.12. TPS growth of lithium niobate crystals [59] . . . 1065
4.12.1. Technique description for growing shaped $LiNbO_3$ crystals . . . 1065
4.12.2. Control of domain sizes . . . 1067
4.12.3. Regular domain structure in growth from doped melts . . . 1069
5. The realization of dynamic stability in the Verneuil technique (VT) . . . 1069
5.1. Growth of cylindrical and plate-shaped crystals . . . 1070
5.1.1. Determination of the coefficients A_{RR}, A_{Rl} and A_{Rh} . . . 1071
5.1.2. Determination of the coefficients A_{ll}, A_{lR} and A_{lh} . . . 1072
5.1.3. Determination of the coefficients A_{hR}, A_{hl} and A_{hh} . . . 1073
5.1.4. Analysis of the process-stability conditions . . . 1075
5.1.5. Plate growth . . . 1076
5.2. Stability-analysis-based automation of VT . . . 1076
5.3. Growth of tube-shaped crystals . . . 1079
5.3.1. Theoretical analysis . . . 1079
5.3.2. Experiments on growing single-crystal corundum tubes under stable conditions . . . 1080
6. Radial instability of vapor whisker growth (VWG) . . . 1080
6.1. Problem formulation . . . 1082
6.2. Stability analysis . . . 1083
6.3. Profile curves of initial whisker sections . . . 1083

7. The realization of dynamic stability in the floating zone technique (FZT) 1084
   7.1. The formulation of the problem 1084
   7.2. The heat flow conditions 1085
   7.3. Mass balance 1086
   7.4. The capillary coefficients 1087
   7.5. Crystallization stability 1088
8. Stability of cylindrical pore (negative crystal) growth 1089
   8.1. Formulation of the problem 1089
   8.2. The capillary coefficients A_{RR} and A_{RC} 1090
   8.3. The concentration coefficients A_{CR} and A_{CC} 1090
   8.4. Analysis of the process-stability conditions 1091
9. The two-shaping-elements technique (TSET) 1092
   9.1. Capillary shaping 1093
   9.2. Hydrodynamic conditions 1094
   9.3. Stability investigation 1095
   9.4. Impurity distribution 1096
   9.5. Silicon as a component of the SOC material 1098
10. The variable shaping technique (VST) 1098
11. The local shaping technique (LST) 1100
    11.1. Formulation of the problem 1101
    11.2. Capillary coefficients A_{ll} and $A_{l\psi}$ 1102
    11.3. Thermal coefficients $A_{\psi l}$ and $A_{\psi\psi}$ 1103
    11.4. Analysis of stability 1104
12. The stability of crystal cross-sectional shapes 1104
13. Crystallization by CST under microgravity conditions 1105
References 1108

1. Introduction: capillary shaping techniques (CST)

Modern engineering needs plate-, rod- and tube-shaped crystals. Therefore, crystals of specified size and shape with controlled defect and impurity structure that allow one to use them as final products without additional machining are to be grown. This problem seems to be simply solved by profiled container crystallization just as in the case of casting. However, such a solution is not always acceptable. Container material should satisfy a whole set of requirements: it should neither react with the melt nor be wetted by it, should be high-temperature and aggressive-medium resistant, etc. Even if all these requirements are satisfied, perfect crystal growth is not secured. Therefore, the techniques of crystal lateral surface shaping without contact with container walls rank high among the melt crystallization techniques, i.e. the Czochralski technique – CzT [1], the Stepanov technique – ST [2], edge-defined film-fed growth – EFG [3], capillary action shaping technique – CAST [4], the Verneuil technique – VT [5], the floating-zone technique – FZT [6] and various modifications thereof. Absence of contact between the crystallizing substance and crucible walls in these techniques allows one to improve crystal structures and to decrease mechanical stress levels. The crucible becomes unnecessary for the VT and for the FZT. For all these techniques the shapes and the dimensions of the crystals grown are controlled by the interface meniscus-shaping capillary forces and by the heat- and mass-exchange conditions in the crystal–melt system. Let the above-mentioned techniques be called capillary shaping techniques – CST.

The absence of rigid shaping as well as the advantages noted, cause a number of problems when growing crystals of specified shapes and controlled cross-sections. Indeed, as the crystal is not restricted by crucible walls, its cross-section depends upon the growth parameters. Any deviations of growth rate, temperature conditions, etc. result in crystal cross-section changes – in pinch formation.

It is experimentally found that an increased number of inclusions and non-uniform impurity distributions are observed at pinch locations. Pinches themselves do not seem to cause defect formation, but crystal-dimension change is indicative of growth conditions (mainly crystallization rate) deviating from the steady-state ones. From this point of view, constant cross-section crystal growth is, to a certain extent, evidence of crystallization process stability. Therefore, the problem of crystal cross-section stabilizations arises. For the CST the stabilization problem proved to be solvable by investigation of crystallization process dynamic stability and by selecting, on the basis of this analysis, crystallization schemes and conditions that provide crystals of specified shape and controlled cross-section.

For the first time this analysis was carried out by the author of the present chapter for the CzT and the ST [7] in 1971 and later by Surek with coauthors [8, 9]. Recently interest in the problem under consideration has increased due to development of crystal-growth computer-aided analysis [10–12]. For the twenty years that have passed since the first publication, dynamic stability in CST has been investigated in detail, steady-state conditions and schemes for crystallization have been selected. Growth of corundum tubes of constant cross-sections by the VT using pre-calculated steady-state conditions can illustrate the success of this approach [13].

This chapter mainly is based on the monograph [14] published in 1988 (in Russian) and translated into English in 1993.

2. *The dynamic stability of crystallization – the basis of shaped crystal growth*

2.1. *Lyapunov set of equations*

Various crystallization schemes and conditions will be analyzed from the point of view of growing crystals of specified cross-section under capillary shaping conditions.

All the crystallization techniques will be characterized by a finite number, n, of the main variables, degrees of freedom, X_i, that can arbitrarily vary in the process of crystallization. Thus for the CzT and ST they will include crystal dimensions, R, and crystallization front position, h, i.e. these techniques have two degrees of freedom ($n = 2$). For a more detailed analysis these techniques can be represented as systems with three degrees of freedom where the third degree of freedom is pressure, d, under which the melt is fed to the shaper for the ST and the melt level, H, in the crucible for the CzT.

The FZT has four degrees of freedom ($n = 4$), they are crystal dimensions, R, melted zone volume, W, and position of crystallization surface, h_c, and melting surface, h_m, relative to the induction heater.

In order to grow a crystal of constant cross-section, stabilization of the above-mentioned parameters should be provided during the crystallization process.

Growth of a crystal of constant cross-section is a particular case of the problem of growing a crystal of controlled cross-section. The latter problem can be solved by changing $n - 1$ of the parameters X_i (naturally, excluding the crystal dimensions) and crystallization conditions (pulling rate, melt temperature, etc.) taking into consideration the functional dependence of all the parameters X_i and the crystallization conditions found beforehand.

To find an explicit function, a set of simultaneous equations should be solved. The set of equations should include: (1) the melt or solution flow equation (the Navier–Stokes equation) with the boundary conditions on the meniscus free surface (the Laplace capillary equation); (2) the continuity equation (the law of crystallizing substance mass conservation); (3) the heat-transfer equation for the liquid and the solid phases with the equation of heat balance at the crystallization front and at the melting front as the boundary conditions (the law of energy conservation); (4) the diffusion equation (impurity mass conservation); and (5) employment of the growth angle constancy condition specific for melt crystallization. The set of equations mentioned above is general for all the crystallization techniques under consideration while specific features of each of the crystallization techniques and schemes are characterized by the set of boundary conditions and concrete values of the parameters included in the equations. The approach proposed here is valid if heat and mass transfer is the limiting stage of the process rather than the crystallization kinetics; this is usually the case for real systems and pulling rates. Then such kinetic effects as the dependence of crystallization front supercooling on the crystallization rate or impurity effect on the crystallization temperature can be taken into account within the model developed here.

The number of equations used should be equal to the number of degrees of freedom, n, in the crystallization scheme under consideration; multiple use of each of the equations mentioned being possible, e.g. the growth angle constancy conditions, for investigation of inner and outer surface shaping when tubular crystals are grown. The following set of differential equations is obtained for the set of X_i:

$$\frac{dX_i}{dt} = f_i\left(X_1, X_2, \ldots, X_n, \frac{dX_1}{dt}, \frac{dX_2}{dt}, \ldots, \frac{dX_n}{dt}, t, C\right), \quad i = 1, 2, \ldots n. \tag{2.1}$$

Here t denotes the time, n is the number of the unknowns that depend on the crystallization technique and the growing crystal cross-section, C denotes a set of controllable crystallization parameters, thermophysical and other constants of the substance being crystallized.

As was mentioned above, stationary values of the variables X_i^0 will be especially interesting for us, i.e. growing crystals of constant cross-sections with stationary crystallization-front position and other parameters of the system. These values satisfy eq. (2.1) with left-hand side zero:

$$f_i(X_1^0, X_2^0, \ldots, X_n^0, t, C) = 0. \tag{2.2}$$

Only those solutions of eq. (2.2) that are stable can be physically realized. According to Lyapunov [15], the solutions of eq. (2.1) are stable if they are stable for the linearized set of equations:

$$\frac{\partial X_i}{\partial t} = \sum_{k=1}^{n} \frac{\partial f_i}{\partial X_k} \delta X_k. \tag{2.3}$$

Here $\delta X_k = X_k - X_k^0$, all partial derivatives are taken with $X_k = X_k^0$. The stability of eq. (2.3) in its turn is observed when all the roots of S in the characteristic equation,

$$\det\left(\frac{\partial f_i}{\partial X_k} - S\delta_{ik}\right) = 0, \tag{2.4}$$

have negative real components, δ_{ik} is the Kronecker delta [15]. Equilibrium will be unstable if eq. (2.4) has at least one root with a positive real component; if an imaginary number can be found among the roots, additional study including allowance for non-linear terms in eq. (2.3) is required.

Calculation of the time-dependent nonstationary functions f_i requires solution of eq. (2.1) and is usually rather difficult. These difficulties can be avoided using the quasi-stationary approach. For example, Mullins and Sekerka [16] applied this approach to the temperature and impurity distribution problem while studying morphological stability of crystallization-front shape, which led to the results that agreed with the experiment. However in each particular case the quasi-stationary approach is to be founded.

2.2. The melt column shaping conditions

For the capillary shaping techniques crystal cross-section results from the melt column section formed by the crystallization surface. Melt column shape can be calculated on

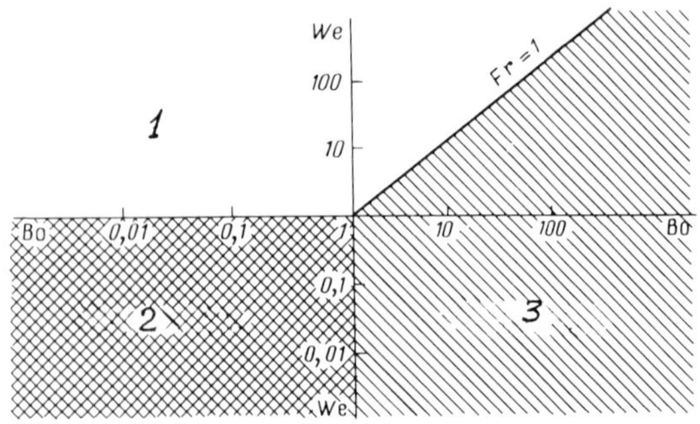

Fig. 1. (1) Inertial, (2) gravitational and (3) capillary force effects on melt column shaping [17] (We, Bo, Fr are the characteristic Weber, Bond and Froude numbers, respectively).

the basis of the Navier–Stokes equation, the Laplace capillary equation being the free-surface boundary condition.

The contribution of various factors of meniscus shaping should be estimated: inertial forces associated with melt flow, capillary forces, gravity forces. The relative effect of the first three factors can be estimated by means of dimensionless numbers: the Weber number $\mathrm{We} = \rho_L V^2 L/\gamma_{LG}$ characterizing comparative action of inertial and capillary forces; the Froude number, $\mathrm{Fr} = V/(gL)^{1/2}$, characterizing comparative action of inertial and gravity forces; the Bond number, $\mathrm{Bo} = \rho_L g L^2/\gamma_{LG}$, characterizing comparative action of gravity and capillary forces. Here ρ_L denotes liquid density, L denotes the liquid meniscus dimensions, V is the liquid flow rate, γ_{LG} denotes the liquid surface tension coefficient, g is the gravitational acceleration. When the Weber and Froude numbers are small, melt flow can be neglected. The Bond number defines the region of capillary or gravity force predominance (fig. 1). The effect of the inertial force as compared to the gravity and capillary ones proves to be negligible, for all the crystallization rates used. Indeed, if we assume that linear dimensions of the meniscus lie within the range of 10^{-3}–10^{-2} m, $\rho_L \approx 10^{-3}$–$10^{-4}\,\mathrm{kg\,m^{-3}}$, $\gamma_{LG} \approx 1\,\mathrm{N\,m^{-2}}$ the liquid column shaping can be examined in hydrostatic approximation up to the fluid speed 0.1–1 $\mathrm{m\,s^{-1}}$.

2.3. The meniscus surface equation

In the hydrostatic approximation the equilibrium shape of liquid surface is described by the Laplace capillary equation [18]

$$\frac{\gamma_{LG}}{R_1(M)} + \frac{\gamma_{LG}}{R_2(M)} + \rho_L g w = \text{constant}. \tag{2.5}$$

Here $R_1(M)$ and $R_2(M)$ denote the main radii of liquid surface curvature at the point M of the liquid surface. Curvature radii are considered to be positive if the normal positive direction goes inside the liquid, and negative if it goes outside the liquid. In this case the normal positive direction towards a curve on the meniscus surface coincides with the direction from the curvature center towards a given point of it. The w-axis is directed vertically upwards. The value of the constant depends upon w-coordinate origin selection and is equal to the pressure on the liquid in the plane $w = 0$. In particular, if the w-coordinate origin coincides with the plane of the liquid surface, constant $= 0$.

We will confine ourselves to the study of a meniscus possessing axial symmetry. Such menisci are obtained during melt pulling of straight circular cylinder-shaped crystals.

The problem of liquid column shape calculation for an axially symmetric meniscus is reduced to finding the shape of a profile curve $w = f(r^*)$, the liquid surface column being obtained by rotation of this curve about the w-axis. Now we will introduce the capillary constant, a, and pass to dimensionless coordinates and parameters.

$$(2\gamma_{LG}/\rho_L g)^{1/2} = a, \qquad w/a = z,$$

$$r^*/a = r, \qquad Pa/(2\gamma_{LG}) = d.$$

This transition means that the capillary constant serves as a linear dimension unit, and the weight of a liquid column one capillary constant high corresponds to the pressure equal to one. This approach allows application of calculation results to any substance and magnitude of gravity.

Then eq. (2.5) takes the form

$$z''r + z'(1 + z'^2) \pm 2(d - z)(1 + z'^2)^{3/2} r = 0. \qquad (2.6)$$

For large Bond numbers (Bo > 1; fig. 1) when gravity prevails (this condition corresponds to growing crystals with a diameter of the order of ten capillary constants) eq. (2.6) is simplified:

$$z'' \pm 2(d - z)(1 + z'^2)^{3/2} r = 0. \qquad (2.7)$$

For small Bond numbers (Bo < 1, fig. 1) when capillarity prevails (this condition corresponds to growing crystals with a diameter smaller than the capillary constant and for large diameters it can be easily achieved under reduced gravity when the capillary constant is high), eq. (2.6) can also be simplified:

$$z''r + z'(1 + z'^2) \pm 2d(1 + z'^2)^{3/2} r = 0. \qquad (2.8)$$

2.4. *Boundary conditions of the capillary problem*

As the Laplace capillary equation is a second-order differential equation, formulation of the boundary problem for melt column shape calculation requires the assignment of two boundary conditions. These boundary conditions are determined by structural features of each specific capillary shaping scheme and will be analyzed in detail below. In this case one of the crystal–melt interface conditions is common for all the melt crystallization techniques. This condition follows from growth angle constancy.

2.5. *Growth angle*

Let the angle ψ_0 (fig. 2) made by the line tangent to the meniscus and the lateral surface of the growing crystal be called the growth angle.

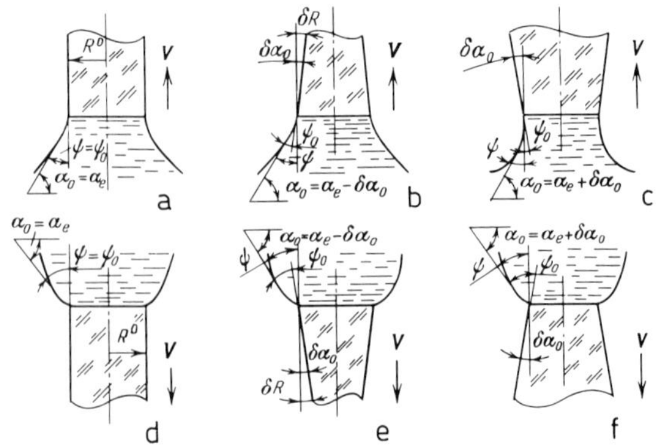

Fig. 2. Crystal growth for the case of capillary shaping: (a, d) $\alpha_0 = \alpha_e$ growth of crystals of constant cross-sections; (b, e) $\alpha_0 < \alpha_e$, growth of crystals of widening cross-sections; (c, f) $\alpha_0 > \alpha_e$, growth of crystals of narrowing cross-sections. V is crystal displacement rate.

In the first works on this subject a crystal of constant cross-section was considered to grow in case the growth angle ψ_0 is equal to zero [19]. Judging by a purely geometric diagram of liquid–solid phase conjugation, such an assumption is quite natural. However, investigation of the crystal growth process showed that the condition $\psi_0 = 0$ is not satisfied while growing crystals of constant cross-sections.

Direct measurements of this angle using photographs of melt columns during crystallization by the CzT were carried out [20, 21]. The technique from [8] is based on determination of crystallized melted disk shape. In our experiments [22] we grew and studied crystallized drops obtained on the bottom of crystals detached from the melt in the process of pulling by the CzT. The value of $\psi_0 = 25° \mp 1°$ for indium antimonide, $\psi_0 = 11° \mp 1°$ for silicon and $\psi_0 = 12° \mp 1°$ for germanium were obtained.

The growth-angle expression was first derived in [23] from the thermodynamic equilibrium condition at the three-phase interface line.

$$\psi_0 = \arccos[(\gamma_{LG}^2 + \gamma_{SG}^2 - \gamma_{SL}^2)/2\gamma_{LG}\gamma_{SG}]. \tag{2.9}$$

Here γ_{LG}, γ_{SG} and γ_{SL} denote interface free energies at the liquid–gas, solid–gas and solid–liquid interfaces, respectively. The expression for ψ_0 obtained in [24] differs from eq. (2.9) in the correction factor, including derivatives of the interface surface energy. Hence, the correction factor results from the anisotropy and for an isotropic case is equal to zero.

If crystallization is referred to as a kinetic process, growth angle dependence on the three-phase line displacement rate V_C and crystallization front supercooling ΔT occurs [25]:

$$\psi_k = \psi_0 - k_\psi V_C^{-1/3}\Delta T. \tag{2.10}$$

Here k_ψ is a kinetic coefficient.

The value of the crystallization front supercooling is strictly dependent upon the growth mechanism. Thus, according to the estimations cited in [26], for silicon the value of ΔT is equal to 3.7° when growing dislocation-free crystals if the crystallization front coincides with the singular face and layer-by-layer growth takes place; $\Delta T \approx 0.3^\circ$ when dislocations are formed in the crystal and $\Delta T \approx 0.05^\circ$ during normal non-singular surface growth. In this case the kinetic correction to the growth angle in eq. (2.10) changes from 20° for dislocation-free crystals when the singular face is exposed at the crystallization front to $\sim 0.3^\circ$ with the normal mechanism taking place.

As known from experiments, in the case of crystal melting, in contrast to crystallization , the angle between the crystal surface and the melt can have any value. As the processes of crystallization and melting differ only by small deviations from the thermodynamic equilibrium, it is natural to try to explain the great difference between them by kinetic effects [27].

2.6. *The equation of crystal dimension change rate*

Proceeding from the condition of growth angle constancy, we can obtain an equation for crystal characteristic dimension change rate, R, from the geometric diagram given in fig. 2. The vector $\boldsymbol{R}$ lies within the diagram plane and represents the radius for a straight circular cylinder shaped crystal. For a plate it is the half-thickness. Now we will introduce the angles made by the line tangent to the meniscus on the three-phase line with the horizontal, α_0, and with vertical, ψ (the crystal grows in the vertical direction; fig. 2). When the angles α_0 and ψ add up to $\pi/2$, a crystal of constant cross-section R^0, grows (fig. 2a, d) and in this case the angle ψ is equal to the growth angle ψ_0. Let α_e denote the values of the angle α_0 corresponding to this stationary-growth process: $\alpha_e = \pi/2 - \psi_0$. If $\alpha_0 \neq \alpha_e$, the crystal changes its dimension R in accordance with the law (fig. 2b, c, e, f)

$$\partial R/\partial t = V_C \operatorname{tg}(\alpha - \alpha_0) = -V_C \operatorname{tg}(\delta\alpha_0). \tag{2.11}$$

Here V_C denotes the crystallization rate. In this case the angle of crystal tapering at any moment is equal to $\delta\alpha_0 = \alpha_0 - \alpha_e$ and the crystallization rate V_C is equal to the difference in rates between pulling and front displacement:

$$V_C = V - \mathrm{d}h/\mathrm{d}t. \tag{2.12}$$

As we have agreed to obtain the qualities characterizing eq. (2.3) near the stationary state, deviations of $\delta\alpha_0$ and $\mathrm{d}h/\mathrm{d}t$ can be regarded as negligible. Hence, by expanding $\operatorname{tg}(\delta\alpha_0)$ into a Taylor series in the vicinity of the stationary point and using the terms of the first-order infinitesimal only, we have:

$$\partial R/\partial t = -(V - \mathrm{d}h/\mathrm{d}t)\operatorname{tg}(\delta\alpha_0) \approx -V\,\delta\alpha_0. \tag{2.13}$$

The angle α_0 as a function of the crystal size, crystallization front position and other factors is found, together with meniscus shape, by solving the Navier–Stokes or Laplace equation (it depends on conditions discussed in section 2.2) with the corresponding system of boundary conditions, depending on the specific crystallization technique. In that case eq. (2.13) can be presented in the form

$$\partial R/\partial t = -V \sum_{k=1}^{n} (\partial\alpha_0/\partial X_k)\delta X_k. \tag{2.14}$$

Here $\delta X_k = X_k - X_k^0$, X_k^0 is the value of the parameter X_k with $\alpha_0 = \alpha_e$. Thus the first equation (for the crystal cross-section change rate) of eq. (2.3) is obtained.

2.7. The equation of the crystallization front displacement rate

Now we will derive another equation from eq. (2.3) that is general for all the capillary shaping techniques, i.e., the equation of the crystallization front displacement rate. At the crystallization front the heat-balance equation should be satisfied:

$$-\lambda_S G_S(h) + \lambda_L G_L(h) = \mathscr{L} V_C. \tag{2.15}$$

Here λ_S and λ_L denote thermal conductivities of solid and liquid phases, respectively, $G_S(h)$ and $G_L(h)$ are the temperature gradients of solid and liquid phases at the crystallization front, h is a crystallization front coordinate; $\mathscr{L}$ denotes the latent heat of melting of a unit volume of the material, V_C is the crystallization rate. In this case, if we use V to denote the pulling rate, then in accordance with eq. (2.12) $V_C = V - dh/dt$. Thus, from eq. (2.15) we obtain a (2.1)-type equation for dh/dt:

$$dh/dt = V - \mathscr{L}^{-1}(\lambda_L G_L - \lambda_S G_S). \tag{2.16}$$

Now expressions for G_L and G_S are to be found, which can be done using the following equation with corresponding boundary conditions:

$$\frac{1}{k_i}\frac{\partial T_i}{\partial t} = \frac{dT_i^2}{dz^2} - \frac{V}{k_i}\frac{dT_i}{dz} - \frac{\mu_i}{\lambda_i}F(T_i - T_e). \tag{2.17}$$

Here $i = $ L, S ($i = $ L for a liquid, $i = $ S for a solid body), T_i denotes the temperature, k_i is the thermal diffusivity coefficient, z is the vertical coordinate, μ_i denotes the coefficient of heat exchange with the environment, F denotes the ratio of the crystal (meniscus) cross-section perimeter to its area, T_e is the environmental temperature, λ_i is the thermal conductivity coefficient.

Then eq. (2.17) describes the temperature distribution of the crystal–melt system in a one-dimensional approximation. This means that the temperature in the crystal (meniscus) cross-section is averaged and real isotherms are replaced with flat ones. In this case the heat exchange with the environment is allowed for not in the form of the boundary condition as in the case of multi-dimensional problems but by introducing heat losses on the lateral surface in the form of an additional term in the heat equation.

Then eq. (2.17) gives a good description of the real temperature distribution for small Biot numbers ($\mathrm{Bi} = \mu_i R/\lambda_i \ll 1$). This can be observed during thin crystal growth (characteristic of the CST), for low coefficients of convective heat transfer from the crystal (melt) surface, μ_i, and high thermal conductivities, λ_i, (characteristic of metals). However, if Biot numbers are not too small, eq. (2.17) gives a qualitatively correct description of the dependence of the temperature gradient in the melt, G_L, and in the crystal, G_S, on crystal dimensions, crystallization front position, pulling rate and some other parameters.

The interface temperature is equal to the crystallization temperature, T_0. In some cases the dependence of T_0 on the crystallization rate, V_C, is to be taken into consideration. This effect will be analyzed below. All the other boundary conditions depend upon the specific crystallization schemes and cooling conditions.

Assume that the solution of the stationary problem (the boundary problem for eq. (2.17) with the left-hand side equal to zero for corresponding boundary conditions) result in calculation of the liquid- and solid-phase temperature distributions. Then the temperature is a function of the boundary conditions, the crystallization front position h^0 being one of them. Liquid- and solid-phase temperature gradients will be functions of the same boundary conditions and the heat balance equation (2.15) should be satisfied at the crystallization front; then $V = V_C$ as the crystallization front displacement rate is equal to zero due to the process stationarity.

Assume that some of the parameters X_k of our system change by a small value $\delta X_k = X_k - X_k^0$. New temperature gradients in eq. (2.15) will be found from the stationary problem as functions of the new parameters X_k, and the crystallization front will move with the rate $\mathrm{d}(h - h^0)/\mathrm{d}t$ towards a new equilibrium position. By expanding the temperature gradients into Taylor series in terms of δX_k in the vicinity of the stationary point and using linear terms δX_k of expansion and taking into consideration that the temperature gradients with $X_k = X_k^0$ are the solutions of eq. (2.15) with $\partial h/\partial t = 0$ (the non-perturbed problem), we obtain:

$$\partial h/\partial t = \mathscr{L}^{-1} \sum_{k=1}^{n} [\lambda_S(\partial G_S/\partial X_k) - \lambda_L(\partial G_L/\partial X_k)]\delta X_k. \tag{2.18}$$

Thus, we have the second equation of eq. (2.3) for the crystallization front displacement rate. It is more convenient to consider the explicit form of the other equations of eq. (2.3) for each crystallization technique separately.

2.8. *Convection influence on the capillarity and liquid-phase heat-transfer processes*

As a rule, crystallization processes are carried out under the conditions of liquid-phase temperature gradient. As is known, the availability of such a gradient leads to convective flow formation, of free convection in particular, associated with the presence of gravity, which is characterized by the Rayleigh number, $\mathrm{Ra} = g\beta\Delta TL^3(\eta k)^{-1}$, and by the Marangoni convection associated with the surface-tension gradient, which is characterized by the thermal Marangoni number, $\mathrm{Ma} = (\mathrm{d}\gamma_{LG}/\mathrm{d}T)\Delta TL/\eta k$. Here β denotes the volume expansion coefficient, ΔT denotes the characteristic temperature difference and η is the coefficient of dynamic viscosity.

The number $(g\beta L)^2(\mathrm{d}\gamma_{LG}/\mathrm{d}T)^{-1}$, that is the ratio of the Rayleigh number to the Marangoni number, characterizes the relative effects of buoyancy and capillary force on convective flow formation. For small characteristic dimensions of the system capillary forces prevail. In [28] it is experimentally shown that Marangoni convection develops in the zone with the characteristic dimensions of $L \approx 5$ mm. As the characteristic dimensions of the menisci in the CST are of the order of a few millimeters, the influence of this very convection on capillarity and heat-transfer processes is discussed here.

It is shown [29] that, from the point of view of capillarity, convection has to change the meniscus shape particularly in microgravity conditions. From the point of view of heat-transfer the convection results in melt stirring and in heat removal increase at the crystallization front for overheated-melt growth. The magnitude of the rates of this flow can be estimated from the meniscus free-surface boundary condition that can have the following form

$$\eta \mathrm{d}V/\mathrm{d}r = -(\mathrm{d}\gamma_{\mathrm{LG}}/\mathrm{d}T)\,\mathrm{grad}\ T. \tag{2.19}$$

Hence,

$$V \approx \frac{\mathrm{d}\gamma_{\mathrm{LG}}}{\mathrm{d}T}\frac{R\Delta T}{\eta h}, \tag{2.20}$$

where R denotes the crystal radius (meniscus cross-section), h denotes the meniscus height. The thermal Marangoni number is characterized by the ratio of the melt-motion-transferred heat flow with the rate V to the diffusion-transferred heat flow. Rate estimation from the expression cited gives values an order of magnitude higher than those observed in [28]. Such a difference is explained by the frictional effect of the crystal solid surface. As is noted in [28], the convective contribution to the heat transfer is considerable for liquid zones more then 1 mm long. Note that the convective heat flow towards the crystallization front increases with R increasing and h decreasing, which can be seen when estimating the magnitudes of the convective flow rates. The behavior of the diffusion-transferred heat flow is the same.

Thus, within crystallization process stability study, convective influence on the heat transfer can be allowed for by formal introduction of the effective coefficient of thermal conductivity and that of lateral-surface heat exchange.

3. *The realization of dynamic stability in the Czochralski technique (CzT)*

By the CzT [1] crystals grow from a meniscus formed on a free surface of melt onto the seed pulled upright (fig. 3). Let the temperature of the meniscus horizontal section be

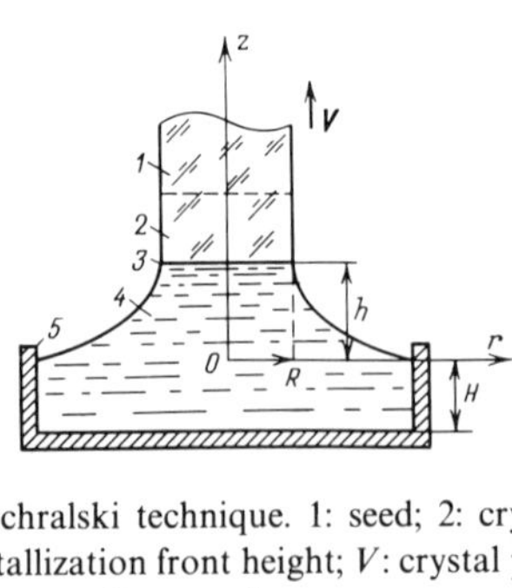

Fig. 3. Crystal growth by the Czochralski technique. 1: seed; 2: crystal; 3: crystallization front; 4: melt; 5: crucible; R: crystal radius; h: crystallization front height; V: crystal pulling rate; H : melt level in crucible; zOr : coordinate system.

T_0, the temperature of liquid crystallization. So, above this plane section the liquid transforms into a solid phase (fig. 3). Now we will set the liquid phase into upward motion with a constant rate, V, keeping the position of the phase transition plane invariable due to proper heat condition selection. When the motion starts, the crystallized portion of the meniscus will continuously form a solid upward or downward tapering off. In the particular case when a line tangent to the liquid meniscus surface makes an angle with the vertical equal to the angle of growth, the crystallized portion of the meniscus will have the form of a cylindrical body whose cross-section will coincide with meniscus section formed by the crystallization plane (fig. 3).

Thus, the seed serves to form a meniscus which later on determines the form of the crystallized product, phase transition surface position being fixed. And now a specimen of specified shape seems to be easy to pull just by selecting a proper seed and meniscus and positioning the phase transition plane into a certain place. However, it is difficult to obtain desired shapes of crystallized products in such a way. You can easily understand why. In the process of pulling, liquid surface and phase transition surface positions are subjected to perturbations that affect the shape of the product being pulled. This, in turn, results in changes in the position of each meniscus point, with a new equilibrium surface of the liquid being formed. These perturbations accumulate with time and the shape of the profile being pulled is distorted.

The idea of steady-state crystal growth is to provide negative feedback presence in the crystallization system due to utilization of inherent system characteristics. In this case perturbations occurring in the system are suppressed, which ensures crystallization process stability and leads to the possibility of growing a crystal of constant cross-sectional shape.

3.1. The Lyapunov set of equations

If the crystal dimension (its radius), R, and the crystallization front position relative to the melt free surface, h, are regarded as variable parameters in the Czochralski technique, for a stability study eq. (2.3) will include eqs. (2.14) and (2.18).

$$\partial R/\partial t = A_{RR}\delta R + A_{Rh}\delta h, \tag{3.1}$$

$$\partial h/\partial t = A_{hR}\delta R + A_{hh}\delta h. \tag{3.2}$$

The following notations are used here:

$$A_{RR} = -V(\partial\alpha_0/\partial R), \tag{3.3}$$

$$A_{Rh} = -V(\partial\alpha_0/\partial h), \tag{3.4}$$

$$A_{hR} = \mathscr{L}^{-1}[\lambda_S(\partial G_S/\partial R) - \lambda_L(\partial G_L/\partial R)], \tag{3.5}$$

$$A_{hh} = \mathscr{L}^{-1}[\lambda_S(\partial G_S/\partial h) - \lambda_L(\partial G_L/\partial h)]. \tag{3.6}$$

For stability of the set of eqs. (3.1) and (3.2) it is necessary and sufficient that the equation coefficients should satisfy the following conditions:

$$A_{RR} + A_{hh} < 0, \tag{3.7}$$

$$A_{RR}A_{hh} - A_{Rh}A_{hR} > 0. \tag{3.8}$$

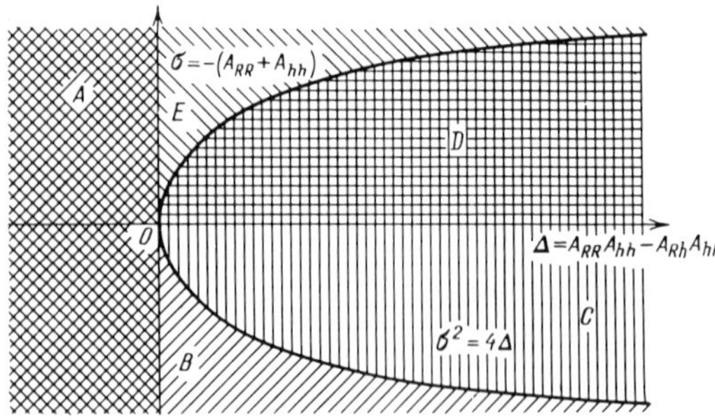

Fig. 4. Areas and types of stability : A : saddles; B : unstable nodes; C : unstable focuses; D : stable focuses; E : stable nodes.

With the conditions of eqs. (3.7) and (3.8) satisfied, two types of stable behavior of the system are possible depending upon the sign of the inequality

$$4A_{Rh}A_{hR} \lesseqgtr (A_{RR} - A_{hh})^2.$$

The sign < corresponds to the point called a node on the phase diagram, while the sign > corresponds to the point called a focus (fig. 4). The focus corresponds to the system absolute stability.

The coefficients A_{RR} and A_{hh} indicate a direct correlation between $\partial R/\partial t$ and δR, $\partial h/\partial t$ and δh, i.e. self-stability of the parameters. The coefficients A_{Rh} and A_{hR} represent the effect of one value change on the other value change rate, i.e. interstability of the parameters. To achieve interstability in the system, the inequality $A_{Rh}A_{hR} < 0$ should be satisfied, i.e. the coefficients A_{Rh} and A_{hR} should have different signs, see eq. (3.8). If $A_{RR} > 0$ and $A_{hh} > 0$ the system will be unstable as follows from eq. (3.7) as self-stabilization of both the parameters is absent. In the case that $A_{RR}A_{hh} < 0$, i.e. self-stabilization of one of the parameters is absent, system stability is possible if interstabilization is available.

If at least one of the inequalities of eqs. (3.7) and (3.8) is replaced by an equality, we cannot judge the system stability by its linear approximation, and a non-linear model should be analyzed.

To define the explicit form of the coefficients A_{jk} specified by eqs. (3.3)–(3.6) the capillary problem and the heat transfer problem for crystal growth by the CzT should be solved.

3.2. Analysis of capillary coefficients

An analytical approximate solution of the Laplace capillary equation or a numerical one allows one to plot the relatives $h(R)|_{\psi_0}$ (fig. 5), $\partial\alpha_0/\partial R(R)$ and $\partial\alpha_0/\partial h(R)$ (fig. 6). The coefficients $A_{RR}(R)$ and $A_{Rh}(R)$ are obtained from $\partial\alpha_0/\partial R(R)$ and $\partial\alpha_0/\partial h(R)$ by eqs. (3.3) and (3.4). From fig. 6 it follows that the value of A_{RR} remains positive for all R-values, approaching zero for crystals of large diameter, and the value of A_{Rh} remains negative for all R-values. In order that the stability conditions, eqs. (3.7) and (3.8), could be satisfied with the sign of the coefficients A_{RR} and A_{Rh} obtained, the following unequalities should be satisfied:

$$A_{hh} < 0, \qquad |A_{hh}| > |A_{RR}|, \tag{3.9}$$

$$A_{hR} > 0, \qquad |A_{Rh}||A_{hR}| > |A_{RR}||A_{hh}|. \tag{3.10}$$

3.2.1. Capillary stability

Consider the case when a change in crystal dimensions and crystallization front position does not lead to a change in liquid- and solid-phase temperature gradients at the crystallization front. In this case the stability of our system can be provided only by the capillary effects (capillary stability). Practically a crystallization system of one degree of freedom is considered in this case and the set of eq. (2.3) describing stability will reduce to the simple relation $\partial R/\partial t = A_{RR}\delta R$. The condition $A_{RR} < 0$ is a condition necessary and sufficient to attain capillary stability. In this case any crystal dimension perturbations will attenuate.

A positive value of the coefficient A_{RR} is indicative of capillary stability absence during crystal pulling by the CzT. Figure 5 gives a good illustration of it. Let point A characterize constant-diameter crystal growth ($\alpha_0 = \alpha_e$). Arbitrary change in crystal dimensions, upward tapering A', downward tapering A'', or in crystallization-front position, rise $\dot{A}$, fall $\ddot{A}$, will cause such a change of the angle α_0 that it will lead to further crystal tapering according to fig. 2.

Crystallization stability can be ensured here only at the expense of heat transfer effects if the conditions of eqs. (3.9) and (3.10) are satisfied. In this case, crystal dimension change (points B' and B'', fig. 5) results in such crystallization front change (points $\dot{B}$ and $\ddot{B}$, fig. 5) that provides growth of crystals of slightly changed dimensions with the liquid–solid phase conjugation angle equal to the growth angle.

3.3. The heat transfer problem

The stationary state temperature distribution in liquid and solid phases in the coordinate system shown in fig. 3 will be described by eq. (2.17) with left-hand side zero.

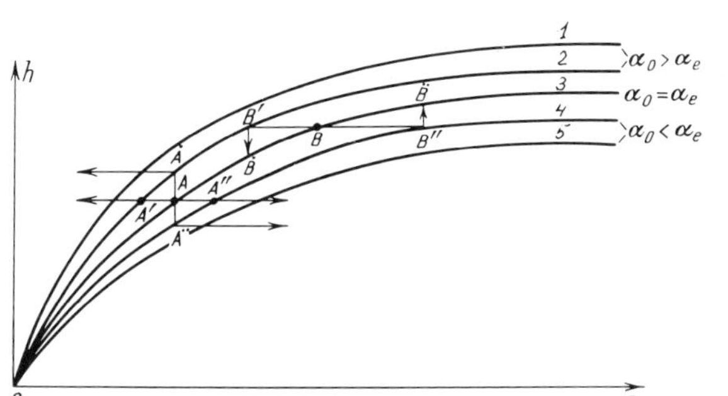

Fig. 5. Profile curve height versus crystal radius for various growth angles ψ_0: (1) – 10°; (2) 0; (3) 10°; (4) 20°; (5) 30°; A: scheme of capillary instability; B: scheme of thermal stability.

If the true meniscus shape is referred to as a right circular cylinder of radius R as the first approximation while solving this equation for the liquid phase, the ratio of the cross-section perimeter to its area (F) both for the meniscus and for the crystal will be equal to $2/R$.

The explicit form of the coefficients A_{hh} and A_{hR} obtained from eqs. (3.5) and (3.6) as a result of solving the heat transfer problem will be given.

3.3.1. Thermal stability

The coefficient A_{hh} is associated with crystallization-front displacement and indicates self-stabilization of the parameter h. Indeed, if the crystal radius is considered to be fixed (e.g. Bridgman growth) the sign of the coefficient A_{hh} determines crystallization stability. We will say that the system is thermally stable if $A_{hh} < 0$ (thermal stability by analogy with capillary stability).

The coefficient A_{hR} indicates interstabilization of the parameters, i.e. crystallization front position and crystal or melt column dimensions. In this case the liquid column dimensions change all over its length, while the crystal dimensions change only near the crystallization front. Accurate calculation of the contribution of crystal-dimension changes near the crystallization front to the coefficient A_{hR} is rather a lengthy problem, therefore only two limiting cases will be analyzed: allowance for crystal dimension changes all over its length and absence of a contribution associated with crystal dimension changes, the latter being easily obtained from the former if all the terms relating to the solid phase and dependent upon $\partial/\partial R$ are equated to zero in the expressions for A_{hh} and A_{hR}.

3.3.2. A semi-infinite crystal where heat removal from the crystal lateral surface is allowed for, while that from the melt is not taken into consideration, i.e. $\mu_S \neq 0$; $\mu_L = 0$

This approximation is valid for long-crystal growth (the case of continuous pulling), with good thermal screening of the melt column provided. For A_{hh} and A_{hR} we obtain

$$A_{hh} = -\frac{\lambda_L}{\mathscr{L}}\frac{(T_p - T_0)}{h^2}, \tag{3.11}$$

$$A_{hR} = \frac{\mu_S}{\mathscr{L}R^2}\frac{(T_0 - T_e)}{\zeta_S} > 0. \tag{3.12}$$

Here

$$\zeta_S = \sqrt{\frac{V^2}{4k_S^2} + \frac{2\mu_S}{\lambda_S R}}$$

where T_p is the melt temperature on a shape level.

In the case under consideration, thermal conditions stabilize the crystallization front position ($A_{hh} < 0$) if the melt is overheated ($T_p > T_0$). Indeed, if an arbitrary perturbation leads to an increase in h, the liquid-phase temperature gradient decreases, and hence the heat supply to the front decreases. In this case the true crystallization rate should increase, which will result in downward front displacement, and hence, in a decrease in h.

The unequality $A_{hR} > 0$ can be explained from the physical point of view in the following way. Heat flow $\pi R^2 G_S$, removed via the crystal lateral surface, is proportional to the lateral surface area, i.e. to R. Therefore, with crystal radius increasing, G_S decreases, which leads to an increase in h.

As follows from section 3.2, for CzT, $A_{Rh} < 0$ and hence $A_{Rh}A_{hR} < 0$, which contributes to satisfying eq. (3.8). Hence, it follows that interstabilization of the parameters is available in the system. Indeed, in the case where the crystal diameter uncontrollably increases, crystallization front height will increase due to $A_{hR} > 0$. This will result in an increase in α_0 as $\partial\alpha_0/\partial h > 0$, and hence in crystal contraction according to the diagram given in fig. 2.

As interstabilization exists, stable crystal pulling by CzT is possible if capillary stability is absent. To achieve this, it is necessary and sufficient that the following inequalities obtained from eqs. (3.9) and (3.10) should be satisfied:

$$\left|\frac{\partial\alpha_0}{\partial R}\right| < \frac{\lambda_L}{\mathscr{L}Vh^2}(T_p - T_0), \tag{3.13}$$

$$\left|\frac{\partial\alpha_0}{\partial h}\right| > \left|\frac{\partial\alpha_0}{\partial R}\right| \frac{(T_p - T_0)R^2\lambda_L\zeta_S}{(T_0 - T_e)h^2\mu_S}. \tag{3.14}$$

3.3.3. A crystal of finite length, l, where the temperature of the crystal upper end is equal to the environmental temperature, i.e. $\mu_L = 0$, $\mu_S \neq 0$, $T_l = T_e$

Most of all the approximation fits growth of finite-length crystals with the seed temperature fixed, e.g. a water-cooled seed, and melt column heat-screened:

$$A_{hh} = -\frac{1}{\mathscr{L}}\left\{\frac{\lambda_L(T_p - T_0)}{h^2} + \frac{\lambda_S\zeta_S^2(T_p - T_e)}{\mathrm{sh}^2[\zeta_S(l-h)]}\right\}, \tag{3.15}$$

$$A_{hR} = \frac{(T_p - T_e)\mu_S}{\mathscr{L}R^2\zeta_S}\left\{\mathrm{cth}[\zeta_S(l-h)] - \frac{\zeta_S(l-h)}{\mathrm{sh}^2[\zeta_S(l-h)]}\right\}. \tag{3.16}$$

In principle this case differs from the previous one in the fact that the solid-phase temperature gradient depends upon the crystallization front position, the gradient modulus increasing with h. This has its effect on A_{hh}, the second term on the right-hand side of eq. (3.15) being negative, and hence improves the process stability. For small values of $(l-h)$, we can expand $\mathrm{sh}^2[\zeta_S(l-h)]$ into a series, restrict ourselves to the first term of the expansion, and then the second term on the right-hand side of eq. (3.15) is found to be similar in structure to the first one: $\lambda_S(T_0 - T_e)/(l-h)^2$. The coefficient A_{hR} in eq. (3.16) differs from that in eq. (3.12) in the factor within the parentheses that tends to one with increasing l. If l is small (approaches h in the order of magnitude), the factor tends to $2\zeta_S(l-h)/3$.

3.3.4. A semi-infinite crystal where heat removal from the melt-column lateral surface is allowed for, i.e. $\mu_L \neq 0$, $\mu_S \neq 0$

If the melt-column screening is not effective enough to ignore heat removal from its surface, we obtain instead of eqs. (3.11) and (3.12) for A_{hh} and A_{hR}, rather awkward expressions including hyperbolic functions. Regarding h as a small value, we can restrict ourselves to the two first terms when expanding the hyperbolic functions into a series:

$$A_{hh} = -\frac{\lambda_L}{\mathscr{L}}\left[\frac{(T_p - T_0)}{h^2} + \tfrac{1}{2}(T_p - T_e)\zeta_L^2\left(1 - \frac{Vh}{3k_L}\right)\right], \tag{3.17}$$

$$A_{hR} = \frac{1}{\mathscr{L}R^2}\left[\frac{(T_0 - T_e)\mu_S}{\zeta_S} - \frac{\mu_L(T_0 - T_e)(1 + Vh/2k_L)h}{3} - \frac{\lambda_L(T_0 - T_e)\zeta_L h}{2}\right]. \tag{3.18}$$

The coefficient A_{hh} in eq. (3.17) is different as compared to eq. (3.11). If not-overheated-melt pulling is carried out ($T_p = T_0$), $A_{hh} = 0$ in eq. (3.11), while in eq. (3.17) $A_{hh} < 0$, i.e. the system stability is improved. It is attributed to the fact that the melt gets cooled while moving along the column, creating a temperature gradient. Changes in melt-column height cause changes in the gradient value, thus stabilizing the system.

As compared with eq. (3.12), the value of the positive coefficient A_{hR} decreases (as was mentioned above, $A_{hR} > 0$ in eq. (3.12) increases due to crystal surface cooling, while melt-column cooling results in an opposite effect).

When melt-column cooling is allowed for, quite similar results are obtained for a finite crystal with specified temperature at its end, which is shown by additional terms appearing in eqs. (3.15) and (3.16).

3.3.5. *Quantitative stability criteria*

Using eqs. (3.13) and (3.14) quantitative values of the parameters ensuring stability of CzT will be found. When growing metal and semiconductor single crystals, it can be assumed that $V \approx 10^{-3}$–10^{-4} m/s, $\mu_i \approx 10^2\,\mathrm{J\,m^{-2}\,K^{-1}}$, $\mathscr{L} \approx 5 \times 10^9\,\mathrm{J\,m^{-3}}$, $\lambda_i \approx 10^2\,\mathrm{J\,m^{-1}s^{-1}K^{-1}}$, $G_L|_{z=h} \approx 10^3$–$10^4\,\mathrm{K\,m^{-1}}$, $h \approx 10^{-3}$ m, $R \approx 10^{-2}$ m, $T_p - T_0 \approx 10$ K, heat capacity $c \approx 1\,\mathrm{J\,kg^{-1}K^{-1}}$, $a \approx 5 \times 10^{-3}$ m. Hence, it follows that for overheated-melt growth $A_{hR} \approx 0.1$–$1\,\mathrm{s^{-1}}$, $A_{hh} \approx -(1$–$10)\,\mathrm{s^{-1}}$ and on the basis of eqs. (3.13) and (3.14) stable-growth conditions for CzT are obtained.

$$\left|\frac{\partial\alpha_0}{\partial R}\right| a \leq 5, \qquad \frac{\partial\alpha_0}{\partial h} \geq 10\left|\frac{\partial\alpha_0}{\partial R}\right|. \tag{3.19}$$

Comparison of eq. (3.19) with the plots (fig. 6) allows to conclude that only for $R/a \geq 1$ satisfying both the inequalities is possible, i.e. stable growth of cylindrical crystals by CzT is possible.

Based on the analysis carried out, the main data obtained on the stability of crystallization by CzT can be summarized. Here CzT is referred to as a system with two degrees of freedom.

(1) Capillary stability is absent: $A_{RR} > 0$.

(2) Interstabilization of the parameters $A_{Rh}A_{hR} < 0$ is observed.

(3) As a result, stable growth is possible only in the case that thermal stability is provided, $A_{hh} < 0$, and the condition $|A_{hh}| > A_{RR}$ is satisfied.

(4) The condition $A_{hh} < 0$ is easily satisfied by providing overheated-melt growth.

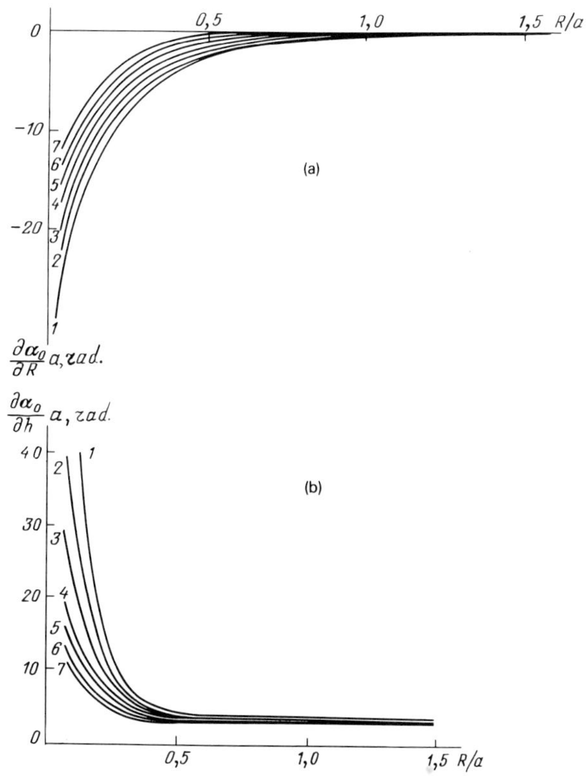

Fig. 6. (a) $\partial\alpha_0/\partial R$ and (b) $\partial\alpha_0/\partial h$ versus crystal radius for various growth angles ψ_0: (1) – 15°; (2) – 5°; (3) 0; (4) 10°; (5) 20°; (6) 30°; (7) 40°.

(5) The condition $|A_{hh}| > A_{RR}$ is easily satisfied by providing growth of crystals of large diameters, $R > a$, for which $A_{RR} \to 0$.

(6) Providing process stability by means of overheating increase is associated with the difficulties of heat removal from the crystallization front. Therefore, a technique increasing the coefficient of convective heat transfer from the crystal surface, e.g. by blowing gas on the crystal localized zone above the crystallization front, can be recommended.

4. *Techniques of pulling from a shaper (TPS)*

TPS differ from CzT by the presence of a shaper on the melt surface (fig. 7). It should be noted that the priority in applying holes in plates placed onto the melt surface for melt pulled crystal shaping belongs to Gomperz [30] and Kapitza [31] but only due to methodical work carried out by Stepanov and his collaborators was the technique developed and is it now widely used for crystallization of various materials. Some conferences were carried out in Russia [32] and all techniques of crystal growth from a shaper were named Stepanov techniques. In recent years the EFG, CAST and some other techniques are opposed to the ST [33–35]. We suggest to combine all these techniques by the common abbreviation TPS. In fact, as was shown in our analysis [36–38], only two possibilities can be used for shaping under these conditions – edges or walls of a shaper. From a mathematical point of view this corresponds to two boundary conditions of the capillary problem – the fixing of a line on the surface of the meniscus or the fixing of the angle between the melt and the wall of the shaper. All mentioned techniques use this principle of shaping, but each of them is characterized by special design features, e.g. using external (fig. 7e) or internal (fig. 7b, d) free edges, surfaces (fig. 7a) or a combination thereof (fig. 7c) for solid bodies in contact with a melt. Some positive or negative pressure, d, can be applied to the melt. This can be achieved by positioning the shaper free edge relative to the liquid free surface level (fig. 7).

Based on this description the conclusion can be drawn that the dimensions and the shapes of the specimens being pulled by TPS depend on the following factors: *firstly*, shaper geometry; *secondly*, the pressure of feeding the melt to the shaper; *thirdly*, the crystallization surface position and shape; *fourthly*, the seed shape. Seed shape is important only for stationary pulling; in this case its cross-section should coincide with

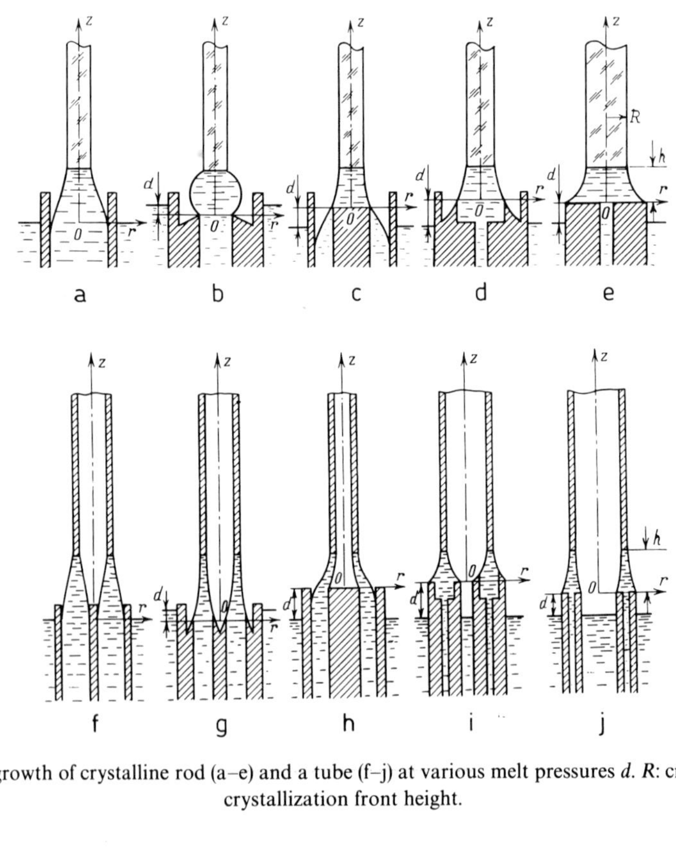

Fig. 7. Melt growth of crystalline rod (a–e) and a tube (f–j) at various melt pressures d. R: crystal radius; h: crystallization front height.

the desired product cross-section. Often, especially when complicated profiles are grown, a wettable rod can be used as a seed, its main function being to raise the melt column above the shaper level. Furthermore, a non-stationary pulling process can be introduced by lowering the crystallization surface, which leads to enhancing the shaper dependence of the crystal cross-section. With such an approach applied to the pulling process, the dimensions and the shape of the crystal grown are determined by the above-mentioned factors and by the ratio of the pulling rate to the crystallization front displacement rate.

As follows from the process description, a shaper in TPS does not serve as a gauged orifice that strictly determines the dimensions and the shape of the crystal being pulled. For this reason it is not advisable to use the term "die" instead of the term "shaper" as is often done, for the extrusion die strictly determines the profile cross-section. The shaper is one of the main elements of the crystallization heat zone in TPS. To demonstrate the problems that are to be solved in the process of designing it, we will analyze a very simple example – melt pulling of a circular cylindrical rod. Even for the case that one shaping scheme is used, this rod can be grown using shapers of various dimensions as well as various pressures (fig. 7a–e). That is why criteria allowing proper selection thereof have to be developed. The problem becomes still more complicated if the crystal shape differs from a circular one. In this case the shaper can compensate for the difference in melt-rise height caused by crystal surface curvature variations to provide a flat crystallization front during pulling (fig. 8).

The above-given description does not completely show the shaper role in TPS. In essence the description does not answer the question as to what differences exist between TPS and CzT. Indeed, in TPS, shaper walls can be moved off the crystal (fig. 7a). In the CzT, crystal size can be commensurate with the crucible size (fig. 3). Under what conditions can the effect of the crucible wall on shaping be ignored? Where is the end of CzT and the beginning of TPS?

The results of a crystallization process stability study have answered these questions [7]. The shaper proved not only to limit the area of melt perturbations as was supposed above but in many cases stabilizes the crystallization process. Therefore, we will give

Fig. 8. Growth of crystal of complicated shape with the difference in melt rise height at the sections of various curvature compensated by the shaper.

considerable attention to crystallization process stability with various schemes of shaping applied.

Profiles of complicated cross-sectional shape can be grown by TPS. Figure 9 shows aluminum alloy profiles, grown in the Stepanov laboratory in 1964–1993. The left one is a heat-exchanger. It consists of an external shell, an internal eight-tube panel and six cross pieces which are removed. The accuracy of maintaining profile dimensions can be judged by such a fact: the heat-exchanger is more than one meter long, it was assembled without any mechanical treatment of the profiles.

Below, we analyze the stability conditions of TPS and give some examples of shaped crystal growth – sapphire, silicon and lithium niobate. These crystals demonstrate the unusual peculiarity of structure typical of TPS.

While analyzing stability of TPS as a system with two degrees of freedom, the data of section 3.1 can be used. The set of eqs. (3.1) and (3.2), the expressions given in

Fig. 9. Aluminum alloy profiles melt-grown by TPS.

eqs. (3.3)–(3.6) for the coefficients appearing in the inequalities (3.7) and (3.8) characterizing stability conditions will be the same for CzT and for TPS. Therefore we can proceed with the analysis of melt-column shaping conditions in TPS.

4.1. Boundary conditions of the capillary problem [37, 38]

As was mentioned above, when formulating the boundary problem for the Laplace capillary equation, two boundary conditions should be specified. As follows from the description of the pulling techniques, one of them will be specified at the lower boundary of the melt column, the other one being specified at the upper boundary, i.e. along the liquid-pulled crystal contact line. A shaper is used for melt-column shaping at its lower boundary in TPS (fig. 10). In the present section the shaper is referred to as a device to control melt-column shape only. Shaping is accomplished either on the walls (fig. 10b, c) or on sharp edges (fig. 10a) of the shaper.

The shaper (fig. 10) will be characterized by its depth d_0, its wall or free edge curvature radius r_0 in the horizontal plane and the angle B_0 made by its wall with the horizontal. The wetting angle θ formed by the melt and the shaper wall is a very important shaper characteristic. If this angle exceeds 90°, the shaper material is not wetted by the melt. If it is smaller, the melt wets the shaper material.

The following boundary conditions on the shaper walls and at its free edges are possible.

4.1.1. The catching conditions

In the case that the shaper material is wetted by the melt, the liquid is easily caught by its sharp edge. This boundary condition will be called the "catching condition". For sharp shaper contours Γ this condition will take the following form (fig. 10a):

$$z|_{\Gamma} = d(r, \varphi). \tag{4.1}$$

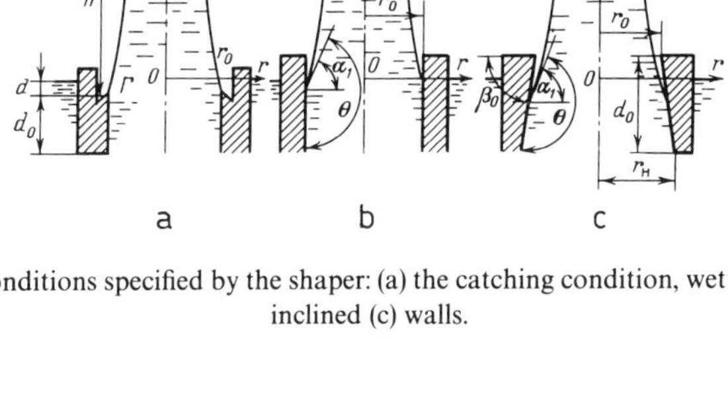

Fig. 10. Boundary conditions specified by the shaper: (a) the catching condition, wetting on vertical (b) and inclined (c) walls.

Here d is the distance from the shaper edge to the plane coordinate that coincides with the liquid surface plane. In the case that the shaper is flat and is placed parallel to the liquid plane, d is constant and represents the pressure being positive, the shaper is placed below the melt free-surface level and vice versa.

4.1.2. *The angle fixing condition (the wetting condition)*

If the liquid does not reach the shaper sharp edges, its surface makes the wetting angle with the shaper walls (fig. 10b, c). This boundary condition, in a general case, will be called the "angle fixing condition" and for the shaper it will be called the "wetting condition". It can be written down in the following form:

$$z'|_{r=r_0} = -\operatorname{tg}(\theta - \tfrac{1}{2}\pi) = -\operatorname{tg}\alpha_1. \tag{4.2}$$

Here, the case of an inclined-wall shaper should be considered separately (fig. 10c):

$$z'|_{r=r_0-\operatorname{ctg}B_0 z} = \operatorname{tg}(\theta + B_0 - 180°) = -\operatorname{tg}\alpha_1, \tag{4.3}$$

i.e. with the wetting boundary condition satisfied by changing the angle of the shaper free edges B_0. The angle α_1 between the horizontal and the line tangent to the liquid surface at the point of its contact with the shaper wall can be changed.

4.1.3. *The wetting-to-catching transition*

With melt pressure increasing, the catching condition at non-wettable shaper free-edges can be obtained. Figure 11a shows a diagram of this transition.

The same result can be obtained by changing the seed-to-shaper hole dimension ratio alone (fig. 11b).

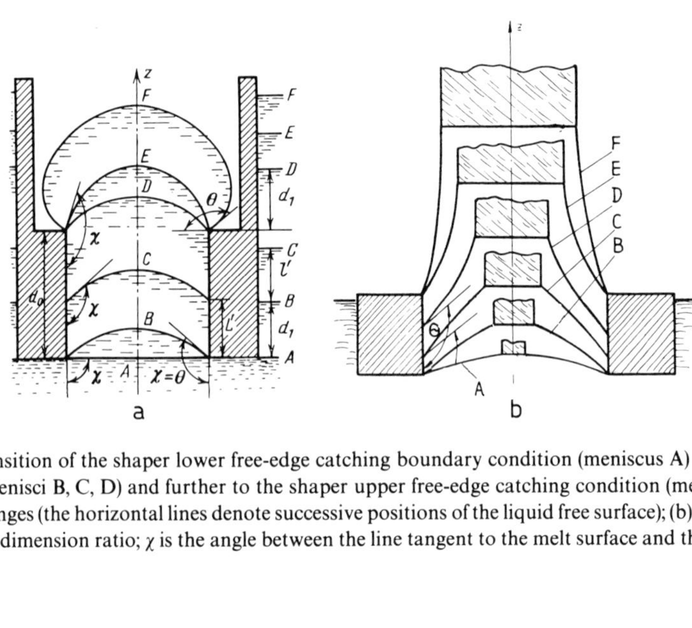

Fig. 11. Transition of the shaper lower free-edge catching boundary condition (meniscus A) to the wetting condition (menisci B, C, D) and further to the shaper upper free-edge catching condition (menisci E, F). (a) Pressure changes (the horizontal lines denote successive positions of the liquid free surface); (b) changes of the seed–shaper dimension ratio; χ is the angle between the line tangent to the melt surface and the shaper wall.

Hereafter it will be shown that the catching boundary condition allows one, in most cases, to provide capillary stability of the process. Therefore, the ways of achieving the catching condition at the shaper free edges in TPS should be specified.

Firstly, a melt-wettable material should be used for the shaper, and the latter should be designed in such a way that the melt could raise up to the shaper free edges due to capillary forces (fig. 7d, e, i, j).

Secondly, a melt-wettable material should be used, the melt column should be supported from inside (fig. 7c).

Thirdly, while using a melt-nonwettable material, the melt column should be embraced from outside providing additional pressure on the liquid to make the melt-shaper contact point touch the shaper sharp edge (figs. 7b and 11a).

Fourthly, for poorly wettable shaper materials, a seed of corresponding dimensions should be used in order that it can ensure melt column contact with the shaper sharp edges (fig. 11b).

4.1.4. *Influence of wetting-angle hysteresis on the capillary effects*

While analyzing all capillary effects, existence of wetting-angle hysteresis should be taken into consideration. Wetting-angle hysteresis reveals itself in the fact [39] that the wetting angle of a liquid run on a solid body θ_{on} is larger than that of liquid run off a solid body θ_{of}. This means that the stationary wetting angle depends upon the process of meniscus formation (running on or off). This results, e.g., in the fact that higher pressure needs to be applied to create the catching condition at the shaper free edge than to keep it unchanged. In the case that the catching condition is created by a seed, this condition can remain unchanged in the process of growth, with the clearance between the shaper free edge and the growing crystal changing.

4.1.5. *The boundary conditions at the crystal free edges*

As has already been mentioned, while pulling a crystal of constant cross-section, the angle fixing condition should be satisfied over the whole contour $\tilde{\mathscr{L}}$ (fig. 8). In this case the $\tilde{\mathscr{L}}$-contour position relative to the shaper free edges or to the melt surface can be calculated by solving the boundary problem.

Besides, meniscus catching on the crystal edge occurs; hence the $\tilde{\mathscr{L}}$-contour position relative to the shaper free edges or the melt surface can be specified. Then, as a result of the boundary problem solution, the angle made by the line tangent to the melt along the contour $\tilde{\mathscr{L}}$ with any fixed direction can be calculated.

Hereafter, these boundary conditions will be specified depending upon the particular problem.

4.2. *Liquid meniscus heights and shapes*

When growing crystals by TPS, melt columns of three types can be formed: concave, convex and convexo-concave ones (figs. 12 and 13).

In a number of works, e.g. [40], to calculate the crystallization front height it is proposed that the approximate formula obtained in [41] should be used. In the

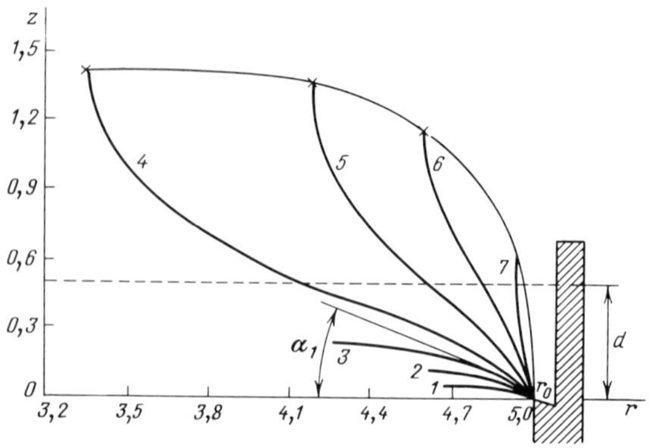

Fig. 12. Convex (1, 2, 3), convexo-concave (4, 5, 6) and concave (7) profile curves plotted by computer-aided calculation of eq. (2.6) for $r_0 = 5.0$, $d = 0.5$. Crosses mark the profile-curve heights calculated from eq. (2.7).

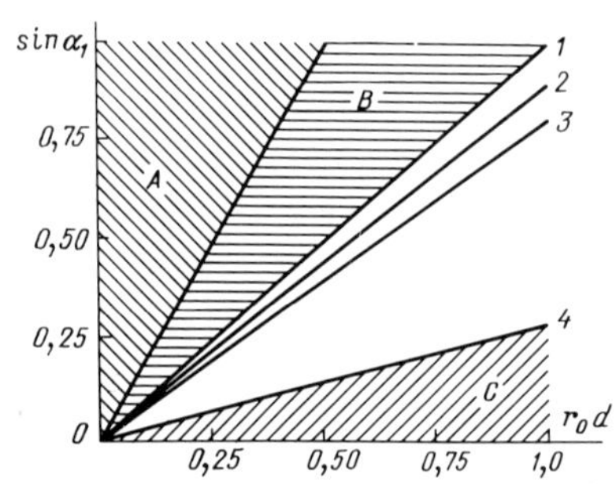

Fig. 13. Ranges of existence of (A) concave, (B) convexo-concave and (C) convex profile curves. Lines 1, 2, 3, 4 separate ranges B and C for various r_0 values: 1: corresponds to $r_0 = 0.1$; 2: $r_0 = 0.5$; 3: $r_0 = 1.0$; 4: $r_0 = 5.0$.

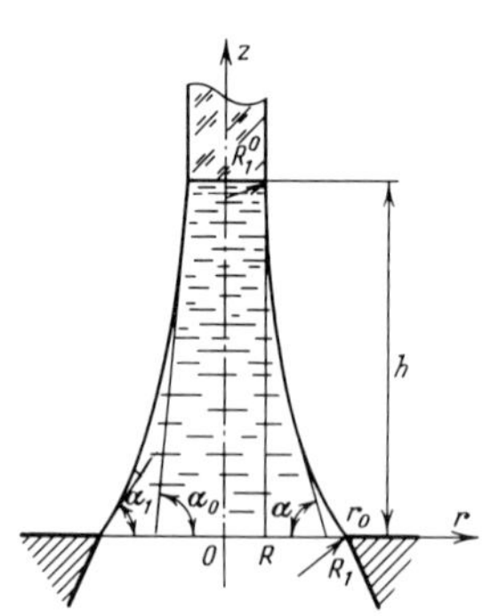

Fig. 14. Liquid column parameters for eq. (4.4).

notation shown on fig. 14 this has the following form:

$$\frac{h}{a} = \pm\left\{\pm(\cos\alpha_0 - \cos\alpha_1) + \left[\frac{(A+1)\cos\alpha_0}{4RA} - d\right]^2\right\}^{1/2} - \left[\frac{(A+1)\cos\alpha_0}{4RA} - d\right], \tag{4.4}$$

where $A = R_1/R_1^0$. However, application of eq. (4.4) to TPS is impossible because of the following reasons:

Firstly, when calculating the melt column height from eq. (4.4), one should know the angles α_0 and α_1, which is equivalent to specifying the derivative $\mathrm{d}z/\mathrm{d}r$ at the crystallization front and the shaper free edge. Besides, specification of pressure is equivalent to specification of the coordinate of the melt column-base point. Thus, specification of three boundary conditions instead of two is required for a second-order differential equation. In practice these conditions cannot be independent. For example, if the pressure d and the angle $\alpha_0|_{z=R}$ are specified, the angle $\alpha_1|_{z=r_0}$ should be calculated by solving the boundary problem. And vice versa, only one value of d corresponds to the specified angles $\alpha_0|_{z=R}$ and $\alpha_1|_{z=r_0}$.

Secondly, as the formula is applied to either concave or convex melt columns, a very wide class of convexo-concave columns for which the coordinates of the inflection points are unknown beforehand is excluded from consideration.

The possibility of applying the approximation from [41] to the CzT can be accounted for by the fact that three boundary conditions are nominally specified for the CzT:

$$\alpha|_{r=R} = \alpha_0, \qquad z|_{r\to\infty} = 0, \qquad \alpha|_{r\to\infty} = 0.$$

This specific feature is associated with the fact that the conditions at the melt-column lower end are specified for $r \to \infty$, where, with the z-coordinate tending to zero, the derivative $\mathrm{d}z/\mathrm{d}r$ also tends to zero.

4.3. The capillary problem for systems with large Bond number [14]

As an example of an analytical investigation of a capillary problem we consider the boundary problem for eq. (2.7) that describes the shape of a liquid column formed when pulling a crystal in the form of a plate or a rod of a large diameter compared to the capillary constant (large Bond number).

The boundary problem will be formulated using the following boundary conditions (fig. 10b). At the point where $r = r_0$ the shaper wall is situated. The line tangent to the melt surface in the liquid–shaper contact point and the negative direction of the r-axis make the angle α_1. Here $\alpha_1 = \theta - \frac{1}{2}\pi$ (θ is the angle of shaper wall wetting by the melt). Where $r = R$ the edge of a melt-grown flat crystalline tape or of a circular cylindrical crystal of a large diameter is situated. Let l_0 denote the clearance between the crystal edge and the shaper: $l_0 = r_0 - R$. Let the angle α_0 made by the line tangent to the melt surface at the melt/crystal contact point and the negative direction of the r-axis be specified. While growing crystals of constant cross-sections, the angle $\alpha_0 = \alpha_e = \frac{1}{2}\pi - \psi_0$ is the complement of the growth angle (*the boundary problem with the angle-fixing boundary conditions at both the ends of the r-variation interval*). Consider what process parameter data can be obtained with such a problem formulation. The boundary conditions can be written down as follows:

$$\left.\frac{dz}{dr}\right|_{r=r_0} = -\operatorname{tg}\alpha_1, \qquad \left.\frac{dz}{dr}\right|_{r=R} = -\operatorname{tg}\alpha_0. \tag{4.5}$$

An expression in the form of a set of Legendre elliptical functions of the first and the second kinds will be obtained for the profile curve shape and height as the result of solving this problem. Figure 15 shows a family of profile curves for various crystal dimension (curves 3–12). Therein, boundary curves representing loci of the ends of the profile curves characterized by the same α_0-values, i.e. providing the growth angle value $\psi_0 = \frac{1}{2}\pi - \alpha_0$, are plotted. Based on this boundary problem solution, the following conclusions can be drawn:

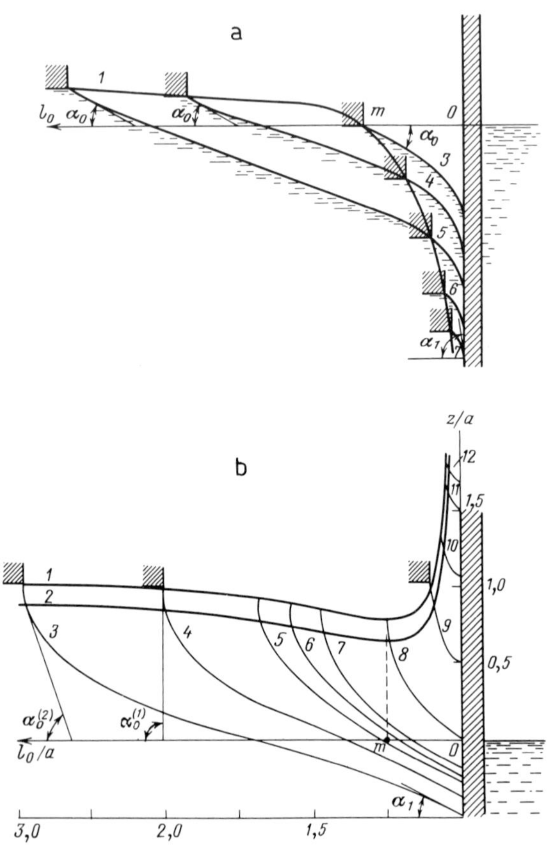

Fig. 15. Boundary (1, 2) and profile (3–12) curves: (a) $\alpha_0 < \alpha_1$; (b) $\alpha_0 > \alpha_1$ with $\alpha_1 = 45°$, $\alpha_0^1 = 90°$, $\alpha_0^2 = 75°$.

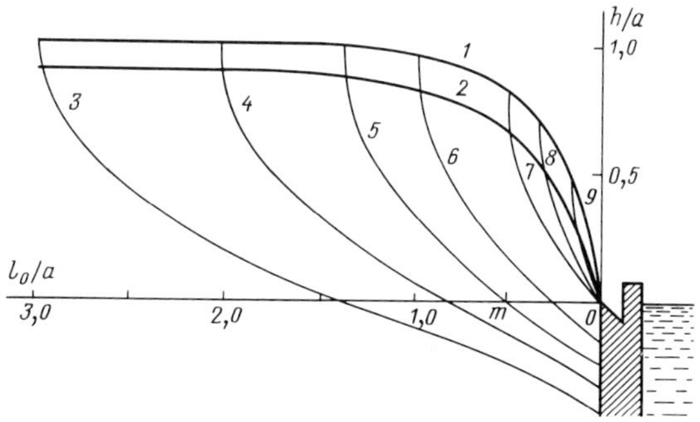

Fig. 16. Boundary (1, 2) and profile (3–9) curves for transition from the wetting condition (3–7) to the catching condition (7–9) : $\alpha_1 = 45°$; (1) $\alpha_0 = 90°$; (2) $\alpha_0 = 75°$.

There exists a minimum value m of the clearance l_0 between the shaper and the crystal being pulled when the meniscus lies both above and below the melt free surface. With $l_0 < m$, the whole meniscus lies above the liquid free surface if $\alpha_0 > \alpha_1$ (fig. 15b); the whole meniscus lies below the liquid free surface if $\alpha_0 < \alpha_1$ (fig. 15a). It is as if the larger angle overpulled the meniscus to its side. Note that this effect, analytically obtained by us, was pointed out by Poincaré [42].

Thus, the vertical coordinates of the liquid–solid phase contact points with respect to the melt free surface are not fixed but depend on the relation between the angles at both ends of the interval and on the value of the clearance between the shaper and the crystal being pulled.

The transition from the wetting boundary condition on the shaper wall to the catching condition at the shaper free edge is possible in the case where the shaper sharp edge is placed below the melt–shaper wall contact point and its coordinate l_0 can be found by solving this boundary problem (fig. 16).

A similar analysis was carried out for other schemes of capillary shaping and Bond numbers [38, 14].

4.4. *The capillary coefficients A_{RR} and A_{Rh}*

To define the character of TPS growth depending on the values of the system parameters as follows from eqs. (3.3) and (3.4), the behavior of the quantities $\partial\alpha_0/\partial R$ and $\partial\alpha_0/\partial h$ should be analyzed. In this case, *firstly*, the character of the boundary condition specified by the shaper (catching or wetting); *secondly*, the pressure value when using the catching boundary condition (it is controlled by changing the shaper position relative to the level of the melt free surface) or the relation between the growth angle and the shaper wall angle when using the wetting boundary condition (it is controlled either by the wall slope or by selecting a proper material for the shaper); *thirdly*, the relation between the cross-sectional dimensions of the crystal being grown and the shaping contour can be used as variable parameters. The meniscus is assumed to make an angle equal to the growth angle at the crystal free edge.

4.4.1. Small Bond numbers, the catching boundary condition, zero pressure

In this case not only the sign of the derivatives $\partial\alpha_0/\partial R$ and $\partial\alpha_0/\partial h$ can be defined but also an analytical expression for them can be derived:

$$\frac{\partial\alpha_0}{\partial R} = -\frac{F'_R}{F'_{\alpha_0}}, \tag{4.6}$$

$$\frac{\partial\alpha_0}{\partial h} = -\frac{F'_h}{F'_{\alpha_0}}, \tag{4.7}$$

Here

$$F'_h = -1,$$

$$F'_R = \sin\alpha_0\left(\operatorname{arch}\frac{r_0}{R\sin\alpha_0} - \operatorname{arch}\frac{1}{\sin\alpha_0}\right) - \frac{r_0\sin\alpha_0}{\sqrt{r_0^2 - R^2\sin^2\alpha_0}},$$

$$F'_{\alpha_0} = R\cos\alpha_0\left(\operatorname{arch}\frac{r_0}{R\sin\alpha_0} - \operatorname{arch}\frac{1}{\sin\alpha_0}\right) - \frac{Rr_0\cos\alpha_0}{\sqrt{r_0^2 - R^2\sin^2\alpha_0}} + R.$$

Figure 17 shows the behavior of the functions $\partial\alpha_0/\partial R$ and $\partial\alpha_0/\partial h$ in the range of $0 \le R \le r_0$ for two values of α_0 $(\alpha_0^{(1)} > \alpha_0^{(2)})$.

4.4.2. Small Bond numbers, the catching boundary condition, non-zero pressure

$$\frac{\partial\alpha_0}{\partial h} = \left\{R\cos\alpha_0\int_R^{r_0} r^2[r^2 - (r^2d - R^2d + R\sin\alpha_0)^2]^{-3/2}\mathrm{d}r\right\}^{-1}, \tag{4.8}$$

$$\frac{\partial\alpha_0}{\partial R} = -\frac{\partial h/\partial R}{\partial h/\partial\alpha_0}. \tag{4.9}$$

Here

$$\partial h/\partial R = -\operatorname{tg}\alpha_0 + (\sin\alpha_0 - 2r_0 d)\int_R^{r_0} r^2[r^2 - (r^2d - R^2d + R\sin\alpha_0)^2]^{-3/2}\,\mathrm{d}r,$$

$$\partial h/\partial\alpha_0 = (\partial\alpha_0/\partial h)^{-1}.$$

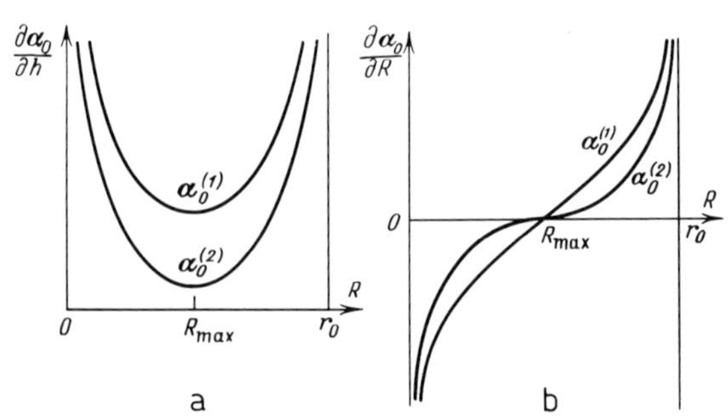

Fig. 17. The functions $\partial\alpha_0/\partial h$ (a) and $\partial\alpha_0/\partial R$ (b) within the range of R-values from 0 to r_0.

From eq. (4.8) it follows that $\partial\alpha_0/\partial h > 0$ within the whole range of $0 \le R \le r_0$, i.e. the behavior in the case of $d = 0$. (fig. 17a). The behavior of the function $\partial\alpha_0/\partial R(R)$ depends on the profile curve type. For convex profile curves $\partial\alpha_0/\partial R > 0$. For concave and convexo–concave profile curves the behavior of this function is similar to its behavior for $d = 0$ (fig. 17b).

4.4.3. Small Bond numbers, the angle-fixing boundary condition

$\partial\alpha_0/\partial h > 0$ within the whole range of R-values. The sign of the function $\partial\alpha_0/\partial R$ depends on the ratio of α_0 and α_1 values:

$\alpha_0 < \alpha_1$. Profile curves are either convex or convexo–concave in this case and $\partial\alpha_0/\partial R > 0$ in the whole range of R-values.

$\alpha_0 > \alpha_1$. Profile curves are either concave or convexo–concave. The most interesting case is $R/r_0 \to 1$. The range of existence of convexo–concave columns is very small (fig. 13) in this case. Therefore concave columns can exist in the main, for which $\partial\alpha_0/\partial R < 0$.

4.4.4. Large Bond numbers, the catching boundary condition

$$\frac{\partial\alpha_0}{\partial R} = \frac{1}{R'_{\alpha_0}} = \left\{\int_0^h \sin\alpha_0\{1 - [\cos\alpha_0 + (h-d)^2 - (z-d)^2]^2\}^{-3/2}\mathrm{d}z\right\}^{-1}, \tag{4.10}$$

$$\frac{\partial\alpha_0}{\partial h} = -\frac{\partial R/\partial h}{\partial R/\partial\alpha_0}. \tag{4.11}$$

Here

$$\partial R/\partial h = -\operatorname{ctg}\alpha_0 + 2\int_0^h (d-h)\{1 - [\cos\alpha_0 + (h-d)^2 - (z-d)^2]^2\}^{-3/2}\,\mathrm{d}z.$$

In this case $\partial\alpha_0/\partial R > 0$. The sign of the derivative $\partial\alpha_0/\partial h$ depends on the sign of $\partial h/\partial R$. Practically, for all the shaping conditions $\partial h/\partial R < 0$, i.e. $\partial\alpha_0/\partial h > 0$. The exception where $\partial h/\partial R > 0$ is characterized by availability of menisci with an ambiguous projection on the r-axis that can exist under positive pressures only.

4.4.5. Large Bond numbers, the angle-fixing boundary condition

Figure 15 shows the crystallization-front $h|_{\alpha_0 = \mathrm{const}}$ height versus the crystal-to-shaper clearance $l_0 = r_0 - R$ for various relations between the growth angle and the wetting angle. The figures illustrate the following.

If $\alpha_1 > \alpha_0$ the following relations are obtained: $A_{RR} < 0$, $A_{Rh} < 0$.

If $\alpha_1 < \alpha_0$, as in the previous case, $A_{Rh} < 0$ and the coefficient A_{RR} can be of either sign. For small clearances ($l_0 < m$) $A_{RR} > 0$, for large clearance ($l_0 > m$) $A_{RR} < 0$.

The results of the analysis of the capillary coefficients A_{RR} and A_{Rh} for all the cases considered are given in table 1.

Analysis of the data of table 1 shows that the stability range is determined by the character of the boundary conditions and the shaper-to-crystal clearance value. The catching boundary condition and a small shaper-to-crystal clearance are preferable. Now it is possible to answer the question where CzT "ends" and TPS "begins". *CzT*

Table 1
The evidence of capillary coefficients as a function of parameters of capillary shaping.

Conditions of crystallization				A_{RR}	A_{Rh}
method[1]	Bond number	boundary condition	parameter values		
CzT				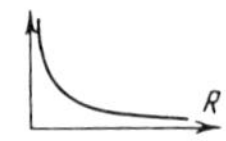	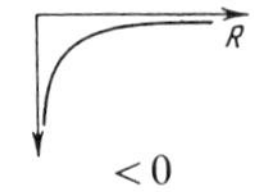
CST	small	catching	$\sin\alpha_0 < 2r_0d$	<0	<0
			$\sin\alpha_0 > 2r_0d$	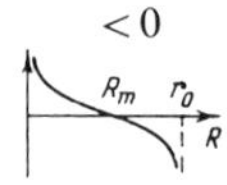	<0
		angle fixing	$\alpha_0 < \alpha_1$	<0	<0
			$\alpha_0 > \alpha_1$	>0	<0
	large	catching	single- and double-valued meniscus	<0	<0
			double-valued mensicus $R \approx r_0$	<0	>0
		angle fixing	$\alpha_0 < \alpha_1$	<0	<0
			$\alpha_0 > \alpha_1, r - R > m$	<0	<0
			$\alpha_0 > \alpha_1, r - R > m$	>0	<0

[1]CzT: Czochralski method; CST: capillary shaping technique.

differs from TPS in the absence of ranges of capillary stability ($A_{RR} > 0$). Hence, in such cases when shaper presence leads to capillary stability, we deal with TPS. Thus, when pulling a circular cylindrical rod with the catching boundary condition satisfied, the range of capillary stability exists only in the case where the crystal-to-shaper dimension ratio $R/r_0 \geq \frac{1}{2}$. This ratio is the condition of CzT-to-TPS transition.

4.5. *The conditions for crystallization stability*

Formulation of the heat transfer problem for TPS does not differ from that for CzT. The only difference is that the one-dimensional heat-conductivity equation can more reasonably be applied to TPS since the crystals grown have smaller cross-sections and the shape of the melt column is closer to the cylindrical one.

Now crystallization process stability of TPS can be investigated by analyzing the conditions that satisfy the inequalities (3.7), (3.8). The capillary coefficients A_{RR} and A_{Rh} are given in table 1. Analysis of the results of the heat flow problem solution, eqs. (3.11)–(3.18), shows that $A_{hR} > 0$ and the conditions that make $A_{hh} < 0$ can be provided.

For *small Bond numbers* in the case where the *catching condition* is satisfied at the shaper edge, interstabilization of the parameters takes place in the system ($A_{Rh}A_{hR} < 0$). If $R > R_m$ is the region of capillary stability ($A_{RR} < 0$), in this case stable growth is observed in the range of $R > R_b$, where R_b is the stability limit calculated from the conditions of eqs. (3.7) and (3.8). In this case $R_b > R_m$ (table 1) for supercooled melts and $R_b \leq R_m$ for overheated ones. In other words, when $A_{hh} < 0$ the range of overall stability is wider than that of capillary stability. With capillary stability available, supercooled-melt stable growth is possible ($A_{hh} > 0$), which allows pulling rate increase due to crystallization heat removal from the front not only through the crystal but also through the supercooled melt.

When the *angle-fixing boundary condition* is satisfied (*small Bond number*) on the shaper wall the availability of capillary stability depends on the growth angle-wetting angle relation. If the shaper is absolutely non-wettable ($\Theta = 180°$, $\alpha_1 = \frac{1}{2}\pi$), capillary stability can exist for small growth angles ($\psi_0 > 0$, $\alpha_0 < \alpha_1 = \frac{1}{2}\pi$). Otherwise, stable growth is possible only for $A_{hh} < 0$ if the inequalities (3.7) and (3.8) hold.

For a *large Bond number* (growing crystals in the shape of a plate or a cylinder of large diameter) using the *catching boundary condition*, capillary stability is always present ($A_{RR} < 0$). Besides, for a double-valued meniscus with $R \approx r_0$ (table 1) interstabilization exists ($A_{Rh}A_{hR} < 0$). In this case stable growth is possible both for $A_{hh} < 0$ and for $A_{hh} > 0$ if the conditions of eqs. (3.7) and (3.8) are satisfied.

For crystals whose parameters lie out of the interval: double-valued meniscus with $R \approx r_0$ (table 1) interstabilization cannot exist and stable growth is possible only from an overheated melt if the condition of eq. (3.8) is satisfied.

When the *angle-fixing boundary condition* is satisfied for a *large Bond number*, capillary stability can exist under the condition $\alpha_0 < \alpha_1$ (the growth angle ψ_0 should be larger than the complement of the wetting angle Θ). With $\alpha_0 > \alpha_1$, capillary stability is possible only for large crystal-to-shaper wall clearances ($r_0 - R > m$). With the clearance and the crystal dimension decreasing ($r_0 - R < m$, $R/a \approx 1$), the coefficient A_{RR} becomes positive and capillary stability is absent.

The possibility of using capillary stability is an important advantage of TPS over the CzT. In this case both overheated- and supercooled-melt stable growth of crystals of constant cross-sections are possible without application of automatic-control systems.

To numerically estimate the degree of stability a quantity inverse to the characteristic time of perturbation relaxation should be introduced.

When stability is present, the roots S of the characteristic equation (2.4) have negative real parts. Then min ($|\text{Re}\, S_1|$, $|\text{Re}\, S_2|$), where $|\text{Re}\, S_1|$ is the real part of the root S_1 of the characteristic equation ($i = 1, 2$), will be regarded as the degree of stability S.

As was mentioned above, the degree of crystallization process stability can be improved and the range of stable growth can be extended by applying forced blowing in the zone above the crystallization front and by increasing the degree of melt overheating as well as by decreasing the pressure d in the meniscus.

From eq. (3.11) it is obvious that the degree of thermal stability (the modulus of the coefficient A_{hh}) is higher for the substances possessing higher thermal conductivity coefficients and lower latent melting heat per unit substance volume.

4.6. *Stationary values of R^0 and h^0 as simultaneous solutions of the capillary and the heat transfer problems*

To find stationary values of the crystal dimension R^0 and the meniscus height h^0, the function $h(R)$ calculated while solving the capillary and the heat flow problems should be simultaneously analyzed. The case of overheated-melt growth of a thin rod-shaper crystal with the catching boundary condition satisfied under zero pressure will be used as an example of this procedure in TPS.

In this case the function $h(R)|_{\alpha_0=\alpha_e}$ calculated while solving the capillary problem is described by analytical expression and is plotted in fig. 18 (curve 1: $r_0 = 6 \times 10^{-2}$ cm, $\psi_0 = 11°$ – typical for Si).

As a result of the solution of the heat flow problem, the function $h(R)$ can be represented in the following way: For small pulling rates heat transfer caused by crystal and melt movement can be neglected. For this purpose the inequality $V^2/4k_i^2 \ll 2\mu_i/\lambda_i R$ should be satisfied in the heat flow problem of section 3.3.2. For $\zeta_S L \gg 1$, the crystal can be considered as an infinitely long one. For $\zeta_L L \ll 1$ the heat removal from the liquid-column lateral surface can be neglected. Under such constraints imposed for the temperature gradients G_L and G_S the following estimates are obtained:

$$G_L \approx \frac{T_p - T_0}{h}, \qquad G_S = -(T_0 - T_e)\sqrt{\frac{2\mu_S}{\lambda_S R}}.$$

Then the equation of crystallization-front heat balance can be written as follows:

$$\sqrt{\frac{\beta}{R}} - \frac{1}{h} = \eta V. \tag{4.12}$$

Here

$$\beta = \frac{2\mu_S \lambda_S^2 (T_0 - T_e)^2}{\lambda_L^2 (T_p - T_0)^2}, \qquad \eta = \frac{\mathscr{L}}{\lambda_L (T_p - T_0)}.$$

Figure 18 shows the functions $h(R)$ calculated from eq. (4.12) (curves 2–6) for five various pulling rates V. To plot the function $h(R)$ from eq. (4.12) the values characteristic of silicon growth were taken, i.e., $\beta = 5 \times 10^4\ \mathrm{m}^{-1}$, $\eta = 10^7\ \mathrm{s\,m}^{-2}$.

The maximum possible crystal dimensions for fixed β and η values correspond to simultaneous solutions of the capillary and heat flow problems with $V = 0$ (point A, fig. 18). Depending on the parameter values, cases of two, one or no simultaneous

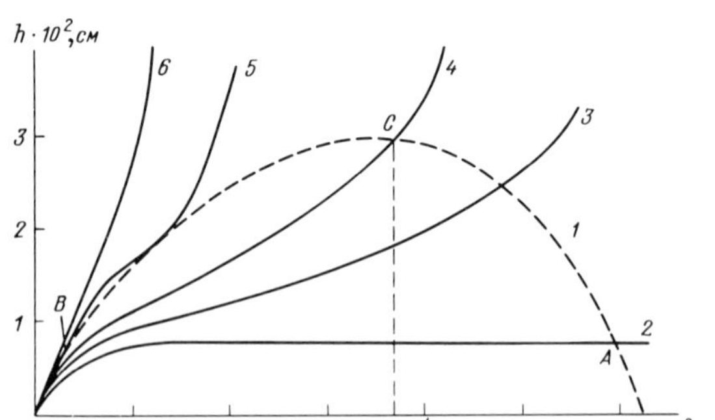

Fig. 18. Crystallization front height versus crystal dimension for cylindrical silicon crystals from (1) the capillary and (2–6) the heat flow problem for various growth rates V_i ($V_{i+1} > V_i$, where $i = 2, 3, \ldots, 6$).

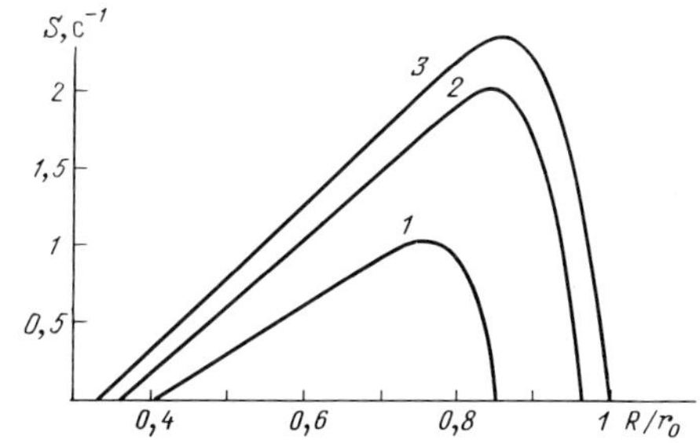

Fig. 19. Stability versus crystal-to-shaper dimension ratio for : (1) Al; (2) Si; (3) Al_2O_3.

solutions exist. In the case that two solutions exist, one of them, corresponding to the smaller radius (point *B*), is always unstable as the inequality (3.8) is violated. Indeed, the inequality (3.8) can be represented in the following form:

$$A_{hh}A_{Rh}\left[\left(\frac{\mathrm{d}h}{\mathrm{d}R}\right)_{\mathrm{T}}-\left(\frac{\mathrm{d}h}{\mathrm{d}R}\right)_{\mathrm{C}}\right]>0. \tag{4.13}$$

Here $(\mathrm{d}h/\mathrm{d}R)_{\mathrm{T}}$ and $(\mathrm{d}h/\mathrm{d}R)_{\mathrm{C}}$ are derivatives of the function $h(R)$ from heat flow and capillary problems, respectively. It is obvious from fig. 18 that for the solution corresponding to the smaller radius $(\mathrm{d}h/\mathrm{d}R)_{\mathrm{T}}-(\mathrm{d}h/\mathrm{d}R)_{\mathrm{C}}<0$ and inequality (3.8) is violated (instability of the saddle-type). With the pulling rate increasing, the simultaneous solutions close in and for some critical pulling rate V_{ex} only one simultaneous solution exists (curve 5, fig. 18) and for $V>V_{\mathrm{ex}}$ simultaneous solutions do not exist (curve 6).

For the simultaneous solution corresponding to the larger radius (point C) the inequality (3.8) always holds. If this R-value lies within the range of capillary stability $(R>R_{\mathrm{m}})$ it is always stable. If $R<R_{\mathrm{m}}$, for the solution to be stable, inequality (3.7) should be satisfied.

Figure 19 shows the degree of stability S versus the value of the equilibrium crystal radius R^0. The right stability limit (the point corresponding to $S=0$ in the range of large radii) satisfies the simultaneous solution with $V=0$. The left stability limit satisfies the condition $A_{RR}+A_{hh}=0$. From fig. 19 it follows that there exists an optimal ratio of the dimension of the crystal being pulled, R^0, to the shaper dimension r_0 (with the values of β and η fixed) when the maximum degree of stability can be achieved.

With the radius of the growing crystal increasing from zero to r_0 (the radius of the shaper), the type of stability changes in the following way: "unstable saddle"–"unstable node"–"unstable focus"–"stable focus"–"stable node" (fig. 4).

4.7. *Influence of melt pressure variations on crystallization stability*

In previous investigations on the stability of TPS the crystal dimension R and the crystallization front height h were used as the variable parameters. Another parameter, the pressure d, is added to them and our system becomes a system of three degrees of freedom. The coefficients of the third equation are found from the mass-balance condition.

Conclusion: Taking account of pressure in the investigation of stability shows [43, 14] that the crystallization process is stable if its stability is ensured by capillary and heat transfer phenomena. Therefore, all the earlier conclusions regarding the influence of capillary and heat flow effects on stability remain valid. Thus, pressure variations without compensating the thermal or capillary instability reinforce inequalities (3.7) and (3.8), i.e. make stable processes even more stable and unstable processes even more unstable.

4.8. *Growth of tubular crystals*

Cross-sections of tubular crystals are binary-connected feasible areas and are characterized by two dimensions: the outer radius of the tube R_1 and the inner one R_2. Thus, in this case a system with three degrees of freedom is obtained, i.e. in eq. (2.1) $n = 3$, $x_1 = R_1$, $x_2 = R_2$, $x_3 = h$ (taking account of pressure gives four degrees of freedom).

The main *conclusions* of the stability analysis [44, 14] are:

When *tubes of large diameters* ($R_1 \gg 1$) are crystallized, a stable solution always exists. Apparently, this situation accounts for the relative simplicity of such tube production as compared with the growth of plates, rods and tubes of small diameters.

When *tubes of small diameters* ($R_1 \ll 1$) are crystallized the process can be stable or unstable. It depends on the scheme adopted for shaping and the values of the process parameters.

4.9. *Impurity distribution*

The present section gives a model of impurity transfer for TPS that allows one to relate the effective coefficient K of impurity distribution to the parameters of capillary shaping [45], because extension of the mechanisms known to thin-profile growth using capillary fed shapes can lead to the wrong conclusions. So application of the Burton equation [46] to calculate K requires specification of the thickness of the diffusion boundary layer. In the case where this thickness is assumed to be equal to the total height of the meniscus and the shaper capillary channel, $K = 1$ for any impurity [47].

Figure 20a illustrates the case of silicon-tape growth. A stationary process is considered; it is assumed that the melt in the zone of the meniscus and the capillary channel is not stirred and the conditions of complete stirring are maintained in the crucible. Under the assumptions made, impurity transfer in the meniscus is described by the equation:

$$D\left(\frac{\mathrm{d}^2C}{\mathrm{d}r^2}+\frac{1}{r}\frac{\mathrm{d}C}{\mathrm{d}r}\right)=-V\frac{\mathrm{d}C}{\mathrm{d}r}, \tag{4.14}$$

with the following boundary conditions at the crystallization front:

$$-D\left.\frac{\mathrm{d}C}{\mathrm{d}r}\right|_{r=r_0}=V_0(1-K_0)C(r_0). \tag{4.15}$$

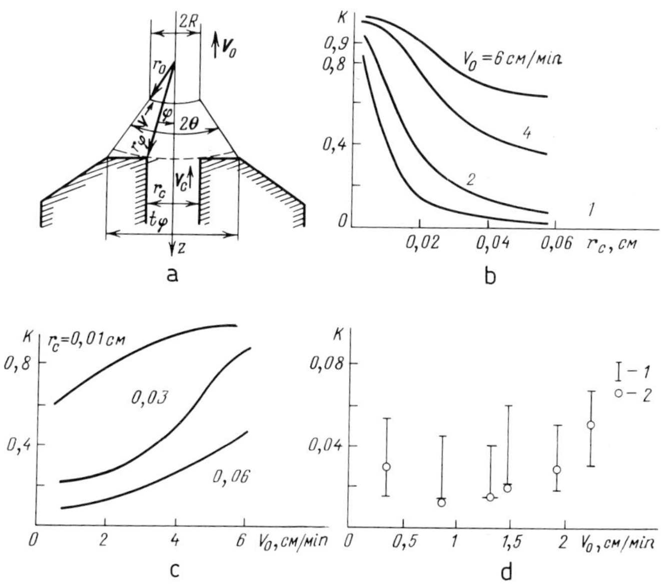

Fig. 20. On calculating the effective coefficient of impurity distribution: (a) diagram; (b) $K = f(r_c)$ – calculated; (c) $K = f(V_0)$ – calculated; (d) comparison of (1) experimental values and (2) values calculated from eq. (4.19) of the effective coefficient of In distribution for thin-walled profiled Si crystal growth.

Here C denotes the impurity concentration in the melt, D is the diffusion coefficient, K_0 is the equilibrium coefficient of impurity distribution; other notation is given in fig. 20a.

The polar coordinate system chosen allows easy specification of the melt-flow rate distribution in the meniscus $V(r) = V_0 r_0 / r$. Then eq. (4.14) is rearranged into the following form:

$$\frac{\mathrm{d}^2 C}{\mathrm{d}r^2} + \frac{1}{r}\left(1 + \frac{V_0 r_0}{D}\right)\frac{\mathrm{d}C}{\mathrm{d}r} = 0. \tag{4.16}$$

Impurity transfer in the capillary channel is described by the equation

$$D\frac{\mathrm{d}^2 C}{\mathrm{d}z^2} = -V_c\frac{\mathrm{d}C}{\mathrm{d}z}, \tag{4.17}$$

with the following boundary conditions:

$$C \to C_\infty, \quad \text{if } z \to \infty. \tag{4.18}$$

It is assumed that the impurity concentration in the bulk of the melt C_∞ is constant in the process of tape growth. The solution of the problem relates the value of the effective coefficient of impurity distribution K to the parameters of the crystallization conditions (V_0 and $2R$), with the conditions of capillary shaping (t_φ, r_C, θ) and the impurity characteristics (D and K_0).

$$K = \frac{K_0}{K_0[1 + (\theta r_C/\varphi t_\varphi - 1)\cos\varphi] + (1 - K_0)(1 - \cos\varphi)(2R/t_\varphi)^{V_0 R/D \sin\theta}}. \tag{4.19}$$

In the case where the width of the capillary channel reaches its maximum value $r_C = t_\varphi$, eq. (4.19) is reduced to

$$K = \frac{K_0}{K_0 + (1 - K_0)(1 - \cos\varphi)(2R/t_\varphi)^{V_0 R/D \sin\theta}}. \tag{4.20}$$

Figure 20b shows $K = f(r_C)$ calculated for aluminum impurities in silicon. The following values are used: $K_0 = 0.002$, $D = 5.3 \times 10^4\,\mathrm{cm^2\,s^{-1}}$, $2R = 0.03\,\mathrm{cm}$, $t_\varphi = 0.06\,\mathrm{cm}$.

Figure 20c shows $K = f(V_0)$ calculated from eq. (4.19) for various r_C values.

It is obvious that the effective coefficient of impurity distribution increases with the growth rate increasing and the melt flow rate in the shaper capillary channel increasing. Under the actual conditions of growing thin-walled profiled silicon crystals $K_{Al} \ll 1$.

Processing of the data of [48] gives a value of $K = 0.39$ for aluminum impurity in silicon, which agrees very well with the value of $K = 0.3$–0.4 calculated from eq. (4.19) in accordance with the initial data of [48]. The results obtained using mass spectrographic analysis of impurity content in tapes grown from refined metallurgical silicon are given in table 2. They confirm that the effective coefficient of distribution of the impurities analyzed is much less than one.

Figure 21 gives data on sulphur distribution along the axis of a silicon tape obtained by laser emission microanalysis (LEM). The ratio of the sulphur spectral line strength to that of silicon defined from the lines of Si ($\lambda = 288.1$ nm, $\lambda = 363.2$ nm) is plotted on the ordinate (the J_{el}/J_{si} ratio is proportional to the concentration of the element analyzed in the silicon matrix). Increase of sulphur concentration in the silicon tape in the process of its pulling shows that $K_S \ll 1$. (It is assumed that for sulphur impurity in

Table 2
Impurity content in silicon precursor and tapes.

Material	Impurity, 10^4 mass%					
	Cu	Mg	B	Ni	Mn	Fe
silicon precursor	1	0.5	1	0.1	1	5
silicon tape	1	5	1	1	1	1

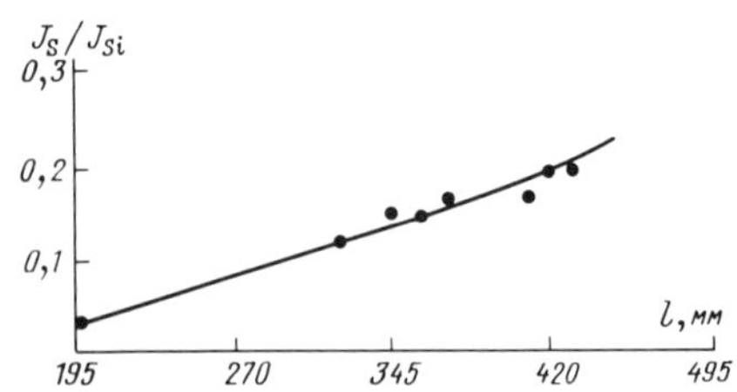

Fig. 21. Sulphur distribution along the length of a silicon tape (LEM data).

silicon $K_0 = 1.0 \times 10^{-5}$.) The rate of tape pulling was equal to 12 mm/min. While analyzing other impurities, no explicit regularities were observed.

To verify the main eq. (4.19), a series of 0.01 mass indium-doped silicon tapes was grown. The mass of each tape-shaped crystal did not exceed 7% of that of silicon charged into the crucible. The tape width $2R$, the shaper transverse dimension t_φ, the capillary slot dimension r_C and the growth rate V_0 were measured. Then photometrical spectral lines of indium in the crystals grown and in the crucible residue were drawn. The measurement accuracy of the effective coefficient of indium distribution was equal to 40–50%.

The results obtained are depicted in fig. 20d. Vertical arrows indicate experimental results, circles show the values calculated from eq. (4.19) in accordance with the above-mentioned parameters. The values of $K_0 = 4.0 \times 10^{-4}$, $D = 5.2 \times 10^{-4}\,\text{cm}^2\,\text{s}^{-1}$ are assumed for indium impurity in silicon.

The results shown in fig. 20d do not allow one to plot K versus the growth rate (V_0) since the thickness of a silicon tape, $2R$, decreases with increasing V_0 (the value of $2V_0R$ practically did not change in the experiments), and they can only demonstrate satisfactory agreement between calculated and experimental values. The values of K calculated from the Burton–Prim–Slichter equation are equal to 0.8–0.9, i.e. they are some orders of magnitude greater than the experimental values.

Impurity distribution across the width of profiled crystals is to a great extent determined by the technique used to feed the melt to the growth meniscus. In fig. 22 resistivity distributions ρ across the width of silicon tapes grown under the conditions of the two versions of meniscus melt replenishment are compared. The following explanation of the observed non-uniformity of impurity distribution across the crystal width (fig. 22b) can be offered: In the case of horizontal melt flow in the meniscus from the shaper edges towards its center, impurities with $K < 1$ driven off by the growing crystal accumulate in the central part of the meniscus. Hence, the corresponding distribution of capillary channels in the shaper permits control of the impurity distribution across the width of the crystal being grown.

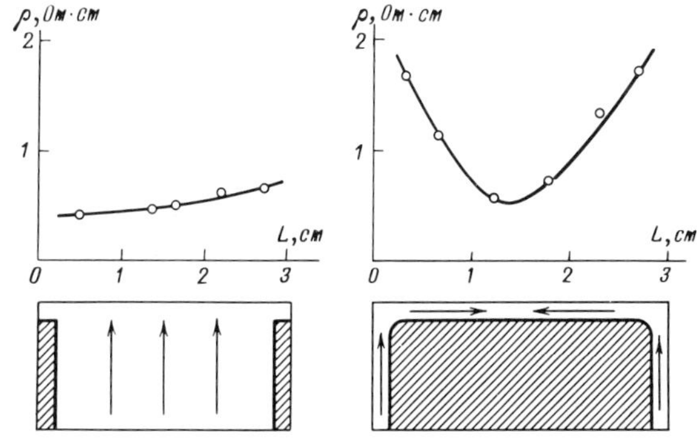

Fig. 22. Resistivity distribution across the width of silicon tapes grown using two different versions of liquid-phase replenishment of the meniscus.

4.10. TPS growth of sapphire crystals

A possibility of sapphire shaped-crystal growth by the different variants of CST are widely demonstrated in this chapter. The main aim of this section is to describe the most significant characteristics of the defect structure of sapphire tubes manufactured by TPS. We used tube samples which enabled us to characterize most clearly the defects formed during the growth process [49]. In the experiments which were carried out, the growth of single sapphire tubes with diameters of 4 to 40 mm and wall thicknesses in the range of 0.5 to 3 mm was investigated.

4.10.1. Morphological features of sapphire tubes

The geometrical form of a tube can differ slightly from the ideal for reasons of a crystallographic nature. According to the Curie theorem, in the process of growth, there is an interaction between the crystal symmetry and that of the medium (i.e. the thermal environment) in which the crystal grows. In the grown crystal, only those elements of symmetry are exhibited which are common both to the crystal and to the medium. That is the reason why grown crystals are faceted depending on the orientation. If a sapphire tube is pulled in the directions $\langle 10\bar{1}0\rangle$, $\langle 11\bar{2}0\rangle$ or intermediate ones with $\rho = 90°$ (where ρ is the angle between the c-axis and the pulling direction), the close-packed basal plane (0001) becomes parallel to the growth direction and facets the tube (fig. 23a). If ρ differs slightly from 90°, a step-like faceting by the basal plane appears on the lateral surface of the tube (fig. 23b). Tubes grown in the [0001] direction are faceted on the outside by $\{11\bar{2}0\}$ planes and on the inside by $\{10\bar{1}0\}$ planes (fig. 23c). Faceting is reduced if the temperature gradient at the crystallization front is increased.

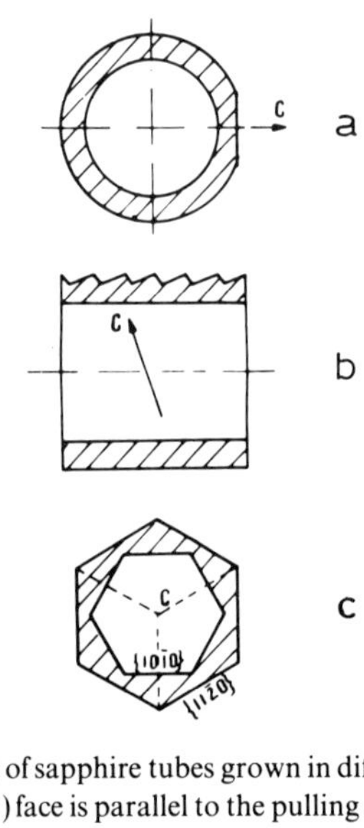

Fig. 23. Faceting of the lateral surface of sapphire tubes grown in different orientations: (a) The cross-section of a tube grown with $\rho = 90°$; the (0001) face is parallel to the pulling direction. (b) The longitudinal section of a tube grown with $\rho < 90°$; a step-like faceting occurs by the (0001) face. (c) The cross-section of a tube grown with $\rho = 0°$; the faceting on the outside occurs by the $\{11\bar{2}0\}$ faces and on the inside by the $\{10\bar{1}0\}$ faces.

4.10.2. Voids

Voids are one of the major defects in shaped sapphire crystals. Among the causes producing voids in sapphire grown from the melt, one can note the thermal dissociation of the melt, the liberation of dilute gases, and the shrinkage of the melt during crystallization. Although the problem is not completely understood at this time, the thermal dissociation of the melt in contact with the crucible material seems to be the most probable cause of the voids.

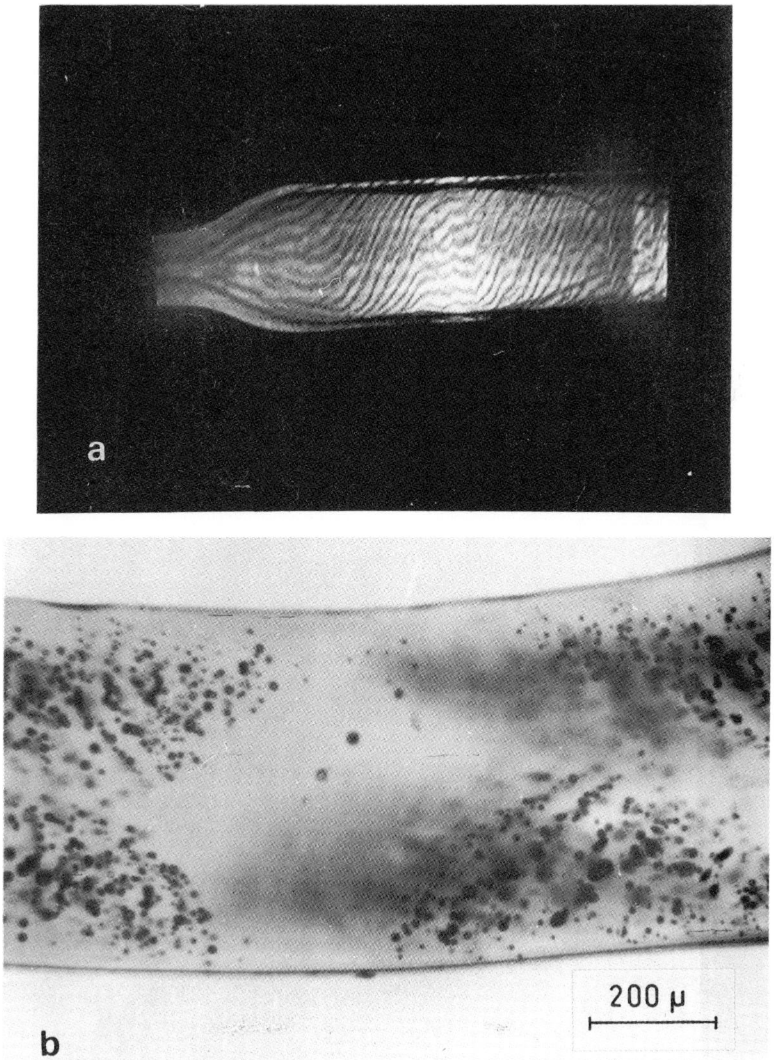

Fig. 24. Tree-like void distribution in a sapphire tube grown at low rates (< 1 mm/min): (a) general view of the tube; (b) void distribution over the cross-section of the tube.

Void distribution in the crystal is not uniform. At low crystallization rates (< 1 mm min^{-1}) the voids are absent. The void distribution is at times in the shape of a tree-like pile-up with a branch diameter of up to 1 mm. Figure 24a shows an example of this tree-like void distribution in a tube. The distribution of voids over the tube section normal to the growth direction is shown in fig. 24b. The characteristic feature of the void distribution over the cross section at growth rates of 1 to 3 mm min^{-1} is the

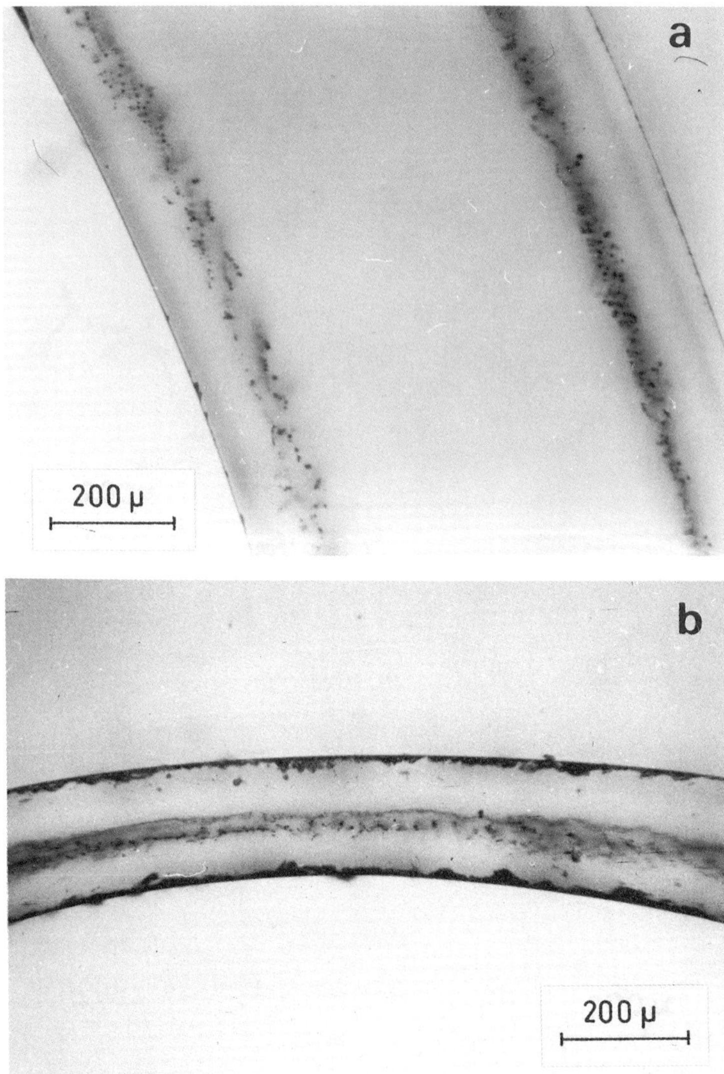

Fig. 25. The void distribution over the cross-section at growth rates of 1–3 mm/min.

presence of a 50 to 200 μm wide void-free zone near the crystal surface and a large void-free region in the center (fig. 25a). As the growth rate increases, the near-surface layer remains free of voids, but the rest of the cross section may contain voids (fig. 25b). These voids are located primarily in planes making an angle $\leq 90°$ with the crystallization front. As the crystallization rate increases further, this mode of void distribution remains in the center of the crystal, but the void-free zones near the surface disappear (fig. 26).

The above features of void distribution in shaped sapphire crystals are not strictly connected only to the growth rate. The void distribution over the crystal cross section is determined by the shape of the crystallization front. The latter, in turn, depends on whether or not the conditions determining stability or instability of a plane crystal–melt interface are satisifed.

It is shown [16] that, in the stable state there is a critical frequency ω_c for which the system exhibits the smallest resistance to an instability; this enables one to obtain the size of cells of an instability front:

$$\lambda = 2\pi/\omega_c = 2\pi(T_m\gamma/LG)^{1/2},$$

where T_m is the melting temperature, γ the specific interface free energy, L the heat of fusion and G the generalized temperature gradient. For real crystallization conditions of sapphire: $T_m = 2320$ K, $\gamma \approx 300$ erg/cm, $L = 3.4$ erg/cm^3, $G = 5$–50 K/cm, and the size of the cells is approximately 40–150 μm.

In [50] it has been shown that the radius of curvature of the interface is, to a first approximation, the perturbation amplitude of the perturbed surface.

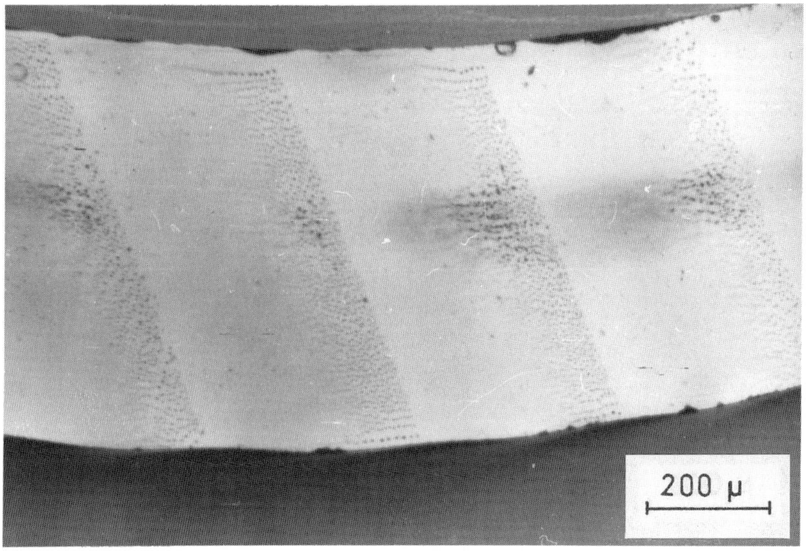

Fig. 26. The cross-section of a tube pulled at a rate greater than 5 mm/min.

The following explanation is proposed, therefore, for the void distribution observed in shaped sapphire crystals. At very low crystallization rates, the interphase boundary is plane and voids are not trapped. In [51] the critical growth rate has been calculated below which a foreign particle is not trapped by the grown crystal. The critical growth rate for sapphire is approximately 0.5 mm/min.

The plane interface may become unstable in some regions; convective currents in the melt may stimulate this process. Macroscopic concavities, with a diameter of up to 1 mm, arise on the crystallization front; these concavities are characterized by an enhanced trapping of voids. We suppose that the change of position of the concavities on the crystallization front with time depends on the pattern of the convective flows in the melt which leads to the void distribution shown in fig. 24a. This process is in accordance with [16] for the lowest frequencies.

When growing shaped sapphire crystals at rates of 1 to 3 mm min^{-1}, the interface has, as a rule, a plane central part and convex peripheral sections. In this case, the curvature of the crystallization front plays the role of a perturbation, and conditions of instability occur primarily at regions of maximum curvature at a distance $\lambda_c = 2\pi/\omega_c$ from the interface edge, where ω_c is the critical frequency. If the crystal cross-sectional dimension $d > 2\lambda_c$, then two bands are observed in the void pattern corresponding to the regions where instability occurs. If the cross-sectional dimension is small ($d \leq 2\lambda_c$), then the interface does not contain the plane part; the instability occurs in the central part of the cross-section and the void distribution is of the type shown in fig. 25b. Figure 27 shows the decanted crystallization front obtained at the same growth rates. The unstable region of the interface is the system of hollows and crests which is faceted by close-packed faces as the instability proceeds. On occasion, smaller cells can be

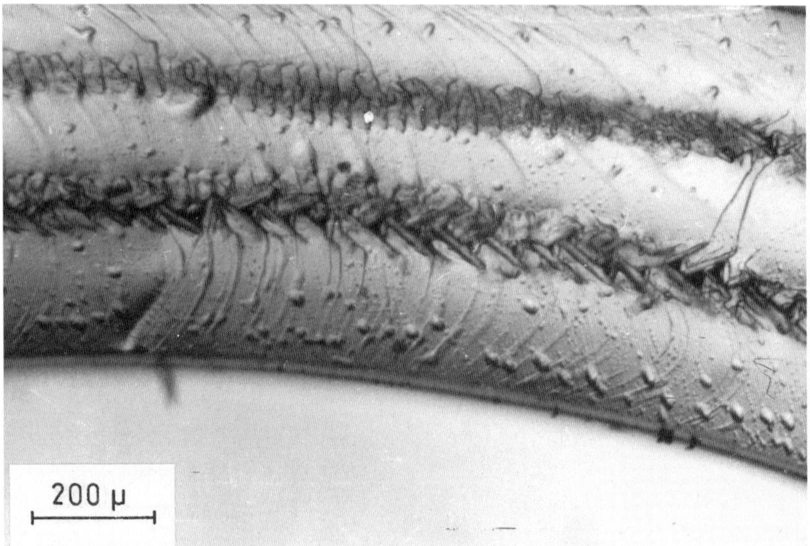

Fig. 27. The decanted crystallization front of a tube grown at rates of 1–3 mm/min.

observed at the interface (fig. 28); these probably correspond to perturbations of higher harmonics.

As the growth rate is increased further, the entire solid–liquid interface becomes unstable (fig. 26). The crystallization front becomes faceted, with the new interface shape becoming stabilized by the faceting. Figure 29 shows the decanted crystallization front of a tube grown in the [0001] direction; the front is faceted by rhombohedral planes, and the void distribution in the crystal corresponds to the positions of hollows on the front. If the crystallization front is completely faceted, the void distribution in the

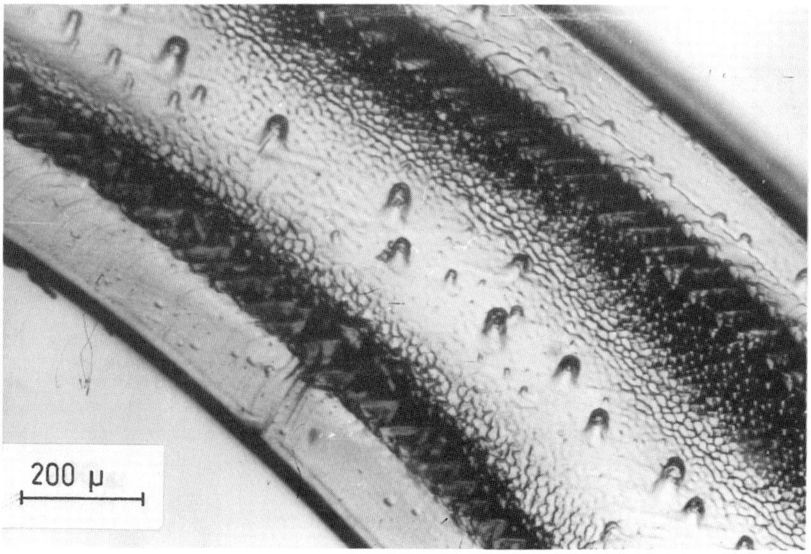

Fig. 28. The crystallization front of a tube grown at 1–3 mm/min showing less commonly observed fine cell structure.

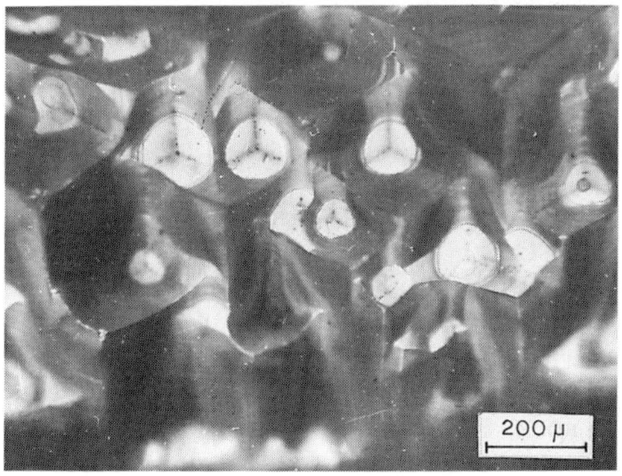

Fig. 29. The crystallization front of a tube grown in the [0001] direction showing rhombohedral faceting.

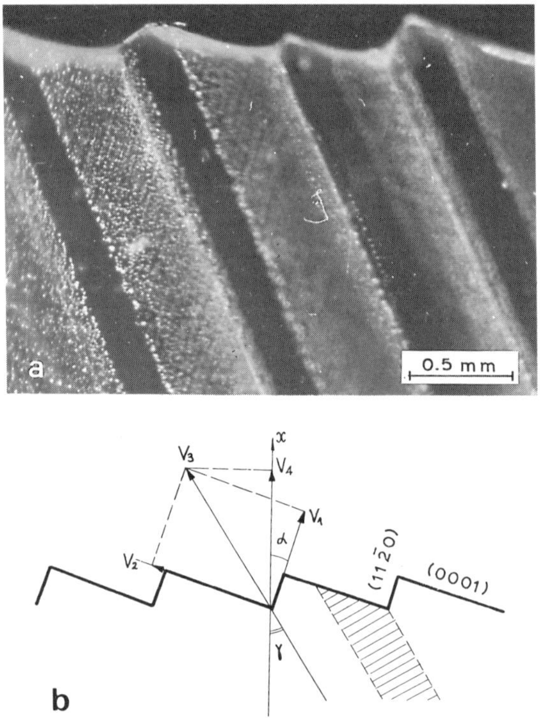

Fig. 30. (a) The decanted crystallization front of a tube pulled at a rate greater than 5 mm/min. (b) Schematic of a faceted crystallization front in (a) used to calculate the ratio of crystallization rates of facets.

crystal may reflect the relationship between the growth rates of the faceting planes. The decanted crystallization front of a tube faceted by the $\{11\bar{2}0\}$ and $\{10\bar{1}1\}$ planes is shown in fig. 30a. From geometrical considerations (fig. 30b), it is apparent that the angle γ between the crystal pulling direction and the direction of the band of void pile-up is given by $\operatorname{tg}\gamma = V_3/V_4$, where $V_3 = V_1 + V_2$ is the vector sum of growth rates of faces 1 and 2, and V_4 is the pulling rate. Knowing the direction of the void pile-up band, one can estimate the ratio of crystallization rates of faces 1 and 2.

$$V_2/V_1 = \sin(\alpha + \gamma)/\cos(\alpha + \gamma),$$

where α is the angle between the normal to face 1 and the pulling direction. For the case of fig. 30, we have

$$V_2/V_1 = V_{\{11\bar{2}0\}}/V_{\{10\bar{1}1\}} = 0.965.$$

The actual growth rates of the facets can be calculated from the known pulling rate. It is noteworthy that the rhombohedral planes are most frequently found in interface faceting.

4.10.3. *Low-angle grain boundaries*

The dislocation defect structure of shaped sapphire crystals was studied both by the optical polarization method and by the technique of widely diverging X-ray beams. It

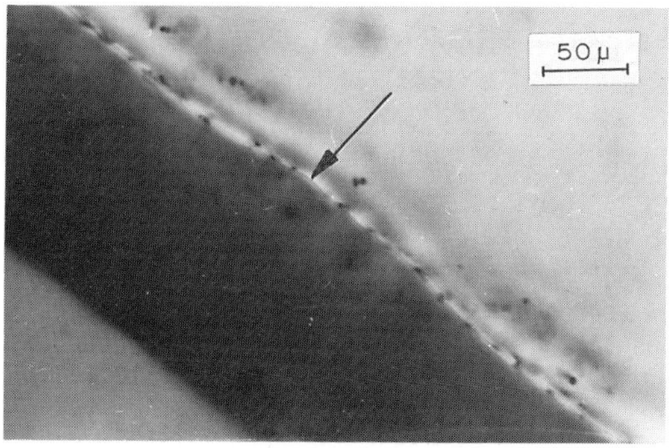

Fig. 31. The cross-section of a tube with a subgrain boundary passing along the void pile-up plane (shown by arrow).

has been established that, in the absence of low-angle grain boundaries propagating from the seed, the initial part of the crystal does not contain subgrain (or low-angle grain) boundaries. Then, as the crystal grows, the dislocation density increases and subgrain boundaries are formed. The misorientation of adjacent subgrains increases with distance from the seed, reaches a certain limit, and then decreases slightly with further growth. The decrease is probably associated with a 'rearrangement of the subgrain structure; i.e. subgrains with large misorientations branch into a series of subgrain boundaries with smaller misorientations. In addition, there is a decrease in the density of subgrain boundaries which are at large angles to the growth direction. The subgrain boundaries which remain are approximately parallel to the growth direction; the density of these boudaries (with misorientations as high as 5° to 10°) stays constant.

It should be noted that, at high crystallization rates, crystals which are free of low-angle grain boundaries can sometimes be grown. This is probably explained by the fact that the time spent by the growing crystal in the plastic zone is not long enough for polygonization processes to occur.

The presence of voids is a source of additional stresses which gives rise to dislocation generation and boundary formation. Subgrain boundaries are frequently observed to form along planes of void pile-up, as shown in fig. 31.

Thus, the defect structure in sapphire tube crystals is related to the distortion of the planar solid–liquid interface, which, in turn, is controlled by the growth conditions.

4.11. TPS growth of silicon crystals

4.11.1 Silicon tapes

These were grown from a graphite shaper using equipment with both resistance furnaces and induction heating [52]. In the former case the hot zone was made to the shape of a rectangle so that it should follow the shape of the pulled crystal cross-section as closely as possible. This hot zone was used to grow silicon tapes up to 40 mm wide. The hot zone of the induction heater was circular in shape. This zone was used to grow silicon tapes 25 mm wide. In both versions, tapes were grown in an argon flow. Six percent of hydrogen was added to the argon used in the equipment with induction heating used to avoid break-downs.

Table 3
The parameters characterizing stability of silicon tape growth.

Parameter	Meniscus flat section	Meniscus end
$A_{RR}\,(s^{-1})$	−0.04	−0.10
$A_{Rh}\,(s^{-1})$	−0.07	−0.19
$A_{hR}\,(s^{-1})$	3.37	2.17
$A_{hh}\,(s^{-1})$	−0.46	−0.42
$(A_{RR}+A_{hh})\,(s^{-1})$	−0.51	−0.51
$\tau_{p\phi}\,(s)$	4.0	3.8

Table 3 gives the coefficients A_{ik} characterizing stability and the combinations $A_{RR}+A_{hh}$ and the time of relaxation of crystallization front $\tau_{p\phi}$ for meniscus end faces and midparts for the case of a silicon tape 0.6 mm grown at a rate of 2 cm/min. It can be seen that the conditions of process self-stabilization hold, the relaxation time for the perturbations δR and δh is equal to 4 s. Thus, stability exists and practical problems of thin-walled silicon tape grown using a shaper are caused by the small meniscus height and by chemical interaction between the melt, the shaper and the environment.

To stabilize the width of the tape being pulled, its ends could be blown with gas flows through special nozzles. Independent flow control allows effective control of tape widening to the width specified.

4.11.2. Silicon tube growth

A circular hot zone with resistance heating was used to grow silicon tubes [53]. They were grown in an argon flow at a rate of up to 6 cm/min. While growing silicon tubes by the semi-continuous technique, the crystal is withdrawn into the air through a special outlet device preventing ingress of air into the chamber. A contactless water-cooled refrigerator can serve as an outlet device, which allows additional increase of process output. Now the semi-continuous technique is used to grow silicon tubes with diameters of up to 12 mm.

Depending on the field of application, silicon profiles of various wall thicknesses are required. The wall thickness of a growing tube depends on the shaper design and on the process parameters. However, there exist principal restrictions that limit one's ability to grow crystals with a specified wall thickness using rates not exceeding the maximum growth rate. For example, a tube with the wall thickness of 0.3 mm could be grown with a growth rate of up to 32 mm/min. Tubes with diameters of 50 mm had walls 3 mm thick in some particular cases (fig. 32).

Overall stability of the process of silicon tube growth is so high that several tubes can simultaneously grow from shapers disposed in the crucible. Seeds are also fixed independently in accordance with the shaper arrangement.

4.11.3. Growth of thin-walled polyhedral prism-shaped crystals

A thin-walled polyhedron-shaped crystal can be considered as a number of tapes (depending on the number of its faces) connected to each other. The procedure for

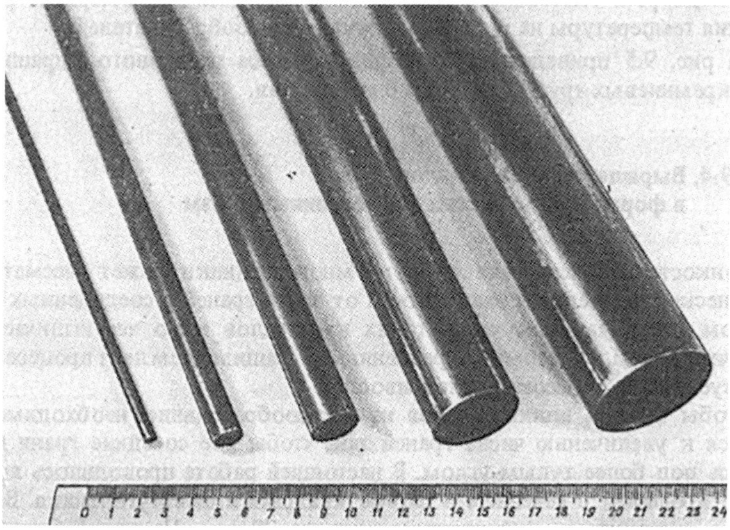

Fig. 32. Specimens of silicon tubes grown by TPS.

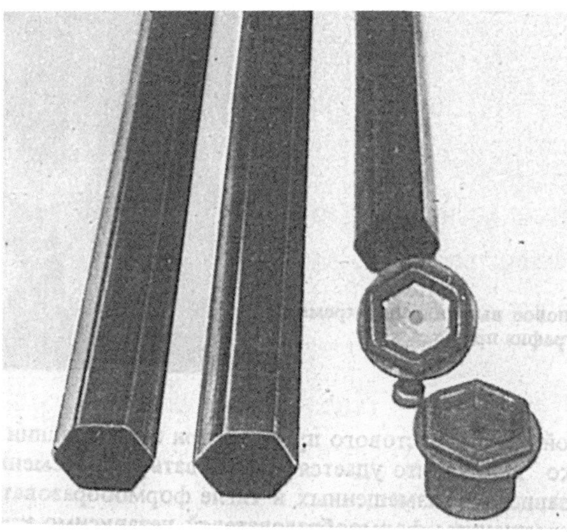

Fig. 33. Hexahedral prism-shaped silicon profiles and shapers to grow them.

growing such crystals differs little from that of growing tubes; therefore the specific feature of the given process is its higher stability as compared with tape growth.

To reduce the influence of the angle on shaping, it is necessary to try to increase the number of faces so that adjacent faces should intersect at more obtuse angles. Figure 33 gives a photograph of hexahedrons and shapers used to grow these crystals.

4.11.4. Profiled silicon structure

Profiled silicon crystals possess characteristic defects in their crystalline structure including planar boundaries (most of those of the twinning type), dislocations and pile-ups thereof, as well as SiC particles and particle aggregations. As a result of twinning along intersecting planes, which can be observed at the initial stage of growth, a stable and quasi-equilibrium structure characterized by the existence of defect areas (planar boundaries) perpendicular to the tape plane and parallel to the direction of growth [54, 55]. In this case the silicon tape surface orientation is $\{110\}$ and the direction of pulling coincides with the axis $\langle 211 \rangle$. Such orientation is formed and maintained irrespective of the seed orientation and can be attained directly at the seed-crystal contact point when the seed orientation is $\{110\}$ $\langle 211 \rangle$ or a section of the silicon tape with a stable structure is used as a seed. Figure 34 shows a photograph of the surface of a silicon tape of a stable structure: SiC inclusions surrounded with dislocation pile-ups can be observed on the tape surface.

On the basis of X-ray structure analysis it was established that the orientation of silicon tape surfaces can vary within a range of 15° from $\{110\}$. Considerable dimensional fragments of defect structures identical to those of tapes are observed in silicon tubes (fig. 35a). The tube surface orientation in the vicinity of such a fragment can differ

Fig. 34. Defect structure of a silicon tape. A SiC inclusion can be observed in its center (a pattern of selective chemical etching; × 120; H.p.: the direction of crystal growth).

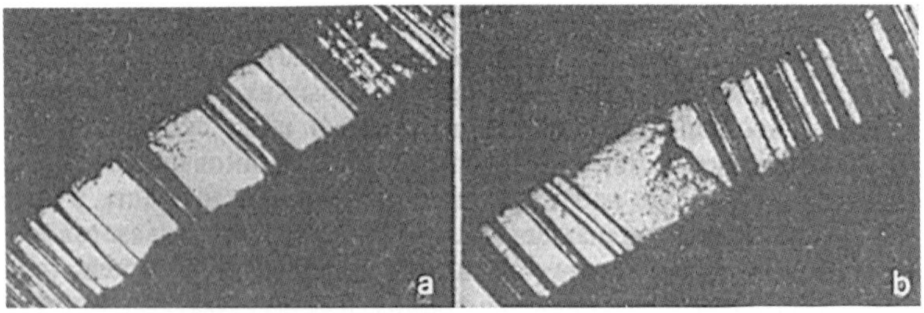

Fig. 35. Defect structures of silicon tubes (cross-sectional fragments): (a) typical structure; (b) defect area providing rotation of the system of twin-boundary planes. (× 200; pattern of selective chemical etching).

from {110} due to surface curvature. This change in orientation does not cause additional structure defects up to values equal to 15° and as soon as that value reaches 15° the surface orientation abruptly changes because there appears "a severely defective spot" that provides an indispensable turn of the system of twin-boundary planes in such a way that the boundaries should again be approximately perpendicular to the tube surface. Either grain boundaries of a general type or dislocation pile-ups can act as such defect areas (fig. 35b).

Apparently, the number of "severely defective spots" and, correspondingly, the number of fragments of steady structures is determined by the necessity of closing the system of fragments into a cylinder. Since the angular magnitude of an arc of 30° corresponds to the variation of the tube surface orientation of 15° from {110}, the overall number of defect spots is equal to at least twelve, which has been proved experimentally. It should also be noted that this linear length of "severely defective spots" along the perimeter of a tube depends but little on its diameter; therefore the relation of the volume of "severely defective spots" to the crystal volume should quickly decrease with increasing tube diameter.

Defect structures of silicon tapes were studied by the TEM technique [56]. With a ~ 200 magnification, the defect structure pattern proved to be similar to that of selective etching. It was observed that the width of defective areas is equal to 3–5 μm and the width of monocrystalline regions between them varies from 10 to 500 μm. Interpretation of the fine structures of the defect areas, that look like dark lines parallel to the direction of pulling in the pattern of selective chemical etching (fig. 34), revealed that each area represents a set of microtwins 40–200 Å wide. Small-angle disorientation of monocrystalline sections adjacent to the defect area is caused by such a set of microtwins. Disorientation measured by the kikuchi-line technique varied from 40° to 4°. Beside the defects, areas of a twinning nature, multilayer lattice defects, small-angle dislocation boundaries and separate dislocations and inclusions were observed in 2% of the cases.

To interpret the defect structure and macroscopic pattern of profiled silicon crystal growth, a model according to which the crystallization front tends to be shaped by the most slowly growing crystal faces {111} while high rates are provided by the availability of acute angles with their vertexes coinciding with twin boundaries was offered in [57] (fig. 36).

To check the proposed model, experiments to investigate the crystallization front shape visualized by impulse changes in the crystal pulling rate were carried out. In this case local changes in crystal thickness follow the changes in crystallization front shape, as does the horizontal hatching observed on the surface of profiled silicon crystals (fig. 37).

Crystallization front shaping by {111} faces proved to be observed only in the vicinity of high-energy boundaries of a general type and at twin boundaries of higher orders, e.g. {111}–{115}. Crystallization front deflection from the planar state does not exceed 2 μm in the coherent-twin region. The overall crystallization front area made by {111} faces is not large; the major part of the crystallization front follows the crystallization isotherm and corresponds to a face close to (112) and in separate cases to (110) (fig. 36) each of which can grow according to the normal mechanism, i.e. both

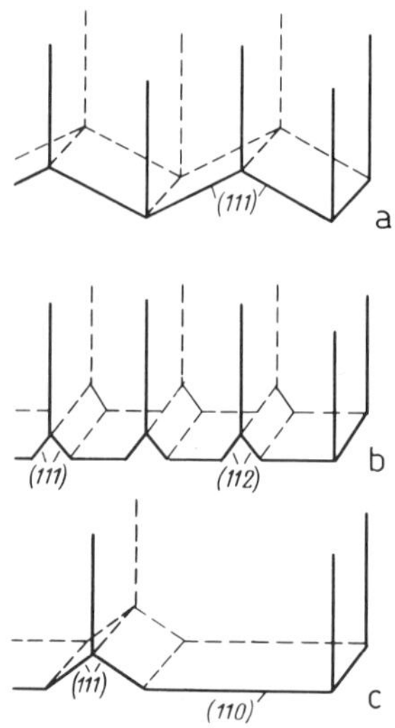

Fig. 36. (a) Crystallization front intersected by planes (111) in accordance with the model of [57]; (b, c) experimentally observed crystallization front shapes.

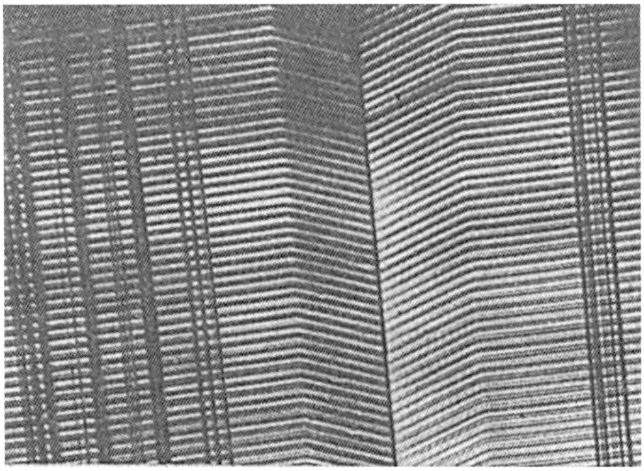

Fig. 37. Agreement between the crystallization front shape and the horizontal hatching on the surface of profiled crystals.

layer-by-layer and normal mechanisms of growth take place when profiled crystals grow.

The specific feathers the crystallization front revealed might be associated with the influence of the impurity on the face free energy or on the growth kinetics. Therefore, it can be assumed that front shaping and perhaps inlet angle formation at twin boundaries are not necessary for high rates of profiled silicon crystal growth to be put into practice. It can be confirmed by the fact that in separate cases silicon tapes with wide (up to 10–15 mm) monocrystalline areas can be grown.

It was noticed that, in the process of silicon tube growth, the melt meniscus height strongly influences the character of defect formation in crystals. With the crystallization front in a tube sinking, the number of boundaries of a general type increases, i.e. profile defect structure depends on the interaction efficiency in the crystallization-front–shaper system. The influence of the shaper is mainly determined by the distortions (thermal and capillary) introduced by the silicon carbide layer formed on the operating free edges of a graphite shaper when it interacts with the silicon melt.

Besides such indirect influence on the structure of profiles, silicon carbide entering the crystal sub-surface layer in the form of SiC particles greatly affects the quality of the material produced.

SiC microcrystals measuring 25–30 μm produced in the melt meniscus form a carbide layer growing on the shaper free edges. When bulges appear on the layer, the meniscus becomes distorted and on the crystal surface there appear distinctive furrows and, in separate cases, bulges along the direction of pulling that break when SiC particles and aggregations thereof result from random fluctuations of the crystallization front when particles reaching the size of the meniscus escape the carbide layer [58]. The number and the size of simultaneously entrapped particles depends on the amplitude of front fluctuations and on the mean height of the meniscus.

The mechanism of carbide inclusion entrapment sharply decreases for a high meniscus since the number of particles whose dimensions reach the meniscus height decreases.

The number of carbide inclusions in profiled silicon crystals can be decreased in several ways: by using crucibles made of graphite of high density; by replacing the graphite container for crucibles with a container coated with a protective layer; by maintaining a high melt meniscus (easy to secure when growing closed profiles). Besides, experiments on alloying profiled silicon with rare-earth elements showed that no SiC inclusions can be found on profile surfaces for any meniscus height for the case where at least 0.05 mass% of gadolinium dopant is introduced into the silicon melt.

4.12. TPS growth of lithium niobate crystals [59]

4.12.1. Technique description for growing shaped $LiNbO_3$ crystals

Shaped $LiNbO_3$ crystals were grown in a modified CzT crystal grower provided with an induction heater (8000 Hz). A special thermal zone was created (fig. 38). An important feature of this zone is that the shaper can be adjusted with respect to the seed crystal, thus enabling one to grow crystals of desired orientation. The crystals were grown in air from platinum crucibles of various dimensions and shapes. The shapers for growing the shaped $LiNbO_3$ crystals were also manufactured from platinum and placed directly into the crucible.

$LiNbO_3$ powder of a congruent-melt composition or CzT-grown crystal scrap was used as a charge. The growth rate of the shaped crystals was 5 to 60 mm/h.

In order to reduce thermoelastic stresses that result from growth and cooling, a platinum shield was placed in the thermal zone. This shield was heated by the same

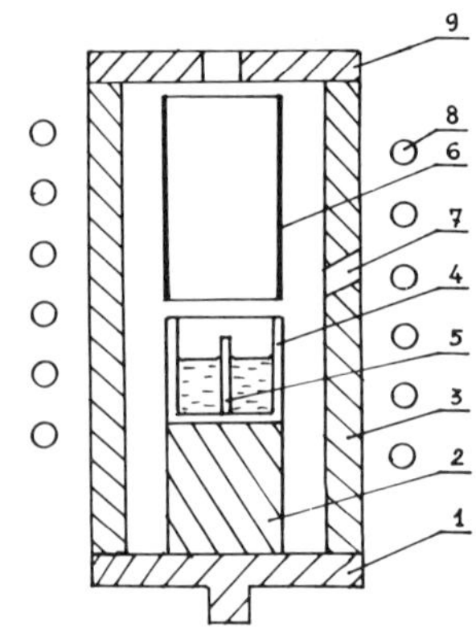

Fig. 38. Thermal zone for growing shaped lithium niobate crystals: (1) table; (2) ceramic support; (3) ceramic tube; (4) platinum crucible; (5) platinum shaper; (6) platinum shield; (7) sapphire window; (8) inductor; (9) ceramic cover.

inductor as the crucible. In addition, passive shields of sintered aluminum oxide tubes were used. The crystallization process at the shaper end surface was monitored by means of a long-focal-length microscope through small openings in the tubes. These openings in the thermal zone were later closed by a sapphire window because of undesirable convective gas flows.

The crystals were cooled directly in the thermal zone in two stages in the following way: (1) From the melting point to 500°C, the temperature in the zone was decreased linearly at a rate of 50 to 250°C/h by means of a programmed controller. (2) The final

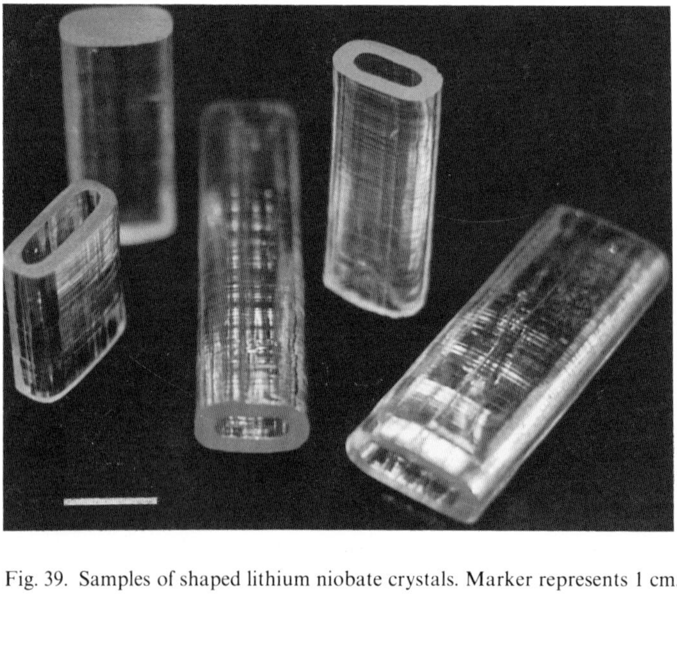

Fig. 39. Samples of shaped lithium niobate crystals. Marker represents 1 cm.

Fig. 40. Lithium niobate ribbon crystals grown simultaneously. Marker represent 1 cm.

cooling from 500°C down to room temperature was obtained by turning off the RF generator and the cooling water.

The overall cooling process took 6 to 12 h depending on the crystal dimensions.

The above technique was employed to obtain ribbons, tubes and other profiles up to 150 mm long (fig. 39). A simultaneous growth of several $LiNbO_3$ ribbons was successfully performed (fig. 40).

4.12.2. *Control of domain sizes*

The presence of a spontaneous dipole moment causes the formation of a domain structure of non-polarized ferroelectric $LiNbO_3$ crystals. Because the point group symmetry is 3 m with a special direction C, the domains are 180°. There are several techniques to obtain mono-domain $LiNbO_3$ crystals. CzT-grown $LiNbO_3$ crystals are normally polarized in special annealing furnaces. In contrast, shaped lithium niobate crystals are polarized directly in the process of growth.

The polarization technique for shaped $LiNbO_3$ crystals is demonstrated in fig. 41. The system for obtaining a single domain was turned on immediately after the crystal was enlarged to the desirable dimensions. At the end of the process, in order to prevent dielectric breakdown, the crystal neck was reduced to 2 to 3 mm in cross-section, after which pulling was stopped. The polarization system was turned off as the shaper temperature was decreased below the Curie temperature, $T_c = 100$°C. The magnitude of the current density was 1–2.5 mA/cm^2, the polarity being (+) for the seed and (−) for the crucible.

It has been shown [60] that if, in the process of the CzT growth of $LiNbO_3$, the direction of the applied electrical field is reversed, 180° domains are formed. Analogous experiments were carried out during growth of shaped $LiNbO_3$ crystals. It should be noted that, in this case, the sign-variable electric field can be applied to the crystal not

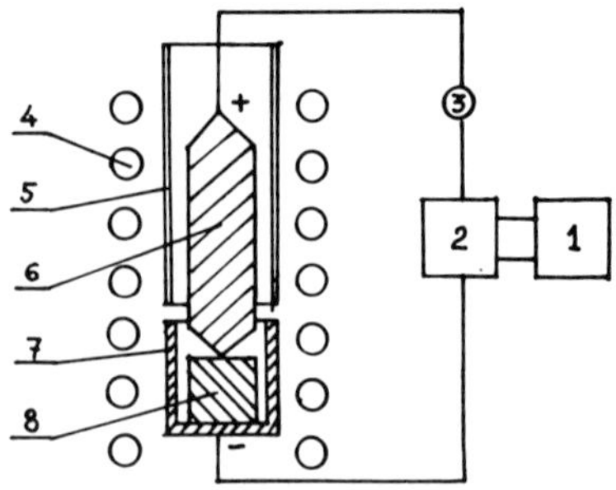

Fig. 41. Process of polarization of shaped $LiNbO_3$ crystals: (1) power supply; (2) current stabilizer; (3) milliammeter; (4) inductor; (5) platinum shield; (6) crystal; (7) platinum crucible; (8) shaper.

only in the growth process but during cooling was well; i.e. one can control the domain structure in the part of the crystal which has the temperature above T_c immediately after pulling is stopped. By varying the growth rate, cooling rate and the period of the applied sign-variable electric field, we were able to obtain domains from several cm to several tens of μm and smaller in size. Figure 42 demonstrates one of the domain structures obtained.

The shape of the domains is related to the Curie isotherm, therefore it is possible to select optimal thermal conditions for the crystallization process.

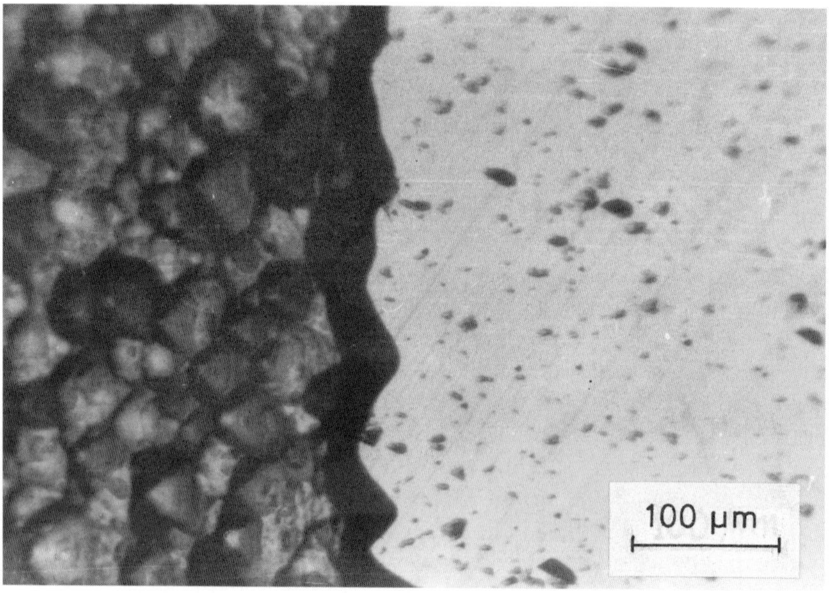

Fig. 42. Domain structure formed in a sign-variable electric field during the crystal cooling.

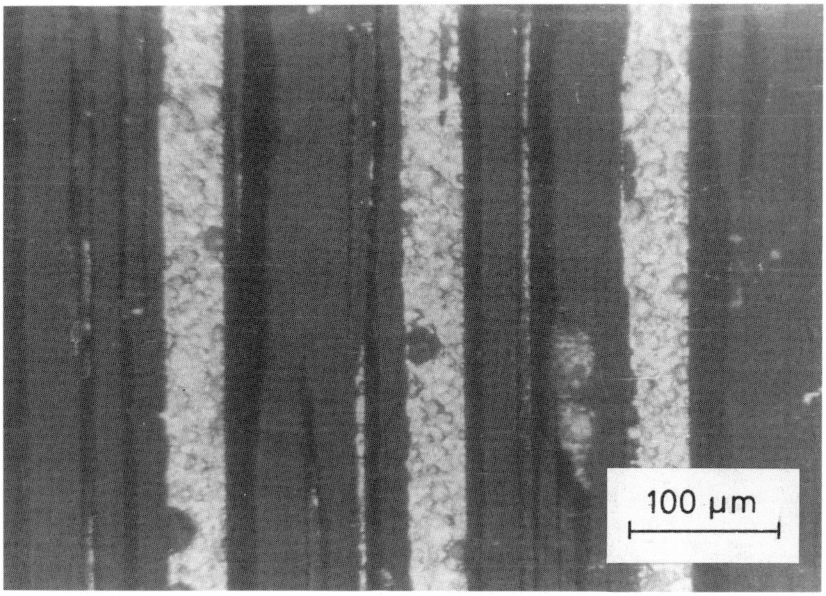

Fig. 43. Domain structure formed at a discontinuous alteration of the pulling rate.

4.12.3. Regular domain structure in growth from doped melts
A periodic domain structure can be formed in lithium niobate crystals grown by the CzT from the melt with scandium, indium and yttrium as dopants [61]. Shaped lithium niobate ribbon crystals were doped with 1 wt% $LiTaO_3$. The growth direction coincided with the [0001] direction, the $(10\bar{1}0)$ plane being the ribbon plane.

The impurity distribution in the crystals and hence their domain structure was controlled by varying the pulling rate. Under these growth conditions, there occurs a periodic alteration of the shaped crystal cross-section, which leads to a change of the effective coefficient of the impurity distribution. Thus, domains measuring 200 to 50 μm were obtained (fig. 43). In as much as the shape of the domains reproduced that of the shaper end surfaces, we obtained horizontal, spherical, conical as well as some other shapes of domain structure by varying the shaper geometry.

5. *The realization of dynamic stability in the Verneuil technique (VT)*

Figure 44 shows a diagram of crystal growth by VT. Fine powder of the material to be crystallized is put into the burner flame. The powder melts partially or completely, forming a melted layer on the surface of the seed crystal. The crystal grows in the process of consecutive melt crystallization as a result of lowering the crystal into the colder zone.

In the classical version of VT, gaseous hydrogen and oxygen are fed into the burner. Later versions using plasma, radiation heating, etc. appeared. However, they did not change the principal scheme.

At the beginning of the century this technique was exclusively used to grow precious crystals from refractory oxides. Considerable changes of the equipment resulted from its application to growing ruby single crystals for laser elements. However, no ideas about optimal temperature distribution for this technique could be found in the literature, only very general calculations according to simplified schemes had been made [62]. Attention was turned to the fact that thin ruby crystals (with a diameter of 4 mm) had a smooth surface and cylindrical shape. This phenomenon was explained in our work investigating crystallization process stability [63].

5.1. *Growth of cylindrical and plate-shaped crystals*

Stability of cylindrical crystal growth by VT will be analyzed with an allowance for three degrees of freedom: (1) the crystal radius R; (2) the distance between the burner face and the crystallization front l_1; and (3) the liquid–gas interface position l (fig. 44). Let the melt meniscus height $h = l - l_1$ measured from the interface to the melt peak point and representing the thickness of the melted layer for a planar crystallization front be introduced into the considerations.

The linearized set of eq. (2.3) defined from the Lyapunov criteria of growth process stability takes the following form in the present case:

$$\partial R/\partial t = A_{RR}\delta R + A_{Rl}\delta l + A_{Rh}\delta h, \tag{5.1}$$

$$\partial l/\partial t = A_{lR}\delta R + A_{ll}\delta l + A_{lh}\delta h, \tag{5.2}$$

$$\partial h/\partial t = A_{hR}\delta R + A_{hl}\delta l + A_{hh}\delta h. \tag{5.3}$$

The explicit form of the coefficients of the system require to be found.

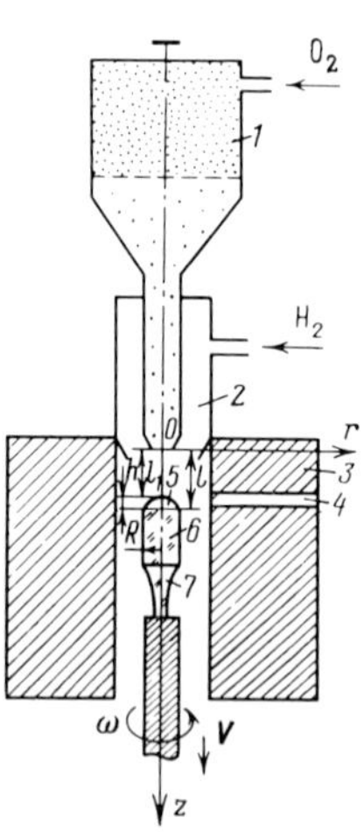

Fig. 44. Cylindrical crystal growth by the Verneuil technique: 1: stock bin; 2: burner; 3: furnace thermal insulation; 4: peephole; 5: melt layer; 6: crystal; 7: seed; R: crystal radius; l_1: melt surface position relative to the burner; $h = l - l_1$: melt meniscus height; zOr: coordinate system; V: displacement rate; ω: crystal rotation rate.

5.1.1. Determination of the coefficients A_{RR}, A_{Rl} and A_{Rh}

Comparison of eqs. (5.1) and (2.14) gives:

$$A_{RR} = -V\frac{\partial \alpha_0}{\partial R}, \qquad A_{Rl} = -V\frac{\partial \alpha_0}{\partial l}, \qquad A_{Rh} = -V\frac{\partial \alpha_0}{\partial h}. \tag{5.4}$$

$A_{Rl} = 0$ as the meniscus shape and angle α_0 do not explicitly depend on the liquid–gas interface position.

The functional relation between the angle α_0, the crystal radius R and the meniscus height h is defined from the Laplace capillary equation (2.6).

The specificity of the capillary shaping by VT formally reveals itself in the form of the following boundary condition: the line tangent to the melt surface is horizontal at the peak point of the melted layer (fig. 45):

$$\mathrm{d}z/\mathrm{d}r\,|_{r=0} = 0. \tag{5.5}$$

The second boundary condition is specified at the interface and follows from the condition of growth angle constancy:

$$\mathrm{d}z/\mathrm{d}r\,|_{r=R} = -\operatorname{tg}\alpha_0. \tag{5.6}$$

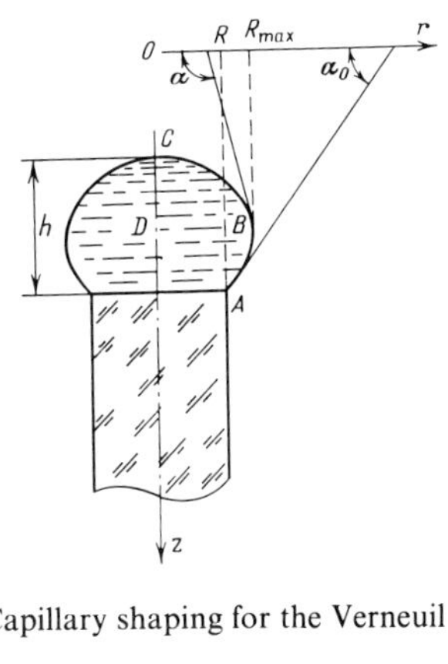

Fig. 45. Capillary shaping for the Verneuil technique.

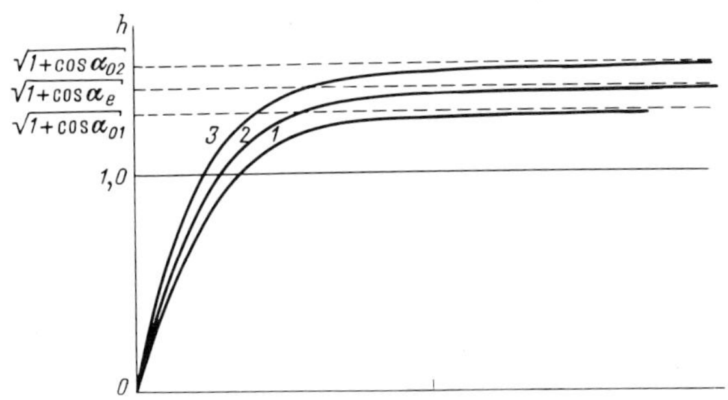

Fig. 46. Melt meniscus height versus crystal diameter for various angles α_0: $\alpha_{01} > \alpha_e > \alpha_{02}$.

For $R \geq 1$ the Laplace equation (2.6) is not solvable by quadrature, and numerical calculation is required. However, qualitatively the behavior of the relation can be estimated in the following way. For the meniscus parts where $R \ll 1$ and $R \gg 1$ analytical solutions can be obtained. Then, by joining the functions and their derivatives at the point $R = 1$, a common solution can be obtained. The meniscus height tends asymptotically to $(1 + \cos \alpha_0)^{1/2}$ with increasing R. Figure 46 shows the qualitative behavior of the relation $h(R)$ for various α_0 values. The initial portion of each curve is a straight line defined from the equation

$$h = R(1 + \cos \alpha_0) \sin^{-1} \alpha_0. \tag{5.7}$$

The capillary coefficients have the following forms:

$$A_{RR} = -VR^{-1} \sin \alpha_0 < 0, \tag{5.8}$$

$$A_{Rh} = VR^{-1} \sin \alpha_0 (1 + \cos \alpha_0)^{-1} > 0. \tag{5.9}$$

Within the whole range of crystal dimensions of $\partial \alpha_0 / \partial R > 0$, $\partial \alpha_0 / \partial h < 0$ and, correspondingly, the signs of the capillary coefficients do not change: $A_{RR} < 0$ and $A_{Rh} > 0$.

5.1.2. *Determination of the coefficients A_{ll}, A_{lR} and A_{lh}*

The mass balance conditions in the crystal–melt system can be represented in the following form:

$$W_0 = \pi \rho_s R^2 \left(V - \frac{dl_1}{dt} \right) + \pi \rho_L R^2 \left(\frac{dl_1}{dt} - \frac{dl}{dt} \right). \tag{5.10}$$

Here W_0 is the charge mass fed into the melted layer per unit time, ρ_s and ρ_L are solid- and liquid-phase densities, respectively. The first component on the right-hand side of eq. (5.10) denotes the substance mass crystallizing per unit time, while the second component is the change in mass of the melted layer. The assumption that $\rho_s = \rho_L = \rho$, made for simplicity, gives:

$$\frac{dl}{dt} = -W(\pi \rho R^2)^{-1} + V. \tag{5.11}$$

Under stationary conditions $dl/dt = 0$ and $V = W_0[\pi \rho (R^0)^2]^{-1}$. When equilibrium is departed from, linearization of eq. (5.11) gives:

$$\partial l / \partial t = 2W_0 (\pi \rho R^3)^{-1} \delta R. \tag{5.12}$$

Comparison of this equation with eq. (5.2) gives:

$$A_{lR} = 2W_0 (\pi \rho R^3)^{-1} > 0, \quad A_{ll} = 0, \quad A_{lh} = 0.$$

When calculating the coefficient A_{ll} the charge-mass distribution in the falling flow was assumed to have the form of a delta function (fig. 47, curve 1) and, with the crystal radius changing, the amount of the substance fed to the melt remains unchanged. In a real situation the charge-mass distribution in the flow along the muffle cross-section in the crystallization area $F(r)$ depends on the design of the equipment used in particular (fig. 47, curves 2–4) and the substance mass fed into the melt per unit time depends on

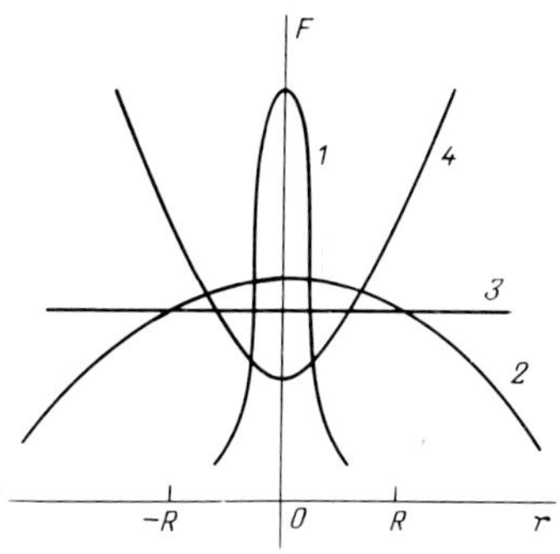

Fig. 47. The function $F(r)$ (charge distribution in the gas flow) depending on the burner design.

the crystal radius:

$$W_0 = 2\pi\beta \int_0^R F(r)r \, \mathrm{d}r. \qquad (5.13)$$

Here β denotes the coefficient of mass flow rate prescribed by the feeder. Substitution of eq. (5.13) into eq. (5.11) and linearization give:

$$A_{lR} = \beta R^{-3}\left[2\int_0^R F(r)r \, \mathrm{d}r - R^2 F(R)\right]. \qquad (5.14)$$

It is obvious that if $F(r) = \text{const} = F(R)$, $A_{lR} = 0$. If the function $F(r)$ reaches its extremum when $r = 0$, the sign of A_{lR} will not change. If the extremum is the minimum, $A_{lR} < 0$. When the charge is fed from the central nozzle of the burner (a most widely used case in practice) the maximum of the charge distribution in the flow lies in the center of the burner muffle (curve 2, fig. 47), therefore $A_{lR} > 0$.

5.1.3. Determination of the coefficients A_{hR}, A_{hl} and A_{hh}

The rate of meniscus height change $\partial h/\partial t$ is specified by eq. (2.18). Liquid- and solid-phase temperature gradients can be defined from eq. (2.17). When formulating the heat flow problem for CzT and TPS the liquid-phase temperature on the fixed level was specified as a boundary condition. This boundary condition for VT does not correspond to the real situation. Crystal displacement in the furnace muffle results in melt temperature change on its surface. Specificity of the thermal conditions of the technique under consideration will be allowed for by specifying the density Q of the heat flow fed from the burner onto the surface of the melted layer. With the gas flow specified, the density of the heat flow depends on the distance between the burner and the level of the melt surface. Figure 48 demonstrates the validity of this conclusion. The function $Q(l)$ is determined by the burner design and the gas rate.

Some part of the heat flow fed onto the melt is used for charge melting directly on the melt surface. It is equal to $\eta \mathscr{L} V$, where $\mathscr{L}$ is the melting heat per unit charge volume, η is the coefficient allowing for the state of the charge ($0 \le \eta \le 1$). When $\eta = 0$ the charge is fed into the melt in completely liquid state. When $\eta = 1$ the charge surface starts melting

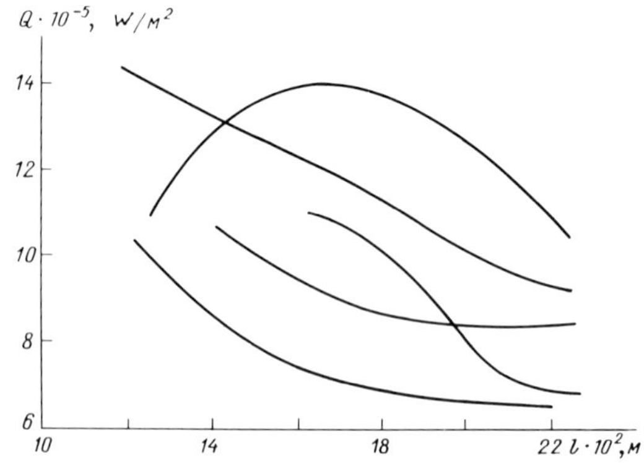

Fig. 48. Heat flow density versus the distance between the burner and the crystal face for various burner designs [64].

on the melt surface. Thus, the boundary conditions for liquid-phase temperature calculation have the following form:

$$\lambda_{\mathrm{L}} \frac{\mathrm{d}T_{\mathrm{L}}^*}{\mathrm{d}z}\bigg|_{z=l} = -Q + \eta\left(V - \frac{\mathrm{d}l}{\mathrm{d}t}\right), \tag{5.15}$$

$$T_{\mathrm{L}}^*|_{z=l+h} = T_0^*.$$

Here $T_i^* = T_i - T_{\mathrm{e}}$ is the reduced temperature, the index $i = \mathrm{L}$ relates to the liquid phase; $i = \mathrm{S}$ relates to the solid phase, $T_0^* = T_0 - T_{\mathrm{e}}$ is the reduced crystallization temperature, T_0 denotes the crystallization temperature, T_{e} is the environmental gas temperature.

Solid-phase temperature distribution is calculated using the following conditions:

$$T_{\mathrm{S}}^*|_{z=l+h} = T_0^*, \tag{5.16}$$

$$\frac{\mathrm{d}T_{\mathrm{S}}^*}{\mathrm{d}z}\bigg|_{z\to\infty} = 0. \tag{5.17}$$

The solution of the problem gives the following expressions for the thermal coefficients:

$$A_{hR} = \frac{1}{\mathscr{L}}\left[\lambda_{\mathrm{L}} T_0^* \zeta_{\mathrm{L}}^2 h \frac{1}{R} + \frac{(Q - \eta \mathscr{L} V)\zeta_{\mathrm{L}}^2 h^2}{2R}\right] - A_{lR}(1-\eta) > 0, \tag{5.18}$$

$$A_{hl} = \frac{1}{\mathscr{L}} \frac{\mathrm{d}Q}{\mathrm{d}l}, \tag{5.19}$$

$$A_{hh} = -\frac{1}{\mathscr{L}}[\lambda_{\mathrm{L}} T_0^* \zeta_{\mathrm{L}}^2 + (Q - \eta \mathscr{L} V)\zeta_{\mathrm{L}}^2 h] < 0. \tag{5.20}$$

Here,

$$\zeta_i = \sqrt{2\mu_i(\lambda_i R)^{-1}}, \quad i = \mathrm{L}, \mathrm{S}. \tag{5.21}$$

5.1.4. Analysis of the process-stability conditions

For the system of eqs. (5.1)–(5.3) to be stable it is necessary and sufficient that the Routh–Hurvitz conditions [15] should be satisfied. For this system they take the following form:

$$A_{RR}+A_{hh}<0, \tag{5.22}$$

$$A_{Rh}A_{lR}A_{hl}<0, \tag{5.23}$$

$$-(A_{RR}+A_{hh})(A_{hh}A_{RR}-A_{Rh}A_{hR})+A_{Rh}A_{lR}A_{hl}>0. \tag{5.24}$$

Since A_{RR} and A_{hh} are negative, eq. (5.22) holds.

The coefficient A_{Rh} is positive and therefore for eq. (5.23) to hold, the coefficients A_{lR} and A_{hl} should be opposite in sign. As was shown above, when the charge is fed from the central nozzle of the burner $A_{lR}>0$, then the coefficient A_{hl} should be negative. It means that in the vicinity of the crystallization zone the density of the heat flow fed onto the melt surface should satisfy the condition $\mathrm{d}Q/\mathrm{d}l<0$, i.e. the heat-flow density should decrease with the distance from the burner face increasing. For eq. (5.24) to be satisfied the necessary and sufficient conditions are the following:

$$A_{RR}A_{hh}-A_{Rh}A_{hR}>0, \tag{5.25}$$

$$-(A_{RR}+A_{hh})(A_{hh}A_{RR}+A_{Rh}A_{hR})>|A_{Rh}A_{lR}A_{hl}|. \tag{5.26}$$

Taking into consideration that $A_{RR}=-(\partial h/\partial R)A_{Rh}$, where the value of $\partial h/\partial R$ is calculated from the capillary problem (fig. 46) the following expression can be derived:

$$A_{RR}A_{hh}-A_{Rh}A_{hR}=\frac{\partial h}{\partial R}\frac{1}{\mathscr{L}}[\lambda_{\mathrm{L}}T_0^*\zeta_{\mathrm{L}}^2+(Q-\eta\mathscr{L}V)\zeta_{\mathrm{L}}^2h]$$

$$-\frac{1}{\mathscr{L}}\left[\frac{\lambda_{\mathrm{L}}T_0^*\zeta_{\mathrm{L}}^2h}{l}+\frac{(Q-\eta\mathscr{L}V)\zeta_{\mathrm{L}}^2h^2}{2R}\right]+A_{lR}(1-\eta). \tag{5.27}$$

Let $R\ll 1$, then taking into account that $\partial h/\partial R\approx h/R$ (in accordance with eq. (5.7) h is linearly dependent on R when small radii are used) from eq. (5.27), it can be obtained that $A_{RR}A_{hh}-A_{Rh}A_{hR}>0$. From here for eq. (5.26) to hold it is sufficient that

$$\left|\frac{\mathrm{d}Q}{\mathrm{d}l}\right|<\frac{V}{R}. \tag{5.28}$$

It is an approximate evaluation that under-estimates $|\mathrm{d}Q/\mathrm{d}l|$ calculated from eq. (5.26). However, this value of $|\mathrm{d}Q/\mathrm{d}l|$ deliberately satisfies eq. (5.24) and besides eq. (5.28) makes clear physical sense: for the process to be stable, the change in the heat flow density Q along the furnace muffle in the vicinity of the growth zone at the distance of an order of R should not exceed the crystallization heat. When this requirement is met and the condition $|\mathrm{d}Q/\mathrm{d}l|<0$ is satisfied the process stability is provided for $R\ll 1$.

If $R>1$, then $\partial h/\partial R$ decreases much faster than h/R does (fig. 46) and starting with some values in the range of $R>1$ crystallization process stability is lost. From eq. (5.27) it follows that the stability range can be slightly widened by decreasing η, i.e. the charge should be melted before feeding it into the melt. This can be achieved by decreasing the size of powder particles, e.g. by means of its careful separation. Decrease of $|dQ/dl|$ contributes to stability-range widening; however, all these measures do not ensure process stability when growing crystals of large diameters.

For corundum the capillary constant $a = 6$ mm. Proceeding from the analysis carried out, one can state that, when corundum crystals are grown, crystallization process stability will exist up to diameters of 10–12 mm. Experience in commercial growth of long thin corundum rods of constant diameters of 3–6 mm for the watch industry is in a good agreement with this estimate.

5.1.5. Plate growth

When growing plates whose thickness does not exceed the capillary constant a by VT, the meniscus height h is linearly dependent on the thickness, i.e. it is described by eq. (5.7), where R is the plate half-thickness, which is similar to the case of thin-rod growth. If the hot zone matches the plate geometry, the thermal problem also remains unchanged and hence the conclusion regarding the stability of thin rod crystallization also holds for thin plate crystallization.

5.2. Stability-analysis-based automation of VT

As was shown above, when passing to crystals with diameters exceeding the capillary constant crystallization stability is lost. It means that to grow large crystals of constant cross-sections in situ control of the parameters is required. When developing systems for the automatic control of growing crystal diameters, a problem of the laws of automatic control of the process parameters under some changes in crystal dimension arises. It will be shown that theoretical derivation of the laws of process parameter control ensuring maintenance of the crystal cross-section specified can be performed on the basis of the process stability analysis. In this case the controllable parameters along with the crystal radius, the liquid–gas interface position and the meniscus shape can be required as the degrees of freedom of the process. For VT the density of the heat flow from the burner Q, the rate of crystal powering V and the charge flow rate W_0 can be used as controllable parameters. Usually, after the crystal has already widened from the seed size to the desired diameter, control is provided by changing the charge flow rate. The controllable charge flow rate W will be regarded as an additional degree of freedom. In this case the linearized set of equations defining stability of the crystallization process with variable charge flow rate takes the following form:

$$\partial R/\partial t = A_{RR}\delta R + A_{Rl}\delta l + A_{Rh}\delta h + A_{RW}\delta W, \tag{5.29}$$

$$\partial l/\partial t = A_{lR}\delta R + A_{ll}\delta l + A_{lh}\delta h + A_{lW}\delta W, \tag{5.30}$$

$$\partial h/\partial t = A_{hR}\delta R + A_{hl}\delta l + A_{hh}\delta h + A_{hW}\delta W, \tag{5.31}$$

$$\partial W/\partial t = A_{WR}\delta R + A_{Wl}\delta l + A_{Wh}\delta h + A_{WW}\delta W, \tag{5.32}$$

Equation (5.32) is the law of charge flow rate control in its general form. It is quite obvious that in this case a linear law should be selected among all the variety of possible laws of control. The law of control provides maintenance of the constant crystal cross-section if the set of eqs. (5.29)–(5.32) is stable. Hence, the problem is to define the coefficients of eq. (5.32) proceeding from the necessity for the set of eqs. (5.29)–(5.32) to be stable.

The results obtained in the previous paragraph where VT was analyzed for stability will be used to define the coefficients A_{ij}. Let growth of crystals of large diameter ($R \gg 1$) be considered. Then $\partial\alpha_0/\partial R \to 0$ and $\partial\alpha_0/\partial h < 0$ and does not depend on the crystal diameter. Therefore it is assumed that $A_{RR} = 0$, $A_{Rh} > 0$. The coefficients A_{Rl} and A_{RW} are equal to zero as the meniscus shape and the angle α_0 do not depend on the liquid–gas interface position and on the charge flow rate.

The coefficients of eq. (5.30) are defined from the law of mass conservation in the crystal–melt system. The mass-balance equation takes the form of eq. (5.10). Substitution of the variable flow rate W into eq. (5.11) instead of the constant charge flow rate W_0 and linearizing gives:

$$\partial l/\partial t = 2W(\pi\rho R^3)^{-1}\delta R - (\pi\rho R^2)^{-1}\delta W, \qquad (5.33)$$

from which

$$A_{lR} = \frac{2W}{\pi\rho R^3} > 0, \qquad A_{ll} = 0, \qquad A_{lh} = 0, \qquad A_{lW} = -\frac{1}{\pi\rho R^2} < 0.$$

In contrast to the case when the charge flow rate is constant, $A_{lW} \neq 0$.

The coefficients of eq. (5.31) coincide with those from eq. (5.3): $A_{hR} > 0$, $A_{hl} < 0$, $A_{hh} < 0$. $A_{hW} = 0$, as the meniscus height is not explicitly dependent on the charge flow rate. To distinguish the coefficients included into the law of charge flow rate control from the coefficients of eqs. (5.29)–(5.31), the following notation is introduced: $A_{RW} = K_R$, $A_{lW} = K_l$, $A_{hW} = K_h$, $A_{WW} = K_W$. $K_h = 0$ will be selected since the law of charge flow rate control into which K_h is included is of no practical interest because of the difficulties of monitoring the change in the meniscus height.

The set of coefficients K_j that provide stability of the set of eqs. (5.29)–(5.32) and hence growth of crystals of constant cross-sections are obtained (table 4). The laws of charge flow rate control that allow maintenance of the specified crystal diameter are called stable ones.

The law of control $\partial W/\partial t = K_l\delta l + K_i\partial l/\partial t$ can be obtained if it is taken into account that $\partial l/\partial t = A_{lR}\delta R + A_{lW}\delta W$.

When experimentally checking the theoretically derived laws of control, two series of crystals were grown [65]. The first series was grown using the stable law $\partial W/\partial t = K_l\delta l + K_i\partial l/\partial t$, the second one was grown using the unstable law $\partial W/\partial t = K_l\delta l$ (fig. 49).

Table 4
The laws of charge flow rate control provided stable and unstable crystal growth.

Stable laws of control	Unstable laws of control
(1) $\partial W/\partial t = K_i\partial l/\partial t + K_W\delta W$, for $K_i > 0$, $K_W < 0$	(1) $\partial W/\partial t = K_R\delta R + K_i\partial l/\partial t$, for K_R and K_i of any sign
(2) $\partial W/\partial t = K_R\delta R + K_i\partial l/\partial t + K_W\delta W$, $K_i > 0$, $K_W < 0$, $K_R < 0$	(2) $\partial W/\partial t = K_i\delta l$
(3) $\partial W/\partial t = K_i\partial l/\partial t + K_l\delta l$, $K_i < 0$, $K_l > 0$	(3) $\partial W/\partial t = K_R\delta R$
	(4) $\partial W/\partial t = K_R\delta R + K_W\delta W$

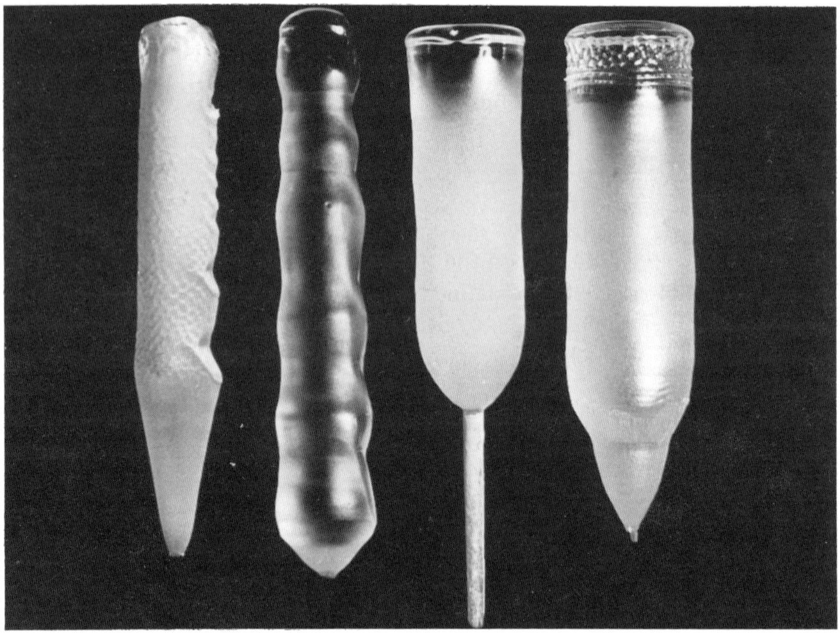

Fig. 49. Corundum crystals grown by stable (to the right) and unstable (to the left) laws of charge control.

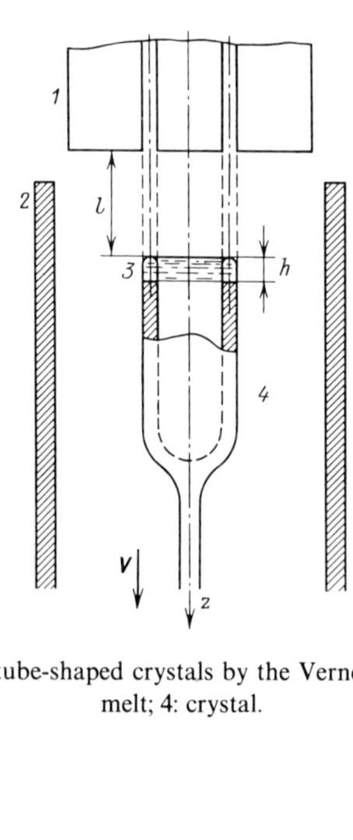

Fig. 50. Growth of one-end-sealed tube-shaped crystals by the Verneuil technique: 1: burner; 2: muffle; 3: melt; 4: crystal.

5.3. Growth of tube-shaped crystals

5.3.1. Theoretical analysis

Figure 50 shows a diagram of growing tubular crystals shaped by tubular powder flow in the oxygen–hydrogen flame also possessing tubular shape. In contrast to rod-shaped crystal growth, the process of tube growth exhibits not three but four degrees of

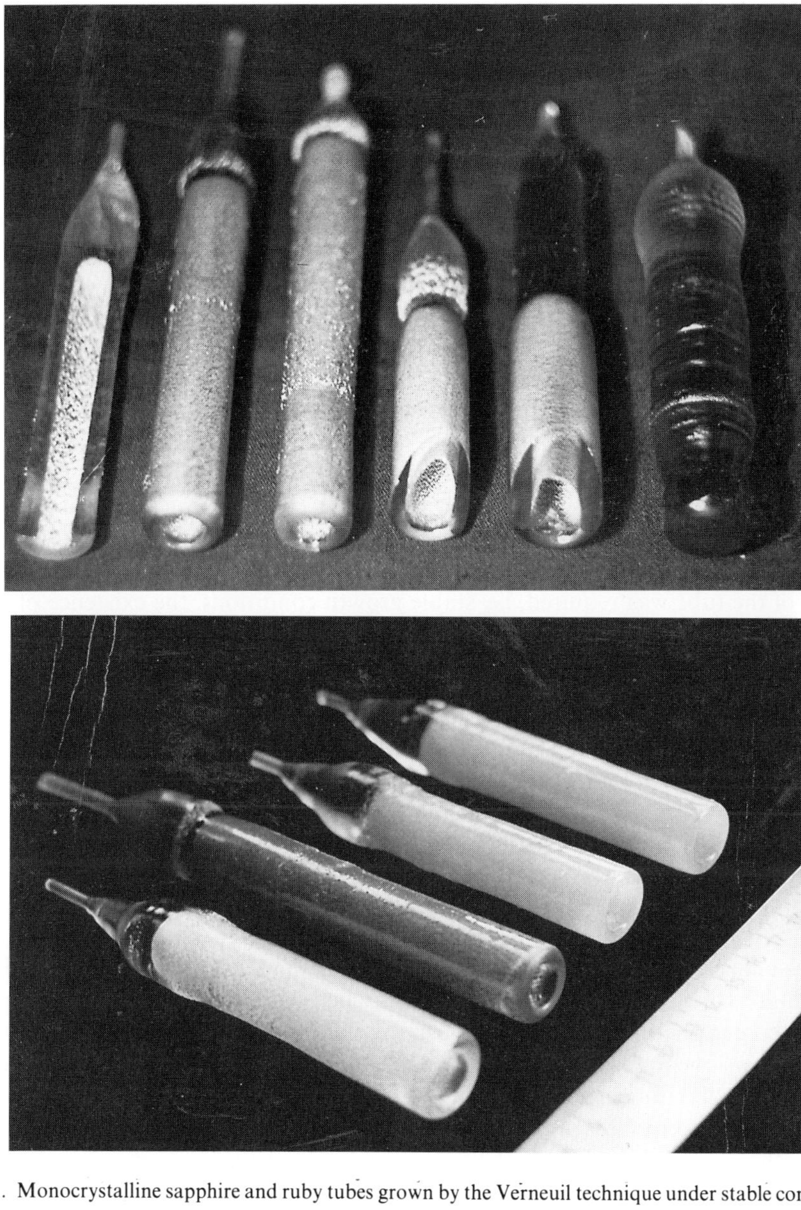

Fig. 51. Monocrystalline sapphire and ruby tubes grown by the Verneuil technique under stable conditions.

freedom. The other tube radius R_1, its inner radius R_2, the liquid–gas interface position l and the crystallization front position l_1 can all vary independently.

The analysis of stability carried out [13] shows: Crystallization of tubes of arbitrary outer diameter is stable if the tube wall thickness is smaller than some critical thickness. This thickness is significantly smaller than the capillary constant and depends both on the thermal conditions of the process and on the outer radius of the tube. This crystal thickness of the tube wall increases with the outer diameter increase. The crystallization stability is lost when the tube-wall thickness is larger than the capillary constant.

5.3.2. *Experiments on growing single-crystal corundum tubes under stable conditions*

In our experiments tubes with an outer diameter of 17 to 25 mm and with the walls 3 to 4 mm thick were grown. A crystallization apparatus fitted with a four-channel burner providing charge supply via the central and the periphery channels was used. Crystal growth was initiated from a seed 3 to 4 mm in diameter. It was experimentally determined that the optimal gas distribution in the burner channels is as follows: oxygen–hydrogen–oxygen–hydrogen. Firstly, a seed cone was grown. The cone was widened by feeding the charge through the central channel, the peripheral oxygen flow rate being increased. The base diameter of the seed core reached 20–22 mm . As soon as the crystal diameter reached the specified value the charge was fed through the peripheral channel. 10–15 min later charge supply from the central tank was cut off and within 30–60 min the rate of the central oxygen flow was reduced. The latter provides a smooth transition from a solid crystal to a tube. The lowering rate was smoothly increased. After that the process of stationary growth was attained. Tubes up to 120 mm long were grown. Usually no parameter control to maintain constant cross-section of the tube was required, i.e. stable growth conditions, the existence of which had been theoretically predicted, could be attained. Figure 51 shows some of the tubes grown.

6. *Radial instability of vapor whisker growth (VWG)*

The main whisker growth by vapor–liquid–crystal (VLC) transition is considered to be proved now [66]. This mechanism (vapor–liquid–solid–VLS) was described by Wagner and Ellis [67,68] to explain unidirectional whisker growth involving impurities (fig. 52). Vapor-phase substance is absorbed by the melt at the crystal end face and then crystallizes according to the scheme resembling the VT. The only difference is that solution growth takes place as the melt is enriched with a great amount of impurity used for controlled whisker nucleation on the substrate. This whisker crystallization scheme allows one to regard this process as a capillary shaping technique and study its stability according to general procedure.

The main purpose of this investigation is to explain a radial instability (fig. 53) and to describe the shape of initial parts of whiskers grown by the VLC mechanism. However, in contrast to the previous sections, kinetic effects that influence the magnitude of the

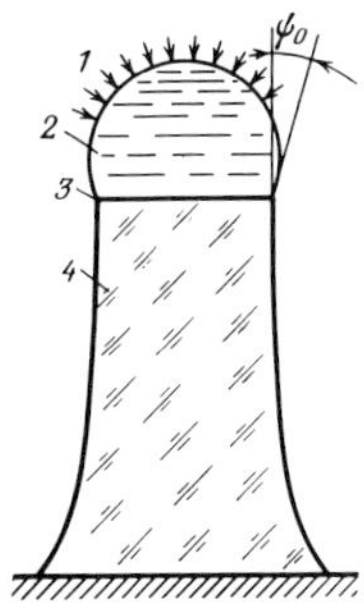

Fig. 52. Whisker growth according to the VLC mechanism: 1: vapor; 2: flux; 3: crystallization front; 4: crystal; ψ_0: growth angle.

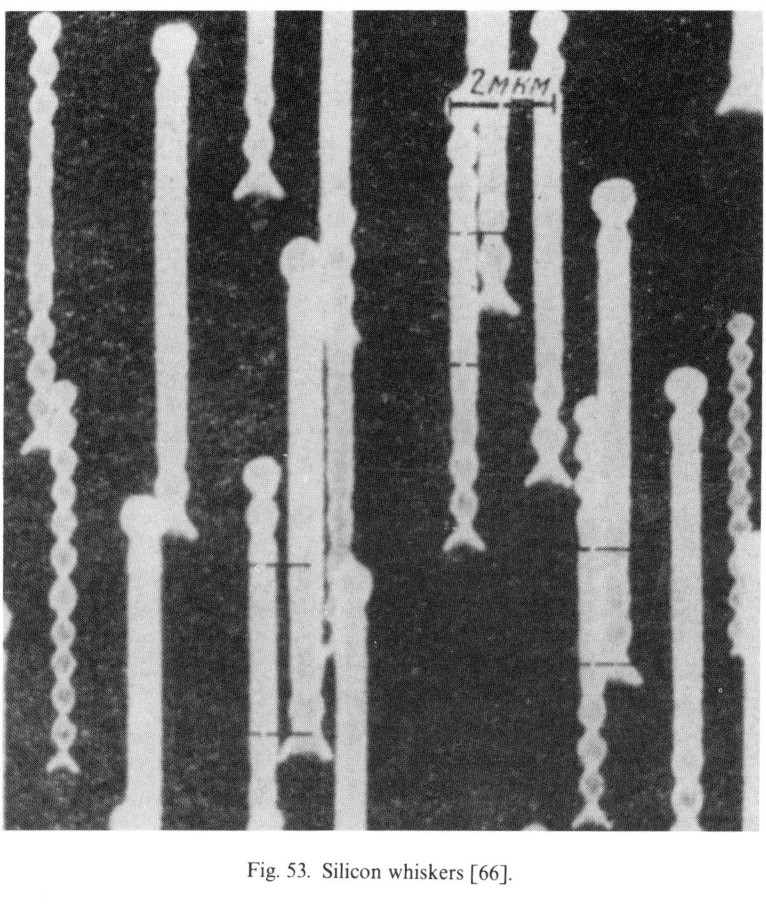

Fig. 53. Silicon whiskers [66].

growth angle will play an important part here. Periodic radial instability of VWG is described in detail in [66]. At present some attempts to explain the causes of the self-exited oscillations exist [66, 69]. Here it will be shown that radial instability of surfactants is a natural consequence of Gibbs–Thompson dimensional effects when whisker diameters reduce to submicron dimensions [70]. A real experimental silicon gold system investigated most thoroughly is analyzed with an allowance for its specific features using the chloride growth process in a $SiCl_4 + H_2$ gaseous mixture in a flowing-gas reactor [66].

6.1. Problem formulation

If the whisker radius R and the main component concentration C in the solution–melt system are selected as independent variables, a linearized set of differential equations (2.3) describing the rate of change in the concentration and in the whisker radius for small deviations from the stationary state will assume the following form:

$$\partial R/\partial t = A_{RR}\delta R + A_{RC}\delta C, \tag{6.1}$$

$$\partial C/\partial t = A_{CR}\delta R + A_{CC}\delta C.$$

The Routh–Hurvitz inequalities for this system take the form of eqs. (3.7) and (3.8). The values of the coefficients A_{ij} depend on the selected stationary-state parameters of a real system (whisker diameter, gas-mixture composition and pressure, temperature, etc.). If eqs. (3.7) and (3.8) hold, the stationary state is of the stable focus type, in the case where the sign of the inequality for the diagonal coefficients is reversed ($A_{RR} + A_{CC} > 0$) while the sign of the inequality $A_{RR}A_{CC} + A_{RC}A_{CR} > 0$ remains unchanged, the dynamic system loses its stability and passes to an unstable focus type state (fig. 4). It means that limit cycles corresponding to self-excited oscillations of a real dynamic system can appear on the phase diagram [66]. The next problem of the analysis is to concretize eq. (6.1) and to investigate the conditions satisfying eqs. (3.7) and (3.8).

The coefficients of the first equation are found, as in previous cases, from the condition that the final lateral surface of the crystal is shaped at the growth angle ψ_k:

$$\psi_k = \psi_0 - \psi_1 - \psi_2 + \psi_3. \tag{6.2}$$

In eq. (6.2) the first term is the equilibrium angle ψ_0 determined by the values of the free surface energy according to eq. (2.9). The second term is identical to eq. (2.10) and, in this case, the contribution of the concentrational supercooling $\Delta T = \eta\Delta C$ at the crystallization front to whisker lateral surface curvature is taken into consideration. The third and the fourth forms appear as a result of the allowance for liquid- and solid-phase surface curvatures.

The coefficients of the second equation of the set of eqs. (6.1) are obtained from the balance of flow of main substance fed from the gas phase through the liquid-drop surface towards the whisker top and removed from the drop in the process of crystallization.

6.2. Stability analysis [70, 74]

From the analysis carried out [70] it follows that, for systems of the Si–Au type when thick whiskers are crystallized, the natural equilibrium state is that of the stable focus type. When passing to small whisker diameters, the corrections for the dimensional effects reverse the signs of all the coefficients in eq. (6.1). In this case the new equilibrium state will be the state of the unstable focus type.

Nonformally the mechanism of positive feedback occurrence in the system can be explained by the dimensional dependence of the growth angle. For the section of thin whisker contraction the growth angle ψ_k increases a little faster than the slope of the line tangent to the melt meniscus. In this case, the smaller the whisker diameter at the drop base, the faster it should contract. However, simultaneously a decrease of the oversaturation occurs at the crystallization front, which results in decrease of the crystallization rate and the rate of whisker contraction. The dynamic equilibrium is unstable and develops in the opposite direction as soon as the process is stopped.

6.3. Profile curves of initial whisker sections

The process of whisker growth is as follows: when a metal particle is melted into the substrate a melt drop assumes the shape of a lens. The lens geometry is determined by the condition of surface-tension force equilibrium at some point on the three-phase line (fig. 54b). Since the gas mixture flows above the melt drop and the oversaturation required is achieved in the melt, crystallization initiation is observed at the front, the drop base representing this front. If, in an isotropic approximation, the meniscus makes an angle ψ with the horizontal surface of the substrate (fig. 54b1), in the process of crystallization the angle between the line tangent to the meniscus and the horizontal increases, the crystallization front coincides with the substrate surface at some instant. In this case $\psi = \psi_0$ (fig. 54b2). From this instant the initial cone-like whisker section crystallizes (fig. 54b3).

Ignoring the dimensional effects for rather thick whiskers, the growth angle can be considered as a constant at any instant of crystallization. Then the process of whisker growth follows the scheme of small-drop crystallization described in [22]. The only difference is that, during drop crystallization, the overall crystal and melt mass is constant while, during whisker crystallization, the volume of the meniscus on the whisker end-face can be regarded as constant.

The profile curves of the initial sections were numerically calculated and are given in (fig. 54a). The initial data are the experimentally found angle $\psi_0 \approx 54^\circ$ and the drop volume W selected arbitrarily. The initial dimension R_0 of the whisker base is calculated from the formula $R_0 = [6W\pi D(3 + D^2)]^{1/3}$, where $D = (1 - \cos\psi_0)\sin\psi_0$. Some discrepancy between the experimental and calculated curves can be attributed to the inaccuracy of growth-angle measurements as well as to the simplified scheme of calculation ignoring changes in the drop volume in the process of crystallization front straightening as the initial whisker section grows.

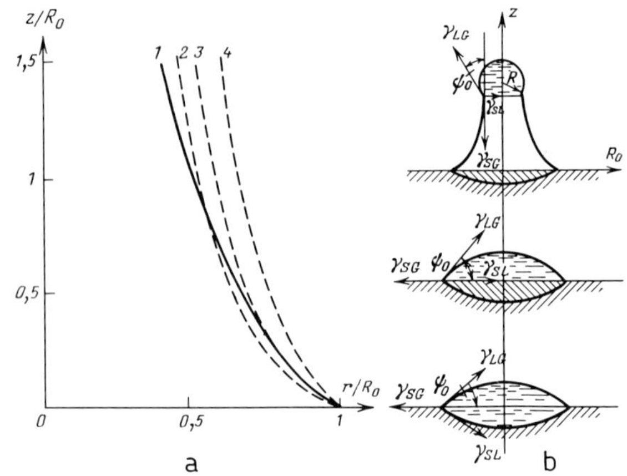

Fig. 54. (a) On calculating the initial whisker section: 1: a whisker profile curve taken from a photograph (fig. 53) in [66]; 2, 3, 4: calculated curves for $\psi_0 = 54°$, $65°$, and $70°$, respectively; (b) diagram illustrating formation of the initial cone-shaped section.

Note that the initial cone formation results in the first necking both for thick and for thin silicon whiskers (fig. 53). This fact is attributed to the first oscillation in the process of transition to the stationary state of the stable focus type for thick whiskers or to the conditions of self-excited oscillations for thin whiskers.

In conclusion it should be noted that the mechanism offered here does not reject other possible models interpreting radial instability in the process of whiskers growth. It is quite obvious that, to finally solve this problem, a study quantitatively verifying the conditions under which self-excited oscillations occur – and not for silicon alone – is required.

7. *The realization of dynamic stability in the floating zone technique (FZT)*

FZT was first analyzed for crystallization stability in [9]. However, this analysis is not complete, as the thermal conditions of crystallization are not taken into account and the capillary part of the problem is violently simplified, i.e. in fact, only capillary stability is studied while thermal fluctuations are considered to produce no effect on the process stability. Besides, from the point of view of the capillary phenomenon analysis, only the simplest versions were considered. Capillary stability proved to exist in the cases under consideration and this conclusion was extended to all other versions of the floating-zone technique. In this section it will be shown that capillary stability does not exist for some crystal dimensions and pressure in the meniscus [71].

7.1. *The formulation of the problem*

A diagram of crystallization by FZT is depicted in fig. 55 (upward pulling). At present, the versions of downward pulling are widely used. Irrespective of the heater design and

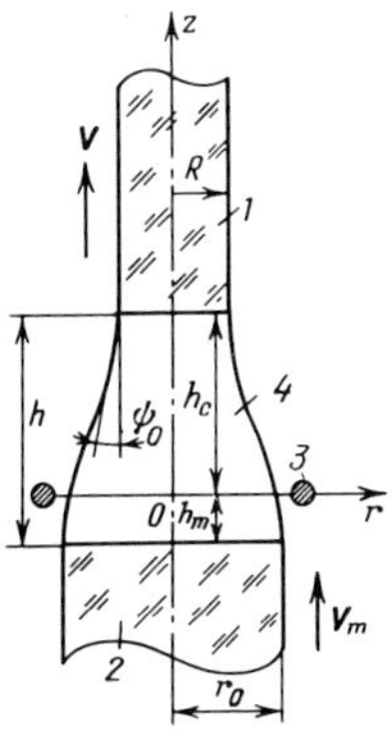

Fig. 55. Cylindrical crystal growth by the floating zone technique: 1: growing crystal with the radius R; 2: feeding rod with the radius r_0; 3: heater; 4: melted zone; h_c, h_m: positions of crystallization front and melting front relative to the heater, respectively.

geometry a region of maximum heating exist in the middle part of the zone that is a convenient physical point to start measurements when analyzing heat flow phenomena of this crystallization technique (fig. 55). In contrast to TPS here the pressure d in the meniscus is determined by melt-zone curvature and is unambiguously related to the zone volume W. In view of this, it is convenient to introduce the melt volume W as one of the degrees of freedom when analyzing stability of crystallization by FZT, its variation in time being easy to find on the basis of the condition of mass balance. In this case both W and d can be found by simultaneously solving the capillary and the heat flow problems with an allowance for the condition of mass balance.

Beside the crystallization front position, h_c, the melting front position h_m can vary independently in the floating-zone technique; variation h_m in time can be found on the basis of the heat balance condition at the melting boundary. Thus, in the floating-zone technique, four quantities can vary independently: they are the crystal dimension R, the crystallization front position h_c, the melting front position h_m and the zone volume W. The set of equations defining crystallization stability in a general case takes the following form:

$$\begin{aligned}
\partial R/\partial t &= A_{RR}\delta R + A_{Rh_c}\delta h_c + A_{Rh_m}\delta h_m + A_{RW}\delta W,\\
\partial h_c/\partial t &= A_{h_cR}\delta R + A_{h_ch_c}\delta h_c + A_{h_ch_m}\delta h_m + A_{h_cW}\delta W,\\
\partial h_m/\partial t &= A_{h_mR}\delta R + A_{h_mh_c}\delta h_c + A_{h_mh_m}\delta h_m + A_{h_mW}\delta W,\\
\partial W/\partial t &= A_{WR}\delta R + A_{Wh_c}\delta h_c + A_{Wh_m}\delta h_m + A_{WW}\delta W.
\end{aligned} \tag{7.1}$$

7.2. *The heat flow conditions*

The heat flow conditions for a crystal being grown and the zone part between the origin of coordinates coinciding with the point of maximum heating and the crystallization front are absolutely identical to those discussed above for CzT. Hence, the coefficients

A_{h_cR} and $A_{h_ch_c}$ coincide with the corresponding heat transfer coefficients A_{hR} and A_{hh}. The coefficient $A_{h_ch_m} = 0$ as the temperature gradients in the zone part under consideration do not depend on the melting front position h_m. The rate of melting front displacement can be defined from the heat balance condition at the melting boundary given below:

$$\lambda_L G_{Lm} - \lambda_S G_{Sm} = \left(V_m - \frac{dh_m}{dt} \right). \tag{7.2}$$

Here V_m is the rate of translation of the melting crystal. The heat flow from the liquid phase to the melting front should provide melting of the crystal fed with a rate V_m. This leads to a higher liquid-phase temperature gradient at the melting front than the corresponding gradient at the crystallization front. Since the temperature difference $T_m - T_0$ relative to the point of maximum heating is the same in both the cases it appears that $h_m < h_c$, this inequality being enhanced with the crystallization rate V_c increasing and, correspondingly, with the melting rate V_m increasing. For small h_m values heat removal from the lateral surface of this meniscus part can be neglected in the first approximation and the liquid-phase gradient at the melting front assume the following form: $G_{Lm} = -(T_m - T_0)/h_m$. Substitution of G_{Lm} into eq. (7.2) gives:

$$A_{h_mh_m} = \frac{\lambda_L}{\mathscr{L}} \frac{(T_m - T_0)}{h_m^2}.$$

Here $h_m/h_c \ll 1$ serves as an applicability criterion of the approximation mentioned for which A_{h_mR}, $A_{h_mh_c}$ and A_{h_mW} can be considered as equal to zero.

Usually, the true meniscus shape is not taken into consideration when heat flow phenomena are analyzed. It is replaced with a cylinder of radius R. In the case of the floating-zone technique, an independent change of W results in a change in the mean radius of the meniscus and in a corresponding change in the liquid–phase temperature gradient. The sign of the coefficient A_{h_cW} coincides with the sign of A_{h_cR}. Indeed, increasing both W and R leads to increases of the heat flow from the liquid phase towards the crystallization front and to a corresponding decrease of the crystallization rate. The influence of the effect mentioned on the crystallization stability will be discussed below and, in the first approximation, it will be assumed that $A_{h_cW} = 0$.

7.3. Mass balance

The rate of melt-zone volume change can be determined on the basis of the heat balance condition:

$$\partial W/\partial t = \pi \rho_S \rho_L^{-1} [r_0^2 (V_m - \partial h_m/\partial t) - R^2 (V_C - \partial h_c/\partial t)]. \tag{7.3}$$

Here ρ_S and ρ_L are the densities of the solid and the liquid phases, respectively. Under stationary-growth conditions $\partial W/\partial t = 0$ and $r_0^2 V_m = R^2 V_C$. From the radius of the crystal being melted r_0, V_m and V_c, the dimension of the stationary growing crystal R^0 can be calculated. When equilibrium is violated $\partial W/\partial t \neq 0$, $\partial R/\partial t \neq 0$, $\partial h_m/\partial t \neq 0$, $\partial h_c/\partial t \neq 0$. Application of eqs. (7.1) and (7.3) for the corresponding coefficients gives:

$$A_{WR} = B + CA_{h_cR}, \qquad A_{Wh_c} = CA_{h_ch_c}, \qquad A_{Wh_m} = -Cr_0^2R^{-2}A_{h_mh_m},$$

$$A_{WW} = CA_{h_cW}, \qquad B = -2\pi\rho_S\rho_L^{-1}RV, \qquad C = \pi\rho_S\rho_L^{-1}R^2.$$

It should be remembered that A_{hW} is assumed to be equal to zero.

7.4. *The capillary coefficients*

As is shown in [27], the boundary condition of meniscus catching on the edge of the crystal being melted is satisfied at the melting front. The meniscus slope towards the lateral surface of this crystal can be arbitrary in contrast to the growth angle constancy during crystallization.

For these boundary conditions, i.e. angle fixing at the crystallization front, catching at the melting front, the behavior of the capillary coefficients as a function of the

a
b
c
d

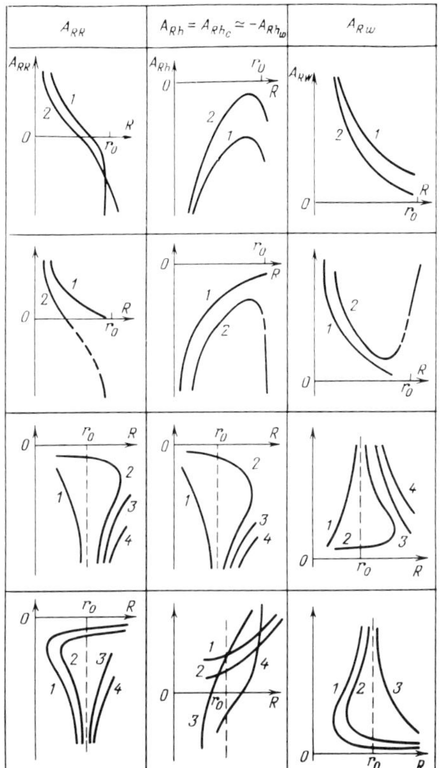

Fig. 56. The coefficients A_{ik} versus R ($\psi_0 = 12$) menisci of various shapes (a–d) under various pressures (1–4) (the value of r_0d increases with the curve number): (a) one-valued and (b) two-valued menisci for small Bond numbers; crystals are pulled upwards (c), or downwards (d) for large Bond number.

relation between the dimensions of the crystal being grown and of the crystal being melted was investigated in [70, 71, 14]. Figure 56 gives the summarized results.

7.5. Crystallization stability

With the results of the analysis of the coefficients A_{ik} carried out above taken into consideration, according to the Routh–Hurvitz criterion for the set of eq. (7.1) to be stable it is necessary and sufficient that the following inequalities should hold:

$$A_{h_m h_m} < 0, \tag{7.4}$$

$$A_{RR} + A_{h_c h_c} < 0, \tag{7.5}$$

$$BA_{RW} + A_{h_c h_c} > 0, \tag{7.6}$$

$$BA_{RR}A_{RW} - A_{RR}A_{h_c h_c}(A_{RR} + A_{h_c h_c}) + A_{h_c R}(A_{RR} + A_{h_c h_c})(A_{Rh_c} + CA_{RW}) > 0. \tag{7.7}$$

If, in the inequalities written above, the heat transfer coefficients $A_{h_c R}$ and $A_{h_c h_c}$ are formally assumed to be equal to zero, the stability criteria $\partial\alpha_0/\partial R > 0$ and $\partial\alpha_0/\partial W < 0$ coinciding with those from [9] can be obtained. The conditions of eqs. (7.4)–(7.7) are more general. If $T_m > T_0$ (the zone is overheated) the coefficients $A_{h_m h_m}$ and $A_{h_c h_c}$ are negative, eqs. (7.4) and (7.5) always hold, and, only under such conditions, is stable growth by FZT possible. As was shown above, for all the parameters of the zone $A_{RW} > 0$ and since $B < 0$, eq. (7.6) always holds; eq. (7.7) is satisfied outside the range of the boundary curves where $\partial R/\partial h > 0$.

By comparing the versions of upward and downward crystallization by FZT it can be found for upward crystallization a larger portion of the boundary curve (and then the majority of possible crystal dimensions) corresponds to the range of stable growth but in this case the diameter of the crystal being pulled is less than one being melted and, on the contrary, for downward crystallization a larger portion of the boundary curve corresponds to the range of possible instability $\partial R/\partial h > 0$. Nevertheless in practice the downward version is used more frequently when growing crystals by FZT since only in this case can crystals of large diameters be pulled under stable conditions using a corresponding selection of the diameter of the crystal being melted and the zone volume.

From the point of view of stability, for FZT, menisci of large curvature are identical to those of TPS in the sense that A_{RR} reverses its sign. In this case some stability limit exists and it is such that crystals of dimensions smaller than the critical one cannot grow in a stable way. However, it can be strictly shown that for FZT the limit of capillary stability ($\partial\alpha_0/\partial R = 0$) is displaced towards the range of smaller crystal diameters as compared with TPS.

As has already been mentioned, changes in the zone volume W with fixed R and h cause changes in the liquid-phase temperature gradient. Up to now the influence of this effect on crystallization stability has not been analyzed. However, it can easily be noticed that the effect decreases stability. Indeed, as was shown above, it results in an increase in the heat flow towards the crystallization front and to a decrease of the crystallization rate, which in turn leads to a further increase of W.

Absence of such an effect in TPS is a certain advantage thereof.

8. Stability of cylindrical pore (negative crystal) growth

The nature of the porosity of crystals depends on many factors such as the material of the crystals, tendency of the melt to dissociate and to interact with material of the container, the morphology of the crystallization front, the hydrodynamics of the various flows in the system, and a number of other physical and technological factors. In this section the dynamic stability approach is used to investigate the conditions of elongated cylindrical pore (very serious defects – fig. 57) formation and steady growth.

8.1. Formulation of the problem

Figure 58 shows stationary growth of a cylindrical pore. Here the meniscus can be considered as a sphere since the pore diameter is equal to a few micrometers – small

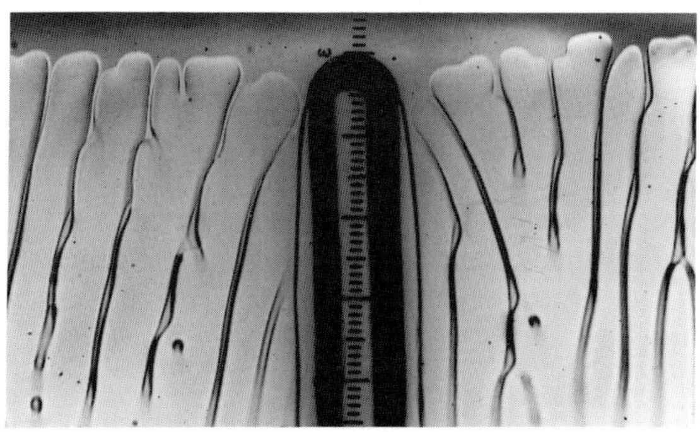

Fig. 57. Optical pattern of gas inclusions in a crystal [73].

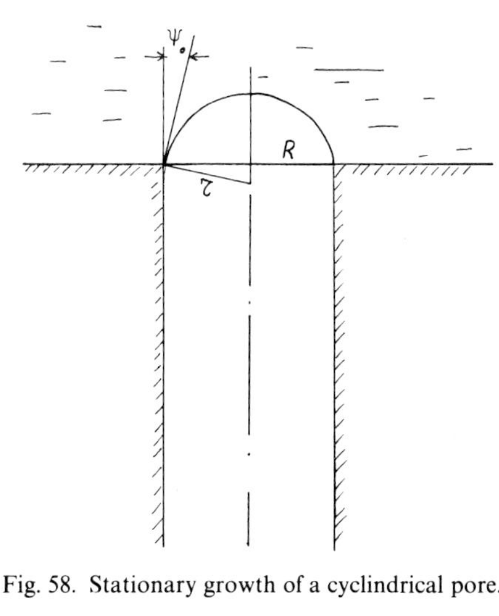

Fig. 58. Stationary growth of a cyclindrical pore.

Bond number. The pore growth is stationary if a line tangent to the meniscus makes the angle of growth ψ_0 with the normal to the crystallization front. If the pore diameter $2R$ and the gas concentration C within a pore are regarded as variable parameters, the set of equation (2.3) will include eq. (2.14) and the equation of the concentration change rate

$$\partial R/\partial t = A_{RR}\delta R + A_{RC}\delta C, \tag{8.1}$$

$$\partial C/\partial t = A_{CR}\delta R + A_{CC}\delta C. \tag{8.2}$$

The pressure P within the pore can be considered equal to CBT, where B is the universal gas constant and T is the absolute temperature.

8.2. *The capillary coefficients A_{RR} and A_{RC}*

From eq. (2.14): $A_{RR} = -V\partial\alpha_0/\partial R$ and $A_{RC} = -V\partial\alpha_0/\partial C$, where $\alpha_0 = \pi/2 - \psi_0$ and V is the rate of crystal growth.

From fig. 58 the following ratio can be derived:

$$\frac{R}{r} = \sin\alpha_0. \tag{8.3}$$

Here r is the radius of curvature of the spherical meniscus.

The Laplace capillary equation has the following form in this case:

$$\frac{2\sigma}{r} = CBT. \tag{8.4}$$

Simultaneous solution of eqs. (2.14), (8.3) and (8.4) gives the explicit functions A_{RR} and A_{RC}:

$$A_{RR} = -\frac{V}{r\cos\alpha_0} = -\frac{V}{R}\,\mathrm{tg}\,\alpha_0 = -\frac{VCBT}{(4\sigma^2 - R^2C^2B^2T^2)^{1/2}}, \tag{8.5}$$

$$A_{RC} = -\frac{VRBT}{2\sigma\cos\alpha_0} = -\frac{VRBT}{(4\sigma^2 - R^2C^2B^2T^2)^{1/2}}. \tag{8.6}$$

8.3. *The concentration coefficients A_{CR} and A_{CC}*

The rate of gas concentration change in a pore depends on the rate of the gas flow through the surface of the melt meniscus and the rate of pore volume change:

$$\frac{\mathrm{d}C}{\mathrm{d}t} = \frac{1}{W}\left(\frac{\mathrm{d}m}{\mathrm{d}t} - C\frac{\mathrm{d}W}{\mathrm{d}t}\right). \tag{8.7}$$

Here m is the mass of the gas in the pore, W is the pore volume: $W = \pi R^2 l$, where l is the pore length. The explicit function of the rate of gas inflow through the surface of the melt meniscus can be written as follows:

$$\frac{\mathrm{d}m}{\mathrm{d}t} = \beta(C_{\mathrm{m}} - C)S. \tag{8.8}$$

Here β is a kinetic coefficient that will be discussed below, C_{m} the concentration of gas in the melt on the surface of the liquid meniscus, S is the area of the liquid meniscus (a part of the spherical segment):

$$S = 2\pi r[r - (r^2 - R^2)^{1/2}]. \tag{8.9}$$

The rate of pore volume change can be given by the following equation:

$$\frac{\mathrm{d}W}{\mathrm{d}t} = \pi R^2 V. \tag{8.10}$$

In accordance with eq. (2.3) the coefficients A_{CR} and A_{CC} are defined by differentiating the right-hand side of eq. (8.7):

$$A_{CR} = \frac{\pi R}{W}\left[\frac{4\beta\sigma(C_{\mathrm{m}} - C)}{(4\sigma^2 - C^2B^2T^2R^2)^{1/2}} - 2CV\right], \tag{8.11}$$

$$A_{CC} = \frac{\pi}{W}\left\{(C_{\mathrm{m}} - C)\left(\frac{4\beta R^2\sigma}{C(4\sigma^2 - C^2B^2T^2R^2)^{1/2}} - \frac{32\beta V\sigma^2}{B^2C^2T^2}\left[1 - \left(1 - \frac{R^2C^2B^2T^2}{4\sigma^2}\right)^{1/2}\right]\right) - \frac{8\beta\sigma^2}{B^2T^2C^2}\left[1 - \left(1 - \frac{R^2C^2B^2T^2}{4\sigma^2}\right)^{1/2}\right] - R^2V\right\} \tag{8.12}$$

8.4. *Analysis of the process-stability conditions*

The process-stability conditions will be analyzed for $\psi_0 = 14°$, which is characteristic for some insulators, e.g. sapphire:

$$A_{RR} = -\frac{4V}{R}, \tag{8.13}$$

$$A_{RC} = -VR\frac{2BT}{\sigma} = -\frac{4V}{C}, \tag{8.14}$$

$$A_{CC} = \frac{1}{l}\left[\beta\left(5\frac{C_{\mathrm{m}}}{C} - 6.6\right) - V\right], \tag{8.15}$$

$$A_{CR} = \frac{2C}{Rl}\left[4\beta\left(\frac{C_{\mathrm{m}}}{C} - 1\right) - V\right]. \tag{8.16}$$

Now from the Routh–Hurvitz conditions ($A_{RR} + A_{CC} < 0$; $A_{RR}A_{CC} - A_{RC}A_{CR} > 0$) the necessary and sufficient conditions of the growth of cylindrical ($l \gg R$) pores can be defined:

$$3\frac{C_{\mathrm{m}}}{C}-1.5>\frac{V}{\beta}>\tfrac{5}{4}\frac{R}{l}\left(\frac{C_{\mathrm{m}}}{C}-1.3\right). \tag{8.17}$$

In particular, for the beginning of the process $l \sim 2R$ and the conditions are as follows:

$$3\frac{C_{\mathrm{m}}}{C}-1.5>\frac{V}{\beta}>\tfrac{5}{8}\frac{C_{\mathrm{m}}}{C}-0.81. \tag{8.18}$$

From [74] it follows that $\beta \sim D_l/R$, where D_l is the diffusion coefficient of the gas in the melt, which is equal to $10^{-5}\,\mathrm{cm^2s^{-1}}$. It allows us to rewrite eq. (8.18) and to obtain the range of rate growth for cylindrical pore formation:

$$10^{-5}\left(\tfrac{3}{2}\frac{C_{\mathrm{m}}BT}{\sigma}-\frac{1.5}{R}\right)>V>10^{-5}\left(\tfrac{1}{4}\frac{C_{\mathrm{m}}BT}{\sigma}-\frac{0.72}{R}\right). \tag{8.19}$$

We did not take into account external pressure and did not use the Gibbs–Thompson dimensional dependence of growth angle. But, as a first approximation, this model allows one to understand the conditions of sufficiently large diameter (more than some micrometry) cylindrical pore formation and growth.

9. *The two-shaping-elements technique (TSET)*

Manufacture of cheap silicon sheets is one of the most important problems of modern semiconductor metallurgy. Its solution would, to a considerable extent, determine the ways of further development of solar power engineering and microelectronics. Silicon profiling by TPS allows the manufacture of thin-walled crystals suitable for solar cell production. Investigation of capillary shaping during such crystal growth reveals that growth of the meniscus length is of the order of a hundred micrometers. Owing to this, the professional level of an operator needs to be high and the process is difficult to automate.

A new melt crystallization technique allowing more than an order of magnitude increase of growth meniscus length was developed [75, 76]. This is achieved by using two shaping elements, a melt wettable support substrate and a feeder, a device for controlled growth meniscus formation, placed between the main melt volume and the substrate and connected with the main melt volume by capillary channels. A diagram of a horizontal version of the technique is given in fig. 59.

The feeder provides control of a number of crystallization process parameters. When comparing TSET with the technique of melt crystal growth on substrates known up to now [77] we can notice that the difference between them is very much like that between TPS and CzT. The feeder can be regarded as a shaper. Unlike the shaper, the feeder is not the only shaping element as the geometry of the crystal being grown is also determined by the substrate. In the case where graphite cloth is used as a substrate, the technique is usually called the SOC technique ("silicon-on-cloth") [75, 76]. We will consider the SOC technique as a particular case of TSET.

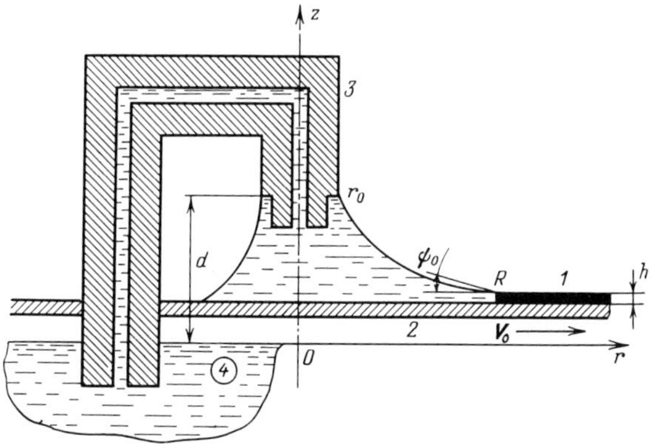

Fig. 59. Crystal growth by the two-shaping-elements technique TSET: 1: tape; 2: substrate; 3: feeder; 4: melt volume; V_0: substrate displacement velocity; h: crystallized layer thickness; ψ_0: growth angle; r_0: feeder free-edge position relative to the feeder axis; d: melt pressure in the meniscus; R: crystallization front position relative to the feeder axis.

9.1. *Capillary shaping*

To define specific features of capillary shaping of plates in TSET it is necessary to solve the capillary problem for eq. (2.7) with the catching boundary conditions at the feeder free edge and the angle fixing condition at the crystal free edge (fig. 59). Profile curves for various pressure values for crystals 0.25 mm thick are given in fig. 60.

The fact that in the growth of thin-walled crystals the meniscus length is an order or magnitude higher than that when one shaping element, a shaper, is used, attracts attention at once. The second specific feature is that the meniscus length does not decrease with the growing crystal thickness decreasing as in the case where a shaper is used but on the contrary it increases.

The shaping function of the substrate is in the first place to "support" the meniscus end-face sections where its length is smaller than in its main part. Besides, the substrate presence allows large deviations of the crystallization front from a straight line, which extends the technological potentialities of the technique.

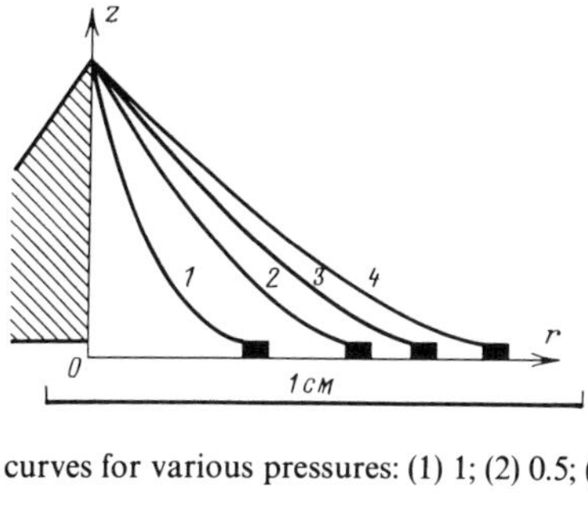

Fig. 60. Profile curves for various pressures: (1) 1; (2) 0.5; (3) 0.25; (4) 0.1.

9.2. *Hydrodynamic conditions*

At first the process of a flat horizontally pulled substrate wetted by the melt will be considered without an allowance for crystallization. The solution of this probelm will give data on minimum film thicknesses. A similar problem was analyzed in [78]. Let the liquid surface on the substrate be divided into two sections: (I) static meniscus limited by the feeder free edge and (II) substrate-entrained liquid film (fig. 61).

Range I is described by the Laplace capillary equation (2.7), range II is described by the Navier–Stokes equation as a boundary-layer approximation. From the condition of solution joining for ranges I and II (continuity of d^2h/dr^2) the amount of substrate-entrained liquid and the liquid film thickness h_f can be determined.

So, range II is described by the following equation:

$$\eta \frac{d^2V}{dz^2} = \frac{1}{\rho}\frac{dP}{dr}, \tag{9.1}$$

with the following boundary conditions:

$$V = V_0, \quad z = 0, \tag{9.2}$$

$$\frac{dV}{dz} = 0, \quad z = h. \tag{9.3}$$

Here η is the kinematic melt viscosity, V is a component of pulling rate parallel to the pulling direction, P denotes the capillary pressure, V_0 is the substrate rate, z is the vertical coordinate measured from the substrate free edge, $h(r)$ is the melt surface profile.

The solution of the problem gives, for silicon (the values of the silicon melt constants are given in table 5):

$$h_f = 0.39\, V_0^{2/3}/P. \tag{9.4}$$

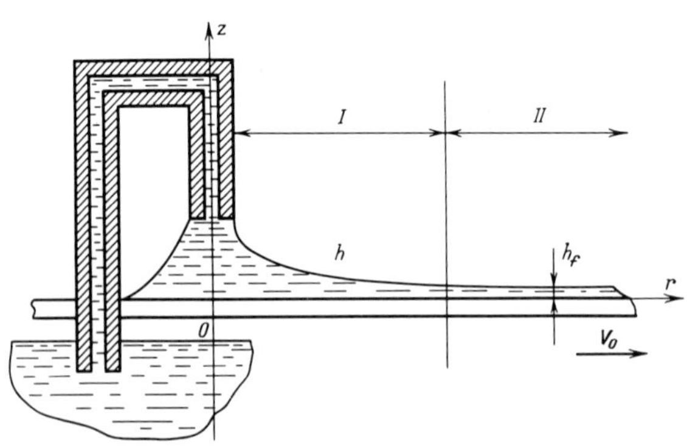

Fig. 61. Hydrodynamic conditions of crystallization using a feeder and substrate: I: the range of capillary shaping; II: the range of viscous flow; h_f: film thickness; V_0: the rate of substrate motion.

Taking into consideration the substrate displacement rates within the range used in growth practice (1–5 cm/min) and pressures providing spontaneous filling of capillary channels, eq. (9.4) gives values of liquid film thickness h_f within the interval of 0.1 to 3 μm. The calculations made allow one to consider two possible regimes of melt crystallization on the substrate. In the case where the thermal conditions specify the crystallization front position within liquid-film range II (fig. 61), the thickness of the layer obtained does not depend on r, the front coordinate, and hence on its shape. Such crystallization conditions are attractive from the point of view of growth procedure because the thermal conditions need not be controlled to produce a crystalline layer uniform along the substrate width. However, such conditions can be used only in cases when very thin layers are to be grown. To produce various devices, photoconverters in particular, a silicon layer ~ 100 μm thick is required; in this case the crystallization front should lie within the range of static menisci as is shown in fig. 59. Under such conditions the thickness of the growing crystal depends on the meniscus height at the point of its intersection with the crystallization front. Thus, crystallization in range I requires controllable heat transfer conditions, i.e. it states the problem of matching the thermal and the capillary conditions of crystallization usual for the melt growth of crystals.

9.3. Stability investigation

A model of the TSET regarded as a system of two degrees of freedom, R and h, can be constructed (fig. 59). While determining the coefficient A_{hR}, heat removal is considered to take place from the upper surface of the crystallizing layer while the carbon-fiber-reinforced substrate impregnated with the melt is a thermal insulator.

Table 5 gives the values of the coefficients of the stability equations. Figure 62 gives calculated curves of the attenuation of crystallization front position perturbations $\delta h(f)/\delta h(0)$ for silicon tapes grown by TPS and by TSET under similar conditions. The behavior of the attenuation curves and the values of the main stability parameter τ_{rp} (the time of perturbation relaxation) show higher stability for the TSET. This result agrees with experiment.

Table 5
The values of the coefficients of stability and of the silicon melt constants.

Crystallization technique	$A_{RR}(s^{-1})$	$A_{Rh}(s^{-1})$	$A_{hR}(s^{-1})$	$A_{hh}(s^{-1})$	$\tau_R(s)$
TPS	−0.044	−0.070	0.970	−0.085	15.5
TSET	−0.230	−0.006	90.45	−0.112	5.8

$\psi_0 = 11°$, $V = 3.3 \times 10^{-4}\,ms^{-1}$, $P = 10^{-2}\,m$, $\mathcal{L} = 4.16 \times 10^9\,Jm^{-3}$, $\rho = 2560\,kg\,m^{-3}$, $\lambda = 60\,vt\,m^{-1}$, $\lambda_s = 21.6\,vt\,m^{-1}$, $T_0 = 1685\,K$, $T_1 = 1385\,K$, $\mu_{rad} = 414\,vt\,m^{-2}$, $\mu_{conv} = 10\,vt\,m^{-2}\,K^{-1}$.

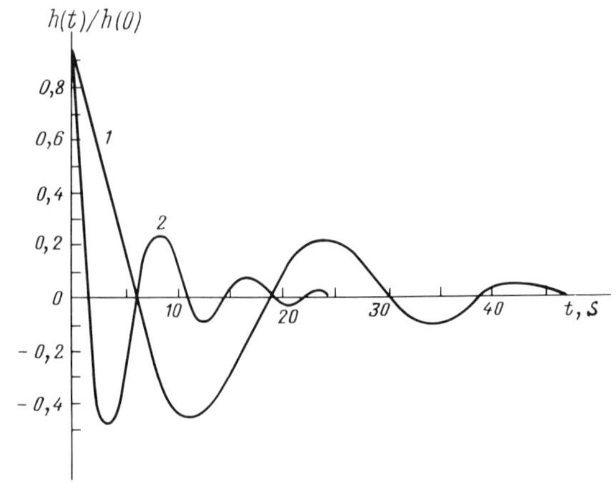

Fig. 62. Calculated curves of attenuation of crystallization front perturbation for a silicon tape: 1: TPS; 2: TSET.

9.4. Impurity distribution

The process of impurity transfer from the bulk of the melt towards the growth meniscus and impurity transfer within the meniscus will be considered in accordance with the diagram given in fig. 63. Melt contamination of the substrate material by impurities is not taken into consideration here.

Let us assume that, in the melt zone between the feeder and the substrate, complete stirring takes place, while in the meniscus a laminar liquid flow is observed for which impurity transfer is described by the one-dimensional diffusion equation.

$$D\left(\frac{d^2C}{dr^2}+\frac{1}{r}\frac{dC}{dr}\right)=-V\frac{dC}{dr} \tag{9.5}$$

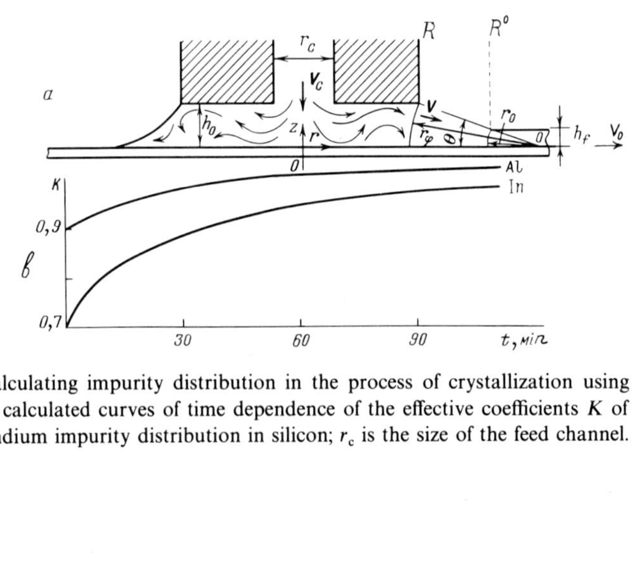

Fig. 63. (a) On calculating impurity distribution in the process of crystallization using a feeder and a substrate, and (b) calculated curves of time dependence of the effective coefficients K of aluminium and indium impurity distribution in silicon; r_c is the size of the feed channel.

with the following boundary conditions at the crystallization front:

$$-D\frac{dC}{dr}\bigg|_{r=R^0} = -V_0(1-K_0)C(R), \tag{9.6}$$

and at the left meniscus boundary

$$C|_{r=R} = C_m. \tag{9.7}$$

Here D is the coefficient of impurity diffusion in the melt, C_m denotes impurity concentration in the stirring zone; other notation is given in fig. 59.

The melt flow in the meniscus is specified by the relation $V = V_0 r_0/r$; with eq. (9.5) taking the following form:

$$\frac{d^2C}{dr^2} + \frac{1}{r}\left(1 + \frac{V_0 r_0}{D}\right)\frac{dC}{dr} = 0. \tag{9.8}$$

Solution of this equation with the boundary conditions formulated above allows one to derive a ratio of the impurity concentration in the crystal, C_S, to its concentration in the stirring zone, C_m ($K^* = C_S/C_m$):

$$K^* = \frac{K_0}{K_0 + (1-K_0)(h_f/h_0)^{h_f V_0/D\sin\theta}}. \tag{9.9}$$

Here K_0 is the equilibrium coefficient of impurity distribution.

Since for the crystallization scheme under consideration the length of the feeder capillary channel l_c is large in comparison with the length of the shaper capillary channel of TPS, the condition $V_c \gg D/l_c$ (V_c is the rate of the laminar melt flow through a capillary channel of the width r_c) (fig. 63) holds for the most part of the impurities in silicon melt and the reverse impurity diffusion into the bulk of the melt can be ignored.

Impurity concentration in the stirring zone C_m depends on time in the following manner:

$$C_m(t) = C_\infty + \frac{V_c r_c C_\infty - K_0 C(R_0)}{W_m} t, \tag{9.10}$$

where W_m denotes the stirring zone volume, t is the time. With an allowance for the effective coefficient of impurity distribution $K = C_S/C_\infty$, eq. (9.10) allows to derive an expression for $K(t)$:

$$K(t) = \frac{1 + V_c r_c t/W_m}{1/K^* + V_0 h_f t/W_m}. \tag{9.11}$$

For the most part of impurities in silicon melt $K \approx 1$ irrespective of the growth time, however for impurities characterized by small K_0 values and large D values the effective coefficient of impurity distribution reaches the value of one only for a long period of time.

Figure 63b shows the values of $K(t)$ for Al ($K_0 = 0.002$, $D = 5.3 \times 10^{-8}\ m^2 s^{-1}$) and In ($K_0 = 0.0004$, $D = 5.2 \times 10^{-8}\ m^2 s^{-1}$). The values of $h_f = 2.5 \times 10^{-4}$ m, $r_c = 8 \times 10^{-4}$ m, $W_m = 1.16\ cm^2$, $h_0 = 3.5 \times 10^{-3}$ m, $V_0 = 3 \times 10^{-4}\ m\,s^{-1}$ used in practice are taken.

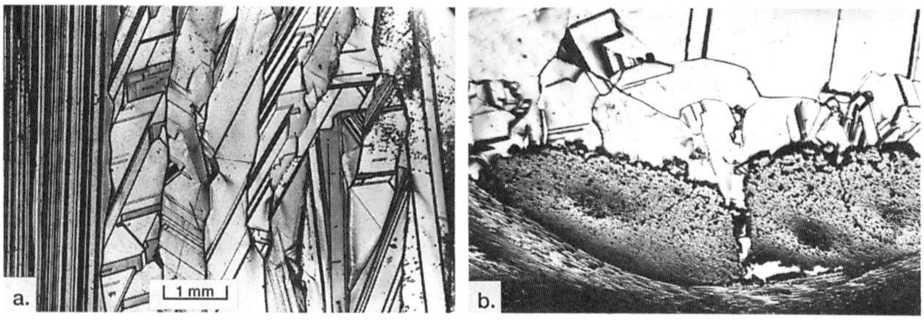

Fig. 64. (a) Etched surface and (b) cross-section of SOC specimen.

The analysis demonstrates that macro-uniform crystalline layers can be grown by silicon crystallization using a feeder and a substrate. The effect of the gradual approach of K to unity becomes important only for thin-layer growth when crystallization takes place in the liquid film range II (fig. 61).

9.5. *Silicon as a component of the SOC material*

Crystalline structure of silicon is of greatest importance since the grain side of polysilicon affects photoconverter efficiency. Figure 64a shows that the silicon layer is formed of elongated grains 0.5–3 mm wide and 5–30 mm long stretched in the direction of growth. Here and there, multi-twinned sections whose structure is similar to that of thin-walled profiled silicon crystals grown by TPS can be observed but their share does not exceed 3–5% of the overall area.

In the immediate vicinity of the carbon-fabric filaments, fine practically equiaxial grains 20–50 μm in size can be observed (fig. 64b). The patterns of transverse microsection metallographic specimen given show that the boundaries of coarse grains forming the surface silicon layer are preferentially oriented perpendicular to the surface.

The materials are designed for photoconverter substrate application. Usually, the silicon layer resistivity does not exceed $1\,\Omega \cdot \mathrm{cm}$, resistivity of the ternary Si–SiC–C base is $0.02\text{–}0.04\,\Omega \cdot \mathrm{cm}$. Thus, the SOC technique allows one to produce a semiconducting crystalline layer on an electro-conductive substrate directly from the melt in the course of one production cycle.

10. *The variable shaping technique (VST)*

Growth of profiled crystals with variable cross-sectional shape is a very important problem. While growing crystals by TPS, the profile cross-sectional shape is mainly controlled by the shaper design, and its cross-section dimensions can be varied only within very narrow limits restricted by the meniscus existence zone. Therefore, the first attempt to widen the limits of varying pulled profile cross-section dimensions in the process of growing was to displace movable shaper elements [79].

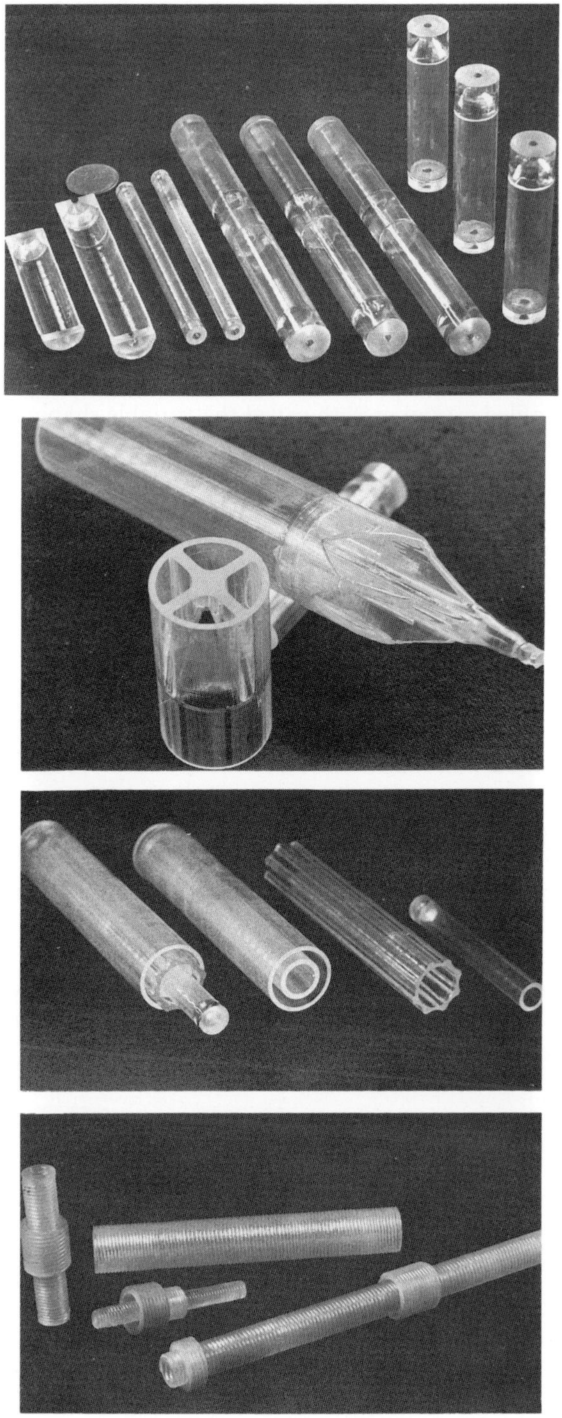

Fig. 65. Monocrystalline sapphire profiles with variable cross-section shapes grown by VST [82].

The second technique of varying profile dimension (e.g. tube inner diameter) is based on melt spreading from a circular capillary channel over the shaper solid surface with proper melt temperature and pulling rate control provided [80].

Experience in application of shaper movable element displacement techniques and control of melt spreading over the shaper solid surface revealed a number of disadvantages thereof. Even in the case of precise manufacture of shaper movable elements their catching occurred. Practically, a shaper could not be used for more than two growing procedures because of its deformation caused by temperature changes. The second technique is difficult to reproduce when tubes with diameter exceeding 15 mm are grown because of process instability caused by absence of the inner meniscus catching on the shaper edge.

A technique of profiled crystal growth called the variable shaping technique (VST) was developed [81–83]. It possesses none of the disadvantages mentioned above and allows one to change within wide limits both the dimensions and the shape of the crystal cross-section during the growing process. VST is based on controlling the melt mass flow towards the crystallization interface when changing for a new specified cross-section shape. In this case different shaping elements of the same shaper, its free edges or walls, operate by turns. Periodic changes of tube diameter and the attachment and breaking of side ribs of the growing tube at a moment specified offer no difficulties (fig. 65).

Transition from a rectangular cross-section profile to a tube and back is also easy to perform. Complicated shape corundum profiles and crucibles of various diameters, 16, 25 and 30 mm, with a wall thickness of 2.5 mm have been produced (fig. 65).

The VST allows improvement of multiple-growth techniques, i.e. growing a group of profiles using one seed crystal. A profile with a cross-section containing all the elements of the future profile group and cross-pieces between them is formed in the process of growing. When growth is completed, change for a new cross-sectional shape without cross-pieces is carried out and the group of profiles grows independently. VST can be applied to growing profiled crystals of various materials.

The investigation of the dynamic stability of VST does not differ from that of TPS. It is necessary only to analyze the stability of each part of the crystal unchanged in cross section. The scheme of analysis described in section 4 can be used.

11. The local shaping technique (LST)

The local shaping technique is one of the many versions of TPS. In contrast to conventional TPS, in this process, the shaper forms only an element of the predetermined cross-section of the crystal being pulled, while the whole cross-section is obtained by means of crystal rotation (fig. 66). This technique was applied to growth of lithium fluoride tubes with 20 mm external diameter and 4 mm thick walls [84], sapphire darks of 60 mm in diameter [85] and complex sapphire products [83, 86, 87], with continuous alteration of the side surface of hollow cone-shaped tubes (fig. 67).

In fact LST is the combination of TPS and TSET, where the bottom of the grown crystal plays the role of substrate.

Let us analyze the dynamic stability of LST for the example of tube growth [87].

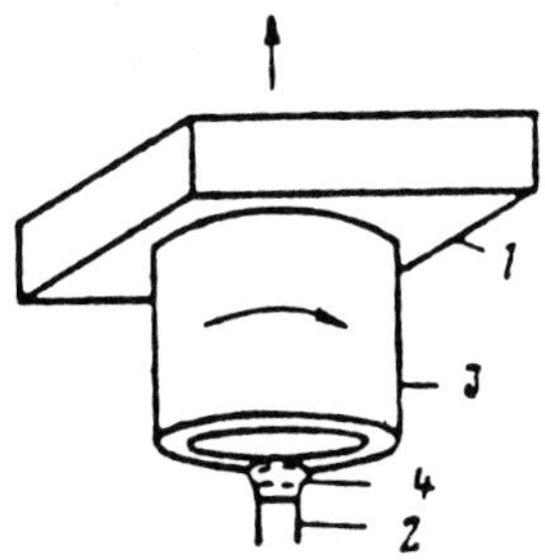

Fig. 66. Schematic drawing of LST realization tube-growth [84]: 1: seed; 2: shaper; 3: grown crystal; 4: meniscus.

Fig. 67. Some complex sapphire products grown by LST [83].

11.1. Formulation of the problem

In steady-state growth, the length of the crystal increases by $l = V/n$ per revolution. From the viewpoint of mass balance, this is equivalent to a situation as though, at the interface, there were a macrostep normal to the growing surface (fig. 68a). The height to the step is $l = V/n$, its growth occurs in the plane of the step only, and there is no mass transfer over the rest of the melt–crystal interface. Therefore, as in the analysis of other processes of growing crystals from the melt, the real shape of the crystallization front can be replaced by a planar front: in our case, the flat surface of the step is assumed to serve as the crystallization front. The position of the crystallization front relative to the edge of the shaper will be given by the displacement $u = R\psi$, where ψ is the central angle, measured with respect to the shaper's edge in the direction of the crystal rotation (fig. 68b).

If the level of melt in the crucible is assumed to be constant during the course of crystallization, the process will have three degrees of freedom: the height of the step l;

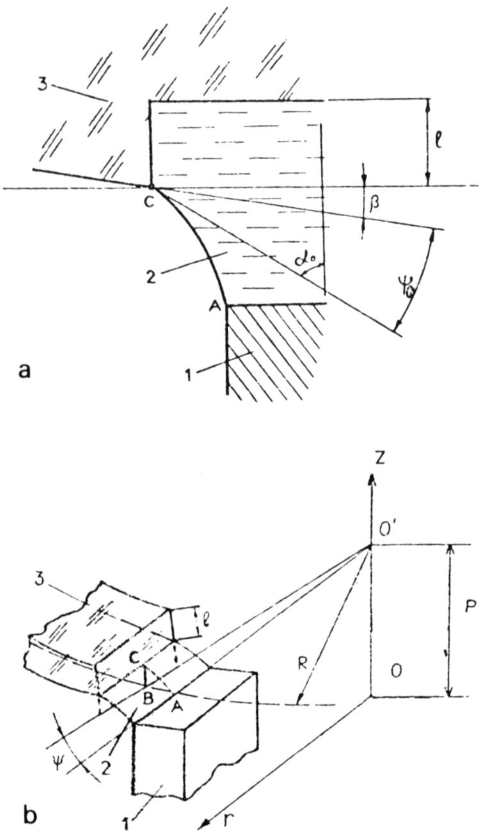

Fig. 68. (a) Schematic representation of crystal formation and (b) degrees of freedom: 1: shaper; 2: meniscus; 3: crystal.

the angle ψ, which determines the position of the crystallization front relative to the shaper's edge; and the wall thickness, m, of the tube. However, since the pulling rate of the tube is much less than its linear rate of rotation ($V/2\pi Rn \sim 10^{-3}$ to 10^{-4}), then, for perturbations of steady-state growth, the thickness of the wall will vary much more slowly than the height of the step l. This allows an analysis with two degrees of freedom only: l and ψ. The linearized set of eq. (2.3) is of the form:

$$\partial l/\partial t = A_{ll}\delta l + A_{l\psi}\delta\psi, \tag{11.1}$$

$$\partial\psi/\delta t = A_{\psi l}\partial l + A_{\psi\psi}\delta\psi. \tag{11.2}$$

11.2. Capillary coefficients A_{ll} and $A_{l\psi}$

The estimates made in section 2.2 have shown that, at rates of crystallization below 100 cm/s, the effects of inertial forces can be neglected. The rate of crystallization of sapphire tubes of 60 mm diameter does not exceed 2 cm/s, which makes it possible to consider the formation of the melt column in a hydrostatic approximation.

In the case where stationary growth is disturbed, the rate of variation of the step height is similar to that of the characteristic transverse size of crystals in other techniques of crystallization from the melt and may be given by the relation of eq. (2.13) $dl/dt = V\delta\alpha_0 = 2\pi Rn\,\delta\alpha_0$, where $\delta\alpha_0 = \alpha_0 - \alpha_e$, is the angle between the tangent to the meniscus at point C and the vertical (fig. 68a), and α_e is the equilibrium value of the angle α_0. In steady-state growth $\alpha_0 = \frac{1}{2}\pi - \psi_0 - \beta$, where ψ_0 is the growth angle and β is the angle of the spiral line slope. Since $\mathrm{tg}\,\beta \approx \beta \approx V/2\pi Rn = 10^{-3}$ to 10^{-4}, one may neglect its value and assume that $\alpha_e = \frac{1}{2}\pi - \psi_0$. The linearized equation for the rate of step height variation is

$$\partial l/\partial t = -2\pi Rn\frac{\partial\alpha_0}{\partial l}\delta l - 2\pi Rn\frac{\partial\alpha_0}{\partial\psi}\delta\psi. \tag{11.3}$$

Where $A_{ll} = -2\pi Rn(\partial\alpha_0/\partial l)$ and $A_{l\psi} = 2\pi Rn(\partial\alpha_0/\partial\psi)$. The values of the derivatives $\partial\alpha_0/\partial l$ and $\partial\alpha_0/\partial\psi$ are determined from the solution of the Laplace capillary equation. In the present case, the meniscus is not symmetrical and is rather complex in shape. However, to analyze the stability of the crystallization process (as was done in TSET), the sign of the derivatives can be determined from the shape of the profile curve AC (fig. 68a), without considering the bulk of the melt column. In solving the Laplace equation, the boundary condition of catching of the meniscus at the edge of the shaper and the condition of fixing of the growth angle are used.

We consider two limiting cases: (1) AC is the profile curve of the meniscus formed in the process of growing a plate; and (2) AC is the profile curve of the meniscus formed in the process of growing a thin rod with radius $r_c < a$, where a is the capillary constant. Using the results of detailed studies of shapes of different menisci, it can be easily shown that $A_{ll} < 0$ and $A_{l\psi} < 0$ for both limiting cases considered.

11.3. Thermal coefficients $A_{\psi l}$ and $A_{\psi\psi}$

The values of the thermal coefficients can be found from the condition of heat balance at the crystallization front, eq. (2.15).

In this case, $V_c = V - du/dt = 2\pi Rn - R(d\psi/dt)$. The temperature gradients G_i at the crystallization front, as in the analysis of other techniques of crystallization form the melt, are determined from the solution of eq. (2.17) by assuming the tube wall thickness $m \gg l$. The boundary conditions involve the given temperatures at the crystallization front, T_0, and that of the melt, T_m, at a distance of $u = R\psi^0$ from the crystallization front, where ψ^0 is the value of the angle determining the position of the step l_0 under steady-state conditions. Far from the crystallization front the temperature gradient in the solid phase is taken to be zero. It follows from the solution that $G_s = -T_0\xi$ and $G_l = (T_0 - T_m)/R\psi^0$, where

$$\xi = \frac{VR}{2k_s} - \left[\left(\frac{VR}{2k_s}\right)^2 + \frac{\mu_s R^2}{\lambda_s l}\right]^{1/2}.$$

k_s and μ_s are the coefficients of thermal conductivity and heat exchange, respectively, of the solid phase. Taking into account the fact that

$$VR/2k_s \ll \mu_s R^2/\lambda_s l,$$

we obtain

$$\xi = (\mu_s R^2/\lambda_s l)^{1/2}.$$

Substituting the values of temperature gradients and crystallization velocity into eq. (2.15) and linearizing the latter, we obtain

$$A_{\psi l} = \frac{\lambda_s T_0}{2L\rho R} \frac{\xi}{l_0} > 0, \qquad A_{\psi\psi} = \frac{\lambda_e}{L\rho} \frac{(T_0 - T_m)}{R^2 \psi_0^2}. \tag{11.4}$$

It follows that, when the melt is overheated, $T_0 < T_m$ and $A_{\psi\psi} < 0$. When the melt is supercooled, $A_{\psi\psi} > 0$.

11.4. Analysis of stability

In order for the set of eqs. (11.1) and (11.2) to be stable, it is necessary and sufficient that (Routh–Hurvitz conditions):

$$A_{ll} + A_{\psi\psi} < 0, \tag{11.5}$$

$$A_{ll} A_{\psi\psi} - A_{l\psi} A_{\psi l} > 0. \tag{11.6}$$

From the values of the coefficients obtained in the foregoing analyses, we can conclude that the conditions of eqs. (11.5) and (11.6) are satisified when the melt is overheated. The experimental data indicate that the surface of the crystallized layer melts whenever it enters the melt column, thus indicating overheating of the melt. Therefore, we may conclude that the growth of shaped crystals by LST is realized in practice as a result of the stability of the process.

12. Stability of crystal cross-sectional shapes

In the previous paragraphs *uniform perturbations* (change in the radius of a cylindrical crystal, tube or plate thickness) were superimposed on the cross-section of a growing crystal when stability was analyzed. Such a consideration solves the problem of characteristic profile dimension stability; however the problem of maintaining the shape of a crystal being pulled remains unsolved.

To analyze shape stability, an *arbitrary small perturbation* that can be expanded into a series in respect to corresponding set of orthogonal functions depending on the shape of the perturbed contour should be superimposed on the contour limiting the profile cross-section and then the time dependence of the coefficients of this expansion should be calculated [88–90]. In this case the cross-sectional shape is stable, provided the coefficients of all the terms of the expansion decrease with time. The first term of the expansion corresponds to the uniform perturbation, i.e. to some change in the cross-section dimension while its shape remains unchanged. The unperturbed meniscus shape described by the function z_0 satisfying the Laplace capillary equation with corresponding boundary conditions and the equilibrium angle α_e correspond to an unperturbed contour. The solution $z = z_0 + z_1$ satisfying the two-dimensional eq. (2.5) linearized in respect to a small correction factor z_1 and the corresponding angle $\alpha_0 = \alpha_e + \delta\alpha_0$ will correspond to the small perturbation of the contour shape. Since all the equations obtained in this case are linear with respect to z_1, the behavior of an arbitrary small and rather smooth distortion can be investigated by analyzing the behavior of the superposition of the components of the z_1 expansion with respect to the corresponding set of orthogonal functions.

The solution of this problem [88, 90] allows one to conclude that *if the system is stable in a capillary sense relatively to uniform perturbations* ($A_{RR} < 0$) *it is stable relatively to arbitrary perturbation* for both types of boundary conditions satisfied on the shaper.

The allowance for morphological instability [89] of the crystallization front gives an additional condition of crystal cross-sectional shape stability. *In the case where conditions of morphological stability are provided, the shape of the crystal cross-section is also stable. However, in the case where morphological instability exists, capillary instability of the crystal shape will be observed.*

The results of experiments [91] on growing sapphire filaments by TPS can be explained by this statement. If the growth rate was high enough ($V \sim 0.2$ cm/s) and the filament radius $R \leq 0.01$ cm, the surface of the crystallization front represented two bulges (dendrites) towards the melt while the filament cross-section represented an ellipse. Thus, morphological instability led to second-harmonic distortion of the front, which resulted in distortions of the crystal cross-sectional shape.

Our experiments on growing sapphire rods with the crystallization front inclined [92] can also be described by the model discussed here. The distortion of the crystallization front in that experiment was achieved by inclining it and one-sided crystal contraction or widening naturally leads to distortions of initially circular crystals.

13. *Crystallization by CST under microgravity conditions*

Gravity is an important factor affecting the crystallization process when the CST are used. Therefore, with technical facilities to perform microgravity crystallization available, it is important to study the effect of gravity on the stability of crystallization carried out using these techniques [93, 94]. The above theoretical data for small Bond number can be used.

Application of the stable growth concept to crystallization using the VT is in particular an illustrative example. As was shown above, the stable–unstable process transition is first of all determined by capillary effects. With the meniscus height being linearly dependent on the crystal diameter (small Bond numbers), i.e. $h = R(1 + \cos\alpha_0)\sin^{-1}\alpha_0$, the process is stable. Under gravity conditions it is possible only for a small diameter, i.e. for sapphire $2R \ll 6$ mm (capillary constant for sapphire).

However, under microgravity conditions, e.g. on board a space station for the level of microgravity of $(10^{-4}–10^{-6})g_0$, the capillary constant $a_s \approx (10^{-2}–10^{-3})a_0$. Hence, the region of stability is two–three orders of magnitude larger: $2R \ll 600–6000$ mm. This is a very important conclusion.

It is of great interest due to the experiments on crystal growth conducted in space. An oxidic burner should not necessarily be used as a heat source. Therefore, any heat source which satisfies the requirement of eq. (5.28) can be used. In addition to process stability, large-dimensional crystal growth by VT under microgravity conditions can offer additional positive effects. *Firstly*, the melted-layer thickness significantly increases, which will allow one to decrease the temperature gradients at the crystallization front and, correspondingly, to decrease thermal stresses. *Secondly*, for thick melt layers the size of the feed particles can be increased. This significantly reduces the requirements on the initial charge thus simplifying its preparation.

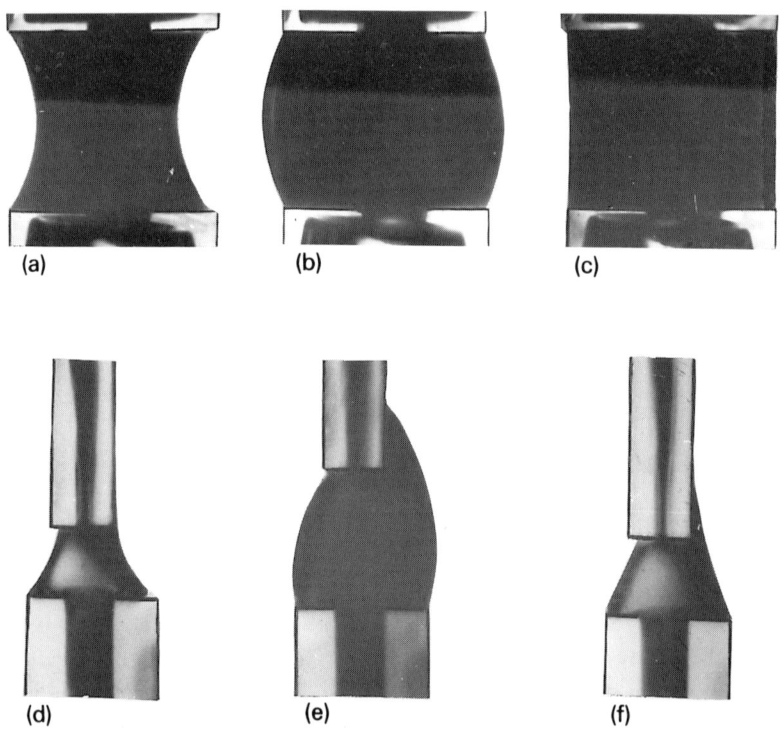

Fig. 69. Meniscus model appearing on pulling a circular rod (a, b, c) and a plate (d, e, f) under zero-g conditions: (a, d) concave meniscus; (b, e) convex meniscus; (c) meniscus in the form of a right circular cylinder; (f) meniscus of a straight profile.

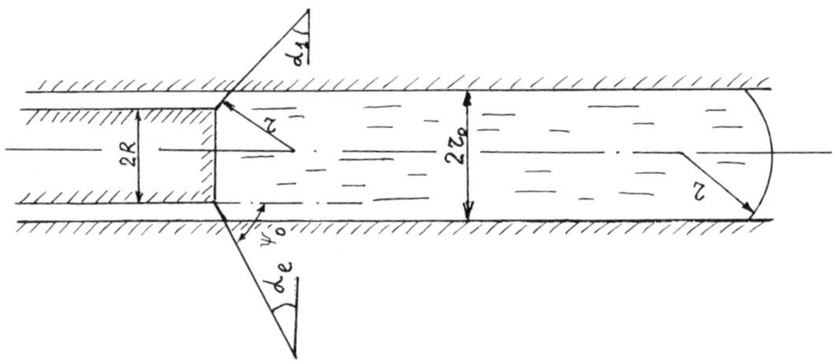

Fig. 70. Schematic drawing of a tape-growth space experiments.

Crystal growth experiments under space conditions were preceded by simulating the liquid column shape using immiscible liquids of equal densities (fig. 69) [94, 95] and the crystallization of copper under the conditions of capillary shaping on a high-altitude rocket [94, 96].

The solution of the capillary problem for small Bond number [38, 14] shows that a cylindrical melt column ($R = r_0; \alpha_0 = \frac{1}{2}\pi$) can be realized in microgravity conditions and we managed to obtain it in simulation experiments (fig. 69). In the conditions of microgravity on the "Salyut" orbital space probe, crystallization of indium was carried out by TPS from the cylindrical melt column [94, 97].

The crystallization of ribbons from the melt of Ge and GaAs in the orbital station "Mir" is the next example of CST using microgravity conditions [14]. The melt is placed in a fixed gap between two flat walls. It is the realization of TPS with the boundary conditions of wetting on the walls of a shaper. The employment of unwetted walls was successful, if the conditions $\alpha_1 > \alpha_0$ and $\theta - \frac{1}{2}\pi > \frac{1}{2}\pi - \psi_0$ were satisfied.

The meniscus projection is part of a circle and the dependence between the ribbon thickness $2R$ and the gap $2r_0$ is (fig. 70):

$$R = r_0 \frac{\sin \alpha_0}{\sin \alpha_1}. \tag{13.1}$$

The capillary coefficients of eq. (3.1) are:

$$A_{RR} = -V \frac{\sin \alpha_1}{r_0 \cos \alpha_0}, \tag{13.2}$$

$$A_{Rh} = -V \frac{\sin\alpha_1}{r_0 \sin\alpha_0}. \tag{13.3}$$

The coefficient A_{RR} is negative and there is capillary stability from a theoretical point of view.

References

[1] J. Czochralski, Z. Phys. Chem. 92 (1917) 219.
[2] A.V. Stepanov, Zh. Tech. Fiz. 29 (1959) 339 [Soviet Phys.-JETP].
[3] B. Chalmers, H.E. LaBell and A.J. Mlavsky, J. Cryst. Growth 13/14 (1972) 84.
[4] G.H. Schwuttke, Phys. Status Solidi (a) 43 (1977) 43.
[5] M.A. Verneuil, Ann. Chim. Phys., Huitième ser. 3 (1904) 20.
[6] W.G. Pfann, Zone Melting (Wiley, New York, London, 1966).
[7] V.A. Tatarchenko, Fiz. Khim. Obrab. Mater. 6 (1973) 47 [Sov. Phys. Chem. Mater. Treatment].
[8] T. Surek and B. Chalmers, J. Cryst. Growth 29 (1975) 1.
[9] T. Surek and S.R. Coriell, J. Cryst. Growth 37 (1977) 253.
[10] D.T.J. Hurle, J. Cryst. Growth 42 (1977) 473.
[11] D.T.J. Hurle, G.C. Joyce, G.C. Wilson et al., J. Cryst. Growth 74 (1986) 480.
[12] D.T.J. Hurle, G.C. Joyce, M Ghassempoory et al., J. Cryst. Growth 100 (1990) 11.
[13] V.A. Borodin, E.A. Brener, T.A. Steriopolo and V.A. Tatarchenko, Cryst. Res. Technol. 17 (1982) 1199.
[14] V.A. Tatarchenko, Ustoichiviy Rost Kristallov (Nauka, Moscow, 1988) p. 240; Engl. transl.: Shaped Crystal Growth (Kluwer, Dordrecht, 1993).
[15] G.A. Korn and T.M. Korn, Mathematical Handbook for Scientists and Engineers (McGraw-Hill, New York, 1961) p. 282.
[16] W.W. Mullins and R.F. Sekerka, J. Appl. Phys. 35 (1964) 444.
[17] V.A. Likov, Handbook on Heat and Mass Transfer (Energia, Moscow, 1978) p. 132.
[18] L.D. Landau and E.M. Lifshitz, Mecanique des Fluids (Mir, Moscow, 1971) p. 289.
[19] G.K. Gaule and J.R. Pastore, in: Metallurgy of Elemental and Compound Semiconductors (Interscience, New York, 1961) p. 201.
[20] Ju. M. Shashkov and E.V. Melnikov, Zh. Fiz. Khim. 39 (1965) 1364 (Sov. J. Phys. Chem.).
[21] P.I. Antonov, Growth Cryst. 6 (1965) 158.
[22] G.A. Satunkin, V.A. Tatarchenko and V.Y. Shaitanov, J. Cryst. Growth 50 (1980) 133.
[23] V.V. Voronkov, Fiz. Tverd. Tela 5 (1963) 571 [Sov. Solid St. Phys.].
[24] W. Bardsley, F.C. Frank, G.W. Green and D.T.J. Hurle, J. Cryst. Growth 10 (1971) 263.
[25] V.V. Voronkov, Kristallogr. 23 (1978) 249 [Sov. Crystallogr.].
[26] V.V. Voronkov, Kristallogr. 19 (1974) 927 [Sov. Crystallogr.].
[27] E.A. Brener, V.A. Tatarchenko and B.E. Fradkov, Kristallogr. 27 (1982) 205 [Sov. Crystallogr.].
[28] D. Schwabe, A. Scharmen, F. Presser and R. Ocder, J. Cryst. Growth 43 (1978) 305.
[29] V.A. Tatarchenko and E.A. Brener, in: Materials and Processes of Space Technology (in Russian) (Nauka, Moscow, 1980) p. 75.
[30] E.V. Gomperz, Z. Phys. 8 (1922) 184.
[31] P. Kapitza, Proc. Roy. Soc. Ser. A. 119 (1928) 358.
[32] Bull. Acad. Sci. USSR, Phys. Ser. 3 (1969) No. 12; 35 (1971) No. 3; 36 (1972) No. 3; 37 (1973) No. 11; 40 (1976) No. 7; 44 (1980) No. 2; 47 (1983) No. 2; etc.
[33] H.E. LaBelle, US Patent No. 3915662 (1975).
[34] T. Toshiba, J. Electron. Eng. 5 (1976) 46.
[35] S.K. Brantov, B.M. Epelbaum and V.A. Tatarchenko, Mater. Lett. 2 (1984) 274.
[36] V.A. Tatarchenko, A.I, Saet and A.V. Stepanov, in: Proc. 1st Conf. on Stepanov's Growth of Semiconductor Single Crystals (in Russian) (Ioffe Phys.-Tech. Inst, Leningrad, 1968) p. 83.
[37] V.A. Tatarchenko, A.I. Saet and A.V. Stepanov, Bull. Acad. Sci. USSR, Phys. Ser. 33 (1969) 1782.
[38] V.A. Tatarchenko, J. Cryst. Growth 37 (1977) 272.
[39] N.K. Adam, The Physics and Chemistry of Surfaces (Cambridge University Press, Cambridge, 1941) p. 238.
[40] S.V. Tsivinskiy, P.I. Antonov and A.V. Stepanov, Zh. Techn. Fiz. 40 (1970) 372 [Soviet Phys.-JETP].
[41] S.V. Tsivinskiy, Inzh. Fiz. Zh. 5 (1962) 59 [Sov. J. Eng. Phys.].
[42] G.A. Poincaré, Course de Physique Mathem, Capillarite (Paris, 1885) p. 163.
[43] V.A. Tatarchenko, Bull. Acad. Sci. USSR, Phys. Ser. 47 (1983) 20.
[44] E.A. Brener and V.A. Tatarchenko, Bull. Acad. Sci. USSR, Phys. Ser. 47 (1983) 25.

[45] S.K. Brantov and V.A. Tatarchenko, Cryst. Res. Technol. 18 (1983) K59.
[46] J.A. Burton, P.C. Prim and W.P. Slichter, J. Chem. Phys. 21 (1953) 1987.
[47] J.C. Swartz, T. Surek and B. Chalmers, J. Electron. Mater. 4 (1975) 255.
[48] J.P. Kalejes, J. Cryst. Growth 44 (1978) 329.
[49] V.A. Tatarchenko, T.N. Jalovets, G.A. Satunkin et al., J. Cryst. Growth 50 (1980) 335.
[50] S. Coriell and R. Sekerka, J. Cryst. Growth 34 (1976) 157.
[51] A.A. Chernov, D.T. Temkin and A.M. Melnikova, Kristallogr. 21 (1976) 652 [Sov. Crystallogr.].
[52] V.A. Tatarchenko, S.K. Brantov and N.V. Abrosimov, Fiz. Khim. Obrab. Mater. (1978) 83 [Sov. Phys. Chem. Mater. Treatm].
[53] N.V. Abrosimov, S.K. Brantov, B. Ljuks and V.A. Tatarchenko, Bull. Acad. Sci. USSR, Phys. Ser. 47 (1983) 351.
[54] J. Shi, G. Geitos, N.V. Abrosimov, S.K. Brantov, S.A. Erofeeva and V.A. Tatarchenko, Bull Acad. Sci. USSR, Phys. Ser. 43 (1979) 174.
[55] N.V. Abrosimov, A.V. Bazhenov and V.A. Tatarchenko, J. Cryst. Growth 82 (1987) 203.
[56] N.V. Abrosimov, A.V. Bazhenov, V.A. Goncharov and S.A. Erofeeva, Bull. Acad. Sci. USSR, Phys. Ser. 47 (1983) 134.
[57] M.N. Leipold, in: Proc. Photovolt. Solar Energy Conf., Luxembourg (1977) p. 872.
[58] N.V. Abrosimov, S.K. Brantov, V.A. Tatarchenko and B.M. Epelbaum, Inorg. Mater. 28 (1982) 181.
[59] B.S. Red' kin, V.N. Kurlov and V.A. Tatarchenko, J. Cryst. Growth 82 (1987) 106.
[60] K. Nassau, H.J. Levinstein and G.H. Loiacano. J. Phys. Chem. Solids 27 (1966) 983.
[61] A. Räuber, in: Current Topics in Materials Science, Vol. 1, ed. E. Kaldis (North-Holland, Amsterdam, 1978) p. 481.
[62] T. Arizumi and N. Kabayashi, J. Cryst. Growth 13/14 (1972) 615.
[63] V.A. Borodin, E.A. Brener and V.A. Tatarchenko, Cryst. Res. Technol. 17 (1982) 1187.
[64] Sh. O. Arzumanian, L.A. Litvinov and S.N. Shorin, Monokrist. Tekh. 2 (1973) 201 [Sov. Single Crystals Techn., Kharkov].
[65] V.A. Borodin, E.A. Brener, V.A. Tatarchenko et al., J. Cryst. Growth 52 (1981) 505.
[66] E.I. Givargizov, The Growth of Whiskers and Plate Crystals from Vapor (in Russian) (Nauka, Moscow, 1977).
[67] R.S. Wagner and W.C. Ellis. Appl. Phys. Lett. 4 (1964) 89.
[68] R.S. Wagner and W.C. Ellis, Trans. Met. Soc. AIME 233 (1965) 1053.
[69] S.A. Ammer and V.B. Fursov, Fiz. Khim. Obrab. Mater. 4 (1980) 71 [Sov. Phys. Chem. Mater. Treatm.].
[70] G.A. Satunkin and V.A. Tatarchenko. Kristallogr. 30 (1985) 772 [Sov. Crystallogr.].
[71] E.A. Brener, G.A. Satunkin and V.A. Tatarchenko, Acta Phys. Sci. Hung. 47 (1978) 159.
[72] E.A. Brener, G.A. Satunkin and V.A. Tatarchenko, Growth Cryst. 14 (1983) 153.
[73] H. Jamgotchian, R. Trivedi and B. Billia, J. Cryst. Growth, to be published.
[74] J.E. Geguzin and A.S. Dzuba, J. Cryst. Growth 52 (1981) 337.
[75] S.K. Brantov, V.A. Tatarchenko and B.M. Epelbaum, Poverkhn. Fis. Chim. Mech. 2 (1985) 139 [Sov. Surf. Phys. Chem. Mech.]
[76] S.K. Brantov, B.M. Epelbaum and V.A. Tatarchenko, J. Cryst. Growth 82 (1987) 122.
[77] J.D. Zook, B.G. Hoepke, B.L. Grung and M.H. Leipold, J. Cryst. Growth 50 (1980) 260.
[78] L.D. Landau and V.G. Levich, Acta Phys. Chim. URSS 17 (1942) 42.
[79] A. Mlavsky and N.A. Panidscic, US Patent No. 3868228 (1975).
[80] H.E. LaBelle, US Patent No. 3915662 (1975).
[81] V.A. Borodin, T.A. Steriopolo, V.A. Tatarchenko and T.N. Jalovets, Bull. Acad. Sci. USSR, Phys. Ser. 47 (1983) 151.
[82] V.A. Borodin, T.A. Steriopolo and V.A. Tatarchenko, Cryst. Res. Technol. 20 (1985) 833.
[83] V.A. Borodin, V.V. Sidorov, T.A. Steriopolo and V.A. Tatarchenko, J. Cryst. Growth 82 (1987) 89.
[84] P.I. Antonov, J.G. Nosov and S.P. Nikanorov, Bull. Acad. Sci. USSR, Phys. Ser. 49 (1985) 6.
[85] I.V. Alyab'ev, S.V. Artemov, N.I. Bletzkan and V.S. Papkov, Bull. Acad. Sci. USSR, Phys. Ser. 49 (1985) 9.
[86] V.A. Borodin, V.V. Sidorov, T.A. Steriopolo, V.A. Tatarchenko and T.N. Jalovets, Bull. Acad. Sci. USSR, Phys. Ser. 52 (1988) 2009.

[87] V.A. Borodin, V.V. Sidorov, S.N. Rossolenko, T.A. Steriopolo, V.A. Tatarchenko and T.N. Jalovets, J. Cryst. Growth 104 (1990) 69.
[88] G.I. Babkin, E.A. Brener and V.A. Tatarchenko, J. Cryst. Growth 50 (1980) 45.
[89] B.L. Timan and O.D. Kolotiy, Rost Svoistva Krist. 6 (1980) 134 [Sov. Growth Properties Cryst., Charkov].
[90] B.L. Timan and O.D. Kolotiy, in: Proc. 9th Conf. on Stepanov's Growth of Single Crystals (in Russian), (Ioffe Phys.-Tech. Inst, Leningrad, 1982) p. 66.
[91] J.T.H. Pollock, J. Mater. Science 7 (1972) 632.
[92] V.A. Tatarchenko and G.A. Satunkin, J. Cryst. Growth 37 (1977) 285.
[93] V.A. Tatarchenko, Adv. Space Res. 11 (1991) 307.
[94] Ju. A. Osip'ian and V.A. Tatarchenko, Adv. Space Res. 8 (1988) (12) 17.
[95] V.A. Tatarchenko and S.K. Brantov, Bull. Acad. Sci. USSR, Phys. Ser. 40 (1976) 116.
[96] M.S. Agafonov, G.A. Gavrilov, L.N. Gubina, V.A. Tatarchenko et al., Bull. Acad. Sci. USSR, Phys. Ser. 43 (1979) 124.
[97] V.A. Tatarchenko, S.K. Brantov, V.L. Levtov et al., Bull. Acad. Sci. USSR, Phys. Sci. 49 (1985) 77.

CHAPTER 17

Crystallisation of Eutectics, Monotectics and Peritectics

J.D. HUNT

Department of Materials, University of Oxford, Oxford, United Kingdom OX1 3PH

SHU-ZU LU

Faculty of Metallurgy, University of Science and Technology, Beijing, China

Contents

1. Introduction . . . 1113
2. The eutectic reaction . . . 1115
    2.1. Regular lamellar growth: both phases non-faceted . . . 1117
        2.1.1. Lamellar spacings . . . 1117
        2.1.2. The extent of the eutectic range . . . 1137
        2.1.3. The addition of a third component . . . 1142
    2.2. Rod eutectics . . . 1144
    2.3. Irregular structures: faceted/nonfaceted eutectics . . . 1146
        2.3.1. Modification of Al–Si . . . 1151
        2.3.2. Cast iron . . . 1152
3. The monotectic reaction . . . 1153
4. The peritectic reaction . . . 1158
5. Concluding remarks . . . 1163
References . . . 1164

HANDBOOK OF CRYSTAL GROWTH, VOL. 2
edited by D.T.J. Hurle

1. Introduction

Examples of eutectic, monotectic and peritectic phase diagrams are shown in fig.1. The eutectic and monotectic are very similar reactions. On cooling a eutectic or monotectic, two phases are simultaneously deposited at one temperature. For the eutectic, two solid phases are formed but for the monotectic one solid phase and a second liquid phase are deposited. The peritectic is a very different type of reaction and could be considered to be an inverse eutectic reaction. The difference is most easily seen by noticing the single phase "V" above the horizontal line for the eutectic and monotectic whereas an inverted "V" occurs below the line for the peritectic (fig. 1). What this means is that one phase goes to two phases on *cooling* for the eutectic and monotectic, and one phase goes to two phases on *heating* for the peritectic. On cooling, the crystal grower expects and finds two-phase structures for the eutectic and monotectic reactions. On cooling, the crystal grower would often like to get a single phase from the peritectic reaction but usually does not.

In the eutectic-like reactions beautiful regular structures can often be formed. The reasons for the formation of the different structures will be discussed and the theory behind the scale of the microstructure will be presented. In the case of the peritectic the

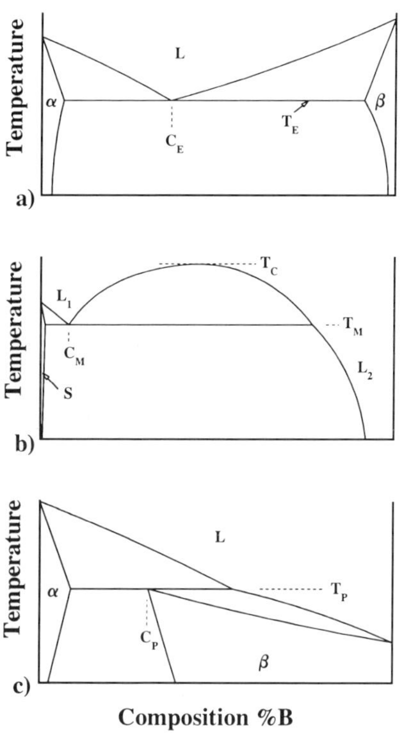

Fig. 1. (a) Eutectic, (b) monotectic and (c) peritectic phase diagrams.

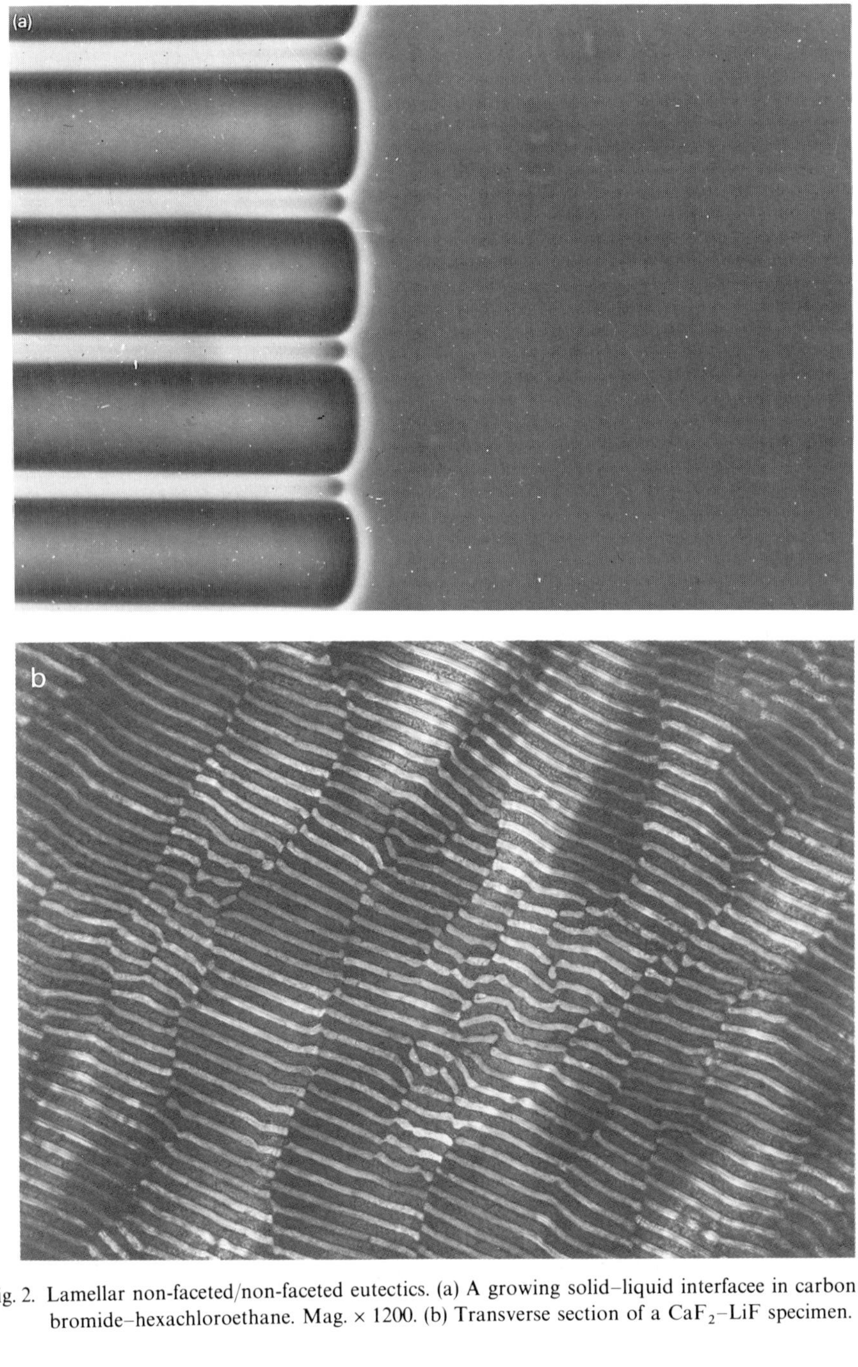

Fig. 2. Lamellar non-faceted/non-faceted eutectics. (a) A growing solid–liquid interfacee in carbon tetrabromide–hexachloroethane. Mag. × 1200. (b) Transverse section of a CaF_2–LiF specimen.

course of the reaction and the problems associated with the formation of a single phase will be discussed.

2. The eutectic reaction

Eutectics have long been used as low-temperature soldering or brazing materials and are the basis of the many widely used casting alloys such as cast iron and Al–Si. The eutectic reaction is potentially a method of producing in situ aligned composites. Attempts have been made to develop eutectic materials with high-strength fibres for turbine blades (see, e.g. [1]). Although possible materials were developed and tested in engines none appear to have been used commercially. Other uses have been in magneto-resistors [2] and it has been suggested the pairs of materials with different piezo properties could be coupled mechanically in a eutectic structure [3].

During directional growth a wide variety of microstructures can be formed. Examples are shown in figs. 2–7. A useful method of classification is to divide eutectics into those which can form regular lamellar or rod-like structures, and those which do not [4, 5]. It is found that if certain eutectics are pure enough and are grown directionally lamellar or rod-like structures are produced (see, e.g. [4, 6–9]) (figs. 2, 3). For other systems it is impossible to form regular structures over large areas (figs. 4, 5). In the latter systems, small parts of the specimen can have a regular structure (figs. 6, 7). The most important feature that distinguishes the two groups is whether or not a facet

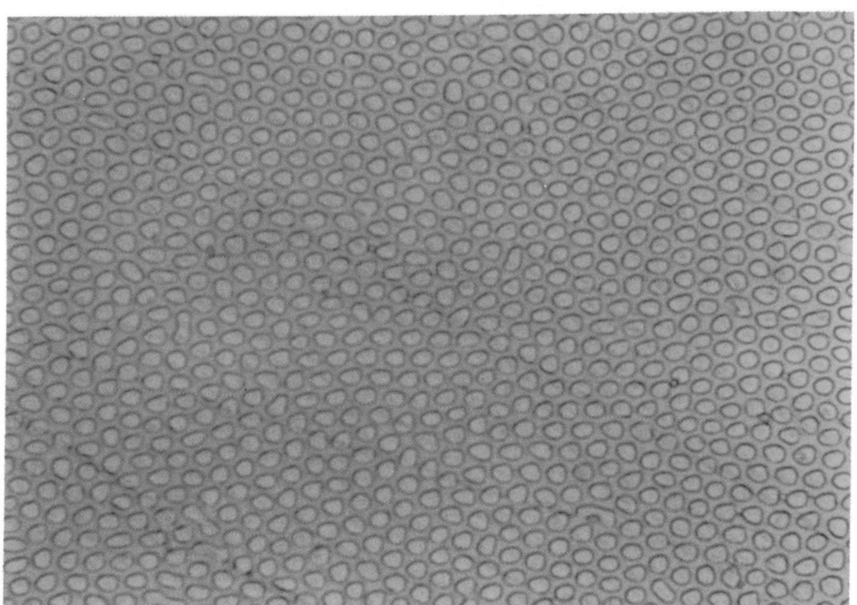

Fig. 3. Rod-like structure in off eutectic Sn–Cd. Mag. × 900.

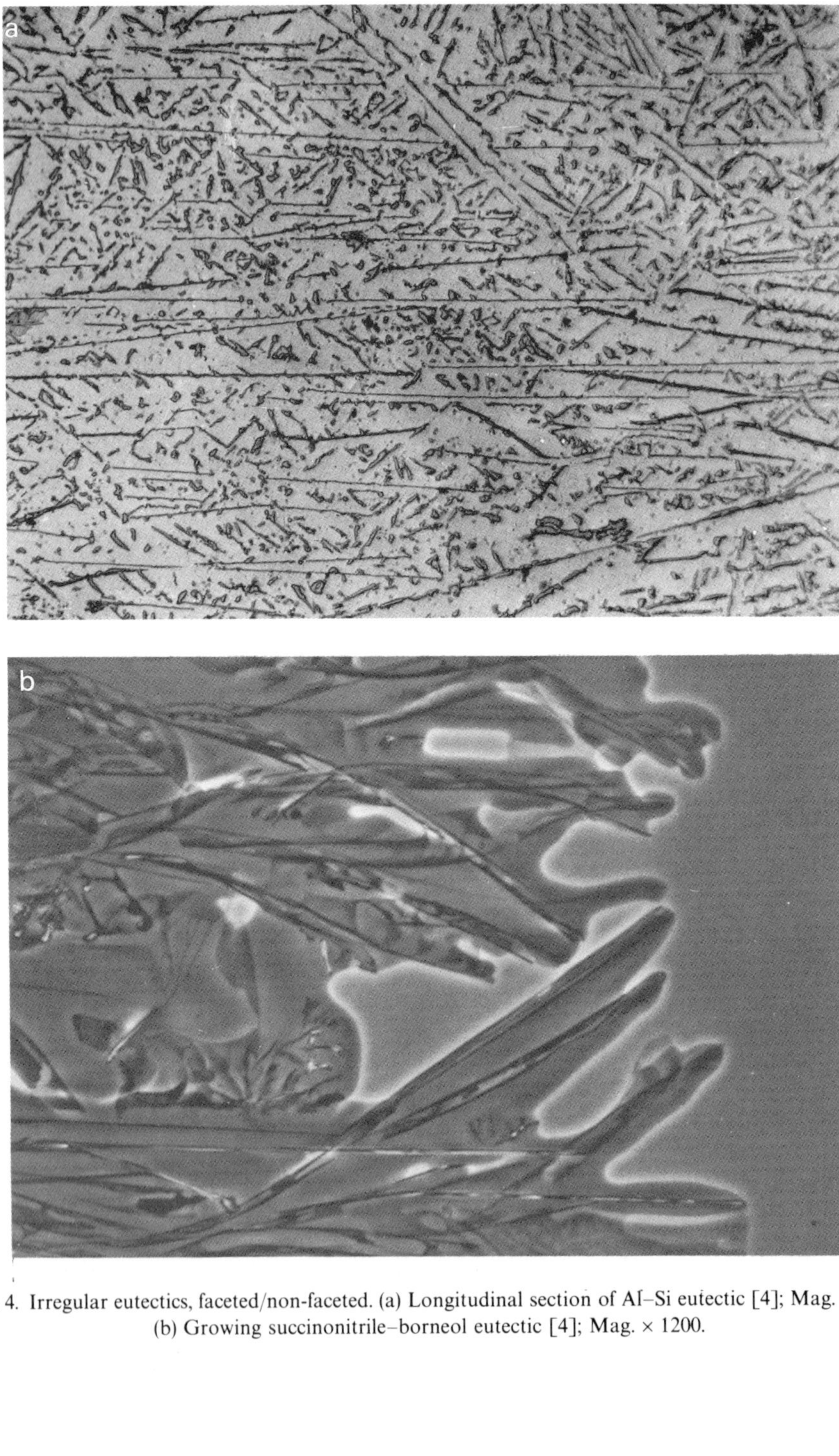

Fig. 4. Irregular eutectics, faceted/non-faceted. (a) Longitudinal section of Al–Si eutectic [4]; Mag. × 480. (b) Growing succinonitrile–borneol eutectic [4]; Mag. × 1200.

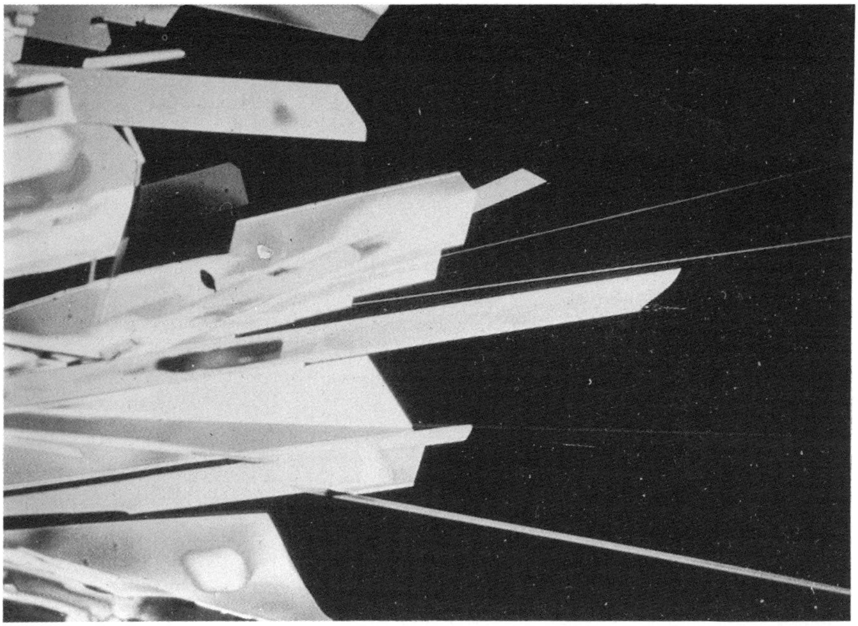

Fig. 5. Faceted/faceted azobenzene–benzil eutectic [4]. Mag. × 240.

forms at the interface during growth [4, 10]. The eutectics where both phases are non-faceted grow with lamellar or rod-like structures, whereas those with one or more faceted phases grow as irregular structures with or without small regular regions [4]. The presence of a faceting phase can be reasonably well predicted by the Jackson model [11] possibly modified to allow for the other component [12, 13]. It seems as if the presence of a facet restricts the ability of the eutectic interface to gradually accommodate change rather than catastrophically. In nature regularity is usually the result of the ability to change. A flock of geese flying in a "V" formation is possible because each is able to easily adjust its own speed and direction and the same appears to be true for eutectics.

2.1. Regular lamellar growth: both phases non-faceted

2.1.1. Lamellar spacings

During the growth of a single phase as a dendrite or cell, one of the components will be rejected by the solid into the liquid. For a eutectic, as the two phases grow each rejects the component used by the other solid phase. The solidification process is thus an example of cooperative growth, each solid phase growing more easily next to the other. The problem to be modeled can be illustrated by considering an α and β lamella growing under steady state (see fig. 8). The concentration of B atoms in the liquid ahead of the α phase must build up until sufficient solute is transferred laterally across the interface as the interface moves forward. This lateral transport of solute leads to a

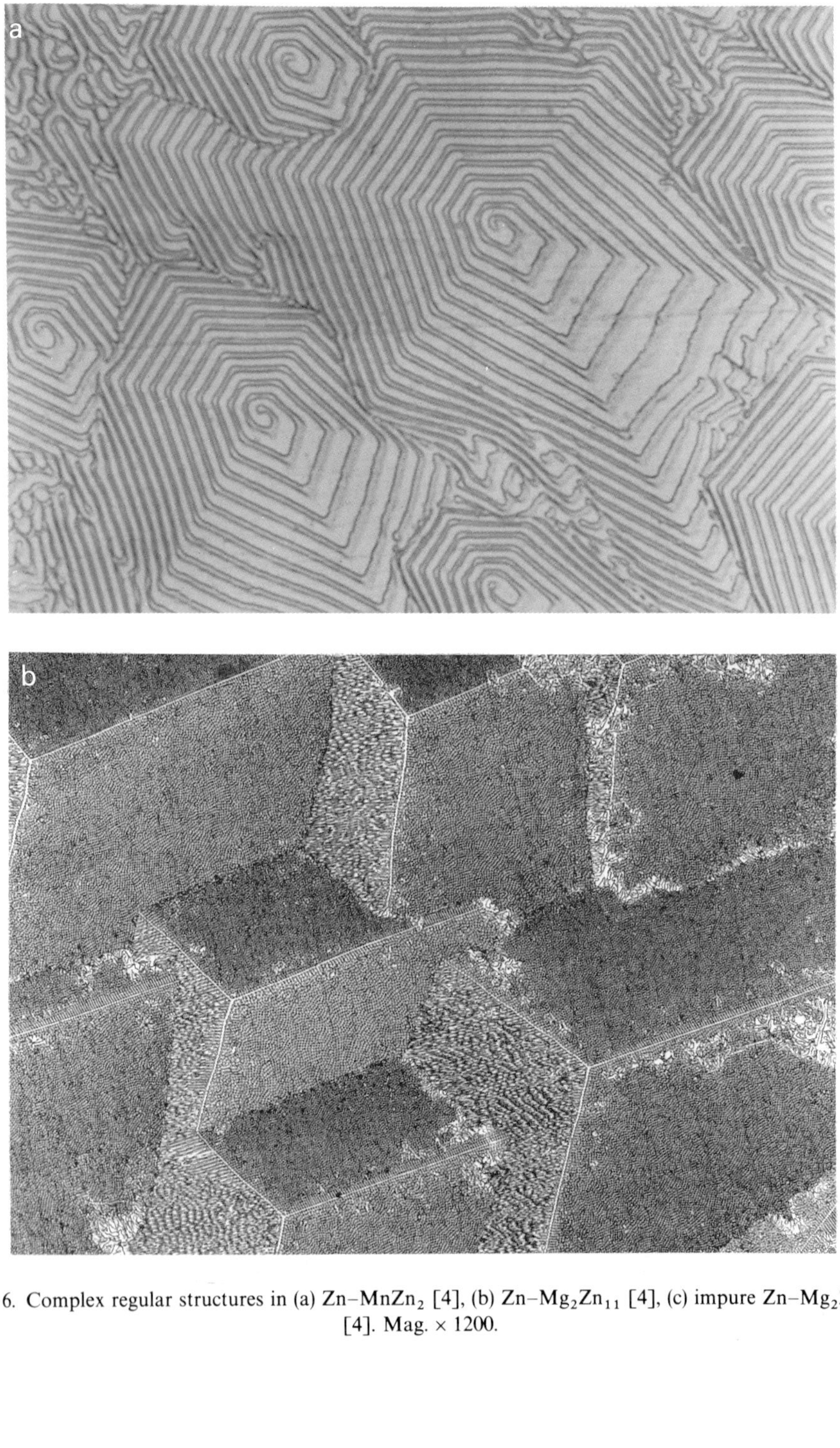

Fig. 6. Complex regular structures in (a) $Zn–MnZn_2$ [4], (b) $Zn–Mg_2Zn_{11}$ [4], (c) impure $Zn–Mg_2Zn_{11}$ [4]. Mag. × 1200.

(Fig. 6. cont'd)

concentration profile similar to that shown in fig. 8. The different solute concentrations along the interface for a planar interface would lead to different melting temperatures. This is illustrated as a reduced temperature or solute undercooling ΔT_C on the figure. The actual temperature of the interface will be the result of heat flow and will include the different conductivities of the three phases and the latent heat evolved as they solidify. Because thermal diffusivities are usually four to five orders of magnitude greater than solute diffusivities there will generally be little variation in the temperature, T_I, along the solid–liquid interface. The very small variation in T_I or the total undercooling, ΔT, is illustrated in the figure. Assuming that the kinetic undercooling is small, the curvature undercooling, ΔT_σ, is the only other major contribution to the total undercooling which means that the interface shape must vary so that the sum of $\Delta T_C + \Delta T_\sigma$ gives T_I. An additional requirement at steady state is that the volume fraction of the phases must be adjusted to deposit the bulk alloy composition. The problem of steady-state eutectic growth is thus one of solving the solute and heat flow for an interface shape that is consistent with the undercooling equation. This is a difficult problem because the interface shape and the fraction of the two solid phases must be a result of the calculation and are not known a priori.

Numerous workers [14–24] have treated the lamella eutectic or the analogous eutectoid problem. Zener [14] perhaps made the most important contribution. In this work it was first appreciated that both diffusion and the incorporation of surface energy must be considered and a crude order of magnitude analysis was carried out which correctly showed that steady state solutions were possible for a range of spacings

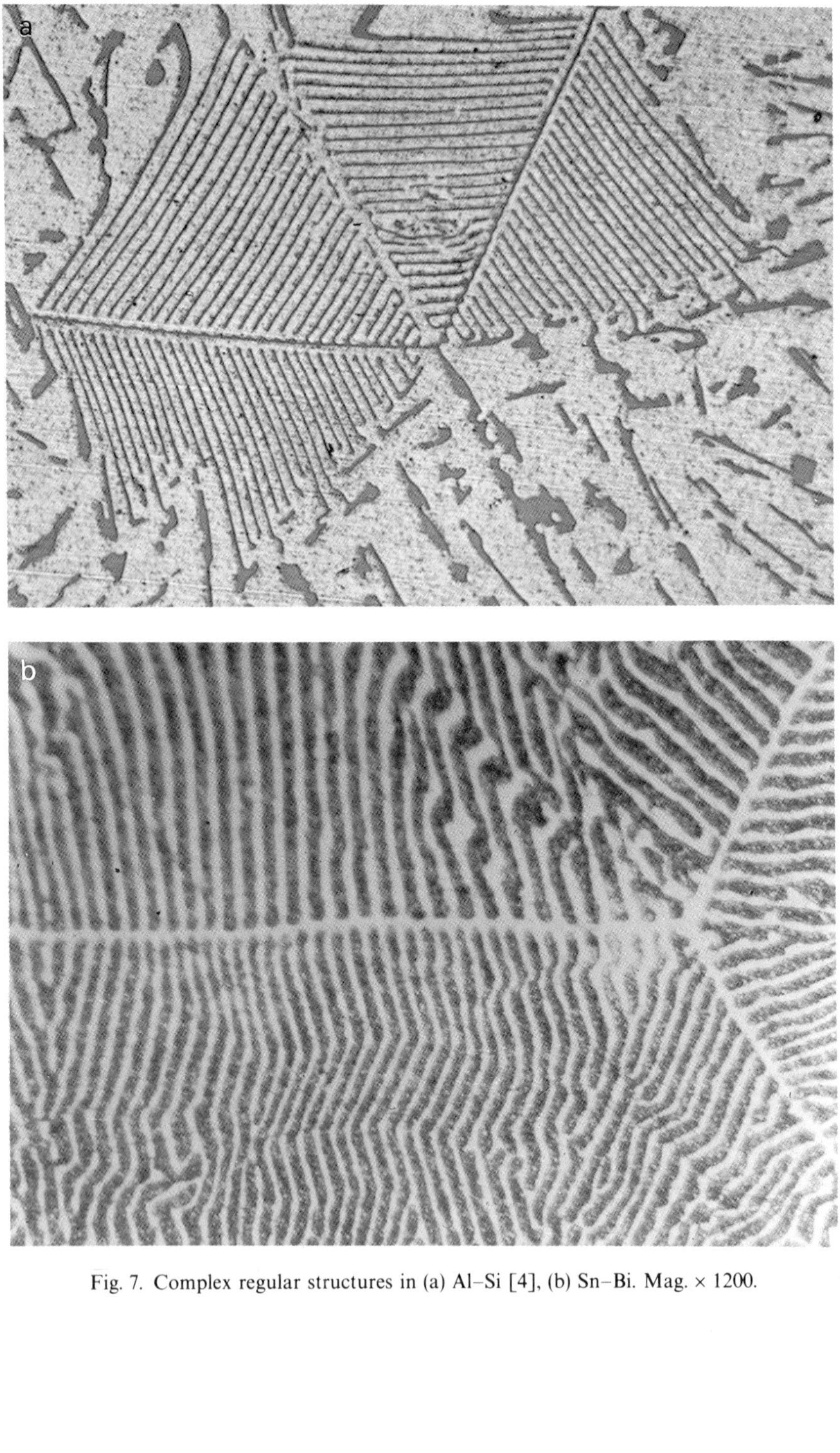

Fig. 7. Complex regular structures in (a) Al–Si [4], (b) Sn–Bi. Mag. × 1200.

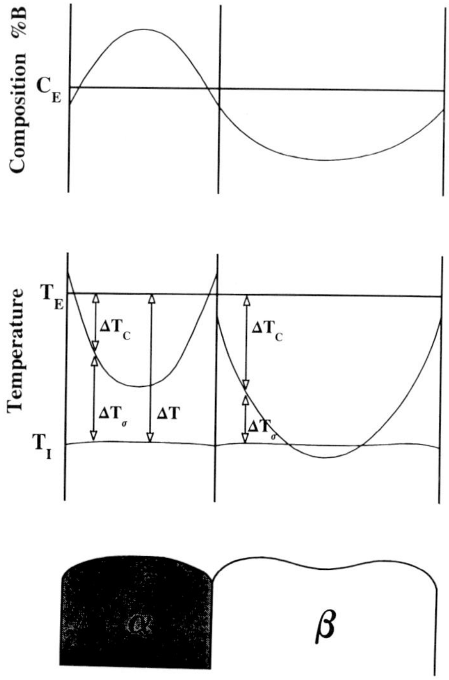

Fig. 8. Schematic eutectic interface showing a plot of interface composition and a plot showing the solute, ΔT_C, and curvature, ΔT_σ, undercooling along the interface.

at a particular velocity. This work was the first to suggest that the practical spacing was that which grew near the minimum undercooling. Most of the work since that time has set out to increase the accuracy and to consider spacing selection in more detail. Hillert [17] and Jackson and Hunt [18] produced more accurate diffusion solutions. It is the latter treatment which will be presented in more detail later in this section. Important contributions have been made by Donaghey and Tiller [19] and Nash [21]. The stability of the eutectic front has been considered by Cline [25], Hurle and Jakeman [26], Strässler and Schneider [20], Datye and Langer [22]. More recently, Trivedi, Magnin and Kurz [23] and Kurz and Trivedi [24] have produced analytic models which relax some of the assumptions made in the earlier work. Recent numerical work on tilted lamellae have been carried out by Kassner and Misbah [27] and Brattkus et al. [28].

In the theoretical models it is generally assumed that convection does not occur and that solute flow occurs in the liquid by diffusion. This assumption is usually valid because the important region for solute diffusion is for about one lamellar spacing ahead of the solid (typically about a micron). It is also usually assumed that the atomic volume of each phase is identical, or at least if they are not, that the atomic volume of the two solid phases are equal [29, 30]. In the Jackson and Hunt [18] (JH) model and most of the other models it is assumed that heat flow is sufficiently rapid when compared with solute diffusion, for a good approximate solution to the heat flow

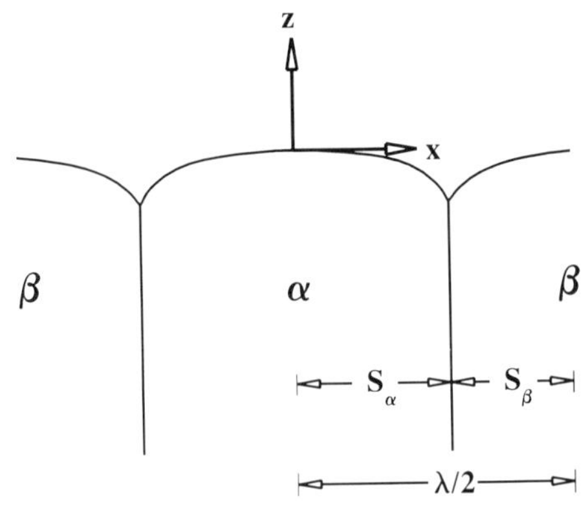

Fig. 9. A diagram showing the definition of x, z, λ, S_α and S_β.

problem to be a constant solid–liquid interface temperature. A first-order correction to this assumption is to suggest [31] there is a linear temperature gradient in the z-direction (fig. 9) and that the interface temperature depends only on the position in this gradient. Either of these assumptions allows the solute and heat flow problems to be decoupled.

In the JH model the problem is further simplified by solving the liquid diffusion problem for a planar front rather than for a curved front. The effect of the curved front is only brought in when the effect of curvature on the melting point is considered. The approximation of using a planar interface diffusion solution has been examined e.g. by Series et al. [29] and Nash [21], where it is found that it is usually not sufficiently serious to offset the value of an analytic solution. The diffusion equation transformed to coordinates moving at a constant velocity v in the z-direction,

$$\nabla^2 C + \frac{v}{D}\frac{\partial C}{\partial z} = 0, \tag{1}$$

is solved, where C is the composition and D the liquid diffusion coefficient. The boundary conditions are

$$C = C_{\mathrm{E}} + C_\infty \quad \text{at } x = \infty, \tag{2}$$

$$\frac{\partial C}{\partial x} = 0 \quad \text{at } x = 0 \text{ and } x = \tfrac{1}{2}\lambda = S_\alpha + S_\beta,$$

where C_{E} is the eutectic composition, λ is the spacing and S_α and S_β are the half-width of the α and β phases, respectively (defined in fig. 9). Because the interface is planar, the conservation equations become

$$\left(\frac{\partial C}{\partial z}\right)_{z=0} = \begin{cases} -\dfrac{vC_0^\alpha}{D}, & 0 < x < S_\alpha, \\[2mm] \dfrac{vC_0^\beta}{D}, & S_\alpha < x < S_\alpha + S_\beta, \end{cases} \tag{3}$$

when $C_0^\alpha = C_{\mathrm{L}\alpha} - C_{\mathrm{S}\beta}$ and $C_0^\beta = C_{\mathrm{S}\beta} - C_{\mathrm{L}\beta}$ are both positive and are the amounts of A and B rejected by the α and β phases as a unit volume freezes and depend on the local liquid composition [for a nonplanar interface the velocity and gradient in eq. (3) would be normal to the interface].

At the solid–liquid interface it is also necessary to satisfy the undercooling equations [18]

$$\Delta T = T_E - T_I = \Delta T_C + \Delta T_\sigma + \Delta T_k, \tag{4}$$

where T_E is the eutectic temperature and T_I is local interface temperature. The solute undercooling $\Delta T_C = m(C_E - C)$ allows for the variation in composition along the interface and curvature undercooling $\Delta T_\sigma = a/r$ allows for the different radius of curvature at different positions across the interface. Both ΔT_C and ΔT_σ are calculated assuming local equilibrium, the final term ΔT_k is the kinetic undercooling and includes all the non-equilibrium contributions to the interface temperature. Jackson and Hunt assume that at low growth rates, ΔT_k is small compared with ΔT for nonfaceting materials. At very high growth rates, ΔT_k may become significant and the C_0^α and C_0^β of eq. (3) would then depend on both liquid composition and growth rate.

The solution to eq. (2) by separation of variables is

$$C = C_E + C_\infty + \sum_{n=0}^{\infty} B_n \cos\left(\frac{2n\pi x}{\lambda}\right) \exp\left[\left(-\frac{v}{2D} - \sqrt{\left(\frac{v}{2D}\right)^2 + \left(\frac{2n\pi}{\lambda}\right)^2}\right) z\right], \tag{5}$$

which reduces to

$$C = C_E + C_\infty + B_0 \exp\left(-\frac{v}{2D}\right) + \sum_{n=1}^{\infty} B_n \cos\left(\frac{2n\pi x}{\lambda}\right) \exp\left(-\frac{2n\pi z}{\lambda}\right), \tag{6a}$$

provided

$$\frac{2n\pi}{\lambda} \gg \frac{v}{2D} \quad \text{for } n > 0, \tag{6b}$$

i.e. when the spacing is smaller than the diffusion distance D/v. This assumption is valid except for the highest velocities (typically greater than $10\,\mathrm{mm\,s^{-1}}$. Trivedi et al. [23] have relaxed this restriction so that rapid solidification processes may be treated and this work will be discussed later. Except for the very important B_0 in eq. (6) the expression is the solution to the Laplace equation rather than eq. (1).

The Fourier coefficients were evaluated by JH using eq. (6) to give the gradients in eq. (3), then multiplying by $\cos(2m\pi/\lambda)$ and integrating across the phases. They assumed that C_0^α and C_0^β could be approximated by the eutectic values C_{0e}^α and C_{0e}^β (see fig. 10a). This gave

$$B_0 = \frac{2(C_{0e}^\alpha S_\alpha - C_{0e}^\beta S_\beta)}{\lambda} \tag{7}$$

and

$$B_n = \frac{\lambda v C_0}{(n\pi)^2 D} \sin\left(\frac{2n\pi S_\alpha}{\lambda}\right), \tag{8}$$

where $C_{0e} = C_{0e}^\alpha + C_{0e}^\beta \simeq C_0$ is shown in fig. 10a.

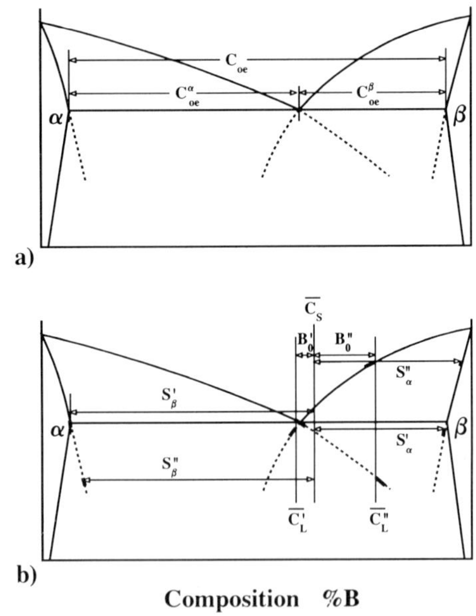

Fig. 10. A eutectic phase diagram showing the definition of C_{0e}, C^{α}_{0e} and C^{β}_{0e}. (b) A eutectic phase diagram showing two possible average liquid interface compositions $\bar{C}'_{L}$ and $\bar{C}''_{L}$ for an alloy of composition $\bar{C}_{s}$. Assuming the atoms have equal atomic volume in each phase the relative sizes of the phases are given approximately by S'_{α}, S'_{β}, S''_{α}, S''_{β}, for $\bar{C}'_{L}$ and $\bar{C}''_{L}$.

Unfortunately eq. (7) is not a good approximation for B_0; the two terms in the numerator are large and are of almost equal size, so that errors in assuming C^{α}_{0} and C^{β}_{0} to be the eutectic values are significant [29]. In general no great error arises [29] for the B_n terms at low velocities because it is the sum of C^{α}_{0} and C^{β}_{0} which appears in eq. (8). It will be seen later that because of the way that (JH) eliminate B_0, the error in eq. (7) is not important. However, the expression for B_0 is wrong and misleading.

Arguably the B_0 term is the most important aspect of eutectic growth. Crudely the B_n terms provide the gradient to transfer solute across the interface and the B_0 term determines the average composition. If too much of the A component is present at the eutectic front, as steady state is approached, the α phase grows at the expense of β at the α–β boundary. The average liquid composition is thus adjusted, so that eventually the average solid composition equals the bulk composition and in addition both phases grow at the same temperature (or satisfy the local undercooling equation). The B_0 term is the term which specifically allows asymmetric eutectics, off-eutectic compositions or eutectics with large kinetic undercoolings on one phase to grow at the same temperature.

A better physical meaning for the term is obtained by averaging eq. (7) across the interface and evaluating at $z = 0$. This shows that

$$B_0 = \bar{C}_{L} - \bar{C}_{S} \tag{9}$$

and is thus just the difference between average liquid composition $\bar{C}_{L}$ and the average solid composition $\bar{C}_S = C_E + C_\infty$. In a eutectic, in contrast to a single phase, there is no simple relationship between the average composition of the solid and that in the liquid. If only the diffusion process is considered [i.e. eqs. (1)–(3) are applied but eq. (4) is not], the average liquid composition at the interface can take any value. This is illustrated in fig. 10b for the alloy $\bar{C}_S = C_E + C_\infty$. Two average liquid compositions $\bar{C}'_L$ (and B'_0) and $\bar{C}''_L$ (and B''_0) are shown together with the range of liquid compositions needed to transport material across the interface. The main difference between the two is that for the second case, $\bar{C}''_L$, the two phases grow at very different temperatures. If the atoms have equal atomic volume in all phases, a good estimate of the relative size of S'_α, S'_β and S''_α, S''_β can be read from the atom fraction phase diagram. The volume ratio $\zeta = S_\beta/S_\alpha$ is different for the two cases (fig. 10b). By considering the rejection of solute on each phase, Series et al. [29] and Magnin and Trivedi [30] derive a better expression for B_0 in terms of the unknown ζ. It is of interest to note that if no solid solubility exists, the fraction of solid phases is fixed and no relationship exists between the average liquid and solid compositions until the undercooling equation [eq. (4)] is applied.

Jackson and Hunt [18] proceed by calculating average undercoolings across the interface. This is done because the effect of curvature can be brought in without considering the detailed shape of the curved lamellar surface. The average radius of curvature depends only on the width of the phase and the angle it makes at the α–β boundary [18, 32]. It should be noted that, because of the equilibrium condition at the α–β liquid cusp, the average curvature undercooling is identical [30, 32] to that which would be calculated by considering the incorporation of solid–solid surface energy (the approach used by Zener [14] and Tiller [15]).

Substituting for the B_n terms and integrating across the α and β phases separately gives

$$\bar{C}_\alpha = C_E + C_\infty + B_0 + \frac{2(S_\alpha + S_\beta)^2}{S_\alpha}\frac{v}{D}C_0 P,$$
$$\bar{C}_\beta = C_E + C_\infty + B_0 - \frac{2(S_\alpha + S_\beta)^2}{S_\beta}\frac{v}{D}C_0 P, \tag{10}$$

where

$$P = \sum_{n=1}^{\infty}\left(\frac{1}{n\pi}\right)^3 \sin^2\left(\frac{n\pi S_\alpha}{S_\alpha + S_\beta}\right). \tag{11}$$

P depends only on the ratio S_α/S_β and is tabulated in [18].

The average undercooling on each phase become

$$\Delta T_\alpha = m_\alpha\left[C_\infty + B_0 + \frac{2v}{D}C_0\frac{(S_\alpha + S_\beta)^2}{S_\alpha}P\right] + \frac{a^L_\alpha}{S_\alpha},$$
$$\Delta T_\beta = m_\beta\left[-C_\infty - B_0 + \frac{2v}{D}C_0\frac{(S_\alpha + S_\beta)^2}{S_\beta}P\right] + \frac{a^L_\beta}{S_\beta}, \tag{12}$$

where m_α and m_β are the slopes of the liquidus lines both defined positive and a^L_α and a^L_β are obtained from the Gibbs–Thomson coefficient and are defined in [18].

As discussed earlier, because heat flow is much more rapid than solute flow the solid–liquid interface was assumed to be approximately isothermal. Thus the average undercooling on each phase must be equal. It should be stressed that this is not an arbitrary assumption but is an approximate solution to the heat flow problem. If an isothermal interface is not used, the heat flow problem must be solved or approximated in some other way.

In general eq. (12) contains two unknowns B_0 and the volume ratio $\zeta = S_\beta/S_\alpha$, where $S_\alpha = \lambda/2(1+\zeta)$ and $S_\beta = \zeta\lambda/2(1+\zeta)$. For the particular case of no solid solubility in either solid phase, ζ is fixed a "priori" by the alloy composition and B_0 should be eliminated using eq. (12), thus ensuring that both phases grow at the same temperature. When solid solubility exists, in principle an expression for B_0 should be obtained [18, 29] in terms of ζ and substituted into eq. (12). Then the two equations should be used to eliminate ζ. Jackson and Hunt proceeded by noting that only B_0 and the average undercoolings were sensitive function of ζ so that the two eqs. (12) were used to eliminate B_0 and the value of ζ in the other terms taken to be the equilibrium value. This procedure thus picks the correct value of B_0 that ensures that both phases grow at the same temperature.

Assumptions about the composition at the solid–liquid interface should not be used to obtain B_0 prior to the use of eq. (12). The assumption that the average liquid composition exactly equals the eutectic composition prevents the two average undercoolings becoming equal. The assumption that the composition at the α–β liquid groove exactly equals the eutectic composition prevents the undercooling equation [eq. (4)] being satisfied at the groove, since, in general, the curvature undercooling will be different on either side of the groove. Both of these assumptions are approximately correct but should be regarded as a result of applying the undercooling equation rather than a procedure to give an expression for B_0.

Eliminating B_0 as described above gives

$$\frac{\Delta T}{m} = v\lambda Q^{\mathrm{L}} + \frac{a^{\mathrm{L}}}{\lambda}, \tag{13}$$

where

$$\frac{1}{m} = \frac{1}{m_\alpha} + \frac{1}{m_\beta}, \qquad Q^{\mathrm{L}} = \frac{P(1+\zeta)^2 C_0}{\zeta D}, \qquad a^{\mathrm{L}} = 2(1+\zeta)\left(\frac{a_\alpha^{\mathrm{L}}}{m_\alpha} + \frac{a_\beta^{\mathrm{L}}}{\zeta m_\beta}\right). \tag{14}$$

Equation (13) is identical in form to that obtained by Zener [14] and Tiller [15]. The undercooling is a function of both velocity and spacing. A plot of undercooling against spacing for a fixed velocity is shown in fig. 11. In practice a well-defined average spacing can be measured so that an additional condition is now needed to define the problem. Zener [14] and Tiller [15] arbitrarily proposed that eutectics grow at the extremum condition; that is with the spacing of the minimum undercooling of fig. 11 at fixed velocity (this is identical to the spacing of the maximum growth rate for a fixed undercooling). Minimising eq. (13) gives:

$$\lambda^2 v = \frac{a^{\mathrm{L}}}{Q^{\mathrm{L}}}, \qquad \frac{\Delta T^2}{v} = 4m^2 a^{\mathrm{L}} Q^{\mathrm{L}}, \qquad \Delta T\lambda = 2ma^{\mathrm{L}}. \tag{15}$$

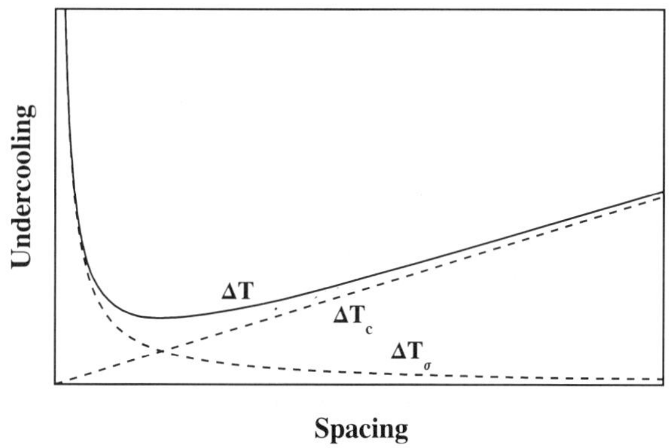

Fig. 11. A plot of total undercooling ΔT, solute undercooling ΔT_C, and curvature undercooling ΔT_σ against spacing.

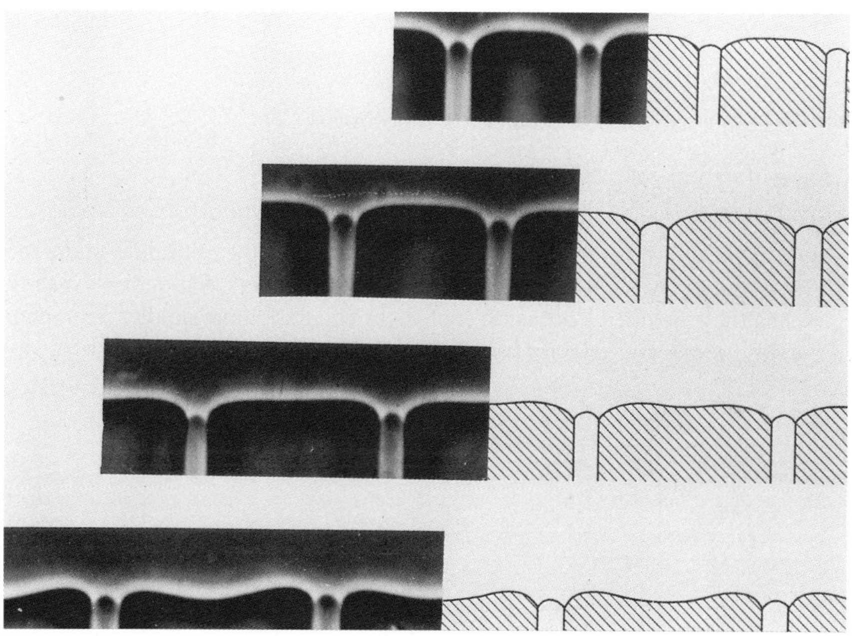

Fig. 12. Calculated and observed interface shapes in carbon tetrabromide–hexachloroethane [18]. Mag. × 1200.

The diffusion solution and the assumption of growth near the extremum allow interface shapes to be calculated [18]. Calculated and actual interface shapes for the transparent carbon tetrachloride–hexachloroethane system are shown in fig. 12. The agreement is remarkably good.

A number of workers have improved on the accuracy of the simple model relaxing various assumptions. Donaghey and Tiller [19] showed how models could be extended to conditions where eq. (6b) was no longer valid. Strässler and Schneider [20] used potential theory to formulate the problem and obtain numerical solutions. Series et al. [29] provided more accurate expressions for B_0 and B_n in the JH model and showed that the changes in B_n became significant at high velocities. They showed that their modified model using the planar front approximation agreed well with actual interface shapes obtained using an electrical analogue. Nash [21] obtained self-consistent interface shapes using an integro-differential equation for the shape of the interface. Composition differences were significant when compared with the simple model in very nonsymmetric eutectics. However, the work restricts the average liquid composition to the eutectic composition which could account for much of the difference. Trivedi, Magnin and Kurz [23] extended the (JH) model so that it could be used at high velocities where the assumption $2n\pi/\lambda \gg V/2D$ is no longer valid for small values of n. This occurs typically when the velocity is greater than about 10 mm s^{-1}. They obtained analytic results for the cases where the phase diagram is cigar-shaped (i.e. C_0^α and C_0^β do not change with undercooling) and also for the case where $k_\alpha = k_\beta$. The term for P in eq. (11) for the cigar-shaped phase diagram becomes

$$P = \sum_{n=1}^{\infty} \left(\frac{1}{n\pi}\right)^3 \sin^2\left(\frac{n\pi S_\alpha}{S_\alpha + S_\beta}\right) \frac{p_n}{1 + (1 + p_n^2)^{0.5}}, \tag{16}$$

and for the equal distribution coefficients P becomes

$$P = \sum_{n=1}^{\infty} \left(\frac{1}{n\pi}\right)^3 \sin^2\left(\frac{n\pi S_\alpha}{S_\alpha + S_\beta}\right) \frac{p_n}{(1 + p_n^2)^{0.5} - 1 + 2k}, \tag{17}$$

where $p_n = 4Dn\pi/(v\lambda)$. Assuming that the eutectic grows at the minimum undercooling they showed that $v\lambda^2$ was not constant at large Péclet numbers. A plot of $v\lambda^2$ is shown in fig. 13. When they included a diffusion coefficient which became smaller with temperature a maximum eutectic velocity was obtained (see fig. 14). This type of behaviour is

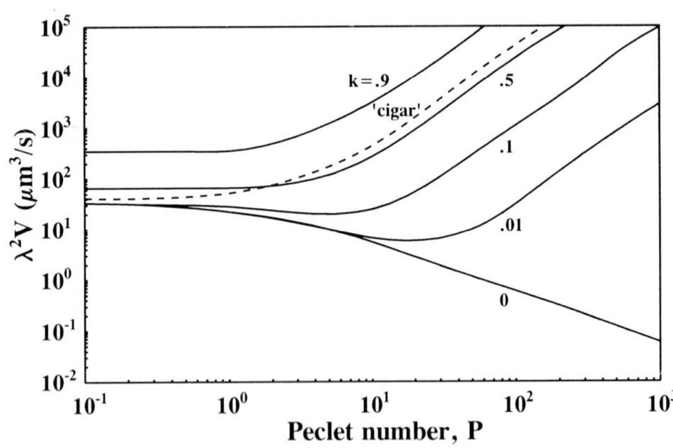

Fig. 13. Plots of $v\lambda^2$ (at the extremum condition) against the Péclet number, $v\lambda/2D$, showing a deviation from a constant value for Péclet numbers above one; from Trivedi, Magnin and Kurz [23]. The "cigar" plot refers to a eutectic where C_0^α and C_0^β do not change with temperature.

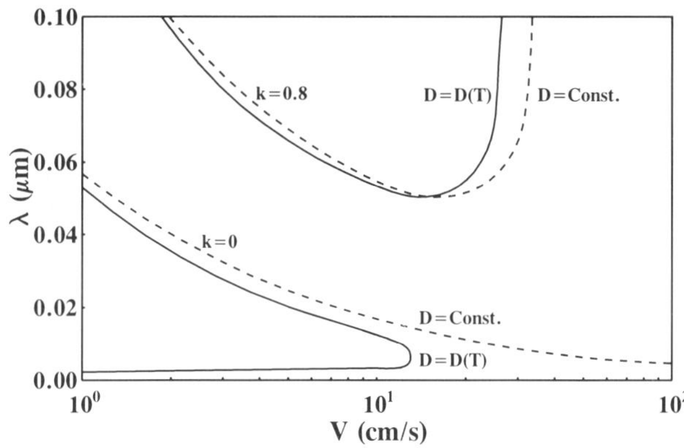

Fig. 14. A plot of extremum spacing against velocity, showing that a maximum velocity exists, from Trivedi Magnin and Kurz [23].

obtained in practice [33, 34]. Series et al. [29] pointed out that a different density in the liquid and solid could be allowed for by a different steady-state velocity in the solid and liquid. Magnin and Trivedi [30] considered the effect of density difference in all three phases and obtained modified expressions which, for certain materials, became significant, but these did not include the differential fluid flow that would be needed. Kurz and Trivedi [24] have recently extended the very-high-velocity model [23] to include non-equilibrium interface kinetics and non-equilibrium partitioning of solute at the interface. Treatments for lamellae tilted to the interface have been carried out by Kassner and Misbah [27] and recently Caroli et al. [35] have considered the effect of anisotropy.

Without further justification growth at the extremum is an arbitrary assumption. From eq. (13) it might be suggested that growth could occur at any spacing. In the early work [18] and later [20, 22, 25], it has been recognised that the complete range of spacings shown could not occur in practice. It was pointed out that there should be a maximum spacing λ_{max} and a minimum spacing λ_{min}.

Experiments on thin specimens of transparent materials have allowed the mechanism limiting the range of stable spacings to be examined. In thin specimens spacing changes occur much more slowly than in bulk specimens. Below a critical spacing a slightly smaller lamella will drop back and be overgrown by its neighbours. Practically this can be induced by gradually decreasing the velocity. Jackson and Hunt [18] argued that the limiting spacing occurred at the minimum undercooling, λ_{ext}. Their argument is illustrated in fig. 15. Figure 15a represents the case where the spacing is larger than that of the minimum undercooling. A fluctuation in spacing produces a slightly narrower spacing at the centre of the diagram. Since the average spacing is larger than the minimum undercooling spacing, the narrower lamellae grow at a higher temperature (smaller undercooling) so that they will grow ahead of the others and since the boundary between the phases tends to be normal to the interface the narrower lamellae will get wider. Figure 15b shows the situation where the average spacing is less than the spacing of the minimum undercooling. Again a fluctuation in spacing occurs

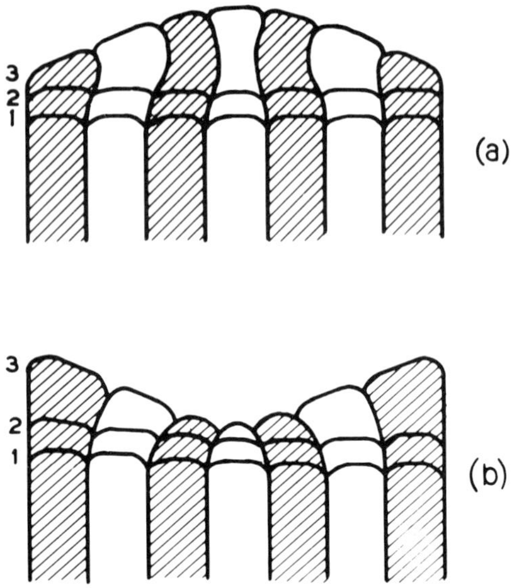

Fig. 15. The lamellae in (a) are wider than the extremum spacing, in (b) narrower than the extremum spacing. Interface position 1 shows the originally regular lamellae, position 2 the formation of narrower lamellae at the centre, position 3 the change in interface shape due to the change in local undercooling. For (a) the structure is restablised by the change in interface shape; in (b) it is not. It the latter case a lamella will grow out.

with narrower lamellae at the centre of the diagram. The narrower lamellae grow at a lower temperature (larger undercooling) so that they drop back. The shape of the interface in this case leads to the lamellae in the centre becoming narrower and eventually a lamella will grow out. The minimum spacing is thus the smallest spacing which is just stable to a perturbation of the array.

As the lamella spacing is increased, at a given velocity a hollow appears at the centre of one of the phases, see fig. 16. Practically, in transparent materials, it is easier to observe the change in interface shape as the velocity is increased. Eventually the hollow becomes sufficiently deep so that after any fluctuation the interface does not return to the steady-state shape. The largest spacing which is just stable is defined as λ_{max}. A good model of the problem needs to include the effect of the increased diffusion length as the hollow gets deeper. Clearly the assumption that the diffusion solution can be approximated by that of a planar interface made in the (JH) model means that the analysis cannot really be used for this problem. Depite this reservation they suggested that the maximum spacing should occur approximately when the slope in the hollow became vertical when the interface shape was calculated from the diffusion solution for a planar interface. Recently, Seetharaman and Trivedi [36] and Trivedi et al. [37] have shown that the spacing predicted is much larger than that found in practice.

A number of theoretical treatments of the stability of eutectic fronts have been made [20, 22, 25, 26, 38]. The difficulty with these analyses is to ensure that unforeseen effects which result from the simplifying assumptions that have to be made do not arise. As has been discussed earlier, the critical feature of steady-state growth is the build-up of the

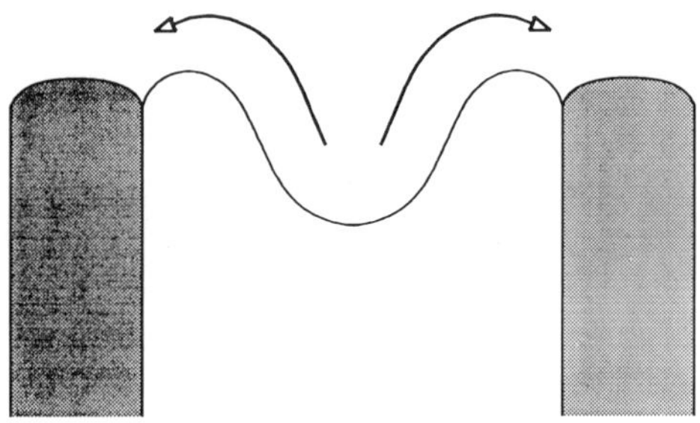

Fig. 16. For wide spacings satisfying the undercooling condition means that a hollow appears in the widest phase. Eventually the increased diffusion length from the hollow prevents a steady-state shape being formed.

average composition ahead of the eutectic, or, implicity, the composition at the $\alpha-\beta$ boundary. Physically this layer is the result of competitive growth at the $\alpha-\beta$ boundary. The immediate response of the eutectic to any change is for the α and β phases to grow at different rates and thus for the boundary to move laterally. The volume fraction of the phases changes and thus the average liquid composition changes. Ideally any perturbation treatment should not restrict the immediate local change in growth rate at the $\alpha-\beta$ boundary, i.e. it should allow changes on a length scale smaller than the spacing. The treatment should also include the coupled interaction on a scale of many lamellae. For problems such as calculating the maximum steady-state spacing or the extent of the eutectic range, effects on the lamella scale will be more important. For problems such as the analysis of array stability needed to calculate λ_{min} and the planar to celluar breakdown with a third component, averaged effects over many lamellae will be the more important. Cline [25], and Hurle and Jakeman [26] treated the problem as if the eutectic was a single phase, averaging properties across the interface. Hurle and Jakeman pointed out that for the pure eutectic instability occurred first at short wavelengths and that this was not really covered by their analysis. Strässler and Schneider [20], and Cline [38] treated lamellae on a scale comparable with the spacing but assumed that the macroscopic interface remained planar. These authors concluded that the minimum stable spacing occurred at the minimum undercooling and that there was also a maximum spacing when a hollow appears in the centre of one lamellae. Datye and Langer [22] treated a nonplanar macroscopic interface and assumed that lamellae grew perpendicular to the local interface. They implicitly assume that local changes in composition are accommodated rapidly by changes in solid fraction so that effectively the lamellae grow perpendicular to the interface. This model provides the most satisfactory description of the situation shown in fig. 15 and again the authors conclude that the smallest stable spacing occurs at the minimum undercooling spacing. Datye and Langer [22] also reproduced the oscillatory solutions in off-eutectic compositions similar to those observed experimentally in transparent materials (fig. 17). Karma [39], in elegant numerical work, which coupled the short- and longer-range effects, reproduced the oscillatory solutions and also lamellae which grew at an angle to the interface such as shown in fig. 18.

The general conclusion appears to be that lamellae are unstable to a perturbation in the array spacing at spacings smaller than that of the minimum undercooling. For large

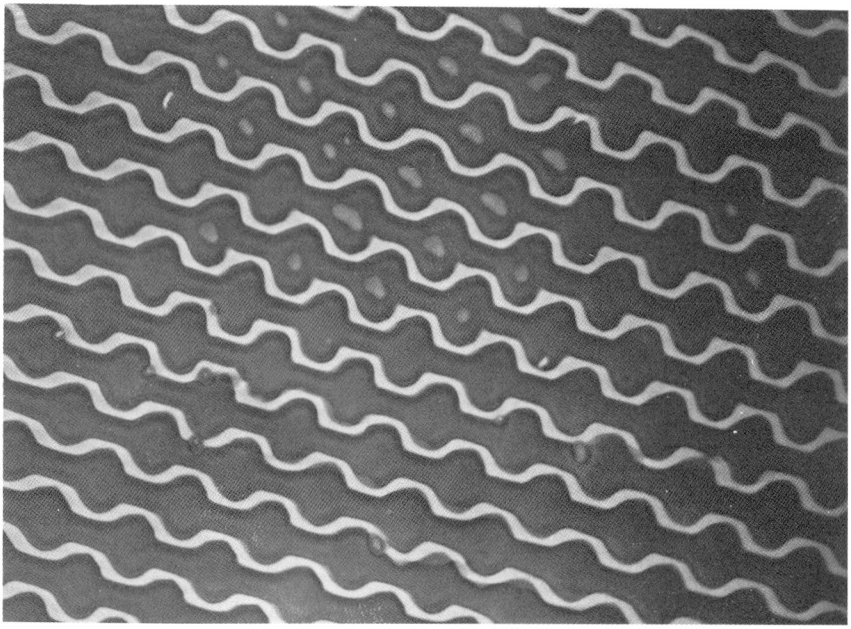

Fig. 17. Oscillatory lamellar structure formed in wide carbon tetrabromide–hexachloroethane eutectic. Mag. × 1200.

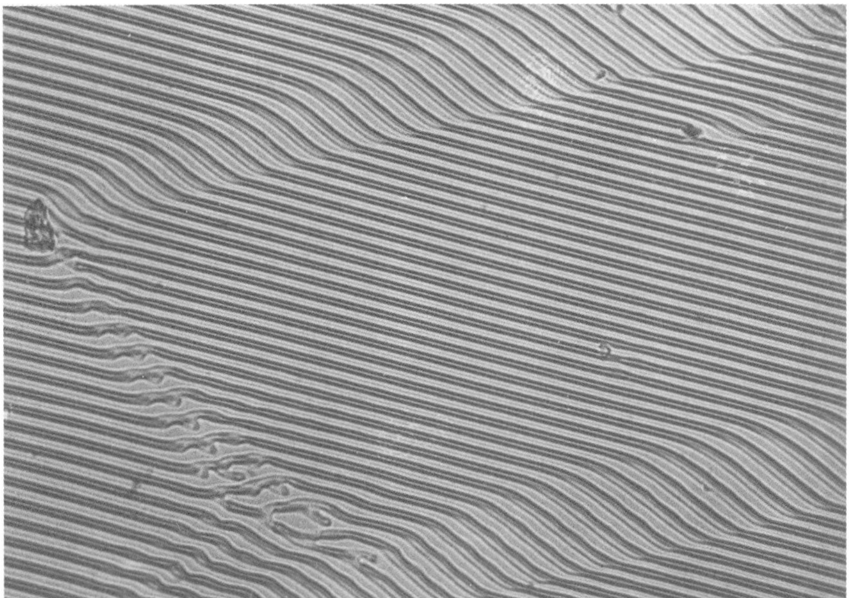

Fig. 18. Lamellae in carbon tetrabromide–hexachloroethane growing at an angle to the interface after a disturbance. Mag. ×900.

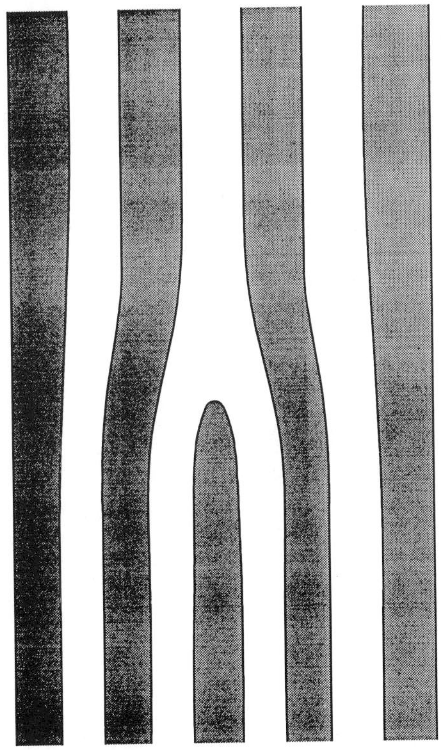

Fig. 19. A schematic illustration of a lamellar fault. The growth direction is out of the paper (from [18]).

enough spacings lamellae are unstable to a perturbation on a lamellar scale where a hollow in one lamella continues to grow deeper. The latter limit has been less well characterised because of the difficulty in using models which assume a planar interface to get the diffusion solution. The problem will be discussed further in the section on faceted/non-faceted eutectics.

Practically, it is found [36, 37, 40–43] that lamella eutectics grow with average spacings slightly larger than that of the minimum undercooling; that is much nearer λ_{min} than λ_{max}. After a change in velocity the change to the steady-state spacing occurs much more rapidly in bulk specimens than in thin specimens such as the glass cells used for observing transparent materials. Jackson and Hunt [18] proposed that small-scale space changes took place by lamellar faults such as that shown schematically in fig. 19. Double [44] showed, by careful sectioning, that faults such as the one shown in fig. 19 did not exist in sufficient quantities to explain the changes. Instead gradual spacing changes took place by the movement of fault lines (see fig. 20). These fault lines are always present in bulk specimens. Double proposed that, when there was a different spacing on either side of the line, the line would migrate across the interface overgrowing the subgrain furthest from the minimum undercooling. This mechanism should thus produce average spacings near the minimum undercooling. He showed that new

Fig. 20. A series (1 to 6) of transverse sections showing the gradual formation of a fault line. The fault line has the lamellae numbered in 5. The fault line is not apparent in 1, is becoming apparent in 2–3 and is beginning to rotate in 4–6. From Double [44].

fault lines could be produced from low-angle boundaries within each lamella combining to give a fault line (see figs. 21b). The relaxation to form triple junctions leads to different spacing at either end of a fault line (figs. 21c and d) and the consequent rotation of the fault line. Eventually the rotation leads to the local elimination or formation of lamellae. The rotation mechanism is shown schematically in fig. 22 and is shown on sections in fig. 20.

At low velocities the functional form of the variation of spacing or undercooling with velocity agrees well with the expressions obtained by minimising the undercooling

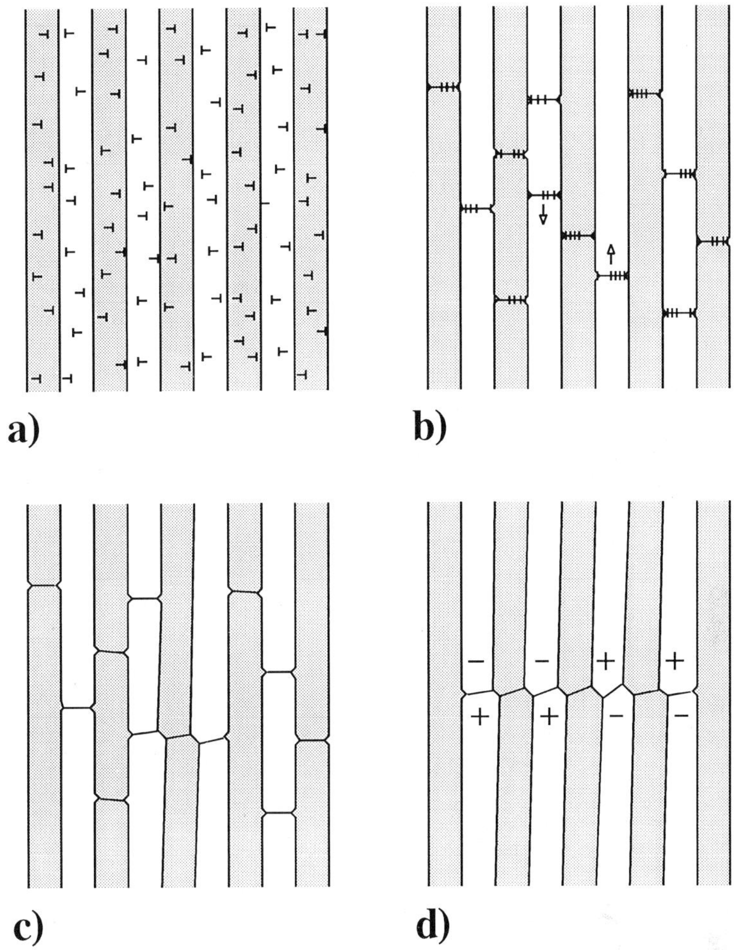

Fig. 21. A schematic illustration of the formation of a fault line by the accumulation of dislocations (a) forming a sub-boundary (b), the interaction of the lamellar sub-boundaries to form a fault line (c) and (d). Relaxation of the triple junctions leads to regions of slightly larger and smaller spacing on the two sides of the fault shown by + and − in (d). After Double [44].

equation. A more detailed comparison of theory with experiment is difficult because generally the surface energies in eq. (14) are not known. A way round this difficulty used by Jordan and Hunt [40], Tassa and Hunt [42], and Clarke and Elliott [41, 43] was to obtain values for the experimental constants $v\lambda^2$ and $\Delta T/v^{0.5}$ and to use these in eq. (13) to give a value for a^{L}. Using the calculated value of a^{L} it is then possible to see that growth occurred at slightly larger spacings than the minimum undercooling [40–43]. An example is shown in fig. 23. Recently, surface energy measurements have become available and again it has been shown by Seetharaman and Trivedi [36] and Trivedi et al. [37] that the average spacing occurs just to the right of the minimum undercooling and that the minimum observable spacing is very nearly the minimum undercool-

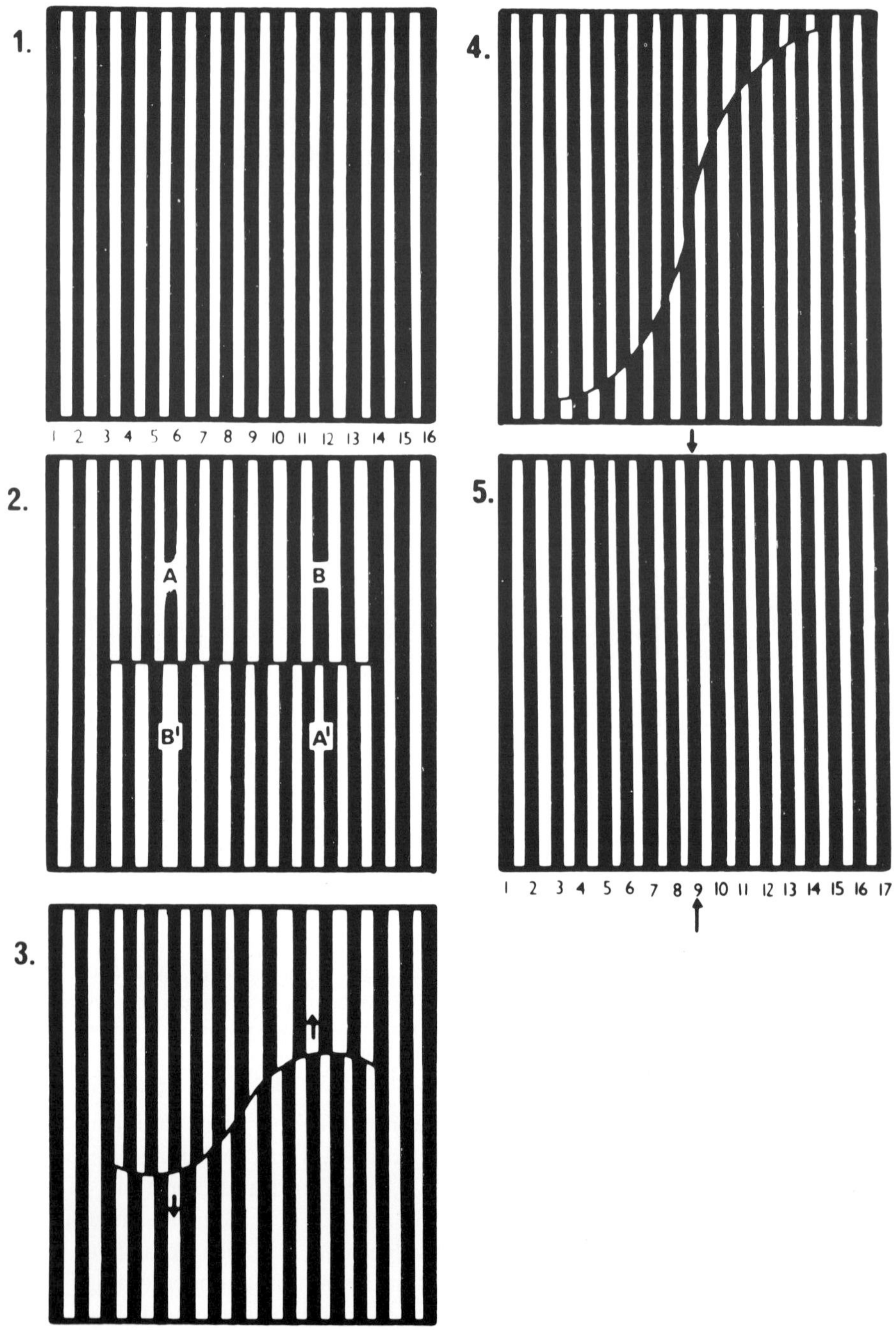

Fig. 22. The localised smaller and larger spacings can lead to a rotating fault line and the formation (or loss) of a lamella. In region AA′ relaxation of the boundary leads to narrower lamellae which grow at the expense of BB′ regions of 2. The fault line rotates in 3–4 leading to an extra lamellae in 5. After Double [44].

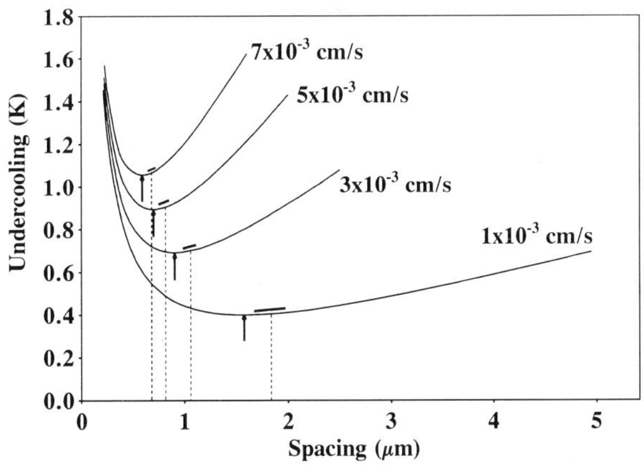

Fig. 23. A diagram showing the relationship between undercooling and spacing indicating the range of observed spacings, the average spacing (dotted line) and the theoretical minimum spacing (arrows) for four different velocities in Pb–Sn. It can be seen that growth occurs slightly to the right of the extremum spacing. From Jordan and Hunt [40].

ing spacing. The maximum observable spacing did not however correspond to that predicted by the JH model [18]. Possible reasons for this discrepancy were discussed earlier.

In most of the theoretical treatments of lamellar growth, transport of solute by convection is neglected. As mentioned earlier, the justification for doing this is that solute is transported across the solid–liquid interface in a layer typically about one lamellar spacing thick (about 1 μm). This layer is thus usually thin enough to be undisturbed by fluid flow. In the analysis it was, however, pointed out that an essential feature of eutectic growth is the B_0 term which allows the phases to grow at the same temperature and decays in a distance of order D/v and will be much larger (typically 100 μm). This layer will clearly often be disturbed by convection. One consequence of this convection will be that the average solid composition will not equal the bulk alloy composition and will change as the fraction liquid and extent of convection changes. The size of this effect will depend on the size of the B_0 term which will in turn depend on the asymmetry of the eutectic and alloy composition. Drevet et al. [45] used this phenomena to explain the difference in measure $v\lambda^2$ on earth and in space. For some alloys the change was small but for others it was much larger and this could be correlated to the size of the B_0 term.

2.1.2. The extent of the eutectic range

Lamellar eutectics can be grown not just at the eutectic composition but over a range of compositions. Outside this range single-phase dendrites or cells grow followed by a eutectic front (see fig. 24). The extent of the range has been found to depend on the growth rate and on the temperature gradient. Perturbation treatments of the problem are difficult because instability first occurs at short wavelengths [26].

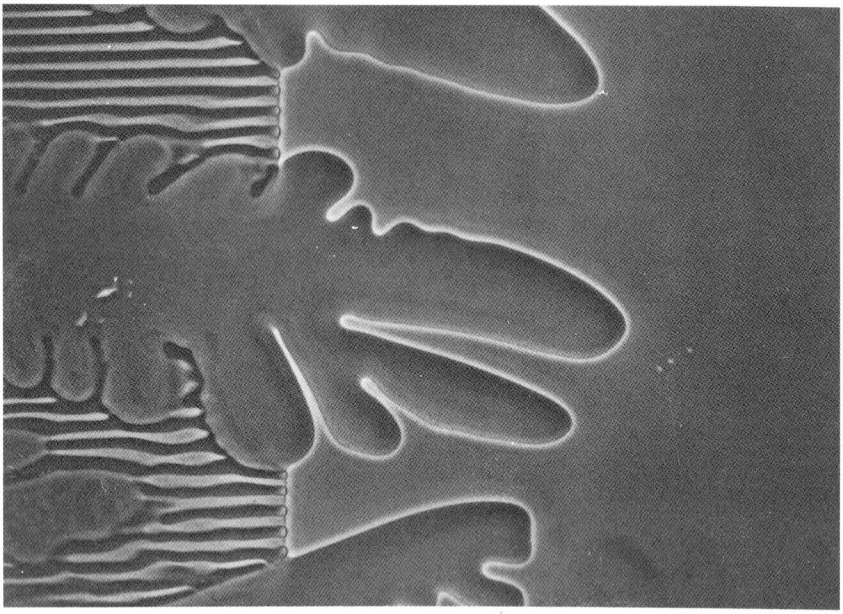

Fig. 24. A growing dendrite and eutectic growth front, carbon tetrabromide–hexachloroethane. Mag. × 1200.

An alternative approach to the problem is competitive growth. It is clear that cells or dendrites cannot be present unless they grow with a tip temperature greater than the eutectic growth temperature [46]. A similar approach based on maximum growth rate at a given temperature was proposed much earlier by Tamman and Botschwar [47]. Hunt and Jackson [46] used the JH model and experimental measurements to predict the eutectic range in the Pb–Sn system. Burden and Hunt [48] put forward a simple model for dendritic growth based on experimental measurements and combined this [49] with JH eutectic model. They suggested that the dendrite tip undercooling was given by

$$\Delta T_{\mathrm{D}} = T_0 - T_{\mathrm{D}} = \frac{GD}{v} + B\sqrt{v}, \tag{18}$$

where T_0 is the alloy liquidus temperature, T_{D} is the tip temperature defined in [48] and, for eutectic, eq. (15) becomes

$$\Delta T = T_{\mathrm{E}} - T_{\mathrm{I}} = A\sqrt{v}, \tag{19}$$

where A is a constant. At the critical condition $T_{\mathrm{D}} = T_{\mathrm{I}}$ so that the eutectic range on the α side of the eutectic becomes

$$\Delta C = C_{\mathrm{E}} - C_0 = \frac{T_0 - T_{\mathrm{E}}}{m_\alpha} = \frac{GD}{m_\alpha v} + \frac{B - A}{m_\alpha}\sqrt{v}. \tag{20}$$

The approach is shown schematically in fig. 25. Eutectic would be formed at both low

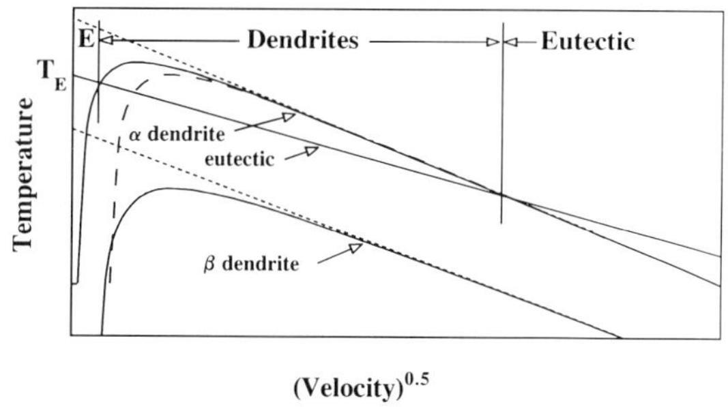

Fig. 25. A schematic diagram of the α and β dendrite tip temperature (the dotted line shows plots for a zero gradient, the filled line for a larger gradient) and the eutectic interface temperature plotted against the square root of velocity for a particular temperature gradient. It is assumed that the eutectic structure occurs without dendrites when the eutectic has the larger growth temperature (i.e. at high and low velocities). Making the alloy B-rich will raise the β but lower α line leading to a change in the transition velocities.

and high velocities for the composition shown. As the composition is changed by making the alloy B-rich, the β liquidus temperature moves up and the α down relative to the eutectic so that the transition velocities change. The results can be summarised on a plot of velocity against composition, or, alternatively, since for the eutectic undercooling is proportional to the square root of the velocity, the results can be

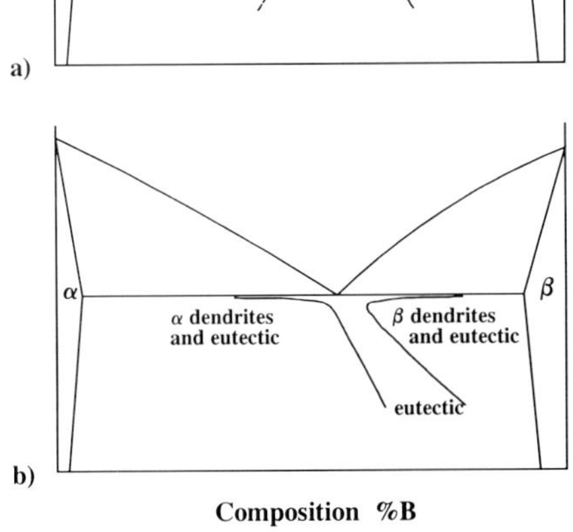

Fig. 26. (a) The results from diagrams (fig. 25) for different compositions can be summarised on the phase diagram by plotting the eutectic growth temperature against composition. The dashed line shows the boundary for a larger gradient. The effect of a steeper β dendrite line on the eutectic range is shown schematically in (b).

superimposed on the phase diagram. An example for two different gradients is shown in fig. 26a. If the primary phases have very different growth kinetics the eutectic can be skewed, fig. 26b. From eq. (20) it can be seen that the eutectic range is dependent on GD/v at high gradients and low velocities but only on velocity at high velocities. A good fit was obtained [49] using the experimental results of Mollard and Flemings [50] and Cline and Livingston [51] and this is shown in fig. 27. A similar good fit was obtained by Tassa and Hunt [42] in Al–Cu. Hughes and Jones [52] used the above competitive growth model to explain their results for Al–Fe. Juarez-Islas and Jones [53] investigated the Al–Mn system and used a treatment of dendritic growth by Kurz et al. [54] obtaining reasonable agreement. Recently, the model for rapid eutectic solidification developed by Trivedi, Magnin and Kurz [23] and Kurz and Trivedi [24] has been combined with their dendrite model [55] to describe rapid solidification produced by laser treatment. Again reasonably good agreement is obtained [56, 57]. Trivedi and Kurz [58] produce phase selection diagrams based on the competitive growth model. A schematic illustration from their work is shown in fig. 28. The lines are similar to fig. 25 except that the eutectic now has a maximum velocity as predicted from their high-velocity eutectic analysis and the dendrite line becomes cellular and then planar again at very high velocities. The results may again be summarised by superimposing them on a phase diagram (see fig. 29a). An example for calculated results [58] also showing a metastable phase is shown in fig. 29b. In conclusion it appears that the competitive growth model describes the structure selection rather well. Probably the

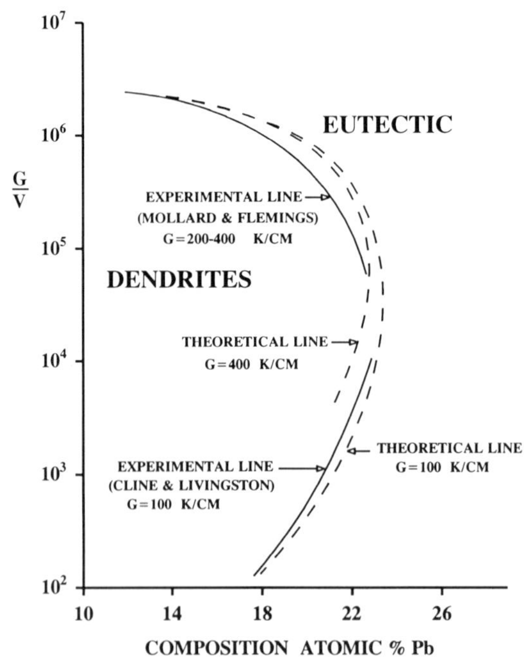

Fig. 27. A comparison of the predicted and observed eutectic range from Burden and Hunt [48].

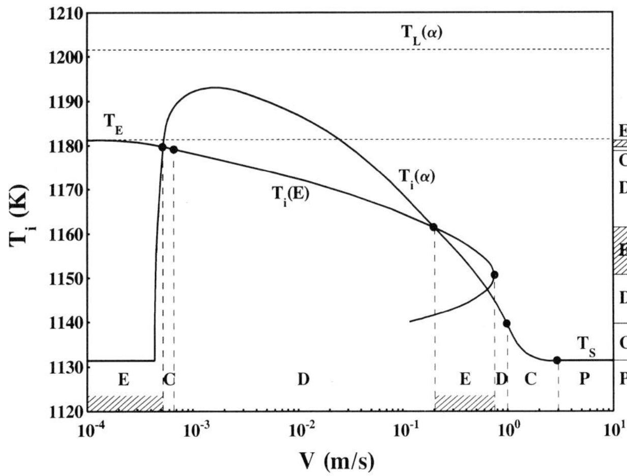

Fig. 28. A plot similar to fig. 25 except that the eutectic line has a maximum velocity and the transition between eutectic (E), planar (P), cells (C), and dendrites (D) is included, from Trivedi and Kurz [58].

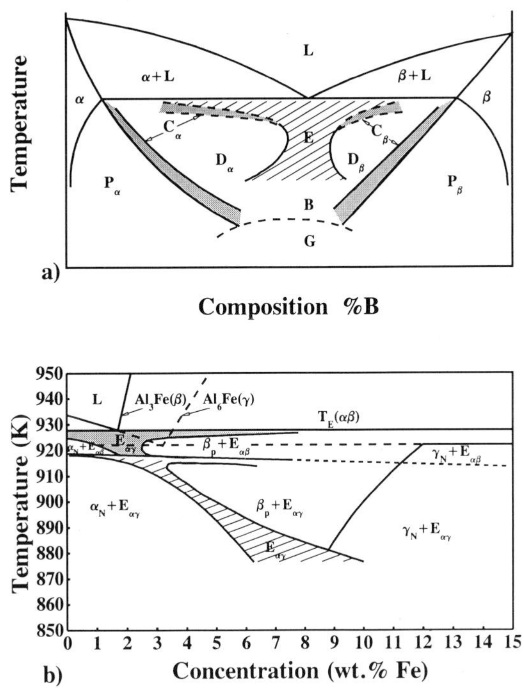

Fig. 29. (a) A schematic phase selection may produced from fig. 28 for a particular gradient. E: eutectic, P: planar, C: cellular, D: dendritic, B: banded, and G: glass. (b) A calculated phase selection map for the Al–Fe system. From Trivedi and Kurz [58].

weakest link in the analyses is the calculation of the array cell or dendrite growth temperature [59].

2.1.3. The addition of a third component

When a third component is added to a two-component eutectic or when the eutectic is impure, the macroscopic eutectic interface behaves in an analogous way to a single phase. For low velocities and high gradients the macroscopic interface remains planar but at higher velocities first two-phase eutectic cells such as those shown in fig. 30 are formed. Further additions or higher velocities can lead to two-phase dendrite-like structures (fig. 31).

During planar growth the third component may partition differently in the two phases but will nonetheless build up ahead of the eutectic, until it is deposited (in the absence of convection) with an average solid composition equivalent to the bulk liquid composition. Since the third component may be deposited in one phase rather than the other a lateral gradient across the interface will be needed and the combined B_0 terms for both components will adjust themselves so that both phases grow at the same temperature. In principal, on a lamellar scale, the addition of third component should be little different from any other term giving asymmetry. On a larger length scale the average solid composition and the average interface temperature will depend on the average liquid composition in much the same way as a single phase.

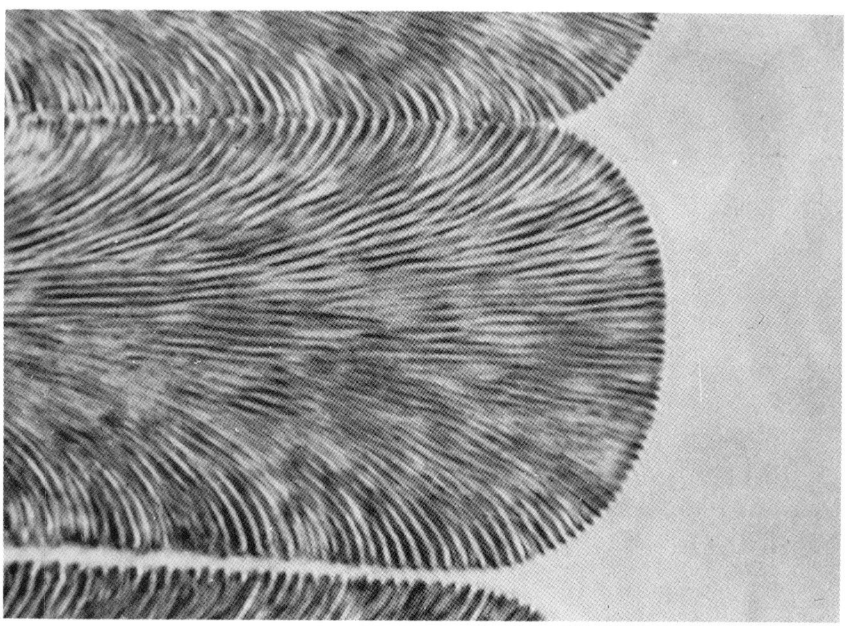

Fig. 30. Cell structure produced in camphor-succinonitrile eutectic containing an unidentified impurity showing growing interface shape. Mag. × 900. From [4].

Fig. 31. Eutectic dendrites succinonitrile–25 wt% camphor with impurity. Mag. × 90.

The problem has been treated by McCartney et al. [60] who derive an expression for planar two-phase eutectic growth when a third component is present. Using a perturbation treatment, they obtain a constitutional undercooling-like expression, to describe the breakdown from a planar eutectic front to eutectic cells. An expression for the growth of two-phase cells or dendrites is also derived based on the Burden and Hunt dendrite model. McCartney et al. [60] show how the limits of the different growth regions in a three-component system can be analysed using stability and competitive growth. This is illustrated in fig. 32. The diagram shows the regions for different structures in a simple system with three binary eutectics and one ternary eutectic. The diagram would apply to directional growth at one velocity and at a particular temperature gradient.

Regions 1: Near the pure component corners planar single-phase growth occurs.
Regions 2: Near the two-component eutectic compositions two-phase eutectic structures grow with a macroscopically planar front.
Regions 3: Near the eutectic valleys two-phase eutectic cells or dendrites are formed with the third component appearing at the ternary eutectic temperature as a thin layer of the third phase or in ternary eutectic depending on the volume fraction of the ternary eutectic.
Regions 4: Near the ternary eutectic composition a three-phase eutectic is formed and this can grow with a macroscopically planar front.
Regions 5: Primary cells or dendrites are formed followed by a planar eutectic front for

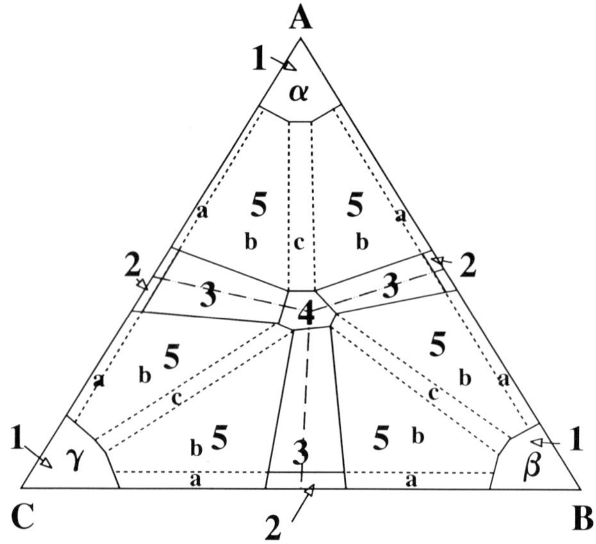

Fig. 32. A schematic diagram of a ternary eutectic system showing the different structural regions that would be expected for different alloy compositions. 1: planar single-phase growth; 2: planar two-phase eutectic; 3: two-phase eutectic cells or dendrites; 4: planar three-phase eutectic; 5a: single-phase cells or dendrites followed by planar two-phase eutectic; 5b: single-phase cells or dendrites followed by two-phase cells or dendrites; and 5c: single-phase cells or dendrites followed by planar three-phase eutectic.

regions 5a; followed by two-phase eutectic cells and ternary eutectic for regions 5b; followed by a ternary eutectic front for regions 5c.

The boundary between 1 and 5 is given by the usual stability treatment for a planar single phase modified to allow for a third component. The boundary between 2 and 3, planar to cellular eutectic, is given by the averaged constitutional undercooling condition. The boundaries between 2, 3 and 4 with 5 can be given by modified competitive growth, as can the boundary between 3 and 4.

2.2. Rod eutectics

Two nonfaceted phases form either lamellar or rod-like structures. Rod-like structures tend to be formed when the volume fraction of one phase is much less than the other. In other systems rod-like grains are present soon after nucleation but grow out during steady-state directional growth.

The growth may be analysed in much the same way as lamellar eutectic [18]. The typical hexagonal repeat unit is replaced by an axisymmetric cylinder of the same area with a rod at its centre. This is shown in fig. 33 which also shows the definition of r_α, r_β and R. The diffusion solution is obtained as for the lamellar problem, eq. (6), but here the cosine series is replaced by a Bessel function series. Proceeding as for lamellar eutectics by calculating average interface undercoolings and using the total undercool-

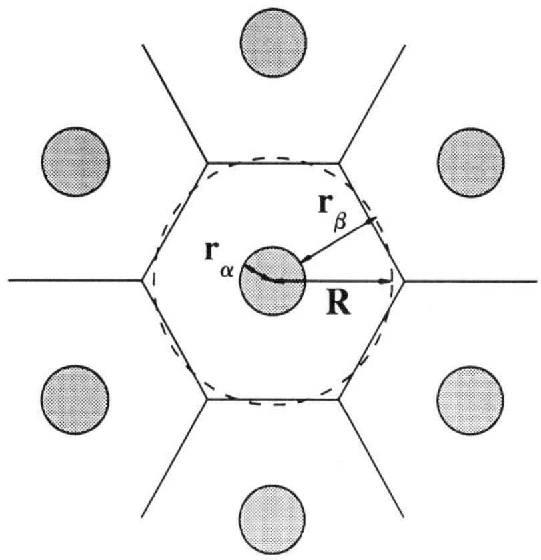

Fig. 33. A diagram showing the definitions of R, r_α and r_β for a rod eutectic.

ings to eliminate the equivalent to the B_0 term gives

$$\frac{\Delta T}{m} = vRq^{\mathrm{R}} + \frac{a^{\mathrm{R}}}{R}, \tag{21}$$

where

$$a^{\mathrm{R}} = 2\sqrt{1+\zeta}\left(\frac{a_\alpha^{\mathrm{R}}}{m_\alpha} + \frac{a_\beta^{\mathrm{R}}}{\zeta m_\beta}\right), \qquad Q^{\mathrm{R}} = \frac{4(1+\zeta)}{D\zeta} C_0 M. \tag{22}$$

M is a sum of Bessel function terms similar to P, defined and tabulated in [18]. Assuming that growth occurs near the minimum undercooling gives relationships indentical to eq. (15) except that the L superscript is replaced by R. Experimental work, e.g. by Walter and Cline [61], shows that the form of the expressions is correct; however detailed comparison does not appear to have been carried out.

It has been suggested by a number of authors [15, 18, 62–64] that the structure formed, lamellar or rod-like, should depend on which phase grows at the lower undercooling or higher temperature. Although this is similar to the competitive growth arguments used for the extent of the eutectic range there is much less physical justification in this case. By combining eq. (15) with the equivalent ones for rods gives the condition that the undercooling for rods will be less than that for lamellar when

$$\frac{a_\alpha^{\mathrm{L}}/m_\alpha + a_\beta^{\mathrm{L}}/(\zeta m_\beta)}{a_\alpha^{\mathrm{R}}/m_\alpha + a_\beta^{\mathrm{R}}/(\zeta m_\beta)} > \frac{4M}{P}\frac{1}{(1+\zeta)^{3/2}}. \tag{23}$$

If the surface energies for rods and lamellae are equal (that is isotropic) the left-hand side becomes equal to one. After evaluating the terms for M and P it is found that the critical volume fraction for the minor phase to give rods is [62, 18]:

$$\frac{S_\alpha}{S_\alpha + S_\beta} = \frac{1}{1+\zeta} < \frac{1}{\pi}. \tag{24}$$

Practically, it is found that rod-like structures are not found for volume fractions greater than this value [62]. Lamellar structures, however, are found for fractions slightly less than this value and it has been suggested that this is the result of low-energy boundaries being produced between the solid phases for lamellae but not for rods. Gradual elimination of rod-like grains can be explained by the selection of lamellar grains with low-energy boundaries between the solid phases. The fraction solid effect can also be shown in a single eutectic grown over a large range of compositions. Rod-like structures are present at low volume fractions but lamellae are found when the fraction of the phases are similar [50, 65].

2.3. *Irregular structures: faceted/nonfaceted eutectics*

Faceted/nonfaceted (f–nf) eutectic growth occurs when one of the product phases solidifies with a faceted and the other with a nonfaceted interface. As was discussed earlier, irregular structures are generally formed in these systems. Examples are shown in figs. 4, 34, 35; sometimes small regions of a complex regular structure [4] can be produced (figs. 5, 6a, b). The presence of a facet can be predicted using Jackson's α factor [11] or extensions of this model [12, 13]. Generally the facets lead to highly anisotropic growth. Most metal/nonmetal eutectic systems belong to this group of eutectics [5]. The best examples are the industrially important cast alloys of the Fe–C and Al–Si systems.

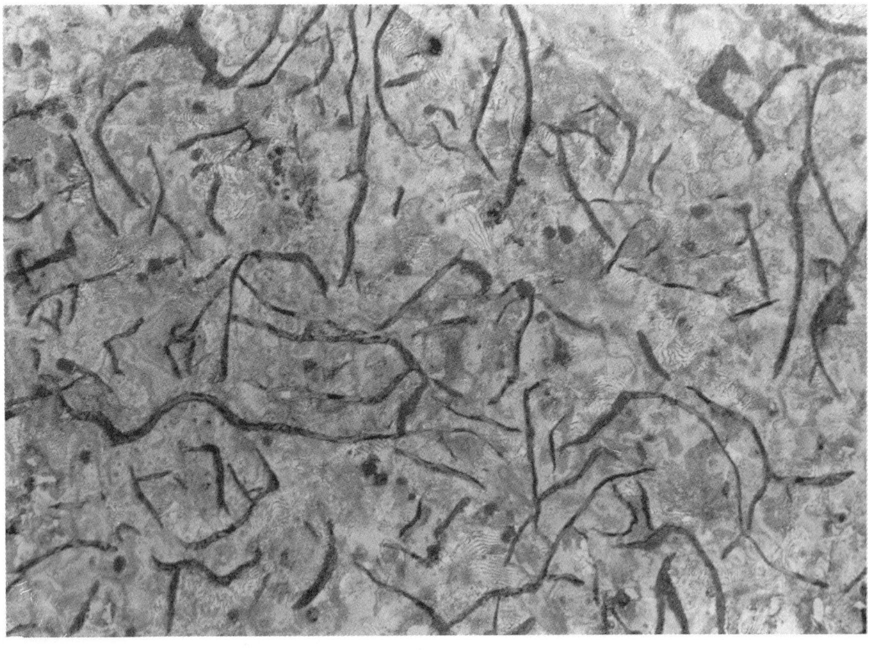

Fig. 34. Grey cast iron. Mag. × 180.

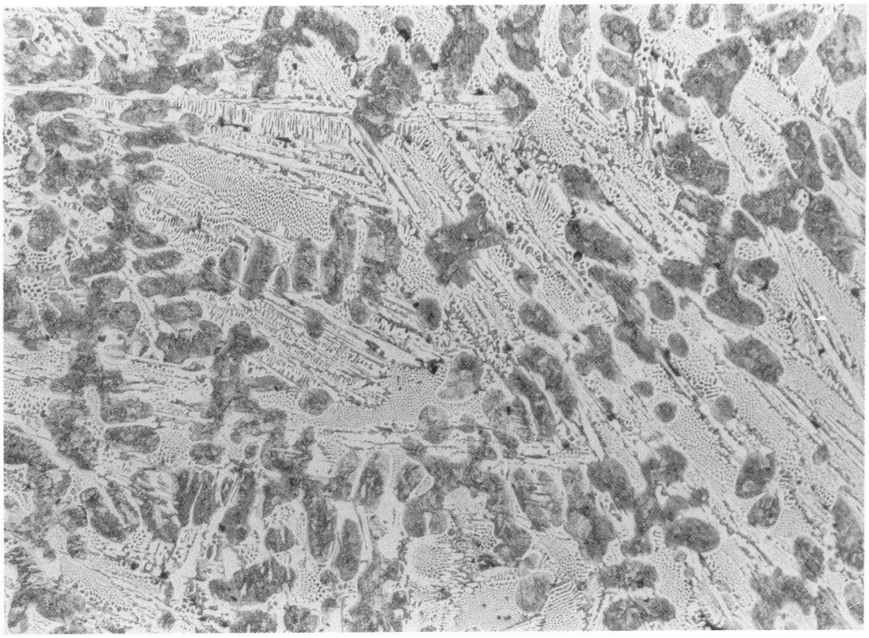

Fig. 35. White cast iron with primary austenite dendrites. Mag. × 90.

Faceted/nonfaceted eutectics have a number of features which distinguish them from the nonfaceted/nonfaceted eutectics.

- The growing solid/liquid interface is non-isothermal (fig. 4a).
- The microstructure is irregular due to growth anisotropy of the faceted phase.
- Quantitative measurements show that generally f–nf eutectics have larger undercooling and larger spacings [66–78] than nf–nf eutectics. The results do not fit simple lamellar theory for growth at the minimum undercooling condition.
- Unlike nf–nf eutectics, the f–nf average spacing depends on the temperature gradient [69] as well as on growth rate.
- Because of the larger growth undercoolings the nucleation of new eutectic grains can often occur ahead of the freezing eutectic front.
- The addition of very small quantities of a third component can lead to dramatic changes in microstructure, e.g. Na to Al–Si, FeSi to cast iron, FeSi and Mg to cast iron (see, e.g. [79–82]).
- The severity of faceting on the faceted phase is though to decrease with increasing growth velocity [83, 84].

The key feature which leads to irregular growth is the inability of the α–β boundary to change its position smoothly because of the presence of a facet in or near the boundary. This feature is illustrated in fig. 36a, b. For two nonfaceted phases a slight change in composition ahead of the solid phases leads to a slight change in composition at the triple junction so that one phase grows wider. This change accommodates the

a)

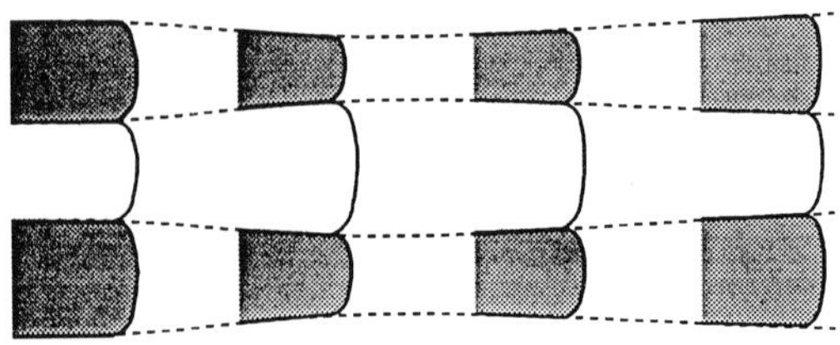

b)

c)

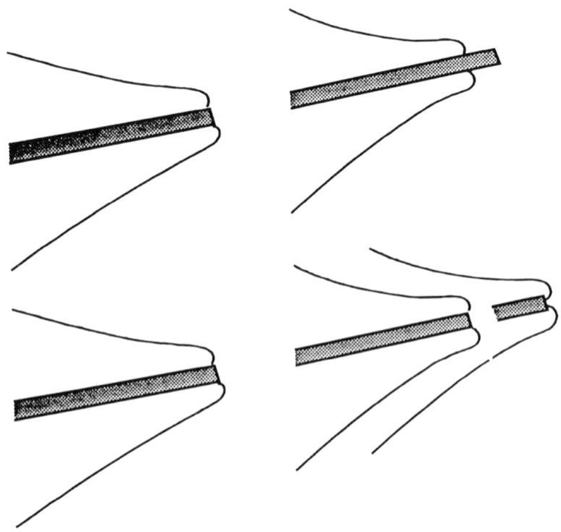

Fig. 36. (a) A diagram showing how a local fluctuation in composition in a nf/nf systems can be easily accommodated by a local change in the position of the α–β boundary. (b) for a f/nf eutectic an increase in B composition leads to increased growth rate of the β faceted phase but the β plate does not get wider. (c) An increase in A composition will eventually lead to the overgrowth of the faceted plate.

fluctuation in composition as a local change in fraction of the two solid phases (see fig. 36a). When a faceting phase is present and a slow growing facet exits at the α–β boundary, an increase in the B composition will lead to more rapid growth of the β plate but little sideways growth will occur on the slow growing facet (see fig. 36b). Thus the fraction of each solid deposited will not change. An increase in A will eventually lead to the overgrowth of the β plate but this response may be delayed (see fig. 36c) so that even in this case a smooth change may not occur.

Another consequence of the lack of response of the boundary is that a faceted plate will grow in a direction determined by its crystallography and not by heat flow. Since many orientations of plate will generally be present, some faceted plates will grow towards one another and thus one of them will be overgrown (see fig. 37). This will mean that there will always be a tendency for the average spacing to increase. The average spacings found in these systems can thus be understood in terms of the ease or otherwise for new plates to be formed by branching (or by some other mechanism). In Bi–Sn or Bi–Pb systems average spacings are similar to those found in nf–nf eutectics whereas the Fe–C and Al–Si the average spacings are very much larger [33].

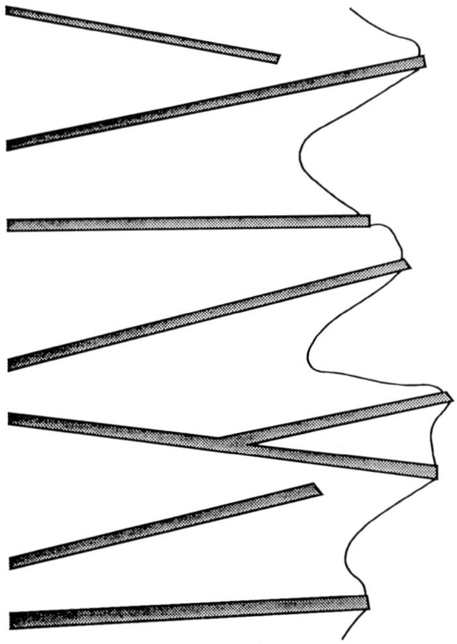

Fig. 37. Schematic illustration of a f–nf eutectic front showing that when plates are not parallel overgrowth will lead to a coarsening structure.

Because of the irregular nature of f–nf eutectics the process is very difficult to model. The usual approach [32, 85, 86] has been to assume that the structure can be approximated by the steady-state lamellar problem considered earlier and some condition other than minimum undercooling is then used to predict the average spacing. The validity of using the steady-state analysis to describe the transient process of irregular growth is difficult to justify. Transient spacings could clearly be greater than the maximum steady-state spacing in a uniform structure. Despite these reservations no realistic alternative approach can be suggested which could give analytic results.

The theoretical models for f–nf growth assume steady state and attempt to calculate a better solution for the lamellar problem at large spacings. As was discussed earlier, as the lamellar spacing increases a hollow begins to form at the centre of the larger volume fraction phase. Eventually steady-state solutions become impossible because of the increasing diffusion length (see fig. 16). The increased diffusion length will tend to make the hollow deeper; the curvature at the base of the hollow and any imposed temperature gradient will tend to make the hollow less deep. Ideally a model for larger spacings should allow for the increased difficulty of diffusion and also for the change in temperature of the interface with position in the temperature gradient.

Sato and Sayama [85] carried out an analysis which assumed only part of the interface was isothermal but the fraction chosen was obtained from experiment. Fisher and Kurz [31] tackled the non-isothermal interface but since they used the plane front diffusion solution, did not allow for the increased diffusion length. They suggested that the spacing was controlled by the splitting of the minor (faceted) phase and proposed that this occurred at a marginal stability condition. Magnin and Kurz [86] carried out an analysis similar to the earlier paper [31] but imposed a condition related to the depth of the hollow produced in the minor phase. This approach is open to the criticism that the largest volume fraction will become unstable first. Liu et al. [87] carry out an analysis similar to Kurz and Fisher but again did not correctly allow for diffusion from the hollow.

Magnin and Kurz [86] and Magnin et al. [78] point out that the correction for a non-isothermal interface is very small for the spacing typically found in the Al–Si system. It should also be noted that solutions obtained by using averaged quantities rather than the local undercooling do not imply that a steady-state solution exists; this is particularly true for large spacings. The JH upper spacing limit will be increased by the inclusion of temperature gradient but decreased by a better solution to the diffusion from the hollow in the lamellar. Because of the small size of the non-isothermal correction, the JH maximum spacing should be regarded as an upper bound. This is

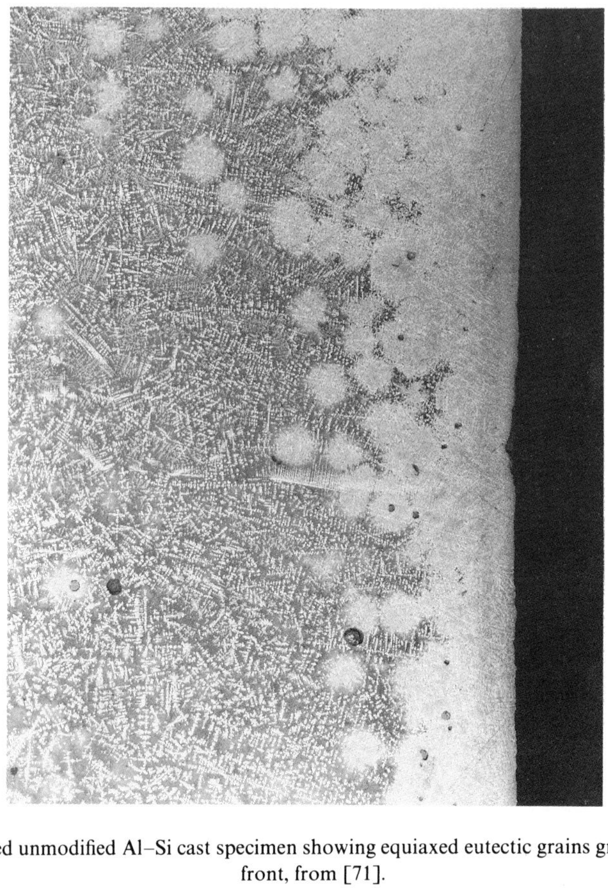

Fig. 38. Quenched unmodified Al–Si cast specimen showing equiaxed eutectic grains growing ahead of the front, from [71].

typically two to three times the extremum spacing and was shown to considerably overestimate the experimental maximum spacing found in nf–nf eutectics by Seetharaman and Trivedi [36]. Clearly if f–nf growth occurs with local spacings greater than the JH upper spacing limit they cannot realistically be described by a model based on steady-state grown.

Magnin et al. [78] approached the problem in a phenomenological manner [72, 86]. They showed the experimental undercoolings and average spacings fitted the nf–nf equation (13) and that the average spacings were always given by $\lambda = \phi\lambda_{ext}$. Equations (15) become

$$\lambda^2 v = \phi^2 \frac{a^L}{Q^L}, \qquad \frac{\Delta T^2}{v} = \left(\phi + \frac{1}{\phi}\right)^2 m^2 a^L Q^L, \qquad \Delta T\lambda = (\phi^2 + 1) m a^L. \tag{25}$$

For the unmodified Al–Si system $\phi = 3.2$, for grey cast iron $\phi = 5.4$ [58]. In conclusion the average spacing and undercooling appear to follow eq. (25) (even though the spacings are probably too large for steady-state growth). No criterion appears to be available to predict the value of ϕ.

Another aspect of growth in these f–nf systems is often important. Because of the large interface temperatures found during directional growth, equiaxed eutectic growth can occur. During casting, instead of growth occurring as a columnar eutectic front growing away from the mould wall, growth occurs as equiaxed grains expanding and filling the volume. Equiaxed growth occurs when the temperature gradients are low, such as in sand casting, and when efficient nucleation sites for the second phase are present. It is apparent (e.g. [88, 71]) that the average interface growth rate is very much smaller when equiaxed rather than columnar growth occurs. Equiaxed growth of this type can occur in grey iron castings and in Al–Si. A quenched sample in Al–Si is shown fig. 38.

2.3.1. *Modification of Al–Si*

When small quantities of Na (0.01 wt%) are added to Al–Si and the material is cast the microstructure changes from an irregular plate-like structure to a fibrous structure with a much finer particle spacing (see, e.g. [89]) (typically 20% of the average plate spacing). The Na-modified structures have better mechanical properties, in particular toughness. Extensive work has been carried on the modification process. Early work by Day and Hellawell [90] showed that both phases were continuous at the eutectic front indicating that the mechanism did not involve repeated nucleation of one phase. This observation suggested that modification involved a change in the growth of the Si phase.

In castings, although Na refines the structure it also increases the undercooling (see, e.g. [71, 74, 75]). If the action of Na was merely to increase the ease of branching in the Si, it should then grow nearer the extremum spacing. This would mean that both the spacing and the undercooling should be reduced. Flood and Hunt [71] showed that during casting, equiaxed-eutectic growth occurred in unmodified castings but columnar-eutectic growth occurred in modified castings. They estimated that the average growth rate was as much as eighteen times faster for the modified structure. They proposed that this could explain the larger undercoolings even though the structure was finer. Hanna et al. [74] showed that during furnace cooling the differences in the undercoolings were still present even when equiaxed growth did not occur. Thus differences in average growth rate could thus not be responsible for the larger undercooling. This result is in agreement with Hogan and Song [75] who measured the undercooling during the directional growth of modified and unmodified structures and again found larger undercooling for modified structures. Lu and Hellawell [91–93] found that the addition of Na and other elements dramatically increased the number of twins in the modified Si phase and proposed that the larger density of twins led to easier branching. They pointed out that Na had just the right atomic radius to be adsorbed onto a faceted step, such that as material grew around the Na, a twin was formed in the Si. Other materials such Sr, Ba, Yb and Ca having a similar atomic radius were also found to have an increased twin density and tendency for modified structure to be produced. Hogan and Song [75] pointed out that the larger undercoolings could be a result of the much larger amount of surfacce energy incorporated into the solidified structure. As was pointed out earlier the undercooling due to curvature in nf–nf eutectics exactly allows for the incorporation of surface energy. This will be true even for faceted phases provided the torque terms are correctly allowed for. The increased surface energy undercooling may be estimated by noting that a Si plate might have $\{111\}$ twins each with an energy of $\sim 30\,\mathrm{erg/cm^2}$ giving an additional undercooling of ~ 1–2 K. In conclusion the mechanism of Na modification appears to be the result of a much larger twin density leading to easier branching. The larger undercoolings are probably the result of incorporating more surface energy. It should be noted, however, that these difference in scale of the microstructure will be increased if equiaxed growth occurs in the unmodified but not in the modified structure.

2.3.2. Cast iron

Cast iron is the most widely used cast material. For this reason, considerable work has been carried on this very complicated alloy system and the reader is referred to more detailed reviews [79, 94–96]. Cast iron of approximately eutectic composition can be solidified as the equilibrium Fe–graphite (grey iron fig. 34) or with the metastable Fe–Fe_3C (white iron fig. 35). During directional growth the structure formed depends on the growth rate, grey iron forming at the slower rates. During casting the structure depends in much the same way on cooling rate. A wide variety of grey-iron structure are formed depending on the presence of other elements and on cooling rate [94]. Only two topics will be briefly covered in the present work, inoculation and the use of Mg to form nodular cast iron (spheroidal graphite SG iron).

Certain elements such as Si tend to promote grey iron by lowering the Fe_3C eutectic and by raising the Fe–graphite eutectic temperature. Commercially grey irons are produced even in thin sections by adding an inoculant containing Si just before pouring. The ferro-silicon addition usually contains small amounts of materials which act as nucleation sites for graphite. A possible explanation for the operation of the inoculant is as follows [97, 98]: As the ferro-silicon melts the Si-rich liquid around, the ferro-silicon is supersaturated in carbon so that graphite nucleates even when the bulk liquid is above the eutectic temperature. The small pockets of Si-rich liquid take some time to disperse and the resolution of the graphite is slow. When a hypoeutectic iron is cast Fe dendrites grow rapidly through the casting; if graphite particles are present,

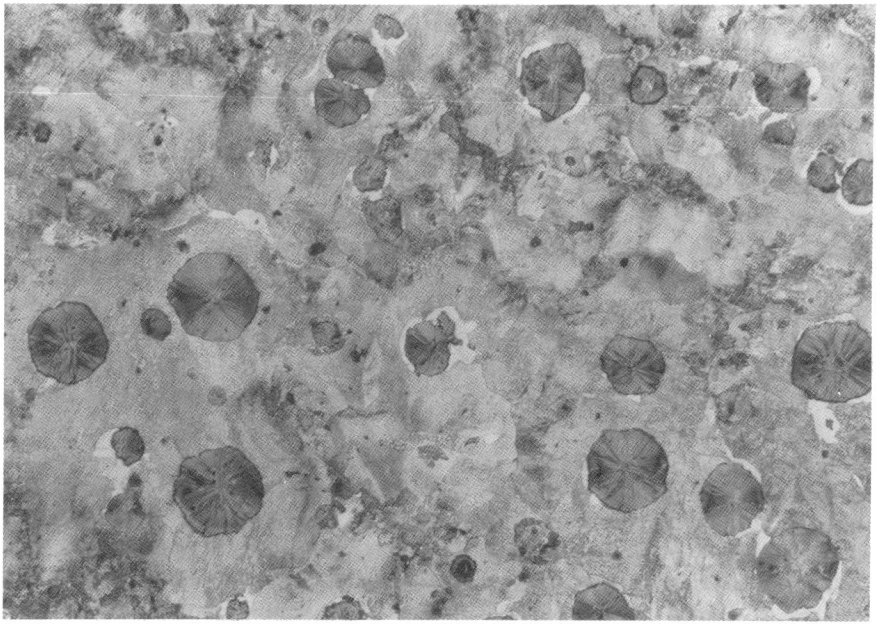

Fig. 39. Pealitic nodular iron. Mag. × 180.

separate cells or eutectic grains will be formed around each particle. The large number of equiaxed grains means that the growth rate on each grain will be kept low for the reasons discussed above. In contrast, the absence of graphite particles means that Fe_3C will nucleate near the mould wall and rapid columnar eutectic growth will take place leading to the white eutectic structure.

When about 0.05% Mg is added to a cast iron and the alloy is inoculated with ferro-silicon, nodular irons are formed [96]. In these alloys the graphite is deposited as spheres (fig. 39) rather than as a graphite flakes. The change in graphite morphology leads to a considerable increase in ductility. The action of Mg in promoting graphite spheres rather than flakes is not well understood, no model appears to be generally accepted for this transition, but there appears to be evidence for spheres being formed in the absence of S [99]. Cooperative growth appears not to have taken place in these alloys. The graphite nodule would be formed in the liquid and would grow until surrounded by iron. Once the graphite is surrounded it can only grow by solid state diffusion. This growth, however, can be appreciable because of the very high solid-state diffusion coefficient of interstitial carbon in iron.

3. *The monotectic reaction*

A monotectic reaction is one in which a liquid phase L_1 at a temperature T_M decomposes to a solid phase S and a liquid phase L_2 simultaneously:

$$L_1 \rightarrow S + L_2.$$

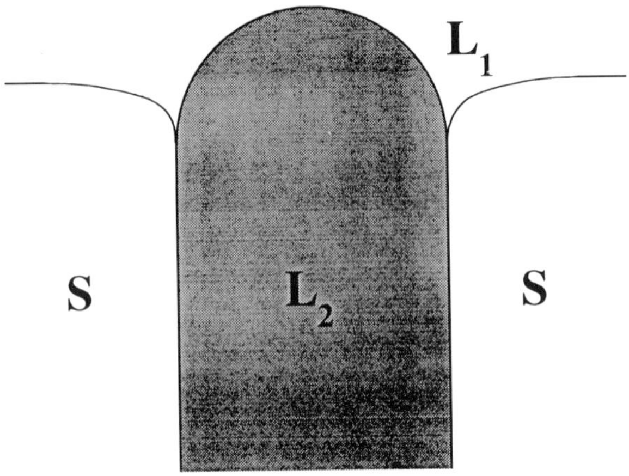

Fig. 40. A schematic monotectic interface shape for the limiting case where the cusp angle in L_1 is zero.

The reaction is similar to the eutectic except that one of the product phases is a liquid. The phase systems containing monotectic reactions show a liquid miscibility gap or dome as can be seen in fig. 1b, a typical binary system containing a monotectic reaction. C_M is the monotectic composition, T_M is the monotectic horizontal and T_C is the upper consolute temperature where the two liquids become miscible.

There appears to have been no detailed theoretical model for steady-state monotectic growth. The problem is very similar to the rod-like eutectic problem discussed earlier. One difference, however, is the fact that the L_1–L_2 interface will not vary in curvature and must be part of a sphere (see fig. 40). This means that the liquid compositions will be constant over the surface and most of the phase change, L_1 to L_2, will occur near the triple junction. The steady-state shape will be maintained by fluid flow. An approximate model similar to that for the eutectic [18] could be carried out neglecting the fluid flow but including the revised interface condition. Spacing measurements show that $R^2 v = K$ for the eutectic is still valid for the monotectic reaction, but the constant, K, is slightly larger than in comparable eutectic systems [100–102]. This observation is consistent with most of the L_2 being deposited near the triple junction.

Chadwick [103] suggested that there are three types of monotectic solidification depending on the relative surface energies of the solid and the two liquid phases:

Case 1: $\sigma_{S_1L_2} > \sigma_{S_1L_1} + \sigma_{L_1L_1}$ (L_2 does not wet the solid, wetting angle $\theta = 180°$, fig. 41a). The liquid phase L_2will tend not to nucleate on the solid substrate so that isolated droplets of L_2 will nucleate within L_1. The droplets will float or sink depending upon the density difference between the two liquids. There may then be a critical growth velocity above which the L_2 droplets in suspension in the melt are taken into the solid, and below which the droplets in suspension float away or are pushed along by the moving solid/liquid interface. Chadwick suggests that regular rod-like structures would not be formed in this type of system.

Case 2: $\sigma_{S_1L_2} < \sigma_{S_1L_1} + \sigma_{L_1L_2}$ (L_2 wets the solid, $0° < \theta < 180°$, see fig. 41 b). This situation is illustrated in fig. 41b. The L_2 droplet can nucleate on the solid substrate and would

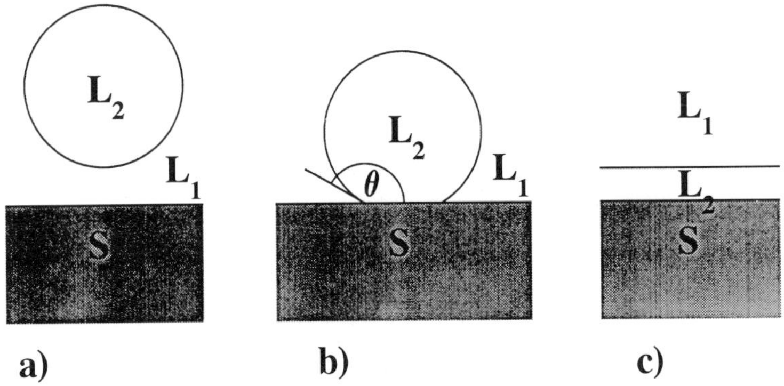

Fig. 41. A schematic diagram showing (a) L_2 does not wet the solid, wetting angle $\theta = 180°$; (b) L_2 wets the solid, $0° < \theta < 180°$; (c) L_2 wets the solid perfectly, $\theta = 0°$. From [104].

then elongate in the growth direction as the solid grows. A surface energy balance could then be established at the triple phase junction and steady-state growth similar to the eutectic could occur.

Case 3: $\sigma_{S_1L_1} > \sigma_{S_1L_2} + \sigma_{L_1L_2}$ (L_2 wets the solid perfectly, $\theta = 0°$, fig. 41c). As can be seen in fig. 41c, perfect wetting between the L_2 and the solid occurs so that a layer of L_2 seals the solid from the L_1 and steady-state growth will then be impossible. Cahn [104, 105] suggests that this surface energy relationship is unlikely to occur.

Cahn [104, 105] extended Chadwick's hypothesis and pointed out that near the critical temperature, T_C (fig. 1b), the relative surface energies will have different temperature dependencies: at the critical temperature the two solid–liquid surface energies will be equal and the L_1–L_2 surface energy will be zero (see fig. 42). As the temperature decreases σ_{SL_2} gets larger and σ_{SL_2} smaller. It is thus suggested that:

- at high temperatures: $\sigma_{S_1L_2} > \sigma_{S_1L_1} + \sigma_{L_1L_2}$;
- at T_w (the wetting transition temperature) $\sigma_{S_1L_2} = \sigma_{S_1L_1} + \sigma_{L_1L_2}$;
- for lower temperatures where $T < T_w$: $\sigma_{S_1L_2} < \sigma_{S_1L_1} + \sigma_{L_1L_2}$.

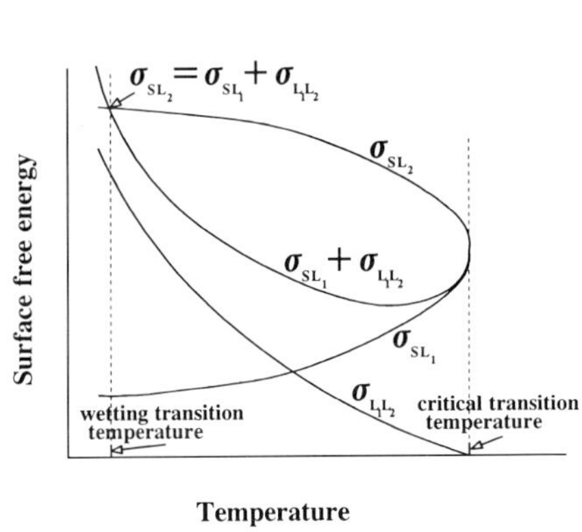

Fig. 42. A schematic plot of σ_{SL_1}, σ_{SL_2} and $\sigma_{L_1L_2}$ against temperature showing that $\sigma_{SL_1} = \sigma_{SL_2}$ and $\sigma_{L_1L_2} = 0$ at the critical temperature. At lower temperature σ_{SL_2} initially gets larger and σ_{SL_1} smaller because of the composition change. The wetting transition temperature occurs when $\sigma_{SL_2} = \sigma_{SL_1} + \sigma_{L_1L_2}$.

Chadwick's case 1 should occur when the monotectic temperature $T_M > T_w$ and case 2 when $T_M < T_w$.

Cahn pointed out that, from a growth point of view, there appears to be no a priori reason why steady-state structure should not be formed as the angle in L_1 approaches and equals zero. For case 1 the angle in L_1 should be zero and the boundary between the solid and L_2 should consist of two surfaces sufficiently close together so that their combined energy is eventually raised to σ_{SL_2}. A possible monotectic interface shape where the angle in L_1 is zero is shown in fig. 40. It is not clear that the steady-state structures should be less stable as the angle in L_2 approaches zero although as will be seen below there is some experimental evidence that this is the case.

Gruegel et al. [100] carried out extensive work on the directional solidification of monotectic systems. They divided the monotectic systems into two structural categories which could later be related to Chadwick's case 1 and 2.

Type A. The L_2 phase grows as well-aligned fibres at low velocities and high gradients (fig. 43). At higher velocities, fluctuations at the growth front lead to the fibre of L_2 breaking down into an array of droplets. At still higher growth rates irregular fibres were formed. These structures were found in succinonitrile–glycerol, succinonitrile–water, Al–Bi, Al–In and CuAl–Pb.

Type B. The L_2 phase grows as coarse elongated globules (see fig. 44) at all velocities. This structure was found in the systems succinonitrile–ethanol, Cu–Pb and Cd–Ga.

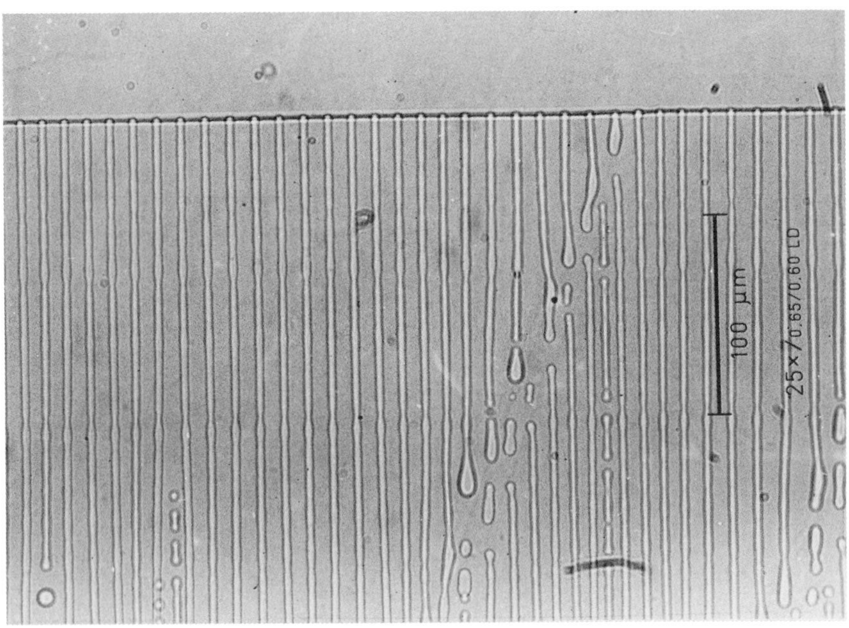

Fig. 43. Type A monotectic. Steady-state growth in succinonitrile–7.5 wt% glycerol. Courtesy A. Hellawell et al. [106].

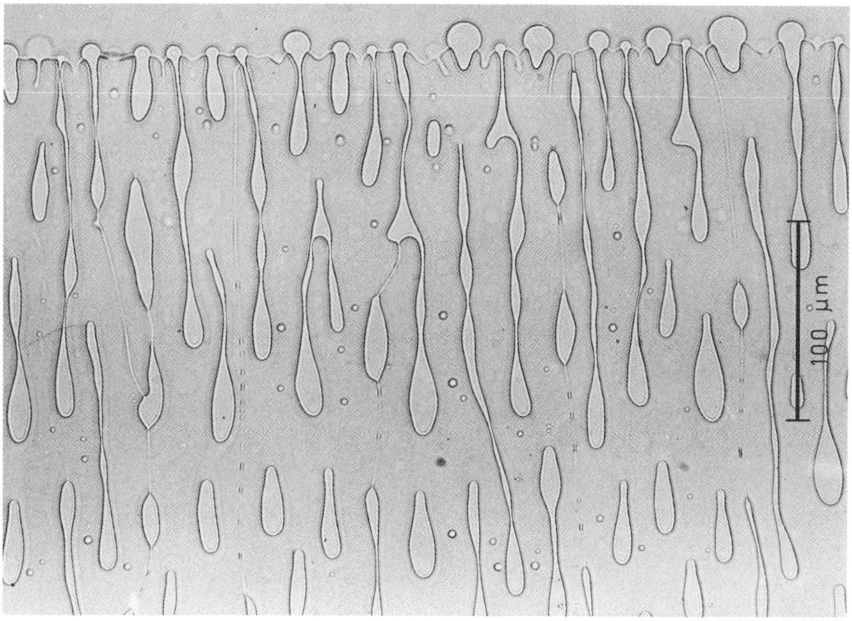

Fig. 44. Type B monotectic. Non-steady-state growth front in succinonitrile–20 wt% ethanol. Courtesy A. Hellawell et al. [100].

Hellawell and his co-workers [100, 101, 106] pointed out that a correlation exists between these two types of system and the phase diagrams: type A was associated with a high liquid miscibility gap, while type B occurred in systems with a low miscibility gap. They showed that by adding a third component to a monotectic the type of structure changed in a manner consistent with the height of the miscibility gap [106]. The ratio of T_M/T_C is shown in table 1 for different monotectics. It can be seen that the transition

Table 1
Monotectic alloys–directionally solidified.

System	(T_M/T_C)	Type
Ga–Pb	~ 0.5	aligned "A"
$Sb–Sb_2S_3$	~ 0.5	aligned "A"
Al–Bi	0.59	aligned "A"
Al–In	0.75	aligned "A"
$(CH_2CN)_2–H_2O$ (S–W)	0.887	aligned "A"
$(CH_2CN)_2–C_3H_5(OH)_3$ (S–G)	0.986	aligned "A"
Cu–16wt.%Pb–3wt.%Al	~ 0.9*	transition
$(CH_2CN)_2$–7.5wt.%E–6.9wt.%G	0.937	transition
$(CH_2CN)_2–C_2H_5OH$ (S–E)	0.94	irregular "B"
Cu–Pb	0.97	irregular "B"
Cd–Ga	0.98	irregular "B"
$C_7F_{14}–C_3H_7OH$	0.857	surface wetting

*Estimated.

occurs at about 0.9. It is suggested that the transition occurs when the monotectic temperature equals the temperature where $\sigma_{S_1L_2} = \sigma_{S_1L_1} + \sigma_{L_1L_2}$. They A is thus case 1 and type B case 2.

Grugel et al. [100] point out the formation of regular droplets at high growth rates in type A monotectics is a result of an oscillatory motion of the triple junction. This oscillatory motion can occur in eutectics [39] and similarly fibres can break up into particles [107]. It is possible that degenerate lamellae found in some eutectic [32] at high growth rates form for similar reasons to the irregular structure formed in type A monotectics at the highest velocities.

Derby and Favier [102] suggest that the types of monotectic depends on the size of the undercooling on the solid and liquid phases but specifically neglect the B_0 term which would be responsible for adjusting the relevant undercooling on the two phases.

4. The peritectic reaction

A peritectic phase digram is shown in fig. 45. On cooling past the peritectic temperature (T_P) liquid will react with α to give β. That is

$$L + \alpha \rightarrow \beta.$$

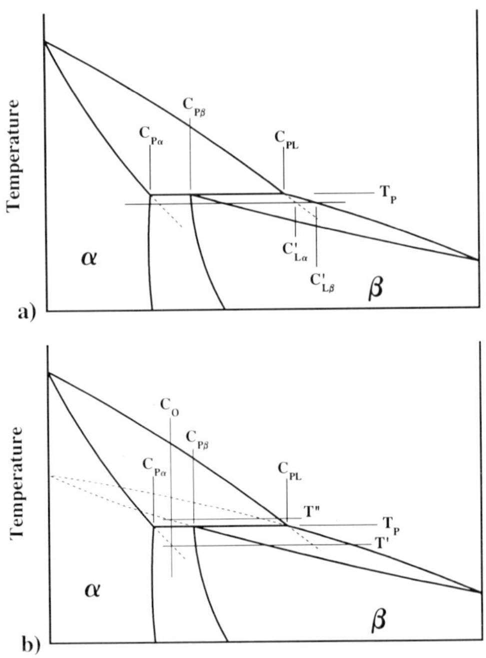

Fig. 45. (a) A peritectic diagram showing the definition of $C_{P\alpha}$, $C_{P\beta}$ and C_{PL}; also showing the liquid composition at the α and β interface below peritectic temperature, $C'_{L\alpha}$ and $C'_{L\beta}$. (b) If the alloy C_0 grows with a planar front it would grow at T' if single-phase α is deposited and at T'' if single-phase β is deposited. T' is below the peritectic temperature, T'' above the peritectic temperature.

This is in contrast to the eutectic-like reactions where, on cooling, one phase goes to two:

$$L \to \alpha + \beta.$$

As pointed out earlier, the transformation involved in heating a eutectic has much in common with a cooled peritectic.

The usual way to discuss a peritectic reaction is to consider the uniform cooling of a sample with a composition between C_{PL} and $C_{P\alpha}$ (fig. 45a). When the α liquidus is crossed, the α phase nucleates and grows dendritically throughout the sample. As the temperature drops, the dendrites thicken and eventually, below T_P (fig. 45a), the β phase nucleates and grows rapidly around the α phase. The rapid growth of β occurs on the α dendrites because it is here that liquid can react with α to deposit β on existing solid β. Once the α phase is covered the peritectic reaction effectively ceases because now the peritectic reaction can only take place by the much slower solid-state diffusion through the layer of β. The solidificiation process then continues by the thickening of the solid β as the temperature is further reduced. Equilibrium quantities of the phases will not be present; in general, some β will be present for compositions with less B than $C_{P\alpha}$ and some α will be present for alloys with more B than $C_{P\beta}$. As the alloy composition approaches C_{PL} the volume fraction of the α phase will be reduced and will eventually disappear. Approximate equilibrium volume fraction of the phases will only occur after extensive annealing at one temperature and not during continuous cooling.

It is of value to examine in more detail the kinetic reason for the rapid growth of the β over the α phase. The rate of growth of both α and β depends on how easily the rejected B atoms can be removed from the interface. Below the peritectic temperature and near the triple αβL junction B atoms can diffuse from the β phase to the α phase ($C'_{L\beta} > C'_{L\alpha}$ see fig. 45a). At the α front, the arrival of B atoms leads to melting of the α phase whereas at the β front the removal of B atoms leads to more rapid growth. The whole of the difference $C'_{L\beta} - C'_{L\alpha}$ will not be available because of curvature consideration [108]. A possible interface shape is shown schematically in fig. 46. The same arguments can be used to show that the β phase will automatically grow towards the α phase even if separated.

Similar processes occur during directional growth and these have been discussed by a number of authors (see, e.g. [109–112]). Consider first alloys and growth conditions

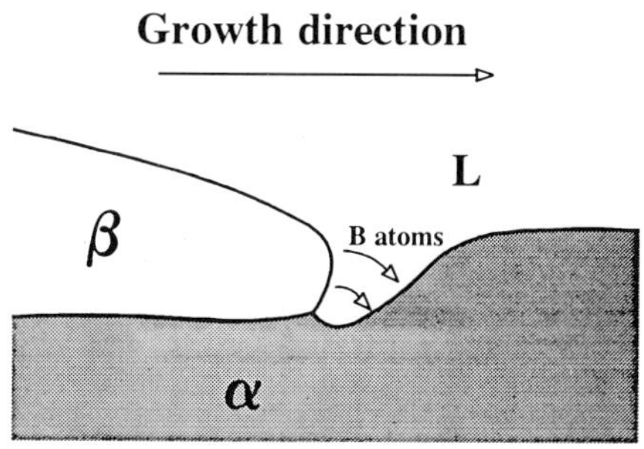

Fig. 46. A schematic view of the α–β–liquid junction during a peritectic reaction, from Hillert [108].

such that the α phase grows initially as primary cells or dendrites. The cell or dendrite tip temperature can be estimated from the models referred to earlier [49, 54, 55, 59]. Assuming that this temperature is above the peritectic temperature α cells or dendrites will form. The liquid between the cells or dendrites becomes richer in B as the temperature drops. Below T_P, the β phase will form, following and dissolving the primary α. A possible steady-state interface shape is shown in Fig. 46. It is interesting to note that, because of the triple junction, curvature leads to freezing of α just before the triple junction even though considerable dissolution occurs further away. As before, at temperatures where the α phase is covered, the peritectic reaction will virtually cease. During further cooling, the β phase will thicken in the usual way. The fraction of the solid phases will not be given by the equilibrium values but will be qualitatively as discussed above. The regularity of the two phase structure will depend on the regularity of the original α cells or dendrites. Typical structures for transformed cells and dendrites are shown in fig. 47. When the α cells or dendrites grow with a tip temperature below the peritectic temperature banded structures similar to those discussed below can form.

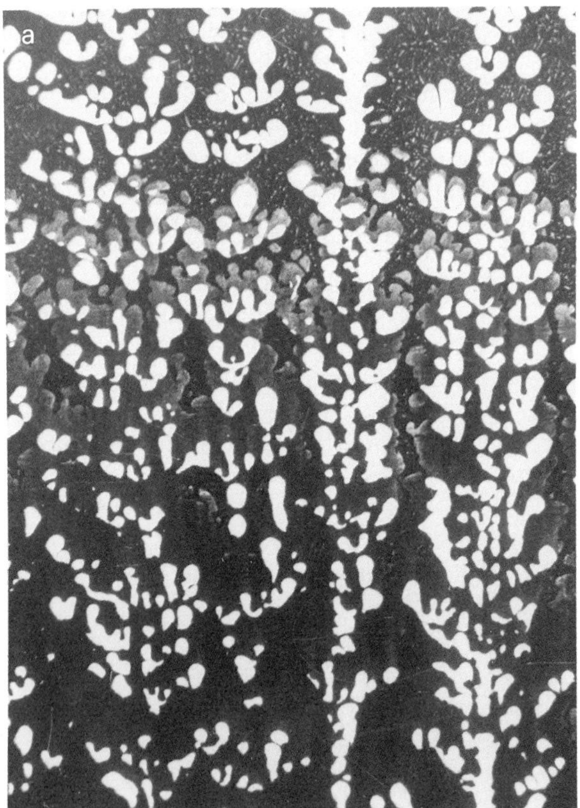

(Fig. 47. cont'd)

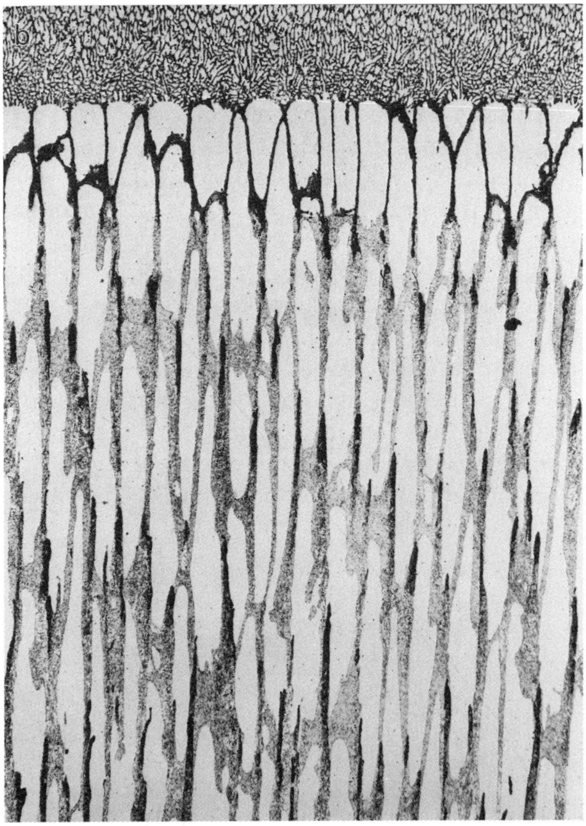

Fig. 47. Quenched interface for cellular and dendritic peritectic (growth upwards). (a) Longitudinal section Cd–8 wt.% Ag. The quenched interface is just visible as a lighter grey phase between the dendrites. Mag. × 28.2. Courtesy D. St. John [116]. (b) Longitudinal section of Sn–1 wt.% Cd. The quenched liquid is apparent as a darker phase between the cells near the tip. Mag. × 48. Courtesy W.J. Boettinger [110].

When the α phase grows as a planar front, the transformations occur in a different manner [104–107]. The constitutional undercooling condition equation for steady-state planar growth is

$$G > m\frac{\mathrm{d}c}{\mathrm{d}x} = \frac{mv}{D}\left(1 - \frac{1}{k}\right)C_0,$$

or

$$\frac{GD}{v} > m\left(1 - \frac{1}{k}\right)C_0 = \Delta T_{\text{Planar}} \tag{26}$$

where ΔT_{Planar} is the difference temperature between the liquidus and single-phase solidus at the bulk alloy composition. In many peritectic systems this difference ΔT_{Planar} can be quite small so that planar front growth can often be achieved in practice. Under these conditions, alloys with bulk compositions $C_0 < C_{P\alpha}$ grow as single-phase α. The β phase will not be formed because the alloy will be completely solid before the peritectic temperature is reached. Alloys with compositions greater than $C_{P\beta}$ can grow as single-phase β without any α, provided the β phase nucleates. Initially α will be formed; as the sample approaches steady state the interface temperature will decrease below T_P and, eventually, if β nucleates it will grow rapidly over the α phase. There will be no tendency for a hole in the β layer to remain since the transformation to β will be more rapid in those regions.

For alloys with bulk composition between $C_{P\alpha}$ and $C_{P\beta}$ (fig. 45) the behaviour is potentially more complicated, and this has led to speculation [113–115] that coupled eutectic-like growth could occur, but this has not been found [109–112] and would not be expected for theoretical reasons to be discussed below. For alloys with compositions slightly greater than $C_{P\alpha}$, planar α growth would occur. For these alloys the steady-state growth temperature would be below the peritectic temperature. For a large enough undercooling below T_P the β phase will nucleate at the growth front. As before, provided the interface temperature is below the peritectic temperature, the β phase will

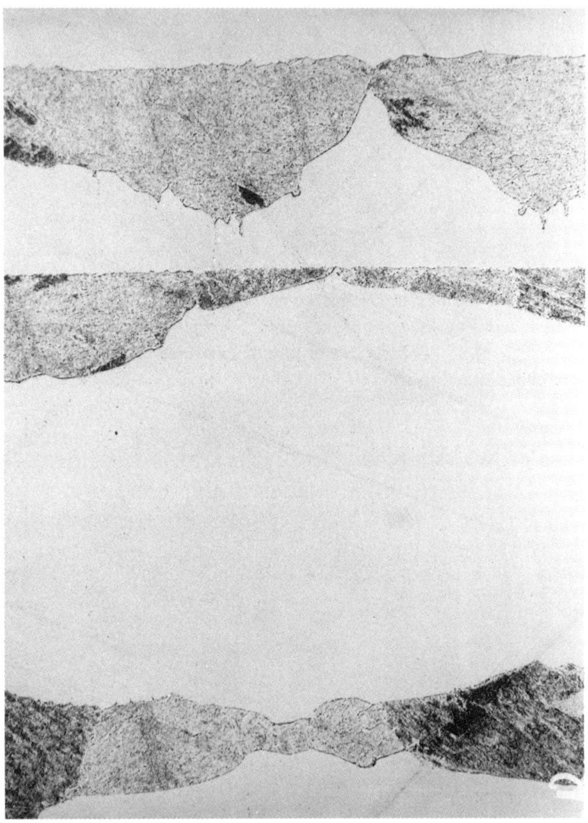

Fig. 48. Banded peritectic structure, Sn–1 wt.% Cd. Mag. × 48. Courtesy W.J. Boettinger [110].

grow more rapidly in contact with α and liquid so that the α phase should be rapidly covered by the β phase. The problem now is that a planar steady-state β front at this composition should grow at a temperature above the peritectic temperature T'' (see fig. 45b). This will mean that, as steady state is approached, the temperature of the front will rise above the peritectic temperature and eventually α will be nucleated. Above the peritectic temperature $C'_{P\beta} < C'_{P\alpha}$ so that here α will grow at the expense of β again and a planar α front will be formed. As before, the temperature will go down until β is formed and the cycle will be repeated. This type of banding behaviour is observed in practice (see, e.g. [109–112] and fig. 48). For coupled growth of peritectic α and β together, B atoms need to be taken from the liquid ahead of both the α and β phases but *much* more B need to be removed from ahead of the α phase. An examination of fig. 45 shows that below T_P the changes in liquid composition with temperature do not assist transfer across the interface. The composition ahead of the α, $C'_{P\alpha}$ (fig. 45a), will be lower in B than that ahead of the β phase, $C'_{P\beta}$. This means that the β phase will grow more easily near the α but the α phase grows with more difficulty near β, if it grows at all. This is in contrast to the eutectic-like reactions where each grows more easily near the other.

The key feature for cooperative growth is, as discussed earlier, the competitive growth of the two new phases at the triple junction. Too much of one new phase should automatically lead to more of the other phase being formed. For this to occur, the initial single phase needs to have an equilibrium composition between the compositions of the two new phases since this allows potential competitive growth at the triple phase boundary. When a peritectic is heated, cooperative formation of liquid and α should be possible from single-phase β but would not be expected when a peritectic is cooled.

In conclusion, during directional growth in peritectic systems a number of structures are expected and have been observed (see, e.g. [109–112]). If planar interface growth does not occur, irregular fibrous structures consisting of partially decomposed α primary structures will be formed for compositions near $C_{P\beta}$ (fig. 47) [116]. At the composition extremes, highly segregated single-phase cells or dendrites will be formed (either α or β). For planar interface growth conditions single-phase α will be formed for alloys where $C_0 < C_{P\alpha}$ and single-phase β can be formed if $C_0 > C_{P\beta}$. For intermediate compositions either a single phase or alternate bands of each phase will be formed (fig. 48). Banding may also occur when cells are present for one or both of the phases. Clearly, the structures formed may be modified by solid-state processes occurring at lower temperatures. If uniform-composition single-phase grains of β are required it would seem that samples of a high enough B composition should be grown with a planar front or, alternatively, the samples should be annealed after solidification in a single-phase region until almost equilibrium conditions are achieved.

5. *Concluding remarks*

Although some progress has been made in the understanding of eutectic, monotectic and peritectic growth, much of the work is rather qualitative. Steady-state lamellar eutectic growth appears to be relatively well described by the basic theoretical models. Clearly, improvements in accuracy and the removal of restrictive assumptions have been made and will need to be made in the future. One particular area which may need more attention is the interaction of fluid flow and solidification, and another the correct handling of the different atomic volumes in the different phases. Although transitions such as the lamelar-to-rod, and dendrite-to-eutectic have been analysed, phenomenological assumptions such as growth of the structures with the lower undercooling have been used. These assumptions need to be examined in more detail. In the case of faceted–nonfaceted eutectics it would appear that, although steady-state analysis is used, the average spacings are larger than the maximum possible spacing. It seems probable that further progress will only be made when an attempt is made to model the non-steady-state process. As mentioned earlier it should be possible to produce a simple analytic model for monotectic growth and eventually to examine the stability of the phases. A good description of peritectic growth must await a better description of the underlying cellular and dendritic growth.

References

[1] F.D. Lemkey and G.P. McCarthy, Conference on In Situ Composites – II, Bolton Landing, Lake George, NY, Xerox publishing (1975) p. 77.
[2] H. Weiss, Conference on In Situ Composites – II, Bolton Landing, Lake George, NY, Xerox publishing (1975) p. 377.
[3] W. Albers, Proc. of the Conference of In Situ Composites I, Lakeville, CT (1972) p.1.
[4] J.D. Hunt and K.A. Jackson, Trans. Metall. Soc. AIME 236 (1966) 843.
[5] E. Scheil, Z. Metallk. 37 (1946) 1.
[6] J.P. Chilton and W.C. Winegard, J. Inst. Met. 89 (1961) 962.
[7] A.S. Yue, Trans. Metall. Soc. AIME 224 (1962) 1010.
[8] J.D. Hunt and J.P. Chilton, J. Inst. Met. 92 (1963) 21.
[9] F.D. Lemkey, R.W. Hertzberg and J.A. Ford, Trans. Metall. Soc. AIME 233 (1965) 334.
[10] D.T.J. Hurle and J.D. Hunt, Proc. Solidification of Metals, Brighton, 1967, ISI. P110 (1968) p. 162.
[11] K.A. Jackson, Liquid Metals and Solidification (ASM, Cleveland, OH, 1958) p. 130.
[12] H.W. Kerr and W.C. Winegard, Crystal Growth, ed. H.A. Peiser (Pergamon, Oxford, 1967) p. 179.
[13] R.S. Fidler, M.N. Crocker and R.W. Smith, J. Cryst. Growth 13/14 (1972) 739.
[14] C. Zener, AIME Trans 167 (1946) 550.
[15] W.A. Tiller, Liquid Metals and Solidification (ASM, Cleveland, OH, 1958) p. 276.
[16] W.H. Brandt, J. Appl. Phys. 16 (1945) 139.
[17] M. Hillert, Jernkontorets Ann. 144 (1960) 520.
[18] K.A. Jackson and J.D. Hunt, Trans. Metall. Soc. AIME 236 (1966) 1129.
[19] L.F. Donaghey and W.A. Tiller, Mater. Sci. Eng. 3 (1968/9) 231.
[20] S. Strässler and W.R. Schneider, Phys. Condens. Matter 17 (1974) 153.
[21] G.E. Nash, J. Cryst. Growth 38 (1977) 155.
[22] V. Datye and J.S. Langer, Phys. Rev. B24 (1981) 4155.
[23] R. Trivedi, P. Magnin and W. Kurz, Acta Metall. 35 (1987) 971.
[24] W. Kurz and R. Trivedi, Met. Trans. 22A (1991) 3051.
[25] H.E. Cline, Trans. Metall. Soc. AIME 242 (1968) 1613.
[26] D.T.J. Hurle and E. Jakeman, J. Cryst. Growth 3/4 (1968) 574.
[27] K. Kassner and C. Misbah, Phys. Rev. Lett. 66 (1991) 1458; 522.
[28] K. Brattkus, B. Caroli, C. Caroli and B. Roulet, J. de Physique 51 (1990) 833.
[29] R.W. Series, J.D. Hunt and K.A. Jackson, J. Cryst. Growth 40 (1977) 221.
[30] P. Magnin and R. Trivedi, Acta Metall. Mater. 39(4) (1991) 453.
[31] D.J. Fisher and W. Kurz, Acta Metall. 28 (1980) 777.

[32] J.D. Hunt, Ph.D. Thesis, Cambridge (1963).
[33] M. Zimmerman, M. Carrard and W. Kurz, Acta Metall. 37 (1991) 3305.
[34] M. Zimmerman, M. Carrard, M. Gremaud and W. Kurz, Mater. Sci. Eng. A134 (1991) 1278.
[35] B. Caroli, C. Caroli, G. Faivre and J. Mergy, J. Cryst. Growth 118 (1992) 135.
[36] V. Seetharaman and R. Trivedi, Met. Trans. 19A (1988) 2955.
[37] R. Trivedi, J.T. Mason, J.D. Verhoeven and W. Kurz, Met. Trans. 22A (1991) 2523.
[38] H.E. Cline, J. Appl. Phys. 50 (1979) 4780.
[39] A. Karma, Phys. Rev. Lett. 59 (1987) 71.
[40] R.M. Jordan and J.D. Hunt, Met. Trans. AIME 3 (1972) 1385.
[41] J.N. Clark and R. Elliott, J. Cryst. Growth 33 (1976) 169.
[42] M. Tassa and J.D. Hunt, J. Cryst. Growth 34 (1976) 38.
[43] J.N. Clark and R. Elliott, Met. Trans. AIME 7A (1976) 1197.
[44] D.D. Double, Mater. Sci. Eng. 11 (1973) 325.
[45] B. Drevet, D. Camel and J.J. Favier, Proc. 8th European Symposium on Materials and Fluid Sciences in Microgravity, Oxford, UK, ESA (1989) p. 101.
[46] J.D. Hunt and K.A. Jackson, Trans. Metall. Soc. AIME 239 (1967) 864.
[47] G. Tammann and A.A. Botschwar, Z. Anorg. Chem. 157 (1926) 26.
[48] M.H. Burden and J.D. Hunt, J. Cryst. Growth 22 (1974) 99; 109.
[49] M.H. Burden and J.D. Hunt, J. Cryst. Growth 22 (1974) 328.
[50] F.R. Mollard and M.C. Flemings, Trans. Metall. Soc. AIME 239 (1967) 1526.
[51] H.E. Cline and J.D. Livingston, Trans. Metall. Soc. AIME 245 (1969) 1987.
[52] I.R. Hughes and H. Jones, J. Mater. Sci. 11 (1976) 1781.
[53] J.A. Juarez-Islas and H. Jones, Acta Metall. 35 (1987) 499.
[54] W. Kurz, B. Giovanola and R. Trivedi, Acta Metall. 34 (1986) 823.
[55] R. Trivedi, J. Lipton and W. Kurz, Acta Metall, 35 (1987) 965.
[56] M. Zimmermann, M. Carrard and W. Kurz, Acta Metall. 37 (1989) 3305.
[57] M. Carrard, M. Gremaud, M. Zimmermann and W. Kurz, Acta Metall. Mater. 40 (1992) 983.
[58] R. Trivedi and W. Kurz, in: Solidification Processing of Eutectic Alloys, eds. D.M. Stefanescu, G.J. Abbaschian and R.J. Bayuzick (Met. Soc., 1988).
[59] S.-Z. Lu and J.D. Hunt, J. Cryst. Growth 123 (1992) 17.
[60] D.G. McCartney, J.D. Hunt and R.M. Jordan, Met. Trans. 11A (1980) 1243.
[61] J. Walter and H.E. Cline, Met. Trans. 1 (1969) 1221.
[62] J.D. Hunt and J.P. Chilton, J. Inst. Met. 91 (1963) 338.
[63] F.C. Frank and K.E. Puttick, Acta Metall. 4 (1956) 206.
[64] D.J.S. Cooksey, D. Munsen, M. P. Wilkinson and A. Hellawell, Philos. Mag. 10 (1964) 745.
[65] R.M. Jordan and J.D. Hunt, Met. Trans. 2 (1971) 3401.
[66] H.A.H. Steen and A. Hellawell, Acta Metall. 20 (1972) 363.
[67] H. Fredricksson, M. Hillert and N. Lange, J. Inst. Met. 101 (1973) 285.
[68] D.C. Jenkinson and L.M. Hogan, J. Cryst. Growth 28 (1975) 171.
[69] B. Toloui and A. Hellawell, Acta Metall. 24 (1976) 565.
[70] R. Elliott and S.M.D. Glenister, Acta Metall. 28 (1980) 1489.
[71] S.C. Flood and J.D. Hunt, Met. Sci. 15 (1981) 287.
[72] H. Jones and W. Kurz, Met. Trans. 11A (1980) 1265.
[73] H. Jones and W. Kurz, Z. Metallk. 72 (1981) 792.
[74] M.D. Hanna, Shu-Zu Lu and A. Hellawell, Met. Trans. 15A (1984) 459.
[75] L.M. Hogan and H. Song, Met. Trans. 18A (1987) 707.
[76] P. Magnin and W. Kurz, Met. Trans. 19A (1988) 1955.
[77] P. Magnin and W. Kurz, Met. Trans. 19A (1988) 1965.
[78] P. Magnin, J.T. Mason and W. Kurz, Acta Metall. 39 (1991) 469.
[79] R. Elliott, Eutectic Solidification Processing. (Butterworths, London, 1983).
[80] I. Minkoff, The Physical Metallurgy of cast Iron. Wiley, Chichester (1983).
[81] A. Hellawell, Prog. Mater. Sci. 15 (1973) 1.
[82] M.N. Croker, R.S. Fidler and R.W. Smith, Proc. Roy. Soc. A335 (1973) 15.
[83] K.A. Jackson, Crystal Growth'74 (North-Holland, Amsterdam, 1974) p. 130.

[84] K.A. Jackson, Mater. Sci. Eng. 65 (1984) 7.
[85] T. Sato and Y. Sayama, J. Cryst. Growth 22 (1974) 529.
[86] P. Magnin and W. Kurz, Acta Metall. 35 (1987) 1119.
[87] J. Liu, Y. Zhou and B. Shang, Acta Metall. 38 (1990) 1625.
[88] J.D. Hunt, Mater. Sci. Eng. 65 (1984) 75.
[89] D.A. Granger and R. Elliott, in: Metals Handbook 15 (ASM, Cleveland, OH, 1988) 159.
[90] M.G. Day and A. Hellawell, J. Inst. Met. 95 (1967) 377.
[91] S.-Z. Lu and A. Hellawell, J. Cryst. Growth 73 (1985) 316.
[92] S.-Z. Lu and A. Hellawell, Solidification Processing'87 (Inst. Met., London, 1988) p. 44.
[93] S.-Z. Lu and A. Hellawell, Met. Trans. 18A (1987) 1721.
[94] D.M. Stefanescu, in: Metals Handbook 15 (ASM, Cleveland, OH, 1988) 168.
[95] D.B. Craig, M.J. Hornung and T.K. McCluhan, in: Metals Handbook 15 (ASM, Cleveland, OH, 1988) 629.
[96] I.C.H. Hughes, in: Metals Handbook 15 (ASM, Cleveland, OH, 1988) 647.
[97] H. Fredriksson, Mater. Sci. Eng. 65 (1984) 137.
[98] G.A. Feest, G. McHugh, D.O. Morton, L.S. Welch and I.A. Cook, Proc. of Solidification Technology in the Foundry and Casthouse (Metals Society, London, 1980).
[99] W.C. Johnson and H.B. Smartt, Solidification Processing'87 (Inst. Met., London, 1988) p. 125.
[100] R.N. Grugel, T.A. Lograsso and A. Hellawell, Met. Trans. 15A (1984) 1003.
[101] R.N. Grugel and A. Hellawell, Met. Trans. 12A (1981) 669.
[102] B. Derby and J.J. Favier, Acta Metall. 31 (1983) 1123.
[103] G.A. Chadwick, Brit. J. Appl. Phys. 16 (1965) 1095.
[104] J.W. Cahn, J. Chem. Phys. 66 (1977) 3667.
[105] J.W. Cahn, Met. Trans. 10A (1979) 119.
[106] R.N. Grugel and A. Hellawell, Met. Trans. 15A (1984) 1626.
[107] S. Marich and D. Jaffrey, Met. Trans. 2 (1971) 2681.
[108] M. Hillert, Solidification and Casting of Metals (Metals Society, London, 1979) p. 81.
[109] D.R. Uhlmann and G.A. Chadwick, Acta Metall. 9 (1961) 835.
[110] W.J. Boettinger, Met. Trans. 5 (1974) 2023.
[111] H.D. Brody and S.A. David, Solidification and Casting of Metals (Metals Society, London, 1979) p. 144.
[112] A. Ostrowski and E.W. Langer, Solidification and Casting of Metals (Metals Society, London, 1979) p. 139.
[113] B. Chalmers, Principles of Solidification (Wiley, New York, 1964) pp. 224–227.
[114] J.D. Livingston, Mater. Sci. Eng. 7 (1971) 61.
[115] M.C. Flemings, Solidification Processing (McGraw-Hill, New York, 1974) pp. 117–178.
[116] D.H. St. John and L.M. Hogan, J. Mater. Sci. 19 (1984) 939.

CHAPTER 18

Spheralitic Crystallization in Macromolecules

PAUL J. PHILLIPS

Department of Materials Science and Engineering, University of Tennessee, Knoxville, TN 37996-2200, USA

Contents

1. Introduction . . . 1169
2. Bulk crystallization behavior . . . 1173
   2.1. Avrami analyses . . . 1173
   2.2. Half-time analyses . . . 1177
   2.3 Nucleating agents . . . 1177
3. The Lamellar crystal . . . 1179
   3.1. The molecular trajectory . . . 1187
   3.2. The nature of the molecular re-entry process . . . 1189
4. The crystal growth process . . . 1192
   4.1. The influence of molecular structure . . . 1199
   4.2. The influence of pressure . . . 1203
5. Spherulitic morphogenesis . . . 1205
6. Summary . . . 1214
Acknowledgments . . . 1214
References . . . 1214

HANDBOOK OF CRYSTAL GROWTH, VOL. 2
edited by D.T.J. Hurle

1. Introduction

Since the earliest studies of crystallization, spherulitic growth has been recognized. Well-known are the geological specimens which are eagerly purchased by tourists in desert places like Arizona, where "rock roses" and pyritic "suns" (imported from Illinois!) [1, 2] are found in large numbers in souvenir shops (fig. 1). Approximately spherical polycrystalline entities are also known to metallurgists, e.g. in cast iron. However, spherulites par excellence are usually encountered in organic materials, especially in macromolecules, both synthetic and natural. Indeed, the paradigm of bulk macromolecular morphology is the spherulite [3]. Because of its special importance in controlling the physical and mechanical properties of semi-crystalline thermoplastics, it is in macromolecules that the study of spherulitic growth has become paramount. Much of our understanding of the details of spherulitic crystallization comes from studies of these man-made materials. Spherulites can also be encountered in natural polymers, such as DNA and other polypeptides (fig. 2). For an excellent descriptive review on the phenomenon of spherulites and optical behavior the paper by Morse and Donnay [4] is strongly recommended.

As can be seen in figs. 1 and 2, spherulites are approximately radial arrays of long thin crystals (lamellae), the first crystals having been nucleated at a central point. However, in order for the fully formed disc or sphere to be generated it is essential that new crystals be continuously nucleated as the radius increases. From the earliest studies of spherulites [2] it has been recognized that there are two essential forms of spherulite. The first is that already described, of tight radial arrays of crystals (fig. 1a). The second is represented in fig. 1b and is a less compact form in which interstices between the crystals are filled with some other materials. In the case of macromolecules this form corresponds closest to what are known as *axialites*. It has also been recognized for a long time that spherulites are not dendritic in nature as they are not usually generated by crystallographic branching. In the majority of cases the branching process is non-crystallographic. There can be exceptions where crystallographic branching occurs as a contributory process. An understanding of spherulitic growth requires that the three basic processes of initial nucleation, crystal growth and continuous crystal generation be understood separately. Furthermore, the reasons why crystals that are initially growing parallel to one another should splay apart and grow in a nonparallel fashion need to be determined.

In addition, a fundamental question which needs to be answered in most cases is why the spherulite forms in the first place, instead of a single grain growing in the conventional manner to generate a polycrystalline material. Early speculations on the reasons for the formation of mineral spherulites (such as lithoidal rhyolite) were based on their being encountered in obsidian, or other glasses, and it was suggested that the diffusion of vapours into the glass caused the crystallization to occur [5]. Such explanations, of course, were not based on scientific studies of the process and now that such studies are possible, but difficult, interpretations have changed [6]. In the cases of inorganic materials, whether mineral or metallic, the basic reasons may vary from one material to another [7]. In many cases the primary cause lies with chemical impurities and the demonstration of that source for some materials has been quite graphic (see, e.g.

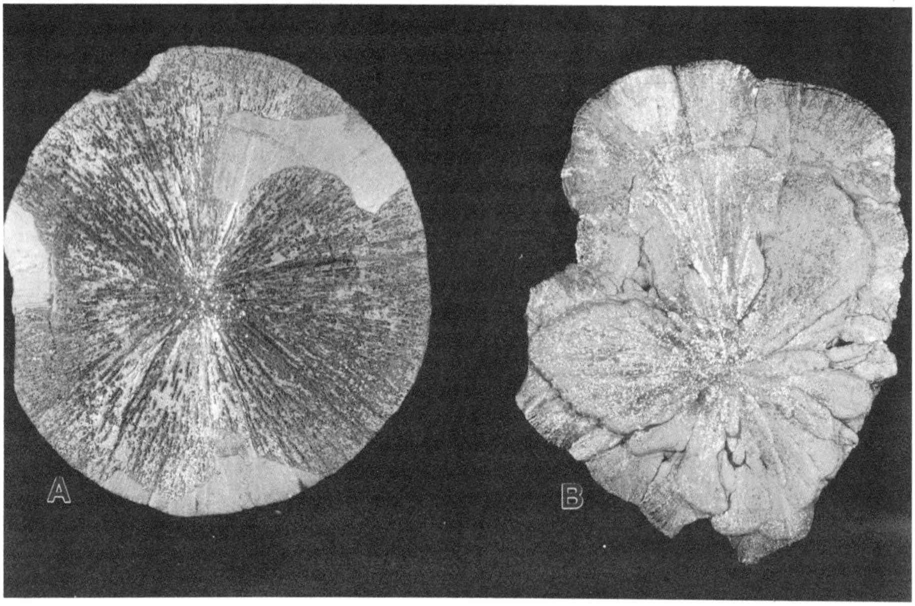

Fig. 1. The two basic types of aggregation found in spherulites, illustrated using pyritic suns. (a) Tight arrays of narrow radial lamellae; (b) open arrays of wide lamellae.

Fig. 2. Spherulitic growth of DNA from the gel state [119].

[8]). Indeed, it has generally been assumed that impurities need to be present to generate spherulites, and that indeed their formation may be induced by the addition of impurities [9]. It now seems clear that the presence of heterogeneous nuclei play a critical role in the generation of spherulitic and related textures in minerals [10]. However, one study of spherulites in antimony concluded that stress was necessary to the nucleation of groups of crystals, rather than single crystals or conventional polycrystals [11]. These authors also showed that nucleation involved the presence of a hole, an occurrence that is well established in polymer thin films [12], but not essential for spherulitic formation in macromolecules. In the case of graphite spherulites in cast irons and related alloys, it was suggested that spherulites could only form when sulphur was absent [13]. In this case stress was also invoked as a mechanism to generate the folded fan-like crystals which were found to compose the spherulite. So it does need to be recognized that the reasons for the occurrence of spherulites may vary considerably from material to material.

When macromolecules and hydrocarbons are considered, intensive studies using polarized microscopy of the growing spherulites have made it a little easier to obtain quantitative definitive data that stand up to the test of theory. Macromolecules and similar shorter molecules do have the unique common feature that the lamellar crystals which make up the spherulite are a natural form of the crystal caused by *inherent limitations in the molecules* themselves. Any real understanding of the phenomenon, as in all cases of crystal growth, comes from observation of the growth kinetics of the individual crystals and the development of models that predict all aspects of the observed behavior, including the crystal shape, the growth faces, the rates of linear growth and the dependence of all these parameters on temperature and pressure. Why lamellar crystals of macromolecules should be found in a spherulitic aggregate is another matter. The early explanations, which held up for over two decades [14, 15] invoked the effect of impurity concentrations on diffusion. This very sound theory was based on constitutional supercooling (see [16], for a review) which had been derived for metallic systems. Expressed very simply it comes down to the suppression of melting point by impurities, or in other words the very well known principle from physical chemistry of the depression of freezing point in mixtures. However, the advent of new methods of preparation for electron microscopy (see, e.g. [3, 17]) has shown that although diffusion of impurities is undoubtedly a factor in the generation of spherulitic textures, it does not account for the detailed mode of formation of a spherulite. It is somewhat ironic that a well-developed theory based on solid scientific logic should be proven not to apply to polymers at a time when it has been found to apply well to minerals [6, 7, 18]. However, there can be little doubt that when impurities are present in a material, and of sufficient concentration at the growth front, then the Keith and Padden theory is perfectly appropriate.

In this chapter the general phenomenology of bulk spherulitic growth will be discussed first, together with an analysis of the bulk growth kinetics, including nucleation and growth processes. Then the actual form of the individual crystals that comprise the spherulite will be discussed and the reasons for the known dependences. This will be followed by an analysis of the growth kinetics and the models which account best for the behavior. The mechanisms by which new, or "daughter" crystals are generated will then be considered in the context of the spherulitic morphogenesis. Throughout, the emphasis will be on the crystals formed by synthetic macromolecules and how the behavior is influenced by the molecular structure and by the length of the molecule.

Many of the concepts that apply, and are understood best, simply do not translate to the non-macromolecular world of elemental and inorganic substances. The reason for this is simple, the atoms that make up macromolecules are joined together to form *strings which survive the melting process.* Hence, unlike the metallic and mineral systems, the generation of a crystal involves the rearrangement of the pre-formed strings to form the crystalline state. Naturally, this process is much more complicated than that of packing together spheres or sphere-like objects. The string has its own characteristics, which can range from wire-like stiffness to rope-like flexibility to wet spaghetti-like tangled mattedness. Additionally, the strings can have both short or long branches, or even may be joined together into one continuous three-dimensional grid [19, 20]. Each string may have a chemical microstructure which varies along its length in a highly ordered or a random fashion, all determined by the details of the process by which it was synthesized. Despite this richness of diversity the basic concepts of nucleation and growth apply ubiquitously to all systems. Of paramount importance to the commonality of behavior is the presence of the lamellar crystal and the manner in which the molecule adds to the crystal.

A first glance at any table of the unit cells formed by polymers tends to generate a phobia for all things polymeric because of the complexity and diversity of the structures generated. However, it should be recognized from the outset that the complexity is a result of the rules of crystallography. In fact polymer crystals have a remarkably uniform method of general packing, which when understood, by even the novice, cures the phobia. The simplest way to envision a polymer crystal is to recognize that each chain has a lowest-energy form, which is normally only attained in the crystal. These lowest-energy forms invariably generate cylindrical shapes, which then pack together laterally to form the crystal. Of course, this is an oversimplification since the actual chain shape will also be influenced by the lateral packing, however, in most cases this is a very useful concept.

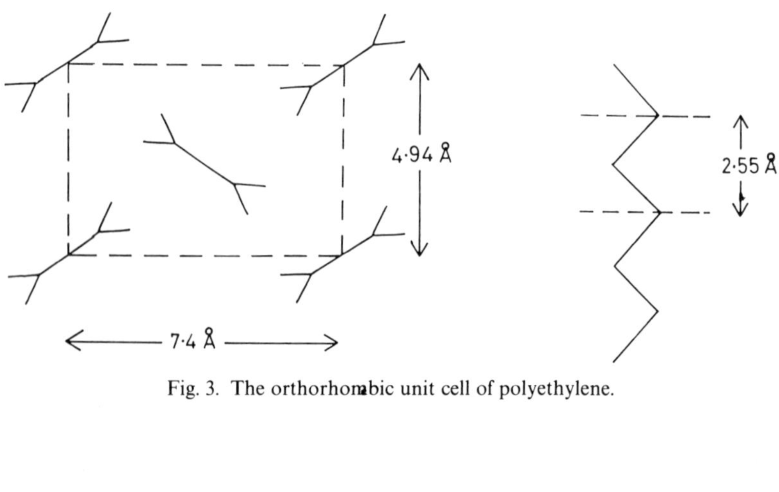

Fig. 3. The orthorhombic unit cell of polyethylene.

Polymer chains tend to form their own stable conformations (or shapes) determined by the energetics of rotation about the chain backbone atoms. The most intensively studied system is the highly flexible chain, polyethylene, in which the chains first form an extended structure, usually refferred to as a linear zig–zag (see later). The extended forms then pack together laterally, parallel to one another, giving an orthorhombic unit cell (fig. 3). The other general form of chain shape is the helix, which again gives rise to an approximately cylindrical shape for the chain. Again the cylinders pack together efficiently to form the crystal. Unless mentioned otherwise, it should be assumed that polymers generally form crystals in which the chains are stacked parallel to one another and that the approximate shape of the units that pack together laterally is cylindrical.

2. *Bulk crystallization behavior*

The measurement of bulk changes can be followed by any method which accurately responds to the changes occurring in any fundamental property, such as volume or heat capacity. The earlier studies generally used some dilatometric method to measure volume changes [21]. However, dilatometry is a laborious technique and more utilitarian methods are now available. Two techniques are popular at the present time: (a) transmitted polarized light intensity [22, 23]; and (b) the most commonly used technique, differential scanning calorimetry (DSC) [24], which essentially measures the energy released by the crystallization process as a function of time (fig. 4a). When such data are plotted as a fraction of the crystallinity achieved at completion of the crystallization process versus the logarithm of time, characteristic sigmoidal curves are obtained (fig. 4b).

In most experimentation it is usual to study the bulk growth rate under isothermal conditions and to repeat the experiment for the widest range of supercoolings over which accurate data can be obtained. Data may also be obtained as a function of both pressure and temperature for the case of transmitted light intensity studies [25].

2.1. *Avrami analyses*

Experimental data are usually analyzed first using the Avrami equation [26]:

$$\ln(1 - X_c) = kt^n, \tag{1}$$

where X_c is the volumetric crystallinity, t the time and n is known as the Avrami exponent. k is simply a scaling constant which contains information on the individual rates of nucleation and linear growth. This generalized equation results from several models of the crystallization process that utilize combinations of the type of nucleation process with the shape of the growing entity. The use of instantaneous (heterogeneous) nucleation or sporadic (homogeneous) nucleation gives different time dependencies because of the time dependence of the sporadic process. Similarly, the assumptions of a disc or a sphere as the growing entity give different powers of time dependence because of the different dimensionalities of the objects. All the versions of the model use the concept of expanding wavefronts governed by a Poisson distribution of nucleation

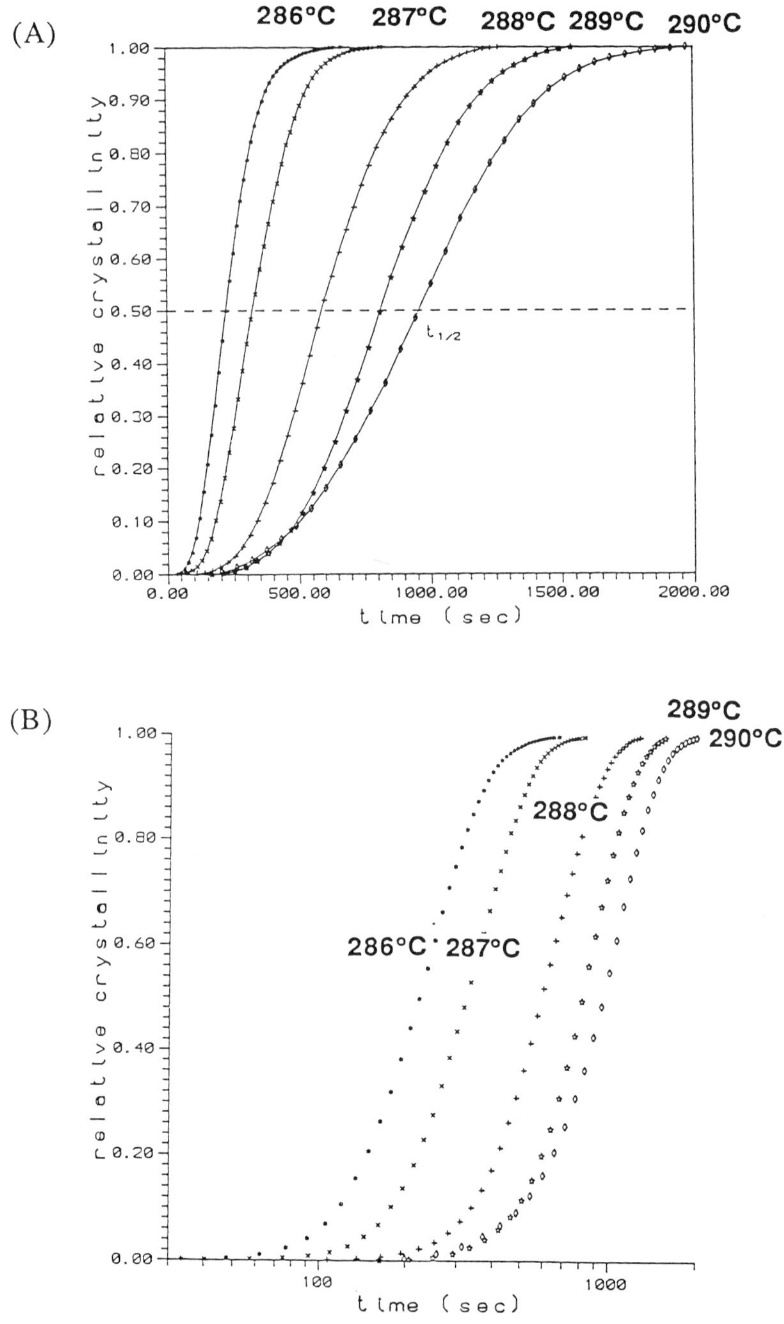

Fig. 4. (a) Fractional crystallinity versus time, obtained as the energy released in a DSC experiment, during the conversion of a polymer liquid crystal from the nematic to the crystalline state [27]. (b) Fractional crystallinity versus logarithm of crystallization time for the data of (a) [27].

sites. Since all of the models use integral growth processes, rather than a mixture of the possible growth processes, the volumetric growth rate always depends on an integral power of time. The values of the time exponent (usually referred to as the Avrami exponent) and the combinations of nucleation type and growth morphology giving rise to those values are summarized for principal types in table 1.

Although, very simple in approach, this type of analysis was for many years the only way of obtaining an estimate of the morphology occurring in a system. For opaque systems and for systems in which the morphological units are too small to be seen in an optical microscope, it still remains a valuable tool.

One example is that of fig. 4, a liquid crystalline macromolecule which forms crystals from its nematic liquid crystalline state with no change of the morphology being

Table 1
Avrami exponents and morphologies

Morphology	Instantaneous	Sporadic
disc	2	3
sphere	3	4
rod	1	2
sheaf	5	6

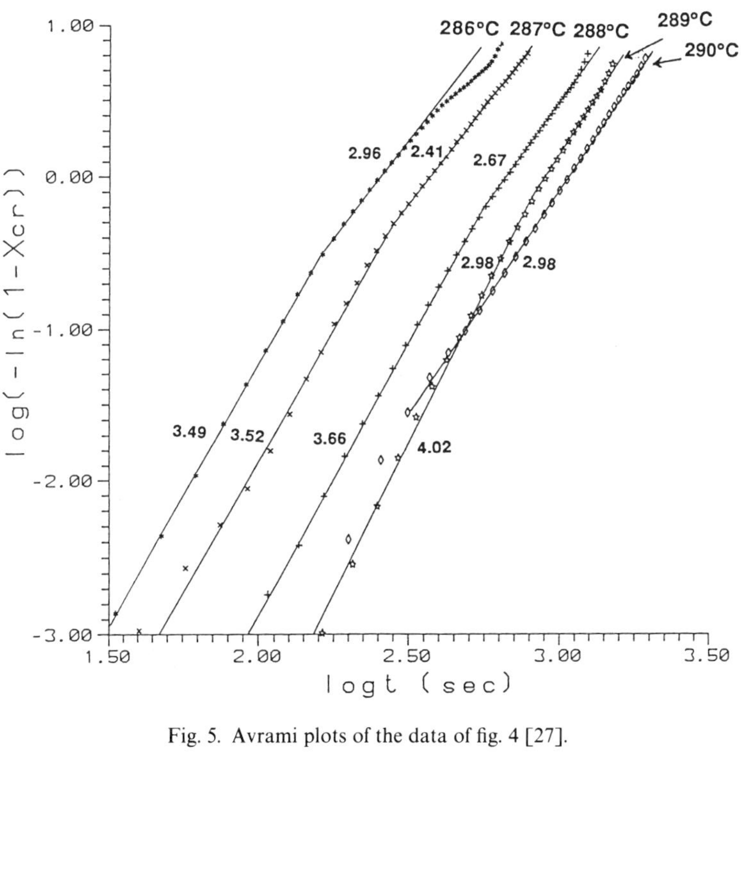

Fig. 5. Avrami plots of the data of fig. 4 [27].

apparent in the optical microscope [27, 28]. The typical Avrami curve is shown in fig. 5, again for this liquid crystalline polymer. As can be seen, the Avrami exponent (the slope) lies between 3.5 and 4.0. Hence, its morphology approximates a sphere with considerable disc-like character (it being a fair assumption that the nucleation process is homogeneous for this unusual substance). The values were surprising, since the process is one of condensation of pre-aligned rods into a crystalline form, and disc-like crystals were expected. The data resulted in the formulation of a model which assumes that one disc can easily nucleate others on its upper and lower faces, because of molecules protruding from the surfaces (fig. 6). The model generated values of the Avrami exponent which correlated well with the experimental values and also explained other types of data.

Because transmission optical microscopy of many substances permits the observation of the nucleation of crystals and the quantitative study of the associated growth rates, the use of the Avrami analysis has decreased significantly. However, in industrial solidification processes, unlike in basic physics, it is important to know the time it will take for a given process to be completed. Because of this fact, studies of bulk transformations, using especially DSC, are a very popular analytical tool in processing applications. Whereas, in basic studies it is important to carry out a study isothermally, in commercial operations the processes occurring are dynamic, so modelling of processes using the Avrami approach are of importance [29].

In general, it should be recognized that a mixture of instantaneous and sporadic nucleation occur and that Avrami exponents should be non-integral. Additionally, a very useful way of obtaining bulk growth kinetics, whilst at the same time observing and measuring the linear growth rates, is to use a trinocular eyepiece in a polarizing microscope [30]. A light intensity detector is set in the monocular to monitor the

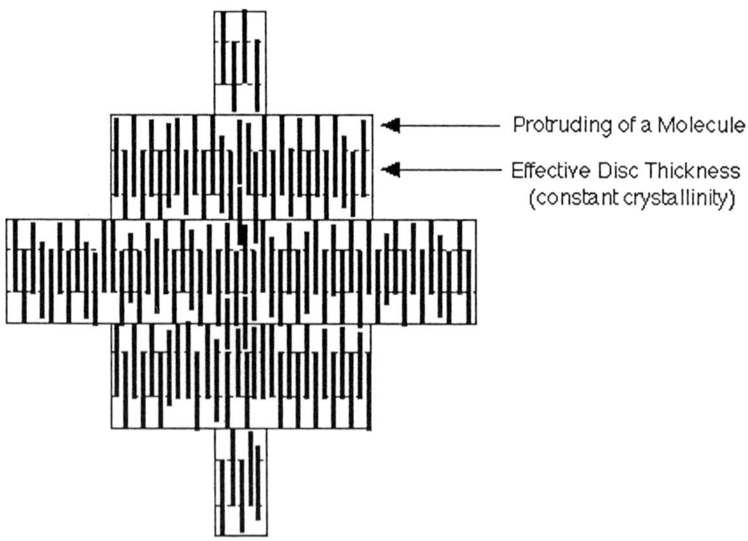

Fig. 6. Model for the growth morphology of the crystalline state of a liquid crystalline polymer [27].

increase in polarized light intensity as the crystallization progresses. In this case, as in any experiment using thin films on microscope slides, the spherulites may indeed simply occur as discs because of lateral size constraints. This fact has been allowed for in studies of bulk growth using this approach, Avrami exponents being approximately 1.0 less than they would be for true bulk studies.

The Avrami equation, as derived, considers the material ahead of the growth front to be liquid and the material behind it to be crystalline. In practice, the latter condition does not occur in any macromolecular material. The spherulites contain both *amorphous and crystalline material* at the completion of the solidification process. Many macromolecular materials are only 30% crystalline, and the highest observed crystallinities under most circumstances are in the range 80–90%. In practice, this seemingly insurmountable complication is easily handled by defining the reference "completely crystalline material" as being the "*completely spherulitic material*". This procedure permits the estimation of the initial slope, as in fig. 5, and hence the Avrami exponent appropriate to the growth of the spherulites prior to impingement. There is always an additional complication in the form of crystallization within the spherulites behind the growth front. Hence a material composed of fully impinged spherulites, which when viewed through a microscope would resemble a polycrystalline metal, continues to crystallize and hence densify. This process, usually referred to as *secondary crystallization*, is responsible for the second lower slope observable in fig. 4. (In the case of the liquid crystal the mechanism is different from that of flexible macromolecules and believed to be the incorporation of the protruding parts of the molecules into the crystal by a thickening process.)

2.2. *Half-time analyses*

The Avrami analyses give indications of the morphologies being developed and of the nucleation processes occurring, but they give no analysis of the actual growth rates of the crystals. A simple approach which can be used to give broad estimates of relative growth rates that is especially useful for a variable temperature study is to use the concept of half-time. Any basic growth curve, such as those shown in fig. 4b, can be converted into a rate study by taking the reciprocal of the time it takes for the process to be half complete. If such data are plotted versus reciprocal supercooling, a linear relation occurs for instantaneous nucleation, whereas a linear relation will be found when plotted versus the reciprocal of the square of the supercooling if sporadic nucleation results [31]. Much more in-depth analyses of the crystal growth process can also be carried out using this approach as will be seen later.

2.3. *Nucleating agents*

Commercial polymers often contain additives to enhance the rate of crystallization and so reduce the processing time during molding. This practice is especially prevalent in the higher-temperature plastics such as PET and PEEK. For many years it was thought that such agents functioned simply through providing heterogeneous surfaces on which crystals could nucleate, despite the fact that there were significant variations in the effectiveness of many of these additives. The reasons for this belief lay in the observation that many different nucleating agents of widely differing crystal structures could nucleate many polymers which also had widely differing crystal structures. Known as the Binsbergen hypothesis [32, 33], this concept was proven totally erroneous in a series of elegant experiments by Wittmann and co-workers [34–36].

They were able to demonstrate that the effective nucleating agents functioned through epitaxy! The key to obtaining a true understanding was the recognition that all the agents contained benzene rings in their chemical structures. These could occur as single or fused rings or as polyphenyls. In all cases there is a set of lattice spacings in their crystals determined by the stacking periodicity of the ringed structures, which is essentially a constant. All polyphenyls also have a constant stacking periodicity, but it is different from that for the single and fused ring systems. Many polymers fall into two general classes of crystal packing. First, there is the polyethylene type already discussed. The lateral packing distances of chains of aliphatic polyamides, polyesters and other polymers are essentially those of polyethylene. Lotz and Wittmann were able to demonstrate that some of the lateral packing lattice parameters of polyethylene matched those of the aromatic stacking lattice parameters of the nucleating agents. In all cases considered there existed the two lattice matches needed for these orthorhombic crystals that were well within the 13% tolerance limits needed for epitaxy. They also clearly demonstrated experimentally that polyethylene could be grown epitaxially on large single crystals of chosen aromatic molecules. It was later shown that more than one set of lattice matches functions in practice for polyethylene [37].

The second general type of polymer crystal is that exemplified by polypropylene in which the equilibrium chain structure is a 3:1 helix [this convention gives the number of chemical repeat units (in this case 3) found in a given number of turns of the helix (in this case 1)]. The chain helices then pack together laterally to form the crystal. It was also demonstrated by Lotz and Wittmann [38] that polyethylene and polypropylene have an effective epitaxial relationship, polyethylene growing on polypropylene because of its lower melting point. So an epitaxial match between polypropylene and the nucleating agents was a given. The concept has also been extended to polymers containing benzene rings in the main chain such as PEEK and PET, it having been demonstrated that the stacking of rings in these polymers generates epitaxial relations between them, nucleating agents and the commoner aliphatic polymers such as the polyethylene and polypropylene already mentioned [39].

The most effective nucleating agents known are generally referred to as *clarifiers* [40] since their nucleating ability is so great that the size of the resultant spherulites is so small that the haze or opacity of crystalline polymers, which is a direct result of light scattering by these optically anisotropic spheres (see later), is essentially eliminated. These clarifiers are based on sorbitol derivatives and are known to form single string-like crystals only about 10 nm in diameter [41, 42]. A small amount of the material produces a dense web of these crystalline strings with a very high surface-to-volume ratio. The web appears to function as an epitaxial agent for polymers perhaps because of the six-membered rings of the sorbitol, or because of the benzene rings attached to them, forming stacks.

3. The lamellar crystal

For certain macromolecules it is possible to observe the individual crystals that form the spherulitic aggregates and even to observe the nucleation process. The greatest degree of resolution has been possible for certain polymers known as polydienes. These materials contain reactive double bonds in the main chain of the polymer and can be reacted with osmium tetraoxide vapour. The reaction is rapid if the polymer is in the form of a thin film (i.e. 80 nm thick) and occurs in 10–15 s. The gas penetrates all of the amorphous material and none of the crystalline material. It adds onto the double bond, thereby generating electron-dense amorphous regions, which can be observed with high levels of resolution in a transmission electron microscope (fig. 7). Since the crystals tend be of the order of 10 nm thick, transmission electron microscopy is the only way of

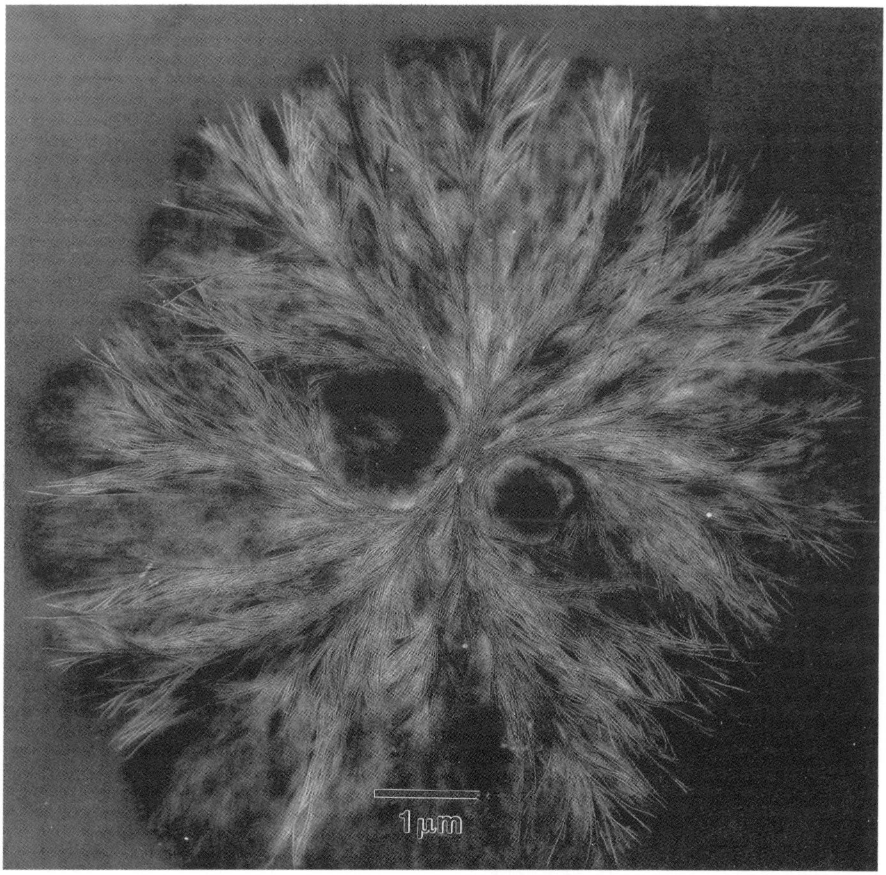

Fig. 7. Lamellar crystals in a spherulite of a film of cis-polyisoprene stained using osmium tetraoxide (crystallized at $-12°C$).

conducting a high-resolution study of such crystals. This technique was first used by Andrews [12, 43] to directly visualize the lamellar crystals formed in bulk polymer.

The technique has an added advantage in that the sections of chain to which the osmium tetraoxide have added are no longer mobile and cannot crystallize. It therefore stops the crystallization process permitting the earliest stages of growth to be observed in a specimen [44]. Schematics are shown in fig. 8 of the centers of spherulites shortly after nucleation. In the schematic A an example is shown of heterogeneous necleation where several of the lamellar crystals have nucleated on a heterogeneous impurity. The schematic B shows the most commonly observed nucleation process in highly purified material, in which no heterogeneous objects can be observed in the centers of the crystals. Generally regarded as a form of homogeneous nucleation, this form gives rise to spherulitic textures which are quite different from those generated from obviously heterogeneous nuclei. The aggregates grow as *sheaves* (fig. 9) and late in their growth process take on the characteristic spherulitic texture (fig. 7). Those species growing from a heterogeneous nucleus always have a spherulitic appearance. It should be noted that commercial polymers usually have a predominantly heterogeneous nucleation process because of the high levels of dust and other particles present from processing operations. However, if properly cleaned and filtered from dilute solutions they will also give rise to the sheaf-like textures in the early stages of growth.

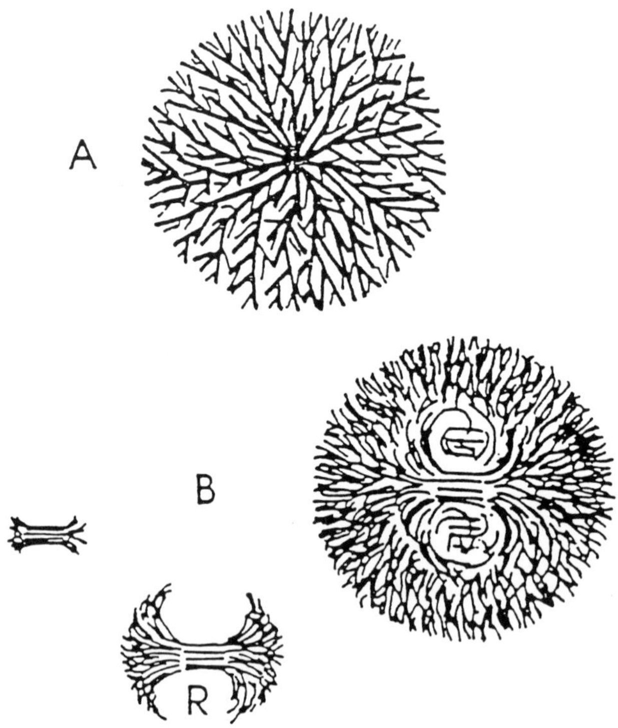

Fig. 8. Schematics of (a) heterogeneous and (b) homogeneous nucleation in a high polymer.

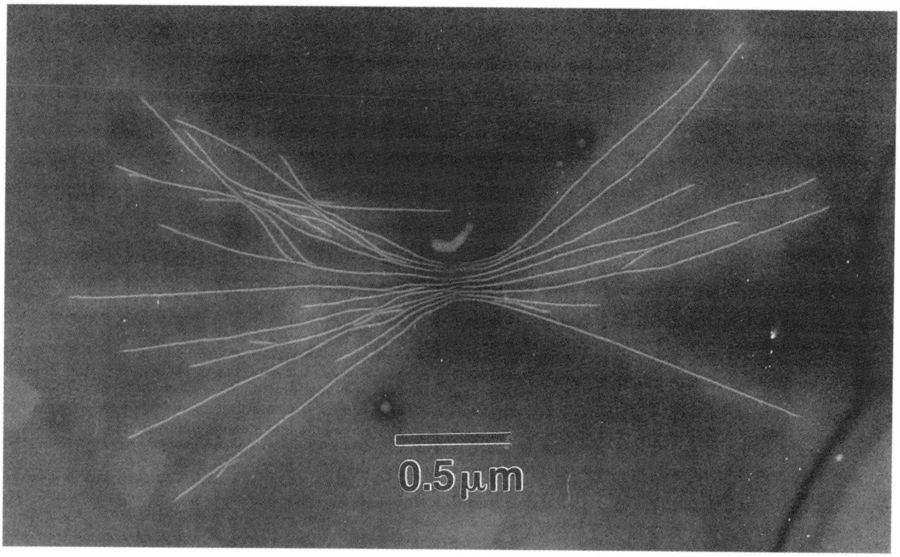

Fig. 9. Sheaf-like growth characteristic of the early stages of spherulitic growth in polymers (cis-polyisoprene crystallized at $-9.2°C$).

In the pictures shown of cis-polyisoprene occasional crystals can be observed having a wide profile and of intermediate contrast. These are the same crystals but in a different orientation from normal. In fact the crystals of cis-polyisoprene are *ribbon-like* in shape. Dependent on the growth regime (see later) they may also grow with a wider *leaf-shaped* profile [45]. Under such circumstances the aggregates consist of stacks of leaves which have all nucleated at about the same time at the center of the approximately elliptical leaf. However, the crystals are not directly superposable on each other as would be expected of crystalline stacks grown on a screw dislocation (fig. 10).

In the case of ribbon-like crystals the growth face can be seen directly in micrographs (fig. 11) and is flat, corresponding to the (010) plane. For the leaf-shaped crystals there is not a well-defined growth plane visible. The tips of the crystals do appear to be pointed in shape and the growth face may be the (110) plane. It should be noted that in these two types of crystal the growth direction is the same and is the [100] direction of the crystal.

When other polymers are considered, the general phenomenology often corresponds to that of cis-polyisoprene. The most commonly studied polymer is linear polyethylene. It is from solutions of this polymer that the first in-depth studies of single crystals of a polymer were conducted [46–48]. These crystals were rhombus-shaped and very thin (fig. 12). They were nicely faceted and the growth faces could be identified easily as predominantly $\{110\}$. Later studies showed that they were actually *hollow pyramids*, which had collapsed to give the flat rhomboid crystals during specimen collection [49].

When directly studied using electron diffraction it was clear that the polymer molecules were oriented with their axes across the smallest dimension, i.e. the crystal thickness. This was directly contrary to intuition, which would lead one to assume that the molecules would be oriented along one of the major dimensions of the crystal,

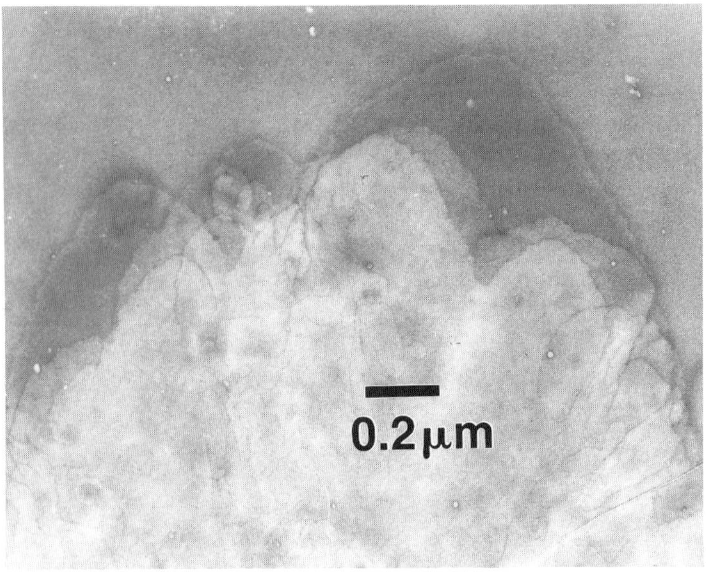

Fig. 10. Leaf-like crystals growing in axialites of cis-polyisoprene.

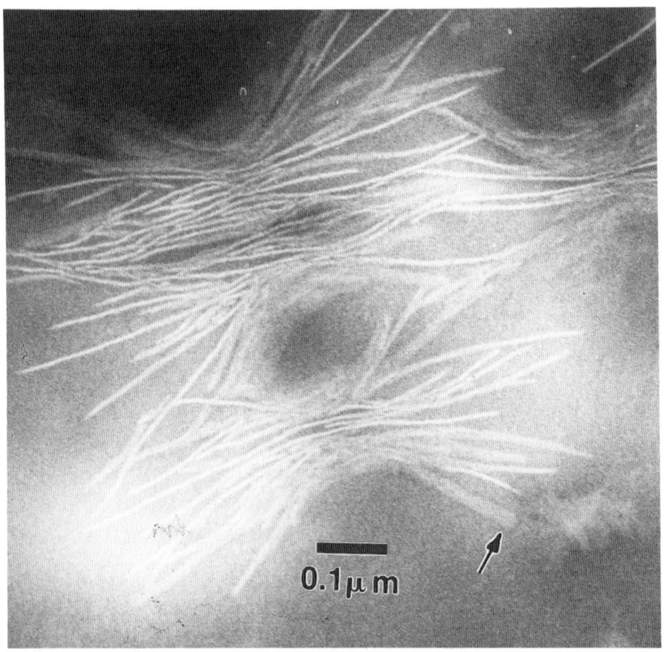

Fig. 11. Ribbon-like crystal growth in spherulites of cis-polyisoprene showing the (010) growth face (arrow).

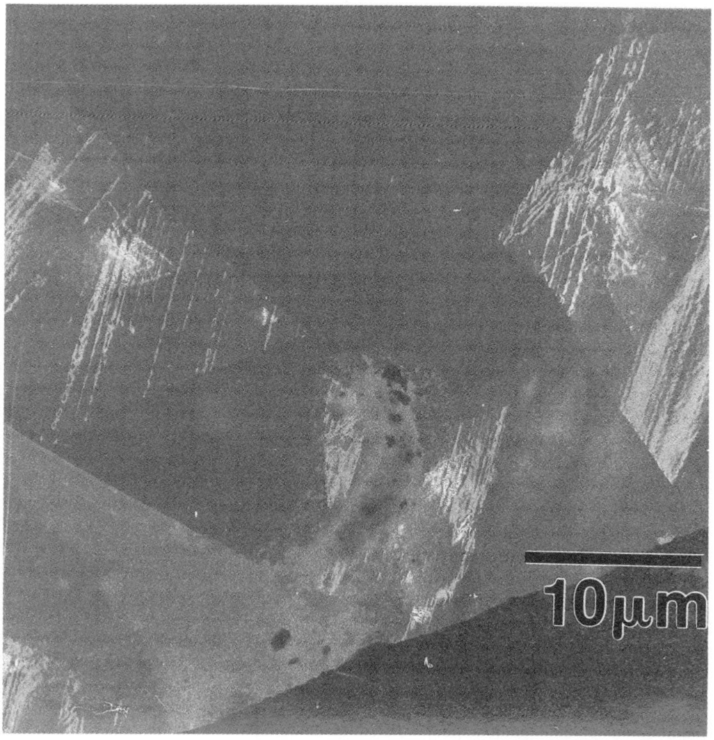

Fig. 12. Single crystals of polyethylene grown from dilute solution. Dark-field TEM using (110) and (200) rings. (Photomicrograph supplied by J.R. White).

indeed the molecular length is about the same size as the edge lengths of typical crystals. This simple and irrefutable piece of evidence led to one of the greatest controversies in macromolecular science. *Simply, the length of the molecule is many times greater than the thickness of the crystal.* For example, a typical polyethylene molecule of molecular weight 100 000 is about one micron in length yet its crystals are only one hundredth of that value in thickness. Since the crystals were about 90% crystalline, it was obvious that the molecule had to fold back and forth re-entering the crystal many times. The *nature* of the folding has been the controversy.

Crystals of linear hydrocarbons have the same molecular structure as polyethylene but are much shorter in length. Typical molecules are 3–10 nm long. As such, the lengths of the longest ones are similar to the crystal thicknesses of polyethylene crystals. They have been studied for many years and much of the initial understanding of polyethylene crystallization came from knowledge of their behavior. Lamellar crystals are formed which have the same crystal structure as polyethylene crystals [50, 51]. They differ from polyethylene in that the molecular length and the crystal thickness are the same. Hence they are *extended-chain* crystals.

Here it will be necessary to diverge from the general narrative in order to acquaint the general reader with the basic nature of polymer molecules and the necessity for a molecular shape change before crystals can be formed. The majority of synthetic macromolecules are flexible in nature. This flexibility is a direct result of the ability of individual links, or groups of links, in the chain to rotate into three major positions (fig. 13). For polyethylene the increase in energy between the *trans* and *gauche* positions is only 800 cal/mol and hence changes can occur easily at room temperature or above. The molecule hence exists in the form of a random coil, whose properties can be modelled using Gaussian statistics. It does however have a lowest-energy configuration which corresponds to every bond being in the trans state (fig. 13). Such a chain is known as the all-trans or *linear zig–zag* chain. It is this lowest-energy shape which must be achieved to make crystalline packing possible. Crystallization in long-chain systems therefore includes as an essential component *the achievement of the lowest-energy state*

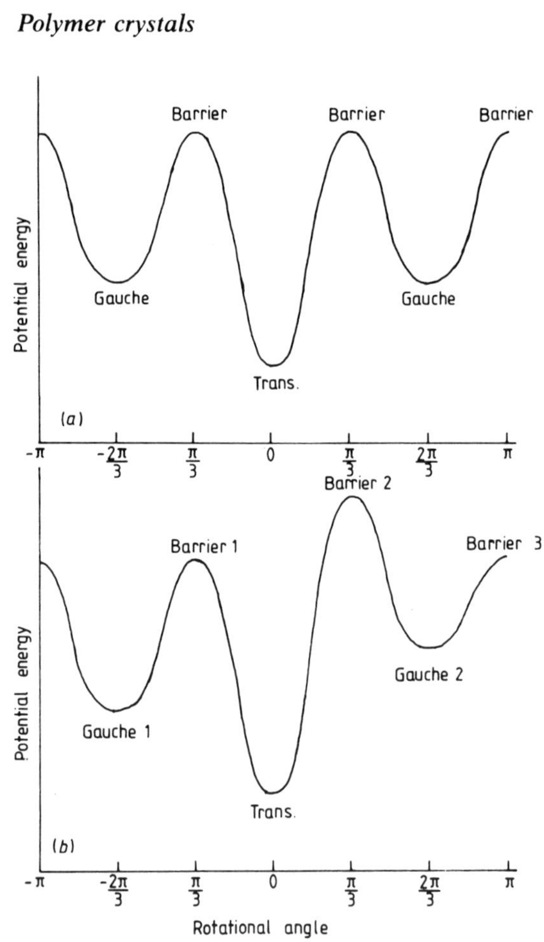

Fig. 13. Schematic potential energy diagrams for (a) polyethylene and (b) a vinyl polymer (first published in [3]).

of the isolated chains. This is a very slow process as it requires cooperation from all units present in the chain and also those other molecules with which the crystallizing molecule is intimately entangled. For a more extensive basic discussion of this subject the reader is referred to Boyd and Phillips [20].

Similar considerations apply to the shorter hydrocarbon systems, however, their shorter length precludes the major problems associated with chain entanglements, and full extension is readily achieved. They do comprise a valuable corroborative tool for the understanding of polyethylene crystallization, as they possess all the same features, but without the complication of chain folding. Studies of their crystal growth have provided values of surface free energies and other parameters for use in analyses of polyethylene crystal growth [52]. In all of the cases considered so far, the polymer chains in their extended linear zig–zag form *stack parallel to one another* like rods to form their crystals. The packings are simply achieved by obtaining the tightest lateral packing of the non-circular cross-section rods. In practice, this results most commonly in the form of an orthorhombic crystal (fig. 3). At elevated temperatures where the rods can rotate and twist, the orthorhombic structure adjusts and approaches a hexagonal state.

When crystallization from the bulk polymer is considered, most of the behavior characteristic of dilute solution growth is retained, with the exception that isolated single crystals cannot occur. This is an essential result of the entangled nature of these long-chain molecules, since it is possible for any given molecule to crystallize in *several crystals at the same time.* The most essential characteristic that is retained is the dependence of the lamellar (crystal) thickness on temperature. The crystal thickness is in fact controlled by thermodynamic variables and is a constant at any temperature. It

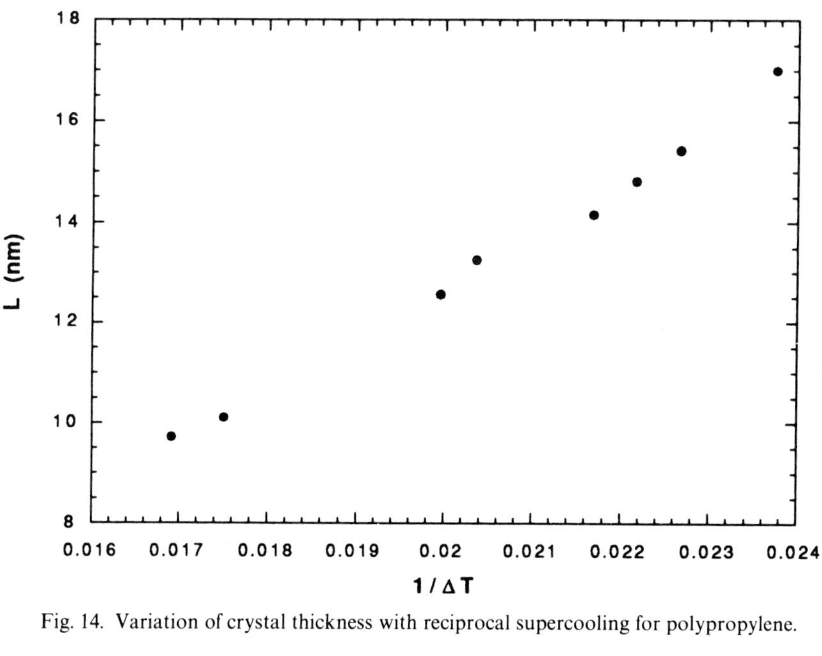

Fig. 14. Variation of crystal thickness with reciprocal supercooling for polypropylene.

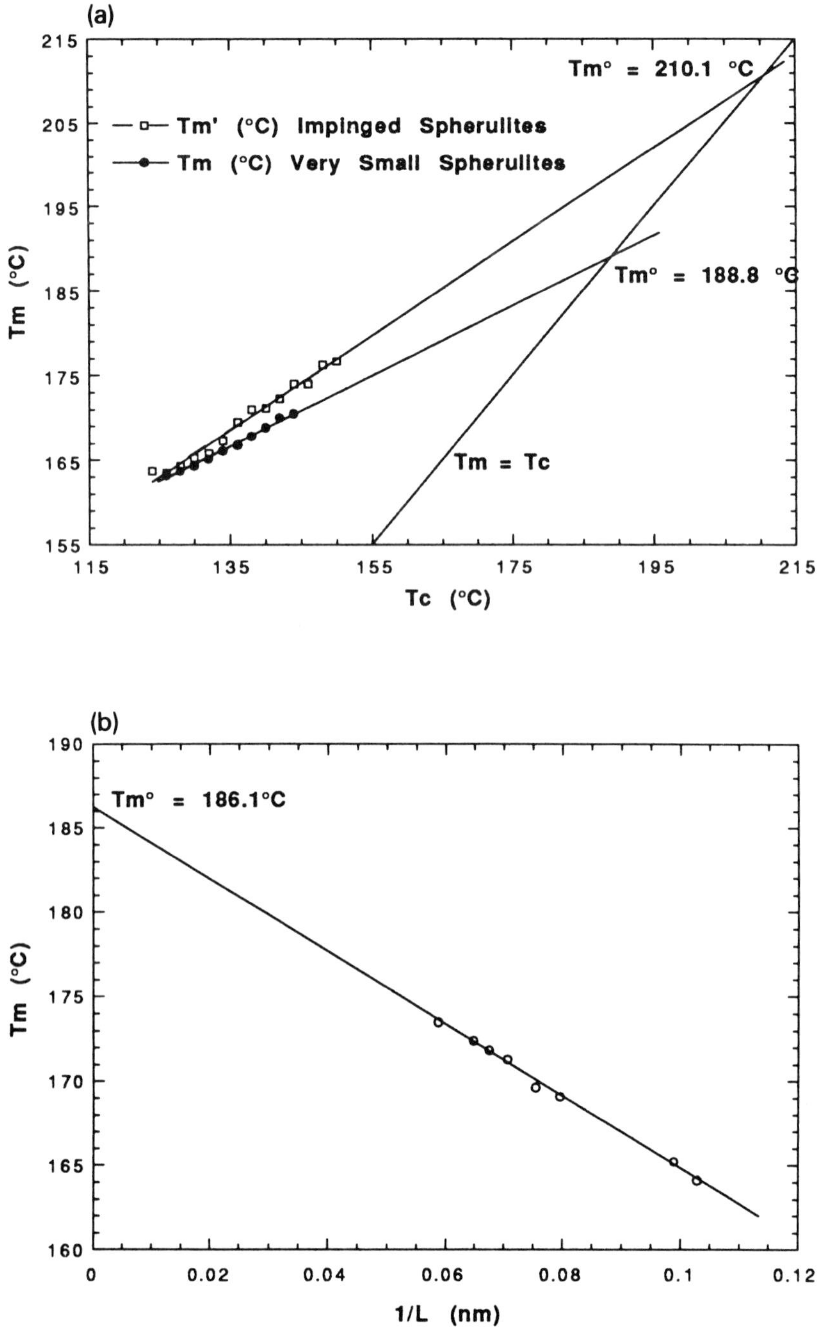

Fig. 15. (a) Hoffman–Weeks plot of melting temperature versus crystallization temperature of polypropylene obtained for small spherulites and well-formed annealed spherulites (Mezghani and Phillips, unpublished data). (b) Melting temperature versus reciprocal lamellar thickness of polypropylene [54].

does not vary with time of crystallization, unlike some other crystal dimensions. The dependence is a simple one, namely the reciprocal of the supercooling (fig. 14). Because of the thinness of the crystals the melting point is a function of crystal thickness and, hence, also a function of the supercooling. This leads to the well-known Hoffman–Weeks plot of melting point versus crystallization temperature (fig. 15) which can be used with caution to estimate the thickness-free or *equilibrium melting point* of the polymer [53]. This task is achieved by extrapolating the $T_M - T_C$ straight line relation to $T_M = T_C$. It does incorporate several assumptions regarding the annealing behavior of the crystals as a function of temperature and has been misused. The accurate extrapolation of this type of curve requires a knowledge of the crystal thickening behavior of the polymer. It has generally been assumed that the behavior reported for polyethylene applies to all systems. It is characterized by an increase in lamellar thickness which is greatest for the thinnest crystals and results in a lower slope for the Hoffman–Weeks plot than would be obtained without significant thickening having occurred. In this case the extrapolation always leads to about the same value. It has recently been demonstrated that polypropylene exhibits a quite different behavior (fig. 15a). In this case it is the crystals produced at lowest supercoolings (i.e. the thicker crystals) that are able to thicken most. Enhanced thickening therefore leads to an erroneously high estimate of the equilibrium melting point and studies have to be conducted using specimens produced at very short crystallization times [54].

In reality the only definitive method of obtaining an estimate of the equilibrium melting point is to perform an extrapolation to infinite thickness of a plot of melting point versus reciprocal lamellar thickness (fig. 15b). It can be seen that the estimate of the T_M value from thickness studies correlates well with that obtained from the Hoffman–Weeks plot for small spherulites.

Appropriate points to consider at this time are the driving force for the thickening process and the true state of an equilibrium crystal. Undoubtedly, the lamellar crystals described are not equilibrium structures, since they are of finite size and only part of the molecule is in the crystalline state. The equilibrium crystal form must be the *fully-extended-chain crystal* characteristic of the hydrocarbons. All lamellae must tend to increase the length of extended chain present and this can only be done by thickening. Hence as a crystal is heated it will tend to thicken. For the purposes of characterization it is fortunate that the thickening process is slow and it is possible to carry out melting studies without appreciable thickening occurring if the rate of heating is high enough. An additional factor always present is the low thermal conductivity of polymer crystals which can cause superheating to occur and hence the acquisition of inaccurately high melting points for a different reason.

The ability of a crystal to thicken is controlled by the detailed manner of re-entry of the molecules into the crystal and its neighbours. Before it is possible to proceed further it will be necessary to explore that problem in detail.

3.1. *The molecular trajectory*

There are only two methods available for the determination of the shape attained by the molecule in the bulk state. Both involve the "dyeing" of the molecule through replacement of hydrogen atoms by deuterium atoms. Once this change has been made, infra-red spectroscopy [55, 56] and neutron scattering [57] can be carried out. A blend is then made of the deuterated polymer with the hydrogenated (i.e. regular) polymer and, provided there is no segregation into isotopic species during crystallization, the deuterated polymer can be regarded as an indicator of general molecular behavior.

In the case of infra-red spectroscopy (FTIR) it is possible to study changes in spectra produced by changes in the nearest-neighbor interactions. So such a study may provide proof as to whether or not adjacent chains in a crystal are from the same molecule. As will be seen later, this microscopic scale view of the detailed molecular placement is extremely important in understanding the manner in which a molecule adds to a growing crystal.

Elastic neutron scattering studies take advantage of the fact that deuterium has a much larger scattering cross-section than hydrogen. The scattering of neutrons by the deuterium is much greater than by hydrogen and so the blend behaves as if the deuterated molecule were dissolved in a transparent solvent. Indeed the manner of the scattering and the molecular analyses that follow are directly derived from the well-known scattering of light by polymer molecules in an organic solvent [20, 58]. Two parameters extractable from the data are the molecular weight and the radius of gyration of the molecule. From the earliest studies it became apparent that the radius of gyration of the molecule in the crystalline state remained close to its value in the randomly coiled melt. The obvious conclusion to be drawn is that the molecule crystallizes substantially in place without any major change of shape.

This result was substantially different from the result obtained from studies of solution-grown single crystals where it was clear that the molecule is segregated into two or three adjacent growth planes of the crystal (in the case of polyethylene that corresponded to the (110) [59]). Hence although the lamellar crystals grown from the melt and from dilute solution show similar thickness dependences on supercooling the detailed molecular placements in the two forms are quite different. Such observations led to several attempts being made at regenerating the scattering intensity curves using computer simulations of different models. Some of the modelling studies were ultimately shown to be seriously flawed after considerable debate [60–62]. None of the models generated is sufficiently definitive for a molecular scale picture to be developed on the scale of chain adjacency. However, the broader scale picture of the molecular chain retaining essentially spherical overall shape is undisputed.

The use of FTIR has not resolved the problem either for melt-grown systems, although data for solution-grown crystals strongly support the interpretations of the neutron scattering data and indeed pre-dated those studies [55, 56]. It is possible to move towards a more detailed picture from studies of the crystallization kinetics since any model proposed must be consistent with and capable of explaining the growth behavior in a kinetic sense and not simply a morphological sense. In order to discuss further the detailed molecular trajectory it is necessary to consider the ways in which the molecule may re-enter the crystal and/or enter an adjacent crystal whilst maintaining its molecular integrity. It is also important to come to some conclusions regarding the conditions under which various types of re-entry would be predicted to predominate.

3.2. The nature of the molecular re-entry process

Many studies have been conducted using neutron scattering, especially on polyethylene. In all cases of bulk crystallization, because of isotopic segregation problems, they have been conducted on crystals produced by rapid quenching [57]. Therefore, they have contained the greatest amount of disorder that can be generated in the system. The retention of the approximately spherical molecular shape should not therefore constitute a major surprise. Prior to these studies the preponderance of experimental evidence was in favor of a molecular trajectory similar to that believed to be present in single crystals. This picture is generally referred to as the adjacent re-entry model (fig. 16) since it assumes that the molecule folds regularly along the growth plane in an adjacent manner [47, 49]. Clearly, this precise picture is inconsistent with the neutron scattering evidence. The predominant alternative model assumes total non-adjacency of re-entry and is usually referred to as the switchboard model [63] (fig. 17). It is now recognized by most researchers that this model generates an unacceptably high density in the crystal surface when the emerging chains are regarded as Gaussian at the crystal interface.

Another model is the interzonal model (fig. 18) which assumes a total lack of adjacency, but combines non-Gaussian chains in the vicinity of the surface with

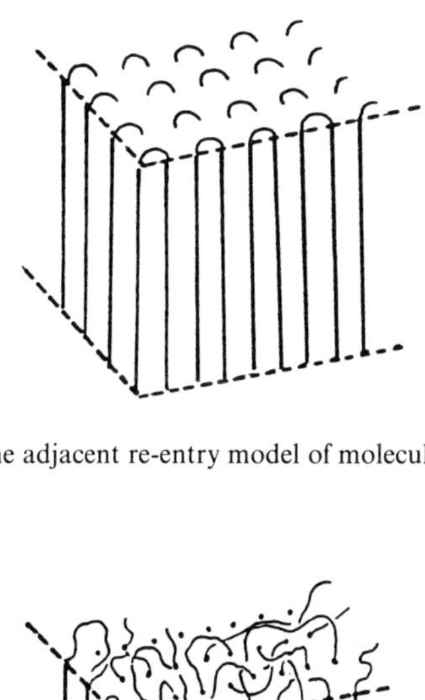

Fig. 16. The adjacent re-entry model of molecular re-entry.

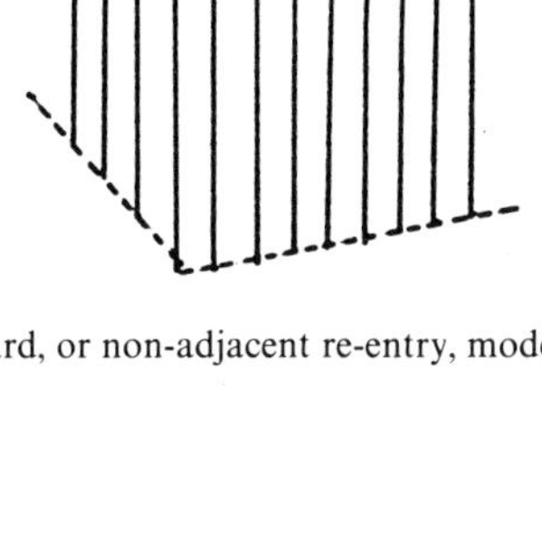

Fig. 17. The switchboard, or non-adjacent re-entry, model of molecular re-entry.

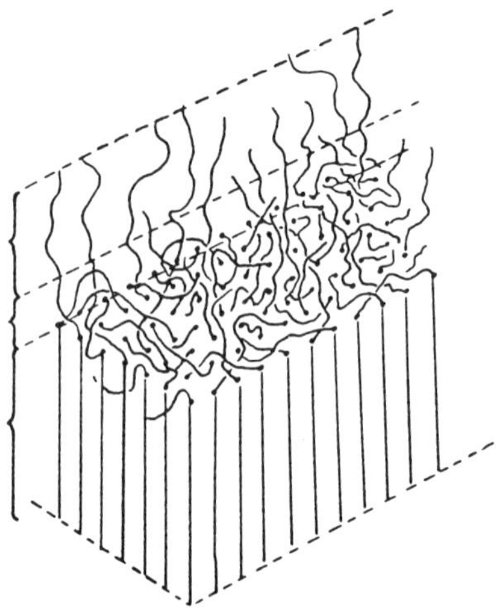

Fig. 18. The interzonal model of molecular re-entry.

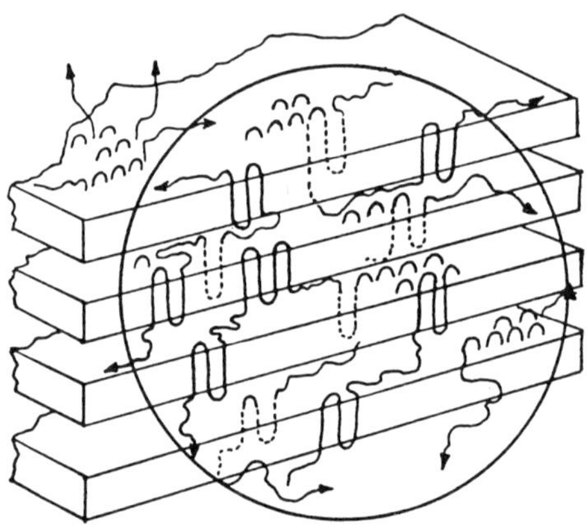

Fig. 19. The synoptic, or comprehensive, model of molecular re-entry. The circle represents the radius of gyration of a typical high-molecular-weight polymer.

switchboard folding [64]. The picture which is closest to the truth would incorporate features of all three models, but would recognize that the frequency with which each type of emergence and re-entry occurs would depend very much on the detailed conditions of any given crystallization experiment and on the chemical structure of the polymer. Such a model is presented in fig. 19 and is believed to apply to all systems that have been studied.

This *synoptic model* combines the following essential features:

(1) adjacent re-entry folds;
(2) non-adjacent re-entry folds, which

(a) re-enter along the same growth face a few stems away, dependent on internal chain stiffness,

or

(b) re-enter on a subsequent growth face parallel to the first, dependent on chain stiffness;

(3) emergent non-Gaussian chains which

(a) remain in the amorphous phase, or

(b) re-enter at a distant location, or

(c) re-enter only in other crystals, having in each case gone through a Gaussian form some distance away from the crystal surface.

Clearly, solution-grown single crystals are composed of predominantly type 1 with some type 2b re-entry. Melt-grown crystals produced at low supercoolings of a flexible polymer such as polyethylene would be expected to contain predominantly type 1 with some type 2a, 2b, 3a, 3c and smaller amounts of the other forms. When grown at high supercoolings, lesser amounts of type 1 would be expected as would larger amounts of types 2 and 3.

The semi-rigid backbone polymers, such as PEEK and others containing aromatic rings in their main chains are incapable of adjacent reentry as they cannot turn through 360° within the distance available and hence can only re-enter the crystal several chain stems away from the point of emergence. They will therefore contain predominantly types 2a, 2b, 3a and 3c.

Very rigid systems, such as the liquid crystalline polymer referred to earlier cannot fold at all and their crystal surfaces will contain only types 3a and 3c.

It is, of course, the flexible systems that have received greatest attention, and it is there that major variations in relative proportions of the different types of surface features can occur, dependent on the crystallization conditions employed. For the semi-rigid systems variations in the proportions of the different types there can also be expected, again dependent on preparation conditions. For these latter materials the term "rigid amorphous phase" was coined to describe the region near the surface where the chains are non-Gaussian and hence have a degree of rigidity that is higher than the equilibrium conformation would dictate [65, 66].

Chains can also have branches on them and this feature also influences the character of the surface. Branches such as butyl and hexyl are commonly found in low-density and linear low-density polyethylenes (readers unfamiliar with this terminology are referred to Phillips [3], or to Boyd and Phillips [20] for a discussion). Such groups cannot be incorporated into the crystal structure because of their size and are rejected to the surfaces. The chain section to which they are attached must either be incorporated into a fold or remain in the amorphous phase. However, incorporation of the branches into the folds in large numbers may cause packing problems and reduce the number of tight adjacent re-entry folds. One possibility is to make the fold into a loop or to cause non-adjacent re-entry. These possibilities themselves put more material into the interface and it must expand in thickness, hence many of the ideas incorporated into the interzonal model become operative.

Any fold, whether adjacent re-entry or not, must attain its own lowest energy state. In the case of an adjacent re-entry fold in polyethylene this was shown to require five main chain bonds. For this fold shape to be achieved it was necessary for several bonds to be in the higher-energy gauche configuration and hence, by its very nature, a fold has energy in excess of a molecular section of equal length in the crystal. It was recently demonstrated in simulations that the incorporation of a branch into the fold changes the conformation of the fold and hence the amount of *excess energy* stored in it. Such details of the fold structure have a major bearing on the crystallization process since the excess energy stored in a fold contributes to the surface free energy of the crystal and also to the *surface free energy of the critical nucleus*. Hence, what appear to be minor details of the molecular trajectory are in fact of major fundamental importance in the formulation of any theory of crystal growth. It is the folded surface of the lamellar crystals that is directly responsible for the thinness of the crystals and hence for the spherulitic habit.

4. *The crystal growth process*

The theory of crystal growth must account for the thickness behavior of the crystals in addition to the growth habits, temperature dependence and other features that are common to all materials. It must recognize that growth does not occur at the highest-energy surface, unlike in other materials, and that the crystal thickness is essentially constant with time of crystallization. It must also recognize the possible presence of all of the potential surface features of the crystals and how they will influence the growth process.

The general phenomenology of supercooling dependence of linear growth rates and of the lamellar thickness has been incorporated from the beginning [67, 68]. Indeed, they can be incorporated without recourse to the detailed structure of the surface. The simple fact that the folded surfaces, regardless of their detailed structures, will have a much higher surface free energy than the lateral surfaces of the crystals explains the behavior. All successful approaches are based on the secondary nucleation approach of Turnbull and Fisher [69], adapted for the chain molecule.

Other approaches are, of course, possible, such as the surface roughening model originally proposed by Burton et al. [70] and later applied to polymeric systems [71–73]. Although this model does indeed apply to many elemental systems the basic concept of atoms being able to jump out of the uppermost crystal layer to sit on the surface, simply does not transfer across to macromolecular systems. Entire sections of a chain would have to jump out and the activation energy required would be prohibitively high for it to occur with any reasonable frequency. It most certainly is possible, and indeed highly probable, that chain sections are sitting on the surface and mobile (see later). Since they are not jumping out of the crystal to do so, the basic concept of a low-activation-energy surface defect model as the controlling process is not applicable. The necessity for the atoms to be permanently connected together in the form of long strings effectively eliminates the possibility of applying conventional atomic mobility statistics to a macromolecular problem. It is a surface adsorption process, followed by the rearrangement of the chain into a partially extended form, that is generally recognized as the initial step in the addition of a flexible polymer molecule to a crystal.

It still remains to be seen if a version of roughening theory applicable to polymers can be developed and tested.

The version of secondary nucleation theory most commonly used is that originated by Hoffman and Lauritzen in 1961 [74], modified several times to incorporate complicating factors that were omitted from the original version for simplicity. The current version [75] incorporates an essential feature missing from all earlier versions, namely the change of shape of the molecule. The most fundamental of all aspects of polymer structure, the conformation of the molecule and the statistical entropy associated with it, now a part of crystallization theory, provides major new insights into polymer crystallization. Of considerable significance is the unification of secondary nucleation theory for polymers with classical physical chemistry of polymer molecules [76].

Secondary nucleation theory of polymers has always embodied the fundamental concept that the critical nucleus must be formed by condensation of stems of approximately equal length on the substrate crystal (fig. 20). In this respect it is very different from classical nucleation theory applied to elemental systems. The freedom of action of individual molecules or atoms is not present, and is a general consequence of the fact that indissoluble strings of atoms are considered. It is this aspect that partially unites the application of the theory to polymers with the application to hydrocarbon chains.

For the first chain to deposit there is an energy penalty of

$$\Delta\Phi = 2abl\sigma + abl\Delta G_f, \tag{2}$$

if it deposits in crystallographic register, which is very unlikely in the case of flexible chain molecules. a, b and l are the dimensions of the nucleus as shown in fig. 20. σ is the

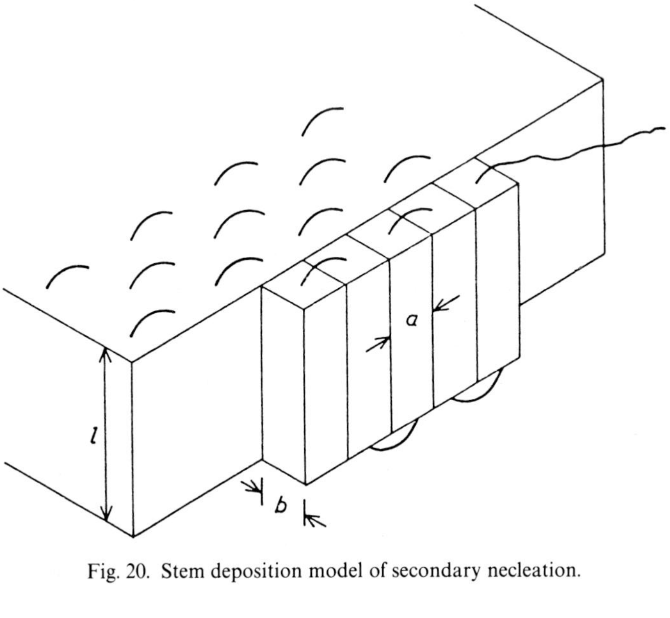

Fig. 20. Stem deposition model of secondary necleation.

lateral surface free energy and ΔG_f the volumetric free energy of fusion. For non-crystallographic register the energy penalty will be at least as high, if not greater due to the extra length of chain adsorbed on the surface.

In practice, chains do not adsorb on the surface in register, they must rearrange their shape on the surface in order to achieve register. This means that the free energy of fusion is released in two steps, whereas the lateral free energy penalty is incurred immediately on adsorption. It is necessary for the process to be separated into two steps, as shown in eq. (3). Sections of the chain that attain crystallographic register release free energy of fusion which partially compensates for the lateral free energy penalty of eq. (3a). At some later stage the stem attains complete crystallographic register, this second stage being handled by the second expression [eq. (3b)]. The symbol Ψ is used to denote the fraction of the free energy of fusion released during the adsorption step. During the formal derivation of the theory it automatically apportions the activation energy of the Eyring–Kauzman equation into its forward and backward reaction components.

$$\Delta\Phi_1 = 2abl\sigma + \Psi abl\Delta G_f, \tag{3a}$$

$$\Delta\Phi_2 = (1 - \Psi)abl\Delta G_f. \tag{3b}$$

The lateral free energy penalty of eq. (3) in fact does not simply represent a lateral free-energy term as traditionally represented. Attainment of deposition on the surface requires the loss of a major fraction of the entropy of the chain and it is implicitly incorporated into the value of the surface free energy, σ. So, in the case of polymers the surface free energy, σ, represents more than simply the mismatched bonding at an interface. It is the most recent adaptation of the Lauritzen-Hoffman theory that *explicitly* relates σ to the change in shape of the polymer chain. This approach writes σ as equal to a constant multiplied by the reciprocal of the *characteristic ratio* of the polymer chain. This ratio, C_N, basically represents the stiffness of the polymer chain as a factor which scales the random coil dimensions, calculated using the number, N, and length, l, of bonds in the chain, to reach the dimension of the real chain, end-to-end distance, R:

$$C_N = \langle R^2 \rangle_0 / Nl^2. \tag{4}$$

The characteristic ratio can be computed with reasonable accuracy, using approaches developed by Flory [77], which involve details of the molecular structure of the chain and the barriers to rotation of the chain. For an ideal flexible chain the characteristic ratio is 1.0, since it is a completely random coil. For the totally stiff chain the characteristic ratio is simply the number of bonds in the chain. Clearly, a totally stiff chain does not need to lose the statistical entropy that must be lost by a random coil. It simply needs to lose translational and rotational entropy and should crystallize much more easily than a random chain. It would therefore be expected that the value of σ would directly reflect the magnitude of the entropy loss required, since it represents the inhibition to secondary nucleus formation.

This effect can be understood more easily for semi-rigid, as opposed to totally rigid, chains through the concept of the *persistence length*, which is related to the characteristic ratio. It basically represents the minimum length of chain needed for a given polymer to develop a coil-like nature. In such a case it might be expected that if the lamellar thickness under a specific crystallization condition were approximately equal to the persistence length that crystallization would be easy because of small loss of statistical entropy needed for incorporation of a persistence length in the crystal.

For flexible chain polymers such as polyethylene, however, the persistence length is about the size of the bond and no unusual effects would be expected. Equation (2) can be used in a continuous manner, as in classical secondary nucleation theory to predict the size of the critical nucleus. In the case of a polymer it has been assumed that the chain is added one stem at a time and so the only dimensional variable is the length of the chain in the first stem. It is this dimension which therefore becomes constant as a function of time. It does, however, vary with temperature, or more accurately supercooling, and the final equation [52] predicts the experimental behavior.

$$l = 2\sigma_e T_m^0/\Delta H_f \Delta T + \delta l, \tag{5}$$

where σ_e is the free energy of the folded surface. In practice, the value of l produced from the equation is not immediately identifiable with the measured value of the lamellar thickness. This occurs because there exists in most cases a thickening process which appears shortly after deposition of the stem as the crystal reduces its total free energy. However, the equation can be used accurately to describe the experimental behavior if a thickening coefficient is included in the equation.

Complete development of the theory requires the generation of a series of equations for the forward and backward steps of each stem addition. This is then followed by the formulation of a flux equation from which an equation for the rate of deposition of critical first stems, and hence, the rate of linear growth can be predicted. In the early versions of this theory it was assumed, as in classical nucleation theory, that the deposition of the first stem was the rate-controlling step. This assumption made it possible to easily generate the equation.

$$G = G_0 \exp(-\Delta F^*/kT)\exp(-\Delta U^*/kT), \tag{6}$$

where G is the linear growth rate, G_0 the pre-exponential factor, ΔF^* the free energy of formation of the critical nucleus and ΔU^* the activation energy of the elementary jump process. For the classical case the value of ΔF^* is given by

$$\Delta F^* = 4\sigma\sigma_e T_m^0/\Delta H_f \Delta T. \tag{7}$$

The theory therefore predicted the dependence of lammellar thickness and of linear growth rate on supercooling, the most important experimental dependences.

The most important fundamental revision to be made to the theory until recently was the introduction of regimes in which the formation of a single critical nucleus need not be the rate-controlling step [78, 52, 79]. One of the simplest ways to envision regimes to think of the crystal growth process as being controlled by two competing rates, the first being the deposition of the critical nucleus (i.e. first stem) and the second being the subsequent surface spreading process (i.e. second and other stems). Regime I consists of the classical picture in which the spreading rate is much greater than the nucleation rate. Regime II occurs when the two rates are comparable, and regime III when the rate of nucleation is much greater than the rate of spreading (see fig. 21).

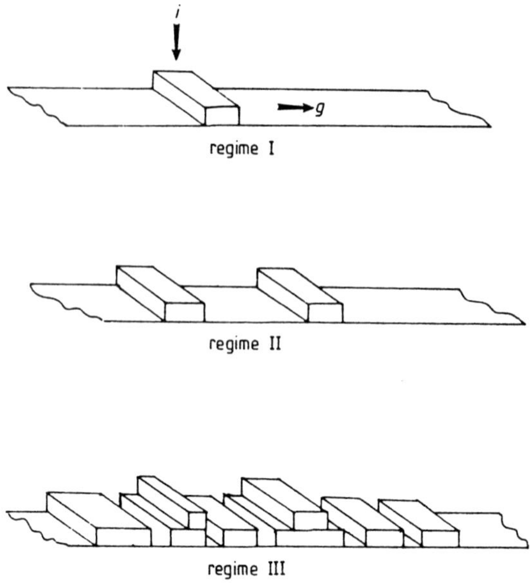

Fig. 21. Schematic diagram of secondary nucleation profiles in the three regimes of polymer crystallization (first published in [3]).

The original experiments which spurred this development were reported by Hoffman et al. in 1975 [80] and were carried out on polyethylenes having very narrow molecular weight distributions. Intermediate molecular weight fractions were found to exhibit a change of slope in the growth rate versus temperature curve at a supercooling of 17 K. The slope in the lower temperature range studied was about a half of that for the upper temperature range. Regime theory, as published by Lauritzen and Hoffman in 1973 [78], uses the symbols i for the rate of secondary nucleation and g for the rate of lateral spreading. The theory predicts that in regime II, where the two rates are comparable, that the linear growth rate is proportional to the square root of the product ig. The immediate result of this assumption is that the free-energy term, ΔF^*, in regime II is a half of that in regime I, explaining the observed change of slope. The observed regime behavior is strongly dependent on molecular weight. Low-molecular-weight fractions crystallized exclusively in regime I and high-molecular-weight fractions exclusively in regime II. Although not yet understood in detail, the difference in behavior as a function of molecular weight is undoubtedly related to ability of the molecule to translate across the crystal–amorphous interface, since higher-molecular-weight molecules are unable to move as rapidly as low-molecular-weight molecules.

Following studies on polymers other than polyethylene, Phillips [81] in 1979 suggested that there might be a third regime occurring at even lower supercoolings than the second regime. This third regime would require that the rate of nucleation would exceed the rate of spreading. The theory for regime III was formulated by Hoffman and published in 1983 [79]. The concept of three regimes has received broad acceptance and it is now recognized that the commonly encountered regimes are regime II and regime III. The classical approach, regime I, is rarely encountered [82]. It is also important to recognize that regime transitions can be observed in unfractionated polymers, and indeed most of the reported studies use unfractionated systems. The effect of polydispersity is to broaden the transitions and they can usually be identified with care in the analyses.

In practical terms, the most appropriate test for regimes is to plot eq. (8) in the form $\log G + \Delta U^*/kT$ versus $1/T\Delta T$ where ΔT is the supercooling. The major problem encountered with this approach is the need to factor out the effect of the elementary jump process, represented as ΔU^*. It is preferable for some form of independent estimate of this parameter to be available. The most useful is to use the Williams–Landel–Ferry (WLF) relation, which is well established and for which calibration constants are available for many systems. Originally suggested by Hoffman and Weeks [53], the approach has been found to work well. Its major problem is that it is an empirical relation usually derived from studies of relaxation or of viscous flow. Bulk studies may not provide an appropriate estimate of the behavior of a section of crystallizing chain. However, since in many cases it appears that the large-scale mobility of the chain, including the effects of entanglements, may be an important controlling variable, the WLF equation may be adventitiously quite appropriate. For polymers which crystallize over a narrow range of temperatures close to the melting point, the ΔU^* factor has little influence on the analyses [25] and the way ΔU^* is treated is not critical. Because of this observation a simplified approach using one "universal" value has been popular for those systems [83, 52]. For polymers having a wide crystallization range, this approach is inadequate and the WLF equation using the values of the constants for the polymer being investigated has proved to be the most reliable method (see, e.g. [23, 45]).

The essential problem facing theorists is that the molecule must be extracted from the entangled melt at high speed and sufficiently fast not to inhibit the very rapid surface spreading process. The process does not require that the entire molecule translate as a whole, but that it be reeled in from one end, or perhaps from the middle. The introduction of the concept of reptation by De Gennes [84] has been of assistance. This different approach to polymer mobility assumes that the molecule has to snake its way through a "hole" defined by its neighbours without their behavior being significantly modified. Experimental values for the activation energy of such a process are hard to come by and are available for very few polymers. Hence, the continuing need for some sort of reliable semi-empirical approach. A value for the activation energy of reptation has been obtained for polyethylene, using studies of self-diffusion in isotopic mixtures [85]. A major achievement of the introduction of this approach to crystallization was the demonstration by Hoffman [86] that the value of the rate of reptation in polyethylene was just adequate for the rate of spreading to occur at its known rate, whereas estimates of the rate of translation of the entire molecule were inadequate by orders of magnitude.

A major problem arising from the lack of real knowledge of the correct method of handling the mobility exponential is the possibility of "generating" a regime transition where none exists. The reverse is also possible, in that a single regime can disappear in performing the plot. The latter had occurred several times prior to the recognition of

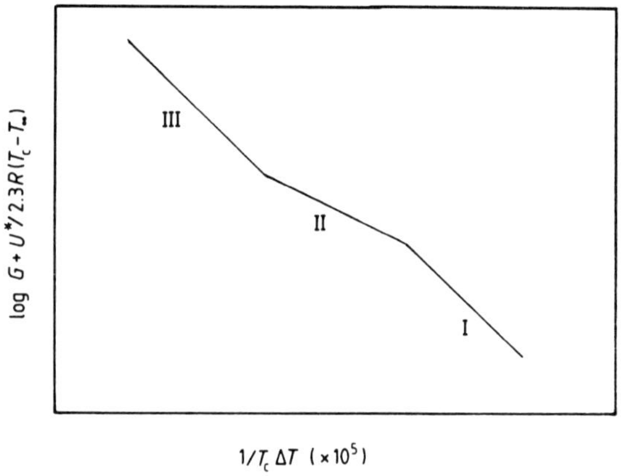

Fig. 22. Schematic diagram of a regime plot of linear growth rate data (first published in [3]).

the existence of regimes, where the objective was to produce a single slope line through appropriate choice of ΔU^* [87]. The true test of regime theory is the existence of all three regimes in a given polymer, since no amount of inappropriate curve fitting can generate the expected two changes of slope (fig. 22). The first such incontrovertible study was for studies of cis-polyisoprene [45] (fig. 23), where all three regimes were observed in a polymer of molecular weight 314 000. Higher molecular weights showed either a combination of regimes II and III or regime III alone. These results confirmed the conclusions regarding the molecular weight dependence of the regime I–regime II transition obtained from polyethylene studies and demonstrated the molecular weight dependence of the regime II–regime III transition. All three transitions have now been reported for an additional three polymers [88–90].

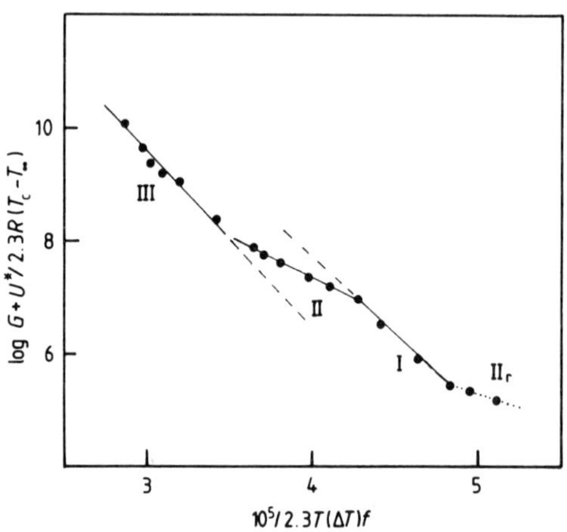

Fig. 23. Crystallization regimes in cis-polyisoprene determined from transmission electron microscopy of osmium tetraoxide stained films (first published in [45]).

It is important at this point to discuss the influence of regime on the detailed structure of the fold surface and on mechanical properties. Regime I corresponds to type 1 folds almost exclusively, i.e. adjacent re-entry, which is possible because of the reptation of the relatively short molecules which normally give rise to this regime. Any one molecule lies almost exclusively within a single lamella, its ends being either buried in the surface or dangling from the surface as cilia. There are therefore very few molecules which add to two or more crystals concurrently and hence bridge the amorphous phase. The material therefore consists of crystals suspended in a highly viscous fluid (the amorphous phase). Such materials are usually of the highest crystallinities possible, ca. 80%. It is well known that such materials have very poor mechanical properties and usually fail in a brittle manner at low stresses [91].

Regime II crystallization will give rise to many molecules which bridge the gaps between crystals. This occurs largely because the effective diameter of a random coil molecule of medium to high molecular weight is several times the lamellar thickness. As demonstrated by neutron scattering, this diameter is approximately retained during crystallization. The chains can be envisioned as crystallizing in many crystals concurrently and also at many sites in each lamella being formed. Adjacent re-entry folding is probably restricted to less than ten contiguous stems, dependent on crystallization temperature. The large concentration of tie-molecules generated by the process form effective mechanical links between the crystals, the material failing by the plastic deformation processes for which polymers are well-known.

In regime III the detailed surface structure is more difficult to predict. The situation is similar to that of regime II with the exception that the number of chain sections depositing as secondary nuclei is much greater. There are limits to the concentration of emerging non-folded chains that can exist in a surface because of packing constraints [62, 63]. Such limitations have been estimated on the assumption that the chains are Gaussian immediately on emergence. It seems likely that the chains will be non-Gaussian on emergence and become Gaussian some distance from the crystal, as suggested in Mandelkern's interzonal model [64] and in Wunderlich's rigid amorphous phase model [65, 66]. Such structures correspond to type 3 chains of the synoptic model given earlier. It has been well known for generations to plastics fabricators that the best mechanical properties are obtained when crystalline thermoplastics are quench-crystallized. Hence they are trying to crystallize the polymers in regime III or deep regime II in order to maximize the number of tie-molecules present.

There is a misconception prevalent with regard to this effect. It is often thought that brittleness is a result of too large a spherulite size, whereas it is really the conditions necessary to generate a large spherulite that are those that produce regime I crystallization, with its paucity of tie-molecules. Additionally, at such high crystallization temperatures in some polymers impurities tend to be effectively segregated to the spherulite boundaries, intensifying the brittleness problem.

4.1. The influence of molecular structure

In the above discussion it was mentioned a number of times that the molecular weight of the polymer was of significance in influencing regime behavior. Recognizing that the regime transitions are determined by the relative values of the rate of secondary nucleation and the rate of surface spreading, it should be possible to change the transition temperatures through appropriate changes of variables that might influence one rate more than the other. It is here that the synthetic chemistry can be used in a very useful manner. The major principles are still being elucidated; however, several important relations have already been demonstrated.

The first clear demonstration of the influence of structure on regimes was demonstrated by crosslinking polyethylene [92]. First, it was predicted that crosslinking should eliminate reptation as a mechanism for transfer of the molecule from the amorphous to the crystalline state. As such, it should significantly reduce the rate of spreading, g, and switch the regime from I or II to III. This was seen to occur clearly (fig. 24). This effect would be expected to be unique to crosslinking. A second effect observed should also be seen in random copolymers, but requires more explanation.

When a polymer chain contains a defect (e.g. comonomer units, branches or isomers) which cannot be incorporated into the crystal, the exclusion process slows down the rate of secondary nucleation. This is largely because the first stem that deposits as the critical nucleus requires a finite number of crystallizable units in a continuous uninterrupted sequence. There is therefore a limited probability of locating the appropriate string needed and this probability has to be incorporated into the theory. This was first done by Andrews et al. [93] and applied to isomerized cis-polyisoprene. The theory predicts that the logarithm of the rate of secondary nucleation should depend on the mole fraction of chain impurities in an inverse fashion. Hence a few

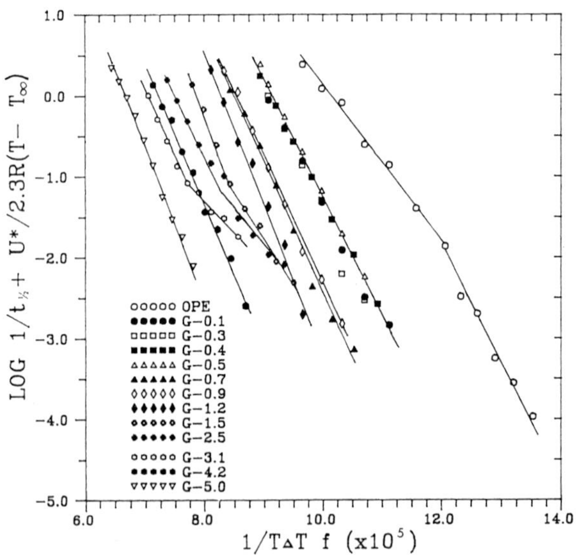

Fig. 24. Crystallization regimes in crosslinked linear polyethylene as a function of the molecular weight between crosslinks (MWX), determined from bulk growth rates. G–0.1: MWX 17 300; G–0.3: MWX 12 700; G–0.4: MWX 9400; G–0.5: MWX 5700; G–0.7: MWX 4500; G–0.9: MWX 3600; G–1.2: MWX 3100; G–1.5: MWX 2200; G–2.5: MWX 1900; G–3.1: MWX 1300; G–4.2: MWX 960 and G–4.5: MWX 560 [94].

percent of chain defects cause the rate of linear growth to drop by orders of magnitude. The Andrews theory of course pre-dated regime theory by several years. As the degree of crosslinking is increased the rate of secondary nucleation, i, must decrease and ultimately must become comparable to the rate of spreading, g, and the regime III induced by the first crosslinks must revert to regime II. This effect is demonstrated in fig. 24.

The second effect, that of a reducing rate of secondary nucleation, is to be expected of all polymers containing chain defects, including copolymers. Hence copolymers such as branched polyethylenes, where the branch is excluded from the crystal, are expected to have regime transitions different from those of linear polyethylene and dependent on the level of branching. Such effects have recently been studied for cross-fractionated polyethylenes (i.e. they have been fractionated according to molecular weight and then according to branching frequency). The effect is quite clear for relatively low-molecular-weight fractions (i.e. those that show a regime I–regime II transition). As the level of branching increases the regime I–regime II transition temperature translates to successively lower temperatures (fig. 25). In the case of higher-molecular-weight fractions, where only regime II is observed for the linear polymer, a regime II–regime III transition is induced for high levels of branching [94].

All recent versions of secondary nucleation theory start the first stem at a chain end and then reel in additional stems in one direction along the growth face. The Andrews theory contains as one of its predictions that the number of crystallizable units in a critical nucleus can be calculated from the slope of a plot of the logarithm of growth rate at a given temperature versus the defect content. When this calculation was performed for cis-polyisoprene it was found that the number corresponded not to a single stem but to three stems. A similar conclusion was reached in the aforementioned studies of

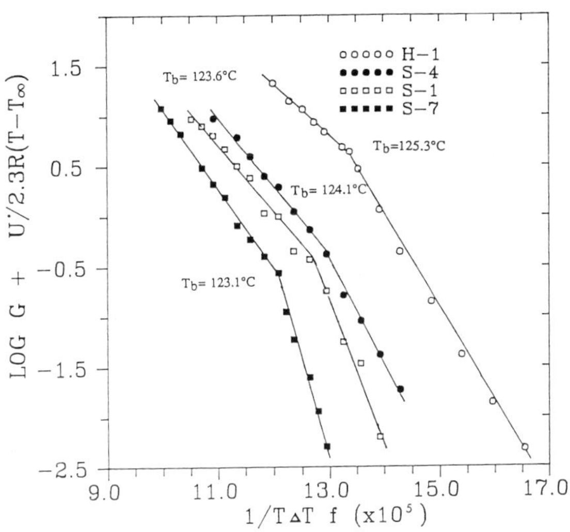

Fig. 25. Crystallization regimes in polyethylene fractions of molecular weight 20000 as a function of hexyl branch content per 1000 C atoms. H–1: linear; S–4: 8.8/1000; S–1: 16.7/1000 and S–7: 20.7/1000 [94].

crosslinked polyethylene. In both cases free ends are not easily available. In the case of cis-polyisoprene this is because of its high molecular weight giving rise to "permanent" entanglements, and in the case of crosslinked polyethylene due to the crosslinks having eliminated chain ends. Hence, it was postulated that the three-stem nucleus is a natural consequence when there are major constraints on chain mobility [92].

There is, however, an additional major change that can be generated by increasing the chain defect content. This effect was seen first in the crosslinked polyethylenes [94]. The Andrews theory assumes that the basic tenets of secondary nucleation theory remain in effect when the chain defects are present. One of the major tenets is that embodied in eq. (4), namely that the lamellar thickness is determined by the supercooling. In the cases under consideration currently, this corresponds to the combined length of three stems and the folds linking them together. There is a point reached, as defect level is increased, where the total length required at a given temperature of crystallization by eq. (4) simply is not attainable. The maximum length achievable under such circumstances corresponds to the average separation of chain defects, which is less than the total length required by the three-stem nucleus. The system therefore reverts to a single-stem nucleus with a lamellar thickness determined by the chain microstructure. This results in a switch to regime III as there is a higher probability of finding these less stable structures than there is of finding the preferred three-stem nuclei. This type of behavior is to be expected in all copolymer systems containing excluded units.

There is another type of effect involving the polymer structure in a more subtle fashion. As mentioned earlier, polypropylene forms a 3:1 helix and these helices pack together to form the crystal. Because the methyl side groups of the polymer stick out from the helical cylinder there exists a helical groove in the surface of the cylinder. This unique geometry (fig. 26) makes it possible for additional helices to lay down at an angle of 80° to the helices in the layer below [95, 96]. When the chains in successive layers are all parallel to one another the common alpha crystal results. A layer of chains at 80° to the first generates branches of alpha if the chains in succeeding layers are all parallel to those on the 80° layer. A unique situation occurs in the crystallization of low-molecular-weight polymers and in high-pressure crystallization of high-molecular-weight polymers. Bilayers are formed involving parallel chains, but each succeeding set of bilayers is oriented at 80° to the previous one. This results in a new type of crystal, the gamma form [96]. For reasons not yet understood, the alpha form is encountered at atmospheric pressure rather than the more dense gamma form. When crystallization is carried out at elevated pressures the gamma starts to form, but does not dominate unless a pressure in excess of 2 kbar is used.

There exists a wide range of crystallization pressures (from 1 bar to 2 kbar) over which the alpha and gamma forms are generated concurrently. Here unusual effects on regimes are observed because of one type of secondary nucleus trying to deposit on the other type of crystal face. Hence at low pressures of crystallization the gamma secondary nuclei deposit on the alpha growth face and have to be removed for alpha growth to proceed. This results in a lowering of the rate of secondary nucleation, i, and hence a movement from the normal regime II–regime III system found at atmospheric pressure to a regime I–regime II system. At higher pressures where the gamma crystals

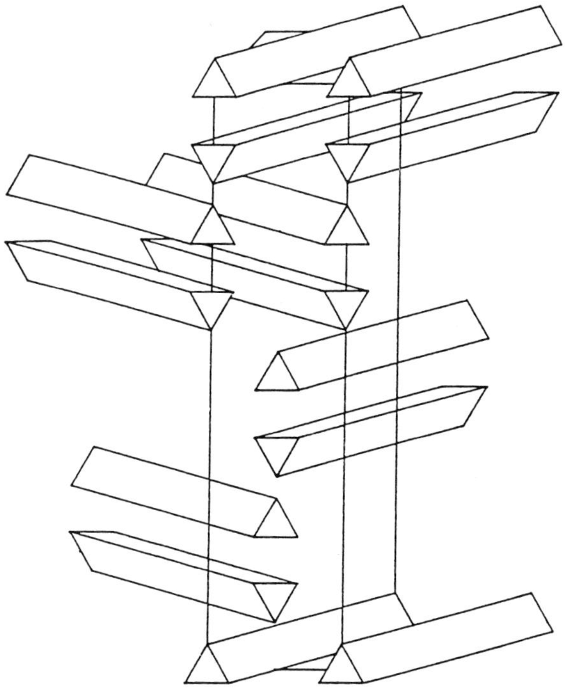

Fig. 26. Chain packing in a gamma crystal of polypropylene [30].

are growing, an analogous effect occurs and a regime I–regime II gamma system is observed. At even higher pressures, where only gamma occurs, a regime II–regime III system is found. The two forms have similar values of stem lengths and fold surface free energies at equivalent supercoolings. This tends to indicate that the differences between the two forms are somewhat analogous to the well-known differences between bcc and hcp crystals, namely different stacking sequences between essentially identical layers [30, 97, 98].

This unusual behavior of polypropylene is the only known example of two competing crystalline forms poisoning each others growth surfaces with alien secondary nuclei and is a direct result of the molecular structure of polypropylene.

4.2. The influence of pressure

The majority of the research carried out on spherulitic growth has considered temperature alone as the variable. Less than one percent of the effort has considered both primary thermodynamic variables, pressure and temperature. Pressure is routinely used in plastics moulding operations and in many cases the pressures used do cause significant changes in growth behavior, an excellent example being that of polypropylene given above. Yet pressure is not used as an active variable in the control of properties, simply being there to obviate problems caused by the large thermal expansion coefficient of polymers.

Both the melting point and the glass transition temperature are strong functions of pressure, both being raised by about 20°C per kbar. Generally the two parameters have different pressure coefficients and so the temperature range available for crystallization may be extended or contracted by the imposition of an elevated pressure. The only other major parameter that tends to be influenced by pressure is the folded structure. Elevated pressure tends to favor higher-density forms and so fold conformations may be changed by the use of elevated pressures. This in turn will have a major effect on the value of the fold surface free energy, and hence on the rates of growth of the lamellae and also on the lamellar thickness. In the case of cis-polyisoprene [87, 99, 100] this effect was found to be of major importance, it being a polymer where the adjacent re-entry folds are loopy and able to change conformation easily. In an analogous study of polyethylene [25] no change of fold surface free energy could be found in this polymer where the adjacent re-entry folds are tight and relatively incapable of forming more constricted forms.

In another polymer, polyethylene terephthalate, there was found to be a major continuous increase in the fold surface free energy with applied pressure during crystallization [23]. The chain structure of PET contains phenyl rings in the backbone which cause a major change in potential folded structures. Only one type of adjacent re-entry structure is possible, involving the aliphatic groupings. Any other surface structures must use nonfolded structures such as tie-chains or non-Gaussian cilia. It is possible that the studies of PET at elevated pressures have involved the pressure

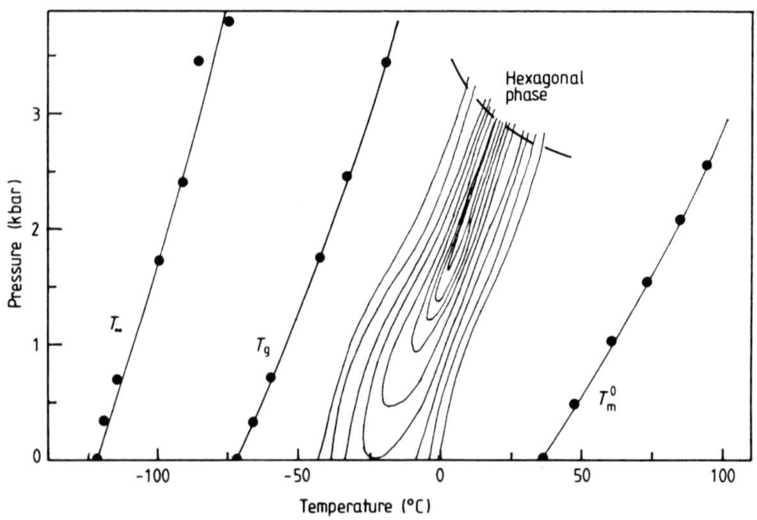

Fig. 27. The phase diagram of cis-polyisoprene showing the equilibrium melting point, glass transition temperature, WLF T_{00} temperature, and the lamellar growth rates in contour form. The contour values are (inward) 0.05, 0.1, 0.25, 0.5, 1, 1.5, 2, 2.5 and 3 microns per hour (first published in [99]).

dependence of the conformations of non-Gaussian chains, but theoretical approaches are not available to test such a hypothesis.

Polymers which can be crystallized over a complete range of temperatures from the glass transition to the melting point are expected to show combined phase-kinetic diagrams such as those of cis-polyisoprene (fig. 27). There are four basic forms of fold surface–pressure dependence to be expected:

(1) insensitivity due to tight adjacent re-entry folding,
(2) step changes due to loose adjacent re-entry folding,
(3) small continuous increase due to switchboard folding,
(4) large continuous or discontinuous increase due to nonfolded forms.

Only type 3 has not yet been observed in experimentation.

There is an additional effect which was once though to be general, but is now known to be restricted to a small number of polymers that crystallize in regime I where continuous adjacent re-entry folding occurs and where activation energies for "snaking" through the crystal are relatively low. This is the formation of chain-extended crystals in which the lamellar thicknesses approach those characteristic of the fully extended chain [101–104]. The effect is only seen in molecules of low and intermediate length.

In the case of polyethylene it has been established that the mechanism for the formation of these unusual species involves first the growth of hexagonal crystals, followed by their transformation to the orthorhombic form [105,106]. Thickening occurred in two stages: first, the lower heat of fusion of the hexagonal phase causes thicker crystals to form; second, during the transformation more thickening can occur. It should be recalled that when crystallization of orthorhombic polyethylene occurs at very low supercoolings at atmospheric pressure (i.e. in regime I), crystals as thick as 50 nm are easily produced. Only two additional polymers exhibit major chain extension, poly(tetrafluoroethylene) and poly(chlorotrifluoroethylene).

5. *Spherulitic morphogenesis*

The mechanisms whereby the lamellar crystals generate spherulitic textures will now be considered. It has to be recognized at the outset that lamellae are single crystals and differ only from those large species grown from solution in size, degree of faceting and in the mix of fold surface structures present. The equations given earlier that describe the relation between lamellar thickness and temperature and between crystal growth rate and temperature apply to all conditions. In metals the differences between single-crystal growth and polycrystalline growth are largely due to changes in primary nucleation. In macromolecules many of the differences are endemic to their length and entangled nature. Placing macromolecules in solution serves to isolate the molecules and mitigates complications due to their long-chain nature. A graphic example of this effect is the ability to draw out the macromolecules from a gel in such a way that the molecules crystallize from an essentially fully extended form to generate fibers which have a modulus approaching that of a carbon–carbon bond [107].

One important, but poorly understood, aspect of spherulitic generation is the effect of molecular variables on the lamellar width. Although the lamellar thickness has been studied quantitatively for many years (see [3]) and so also has the lamellar length through spherulitic growth, the lamellar width is a much more elusive quantity. Its study requires some form of electron microscopy combined with some form of specimen orientation. It also requires that the lamellae be relatively loosely packed so that individual crystal shapes can be discerned. Sometimes it is possible to discern the width if some crystals at the growth front protrude into the melt. It is generally accepted that the width increases with crystallization temperature, but the detailed dependence is not understood in general.

In the case of cis-polyisoprene, because of the afore-mentioned ability of osmium tetraoxide to terminate the crystallization process at will, it was possible to observe clearly that there was a major change in the width dependence on temperature at the regime I–regime II transition temperature [108, 45]. In both regimes III and II the

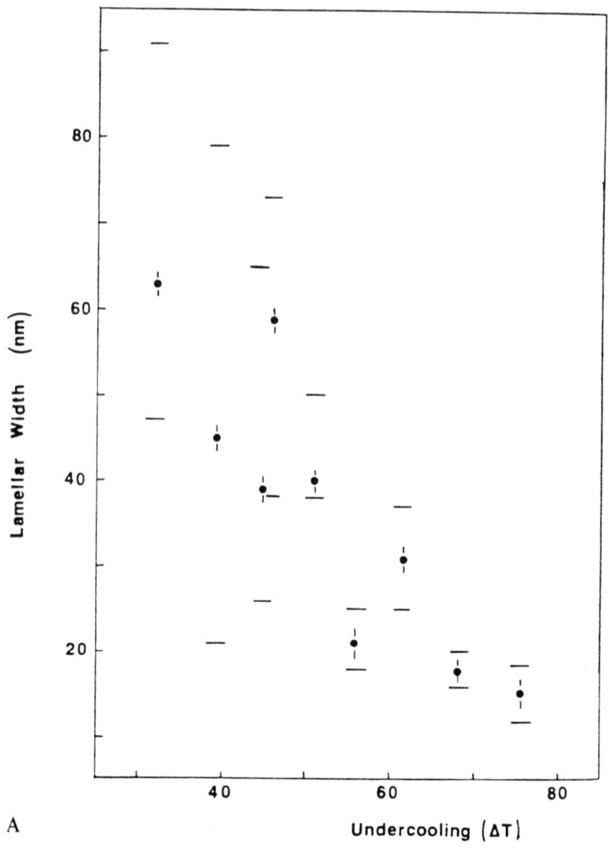

A

Fig. 28. (A) Distribution of lamellar widths in a fraction of cis-polyisoprene of molecular weight 313 000, and (B) dependence of lamellar width on supercooling at (a) − 20°C and (b) − 0.6°C (molecular weight 543 000) [108].

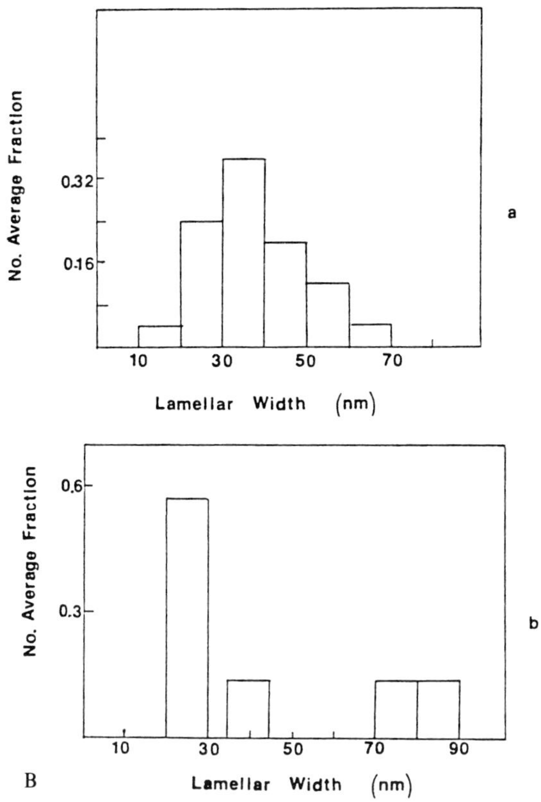

B

(Fig. 28 cont'd)

crystal grew as a ribbon of constant width. Although there was a distribution of widths present (fig. 28a) the peak width clearly was inversely proportional to supercooling in a similar way to lamellar thickness (fig. 28b). In regime I leaf-shaped crystals resulted (fig. 17) and the constant width relation disappeared. In polyethylene and some other highly crystalline polymers the shapes of the dominant lamellae appear to be leaf-shaped in both regimes I and II (personal communication from D.C. Bassett). The ribbon-like lamellae observed in regimes II and III in cis-polyisoprene and other low-crystallinity polymers may be characteristic of only polymers where for some reason or other the lamellae are separated by large amounts of inherently non-crystallizable material. It may be a direct result of the very long molecules, long-chain branches or even gels present in such systems.

In other polymers studied, researchers have of necessity considered the shapes present at the growth tips. It has also been necessary to carry out the studies at low supercoolings in order to obtain reasonable descriptions, and so studies of polyethylene have been primarily in the regime I region. They have tended to confirm a wide lamellar profile of approximately leaf-shaped entities, often with relatively well-defined growth tips.

This aspect of lamellar growth is very important to the generation of a real understanding of the spherulitic formation process, simply because the constraints placed on packing together lozenge or leaf-shaped crystals into a spherical entity are much greater than for packing together ribbon-like crystals. The most successful models that have been used to describe the light scattering [109] from spherulites have assumed radiating arrays of ribbons and are very effective for polymers crystallized in regime II and III, but have major problems when applied to polyethylene grown in regime I [110]. It is therefore important that features defined for one regime of growth are not taken as characteristic of all conditions. This is very important when polyethylene is considered, and perhaps taken as the paradigm for all polymers, since the most definitive experiments have been made at low supercoolings in regime I, where axialites, not spherulites, are formed.

Undoubtedly spherulites or axialites in the vast majority of polymers begin from either a single lamellar crystal or a cluster of such lamellae, unless a particulate nucleus large enough to generate many lamellae is operative. The initial lamellae are grown from the microstructurally purest molecules present and they tend to grow in a linear fashion. Termed either leading lamellae or dominant lamellae, they form the superstructure of whatever morphological species is being formed. They tend to generate additional lamellae in their immediate vicinity, presumably by a branching process. The branching process may have a crystallographic contribution to it, but more often than not in the case of high-molecular-weight polymers, it may be non-crystallographic. The latter process has been demonstrated to occur in polymers such as cis-polyisoprene (fig. 29) where lamellar branches are seen to have no physical attachment to the leading lamellae and to splay away shortly after formation. After a short distance they tend to propagate in the same radial fashion as the first lamellae created.

In the early stages of growth the morphological units do not have a spherulitic appearance, but are sheaves. In the case of axialites, where the lamellae are very wide,

Fig. 29. Non-crystallographic branching in cis-polyisoprene crystallized at −15°C.

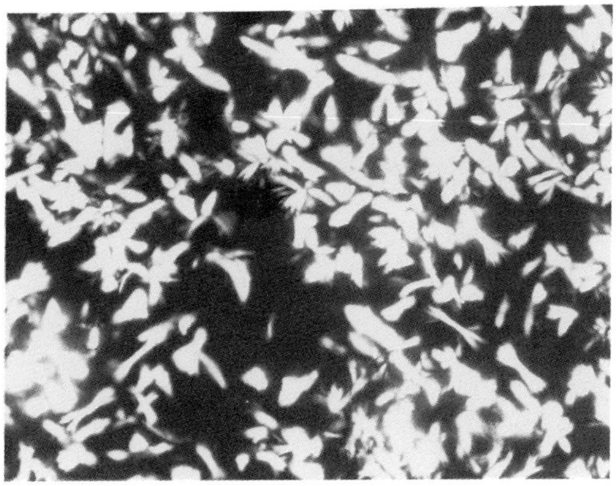

Fig. 30. Axialites growing in a branched fraction of polyethylene of molecular weight 35700 and 15.5 hexyl branches per 1000 C atoms [94].

the growing entities appear as single, or more often as double, propellers (fig. 30) in an optical microscope. In the case of polyethylene grown in regime I, the dominant lamellae tend to grow in cells and are very wide and ridged (see [17] for a review). The ridges are a reflection in the melt of the tendency of the solution-grown single crystals [111] to form hollow pyramids and ridged pyramids (fig. 31). The chain stem is inclined to the surface of the fold surface in these ridged crystals at an angle consistent with that in single-crystal growth. The ridging disappears when lower temperatures of crystallization consistent with regime II are used.

There can be little doubt that the leading lamellae are formed from the microstucturally purest molecules and they have a higher probability of secondary nucleation

Fig. 31. Ridged crystals in linear polyethylene (photograph supplied by D.C. Bassett; first published in [3]).

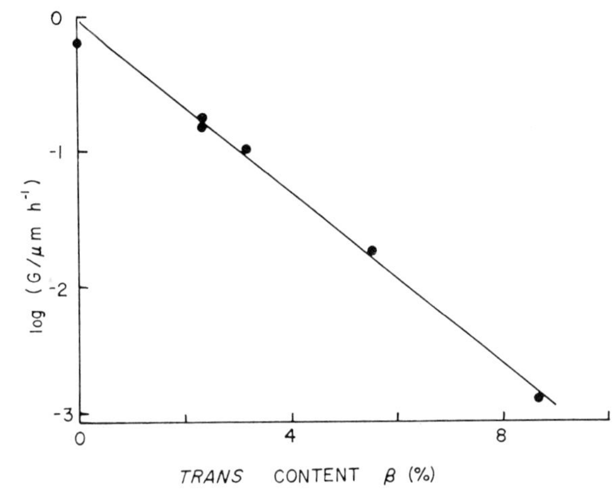

Fig. 32. The growth rate of cis-polyisoprene lamellae at $-26°C$ as a function of microstructural impurity content (first published in [120]).

than do microstructurally impure molecules. The research referred to earlier, in which the effect of branches and of comonomers was considered, predicts that this must be the case. It has been recognized for many years that impure molecules are rejected at the growth face and crystallize later. The reason did not become clear until Andrews et al. [93] clearly demonstrated that this rejection was due to their lower probability of secondary nucleation (fig. 32) and was a logarithmic dependence on impurity content. Although the effects of chain impurity content on crystallization are now beginning to be understood in terms of probability of rejection and of regimes for narrow fractions, the details of blending of fractions together defines the ultimate events and are only now being investigated.

The recognition of the presence of leading lamellae and their control over the spherulitic superstructure and the growth front has emphasized the lack of understanding that exists with regard to subsequent events. Prior to this realization it had been generally assumed that the branching was sufficiently rapid for the newer crystals to be growing at the front. It is now recognized that a large amount of the crystallization occurs behind the growth front and the degree of retardation will depend on the polymer. In the case of axialites in polyethylene, the filling in between cells of the leading lamellae may not take place until the temperature is lowered. In the case of spherulitic growth in polyethylene in regime II many of the branched crystals growing from relatively pure microstructurally defective molecules may be only slightly behind the growth front of the dominant lamellae and impossible to resolve optically. In the case of PEEK the secondary crystallization can be seen as a second spherulitic growth front emanating from the center of the spherulite due to an unusually severe retardation of the branching processes [112].

It has to be recognized from fig. 32 that in any system there will be a whole range of molecules of differing levels of microstructural impurity and molecular weight that simply will crystallize on different time scales at a given isothermal condition. The conditions, both environmental and molecular, that are needed to convert axialitic growth into spherulitic growth are simply not understood in any detail. It involves a change of lamellar shape from leaf to ribbon, probably the infilling taking place as ribbons. It is known [95] that the tendency to form axialites initially is controlled by molecular weight and temperature alone in fractions and not by impurity levels. However, high molecular weights combined with high levels of branching ensure spherulitic textures.

A still poorly understood feature of spherulitic growth in polyethylene is the formation of banding. This feature has been recognized from the earliest studies using optical microscopy (see, e.g. [14]). Early interpretations involved a regular helicoidal twisting of the lamellae as they radiated outwards and there is no doubt that such an explanation is absolutely consistent with polarized microscopy. It is a feature that is associated with regime II growth. More recent studies using permanganic etching and

Fig. 33. Banding in linear polyethylene in a TEM photograph of a chromium shadowed permanganically etched surface (first published in [3]).

chlorosulphonation show that individual lamellae do not twist continuously (see [17] for a review), but that the regular banding is the result of an aggregate effect. Permanganic etching studies suggest that screw-dislocation-generated branching is an integral feature of regular banding (fig. 33). Such a mechanism generates six to twelve new crystals in a stack each time it functions. Under such circumstances the new lamellae will rapidly outweigh their progenitors and will follow very closely behind the original lamellae as they grow. It is also clear from micrographs such as that in fig. 33 that twisting of the lamellae remains an integral part of the process. It is possible that the new stacks are generated whenever there is space generated by the radial growth process for their generation [113]. Why such a process should generate so many crystals at one time rather than simply one or two at a time has not been established.

It has also been demonstrated [94] that banding is rarely present in crossfractioned specimens of polyethylenes. Materials which had been subjected to fractionation for both branching and molecular weight in two separate consecutive treatments did not show banding when crystallized under conditions which generated spherulites. Indeed, the spherulites formed were very poorly developed. However, specimens which had only been subjected to TREF (temperature rising elution fractionation), which essentially separates largely by branching, formed classical spherulites. The complicated molecular features which control the generation of spherulites and also the generation of banding in bulk unfractionated materials are quite poorly understood. It is something of an anachronism that the best known morphological unit manifest by macromolecules is generated most easily and most perfectly by polymers showing a wide range of molecular weights combined with a wide range of chemically microstructural imperfections.

In less densely packed spherulites it is possible for the number of branch stacks being generated by screw dislocations, or some other process, to be estimated. The results for cis-polyisoprene are shown in fig. 34, where it can be seen that the process is controlled by the regime operating. In this case there is a discontinuous jump as regime II is entered by lowering the temperature and that the maximum generation is

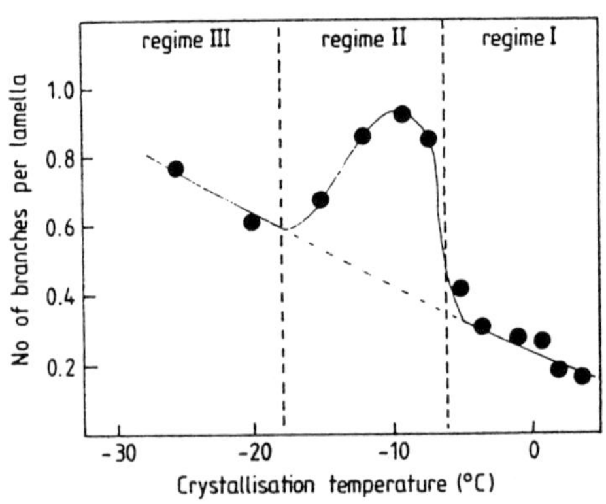

Fig. 34. Average number of branches per lamellar crystal in cis-polyisoprene as a function of crystallization temperature for a fraction of molecular weight 313 000 (first published in [45]).

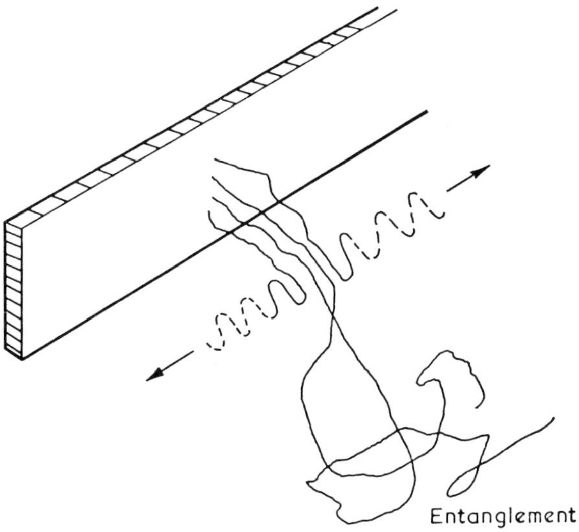

Fig. 35. Early view of the mechanism of non-crystallographic branching in a high polymer (first published in [121]).

somehow associated with the regime change. Since regime II occurs when multiple nucleation becomes prevalent it seems reasonable to speculate that the occurrence is related to the formation of tie-molecules and to stress being generated at the crystal–amorphous interface by tie-chain extension. An early view of the process is depicted in fig. 35 but it does not involve the initiation of screw-dislocation-type growth. Whether or not such a type of stress generation could produce a screw dislocation and the growth that can be generated by it remains to be explored theoretically.

It should also be noted that, although lamellae have been referred to as ribbons or leaves throughout, they are not necessarily planar entities. A glance at any of the micrographs of cis-polyisoprene used in this chapter shows that they tend to bend and curve, especially near the sites of their inception. This comment is true of the leaf-shaped lamellae found in axialites. Clearly, for a leaf-shape to be curved the object itself must have some overall convex or concave shape. The existence of such "bowl-shaped" lamellae has been investigated in some detail for a small number of polymers largely by Khoury and Barnes [114–116]. Although circumstances leading to their formation have been well documented the detailed explanations for the behavior have not yet been resolved. It is unlikely that they are quirks, since similar objects in the form of multilayered aggregates have been seen in several systems (see, e.g. [115]). Additionally, it is worth noting that solution-grown single crystals of polyethylene are generally reported as hollow pyramids, but that in the early work of Bassett [111, 118] it was noted that a substantial number of double-cupped crystals were found.

6. Summary

Although considerable progress has been made in understanding the essentials of lamellar growth in macromolecules, as a community we are well away from a complete understanding of all the detailed processes that lead to the generation of the classical spherulite in macromolecular systems. It is clear that much additional work is needed on the mechanisms by which lamellae branch and splay apart, and how these processes are influenced by molecular weight and molecular microstructure, before any overall understanding can be achieved. This is despite the volume of excellent research that has been generated on these systems in the past forty years. Even for polyethylene, the "water" of polymer science, the situation is far from being resolved.

Acknowledgments

Much of the original research reported in this chapter and the time for its writing have been supported by the Polymers Program of the National Science Foundation and by the Dow Chemical Company.

References

[1] C. Darwin, Geological Observations, etc., Part II, London (1851).
[2] J.P. Iddings, Bull. Phil. Soc. Wash. 11 (1891) 445.
[3] P.J. Phillips, Rep. Prog. Phys. 53 (1990) 549.
[4] H.W. Morse and J.D.H. Donnay, Amer. Mineral. 7 (1936) 391.
[5] J.P. Iddings. Amer. J. Sci. 33 (1887) 45.
[6] G.E. Lofgren, Bull. Geol. Soc. Am. 82 (1971) 111.
[7] G.E. Lofgren, J. Geophys. Res. 76 (1971) 5635.
[8] S.E. Kesler and P.W. Weiblen, Amer. Mineral. 53 (1968) 2025.
[9] F. Bernauer, Gedrillte Kristalle (Borntrager, Berlin, 1929).
[10] G.E. Lofgren, J. Petrol. 24 (1983) 229.
[11] L.S. Palatnik and V.M. Kosevich, in: Growth of Crystals, Vol. 3, eds. A.V. Shubnikov and N.N. Sheftal (Consultants Bureau, New York, 1962). p. 121.
[12] E.H. Andrews, Proc. R. Soc. A 270 (1962) 332.
[13] I.E. Bolotov, in: Growth of Crystals, Vol. 5b, ed. N.N. Sheftal (Consultants Bureau, New York, 1968) p. 3.
[14] H.D. Keith and F.J. Padden, J. Polym. Sci. 51 (1961) S4.
[15] H.D. Keith and F.J. Padden, J. Appl. Phys. 35 (1964) 1270.
[16] C. Elbaum, Prog. Metal Phys. 8 (1959) 203.
[17] D.C. Bassett, CRC Crit. Rev. Solid State Mater. Sci. 12 (1984) 97.
[18] G.E. Lofgren, Amer. J. Sci. 274 (1974) 243.
[19] F.W. Billmeyer, Textbook of Polymer Science, 2nd ed. (Wiley, New York, 1984).
[20] R.H. Boyd and P.J. Phillips, The Science of Polymer Molecules (Cambridge University Press, Cambridge, 1992).
[21] H.-G. Kim and L. Mandelkern, J. Polym. Sci. A-2 10 (1972) 1125.
[22] J.H. Magill, Polymer 2 (1961) 221.
[23] P.J. Phillips and H.T. Tseng, Macromol. 22 (1979) 1649.
[24] B. Wunderlich, Thermal Analysis (Academic Press, San Diego, 1990).
[25] H.T. Tseng and P.J. Phillips, Macromol. 18 (1985) 1565.

[26] M. Avrami, J. Chem. Phys. 7 (1939) 1103.
[27] A. Fukada, M.Sc. Thesis, University of Tennessee, Knoxville, TN (1991).
[28] A. Fukada and P.J. Phillips, in: Proc. ANTEC 92 (SPE, Brookfield, CT, 1992).
[29] R.J. Patel and J.E. Spruiell, Polym. Eng. Sci. 31 (1991) 730.
[30] R.A. Campbell, Ph.D. Dissertation, University of Tennessee, Knoxville, TN (1991).
[31] G. Ross and L. Frolen, J. Res. NBS 79A (1975) 701.
[32] F.L. Binsbergen, Polymer 11 (1970) 253.
[33] F.L. Binsbergen, J. Polym. Sci. Symp. 59 (1977) 11.
[34] J.C. Wittmann and B. Lotz, J. Polym. Sci. Phys. 19 (1981) 1837.
[35] J.C. Wittmann and B. Lotz, J. Polym. Sci. Phys. 19 (1981) 1853.
[36] J.C. Wittmann, A.M. Hodge and B. Lotz, J. Polym. Sci. Phys. 21 (1983) 2495.
[37] D. Parikh and P.J. Phillips, J. Polym. Sci. Polym. Lett. Ed. 28 (1990) 167.
[38] B. Lotz and J.C. Wittmann, J. Polym. Sci. Polym. Phys. Ed. 24 (1986) 1559.
[39] P.J. Phillips and A.J. Greso, in: Proc. ANTEC 1992 (SPE, Brookfield, CT, 1992).
[40] Milliken, Millad – Clear Thinking (Milliken Chemical, Spartanburg, SC, 1991).
[41] A. Thierry, C. Straupe, B. Lotz and J.C. Wittmann, Polym. Commun. 31 (1990) 299.
[42] T. Kobayashi, H. Hasegawa and T. Hashimoto, Nih. Reor. Gakk. 17 (1989) 86.
[43] E.H. Andrews, Proc. R. Soc. A 272 (1964) 562.
[44] B.C. Edwards and P.J. Phillips, J. Polym. Sci. Polym. Phys. Ed. 13 (1975) 2117.
[45] P.J. Phillips and N. Vatansever, Macromol. 20 (1987) 2138.
[46] P.H. Till, J. Polym. Sci. 24 (1957) 30.
[47] A. Keller, Phil. Mag. 2 (1957) 1171.
[48] E.W. Fischer, Z. Naturf. 12a (1957) 753.
[49] P.H. Geil, Polymer Single Crystals (Wiley, New York, 1963).
[50] C.W. Bunn, Trans. Farad. Soc. 35 (1939) 482.
[51] D. Turnbull and R.L. Cormia, J. Chem. Phys. 34 (1961) 820.
[52] J.D. Hoffman, G.T. Davis and J.I. Lauritzen, in: Treatise on Solid State Chemistry, Vol. 3, ed. N.B. Hannay (Plenum, New York, 1976) Ch.7.
[53] J.D. Hoffman and J.J. Weeks, J. Chem. Phys. 37 (1962) 1723.
[54] R.A. Campbell, K. Mezghani and P.J. Phillips (1992) submitted to Macrom.
[55] I.M. Bank and S. Krimm, J. Polym. Sci. A2 7 (1969) 1785.
[56] I.M. Bank and S. Krimm, J. Polym. Sci. B 8 (1970) 143.
[57] D.M. Sadler, in: The Structure of Crystalline Polymers, ed. I. Hall (Elsevier Applied Science, Barking, UK, 1984).
[58] K.S. Schmitz, An Introduction to Dynamic Light Scattering by Macromolecules (Academic Press, San Diego, 1990).
[59] S.J. Spells, D.M. Sadler and A. Keller, Polymer 21 (1980) 1121.
[60] D.Y. Yoon and P.J. Flory, Polymer 18 (1977) 509.
[61] E.A. DiMarzio and C.M. Guttman, Polymer 21 (1980) 733.
[62] C.M. Guttman, E.A. DiMarzio and J.D. Hoffman, Polymer 22 (1981) 597.
[63] P.J. Flory, J. Am. Chem. Soc. 84 (1962) 2857.
[64] M. Glotin and L. Mandelkern, Coll. Pol. Sci. 260 (1982) 182.
[65] J. Grebowitz, S.F. Lau and B. Wunderlich, J. Polym. Sci. Symp. 71 (1984) 19.
[66] B. Wunderlich and S.Z.D. Cheng, Gazz. Chim. Ital. 116 (1986) 345.
[67] L. Mandelkern, Growth and Perfection of Crystals (Wiley, New York, 1958).
[68] L. Mandelkern, Crystallization of Polymers (McGraw-Hill, New York, 1964).
[69] D. Turnbull and J.C. Fisher, J. Chem. Phys. 17 (1949) 71.
[70] W.K. Burton, N. Cabrera and F.C. Frank, Phil. Trans. R. Soc. A 243 (1951) 299.
[71] D.M. Sadler and G.H. Gilmer, Polymer 25 (1984) 1446.
[72] D.M. Sadler and G.H. Gilmer, Phys. Rev. Lett. 56 (1986) 2708.
[73] D.M. Sadler, Polym. Commun. 27 (1986) 140.
[74] J.D. Hoffman and J.I. Lauritzen, J. Res. NBS A 65 (1961) 297.
[75] J.D. Hoffman, R.L. Miller, H. Marand and D.B. Roitman, Macrom. 25 (1992) 2221.
[76] P.J. Flory, Principles of Polymer Chemistry (Cornell University Press, Ithaca, NY, 1953).

[77] P.J. Flory, Statistical Mechanics of Chain Molecules (Wiley, New York, 1969).
[78] J.D. Hoffman and J.I. Lauritzen, J. Appl. Phys. 44 (1973) 4340.
[79] J.D. Hoffman, Polymer 24 (1983) 3.
[80] J.D. Hoffman, G. Ross, L. Frolen and J.I. Lauritzen, J. Res. NBS A 79 (1975) 671.
[81] P.J. Phillips, Polym. Prepr. (Am. Chem. Soc., Div. Polym. Chem.) 20 (1979) 438.
[82] A.J. Lovinger, D.D. Davis and F.J. Padden, Polymer 26 (1985) 1595.
[83] T. Suzuki and A. Kovacs, Polym. J. 1 (1970) 82.
[84] P.G. De Gennes, Scaling Effects in Polymer Physics (Cornell University Press, Ithaca, NY, 1979).
[85] J. Klein and B.J. Briscoe, Proc. R. Soc. A 365 (1979) 53.
[86] J.D. Hoffman, Polymer 23 (1982) 656.
[87] E.N. Dalal and P.J. Phillips, J. Polym. Sci. Polym. Lett. Ed. 22 (1984) 7.
[88] S. Lazcano, J.G. Fatou, C. Marco and A. Bello, Polymer 29 (1988) 2076.
[89] S.Z.D. Cheng, J. Chen and J.J. Janimak, Polymer 31 (1989) 1018.
[90] J.J. Janimak, Ph.D. Dissertation, University of Akron, OH (1991).
[91] P.J. Phillips and J. Patel, Polym. Eng. Sci. 18 (1978) 943.
[92] P.J. Phillips and W.S. Lambert, Macromol. 23 (1990) 2075.
[93] E.H. Andrews, P.J. Owen and A. Singh, Proc. R. Soc. A 324 (1971) 79.
[94] W.S. Lambert, Ph.D. Dissertation, University of Tennessee, Knoxville, TN (1991).
[95] B. Lotz and J.C. Wittman, J. Polym. Sci. Polym. Phys. Ed. 24 (1986b) 1541.
[96] S. Bruckner and S.V. Meille, Nature 340 (1989) 455.
[97] P.J. Phillips and R.A. Campbell, in: Proc. ANTEC 1991 (SPE, Brookfield, CT, 1991) p. 896.
[98] R.A. Campbell, P.J. Phillips and J.S. Lin (1992) Polymer, in press.
[99] E.N. Dalal and P.J. Phillips, Macromol 17 (1984) 248.
[100] P.J. Phillips and B.C. Edwards, J. Polym. Sci. Phys. 13 (1975) 1819.
[101] P. Geil, F.R. Anderson, B. Wunderlich and T. Arakawa, J. Polym. Sci. A 2 (1964) 3707.
[102] B. Wunderlich and T. Davidson, J. Polym. Sci. A2 7 (1969) 2043.
[103] D.C. Bassett, Principles of Polymer Morphology (Cambridge University Press, Cambridge, 1981).
[104] D.C. Bassett, in: Developments in Crystalline Polymers, Vol. 3 (Elsevier Applied Science, Barking, UK, 1985) Ch. 3.
[105] D.C. Bassett and B. Turner, Phil. Mag. 29 (1974) 925.
[106] D.C. Bassett, S. Block and G.J. Piermarini, J. Appl. Phys. 45 (1974) 4146.
[107] P.J. Lemstra, N.A.J.M. van Earle and C.W.M. Bastiaansen, Polym. J. 19 (1987) 85.
[108] N. Vatansever, Ph.D. Dissertation, University of Tennessee, Knoxville, TN (1987).
[109] M.B. Rhodes and R.S. Stein, J. Polym. Sci. A-2 7 (1969) 1539.
[110] J. Maxfield and L. Mandelkern, Macromol. 10 (1977) 1141.
[111] D.C. Bassett, F.C. Frank and A. Keller, Phil. Mag. 8 (1963) 1739; 1753.
[112] F.J. Medellin-Rodriguez, in: Proc. ANTEC 1991 (SPE, Brookfield, CT, 1991) p. 893.
[113] P.J. Phillips and H.P. Sien, Polym. Eng. Sci. 18 (1978) 299.
[114] F. Khoury and J.D. Barnes, J. Res. NBS A 76 (1972) 225.
[115] F. Khoury and J.D. Barnes, J. Res. NBS. A 78 (1974) 95.
[116] J.D. Barnes and F. Khoury, J. Res. NBS. A 78 (1974) 363.
[117] A.E. Woodward, Atlas of Polymer Morphology (Hanser, New York, 1989).
[118] D.C. Bassett and A. Keller, Phil. Mag. 7 (1962) 1553.
[119] K.L. Monar, M.Sc. Thesis, University of Tennessee, Knoxville, TN (1992).
[120] G.J. Rensch, P.J. Phillips, A. Vatansever and V.A. Gonzalez, J. Polym. Sci. Polym. Phys. 24 (1986) 1943.
[121] B.C. Edwards and P.J. Phillips, Polymer 15 (1974) 351.

CHAPTER 19

Fundamentals of Aqueous Solution Growth

SARA SARIG

Casali Institute, Hebrew University, Jerusalem 91904, Israel

Contents

1. The medium . . . 1219
   1.1. Structure of water . . . 1219
   1.2. Aqueous solutions . . . 1219
   1.3. The thermodynamic states of the medium . . . 1222
      1.3.1. Solubility curves . . . 1222
      1.3.2. Supersaturation . . . 1223
      1.3.3. Measurement of supersaturation . . . 1224
      1.3.4. The structure of supersaturated solutions . . . 1225
2. Nucleation. . . . 1225
   2.1. Homogeneous nucleation in gases . . . 1226
   2.2. Homogeneous nucleation in solutions . . . 1227
   2.3. Heterogeneous nucleation in solutions . . . 1230
3. Crystal growth . . . 1234
   3.1. The various approaches to theories of crystal growth . . . 1234
   3.2. Electrolytic growth kinetics . . . 1235
   3.3. Growth rate isograms and isotachs . . . 1238
   3.4. Electrolytic crystal growth mechanisms . . . 1239
   3.5. Some chemical factors in crystal growth mechanisms . . . 1240
   3.6. Rate laws and rate constants in crystal growth . . . 1242
   3.7. Unconventional approaches to growth mechanisms . . . 1243
   3.8. Precipitation . . . 1244
   3.9. Ripening . . . 1246
   3.10. Growth kinetics and agglomeration . . . 1247
4. The effect of impurities . . . 1249

HANDBOOK OF CRYSTAL GROWTH, VOL. 2
edited by D.T.J. Hurle

4.1. Definition of an impurity 1249
4.2. Modification of crystal habit 1251
4.2.1. Crystal habit 1251
4.2.2. The effect of impurities on crystal habit 1251
4.3. Effect of impurities on nucleation 1259
4.4. Effect of impurities on growth rate 1262
5. Growth of single crystals 1264
References 1266

1. The medium

1.1. Structure of water

Most of the crystallization processes of ionic salts are conducted in aqueous solutions or, in some cases, in solvents which are mixtures of water and miscible organic solvents. The organic solvent/water mixtures serve also as the medium from which many organic crystals are produced. Water is non-toxic, the most abundant and cheapest solvent which remains liquid over a wide range of temperatures, with a very high boiling point and thus also the most convenient for use. The almost limitless availability of water gives the misleading impression that it is a simple compound and that it may act in dissolution and crystallization of materials as an inert medium, its participation in those processes being non-specific.

The existence of several points of anomaly in water indicates the complexity of the compound. Water does not fit into the series of hydrides of the sixth group of elements especially with respect to boiling point, it is denser in the liquid than in the solid state, its density increases with rising temperature in the range 0°C to 4°C. The anomaly of water as well as the radial distribution curves, which indicate almost fixed near neighbors may be accounted for by postulating that liquid water possesses a defined structure.

The structure of water may be explained on the molecular level. The water molecule is a slightly distorted tetrahedron with oxygen bonded to two hydrogens and having two lone electron pairs, thus being able to form four hydrogen bonds, two of them with hydrogens of other molecules using the lone electron pairs and two by its own hydrogens bonding to oxygen lone pairs of other molecules. The expected geometry would be a tetrahedral unit, hydrogen bonding being postulated between the oxygen atom at the center of the tetrahedron and those at the corners. This arrangement is in good agreement with the model of the sp^3 hybrid orbitals. Owing to the rapidity of proton motion it is not possible to distinguish between the covalently and the hydrogen bonded positions. Thus, a tri-dimensional quartz-like structure, characteristic of ice is formed. There are also other possible arrangements like clathrate structures built by dodecahedra with pentagonal faces and some more complex structures.

It is believed that when ice melts the above described structures are maintained on short range basis. However, because the short range order is not perfect, the long range order characteristic of ice is destroyed. As the temperature is raised, the number of near neighbors increases. At the same time the distance between adjacent molecules increases. For each temperature there is a ratio between the water which remains in the bonded network and that for which the order is destroyed.

1.2. Aqueous solutions

When a substance is dissolved in water it disturbs the pattern. Around a non-polar molecule a hydrophobic atmosphere is formed, whereas around polar molecules and ions a Coulombic sphere is formed. The complexity of the Coulombic hydration sphere is shown in fig. 1 which is a two-dimensional drawing, representing a real three-

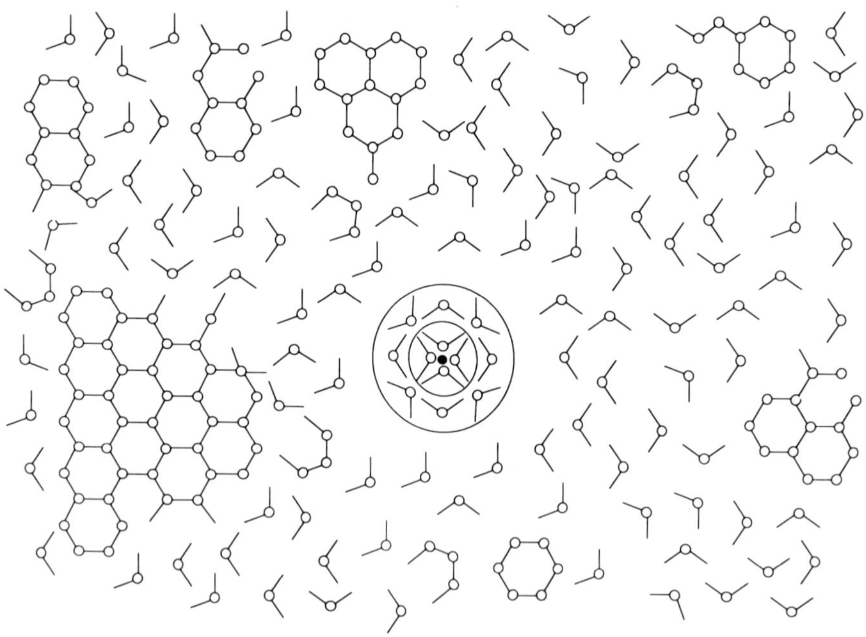

Fig. 1. A two-dimensional representation of a model of water structure surrounding a cation. Note the primary and secondary hydration circles (representing hydration spheres), structured regions and unbonded water molecules.

dimensional situation, using Mg^{2+} ion as an example. The Mg^{2+} ion has a primary hydration sphere of six water molecules which are bound by coordinative bonds, forming $Mg(H_2O)_6$ complex. Ions having lower electrostatic charge density form a primary hydration sphere of four water molecules. Around this hydration sphere the secondary hydration sphere exists, constituted evidently of "flickering clusters", or Frank–Wen clusters [1] with lifetimes of about 10^{-11} s. Beyond this, both broken water structure, i.e. unbonded, unaligned water molecules and fragments of structured water, surround the cluster. A neighboring ion forms a similar structured pattern at some distance.

The water which is not engaged in the primary and secondary hydration spheres, partly structured and partly unassociated with relative quantities strongly depending on temperature and with very fast rates of exchange between the two states, is termed "free water". The amount of free water in solutions of electrolytes depends on the solubility of the salt. For sparingly soluble salts with K_{SP} of about 10^{-10} ($BaSO_4$, $CaCO_3$), the equilibrium concentration is $\approx 10^{-5}$ M. If we assume that the hydration spheres of the cation and the anion and the neighboring water consists of 200 molecules it will mean that 2×10^{-3} mol of water in a liter are engaged in hydration. This is a negligible amount compared to the 55.5 mol of water contained in one liter of water. However, for soluble electrolytes such as NaCl and KCl the equilibrium solubility is about 5 M. It allows about 11 molecules of water for the hydration of each ion pair, hardly leaving any free water. This point will be discussed further with connection to supersaturation

There are few instances in which the structure of a concentrated electrolyte solution was worked out in detail. Such an illustrative example is the system of Dead Sea water, a concentrated solution of ionic strength 7.874 and total chloride concentration 5.891 M with the main accompanying cations Mg, Na, Ca and K. Marcus calculated the activities of the components and the changes of activities on isothermal evaporation. The structural aspects of Dead Sea brines are cited in the following paragraphs, with few omissions, from his contribution to *Ionic Liquids* [2].

Dead Sea water (DSW), has a density of 1.207 kg/dm^3 and contains 9.8 mol of ions, hence 5.29 ions per nm^3. The mean distance between two ions in DSW is therefore 0.57 nm. The end brine (from Dead Sea Works) contains 9.01 ions per nm^3 but the mean distance apart shrunk only to 0.48 nm. The diameters of the ions themselves are considerably smaller than the distances above, the largest chloride anion having a Pauling diameter of 0.362 nm. The solution is not close packed in terms of bare ions.

Consider, however, the hydrated ions (fig. 2). If only primary hydration is taken into account, then 6 mol of Mg and Ca ions must be considered bound. In DSW these are altogether 11.9 and in End Brine 33.2 mol of water out of the 55.5 mol present in 1 kg of water. The pimary hydration, in the above sense, does not use up the free water in the

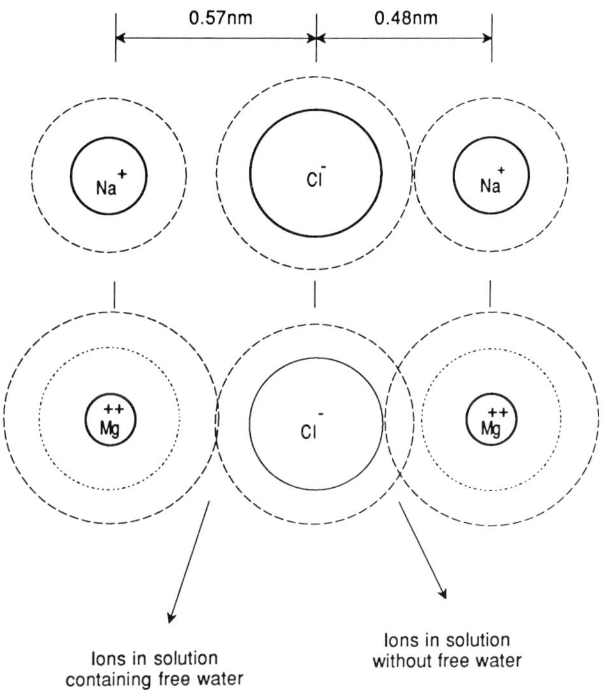

Fig. 2. Ions in solutions containing free water and without free water. A model of Mg^{2+}, Na^+ and Cl^- ions (solid lines), their first hydration spheres (Na^+, Cl^-), first and second hydration spheres (Mg^{2+}) drawn to approximate relative sizes. In a solution devoid of free water the hydration spheres are forced to interpenetrate.

solution. If, however, secondary hydration is also considered [3], the diameters of Cl, Mg and Ca ions, 0.51, 0.62 and 0.62 are commensurate with the mean distance apart of the ions. The water bound in these ions, together with that in the ions Na and K, becomes 39.7 mol in Dead Sea water and 82.1 mol in End Brine water. In the former there is still 28% free water, but there is a deficit of 31% in the latter. It means that in this case all the hydrated ions must be in intimate contact, so that the End Brine is a *molten hydrated salt*.

What kind of order, if any, prevails in these concentrated solutions? The interpretation of the radial distribution in terms of pair correlation functions is virtually impossible in the case of X-ray diffraction. There is no conclusive evidence for ion–ion contacts or ion pairs.

1.3. The thermodynamic states of the medium

1.3.1. Solubility curves

Solubility depends upon the difference between the energetic levels of ions in a crystal and their respective energetic levels as hydrated ions in a solution. This difference, on the whole, is small as compared to the large terms of crystal bond energy and hydration energy. Thus chemists, with all the information at their command, cannot explain easily and qualitatively to first year students why AgCl and $BaSO_4$ are sparingly soluble while $AgNO_3$ and $BaCl_2$ are soluble.

Solubility constants for sparingly soluble salts are determined experimentally. The experimental set up is simple: an excess of the salt in question is dispersed in a measured quantity of water. The conditions to be considered are: no change in water volume, therefore the process is carried out under reflux, no change of temperature, therefore use of thermostated bath and intimate, prolonged contact between the substance and the solvent, therefore satisfactory stirring and use of finely ground powders of the salt. To ascertain the attainment of equilibrium, consecutive samples of the solution are tested at different times for the concentration of the salt until two subsequent measurements are identical. As withdrawal of a sample for analysis from an investigated solution may affect the accuracy, several samples, kept under identical conditions with respect to temperature and stirring are used, each one corresponding to a specified time of reaction. Precaution must be observed to allow adequate time for settling of the dispersed powder after the stirring is stopped.

It would seem that mixing of the constituents of the salts to allow precipitation and determining the point of turbidity appearance would allow one an alternative way to determine solubility. Such an approach is, however, totally wrong, as will be shown in the discussion concerning supersaturation.

The solubility determinations of soluble salts is conducted essentially in the same way. The time needed to reach equilibrium is usually shorter than with sparingly soluble salts, but the settling times before analyses may be longer. In many soluble salts the solubility is strongly temperature dependent, the function being either linear (fig. 3) or exponential, close to linear. Therefore solubility determination must be performed in a range of temperatures. In this case also, evaporation of a dilute solution until the appearance of crystals cannot serve as an alternative method.

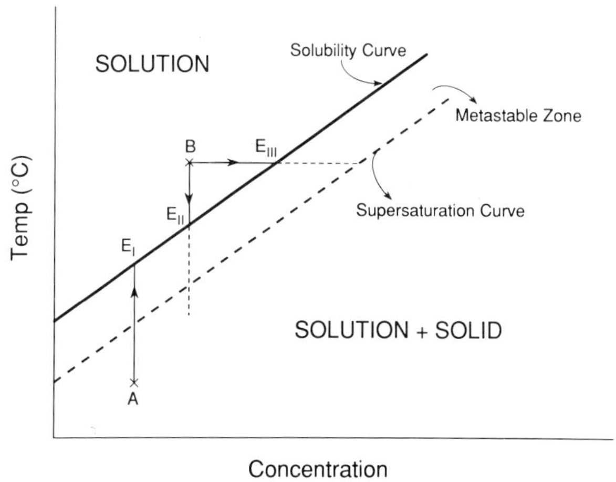

Fig. 3. Generalized solubility and supersaturation curves as a function of concentration versus temperature. E_I a composition attained by heating a solution containing solids (composition A). E_{II} composition attained by cooling clear solution of composition B. E_{III} composition attained by isothermal evaporation of solution B.

Several salts have solubility curves without slope, i.e. the solubility is not dependent on temperature, or even with inverted slope. This phenomenon has no simple explanation.

1.3.2. Supersaturation

Supersaturation is a measure of metastability. Its importance in physical processes was first observed, appreciated and used for theoretical analysis of the phenomenon of liquid droplet formation in the gaseous phase.

The experimental observation was that liquid droplets never form in vapor which is in equilibrium with its surroundings, at a given temperature. The actual vapor pressure P has to be significantly higher from the equilibrium vapor pressure P_∞ to induce the formation of droplets. The quotient P/P_∞ was termed supersaturation and is usually designed by S. The value of S which is needed to start the formation of droplets depends on the purity of the gaseous system. The purer is the system, the higher S is needed. Impurities, as rule, decrease the degree of S which induces the formation of droplets.

Supersaturation in solutions expresses also the degree of metastability of the system with respect to concentration of salts dissolved in water. If the equilibrium concentration of a salt in solution is designed C^* and the actual concentration by C and if $C > C^*$, the solution is supersaturated with respect to this salt.

The solid line in fig. 3, is designating the solubility curve, the region above the curve describes the conditions for the solution's existence, whereas the region below the curve corresponds to a solution co-existing with solid. If a mixture of solid and solution, having the composition of point A is heated, the solid will dissolve gradually. The last crystal will disappear at the temperature corresponding to point E_I. It would seem that, according to the definition of equilibrium, if a true solution having the composition of B is cooled, crystallization should start at point E_{II}, which lies on the solubility curve. However, only in very few cases this will occur. In the great majority of cases a considerable undercooling precedes the onset of nucleation of the solid phase. Such points, at the various concentrations and temperatures form a set of supersolubility curves, one of them represented by the dashed line in fig. 3. The zone between the solubility and supersolubility curves is called the metastable zone. Its width depends upon the chemical nature of the system, the flow regime, the impurity content and other operational parameters. The zone below the supersolubility curve is called the labile zone. The true solution zone, above the solubility curve, is of course stable.

Another way to define the zones on the concentration versus temperature diagram is to follow what happens to a solute crystal introduced to the solution in the three zones delineated in fig. 3. A crystal introduced to a solution of composition above the solubility zone will dissolve as long as the new composition remains above the equilibrium line. If a particle is added at the labile zone it will continue to grow and moreover it will induce nucleation of other crystals. If it is introduced into the metastable zone, it will continue to grow but it will not cause spontaneous nucleation.

Supersaturation can be attained not only by cooling, but by removing some solvent by isothermal evaporation as indicated by line BE_{III}. Whereas penetration into the metastable zone is natural and easy, penetration into the labile zone rarely occurs. The reason is that the surface at which the evaporation takes place, becomes more supersaturated than the bulk of the solution. Thus the first crystals are formed at the surface; eventually they fall into the solution acting as seed crystals.

1.3.3. Measurement of supersaturation

The values of equilibrium solubility, C^*, are either tabulated in handbooks or can be experimentally determined. If the actual concentrations, C, in a supersaturated solutions could be measured directly or by any physical method, the value S could have been determined. However, in almost all cases the measurement of ion concentration in a metastable solution, tends to destroy the parameter which the measurement seeks to determine. Any slight disturbance introduced into the metastable state causes its immediate collapse.

Thus only non-invasive methods which do not affect the metastable state may be used. Among those are conductivity measurements and refractive index determinations. Soluble salts i.e. salts which have high saturation concentration and very little free water content, can support only low supersaturations of few tenths of percent. Neither conductivity nor refractive index are sensitive enough to measure changes with the required degree of precision needed for the calculation of the supersaturation values in ordinary ionic salts. Refractive index was used to determine supersaturation in solution of citric acid which can form large clusters stabilized by hydrogen bonds [4]. Similar results were found later for sucrose [5]. Hussman et al. carried out Raman spectroscopic studies in $NaNO_3$ aqueous solutions and found evidence for the existence of solute clusters [6]. In 1986 Larson and Garside repeated the observations with respect to citric acid and $NaNO_3$, extending them to urea and K_2SO_4 [7].

1.3.4. The structure of supersaturated solutions

There cannot be a fundamental difference between the structure of a very concentrated solution of calcium and magnesium salts in a medium devoid of free water, as described above and between the structure of a supersaturated solutions of a soluble salt in which there is very little free water, as calculated from the hydration spheres of the ions and the total amount of water available in the solution. In calcium and magnesium solutions, above a certain temperature, these structures are actually molten hydrated salts and crystallization can be effected by cooling or evaporation. Salts which are less soluble like chlorides, nitrates and sulfates of alkali elements, form quasi-crystalline nets of hydrated ions. The formation of a crystal nucleus is an occurrence of partial dehydration, followed by a shift of partly bared ions toward each other. Such nuclei resemble closely the electrostatic charges, dipole moments and other properties of the solutions around them. Hence the surface tension between the continuous solution phase and the separating solid phase has a low value.

On the other hand, the solutions of sparingly soluble salts retain only a negligible fraction of the total water content of the solution in the hydration spheres. When nuclei of the solid phase are formed, they are surrounded by water possessing the characteristics of pure water: structured quartz-like fragments, unassociated molecules and flickering clusters. The surface tension of an ionic entity and an unsimilar medium necessitates very high surface tension to ensure the survival of the nucleus.

2. Nucleation

The formation of a condensed phase, solid out of liquid or liquid out of gaseous phases, proceeds through multicenter nucleation. In a one-phase system the nucleation can be homogeneous i.e. the chemical nature of the condensed phase and the remaining non-condensed one being chemically identical, with only new bonds being formed and boundary surfaces being created. Homogeneous nucleation does not involve catalytic activity of foreign materials or surfaces. The formation of mercury droplets from a pure gaseous phase, without interaction with the walls of a container is a good example. In aqueous solutions, the solvent is ever present. In case when there is only one salt dissolved the nucleation may be considered homogeneous in one of two situations: (a) the solution is extremely pure, not containing impurities which may induce nucleation, acting as catalysts to reduce the activation energy for the process, (b) the concentration of the condensing phase being so much greater than the equilibrium concentration that the effect of the foreign agents becomes negligible in the catastrophic occurrence of the nucleation. Also, nucleation on the walls of the reaction vessel, on the stirrers or on the solution surface is negligible.

Most crystallization processes in geological, biological, industrial and laboratory processes start by heterogeneous nucleation i.e. the formation of the condensed phase starts on foreign substrates or templates. The templates may be of similar chemical character but in many cases they are very different, like organic amino acids and mineral salts. Nevertheless they have chemical affinity such as electrostatic attraction and most importantly they possess structural compatibility in two dimensions or even one dimension. Foreign materials may also suppress nucleation by adsorption on the growing nuclei. Small fragments of crystals separated through breakage or attrition can also serve as nuclei for continued growth of new crystals. In this case the nucleation is termed secondary. It is most common in industrial processes.

There are solutions in which spontaneous nucleation does not occur before they pass to a viscous stage, as for instance in sugar solutions, and nucleation is usually achieved by seeding the solution with preformed nuclei.

The various mechanisms of nucleation in aqueous solution will be reviewed in the following sections.

2.1. Homogeneous nucleation in gases

The formal theory of nuclei formation in a homogeneous medium was developed with regard to droplet formation in a gaseous medium. When a spherical nucleus of radius r is formed, the change in the energy of the system relates to the new bonds formed within the droplet and the energy σ required to create the surface.

The standard free energy change is therefore represented by two terms having opposite contributions.

$$\Delta G^0 = -\tfrac{4}{3}\pi r^3 \Delta G_v + 4\pi r^2 \sigma, \quad (1)$$

where ΔG_v is the volume free energy resulting from bond formation.

For very small droplets the surface energy term predominates. In this region of sizes the nuclei are formed and dispersed, without starting growth. For large drops the volume free energy term predominates. The maximum value ΔG^* for a spherical cluster is

$$\Delta G^* = 16\pi\sigma^3 / 3\Delta G_v^2. \quad (2)$$

ΔG^* may be viewed as the activation energy barrier for the nucleation process to take place. The nucleus corresponding to this energy is termed the "critical nucleus" i.e. a nucleus which will grow spontaneously, releasing energy.

The volume free energy is expressed by the Gibbs–Kelvin equation

$$\Delta G_v = KT \ln(P/P_\infty)/v = KT(\ln S)/v. \quad (3)$$

The theory, introduced by Becker and Doring [8], based on the ideas of Volmer and Weber [9] can be tested experimentally by measuring the rate of nucleation J as a function of supersaturation.

$$J = \mathrm{d}N/\mathrm{d}t = A \exp(-\Delta G^\circ/KT). \quad (4)$$

Combining eqs. (1)–(4) gives

$$J = A \exp[-16\pi\sigma^3 v^2 / 3(KT)^3 (m \ln S)^2]. \quad (5)$$

The curve of J versus S shows negligible amounts of nucleation at low supersaturations and above a critical value of S a sudden very fast surge of nuclei.

The first experimental examinations of the theory [10] were successful enough. The attempts to calculate the value of the pre-exponent A by thermodynamical consideration were less satisfying [11]. The theory was scrutinized and remarkably large number of modifications were suggested and tested [12–15], to mention just a few, developing new techniques which could shorten the time between the nuclei formation and detection. It seems doubtful whether any theory, based on pure physical concepts, without taking into account the chemical parameters of the condensing molecules will have a reliable predictive power of general validity.

2.2. Homogeneous nucleation in solutions

The set of eqs. (1)–(5) may be used to describe the energetics of crystal nucleation from solution. The main and most important difference is that, instead of using the surface tension value σ, one has to use the function of the interfacial tension between the crystal face and the mother liquor σ_{CL}. This difference is not trivial and has a deep physical significance. There are no experimental techniques to measure the interfacial tension between a growing crystal and its mother liquor. The nucleus is supposedly not spherical and the polyhedral edges may have edge energy quite different from that of flat faces, and this contribution may not be negligible. The notion that σ_{CL} may be size dependent is even stronger in the case of polyhedral nuclei than in the case of spherical droplets, for which it had been expressed.

The uncertainty with respect to solid nuclei extends beyond its shape, because it includes the degree of desolvation. Molecular or ionic attraction forces at the very early stages of nucleus formation must be sufficient to overcome the bonding forces in the solvation spheres. It remains quite uncertain whether the differences in σ_{CL} between mature crystals and embryonic clusters can be disregarded.

There exists another fundamental difference between nucleation in vapor and in solution. The vapor phase can be effectively purified, however one cannot completely purify either the solvent or the solute used for nucleation experiments to sufficient degree.

In spite of careful and meticulous purifying procedures [16], homogeneous nucleation was not easily attained in the investigated systems. It was found [17] that in order to obtain reproducible precipitation of a sparingly soluble salt, the reactants must be mixed rapidly. In spite of numerous studies, for a long period, there was no accepted criterion for the occurrence of a homogeneous nucleation in solution.

In 1961, Nielsen reported a series of results in precipitation of $BaSO_4$ [18]. The initial concentrations ranged from 10^{-4} to 4×10^{-2} M. The experiments were performed by mixing equivalent amounts of solutions of $BaCl_2$ and Na_2SO_4. The induction period t_i was measured as the time between the mixing and the appearance of visible turbidity. At the highest concentrations the solutions were mixed in a Roughton T-tube and the induction period calculated from the rate of flow and the distance between the place where the solutions met and the site of just visible turbidity. In the end of each experiment the number of crystals formed per unit volume was determined by counting under an optical microscope.

The results are summarized in fig. 4. The upper curve shows that the number of crystals is independent of the initial concentration when the latter is less than 10^{-2} M. Then, in a very steep function, t_i jumps from about 10^5 to above 10^{10} within a very

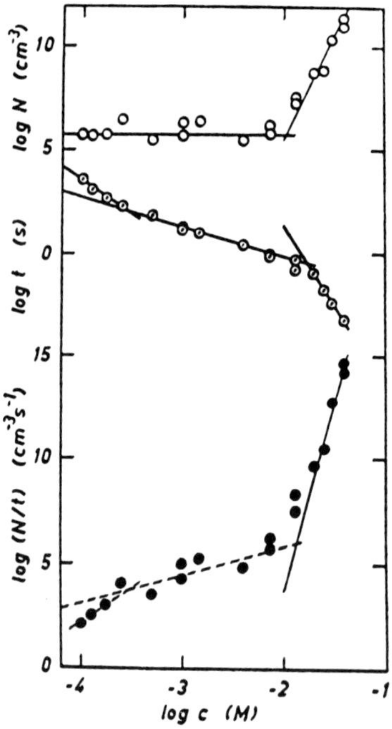

Fig. 4. Study of $BaSO_4$ precipitation: numbers of crystals, N, per unit volume, induction periods, t_i, and the number of crystals per unit volume divided by the induction period – as function of $BaSO_4$ concentration. Note the change of slope at 2×10^{-2} M concentration which corresponds to supersaturation of 2000 [18].

narrow range of concentrations. The second curve of induction period versus concentration consists of three slopes. The pronounced change of slope coincides with the change of slope in the upper curve, namely at the concentration of 10^{-2} M. The bottom curve was calculated as the number of crystals formed per unit volume, divided by the induction period.

The functions of induction periods, numbers of crystals and nucleation rates have a common sharp inflection point at log $c = -2$ i.e. for 0.01 M solution. This determination of the critical supersaturation, at which the homogeneous nucleation starts, allowed to calculate the interfacial tension in the case of $BaSO_4$ crystals. This value, $90 \pm 10\,\mathrm{erg/cm^2}$ is within the range of σ_{CL} for $SrSO_4$ namely $84 \pm 8\,\mathrm{erg/cm^2}$ derived by different reasoning based on solubility data [19].

Later on, the technique of observing the turbidity onset and measuring the time by taking motion pictures of the outlet tube, was improved [20]. This "stop flow" technique was used in many subsequent studies.

When a precipitate is detected, the nuclei had already grown to appropriate sizes. Therefore for measuring the induction periods of crystallite appearance one has to take into account the kinetics of growth of nuclei into detectable particles. Nielsen [21] solved a combined set of nucleation equations and precipitation rates by diffusion and surface controlled mechanisms.

For diffusion controlled growth, the number of crystals formed, N, is

$$N = 10^{22} \exp(-3\Delta G^*/5kT) \tag{6}$$

and the equation representing the induction period is

$$t_i = 10^{-10} \exp(2\Delta G^*/5kT) \tag{7}$$

Plots of $\log t_i$ versus $(\log S)^{-2}$ yield straight lines intersecting the y-axis at -10. If in a system such line or part of it has this predicted slope, it may be accepted as a proof that homogeneous nucleation occurred in this system i.e. that one may use the set of equations and calculate the σ_{CL} values for the salt in question. And indeed, though the equations of homogeneous nucleation could not be used for solving the parameters of nucleation energetics, because the values of σ_{CL} were not known and not measurable, they were used in a reversed way to estimate the values of interfacial tension for numerous salts.

These results were correlated by Sohnel [22] with solubility of the respective materials. Figure 5 shows that there exists, in a broad sense, a correlation between the

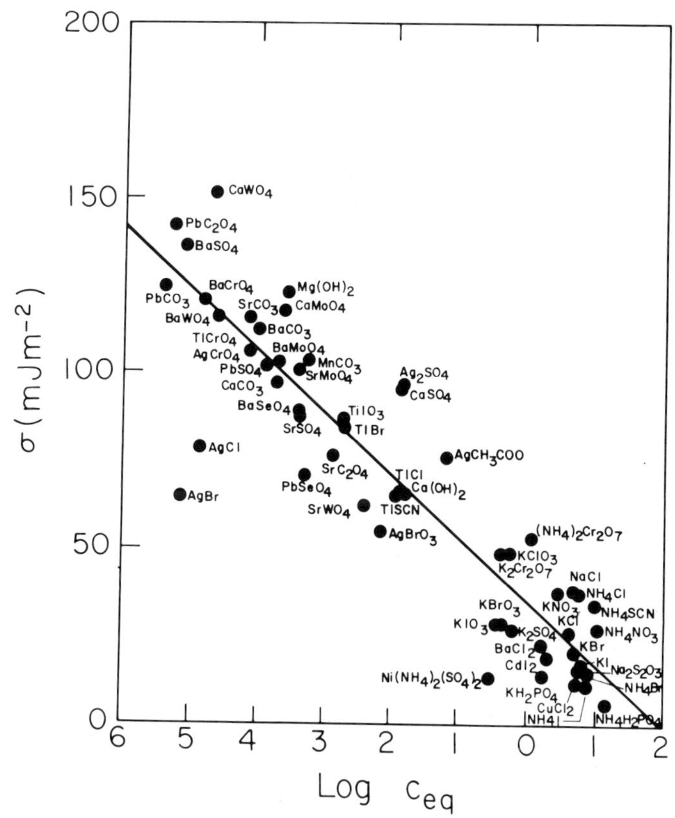

Fig. 5. Correlation between solubility of various salts and the interfacial tension of their nuclei and the mother liquor in which they are formed [22].

surface tension and solubility of many diverse salts. The sparingly soluble salts have high surface tension while the soluble and very soluble salts have low surface tensions. The few exceptions, like silver salts may be explained by their colloidal nature.

It is interesting to point out the profound difference between the nature of the investigations of homogeneous nucleation in the vapor and in solution. The research studies in the vapor developed the basic Classical Theory of Becker–Doring by including thermodynamic considerations, partition functions, equilibrium, steady state and kinetic approaches. Each investigation envisions the nucleation process from a special viewpoint: stressing the importance of the rotational and translational partition functions, the size distribution of clusters, considering the heat released in cluster formation or calculating the bond energy from first principles. There are no such attempts in the theory of homogeneous nucleation in solution. Actually most research studies are committed to the determination of interfacial tensions in specific systems. Therefore the correlation study between interfacial tension and saturation solubility [22] may be seen as a summit of this research effort.

In the light of the latest research results in the field of water structure and of the probable structure of solutions containing ions, it seems reasonable that the very low interfacial tensions of very soluble salts is due to the similarity between the structures of the forming nuclei and the structure of the supersaturated mother liquor. In saturated ≈ 5 M solution only 11 molecules are available to solvate the ions, in many cases several molecules entering the forming crystal. There is hardly a surplus of water molecules to form the structure of free water. The nucleus separates itself from a matrix which has similar characteristics to its own structure. It is expected that the interfacial tension between similar structures would be low.

On the other hand, when the equilibrium concentration of a sparingly soluble salt is $\approx 10^{-5}$ M and the supersaturation for homogeneous nucleation is $\approx 10^{-2}$ M the ratio of 55 mol H_2O/ 0.01 mol salt in one liter, each nucleus of ≈ 50 ion pairs is surrounded by $\approx 275\,000$ molecules of water. It means that the nuclei must form a surface which differentiates between them and water and which must have high energy.

2.3. *Heterogeneous nucleation in solutions*

The equations related to the energetics of heterogeneous nucleation can be derived, in principle, following the same pattern as was used for the process of homogeneous nucleation (eq. (1)), using for a model a cubic nucleus. The energy balance for the formation of such a nucleus having σ_{CL} interfacial tension between the nucleus and the mother liquor in homogeneous nucleation will be

$$\Delta G^0_{Hom} = -l^3\Delta G_v + 6l^2\sigma_{CL}. \tag{8}$$

From this equation, the energy barrier for nucleation, in analogy to eq. (2) will be

$$\Delta G^*_{Hom} = 32\sigma^3_{CL}/\Delta G^2_v. \tag{9}$$

Let us assume that a preformed particle of volume $\frac{1}{2}l^3$ exists in the solution and that a nucleus of the precipitating phase forms on it, completing it to a cube. The energy released by the formation of bonds will be only half of that specified by eq. (8), but the interface with the liquor will be halved too; instead a new interface, l^2, between the crystal and substrate is formed having the interfacial tension of σ_{CS}. This will constitute heterogeneous nucleation and the energy balance will be

$$\Delta G^0_{Het} = -l^3\Delta G_v/2 + 3l^2\sigma_{CL} + l^2\sigma_{CS}. \tag{10}$$

The energy barrier calculated for the maximum value of ΔG^*

$$\Delta G^*_{Het} = 16(3\sigma_{CL} + \sigma_{CS})^3/27\Delta G_v^2. \tag{11}$$

If the interfacial tension σ_{CS} is zero, as for a nucleus growing on its own kind of crystal, then

$$\Delta G^*_{Het} = \tfrac{1}{2}\Delta G^*_{Hom}.$$

The smaller the interfacial tension between the substrate and the forming nucleus, the lower will be the energy barrier for the nucleus formation. For instance if σ_{CS} is half σ_{CL} the energy barrier will be $25.4\sigma^3_{CL}/\Delta G_v^2$, still lower than that needed for homogeneous nucleation.

The interfacial tension between the nucleus and the substrate depends upon the registry between them, i.e. depends upon the degree of structural fit.

This seemingly simple concept developed slowly from many studies. Packter et al. [23] found that the number of crystallites in a series of precipitated salts (N_{Het}) was ten to about hundred times larger in glass beakers than it was in propylene beakers. N_{Het} increased only slightly with increased salt concentrations and supersaturation. It was concluded that heterogeneous nucleation probably occurred onto microcrystalline particles of some siliceous mineral dispersed in the solution.

House and Tutton [24] also found that pyrex seeds induced the nucleation of calcite from a metastable solution. Increase in the mass of seeds produced an increase in the number of crystals. The nucleation potential of the seed material was significantly reduced by alkali treatment – the number was reduced by three orders of magnitude for unchanged mass of seeds.

Studies of this kind showed that the surfaces of the substrates were important in lowering the energy barrier to nucleation, but they could not supply enough data for quantitative formulation and for predictions. For instance Sohnel [25] maintains that during ageing of suspensions formed by heterogeneous nucleation, the primary particles aggregate and subsequent ripening by recrystallization takes place.

In studies of heterogeneous nucleation on unspecified substrates, any slight deviation from the experimental set-up, even such seemingly trivial ones as a change in stirring rate can affect the nucleation rate.

When the substrates were well defined, it was found that it was necessary to have a substrate with known surface structure, a depositing phase of known solubility and crystal structure, a known epitaxial relationship between deposit and substrate and methods with which one could measure the supersaturation in the solution.

It was found that the degree of misfit affects the energy barrier to heterogeneous nucleation. Experimental data for the nucleation of KCl, KBr, KI, NH_4I, RbCl and RbBr onto cleaved mica substrate is a function of lattice registry and temperature. The mismatch of lattice between deposit and substrate was expressed [26] as: $1 - \delta = \frac{1}{2} \times$ (111 unit cell dimension of alkali halide/a_0 dimension of mica).

KCl which has the largest mismatch with mica, in this series, namely -14%, needs a critical supersaturation of 1.081 to start nucleation at 19.8°C, whereas NH_4I which has the lowest mismatch of -1%, at the same temperature, needs a supersaturation of only 1.002. The general trend in this series is: the better the structural match, the lower the supersaturation needed for nucleation i.e. that the energy barrier for nucleation is lowered by good structural fit between the nucleating crystal and the substrate.

A convincing demonstration of heterogeneous nucleation of one crystallographic species upon another was presented by Rinaudo, Abbona and Boistelle using the system of newberyite and uric acid dihydrate [27]. If a solution is highly supersaturated both with respect to newberyite ($MgHPO_4 \cdot 3H_2O$) and uric acid both hydrated phases may nucleate independently. From weakly supersaturated solutions, anhydrous uric acid precipitates first; upon its (100) faces heterogeneous nucleation of newberyite was observed. A second generation of anhydrous uric acid grows epitaxially on the (010) of newberyite crystals. Some preferential orientations were determined on the contact planes between the lattice rows which correspond to periodic bond chain directions for both phases. This study proves that heterogeneous nucleation has a relatively lower energy barrier to overcome (occurs at low supersaturation) than homogeneous nucleation (needs high supersaturation).

However, some important queries remain unanswered. First, what is the extent of the energy decrease in heterogeneous nucleation and second, is this decrease dependent on the size of the substrate, or, in other words, what would be the smallest fragment of the foreign phase which may induce efficient heterogeneous nucleation.

The second question was partly referred to in a work studying the effect of water soluble polymers, polyvinyl sulfonate (PVS) and polyglutamic acid (PGA) on the carbonates of Mg, Ca, Sr, Ba and Pb [28]. In this study the minimal labile supersaturation c/c^* was defined as the ratio between the concentration c which will induce visible precipitation in ≈ 1 min, and between the equilibrium concentration c^*. In the family of carbonates studied c/c^* was inversely proportional to the respective solubility product. The soluble polymers had retarding effect on the appearance of the turbidity. These retarding effects were graded into: slight, medium, good, very good and excellent. They were correlated with the dimensional matching between the intercationic distances in the respective carbonate lattices and the distances between acidic functional groups in the polymeric chains. The idea was simple – the better the fit between the positive charges on the crystal faces and the distance between the negatively charged functional groups on the polymers, the more effective would be the adsorption of the chains on the growing nuclei and hence the stronger the retarding effect. This correlation made clear why PVS had no effect on magnesium carbonate and calcite. Both had distances between the cations shorter than the distances in the polymer. On the other hand both polymers have excellent retarding effect on barium carbonate, the intercationic distances of which are in the range of the functional groups distances.

As frequently encountered in research this orderly picture of good correlations was marred by one gross inconsistency. The effect of PGA on basic lead carbonate

$2PbCO_3 \cdot Pb(OH)_2$ was unexpected: instead of showing retardation, the system showed speeding up of precipitation. In basic lead carbonate the intercationic distances range from 3.209 to 7.514 Å; the distances on the polypeptide chain of PGA are 7.5–8.5 Å. Whereas retardation of precipitation could be explained by adsorption on growing faces interrupting growth, it can hardly explain acceleration of turbidity formation.

The formation of carbonates occurs at the supersaturation of 7, 50, 50 and 53 for Mg, Ca, Sr and Ba respectively. It may be deduced from the experiments of Nielsen [18] that the nucleation in these systems is heterogeneous. Basic lead carbonate, without additives nucleates at supersaturation of 1250, which may suggest homogeneous or partly homogeneous nucleation.

Because of the unusual effect of PGA on basic lead carbonate, the precipitates from the pure system and those with the additive were analyzed by three different methods: X-ray diffraction, scanning electron microscope and thermal analysis. All three showed that the precipitate from the system containing PGA was nearly amorphous, whereas the pure basic lead carbonate was well crystallized. The amorphous character and the accelerated formation indicated possible heterogeneous nucleation on the polymeric chains, the number of which exceeded the number of nuclei in usual homogeneous nucleation thus causing formation of an enormous number of very small crystallites.

Another study [29] , using PVS and a sparingly soluble salt having a simple crystallographic cell, showed that the assumption drawn from the former study was feasible. This study attempted to focus the perception of heterogeneous nucleation onto micro-substrates by means of size measurements using light beating spectroscopy apparatus equipped with a specially designed multifilter spectrum analyzer. The concentrations of $SrSO_4$ and of PVS were adjusted to fulfill essential requirements which are almost contradictory: the particles of $SrSO_4$ had to be large enough to be detectable by the available equipment, but small enough to remain suspended in solution for a period sufficient to carry out the measurements. The range of concentrations was rather narrow, allowing 18–19 fold supersaturation ratios, corresponding to 0.0156–0.0168 M $SrSO_4$ solutions. The employed concentrations of PVS were 5.8×10^{-7} M and 1.17×10^{-6} M solutions corresponding to 10 and 20 ppm respectively.

The striking result was that the $SrSO_4$ particles in the 10 ppm PVS solution were about 1.3 times larger than those in the 20 ppm solution. This means that the ratio of the volumes and consequently of the weights of the particles is 2.2, close to the inverse ratio of the polymer concentrations. It means that by raising the polymer concentration by a factor of two, the number of the particles was doubled from 3×10^{14} to 6×10^{14} per ml. At a constant supersaturation i.e. the same solute supply, the particle weights should be halved being distributed on a doubled number of substrates.

The systematic studies of heterogeneous nucleation were carried out on substrates with well developed, close to perfect, faces. However, no crystal surface can be perfect on the atomic scale, and photomicrographs of crystals grown on substrates show that only certain sites on the surface are used for nucleation. Nucleation would probably also occur if those sites were separated from the bulk of the solid and isolated from each other, provided they retained their special structural properties. It follows that very small, active fragments of matter can be prepared which will possess the required degree of matching with the crystal lattice, on which heterogeneous nucleation can take place. A polymer, several tens of thousands daltons in weight, with hundreds of functional groups can serve as such a microsubstrate.

When polymeric substrates in the ppm range, are introduced to a supersaturated solution of a sparingly soluble salt, the ordinary nucleation mechanism may be overcome. The ordinary unspecified hetero-centers, estimated as 10^6 to 10^9 per ml are negligible with respect to the overwhelming number of 10^{14} macromolecules per ml contained in an ordinary solution of a soluble 10 000 to 20 000 dalton species. Even the massive homogeneous nucleation with estimated 10^{12} nuclei per ml seems to be negligible.

PVS polymeric molecules are especially suitable to act as microsubstrates for $SrSO_4$. The polymer has negatively charged functional groups 0.4–0.5 nm apart on the free flexing chain. The intercationic distances in $SrSO_4$ are 0.432 nm and this makes the two compounds an almost perfect dimensional match.

In another study, the effect of two naturally occurring polysaccharidic electrolytes, carragheenan and sodium alginate on the crystallization of sodium chloride was studied. In this study [30] a different interesting idea concerning heterogeneous nucleation was presented. It was assumed that normally, in pure NaCl solutions crystals nucleate heterogeneously on motes present in the solution. As the polysaccharides interfered with the nucleation process, it was suggested that the polyelectrolyte molecules adsorbed onto the motes rendering them inactive.

It is very difficult to isolate the stage of nucleation from the overall process of crystal formation, and therefore the studies concerning heterogeneous nucleation do not seem straightforward allowing one to draw clear cut conclusions. In forthcoming sections describing the overal effects of impurities on the crystallization processes some ideas concerning nucleation will be considered again.

3. *Crystal growth*

3.1. *The various approaches to theories of crystal growth*

The process of crystal growth is being studied by three very different approaches: First, the thermodynamic approach, considering the free energies of the growing faces [31]. Second, the various mechanisms governing the incorporation of the building elements into the growing crystal. This is the most widely researched and the most fertile field of research extending from two-dimensional nucleation on a growing surface [32], through the brilliant theory of Burton, Cabrera and Frank [33] envisaging the self perpetuating dislocation with the ensuing spirals, their curvatures, step velocity, step bunching [34] and the computer simulations of Monte Carlo principles [35]. Third, the approach which took into account the phenomena occurring in the supersaturated solution which surrounds the growing crystal. This approach uses in a very general way the principles embodied in the theories of surface mechanisms, in cases when the growth is controlled by surface reaction, but mostly deals with kinetics of growth when it is diffusion controlled.

The thermodynamic approach had a renaissance in the theory of Hartman and Perdok [36]. As happens with some attractive theories which supply a wealth of mathematical detail and hence giving the feeling of almost infinite potential, it was difficult to discern a weakness in the theory. But the weakness existed and it was the disregard of the role of the solution in the process of growth and the disregard of the solvent–solute interactions.

The thermodynamical and mechanical statistical approaches are presented in detail in the first volume of this Handbook. This section deals primarily with the aspects of crystal growth from aqueous solutions.

In 1974 Bennema [37] published a very thorough critique entitled *Crystal Growth from Solution – Theory and Experiment*. He concluded that a unified growth theory had not been developed. Instead, a set of complementary crystal growth theories dealt each with a certain aspect of the crystal growth process. Specific growth theories from solution were not developed.

Three major theories were reviewed and confronted with experimental results.

(1) *Hartman – Perdok theory*. This theory had a predictive power with respect to crystal habit, when the crystal is grown either from vapor or from pure melt. If the crystal is grown from solution, the predictive power may be preserved, provided there are no specific interactions between the solvent and the crystallizing salt and that there are no impurities.

In retrospect, the restriction concerning lack of interaction with the solvent and requirement for the absence of impurities excludes the majority of aqueous systems from feasibility of being analyzed by the H–P theory.

(2) *Statistical mechanical surface models*. Crystal–fluid models are block models in which the whole interface is divided into blocks of the same shape and size and each block is either solid or fluid. In these models energy and configurational entropy terms are involved. The rigorous simplification of the interface to a system of solid in contact with a fluid continuum, raises problems when confronted with a complicated real solid–solution interface.

Though nearly twenty years have passed since this criticism had been formulated, no studies utilizing the statistical mechanical approach have been reported with respect to crystal growth from aqueous solutions. Until the expected breakthrough, this approach does not seem to be fruitful.

(3) *Burton–Cabrera–Frank theory*. The most direct way to check the theory is to look for mono-atomic Frank spirals. Such spirals have been observed on mineral crystals grown under hydrothermal conditions and for crystals of alcohols and paraffins. For growth from solutions at moderate temperatures as far as was known to Bennema [37], and as far as is known up to 1991, no such spirals have been reported, though rotating macro-spirals have been observed.

On the other hand, the confrontation of calculated BCF curves with observed ones yields good agreement [38].

3.2. Electrolytic growth kinetics

It is difficult to ascribe growth mechanisms to very soluble electrolytes because they grow from aqueous solutions by a linear rate law – the rate being proportional to the difference between the concentration and the solubility. The rate is controlled by the transport of the ions in the bulk solution and by the adsorption process. On the other hand, sparingly soluble electrolytes mostly grow by parabolic or exponential rate laws which can be accounted for in terms of the classical theories of surface nucleation, screw dislocation and surface spirals.

In order to obtain reliable data concerning growth rates, the investigations should include only the stage of ion transport to growing surfaces, excluding nucleation. Also chemical phenomena had to be taken into consideration. This objective could be attained by using well defined "seed populations" in the investigated systems and evaluating their growth at given supersaturations. Numerous studies of this kind were carried out, a distinguished part of them in Buffalo State University of New York in the group headed by Nancollas.

An example is presented by a study of crystallization of dicalcium phosphate dihydrate (DCPD) [39]. A supersaturated solution was inoculated with seed crystals of the salt to ensure follow up of the growth process alone. The rate of growth was followed by the change of concentration of the calcium, phosphate and hydrogen ions. It was found that, after an initial surge, the rate of growth could be described by a 2nd order equation with respect to concentrations of $[Ca^{2+}][HPO_4^{2-}]$ over a wide range of calcium/phosphate ratios, suggesting a predominantly surface reaction controlled process. The experiments were carried out at 15, 25 and 37°C to enable the computation of the activation energy, which in this case was 10.3 kcal per mol.

Table 1 is a list of studies of similar type performed on systems of sparingly soluble salts. It shows the order of reaction, the suggested mechanism of the reaction and the reference number. This impressive list of studies illustrates the wide range of chemical systems which may serve as models for the determination of crystal growth kinetics and mechanisms.

It is interesting to note that in some of the studies the growth obeys the second order equation with respect to ion concentration whereas in others with respect to supersaturation. The studies were not compared or confronted with each other and therefore it is difficult to see why, for instance, the growth of calcium oxalate monohydrate [45] is

Table 1
Kinetic studies of rates and mechanisms in crystallization of sparingly soluble salts (Nancollas).

Substance	Order of reaction	Mechanism	Reference
Ca phosphate	2nd	Surface reaction	[39]
Ca sulfate	2nd	No. of growth sites	[40]
Hydroxyapatite		Precursor phase	[41]
Pb sulfate	> 2nd	Surface nucleation	[42]
Ca carbonate	2nd	Surface reaction	[43]
Mg hydroxide	1st	Diffusion	[44]
CaOx monohydrate	2nd	Surface reaction	[45]
Ba oxalate	2nd	Surface controlled	[46]
CaOx trihydrate	1st	Diffusion	[46]
Ba sulfate	2nd	Surface controlled	[47]
Ca fluoride		Surface reaction	[48]
Mg fluoride		Surface reaction	[48]

surface reaction controlled whereas of calcium oxalate trihydrate is diffusion controlled [46], though chemical reasons must exist. It seems that phenomena perceived from the chemical viewpoint are more complex than when viewed from the side of the mathematical-physical abstractions. However, the rules of crystal growth from solutions have to be derived from real systems because they are expected to serve for understanding processes and designing operations.

Different salts have different solubilities and therefore can support differing supersaturations. Their ions have different spheres of hydration and hence different frequencies of dehydration. The crystal lattices have widely diverse energies of formation and their velocity of growth may vary over several orders of magnitude.

There was an aspect in the technique of "seeded growth" which aroused some criticism. When growth rates are determined in a seeded supersaturated solution, kept in a metastable state, very slight but nevertheless measurable decrease in the concentration of the precipitating ions occurs due to depletion, which affects the supersaturation and hence the growth rate. This flaw was amended by the introduction of the "Constant Composition Method" [49, 50]. If the concentration of the relevant ions can be measured with a high degree of precision, and the depletion automatically compensated, the growth rates will be determined under constant composition conditions. In one of the first studies of this type [49], a metastable supersaturated solution was prepared by mixing solutions of $CaCl_2$ and K_2HPO_4 and bringing it to the desired pH by addition of dilute KOH. Precipitation started immediately upon addition of seed material such as calcium phosphate crystallites, dentin or tooth enamel. Any pH change triggered the automatic release of titrants, thus keeping the solution composition constant. Additions of calcium, phosphate and base were made from mechanically coupled burets.

The advantages of the Constant Composition Method are the possibility to attain higher precision and to follow growth processes for prolonged periods at very low but steady supersaturation. Thus, for instance, in calcium phosphate solutions saturated only with respect to the most thermodynamically stable hydroxyapatite (HAP), the phase could be grown on HAP seeds without the complications involved in the formation of a precursor phase [50]. Even more sophisticated application of this method was to examine the kinetics of growth of HAP and calcium oxalate on the seeds of crystallographically compatible uric acid [51].

This method was used to determine the overall growth rates of varying amounts of $CaSO_4 \cdot 2H_2O$ crystals. The analysis of the results showed that the rate of growth in aqueous suspension was screw dislocation controlled in a supersaturation range $1.03 < c/c^* < 1.15$ [52]. On one hand, the result shows the great sensitivity of the method, but on the other it exposes the inherent difficulties in formulating theories of general validity concerning crystallization of chemical materials in aqueous solutions. One may wonder what happens at supersaturations higher than 1.15. The growth mechanisms that could be identified included a combination of normal spiral growth, growth through cooperating spirals with non-parallel Burger's vectors and growth of grain boundary spirals together with partial outgrowths of concave parts of crystals. These could explain the different rates of growth found for different preparations of crystals.

3.3. Growth rate isograms and isotachs

Nielsen and Toft [53] present data for a number of electrolytes expressed by means of empirical kinetic formulas and rate constants which may be compared with growth models. The systems included solutions in which indifferent ions might also be present.

A binary electrolyte $A_\alpha B_\beta$ dissolved in pure water would yield strictly equivalent ratio of ions [A] and [B]. However, their concentrations can be varied independently as for instance $[Ba^{2+}]$ and $[SO_4^{2-}]$ in solutions of $BaCl_2$ and Na_2SO_4. Systems in which non-equivalent concentrations of the salt forming ions are present are quite common in industry and biology, but one must admit that this particular configuration has not been considered in any physical models of crystal growth.

The effects of any set of values of the two independent variables [A] and [B] may be plotted in a pA, pB diagram. The contour lines (isograms) for constant values of the growth rate are called isotachs [53]. The isotach for zero growth rate is the solubility curve, separating the areas of supersaturation and undersaturation. For a binary electrolyte $A_\alpha B_\beta$

$$a_A^\alpha a_B^\beta = K_{SP} \tag{12}$$

where

a_A, a_B = activities of A and B resp.
α, β = stoichiometric coefficients,
K_{SP} = solubility product.

Hence

$$\alpha \mathrm{p}A + \beta \mathrm{p}B = \mathrm{p}K_{SP}, \tag{13}$$

and

$$\mathrm{dp}B/\mathrm{dp}A = -\alpha/\beta, \tag{14}$$

i.e. that the solubility curve in the pA, pB diagram is a straight line with a slope $-\alpha/\beta$. However in cases in which ion pairing or complex formation occur this line will be a bent curve. Isotachs for diffusion controlled growth rates are always bent curves. At high, non-equivalent concentrations the rate depends strongly on the concentration of the deficient component.

Electrolytes growing by a surface controlled rate usually have approximately linear isotachs with slopes close to $-\alpha/\beta$. In those cases the growth rate, v_g, is a function of the ionic product:

$$v_g = f([A]^\alpha [B]^\beta). \tag{15}$$

The rate is controlled by the integration jump of cations into the kinks from an adsorption layer which is in equilibrium with the solution, and which is electroneutral. It follows that $\alpha[A]_{ad} = \beta[B]_{ad}$ in the adsorption layer even when those concentrations are non-equivalent in the bulk.

The electrolytes investigated by the "isotach approach" yielded the following rate laws. Diffusion control: AgCl and $BaSO_4$ at relatively high concentration typically when $dr/dt > 10\,nm/s$. Parabolic rate law: AgCl, $Ag_2C_2O_4$, $BaSO_4$, $CaCO_3$ and $CaC_2O_4 \cdot H_2O$ at moderate supersaturation where $dr/dt < 10\,nm/s$ and CaF_2 at rather small supersaturation where $dr/dt < 0.02\,nm/s$. Exponential rate law: CaF_2 and SrF_2.

Only several of the investigated materials in the studies of Nancollas and co-workers [39–50] are included in the study of Nielsen and Toft [53] and vice versa, therefore the interesting comparison is limited to few materials. The comparison is interesting because the experimental techniques are basically different. Whereas Nancollas used seeded solutions, avoiding the complication of nucleation process, Nielsen used the technique of mixing solutions and checking the particles populations for a defined "good behaviour": they should not aggregate, split or show nucleation events. Experiments which did not comply with these assumptions were discarded. In spite of the very different experimental approach the comparison is favorable for calcium carbonate, calcium oxalate monohydrate and partially for barium sulphate with the parabolic rate law. This agreement is highly encouraging emphasizing that though the rate laws are dependent on experimental conditions, they are nevertheless mainly dependent on the chemical character of the growing crystals.

3.4. *Electrolytic crystal growth mechanisms*

The mechanisms of growth in solution are expressed in terms of supersaturation, the mathematical expression of which can take various forms derived from the two basic quantities: c is the concentration of the crystallizing solute and c^* the equilibrium concentration or solubility. For soluble salts the supersaturation is expressed as the excess solute, namely, $c - c^*$ is the supersaturation, or the driving force. For sparingly soluble salts, for which large supersaturations can be built up $c/c^* = S$ is the supersaturation ratio. $S - 1 = \sigma = (c - c^*)/c^*$ is the relative supersaturation used in many equations. $R = dr/dt$ is the linear growth rate.

A widely accepted classification of growth mechanisms is based on the law rate when they are, one at a time, fully controlling the growth rate. The classes were set forth according to the rate determining mechanisms:

(1) Diffusion or transport followed by adsorption at the crystal face. The rate expression is

$$v_g = k_1(S - 1) = k_1\sigma \tag{16}$$

i.e. a linear function.

(2) Surface reaction, such as spiral growth with rate expression

$$v_g = k_2(S - 1)^2 = k_2\sigma^2 \tag{17}$$

i.e. a parabolic function.

(3) Surface nucleation (especially polynuclear) which is expressed by an exponential function.

The diffusion controlled rate may be safely estimated within half an order of magnitude when the concentrations and the particle dimensions are known [54]. By comparing the calculated diffusion controlled rate and the experimentally measured one it is then possible to determine whether the particular process is either diffusion or surface controlled.

In many practical cases two or even more elementary mechanisms are simultaneously affecting the growth rate. When their influences drive the growth in the same direction, such as spirals and surface nucleation, they are additive. When they are inherently different but simultaneous, the faster one may be rate determining. However, if they are consecutive like for instance bulk diffusion and surface spiral growth, the slower one will be rate determining [54].

When mechanisms change following changes in supersaturation no abrupt switches can be discerned, the transition being rather gradual.

The most important parameter in the model of two-dimensional surface nucleation is the edge energy σ_{ed} (designated by many investigators as τ). This quantity cannot be measured by independent methods on macroscopic crystals. Like interfacial energy in homogeneous nucleation it can be derived from data on induction periods (eq.(7)) and from data on two-dimensional surface nucleation controlled growth such as polynuclear growth. Nielsen and Sohnel [55] collected data on this parameter expressed as surface tension ($\sigma = \tau/a^2$) and found a good correlation between σ and the solubility of the electrolytes. The few deviations from the main trend, that were encountered, could be explained by chemical reasoning. The general correction may be used for prediction of τ values still not measured and for "smoothing" out empirical values [54]

$$\tau/kT = 2.82 - 0.272\ln(c^*/\text{mol}\,\text{m}^{-3}) \tag{18}$$

Nielsen also [56] tried to estimate the adsorption coefficient K_{ad} from the ion pair association constants. If the electrolyte consists of A and B ions, an adsorbed A ion will be attracted to a B ion on the crystal surface with a strength that may be correlated with the attraction between the two ions if they were forming an ion pair in the solution. The orders of magnitude of K_{ad} for various types of ionic electrolytes are as follows; If for 1:1 type electrolyte K_{ad} is given the value of 10 then for 1:2 or 2:1 type it is 30 and for 2:2 type it is 200.

3.5. *Some chemical factors in crystal growth mechanisms*

Crystal growth theories are, as a rule, presented from the viewpoint of physical phenomena. The investigated parameters are bulk diffusion, surface migration, dislocations, flow regime and, because of the wealth of theories and data on growth from vapor and melt, the surface structure. The molecular processes and chemical changes are hardly in the focus of attention.

Most of the ions in aqueous solutions, especially cations, have a coordination sphere of water molecules. On entering into a crystal structure the ions release some, or all, of their hydration water. It was suggested that the activation energy of adsorption and of integration depends strongly on the energy needed for the dissociation of a water molecule from the hydrated ion [57]. These energies relate to the corresponding dehydration frequencies by the Eyring equation

$$\nu_{dh} = \nu_0 \exp(\Delta G^{\nu}_{dh}/kT), \tag{19}$$

where

$\nu_0 = kT/h$.

ΔG^{ν}_{dh} = activation Gibbs energy of dehydration for one water molecule from the inner coordination sphere.

Dehydration frequencies have been reported for a great number of important cations. Nielsen [54] presented them on the basis of Petrucci and Burgess data; the numbers are $\log(\nu_{dh}/s)$. For monovalent cations the values range from 8.2 to 10, for divalent from 3.5 for Be^{2+} to 9.2 for Ba^{2+}. Transition metal ions and rare earths also fall within these ranges. The frequencies were measured by kinetic reaction methods such as pressure jump, temperature jump, ultrasonic absorption and NMR spectrometry. The dehydration frequencies for anions are too large to be measured by known methods.

Various methods for determining the hydration numbers gave estimated values of 4–6 water molecules for cations whereas only 0–2 were found for most anions. The numbers are not exact or absolute because they depend upon the method of measurement and upon the determination of the nature of the hydration sphere but they express the general trend i.e. that the water–cation bonds are stronger than those of water–anion. Nielsen maintains that both kinds of bonds are essentially electrostatic [54]; water molecules are very polar, so this statement may be true especially when taking into account the polarizing properties of the cations. However, the dominant character of water–anion association is the hydrogen bond, whereas the bonds with transition metal cations are coordinative. The bonding strength of alkali metal and alkaline earth ions depends upon their charge density i.e. upon their polarizing effect on the water molecules.

The spread of dehydration frequency values for various cations with close dimensions and the same charge is due to electronic structures. Let us look for instance at three monovalent cations; the order of increasing radii is $Na^+(0.102\,nm) < Ag^+(0.115\,nm) < K^+(0.138\,nm)$. If the size were the only determining factor of attraction, one could expect potassium salts would be less likely to contain water of crystallization than silver salts. Actually the numbers of water molecules in salts containing water of crystallization for those cations decreases in the order: Na > K > Ag.

For hydrated ions entering the crystal lattice with all or most of their hydration water the ν_{dh} is close to $10^{10}\,s^{-1}$ which means that there is practically no effect of the dehydration step on the growth rate. However, in the same sense it means that if a naturally hydrated cation in the solution is nearly or completely dehydrated on entering into the lattice, it will affect growth.

Calcium is strongly hydrated in solution. Calcium phosphates are the most important deposits in biomineralization; some of them, dicalcium phosphate dihydrate and octacalcium phosphate are hydrated salts whereas hydroxyapatite does not contain water of crystallization. Possible implications of the dehydration frequency of calcium cation on the formation of calcium phosphates in biological systems may be important in biomineralization phenomena.

3.6. Rate laws and rate constants in crystal growth

In 1987 Nielsen confronted rate constants and rate laws of crystal growth, calculated for processes in solutions with experimental measurements. He did not consider the prestigeous mathematical physical models developed for growth from vapor and growth in melt, and their subsequent troublesome adjustment to solution processes, but started with realistic rate determining steps actually occurring in solutions [58]. The following steps were considered:

(1) Transport by convection in the solution.
(2) Transport by diffusion through the solution.
(3) Adsorption at the crystal/solution interface.
(4) Migration within the adsorption layer.
(5) Adsorption at a step (ledge) in the crystal surface.
(6) Migration along the step.
(7) Integration at a kink in a step.
(8) Partial or total dehydration of ions.

The symbols used in the equations are in accordance with previous designations. In order to simplify the equations activity coefficients are not included. Corresponding to elementary mechanisms (1)–(8) the rate laws will be:

(1) Convection $R = k_C(S-1) = k_C\sigma$.
(2) Diffusion $R = k_D(S-1) = k_D\sigma$.
(3) Adsorption $R = k_I(S-1) = k_I\sigma$.
(4) Two-dimensional diffusion; R depends on surface geometry.
(5) Adsorption at a step: $R = k' \times$ (step density) $(S-1)$.
(6) One-dimensional diffusion: R depends on step geometry.
(7) Integration: $R = k'' \times$ (kink density) $(S-1)$.
(8) Dehydration: R depends on surface geometry.

Linear rate laws are probably rate controlled by adsorption processes that need activation energy. The coefficient k_I can be calculated as

$$k_I = R/\sigma = DV_m c^*/\delta, \tag{20}$$

where

D = diffusion coefficient,
δ = diffusion layer thickness of about 100 μm.

Parabolic rate laws are assumed to be caused by integration rate control. The rate constant $k_2 = R/\sigma^2$ can be expressed in terms of parameters that may be calculated or estimated independently of growth rate experiments:

$$k_2 = 10^{-4} a v_{dh} K_{ad} V_m c^*/(\tau/kT)\exp(\tau/kT). \tag{21}$$

Exponential rate laws are attributed to surface nucleation growth. The constant k_c will be

$$k_c = 0.002 a v_{dh}(K_{ad} c^* V_m)^{4/3}\exp(-\tau/kT). \tag{22}$$

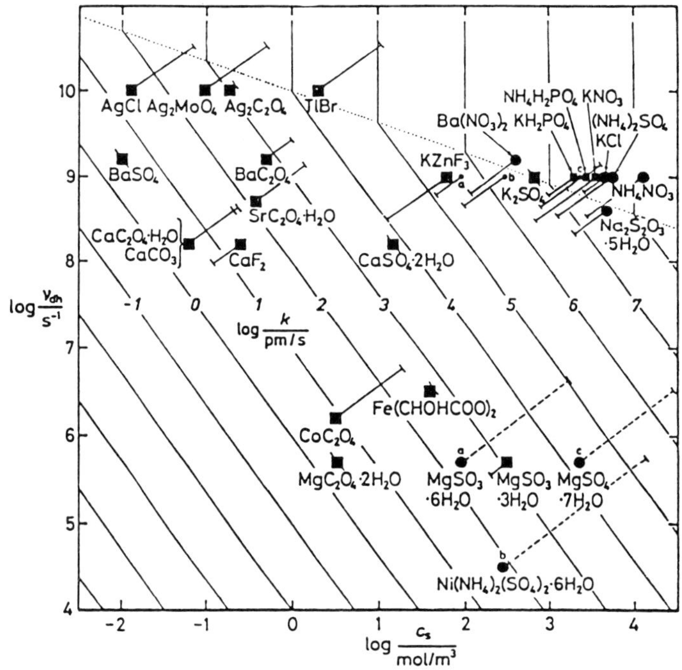

Fig. 6. Generalized crystal growth rate diagram: dehydration frequency ν_{dh}, versus solubility, c_s. The vertical and sloping lines are isograms [58].

The theory comprises several parameters that can only be measured with considerable uncertainty, but which may be estimated from general correlations. Nielsen maintains that the best way to compare theoretical and experimental growth rates may be to construct a generalized rate diagram such as fig. 6 where the theoretical rates are plotted in form of isograms as a function of experimental parameters: solubility and dehydration frequency. In fig. 6 the vertical lines indicate the theoretical values of k for linear rate law and the sloping lines the parabolic rate law. For electrolytes for which the growth rates measured those values are indicated by black circles for linear and black squares for parabolic rate laws. The lengths of the bars indicate the deviations. In many cases the deviation is of an order of magnitude, though some are more extended.

3.7. *Unconventional approaches to growth mechanisms*

All the theories presented, in spite of the significant differences, favor a common axiom: the molecules or the atoms or the ions arrive to the growth sites as individual moieties, one by one. The supersaturated solution is a reservoir of separate species. If it has any structure, it is inconsequent in relation to growth mechanisms.

Glasner [59] suggested that the supersaturated solution is a solution of "block nuclei", too small to be detected by available physical methods. If such preformed block nuclei exist, growth may be envisioned as incorporation of the blocks in the growing crystal. As experimental evidence, he brought forward heat effects accompanying the crystallization of alkali halides, especially in the presence lead complexes [60]. This approach had been presented as an explanation to observed growth rates of KCl in the presence of Pb^{2+} ions [61, 62] but was not accepted in the mainstream of growth theories. Structures of three-dimensional mosaics, were observed in potassium chloride crystals when cut to expose inner surfaces [63].

Recently, a new theoretical approach of Sohnel and Garside [64] provided a basis for the assumption that the solute can form stable clusters of a subcritical size. The proof lies in the theoretical functions concerning dependence of surface tension on the size of the nuclei as was suggested by Tolman [65]. From available experimental evidence it was deduced that these clusters behaved as a new phase displaying properties of the bulk solid [64, 66]. Recently Veverka et al. provided a rigorous version on non-equilibrium thermodynamic grounds [67]. Chang and Myerson [68] measured solute diffusivity as a function of concentration. Their results are also indicative of the presence of solute clusters. It will be interesting to follow up the ramification extending from the notion that subcritical size clusters are formed in supersaturated solutions. One of them may be the renewed examination of the possibility that growth may occur by incorporation of preformed clusters (block nuclei) along with the addition of monomeric species as suggested by Glasner and Skurnik (Sarig) [59, 61, 62].

3.8. Precipitation

Precipitation occurs on mixing two solutions, each of them containing one ionic species of the sparingly soluble salt. Mixing which allows formation of a temporary local excess of one of the ions anywhere in the solution results in non-reproducible precipitation kinetics. In spite of these difficulties several studies of this type of precipitation were published. Probably the first one was that of Turnbull, analyzing the precipitation of $BaSO_4$, in 1954 [69].

All precipitation processes are characterized by an induction period which depends upon concentration of the reactants, the impurity content and upon the sensitivity of the detection technique used.

It is an accepted procedure to express results in terms of the fraction of the total precipitate formed during a specified period: $\alpha = (c - c_f)/(c_f - c_i)$, where c_i and c_f are the initial and final concentrations of reactants and c the solubility. Turnbull deduced the radius of a spherical growing particle from the diffusion equations:

$$r^2/2D + r/G = mKt, \tag{23}$$

where

D = diffusion coefficient,
G = interface transfer coefficient,
m = concentration of the precipitant in the bulk solution,
K = rate constant.

For small radii, r^2 becomes negligible:

$$r = mKGt, \tag{24}$$

which indicates that the growth process is controlled by the interface transfer G.

For large radii, r^2 becomes dominant and r is negligible:

$$r^2 = 2mKDt, \tag{25}$$

meaning that the process is diffusion controlled.

The most significant point brought forward by this analysis is that $BaSO_4$ and numerous other sparingly soluble inorganic salts grow by an interface controlled process at the early stages of growth, when the crystals are small.

It was shown by conductometry that the precipitation of $BaSO_4$ could be represented by an empirical equation [70]:

$$\mathrm{d}\alpha/\mathrm{d}t = k\alpha^{2/3}(1-\alpha)^p. \tag{26}$$

The value of the exponent p was found to be 3 or 4 for $BaSO_4$, $SrSO_4$, Ag_2CrO_4 and CaC_2O_4. Other investigators confirmed these determinations but they were shown to be an artifact of the experimental technique . The measurements were made using a platinum electrode which catalyzed the precipitation. Under similar conditions surface reaction controlled growth was reported with the order of reaction varying from 0–4. It seems that it is extremely difficult to follow precipitation and growth of particles quantitatively because conductivity gives no information relating to size, shape or numbers of precipitated particles. In an effort to circumvent the difficulties with the conductivity technique, Walton [71] attempted an interpretation of turbidity data for AgCl precipitation and found a reaction order of ≈ 2. Later study on the more complex $BaSO_4$ also yielded a kinetic order close to 2.

Walton [72] summarized the precipitation studies as known in 1967:

In general the appropriate equation for the rate of deposition was supposed to be

$$R = kn(c-c^*)^x, \tag{27}$$

where

n = number of seed crystals,
$(c-c^*)$ = excess solute concentration,
x = order of reaction.

In the cases examined prior to this summary for AgCl $x=2$, for $AgIO_3$ $x=2$, for (Ba, Sr, Pb) SO_4 $x=2$ and for Ag_2CrO_4 $x=3$. As far as such a small sample allows generalization x is equal to the number of ions in the formula molecule.

It has not yet been established whether the number of seed crystals n affects the growth rate through the total surface area as in surface reaction processes, or through surface active sites with screw dislocations which are not necessarily dependent upon the total surface area.

Moreover, growth of inorganic crystals may be limited by transport across the boundary layer surrounding a crystal by the incorporation process. Surface reaction theories yield the correct form of the rate equation and may be related to surface nucleation, but they are generally based upon the concept of equilibrium adsorption.

Precipitation kinetics are believed to follow equations similar to those describing seed-crystal kinetics, but the considerable experimental difficulties did not allow a thorough analysis.

The factors affecting precipitation kinetics were summarized:

(1) Numbers of initial impurities present – large numbers of impurities act as nuclei and produce large numbers of growing precipitate particles. The total surface area of the precipitate particles may affect the kinetic order of the process.

(2) The basic mechanism is probably a mixture of screw dislocation, surface nucleation, dendritic growth and agglomeration. In many systems an exact kinetic order cannot be expected to hold.

(3) Catalysis by active sites in container walls, electrodes, stirrer etc. often occurs.

(4) Methods of mixing reactants are very important in controlling the rate of precipitation.

(5) Traces of impurities may inhibit both growth and dissolution processes.

(6) Stirring usually disturbs the boundary layer at the crystal surfaces and hence the growth rate.

(7) Small activation energies indicates relatively little direct effect of increased temperature.

Finally Walton states that increased agglomeration will slow down the precipitation rate [72]. This statement might have been brought forward by his insight into precipitation processes, but no specific study could be found to corroborate it.

Most of the available data indicate that precipitation is too slow to be diffusion controlled and, with few exceptions, interface processes are involved to account for the kinetics of growth and precipitation. Dissolution and ripening may be believed to occur mainly via diffusion controlled processes. This would justify the assumption that dissolution, being non-symmetric with growth, cannot teach us about the mechnisms of growth.

3.9. Ripening

In a two-phase system of solution and disperse solid with positive interfacial tension, the free energy will decrease when the surface diminishes under constant conditions. This implies that the formation of a finely divided precipitate is not concurrent with attaining thermodynamical equilibrium. True thermodynamical equilibrium will be reached in the idealized case when all the disperse phase is incorporated into one single crystal. This crystal will be shaped according to the principle of minimum total surface free energy. The crystal will contain only the number of dislocations which are allowed, at the temperature of its existence, by Boltzmann's distribution law.

The processes prompted by the momentum towards equilibrium is called "ripening". There are two main routes by which a system of many small particles can be transformed into a more "advanced" system of larger particles in smaller number: one by which the smaller particles are dissolved and the larger ones grow (Ostwald ripening), and the other by which the particles combine by coagulation or agglomeration.

Nielsen presents the kinetics of Ostwald ripening in his fundamental book *Kinetics of Precipitation* [73]. According to his assumptions a strictly monodisperse suspension cannot undergo ripening; it is in metastable equilibrium. However, the sizes of the particles participate in thermal fluctuations, and therefore, even if an ideally monodisperse system could be established, it would show in time a distribution of sizes.

According to Thomson the (Gibbs–Kelvin) equation, $r^* = 2\sigma v/kT \ln S$, only particles of one size can be in equilibrium with the solution for each value of S. Also, as the equilibrium value r^* increases with decreasing S, a particle with $r > r^*$ will find the solution supersaturated and hence tend to grow larger, whereas a particle with $r < r^*$ will dissolve and ultimately disappear. Ripening can only occur close to saturation i.e. when S is $\sim$ unity, therefore surface reaction may be more important than diffusion. Nielsen maintains that it is of great interest to know the limiting velocity [73] and therefore he developed the theory of diffusion controlled Ostwald ripening.

Nielsen calculated that for purely diffusional growth and dissolution a particle of 50 nm will grow to 5000 nm in 3 h and to 50 μm in 12 days. As most probably any precipitate particle will actually grow either by surface nucleation or by the screw dislocation spiral step mechanism, the rate of change will be inevitably smaller by several orders of magnitude.

3.10. Growth kinetics and agglomeration

When growth kinetics of calcium oxalate monohydrate was investigated in relationship to variation of seed concentration, it was found that it could be described by a simple orthogonal hyperbolic formula [74]. The two main parameters were U_∞, the predicted uptake of calcium ions from the solution as time approaches infinity and t_m, the time at which uptake is half maximum ($U_\infty/2$). Whereas it is clear that there exists a relationship of these parameters and the solubility product, the study intended to find out how U_∞ and t_m were affected by changing the quantity of seed. It was assumed that the variation in seed concentration would affect the mass growth rate in a simple way if agglomeration were absent but it would have a more complex effect in the presence of agglomeration.

A special, sophisticated method was developed for this study, based on the direct uptake of ^{45}Ca tracer into the whewhelite crystals. It allowed direct measurement of crystal growth. Using seeds obviated the notorious difficulty of systems with spontaneous nucleation. The reliability of the isotope measurements was confirmed by isotope balances and conductivity method. The most prominent result of the study [75] was the substantial deviation of the best-fit growth parameter U_∞ from U_{eq}. The probable cause for this is the influence of the agglomeration of crystals on growth, suggesting that a diffusion process limits the influence of the thermodynamic driving force. The authors conclude that all growth techniques in suspension measure the resultant effect of growth, nucleation and agglomeration.

Though the analysis is quite impressive, the techniques employed are complicated and cannot be expected to be used as standard tools to explore aggregation. A somewhat similar approach to the analysis of the combined processes in precipitation, based on simpler experimental techniques was reported by the group headed by Furedi–Milhofer in Zagreb [76]. The samples were prepared by mixing equimolar solutions of sodium oxalate and calcium chloride; the concentrations of calcium ions in solution were measured with selective electrode, particle number and size analysis were performed by a Coulter counter. For the quantitative analysis of the experimental data the degree of reaction, α, was used:

$$\alpha = (c_0 - c_t)/(c_0 - c_S), \tag{28}$$

where

c_0 = initial total calcium concentration,
c_t, c_S = concentrations at time t and at equilibrium, resp.

For self-preserving distribution of sizes and constant shape of particles

$$(\mathrm{d}\alpha/\mathrm{d}t)\alpha^{-2/3} = KN_t^{1/3}(c_t - c_S)^n. \tag{29}$$

N_t denotes the total number of particles.

Equation (29) generally describes the kinetics of spontaneous (unseeded) precipitation and may serve for an analysis to distinguish between different successive precipitation stages. A log (rate) versus log (supersaturation) plot should give an upward surge in case of nucleation (heterogeneous) and a straight line if crystal growth is rate controlling. In the case of decreasing N_t, due to aggregation or recrystallization, the rate should decrease faster and by a different rate law.

Aggregation as a rate controlling process was investigated by the above presented analysis in the system of precipitating calcium oxalate trihydrate [77]. It was found that the rate of crystal growth followed the equation:

$$R = (\mathrm{d}\alpha/\mathrm{d}t)\alpha^{-2/3} = 19^{9.81}(c_t - c_S)^{3.6}.$$

Figure 7 shows in the upper part the experimental determinations of crystal number N_t and corresponding volume V_t as functions of time and in the lower part the growth rate R versus supersaturation. The rate versus supersaturation curve evidently rises steeply (dotted line in fig. 3) which would correspond to nucleation, then decreases linearly as crystal growth becomes dominating. The change of slope indicates aggregation. The change in N precedes the change in growth rate which can be effectively affected only after the shielding of the crystal surfaces become sufficient to impair crystal growth [77].

The authors performed an analysis of the transport mechanisms of aggregation. It showed that, in the early stages, when particles of sizes $d < 3$ μm prevail in suspension, the collision rates are controlled by Brownian motion, laminar flow and/or differential settling. In later periods when particles of $d > 3$ μm prevail in suspension, differential settling becomes the dominant rate controlling mechanism.

Both sets of studies [74, 75] and [76, 77] investigated the aggregation processes of calcium oxalate which is the main constituent of human kidney stones. The independent assessment of growth and aggregation of calcium oxalate crystals was reported in *Investigative Urology* by Ryall et al. in 1981 [78]. A Coulter counter was used to

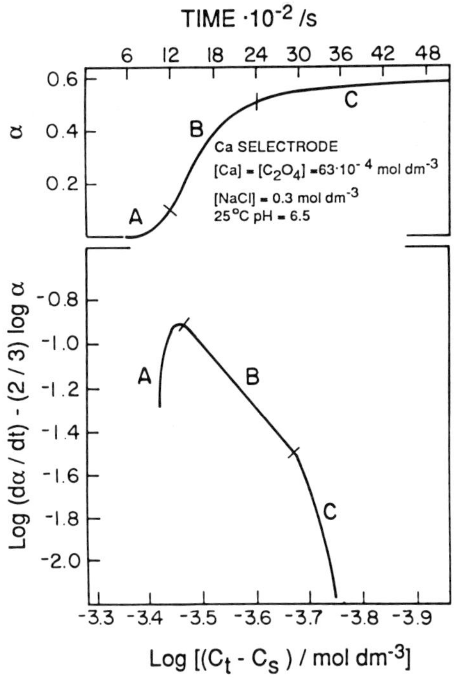

Fig. 7. Aggregation in a system of precipitating calcium oxalate trihydrate. Upper curve: The fraction of completed reaction, α, versus time curve obtained by selectrode. Lower curve: growth rate versus supersaturation. Sections A, B, and C correspond in time. Change of slope indicates aggregation.

measure the net increase in total crystal volume which indicated the extent of growth and the percentage of change in total crystal number, which indicated the extent of aggregation. A simple theoretical model was used to simulate the changes in particle size distribution that occur when crystals grow, aggregate or disaggregate to a define extent [79].

4. *The effect of impurities*

4.1. *Definition of an impurity*

An impurity in a crystallizing system is a foreign material, chemically different from the main component and present at low concentration. There is no clear cut borderline distinguishing betweeen two- or multi-component system and a mono-component system with impurities, but usually impurities are considered to be under a one percent limit. In most chemical processes foreign substances below one percent concentration have little effect on the process. However in crystallization processes some substances in the ppm range exert a very pronounced effect on crystallization parameters: nucleation, growth, crystal habit, agglomeration, polymorphic transitions and size distribution.

An important aspect of impurity effects is the specificity. Some cations, or more correctly their complexes, have a very strong effect on certain salts, e.g. Pb on KCl or Co on KNO_3. Organic molecules have an effect on the crystallization of inorganic salts, e.g. urea on NaCl or glutamic acid on calcium oxalate. Macromolecules and soluble polymeric species affect the precipitation of sparingly soluble and soluble minerals, a phenomenon which is of the utmost importance in biological mineralization and has application in technology.

Though impurities affect each stage of crystallization, the most discernible influence is on the growth rate of soluble salts and on the crystal habit, which will be described in the following section. These effects were the first reported in the literature and were the first studied. The classic example of crystal growth modification by foreign substances is the formation of octahedral crystals, instead of the normal cubes, when common table salt is grown in the presence of urine [80]. Systematic observations were carried out in studies of the effects of impurities upon the crystallization velocity of supercooled melts. It was established that many soluble substances lowered the crystallization rate. Freundlich [81] suggested that the pronounced effect of the impurities may be due to the formation of an adsorbed layer which retarded the normal course of crystallization. Application of his famous adsorption equation to the case of partial coverage of the growing face yielded calculated results in satisfactory agreement with the observations.

The prominent role of the surface layer was emphasized in the studies of Marc [82] on the rates of crystal growth from solutions. His study of crystallization of potassium sulfate from solution showed that the rate was markedly retarded when some organic substances such as quinoline yellow, Bismark brown or ponceau red were added to the solution. The interesting fact is that these dyes tend to color the normally white potassium sulfate crystals, whereas methylene blue which does not impart any color to the crystals, does not affect the rate of crystallization either. It might have been concluded that the inhibition of growth was due to the strong adsorption of the dyestuffs on the growing crystals.

Adsorption is a phenomenon which occurs on an atomic level and therefore should be explained by ionic or molecular interactions. It follows that the adsorption of foreign substances on a growing crystal surface must be highly preferential. Crystal faces may contain alternating positive and negative ions as for instance the (100) face of NaCl or ions of one kind as the (111) face of the same mineral. Therefore, the latter charged face would attract polar molecules preferentially. The covered face will grow more slowly than the uncovered ones thus coming into prominence in the crystal habit. This could be the explanation for the octahedral shape of NaCl crystals formed in the presence of urea [83].

Although the chain of reasoning was quite logical, a more direct proof was needed to firmly establish the adsorption theory. One of the experimental techniques which helped was the observation that radioactive elements were adsorbed on precipitated particles. Soon these "tagged" atoms were used to measure the total surface of the crystals in the precipitate [84].

Numerous reviews are available. Up to 1959 mostly observations of habit modifications due to impurity adsorption were considered [85].

4.2. Modification of crystal habit

4.2.1. Crystal habit

Crystals grown under conditions close to equilibrium are generally well formed, displaying well defined faces. The concept, as expressed by Curie, that the shape of a crystal will be such as to make the total surface energy to be a minimum was subsequently broadened by the principle of Bravais. This approach states that the most important faces of the crystal will be those which have the highest reticular densities and the greatest interplanar distances. In crystals grown at high supersaturations and having interactions with a solvent or under the influence of impurities, these principles seldom apply. In alkali halides grown from vapor, or very slowly from aqueous solutions, the low surface energy (100) planes predominate, thus resulting in the formation of perfect cubes. Stranski et al. maintained that the equilibrium shape for NaCl-type structures were cubic [86], rhombic dodecahedral for CsCl, octahedral for CaF_2 [87] and rhombohedral for $CaCO_3$ and $NaNO_3$. It was concluded that the form should favor the following surface planes: NaCl (100), CsCl (110), CaF_2 (111) and $CaCO_3$ (calcite) (100). In many cases the low energy planes are those which correspond to the unit cell surfaces so that a relation between crystallographic structure and morphology exists. However, the crystal habit of some of the soluble salts are not simply related to their unit cell structure. For example K_2SO_4, K_2CrO_4 and $(NH_4)_2SO_4$ are all of the same rhombic structure but crystallize with distinctly different habits from pure aqueous solutions. Instead of taking into account only the simple thermodynamic considerations, the kinetic factors should also be considered. If the accommodation coefficients of various faces are fixed, one would expect that the rates of growth of these faces would depend upon the frequency of lattice points in the various planes, that is on their reticular density and on the interplanar spacing along the surface. Bravais [88] demonstrated that the preferred planes were those that progressed at the slowest rates extending themselves to the exclusion of the more rapidly advancing planes. This fundamental principle is illustrated in an elementary way by the geometric scheme shown in fig. 8.

4.2.2. The effect of impurities on crystal habit

In many cases of habit modification, evidence may be found of structural compatibility between the modifying agent and the crystal. As early as in 1949, the growth inhibition of $CaCO_3$ (111) faces in solutions containing metaphosphates was attributed to the similar orientation of Ca^{2+} in those faces and the oxygen atoms in the metaphosphate

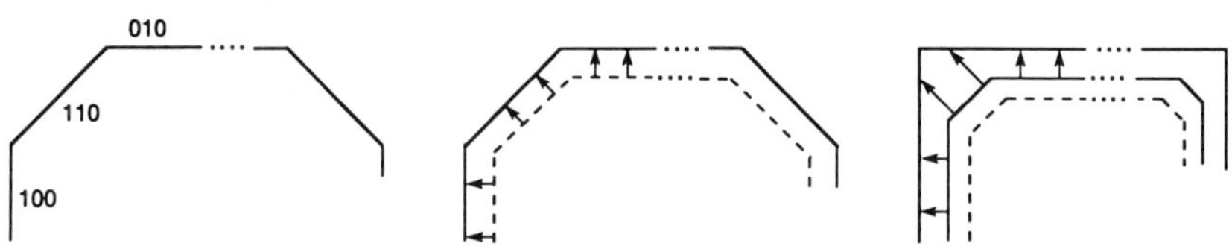

Fig. 8. Elementary explanation of crystal habit formation. Different faces have different growth rates, as indicated by vectors. The faster growing face grows out of existence.

chain [89]. A more detailed study of gypsum, showed that the growth of (111) faces was inhibited by traces of discarboxylic acids in which the COOH groups were separated by two carbon atoms in aliphatics, occupy α-position in aromatics and cis-position in olefins [90].

The spectacular specificity of the effects of particular molecules on the crystal habit of the two common species of calcium sulfate is illustrated in a study of sea water desalination [91]. Calcium sulfate hemihydrate crystals, deposited from boiling sea water have the distinct habit of very elongated hexagonal rods (fig. 9a). In the presence of 20 ppm of polyvinyl sulfonate (PVS) the crystals become deformed, making up a sponge-like mass (fig. 9b). In technological terms, as verified by scale formation on an aluminum pipe, it meant that whereas a firm gypsum film is formed in the absence of

Fig. 9. Scanning electron microscope photomicrographs of calcium sulfate hemihydrate obtained by precipitation from boiling sea water. (a) Crystal grown without additives. Magn: × 490. (b) Crystals grown in the presence of 20 ppm of polyvinyl sulfonate. Note the drastic change of habit. Magn: × 500. (c) Crystals grown in the presence of 20 ppm of poly-glutamic acid. Magn: × 500.

any additive, in the presence of PVS a soft deposit was formed, which could be easily washed away. Another polyelectrolyte, polyglutamic acid (PGA), exerts the opposite effect – the crystals presented in fig. 9c at the same magnification as the other photomicrographs in this figure, are huge in comparison to those of fig. 9a. In technological terms it meant that the scale formed in the presence of PGA was harder and more persistent than that formed in the absence of any additive.

The molecular model of PVS confirms that in the molecule, a simple polymeric carbon chain with sulfonic groups on every alternating carbon, the distances between the functional groups are about 4.5–5 Å. PGA has carboxylic groups spaced at about 8 Å.

Fig. 10. Scanning electron microscope photomicrographs of calcium sulfate dihydrate obtained by evaporation of sea water at ambient temperature. (a) Plates grown without additives. Magn: × 2000. (b) Crystals grown in the presense of 20 ppm of polyglutamic acid. Note the drastic change of habit. Magn: × 470. (c) Crystals grown with 20 ppm of polyvinyl sulfonate. Magn: × 2000.

The intercationic distances between calcium ions in the hemihydrate structure are smaller than 8 Å, but some of them are in the shorter 4.5–6 Å range, which constitutes a perfect fit with PVS. On the other hand, the calcium sulfate dihydrate has cell dimensions: $a_0 = 5.68$ Å; $b_0 = 15.18$; $c_0 = 6.52$ Å and $\beta = 118°23'$. The positions of two of the calcium ions are $(0, u, 1/2)$ and $(1/2, u + 1/2, 3/4)$ with $u = 0.4200$. The distance between these two particular ions is 8.1 Å. This makes PGA a perfect partner for interaction with calcium sulfate dihydrate. Whether viewed as a prediction or an explanation of an observation, the plain fact is that the same relation that exists between PVS and the hemihydrate, exists also between PGA and the dihydrate. This can be seen in fig. 10. Calcium sulfate dihydrate without additives grows from sea water evaporated under reduced pressure at 45°C, as plates (fig. 10a); the addition of 20 ppm PGA completely disrupts this habit, causing a sponge-like appearance (fig. 10b). The addition of the non-fitting PVS causes growth of quite thick plates (fig. 10c) instead of the thin plates which were formed without additives.

Another graphical example of a specific interaction is the case of sodium chloride [92] grown in the presence of ppm traces of the surfactant nitrilotriacetamide (NTAA), which is an excellent anti-caking agent. Nitrilomonoacetic acid and nitrilotriacetic acid, closely related in structure to NTAA, but each having less functional groups, were quite ineffective as anti-caking agents.

The residual electrostatic charges on the surface of the NaCl crystal and those of the functional groups, CO and NH_2 are weak: even three points of contact are insufficient to anchor a relatively large organic molecule on the surface and keep it adsorbed. Six points of contact are needed to make an efficient interaction.

Figure 11a in which a molecular model of NTAA is placed on a diagram of a (100) NaCl face, drawn to the same scale, explains the specific interaction. The CO group, which has a slight negative charge, is positioned above the cation whereas the slightly positive amide orients itself towards the anion. The six points of attraction are evidently sufficient to induce a distinct mode of growth: cubes with pits in the presence of 50 ppm (fig. 11b) and unusual, dendritic crystals resembling snow flakes, but with four-fold symmetry, in the presence of 200 ppm NTAA (fig. 11c).

Figure 12a is the same NTAA model placed on a (100) KCl face. As the interionic distances are larger, there is a lack of fit. KCl crystals grown with NTAA developed non-planar surfaces with no right angles. A typical crystal (fig. 12b) looks rather like a piece of modern sculpture. An intriguing query posed itself: could it be restored to its regular habit if the modifying molecule was properly adsorbed? A compound related to NTAA, namely nitrilotripropamide (NTPA), with side chains extended by a CH_2 group, can be properly aligned on the KCl surface. Indeed, KCl crystals grown in the presence of NTPA resemble NaCl crystals grown with NTAA (fig. 12c).

A special habit modification of basic lead carbonate by polyglutamic acid was associated with a quite unique set of phenomena: whereas polyglutamic acid retarded the precipitation of Mg, Ca, Sr, Ba and normal Pb carbonates, it accelerated the precipitation of basic lead carbonate. In the X-ray diffraction pattern of pure $2PbCO_3 \cdot Pb(OH)_2$, the peaks were sharp, but the crystals formed in the presence of polyglutamic acid yielded flattened, smeared out, diffraction peaks which could be obtained from an almost amorphous solid [28]. The thermal gravimetry analysis curves for both

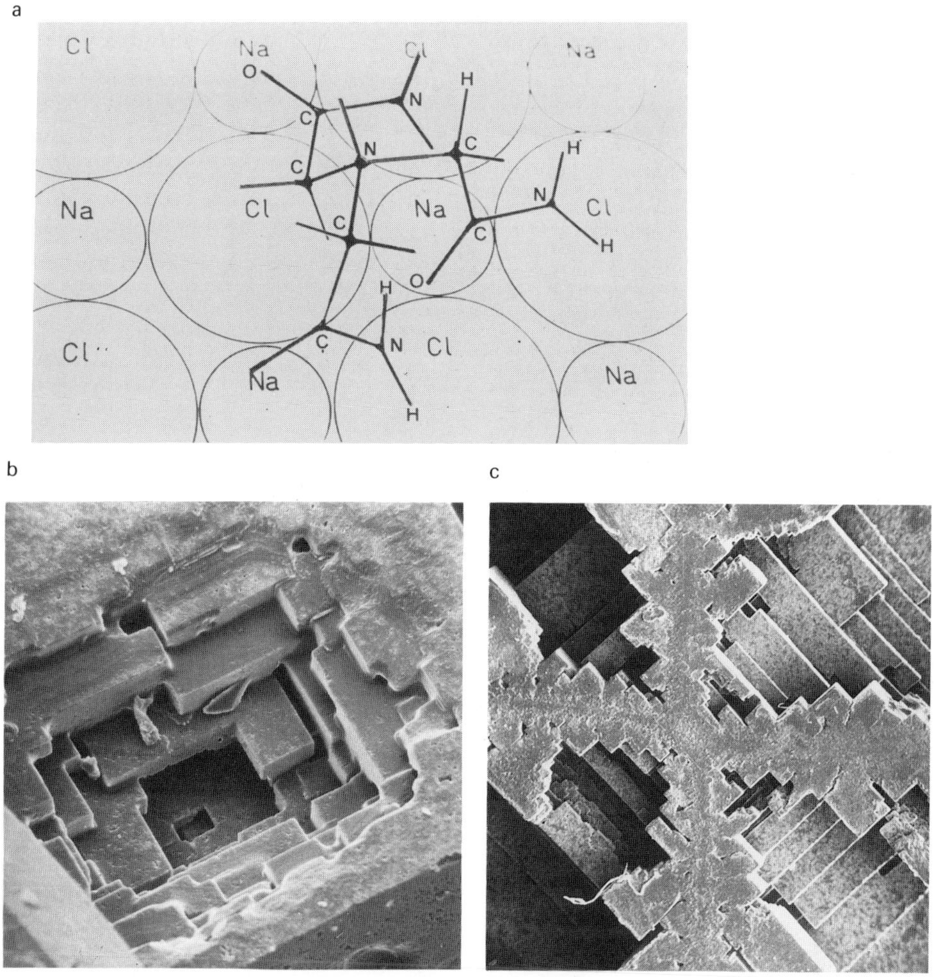

Fig. 11. Dimensional fit of opposite electrostatic charges of an impurity and interionic distances on a crystal face. (a) Molecular model of nitrilotriacetamide (NTAA) placed on a NaCl diagram drawn to the same scale. The COO^- groups are attracted to the positive cation and the slightly positive NH_2 to the anion. (b) A skeleton cube of NaCl grown in the presence of 50 ppm of NTTAA. Magn: × 200. (c) A pseudo-dendritic crystal grown in the presence of 200 ppm of NTAA. Magn: × 80.

the pure and the lead carbonate with additive were quite different, but the most convincing evidence of the special effect of polyglutamic acid on the crystallization of basic lead carbonate was the change in the shape of the crystals. Whereas the pure crystals were nicely shaped, even if very small, hexagonal plates, the presence of 10 ppm of polyglutamic acid yielded a glassy mass. Similar accelerating effect of polyglutamic acid on hydroxyapatite, also attributed to structural factors, revealed an unusual habit of hydroxyapatite precipitate which resembled remarkably the habit of this mineral in renal stones [93].

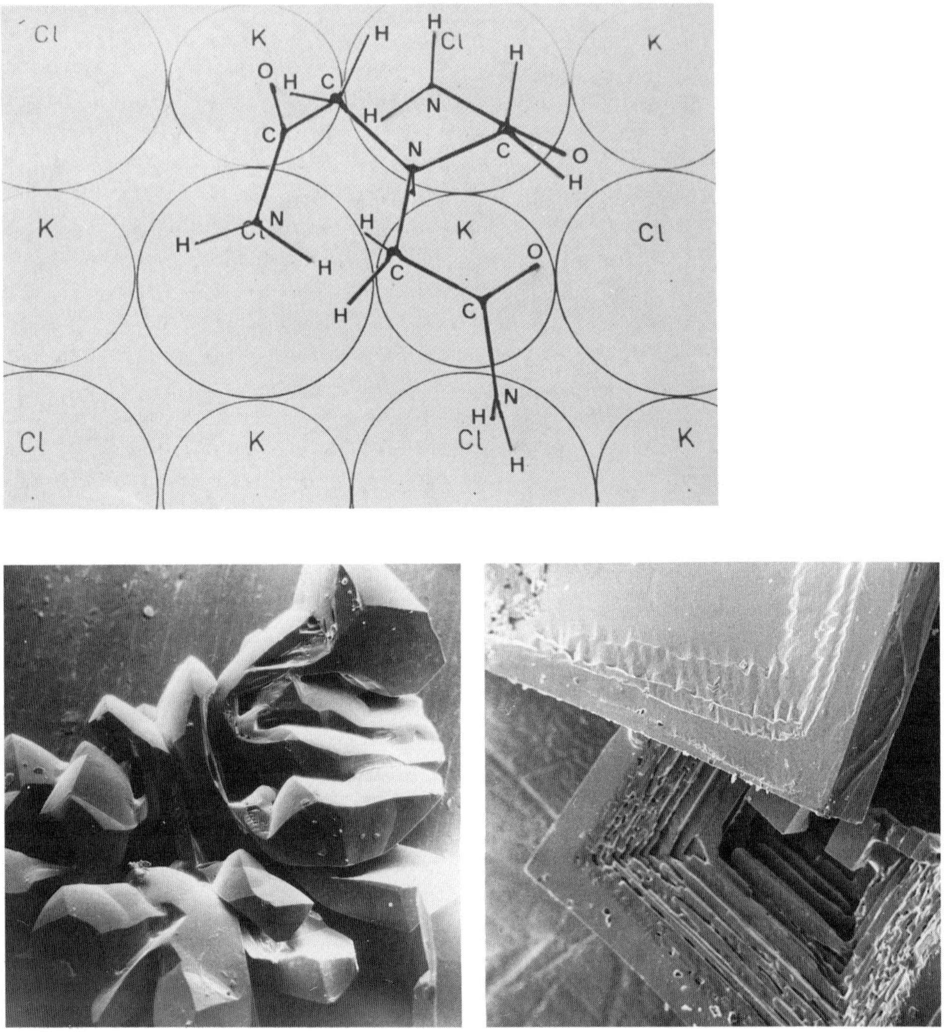

Fig. 12. (a) Molecular model of nitrilotriacetamide (NTAA) placed on a KCl diagram drawn to the same scale. There is a lack of the structural fit shown in fig. 11. When the NH_2 groups are accommodated on the anions, the side chains of NTAA are too short to allow the accommodation of COO^- groups on the cations. (b) Distorted KCl crystal grown in the presence of 50 ppm of NTAA. Magn: × 92. (c) KCl grown in the presence of 50 ppm nitrilotripropamide (NTPA) which is similar to NTAA, but has side chains longer by one CH_2 unit. The longer chain enable NTPA to be placed on the KCl surface in the same mode as NTAA on NaCl. Note that this crystal modification is similar to that in fig. 11b.

New developments in the theoretical approach to the requirements for interactions between the crystal faces and the modifying organic molecules came from the field of biomineralization [94, 95]. In vitro experiments in which crystals of calcium dicarboxylate salts were grown in the presence of aspartic acid-rich protein, revealed a stereochemical property common to all the interacting faces.

One of the illuminating examples is the calcium malonate/aspartic acid-rich protein system. The pure calcium malonate crystals are prismatic and elongated along the *c*-axis. The predominant faces are (100), (110) and (010). The crystals grown in the presence of protein are elongated along the *b*-axis and the predominant faces are (101), (101) and (110). An analysis of the crystal morphologies shows that acidic proteins specifically influenced the growth of the (101) and the (101) faces, which developed at the expense of the (100), (210) and (001) faces. The characteristic feature of the (101) and (101) faces is the carboxylate orientation of the malonate molecules with respect to calcium. Complexation motifs of these faces define Ca–Ca and carboxylate–carboxylate repeat distances of 6.1 Å. This "match" of repeat distances is not sufficient, however, to explain the selectivity of the proteins for certain faces, as any face parallel to the *b*-axis, or for that matter to the *c*-axis, has a repeating Ca–Ca distance of 6.8 Å.

After ascertaining that similar morphological changes were induced by aspartic acid-rich protein in calcium fumarate and calcium maleate crystals, an important recognition of the "stereochemical effect" emerged [94]. These experiments showed that rigid, calcium loaded, aspartic acid-rich matrix proteins, which adopt the β-sheet conformation, can co-operatively bind specifically to those crystal faces in which the carboxylate groups are approximately perpendicular to the plane of the face. This effect may be attributed to the requirement of the protein-bound calcium ions to optimally complete their coordination polyhedrons with the carboxylate groups at the crystal surface. It would imply that the carboxylate and calcium ions of the protein would have occupied lattice sites of the substrate ions had the crystal layer been extended.

The serious difficulties encountered when attempting to predict the interaction between an impurity and a growing crystal, even having in command the crystallographic data, may be well illustrated by the adaption, by Addadi and Weiner [94], of their own "stereochemical effect" for the pre-evaluation of the effect of the aspartic acid-rich protein on the modification of the crystal habit of calcium carbonate. According to this stereochemical effect, the optimal plane for the interaction should be the (001) plane, in which the planar carbonate ions are well aligned. However, it was found that the growth of calcite was inhibited only in a non-specific manner at protein concentrations $> 0.5\ \mu g/ml$ not inducing the expected habit change. There was nevertheless a very rewarding compensation: crystals grown in the presence of smaller amounts of protein developed very small (001) faces. The experimental evidence that calcite crystals developed on these faces as a result of being nucleated on immobilized protein, previously adsorbed on the container surface, may be presented as an outstanding example of heterogeneous nucleation.

The recognition that the morphological changes can be explained by the adsorption-inhibition process on the molecular level, brought Addadi et al. [89] to explore systematically the possibility to modify the habit of crystals by tailoring additives which bind at a preselected face and thus inhibit growth in a predictable manner.

An instructive and illustrative example is the modification of benzamide morphology by the "tailor made" additives. Benzamide crystallizes from ethanol in the form of plate-like crystals elongated along the *b*-axis. The main packing feature of the structure (fig. 13a) consists of a ribbon motif of hydrogen bonded cyclic dimers interlinked by NH···O bonds along the *b*-axis. Retardation of growth of the (011) faces along the *b*-axis was achieved with benzoic acid as additive. The substitution of a molecule of

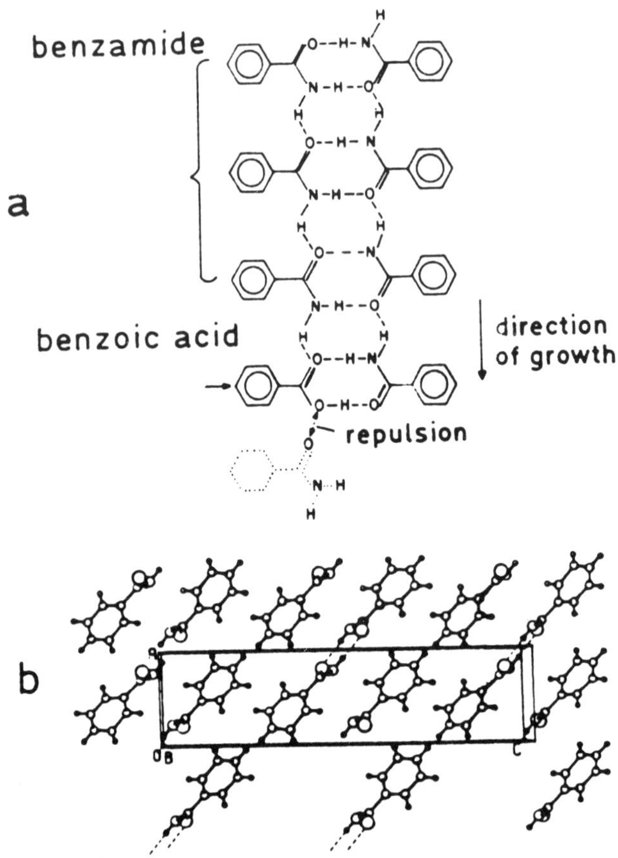

Fig. 13. (a) Schematic representation of part of the benzamide structure showing a benzoic acid molecule inserted at the end of a ribbon motif of hydrogen bonded dimers along the *b*-axis. (b) Packing arrangement of benzamide viewed along the *b*-axis.

benzamide at the end of the ribbon motif by benzoic acid results in replacement in the *b*-direction of an NH···O bond by a repulsive O···O interaction, thus retarding growth in this direction and inducing the growth of bar-shaped crystals elongated in the "*a*"-axis direction. Inhibition of growth along *a* was accomplished by adding *o*-toluamide to the crystallization solution. This additive can be easily adsorbed in the hydrogen bonding chain without interfering with the growth in the "*b*"-direction. The ortho-methyl group emerges from the (104) face, and thus interferes with growth along the *a*-direction, in which the dimers are stacked. Finally, thinner and thinner plates are obtained by adding increasing amounts of *p*-toluamide, the methyl substituent of which perturbs the weak van der Waals interactions between the phenyl layers in the *c*-direction.

Similar morphological changes were induced in crystals of other primary amides and other organic molecules including carboxylic acids, amino acids, dipeptides and steroids [95]. Though these studies introduced a new insight to the complex host-additive relationship, they were, up to 1989, limited to the realm of organic crystals. The most extensively investigated mineral systems are calcium phosphate, oxalate and carbonate with organic admixtures, in view of their importance in biology. Going through the recent studies one has the distinct impression that the scientists are on the verge of accomplishing the ability to predict the effect of any additive on any inorganic crystal with respect to habit modification and probably rate of growth. The needed data of computer modeling are almost within reach; it does not seem that there are insurmountable difficulties to attain this aim and there is a reasonable hope that with suitable investment of work the aim will be attained within a few years.

4.3. Effect of impurities on nucleation

When the interaction between an impurity molecule and the crystallizing main phase is as strong or nearly as strong as the interaction between the molecules or ions of the crystallizing phase, the effects on nucleation may be substantial.

While trying to study the quantitative effect of foreign ions on nucleation rates, investigators used the available equations of the description of homogeneous nucleation in solution. The effect of foreign ions on precipitation could be obtained through the evaluation of their effect on crystallization parameters expressed in these equations, especially on the critical supersaturation. Thus, for instance, when the effect of iron and thorium ions on the precipitation of barium sulfate was studied in 1969 [96], a simple equation was derived from the classical Volmer–Becker–Doring homogeneous nucleation theory which relates the critical radius and accordingly number of ions in the critical nucleus to the critical supersaturation ratio (eqs. (4) and (6)). Small concentrations of Th raise the critical supersaturation ratio and affect the growth rate. Similar effects were found for ferric ions on the rates of nucleation and growth. Since then many systems have been analyzed according to this point of view and many additional examples allow one to form a generalization: the presence of a multivalent cation in a system of a sparingly soluble, precipitating salt of alkaline earths may affect the course of nucleation and growth, modifying the size distribution in the precipitate.

Another fruitful way to use classical nucleation theory is to measure the "induction period" by the stopped-flow method introduced by Nielsen [18, 20]. Induction periods were measured utilizing the changes in the conductivity of $CaSO_4 \cdot 2H_2O$ solution after nucleation and growth [97]. The measured induction periods were insensitive to calcium sulfate concentrations, indicating that the nucleation process was heterogeneous. Low additions of Al^{3+} increasingly prolonged the induction period. The explanation was based on the assumption that $[Al(H_2O)_6]$ complexes poisoned the growth sites. The straightforward observation and its simple explanation is, as in many other studies, obscured by additional occurrences: above a certain concentration of Al^{3+}, further addition caused the induction period to reduce again. This effect may be due to the formation of precursors of aluminum hydroxide poly-complexes which are less effective as growth site inhibitors.

It seems that the possibility to detect the effect of foreign ions on the nucleation rate through the use of homogeneous nucleation parameters is limited to very sparingly soluble salts, with K_{SP} of 10^{-10} or lower, for which high supersaturation ratios can be easily attained by appropriate mixing of solutions containing the salt forming ions.

The sequence of events in the nucleation process of soluble salts such as NaCl, KCl or KNO_3 could not have been detected and analyzed in aqueous solutions by the approaches and methods developed for the study of homogeneous nucleation in which sparingly soluble salts were used. In this case the rate of nucleation was inferred from the analysis of the polycrystalline mass produced. One of the most sophisticated approaches was the "population balance" analysis [98]. Shor and Larson [99] used the population balance principle to analyze potassium nitrate crystals produced in a mixed suspension mixed product removal (MSMPR) crystallizer. The experimental procedure and the mathematical analyses are presented in detail in [98] and in chapter 7 of this Volume of the Handbook. The possibility to estimate the number N^0 of "zero size" crystals (i.e. nuclei) which would keep a constant population of crystals throughout a steady state process was used to investigate the effect of polyvalent cations Co^{2+} and Cr^{3+} on this particular number as well as on other crystallization parameters. The investigators found that the cations suppressed nucleation, the inhibiting effect of Cr^{3+} being stronger than that of Co^{2+}. Though the verbal conclusion of the study was clear cut and unambiguous, careful scrutinization of the experimental results showed that the correlation between the nuleation rate $B^0(=N^0$ per unit of time) and the impurity concentration existed only for normalized values, i.e. for nucleation rates divided by suspension density. Actually both quantities varied with inconsistent trends, but nevertheless parallel to each other, with impurity concentration. The rather irregular dependence of an impurity effect on impurity concentration, leading actually to reversal of trends in impurity effects on nucleation and other crystallization parameters, has been recently pointed out [100].

The difference between impurity effect at low and high concentration was easily discerned with respect to surface active agents: at low concentrations methylamine and dodecylaminehydrochlorides enhanced nucleation rates of potassium nitrate [99] but above certain concentration the effect was weakened. In this specific case a very reasonable explanation was found. When the "critical micelle concentration" was reached, the molecules of the surfactant reacted with each other to form micelles and were not available to interact with the mineral phase.

Liu and Botsaris [101] employed the same MSMPR configuration to investigate the crystallization of sodium chloride in this presence of Pb^{2+} ions. They found that the nucleation rate increased with increasing impurity concentration, just the opposite to what was observed for potassium nitrate in the presence of Co^{2+} and Cr^{3+} [99]. The authors [101] maintain that in the MSMPR crystallizer the increase in impurity concentration may shift the steady state supersaturation to a different value. At the new supersaturation, the nucleation rate may be lower or higher. Under the operating conditions of the MSMPR crystallizer, used in both of the above studies, the growth and nucleation rates should shift to higher or lower rates correspondingly. It seems that no theoretical approach has been developed which could predict the direction of the supersaturation shift. Even, there was no way to measure the supersaturation, when the shift occurred, by an independent, direct method.

The important conclusion of the study [101], is that the relative kinetic order *i* of NaCl crystallizing in the MSMPR crystallizer is independent of the concentration of the Pb^{2+} impurity. The applicability of this conclusion to the system of KNO_3–Co^{2+} was also demonstrated and concluded to be of general validity.

With improving techniques of crystal size distribution measurement, laser light-scattering instruments were connected on-line to the MSMPR crystallizers. Such a configuration was used in 1986 to study the effect of selected cationic and anionic impurities on the precipitation of gypsum in aqueous brine [102]. As the size distributions were bimodal, the existence of agglomerates was suspected and subsequently confirmed by microscopy. Both aluminum and maleate ions enhanced the degree of agglomeration but reduced the overall growth and nucleation rates. It is interesting to note that at this state of the art, it was assumed that an intelligent guess might have been made concerning the effect of additional ions in a known system. The authors tried their thesis and predicted fumarate ions to be rather ineffective. Contrary to expectations, however, the fumarate ions also exerted an observable effect on gypsum precipitation parameters.

The use of a MSMPR crystallizer and the analysis by the population balance principle is very suitable for the investigation of systems of soluble salts such as NaCl, KCl, and KNO_3. Actually in the "mixed suspension" systems, the formation of new particles is almost totally of the *secondary* type. The character of *secondary nucleation* is not well defined. It consists of a number of mechanisms including dendritic growth fracture, surface nucleation, impact phenomena and mechanical attrition i.e. it is related to the conditions existing at the surface of the particles in suspension.

The study of Botsaris and Sutwala [103] demonstrates the idea that secondary nucleation could become a tool for the study of other aspects of crystallization theory, like growth and primary nucleation. The Strickland–Constable experiments on contact nucleation of $NaClO_3$ by sliding of crystals were repeated in the presence of small amounts of an impurity (borax). The number of nuclei produced by the seed at various supersaturations and impurity concentrations were counted. The results provide information about the effect of impurities or additives, not only on secondary nucleation but also on the process of primary nucleation. The effect of impurities on primary nucleation was also investigated by the modified droplet technique [104]. The concentrations of Ca^{2+}, Sr^{2+}, Ba^{2+}, Ag^{+}, Pb^{2+}, Cr^{3+} and Ni^{2+} directly affect the nucleation characteristics of K_2SO_4. The effects ranged from complete suppression of nucleation by Cr^{3+} to no effect at all by Ni^{2+}. As in other systems, increasing the impurity concentration brings the effect, at a certain level, to an abrupt change. In this case the discontinuity corresponded to the solubility of the impurity sulfates. When the data were analyzed by other investigators [105], using their own model, they did not confirm the existence of the discontinuity.

However, the dependence of the impurity effect on concentration was repeatedly reported in crystallization studies: the effects of small amounts of impurities on nucleation and crystallization were observed to be often stronger than for large amounts [106]. This study deals with the crystallization of KCl from saturated aqueous solutions when solid $PbCl_2$, $K_4[Fe(CN)_6]\cdot 3H_2O$ or $K_3[Fe(CN)_6]$ were added to 100 ml saturated KCl solution. The kinetics was followed by ^{42}K tracer nuclei. The surprising finding was that when $K_4[Fe(CN)_6] \cdot 3H_2O$ was added the minimum of crystallization occurred at ~5 mg and the maximum at ~20 mg. The question is: how a substance which acts as a potent *inhibitor*, believed to poison growth sites, is able to change its "basic nature" in the same system and cause quite an opposite effect? No explanation was attempted. The authors expressed a somewhat bland statement that "The effects of small amounts of impurities on the crystallization and nucleation are often stronger than the large ones." The acute question about the role of ferrocyanide suppressing crystal formation of KCl at the concentration ratio of about 150 µg of impurity to 1 g of crystallizing salt and not being able to do so at the concentration ratio of 600 µg to 1 g of salt did not trigger a surge of investigations in the international scientific community or discussions at the prestigious international conferences.

4.4. *Effect of impurities on growth rate*

Literally thousands of scientific reports showed that the presence of impurities in a crystallizing system can drastically change the course of crystal formation altering the nature of nucleation, usually decreasing but sometimes increasing the growth rate, promoting or preventing aggregation, directing the preferential formation of a polymorphic form and perfecting or distorting the crystal habit. Usually all these effects are concurrent; following any single process in this maze is very difficult. In the previous section some of the investigations which observed the effect of impurities on nucleation were presented. The inherent difficulty that all of them faced was that the nucleation is immediately followed by growth and nuclei can be detected only after they grew to a certain size.

The separation of the growth stage from all others is easier to achieve than the isolation of the nucleation stage. This was carried out in seeded systems of all kinds, especially with the "constant composition" approach. Thus the determination of the effect of impurities on the growth rates was successful in numerous cases in a quantitative manner. Still, it is rarely possible to find a simple explanation of the mechanism by which a certain impurity affects the growth of a given crystal. The presentation of the studies carried out in this area in a roughly chronological order may help to organize the body of knowledge toward the formulation of some generalization.

Especially illuminating is one of the early papers by Bunn [107] in which the properties of crystals built up by the continual adsorption and incorporation of impurity, termed by him "adsorption bodies", during growth were compared with those of mixed crystals. He concluded that the two types of crystal differ only in degree: adsorption bodies may be regarded as unstable mixed crystals. It was suggested, and this basic assumption remained valid until the present time, that the condition for strong adsorption is similarity of lattice structure and interatomic distances on specific planes only; the rest of the structure may be quite different. This condition is the same as required for oriented epitaxial growth; in fact oriented overgrowth may be regarded as an adsorbed layer developed to macroscopic size.

A comprehensive summary of the effects of small amounts of soluble impurities on both the growth rates and the habit of crystals formed in solutions, as known up to 1951, could be found in the excellent book of Buckley [108]. The added impurities were classified by him into two main types: large organic molecules and complex inorganic ions. The organic molecules exerted strong inhibiting effect on growth rates at concentrations of 10^{-3} to 10^{-4} M, whereas dissolution was virtually unaffected. Though the inhibiting effect has been confirmed in principle in subsequent studies, up to very recent investigations, it seems that the total lack of effect on dissolution did not withstand the test of many modern studies. The class of the inorganic complexes was investigated mainly by Sears [109]. These additives are active at the minute concentrations of 10^{-5} to 10^{-6} M. Sears maintained that the most probable mechanism of growth poisoning involved adsorption at the growth sites, i.e. at kinks or growth steps, rather than monolayer adsorption.

One of the first examples of a systematic study examining the effects of foreign ions on the growth of seed suspension was that of Davies and Nancollas [110]. They found that the potassium salt of eosin completely prevented crystal growth of silver chloride. Some retardation of dissolution of seed crystals was also observed in the presence of this material. Other surfactants inhibited growth but not dissolution. The modification of the growth rate was observable, but it was still possible to express the growth rates by second order equations, with the exception of eosin poisoning. It was also established that eosin and dodecyl sulfate had the same effect on the growth kinetics of magnesium oxalate [111].

The studies of Russian scientists were summarized by Khamskii [112]. Some of them suggested that the action of impurities was caused by the electric field in the adsorption layer, but those explanations were not considered satisfactory in following studies. Mullin et al. [113] considered the ionic influence associated with the hydration of the ions. Kern [114] held the opinion that impurities can modify the different crystallographic faces and thus affect their relative growth rates.

Though the effect of impurities on growth was analyzed by kinetic theories and by surface energy approach, Mullin [115] concluded that it was rarely possible to arrive at a well founded explanation of how a given impurity affected the growth of a given crystal. It would rather seem that more than one mechanism might apply to any particular case.

The series of "seeded" growth studies, and especially those which used the "constant composition" method, greatly contributed to the systematization of the impurity effects on growth. The main purpose of these studies was to determine the order of the growth process, thus indicating the most probable growth mechanism. However, the majority of these studies were engaged concurrently with the effect of impurities on the growth rates. The investigated salts were sparingly soluble salts and the impurities were pyrophosphate and a series of phosphonates. The characteristic feature of phosphonates is a backbone of two or more carbon atoms to which PO_3^{2-} functional groups are bonded imparting the molecule reactivity with respect to adsorption, whereas the carbon backbone makes them inert toward hydrolysis. The molecules are effective in inhibiting growth of mineral deposits.

The indirect evidence that the phosphonates retard growth rates of mineral deposits by adsorbing on growing crystal faces may be deduced from comparison of the effects of the common chelating agent, EDTA, and its phosphonate analog ENTMP. When represented in writing the chemical formulae of the two substances look very similar.

The difference is that EDTA possesses four carbonate terminal groups, whereas ENTMP has four phosphonic groups. In space they display different properties: because of the planar structure of carbonate the EDTA molecule is flexible and can bend over whereas the more bulky tetrahedral phosphonate make this molecule less able to bend. The free bending capacity makes EDTA a six-dentate chelating agent, wherein one molecule of EDTA reacts with only one calcium ion. The sluggish ENTMP molecule is inferior with respect to complexing capacity but 0.1 mol% of this additive is sufficient to inhibit growth of calcium sulphate or phosphate crystals [116] whereas a relatively high proportion (almost 1:1 molar ratio) of EDTA is needed for the same effect. ENTMP is able to adsorb on the faces of the mineral phases, whereas the prefered state for EDTA is in the form of a ligand to a single ion.

This example highlights the importance of structural factors in the interactions between the impurity and the affected crystal. If an organic molecule has the proneness to adsorb on a crystal face because of charge affinity and dimensional fit, it can adsorb on growth sites and retard growth. This topic will also be researched and hopefully solved by computer modeling.

5. *Growth of single crystals*

"Solution growth" of single crystals includes many types of techniques which are dealt with in the appropriate chapters of this Handbook-growth from molten salt, flux, liquid metal etc. In this chapter, only the comparatively narrow aspect of single crystal growth from aqueous solutions is presented. Though the solution growth techniques are different with respect to range of technological application, industrial and economic importance and extent of research, all of them have some common scientific background.

A comprehensive review on aqueous solution growth up to 1950 is included in the classic book of Buckley [108].

There are several salts easily grown from each large crystal that are reasonably free of imperfections. For instance crystals of potassium hydrogen phosphate are grown in simple tanks by evaporation and cooling without engaging a particular mode of control. However, the growth of the majority of crystals necessitate the employment of more careful and sophisticated techniques.

Hurle [117] classified the various techniques into those which have a constant (time-invariant) growth temperature and those for which the interface temperature is varying with time.

A simple aqueous solution crystallizer commonly consists of a rotating shaft with perpendicular branches (rotating tree) with a seed crystal attached to each arm [117]. The fluid flow so provided breaks up the boundary layer of rejected solvent and leads to somewhat faster growth and allows one to obtain more perfect crystals than those grown in quiescent solutions. Supersaturation is maintained by slow cooling, the rates being indeed very slow: 0.1 to 1°C per day. This requires a very careful and reliable temperature control.

More elaborate crystallizers use the possibility of easily pumping aqueous solutions. Multiple tanks are employed, a "saturator" and "growth" tank; the saturator is kept at a temperature slightly higher than the growth tank with solution being loaded with the solute at the saturator, pumped to the growth tank, depleted there by growth and recycled to the saturator.

A major problem is to attain sufficient supersaturation to maintain an adequate growth rate, while avoiding spurious nucleation of undesirable embryos. Because keeping mechanical control of the supersaturation on the very narrow edge of economic demands is difficult, chemical means were sought after. Complexing agents were added to such solutions, lowering the concentration of the ions or molecules in the vicinity of the growing crystal, but releasing them after the depletion caused by growth.

Another major problem is poisoning of the growing faces by trace impurities. The growth rates and the crystal habit may be extremely sensitive to impurities, as mentioned in the former sections of this chapter. However, in agreement with the multi-faced effect of impurities, also mentioned, in some cases the growth rates can be enhanced by impurities.

The technique has, nevertheless, many advantages. The main advantage is that it usually is used at a temperature near to room temperature, thus lowering energy consumption. Frequently the growth proceeds layer by layer, the crystal showing vapor growth habit. This results in high purity single crystals, relatively free of stresses and strains.

Unfortunately, the method is best suited for the growth of easy soluble ionic or molecular crystals, having a relative small technical importance. Nicolau [118] analyzed in detail the factors which could allow one to extend the solution growth method to the growth of iono-covalent crystals.

Iono-covalent compounds are commonly sparingly soluble, because the iono-covalent bond is stronger than the ionic or van der Waals bond owing to a partial overlapping of the molecular orbitals. Liberation of ions or molecules in solution would need the supply free energy to be higher than that supplied by solvation. One way to circumvent this difficulty is to use a double exchange chemical reaction involving electron donors and acceptors:

$$C_1A_1 + C_2A_2 \longrightarrow C_1A_2 + C_2A_1,$$

giving the desired iono-covalent compound C_1A_2.

As the compound is sparingly soluble, the solution can hardly support any supersaturation and the product precipitates. To impede precipitation, dilute solutions were used and the ions of the crystal to be grown were brought together by counter-diffusion. However, in practical cases, in order to improve control a gel is used as the reaction medium.

Another way to use aqueous solution growth for iono-covalent crystal growth is to use a complexing agent. The free energy needed to break up the iono-covalent bonds is supplied, in this situation, by the energy released through the formation of the complexing bonds.

The complexing agent must be carefully selected, fulfilling two main requirements:

(a) It must interact chemically with the iono-covalent compound and yield an easily soluble compound. The shift of the reaction equilibrium constant from unity must not be significant.

(b) The reaction must be strong enough to dissolve adequate quantities of the complexed component, but on the other hand must be weak enough to allow the incorporation of this component in the growing crystal.

For a given system the crystallization kinetics depends not only on the temperature and supersaturation but also on the solvent/complexing agent mol ratio. Optimization of the growth conditions involves: a physico-chemical characterization of the solution, a study of the crystal habit and a study on crystallization kinetics in connection with the solution properties and crystal structure.

Nicolau [118] listed the single crystals that were grown by this method up to 1980: AgCl, AgBr, β-AgI, Se, α-HgS, CuCl, PbO, PbI_2 and α-HgI_2 using respectively, as complexing agents NH_3, KI · Na_2S · HCl, NaOH, NaI and HCl.

Nicolau remarks that due to the complexity of the coordination chemistry, it is not possible to make valid estimates about the ease with which different classes of iono-covalent compounds might be grown. However, general estimates within a class may be made. Thus the compounds derived from a mono-valent acid radical such as the halides or pseudo-halides of transition or heavy metals are expected to grow easily. Then, some hydroxides, sulfides, oxysulfides, oxyhalides, inorganic anhydrides, amphoteric oxides, metalloides, intermetalloidic compounds, metals and some salts derived from organic or inorganic polybasic oxyacids may be grown.

A practical example is presented [119] by the growth procedure of α-HgI_2 single crystals from dimethylsulfoxide complexes.

A water solution method with a controlled chemical reaction was used for growth of ferroelectric $PbHPO_4$ [120]. First, the hydrogen lead phosphate was precipitated from lead nitrate and disodium hydrogen phosphate. Then it was dissolved in dilute nitric acid and the driving force was supplied by absorption of gaseous ammonia. A flat bottomed vessel was used with a slow stirrer located in a closed system. Plate-like crystals with average dimensions of $5 \times 1.5 \times 4\,mm^3$ were grown during 10 days.

Actual aqueous solution growth is presently of rather limited use. Most crystals of technological interest are grown from the melt. Few compounds such as for instance vanillin [121] were reported to be grown from methanol, ethanol and chloroform which strongly and variably affected its crystal habit.

References

[1] R.H. Horne, Water and Aqueous Solutions (Wiley–Interscience, New York, 1972) p. 245.

[2] Y. Marcus, Ionic Liquids (Plenum Press, New York, 1981) p. 108.

[3] V.I. Voznesenskaya, Vop. Fiz. Khim. Rastvorov Electrolit, ed. G.I. Mikulin (Khimia, Leningrad, 1968) p. 172.

[4] C.L. Leci and J.W. Mullin, Phil. Mag. 19 (1969) 1075.

[5] A.T. Allen, M.P. McDonald, W.M. Nicol and R.M. Wood, Nature 235 (1972) 36.

[6] G.A. Hussmann, M.A. Larson and K.A. Berglund, in: Industrial Crystallization 84, eds. S.J. Jancic and E.J. De Jong (Elsevier, Amsterdam, 1984) p. 21.
[7] M.A. Larson and J. Garside, Chem. Eng. Sci. 41 (1986) 1285.
[8] R. Becker and W. Doring, Ann. Physik 24 (1935) 719.
[9] M. Volmer and A.Z. Weber, Z. Physik. Chem. 119 (1925) 277.
[10] M. Volmer and H. Flood, Z. Physik. Chem. A 170 (1934) 272.
[11] R.F. Strickland-Constable, Kinetics and Mechanism of Crystallization (Academic Press, New York, 1968).
[12] J. Lothe and G.M. Pound, J. Chem. Phys. 36 (1962) 2080.
[13] J.L. Katz, Ph. Mirabel, C.J. Scoppa and T.R. Virkler, J. Chem. Phys. 65 (1976) 382.
[14] H.A. Sharaf and R.A. Dobbins, J. Chem. Phys. 77 (1982) 1517.
[15] R. Strey, T. Shmeling and P.E. Wagner, J. Chem. Phys. 85 (1986) 6192.
[16] F.C. Collins and J.P. Leinweber, J. Phys. Chem. 60 (1956) 389.
[17] I.M. Kolthoff and B. van't Riet, J. Phys. Chem. 63 (1959) 817.
[18] A.E. Nielsen, Acta Chem. Scand. 15 (1961) 441.
[19] B.V. Enustun and Turkevitz, J. Am. Chem. Soc. 82 (1960) 4502.
[20] A.E. Nielsen, Krist. Tech. 4 (1969) 17.
[21] A.E. Nielsen, Proc. Int. Symp. Nucleation Phenomena (Cleveland, 1965).
[22] O. Sohnel, J. Cryst. Growth 57 (1982) 101.
[23] A. Packter, A. Alleen, P. Chauhan and S.C. Uppaladini, Cryst. Res. Technol. 16 (1981) 1419.
[24] W.A. House and J.A. Tutton, J. Cryst. Growth 56 (1982) 699.
[25] O. Sohnel, Cryst. Res. Technol. 16 (1981) 651.
[26] M.C. Uppreti and A.G. Walton, J. Chem. Phys. 44 (1966) 1936.
[27] C. Rinaudo, F. Abbona and R. Boistelle, J. Cryst. Growth 66 (1984) 607.
[28] S. Sarig and F. Kahana, J. Cryst. Growth 35 (1976) 145.
[29] S. Sarig and O. Ginio, J. Phys. Chem. 80 (1976) 256.
[30] J.D. Birchall and R.J. Davey, J. Cryst. Growth 54 (1981) 323.
[31] G. Wulf, Z. Kristallogr. 34 (1901) 449.
[32] N.F. Mott, Nature 165 (1950) 295.
[33] W.K. Burton, N. Cabrera and F.C. Frank, Phil. Trans. A 243 (1951) 299.
[34] A.A. Chernov, Sov. Phys. Usp. 4 (1961) 116.
[35] K.A. Jackson, J. Cryst Growth 24 & 25 (1974) 130.
[36] P. Hartman and W.G. Perdok, Acta Cryst. 8 (1955) 49.
[37] P. Bennema, J. Cryst. Growth 24 & 25 (1974) 76.
[38] P. Bennema, J. Boon, C. van Leeuwen and G.H. Gilmer, Krist. Technik 8 (1973) 659.
[39] W.R. Marshall and G.H. Nancollas. J. Phys. Chem. 73 (1969) 3838.
[40] S.T. Liu and G.H. Nancollas, J. Cryst. Growth 6 (1970) 281.
[41] G.H. Nancollas and M.S. Mohan, Arch. Oral. Biol. 15 (1970) 731.
[42] D.M.S. Little and G.H. Nancollas, Trans. Faraday Soc. 66 (1970) 3103.
[43] M.M. Reddy and G.H. Nancollas, J. Colloid Interface Sci. 36 (1971) 166.
[44] S.T. Liu and G.H. Nancollas, Desalination 12 (1973) 75.
[45] G.H. Nancollas and G.L. Gardner, J. Cryst. Growth 21 (1974) 267.
[46] S.T. Liu and G.H. Nancollas, J. Inorg. Nucl. Chem. 38 (1976) 515.
[47] S.T. Liu, G.H. Nancollas and E.A. Gasiecki, J. Cryst. Growth 33 (1976) 11.
[48] G.L. Gardner and G.H. Nancollas, J. Dent. Res. 55 (1976) 342.
[49] Z. Amjad, P. Koutsoukos, M.B. Tomson and G.H. Nancollas, J. Dent. Res. 55 (1976) 342.
[50] P.G. Koutsoukos, Z. Amjad, M.B. Tomson and G.H. Nancollas, J. Am. Chem. Soc. 102 (1980) 1553.
[51] P.G. Koutsoukos, C.Y. Lam-Erwin and G.H. Nancollas, Invest. Urol. 18 (1980) 178.
[52] M.R. Christoffersen, J. Christoffersen, M.P.C. Weijnen and G.M. Van Rosmalen, J. Cryst. Growth 58 (1982) 585.
[53] A.E. Nielsen and J.M. Toft, J. Cryst. Growth 67 (1984) 278.
[54] A.E. Nielsen, J. Cryst. Growth 67 (1984) 289.
[55] A.E. Nielsen and O. Sohnel, J. Cryst. Growth 11 (1971) 233.
[56] A.E. Nielsen, Pure Appl. Chem. 53 (1981) 2025.

[57] R. Reich and M. Kahlweit, Ber. Busenges. 72 (1968) 66.
[58] A.E. Nielsen, Croat. Chem. Acta 60 (1987) 531.
[59] A. Glasner, Mater. Res. Bull. 8 (1973) 413.
[60] A. Glasner, Israel J. Chem. 12 (1974) 817.
[61] A. Glasner and S. Skurnik (Sarig), Israel J. Chem. 3 (1965) 143.
[62] A. Glasner and S. Skurnik (Sarig), J. Chem. Phys. 47 (1967) 3687.
[63] S. Sarig, C.L. Leci and N. Eidelman, J. Cryst. Growth 47 (1979) 365.
[64] O. Sohnel and J. Garside, J. Cryst. Growth 89 (1988) 202.
[65] R.C. Tolman, J. Chem. Phys. 16 (1948) 758; J. Chem. Phys. 17 (1949) 333.
[66] M.A. Larson, and J. Garside, J. Cryst. Growth 76 (1986) 88.
[67] V. Veverka, O. Sohnel, P. Bennema and J. Garside, AIChE J. 37 (1991) 490.
[68] Y.C. Chang and A.S. Myerson, AIChE J. 32 (1986) 1567.
[69] D. Turnbull, Acta Met. 1 (1954) 684.
[70] R.A. Johnson and J.D. O' Rourke, J. Am. Chem. Soc. 76 (1954) 2124.
[71] A.G. Walton, Z. Physik. Chem. 35 (1962) 326.
[72] A.G. Walton, The Formation and Properties of Precipitates (Interscience, New York, 1967).
[73] A.E. Nielsen, Kinetics of Precipitation (Pergamon Press, New York, 1964).
[74] L.J.M. Blomen, E.J. Will, O.L.M. Bijvoet and H. Van Der Linden, J. Cryst. Growth 64 (1983) 306.
[75] E.J. Will, O.L.M. Bijvoet, L.J.M. Blomen and H. Van Der Linden, J. Cryst. Growth 64 (1983) 297.
[76] D. Skrtic, M. Markovic, L.J. Komunjer and H. Furedi-Milhofer, J. Cryst. Growth 66 (1984) 431.
[77] M. Markovic, D. Skrtic and H. Furedi-Milhofer, J. Cryst. Growth 67 (1984) 645.
[78] R.L. Ryall, C.J. Bagley and V.R. Marshall, Invest. Urol. 18 (1981) 401.
[79] R.L. Ryall, R.G. Ryall and V.R. Marshall, Invest. Urol. 18 (1981) 396.
[80] Rome de l'Isle, Crystallographie (Paris) 1703.
[81] H. Freundlich, Colloid and Capilary Chemistry (Dutton, New York, 1922).
[82] K. Marc, Z. Physik. Chem. 61 (1900) 305; 67 (1909) 70; 68 (1909) 104; 73 (1910) 605.
[83] J.J.P. Valston, Physik. Z. 21 (1920) 606; 59 (1924) 335.
[84] P. Paneth and Z. Horowitz, Z. Physik. Chem. 89 (1915) 513.
[85] G. Bliznakov, Fortschr. Mineral. 36 (1959) 149.
[86] J.N. Stranski, Z. Krist. 105 (1943) 287.
[87] G. Bradistilov and J.N. Stranski, Z. Krist. 103 (1940) 1.
[88] A. Bravais, J. L'Ecole Polytechnique 167 (1851).
[89] D. Raistrick, Discuss. Faraday Soc. 5 (1949) 234.
[90] M.E. Tadros and I. Mayers, J. Colloid Interface Sci. 72 (1979) 245.
[91] S. Sarig, F. Kahana and R. Leshem, Desalination 17 (1975) 215.
[92] S. Sarig, A. Glasner and J. Epstein, J. Cryst. Growth 28 (1975) 295.
[93] J.V. Garcia-Ramos and P. Carmora, J. Cryst. Growth 57 (1982) 336.
[94] L. Addadi and S. Weiner, Proc. Natl. Acad. Sci. USA 82 (1985) 4110.
[95] L. Addadi, Z. Berkovitch-Yellin, I. Weisbush, J. van Mil, L.J.W. Shimon, M. Lahav and L. Leiserowitz, Angew. Chem. Int. Ed. 24 (1985) 466.
[96] C.E. Bricker and D.G. Myers, US Clearinghouse Fed. Sci. Tech. Inf. PB Report No. 190 (1969) p. 570.
[97] S. Sarig and J.W. Mullin, J. Chem. Tech. Biotech. 32 (1982) 525.
[98] A.D. Randolph and M.A. Larson, Theory of Particulate Processes (Academic Press, New York, 1971).
[99] S.M. Shor and M.A. Larson, Chem. Eng. Prog. Symp. Ser. 87 (1971) 32.
[100] N. Eidelman, R. Azoury and S. Sarig, J. Cryst. Growth 74 (1986) 1.
[101] Y. Liu and G.D. Botsaris, AIChE J. 19 (1973) 510.
[102] J. Budz, A.G. Jones and J.W. Mullin, J. Chem. Tech. Biotech. 36 (1986) 153.
[103] G.D. Botsaris and C. Sutwala, AIChE Symp. Ser. 72 (1976) 7.
[104] J. Garside and V.F. Komarov, Krist, Tech. 11 (1976) 699.
[105] K. Shimizu, N. Kubota and T. Kawakami, Kagaku Kogaku Rombashu 5 (1979) 433.
[106] A. Winzer, H.H. Emos and V. Buerger, Freiberg Forsch. A 600 (1979) 61.
[107] C.W. Bunn, Proc. R. Soc. 141 (1933) 567.
[108] H.E. Buckley, Crystal Growth (Wiley, New York, 1951).

[109] G.W. Sears, in: Growth and Perfection of Crystals, eds. R.H. Doremus, B.W. Roberts and D. Turnbull (Wiley, New York, 1958).
[110] C.W. Davies and G.H. Nancollas, Trans. Faraday Soc. 51 (1955) 823.
[111] G.H. Nancollas and N. Purdie, Trans. Faraday Soc. 57 (1961) 1.
[112] E.V. Khamskii, Crystallization from Solutions (Consultants Bureau, New York, 1969).
[113] J.W. Mullin, A. Amatavivadhana and M. Chakraborty, J. Appl. Chem. 20 (1970) 159.
[114] R. Kern, Bull. Soc. France Mineral. Cristallogr. 91 (1968) 247.
[115] J.W. Mullin, Crystallisation (Butterworth, London, 1972).
[116] S. Sarig, Report to Stauffer Chem. Co. (Dobbs Ferry, New York, 1986).
[117] D.T.J. Hurle, Crystal Growth: A Tutorial Approach, eds. W. Bardesley, D.T.J. Hurle and J.B. Mullin (North-Holland, Amsterdam, 1979).
[118] I.F. Nicolau, J. Cryst. Growth 48 (1980) 45.
[119] I.F. Nicolau and J.P. Joly, J. Cryst. Growth 48 (1980) 61.
[120] B. Brezina and J. Horvath, J. Cryst. Growth 52 (1981) 858.
[121] N.B. Sing, T. Henningsen, E.P. Metz, R. Hammacher, E.E. Cumberledge, R.H. Hopkins and R. Mazelsky, Mater. Lett. 12 (1991) 270.

CHAPTER 20

Crystal Growth in Gels

F. LEFAUCHEUX and M.C. ROBERT

Laboratoire de Minéralogie-Cristallographie, associé au CNRS, Universités PVI et PVII, Paris, France

Contents

1. Introduction . . . 1273
2. Techniques and new materials grown in gel . . . 1273
  2.1. Gel growth techniques . . . 1273
    2.1.1. Insoluble crystals . . . 1273
    2.1.2. Soluble crystals . . . 1274
  2.2. Materials grown in gel . . . 1276
    2.2.1. Inorganic materials . . . 1276
    2.2.2. Organic materials . . . 1276
3. Gel structure . . . 1279
  3.1. Different types of hydrogels . . . 1280
    3.1.1. Lamellar gels . . . 1280
    3.1.2. Chemical gels . . . 1280
    3.1.3. Physical gels . . . 1280
    3.1.4. Aggregation of particles . . . 1280
  3.2. Structure of hydrogels . . . 1280
    3.2.1. Silica gels . . . 1280
    3.2.2. Agarose gels . . . 1285
  3.3. Gel properties related to crystal growth . . . 1285
4. Mass transfer in gel . . . 1286
  4.1. Diffusion coefficient . . . 1286
  4.2. Mass transfer associated with different diffusion systems . . . 1288
    4.2.1. Numerical simulations . . . 1288
    4.2.2. Experimental determination by optical methods . . . 1288
    4.2.3. Experimental determination by chemical methods . . . 1290
  4.3. Application of the diffusion properties of gels to microgravity simulations . . . 1291
5. Nucleation and growth in gel media . . . 1292
  5.1. Nucleation . . . 1293
    5.1.1. Some definitions . . . 1293
    5.1.2. An experimental study . . . 1293
    5.1.3. Influence of the gel upon nucleation . . . 1294
  5.2. Crystal growth . . . 1294

HANDBOOK OF CRYSTAL GROWTH, VOL. 2
edited by D.T.J. Hurle

5.3. Analysis of gel behaviour during nucleation and growth 1296
6. Crystalline quality and physical properties of gel-grown crystals 1297
6.1. Crystal habit 1297
6.1.1. Dendrites 1297
6.1.2. Twins 1297
6.2. Impurity 1298
6.2.1. Homogeneous incorporation 1298
6.2.2. Doping levels in gels 1298
6.3. Lattice defects (X-ray diffraction) 1298
6.3.1. Global measurements 1298
6.3.2. Local measurements 1299
6.4. Physical measurements on gel-grown crystals 1300
6.4.1. Phase transition 1300
6.4.2. Second harmonic generation in non-linear optics 1300
7. Conclusions 1301
References 1301

1. Introduction

Crystalline quality and physical properties of materials are highly dependent on the growth method used. High-temperature growth often leads to strained crystals; so when possible, it is preferable to use soft methods such as low-temperature solution growth and among them, gel growth techniques. Anyway, these methods must be used in the case of mechanically fragile or thermally unstable crystals.

Gel growth is, in fact, a particular case of solution growth where the solution is trapped in a polymeric structure; the presence of the gel, on the one hand induces particular features concerning nucleation and growth and, on the other hand, needs adapted growth techniques.

During these last ten years, many papers have been published concerning particular points of gel growth but very few general reviews on this subject appeared [1–3] and the reference in this domain remains Henisch's book, recently re-published [4]. In the review presented here, we restrict the references to those published after 1986, the latest that have been quoted in Henisch's work.

This chapter will cover the following topics: first an outlook on different techniques which can be applied to grow crystals in gels; then some examples of crystals which draw benefit of this technique will be given. Characteristics of nucleation and growth depend on the gel structure which will be related in detail for silica gels. Mass transfer in a gel occurs only by diffusion, no convection currents are present so that gel can be used as a model of a diffusive medium, which, e.g. allows one to simulate on earth microgravity experiments. The supersaturation which can be attained at nucleation or during growth depends on the gel structure, as will be shown, but it is always higher than in free solution. Finally, physical properties of gel-grown crystals will be presented in comparison with those of solution-grown crystals.

2. Techniques and new materials grown in gel

2.1. Gel growth techniques

Crystallization takes place in a gel obtained by polymerization just before the growth. In free solution growth, crystals can be grown either from a chemical equilibrium (the case of "insoluble" crystals) or from a physical equilibrium (the case of "soluble" crystals). Let us examine these two possibilities.

2.1.1. Insoluble crystals

As an example, growth of lead monetite crystals can be made by mixing two solutions, one containing Pb^{2+}, the other phosphate ions. According to the following chemical reaction,

$$Pb^{2+} + HPO_4^{2-} \rightleftharpoons PbHPO_4,$$

precipitation takes place provided that the solubility product of $PbHPO_4$ is exceeded.

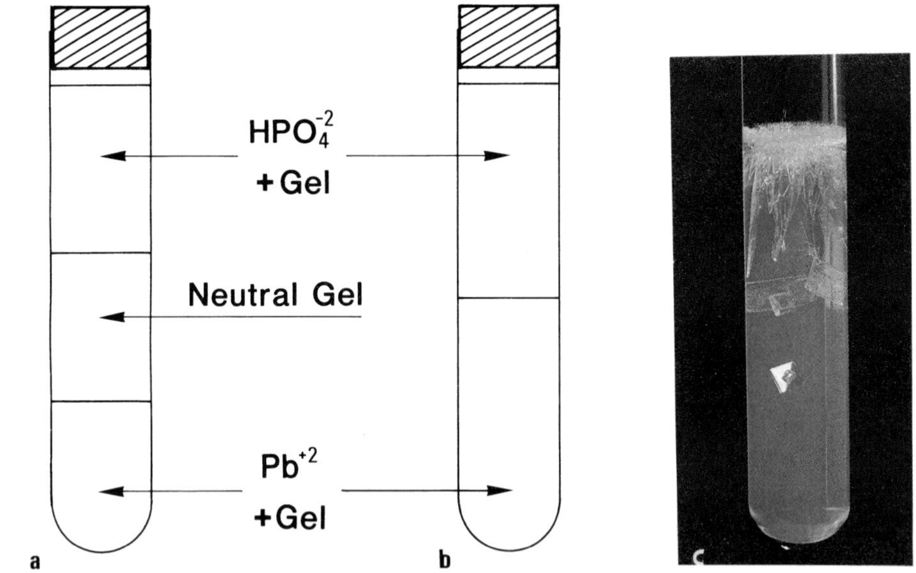

Fig. 1. Gel growth by double diffusion and precipitation. (a) Three-layer technique. (b) Two-layer technique. (c) $PbHPO_4$ crystals seen in their growth medium obtained by the two-layer technique in a 7% TMOS gel. Inner diameter of the test tube: 2 cm.

Two techniques can be used in gel (fig. 1): the three-layer technique (a) separates initially the two considered ions; after diffusion towards the central part, nuclei appear where the needed supersaturation is attained, generally in the central part if the diffusion rate of both ions is about the same. The two-layer technique (b) is rapid to implement but, at the interface, a spurious nucleation appears and a polycrystalline crust often hinders further diffusion and crystallization. However, for crystals whose nucleation is easy, only a few crystals appear within the gel as shown in fig. 1c, which shows lead monetite crystals grown in a silica gel by this technique.

2.1.2. Soluble crystals

In the case of soluble crystals, one can make use of the variation of solubility of the material in a given solvent as a function of parameters such as temperature or addition of other solvents. If the solubility increases with temperature, crystals can be grown by T decrease in free solution: this method is easily transposed to gel growth as shown in fig. 2a, where the gel is made in the presence of a saturated solution of the material at the initial temperature. When the gel is set, the supersaturation is established by cooling the test tube (initially or continuously during growth). Crystals appear everywhere in the gel as can be seen in fig. 2c which shows KH_2PO_4 crystals (KDP) in their growth medium.

A second method which can be proposed is shown in fig. 2b. After the gel with the saturated solution is set, a liquid layer of a solvent in which the material is insoluble is added (e.g. acetone in the case of KDP crystals, because KDP is insoluble in acetone).

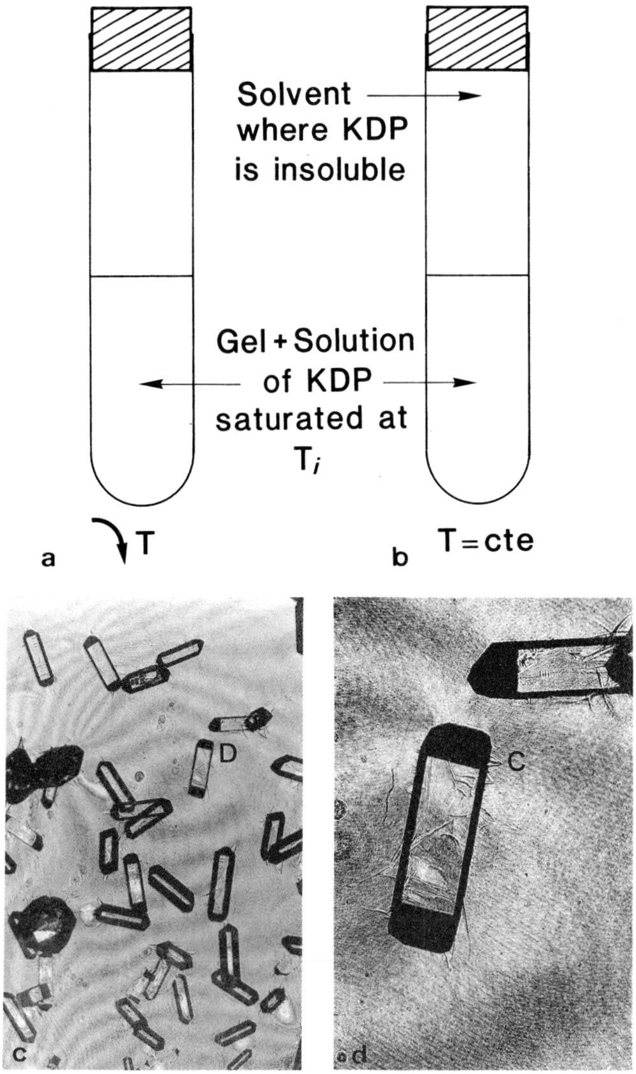

Fig. 2. Gel growth of soluble crystals. Principle of the experiment: (a) by temperature decrease; (b) by the addition of a liquid in which the material is insoluble. (c) KH_2PO_4 crystals grown by temperature decrease. (d) Enlargement of crystal D showing modification of the gel due to crystal growth. Crystal D has a length of 500 μm. C: cusp-like cavities.

Diffusion of this liquid in the gel increases the supersaturation and can induce crystallization. The crystal distribution differs from the preceding case, due to a concentration profile along the gel column depending on diffusion of the solvent: the number of crystals decreases rapidly with distance from the interface. Different modifications of these basic techniques can be achieved according to the material to be grown and a full review of these techniques is given in [2]. Note that the described techniques lead to spontaneously nucleated crystals formed within these gels, contrary to free solution-grown crystals which are usually obtained from a seed in a stirred bath: growth conditions are then completely different. Seeding in a gel can nevertheless be performed in particular cases that will be discussed later.

2.2. *Materials grown in gel*

2.2.1. *Inorganic materials*

Crystals which have been grown in gel are almost always sparingly soluble because of the difficulty to get large crystals in this case. Table 1 presents a non-exhaustive list of inorganic materials which have been obtained in recent years. Other examples can be found in [1–4].

2.2.2. *Organic materials*

Table 2 shows some organic crystals grown, in most cases, in silica gel.
Two particular cases will be discussed here depending on the solvent.

2.2.2.1. Proteins. Protein crystal growth in gel presents some particular features which makes crystallization more difficult than for other crystals. First, crystals of proteins are rather fragile, mechanically, as they contain a large percentage of water (15–80%) so that the required gel must be broken easily by the growing crystal. This is possible with low-concentrated gels which present a low stiffness.

Second, the protein to be crystallized is available in small quantities; gel volumes have to be only a few microliters contrary to a, more usual, few milliliters.

Table 1
Inorganic materials newly grown in gels.

Substance	Gel	Reference	Year
$UO_2HPO_4, 4H_2O$	TMOS gel (silica)	[5]	1986
$PbHPO_4$	silica	[6]	1990
$CaHPO_4, 2H_2O$	agar	[7]	1991
$Ca_8H_2(PO_4)_6, 5H_2O$	silica	[8]	1990
$PbCrO_4$	film	[9]	1987
$CaSO_4, \frac{1}{2}H_2O$	silica (epitaxy)	[10]	1991
$CaSO_3, \frac{1}{2}H_2O$	silica	[11]	1991
$RbClO_4$	ion-exchanged silica gel	[12]	1987
BiSI	silica	[13]	1990
BiSCl	silica	[14]	1990
SbSeCl	silica	[15]	1991
As_2S_3	silica	[16]	1986
γ-Al_2O_3	alumina	[17]	1987
sapphire	boehmite	[18]	1987
$Pb(N_3)_2$	agar	[19]	1989
MIO_4 (M = K, Rb, Cs)	silica	[20]	1992

Table 2
Organic materials newly grown in gels.

Substance	Gel	Reference	Year
Metal II carboxylates			
formates			
acetates	TMOS gel	[21]	1990
propionates			
Oxalates			
CaC_2O_4, H_2O	silica	[22]	1987
$La_{1-x}Cu_x$, $3C_2O_4$, nH_2O	silica	[23]	1990
CdC_2O_4, $3H_2O$			
oxamic acid	silica	[24]	1989
complex Pb^{2+}, Ca^{2+}, Cd^{2+}			
α-glycine	silica	[25]	1988
$C_{27}H_{46}O$ cholesterol	silica	[26]	1987
CoANP and PNP	TMOS gel	[27]	1991
hel egg white lyzozyme	agarose, TMOS gel	[28]	1991

Finally, proteins must be grown at a given pH; this recommendation is fulfilled by adding a buffer to the growth medium.

The choice of the gel depends mainly on the type of technique used for growing the protein. Silica gel can be applied, as shown previously, to liquid–liquid diffusion but it is not well adapted to a vapour phase because the gel forms a crust by dehydration. These two techniques will be examined successively.

2.2.2.1.1. Hanging drop technique in gel. Agarose is a powder extracted from seaweed; it consists of polysaccharide chains carrying substituents (methyl, sulphate, etc.) which vary according to the origin of agarose. Agarose gel is obtained by decreasing the temperature of an aqueous solution of agarose, down to the gelation temperature which is about 36°C according to the substituents and the gel concentration.

A possible experiment protocol is minutely described in [28]. Let us summarize the main features. In this technique (fig. 3, [28]) a drop of gelled growth solution is set

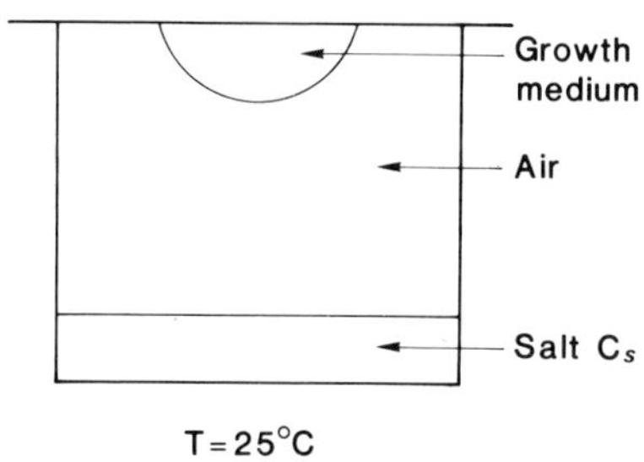

Fig. 3. Principle of the hanging drop technique. Composition of the growth medium (example of hen egg white lysozyme): agarose: 0.1%, NaCl: $\frac{1}{2}C_s$ acetate buffer: 0.1 M, protein: 40 g/l. Volume of the droplet: 20 μm.

above a salt (e.g. NaCl if the protein under study is insoluble in concentrated NaCl solutions) reservoir at a given concentration C_S. The drop is formed with a rather complicated medium whose initial composition for a particular protein is given in the caption of fig. 3. In fact, the mixing cannot be made directly because of the thermal fragility of the protein: a first mixing is made with the first three constituents at 40°C and is added to an equal volume droplet of protein solution at 20°C just before turning it over the reservoir. The whole set-up is then maintained at a constant temperature. The equilibration of vapour pressure (which depends on salt concentration) between droplet and reservoir leads to a dehydration of the droplet, whose volume decreases, inducing supersaturation and consequently nucleation. Vapour diffusion techniques can be performed with other reactants in the reservoir such as volatile acids or bases if a pH variation is needed for supersaturation.

2.2.2.1.2. Batch crystallization. This technique consists in preparing directly a supersaturated solution in a closed cuvette at a given temperature [29]. The delay of nucleation of crystals depends only on nucleation kinetics while in the preceding technique the equilibration time must also be taken into account. Hen egg white lysozyme crystals obtained by this technique are shown in fig. 4 after two weeks. They are obtained in an agarose gel (0.1%) from a growth solution whose initial concentrations are the following agarose 0.1% weight, protein 20 mg/ml, NaCl0.4M, acetic buffer 0.1M. The crystallization performed at 25°C allowed to grow crystals of about 500 μm, well isolated in the gel.

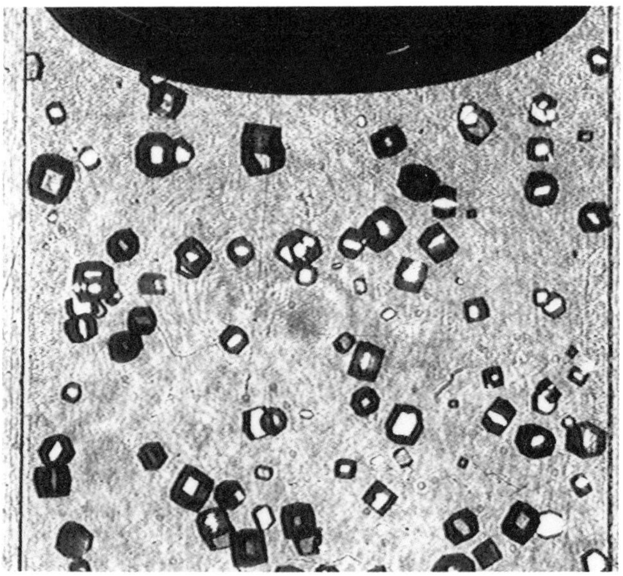

Fig. 4. Batch crystallization of HEWL in agarose gel. Growth conditions: agarose: 0.1%; HEWL; 20 mg/ml; NaCl: 0.4M; acetate buffer: 0.1M. Crystals were grown in two weeks by temperature decrease from 40 to 25°C. Width of the cuvette 1 cm.

Fig. 5. A gel-grown crystal of POM (3-methyl-4-nitro-pyridine-1-oxide). Growth conditions: ACN: 60%; POM: 350 g/l; TMOS; 5%. Temperature decrease from 40 to 25°C. Inner diameter of the test tube: 2 cm.

2.2.2.2. Organic crystals for non-linear optics. Organic crystals are good candidates for frequency doublers in the visible range, but free-solution techniques fail for growing crystals of good optical quality with a rather large size. Growth in TMOS gels has been tested with success for a few crystals: POM [30], NPP [31], CoANP and PNP [27]. These organic materials are insoluble in water but soluble in organic solvents so that different mixtures of both allow one to vary the solubility at a given temperature. For POM and NPP crystals, acetonitrile is the chosen solvent; it forms with water an azeotrope at 83% weight in ACN for which the solubility of POM or NPP is considerably enhanced; furthermore, this solubility varies significantly with temperature for the azeotropic mixture. So it is possible to perform growth either by T-decrease or by progressive addition of water. A unique POM crystal (fig. 5) was grown in a gel by T-decrease (40–25°C). The nucleation rate was very low due to the growth conditions: gel, temperature and growth-medium composition. This composition was the following: ACN 60% – water 40% (percentage in volume measured at 20°C); solute: POM, 350 g/l; TMOS 5% in volume.

Water plays a double role here; firstly by making hydrolysis of TMOS possible, secondly by tuning POM solubility which must be as high as possible. But prior to the crystal growth, the gel must be obtained in a reasonable time (within one week); in fact, gelation time, which depends on several parameters such as TMOS content, pH, additives, is an important criterion for growth medium selection.

For instance, acid must be added in the case of NPP crystallization for lowering the pH to a value inducing a rapid polymerization of the gel.

3. *Gel structure*

Hydrogels used in crystal growth are to be distinguished from xerogels used, e.g. in catalysis, the latter being often obtained by dehydration of the former. Hydrogels mean only a biphasic medium with a solid phase trapping the liquid phase (organic or aqueous solution).

Some examples of hydrogels will be given to explain some details of their structure.

3.1. Different types of hydrogels

A hydrogel can be defined as a colloidal system with at least two components, whose microscopic structure spreads continuously over a macroscopic scale. Gel behaviour (especially mechanical properties) depends on the type of bonding responsible for the gel coherence. Flory [32] distinguishes four classes of gels:

3.1.1. Lamellar gels

An example of lamellar gels are clays gels. These systems are well ordered with the lamellae all parallel. Natural materials such as bentonite or montmorillonite have been used to grow calcium oxalates [33] or gypsum [10, 34] .

3.1.2. Chemical gels

If the polymer which forms the gel is obtained by chemical bonding, gels are called chemical gels. They can be obtained by condensation or copolymerization. Polyacrylamide gels which have been used to grow $PbHPO_4$ crystals [35] are an example.

3.1.3. Physical gels

By contrast physical gels are weakly bound and gelation is obtained by variation of physical parameters such as temperature or pressure. Furthermore, the sol–gel transition is reversible which is not the case for chemical gels. Gelatine [36] and agarose [28, 37] have been used for growing mineral and organic materials.

3.1.4. Aggregation of particles

An intermediate case between chemical and physical gels consists of an aggregation of particles. These particles, formed by chemical reaction (strong bonding), aggregate to form a gel because of electrostatic charges deposited on their surface (weak bonding). This is the case for silica gels widely used in crystal growth.

Anyway the structure of the hydrogel is greatly dependent on the gel formation whose conditions vary with growth conditions of the crystals. We will detail it for silica gel (namely TMOS gel) and give some information about agarose gels.

3.2. Structure of hydrogels

3.2.1. Silica gels (the case of TMOS gels)

Silica gels are obtained by polymerization of silicic acid; the starting material can be sodium metasilicate or silanes: tetramethoxysilane (TMOS) and tetraethoxysilane (TEOS); the chemical reactions are neutralization in the first case and hydrolysis for silanes according to the global schemes:

$$Na_2SiO_3 + H_2O + 2HCl \longrightarrow Si(OH)_4 + 2NaCl,$$

$$(CH_3O)_4\,Si + 4H_2O \longrightarrow Si(OH)_4 + 4CH_3OH.$$

The use of TMOS or TEOS provides several advantages in comparison to metasilicate use. Silicic acid is formed only by addition of water, so the pH can be fixed by the growth conditions over a wide range ($3 < pH < 8$). Water can be added in a water-to-TMOS volume ratio of 1:4 so that the concentration of the growth solution does not change in a large way. Lastly, TMOS gels do not contain Na ions which can be undesirable. But CH_3OH present in the growth solution can damage crystals (in the case of some proteins) and often modifies the solubility of the material. If ethanol is preferable for the growth medium, TEOS can replace TMOS. Both are insoluble in water and an emulsion must be made by stirring vigorously the solution, the hydrolysis time being shortened at low pH (instantaneous at pH:1; infinite at pH:8).

After hydrolysis, polymerization of silicic acid takes place, depending on various parameters (silica content, additives, pH, T) which fix the structure and the kinetics at which this structure is attained. Gelation time, which measures the kinetics of both reactions (hydrolysis and polymerization), can be deduced from a mechanical study of the sol–gel transition. The structure of the gel in growth conditions can be inferred either by a direct visualization of the hydrogel or by light scattering.

3.2.1.1. Sol–gel transition – a mechanical study. The polymerization of silicic acid has been extensively studied [38]: it leads to silica particles whose diameter increases with pH. At low pH or at large ionic strength, the charges deposited at the surface of silica particles are screened : aggregation takes place and chains of polymers are formed. By increasing the pH, individual particles grow in size before aggregation and at high pH, gel cannot be formed.

The sol–gel transition is a critical phenomenon at which many physical variables diverge: in the sol phase, viscosity of the sol increases as the transition is approached, while the gel phase formation is marked by a sudden increase of elasticity. Viscosity and elasticity have been measured with a sphere-rheometer for different gels as a function of time. Gelation times t_G [39] are given below for 0.5M NaCl solutions.

TMOS:	2%	5%	10%
t_G(23°C):	> 1 day	22 h	5h

NaCl was added in order to simulate a high ionic strength. It must be noted that gelation time corresponds to the time of spreading a polymeric chain from one extremity of the tube to the other, but the gel continues to grow as will be shown later.

For crystal growers, determination of the gel structure is needed when the gel can be set in a supersaturated medium; that can be made when the gel is firm enough so that it resists pouring (t:$3t_G$). A structure determination by direct visualization has been made in these conditions.

3.2.1.2. Direct visualization of a TMOS gel. Small pieces of TMOS gels were submitted to ultra-rapid freezing followed by a deep-etch and a rotary replication. This technique,

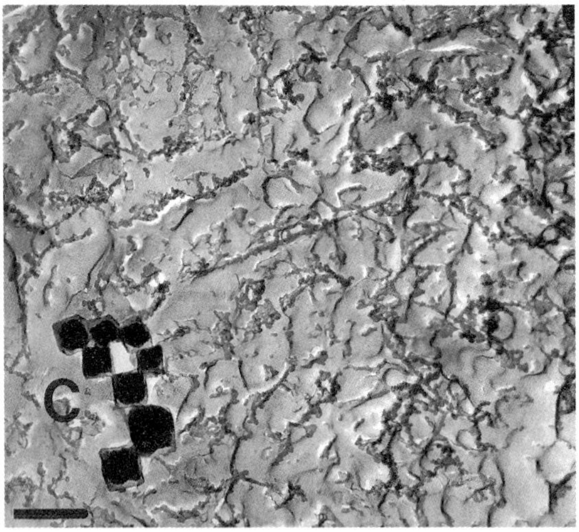

Fig. 6. Micrograph of a replica of a 5% TMOS gel. MgO crystals (C) indicate the surface of the sample. Polymer spheres (P), more or less connected, form chains limiting pores where growth solution is trapped. Bar represents 200 nm [40].

already used for fragile biological specimens, has been retained for TMOS which present a low mechanical and thermal stability, so that artefact formation is prevented at each preparation step.

Figure 6 [40] presents a micrograph obtained for a 5% TMOS gel where the polymer structure is well visible: silica spheres of about 300 Å, slightly connected and forming polymer chains. The cavities in which the solution is trapped are not well defined because the etching was not deep enough. For deep-etching, a three-dimensional visualization is needed which is obtained by tilting the sample. This has been done for 5 and 10% TMOS gels and a statistical treatment provided the following results.

TMOS	5%	10%
Mean diameter of macropores	2100 Å	600 Å
Diameter of polymer spheres	300 Å	100 Å

The visualization can give only the distribution of the largest cavities or macropores but they are the most important ones for nucleation and growth of a new solid phase. A systematic study was made on these gels by light scattering.

3.2.1.3. Mesh size in TMOS gels determined by light scattering. As previously shown, transmission electron microscopy images of frozen gel slices lead to the macropore distribution of gels ready to use. Light scattering techniques can give the evolution of polymer and cavity size during the gelation; in fact, the coherence length deduced from data analysis measures the largest distance between two polymer segments (maximum pore size).

Figure 7 summarizes the main results obtained for TMOS gels formed in an aqueous solution of $NH_4H_2PO_4$ saturated at 20°C [41]. Gelation time was estimated for these

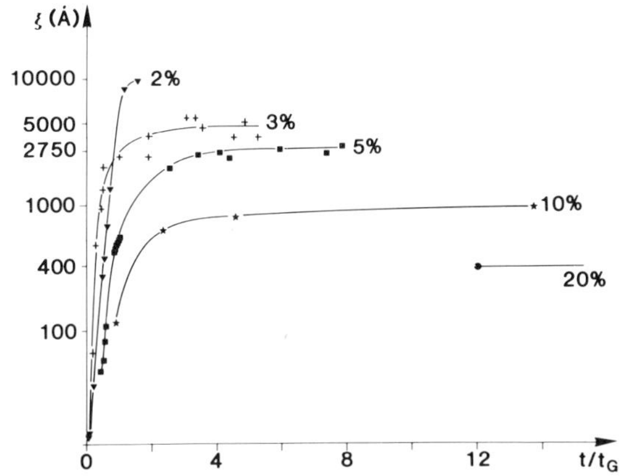

Fig. 7. Time dependence for the coherence length ξ measured for TMOS gels of different density.

experiments as the time before the gel resists pouring. It is clear that the polymer keeps growing beyond t_G, the plateau being attained for about $3t_G$. Mesh size calculated for different TMOS contents allowed one to fix a fractal dimension of 2.3 for polymer building; this fractal dimension permits extrapolation of the pore size to any TMOS content.

3.2.1.4. Mechanical properties of TMOS gels used in crystal growth. For these measurements, a gel cyclinder is strained and stress measurements are made with a balance [42]. By increasing strain, the elastic modulus increases and the elastic limit increases with the density of the gel. For further strain, gel is reduced to a dusty material according to the structure of weakly bound silica particles.

Fig. 8. KDP crystals growing in a TMOS gel. The gel is strained, due to crystal growth. Crystal size is about 3 mm.

These measurements can afford some information on gel behaviour during crystal growth; growing crystals induce strain as is well seen in fig. 8. Some cusp-like cavities have been formed starting from crystal edges.

When crystal growth lasts a long time (a few months in the case of precipitation), one must take aging and drying of the gel [43] into account. In fact, polymerization and syneresis (gel shrinkage with expulsion of solvent) are the most striking features which can induce modifications to the growth conditions.

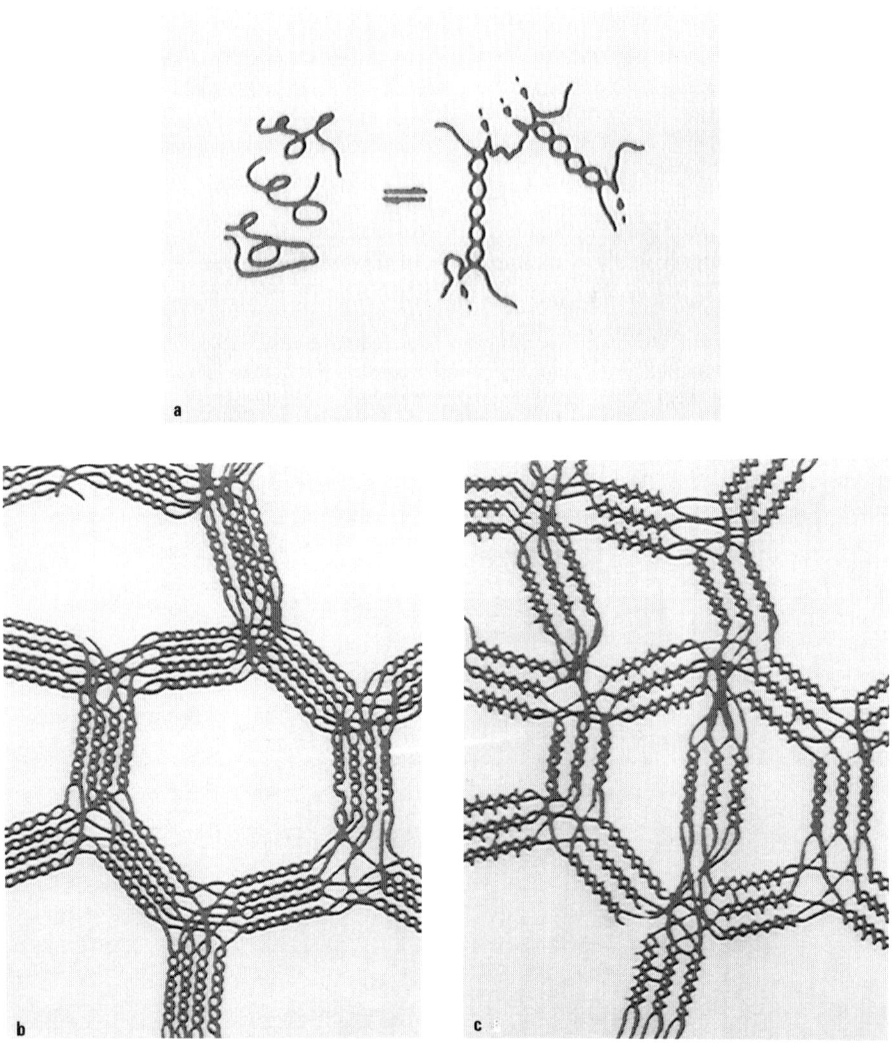

Fig. 9. Gel formation of agarose. By decreasing the temperature of an agarose solution, strands form a double helix (a) which aggregate at lower temperature to constitute suprafibres, more or less separated (b, c).

3.2.2. Agarose gels

Agarose is a polysaccharide isolated from agar, the basic disaccharide is known as agarobiose [44] whose substituents vary with the origin of the agarose.

When an agarose solution is isolated (fig. 9), gel is formed in two steps: first, single strands of agarose aggregate to form a double helix, second, a lateral aggregation of these helices produce fibers that are more or less separated depending on the agarose content [45]. The temperature of each transition varies according to agarose origin from 15 to 40°C; the resulting structure is greatly dependent on the whole thermal treatment.

Agarose gels are physical gels, weakly bound, particularly suitable for fragile materials.

We have presented only two particular gels but many others have been characterized, e.g. gelatin [46], polyacrylamide [47], by direct visualization, or TEOS gels by thermoporometry [48].

3.3. Gel properties related to crystal growth

Gel structure influences crystal growth at every step of the crystal formation.

The solid phase supports the growing crystal which remains where it has nucleated; no holder is needed and no sudden supersaturation changes occur, which often cause defects in the crystal.

The solid phase partitions the liquid phase so that convection currents cannot occur. Mass transfer is then ensured only by diffusion.

Nucleation takes place in a confined space so that it is strongly reduced in comparison to a free medium. In other words, one has to establish a high supersaturation in order to obtain nucleation.

The main characteristics of gel growth will be examined: first diffusion properties, then nucleation and growth in a gel, lastly crystalline quality of the grown material.

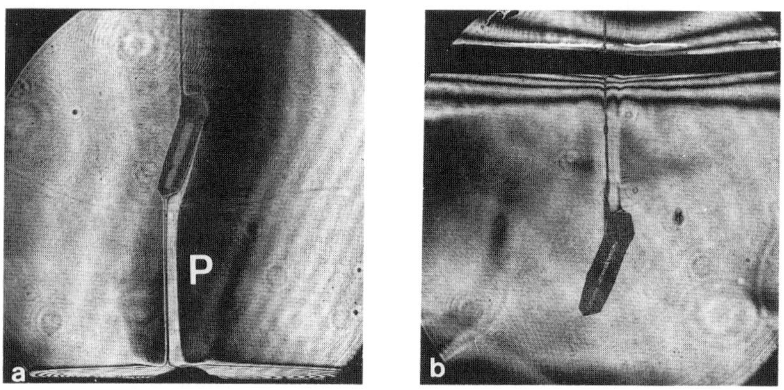

Fig. 10. Interferograms showing dissolution (a) and growth (b) of an $NH_4H_2PO_4$ (ADP) in solution. P marks convection movements due to concentration changes around the crystal.

The difference between free and gelled solutions can be seen by holographic interferometry which is sensitive to local concentration changes between two instants. Figure 10 concerns free solution where an ADP crystal was allowed to dissolve (a) or to grow (b). Dissolution enriches the layers around the crystal; they are then heavier than the surrounding ones and are seen as a plume P below the crystal. Crystal growth leads to the opposite: light layers which move upwards are seen as fringes on the top. In both cases buoyancy-driven convection currents appear because the experimental cell is large. This is not the case in a gel: fig. 13 presents some gel-grown crystals seen by holographic interferometry: concentration changes are seen around each crystal, no convection is present and mass transfer takes place by diffusion.

4. *Mass transfer in gel*

4.1. *Diffusion coefficient*

The evolution of concentration gradients of a species in a gel column elongated along an x-axis is usually represented by solutions of a Fick's equation

$$\frac{\partial c}{\partial t} = D \frac{\partial^2 c}{\partial x^2}.$$

In this representation the diffusion coefficient D is assumed to be a constant. As for diffusion in free solutions, D varies with the concentration C [49] but also with grad C. Experimentally, it means that for the determination of D, one has to choose an initial concentration gradient small enough to get a significant value of D and high enough to have a precise determination of further concentration profiles.

As a matter of fact all these diffusion coefficient determinations might be made at constant chemical potential, e.g. by using radioisotope tracers [50], but practically most of the studies have been carried out using variable regime methods which correspond to the actual situation of mass transfer in crystal growth. Fortunately these situations are rather well approximated by simple numerical solutions of Fick's equation [51].

D generally decreases when the gel content increases; several reasons are put forward to explain such a variation: lengthening of diffusion paths and/or decreasing of particle mobility due to either a mechanical obstruction by gel pores or changes in water viscosity for solution trapped in a gel structure [52].

The data on diffusion coefficients in gels found in the literature concern mainly dilute solutions of mineral species. For example, in the concentration range around 0.5M, the diffusion coefficients of NaCl in 5% and 10 vol.% TMOS gels are, respectively, 80% and 65% of the value obtained in free solutions [51]. In the same concentration range, the diffusion coefficient of chloride ions in 0.04 wt.% agarose is 90% of D_{solution} [52]. All these gels have been used fruitfully for crystal growth. These papers present some diffusion-coefficients-versus-gel-content curves $D(C_g)$. It happens that extrapolation to zero gel concentration $D(0)$ does not coincide with the diffusion coefficient measured in solution D_s in similar conditions [52, 53]. This may indicate that the diffusing species are adsorbed on or react with the gel walls.

Only a few measurements have been done with organic compounds [54] or on biological macromolecules [55].

As a rule, for a growth system allowing one to get good monocrystals, very light gels are used and consequently the diffusion coefficients in gels remain of the same order of magnitude as they are in free solution.

The determinations of D are generally made by measuring the rate at which the concentration of a given species increases or decreases in a gel column according to the conditions imposed for the solutions bathing one or both ends of the column. The measurements are made continuously in the solution by chemical or electrochemical techniques or, after a certain time of diffusion, in the successive slices of the gel column. When the gel is transparent, a very efficient way to follow in real time the diffusion profiles and to keep it free from artefacts is to determine the refractive index variations by optical techniques such as holographic interferometry [51].

It consists of recording a hologram of a first layer of gel filling the half of a parallel-sided cell. This gel contains the species under study at concentration C_0. This hologram memorizes the optical waves emitted by the cell at the hologram recording time. Then a solution containing the same species at concentration $C_0 - \Delta C$ is poured over the first layer at time t_0. A diffusion of species occurs from the bottom layer to the upper one; it results in concentration variation and correlative refractive index changes as a function of time. If the cell is kept on the holographic bench at the same position as for hologram recording, a fringe pattern (interferogram) due to interference of the present (t) and reconstructed (t_0) waves is created, permitting one to calculate optical path changes between the two times.

For a planar initial interface a system of fringes parallel to this interface is obtained [56].

Figure 11 shows two interferograms corresponding to the diffusion of a nearly saturated ($C_0 = 2.2$ M) solution of ammonium dihydrogen phosphate (ADP) in a 5 vol.% TMOS gel at two times, $t_1 = 23\,700$ s and $t_2 = 83\,400$ s. The initial concentration difference between the two layers is 0.27 M. The distance from one fringe to the following one corresponds to a variation of the optical path equal to one wavelength ($\lambda = 0.6328$ μm). The thickness of the cell $e = 1$ cm, one can calculate the corresponding variation of the refractive index ($\Delta n = \lambda/e$) and of the concentration ($\Delta n = K\Delta C$ with $K = 140 \times 10^{-4}\,\mathrm{M}^{-1}$).

So, from the flat contrast on the right-hand side of the interferogram which corresponds to the volume of the gel layer in which the concentration is unchanged, one can calculate the concentration decrease on approaching the interface, each fringe corresponding to an additional decrease ΔC.

Each interferogram enables one to draw the concentration profile at a given time. All the profiles are fitted with calculated ones [56, 57]. In the present case, the best fit for the whole set of curves for this experiment is $D_{\mathrm{gel}} = 3 \times 10^{-6}\,\mathrm{cm^2\,s^{-1}}$ (instead of $6 \times 10^{-6}\,\mathrm{cm^2\,s^{-1}}$ for the same system in free gel solution).

Such measurements could be made more systematically when knowledge of a diffusion coefficient is required: not only are these data very few but, even when existing

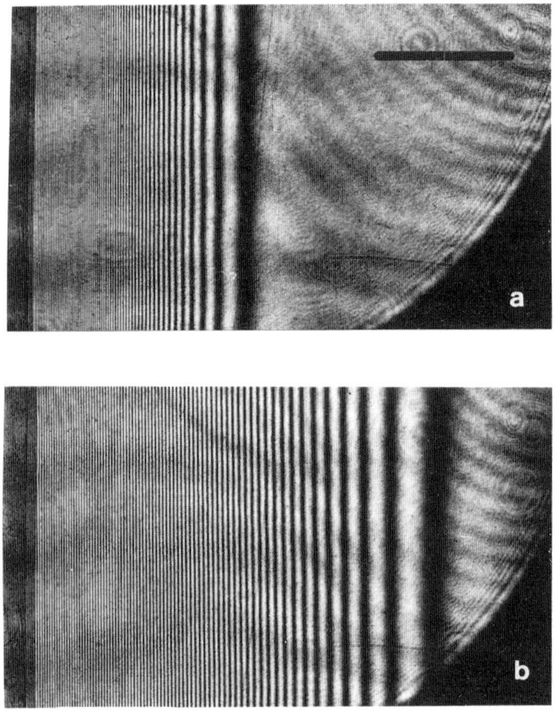

Fig. 11. Interferograms $(0, t)$ showing the diffusion of an ADP solution in a 5% TMOS gel. (a) $t_1 = 23\,700$ s, $t_2 = 83\,400$ s. Bar represents 0.5 mm.

for the concerned species, they have not been measured in the required conditions (concentration range, presence of other constituents in the solution, counter-diffusion of other species).

4.2. *Mass transfer associated with different diffusion systems*

4.2.1. *Numerical simulations*

Several attempts [57] have been made to achieve mathematical simulations of what occurs in growth systems using simple or double diffusions of two reactants A and B through a gel column to form the reaction product AB. In order to simplify the calculations, the diffusion coefficients of the species, D_A and D_B, are taken to be constant. This allows one to examine what happens with inexhaustible or finite reservoirs containing either gelled or free solutions. It is also interesting to test the effect of concentration programming.

From these calculations, it is possible to determine the locations and times at which some specific or several conditions are fulfilled.

4.2.2. *Experimental determination by optical methods*

As the gels used for growth are mostly transparent, the changes occurring in a gel column submitted to counter-diffusion of two reactants can be followed by holographic interferometry (see section 4.1) which enables one to measure refractive index variations as a function of time [5].

Figure 12 corresponds to two interferograms of a three-layer column of gel contained in a parallel-sided cell. The positions of interfaces I_1 and I_2 between the three layers are marked by arrows. The bottom layer is the reservoir of the first reactant, a 1 M solution of ammonium hydrogen phosphate (ADP) trapped in the gel; the middle part is a reactant-free gel layer; the upper layer is the reservoir of the second reactant, a 1 M free solution of calcium nitrate.

A first reference hologram has been taken at t_0, just after pouring the middle layer onto the bottom layer. The interferogram (fig. 12a) has been taken at t_1, 24 h later, which is the time required for gelling the second gel layer. The pattern of thin fringes shows the upwards diffusion of ADP whose concentration progressively increases at the bottom of the middle layer: it results in a density gradient parallel to the gravity vector which gives a stable configuration, even when the gel is not yet set. No convection is generated and mass transfer proceeds by diffusion on both sides of the interface I_2.

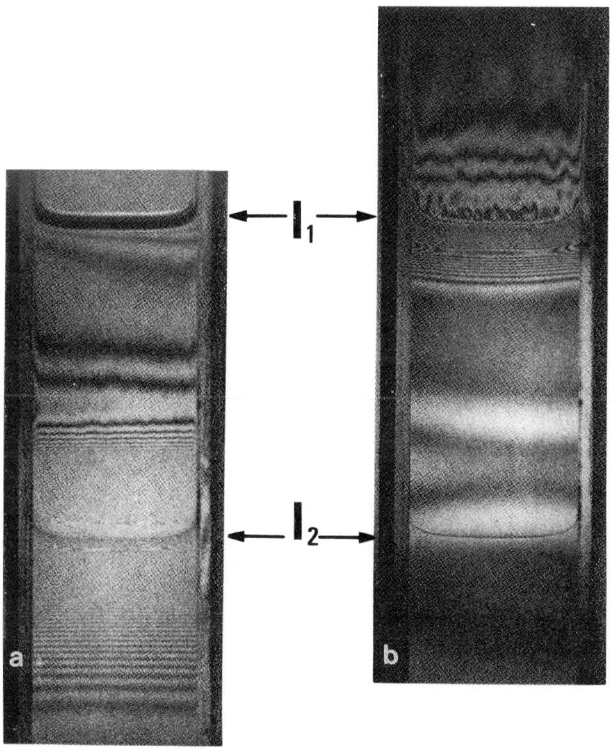

Fig. 12. Interferograms showing the double diffusion of the two reactants towards the middle layer. Arrows show the interfaces between the three layers (I). Bottom and middle layers are gelled. (a) Reference hologram was taken just after pouring the middle layer. Interferogram is made 24 h later. (b) Reference hologram was taken immediately after the first interferogram. Interferogram (b) is made 3.5 h later. Width of the cuvette: 2.2 cm.

A second reference hologram is then taken (at t_2, a short time after t_1) just after pouring the upper liquid layer. The interferogram (fig. 12b) has been taken at t_3, 3.5 h after t_2. The concentration changes related to diffusion of ADP between t_2 and t_3 are faint (the concentration gradients vanish with time) and associated refractive index changes give rise to the formation of only a few thick fringes. By contrast, at the same time, the downwards diffusion of $Ca(NO_3)_2$ is just beginning, the diffusion profile in the middle layer is still rather sharp as shown by the system of thin fringes close to the upper interface. Above this interface, no diffusion fringes are visible in the liquid layer but only some flame-shaped contrasts which correspond to convection rolls; as a matter of fact, the liquid layers close to the interface are depleted due to diffusion of $Ca(NO_3)_2$ towards the gel; the depleted layers, being lighter, are driven by buoyancy. Three gel layers would be required to have a purely diffusive system.

It is rather simple to correlate refractive index changes and composition of reactants in the different slices of gel when there are only a few diffusing species; with more than three reactants, it is not easy to de-correlate refractive index changes related to each reactant and additional characterization methods are required. Chemical ones are especially convenient.

4.2.3. Experimental determination by chemical methods

In the recent past, Prieto et al. [59–61] have developed an impressive amount of work devoted to characterization of precipitation growth systems. The purpose of these experiments was first to follow mass transfer and consequently supersaturation evolution (space and time) in a classical double-diffusion system allowing crystal growth by precipitation.

A series of identical gel columns, fed at both ends by two reactants reservoirs, is prepared. A reference column contains a pH indicator allowing one to know the pH variations. Another one is connected to water reservoirs in order to follow the "unloading" of waste products of the gelling reaction; in the present case NaCl (for silica gel, see section 3.2.1).

After the previously stated diffusion periods, each gel column is sliced and the global composition of each slice is determined by chemical analysis. As the corresponding pH profiles are known, one can calculate, by applying the Debye–Hückel formula, the concentration of different chemical species present in the slice, free ions as well as ionic associations. Knowing the ionic strength and the activity coefficient of the free ions involved in the precipitation reaction one can calculate the actual supersaturation. It appears that the location of the maximum supersaturation determined that way is quite different from the one assumed from the simple hypothesis of complete ionization of reactants.

Furthermore, all these experiments clearly demonstrate that diffusion coefficients cannot be considered as constant in these precipitation systems: the simplification commonly made in models of mass transfer in a gel leads to important departures from reality.

All these studies naturally lead to correlation of first precipitate location and supersaturation evolution. In addition to the condition of exceeded critical supersaturation (minimum supersaturation required for nucleation), the condition of "equality range" of reactant concentration (for equimolar reactions) suggested by the theoretical work of Henisch and Garcia Ruiz [57] is experimentally demonstrated. The nucleation density drastically decreases when the ratio of reacting species significantly departs from unity. The critical supersaturation values, expressed as the ratio of precipitation concentration over the equilibrium concentration, vary widely inside the same growth system but remain high; e.g. it varies from 457 to 3742 for silica gel growth of $SrCO_3$ and from 82 to 889 for $BaCO_3$; an earlier work [62] already mentioned very high values of critical supersaturation: 200 to 700 for tricalcium phosphate and hydroxyapatite in gelatine gels.

As a matter of fact, the nucleation rate depends not only on the supersaturation value but also on the supersaturation rate, the rate at which the supersaturation increases with time [61].

Several experimental [62, 63] or theoretical [64, 65] studies have been made in order to establish good models and crystallization criteria for nucleation in diffusing systems. Up to now, no clear and simple nucleation studies could be made in these evolutionary and multiparameter systems. For example, when working with pH variations, not only are the gel properties affected but also the precipitation reaction. Nucleation studies in gels have been done on simpler systems (see section 5). However, due to the widespread utilization of double-diffusing systems to grow valuable crystals, further investigations are required, not only for determining nucleation and further growth conditions but also dissolution conditions which seem to be involved in the very interesting phenomenon of periodic precipitation leading to Liesegang ring formation.

4.3. Application of the diffusion properties of gels to microgravity simulations

At the beginning of the seventies it was proposed to the scientific community and in particular to crystal growers, to prepare some space experiments in order to evaluate to what extent a microgravity environment could be beneficial in the materials science domain. Two advantages were put forward: absence of buoyancy-driven convection and absence of sedimentation; it was obvious that growth systems requiring either a slow supply of solute, like low-solubility substances, or a protective surrounding medium, like biological macromolecules, could be good candidates.

The counterpart was that very little was known at that time about the actual space conditions, e.g. on the effect of a residual gravity, or on the way to design these new types of experiments.

Concerning the first question about the absence of convection, a first answer was given in 1985 with the first interferograms of a growth cell showing characteristic diffusion patterns around a crystal growing in solution being obtained; these patterns are quite similar to those obtained with gel growth (fig. 14). Concerning the absence of sedimentation, the question is still pending, but one can say that important drifts of growing crystals have not yet been pointed out [67].

Considering that space experiments are rare and expensive, it was of primary interest to develop different ways of simulating space experiments. In order to prevent convection, one has to use systems whose Rayleigh number ($\mathrm{Ra} = g\Delta\rho L^3/\eta D$) is smaller than a critical value Ra_c.

As it is not possible on earth to reduce the gravity vector, or to decrease the diffusion coefficients, one can either reduce the density differences $\Delta\rho$ by using iso-density systems or by reducing characteristic lengths L such as in microscopic devices, or still by increasing the viscosity η with highly viscous solutions.

All these means are valuable to reduce convection but, except in the case of highly viscous solutions, none is efficient in reducing sedimentation effects.

It appears that gel growth is the easiest and the most consistent way to simulate solution growth in space (the viscosity is infinite and the characteristic length, equated to pore diameter is quite small). The discussion here is restricted to experimental simulations and does not concern numerical simulations which do not supply real crystals.

As a matter of fact, most of the first space experiments [67] were conceived as transpositions of experiments normally performed on earth in gel media: space growth conditions were tested on earth by a series of gel experiments using decreasing amounts of gelling substances [68]. Indeed, besides diffusion properties, a gelled solution has some specific properties (e.g. nucleation inhibition) which are not representative of a free solution in a microgravity environment. By decreasing the gel content one can extrapolate to free solution: the results of some space experiments prepared that way were in good agreement with the results of simulation experiments [69].

It is obvious that microgravity crystal growth activities have given fresh impetus to gel growth activities. This is especially true concerning the protein crystal growth domain. The very first space experiments were conceived as substitutes to gel experiments considered as unrealistic due to gel rigidity. However, these experiments could be simulated some time later, after the space mission, when new gel media [55] were developed. Protein crystal growth in gels [28] is now a very open field.

The question is often asked as to whether it would be advisable to use gel growth media for microgravity experiments. At first sight, it could appear redundant to proceed that way but two special cases need to be considered; first, when perturbations in the gravity level are feared, one could add a very low amount of gel. Second, when the growth solutions present gas–liquid interfaces, superficial tension (Marangoni) convection can be generated; this is due to superficial tension gradients induced by temperature or concentration gradients. As gel networks prevent all types of convection, small amounts of gel can be added. This is the case for zeolite crystal growth in space [70, 71]. Several types of zeolites have been grown under hydrothermal conditions; the autoclaves consist of two communicating compartments originally filled with liquid and gas. Different mother solutions have been used: they are either clear or gelled, according to the composition. Obvious differences appeared in the crystal distributions obtained for the two types of mother solution.

To sum up, one can say that gel growth is certainly indebted to space activities for its recent and rapid development but, in turn, gels have afforded a big contribution to preparing and interpreting the results of space experiments.

5. *Nucleation and growth in gel media*

Gels have been shown to promote a diffusive process for mass transfer, so growth or precipitation are obviously dependent on the gel structure but it is also the case of nucleation by solubility decrease because this last phenomenon takes place in a confined medium.

5.1. Nucleation

5.1.1. Some definitions

Nucleation of a solid phase from the liquid can take place heterogeneously (on walls or dust) or homogeneously if a sufficient supersaturation is set up. Supersaturation is defined either by $\beta = C/C_{eq}$ or by $\sigma = (C - C_{eq})/C_{eq}$, where the actual concentration of the material is compared to the equilibrium concentration.

This solid phase appears at a critical threshold β_{crit}, after a latence time; at higher β, the number N of nuclei increases linearly with time: this allows one to fix the nucleation frequency J as the slope of the graph $N(t)$.

Nucleation studies lead to the determination of β_{crit} and J at a given supersaturation. In gels the parameter to be used is the pore size which is essentially dependent on the gel content.

5.1.2. An experimental study

Nucleation in a gel medium can be followed by holographic interferometry as seen in fig. 13 which shows three interferograms taken at successive times during a nucleation experiment of ADP induced by T decrease in a gel. Crystals are seen with concentric fringes around them, which correspond to mass transfer from the gel to the crystal. From one interferogram to the following one, some nuclei appear just visible by their diffusion envelope. So long as these diffusion envelopes do not interact, the analysis of nucleation is still valid though nuclei exceed the critical size. This experiment shows clearly how unique this medium is for nucleation studies in solution: indeed, in free solution, a homogeneous nucleus falls under gravity forces and grows in a region where the supersaturation is different. In a gel, nuclei are sustained by the solid phase and grow where they have nucleated.

By varying the initial supersaturation, one can determine the critical threshold and the nucleation frequency versus supersaturation for different gel media.

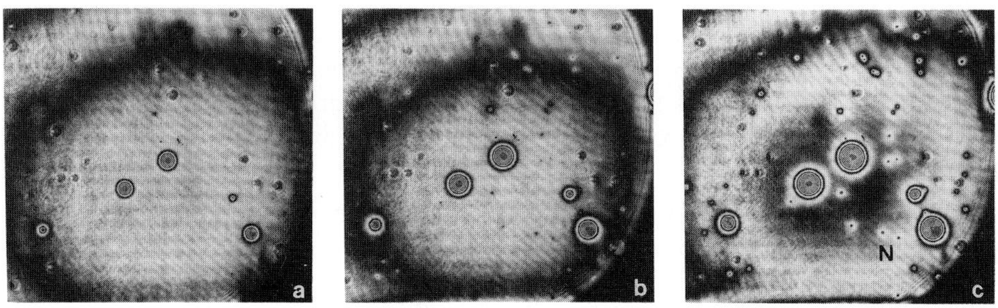

Fig. 13. Nucleation of ADP crystallites. Interferograms $(0, t)$ are taken at different times $t_1 = 40$ min (a), $t_2 = 46$ min (b), $t_3 = 52$ min (c). New nuclei (some of them are marked N) appear from one interferogram to the following one.

5.1.3. *Influence of the gel upon nucleation*

Two gels have been tested: they show an opposite behaviour.

5.1.3.1. Silica gel. The nucleation threshold has been measured for different concentrations of TMOS gels and the results are the following [72]:

TMOS (vol.%):	0	3	5	7
β_{crit}	1.30	1.74	1.82	1.91

$\beta = 1.30$ measured in pure aqueous solution must be lower than for homogeneous nucleation because of the impossibility of eliminating foreign particles which favour heterogeneous nucleation. This is not the case for a gel because polymerization of the gel has embedded these foreign particles.

Holographic interferometry allows one to determine the actual supersaturation at the nucleation step. It is generally lower than the previous ones, essentially because of non-pure homogeneous nucleation [58].

From these experimental data, it is clear that the gel reduces strongly the nucleation density as the critical supersaturation increases with the silica content, and is, in any event much higher than in free solution. These results agree with the hypothesis of homogeneous nucleation which depends on the cavity size. A theoretical study [72] has shown that the confined medium which is the gel, affects more the kinetics of nucleation than the thermodynamics.

5.1.3.2. Agarose gel. The nucleation density of hen egg white lysozyme has been measured in gelled hanging drops (see section 2.2.2.1) made by varying agarose concentrations.

Results obtained at 20°C with a given salt concentration (NaCl: 1.2 M) are the following [73]; the counting has been done after eight days on at least four droplets.

Agarose wt/V (%):	0	0.05	0.1	0.125
Mean number of nuclei in a 20 μl droplet: ($C_{protein} = 50$ mg/ml).	4	8	12	15

The same increase of nucleation density with agarose content has been found for another salt concentration. These rather surprising results can be explained by taking into account the structure of agarose. In fact, for this increase of content, the porosity decreases by about 20% [74] but forming chemical weak bonds between double helices needs water which increases the supersaturation in the droplet. This last effect must predominate over the decrease of nucleation density expected with the porosity decrease.

5.2. *Crystal growth*

Figure 14 shows four interferograms taken during growth of an ADP crystal in a 5% TMOS gel by T decrease from 34 to 22°C. Only one crystal has nucleated and growth could be followed over the whole growth duration [56].

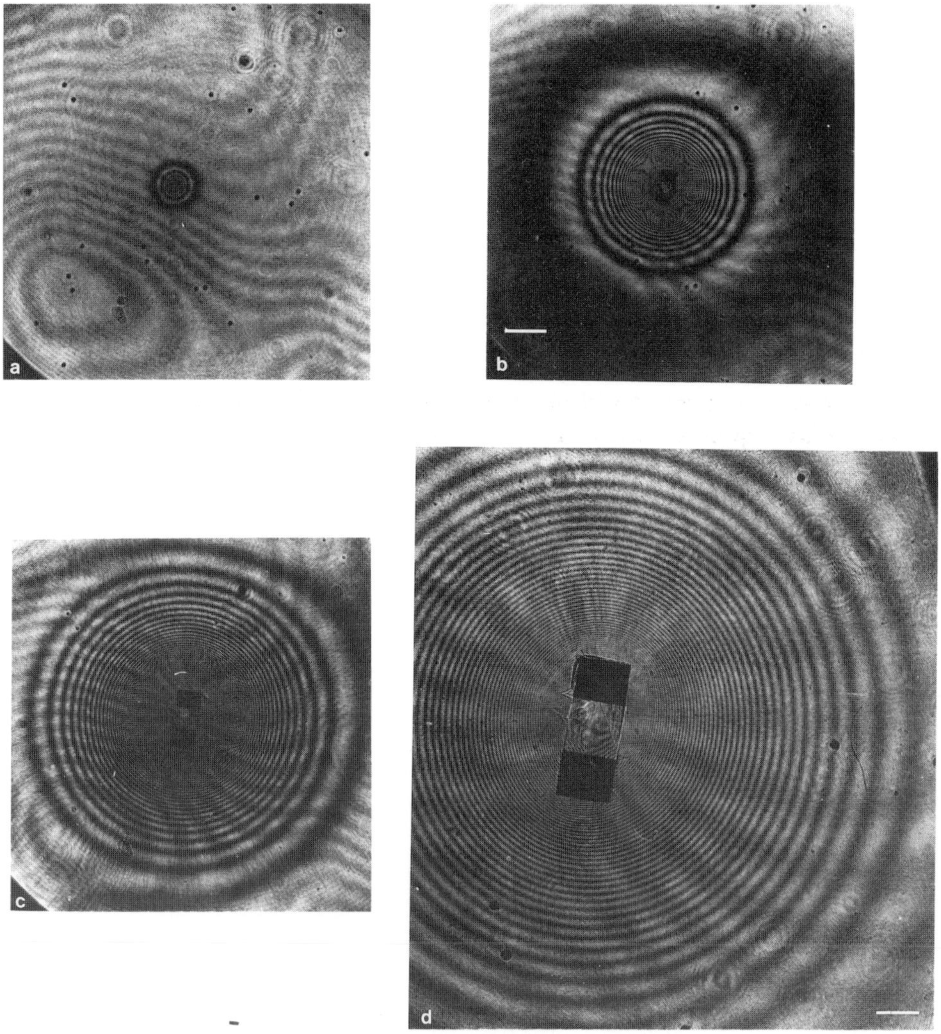

Fig. 14. Interferograms (0, t) of an ADP crystal, taken at the following times; (a) $t = 1.5$ h, (b) $t = 8$ h, (c) $t = 28$ h, (d) $t = 48$ h. For (a, b, c), the bar represents 5 mm; for (d), the bar represents 2 mm.

Interferograms can provide data on mass transfer from the gel towards the crystal: change of optical path between reference time (0) and interferogram time (t) has to be calculated for a given point on the interferogram; furthermore from one fringe to the following one, this change varies by one wavelength. This allows one to plot the concentration profile in a given direction for each interferogram. If direct views obtained by using only the object beam are registered simultaneously, the moving forward of the growth front and hence the growth rate can be measured. From these data which provide growth rate R at various supersaturations σ, the growth mechanism can be deduced. In the case of ADP, pyramidal faces follow an exponential law $R(\sigma)$ corresponding to a two-dimensional growth mechanism.

5.3. Analysis of gel behaviour during nucleation and growth

Let us summarize some features already developed in preceding sections:

At the nucleation step, some nuclei appear in macropores filled with the growth solution. As growth proceeds, two processes can happen; if the gel is too rigid with respect to the crystal, it is incorporated progressively and the crystal is contaminated: that is the case for calcite growing in silica gels [1], or it often arises with chemical gels such as polyacrylamide. But if the gel is weakly bound (physical gels or gels of particles), it can be easily rejected and broken forming around the crystal cusp-like cavities (fig. 15). These cavities filled with growth solution surround the crystal so that the crystal grows in a liquid pocket preventing polynucleation due to sticking of the gel. So a regular mass transfer is obtained by diffusion through the gel towards this liquid pocket.

Spontaneously nucleated crystals are then grown in this way; indeed re-implantation of the crystal in a gel often leads to polynucleation induced by sticking of the gel to the seed. Nevertheless, feeding experiments have been performed with success in the case of HEW lysozyme grown by the hanging drop technique [75].

Supersaturation at nucleation is always higher in a gel than in free solution (σ is then about 25% against 3–5%) so that initial growth rates can be very large (0.5 to 1 mm/h) [56]. So growth duration is of the same order of magnitude in gel and in free solution.

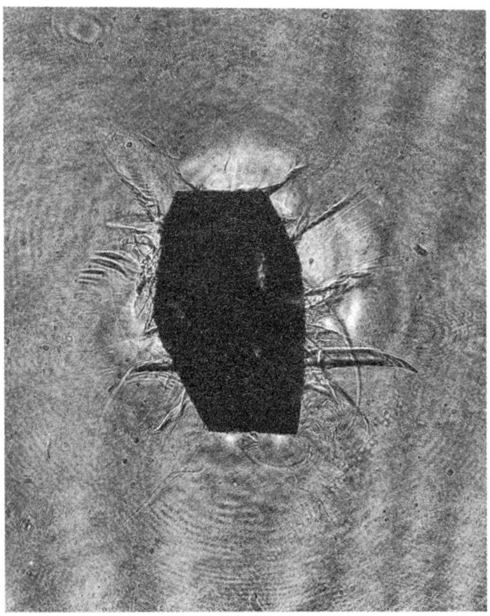

Fig. 15. Cusp-like cavities around a gel-grown crystal. Gel is broken from crystal edges.

6. Crystalline quality and physical properties of gel-grown crystals

Such a study will not be complete without characterization of grown crystals. The main features of gel-grown crystals will be underlined in comparison to free solution-grown crystals.

6.1. Crystal habit

Much work has been done on crystal habit modification especially by varying different parameters controlling the gel structure (*T*, pH, impurity, gel density) [1–4]. Changes in morphology can be due to face blocking by impurity but more generally to growth-rate variations with supersaturation.

Because the nucleation is often difficult, gels promote the growth of the only a few crystals which have nucleated. For instance, hexacyanoferrates are always obtained in solution as colloidal particles and a structure determination is impossible. Some of them (Zn–Cs hexacyanoferrates) have been grown in gel as micrometric crystals large enough for crystalline structure determination [76].

As nucleation often occurs at high supersaturation, it is easy to develop dendritic growth.

6.1.1. Dendrites

Dendritic growth is often considered as unlucky because it develops thin crystals unusable for any application. But gel can be useful for the study of dendritic growth itself as shown in [77] where the three-dimensional structure of NH_4Br dendrites could be described thanks to gel growth. Dendrites develop a tip with a parabolic shape varying with supersaturation and secondary arms. For the first crystallization, the dendrite which is in direct contact with the gel is not very regular, so it is remelted and further crystallization leads to a dendrite with a well-defined shape at the tip; the growth then takes place in a tube of liquid for which mass transfer occurs through the gel by diffusion but without being hampered by gel. Results in gels are particularly convincing with reference to free solution results.

6.1.2. Twins

Gel growth allows one to grow gypsum crystals large enough for characterization. By varying the nature of the gel, different habits have been obtained: needle-shaped crystals in silica and agar gels [78], never-twinned crystals in bentonite [79]. Large crystals obtained in silica gels permitted a study of twinning by X-ray topography [80]. In this paper, comparisons with free-solution-grown crystals have been made; twinning is more frequently observed in free solution than in gelled solution. If the used gel is dense enough (TMOS 10%) to give small pores, nucleation is also reduced for twinning.

But twinning can be favoured by the gel, depending on the growth conditions. NPP, an organic crystal previously mentioned, has been grown in gel as a twinned crystal [31] while single crystals are obtained in free solution. This must be due to the high initial supersaturation needed for nucleation.

6.2. Impurity

Impurities can change the habit by preventing crystal growth of some faces but moreover they can be incorporated, changing the crystalline quality.

6.2.1. Homogeneous incorporation

Some authors [4] have reported that crystals grown in gels are purer than the reagents, probably due to the preliminary polymerization which incorporates dust and some impurities, producing a pure growth medium. But gel growth can improve crystalline quality even for high doping levels because the doping is homogeneous in every growth sector: no segregation appears such as is usually observed in solution. That is the case for ADP crystals doped with chromium whose crystalline quality is much better than in free solution [81]; nevertheless the doping level is higher but chromium is equally distributed in the crystal as shown by secondary ion mass spectrometry [82]. Furthermore, no habit change is observed in ADP gel-grown crystals, while free-solution-grown crystals present habit modifications (tapering) depending on chromium content.

Homogeneous distribution of impurity takes place at high growth rate (due to high initial supersaturation) for which incorporation occurs continuously without impurity diffusion. Because of this homogeneous distribution, no strain appears at the growth sector boundary, reducing the dislocation density.

6.2.2. Doping levels in gels

Doping proteins can be easier in a gel than in free solution [83]. Turkey egg white lysozyme (TEWL) is a protein whose molecule differs from hen egg white lysozyme (HEWL) only by seven amino acids of the 129 of the whole sequence; lattice parameters differ of about 20%, but both crystallize in the same space group. Growth of TEWL has been performed in the presence of increasing amounts of HEWL both in free solution and in agarose gels (0.15%). Habit change is observed in both cases: by increasing HEWL content, hexagonal crystals of TEWL, elongated along the *c*-axis, first become shorter, then rounded. The latter correspond to a central part of TEWL surrounded by a layer of both intergrown materials: the contamination is effective for lower HEWL contents in free solution. For instance, if the mother solution contamination rate is 50%, no TEWL crystals appear in free solution while in agarose 80% of gel-grown crystals show a well-faceted habit and are TEWL pure crystals as shown by X-ray diffraction.

6.3. Lattice defects (X-ray diffraction)

Crystal quality of gel-grown crystals can be tested by X-ray diffraction.

6.3.1. Global measurements

When crystals are large enough (about 50 μm) structure determination becomes possible and the resolution of diffraction patterns can be taken as a criterion of crystalline quality. For instance, with gel-grown crystals of HEWL [84], a resolution of 1.5 Å has been attained, which represents the limit obtained with free-solution crystals.

Global diffracted intensity decreases with decreasing lattice plane spacing (d) for a given wavelength but it is also dependent on crystal quality: the better the crystal is, the larger is the intensity collected for low values of d. By comparing intensity diagrams versus d, human albumine has been found to be of better quality when grown in agarose gel than in free solution [85]. In this study, gel was used as a model growth medium for preparing microgravity experiments (see section 4.3). So space crystals have been compared to gel-grown crystals and found worse, but this comparison is not quite appropriate because differences are also due to crystal size which is proportional to growth duration; up to now the growth durations are shorter in space than on earth.

6.3.2. Local measurements

When crystal size is more than $5 \times 5\,mm^2$, X-ray topographs can give the defect distribution within the crystal. Monochromatic X-rays diffracted by a given reflector plane show lattice defects as contrast variations. Figure 16 presents a Lang X-ray topograph of a gel-grown $NaClO_3$ crystal: a few defects are present, showing a black contrast in a rather uniform background. Two groups of dislocations (d) starting from the nucleus propagate in opposite (1 0 0) and ($\bar{1}$ 0 0) growth sectors. Because the reflector plane (1 0 0) gives a strong contrast, they are screw dislocations. Furthermore, a topograph taken with 0 2 0 reflection reveals two other groups of screw dislocations in perpendicular growth sectors (0 1 0); then the development of the four homologous

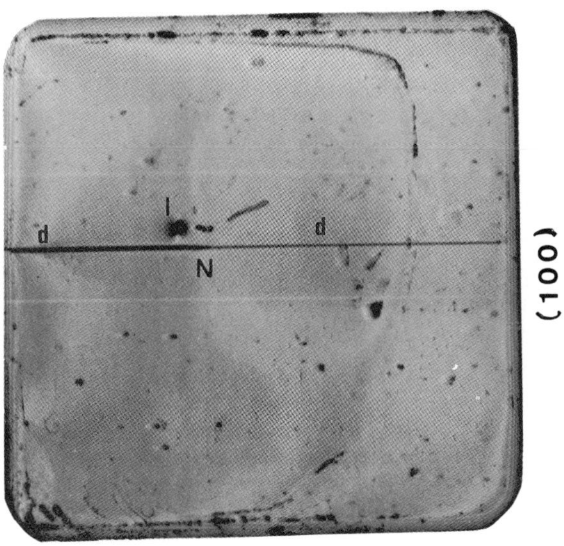

Fig. 16. X-ray translation topography of a $NaClO_3$ (001) slice. Mo Kα Ref 200; N: nucleus; d: dislocation; I: inclusion. Bar represents 0.8 mm.

growth sectors is about the same. Some solid inclusions (I) are visible but this large crystal case is considered to be of high quality.

A series of ADP crystals grown in different TMOS gels have been characterized by X-ray topography and compared with gel-grown crystals. At high gel content (10%), numerous dislocations start from the nucleus and propagate in different growth sectors while at low content (2%), no dislocations appear; only a weak contrast at growth sector boundaries indicates a slight distortion. Free solution-grown crystals [81] usually have dislocations at the seed–new crystal interface depending on convection currents.

So the only defects which appear in gel-grown crystals are produced at the nucleation step and are due to gel incorporation before the formation of a liquid pocket around the crystal. This type of defect can be eliminated by using lighter gels which must, nevertheless, be firm enough to support the crystal.

Lattice parameter mapping has been obtained by plane wave reflection topography for ADP crystals [82]. The experiment consists in recording topographs for various positions along the rocking curve. Lattice parameter difference between pyramidal and prismatic sectors is estimated to be 1×10^{-5} for gel-grown crystals while it is 2.2×10^{-5} for solution-grown crystals.

6.4. *Physical measurements on gel-grown crystals*

Characterization of some peculiar physical properties complements X-ray topography results and can lead to a criterion of crystalline quality, more directly usable than X-ray measurements. Let us consider some examples.

6.4.1. *Phase transition*

Lead phosphate ($PbHPO_4$) displays a ferroelectric–paraelectric transition which is characterized by a transition temperature T_c at which the dielectric constant is a maximum. Furthermore, the T_c value is highly sensitive to the crystal quality; the better the crystal is, the lower is T_c and the more nearly intrinsic properties are obtained. By using this criterion, space-grown crystals [68] appear to be of a better quality than gel-grown crystals, which themselves are better than solution-grown crystals.

$(TMA)_2ZnCl_4$(TMA = trimethylammonium) presents an incommensurate phase between two commensurate phases, spreading at ambient temperatures. The quality of crystals grown by different methods has been studied by X-ray topography and by dielectric measurements [87]: study of phase transitions was possible only with gel-grown crystals which showed a few defects [88].

6.4.2. *Second harmonic generation in non-linear optics*

Organic molecular crystals of the nitroaniline family are highly attractive because of their application in non-linear optics, surpassing the use of inorganic materials such as KH_2PO_4 or $LiNbO_3$. Among these organic crystals, POM and NPP crystals are efficient for generating second harmonic radiation in the visible range. The experiment consists of illuminating such a crystal (set in phase-matching conditions) by an infra-red laser beam and recording only second harmonic intensity as a function of incidence angle of the laser beam. The width at half height of the rocking curve allows one to calculate the effective interaction length (l_{eff}), which has to be compared to the actual length (l): the more nearly l_{eff} approaches l, the better is the crystal. For POM crystals [30], l_{eff} is about the actual length if only one growth sector is involved; on the other hand, l_{eff} decreases if the laser beam crosses defects at growth sector boundaries. For NPP crystals [31], $l_{eff}/l = 0.77$ which qualifies NPP gel-grown crystals as the most efficient grown so far.

7. *Conclusion*

Although gel growth is just leaving its primary stage when compared to other growth techniques, it is clear that a large development is expected. The main characteristics have been pointed out.

A few growth techniques have been described here because they correspond to basic configurations. Each of them may give rise to numerous variations.

Among the gel-grown materials which have been quoted, organic ones are now raising a particular interest. This has been made possible by improving the knowledge about the gel structure and the way in which it interacts with the growth processes.

Several features are well established, but some of them are not fully understood, i.e. the influence of the gel on nucleation or on the incorporation of impurities.

An interpretation will probably be given when many other examples, allowing comparative studies, become available; up to now, gel growth has been applied in a rather limited domain, i.e. low temperatures and normal pressure. A wider range could be explored such as has been done in one case of hydrothermal gel growth [89].

In the same way, only a few non-aqueous solvents have been tested. It appears that mixtures of solvents could give access to a wide variety of growth conditions.

It is now absolutely necessary to gather quantitative data and consequently to develop many other characterization techniques. Those which allow real-time control and reprogramming of growth conditions seem expecially convenient.

The goal of such studies is not only to obtain high-quality crystals but also an improved knowledge of phenomena governing growth processes.

References

[1] I. Karniewicz and B. Wiktorowska, Krist. Tech. 15 (1980) 12.
[2] S.K. Arora, Prog. Cryst. Growth Charact. 4 (1981) 345.
[3] A.R. Patel and A. Venkateswara Rao, Bull. Mater. Sci. 4 (5) (1982) 527.
[4] H.K. Henisch, Crystals in Gels and Liesegang Rings (Cambridge University Press, Cambridge, 1988).
[5] E. Manghi and G. Polla, J. Cryst. Growth 74 (1986) 380.
[6] C.C. Desai and M.S.V. Ramana, J. Cryst. Growth 102 (1990) 191.
[7] R.H. Plovnick, J. Cryst. Growth 114 (1991) 22.
[8] T. Sokolowski and B. Wiktorowska, Cryst. Res. Technol. 25 (1990) 481.
[9] I. Das, A. Pushkarna and R.S. Lall, J. Cryst. Growth 82 (1987) 361.
[10] A.M. Cody and R.D. Cody, J. Cryst. Growth 113 (1991) 508.

[11] S.N. Patil and A. Venkatewara Rao, Cryst. Res. Technol. 26 (1991) 841.
[12] B. Wiktorowska, Cryst. Res. Technol. 22 (1987) K83.
[13] R. Roopkumar, G. Raman and F.D. Gnamam, Cryst. Res. Technol. 25 (1990) K289.
[14] R. Roopkumar, G. Raman and F.D. Gnanam, Cryst. Res. Technol. 25 (1990) 7; K166.
[15] G. Raman, R. Roopkumar and F.D. Gnanam, Cryst. Res. Technol. 26 (1991) K160.
[16] G. Raman, F.D. Gnanam and P. Ramasany, J. Cryst. Growth 75 (1986) 471.
[17] V. Saraswati and G.V.N. Rao, J. Cryst. Growth 83 (1987) 606.
[18] S. Jagota and R. Raj, J. Cryst. Growth 85 (1987) 527.
[19] V. Lopez-Acevedo and C. Viedma, J. Cryst. Growth 98 (1989) 746.
[20] T.A. Al-Dhahir, G. Dhanaraj and H.L. Bhat, J. Cryst. Growth 121 (1992) 132.
[21] A. Lentz, Th. Fetzer, I. Voitenleitner, Th. Haner and O. Abou E1-Wafa, Cryst. Res. Technol. 25 (1990) 391.
[22] A.M. Cody and R.D. Cody, J. Cryst. Growth 83 (1987) 485.
[23] G. Varghese, M.A. Ittyachen and J. Issac, Cryst. Res. Technol. 25 (1990) K153.
[24] A. Michaelides and S. Skoulika, J. Cryst. Growth 94 (1989) 208.
[25] S. Natarajan, H.A. Devadoss and K. Alwan, Cryst. Res. Technol. 23 (1988) 1343.
[26] S. Narayana-Kalkura and S. Devanarayanan, J. Cryst. Growth 83 (1987) 446.
[27] B. Brezina and J. Hulliger, Cryst. Res. Technol. 26 (1991) 155.
[28] M.C. Robert, K. Provost and F. Lefaucheux, in: Crystallization of Nucleic Acids and Proteins, eds. A. Ducruix and R. Giégé, The Practical Approach Series (IRL Press, Oxford, 1992) 127.
[29] A. Ducruix and R. Giégé, in: Crystallization of Nucleic Acids and Proteins, eds. A. Ducruix and R. Giéré, The Practical Approach Series (IRL Press, Oxford, 1992) 73.
[30] P. Andreazza, F. Lefaucheux, M.C. Robert, D. Josse and J. Zyss, J. Appl. Phys. 68 (1990) 8.
[31] P. Andreazza, D. Josse, F. Lefaucheux, M.C. Robert and J. Zyss, Phys. Rev. B 45 (1992) 7640.
[32] P.J. Flory, Disc. Faraday Soc. 57 (1974) 7.
[33] A.M. Cody, H.T. Horner and R.D. Cody, Scanning Electron Microsc. I (1982) 185.
[34] R.D. Cody, Geol. Soc. Am. Bull. 87 (1976) 1163.
[35] B. Brezina and M. Havrankova, J. Cryst. Growth 34 (1976) 248.
[36] E. Banks, R. Chianelli and F. Pintchovsky, J. Cryst. Growth 18 (1973) 185.
[37] R.H. Plovnick, J. Cryst. Growth 114 (1991) 22.
[38] R.K. Iler, The Chemistry of Silica (Wiley Interscience, New York, 1979).
[39] B. Gauthier-Manuel, E. Guyon, S. Roux, S. Gits and F. Lefaucheux, J. Physique 48 (1987) 869.
[40] P. Favard, J.P. Lechaire, M. Maillard, N. Favard, P. Andreazza, F. Lefaucheux and M.C. Robert, Colloid Polym. Sci. 270 (1992) 584.
[41] B. Cabane, M. Dubois, F. Lefaucheux and M.C. Robert, J. Non-Cryst. Solids 119 (1990) 121.
[42] F. Lefaucheux, M.C. Robert, S. Gits, Y. Bernard and B. Gauthier-Manuel, Rev. Int. Hautes Temp. Réfract. 23 (1986) 57.
[43] G.W. Scherer, J. Non-Cryst. Solids 100 (1987) 77.
[44] C. Vincent, Technoscope de Biofutur 35 (1990) 2.
[45] P.G. Righetti, J. Biochem. Biophys. Meth. 19 (1989) 1.
[46] P. Favard, J.P. Lechaire, M. Maillard, N. Favard, M. Djabourov and J. Leblond, Biol. Cell 67 (1989) 201.
[47] R. Rüchel, R. Sterre and E. Erbe, J. Chromatography 166 (1978) 563.
[48] J.F. Quinson, J. Dumas and J. Serughetti, J. Non-Cryst. Solids 79 (1986) 397.
[49] Y.C. Chang and A.S. Myerson, AICHE J. Phys. Chem. 31(6) (1985) 890.
[50] J. Salvinien, R. Marignan and S. Cordier, J. Chim. Phys. 48 (1951) 47.
[51] S. Gits-Léon, F. Lefaucheux and M.C. Robert, J. Cryst. Growth 84 (1987) 155.
[52] A.G. Langdon and H.C. Thomas, J. Phys. Chem. 75(12) (1971) 1821.
[53] J. Salvinien, J. Chim. Phys. 48 (1951) 465.
[54] F. Gonzalez, M. Cunisse, A. Schan, M.C. Robert, F. Lefaucheux, P. Andreazza, A. Perigaud and J. Zyss, Proc. 7th European Symp. on Materials and Fluid Sciences in Microgravity, Oxford, UK, ESA-SP 295 (1989) 581.
[55] M.C. Robert and F. Lefaucheux, J. Cryst. Growth 90 (1988) 358.
[56] F. Lefaucheux, M.C. Robert and Y. Bernard, J. Cryst. Growth 88 (1988) 97.

[57] H.K. Henisch and J.M. Garcia-Ruiz, J. Cryst. Growth 75 (1986) 195; 203.
[58] F. Lefaucheux, M.C. Robert, Y. Bernard and S. Gits, Cryst. Res. Technol. 19 (1984) 1541.
[59] M. Prieto, C. Viedma, V. Lopez Acevedo, J.L. Martin Vivaldi and S. Lopez Andres, J. Cryst. Growth 92 (1988) 61.
[60] M. Prieto, L. Fernandez Dias and S. Lopez Andres, J. Cryst. Growth 98 (1989) 447.
[61] M. Prieto, L. Fernandez Dias and S. Lopez Andres, J. Cryst. Growth 108 (1991) 770.
[62] A. Packter and A. Chatterjee, Cryst. Res. Technol. 20 (1985) 663.
[63] S.N. Patil and A. Venkateswara Rao, Cryst. Res. Technol. 6 (1991) 847.
[64] O.A. Brevnova, L.L. Goshka and V.P. Ruzov, Cryst. Res. Technol. 27 (1992) 301.
[65] A.V. Cipanov, L.L. Goshka and V.P. Ruzov, Cryst. Res. Technol. 7 (1990) 737.
[66] A.V. Cipanov, L.L. Goshka and V.P. Ruzov, Cryst. Res. Technol. 2 (1990) 119(b).
[67] A. Authier, K.W. Benz, M.C. Robert and F. Wallrafen, in: Fluid Sciences and Materials Science in Space, ed. H.U. Walter (Springer, Berlin, 1987) 405.
[68] F. Lefaucheux, M.C. Robert, E. Manghi and H. Arend, J. Cryst. Growth 51 (1981) 551.
[69] M.C. Robert, F. Lefaucheux, B. Jannot, G. Godefroy and E. Garnier, J. Cryst. Growth 88 (1988) 499.
[70] A. Sacco, Jr., SPIE 1557 (1991) 6.
[71] F. di Renzo, R. Dutartre, P. Espiau, F. Fajula and M.A. Nicolle Proc. 8th European Symp. on Material and Fluid Sciences in Microgravity, Brussels, Belgium, 1992, ESA SP-333 (1992) 691.
[72] P. Andreazza, F. Lefaucheux and B. Mutaftschiev, J. Cryst. Growth 92 (1988) 415.
[73] K. Provost and M.C. Robert J. Cryst. Growth 110 (1991) 258.
[74] P. Serwer, E.T. Moreno and G.A. Griess, in: Electrophoresis 88, ed. C. Schafer-Nielsen (VCH, Weinhein, 1988) p. 26.
[75] Y. Bernard, S. Degoy, F. Lefaucheux and M.C. Robert, Acta Cryst. D, submitted.
[76] P. Gravereau and E. Garnier, Rev. Chim. Miner. 20 (1983) 68, A.2.
[77] J. Maurer, B. Perrin and P. Tabeling, Europhys. Lett. 14(6) (1991) 575.
[78] G.M. van Rosmalen, W.G.J. Marchee and P. Bennema, J. Cryst. Growth 35 (1976) 169.
[79] R.D. Cody and H.R. Shanks, J. Cryst. Growth 23 (1974) 275.
[80] C. Rinaudo, M.C. Robert and F. Lefaucheux, J. Cryst. Growth 71 (1985) 803.
[81] S. Gits, M.C. Robert and F. Lefaucheux, J. Cryst. Growth 71 (1985) 203.
[82] S. Léon, F. Lefaucheux, M.C. Robert, C. Malgrange and J.C. Lorin, J. Cryst. Growth 97 (1989) 631.
[83] K. Provost., Thèse de Doctorat, Université Pierre et Marie Curie, Paris (1992).
[84] M.C. Robert and J. Berthou, C.R. Acad. Sci. Paris 305, Série II (1987) 847.
[85] T:Y. Miller, X.M. He and D.C. Carter, J. Cryst. Growth 122 (1992) 305.
[86] F. Lefaucheux, M.C. Robert and E. Manghi, J. Cryst. Growth 56 (1982) 141.
[87] B. Jannot, M. Benhmida, M. Fahli, C. Dumas, M. Jannin, S. Gits, M. Ribet and G. Godefroy, Jpn. J. Appl. Phys. 24 (suppl. 24–2) (1985) 753.
[88] M. Ribet, S. Gits-Léon, F. Lefaucheux and M.C. Robert, Ferroelectrics 66 (1986) 259.
[89] W.A. Franke and N.A. Mehran, Cryst. Res. Technol. 27 (1992) 295.

Subject Index

absolute solubility 477
absorbed flux 935
absorption 930, 931
– coefficient 930, 938, 958
absorptivity 931
accelerated crucible rotation 731
– technique 56, 576
acid digestion bombs 491, 510
acid solvents 485
acidic solvents 529
acmite 492
acoustic spectrosopy 357
acousto-optic 536
ACRT 576–579, 584, 602, 603
activated complex 478
activation barriers 433
activation energy 419
active slip system 862
activity 479
– coefficient 472, 478, 481
adaptive control 205
additive 346, 1258
ADI methods 917
adjacent re-entry 1189
advection 616
advective flow 617
after-heaters 115
agarose 1285
agglomeration 343
aggregation 343, 1247, 1248
– kernel 344
– state 453
air-cooled finger 578
alkali halides 203
alkali or rare-earth ions 595
alkali–zirconium silicates 541
alkaline solutions 545
allogenic minerals 11
alloy single crystals 178
alpha amylase 426, 427, 429
α-berlinite 467
α-quartz 467, 469, 511
aluminium antimonide 167
aluminium garnets 605
aluminium–copper alloy 279
alumino-silicates 484
ambient gas, heat transfer in- 963
amethyst 544
amino acid 420, 447
ammonium sulfate 430, 444, 447
amorphous precipitate 430, 441, 449
amorphous silicon monoxide 511
analcime 484
analsite 468
angle fixing condition 1036
anisotropy of the structure 653
anomalous grain growth 595
anomaly effect 121
anorthide 596
antisotropic layer structure 639
apparatus for solubility measurements 509
aqueous chemistry 424
aqueous solutions 520
aqueous solvents 467
As injection 172
aspartic acid-rich protein 1257
attrition 345
authigenic minerals 11
autoclave for composite gradient method 522
autoclave materials 491
autoclaves 468, 469, 491
automated pulling 203
automatic control 103
automatic diameter control 116, 122, 126, 144
Avrami analyses 1173
axial field 263, 268
axial gradients 832
axial segregation 676
axial solute concentration 783
axial temperature profile 79, 84
axialites 1169, 1208
axisymmetric view factor symmetry 946

back-melting 895, 907, 991
back-stress parameter 846
backward light scattering 356
$BaFe_{12-2x}Co_xTi_xO_{19}$ 595
$BaFe_{12}O_{19}$ 586, 593
$BaFe_{12-x}M_xO_{19}$ 586, 587

baffle 186, 495, 520
– opening 527
$BaMoO_4$ 590
band-energy method 933, 936, 940
banding 1211
barium carbonates 468
barium hexaferrite 593, 596
baroclinic instability 724
batch crystallisation model 379
batch crystallization 433
batch diffusion 436
batch effects 985
batchwise simulations 913, 984
$BaTiO_3$ 584, 586, 587
$Ba_{1-x}SrTiO_3$ 586
$Ba_{1-x}Sr_xTiO_3$ 587
Belt apparatus 491, 508
benzil 203
benzophenone 203
BeO 590
Berg effect 40
berlinite 484, 486, 495, 497, 510, 520
beryl 467, 544
beryllia 592
$Bi_{40}Fe_2O_{63}$ 588
$Bi_{40}Ga_2O_{63}$ 588
$Bi_{40}Ga_2TiO_{63}$ 586
$Bi_2Mo_3O_{12}$ 590
Biot numbers 1022
Biron emerald 544
birth rate function 360
bismuth 123
– germanate 118
bismuth–manganese eutectic 280
$Bi_{12}TiO_{20}$ 586–588
$Bi_4Ti_3O_{12}$ 586
block casting 753
block nuclei 1243
Bond number 736, 1018
borate systems 492
borates 531
boric oxide 119, 123, 159
– encapsulation 64
boron nitride 114
bottle neck 231, 233
bottom growth 575
– with a nutrient 590
boundary layer 781, 884
– thickness 271
Boussinesq approximation 266, 621, 890
breeding 326
– dendritic 328
– fluid shear 328, 333
– initial 327
– mechanical 329
– surface 329
Brembelt 299
Bridgman growth 280
Bridgman technique 53, 55
Bridgman unsupported 498
Bridgman–Stockbarger technique 201
bright ring 117, 144
bromelite 520
bubble-column 303
bulk acoustic wave devices 531
bulk flow simulations 885
bulk growth kinetics 1176
buoyancy 712
– convection 716, 774
buoyancy-driven convection 34, 38–40
Burton–Cabrera–Frank 1235

$CaAl_2Si_2O_8$ 596
$CaCO_3$ 586, 594
cadmium sulphide 540
cadmium telluride 180
calcite 594
– single crystals 535
calcium fumarate 1257
calcium maleate 1257
calcium sulfate dihydrate 1254
calcium sulfate hemihydrate 1252
calcium tungstate 182
calculated plastic strain 868
$CaMgSi_2O_6$ 596
capillary coefficients 1026
capillary constant 1019
capillary convection 713, 730
capillary pressure 227
capillary shaping techniques (CST) 1015
capillary stability 1027
carbonates 468, 492, 534
cascade of evaporative crystallisers 394
cascaded control system 76
CAST 1032
cast iron 1152
catapleite 541
catching conditions 1035
$CaTiO_3$ 584
$CdTiO_3$ 584
cell boundaries 136
cellular interface 135
cellular structure 130, 132, 166, 181, 187, 189, 197

centrifugal forces 803
ceramic 467
chain-extended crystals 1205
chemical bond 477
chemical potential 471
chemical zoning 33
χ-Al_2O_3 584
chloresodalite 541
chlorosulphonation 1212
chondrules 13
chuck 102
cinnabar 539, 540
circular projections 943
cis-polyisoprene 1181, 1198
clarifiers 1178
Classical Theory of Becker–Doring 1230
classified product removal 367
cluster formation 34
Co-Ti:$YAlO_3$ 584
cobalt nickel iron spinels 593
Cochran flow 267
cocurrent cascade of crystallizers 304
coefficient of thermal expansion 481
coenzyme 443, 454
cold-cone seal 491, 498, 606
cold crucible 101, 125
colour centre 185, 189
coloured quartz 544
comonomer 1200
complex oxides 584
complex silicates 467
composition diagrams 476
compound semiconductors 155
compressibility 481
computational cost 910
computer control 194
concanavalin B 427
concentration boundary layer 782
concentrators 240
conditions for crystallization stability 1044
cone closure 499
conformational dynamics 428
conformational equilibrium states 443
conformations 1173
congruent sublimation 623
conservative dopant 149, 276
Constant Composition Method 1237
constitutional supercooling 130, 134, 189, 481, 632, 1171
constitutional supersaturation 294, 632
contact line 908
contact metasomatism 11, 32
contact nucleation 328, 1261
contacting angle 106, 119, 120
continuous crystallisation models 358
continuous-wave laser 181
control of convection 795
control of domain sizes 1067
control of nucleation 640
control (proportional/integral) 987
control strategies of batch processes 382
control systems 116
control volume methods 916, 950
controlled growth 503
controller 914
convection 437, 450, 621, 711
– cells 723
convective contribution 645
convective currents 436
convective flows 749
convective mass transfer 471
convective motions 646
convective turbulence 205, 283
cool-down stage 1002
cooled tip 582
cooling crystallisation 289, 323
cooling disc crystalliser 304, 394
copolymers 1201
copper 123, 124
– lining 519
coracle 123, 173
corrugations 132
corundum 518, 545–547
Coulter counter 350
counter-diffusion 1265
countercurrent cascade of crystallizers 305
$Co_{1-x}Ni_xFe_2O_4$ 593
Cr–Al 497
cracks 185
creep 838
– mechanism 838
cristabolite 518
critical density 499
critical Marangoni number 771, 799
critical nucleus 1192
critical resolved shear stress 71, 838, 888, 952, 974
critical supersaturation 1228
critical temperature 483
– of water 499
crosslinking 1200
crucible 102, 103, 113, 114, 118, 143, 571
– charging 184
– rotation 103, 108, 109, 270, 273, 279

crucible (*cont'd*)
– thermal conductivity 90
crystal growth by composite gradient method 522
crystal growth by reverse temperature gradient method 523
crystal growth by slow heating method 521
crystal growth mechanisms 336
crystal habit 1251
crystal pulling 101, 204
crystal quality 823, 887, 972
crystal radius oscillations 985
crystal rotation 103, 104, 266, 919
crystal shape 883, 908, 989, 997
crystal size distribution 10, 346
crystal thickening 1187
crystal thickness 1185
crystal weighing 119
crystal–melt interface 116
crystalline quality 1297
crystallization front: the heat-balance equation 1022
crystallization kinetics 331
crystallization methods 322
crystallization techniques 321
crystallization temperature 526
CSD measuring techniques 349
cuprite 520
curvature of a temperature field 835
cusped field 263, 278
cylindrical growth 105
cylindrical pore 1089
Cz silicon 143
Czochralski 101, 105, 753, 803, 825
– configuration 720, 775, 799
– growth 108, 109, 123, 180, 263, 267, 268, 278
– method 536
– technique 182, 1024

Dauphine twin quartz 545
Dead Sea water 1221
death rate function 360
Debye–Hückel equation 481
decoupled schemes 964
decrease of growth rate 643
defects 951
deformable gasket 501
deformation behaviour of InP 843
deformation behaviour of Si 843
deformed copper ring 502
degree of filling 483
degree of supersaturation 472, 529
dehydration frequencies 1240
denaturation 445, 449
denature 422
dendritic crystals 17
dendritic growth 481, 1297
dendritic web growth 155
depolymerization 541
detergent 453, 454, 456
determination of supercooling 603
deuterium 1188
Dewar flask 443
diagenesis 5
dialysis 431, 434–436, 443, 447
– buttons 435
dialyzed 445
diameter control 113
diamond 7, 18, 23, 26, 27, 31, 469
– morphology 18
dielectric constant 477, 481
dielectric effects 445
dielectric properties 446
dielectric screening 432
dies 102, 121, 122
differential scanning calorimetry 1173
differential thermal analysis 573
diffraction pattern 428
diffuse radiation 931
diffusion 481, 1239
– approximation 932
– boundary layer 34
– coefficient 813, 1286
– equation 891
– limitation 631
– of elements 32
– profile 275
– rate 436, 448
diffusive flow 617
dilatometry 1173
dilute alloys 104
diopside 596
discretization 916
– coupled space time 907
– space 901, 906
– time 902
dislocation density 649, 650, 856, 857
– in GaAs 62, 67, 71, 72
– in GaP 82
– in InP 85, 87
dislocation etch pits 140

dislocation formation 128, 129, 862
dislocation-free crystals 124
dislocation-free growth 148
dislocation loops 140
dislocation pattern 823, 861
dislocation velocity 840
dislocations 888, 976, 1000
dissociative sublimation 629
dissolution 27, 29, 538
dissolution striations 604
distribution coefficient 280, 283, 664
distribution moment 347
division lines 946
division points 945
DNA 1169
dominant lamellae 1208
dopant 545
– segregation 270
– striation 270
doping 1298
double container technique 194
draft-tube baffle crystalliser 388
driving force 321
drop doping 235
dynamic layer growth 299
dynamic operation 372
dynamic similarity 737
dynamic simulations 976
– with melt convection 998
dynamic stability of crystallization 1016
dynamical crystallization experiments 41
$DyPO_4$ 588

eddy current techniques 743
eddy diffusivity 811
edge-defined film-fed growth 123, 154, 198
edge energy 1240
edge facets 94
EDTA 1263
effective ambient temperature 947
effective distribution coefficient 271, 293
effective partition coefficient 682
effective segregation coefficient 785, 786
effective stress 840
EFG 1032
Ekman layer 108, 270
Ekman number 737
elastic stresses near interface 850
elastic–plastic interaction 854
electro-magnetic stirring 161
electro-optical 538
electrodiffusion 518
electrodynamic forces 226
electrodynamic gradient freeze (EDGF) 59
electrodynamic pressure 226
electrolytes 481
electromagnet 263
electron beam heating 101
electron bombardment 246
emerald 469, 542
Emerita 543
emission coefficient 929
emissive power of the blackbody 928, 932
emissivity 931, 958, 965
emitted flux 935
encapsulant 157
end brine 1221
energy barrier 1231
energy equation 900
enhanced dislocation multiplication 858
enthalpy 477, 479
ENTMP 1263
entropy 477
environmental conditions 5
– natural crystallization 5
epitaxial nucleants 458
epitaxy 1178
equation of state 840
equation of transfer 937, 956
equilibrium condition 476
equilibrium constant 477
equilibrium distribution coefficient 292
equilibrium melting point 1187
equilibrium segregation coefficient 813
equilibrium shape 1251
etch channels 517
etch figures 22
etch pit density 854
etch tunnels 518
Eulerian models for mixing 408
eutectic cells 1142
eutectic phase 1113
eutectic reaction 1115
eutectic system 291
evaporative crystallisation 323
exoclave 508
expansion coefficient 826
exponential rate laws 1242
exsolution lamellae 7
extended-chain crystals 1183
extent of the eutectic range 1137
external heating 239
extra-terrestrial materials 6, 21, 47

faces 1234
facet 129, 193, 236
– effect 161
faceted eutectics 1146
falling film crystallizer 300
fastening of the seed 580
$FeBO_3$ 584
feldspar 468
ferrielectrics 469
ferrite 586
ferrofluidic seal 111
ferromagnetic chromium (IV) oxide 467
ferromagnets 469
Fickian diffusion 616
filtration 397
fines destruction 365, 385
finite difference methods 916
finite element method 901, 941
flange type seals 503
flash evaporative crystallisation 324
flat plate closure autoclave 498
flat plate closures 498
flat plate seal 491
floating crucible 127
floating technique 215
floating-type insert 503
floating-zone 263, 279, 280, 807
– melting 198
– technique 143, 215, 1084
flow oscillations 878
flow pressure 227
fluid salinity 476
fluoride crystals 201
fluorides 468
flux growth 519
flux techniques 525
fluxes for oxide materials 567
forced circulation crystalliser 391
forced convection 266
forward light scattering 352
fractal patterns 17
fractional crystallisation 289
framework phosphates 528, 529
framework structures 528
Frank–Read mechanism 870
free energy 470, 1234
free interface diffusion 435, 437
free oxygen 568
free water 1220
Froude number 1018
Furedi–Millhofer 1248
furnace skeleton 962
g-jitter 283
gadolinium gallium garnet 599
galena 541
Galerkin method 899, 916, 941, 944, 980
gallium 605
– antimonide 167, 282
– arsenide 123, 167, 267, 271, 276, 282, 283
– garnets 189
– phosphide 118, 173
– substituted iron garnets 590
γ-ray rocking curves 650
garnets 185, 193, 566, 575, 576, 595, 599, 603, 605–607
gas-phase doping 235
$GdAlO_3$ 582, 584
$Gd_3Fe_5O_{12}$ 575
Gebhart factors 941, 947
Gebhart methods 980
gel 423
– growth techniques 1273
gem identification 32
gem-quality synthetic diamonds 31
gemstones 542
geological processes 5, 7, 23
geometric similarity 735
germanates 467
germanium 101, 119, 126, 280
– antimonide 121
GGG 599–601, 603, 604
Gibbs free energy 476
Gibbs–Thomson effect 341
glass crystallization 595
glass liners 468
glass tube 493
glass vessels 497
glasses 1169
global modelling 796
global problem 959
global simulations 965
gold 114
– capsules 511
– liners 527
gradient freeze 753, 755
graphite 114
– grease 510
– lubricant 510
Grashof number 116, 621, 673, 736, 737
gravitational pressure 227
gravity 436, 437, 450, 824, 825
– wash column 306
gray materials 932
Gray–Loc seals 503

grow-out 104
growth 336
– angle 894, 905, 981, 1019
– banding 30
– by a travelling solvent zone 592
– by flux evaporation 572, 594
– by slow cooling 572
– by vertical temperature gradient transport 572
– chamber 110, 111
– chamber ambient 115
– conditions 5, 831
– – natural crystallization 5
– dynamics 911
– histories 3, 7, 14, 27, 31
– interface 849
– kinetics 546
– material 517
– of α-quartz 512
– of corundum 546
– of garnets 511
– of quartz 491
– of zincite 519
– phenomena 596
– rate 104, 481, 514, 516, 519, 542, 543, 642, 855
– – dispersion 342
– – measurements 46
– speed 891, 902
– spiral 22, 23, 26, 596
– steps 1263
– striations 81, 88, 135, 185, 603
– temperature 514
– twinning 164
– twins 129
growth-induced dislocations 30

habit 643
– modification 1251
– variation 20
hackmanite 541, 542
half-time analysis 1177
halides 101
hanging drop 438–440
Hartman–Perdok 1235
Hartmann number 266, 727
heat balance 359
heat flux balance 830
heat of dissolution 479
heat pipe 588
heat shields 951, 974
heat supply distribution 896, 900, 969, 971
heat transfer 340, 1027
heater power 103
– calculated 989, 999
– smoothing 982, 990
heating system 895, 974
hegemonic ligand 489
helical dislocations 191
helicoidal twisting 1211
Helmholtz coils 278
hematite 23
Herring condition 106
heterogeneous nucleation 326, 458, 1173, 1230
hexaferrite 595
hexagonal ferrites 588
high molar acid solutions 497
high pressure 467, 595
– crystallization 289
– puller 168, 169
high quality quartz 514, 517
– growth 518
high-temperature creep 839
high-temperature vapour growth 654
hollow pyramids 1181
holm 107
holographic interferometry 1287
homogeneity 422
homogeneous nucleation 1173, 1225, 1227
Hooke's law 189, 834
hoop stresses 190
horizontal autoclave 503, 504
horizontal Bridgman growth of InP 83
horizontal Bridgman technique 774
horizontal zone-melting 282
hornblende 468
hot wall technology 156
hyalopyritic 41
hydrated ions 1241
hydration layers 430
hydration numbers 1241
hydration sphere 1219
hydraulic wash column 307
hydrocarbons 1171
hydrodynamical principles 470
hydrodynamics 507
hydrogel 1280
hydrogen bonding 478
hydrohalogenic acids 536
hydrolysis 538
– process 533
hydromagnetic flow 269
hydrosodalite 541

hydrothermal apparatus for visual observation 506
hydrothermal chemistry 470, 480
hydrothermal crystal growth 467, 490
hydrothermal dehydration process 512
hydrothermal epitaxial growth 511
hydrothermal experiments 469
hydrothermal metasomatism 11, 26
hydrothermal mineralizer 477
hydrothermal reaction sintering 511
hydrothermal reactions 467
hydrothermal synthesis 531
hydrothermal treatment 467
hydrothermally grown berlinite crystals 524

ideal solutions 476
igneous process 5
II-VI compounds 654
impurity 1171, 1259
– distribution 1048, 1096
– effects 1250
– hardening 87
in situ observation 36
– aqueous solution 34, 36
– high temperature 14, 36
– method 8, 45
in situ observation method at high temperatures 41, 42, 46
in situ synthesis 69
inclusion of impurities 294
inclusions 30, 829
inconel 499
– vessel 498
incongruent melting 525
incorporation of impurity 292, 1262
indium antimonide 121, 159, 279, 280
indium antimonide–nickel antimonide eutectic 280
indium arsenide 173
indium-gallium antimonide 280
indium phosphide 174, 270, 274, 282
induced striations 582, 586, 589, 602, 607
induction heating 225, 879, 895, 968
induction period 1227, 1259
induction time 335
inductor 232
industrial crystallisers 385
industrial layer growth equipment 299
industrial mass crystallisation 315
industrial pressure crystallisation equipment 308
industrial suspension growth equipment 303
inelastic light scattering 433
infra–red striagraphs 90
inhibitor 454, 1262
inhomogeneities 779, 827
initial whisker sections 1083
InP 832
insolubility point 445
instabilities 646
integrated analysis of crystal growth 865
intensity of the blackbody 938
intensity (polarized spectral) 929
intensity (spectral) 930
interface behaviour 965, 992, 994
interface curvature 689
interface deflection 959
interface inertia 983
interface instabilities 745
interface inversion 918, 919
interface shape 730, 880, 925
interfacial tension 925, 1227, 1230
interferometry 22
interionic distances 1254
internal heating 239
internal morphology 16, 26
internal order 428
internal radiation 881, 937, 950, 953, 956
interplanetary dust particles 47
interstability 1026
inverse problems 960, 971, 978, 987, 990
ionic conductivity 529
ionization 481
iono-covalent compounds 1265
iridium 114
– platelets 188
iron garnets 574, 585, 590, 605
irradiance 939
irradiation 545
isochores 499
isoelectric focusing 423
isograms 1238
isomers 1200
isomorphous heavy-atom derivatives 446
isotachs 1238
isotherm method 897
isothermal hydrothermal autoclave 484
isotherms 832, 833
isotope doping 512

Jacobian matrices 909, 978

Kennedy's *PVT* diagram 497
kinematic similarity 735
kinetic instability 647

kinetic model comet tails 47
kinetic studies 503, 514
kinetics 440
– of crystallization 476
kinks 1263
$KNbO_3$ 586, 587
$KTa_{1-x}NbO_4$ 590
$KTa_{1-x}Nb_xO_3$ 584, 586, 588, 590, 592
$KTiOPO_4$ 584, 586, 588
KTP 588, 592
Kyropulos 753

labile regions 422
labile zone 1224
$LaGaO_3$ 584
Lambert's cosine law 935
lamellae 1205
lamellar crystal 1179
lamellar width 1206
lanthanum 126
Laplace capillary equation 1018
Laplace constant 107, 118
Laplace–Young equation 107, 117, 217
largest autoclave 512, 513
laser-heated growth 576, 594
laser-heated pedestal growth 200
lasers 469
latent heat 103, 104
lateral free energy 1194
lateral inhomogeneity 782
lattice hardening 87, 827
lattice interactions 423
$La_{2-x}Ba_xCuO_4$ 587
layer growth 295, 299, 311, 337
layered structure 653
lazurite 541
lead germanate 182
lead telluride 181
leaf-shaped crystals 1207
LEC technique 274
Lechleitner emerald 543, 544
lever rule 292
ligand 443, 453
ligand binding 430
light cut technique 743
light radiation 242
light scattering 1282
linear rate laws 1242
linear systems 910
linear zig–zag 1173
liners 492
lipophilic 454
liquid channels 428
liquid crystal 1174
liquid encapsulated growth 879, 973, 986
liquid encapsulated vertical Bridgman growth (LEVB) of GaAs 72
liquid encapsulated vertical zone melting (LEVZM) 72
liquid encapsulation 64
– Czochralski technique 156, 157
– technique 101
liquid meniscus heights 1037
liquid meniscus shapes 1037
liquid phase epitaxy 565
liquid phosphorus 175
liquid separation 396
liquid–liquid dialysis 438
liquid–liquid diffusion 435, 436, 440
liquid–solid interface shape 81, 91, 94, 763
liquid–solid transformation 849
liquidus temperature 104
lithium niobate 195, 531, 1065
lithium tantalate 196
lithium tetraborate 531
local heating 593
local shaping technique (LST) 1100
local simulations 885
local–global iterations 964
long-chain polymers 444
long-range stress 849
Lorentz force 225, 261, 265, 283
low-angle grain boundaries 1058
low-carbon steel 501
– autoclaves 491
low dislocation density 859
low dislocation quartz 514
low-temperature creep 839
lower yield point 841
LPE using PbO/B_2O_3-based fluxes 574, 600
Lyapunov set of equations 1016

MAC method 917
macro-elements 959
– linear 961
– non-linear 961
– twin 963
macromixing 402
macromolecules 1167
magma mixing 41
magmas 5
magmatic crystallization 34
magnesium aluminate 594
– spinel 182, 192

magnetic field 151, 173, 205, 237, 797, 799, 881, 891, 896, 915, 969
magnetic interaction parameter 269
magnetic Prandtl number 265
magnetic Reynolds number 265
magnetic viscosity 270
magnetite 511
magnetohydrodynamics 748
malachite 484, 489
– solubility 490
Marangoni convection 713, 765, 774, 807, 893
Marangoni effect 151, 224, 268
Marangoni number 224, 736, 737, 809, 1023
mass balance 359, 382
mass crystallization 536
mass transfer 339, 537, 661
mass transport 472, 538, 616
material parameter 814, 842
maximum zone length 220
Maxwell's equations 225, 264, 969
mechanical design 110
Mechanical Wash Column 308
melt column shaping conditions 1017
melt convection 877, 914, 971
– dynamic simulations with- 982
melt flow patterns 883, 918, 919, 955
melt growth 711
– configurations 712
melt replenishment 205, 986
melt stability 925
melt techniques 531
melt turbulence 143
melt–solid interface 56, 88, 90
– shape 73
melting points of metastably nucleated phase 45
melting temperature 904
membrane 454
meniscus 105, 116, 119, 217
– angle 218
– height 123
– shape 775, 893, 905, 908, 995
– surface equation 1018
mercury iodide 638
mesh deformation 906
metal complex stability 476
metal ions 454
metal lining 492
metals 101
metasomatic stages 467
metastable nucleation 13, 14, 43, 44
metastable phases 484
metastable regions 422
metastable solutions 484
metastable zone 1224
– width 334
$MgAl_2O_4$ 595
$MgGa_2O_4$ 595
MHD equations 261, 264
mica 12, 25
micro-facets 131
micro-precipitates 153
micro-substrates 1233
microbes 447, 454
microdefects 238
microdialysis 440
– buttons 445
microgravity 283, 436, 437, 628, 787, 788, 800, 802
– conditions 625, 628, 1105
– simulations 1291
microheterogeneity 422, 423
micromixing 403
– models 405
microprocessor control 144
microsegregation 140, 696, 789, 790
microstructural impurity content 1210
microtechniques 435
mild steel liner 501
Milne–Eddington approximation 939, 954
mineral parageneses 7
mineralizer 467, 526, 527
– solution 481, 533
minerals like quartz 468
mirror furnace 593
mixed fluorides 201
mixed solvents 497
mixing 402
– times 406
$Mn_{0.5}Zn_{0.5}Fe_2O_4$ 593
$Mn_{0.5}Zn_{0.7}Fe_2O_4$ 593
$Mn{:}Y_3Fe_5O_{12}$ 575, 586
model of plastic deformation 839
modelling of dislocation formation 853
modelling of mechanical breeding 332
modelling of secondary contact nucleation 331
modelling of surface breeding 333
modification of Al–Si 1151
modified Bridgman 491, 494
– autoclave 501, 502, 512, 513
– seal 503
modified directional freezing 62
modified growth 54
modified stress field 859

molar volume 480
molds 447
molecular shape 1184
molten charge 104
molybdenum 114
momentum boundary layer 739, 782
– thickness 275
monophase crystallization 476
monotectic phase 1113
monotectic reaction 1153
Morey autoclave 491, 497, 498, 529
Morey flat plate 494
morphological evolution 32
morphological features 1052
morphological instability 282
morphological stability 632
morphologies of crystals 16
morphology of mineral crystals 14
mosaic spread 342
mosaicity 653
mother liquid inclusion 40
mother liquor 424, 432, 433, 438, 439, 441, 448, 454
moving mesh 897
– explicit techniques 898
– implicit techniques 899
MSMPR crystallizer 1260
MSMPR model 362
multi-grid technique 778
multi-variate control 205
multi-zone furnace 59, 61, 94, 496
multiplication of dislocations 840
M_xWO_3, M = Cu 595

n-type semiconductor 518
Na-β-alumina 528
NaFe–silicate 491
NASICON 528, 529
native mineral formation 499
natural convective flow 268
natural emerald 544
natural emission 929, 931
natural remnant magnetism of chondrules 47
Navier–Stokes equation 264, 265
NC diagrams 1046
– of crystallization 1253
$NdGaO_3$ 584
$Nd_xR_{1-x}P_5O_{14}$, R = La, Y 595
$Nd{:}Y_3Al_5O_{12}$ 575
nearly-sealed ampoule 635
needle-eye technique 223
negative crystal 1089
negative temperature coefficient of solubility 495, 520, 521
nematic 1175
net flux 936
neutron scattering 1188
newberyite 1232
nickel 126
nickel linings 492
niobium 126
nitrilotriacetamide (NTAA) 1254
nitrogen platelets 30
noble metal lining 468, 501
non-adjacent re-entry 1189
non-aqueous solvents 484, 485
non-conservative dopant 150, 276
non-conventional phase diagrams 475, 476
non-crystallographic branching 1213
non-dilute alloys 922
non-linear optics 1300
non-specific solvation 478
nonfaceted eutectics 1146
nonspecific aggregation 454
normal freezing 677
novel equipment design 503
nuclear transmutation 143
nucleating agents 1177
nucleation 8, 325, 1293
nucleation control 634
nucleation energetics 1229
nucleus 1226
numerical integration 845
numerical modelling 105
numerical remelting 981, 992
numerical viscosity 916
nutrient (dissolving) temperature 526
nutrient effect 517

octahedral slip 128
Ohm's law 265
ontogeny 21
opacity coefficients 954
opaque bodies 930
opposed diamond anvil 491
optic 467
optical fibres 198
optical furnaces 242
optical properties 424
optical thickness 932, 937, 955
ordered aggregates 420
organic crystals 203
organic solvent 432, 439, 444, 446–448, 497
orientation 643

oriented overgrowth 1262
origin of stress 824
Orowan equation 839
orthokinetic aggregation 344
Oslo crystalliser 395
Ostwald ripening 11, 398, 1247
Oswald stage law 640
oxidation potential 476
oxide crystals 119, 881, 895, 918, 956, 967
oxides 101
oxygen in silicon 268
oxygen isotope doping 512

P–V–T apparatus 507, 508
parabolic rate laws 1242
parakeldyshite 541
paratellurite 536, 537
partial mixing 682, 691
partial molar volume 480
partition coefficient 516, 664
$PbFe_{7.8}Al_{4.2}O_{19}$ 584
$PbFe_{0.5}Nb_{0.5}O_3$ 584
$PbFe_{12-x}Ga_xO_{19}$ 584
$PbMn_{0.5}Nb_{0.5}O_3$ 584
PBN crucible 78
Péclet number 618, 674, 736
pedestal 113
pedestal technique 217
PEEK 1191
pegmatite 26, 32
pendulum apparatus 504, 505
percent fill 515, 546
percent of baffle 514
percent opening of baffle 537
perikinetic aggregation 344
periodic bond chain 599
peritectic phase 1113
peritectic reaction 1158
permanganic studies 1212
perovskites 197, 584
persistence length 1194
perturbation techniques 104, 106, 949, 950
pH-induced crystallization 443
phase diagrams 291, 422, 473, 923
– in crystal growth 567
phase equilibria 491, 507
phase-kinetic diagrams 1205
phase matching 197
phase relationships 473
phase separation 432
phase transition 471, 1300
phase transition temperature 538
phenakite 544
phosphates 467, 484, 520, 528
phosphonates 1263
photo-semiconductors 469
photochromic 541
photon correlation spectroscopy 355
physical vapour transport 616
physico-chemical investigations 503
physico-chemical principles 470, 471, 477
physico-chemical processes 507
picket heaters 112
piezoelectric 467, 518, 536, 541
– crystal 512
piezoelectrics 469
pill doping 235
piston cylinder 491
Planck's law 928
plastic deformation 149
plastic strain, scalar 890
plasticity 189, 838
platinum 126
Poiseuille flow 128
polarization 929
polarized devices 534
polarized light intensity 1173
polydispersity 1197
polyethylene 1181
polyethylene glycol 432, 434, 444–446, 448, 454, 458, 459
polyethylene terephthalate 1204
polyethylenes 1191
polyglutamic acid (PGA) 1253
polygonal shapes 102
polyhedral prism-shaped crystals 1060
polymerization 35, 38
polymorphism 430
polynuclear complexes 489
polypropylene 1202
polysilicon 143, 148
polyvinyl sulfonate (PVS) 1252
population balance 360, 381, 1260
population density 347
post-growth histories 3, 7, 14, 27, 31
postmagmatic stages 467
potash alum 588
potassium chloride 118
potassium niobate crystals 504
potassium silicates 485
potassium titanyl phosphate 523
Prandtl number 266, 736
precipitant 440, 444, 445, 447, 448, 453
precipitates 829

precipitating agent 430, 432, 443, 454
precipitation 324, 397
precipitation kinetics 1246
precipitation point 432
prediction of the dislocation density 864
predictor–corrector technique 903
pressure chamber 111, 118
pressure crystallisation 297, 309, 311
$PrGaO_3$ 584
primary nucleation 326
principle of phase formation 473
Proabd refiner 299
process
– control 913, 983
– Czochralski 878
– dynamic gradient freeze 881, 970
– floating zone 882
– horizontal Bridgman 882
– simulation 372
– – discretisation of CSD 375
– – method of moments 374
– stability 877, 883, 911, 953, 983
– vertical Bridgman 881
profiled silicon structure 1062
protective gas 233
protein crystal growth 1276
protein–ligand complexes 446
proteolysis 454
pseudo-steady model 884, 903, 910
Pt liners 494
Pt–Rh 497
pull rate 236, 856, 902, 903
– critical 984
pull rod 110
puller design 144
pulling length 230
pulling mechanism 115
pulling speed 105
pulsated column 303
purification 423
– mechanisms 297
PVT curves 481
PVT studies 503
pyrite 21
pyrolytic boron nitride 174

quadrature 945
quartz 13, 38, 104, 113, 148, 469, 484, 512
– crystal 467
– glass ampoules 537
– reactor 506
– vessel 507
– windows 505
quasi-elastic light scattering 355
quasi-steady assumption 914
quasi-steady model 884, 903, 910, 959

R–z model 369
radial mass transfer 625
radial segregation 688
radial temperature differences 860
radiation 643
– furnaces 242
– shields 113
radiative enclosures 962
radiative flux vector 938–940
radiative heat transfer 654, 878
radiative integral equation 936, 941
radio frequency induction heating 101
radiosity 935
random coil 1184
Raoult's law 476
rapid screens 439
rare earth disilicates 585
rare earth gallates 197
rare earth metals 125
rare earth orthoferrite 583, 596
rare earth phosphates 529
rare earth silicates 476
rate constant 478
rate of dissolution of the nutrient 472
rate of growth 453, 1236
rate of nucleation 1226
rates 45
– dissolution 45
– growth 45
– nucleation 45
Rayleigh number 736, 1023
reactants 477
real hydrothermal solutions 476
recrystallization 298, 308, 400, 470
recrystallized glass 517
red phosphorus 175
redox conditions 540
redox potential 477, 511
reduced interface concentration 626
reduction of the dislocation density 858
reflectivity 930
reflector 242
Regency emerald 544
regional metamorphism 5, 26, 32
regular domain structure 1069
relative supersaturation 472
relaxation during cooling 867

released plastic strain 857
reptation 1197
residence
– time distribution 402
residence time 359
residual elastic stress 855, 867
resistive heating elements 112
Reynolds number 736, 737
rf induction heating 113, 239
$RFe_3B_4O_{12}$ 584
rhodium 114
ribbon-like lamellae 1207
ripening 1246
RNase 428
rod eutectics 1144
rod-like 1117
Rosseland approximation 932, 940
rotary-drum crystalliser 299
rotation rate 236
rough growth 337
roughening temperature 129, 337
RPO_4, R = Y, Dy 586
ruby 181, 469, 542, 547
rupture discs 509
ruptured autoclaves 510
Russian emerald 544
RVO_4, R = Tb, Sy, Tm 590

salt 432, 444, 445
– glycol 434
– solvent 439
salting-in 431, 434, 441
salting-out 430, 441, 442, 445
– crystallisation 325
sampling technique 483
sapphire 192, 469
– crystals 1052
– windows 111, 505
satellite tobacco 426
saturated solutions 470
saturated vapour pressure 476
scale-up 377
scaling analysis 671, 884, 977
scaling laws 671
Schmidt number 109, 273, 275, 674, 736
screening technique 433
sealed ampoule 634
second law of thermodynamics 931, 940
secondary crystallization 1177
secondary nucleation 10, 39, 40, 327, 1261
– theory 1193
sedimentation 396, 436, 437
sedimentgeness 5
sedimentogenetic crystallization 10
sedimentogenetic-diagenetic processes 35
seed crystal 103, 115, 231, 233
seed effect 516
seed induced growth 470
seed orientation 513, 514, 526
seed plate 469
seed populations 1236
seeded batch 379, 383
seeding 457, 458
seeding on 104
segregation 661, 880, 892, 919
– coefficient 664, 779
selection of crystalliser 387
self-energized closures 503
self perpetuating dislocation 1234
self-stability 1026
self-transparent crystals 105
semi-permeable membrane 434, 435
semi-transparent materials 936, 949, 973
semiconducting piezoelectric 538
separated charge method 511
separation process 289
sequential countercurrent staging 301
sequential extraction 443
servo-controller 120
shadow algorithm 944, 948
shear stress 850
sheaves 1180
shouldering process 988
sieve analysis 350
silica autoclaves 485
silica crucible 150
silica gels 1280
silica polymerization 13
silicate systems 41, 43
silicates 484
silicon 101, 109, 110, 118, 125, 140, 205, 228, 261, 267, 273, 276, 278, 280, 281
– crystal puller 146
– crystals 1059
– ribbons 154
– tapes 1059
– tube 1060
silver 114, 124, 519
– boat technique 125
– lining 492
simulation of deformation experiments 845
sitting drop 440
size-dependent growth 341
slip systems 843, 850

slow cooling 573, 575
– bottom growth 574
slurry handling 303–305
$Sm_{0.55}Tb_{0.45}FeO_3$ 596
smoky quartz 544
SOC technique (silicon-on-cloth-) 1092
sodalite 511, 541–543
sodium chloride 119
sodium iron titanium bronzes 511
sodium silicates 485
solar ponds 387
solenoids 261
solid electrolytes 541
solid separation 396
solid solution system 291, 295
solid state diffusion 298
solid state electrolysis 518
solidification front 891
solidification heat 892, 901, 904, 993, 995
solid/liquid separation 303, 306
solubility 419, 476, 479, 483, 484, 507, 514, 1222
– curve 511, 529, 531, 532, 539, 542, 1223
– determinations 1222
– measurements 503
– minimum 431, 432, 442, 443
– of albite 484
– of berlinite 466–488, 521
– of calcite 534
– of cristabolite 518
– of crystallization 476
– of malachite 489
– of quartz 484–486, 1688
– of ruby 577
– of tellurium dioxide 536
– of zincite 518
solutal convection 628
solute 103
– boundary layer 670, 705
– – thickness 275
– distribution 893, 915, 923
– segregation 779
– trail 138, 141, 187
solution density 484
solutions of crystallization 476
solvates 477
solvation number 477
solvation process 477
solvent viscosity 481
solvent–solute interaction 476
sources of stress 835
space processing 283
Special Construction Bureau 503
special growth methods 595
specific protein–ion interactions 445
specific solvation 478
spectral methods 917
sphalerite 538, 539
spherulite 1169
spherulitic morphology 17
spiral growth 338
spoke patterns 723
spontaneous nucleation 529, 536
spurious nucleation 1265
$Sr_{1.03}Ga_{10.81}Mg_{0.58}Zr_{0.58}O_{18.99}$ 586
$SrFe_{12}O_{19}$ 593
$SrFe_{12-x}M_xO_{19}$ 587
$SrZrO_3$ 584
stabilised zirconia 194
stabilities of metastably nucleated phases 44
stability of crystal cross-sectional shapes 1104
stable clusters 1244
stable growth 633
stable laws of control 1077
stable nucleus 420
stacking fault energy 93
stagnant film model 682, 684
stainless steel acid digestion bombs 509
static condensation 961
static layer growth 299
static magnetic field 703
statistical entropy 1193
steady-state operation 362
steel autoclaves 497
Stefan flow 617
Stefan problem 760, 877, 896
stellite 499
stellite Tuttle-cold cone sealed autoclaves 509
step change in heater power 993
step change in heater temperature 986
step change in pull rate 912, 984, 993
step growth 600
step separation 22
Stepanov techniques 1032
stepwise perturbation 977, 997
stereochemical effect 1257
stick-slip effect 229
Stockbarger method 53, 57
stoichiometry 528
– in the vapour 630
Stokes stream function 982, 998, 999
strain 836
– rate 847
– rates during crystal growth 856

stress 834, 836
– at room temperature 867
– distribution 869
– exponent 842
– in the cooling crystal 866
– induced twinning 94
stresses 138, 262
– thermoelastic 888, 950
– thermoplastic 888
striations 136, 186, 235, 648, 789, 790, 792, 794, 803, 827, 828
strip heater 241
strontium 468
– hexaferrite 593
– hexagallate 586
structural defects 650
structural factors 1264
structural fit 1231
structural match 1232
structure of hydrogels 1280
structure of water 1219
sub-critical system 483
subiodides of tellurium 467
Submerged Heater Method 753
substrate 443, 454, 1231
sulphates 468, 492
sulphides 468, 492, 537
super-critical system 483
super-solubility 1224
superconducting magnet 261
superconducting solenoid 277
supercooling 1187
superionic borates 532
superionic conductor 528, 533
superionic phosphates 528, 529
supersaturated state 419, 433, 435, 442, 457
supersaturation 321, 399, 430, 432, 440, 443, 458, 542, 543, 1223
surface acoustic wave 531
surface energy 1226
surface free energy 1192
surface kinetics 623
surface layer 1250
surface microtopographs 12, 22, 39
surface microtopography 16, 40
surface nucleation 1239
surface radiation 932, 933
surface reaction 1239
surface roughening 1192
surface tension 217, 814, 1225
susceptor 113, 241
suspension growth 296, 303, 311
sweating 300, 301, 310, 318, 319
sweeping 516, 518
swirl defects 152
swirls 238
Symerald 543
synoptic model of molecular re-entry 1190
synthetic macromolecules 1172

tantalate 531
Taylor–Proudman theorem 266
Teal and Little 101
techniques of pulling from a shaper (TPS) 1031
teflon beakers 494
teflon liner 491, 494, 529, 531
tellurium dioxide 473, 536
temperature control system 76
temperature differential 526, 546
temperature distribution 579
temperature fluctuations 267
temperature gradient 471, 513, 514, 835
– zone-melting 137
temperature jump 932, 941
temperature oscillation method 636
temperature stability 91, 577, 578
temperature–density diagram of water 482
ternary phase diagram 526
ternary system 476
test tube pressure vessel 469
textures 8, 41
– holocrystalline 8
– hyalopyritic 8, 9
– igneous rock 8, 41
– porphyritic 8
thermal balance 104
Thermal Boundary Layers 739
thermal conductivity 814
thermal history 831
thermal radiation 928
thermal resistance of the crystal 642
thermal shock 103
thermal stability 1028
thermal stress 105
thermally induced stress 830
thermo-solutal convection 925, 927
thermo-syphon crystalliser 388
thermocapillary convection 765, 807
thermocouples 497
thermodynamic principles 478
thermodynamic properties 481
thermographic method 584
thermogravimetric analysis 573, 585

three-dimensional high ionic conductivity 528
three-phase boundary 57, 105, 106
III–V compounds 261
thrichlorosilane 143
tie-molecules 1199
tilt mosaics 650
time constants 986
time integration scheme 980, 989
time-lapse photography 433
time step 907, 981
tissue culture plates 439
$TmVO_4$ 590
TNO-Thijssen wash column 307
top-seeded solution growth 584
top-seeded VTGT growth with a nutrient 589
top-seeding solution growth 575, 590
topaz 469
total encapsulation 64, 65
toxicity 477
transfer function 120
transient calculations 913
transient effects 976, 990
transitional metal compounds 467
transmission cathodo-luminescence (TCL) 89
transmutation doping 238
transverse field 267, 281
travelling heater method 803
tri-junction 894, 905
– locus 981
trichlorsilane 231
12-tricosanone (laurone) 203
trigons 27, 28
true solubility 484
TSK-4c 305
tungsten 114
turbidity data 1245
turbulence 915
– model 775, 918
turbulent flow 811
turbulent melt convection 810
Tuttle autoclave 499, 510
Tuttle cold cone seal autoclave 494, 497, 499, 507
Tuttle cold-seal vessel 606
twinning 93, 104, 177, 190, 450
twins 185, 1297
two-dimensional nucleation 338
two-shaping-elements technique (TSET) 1092
two-stage crystalliser 370
II–VI compound semiconductors 94
typomorphology 21
TZM autoclaves 492
TZM vessels 606

udimett 499
ultrapurification of organic compounds 289
undercooling 892
unified growth theory 1235
unit cells 1172
unseeded batch 380, 384
unstable growth morphologies 294
unsteady convection 696, 758, 789
unsteady flow 718
unsupported area closures 498
uric acid dihydrate 1232

valsovite 541
van 't Hoff equation 472, 484
vapor diffusion 432, 436–438, 440, 443
vapor whisker growth (VWG) 1080
vapor–liquid–crystal (VCL) 1080
vapor–liquid–solid–VLS 1080
variable shaping technique (VST) 1098
variable-volume insert 503
Velocity Boundary Layers 739
Verneuil technique (VT) 1069
vertical Bridgman 753, 755, 758, 804, 826
vertical gradient freeze growth of GaP 74
vertical gradient freeze growth of InP 83
vertical gradient freeze (VGF) 58
vertical temperature gradient transport 575, 589
VGF of GaAs 70
view factor algebra 949
view factors 935, 942, 947, 978
viewed parts 943
viscosity 477, 478, 481, 814
voids 187, 192, 1053
volume change 830
volume free energy 1226
von Mises invariant 888, 1061

warpage 829
wash columns 306
washing 298, 308
weak formulation 900, 905, 915, 939, 944
Weber number 1018
weight loss method 484
weight signal 120
welded closure 498, 501, 606
welded Walker–Nuehler closure 491
wetted areas 607
wetting 599, 826
– angle 894, 905
– condition 1036
wetting-to-catching transition 1036
wollastonite 468
work hardening 841

X-ray damage 428
X-ray diffraction 1298
X-ray topographs 606
X-ray transmission topographs 649

YAG 589, 599, 603
$Y_3Al_5O_{12}$ 575
$YAlO_3$ 584
$Yb_3Fe_5O_{12}$ 569, 570
yeast phenylalanine tRNA 429
$Y_3Fe_{4.2}Ga_{0.8}O_{12}$ 606, 607
$Y_3Fe_{4.5}Ga_{0.5}O_{12}$ 607
$Y_3Fe_4Ga_1O_{12}$ 576, 593
$Y_3Fe_5O_{12}$ 570, 574–576, 606
$YFeO_3$ 574, 582, 584
$Y_3Fe_{5-x}Ga_xO_{12}$ 575, 576, 590, 591
$Y_3Fe_{4-x}Ga_xO_{12}$, $0 < x < 1$ 593
$Y_3Fe_{5-x}M_xO_{12}$ 589
$Y_3Ga_5O_{12}$ 575, 606
YIG 492, 569–571, 574, 580–582, 589, 593–595, 598–603, 605, 607
YPO_4 586
yttrium aluminium garnet 582, 589, 599
yttrium iron garnet 570
yttrium orthoferrite 574
$Y_{3-x-z}Bi_xPb_zFe_{5-y}Al_yO_{12}$ 604
$Y_{3-x-z}La_xPb_zFe_{5-y}Ga_yO_{12}$ 605
$Y_{3-z}Pb_zFe_{5-y}Ga_yO_{12}$ 605

zincite 518, 519
zirconia 114
zirconium minerals 541
zirconium silicates 541
$Zn_{1.9}Fe^{2+}_{0.1}Ba_2Fe_{12}O_{22}$ 590
$Zn_{1.9}Mn^{2+}_{0.1}Ba_2Fe_{12-x}Mn_xO_{22}$ 590
$ZnAl_2O_4$ 595
$Zn_2Ba_2Fe_{12}O_{22}$ 586, 588
zone-melting 53, 678
zone-refining 202